TRIGONOMETRIC FORMULAS

FUNDAMENTAL FORMULAS

$$\tan\theta = \frac{\sin\theta}{\cos\theta}$$

$$\cot\theta = \frac{\cos\theta}{\sin\theta}$$

$$\sec\theta = \frac{1}{\cos\theta}$$

$$\csc\theta = \frac{1}{\sin\theta}$$

$$\sin^2\theta + \cos^2\theta = 1$$

$$\tan^2\theta + 1 = \sec^2\theta$$

$$1 + \cot^2\theta = \csc^2\theta$$

NEGATIVE ANGLE FORMULAS

$$\sin(-\theta) = -\sin\theta$$

$$\cos(-\theta) = \cos\theta$$

$$\tan(-\theta) = -\tan\theta$$

ADDITION FORMULAS

$$\sin(\theta \pm \phi) = \sin\theta\cos\phi \pm \cos\theta\sin\phi$$

$$\cos(\theta \pm \phi) = \cos\theta\cos\phi \mp \sin\theta\sin\phi$$

$$\tan(\theta \pm \phi) = \frac{\tan\theta \pm \tan\phi}{1 \mp \tan\theta\tan\phi}$$

DOUBLE ANGLE FORMULAS

$$\sin 2\theta = 2\sin\theta\cos\theta$$

$$\cos 2\theta = \cos^2\theta - \sin^2\theta$$

$$\tan 2\theta = \frac{2\tan\theta}{1 - \tan^2\theta}$$

$$\sin^2\theta = \frac{1 - \cos 2\theta}{2}$$

$$\cos^2\theta = \frac{1 + \cos 2\theta}{2}$$

PRODUCT FORMULAS

$$\sin\theta\sin\phi = \frac{1}{2}\left(\cos(\theta - \phi) - \cos(\theta + \phi)\right)$$

$$\sin\theta\cos\phi = \frac{1}{2}\left(\sin(\theta - \phi) + \sin(\theta + \phi)\right)$$

$$\cos\theta\cos\phi = \frac{1}{2}\left(\cos(\theta - \phi) + \cos(\theta + \phi)\right)$$

COMMON TRIANGLES

RIGHT TRIANGLE

Pythagorean Theorem

$$c^2 = a^2 + b^2$$

Trigonometric functions

$$\sin\theta = \frac{a}{c} \qquad \cot\theta = \frac{b}{a}$$

$$\cos\theta = \frac{b}{c} \qquad \sec\theta = \frac{c}{b}$$

$$\tan\theta = \frac{a}{b} \qquad \csc\theta = \frac{c}{a}$$

ANY TRIANGLE

Law of Cosines

$$c^2 = a^2 + b^2 - 2ab\cos\gamma$$

Law of Sines

$$\frac{\sin\alpha}{a} = \frac{\sin\beta}{b} = \frac{\sin\gamma}{c}$$

Ratios of corresponding sides of similar triangles

$$\frac{a}{b} = \frac{a'}{b'}, \quad \frac{a}{c} = \frac{a'}{c'}, \quad \frac{b}{c} = \frac{b'}{c'}$$

CALCULUS

Second Edition

Richard A. Hunt
Purdue University

■ HarperCollins*CollegePublishers*

Sponsoring Editor: George Duda
Developmental Editor: Louise Howe
Project Coordination: Lifland et al., Bookmakers
Electronic Art: Tech-Graphics Corporation
Cover Design: Lesiak/Crampton Design: Cynthia Crampton
Cover Photo: Dominique Sarraute/The Image Bank
Design and Composition: ATLIS Graphics & Design
Printer and Binder: R.R. Donnelley & Sons Company
Cover Printer: R.R. Donnelley & Sons Company

Calculus, Second Edition
Copyright © 1994 by HarperCollins College Publishers

Library of Congress Cataloging-in-Publication Data

Hunt, Richard A., 1937–
    Calculus / Richard A. Hunt.—2nd ed.
        p.   cm.
    Includes index.
    ISBN 0-06-043046-X (student ed)
    ISBN 0-06-501449-9 (free copy)
    1. Calculus.   I. Title.
QA303.H8513 1994   93-32071
515—dc20   CIP

93 94 95 96   9 8 7 6 5 4 3 2 1

Brief Contents

Contents

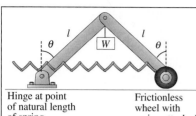

Hinge at point
of natural length
of spring

Frictionless
wheel with
spring attached

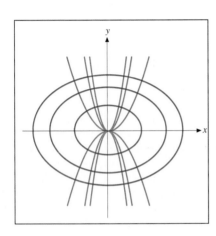

To the Instructor

The purpose of this text is to illustrate the concepts, techniques, and applications of calculus. Intended primarily for science and engineering majors, the text contains all the topics of a standard three-semester introductory calculus sequence. A two-semester version, *Calculus of a Single Variable, Second Edition,* is also available. The second edition is a student-oriented calculus text that is easily integrated into a calculator/computer environment.

The major focus of the second edition is problem solving. The teaching strategy of isolating and emphasizing basic prerequisite skills required for the successful solution of calculus problems has been retained from the first edition. The role of concepts in problem solving is emphasized throughout.

The examples and exposition are again major strengths. Numerous detailed examples relate particular problems to general problem solving strategies. Many counterexamples are used to clarify concepts.

As in the first edition, the theory is mathematically honest, but not heavy-handed. Important concepts are introduced intuitively and then developed more analytically. Proofs illustrate the role of concepts in the theory that forms the basis of problem solving. The analytic development is not intended to be complete and may be omitted by those who prefer to use only the intuitive approach.

The second edition contains many new pedagogic features to help students learn. More challenging exercises, conceptual exercises, writing exercises, exploratory and open-ended exercises, and the *Extended Applications* found at chapter endings have been added. There are also exercises designed to utilize calculator and computer technology to improve understanding. These include *Calculus Explorer & Tutor* exercises.

FOCUS ON PROBLEM SOLVING

The underlying goal of the text is to help students solve problems. The teaching strategy used to accomplish this goal is based on the premise that students have a much better chance to solve a complicated problem if they understand the concepts involved and have developed the skills required to do each component of the problem.

Every calculus teacher is aware of content that students find difficult. This text contains several sections that help students in these content areas. It was the success of these special help sections, which I first wrote for my students to supplement other texts, that led me to write the first edition of my calculus book.

- Section 4.5 helps students translate word problems into mathematical expressions. The systematic study of methods for finding equations that relate variables forms a basis for problem solving strategies. This unique presentation is particularly useful for helping students set up problems involving related rates and extrema.
- Section 6.1 develops student skills in using analytic geometry to describe physical quantities such as the length and midpoint of a variable vertical line segment between two curves. These are essential basic skills required to set up definite integrals that represent physical quantities such as area between curves, volumes, and moments.
- Unique "sketch and describe" problems are introduced in Section 13.1. These problems are designed especially to help students understand the relation of limits of integration of iterated multiple integrals to corresponding regions.

- Students are taught how to choose parametric equations for some common curves in Section 11.1. This basic skill is essential for setting up line integrals for applications of the vector calculus covered in Chapter 14.

EMPHASIS ON CONCEPTS

The text emphasizes understanding ideas and relating concepts to problem solving. An important objective is for students to see how the fundamental concepts of calculus, including limits, continuity, differentiation, and Riemann sums, relate to problem solving. The role of proofs in this text is to emphasize concepts in the theory, not to provide mathematical completeness. The proofs are honest with regard to concepts, although not complete in detail.

The conceptual focus and problem solving philosophy of the text are well illustrated by the treatment of volumes of revolution in Section 6.4. First, the relevant prerequisite analytic geometry is presented in Section 6.1. The students are then taught to determine an integral that represents a volume of revolution by considering the result of rotating a representative area strip. The integral may represent a sum of either disks, washers, or cylindrical shells, depending on the type of area strip used. Thus, students are not taught the "disk method" and the "shell method" as separate methods; they are seen as examples of the same concept. The concept of applications of the definite integral is further emphasized by the pedagogic device of annotating the components of definite integrals so that each part has a clear physical meaning.

Concepts are presented graphically, as well as analytically. For example, limits and continuity are introduced graphically in Section 2.3, while Section 2.4 contains a unique analytical presentation of limits and the relation of the $\epsilon-\delta$ process to applications. The availability of the concept of bounds of functions and the simple device of giving the students a preliminary assumption make the analytic approach accessible to students.

Interpretation of graphs is emphasized. The text relates calculus concepts to the graph of a function, as opposed to using calculus to obtain the graph. Many new exercises challenge students to determine analytic properties of a function from either a given graph or a calculator-generated graph. This approach is consistent with the use of graphics calculators.

The study of infinite series is motivated in Section 9.2 by asking a question: if a ball always bounces back to one-half the height of its previous bounce, will it (theoretically) bounce forever? The difficult concept of infinite series has been made more accessible to students by the inclusion of suggestions for which convergence test should be applied to a particular type of series and the frequent review of these suggestions throughout Chapter 9.

EXAMPLES THAT ILLUSTRATE CONCEPTS AND APPLICATIONS

The selection and presentation of the examples are a major strength of the text. The examples reflect my many years of teaching calculus. They are carefully chosen to simultaneously illustrate the concepts, problem solving strategy, and all technical skills required.

Algebraic details are provided throughout to increase readability and promote understanding. Students cannot follow the flow of ideas in a solution if they cannot follow the algebraic details.

Example solutions are related to general problem solving strategies. For example, Example 1 of Section 9.4 first investigates why the convergence or divergence of the series $\sum_{n=2}^{\infty} 1/(n \ln n)$ cannot be determined by comparison with a p-series, and then illustrates how the Integral Test is used to determine divergence.

Many counterexamples clarify the concepts. These counterexamples range from simple geometric examples of discontinuities in Section 2.3 and illustrations of the Extreme Value Theorem in Section 3.8 to more sophisticated examples of discontinuities of functions of two variables in Section 12.2.

The use of integration tables is integrated throughout Chapter 8. The basic integration techniques of substitution, integration by parts, and partial fractions are presented and used to develop formulas. Thereafter, complicated integrals that occur in applications are evaluated by referring to a table of integrals. This approach is consistent with using either a computer or a calculator as an electronic table of integrals.

The development of the concept of extrema of functions of several variables has been sharpened in Sections 12.8 and 12.9. Reviewers were impressed by the step-by-step progression of ideas, the depth of coverage, and the geometric interpretation of the Lagrange method.

Fluid dynamics is used in Section 14.3 to give a physical sense to Green's Theorem. Example 6 of Section 14.3 illustrates the fact that Green's Theorem cannot always be used to evaluate the flux of a vector field across a given closed curve. Students are encouraged to verify hypotheses before applying a theorem to the solution of a particular problem.

MORE CHALLENGING CONCEPTUAL EXERCISES

This second edition contains over 1700 new exercises. These include more real-life applications, conceptual and theoretical exercises, graphics calculator exercises, writing exercises, *Extended Applications*, and exercises from *Calculus Explorer & Tutor*.

- I have searched extensively for meaningful and realistic applications. The text contains many new exercises taken from upperclass science and engineering textbooks. For example, Doppler echocardiography is related to the $\epsilon-\delta$ process in Section 2.4, Exercise 22.
- Although most exercises can be done with pencil and paper, this edition contains many exercises, marked with a graphics calculator icon, that require a graphics calculator. These exercises are designed to encourage student exploration of the richness of calculus.
- Appearing throughout the text are more than 100 problems taken from *Calculus Explorer & Tutor*, developed by Karl Petersen. These problems can be worked with pencil and paper and then investigated further with a computer.
- *Extended Applications* are given at the end of most chapters. For example, see "Roller Coaster" after Chapter 3 and "Highway Construction" after Chapter 5. These are designed for individual or group work. Many of the projects have a discovery component in addition to requiring a written or verbal report of the type expected in a real-world workplace.

A VISUAL APPROACH

The use of graphics is an important feature of the text. This edition contains approximately 2390 figures, including those in the Answer Section. Concepts are introduced graphically, and questions about graphs are used to challenge student comprehension of concepts. Interpretation of graphs is stressed.

The visual approach is emphasized by the inclusion of many new exercises about given graphs. For example, in Section 4.3 students are asked to relate the graph of a function and the graph of its derivative. In Section 4.4 students are asked to determine the approximate position of extrema and points of inflection, and to determine the sign of $f(x)$, $f'(x)$, and $f''(x)$ from a given graph. Problems of this type reflect the fact that it is now easy for students to obtain a calculator-generated graph and thus the learning focus should be on the interpretation of graphs.

The ability to sketch helps students to visualize and solve many problems, so I have made a special effort to develop this skill. Suggestions for sketching graphs of lines, parabolas, basic trigonometric functions, and some other simple curves are given in Sections 1.4, 1.5, and 1.6. Graphs of polar equations are studied in Section 10.5, where the emphasis is on the dynamics of how positive and negative loops of polar graphs are generated.

The three-dimensional sketching instructions in the text have been praised unanimously by reviewers. Detailed instructions for sketching three-dimensional surfaces in the axonometric style of computer drawings are given in Sections 10.7 and 10.8. The development includes explanations of how to change the orientation and scale of the axes to obtain different views. The instructions for sketching surfaces begin with sketching curves. This sequencing is consistent with the basic approach of the text in that key basic elements of a subject are studied before the more complicated problems are attacked.

All of the figures in the text are drawn in the simple style that the students are taught to use, so each figure reinforces the sketching instructions. Systematic use of color helps students to distinguish different components of figures. Most notably, color is used to distinguish the "top," "sides," and "front" of three-dimensional regions in Section 13.1 and sections that follow.

EMPHASIS ON NUMERICAL SOLUTIONS AND ERROR ANALYSIS

Numerical solutions and error analysis are essential components of real-life applications of calculus. Numerical solutions to problems are enhanced in this text by a carefully coordinated error analysis track, which builds on bounds of functions and relates several different types of errors.

- Section 2.2 contains a simple development of the concept of bounds of functions. Upper and lower bounds are introduced in terms of the graph of a function and related to calculator-generated graphs. The concept of bounds is the foundation for various estimates of error throughout the text.
- Linear approximation and error analysis of functions of one real variable are introduced in Section 3.6. The corresponding concepts for functions of several variables are related to the one-dimensional case in Section 12.4.
- Section 3.9 introduces Taylor polynomials and compares the error in using linear and quadratic approximations of function values. These results are then used to motivate the Trapezoidal Rule and Simpson's Rule in Section 5.6 and the study of Taylor series in Section 9.9. Polynomial approximations of solutions of differential equations are given in Section 15.8. The early introduction and use of Taylor polynomials for motivation is based, in part, on the availability of calculator-generated graphs of Taylor polynomials.
- Numerical solutions of first-order differential equations are studied in Section 15.9.

STUDENT-ORIENTED PEDAGOGY

The pedagogic features of this text greatly enhance student learning. Some of these features are indicated below.

- The *Pretest* at the beginning of Chapter 1 helps pinpoint possible student weaknesses in prerequisite skills.

- *Connections* at the beginning of each section alert students and instructors to prerequisite knowledge for the section, pinpoint where the background material was developed, and emphasize interconnections among concepts.
- Theorems are expressed formally and then restated in plain talk to emphasize the essential idea of the result. Important formulas are expressed in words in order to clarify their meaning.
- Systematic use of color allows students to distinguish different components of the figures.
- Figure legends emphasize the pedagogic point of each figure and reinforce the textual explanations.
- Algebraically detailed examples throughout enhance readability for students.
- Designated *Notes* provide additional insight into the subject matter by offering words of advice or cautions with regard to common student difficulties.
- Solution *strategies* help students to organize their thoughts and form a plan of attack for problem solving.
- *Summaries* of formulas aid students in reviewing what they have learned.
- The *Review of Chapter Concepts* provides a bird's-eye view of the whole chapter through a narrative summary of results, formulas, and concepts, with an emphasis on the interconnection of results.

ADAPTABILITY IN A CALCULATOR/COMPUTER ENVIRONMENT

The text can be used to teach calculus either in a traditional manner or with elements of calculus reform. The examples, explanations of concepts, and exercises in this text have been carefully designed to help students to understand calculus. Use of a calculator/computer to carry out calculations, generate graphs, or engage in exploratory learning could enhance this understanding.

The text has been written with an awareness that it may be used in connection with either graphics calculators or a computer algebra system. This awareness is reflected in a change in emphasis in the treatment of topics such as graphing and techniques of integration and in the inclusion of exercises and remarks that relate to technology. Exercises that encourage students to explore and write and the *Extended Applications* reflect calculus reform and NCTM recommendations.

I believe the numerical, graphical, and analytical presentations of the text are very appropriate for contemporary calculus. A judicious selection of traditional topics, combined with supplemental reform-type projects and the option of using calculators and/or computers as you feel appropriate, provides the flexibility you need to structure a calculus course that meets the needs of your students.

SUPPLEMENTS FOR THE INSTRUCTOR/ADOPTER

Many supplements for use in connection with this text are available to qualified adopters. Some restrictions do apply. For details about these supplements and others, please contact your local HarperCollins College sales representative.

PRINTWARE: *Instructor's Complete Solutions Manual, Volumes I and II* • *Printed Test Forms* • *Transparency Masters* • Demko, *A Primer for Linear Algebra* • *Using Software: A Guide to the Ethical and Legal Use of Software for Members of the Academic Community* © Educom; Adapso • Lockard et al., *Microcomputer for Educators, 2e* • Kemp/Smellie, *Planning, Producing, and Using Instructional Media, 6e*

AUDIOVISUAL: Access to HarperCollins College Publisher's VCR (Videos in Calculus Reference) files and library. HarperCollins is a disability-friendly corporation. For more information contact your local sales representative.

TESTING SYSTEM: HarperCollins Test Generator for both IBM and Mac platforms

SOFTWARE: *Graph Explorer* (IBM and Mac) • *Math Utilities* (Curves and Surfs)(Version 4.0, IBM) • *Calculus Explorer and Tutor* • *Visual Calculus* to accompany Hunt, *Calculus, 2e* • *Interactive Mathematics for the Macintosh* (The Math Lab) • MATHCAD®, CONVERGE®, Theorist®, MACSYMA®, and MAPLE® available to qualified adopters • Information on several public domain packages • Several sofware packages available from Conduit®

SUPPLEMENTS FOR THE STUDENT

STUDYWARE: Hunt, *Student's Solutions Manual* • Turner, *Study Guide to Accompany Hunt, 2e* • Turner, *A Calculus Workbook* • Demko, *A Primer for Linear Algebra* • Benbury/Lee, HarperCollins College Outline, *Calculus with Analytic Geometry* • Minnick/Gerber, *An Outline for the Study of Calculus, Volumes I, II, and III* • *Study Guide for College Algebra* • *Study Guide for Trigonometry*

SOFTWARE: *Calculus Explorer and Tutor* • *Visual Calculus* • *Calculus* (Kemery/Kuriz Software, $39.94—student discount card in book) • *Are You Ready for Calculus?*

LABWARE: Sparks/Davenport/Braselton, *Calculus Labs Using Mathematica®* • Hundhausen/Yeatts, *Laboratory Explorations in Calculus with Applications to Physics* • Shenk, *Graphing Calculator Workbook for Calculus: An Exploratory Approach* • Mathews/Eidswick, *An HP 48G Calculus Companion* • Spero, *The Electronic Spreadsheet and Elementary Calculus* • Moody, *Computer Experiments for Calculus: A Laboratory Workbook* • DERIVE® and MAPLE V® laboratory products also available

ACCURACY

I am very proud of the record of mathematical and technical accuracy of the first edition, and every effort has been made, in a continuing commitment, to ensure the accuracy of the second edition. I have justifiable confidence in the accuracy of this textbook.

I was most fortunate to have an excellent group of twenty-two mathematics professors, one or two for each chapter, to read page proofs of the manuscript. I thank each of them for their help:

George Boros, University of New Orleans and Tulane University; Stephen W. Brady, Wichita State University; Gary Buls, St. Cloud State University; Stephen D. Casey, American University; Christopher Cotter, University of Northern Colorado; Donald Goral, Northern Virginia Community College; Harvey Greenwald, California Polytechnic State University; William R. Hintzman, San Diego State University; James Jordan, Washington State University; Andrew Karantinos, University of South Dakota; Hidefumi Katsuura, San Jose State University; Robert Lax, Louisiana State University; Sarah Mabrouk, Boston University; Walter F. Martens, University of Alabama at Birmingham; Daniel Moak, Michigan Technological University; Donald E. Myers, University of Arizona; Walter Potter, Southwestern University; R. K. RaiChoudhary, Miami Dade Community College; Sahara Schiavone, University of Alberta; Charles Waters, Mankato State University; Mary Wilson, Austin Community College; Graham Zelmer, Carleton University, Ottawa

®All trademarks are the property of their respective owners.

As every teacher knows, it is essential that answers to exercises be correct. Checking the accuracy of exercise solutions in this second edition involved several stages. First, I personally worked all exercises and prepared a first draft of the complete *Solutions Manual*. Personally working the exercises allowed me to verify that the exercises actually fulfilled their intended educational purposes and included a wide variety of interesting and workable problems. A group of twenty-four mathematics professors, one or two for each chapter, then reviewed the exercises and checked the accuracy of answers to the odd-numbered exercises, which appear in the back of the book. The next step was for two Purdue mathematics graduate students to provide a complete set of solutions that were used to check the typed manuscript of the *Solutions Manuals*. Calculations were verified by computer for problems for which that was appropriate. I thank each of the solutions checkers:

Richard Bagby, New Mexico State University; Denis Bertholf, Oklahoma State University; George Boros, University of New Orleans and Tulane University; Stephen W. Brady, Wichita State University; Stephen D. Casey, American University; Christopher Cotter, University of Northern Colorado; Audrey Douthit, Pennsylvania State University; Stuart Goldenberg, California State Polytechnic University; Donald Goral, Northern Virginia Community College; William R. Hintzman, San Diego State University; Arthur Hobbs, Texas A&M University; Hidefumi Katsuura, San Jose State University; Walter F. Martens, University of Alabama at Birmingham; Paul E. McDougle, University of Miami; Daniel Moak, Michigan Technological University; Donald E. Myers, University of Arizona; Barbara Osofsky, Rutgers University; Emily Mann Peck, University of Illinois, Urbana; Thomas Rishel, Cornell University; John Scheick, Ohio State University; Charles Waters, Mankato State University; Mary Wilson, Austin Community College; James Wooland, Florida State University; Graham Zelmer, Carleton University, Ottawa

Paul Hurst and Sandya Liyanarachchi, mathematics graduate students at Purdue, provided a complete set of exercise solutions.

ACKNOWLEDGMENTS

Many people have made significant and valuable contributions to this book. I thank them for their help in this project.

I want to especially thank users of the first edition and the following professors, who reviewed either the first edition or the manuscript of the second edition and suggested improvements:

Robert D. Adams, University of Kansas; Paul J. Allen, University of Alabama; Thomas E. Armstrong, University of Maryland; Holly J. Ashton, Brevard Community College; Nazanin Azarnia, Miami University, Hamilton Campus; Sitadri N. Bagchi, University of Nevada at Reno; Maurino Bautista, Rochester Institute of Technology; John G. Bergman, University of Delaware; Armand Berliner, New Jersey Institute of Technology; Andreas Blass, University of Michigan; George Boros, University of New Orleans and Tulane University; Therlene Boyett, Valencia Community College; Stephen W. Brady, Wichita State University; Stephen D. Casey, American University; Daniel S. Chess, CUNY, Hunter College; Kevin F. Clancey, University of Georgia; Gabriel B. Costa, Seton Hall University; Sami Deek, Richland College; Thomas Dence, Ashland University; John F. Detlef, University of Minnesota–Morris; William D. Emerson, Metropolitan State College of Denver; William P. Francis, Michigan Technological University; Jim Franklin, Valencia Community College; Karen Graham, University of New Hampshire; John Gregory, Southern Illinois University; C. W. Groetsch, University of Cincinnati; Frederick Hoffman, Florida Atlantic University; Calvin V. Holmes, San Diego State University; Dale W. Hughes, Johnson County Community College; Clark "Bubba" Jeffries,

Clemson University; Thomas Kelley, Metropolitan State College of Denver; Daniel Kemp, South Dakota State University; Otis Kenny, Boise State University; Arthur Lieberman, Cleveland State University; Benny P. Lo, Ohlone College; Joyce Longman, Villanova University; Jorge M. Lopez, University of Puerto Rico–Ro Piedras; M. A. Malik, Concordia University (Quebec); Martin J. Marsden, University of Pittsburgh; Joan McCarter, Arizona State University; Thomas A. Metzger, University of Pittsburgh; Aaron Meyerowitz, Florida Atlantic University; Betty Louise Miller, West Virginia University; Art Moore, Orange Coast College; Weston I. Nathanson, California State University–Northridge; James Osterburg, University of Cincinnati; Robert Piziak, Baylor University; Srinivasa G. Rajalakshmi, Southwest Missouri State University; Andrew Rich, Kansas State University; Richard C. Roberts, University of Maryland, Baltimore County; Charles Roth, McGill University, Montreal; David Royster, University of North Carolina–Charlotte; Laurence Small, Los Angeles Pierce College; Marvin Solberg, U.S. Naval Academy; David C. Sutherland, Middle Tennessee State University; Willie E. Taylor, Texas Southern University; Charles Traina, St. John's University; Dale E. Walston, University of Texas at Austin; Charles W. Waters, Mankato State University; Tony Werckman, University of Colorado; Graham K. Zelmer, Carleton University–Ottawa

I thank the following individuals and groups for their contributions of novel exercises that added to the interest and diversity of the book: David Wells, who provided the *Extended Applications* at chapter endings; Karl Petersen, for the *Calculus Explorer & Tutor* exercises that are in the text; and Sandra Keith, who wrote some of the writing exercises.

I am grateful to all of the people who contributed to the production of the book: Carl Cowen, my colleague and general consultant at Purdue; Judy Stout, who typed the first drafts of the text and *Solutions Manual* in TEX; Shirley Cicero, who helped coordinate the reviewing process and skillfully handled many technical details; Sally Lifland, who coordinated production for ATLIS Graphics; Ann Buesing, who coordinated production from HarperCollins. HarperCollins Mathematics Editor George Duda and Developmental Editor Louise Howe contributed many useful ideas; it was a pleasure to work with them.

Finally, I would like to thank my wife, Ann, for her continuing support, patience, and devotion.

To the Student

This book is written to help you learn calculus. Based on thousands of student evaluations of the first edition, I can assure you that you will find the book readable and useful for learning to solve calculus problems.

SPECIAL LEARNING FEATURES

The book contains many special features to help you learn. Chapter 1 reviews precisely the algebra, analytic geometry, and trigonometry that you will need to solve calculus problems. There are unique sections that help you translate word problems into equations and use analytic geometry effectively in calculus problems.

The text contains many figures to help you to visualize the mathematics. The examples are worked out in detail to illustrate how concepts, general strategies, and basic algebra/trigonometry skills relate to problem solving. Each section begins with a list of *Connections*, which refer to prior material that relates to the section. Each chapter ends with a review of concepts and important formulas. I urge you to take advantage of these special features.

A BRIEF PREVIEW OF CALCULUS

Calculus can be described as a tool for analyzing functions. This analysis gives us information about the physical quantities functions represent and allows us to solve a variety of practical problems that require techniques beyond those of algebra, trigonometry, and analytic geometry. Some of these problems are listed below:

- If a ball always bounces back to three-fourths the height of the previous bounce, it will (theoretically) bounce an infinite number of times, but will it (theoretically) continue to bounce forever? (Most students who have not yet studied calculus answer this question incorrectly.) The solution involves infinite series. (Section 9.2)
- What can you say about values of $(\sin x)/x$ and $\sin(1/x)$ for small, nonzero values of x? This question involves the fundamental concept of limits. (Section 2.3)
- How does your calculator evaluate $\sqrt{2}$? This question involves approximation techniques and error analysis. (Section 3.6)
- What is the maximum speed of a piston attached by a rod to a flywheel as the flywheel rotates at a constant rate? Finding extreme values is an important application of calculus. (Section 4.7)
- Is your speed along a straight road less than, equal to, or greater than the rate of change of the distance, as measured by a radar gun, between your car and a police officer who is parked some distance from the road? The study of rates of change is especially well suited for calculus. (Section 4.6)
- How can you determine the balance point of an irregular-shaped plate? Calculus can be used to study many physical quantities such as area, volume, mass, and centers of mass. (Section 6.7)

From Concepts to Applications

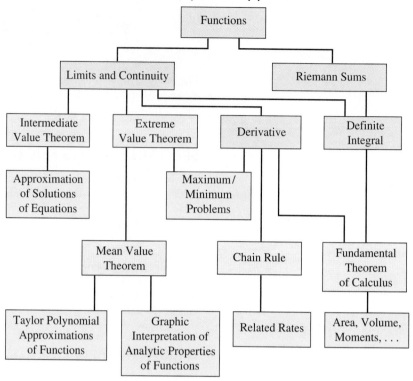

You will see many applications of calculus in this text, but you should realize that our goal includes more than the examples and exercises given here. When you understand the concepts of calculus, you will be able to apply this tool to problems you encounter as you continue your professional advancement. (See the chart above for an indication of how some basic concepts lead to applications.)

You can learn calculus by following the same steps you might take in learning to use any new tool. First, you see in general terms what the tool is and what it can do, perhaps by reading the instruction manual and learning about the important parts of the tool. In this step, you study *concepts*. Second, you practice the basic function of the tool until you feel confident with it. You learn *techniques*. Finally, you use the tool to create something useful. You use the tool for *applications*. This text explains and illustrates the concepts, techniques, and applications of calculus. I hope you will become comfortable and skillful with this powerful tool.

Richard A. Hunt

1 Some Preliminary Topics

This chapter contains a review of topics of algebra, analytic geometry, and trigonometry that are needed for the study of calculus.

Manipulation of real numbers and algebraic expressions is fundamental to the solution of calculus problems. Our attention is restricted to algebra problems that will appear later as parts of calculus problems. Graphs of lines, circles, parabolas, and other standard curves are reviewed so they will be available to graphically illustrate concepts of calculus. The properties of trigonometric functions that are reviewed here are used throughout the text.

The chapter begins with a pretest that is intended to indicate which of the sections in Chapter 1 require more careful study. Each section contains relevant definitions, explanations, additional examples, and exercises. The chapter concludes with a summary and review exercises.

PRETEST

This pretest consists of some of the basic examples from Chapter 1. Answers and complete solutions can be found in the indicated sections where the examples appear.

The problems are not intended to be challenging, but to indicate minimal skills required to begin Chapter 2. Any difficulty with these exercises indicates a serious need for more careful study of the appropriate sections of Chapter 1. More challenging exercises are found at the end of each section and in the review exercises.

1. Solve the inequality $|2x - 3| < 2$. (See §1.1, Example 5.)

2. Express $\dfrac{(x^{-2})^{-3}}{x \cdot x^3}$ in the form x^r. (See §1.1, Example 6f.)

3. Evaluate $(-8)^{1/3}$. (See §1.1, Example 7c.)

4. Simplify $\dfrac{\dfrac{1}{x+1} - \dfrac{1}{3}}{x - 2}$. (See §1.2, Example 4.)

5. Simplify $(x^3)\left(\frac{1}{3}\right)(1 - x^2)^{-2/3}(-2x) + (1 - x^2)^{1/3}(3x^2)$. (See §1.2, Example 6.)

6. Find all values of x for which $(x^2 + 1)(2x - 1) - (x^2 - x)(2x)$ is zero. (See §1.2, Example 8.)

7. Find all intervals on which $\dfrac{x(x - 2)}{x + 1}$ is (a) positive and (b) negative. (See §1.2, Example 9.)

8. Find points (x, y) on the graph of $2x + 3y = 6$ that (a) are intercepts, (b) have $x = -3$, (c) have $y = -2$. (See §1.3, Example 2.)

9. Find the length, slope, and midpoint of the line segment between the point $(4, 3)$ and a variable point $(x, (6 + 2x)/3)$ on the graph of $-2x + 3y = 6$. Simplify. (See §1.3, Example 5.)

10. Graph $2x + 3y = 6$. (See §1.4, Example 1.)

11. Graph $x + 1 = 0$. (See §1.4, Example 4.)

12. Find an equation of the line that has slope $m = 6$ and contains the point on the graph of $y = x^2$ with $x = 3$. (See §1.4, Example 5.)

13. Find an equation of the line through the point $(-2, 3)$ and (a) parallel to the line $4x - 3y + 7 = 0$, (b) perpendicular to the line $4x - 3y + 7 = 0$. (See §1.4, Example 9.)

14. Sketch the graph of the equation $y^2 + 8y = 6x - x^2$. (See §1.5, Example 1.)

15. Sketch the graph of the equation $y = 2x - x^2$. (See §1.5, Example 2.)

16. Sketch the graph of the equation $y = \sqrt{1 - x}$. (See §1.5, Example 4.)

17. Sketch the graph of $y = \dfrac{1}{x + 1}$. (See §1.5, Example 6.)

18. Find an equation that relates the variables b and θ in the figure. (See §1.6, Example 1a.)

19. Express $\cos\theta$ in terms of x, if $\sin\theta = x/3$ and $-\pi/2 \le \theta \le \pi/2$. (See §1.6, Example 2c.)

20. Find all θ, $0 \le \theta \le 2\pi$, that satisfy (a) $\sin\theta = 0.79$ and (b) $\cos\theta = -0.32$. (See §1.6, Example 3a,c.)

21. Find an equation that relates the variables b and β in the figure. (See §1.6, Example 4a.)

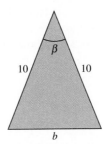

1.1   PROPERTIES OF REAL NUMBERS

In this section we will review solving inequalities, properties of absolute value, and rules of exponents. The development presupposes a working knowledge of the arithmetic properties of the real number system. We also assume familiarity with the association of real numbers with points on a number line, an association that includes the ideas of length, direction, and order. See Figure 1.1.1. We will use the terms "real number" and "point on a line" interchangeably.

Connections

Geometric interpretations of real numbers, inequalities, and absolute value.

Linear inequalities.

Rules of exponents.

A number line

FIGURE 1.1.1 Points on a number line are associated with real numbers. Each real number is either positive, negative, or zero. Real numbers may be either integers, rational numbers, or irrational numbers.

Inequalities

To **solve** an inequality for a variable means to find an equivalent inequality in which the variable is isolated on one side of the inequality; numbers or other known quantities are on the opposite side of the inequality. Recall that the same quantity can be either added to or subtracted from both sides of an inequality without changing the sense (or direction) of the inequality. We can also multiply or divide both sides of an inequality by a positive number without changing the sense. However, multiplication or division by a negative number changes the sense (reverses the direction) of an inequality.

EXAMPLE 1

Solve the inequality $2x - 3 \leq 4 + 5x$.

SOLUTION   We gather terms that involve x on the left-hand side of the inequality and constant terms on the right. To do this we add 3 to and subtract $5x$ from both sides of the original inequality and then combine like terms.

$$2x - 3 \leq 4 + 5x,$$
$$2x - 5x \leq 4 + 3,$$
$$-3x \leq 7.$$

Dividing by -3 we obtain the solution

$$x \geq -\frac{7}{3}.$$

Note that the sense of the inequality was changed when we divided by the negative number -3. ■

If $a < b$, the **double inequality** $a < x < b$ means x is between a and b. The inequalities $a < x$ and $x < b$ are *both* true. See Figure 1.1.2.

We can solve certain types of double inequalities by isolating the variables in the middle position. We follow the usual rules of inequalities, except we must remember to add, subtract, multiply, and divide all three terms by the same number. The senses of both inequalities are changed when we multiply or divide by a negative number.

FIGURE 1.1.2 $a < x < b$ means $a < x$ and $x < b$ are *both* true.

EXAMPLE 2

Solve the double inequality $-2 < 4 - 3x < 13$.

SOLUTION   We first subtract 4 from each of the three terms of the original double inequality to obtain

$$-2 - 4 < -3x < 13 - 4,$$
$$-6 < -3x < 9.$$

We now divide each term by -3, being careful to change the sense of both inequalities, to obtain

$$2 > x > -3.$$

We can rewrite this as $-3 < x < 2$. The solution is sketched in Figure 1.1.3. ■

FIGURE 1.1.3 $-2 < 4 - 3x < 13$ is equivalent to $-3 < x < 2$.

Absolute Value

> The **absolute value** of a real number x is
>
> $$|x| = \begin{cases} x, & x \geq 0, \\ -x, & x < 0. \end{cases}$$

The absolute value of a real number is always nonnegative, so $|x| \geq 0$ for any real number x.

Note that absolute value is defined by two expressions, one the negative of the other. In order to determine which expression to use for the absolute value of a number, we must check whether the number is positive or negative.

Example 3

Evaluating the absolute values, we have

(a) $5 > 0$, so $|5| = 5$,

(b) $-3 < 0$, so $|-3| = -(-3) = 3$,

(c) $1 - \sqrt{2} < 0$, so $|1 - \sqrt{2}| = -(1 - \sqrt{2}) = \sqrt{2} - 1$. ◼

For some of our work, it will be useful to express the absolute value of a variable expression in a form that does not involve absolute value notation. We can do this by first determining values of the variable for which the expression is nonnegative and values of the variable for which the expression is negative. We then have

$$|\text{Expression}| = \begin{cases} \text{Expression}, & \text{intervals where expression is nonnegative,} \\ -(\text{Expression}), & \text{intervals where expression is negative.} \end{cases}$$

Example 4

We can write $|2x - 6|$ without absolute values by noting that $2x - 6 \geq 0$ for $x \geq 3$ and $2x - 6 < 0$ for $x < 3$, so

$$|2x - 6| = \begin{cases} 2x - 6, & x \geq 3, \\ -(2x - 6), & x < 3. \end{cases}$$ ◼

We have the following properties of absolute value:

$$|-x| = |x|, \quad |xy| = |x| \cdot |y|,$$

$$\left|\frac{x}{y}\right| = \frac{|x|}{|y|},$$

$$|x + y| \leq |x| + |y|. \qquad \textbf{(The Triangle Inequality)}$$

The Triangle Inequality can be verified by considering all cases where x and y have like or unlike signs. Equality holds whenever x and y have like signs; inequality holds for unlike signs.

The **distance** between two points x_0 and x_1 on a number line is $|x_1 - x_0|$. See Figure 1.1.4. *Distance is nonnegative* and

$$|x_1 - x_0| = |x_0 - x_1|.$$

The absolute value of x can be interpreted as the distance of x from the origin. It is then easy to see that

$$|x| < d \qquad \text{if and only if } -d < x < d. \qquad (1)$$

See Figure 1.1.5.

We will need to solve inequalities of the form $|u| < d$, where u is an expression that involves the variable x. To do this, we use Statement 1 to write $|u| < d$ as the equivalent double inequality $-d < u < d$, and then we solve the double inequality as before.

FIGURE 1.1.4 The distance between points x_0 and x_1 is $|x_1 - x_0| = |x_0 - x_1|$.

FIGURE 1.1.5 $|x| < d$ means $-d < x < d$.

EXAMPLE 5

Solve the inequality $|2x - 3| < 2$.

SOLUTION   Using Statement 1, we rewrite $|2x - 3| < 2$ as the double inequality

$$-2 < 2x - 3 < 2.$$

Then

$$-2 + 3 < 2x < 2 + 3,$$
$$1 < 2x < 5,$$
$$\frac{1}{2} < x < \frac{5}{2}.$$

FIGURE 1.1.6 $|2x - 3| < 2$ is equivalent to $\frac{1}{2} < x < \frac{5}{2}$.

Note that the senses of the inequalities were not changed when we divided by the positive number 2. The solution is sketched in Figure 1.1.6. ■

The inequality opposite to that in Statement 1 describes points x that are a distance of more than d from the origin.

$$\boxed{|x| > d \qquad \text{if and only if either } x > d \text{ or } x < -d.}$$

If d is nonnegative, a number x cannot satisfy both $x > d$ and $x < -d$; $|x| > d$ means x is in either the interval $x > d$ or the interval $x < -d$, but not both. See Figure 1.1.7.

FIGURE 1.1.7 $|x| > d$ means *either* $x > d$ or $x < -d$.

Exponential Expressions

We begin by setting the notation.

- If n is a positive integer, the nth power of x is $x^n = \underbrace{x \cdots x}_{n \text{ factors}}$. A real number y is an nth root of x if $y^n = x$.
- If x is nonnegative and n is a positive integer, then x has exactly one non-negative nth root, which we denote by either $x^{1/n}$ or $\sqrt[n]{x}$. (If x is nonnegative and n is an *odd* integer, then the nonnegative nth root is the only real number root of x. If x is positive and n is an *even* integer, then x has two real number nth roots, one positive and one negative. The negative nth root of x is then expressed as either $-x^{1/n}$ or $-\sqrt[n]{x}$.)
- If x is negative and n is an *odd* positive integer, then x has only one real number nth root, denoted either $x^{1/n}$ or $\sqrt[n]{x}$. (A real number odd root of a negative number is always negative.)
- Negative numbers have no real number even roots. The expressions $x^{1/n}$ and $\sqrt[n]{x}$ are undefined if x is negative and n is even.
- If m and n are positive integers and x is nonnegative, then $(x^m)^{1/n} = (x^{1/n})^m$ and their common value is denoted $x^{m/n}$, so $x^{m/n} = (x^m)^{1/n} = (x^{1/n})^m$. If x is negative and n is *odd*, then $x^{m/n} = (x^m)^{1/n} = (x^{1/n})^m$. The expression $x^{m/n}$ is undefined if x is negative and n is even.
- $x^0 = 1$ for $x \neq 0$. 0^0 is undefined.
- Expressions with negative exponents are defined by $x^{-r} = 1/x^r$.

Since $\sqrt{x^2}$ is the *nonnegative* square root of the nonnegative number x^2, we have

$$\sqrt{x^2} = |x|.$$

For example, $\sqrt{(2)^2} = \sqrt{4} = 2 = |2|$ and $\sqrt{(-3)^2} = \sqrt{9} = 3 = |-3|$.

A working knowledge of the following rules of exponents is required:

$$x^m \cdot x^n = x^{m+n}, \qquad (x^m)^n = x^{mn}, \qquad (xy)^m = x^m \cdot y^m,$$
$$\left(\frac{x}{y}\right)^m = \frac{x^m}{y^m}, \qquad x^{-m} = \frac{1}{x^m}, \qquad \frac{x^m}{x^n} = x^{m-n}.$$

When dealing with radical expressions it is generally a good idea to express the radicals in exponential notation and use the rules of exponents to simplify the expressions.

EXAMPLE 6

Writing the exponential expressions in the form x^r, we have

(a) $1/x = x^{-1}$,   (b) $x^{1/3} \cdot x^{1/2} = x^{1/3+1/2} = x^{2/6+3/6} = x^{5/6}$,

(c) $x\sqrt{x} = x^1 x^{1/2} = x^{1+1/2} = x^{3/2}$,   (d) $(x^3)^2 = x^{3(2)} = x^6$,

(e) $\dfrac{x}{\sqrt{x}} = \dfrac{x}{x^{1/2}} = x^{1-1/2} = x^{1/2}$,

(f) $\dfrac{(x^{-2})^{-3}}{x \cdot x^3} = \dfrac{x^{-2(-3)}}{x^{1+3}} = \dfrac{x^6}{x^4} = x^{6-4} = x^2$. ∎

We can use a calculator to find approximate values of exponential expressions. We will use the symbol $a \approx b$ to indicate that a and b are **approximately equal**.

You do need to be careful if you try to use a calculator to evaluate a root of a negative number. For example, $(-8)^{1/3} = -2$, but $(-8)^{0.33333333}$ is undefined since it involves an even root of a negative number. To use a calculator to find a root of a negative number you should first decide if the expression is defined. If it is, you should determine its sign and then use your calculator to evaluate the root of the corresponding positive number.

EXAMPLE 7

Using a calculator, we have

(a) $10^{1/2} = \sqrt{10} \approx 3.1622777$,   (b) $3^{-6} \approx 0.00137174$,

(c) $(-8)^{1/3} = -(8^{1/3}) \approx -(8^{0.33333333}) \approx -1.9999999 \approx -2$,

(d) $(-7)^{1/4}$ is undefined. ∎

Scientific Notation

Numbers are said to be expressed in **scientific notation** when they are written as the product of a number between one and ten multiplied by an integer power of ten.

EXAMPLE 8

Using scientific notation, we have

(a) $1234 = 1.234 \times 10^3$,

(b) $0.0056 = 5.6 \times 10^{-3}$. ∎

You should know how to use scientific notation on your calculator.

Interval Notation

We conclude this section by establishing a notation for intervals. This is done in Table 1.1.1. Parentheses indicate that the corresponding endpoints are not included, while square brackets indicate that the corresponding endpoints are included. The *symbols* ∞ (**infinity**) and $-\infty$ do not represent real numbers.

Intervals for which both endpoints are real numbers are called **bounded** (or **finite**) **intervals**. Intervals that are not bounded are called **unbounded** (or **infinite**) **intervals**.

Intervals of the form (a, b) are called **open intervals**. Intervals of the form $[a, b]$ are called **closed intervals**.

TABLE 1.1.1

Interval notation	Inequality notation	Graph
x in: (a, b)	$a < x < b$	
$[a, b]$	$a \leq x \leq b$	
$[a, b)$	$a \leq x < b$	
$(a, b]$	$a < x \leq b$	
$(-\infty, b)$	$-\infty < x < b$ or simply $x < b$	
$(-\infty, b]$	$-\infty < x \leq b$ or simply $x \leq b$	
(a, ∞)	$a < x < \infty$ or simply $x > a$	
$[a, \infty)$	$a \leq x < \infty$ or simply $x \geq a$	
$(-\infty, \infty)$	$-\infty < x < \infty$ or "all real x"	

EXERCISES 1.1

Solve the inequalities in Exercises 1–18.

1. $4x - 3 < 2x + 5$
2. $3x + 2 < 10 - x$
3. $x - 1 \geq 3x + 4$
4. $2x + 3 \leq 5x - 2$
5. $1 < 2x - 1 < 3$
6. $-1 < 3x + 5 < 11$
7. $-4 \leq 2 - 3x \leq -1$
8. $-5 \leq 3 - 2x \leq -1$
9. $|2x - 3| < 1$
10. $|2x + 1| < 3$
11. $|3x - 4| < 2$
12. $|3x + 1| < 1$
13. $|3 - 2x| \leq 1$
14. $|1 - 3x| \leq 2$
15. $|x - 3| > 2$
16. $|3 - 2x| > 5$
17. $|2x - 1| \geq 5$
18. $|5x - 3| \geq 7$

Evaluate the expressions in Exercises 19–26.

19. $|3 - 4|, \quad |3| + |-4|$
20. $|5 - 2|, \quad |5| + |-2|$
21. $|-3 - 2|, \quad |-3| + |-2|$
22. $|5 + 2|, \quad |5| + |2|$
23. $|1 - \sqrt{2}| + |2 - \sqrt{2}|$
24. $|\sqrt{3} - 2| + |\sqrt{3} - 1|$
25. $\sqrt{(-3)^2}$
26. $\sqrt{(-2)^2}$

Write the expressions in Exercises 27–36 in the form x^r.

27. $1/x^2$
28. $1/x^{-1}$
29. $x^2 \cdot x^3$
30. $x^{1/3} \cdot x^{1/4}$

31. $(x^{-3})^2$ **32.** $(\sqrt{x})^3$

33. $1/\sqrt{x}$ **34.** $x^2/\sqrt{x}$

35. $\dfrac{(x^{-1})^2}{x \cdot x^{-4}}$ **36.** $\dfrac{x^2 \cdot x^3}{(x^3)^2}$

Write the expressions in Exercises 37–40 in the form $a^r b^s$.

37. $\dfrac{\sqrt{ab}}{b}$ **38.** $\dfrac{\sqrt[3]{ab}}{b}$

39. $\dfrac{a(ab)^{-2}}{b^{-3}}$ **40.** $\dfrac{a^2(ab)^{-1}}{b^{-2}}$

Evaluate the expressions in Exercises 41–54.

41. $9^{3/2}$ **42.** $(-27)^{2/3}$

43. $16^{1/4}$ **44.** $(-32)^{1/5}$

45. $(-8)^{2/3}$ **46.** $(1/8)^{4/3}$

47. $10^{1/2} \cdot 125^{-1/6} \cdot 2^{3/2}$ **48.** $27^{1/6} \cdot 12^{1/2} \cdot 8^{1/3}$

49. $27^{4/3}/9$ **50.** $9^{2/3}/3^{1/3}$

51. $8^{-2/3} - 9^{-1/2}$ **52.** $5^{-1} - 3^{-2}$

53. $\dfrac{5^{-2} - 3^{-2}}{5^{-1} - 3^{-1}}$ **54.** $\dfrac{2^{-1} - 3^{-1}}{2^{-2} + 3^{-2}}$

Use a calculator to find approximate values of those expressions in Exercises 55–64 that represent real numbers.

55. $(4.7)^{2/3}$ **56.** $(2.4)^{1/4}$

57. $3^{1.7}$ **58.** $2^{1.6}$

59. 2^{-7} **60.** 3^{-5}

61. $(-7)^{2/3}$ **62.** $(-1.4)^{-2/3}$

63. $(-5)^{3/2}$ **64.** $(-0.03)^{3/4}$

Express the numbers in Exercises 65–68 in scientific notation.

65. $34,027$ **66.** $3,209,527$

67. 0.965 **68.** 0.00000023

Express the numbers in Exercises 69–72 as decimal numbers.

69. 7.32×10^4 **70.** 9.45×10^6

71. 7.361×10^{-4} **72.** 2.001×10^{-5}

Complete the tables in Exercises 73–74.

73.

Interval notation	Inequality notation	Graph
(0, 1)		
[1, 3]		
	$x > 0$	
	$x \le 1$	

74.

Interval notation	Inequality notation	Graph
(−1, 1)		
[0, 2]		
	$x \le 1$	
	$x > -1$	

75. Express the inequality $-1 < x < 3$ in the form $|x - c| < d$.

76. Express the inequality $2 < x < 5$ in the form $|x - c| < d$.

77. Show that $a \le x \le b$ implies $|x| \le \max(|a|, |b|)$.

78. (a) Use the triangle inequality to show that $|a - b| \ge |a| - |b|$. (*Hint:* Consider $(a - b) + b$.)

   (b) Show that $|a - b| \ge |b| - |a|$.

   (c) Conclude that $|a - b| \ge ||a| - |b||$.

79. The **Coulomb force** F in newtons acting on the electron of a hydrogen atom as it moves in a circular orbit with radius R meters about the proton with speed v meters per second is given by

$$F = mv^2/R,$$

where $m = 9.1 \times 10^{-31}$ kilogram is the mass of the electron. What is the speed if the Coulomb force is 8.2×10^{-3} newton and the radius is 5.3×10^{-11} meter?

80. An electron of mass m grams is accelerated with E ergs of energy to a velocity of v centimeters per second in an **electrical diffraction experiment**, where

$$E = mv^2/2.$$

Find the velocity if the electron has mass 9.109×10^{-28} gram and the energy is 6.408×10^{-8} erg.

81. If the **minimum stopping distance** of an automobile traveling at a speed of v_1 is d_1 and its minimum stopping distance at speed v_2 is d_2, then

$$\frac{d_2}{d_1} = \left(\frac{v_2}{v_1}\right)^2.$$

If a certain automobile can be stopped in a minimum of 120 ft when traveling at 30 mph, what is the shortest distance in which it can be stopped when it is traveling 70 mph?

82. The initial pressure P_i, the initial volume V_i, the terminal pressure P_t, and the terminal volume V_t of a gas that has been **compressed adiabatically** (with no heat exchange) are related by the equation

$$P_i V_i^{1.4} = P_t V_t^{1.4}.$$

What is the pressure of the gas after it has been compressed adiabatically from a volume of $200 \, \text{cm}^3$ at a pressure of 1 atmosphere to a volume of $20 \, \text{cm}^3$?

1.2 ALGEBRA FOR CALCULUS

The algebra problems in this section are parts of actual calculus problems. The algebra we do here is necessary to complete the solutions of the calculus problems. The better we can handle the algebra, the more time we will have to concentrate on the calculus aspects of the problems.

Connections

Multiplying polynomials.

Using fractional and negative exponents, 1.1.

Using the Distributed Law to factor.

Factoring simple polynomials.

The Quadratic Formula.

Simplifying Expressions

Calculus requires the simplification of many types of algebraic expressions. The simplest of these involve carrying out multiplications, removing parentheses, and combining like terms.

EXAMPLE 1

Simplifying, we have

$$
\begin{aligned}
(x^2 + 1)&(2x - 1) - (x^2 - x + 1)(2x) \\
&= (2x^3 - x^2 + 2x - 1) - (2x^3 - 2x^2 + 2x) \\
&= 2x^3 - x^2 + 2x - 1 - 2x^3 + 2x^2 - 2x \\
&= x^2 - 1. \quad \blacksquare
\end{aligned}
$$

It is worthwhile to look for factors that are common to all terms. The factored form of the simplified expression is the desired form for most applications.

EXAMPLE 2

Simplifying, we have

$$
\begin{aligned}
(2x - 3)^2(4) &+ (4x - 1)(2)(2x - 3)(2) \\
&= 4(2x - 3)[(2x - 3) + (4x - 1)] \\
&= 4(2x - 3)(6x - 4) \\
&= 4(2x - 3)(2)(3x - 2) \\
&= 8(2x - 3)(3x - 2). \quad \blacksquare
\end{aligned}
$$

Since division by zero is undefined, quotients are undefined at points where the denominator is zero.

Canceling a common variable factor from the numerator and denominator of an expression can give a simplified expression that is defined at a point where the original expression is not defined. We should indicate when this is the case.

EXAMPLE 3

Simplifying, we have

$$
\frac{x^2 - x - 6}{x - 3} = \frac{(x - 3)(x + 2)}{x - 3} = x + 2, \qquad x \neq 3.
$$

The statement $x \neq 3$ indicates that the simplified expression $x + 2$ is equal to the original expression only for $x \neq 3$. The original expression is not defined for $x = 3$, but the simplified expression is defined for all x. $\blacksquare$

EXAMPLE 4

Simplifying, we have

$$
\frac{\dfrac{1}{x + 1} - \dfrac{1}{3}}{x - 2} = \frac{\dfrac{1}{x + 1}\left(\dfrac{3}{3}\right) - \dfrac{1}{3}\left(\dfrac{x + 1}{x + 1}\right)}{x - 2}
$$

$$= \frac{\dfrac{3-(x+1)}{3(x+1)}}{x-2} = \frac{3-x-1}{3(x+1)(x-2)}$$

$$= \frac{-x+2}{3(x+1)(x-2)} = \frac{-(x-2)}{3(x+1)(x-2)}$$

$$= -\frac{1}{3(x+1)}, \qquad x \neq 2.$$

Since the simplified expression is defined for $x = 2$ and the original expression is not, we have indicated that the expressions are unequal for $x = 2$. Neither expression is defined for $x = -1$. ∎

Adding Expressions That Contain Radicals

The nature of some calculus formulas gives us expressions involving the sum or difference of powers where the powers differ by one. We must be prepared to handle these expressions. One method is to factor out the term having the smaller exponent. This requires special care when negative exponents are involved.

EXAMPLE 5
Simplify

$$\left(\frac{3}{2}\right)x^{1/2} - \left(\frac{1}{2}\right)x^{-1/2}.$$

SOLUTION   Each term contains a power of x. Since $-1/2 < 1/2$, the smaller exponent of x is $-1/2$, so we factor $x^{-1/2}$ from each term. We also factor $1/2$ from each term.

$$\left(\frac{3}{2}\right)x^{1/2} - \left(\frac{1}{2}\right)x^{-1/2} = \left(\frac{3}{2}\right)x^{1-1/2} - \left(\frac{1}{2}\right)x^{-1/2}$$

$$= \frac{x^{-1/2}}{2}(3x^1 - 1)$$

$$= \frac{3x-1}{2x^{1/2}}. \quad ∎$$

An alternative method of handling a sum or difference of expressions that involve negative exponents is to write the expressions with positive exponents and obtain a common denominator in order to carry out the addition or subtraction.

EXAMPLE 6
Simplifying, we have

$$(x^3)\left(\frac{1}{3}\right)(1-x^2)^{-2/3}(-2x) + (1-x^2)^{1/3}(3x^2)$$

$$= x^2\left[\frac{-2x^2}{3(1-x^2)^{2/3}} + 3(1-x^2)^{1/3}\left(\frac{3(1-x^2)^{2/3}}{3(1-x^2)^{2/3}}\right)\right]$$

$$= x^2\left[\frac{-2x^2 + 9(1-x^2)}{3(1-x^2)^{2/3}}\right] = x^2\left[\frac{9-11x^2}{3(1-x^2)^{2/3}}\right]. \quad ∎$$

EXAMPLE 7

Simplifying, we have

$$(x)\left(\frac{1}{2\sqrt{1-2x}}\right)(-2) + (\sqrt{1-2x})(1)$$

$$= \frac{-x}{\sqrt{1-2x}} + \sqrt{1-2x}\left(\frac{\sqrt{1-2x}}{\sqrt{1-2x}}\right)$$

$$= \frac{-x+(1-2x)}{\sqrt{1-2x}} = \frac{1-3x}{\sqrt{1-2x}}. \quad\blacksquare$$

Zeros of Expressions

Values of the variable for which certain expressions are zero are of particular interest for applications of calculus.

It is easy to determine values of x for which a polynomial is zero if the polynomial is factored into linear terms. For the purpose of illustrating calculus concepts, we will deal often with polynomials that can be factored easily.

If a quadratic polynomial cannot be factored easily, the **Quadratic Formula** can be used to find values for which it is zero. Recall that the Quadratic Formula tells us that

$$ax^2 + bx + c = 0 \text{ has solution } x = \frac{-b \pm \sqrt{b^2 - 4ac}}{2a}.$$

From the Quadratic Formula we see that

$$ax^2 + bx + c \text{ has no real number zeros if } b^2 - 4ac < 0.$$

EXAMPLE 8

Find all values of x for which

$$(x^2 + 1)(2x - 1) - (x^2 - x)(2x)$$

is zero.

SOLUTION   Setting the expression equal to zero and then simplifying, we obtain

$$(x^2 + 1)(2x - 1) - (x^2 - x)(2x) = 0,$$
$$2x^3 - x^2 + 2x - 1 - 2x^3 + 2x^2 = 0,$$
$$x^2 + 2x - 1 = 0.$$

From the Quadratic Formula, we see this is true for

$$x = \frac{-(2) \pm \sqrt{(2)^2 - 4(1)(-1)}}{2(1)}$$

$$= \frac{-2 \pm \sqrt{8}}{2} = \frac{-2 \pm 2\sqrt{2}}{2} = \frac{2(-1 \pm \sqrt{2})}{2}$$

$$= -1 \pm \sqrt{2}.$$

That is, the expression is zero if either $x = -1 + \sqrt{2}$ or $x = -1 - \sqrt{2}$. $\blacksquare$

Sign of Expressions

In calculus it is important to know the sign of certain expressions. We will illustrate a scheme for solving this problem in simple cases. The scheme is based on the following two facts:

Most of the expressions we will be dealing with will be either always positive or always negative on intervals between successive values where they are either zero or undefined.

The product or quotient of an odd number of negative factors is negative and the product or quotient of an even number of negative factors is positive.

We will determine the sign of an expression from a **table of signs.** The idea of a table of signs is to determine the sign of a complicated expression by determining the sign of each of its simpler factors. Note that it is easy to determine the sign of a power of a linear factor in intervals that do not contain its zero.

TABLE 1.2.1

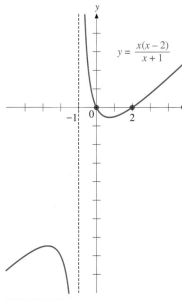

	$-\infty$	-1	0	2	∞
$x + 1$	$-$	0 $+$	$+$	$+$	
x	$-$	$-$	0 $+$	$+$	
$x - 2$	$-$	$-$	$-$	0 $+$	
$\dfrac{x(x-2)}{x+2}$	$-$	$+$	$-$	$+$	

To construct a table of signs

- Write the expression in factored form and list the factors in the left-hand column of the table.
- List the zeros of the factors in order along the top of the table and mark the zeros of each factor in the row of that factor.
- Determine and mark the sign of each factor in each of the intervals determined by the zeros of the factors. Note that each factor can change sign only where that factor is zero.
- Determine the sign of the expression in each interval by counting the number of factors that are negative in the interval.

The zeros and sign of an expression can be determined from a calculator-generated graph of the expression. The zeros are the x-coordinates of the points where the graph intersects the x-axis. The expression is positive where the graph is above the x-axis and the expression is negative where the graph is below the x-axis.

EXAMPLE 9

Find all intervals on which

$$\frac{x(x-2)}{x+1}$$

is **(a)** positive and **(b)** negative.

SOLUTION   The expression has factors x, $x-2$, and $x+1$, which are listed in the left-hand column of Table 1.2.1. The zeros of the factors are 0, 2, and -1. These are listed in order along the top of the table and the zero of each factor is marked at the appropriate spot in the table. The sign of each factor is determined by noting that linear factors are positive on one side of their zero and negative on the other. It is easy to determine which is which by, for example, substitution of one number other than the zero into the expression. The sign of each factor in each interval is marked in the row of that factor. The sign of the expression in each interval is then determined by counting the number of negative signs of factors in the column of that interval.

$$y = \frac{x(x-2)}{x+1}$$

FIGURE 1.2.1   The expression $x(x-2)/(x+1)$ is positive on the intervals $-1 < x < 0$ and $x > 2$; it is negative on the intervals $x < -1$ and $0 < x < 2$.

We conclude that $x(x-2)/(x+1)$ is positive on the intervals $-1 < x < 0$ and $x > 2$; the expression is negative on the intervals $x < -1$ and $0 < x < 2$. This information can also be obtained from the graph given in Figure 1.2.1. ■

EXAMPLE 10

Find all intervals on which

$$\frac{(1-x)^{1/3}}{x^{2/3}}$$

is **(a)** positive and **(b)** negative.

SOLUTION   The factors are $(1-x)^{1/3}$ and $x^{2/3}$. The zeros of the factors are 1 and 0.

Odd roots of any number have the same sign as the number. Hence, $(1-x)^{1/3}$ has the same sign as $1-x$. The expression $x^{2/3} = (x^{1/3})^2$ is positive for $x \neq 0$. The corresponding sign table is given in Table 1.2.2.

We see from the table that the given expression is positive on the intervals $x < 0$ and $0 < x < 1$, and negative on the interval $x > 1$. See Figure 1.2.2.

Note that we cannot say that the expression is positive on the interval $x < 1$ because it is not defined for $x = 0$. That is why the factor $x^{2/3}$ and its zero were included in the table even though $x^{2/3}$ does not contribute any negative signs. ■

The method used in Examples 9 and 10 can be used to compare the values of two given expressions.

EXAMPLE 11

Find all values of x for which $x + \dfrac{1}{x} > 2x$.

SOLUTION   We first rewrite the inequality as an equivalent inequality with zero on one side and simplify the expression on the other side. (We avoid multiplying the inequality by an expression of variable or unknown sign since that could change the sense of the inequality.)

$$x + \frac{1}{x} > 2x,$$

$$-x + \frac{1}{x} > 0,$$

$$\frac{-x^2 + 1}{x} > 0,$$

$$\frac{(1-x)(1+x)}{x} > 0.$$

It follows that the solution of the original inequality consists of those x for which the expression

$$\frac{(1-x)(1+x)}{x}$$

is positive. From Table 1.2.3 we see that this is true on the intervals $x < -1$ and $0 < x < 1$.

TABLE 1.2.2 —————

	$-\infty$	0	1	∞
$x^{2/3}$	$+$	0	$+$	$+$
$(1-x)^{1/3}$	$+$	$+$	0	$-$
$\dfrac{(1-x)^{1/3}}{x^{2/3}}$	$+$	$+$		$-$

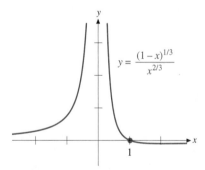

$$y = \frac{(1-x)^{1/3}}{x^{2/3}}$$

FIGURE 1.2.2   The expression $(1-x)^{1/3}/x^{2/3}$ is positive on the intervals $x < 0$ and $0 < x < 1$; it is negative on the interval $x > 1$.

TABLE 1.2.3 —————

	$-\infty$	-1	0	1	∞
$1-x$	$+$	$+$	$+$	0	$-$
$1+x$	$-$	0	$+$	$+$	$+$
x	$-$	$-$	0	$+$	$+$
$\dfrac{(1-x)(1+x)}{x}$	$+$	$-$		$+$	$-$

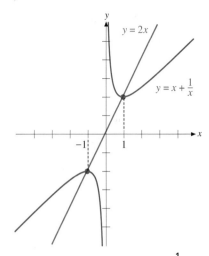

FIGURE 1.2.3 We see that $x + \dfrac{1}{x}$ is greater than $2x$ on the intervals $x < -1$ and $0 < x < 1$.

Note that the graphs of $y = x + \dfrac{1}{x}$ and $y = 2x$ given in Figure 1.2.3 indicate that $x + \dfrac{1}{x} > 2x$ on the intervals $x < -1$ and $0 < x < 1$. ∎

NOTE   *You must be very careful to analyze cases if you multiply an inequality by an expression of variable or unknown sign.*

For example, multiplying the inequality $x + \dfrac{1}{x} > 2x$ by the variable x gives

$$x^2 + 1 > 2x^2 \text{ if } x > 0 \text{ and } x^2 + 1 < 2x^2 \text{ if } x < 0.$$

The solution of $x + \dfrac{1}{x} > 2x$ is then all x that satisfy either

$$x^2 + 1 > 2x^2 \text{ and } x > 0$$

or

$$x^2 + 1 < 2x^2 \text{ and } x < 0.$$

You may verify that correctly solving the above inequalities for x gives the results of Example 11.

EXERCISES 1.2

Simplify the expressions in Exercises 1–16.

1. $(2 - x)(3) + (1 + 3x)(-1)$

2. $(5 - 3x)(2) + (3 + 2x)(-3)$

3. $(2x + 1)^2(5) + (5x - 1)(2)(2x + 1)(2)$

4. $\dfrac{(2x - 1)(2)(3x - 2)(3) - (3x - 2)^2(2)}{(2x - 1)^2}$

5. $\left(-\dfrac{1}{2}x^{-3/2}\right) - \left(\dfrac{3}{2}x^{-5/2}\right)$

6. $\sqrt{x} - \dfrac{1}{\sqrt{x}}$

7. $(x^2)\left(\dfrac{1}{3}\right)(x + 1)^{-2/3}(1) + (x + 1)^{1/3}(2x)$

8. $\dfrac{(\sqrt{x^2 + 1})(1) - (x)\left(\dfrac{1}{2\sqrt{x^2 + 1}}\right)(2x)}{x^2 + 1}$

9. $\dfrac{x^2 - 4}{x - 2}$

10. $\dfrac{x^2 + 3x + 2}{x + 1}$

11. $\dfrac{\dfrac{1}{x^2} - \dfrac{1}{4}}{x - 2}$

12. $\dfrac{\dfrac{1}{x - 3} + \dfrac{1}{6}}{x + 3}$

13. $\dfrac{x^3 + 8}{x + 2}$

14. $\dfrac{x^4 - 16}{x - 2}$

15. $\dfrac{x^2 - x - 2}{x + 1}$

16. $\dfrac{2x^2 - 5x + 3}{x - 1}$

In Exercises 17–28, find all values of x for which the given expressions are zero.

17. $x^2 - 5x - 6$

18. $x^2 - 5x + 6$

19. $3x^2 - 4x + 2$

20. $2x^2 + 3x + 2$

21. $(x^2 - 2x - 2)(-1) + (1 - x)(2x - 2)$

22. $(3x^2 - 2x + 1)(2) + (2x - 3)(6x - 2)$

23. $(2x - 1)(2)(x + 2)(1) + (x + 2)^2(2)$

24. $(x^2 - x)(2) + (2x + 2)(2x - 1)$

25. $x - \dfrac{1}{x}$

26. $1 + \dfrac{3}{2x - 3}$

27. $x^2 + 2x - 2$

28. $x^2 - 3x + 1$

In Exercises 29–34, find all values of x for which the given expressions are (a) undefined and (b) zero.

29. $\dfrac{(x - 1)(2)(3x + 1)(3) - (3x + 1)^2(1)}{(x - 1)^2}$

30. $\dfrac{(x^2 - 1)(1) - (x)(2x)}{(x^2 - 1)^2}$

31. $x - \dfrac{8}{x^2}$

32. $\sqrt{x - 1} - \dfrac{1}{2\sqrt{x - 1}}$

33. $\dfrac{\sqrt{2x + 1}(3) - (3x + 1)\left(\dfrac{1}{2\sqrt{2x + 1}}\right)(2)}{2x + 1}$

34. $\dfrac{\sqrt{3x-2}\,(2) - (2x-1)\left(\dfrac{1}{2\sqrt{3x-2}}\right)(3)}{3x-2}$

In Exercises 35–46, find all intervals on which the given expressions are (a) positive and (b) negative.

35. $(x-1)(x+2)$

36. $x^2 - 3x + 2$

37. $\dfrac{x(x+1)}{x-2}$

38. $\dfrac{2x^2 - 7x + 3}{x-3}$

39. $\dfrac{x+1}{x^2 - x + 1}$

40. $\dfrac{x-2}{x^2 - 2x + 2}$

41. $\dfrac{x}{(x-1)^3}$

42. $\dfrac{x-1}{(x-2)^2}$

43. $\dfrac{(x+1)^{1/2}}{x^{1/3}}$

44. $\dfrac{(x-1)^{1/3}}{x^{3/4}}$

45. $\sqrt{x} - \dfrac{1}{2\sqrt{x}}$

46. $x^{1/4} - x^{-3/4}$

Solve the inequalities in Exercises 47–54.

47. $x^3 < x$

48. $x^3 < x^2$

49. $x^2 < 2 - x$

50. $x > x^2$

51. $\dfrac{1}{x} > \dfrac{1}{x-1}$

52. $\dfrac{2}{x+1} > \dfrac{1}{x}$

53. $\sqrt{x} > x/2$

54. $\sqrt{x} > 2x^2$

Use a calculator graph of the expressions given in Exercises 55–58 to find (a) approximate values of all x for which the given expressions are zero, (b) intervals where the expressions are positive, and (c) intervals where the expressions are negative.

55. $x^3 + 2x + 1$

56. $x^3 + 3x - 5$

57. $x^3 - 4x + 1$

58. $x^3 - x^2 - 2x + 1$

1.3 THE COORDINATE PLANE

Connections

Coordinates and points in a coordinate plane.

Graph of an equation in x and y.

Length, slope, and midpoint formulas for line segments.

We will review the terminology of the coordinate plane and the fundamental concepts of the graph of an equation. We will then introduce the idea of a variable point on a graph, where the coordinates of the point are given in terms of a variable. Formulas for the length, slope, and midpoint of a line segment will be reviewed in terms of variable points, as they are used in calculus.

A Rectangular Coordinate System

A **rectangular**, or **Cartesian**, **coordinate system** for points in a plane is determined by any two number lines that are in the plane and are perpendicular to each other. The number lines are called **coordinate axes**. The **coordinates** of a point in the plane are the unique **ordered pair** of numbers determined by the points of intersection of the coordinate axes with lines through the given point and parallel to the axes. Coordinates are written in the form (a, b). The order of the coordinates indicates which of the coordinates is associated with which axis. The coordinate axes can be used to locate the unique point in the plane that corresponds to a given ordered pair.

It is customary to draw one coordinate axis horizontal, with positive direction to the right, and the other axis vertical, with positive direction upward. The coordinate axes then divide the plane into four **quadrants**, which are numbered as indicated in Figure 1.3.1. Unless there are specific reasons not to, the same scale is chosen for each axis and the intersection of the axes is zero on both scales. The intersection of the axes then has coordinates $(0, 0)$ and is called the **origin**. The first coordinate of a pair (a, b) corresponds to the horizontal axis, and the second coordinate corresponds to the vertical axis.

In Figure 1.3.1, we have located (or plotted) several points and indicated their coordinates. It is important to be able to locate a point from its coordinates and to determine the coordinates of a point from its position. We will not

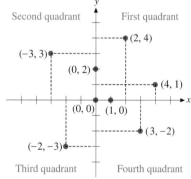

FIGURE 1.3.1 The axes of a rectangular coordinate system divide the plane into four quadrants. The location of a point in the plane is determined by the coordinates of the point.

necessarily distinguish between a point and its coordinates. That is, we may refer to a point P with coordinates (a, b) as either P, (a, b), or $P(a, b)$.

We will most often use x as the variable for the horizontal axis and y as the variable for the vertical axis. The first coordinate is then called the x-coordinate and the second coordinate is called the y-coordinate. The plane is called the xy-plane. Other choices of variables are used when that is appropriate for a particular problem.

Graphs of Equations

The **graph** of an equation that relates two variables x and y is the collection of all points with coordinates (x, y) that satisfy the equation.

EXAMPLE 1

Consider the equation $2x + 3y = 6$. **(a)** Is $(6, -2)$ on the graph? **(b)** Is $(-2, 6)$ on the graph?

SOLUTION

(a) Substitution of 6 for x and -2 for y in the equation gives $2(6) + 3(-2) = 6$, which is true, so $(6, -2)$ is on the graph.

(b) Substitution of the coordinates $(-2, 6)$ yields $2(-2) + 3(6) = 6$, which is not true, so $(-2, 6)$ is not on the graph. See Figure 1.3.2. ■

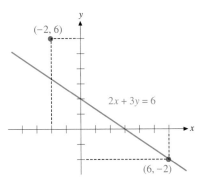

FIGURE 1.3.2 The point $(6, -2)$ is on the graph of $2x + 3y = 6$; $(-2, 6)$ is not.

Points on the graph of an equation can be found by substituting a value of one variable into the equation and then solving the equation for the corresponding value or values of the other variable.

Points of intersection of a graph with the coordinate axes are called **intercepts**. Intercepts are often easy to determine and should be plotted. Points (x, y) on the x-axis have $y = 0$; points on the y-axis have $x = 0$.

EXAMPLE 2

Find points (x, y) on the graph of $2x + 3y = 6$ that **(a)** are intercepts, **(b)** have $x = -3$, **(c)** have $y = -2$.

SOLUTION

(a) Intercepts on the x-axis have $y = 0$. When $y = 0$, $2x + 3(0) = 6$ implies $x = 3$. $(3, 0)$ is the x-intercept. Intercepts on the y-axis have $x = 0$. When $x = 0$, $2(0) + 3y = 6$ implies $y = 2$. $(0, 2)$ is the y-intercept.

(b) Substitution of $x = -3$ gives $2(-3) + 3y = 6$, which implies $y = 4$, so $(-3, 4)$ is a point on the graph with $x = -3$.

(c) Substitution of $y = -2$ gives $2x + 3(-2) = 6$, which implies $x = 6$, so $(6, -2)$ is a point on the graph with $y = -2$.

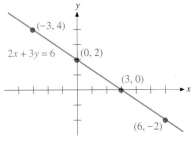

FIGURE 1.3.3 The graph of $2x + 3y = 6$ has intercepts $(3,0)$ and $(0,2)$.

See Figure 1.3.3. ■

Variable Points on a Graph

In many calculus problems we will deal with points that have coordinates expressed in terms of a variable. We will refer to such points as **variable points**. A variable point on the graph of an equation in x and y can be expressed in terms of either the variable x or the variable y. If x is the variable, then the first coordinate of the variable point is x, and the second coordinate is the expression

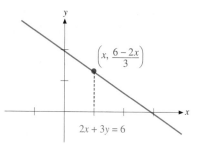

FIGURE 1.3.4a The point $\left(x, \dfrac{6-2x}{3}\right)$ is a variable point on the graph of $2x + 3y = 6$, in terms of x.

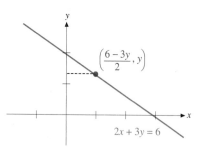

FIGURE 1.3.4b The point $\left(\dfrac{6-3y}{2}, y\right)$ is a variable point on the graph of $2x + 3y = 6$, in terms of y.

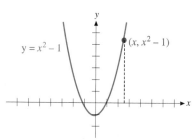

FIGURE 1.3.5a The point $(x, x^2 - 1)$ is a variable point on the graph of $y = x^2 - 1$, in terms of x.

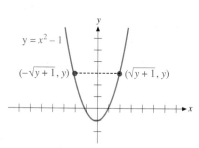

FIGURE 1.3.5b The points $(\sqrt{y+1}, y)$ and $(-\sqrt{y+1}, y)$ are variable points on the graph of $y = x^2 - 1$, in terms of y.

obtained by solving the equation for y in terms of x. If y is the variable, the second coordinate is y, and the first coordinate is the expression obtained by solving the equation for x in terms of y.

EXAMPLE 3

Find the coordinates of a variable point on the graph of $2x + 3y = 6$ in terms of (a) x and (b) y.

SOLUTION

(a) The point (x, y) is on the graph if $2x + 3y = 6$. Solving this equation for y, we obtain

$$y = \frac{6 - 2x}{3}.$$

The point $(x, (6 - 2x)/3)$ is a variable point on the graph, in terms of x. The points $(x, (6 - 2x)/3)$ vary over the graph as x varies. See Figure 1.3.4a.

(b) $2x + 3y = 6$ implies $x = (6 - 3y)/2$. The point $((6 - 3y)/2, y)$ is a variable point on the graph, in terms of y. See Figure 1.3.4b. ∎

EXAMPLE 4

Find the coordinates of a variable point on the graph of $y = x^2 - 1$ in terms of (a) x and (b) y.

SOLUTION

(a) A point (x, y) is on the graph if $y = x^2 - 1$. This means $(x, x^2 - 1)$ is a variable point on the graph, in terms of x. See Figure 1.3.5a.

(b) Solving the equation $y = x^2 - 1$ for x, we obtain $x^2 = y + 1$, so $x = \sqrt{y+1}$ or $x = -\sqrt{y+1}$. If $y > -1$, both points $(\sqrt{y+1}, y)$ and $(-\sqrt{y+1}, y)$ are on the graph. When $y = -1$, $x = 0$, so $(0, -1)$ is on the graph. There is no point on the graph with $y < -1$. The points $(\sqrt{y+1}, y)$ and $(-\sqrt{y+1}, y)$ are variable points on the graph for $y \geq -1$. See Figure 1.3.5b. ∎

Length, Slope, and Midpoint Formulas

> The **line segment** between two points $P_1(x_1, y_1)$ and $P_2(x_2, y_2)$ has
>
> **Length = Distance between points** $= \sqrt{(x_2 - x_1)^2 + (y_2 - y_1)^2}$,
>
> **Slope** $= \dfrac{y_2 - y_1}{x_2 - x_1}$, and
>
> **Midpoint** $= \left(\dfrac{x_1 + x_2}{2}, \dfrac{y_1 + y_2}{2}\right)$.

The distance formula is a consequence of the Pythagorean Theorem. Notice that the points $P_1(x_1, y_1)$, $P_2(x_2, y_2)$, and (x_2, y_1) form a right triangle in Figure 1.3.6. We will use the notation $|P_1 P_2|$ to denote the distance between the points P_1 and P_2. Note that the distance between points x_1 and x_2 on a number line can be expressed as $\sqrt{(x_2 - x_1)^2} = |x_2 - x_1|$.

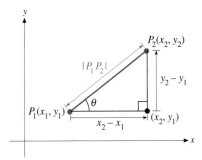

FIGURE 1.3.6 The Pythagorean Theorem implies the line segment between the points $P_1(x_1, y_1)$ and $P_2(x_2, y_2)$ has length $\sqrt{(x_2 - x_1)^2 + (y_2 - y_1)^2}$. The slope is $\dfrac{y_2 - y_1}{x_2 - x_1}$.

We must be careful when using the formula for slope. Choose one point for P_1 and the other for P_2, but do not interchange P_1 and P_2 in mid-formula. It may be helpful to remember

$$\text{Slope} = \frac{\text{Rise}}{\text{Run}}.$$

Also, note that the slope is determined by the angle θ in Figure 1.3.6.

A line segment slopes upward from left to right if the slope is positive. It slopes downward when the slope is negative. Horizontal line segments have slope zero. Slope is undefined for vertical line segments. See Figure 1.3.7.

The coordinates of the midpoint are the averages of the corresponding coordinates of the endpoints. See Figure 1.3.8.

We must be able to use the formulas for length, slope, and midpoint with variable points.

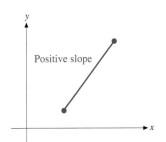

FIGURE 1.3.7a A line segment slopes upward from left to right if the slope is positive.

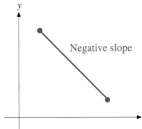

FIGURE 1.3.7b A line segment slopes downward from left to right if the slope is negative.

FIGURE 1.3.7c Horizontal line segments have slope zero.

FIGURE 1.3.7d Slope is undefined for vertical line segments.

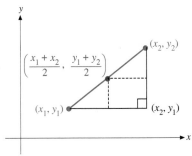

FIGURE 1.3.8 The coordinates of the midpoint of a line segment are the averages of the corresponding coordinates of the endpoints.

EXAMPLE 5

Find the length, slope, and midpoint of the line segment between the point $(4, 3)$ and a variable point on the graph of $-2x + 3y = 6$, in terms of x. Simplify.

SOLUTION Solving the equation $-2x + 3y = 6$ for y, we obtain

$$y = \frac{6 + 2x}{3}.$$

This implies $(x, (6 + 2x)/3)$ is a variable point on the graph, in terms of x. Using $P_1 = (4, 3)$ and $P_2(x, (6 + 2x)/3)$, we have

$$\begin{aligned}
\text{Length} &= \sqrt{(x - 4)^2 + \left(\frac{6 + 2x}{3} - 3\right)^2} \\
&= \sqrt{(x - 4)^2 + \left(\frac{6 + 2x - 9}{3}\right)^2} \\
&= \sqrt{\frac{9(x - 4)^2 + (2x - 3)^2}{9}} \\
&= \frac{\sqrt{9x^2 - 72x + 144 + 4x^2 - 12x + 9}}{3}
\end{aligned}$$

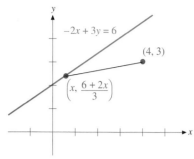

FIGURE 1.3.9 The line segment between (4,3) and a variable point on the graph of $-2x + 3y = 6$ has

length $= \sqrt{\dfrac{13x^2 - 84x + 153}{3}}$,

slope $= \dfrac{2x - 3}{3(x - 4)}$,

and

midpoint $= \left(\dfrac{4 + x}{2}, \dfrac{15 + 2x}{6} \right)$.

$$= \sqrt{\frac{9(x - 4)^2 + (2x - 3)^2}{9}}$$

$$= \frac{\sqrt{9x^2 - 72x + 144 + 4x^2 - 12x + 9}}{3}$$

$$= \frac{\sqrt{13x^2 - 84x + 153}}{3}.$$

The formula for slope gives

$$\text{Slope} = \frac{\dfrac{6 + 2x}{3} - 3}{x - 4} = \frac{\dfrac{6 + 2x - 9}{3}}{x - 4} = \frac{2x - 3}{3(x - 4)}.$$

When $x = 4$, the line segment is vertical and the slope is undefined.

$$\text{Midpoint} = \left(\frac{4 + x}{2}, \frac{3 + \dfrac{6 + 2x}{3}}{2} \right)$$

$$= \left(\frac{4 + x}{2}, \frac{15 + 2x}{6} \right).$$

See Figure 1.3.9.   ■

EXERCISES 1.3

In Exercises 1–6, indicate which of the given points are on the graph of the given equations.

1. $3x - 4y = 7$; $(1, -1)$, $(4, -3)$

2. $2x - 3y = 5$; $(4, -1)$, $(-2, -3)$

3. $y = 2x^2 - 4x + 1$; $(2, 1)$, $(0, 1)$

4. $y = 3x^2 + 6x + 5$; $(-1, 2)$, $(1, 10)$

5. $x^2 + y^2 + 4y - 21 = 0$; $(-3, 2)$, $(2, -3)$

6. $x^2 + y^2 - 6x - 16 = 0$; $(4, -3)$, $(7, -3)$

In Exercises 7–12, plot the intercepts, points with the given x-coordinates, and points with the given y-coordinates on the graphs of the equations.

7. $3x + 2y = 6$; $x = -2$, $y = -3$

8. $y = 2x$; $x = 2$, $y = -4$

9. $y = x^2 - 1$; $x = 2$, $y = -2$

10. $y = x^3$; $x = 1, -1$, $y = 8, -8$

11. $x = y^2$; $x = 4, -4$, $y = 1, -1$

12. $x = y(2 - y)$; $x = -3$, $y = 1$

In Exercises 13–18, find the coordinates of a variable point on the graph of the given equation in terms of (a) x and (b) y.

13. $y = 2x$

14. $3x + 2y = 6$

15. $y = x^3 - 1$

16. $y = 2(x + 1)^2 - 1$

17. $x = y^2 - 1$

18. $x^2 + y^2 = 4$

In Exercises 19–28, find the length, slope, and midpoint of the line segment between the two given points.

19. $(-1, 2)$, $(3, -1)$

20. $(2, 2)$, $(4, 2)$

21. $(1, -2)$, $(1, 3)$

22. $(3, -1)$, $(3, 2)$

23. $(x, 2x)$, $(3, 0)$

24. (x, x^2), $(0, 1)$

25. $(x, 1 - x)$, $(x, x^2 - 1)$

26. $(x, 1)$, $(x, x^2 + 1)$

27. $(0, y)$, $(4 - y^2, y)$

28. $(-\sqrt{4 - y^2}, y)$, $(\sqrt{4 - y^2}, y)$

In Exercises 29–40, find the slope of the line segment between the given point and a variable point on the graph of the given equation, in terms of x.

29. $2x + 3y + 6 = 0$; $(-6, 2)$

30. $3x - 2y + 6 = 0$; $(4, 9)$

31. $y = x^2$; $(0, 0)$

32. $y = x^3$; $(0, 0)$

33. $y = x^3$; $(1, 1)$

34. $y = x^2$; $(2, 4)$

35. $y = x^2 - 3x + 1$; $(2, -1)$

36. $y = x^2 + 2x - 3$; $(1, 0)$

41. Each point on the graph $y = \dfrac{x}{1+x^2}$ is joined by a vertical line segment to a corresponding point on the line $y = x$. Find the length of each vertical segment and find the coordinates of the variable midpoints of these segments in terms of x.

42. Each point on the graph $x = \dfrac{1}{1+|y|}$ is joined by a horizontal line segment to a corresponding point on the line $y = x - 1$. Find the length of each horizontal segment and find the coordinates of the variable midpoints of these segments.

43. Use slopes to determine if the points $A(-1, -1)$, $B(1, 2)$, $C(3, 0)$ and $D(1, -3)$ form a parallelogram.

44. Use slopes to determine if the points $A(-1, -1)$, $B(-2, 2)$, $C(1, -2)$ and $D(2, -5)$ form a parallelogram.

45. Three points in a plane are said to be **collinear** if they are all three on a single line segment. Express the fact that three distinct points are collinear in terms of the distances between the points.

46. Express the fact that three distinct points are collinear in terms of slopes of the line segments between the points.

47. Determine if the points $A(-3, -2)$, $B(2, 0)$, and $C(7, 1)$ are collinear.

48. Determine if the points $A(-3, 2)$, $B(0, -1)$ and $C(4, -4)$ are collinear.

49. Find a formula for the distance between $(1, 0)$ and a variable point on the graph of $y = x^2$. Use a graph of the distance in terms of x to find an approximate value of x for which the distance is smallest.

50. Find a formula for the distance between a variable point on the graph of $y = \dfrac{x^2}{4} - 1$ and the origin. Use a graph of the distance in terms of x to determine if there are any points on the graph of $y = \dfrac{x^2}{4} - 1$ that are closer to the origin than is $(0, -1)$.

51. Find a formula for the distance between a variable point on the graph of $y = x^2 - 1$ and the origin. Use a graph of the distance in terms of x to determine if there are any points on the graph of $y = x^2 - 1$ that are closer to the origin than is $(0, -1)$.

52. Use a calculator to evaluate the slope of the line segment between $(0, 1)$ and a variable point on the graph of $y = e^x$ for $x = 10^{-n}, n = 1, 2, \ldots$. What happens to the slope as n becomes larger?

53. Use a calculator to evaluate the slope of the line segment between $(0, 1)$ and a variable point on the graph of $y = e^{2x}$ for $x = 10^{-n}, n = 1, 2, \ldots$. What happens to the slope as n becomes larger?

54. For several values of a, verify that the slope of the line segment between $(0, 1)$ and a variable point on the graph of $y = e^{ax}$ is approximately a if x is near zero.

1.4 LINES

Connections

Concept of a line.
Graphing lines.
Finding equations of lines.

One of the fundamental applications of calculus is the use of lines to find approximate values of expressions. In this section we will review how to recognize a line from its equation, how to sketch the graph of a line, and how to determine an equation of a line from appropriate data.

Graphing Lines

The **general equation of a line** has form
$$ax + by + c = 0, \quad a \text{ and } b \text{ not both zero.}$$

When sketching the graph of an equation of this form, we should use the fact that the graph is a line. Since any two points on a line determine the line, we can sketch the graph by plotting only two points and then drawing the line through the points. It is particularly easy to locate the intercepts, and these should be plotted.

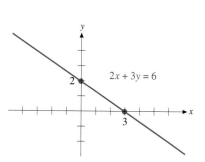

FIGURE 1.4.1   The graph of $2x + 3y = 6$ is a line with intercepts $(3, 0)$ and $(0, 2)$.

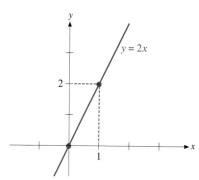

FIGURE 1.4.2   The graph of $y = 2x$ is a line with intercept $(0, 0)$. The point $(0, 0)$ and any other point on the graph, such as $(1, 2)$, determine the graph.

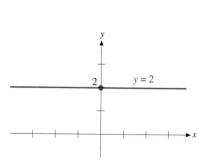

FIGURE 1.4.3   The graph of $y = 2$ is a horizontal line with y-intercept $(0, 2)$.

EXAMPLE 1

Graph $2x + 3y = 6$.

SOLUTION   First, we recognize that this equation can be written in the form $2x + 3y - 6 = 0$, the general form of the equation of a line. This tells us the graph is a line. The intercepts are easily determined from the original form of the line. When $y = 0$, $2x = 6$, so $x = 3$. When $x = 0$, $3y = 6$, so $y = 2$. We have plotted the points $(3, 0)$ and $(0, 2)$, and then sketched the graph in Figure 1.4.1.   ■

EXAMPLE 2

Graph $y = 2x$.

SOLUTION   We recognize that this is the equation of a line, since it can be written in the general form of the equation of a line. We see that $(0, 0)$ is the only intercept of this line. The point $(0, 0)$ and any other point on the graph will determine the graph. For example, we may choose $x = 1$, so $y = 2$. See Figure 1.4.2.   ■

EXAMPLE 3

Graph $y = 2$.

SOLUTION   This is the equation of a line. When $x = 0$, $y = 2$. In fact, the coordinates $(x, 2)$ satisfy the equation for all values of x and the graph is a horizontal line. See Figure 1.4.3.   ■

EXAMPLE 4

Graph $x + 1 = 0$.

SOLUTION   This is the equation of a line. When $y = 0$, $x = -1$. The graph contains all points $(-1, y)$ and is a vertical line. See Figure 1.4.4.   ■

Point-Slope Form

FIGURE 1.4.4   The graph of $x + 1 = 0$ is a vertical line with x-intercept $(-1, 0)$.

It is characteristic of a nonvertical line that the line segments between any two points on the line have a common slope, called the **slope of the line**. Note that the slope of the line is determined by the angle θ in Figure 1.4.5. If (x_1, y_1) is

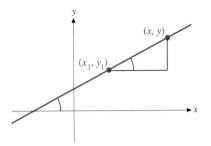

FIGURE 1.4.5 Line segments between any two distinct points on a nonvertical line have a common slope, called the slope of the line. The slope is determined by the angle θ.

a fixed point on a line with slope m, then a point (x, y) is on the line exactly when

$$\frac{y - y_1}{x - x_1} = m \text{ or } y - y_1 = m(x - x_1).$$

This gives us an important characterization of the equation of a nonvertical line.

The **point-slope** form of the equation of the line through the point (x_1, y_1) with slope m is

$$y - y_1 = m(x - x_1).$$

Horizontal lines have slope zero and the point-slope form of the equation of the horizontal line through the point (x_1, y_1) reduces to $y = y_1$.

Slope is undefined for vertical lines, so we cannot use the point-slope form. The equation of the vertical line through the point (x_1, y_1) is $x = x_1$.

Later, we will use calculus to determine the slope of a particular line through a point on a graph. It will then be convenient to use the point-slope form of the equation of a line to *write* an equation of the line.

Strategy for finding an equation of a line

- Find a point on the line.
- Find the slope of the line.
- Use the point-slope form to write an equation of the line.

EXAMPLE 5

Find an equation of the line that has slope $m = 6$ and contains the point on the graph of $y = x^2$ with $x = 3$.

SOLUTION   The point on the graph of $y = x^2$ with $x = 3$ has $y = 3^2$. We want an equation of the line through $(3, 9)$ with slope $m = 6$. The point-slope form of the equation is

$$y - 9 = 6(x - 3).$$

Simplifying, we have

$$y - 9 = 6x - 18,$$
$$y = 6x - 9.$$

The graph is sketched in Figure 1.4.6. ∎

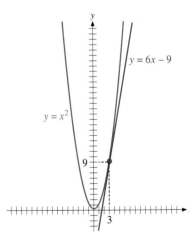

$y = 6x - 9$

$y = x^2$

FIGURE 1.4.6 The point on the graph of $y = x^2$ with $x = 3$ is $(3, 9)$. The line through $(3, 9)$ with slope $m = 6$ has equation $y - 9 = 6(x - 3)$, which simplifies to $y = 6x - 9$.

EXAMPLE 6

Find an equation of the line through the points $(2, -1)$ and $(4, 3)$.

SOLUTION   We have two points on the line; we need the slope. To find the slope, we use the fact that any two points on the line determine the slope. Hence,

$$\text{Slope } = \frac{3 - (-1)}{4 - 2} = \frac{4}{2} = 2.$$

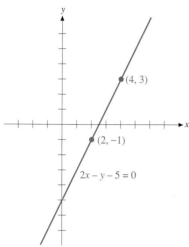

FIGURE 1.4.7 The line through the points $(2, -1)$ and $(4, 3)$ has slope $= \frac{3-(-1)}{4-2} = 2$ and equation $y - 3 = 2(x - 4)$, which simplifies to $2x - y - 5 = 0$.

We can now use the point $(4, 3)$ and slope $m = 2$ in the point-slope form and write the desired equation,

$$y - 3 = 2(x - 4).$$

If we used the point $(2, -1)$, we would write the equation

$$y - (-1) = 2(x - 2).$$

Both of the above equations simplify to $2x - y - 5 = 0$. We can use either point to find the equation. The graph is sketched in Figure 1.4.7.   ∎

EXAMPLE 7

Find an equation of the line through the points $(-1, 4)$ and $(-1, 3)$.

SOLUTION   We might notice immediately that this is a vertical line with equation $x = -1$. If we did not notice that this is a vertical line, we would proceed as in the previous example to find the slope. That is,

$$\text{Slope} = \frac{4 - 3}{-1 - (-1)} = \frac{1}{0}.$$

The fact that the slope is undefined indicates a vertical line. See Figure 1.4.8.   ∎

Slope-Intercept Form

It is not difficult to check that the slope of the line $y = mx + b$ is m; its y-intercept is the point $(0, b)$. See Figure 1.4.9. This gives us the following characterization.

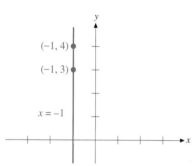

FIGURE 1.4.8 The line through the points $(-1, 4)$ and $(-1, 3)$ is a vertical line with equation $x = -1$. The slope of vertical line is undefined.

> The **slope-intercept** form of the equation of the line with slope m and y-intercept $(0, b)$ is
>
> $$y = mx + b.$$

The slope-intercept form can be used to *read* the slope of a line from its equation.

To find the slope of a line

- Write the equation of the line in slope-intercept form.
- The slope of the line is then the coefficient of x.

EXAMPLE 8

Find the slope of the line $3x + 2y - 6 = 0$.

SOLUTION   We first write the equation in slope-intercept form. That is, we solve the equation for y.

$$3x + 2y - 6 = 0,$$
$$2y = -3x + 6,$$
$$y = -\frac{3}{2}x + 3.$$

FIGURE 1.4.9 The slope-intercept form of the equation of the line with slope m and y-intercept b is $y = mx + b$.

The coefficient of x then gives the slope, $m = -3/2$.   ∎

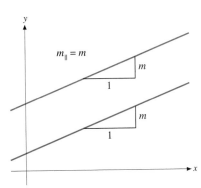

FIGURE 1.4.10a Parallel lines have equal slopes, $m_\parallel = m$.

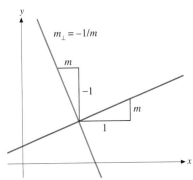

FIGURE 1.4.10b Perpendicular lines have slopes that are negative reciprocals, $m_\perp = -1/m$.

Parallel and Perpendicular Lines

If a line has nonzero slope m, then all parallel lines have slope $m_\parallel = m$ and all perpendicular lines have slope $m_\perp = -1/m$. (These statements can be proved by showing that the right triangles in Figure 1.4.10 are congruent. Corresponding sides then have equal lengths, as is indicated by the labeling.) All vertical lines are parallel to each other and are perpendicular to all horizontal lines.

EXAMPLE 9

Find an equation of the line through the point $(-2, 3)$ and (**a**) parallel to the line $4x - 3y + 7 = 0$, (**b**) perpendicular to the line $4x - 3y + 7 = 0$.

SOLUTION   We have a point on the lines. We need their slopes. Their slopes are found from the slope of the given line.

$$4x - 3y + 7 = 0,$$
$$-3y = -4x - 7,$$
$$y = \frac{4}{3}x + \frac{7}{3}.$$

We see the given line has slope $m = 4/3$.

(**a**) Lines parallel to the given line have slope $m_\parallel = m = 4/3$. This slope and the point $(-2, 3)$ give the equation

$$y - 3 = \frac{4}{3}(x - (-2)),$$
$$3(y - 3) = 4(x + 2),$$
$$3y - 9 = 4x + 8,$$
$$4x - 3y + 17 = 0.$$

(**b**) Lines perpendicular to the given line have slope $m_\perp = -1/m = -3/4$. We can then write the equation

$$y - 3 = -\frac{3}{4}(x - (-2)),$$
$$4(y - 3) = -3(x + 2),$$

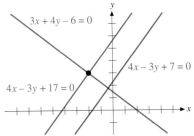

FIGURE 1.4.11 The line $4x - 3y + 17 = 0$ contains the point $(-2, 3)$ and is parallel to the line $4x - 3y + 7 = 0$. The line $3x + 4y - 6 = 0$ contains the point $(-2, 3)$ and is perpendicular to the line $4x - 3y + 7 = 0$.

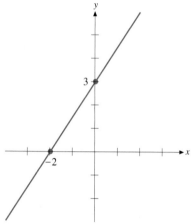

FIGURE 1.4.12 The intercept form of the equation of the line with x-intercept $(a, 0)$ and y-intercept $(0, b)$ is

$$\frac{x}{a} + \frac{y}{b} = 1.$$

FIGURE 1.4.13 The intercept form of the equation of the line with x-intercept $(-2, 0)$ and y-intercept $(0, 3)$ is

$$\frac{x}{-2} + \frac{y}{3} = 1.$$

$$4y - 12 = -3x - 6,$$
$$3x + 4y - 6 = 0.$$

The lines are sketched in Figure 1.4.11. ■

Intercept Form

If a and b are nonzero, it is not difficult to determine that the equation of the line that contains the points $(a, 0)$ and $(0, b)$ can be written in the form

$$\frac{x}{a} + \frac{y}{b} = 1.$$

See Figure 1.4.12. This gives us the following characterization.

The **intercept** form of the equation of the line with x–intercept $(a, 0)$ and y-intercept $(0, b)$ is

$$\frac{x}{a} + \frac{y}{b} = 1.$$

The intercept form is convenient for writing the equation of a line if we know its intercepts.

EXAMPLE 10

Find an equation of the line pictured in Figure 1.4.13.

SOLUTION We see from the picture that the x-intercept is $(a, 0) = (-2, 0)$ and the y-intercept is $(0, b) = (0, 3)$. We then write the equation

$$\frac{x}{-2} + \frac{y}{3} = 1.$$

We can then multiply both sides of the equation by -6 to obtain

$$3x - 2y = -6. \quad ■$$

EXERCISES 1.4

Sketch the graphs of the equations in Exercises 1–14.

1. $2x - 3y = 6$ **2.** $3x - 2y = 6$

3. $x - 2y = 4$ **4.** $2x - y = 2$

5. $2x - y = 0$ **6.** $x - 2y = 0$

7. $x + 3y = 0$ **8.** $3x + y = 0$

9. $x + 2 = 0$ **10.** $x - 1 = 0$

11. $x = 0$ **12.** $y = 0$

13. $y = -1$ **14.** $y - 1 = 0$

In Exercises 15–44, find an equation of the line that satisfies the given conditions.

15. Through $(2, 3)$, slope -1

16. Through $(1, -2)$, slope 3

17. Through $(-1, 2)$, slope $1/3$

18. Through $(2, 4)$, slope $1/2$

19. Through $(2, 1)$ and $(-1, 4)$

20. Through $(-1, 2)$ and $(-3, -1)$

21. Through $(2, 3)$, parallel to the line $2x - y = 4$

22. Through $(-1, 2)$, parallel to the line $6x + 2y = 5$

23. Through $(0, 2)$, perpendicular to the line $x + y = 2$

24. Through $(2, 0)$, perpendicular to the line $x = y$

25. Through $(3, 2)$, parallel to the x-axis

26. Through $(-1, 4)$, perpendicular to the y-axis

27. Through $(-1, -3)$, perpendicular to the x-axis

28. Through $(2, -3)$, parallel to the y-axis

29. Slope -6, through the point on the graph of $y = x^2$ that has $x = -3$

30. Slope -6, through the point on the graph of $y = -x^2$ that has $x = 3$

31. Slope 0, through the point on the graph of $y = x^2 - 2x$ that has $x = 1$

32. Slope 0, through the point on the graph of $y = x^2 + 4x + 2$ that has $x = 2$

33. Through the points on the graph of $y = 1 - x^2$ that have $x = 0$ and $x = 1$

34. Through the points on the graph of $y = x^2 + 4x$ that have $x = -1$ and $x = 1$

35. Through the points on the graph of $y = x^3$ that have $x = 2$ and $x = -1$

36. Through the points on the graph of $y = x^3 - x$ that have $x = 1$ and $x = 2$

37. Graph is

38. Graph is

39. Graph is

40. Graph is

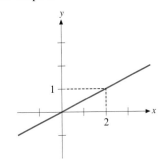

41. Points on the graph are equidistant from $(0, 1)$ and $(2, 0)$.

42. Points on the graph are equidistant from $(-2, 1)$ and $(3, -2)$.

43. Points on the graph are equidistant from $(-1, 3)$ and $(-1, -1)$.

44. Points on the graph are equidistant from $(-1, 2)$ and $(5, 2)$.

45. Find the minimum distance between the origin and points on the line $x + 2y = 2$.

46. Find an equation of each line that contains $(13, 0)$ and is tangent to the circle $x^2 + y^2 = 25$.

47. Let M be the midpoint of the line segment between $A(-3, -1)$ and $B(1, 1)$ and let N be the midpoint of the line segment between B and $C(2, -2)$. Show that the segment MN is parallel to the segment AC.

48. (a) For what values of k is $5x + ky = 3$ parallel to $2x - 3y = 5$? (b) For what values of k is $5x + ky = 3$ perpendicular to $2x - 3y = 5$?

49. For what values of k is the point $(2, 3)$ on the graph of $3x - ky = 2$?

50. If $b \neq 0$, show that the line segment between any two points on the line $ax + by + c = 0$ has slope $-a/b$.

51. Use a graphics calculator to find an approximate value of the m for which the line $y = mx + 1$ seems to best approximate the graph of $y = e^x$ for x near zero.

52. Use a graphics calculator to find an approximate value of the m for which the line $y = mx + 1$ seems to best approximate the graph of $y = 2^x$ for x near zero.

Connections

Recognizing the form of equations of lines, circles, parabolas, and ellipses.

Using the form of an equation to determine key points and general character of its graph, as was done with linear equations, 1.4

Completing a square.

1.5 GRAPHS OF SOME SIMPLE EQUATIONS

In this section we will review graphing techniques for a representative collection of simple equations, so we can use the figures to graphically illustrate calculus concepts. Of course, these graphs can be generated on some calculators, but it is still desirable (and possibly quicker) to hand-sketch the graphs of some simple equations.

General Strategy for Graph Sketching

Let us first note that we should not attempt to sketch a graph by plotting a large number of randomly chosen points on the graph. Locating points on a graph is a basic and important idea, but we must realize that any finite number of points on the graph cannot determine the entire graph. Unless we have additional information about the graph, we can never be certain what happens between any two points on the graph. That is, we need to know something of the general character of the graph. Once we know the general character of a graph, the graph can be obtained by plotting only a few key points on the graph and then sketching in the remainder of the graph. For example, if we recognize that an equation is the equation of a line, we know the general character of the graph. We can then use any two points to determine the graph, but we would ordinarily use the intercepts as key points.

In this section we will consider some examples for which the general character and key points can be determined from the *form* of the equation. Later, we will use calculus to determine the general character and key points of a graph. We always use the following strategy for sketching graphs.

> **Strategy for sketching graphs**
>
> - Determine the general character of the graph.
> - Determine and plot key points.
> - Sketch the remainder of the graph.

Circles and Ellipses

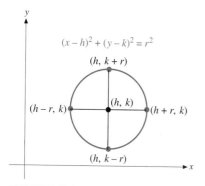

FIGURE 1.5.1 The standard equation of the circle with center (h, k) and radius r is $(x - h)^2 + (y - k)^2 = r^2$. The graph contains the key points $(h + r, k), (h - r, k), (h, k + r)$, and $(h, k - r)$.

> The **standard form** of the equation of the **circle** with center (h, k) and radius r is
> $$(x - h)^2 + (y - k)^2 = r^2. \tag{1}$$

The distance formula tells us that this equation is satisfied by all points (x, y) whose distance from (h, k) is r, and that only these points satisfy the equation.

The center of a circle is read easily from the standard form of its equation. To graph a circle, the center should be located and used to plot the four key points $(h + r, k), (h - r, k), (h, k + r)$, and $(h, k - r)$. See Figure 1.5.1.

Equations of the form

$$x^2 + y^2 + Dx + Ey + F = 0 \tag{2}$$

can be put into the standard form of a circle by *completing the squares*. Note that the graph of $(x - h)^2 + (y - k)^2 = 0$ is the single point (h, k). No points satisfy $(x - h)^2 + (y - k)^2 = $ (a negative number).

EXAMPLE 1
Sketch the graph of the equation $y^2 + 8y = 6x - x^2$.

SOLUTION   We must first recognize that the equation can be written in the form of (2), so the graph is a circle, or a single point, or no points satisfy the equation. In particular, we have

$$y^2 + 8y = 6x - x^2, \text{ so } x^2 + y^2 - 6x + 8y = 0.$$

To write the equation in standard form, we need to complete the squares. To do this, we group the x-terms and the y-terms. Planning ahead, we leave spaces to complete the square inside each pair of parentheses. Also, we write blank parentheses on the right corresponding to each set of parentheses on the left. Equality is then preserved when we fill in corresponding parentheses with equal numbers. We thus obtain

$$(x^2 - 6x \quad) + (y^2 + 8y \quad) = 0 + (\quad) + (\quad).$$

Recalling that the square of one-half the coefficient of the first-degree term is added to complete the square, we obtain

$$\left(x^2 - 6x + \left(\frac{-6}{2}\right)^2\right) + \left(y^2 + 8y + \left(\frac{8}{2}\right)^2\right) = 0 + (9) + (16),$$
$$(x^2 - 6x + 9) + (y^2 + 8y + 16) = 25,$$
$$(x - 3)^2 + (y + 4)^2 = 5^2.$$

From the standard form, we see that the circle has center $(3, -4)$ and radius 5. Key points are plotted and the graph is sketched in Figure 1.5.2. ∎

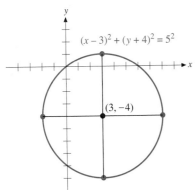

FIGURE 1.5.2   The standard form of the equation $y^2 + 8y = 6x - x^2$ is $(x - 3)^2 + (y + 4)^2 = 5^2$. The graph is a circle with center $(3, -4)$ and radius 5.

The **standard form** of the equation of an **ellipse** with center (h, k) is

$$\frac{(x - h)^2}{a^2} + \frac{(y - k)^2}{b^2} = 1, \quad a \text{ and } b \text{ positive}, a \neq b. \tag{3}$$

The key points $(h - a, k)$ and $(h + a, k)$ are found by moving a distance a to the left and right of the center. The number a is the positive square root of the positive number below $(x - h)^2$, and we move a units parallel to the x-axis. The key points $(h, k + b)$ and $(h, k - b)$ are found by moving b units up and down from the center. The number b is the positive square root of the positive number below $(y - k)^2$, and we move b units parallel to the y-axis. See Figure 1.5.3.

Parabolas

Let us further illustrate the idea of using the general character and key points to sketch a graph by considering equations of the form

$$y - k = a(x - h)^2, \quad a \neq 0. \tag{4}$$

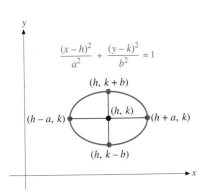

FIGURE 1.5.3   The standard form of the equation of an ellipse with center (h, k) is

$$\frac{(x - h)^2}{a^2} + \frac{(y - k)^2}{b^2} = 1.$$

The graph contains the key points $(h - a, k)$, $(h + a, k)$, $(h, k + b)$, and $(h, k - b)$.

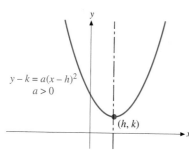

FIGURE 1.5.4a If $a > 0$, the graph of $y - k = a(x - h)^2$ is a parabola that opens upward from the vertex, (h, k).

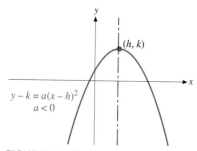

FIGURE 1.5.4b If $a < 0$, the graph of $y - k = a(x - h)^2$ is a parabola that opens downward.

FIGURE 1.5.5 The graph of $y = 2x - x^2$ is a parabola that opens downward. The intercepts are (0,0) and (2,0). The vertex is (1,1).

The graph is a **parabola** with **vertex** at the point (h, k). It is symmetric about the vertical line $x = h$. If $a > 0$, then points (x, y) on the graph satisfy $y - k = a(x - h)^2 \geq 0$, so $y \geq k$; the parabola opens upward from the vertex if $a > 0$. If $a < 0$, then $y - k = a(x - h)^2 \leq 0$, so $y \leq k$; the parabola opens downward from the vertex if $a < 0$. See Figure 1.5.4.

An equation of the form

$$y = ax^2 + bx + c, \qquad a \neq 0, \tag{5}$$

can be put into the form of (4) by completing the square. The graph is a parabola that opens upward if $a > 0$ and opens downward if $a < 0$.

To sketch the graph of an equation of the type of (5), we must first recognize that it is a parabola. The sign of the coefficient of x^2 then gives us the direction the parabola opens. Next, we determine and plot the intercepts. The graph can then be sketched by drawing a parabola that opens in the proper direction, through the intercepts. If a more accurate sketch is desired, we can determine and plot the vertex before sketching the graph.

EXAMPLE 2

Sketch the graph of the equation $y = 2x - x^2$.

SOLUTION We recognize that the graph is a parabola that opens downward.
When $x = 0$, $y = 2(0) - (0)^2$ implies $y = 0$. The y-intercept is $(0, 0)$.
When $y = 0$, we have $0 = 2x - x^2$, or $x(x - 2) = 0$. The solutions are $x = 0$ and $x = 2$. The x-intercepts are $(0, 0)$ and $(2, 0)$.
Any sketch of a parabola that opens downward, through the intercepts $(0, 0)$ and $(2, 0)$, would look similar to the actual graph given in Figure 1.5.5. We know from symmetry that the vertex must have x-coordinate 1, so its y-coordinate is $2(1) - (1)^2 = 1$. We could also determine the vertex by completing the square and writing the equation in the form of (4). ∎

EXAMPLE 3

Sketch the graph of the equation $y = 2x^2 + 4x + 3$.

SOLUTION We recognize that the graph is a parabola that opens upward. When $x = 0$, $y = 3$. The y-intercept is $(0, 3)$. When $y = 0$, we have

$$2x^2 + 4x + 3 = 0.$$

This quadratic equation does not factor easily, so we try to solve it by using the Quadratic Formula. That is,

$$x = \frac{-(4) \pm \sqrt{4^2 - 4(2)(3)}}{2(2)}.$$

Since the formula for the solution contains $\sqrt{-8}$, there is no real number solution. This means the graph has no x-intercept.

We can determine the vertex and line of symmetry of the parabola by completing the square. To complete the square, we first rewrite the equation with the x-terms on one side and all other terms on the other side. We then factor the coefficient of x^2 from the x-terms. We obtain

$$y - 3 + 2(\quad) = 2(x^2 + 2x + \quad),$$

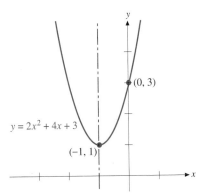

FIGURE 1.5.6 The graph of $y = 2x^2 + 4x + 3$ is a parabola that opens upward. The only intercept is $(0,3)$. The vertex is $(-1, 1)$.

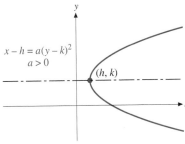

FIGURE 1.5.7a If $a > 0$, the graph of $x - h = a(y - k)^2$ is a parabola that opens to the right from the vertex, (h, k).

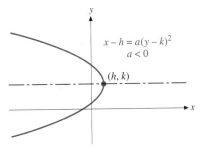

FIGURE 1.5.7b If $a < 0$, the graph of $x - h = a(y - k)^2$ is a parabola that opens to the left.

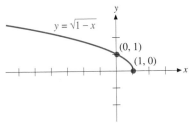

FIGURE 1.5.8 The graph of $y = \sqrt{1-x}$ is the upper half of the parabola $y^2 = 1 - x$.

where we have left space inside the parentheses on the right for the term that will complete the square. Also, we have written blank parentheses on the left to match the parentheses on the right. We have given the pair of parentheses on the left the same coefficient as that on the right, namely, 2. Equality is then retained when equal numbers are written inside the two pairs of parentheses. We have

$$y - 3 + 2(1) = 2\left(x^2 + 2x + \left(\frac{2}{2}\right)^2\right),$$
$$y - 1 = 2(x + 1)^2.$$

From this form of the equation we see that the vertex is $(h, k) = (-1, 1)$.

In Figure 1.5.6 we have plotted the vertex and y-intercept and then used symmetry about the vertical line $x = -1$ to sketch the graph. The y-coordinate of the vertex is positive and the parabola opens upward, so there is no x-intercept. This agrees with the information we obtained earlier. ■

The roles of x and y can be interchanged in the equation of a parabola. That is, the graph of an equation of the form

$$x = ay^2 + by + c, \qquad a \neq 0, \tag{6}$$

is also a parabola. The vertex can be found by completing the square to write the equation in the form

$$x - h = a(y - k)^2. \tag{7}$$

The vertex is then seen to be (h, k). The graph is symmetric about the horizontal line $y = k$. The parabola opens to the right if $a > 0$ and to the left if $a < 0$. See Figure 1.5.7.

Graphs of Equations That Involve Square Roots

The graphs of some equations that involve a square root may be recognized after the equation is squared. Since squaring an equation can introduce extraneous solutions, we must check that the original equation is satisfied.

EXAMPLE 4
Sketch the graph of the equation $y = \sqrt{1-x}$.

SOLUTION Squaring the equation, we obtain

$$y^2 = 1 - x \text{ or } x - 1 = -y^2.$$

This equation is recognized as the equation of a parabola that opens to the left and has vertex at $(1, 0)$. From the original equation, we have that

$$y = \sqrt{1-x} \geq 0,$$

since $\sqrt{1-x}$ is the nonnegative square root of the nonnegative number $1 - x$. It follows that the graph of the original equation is the upper half of the parabola.

In Figure 1.5.8 we have plotted the vertex $(1, 0)$ and the intercept $(0, 1)$, and then sketched the graph. ■

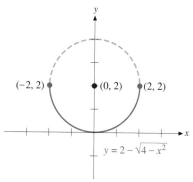

FIGURE 1.5.9 The graph of $y = 2 - \sqrt{4 - x^2}$ is the lower half of the circle $x^2 + (y - 2)^2 = 2^2$.

EXAMPLE 5

Sketch the graph of the equation $y = 2 - \sqrt{4 - x^2}$.

SOLUTION Writing the equation with the radical by itself on one side of the equation and then squaring, we have

$$\sqrt{4 - x^2} = 2 - y,$$
$$4 - x^2 = (2 - y)^2,$$
$$4 - x^2 = (y - 2)^2,$$
$$x^2 + (y - 2)^2 = 2^2.$$

This is the equation of the circle with center at $(0, 2)$ and radius 2. From the original equation, we have

$$y - 2 = -\sqrt{4 - x^2} \leq 0, \text{ so } y \leq 2.$$

Points on the graph of the original equation must satisfy $y \leq 2$, so the graph is the lower half of the circle. The graph is sketched in Figure 1.5.9. ∎

Graphs with Asymptotes

The general character of the graph of

$$y = \frac{1}{x - a}$$

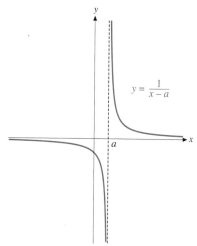

FIGURE 1.5.10 The graph of $y = \frac{1}{x - a}$ has vertical asymptote $x = a$ and horizontal asymptote $y = 0$; y is negative for $x < a$ and y is positive for $x > a$.

is illustrated in Figure 1.5.10. Note that $|y|$ increases without bound as the corresponding values of x approach a from either side. The line $x = a$ is called a **vertical asymptote** of the graph. For $x < a$, the corresponding values of y are negative and the graph is below the x-axis. For $x > a$, the corresponding values of y are positive and the graph is above the x-axis. There is no point on the graph that has x-coordinate a. As $|x|$ becomes large the corresponding values of y approach zero. The line $y = 0$ is called a **horizontal asymptote** of the graph.

EXAMPLE 6

Sketch the graph of $y = \frac{1}{x + 1}$.

SOLUTION The graph has vertical asymptote $x = -1$ and horizontal asymptote $y = 0$; y is negative for $x < -1$ and positive for $x > -1$. The y-intercept $(0, 1)$ is plotted and the graph is sketched in Figure 1.5.11. ∎

The general character of the graph of

$$y = \frac{1}{(x - a)^2}$$

is illustrated in Figure 1.5.12. The graph has vertical asymptote $x = a$ and horizontal asymptote $y = 0$. All points on the graph have positive y-coordinates. There is no point on the graph with x-coordinate a.

Be sure you know how your calculator handles graphs that have asymptotes.

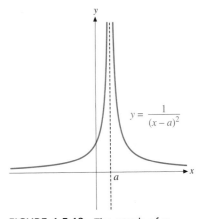

FIGURE 1.5.11 The graph of $y = \dfrac{1}{x+1}$ has vertical asymptote $x = -1$ and horizontal asymptote $y = 0$; y is negative for $x < -1$ and y is positive for $x > -1$. The only intercept is $(0,1)$.

FIGURE 1.5.12 The graph of $y = \dfrac{1}{(x-a)^2}$ has vertical asymptote $x = a$ and horizontal asymptote $y = 0$. All points on the graph have positive y-coordinates. There is no point on the graph with x-coordinate a.

EXERCISES 1.5

Sketch the graphs of the equations given in Exercises 1–50. Illustrate the general character of the graphs; only key points and intercepts need be located accurately.

1. $2x - y = 4$

2. $4x + 3y = 12$

3. $x = 2$

4. $y = -2$

5. $y = 2x$

6. $y = -x/3$

7. $(x-1)^2 + y^2 = 1$

8. $x^2 + (y+1)^2 = 1$

9. $x^2 + y^2 + 4y = 0$

10. $x^2 + y^2 - 6x = 0$

11. $x^2 + y^2 - 2x + 2y = 0$

12. $x^2 + y^2 + 4x - 4y + 6 = 0$

13. $x^2 + y^2 - 3x - y + 9/4 = 0$

14. $x^2 + y^2 + x + 3y + 9/4 = 0$

15. $\dfrac{x^2}{3^2} + \dfrac{y^2}{2^2} = 1$

16. $\dfrac{x^2}{1^2} + \dfrac{(y-2)^2}{2^2} = 1$

17. $\dfrac{(x-2)^2}{2^2} + \dfrac{(y+2)^2}{1^2} = 1$

18. $\dfrac{(x+1)^2}{1^2} + \dfrac{(y-3)^2}{3^2} = 1$

19. $4x^2 + 9y^2 + 24x = 36$

20. $x^2 + 4y^2 - 4x - 8y = 0$

21. $y = 2x^2 - 4x + 3$

22. $y = 2x^2 + 4x + 2$

23. $y = x^2 + 1$

24. $y = x^2 + 2x + 2$

25. $y = -x^2 + 4$

26. $y = x^2 - 4$

27. $y = x(x-2)$

28. $y = x(1-x)$

29. $x = y^2$

30. $x = -y^2$

31. $x = -y^2 + 3y - 2$

32. $x = y^2 + 4y + 3$

33. $x = y^2 - y$

34. $x = (y-1)(y-3)$

35. $y = \sqrt{x}$

36. $y = -\sqrt{x}$

37. $x = -\sqrt{y}$

38. $x = \sqrt{y}$

39. $y = \sqrt{x+4}$

40. $y = \sqrt{4-3x}$

41. $x = \sqrt{y-1} + 1$

42. $x = \sqrt{y+1} - 1$

43. $y = -\sqrt{9-x^2}$

44. $y = \sqrt{9-x^2}$

45. $x - \sqrt{2y - y^2} = 0$

46. $x + \sqrt{-2y - y^2} = 0$

47. $y = \dfrac{1}{x-2}$

48. $y = \dfrac{1}{x+3}$

49. $y = \dfrac{1}{(x+1)^2}$

50. $y = \dfrac{1}{(x-1)^2}$

51. The height above ground in feet of an object t seconds after it is thrown vertically from ground level at 48 ft/s is given by the equation $s = -16t^2 + 48t$. What is the maximum height obtained by the object?

52. The area of a rectangular field with one side of length x and perimeter 100 is $A = x(50 - x)$. What are the dimensions of the field that has the largest area?

53. Find an equation that describes the points whose distance from $(2, 2)$ is twice as great as the distance from $(-1, -1)$. (For example, the origin is such a point since the distance of the origin from $(2, 2)$ is $2\sqrt{2}$ which is twice the distance from $(-1, -1)$ to the origin.) Identify the graph.

54. Find an equation satisfied by all points (x, y) that are equidistant from $(0, 3)$ and $(2, 0)$.

55. Sketch the graph of $y = \dfrac{1}{x} + \dfrac{1}{x - 2}$. What are the x-intercepts? What are the asymptotes?

56. Sketch the graph of $y = \dfrac{1}{x^2} + \dfrac{1}{x - 2}$. What are the x-intercepts? What are the asymptotes?

57. Graph $y = \dfrac{1}{x - 1} - \dfrac{3}{x - 3}$. What are the approximate coordinates of the lowest point on the graph between $x = 1$ and $x = 3$? What are the approximate coordinates of the highest point on the graph for $x < 1$?

58. Graph $y = \dfrac{2}{x} - \dfrac{1}{x - 2}$. What are the approximate coordinates of the lowest point on the graph between $x = 0$ and $x = 2$? What are the approximate coordinates of the highest point on the graph for $x > 2$?

59. Compare the length of a variable vertical line segment between $y = 2 - x$ and $y = \sqrt{2 - x^2}$ with the distance between x and one for values of x near one. What happens to the ratio of the length over the distance as x approaches one?

60. Graph $y = 2 - x$ and $y = -x^2$. What is the apparent slope of the shortest line segment that has one endpoint on each graph? What is the relation between this shortest line segment and the line $y = 2 - x$?

1.6 REVIEW OF TRIGONOMETRY

This section contains a review of the basic definitions and properties of trigonometric functions. We will be dealing with these functions throughout our study of calculus.

Connections

Basic concepts of angles.

Right triangle trigonometry.

Trigonometric functions of a real variable.

Common values and graphs of the sine and cosine functions.

Basic trigonometric identities.

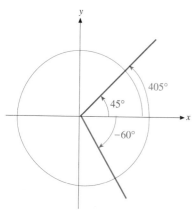

FIGURE 1.6.1 An angle that is positioned at the origin with initial side the positive x-axis is said to be in standard position.

Angles

We may think of an **angle** as a measure of the rotation from an initial side of the angle to a terminal side. Counterclockwise rotations are associated with positive angles, and clockwise rotations are associated with negative angles. Each rotation is associated with one and only one angle. Unequal angles (such as $45°$ and $405°$) may have the same initial and terminal sides if the angles differ by an integer multiple of $360°$. See Figure 1.6.1. An angle that is positioned at the origin with initial side the positive x-axis is said to be in **standard position.**

Many of the calculus formulas we will be using are greatly simplified by using **radian measure** of angles. Since we will be using radian measure for calculus problems, it will be useful to know the radian measure of some of the common angles. Recall that the radian measure of an angle is the ratio of the length s of the arc of a circle subtended by the angle and the radius r of the circle. That is,

$$\theta = \frac{s}{r} \text{ radians.}$$

See Figure 1.6.2. This ratio does not depend on the size of the circle. For the purpose of determining the radian measure of an angle, we take the length of an arc of a circle to be positive in the counterclockwise direction and negative

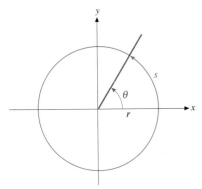

FIGURE 1.6.2   The radian measure of θ is the ratio s/r.

in the clockwise direction. This corresponds to the convention established for positive and negative angles.

The radian measure of one revolution is given by

$$\frac{\text{Circumference of circle}}{\text{Radius}} = \frac{2\pi r}{r} = 2\pi.$$

It follows that $360° = 2\pi$ radians (rad), or

$$180° = \pi \text{ rad}.$$

We can change the units of measure of an angle from degrees to radians by multiplying by the unit factor π rad$/180°$. For example,

$$30° = 30° \left(\frac{\pi \text{ rad}}{180°}\right) = \frac{\pi}{6} \text{ rad} \approx 0.52 \text{ rad},$$

$$45° = 45° \left(\frac{\pi \text{ rad}}{180°}\right) = \frac{\pi}{4} \text{ rad} \approx 0.79 \text{ rad},$$

$$60° = 60° \left(\frac{\pi \text{ rad}}{180°}\right) = \frac{\pi}{3} \text{ rad} \approx 1.05 \text{ rad},$$

$$90° = 90° \left(\frac{\pi \text{ rad}}{180°}\right) = \frac{\pi}{2} \text{ rad} \approx 1.57 \text{ rad}.$$

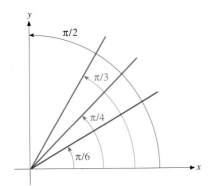

FIGURE 1.6.3   Some common angles can be expressed as fractional multiples of π.

See Figure 1.6.3. We can change from radians to degrees by multiplying by the unit factor $180°/(\pi \text{ rad})$. That is,

$$1 \text{ radian} = (1 \text{ rad}) \left(\frac{180°}{\pi \text{ rad}}\right) \approx 57°,$$

$$2 \text{ radians} = (2 \text{ rad}) \left(\frac{180°}{\pi \text{ rad}}\right) \approx 115°.$$

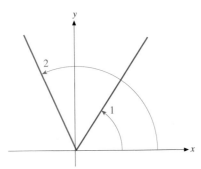

FIGURE 1.6.4   One radian is approximately 57 deg.

See Figure 1.6.4. The radian measure of an angle can be any real number. It is not necessary that radian measure be expressed as a fractional multiple of π, although some common angles can be expressed conveniently in this form.

Right Triangle Trigonometry

Trigonometric functions of an **acute angle** $\theta (0 < \theta < \pi/2)$ can be described in terms of ratios of lengths of the sides and hypotenuse of a right triangle. Recall (see Figure 1.6.5) that

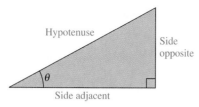

FIGURE 1.6.5 Trigonometric functions of an acute angle θ can be expressed in terms of the ratios of lengths of the side opposite, the side adjacent, and the hypotenuse of a right triangle.

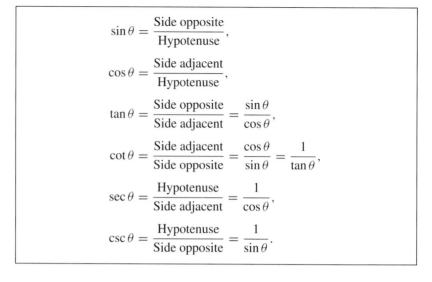

$$\sin\theta = \frac{\text{Side opposite}}{\text{Hypotenuse}},$$

$$\cos\theta = \frac{\text{Side adjacent}}{\text{Hypotenuse}},$$

$$\tan\theta = \frac{\text{Side opposite}}{\text{Side adjacent}} = \frac{\sin\theta}{\cos\theta},$$

$$\cot\theta = \frac{\text{Side adjacent}}{\text{Side opposite}} = \frac{\cos\theta}{\sin\theta} = \frac{1}{\tan\theta},$$

$$\sec\theta = \frac{\text{Hypotenuse}}{\text{Side adjacent}} = \frac{1}{\cos\theta},$$

$$\csc\theta = \frac{\text{Hypotenuse}}{\text{Side opposite}} = \frac{1}{\sin\theta}.$$

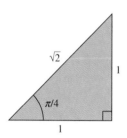

FIGURE 1.6.6 Trigonometric functions of $\pi/6$ and $\pi/3$ are read easily from this right triangle.

It is often more convenient to use the exact value of the trigonometric functions of the angles $\pi/6, \pi/4$, and $\pi/3$ than to use the multiple-digit approximate values given by a calculator. The values of the trigonometric functions for these common angles are easily read from a sketch of a representative right triangle.

To sketch a $\pi/6 - \pi/3 - \pi/2$ triangle, it is only necessary to remember that, since this triangle is half of an equilateral triangle, the side opposite the $\pi/6$ angle is one-half the length of the hypotenuse. That is, $\sin(\pi/6) = 1/2$. If the hypotenuse is 2, then the side opposite the $\pi/6$ angle is 1. The Pythagorean Theorem implies the remaining side has length $\sqrt{3}$. From Figure 1.6.6 we can see, for example, that

$$\cos\frac{\pi}{6} = \frac{\text{Side adjacent } \pi/6}{\text{Hypotenuse}} = \frac{\sqrt{3}}{2} \qquad \text{and}$$

$$\tan\frac{\pi}{3} = \frac{\text{Side opposite } \pi/3}{\text{Side adjacent } \pi/3} = \frac{\sqrt{3}}{1} = \sqrt{3}.$$

The sides of a $\pi/4 - \pi/4 - \pi/2$ triangle are equal, so $\tan(\pi/4) = 1$. If the length of each of its two equal sides is 1, then the hypotenuse is $\sqrt{2}$. We can read the values of the trigonometric functions of $\pi/4$ from the representative right triangle in Figure 1.6.7. For example, we see that

$$\sin\frac{\pi}{4} = \frac{\text{Side opposite } \pi/4}{\text{Hypotenuse}} = \frac{1}{\sqrt{2}}.$$

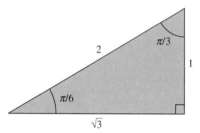

FIGURE 1.6.7 Trigonometric functions of $\pi/4$ are read easily from this right triangle.

We will often encounter problems where the sides and/or angles of a right triangle are variables. The trigonometric functions can then be used to find relations between these variables. It is worth noting that the triangles we will encounter may not appear in a standard or convenient position.

EXAMPLE 1

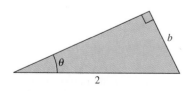

FIGURE 1.6.8 The side opposite θ has length b and the hypotenuse is 2, so $\sin\theta = b/2$.

(a) From Figure 1.6.8 we see that $\sin\theta = b/2$, so $b = 2\sin\theta$.

(b) From Figure 1.6.9 we see that $\cot\phi = d/5000$, so $d = 5000\cot\phi$. ∎

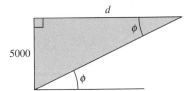

FIGURE 1.6.9 The side opposite ϕ has length 5000 and the side adjacent has length d, so $\cot\phi = d/5000$.

Trigonometric Functions of Any Angle

We need to extend the definition of the trigonometric functions to include all angles. Trigonometric functions of an angle that is in standard position can be defined in terms of the coordinates (x, y) of the point of intersection of the terminal side of the angle and the circle $x^2 + y^2 = 1$. In particular, we define

$$\sin\theta = y \text{ and } \cos\theta = x.$$

See Figure 1.6.10. The other trigonometric functions are then defined by

$$\tan\theta = \frac{y}{x}, \quad \cot\theta = \frac{x}{y}, \quad \sec\theta = \frac{1}{x}, \quad \csc\theta = \frac{1}{y}.$$

These definitions agree with the previous definitions of the trigonometric functions of an acute angle in terms of the sides and hypotenuse of a right triangle.

It is clear that the values of the trigonometric functions are equal for angles that differ by integer multiples of 2π.

It is possible to use the notion of side opposite, side adjacent, and hypotenuse to determine the trigonometric functions of any angle θ. To do this, we first draw the angle in standard position and then form a **representative right triangle** with one side along the x-axis, one side vertical, and hypotenuse along the terminal side. This is done by drawing a vertical line segment from a point (x, y) on the terminal side of the angle to the x-axis. The base of the triangle is labeled x, the height is labeled y, and the hypotenuse is labeled $\sqrt{x^2 + y^2}$. The signs of x and y depend on the quadrant that contains the terminal side of θ. The hypotenuse is always labeled positive. See Figure 1.6.11. The terminal side of θ intersects the unit circle at the point

$$\left(\frac{x}{\sqrt{x^2 + y^2}}, \frac{y}{\sqrt{x^2 + y^2}} \right).$$

It follows that, for example,

$$\sin\theta = \frac{y}{\sqrt{x^2 + y^2}}.$$

This can be interpreted as the side opposite the acute angle at the origin over the hypotenuse. The other trigonometric functions can also be read from the sketch.

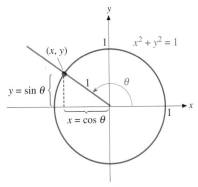

FIGURE 1.6.10 Trigonometric functions of an angle that is in standard position are determined by the coordinates (x, y) of the intersection of the terminal side of the angle and the circle $x^2 + y^2 = 1$.

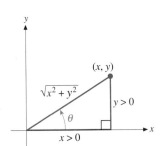

FIGURE 1.6.11a If $0 < \theta < \pi/2$, both sides of a representative right triangle are positive.

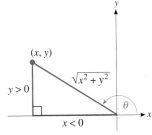

FIGURE 1.6.11b If $\pi/2 < \theta < \pi$, a representative right triangle has side opposite positive and side adjacent negative.

FIGURE 1.6.11c If $\pi < \theta < 3\pi/2$, both sides of a representative right triangle are negative.

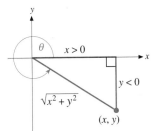

FIGURE 1.6.11d If $3\pi/2 < \theta < 2\pi$, a representative right triangle has side opposite negative and side adjacent positive.

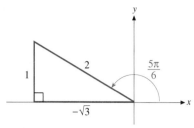

FIGURE 1.6.12 The terminal side of $5\pi/6$ is in the second quadrant and the reference angle is $\pi/6$. A representative right triangle with angle $\pi/6$ shows that $\cos(5\pi/6) = -\sqrt{3}/2$.

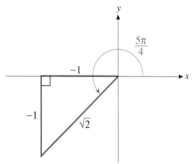

FIGURE 1.6.13 The terminal side of $5\pi/4$ is in the third quadrant and the reference angle is $\pi/4$. A representative right triangle with angle $\pi/4$ shows that $\sec(5\pi/4) = -\sqrt{2}$.

The acute angle between the terminal side and the x-axis of an angle in standard position is called the **reference angle** of the angle. Except for sign, the values of the trigonometric functions of an angle are equal to those of its reference angle.

EXAMPLE 2

Draw representative right triangles and evaluate **(a)** $\cos(5\pi/6)$, **(b)** $\sec(5\pi/4)$, **(c)** $\cos\theta$, if $\sin\theta = x/3$ and $-\pi/2 \leq \theta \leq \pi/2$.

SOLUTION

(a) A representative right triangle corresponding to an angle of $5\pi/6$ is given in Figure 1.6.12. Note that the terminal side of the angle is in the second quadrant and the reference angle is $\pi - 5\pi/6 = \pi/6$. The side opposite the reference angle is labeled 1, the hypotenuse is labeled 2, and the side adjacent is labeled $-\sqrt{3}$. This labeling reflects both the lengths of the sides of a representative $\pi/6 - \pi/3 - \pi/2$ triangle and the sign of x- and y-coordinates in the second quadrant. We can then see that $\cos(5\pi/6) = -\sqrt{3}/2$. (A calculator, set to *radians*, gives $\cos(5\pi/6) = -0.8660$.)

(b) A representative right triangle corresponding to an angle of $5\pi/4$ is drawn in Figure 1.6.13. The terminal side of the angle is in the third quadrant and the reference angle is $5\pi/4 - \pi = \pi/4$. The side opposite the reference angle and the side adjacent are labeled -1, and the hypotenuse is labeled $\sqrt{2}$. This labeling reflects both the lengths of the sides of a representative $\pi/4 - \pi/4 - \pi/2$ triangle and the sign of x- and y-coordinates in the third quadrant. We then see that $\sec(5\pi/4) = -\sqrt{2}$. (A calculator gives $\sec(5\pi/4) = -1.4142$.)

(c) Representative right triangles corresponding to $\sin\theta = x/3$ and $-\pi/2 \leq \theta \leq \pi/2$ are drawn in Figure 1.6.14. The Pythagorean Theorem implies that a right triangle with hypotenuse 3 and one side x, must have the other side of length $\sqrt{9 - x^2}$. If x is either positive as in Figure 1.6.14a or negative as in Figure 1.6.14b, we have $\cos\theta = \sqrt{9 - x^2}/3$. ■

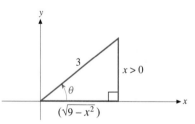

FIGURE 1.6.14a If $0 < \theta < \pi/2$ and $\sin\theta = x/3$, then x is positive and $\cos\theta = \sqrt{9 - x^2}/3$.

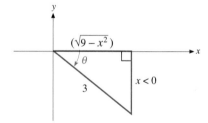

FIGURE 1.6.14b If $-\pi/2 < \theta < 0$ and $\sin\theta = x/3$ then x is negative and $\cos\theta = \sqrt{9 - x^2}/3$.

Using Calculator Values of Inverse Trigonometric Functions

The appropriate inverse function on a calculator can be used to obtain the solutions

$$\sin^{-1} d, \qquad \cos^{-1} d, \qquad \text{and} \qquad \tan^{-1} d$$

of the basic equations

$$\sin \theta = d, \qquad \cos \theta = d, \qquad \text{and} \qquad \tan \theta = d, \qquad \text{respectively.}$$

These solutions satisfy the inequalities

$$-\pi/2 \le \sin^{-1} d \le \pi/2,$$
$$0 \le \cos^{-1} d \le \pi, \quad \text{and}$$
$$-\pi/2 < \tan^{-1} d < \pi/2.$$

Representative right triangles can be used to determine the quadrants of the terminal sides of other solutions and the relation of other solutions to the above solutions. In particular:

$\sin \theta = d$ has solutions $\sin^{-1} d$ and $\pi - \sin^{-1} d$.   See Figure 1.6.15.
$\cos \theta = d$ has solutions $\cos^{-1} d$ and $-\cos^{-1} d$.   See Figure 1.6.16.
$\tan \theta = d$ has solutions $\tan^{-1} d$ and $\tan^{-1} d + \pi$.   See Figure 1.6.17.

All other solutions of these equations are obtained by adding an integer multiple of 2π to the above solutions.

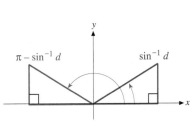

FIGURE 1.6.15a If $d > 0$, the solutions of $\sin \theta = d$ have terminal sides in either the first or second quadrants.

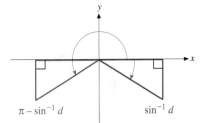

FIGURE 1.6.15b If $d < 0$, the solutions of $\sin \theta = d$ have terminal sides in either the third or fourth quadrants.

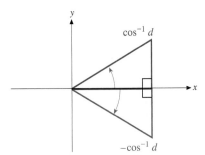

FIGURE 1.6.16a If $d > 0$, the solutions of $\cos \theta = d$ have terminal sides in either the first or fourth quadrants.

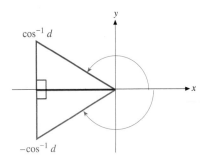

FIGURE 1.6.16b If $d < 0$, the solutions of $\cos \theta = d$ have terminal sides in either the second or third quadrants.

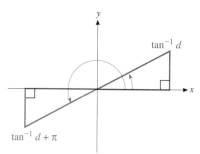

FIGURE 1.6.17a If $d > 0$, the solutions of $\tan\theta = d$ have terminal sides in either the first or third quadrants.

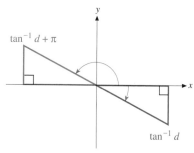

FIGURE 1.6.17b If $d < 0$, the solutions of $\tan\theta = d$ have terminal sides in either the second or fourth quadrants.

EXAMPLE 3

Find all θ, $0 \le \theta \le 2\pi$, that satisfy **(a)** $\sin\theta = 0.79$, **(b)** $\sin\theta = -0.82$, **(c)** $\cos\theta = -0.32$, **(d)** $\tan\theta = \sqrt{3}$.

SOLUTION

(a) In Figure 1.6.18 we have sketched representative right triangles corresponding to $\sin\theta = 0.79$. Note that angles in standard position with positive sines must have terminal sides in either the first or second quadrant. We then use a calculator, *set to radian measure,* to obtain the solution

$$\theta = \sin^{-1} 0.79 \approx 0.910808997.$$

FIGURE 1.6.18 The equation $\sin\theta = 0.79$ has solutions $\theta = \sin^{-1}(0.79)$ and $\theta = \pi - \sin^{-1}(0.79)$.

This angle is in the first quadrant. From Figure 1.6.18 we see that an angle in the second quadrant that has the same sine is

$$\theta = \pi - \sin^{-1} 0.79 \approx 2.230783656.$$

The two solutions of $\sin\theta = 0.79$ that satisfy $0 \le \theta \le 2\pi$ are $\theta = 0.91$ and $\theta = 2.23$, where we have rounded the solutions to two decimal places.

(b) Representative right triangles corresponding to $\sin\theta = -0.82$ are sketched in Figure 1.6.19. Note that the terminal sides are in the third and fourth quadrants. A calculator gives the solution

$$\theta = \sin^{-1}(-0.82) \approx -0.961411018.$$

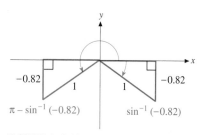

FIGURE 1.6.19 The equation $\sin\theta = -0.82$ has solutions $\theta = \sin^{-1}(-0.82)$ and $\theta = \pi - \sin^{-1}(-0.82)$.

This angle is in the fourth quadrant. We can obtain the solution with terminal side in the fourth quadrant that satisfies $0 \le \theta \le 2\pi$ by adding 2π to $\sin^{-1}(-0.82)$ to obtain

$$\theta = \sin^{-1}(-0.82) + 2\pi \approx 5.321774288.$$

A solution with terminal side in the third quadrant is given by

$$\theta = \pi - \sin^{-1}(-0.82) \approx 4.103003672.$$

The solutions that satisfy $0 \le \theta \le 2\pi$, rounded to two decimal places, are

$$\theta = 4.10 \quad \text{and} \quad \theta = 5.32.$$

(c) Representative right triangles corresponding to $\cos\theta = -0.32$ are sketched in Figure 1.6.20. The angles have terminal sides in the second and third quadrants. A calculator gives the solution

$$\theta = \cos^{-1}(-0.32) \approx 1.896525814.$$

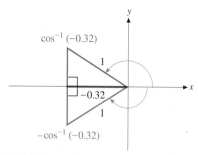

FIGURE 1.6.20 The equation $\cos\theta = -0.32$ has solutions $\theta = \cos^{-1}(-0.32)$ and $\theta = -\cos^{-1}(-0.32)$.

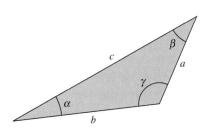

FIGURE 1.6.21 The equation $\tan\theta = \sqrt{3}$ has solutions $\theta = \tan^{-1}(\sqrt{3})$ and $\theta = \tan^{-1}(\sqrt{3}) + \pi$.

This angle is in the second quadrant and satisfies $0 \le \theta \le 2\pi$. The angle

$$\theta = -\cos^{-1}(-0.32) \approx -1.896525814$$

in the third quadrant has the same cosine, but does not satisfy $0 \le \theta \le 2\pi$. We can obtain a coterminal angle that does satisfy $0 \le \theta \le 2\pi$ by adding 2π to $-\cos^{-1}(-0.32)$. We obtain

$$\theta = -\cos^{-1}(-0.32) + 2\pi \approx 4.386659493.$$

The solutions of $\cos\theta = -0.32$ that satisfy $0 \le \theta \le 2\pi$, rounded to two decimal places, are

$$\theta = \cos^{-1}(-0.32) = 1.90 \qquad \text{and} \qquad \theta = -\cos^{-1}(-0.32) + 2\pi = 4.39.$$

(d) Representative right triangles corresponding to $\tan\theta = \sqrt{3}$ are sketched in Figure 1.6.21. These angles have terminal sides in the first and third quadrants. We may recognize that these angles have reference angles $\pi/3$, so the solutions are

$$\theta = \pi/3 \qquad \text{and} \qquad \theta = 4\pi/3.$$

If we did not recognize the reference angle, then we could use a calculator to determine the solution

$$\theta = \tan^{-1}(\sqrt{3}) \approx 1.047197551.$$

($\tan^{-1}(\sqrt{3})$ is exactly $\pi/3$.) We could then calculate the solution

$$\theta = \tan^{-1}(\sqrt{3}) + \pi \approx 4.188790205.$$

($\tan^{-1}(\sqrt{3}) + \pi$ is exactly $4\pi/3$.) ■

Law of Cosines and Law of Sines

The Law of Cosines and the Law of Sines can be used to find relations between the angles and sides of a triangle. These formulas hold for any triangle, not just for right triangles. See Figure 1.6.22.

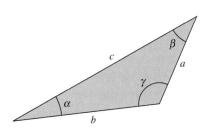

FIGURE 1.6.22 The Law of Cosines and the Law of Sines relate angles and lengths of sides of triangles that are not necessarily right triangles.

Law of Cosines:	$c^2 = a^2 + b^2 - 2ab\cos\gamma.$
Law of Sines:	$\dfrac{\sin\alpha}{a} = \dfrac{\sin\beta}{b} = \dfrac{\sin\gamma}{c}.$

If $\gamma = \pi/2$, we have $\cos\gamma = 0$ and the Law of Cosines gives the familiar result of the Pythagorean Theorem. If $\pi/2 < \gamma < \pi$, $\cos\gamma$ is negative and the Law of Cosines gives a value of c that is greater than $\sqrt{a^2 + b^2}$.

We can obtain relations between variables in a triangle from a sketch of the triangle.

EXAMPLE 4

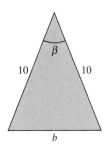

FIGURE 1.6.23 The Law of Cosines relates one angle and three sides of a triangle.

(a) From Figure 1.6.23 we see that the side of length b is the side opposite the angle β in the triangle. Since the other two sides of the triangle are known, we can use the Law of Cosines to obtain the equation

$$b^2 = 10^2 + 10^2 - 2(10)(10)\cos\beta,$$
$$b^2 = 200 - 200\cos\beta.$$

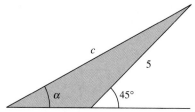

FIGURE 1.6.24 The Law of Sines relates two angles and the two sides opposite the angles in a triangle.

(b) From Figure 1.6.24 we see that the side opposite the angle α has length 5. The angle opposite the side of length c is $\pi - \pi/4 = 3\pi/4$. We can then use the Law of Sines to obtain the equation

$$\frac{\sin(3\pi/4)}{c} = \frac{\sin\alpha}{5} \quad \text{or} \quad \frac{1}{\sqrt{2}c} = \frac{\sin\alpha}{5}. \quad \blacksquare$$

Graphs of the Trigonometric Functions

Since the radian measure of an angle is a real number that is a ratio of lengths, any real number—positive, negative, or zero—may be considered to be the radian measure of an angle. We may then consider the trigonometric functions to be functions of a real variable. That is, $\sin x$ represents the value of the sine function for the angle that has radian measure x. This interpretation is particularly useful for applications of calculus.

The graphs of the trigonometric functions are sketched in Figures 1.6.25–1.6.30. You should know the key points and general character of these graphs, especially those of the sine and cosine functions, since these functions determine values of the others. The graphs contain important information about the corresponding functions. For example, we see that the sine and cosine functions have extreme values of ± 1 and that

$$\sin(k\pi) = 0 \quad \text{and} \quad \cos\left(\frac{\pi}{2} + k\pi\right) = 0, \quad k \text{ any integer.}$$

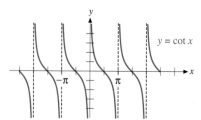

FIGURE 1.6.25 $\sin x = 0$ when x is an integer multiple of π. The values of $\sin x$ range between 1 and -1 and repeat in intervals of 2π.

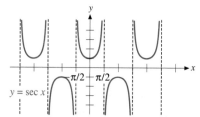

FIGURE 1.6.26 $\cos x = 0$ when x is $\pi/2$ plus an integer multiple of π. The values of $\cos x$ range between 1 and -1 and repeat in intervals of 2π.

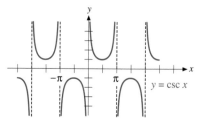

FIGURE 1.6.27 $\tan x$ is zero where $\sin x = 0$ and has vertical asymptotes where $\cos x = 0$. The values of $\tan x$ repeat in intervals of π.

FIGURE 1.6.28 $\cot x$ is zero where $\cos x = 0$ and has vertical asymptotes where $\sin x = 0$. The values of $\cot x$ repeat in intervals of π.

FIGURE 1.6.29 $\sec x$ has vertical asymptotes where $\cos x = 0$. The values of $\sec x$ are greater than one in absolute value and repeat in intervals of 2π.

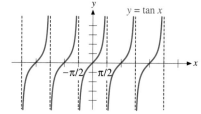

FIGURE 1.6.30 $\csc x$ has vertical asymptotes where $\sin x = 0$. The values of $\csc x$ are greater than one in absolute value and repeat in intervals of 2π.

The zeros of $\sin x$ give the vertical asymptotes of the graphs $\csc x = \dfrac{1}{\sin x}$ and $\cot x = \dfrac{\cos x}{\sin x}$; the zeros of $\cos x$ give the vertical asymptotes of the graphs $\sec x = \dfrac{1}{\cos x}$ and $\tan x = \dfrac{\sin x}{\cos x}$. The values of $\sin x$, $\cos x$, $\sec x$, and $\csc x$ repeat in intervals of 2π. These functions are said to be **periodic** with period 2π. The values of $\tan x$ and $\cot x$ repeat in intervals of π. These functions are periodic with period π.

Functions of the form

$$y = a\cos(bx - c)$$

occur in applications from physics and electrical engineering. Let us utilize the fact that the graph has the general character of the cosine curve to sketch its graph.

EXAMPLE 5

Sketch the graph of $y = 3\cos(2x + \pi/2)$.

SOLUTION   We use the facts that $\cos 0 = 1$, $\cos(\pi/2) = 0$, $\cos \pi = -1$, $\cos(3\pi/2) = 0$, and $\cos 2\pi = 1$ to determine key points on the graph.

When $2x + \pi/2 = 0$, $x = -\pi/4$ and $y = 3\cos 0 = 3$. The point $(-\pi/4, 3)$ is a key point on the graph of $y = 3\cos(2x + \pi/2)$.

When $2x + \pi/2 = \pi/2$, $x = 0$ and $y = 3\cos(\pi/2) = 0$. The point $(0, 0)$ is a key point on the graph.

When $2x + \pi/2 = \pi$, $x = \pi/4$ and $y = 3\cos(\pi) = -3$. The point $(\pi/4, -3)$ is a key point on the graph.

When $2x + \pi/2 = 3\pi/2$, $x = \pi/2$ and $y = 3\cos(3\pi/2) = 0$. The point $(\pi/2, 0)$ is a key point on the graph.

When $2x + \pi/2 = 2\pi$, $x = 3\pi/4$ and $y = 3\cos(2\pi) = 3$. The point $(3\pi/4, 3)$ is a key point on the graph. In Figure 1.6.31 we have plotted the key points determined above and then sketched a cosine curve through these points. ■

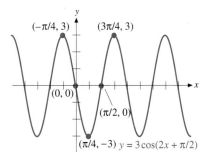

FIGURE 1.6.31 The graph of $y = 3\cos(2x + \pi/2)$ is sketched by plotting key points where the argument $2x + \pi/2 = 0, \pi/2, \pi, 3\pi/2, 2\pi$ and then sketching a cosine curve through these points.

Using Trigonometric Identities

For some calculus problems we will need to simplify or change the form of a trigonometric expression. The trigonometric formulas that we will need are listed inside the front cover for reference. Verification of these formulas can be found in any standard algebra-trigonometry text.

The examples below illustrate how we will use trigonometric identities to change certain types of trigonometric expressions into a form that is convenient for applications of calculus.

We can use the formula

$$\sin^2 x + \cos^2 x = 1$$

to express either *even* powers of $\cos x$ as a polynomial in $\sin x$ or *even* powers of $\sin x$ as a polynomial in $\cos x$. Similarly, we can use the formula

$$\tan^2 x + 1 = \sec^2 x$$

to express either *even* powers of $\tan x$ as a polynomial in $\sec x$ or *even* powers of $\sec x$ as a polynomial in $\tan x$.

EXAMPLE 6

Express $\sin^3 x$ as either $\cos x$ times a polynomial in $\sin x$ or $\sin x$ times a polynomial in $\cos x$.

SOLUTION   We first note that the power of $\sin x$ is odd. We can then factor one power of $\sin x$ and use $\sin^2 x + \cos^2 x = 1$ to express the remaining *even* power of $\sin x$ as a polynomial in $\cos x$. We have

$$\sin^3 x = (\sin^2 x)\sin x = (1 - \cos^2 x)\sin x. \quad \blacksquare$$

The formulas

$$\sin^2 x = \frac{1}{2} - \frac{\cos 2x}{2} \quad \text{and} \quad \cos^2 x = \frac{1}{2} + \frac{\cos 2x}{2}$$

can be used to express products of *even* powers of $\sin x$ and $\cos x$ as a sum of terms of the form $a \cos kx$, k an integer.

EXAMPLE 7

Express $\sin^4 x$ as a sum of terms of the form $a \cos kx$, k an integer.

SOLUTION

$$\sin^4 x = (\sin^2 x)^2 = \left(\frac{1}{2} - \frac{\cos 2x}{2}\right)^2$$

$$= \frac{1}{4} - \frac{1}{2}\cos 2x + \frac{1}{4}\cos^2 2x$$

$$= \frac{1}{4} - \frac{1}{2}\cos 2x + \frac{1}{4}\left(\frac{1}{2} + \frac{\cos 4x}{2}\right)$$

$$= \frac{3}{8} - \frac{1}{2}\cos 2x + \frac{1}{8}\cos 4x. \quad \blacksquare$$

EXERCISES 1.6

In Exercises 1–6 draw representative right triangles and determine the indicated trigonometric function of the acute angle θ that satisfies the given condition.

1. $\cos\theta$, if $\sin\theta = 3/5$

2. $\sin\theta$, if $\cos\theta = 5/13$

3. $\tan\theta$, if $\sin\theta = x/2$

4. $\cos\theta$, if $\sin\theta = \sqrt{9 - x^2}/3$

5. $\sin\theta$, if $\tan\theta = x/2$

6. $\sin\theta$, if $\sec\theta = x/2$

In Exercises 7–12, express the indicated variable in terms of a trigonometric function of θ.

7.

8.

9.

10.

11.

12.

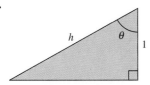

Evaluate the trigonometric functions in Exercises 13–20. (These angles have reference angles of 0, $\pi/6$, $\pi/4$, $\pi/3$, or $\pi/2$, so the exact values can be expressed in terms of $\sqrt{2}$ and $\sqrt{3}$.)

13. (a) $\sin \pi$, (b) $\cos \pi$, (c) $\tan \pi$

14. (a) $\sin(3\pi/2)$, (b) $\cos(3\pi/2)$, (c) $\tan(3\pi/2)$

15. (a) $\sin(5\pi/6)$, (b) $\cos(5\pi/6)$, (c) $\tan(5\pi/6)$

16. (a) $\sin(4\pi/3)$, (b) $\cos(4\pi/3)$, (c) $\tan(4\pi/3)$

17. (a) $\sin(7\pi/4)$, (b) $\cos(7\pi/4)$, (c) $\tan(7\pi/4)$

18. (a) $\cot(2\pi/3)$, (b) $\sec(2\pi/3)$, (c) $\csc(2\pi/3)$

19. (a) $\cot(5\pi/4)$, (b) $\sec(5\pi/4)$, (c) $\csc(5\pi/4)$

20. (a) $\cot(11\pi/6)$, (b) $\sec(11\pi/6)$, (c) $\csc(11\pi/6)$

In Exercises 21–26, draw representative right triangles and determine the indicated trigonometric function of the angle θ that satisfies the given conditions.

21. $\cos \theta$, if $\sin \theta = -3/5$ and $\cos \theta > 0$

22. $\sin \theta$, if $\cos \theta = -5/13$ and $\sin \theta < 0$

23. $\cos \theta$, if $\tan \theta = -1/2$ and $\sin \theta > 0$

24. $\tan \theta$, if $\cos \theta = -1/3$ and $\sin \theta > 0$

25. $\cos \theta$, if $\sin \theta = x$ and $-\pi/2 < \theta < \pi/2$

26. $\sin \theta$, if $\tan \theta = x$ and $-\pi/2 < \theta < \pi/2$

In Exercises 27–42, find all θ, $0 \leq \theta \leq 2\pi$, that satisfy the given equations. Use radian measure.

27. $\sin \theta = 1/2$ **28.** $\tan \theta = -1$

29. $\tan \theta = \sqrt{3}$ **30.** $\sin \theta = -\sqrt{3}/2$

31. $\sin(2\theta) = 0$ **32.** $\sin(3\theta) = 0$

33. $\cos(3\theta) = 0$ **34.** $\cos(4\theta) = 0$

35. $\sin x = 1/3$ **36.** $\sin x = -1/4$

37. $\cos x = -.71$ **38.** $\cos x = 1/3$

39. $\tan x = -1/4$ **40.** $\tan x = 2$

41. $2\sin^2 x - \sin x = 0$

42. $2\cos^2 x + \cos x - 1 = 0$

In Exercises 43–48, use either the Law of Cosines or the Law of Sines to find relations between the variables in the pictured triangles.

43.

44.

45.

46.

47.

48.

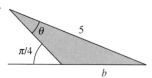

Find all unknown lengths of sides and angles of the triangles described in Exercises 49–52.

49. Lengths of sides 3, 4, and 6.

50. Lengths of sides 4 and 6, included angle 40°.

51. Angles 65° and 35°, length of included side 5.

52. Angle 40°, side adjacent of length 7, side opposition of length 5.

Sketch the graphs of the equations given in Exercises 53–64.

53. $y = \sin x, -\pi/2 \le x \le \pi/2$

54. $y = \cos x, 0 \le x \le \pi$

55. $y = \tan x, -\pi/2 < x < \pi/2$

56. $y = \sec x$

57. $y = 1 - \sin x$ **58.** $y = 2 + 2\cos x$

59. $y = \sin x + \cos x$ **60.** $y = \cos x - \sin x$

61. $y = \cos\left(x - \dfrac{\pi}{2}\right)$ **62.** $y = \sin\left(x - \dfrac{\pi}{2}\right)$

63. $y = \sin 2x$ **64.** $y = \sin\dfrac{x}{2}$

In Exercises 65–68, express the given expressions as either $\cos x$ times a polynomial in $\sin x$ or $\sin x$ times a polynomial in $\cos x$.

65. $\cos^3 x$ **66.** $\sin^5 x$

67. $\sin^3 x \cos^2 x$ **68.** $\sin^3 x \cos^3 x$

In Exercises 69–72, express the given expressions as either $\sec^2 x$ times a polynomial in $\tan x$ or $\sec x \tan x$ times a polynomial in $\sec x$.

69. $\sec^4 x$ **70.** $\tan^3 x \sec x$

71. $\tan^3 x \sec^3 x$ **72.** $\tan^3 x \sec^4 x$

In Exercises 73–74, express the given expressions as a sum of terms of the form $a \cos kx$, k an integer.

73. $\cos^4 x$ **74.** $\sin^2 x \cos^2 x$

75. A 12-ft ladder is leaning against a vertical wall at an angle of 70° from horizontal. See figure. What is the height of the top of the ladder?

76. The angle of elevation to a balloon from a point at ground level that is 30 m from the point on the ground that is directly below the balloon is 82°. See figure. How high is the balloon?

77. A wheel with diameter 14 in. rolls 12 ft along a level surface. How many revolutions does it make?

78. (a) A **nautical mile** is approximately the length of the arc subtended at the surface of the earth by an angle of $1'$ at the center of the earth. Use 4000 miles as the approximate radius of the earth to find the length in miles of one nautical mile.

(b) A **knot** is one nautical mile per hour. What is the speed in miles per hour of a ship that is traveling at 20 knots?

79. The front sprocket of a bicycle has radius 7 in. and the rear high gear sprocket has radius 2 in. The diameter of the rear wheel of the bicycle is 27 in. The front and rear sprockets are connected by a chain and the rear wheel rotates with the rear sprocket. See figure. How fast, in miles per hour, will the bicycle travel in high gear if the front sprocket rotates at 2 rev/s?

$r = 2$ in.
$r = 7$ in.
$d = 27$ in.

80. Graph $y = a \sin x$ for several values of a and compare the graphs with that of $y = \sin x$.

81. Graph $y = \sin bx$ for several values of b and compare the graphs with that of $y = \sin x$.

82. Graph $y = \cos(x - c)$ for several values of c and compare the graphs with that of $y = \cos x$.

83. Graph $y = a \cos(bx - c)$ for several values of a, b, and c and discuss how the value of each of a, b, and c affects the graph.

84. Graph $y_1 = x$, $y_3 = x - \dfrac{x^3}{3!}$, $y_5 = x - \dfrac{x^3}{3!} + \dfrac{x^5}{5!}$, and

$y_7 = x - \dfrac{x^3}{3!} + \dfrac{x^5}{5!} - \dfrac{x^7}{7!}$ and compare the graphs with that of $\sin x$ for $0 \le x \le \pi$.

85. Graph $y_0 = 1$, $y_2 = 1 - \dfrac{x^2}{2!}$, $y_4 = 1 - \dfrac{x^2}{2!} + \dfrac{x^4}{4!}$, and

$y_6 = 1 - \dfrac{x^2}{2!} + \dfrac{x^4}{41} - \dfrac{x^6}{6!}$ and compare the graphs with that of $\cos x$ for $-\pi/2 \le x \le \pi/2$.

REVIEW OF CHAPTER 1 CONCEPTS

The algebra and trigonometry reviewed in this chapter is very specific to calculus. Mastery of this material will give you the background necessary for success in calculus. Some highlights are mentioned below, but you should make every effort to master all the material in this chapter.

- The inequality $|x - c| < d$ says that the distance between x and c is less than d. The inequality $|x - c| < d$ is equivalent to the double inequality $-d < x - c < d$. Inequalities of the form $|x - c| < d$ will be used in Chapter 2 to develop fundamental calculus concepts.
- Many calculus problems involve exponents. Use of the rules of exponents are essential for dealing with these problems. Be sure you know how to simplify expressions that involve fractional powers.
- We will see in Chapter 4 that determining the sign of certain expressions is an important part of using calculus. We will use a table of signs, as introduced in Section 1.2, to determine intervals on which expressions are either always positive or always negative. This problem can also be solved by using calculator-generated graphs.
- The line segment between the points (x_1, y_1) and (x_2, y_2) has

Length = distance between points

$$= \sqrt{(x_2 - x_1)^2 + (y_2 - y_1)^2},$$

Slope $= \dfrac{y_2 - y_1}{x_2 - x_1}$, and

Midpoint $= \left(\dfrac{x_1 + x_2}{2}, \dfrac{y_1 + y_2}{2} \right).$

We will use these formulas in cases where the coordinates are variables. The slope of a line segment will be used in Chapter 2 in the development of the derivative, one of the basic tools of calculus.

- The point-slope form of the equation of the line through the point (x_1, y_1) with slope m is $y - y_1 = m(x - x_1)$. You will be using this form often and you should know it well.
- You should know the trigonometric functions in terms of side adjacent, side opposite, and hypotenuse of a right triangle.
- You should be very familiar with the graphs of $\sin x$ and $\cos x$. In particular, you should know the values of $\sin x$ and $\cos x$ for $x = 0, \pi/2, \pi, 3\pi/2$, and 2π.
- Trigonometric formulas that we will use are listed inside the front cover. You should know at least the fundamental formulas and where to find the others.

CHAPTER 1 REVIEW EXERCISES

Solve the inequalities in Exercises 1–8.

1. $-2x > 6$

2. $2x - 3 < 5x + 9$

3. $-3 < 2x - 3 < 3$

4. $-1 < 2 - 3x < 5$

5. $|2x - 1| < 3$

6. $|3x - 2| < 4$

7. $|x + 2| > 1$

8. $|2x - 3| > 1$

9. Evaluate $|\sqrt{3} - 3| - |\sqrt{3} - 2|$.

10. Evaluate $|2\pi - 9| - |2\pi - 6|$.

Write the expressions in Exercises 11–14 in the form x^r.

11. $\dfrac{x(x^3)^2}{x^5 x^{-2}}$

12. $\dfrac{x(\sqrt{x})^3}{(\sqrt[3]{x})^5}$

13. $\sqrt{\sqrt{x}}$

14. $\sqrt{1/x}$

Write the expressions in Exercises 15–16 in the form $a^r b^s$.

15. $\dfrac{(a^2 b^{-1})^3}{a(a^{-1} b^2)^2}$

16. $\dfrac{\sqrt{ab^2}}{\sqrt[3]{a^2 b}}$

17. Evaluate $(-8)^{-1/3}(18)^{3/2}\sqrt{2}$.

18. Evaluate $\dfrac{6^{1/2}(\sqrt{3})^3}{27^{2/3} 8^{5/6}}$.

Express the numbers in Exercises 19–20 in scientific notation.

19. $4,371,000$

20. 0.002591

Express the numbers in Exercises 21–22 as decimal numbers.

21. 2.735×10^{-3}

22. 7.59×10^4

Simplify the expressions in Exercises 23–26.

23. $\dfrac{1}{a^{-1} + b^{-1}}$

24. $\dfrac{a - b}{a^{-2} - b^{-2}}$

25. $\dfrac{\sqrt{1 - x^2}(1) - (x)\left(\dfrac{1}{2\sqrt{1 - x^2}}\right)(-2x)}{(\sqrt{1 - x^2})^2}$

26. $\dfrac{(x^2 + 1)^{2/3}(1) - (x)(2/3)(x^2 + 1)^{-1/3}(2x)}{(x^2 + 1)^{4/3}}$

In Exercises 27–30, find all values of x for which the given expressions are zero.

27. $(3x - 1)(2)(2x - 3)(2) + (2x - 3)^2(3)$

28. $(3x - 1)^2(2) + (2x + 1)(2)(3x - 1)(3)$

29. $2x^2 - 3x - 3$

30. $x^2 + 4x - 2$

In Exercises 31–34, find all values of x for which the given expressions are (a) undefined and (b) zero.

31. $\dfrac{(x + 1)(3x^2) - (x^3)(1)}{(x + 1)^2}$

32. $\dfrac{(3x + 2)^2(2) - (2x - 1)(2)(3x + 2)(3)}{(3x + 2)^4}$

33. $(x)\left(\dfrac{1}{3}\right)(2x - 1)^{-2/3}(2) + (2x - 1)^{1/3}(1)$

34. $\sqrt{x - 2} - \dfrac{1}{2\sqrt{x - 2}}$

In Exercises 35–38, find all intervals on which the given expressions are (a) positive and (b) negative.

35. $(x - 1)(x + 2)$

36. $x(x + 1)(x - 2)$

37. $x(x - 1)(x - 2)^2$

38. $\dfrac{x^2(x + 1)}{x - 2}$

39. Find the length, slope, and midpoint of the line segment between the points $(2, 3)$, and $(0, y)$.

40. Find the length, slope, and midpoint of the line segment between the points $(2, 3)$, and $(x, 0)$.

41. Find the slope of the line segment between the point $(2, -3)$ and a variable point on the graph of $f(x) = 1 - x^2$, in terms of x.

42. Find the slope of the line segment between the point $(2, 2)$ and a variable point on the graph of $f(x) = x^2 - x$, in terms of x.

43. Find the slope of the line segment between the point $(4, 2)$ and a variable point on the graph of $f(x) = \sqrt{x}$, in terms of x.

44. Find the slope of the line segment between the point $(1, 1)$ and a variable point on the graph of $f(x) = 1/x$, in terms of x.

45. Find an equation that is satisfied by all points (x, y) that are twice as far from $(3, 0)$ as they are from $(0, 0)$.

46. Find an equation that is satisfied by all points (x, y) that are equidistant from the point $(0, 0)$ and the line $y = -2$.

47. Find an equation that is satisfied by all points (x, y) with the sum of their distance from $(4, 0)$ and their distance from $(-4, 0)$ equal 10.

48. Find an equation that is satisfied by all points (x, y) with the difference of their distance from $(0, 5)$ and their distance from $(0, -5)$ equal 8.

49. Find an equation of the line that contains the point $(1, -2)$ and has slope 3.

50. Find an equation of the line that contains the points $(4, 3)$ and $(2, -1)$.

51. Find an equation of the line that contains the point $(5, 3)$ and is parallel to the line $3x - 4y = 12$.

52. Find an equation of the line that contains the point $(2, 3)$ and is perpendicular to the line $2x + 3y = 1$.

53. Find an equation of the line that has graph

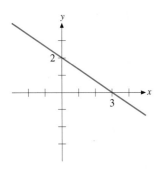

54. Find an equation of the line that has graph

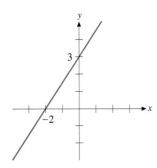

Sketch the graphs of the equations in Exercises 55–74.

55. $x + 2y = 2$ **56.** $2x - 3y + 6 = 0$

57. $2y = 3x$ **58.** $x = 2y$

59. $x^2 + y^2 = 4$ **60.** $x^2 + y^2 = 4x$

61. $\dfrac{x^2}{4} + \dfrac{y^2}{9} = 1$

62. $4x^2 + y^2 - 8x - 4y + 4 = 0$

63. $y = x(x - 3)$ **64.** $y - 1 = 2(x - 1)^2$

65. $y = \sqrt{1 - x}$ **66.** $y = \sqrt{1 - x^2}$

67. $y = \dfrac{1}{2 - x}$ **68.** $y = \dfrac{1}{(x - 1)^2}$

69. $y = \dfrac{1}{x - 1} - \dfrac{1}{x + 1}$ **70.** $y = \dfrac{1}{x - 1} + \dfrac{1}{x + 1}$

71. $y = 1 - \cos x$ **72.** $y = 1 + \sin x$

73. $y = \cos(2x)$ **74.** $y = \cos(x/2)$

75. Express x in terms of a trigonometric function of θ.

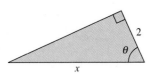

76. Express x in terms of a trigonometric function of θ.

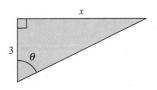

77. Express x in terms of $\cos \theta$.

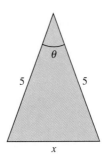

78. Express x in terms of $\sin \theta$.

79. Find $\tan \theta$, if $\sin \theta = -3/5$ and $\cos \theta < 0$.

80. Find $\sin \theta$, if $\sec \theta = 5/4$ and $\tan \theta < 0$.

81. Find $\sin \theta$ if $\cos \theta = 1/2$ and $\tan \theta < 0$.

82. Find $\cos \theta$, if $\tan \theta = 3$ and $\sin \theta > 0$.

In Exercises 83–94, find all θ, $0 \leq \theta \leq 2\pi$, that satisfy the given equations. Use radian measure.

83. $\cos \theta = -1/2$ **84.** $\sin \theta = -\sqrt{3}/2$

85. $\sec \theta = -\sqrt{2}$ **86.** $\csc \theta = 2$

87. ▦ $\sin x = 0.6$ **88.** ▦ $\sin x = -0.4$

89. ▦ $\cos x = -0.4$ **90.** ▦ $\cos x = 0.6$

91. ▦ $\tan x = 3.3$ **92.** ▦ $\tan x = -2.4$

93. $\cos(2\theta) = 0$ **94.** $\tan(2\theta) = 0$

Verify the identities in Exercises 95–98.

95. $\dfrac{\cos \theta}{\tan \theta} + \sin \theta = \csc \theta$

96. $\tan \theta + \cot \theta = \sec \theta \csc \theta$

97. $\dfrac{1}{1 - \tan\theta} + \dfrac{1}{1 - \cot\theta} = 1$

98. $2\cot 2\theta = \cot\theta - \tan\theta$

99. Show that
$$\frac{\sin(\theta + h) - \sin\theta}{h}$$
$$= \cos\theta\left(\frac{\sin h}{h}\right) + \sin\theta\left(\frac{\cos h - 1}{h}\right).$$

100. Show that
$$\frac{\cos(\theta + h) - \cos\theta}{h}$$
$$= -\sin\theta\left(\frac{\sin h}{h}\right) + \cos\theta\left(\frac{\cos h - 1}{h}\right).$$

101. Show that $B(\cos^2\theta - \sin^2\theta) + 2(C - A)\sin\theta\cos\theta = 0$ implies
$$\cot 2\theta = \frac{A - C}{B}.$$

102. Show that $\tan 2\theta = -d/h$, $0 < \theta < \pi/2$, d and h positive, implies
$$\tan\theta = \frac{h + \sqrt{h^2 + d^2}}{d}.$$

103. A baseball of mass m is moving toward home plate at a speed of v when it meets a bat that applies a force F for a brief time t and then the ball returns toward the outfield with speed u. The **change in momentum** due to the **impulse** is expressed by the equation
$$Ft = mv - mu.$$
Solve this equation for u.

104. The speed of radio waves v is related to their frequency f and their wavelength λ by the equation
$$v = f\lambda.$$
Radio waves travel in air at 3×10^8 m/s. What is the wavelength of an FM station if the waves used have a frequency of 105 MHz? (1 MHz $= 10^6$ hertz $= 10^6$/s.)

105. An object of mass m_1 is moving with speed v_1 when it meets head-on with an object of mass m_2 that is moving with speed v_2 in the opposite direction. If the collision is **inelastic**, so the two objects move together

with speed v after the collision, the **Law of Conservation of Momentum** implies
$$m_2 v_2 - m_1 v_1 = (m_1 + m_2)v,$$
where we have assumed that the objects move together in the direction of the second object after the collision. Solve this equation for m_2.

106. A winch with diameter 8 inches is used to lift a weight. See figure. How many revolutions must the winch be rotated to lift the weight 2 feet?

$d = 8$ in.

107. The front sprocket of a bicycle has radius 7 inches and the rear high-gear sprocket has radius 2 inches. The diameter of the wheels of the bicycle are 27 inches. The front and rear sprockets are connected by a chain and the rear wheel rotates with the rear sprocket. How many revolutions per minute must the front sprocket make for the bicycle to go 25 MPH?

108. A triangle has length of base 12. The angles at the ends of the base are 30° and 40°. What is the height of the triangle?

109. The angle of elevation from one point at ground level to the top of a building is measured to be 32°. From a point that is 100 feet closer to the building, the angle of elevation is 38°. See figure. How tall is the building?

32° 38°

100 ft

2 Limits, Continuity, and Differentiability at a Point

Using mathematical models to obtain information about real-life systems is a central idea of mathematics. The concept of expressing relations between physical quantities as functions is fundamental to many mathematical models.

Calculus is used to study functions. By using the techniques of calculus, we can obtain information about functions and, hence, the physical quantities represented by the functions. In order to understand calculus, we must understand the concepts on which it is based. Some of the fundamental concepts of calculus are introduced in this chapter.

A systematic development of the concept of bounds of functions enables us to obtain information about the size of the values of functions. Such information can be used, for example, to obtain estimates of error in several types of calculations we will be using.

The concept of the limit of a function at a point allows us to describe the behavior of the values of a function near the point. This is one of the key ideas of calculus and distinguishes calculus from other areas of mathematics, such as algebra and trigonometry. The concept of limit is the foundation for the concepts of continuity and differentiability, which are important characteristics of functions. The derivative is one of the central tools in the use of calculus to study functions.

2.1 FUNCTIONS

The application of calculus to practical problems depends on expressing physical quantities in terms of functions. Analysis of the functions then gives us information about the physical quantities of interest. In order to use calculus, we must understand the basic ideas of functions.

Relation Between a Function and Its Graph

A formal definition of function can be given in terms of what we would ordinarily think of as the graph of the function.

DEFINITION

A **function with domain** D is a collection of ordered pairs with the property that, for each x in D, there is exactly one pair (x, y) in the collection with first element x. The set of all second elements y is called the **range** of the function.

Thus, the graph in Figure 2.1.1 represents a function. The graph in Figure 2.1.2 cannot represent a function, because the graph contains more than one pair (x, y) that have the same first element x.

A function is specified by a **rule** that determines, for each x in the domain, exactly one corresponding value y in the range. It is customary to use a letter for the name of a function. For example, we may speak of the function f. The value y that is determined by a function f for a particular x is denoted $f(x)$—read "f of x."

If f is a function and $y = f(x)$, we say y is a function of x. The variable x is called the **independent variable**. The variable y is called the **dependent variable**. The value $y = f(x)$ *depends* on the particular choice of x in the domain of f. The choice of x determines the corresponding value $y = f(x)$. Until stated otherwise, the functions we will study will involve only variables that represent real numbers.

Let us illustrate the concept of function and graph with an example.

EXAMPLE 1

The graph in Figure 2.1.3 is the graph of a function f. Find the domain and range of f. Find $f(-2)$, $f(1)$, and $f(4)$.

SOLUTION   The domain of f consists of the set of x-coordinates of points (x, y) on the graph. This set is given by the x-coordinates of vertical lines that intersect the graph. Since vertical lines intersect the graph if and only if their x-coordinates satisfy $0 \leq x < 5$, the domain of f is $0 \leq x < 5$. (The solid dot at $(0, 1)$ emphasizes that the point $(0, 1)$ is on the graph, while the open dot at $(5, 3)$ indicates that point is not on the graph.)

The range of f is given by the set of y-coordinates of points (x, y) on the graph. This set is given by the y-coordinates of horizontal lines that intersect the graph. Horizontal lines intersect the graph if and only if their y-intercepts are in the interval $1 \leq y \leq 3$, so the range of f is $1 \leq y \leq 3$.

There is no point on the graph with x-coordinate -2. The number -2 is not in the domain of f; $f(-2)$ is undefined.

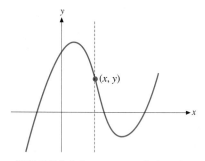

FIGURE 2.1.1 We can tell that this graph is the graph of a function, because each vertical line that intersects the graph intersects it in exactly one point.

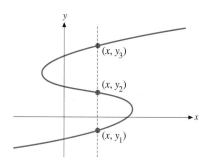

FIGURE 2.1.2 This graph is not the graph of a function, because some vertical lines intersect the graph in more than one point.

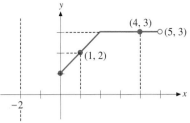

FIGURE 2.1.3 This graph is the graph of a function f that has domain $0 \le x < 5$ and range $1 \le y \le 3$. The point $(1, 2)$ is on the graph, so $f(1) = 2$; $(4, 3)$ is on the graph, so $f(4) = 3$. There is no point on the graph with x-coordinate -2, so $f(-2)$ is undefined.

The point $(1, 2)$ is on the graph, so $f(1) = 2$.
The point $(4, 3)$ is on the graph, so $f(4) = 3$. ∎

Substitution and Simplification

We will most often deal with the case where the values of a function are given by a *formula*. The values of the function are then obtained by *substitution* into the formula. We will see that:

Substitution is copying the right thing into the right place.

Each component of the formula must be copied into its proper position. After the substitution has been carried out, we can perform whatever algebra is necessary to simplify the expression. It is important that substitution and simplification be considered as separate steps.

EXAMPLE 2
$f(x) = 2x^2 - x - 3$. Find (a) $f(3)$ and (b) $f(x - 2)$.

SOLUTION   We first note that the formula $f(x) = 2x^2 - x - 3$ really *has nothing to do with x*. That is, the same function f could be defined with any variable in place of x. For example, we could write $f(u) = 2u^2 - u - 3$. The variable serves only as a place holder. In fact, it may be useful to use the **blank form** of the formula,

$$f(\ ) = 2(\ )^2 - (\ ) - 3.$$

We can then obtain the value that corresponds to any choice of independent variable simply by *copying* the variable inside each set of parentheses. In each case we will substitute and then simplify.

(a) Substitute: $f(3) = 2(3)^2 - (3) - 3$
     Simplify: $= 2(9) - 3 - 3$
              $= 12.$
(b) Substitute: $f(x - 2) = 2(x - 2)^2 - (x - 2) - 3$
     Simplify: $= 2(x^2 - 4x + 4) - x + 2 - 3$
             $= 2x^2 - 8x + 8 - x - 1$
             $= 2x^2 - 9x + 7.$

Note that parentheses were useful in both the substitution and simplification stages. ∎

The idea of substitution applies to complicated expressions that involve several components. We copy each component into its proper place. Parentheses should be used to organize our work.

EXAMPLE 3
$f(x) = x^2 - 4x$, $g(x) = 2x - 4$. Find

(a) $f(x) + g(x)$,
(b) $f(x) \cdot g(x)$,
(c) $f(x) - xg(x)$,
(d) $\dfrac{f(x) - f(3)}{x - 3}$,

(e) $f(g(x))$,

(f) $g(f(x))$.

SOLUTION   In each case we will first substitute, then simplify:

(a) $f(x) + g(x) = (x^2 - 4x) + (2x - 4)$

$$= x^2 - 4x + 2x - 4$$

$$= x^2 - 2x - 4.$$

(b) $f(x) \cdot g(x) = (x^2 - 4x)(2x - 4)$

$$= 2x^3 - 12x^2 + 16x.$$

(c) $f(x) - xg(x) = (x^2 - 4x) - x(2x - 4)$

$$= x^2 - 4x - 2x^2 + 4x$$

$$= -x^2.$$

(d) $\dfrac{f(x) - f(3)}{x - 3} = \dfrac{(x^2 - 4x) - ((3)^2 - 4(3))}{x - 3}$

$$= \dfrac{x^2 - 4x + 3}{x - 3}$$

$$= \dfrac{(x - 3)(x - 1)}{x - 3}$$

$$= x - 1, \quad x \neq 3.$$

(We have indicated that $(f(x) - f(3))/(x - 3) = x - 1$ only for $x \neq 3$.)

(e) $f(g(x))$ is read as "f of g of x." Taking one step at a time, we first substitute $2x - 4$ in place of $g(x)$ and then evaluate $f(2x - 4)$.

$$f(g(x)) = f(2x - 4)$$

$$= (2x - 4)^2 - 4(2x - 4)$$

$$= 4x^2 - 16x + 16 - 8x + 16$$

$$= 4x^2 - 24x + 32.$$

(f) $g(f(x)) = g(x^2 - 4x)$

$$= 2(x^2 - 4x) - 4$$

$$= 2x^2 - 8x - 4. \quad \blacksquare$$

Combinations of Functions

If f and g are functions, the **sum** $f + g$ is the function whose values are given by the formula $(f + g)(x) = f(x) + g(x)$. The domain of $f + g$ consists of those x values for which both $f(x)$ and $g(x)$ are defined. The **product** fg, **quotient** f/g, and other arithmetic combinations of functions are defined as suggested by the notation. The domains consist of those x for which every component function and the combined arithmetic expression are defined. For example, the domain of f/g consists of those x for which both $f(x)$ and $g(x)$ are defined, except for those x with $g(x) = 0$. The **composite function** $f \circ g$ is defined by the equation

$$f \circ g(x) = f(g(x)).$$

The domain of $f \circ g$ consists of those x in the domain of g for which $g(x)$ is in the domain of f. From Examples 3e and 3f we see that it is possible that $f \circ g$ and $g \circ f$ are different functions.

NOTE   *The compositions $f \circ g$ and $g \circ f$ are usually unequal.*

Later, we will develop calculus formulas related to basic functions such as power functions, root functions, and the basic trigonometric functions. We will then develop formulas for handling combinations of these basic functions. In order to apply these formulas, you will need to be able to recognize products, quotients, and compositions of the basic functions. We will treat composite functions informally, but you must recognize them when you see them. The key is to read the functions properly. For example:

- $\sqrt{x^2 + 1}$ is the *square root of* $x^2 + 1$, a composition of the square root function and the function $x^2 + 1$. The function $\sqrt{x^2 + 1}$ is not the basic square root function, which has values $\sqrt{x}$. Formulas that we will develop for the square root function will not apply directly to the composition.
- $(2x - 1)^3$ is the *third power of* $2x - 1$, a composition of the power function and the function $2x - 1$. This function is different from the basic power function that has values x^3.
- $\sin(x^2)$ is the *sine of* x^2, a composition of the sine function and the square function.
- $\sin^2 x = (\sin x)^2$ is the *square of* $\sin x$, a composition of the square function and the sine function.

NOTE $\sin(x^2)$ *is read "sine of x^2" and* $\sin^2 x$ *is read "sine squared of x." Be sure that you understand the difference between these two different compositions of the sine function and the square function.*

Restrictions on the Domain of a Function

The domain is an important component of a function. In particular:

DEFINITION

Two functions f and g are **equal** if they have the same domain and $f(x) = g(x)$ for every x in that domain.

This means that two functions are considered to be unequal if they have different domains, even if they have the same values at points where both are defined.

Let us consider some factors that may restrict the domain of a function.

The domain of a function may be restricted by an explicit statement.

EXAMPLE 4

Sketch the graph of the function

$$S(x) = \sin x, \qquad -\frac{\pi}{2} \le x \le \frac{\pi}{2}.$$

Evaluate $S(\pi/6)$, $S(\pi)$, and $S(1)$.

SOLUTION   Although $\sin x$ is defined for all x, it is stated that the domain of S is restricted to $-\pi/2 \le x \le \pi/2$. The graph of S consists of that part of the graph of $y = \sin x$ with $-\pi/2 \le x \le \pi/2$. See Figure 2.1.4, where we have emphasized that the points $(-\pi/2, -1)$ and $(\pi/2, 1)$ are included in the graph.

We are told that the values of the function S are given by the formula $\sin x$ only for $-\pi/2 \le x \le \pi/2$. The function S is undefined for other values of x.

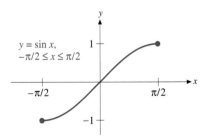

FIGURE 2.1.4 The domain of the function S is $-\pi/2 \le x \le \pi/2$. For these values of x, $S(x) = \sin x$; $S(x)$ is undefined for other values of x.

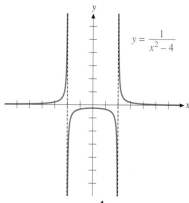

FIGURE 2.1.5 $\dfrac{1}{x^2 - 4}$ is undefined for $x = \pm 2$, where the denominator is zero.

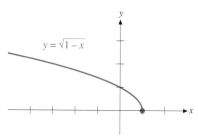

FIGURE 2.1.6 The domain of $g(x) = \sqrt{1 - x}$ consists of all $x \leq 1$.

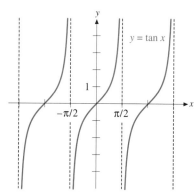

FIGURE 2.1.7 The function $h(x) = \tan x$ is undefined at all odd integer multiples of $\pi/2$.

$S(\pi/6) = \sin(\pi/6) = 1/2.$
$S(\pi)$ is undefined.
$S(1) = \sin 1 \approx 0.8415.$ ∎

Unless stated otherwise, the domain of a function that is given by a formula includes all points where the formula makes sense.

Points where the formula does not make sense are excluded from the domain without comment.

EXAMPLE 5

Indicate the domains of the functions

    (a) $f(x) = \dfrac{1}{x^2 - 4}$,

    (b) $g(x) = \sqrt{1 - x}$,

    (c) $h(x) = \tan x$.

SOLUTION

    (a) Since we cannot divide by zero, we must avoid $x = \pm 2$. The domain of f consists of all real numbers except $x = 2$ and $x = -2$. The graph of f is given in Figure 2.1.5.

    (b) Since the square root of a negative number is undefined, we need $1 - x \geq 0$. Solving this inequality, we find the domain of g consists of all real numbers x with $x \leq 1$. See Figure 2.1.6.

    (c) The tangent function is undefined at all odd integer multiples of $\pi/2$. The domain of h consists of all real numbers except odd integer multiples of $\pi/2$. See Figure 2.1.7. ∎

The domain of a function may be restricted by the interpretation of the values of the function as a physical quantity.

EXAMPLE 6

Express the slope of the line segment between the origin and a variable point (x, x^2) on the graph of $y = x^2$, as a function of x.

SOLUTION   Let $m(x)$ denote the desired slope. Substitution into the formula for the slope of the line segment between the points (x, x^2) and $(0, 0)$ gives

$$m(x) = \frac{x^2 - 0}{x - 0}.$$

Simplifying, we obtain the function

$$m(x) = x, \qquad x \neq 0.$$

The restriction of the domain of m reflects the fact that $m(x)$ does not represent the slope of the line segment between $(0, 0)$ and (x, x^2) if $x = 0$. When $x = 0$, (x, x^2) and $(0, 0)$ are the same point and there is no line segment between them. See Figure 2.1.8. ∎

The domain of a function may be restricted by the interpretation of the independent variable as a physical quality.

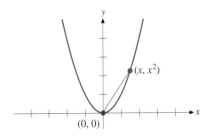

FIGURE 2.1.8 The slope of the line segment between $(0, 0)$ and (x, x^2) is $m(x) = x$, $x \neq 0$. When $x = 0$, $(0, 0)$ and (x, x^2) are the same point and there is no line segment between them.

FIGURE 2.1.9 The area of the rectangle is $A(x) = x(2 - x)$, $0 \leq x \leq 2$. Since x and $2 - x$ represent lengths, they must be nonnegative. We include the endpoints, $x = 0$ and $x = 2$, in the domain, even though a rectangle with two sides of length zero is not really a rectangle.

In particular:

Length is nonnegative.

Variable expressions that represent length must be nonnegative. (Length zero gives degenerate cases of geometric figures such as rectangles, but it will be convenient for future applications of calculus to include these cases. We will need to check that we do not get a degenerate case as the solution of a real problem.)

EXAMPLE 7

A rectangle has length x and width $2 - x$. Express the area of the rectangle as a function of x. Indicate the domain.

SOLUTION   The rectangle is sketched and labeled in Figure 2.1.9. The area of a rectangle is given by

$$\text{Area} = (\text{Length})(\text{Width}).$$

Substitution then gives the formula

$$\text{Area} = x(2 - x).$$

The formula makes sense for all x, but it will represent the area of a rectangle only if the values of x are restricted. Since x and $2 - x$ represent the length of sides of a rectangle, we must have $x \geq 0$ and $2 - x \geq 0$. These two inequalities are equivalent to the double inequality $0 \leq x \leq 2$. We include the endpoints, $x = 0$ and $x = 2$, in the domain, even though a rectangle with the length of two sides zero is not really a rectangle. Thus, we obtain the function

$$A(x) = x(2 - x), \qquad 0 \leq x \leq 2. \quad \blacksquare$$

Functions Defined by More Than One Formula

The most familiar function that is defined by different formulas for different values of x is the absolute value function, $|x|$, which we express in the form

$$|x| = \begin{cases} x, & x \geq 0, \\ -x, & x < 0. \end{cases}$$

Generally, we will use the notation

$$f(x) = \begin{cases} \text{First formula,} & \text{Interval where first formula holds,} \\ \vdots & \vdots \\ \text{Last formula,} & \text{Interval where last formula holds,} \end{cases}$$

where the intervals where different formulas hold are disjoint. In order to evaluate a function defined in this form, it is important to check which interval a particular value of x is in, so we know which formula gives the value $f(x)$.

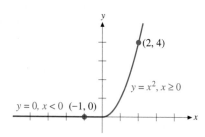

FIGURE 2.1.10   Since $-1 < 0$ and $f(x) = 0$ for $x < 0$, we have $f(-1) = 0$. Since $2 \geq 0$ and $f(x) = x^2$ for $x \geq 0$, we have $f(2) = (2)^2 = 4$.

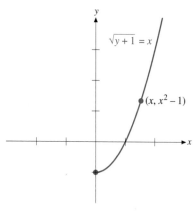

FIGURE 2.1.11   The equation $\sqrt{y+1} = x$ does define y as a function of x. The function is $f(x) = x^2 - 1$, $x \geq 0$.

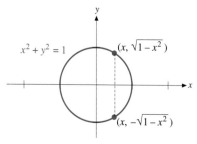

FIGURE 2.1.12   The equation $x^2 + y^2 = 1$ does not define y as a function of x. Some vertical lines intersect the graph at more than one point.

EXAMPLE 8

Indicate the domain and sketch the graph of the function

$$f(x) = \begin{cases} 0, & x < 0, \\ x^2, & x \geq 0. \end{cases}$$

Evaluate $f(-1)$ and $f(2)$.

SOLUTION   The function is defined for all x, so the domain is the set of all real numbers. For $x < 0$, the graph coincides with the graph of the line $y = 0$. For $x \geq 0$, the graph coincides with the parabola $y = x^2$. The graph is sketched in Figure 2.1.10.

In order to find the value $f(x)$ for a particular value of x, we must check the definition of f to determine which formula applies for the interval of definition that contains x.

Since $x = -1$ satisfies $x < 0$ and $f(x) = 0$ for $x < 0$, we have $f(-1) = 0$.

Since $x = 2$ satisfies $x \geq 0$ and $f(x) = x^2$ for $x \geq 0$, we have $f(2) = (2)^2 = 4$.   ∎

Functions Defined by Equations

An *equation* in x and y determines y as a function of x if and only if the *graph of the equation* determines y as a function of x. In many cases we can obtain a formula for a function that is given by an equation in x and y by solving the equation for y in terms of x.

EXAMPLE 9

Sketch the graphs of the following equations and determine which define y as a function of x. Find a formula for the values and indicate the domain and range of each function.

(a) $\sqrt{y+1} = x$,
(b) $x^2 + y^2 = 1$.

SOLUTION

(a) The graph of $\sqrt{y+1} = x$ is given in Figure 2.1.11. Note that the square root function has nonnegative values, so points (x, y) on the graph satisfy $x \geq 0$. From the graph we see that the equation does define y as a function of x for $x \geq 0$. We also see that the range consists of all $y \geq -1$.

Solving the equation for y in terms of x, we have

$$\sqrt{y+1} = x,$$
$$y + 1 = x^2,$$
$$y = x^2 - 1.$$

This formula gives the values of the function, but the domain must be restricted to $x \geq 0$, since the original equation cannot be satisfied for $x < 0$. The function is

$$f(x) = x^2 - 1, \qquad x \geq 0.$$

(b) The graph is sketched in Figure 2.1.12. Since some vertical lines intersect the graph at more than one point, this is not the graph of a function.

If we solve the equation $x^2 + y^2 = 1$ for y in terms of x, we obtain $y^2 = 1 - x^2$, so $y = \pm\sqrt{1 - x^2}$. Thus, the equation does not have a *unique* solution for y in terms of x. This equation does not define y as a function of x. ∎

Symmetry; Even and Odd Functions

When sketching graphs, it is sometimes useful to take advantage of certain simple symmetries.

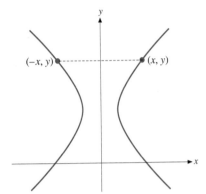

FIGURE 2.1.13a   A graph is symmetric with respect to the *y*-axis if reflecting the graph about the *y*-axis does not change the graph.

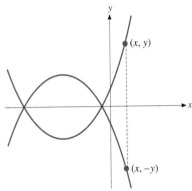

FIGURE 2.1.13b   A graph is symmetric with respect to the *x*-axis if reflecting the graph about the *x*-axis does not change the graph.

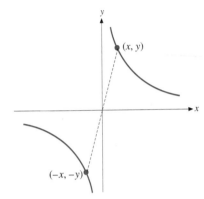

FIGURE 2.1.13c   A graph is symmetric with respect to the origin if reflecting each point on the graph about the origin does not change the graph.

DEFINITION

A graph is said to be **symmetric with respect to the y-axis** if $(-x, y)$ is on the graph whenever (x, y) is. See Figure 2.1.13a.

A graph is **symmetric with respect to the x-axis** if $(x, -y)$ is on the graph whenever (x, y) is. See Figure 2.1.13b.

A graph is **symmetric with respect to the origin** if $(-x, -y)$ is on the graph whenever (x, y) is. See Figure 2.1.13c.

A function f is called **even** if $f(-x) = f(x)$ for all x in the domain of f; f is called **odd** if $f(-x) = -f(x)$ for all x in the domain.

Note that if f is either odd or even, then $-x$ must be in the domain of f whenever x is.

It is easy to check that the graph of an even function is symmetric with respect to the y-axis. The graph of an odd function is symmetric with respect to the origin. Except for the trivial case $f(x) = 0$, the graph of a *function* cannot be symmetric with respect to the x-axis.

EXAMPLE 10

(a) $f(x) = x^3 - x$ is an odd function, because

$$f(-x) = (-x)^3 - (-x)$$
$$= -(x^3 - x) = -f(x).$$

The graph is symmetric with respect to the origin. See Figure 2.1.14.

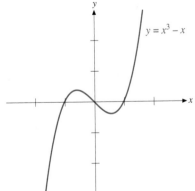

FIGURE 2.1.14 $f(x) = x^3 - x$ is an odd function. Its graph is symmetric with respect to the origin.

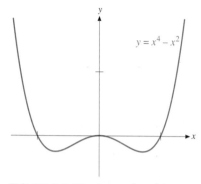

FIGURE 2.1.15 $f(x) = x^4 - x^2$ is an even function. Its graph is symmetric with respect to the *y*-axis.

(b) $f(x) = x^4 - x^2$ is an even function, because

$$f(-x) = (-x)^4 - (-x)^2 = x^4 - x^2 = f(x).$$

The graph is symmetric with respect to the *y*-axis. See Figure 2.1.15.

(c) $f(x) = x^2 - x$ is neither an odd function nor an even function. The expression

$$f(-x) = (-x)^2 - (-x) = x^2 + x$$

is different from both $f(x)$ and $-f(x)$. The graph is symmetric with respect to neither the *y*-axis nor the origin. See Figure 2.1.16.

(d) $f(x) = \tan x$ is an odd function, because

$$f(-x) = \tan(-x) = -\tan x = -f(x).$$

The graph is symmetric with respect to the origin. See Figure 2.1.17.

(e) $f(x) = 1/x^2$ is an even function, because

$$f(-x) = \frac{1}{(-x)^2} = \frac{1}{x^2} = f(x).$$

The graph is symmetric with respect to the *y*-axis. See Figure 2.1.18. ■

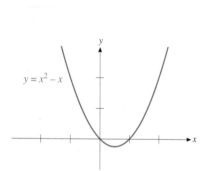

FIGURE 2.1.16 $f(x) = x^2 - x$ is neither an odd function nor an even function. The graph is symmetric with respect to neither the *y*-axis nor the origin.

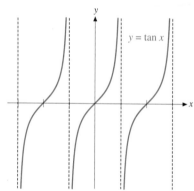

FIGURE 2.1.17 $f(x) = \tan x$ is an odd function. Its graph is symmetric with respect to the origin.

FIGURE 2.1.18 $f(x) = 1/x^2$ is an even function. Its graph is symmetric with respect to the *y*-axis.

EXERCISES 2.1

1. $f(x) = 2x^2 - 3$. Find $f(3)$ and $f(\sqrt{2})$.

2. $f(x) = \dfrac{x}{1-x}$. Find $f(5)$ and $f\left(\dfrac{2}{3}\right)$.

3. $f(x) = 2x - 1$. Find $f\left(\dfrac{1}{x}\right)$ and $\dfrac{1}{f(x)}$.

4. $f(x) = 2x - 3$. Find $f(x^2)$ and $[f(x)]^2$.

5. $f(x) = 3x^2 - 2$. Find $f(x-1)$ and $f(x) - 1$.

6. $f(x) = 3x - 2$. Find $\dfrac{f(x) - f(-3)}{x+3}$.

7. $f(x) = x^2 + x$. Find $\dfrac{f(x) - f(-2)}{x+2}$.

8. $f(x) = \dfrac{x}{x-3}$. Find $\dfrac{f(x) - f(1)}{x-1}$.

In Exercises 9–12, evaluate the given expressions for the functions $f(x) = 3x^2 - 2x + 1$ and $g(x) = 2x - 1$.

9. $f(x) + g(x)$, $f(x) + (x+1)g(x)$

10. $f(x) \cdot g(x)$, $f(x) - [g(x)]^2$

11. $f(g(x))$

12. $g(f(x))$

Sketch the graphs of the functions in Exercises 13–20. Find the values indicated.

13. $f(x) = x^2, x \geq 0$; $f(-1)$, $f(0)$, $f(2)$

14. $f(x) = 2x + 4, x \geq -2$; $f(-3)$, $f(0)$, $f(2)$

15. $f(x) = \cos x, 0 \leq x \leq \pi$; $f(-\pi/2)$, $f(\pi/2)$, $f(2\pi/3)$

16. $f(x) = \tan x, \quad -\pi/2 < x < \pi/2$; $f(-\pi/2)$, $f(\pi/4)$, $f(3\pi/4)$

17. $f(x) = \begin{cases} 1, & x < 0, \\ 1 - x, & x > 0; \end{cases}$ $f(-2)$, $f(0)$, $f(2)$

18. $f(x) = \begin{cases} 1 - x, & x < 1, \\ x - 1, & x > 1; \end{cases}$ $f(0)$, $f(1)$, $f(3)$

19. $f(x) = \begin{cases} x, & 0 \leq x \leq 2, \\ 5 - x, & x > 2; \end{cases}$ $f(-1)$, $f(2)$, $f(3)$

20. $f(x) = \begin{cases} x^2, & x < 0, \\ x + 1, & 0 \leq x \leq 2; \end{cases}$ $f(-2)$, $f(0)$, $f(3)$

Indicate the domain of the functions in Exercises 21–26.

21. $f(x) = \dfrac{x}{x^2 - 1}$

22. $f(x) = \dfrac{x + 1}{x^2 + 5x + 4}$

23. $f(x) = \sqrt{3 - 2x}$

24. $f(x) = \dfrac{1}{\sqrt{x^2 + 3x - 4}}$

25. $f(x) = \sec x$

26. $f(x) = \csc x$

27. Find the slope of the line segment between the point $(1, 2)$ and a variable point on the line $2x + y = 4$, as a function of x. Indicate the domain.

28. Find the slope of the line segment between the point $(-2, 4)$ and a variable point on the parabola $y = x^2$, as a function of x. Indicate the domain.

29. Find the slope of the line segment between the point on the parabola $y = x^2$ that has x-coordinate 1 and a variable point on the graph, as a function of x. Indicate the domain.

30. Find the slope of the line segment between the point on the parabola $y = 1 - x^2$ that has x-coordinate 2 and a variable point on the graph, as a function of x. Indicate the domain.

31. A rectangular box has length x, width $2 - x$, and height $3 - x$. Express the volume of the box as a function of x. Indicate the domain.

32. Express the total surface area of the six sides of the box in Exercise 31 as a function of x. Indicate the domain.

33. A rectangular box has a square base and no top. The length of one side of the square base is x. The height of the box is $5 - x$. Express the total surface area of the five sides of the box as a function of x. Indicate the domain.

34. Express the volume of the box in Exercise 33 as a function of x. Indicate the domain.

35. A container with no top is made in the shape of a cube by cutting five squares whose sides have length s and soldering them along the edges. If the metal costs 0.5 cent a square centimeter and soldering costs 6 cents a centimeter, write the total cost C in cents of the metal and soldering as a function of s. Indicate the domain.

36. As many as possible square pieces with length of sides s are to be cut from a square sheet that is 36 cm on a side, with cuts parallel to the sides of the sheet. (See figure.) Express the cost per piece as a function of the length of sides, s, if the pieces must have sides at least 9 cm and the sheet costs 72 cents.

37. A variable point $(x, \sqrt{x})$ on the graph of $y = \sqrt{x}$ is revolved about the x-axis. Express the (a) area and (b) circumference of the resulting circle as a function of x.

38. A variable point $(x, \sqrt{x})$ on the graph of $y = \sqrt{x}$ is revolved about the y-axis. Express the (a) area and (b) circumference of the resulting circle as a function of x.

39. Express the length of a vertical line segment between the x-axis and a variable point (x, x^2) on the graph of $y = x^2$ as a function of x.

40. Express the length of a vertical line segment between a variable point $(x, x - 1)$ on the graph of $y = x - 1$ and a variable point (x, x^2) on the graph of $y = x^2$ as a function of x.

Each graph in Exercises 41–44 is the graph of a function. Find the domain, range, and values indicated.

41. $f(-4)$, $f(1)$, $f(3)$

42. $f(-2), f(0), f(2)$

43. $f(-1), f(0), f(2)$

44. $f(1), f(3/2), f(3)$

Sketch the graphs of the equations in Exercises 45–54. Determine which of them define y as a function of x. Find a formula for the values and indicate the domain and range of each function.

45. $2x + 3y + 6 = 0$ **46.** $x^2 - 4x + y = 0$

47. $x - y^2 + 1 = 0$ **48.** $x - \sqrt{y+1} = 0$

49. $\sqrt{1-x} + y = 0$ **50.** $(x-1)^2 + y^2 = 1$

51. $x^2 y = 4$ **52.** $xy^2 = 4$

53. $x = \sin y$ **54.** $x = \cos y$

55. Does the equation $4x^2 + 4xy + y^2 = 0$ define y as a function of x? Explain.

56. Does the equation $2x^2 + 3xy + y^2 = 0$ define y as a function of x? Explain.

57. Find a function of x with a graph that is the upper half of the circle with center the origin and radius 2.

58. Explain why there is no function of x with a graph that is the right half of the circle with center the origin and radius 2.

Express the functions in Exercises 59–62 without absolute values by writing them in terms of different formulas for different values of x.

59. $f(x) = |x - 2|$ **60.** $f(x) = |2x - 1|$

61. $f(x) = \dfrac{|x-2|}{x-2}$ **62.** $f(x) = \dfrac{|x+1|}{x+1}$

63. Verify that $\dfrac{|x|+x}{2} = \begin{cases} x, & \text{if } x \geq 0, \\ 0, & \text{if } x < 0. \end{cases}$

64. Use the absolute value function in a way that is similar to the way it is used in Exercise 63 to write a single formula for the function that has value $f(x)$ if $f(x) \geq g(x)$, and value $g(x)$ if $g(x) > f(x)$. (This function is called the **maximum** of f and g.)

In Exercises 65–66, find formulas for and indicate the domains of the functions (a) $f \circ g$ and (b) $g \circ f$.

65. $f(x) = \sqrt{x-1}$, $g(x) = x^2 - 1$

66. $f(x) = \sqrt{1-x^2}$, $g(x) = \sqrt{2-x^2}$

67. If $f(x) = x^2$, $g(x) = \sqrt{x}$, and $h(x) = x$, explain why $f \circ g \neq h$.

68. If $f(x) = \sqrt{x}$, $g(x) = x^2$, and $h(x) = x$, explain why $f \circ g \neq h$.

69. $f(x) = 2x - 1$ and $h(x) = x$. Find a function g such that $f \circ g = h$.

70. $f(x) = x^3 + 1$ and $h(x) = x$. Find a function g such that $f \circ g = h$.

Determine whether the graphs given in Exercises 71–76 appear to be symmetric with respect to the x-axis, the y-axis, or the origin.

71.

72.

73.

74.

75.

76.

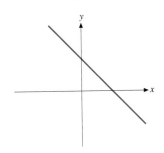

Determine if the functions given in Exercises 77–84 are odd, even, or neither.

77. $f(x) = x^4 + 2x^2 + 1$ **78.** $f(x) = 5x^4 + 3x^2 + 1$

79. $f(x) = 2x^3 - 3x + 1$ **80.** $f(x) = x^3 + 2x$

81. $f(x) = \dfrac{x^2 + 4}{x^3 - x}$ **82.** $f(x) = \dfrac{x^3 + 5}{x^2 + 9}$

83. $f(x) = x \sin x$ **84.** $f(x) = x \cos x$

85. For which integers n is $f(x) = x^n$ an even function? For which integers n is it an odd function?

86. Is a product or quotient of two even functions odd or even?

87. Is a product or quotient of two odd functions odd or even?

88. Is a product or quotient of an odd function and an even function odd or even?

89. Use the given graph of $f(x)$ to sketch the graphs of
   (**a**) $f(-x)$,   (**b**) $-f(x)$,
   (**c**) $f(|x|)$, and   (**d**) $|f(x)|$.

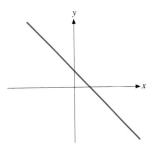

90. Use the given graph of $f(x)$ to sketch the graphs of
   (**a**) $f(-x)$,   (**b**) $-f(x)$,
   (**c**) $f(|x|)$, and   (**d**) $|f(x)|$.

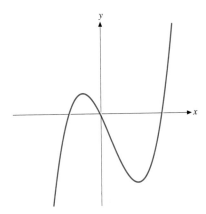

91. A function of the form $f(x) = ax + b$, where a and b are constants, is called a **linear function**.
   (**a**) Show that the composition of two linear functions is a linear function.
   (**b**) Express the slope of the line that is the graph of the composition in terms of the slopes of the graphs of the composing functions.

2.2 A GRAPHIC APPROACH TO BOUNDS OF FUNCTIONS

Connections

Inequalities and absolute value, 1.1.

Interpretation of graphs, 1.4–5.

Graphs of the sine and cosine functions, 1.6.

FIGURE 2.2.1 The function *f* is bounded from below by *m* and *m'*; *f* is bounded from above by *M* and *M'*.

The concept of bounds of functions is basic to the study of calculus. In this section we will introduce the concept in terms of graphs and show that it is easy to determine a bound of a function from its graph. We will then illustrate some analytic techniques for finding bounds of certain types of functions. In later sections we will use bounds of functions to facilitate understanding of the concept of the limit of a function. We will also use bounds in several types of error analyses.

The Relation of Upper and Lower Bounds to Graphs

DEFINITION

If $f(x) \geq m$ for all x in the domain of f, then m is a **lower bound** of f and f is said to be **bounded from below by *m*.** If $f(x) \leq M$ for all x in the domain of f, then M is an **upper bound** of f and f is said to be **bounded from above by *M*.**

Geometrically, a function f is bounded from below by m if the graph of f is above or touching the horizontal line $y = m$. Note that if m is a lower bound of f and $m' \leq m$, then m' is also a lower bound of f. A function f is bounded from above by M if the graph of f is below or touching the horizontal line $y = M$. If M is an upper bound of f and $M' \geq M$, then M' is also an upper bound of f. See Figure 2.2.1.

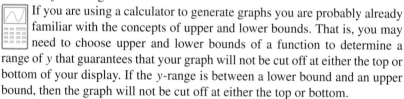 If you are using a calculator to generate graphs you are probably already familiar with the concepts of upper and lower bounds. That is, you may need to choose upper and lower bounds of a function to determine a range of y that guarantees that your graph will not be cut off at either the top or bottom of your display. If the y-range is between a lower bound and an upper bound, then the graph will not be cut off at either the top or bottom.

It is easy to determine upper and lower bounds of a function from the graph of the function. We need only choose numbers m and M so that the graph of the function is between (possibly touching) the horizontal lines $y = m$ and $y = M$.

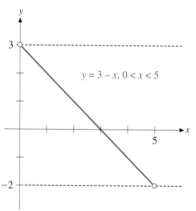

FIGURE 2.2.2 The graph of $f(x) = 3 - x$, $0 < x < 5$ is between the horizontal lines $y = -2$ and $y = 3$.

EXAMPLE 1

The graph of $f(x) = 3 - x$, $0 < x < 5$, is sketched in Figure 2.2.2. From the graph we see that $m = -2$ is a lower bound and $M = 3$ is an upper bound. ∎

EXAMPLE 2

From the graph in Figure 2.2.3 we see that $f(x) = 2 \sin(3x)$ has lower bound $m = -2$ and upper bound $M = 2$. ∎

A sketch of the graph of a function can also be used to determine that a function is unbounded from either above or below. Functions become unbounded as points on the graph approach a vertical asymptote.

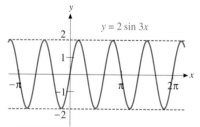

FIGURE 2.2.3 The graph of $f(x) = 2\sin(3x)$ is between the horizontal lines $y = 2$ and $y = -2$.

EXAMPLE 3

From the graph in Figure 2.2.4 we see that the function $f(x) = 1/(2x - 1)$, $x > 1/2$, is unbounded from above. Note that no matter how large M is chosen,

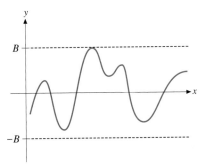

FIGURE 2.2.4 The function $f(x) = \dfrac{1}{2x - 1}$, $x > 1/2$, is unbounded from above.

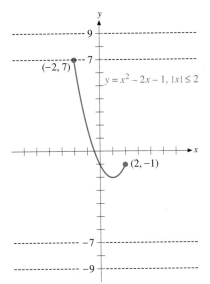

FIGURE 2.2.5 A function is bounded by B if its graph is between the horizontal lines $y = B$ and $y = -B$.

FIGURE 2.2.6 The Triangle Inequality can be used to show that $B = 9$ is a bound of $f(x) = x^2 - 2x - 1$, $|x| \leq 2$. From the graph we see that $B = 7$ is also a bound.

the graph of $y = f(x)$ is above the line $y = M$ for all values of $x > 1/2$ that are near enough to $1/2$, so no number M can be an upper bound of f. ∎

We can determine upper and lower bounds of a function simultaneously by investigating the absolute value of a function.

DEFINITION

If $|f(x)| \leq B$ for all x in the domain of f, then B is called a **bound** of f and f is said to be **bounded by B.**

Geometrically, f is bounded by B if the graph of f is between (possibly touching) the horizontal lines $y = -B$ and $y = B$. See Figure 2.2.5. It is geometrically clear that a function is bounded if and only if it is both bounded from below and bounded from above.

The following strategy can be used to determine a bound of a function from its graph. Of course, this method is convenient only if you have easy access to the graph of the function.

To find a bound of a function from its graph

- Graph the function.
- Find a number B such that the graph of the function is between the horizontal lines $y = -B$ and $y = B$.
- The number B is then a bound of the function.

Using Absolute Values and the Triangle Inequality

It is worthwhile to be able to determine bounds of functions in a systematic and *easy* way from the formulas that define the functions. To do this we will use the following properties of absolute value to separate complicated problems into simpler parts:

$$|f(x) + g(x)| \leq |f(x)| + |g(x)|,$$
$$|f(x)g(x)| = |f(x)||g(x)|,$$
$$\left|\frac{f(x)}{g(x)}\right| = \frac{|f(x)|}{|g(x)|}.$$

The **Triangle Inequality** $|f + g| \leq |f| + |g|$ allows us to find a bound of a sum by finding bounds of each term separately. That is:

If $|f(x)| \leq M$ and $|g(x)| \leq N$ for all x in the domain of $f + g$, then

$$|f(x) + g(x)| \leq |f(x)| + |g(x)| \leq M + N,$$

so $M + N$ is a bound of $f + g$.

EXAMPLE 4
Find a bound of the function

$$f(x) = x^2 - 2x - 1, \qquad |x| \leq 2.$$

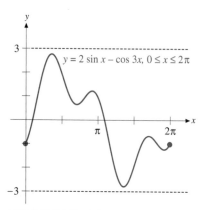

FIGURE 2.2.7 The graph of $\sin x$ is between the horizontal lines $y = -1$ and $y = 1$, so $|\sin x| \le 1$. We have $|\cos x| \le 1$, so $B = 1$ is a bound of $\cos x$.

FIGURE 2.2.8 It is very easy to use the Triangle Inequality to determine that $B = 3$ is a bound of $f(x) = 2\sin x - \cos 3x$, $0 \le x \le 2\pi$.

SOLUTION Using the Triangle Inequality, we have

$$|f(x)| = |x^2 - 2x - 1| \le |x^2| + |-2x| + |-1|$$
$$= |x|^2 + |-2||x| + |-1|$$
$$= |x|^2 + 2|x| + 1.$$

Since $|x| \le 2$, we then obtain

$$|f(x)| \le (2)^2 + 2(2) + 1 = 4 + 4 + 1 = 9,$$

so $B = 9$ is a bound of f on the set $|x| \le 2$. See Figure 2.2.6. ∎

Note that use of the Triangle Inequality results in a *sum of nonnegative terms*. Each term is positive or zero and there are plus signs between each of the terms.

The bound $B = 9$ obtained in Example 4 is not the smallest bound of the function. From the graph in Figure 2.2.6 we see that $B = 7$ is the smallest choice for bounds of f. For many problems the exact value of the smallest bound is not important. The important point is that we obtained the bound $B = 9$ in a systematic (and easy) way.

Bounds of the Sine and Cosine Functions

We know that the values of the sine and cosine functions range between -1 and 1. See Figure 2.2.7. It follows that

$$|\sin u| \le 1 \quad \text{and} \quad |\cos u| \le 1,$$

where u represents any real number or expression.

EXAMPLE 5

Find a bound of the function

$$f(x) = 2\sin x - \cos 3x, \qquad 0 \le x \le 2\pi.$$

SOLUTION Using the Triangle Inequality, we have

$$|f(x)| = |2\sin x - \cos 3x| \le |2\sin x| + |-\cos 3x|$$
$$= 2|\sin x| + |\cos 3x| \le 2(1) + (1) = 3,$$

so $B = 3$ is a bound of f. See Figure 2.2.8. ∎

We see from the graph in Figure 2.2.8 that $B = 3$ is not the smallest bound of the function in Example 5. In this example it is somewhat difficult to determine the smallest bound, but it is very easy to establish that $B = 3$ is a bound.

Bounds of Products

We can use $|fg| = |f||g|$ to find a bound of a product by finding a bound for each factor separately. That is:

If $|f(x)| \le M$ and $|g(x)| \le N$ for all x in the domain of fg, then

$$|f(x)g(x)| = |f(x)||g(x)| \le MN,$$

so MN is a bound of fg.

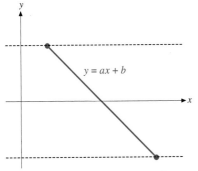

FIGURE 2.2.9   We can find a bound of $f(x) = x \sin x$, $0 \leq x \leq 2\pi$, by finding a bound for each factor separately.

FIGURE 2.2.10   The values of $ax + b$ for x in an interval are contained between its values at the endpoints of the interval.

EXAMPLE 6

Find a bound of the function

$$f(x) = x \sin x, \qquad 0 \leq x \leq 2\pi.$$

SOLUTION   We see that $f(x)$ is a product of the two factors x and $\sin x$. We have $|x| \leq 2\pi$ for $0 \leq x \leq 2\pi$ and we know that $|\sin x| \leq 1$. It follows that

$$|f(x)| = |x \sin x| = |x||\sin x| \leq (2\pi)(1) = 2\pi.$$

Therefore, $B = 2\pi$ is a bound of f on the set $0 \leq x \leq 2\pi$. See Figure 2.2.9. ■

Bounds of Linear Factors

The graph of $y = ax + b$ is a line, so the values of $ax + b$ for x in an interval are contained between its values at the endpoints of the interval. See Figure 2.2.10.

A bound of the linear factor $ax + b$ for x in an interval is given by the larger of the absolute value of $ax + b$ at each endpoint of the interval.

If a function is defined on a bounded interval and has values that are products of linear factors, we can use the endpoints of the interval to find bounds of each linear factor. It does not matter whether or not the domain of the function includes the endpoints of the interval. Also, bounds of different linear factors may be given by different endpoints of the interval.

EXAMPLE 7

Find a bound of the function

$$f(x) = (x + 1)(2x - 5), \qquad 0 < x < 2.$$

SOLUTION   Since the two factors of f are linear, we can obtain a bound for each of them by evaluating their absolute values at 0 and 2, the endpoints of the interval of definition of f. It does not matter that these points are not included in the domain of f.

Substitution of 0 and 2 for x in the linear factor $x + 1$ gives

$$|(0) + 1| = 1 \text{ and } |(2) + 1| = 3.$$

The larger of these values gives a bound of 3 for the factor $x + 1$.

Substitution of 0 and 2 for x in the linear factor $2x - 5$ gives

$$|2(0) - 5| = 5 \text{ and } |2(2) - 5| = 1,$$

so the factor $2x - 5$ is bounded by 5 on the interval $0 < x < 2$.

Combining results we have

$$|f(x)| = |x + 1||2x - 5| \leq (3)(5) = 15,$$

so $B = 15$ is a bound of f. See Figure 2.2.11. ■

From the graph in Figure 2.2.11 it appears that the function in Example 7 is bounded by a number much less than fifteen. Also, we see that values of the product of the two linear terms at endpoints of the interval of definition do not give a bound of the product. We can use endpoint values to obtain bounds for each linear factor, but we cannot use endpoint values of the (nonlinear) product to obtain a bound of the product.

NOTE   *Do not try to obtain a bound of a nonlinear function by using values at the endpoints of the interval where the function is defined.*

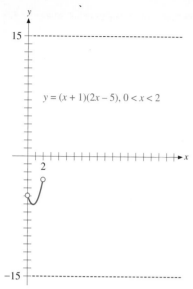

FIGURE 2.2.11 We can find a bound of $f(x) = (x + 1)(2x - 5)$, $0 < x < 2$, by finding a bound for each linear factor separately.

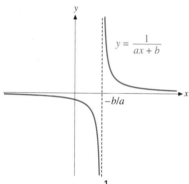

FIGURE 2.2.12 $\dfrac{1}{ax + b}$ is bounded on an interval unless the interval either contains $x = -b/a$ or has $x = -b/a$ as an endpoint.

FIGURE 2.2.13 We can see from the graph that $1/(2x - 1)$ is bounded on the interval $x > 2$. The absolute value of the expression at the endpoint closer to the vertical asymptote gives a bound.

Linear Factors in the Denominator

It is easy to determine a bound of a factor of the form $1/(ax + b)$ for x in an interval by sketching a graph of $1/(ax + b)$. From Figure 2.2.12 we see that the graph has vertical asymptote $x = -b/a$, but is bounded on intervals that neither contain $-b/a$ nor have $-b/a$ as an endpoint. Also, the value of $1/(ax + b)$ at the endpoint of the interval that is closer to $-b/a$ gives a bound of $1/(ax + b)$ for x in the interval.

To find a bound of 1 / (ax + b) on an interval

- Sketch the graph of $1/(ax + b)$ with its vertical asymptote $x = -b/a$.
- Check that the interval neither contains $-b/a$ nor has $-b/a$ as an endpoint.
- The value of $1/|ax + b|$ at the endpoint of the interval that is closer to $-b/a$ then gives a bound of $1/(ax + b)$ on the interval.

Note that $1/(ax + b)$ is unbounded on intervals that either contain $-b/a$ or have $-b/a$ as an endpoint.

NOTE   *Do not use the value of $1/|ax + b|$ at an endpoint of an interval as a bound of $1/(ax + b)$ for x in the interval without checking that the interval neither contains $-b/a$ nor has $-b/a$ as an endpoint.*

EXAMPLE 8
Find a bound of the function

$$f(x) = \frac{1}{2x - 1}, \qquad x > 2.$$

SOLUTION   A sketch of the graph of $1/(2x - 1)$ is given in Figure 2.2.13. The denominator $2x - 1$ is zero when $x = 1/2$, so the graph has vertical asymptote $x = 1/2$. It is clear from the figure that $1/(2x - 1)$ is bounded on the interval $x > 2$ and that the value of $1/|2x - 1|$ at $x = 2$ gives a bound of f on the interval. We have

$$|f(x)| = \frac{1}{|2x - 1|} \leq \frac{1}{|2(2) - 1|} = \frac{1}{3} \quad \text{for} \quad x > 2.$$

The function f is bounded by $B = 1/3$ on the set $x > 2$. ∎

Recall from Example 3 that $1/(2x - 1)$ is not bounded on the interval $x > 1/2$. The vertical asymptote of $1/(2x - 1)$ is at an endpoint of that interval. Also, $1/(2x - 1)$ is unbounded on any interval that contains $x = 1/2$ in its interior. Don't forget to check that an interval does not contain an asymptote of $1/(ax + b)$ before using values at the endpoint of the interval to obtain a bound.

EXAMPLE 9
Find a bound of

$$f(x) = \frac{(x + 1)^2}{x - 1}, \qquad |x| < 0.2.$$

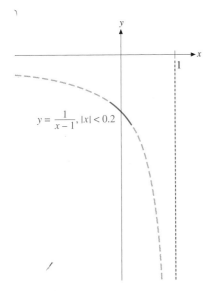

$$y = \frac{1}{x-1}, |x| < 0.2$$

FIGURE 2.2.14 We see from the graph that $1/(x-1)$ is bounded for $-0.2 < x < 0.2$. The absolute value of the expression at the endpoint closer to the vertical asymptote gives a bound.

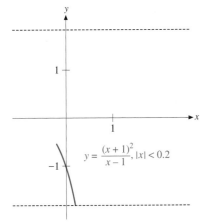

$$y = \frac{(x+1)^2}{x-1}, |x| < 0.2$$

FIGURE 2.2.15 We can find a bound of $f(x) = (x+1)^2/(x-1)$, $|x| < 0.2$, by finding a bound for each of the factors $(x+1)^2$ and $1/(x-1)$.

SOLUTION We need bounds of $(x+1)^2$ and $1/(x-1)$ for $|x| < 0.2$. These factors involve linear terms, so we can obtain bounds by using values at endpoints of the interval $|x| < 0.2$. Rewriting $|x| < 0.2$ as the double inequality

$$-0.2 < x < 0.2,$$

we see that the endpoints are 0.2 and -0.2.

Substitution of 0.2 and -0.2 for x in the linear factor $x + 1$ gives

$$|(0.2) + 1| = 1.2 \text{ and } |(-0.2) + 1| = 0.8.$$

The larger of these values, 1.2, gives a bound of $x + 1$.

From the sketch of $1/|x - 1|$ in Figure 2.2.14 we see that $1/(x - 1)$ is bounded for $-0.2 < x < 0.2$ and that a bound is given by the value of $1/|x - 1|$ at the endpoint $x = 0.2$. Substitution of $x = 0.2$ gives

$$\frac{1}{|(0.2) - 1|} = \frac{1}{0.8}.$$

Combining results, we obtain

$$|f(x)| = |x + 1|^2 \frac{1}{|x - 1|} \leq (1.2)^2 \left(\frac{1}{0.8}\right) = 1.8.$$

The function is bounded by $B = 1.8$. See Figure 2.2.15. ■

Factors of the Form $\dfrac{1}{u^2 + a^2}$

If u is any function of x and a is a positive number, then we have

$$0 \leq u^2,$$
$$a^2 \leq u^2 + a^2,$$
$$\frac{1}{u^2 + a^2} \leq \frac{1}{a^2},$$

so $1/a^2$ is a bound of $1/(u^2 + a^2)$ on any interval.

Similarly, we have

$$\frac{1}{\left(\begin{array}{c}\text{a nonnegative}\\\text{variable term}\end{array}\right) + \left(\begin{array}{c}\text{a positive}\\\text{constant}\end{array}\right)} \leq \frac{1}{0 + \left(\begin{array}{c}\text{the positive}\\\text{constant}\end{array}\right)}.$$

It follows that we can obtain bounds for functions that have factors in the denominator that consist of a nonnegative variable term plus a positive constant by substitution of zero for the nonnegative variable term.

EXAMPLE 10

Find a bound of the function

$$f(x) = \frac{1}{\cos^2 x + 4}.$$

SOLUTION We see that the denominator consists of the nonnegative variable term $\cos^2 x$ plus the positive constant four. We then have

$$0 \leq \cos^2 x$$
$$0 + 4 \leq \cos^2 x + 4,$$

FIGURE 2.2.16 The denominator consists of the nonnegative variable term $\cos^2 x$ plus the positive constant 4. We can obtain a bound by substituting zero for the nonnegative variable term.

so

$$|f(x)| = \frac{1}{\cos^2 x + 4} \le \frac{1}{0 + 4} \le \frac{1}{4}.$$

$B = 1/4$ is a bound of f. See Figure 2.2.16. ∎

Using Inequalities to Determine Bounds

We have seen how to find a bound of $1/g$ in case g is either a linear function or a sum of a nonnegative variable term and a positive constant. These are special cases of the general result that a function of the form $1/g(x)$ is bounded if $|g(x)|$ is bounded from below by a positive number. That is:

If there is a positive number d such that $|g(x)| \ge d > 0$ for all x in the domain of g, then

$$\frac{1}{|g(x)|} \le \frac{1}{d},$$

so $1/d$ is a bound of $1/g$.

In order that $1/g$ be bounded, it is important that there is a fixed positive number d such that $|g(x)| \ge d$ for all x in the domain. It is not enough merely that $|g(x)|$ is positive for each x in the domain. For example, $|2x - 1| > 0$ for $x > 1/2$, but it follows from Example 3 that $1/(2x - 1)$ is unbounded on the interval $x > 1/2$.

EXAMPLE 11

$|f(x) - 7| < 2$. Find a bound of **(a)** f and **(b)** $1/f$.

SOLUTION $|f(x) - 7| < 2$ implies the double inequality

$$-2 < f(x) - 7 < 2.$$

Adding 7 to the double inequality we obtain

$$5 < f(x) < 9, \text{ so } 5 < |f(x)| < 9.$$

(a) $|f(x)| < 9$ implies 9 is a bound of f.

(b) $5 < |f(x)|$ implies $\dfrac{1}{|f(x)|} < \dfrac{1}{5}$, so $1/5$ is a bound of $1/f$. ∎

EXERCISES 2.2

In Exercises 1–8 use the given graphs to find lower and upper bounds of the indicated functions.

1.

2.

3.

4.

5.

6.

7.

8.

In Exercises 9–48 find bounds of the given functions or indicate they are not bounded.

9. $f(x) = 3x^2 - 4x - 5, |x| \le 2$

10. $f(x) = 2x^2 - 3x - 7, |x| \leq 3$

11. $f(x) = x^3 + 2x - 3, |x| \leq 2$

12. $f(x) = x^4 - 5x^2 - 8, |x| \leq 2$

13. $f(x) = 1 - 2x$, all x

14. $f(x) = x^2 - 3x - 5$, all x

15. $f(x) = 2\sin x - 3\cos x, 0 \leq x \leq 2\pi$

16. $f(x) = 3\sin x + 2\cos x, 0 \leq x \leq 2\pi$

17. $f(x) = \sin 2x - \cos 3x, 0 \leq x \leq 2\pi$

18. $f(x) = \sin(x/2) + \cos(x/3), 0 \leq x \leq 2\pi$

19. $f(x) = \dfrac{1}{x+3}, x > 2$

20. $f(x) = \dfrac{1}{x+2}, x > 3$

21. $f(x) = \dfrac{1}{x-2}, x > 2$

22. $f(x) = \dfrac{1}{3-x}, x > 3$

23. $f(x) = \dfrac{1}{3x-2}, x > 4$

24. $f(x) = \dfrac{1}{2x-1}, x > 3$

25. $f(x) = \dfrac{1}{1-2x}, |x| < 0.3$

26. $f(x) = \dfrac{1}{1-3x}, |x| < 0.3$

27. $f(x) = \dfrac{1}{x}, |x-4| < 1$

28. $f(x) = \dfrac{1}{x}, |x-5| < 1$

29. $f(x) = \dfrac{1}{x^2+4}$, all x

30. $f(x) = \dfrac{1}{x^2+5}$, all x

31. $f(x) = \dfrac{1}{x^4+2x^2+5}$, all x

32. $f(x) = \dfrac{1}{x^4+5x^2+2}$, all x

33. $f(x) = x\cos x, |x| < 3$

34. $f(x) = x^2 \sin x, |x| < 3$

35. $f(x) = x^2 \sin^2 x, |x| \leq \pi$

36. $f(x) = x\cos(x^2), |x| \leq \pi$

37. $f(x) = \dfrac{x}{2x-6}, |x| < 2$

38. $f(x) = \dfrac{4x}{3x-11}, |x| < 3$

39. $f(x) = \dfrac{1+x}{1-x}, |x| < 0.2$

40. $f(x) = \dfrac{1-2x}{1+2x}, |x| < 0.3$

41. $f(x) = \dfrac{x+4}{x^2}, |x-4| < 1$

42. $f(x) = \dfrac{x+5}{x^2}, |x-5| < 1$

43. $f(x) = \dfrac{x^3}{x^2+4}, |x| < 2$

44. $f(x) = \dfrac{x^2}{x^2+5}, |x| < 3$

45. $f(x) = \dfrac{x^3}{(1-x)^2}, |x| < 0.2$

46. $f(x) = \dfrac{x^2}{(1+x)^2}, |x| < 0.3$

47. $f(x) = \dfrac{\sin x + \cos x}{\sin^2 x + 1}$, all x

48. $f(x) = \dfrac{\sin x + \sin 2x}{\tan^2 x + 1}, -\pi/2 < x < \pi/2$

49. $|f(x) - 3| < 1$. Find a bound of f.

50. $|f(x) - 2| < 5$. Find a bound of f.

51. $|g(x) - 2| < 1$. Find a bound of $1/g$.

52. $|g(x) - 5| < 3$. Find a bound of $1/g$.

53. $|f(x) - 3| < 0.005$ and $|g(x) - 2| < 0.005$. Find a bound of $|f(x) + g(x) - 5|$.

54. $|g(x) - M| < M/2$, where M is a fixed positive number. Find a bound of $1/g$ in terms of M.

55. Show that

$$\frac{a^2}{a^2 + b^2} \leq 1.$$

56. Show that

$$\frac{2|ab|}{a^2 + b^2} \leq 1.$$

(*Hint:* $0 \leq (|a| - |b|)^2$.)

Use the results of Exercises 55 and 56 to verify that the given numbers B are bounds of the functions in Exercises 57–62.

57. $f(x) = \dfrac{2x^2}{x^2+4}, B = 2$

58. $f(x) = \dfrac{18}{x^2+9}, B = 2$

59. $f(x) = \dfrac{8x}{x^2 + 4}, B = 2$

60. $f(x) = \dfrac{3x}{x^2 + 9}, B = \frac{1}{2}$

61. $f(x) = \dfrac{x^3}{x^2 + 9}, |x| \leq 3, B = \dfrac{3}{2}$

62. $f(x) = \dfrac{4x^3}{x^2 + 4}, |x| \leq 2, B = 4$

In Exercises 63–68 graph the given functions and use the graphs to determine bounds.

63. $f(x) = xe^{-x}, x \geq 0$ **64.** $f(x) = \dfrac{\sin x}{x}, x > 0$

65. $f(x) = x \ln x, 0 < x \leq 1$

66. $f(x) = \dfrac{1 - \cos x}{x^2}, x > 0$

67. $f(x) = \dfrac{\ln x}{x}, x \geq 1$ **68.** $f(x) = \dfrac{\sin 3x}{x}, x > 0$

2.3 AN INTUITIVE INTRODUCTION TO LIMITS AND CONTINUITY

Connections

Interpretation of graphs, 1.4–5.

Real-world problems involve error and approximation. For example, a piston and cylinder cannot be manufactured to exact specifications; some tolerance is necessary. This leads to the question, *how close* to the ideal must the cylinder and piston be to insure that they operate at a prescribed degree of efficiency? Questions of this type involve the mathematical concepts of limit and continuity.

In this section we will illustrate the idea of the limit of a function at a point intuitively, in terms of approximation and graphs. We will also introduce the concept of continuity of a function at a point and see how the concepts of limit and continuity are related.

Limits and Graphs

We will use the following informal definition in terms of the graph of a function to illustrate the ideas involved in the concept of the limit of a function at a point.

INFORMAL DEFINITION

We say that **the limit of $f(x)$ as x approaches c is L**, written $\lim\limits_{x \to c} f(x) = L$, if the y-coordinate of a point $(x, f(x))$ on the graph of f approaches the single number L as x approaches c from each side of c. We will sometimes write $f(x) \to L$ as $x \to c$ to indicate that $\lim\limits_{x \to c} f(x) = L$.

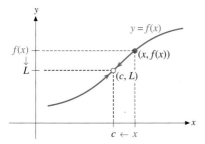

FIGURE 2.3.1 The y-coordinate of the point $(x, f(x))$ approaches L as x approaches c, so $\lim\limits_{x \to c} f(x) = L$.

See Figure 2.3.1. The idea is that $f(x)$ is approximately L for all x that are close enough to c and that the approximation can be made as accurate as is desired. The formal definition of limit (and the related real-life problems) require consideration of *how close x* must be to c to guarantee that the approximation is of a prescribed accuracy. We will consider these quantitative aspects of limit in the next section.

If $\lim_{x \to c} f(x) = L$, then the values $f(x)$ are approximately L for all x that are close enough to c. We can use calculator values of $f(x)$ for x close to c to indicate the existence and value of a limit, but calculator evaluation of function values does not constitute formal verification of a limit.

EXAMPLE 1

(a) From the graph in Figure 2.3.2a we see that the y-coordinate of a point $(x, 4 - 2x)$ on the graph of $f(x) = 4 - 2x$ approaches 2 as x approaches 1, so $\lim\limits_{x \to 1}(4 - 2x) = 2$. Calculator values of $4 - 2x$ are clearly near 2 whenever x is near 1.

(b) From the graph of $f(x) = x^2$ in Figure 2.3.2b we see that the y-coordinate of a point (x, x^2) on the graph of $f(x) = x^2$ approaches 4 as x approaches 2, so $\lim\limits_{x \to 2} x^2 = 4$. It is clear that calculator values of x^2 are near 4 whenever x is close to 2.

(c) From the graph in Figure 2.3.2c we see that the values of $f(x) = |x|$ approach 2 as x approaches -2, so $\lim\limits_{x \to -2} |x| = 2$.

(d) From the graph in Figure 2.3.2d we see that the values of $f(x) = \sqrt{x}$ approach 3 as x approaches 9, so $\lim\limits_{x \to 9} \sqrt{x} = 3$. Calculator values of $\sqrt{x}$ are near 3 whenever x is near 9.

(e) From the graph in Figure 2.3.2e we see that the values of $f(x) = \sin x$ approach $\sin(\pi/6) = 1/2$ as x approaches $\pi/6$, so $\lim\limits_{x \to \pi/6} \sin x = 1/2$. Calculator values of $\sin x$ are near $1/2$ whenever x is near $\pi/6$ *radians*. ■

FIGURE 2.3.2a The y-coordinate of the point $(x, 4 - 2x)$ approaches 2 as x approaches 1, so $\lim\limits_{x \to 1}(4 - 2x) = 2$.

FIGURE 2.3.2b The y-coordinate of the point (x, x^2) approaches 4 as x approaches 2, so $\lim\limits_{x \to 2} x^2 = 4$.

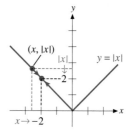

FIGURE 2.3.2c The y-coordinate of the point $(x, |x|)$ approaches 2 as x approaches -2, so $\lim\limits_{x \to -2} |x| = 2$.

FIGURE 2.3.2d The y-coordinate of the point $(x, \sqrt{x})$ approaches 3 as x approaches 9, so $\lim\limits_{x \to 9} \sqrt{x} = 3$.

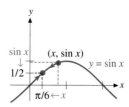

FIGURE 2.3.2e The y-coordinate of the point $(x, \sin x)$ approaches 1/2 as x approaches $\pi/6$, so $\lim\limits_{x \to \pi/6} \sin x = 1/2$.

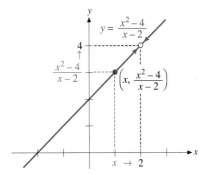

FIGURE 2.3.3a The y-coordinate of the point $\left(x, \dfrac{x^2 - 4}{x - 2}\right)$ approaches 4 as x approaches 2, so $\lim\limits_{x \to 2} \dfrac{x^2 - 4}{x - 2} = 4$.

FIGURE 2.3.3b The y-coordinate of the point $\left(x, \dfrac{\sin x}{x}\right)$ approaches 1 as x approaches 0, so $\lim\limits_{x \to 0} \dfrac{\sin x}{x} = 1$.

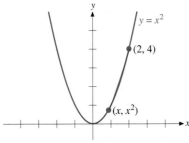

FIGURE 2.3.4 The slope of the line segment between (x, x^2) and $(2, 4)$ is $\dfrac{x^2 - 4}{x - 2}$. If $x = 2$, the points are identical and there is no line segment between them.

Limits at Points Where a Function is Undefined

The value of the limit of each function in Example 1 agrees with the value of the function at the point of the limit. This is the case with many of the functions with which we will be dealing. However, as we shall see in Section 2.6, a fundamental problem of calculus involves limits of functions at points where the functions are undefined. Thus, we exclude the point $x = c$ from consideration of the limit at c. It is not necessary that f be defined at c in order to have a limit at c.

The limit of $f(x)$ as x approaches c depends on the values of $f(x)$ as x becomes close to c, but we exclude $x = c$.

EXAMPLE 2

(a) From the graph of $f(x) = \dfrac{x^2 - 4}{x - 2}$ in Figure 2.3.3a we see that $\lim\limits_{x \to 2} \dfrac{x^2 - 4}{x - 2} = 4$. Note that

$$f(x) = \frac{x^2 - 4}{x - 2} = \frac{(x - 2)(x + 2)}{x - 2} = x + 2, x \neq 2.$$

Hence, the graph of f is the same as the graph of the line $y = x + 2$, except $f(x)$ is undefined at $x = 2$. Calculator values of $\dfrac{x^2 - 4}{x - 2}$ will give values $x + 2$ for $x \neq 2$.

(b) From the graph in Figure 2.3.3b we see that the values $\dfrac{\sin x}{x}$ approach one as x approaches zero, so $\lim\limits_{x \to 0} \dfrac{\sin x}{x} = 1$. If x is measured in *radians*, then calculator values of $\dfrac{\sin x}{x}$ will be approximately one for small, nonzero values of x. ■

In Example 2a, there may be a reason for writing $f(x) = \dfrac{x^2 - 4}{x - 2}$ instead of the simpler expression $x + 2$. For example, $f(x)$ may represent the slope of the line segment between the point $(2, 4)$ and a variable point (x, x^2) on the graph of $y = x^2$. See Figure 2.3.4. If $x = 2$, the points $(2, 4)$ and (x, x^2) are identical and it does not make sense to speak of the line segment between them. Thus, the fact that $f(x)$ is not defined for $x = 2$ reflects the more important fact that the physical interpretation of $f(x)$ does not make sense for $x = 2$.

Some Examples of Limits That Do Not Exist

If $f(x)$ does not approach a single *number as x approaches c, then $\lim\limits_{x \to c} f(x)$ does not exist.*

EXAMPLE 3

(a) From Figure 2.3.5a we see that $\lim\limits_{x \to 0} |x|/x$ does not exist. The y-coordinates of points on the graph approach -1 as x approaches 0 from the left and they approach 1 as x approaches 0 from the right. This means that some values $f(x)$ approach 1 and other values of $f(x)$ approach -1 as x approaches 0. Thus, the values of $f(x)$ do not approach a *single* number as x approaches 0.

(b) From the graph in Figure 2.3.5b we see that $\lim_{x\to 0} \sin(1/x)$ does not exist. The function $\sin(1/x)$ repeatedly assumes all values between 1 and -1 as x approaches 0. Calculator values of $\sin(1/x)$ for small, $x \neq 0$, may be any number between one and negative one. ■

If $f(x)$ becomes unbounded as x approaches c, then $\lim_{x\to c} f(x)$ does not exist.

If $f(x)$ becomes unbounded as x approaches c, then $f(x)$ cannot approach any real number.

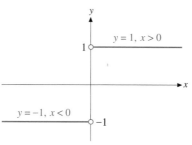

FIGURE 2.3.5(a)   The values $|x|/x$ approach different numbers as x approaches 0 from different sides, so $\lim_{x\to 0} |x|/x$ does not exist.

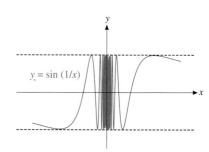

FIGURE 2.3.5(b)   The values $\sin(1/x)$ repeatedly assume all values between 1 and -1 as x approaches 0, so $\lim_{x\to 0} \sin(1/x)$ does not exist.

EXAMPLE 4

From the graph in Figure 2.3.6 we see that $\lim_{x\to 0} \dfrac{1}{x^2}$ does not exist. The function $1/x^2$ becomes unbounded as x approaches 0. ■

One-Sided Limits

The limit of $f(x)$ as x approaches c involves the behavior of $f(x)$ for x on both sides of c. The same ideas and theory hold if we restrict attention to x on only one side of c.

FIGURE 2.3.6 $\dfrac{1}{x^2}$ becomes unbounded as x approaches 0, so $\lim_{x\to 0} \dfrac{1}{x^2}$ does not exist.

INFORMAL DEFINITION

   The limit of $f(x)$ as x approaches c from the left is L (written $\lim_{x\to c^-} f(x) = L$) if the y-coordinate of a point $(x, f(x))$ on the graph of f approaches L as x approaches c from the left of c.
   The limit of $f(x)$ as x approaches c from the right is L (written $\lim_{x\to c^+} f(x) = L$) if the y-coordinate of a point $(x, f(x))$ on the graph of f approaches L as x approaches c from the right of c.

Values to the left of c are less than c. This is indicated by the negative sign in the symbol $x \to c^-$. The positive sign in the symbol $x \to c^+$ indicates values of x that are to the right of c, greater than c.

The expression $\lim_{x\to c} f(x) = L$ means that $f(x)$ approaches L as x approaches c from each side of c. This gives the following relations between a limit and one-sided limits at a point.

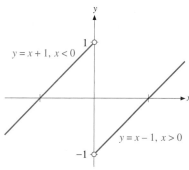

FIGURE 2.3.7a $\lim\limits_{x\to 0^-} f(x) = 1$, $\lim\limits_{x\to 0^+} f(x) = -1$, and $\lim\limits_{x\to 0} f(x)$ does not exist.

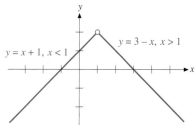

FIGURE 2.3.7b $\lim\limits_{x\to 1^-} g(x) = 2$, $\lim\limits_{x\to 1^+} g(x) = 2$, and $\lim\limits_{x\to 1} g(x) = 2$.

If $\lim\limits_{x\to c} f(x) = L$, *then* $\lim\limits_{x\to c^-} f(x) = \lim\limits_{x\to c^+} f(x) = L$.

If $\lim\limits_{x\to c^-} f(x) = \lim\limits_{x\to c^+} f(x) = L$, *then* $\lim\limits_{x\to c} f(x) = L$.

If $\lim\limits_{x\to c^-} f(x) \neq \lim\limits_{x\to c^+} f(x)$, *then* $\lim\limits_{x\to c} f(x)$ *does not exist.*

In the latter case, $f(x)$ does not approach a *single* number L as x approaches c.

EXAMPLE 5

(a) From the graph of

$$f(x) = \begin{cases} x + 1, & x < 0, \\ x - 1, & x > 0, \end{cases}$$

in Figure 2.3.7a we see that $\lim\limits_{x\to 0^-} f(x) = 1$, $\lim\limits_{x\to 0^+} f(x) = -1$, and $\lim\limits_{x\to 0} f(x)$ does not exist.

(b) From the graph of

$$g(x) = \begin{cases} x + 1, & x < 1, \\ 3 - x, & x > 1, \end{cases}$$

in Figure 2.3.7b we see that $\lim\limits_{x\to 1^-} g(x) = 2$, $\lim\limits_{x\to 1^+} g(x) = 2$, and $\lim\limits_{x\to 1} g(x) = 2$. ∎

Continuity at a Point

We have seen that it is not necessary that a function f be defined at a point c in order to have a limit at c. However, in many cases the limit of f at c is $f(c)$. This leads to the following definition, which must be considered to be informal until limit is defined formally.

> **DEFINITION**
>
> The function f is **continuous** at c if $\lim\limits_{x\to c} f(x) = f(c)$.

The definition means that for f to be continuous at c,

  (i) f must have a limit at c,

  (ii) f must be defined at c, and

  (iii) the value of the limit at c must be equal to the value of the function at c.

If one (or more) of the above three conditions is not satisfied, f is said to be **discontinuous at c.**

The statement $\lim\limits_{x\to c} f(x) = f(c)$ says that the values $f(x)$ are close to $f(c)$ for x close to c. This means that small changes, or errors, in x will result in only small changes in the values of a *continuous* function f. Fortunately, most of the functions that occur in science and engineering are continuous, so small errors are negligible. For example, a cylinder and piston will operate efficiently, even though they cannot be manufactured to exact specifications.

As an example of a function that is not continuous, consider the amount of air enclosed by a balloon that is being filled at a constant rate as a function of

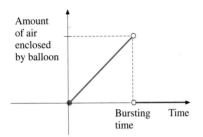

FIGURE 2.3.8 The amount of air enclosed by the balloon shortly before it bursts is significantly different from the amount it encloses shortly after it bursts.

time. The amount of air varies continuously up to the time the balloon bursts, at which time the amount of air enclosed by the balloon is undefined. The amount enclosed could be taken to be zero after the balloon bursts. In this example, a small variation in time near the bursting point has a very significant effect on the amount of air enclosed by the balloon. See Figure 2.3.8.

As in Example 1, we could use graphs to determine that linear functions $ax + b$, power functions x^n, the absolute value function $|x|$, and the trigonometric functions $\sin x$ and $\cos x$ are continuous at each point. The root functions $x^{1/n}$ are continuous at each point c where they are defined on both sides of c. We will see in Section 2.5 how to obtain continuity of sums, products, quotients, and compositions of these basic functions.

If $\lim_{x \to c} f(x)$ exists, but f fails to be continuous at c because either $f(c)$ is undefined or $f(c)$ is defined to be a number other than $\lim_{x \to c} f(x)$, then f is said to have a **removable discontinuity at** c. In this case, we can obtain a *new function* that is continuous at c by defining (or redefining) the value at c to be $\lim_{x \to c} f(x)$. If $\lim_{x \to c} f(x)$ does not exist, then no choice of value at c will give a new function that is continuous at c.

EXAMPLE 6

(a) From Figure 2.3.9a we see that

$$f(x) = \begin{cases} 1, & x < 0, \\ 1 - x, & x \geq 0, \end{cases}$$

is continuous at every point.

FIGURE 2.3.9a This function is continuous at every point. The limit of f exists at zero, f is defined at zero, and the limit at zero is equal to the value at zero.

FIGURE 2.3.9b The function g is not continuous at zero, because $\lim_{x \to 0} g(x)$ does not exist.

FIGURE 2.3.9c The function h is not continuous at zero, because h is not defined at zero. A new function that is continuous at zero could be obtained by defining $h(0)$ to be one, the value of the limit at zero.

FIGURE 2.3.9d The function k is not continuous at zero, because the limit at zero is unequal its value at zero. A new function that is continuous at zero could be obtained by redefining $k(0)$ to be one, the value of the limit at zero.

b. From Figure 2.3.9b we see that

$$g(x) = \begin{cases} 1, & x < 0, \\ 0, & x \geq 0, \end{cases}$$

is not continuous at zero, because $\lim_{x \to 0} g(x)$ does not exist.

c. From Figure 2.3.9c we see that

$$h(x) = \begin{cases} 1, & x < 0, \\ 1 - x, & x > 0, \end{cases}$$

is not continuous at zero, because h is undefined at zero. Since $\lim_{x \to 0} h(x)$ exists, the discontinuity is removable. We could obtain a new function that is continuous at zero by defining $h(0)$ to be one, the value of the limit at zero.

d. From Figure 2.3.9d we see that

$$k(x) = \begin{cases} 1, & x \neq 0, \\ 0, & x = 0, \end{cases}$$

is not continuous at zero, because $\lim_{x \to 0} k(x) = 1$, while $k(0) = 0$. Since the limit at zero exists, the function has a removable discontinuity at zero. We could obtain a new function that is continuous at zero by redefining $k(0)$ to be one, the value of the limit at zero. ■

We can use one-sided limits to define continuity from each side separately.

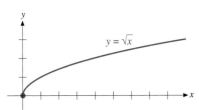

FIGURE 2.3.10 The square root function is continuous at all points $c > 0$; it is continuous from the right at zero.

DEFINITION

The function f is **continuous from the left at** c if $\lim_{x \to c^-} f(x) = f(c)$.
The function f is **continuous from the right at** c if $\lim_{x \to c^+} f(x) = f(c)$.

A function f is continuous at c if and only if it is continuous from both the left and right at c, so $\lim_{x \to c} f(x) = f(c)$.

From Figure 2.3.10 we see that the square root funtions is continuous from the right at zero, because $\lim_{x \to 0^+} \sqrt{x} = 0 = \sqrt{0}$. It is not continuous from the left at zero because it is undefined for $x < 0$.

The greatest integer less than or equal to a real number x is denoted $[x]$. For example, $[1.95] = 1$, $[3] = 3$, and $[-1.3] = -2$. The **greatest integer function** can be used in computer science to **truncate** decimal numbers. From the graph in Figure 2.3.11 we see that the function is continuous at every noninteger x, continuous from the right at every integer, but not continuous from the left at any integer. It is discontinuous at each integer.

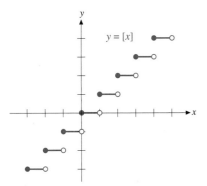

FIGURE 2.3.11 The greatest integer function is continuous at every noninteger, continuous from the right at every integer, but not continuous from the left at any integer.

EXERCISES 2.3

For each of the limits in Exercises 1–18, use the given graph to find the value or determine that the limit does not exist.

1. $\lim\limits_{x \to 0} \dfrac{x^2}{x}$

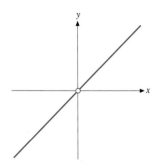

2. $\lim\limits_{x \to 0} \dfrac{x^3}{x}$

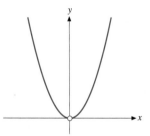

3. $\lim\limits_{x \to 0} \dfrac{|x| + x}{x}$

4. $\lim\limits_{x \to 0} \dfrac{|x| - x}{x}$

5. $\lim\limits_{x \to 0} \dfrac{1}{x^{1/3}}$

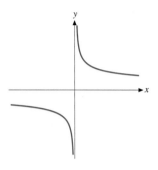

6. $\lim\limits_{x \to 1} \dfrac{1}{(x - 1)^2}$

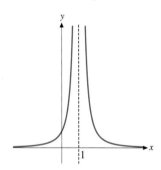

7. $\lim\limits_{x \to 1} \dfrac{x^2 - 1}{x - 1}$

8. $\displaystyle\lim_{x\to-1}\frac{x^2-1}{x+1}$

9. $\displaystyle\lim_{x\to1}\frac{x^3-1}{x-1}$

10. $\displaystyle\lim_{x\to-1}\frac{x^3+1}{x+1}$

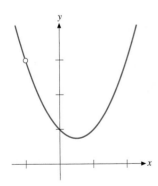

11. (a) $\displaystyle\lim_{x\to-4}\frac{x^2}{4(x+2)}$

   (b) $\displaystyle\lim_{x\to-2}\frac{x^2}{4(x+2)}$

(c) $\displaystyle\lim_{x\to0}\frac{x^2}{4(x+2)}$

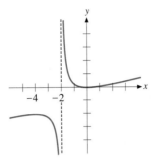

12. (a) $\displaystyle\lim_{x\to-2}\frac{8x}{(x-2)^2}$

   (b) $\displaystyle\lim_{x\to0}\frac{8x}{(x-2)^2}$

   (c) $\displaystyle\lim_{x\to2}\frac{8x}{(x-2)^2}$

13. $\displaystyle\lim_{x\to0}\frac{\sin(1/x)}{x}$

14. $\lim\limits_{x \to 0}(1 - \cos(1/x))$

15. $\lim\limits_{x \to 0} x^2 \cos^2(1/x)$

16. $\lim\limits_{x \to 0} x^{1/3} \sin(1/x)$

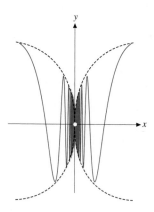

17. $\lim\limits_{x \to 0} \dfrac{\sin x}{x}$

18. $\lim\limits_{x \to 0} \dfrac{1 - \cos^2 x}{x}$

In Exercises 19–24 use the given graphs to evaluate or determine "does not exist" for each of the limits (a) $\lim\limits_{x \to c^-} f(x)$*, (b)* $\lim\limits_{x \to c^+} f(x)$*, and (c)* $\lim\limits_{x \to c} f(x)$ *for the indicated number c.*

19. $f(x) = \begin{cases} -1, & x < 0, \\ 1, & x > 0; \end{cases} \quad c = 0$

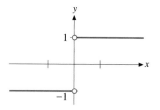

20. $f(x) = \begin{cases} -1 - x, & x < 0, \\ 1 - x, & x > 0; \end{cases} \quad c = 0$

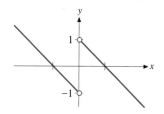

21. $f(x) = \begin{cases} 1 + x, & x < 0, \\ 0, & x = 0, \\ 1 - x, & x > 0; \end{cases} \quad c = 0$

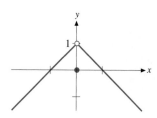

22. $f(x) = \begin{cases} 1, & x \neq 0, \\ 0, & x = 0; \end{cases} \quad c = 0$

23. $f(x) = |x^2 - 1|; c = 1$

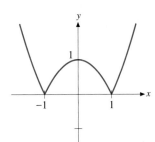

24. $f(x) = \begin{cases} x^2, & x \le 1, \\ 2 - x, & x > 1; \end{cases}$ $c = 1$

In Exercises 25–28 indicate whether the functions whose graphs are given are continuous at each of the points a, b, and c.

25.

26.

27.

28.

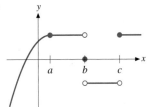

In each of Exercises 29–32, either find a value of a so that f is continuous at 0, or indicate that this is impossible. (Hint: Sketch the graph of f.)

29. $f(x) = \begin{cases} 1 + x, & x < 0 \\ a, & x = 0 \\ 1 - x, & x > 0 \end{cases}$

30. $f(x) = \begin{cases} 0, & x < 0 \\ a, & x = 0 \\ x \cos x, & x > 0 \end{cases}$

31. $f(x) = \begin{cases} \dfrac{|x| + x}{2x}, & x \ne 0 \\ a, & x = 0 \end{cases}$

32. $f(x) = \begin{cases} 0, & x < 0 \\ a, & x = 0 \\ \cos x, & x > 0 \end{cases}$

In Exercises 33–34 find numbers a and b so that f is continuous at every point. (Hint: Use the fact that the graph of $y = ax + b$ is a line to sketch the graph of f.)

33. $f(x) = \begin{cases} 2, & x < 0 \\ ax + b, & 0 \le x \le 4 \\ 0, & x > 4 \end{cases}$

34. $f(x) = \begin{cases} x, & x < -1 \\ ax + b, & -1 \le x \le 2 \\ 2x - 7, & x > 2 \end{cases}$

In Exercises 35–38 sketch the graphs and determine if the functions are continuous from the left at zero or continuous from the right at zero.

35. $f(x) = \dfrac{|x| + x}{2}$ **36.** $f(x) = [x] + x$

37. $f(x) = -[-x]$ **38.** $f(x) = \sqrt[3]{x}$

39. Sketch the graph of the function $f(x) = [10x]/10$.

40. The function of Exercise 39 truncates a positive decimal number x to one decimal place. Find a function that truncates a positive decimal number to five decimal places.

41. Sketch the graph of $f(x) = [x + 0.5]$.

42. The function of Exercise 41 rounds a decimal number x to the nearest integer. Find a function that rounds a decimal number to the nearest hundredth.

43. Sketch the graph of the **truncation function**, $T(x) = (\text{sign } x) \cdot [\![|x|]\!]$, where

$$\text{sign } x = \begin{cases} 1, & \text{if } x > 0, \\ 0, & \text{if } x = 0, \\ -1, & \text{if } x < 0. \end{cases}$$

This function truncates the decimal part from any decimal number, positive or negative.

44. Graph $f(x) = x \ln x$. What is the apparent value of $\lim\limits_{x \to 0^+} f(x)$?

45. Graph $f(x) = e^{-1/x^2}$. What is the apparent value of $\lim\limits_{x \to 0} f(x)$?

In Exercises 46–48 use calculator values to indicate the existence and value of the limits.

46. $\lim\limits_{x \to 0} \dfrac{1 - \cos x}{x}$. Use radian measure. (**Calculus Explorer & Tutor I, Limits, 14.**)

47. $\lim\limits_{x \to 0} \dfrac{\tan x - x}{x^3}$. Use radian measure. (**Calculus Explorer & Tutor I, Limits, 15.**)

48. $\lim\limits_{h \to 0}(1 + h)^{1/h}$. (**Calculus Explorer & Tutor I, Limits, 16.**)

2.4   THE QUANTITATIVE ASPECT OF LIMITS; APPLICATIONS

Connections

Inequalities and absolute value, 1.1.

Factoring polynomials, 1.2

Bounds of functions, 2.2.

In this section we will investigate a basic quantitative problem and show how it relates to some applications that involve accuracy of measurement. We will also show how the problem relates to a formal definition of the limit of a function at a point and illustrate how the definition can be used to obtain formulas for evaluating limits.

Using ϵ–δ to Quantify Closeness

The intuitive idea of values $f(x)$ approaching L as x approaches c is related to the formal statement

$$|f(x) - L| < \epsilon \text{ whenever } 0 < |x - c| < \delta, \tag{1}$$

where the Greek letters ϵ (epsilon) and δ (delta) represent positive numbers. The number ϵ measures how close $f(x)$ is to L; δ measures how close x is to c. The inequality $0 < |x - c|$ excludes $x = c$.

In order to interpret Statement 1 graphically, we note that the inequality $|f(x) - L| < \epsilon$ can be rewritten as

$$-\epsilon < f(x) - L < \epsilon,$$
$$L - \epsilon < f(x) < L + \epsilon.$$

This means that the point $(x, f(x))$ on the graph of f is between the horizontal lines $y = L - \epsilon$ and $y = L + \epsilon$. The inequality $0 < |x - c| < \delta$ means

$$c - \delta < x < c + \delta, \quad c \neq x.$$

We can then interpret Statement 1 as saying that the points $(x, f(x))$ are between the horizontal lines $y = L - \epsilon$ and $y = L + \epsilon$ whenever x satisfies $c - \delta < x < c + \delta$, $x \neq c$. See Figure 2.4.1. If the y-coordinates of points $(x, f(x))$ approach the point L as x approaches c, then for each fixed positive number ϵ, the points must be between the horizontal lines $y = L - \epsilon$ and $y = L + \epsilon$ for all x that are close enough to (but unequal to) c. This means that

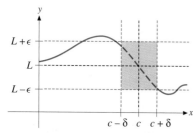

FIGURE 2.4.1 The statement, $|f(x) - L| < \epsilon$ whenever $0 < |x - c| < \delta$, means that the points $(x, f(x))$ are between the horizontal lines $y = L - \epsilon$ and $y = L + \epsilon$ whenever x satisfies $c - \delta < x < c + \delta, x \neq c.$

for each positive number ϵ, there must be some positive number δ for which Statement 1 is true. We are thus led to the following problem.

Quantitative Problem: *Given a positive number ϵ, can we find a positive number δ such that*

$$|f(x) - L| < \epsilon \ whenever \ |x - c| < \delta?$$

There is not a single method that solves this problem for all functions. In fact, the problem does not always have a solution. However, in many important cases we can use the following procedure.

One solution strategy for ϵ–δ problems

- Use algebra and factoring to express $|f(x) - L|$ in terms of $|x - c|$.
- Use a preliminary assumption, if necessary, to find bounds of the factors of $|f(x) - L|$ other than $|x - c|$.
- Use the results of the above steps to obtain an inequality of the form $|f(x) - L| \le B|x - c|$. It follows that

$$|f(x) - L| < \epsilon \ whenever \ |x - c| < \epsilon/B,$$

so $\delta = \epsilon/B$ is a solution of the problem.

EXAMPLE 1
Find a positive number δ such that

$$|12x - 120| < 0.3 \ whenever \ |x - 10| < \delta.$$

SOLUTION Notice that the expression $12x - 120$ is zero when $x = 10$. We should expect that $|12x - 120|$ is less than 0.3 whenever $|x - 10|$ is small enough. We want to know *how small* $|x - 10|$ should be to guarantee that $|12x - 120| < 0.3$.

The first step is to factor. We have

$$|12x - 120| = 12|x - 10|.$$

This gives us an expression on the right side of the above equality of the form $B|x - c|$, where B is a constant. We can then solve the inequality $12|x - 10| < 0.3$ for $|x - 10|$ to determine that

$$|12x - 120| < 0.3 \ whenever \ |x - 10| < 0.3/12 = 0.025.$$

That is, if we choose $\delta = 0.025$, then

$$|x - 10| < \delta \ implies \ |12x - 120| = 12|x - 10| < 12\delta = 12(0.025) = 0.3.$$

This shows that $\delta = 0.025$ is a solution of our problem. See Figure 2.4.2. ■

FIGURE 2.4.2 $|12x - 120| < 0.3$ whenever $|x - 10| < 0.025$.

EXAMPLE 2
Find a positive number δ such that

$$|\pi r^2 - 9\pi| < 0.01 \ whenever \ |r - 3| < \delta.$$

Assume $|r - 3| < 0.5$.

SOLUTION We want

$$|\pi r^2 - 9\pi| < 0.01,$$

but it is not necessary to solve this inequality for exact values of r. We will follow our procedure and reduce the problem to a much simpler inequality.

The first step is to factor $|\pi r^2 - 9\pi|$. This allows us to express $|\pi r^2 - 9\pi|$ in terms $|r - 3|$. We have

$$|\pi r^2 - 9\pi| = \pi |r + 3| \, |r - 3|. \tag{2}$$

NOTE   *Do not try to solve $\pi |r + 3| \, |r - 3| < 0.01$ by writing*

$$|r - 3| < \frac{0.01}{\pi |r + 3|}.$$

It is very difficult to draw useful conclusions from this type of inequality.

The next step is to find a bound for the factors other than $|r - 3|$. That is, we need a bound of the factor $|r + 3|$. Note that $|r + 3|$ is not bounded unless r is restricted. However, we can obtain a bound by using the preliminary assumption $|r - 3| < 0.5$. Thus, we need a bound of the function

$$r + 3, \qquad |r - 3| < 0.5.$$

Following the technique of Section 2.2, we write $|r - 3| < 0.5$ as the double inequality

$$-0.5 < r - 3 < 0.5$$

and solve for r. We obtain

$$2.5 < r < 3.5.$$

The linear factor $r + 3$ is bounded on the interval by the value of its absolute value at the endpoint $r = 3.5$, so

$$|r + 3| < |(3.5) + 3| = 6.5.$$

Using the bound of 6.5 for $|r + 3|$ in (2), we obtain

$$|\pi r^2 - 9\pi| \leq \pi(6.5)|r - 3|. \tag{3}$$

It is clear from (3) that $|\pi r^2 - 9\pi| < 0.01$ whenever $\pi(6.5)|r - 3| < 0.01$. It is easy to solve the latter inequality for $|r - 3|$ to obtain $|r - 3| < 0.01/(6.5\pi)$. (Note that $|r - 3| < 0.01/(6.5\pi) \approx 0.00048971$ implies that $|r - 3| < 0.5$, so our preliminary assumption is valid.) We then have

$$|\pi r^2 - 9\pi| < 0.01 \text{ whenever } |r - 3| < 0.01/(6.5\pi).$$

The positive number $\delta = 0.01/(6.5\pi)$ is a solution of our problem. See Figure 2.4.3. ∎

In Example 2, we discovered that $|r - 3| < 0.01/(6.5\pi)$ ensures $|\pi r^2 - 9\pi| < 0.01$. That is, $0.01/(6.5\pi)$ is an "answer" to our problem. It would also be a correct solution of our problem to require r to be even closer to 3. For example, we could *round down* the calculator value of $0.01/(6.5\pi) \approx 0.00048971$ and say that

$$|\pi r^2 - 9\pi| < 0.01 \text{ whenever } |r - 3| < 0.00048.$$

That is, if $|r - 3| < 0.00048$, then $|r - 3|$ is also less than the larger number $0.01/(6.5\pi)$, so we know that we must have $|\pi r^2 - 9\pi| < 0.01$.

The "answer" $0.01/(6.5\pi)$ obtained in Example 2 was determined in a systematic manner by using the preliminary assumption that $|r - 3| < 0.5$.

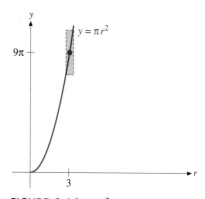

FIGURE 2.4.3 $|\pi r^2 - 9\pi| < 0.01$ whenever $|r - 3| < 0.01/(6.5\pi)$.

If we used the same procedure with a different preliminary assumption we would get a number different from $0.01/(6.5\pi)$. For example, if we used the preliminary assumption $|r - 3| < 2$, we would have

$$-2 < r - 3 < 2,$$
$$1 < r < 5, \text{ so}$$
$$|r + 3| < 8.$$

We would then obtain

$$|\pi r^2 - 9\pi| = \pi|r + 3||r - 3| < 8\pi|r - 3| < 0.01$$

whenever $|r - 3| < 0.01/(8\pi) \approx 0.00039$, so 0.00039 would be an "answer" to our problem. The difference in the "answers" obtained is not significant for this type of problem. The important issue is whether *some* answer can be determined.

EXAMPLE 3

Find a positive number δ such that

$$|x^3 - 1| < 0.5 \text{ whenever } |x - 1| < \delta.$$

Assume $|x - 1| \leq 1$.

SOLUTION   We begin by factoring to express $|x^3 - 1|$ in terms of $|x - 1|$. We obtain

$$|x^3 - 1| = |x^2 + x + 1||x - 1|. \tag{4}$$

We then see that we need a bound of the factor $|x^2 + x + 1|$. The preliminary assumption $|x - 1| \leq 1$ implies $-1 \leq x - 1 \leq 1$. Solving for x we obtain $0 \leq x \leq 2$, so $0 \leq |x| \leq 2$. Using $|x| \leq 2$ and the Triangle Inequality, we then have

$$|x^2 + x + 1| \leq |x|^2 + |x| + 1 \leq (2)^2 + (2) + 1 = 7.$$

Using this bound in (4), we obtain

$$|x^3 - 1| \leq 7|x - 1|.$$

It follows that $|x^3 - 1| < 0.5$ whenever $7|x - 1| < 0.5$. Solving the latter inequality for $|x - 1|$, we obtain

$$|x^3 - 1| < 0.5 \text{ whenever } |x - 1| < 0.5/7,$$

so $\delta = 0.5/7$ is a solution to the problem. See Figure 2.4.4. ■

FIGURE 2.4.4 $|x^3 - 1| < 0.5$ whenever $|x - 1| < 0.5/7$

The positive number δ that corresponds to a given positive number ϵ usually depends on ϵ, but it is not necessarily a constant multiple of ϵ.

EXAMPLE 4

For each positive number ϵ, find a positive number δ such that $|x^{1/3}| < \epsilon$ whenever $|x| < \delta$.

SOLUTION   In this case we can solve the inequality $|x^{1/3}| < \epsilon$ for $|x|$ to see that

$$|x^{1/3}| < \epsilon \text{ whenever } |x| < \epsilon^3.$$

It follows that $\delta = \epsilon^3$ is a solution to our problem. See Figure 2.4.5. ■

Note that the use of ϵ in place of a particular decimal number did not cause any difficulty in Example 4.

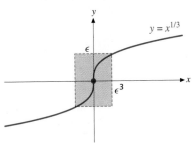

FIGURE 2.4.5 $|x^{1/3}| < \epsilon$ whenever $|x| < \epsilon^3$.

FIGURE 2.4.6 How close to 10 should x be to insure that 12x is within 0.3 of 120?

FIGURE 2.4.7 How close to 3 should r be to insure that the area of the circle is within 0.01 of 9π?

FIGURE 2.4.8 The current in a simple RV circuit is given by $I = V/R$.

FIGURE 2.4.9 The values of $1/|R|$ for $1.5 \leq R \leq 2.5$ are bounded by its value at the endpoint $R = 1.5$.

Applications to Accuracy of Measurement

In order to see how our quantitative problem relates to practical applications, let us restate the problem in terms of input-output:

If it is required that an output $f(x)$ be within ϵ of L, how close should the input x be to c?

- For example, suppose a 120-yard football field (including both endzones) is to be marked by using a measuring tape that is approximately 10 yards long. We may ask how close to 10 yards must the measuring tape be to ensure that the actual length of the field is within 0.3 yard of 120 yards? Let x denote the actual length in yards of the measuring tape. Since the length of the tape is essentially 10 yards, 12 lengths of the tape would be used to mark a field 120 yards long. That is, the actual length of the field is $12x$ yards. See Figure 2.4.6. The absolute value of the difference between the actual length and 120 is $|12x - 120|$. We want the actual length to be within 0.3 of 120, or $|12x - 120| < 0.3$. The result of Example 1 shows that

$$|12x - 120| < 0.3 \text{ whenever } |x - 10| < 0.025.$$

That is, the actual length is within 0.3 yard of 120 yards whenever the length of the measuring tape is within 0.025 yard of 10 yards (0.025 yard = 0.9 inch).

- Example 2 tells us how close to 3 the radius of a circle should be in order to ensure that the area of the circle be within 0.01 of 9π, where we have assumed the radius is within 0.5 of 3. See Figure 2.4.7.

EXAMPLE 5

How close to 2 ohms should the resistance in the 12-volt electrical circuit indicated in Figure 2.4.8 be in order for the current to be within 0.05 amp of 6 amps? Assume the resistance is within 0.5 ohm of 2 ohms. $I = V/R$, where I is the current in amps, V is the voltage in volts, and R is the resistance in ohms.

SOLUTION   We are told that $V = 12$, so the current in the circuit is $I = 12/R$. We want to determine $|R - 2|$ so that $\left|\dfrac{12}{R} - 6\right| < 0.05$. Carrying out the arithmetic, we have

$$\left|\frac{12}{R} - 6\right| = \frac{|12 - 6R|}{|R|} = \frac{6}{|R|}|2 - R| = \frac{6}{|R|}|R - 2|. \tag{5}$$

We need a bound of the factor $1/R$. We have assumed that the resistance is within 0.5 ohm of 2 ohms. That is,

$$|R - 2| \leq 0.5, \text{ so } -0.5 \leq R - 2 \leq 0.5.$$

The double inequality implies

$$1.5 \leq R \leq 2.5.$$

From the graph in Figure 2.4.9 we see that the values of $1/|R|$ in this interval are bounded by its value at $R = 1.5$, so

$$\frac{1}{|R|} \leq \frac{1}{1.5}.$$

Using this bound in (5), we obtain

$$\left|\frac{12}{R} - 6\right| \le \frac{6}{1.5}|R - 2|.$$

It follows that $\left|\dfrac{12}{R} - 6\right| < 0.05$ whenever $\dfrac{6}{1.5}|R - 2| < 0.05$, or $|R - 2| < \dfrac{(1.5)(0.05)}{6} = 0.0125$. (Note that $|R - 2| < 0.0125$ implies that R is within 0.5 of 2, so our preliminary assumption holds.) The current is within 0.05 amp of 6 amps whenever the resistance is within 0.0125 ohm of 2 ohms. ∎

EXAMPLE 6

The horizontal distance traveled by an arrow that is shot with initial velocity 192 ft/s at an angle θ from horizontal is given by the formula $L = 1152 \sin 2\theta$. See Figure 2.4.10. The arrow travels 500 ft when $\theta = 0.2245$ radian. How close to 0.2245 radian does θ need to be to ensure that the arrow will travel within 12 ft of 500 ft?

SOLUTION   We see from Figure 2.4.11 that the angle θ must be between the values for which $L = 488$ and $L = 512$.

When $L = 488$, we have $488 = 1152 \sin 2\theta$. Using a calculator to solve this equation for the value of θ that is near 0.2245, we have

$$\theta = \frac{1}{2}\sin^{-1}\frac{488}{1152} = 0.2187 \text{ radian.}$$

FIGURE 2.4.10   The horizontal distance in feet traveled by an arrow that is shot from a certain bow at angle θ is $L = 1152 \sin 2\theta$.

The equation $512 = 1152 \sin 2\theta$ has solution

$$\theta = \frac{1}{2}\sin^{-1}\frac{512}{1152} = 0.2303 \text{ radian.}$$

FIGURE 2.4.11 L is between 488 and 512 whenever θ is between the values for which $L = 488$ and $L = 512$.

We conclude that θ must satisfy $0.2187 < \theta < 0.2303$, so $-0.0058 < \theta - 0.2245 < 0.0058$. The arrow will travel within 12 ft of 500 ft if the angle is within 0.0058 radian of 0.2245 radian. (0.0058 radian $\approx 1/3$ deg). ∎

The Formal Definition of Limit

We have now seen how the concept of limit is related to graphs and how the quantitative aspect of the concept relates to applications that involve accuracy of measurement. Let us now state a formal and precise definition of the limit of a function at a point and see how it reflects the ideas we have discussed.

DEFINITION

The **limit of $f(x)$ as x approaches c is L** (written $\lim\limits_{x \to c} f(x) = L$) if, for each positive number ϵ, there is a positive number δ such that

$$|f(x) - L| < \epsilon \text{ whenever } 0 < |x - c| < \delta.$$

If the condition of the definition is not satisfied for any number L, we say $\lim\limits_{x \to c} f(x)$ **does not exist**.

Let us see how the formal definition of limit forces the y-coordinate of points $(x, f(x))$ to approach L as x approaches c. The condition of the definition asserts that, for each particular positive number ϵ, there is a positive number δ such that $|f(x) - L| < \epsilon$ whenever $0 < |x - c| < \delta$. As we have seen earlier, this means the points $(x, f(x))$ are between the horizontal lines $y = L - \epsilon$ and $y = L + \epsilon$ for all x that satisfy $c - \delta < x < c + \delta$, except possibly for $x = c$, which is excluded. As x approaches c from either side, x will eventually be within δ of c, so $(x, f(x))$ eventually stays between the horizontal lines. This means that $f(x)$ can approach only numbers L' with $L - \epsilon \leq L' \leq L + \epsilon$ as x approaches c. See Figure 2.4.12a. Finally, we use the fact that the condition of the definition holds for each positive number ϵ to exclude the possibility that $f(x)$ approaches a number L' with $L' \neq L$. That is, if $L' \neq L$, we can choose the positive number ϵ so small that the point (c, L') is not between the horizontal lines. See Figure 2.4.12b. We conclude that $f(x)$ must approach the single number L as x approaches c, as indicated in Figure 2.4.12c.

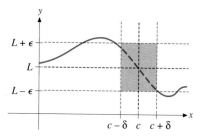

FIGURE 2.4.12a If the points $(x, f(x))$ are between the horizontal lines $y = L - \epsilon$ and $y = L + \epsilon$ for $c - \delta < x < c + \delta, x \neq c$, then the values $f(x)$ can approach only numbers L' with $L - \epsilon \leq L' \leq L + \epsilon$ as x approaches c.

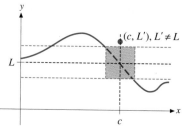

FIGURE 2.4.12b If $L' \neq L$, we can choose ϵ so small that (c, L') is not between the horizontal lines $y = L - \epsilon$ and $y = L + \epsilon$. This implies that $f(x)$ cannot approach a number L' with $L' \neq L$ as x approaches c.

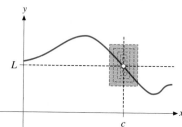

FIGURE 2.4.12c The fact that ϵ may be as small as we please forces $f(x)$ to approach the single number L as x approaches c.

It should be noted that the problem of verifying that $\lim_{x \to c} f(x) = L$ is the same quantitative problem that we encountered in Examples 1–6. Namely, given a positive number ϵ, we must find a positive number δ such that

$$|f(x) - L| < \epsilon \text{ whenever } 0 < |x - c| < \delta.$$

Let us see how the formal definition of limit can be used to verify some simple limits.

EXAMPLE 7

Use the definition to verify that $\lim_{x \to 3} (2x - 5) = 1$.

SOLUTION We must show that, for each positive number ϵ, there is a positive number δ such that

$$|(2x - 5) - 1| < \epsilon \text{ whenever } 0 < |x - 3| < \delta.$$

We have

$$|(2x - 5) - 1| = |2x - 6| = 2|x - 3| < \epsilon$$

whenever $|x - 3| < \epsilon/2$. The condition of the definition is satisfied with $\delta = \epsilon/2$. See Figure 2.4.13. ∎

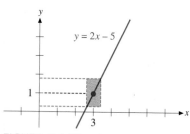

FIGURE 2.4.13 $|(2x - 5) - 1| < \epsilon$ whenever $|x - 3| < \epsilon/2$, so $\lim_{x \to 3} (2x - 5) = 1$.

EXAMPLE 8

Use the definition to verify that $\lim\limits_{x \to -1} x^2 = 1$.

SOLUTION   We must show that, for each positive number ϵ, there is a positive number δ such that $|x^2 - 1| < \epsilon$ whenever $0 < |x + 1| < \delta$. We make the preliminary assumption that $|x + 1| \leq 1$. (Any preliminary assumption $|x + 1| \leq k$, k a positive number, would do.)

Factoring, we have

$$|x^2 - 1| = |x - 1||x + 1|,$$

so we see that we need a bound of the factor $|x - 1|$. Using the preliminary assumption $|x + 1| \leq 1$, we obtain

$$-1 \leq x + 1 \leq 1,$$
$$-2 \leq x \leq 0.$$

The values of $|x - 1|$ at the endpoints of this interval are

$$|(-2) - 1| = 3 \text{ and } |(0) - 1| = 1,$$

so $|x - 1| \leq 3$ in the interval.

Using the inequality $|x - 1| \leq 3$, we obtain

$$|x^2 - 1| = |x - 1||x + 1| \leq 3|x + 1| < \epsilon$$

whenever $0 < |x + 1| < \epsilon/3$ and $|x + 1| \leq 1$, so our preliminary assumption is satisfied. The condition of the definition is satisfied with δ equal to the smaller of $\epsilon/3$ and 1. See Figure 2.4.14.   ∎

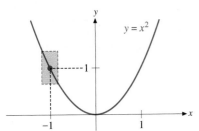

FIGURE 2.4.14 $|x^2 - 1| < \epsilon$ whenever $|x + 1| \leq 1$ and $|x + 1| < \epsilon/3$, so $\lim\limits_{x \to -1} x^2 = 1$.

EXAMPLE 9

Use the definition to verify that $\lim\limits_{x \to 4} \sqrt{x} = 2$.

SOLUTION   We must show that, for each positive number ϵ, there is a positive number δ such that

$$|\sqrt{x} - 2| < \epsilon \text{ whenever } 0 < |x - 4| < \delta.$$

We make the preliminary assumption that $x \geq 0$.

In order to verify the condition, we first use the factorization formula

$$x - 4 = (\sqrt{x} - 2)(\sqrt{x} + 2)$$

to express $|\sqrt{x} - 2|$ in terms of $|x - 4|$. We obtain

$$|\sqrt{x} - 2| = \frac{|x - 4|}{\sqrt{x} + 2}.$$

(We may think of the above step as rationalization of the numerator.) The next step is to find a bound of the factor $1/(\sqrt{x} + 2)$. The denominator is a sum of a nonnegative variable term $\sqrt{x}$ and the positive constant 2. We can obtain a bound by substituting zero for the nonnegative term $\sqrt{x}$. We obtain

$$\frac{1}{\sqrt{x} + 2} \leq \frac{1}{0 + 2} = \frac{1}{2}.$$

Combining results, we have

$$|\sqrt{x} - 2| \leq \frac{1}{2}|x - 4|.$$

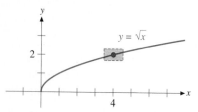

FIGURE 2.4.15 $|\sqrt{x} - 2| < \epsilon$ whenever $x \geq 0$ and $|x - 4| < 2\epsilon$, so $\lim\limits_{x \to 4} \sqrt{x} = 2$.

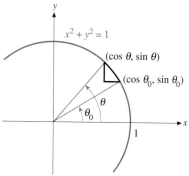

FIGURE 2.4.16 The terminal side of an angle θ in standard position intersects the circle $x^2 + y^2 = 1$ at the point with coordinates $x = \cos\theta$ and $y = \sin\theta$.

FIGURE 2.4.17 $|\cos\theta - \cos\theta_0| < \epsilon$ whenever $|\theta - \theta_0| < \epsilon$, so $\lim\limits_{\theta \to \theta_0} \cos\theta = \cos\theta_0$.

FIGURE 2.4.18 $|\sin\theta - \sin\theta_0| < \epsilon$ whenever $|\theta - \theta_0| < \epsilon$, so $\lim\limits_{\theta \to \theta_0} \sin\theta = \sin\theta_0$.

It follows that $|\sqrt{x} - 2| < \epsilon$ whenever $\frac{1}{2}|x - 4| < \epsilon$, or $|x - 4| < 2\epsilon$. The condition of the definition is satisfied for $\delta = 2\epsilon$. See Figure 2.4.15. ■

It is not always possible to use only algebraic methods to verify the condition in the definition of limit.

EXAMPLE 10

Use the definition to verify that $\lim\limits_{\theta \to \theta_0} \cos\theta = \cos\theta_0$ and $\lim\limits_{\theta \to \theta_0} \sin\theta = \sin\theta_0$.

SOLUTION We will use a geometric argument to verify the condition of the definition for the limits. Recall that the terminal side of an angle of θ radians in standard position intersects the circle $x^2 + y^2 = 1$ at the point with coordinates $x = \cos\theta$ and $y = \sin\theta$. In Figure 2.4.16 we have labeled the coordinates of points that correspond to angles θ and θ_0. It follows from the definition of radian measure (Section 1.6) that the length of the arc of the circle between the points is $|\theta - \theta_0|$. The length of the horizontal side of the "triangle" indicated in Figure 2.4.16 is $|\cos\theta - \cos\theta_0|$, while the length of the vertical side is $|\sin\theta - \sin\theta_0|$. Since the arc is longer than the length of either side, we have

$$|\cos\theta - \cos\theta_0| \leq |\theta - \theta_0|$$

and

$$|\sin\theta - \sin\theta_0| \leq |\theta - \theta_0|.$$

It is clear from the above inequalities that, given any positive number ϵ, we have

$$|\cos\theta - \cos\theta_0| < \epsilon \text{ whenever } |\theta - \theta_0| < \epsilon$$

and

$$|\sin\theta - \sin\theta_0| < \epsilon \text{ whenever } |\theta - \theta_0| < \epsilon.$$

In both cases, the condition of the definition is satisfied for $\delta = \epsilon$. See Figure 2.4.17 and Figure 2.4.18. ■

We conclude this section with a formal definition of one-sided limits.

DEFINITION

    The limit of $f(x)$ as x approaches c from the right is L (written $\lim\limits_{x \to c^+} f(x) = L$) if, for each $\epsilon > 0$, there is a $\delta > 0$ such that $|f(x) - L| < \epsilon$ whenever $c < x < c + \delta$.

    The limit of $f(x)$ as x approaches c from the left is L (written $\lim\limits_{x \to c^-} f(x) = L$) if, for each $\epsilon > 0$, there is a $\delta > 0$ such that $|f(x) - L| < \epsilon$ whenever $c - \delta < x < c$.

EXERCISES 2.4

In Exercises 1–12 find a positive number δ such that the given statement is true.

1. For each positive number ϵ, $|(2x - 3) - 1| < \epsilon$ whenever $|x - 2| < \delta$.

2. For each positive number ϵ, $|(4 - 3x) - 7| < \epsilon$ whenever $|x + 1| < \delta$.

3. For each positive number ϵ, $|x^3| < \epsilon$ whenever $|x| < \delta$.

4. For each positive number ϵ, $|x^5| < \epsilon$ whenever $|x| < \delta$.

5. For each positive number ϵ, $|x^{1/3} \sin 3x| < \epsilon$ whenever $|x| < \delta$.

6. For each positive number ϵ, $|x^{1/5} \cos 2x| < \epsilon$ whenever $|x| < \delta$.

7. $|x^2 - 4| < 0.001$ whenever $|x + 2| < \delta$. Assume $|x + 2| \leq 1$.

8. $|(x^2 + 5x - 1) + 5| < 0.005$ whenever $|x + 1| < \delta$. Assume $|x + 1| \leq 1$.

9. $\left|\dfrac{1}{x} + \dfrac{1}{4}\right| < 0.001$ whenever $|x + 4| < \delta$. Assume $|x + 4| \leq 1$.

10. $\left|\dfrac{x}{x + 2} - \dfrac{1}{2}\right| < 0.001$ whenever $|x - 2| < \delta$. Assume $|x - 2| \leq 1$.

11. For each positive number ϵ, $\left|\dfrac{x^2 - 4}{x - 2} - 4\right| < \epsilon$ whenever $0 < |x - 2| < \delta$.

12. For each positive number ϵ, $\left|\dfrac{2x^2 - 3x + 1}{x - 1} - 1\right| < \epsilon$ whenever $0 < |x - 1| < \delta$.

13. A 100-m field is to be marked by using a measuring tape that is approximately 10 m long. How close to 10 m must the tape be in order for the actual length of the field to be within 0.01 m of 100 m?

14. A 12-ft board is to be marked by using a measuring stick that is approximately 3 ft long. How close to 3 ft must the stick be in order for the actual length of the board to be within 0.02 ft of 12 ft?

15. How close to 5 m should each side of a square be in order for the area of the square to be within 0.01 m² of 25 m²? Assume each side is within 0.01 m of 5 m.

16. How close to 4 m should be radius of a circle be in order for the area of the circle to be within 0.01 m² of 16π m²? Assume the radius is within 0.01 m of 4 m.

17. How close to 2 should each edge of a cube be in order for the volume of the cube to be within 0.01 of 8? Assume each edge is within 0.01 of 2.

18. How close to 3 should the radius of a sphere be in order for the volume of the sphere to be within 0.01 of 36π? Assume the radius can be measured within 0.01 of 3. $V = 4\pi r^3/3$.

19. How close to 3 Ω must the resistance in a 12-V electrical system be in order for the current to be within 0.01 A of 4 A? Assume the resistance is within 1 Ω of 3 Ω. $I = V/R$, where I is the current in amperes, V is the voltage in volts, and R is the resistance in ohms.

20. How close must x be to 3 to guarantee that x^2 is within 0.4 of 9? (**Calculus Explorer & Tutor I, Limits, 8.**)

21. Resistors are available in a variety of standard resistances with 5% tolerance. For example, a 500 Ω resistor with a 5% tolerance is guaranteed to have a resistance between 475 Ω and 525 Ω. An engineer is designing a 110-V circuit that should carry between 0.215 A and 0.225 A. If $I = V/R$, where I is the current in amperes, V the voltage in volts, and R the resistance in ohms, will a resistor rated at 500 Ω with a 5% tolerance be adequate? If not, what tolerance should the engineer insist upon in the specifications?

22. **Doppler echocardiography** is a noninvasive method of determining blood velocity within the heart by measuring how the frequency of reflected sound waves differs from the frequency of transmitted sound waves. The velocity is given by $V = \dfrac{kFd}{\cos\theta}$, where Fd is the frequency shift, θ is the angle between the ultrasound beam and the direction of the blood cells sampled, and k is a constant. Usually, the ultrasound beam is parallel to the blood velocity and the cosine of the Doppler angle is assumed to be 1.00. The velocity is underestimated if the Doppler angle is greater than zero. How near to 0° should the Doppler angle be to guarantee that the error in velocity is less than 10%? See figure. (*Hint:* Which values of θ near zero guarantee that $\dfrac{kFd}{\cos\theta} - kFd < 0.10kFd$?)

Blood flows through an open valve.

There is no regurgitant flow when the valve closes sufficiently.

Insufficient valve closure results in high velocity and turbulent regurgitant flow.

Verify the condition of the formal definition for each of the limits in Exercises 23–35.

23. $\lim\limits_{x \to 3} 2x = 6$

24. $\lim\limits_{x \to 2} (3x - 1) = 5$ **(Calculus Explorer & Tutor I, Limits, 7.)**

25. $\lim\limits_{x \to 0} x^{1/3} = 0$

26. $\lim\limits_{x \to 0} x^3 = 0$

27. $\lim\limits_{x \to 16} \sqrt{x} = 4$

28. $\lim\limits_{x \to 9} \sqrt{x} = 3$

29. $\lim\limits_{x \to 27} x^{1/3} = 3$
(*Hint:* $x - c = (x^{1/3} - c^{1/3})(x^{2/3} + x^{1/3}c^{1/3} + c^{2/3})$.)

30. $\lim\limits_{x \to -8} x^{1/3} = -2$

31. $\lim\limits_{x \to \pi/4} \sin 2x = 1$

32. $\lim\limits_{x \to \pi/3} \cos 3x = -1$

33. $\lim\limits_{x \to 0} x^2 \cos^2 x = 0$

34. $\lim\limits_{x \to 0} x^{1/3} \sin(1/x) = 0$

35. $\lim\limits_{x \to c} x^{1/3} = c^{1/3}, c \neq 0$

36. (a) Starting with $x = 2$, repeatedly press the $\sqrt{}$ key on your calculator and explain what happens to the display.
(b) Repeat (a), except start with $x = 0.5$.

37. (a) Starting with $x = 1.1$, repeatedly press the x^2 key on your calculator and explain what happens to the display.
(b) Repeat (a), except start with $x = 0.9$.

38. Use a calculator to evaluate $\sqrt{5}$ and $10^{\sqrt{5}}$. For $n = 1, 2, 3, 4, \ldots$, let x_n be the decimal representation of $\sqrt{5}$ truncated to n decimal places, so $x_1 = 2.2$, $x_2 = 2.23$, etc. Use your calculator to evaluate 10^{x_n} for some values of n to determine the smallest n for which 10^{x_n} is within 0.1 of $10^{\sqrt{5}}$.

39. If $\epsilon > 0$, then $\epsilon^0 = 1$ and $0^\epsilon = 0$. It should then seem reasonable that $\epsilon^x \approx 1$ for $x \approx 0$ and $x^\epsilon \approx 0$ for $x \approx 0$, $x > 0$.
(a) Use your calculator to evaluate $(0.1)^{10^{-n}}$ for some values of n to determine the smallest positive integer n for which $(0.1)^{10^{-n}} \geq 0.99$.
(b) Use your calculator to evaluate $(10^{-n})^{0.1}$ for some values of n to determine the smallest positive integer n for which $(10^{-n})^{0.1} \leq 0.01$.

40. Let $x_n = 2 + 10^{-n}$. What is the smallest n such that $|e^{x_n} - e^2| < 0.0001$?

41. Let $x_n = 10 + 10^{-n}$. What is the smallest n such that $|\ln x_n - \ln 10| < 0.0001$?

42.
(a) Let x_n be the decimal expansion of $\sqrt{3}$ truncated at the nth decimal place, so $x_1 = 1.7$, $x_2 = 1.73$, and $x_3 = 1.732$, etc. What is the smallest n for which x_n^2 is within 0.0001 of 3?
(b) Use the preliminary assumption $0 < x < \sqrt{3}$ to determine a positive number δ such that $|x^2 - 3| < 0.0001$ whenever $|x - \sqrt{3}| < \delta$. Compare the result with that of (a).

43. If
$$f(x) = \begin{cases} 1 - x, & x < 0, \\ 0, & x = 0, \\ -1 - x, & x > 0, \end{cases}$$
for what positive values of ϵ is it possible to find some $\delta > 0$ such that $|f(x) - f(0)| < \epsilon$ whenever $|x| < \delta$?

2.5 EVALUATION OF LIMITS

Connections

Arithmetic combinations and composition of functions, 2.1.

An intuitive idea of limits and continuity, 2.3.

It is necessary to evaluate limits in order to derive many of the formulas that we will use to solve calculus problems. In this section we will illustrate how continuity can be used to evaluate many limits easily. We conclude the section by developing some important general results about limits.

Limits of Functions Known to Be Continuous

We begin by noting that it is easy to evaluate the limit at c of a function that is known to be continuous at c. That is, if f is continuous at c, then $\lim\limits_{x \to c} f(x) = f(c)$. This means:

The limit at c of a function that is continuous at c is the value of the function at c.

We could use either the informal techniques of Section 2.3 or the formal techniques of Section 2.4 to verify that many familiar functions are continuous at each point where they are defined. For example, we can verify that linear functions $ax + b$, power functions x^n, the sine function, the cosine function, and the absolute value function are continuous at every point; if n is any positive integer, the root function $x^{1/n}$ is continuous at those c where it is defined for all x near c.

EXAMPLE 1

Since the functions involved are known to be continuous at the points of interest, we have

(a) $\displaystyle\lim_{x \to 2}(3x - 1) = 3(2) - 1 = 5,$

(b) $\displaystyle\lim_{x \to 3} x^2 = (3)^2 = 9,$

(c) $\displaystyle\lim_{x \to \pi/6} \sin x = \sin(\pi/6) = 1/2,$

(d) $\displaystyle\lim_{x \to \pi/2} \cos x = \cos(\pi/2) = 0,$

(e) $\displaystyle\lim_{x \to 16} \sqrt{x} = \sqrt{16} = 4.$ ∎

Arithmetic Combinations of Continuous Functions

In order to use continuity to evaluate a limit, we need to know that the function involved is continuous. The following theorem is useful for determining the continuity of many functions.

THEOREM 1

If f and g are continuous at c, then af, $f + g$, and fg are continuous at c. If $g(c) \neq 0$, then $1/g$ and f/g are also continuous at c.

Theorem 1 is an easy consequence of a corresponding result for limits, which we will develop later in this section. For now, let us see how we can use Theorem 1.

A polynomial $P(x) = a_0 + a_1 x + \cdots + a_n x^n$ is the sum of a linear function and terms that are a constant multiples of power functions. Since we know that each of these functions is continuous at each point, Theorem 1 implies that polynomials are continuous at each point. Since Theorem 1 also implies that quotients of continuous functions are continuous at points where the denominator is nonzero, we obtain the following:

Polynomials are continuous at each point. If P and Q are polynomials, the rational function P/Q is continuous at each point where it is defined.

EXAMPLE 2

Since polynomials and rational functions are continuous at each point where they are defined, we have

(a) $\displaystyle\lim_{x \to 2}(x^2 - 3x + 2) = (2)^2 - 3(2) + 2 = 0,$

(b) $\displaystyle\lim_{x \to -3}(x + 5)(x - 3) = (-3 + 5)(-3 - 3) = -12,$

(c) $\lim\limits_{x \to 3} \dfrac{x^2}{x - 5} = \dfrac{3^2}{3 - 5} = -\dfrac{9}{2}$. ■

Since

$$\tan x = \frac{\sin x}{\cos x}, \cot x = \frac{\cos x}{\sin x}, \sec x = \frac{1}{\cos x}, \text{ and } \csc x = \frac{1}{\sin x},$$

we can use the Theorem 1 and the continuity of the sine and cosine functions to conclude the following:

The six basic trigonometric functions are continuous at each point where they are defined.

EXAMPLE 3

Since the tangent function is continuous at $\pi/4$, $\lim\limits_{x \to \pi/4} \tan x = \tan \dfrac{\pi}{4} = 1$. ■

Theorem 1 also allows us to evaluate limits of sums, products, and quotients of trigonometric and root functions.

EXAMPLE 4

(a) $\lim\limits_{x \to 0} x^{1/3} \sec x = (0)^{1/3}(\sec 0) = (0)(1) = 0$.

(b) $\lim\limits_{x \to 0} \dfrac{\tan x + \sec x}{\cos x} = \dfrac{\tan 0 + \sec 0}{\cos 0} = \dfrac{0 + 1}{1} = 1$. ■

Continuity of Composite Functions

Recall that the composition of two functions is defined by the equation $f \circ g(x) = f(g(x))$. The following result on the continuity of composite functions is an easy consequence of the corresponding result for limits, which we will consider later in this section.

COMPOSITION THEOREM

If g is continuous at c and f is continuous at $g(c)$, then $f \circ g$ is continuous at c.

EXAMPLE 5

Evaluate

(a) $\lim\limits_{x \to 3} |1 - 2x - 3x^2|$,

(b) $\lim\limits_{x \to \pi/3} \sin^2 x$,

(c) $\lim\limits_{x \to \pi/6} \cos 2x$.

SOLUTION

(a) $|1 - 2x - 3x^2|$ is the "absolute value of $1 - 2x - 3x^2$." Since the absolute value function and the polynomial are continuous, the composition is continuous.

$$\lim\limits_{x \to 3} |1 - 2x - 3x^2| = |1 - 2(3) - 3(3)^2| = |-32| = 32.$$

(b) $\sin^2 x = (\sin x)^2$ is the "square of $\sin x$." The square function and the sine function are continuous at every point, so

$$\lim_{x \to \pi/3} \sin^2 x = \sin^2 \left(\frac{\pi}{3}\right) = \left(\frac{\sqrt{3}}{2}\right)^2 = \frac{3}{4}.$$

(c) $\cos 2x$ is the "cosine of $2x$." Since the cosine function and the function $2x$ are continuous at every point, the composition is continuous.

$$\lim_{x \to \pi/6} \cos 2x = \cos 2 \left(\frac{\pi}{6}\right) = \cos \frac{\pi}{3} = \frac{1}{2}. \quad \blacksquare$$

Limits of Difference Quotients

One of the fundamental concepts of calculus, which will be introduced in the next section, involves limits of the form

$$\lim_{x \to c} \frac{f(x) - f(c)}{x - c}.$$

Thus, we are interested in the behavior of the quotient

$$\frac{f(x) - f(c)}{x - c}$$

for x near the point c where the denominator is zero. This important type of limit cannot be evaluated simply by substitution of c for x, since the expression of interest is undefined at $x = c$. Some limits of this type can be evaluated by using the following result.

THEOREM 2

If $\lim_{x \to c} g(x)$ exists and there is some positive number δ_1 such that $f(x) = g(x)$ for $0 < |x - c| < \delta_1$, then $\lim_{x \to c} f(x) = \lim_{x \to c} g(x)$.

Theorem 2 is true because limits at c depend only on values of x near c, $x \neq c$, and $f(x) = g(x)$ for those x. Limits at c do not depend on values at c.

EXAMPLE 6

Evaluate $\lim_{x \to 1} \dfrac{x^2 - 1}{x - 1}$.

SOLUTION   We first note that we cannot evaluate the limit at 1 of $(x^2 - 1)/(x - 1)$ by substitution of 1 for x, because the denominator is zero when $x = 1$. However,

$$f(x) = \frac{x^2 - 1}{x - 1} = \frac{(x - 1)(x + 1)}{x - 1} = x + 1, \quad x \neq 1.$$

That is, if $g(x) = x + 1$, we have $f(x) = g(x)$, $x \neq 1$. Compare the graph of f in Figure 2.5.1a with the graph of g in Figure 2.5.1b. The graphs are the same,

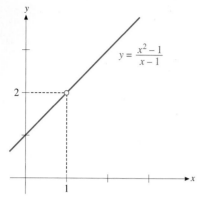

FIGURE 2.5.1a The functions $f(x) = \frac{x^2 - 1}{x - 1}$ and $g(x) = x + 1$ satisfy $f(x) = g(x)$, except for $x = 1$, so the two functions have the same limits at one.

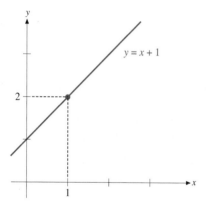

FIGURE 2.5.1b The linear function $g(x) = x + 1$ is continuous, so $\lim_{x \to 1} g(x) = g(1) = 2$.

except for the point $(1, 2)$. Since $g(x) = x + 1$ is a polynomial, we already know how to evaluate limits of g. In particular,

$$\lim_{x \to 1} g(x) = \lim_{x \to 1} (x + 1) = (1) + 1 = 2.$$

Since f and g have the same limit at 1, it follows that $\lim_{x \to 1} f(x) = 2$. ∎

We will ordinarily evaluate limits of difference quotients without referring specifically to the functions f and g of Theorem 2. For example, the limit in Example 6 can be evaluated by writing simply

$$\lim_{x \to 1} \frac{x^2 - 1}{x - 1} = \lim_{x \to 1} \frac{(x - 1)(x + 1)}{x - 1} = \lim_{x \to 1} (x + 1) = (1) + 1 = 2.$$

One-Sided Limits

We can use what we know about limits of continuous functions to evaluate one-sided limits. Recall from Section 2.3 that

> If $\lim_{x \to c} f(x) = L$, then $\lim_{x \to c^-} f(x) = \lim_{x \to c^+} f(x) = L$.
>
> If $\lim_{x \to c^-} f(x) = \lim_{x \to c^+} f(x) = L$, then $\lim_{x \to c} f(x) = L$.
>
> If $\lim_{x \to c^-} f(x) \neq \lim_{x \to c^+} f(x)$, then $\lim_{x \to c} f(x)$ does not exist.

Suppose that $f(x)$ is given by a particular formula for x in an interval. The formula may then be used to determine the limit of f at interior points of the interval and the one-sided limits from inside the interval at each endpoint of the interval. The values of the one-sided limits of f at the endpoints are not affected by whether or not the formula gives the values of f at the endpoints of the interval, since limits at a point do not depend on the value of the function at the point.

EXAMPLE 7

$f(x) = x^2$, $-1 < x \leq 2$. Evaluate **(a)** $\lim_{x \to 0} f(x)$, **(b)** $\lim_{x \to -1^+} f(x)$, and **(c)** $\lim_{x \to 2^-} f(x)$.

SOLUTION

(a) Zero is an interior point of the interval $-1 < x \leq 2$, so the values of f are given by the formula $f(x) = x^2$ as x approaches 0 from each side. We then have

$$\lim_{x \to 0} f(x) = \lim_{x \to 0} x^2 = 0.$$

(b) The values of f are given by the formula $f(x) = x^2$ as x approaches -1 from the right, so

$$\lim_{x \to -1^+} f(x) = \lim_{x \to -1^+} x^2 = (-1)^2 = 1.$$

Note that we have used the formula x^2 to evaluate the limit from the right of f at -1. The value of the limit is the value of x^2 at $x = -1$, but this is not the value of f at $x = -1$; $f(-1)$ is undefined.

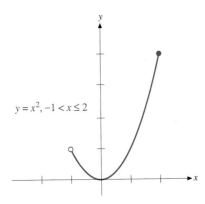

$y = x^2, -1 < x \leq 2$

FIGURE 2.5.2 The formula x^2 gives the values of the limit of f at interior points of the interval and the value of the one-sided limits from within the interval at each endpoint. It does not matter whether or not the formula gives the values of the function at the endpoints, since limits at a point do not depend on the value of the function at the point.

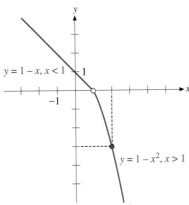

$y = 1 - x, x < 1$

$y = 1 - x^2, x > 1$

FIGURE 2.5.3 The formula $1 - x^2$ gives the limit of f at two and the limit from the right at one. The formula $1 - x$ gives the limit from the left at one. Since both one-sided limits at one have value zero, $\lim_{x \to 1} f(x) = 0$.

(c) The values of f are given by the formula $f(x) = x^2$ as x approaches 2 from the left, so

$$\lim_{x \to 2^-} f(x) = \lim_{x \to 2^-} x^2 = (2)^2 = 4.$$

Again, we have used the formula x^2 to evaluate the limit from the left at 2. It is coincidental that this value is also the value of f at 2. The graph of f is given in Figure 2.5.2. ∎

If the values of a function f are given by different formulas for x in different intervals, we must use one-sided limits to evaluate or determine the existence of limits at the endpoints of the intervals.

EXAMPLE 8

$$f(x) = \begin{cases} 1 - x, & x < 1, \\ 1 - x^2, & x > 1. \end{cases}$$

Evaluate **(a)** $\lim_{x \to 2} f(x)$ and **(b)** $\lim_{x \to 1} f(x)$.

SOLUTION

(a) The formula $f(x) = 1 - x^2$ holds as x approaches 2 from each side of 2, so we have

$$\lim_{x \to 2} f(x) = \lim_{x \to 2} (1 - x^2) = 1 - (2)^2 = -3.$$

(b) Different formulas for x hold as x approaches 1 from different sides of 1, so we must look at each one-sided limit separately. The limit from the left at one involves only $x < 1$. For $x < 1$ we use the formula $f(x) = 1 - x$, so

$$\lim_{x \to 1^-} f(x) = \lim_{x \to 1^-} (1 - x) = 1 - (1) = 0.$$

The limit from the right at one involves only $x > 1$. For $x > 1$, we use the formula $f(x) = 1 - x^2$, so

$$\lim_{x \to 1^+} f(x) = \lim_{x \to 1^+} (1 - x^2) = 1 - (1)^2 = 0.$$

Since both one-sided limits at one have value zero, $\lim_{x \to 1} f(x) = 0$. See Figure 2.5.3. ∎

EXAMPLE 9

Evaluate $\lim_{x \to 2} \dfrac{|x - 2|}{x - 2}$.

SOLUTION Since $x - 2 \geq 0$ for $x \geq 2$ and $x - 2 < 0$ for $x < 2$, we know that

$$|x - 2| = \begin{cases} x - 2, & x \geq 2, \\ -(x - 2), & x < 2, \end{cases}$$

so $|x - 2|$ is a function that is defined by different formulas for different values of x. We then have

$$\frac{|x - 2|}{x - 2} = \begin{cases} 1, & x > 2, \\ -1, & x < 2. \end{cases}$$

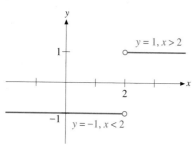

FIGURE 2.5.4 $\lim\limits_{x \to 2^-} \dfrac{|x-2|}{x-2} = -1$ and $\lim\limits_{x \to 2^+} \dfrac{|x-2|}{x-2} = 1$. Since the two one-sided limits are unequal, $\lim\limits_{x \to 2} \dfrac{|x-2|}{x-2}$ does not exist.

Since the values of $\dfrac{|x-2|}{x-2}$ are given by different formulas as x approaches 2 from different sides, we must look at the one-sided limits at 2. The limit from the left at two involves only $x < 2$, so we use the formula $\dfrac{|x-2|}{x-2} = -1$ to obtain

$$\lim_{x \to 2^-} \frac{|x-2|}{x-2} = \lim_{x \to 2^-} -1 = -1.$$

The limit from the right at two involves only $x > 2$, so we use the formula $\dfrac{|x-2|}{x-2} = 1$ to obtain

$$\lim_{x \to 2^+} \frac{|x-2|}{x-2} = \lim_{x \to 2^+} 1 = 1.$$

Since the one-sided limits have unequal values, $\lim\limits_{x \to 2} \dfrac{|x-2|}{x-2}$ does not exist. See Figure 2.5.4. ■

Development of Results for Limits

We have seen how to use results for continuous functions to evaluate many limits easily. The continuity results are easy consequences of the corresponding results for limits, which we now develop.

We have seen in Section 2.4 that the following formal definition of limit reflects the ideas that were illustrated graphically in Section 2.3.

DEFINITION

The **limit of f at c is L** if, for each $\epsilon > 0$, there is a $\delta > 0$ such that

$$|f(x) - L| < \epsilon \text{ whenever } 0 < |x - c| < \delta.$$

It follows from the definition that the statement "$\lim\limits_{x \to c} f(x) = L$" is equivalent to the statement "$\lim\limits_{x \to c} |f(x) - L| = 0$." That is, we can show that $f(x) \to L$ as $x \to c$ by showing that $|f(x) - L| \to 0$ as $x \to c$.

Strategy for showing that $\lim\limits_{x \to c} f(x) = L$:

- Use inequalities and bounds of functions to obtain an inequality of the form $|f(x) - L| \leq |z(x)|$, where $z(x)$ is a function that is known to approach zero as x approaches c.

The following result is the formal basis of our strategy.

If $\lim\limits_{x \to c} z(x) = 0$ and there is some positive number δ_1 such that $|f(x) - L| \leq |z(x)|$ on the set $0 < |x - c| < \delta_1$, then $\lim\limits_{x \to c} f(x) = L$. (1)

If $|f(x) - L| \leq |z(x)|$, then it is clear that $|f(x) - L|$ is less than a positive number ϵ whenever $|z(x)|$ is.

Statement 1 allows us to verify that $\lim_{x \to c} f(x) = L$ without carrying out the details of verifying the ϵ–δ definition of limit. We will also use the following two facts.

$$\text{If } \lim_{x \to c} z_1(x) = \lim_{x \to c} z_2(x) = 0, \text{ then } \lim_{x \to c} (z_1(x) + z_2(x)) = 0. \tag{2}$$

For example, if $|z_1(x)| < \epsilon/2$ whenever $0 < |x - c| < \delta_1$ and $|z_2(x)| < \epsilon/2$ whenever $0 < |x - c| < \delta_2$, then $|z_1(x) + z_2(x)| < \epsilon$ whenever $0 < |x - c| < \delta$, where δ is the smaller of δ_1 and δ_2. Statement 2 says:

The sum of two functions with limit zero has limit zero.

If $\lim_{x \to c} z(x) = 0$ and there are positive numbers B and δ_1 such that $|b(x)| \leq B$ whenever $0 < |x - c| < \delta_1$, then $\lim_{x \to c} b(x)z(x) = 0$. $\tag{3}$

The product $|b(x)z(x)|$ is less than a positive number ϵ whenever $|z(x)|$ is less than ϵ/B and x is within δ_1 of c, $x \neq c$. Statement 3 says:

The product of a bounded function and a function with limit zero has limit zero.

The following two results relate limits and bounds of functions.

If $\lim_{x \to c} f(x) = L$, then there is some positive number δ_1 such that $f(x)$ is bounded on the set $0 < |x - c| < \delta_1$. $\tag{4}$

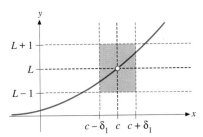

For example, if $f(x)$ is within one of L, then f is bounded. See Figure 2.5.5. Statement 4 says:

FIGURE 2.5.5   If $\lim_{x \to c} f(x) = L$, then f is bounded near c.

Functions are bounded near points where they have limits.

If $\lim_{x \to c} g(x) = M$ and $M \neq 0$, then there is some positive number δ_1 such that $1/g(x)$ is bounded on the set $0 < |x - c| < \delta_1$. $\tag{5}$

We know that $1/g(x)$ is bounded on sets where $g(x)$ stays away from zero. If $g(x)$ is near M and $M \neq 0$, then $g(x)$ stays away from zero. See Figure 2.5.6. Statement 5 says:

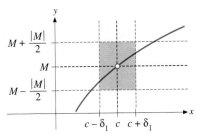

$\dfrac{1}{g(x)}$ *is bounded near points where g has nonzero limit.*

FIGURE 2.5.6   If $\lim_{x \to c} g(x) = M$ and $M \neq 0$, then $1/g(x)$ is bounded near c.

We now consider the main theorem on limits. This theorem allows us to evaluate the limit of a complicated expression by finding the limits of each of its less complicated parts.

LIMIT THEOREM

If $\lim\limits_{x \to c} f(x) = L$ and $\lim\limits_{x \to c} g(x) = M$, then

(a) $\lim\limits_{x \to c}(af(x)) = a \lim\limits_{x \to c} f(x) = aL$, a any constant,

(b) $\lim\limits_{x \to c}(f(x) + g(x)) = \lim\limits_{x \to c} f(x) + \lim\limits_{x \to c} g(x) = L + M$,

(c) $\lim\limits_{x \to c}(f(x)g(x)) = \lim\limits_{x \to c} f(x) \cdot \lim\limits_{x \to c} g(x) = LM$,

(d) $\lim\limits_{x \to c} \dfrac{1}{g(x)} = \dfrac{1}{\lim\limits_{x \to c} g(x)} = \dfrac{1}{M}$, $M \neq 0$,

(e) $\lim\limits_{x \to c} \dfrac{f(x)}{g(x)} = \dfrac{\lim\limits_{x \to c} f(x)}{\lim\limits_{x \to c} g(x)} = \dfrac{L}{M}$, $M \neq 0$.

Proof   We will use the facts that $\lim\limits_{x \to c} f(x) = L$ implies $|f(x) - L| \to 0$ as $x \to c$ and that $\lim\limits_{x \to c} g(x) = M$ implies $|g(x) - M| \to 0$ as $x \to c$.

(a) Factoring, we have

$$|af(x) - aL| = |a||f(x) - L|.$$

Since we know that $|f(x) - L|$ approaches zero as x approaches c, the product with the constant $|a|$ also approaches zero. Statement 1 then implies $\lim\limits_{x \to c}(af(x)) = aL$.

(b) Rearranging the terms and using the Triangle Inequality, we have

$$|(f(x) + g(x)) - (L + M)| = |(f(x) - L) + (g(x) - M)|$$
$$\leq |f(x) - L| + |g(x) - M|.$$

Since each of the two terms on the right approach zero, Statement 2 implies the sum also approaches zero. Statement 1 then implies $\lim\limits_{x \to c}(f(x) + g(x)) = L + M$.

(c) The algebra trick of adding and subtracting $f(x)M$ and the Triangle Inequality are used to verify this statement.

$$|f(x)g(x) - LM| = |f(x)g(x) - f(x)M + f(x)M - LM|$$
$$= |f(x)(g(x) - M) + (f(x) - L)M|$$
$$\leq |f(x)||g(x) - M| + |f(x) - L||M|.$$

Since f has a limit at c, Statement 4 implies that $f(x)$ is bounded near c, $x \neq c$, so the product $|f(x)||g(x) - M| \to 0$ as $x \to c$, as in Statement 3. Since, also, $|f(x) - L||M| \to 0$ as $x \to c$, Statements 2 and 1 then imply $\lim\limits_{x \to c} f(x)g(x) = LM$.

(d) $\left| \dfrac{1}{g(x)} - \dfrac{1}{M} \right| = \dfrac{|M - g(x)|}{|g(x)M|}$. Since g has a nonzero limit, Statement 5 implies that $1/g(x)$ is bounded near c, $x \neq c$. Statement 2 then implies

$$\left| \dfrac{1}{g(x)} - \dfrac{1}{M} \right| = \dfrac{1}{|g(x)|} \dfrac{1}{|M|} |M - g(x)| \to 0 \text{ as } x \to c,$$

so Statement 1 implies $\lim\limits_{x \to c} \dfrac{1}{g(x)} = \dfrac{1}{M}$.

(e) From (d), we have

$$\frac{1}{g(x)} \rightarrow \frac{1}{M} \quad \text{as } x \rightarrow c.$$

From (c), applied to the product

$$f(x) \cdot \frac{1}{g(x)},$$

we have

$$\lim_{x \rightarrow c} \frac{f(x)}{g(x)} = \lim_{x \rightarrow c} f(x) \cdot \frac{1}{g(x)} = L \cdot \frac{1}{M} = \frac{L}{M}, \quad M \neq 0. \quad \blacksquare$$

Theorem 1 is an easy consequence of the Limit Theorem. For example, if f and g are continuous at c, so $\lim_{x \rightarrow c} f(x) = f(c)$ and $\lim_{x \rightarrow c} g(x) = g(c)$, the Limit Theorem implies that

$$\lim_{x \rightarrow c}(f+g)(x) = \lim_{x \rightarrow c}(f(x) + g(x)) = \lim_{x \rightarrow c} f(x) + \lim_{x \rightarrow c} g(x)$$
$$= f(c) + g(c) = (f+g)(c),$$

so $f + g$ is continuous at c.

Note that we cannot apply the Limit Theorem to obtain the limit of a sum, product, or quotient unless both of the functions involved have limits. We can use Statement 3 to evaluate the limit of a product of a bounded function and a function that approaches zero, even though the bounded function may not have a limit.

EXAMPLE 10

Evaluate $\lim_{x \rightarrow 0} x \cos(1/x)$.

SOLUTION   We see this as the limit of the product of x and $\cos(1/x)$. However, we cannot use the Limit Theorem because the factor $\cos(1/x)$ does not have a limit. (The latter statement can be seen from the graph of $\cos(1/x)$ given in Figure 2.5.7.) We do know that $|\cos(1/x)| \leq 1$, $x \neq 0$, so the factor is bounded. Since $\lim_{x \rightarrow 0} x = 0$, we can use Statement 3, which tells us the product of a bounded function and a function with limit zero has limit zero. That is,

$$\lim_{x \rightarrow 0} x \cos(1/x) = 0.$$

The graph of $x \cos(1/x)$ is indicated in Figure 2.5.8. The limit at zero exists, even though the function is not defined at $x = 0$, and we can only indicate the behavior of the graph for x near the origin. $\quad \blacksquare$

The Composition Theorem for continuous functions is an easy consequence of the following result.

If $\lim_{x \rightarrow c} g(x) = M$ *and* f *is continuous at* M, *then* $\lim_{x \rightarrow c} f(g(x)) = f(\lim_{x \rightarrow c} g(x)) = f(M)$. (6)

We note that Statement 6 could be proved by showing that $|f(g(x)) - f(M)|$ is small whenever $|x - c|$ is small enough, $x \neq c$. However, since f is continuous at M, $|f(g(x)) - f(M)|$ is small whenever $|g(x) - M|$ is small enough,

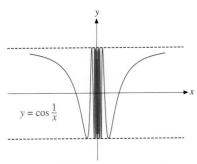

$y = \cos \frac{1}{x}$

FIGURE 2.5.7 $\lim_{x \rightarrow 0} \cos(1/x)$ does not exist, since the graph oscillates infinitely often between one and negative one as x approaches zero.

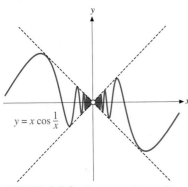

$y = x \cos \frac{1}{x}$

FIGURE 2.5.8 $\lim_{x \rightarrow 0} x \cos(1/x) = 0$, since the factor $\cos(1/x)$ is bounded and the factor x approaches zero as x approaches zero.

including when $g(x) = M$. But, $\lim\limits_{x \to c} g(x) = M$ implies $|g(x) - M|$ is small (possibly zero) whenever $|x - c|$ is small enough, $x \neq c$.

If g is continuous at c, then $\lim\limits_{x \to c} g(x) = g(c)$. If f is continuous at $g(c)$, then Statement 6 implies

$$\lim_{x \to c} f \circ g(x) = \lim_{x \to c} f(g(x)) = f(\lim_{x \to c} g(x)) = f(g(c)) = f \circ g(c).$$

This shows that a composition of continuous functions is continuous, as stated in the Composition Theorem.

The Squeeze Theorem

The following theorem tells us, if the graph of f is squeezed between the graphs of functions g and h that have limit L at c, then f also has limit L at c. The result is geometrically obvious. See Figure 2.5.9.

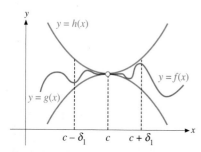

FIGURE 2.5.9 If the graph of f is squeezed between the graphs of two functions that have limit L at c, then f must also have limit L at c.

SQUEEZE THEOREM

If $\lim\limits_{x \to c} g(x) = \lim\limits_{x \to c} h(x) = L$ and there is some positive number δ_1 such that $g(x) \leq f(x) \leq h(x)$ whenever $0 < |x - c| < \delta_1$, then $\lim\limits_{x \to c} f(x) = L$.

The Squeeze Theorem is closely related to our strategy for showing that $\lim\limits_{x \to c} f(x) = L$. In particular, if we show there is a function z such that $|f(x) - L| \leq z(x)$ and $z(x) \to 0$ as $x \to c$, then $-z(x) \leq f(x) - L \leq z(x)$, or $L - z(x) \leq f(x) \leq L + z(x)$, so $f(x)$ is squeezed between the functions $g(x) = L - z(x)$ and $h(x) = L + z(x)$. Since z has limit zero at c, the functions g and h both have limits L at c, and so must f.

EXERCISES 2.5

Evaluate the limits in Exercises 1–42.

1. $\lim\limits_{x \to 0} (x^2 - 3x + 4)$

2. $\lim\limits_{x \to 0} (2 - 4x + x^2)$

3. $\lim\limits_{x \to -2} (2x - 1)(x + 4)$

4. $\lim\limits_{x \to -3} (x - 1)(4x + 2)$

5. $\lim\limits_{x \to 8} (2x^{1/3} + 12x^{-2/3})$

6. $\lim\limits_{x \to 4} (\sqrt{x} + 1/\sqrt{x})$

7. $\lim\limits_{x \to 0} (2 \sin x + 3 \cos x)$

8. $\lim\limits_{x \to \pi/4} (3 \sin x + \cos x)$

9. $\lim\limits_{x \to 3} \dfrac{6x + 3}{2x - 1}$ **(Calculus Explorer & Tutor I, Limits, 1.)**

10. $\lim\limits_{x \to -2} \dfrac{x^2 - 4}{x - 2}$

11. $\lim\limits_{x \to -1} \dfrac{x^2}{x^2 + 1}$

12. $\lim\limits_{x \to 0} \dfrac{2x - 1}{x - 2}$

13. $\lim\limits_{x \to \pi/3} \tan x$

14. $\lim\limits_{x \to \pi/6} \sec x$

15. $\lim\limits_{x \to 0} x \sin x$

16. $\lim\limits_{x \to 0} (x + 1) \cos x$

17. $\lim\limits_{x \to 0} x^3 \cos(1/x)$

18. $\lim\limits_{x \to 0} x \cos(1/x^2)$

19. $\lim\limits_{x \to 4} \dfrac{x^2 - 16}{x - 4}$ **(Calculus Explorer & Tutor I, Limits, 2.)**

20. $\lim\limits_{x \to -5} \dfrac{x^2 - 25}{x + 5}$

21. $\lim\limits_{x \to 2} \dfrac{2x^2 - 3x - 2}{x - 2}$

22. $\lim\limits_{x \to -1} \dfrac{x^2 - x - 2}{x^2 + 4x + 3}$ **(Calculus Explorer & Tutor I, Limits, 3.)**

23. $\lim\limits_{x \to 7} \dfrac{\frac{1}{x} - \frac{1}{7}}{x - 7}$ **(Calculus Explorer & Tutor I, Limits, 5.)**

24. $\lim\limits_{x \to 2} \dfrac{\frac{1}{x^2} - \frac{1}{4}}{x - 2}$

25. $\lim\limits_{x \to 2} \dfrac{\frac{x}{x + 1} - \frac{2}{3}}{x - 2}$

26. $\lim\limits_{x \to 2} \dfrac{\frac{x}{1 - x} - (-2)}{x - 2}$

27. $\lim_{x \to 1}(3x^2 - 7x + 3)^{10}$

28. $\lim_{x \to -1}\left(\dfrac{10x + 2}{x + 2}\right)^{1/3}$

29. $\lim_{x \to 0}\left|\dfrac{x + 4}{x - 2}\right|$

30. $\lim_{x \to 4}\left|\dfrac{x}{2 - x}\right|$

31. $\lim_{x \to 3}\sqrt{x^2 + 16}$

32. $\lim_{x \to 5}\sqrt{x^2 + 144}$

33. $\lim_{x \to 2}(x^2 + 4)^{-1/3}$

34. $\lim_{x \to 3}|3x - 1|^{2/3}$

35. $\lim_{x \to 3}\sqrt{\dfrac{x^2 - 4x + 3}{x - 3}}$

36. $\lim_{x \to 1}\sqrt{\dfrac{x^2 + 2x - 3}{x - 1}}$

37. $\lim_{x \to \pi/4}\tan 3x$

38. $\lim_{x \to \pi/3}\sec 2x$

39. $\lim_{x \to \pi/2}\cos^2(x + \pi/2)$

40. $\lim_{x \to \pi/2}\sin^2(x + \pi/4)$

41. $\lim_{x \to 5}\dfrac{\sqrt{x} - \sqrt{5}}{x - 5}$ **(Calculus Explorer & Tutor I, Limits, 4.)**

42. $\lim_{x \to 0}\dfrac{x^2}{4 - \sqrt{4^2 - x^2}}$. (*Hint:* Multiply numerator and denominator by $4 + \sqrt{4^2 - x^2}$.)

In Exercises 43–54 evaluate each of the limits $\lim_{x \to c^-} f(x)$, $\lim_{x \to c^+} f(x)$, and $\lim_{x \to c} f(x)$, for the given functions f and given numbers c, or determine that the limit does not exist.

43. $f(x) = \begin{cases} 0, & x \le 0, \\ \sin x, & x > 0; \end{cases}$ $\quad c = 0$

44. $f(x) = \begin{cases} 1, & x \le 0, \\ \dfrac{1}{1 + x^2}, & x > 0; \end{cases}$ $\quad c = 0$

45. $f(x) = \begin{cases} x + 1, & x < 0, \\ 0, & x = 0, \\ x - 1, & x > 0; \end{cases}$ $\quad c = 0$

46. $f(x) = \begin{cases} 3x^2, & x < 2, \\ 7, & x = 2, \\ 2x, & x > 2; \end{cases}$ $\quad c = 2$ **(Calculus Explorer & Tutor I, Limits, 10.)**

47. $f(x) = \dfrac{|x - 1|}{x - 1}; c = 1$

48. $f(x) = \dfrac{x - 4}{|x - 4|}; c = 4$ **(Calculus Explorer & Tutor I, Limits, 9.)**

49. $f(x) = \sqrt{x + 1}; c = -1$

50. $f(x) = \sqrt{1 - x}; c = 1$

51. $f(x) = \begin{cases} x, & x < 0, \\ 2x - x^2, & x > 0; \end{cases}$ $\quad c = 0, \quad c = 1$

52. $f(x) = \begin{cases} 1 - x, & x < 0, \\ x^2 - 2x + 1, & x > 0; \end{cases}$ $\quad c = 0, \quad c = 1$

53. $f(x) = \begin{cases} x - 1, & x < 2, \\ \dfrac{x^2}{x - 2}, & x > 2; \end{cases}$ $\quad c = 0, \quad c = 2$

54. $f(x) = \begin{cases} x + 1, & x < 1, \\ x + 3, & x > 1; \end{cases}$ $\quad c = 0, \quad c = 1$

55. Evaluate $\lim_{h \to 0}\dfrac{(a + h)^3 - a^3}{h}$. **(Calculus Explorer & Tutor I, Limits, 6.)**

56. Show that, if $\lim_{x \to c}[f(x)/g(x)]$ exists and $\lim_{x \to c} g(x) = 0$, then $\lim_{x \to c} f(x) = 0$. (This means that $\lim_{x \to c}[f(x)/g(x)]$ cannot exist if $\lim_{x \to c} g(x) = 0$ and $\lim_{x \to c} f(x) \ne 0$.)

57. Use the definition of limit to verify that the sum of two functions with limit zero has limit zero. (*Hint:* If $z_1(x)$ and $z_2(x)$ are each less than $\epsilon/2$ in absolute value, then the absolute value of their sum is less than ϵ. If $\lim_{x \to c} z(x) = 0$, then for each given positive number ϵ, there is a positive number δ such that $|z(x)| < \epsilon/2$ whenever $0 < |x - c| < \delta$.)

58. Use the definition of limit to verify that if $\lim_{x \to c} z(x) = 0$ and there is a positive number δ_1 such that the function b is bounded on the set $0 < |x - c| < \delta_1$, then

$$\lim_{x \to c} b(x)z(x) = 0.$$

59. If $\lim_{x \to c} f(x) = L$, show that there is a positive number δ_1 such that $|f(x)| < 1 + |L|$ whenever $0 < |x - c| < \delta_1$. (*Hint:* $|f(x)| = |f(x) - L + L| \le |f(x) - L| + |L|$. Use $\epsilon = 1$ in the definition of $\lim_{x \to c} f(x) = L$.)

60. If $\lim_{x \to c} g(x) = M$ and $M \ne 0$, show that there is a positive number δ_1 such that $1/|g(x)| < 2/|M|$ whenever $0 < |x - c| < \delta_1$. (*Hint:* $|M| \le |M - g(x)| + |g(x)|$, so $|g(x)| > |M|/2$ if $|M - g(x)| < |M|/2$.)

61. Use the definition of limit to verify that, if f is continuous at M and $\lim_{x \to c} g(x) = M$, then $\lim_{x \to c} f(g(x)) = f(M)$.

62. Is it true that $\lim_{x \to c} g(x) = 0$ implies $\lim_{x \to c}\sqrt{g(x)} = 0$? Explain.

63. If

$$f(x) = \begin{cases} 0, & \text{if } x \text{ is rational}, \\ 1, & \text{if } x \text{ is irrational}, \end{cases}$$

show that f is not continuous at any point.

64. If

$$f(x) = \begin{cases} 0, & \text{if } x \text{ is rational}, \\ x, & \text{if } x \text{ is irrational}, \end{cases}$$

show that f is continuous at 0.

65. Prove the Squeeze Theorem by showing that $g(x) \le f(x) \le h(x)$ implies

$$|f(x) - L| \le \max(|g(x) - L|, |h(x) - L|),$$

so $|f(x) - L| \to 0$ as $x \to c$.

66. In Section 3.3 we will see that

$$\cos x < \frac{\sin x}{x} < 1 \text{ for } 0 < |x| < \frac{\pi}{2}, x \text{ in radians.}$$

What can you conclude about $\lim_{x \to 0} \dfrac{\sin x}{x}$?

2.6 THE DERIVATIVE AT A POINT

Connections

Simplifying expressions, 1.2.

Slope of a line segment, 1.3.

Evaluating limits, 2.5.

Concept of average velocity.

The derivative is one of the fundamental working tools of calculus. The mathematical idea of a derivative has application to a wide range of physical problems. In this section we will see what a derivative is and how the concept relates to a few of these problems. We must understand the idea of the derivative if we are to understand its applications.

Let us state the definition of the derivative of a function at a point. This sets the notation and gives a common point of reference for the examples that illustrate how the concept is used.

DEFINITION

If $\lim_{x \to c} (f(x) - f(c))/(x - c)$ exists, f is said to be **differentiable at c**. The value of the limit is denoted $f'(c)$ (read "f prime of c") and is called the derivative of f at c. That is, the **derivative of f at c** is

$$f'(c) = \lim_{x \to c} \frac{f(x) - f(c)}{x - c},$$

if the limit exists. If the limit does not exist, f is **not differentiable at c**.

The *mathematical meaning* of the existence of the derivative is simply that the ratios $(f(x) - f(c))/(x - c)$ approach a single number $f'(c)$ as x approaches c. The derivative *is* the limit of these ratios. The *physical meaning* of the derivative depends on the *interpretation* of the ratios as physical quantities. We will illustrate some of the possible interpretations in this section.

The Derivative as a Slope, Tangent Lines

The quotient $(f(x) - f(c))/(x - c)$, $x \ne c$, may be interpreted as the slope of the line through the fixed point $(c, f(c))$ and a variable point $(x, f(x))$ on the graph of $y = f(x)$. This slope is not defined when $x = c$. However, if f is differentiable at c, the slope is close to the number $f'(c)$ whenever x is close enough to c, $x \ne c$. See Figure 2.6.1. This leads us to consider the line through the point $(c, f(c))$ with slope $f'(c)$.

If f is differentiable at c, the line through the point $(c, f(c))$ with slope $f'(c)$ is said to be **tangent** to the graph of $y = f(x)$ at the point $(c, f(c))$. The point-slope form of the equation of the line through the point $(c, f(c))$ with slope $f'(c)$ is

$$y - f(c) = f'(c)(x - c).$$

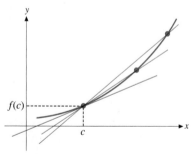

FIGURE 2.6.1 If f is differentiable at c, the slope of the line through the fixed point $(c, f(c))$ and a variable point $(x, f(x))$ approaches $f'(c)$ as x approaches c.

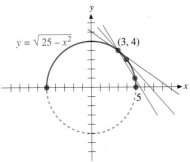

FIGURE 2.6.2 As x approaches 3, the slope of the line through the point $(3, 4)$ and the point $(x, \sqrt{25 - x^2})$ on the graph of $f(x) = \sqrt{25 - x^2}$ approaches the slope of the line that is geometrically tangent to the circle at $(3, 4)$.

This is an **equation of the line tangent to the graph at $(c, f(c))$.**

We should realize that the slope of the line between two *different* points on a graph is a familiar idea, but the idea of a line tangent to a graph needed to be defined.

In later sections we will use the line tangent to the graph of a function to find approximate values of the function. If f is not differentiable at c, then there is no line that will serve this purpose. No line has been defined to be tangent to the graph at $(c, f(c))$ if f is not differentiable at c.

Let us check that the definition of the line tangent to the graph of a function reflects the notion of a line that is geometrically tangent to a circle. That is, the graph of the function $f(x) = \sqrt{25 - x^2}$ is the upper half of the circle $x^2 + y^2 = 25$. We see in Figure 2.6.2 that the slope of the line through the point $(3, 4)$ and another point $(x, \sqrt{25 - x^2})$ on the graph of f approaches the slope of the line geometrically tangent to the circle at $(3, 4)$ as x approaches 3. This means that our definition of the line tangent to the graph of the function in terms of the derivative agrees with the line that is geometrically tangent to the circle.

EXAMPLE 1

(a) Find the slope of the line through the point $(2, 4)$ and a variable point (x, x^2) on the graph of $f(x) = x^2$, in terms of x.

(b) Find the slope of the line tangent to the graph at $(2, 4)$.

(c) Find an equation of the line tangent to the graph at $(2, 4)$.

SOLUTION

(a) If $x \neq 2$, the slope of the line through $(2, 4)$ and (x, x^2) is

$$\frac{f(x) - f(2)}{x - 2} = \frac{x^2 - 4}{x - 2}.$$

Simplifying, we obtain

$$\frac{x^2 - 4}{x - 2} = \frac{(x - 2)(x + 2)}{x - 2} = x + 2, \qquad x \neq 2.$$

(b) The slope of the line tangent to the graph at $(2, 4)$ is

$$f'(2) = \lim_{x \to 2} \frac{f(x) - f(2)}{x - 2} = \lim_{x \to 2} \frac{x^2 - 4}{x - 2}.$$

Since the limits at 2 do not depend on the values at 2, we can replace $(x^2 - 4)/(x - 2)$ by the simplified expression $x + 2$ in the above limit. We obtain

$$f'(2) = \lim_{x \to 2} (x + 2) = (2) + 2 = 4.$$

(c) The point-slope form of the equation of the line tangent to the graph at $(2, 4)$ is $y - 4 = 4(x - 2)$. Simplifying, we obtain

$$y = 4x - 4.$$

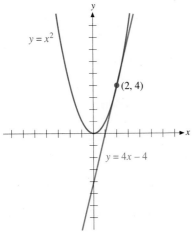

FIGURE 2.6.3 The line through the point $(2, 4)$ with slope $f'(2) = 4$ is tangent to the graph of f at the point.

See Figure 2.6.3. ■

FIGURE 2.6.4 The position of a particle that is moving along a number line with constant velocity v_0 is given by $s(t) = s_0 + v_0 t$, where s_0 is the position of the particle at time $t = 0$.

FIGURE 2.6.5 The height in feet of a particle t seconds after it is thrown vertically from ground level with initial velocity v_0 is $s(t) = -16t^2 + v_0 t$.

The Derivative as Velocity

Let us consider a particle moving along a number line. The position of the particle at time t may be given by a function $s(t)$. For example:

- If a particle moves along a number line with constant velocity v_0, its position is given by the formula $s(t) = s_0 + v_0 t$, where s_0 is the position of the particle at time $t = 0$. See Figure 2.6.4, where we have used a schematic drawing to indicate the motion; the arrows indicate the direction and speed of the particle.
- If a particle is thrown upward from ground level with initial velocity v_0(ft/s), its height (in feet) t seconds later is known to be given by the formula $s(t) = -16t^2 + v_0 t$. See Figure 2.6.5, where we have used a schematic drawing to indicate the direction and speed of the particle at various times.

If the position of a particle on a number line at time t is given by $s(t)$, we can determine the average velocity of the particle over any interval of time. That is, the average velocity over the time interval from time t_0 to another time t is

$$\text{Average velocity} = \frac{\text{Change in position}}{\text{Length of time}} = \frac{s(t) - s(t_0)}{t - t_0}.$$

If the function s is differentiable at t_0, the average velocity of the particle over small intervals of time that either begin at $t_0 (t_0 < t)$ or end at $t_0 (t < t_0)$ will be near the number $s'(t_0)$. This leads to the following definition.

DEFINITION

If the position function $s(t)$ is differentiable at t_0, the derivative

$$s'(t_0) = \lim_{t \to t_0} \frac{s(t) - s(t_0)}{t - t_0}$$

is called the (instantaneous) **velocity** at time t_0.

The sign of the velocity indicates the direction of the object on the number line. The **speed** at time t_0 is given by $|s'(t_0)|$.

Average velocity over a nonzero interval of time is a familiar idea. However, it does not make sense to consider instantaneous velocity to be the average velocity over a time interval of length zero. Instantaneous velocity is a limit of average velocities.

EXAMPLE 2
The position of a particle on a number line at time t is given by $s(t) = 1/(t + 1)$.

(a) Find the average velocity of the particle over the interval of time between 1 and t, $t \neq 1$.

(b) Find the (instantaneous) velocity at time 1.

SOLUTION

(a) For $t \neq 1$, the average velocity over the interval of time between 1 and t is

$$\frac{s(t) - s(1)}{t - 1} = \frac{\dfrac{1}{t+1} - \dfrac{1}{1+1}}{t - 1}.$$

Simplifying, we have

$$\frac{\dfrac{1}{t+1} - \dfrac{1}{2}}{t - 1} = \frac{\dfrac{2 - (t+1)}{2(t+1)}}{t - 1} = \frac{\dfrac{2 - t - 1}{2(t+1)}}{t - 1}$$

$$= \frac{\dfrac{1 - t}{2(t+1)}}{t - 1} = \frac{1 - t}{2(t+1)(t-1)} = \frac{-(t-1)}{2(t+1)(t-1)}$$

$$= \frac{-1}{2(t+1)}, \qquad t \neq 1.$$

(b) The instantaneous velocity at time 1 is the derivative

$$s'(1) = \lim_{t \to 1} \frac{s(t) - s(t)}{t - 1}.$$

Since limits at 1 do not depend on values at 1, we can use the simplified expression for the average velocity in the above limit. We obtain

$$s'(1) = \lim_{t \to 1} \frac{-1}{2(t+1)} = \frac{-1}{2(1+1)} = -\frac{1}{4}.$$

The negative velocity indicates that the particle is moving in the negative direction when $t = 1$. See Figure 2.6.6. ∎

FIGURE 2.6.6 Since $s'(1)$ is negative, the particle is moving in the negative direction when $t = 1$.

The Derivative as a General Rate of Change

Generally, we may interpret the ratio $(f(x) - f(c))/(x - c)$ as the average rate of change of $f(x)$ with respect to the variable x.

If f is differentiable at c, the derivative

$$f'(c) = \lim_{x \to c} \frac{f(x) - f(c)}{x - c}$$

is called the (instantaneous) **rate of change of f with respect to the variable x at c.**

EXAMPLE 3

Find the rate of change of the volume of a cube with respect to the length of one of its edges when each edge is 10 ft.

SOLUTION   It is very important that the length of the edges and the volume of the cube be considered variables. Let x denote the length of each edge of the cube. The volume is then given by $V = x^3$. See Figure 2.6.7.

FIGURE 2.6.7 The volume of a cube with length of edge x is $V = x^3$.

The rate of change of the volume V with respect to the length of each edge x when $x = 10$ is defined to be the derivative

$$V'(10) = \lim_{x \to 10} \frac{V(x) - V(10)}{x - 10}$$

$$= \lim_{x \to 10} \frac{x^3 - 10^3}{x - 10}.$$

Factoring the numerator and simplifying, we have

$$V'(10) = \lim_{x \to 10} \frac{(x - 10)(x^2 + 10x + 10^2)}{x - 10}$$

$$= \lim_{x \to 10} (x^2 + 10x + 10^2)$$

$$= (10)^2 + 10(10) + 10^2 = 300.$$

The volume of the cube is changing at a rate of 300 ft^3 per 1 ft change in the length of its edges at the instant when the edges are 10 ft. ∎

EXAMPLE 4

Find the rate of change of the distance between the points $(0, 3)$ and $(x, 0)$ with respect to x when x is 4.

SOLUTION   Note that $(x, 0)$ is a variable point on the x-axis. See Figure 2.6.8. The Distance Formula gives

$$D(x) = \sqrt{(x - 0)^2 + (0 - 3)^2} = \sqrt{x^2 + 9}.$$

FIGURE 2.6.8 The distance between the points $(0, 3)$ and $(x, 0)$ is $D(x) = \sqrt{x^2 + 9}$.

The desired rate of change is the derivative

$$D'(4) = \lim_{x \to 4} \frac{\sqrt{x^2 + 9} - \sqrt{4^2 + 9}}{x - 4} = \lim_{x \to 4} \frac{\sqrt{x^2 + 9} - 5}{x - 4}.$$

Rationalizing the numerator and simplifying, we have

$$D'(4) = \lim_{x \to 4} \frac{\sqrt{x^2 + 9} - 5}{x - 4} \cdot \frac{\sqrt{x^2 + 9} + 5}{\sqrt{x^2 + 9} + 5}$$

$$= \lim_{x \to 4} \frac{(x^2 + 9) - 25}{(x - 4)(\sqrt{x^2 + 9} + 5)}$$

$$= \lim_{x \to 4} \frac{x^2 - 16}{(x - 4)(\sqrt{x^2 + 9} + 5)}$$

$$= \lim_{x \to 4} \frac{(x - 4)(x + 4)}{(x - 4)(\sqrt{x^2 + 9} + 5)}$$

$$= \lim_{x \to 4} \frac{x + 4}{\sqrt{x^2 + 9} + 5}$$

$$= \frac{4 + 4}{\sqrt{4^2 + 9} + 5} = \frac{8}{5 + 5} = \frac{4}{5}.$$

The distance is changing at a rate of 4/5 units per unit change in x when $x = 4$.

∎

Geometric Aspects of Differentiability

Recall that f is continuous at c if $\lim_{x \to c} f(x) = f(c)$. We now show that differentiability implies continuity.

If f is differentiable at c, then f is continuous at c.   (1)

Proof   To verify Statement 1 we must show $\lim_{x \to c} f(x) = f(c)$. Using some algebraic manipulation to write $f(x)$ as an expression that is equivalent for $x \neq c$ and then using properties of limits, we have

$$\lim_{x \to c} f(x) = \lim_{x \to c} \left[\frac{f(x) - f(c)}{x - c}(x - c) + f(c) \right]$$

$$= \lim_{x \to c} \frac{f(x) - f(c)}{x - c} \cdot \lim_{x \to c}(x - c) + \lim_{x \to c} f(c)$$

$$= f'(c) \cdot (c - c) + f(c) = f(c).$$

Note that we needed to know that $\lim_{x \to c}(f(x) - f(x))/(x - c)$ exists in order to use the Limit Theorem. The fact that f is differentiable at c means that this limit exists. ∎

It follows from Statement 1 that

f is not differentiable at points where it is not continuous.

EXAMPLE 5
Is

$$f(x) = \begin{cases} x + 1, & x < 0, \\ 0, & x = 0, \\ x - 1, & x > 0, \end{cases}$$

differentiable at 0?

SOLUTION   We see from the graph in Figure 2.6.9a that f is not continuous at 0, so f is not differentiable at 0. We could also verify directly that

$$f'(0) = \lim_{x \to 0} \frac{f(x) - f(0)}{x - 0}$$

does not exist by investigating the corresponding one-sided limits of the difference quotient, neither of which exists. For example, using the formula $f(x) = x - 1$ for $x > 0$, we have

$$\lim_{x \to 0^+} \frac{f(x) - f(0)}{x - 0} = \lim_{x \to 0^+} \frac{(x - 1) - (0)}{x - 0} = \lim_{x \to 0^+} \frac{x - 1}{x}.$$

This limit does not exist, because $(x - 1)/x$ becomes unbounded as x approaches 0 from the right. This corresponds to the fact that the line segment between $(0, 0)$ and a variable point $(x, f(x))$ on the graph becomes vertical as x approaches 0 from the right. See Figure 2.6.9b. ∎

f is not differentiable at c if $(f(x) - f(c))/(x - c)$ becomes unbounded as x approaches c.   (2)

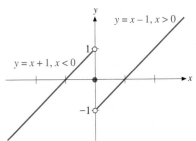

FIGURE 2.6.9a The function f is not continuous at 0, so it is not differentiable at 0.

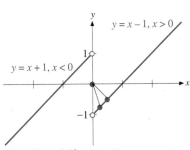

FIGURE 2.6.9b The line segment between $(0, 0)$ and $(x, f(x))$ becomes vertical as x approaches 0 from the right.

This can happen, even though f is continuous at c. In simple cases this means that the graph either passes smoothly through $(c, f(c))$ vertically or has a vertical sharp point at $(c, f(c))$.

EXAMPLE 6

Is $f(x) = x^{1/3}$ differentiable at 0?

SOLUTION

$$f'(0) = \lim_{x \to 0} \frac{f(x) - f(0)}{x - 0} = \lim_{x \to 0} \frac{x^{1/3} - 0}{x} = \lim_{x \to 0} \frac{1}{x^{2/3}}.$$

Since $1/x^{2/3}$ becomes unbounded as x approaches zero, $f'(0)$ does not exist. This means that f is not differentiable at 0. We see from Figure 2.6.10 that the graph passes smoothly through the origin vertically. ∎

EXAMPLE 7

Is $f(x) = x^{2/3}$ differentiable at 0?

SOLUTION

$$f'(0) = \lim_{x \to 0} \frac{f(x) - f(0)}{x - 0} = \lim_{x \to 0} \frac{x^{2/3} - 0}{x} = \lim_{x \to 0} \frac{1}{x^{1/3}}.$$

The limit of $1/x^{1/3}$ at zero does not exist, because $1/x^{1/3}$ becomes unbounded as x approaches zero. This means that $f(x) = x^{2/3}$ is not differentiable at zero. From Figure 2.6.11 we see that the slope of the line through the origin and a point $(x, x^{2/3})$ becomes increasingly large and positive as x approaches zero from the right. The slope becomes increasingly large and negative as x approaches zero from the left. The graph of f has a sharp point at $(0, 0)$. ∎

Let us emphasize that a function can be continuous but not differentiable at a point c. However, a function cannot be differentiable but not continuous at c.

If f is differentiable at c, the slope of the line through $(c, f(c))$ and $(x, f(x))$ approaches $f'(c)$, the slope of the line tangent to the graph at $(c, f(c))$, as x approaches c. This means the angle between the two lines approaches zero as x approaches c. See Figure 2.6.12. In this sense we have the following.

If f is differentiable at c, then the graph of f passes through the point $(c, f(c))$ in the same direction as the line tangent to the graph at that point.   (3)

The graph may touch or cross the tangent line at any number of points in addition to the point $(c, f(c))$. See Figure 2.6.13.

If both one-sided limits of the difference quotient exist and

$$\lim_{x \to c^+} \frac{f(x) - f(c)}{x - c} \neq \lim_{x \to c^-} \frac{f(x) - f(c)}{x - c},$$ (4)

then f is not differentiable at c.

The graph has a corner at $(c, f(c))$.

FIGURE 2.6.10 $\dfrac{f(x) - f(0)}{x - 0}$ becomes unbounded as x approaches zero, so f is not differentiable at zero. The graph passes smoothly through the origin vertically.

FIGURE 2.6.11 The slope of the line segment between $(0, 0)$ and the point $(x, x^{2/3})$ becomes unbounded as x approaches zero. The graph of f has a sharp corner at zero.

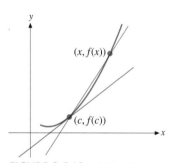

FIGURE 2.6.12 If f is differentiable at c, the graph of f passes through the point $(c, f(c))$ in the same direction as the line tangent to the graph at that point.

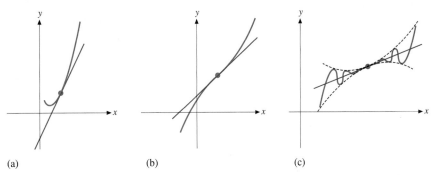

(a)   (b)   (c)

FIGURE 2.6.13   The graph of a function may touch or cross a tangent line at any number of points in addition to the point $(c, f(c))$.

EXAMPLE 8

Is

$$f(x) = \begin{cases} x^2, & x \le 1, \\ x, & x > 1, \end{cases}$$

differentiable at 1?

SOLUTION   One-sided limits or a sketch of the graph can be used to show that $\lim_{x \to 1} f(x) = 1 = f(1)$, so f is continuous at 1.

To evaluate

$$f'(1) = \lim_{x \to 1} \frac{f(x) - f(1)}{x - 1}$$

we must evaluate each one-sided limit, since different formulas for $f(x)$ hold for x on different sides of 1. The limit from the right at 1 involves $x > 1$, so we use the formula $f(x) = x$ to obtain

$$\lim_{x \to 1^+} \frac{f(x) - f(1)}{x - 1} = \lim_{x \to 1^+} \frac{x - 1}{x - 1} = \lim_{x \to 1^+} 1 = 1.$$

The limit from the left at 1 involves $x < 1$, so we use the formula $f(x) = x^2$ to obtain

$$\lim_{x \to 1^-} \frac{f(x) - f(1)}{x - 1} = \lim_{x \to 1^-} \frac{x^2 - 1}{x - 1}$$

$$= \lim_{x \to 1^-} \frac{(x - 1)(x + 1)}{x - 1}$$

$$= \lim_{x \to 1^-} (x + 1) = 1 + 1 = 2.$$

Since the one-sided limits are unequal,

$$f'(1) = \lim_{x \to 1} \frac{f(x) - f(1)}{x - 1}$$

does not exist, so f is not differentiable at 1. In Figure 2.6.14 we see that the graph has a corner at $(1, 1)$. ∎

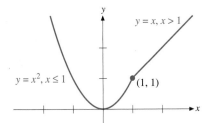

FIGURE 2.6.14   The function f is not differentiable at one, where the graph has a corner.

A calculator generated graph can be used to indicate points where a function is not differentiable because of a discontinuity or because the graph either becomes vertical or has a sharp corner.

Approximating $f'(c)$ from the Graph of f

Since the value of the derivative $f'(c)$ is the slope of the line tangent to the graph of f at the point $(c, f(c))$, we can use the graph of f to determine an approximate value of $f'(c)$.

NOTE *The apparent slope of the line tangent to the graph of a function f at a point $(c, f(c))$ depends on the relative scales used on the x-axis and the y-axis. The apparent slope will be the value $f'(c)$ only if equal scales are used on the two axes.*

 If you are using a calculator generated graph to approximate the values of a derivative, be sure to account for the relative scales on the coordinate axes.

FIGURE 2.6.15 The slope of the line tangent to the graph of f at $(c, f(c))$ is the value of the derivative $f'(c)$.

EXAMPLE 9

From the graph of $f(x) = \sin x$ in Figure 2.6.15 it appears that $f'(0) = 1$, $f'(\pi/2) = 0$, $f'(\pi) = -1$, and $f'(3\pi/2) = 0$. ∎

EXERCISES 2.6

In Exercises 1–8: (a) Find the slope of the line between the point $(c, f(c))$ and a variable point $(x, f(x))$ on the graph of $y = f(x)$ in terms of x. (b) Find the slope of the line tangent to the graph at the point $(c, f(c))$. (c) Find the equation of the line tangent to the graph at $(c, f(c))$.

1. $f(x) = x^2 - 3x; c = 1$ **2.** $f(x) = 4x - x^2; c = 1$

3. $f(x) = 3x + 4; c = -3$ **4.** $f(x) = 2x + 5; c = -3$

5. $f(x) = x^3; c = 2$ **6.** $f(x) = 1 - x^3; c = 2$

7. $f(x) = 1/x; c = 4$ **8.** $f(x) = 1/(x + 1); c = 4$

In Exercises 9–16 the position of a particle on a number line at time t is given by $s(t)$. (a) Find the average velocity of the particle as t ranges over the interval of time between t_0 and t, $t \neq t_0$. (b) Find the (instantaneous) velocity at time t_0.

9. $s(t) = t^2 - 2; t_0 = 2$ **10.** $s(t) = 3 - t^2; t_0 = 2$

11. $s(t) = 6 - 2t; t_0 = 3$ **12.** $s(t) = 4 + 5t; t_0 = 3$

13. $s(t) = \sqrt{t + 1}; t_0 = 8$ **14.** $s(t) = \sqrt{t}; t_0 = 4$

15. $s(t) = \dfrac{t}{t + 1}; t_0 = 3$ **16.** $s(t) = \dfrac{t}{t + 2}, t_0 = 8$

17. If a particle is thrown vertically from ground level with initial velocity v_0 ft/s, its height in feet t seconds later is known to be given by $s(t) = -16t^2 + v_0 t$. Find the average velocity of the particle over the interval of time between 0 and t. What is the approximate value of these average velocities for small values of t?

18. Use the formula for $s(t)$ given in Exercise 17 to find the average velocity of the particle over the interval of time between $v_0/32$ and t. What is the approximate value of these average velocities for values of t near $v_0/32$?

19. Find the rate of change of the area of a circle with respect to its radius when the radius is 3.

20. Find the rate of change of the area of a square with respect to the length of one side when the side is 5.

21. Find the rate of change of the distance between the points $(0, y)$ and $(-3, 0)$ with respect to y when y is -4.

22. Find the rate of change of the distance between the points $(0, y)$ and $(12, 0)$ with respect to y when y is 5.

Determine whether the functions in Exercises 23–38 are differentiable at zero by investigating $\lim_{x \to 0}[f(x) - f(0)]/(x - 0)$. Use one-sided limits where appropriate.

23. $f(x) = |x|$

24. $f(x) = \begin{cases} 1 + x, & x \leq 0 \\ 1 - x, & x > 0 \end{cases}$

25. $f(x) = \begin{cases} x^2, & x \leq 0 \\ 0, & x > 0 \end{cases}$

26. $f(x) = \begin{cases} x^2 + x, & x \leq 0 \\ x, & x > 0 \end{cases}$

27. $f(x) = \begin{cases} 1 + x + x^2, & x \leq 0 \\ 2 - x - x^2, & x > 0 \end{cases}$

28. $f(x) = \begin{cases} |x|/x, & x \neq 0 \\ 0, & x = 0 \end{cases}$

29. $f(x) = x^{4/5}$

30. $f(x) = x^{4/3}$

31. $f(x) = x^{5/3}$

32. $f(x) = x^{3/5}$

33. $f(x) = |x^2 - 2x|$

34. $f(x) = |x^3 - 3x^2|$

35. $f(x) = \begin{cases} x^2 \sin(1/x), & x \neq 0 \\ 0, & x = 0 \end{cases}$

36. $f(x) = \begin{cases} x \sin(1/x), & x \neq 0 \\ 0, & x = 0 \end{cases}$

37. $f(x) = \begin{cases} 0, & x \text{ rational} \\ x, & x \text{ irrational} \end{cases}$

38. $f(x) = \begin{cases} 0, & x \text{ rational} \\ x^2, & x \text{ irrational} \end{cases}$

39. At what points is the function $f(x) = |x^2 - 3x + 2|$ not differentiable?

40. At what points is the function $f(x) = |x^3 - 4x^2 + 4x|$ not differentiable?

41. $f(x) = 5x - 3$. Use the definition of the derivative to show that $f'(2) = 5$.

42. $f(x) = 7 - 5x$. Use the definition of the derivative to show that $f'(3) = -5$.

43. $f(x) = x^2$. Use the definition of the derivative to show that $f'(3) = 6$.

44. $f(x) = 2x^2 - 3x - 4$. Use the definition of the derivative to show that $f'(3) = 9$.

45. $f(x) = 1/x$. Use the definition of the derivative to show that $f'(2) = -1/4$.

46. $f(x) = \sqrt{x}$. Use the definition of the derivative to show that $f'(4) = 1/4$.

In Exercises 47–52, use the given graph of f to find approximate values of (a) $f'(-1)$, (b) $f'(0)$, and (c) $f'(1)$.

47.

48.

49.

50.

51.

52.

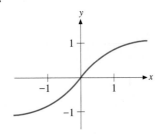

53. Use graphs of (a) $f(x) = e^x$, (b) $f(x) = e^{2x}$ and (c) $f(x) = e^{x/2}$ to find approximate values of $f'(0)$.

54. Use the graph of $f(x) = \ln x$ to find approximate values of (a) $f'(1/2)$, (b) $f'(1)$, and (c) $f'(2)$.

REVIEW OF CHAPTER 2 CONCEPTS

In this chapter we have studied functions, bounds, limits, continuity, and differentiability. Functions are basic to mathematics. We have seen that bounds play an important role in the theory of limits. Continuity and differentiability are defined in terms of limits.

- You should be able to determine bounds, limits, continuity, and differentiability of a function from the graph of the function.
- The concept of bounds of functions will be important for several types of error analyses that we will be doing in later chapters. You should know how to use graphs and inequalities to find bounds of simple functions.
- If a function has a limit at a point c, then the function must be bounded near the point. You should know examples of functions that do not have a limit at a point, even though they are bounded near the point.
- If a function is continuous at a point, then the function must have a limit at the point. You should know examples of functions that are not continuous at a point, even though they have a limit at the point.

- If a function f is differentiable at c, then f must be continuous at c. You should know examples of functions that are not differentiable at a point, even though they are continuous at the point.
- The ability to recognize compositions of basic functions is a skill that will prove most useful for working with the differentiation formulas that we will derive in the next chapter.
- You should know how to use the Limit Theorem to evaluate limits of sums, products, and quotients. It is also useful to remember that the product of a bounded function and a function with limit zero has limit zero.
- The derivative at a point is defined to be the limit of the difference quotient, $f'(c) = \lim\limits_{x \to c} \dfrac{f(x) - f(c)}{x - c}$. The physical meaning of the derivative depends on the interpretation of the quotient as a physical quantity such as the slope of a line segment, an average velocity, or a general rate of change.

CHAPTER 2 REVIEW EXERCISES

1. $f(x) = 2x^2 - 3x + 1$, $g(x) = 3x - 1$. Find
   (a) $f(x - 2) - 2g(x)$ and (b) $5f(x) - [g(x)]^2$.

2. $f(x) = 1 - 2x - x^2$, $g(x) = x - 1$. Find
   (a) $f(g(x)) + [g(x)]^2$ and (b) $g(f(x)) + xg(x)$.

3. $f(x) = 3x^2 - 2x + 5$, $g(x) = 2x - 3$. Find
   (a) $[f(x) - f(2)]/(x - 2)$ and (b) $f(g(x)) - 2g(f(x))$.

4. $f(x) = 1 - 2x - 3x^2$, $g(x) = 2x + 1$. Find
   (a) $\dfrac{f(x) - f(1)}{x - 1}$ and (b) $f(g(x)) - 2g(f(x))$.

5. $f(x) = \sqrt{x - 1}$, $g(x) = \sqrt{4 - x}$. Find the domain of
   (a) $f + g$, (b) f/g, and (c) $f \circ g$.

6. $f(x) = \sqrt{1 - x^2}$, $g(x) = \sqrt{1 - x}$. Find the domain of
   (a) $f + g$, (b) f/g, and (c) $f \circ g$.

7. Find the domain of the function
$$f(x) = \frac{1}{\sqrt{3 - 2x}}.$$

8. Find the domain of the function
$$f(x) = \frac{x^2 - 3x - 4}{x - 4}.$$

Sketch the graphs of the equations in Exercises 9–14 and determine which of them define y as a function of x. Find a formula for the values and indicate the domain and range of each function.

9. $3x - 4y = 12$

10. $x + \sqrt{y - 1} = 0$

11. $x^2 + y^2 = 4$

12. $|x| + |y| = 1$

13. $y^2 = 4x^2$

14. $x + \sqrt{y} = 1$

15. A rectangular box has length $11 - 2x$, width $8 - 2x$, and height x. Express the volume of the box as a function of x. Indicate the domain.

16. A right circular cylinder has radius r and height $2 - r$. Express the total area of the lateral surface, the circular top, and the circular bottom of the cylinder as a function of r. Indicate the domain.

17. A variable point on the graph of $y = x^2$ is revolved about the x-axis. Express the (a) area and (b) circumference of the resulting circle as a function of x.

18. A variable point on the graph of $y = 1 + \sin x$ is revolved about the x-axis. Express the (a) area and (b) circumference of the resulting circle as a function of x.

19. Express the length of a vertical line segment between a variable point on the graph of $y = 1 + \cos x$ and the x-axis as a function of x.

20. Express the length of a vertical line segment between a variable point on the graph of $y = \dfrac{x^2}{1 + x^2}$ and $y = -1$ as a function of x.

21. Express the function $f(x) = |x^2 - 4|$ without absolute values.

22. Express the function $f(x) = |\sin x - 0.5|,\ 0 \le x \le \pi/2$ without absolute values.

23. $f(x) = 3x - 2$. Find a function g such that $f \circ g = x$.

24. $f(x) = x^2$. Explain why there is no function g such that $f \circ g = x$ for all x.

Find a bound or indicate unbounded for each of the functions in Exercises 25–38.

25. $f(x) = 3x^2 - x \cos 2x,\ |x| < 3$

26. $f(x) = 2x^2 - 3x + 1,\ |x| < 2$

27. $f(x) = x \sin 3x - x^2 \cos 5x,\ |x| < 2$

28. $f(x) = \sin x - \sin 2x + \sin 3x$

29. $f(x) = \dfrac{1}{3x - 1},\ 1 < x < 3$

30. $f(x) = \dfrac{x}{1 - x},\ |x| < 0.3$

31. $f(x) = \dfrac{x}{1 + x^2},\ 0 < x < 1$

32. $f(x) = \dfrac{1}{2 + \cos x}$

33. $f(x) = \dfrac{x}{2x - 0.8},\ |x - 1| < 0.2$

34. $f(x) = \dfrac{x - 1}{x + 1},\ |x| < 0.2$

35. $f(x) = \dfrac{x^3}{x^2 + 4},\ |x - 2| < 2$

36. $f(x) = \dfrac{\cos x}{2 + \sin^2 x},\ |x| < \pi$

37. $f(x) = \dfrac{6}{x - 2},\ |x - 3| < 1$

38. $f(x) = \dfrac{1}{1 + \cos x},\ |x| < \pi$

Use the graphs of $|f|$ given in Exercises 39–42 to find bounds of f.

39.

40.

41.

42.

Use the graphs of the function f given in Exercises 43–44 to evaluate (a) $f(a)$, (b) $\lim\limits_{x \to a^-} f(x)$, (c) $\lim\limits_{x \to a^+} f(x)$, (d) $f(b)$, (e) $\lim\limits_{x \to b^-} f(x)$, and (f) $\lim\limits_{x \to b^+} f(x)$.

43.   **44.**

45. For the tax year 1989, the U.S. Federal Income Tax T for a single person with taxable income i taking the standard deduction was

$$T(i) = \begin{cases} 0 & 0 \leq i < 5100 \\ .15(i - 5100) & 5100 \leq i < 23650 \\ .28(i - 23650) \\ \quad + 2782.50 & 23650 \leq i < 50000 \\ .33(i - 50000) \\ \quad + 10160.50 & 50000 \leq i \end{cases}$$

(Assume that all nonnegative real numbers are possible values of i.)

(a) Is the Federal Income Tax function discontinuous at any point $i > 0$?
(b) What is the derivative of T at $i = 29500$?
(c) At which points is the Federal Income Tax function not differentiable?
(d) Is there any value of i for which the tax on additional income is more than the additional income?

46. $f(x) = \begin{cases} 1 - \cos x, & -2\pi \leq x < 0 \\ 0, & x = 0 \\ x^2 - 1, & 0 < x \leq 2 \end{cases}$

(a) Sketch the graph of f.
(b) Find the domain and range of f.
(c) Evaluate $f(-\pi)$, $f(1)$, and $f(3)$.
(d) Evaluate $\lim_{x \to 0^-} f(x)$ and $\lim_{x \to 0^+} f(x)$.
(e) Is f continuous from the left at 0? Is f continuous from the right at 0? Is f continuous at 0?

47. $f(x) = \dfrac{|x| + x}{2}$.

(a) Sketch the graph of f.
(b) Find the domain and range of f.
(c) Evaluate $f(-1)$, $f(0)$ and $f(1)$.
(d) Evaluate $\lim_{x \to 0^-} f(x)$ and $\lim_{x \to 0^+} f(x)$.
(e) Is f continuous from the left at $x = 0$? Is f continuous from the right at $x = 0$? Is f continuous at $x = 0$?

48. $f(x) = \begin{cases} \dfrac{|x - 2|}{x - 2}, & x \neq 2, \\ 0, & x = 2. \end{cases}$

(a) Sketch the graph of f.
(b) Find the domain and range of f.
(c) Evaluate $f(0)$, $f(2)$, and $f(4)$.
(d) Evaluate $\lim_{x \to 0^-} f(x)$ and $\lim_{x \to 0^+} f(x)$.
(e) Is f continuous from the left at $x = 0$? Is f continuous from the right at $x = 0$? Is f continuous at $x = 0$?

49. $f(x) = \begin{cases} \cos x, & -\pi/2 < x < 0, \\ 1, & x = 0 \\ \sec x, & 0 < x < \pi/2. \end{cases}$

(a) Sketch the graph of f.
(b) Find the domain and range of f.
(c) Evaluate $f(-\pi/4)$, $f(0)$, and $f(\pi/4)$.

(d) Evaluate $\lim_{x \to 0^-} f(x)$ and $\lim_{x \to 0^+} f(x)$.
(e) Is f continuous from the left at $x = 0$? Is f continuous from the right at $x = 0$? Is f continuous at $x = 0$?

50. $f(x) = \begin{cases} x^2 - 1, & -2 \leq x < 1, \\ 2x - 1, & 1 \leq x \leq 2. \end{cases}$

(a) Sketch the graph of f.
(b) Find the domain and range of f.
(c) Evaluate $f(-1)$, $f(1)$, and $f(2)$.
(d) Is f continuous at $x = 1$?
(e) Evaluate $\lim_{x \to 1^-} \dfrac{f(x) - f(1)}{x - 1}$ and $\lim_{x \to 1^+} \dfrac{f(x) - f(1)}{x - 1}$.
(f) Is f differentiable at $x = 1$?

51. $f(x) = \begin{cases} x^2, & -2 \leq x < 1, \\ 2x - 1, & 1 \leq x \leq 2. \end{cases}$

(a) Sketch the graph of f.
(b) Find the domain and range of f.
(c) Evaluate $f(-1)$, $f(1)$, and $f(2)$.
(d) Is f continuous at $x = 1$?
(e) Evaluate $\lim_{x \to 1^-} \dfrac{f(x) - f(1)}{x - 1}$ and $\lim_{x \to 1^+} \dfrac{f(x) - f(1)}{x - 1}$.
(f) Is f differentiable at $x = 1$?

52. $f(x) = \begin{cases} -x^2 - x, & -2 \leq x < 0, \\ 0, & x = 0 \\ x^2 - x, & 0 < x \leq 2. \end{cases}$

(a) Sketch the graph of f.
(b) Find the domain and range of f.
(c) Evaluate $f(-1)$, $f(0)$, and $f(1)$.
(d) Is f continuous at $x = 1$?
(e) Evaluate $\lim_{x \to 1^-} \dfrac{f(x) - f(1)}{x - 1}$ and $\lim_{x \to 1^+} \dfrac{f(x) - f(1)}{x - 1}$.
(f) Is f differentiable at $x = 1$?

53. $f(x) = \begin{cases} x^2 - 1, & -2 \leq x \leq 0 \\ x - 1, & 0 < x < 3 \end{cases}$

(a) Sketch the graph of f.
(b) Find the domain and range of f.
(c) Evaluate $f(-1)$, $f(2)$, and $f(4)$.
(d) Is f continuous at 0?
(e) Evaluate $\lim_{x \to 0^-} \dfrac{f(x) - f(0)}{x - 0}$ and $\lim_{x \to 0^+} \dfrac{f(x) - f(0)}{x - 0}$.
(f) Is f differentiable at 0?

In Exercises 54–63, find a positive number δ such that the given statement is true for each positive number ϵ.

54. $|(2x - 1) - 5| < \epsilon$ whenever $|x - 3| < \delta$.

55. $|(3x + 2) - 8| < \epsilon$ whenever $|x - 2| < \delta$.

56. $\left| \dfrac{x^2 - 3x + 2}{x - 2} - 1 \right| < \epsilon$ whenever $0 < |x - 2| < \delta$.

57. $\left| \dfrac{x^2 - 1}{x - 1} - 2 \right| < \epsilon$ whenever $0 < |x - 1| < \delta$.

58. $|3x^{1/3}| < \epsilon$ whenever $|x| < \delta$.

59. $|x^4| < \epsilon$ whenever $|x| < \delta$.

60. $|(2x^2 - 3x + 5) - 14| < \epsilon$ whenever $|x - 3| < \delta$. Assume $|x - 3| \leq 1$.

61. $|(x^2 - 3x + 1) - (-1)| < \epsilon$ whenever $|x - 2| < \delta$. Assume $|x - 2| \leq 1$.

62. $\left| \dfrac{1}{x} - \dfrac{1}{4} \right| < \epsilon$ whenever $|x - 4| < \delta$. Assume $|x - 4| \leq 1$.

63. $|\sqrt{x} - 3| < \epsilon$ whenever $|x - 9| < \delta$. Assume $x \geq 0$.

64. A 12-ft length of wood is to be measured by using a measuring stick that is approximately 3 ft long. How close to 3 ft must the measuring stick be to ensure that the measured length is within 0.1 ft of 12 ft?

65. How close to 4Ω should the resistance in a 12-V electrical circuit be in order for the current to be within 0.05 A of 3 A? Assume the resistance is within 0.5Ω of 4Ω. $I = V/R$, where I is the current in amperes, V is the voltage in volts, and R is the resistance in ohms.

66. Evaluate $\lim\limits_{x \to 0} \cos(1/x)$.

67. Evaluate $\lim\limits_{x \to 0} x \cos(1/x)$.

68. Find the slope of the line segment between the point $(3, 9)$ and a variable point (x, x^2) on the graph of $f(x) = x^2$. Find the equation of the line tangent to the graph at the point $(3, 9)$.

69. The position of a particle on a number line at time t is given by $s(t) = -16t^2 + 32t$. Find the average velocity of the particle over the interval of time from 2 and t. Find the velocity of the particle at time 2.

70. Use the definition of the derivative to show that $f(x) = x^{1/5}$ is not differentiable at 0. Explain what happens to the slope of the line segment between $(0, 0)$ and $(x, x^{1/5})$ as x approaches 0 from each side. Sketch the graph.

71. Use the definition of the derivative to show that $f(x) = x^{2/5}$ is not differentiable at 0. Explain what happens to the slope of the line segment between $(0, 0)$ and $(x, x^{2/5})$ as x approaches 0 from each side. Sketch the graph.

72. $f(x) = 2x - 3$. Use the definition of the derivative to show that $f'(c) = 2$ for each c.

73. $f(x) = x^2$. Use the definition of the derivative to show that $f'(c) = 2c$ for each c.

74. $f(x) = 2x^2 + 3x - 5$. Use the definition of the derivative to show that $f'(c) = 4c + 3$ for each c.

75. $f(x) = 1/x$. Use the definition of the derivative to show that $f'(c) = -1/c^2$ for each nonzero c.

76. (a) Observe that $f(x) = |x|$ is continuous (but not differentiable) at $x = 0$ and observe that $g(x) = x^2$ is differentiable with $g(0) = g'(0) = 0$. Show that $(fg)(x) = |x|x^2 = |x^3|$ is differentiable at $x = 0$.

(b) More generally, suppose f is continuous at $x = 0$ and suppose that g is differentiable at $x = 0$ with $g(0) = g'(0) = 0$. Prove that fg is differentiable at $x = 0$ and $(fg)'(0) = 0$.

Determine if the graphs in Exercises 77–80 are symmetric with respect to either the x-axis, the y-axis, or the origin.

77.

78.

79.

80.

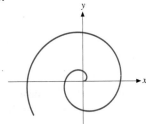

Determine if the functions given in Exercises 81–84 are odd or even.

81. $f(x) = 2x^4 - 1$

82. $f(x) = 3x^3 + 5x$

83. $f(x) = x^3 - 2x + 1$

84. $f(x) = x \tan x$

85. Use the given graph of $f(x)$ to sketch the graphs of
(a) $f(-x)$, **(b)** $-f(x)$, **(c)** $f(|x|)$, and **(d)** $|f(x)|$.

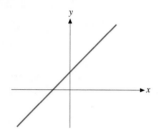

86. Use the given graph of $f(x)$ to sketch the graphs of
(a) $f(-x)$, **(b)** $-f(x)$, **(c)** $f(|x|)$, and **(d)** $|f(x)|$.

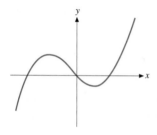

87. Give some reasons why a limit of a function at a point may not exist. Illustrate each reason with a sketch.

88. Discuss some possibilities for "tangent lines" at a point $(c, f(c))$ on the graph of a function f that is not differentiable. Explain why no such "tangent line" satisfies the condition that points $(x, f(x))$ on the graph approach $(c, f(c))$ at the angle of the "tangent line" as x approaches c.

EXTENDED APPLICATION

HIGHWAY CONSTRUCTION

As a civil engineer working for your state's Department of Transportation, you have been assigned responsibility for designing a stretch of highway that runs through a valley. Along this stretch, a long 8% downgrade is followed by a long 10% upgrade. (Grade is another word for slope, so an 8% downgrade has a slope of −0.08.) You must design a transitional segment between the two linear segments, to consist of an arc of a parabola. See Figure 1.

FIGURE 1

A suitable parabolic arc must be tangent to the two linear segments at two points, say P and Q. In order to provide a smooth transition for drivers, P and Q must be separated by a horizontal distance of 1000 feet.

If you superimpose a coordinate system on Figure 1, the two linear segments will become portions of the graphs of two linear functions, say $y = L_1(x)$ and $y = L_2(x)$. The points P and Q will have coordinates $(p, L_1(p))$ and $(p + 10000, L_2(p + 1000))$ for some value of p. The parabola can be represented by an equation of form $f(x) = ax^2 + bx + c$. If the origin is placed at P, then the coefficients a, b, and c must be chosen so that $f(0) = 0$, $f'(0) = -0.08$, and $f'(1000) = 0.10$. (It may appear that a fourth condition $f(1000) = L_2(1000)$ should also be imposed. However, if an arc satisfying the other three conditions can be found, the additional condition can be satisfied by choosing the upgrade to pass through Q.)

Tasks:

1. Explain why each of the three conditions on the coefficients is reasonable, and tell what will happen to vehicles on the highway if each is not met.
2. Write a system of linear equations in a, b, and c to express the three conditions.
3. Solve the system, and write the equation of the parabola. Also write the equations of the two linear segments, and verify analytically that the three conditions are met.
4. Graph the two lines and the parabola, either manually or on a graphing utility. Verify graphically that the three conditions are met.
5. What is the difference in elevation between P and Q? If the point R in Figure 1 has an elevation of 942 feet, what is the elevation of P?

Writing Assignment:

Send the specifications of your proposed transitional segment to your supervisor, accompanied by an accurate sketch and all calculations needed to support your design. Your supervisor knows calculus thoroughly and expects accurate usage of its language and symbols.

Source: Marvin Keener and Jeanne Agnew, Oklahoma State University, *Industry Related Problems for Mathematics Students.*

3 Continuity and Differentiability on an Interval

In Chapter 2 we emphasized that the derivative $f'(c)$ is a number that is obtained as the limit of the difference quotient $(f(x) - f(c))/(x - c)$ as x approaches c. This point of view is important for establishing what a derivative is and how it relates to physical quantities. However, in order to use derivatives effectively in practical problems, we need a method of determining values of a derivative that is easier than using the definition as a limit in each individual case. The idea is to use the definition of the derivative to establish general differentiation formulas. We can then use the formulas for future work.

In this chapter, we will develop formulas for the derivatives of power functions and trigonometric functions. Also, we will develop formulas that allow us to determine the derivatives of sums, products, quotients, and compositions of functions. These formulas are essential for the efficient application of the derivative to practical problems.

We will also see how continuity and differentiability at each point of an interval can be used to obtain solutions to some of the fundamental problems of calculus. In particular, we will investigate the problems of finding approximate values of solutions of equations and finding largest and smallest values of functions. We will also see an example of how information about the derivative of a function can be used to obtain other information about the function. These are all problems that we will return to in later chapters.

3.1 POWER AND SUM RULES; HIGHER-ORDER DERIVATIVES

Connections

Continuity of power functions, 2.5.

Definition of derivative at a point c, 2.6.

In this section we will develop formulas for the derivatives of the power functions $f(x) = x^r$, where r is a rational number. We will also see how to differentiate constant multiples of functions and sums of functions.

The Derivative as a Function

Before we begin to develop differentiation formulas, let us make the observation that, as c varies over the set where f is differentiable, the values $f'(c)$ determine a *function* f'.

We can express the values of f' in terms of the independent variable of f simply by substituting that variable for c in the formula for $f'(c)$. For example, if f is a function of x, we express the values of the derivative as $f'(x)$. It is convenient for the application of differentiation formulas to express f and f' in terms of the same independent variable.

Later in this section we will introduce a formulation of the definition of the derivative of a function that gives the derivative directly in terms of the independent variable of the function. However, the formula

$$f'(c) = \lim_{x \to c} \frac{f(x) - f(c)}{x - c}$$

is (arguably) easier to use and better reflects the concept of the derivative at a point. We will ordinarily develop differentiation formulas by using the above formula for $f'(c)$ as in Section 2.6 and then substituting x for c.

The symbols

$$\frac{df}{dx} \text{ and } Df$$

are commonly used to denote the derivative f'. When the variable of differentiation represents time, the symbol $\dot{f}$ is sometimes used. Both expressions

$$(\text{Formula})' \text{ and } \frac{d}{dx}(\text{Formula})$$

represent the derivative of the function whose Formula is indicated. For functions given in the form $y = (\text{Formula})$, we use either y' or dy/dx to indicate the derivative.

Derivative of a Constant Function

We begin a step-by-step development of the derivatives of the power functions by deriving the formula for the differentiation of constant functions.

DIFFERENTIATION OF CONSTANT FUNCTIONS

If $f(x) = k$, a constant function, then $f'(c) = 0$ for every c.

Proof We will use the definition of the derivative to verify this result. That is, if $f(x) = k$, then

$$f'(c) = \lim_{x \to c} \frac{f(x) - f(c)}{x - c} = \lim_{x \to c} \frac{k - k}{x - c}$$

$$= \lim_{x \to c} \frac{0}{x - c} = \lim_{x \to c} 0 = 0.$$

(We have used the fact that $0/(x - c) = 0$ for $x \ne c$. Since limits at c do not depend on values at c, it follows that $0/(x - c)$ and 0 have equal limits at c.) ∎

Positive Integer Powers

We next derive a formula for the derivative of power functions x^n, where n is a positive integer.

If $f(x) = x^n$, n a positive integer, then $f'(c) = nc^{n-1}$ for every c. (1)

Proof We must show that, if $f(x) = x^n$, where n is a positive integer, then $f'(c) = nc^{n-1}$ for every c. To do this, we will use the factorization formula

$$x^n - c^n = (x - c)(x^{n-1} + cx^{n-2} + c^2 x^{n-3} + \cdots + c^{n-2} x + c^{n-1}),$$

which can be verified by carrying out the multiplication and combining like terms on the right-hand side. We then have

$$f'(c) = \lim_{x \to c} \frac{f(x) - f(c)}{x - c} = \lim_{x \to c} \frac{x^n - c^n}{x - c}$$

$$= \lim_{x \to c} \frac{(x - c)(x^{n-1} + cx^{n-2} + c^2 x^{n-3} + \cdots + c^{n-2} x + c^{n-1})}{x - c}$$

$$= \lim_{x \to c} (x^{n-1} + cx^{n-2} + c^2 x^{n-3} + \cdots + c^{n-2} x + c^{n-1}).$$

Since polynomials are continuous, the latter limit is the value of the polynomial at c. We then obtain

$$f'(c) = c^{n-1} + cc^{n-2} + c^2 c^{n-3} + \cdots + c^{n-2} c + c^{n-1}.$$

Each of the n terms on the right equals c^{n-1}, so

$$f'(c) = nc^{n-1}. \quad \blacksquare$$

We can express the results of the theorem on the differentiation of constant functions and Statement 1 as differentiation formulas in terms of the variable x by writing

$$\frac{d}{dx}(k) = 0, \qquad k \text{ a constant,} \tag{2}$$

and

$$\frac{d}{dx}(x^n) = nx^{n-1}, \qquad n \text{ a positive integer.} \tag{3}$$

Formulas 2 and 3 are used to write the derivatives of constant functions and power functions that have exponents that are positive integers.

EXAMPLE 1

Using Formulas 2 and 3, we have

(a) $\dfrac{d}{dx}(3) = 0$,

(b) $\dfrac{d}{dx}(-7) = 0$,

(c) $\dfrac{d}{dx}(x^2) = 2x^{2-1} = 2x^1 = 2x$,

(d) $\dfrac{d}{dx}(x^{15}) = 15x^{15-1} = 15x^{14}$,

(e) $\dfrac{d}{dx}(x) = \dfrac{d}{dx}(x^1) = 1x^{1-1} = 1x^0 = 1 \cdot 1 = 1$. ∎

Positive Rational Powers

To extend our results for the differentiation of power functions to include rational exponents, we use Statement 1 to verify the following.

If $f(x) = x^{m/n}$, m and n positive integers, then $f'(c) = \dfrac{m}{n}c^{m/n-1}$ for every $c > 0$. $\qquad$ (4)

Proof The derivative of $x^{m/n}$ at c is

$$f'(c) = \lim_{x \to c} \frac{x^{m/n} - c^{m/n}}{x - c}.$$

To evaluate this limit, let us set

$$u = x^{1/n} \text{ and } d = c^{1/n}.$$

If $x \neq c$, then $u \neq d$, and we can write

$$
\begin{aligned}
\frac{x^{m/n} - c^{m/n}}{x - c} &= \frac{(x^{1/n})^m - (c^{1/n})^m}{(x^{1/n})^n - (c^{1/n})^n} \\
&= \frac{u^m - d^m}{u^n - d^n} \\
&= \frac{u^m - d^m}{u - d} \cdot \frac{u - d}{u^n - d^n} \\
&= \frac{u^m - d^m}{u - d} \cdot \frac{1}{\dfrac{u^n - d^n}{u - d}}.
\end{aligned}
$$

The first factor above is the difference quotient for the derivative of the function u^m at the point d. From Statement 1 we know that this approaches the derivative md^{m-1} as u approaches d. The quotient must approach the same value as x approaches c, because $u = x^{1/n} \to c^{1/n} = d$ as $x \to c$. The latter fact is a consequence of the continuity at c of the function $x^{1/n}$. Similarly, the quotient in the denominator of the second factor approaches the derivative nd^{n-1} as x approaches c. Collecting results, we have

$$f'(c) = md^{m-1} \cdot \frac{1}{nd^{n-1}} = \frac{m}{n}d^{m-n} = \frac{m}{n}(c^{1/n})^{m-n} = \frac{m}{n}c^{m/n-1}. \quad ∎$$

With some restrictions, the formula for the derivative of $x^{m/n}$ given in Statement 4 can be extended to nonpositive values of c. We first note that $x^{m/n}$ is defined for negative x only if n is an odd integer. Since the derivative at a point $c \leq 0$ involves negative values of x, it follows that $x^{m/n}$ can be differentiable at $c \leq 0$ only if n is odd. If n is an odd integer, the proof of Statement 4 given above holds for $c < 0$. If n is odd and $m/n \geq 1$, it is easy to use the definition of derivative to show that $x^{m/n}$ is differentiable at zero and that the formula holds. If $m/n < 1$, we can verify directly as in Examples 6 and 7 of Section 2.6 that $x^{m/n}$ is not differentiable at zero. Note that $c^{m/n-1}$ is undefined if $c = 0$ and $m/n < 1$.

We can express Statement 4 in the form

$$\frac{d}{dx}(x^{m/n}) = \frac{m}{n}x^{m/n-1}, \quad m \text{ and } n \text{ positive integers.} \tag{5}$$

EXAMPLE 2

From Formula 5 we have

(a) $\dfrac{d}{dx}(x^{2/3}) = \dfrac{2}{3}x^{2/3-1} = \dfrac{2}{3}x^{-1/3},$

(b) $\dfrac{d}{dx}(x^{1.7}) = 1.7x^{0.7},$

(c) $\dfrac{d}{dx}(\sqrt{x}) = \dfrac{d}{dx}(x^{1/2}) = \dfrac{1}{2}x^{1/2-1},$

$\qquad = \dfrac{1}{2}x^{-1/2} = \dfrac{1}{2x^{1/2}} = \dfrac{1}{2\sqrt{x}}.$ ∎

The square root function occurs so often that it is useful to remember that

$$\frac{d}{dx}(\sqrt{x}) = \frac{1}{2\sqrt{x}}.$$

Negative Rational Powers

Formula 5 gives us the formula for the derivative of x^r, where r is a positive rational number. It is not difficult to remove the requirement that r be positive.

If $f(x) = x^{-r}$, r a positive rational, then $f'(c) = -rc^{-r-1}$ for every $c > 0$. (6)

Proof   We use Statement 5 to prove Statement 6. We have

$$f'(c) = \lim_{x \to c}\frac{f(x) - f(c)}{x - c} = \lim_{x \to c}\frac{x^{-r} - c^{-r}}{x - c}$$

$$= \lim_{x \to c}\frac{\dfrac{1}{x^r} - \dfrac{1}{c^r}}{x - c} = \lim_{x \to c}\frac{\dfrac{c^r - x^r}{x^r \cdot c^r}}{x - c}$$

$$= \lim_{x \to c}\frac{c^r - x^r}{(x - c)x^r c^r}$$

$$= \lim_{x \to c}\left(-\frac{x^r - c^r}{x - c} \cdot \frac{1}{x^r c^r}\right).$$

Formula 5 implies that the limit at c of the first factor above is rc^{r-1}. The continuity at c of x^r gives the limit of the second factor. We obtain

$$f'(c) = -rc^{r-1} \cdot \frac{1}{c^r c^r} = -rc^{-r-1}. \quad \blacksquare$$

Let us summarize our results for the derivatives of power functions in terms of the derivative as a function.

THE POWER RULE

$$\frac{d}{dx}(x^r) = rx^{r-1}, \quad r \text{ a rational number.}$$

We can use the Power Rule to find the derivative of any power function that has a nonzero rational number exponent.

EXAMPLE 3

Using the Power Rule, we have

(a) $\dfrac{d}{dx}\left(\dfrac{1}{x}\right) = \dfrac{d}{dx}(x^{-1}) = (-1)x^{-1-1} = -x^{-2},$

(b) $\dfrac{d}{dx}\left(\dfrac{1}{\sqrt{x}}\right) = \dfrac{d}{dx}(x^{-1/2}) = -\dfrac{1}{2}x^{-3/2}. \quad \blacksquare$

Linear Combinations

We will need formulas that tell us how to differentiate various combinations of functions. Our first result of this type is the following:

THE LINEAR SUM RULE

If f and g are differentiable at c and a is a constant, then af and $f + g$ are differentiable at c. Also,

$$(af)'(c) = af'(c) \text{ and } (f + g)'(c) = f'(c) + g'(c).$$

Proof Using the definition of the derivative and the Limit Theorem to verify the first statement of the theorem, we have

$$\begin{aligned}
(af)'(c) &= \lim_{x \to c} \frac{(af)(x) - (af)(c)}{x - c} \\
&= \lim_{x \to c} \frac{af(x) - af(c)}{x - c} \\
&= \lim_{x \to c} a\frac{f(x) - f(c)}{x - c} \\
&= a\left(\lim_{x \to c} \frac{f(x) - f(c)}{x - c}\right) = af'(c).
\end{aligned}$$

The second statement of the theorem is verified by rearranging terms and using the Limit Theorem.

$$(f+g)'(c) = \lim_{x \to c} \frac{(f+g)(x) - (f+g)(c)}{x-c}$$

$$= \lim_{x \to c} \frac{(f(x)+g(x)) - (f(c)+g(c))}{x-c}$$

$$= \lim_{x \to c} \left(\frac{f(x)-f(c)}{x-c} + \frac{g(x)-g(c)}{x-c} \right)$$

$$= \lim_{x \to c} \frac{f(x)-f(c)}{x-c} + \lim_{x \to c} \frac{g(x)-g(c)}{x-c}$$

$$= f'(c) + g'(c). \quad \blacksquare$$

The Linear Sum Rule tells us that the derivative of a constant multiple of a differentiable function is the constant times the derivative of the function, and that the derivative of a sum of differentiable functions is the sum of the derivatives of the functions. The second statement can be extended to apply to the sum of any finite number of terms.

EXAMPLE 4

Using the theorems we have developed, we have

(a) $\frac{d}{dx}(\ x^2\ -3\ x\ +7\ )$
$= 2x\quad -3\quad (1)\quad +0$

$\left(\begin{array}{c}\text{Power}\\\text{function}\end{array}\right) + \text{(Constant)}\left(\begin{array}{c}\text{Power}\\\text{function}\end{array}\right) + \left(\begin{array}{c}\text{Constant}\\\text{function}\end{array}\right)$

$\left(\begin{array}{c}\text{Derivative}\\\text{of power}\\\text{function}\end{array}\right) + \text{(Constant)}\left(\begin{array}{c}\text{Derivative}\\\text{of power}\\\text{function}\end{array}\right) + \left(\begin{array}{c}\text{Derivative}\\\text{of constant}\\\text{function}\end{array}\right)$

(b) $\frac{d}{dx}(2x^4 - 3x^3 + x^2 - x + 4) = 2(4x^3) - 3(3x^2) + 2x - 1 + 0$
$$= 8x^3 - 9x^2 + 2x - 1. \quad \blacksquare$$

Some products and quotients of functions can be differentiated easily by carrying out a multiplication or division and then using the formulas of this section.

EXAMPLE 5

(a) $\frac{d}{dx}(x(x-1)^2) = \frac{d}{dx}(x(x^2-2x+1))$
$$= \frac{d}{dx}(x^3 - 2x^2 + x)$$
$$= 3x^2 - 2(2x) + 1$$
$$= 3x^2 - 4x + 1,$$

(b) $\frac{d}{dx}\left(\frac{x^2-3x+5}{x}\right) = \frac{d}{dx}\left(\frac{x^2}{x} - \frac{3x}{x} + \frac{5}{x}\right)$
$$= \frac{d}{dx}(x - 3 + 5x^{-1})$$
$$= 1 - 0 + 5(-1)x^{-1-1}$$
$$= 1 - 5x^{-2}. \quad \blacksquare$$

EXAMPLE 6

Find the equation of the line tangent to the graph of $f(x) = x^2 - 5x + 1$ at the point on the graph with $x = 2$.

SOLUTION   We need the coordinates of the point on the graph that has $x = 2$ and the slope of the line tangent at that point.

$$f(2) = (2)^2 - 5(2) + 1 = -5,$$

so $(2, -5)$ is the point on the graph with $x = 2$.

The slope of the line tangent to the graph of f at $(2, -5)$ is $f'(2)$, the value of the derivative at 2.

$$f(x) = x^2 - 5x + 1, \text{ so}$$
$$f'(x) = 2x - 5.$$

Substituting 2 for x in the formula for $f'(x)$, we have

$$f'(2) = 2(2) - 5 = -1.$$

Note that we used a differentiation formula to evaluate $f'(2)$. We no longer need to determine $f'(2)$ by evaluating the limit as x approaches 2 of the slope of the line segments between $(2, -5)$ and a variable point on the graph, as we did in Section 2.6.

The point-slope form of the equation of the line through $(2, -5)$ with slope -1 is

$$y - (-5) = (-1)(x - 2).$$

Simplifying, we have

$$y + 5 = -x + 2,$$
$$x + y + 3 = 0.$$

The graph is sketched in Figure 3.1.1.   ■

In Example 6 we used the formula for $f(x)$ to determine the y-coordinate of the point on the graph with $x = 2$. The formula for $f'(x)$ was used to determine the slope of the line tangent to the graph at $(2, f(2))$.

NOTE   *In many problems we will deal with both a function f and its derivative f'. Always label your functions so you can tell which function is which. Don't just write the formula for $f'(x)$, write "$f'(x) =$" followed by the formula.*

Higher-Order Derivatives

Since the derivative of a function is again a function, we may consider successive derivatives of functions. The derivative of a function is then called the first derivative. The derivative of the first derivative is called the second derivative.

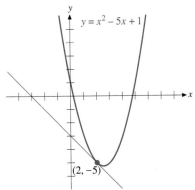

FIGURE 3.1.1   The y-coordinate of the point on the graph of $f(x) = x^2 - 5x + 1$ with $x = 2$ is $f(2) = -5$. The slope of the line tangent to the graph at $(2, -5)$ is $f'(2) = -1$.

The derivative of the second derivative is called the third derivative, and so forth. The following notation is used:

First derivative: $\dfrac{df}{dx}$ f' $f^{(1)}$

Second derivative: $\dfrac{d}{dx}\left(\dfrac{df}{dx}\right)=\dfrac{d^2 f}{dx^2}$ $(f')'=f''$ $f^{(2)}$

Third derivative: $\dfrac{d}{dx}\left(\dfrac{d^2 f}{dx^2}\right)=\dfrac{d^3 f}{dx^3}$ $(f'')'=f'''$ $f^{(3)}$

$\vdots$

nth derivative: $\dfrac{d}{dx}\left(\dfrac{d^{n-1} f}{dx^{n-1}}\right)=\dfrac{d^n f}{dx^n}$ $f^{(n)}$

Both $d^n f/dx^n$ and $f^{(n)}$ indicate derivatives of order n. The function f is sometimes denoted $f^{(0)}$.

The prime notation f', f'', f''' becomes awkward for derivatives of order higher than three, so we will usually use either $d^n f/dx^n$ or $f^{(n)}$ to denote such derivatives.

EXAMPLE 7
Find the first four derivatives of

(a) $f(x)=3-x+5x^2-7x^3$,
(b) $g(x)=1/x$.

SOLUTION

(a) $f(x)=3-x+5x^2-7x^3$,

$$\frac{df}{dx}(x)=-1+10x-21x^2,$$

$$\frac{d^2 f}{dx^2}(x)=\frac{d}{dx}\left(\frac{df}{dx}\right)(x)=10-42x,$$

$$\frac{d^3 f}{dx^3}(x)=\frac{d}{dx}\left(\frac{d^2 f}{dx^2}\right)(x)=-42,$$

$$\frac{d^4 f}{dx^4}(x)=\frac{d}{dx}\left(\frac{d^3 f}{dx^3}\right)(x)=0.$$

(b) $g(x)=x^{-1}$,

$$g'(x)=(-1)x^{-2},$$
$$g''(x)=(g')'(x)=(-1)(-2)x^{-3},$$
$$g'''(x)=(g'')'(x)=(-1)(-2)(-3)x^{-4},$$
$$g^{(4)}(x)=(g''')'(x)=(-1)(-2)(-3)(-4)x^{-5}. \quad\blacksquare$$

Velocity and Acceleration

Recall that if the **position** of a particle along a line at a time t is given by the function $s(t)$, then its **velocity** at time t is given by the formula $v(t)=s'(t)$. We now define the (instantaneous) **acceleration** of the particle to be the rate of change of the velocity with respect to time,

$$a(t)=v'(t)=s''(t).$$

EXAMPLE 8

The height in feet above ground of an object t seconds after it is thrown vertically from a height of 64 ft with initial velocity 48 ft/s is given by the formula

$$s(t) = -16t^2 + 48t + 64, \quad t \geq 0.$$

The formula is valid until the object returns to ground level. Find formulas for the velocity $v(t)$ and acceleration $a(t)$ of the object. At what time t is $v(t) = 0$? What is the height when $v(t) = 0$? At what times is the height zero? What is the velocity at the times when the height is zero?

SOLUTION   We have

$$s(t) = -16t^2 + 48t + 64,$$
$$v(t) = s'(t) = -32t + 48,$$
$$a(t) = v'(t) = s''(t) = -32.$$

(Note that the acceleration is constant. The negative value indicates the acceleration is downward.)

$$v(t) = 0 \text{ when } -32t + 48 = 0, \text{ so } t = \frac{48}{32} = \frac{3}{2}.$$

When $t = 3/2$, the height is

$$s\left(\frac{3}{2}\right) = -16\left(\frac{3}{2}\right)^2 + 48\left(\frac{3}{2}\right) + 64 = 100.$$

The height is zero when $s(t) = 0$, or

$$-16t^2 + 48t + 64 = 0,$$
$$-16(t^2 - 3t - 4) = 0,$$
$$(t - 4)(t + 1) = 0,$$
$$t = 4, \quad t = -1.$$

When $t = 4$, the velocity is $v(4) = -32(4) + 48 = -80$. The value $t = -1$ is not in the domain of the height function. See Figure 3.1.2. ■

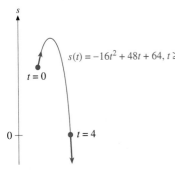

FIGURE 3.1.2 The height in feet of an object t seconds after it is thrown vertically from a height of 64 ft with initial velocity 48 ft/s is given by the formula $s(t) = -16t^2 + 48t + 64$, $t \geq 0$.

An Equivalent Definition of Derivative

We have used the definition $f'(c) = \lim\limits_{x \to c} \dfrac{f(x) - f(c)}{x - c}$ in order to emphasize that the derivative at a point is the limit of the difference quotient and to simplify the calculations in some of the examples. In some cases it is more convenient to use a form of the definition that allows us to obtain a formula for the derivative in terms of the variable x, without referring to a point c. That is, if we set $x - c = h$, so $x = c + h$, then $x \to c$ is equivalent to $h \to 0$, so

$$f'(c) = \lim_{x \to c} \frac{f(x) - f(c)}{x - c} = \lim_{h \to 0} \frac{f(c + h) - f(c)}{h}.$$

In the latter formulation we can replace the point c by the variable x to obtain

$$f'(x) = \lim_{h \to 0} \frac{f(x + h) - f(x)}{h}. \tag{7}$$

We will use Formula 7 when that seems convenient.

EXAMPLE 9

Use Formula 7 to evaluate the derivative of $f(x) = x^3$.

SOLUTION   We have

$$
\begin{aligned}
f'(x) &= \lim_{h \to 0} \frac{f(x+h) - f(x)}{h} \\
&= \lim_{h \to 0} \frac{(x+h)^3 - x^3}{h} \\
&= \lim_{h \to 0} \frac{x^3 + 3x^2h + 3xh^2 + h^3 - x^3}{h} \\
&= \lim_{h \to 0} (3x^2 + 3xh + h^2) = 3x^2. \quad \blacksquare
\end{aligned}
$$

Generally, using Formula 7 to evaluate the derivative of x^n requires the binomial formula to expand $(x+h)^n$.

EXERCISES 3.1

In Exercises 1–20 find the derivatives of the given functions.

1. $f(x) = x^2 - 7x + 3$

2. $f(x) = 3x^2 + 4x - 2$

3. $f(x) = 2x^3 - x + 1$

4. $f(x) = x^4 - 13x^2 + 9x + 7$
   (Calculus Explorer & Tutor I, Differentiation, 1.)

5. $f(x) = 2x - 3 + x^{-1}$

6. $f(x) = 1 - \dfrac{2}{x} + \dfrac{4}{x^2}$ **(Calculus Explorer & Tutor I, Differentiation, 3.)**

7. $f(x) = x^{1/2} + x^{-1/2}$

8. $f(x) = x^{4/3} - 4x^{1/3}$

9. $f(x) = 6x^{2/3} - 9x^{-1/3}$

10. $f(x) = 5x^{-1.6} + 10x^{-0.6}$

11. $f(x) = \sqrt[3]{2x^2}$

12. $f(x) = \sqrt{3x}$

13. $f(x) = \sqrt{x}(2x + 1)$

14. $f(x) = \dfrac{x^3 - x}{x^{1/3}}$ **(Calculus Explorer & Tutor I, Differentiation, 9.)**

15. $f(x) = \dfrac{(x+1)^2}{4x}$

16. $f(x) = \sqrt{x}(\sqrt{x} - 1)^2$

17. $f(t) = 3(t^2 + 1)^2$

18. $f(t) = (3 - 2t)(1 + t)$

19. $f(t) = \dfrac{1 + \sqrt{t}}{2t}$

20. $f(t) = \dfrac{4 + \sqrt{t}}{3t^2}$

Find the first three derivatives of the functions given in Exercises 21–26.

21. $f(x) = 3x^2 - 7x + 2$

22. $f(x) = x^3 - 2x^2 + x - 7$

23. $f(x) = x^{1/3} + x^{-1/3}$

24. $f(x) = x^2 - \dfrac{1}{x^2}$

25. $f(x) = x^9$

26. $f(x) = \sqrt{x}$

In Exercises 27–32 find the equation of the line tangent to the graph of $y = f(x)$ at the point on the graph with $x = c$.

27. $f(x) = 3x + 1, c = -2$

28. $f(x) = x^6 - x^5, c = 1/2$
   (Calculus Explorer & Tutor I, Differentiation, 17.)

29. $f(x) = x^2 - 2, c = 2$

30. $f(x) = 3x^2 - 4x + 1, c = 2$

31. $f(x) = 1/x, c = -2$

32. $f(x) = 1/x^2, c = -2$

In Exercises 33–38 the given function represents the position on a line of a particle at time t. Find the velocity of the particle at time t. Evaluate the position and velocity at the given time t_1.

33. $s(t) = 50t - 16t^2, t_1 = 3$

34. $s(t) = 60t - 16t^2, t_1 = 2$

35. $s(t) = t + 4t^3, t_1 = 1$

36. $s(t) = t^3 - t, t_1 = 2$

37. $s(t) = \sqrt{t}, t_1 = 4$

38. $s(t) = t^{1/3}, t_1 = 8$

In Exercises 39–42 the position of a particle on a line at time t is given by the formula for $s(t)$. Find (a) the velocity $v(t) = s'(t)$ and (b) the acceleration $a(t) = v'(t) = s''(t)$ at time t. (c) Determine the position at each time the velocity is zero. (d) Determine the velocity at each time the position is zero.

39. $s(t) = -16t^2 + 32t + 48$

40. $s(t) = -16t^2 + 16t + 32$

41. $s(t) = t - \dfrac{1}{t^2}$

42. $s(t) = 1 - t^2$

In Exercises 43–48 find all points $(x, f(x))$ on the graph of f for which (a) $f'(x) = 0$ and (b) $f'(x)$ does not exist.

43. $f(x) = x^2 - 6x + 1$

44. $f(x) = x^3 + x^2 - x - 1$

45. $f(x) = x + 4x^{-1}$

46. $f(x) = 2x^{1/3} + x^{-2/3}$

47. $f(x) = x^{4/3} + 4x^{1/3}$

48. $f(x) = x^{7/3} - 7x^{1/3}$

In Exercises 49–52 find all points on the graph of the given function f where (a) $f'(x) = 0$ and (b) $f''(x) = 0$.

49. $f(x) = 4 + 12x - 3x^2$

50. $f(x) = x^3 - 3x^2$

51. $f(x) = x + \dfrac{1}{x}$

52. $f(x) = \sqrt{x} + \dfrac{1}{\sqrt{x}}$

53. Find all points on the graph of $f(x) = x^3 - x$ that have tangent lines with slope 2.

54. Find all points on the graph of $f(x) = x^2 - 2x$ that have tangent lines parallel to the line $8x - 2y = 5$.

55. Find all points on the graph of $f(x) = x^2$ that have tangent lines that contain the point $(5, 9)$. (*Hint:* The equation of the line tangent to the graph of $f(x) = x^2$ at a point (x_0, y_0) on the graph is $y - x_0^2 = 2x_0(x - x_0)$. If this line contains the point $(5, 9)$, then $(x, y) = (5, 9)$ must satisfy the equation of the line. Substitute 5 for x and 9 for y, and solve the resulting equation for x_0.)

56. Find all points on the graph of $f(x) = x^3$ that have tangent lines that intersect the x-axis at the single point $(4, 0)$.

57. Find all points of intersection of the parabola $y = x^2$ and the line $x + y = 2$. At each point of intersection, find the acute angle of intersection of the line $x + y = 2$ and the line

tangent to the parabola at the point. (*Hint:* Use the formula for the tangent of the difference of two angles. The formula is given inside the front cover.)

58. Find all points of intersection of the graphs of $y = x^3$ and $y = 3x^2 - 2x$. At each point of intersection, find the acute angle of intersection of the lines tangent to the graphs at the point.

59. If $P(x) = a_0 + a_1 x$, show that $a_0 = P(0)$ and $a_1 = P'(0)$.

60. If $P(x) = a_0 + a_1 x + a_2 x^2$, show that $a_0 = P(0), a_1 = P'(0)$, and $a_2 = P''(0)/2$.

Use Formula 7 to evaluate the derivatives of the functions given in Exercises 61–65.

61. $f(x) = x^2$

62. $f(x) = 1/x$

63. $f(x) = 1/x^2$

64. $f(x) = \sqrt{x}$

65. $f(x) = 1/\sqrt{x}$

66. Use Mathematical Induction and the Linear Sum Rule to show that $(f_1 + f_2 + \cdots + f_n)' = f_1' + f_2' + \cdots + f_n'$ for every positive integer n.

67. Compare the graphs of $f(x) = e^x$, $P_1(x) = 1 + x$, $P_2(x) = 1 + x + x^2/2$, $P_3(x) = 1 + x + x^2/2 + x^3/6$, and $P_4(x) = 1 + x + x^2/2 + x^3/6 + x^4/24$.

68. Compare the graphs of $f(x) = \sin x$, $P_1(x) = x$, $P_3(x) = x - x^3/6$, and $P_5(x) = x - x^3/6 + x^5/120$.

3.2 PRODUCT AND QUOTIENT RULES

Connections

Simplifying expressions, 1.2.

Definition of derivative at a point c, 2.6.

Continuity of differentiable functions, 2.6.

Power and Linear Sum Rules, 3.1.

Functions defined by more than one formula, 2.1.

We develop formulas that are used to find the derivatives of products and quotients.

The Product Rule

The **Product Rule** is used to differentiate products of differentiable functions.

THE PRODUCT RULE

If f and g are differentiable at c, then fg is differentiable at c and

$$(fg)'(c) = f(c) \cdot g'(c) + g(c) \cdot f'(c).$$

Proof We will use the algebraic trick of adding and subtracting $f(x)g(c)$ in the numerator to evaluate the limit in the definition of the derivative.

$$(fg)'(c) = \lim_{x \to c} \frac{(fg)(x) - (fg)(c)}{x - c}$$

$$= \lim_{x \to c} \frac{f(x)g(x) - f(c)g(c)}{x - c}$$

$$= \lim_{x \to c} \frac{f(x)g(x) - f(x)g(c) + f(x)g(c) - f(c)g(c)}{x - c}$$

$$= \lim_{x \to c} \left[f(x)\frac{g(x) - g(c)}{x - c} + g(c)\frac{f(x) - f(c)}{x - c} \right]$$

$$= \lim_{x \to c} f(x) \cdot \lim_{x \to c} \frac{g(x) - g(c)}{x - c} + g(c) \cdot \lim_{x \to c} \frac{f(x) - f(c)}{x - c}$$

$$= f(c) \cdot g'(c) + g(c) \cdot f'(c).$$

The above limits were evaluated by using the definition of the derivatives of f and g, and the continuity of f. Since f is differentiable at c, f is continuous at c, so $\lim_{x \to c} f(x) = f(c)$. ■

The Product Rule does not depend on the names of the functions involved. Think of the Product Rule as saying

$$\begin{pmatrix} \text{Derivative} \\ \text{of a product} \end{pmatrix} = \begin{pmatrix} \text{First} \\ \text{factor} \end{pmatrix} \begin{pmatrix} \text{Derivative} \\ \text{of second} \end{pmatrix} + \begin{pmatrix} \text{Second} \\ \text{factor} \end{pmatrix} \begin{pmatrix} \text{Derivative} \\ \text{of first} \end{pmatrix}.$$

When the functions involved are given by formulas, we should write the derivative directly, in one step. It is helpful to use the above form every time you differentiate a product. The use of all four sets of parentheses on the right-hand side helps to organize your work.

EXAMPLE 1
Using the Product Rule, we have

(a) $\dfrac{d}{dx}(\underbrace{(x^2 + 1)}_{\substack{\uparrow \\ \left(\begin{smallmatrix}\text{First}\\\text{factor}\end{smallmatrix}\right)}} \underbrace{(3x - 2)}_{\substack{\uparrow \\ \left(\begin{smallmatrix}\text{Second}\\\text{factor}\end{smallmatrix}\right)}})$

$= \underbrace{(x^2 + 1)}_{\substack{\uparrow \\ \left(\begin{smallmatrix}\text{First}\\\text{factor}\end{smallmatrix}\right)}} \underbrace{(3)}_{\substack{\uparrow \\ \left(\begin{smallmatrix}\text{Derivative}\\\text{of second}\end{smallmatrix}\right)}} + \underbrace{(3x - 2)}_{\substack{\uparrow \\ \left(\begin{smallmatrix}\text{Second}\\\text{factor}\end{smallmatrix}\right)}} \underbrace{(2x)}_{\substack{\uparrow \\ \left(\begin{smallmatrix}\text{Derivative}\\\text{of first}\end{smallmatrix}\right)}}$,

(b) $\dfrac{d}{dx}((x - 2)(x^2 + 3x + 1)) = (x - 2)(2x + 3) + (x^2 + 3x + 1)(1),$

(c) $\dfrac{d}{dx}((3x - 2)(x^2 + x^{-2})) = (3x - 2)(2x - 2x^{-3}) + (x^2 + x^{-2})(3).$

■

The derivatives in Example 1 have not been simplified. Simplifying is an additional step, distinct from using the Product Rule to determine the derivatives. Simplification of expressions that arise from the application of differentiation formulas is discussed in Section 1.2.

The Quotient Rule

The **Quotient Rule** is used to differentiate quotients of differentiable functions.

> **THE QUOTIENT RULE**
>
> If f and g are differentiable at c and $g(c) \neq 0$, then f/g is differentiable at c and
>
> $$\left(\frac{f}{g}\right)'(c) = \frac{g(c)f'(c) - f(c)g'(c)}{(g(c))^2}.$$

Proof   The proof of the Quotient Rule is similar to that of the Product Rule. We use the fact that g is continuous at c, so $g(c) \neq 0$ implies $g(x) \neq 0$ for x near c.

$$
\begin{aligned}
\left(\frac{f}{g}\right)'(c) &= \lim_{x \to c} \frac{\left(\dfrac{f}{g}\right)(x) - \left(\dfrac{f}{g}\right)(c)}{x - c} = \lim_{x \to c} \frac{\dfrac{f(x)}{g(x)} - \dfrac{f(c)}{g(c)}}{x - c} \\[2ex]
&= \lim_{x \to c} \frac{\dfrac{g(c)f(x) - f(c)g(x)}{g(x)g(c)}}{x - c} \\[2ex]
&= \lim_{x \to c} \frac{\dfrac{g(c)f(x) - g(c)f(c) + g(c)f(c) - f(c)g(x)}{g(x)g(c)}}{x - c} \\[2ex]
&= \lim_{x \to c} \frac{g(c)(f(x) - f(c)) - f(c)(g(x) - g(c))}{(x - c)g(x)g(c)} \\[2ex]
&= \lim_{x \to c} \frac{\dfrac{g(c)(f(x) - f(c)) - f(c)(g(x) - g(c))}{x - c}}{g(x)g(c)} \\[2ex]
&= \lim_{x \to c} \frac{g(c)\dfrac{f(x) - f(c)}{x - c} - f(c)\dfrac{g(x) - g(c)}{x - c}}{g(x)g(c)}.
\end{aligned}
$$

The Limit Theorem then gives

$$\left(\frac{f}{g}\right)'(c) = \frac{g(c)f'(c) - f(c)g'(c)}{(g(c))^2}. \quad \blacksquare$$

Think of the Quotient Rule as saying

$$\left(\begin{array}{c}\text{Derivative}\\ \text{of a quotient}\end{array}\right) = \frac{\left(\begin{array}{c}\text{Bottom}\\ \text{factor}\end{array}\right)\left(\begin{array}{c}\text{Derivative}\\ \text{of top}\end{array}\right) - \left(\begin{array}{c}\text{Top}\\ \text{factor}\end{array}\right)\left(\begin{array}{c}\text{Derivative}\\ \text{of bottom}\end{array}\right)}{(\text{Bottom factor})^2}.$$

We must be especially careful when using the Quotient Rule. Be sure each term is in its proper place. Using all five sets of parentheses on the right-hand side helps to organize your work.

EXAMPLE 2

From the Quotient Rule, we have

(a)
$$\frac{d}{dx}\left(\frac{3x-2}{x^2+1}\right) = \frac{\overbrace{(x^2+1)}^{\left(\substack{\text{Top}\\\text{factor}}\right)}\overbrace{(3)}^{\left(\substack{\text{Bottom}\\\text{factor}}\right)\left(\substack{\text{Derivative}\\\text{of top}}\right)} - \overbrace{(3x-2)}^{\left(\substack{\text{Top}\\\text{factor}}\right)}\overbrace{(2x)}^{\left(\substack{\text{Derivative}\\\text{of bottom}}\right)}}{\underbrace{(x^2+1)^2}_{\left(\substack{\text{Bottom}\\\text{factor,}\\\text{squared}}\right)}},$$

where the $\frac{3x-2}{x^2+1}$ has top factor $3x-2$ and bottom factor x^2+1 (Bottom factor).

(b)
$$\frac{d}{dx}\left(\frac{x-2}{2x+1}\right) = \frac{(2x+1)(1)-(x-2)(2)}{(2x+1)^2},$$

(c)
$$\frac{d}{dx}\left(\frac{x^3}{x^2+4}\right) = \frac{(x^2+4)(3x^2)-(x^3)(2x)}{(x^2+4)^2}. \quad \blacksquare$$

It is not efficient to use the Product Rule or the Quotient Rule when one of the factors is a constant. That would give the correct derivative, but it is needlessly complicated. It's easier to use

$$(af)' = a \cdot f', \quad \left(\frac{f}{b}\right)' = \left(\frac{1}{b} \cdot f\right)' = \frac{1}{b} \cdot f' = \frac{f'}{b}, \text{ and}$$

$$\left(\frac{a}{x^r}\right)' = (ax^{-r})' = a(-rx^{-r-1}).$$

In Section 3.4 we will learn to differentiate functions of the form $a/g = a \cdot g^{-1}$ without using the Quotient Rule.

EXAMPLE 3

(a)
$$\frac{d}{dx}\left(\frac{4x-3}{7}\right) = \frac{d}{dx}\left[\frac{1}{7}(4x-3)\right] = \frac{1}{7}(4) = \frac{4}{7},$$

(b)
$$\frac{d}{dx}\left(\frac{4}{x^2}\right) = \frac{d}{dx}(4x^{-2}) = 4(-2x^{-3}) = -\frac{8}{x^3}. \quad \blacksquare$$

In some cases you may find it easier to carry out a multiplication or division instead of using the Product or Quotient Rules. For example, we may write either

$$\frac{d}{dx}(x(x^2-3x+2)) = \frac{d}{dx}(x^3-3x^2+2x)$$
$$= 3x^2-6x+2$$

or

$$\frac{d}{dx}(x(x^2-3x+2)) = (x)(2x-3)+(x^2-3x+2)(1)$$
$$= 2x^2-3x+x^2-3x+2$$
$$= 3x^2-6x+2.$$

EXAMPLE 4

Find the equation of the line tangent to the graph of

$$f(x) = (2x^2-3x+1)(3x^2+5x-4)$$

at the point where $x = 2$.

SOLUTION   We need the coordinates of the point on the graph where $x = 2$ and the slope of the line tangent at that point.

$$f(2) = (2(2)^2 - 3(2) + 1)(3(2)^2 + 5(2) - 4)$$
$$= (3)(18) = 54,$$

so the point $(2, 54)$ is on the graph. The slope of the line tangent to the graph at $(2, 54)$ is the value of the derivative $f'(2)$. It is not convenient to carry out the multiplication of the factors of $f(x)$, so we use the Product Rule to find the derivative.

$$f(x) = (2x^2 - 3x + 1)(3x^2 + 5x - 4), \text{ so}$$
$$f'(x) = (2x^2 - 3x + 1)(6x + 5) + (3x^2 + 5x - 4)(4x - 3).$$

Substitution into the formula for f' gives

$$f'(2) = (2(2)^2 - 3(2) + 1)(6(2) + 5) + (3(2)^2 + 5(2) - 4)(4(2) - 3)$$
$$= (3)(17) + (18)(5) = 141.$$

The point-slope form of the equation of the line through the point $(2, 54)$ with slope 141 is

$$y - 54 = 141(x - 2).$$

Simplifying, we obtain

$$y = 141x - 282 + 54,$$
$$y = 141x - 228. \quad \blacksquare$$

EXAMPLE 5

$f(x) = x^{1/3}/(3x - 2)$. Find all points $(x, f(x))$ on the graph of f for which (a) $f'(x) = 0$ and (b) $f'(x)$ does not exist.

SOLUTION   Since

$$f(x) = \frac{x^{1/3}}{3x - 2},$$

the Quotient Rule gives

$$f'(x) = \frac{(3x - 2)\left(\frac{1}{3}x^{-2/3}\right) - (x^{1/3})(3)}{(3x - 2)^2}.$$

To find values of x for which $f'(x) = 0$ and values of x for which $f'(x)$ does not exist, we simplify the expression for $f'(x)$. Each factor in the numerator contains a power of x. Factoring the smaller power of x from each term in the numerator, we have

$$f'(x) = \frac{(3x - 2)\left(\frac{1}{3}x^{-2/3}\right) - (x^{1/3})(3)}{(3x - 2)^2}$$

$$= \frac{(3x-2)\left(\dfrac{x^{-2/3}}{3}\right) - \left(\dfrac{9x^{1-2/3}}{3}\right)}{(3x-2)^2}$$

$$= \frac{\left(\dfrac{x^{-2/3}}{3}\right)((3x-2)-9x)}{(3x-2)^2}$$

$$= \frac{-6x-2}{3x^{2/3}(3x-2)^2}.$$

(a) $f'(x) = 0$ when $-6x - 2 = 0$, or $x = -1/3$. When $x = -1/3$,

$$y = f\left(-\frac{1}{3}\right) = \frac{\left(-\dfrac{1}{3}\right)^{1/3}}{3\left(-\dfrac{1}{3}\right)-2} = \frac{-\left(\dfrac{1}{3}\right)^{1/3}}{-3} = \left(\frac{1}{3}\right)^{4/3} \approx 0.2311.$$

Therefore, $(-1/3, (1/3)^{4/3})$ is a point on the graph with $f'(x) = 0$. The graph has a horizontal tangent at this point.

(b) We see that $f'(x)$ does not exist when $3x^{2/3}(3x-2)^2 = 0$; that is, when $x = 0$ or $x = 2/3$. When $x = 0$, $y = f(0) = 0$. The coordinates $(0, 0)$ represent a point on the graph where $f'(x)$ does not exist. The graph has a vertical tangent at $(0, 0)$. There is no point on the graph of f with $x = 2/3$. See Figure 3.2.1. ∎

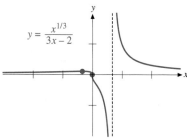

$$y = \frac{x^{1/3}}{3x-2}$$

FIGURE 3.2.1 The function f has a horizontal tangent at $(-1/3, (1/3)^{4/3})$, where $f'(x) = 0$. The graph becomes vertical as it passes through the origin, where $f'(x)$ does not exist.

EXAMPLE 6

$f(x) = x^2 F(x)$, where $F(3) = 5$ and $F'(3) = 2$. Find $f'(3)$.

SOLUTION   The Product Rule gives

$$f'(x) = (x^2)(F'(x)) + (F(x))(2x).$$

Substituting known values when $x = 3$ gives

$$f'(3) = ((3)^2)(F'(3)) + (F(3))(2(3)) = (9)(2) + (5)(6) = 48. \quad ∎$$

Functions Defined by More Than One Formula

If a function f is defined by a formula for x in an interval I, the formula can be used to evaluate the derivative of f at points interior to I. If f is defined at an endpoint of I and the value of f at that endpoint agrees with the value given by the formula, then the formula can be used to evaluate the one-sided derivative of f from within I at the endpoint.

If f is defined by different formulas on each side of a point c, then the one-sided derivatives from each side of c must be investigated to determine the existence and value of the derivative at c. The function is differentiable at c if both one-sided derivatives at c have the same value. This can happen only if f is continuous at c. Recall that a function that is not continuous at a point cannot be differentiable at the point. If f is not continuous at c, then at least one of the two one-sided derivatives of f at c will not exist. (See Section 2.6, Example 5.)

NOTE   *Do not use the values given by the derivatives of the formulas that hold on each side of c without checking that f is continuous at c.*

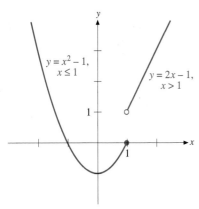

FIGURE 3.2.2 The function f is differentiable at all points, except for $x = 1$. It is discontinuous at $x = 1$, so it is not differentiable there.

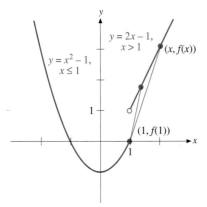

FIGURE 3.2.3 The slopes of the line segments between $(1, f(1))$ and $(x, f(x))$ increase without bound as x approaches one from the right.

EXAMPLE 7

$$f(x) = \begin{cases} x^2 - 1, & x \le 1, \\ 2x - 1, & x > 1. \end{cases}$$

Evaluate (a) $f'(0)$ and (b) $f'(1)$.

SOLUTION   The derivative of f for x in the interior of each of the intervals $x \le 1$ and $x > 1$ is given by the derivative of the formula that defines f in that interval. Differentiation then gives

$$f'(x) = \begin{cases} 2x, & x < 1, \\ 2, & x > 1. \end{cases}$$

Note that we have excluded the point $x = 1$ from the formula for $f'(x)$; $x = 1$ is not an interior point of an interval where the values of f are given by a single formula.

(a) Since $x = 0$ is an interior point of an interval where the values of f are given by a single formula, we can use the formula for f' given above to determine that

$$f'(0) = 2(0) = 0.$$

(b) The formulas we derived for $f'(x)$ do not include $x = 1$. To determine the existence and value of f' at $x = 1$ we first check that f is continuous at $x = 1$.

The values of $f(x)$ for $x \le 1$ are given by the formula $x^2 - 1$, so $f(1) = (1)^2 - 1 = 0$ and f is continuous from the left at $x = 1$. The values of $f(x)$ for $x > 1$ are given by the formula $2x - 1$, so $f(x)$ approaches $2(1) - 1 = 1$ as x approaches 1 from the right. It follows that f is not continuous at $x = 1$, so f is not differentiable at $x = 1$. See Figure 3.2.2.

Note that both formulas for $f'(x)$ give the value two for $x = 1$. Since f is continuous from the left at $x = 1$, two is the value of the derivative from the left at $x = 1$. Since f is not continuous from the right at $x = 1$, f is not differentiable from the right at $x = 1$. It can be verified directly that

$$\lim_{x \to 1^+} \frac{f(x) - f(1)}{x - 1} = \lim_{x \to 1^+} \frac{(2x - 1) - (0)}{x - 1}$$

does not exist. From Figure 3.2.3 we see that the slopes of the line segments between $(1, f(1)) = (1, 0)$ and $(x, f(x))$ increase without bound as x approaches one from the right. ∎

EXERCISES 3.2

Use the Product Rule or the Quotient Rule to evaluate the derivatives in Exercises 1–8. Use parentheses to indicate each factor in the formula you are using and then simplify your answers.

1. $\dfrac{d}{dx}\left((x^2 + 1)(2x^2 - 3x + 1)\right)$

2. $\dfrac{d}{dx}\left((9x^2 + 6x + 5)(x^4 - 7x^3 + 8x^2 + 5x + 2)\right)$

(**Calculus Explorer & Tutor I, Differentiation, 2.**)

3. $\dfrac{d}{dx}\left((x + x^{-1})(x - x^{-1})\right)$ **4.** $\dfrac{d}{dx}\left((1 + \sqrt{x})(x + \sqrt{x})\right)$

5. $\dfrac{d}{dx}\left(\dfrac{3x}{x^2 + 1}\right)$

6. $\dfrac{d}{dx}\left(\dfrac{x^3-1}{x^4+1}\right)$ (Calculus Explorer & Tutor I, Differentiation, 4.)

7. $\dfrac{d}{dx}\left(\dfrac{2x-1}{4x+1}\right)$

8. $\dfrac{d}{dx}\left(\dfrac{x^{1/3}}{x^2-x}\right)$ (Calculus Explorer & Tutor I, Differentiation, 8.)

In Exercises 9–14 find the equation of the line tangent to the graph of $y = f(x)$ at the point on the graph where $x = c$.

9. $f(x) = (x^2+3x+1)(x^3-4x+1), c = 0$

10. $f(x) = (3x-5)(x^3+2x^2+3x+4), c = 0$

11. $f(x) = \dfrac{x^2+9}{5}, c = 4$ **12.** $f(x) = \dfrac{x^2+4}{x}, c = 2$

13. $f(x) = \dfrac{x+1}{x-1}, c = 2$ **14.** $f(x) = \dfrac{x^2-1}{x^2+1}, c = 1$

In Exercises 15–18 find all points $(x, f(x))$ on the graph of f for which (a) $f'(x) = 0$ and (b) $f'(x)$ does not exist.

15. $f(x) = \dfrac{x-1}{x+1}$ **16.** $f(x) = \dfrac{x}{x^2+1}$

17. $f(x) = \dfrac{x^{1/3}}{x^2+5}$ **18.** $f(x) = \dfrac{x^{2/3}}{2x+1}$

In Exercises 19–24 use the Product Rule or the Quotient Rule to find formulas for the derivatives of functions in the given form.

19. $[f(x)]^2$ **20.** $[f(x)]^3$

21. $f(x) \cdot g(x) \cdot h(x)$ **22.** $\dfrac{1}{g(x)}$

23. $\dfrac{1}{[g(x)]^2}$ **24.** $\dfrac{f(x) \cdot g(x)}{h(x)}$

25. Find a formula for $(fg)''$. **26.** Find a formula for $(fg)'''$.

27. $f(x) = x \cdot F(x)$, where $F(2) = 3$ and $F'(2) = 5$. Find $f'(2)$.

28. $f(x) = F(x)/x^2$, where $F(2) = 16$ and $F'(2) = 12$. Find $f'(2)$.

29. $f(x) = xE(x)$, where $E(x) > 0$ and $E'(x) = E(x)$. Find x for which $f'(x) = 0$.

30. $f(x) = E(x)/x$, where $E(x) > 0$ and $E'(x) = E(x)$. Find x for which $f'(x) = 0$.

31. $f(x) = L(x)/x$, $x > 0$, where $L'(x) = 1/x$. Show that $f'(x) = 0$ implies $L(x) = 1$.

32. $f(x) = x^2 L(x)$, $x > 0$, where $L(e) = 1$ and $L'(x) = 1/x$. Evaluate $f'(e)$.

In Exercises 33–42 either evaluate the indicated derivatives of the given functions or show the derivatives do not exist.

33. $f(x) = \begin{cases} 0, & x < 0, \\ x^2, & x \geq 0. \end{cases}$ (a) $f'(-1)$, (b) $f'(0)$, and (c) $f'(1)$.

34. $f(x) = \begin{cases} -x^2, & x < 0, \\ x^2, & x \geq 0. \end{cases}$ (a) $f'(-1)$, (b) $f'(0)$, and (c) $f'(1)$.

35. $f(x) = \begin{cases} 0, & x \leq 0, \\ x, & x > 0. \end{cases}$ (a) $f'(-1)$, (b) $f'(0)$, and (c) $f'(1)$.

36. $f(x) = \begin{cases} x, & x \leq 1, \\ 1/x, & x > 1. \end{cases}$ (a) $f'(0)$, (b) $f'(1)$, and (c) $f'(2)$.

37. $f(x) = \begin{cases} x^2-1, & x \neq 0, \\ 0, & x = 0. \end{cases}$ (a) $f'(-1)$, (b) $f'(0)$, and (c) $f'(1)$.

38. $f(x) = \begin{cases} 1-x^2, & x < 0, \\ 0, & x = 0, \\ x^2-1, & x > 0. \end{cases}$ (a) $f'(-1)$, (b) $f'(0)$, and (c) $f'(1)$.

39. $f(x) = \begin{cases} -x^{4/3}, & x < 0, \\ x^{4/3}, & x \geq 0. \end{cases}$ (a) $f'(-1)$, (b) $f'(0)$, and (c) $f'(1)$.

40. $f(x) = \begin{cases} -\sqrt{-x}, & x < 0, \\ \sqrt{x}, & x \geq 0. \end{cases}$ (a) $f'(-1)$, (b) $f'(0)$, and (c) $f'(1)$.

41. $f(x) = \begin{cases} |x|, & x \leq 1, \\ 2-x, & x > 1. \end{cases}$ (a) $f'(0)$, (b) $f'(1)$, and (c) $f'(2)$.

42. $f(x) = \begin{cases} 0, & x < 0, \\ \dfrac{x}{x^2+1}, & 0 \leq x \leq 1, \\ 1, & x > 1. \end{cases}$ (a) $f'(-1)$, (b) $f'(0)$, and (c) $f'(1)$.

43. Use Mathematical Induction and the Product Rule to show that $\dfrac{d}{dx}(x^n) = nx^{n-1}$ for every positive integer n.

44. Use the fact that $\dfrac{d}{dx}(x^r) = rx^{r-1}$ for positive rational numbers r to verify that $\dfrac{d}{dx}(x^{-r}) = -rx^{-r-1}$, by applying the Quotient Rule to find the derivative of $1/x^r$.

3.3 DIFFERENTIATION OF TRIGONOMETRIC FUNCTIONS

Connections

$\lim\limits_{h\to 0}\dfrac{f(x+h)-f(x)}{h}$ *form of the derivative,*
3.1.

Formulas for the sine and cosine of a sum of angles, inside front cover.

Power, Sum, Product, and Quotient Rules, 3.1–2.

In this section we will develop differentiation formulas for the six basic trigonometric functions. Radian measure of angles will be used to develop the formulas. (Properties of trigonometric functions and radian measure are reviewed in Chapter 1.)

Some Basic Limits

We will need the following limit result to verify the formulas for the differentiation of the sine and cosine functions:

$$\lim_{\theta \to 0} \frac{\sin \theta}{\theta} = 1, \quad \theta \text{ in radians.} \qquad (1)$$

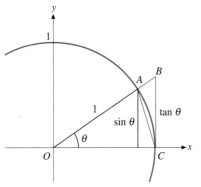

FIGURE 3.3.1 The sector *OAC* contains the triangle *OAC* and is contained in the triangle *OBC*.

Proof We see that the limit in (1) cannot be evaluated by factoring and then cancelling θ. We will use a geometric argument to evaluate the limit. We first assume that $0 < \theta < \pi/2$.

From Figure 3.3.1 we see that

$$(\text{Area triangle } OAC) < (\text{Area sector } OAC) < (\text{Area triangle } OBC). \quad (2)$$

Since the area of a sector is proportional to its central angle, we have

$$\frac{(\text{Area of sector } OAC)}{\theta} = \frac{(\text{Area of circle})}{2\pi} = \frac{\pi(1)^2}{2\pi}.$$

It follows that $(\text{Area of sector } OAC) = \theta/2$. Triangle OAC has height $\sin\theta$ and base 1. Triangle OBC is a right triangle with height $\tan\theta$ and base 1. Substitution into (2) then gives

$$\frac{1}{2}(1)(\sin\theta) < \frac{\theta}{2} < \frac{1}{2}(1)(\tan\theta) \text{ or } \frac{\sin\theta}{2} < \frac{\theta}{2} < \frac{\sin\theta}{2\cos\theta}.$$

Since $\sin\theta > 0$ for $0 < \theta < \pi/2$, we can divide the double inequality by $(\sin\theta)/2$ to obtain

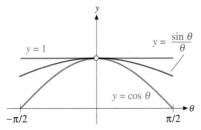

FIGURE 3.3.2 The graph of $\dfrac{\sin\theta}{\theta}$, $0 < |\theta| \le \pi/2$, is squeezed between the graphs of $\cos\theta$ and the constant function one.

$$1 < \frac{\theta}{\sin\theta} < \frac{1}{\cos\theta} \text{ or } \cos\theta < \frac{\sin\theta}{\theta} < 1.$$

The latter inequality also holds for $-\pi/2 < \theta < 0$, because

$$\frac{\sin(-\theta)}{-\theta} = \frac{\sin\theta}{\theta} \text{ and } \cos(-\theta) = \cos\theta.$$

See Figure 3.3.2. Since $\cos\theta$ and the constant function 1 both have limit 1 at 0, it follows from the Squeeze Theorem of Section 2.4 that

$$\lim_{\theta \to 0} \frac{\sin\theta}{\theta} = 1. \quad \blacksquare$$

We will also need

$$\lim_{\theta \to 0} \frac{1 - \cos \theta}{\theta} = 0, \quad \theta \text{ in radians.} \tag{3}$$

Proof We can use (1) to verify (3). That is,

$$\lim_{\theta \to 0} \frac{1 - \cos \theta}{\theta} = \lim_{\theta \to 0} \left(\frac{1 - \cos \theta}{\theta} \right) \left(\frac{1 + \cos \theta}{1 + \cos \theta} \right)$$

$$= \lim_{\theta \to 0} \frac{1 - \cos^2 \theta}{\theta(1 + \cos \theta)}$$

$$= \lim_{\theta \to 0} \frac{\sin^2 \theta}{\theta(1 + \cos \theta)}$$

$$= \lim_{\theta \to 0} \left(\frac{\sin \theta}{\theta} \right) \left(\frac{\sin \theta}{1 + \cos \theta} \right)$$

$$= \left(\lim_{\theta \to 0} \frac{\sin \theta}{\theta} \right) \left(\lim_{\theta \to 0} \frac{\sin \theta}{1 + \cos \theta} \right)$$

$$= (1) \left(\frac{\sin 0}{1 + \cos 0} \right) = 0. \quad \blacksquare$$

We can also use the technique of the proof of (3) to prove

$$\lim_{\theta \to 0} \frac{1 - \cos \theta}{\theta^2} = \frac{1}{2}, \quad \theta \text{ in radians.} \tag{4}$$

The Differentiation Formulas

We can now determine derivatives of the six basic trigonometric functions. We begin with the derivative of the sine function. We will use the limits (1) and (3) that we developed above to evaluate the limit that gives the derivative.

Since the limits in (1) and (3) were evaluated for θ in radians, radian measure must also be used for the differentiation formulas.

$$\frac{d}{dx}(\sin x) = \cos x \text{ for every } x. \tag{5}$$

Proof It is convenient to use the formulation of the definition of the derivative given at the end of Section 3.1. Using the trigonometric identity for the sine of a sum, we have

$$\frac{d}{dx}(\sin x) = \lim_{h \to 0} \frac{\sin(x + h) - \sin x}{h}$$

$$= \lim_{h \to 0} \frac{\sin x \cdot \cos h + \cos x \cdot \sin h - \sin x}{h}$$

$$= \lim_{h \to 0} \left[(\sin x) \left(\frac{\cos h - 1}{h} \right) + (\cos x) \left(\frac{\sin h}{h} \right) \right].$$

Formulas (1) and (3) and the Limit Theorem then imply

$$\frac{d}{dx}(\sin x) = (\sin x)(0) + (\cos x)(1) = \cos$$

Similarly, by using the formula for $\cos(x + h)$, w
following:

$$\frac{d}{dx}(\cos x) = -\sin x \text{ for every } x. \qquad (6)$$

Let us list the derivatives of the six basic trigonometric functions.

$$\frac{d}{dx}(\sin x) = \cos x,$$

$$\frac{d}{dx}(\cos x) = -\sin x,$$

$$\frac{d}{dx}(\tan x) = \sec^2 x,$$

$$\frac{d}{dx}(\cot x) = -\csc^2 x,$$

$$\frac{d}{dx}(\sec x) = \sec x \cdot \tan x,$$

$$\frac{d}{dx}(\csc x) = -\csc x \cdot \cot x.$$

The formulas for the derivatives of $\sin x$ and $\cos x$ are from (5) and (6). The other formulas may then be derived by using the Quotient Rule. For example,

$$\frac{d}{dx}(\tan x) = \frac{d}{dx}\left(\frac{\sin x}{\cos x}\right)$$

$$= \frac{(\cos x)\left(\frac{d}{dx}(\sin x)\right) - (\sin x)\left(\frac{d}{dx}(\cos x)\right)}{\cos^2 x}$$

$$= \frac{(\cos x)(\cos x) - (\sin x)(-\sin x)}{\cos^2 x}$$

$$= \frac{\cos^2 x + \sin^2 x}{\cos^2 x} = \frac{1}{\cos^2 x} = \sec^2 x.$$

Similarly,

$$\frac{d}{dx}(\sec x) = \frac{d}{dx}\left(\frac{1}{\cos x}\right)$$

$$= \frac{(\cos x)(0) - (1)(-\sin x)}{\cos^2 x}$$

$$= \frac{\sin x}{\cos^2 x} = \left(\frac{1}{\cos x}\right) \cdot \left(\frac{\sin x}{\cos x}\right) = \sec x \cdot \tan x.$$

(In Section 3.4 we will learn a method that is more convenient than using the Quotient Rule to differentiate quotients, such as $1/\cos x$, that have constant numerators.)

Formulas for the derivatives of $\cot x$ and $\csc x$ can be obtained in the same way as those for $\tan x$ and $\sec x$.

Be sure you know the formulas for the derivatives of the sine and cosine functions. It is worthwhile to memorize the differentiation formulas for all six trigonometric functions, even though they all can be derived from the formulas for the derivatives of the sine and cosine functions.

We can now differentiate products and quotients that involve trigonometric functions.

EXAMPLE 1

(a) $\dfrac{d}{dx}(x^2 \sin x) = (x^2)(\cos x) + (\sin x)(2x),$

(b) $\dfrac{d}{dx}\left(\dfrac{\cos x}{x}\right) = \dfrac{(x)(-\sin x) - (\cos x)(1)}{x^2},$

(c) $\dfrac{d}{dx}(\sec x \tan x) = (\sec x)(\sec^2 x) + (\tan x)(\sec x \tan x).$ ■

Differentiation of functions with more than two factors may require various combinations of the Product and Quotient Rules.

EXAMPLE 2

Find the derivative of

(a) $f(x) = x \sin x \cos x$ and

(b) $g(x) = \dfrac{x \sin x}{1 + \cos x}.$

SOLUTION

(a) Let us read $x \sin x \cos x$ as the product of the function x and the function $\sin x \cos x$. That is,

$$f(x) = \underset{\substack{\uparrow \\ \left(\substack{\text{First} \\ \text{factor}}\right)}}{(x)} \ \underset{\substack{\uparrow \\ \left(\substack{\text{Second} \\ \text{factor}}\right)}}{(\sin x \cos x)}.$$

Then

$$f'(x) = \underset{\substack{\uparrow \\ \left(\substack{\text{First} \\ \text{factor}}\right)}}{(x)} \ \underset{\substack{\uparrow \\ \left(\substack{\text{Derivative} \\ \text{of second}}\right)}}{((\sin x)(-\sin x) + (\cos x)(\cos x))} + \underset{\substack{\uparrow \\ \left(\substack{\text{Second} \\ \text{factor}}\right)}}{(\sin x \cos x)} \ \underset{\substack{\uparrow \\ \left(\substack{\text{Derivative} \\ \text{of first}}\right)}}{} \quad (1).$$

The Product Rule was used to determine the derivative of the second factor. The derivative simplifies to

$$f'(x) = -x \sin^2 x + x \cos^2 x + \sin x \cos x.$$

We can obtain the same derivative by reading $x \sin x \cos x$ as the product of $x \sin x$ and $\cos x$, so

$$f(x) = (x \sin x)(\cos x)$$

and

$$f'(x) = (x \sin x)(-\sin x) + (\cos x)((x)(\cos x) + (\sin x)(1)).$$

This simplifies to the same expression for $f'(x)$ that we obtained earlier.

(b) We read $g(x)$ as a quotient whose top factor is a product. That is,

$$g(x) = \frac{\overset{\text{(Top factor)}}{\overset{\downarrow}{(x \sin x)}}}{\underset{\underset{\text{(Bottom factor)}}{\uparrow}}{(1 + \cos x)}}.$$

Applying the Quotient Rule, and using the Product Rule to find the derivative of the top factor, we obtain

$$g'(x) = \frac{\overbrace{(1 + \cos x)}^{\left(\substack{\text{Bottom} \\ \text{factor}}\right)} \overbrace{((x)(\cos x) + (\sin x)(1))}^{\left(\substack{\text{Derivative} \\ \text{of top}}\right)} - \overbrace{(x \sin x)}^{\left(\substack{\text{Top} \\ \text{factor}}\right)} \overbrace{(-\sin x)}^{\left(\substack{\text{Derivative} \\ \text{of bottom}}\right)}}{\underset{\left(\substack{\text{Bottom} \\ \text{factor,} \\ \text{squared}}\right)}{(1 + \cos x)^2}}.$$

Simplifying, we have

$$g'(x) = \frac{x \cos x + \sin x + x \cos^2 x + \cos x \sin x + x \sin^2 x}{(1 + \cos x)^2}$$

$$= \frac{x(\cos x + \cos^2 x + \sin^2 x) + (\sin x + \cos x \sin x)}{(1 + \cos x)^2}$$

$$= \frac{x(\cos x + 1) + (\sin x)(1 + \cos x)}{(1 + \cos x)^2}$$

$$= \frac{(x + \sin x)(1 + \cos x)}{(1 + \cos x)^2}$$

$$= \frac{x + \sin x}{1 + \cos x}. \quad \blacksquare$$

Applications

EXAMPLE 3

Find the equation of the line tangent to the graph of $f(x) = 2 \sin x \cos x$ at the point where $x = \pi/3$.

SOLUTION   When $x = \pi/3$,

$$y = f\left(\frac{\pi}{3}\right) = 2 \sin \frac{\pi}{3} \cos \frac{\pi}{3} = 2\left(\frac{\sqrt{3}}{2}\right)\left(\frac{1}{2}\right) = \frac{\sqrt{3}}{2},$$

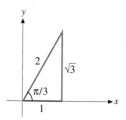

FIGURE 3.3.3 A representative right triangle can be used to determine that $\sin(\pi/3) = \sqrt{3}/2$ and $\cos(\pi/3) = 1/2$.

so $(\pi/3, \sqrt{3}/2)$ is the point on the graph where $x = \pi/3$. (The exact values of $\sin(\pi/3)$ and $\cos(\pi/3)$ can be determined from the representative triangle in Figure 3.3.3. A calculator could be used to determine approximate values.) The slope of the line tangent to the graph at the point $(\pi/3, \sqrt{3}/2)$ is the value of the derivative at $x = \pi/3$. Using the Product Rule, we have

$$f'(x) = 2((\sin x)(-\sin x) + (\cos x)(\cos x))$$
$$= 2(-\sin^2 x + \cos^2 x).$$

Hence,

$$f'\left(\frac{\pi}{3}\right) = 2\left(-\sin^2\frac{\pi}{3} + \cos^2\frac{\pi}{3}\right)$$
$$= 2\left(-\left(\frac{\sqrt{3}}{2}\right)^2 + \left(\frac{1}{2}\right)^2\right) = -1.$$

The point-slope form of the equation of the line through the point $(\pi/3, \sqrt{3}/2)$ with slope -1 is

$$y - \frac{\sqrt{3}}{2} = (-1)\left(x - \frac{\pi}{3}\right)$$
$$x + y = \frac{\pi}{3} + \frac{\sqrt{3}}{2}.$$

See Figure 3.3.4. ∎

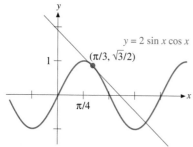

FIGURE 3.3.4 The y-coordinate of the point on the graph of $f(x) = 2\sin x \cos x$ with $x = \pi/3$ is $f(\pi/3) = \sqrt{3}/2$. The slope of the line tangent to the graph at $(\pi/3, \sqrt{3}/2)$ is $f'(\pi/3) = -1$.

EXAMPLE 4

Find points on the graph of

$$f(x) = \sin x + \cos x, \qquad 0 \le x \le 2\pi,$$

where the tangent line is horizontal.

SOLUTION   We know that the slope of the tangent line at the point $(x, f(x))$ is given by the value of the derivative $f'(x)$. Horizontal lines have slope zero, so we need to find values of x for which $f'(x) = 0$.

$$f(x) = \sin x + \cos x, \text{ so } f'(x) = \cos x - \sin x.$$

Setting $f'(x) = 0$, we have

$$\cos x - \sin x = 0.$$

To solve this equation, we express it in terms of $\tan x$. We have

$$-\sin x = -\cos x,$$
$$\frac{\sin x}{\cos x} = 1,$$
$$\tan x = 1.$$

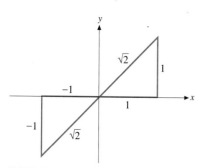

FIGURE 3.3.5 Representative right triangles can be used to determine that the equation $\tan x = 1$ has solutions $x = \pi/4$ and $x = 5\pi/4$.

There are two solutions of the equation $\tan x = 1$ in the interval $0 \le x \le 2\pi$. Both solutions have reference angle $\pi/4$. Representative right triangles are sketched in Figure 3.3.5. The solutions are $x = \pi/4$ and $x = 5\pi/4$. These are

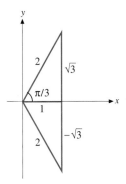

FIGURE 3.3.6 The graph of $f(x) = \sin x + \cos x$ has horizontal tangents at points with $x = \pi/4$ and $x = 5\pi/4$, where $f'(x) = 0$.

FIGURE 3.3.7 Representative right triangles can be used to determine that $\cos(\pi/3) = \cos(5\pi/3) = 1/2$.

the values of x that correspond to points on the graph with horizontal tangents. The y-coordinates are given by the corresponding values $f(x)$.

$$f\left(\frac{\pi}{4}\right) = \sin\frac{\pi}{4} + \cos\frac{\pi}{4} = \frac{1}{\sqrt{2}} + \frac{1}{\sqrt{2}} = \frac{2}{\sqrt{2}} = \sqrt{2} \text{ and}$$

$$f\left(\frac{5\pi}{4}\right) = \sin\frac{5\pi}{4} + \cos\frac{5\pi}{4} = \frac{-1}{\sqrt{2}} + \frac{-1}{\sqrt{2}} = -\frac{2}{\sqrt{2}} = -\sqrt{2}.$$

$(\pi/4, \sqrt{2})$ and $(5\pi/4, -\sqrt{2})$ are points on the graph that have horizontal tangents. The graph is sketched in Figure 3.3.6. ■

EXAMPLE 5

The position of a particle on a number line at time t is given by $s(t) = t - 2\sin t$, $0 \le t \le 2\pi$. Find intervals of time for which the particle is moving in the positive direction.

SOLUTION   The particle is moving in the positive direction when the velocity is positive. The velocity is given by the derivative $s'(t)$.

$$s(t) = t - 2\sin t, \text{ so } s'(t) = 1 - 2\cos t.$$

We need to find values of t, $0 \le t \le 2\pi$, such that

$$s'(t) > 0,$$
$$1 - 2\cos t > 0.$$

The expression $1 - 2\cos t$ can change sign only at points where $1 - 2\cos t = 0$, or $\cos t = 1/2$. From the representative right triangles drawn in Figure 3.3.7, we see that $\cos t = 1/2$ for $t = \pi/3$ and $t = 5\pi/3$. The table of signs given in Table 3.3.1 can then be determined by calculating a value of $1 - 2\cos t$ in each interval. We see that $s'(t)$ is positive and the particle is moving in the positive direction for $\pi/3 < t < 5\pi/3$. (We could also determine that $s'(t)$ is positive for $\pi/3 < t < 5\pi/3$ from the graph of s' given in Figure 3.3.8.) ■

TABLE 3.3.1

	0		$\pi/3$		$5\pi/3$		2π
$s'(t) = 1 - 2\cos t$		$-$		$+$		$-$	

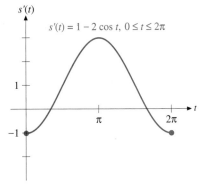

FIGURE 3.3.8 The derivative $s'(t)$ is positive for $\pi/3 < t < 5\pi/3$.

EXERCISES 3.3

Find the derivatives in Exercises 1–12. Use parentheses to indicate each factor when using the Product or Quotient Rules, and then simplify your answers.

1. $\dfrac{d}{dx}(\sin x + \cos x)$

2. $\dfrac{d}{dx}(2\sin x - \cos x)$

3. $\dfrac{d}{dx}(\sec x - 2\tan x)$

4. $\dfrac{d}{dx}\left(\dfrac{2\cot x - \sec x}{3}\right)$

5. $\dfrac{d}{dx}(3x^{1/3} \cdot \sin x)$

6. $\dfrac{d}{dx}(\sec x \cdot \csc x)$ **(Calculus Explorer & Tutor I, Trigonometric Differentiation, 12.)**

7. $\dfrac{d}{dx}(2x\sin x)$

8. $\dfrac{d}{dx}(x^2 \sin x)$

9. $\dfrac{d}{dx}\left(\dfrac{\sin x}{x}\right)$ **(Calculus Explorer & Tutor I, Trigonometric Differentiation, 2.)**

10. $\dfrac{d}{dx}\left(\dfrac{3x^2}{1+\sin x}\right)$

11. $\dfrac{d}{dx}\left(\dfrac{\cos x}{x^2+4}\right)$

12. $\dfrac{d}{dx}\left(\dfrac{2\cos x}{\sin x + \cos x}\right)$

In Exercises 13–16 find the equation of the line tangent to the graph at the point where $x = c$.

13. $f(x) = x + \sin x,\ c = 0$

14. $f(x) = x\cos x,\ c = 0$

15. $f(x) = \sin x \cdot \tan x,\ c = \pi/6$

16. $f(x) = \dfrac{\sin x + \cos x}{\sin x - \cos x},\ c = \pi/2$

In Exercises 17–20 find all points on the graphs of the given functions where the tangent line is horizontal.

17. $f(x) = x + 2\sin x,\ 0 \le x \le 2\pi$

18. $f(x) = x - \sqrt{2}\cos x,\ 0 \le x \le 2\pi$

19. $f(x) = \sqrt{3}\sin x + \cos x,\ 0 \le x \le 2\pi$

20. $f(x) = \sqrt{3}\sin x - \cos x,\ 0 \le x \le 2\pi$

In Exercises 21–24 the position of a particle on a number line at time t is given by the function $s(t)$. Find intervals of time for which the particle is moving in the positive direction.

21. $s(t) = \sqrt{3}t + 2\cos t,\ 0 \le t \le 2\pi$

22. $s(t) = \sqrt{3}t + 2\sin t,\ 0 \le t \le 2\pi$

23. $s(t) = 2t - \tan t,\ 0 \le t < \pi/2$

24. $s(t) = 2\sec t - \tan t,\ 0 \le t < \pi/2$

25. Show that $\displaystyle\lim_{\theta \to 0}\frac{1-\cos\theta}{\theta^2} = \frac{1}{2}$.

26. Derive the formula $\dfrac{d}{dx}(\cos x) = -\sin x$.

27. Derive the formula for the derivative of $\cot x$ by expressing $\cot x$ in terms of $\sin x$ and $\cos x$ and then using the Quotient Rule.

28. Derive the formula for the derivative of $\csc x$ by expressing $\csc x$ in terms of $\sin x$ and then using the Quotient Rule.

29. Show that $x_0 = \tan x_0$ whenever the line tangent to the graph of $y = \sin x$ at the point $(x_0, \sin x_0)$ intersects the origin. See figure.

30. The formula $\dfrac{d}{dx}(\sin x) = \cos x$ was derived from the formula $\displaystyle\lim_{x \to 0}\frac{\sin x}{x} = 1$, where $\sin x$ represents the sine of an angle with *radian* measure x. If $\sin x°$ represents the sine of an angle of x degrees, then $\displaystyle\lim_{x \to 0}\frac{\sin x°}{x} = \frac{\pi}{180}$. (You can verify that this limit seems correct by using your calculator to evaluate $\dfrac{\sin x°}{x}$ for small values of x.) What would be the derivative at c of the sine function if degrees were used instead of radians? That is, what is the value of $\displaystyle\lim_{h \to 0}\frac{\sin(x° + h°) - \sin x°}{h}$?

Use Formulas 1, 3, and 4 to evaluate the limits in Exercises 31–36.

31. $\displaystyle\lim_{x \to 0}\frac{\tan x}{x}$

32. $\displaystyle\lim_{x \to 0}\frac{1-\sec x}{x}$

33. $\displaystyle\lim_{x \to 0}\frac{\sin(2x)}{x}$

34. $\displaystyle\lim_{x \to 0}\frac{\tan(4x)}{\sin(2x)}$

35. $\lim\limits_{x\to 0} \dfrac{1 - \cos(3x)}{x^2}$

36. $\lim\limits_{x\to 0} \dfrac{1 - \cos(3x)}{\sin^2(2x)}$

37. $f(x) = \begin{cases} \dfrac{x^2}{\sin x}, & x \neq 0, \\ 0, & x = 0. \end{cases}$

    (a) Find a formula for $f'(x)$, $x \neq 0$.

    (b) Does $f'(x)$ approach a limit as x approaches zero?

    (c) Is f differentiable at zero?

    (d) Is f' continuous at zero?

38. $f(x) = \begin{cases} \dfrac{1 - \cos x}{x}, & x \neq 0, \\ 0, & x = 0. \end{cases}$

    (a) Find a formula for $f'(x)$, $x \neq 0$.

    (b) Does $f'(x)$ approach a limit as x approaches zero?

    (c) Is f differentiable at zero?

    (d) Is f' continuous at zero?

39. Graph $f(x) = \sec x - x$, $-\pi/2 < x < \pi/2$ and use the graph to determine an approximate value of x for which $f'(x) = 0$.

40. Graph $f(x) = x^2 - \sin x$, $0 \leq x \leq \pi/2$ and use the graph to determine an approximate value of x for which $f'(x) = 0$.

Connections

Composition of functions, 2.1.

Power, Sum, Product, and Quotient rules, 3.1–2.

Differentiation formulas for the trigonometric functions, 3.3.

3.4 THE CHAIN RULE

In previous sections we have developed formulas for the differentiation of sums, products, and quotients of functions. In this section we will develop a formula for the differentiation of composite functions and see how to apply the formula to compositions of power functions and the six basic trigonometric functions.

The Chain Rule Formula

The **Chain Rule** is used to differentiate composite functions.

THE CHAIN RULE

If g is differentiable at c and f is differentiable at $g(c)$, then the composite function $f \circ g(x) = f(g(x))$ is differentiable at c and

$$(f \circ g)'(c) = f'(g(c)) \cdot g'(c).$$

Proof It follows from the definitions that

$$(f \circ g)'(c) = \lim_{x\to c} \frac{f \circ g(x) - f \circ g(c)}{x - c} = \lim_{x\to c} \frac{f(g(x)) - f(g(c))}{x - c}.$$

We must show that the limit on the right has value $f'(g(c)) \cdot g'(c)$.

Let us first note that if we knew that $g(x) - g(c) \neq 0$ for all x near c, $x \neq c$, we could write

$$\frac{f(g(x)) - f(g(c))}{x - c} = \frac{f(g(x)) - f(g(c))}{g(x) - g(c)} \cdot \frac{g(x) - g(c)}{x - c}$$

and show that the first factor on the right has limit $f'(g(c))$ and that the second factor has limit $g'(c)$, so we would be done. However, it is possible that $g(x) - g(c) = 0$, even though $x \neq c$. In order to cover this possibility, we introduce the function

$$\epsilon(u) = \begin{cases} \dfrac{f(u) - f(g(c))}{u - g(c)} - f'(g(c)), & u \neq g(c), \\ 0, & u = g(c). \end{cases}$$

Since f is differentiable at $g(c)$, we have

$$\lim_{u \to g(c)} \epsilon(u) = f'(g(c)) - f'(g(c)) = 0 = \epsilon(g(c)),$$

so ϵ is continuous at $g(c)$. We know that g is continuous at c, since it is differentiable at c. It follows that the composition $\epsilon \circ g$ is continuous at c, so

$$\lim_{x \to c} \epsilon(g(x)) = \epsilon(g(c)) = 0.$$

If $g(x) \neq g(c)$, the definition of the function ϵ gives

$$\frac{f(g(x)) - f(g(c))}{g(x) - g(c)} - f'(g(c)) = \epsilon(g(x)),$$

so

$$f(g(x)) - f(g(c)) = f'(g(c))(g(x) - g(c)) + \epsilon(g(x))(g(x) - g(c)). \quad (*)$$

Note that $(*)$ is also true if $g(x) = g(c)$. If $x \neq c$, we then use $(*)$ to obtain

$$\lim_{x \to c} \frac{f(g(x)) - f(g(c))}{x - c}$$

$$= \lim_{x \to c} \left(f'(g(c)) \frac{g(x) - g(c)}{x - c} + \epsilon(g(x)) \frac{g(x) - g(c)}{x - c} \right)$$

$$= f'(g(c))g'(c) + 0 \cdot g'(c) = f'(g(c)) \cdot g'(c). \quad \blacksquare$$

The Chain Rule may also be expressed in the following form. If $y = y(u)$ is a differentiable function of u and $u = u(x)$ is a differentiable function of x, then $y = y(u(x))$ is a differentiable function of x and

$$\frac{dy}{dx} = \frac{dy}{du} \cdot \frac{du}{dx}.$$

In the case $y = u^r$ and $u = u(x)$, the Chain Rule gives

$$\frac{d}{dx}(u^r) = \frac{dy}{dx} = \frac{dy}{du} \cdot \frac{du}{dx} = ru^{r-1} \cdot \frac{du}{dx}.$$

In the formulas

$$\frac{d}{dx}(x^r) = rx^{r-1} \text{ and } \frac{d}{du}(u^r) = ru^{r-1},$$

we are differentiating a constant power of the variable of differentiation. We know from Section 3.1 how to differentiate a constant power of the variable of differentiation; the Chain Rule is not required. The Chain Rule is required to differentiate a power of a *function* that is not simply the variable of differentiation. This version of the Chain Rule should be thought of as saying

Derivative of (Function)^r = r(Function)$^{r-1}$(Derivative of function).

Using the Chain Rule

EXAMPLE 1
Find the derivatives.

    (a) $\dfrac{d}{dx}((x^2 + 7x - 3)^4)$,

    (b) $\dfrac{d}{dx}\left(\dfrac{1}{x^2 + 1}\right)$,

    (c) $\dfrac{d}{dx}(\sqrt{x^2 + x + 4})$,

    (d) $\dfrac{d}{dx}(\sin^2 x)$.

SOLUTION

    (a) $(x^2 + 7x - 3)^4$ is the fourth power of the function $x^2 + 7x - 3$. We can use the Chain Rule to write the derivative in one step:

$$\frac{d}{dx}\underbrace{(x^2 + 7x - 3)}_{\text{(Function)}}{}^{\underset{\text{(Exponent } r)}{4}}) = 4 \underset{(r)}{} \underbrace{(x^2 + 7x - 3)}_{\text{(Function)}}{}^{\underset{\text{(Exponent } r-1)}{3}} \underbrace{(2x + 7)}_{\binom{\text{Derivative}}{\text{of function}}} \cdot$$

    (b) Let us write $1/(x^2 + 1)$ as $(x^2 + 1)^{-1}$, the -1 power of the function $x^2 + 1$. The Chain Rule then gives the derivative.

$$\frac{d}{dx}\left(\frac{1}{x^2 + 1}\right) = \frac{d}{dx}(\underbrace{(x^2 + 1)}_{\text{(Function)}}{}^{\underset{\text{(Exponent } r)}{-1}})$$

$$= -1 \underset{(r)}{} \underbrace{(x^2 + 1)}_{\text{(Function)}}{}^{\underset{\text{(Exponent } r-1)}{-2}} \underbrace{(2x)}_{\binom{\text{Derivative}}{\text{of function}}} \cdot$$

We could have used the Quotient Rule to differentiate $1/(x^2 + 1)$, but it is much easier to use the Chain Rule. In general, don't use the Quotient Rule to differentiate functions of the form $1/g(x)$. Instead, write $1/g(x)$ as $[g(x)]^{-1}$ and use the Chain Rule.

    (c) The square root function can be written as the half-power function. The Chain Rule for powers of functions can then be used to differentiate square roots of functions.

$$\frac{d}{dx}(\sqrt{x^2 + x + 4}) = \frac{d}{dx}((x^2 + x + 4)^{1/2})$$

$$= \frac{1}{2}(x^2 + x + 4)^{-1/2}(2x + 1).$$

Alternatively, we could use the fact that

$$\frac{d}{du}(\sqrt{u}) = \frac{1}{2\sqrt{u}}$$

and write

$$\frac{d}{dx}(\sqrt{x^2 + x + 4}) = \frac{1}{2\sqrt{x^2 + x + 4}}(2x + 1).$$

(d) The function $\sin^2 x$ is read as "sine squared of x," but is important to recognize that the composition is the *square* of the function $\sin x$, since $\sin^2 x = (\sin x)^2$. The Chain Rule applied to the square of a function then gives the derivative.

$$\frac{d}{dx}(\sin^2 x) = \frac{d}{dx}((\sin x)^2) = 2(\sin x)(\cos x). \quad \blacksquare$$

It is important to know how to use the Chain Rule for each function we can differentiate. For example, we know that

$$\frac{d}{du}(\sin u) = \cos u.$$

Hence, in the case $y = \sin u$ and $u = u(x)$, the Chain Rule gives

$$\frac{d}{dx}(\sin u) = \frac{dy}{dx} = \frac{dy}{du} \cdot \frac{du}{dx} = \cos u \cdot \frac{du}{dx}.$$

This formula should be thought of as saying

Derivative of sin(Function) = cos(Function) · (Derivative of Function).

The versions of the Chain Rule that correspond to the differentiation formulas we know are listed below.

$$\frac{d}{dx}(u^r) = ru^{r-1} \cdot \frac{du}{dx}, r \text{ rational}$$

$$\frac{d}{dx}(\sin u) = \cos u \cdot \frac{du}{dx} \qquad \frac{d}{dx}(\cot u) = -\csc^2 u \cdot \frac{du}{dx}$$

$$\frac{d}{dx}(\cos u) = -\sin u \cdot \frac{du}{dx} \qquad \frac{d}{dx}(\sec u) = \sec u \cdot \tan u \cdot \frac{du}{dx}$$

$$\frac{d}{dx}(\tan u) = \sec^2 u \cdot \frac{du}{dx} \qquad \frac{d}{dx}(\csc u) = -\csc u \cdot \cot u \cdot \frac{du}{dx}$$

The above formulas should be read with "function" in place of u. In most of the examples we will be dealing with, the "function" will be given by a formula. We should then use the above formulas to write the formulas for the derivatives in one step. It is helpful to use parentheses when doing this.

EXAMPLE 2
Find the derivatives.

(a) $\dfrac{d}{dx}(\sin(x^2 + 1))$,

(b) $\dfrac{d}{dx}(\tan(2x))$,

(c) $\dfrac{d}{dx}(\sec \sqrt{x})$.

SOLUTION

(a) $\sin(x^2 + 1)$ is the sine of $x^2 + 1$. The Chain Rule gives the derivative.

$$\frac{d}{dx}(\underbrace{\sin(x^2 + 1)}_{(\sin(\text{Function}))}) = (\underbrace{\cos(x^2 + 1)}_{(\cos(\text{Function}))})\ \underbrace{2x}_{\left(\substack{\text{Derivative} \\ \text{of function}}\right)}.$$

(b) $\tan(2x)$ is the tangent of the function $2x$. The Chain Rule gives the derivative:

$$\frac{d}{dx}(\underbrace{\tan(2x)}_{(\tan(\text{Function}))}) = (\underbrace{\sec^2(2x)}_{(\sec^2(\text{Function}))})\ \underbrace{(2)}_{\left(\substack{\text{Derivative} \\ \text{of function}}\right)}.$$

(c) $\sec \sqrt{x}$ is the secant of the square root of x. The Chain Rule gives the derivative:

$$\frac{d}{dx}(\sec \sqrt{x}) = (\sec \sqrt{x} \cdot \tan \sqrt{x})\left(\frac{1}{2\sqrt{x}}\right). \quad\blacksquare$$

Some problems require that we use the Chain Rule more than once, or that we use the Chain Rule with either or both of the Product and Quotient Rules.

EXAMPLE 3
Find the derivatives.

(a) $\dfrac{d}{dx}\left(x \sin \dfrac{1}{x}\right)$,

(b) $\dfrac{d}{dx}\left(\sqrt{\dfrac{1-x}{2x+1}}\right)$,

(c) $\dfrac{d}{dx}\left((1 + \sec^2(3x))^5\right)$.

SOLUTION

(a) We see that $x \sin(1/x)$ is the product of x and $\sin(1/x)$, so the Product Rule is used to find its derivative. Since $\sin(1/x)$ is the sine of the function $1/x$, the Chain Rule is used to differentiate this factor.

$$\frac{d}{dx}\left(\underbrace{x}_{\left(\substack{\text{First} \\ \text{factor}}\right)}\ \underbrace{\sin\frac{1}{x}}_{\left(\substack{\text{Second} \\ \text{factor}}\right)}\right)$$

$$= \underbrace{(x)}_{\left(\substack{\text{First} \\ \text{factor}}\right)}\ \underbrace{\left(\left(\cos\frac{1}{x}\right)(-x^{-2})\right)}_{\left(\substack{\text{Derivative} \\ \text{of second}}\right)} + \underbrace{\left(\sin\frac{1}{x}\right)}_{\left(\substack{\text{Second} \\ \text{factor}}\right)}\ \underbrace{(1)}_{\left(\substack{\text{Derivative} \\ \text{of first}}\right)}.$$

(b) The expression

$$\sqrt{\frac{1-x}{2x+1}} = \left(\frac{1-x}{2x+1}\right)^{1/2}$$

is the half-power of the quotient $(1-x)/(2x+1)$. To find the derivative we use the Chain Rule and then the Quotient Rule.

$$\frac{d}{dx}\left(\left(\frac{1-x}{2x+1}\right)^{1/2}\right)$$

$$= \frac{d}{dx}\left(\underbrace{\left(\frac{1-x}{2x+1}\right)}_{\text{(Function)}} \overset{\text{(Exponent } r)}{1/2}\right)$$

$$= \underbrace{\frac{1}{2}}_{(r)} \underbrace{\left(\frac{1-x}{2x+1}\right)}_{\text{(Function)}} \overset{-1/2}{\underset{\text{(Exponent } r-1)}{}} \underbrace{\left(\frac{(2x+1)(-1)-(1-x)(2)}{(2x+1)^2}\right)}_{\text{(Derivative of function)}}.$$

(c) This example requires the Chain Rule more than once. We will take the derivative one step at a time. We first note that $(1+\sec^2(3x))^5$ is the fifth power of the function $1+\sec^2(3x)$. The Chain Rule tells us that

$$\frac{d}{dx}((1+\sec^2(3x))^5) = 5(1+\sec^2(3x))^4\frac{d}{dx}(1+\sec^2(3x)).$$

We now read $1+\sec^2(3x)$ as the sum $1+(\sec(3x))^2$. The second term is the square of the function $\sec(3x)$, so we use the Chain Rule to find the derivative of that term. That is,

$$\frac{d}{dx}(1+\sec^2(3x)) = 0 + 2(\sec(3x))\frac{d}{dx}(\sec(3x)).$$

Since $\sec(3x)$ is the secant of the function $3x$, the Chain Rule is used to find the derivative.

$$\frac{d}{dx}(\sec(3x)) = (\sec(3x)\tan(3x))(3).$$

Combining results, we have

$$\frac{d}{dx}\left((1+\sec^2(3x))^5\right)$$

$$= 5(1+\sec^2(3x))^4\{0 + 2(\sec(3x))(\sec(3x)\tan(3x))(3)\}.$$

Let us note that the factors in the above derivative were determined in order from left to right. Also, each term is determined by the previous term. Hence, the factors could have been written in order on one line without writing the component derivatives on separate lines and combining the results. That is, the first factor of the derivative is $5(1+\sec^2(3x))^4$, which is the first term of the derivative of (Function)^5, where Function $= 1+\sec^2(3x)$. After we have written this term, we write $0 + 2\sec(3x)$, which is the derivative of the constant function 1 plus the first term of the derivative (Function)^2, where Function =

$\sec(3x)$. We then multiply $2\sec(3x)$ by $\sec(3x)\tan(3x)$, which is the first term of the derivative of $\sec(\text{Function})$, where Function $= 3x$. The chain stops with the derivative of $3x$, which we know is 3 and does not require the Chain Rule. Pictorially, we have

$$\frac{d}{dx}\overbrace{\left((1+\sec^2(3x))^5\right)}^{\binom{\text{Derivative of}}{(\text{Function})^5}}$$

$$= \underbrace{5(1+\sec^2(3x))^4}_{\binom{\text{First term of derivative}}{\text{of }(\text{Function})^5,\ \text{Function}=1+\sec^2(3x)}} \times$$

$$\left\{ \underbrace{0}_{\binom{\text{Derivative}}{\text{of }1}} + \underbrace{2(\sec(3x))}_{\binom{\text{First term}}{\text{of derivative of}\ (\text{Function})^2,\ \text{Function}=\sec(3x)}} \underbrace{(\sec(3x)\tan(3x))}_{\binom{\text{First term}}{\text{of derivative of}\ \sec\ (\text{Function}),\ \text{Function}=3x}} \underbrace{(3)}_{\binom{\text{Derivative of}}{\text{Function},\ \text{Function}=3x}} \right\}.$$

EXAMPLE 4

Find all points on the graph of $f(x) = x(1-x)^{1/3}$ for which (a) $f'(x) = 0$ and (b) $f'(x)$ does not exist.

SOLUTION $f(x)$ is the product of x and $(1-x)^{1/3}$, so we use the Product Rule to find $f'(x)$. Since $(1-x)^{1/3}$ is a power of a function, we use the Chain Rule to find the derivative of this factor. We have

$$f(x) = x(1-x)^{1/3}, \quad \text{so}$$

$$f'(x) = (x)\left(\frac{1}{3}(1-x)^{-2/3}(-1)\right) + ((1-x)^{1/3})(1).$$

We simplify by factoring $(1-x)^{-2/3}/3$ from each term. We have

$$f'(x) = -\frac{x(1-x)^{-2/3}}{3} + \frac{3(1-x)^{1-(2/3)}}{3}$$

$$= \left(\frac{(1-x)^{-2/3}}{3}\right)(-x + 3(1-x))$$

$$= \frac{3-4x}{3(1-x)^{2/3}}.$$

(a) From the simplified form of $f'(x)$, we see that $f'(x) = 0$ when $x = 3/4$. When $x = 3/4$, $y = f(3/4) = (3/4)(1/4)^{1/3} = 3(1/4)^{4/3}$, or $y \approx 0.47$. Then $(3/4, 3(1/4)^{4/3})$ is a point on the graph of f with $f'(x) = 0$. The graph has a horizontal tangent at that point. See Figure 3.4.1.

(b) We see that $f'(x)$ does not exist when $x = 1$. When $x = 1$, $y = f(1) = 0$. The point $(1, 0)$ is a point on the graph where $f'(x)$ does not exist. The graph passes through the point $(1, 0)$ vertically. See Figure 3.4.1. ∎

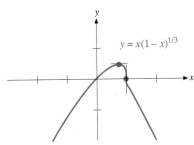

$y = x(1-x)^{1/3}$

FIGURE 3.4.1 The function f has a horizontal tangent at $(3/4, 3(1/4)^{4/3})$, where $f'(x) = 0$. The graph becomes vertical as it passes through the point $(1, 0)$, where $f'(x)$ does not exist.

EXAMPLE 5

$f(x) = x/\sqrt{x^2+1}$. Find $f''(x)$.

SOLUTION   We must first find f'. Using the Quotient Rule, we have

$$f'(x) = \frac{(\sqrt{x^2+1})(1) - (x)\left(\dfrac{1}{2\sqrt{x^2+1}}(2x)\right)}{(\sqrt{x^2+1})^2}.$$

It is worthwhile to simplify $f'(x)$ before calculating $f''(x)$.

We have

$$f'(x) = \frac{\sqrt{x^2+1} - \dfrac{x^2}{\sqrt{x^2+1}}}{x^2+1}$$

$$= \frac{\sqrt{x^2+1}\cdot\dfrac{\sqrt{x^2+1}}{\sqrt{x^2+1}} - \dfrac{x^2}{\sqrt{x^2+1}}}{x^2+1}$$

$$= \frac{(x^2+1) - x^2}{(x^2+1)^{3/2}}$$

$$= \frac{1}{(x^2+1)^{3/2}} = (x^2+1)^{-3/2}.$$

The Chain Rule then gives

$$f''(x) = -\frac{3}{2}(x^2+1)^{-5/2}(2x) = \frac{-3x}{(x^2+1)^{5/2}}. \quad \blacksquare$$

The Chain Rule can be used even if one of the functions involved is not completely known. It is only necessary that we know the values of the functions and derivatives at the appropriate points.

EXAMPLE 6

$V(t) = (s(t))^3$, where $s(2) = 4$ and $s'(2) = 5$. Find $V'(2)$.

SOLUTION   Using the Chain Rule to differentiate the power of the function s, we obtain

$$V'(t) = 3(s(t))^2(s'(t)).$$

When $t = 2$, we have

$$V'(2) = 3(s(2))^2(s'(2))$$
$$= 3(4)^2(5) = 240. \quad \blacksquare$$

EXAMPLE 7

$F(x) = f(3x-1)$, where $f'(5) = 2$. Find $F'(2)$.

SOLUTION   The function F is the composition of the function f and the function $3x - 1$. The Chain Rule gives the derivative

$$F'(x) = f'(3x-1)\cdot\frac{d}{dx}(3x-1)$$

$$= f'(3x-1)\cdot(3).$$

Substituting $x = 2$ and $f'(5) = 2$, we obtain

$$F'(2) = f'(3(2) - 1) \cdot (3)$$
$$= f'(5) \cdot (3)$$
$$= (2)(3) = 6. \quad \blacksquare$$

EXERCISES 3.4

Evaluate the derivatives in Exercises 1–24. Use parentheses to indicate each factor and then simplify your answers.

1. $\dfrac{d}{dx}((7x + 1)^6)$ **(Calculus Explorer & Tutor I, Differentiation, 5.)**

2. $\dfrac{d}{dx}((3x^2 - 4x + 5)^7)$

3. $\dfrac{d}{dx}((2x^3 - 1)^{-2/3})$

4. $\dfrac{d}{dx}((x^2 - 2x + 4)^{-1/3})$

5. $\dfrac{d}{dx}\left(\dfrac{1}{2\sin x + 3}\right)$

6. $\dfrac{d}{dx}\left(\dfrac{1}{3\sec x - 1}\right)$

7. $\dfrac{d}{dx}(\sin(x^3 - x))$

8. $\dfrac{d}{dx}(\sin \sqrt{x})$ **(Calculus Explorer & Tutor I, Trigonometric Differentiation, 4.)**

9. $\dfrac{d}{dx}(\sqrt{\sin x})$ **(Calculus Explorer & Tutor I, Trigonometric Differentiation, 6.)**

10. $\dfrac{d}{dx}(\cos(1 - x))$

11. $\dfrac{d}{dx}\left(\sec\dfrac{1}{x}\right)$

12. $\dfrac{d}{dx}\left(\csc\dfrac{1}{x^2}\right)$

13. $\dfrac{d}{dx}(x\sqrt{x^2 + 1})$

14. $\dfrac{d}{dx}(x^2(1 + 2x)^3)$

15. $\dfrac{d}{dx}\left(\left(\dfrac{x}{1 + x^2}\right)^7\right)$ **(Calculus Explorer & Tutor I, Differentiation, 14.)**

16. $\dfrac{d}{dx}\left(\sqrt{\dfrac{1 - x^2}{1 + x^2}}\right)$ **(Calculus Explorer & Tutor I, Differentiation, 11.)**

17. $\dfrac{d}{dx}\left(\dfrac{\sin^2 x}{1 + \tan^4 x}\right)$ **(Calculus Explorer & Tutor I, Trigonometric Differentiation, 18.)**

18. $\dfrac{d}{dx}\left(\dfrac{\tan x}{\sqrt{\sec^2 x + 1}}\right)$

19. $\dfrac{d}{dx}(\tan^2(2x))$

20. $\dfrac{d}{dx}((\sin 2x)(\cos 2x))$

21. $\dfrac{d}{dx}(\sqrt{x^2 + \sin^2 x})$

22. $\dfrac{d}{dx}(\sqrt{1 + \cos^2 x})$

23. $\dfrac{d}{dx}(\sqrt{x + \sqrt{x + x^2}})$

24. $\dfrac{d}{dx}([1 + (1 + x^2)^2]^2)$

In Exercises 25–32 find the equation of the line tangent to the graph of $y = f(x)$ at the point on the graph with $x = c$.

25. $f(x) = \sqrt{\dfrac{x^3}{x + 3}}, c = 1$

26. $f(x) = \left(\dfrac{x}{x^2 + 16}\right)^{2/3}, c = 4$

27. $f(x) = \dfrac{1}{\sqrt{x^2 + 9}}, c = 4$

28. $f(x) = \dfrac{1}{3x - 2}, c = 2$

29. $f(x) = \cot^2 x, c = \pi/4$

30. $f(x) = \csc^2 x, c = \pi/3$

31. $f(x) = x - \sin(2x), c = \pi/2$

32. $f(x) = x + \cos(2x), c = \pi/4$

In Exercises 33–38 find all points $(x, f(x))$ on the graph of f for which (a) $f'(x) = 0$ and (b) $f'(x)$ does not exist.

33. $f(x) = x(3x - 4)^{1/3}$

34. $f(x) = x(3x - 20)^{2/3}$

35. $f(x) = \dfrac{x}{(6x + 4)^{2/3}}$

36. $f(x) = \dfrac{x}{(3x - 2)^{2/3}}$

37. $f(x) = x - \tan\dfrac{x}{2}, 0 < x < \pi$

38. $f(x) = \sin x - \sin^2 x, 0 \le x \le 2\pi$

39. $F(t) = f(4t + 3)$, where $f'(7) = 5$. Find $F'(1)$.

40. $F(t) = f(t^2 + 1)$, where $f'(5) = 3$. Find $F'(2)$.

41. $F(t) = \sqrt{g(t)}$, where $g(2) = 9$ and $g'(2) = 4$. Find $F'(2)$.

42. $F(t) = (g(t))^{1/3}$, where $g(3) = 8$ and $g'(3) = 6$. Find $F'(3)$.

43. $F(x) = f(a + bx) + f(a - bx)$, where f is differentiable at a. Find $F'(0)$.

44. Find $f'(3)$ if $f(t) = u(x(t)), x(3) = 2, u(2) = 5, x'(3) = -1, u'(3) = 4, u'(2) = 6$, and $x(2) = 10$. **(Calculus Explorer & Tutor I, Differentiation, 12.)**

45. Use the formula $|x| = \sqrt{x^2}$ and the Chain Rule to show that

$$\frac{d}{dx}(|x|) = \frac{x}{|x|}, \qquad x \neq 0.$$

The Chain Rule then implies

$$\frac{d}{dx}(|u|) = \frac{u}{|u|}\frac{du}{dx}, \qquad u \neq 0.$$

46. Find the equation of the line tangent to the graph of $f(x) = |1 - x^2|$ at the point where $x = 2$.

47. Find the equation of the line tangent to the graph of $f(x) = |\sin(5x)|$ at the point where $x = \pi/4$.

48. Find the equation of the line tangent to the graph of $f(x) = |(1-x)^{1/3}|$ at the point where $x = 9$.

In Chapter 7 we will introduce a function L with derivative $L'(x) = 1/x$. Use this formula to evaluate the derivatives in Exercises 49–52.

49. $\dfrac{d}{dx}(L(x^2))$

50. $\dfrac{d}{dx}(L(3x))$

51. $\dfrac{d}{dx}(L(x^2 + 1))$

52. $\dfrac{d}{dx}(L(\sec x + \tan x))$

53. The radius of a balloon is observed to be increasing at a rate of 3 in./s when the radius is 2 in. The radius is increasing at a rate of 1.5 in./s when the radius is 3 in. Is the rate of change of the volume of the balloon when the radius is 2 in. greater than, less than, or equal to the rate of change of the volume when the radius is 3 in.? (*Hint:* $V = 4\pi r^3/3$, where the volume and radius are functions of time. Differentiation of this equation with respect to time gives an equation that relates the rates of change of the volume and the radius.)

54. Find the rate of change of the volume of a weather balloon when its radius is 10 cm if its radius is increasing at a rate of 0.3 cm/s at that time. $V = 4\pi r^3/3$, where V and r are functions of time.

55. If a balloon is rising vertically from a point on the ground that is 50 ft from a ground-level observer, the height of the balloon and the angle of elevation from the observer to the balloon are related by the equation $h = 50\tan\theta$, where h and θ are functions of time. The angle of elevation is observed to be increasing at a rate of $2°$/s when the angle of elevation is $45°$. Find the rate at which the balloon is rising at that time. (*Hint:* Angles must be expressed in radian measure in order to apply the formulas for the differentiation of the trigonometric functions.)

56. The electrical resistance R of a wire 100 m long varies with its radius r according to

$$R = \frac{0.001}{r^2}.$$

The radius varies with the absolute temperature T according to

$$r = 0.2000 + 0.0000003T.$$

What is the rate of change of R with respect to T when $T = 300°$?

57. In Chapter 7 we will introduce a function E with $E(0) = 1$ and $E'(x) = E(x)$. Use these formulas to show that the line tangent to the graph of $y = E(ax)$ at the point where the graph intersects the y-axis has slope a.

58. Show that

$$\frac{d^2}{dx^2}(f(g(x))) = f'(g(x))g''(x) + f''(g(x))(g'(x))^2.$$

59. If A, B, and ω are constants, show that $y = A\sin\omega x + B\cos\omega x$ satisfies the equation $y'' + \omega^2 y = 0$.

60. The displacement from equilibrium at time t of a particle moving along a line with **simple harmonic motion** is

$$s(t) = A\cos(\omega t + \delta).$$

Let s_0 be the displacement when $t = 0$ and let v_0 be the velocity when $t = 0$. Show that (a) $\tan\delta = -\dfrac{v_0}{\omega s_0}$ and (b) $A^2 = s(t)^2 + \left(\dfrac{v(t)}{\omega}\right)^2$, where $v(t)$ is the velocity at time t.

61. If the velocity of a particle moving on a number line is proportional to the square root of its position, so $v = k\sqrt{s}$, show that the acceleration is constant.

62. The period of a simple pedulum of length L is $T = 2\pi\sqrt{\dfrac{L}{g}}$. If the rate of change of the length with respect to temperature is proportional to the length, so $\dfrac{dL}{d\tau} = kL$, show that the rate of change of the period with respect to temperature is proportional to the period.

63. If $P(x)$ is a polynomial of the form

$$P(x) = a_0 + a_1(x - c) + a_2(x - c)^2 + \cdots + a_n(x - c)^n,$$

show that

$$a_j = \frac{P^{(j)}(c)}{j!}, \qquad j = 0, 1, 2, \ldots, n.$$

(*Hint:* Use the fact that positive powers of $x - c$ are zero when $x = c$ to evaluate $P(c), P'(c), P''(c), \ldots, P^{(n)}(c)$.)

64. If $P(x) = a_0 + a_1 x^2 + \cdots + a_n x^n$ is a polynomial and $P'(x) = 0$ for all x, show that $P(x) = a_0$, a constant. (*Hint:* Use the result of Exercise 63 and the fact that higher-order derivatives of P are derivatives of the constant function $P'(x) = 0$ and, hence, are zero.)

65. $f(x) = \begin{cases} x^2\sin\left(\dfrac{1}{x}\right), & x \neq 0, \\ 0, & x = 0. \end{cases}$

(a) Find a formula for $f'(x)$, $x \neq 0$.
(b) Does $f'(x)$ approach a limit as x approaches zero?
(c) Is f differentiable at zero?
(d) Is f' continuous at zero?

66. $f(x) = \begin{cases} x \sin\left(\dfrac{1}{x}\right), & x \neq 0, \\ 0, & x = 0. \end{cases}$

    (a) Find a formula for $f'(x)$, $x \neq 0$.

    (b) Does $f'(x)$ approach a limit as x approaches zero?

    (c) Is f differentiable at zero?

    (d) Is f' continuous at zero?

67. $f(x) = \begin{cases} x^2 \sin\left(\dfrac{1}{x^2}\right), & x \neq 0, \\ 0, & x = 0. \end{cases}$

Show that $f'(x)$ becomes unbounded as x approaches zero, but $f'(0) = 0$.

3.5   IMPLICIT DIFFERENTIATION

Connections

Concept of function, 2.1

Rules of differentiation, especially the Chain Rule, 3.1–4.

It is possible to evaluate the derivatives of some functions even though we do not have a formula for their values.

Functions Defined Implicitly

Recall from Section 2.1 that a graph defines y as a function of x if every vertical line that intersects the graph intersects it at exactly one point. Various parts of a graph may define functions of x, even though the entire graph does not. For example, the graph in Figure 3.5.1 does not define y as a function of x. However, if we restrict our attention to any one of the rectangles in Figure 3.5.2, the graph does define y as a function of x.

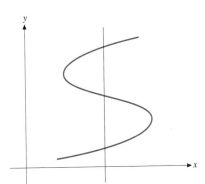

FIGURE 3.5.1 This graph does not define y as a function of x, because some vertical lines intersect the graph at more than one point.

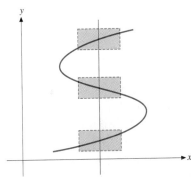

FIGURE 3.5.2 The graph defines y as a function of x inside each of the indicated rectangles, since each vertical line that intersects any one of the rectangles intersects the graph at exactly one point inside that rectangle.

DEFINITION

A graph **defines y as a function of x in a rectangle with center** (x_0, y_0) on the graph if each vertical line that intersects the rectangle intersects the graph at exactly one point (x, y) inside the rectangle. In that case, y is the value of the function at x.

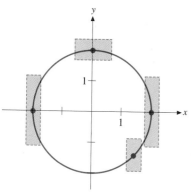

FIGURE 3.5.3 The graph defines y as a function of x in a rectangle with center (x_0, y_0) at every point on the graph, except for $(-2, 0)$ and $(2, 0)$.

EXAMPLE 1

Find all points (x_0, y_0) on the graph of the equation $x^2 + y^2 = 4$ such that the graph defines y as a function of x in some rectangle with center (x_0, y_0).

SOLUTION   The graph of the equation $x^2 + y^2 = 4$ is a circle withcenter $(0, 0)$ and radius 2. We see from Figure 3.5.3 that the graph defines y as a function of x in some rectangle with center (x_0, y_0) at every point (x_0, y_0) on the graph, except for $(2, 0)$ and $(-2, 0)$. Different points (x_0, y_0) may require rectangles of different dimensions to satisfy the condition of the definition. No rectangle with center at either $(2, 0)$ or $(-2, 0)$ can satisfy the desired condition, since some vertical lines that intersect the rectangle would intersect the graph at two points within the rectangle and some vertical lines that intersect the rectangle would not intersect the graph at any point. ∎

Equations of the form $y = f(x)$ give y **explicitly** as a function of x. Various parts of the graph of an equation in x and y may define y as a function of x, even though we cannot solve the equation for y and find a formula for the function. Such functions are said to be given **implicitly** by the equation. When these functions are differentiable, it is possible to find their derivatives without solving the equation for y. The method is called **implicit differentiation.**

Implicit Differentiation

The **Method of Implicit Differentiation** is used to find derivatives of functions that are defined implicitly by an equation.

Method of implicit differentiation

- Assume an equation defines y as a differentiable function of x.
- Differentiate both sides of the equation with respect to x. Remember that y is a function of x. (The equation will ordinarily involve a function of y, so the Chain Rule will be required for this differentiation.)
- Solve the differentiated equation for dy/dx in terms of x and y. (This is always easy to do, because the differentiated equation is linear in dy/dx.)

EXAMPLE 2

Use the Method of Implicit Differentiation to find dy/dx for the functions of x defined implicitly by the equation $x^2 + y^2 = 4$.

SOLUTION   We assume the equation $x^2 + y^2 = 4$ defines y as a differentiable function of x. The equal sign in the equation tells us both sides of the equation represent the same function of x. Since the derivatives of equal functions are equal, we can obtain another equation by differentiating both sides of the equation with respect to x.

$$x^2 + y^2 = 4,$$

$$\frac{d}{dx}(x^2 + y^2) = \frac{d}{dx}(4),$$

$$\frac{d}{dx}(x^2) + \frac{d}{dx}(y^2) = \frac{d}{dx}(4).$$

We know that

$$\frac{d}{dx}(x^2) = 2x \text{ and } \frac{d}{dx}(4) = 0.$$

Since y is a function of x, y^2 is of the form (Function)². The Chain Rule then gives

$$\frac{d}{dx}(y^2) = (2y)\frac{dy}{dx}.$$

Substitution in the above formula then gives

$$2x + (2y)\frac{dy}{dx} = 0 \text{ or } \frac{dy}{dx} = -\frac{x}{y}. \quad \blacksquare$$

In Example 2 we can find formulas for the implicitly defined functions by solving the equation $x^2 + y^2 = 4$ for y. We obtain

$$y = \sqrt{4 - x^2} \text{ and } y = -\sqrt{4 - x^2}.$$

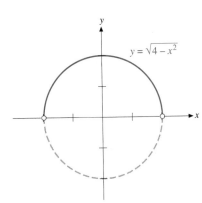

FIGURE 3.5.4 The graph of $y = \sqrt{4 - x^2}$ is the upper half of the graph of the circle $x^2 + y^2 = 4$.

The function defined by the upper part of the graph ($y > 0$) is given by the formula $y = \sqrt{4 - x^2}$. See Figure 3.5.4. Using the Chain Rule, we see that the derivative of this function is

$$\frac{dy}{dx} = \frac{1}{2\sqrt{4 - x^2}}(-2x) = -\frac{x}{\sqrt{4 - x^2}}.$$

Since $y = \sqrt{4 - x^2}$, we can write $dy/dx = -x/y$. Similarly, the function defined by the lower part of the graph ($y < 0$) is given by the formula $y = -\sqrt{4 - x^2}$. See Figure 3.5.5. The derivative is

$$\frac{dy}{dx} = -\frac{1}{2\sqrt{4 - x^2}}(-2x) = -\frac{x}{-\sqrt{4 - x^2}}.$$

Since $y = -\sqrt{4 - x^2}$, we can write $dy/dx = -x/y$. Thus, the formula $dy/dx = -x/y$ holds in either case, as we had previously discovered by using the (easier) method of Implicit Differentiation.

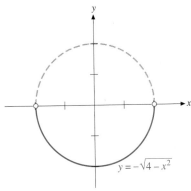

FIGURE 3.5.5 The graph of $y = -\sqrt{4 - x^2}$ is the lower half of the graph of the circle $x^2 + y^2 = 4$.

NOTE *When we are using the method of Implicit Differentiation it is important to remember that y is a function of x and that we are differentiating with respect to x. Hence, powers of y are powers of functions and the Chain Rule is required for their differentiation. We should also be prepared to use combinations of the Chain Rule, Product Rule, and Quotient Rule.*

EXAMPLE 3
Use the Method of Implicit Differentiation to find dy/dx for the functions of x defined implicitly by the equation $x^{2/3} + y^{2/3} = 1$.

SOLUTION We assume y is a differentiable function of x. Differentiating the equation

$$x^{2/3} + y^{2/3} = 1$$

with respect to x, and using the Chain Rule to find the derivative of the 2/3 power of the function y, we have

$$\frac{2}{3}x^{-1/3} + \frac{2}{3}y^{-1/3}\frac{dy}{dx} = 0.$$

(The above expression does not make sense if either $x = 0$ or $y = 0$.) Solving for dy/dx, we have

$$\frac{2}{3}y^{-1/3}\frac{dy}{dx} = -\frac{2}{3}x^{-1/3},$$

$$\frac{dy}{dx} = -\frac{x^{-1/3}}{y^{-1/3}},$$

$$\frac{dy}{dx} = -\frac{y^{1/3}}{x^{1/3}} = -\left(\frac{y}{x}\right)^{1/3}.$$

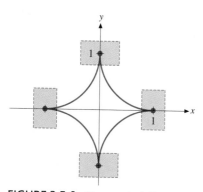

FIGURE 3.5.6   The graph defines y as a function of x at every point on the graph, except for $(1, 0)$ and $(-1, 0)$. The functions with graphs through $(0, 1)$ and $(0, -1)$ are not differentiable at $x = 0$.

From Figure 3.5.6 we see that the graph of the equation $x^{2/3} + y^{2/3} = 1$ does not define y as a function of x in a rectangle with center at either of the points $(1, 0)$ or $(-1, 0)$ on the graph. The graph defines y as a function of x in rectangles with centers $(0, 1)$ and $(0, -1)$, but the functions are not differentiable at $x = 0$. ∎

EXAMPLE 4

Use the Method of Implicit Differentiation to find dy/dx for the functions of x defined implicitly by the equation $\cos(x + y) = x$.

SOLUTION   We assume y is a differentiable function of x. Differentiating the equation

$$\cos(x + y) = x$$

with respect to x, and using the Chain Rule to differentiate the cosine of the function $x + y$, we have

$$-\sin(x + y) \cdot \left(1 + \frac{dy}{dx}\right) = 1.$$

Solving for dy/dx, we have

$$-\sin(x + y) - \sin(x + y) \cdot \frac{dy}{dx} = 1,$$

$$-\sin(x + y) \cdot \frac{dy}{dx} = 1 + \sin(x + y),$$

$$\frac{dy}{dx} = \frac{1 + \sin(x + y)}{-\sin(x + y)}.$$

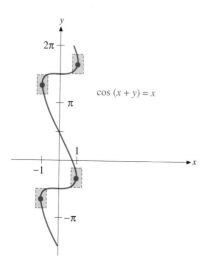

FIGURE 3.5.7   The graph does not define y as a function of x at points with $x = \pm 1$.

We see that the formula for dy/dx does not make sense whenever $\sin(x + y) = 0$—that is, whenever $x + y = k\pi, k = 0, \pm 1, \pm 2, \ldots$. For those values of (x, y), the defining equation $\cos(x + y) = x$ becomes $\cos(k\pi) = x$, so x must be 1 when k is even and -1 when k is odd. The graph becomes vertical and the equation does not define y as a function of x in rectangles with centers at the corresponding points on the graph. See Figure 3.5.7. ∎

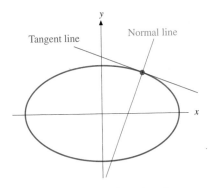

FIGURE 3.5.8   The line that is perpendicular to the line tangent to the graph at a point on the graph is called the line normal to the graph at the point.

Normal Lines

For some applications of calculus, we are interested in the direction that is perpendicular to the line that is tangent to the graph of an equation in x and y at a point (x_0, y_0) on the graph. The line that is perpendicular to the tangent line is called the **line normal to the graph** at the point (x_0, y_0). See Figure 3.5.8. We know that the slope of the line tangent to the graph at a point (x_0, y_0) is the value of the derivative dy/dx at the point.

If dy/dx is nonzero at (x_0, y_0), the line normal to the graph at (x_0, y_0) has slope that is the negative reciprocal of dy/dx.

If dy/dx is zero at (x_0, y_0), the line normal to the graph is vertical and has equation $x = x_0$.

EXAMPLE 5

Find the equation of the line normal to the graph of $x^3 + y^3 - 4.5xy = 0$ at the point $(2, 1)$ on the graph.

SOLUTION   We need the value of the derivative dy/dx at the point $(2, 1)$. It is not convenient to solve the equation for y in terms of x, so we use the Method of Implicit Differentiation to find dy/dx. We assume the equation defines y as a differentiable function of x and differentiate the equation

$$x^3 + y^3 - 4.5xy = 0$$

with respect to x. Using the Chain Rule to differentiate the third power of the function y and the Product Rule to differentiate the product xy, we obtain

$$3x^2 + 3y^2\frac{dy}{dx} - 4.5\left((x)\left(\frac{dy}{dx}\right) + (y)(1)\right) = 0.$$

Solving for dy/dx, we have

$$3x^2 + 3y^2\frac{dy}{dx} - 4.5x\frac{dy}{dx} - 4.5y = 0,$$

$$(3y^2 - 4.5x)\frac{dy}{dx} = 4.5y - 3x^2,$$

$$\frac{dy}{dx} = \frac{4.5y - 3x^2}{3y^2 - 4.5x} = \frac{1.5(3y - 2x^2)}{1.5(2y^2 - 3x)} = \frac{3y - 2x^2}{2y^2 - 3x}.$$

When $(x, y) = (2, 1)$,

$$\frac{dy}{dx} = \frac{3(1) - 2(2)^2}{2(1)^2 - 3(2)} = \frac{-5}{-4} = \frac{5}{4}.$$

The slope of the line normal to the graph at the point $(2, 1)$ is then $-1/(5/4) = -4/5$. The point-slope form of the equation of the line through $(2, 1)$ with slope $-4/5$ is

$$y - 1 = -\frac{4}{5}(x - 2).$$

Simplifying, we obtain

$$5(y - 1) = -4(x - 2),$$
$$5y - 5 = -4x + 8,$$
$$4x + 5y = 13.$$

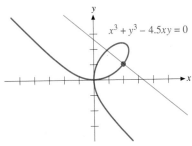

FIGURE 3.5.9   The line normal to the graph at the point $(2, 1)$ has slope $-4/5$, which is the negative reciprocal of the value of $\frac{dy}{dx}$ at that point.

The graph of the equation and the normal line are sketched in Figure 3.5.9. ∎

Other Variables

The Method of Implicit Differentiation can be used to find the derivative of any variable with respect to another from an equation that relates the two variables. We must distinguish between the independent variable of differentiation and the dependent variable that is considered to be the function.

EXAMPLE 6

If a rocket is rising vertically from a point on the ground that is 200 ft from an observer at ground level, the height of the rocket and the angle of elevation from the observer to the rocket are related by the formula $h = 200 \tan \theta$. See Figure 3.5.10. Find $d\theta/dh$, the rate of change of the angle of elevation with respect to the height, when the height is 600 ft.

SOLUTION   We first note that we want the derivative of the *function* θ with respect to the *independent variable* h. We assume that the equation $h = 200 \tan \theta$ defines θ as a differentiable function of h and differentiate the equation with respect to h. Since $\tan \theta$ is the tangent of a function, we need to use the Chain Rule to differentiate that term. We have

FIGURE 3.5.10   The angle of elevation θ is related to the height h by the equation $h = 200 \tan \theta$.

$$h = 200 \tan \theta,$$
$$1 = 200 \sec^2 \theta \frac{d\theta}{dh},$$
$$\frac{d\theta}{dh} = \frac{\cos^2 \theta}{200}.$$

When $h = 600$, we have

$$600 = 200 \tan \theta, \text{ so } \tan \theta = 3.$$

The exact value of $\cos \theta$ can then be determined from the representative triangle in Figure 3.5.11, or a calculator can be used to find an approximate value. We obtain $\cos \theta = 1/\sqrt{10}$, so

$$\frac{d\theta}{dh} = \frac{(1/\sqrt{10})^2}{200} = 0.0005 \text{ rad/ft}.$$

Recall that the formulas we use for the differentiation of trigonometric functions hold only if the angles are measured in radians. ∎

FIGURE 3.5.11   We can use a representative right triangle to determine that an acute angle θ with $\tan \theta = 3$ satisfies $\cos \theta = 1/\sqrt{10}$.

Higher-Order Derivatives

We can find higher-order derivatives of functions that are defined implicitly.

To evaluate second derivatives of implicit functions
• Evaluate $\dfrac{dy}{dx}$ in terms of x and y.   • Differentiate the equation for $\dfrac{dy}{dx}$ with respect to x to obtain $\dfrac{d^2 y}{dx^2} = \dfrac{d}{dx}\left(\dfrac{dy}{dx}\right)$ in terms of x, y, and $\dfrac{dy}{dx}$. Remember that y is a function of x and you are differentiating with respect to x.   • Replace $\dfrac{dy}{dx}$ with its value in terms of x and y to obtain $\dfrac{d^2 y}{dx^2}$ in terms of x and y.

EXAMPLE 7

$y^2 - 2xy = 1$. Find formulas for dy/dx and d^2y/dx^2 in terms of x and y.

SOLUTION   We assume the equation defines y as a differentiable function of x. Differentiating the equation $y^2 - 2xy = 1$ with respect to x and solving the resulting equation for dy/dx, we have

$$2y\frac{dy}{dx} - 2\left((x)\left(\frac{dy}{dx}\right) + (y)(1)\right) = 0,$$

$$2y\frac{dy}{dx} - 2x\frac{dy}{dx} - 2y = 0,$$

$$(2y - 2x)\frac{dy}{dx} = 2y,$$

$$\frac{dy}{dx} = \frac{y}{y-x}.$$

We can obtain d^2y/dx^2 by differentiating the above equation with respect to x. We must remember that both y and dy/dx are functions of x. Using the Quotient Rule, we have

$$\frac{d^2y}{dx^2} = \frac{(y-x)\left(\frac{dy}{dx}\right) - (y)\left(\frac{dy}{dx} - 1\right)}{(y-x)^2}.$$

Simplifying, we have

$$\frac{d^2y}{dx^2} = \frac{y\frac{dy}{dx} - x\frac{dy}{dx} - y\frac{dy}{dx} + y}{(y-x)^2}$$

$$= \frac{y - x\frac{dy}{dx}}{(y-x)^2}.$$

Using the formula $dy/dx = y/(y-x)$ found above, we can express the second derivative in terms of x and y.

$$\frac{d^2y}{dx^2} = \frac{y - x\left(\frac{y}{y-x}\right)}{(y-x)^2}$$

$$= \frac{y\frac{y-x}{y-x} - \frac{xy}{y-x}}{(y-x)^2}$$

$$= \frac{y(y-x) - xy}{(y-x)^3}$$

$$= \frac{y^2 - 2xy}{(y-x)^3}.$$

Since the original equation is $y^2 - 2xy = 1$, we can further simplify to obtain

$$\frac{d^2y}{dx^2} = \frac{1}{(y-x)^3}. \blacksquare$$

EXERCISES 3.5

In Exercises 1–8 find all points (x_0, y_0) on the graphs where the graphs define y as a function of x in a rectangle with center (x_0, y_0).

1.

2.

3.

4.

5.

6.

7.

8.

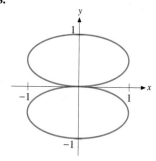

Use the method of Implicit Differentiation to find dy/dx for the functions of x that are defined implicitly by the equations in Exercises 9–22.

9. $x^2 - y^2 = 4$ **10.** $y^2 - x^2 = 49$

11. $y^3 = x$ **12.** $y^3 = x^2$

13. $\sqrt{x} + \sqrt{y} = 4$

14. $x^{1/3} + y^{1/3} = 2$

15. $\sin(x + y) = x$

16. $\sin(x + y) = y$

17. $xy^2 - yx^2 = 16$

18. $xy^3 - yx^3 = 4$

19. $x^3 + y^3 - 3xy = 0$

20. $2x^2y - 3x^4y^2 = -1$ **(Calculus Explorer & Tutor I, Differentiation, 19.)**

21. $(\sin x)(\cos y) = y$

22. $(\sin x)(\cos y) = x$

In Exercises 23–28 find the equation of the line tangent to the graph at the given point.

23. $y^2 = 2x^3$, $(2, -4)$

24. $2y^3 = x^2$, $(-4, 2)$

25. $(x + y)^3 = x + y + 6$, $(3, -1)$

26. $\sqrt{x + y} = x + y - 6$, $(5, 4)$

27. $xy^3 - x^3y = 30$, $(2, 3)$

28. $xy^2 + yx^2 = 6$, $(2, -3)$

In Exercises 29–34 find the equation of the line normal to the graph at the given point.

29. $\dfrac{1}{x^2} + \dfrac{1}{y^2} = 20$, $\left(\dfrac{1}{2}, -\dfrac{1}{4}\right)$

30. $\dfrac{1}{x} + \dfrac{1}{y} = 1$, $\left(\dfrac{1}{4}, -\dfrac{1}{3}\right)$

31. $x = \sin y$, $(1/2, \pi/6)$

32. $x = \sec y$, $(2, \pi/3)$

33. $x = \tan y$, $(1, \pi/4)$

34. $x = y - \cos y$, $(\pi/2, \pi/2)$

35. Show that the line normal to the circle $x^2 + y^2 = r^2$ at any point (x_0, y_0) on the circle, passes through the origin.

36. Find the equation of each line that contains the origin and is tangent to the circle $(x - 4)^2 + y^2 = 4$.

37. The volume V and the surface area S of a sphere are related by the equation $36\pi V^2 = S^3$. Find dS/dV, the rate of change of the surface area with respect to the volume when the volume is $\pi\sqrt{6}\,\text{ft}^3$.

38. The volume V of water in a certain conical tank is given by the equation $V = \pi h^3/3$, where h is the depth of the water above the vertex of the cone. Find dh/dV, the rate

of change of the depth with respect to the volume when the volume is $9\pi\,\text{ft}^3$.

39. Find and compare dv/du and du/dv, if $u = 1 + 2v + 3v^2$.

40. Find and compare ds/dr and dr/ds, if $r = 4 - s + 2s^2$.

In Exercises 41–46 use the Method of Implicit Differentiation to find formulas for dy/dx and d^2y/dx^2 in terms of x and y.

41. $x^2 + y^2 = 4$

42. $y^3 = x^2$

43. $y^2 - 2xy + 4 = 0$

44. $\dfrac{1}{x} + \dfrac{1}{y} = 1$

45. $x = \tan y$

46. $x = \sin y$

In Exercises 47–50, (a) sketch the graphs of the given equations for several values of the constants C and C'. (Graphs of the first equation intersect graphs of the second equation orthogonally at each point of intersection.) (b) Find the slope of the tangent to each curve at a point (x, y) on the curve, in terms of x and y. (Slopes from the first equation are negative reciprocals of the slopes from the second equation at each point of intersection.)

47. $y = x^3 + C$; $y = \dfrac{1}{3x} + C'$

48. $x^2 + y^2 = C$; $y = C'x$

49. $\dfrac{x^2}{2} + y^2 = C$; $y = C'x^2$

50. $\dfrac{x^2}{3} + y^2 = C$; $y = C'x^3$

In Exercises 51–52 graph the given equations for several values of the constants C and C'. Observe that graphs of the first equation intersect graphs of the second equation orthogonally at each point of intersection.

51. $y = \sin x + C$; $y = -\ln|\sec x + \tan x| + C'$

52. $y = \tan^{-1} x + C$; $y = -x - x^3/3 + C'$

53. **Bessel's function of order zero,** which we denote $J(x)$, is continuous and has continuous derivatives of all orders for all x. It satisfies $xJ''(x) + J'(x) + xJ(x) = 0$ and $J(0) = 1$.
 (a) Find the value of $J'(0)$.
 (b) Use the method of implicit differentiation to find $J''(0)$.

3.6 LINEAR APPROXIMATION; DIFFERENTIALS

Connections

Function notation, 2.1.

Equation of tangent line, 2.6.

Rules of differentiation, 3.1–5.

You are already familiar with applications of the mathematical idea of approximation of values of functions. For example, when you use a calculator to determine the value of $\sqrt{2}$, the value given by your calculator is not the exact value, but an approximate value of the function $\sqrt{x}$ for $x = 2$. In this section, we will illustrate one of the most simple and important types of approximation methods, **linear approximation.** The idea is to use the values of a function and its derivative at a single point to find approximate values of the function at nearby points.

Approximation by Tangent Lines

Let us assume that f is differentiable at c, and that we are interested in approximate values of $f(x)$ for x near c. We know from Section 2.6 that the graph of f passes through the point $(c, f(c))$ in the same direction as the line tangent to the graph at that point. We also note that the tangent line is the graph of the *linear function*

$$L(x) = f(c) + f'(c)(x - c).$$

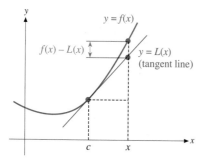

FIGURE 3.6.1 The value $L(x)$ on the line tangent to the graph of f at $(c, f(c))$ can be used as an approximate value of $f(x)$ for x near c.

See Figure 3.6.1.

Let us see that the difference $f(x) - L(x)$ becomes small relative to $x - c$ as x approaches c. That is, for x near c, $x \neq c$, we have

$$\frac{f(x) - L(x)}{x - c} = \frac{f(x) - (f(c) + f'(c)(x - c))}{x - c}$$

$$= \frac{f(x) - f(c)}{x - c} - \frac{f'(c)(x - c)}{x - c}$$

$$= \frac{f(x) - f(c)}{x - c} - f'(c).$$

Since

$$\lim_{x \to c} \frac{f(x) - f(c)}{x - c} = f'(c),$$

we see that the above ratio approaches zero as x approaches c. Thus, if x is near c, $x \neq c$, we have

$$\frac{f(x) - L(x)}{x - c} = \text{(A small number)},$$
$$f(x) - L(x) = \text{(A small number)}(x - c).$$

If x is near c, then $x - c$ is small and the product (A small number)$(x - c)$ is very small. That is, $L(x)$ is very near $f(x)$, and the difference between $f(x)$ and $L(x)$ becomes much smaller than $x - c$ as x approaches c. This means that the value $L(x)$ on the tangent line can be used as an approximate value of $f(x)$ for x near c. We write

$$f(x) \approx f(c) + f'(c)(x - c) \text{ whenever } x \approx c. \tag{1}$$

Let us illustrate how to use Formula 1 to find approximate values of a function f. To do this we must know the value $f(c)$ and the value of the derivative $f'(c)$, both at the point c. We can then use that information to find approximate values of $f(x)$ for x near c.

EXAMPLE 1

Use a linear approximation of $\sqrt{x}$ for x near 1 to find an approximate value of $\sqrt{1.02}$.

SOLUTION   We use Formula 1 with $f(x) = \sqrt{x}$ and $c = 1$. We have $f(1) = \sqrt{1} = 1$ and $f'(x) = 1/(2\sqrt{x})$, so $f'(1) = 1/2 = 0.5$. Formula 1 then gives

$$f(x) \approx f(c) + f'(c)(x - c),$$
$$\sqrt{x} \approx 1 + (0.5)(x - 1).$$

Substitution of $x = 1.02$ gives

$$\sqrt{1.02} \approx 1 + (0.5)(1.02 - 1) = 1.01.$$

The approximate value of $\sqrt{1.02}$ obtained by linear approximation is 1.01. (A calculator gives $\sqrt{1.02} \approx 1.0099505$.) The graph of $f(x) = \sqrt{x}$ and the linear function $L(x) = 1 + (0.5)(x - 1)$ are given in Figure 3.6.2. ∎

Note that the purpose of Example 1 is to illustrate the concept of linear approximation, not to suggest a new way to find square roots without using a calculator!

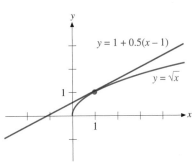

FIGURE 3.6.2 The line tangent to the graph of $\sqrt{x}$ at (1, 1) gives a linear approximation of $\sqrt{x}$ for x near 1.

EXAMPLE 2

Use a linear approximation of $\sin x$ for x near zero to find an approximate value of $\sin(0.2)$.

SOLUTION   We are interested in a linear approximation of the function $f(x) = \sin x$ for x near $c = 0$. We have $f(0) = \sin(0) = 0$ and $f'(x) = \cos x$, so $f'(0) = \cos(0) = 1$. Using Formula 1 with $f(x) = \sin x$ and $c = 0$, we have

$$f(x) \approx f(c) + f'(c)(x - c),$$
$$\sin x \approx 0 + (1)(x - 0) = x.$$

Substitution of 0.2 for x gives

$$\sin(0.2) \approx 0.2.$$

The desired approximate value of $\sin(0.2)$ is 0.2. (A calculator value of the sine of 0.2 rad is $\sin(0.2) \approx 0.19866933$.) The graph of $f(x) = \sin x$ and the linear function $L(x) = x$ are given in Figure 3.6.3. ∎

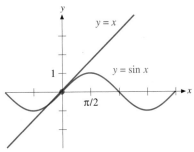

FIGURE 3.6.3 For x near zero, $\sin x$ can be approximated by the linear function x.

Linear Approximation of Implicit Functions

It is not necessary to have an explicit formula for the values of a function in order to use linear approximation. In particular, we can use the method of linear approximation to find approximate values of functions that are defined implicitly.

EXAMPLE 3

Use the line tangent to the graph of the equation $(x^2 + y^2)^2 = 16xy$ at the point $(2, 2)$ to approximate the value of y near 2 that satisfies the equation when $x = 1.85$.

SOLUTION   We assume that the equation $(x^2 + y^2)^2 = 16xy$ defines y as a differentiable function of x in some rectangle with center $(2, 2)$. We want a linear approximation of this function $y(x)$. Since the graph of this function passes through the point $(2, 2)$, we have $y(2) = 2$. We need the value of $y'(2)$, which we obtain by implicit differentiation. Using the Chain Rule and Product Rule to differentiate both sides of the equation

$$(x^2 + y^2)^2 = 16xy$$

with respect to x, we obtain

$$2(x^2 + y^2)(2x + 2yy') = 16((x)(y') + (y)(1)),$$
$$(x^2 + y^2)(2x) + (x^2 + y^2)(2yy') = 8xy' + 8y,$$
$$(2y(x^2 + y^2) - 8x)y' = 8y - 2x(x^2 + y^2),$$
$$y' = \frac{8y - 2x(x^2 + y^2)}{2y(x^2 + y^2) - 8x}.$$

When $(x, y) = (2, 2)$,

$$y'(2) = \frac{8(2) - 2(2)(4 + 4)}{2(2)(4 + 4) - 8(2)} = \frac{-16}{16} = -1.$$

Substitution of $c = 2$ in Formula 1 then gives

$$y(x) \approx y(2) + y'(2)(x - 2)$$
$$= 2 - (x - 2).$$

When $x = 1.85$, we obtain

$$y(1.85) \approx 2 - (1.85 - 2) = 2.15.$$

This is the desired approximate value. The graph of the equation $(x^2 + y^2)^2 = 16xy$ and the line tangent to the graph at $(2, 2)$ are sketched in Figure 3.6.4. ∎

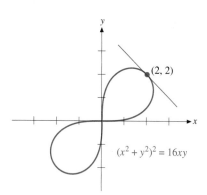

FIGURE 3.6.4 The method of linear aproximation can be used to find approximate values of implicitly defined functions.

Differentials

In many problems we are interested in the change, or approximate change, of values $f(x)$ that correspond to a change in x. If $\Delta x = x - c$ represents a change in x, the corresponding change in $f(x)$ is denoted $\Delta f = f(x) - f(c)$. If f is differentiable at c, then we know from (1) that

$$f(x) \approx f(c) + f'(c)(x - c), \text{ so}$$
$$f(x) - f(c) \approx f'(c)(x - c).$$

Let us rewrite this last statement as

$$\Delta f \approx f'(c) \cdot \Delta x. \tag{2}$$

It is convenient to introduce a notation that suggests the use of the derivative in the linear approximation of the change of values of a function.

DEFINITION

If f is a differentiable function of x, the **differential** of f, denoted df, is defined in terms of the real variables x and dx by the equation $df = f'(x) \cdot dx$.

If we choose $x = c$ and $dx = \Delta x = x - c$ in the definition of df, then $df = f'(c) \cdot \Delta x$. The differential then represents the change of values along the line tangent to the graph of f at the point $(c, f(c))$, so df is a linear approximation of the difference $\Delta f = f(x) - f(c)$; df is also called a **differential approximation** of Δf. See Figure 3.6.5. From Formula 2, we then have

$$\Delta f \approx f'(c) \cdot \Delta x = f'(c) \cdot dx = df. \qquad (3)$$

FIGURE 3.6.5 The value $df = f'(c) \cdot \Delta x$ is called the differential approximation of Δf.

EXAMPLE 4

Find the differentials du in terms of the variables x and dx.

(a) $u = 3x^2 - 2$,
(b) $u = x^{1/3}$,
(c) $u = \cos 2x$.

SOLUTION   In each case $du = u'(x) \cdot dx$.

(a) $u = 3x^2 - 2$ implies $u'(x) = 6x$, so $du = 6x\,dx$.
(b) $u = x^{1/3}$ implies $u'(x) = \dfrac{1}{3}x^{-2/3}$, so $du = \dfrac{1}{3}x^{-2/3}dx$.
(c) $u = \cos 2x$ implies $u'(x) = (-\sin 2x)(2)$, so $du = -2\sin 2x\,dx$.

(Note that we used the Chain Rule to find the derivative with respect to x of $\cos 2x$.) ∎

EXAMPLE 5

Find the change in area of a circle due to a change of dr in its radius. Find the differential approximation of the change in area.

SOLUTION   The area of a circle in terms of its radius is given by the formula

$$A = A(r) = \pi r^2.$$

The actual change in area of the circle due to change of dr in its radius is

$$
\begin{aligned}
\Delta A &= A(r + dr) - A(r) \\
&= \pi (r + dr)^2 - \pi r^2 \\
&= \pi r^2 + 2\pi r(dr) + \pi (dr)^2 - \pi r^2 \\
&= 2\pi r\,dr + \pi (dr)^2.
\end{aligned}
$$

The differentiational approximation of the change in area is the differential dA. We have $A'(r) = 2\pi r$, so

$$dA = A'(r)dr = 2\pi r\,dr.$$

The approximate change in area is given by the circumference of the circle times the change in radius. We can think of dA as the area of the "rectangle" obtained by cutting and unrolling a band of width dr around the circle of radius r. See Figure 3.6.6. If dr is small, then

$$\Delta A - dA = 2\pi r\,dr + \pi (dr)^2 - 2\pi r\,dr = \pi (dr)^2$$

is very small and the differential approximation dA is very close to the actual change ΔA. ∎

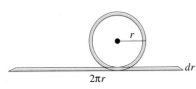

FIGURE 3.6.6 The differential $dA = 2\pi r\,dr$ can be thought of as the area of the "rectangle" obtained by cutting and unrolling a band of width dr around the circle of radius r. Note that $2\pi r$ is the circumference of a circle with radius r.

Approximation of Error

We may think of the function value $f(x)$ as the output of a process corresponding to an input x. In a real-world process we may be able to control the input x only within a certain tolerance of the desired input c. The difference $\Delta f = f(x) - f(c)$ may then be interpreted as the **error** in the output that corresponds to an error of $\Delta x = x - c$ in the input. Since the corresponding differential df is approximately Δf for small values of Δx, we may use df as an **approximate value of the error** in f due to a possible error of Δx in x. The ratio df/f is an approximate value of the **relative error** $\Delta f/f$.

EXAMPLE 6

Find an approximate value of the error in the volume of a cube if each edge of the cube is intended to be 2 ft with a possible error of 0.005 ft. Find an approximate value of the relative error.

SOLUTION   The volume of a cube with length of each edge s is given by the formula $V = V(s) = s^3$. An approximate value of the error in volume is given by the differential

$$dV = V'(s)\,ds = 3s^2\,ds.$$

We want the value of the differential dV when the intended length of edge is $s = 2$ and the possible error in s is $ds = \pm 0.005$. Substitution then gives

$$dV = 3(2)^2(\pm 0.005) = \pm 0.06 \text{ ft}^3.$$

This is the approximate value of the error in volume.
  When $x = 2$, we have $V = 2^3$ and the relative error is approximately

$$\frac{dV}{V} = \pm\frac{0.06}{2^3} = \pm 0.0075 = \pm 0.75\%. \quad \blacksquare$$

Differential Formulas

We can use the rules for differentiation of products, quotients, and composite functions to find differentials.

THEOREM

If f and g are differentiable functions of x, then

$$d(fg) = f\,dg + g\,df, \qquad (4)$$

$$d\left(\frac{f}{g}\right) = \frac{g\,df - f\,dg}{g^2}, \qquad (5)$$

and

$$d(f \circ g) = (f' \circ g)\,dg. \qquad (6)$$

Proof   Let us verify Formula 4. The other formulas are verified similarly.

$$d(fg) = \frac{d}{dx}(fg)\,dx$$

$$= \left(f \frac{dg}{dx} + g \frac{df}{dx} \right) dx$$

$$= f \left(\frac{dg}{dx} dx \right) + g \left(\frac{df}{dx} dx \right)$$

$$= f \, dg + g \, df. \quad \blacksquare$$

If several functions of the same independent variable are related by an equation, then their differentials are also related. We can determine the relation between the differentials by using the usual rules of differentiation with differentials in place of derivatives. For example, if V, x, and y are all functions of the same variable and

$$V = x^2 y,$$

then

$$dV = (x^2)(dy) + (y)(2x \, dx).$$

Note that we used the Product Rule to differentiate the product of x^2 and y. The differential of x^2 involves using the Chain Rule to differentiate the square of the function x.

EXERCISES 3.6

1. Use a linear approximation of $x^{2/3}$ for x near 8 to find an approximate value of $(7.97)^{2/3}$.

2. Use a linear approximation of $x^{3/4}$ for x near 16 to find an approximate value of $(15.96)^{3/4}$.

3. Use a linear approximation of x^{17} for x near 1 to find an approximate value of $(1.03)^{17}$.

4. Use a linear approximation of x^{13} for x near 1 to find an approximate value of $(1.02)^{13}$.

5. Use a linear approximation of $\cos x$ for x near 0 to find an approximate value of $\cos(0.05)$.

6. Use a linear approximation of $\sin x$ for x near 0 to find an approximate value of $\sin(-0.15)$.

7. Use a linear approximation of $\sec x$ for x near 0 to find an approximate value of $\sec(\pi/12)$.

8. Use a linear approximation of $\tan x$ for x near 0 to find an approximate value of $\tan(\pi/12)$.

9. Use the line tangent to the graph of the equation $x^2 + y^2 = 25$ at the point $(3, 4)$ to approximate the value of y near 4 that satisfies the equation when $x = 2.9$.

10. Use the line tangent to the graph of the equation $x^2 - y^2 = 16$ at the point $(5, 3)$ to approximate the value of y near 3 that satisfies the equation when $x = 5.1$.

11. Use the line tangent to the graph of the equation $x = \tan y$ at the point $(1, \pi/4)$ to approximate the value of y near $\pi/4$ that satisfies the equation when $x = 1.2$.

12. Use the line tangent to the graph of the equation $x = \sin y$ at the point $(0.5, \pi/6)$ to approximate the value of y near $\pi/6$ that satisfies the equation when $x = 0.45$.

13. Use the line tangent to the graph of the equation $(x^2 + y^2)^{3/2} = 2^{3/2} xy$ at the point $(1, 1)$ to approximate the value of y near 1 that satisfies the equation when $x = 1.1$.

14. Use the line tangent to the graph of the equation $(x^2 + y^2)^2 = 3x^2 y - y^3$ at the point $(\sqrt{3}/2, 1/2)$ to approximate the value of y near $1/2$ that satisfies the equation when $x = 0.9$.

In Exercises 15–26 express the differentials du in terms of the variables x and dx.

15. $u = 3x - 1$ **16.** $u = 2x + 7$

17. $u = x^2 + 1$ **18.** $u = x^2 + 2x + 2$

19. $u = x^3 + 4x^2 - 3$ **20.** $u = x^3 - x$

21. $u = \sqrt{x}$ **22.** $u = \sqrt{2x + 1}$

23. $u = \cos(3x)$ **24.** $u = \sin(2x)$

25. $u = \tan x$ **26.** $u = \sec x$

27. A right circular cylinder has fixed height h_0 and variable radius r. Find the change in its volume due to a change of dr in its radius. Find the differential approximation of the change in volume.

28. Find the change in volume of a sphere due to a change of dr in its radius. Find the differential approximation of the change in volume.

29. Find the change in the area of a square due to a change of ds in the length of each of its sides. Find the differential approximation of the change in the area.

30. Find the change in the area of an equilateral triangle due to a change of ds in the length of each of its sides. Find the differential approximation of the change in the area.

31. The height of a right circular cylinder is twice the radius of its base. Express the volume of the cylinder in terms of its radius. Use differentials to find the approximate error in volume if the radius is 4 m with a possible error of ± 0.03 m. Find the approximate relative error.

32. The height of a right circular cone is one-half the radius of its base. Express the volume of the cone in terms of its radius. Use differentials to find the approximate error in volume if the radius is 2 m with a possible error of ± 0.02 m. Find the approximate relative error.

33. The height of a tree is determined by measuring the angle of elevation to the top of the tree from a point on the ground that is 40 ft from the base of the tree. Find the height of the tree and the approximate error in the calculation if the angle of elevation is measured to be $45°$ with a possible error of $1°$.

34. The angle of elevation between an observer at ground level and an airplane is measured to be $30°$ with a possible error of $1°$ as the plane passes over a point on the ground that is 5000 ft from the observer. Find the height of the plane and the approximate error in the calculation.

In Exercises 35–40 assume that the variables in each equation represent functions of the same independent variable and find equations that relate their differentials.

35. $A = \pi r^3$

36. $V = \dfrac{4}{3}\pi r^3$

37. $V = \pi r^2 h$

38. $V = \ell w h$

39. $S = \pi r \sqrt{r^2 + h^2}$

40. $\dfrac{1}{R} = \dfrac{1}{R_1} + \dfrac{1}{R_2}$

41. A **standard American golf ball** weighs $W = 1.62 \pm 0.01$ oz and has diameter $D = 1.68 \pm 0.01$ in. (a) Find a formula for the weight-density, δ, of the ball in terms of its weight and its diameter. (Weight-density = weight/volume.) (b) Evaluate the weight-density when the weight is 1.62 oz

and the diameter is 1.68 in. (c) Express the differential of the weight-density in terms of the differentials of the weight and diameter. (d) Use differentials to determine the approximate maximum error of the weight-density.

42. The circumference of a Rawlings **NCAA men's tournament basketball** must be 75.56 ± 0.64 cm. Use differentials to find the approximate percent error in the surface area of the ball. (Errors of approximately 5% are allowed in weight and bounce.)

43. The **gravitational force on the surface of the moon** can be measured by observing the period of a simple pendulum and using the formula

$$g = 4\pi^2 \frac{L}{T^2}.$$

What is the maximum percentage error in g if there are possible errors of 2% in measuring L and T?

44. The radius, height, and volume of a right circular cylinder are related by the equation $v = \pi r^2 h$. (a) Express the differential of the volume in terms of the differentials of the radius and height. When the radius is 10 and the height is (b) 4 and (c) 6, will the volume be affected more by error in the radius or the same error in the height?

45. The hypotenuse and the two sides of a right triangle are related by the equation $h^2 = x^2 + y^2$. Express the differential of h in terms of x, y, and the differentials dx and dy.

46. Specific volume, v, pressure, p, absolute temperature T, internal energy, u, and entropy, s, are fundamental properties of thermodynamics. These properties are related by **Gibbs equation**, $T\,ds = du + p\,dv$. Specific enthalpy is defined to be $h = u + pv$. Show that Gibbs equation can be expressed in the form $T\,ds = dh - v\,dp$.

47. Prove Formula 5.

48. Prove Formula 6.

3.7 THE INTERMEDIATE VALUE THEOREM; NEWTON'S METHOD

Connections

Intuitive idea of continuity, 2.3.

Approximation by tangent lines, 3.6.

Continuity is the key to the problem of finding approximate values of solutions of equations that cannot be solved by methods of algebra. The Intermediate Value Theorem provides the theoretical basis for the solution of this problem.

Continuity on an Interval

The Intermediate Value Theorem expresses one of several important and useful properties of functions that are continuous at each point of an interval. Let us begin our study by defining the notion of continuity on an interval.

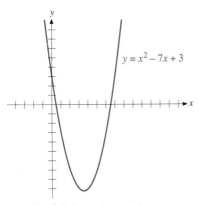

FIGURE 3.7.1 Polynomials are continuous at every point.

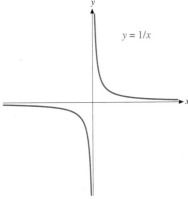

FIGURE 3.7.2 Rational functions are continuous at every point where they are defined.

> **DEFINITION**
>
> A function f is said to be **continuous on an interval** if (i) f is continuous at each point of the interval that is not an endpoint of the interval; (ii) f is continuous from the right at the left endpoint of the interval, if that point is included in the interval; and (iii) f is continuous from the left at the right endpoint of the interval, if that point is included in the interval.

Continuity on an interval depends only on the values of the function at points of the interval. At endpoints that are included in the interval, the function is required to be continuous only from the direction of points inside the interval; no requirements are made at endpoints of the interval that are not included in the interval. If f is continuous on an interval I, then f is continuous on any interval contained in I.

We can use what we learned in Section 2.5 about continuity at a point to determine continuity on intervals. For example:

- The function $f(x) = x^2 - 7x + 3$ is continuous on any interval, because *polynomials are continuous at every point.* See Figure 3.7.1.
- The function $f(x) = 1/x$ is continuous on any interval that does not contain the origin. In particular, f is continuous on each of the intervals $x > 0$ and $x < 0$. See Figure 3.7.2. We know that *rational functions are continuous at every point where they are defined.*
- The function $f(x) = \tan x$ is continuous on intervals on which it is defined, such as $-\pi/2 < x < \pi/2$. The *trigonometric functions are continuous at every point where they are defined.* See Figure 3.7.3.
- The function $f(x) = \sqrt{x}$ is continuous at every $x > 0$ and continuous from the right at $x = 0$, so it is continuous on the interval $x \geq 0$ and all of its subintervals. See Figure 3.7.4.

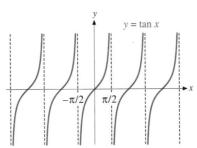

FIGURE 3.7.3 The trigonometric functions are continuous at every point where they are defined.

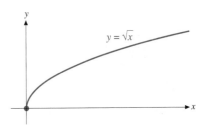

FIGURE 3.7.4 The square root function is continuous on the interval $x \geq 0$.

The Intermediate Value Theorem

The Intermediate Value Theorem tells us that the graph of a continuous function cannot pass from one side of a horizontal line $y = L$ to the other without intersecting the line at least once. See Figure 3.7.5.

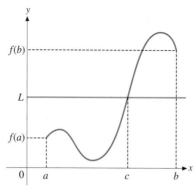

FIGURE 3.7.5 The graph of a continuous function cannot pass from one side of a horizontal line $y = L$ to the other without intersecting the line at least once.

INTERMEDIATE VALUE THEOREM

If f is continuous on the interval $a \leq x \leq b$, and L is any number between $f(a)$ and $f(b)$, then there is at least one number c such that $a \leq c \leq b$ and $f(c) = L$.

The Intermediate Value Theorem is a simple consequence of the following special case, which we will use to find approximate values of solutions of equations $f(x) = 0$. (See Exercise 16.)

SPECIAL CASE OF THE INTERMEDIATE VALUE THEOREM

If f is continuous on the interval $a \leq x \leq b$, and either $f(a) < 0 < f(b)$ or $f(b) < 0 < f(a)$, then there is at least one number c such that $a < c < b$ and $f(c) = 0$.

FIGURE 3.7.6 Repeatedly cutting each interval in half gives an unending sequence of intervals that shrink down to a single point c.

Proof   Let us verify the theorem in case $f(a) < 0 < f(b)$. The other case is similar. The idea is to locate a particular point c and then show $f(c) = 0$.

We begin by cutting the interval $a \leq x \leq b$ at its midpoint $(a + b)/2$ and selecting one of the two half-intervals: select the left half-interval if $f((a + b)/2) \geq 0$; select the right half-interval if $f((a + b)/2) < 0$. If $a_1 \leq x \leq b_1$ is the selected half-interval, we will have $f(a_1) < 0 \leq f(b_1)$.

We repeat the procedure of cutting in half and selecting one of the halves, with the interval $a \leq x \leq b$ replaced by the interval $a_1 \leq x \leq b_1$. We obtain a new interval $a_2 \leq x \leq b_2$ with $f(a_2) < 0 \leq f(b_2)$. See Figure 3.7.6.

Theoretically, this procedure of cutting in half and selecting can be continued indefinitely to obtain an unending sequence of closed intervals. These intervals are nested, each one inside the previous one, and the lengths of the intervals approach zero. It is an axiom of the real number system that there is a single point c that is contained in the intersection of all these intervals. It should seem reasonable that the intervals do shrink down to a single point on the real number line.

As our selected intervals shrink down to c, their endpoints approach c. Since the endpoints of our selected intervals and the point c are in the interval $a \leq x \leq b$ and f is continuous on this interval, it follows that the values of f at the endpoints of our selected intervals must approach $f(c)$. Since we have selected the intervals in such a way that the values of f at their left endpoints are less than zero, these values cannot approach any positive number, so $f(c) \leq 0$. Since the values of f at the right endpoints of our selected intervals are greater than or equal to zero, these values cannot approach any negative number, so $f(c) \geq 0$. It follows that $f(c) = 0$ and $a < c < b$. ■

The procedure used in the above proof is called the **bisection method** of locating the zeros of a function. We will return to this method shortly. Let us now see how the special case of the Intermediate Value Theorem can be used to determine the existence and approximate location of solutions of $f(x) = 0$, where f is a continuous function.

EXAMPLE 1

Use the Intermediate Value Theorem to show that the equation $x^3 - 4x^2 + 2x + 2 = 0$ has three solutions. Find intervals between successive integers that contain the solutions.

SOLUTION We want to determine the existence and location of solutions of the equation $f(x) = 0$, where $f(x) = x^3 - 4x^2 + 2x + 2$. The idea is to evaluate $f(x)$ at various integers until we find successive integers for which the sign of $f(x)$ changes. Since f is continuous, the Intermediate Value Theorem then implies that there is a value c in each such interval with $f(c) = 0$. A table of values is given in Table 3.7.1. Note that if $|x|$ is large $f(x)$ will have the same sign as the highest-order term, x^3.

From Table 3.7.1 we see that $f(-1) < 0 < f(0)$, $f(1) > 0 > f(2)$, and $f(3) < 0 < f(4)$, so the intervals $(-1, 0)$, $(1, 2)$, and $(3, 4)$ each contain a solution of the equation $f(x) = 0$. See Figure 3.7.7. ■

TABLE 3.7.1

x	$f(x)$
-3	-67
-2	-26
-1	-5
0	2
1	1
2	-2
3	-1
4	10

Let us emphasize that the Intermediate Value Theorem depends on the continuity of the function on the interval $[a, b]$. If there is even one point between a and b where f is discontinuous, the conclusion of the Intermediate Value Theorem may fail. For example, from the graph in Figure 3.7.8, we see

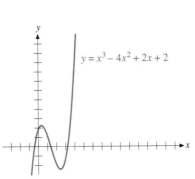

$y = x^3 - 4x^2 + 2x + 2$

FIGURE 3.7.7 Each of the intervals $(-1, 0)$, $(1, 2)$, and $(3, 4)$ contain a solution of $f(x) = 0$.

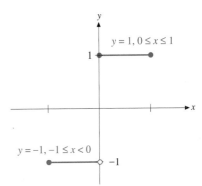

$y = 1, 0 \leq x \leq 1$

$y = -1, -1 \leq x < 0$

FIGURE 3.7.8 A function that is not continuous need not be zero between a point where it is positive and a point where it is negative.

that the function

$$f(x) = \begin{cases} -1, & -1 \le x < 0, \\ 1, & 0 \le x \le 1, \end{cases}$$

is not continuous at $x = 0$. We have $f(-1) < 0 < f(1)$, but there is no point c that satisfies $-1 < c < 1$ and $f(c) = 0$. The graph jumps over the horizontal line $y = 0$ at the point where the function is discontinuous.

The Bisection Method

Let us now see how we can use the Bisection Method to find approximate values of solutions of $f(x) = 0$.

The essential feature of the Bisection Method is that the selection process produces a sequence of nested intervals, each contained in the previous interval, with the lengths of the intervals approaching zero, and such that values of the continuous function change sign on each interval. This feature can be retained without using bisection strictly. We can modify the selection process by choosing points other than the midpoint of each successive interval if that is convenient for a particular problem.

The Bisection Method of approximating zeros of a continuous function:

- Evaluate $f(x)$ for several values of x. The continuous function will have at least one zero between x_1 and x_2 if $f(x_1)$ and $f(x_2)$ are of opposite sign.
- Evaluate $f(x_3)$ for a point x_3 in an interval where $f(x)$ changes sign to obtain a smaller interval where $f(x)$ changes sign.
- Continue cutting down the size of the interval where $f(x)$ changes sign until you have an interval that is the desired size.

The selection process of the Bisection Method is easily programmed for use with a computer. A computor/calculator-generated graph of a continuous function f can also be used to determine approximate values of solutions of $f(x) = 0$.

EXAMPLE 2

Determine that the equation $\sin x = 1 - x$ has a solution in the interval $0 \le x \le \pi/2$, and then find an interval of length 0.1 that contains the solution.

SOLUTION We begin by writing the given equation as an equivalent equation in the form $f(x) = 0$. We have

$$\sin x = 1 - x,$$
$$x - 1 + \sin x = 0.$$

We then see that the solution of the original equation satisfies $f(x) = 0$, where

$$f(x) = x - 1 + \sin x.$$

Evaluating the function f at 0 and $\pi/2$, we obtain

$$f(0) = 0 - 1 + \sin(0) = -1 \text{ and } f\left(\frac{\pi}{2}\right) = \frac{\pi}{2} - 1 + \sin\left(\frac{\pi}{2}\right) = \frac{\pi}{2}.$$

Since f is a continuous function that has a negative value at 0 and a positive value at $\pi/2$, the special case of the Intermediate Value Theorem tells us that there is at least one point c between 0 and $\pi/2$ with $f(c) = 0$. This means that c is a solution of our original equation.

To locate the solution c more accurately, we evaluate the function f at some point between 0 and $\pi/2$. Let us try the value at $x = 0.8$, a point that is near the midpoint of the interval between 0 and $\pi/2 \approx 1.57$. We must evaluate the sine of 0.8 *radian.* We have

$$f(0.8) = 0.8 - 1 + \sin(0.8) \approx 0.517.$$

Since $f(0) < 0 < f(0.8)$, we know that there is a solution c between 0 and 0.8. Continuing to cut down the size of the interval that contains a solution, we evaluate $f(0.4)$:

$$f(0.4) = 0.4 - 1 + \sin(0.4) \approx -0.210.$$

Since $f(0.4) < 0 < f(0.8)$, we know there is a solution c between 0.4 and 0.8. Evaluating f at 0.6, we have

$$f(0.6) = 0.6 - 1 + \sin(0.6) \approx 0.164.$$

We now have $f(0.4) < 0 < f(0.6)$. Evaluating f at 0.5, we obtain

$$f(0.5) = 0.5 - 1 + \sin(0.5) \approx -0.020.$$

Since $f(0.5) < 0 < f(0.6)$, we know that the interval $[0.5, 0.6]$ contains a solution of our equation. See Figure 3.7.9. ∎

In Example 2 we determined that the interval $[0.5, 0.6]$ contains a solution of the equation $\sin x = 1 - x$. This means that any point of the interval $[0.5, 0.6]$ can be used as an approximate value of the solution that is accurate to within 0.1 of the exact solution. We could obtain approximate values that are more accurate by continuing to cut down the size of the interval that contains the solution. However, more efficient methods are generally used if accurate approximate values are required.

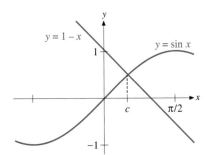

FIGURE 3.7.9 The solution of the equation $\sin x = 1 - x$ is the x-coordinate of the point of intersection of the graph of $\sin x$ and the graph of $1 - x$. We see that there is a solution between zero and $\pi/2$.

Newton's Method

If $f(x_0)$ and $f(x_1)$ have opposite signs and f is continuous on the interval $[x_0, x_1]$, we know from the Intermediate Value Theorem that there is at least one c between x_0 and x_1 such that $f(c) = 0$. That is, c is a solution of the equation $f(x) = 0$. See Figure 3.7.10. We could use the Bisection Method to find approximate values of c, but that method is not efficient for problems that require high accuracy. We now introduce a much more efficient method for finding approximate values of solutions of $f(x) = 0$ in the case that f is differentiable. The method is called **Newton's Method**.

Newton's Method is based on the fact that tangent lines can be used to approximate values of a *differentiable* function. For example, suppose x_1 is a point that is near the exact solution c of the equation $f(x) = 0$. The equation of the line tangent to the graph of f at the point $(x_1, f(x_1))$ is

$$y - f(x_1) = f'(x_1)(x - x_1).$$

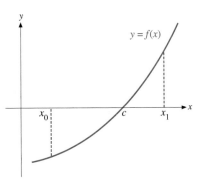

FIGURE 3.7.10 Solutions c of the equation $f(x) = 0$ correspond to points where the graph of f intersects the x-axis.

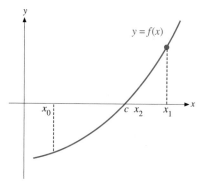

FIGURE 3.7.11 The line tangent to the graph of f at $(x_1, f(x_1))$ intersects the x-axis near the point where the graph of f intesects the x-axis.

Since the graph of f is near the graph of the tangent line, we should expect that the graph would cross the x-axis near the point where the tangent line crosses the x-axis. If $f'(x_1) \neq 0$, this line will intersect the x-axis at a point $(x_2, 0)$. Substituting these coordinates into the equation of the tangent line and solving for x_2, we have

$$0 - f(x_1) = f'(x_1)(x_2 - x_1),$$
$$-f(x_1) = f'(x_1) \cdot x_2 - f'(x_1) \cdot x_1,$$
$$f'(x_1) \cdot x_2 = f'(x_1) \cdot x_1 - f(x_1),$$
$$x_2 = x_1 - \frac{f(x_1)}{f'(x_1)}.$$

This value of x_2 should be better than x_1 as an approximate value of the unknown exact solution c. See Figure 3.7.11. We could repeat the process with x_2 in place of x_1 to obtain $x_3 = x_2 - f(x_2)/f'(x_2)$. Generally, we could repeat the process as often as we please and obtain successively more accurate approximate values of the exact solution c.

Let us summarize:

> ### Newton's Method for finding approximate values of solutions c of the equation $f(x) = 0$
>
> - Use values of $f(x)$ to find an interval such that $f(x)$ has opposite signs at the endpoints.
> - Choose a point x_1 in the interval and then use the formula
>
> $$x_{n+1} = x_n - \frac{f(x_n)}{f'(x_n)}, \qquad n = 1, 2, 3, \ldots,$$
>
> to find the successive approximate values $x_2, x_3, x_4, \ldots$.
> - Choose x_{n+1} (appropriately rounded) to be the final approximate value of c if the difference between x_{n+1} and x_n is within the desired accuracy.

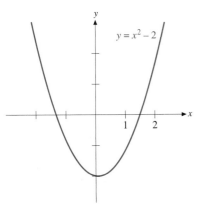

$y = x^2 - 2$

FIGURE 3.7.12 The graph of $x^2 - 2$ intersects the x-axis at a point between one and two.

EXAMPLE 3

Use Newton's Method to find an approximate value of the positive solution of the equation $x^2 - 2 = 0$. Obtain accuracy to five decimal places. (The exact solution is $\sqrt{2}$, so we can check the accuracy of the method with a calculator.)

SOLUTION   We seek the positive solution of $f(x) = 0$, where

$$f(x) = x^2 - 2.$$

This is the point on the positive x-axis where the graph of f crosses the x-axis. See Figure 3.7.12.

Since $f(1) = 1^2 - 2 = -1 < 0$ and $f(2) = 2^2 - 2 = 2 > 0$, we know there is a solution c between 1 and 2. Since the value $f(1) = -1$ is closer to zero than the value $f(2) = 2$, let us choose x_1 to be closer to 1 than 2. We choose $x_1 = 1.3$ to be our first approximation of c.

We need to determine the formula for successive approximations. We have $f(x) = x^2 - 2$, so $f'(x) = 2x$. Substitution into the formula for successive approximation gives

$$x_{n+1} = x_n - \frac{f(x_n)}{f'(x_n)} = x_n - \frac{x_n^2 - 2}{2x_n}.$$

Simplifying, we have

$$x_{n+1} = x_n - \frac{x_n^2}{2x_n} + \frac{2}{2x_n}$$

$$= x_n - \frac{x_n}{2} + \frac{1}{x_n}$$

$$= \frac{x_n}{2} + \frac{1}{x_n}.$$

We have chosen $x_1 = 1.3$, so the formula (with $n = 1$) gives

$$x_2 = \frac{1}{2}x_1 + \frac{1}{x_1} = \frac{1}{2}(1.3) + \frac{1}{(1.3)} = 1.4192308.$$

Continuing until there is no change in the first five decimal places, we obtain

$$x_3 = \frac{1}{2}x_2 + \frac{1}{x_2} = \frac{1}{2}(1.4192308) + \frac{1}{(1.4192308)} = 1.4142224.$$

$$x_4 = \frac{1}{2}x_3 + \frac{1}{x_3} = \frac{1}{2}(1.4142224) + \frac{1}{(1.4142224)} = 1.4142136.$$

$$x_5 = \frac{1}{2}x_4 + \frac{1}{x_4} = \frac{1}{2}(1.4142136) + \frac{1}{(1.4142136)} = 1.4142136.$$

These calculations are arranged in Table 3.7.2.

TABLE 3.7.2

n	x_n		
1	1.3	←	Our choice
2	1.4192308	←	Calculated from $x_1 = 1.3$
3	1.4142224	←	Calculated from $x_2 = 1.4192308$
4	1.4142136	←	Calculated from $x_3 = 1.4142224$
5	1.4142136	←	Calculated from $x_4 = 1.4142136$

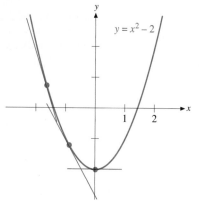

FIGURE 3.7.13 A line tangent to the graph of $x^2 - 2$ at a point with x negative intersects the x-axis at a point with negative x-coordinate.

We have used a calculator to evaluate the successive approximations. It is convenient to store x_n and then recall it as it is needed in the calculation of x_{n+1}. A programmable calculator or a computer would be very appropriate for this work.

The first five decimal digits did not change when x_5 was calculated from x_4. We use 1.41421 as our approximate value of $\sqrt{2}$. (A calculator gives $\sqrt{2} \approx 1.4142136$, so our approximate value does seem to be accurate to five decimal places.) ■

Any choice of x_1 that is close enough to a particular solution c of the equation $f(x) = 0$ will give values x_n that approach the same number c. For example, the calculations that correspond to the choice $x_1 = 2$ for the positive solution of the equation $x^2 - 2 = 0$ considered in Example 3 are given in Table 3.7.3. We see that this choice also gives the approximate value 1.41421.

Different choices of x_1 may approach different solutions of the equation $f(x) = 0$. For example, we see from Figure 3.7.13 that the line tangent to the curve $f(x) = x^2 - 2$ at a point with negative x intersects the x-axis at a point with negative x-coordinate; a negative choice of x_1 would give approximate values of the negative solution of the equation $f(x) = 0$. The choice $x_1 = 0$ would not work at all, because the line tangent to the graph of $f(x) = x^2 - 2$ at the point $(0, -2)$ is horizontal and does not intersect the x-axis; the derivative of $f(x) = x^2 - 2$ is zero at zero and the formula for x_2 would be invalid.

TABLE 3.7.3

n	x_n		
1	2	←	Our choice
2	1.5	←	Calculated from $x_1 = 2$
3	1.4166667	←	Calculated from $x_2 = 1.5$
4	1.4142157	←	Calculated from $x_3 = 1.4166667$
5	1.4142136	←	Calculated from $x_4 = 1.4142157$

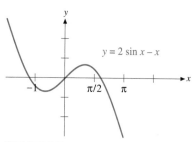

FIGURE 3.7.14 The equation $f(x) = 0$ has a positive solution between $\pi/2$ and π, where the graph intersects the x-axis.

EXAMPLE 4

Find an approximate value of the positive solution of $2 \sin x = x$. Obtain accuracy to five decimal places.

SOLUTION   We want the positive solution of $f(x) = 0$, where

$$f(x) = 2 \sin x - x.$$

This is the point on the positive x-axis where the graph of f crosses the x-axis. See Figure 3.7.14. Note that $f(\pi/2) = 2\sin(\pi/2) - \pi/2 = 2(1) - \pi/2 > 0$ and $f(\pi) = 2 \sin \pi - \pi = -\pi < 0$, so there is a solution between π and $\pi/2$.

We have $f(x) = 2 \sin x - x$, so $f'(x) = 2 \cos x - 1$. Substituting in the formula for successive approximations, we have

$$x_{n+1} = x_n - \frac{f(x_n)}{f'(x_n)} = x_n - \frac{2 \sin x_n - x_n}{2 \cos x_n - 1}.$$

TABLE 3.7.4

n	x_n
1	1.8
2	1.9015504
3	1.8955153
4	1.8954943
5	1.8954943

Since the value $f(\pi/2) = 2 - \pi/2$ is much closer to zero than the value $f(\pi) = -\pi$, let us choose x_1 to be near $\pi/2$. We choose $x_1 = 1.8$ and calculate the successive approximate values from the above formula. The results are arranged in Table 3.7.4.

NOTE *Don't forget that your calculator must be set for radian measure to calculate the values at x_n of the sine and cosine functions.*

Since x_4 and x_5 agree to five decimal places, we use 1.89549 as our final approximate value. ■

EXAMPLE 5
Show that the equation $x^3 - 3x + 1 = 0$ has three solutions and find approximate values for each of them. Obtain accuracy to three decimal places.

SOLUTION To show that the equation has three solutions, we need to find three intervals on which the continuous function $f(x) = x^3 - 3x + 1$ changes sign. We can do this by evaluating $f(x)$ at various points until we obtain three changes of sign. For example, we have

$$f(-2) = -1,$$
$$f(-1) = 3,$$
$$f(0) = 1,$$
$$f(1) = -1, \text{ and}$$
$$f(2) = 3.$$

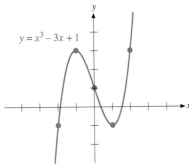

$y = x^3 - 3x + 1$

FIGURE 3.7.15 The equation $f(x) = 0$ has solutions in each of the intervals $(-2, -1)$, $(0, 1)$, and $(1, 2)$.

In Figure 3.7.15 we have plotted the corresponding points and then sketched the graph of f through these points. There is a solution of $f(x) = 0$ on each of the intervals $(-2, -1)$, $(0, 1)$, and $(1, 2)$, where the values $f(x)$ change sign.

To apply Newton's Method, we need to find the derivative of $f(x) = x^3 - 3x + 1$. We have

$$f'(x) = 3x^2 - 3.$$

The formula for the successive approximations is

$$x_{n+1} = x_n - \frac{f(x_n)}{f'(x_n)} = x_n - \frac{x_n^3 - 3x_n + 1}{3x_n^2 - 3}.$$

Simplifying, we have

$$x_{n+1} = x_n \frac{3x_n^2 - 3}{3x_n^2 - 3} - \frac{x_n^3 - 3x_n + 1}{3x_n^2 - 3}$$
$$= \frac{3x_n^3 - 3x_n - (x_n^3 - 3x_n + 1)}{3x_n^2 - 3}$$
$$= \frac{2x_n^3 - 1}{3x_n^2 - 3}$$

For each of the three solutions, we need to choose a point x_1 that is near the solution. Let us begin with the choices -1.8, 0.5, and 1.5. The results of the calculations are given in Table 3.7.5. The approximate values of the solutions are -1.879, 0.347, and 1.532. ■

NOTE *Newton's Method does not always work. In some cases the numbers x_n may wander around and not approach any particular number.*

TABLE 3.7.5

n	x_n	x_n	x_n
1	-1.8	0.5	1.5
2	-1.8845238	0.3333333	1.5333333
3	-1.8794047	0.3472222	1.5320906
4	-1.8793852	0.34729635	1.5320889

EXAMPLE 6

Suppose we try to use Newton's Method to find the solution of $f(x) = 0$, where $f(x) = x^{1/3}$. We have $f'(x) = (1/3)x^{-2/3}$, so

$$x_{n+1} = x_n - \frac{f(x_n)}{f'(x_n)} = x_n - \frac{x_n^{1/3}}{(1/3)x^{-2/3}} = x_n - 3x_n = -2x_n.$$

This gives

$$x_2 = -2x_1,$$
$$x_3 = -2x_2 = -2(-2x_1) = 4x_1,$$
$$x_4 = -2x_3 = -2(4x_1) = -8x_1,$$
$$\vdots$$

FIGURE 3.7.16 The graph opens upward on the interval $x < 0$ and opens downward on the interval $x > 0$.

Hence, we see that for any choice of $x_1 \neq 0$, the successive approximations are on the opposite side of zero (the exact solution) and twice as far away from zero as the previous approximation. This is illustrated in Figure 3.7.16. ■

It is possible to show that, if f' and f'' are each of constant sign on an open interval that contains a solution of $f(x) = 0$, then any starting point that is close enough to the solution will give successive Newton approximations that converge to the solution. Also, the difference between the exact solution and a particular Newton approximate value is less than the difference between that approximate value and the previous approximate value. If your successive approximations appear to jump around, you may need to start over with a point that is closer to the solution. You can also evaluate $f(x_n)$ to see if the values are approaching zero.

EXERCISES 3.7

Use the Intermediate Value Theorem to show that each of the equations given in Exercises 1–4 has three solutions. Find intervals between successive integers that contain the solutions.

1. $x^3 - 3x + 1 = 0$

2. $x^3 - 12x + 8 = 0$

3. $x^3 - 3x^2 + 1 = 0$

4. $x^3 - 6x^2 + 12 = 0$

In Exercises 5–12, determine that the given equation has a solution in the given interval and then find an interval of length 0.1 that contains the solution.

5. $x^3 - 4x^2 + x + 3 = 0, 1 \leq x \leq 2$

6. $x^3 + 2x^2 - 8x + 2 = 0, 1 \leq x \leq 2$

7. $x^4 - 3x^3 + 1 = 0, 0 \leq x \leq 1$

8. $2x^4 - 6x + 3 = 0, 0 \leq x \leq 1$

9. $x = 2\sin x, \pi/2 \leq x \leq \pi$

10. $x = \cos x, 0 \leq x \leq \pi/2$

11. $\cos x = \tan x, 0 \leq x \leq \pi/4$

12. $\sin x = 2\tan x - 1, 0 \leq x \leq \pi/4$

13. Show that for any constants b, c, d the equation $x^3 + bx^2 + cx + d = 0$ has at least one solution.

14. Find a function f such that $f(-1) = -1$ and $f(1) = 1$, where f is continuous on the interval $-1 < x < 1$, but there is no point c between -1 and 1 for which $f(c) = 0$.

15. The function $f(x) = |x|/x$ satisfies $f(-1) = -1$ and $f(1) = 1$, but there is no point c between -1 and 1 with

$f(c) = 0$. Why doesn't the Intermediate Value Theorem apply to this function on the interval $-1 \le x \le 1$?

16. Show that the Intermediate Value Theorem is a consequence of the special case we proved by applying the special case to the function $g(x) = f(x) - L$.

In Exercises 17–26 use Newton's Method to find an approximate value of the equation $f(x) = 0$ in the given interval. Obtain accuracy to four decimal places. (The solutions may be checked by direct calculation on a calculator.)

17. $f(x) = x^2 - 3, [1, 2]$

18. $f(x) = x^2 - 5, [2, 3]$

19. $f(x) = x^2 - 8, [-3, -2]$

20. $f(x) = x^2 - 10, [-4, -3]$

21. $f(x) = x^3 - 9, [2, 3]$

22. $f(x) = x^3 - 12, [2, 3]$

23. $f(x) = x^4 - 8, [1, 2]$

24. $f(x) = x^4 - 12, [1, 2]$

25. $f(x) = \cos x - 0.4, [0, \pi/2]$

26. $f(x) = \sin x - 0.4, [0, \pi/2]$

Each of the equations in Exercises 27–36 has one positive solution. Use Newton's Method to approximate its value to four decimal places.

27. $x^2 - x - 1 = 0$

28. $x^2 + x - 1 = 0$

29. $3x^2 + 3x - 1 = 0$

30. $3x^2 + 6x - 2 = 0$

31. $x^3 = 1 - 3x$

32. $x^3 = 3 - x$

33. $x = \cos x$

34. $x = 3 \sin x$

35. $x^2 = \cos x$

36. $x^2 = \sin x$

Show that each of the equations in Exercises 37–42 has three solutions. Use Newton's Method to approximate their values to three decimal places.

37. $x^3 - 3x + 1 = 0$

38. $x^3 - 3x - 1 = 0$

39. $x^3 - 3x^2 + 2 = 0$

40. $x^3 - 3x^2 + 3 = 0$

41. $x^3 - 3x^2 - 9x + 10 = 0$

42. $x^3 - 3x^2 - 9x + 8 = 0$

43. Newton's Method can give different results when applied to different forms of the same equation. If $c > 0$, for what values of x_1 does Newton's Method give x_n that approach c when applied to find the positive zero of the function (a) $f(x) = x - c$, (b) $g(x) = x^2 - c^2$, and (c) $h(x) = \sqrt{x} - \sqrt{c}$? (*Hint:* Sketch the graphs of each function and check where the tangent line at $(x_1, f(x_1))$ intersects the x-axis.)

44. Suppose we use Newton's Method with $f(x) = \sin x$ to find an approximate solution of $\sin x = 0$, $-\pi/2 < x < \pi/2$.
(a) If $-\pi/2 < x_1 < \pi/2$ satisfies $\tan x_1 = 2x_1$, show that $x_n = \pm x_1$ for all n, so Newton's Method does not give approximate values of the solution $c = 0$ unless $x_1 = 0$.
(b) Show that there is at least one point x_0 with $0 < x_0 < \pi/2$ and $\tan x_0 = 2x_0$. (c) Estimate the value of the number x_0 in part (b). From the figure we see that $|x_1| < x_0$ implies that Newton's Method gives x_n that approach zero.

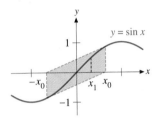

45. The **Mach number** M of a certain ramjet satisfies the equation

$$\left(\frac{5}{6}\right)^3 (1 + 0.2M^2)^3/M - 2 = 0.$$

(a) Show that the function

$$f(M) = \left(\frac{5}{6}\right)^3 (1 + 0.2M^2)^3/M - 2$$

has two zeros, one subsonic $(M < 1)$ and one supersonic $(M > 1)$.
(b) Use Newton's Method to approximate the two zeros of $f(M) = 0$.

In Exercises 46–49, use a calculator generated graph to determine approximate values of x_1 for which Newton's Method will give values x_n that approach the zero of the function.

46. $y = (x + 1)e^{-x}$

47. $y = \cos(\pi x/2), -1 < x < 3$

48. $y = (-x + 2)/(x - 1)$

49. $y = \tan^{-1} x$

3.8 THE EXTREME VALUE THEOREM; THE MEAN VALUE THEOREM

Connections

Finding zeros of expressions, 1.2.

Interpretation of graphs, 1.4–5.

Concept of the derivative as a slope, 2.6.

The Extreme Value Theorem provides the theoretical basis for the solution of the problem of finding the largest and smallest values of a function. In this section we will see how continuity is related to this problem and how the derivative can be used in its solution.

The Mean Value Theorem is an important tool for using information about the derivative of a function to obtain information about the function.

The Extreme Value Theorem

We begin our development of the Extreme Value Theorem with a formal definition of the largest and smallest values of a function on a set.

DEFINITION

The value $f(c)$ is called the **maximum value of f on a set S** if c is in S and $f(x) \leq f(c)$ for all x in S; $f(d)$ is called the **minimum value of f on a set S** if d is in S and $f(d) \leq f(x)$ for all x in S. Maximum and minimum values of a function on a set are called **extreme values** or **extrema** of the function.

The following theorem guarantees the existence of extrema of a function that is continuous on an interval that contains both endpoints.

THE EXTREME VALUE THEOREM

If f is continuous on the interval $a \leq x \leq b$, then there are points c and d in the interval such that $f(c)$ is the maximum value of f on the interval and $f(d)$ is the minimum value of f on the interval.

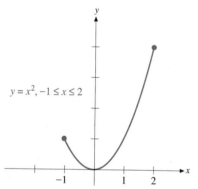

$y = x^2, -1 \leq x \leq 2$

FIGURE 3.8.1 A function that is continuous on an interval that contains both endpoints has both a maximum value and a minimum value.

We will not verify this theorem. We will illustrate the role of continuity in the result with some examples. Note that if f is continuous on an interval that contains both endpoints, then the theorem says f has both a maximum value and a minimum value on the interval. If either the interval does not contain both endpoints or f is not continuous at every point of the interval, then f may or may not have maximum or minimum values.

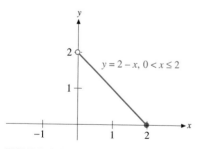

$y = 2 - x, 0 < x \leq 2$

FIGURE 3.8.2 A function that is continuous on an interval that does not contain both endpoints may or may not have maximum or minimum values.

EXAMPLE 1

Sketch the graphs of the given functions and use the graphs to find their maximum and minimum values; indicate "does not exist" where appropriate.

(a) $f(x) = x^2, \qquad -1 \leq x \leq 2,$

(b) $f(x) = 2 - x, \qquad 0 < x \leq 2,$

(c) $f(x) = 1/x, \qquad x \geq 1,$

(d) $f(x) = \begin{cases} x + 1, & -1 \leq x < 0, \\ 0, & x = 0, \\ x - 1, & 0 < x \leq 1. \end{cases}$

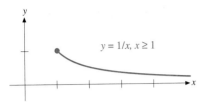

FIGURE 3.8.3 A function that is continuous on an infinite interval may or may not have maximum or minimum values.

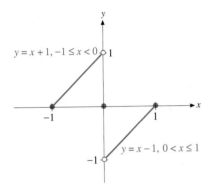

FIGURE 3.8.4 A function that is not continuous on an interval may or may not have maximum or minimum values.

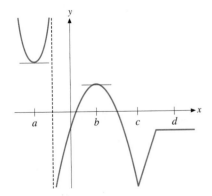

FIGURE 3.8.5 $f(a)$ and $f(c)$ are local minima of f and $f(b)$ is a local maximum of f.

SOLUTION

(a) The function $f(x) = x^2$ is continuous on the interval $-1 \leq x \leq 2$, and the interval contains both endpoints. Hence, the theorem guarantees maximum and minimum values. From the graph in Figure 3.8.1, we see that $f(2) = 4$ is the maximum value and $f(0) = 0$ is the minimum value.

(b) The theorem does not guarantee either a maximum or minimum value of $f(x) = 2 - x$ on the interval $0 < x \leq 2$, because the endpoint 0 is not included in the interval. From Figure 3.8.2, we see that $f(2) = 0$ is the minimum value of f; there is no maximum value. Two is not a *value* of f, because there is no x in the interval $0 < x \leq 2$ with $f(x) = 2$.

(c) The theorem does not guarantee either a maximum or minimum value of $f(x) = 1/x$ on the interval $x \geq 1$, because the interval is infinite. Infinite endpoints are not included in an interval. We see from Figure 3.8.3 that f has a maximum value of $f(1) = 1$; f does not have a minimum value.

(d) The theorem does not guarantee either a maximum value or a minimum value of f on the interval $-1 \leq x \leq 1$, because the function is discontinuous at $x = 0$, a point in the interval. We see from Figure 3.8.4 that f has neither a maximum value nor a minimum value on the interval; 1 and -1 are not values of f. ∎

Local Extrema

The following definition is related to the use of calculus to find extrema.

DEFINITION

The value $f(c)$ is called a **local maximum** of f if there is some open interval I with center c such that $f(x) \leq f(c)$ for all x in I. The value $f(c)$ is called a **local minimum** of f if there is some open interval I with center c such that $f(c) \leq f(x)$ for all x in I. Local maxima and local minima are called **local extrema**.

The value $f(c)$ is a local maximum of f if $f(c)$ is greater than or equal to $f(x)$ whenever x is close enough to c. A local minimum of f is less than or equal to all values of $f(x)$ for nearby x. It is required that $f(x)$ be defined for all x in some interval $|x - c| < \delta$ in order that $f(c)$ be a local maximum or minimum of f.

From the graph in Figure 3.8.5 we see that $f(a)$ and $f(c)$ are local minima of f and that $f(b)$ is a local maximum of f; $f(d)$ is both a local minimum and a local maximum of f. It is possible that a local minimum is greater than a local maximum.

The following theorem is the basis of using calculus for finding maximum and minimum values of a function.

THEOREM 1

If $f(c)$ is either a local maximum or minimum of f, then either (i) $f'(c) = 0$ or (ii) $f'(c)$ does not exist.

Proof Let us verify Theorem 1 in the case where $f(c)$ is a local maximum of f, so $f(x) \le f(c)$ for x near c. Then $f(x) - f(c) \le 0$ for x near c. We now look directly at the definition of the derivative of f at c. For x near c, $x > c$, we have $x - c > 0$, so $f(x) - f(c) \le 0$ implies

$$\frac{f(x) - f(c)}{x - c} \le 0.$$

This implies that

$$\lim_{x \to c^+} \frac{f(x) - f(c)}{x - c} \le 0,$$

if that one-sided limit exists.

For x near c, $x < c$, we have $x - c < 0$, so $f(x) - f(c) \le 0$ implies

$$\frac{f(x) - f(c)}{x - c} \ge 0.$$

This implies that

$$\lim_{x \to c^-} \frac{f(x) - f(c)}{x - c} \ge 0,$$

if that one-sided limit exists.

If $f'(c)$ exists, then both of the above one-sided limits exist and have the same value. The only possible common value of the one-sided limits is zero, so $f'(c) = 0$. Geometrically, if $f(c)$ is a local maximum of f, then the line through the points $(c, f(c))$ and $(x, f(x))$, x near c, $x \ne c$, slopes downward if $x > c$ (Figure 3.8.6a) and upward if $x < c$ (Figure 3.8.6b). If these slopes approach a common value as x approaches c, that common value must be $f'(c) = 0$ (Figure 3.8.6c).

It is not difficult to modify the above argument to show that if the derivative exists at a point whose value is a local minimum, then the derivative must equal zero. ∎

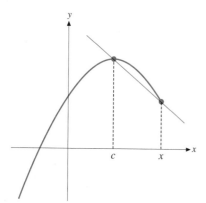

FIGURE 3.8.6a If $f(c)$ is a local maximum, the line through $(c, f(c))$ and $(x, f(x))$ slopes downward if $x > c$.

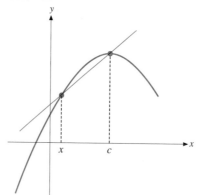

FIGURE 3.8.6b If $f(c)$ is a local maximum, the line through $(c, f(c))$ and $(x, f(x))$ slopes upward if $x < c$.

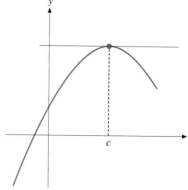

FIGURE 3.8.6c If the derivative of f exists at a point whose value is a local maximum, then the derivative at that point must be zero.

A function can have a local maximum or minimum of $f(c)$ at a point where $f'(c)$ does not exist. In Figure 3.8.7, $f(c)$ is a local minimum, but $f'(c)$ does not exist; $f'(a) = f'(b) = f'(d) = 0$.

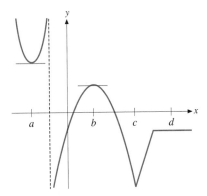

FIGURE 3.8.7 A function can have a local maximum or minimum at a point where either the derivative is zero, as at *a*, *b*, and *d*, or the derivative does not exist, as at *c*.

If f is continuous on the interval $a \leq x \leq b$, we know from the Extreme Value Theorem that f has a maximum value and a minimum value on the interval. If $f(c)$ is either a maximum or minimum value of f and $a < c < b$, then $f(c)$ must also be a local maximum or minimum, respectively. This means either $f'(c) = 0$ or $f'(c)$ does not exist. Noting that the maximum and minimum values of f on the interval $a \leq x \leq b$ can occur at either of the two endpoints of the interval, we obtain the following result:

THEOREM 2

If f is continuous on the closed interval $a \leq x \leq b$, then the maximum and minimum values of f on the interval occur at points x, where either

(i) $a < x < b$ and $f'(x) = 0$;
(ii) $a < x < b$ and $f'(x)$ does not exist; or
(iii) x is an endpoint of the interval ($x = a$ or $x = b$).

Points x for which f is defined in an open interval that contains x and either $f'(x) = 0$ or $f'(x)$ does not exist are called **critical points** of f. Theorem 2 tells us that we can find extrema of a function that is continuous on an interval that contains both endpoints by evaluating $f(x)$ at each critical point inside the interval and at both endpoints of the interval; the largest of these values is the maximum value and the smallest of these values is the minimum value.

EXAMPLE 2

Find the maximum and minimum values of the function

$$f(x) = x^3 - 12x + 3, \quad 0 \leq x \leq 3.$$

SOLUTION   Since f is continuous on the interval $0 \leq x \leq 3$, an interval that contains both endpoints, the Extreme Value Theorem guarantees both maximum and minimum values. To find them we investigate the derivative of f.

$$f(x) = x^3 - 12x + 3, \text{ so}$$
$$f'(x) = 3x^2 - 12.$$

(i) $f'(x) = 0$ implies

$$3x^2 - 12 = 0,$$
$$3(x^2 - 4) = 0,$$
$$3(x - 2)(x + 2) = 0.$$

The derivative is zero when $x = 2$. ($x = -2$ is not in the interval $0 \leq x \leq 3$.)

(ii) The derivative exists for all points in the interval.
(iii) The endpoints are $x = 0$ and $x = 3$.

The values of f at each x determined above, $x = 2, 0, 3$, are found by substitution in the formula $f(x) = x^3 - 12x + 3$. The values are given in Table 3.8.1. We see that

$$f(0) = 3 \text{ is the maximum value of } f;$$
$$f(2) = -13 \text{ is the minimum value of } f.$$

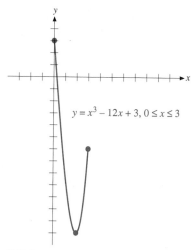

$y = x^3 - 12x + 3, 0 \le x \le 3$

FIGURE 3.8.8 The maximum value of this function occurs at an endpoint of the interval. The minimum occurs at a point where $f'(x) = 0$.

TABLE 3.8.1

	x	$f(x)$	
$f' = 0$:	2	-13	$\leftarrow$ Smallest value of f in table
Endpoints: $\begin{cases} \\ \end{cases}$	0	3	$\leftarrow$ Largest value of f in table
	3	-6	

The table of values of $f(x)$ constructed above has values of $f(x)$ for exactly those x in the interval $0 \le x \le 3$ for which either (i) $f'(x) = 0$, (ii) $f'(x)$ does not exist, or (iii) x is an endpoint of the interval. These values, and only these values, appear in the table.

The graph of $f(x) = x^3 - 12x + 3, 0 \le x \le 3$, is sketched in Figure 3.8.8. It is not necessary to graph the function to determine its maximum and minimum values. ■

EXAMPLE 3

Find maximum and minimum values of the function $f(x) = x - 3x^{1/3}$, $-1 \le x \le 8$.

SOLUTION   Since f is continuous on an interval that contains both endpoints, we know there are maximum and minimum values. We begin by investigating the derivative.

$$f(x) = x - 3x^{1/3}, \text{ so}$$

$$f'(x) = 1 - x^{-2/3}$$
$$= x^{-2/3}(x^{2/3} - 1)$$
$$= \frac{x^{2/3} - 1}{x^{2/3}}.$$

We have simplified $f'(x)$ to find values of f for which either $f'(x) = 0$ or $f'(x)$ does not exist.

(i) $f'(x) = 0$ whenever $x^{2/3} - 1 = 0$, so

$$x^{2/3} = 1,$$
$$x^2 = 1^3,$$
$$x = 1.$$

(-1 is an endpoint of the interval.)

(ii) $f'(x)$ does not exist when $x = 0$.

(iii) Endpoints are $x = -1, 8$.

The values of $f(x) = x - 3x^{1/3}$ at the points determined above are given in Table 3.8.2. We see that

$$f(-1) = f(8) = 2 \text{ is the maximum value of } f;$$
$$f(1) = -2 \text{ is the minimum value of } f.$$

The graph of $f(x) = x - 3x^{1/3}, -1 \le x \le 8$, is shown in Figure 3.8.9. ■

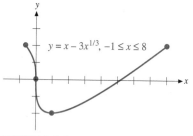

$y = x - 3x^{1/3}, -1 \le x \le 8$

FIGURE 3.8.9 This function obtains its maximum value at both endpoints. Its minimum occurs at a point where the derivative is zero.

TABLE 3.8.2

		x	$f(x)$	
$f' = 0$:		1	-2	← Smallest value of f in table
f' does not exist:		0	0	
Endpoints:	$\left\{\begin{matrix} -1 \\ 8 \end{matrix}\right.$		$\left.\begin{matrix} 2 \\ 2 \end{matrix}\right\}$	← Largest value of f in table

EXAMPLE 4

The formula $s(t) = -16t^2 + 48t$ gives the height in feet of an object t seconds after it is thrown vertically at a speed of 48 ft/s from ground level. The formula is valid until the object returns to ground level. How high above ground level will the object reach?

SOLUTION   Experience suggests that the object leaves ground level at $t = 0$, moves upward to a highest point, and then falls back to ground level, as indicated in Figure 3.8.10. To fit this problem into the mathematical context of this section, we consider the function

$$s(t) = -16t^2 + 48t, \qquad 0 \le t \le t_1,$$

where t_1 is the value of t when the object returns to ground level. This means that $s(t_1) = 0$. (It is easy to see that $s(t) = 0$ implies $t = 0, 3$, so $t_1 = 3$.) The greatest height reached by the object is the maximum value of the function $s(t)$. Since s is continuous on the interval $0 \le t \le t_1$, theory guarantees that s will have a maximum value on this interval. We have

$$s(t) = -16t^2 + 48t, \text{ so}$$
$$s'(t) = -32t + 48.$$

(i) $s'(t) = 0$ when $-32t + 48 = 0$, or $t = 1.5$. The height of the object when $t = 1.5$ is

$$s(1.5) = -16(1.5)^2 + 48(1.5) = 36.$$

(ii) $s'(t)$ exists for all x.

(iii) $s(t) = 0$ at both endpoints.

We conclude that the maximum height is $s(1.5) = 36$ ft. Note that the velocity is zero when the object is at its greatest height. ∎

FIGURE 3.8.10 The height in feet above ground of an object t seconds after it is thrown vertically at a speed of 48 ft/s from ground level is given by $s(t) = -16t^2 + 48t$.

The Mean Value Theorem

The Mean Value Theorem is a consequence of the following basic result, which we will use to establish several other results.

ROLLE'S THEOREM

If f is continuous on the closed interval $a \le x \le b$, $f(a) = f(b) = 0$, and f is differentiable in the open interval $a < x < b$, then there is a point z with $a < z < b$ and $f'(z) = 0$.

Proof   Since f is continuous on an interval that contains both endpoints, we know that f has both a maximum value and a minimum value on the

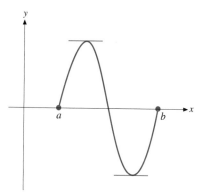

FIGURE 3.8.11 If $f(z)$ is a local extreme and f is differentiable at z, then $f'(z) = 0$.

interval. Let us see that one of these extreme values occurs at a point z with $a < z < b$.

If both extrema are zero, then $f(x) = 0$ for all points in the interval and we can choose z to be any point with $a < z < b$. If one of the extreme values is nonzero, then that extreme value must occur at a point z with $a < z < b$, since $f(a) = f(b) = 0$. It follows that one of the extreme values of f must occur at a point z with $a < z < b$.

If $f(z)$ is an extreme value of f and $a < z < b$, then $f(z)$ is a local extremum of f. Since f is differentiable at z, we must then have $f'(z) = 0$. See Figure 3.8.11. ∎

Let us now state the Mean Value Theorem and see how it can be used.

MEAN VALUE THEOREM

If f is continuous on the closed interval $a \leq x \leq b$ and differentiable on the open interval $a < x < b$, then there is a point z with $a < z < b$ and

$$\frac{f(b) - f(a)}{b - a} = f'(z).$$

Proof To verify the Mean Value Theorem, we apply Rolle's Theorem to the function

$$g(x) = f(x) - f(a) - \left(\frac{f(b) - f(a)}{b - a}\right)(x - a).$$

(The value of g is obtained by subtracting the y-coordinate of points on the line through $(a, f(a))$ and $(b, f(b))$ from the corresponding values of f. See Exercise 59.) The function g satisfies the hypotheses of Rolle's Theorem and

$$g'(z) = f'(z) - \frac{f(b) - f(a)}{b - a}.$$

Thus $g'(z) = 0$ gives the conclusion of the Mean Value Theorem. A representative graph of f is given in Figure 3.8.12. ∎

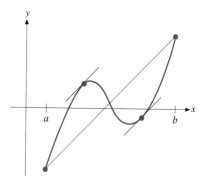

FIGURE 3.8.12 The conclusion of the Mean Value Theorem means there is at least one point between a and b where the tangent line is parallel to the secant line through $(a, f(a))$ and $(b, f(b))$.

The average (or mean) rate of change of f over the interval $a \leq x \leq b$ is

$$\frac{f(b) - f(a)}{b - a}.$$

The Mean Value Theorem tells us that, subject to the conditions of the theorem, the mean rate of change of f over the interval is equal to the instantaneous rate of change of f with respect to x at some point z, $a < z < b$.

We can use the Mean Value Theorem to obtain useful information about f, even though we may not know the particular value of z for which the conclusion holds. In fact, the usefulness of the theorem is that it is not necessary to go to the trouble of finding z. It is enough for many purposes to know that such a z exists.

EXAMPLE 5

Suppose f is continuous on the closed interval $1 \leq x \leq 3$ and $f'(x) \geq 0$ for $1 < x < 3$. Show that $f(3) \geq f(1)$.

SOLUTION   We apply the Mean Value Theorem to the function f on the interval $1 \leq x \leq 3$ to obtain

$$\frac{f(3) - f(1)}{3 - 1} = f'(z)$$

for some z, $1 < z < 3$. Since $f'(x) \geq 0$ for all x, $1 < x < 3$, we must have $f'(z) \geq 0$. Then

$$\frac{f(3) - f(1)}{3 - 1} = f'(z) \geq 0,$$

so $f(3) - f(1) \geq 0(3 - 1) = 0$, or $f(3) \geq f(1)$. ■

EXAMPLE 6

Suppose f is differentiable and $|f'(x) - 2| \leq 0.5$ whenever $|x - 2| \leq 1$. If $f(2) = 1$, find an interval that is certain to contain the value of $f(2.2)$.

SOLUTION   We can apply the Mean Value Theorem to the function f on the interval $2 \leq x \leq 2.2$ to obtain

$$\frac{f(2.2) - f(2)}{2.2 - 2} = f'(z)$$

for some z between 2 and 2.2. Since z is within 1 of 2, we know that

$$|f'(z) - 2| \leq 0.5,$$
$$-0.5 \leq f'(z) - 2 \leq 0.5,$$
$$1.5 \leq f'(z) \leq 2.5.$$

Combining results, we then have

$$1.5 \leq \frac{f(2.2) - f(2)}{2.2 - 2} \leq 2.5,$$
$$(1.5)(0.2) \leq f(2.2) - 1 \leq (2.5)(0.2),$$
$$1.3 \leq f(2.2) \leq 1.5.$$

We see that $f(2.2)$ must be in the interval $[1.3, 1.5]$. Geometrically, the graph of f must be contained between the lines $y = 1 + 1.5(x - 2)$ and $y = 1 + 2.5(x - 2)$ for $|x - 2| \leq 1$. See Figure 3.8.13. ■

We can use the Mean Value Theorem to obtain some results that are fundamental to integral calculus, a topic that we will study at great length in later chapters.

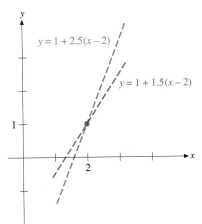

FIGURE 3.8.13   If $|f'(x) - 2| \leq 0.5$ whenever $|x - 2| \leq 1$ and $f(2) = 1$, then the graph of f must be contained between the two lines $y = 1 + 1.5(x - 2)$ and $y = 1 + 2.5(x - 2)$.

THEOREM 3

If $f'(x) = 0$ for all x in an interval, then $f(x)$ is constant on the interval.

Proof Choose any point x_1 in the interval. If x is in the interval, the Mean Value Theorem gives

$$\frac{f(x) - f(x_1)}{x - x_1} = f'(z)$$

for some z between x and x_1. Then z is in the interval, so $f'(z) = 0$. This gives

$$f(x) - f(x_1) = f'(z) \cdot (x - x_1) = 0,$$

so $f(x) = f(x_1)$. That is, all values of $f(x)$, x in the interval, are equal to the constant $f(x_1)$. ∎

It is important in the above theorem that the derivative be zero on an *interval*. A function can have derivative zero at every point in a set and not be constant on the set. For example, the function

$$f(x) = \begin{cases} 1, & 0 < x < 1, \\ -1, & -1 < x < 0, \end{cases}$$

has $f'(x) = 0$ for all x in the set that consists of the two intervals $(-1, 0)$ and $(0, 1)$. The function f is not constant on the *set,* but it is constant on each of the *intervals* where $f'(x)$ exists and is zero. See Figure 3.8.14.

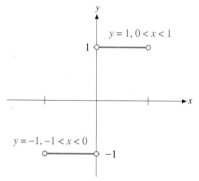

FIGURE 3.8.14 A function can have different values on a set where the derivative is zero if the set is not an interval.

THEOREM 4

If $f'(x) = g'(x)$ for all x in an interval, then there is some constant C such that $g(x) = f(x) + C$ for x in the interval.

Theorem 4 can be proved by applying Theorem 3 to the function $g - f$.

EXERCISES 3.8

In Exercises 1–8, sketch the graphs of the given functions and use the graphs to find their maximum and minimum values; indicate "does not exist" where appropriate.

1. $f(x) = x - 2, 0 \le x \le 3$

2. $f(x) = x^2 + 1, -1 \le x \le 2$

3. $f(x) = 1 - x^2, -2 < x < 1$

4. $f(x) = |x - 1| - 1, -1 < x < 2$

5. $f(x) = \tan x, -\pi/2 < x < \pi/2$

6. $f(x) = \sec x, -\pi/2 < x < \pi/2$

7. $f(x) = \begin{cases} x + 1, & -1 \le x < 0 \\ x - 1, & 0 \le x \le 1 \end{cases}$

8. $f(x) = \begin{cases} \dfrac{|x|}{x}, & x \neq 0 \\ 0 & x = 0 \end{cases}$

Find the maximum and minimum values of the functions given in Exercises 9–24.

9. $f(x) = 2x^2 - 8x + 3, 0 \le x \le 3$

10. $f(x) = 3x^2 + 6x - 10, 0 \le x \le 3$

11. $f(x) = x^3 - 6x^2 + 4, -1 \le x \le 6$

12. $f(x) = x^3 - 12x - 8, -1 \le x \le 2$

13. $f(x) = 3x^{2/3}, -1 \le x \le 8$

14. $f(x) = 3x^{1/3}, -1 \le x \le 8$

15. $f(x) = x(1 - x)^{1/3}, 0 \le x \le 2$

16. $f(x) = x - 3x^{1/3}, -1 \le x \le 27$

17. $f(x) = \dfrac{x}{x^2 + 1}, 0 \le x \le 2$

18. $f(x) = \dfrac{x^2}{x^2 + 1}, -1 \le x \le 2$

19. $f(x) = \sin \dfrac{3x}{5}, 0 \le x \le 2\pi$

20. $f(x) = \cos \dfrac{3x}{5}, 0 \le x \le \pi$

21. $f(x) = \sin x + \cos x, 0 \le x \le 2\pi$

22. $f(x) = \sin x + \cos x, 0 \le x \le \pi$

23. $f(x) = 4 - |2x - 3|, 0 \le x \le 4$

24. $f(x) = |x^3 - 1|, -1 \le x \le 2$

25. The formula $s(t) = -16t^2 + 64t$ gives the height in feet of an object t seconds after it is thrown vertically at a speed of 64 ft/s from ground level. The formula is valid until the object returns to ground level. How high above ground level will the object reach?

26. The formula $s(t) = -2.7t^2 + 64t$ gives the height in feet of an object t seconds after it is thrown vertically at a speed of 64 ft/s from the surface of the moon. The formula is valid until the object returns to the surface. How high above the surface of the moon will the object reach?

27. The formula $s(t) = -4.9t^2 + 98t$ gives the height in meters of an object t seconds after it is thrown vertically at a speed of 98 m/s from ground level. The formula is valid until the object returns to ground level. How high above ground level will the object reach?

28. The formula $s(t) = -4.9t^2 + 49t + 6$ gives the height in meters of an object t seconds after it is thrown vertically from a height 6 m above ground level at a speed of 49 m/s. The formula is valid until the object returns to ground level. How high above ground level will the object reach?

29. Find the maximum area of a rectangle that has sides of lengths x and $6 - x$. (*Hint:* Use the fact that length is nonnegative to establish an appropriate domain for the function that gives the area of the rectangle.)

30. Find the maximum volume of a rectangular box that has edges of lengths x, $3 - 2x$, and $2 - 2x$.

Find the minimum value of the functions in Exercises 31–32.

31. $f(x) = \sec x - x, -\pi/2 < x < \pi/2$

32. $f(x) = x^2 - \sin x, 0 \le x \le \pi/2$

33. Show that among all rectangles that have lengths of sides x and $5 - 2x$, there is none that has minimum area. (It is convenient to use the fact that length is nonnegative to set up the problem mathematically, but a rectangle that has length of one side zero is not actually a rectangle.)

34. If f is continuous on the interval $a \le x \le b$, differentiable on the interval $a < x < b$, with $f(a) \ge 0$, $f(b) \ge 0$, and $f'(x) \ne 0$ for $a < x < b$, show that $f(x) > 0$ for $a < x < b$. (*Hint:* Can f obtain its minimum value at a point x with $a < x < b$?)

35. Show that $\tan x > x, 0 < x < \pi/4$.

36. Complete the proof of Theorem 1 by showing that, if $f(c)$ is a local minimum of f and $f'(c)$ exists, then $f'(c) = 0$.

37. The vertical height at time t of a weight hanging on a spring is $s(t) = \ell + a \cos \omega t + b \sin \omega t$. Find (a) the maximum speed of the weight and (b) the position of the weight when the speed is maximum.

38. As a wheel with radius r rolls one revolution up an incline of angle α, the vertical height of a fixed point on the edge of the wheel is $h(s) = \left(s - r \sin \dfrac{s}{r}\right) \sin \alpha + \left(r - r \cos \dfrac{s}{r}\right) \cos \alpha, 0 \le s \le 2\pi r$. See figure. Find the maximum height of the point. (*Hint:* $\cos \dfrac{s}{r} \cos \alpha + \sin \dfrac{s}{r} \sin \alpha = \cos\left(\dfrac{s}{r} - \alpha\right)$.)

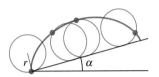

39. Functions that are derivatives satisfy the Intermediate Value Theorem, even though they may not be continuous. In particular, if $f'(x)$ is defined for $a \le x \le b$ with $f'(a) < 0$ and $f'(b) > 0$, there must be a point c between a and b with $f'(c) = 0$. Use the Extreme Value Theorem for f and Theorem 1 of this section to verify this special case. (*Hint:* Can $f(a)$ or $f(b)$ be the minimum value of f on the interval $a \le x \le b$?)

In Exercises 40–47: (a) Determine if the hypotheses of the Mean Value Theorem hold for the given functions on the indicated interval. If not, indicate why not. (b) Find all z in the interval that satisfy the conclusion of the Mean Value Theorem.

40. $f(x) = x^2 - 1, -1 \le x \le 2$

41. $f(x) = x^2 - 2x, -2 \le x \le 2$

42. $f(x) = |x|, -1 \le x \le 1$

43. $f(x) = \begin{cases} 1/x, & 0 < |x| \le 1, \\ 0, & x = 0. \end{cases}$

44. $f(x) = x^{1/3}, -1 \le x \le 1$

45. $f(x) = x^{2/3}, -1 \le x \le 1$

46. $f(x) = \begin{cases} x^2 + x, & -1 \le x \le 0, \\ x, & 0 < x \le 1. \end{cases}$

47. $f(x) = \begin{cases} x^2 + x, & -1 \le x \le 0, \\ 3x/2, & 0 < x \le 1 \end{cases}$

48. Suppose f is continuous for $0 \le x \le 3$ and differentiable for $0 < x < 3$ with $f'(x) \ge 2$ for $0 < x < 3$. If $f(0) \ge 4$, show $f(3) \ge 10$.

49. Suppose f is continuous for $-1 \le x \le 1$ and differentiable for $-1 < x < 1$ with $f'(x) \ge 3$ for $-1 < x < 1$. If $f(1) \le 0$, show $f(-1) \le -6$.

50. Suppose f is differentiable with $f'(x) \le -4$ for all x. If $f(-2) \le 7$, show $f(2) \le -9$.

51. Suppose f is differentiable with $f'(x) \le 5$ for all x. If $f(4) \ge 18$, show $f(1) \ge 3$.

52. Suppose f is differentiable and $|f'(x) - 5| \le 0.5$ whenever $|x - 2| \le 0.4$. If $f(2) = 3$, find an interval that is certain to contain the value of $f(2.3)$.

53. Suppose f is differentiable and $|f'(x) - 5| \le 0.1$ whenever $|x - 2| \le 0.4$. If $f(2) = 3$, find an interval that is certain to contain the value of $f(2.3)$.

54. Suppose f is differentiable and $|f'(x) - 3| \le 0.2$ whenever $|x - 7| \le 0.4$. If $f(7) = 3$, find an interval that is certain to contain the value of $f(6.9)$.

55. Suppose f is differentiable and $|f'(x) - 1.5| \le 0.01$ whenever $|x - 3| \le 0.05$. If $f(3) = 2.25$, find an interval that is certain to contain the value of $f(2.98)$.

56. A motorist tells a police officer that although he drove 84 miles in 1 hour 24 minutes, he never exceeded 55 mph. Explain why the officer (who understands calculus) knows that this was impossible. (Assume that the driver's speed was a differentiable function of time.)

57. Suppose f and g are both differentiable for all x, $f(0) \ge g(0)$, and $f'(x) \ge g'(x)$ for $x > 0$. Show that $f(x) \ge g(x)$ for $x > 0$. (Apply the Mean Value Theorem to the function $h = f - g$ over the interval $[0, x]$.)

58. Find a function f with $f(0) = 0$, $f(1) = 1$, and $f'(x) = 0$ for $0 < x < 1$. Why doesn't the Mean Value Theorem apply to this function on the interval $[0, 1]$?

59. Find the equation of the line through the two points $(a, f(a))$ and $(b, f(b))$. If L denotes the function whose graph is that line, show that the function used in the proof of the Mean Value Theorem is $g = f - L$.

60. Use the Intermediate Value to show that the equation $2x^3 + 3x - 1 = 0$ has exactly one solution.

61. Use Rolle's Theorem to show that a polynomial of degree 2 can have at most two distinct zeros. (*Hint:* The derivative must be zero at some point between two distinct zeros of a polynomial. The derivative of a polynomial of degree 2 is a polynomial of degree 1.)

62. If $f''(x) > 0$ for $a < x < b$, show that f can have at most one critical point in the interval. (Recall that z is a critical point of f if either $f'(z) = 0$ or $f'(z)$ does not exist.)

63. If $f'(x) = a$ for all x in an interval I, show that there is a constant b such that $f(x) = ax + b$ for all x in I. (*Hint:* Consider the derivative of $f(x) - ax$.)

64. Use Theorem 3 to prove Theorem 4.

65. If $f(x) = \sin^2 x$ and $g(x) = -\cos^2 x$, show that $f'(x) = g'(x)$. Find a constant C such that $g(x) = f(x) + C$.

66. If $f(x) = \sec^2 x$ and $g(x) = \tan^2 x$, show that $f'(x) = g'(x)$. Find a constant C such that $g(x) = f(x) + C$.

67. If $f'(x) = \sec^2 x$ and $f(0) = 0$, show that $f(x) = \tan x$, $-\pi/2 < x < \pi/2$. Is it necessarily true that $f(x) = \tan x$ for any values of x that are not in the interval $-\pi/2 < x < \pi/2$?

3.9 ACCURACY OF APPROXIMATION; TAYLOR POLYNOMIALS

Connections

Linear approximation, 3.6.

Higher-order derivatives, 3.1.

Finding bounds of functions, 2.2.

Since the derivative of a function tells us something about the function, we should expect that the second derivative and higher-order derivatives give more information about the function. We will see how to use higher-order derivatives to obtain polynomials that can be used to find approximate values of a given function.

If f is differentiable at c, we have seen in Section 3.6 that the linear function

$$P_1(x) = f(c) + f'(c)(x - c)$$

can be used to approximate values of $f(x)$ for x near c. Note that the graph of P_1 is the line tangent to the graph of f at the point $(c, f(c))$. See Figure 3.9.1. It is easy to check that

$$P_1(c) = f(c) \text{ and } P_1'(c) = f'(c),$$

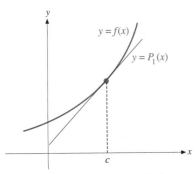

FIGURE 3.9.1 The graph of the linear function P_1 is the line tangent to the graph of $f(x)$ at the point $(c, f(c))$.

so the values of P_1 and its first derivative at c agree with the values of f and f' at c. This suggests that, if we would like to approximate the values of $f(x)$ with a quadratic function, we should try the quadratic function P_2 that satisfies

$$P_2(c) = f(c), \quad P_2'(c) = f'(c), \text{ and } P_2''(c) = f''(c).$$

The unique quadratic function that satisfies these conditions is

$$P_2(x) = f(c) + f'(c)(x - c) + \frac{f''(c)}{2}(x - c)^2.$$

Taylor's Formulas

Let us now introduce some results that allow us to compare the accuracy of linear and quadratic approximations. The results are based on the following generalization of the Mean Value Theorem.

TAYLOR'S FORMULAS

Suppose I is an interval that contains c. If f'' exists on I, then for each point x in I, there is a point z between x and c such that

$$f(x) = f(c) + f'(c)(x - c) + \frac{f''(z)}{2}(x - c)^2.$$

If f''' exists on I, then for each point x in I, there is a point z between x and c such that

$$f(x) = f(c) + f'(c) + \frac{f''(c)}{2}(x - c)^2 + \frac{f'''(z)}{6}(x - c)^3.$$

These formulas can be verified in the same manner in which we verified the Mean Value Theorem in Section 8. That is, we apply Rolle's Theorem to an auxiliary function g. (See Exercises 21 and 22.)

The linear and quadratic approximations of $f(x)$ are given by

$$P_1(x) = f(c) + f'(c)(x - c)$$

and

$$P_2(x) = f(c) + f'(c)(x - c) + \frac{f''(c)}{2}(x - c)^2,$$

respectively. P_1 and P_2 are called the first- and second-order **Taylor polynomials** of f about c. The expressions

$$R_1 = \frac{f''(z)}{2}(x - c)^2 \text{ and } R_2 = \frac{f'''(z)}{6}(x - c)^3$$

are called **remainder terms.** Taylor's Formulas can be expressed as

$$f(x) = P_1(x) + R_1 \text{ and } f(x) = P_2(x) + R_2, \text{ or}$$
$$f(x) - P_1(x) = R_1 \text{ and } f(x) - P_2(x) = R_2.$$

We then see that the remainder term in Taylor's Formula measures the difference between the actual value of the function $f(x)$ and the approximate value given by the corresponding Taylor polynomial. We do not know the value of the remainder term, because we do not know the value of the number z in the formula. However, we can determine a bound of the error in using the Taylor polynomial as an approximate value of $f(x)$ by using the methods of Section 2.2 to obtain a bound of the remainder term that holds for all values of z between x and c.

EXAMPLE 1

Use linear and quadratic approximations of x^{15} for x near 1 to find approximate values of $(0.97)^{15}$. Use the remainder terms in Taylor's Formulas to determine the accuracy of the approximations.

SOLUTION We will need formulas for the first three derivatives of the function $f(x) = x^{15}$.

We have

$$f(x) = x^{15},$$
$$f'(x) = 15x^{14},$$
$$f''(x) = 15(14)x^{13},$$
$$f'''(x) = 15(14)(13)x^{12}.$$

When $x = 1$, we obtain

$$f(1) = (1)^{15} = 1,$$
$$f'(1) = 15(1)^{14} = 15,$$
$$f''(1) = 15(14)(1)^{13} = 15(14).$$

The first- and second-order Taylor polynomials about $c = 1$ are

$$P_1(x) = f(1) + f'(1)(x - 1) = 1 + 15(x - 1) \text{ and}$$

$$P_2(x) = f(1) + f'(1)(x - 1) + \frac{f''(1)}{2}(x - 1)^2$$

$$= 1 + 15(x - 1) + \frac{15(14)}{2}(x - 1)^2.$$

Note that we do not expand the powers of $x - c$ in the Taylor polynomial of f about c.

The desired linear and quadratic approximations of $(0.97)^{15}$ are

$$P_1(0.97) = 1 + 15(0.97 - 1) = 0.55 \text{ and}$$
$$P_2(0.97) = 1 + 15(0.97 - 1) + \frac{15(14)}{2}(0.97 - 1)^2 = 0.6445.$$

To determine the accuracy of the above approximate values, we investigate the remainder terms in Taylor's Formulas. The remainder that corresponds to the linear approximation is

$$R_1 = \frac{f''(z)}{2}(x - c)^2, \text{ where } z \text{ is some number between } x \text{ and } c.$$

Using $x = 0.97$, $c = 1$, and the formula for f'' from above, we obtain

$$R_1 = \frac{15(14)z^{13}}{2}(0.97 - 1)^2, \text{ where } z \text{ satisfies } 0.97 < z < 1.$$

The inequality $0.97 < z < 1$ implies $0.97 < |z| < 1$. Since R_1 involves a positive power of z, we use the inequality $|z| < 1$ to obtain

$$|R_1| = \frac{15(14)|z|^{13}}{2}(0.97 - 1)^2$$
$$< \frac{15(14)(1)^{13}}{2}(0.97 - 1)^2$$
$$= 0.0945.$$

It follows that

$$|f(0.97) - P_1(0.97)| = |R_1| < 0.0945,$$

so the linear approximation $P_1(0.97) = 0.55$ is within 0.0945 of the exact value of $f(0.97) = (0.97)^{15}$.

The accuracy of the quadratic approximation can be determined from the remainder R_2. We have

$$R_2 = \frac{f'''(z)}{6}(x - c)^3 \text{ for some } z \text{ between } x \text{ and } c.$$

Using $x = 0.97$, $c = 1$, and the formula for f''', we obtain

$$R_1 = \frac{15(14)(13)z^{12}}{6}(0.97 - 1)^3, \quad 0.97 < z < 1.$$

As before, $0.97 < z < 1$ implies $|z| < 1$, so

$$|R_2| = \frac{15(14)(13)|z|^{12}}{6}|0.97 - 1|^3$$
$$< \frac{15(14)(13)(1)^{12}}{6}|0.97 - 1|^3$$
$$= 0.012285.$$

We then have

$$|f(0.97) - P_2(0.97)| = |R_2| < 0.012285,$$

so the quadratic approximation $P_2(0.97) = 0.6445$ is within 0.012285 of the actual value of $(0.97)^{15}$. (A calculator gives $(0.97)^{15} \approx 0.63325119$. This is consistent with the results we obtained.) The graphs of x^{15}, the linear function $P_1(x)$, and the quadratic function $P_2(x)$ are given in Figure 3.9.2. ■

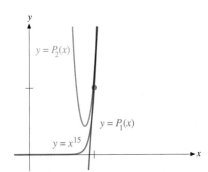

FIGURE 3.9.2 The quadratic function $P_2(x)$ gives more accurate approximate values of x^{15} than the linear function $P_1(x)$.

EXAMPLE 2

Use linear and quadratic approximations of $\cos x$ for x near 0 to find approximate values of $\cos(0.03)$. Use the remainder terms in Taylor's Formulas to determine the accuracy of the approximations.

SOLUTION   We will need the first three derivatives of $f(x) = \cos x$ and the values of f, f', and f'' at $c = 0$:

$$
\begin{aligned}
f(x) &= \cos x, & f(0) &= \cos 0 = 1, \\
f'(x) &= -\sin x, & f'(0) &= -\sin 0 = 0, \\
f''(x) &= -\cos x, & f''(0) &= -\cos 0 = -1, \\
f'''(x) &= \sin x.
\end{aligned}
$$

The first- and second-order Taylor polynomials of f at $c = 0$ are

$$P_1(x) = f(0) + f'(0)(x - 0)$$
$$= 1 + (0)(x - 0) = 1 \text{ and}$$

$$P_2(x) = f(0) + f'(0)(x - 0) + \frac{f''(0)}{2}(x - 0)^2$$

$$= 1 + (0)(x) + \frac{-1}{2}(x)^2 = 1 - 0.5x^2.$$

The desired linear and quadratic approximations are

$$P_1(0.03) = 1 \text{ and}$$
$$P_2(0.03) = 1 - 0.5(0.03)^2 = 0.99955.$$

Using $x = 0.03$, $c = 0$, and the formula for f'' from above, we have

$$R_1 = \frac{f''(z)}{2}(x - c)^2 = \frac{-\cos z}{2}(0.03)^2, \quad z \text{ between } 0.03 \text{ and } 0.$$

Since $|\cos z| \leq 1$ for any value of z, we have

$$|R_1| = \frac{|-\cos z|}{2}|0.03|^2 \leq \frac{1}{2}(0.03)^2 = 0.00045.$$

We conclude that the linear approximation $P_1(0.03) = 1$ is within 0.00045 of the actual value of $\cos(0.03)$.

The accuracy of the quadratic approximation is determined from R_2. Using $x = 0.03$, $c = 0$, and the formula for f''', we obtain

$$R_2 = \frac{f'''(z)}{6}(x - c)^3 = \frac{\sin z}{6}(0.03)^3, \quad z \text{ between } 0.03 \text{ and } 0.$$

Since $|\sin z| \leq 1$, we obtain

$$|R_2| = \frac{|\sin z|}{6}|0.03|^3 \leq \frac{1}{6}(0.03)^3 = 0.0000045.$$

We conclude that the quadratic approximation $P_2(0.03) = 0.99955$ is within 0.0000045 of the actual value of $\cos(0.03)$. (A calculator gives $\cos(0.03) \approx 0.99955003$.) The graphs of $\cos x$, $P_1(x)$, and $P_2(x)$ are given in Figure 3.9.3. ■

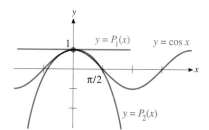

FIGURE 3.9.3   The quadratic function $P_2(x)$ gives more accurate approximate values of $\cos x$ than the linear function $P_1(x)$.

The idea in using Taylor polynomials to find approximate values of functions is to use the values of the function and its derivatives at a single point to determine values at nearby points. It is not necessary to have a formula for the function.

EXAMPLE 3

Suppose that a function f satisfies $f(1) = 0$ and is differentiable for $x > 0$ with $f'(x) = 1/x$. Use linear and quadratic approximation of $f(x)$ for x near 1 to find approximate values of $f(1.2)$. Use the remainder terms in Taylor's Formulas to determine the accuracy of the approximations.

SOLUTION   We will need the first three derivatives. We are given that

$$f'(x) = x^{-1}.$$

Differentiating, we obtain

$$f''(x) = -x^{-2},$$
$$f'''(x) = 2x^{-3}.$$

We are given that $f(1) = 0$. From the above formulas we see that $f'(1) = (1)^{-1} = 1$ and $f''(1) = -(1)^{-2} = -1$, so

$$P_1(x) = f(1) + f'(1)(x - 1)$$
$$= 0 + (1)(x - 1) = x - 1$$

and

$$P_2(x) = f(1) + f'(1)(x - 1) + \frac{f''(1)}{2}(x - 1)^2$$

$$= 0 + (x - 1) - \frac{1}{2}(x - 1)^2.$$

The approximate values of $f(1.2)$ are

$$P_1(1.2) = (1.2 - 1) = 0.2$$

and

$$P_2(1.2) = (1.2 - 1) - \frac{1}{2}(1.2 - 1)^2 = 0.18.$$

Using $x = 1.2$, $c = 1$, and the formula for f'', we obtain

$$R_1 = \frac{f''(z)}{2}(x - c)^2 = \frac{-z^{-2}}{2}(1.2 - 1)^2$$

$$= \left(-\frac{1}{2}\right)\left(\frac{1}{z^2}\right)(0.2)^2, \qquad 1 < z < 1.2.$$

The values of $\frac{1}{z^2}$ for $1 < z < 1.2$ are less than the value at $z = 1$, so

$$\frac{1}{|z|^2} < \frac{1}{(1)^2}.$$

We then have

$$|R_1| = \left(\frac{1}{2}\right)\left(\frac{1}{|z|^2}\right)(0.2)^2 < \left(\frac{1}{2}\right)\left(\frac{1}{(1)^2}\right)(0.2)^2 = 0.02.$$

The linear approximation $P_1(1.2) = 0.2$ is within 0.02 of the actual value of $f(1.2)$.

We have

$$R_2 = \frac{f'''(z)}{6}(x - c)^3 = \left(\frac{1}{6}\right)\left(\frac{2}{z^3}\right)(1.2 - 1)^3, \qquad 1 < z < 1.2.$$

As before, $1 < z < 1.2$ implies

$$\frac{1}{|z|} < \frac{1}{1}, \text{ so } |R_2| < \left(\frac{1}{6}\right)\left(\frac{2}{(1)^3}\right)(0.2)^3 \approx 0.00266667.$$

We conclude that the approximate value $P_2(1.2) = 0.18$ is within 0.00266667

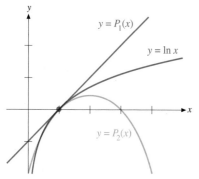

FIGURE 3.9.4 The quadratic function $P_2(x)$ gives more accurate approximate values of $\ln x$ than the linear function $P_1(x)$.

of the actual value of $f(0.02)$. The graphs of f, P_1, and P_2 are given in Figure 3.9.4. ∎

The function described in Example 3 is called the **natural logarithmic function** and is denoted $\ln(x)$. We will study this function more in Chapter 7. You may check with a calculator that $\ln(1.2) \approx 0.18232156$. This is consistent with the results of Example 3.

Higher-Order Taylor Polynomials

If f is any function that has n derivatives at c, we can form the polynomial

$$P_n(x) = f(c) + f'(c)(x - c) + \frac{f''(c)}{2!}(x - c)^2 + \cdots + \frac{f^{(n)}(c)}{n!}(x - c)^n.$$

$P_n(x)$ is called the **nth-order Taylor polynomial of f about c.** $P_n(x)$ is the unique nth-order polynomial for which the values of its derivatives of order up to and including n agree with these of f at the point c. Note that $P_n(x)$ is expressed in terms of powers of $(x - c)$. If $c \neq 0$, these powers should not be expanded in an attempt to express $P_n(x)$ in terms of powers of x. The powers of $(x - c)$ in $P_n(x)$ go from zero to n, where the zero power of $(x - c)$ is taken to be one; the coefficient of $(x - c)^k$ is $f^{(k)}(c)/k!$, $0 \leq k \leq n$, where $f^{(0)}(c) = f(c)$, $0! = 1$, and $k! = 1 \cdot 2 \cdots k$ for $k \geq 1$.

EXAMPLE 4

Find the first-, second-, third-, fourth-, fifth-, sixth-, and seventh-order Taylor polynomials of $\cos x$ about 0. Compare the values of these Taylor polynomials at $x = 0.2$ with a calculator value of $\cos(0.2)$.

SOLUTION We need the values of $f(x) = \cos x$ and the first seven derivatives at 0:

$$
\begin{aligned}
f(x) &= \cos x, & f(0) &= \cos 0 = 1, \\
f'(x) &= -\sin x, & f'(0) &= -\sin 0 = 0, \\
f''(x) &= -\cos x, & f''(0) &= -\cos 0 = -1, \\
f'''(x) &= \sin x, & f'''(0) &= \sin 0 = 0, \\
f^{(4)}(x) &= \cos x, & f^{(4)}(0) &= \cos 0 = 1, \\
f^{(5)}(x) &= -\sin x, & f^{(5)}(0) &= -\sin 0 = 0, \\
f^{(6)}(x) &= -\cos x, & f^{(6)}(0) &= -\cos 0 = -1, \\
f^{(7)}(x) &= \sin x, & f^{(7)}(0) &= \sin 0 = 0.
\end{aligned}
$$

Substitution of values in the defining formulas gives

$$P_1(x) = f(0) + f'(0)(x - 0) = 1,$$

$$P_2 = f(0) + f'(0)(x - 0) + \frac{f''(0)}{2!}(x - 0)^2 = 1 - \frac{1}{2!}x^2,$$

$$P_3 = f(0) + f'(0)(x - 0) + \frac{f''(0)}{2!}(x - 0)^2 + \frac{f'''(0)}{3!}(x - 0)^3 = 1 - \frac{1}{2!}x^2,$$

$$P_4(x) = P_5(x) = 1 - \frac{1}{2!}x^2 + \frac{1}{4!}x^4,$$

$$P_6(x) = P_7(x) = 1 - \frac{1}{2!}x^2 + \frac{1}{4!}x^4 - \frac{1}{6!}x^6.$$

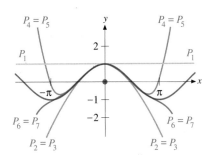

FIGURE 3.9.5 The graphs of the Taylor polynomials $P_n(x)$ become closer to the graph of $\cos x$ as n increases.

Evaluating the above Taylor polynomials at $x = 0.2$, we obtain

$$P_1(0.2) = 1,$$

$$P_2(0.2) = P_3(0.2) = 1 - \frac{1}{2}(0.2)^2 = 0.98,$$

$$P_4(0.2) = P_5(0.2) = 1 - \frac{1}{2}(0.2)^2 + \frac{1}{24}(0.2)^4 \approx 0.98006667,$$

$$P_6(0.2) = P_7(0.2) = 1 - \frac{1}{2}(0.2)^2 + \frac{1}{24}(0.2)^4 - \frac{1}{720}(0.2)^6 \approx 0.98006658.$$

A calculator gives $\cos(0.2) \approx 0.98006658$. (Be sure to use radian measure.) We can see from Figure 3.9.5 how the values of the Taylor polynomials $P_n(x)$ begin to approach $\cos x$ as n increases.

EXERCISES 3.9

In Exercises 1–12, use linear and quadratic approximations of the given function near the given point c to find approximate values of the given expression. Use the remainder terms in Taylor's Formulas to determine the accuracy of the approximations. Compare with a calculator value of the function at c.

1. $f(x) = x^{20}, c = 1, (0.97)^{20}$

2. $f(x) = x^{10}, c = 1, (0.98)^{10}$

3. $f(x) = x^{1/3}, c = 8, (8.04)^{1/3}$

4. $f(x) = x^{1/4}, c = 16, (16.05)^{1/4}$

5. $f(x) = \cos x, c = 0, \cos(0.06)$

6. $f(x) = \sin x, c = 0, \sin(0.02)$

7. $f(x) = 1/(1 - x), c = 0, 1/(1 - 0.2)$

8. $f(x) = 1/(1 - x), c = 0, 1/(1 - (-0.2))$

9. The function f satisfies $f(0) = 1$, $f'(x) = f(x)$, and $|f(x)| < 3$ for $x \le 1$; $c = 0$, $f(0.2)$. [$f(x) = e^x$.]

10. The function f satisfies $f(0) = 1$, $f'(x) = f(x)$, and $|f(x)| \le 1$ for $x \le 0$; $c = 0$, $f(-0.2)$. [$f(x) = e^x$.]

11. The function f satisfies $f(0) = 0$ and has derivative $f'(x) = (x + 1)^{-1}$; $c = 0$, $f(-0.25)$. [$f(x) = \ln(x + 1)$.]

12. The function f satisfies $f(0) = 0$ and has derivative $f'(x) = (1 - x)^{-1}$; $c = 0$, $f(0.25)$. [$f(x) = \ln\dfrac{1}{1 - x}$.]

In Exercises 13–16, find linear and quadratic polynomials that approximate the graph of the given equation near the point (x_0, y_0) on the graph. Use the polynomials to find approximate values of the y near y_0 that satisfy the equation for the given value of x. Compare with a calculator value of that y.

13. $x^2 + y^2 = 25, (x_0, y_0) = (3, 4), x = 3.1$

14. $x^2 - y^2 = 16, (x_0, y_0) = (5, 3), x = 4.9$

15. $x = \sin y, (x_0, y_0) = \left(\dfrac{1}{2}, \dfrac{\pi}{6}\right), x = 0.55$

16. $x = \tan y, (x_0, y_0) = \left(1, \dfrac{\pi}{4}\right), x = 0.94$

17. Show that

$$|\sin x - x| \le \frac{|x|^3}{6}.$$

18. Use the fact that $|\sin z| \le |z|$ to show that

$$\left|\cos x - \left(1 - \frac{x^2}{2}\right)\right| \le \frac{|x|^4}{6}.$$

19. Show that $P_3(x)$, the third-order Taylor polynomial of f about c, satisfies $P_3^{(k)}(c) = f^{(k)}(c), k = 0, 1, 2, 3$.

20. Show that $P_n(x)$, the nth-order Taylor polynomial of f about c, satisfies $P_n^{(k)}(c) = f^{(k)}(c), k = 0, \ldots, n$.

21. Suppose I is an interval that contains c and x, and that f'' exists on I. Let

$$g(t) = f(x) - f(t) - f'(t)(x - t) - R_1 \frac{(x - t)^2}{(x - c)^2},$$

where $R_1 = f(x) - f(c) - f'(c)(x - c)$. Verify the formula for the remainder term R_1 in Taylor's Formula by applying Rolle's Theorem to $g(t)$ on the interval between x and c, and showing that $g'(z) = 0$ implies

$$R_1 = \frac{f''(z)}{2}(x - c)^2.$$

22. Suppose I is an interval that contains c and x, and that f''' exists on I. Let

$$g(t) = f(x) - f(t) - f'(t)(x - t) - \frac{f''(t)}{2}(x - t)^2$$

$$- R_2 \frac{(x - t)^3}{(x - c)^3},$$

where

$$R_2 = f(x) - f(c) - f'(c)(x - c) - \frac{f''(c)}{2}(x - c)^2.$$

Verify the formula for the remainder term R_2 in Taylor's Formula by applying Rolle's Theorem to $g(t)$ on the interval between x and c, and showing that $g'(z) = 0$ implies

$$R_2 = \frac{f'''(z)}{6}(x - c)^3.$$

In Exercises 23–30 find the nth-order Taylor polynomial of f about c and compare the values of $f(x_0)$ and $P_n(x_0)$.

Plot the graphs of f and P_n for x in an interval that contains c to illustrate how the polynomials approximate the function over an interval.

23. $f(x) = x^3 - x^2 + x - 1, n = 1, 2, 3, c = 1, x_0 = 1.2$

24. $f(x) = x^3, n = 1, 2, 3, c = 1, x_0 = 1.1$

25. $f(x) = (x - 2)^3, n = 1, 2, 3, c = 0, x_0 = -0.1$

26. $f(x) = \dfrac{1}{1 - x}, n = 1, 2, 3, c = 0, x_0 = 0.1$

27. $f(x) = \tan x, n = 1, 2, 3, c = 0, x_0 = 0.2$

28. $f(x) = \sin x, n = 2, 4, 6, c = 0, x_0 = -0.1$

29. $f(x) = \sqrt{x}, n = 1, 2, 3, c = 1, x_0 = 0.98$

30. $f(x) = x^{1/3}, n = 1, 2, 3, c = 1, x_0 = 1.02$

REVIEW OF CHAPTER 3 CONCEPTS

In this chapter we developed formulas for the derivatives of power and root functions and the six trigonometric functions. We also developed formulas for the derivatives of sums, products, quotients, and compositions of functions. You should know each of these basic formulas.

- You must recognize compositions of functions and know how to apply the Chain Rule to find their derivatives.

- The method of implicit differentiation can be used to find derivatives of functions that are defined implicitly by an equation. The method involves differentiating each side of the equation with respect to the independent variable. It is important to remember which variable is considered to be a function of the independent variable and to treat that variable as a function when applying the Chain Rule and other rules of differentiation.

- Linear approximation is based on the fact that, if a function f is differentiable at a point c, then

$$\frac{f(x) - f(c)}{x - c} \approx f'(c), \text{ so}$$
$$f(x) \approx f(c) + f'(c)(x - c) \text{ for all } x \text{ near } c.$$

- Newton's Method of approximating the zeros of a function is based on the fact that differentiable functions can be approximated by their tangent lines.

- The Intermediate Value Theorem tells us that the graph of a function that is continuous on an interval cannot jump over a horizontal line without intersecting the line. The conclusion is not necessarily true if the function is discontinuous at some point on the interval.

- The Extreme Value Theorem tells us that a function that is continuous on an interval that contains both endpoints must have both a maximum value and a minimum value on the interval. The conclusion is not necessarily true if either there are points in the interval where the function is discontinuous or the interval does not contain both endpoints.

- The Mean Value Theorem can be used to obtain information about a function from information about its derivative. We have seen how the Mean Value Theorem can be used to obtain bounds of error of linear and quadratic approximations.

CHAPTER 3 REVIEW EXERCISES

Find the derivatives in Exercises 1–30.

1. $\dfrac{d}{dx}(x^3 - 2x^2 + x - 4)$

2. $\dfrac{d}{dx}\left(\sqrt{x} + \dfrac{1}{\sqrt{x}}\right)$

3. $\dfrac{d}{dx}\left(\dfrac{x^2 + 1}{x^2}\right)$

4. $\dfrac{d}{dx}\left(\dfrac{2x^2 - 3x + 5}{x}\right)$

5. $\dfrac{d}{dx}(x^2(3x^2 + 2x - 3))$

6. $\dfrac{d}{dx}(\cos x(\sec x - \tan x))$

7. $\dfrac{d}{dx}(x^2 \cdot \tan x)$

8. $\dfrac{d}{dx}(x \cdot \sin x)$

9. $\dfrac{d}{dx}(\sec x \tan x)$

10. $\dfrac{d}{dx}(\sec^2 x)$

11. $\dfrac{d}{dx}\left(\dfrac{\cos x}{\sqrt{x}}\right)$

12. $\dfrac{d}{dx}(\sqrt{x}\,\sin x)$

13. $\dfrac{d}{dx}\left(\dfrac{1-x}{1+x}\right)$

14. $\dfrac{d}{dx}\left(\dfrac{2x-1}{3x+2}\right)$

15. $\dfrac{d}{dx}\left(\dfrac{x^2}{\cos x}\right)$

16. $\dfrac{d}{dx}\left(\dfrac{\cos x}{1+x^2}\right)$

17. $\dfrac{d}{dx}((5x-1)^{-1/3})$

18. $\dfrac{d}{dx}(\sec\sqrt{x})$

19. $\dfrac{d}{dx}\left(\dfrac{1}{1-\cos 3x}\right)$

20. $\dfrac{d}{dx}\left(\dfrac{1}{1-\csc 2x}\right)$

21. $\dfrac{d}{dx}(\sin^2 x)$

22. $\dfrac{d}{dx}(\sin x^2)$

23. $\dfrac{d}{dx}(\tan^3 3x)$

24. $\dfrac{d}{dx}(\tan\sqrt{x^2+1})$

25. $\dfrac{d}{dx}(\sec^2 2x)$

26. $\dfrac{d}{dx}(\sqrt{1+\cos^2 x})$

27. $\dfrac{d}{dx}(\csc(\cot x))$

28. $\dfrac{d}{dx}((1+\sqrt{x^2+1})^2)$

29. $\dfrac{d}{dx}(x\sqrt{4x+1})$

30. $\dfrac{d}{dx}\left(\sqrt{\dfrac{2x+1}{3x-1}}\right)$

31. $f(x) = \begin{cases} x^3, & x < 0 \\ 0, & x = 0 \\ x^2, & x > 0. \end{cases}$ Evaluate (a) $f'(-1)$, (b) $f'(0)$, and (c) $f'(1)$.

32. $f(x) = \begin{cases} \sin x, & x < 0 \\ 0, & x = 0 \\ x, & x > 0. \end{cases}$ Evaluate (a) $f'(-1)$, (b) $f'(0)$, and (c) $f'(1)$.

33. $f(x) = \begin{cases} 2x - x^2, & x < 0 \\ 0, & x = 0 \\ -2x - x^2, & x > 0. \end{cases}$ Evaluate (a) $f'(-1)$, (b) $f'(0)$, and (c) $f'(1)$.

34. $f(x) = \begin{cases} x+1, & x < 0 \\ 0, & x = 0 \\ x-1, & x > 0. \end{cases}$ Evaluate (a) $f'(-1)$, (b) $f'(0)$, and (c) $f'(1)$.

35. $f(x) = \begin{cases} \sqrt{-x}, & x < 0 \\ 0, & x = 0 \\ \sqrt{x}, & x > 0. \end{cases}$ Evaluate (a) $f'(-1)$, (b) $f'(0)$, and (c) $f'(1)$.

36. $f(x) = \begin{cases} x/2, & x < 0 \\ 0, & x = 0 \\ 2x, & x > 0. \end{cases}$ Evaluate (a) $f'(-1)$, (b) $f'(0)$, and (c) $f'(1)$.

Find the first three derivatives of the given function in each of Exercises 37–40.

37. $f(x) = \dfrac{1}{1-2x}$

38. $f(x) = (1-x)^{1/3}$

39. $f(x) = \sin^2 x$

40. $f(x) = \tan x$

41. Find the equation of the line tangent to the graph of $f(x) = \sqrt{x}$ at the point on the graph with $x = 4$.

42. Find the equation of the line tangent to the graph of $f(x) = \sin^2 x$ at the point where $x = \pi/6$.

43. Find the equation of the line normal to the graph of $x = \sin y$ at the point $(1/2, \pi/6)$.

44. Find the equation of the line normal to the graph of $x = \tan y$ at the point $(1, \pi/4)$.

45. Find all points on the graph of $f(x) = x(x+32)^{1/3}$ where (a) $f'(x) = 0$ and (b) $f'(x)$ does not exist.

46. Find all points on the graph of $f(x) = 3\sin x^{1/3} + 3\cos x^{1/3}$, $-\pi^3 \le x \le \pi^3$, where (a) $f'(x) = 0$ and (b) $f'(x)$ does not exist.

47. Find all points on the graph of $f(x) = x - 2\cos x$, $0 \le x \le 2\pi$, where the graph has a horizontal tangent.

48. Find all points on the graph of $f(x) = \dfrac{x}{(3x+16)^{1/3}}$ where the graph has a horizontal tangent.

49. The position of a particle on a number line at time t is given by $s(t) = \sqrt{t}/(1+t)$. Find all intervals of time for which the particle is moving in the positive direction.

50. The distance from the ceiling at time t of a certain weight that is bouncing up and down on a spring attached to the ceiling is given by $s(t) = 2 + \cos 3t$. Find the position of the weight at times when the velocity is zero.

51. In Chapter 7 we will introduce a function S with the derivative

$$S'(x) = \dfrac{1}{\sqrt{1-x^2}}.$$

Use this formula to evaluate

$$\dfrac{d}{dx}(S(x/a)), \quad a \text{ a positive constant.}$$

52. In Chapter 7 we will introduce a function T with the derivative

$$T'(x) = \dfrac{1}{1+x^2}.$$

Use this formula to evaluate

$$\frac{d}{dx}(T(x/a)), \quad a \text{ a nonzero constant.}$$

53. A ladder 12 ft long is leaning against a wall. The bottom of the ladder is sliding away from the wall at a rate of 3 ft/s, while the top of the ladder slides down the wall. Let x denote the distance of the bottom of the ladder from the wall and let y denote the height of the top of the ladder. Since the ladder is moving, x and y are functions of time. The functions are related by the equation $x^2 + y^2 = 12^2$. We are given that the rate of change of x with respect to time is $dx/dt = 3$. Find the rate of change of y with respect to time when $x = 3$, 6, and 9. What happens to dy/dt as x approaches 12? See figure.

54. If a rocket is rising vertically from a point on the ground that is 60 ft from a ground-level observer, the height of the rocket and the angle of elevation from the observer to the rocket are related by the equation $h = 60 \tan \theta$, where h and θ are functions of time. The rocket is rising at a rate of 120 ft/s when it is 80 ft high. What is the rate of change of the angle of elevation at this time? See figure.

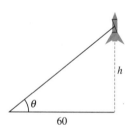

In Exercises 55–60 (a) sketch the graph of the equation, (b) find all points (x_0, y_0) on the graph such that the graph defines y as a function of x in some rectangle with center (x_0, y_0), (c) determine points on the graph where y is a differentiable function of x, and (d) find $\dfrac{dy}{dx}$ at each point where y is differentiable.

55. $4x^2 + y^2 = 4$ **56.** $y^2 = x^2$

57. $4x^2 - y^2 = 4$ **58.** $y^2 - 2xy = 0$

59. $x = |y|$ **60.** $|x| + |y| = 1$

Use the Method of Implicit Differentiation to find dy/dx for the functions of x defined implicitly by the equations in Exercises 61–68.

61. $x^3 - y^3 = 1$ **62.** $\sqrt{x} + \sqrt{y} = 1$

63. $y^2 - 3xy - x^2 = 0$ **64.** $y^2 + 2xy - x^2 = 0$

65. $x = \tan 2y$ **66.** $x = y^2 - 2y$

67. $x = \cos y$ **68.** $x = 2y^2 - 3y + 1$

Find dy/dx and d^2y/dx^2 for the functions of x that are defined implicitly by equations given in Exercises 69–74.

69. $x^2 - y^2 = 3$ **70.** $x^3 + y^2 = 1$

71. $x = \sec y$ **72.** $x = \sin y + \cos y$

73. $\sin y = \cos x$ **74.** $\sin \dfrac{y}{x} = \dfrac{1}{2}$

75. Assume the equation $x = \tan y$ defines y as a differentiable function of x. Show that

$$\frac{dy}{dx} = \frac{1}{1 + x^2}.$$

(*Hint:* $\sec^2 y = 1 + \tan^2 y$.)

76. Assume the equation $x = \sin y$ defines y as a differentiable function of x with $-\pi/2 < y < \pi/2$. Show that $\dfrac{dy}{dx} = \dfrac{1}{\sqrt{1 - x^2}}$. (*Hint:* $\sin^2 y + \cos^2 y = 1$.)

Express the differentials du in terms of the variables x and dx in Exercises 77–80.

77. $u = x^2 + 4$ **78.** $u = \sqrt{x}$

79. $u = \sin x$ **80.** $u = \tan(x/2)$

Find an equation that relates the differentials of the variables in each equation in Exercises 81–86.

81. $x^2 + y^2 = r^2$ **82.** $A = \ell w$

83. $x = r \cos \theta$ **84.** $y = r \sin \theta$

85. $V = \pi r^2 h$ **86.** $S = 2\pi rh + 2\pi r^2$

87. The height of a rectangular box is half the length of one side of its square base. Use differentials to find the approximate error in volume if the length of each side of the base is 2 m with a possible error of ± 0.02 m. Find the approximate relative error.

88. The angle of elevation between an observer at ground level and an airplane is measured to be $20°$ with a possible error of $1°$ as the plane passes over a point on the ground that is 1 mi from the observer. Find the height of the plane and the differential approximation of the error in calculation. Find the approximate relative error.

89. Use differentials to find the approximate maximum error in the volume of a rectangular box if the length is 7 ± 0.05, the width is 5 ± 0.04, and the height is 3 ± 0.03.

90. A rectangular box is to have length 5, width 4, and height 3. If the same error is made in each of the length, width, and height, which will affect the volume most?

91. Determine that the equation $x^3 + \sqrt{3}x - \sqrt{19} = 0$ has a solution in the interval $1 \le x \le 2$, and then find an interval of length 0.1 that contains the solution.

92. Determine that the equation $x^3 = \cos x$ has a solution in the interval $0 \le x \le \pi/2$, and then find an interval of length 0.1 that contains the solution.

93. Use Newton's Method to approximate to four decimal places the solution of the equation $2x = \cos x$.

94. Use Newton's Method to approximate to four decimal places the solution of the equation $x^3 + 3x = 2$.

95. The equation $\tan x = \cos x$ has one solution between 0 and $\pi/2$. Use Newton's Method to approximate the solution to three decimal places.

96. The equation $x^4 + x = 1$ has two solutions. Use Newton's Method to approximate the solutions to three decimal places.

97. The equation $x^3 - 6x - 1 = 0$ has three solutions. Use Newton's Method to approximate the solutions to three decimal places.

98. The equation $x^3 - 3x^2 + 1 = 0$ has three solutions. Use Newton's Method to approximate the solutions to three decimal places.

In Exercises 99–100 use the given graph of f to determine the first approximations x_1 for which successive approximations given by Newton's Method will approach the solution of $f(x) = 0$.

99.   **100.**

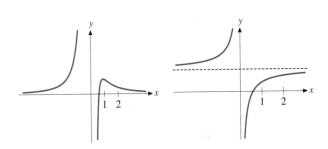

In Exercises 101–104 sketch the graphs and find the maximum and minimum values of the given functions. Indicate "does not exist" where that is appropriate.

101. $f(x) = 3 - 2x, 0 \le x < 2$

102. $f(x) = x^2 - 1, -1 < x < 2$

103. $f(x) = 1 - |x|, -1 < x < 2$

104. $f(x) = \begin{cases} -1 - x, & -1 \le x < 0 \\ 0, & x = 0 \\ 1 - x, & 0 < x \le 1. \end{cases}$

Find the maximum and minimum values of the functions given in Exercises 105–108.

105. $f(x) = x^3 - x^2 - 2, -1 \le x \le 2$

106. $f(x) = \sin \dfrac{2\pi x}{3}, 0 \le x \le 2$

107. $f(x) = (1 - 3x)^{2/3}, 0 \le x \le 3$

108. $f(x) = |x^2 - 4x - 5|, 0 \le x \le 6$

109. Find the maximum volume of a rectangular box that has edges of lengths $x, 6 - 2x$, and $6 - 2x$.

110. The perimeter of a triangle that has sides of lengths 1 and 2 with included angle θ is given by the formula $p(\theta) = 3 + \sqrt{5 - 4\cos\theta}$. Show that among all such triangles there is none for which the perimeter is either maximum or minimum.

111. If $f(2) = 3$ and $f'(x) \ge -1$ for all x, show that $f(7) \ge -2$.

112. If $f(2) = 3$ and $|f'(x) - 2| \le 0.2$ for all x, show that $3 + 2.2(x - 2) \le f(x) \le 3 + 1.8(x - 2)$ for $x \le 2$ and $3 + 1.8(x - 2) \le f(x) \le 3 + 2.2(x - 2)$ for $x \ge 2$, so the graph of f is contained between the two lines $y = 3 + 1.8(x - 2)$ and $y = 3 + 2.2(x - 2)$.

113. If f is continuous on the interval $x \ge c$ and $f'(x) > 0$ for all $x > c$, show that $f(x) > f(c)$ for all $x > c$.

114. If $f'(x) = \cos x$ and $f(0) = 0$, show that $f(x) = \sin x$.

115. Use linear and quadratic approximations of x^{16} for x near 1 to find approximate values of $(0.98)^{16}$. Use the remainder terms in Taylor's Formulas to determine the accuracy of the approximations. Compare with a calculator value.

116. Use linear and quadratic approximations of x^{-8} for x near 1 to find approximate values of $(1.06)^{-8}$. Use the remainder terms in Taylor's Formulas to determine the accuracy of the approximations. Compare with a calculator value.

117. Use linear and quadratic approximations of $\tan x$ for x near $\pi/4$ to find approximate values of $\tan 0.75$. Use the remainder terms in Taylor's Formulas to determine the accuracy of the approximations. Compare with a calculator value.

118. Use a quadratic approximation of $\cos x$ for x near 0 to find an approximate value of $\cos(-0.15)$. Use the remainder term in Taylor's Formula to determine the accuracy of the approximation. Compare with a calculator value.

119. Use the line tangent to the graph of the equation $x = 3y^2 - y^3$ at the point $(2, 1)$ to approximate the value of y near 1 that satisfies the equation when $x = 1.96$.

120. Use the line tangent to the graph of $x^3 = y^2 + y + 6$ at the point $(2, 1)$ to approximate the value of y near 1 that satisfies the equation when $x = 2.04$.

121. Find the fourth-order Taylor polynomial of $f(x) = \cos(2x)$ about 0.

122. Find the third-order Taylor polynomial of $f(x) = 1/(1+x)$ about 0.

123. Find the third-order Taylor polynomial of $f(x) = \sqrt{x}$ about 4.

124. Find the third-order Taylor polynomial of $f(x) = \sin x$ about $\pi/6$.

125. Find the third-order Taylor polynomial of $f(x) = \sqrt{1+x}$ about 0.

126. Find the third-order Taylor polynomial of $f(x) = (1+x)^{-1/2}$ about 0.

127. The function $f(x) = \csc x$ satisfies $f(-\pi/2) = -1$ and $f(\pi/2) = 1$, but there is no point c between $-\pi/2$ and $\pi/2$ with $f(c) = 0$. Why doesn't the Intermediate Value Theorem apply to this function on the interval $-\pi/2 \le x \le \pi/2$?

128. Evaluate $\dfrac{d}{dx}\left(\cos(x^2+3)^4 \sin^2 x\right)$ in several steps and explain what is being done in each step.

129. Use the Intermediate Value Theorem to explain why an irregularly shaped cake such as the one shown in the figure can be divided into two equal parts by a single vertical slice.

130. The formula $A = \pi r^2$ for the area of a circle of radius r gives the differential expression $dA = 2\pi r\, dr$. Use a geometric argument in terms of areas to explain why dA is a good approximation of $\Delta A = A(r + dr) - A(r)$ if dr is small.

EXTENDED APPLICATION

ROLLER COASTER

You have just been contacted by an entrepreneur who plans to build an amusement park in Arizona, to be called Seven Flags over Flagstaff. The park is to feature the world's highest wooden frame roller coaster. You have been asked to assist in its design because of your experience in designing urban rail systems, as well as your love of roller coasters. You are working under the following constraints on the shape of the track:

- In order to secure the coaster's claim to record height, the highest point on the track will be at least 240 feet above ground level. However, the strength of the frame will not permit a height greater than 270 feet.
- Because the car loses speed as a result of friction, the vertical rise on any uphill segment cannot exceed 2/3 of the vertical drop in the preceding downhill segment.
- Because you believe that part of the ride's thrill comes from constant changes in direction, no segment of the track will be linear.
- To keep the ride safe, the slope of the track must not exceed 1.5 in absolute value at any point.

One of your responsibilities is to design the stretch of track immediately beyond the highest point. The track will not curve to the side during this stretch, but will only move up and down. You have sketched the track in a coordinate system, as in Figure 1. The track will have horizontal tangent lines at A, which is between (0, 240) and (0, 270), and at D, which is between (690, 0) and (710, 0). It will contain at least one uphill segment between the two downhills AB and CD. You may elect to include additional uphill segments.

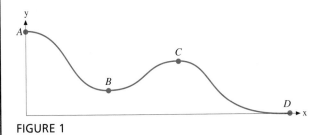

FIGURE 1

You have decided that construction will be simpler if each of the segments AB, BC, etc., has the shape of the graph of some elementary function over an appropriate interval. You are presently exploring the graphs of several types of

functions, in order to determine which ones will meet your requirements.

Tasks:

1. On a piece of graph paper, carefully draw the track as you would like to have it constructed. You need not find a formula for your graph.

2. **a.** Find a function $f(x) = ax^3 + bx^2 + cx + d$ that passes through (0, 1) and (1, 0) and has horizontal tangent lines at both points.

   b. Graph $y = f'(x)$ on a graphing utility, and estimate the maximum value of $|f'(x)|$ on [0, 1].

   c. Let $g(x) = 250f(x)$, and let $h(x) = g(x/300)$. Verify analytically that $g(x)$ has horizontal tangent lines at (0, 250) and (1, 0), and $h(x)$ at (0, 250) and (300, 0).

   d. Graph $y = g'(x)$ and $y = h'(x)$ on a graphing utility. Estimate the maximum value of $|g'(x)|$ on [0, 1], and of $|h'(x)|$ on [0, 300]. Use your graphs to tell how g' is related to f', and h' to g'. Confirm your conclusions analytically.

3. Repeat Task 2 for functions of each of the following forms:
   a. $f(x) = a(x^2 - c^2)^2$
   b. $f(x) = a \cos bx + c$

4. Construct a function to approximate your drawing from Task 1 as closely as possible, as follows:

   a. Choose an elementary function whose graph will represent the segment AB. It may be of one of the types explored in Tasks 2 and 3, or some other type.

   b. Choose an elementary function to represent the segment BC. You may need to use your knowledge of translations and reflections of graphs, as well as your observations from Tasks 2 and 3.

   c. Continue for all other segments in your drawing. Make sure that you have differentiability at each point where two segments join, and that you have met all required conditions.

   d. Graph the function that represents the entire stretch of track. If possible, use a graphing utility and obtain a hard copy. Compare your graph with your drawing from Task 1.

Writing Assignment:
Submit a technical report to your colleagues on the coaster design team, in which you specify your design. Include all graphs and calculations, and discuss the advantages of your design compared to others.

4 Graphs and Applications of the Derivative

In this chapter we will study some applications of the ideas developed in Chapter 3.

One major objective is to see how various calculus concepts are related to graphs of functions. We begin by extending the definition of limit to include infinite-valued limits and limits at infinity and illustrating how these concepts are related to asymptotes. We will also see how the signs and zeros of a function and its first and second derivatives can be used to study properties of the function and its graph.

Another important objective is to illustrate how calculus can be used to solve some practical problems. This involves translating problems that arise in real-world situations into mathematical expressions that involve real variables. We will investigate some representative procedures for doing this and then see how to use the results in some problems that involve rates of change and maximum-minimum values.

4.1 INFINITE-VALUED LIMITS; L'HÔPITAL'S RULE

Connections

Graphs with asymptotes, 1.5.

Evaluation of limits, 2.5.

In this section we will investigate the behavior of graphs where the functions (possibly) become unbounded.

Infinite-Valued Limits

We saw in Chapter 2 that the limit of $f(x)$ as x approaches a point c is useful for describing a particular behavior of the graph of f for x near c. We also know that if $f(x)$ becomes unbounded as x approaches c, then $\lim_{x \to c} f(x)$ does not exist as a real number. It is important for the concepts of continuity and differentiability that a limit be a real number. For other purposes, it is convenient to extend the idea to include cases where the values $f(x)$ approach either infinity or negative infinity as x approaches c. The formal definitions follow.

DEFINITION

We say $\lim_{x \to c} f(x) = \infty$ if, for each number M, there is a number $\delta > 0$ such that $f(x) > M$ whenever $0 < |x - c| < \delta$. See Figure 4.1.1a. We say $\lim_{x \to c} f(x) = -\infty$ if, for each number m, there is a number $\delta > 0$ such that $f(x) < m$ whenever $0 < |x - c| < \delta$. See Figure 4.1.1b.

One-sided limits can also be extended to include the values ∞ and $-\infty$. As before, the limit at c exists if and only if both one-sided limits exist and have the same value. This now includes values of infinity and negative infinity.

We say $x = c$ is a **vertical asymptote** of a function f if either of the one-sided limits of $f(x)$ at c is either ∞ or $-\infty$. The expression $x = c$ is the equation of a vertical line, so a vertical asymptote is a vertical line.

A quotient of continuous functions u/v has a vertical asymptote at points c where the denominator is zero and the numerator is not. The sign of $u(x)/v(x)$ for x on each side of c, x near c, then determines the behavior of the graph near $x = c$.

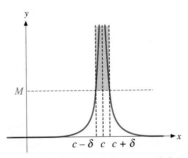

FIGURE 4.1.1a $\lim_{x \to c} f(x) = \infty$ if, for each number M, there is a positive number δ such that the graph of f is above the horizontal line $y = M$ whenever $c - \delta < x < c + \delta$, $\quad x \neq c$.

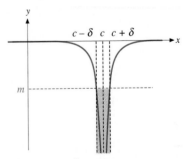

FIGURE 4.1.1b $\lim_{x \to c} f(x) = -\infty$ if, for each number m, there is a positive number δ such that the graph of f is below the horizontal line $y = m$ whenever $c - \delta < x < c + \delta$, $\quad x \neq c$.

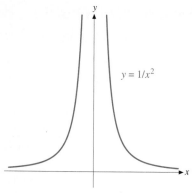

FIGURE 4.1.2 $f(x) = \dfrac{1}{x^2}$ has vertical asymptote $x = 0$. $\lim\limits_{x \to 0} \dfrac{1}{x^2} = \infty$.

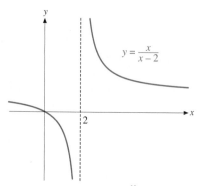

FIGURE 4.1.3 $f(x) = \dfrac{x}{x-2}$ has vertical asymptote $x = 2$. $\lim\limits_{x \to 2^-} \dfrac{x}{x-2} = -\infty$ and $\lim\limits_{x \to 2^+} \dfrac{x}{x-2} = \infty$.

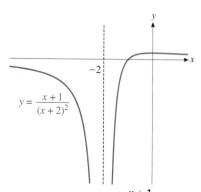

FIGURE 4.1.4 $f(x) = \dfrac{x+1}{(x+2)^2}$ has vertical asymptote $x = -2$. $\lim\limits_{x \to -2} \dfrac{x+1}{(x+2)^2} = -\infty$.

EXAMPLE 1

The function $f(x) = \dfrac{1}{x^2}$ has vertical asymptote $x = 0$, where the denominator is zero. It is clear that $1/x^2$ is positive for $x \neq 0$, so

$$\lim_{x \to 0^-} \frac{1}{x^2} = \lim_{x \to 0^+} \frac{1}{x^2} = \lim_{x \to 0} \frac{1}{x^2} = \infty.$$

That portion of the graph near $x = 0$ is sketched in Figure 4.1.2.   ∎

EXAMPLE 2

The function $f(x) = \dfrac{x}{x-2}$ has vertical asymptote $x = 2$, where the denominator is zero. We can check that $\dfrac{x}{x-2}$ is negative for $0 < x < 2$ and positive for $x > 2$. It follows that

$$\lim_{x \to 2^-} \frac{x}{x-2} = -\infty \quad \text{and} \quad \lim_{x \to 2^+} \frac{x}{x-2} = \infty.$$

Since the one-sided limits are unequal, $\lim\limits_{x \to 2} \dfrac{x}{x-2}$ does not exist. The graph of f for x near 2 is sketched in Figure 4.1.3.   ∎

EXAMPLE 3

The function $f(x) = \dfrac{x+1}{(x+2)^2}$ has vertical asymptote $x = -2$, where the denominator is zero. We can check that $\dfrac{x+1}{(x+2)^2}$ is negative for $x < -2$ and $-2 < x < -1$, so

$$\lim_{x \to -2} \frac{x+1}{(x+2)^2} = -\infty$$

and both one-sided limits at -2 are negative infinity. The graph of f for x near -2 is sketched in Figure 4.1.4.   ∎

EXAMPLE 4

The function $f(x) = \tan x = \dfrac{\sin x}{\cos x}$ has vertical asymptote $x = \pi/2$, because $\cos(\pi/2) = 0$ and $\sin(\pi/2) = 1 \neq 0$. The values $\tan x$ are positive for $0 < x < \pi/2$ and negative for $\pi/2 < x < \pi$, so

$$\lim_{x \to \pi/2^-} \tan x = \infty \quad \text{and} \quad \lim_{x \to \pi/2^+} \tan x = -\infty.$$

Since the one-sided limits are unequal, $\lim\limits_{x \to \pi/2} \tan x$ does not exist. The graph of f for x near $\pi/2$ is sketched in Figure 4.1.5.   ∎

You should know how to use the sign of $f(x)$ for x near a vertical asymptote to determine one-sided limits and sketch the graph of f near the asymptote. You should also be able to use a graph of f to determine the limits of f on each side of a vertical asymptote.

 Note that a calculator/computer graph of a function can be used to indicate vertical asymptotes and the values of one-sided infinite limits. Be sure you know how your calculator/computer handles the graph of a function near points where it has a vertical asymptote.

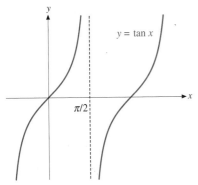

FIGURE 4.1.5 $f(x) = \tan x$ has vertical asymptote $x = \pi/2$. $\lim\limits_{x \to \pi/2^-} \tan x = \infty$ and $\lim\limits_{x \to \pi/2^+} \tan x = -\infty$.

We must be very careful when taking the limit of a sum, product, or quotient, if one of the terms has an infinite limit, since the usual Limit Theorem does not apply to all limits that involve infinite values. The problem is that the rules of arithmetic do not apply to ∞ and $-\infty$, because they are not real numbers. We do have the following results, which can be verified by using the definition of limits with infinite values.

THEOREM 1

If $\lim\limits_{x \to c} u(x) = \lim\limits_{x \to c} v(x) = \infty$, then

$$\lim_{x \to c}(u(x) + v(x)) = \infty \text{ and } \lim_{x \to c} u(x) \cdot v(x) = \infty.$$

If $\lim\limits_{x \to c} u(x) = L$, L a real number, and $\lim\limits_{x \to c} v(x) = \infty$, then

$$\lim_{x \to c}(u(x) + v(x)) = \infty,$$

$$\lim_{x \to c} u(x) \cdot v(x) = \begin{cases} \infty, & \text{if } L > 0, \\ -\infty, & \text{if } L < 0, \end{cases}$$

$$\lim_{x \to c} \frac{u(x)}{v(x)} = 0.$$

Corresponding results hold for limits that have values of negative infinity and for one-sided limits.

Intuitively, if $\lim\limits_{x \to c} u(x) = \lim\limits_{x \to c} v(x) = \infty$, we can think of $u(x)$ and $v(x)$ as large positive numbers. Since the sum and product of two large positive numbers are also large positive numbers, it should then seem reasonable that $\lim\limits_{x \to c}(u(x) + v(x)) = \lim\limits_{x \to c} u(x) \cdot v(x) = \infty$. Similarly, if $u(x) \approx L$ and $v(x)$ is much larger than L, then $u(x) + v(x)$ is large and $\dfrac{u(x)}{v(x)}$ is small. If $u(x) \approx L$, $L \neq 0$, and $v(x)$ is much larger than $1/L$, then $|u(x)v(x)|$ is large.

Theorem 1 can be used to evaluate some limits directly. Although Theorem 1 does not apply to a difference of terms that approach infinity, such expressions can sometimes be written as a product to which Theorem 1 can be applied.

EXAMPLE 5

Evaluate **(a)** $\lim\limits_{x \to 0}\left(\dfrac{1}{x^2} + \dfrac{1}{x^4}\right)$ and **(b)** $\lim\limits_{x \to 0}\left(\dfrac{1}{x^2} - \dfrac{1}{x^4}\right)$.

SOLUTION

(a) We have $\lim\limits_{x \to 0}\dfrac{1}{x^2} = \lim\limits_{x \to 0}\dfrac{1}{x^4} = \infty$. Theorem 1 tells us that a sum of terms that approach infinity also approaches infinity, so

$$\lim_{x \to 0}\left(\frac{1}{x^2} + \frac{1}{x^4}\right) = \infty.$$

(b) Theorem 1 does not apply to a difference of terms that approach infinity, but we can write

$$\frac{1}{x^2} - \frac{1}{x^4} = \frac{1}{x^4}\left(x^2 - 1\right).$$

Since $\lim\limits_{x \to 0} \dfrac{1}{x^4} = \infty$ and $\lim\limits_{x \to 0}\left(x^2 - 1\right) = -1$, Theorem 1 tells us

$$\lim_{x \to 0}\left(\frac{1}{x^2} - \frac{1}{x^4}\right) = \lim_{x \to 0} \frac{1}{x^4}\left(x^2 - 1\right) = -\infty. \quad \blacksquare$$

Indeterminate Forms

Certain algebraic combinations of small, nonzero numbers and large numbers depend on the relative size of the numbers and can be of any magnitude. This leads to the following definition.

DEFINITION

If $\lim\limits_{x \to c} u(x) = \lim\limits_{x \to c} v(x) = 0$, the quotient $u(x)/v(x)$ is said to have **indeterminate form 0/0** at $x = c$.

If $\lim\limits_{x \to c} u(x) = \pm\infty$ and $\lim\limits_{x \to c} v(x) = \pm\infty$, the quotient $u(x)/v(x)$ is said to have **indeterminate form ∞ / ∞** at $x = c$.

If $\lim\limits_{x \to c} u(x) = 0$ and $\lim\limits_{x \to c} v(x) = \pm\infty$, the product $u(x) \cdot v(x)$ is said to have **indeterminate form $0 \cdot \infty$** at $x = c$.

If $\lim\limits_{x \to c} u(x) = \lim\limits_{x \to c} v(x) = \pm\infty$, the difference $u(x) - v(x)$ is said to have **indeterminate form $\infty - \infty$** at $x = c$.

Analogous definitions can be made for one-sided indeterminate forms.

The limit at c of each of the above indeterminate forms may have value infinity, so the corresponding graph may have a vertical asymptote $x = c$. In order to determine whether or not the graph actually has a vertical asymptote, we need to evaluate the limit at c. We will now introduce a theorem that allows us to evaluate limits of some of the above indeterminate forms.

L'HÔPITAL'S RULE

Suppose that the quotient u/v has indeterminate form either 0/0 or ∞/∞ at c. Then

$$\lim_{x \to c} \frac{u'(x)}{v'(x)} = L \text{ implies } \lim_{x \to c} \frac{u(x)}{v(x)} = L.$$

The results also hold for one-sided limits.

Proof We will prove only the 0/0 case. Proof of the ∞/∞ case is left for more advanced texts.

We assume that u and v are continuous at c and $u(c) = v(c) = 0$. We also assume that u and v are differentiable for $0 < |x - c| < \delta_0$ and that neither

$v(x)$ nor $v'(x)$ is zero in that interval. (It is possible, but tedious, to show the general result for indeterminate forms $0/0$ is a consequence of this case.)

The idea of the proof is to obtain a relation between values of the quotients u/v and u'/v'. To do this we fix a point x with $0 < |x - c| < \delta_0$ and apply Rolle's Theorem to the function

$$h(t) = v(x)u(t) - u(x)v(t), \quad t \text{ between } x \text{ and } c.$$

(The function $h(t)$ is continuous on the closed interval between x and c, and is differentiable on the open interval between x and c, because $u(t)$ and $v(t)$ are. Clearly $h(x) = 0$; $h(c) = 0$ because $u(c) = v(c) = 0$.) The conclusion of Rolle's Theorem gives us a point z that is between x and c with $h'(z) = 0$. That is,

$$v(x)u'(z) - u(x)v'(z) = 0, \quad \text{so } \frac{u(x)}{v(x)} = \frac{u'(z)}{v'(z)}.$$

$\lim_{x \to c} \dfrac{u'(x)}{v'(x)} = L$ implies that $\dfrac{u'(z)}{v'(z)}$ approaches L as z approaches c. As x approaches c, the corresponding value z must also approach c, since z is between x and c. It follows that

$$\frac{u(x)}{v(x)} = \frac{u'(z)}{v'(z)} \to L \text{ as } x \to c. \quad \blacksquare$$

L'Hôpital's Rule applies only to *quotients* of indeterminate form either $0/0$ or ∞/∞. Note that $\dfrac{u'(x)}{v'(x)}$ is the quotient of the derivative of $u(x)$ and the derivative of $v(x)$. This is *not* the same as the derivative of the quotient $\dfrac{u(x)}{v(x)}$.

If u and v are continuous at c and $u(c) = v(c) = 0$, then u/v has indeterminate form $0/0$ at c. This means that we can check that a quotient of continuous functions has indeterminate form $0/0$ at c by substitution of the value $x = c$; $u(c)/v(c) = 0/0$ tells us that u/v has indeterminate form $0/0$ at c. The fact that $u(c)/v(c)$ is undefined has no bearing on either the existence or the value of the limit of $u(x)/v(x)$ as x approaches c.

Applications of L'Hôpital's Rule

EXAMPLE 6

Evaluate $\lim\limits_{x \to 3} \dfrac{x^2 - 2x - 3}{x - 3}$.

SOLUTION   We could evaluate this limit by factoring, but let us see how to use l'Hôpital's Rule.

The expression $\dfrac{x^2 - 2x - 3}{x - 3}$ is of the form $\dfrac{u(x)}{v(x)}$, where $u(x) = x^2 - 2x - 3$ and $v(x) = x - 3$. The functions u and v are continuous at $x = 3$ with $u(3) = v(3) = 0$, so u/v has indeterminate form $0/0$ at $x = 3$. Also,

$$\lim_{x \to 0} \frac{u'(x)}{v'(x)} = \lim_{x \to 3} \frac{\dfrac{d}{dx}(x^2 - 2x - 3)}{\dfrac{d}{dx}(x - 3)} = \lim_{x \to 3} \frac{2x - 2}{1} = 2(3) - 2 = 4.$$

L'Hôpital's Rule then tells us

$$\lim_{x\to3}\frac{x^2-2x-3}{x-3}=\lim_{x\to3}\frac{u(x)}{v(x)}=\lim_{x\to3}\frac{u'(x)}{v'(x)}=\lim_{x\to3}\frac{2x-2}{1}=4.$$

We will write the solution in the shorter form

$$\lim_{x\to3}\frac{x^2-2x-3}{x-3}\underset{(0/0,\text{l'H})}{=}\lim_{x\to3}\frac{2x-2}{1}=4.$$

We have written $(0/0, \text{l'H})$ under the equal sign to indicate that we have checked that l'Hôpital's Rule applies to the indeterminate form $0/0$ and that equality holds if the limit of u'/v' exists. The existence of the latter limit then justifies the use of l'Hôpital's Rule. If the limit of u'/v' does not exist, then l'Hôpital's Rule does not apply and the equation we have written is invalid. The limit of u/v may exist, even though the limit of u'/v' does not.

We have indicated in Figure 4.1.6 that the values $\dfrac{x^2-2x-3}{x-3}$ approach four as x approaches three. The expression $\dfrac{x^2-2x-3}{x-3}$ is undefined at $x=3$, but $x=3$ is not a vertical asymptote. ■

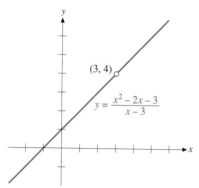

FIGURE 4.1.6 The function $f(x) = \dfrac{x^2-2x-3}{x-3}$ has indeterminate form $0/0$ at $x=3$. L'Hôpital's Rule can be used to show that $\lim\limits_{x\to3}\dfrac{x^2-2x-3}{x-3}=4$.

EXAMPLE 7

Evaluate $\lim\limits_{x\to0}\dfrac{1-\cos2x}{x^2}$.

SOLUTION   It is easy to see that $(1-\cos2x)/x^2$ has indeterminate form $0/0$ at $x=0$. We can then write the solution in the short form.

$$\lim_{x\to0}\frac{1-\cos2x}{x^2}\underset{(0/0,\text{l'H})}{=}\lim_{x\to0}\frac{-(-\sin2x)(2)}{2x}$$

$$\underset{(\text{Simplifying})}{=}\lim_{x\to0}\frac{\sin2x}{x}$$

$$\underset{(0/0,\text{l'H})}{=}\lim_{x\to0}\frac{(\cos2x)(2)}{1}=2.$$

In this example the first application of l'Hôpital's Rule gave us $(\sin2x)/x$, which also has indeterminate form $0/0$. We were able to evaluate the limit after a second application of l'Hôpital's Rule. The graph of $\dfrac{1-\cos2x}{x^2}$ for x near 0 is sketched in Figure 4.1.7. ■

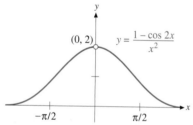

FIGURE 4.1.7 The quotient $\dfrac{1-\cos2x}{x^2}$ has indeterminate form $0/0$ at $x=0$. Two applications of l'Hôpital's Rule can be used to show that $\lim\limits_{x\to0}\dfrac{1-\cos2x}{x^2}=2$.

The limit of some expressions of indeterminate form $\infty-\infty$ can be evaluated by writing the expressions as quotients.

EXAMPLE 8

Evaluate $\lim\limits_{x\to0}\left(\dfrac{1+\sin x}{x}-\dfrac{1}{\sin x}\right)$.

SOLUTION   The expression $\dfrac{1+\sin x}{x}-\dfrac{1}{\sin x}$ is of indeterminate form $\infty-\infty$ at $x=0$. We then have

$$\lim_{x\to0}\left(\frac{1+\sin x}{x}-\frac{1}{\sin x}\right)\underset{(\text{Rewriting})}{=}\lim_{x\to0}\frac{\sin x+\sin^2x-x}{x\sin x}$$

$$= \lim_{(0/0,\text{l'H})x\to 0} \frac{\cos x + 2\sin x\cos x - 1}{(x)(\cos x) + (\sin x)(1)}$$

$$= \lim_{(0/0,\text{l'H})x\to 0} \frac{-\sin x + 2(\sin x)(-\sin x) + 2(\cos x)(\cos x)}{(x)(-\sin x) + (\cos x)(1) + \cos x}$$

$$= \frac{2}{2} = 1.$$

The graph in Figure 4.1.8 indicates that $f(x) = \dfrac{1 + \sin x}{x} - \dfrac{1}{\sin x} \to 1$ as $x \to 0$. ∎

We can sometimes use l'Hôpital's Rule to evaluate limits of indeterminate form $0 \cdot \infty$ by rewriting the expressions in the form of a quotient to which l'Hôpital's Rule applies. In particular, if the product uv has indeterminate form $0 \cdot \infty$, then by writing either

$$u(x)v(x) = \frac{u(x)}{1/v(x)} \text{ or } u(x)v(x) = \frac{v(x)}{1/u(x)}$$

we can express uv in indeterminate form $0/0$ or ∞/∞, respectively.

EXAMPLE 9
Evaluate $\lim\limits_{x\to 0} x\cot x$.

SOLUTION   The product $x\cot x$ has indeterminate form $0 \cdot \infty$ at $x = 0$. Rewriting the product as a quotient, we have

$$\lim_{x\to 0} x\cot x \underset{(\text{Rewriting})}{=} \lim_{x\to 0} \frac{\cot x}{1/x} \underset{(\infty/\infty,\text{l'H})}{=} \lim_{x\to 0} \frac{-\csc^2 x}{-1/x^2}.$$

At this point we see that application of l'Hôpital's Rule to the quotient has given us an even more complicated quotient, so we abandon this effort and try rewriting the product as a different quotient. We then obtain

$$\lim_{x\to 0} x\cot x \underset{(\text{Rewriting})}{=} \lim_{x\to 0} \frac{x}{\tan x} \underset{(0/0,\text{l'H})}{=} \lim_{x\to 0} \frac{1}{\sec^2 x} = 1.$$

The graph of $x\cot x$ for x near zero is given in Figure 4.1.9. ∎

We conclude this section with some suggestions for using l'Hôpital's Rule:

Be sure to check that the expression has indeterminate form either $0/0$ or ∞/∞ before applying l'Hôpital's Rule.

- The expression $\dfrac{\cos 2x}{x^2}$ does not have indeterminate form at $x = 0$, where the denominator is zero. Since $\dfrac{\cos x}{x^2}$ is positive for x near zero, $x \neq 0$, we have

$$\lim_{x\to 0} \frac{\cos 2x}{x^2} = \infty,$$

so the graph has vertical asymptote $x = 0$. See Figure 4.1.10. It is left to the reader to show that two blind applications of l'Hôpital's Rule give the incorrect value of -2 for the limit of the expression as x approaches zero.

FIGURE 4.1.8 $\dfrac{1 + \sin x}{x} - \dfrac{1}{\sin x} = \dfrac{\sin x + \sin^2 x - x}{x\sin x}$, and then two applications of l'Hôpital's Rule show $\lim\limits_{x\to 0}\left(\dfrac{1 + \sin x}{x} - \dfrac{1}{\sin x}\right) = 1.$

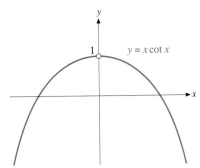

FIGURE 4.1.9 $x\cot x = \dfrac{x}{\tan x}$, so l'Hôpital's Rule can be used to show $\lim\limits_{x\to 0} x\cot x = 1.$

FIGURE 4.1.10 $\dfrac{\cos 2x}{x^2}$ has form 1/0 at $x = 0$. L'Hôpital's Rule does not apply. Since $(\cos 2x)/x^2$ is positive near zero and becomes unbounded as x approaches 0, $\lim\limits_{x\to 0}\dfrac{\cos 2x}{x^2} = \infty.$

FIGURE 4.1.11 The expression $\dfrac{\sin x}{\tan x}$ is equal to $\cos x$, except for integer multiples of $\pi/2$. L'Hôpital's Rule is not needed to determine that $\displaystyle\lim_{x\to 0}\frac{\sin x}{\tan x} = \lim_{x\to 0}\cos x = 1$.

Don't forget to try simplifying before using l'Hôpital's Rule.

- The expression $\dfrac{\sin x}{\tan x}$ has indeterminate form $0/0$ at $x = 0$, but the expression simplifies to $\cos x$, so l'Hôpital's Rule is not needed to evaluate the limit at zero. See Figure 4.1.11. We have

$$\lim_{x\to 0}\frac{\sin x}{\tan x} = \lim_{x\to 0}\frac{\sin x}{\dfrac{\sin x}{\cos x}} = \lim_{x\to 0}\cos x = \cos 0 = 1.$$

We can simplify the differentiation in some problems by excluding factors that have finite, nonzero limits from the application of l'Hôpital's Rule.

- The expression $\dfrac{\sin 3x}{x\sqrt{4 - x^2}}$ has indeterminate form $0/0$ at $x = 0$ and we could apply l'Hôpital's Rule to the quotient. However, since the factor $\sqrt{4 - x^2}$ has finite, nonzero limit two at zero, we can simplify the differentiation by writing

$$\lim_{x\to 0}\frac{\sin 3x}{x\sqrt{4 - x^2}} = \left(\lim_{x\to 0}\frac{1}{\sqrt{4 - x^2}}\right)\left(\lim_{x\to 0}\frac{\sin 3x}{x}\right)$$

and applying l'Hôpital's Rule only to the second limit. The value of the first limit is $1/2$ and one application of l'Hôpital's Rule shows the value of the second limit is 3, so

$$\lim_{x\to 0}\frac{\sin 3x}{x\sqrt{4 - x^2}} = \left(\frac{1}{2}\right)(3) = \frac{3}{2}.$$

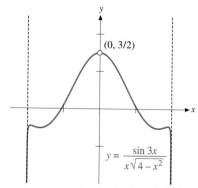

FIGURE 4.1.12 Since $\sqrt{4 - x^2}$ has nonzero, finite limit 2 as x approaches 0, this expression should be factored before l'Hôpital's Rule is used to evaluate the limit of $\dfrac{\sin 3x}{x}$.

The graph of $\dfrac{\sin 3x}{x\sqrt{4 - x^2}}$ is sketched in Figure 4.1.12.

We may need to apply l'Hôpital's Rule more than once to evaluate a limit, but we should check that we are making progress.

- The expression $\cot x / \csc x$ is of indeterminate form ∞/∞ at $x = 0$. If we applied l'Hôpital's Rule we would obtain

$$\lim_{x\to 0}\frac{\cot x}{\csc x} \underset{(\infty/\infty, \text{l'H})}{=} \lim_{x\to 0}\frac{-\csc^2 x}{-\csc x \cot x} = \lim_{x\to 0}\frac{\csc x}{\cot x} = (?).$$

We see that we are not making any progress. If we try simplifying, as we should, we obtain

$$\lim_{x\to 0}\frac{\cot x}{\csc x} = \lim_{x\to 0}\frac{\dfrac{\cos x}{\sin x}}{\dfrac{1}{\sin x}} = \lim_{x\to 0}\cos x = 1.$$

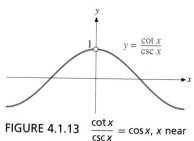

FIGURE 4.1.13 $\dfrac{\cot x}{\csc x} = \cos x$, x near 0, so $\displaystyle\lim_{x\to 0}\frac{\cot x}{\csc x} = 1$.

The graph of $\dfrac{\cot x}{\csc x}$ is sketched in Figure 4.1.13.

EXERCISES 4.1

In Exercises 1–14, sketch the graph of the given function f near the vertical asymptote $x = c$. Indicate the value of the one-sided limits $\lim\limits_{x \to c^-} f(x)$ and $\lim\limits_{x \to c^+} f(x)$.

1. $f(x) = \dfrac{1}{x+1}$

2. $f(x) = \dfrac{1}{x-2}$

3. $f(x) = \dfrac{x}{x+1}$

4. $f(x) = \dfrac{x}{x+2}$

5. $f(x) = \cos x \cot x, c = 0$

6. $f(x) = \sin x \tan x, c = \pi/2$

7. $f(x) = \sec x \tan x, c = \pi/2$

8. $f(x) = \csc x \cot x, c = 0$

9. $f(x) = \dfrac{x-3}{(x-1)^2}$

10. $f(x) = \dfrac{x+1}{(x+3)^2}$

11. $f(x) = \dfrac{x}{1-x}$

12. $f(x) = \dfrac{x+1}{x+2}$

13. $f(x) = \dfrac{\cos x}{x^2}$

14. $f(x) = \dfrac{\sin x}{x^2}$

Evaluate the limits in Exercises 15–46.

15. $\lim\limits_{x \to 0^+} \left(\dfrac{1}{x^2} + \dfrac{1}{x^3} \right)$

16. $\lim\limits_{x \to 0^-} \left(\dfrac{1}{x^2} - \dfrac{1}{x} \right)$

17. $\lim\limits_{x \to 0^-} \left(\dfrac{1}{x^2} + \dfrac{1}{x^3} \right)$

18. $\lim\limits_{x \to 0^+} \left(\dfrac{1}{x^2} - \dfrac{1}{x} \right)$

19. $\lim\limits_{x \to 0^+} \left(\dfrac{1}{x^3} - \dfrac{1}{x^2} \right)$

20. $\lim\limits_{x \to 0^-} \left(\dfrac{1}{x^3} - \dfrac{1}{x^2} \right)$

21. $\lim\limits_{x \to 0^-} \left(\dfrac{1}{x^2} - \dfrac{1}{\sin x} \right)$

22. $\lim\limits_{x \to 0} \left(\dfrac{1}{x} - \dfrac{1}{x \sin x} \right)$

23. $\lim\limits_{x \to 3} \dfrac{x^2 - 9}{x - 3}$

24. $\lim\limits_{x \to 2} \dfrac{x^2 - 3x + 2}{x - 2}$

25. $\lim\limits_{x \to \pi/6} \dfrac{\sin x - (1/2)}{x - (\pi/6)}$

26. $\lim\limits_{x \to \pi/4} \dfrac{\tan x - 1}{x - (\pi/4)}$

27. $\lim\limits_{x \to 0} \dfrac{\sin 3x}{\cos 2x}$

28. $\lim\limits_{x \to 0} \dfrac{\cos x}{\sin^2 x}$

29. $\lim\limits_{x \to 0} \dfrac{\sin 2x}{x}$

30. $\lim\limits_{x \to 0} \dfrac{\sin 3x}{\sin 2x}$

31. $\lim\limits_{x \to 0} \dfrac{\sin x - x}{x^3}$

32. $\lim\limits_{x \to 0} \dfrac{\cos 2x - 1}{x^2}$

33. $\lim\limits_{x \to 0^+} \dfrac{\cos 2x}{\sin 3x}$

34. $\lim\limits_{x \to 0^-} \dfrac{\cos 3x}{\sin 2x}$

35. $\lim\limits_{x \to 0} \dfrac{x^2 + 3x + 2}{x^2 - 1}$

36. $\lim\limits_{x \to 0} \dfrac{4x^2 - 3x + 5}{x^2 + 2x + 3}$

37. $\lim\limits_{x \to 0} \sin x \cot x$

38. $\lim\limits_{x \to 0} \sin x \csc x$

39. $\lim\limits_{x \to 0} x \cot 2x$

40. $\lim\limits_{x \to 0} \sin 2x \cot 3x$

41. $\lim\limits_{x \to 0} \dfrac{\cos^2 x \sin x}{x}$

42. $\lim\limits_{x \to 0} \dfrac{x\sqrt{x+4}}{\sin 2x}$

43. $\lim\limits_{x \to 0} \dfrac{\sin x \sqrt{1 - \sin x}}{x}$

44. $\lim\limits_{x \to 0} \dfrac{\cos x \sin^2 x}{1 - \cos x}$

45. $\lim\limits_{x \to 0} \left(\dfrac{1 + 2x}{\sin x} - \dfrac{1}{x} \right)$

46. $\lim\limits_{x \to 0} \left(\dfrac{\tan x + \sec x}{x} - \dfrac{1}{\sin x} \right)$

47. Use the definition to show that

$$\lim_{x \to 0} \frac{1}{x^2} = \infty.$$

48. Use the definition to show that

$$\lim_{x \to 3} \frac{1}{(2x - 6)^2} = \infty.$$

49. (a) Write a formal definition of the statement

$$\lim_{x \to c^+} f(x) = \infty$$

and (b) use the definition to show that

$$\lim_{x \to 2^+} \frac{2}{3x - 6} = \infty.$$

50. (a) Write a formal definition of the statement

$$\lim_{x \to c^-} f(x) = -\infty$$

and (b) use the definition to show that

$$\lim_{x \to 2^-} \frac{3}{4x - 8} = -\infty.$$

51. If $1/2 \le f(x) \le 3/2$ for all x and $\lim\limits_{x \to c} g(x) = \infty$, use the definition to show that $\lim\limits_{x \to c} f(x)g(x) = \infty$.

52. If $\lim\limits_{x \to c} g(x) = \infty$, use the definition to show that

$$\lim_{x \to c} \frac{1}{g(x)} = 0.$$

53. If $\lim\limits_{x \to c} g(x) = 0$, explain why it is not necessarily true that

$$\lim_{x \to c} \frac{1}{g(x)} = \infty.$$

54. If f is continuous on the interval $a < x < b$, and $\lim\limits_{x \to a^+} f(x) = \lim\limits_{x \to b^-} f(x) = \infty$, show that f has a minimum value on the interval.

Use calculator-generated graphs to determine apparent values of (a) $\lim_{x \to 0^-} f(x)$, *(b)* $\lim_{x \to 0^+} f(x)$, *and (c)* $\lim_{x \to 0} f(x)$ *for the functions given in Exercises 55–58.*

55. $f(x) = \ln|x|$

56. $f(x) = \dfrac{\ln|x|}{x}$

57. $f(x) = x \ln|x|$

58. $f(x) = \dfrac{x}{\ln(1 + x^2)}$

4.2 LIMITS AT INFINITY; ASYMPTOTIC BEHAVIOR

Connections

Graphs with asymptotes, 1.5.

Knowledge of the behavior of $f(x)$ for $|x|$ large can be used to obtain information about the quantity represented by the function. For example, $f(x)$ might represent the temperature of a particular piece of machinery after it has been run continuously for x units of time. The behavior of $f(x)$ for large values of x could indicate if the temperature will approach an acceptable steady-state value or overheat after being run for a long period of time. Limits at infinity are used to investigate such problems.

Limits at Infinity

We begin our study of the behavior of $f(x)$ for $|x|$ large with the following formal definitions of limits at infinity.

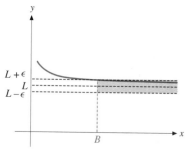

FIGURE 4.2.1a $\lim_{x \to \infty} f(x) = L$ if, for each positive number ϵ, there is a positive number B such that the graph of f is between the horizontal lines $y = L - \epsilon$ and $y = L + \epsilon$ whenever $x > B$.

DEFINITION

We say $\lim_{x \to \infty} \boldsymbol{f(x) = L}$ if, for each positive number ϵ, there is a positive number B such that $|f(x) - L| < \epsilon$ whenever $x > B$.

We say $\lim_{x \to -\infty} \boldsymbol{f(x) = L}$ if, for each positive number ϵ, there is a negative number $-B$ such that $|f(x) - L| < \epsilon$ whenever $x < -B$.

We write $\lim_{x \to \pm\infty} f(x) = L$ if both limits are L.

FIGURE 4.2.1b $\lim_{x \to -\infty} f(x) = L$ if, for each positive number ϵ, there is a negative number $-B$ such that the graph of f is between the horizontal lines $y = L - \epsilon$ and $y = L + \epsilon$ whenever $x < -B$.

See Figure 4.2.1.

The statement $\lim_{x \to \infty} f(x) = L$ means that the values $f(x)$ are near L for all large enough x. The statement $\lim_{x \to -\infty} f(x) = L$ means that $f(x)$ is near L for all negative x with large enough absolute value.

The definitions of limits at ∞ and $-\infty$ can be modified as in Section 4.1 to include infinite values. The usual rules of limits hold for *finite-valued* limits at ∞ and $-\infty$. Theorem 1 of Section 4.1 applies to infinite-valued limits at ∞ and $-\infty$. Also:

THEOREM

L'Hôpital's Rule holds for $c = \pm\infty$.

We will not prove this version of l'Hôpital's Rule.

We will want to investigate the behavior as $x \to \pm\infty$ of functions that are quotients of terms that involve powers of x. We could use l'Hôpital's Rule in this investigation, but we prefer to illustrate a method that indicates the size of the quotient for $|x|$ large. We will find this particular information useful in Chapter 9.

To approximate the size of quotients for x near $\pm\infty$

- Scale the ratio of terms that involve powers of x by factoring the highest power of x from each of the numerator and denominator.
- Combine factored powers of x to obtain a single power of x times a quotient of terms that have finite, nonzero limits at infinity. The limit of expressions of this form can be evaluated easily.

We will use the following facts.

$$\textit{If } r > 0, \textit{ then } \lim_{x \to \infty} \frac{1}{x^r} = 0.$$

$$\textit{If } x^r \textit{ is defined for } x < 0, \textit{ then } r > 0 \textit{ implies } \lim_{x \to -\infty} \frac{1}{x^r} = 0. \tag{1}$$

Statements 1 merely reflect the fact that a positive power or root of a large number is large and that dividing a fixed number by a much larger number gives a small number.

EXAMPLE 1

Evaluate $\displaystyle\lim_{x \to \infty} \frac{2x - 3}{x^2 + 7x - 1}$.

SOLUTION We have a quotient where both the numerator and the denominator approach infinity as x approaches infinity. To investigate the ratio, we scale the ratio by factoring the highest power of x from each of the numerator and denominator. We then obtain

$$\lim_{x \to \infty} \frac{2x - 3}{x^2 + 7x - 1} = \lim_{x \to \infty} \frac{x\left(2 - \dfrac{3}{x}\right)}{x^2\left(1 + \dfrac{7}{x} - \dfrac{1}{x^2}\right)}$$

$$= \lim_{x \to \infty} \left(\frac{1}{x}\right)\left(\frac{2 - \dfrac{3}{x}}{1 + \dfrac{7}{x} - \dfrac{1}{x^2}}\right).$$

The scaling has given us an expression that consists of $1/x$ times a quotient of terms that have finite, nonzero limits at infinity. We can then use rules of limits to evaluate the limit of the expression. Using (1), we obtain

$$\lim_{x \to \infty} \frac{2x - 3}{x^2 + 7x - 1} = \lim_{x \to \infty} \left(\frac{1}{x}\right)\left(\frac{2 - \dfrac{3}{x}}{1 + \dfrac{7}{x} - \dfrac{1}{x^2}}\right) = (0)\frac{2 - 0}{1 + 0 - 0} = 0.$$

Note that we also have

$$\lim_{x \to \infty} \frac{2x - 3}{x^2 + 7x - 1} \underset{(\infty/\infty,\,\text{l'H})}{=} \lim_{x \to \infty} \frac{2}{2x + 7} = 0.$$

Thus, l'Hôpital's Rule gives the same value of the limit, but does not give the information that

$$\frac{2x - 3}{x^2 + 7x - 1} \approx \frac{2}{x} \quad \text{for } |x| \text{ large.} \quad \blacksquare$$

The Method of Example 1 can be used to verify the following fact.

If P and Q are polynomials with the degree of P less than the degree of Q, then

$$\lim_{x \to \pm\infty} \frac{P(x)}{Q(x)} = 0.$$

EXAMPLE 2

Evaluate $\lim\limits_{x \to -\infty} \dfrac{2x^2 - 1}{3x^2 + x - 4}$.

SOLUTION As in the previous example, we have a quotient of functions that approach infinity. We scale by factoring the highest power of x from each of the numerator and the denominator. This gives

$$\lim_{x \to -\infty} \frac{2x^2 - 1}{3x^2 + x - 4} = \lim_{x \to -\infty} \frac{x^2 \left(2 - \dfrac{1}{x^2}\right)}{x^2 \left(3 + \dfrac{1}{x} - \dfrac{4}{x^2}\right)}$$

$$= \lim_{x \to -\infty} \frac{2 - \dfrac{1}{x^2}}{3 + \dfrac{1}{x} - \dfrac{4}{x^2}} = \frac{2}{3}. \quad \blacksquare$$

Example 2 illustrates the following general fact.

If P and Q are polynomials of equal degree, then

$$\lim_{x \to \pm\infty} \frac{P(x)}{Q(x)}$$

is the quotient of the coefficient of the highest-degree term of P and the coefficient of the highest-degree term of Q.

We cannot use the Limit Theorem to evaluate the limit of a sum of two functions that have infinite limits of opposite sign. In some cases, we can evaluate limits of this type by rewriting the expression as a product of a factor with an infinite limit and a factor with *finite nonzero* limit.

EXAMPLE 3

Evaluate $\lim\limits_{x \to -\infty} (x^3 + 2x^2)$.

SOLUTION We note that $x^3 \to -\infty$ and $2x^2 \to \infty$ as $x \to -\infty$. Rewriting the sum by factoring out the highest power of x that is a factor of each term, we have

$$x^3 + 2x^2 = x^3 \left(1 + \frac{2}{x}\right).$$

We then see that the given expression is the product of x^3 and a factor that has limit one as x approaches negative infinity. Since $x^3 \to -\infty$ as $x \to -\infty$, we conclude that

$$\lim_{x \to -\infty} (x^3 + 2x^2) = -\infty. \quad \blacksquare$$

The method of Example 3 can be used to verify the following:

The limits at infinity of any polynomial are the same as the limits of the highest-degree term of the polynomial.

We have seen that the behavior of a polynomial at $\pm\infty$ reflects the behavior of the highest-degree term of the polynomial. The behavior of a rational function at $\pm\infty$ is determined by the highest-degree terms in each of its numerator and denominator. It follows that the behavior at $\pm\infty$ of such functions can be determined by observation of these *dominant,* highest-degree terms. If this is understood, then it is not actually necessary to carry out the factorization to determine the behavior at $\pm\infty$.

If either $\lim_{x \to \infty} f(x) = L$ or $\lim_{x \to -\infty} f(x) = L$, $y = L$ is called a **horizontal asymptote of** f. In that case, points $(x, f(x))$ on the graph f approach the horizontal line $y = L$ as x approaches infinity or negative infinity, respectively. The graph of a function f intersects a horizontal asymptote $y = L$ whenever $f(x) = L$. A graph may touch or cross a horizontal asymptote any number of times, as is illustrated by the graph of $(\sin x)/x$ given in Figure 4.2.2. We see that $\lim_{x \to \pm\infty} (\sin x)/x = 0$, so $y = 0$ is a horizontal asymptote of $(\sin x)/x$. The graph crosses the asymptote at each point on the asymptote that has x-coordinate a nonzero integer multiple of π. ($(\sin x)/x$ is undefined for $x = 0$, but $\lim_{x \to 0}(\sin x)/x = 1$, so $x = 0$ is not a vertical asymptote.)

FIGURE 4.2.2 The graph of $(\sin x)/x$ crosses the horizontal asymptote $y = 0$ at each point where x is a nonzero integer multiple of π.

Graphs with Asymptotes

It is important to know how the analytic properties of domain, intercepts, sign of $f(x)$, continuity, vertical asymptotes, and horizontal asymptotes are reflected in the graph of f. One way to learn the relation between the analytic properties of a function and its graph is to use the properties of the function to sketch its graph. The idea is to sketch the graph in the simplest way that illustrates the desired properties, where "simplest way" means with as few "wiggles" as possible. We will learn more about "wiggles" in the following two sections.

To use properties of f(x) to sketch a graph

- Check the domain, intercepts, and sign of $f(x)$. The graph is above the x-axis where $f(x) > 0$ and the graph is below the x-axis where $f(x) < 0$.
- Check for infinite-valued limits and sketch vertical asymptotes.
- Plot the intercepts and that part of the graph near each vertical asymptote.
- Check limits at $\pm\infty$ and sketch horizontal asymptotes.
- Complete the graph in the simplest way that satisfies the properties that have been determined.

TABLE 4.2.1

	$-\infty$		-2		0		2		∞
x^2		$+$		$+$	0	$+$		$+$	
$x^2 - 4$		$+$	0	$-$		$-$	0	$+$	
$f(x)$		$+$		$-$		$-$		$+$	

↑ Vertical asymptote ↑ Vertical asymptote

A calculator/computer can be used to generate the graph of a function. You can then use the displayed graph to obtain an indication of the properties of the function. You should then verify that the analytic properties correspond to those indicated by the graph.

EXAMPLE 4

Sketch the graph of $f(x) = \dfrac{x^2}{x^2 - 4}$.

SOLUTION   The function f is defined for all x, except for $x = 2$ and $x = -2$. The function is continuous on each interval where it is defined. The point $(0, 0)$ is the only intercept. The lines $x = 2$ and $x = -2$ are vertical asymptotes. We may notice that $f(-x) = f(x)$, so the function is even and the graph is symmetric with respect to the y-axis.

We have used the technique of Section 1.2 to construct a table of signs for $f(x)$ in Table 4.2.1. Note that we have indicated at the bottom of the table that the graph has vertical asymptotes $x = 2$ and $x = -2$, where the denominator is zero.

We see from Table 4.2.1 that $f(x)$ is positive, so the graph is above the x-axis, on the intervals $x < -2$ and $x > 2$; $f(x)$ is negative, and the graph is below the x-axis, on the intervals $-2 < x < 0$ and $0 < x < 2$.

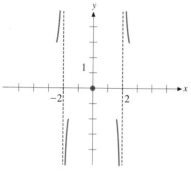

FIGURE 4.2.3a   The graph of $f(x) = \dfrac{x^2}{x^2 - 4}$ has intercept $(0, 0)$ and vertical asymptotes $x = -2$ and $x = 2$, where the denominator is zero. The sign of $f(x)$ determines the behavior of the graph near the vertical asymptotes.

In Figure 4.2.3a we have plotted the intercepts, drawn the vertical asymptotes, and sketched the graph near the vertical asymptotes. Note that the sign of $f(x)$ in intervals to the immediate left and right of each of the vertical asymptotes gives the behavior of the graph near the asymptote.

To complete the graph, we investigate the behavior of $f(x)$ for $|x|$ large. We have

$$f(x) = \frac{x^2}{x^2 - 4} = \frac{x^2}{x^2 \left(1 - \dfrac{4}{x^2}\right)} = \frac{1}{1 - \dfrac{4}{x^2}}.$$

It follows that $\displaystyle\lim_{x \to \pm\infty} f(x) = 1$. The line $y = 1$ is a horizontal asymptote of f; $f(x)$ is close to 1 whenever $|x|$ is large.

We now add the horizontal asymptote to our sketch and then complete the graph in the simplest way that satisfies the conditions we have determined. This is done in Figure 4.2.3b. (We could check that the equation $f(x) = 1$ does not have a solution, so this graph does not cross the horizontal asymptote $y = 1$.) ■

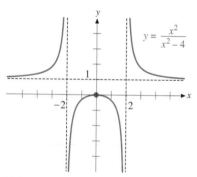

FIGURE 4.2.3b   The horizontal asymptote $y = 1$ is added and the graph is then completed in the simplest way that satisfies the conditions we have determined.

EXAMPLE 5

Sketch the graph of $f(x) = \dfrac{x}{\sqrt{x^2 + 1}}$.

SOLUTION   The function f is defined and continuous for all x. The only intercept is $(0, 0)$. It is clear that $f(x)$ is positive for positive x and negative for negative x. There is no vertical asymptote. We may notice that $f(-x) = -f(x)$, so the function is odd and the graph is symmetric with respect to the origin.

We investigate $f(x)$ for $|x|$ large by scaling. Since $\sqrt{x^2} = |x|$, we have

$$f(x) = \frac{x}{\sqrt{x^2 + 1}} = \frac{x}{\sqrt{x^2 \left(1 + \dfrac{1}{x^2}\right)}} = \frac{x}{|x|\sqrt{1 + \dfrac{1}{x^2}}}$$

$$= \begin{cases} \dfrac{1}{\sqrt{1 + \dfrac{1}{x^2}}}, & x > 0, \\[4mm] \dfrac{-1}{\sqrt{1 + \dfrac{1}{x^2}}}, & x < 0, \end{cases}$$

where we have used the fact that

$$\frac{x}{|x|} = \begin{cases} 1, & x > 0, \\ -1, & x < 0. \end{cases}$$

It follows that $\lim\limits_{x \to \infty} f(x) = 1$ and $\lim\limits_{x \to -\infty} f(x) = -1$. The lines $y = 1$ and $y = -1$ are both horizontal asymptotes of f; $f(x)$ is near 1 for x positive with large absolute value and $f(x)$ is near -1 for negative x with large absolute value. The simplest way to sketch the graph is rising from the asymptote $y = -1$ on the left, through the origin, and rising toward the asymptote $y = 1$ on the right. (We could check that neither of the equations $f(x) = -1$ and $f(x) = 1$ has a solution, so the graph does not touch or cross either asymptote. We might also notice that $\sqrt{x^2 + 1} > |x|$, so $|f(x)| < 1$ for all x. This also implies the graph stays strictly between the lines $y = 1$ and $y = -1$.) The graph is sketched in Figure 4.2.4. ∎

FIGURE 4.2.4 The function $f(x) = \dfrac{x}{\sqrt{x^2 + 1}}$ is defined and continuous for all x. It has intercept $(0, 0)$. The graph approaches the horizontal asymptote $y = -1$ as x approaches negative infinity. The graph approaches the horizontal asymptote $y = 1$ as x approaches infinity.

Oblique Asymptotes

For many functions f that arise as the solutions of science and engineering problems, it is important to know the behavior of $f(x)$ for $|x|$ large. For example, $f(x)$ might represent the temperature of a particular piece of machinery at time x. If $f(x)$ becomes unbounded as x approaches infinity, the machinery will overheat and fail to operate. However, if we knew the rate at which $f(x)$ grows over a long period of time, we could effect design changes in the cooling system to overcome the problem. The study of **asymptotic behavior** allows us to describe and characterize different types of behavior at infinity.

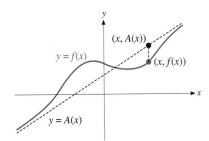

FIGURE 4.2.5 If $\lim\limits_{x \to \pm\infty} (f(x) - A(x)) = 0$, the point $(x, f(x))$ on the graph of f is near the point $(x, A(x))$ on the graph of the asymptote $y = A(x)$ whenever $|x|$ is large enough.

DEFINITION

We say a function A is an **asymptote** of f if either

$$\lim_{x \to \infty} (f(x) - A(x)) = 0 \quad \text{or} \quad \lim_{x \to -\infty} (f(x) - A(x)) = 0.$$

See Figure 4.2.5.

If $A(x) = ax + b$, then $\lim\limits_{x \to \pm\infty} (f(x) - (ax + b)) = 0$ implies the graph of f approaches the line $y = ax + b$ as $x \to \pm\infty$. If $a = 0$, $y = ax + b$ is a horizontal asymptote of f; if $a \neq 0$, the slanted line $y = ax + b$ is called an **oblique asymptote** of f.

TABLE 4.2.2

	$-\infty$	-2	-1	3	∞
$x+2$	$-$	0 $+$	$+$	$+$	$+$
$x+1$	$-$	$-$	0 $+$	$+$	$+$
$x-3$	$-$	$-$	$-$	0 $+$	$+$
$f(x)$	$-$	$+$	$-$	$+$	

↑
Vertical
asymptote

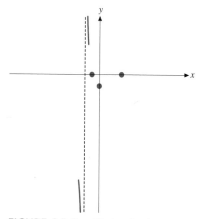

FIGURE 4.2.6a We begin the graph of $f(x) = \dfrac{(x+1)(x-3)}{x+2}$ by plotting the intercepts, drawing the vertical asymptote, and sketching that part of the graph near the vertical asymptote.

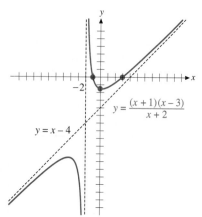

FIGURE 4.2.6b The oblique asymptote $y = x - 4$ is added and the graph is completed in the simplest way that satisfies the conditions we have found.

If P and Q are polynomials with degree $P \geq$ degree Q, we can determine the asymptotic behavior of P/Q by using long division to divide $P(x)$ by $Q(x)$.

EXAMPLE 6

Determine the asymptotic behavior and sketch the graph of

$$f(x) = \frac{(x+1)(x-3)}{x+2}.$$

SOLUTION   The function f is defined for all x, except for $x = -2$. It is continuous on intervals where it is defined. The graph has vertical asymptote $x = -2$ and intercepts $(-1, 0)$, $(3, 0)$, and $(0, -3/2)$. The sign of $f(x)$ is determined from the sign chart given in Table 4.2.2. Note that we have indicated in the table that $x = -2$ is a vertical asymptote. The sign of $f(x)$ on each side of $x = -2$ determines the behavior of the graph near the asymptote.

In Figure 4.2.6a we have plotted the intercepts, sketched the vertical asymptote, and sketched the graph of f near the asymptote.

Let us carry out the division in order to investigate the behavior of $f(x)$ for $|x|$ large. We first expand the numerator to obtain

$$(x+1)(x-3) = x^2 - 2x - 3.$$

Then

$$
\begin{array}{r}
x - 4 \\
x + 2 \overline{\smash{)}\ x^2 - 2x - 3} \\
\underline{x^2 + 2x} \\
-4x - 3 \\
\underline{-4x - 8} \\
5
\end{array}
$$

This gives

$$f(x) = \frac{(x+1)(x-3)}{x+2} = \frac{x^2 - 2x - 3}{x+2} = x - 4 + \frac{5}{x+2}.$$

Since

$$\lim_{x \to \pm\infty} \frac{5}{x+2} = 0,$$

we see that points $(x, f(x))$ on the graph of f are near points $(x, x - 4)$ on the graph of $A(x) = x - 4$ for $|x|$ large. The function $A(x) = x - 4$ is an oblique asymptote of f.

The graph is completed by drawing the oblique asymptote and then sketching the graph in the simplest way that satisfies the conditions we have found. This is done in Figure 4.2.6b. Note that

$$f(x) = x - 4 + \frac{5}{x+2} < x - 4 \qquad \text{whenever } x < -2$$

and

$$f(x) = x - 4 + \frac{5}{x+2} > x - 4 \qquad \text{whenever } x > -2. \ \blacksquare$$

EXERCISES 4.2

Evaluate the limits in Exercises 1–28.

1. $\lim\limits_{x\to\infty}\dfrac{x^2+x}{x^3-x^2}$

2. $\lim\limits_{x\to\infty}\dfrac{(3x-1)^3}{x^4+1}$

3. $\lim\limits_{x\to-\infty}\dfrac{x^2-3x+1}{x^2+27}$

4. $\lim\limits_{x\to-\infty}\dfrac{3-2x-x^2}{1-x-2x^2}$

5. $\lim\limits_{x\to\infty}\dfrac{(x^2+1)(x-3)}{x^3-8}$

6. $\lim\limits_{x\to\infty}\dfrac{2x^2+3x+1}{6x^2-7x-2}$ **(Calculus Explorer & Tutor I, Limits, 11.)**

7. $\lim\limits_{x\to\infty}\sqrt{\dfrac{x^2}{1+4x^2}}$

8. $\lim\limits_{x\to-\infty}\sqrt{\dfrac{x}{9x+4}}$

9. $\lim\limits_{x\to\infty}\dfrac{x^2}{x-1}$

10. $\lim\limits_{x\to\infty}\dfrac{6x^3+2x-3}{4x^2+1}$ **(Calculus Explorer & Tutor I, Limits, 12.)**

11. $\lim\limits_{x\to-\infty}\dfrac{x^3}{1-x}$

12. $\lim\limits_{x\to\infty}\dfrac{1-x}{1-x^3}$

13. $\lim\limits_{x\to-\infty}\dfrac{x}{\sqrt{4x^2+1}}$

14. $\lim\limits_{x\to\infty}\dfrac{x}{\sqrt{9x^2+1}}$

15. $\lim\limits_{x\to\infty}\dfrac{x\sqrt{4x+1}}{(1+\sqrt{x})^3}$

16. $\lim\limits_{x\to-\infty}\dfrac{x\sqrt{1-4x}}{\sqrt{1-9x^3}}$

17. $\lim\limits_{x\to\infty}(x^3-4x^2)$

18. $\lim\limits_{x\to\infty}(x-2x^2)$

19. $\lim\limits_{x\to-\infty}(x-x^2-x^3-4x^4)$

20. $\lim\limits_{x\to-\infty}(2x^5-3x^4+x^3)$

21. $\lim\limits_{x\to\infty}(x^{3/5}-x^{3/4})$

22. $\lim\limits_{x\to\infty}(x-\sqrt{x})$

23. $\lim\limits_{x\to\infty}\left(\dfrac{x^2}{x+1}-\dfrac{x^2}{x-1}\right)$

24. $\lim\limits_{x\to\infty}(\sqrt{x^2+4x}-\sqrt{x^2+1})$
$\left(\textit{Hint: Multiply by } \dfrac{\sqrt{x^2+4x}+\sqrt{x^2+1}}{\sqrt{x^2+4x}+\sqrt{x^2+1}}.\right)$

25. $\lim\limits_{x\to\infty}(\sqrt{x^2+4}-x)$
$\left(\textit{Hint: Multiply by } \dfrac{\sqrt{x^2+4}+x}{\sqrt{x^2+4}+x}.\right)$

26. $\lim\limits_{x\to\infty}(\sqrt{x^2-2x}-x)$
(Calculus Explorer & Tutor I, Limits, 13.)

27. $\lim\limits_{x\to\infty}x\sin(1/x)$

28. $\lim\limits_{x\to\infty}x\sin(1/x^2)$

In Exercises 29–48, sketch the graphs in the simplest way that reflects the domain, intercepts, sign of $f(x)$, continuity, and behavior near vertical and horizontal asymptotes.

29. $f(x)=\dfrac{x}{x-1}$

30. $f(x)=\dfrac{x}{x+1}$

31. $f(x)=\dfrac{x^2}{x^2-1}$

32. $f(x)=\dfrac{(x+1)^2}{x^2+2x}$

33. $f(x)=\dfrac{x}{x^2+x+1}$

34. $f(x)=\dfrac{x}{x^2+1}$

35. $f(x)=\dfrac{x-2}{x^2-4x}$

36. $f(x)=\dfrac{x}{x^2-1}$

37. $f(x)=\dfrac{x^2-4}{x^2-1}$

38. $f(x)=\dfrac{x^2-1}{x^2-4}$

39. $f(x)=\dfrac{x^2-1}{x^3}$

40. $f(x)=\dfrac{x^2-1}{x^4}$

41. $f(x)=\dfrac{1}{\sqrt{x^2+1}}$

42. $f(x)=\dfrac{1}{x^2+1}$

43. $f(x)=\dfrac{|x|}{\sqrt{x^2+1}}$

44. $f(x)=\dfrac{6x}{\sqrt{4x^2+9}}$

45. $f(x)=\dfrac{\sin x}{x^2}$

46. $f(x)=\dfrac{\cos x}{x}$

47. $f(x)=\dfrac{\cos^2 x}{x^2}$

48. $f(x)=\dfrac{\sin^2 x}{x^2}$

In Exercises 49–62, determine the asymptotic behavior and sketch the graphs.

49. $f(x)=x+\dfrac{1}{x}$

50. $f(x)=x-\dfrac{1}{x}$

51. $f(x)=\dfrac{1-4x-x^2}{x+2}$

52. $f(x)=\dfrac{1-2x-x^2}{x+1}$

53. $f(x)=\dfrac{x^2-2x}{x-1}$

54. $f(x)=\dfrac{x^2+4x+5}{x+2}$

55. $f(x)=\dfrac{x^3+8}{x^2}$

56. $f(x)=\dfrac{x^3-1}{x^2}$

57. $f(x)=\dfrac{x^3-1}{x}$

58. $f(x)=\dfrac{x^3-x^2-2x}{x-1}$

59. $f(x)=\dfrac{x^4-1}{x^2}$

60. $f(x)=\dfrac{x^4+1}{x^2}$

61. $f(x)=\dfrac{x^4+1}{x}$

62. $f(x)=\dfrac{x^4-1}{x}$

Write a formal definition for the statements in Exercises 63–66.

63. $\lim\limits_{x\to\infty}f(x)=\infty$

64. $\lim\limits_{x\to\infty}f(x)=-\infty$

65. $\lim\limits_{x\to-\infty}f(x)=-\infty$

66. $\lim\limits_{x\to-\infty}f(x)=\infty$

67. Use the definition to show that $\lim\limits_{x \to \infty} \dfrac{1}{2x - 1} = 0$.

68. Use the definition to show that $\lim\limits_{x \to \infty} \dfrac{x}{x + 1} = 1$.

69. If $\lim\limits_{x \to \infty} f(x) = L$ and $\lim\limits_{x \to \infty} g(x) = M$, where L and M are real numbers, use the definition to show that $\lim\limits_{x \to \infty} [f(x) + g(x)] = L + M$.

In Exercises 70–73, use calculator-generated graphs to determine apparent asymptotes of the given fuctions.

70. $f(x) = \ln \left| \dfrac{x}{x - 1} \right|$

71. $f(x) = \dfrac{\ln |\ln |x||}{x}$

72. $f(x) = \dfrac{e^x}{e^x - 1}$

73. $f(x) = \tan^{-1} \dfrac{x}{x - 1}$

4.3 SIGN OF $f'(x)$; EXTREMA

Connections

Zeros and signs of expressions, 1.2.
Geometric interpretation of derivative, 2.6.
Graphs with asymptotes, 4.1–2.

An important general characteristic of a function involves the rising or falling of the graph as x moves in the positive direction over an interval.

Monotonicity on Intervals

Informally, we say f is **increasing** on an interval if the graph rises as x increases along the interval, as indicated in Figure 4.3.1a. We say f is **decreasing** if the graph falls as x increases along the interval, as illustrated in Figure 4.3.1b. The following formal definition reflects these ideas.

DEFINITION

A function f is **increasing** on an interval I if $f(x_1) < f(x_2)$ whenever x_1 and x_2 are in I and $x_1 < x_2$; f is **decreasing** on I if $f(x_1) > f(x_2)$ whenever x_1 and x_2 are in I and $x_1 < x_2$. If f is either increasing on I or decreasing on I, f is said to be **monotonic** on I.

Recall that the slope of the line tangent to the graph of f at $(c, f(c))$ is given by the value of the derivative $f'(c)$. The tangent line is increasing if $f'(c) > 0$ and decreasing if $f'(c) < 0$. See Figure 4.3.2. The following theorem relates the sign of $f'(x)$ to intervals where the graph of f is either increasing or decreasing.

THEOREM

If $f'(x) > 0$ for all x in an interval I, then f is increasing on I. If $f'(x) < 0$ for all x in I, then f is decreasing on I.

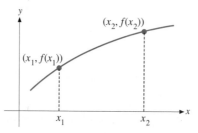

FIGURE 4.3.1a The function f is increasing on an interval if the graph rises as x increases along the interval.

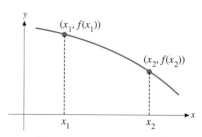

FIGURE 4.3.1b The function f is decreasing on an interval if the graph falls as x increases along the interval.

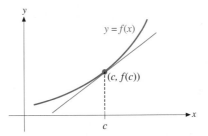

FIGURE 4.3.2a The line tangent to the graph of f at the point $(c, f(c))$ is increasing if $f'(c) > 0$.

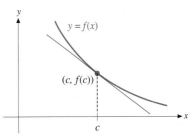

FIGURE 4.3.2b The line tangent to the graph of f at the point $(c, f(c))$ is decreasing if $f'(c) < 0$.

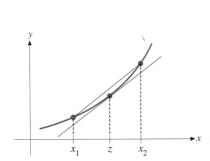

FIGURE 4.3.3 The Mean Value Theorem tells us there is a point z between x_1 and x_2 such that $\dfrac{f(x_2) - f(x_1)}{x_2 - x_1} = f'(z)$.

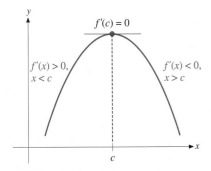

FIGURE 4.3.4a If $f'(c) = 0$, $f'(x) > 0$ for x to the left of c, and $f'(x) < 0$ for x to the right of c, then f has a local maximum at $(c, f(c))$.

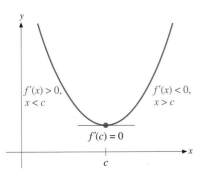

FIGURE 4.3.4b If $f'(c) = 0$, $f'(x) < 0$ for x to the left of c, and $f'(x) > 0$ for x to the right of c, then f has a local minimum at $(c, f(c))$.

Proof Let us verify the first statement of the theorem. Verification of the second statement is similar.

We assume $f'(x) > 0$ for all x in an interval I. We must show that f satisfies the condition of the definition of increasing on I.

If x_1 and x_2 are in I, then the fact that $f'(x)$ exists for all x in I implies that f is continuous and differentiable on the closed interval between x_1 and x_2. The Mean Value Theorem then tells us there is a point z between x_1 and x_2 such that

$$\frac{f(x_2) - f(x_1)}{x_2 - x_1} = f'(z), \text{ or}$$
$$f(x_2) - f(x_1) = f'(z)(x_2 - x_1).$$

Since z is in I, we have $f'(z) > 0$. Then $x_1 < x_2$ implies that the product $f'(z)(x_2 - x_1)$ is positive. That is,

$$f(x_2) - f(x_1) = f'(z)(x_2 - x_1) > 0,$$

so $f(x_2) > f(x_1)$ whenever x_1 and x_2 are in I and $x_1 < x_2$. The condition of the definition is satisfied, so f is increasing on I. See Figure 4.3.3. ∎

Let us see how the monotonicity of f in intervals relates to local extrema of f.

Recall from Section 3.8 that values of x for which f is defined in an open interval that contains x and either

(i) $f'(x) = 0$ or
(ii) $f'(x)$ does not exist

are called **critical points** of f. From Theorem 1 of Section 3.8 we see that if $f(c)$ is either a local maximum or local minimum of a function f, then c is a critical point of f. However, the fact that c is a critical point of f does not guarantee that $f(c)$ is either a local maximum or a local minimum. We can use the monotonicity of f in intervals to determine which critical points correspond to local extrema. The idea is illustrated in Figure 4.3.4 and stated formally in the following theorem.

THE FIRST DERIVATIVE TEST FOR LOCAL EXTREMA

(a) If $f'(c) = 0$ and there is some $\delta > 0$ such that $f'(x) > 0$ for $c - \delta < x < c$ and $f'(x) < 0$ for $c < x < c + \delta$, then f has a local maximum at $(c, f(c))$.

(b) If $f'(c) = 0$ and there is some $\delta > 0$ such that $f'(x) < 0$ for $c - \delta < x < c$ and $f'(x) > 0$ for $c < x < c + \delta$, then f has a local minimum at $(c, f(c))$.

The proof is left for an exercise.

 Of course, monotonicity and local extrema can be determined from either a given or calculator-generated graph.

EXAMPLE 1

From the graph in Figure 4.3.5 we see that f is increasing on the intervals $x \leq 1$ and $x \geq 3$; f is decreasing on the interval $1 \leq x \leq 3$; f has a local maximum at $(1, 3)$ and a local minimum at $(3, -1)$. ∎

Graphs and the Sign of $f'(x)$

Using properties of a function to sketch its graph is a good way to study the properties. Let us see how to use the sign of $f'(x)$, as well as the domain, intercepts, sign of $f(x)$, continuity, and asymptotes, to sketch the graph of f.

Points on the graph of a function f that correspond to critical points of f should be plotted. We should draw horizontal tangents through points on the graph where $f'(x) = 0$. If f is continuous at c and $f'(x)$ approaches either infinity or negative infinity as x approaches c, the graph of f becomes vertical at the point $(c, f(c))$; we should draw vertical tangents through such points. These horizontal and vertical tangents are then used as guides for sketching the graph.

When sketching the graph of a function f, we should sketch the graph increasing on intervals where $f'(x)$ is positive; the graph should be sketched decreasing on intervals where $f'(x)$ is negative.

We will use the fact that $f'(x)$ is of constant sign on the intervals between successive points where either $f'(x) = 0$ or $f'(x)$ does not exist. In some cases the sign of $f'(x)$ in an interval is easily determined by observation of the formula for $f'(x)$. In more complicated examples, the formula for $f'(x)$ should be written in factored form and a table of signs constructed.

EXAMPLE 2

Sketch the graph of $f(x) = 1 + 3x - x^3$. Find local extrema.

SOLUTION   Let us begin by investigating key points and general characteristics that are directly related to the values of the function $f(x)$. The function is defined and continuous for all x. We see that the y-intercept is $(0, 1)$, but the x-intercepts are not easily determined from the equation $f(x) = 0$. Hence, it is not convenient to determine intervals where $f(x)$ is of constant sign. In order to investigate the behavior of $f(x)$ for $|x|$ large, we write

$$f(x) = 1 + 3x - x^3 = x^3 \left(\frac{1}{x^3} + \frac{3}{x^2} - 1 \right).$$

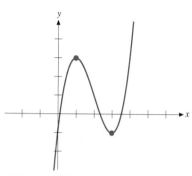

FIGURE 4.3.5 The function is increasing on the intervals $x \leq 1$ and $x \geq 3$; f is decreasing on the interval $1 \leq x \leq 3$; f has a local maximum at $(1, 3)$ and a local minimum at $(3, -1)$.

The second factor approaches -1 as $|x|$ approaches infinity. Since

$$\lim_{x \to -\infty} x^3 = -\infty \text{ and } \lim_{x \to \infty} x^3 = \infty$$

it follows that

$$\lim_{x \to -\infty} f(x) = \infty \text{ and } \lim_{x \to \infty} f(x) = -\infty.$$

Let us now consider information related to the derivative $f'(x)$. We have

$$f(x) = 1 + 3x - x^3, \text{ so}$$

$$\begin{aligned} f'(x) &= 3 - 3x^2 \\ &= 3(1 - x^2) \\ &= 3(1 - x)(1 + x). \end{aligned}$$

TABLE 4.3.1

	$-\infty$		-1		1		∞
$3(1-x)$		$+$		$+$	0	$-$	
$1+x$		$-$	0	$+$		$+$	
$f'(x)$		$-$		$+$		$-$	

From the factored form of $f'(x)$ we see that $f'(x) = 0$ when $x = 1$ or $x = -1$. The corresponding values of y are

$$\begin{aligned} f(1) &= 1 + 3(1) - (1)^3 = 3 \text{ and} \\ f(-1) &= 1 + 3(-1) - (-1)^3 = -1. \end{aligned}$$

The graph has horizontal tangents at $(1, 3)$ and $(-1, -1)$.

The sign of $f'(x)$ can be determined either by direct observation or from the table of signs given in Table 4.3.1. We see that $f'(x)$ is positive on the interval $-1 < x < 1$, so f is increasing on this interval; $f'(x)$ is negative on the intervals $x < -1$ and $x > 1$, so f is decreasing on these intervals.

The graph in Figure 4.3.6 reflects the key points and general characteristics that we have found. The graph decreases to the point $(-1, -1)$, where it levels off and begins increasing. It increases through the y-intercept at $(0, 1)$ to the point $(1, 3)$, where it levels off and begins decreasing. From the graph we see that there are three x-intercepts, one to the left of -1, one between -1 and 0, and one to the right of 1. We also see that the critical point $x = -1$ gives a local minimum and the critical point $x = 1$ gives a local maximum. ■

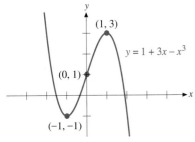

FIGURE 4.3.6 The function $f(x) = 1 + 3x - x^3$ is decreasing on the intervals $x < -1$ and $x > 1$, where the derivative is negative. It is increasing on the interval $-1 < x < 1$, where the derivative is positive. There is a local minimum at $(-1, -1)$ and a local maximum at $(1, 3)$.

EXAMPLE 3

Sketch the graph of $f(x) = \dfrac{1}{9}(x^4 - 4x^3)$. Find local extrema.

SOLUTION   The function is defined and continuous for all x. Factoring $f(x)$, we have

$$f(x) = \frac{x^3}{9}(x - 4).$$

This shows the intercepts are $(0, 0)$ and $(4, 0)$. A sign chart for the function $f(x)$ can be used to see that $f(x) > 0$ for $x < 0$ and $x > 4$, and that $f(x) < 0$ for $0 < x < 4$. Since

$$f(x) = \frac{x^3}{9}(x - 4) = \frac{x^4}{9}\left(1 - \frac{4}{x}\right),$$

we see that $f(x) \to \infty$ as $x \to \pm\infty$.

We could now sketch the graph of f in the simplest way that reflects the properties determined above. The result would be similar to the actual graph given in Figure 4.3.7, although we would not know about the "wiggle" at the

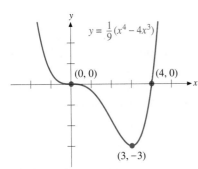

FIGURE 4.3.7   The function $f(x) = \dfrac{1}{9}(x^4 - 4x^3)$ has critical points $(0, 0)$ and $(3, -3)$ where the derivative is zero. The function is decreasing on the interval $x < 3$ and is increasing on the interval $x > 3$. The function has a local minimum at $(3, -3)$; there is no extreme value at the critical point $(0, 0)$.

origin. We would not know for certain on which intervals the graph is *actually* increasing or decreasing, and we would not know the exact location of the local minimum indicated by the sketch. These characteristics are related to the sign of $f'(x)$. We have $f(x) = \dfrac{1}{9}(x^4 - 4x^3)$, so

$$f'(x) = \frac{1}{9}(4x^3 - 12x^2) = \frac{4x^2}{9}(x - 3).$$

We see that $f'(x) = 0$ when $x = 0$ or $x = 3$. We have $f(0) = 0$ and $f(3) = -3$. The graph has horizontal tangents at $(0, 0)$ and $(3, -3)$. A table of signs for $f'(x)$ is given in Table 4.3.2.

TABLE 4.3.2

	$-\infty$		0		3		∞
$\dfrac{4x^2}{9}$		$+$	0	$+$		$+$	
$x - 3$		$-$		$-$	0	$+$	
$f'(x)$		$-$		$-$		$+$	

Theorem 1 tells us that f is increasing on the interval $x > 3$ and decreasing on each of the intervals $x < 0$ and $0 < x < 3$.

To sketch the graph, we plot the points $(0, 0)$ $(4, 0)$, and $(3, -3)$, and draw horizontal tangents through the points $(0, 0)$ and $(3, -3)$. The graph is then sketched through these points in a way that reflects the sign of $f'(x)$. This is done in Figure 4.3.7. The graph decreases to the point $(0, 0)$, levels off, and continues decreasing to the point $(3, -3)$. Hence, the graph is decreasing on the interval $x < 3$. It levels off at $(3, -3)$ and then increases through the point $(4, 0)$.

From the graph we see that the critical point $x = 3$ gives a local minimum. Also, the value of f at the critical point $x = 0$ is neither a local maximum nor a local minimum. ∎

Behavior near Asymptotes

EXAMPLE 4

Sketch the graph of $f(x) = \dfrac{x}{4 - x^2}$. Find local extrema.

SOLUTION The function is defined for all x, except $x = 2$ and $x = -2$. It is continuous on intervals where it is defined. The only intercept is $(0, 0)$. The graph has vertical asymptotes $x = 2$ and $x = -2$. We have

$$\lim_{x \to \pm\infty} \frac{x}{4 - x^2} = 0,$$

so the graph has horizontal asymptote $y = 0$. A table of signs can be used to determine that $f(x)$ is positive on the intervals $x < -2$ and $0 < x < 2$, and that $f(x)$ is negative on the intervals $-2 < x < 0$ and $x > 2$. We may also notice that $f(-x) = -f(x)$, so the function is odd and the graph is symmetric with respect to the origin.

We could now sketch the graph in the simplest way that reflects the above information. The result would probably look like the graph in Figure 4.3.8. However, we would not be certain about the lack of "wiggles" and local extrema unless we investigate the derivative $f'(x)$. We have

$$f(x) = \frac{x}{4 - x^2},$$

FIGURE 4.3.8 The function $f(x) = \dfrac{x}{4 - x^2}$ has intercept $(0, 0)$, vertical asymptotes $x = -2$ and $x = 2$, and horizontal asymptote $y = 0$. The function is increasing on each of the intervals $x < -2$, $-2 < x < 2$, and $x > 2$, where $f'(x) > 0$. There is no local extremum.

so the Quotient Rule gives

$$f'(x) = \frac{(4 - x^2)(1) - (x)(-2x)}{(4 - x^2)^2}.$$

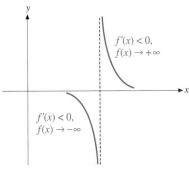

(a)

(b)

FIGURE 4.3.9 The sign of $f'(x)$ as x approaches each side of a vertical asymptote determines if $f(x)$ approaches infinity or negative infinity from that side.

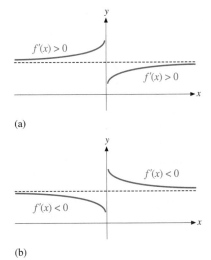

(a)

(b)

FIGURE 4.3.10 The graph of f does not cross a horizontal asymptote in infinite intervals where $f'(x)$ is of constant sign. The graph can cross a horizontal asymptote any number of times if the sign of the derivative changes.

Simplifying, we have

$$f'(x) = \frac{4 - x^2 + 2x^2}{(4 - x^2)^2} = \frac{x^2 + 4}{(4 - x^2)^2}.$$

There are no critical points. ($f'(x)$ does not exist for $x = 2$ and $x = -2$, but these points are not in the domain of f.) There are no local extrema. We see that $f'(x) > 0$ at every point at which it exists. Theorem 1 tells us f is increasing on each of the intervals $x < -2$, $-2 < x < 2$, and $x > 2$.

We can sketch the graph by plotting the intercept $(0, 0)$, drawing the vertical asymptotes $x = -2$ and $x = 2$, drawing the horizontal asymptote $y = 0$, and then sketching the graph increasing on each of the intervals $x < -2, -2 < x < 2$, and $x > 2$. This is done in Figure 4.3.8. Note that f is *not* increasing on the interval $-\infty < x < \infty$. ∎

If $f'(x)$ is of constant sign as $(x, f(x))$ approaches an asymptote, we can infer additional information about the graph. For example, we see in Figure 4.3.9a that if f is increasing on the left side of a vertical asymptote, then $f(x)$ must approach infinity as x approaches the asymptote from the left; if f is increasing on the right side of a vertical asymptote, then $f(x)$ must approach negative infinity as x approaches the asymptote from the right. The behavior of a graph of a function that is decreasing on each side of a vertical asymptote is illustrated in Figure 4.3.9b.

If $f'(x)$ is of constant sign as x approaches either ∞ or $-\infty$, and $(x, f(x))$ approaches a horizontal asymptote, the graph must approach the asymptote from only one side of the asymptote. The possibilities are illustrated in Figure 4.3.10. A graph does not cross a horizontal asymptote in infinite intervals where $f'(x)$ is of constant sign. Of course, the graph can cross a horizontal asymptote any number of times if the sign of $f'(x)$ changes.

Vertical Tangents

EXAMPLE 5
Sketch the graph of $f(x) = x\sqrt{x + 1}$. Find local extrema.

SOLUTION $f(x)$ is defined only for $x + 1 \geq 0$, or $x \geq -1$; it is continuous on the interval $x \geq -1$. The intercepts are $(0, 0)$ and $(-1, 0)$. It is not difficult to check that $f(x)$ is negative for $-1 < x < 0$ and positive for $x > 0$.

The Product Rule is used to find the derivative. Also, the Chain Rule is needed to differentiate the factor that is the square root of the function $x + 1$. We have

$$f(x) = x\sqrt{x + 1}, \text{ so}$$

$$f'(x) = (x)\left(\frac{1}{2\sqrt{x + 1}}\right)(1) + (\sqrt{x + 1})(1).$$

To determine the critical points, we simplify $f'(x)$ to obtain

$$f'(x) = \frac{x}{2\sqrt{x + 1}} + \sqrt{x + 1}\left(\frac{2\sqrt{x + 1}}{2\sqrt{x + 1}}\right)$$

$$= \frac{x + 2(x + 1)}{2\sqrt{x + 1}} = \frac{3x + 2}{2\sqrt{x + 1}}.$$

We see that $f'(x) = 0$ when $x = -2/3$. We have

$$f\left(-\frac{2}{3}\right) = \left(-\frac{2}{3}\right)\sqrt{-\frac{2}{3}+1} = -\left(\frac{2}{3}\right)\sqrt{\frac{1}{3}} = -\frac{2}{3^{3/2}} \approx -0.38.$$

The graph has a horizontal tangent at $(-2/3, -2/3^{3/2})$. We see that $f'(x)$ approaches negative infinity as x approaches -1 from the right, so the graph becomes vertical at $(-1, 0)$. We have $f'(x) < 0$ for $-1 < x < -2/3$, so the graph is decreasing on this interval. We have $f'(x) > 0$ for $x > -2/3$, so the graph is increasing on the interval $x > -2/3$.

To sketch the graph, we plot the points $(-1, 0)$, $(0, 0)$, and $(-2/3, -2/3^{3/2})$, draw a horizontal tangent through $(-2/3, -2/3^{3/2})$ and a vertical tangent through $(-1, 0)$, and then sketch the graph in a way that reflects the sign of $f'(x)$. This is done in Figure 4.3.11. We see that f has a local minimum at $(-2/3, -2/3^{3/2})$. (Recall that we have required f be defined in an open interval that contains c in order that $f(c)$ be a local extremum. This eliminates the point $(-1, 0)$ from consideration as a local maximum.) ∎

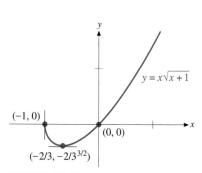

FIGURE 4.3.11 The function $f(x) = x\sqrt{x+1}$ is defined for $x \geq -1$. It is decreasing on the interval $-1 < x < -2/3^{2/3}$ and increasing on the interval $x > -2/3^{2/3}$. The function has a local minimum at $(-2/3, -2/3^{3/2})$.

Relating the Graphs of f and f'

It is possible to use the graph of a function f to determine the graph of the derivative f'. To do this we note that the slope of the line tangent to the graph of f at a point $(x, f(x))$ on the graph is the value of the derivative at x. In particular, if the line tangent to the graph of f at $(x, f(x))$ is horizontal, then $f'(x) = 0$, so the graph of f' intersects the x-axis at the point $(x, 0)$. Also, $f'(x)$ must be positive on intervals where f is increasing and $f'(x)$ must be negative on intervals where f is decreasing.

EXAMPLE 6

Use the graph of f given in Figure 4.3.12a to sketch the graph of f'.

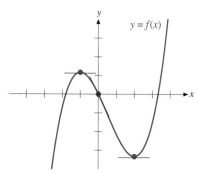

FIGURE 4.3.12a The graph of f has horizontal tangent lines where $x = -1$ and $x = 2$. The slope of the tangent line at the origin appears to be approximately -2.

SOLUTION We see from Figure 4.3.12a that lines tangent to the graph of f at the points where $x = -1$ and $x = 2$ are horizontal, so we must have $f'(-1) = f'(2) = 0$. The graph of f is increasing on the intervals $x < -1$ and $x > 2$, so $f'(x)$ must be positive in those intervals. Since f is decreasing on the interval $-1 < x < 2$, $f'(x)$ must be negative on that interval. Finally, we note that the slope of the line tangent to the graph of f at the origin appears to be approximately negative two, so $f'(0)$ must be approximately negative two. In Figure 4.3.12b we have plotted the points $(-1, 0)$, $(2, 0)$ and $(0, -2)$, and then sketched the graph of f' through these points with $f'(x)$ positive on the intervals $x < -1$ and $x > 2$, and negative on the interval $-1 < x < 2$. ∎

Recall from Section 2.6 that a function is not differentiable at points where either it is not continuous, the graph has a sharp corner, or the graph becomes vertical.

EXAMPLE 7

The graph of f in Figure 4.3.13a becomes vertical where $x = 1$. As x approaches one from either side, the slope of the curve approaches infinity. This means the value of the derivative of f approaches infinity as x approaches one from each side, so $x = 1$ is a vertical asymptote of f'. See Figure 4.3.13b. The function f is continuous, but not differentiable at $x = 1$. ∎

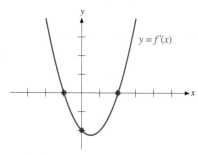

FIGURE 4.3.12b The derivative of f satisfies $f'(-1) = f'(2) = 0$ and $f'(0) \approx -2$. $f'(x)$ is positive in intervals where f is increasing and $f'(x)$ is negative in intervals where f is decreasing.

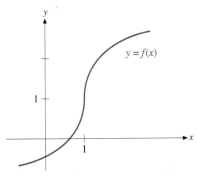

FIGURE 4.3.13a The graph of the continuous function f becomes vertical as x approaches one from each side, so the derivative becomes unbounded.

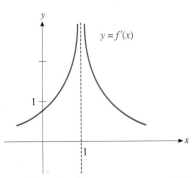

FIGURE 4.3.13b The derivative of f has vertical asymptote $x = 1$; f is not differentiable at $x = 1$.

For some functions g we can use the graph of g to sketchthe graph of a function f such that $f' = g$. The idea is to match intervals where g is positive with intervals where f is increasing and to match intervals where g is negative with intervals where f is decreasing.

Let us note that there is not a unique function f such that $f' = g$. That is, if $f' = g$, then all functions of the form $h = f + C$, C a constant, also satisfy $h' = f' = g$. Theorem 4 of Section 3.8 tells us that all functions h that satisfy $h' = f'$ on an interval are of the form $h = f + C$ on the interval. This means that the graphs of any two functions that have derivative g on an interval differ by a vertical shift.

Statement 1 of Section 2.6 tells us that f must be continuous in intervals where f' exists. This means that if $f' = g$ in an interval, then f must be continuous in the interval.

Finally, we note that not all functions g are derivatives, so there may be no function f with $f' = g$.

EXAMPLE 8

There is no continuous function f such that $f' = g$, where

$$g(x) = \begin{cases} -1, & x < 1, \\ 0, & x = 0, \\ 1, & x > 0. \end{cases}$$

Any function f that satisfies $f' = g$ must be of the form

$$f(x) = \begin{cases} -x + a, & x < 1, \\ b, & x = 0, \\ x + c, & x > 0, \end{cases}$$

where a, b and c are constants. The function f would be continuous at zero if and only if $a = b = c$. However, the graph of f would have a sharp corner where $x = 0$, so f could not be differentiable at zero. See Figure 4.3.14. ∎

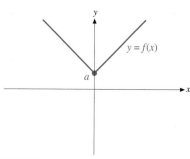

FIGURE 4.3.14a   The function g is
not continuous at zero.

FIGURE 4.3.14b   The continuous
function f is not differentiable at zero,
so we cannot have $f' = g$.

EXERCISES 4.3

Sketch the graphs and find local extrema in Exercises 1–56.

1. $f(x) = x^{4/3}$

2. $f(x) = x^{2/3}$

3. $f(x) = x^3$

4. $f(x) = x^{1/3}$

5. $f(x) = (10 + 6x - x^2)/6$

6. $f(x) = (2x^2 + 8x - 11)/8$

7. $f(x) = (x^3 - 12x)/8$

8. $f(x) = 3(x^3 + 2x^2 + x)$

9. $f(x) = 2 + 3x - x^3$

10. $f(x) = x^3 - 3x^2 + 2$

11. $f(x) = x^3 + x$

12. $f(x) = 2 - x - x^3$

13. $f(x) = 4x^2 - x^4$

14. $f(x) = x^4 + x^2$

15. $f(x) = 3x^4 - 4x^3 + 6x^2$

16. $f(x) = 3x^4 + 4x^3$

17. $f(x) = (12x^5 - 45x^4 + 40x^3)/8$

18. $f(x) = 4x^5 - 5x^4$

19. $f(x) = x^{7/3} - 7x^{1/3}$

20. $f(x) = (16x^{2/3} - x^{8/3})/16$

21. $f(x) = (x^2 - 1)^{3/5}$

22. $f(x) = (x^2 - 1)^{2/5}$

23. $f(x) = \sqrt{x}(x - 3)$

24. $f(x) = x\sqrt{3 - x}$

25. $f(x) = \dfrac{8x}{x^2 + 4}$

26. $f(x) = \dfrac{1}{x^2 + 1}$

27. $f(x) = \dfrac{x^2}{x^2 + 9}$

28. $f(x) = \dfrac{4x^3}{x^4 + 3}$

29. $f(x) = \dfrac{6x^{1/3}}{x^2 + 5}$

30. $f(x) = \dfrac{12x^{2/3}}{x^2 + 8}$

31. $f(x) = x + \dfrac{4}{x}$

32. $f(x) = x + \dfrac{4}{x^2}$

33. $f(x) = x^2 + \dfrac{1}{x^2}$

34. $f(x) = x^3 + \dfrac{3}{x}$

35. $f(x) = \dfrac{1}{x^2 - 1}$

36. $f(x) = \dfrac{x}{x^2 - 1}$

37. $f(x) = \dfrac{8x - 20}{x^2 - 4}$

38. $f(x) = \dfrac{x^2 - 1}{x^2}$

39. $f(x) = \dfrac{(x^2 - 1)^2}{x^4}$

40. $f(x) = 4\left(\dfrac{x^2 - 1}{x^4}\right)$

41. $f(x) = x + \sin x, 0 \le x \le 2\pi$

42. $f(x) = x + \cos x - 1, 0 \le x \le 2\pi$

43. $f(x) = x + \frac{1}{2}\sin 2x, 0 \le x \le 2\pi$

44. $f(x) = x + 2\sin x, 0 \le x \le 2\pi$

45. $f(x) = \sin x + \cos x, 0 \le x \le 2\pi$

46. $f(x) = \cos x - \sin x, 0 \le x \le 2\pi$

47. $f(x) = \tan x - x, -\pi/2 < x < \pi/2$

48. $f(x) = \cot x + x, 0 < x < \pi$

49. $f(x) = \tan x - 2\sec x, -\pi/2 < x < \pi/2$

50. $f(x) = \tan x + 2\sec x, -\pi/2 < x < \pi/2$

51. $f(x) = \dfrac{1}{\sqrt{1 - x^2}}$

52. $f(x) = \dfrac{1}{\sqrt{2x - x^2}}$

53. $f(x) = \sqrt{4x - x^2}$

54. $f(x) = -\sqrt{9 - x^2}$

55. $f(x) = \sqrt{x^2 - 1}$

56. $f(x) = \dfrac{1}{\sqrt{x^2 - 1}}$

*Find intervals on which the functions given in Exercises 57–58
are increasing.*

57. $f(x) = x^3 - 9x^2 - 48x + 25$ (**Calculus Explorer & Tutor I, Shapes of Curves, 1.**)

58. $f(x) = \dfrac{x - 3}{x^2 + 7}$ (**Calculus Explorer & Tutor I, Shapes of Curves, 2.**)

In Exercises 59–62, the position of an object along a number line at time t is given by $s(t)$. Find intervals of time for which the object is moving in the positive direction.

59. $s(t) = -16t^2 + 64t + 64, t \geq 0$

60. $s(t) = -16t^2 + 128t + 32, t \geq 0$

61. $s(t) = \sin t, 0 \leq t \leq 2\pi$

62. $s(t) = \sin 2t, 0 \leq t \leq 2\pi$

In Exercises 63–66, use the given graph of f to **(a)** find values of x for which $f'(x) = 0$, **(b)** find intervals on which $f'(x) > 0$, and **(c)** find intervals on which $f'(x) < 0$.

63.

64.

65.

66.
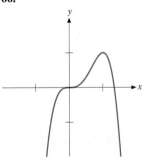

In Exercises 67–70, use the given graphs of f and g to determine whether $g = f'$ or $f = g'$ is true.

67.

68.

69.

70.
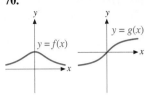

In Exercises 71–78, use the given graph of f to sketch the graph of f'.

71.

72.

73.

74.

75.

76.

77.

78.
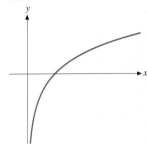

In Exercises 79–82, use the given graph of g to sketch the graph of a continuous function f such that f′ = g and f (0) = 0.

79.

80.

81.

82.

83. The study of thermodynamics and fluid flows related to rocket propulsion leads to the function

$$f(M) = \frac{M^2 + \left(\dfrac{\gamma - 1}{2}\right) M^4}{(1 + \gamma M^2)^2},$$

where the constant $\gamma > 1$ is a ratio of specific heats and the **Mach number** M is a measure of internal energy. Sketch the graph of f as a function of $M \geq 0$. What is the maximum value of f? What is the asymptotic value of $f(M)$ as M approaches infinity?

84. An **ideal voltage source** is a device whose terminal voltage is independent of the current through it. The voltage of any actual voltage device decreases as the current through it increases. For example, if a car is started with the lights on, the lights dim perceptibly while the battery is supplying current to the starter. A **practical voltage source model** consists of an ideal voltage source V_S connected to an ideal resistor R_S. If a practical voltage source is connected to a load resistor R_L, as indicated in the figure below, we have $V_S = i_L(R_S + R_L)$, where i_L, is the current through the load. Since $i_L R_L = V_L$, the latter equation can be written as $V_S = i_L R_S + V_L$. The power delivered to the load is

$$P_L = i_L^2 R_L = \frac{V_S^2 R_L}{(R_S + R_L)^2}.$$

(a) Sketch the graph of $i_L = \dfrac{V_S - V_L}{R_S}$ as a function of V_L with R_S and V_S fixed. What is the load current for a short circuit, when $V_L = 0$? What is the load voltage for an open circuit, when $i_L = 0$?

(b) Show that the maximum power delivered to the load by a practical voltage source with R_S and V_S fixed occurs when the load resistance satisfies $R_L = R_S$. This result is called the **maximum power transform theorem**.

85. Show that the graph of the function $f(x) = x\sqrt{x+1}$ of Example 5 approaches the asymptote

$$A(x) = \left(x + \frac{1}{3}\right)^{3/2}$$

as x approaches infinity.

86. If f is continuous at c and $\lim\limits_{x \to c^+} f'(x) = \infty$, use the Mean Value Theorem to show that

$$\lim_{x \to c^+} \frac{f(x) - f(c)}{x - c} = \infty.$$

87. Use the Mean Value Theorem to prove the **First Derivative Test for Local Extrema**.

(a) If $f'(c) = 0$, $f'(x) > 0$ for $c - \delta < x < c$, and $f'(x) < 0$ for $c < x < c + \delta$, then f has a local maximum at $(c, f(c))$.

(b) If $f'(c) = 0$, $f'(x) < 0$ for $c - \delta < x < c$, and $f'(x) > 0$ for $c < x < c + \delta$, then f has a local minimum at $(c, f(c))$.

4.4 SIGN OF *f″*(*x*); CONCAVITY

We continue the study of functions by introducing another general characteristic, that of concavity. Concavity involves the rate of change of the increase or decrease of a function.

Concavity

In simple terms, a graph is said to be **concave up** if it opens upward. It is said to be **concave down** if it opens downward. The idea of concavity may be expressed more precisely by considering chords drawn between two points on the graph.

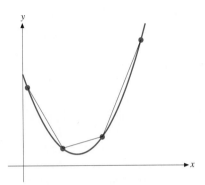

FIGURE 4.4.1a The graph of a function that is concave up opens upward and is below each chord.

> **DEFINITION**
>
> The graph of a function *f* is **concave up** on an interval *I* if it is below or touching each chord between points on the graph with *x*-cordinates in *I*, as indicated in Figure 4.4.1a; it is **concave down** on *I* if it is above or touching each chord, as in Figure 4.4.1b.

The graph of a linear function $f(x) = ax + b$ is the same as the chord between any two points on the line, so linear functions are both concave up and concave down.

It is convenient for the analysis of functions to characterize the concavity of differentiable functions in terms of derivatives. We begin with the following result.

If f′ is increasing, then the graph of f is above lines that are tangent to the graph, as indicated in Figure 4.4.2a; if f′ is decreasing, the graph is below tangent lines, as indicated in Figure 4.4.2b.   (1)

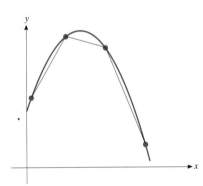

FIGURE 4.4.1b The graph of a function that is concave down opens downward and is above each chord.

FIGURE 4.4.2a If *f′* is increasing on an interval, then *f* is concave up on the interval and the graph of *f* is above the tangent lines. The function *f* is concave up on intervals where $f''(x) > 0$.

FIGURE 4.4.2b If *f′* is decreasing on an interval, then *f* is concave down on the interval and the graph of *f* is below the tangent lines. The function *f* is concave down on intervals where $f''(x) < 0$.

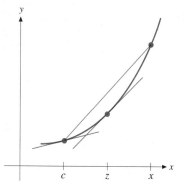

FIGURE 4.4.3 The Mean Value Theorem implies there is some z between c and x such that $\dfrac{f(x) - f(c)}{x - c} = f'(z)$.

Proof This result can be established by comparing the slope of the line tangent to the graph of f at $(c, f(c))$ and the slope of the chord between the points $(c, f(c))$ and $(x, f(x))$, $x \neq c$. The tangent line has slope $f'(c)$ and the chord has slope

$$\frac{f(x) - f(c)}{x - c}.$$

We will use the fact that the Mean Value Theorem implies there is some z between c and x such that

$$\frac{f(x) - f(c)}{x - c} = f'(z),$$

so the slope of the chord is $f'(z)$.

The case where f' is increasing and $c < x$ is illustrated in Figure 4.4.3. Since z is between c and x, $c < x$ implies $c < z < x$. If f' is increasing, $c < z$ implies $f'(c) < f'(z)$. We conclude that the slope of the chord is greater than the slope of the tangent line. The point $(x, f(x))$ is then above the tangent line, as asserted for f' increasing in Lemma 1. The cases for $x < c$ and f' decreasing are similar. ∎

Intuitively, an increasing derivative f' bends the graph upward from its tangent lines and the graph is above its tangent lines. A decreasing derivative f' bends the graph downward, below its tangent lines.

We can use Statement 1 to obtain the following.

The graph of a function f is concave up on intervals where the derivative f' is increasing. It is concave down on intervals where f' is decreasing. (2)

Proof Let us verify the case where f' is increasing. The other case is similar. We will show that, if f' is increasing and $x_1 < c < x_2$, then $(c, f(c))$ is below the chord between $(x_1, f(x_1))$ and $(x_2, f(x_2))$. From Statement 1, we know that both points $(x_1, f(x_1))$ and $(x_2, f(x_2))$ must be above the tangent line at $(c, f(c))$. This is only possible if $(c, f(c))$ is below the chord. See Figure 4.4.4. ∎

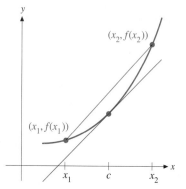

FIGURE 4.4.4 Both points $(x_1, f(x_1))$ and $(x_2, f(x_2))$ can be above the tangent line at $(c, f(c))$ only if $(c, f(c))$ is below the chord.

If the second derivative $f'' = (f')'$ is positive, then the derivative f' is increasing. If f'' is negative, then f' is decreasing. This gives us the following characterization of concavity.

CONCAVITY THEOREM

If the second derivative f'' is positive on an interval I, then f is concave up on I. If f'' is negative on I, then f is concave down on I.

The above characterization of concavity is the one that we will use to determine the concavity of a function.

EXAMPLE 1

Sketch the graphs of $f(x) = x^3$ and $g(x) = x^{1/3}$.

SOLUTION Both functions are defined and continuous for all x. Both have intercept $(0, 0)$. Both are negative for $x < 0$ and positive for $x > 0$.

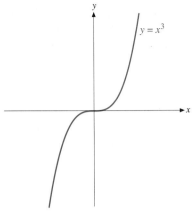

FIGURE 4.4.5a The function $f(x) = x^3$ is concave down on the interval $x < 0$, where $f''(x) < 0$; f is concave up on the interval $x > 0$, where $f''(x) > 0$.

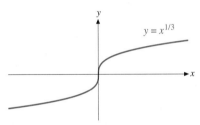

FIGURE 4.4.5b The function $g(x) = x^{1/3}$ is concave up on the interval $x < 0$, where $g''(x) > 0$; g is concave down on the interval $x > 0$, where $g''(x) < 0$.

The derivatives are $f'(x) = 3x^2$ and $g'(x) = (1/3)x^{-2/3}$. The graph of f has a horizontal tangent at $(0, 0)$ and the graph of g has a vertical tangent at $(0, 0)$. Both derivatives are positive on the intervals $x < 0$ and $x > 0$. Both functions are increasing on $-\infty < x < \infty$.

The second derivatives are $f''(x) = 6x$ and $g''(x) = -(2/9)x^{-5/3}$. We see that f'' is negative for $x < 0$ and positive for $x > 0$. It follows that the graph of f is concave down for $x < 0$ and concave up for $x > 0$. See Figure 4.4.5a. We see that g'' is positive for $x < 0$ and negative for $x > 0$. This implies the graph of g is concave up for $x < 0$ and concave down for $x > 0$. See Figure 4.4.5b. ∎

If c is a critical point of f with $f'(c) = 0$ and $f''(c) \neq 0$, the sign of $f''(c)$ can be used to determine if f has a local maximum or minimum at c. We have the following result.

THE SECOND DERIVATIVE TEST FOR LOCAL EXTREMA AT CRITICAL POINTS

If $f'(c) = 0$ and $f''(c) > 0$, then f has a local minimum at c. If $f'(c) = 0$ and $f''(c) < 0$, then f has a local maximum at c. If $f'(c) = 0$ and $f''(c) = 0$, f may have either a local maximum, local minimum, or neither at the critical point c.

Proof Let us verify the first statement. Verification of the second statement is similar. If $f''(c) > 0$, then

$$\lim_{x \to c} \frac{f'(x) - f'(c)}{x - c} = f''(c) > 0.$$

This implies that

$$\frac{f'(x) - f'(c)}{x - c} > 0, \qquad \text{for } x \text{ near } c, \quad x \neq c.$$

For $x < c$, we have $x - c < 0$, so multiplication of the above inequality by $x - c$ gives

$$f'(x) - f'(c) < 0.$$

Since $f'(c) = 0$, we have

$$f'(x) < 0, \qquad \text{for } x < c, \quad x \text{ near } c.$$

This means that f is decreasing on an interval with right endpoint c. Similarly, we can show that

$$f'(x) > 0, \qquad \text{for } x > c, \quad x \text{ near } c,$$

so f is increasing on an interval with left endpoint c. We conclude that $f(c)$ must be a local minimum. This case is illustrated by the graph of $f(x) = x^2$ in Figure 4.4.6a. The second statement is illustrated by the graph of $f(x) = -x^2$ in Figure 4.4.6b.

If $f'(c) = f''(c) = 0$, it may happen that $f'(x)$ does not change sign at $x = c$, so f may not have a local maximum or minimum at the critical point c. This is illustrated in Figure 4.4.7 by the graphs of $f(x) = x^3$ and $f(x) = -x^3$.

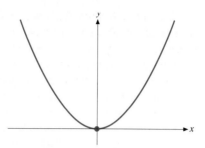

FIGURE 4.4.6a If $f'(c) = 0$ and $f''(c) > 0$, then f has a local minimum at c.

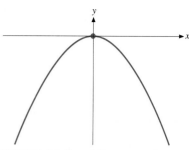

FIGURE 4.4.6b If $f'(c) = 0$ and $f''(c) < 0$, then f has a local maximum at c.

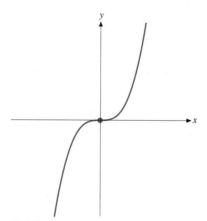

FIGURE 4.4.7a The function $f(x) = x^3$ satisfies $f'(0) = f''(0) = 0$, but $f'(x)$ does not change sign at c; f does not have a local extremum at c.

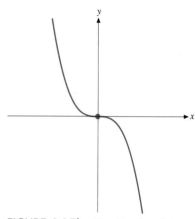

FIGURE 4.4.7b The function $f(x) = -x^3$ satisfies $f'(0) = f''(0) = 0$, but $f'(x)$ does not change sign at c; f does not have a local extremum at c.

It is possible that f has a local minimum or maximum value at a point where $f'(c) = f''(c) = 0$. These cases are illustrated in Figure 4.4.8, where we have sketched the graphs of $f(x) = x^4$ and $f(x) = -x^4$. ■

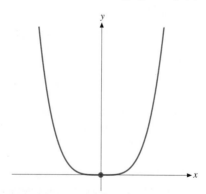

FIGURE 4.4.8a The function $f(x) = x^4$ satisfies $f'(0) = f''(0) = 0$, but f has a local minimum at c.

FIGURE 4.4.8b The function $f(x) = -x^4$ satisfies $f'(0) = f''(0) = 0$, but f has a local maximum at c.

If $f'(c) = f''(c) = 0$, we should use the sign of $f'(x)$ for x near c and a sketch of the graph of f to determine if f has a local extreme value at c.

The second derivative $f''(x)$ is of constant sign on intervals between successive points where it is either zero or undefined. Points on the graph where either $f''(x) = 0$ or $f''(x)$ does not exist should be plotted if it is desired that the graph illustrate concavity.

A point on a graph where the graph changes concavity, from concave up to concave down or from concave down to concave up, is called a **point of inflection** of the graph. If $(d, f(d))$ is a point of inflection of the graph of f, and $f''(x)$ exists for x near d on each side of d, then $f''(x)$ must change sign at $x = d$. This means:

If $(d, f(d))$ is a point of inflection. then either

    (i) $f''(d) = 0$ *or*
    (ii) $f''(d)$ *does not exist.*

Note that this tells us only that $(d, f(d))$ is a *candidate* for a point of inflection if either $f''(d) = 0$ or $f''(d)$ does not exist. We must check that the concavity changes before we know that it is actually a point of inflection.

NOTE   *It is possible that $(d, f(d))$ is not a point of inflection, even though either $f''(d) = 0$ or $f''(d)$ does not exist.*

The Sign of $f(x)$, $f'(x)$, and $f''(x)$

At this point it is worthwhile to review the general characteristics of graphs that we have studied. We can obtain information about a graph from the sign of $f(x)$, the sign of $f'(x)$, and the sign of $f''(x)$. It is important that we understand what information is given by each of f, f', and f''.

If $f(x)$, $f'(x)$, and $f''(x)$ are each of constant sign for x in an interval I of finite length, then all eight combinations of signs are possible; not all combinations of signs of f, f', and f'' are possible on infinite intervals or if the graph has an asymptote.

EXAMPLE 2
Sketch the graph of a function that satisfies the given conditions or indicate that this is impossible.

    (a) $f(x) > 0$, $f'(x) > 0$, $f''(x) > 0$, $0 < x < 1$.
    (b) $f(x) < 0$, $f'(x) > 0$, $f''(x) < 0$, $0 < x < 1$, vertical asymptote $x = 0$.
    (c) $f(x) > 0$, $f'(x) < 0$, $f''(x) < 0$, $0 < x < 1$, vertical asymptote $x = 0$.
    (d) $f(x) > 0$, $f'(x) < 0$, $f''(x) < 0$, $-\infty < x < \infty$.
    (e) $f(x) < 0$, $f'(x) > 0$, $f''(x) < 0$, $-\infty < x < \infty$, horizontal asymptote as $x \to \infty$.

SOLUTION

    (a) The condition $f > 0$ implies the graph is above the x-axis; $f' > 0$ implies the graph is increasing; $f'' > 0$ implies the graph is concave up. See Figure 4.4.9.

    (b) The condition $f < 0$ implies the graph is below the x-axis; $f' > 0$ implies the graph is increasing; $f'' < 0$ implies the graph is concave down. The graph f can have a vertical asymptote $x = 0$. See Figure 4.4.10.

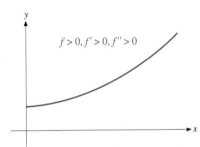

$f > 0, f' > 0, f'' > 0$

FIGURE 4.4.9 The condition $f > 0$ implies the graph is above the x-axis; $f' > 0$ implies f is increasing; $f'' > 0$ implies f is concave up.

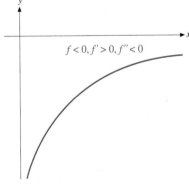

$f < 0, f' > 0, f'' < 0$

FIGURE 4.4.10 The condition $f < 0$ implies the graph is below the x-axis; $f' > 0$ implies f is increasing; $f'' < 0$ implies f is concave down. The graph can have a vertical asymptote $x = 0$.

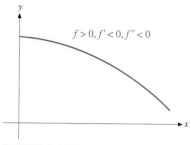

FIGURE 4.4.11 The condition $f > 0$ implies the graph is above the x-axis; $f' < 0$ implies f is decreasing; $f'' < 0$ implies f is concave down. The graph cannot have a vertical asymptote $x = 0$.

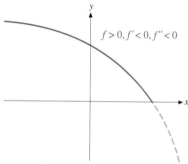

FIGURE 4.4.12 The condition $f > 0$ implies the graph is above the x-axis; $f' < 0$ implies f is decreasing; $f'' < 0$ implies f is concave down. These conditions cannot hold for $-\infty < x < \infty$, because a function that is decreasing and concave down must approach negative infinity as x approaches infinity.

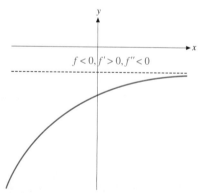

FIGURE 4.4.13 The condition $f < 0$ implies the graph is below the x-axis; $f' > 0$ implies f is increasing; $f'' < 0$ implies f is concave down. The graph can have a horizontal asymptote as x approaches infinity.

(c) The condition $f > 0$ implies the graph is above the x-axis; $f' < 0$ implies the graph is decreasing; $f'' < 0$ implies the graph is concave down. It is impossible for the graph to have a vertical asymptote $x = 0$. See Figure 4.4.11.

(d) The condition $f > 0$ implies the graph is above the x-axis; $f' < 0$ implies the graph is decreasing; $f'' < 0$ implies the graph is concave down. If a function satisfies $f'(x) < 0$ and $f''(x) < 0$ for $-\infty < x < \infty$, then $f(x)$ must approach negative infinity as x approaches infinity. It is impossible that $f(x)$ is positive for all x. See Figure 4.4.12.

(e) The condition $f < 0$ implies the graph is below the x-axis; $f' > 0$ implies the graph is increasing; $f'' < 0$ implies the graph is concave down. The graph can have a horizontal asymptote as $x \to \infty$. In fact, it should seem that a function that is negative and increasing as x approaches infinity should have a horizontal asymptote at infinity. This is true. See Figure 4.4.13. ■

It is useful to note the following:

If $f'(x)$ and $f''(x)$ are both of constant sign as $(x, f(x))$ approaches an asymptote, the graph must open away from the asymptote.

Graph Sketching

We are now ready to sketch the graphs of some specific functions. We have seen that the information we obtain from the values of $f(x)$, $f'(x)$, and $f''(x)$ sometimes overlaps. We should use whatever information is necessary and convenient for illustrating the desired general characteristics. It is not necessary to use all we know for every example, although extra information can be used as a check of our work.

EXAMPLE 3

Sketch the graph of $f(x) = x^3 - 6x^2 + 9x$. Find local extrema and points of inflection.

SOLUTION The function is defined and continuous for all x. Writing $f(x)$ in factored form, we have

$$f(x) = x(x - 3)^2.$$

It follows that f has intercepts $(0, 0)$ and $(3, 0)$. We can check that $f(x)$ is negative for $x < 0$ and positive for $0 < x < 3$ and $x > 3$. Figure 4.4.14a illustrates these characteristics.

The derivative of $f(x) = x^3 - 6x^2 + 9x$ is

$$f'(x) = 3x^2 - 12x + 9.$$

Factoring, we have

$$f'(x) = 3(x^2 - 4x + 3) = 3(x - 1)(x - 3).$$

We see that $f'(x) = 0$ when $x = 1$ and $x = 3$. We have $f(1) = 4$ and $f(3) = 0$. The graph has horizontal tangents at $(1, 4)$ and $(3, 0)$. The first derivative $f'(x)$ is positive and the graph is increasing on the intervals $x < 1$ and $x > 3$. The derivative is negative and the graph is decreasing on the interval $1 < x < 3$.

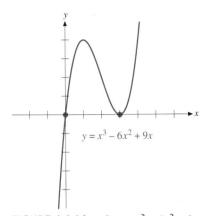

FIGURE 4.4.14a $f(x) = x^3 - 6x^2 + 9x$ has intercepts $(0, 0)$ and $(3, 0)$. The graph is above the x-axis on the intervals $0 < x < 3$ and $x > 3$, where $f(x) > 0$. The graph is below the x-axis on the interval $x < 0$, where $f(x) < 0$.

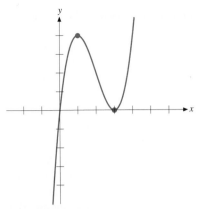

FIGURE 4.4.14b   The function f is increasing on the intervals $x < 1$ and $x > 3$, where $f'(x) > 0$; f is decreasing on the interval $1 < x < 3$, where $f'(x) < 0$; f has a local maximum at $(1, 4)$ and a local minimum at $(3, 0)$.

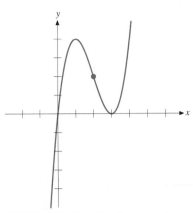

FIGURE 4.4.14c   The function f is concave down on the interval $x < 2$, where $f''(x) < 0$; f is concave up on the interval $x > 2$, where $f''(x) > 0$; the concavity changes at the point $(2, 2)$, so this is a point of inflection.

These characteristics are illustrated in Figure 4.4.14b. The function has a local maximum at $(1, 4)$ and a local minimum at $(3, 0)$. The second derivative is the derivative of

$$f'(x) = 3x^2 - 12x + 9, \text{ so}$$
$$f''(x) = 6x - 12.$$

We see that $f''(x) = 0$ when $x = 2$. We have $f(2) = 2$. The second derivative $f''(x)$ is negative and the graph is concave down on the interval $x < 2$; $f''(x)$ is positive and the graph is concave up on the interval $x > 2$. The concavity of the graph is illustrated in Figure 4.4.14c. The concavity changes from negative to positive at the point $(2, 2)$, so $(2, 2)$ is a point of inflection of the graph. Also, $f'(1) = 0$ and $f''(1) = -6 < 0$, so the Second Derivative Test tells us that f has a local maximum at the critical point $c = 1$. Similarly, $f'(3) = 0$ and $f''(3) = 6 > 0$ imply f has a local minimum at the critical point $c = 3$. Of course, we already discovered these facts by looking at the sign of $f'(x)$ for x near $x = 1$ and $x = 3$.   ■

EXAMPLE 4

Sketch the graph of $f(x) = \dfrac{5x^2 - 5}{x^3}$. Find local extrema and points of inflection.

SOLUTION   We see that $f(x)$ is undefined for $x = 0$. The graph has vertical asymptote $x = 0$. The function is continuous for $x < 0$ and $x > 0$. The intercepts are $(1, 0)$ and $(-1, 0)$. The sign of $f(x)$ can be determined from a table of signs. See Table 4.4.1.

We have

$$\lim_{x \to \pm\infty} \frac{5x^2 - 5}{x^3} = 0,$$

TABLE 4.4.1

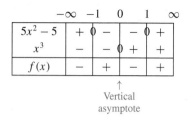

	$-\infty$	-1		0		1	∞
$5x^2 - 5$	$+$	0	$-$		$-$	0	$+$
x^3	$-$		$-$	0	$+$		$+$
$f(x)$	$-$		$+$		$-$		$+$

↑
Vertical
asymptote

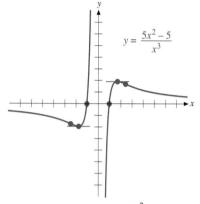

$y = \dfrac{5x^2 - 5}{x^3}$

FIGURE 4.4.15 $f(x) = \dfrac{5x^2 - 5}{x^3}$
has intercepts $(-1, 0)$ and $(1, 0)$; f
has vertical asymptote $x = 0$ and
horizontal asymptote $y = 0$; f has
local maximum at $(\sqrt{3}, 10/3^{3/2})$ and
local minimum at $(-\sqrt{3}, -10/3^{3/2})$;
$(-\sqrt{6}, -25/6^{3/2})$ and $(\sqrt{6}, 25/6^{3/2})$ are
points of inflection.

since the denominator is of higher order than the numerator. The graph has horizontal asymptote $y = 0$. We may also notice that $f(-x) = -f(x)$, so f is an odd function and the graph is symmetric with respect to the origin.

The information we have obtained by investigating the values of $f(x)$ can be used to sketch the graph. The graph in Figure 4.4.15 satisfies the conditions we have determined and is as simple as possible. That is, the graph changes from increasing to decreasing or vice versa only when necessary to satisfy the conditions we have determined.

From the sketch in Figure 4.4.15 it appears that f must have a local minimum somewhere in the interval $x < -1$ and a local maximum somewhere in the interval $x > 1$. By comparing the concavity at the local extrema with that as $|x| \to \infty$, we see that there must be a point of inflection in each of the intervals $x < -1$ and $x > 1$. If we want to locate these key points we must investigate the derivatives of f. The first derivative can be used to determine intervals on which the graph is either increasing or decreasing. The second derivative can be used to determine concavity.

We can easily carry out the division by x^3 to obtain

$$f(x) = 5x^{-1} - 5x^{-3}, \text{ so}$$
$$f'(x) = -5x^{-2} + 15x^{-4} \text{ and}$$
$$f''(x) = 10x^{-3} - 60x^{-5}.$$

Simplifying the first derivative, we obtain

$$f'(x) = \frac{5(3 - x^2)}{x^4}.$$

We see that $f'(x) = 0$ when $x = \sqrt{3}$ and when $x = -\sqrt{3}$. The corresponding y-coordinates are

$$f(\sqrt{3}) = \frac{5(\sqrt{3})^2 - 5}{(\sqrt{3})^3} = \frac{10}{3^{3/2}} \text{ and } f(-\sqrt{3}) = -\frac{10}{3^{3/2}}.$$

The function has a local maximum at $(\sqrt{3}, 10/3^{3/2})$ and a local minimum at $(-\sqrt{3}, -10/3^{3/2})$. Investigation of the sign of $f'(x)$ verifies that the graph is increasing on the intervals $-\sqrt{3} < x < 0$ and $0 < x < \sqrt{3}$; f is decreasing on the intervals $x < -\sqrt{3}$ and $x > \sqrt{3}$.

Simplifying the second derivative, we obtain

$$f''(x) = \frac{10(x^2 - 6)}{x^5}.$$

We have $f''(x) = 0$ when $x = \sqrt{6}$ and when $x = -\sqrt{6}$. The corresponding y-coordinates are

$$f(\sqrt{6}) = \frac{5(\sqrt{6})^2 - 5}{(\sqrt{6})^3} = \frac{25}{6^{3/2}} \text{ and } f(-\sqrt{6}) = -\frac{25}{6^{3/2}}.$$

Investigation of the sign of $f''(x)$ shows that the graph is concave up on the intervals $-\sqrt{6} < x < 0$ and $x > \sqrt{6}$. The graph is concave down on the intervals $x < -\sqrt{6}$ and $0 < x < \sqrt{6}$. The graph has points of inflection at $(\sqrt{6}, 25/6^{3/2})$ and $(-\sqrt{6}, -25/6^{3/2})$, since the graph changes concavity at these points. ■

EXAMPLE 5

Sketch the graph of $f(x) = x^{5/3} + 5x^{2/3}$. Find local extrema and points of inflection.

SOLUTION   The function is defined and continuous for all x. We have

$$f(x) = x^{5/3} + 5x^{2/3} = x^{2/3}(x+5),$$
$$f'(x) = \frac{5}{3}x^{2/3} + \frac{10}{3}x^{-1/3} = \frac{5(x+2)}{3x^{1/3}},$$
$$f''(x) = \frac{10}{9}x^{-1/3} - \frac{10}{9}x^{-4/3} = \frac{10(x-1)}{9x^{4/3}}.$$

From the factored form of $f(x)$, we see that the graph has intercepts $(0,0)$ and $(-5,0)$. $f(x)$ is positive for $-5 < x < 0$ and $x > 0$, and is negative for $x < -5$.

From the factored form of $f'(x)$, we see that $f'(x) = 0$ when $x = -2$, and $f'(x)$ does not exist when $x = 0$. We have $f(-2) = 3 \cdot 2^{2/3}$ and $f(0) = 0$. f is increasing ($f'(x) > 0$) on the intervals $x < -2$ and $x > 0$, decreasing ($f'(x) < 0$) on the interval $-2 < x < 0$. f has a local maximum at $(-2, 3 \cdot 2^{2/3})$ and a local minimum at $(0,0)$.

We see that $f''(x) = 0$ when $x = 1$, and that $f''(x)$ does not exist when $x = 0$. We have $f(1) = 6$ and $f(0) = 0$. The points $(1,6)$ and $(0,0)$ are possible points of inflection. The graph is concave up ($f''(x) > 0$) on the interval $x > 1$. The graph is concave down ($f''(x) < 0$) on the intervals $x < 0$ and $0 < x < 1$. The point $(1,6)$ is a point of inflection, but $(0,0)$ is not, because the graph does not change concavity at $(0,0)$. The graph is given in Figure 4.4.16. Notice that it is difficult to discern the concavity for positive x from this accurate sketch. ∎

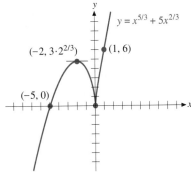

FIGURE 4.4.16 $f(x) = x^{5/3} + 5x^{2/3}$ has intercepts $(-5,0)$ and $(0,0)$; f has a local maximum at $(-2, 3 \cdot 2^{2/3})$, where the derivative is zero; there is a local minimum at $(0,0)$, where the derivative does not exist; $(1,6)$ is a point of inflection and $(0,0)$ is not.

EXERCISES 4.4

In Exercises 1–16, sketch the graph of a function that satisfies the given conditions or indicate that this is impossible.

1. $f(x) > 0, f'(x) > 0, f''(x) < 0, 0 < x < 1$

2. $f(x) > 0, f'(x) < 0, f''(x) > 0, 0 < x < 1$

3. $f(x) < 0, f'(x) > 0, f''(x) > 0, 0 < x < 1$

4. $f(x) < 0, f'(x) < 0, f''(x) > 0, 0 < x < 1$

5. $f(x) > 0, f'(x) > 0, f''(x) < 0, 0 < x < 1$, with vertical asymptote $x = 1$

6. $f(x) > 0, \ f'(x) < 0, \ f''(x) > 0, \ 0 < x < 1$, vertical asymptote $x = 0$

7. $f(x) < 0, \ f'(x) < 0, \ f''(x) < 0, \ 0 < x < 1$, vertical asymptote $x = 1$

8. $f(x) < 0, \ f'(x) < 0, \ f''(x) < 0, \ 0 < x < 1$, vertical asymptote $x = 0$

9. $f(x) > 0, f'(x) < 0, f''(x) > 0, -\infty < x < \infty$

10. $f(x) < 0, f'(x) > 0, f''(x) > 0, -\infty < x < \infty$

11. $f(x) < 0, f'(x) < 0, f''(x) > 0, -\infty < x < \infty$

12. $f(x) < 0, f'(x) < 0, f''(x) < 0, -\infty < x < \infty$

13. $f(x) < 0, f'(x) < 0, f''(x) < 0, -\infty < x < \infty$, horizontal asymptote as $x \to -\infty$

14. $f(x) < 0, f'(x) < 0, f''(x) < 0, -\infty < x < \infty$, horizontal asymptote as $x \to \infty$

15. $f(x) > 0, f'(x) < 0, f''(x) > 0, -\infty < x < \infty$, horizontal asymptote as $x \to -\infty$

16. $f(x) > 0, f'(x) < 0, f''(x) > 0, -\infty < x < \infty$, horizontal asymptote as $x \to \infty$

Sketch the graphs of the functions in Exercises 17–42. Find local extrema and points of inflection.

17. $f(x) = x^3 + 1$

18. $f(x) = (x-1)^{1/3}$

19. $f(x) = -2 + 3x - x^3$

20. $f(x) = x^3 - 3x^2$

21. $f(x) = x^3 + x$

22. $f(x) = 2 - x - x^3$

23. $f(x) = 64x^2 - 16x$

24. $f(x) = 3x^4 + 4x^3$

25. $f(x) = (6x^5 + 20x^3 - 90x)/32$

26. $f(x) = 10x^2 - 4x^5$ **(Calculus Explorer & Tutor I, Shapes of Curves, 10.)**

27. $f(x) = x^{1/3}(x^2 - 7)$

28. $f(x) = x^{2/3}(x^2 - 16)/16$

29. $f(x) = x^{2/3}(x + 40)/4$

30. $f(x) = x(x - 40)^{2/3}/4$

31. $f(x) = \dfrac{8x}{x^2 + 4}$

32. $f(x) = \dfrac{1}{x^2 + 1}$

33. $f(x) = \dfrac{x}{x^2 - 1}$

34. $f(x) = \dfrac{x^3 + 2}{x}$ **(Calculus Explorer & Tutor I, Shapes of Curves, 9.)**

35. $f(x) = \dfrac{x - 1}{x^2}$

36. $f(x) = \dfrac{x^2 - 1}{x^2}$

37. $f(x) = \dfrac{(x^2 - 1)^2}{x^4}$

38. $f(x) = \dfrac{10(x^2 - 1)}{x^5}$

39. $f(x) = x + \sin 2x, 0 \le x \le 2\pi$

40. $f(x) = x - \sin x, 0 \le x \le 2\pi$

41. $f(x) = \sin x + \cos x, 0 \le x \le 2\pi$

42. $f(x) = \cos x - \sin x, 0 \le x \le 2\pi$

In Exercises 43–52, sketch a short arc of the graph of the given equation that contains the given point on the graph. The sketch should indicate the sign of the slope and the concavity of the graph as it passes through the point.

43. $x = y^3 - y, (0, 1)$

44. $x^3 = y^3 + 3y + 1, (1, 0)$

45. $x = \tan y, (1, \pi/4)$

46. $x = \sin y, (1/2, \pi/6)$

47. $x^3 + y^3 = 7, (-1, 2)$

48. $x^{1/3} + y^{1/3} = 1, (-1, 8)$

49. $\sqrt{x} + \sqrt{y} = 5, (4, 9)$

50. $x^2 + 4xy + y^2 + 2 = 0, (1, -1)$

51. $x^2 + y^2 + 2 = xy + 9, (1, -2)$

52. $4x^2 + y^2 = 4xy + 9, (2, 1)$

In Exercises 53–60, use the given graphs to find approximate coordinates of **(a)** *local extrema and* **(b)** *points of inflection.*

53.

54.

55.

56.

57.

58.

59.

60.

61. Some forces, such as the force due to gravity, tend to decrease the potential energy of the system on which they are acting. Therefore, when a system is disturbed from a position of equilibrium, these forces tend to bring it back to its original position if the potential energy V is minimum and move it farther away if V is maximum. If the position is defined by a single independent variable θ, the system is in equilibrium when $\dfrac{dV}{d\theta} = 0$. The system in equilibrium is in **stable equilibrium** if $\dfrac{d^2V}{d\theta^2} > 0$ and in **unstable equilibrium** if $\dfrac{d^2V}{d\theta^2} < 0$. The position of a bar that is hinged from one end is determined by the angle θ, as indicated in the figure below. The potential energy of the bar is $V = w\left(\dfrac{\ell}{2} - \dfrac{\ell}{2}\cos\theta\right)$, where w is the weight of the bar and ℓ is its length. For which values of θ is the bar in **(a)** stable equilibrium and **(b)** unstable equilibrium?

62. The potential energy of the system pictured is $V = 2k\ell^2\sin^2\theta + W\ell\cos\theta$, $0 \le \theta \le \pi$, where k, ℓ, and W are positive constants. If $4k\ell > W$, for which values of θ is the system in **(a)** stable equilibrium and **(b)** unstable equilibrium? (See the previous exercise.)

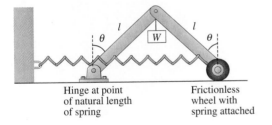

Hinge at point
of natural length
of spring

Frictionless
wheel with
spring attached

63. Suppose **Newton's Method** is used to obtain x_2 from x_1. We can see from the figure that, if $x_2 - c < x_1 - x_2$, then the error in using x_2 as an approximate value of c is less than the difference between x_2 and x_1. Follow the steps indicated below to show that, if $f(x_0) < 0 < f(x_1)$, $f'(x) \ge m > 0$ and $0 \le f''(x) \le M$ for $x_0 < x < x_1$, and $0 < x_1 - x_0 < m/M$, then $x_2 - c < x_1 - x_2$.
(a) Use the Mean Value Theorem and bounds of f'' and $x_1 - c$ to show that $f'(x_1) - f'(c) \le m$.
(b) Use (a) and the fact that $m \le f'(c)$ to show that $f'(x_1) \le 2f'(c)$.
(c) Explain why $f(x_1) \ge f'(c)(x_1 - c)$.
(d) Explain why $f'(x_1) = \dfrac{f(x_1)}{x_1 - x_2}$.
(e) Use the results of (b), (c), and (d) to show $\dfrac{f'(c)(x_1 - c)}{x_1 - x_2} \le 2f'(c)$.
(f) Use (e) to show $x_1 - c \le 2(x_1 - x_2)$.
(g) Use (f) to show $x_2 - c \le x_1 - x_2$.

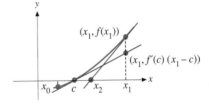

4.5 FINDING EQUATIONS THAT RELATE VARIABLES

Connections

Basic formulas from geometry.

Trigonometric formulas, 1.6.

This section is devoted to the basic skill of translating a problem that is described in words into a mathematical expression that involves real variables. We will deal with the variables as functions in later sections.

No calculus problems are solved (or even stated) in this section. However, the procedures discussed are essential for the solution of the problems considered in ensuing sections.

Strategy for translating problems from words to equations

- Read the problem carefully.
- Identify and name the variables. Draw a picture, if appropriate.
- Find equations that relate the variables.
- Use substitution to eliminate extra variables.

Use as many variables as are convenient to obtain equations that relate the variables of interest. Extra variables can be eliminated after they have served their purpose of making it easier to express the equation that relates the variables of interest.

Formulas from Geometry

Relations between variables are often obtained from *known formulas or given formulas.* In this section we will use the following formulas from geometry:

RECTANGLE
Area $= \ell w$
Perimeter $= 2\ell + 2w$

TRIANGLE
Area $= \frac{1}{2}bh$
Perimeter $= a + b + c$

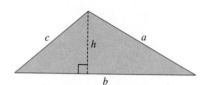

CIRCLE
Area $= \pi r^2$
Circumference $= 2\pi r$

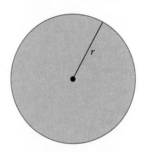

RECTANGULAR BOX
Volume $= \ell w h$
Areas of faces $= \ell w, \ \ell h, \ wh$

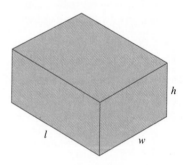

RIGHT CIRCULAR CYLINDER
Volume $= \pi r^2 h$
Area of curved surface $= 2\pi r h$

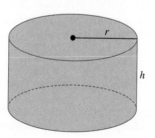

RIGHT CIRCULAR CONE
Volume $= \frac{1}{3}\pi r^2 h$
Area of curved surface
$$= \pi r \sqrt{r^2 + h^2}$$

SPHERE

Volume $= \dfrac{4}{3}\pi r^3$

Surface area $= 4\pi r^2$

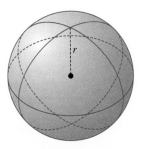

GENERALIZED RIGHT CYLINDER

Volume = (Area of base)(Height)

Lateral surface area
= (Perimeter of base)(Height)

EXAMPLE 1

A rectangular box is to be constructed from a 12-in. square piece of cardboard by cutting equal squares from each corner and then bending up the sides. **(a)** Find an equation that relates the length of one side of the square base of the box and the height of the box. **(b)** Express the volume of the box in terms of its height.

SOLUTION

(a) In Figure 4.5.1a we have sketched a 12-in. square and indicated that a square with length of side h inches is to be cut from each corner. A length of ℓ inches remains of each of the 12-in. sides of the original square. The lengths h and ℓ are variables. After the sides are bent up, we obtain a rectangular box that has length of each side of the square base ℓ and height h, as indicated in Figure 4.5.1b. From Figure 4.5.1a we see that

$$2h + \ell = 12.$$

This is an equation that relates the length of each side of the square base and the height of the box.

(b) From Figure 4.5.1b we see that the volume of the box in cubic inches is given by the equation

$$V = h\ell^2.$$

We want to express the volume in terms of the single variable h. To do this, we use the fact from part (a) that the variables h and ℓ are related by the equation

$$2h + \ell = 12.$$

Solving this equation for ℓ in terms of h, we obtain

$$\ell = 12 - 2h.$$

Substitution in the expression for V then gives the volume in terms of the height. We have

$$V = h\ell^2,$$
$$V = h(12 - 2h)^2.$$

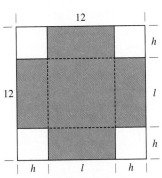

FIGURE 4.5.1a   A square with length of side h inches is cut from each corner of a 12-in. square.

FIGURE 4.5.1b   The volume of the rectangular box with square base is $V = h\ell^2$.

(The variables h and $\ell = 12 - 2h$ cannot represent lengths of sides of a rectangular box unless they are nonnegative, since length is nonnegative. Thus, the above expression for V can represent the volume of the box only if $h \geq 0$ and $12 - 2h \geq 0$, or $0 \leq h \leq 6$.) ∎

EXAMPLE 2

An oil drum in the shape of a right circular cylinder has volume 5.6 ft^3.

(a) Find an equation that relates the height and the radius of the drum.

(b) Express the total amount of material needed to form the top, bottom, and lateral surface of the drum in terms of its radius.

SOLUTION

(a) A sketch of the drum is given in Figure 4.5.2. The radius r and height h have been labeled; these are variables. The volume is not a variable. The radius and height of the cylinder are related because they determine the volume of the cylinder. The equation for the volume of a right circular cylinder tells us the volume is $\pi r^2 h$. We assume that the units of length are feet, so $\pi r^2 h$ gives the volume in cubic feet. Since we know that the volume of the drum is 5.6 ft^3, we must have

$$\pi r^2 h = 5.6.$$

This is the equation that relates the variables.

(b) The total amount of material needed to form the drum is the sum of the areas of the top, bottom, and the lateral surface. The area of each of the top and bottom is πr^2 ft^2, and the area of the lateral surface is $2\pi rh$ ft^2. The total amount of material, in terms of the two variables r and h, is given by the expression

$$M = 2\pi r^2 + 2\pi rh.$$

Since we want M in terms of the single variable r, we use the relation between r and h that was established in part (a) to eliminate the variable h. We have

$$\pi r^2 h = 5.6,$$
$$h = \frac{5.6}{\pi r^2}.$$

Substitution then gives

$$M = 2\pi r^2 + 2\pi rh,$$
$$M = 2\pi r^2 + 2\pi r \left(\frac{5.6}{\pi r^2} \right),$$
$$M = 2\pi r^2 + \frac{11.2}{r}.$$

This gives the total amount of material in terms of the radius of the drum. ∎

Triangles

Many practical problems involve right triangles. For these problems it is common to use the *Pythagorean Theorem* to find relations between variables. It is also helpful to be familiar with the definition of the *trigonometric functions* in terms of side opposite, side adjacent, and hypotenuse. See Figure 4.5.3. (The trigonometric functions are reviewed in Section 1.6.)

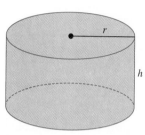

FIGURE 4.5.2 The right circular cylinder has volume $\pi r^2 h$ and total surface area $2\pi r^2 + 2\pi rh$.

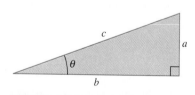

FIGURE 4.5.3 The sides of a right triangle are related by the Pythagorean Theorem. The trigonometric functions of the acute angles are given by ratios of the lengths of the sides.

RIGHT TRIANGLE   (Figure 4.5.3)

Pythagorean Theorem

$$c^2 = a^2 + b^2$$

Trigonometric functions

$$\sin\theta = \frac{a}{c} \quad \cot\theta = \frac{b}{a}$$

$$\cos\theta = \frac{b}{c} \quad \sec\theta = \frac{c}{b}$$

$$\tan\theta = \frac{a}{b} \quad \csc\theta = \frac{c}{a}$$

EXAMPLE 3

A taut rope runs from the bow of a small boat to a windlass that is on the edge of a dock. The windlass is 8 ft higher than the point where the rope is attached to the boat. Find an equation that relates the length of the rope between the boat and the windlass and the horizontal distance of the boat from a point directly below the windlass.

SOLUTION   A sketch is given in Figure 4.5.4. The length of rope between the boat and the windlass is labeled y, and the horizontal distance of the boat from a point directly below the windlass is labeled x; x and y are variables. From the sketch we see that x and 8 are the lengths of the sides of a right triangle that has hypotenuse y. The Pythagorean Theorem then gives

$$x^2 + 8^2 = y^2. \quad \blacksquare$$

FIGURE 4.5.4   The right triangle has hypotenuse of length y and lengths of sides x and 8, so $x^2 + 8^2 = y^2$.

EXAMPLE 4

An airplane is flying horizontally at an altitude of 2000 ft in a direction that will take it directly over an observer at ground level. Find an equation that relates the angle of elevation from the observer to the plane and the distance between the observer and the point at ground level directly below the plane.

SOLUTION   A sketch is given in Figure 4.5.5. The angle of elevation is labeled θ and the distance between the observer and the point at ground level directly below the plane is labeled x; θ and x are variables. From Figure 4.5.5 we see that

$$\tan\theta = \frac{2000}{x} \text{ or } x = 2000\cot\theta. \quad \blacksquare$$

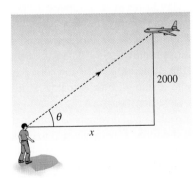

FIGURE 4.5.5   The right triangle has side opposite the angle θ of length 2000 and side adjacent θ of length x, so $\tan\theta = 2000/x$.

For triangles that are not right triangles, we can obtain relations by using the *Law of Cosines,* the *Law of Sines,* and *ratios of corresponding sides of similar triangles.*

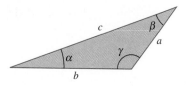

FIGURE 4.5.6 The length of sides and one angle of any triangle are related by the Law of Cosines. The Law of Sines tells us the ratios of the sine of an angle over the length of the side opposite are equal.

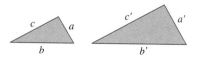

FIGURE 4.5.7 Ratios of corresponding sides of similar triangles are equal.

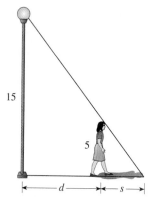

FIGURE 4.5.8 The smaller triangle with base s and height 5 is similar to the larger triangle with base $d + s$ and height 15.

ANY TRIANGLE (Figures 4.5.6 and 4.5.7)

Law of Cosines

$$c^2 = a^2 + b^2 - 2ab \cos \gamma$$

Law of Sines

$$\frac{\sin \alpha}{a} = \frac{\sin \beta}{b} = \frac{\sin \gamma}{c}$$

Ratios of corresponding sides of similar triangles

$$\frac{a}{b} = \frac{a'}{b'}, \quad \frac{a}{c} = \frac{a'}{c'}, \quad \frac{b}{c} = \frac{b'}{c'}$$

EXAMPLE 5

A girl 5 ft tall is walking away from a lamppost that is 15 ft high. Find an equation that relates the length of the girl's shadow and the distance between the girl and the lamppost.

SOLUTION A sketch is given in Figure 4.5.8. The distance between the girl and the lamppost is labeled d and the length of her shadow is labeled s; d and s are variables. Using the ratio of corresponding parts of the larger and smaller similar right triangles, we obtain

$$\frac{d + s}{15} = \frac{s}{5}.$$

Simplifying, we obtain

$$5d + 5s = 15s,$$
$$5d = 10s,$$
$$d = 2s. \quad \blacksquare$$

Variation

Direct or **inverse variation** is a common relation between variables. Either of the statements "y varies directly as x" or, more simply, "y varies as x" is translated into the equation $y = kx$, where k is a constant. In this case x and y are proportional and k is called the constant of proportionality. The statement "y varies inversely as x" is translated into the equation $y = k/x$, where k is a constant. The statement "y varies jointly as x and z" means that y varies as the product of x and z, so $y = kxz$ for some constant k. A variable may simultaneously vary directly as a product of variables and inversely as a product of other variables. For example, the fact that the force F between two masses m_1 and m_2 due to the attraction of gravity varies directly as the product of the masses and inversely as the square of the distance d between them is indicated by the equation

$$F = \frac{km_1 m_2}{d^2}.$$

The value of the gravitational constant k in this equation depends only on the units used to measure mass, distance, and force.

EXAMPLE 6

Hooke's Law asserts that the force required to either stretch or compress a spring from its natural length varies as the displacement. A certain spring is stretched 12 in. beyond its natural length by a 3-lb weight. Express the force in terms of the displacement for this spring.

SOLUTION   Let F denote the force in pounds that corresponds to a displacement of x inches; x and F are variables. Hooke's Law tells us that

$$F = kx \text{ for some constant } k.$$

We are told that a force of 3 lb results in a displacement of 12 in. This means that when the spring is stretched 12 in., the spring force is equal to the force of 3 lb, so $x = 12$ when $F = 3$. See Figure 4.5.9. Substitution of the value 12 for the variable x and the corresponding value 3 for the variable F then allows us to determine the spring constant k. We have

$$F = kx,$$
$$3 = k(12),$$
$$k = \frac{1}{4}.$$

Using this value of k we obtain the equation

$$F = \frac{x}{4}. \quad \blacksquare$$

FIGURE 4.5.9   A 3-lb weight stretches a certain spring 12 in. beyond its natural length.

Linear Relations

Linear relations between variables occur relatively often and we should be prepared to deal with them. We could use the techniques of finding the equation of a line, as in Section 1.4, but it is usually more convenient for word problems to write a linear relation as a **ratio of the changes** of the variables. For example, if $y = y_0$ when $x = x_0$, we can express a linear relation between the variables x and y in the form

$$\frac{y - y_0}{x - x_0} = \frac{\text{Change in } y}{\text{Change in } x}.$$

It doesn't matter which variable is in the numerator, as long as the same variable is in the numerator on each side of the equation. This fact makes the ratio-of-changes form convenient for problems that involve variables other than x and y.

EXAMPLE 7

A farmer wishes to plant a grove of apple trees. He knows from experience that each tree will produce approximately 2 bushels of apples if he plants 100 trees per acre. Also, because of crowding, each tree will produce only 1.85 bushels if he plants 110 trees per acre. **(a)** Find the equation that relates the number of trees per acre, n, and the production of apples per tree, r, if the relation is assumed to be linear. **(b)** Express the total bushels per acre produced in terms of the number of trees per acre.

SOLUTION

(a) We are told that the relation between the variables r and n is linear and that $r = 2$ when $n = 100$. We also know that $r = 1.85$ when $n = 110$. It follows that a change of $1.85 - 2$ in r corresponds to a change of $110 - 100$ in n. We can then express the linear relation between the variables as

$$\frac{\text{Change in } r}{\text{Change in } n} : \quad \frac{r - 2}{n - 100} = \frac{1.85 - 2}{110 - 100},$$

$$\frac{r - 2}{n - 100} = \frac{-0.15}{10},$$

$$10r - 20 = -0.15n + 15,$$

$$10r = -0.15n + 35,$$

$$r = -0.015n + 3.5.$$

FIGURE 4.5.10 $r = 2$ when $n = 100$ and $r = 1.85$ when $n = 110$.

This is an equation that gives the linear relation between r and n. See Figure 4.5.10, where we have plotted the points $(n, r) = (100, 2)$ and $(n, r) = (110, 1.85)$ and sketched a line through the points to indicate the linear relation.

(b) Let T denote the total bushels per acre of apples produced. This is given by the product of the number of trees per acre and the production per tree, so

$$T = nr.$$

We want to express T in terms of n, so we use the linear relation between n and r that was established in part (a) to eliminate the variable r. We have

$$r = -0.015n + 3.5.$$

Substitution then gives us

$$T = nr,$$
$$T = n(-0.015n + 3.5).$$

This is the desired equation.   ∎

Actually, the number of trees per acre planted in Example 7 must be a nonnegative integer. To apply techniques of calculus we will assume that n can be any nonnegative real number. Also, the assumption that the relation between r and n is linear is not likely to hold if either a very large or very small number of trees per acre are planted.

EXERCISES 4.5

1. A rectangular box with no top is to be constructed by cutting equal squares from each corner of an 8.5-in. × 11-in. piece of cardboard. **(a)** Find an equation that relates the height and length. **(b)** Find an equation that relates the height and width. **(c)** Express the volume of the box in terms of its height.

2. A rectangular box with top is to be constructed by cutting and bending a 20-in. × 36-in. piece of cardboard as indicated below. **(a)** Find an equation that relates the height and length. **(b)** Find an equation that relates the height and width. **(c)** Express the volume of the box in terms of its height.

3. A farmer has 600 m of fencing to enclose a rectangular region adjacent to a river. No fencing is required along the river. **(a)** Find an equation that relates the length and width of the region. **(b)** Express the area of the region in terms of the length of fence parallel to the river.

4. A farmer has 800 m of fencing to enclose a rectangular field and divide it in half. **(a)** Find an equation that relates the length and width of the field. **(b)** Express the area of the field in terms of the length of fence that divides the field.

5. **(a)** Find an equation that relates the height h and the length ℓ of one side of an equilateral triangle. **(b)** Express the area of the triangle in terms of the common length of the sides.

6. **(a)** Find an equation that relates the hypotenuse h and the length ℓ of one side of an isosceles right triangle. **(b)** Express the perimeter of the triangle in terms of the hypotenuse.

7. The area of a rectangular field is to be 20,000 m². **(a)** Find an equation that relates the length and width of the field. **(b)** Express the perimeter of the field in terms of the length of one of its sides.

8. The diameter of a conical pile of dry, smooth sand is three times its height. Find an equation that relates the volume of sand and the height of the pile.

9. A tank in the shape of a right circular cylinder with two hemispherical ends is to have volume 600 ft³. **(a)** Find an equation that relates the radius and height of the cylindrical part. **(b)** Express the total surface area of the tank in terms of its radius.

10. A window in the shape of a rectangle with a semicircular top is to have total area 16 ft². **(a)** Find an equation that relates the width and the height of the rectangular part. **(b)** Express the perimeter of the window in terms of the width of the rectangular part.

11. The string between a boy and a kite that is 150 ft above ground is taut. Find an equation that relates the length of string and the distance between the boy and the point at ground level that is directly below the kite.

12. A post 12 ft high is 20 ft from a post 16 ft high. A wire is to be run from the top of one post to the ground at a point between the posts and then to the top of the other post. Find an equation that relates the total length of wire required and the distance between the point where the wire touches the ground and the 16-ft-high post.

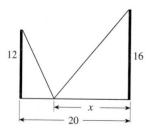

13. A bird feeder is hanging as indicated in the figure below. Express the total length of rope in terms of θ.

14. A bird feeder is hanging as indicated in the figure above. Express the total length of rope in terms of x.

15. A rectangle is inscribed in a right triangle in such a way that one corner of the rectangle is positioned at the right angle of the triangle. The triangle has height 3 and base 4. (a) Find an equation that relates the length and width of the rectangle. (b) Express the area of the rectangle in terms of the length of its side that is along the base of the triangle.

16. A triangular playground is to be constructed at the corner of Lake Street and Elm Street, which meet at right angles. The playground is to be bounded on two sides by the streets and on the other side by a straight wall that runs between the streets. The wall must be built through a lamppost that is 100 ft from Lake Street and 120 ft away from Elm. (a) Find an equation that relates the lengths of the edges of the playground along the two streets. (b) Express the area of the playground in terms of the length of its edge along Lake Street.

17. Express the length $|AB|$ in the figure in terms of the angle θ and the constants L and W.

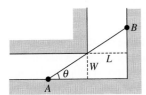

18. Express the length x in the figure in terms of the angle θ and the constants a, b, and c.

19. A rectangle is inscribed in a semicircle of radius 4 with the base of the rectangle along the diameter of the semicircle. Express the area of the rectangle in terms of its height.

20. A right triangle has hypotenuse 5. Express the area of the triangle in terms of one of its acute angles.

21. A trough 12 ft long is partially filled with water. Each vertical end of the trough is an isosceles triangle that is 3 ft across the top and 2 ft deep. Find an equation that relates the depth and volume of the water in the trough.

22. A conical tank that is 12 m across its circular top and 15 m deep is partially filled with water. Find an equation that relates the depth and volume of the water in the tank.

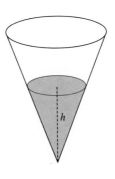

23. The frustrum of a right circular cone has radius of base b, radius of top a, and height h. Express its volume in terms a, b, and h.

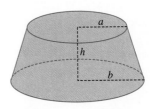

24. A ball is at height s above a point at ground level that is ℓ_0 from the base of a lamppost of height h_0, where $0 \leq s < h_0$. Find an equation that relates s and the distance x of the shadow of the ball from the base of the lamppost.

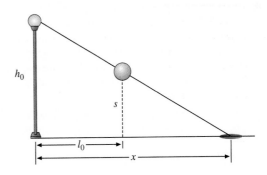

25. A balloon is rising from a point on the ground that is 50 m from an observer. Find an equation that relates the height of the balloon and the angle of elevation from the observer to the balloon.

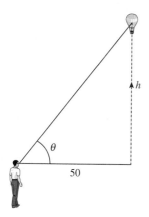

26. A tennis ball is moving in a direction perpendicular to the net, toward a point at the net that is 12 m from a spectator who is sitting in line with the net. Find an equation that relates the distance of the ball from the net and the angle between the net and the line of sight from the spectator to the ball.

27. Hooke's Law asserts that the force required to either stretch or compress a spring from its natural length varies as the length displaced. A certain spring is stretched 6 in. beyond its natural length by a 0.3-lb weight. Express the force in terms of the displacement for this spring.

28. The force on a particular object due to the gravitational attraction of the earth varies inversely as the square of the distance of the object from the center of the earth. Find the equation that relates the force due to gravity and the distance from the center of the earth of the object, if the object weighs 150 lb at the surface of the earth, 4000 miles from the center of the earth.

29. The manager of an apartment building estimates that all 80 apartments can be rented if she charges $300/month, but

one apartment becomes vacant for each $10/month increase in rent. (a) Find an equation that relates the rent and the number of occupied apartments, assuming that the relation is linear. (b) Express the total amount of revenue collected in terms of the unit price.

30. A salesman believes he could sell 100 items at $2 each, but he also believes he could sell an additional 15 items for each $0.10 he decreases the unit price. (a) Find an equation that relates the unit price and the expected number of items sold, assuming that the relation is linear. (b) Express the total amount of expected revenue collected in terms of the unit price.

31. Express the area of the segment of a circle of radius r in terms of the central angle θ subtended by the segment. See figure.

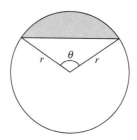

32. Express the area of the segment of a circle of radius r in terms of its width h along the diameter perpendicular to its secant line. See figure.

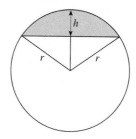

33. Express the area of a rectangle in terms of the length of its diagonals and the acute angle between the diagonals.

34. Find an equation that relates the length of a chord, ℓ, of a circle of radius r and the distance, d, of the chord from the center of the circle.

4.6 RELATED RATES

Connections

Finding equations that relate variables, 4.5.

Chain Rule, 3.4.

In this section we will study quantities that are functions of time. We will study relations between these quantities and their rates of change. The problems involve determining relations between variables that are described in words, finding relations between their derivatives, and then using known data to determine unknown values and rates of change.

We will use the following strategy.

> **Strategy for related rate problems**
>
> - Carefully read the problem.
> - Identify and name the variables. Draw a picture, if appropriate.
> - Determine what you want and what you know.
> - Find equations that relate the variables.
> - Differentiate the equations with respect to the time variable.
> - Substitute the known values of the variables and rates in the original and differentiated equations to solve for what you want.

The strategy is given at this time to indicate the solution pattern we will be using. It is intended to help you organize your work. You should refer back to this strategy as you go through the examples in this section.

Powers of Variables

EXAMPLE 1

A ladder 10 ft long is leaning against a wall. The bottom of the ladder is sliding away from the wall at a rate of 4 ft/s. Find the rate at which the top of the ladder is moving down the wall when the bottom of the ladder is 8 ft from the wall.

FIGURE 4.6.1 The distances x and y are variables, since they are changing as the ladder slides down the wall. The Pythagorean Theorem implies $x^2 + y^2 = 10^2$.

SOLUTION   We must first determine the variables. This is done by sketching and labeling a ladder leaning against a wall, as in Figure 4.6.1. The distance of the bottom of the ladder from the wall, x, and the height of the top of the ladder, y, are variables; they are both changing as time changes. We are interested in the instant when $x = 8$, but we should not label $x = 8$ in our sketch. It is most important to realize that x is a variable and that 8 is one value of this variable. The length of the ladder does not change. We should label the length of the ladder 10.

We next determine what we want and what we know, in terms of the variables. This is done in order to state the problem in mathematical terms. Also, we want to display this information clearly, so it is readily available when needed. We are asked to find the rate of change of y when $x = 8$. The rate of change of y is the derivative of y with respect to time. What we want is displayed as follows:

$$\textit{We want:} \qquad \frac{dy}{dt} \text{ when } x = 8.$$

We are told that the bottom of the ladder is moving away from the wall at a rate of 4 ft/s. Since x is increasing, the rate of change of x with respect to time is positive. We display what we know as follows:

$$\textit{We know:} \qquad \frac{dx}{dt} = +4.$$

The next step is to find an equation that relates the variables. From the Pythagorean Theorem, we obtain the equation

$$x^2 + y^2 = 10^2.$$

We now differentiate the equation that relates the variables to obtain a relation between the rates of change. We must remember that x and y are functions of t, and that we are differentiating with respect to t. Powers of x and y are powers of functions and we must use the Chain Rule for their differentiation. We obtain

$$2x\frac{dx}{dt} + 2y\frac{dy}{dt} = 0.$$

We are now ready to use what we know to determine what we want. When $x = 8$, we know that $dx/dt = 4$. Substitution of these values in the differentiated equation gives

$$2x\frac{dx}{dt} + 2y\frac{dy}{dt} = 0,$$

$$2(8)(4) + 2(y)\left(\frac{dy}{dt}\right) = 0,$$

$$2(y)\left(\frac{dy}{dt}\right) = -2(8)(4).$$

We see that we need to know the value of y when $x = 8$. We determine this by substituting 8 for x in the original equation and then solving for y. Noting that the length y is nonnegative, we have

$$x^2 + y^2 = 10^2,$$
$$8^2 + y^2 = 10^2,$$
$$y^2 = 100 - 64,$$
$$y = \sqrt{36} = 6.$$

That is, $y = 6$ when $x = 8$. Using this value of y in the differentiated equation, we obtain

$$2(6)\left(\frac{dy}{dt}\right) = -2(8)(4),$$

$$\frac{dy}{dt} = -\frac{2(8)(4)}{2(6)} = -\frac{16}{3}.$$

The derivative dy/dt is negative. This indicates y is decreasing. The top of the ladder is sliding down the wall at a rate of 16/3 ft/s when the bottom of the ladder is 8 ft from the wall. ∎

Products of Variables

EXAMPLE 2

The Ideal Gas Law states that $PV = nRT$, where P is the pressure in newtons/m^2, V is the volume in cubic meters, n is the number of moles of the gas, $R = 8.314$ joules per mole per degree Kelvin is the universal gas constant, and T is the temperature in degrees Kelvin. The volume of 0.03 mole of a certain gas is increasing at a rate of 0.0025 m³/s while the temperature is decreasing at a rate of 0.07 K/s. Find the rate of change of the pressure when the volume is 0.06 m³ and the temperature is 295 K. Is the pressure increasing or decreasing at this time?

SOLUTION   The variables are P, V, and T.

$$\text{We want:} \qquad \frac{dP}{dt} \text{ when } V = 0.06 \text{ and } T = 295.$$

We are told that the volume is increasing, so dV/dt is positive. The temperature is decreasing, so dT/dt is negative.

$$\text{We know:} \qquad \frac{dV}{dt} = +0.0025 \text{ and } \frac{dT}{dt} = -0.07.$$

Using $n = 0.03$ and $R = 8.314$ we obtain the relation

$$PV = 0.24942T.$$

Using the Product Rule to differentiate the product PV, we obtain

$$(P)\left(\frac{dV}{dt}\right) + (V)\left(\frac{dP}{dt}\right) = 0.24942\frac{dT}{dt}.$$

When $V = 0.06$ and $T = 295$, the relation $PV = 0.24942T$ gives

$$(P)(0.06) = (0.24942)(295),$$

$$P = \frac{(0.24942)(295)}{0.06} = 1226.315.$$

This is the pressure at the time in which we are interested. The differentiated equation then gives

$$(1226.315)(0.0025) + (0.06)\left(\frac{dP}{dt}\right) = (0.24942)(-0.07),$$

$$\frac{dP}{dt} = \frac{-(0.24942)(0.07) - (1226.315)(0.0025)}{0.06} = -51.387448.$$

When the volume is 0.06 m³ and the temperature is 295 K, the pressure is decreasing at a rate of approximately 51.4 N/m² per second. (The negative value of dP/dt indicates that the pressure is decreasing.) ∎

EXAMPLE 3
A water trough is 10 ft long. The vertical ends are isosceles triangles, 4 ft across the top with sloping sides of length 3 ft. See Figure 4.6.2a. Water is flowing into the trough at a rate of 2 ft³/min. Find the rate of change of the depth of water in the trough when the depth is 1.5 ft.

SOLUTION   The volume of water in the trough, V, and the depth of the water, h, are changing as water flows into the trough; V and h are variables.

$$\text{We want:} \qquad \frac{dh}{dt} \text{ when } h = 1.5.$$

$$\text{We know:} \qquad \frac{dV}{dt} = +2.$$

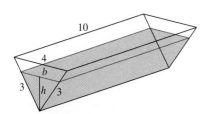

FIGURE 4.6.2a The depth of water, h, is a variable, since the depth changes as water is added to the trough. The variable b is convenient for expressing the volume of water in the trough, $V = \frac{1}{2}(b)(h)(10)$.

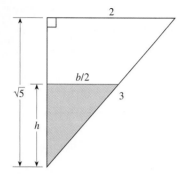

FIGURE 4.6.2b The ratio of corresponding sides of similar triangles gives $\dfrac{b/2}{h} = \dfrac{2}{\sqrt{5}}$.

To find an equation that relates the volume and the depth, it is convenient to introduce an additional variable. In particular, let b denote the distance between the sloping sides of the trough at the uppermost level of the water in the trough. From Figure 4.6.2a we see that the body of water forms a generalized right cylinder with (vertical) triangular base and (horizontal) height 10. The volume of water in the trough is then given by the formula

$$\text{Volume} = (\text{Area of base})(\text{Height}).$$

In this case we have

$$V = \frac{1}{2}(b)(h)(10),$$

$$V = 5bh.$$

Half of the end of the trough forms a right triangle with one side 2 and hypotenuse 3. See Figure 4.6.2b. The Pythagorean Theorem implies that the other side is $\sqrt{3^2 - 2^2} = \sqrt{5}$. We can use the ratio of corresponding sides of similar triangles to obtain the relation

$$\frac{b/2}{h} = \frac{2}{\sqrt{5}},$$

$$b = \frac{4h}{\sqrt{5}}.$$

We now have two equations that relate the variables:

$$V = 5bh \text{ and } b = \frac{4h}{\sqrt{5}}.$$

We could differentiate both of these equations and then work with four equations to determine the unknown rate. However, since the information we have and want does not involve the variable b, it is easier to use the second equation to eliminate b from the first equation. This gives us a relation between the variables V and h, exactly the variables that we are interested in. We obtain

$$V = 5\left(\frac{4h}{\sqrt{5}}\right)h,$$

$$V = 4\sqrt{5}h^2.$$

Differentiating this equation with respect to t, we have

$$\frac{dV}{dt} = 4\sqrt{5}\left(2h\frac{dh}{dt}\right).$$

When $h = 1.5$, the differentiated equation gives

$$2 = 4\sqrt{5}(2)(1.5)\left(\frac{dh}{dt}\right),$$

$$\frac{dh}{dt} = \frac{2}{4\sqrt{5}(2)(1.5)} \approx 0.0745 \,\text{ft/min}.$$

The level of the water is rising at a rate of approximately 0.0745 ft/min when the depth is 1.5 ft. ■

Trigonometric Variables

The formulas we derived for derivatives of the trigonometric functions were based on the change of the trigonometric functions corresponding to a change in the *radian* measure of the angle. This means:

We must use radian measure to measure changes of angles in calculus problems that involve differentiation of trigonometric functions.

EXAMPLE 4

A balloon is rising vertically from a point on the ground that is 200 m from an observer at ground level. The observer determines that the angle of elevation between the observer and the balloon is increasing at a rate of 0.9°/s when the angle of elevation is 45°. How fast is the balloon rising at this time?

SOLUTION A sketch is given in Figure 4.6.3. The angle of elevation θ and the height h are variables.

Since this problem involves the rate of change of a trigonometric function, let us change the given angles from degrees to radians. We have

$$45° = \frac{\pi}{4}\ \text{rad} \quad \text{and} \quad 0.9° = (0.9°)\left(\frac{\pi\ \text{rad}}{180°}\right) = \frac{\pi}{200}\ \text{rad}.$$

We then determine what we want and what we know in the problem.

FIGURE 4.6.3 The right triangle has side opposite θ of length h and side adjacent 200, so $\tan \theta = \dfrac{h}{200}$.

We want: $\dfrac{dh}{dt}$ when $\theta = \dfrac{\pi}{4}$.

We know: $\dfrac{d\theta}{dt} = +\dfrac{\pi}{200}$.

From Figure 4.6.3 we see that θ and h are related by the equation

$$\tan \theta = \frac{h}{200}, \quad \text{so } h = 200 \tan \theta.$$

We then differentiate the latter equation with respect to t to obtain an equation that involves the rates of change of the variables. Since θ is a function of time, the Chain Rule is required for the differentiation of $\tan \theta$ with respect to t. We obtain

$$\frac{dh}{dt} = 200 \sec^2 \theta \frac{d\theta}{dt}.$$

When $\theta = \pi/4$, we have

$$\frac{dh}{dt} = 200 \left(\sec \frac{\pi}{4}\right)^2 \left(\frac{\pi}{200}\right)$$

$$= 200(\sqrt{2})^2 \left(\frac{\pi}{200}\right) = 2\pi.$$

The balloon is rising at a rate of 2π m/s ≈ 6.3 m/s when the angle of elevation is $\pi/4$. ∎

EXAMPLE 5

A connecting rod 60 cm long runs from a point 20 cm from the center of a flywheel to a piston. The piston moves back and forth in line with the center of

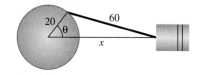

FIGURE 4.6.4   The variables x and θ change as the flywheel rotates.

the flywheel as the flywheel rotates at a rate of 0.5 revolution/s. At what speed is the piston moving when the radial line to the point on the flywheel where the rod is attached makes an angle of 90° with the radial line in the direction of the piston? What is the speed of the piston when the angle is 60°?

SOLUTION   A sketch is given in Figure 4.6.4. The angle θ and the distance x between the center of the flywheel and the piston are variables.

Let us express the angles in terms of radian measure. We know that

$$90° = \frac{\pi}{2}\,\text{rad and } 60° = \frac{\pi}{3}\,\text{rad.}$$

Also, since 1 revolution = 2π rad, we have

$$0.5 \text{ revolution } = (0.5 \text{ revolution}) \left(\frac{2\pi\,\text{rad}}{1\,\text{revolution}} \right) = \pi\,\text{rad,}$$

so 0.5 revolution/s = π rad/s.

We want: $\quad \dfrac{dx}{dt}$ when $\theta = \dfrac{\pi}{2}$ and when $\theta = \dfrac{\pi}{3}$.

Let us assume the flywheel is rotating counterclockwise, so θ is increasing.

We know: $\quad \dfrac{d\theta}{dt} = \pi$ rad/s.

We can use the Law of Cosines to obtain the relation

$$60^2 = 20^2 + x^2 - 2(20)(x)(\cos\theta).$$

To differentiate this equation, we note that x^2 is the square of a function, $\cos\theta$ is the cosine of a function, and $x\cos\theta$ is a product of functions. Using the Chain Rule and the Product Rule, we obtain

$$0 = 0 + 2x\frac{dx}{dt} - 40\left[(x)\left(-\sin\theta\frac{d\theta}{dt} \right) + (\cos\theta)\left(\frac{dx}{dt} \right) \right].$$

When $\theta = \pi/2$, we have $\sin\theta = 1$ and $\cos\theta = 0$. The original equation then gives

$$60^2 = 20^2 + x^2 - 2(20)(x)(0),$$
$$x^2 = 60^2 - 20^2,$$
$$x = \sqrt{60^2 - 20^2} = \sqrt{3200} = 40\sqrt{2}.$$

The differentiated equation then gives

$$0 = 2(40\sqrt{2})\left(\frac{dx}{dt} \right) - 40\left[(40\sqrt{2})(-1)(\pi) + (0)\left(\frac{dx}{dt} \right) \right],$$

$$-2(40\sqrt{2})\left(\frac{dx}{dt} \right) = 40(40\sqrt{2})(\pi),$$

$$\frac{dx}{dt} = -20\pi \approx -62.83.$$

The piston is moving at a speed of 20π cm/s when $\theta = \pi/2$. The sign of the derivative tells us that x is decreasing, so the piston is moving toward the flywheel at this time.

When $\theta = \pi/3$, we have $\sin\theta = \sqrt{3}/2$ and $\cos\theta = 1/2$. The original equation then gives

$$60^2 = 20^2 + x^2 - 2(20)(x)\left(\frac{1}{2}\right),$$
$$x^2 - 20x - 60^2 + 20^2 = 0,$$
$$x^2 - 20x - 3200 = 0.$$

The Quadratic Formula gives

$$x = \frac{-(-20) \pm \sqrt{(-20)^2 - 4(1)(-3200)}}{2(1)}$$
$$= \frac{20 \pm \sqrt{13,200}}{2}$$
$$= \frac{20 \pm 20\sqrt{33}}{2}$$
$$= 10 \pm 10\sqrt{33}.$$

Since x is positive, we have $x = 10 + 10\sqrt{33}$. The differentiated equation then gives

$$0 = 2(10 + 10\sqrt{33})\left(\frac{dx}{dt}\right)$$
$$- 40\left[(10 + 10\sqrt{33})\left(-\frac{\sqrt{3}}{2}\right)(\pi) + \left(\frac{1}{2}\right)\left(\frac{dx}{dt}\right)\right].$$

Multiplying both sides of this equation by 2 and then solving for dx/dt, we obtain

$$0 = 4(10 + 10\sqrt{33})\left(\frac{dx}{dt}\right) + 40(10 + 10\sqrt{33})(\sqrt{3})(\pi) - 40\left(\frac{dx}{dt}\right),$$
$$(40 - 4(10 + 10\sqrt{33}))\left(\frac{dx}{dt}\right) = 40(10 + 10\sqrt{33})(\sqrt{3})(\pi),$$
$$\frac{dx}{dt} = \frac{40(10 + 10\sqrt{33})(\sqrt{3})(\pi)}{-40\sqrt{33}}$$
$$= -\frac{(10 + 10\sqrt{33})(\sqrt{3})(\pi)}{\sqrt{33}}$$
$$= -\frac{10\pi(1 + \sqrt{33})}{\sqrt{11}} \approx -63.89.$$

The piston is moving toward the flywheel at a rate of

$$\frac{10\pi(1 + \sqrt{33})}{\sqrt{11}} \text{ cm/s} \quad \text{when } \theta = \frac{\pi}{3}. \quad \blacksquare$$

The speed of an object that is moving in a circular path is given by the formula

$$\text{Speed} = (\text{Angular velocity})(\text{Radius of circular path}),$$

where the angular velocity is in terms of radian measure of angle per unit of time. Thus, in Example 5, the end of the rod that is attached to the flywheel is moving at a rate of

$$(\pi \text{ rad/s})(20 \text{ cm}) = 20\pi \text{ cm/s}.$$

This is the same as the speed of the piston when $\theta = \pi/2$. When $\theta = \pi/3$, the speed of the piston is $10\pi(1 + \sqrt{33})/\sqrt{11} \approx 63.89$, which is greater than $20\pi \approx 62.83$.

EXERCISES 4.6

1. The radius of a circle is increasing at a rate of 2 ft/s. Find the rate of change of its (a) circumference and (b) area when the radius is 6 ft.

2. Each side of a square is increasing at a rate of 3 ft/s. Find the rate of change of its (a) perimeter and (b) area when each side is 5 ft.

3. Each edge of a cube is increasing at a rate of 3 cm/s. Find the rate of change of its (a) surface area and (b) volume when each edge is 20 cm.

4. The radius of a sphere is increasing at a rate of 2 cm/s. Find the rate of change of its (a) surface area and (b) volume when the radius is 40 cm.

5. Air is being blown into a spherical balloon at a rate of 2 ft³/min. Find the rate of change of its radius when the radius is 1.5 ft.

6. A spherical balloon is being inflated. At a certain instant its radius is 5 inches, and the radius is increasing at a rate of one inch every two seconds. How fast is air entering the balloon at that moment? **(Calculus Explorer & Tutor I, Related Rates, 6.)**

7. A fisherman who hooks a fish notices that the fancy counter attached to his reel indicates that line is being paid out at 2 meters per second. If the tip of the rod is 2 meters above the surface of the water, the fish is at a depth of 1 meter, and there are 6 meters of line out, how fast is the fish swimming away? **(Calculus Explorer & Tutor I, Related Rates, 1.)**

8. Sand is falling into a conical pile at a rate of 2 ft³/s. The height of the cone is always one-third of the diameter of its base. Find the rate of change of the diameter of the pile when it contains 72π ft³ of sand.

9. A water trough is 8 ft long. The back is vertical and the vertical ends are right triangles that are 3 ft across the top and 2 ft down the back. Water is flowing into the trough at a rate of 5 ft³/min. Find the rate of change of the depth of the water in the trough when the depth is 1.5 ft.

10. A water trough is 12 ft long. The vertical ends are trapezoids that are 4 ft across the top, 1 ft deep, and 2 ft across the bottom. Water is pouring into the trough at a rate of 8 ft³/min. Find the rate of change of the depth of water in the trough when the depth is 0.2 ft.

11. A swimming pool has dimensions 10 m × 20 m. The bottom of the pool slants down from a depth of 1 m along one end to a depth of 3 m and then levels off, forming a 10 m × 10 m level square at the other end. Water is being pumped into the pool at a rate of 0.2 m³/min. Find the rate of change of the depth of the water in the pool when the depth is (a) 0.5 m and (b) 2.5 m.

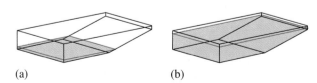

(a)   (b)

12. The bottom of a 10 m × 10 m swimming pool slants from a depth of 1.5 m to a depth of 3 m at the opposite end. Water is being pumped into the pool at a rate of 2 m³/min. Find the rate of change of the depth of water in the pool when the depth is (a) 1 m and (b) 2 m.

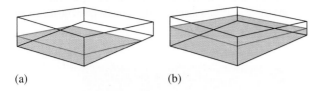

(a)   (b)

13. A right circular cone is 6 m across the top and 8 m deep. Water is flowing into the cone at a rate of 12 m³/min. Find the rate of change of the depth of water in the cone when the depth is 4 m.

14. A water tank has the shape of an inverted right circular cone whose (upper) base radius is equal to one-fourth of its height. Water is being pumped in at 1.75 cubic meters per second. How fast is the water level rising when the water is 2 meters deep? (**Calculus Explorer & Tutor I, Related Rates, 3.**)

15. A girl 5 ft tall walks at a rate of 6 ft/s away from a lamppost that is 15 ft high. Find the rate of change of the length of her shadow when she is (a) 10 ft and (b) 20 ft from the lamppost.

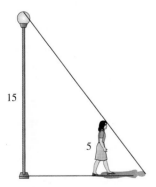

15

5

16. A spy walks away into the gloom of night from a 10-meter lamppost at 1 meter per second. If he is 2 meters tall, how fast is the tip of his shadow moving when he is 30 meters away from the post? (**Calculus Explorer & Tutor I, Related Rates, 2.**)

17. Wind is blowing a kite horizontally at a rate of 6 ft/s, 120 ft above the ground. Find the rate at which the string must be let out to keep the string taut when 130 ft of string is out.

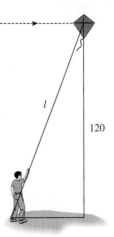

l

120

18. A boat is being towed by a rope that goes from the boat to a pulley on a dock. The pulley is 5 ft above the point on the boat where the rope is attached to the boat. The rope is being pulled in at a rate of 2 ft/s. Find the rate at which the boat is approaching the dock when the boat is 12 ft from the dock.

5

19. A car is traveling north toward an intersection at a rate of 60 mph while a truck is traveling east away from the intersection at a rate of 50 mph. Find the rate of change of the distance between the car and truck when the car is 4 mi south of the intersection and the truck is 3 mi east of the intersection. Is the distance increasing or decreasing at this time?

20. Find the rate of change of the distance between the car and truck in Exercise 19 when the car is 3 mi south of the intersection and the truck is 4 mi east of the intersection. Is the distance increasing or decreasing at this time?

21. Two ships leave from the same point at the same time. The angle between their paths is 60°. One ship travels at 20 mph and the other at 30 mph. Find the rate of change of the distance between the ships two hours after they separated.

22. Find the rate of change of the distance between the ships in Exercise 21 two hours after they separated if the angle between their paths is 120°.

23. At a certain instant a car traveling east at 60 miles per hour is 40 miles west of an intersection, and another car traveling north at 50 miles per hour is 30 miles south of the same intersection. How fast is the distance between the two cars changing one hour later? (**Calculus Explorer & Tutor I, Related Rates, 4.**)

24. The end of rope opposite the weight in the figure is being pulled horizontally away from the fixed pulley at a constant rate. Is the rate at which the weight rises constant, increasing, or decreasing?

25. A rocket is rising vertically from a point on the ground that is 100 m from an observer at ground level. The observer notes that the angle of elevation is increasing at a rate of 12°/s when the angle of elevation is 60°. Find the speed of the rocket at that instant.

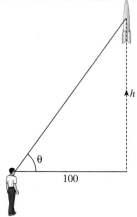

26. A balloon is rising vertically at a rate of 2 ft/s from a point on the ground that is 50 ft from a ground-level observer. Find the rate of change of the angle of elevation between the observer and the balloon when the balloon is 120 ft above ground.

27. A lighthouse that is 200 ft from a straight shoreline contains a light that is revolving at a rate of 0.2 revolution/s. Find the rate at which the beam from the light is moving along the shore at a point that is 100 ft from the point on the shore nearest the lighthouse.

28. An airplane is flying at 150 ft/s at an altitude of 2000 ft in a direction that will take it directly over an observer at ground level. Find the rate of change of the angle of elevation between the observer and the plane when the plane is directly over a point on the ground that is 2000 ft from the observer.

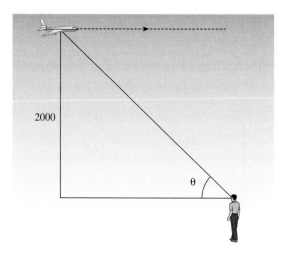

29. An airplane takes off at an angle of 30°, flying at 160 ft/s in a direction that will take it directly over an observer at ground level, 800 ft from the take-off point. Find the rate of change of the angle of elevation between the observer and the plane when the angle of elevation is 60° (a) before and

(b) after the plane passes over the observer. (*Hint:* You can use the Law of Sines.)

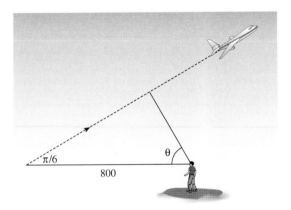

30. A tennis ball is hit at 50 m/s directly toward the net at a point that is 10 m from a spectator sitting in line with the net. Find the rate of change of the angle of the line of sight of the spectator to the ball as the ball passes over the net.

31. A right circular cylinder is being flattened in a manner that does not change its volume. Find the rate of change of its radius when the radius is 3 in. and the height is 4 in., if the height is decreasing at a rate of 0.2 in./s.

32. As a right circular cylinder is being heated, its radius is increasing at a rate of 0.04 m/s and its height is increasing at a rate of 0.15 m/s. Find the rate of change of the volume of the cylinder when the radius is 0.5 m and the height is 0.3 m.

33. Two variable resistors are connected in parallel. The resistance of the first is increasing at a rate of $0.12\,\Omega$/s while the resistance of the second is decreasing at a rate of $0.08\,\Omega$/s. Find the rate of change of the total resistance when the first is $4\,\Omega$ and the second is $2\,\Omega$. Is the total resistance increasing or decreasing at this time? The total resistance R of two parallel resistors R_1 and R_2 is given by the formula $1/R = (1/R_1) + (1/R_2)$.

34. The current in an electrical system is decreasing at a rate of 5 A/s while the voltage remains constant at 9 V. Find the rate of change of the resistance when the current is 18 A.

The voltage V in volts, the resistance R in ohms, and the current I in amperes are related by the formula $V = IR$.

35. The pressure of a fixed amount of a certain gas is increasing at a rate of $500\,(\text{N/m}^2)$/s while the temperature remains constant. Find the rate of change of the volume of the gas when the pressure is $1600\,\text{N/m}^2$ and the volume is $3\,\text{m}^3$. Is the volume increasing or decreasing at this time? PV is constant in this case.

36. The pressure and volume of a fixed quantity of air at a fixed temperature are related by the formula $PV^{1.4} = \text{constant}$. Find the rate of change of the pressure at the instant the pressure is $5\,\text{lb/in.}^2$ and the volume is $45\,\text{in.}^3$, if the volume is increasing at a rate of $3\,\text{in.}^3$/s. Is the pressure increasing or decreasing at this time?

37. The gravitational force of attraction between two masses is inversely proportional to the square of the distance between them. Suppose that when a comet is 600 million kilometers from the earth and receding at 12,000 kilometers per second, the force between the comet and the earth is 3000 newtons. How fast is the force changing at that instant? (**Calculus Explorer & Tutor I, Related Rates, 5.**)

38. The strength of a received radar signal is inversely proportional to the fourth power of the range (distance of the target). If the signal strength meter reads 64 when the range is 1 kilometer, and the signal is decreasing in intensity by two units per minute when the strength reads 20, what is the target doing at that instant? (**Calculus Explorer & Tutor I, Related Rates, 7.**)

39. Water is evaporating from a conical cup at a rate that varies as the area of the surface of the water. Show that the depth of water in the cup decreases at a constant rate that does not depend on the dimensions of the cup.

40. A snowball is melting at a rate that varies as its surface area. Show that its radius decreases at a constant rate.

41. A helicopter is taking off at an angle of $60°$, flying at 20 m/s in a straight line away from an observer at ground level, 100 m away from the take-off point. Find the distance between the helicopter and the observer 5 s after take-off. Find the rate of change of the distance at this time.

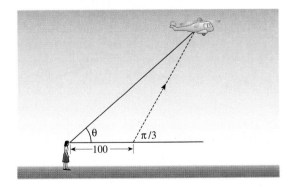

42. For the helicopter in Exercise 41 find the angle of elevation from the observer to the helicopter 5 s after take-off. Find the rate of change of the angle of elevation at this time.

43. A particle is moving along the parabola $y = x^2$ in such a way that its x-coordinate is increasing and

$$\left(\frac{dx}{dt}\right)^2 + \left(\frac{dy}{dt}\right)^2 = 1.$$

Find dx/dt and dy/dt when the particle is at the point $(2, 4)$.

44. A particle moves along the curve $2x^2 + y^3 = 10$ in such a way that its x coordinate is increasing at $1/3$ unit per second. How fast is the y coordinate changing when the particle is at the point $(1, 2)$? **(Calculus Explorer & Tutor I, Related Rates, 8.)**

45. A particle is moving in the counterclockwise direction around the ellipse

$$\frac{x^2}{10^2} + \frac{y^2}{5^2} = 1$$

in such a way that

$$\left(\frac{dx}{dt}\right)^2 + \left(\frac{dy}{dt}\right)^2 = 1.$$

Find dx/dt and dy/dt when the particle is at the point $(6, 4)$.

46. A particle moves counterclockwise around the ellipse $x^2 + 4y^2 = 4$ so that $\left(\frac{dx}{dt}\right)^2 + \left(\frac{dy}{dt}\right)^2 = 1$. What are $\frac{dx}{dt}$ and $\frac{dy}{dt}$ at the point $(\sqrt{2}, \sqrt{2}/2)$?

47. The rate of flow from a hole in the bottom of a spherical tank of radius r is governed by the equation

$$\frac{dV}{dt} = -2\sqrt{y}$$

where V is the volume when the depth is y. For $0 \le y \le r$, $V = \pi \left(ry^2 - \frac{y^3}{3}\right)$. What is $\frac{dy}{dt}$ when $y = r/2$?

48. A cylindrical beaker has radius r and height h. Water is pouring from the beaker, which is tilted. Find a formula that relates the rate of change of the volume of the water in the beaker and the rate of change of the angle the beaker makes with the vertical. Assume that the water covers the entire bottom of the beaker. (*Hint:* Use the average depth of the water from the bottom of the beaker to find the volume of water in the beaker.)

49. A baseball scout is using a radar gun to time a prospective pitcher's throws. The scout is sitting in line with the third base line, 60 ft from home plate. For one pitch the radar gun registers 80 mph when the ball is 50 ft from home plate. How fast is the ball traveling at that time? (*Hint:* The radar gun measures the rate of change of the distance between the ball and the gun.)

50. A policeman stops a motorist after a radar gun indicates the motorist was driving 65 mph. The motorist claims he couldn't have been going that fast and the radar reading must be higher than his actual speed because the policeman was not in line with the direction of motion of the motorist's car when the reading was taken. The policeman admits that he was not in line with the direction of motion, but claims that made the radar reading lower than the actual speed. Who is right and why?

51. A ball is thrown vertically at 48 ft/s from a point on the ground that is 24 ft from the base of a lamppost that is 54 ft high. How fast is the shadow of the ball moving one second after the ball is thrown? (*Hint:* The height of the ball in feet t seconds after it is thrown is given by $s(t) = -16t^2 + 48t$.)

4.7 MAXIMUM-MINIMUM VALUE PROBLEMS

Connections

Domain of a function, 2.1

Finding equations that relate variables, 4.5.

Extreme Value Theorem, 3.8.

Efficient design involves finding maximum and minimum values of quantities subject to some type of constraint. For example, we may want to produce a product in such a way as to maximize profit or minimize cost subject to production and market constraints. To solve this type of problem, we will use the methods of Section 4.5 to express the quantity of interest as a function of a single real variable. We can then use the technique of Section 3.8 to find maximum and minimum values. Let us outline the strategy that we will be following in the examples.

Strategy for maximum-minimum problems

- Carefully read the problem.
- Identify and name the variables; draw a picture, if appropriate.
- Express the maximum-minimum variable of interest in terms of the desired independent variable and any convenient extra variables.
- Find equations that relate the independent variable and any extra variables.
- Use substitution to eliminate all extra variables from the expression for the maximum-minimum variable.
- Determine the domain of the independent variable.
- Determine the maximum-minimum values by checking values of the maximum-minimum variable

  (i) at points in the domain where the derivative is zero,
  (ii) at points in the domain where the derivative does not exist, and
  (iii) at (or near) the endpoints of the domain.

The Importance of the Domain

The domain is an important part of a function. This is particularly true in maximum-minimum value problems. We have seen in Section 2.1 how the domain of a function may be restricted by the interpretation of variables as physical quantities. Other restrictions on the values of variables and the domain of a function may arise as consequences of physical constraints that exist for a particular problem.

   If we obtain the maximum-minimum variable as a function that is continuous on an interval that contains both endpoints, then the Extreme Value Theorem guarantees that the function has both a maximum value and a minimum value. Moreover, the maximum and minimum values must each occur at either a point inside the interval where the derivative is zero, a point inside the interval where the derivative does not exist, or an endpoint of the interval. If a function is not continuous at every point of an interval, or if the interval does not contain both endpoints, then the function may or may not have maximum or minimum values. In these cases a sketch of the graph of the maximum-minimum variable as a function of the independent variable may be used to determine if a solution exists.

   Most of the functions that occur are continuous on an interval. We try to include the endpoints of the interval, if this is possible. Generally speaking, we include the endpoints of the interval in the domain if the formula we obtain makes sense at the endpoints. Care must be taken when interpreting the physical conditions described at the endpoints. Let us illustrate with a simple geometric example.

EXAMPLE 1

A triangle is to have one side of length 3 and another side of length 4. **(a)** Express the area of the triangle as a function of the angle between these sides and find the angle that gives the triangle of maximum area. **(b)** Express the perimeter of the triangle as a function of the angle between the given sides and find the angle that gives the triangle of minimum perimeter.

SOLUTION

(a) A sketch of the triangle is given in Figure 4.7.1. The maximum-minimum variable of interest is the area of the triangle, which we denote by A. The angle between the sides of lengths 3 and 4 is labeled θ; θ is the desired independent variable. We want to express the area as a function of θ, but the *easiest* way to describe the area of a triangle is in terms of its base and height. Hence, we introduce the variable h, which represents the height of the triangle. We then have

$$A = \frac{1}{2}(4)(h).$$

Simplifying, we obtain

$$A = 2h.$$

Let us now use the relation between h and θ to express the area in terms of θ. From Figure 4.7.1 we have

$$\sin \theta = \frac{h}{3}, \text{ so } h = 3 \sin \theta.$$

Substitution then gives us a formula for the area in terms of the angle. We have

$$A = 2h,$$
$$A = 2(3 \sin \theta),$$
$$A = 6 \sin \theta.$$

The expression $6 \sin \theta$ is defined for all θ, but θ can represent an angle of a triangle only if $0 < \theta < \pi$. We include the endpoints $\theta = 0$ and $\theta = \pi$ in order to facilitate the mathematical analysis of the problem. This gives us an area function that is continuous on an interval that contains both endpoints. The Extreme Value Theorem then guarantees that the function has a maximum value and we can use the method of Section 3.8 to find its value. If the maximum value of the area function is obtained for a value of θ that satisfies $0 < \theta < \pi$, that value of θ gives the triangle of maximum area. The area function is

$$A(\theta) = 6 \sin \theta, \qquad 0 \leq \theta \leq \pi.$$

We have

$$A'(\theta) = 6 \cos \theta.$$

(i) Setting $A'(\theta) = 0$, we obtain

$$6 \cos \theta = 0.$$

The solution for $0 < \theta < \pi$ is $\theta = \pi/2$. (ii) The derivative exists for all θ in

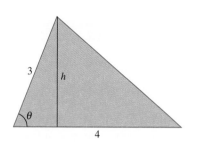

FIGURE 4.7.1 The variable h is convenient for expressing the area of the triangle.

TABLE 4.7.1

θ	$A(\theta)$
$\dfrac{\pi}{2}$	6
0	0
π	0

the interval. (iii) The endpoints of the interval are $\theta = 0$ and $\theta = \pi$. The value of $A(\theta)$ at each θ determined above is given in Table 4.7.1. The maximum value of the function is the largest value of $A(\theta)$ in the table. We see that the area function has a maximum value of 6 when $\theta = \pi/2$. This value of θ gives the triangle of maximum area. The graph of the function $A(\theta)$ is sketched in Figure 4.7.2.

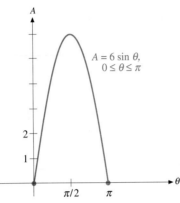

FIGURE 4.7.2 The area of the triangle is $A(\theta) = 6\sin\theta, 0 \leq \theta \leq \pi$. The area is maximum where the derivative is zero.

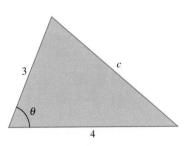

FIGURE 4.7.3 The variable c is convenient for expressing the perimeter of the triangle.

(b) The triangle is sketched in Figure 4.7.3. The maximum-minimum variable is the perimeter P. The desired independent variable is the angle θ. The side opposite θ is labeled c. It is easy to express the perimeter of the triangle in terms of the variable c. We have

$$P = 4 + 3 + c,$$
$$P = 7 + c.$$

Since we wish to express the perimeter as a function of θ, we need to find an equation that relates the variables c and θ. The Law of Cosines gives us

$$c^2 = 4^2 + 3^2 - 2(3)(4)(\cos\theta).$$

Solving for c in terms of θ, we obtain

$$c = \sqrt{25 - 24\cos\theta}.$$

Substitution in the expression for the perimeter then gives

$$P = 7 + \sqrt{25 - 24\cos\theta}.$$

Consideration of the domain as in part (a) gives us the function

$$P(\theta) = 7 + \sqrt{25 - 24\cos\theta}, \qquad 0 \leq \theta \leq \pi.$$

Using the Chain Rule to evaluate the derivative, we have

$$P'(\theta) = \frac{1}{2\sqrt{25 - 24\cos\theta}}(-24)(-\sin\theta).$$

There is no value of θ between 0 and π for which $P'(\theta) = 0$, and there is no point in the interval where the derivative does not exist. This means that

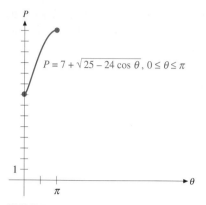

FIGURE 4.7.4 The perimeter of the triangle is $P(\theta) = 7 + \sqrt{25 - 24\cos\theta}$, $0 \le \theta \le \pi$. The perimeter is minimum when $\theta = 0$.

the minimum value of the function must occur at one of the endpoints of the interval and not at any point inside the interval. We have

$$P(0) = 7 + \sqrt{25 - 24\cos 0} = 7 + \sqrt{1} = 8$$

and

$$P(\pi) = 7 + \sqrt{25 - 24\cos \pi} = 7 + \sqrt{49} = 14.$$

It follows that the minimum value of the perimeter function is $P(0) = 8$. However, when $\theta = 0$ (or $\theta = \pi$) the function does not represent the perimeter of an actual triangle. Among all triangles that have sides of length 3 and 4, there is none that has minimum (or maximum) perimeter. The perimeter approaches 8 as the angle θ decreases to zero, but none of the triangles has perimeter 8. See Figure 4.7.4. ■

Area-Perimeter Problems

EXAMPLE 2

A farmer has 200 m of fencing to enclose a rectangular region adjacent to a river. No fencing is required along the river. **(a)** Express the area of the region as a function of the length of the sides of the region perpendicular to the river. Find the dimensions and area of the region that has maximum area. **(b)** Repeat part (a) with the constraint that the region cannot extend more than 40 m from the river.

SOLUTION

(a) The region is sketched in Figure 4.7.5. The maximum-minimum variable of interest is the area of the region, which we denote by A. Note that we have a choice of how much fence to use parallel to the river (x) and how much to use perpendicular to the river (y), so x and y are variables. Also, x and y are exactly the variables that are needed to obtain an expression for the area, the maximum-minimum variable of interest. As a general rule we will use as many variables as is convenient to write down an expression for the maximum-minimum variable. Any extra variables will be eliminated later, when we find relations between variables. It is very important that finding relations between variables should not be confused with the number one task of expressing the maximum-minimum variable in terms of any convenient variables. We have

$$A = xy.$$

We now find a relation between x and y. These variables are related because they determine the amount of fencing, and we are told that the total amount

FIGURE 4.7.5 The area of the rectangular field can be expressed in terms of the two variables x and y. $A = xy$.

of fencing is 200. From our sketch we see that the total amount of fencing is $x + 2y$, since no fencing is required along the river. Hence, consideration of the total amount of fencing gives the relation

$$x + 2y = 200.$$

The variable y is the desired independent variable, so we solve the above equation for x in terms of y. We obtain

$$x = 200 - 2y.$$

Substitution into the expression for the area of the region and then simplifying, we obtain

$$A = xy,$$
$$A = (200 - 2y)y,$$
$$A = 200y - 2y^2.$$

We now need to determine the domain. Since *length is nonnegative,* we must have $x \geq 0$ and $y \geq 0$. Then, substituting $x = 200 - 2y$ into the inequality $x \geq 0$ and solving for y, we obtain

$$x \geq 0,$$
$$200 - 2y \geq 0,$$
$$-2y \geq -200,$$
$$y \leq 100.$$

We include the endpoints $y = 0$ and $y = 100$ in the domain, even though these values of y give a "rectangle" with two sides and area zero. This gives

$$A(y) = 200y - 2y^2, \qquad 0 \leq y \leq 100.$$

The area function A is continuous on an interval that contains both endpoints, so we are guaranteed a maximum (and minimum) value. We use the techniques of Section 3.8 to find the maximum value. We have

$$A'(y) = 200 - 4y.$$

(i) The derivative is zero when $y = 50$, which is a point inside the interval of definition. (ii) The derivative exists for all y in the interval. (iii) The endpoints are $y = 0$ and $y = 100$. The formula for $A(y)$ is used to find the corresponding values of the area. The values are given in Table 4.7.2. We see that $A(50) = 5000$ m^2 is the maximum area of the region described. When $y = 50$, the formula for x in terms of y gives $x = 200 - 2(50) = 100$. The dimensions of the region that has maximum area are 50 m $\times$ 100 m, with the longer side parallel to the river.

Let us note that the area $A(y)$ is clearly positive for $0 < y < 100$ and $A(0) = A(100) = 0$. It follows that the maximum value of $A(x)$ must occur for some y in the interval $0 < y < 100$. Since the derivative exists for all y in the interval and the derivative is zero at only one point, the value at that point must give the maximum value of $A(y)$. The graph of the function $A(y)$ is sketched in Figure 4.7.6.

(b) This problem is the same as that in part (a), except we are told that the region cannot extend more than 40 m from the river. We obtain the same formula for the area in terms of y, but we now have the additional restriction that $y \leq 40$. See Figure 4.7.7. The function is then

TABLE 4.7.2 ──────────

y	$A(y)$
50	5000
0	0
100	0

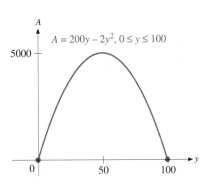

FIGURE 4.7.6 The area of the rectangular field is given by the function $A(y) = 200y - 2y^2$, $0 \leq y \leq 100$. The area is maximum where the derivative of A is zero.

FIGURE 4.7.7 If the region cannot extend more than 40 m from the river, the variable y is restricted to values $0 \le y \le 40$.

$$A(y) = 200y - 2y^2, \qquad 0 \le y \le 40.$$

The derivative $A'(y) = 200 - 4y$ is not equal to zero for any y between 0 and 40, and the derivative exists at all of these points. It follows that the maximum value of the function must occur at one of the endpoints of the interval. We have

$$A(0) = 200(0) - 2(0)^2 = 0$$

and

$$A(40) = 200(40) - 2(40)^2 = 4800.$$

The maximum area of the region subject to the condition that it does not extend more than 40 m from the river is 4800 m². This maximum value is obtained by constructing the fence with the two sides perpendicular to the river 40 m long and the side parallel to the river 120 m long. The graph of the function $A(y)$ is sketched in Figure 4.7.8. ∎

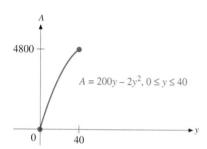

FIGURE 4.7.8 The function $A(y) = 200y - 2y^2, 0 \le y \le 40$, has maximum value when $y = 40$.

EXAMPLE 3

A rectangular field is to have area 60,000 m². Fencing is required to enclose the field and to divide it in half. Express the total amount of fencing required as a function of the length of fence that divides the field. Find the minimum amount of fencing required. What are the outer dimensions of the field that requires the least fencing?

SOLUTION   A sketch is given in Figure 4.7.9. The sides of the field are labeled x and y; these are variables. The maximum-minimum variable of interest is the total amount of fencing, which is given by

$$F = 3x + 2y.$$

We need to find an equation that relates the variables x and y. These variables are related because they determine the area of the field, and the area is given to be 60,000. Consideration of the area of the field gives the relation

$$xy = 60,000.$$

Since the desired independent variable is x, we solve for y in terms of x. We obtain

$$y = \frac{60,000}{x}.$$

Substitution in the formula for the total amount of fencing then gives

$$F = 3x + \frac{120,000}{x}.$$

FIGURE 4.7.9 The outside dimensions of the rectangular field are x and y. The length of fence that divides the field in half is x. The total amount of fencing required is $F = 3x + 2y$.

Since length is nonnegative, we must have $x \geq 0$ and $y \geq 0$. However, neither y nor the total amount of fencing is defined if $x = 0$. (The area of the field cannot be 60,000 if the length of one side is zero.) There is no difficulty (mathematically) for any positive value of x. We obtain the function

$$F(x) = 3x + \frac{120,000}{x}, \qquad x > 0.$$

In this problem, we do not have a function that is continuous on an interval that contains both endpoints, so the Extreme Value Theorem does not guarantee a minimum value. However, we can determine the solution by using information about the graph of the function. This involves investigating the derivative and the behavior of the function near the endpoints of the interval of definition. The derivative is

$$F'(x) = 3 - \frac{120,000}{x^2}.$$

Simplifying, we obtain

$$F'(x) = \frac{3x^2 - 120,000}{x^2}.$$

We see that $F'(x) = 0$ whenever

$$3x^2 - 120,000 = 0,$$
$$3x^2 = 120,000,$$
$$x^2 = 40,000,$$
$$x = 200, \text{ since } x \geq 0.$$

The formula for $F(x)$ gives $F(200) = 1200$. The point $(200, 1200)$ is the only critical point in the interval of definition. It is not difficult to see that the values of

$$F(x) = 3x + \frac{120,000}{x}$$

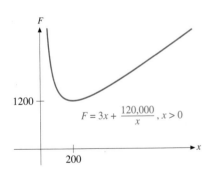

FIGURE 4.7.10 The amount of fencing required is $F(x) = 3x + \frac{120,000}{x}$, $x > 0$. The minimum value is where the derivative is zero.

approach infinity as x approaches zero and as x approaches infinity. Also, the sign of $F'(x)$ can be used to determine that F is decreasing on the interval $0 < x < 200$ and increasing on the interval $x > 200$. We conclude that $F(200) = 1200$ m is the minimum amount of fencing required. See Figure 4.7.10. When $x = 200$, $y = 60,000/200 = 300$. The outer dimensions of the field that requires the least fencing are 200 m × 300 m. ■

Rate Problems

EXAMPLE 4

A cabin is located 2 km directly into the woods from a mailbox on a straight road. A store is located on the road, 5 km from the mailbox. A woman wishes to walk from the cabin to the store. She can walk 3 km/h through the woods and 4 km/h along the road. Express the total time for her walk as a function of the distance between the mailbox and the point on the road toward which

FIGURE 4.7.11 The distance between the mailbox and the point on the road toward which the woman walks is x. The distance walked through the woods is z and the distance walked along the road is y.

she walks. Find the point on the road toward which she should walk in order to minimize the total time for her walk and find the minimum total time.

SOLUTION   A map is sketched in Figure 4.7.11. The variables are the distances labeled x, y, and z. The desired independent variable is x. The maximum-minimum variable of interest is the total time for the walk. We know that

$$\text{Rate} = \frac{\text{Distance}}{\text{Time}}, \quad \text{so} \quad \text{Time} = \frac{\text{Distance}}{\text{Rate}}.$$

The total time T is the sum of time through the woods and the time along the road. Expressing these times in terms of the variables z and y, we obtain the expression

$$T = \frac{z}{3} + \frac{y}{4}.$$

Each of the variables y and z can be expressed in terms of the variable x. The Pythagorean Theorem gives the relation

$$z^2 = x^2 + 2^2, \text{ so } z = \sqrt{x^2 + 4}.$$

Clearly, $x + y = 5$, so $y = 5 - x$. Substitution then gives

$$T = \frac{\sqrt{x^2 + 4}}{3} + \frac{5 - x}{4}.$$

Since the woman is planning to walk toward a point that is between the mailbox and store, we have $0 \le x \le 5$. This gives the function

$$T(x) = \frac{\sqrt{x^2 + 4}}{3} + \frac{5 - x}{4}, \qquad 0 \le x \le 5.$$

The derivative is

$$T'(x) = \frac{1}{3} \left(\frac{1}{2\sqrt{x^2 + 4}} \right)(2x) - \frac{1}{4}$$

$$= \frac{4x - 3\sqrt{x^2 + 4}}{12\sqrt{x^2 + 4}}.$$

(i) $T'(x) = 0$ implies

$$4x = 3\sqrt{x^2 + 4},$$
$$16x^2 = 9(x^2 + 4),$$
$$7x^2 = 36,$$
$$x = \frac{6}{\sqrt{7}}, \text{ since } x \ge 0.$$

Since squaring an equation can give extraneous roots, we must check that we have obtained a solution of the original equation. We have.

$$T\left(\frac{6}{\sqrt{7}}\right) = \frac{\sqrt{(36/7) + 4}}{3} + \frac{5 - (6/\sqrt{7})}{4} \approx 1.69.$$

(ii) $T'(x)$ exists in the interval. (iii) The values at the endpoints are $T(0) = (2/3) + (5/4) \approx 1.92$ and $T(5) = \sqrt{29}/3 \approx 1.80$. We conclude that the minimum total time is $T(6/\sqrt{7}) \approx 1.69$ hours, or approximately 1 hour, 41 minutes.

FIGURE 4.7.12 The time required for the walk is $T(x) = \dfrac{\sqrt{x^2+4}}{3} + \dfrac{5-x}{4}$, $0 \le x \le 5$. The minimum occurs when $x = 6/\sqrt{7}$, where the derivative is zero.

In order to complete the walk this quickly, the woman should walk toward the point on the road that is $x = 6/\sqrt{7} \approx 2.27$ miles, or about $2\frac{1}{4}$ miles from the mailbox. The graph of T is sketched in Figure 4.7.12. ∎

In Example 4, the restriction $x \le 5$ was necessary in the domain of $T(x)$, because we used $5 - x$ to represent a distance. Also, note that $x = 5$ means that the woman walks directly through the woods to the store. For certain combinations of rates and distances this path gives the minimum time. This would be the case in Example 4 if, for example, the store were within 2.26 miles of the mailbox. The endpoint $x = 0$ indicates a path directly toward the road and then along the road to the store. One might verify that the derivative of T at $x = 0$ is negative for any combination of rates and distances, so $x = 0$ can never give a minimum value of T.

Geometric Problems

EXAMPLE 5

The strength of a rectangular beam varies as the product of the width and the square of the height of a cross-section. Find the dimensions of the rectangular beam of maximum strength that can be cut from a cylindrical log with radius 3.

FIGURE 4.7.13a The rectangular beam has width w and height h.

SOLUTION   A sketch of a cross-section of the beam is given in Figure 4.7.13a. The variables are the width w and the height h.

The maximum-minimum variable of interest is the strength of the beam. The statement that the strength of the beam varies as the product of the width and the square of the height translates into the equation

$$S = kwh^2,$$

where S is the strength and k is a constant that depends on the type of wood and the units that are used. Even though we cannot determine the value of the constant k, k is not a variable in this problem. To express S as a function of a single variable, we need to find an equation that relates the variables w and h. To do this, we draw a line from the center of the log to a corner of the beam, as in Figure 4.7.13b. This gives us a right triangle with sides $w/2$ and $h/2$, and hypotenuse 3. The Pythagorean Theorem then gives

$$\left(\frac{w}{2}\right)^2 + \left(\frac{h}{2}\right)^2 = 3^2.$$

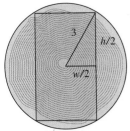

FIGURE 4.7.13b To find an equation that relates w and h, we sketch a radius to one corner of the rectangle and form a right triangle with hypotenuse 3 and sides of lengths $w/2$ and $h/2$.

We could now choose either w or h to be the independent variable and solve the above equation for the other variable. We see that in either case this will result in an expression that involves a square root. However, since the variable h appears squared in the formula for the strength, we can avoid a square root in the formula for the strength if we choose the independent variable to be w. Solving the above equation for h^2 in terms of w, we obtain

$$w^2 + h^2 = 4(9),$$
$$h^2 = 36 - w^2.$$

Substituting in the expression for the strength and then simplifying, we obtain

$$S = kw(36 - w^2),$$
$$S = k(36w - w^3).$$

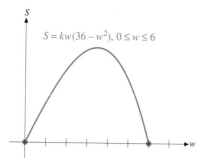

FIGURE 4.7.14 The strength of the beam is $S(w) = kw(36 - w^2)$, $0 \le w \le 6$. The maximum occurs where the derivative is zero.

FIGURE 4.7.15a The volume of the right circular cylinder is $V = \pi r^2 h$.

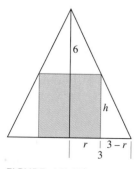

FIGURE 4.7.15b The ratios of corresponding sides of similar triangles give $\dfrac{h}{3-r} = \dfrac{6}{3}$.

It is clear that the width of the beam cannot be greater than the diameter of the log, so we obtain the function

$$S(w) = k(36w - w^3), \qquad 0 \le w \le 6.$$

The derivative is

$$S'(w) = k(36 - 3w^2).$$

(i) $S'(w) = 0$ when

$$36 - 3w^2 = 0,$$
$$-3w^2 = -36,$$
$$w^2 = 12,$$
$$w = \sqrt{12} = 2\sqrt{3}, \text{ since } w \ge 0.$$

(ii) $S'(w)$ exists for all w in the interval of definition. (iii) At the endpoints, we have $S(0) = S(6) = 0$.

Since the function is zero at both endpoints and positive at points between, the value at the single critical number must be the maximum value. When $w = 2\sqrt{3}$, we have

$$h^2 = 36 - w^2,$$
$$h^2 = 36 - 12,$$
$$h = \sqrt{24} = 2\sqrt{6}.$$

The strongest rectangular beam that can be cut from a cylindrical log with radius 3 has dimensions $2\sqrt{2} \times 2\sqrt{6}$. The shape of the strongest beam does not depend on the constant k, although the maximum strength does. The graph of the function S is sketched in Figure 4.7.14. ∎

EXAMPLE 6

A right circular cylinder is inscribed in a right circular cone so that the center-lines of the cylinder and the cone coincide. The cone has height 6 and radius of base 3. Express the volume of the cylinder as a function of its radius. Find the dimensions and volume of the cylinder that has maximum volume.

SOLUTION   A sketch is given in Figure 4.7.15a. The radius r and the height h of the cylinder are variables. The maximum-minimum variable is the volume of the cylinder. We know that

$$V = \pi r^2 h.$$

To find an equation that relates r and h, we sketch a cross-section through the common centerline, as indicated in Figure 4.7.15b. Using the ratio of corresponding sides of the smaller and larger similar triangles, we obtain

$$\frac{h}{3-r} = \frac{6}{3}.$$

The desired independent variable is r. Solving for h in terms of r, we have

$$h = 2(3 - r),$$
$$h = 6 - 2r.$$

Substituting in the equation for the volume of the cylinder, we obtain

$$V = \pi r^2 (6 - 2r),$$
$$V = 2\pi (3r^2 - r^3).$$

Since the radius of the cylinder cannot exceed the radius of the base of the cone, we obtain the function

$$V(r) = 2\pi (3r^2 - r^3), \qquad 0 \le r \le 3.$$

The derivative is

$$V'(r) = 2\pi (6r - 3r^2).$$

We see that $V'(r) = 0$ when

$$6r - 3r^2 = 0,$$
$$3r(2 - r) = 0.$$

The only solution inside the interval of definition is $r = 2$. The derivative exists at all points in the interval. The function is zero at both endpoints and is positive between, so the maximum volume must occur at the single critical point. The maximum volume of the inscribed cylinder is

$$V(2) = 2\pi (3(2)^2 - (2)^3) = 8\pi.$$

When $r = 2$, the formula for h in terms of r gives

$$h = 6 - 2(2) = 2.$$

The cylinder of maximum volume has height 2 and radius 2. The graph of the function V is sketched in Figure 4.7.16. ∎

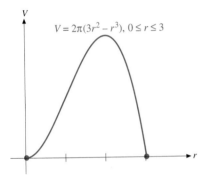

FIGURE 4.7.16 The volume of the cylinder is $V(r) = 2\pi(3r^2 - r^3)$, $0 \le r \le 3$. The volume is maximum where the derivative is zero.

Discrete Variable Problems

Many practical maximum-minimum problems involve variables that are restricted to **discrete** values. For example, the number of items manufactured must be a nonnegative integer and the price charged per item must be restricted to values that are not fractions of a cent. Calculus does not apply directly to problems that involve discrete variables. However, we can sometimes get an indication of the solution of the discrete problem by applying calculus to a real-variable **mathematical model** of the problem. That is, we simply assume that each variable ranges over all real numbers in some interval, even though the physical interpretation of the variable makes sense only for discrete values. This allows us to use calculus to find a solution of the problem for the model. Hopefully, the solution of the discrete problem involves values of the variables that are near the values that give the solution of the problem for the model. It is not within the scope of this text to carefully discuss when this is actually the case. We will simply accept the solution of the model problem as an approximate solution of the discrete problem.

EXAMPLE 7

From experience, a promoter knows that he can sell all 400 tickets to a concert if he charges $6 per ticket. Also, he knows that for each $1 increase in price, he will sell approximately 35 fewer tickets. Express the total revenue collected for tickets as a function of the price per ticket. For what price per ticket would he

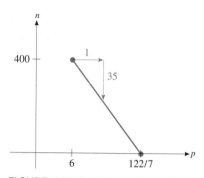

FIGURE 4.7.17 The number of tickets sold, n, decreases by 35 for each $1 increase in the ticket price, p. This indicates a linear relation between the variables n and p.

expect to collect the most money from the sale of tickets and how much money would he expect to collect?

SOLUTION The variables in this problem are the price per ticket, p, and the number of tickets sold, n. In practice, p must be restricted to values that do not involve fractions of a cent and n must be an integer. In order to use calculus, we assume that p and n are real-number variables. The maximum-minimum variable is the total revenue collected, R, which is given by the formula

$$R = np.$$

The statement that the number of tickets sold will decrease by 35 for each $1 increase in ticket price indicates a linear relation between n and p. See Figure 4.7.17. We are told that $n = 400$ when $p = 6$. Also we know that a change of *positive* 1 in p corresponds to a change of *negative* 35 for n, since n *decreases* as p *increases*. We can express the linear relation between n and p as a ratio of changes. We obtain

$$\frac{\text{Change in } n}{\text{Change in } p} : \quad \frac{n - 400}{p - 6} = \frac{-35}{1},$$
$$n - 400 = -35(p - 6).$$

The desired independent variable is p. Solving for n in terms of p, we obtain

$$n = 400 - 35p + 210,$$
$$n = 610 - 35p.$$

Substitution in the formula for the total revenue then gives

$$R = (610 - 35p)(p),$$
$$R = 610p - 35p^2.$$

We assume that $p \geq 6$, since all of the tickets can be sold at this price. Since $n = 610 - 35p$ must be nonnegative, we also have $p \leq 122/7$. This gives the function

$$R(p) = 610p - 35p^2, \qquad 6 \leq p \leq \frac{122}{7}.$$

The derivative is

$$R'(p) = 610 - 70p.$$

We see that $R'(p) = 0$ when $p = 61/7$. This is the only critical point of the function. We have

$$R\left(\frac{61}{7}\right) = 610\left(\frac{61}{7}\right) - 35\left(\frac{61}{7}\right)^2 \approx 2658.$$

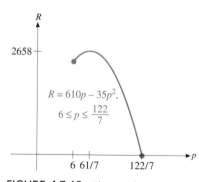

FIGURE 4.7.18 The total revenue collected is $R(p) = 610p - 35p^2$, $6 \leq p \leq \frac{122}{7}$. The maximum revenue is collected at the price for which the derivative is zero.

The values at the endpoints are $R(6) = 610(6) - 35(6)^2 = 2400$ and $R(122/7) = 0$. The maximum value of the revenue function is given by $p = 61/7 \approx 8.71$. The promoter would expect to collect the most money from tickets for the concert if he charges approximately $8.71 per ticket. He would expect to collect approximately $2658. The graph of $R(p)$ is given in Figure 4.7.18. ∎

EXERCISES 4.7

1. A right triangle has hypotenuse 5. Express the area of the triangle as a function of one of its acute angles. Find the angles and the area of the triangle that has maximum area.

2. A right triangle has hypotenuse 5. Express the perimeter of the triangle as a function of one of its acute angles. Find the angles and perimeter of the triangle that has maximum perimeter.

3. A farmer has 900 m of fencing. The fencing is to be used to enclose a rectangular field and to divide it in half. Express the area of the field as a function of the length of fencing used to divide the field in half. Find the outer dimensions and area of the field that has maximum area.

4. A pig farmer has 600 meters with which to enclose and divide four adjacent rectangular pens, as shown in the figure. What dimensions will achieve the maximum possible total area for the four pens? (**Calculus Explorer & Tutor I, Max-min, 7.**)

5. A gutter with vertical sides and no top is to be constructed by bending up equal sides of a rectangular piece of metal that is 10 in. wide. Express the area of the cross-section of the gutter as a function of the length turned up on each side.

Find the dimensions and area of the cross-section that has maximum area.

6. Find the shortest possible perimeter of a rectangle of area 27. (**Calculus Explorer & Tutor I, Max-min, 2.**)

7. A farmer wishes to enclose a rectangular region adjacent to a river. No fencing is required along the river. The area of the region is to be $15,000 \, \text{m}^2$.
   (a) Express the amount of fencing required as a function of the length of fence parallel to the river. Find the dimensions and amount of fencing required for the field that requires the least amount of fencing.
   (b) Repeat part (a) with the constraint that the field cannot extend more than 150 m along the river.

8. A farmer living in Minnesota has 1000 meters of fence with which he wants to bound three sides of a field adjoining the upper Mississippi River. (The river will bound the fourth side.) What is the maximum possible area of a field that he can enclose in this way? Could he do better with a field of any other shape? (**Calculus Explorer & Tutor I, Max-min, 5.**)

9. (a) A rectangular pen is to be constructed so that one side of the pen contains all of a 100-ft straight fence that is already standing and also that the pen will have an area of $16,000 \, \text{ft}^2$. New fencing is not required along the existing fence. Express the amount of new fencing required as a function of the length of the side of the pen that contains the existing fence. Find the dimensions and amount of new fencing required for the pen that require the least new fencing.

(b) Repeat part (a), except assume that one side of the pen must contain all of a 150-ft length of an existing fence.

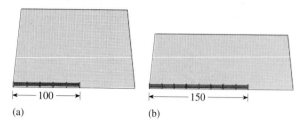

(a)   (b)

10. (a) A 3-m length of wire is to be cut into two pieces. One piece is to be bent to form a circle and the other piece is to be bent to form a square. Find the length of wire that is used for each of the circle and the square, and the total area of the circle and the square that have minimum total area.

(b) Can the wire be cut in such a way that the total area of the circle and square is maximum? What configuration gives the maximum area if it is not required that the wire be cut, so all of the wire may be used to form one of either a circle or a square?

11. Solve Exercise 7 with an equilateral triangle in place of the circle.

12. A rectangular box with no top is to be constructed by cutting equal squares from each corner of a 2 ft × 3 ft rectangular piece of material and bending up the sides. Express the volume of the box as a function of its height. Find the dimensions and volume of the box that has maximum volume.

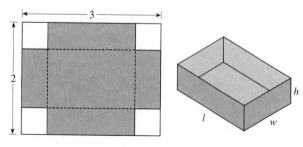

13. A square sheet of cardboard 42 centimeters on a side is to be made into a box open at the top by cutting a square out of

each corner and folding up the sides. What is the maximum possible volume of a box constructed this way? (**Calculus Explorer & Tutor I, Max-min, 3.**)

14. Material for the top and bottom of a cylindrical container costs \$2/ft^2. Material for the lateral surface costs \$1/ft^2. The allotment for the total cost of material is \$48. Express the volume of the cylinder as a function of its radius. Find the dimensions and volume of the cylinder that has maximum volume.

15. Fencing is required to enclose a rectangular field of area 16,000 m^2 and to divide it in half. Fencing to enclose the field costs \$3/m and the fencing used to divide the field costs \$2/m. Express the total cost of fencing as a function of the length of fence that divides the field. Find the dimensions and total cost of fencing for the field that is least expensive to fence.

16. A rectangular field adjacent to a river is to be enclosed. Fencing along the river costs \$5/m and fencing for the other three sides costs \$3/m. The area of the field is to be 9000 m^2. Express the total cost of fencing as a function of the length of fence along the river. Find the dimensions and total cost of fencing for the field that is least expensive to fence.

17. At 2 A.M. the QE2 is 60 miles due north of an oil tanker and sailing south at 15 miles per hour. The tanker is sailing west at 10 miles per hour. At what time are the two ships closest together? (**Calculus Explorer & Tutor I, Max-min, 1.**)

18. (a) A factory is located 100 m downstream and on the opposite side of a river from a power plant. The river is 60 m wide. A power line is to be run from the power plant, diagonally under the river, and then along the bank to the factory. It costs $130/m to lay the line under the river and $50/m along the bank. Express the total cost as a function of the distance between the point across from the power plant and the point where the line emerges from the river. Find the point where the line emerges from the river and the total cost for the cheapest path.

    (b) Repeat part (a), but with the factory 20 m downstream.

(a)   (b)

19. (a) A man is in a boat on the ocean, 300 m directly out from his house on the shore. A lighthouse is on the shore, 200 m from his house. He can row 2 km/h and walk 4 km/h. He plans to row toward a point on the shore between his house and the lighthouse, and then walk to the lighthouse. Express the total time as a function of the distance from his house to the point on the shore toward which he rows. Find the minimum total time this will take.

    (b) Repeat part (a), but with the lighthouse 160 m from the man's house.

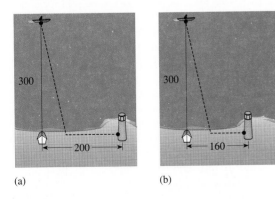

(a)   (b)

20. A post 6 ft high is 21 ft from a post 8 ft high. A line is to be run from the top of one post to the ground between the posts and then to the top of the other post. Express the length of line as a function of the distance between the 6-ft post and the point where the line touches the ground. Find the minimum length of line.

21. The speed of light through air is v_0 ft/s and the speed through water is v_1 ft/s. An object is submerged in water 3 ft below a point that is 10 ft from a point where a boy is standing. The boy's eyes are 5 ft above water level. Express the total time in seconds for a light ray to travel from the object to the boy's eyes, as a function of the distance between the boy and the point where the ray emerges from the water. Show that the total time for a ray of light to travel from the object to water level and then through air to the boy's eyes is minimum when

$$\frac{\sin \theta_0}{v_0} = \frac{\sin \theta_1}{v_1}.$$

(Light travels in a straight line through water and air, but bends sharply as it passes from water to air.)

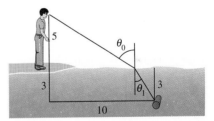

22. According to **Fermat's Principle,** light trying to get from point A to point B will select that path which requires the least amount of travel time. Show that if A is located in a medium in which the velocity of light is a, B is located in a medium in which the velocity of light is b, and α and β are the two angles that a light ray from A to B makes with the perpendicular to the boundary between the mediums, then $\frac{\sin \alpha}{\sin \beta} = \frac{a}{b}$. (**Calculus Explorer & Tutor I, Max-min, 4.**)

23. The velocity in deep water of waves of wavelength λ is $v = k\sqrt{\frac{\lambda}{c} + \frac{c}{\lambda}}$ where k and c are constants. For what value of λ is the velocity minimum?

24. Explain why the point $\left(\frac{\pi}{2a}, 0\right)$ cannot be the point on the curve $y = \cos(ax)$ that is nearest the origin.

25. Find the point on the curve $y = \sqrt{x}$ that is nearest the point $(1, 0)$.

26. Find the point on the curve $y = 2\sqrt{x}$ that is nearest the point $(1, 0)$.

27. A sector of a circle has perimeter 4. Express the area of the sector as a function of the radius. Find the central angle and radius that give the sector of maximum area.

28. An isosceles triangle has perimeter 4. Express the area of the triangle as a function of the length of the two equal sides. Find the length of the equal sides and the included angle of the triangle that has maximum area.

29. (a) A rectangle is inscribed in a right triangle with the corner of the rectangle at the right angle of the triangle. Show that the area of the rectangle is maximum when a corner of the rectangle is at the midpoint of the hypotenuse of the triangle.

(b) A rectangle is inscribed in a triangle with base of the rectangle along the longest side of the triangle. Explain why the rectangle of maximum area has two corners on midpoints of sides of the triangle. (*Hint:* For part (b), use part (a) to show that any other such rectangle has smaller area.)

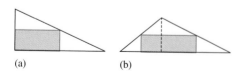

(a)   (b)

30. A ball is thrown vertically at 48 ft/s from a point on the ground that is 64 ft from the base of a lamppost that is 54 ft high. What is the maximum speed of the shadow of the ball along the ground? The height of the ball is $s(t) = -16t^2 + 48t, 0 \le t \le 3.$.

31. What is the minimum length $|AB|$ in the figure, where W and L are constants?

32. Show that the length x in the figure is minimum when θ satisfies $a \sin^3 \theta - b \cos^3 \theta + c \cos^2 \theta - c \sin^2 \theta = 0$.

🖩 Find an approximate value of the minimum length if $a = 7, b = 5$, and $c = 3$.

33. A particle moves along the curve $y = x^2 + 2x$ so that $\dfrac{dx}{dt}$ is constant. At what point on the curve is $\sqrt{\left(\dfrac{dx}{dt}\right)^2 + \left(\dfrac{dy}{dt}\right)^2}$ minimum?

34. At what point on the graph of $y = x^3 + 3x^2 - 2x - 1$ is the slope smallest?

35. An apartment building with 40 apartments can be completely rented if the rent is set at \$400/month. One apartment becomes vacant for every \$10 increase in rent. Express the total amount of rent collected as a function of the unit rent. What rent will give the maximum total amount of rent collected per month?

36. A complete lot of 300 items can be sold at a price of \$2 each. One fewer of the items can be sold for each \$0.02 the price is increased. Express the total amount of money collected as a function of the unit price. What price will give the maximum total amount of money collected?

37. A computer manufacturer can sell 10,000 computers per year if they are priced at \$5000 apiece. The number sold per year increases by 20 for every \$10 decrease in the price. It costs \$20,000,000 per year to operate the plant and \$2500 to make each computer. How should the manufacturer set the price of the computer in order to maximize profit? What is the largest possible yearly profit? (**Calculus Explorer & Tutor I, Max-min, 10.**)

38. A rectangle is inscribed in the ellipse

$$\frac{x^2}{a^2} + \frac{y^2}{b^2} = 1$$

with the centerlines of the rectangle along the axes of the ellipse. Show that the maximum area of the rectangle is $2ab$.

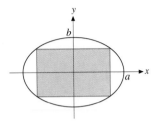

39. A conical cup is to be made by cutting a sector from a circle of radius 6 and joining the radii of the remaining parts. Express the volume of the cup as a function of the height of the cup. Find the central angle of the sector removed that gives the cup of maximum volume.

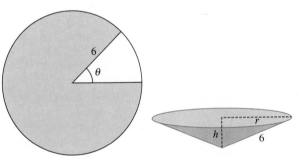

40. A conical cup is to be made by cutting a sector from a semicircle of radius 6 and joining the radii of the remaining parts. Express the volume of the cup as a function of the height of the cup. Find the central angle of the sector removed that gives the cup of maximum volume. (*Hint:* What is the smallest the height can be?)

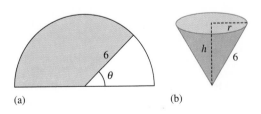

(a)   (b)

41. (a) A gutter is to be formed by bending equal lengths at equal angles from rectangular material that is 12 in. wide. The gutter is to be 3 in. deep. Express the area of a cross-section as a function of the angle that each side is turned up. Find the angle that gives the maximum cross-sectional area.

(b) Repeat (a) in the case that the gutter must be 5.5 in. deep.

42. The base of a bin is 6ft × 10ft. One of the 6-ft sides is 4 ft tall and hinged at the bottom, so it may be lowered away from the bin. The other three sides are vertical and 4 ft tall; two sides extend in quarter circles beyond the hinged end. Express the volume of the bin to the level of the top of the hinged side as a function of the angle that it makes with the horizontal. Find the angle that gives the maximum volume.

43. The cross-section of a parabolic dome is given by the equation $y = (10,000 - x^2)/50$; the floor corresponds to $y = 0$. Find the height at which lights should be installed in order to give the maximum illumination of the floor at its center, if illumination varies directly as the cosine of the angle the rays make with the direction perpendicular to the surface and inversely as the square of the distance between the source and the point of the surface.

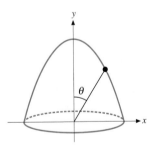

44. Find the dimensions of the right circular cone that has volume 3π and minimum curved surface area.

45. Find the dimemsions of the cylindrical can of volume V of minimal total surface area. (**Calculus Explorer & Tutor I, Max-min, 8.**)

46. Find the dimemsions of the strongest rectangular beam that can be cut from an elliptical log with cross-section given by $4x^2 + y^2 = 4$, if the strength of the beam varies as the product of its width and the square of its height.

47. A right circular cylinder is inscribed in a sphere of radius R. What are the dimensions of the cylinder that has maximum lateral surface area?

48. A solid in the shape of a generalized right cylinder with base an equilateral triangle is to have volume 2. What are the dimensions of the solid that has the total surface area of its five sides minimal?

49. What is the maximum volume of a right circular cone that can be inscribed in a sphere of radius r_0? (**Calculus Explorer & Tutor I, Max-min, 6.**)

50. A rectangle is inscribed in a circle of radius r. **(a)** Express the area of the rectangle in terms of the acute angle between its diagonals. **(b)** Find the angle between the diagonals of the rectangle that has maximum area.

51. Show that the minimum distance of the point (x_1, y_1) to the line $ax + by + c = 0$ is $\dfrac{|ax_1 + by_1 + c|}{\sqrt{a^2 + b^2}}$.

52. Show that the point (x_1, y_1) on the nonhorizontal line $y = mx + b$ that is closest to a point (x_0, y_0) not on the line satisfies

$$\frac{y_1 - y_0}{x_1 - x_0} = -\frac{1}{m}.$$

This means that the line through the points (x_0, y_0) and (x_1, y_1) is perpendicular to the line $y = mx + b$.

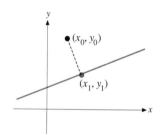

53. Suppose that the minimum distance between a point (x_0, y_0) and points on the graph of a differentiable function f occurs at the point (x_1, y_1) on the graph. Show that

$$\frac{y_1 - y_0}{x_1 - x_0} = -\frac{1}{f'(x_1)}$$

if $f'(x_1) \neq 0$, and that $x_1 = x_0$ if $f'(x_1) = 0$. In either case, the line through (x_0, y_0) and (x_1, y_1) is perpendicular to the

line tangent to the graph f at the point (x_1, y_1), as long as (x_0, y_0) is not on the graph of f.

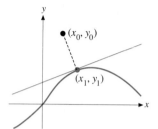

54. ▨ Find an approximate value of the maximum speed of the piston pictured in the figure if the flywheel is rotating at 0.5 revolution/s. The speed of the piston is $\left|\dfrac{ds}{dt}\right|$.

55. ▨ Find approximate values of the coordinates of the point on the parabola $y = x^2$ that is nearest to the point $(4, 1)$.

56. A right circular cylinder with volume 8π is to be constructed from two equal circles and a rectangle that are cut from a rectangular piece of material in one of the three ways indicated below. Determine the radius and height of the cylinder and the method of cutting that requires the least area of the rectangular piece from which the parts are cut. (*Hint:* $h \geq 2r$ in (a) and $h \geq 4r$ in (b).)

(a)

(b)

(c)

REVIEW OF CHAPTER 4 CONCEPTS

- Rational functions have vertical asymptotes at points where the denominator is zero and the numerator is not.
- You should recognize indeterminate forms $0/0$ and ∞/∞ and be prepared to use l'Hôpital's Rule to determine whether or not a function has a vertical asymptote at a point where it may become unbounded.
- Important characteristics of a function that can be obtained directly from the values of the function include its domain, intercepts, continuity, and behavior near asymptotes.
- The sign of the first derivative indicates whether the function is increasing or decreasing and is used to locate local extrema.
- The sign of the second derivative indicates the concavity of the function.
- You should be able to use information about the convexity of graphs near local extrema and asymptotes to approximate points of inflection from the graph of a function.
- It is important to have a strategy for translating word problems into mathematical expressions that involve variables. You should know how to use formulas from geometry and trigonometry, ratios, and linear relations to find equations that relate variables.
- Be sure that any quantity that changes with time is considered to be a variable when setting up problems involving related rates.
- If functions of time are related by an equation, then differentiation of the equation with respect to time gives a new equation that relates the derivatives of the functions. Be sure to use the Chain Rule when differentiating composite functions.
- Remember that the domain of a function is particularly important for extreme value problems. Interpretation of the variables as physical quantities can affect the domain of a function.
- If the function makes mathematical sense, you should try to choose the domain in an extreme value problem to be an interval that contains both endpoints. If the maximum-minimum function is continuous, this guarantees the existence of extreme values. Be sure to check that extreme values at endpoints of the domain are practical solutions of your problem.

CHAPTER 4 REVIEW EXERCISES

Evaluate the limits in Exercises 1–18.

1. $\lim\limits_{x \to 2^+} \dfrac{1-x}{2-x}$

2. $\lim\limits_{x \to -3^-} \dfrac{x}{(x+3)^2}$

3. $\lim\limits_{x \to \pi^+} \dfrac{\cos x}{\tan x}$

4. $\lim\limits_{x \to \pi^-} \dfrac{\sec x}{\sin x}$

5. $\lim\limits_{x \to 0} \dfrac{\tan 2x}{x}$

6. $\lim\limits_{x \to 0} \dfrac{\tan(4x)}{\sin(2x)}$

7. $\lim\limits_{x \to 0} \dfrac{\sin x}{x^{1/3}}$

8. $\lim\limits_{x \to 4} \dfrac{x^{3/2}-8}{x^2-4x}$

9. $\lim\limits_{x \to 0} \sqrt{1+\dfrac{\sin 3x}{x}}$

10. $\lim\limits_{x \to 0} \cos\left(\dfrac{\sin \pi x}{x}\right)$

11. $\lim\limits_{x \to \infty} \dfrac{3x-1}{2x+1}$

12. $\lim\limits_{x \to \infty} \dfrac{x^2+2x+5}{x^3-1}$

13. $\lim\limits_{x \to -\infty} \dfrac{6x+5}{\sqrt{4x^2+9}}$

14. $\lim\limits_{x \to -\infty} (1-2x^2-3x^3)$

15. $\lim\limits_{x \to \infty} \dfrac{x}{(1-x^2)^{1/3}}$

16. $\lim\limits_{x \to \infty} \dfrac{(x+1)^{2/3}}{(x^2+1)^{1/3}}$

17. $\lim\limits_{x \to 0} \left(\dfrac{1}{\tan x}-\dfrac{1}{x}\right)$

18. $\lim\limits_{x \to \infty} (x\sqrt{x^2+1}-x^2)$

Sketch the graphs of the functions in Exercises 19–36. Find local extrema and points of inflection.

19. $f(x) = x(x^2-3)$

20. $f(x) = (4x^3-x^4)/5$

21. $f(x) = \dfrac{x}{x+1}$

22. $f(x) = \dfrac{x^2}{x^2-1}$

23. $f(x) = x - \dfrac{4}{x^2}$

24. $f(x) = \dfrac{20x-40}{x^3}$

25. $f(x) = \dfrac{1}{2-x}$

26. $f(x) = x + \dfrac{1}{x}$

27. $f(x) = 4\sin x - 3\cos x$, $\quad 0 \le x \le 2\pi$

28. $f(x) = \sin^2 x$, $\quad 0 \le x \le 2\pi$

29. $f(x) = \sqrt{1+\sin x}$, $\quad 0 \le x \le 2\pi$

30. $f(x) = \sec x + \tan x$, $\quad -\pi/2 < x < \pi/2$

31. $f(x) = \sec x + 2\tan x$, $\quad -\pi/2 < x < \pi/2$

32. $f(x) = 2\sec x + \tan x$, $\quad -\pi/2 < x < \pi/2$

33. $f(x) = 2\sin x - x$, $\quad 0 \le x \le 2\pi$

34. $f(x) = \tan x - 2x$, $\quad -\pi/2 < x < \pi/2$

35. $f(x) = (x^2-1)^{2/3}$

36. $f(x) = x\sqrt{x+2}$

In Exercises 37–40, sketch a short arc of the graph of the given equation that contains the given point on the graph. The sketch should indicate the sign of the slope and the concavity of the graph as it passes through the point.

37. $x^2 - y^3 = xy - 1;$ $(1, 1)$

38. $\tan y = 2\sin x;$ $(5\pi/6, \pi/4)$

39. $x = \cos y;$ $(-1/2, 2\pi/3)$

40. $x = 2\sin y - \cos y;$ $(-1, 0)$

Find local extrema of the functions y that are defined implicitly as functions of x in Exercises 41–42.

41. $2x^2 + 4xy + 3y^2 = 4$

42. $-3x^2 + 4xy + 4y^2 = 3$

Determine the asymptotic behavior and sketch the graphs of the functions in Exercises 43–46.

43. $f(x) = \dfrac{-2x}{x+2}$

44. $f(x) = x + \dfrac{1}{x^2}$

45. $f(x) = \dfrac{x^2 + 2x + 1}{x}$

46. $f(x) = \dfrac{x^3 - x^2 + x}{x-1}$

In Exercises 47–52 use the given graph of f to sketch the graph of f'.

47.

48.

49.

50.

51.

52.

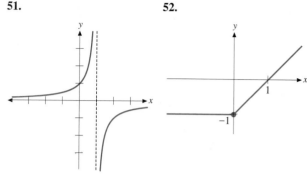

Use the given graphs in Exercises 53–56 to find approximate values of the coordinates of (a) local extrema and (b) points of inflection.

53.

54.

55.

56.

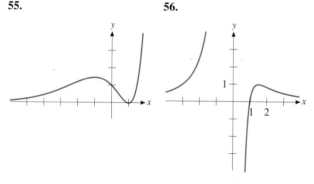

Determine the sign of $f(x)$, $f'(x)$, and $f''(x)$ from the graph of f in each of Exercises 57–60.

57.

58.

59.

60.

In Exercises 61–68 sketch the graph of a function that will satisfy the given conditions or indicate that this is impossible.

61. $f(x) > 0, f'(x) < 0, f''(x) < 0, 0 < x < 1$.

62. $f(x) < 0, \ f'(x) > 0, \ f''(x) < 0, \ 0 < x < 1$, vertical asymptote $x = 0$.

63. $f(x) < 0, \ f'(x) < 0, \ f''(x) > 0, \ 0 < x < 1$, vertical asymptote $x = 1$.

64. $f(x) > 0, \ f'(x) < 0, \ f''(x) < 0, \ 0 < x < 1$, vertical asymptote $x = 0$.

65. $f(x) < 0, f'(x) > 0, f''(x) > 0, x > 0$.

66. $f(x) > 0, f'(x) < 0, f''(x) < 0, x > 0$.

67. $f(x) < 0, f'(x) > 0, f''(x) < 0, -\infty < x < \infty$, horizontal asymptote as x approaches infinity.

68. $f(x) > 0, f'(x) > 0, f''(x) > 0, -\infty < x < \infty$, horizontal asymptote as x approaches infinity.

69. The radius of a right circular cylinder is increasing at a rate of 3 in./s while the height is decreasing at a rate of 4 in./s. Find the rate of change of the volume of the cylinder when the radius is 6 in. and the height is 8 in. Is the volume increasing or decreasing at this time?

70. The volume of a rectangular box remains constant as the length of each side of its square base increases at a rate of 3 cm/s. Find the rate of change of the height of the box when each side of the base is 5 cm and the height is 7 cm.

71. An airplane is flying 200 ft/s at an altitude of 10,000 ft in a line that will take it directly over a person who is observing the plane from the ground. Find the rate of change of the angle of elevation from the observer to the plane when it is directly over a point on the ground that is 10,000 ft from the observer.

72. An oil tank consists of a right circular cylinder of radius 5 m and height 10 m on top of a cone with radius of top 5 m and height 4 m. Oil is being pumped into the tank at a rate of 50 m³/min. What is the rate of change of the depth of the oil in the tank when the depth at the deepest part is (a) 3 m and (b) 7 m?

73. A solution is being poured through a conical funnel into a test tube, as indicated in the figure. The level of the solution in the conical part of the funnel is decreasing 0.5 cm/s when the level is 6 cm. At what rate is the level in the test tube rising at this time?

74. As a boy walks horizontally at 6 ft/s while flying a kite on a 400 ft string, he notices that the angle of elevation is increasing at 5°/s when the angle is 30°. What is the rate of change of the height of the kite at this time?

75. The angle between sides of lengths four and five of a triangle is increasing at a rate of 2°/s. What is the rate of change of the length of the side opposite the angle when the angle is 40°?

76. An angle α of a triangle is increasing at 4°/s while another angle remains fixed at 36°, the length of the side opposite the 36° angle has constant length eight, and the other angle and the lengths of the other two sides are varying. What are the rates of change of the lengths of (a) the side opposite α and (b) the nonconstant side adjacent α when $\alpha = 84°$?

77. The rate of flow from a hole in the bottom of a cylindrical tank of radius two and height three is governed by the differential equation

$$\frac{dV}{dt} = -k\sqrt{y},$$

where V is the volume of fluid when its depth is y and k is a constant. What is the rate of change of the depth when the depth is 1.5?

78. The rate of flow from a hole at the bottom of a conical tank with vertex at the bottom, radius of base two, and height three is governed by the differential equation

$$\frac{dV}{dt} = -k\sqrt{y},$$

where V is the volume of fluid when its depth is y and k is a constant. What is the rate of change of the depth when the depth is 1.5?

79. A farmer has 2000 ft of fencing to enclose a rectangular region adjacent to a river. No fencing is required along the river. (a) Express the area of the region as a function of the length of the fence that is perpendicular to the river. Find the dimensions and area of the region that has maximum area. (b) Repeat part (a) with the constraint that the region cannot extend more than 400 ft from the river.

80. A rectangular box with square base and no top is to have volume of $32\,\text{ft}^3$. (a) Express the total amount of material required for the box as a function of the length of each side of its square base. Find the dimensions and the amount of material required for the box that requires the least material. (b) Repeat part (a) with the restriction that each side of the base must be at least 5 ft.

81. Material for the bottom of a cylindrical container with no top costs three times as much as the material for the cylindrical sides. What are the dimensions of the cylinder for which the cost of material is cheapest for a given volume?

82. A rope is connected to two points A and B that are at the same height and 40 ft apart. A bird feeder hangs from the bottom of a vertical length of rope that is attached to a ring that slides freely on the rope between A and B, so the segments of rope from the ring to A and from the ring to B are essentially straight and of equal length. What are the angles between the rope and the horizontal at A and B and what is the length of rope between the bird feeder and the ring if the amount of rope used is minimum and the bird feeder is to hang (a) 20 ft and (b) 10 ft below the level of A and B?

83. A right circular cone is inscribed in a fixed right circular cone so the cones are coaxial and have vertices in opposite directions. What are the relations between the heights and radii of the cones if the inner cone has maximum volume?

84. A cone is inscribed in a sphere of radius six. What is the maximum volume of the cone?

The **curvature** *of a curve* $y = f(x)$ *at a point on the curve is* $\kappa = \dfrac{|f''|}{(1 + (f')^2)^{3/2}}$. *Find the points on the curves given in Exercises 85–86 where the curvature is maximum.*

85. $y = x^2$

86. $y = \sin x,\ 0 \le x \le 2\pi$

87. A right triangle is formed in the first quadrant by the x-axis, the y-axis, and a line through the point $(2, 1)$. Express the area of the triangle as a function of its base. Find the dimensions and area of the triangle of minimum area.

88. A post 2 m high is 3 m from a post 1 m high. A cable is to be run from the top of one post to the ground at a point between the posts, and then to the top of the other post. Express the length of the cable as a function of the distance between the 2-m post and the point where the cable touches the ground. Find the shortest length of cable that can be used.

89. The **deflection** of a uniformly loaded beam that is fixed at both ends so that $y(0) = y(\ell) = y'(0) = y'(\ell) = 0$ is

$$y = k(x^4 - 2\ell x^3 + \ell^2 x^2).$$

See figure. (a) What is the maximum deflection? (b) Where are the points of inflection of the deflection curve?

90. The horizontal distance traveled by a javelin that is thrown with initial velocity 20 m/s while running at 6 m/s is given by the formula

$$D(\theta) = \left(\frac{40}{4.9}\right)(10 \sin\theta \cos\theta + 3 \sin\theta),$$

where θ is the angle at which it is thrown. Find the angle that gives the longest throw.

91. In statistics, **least square approximation** is an approximation that minimizes the total of the squares of the errors.
    (a) Find the number that is the least square approximation to the three numbers $-1, 4, 6$; that is, find the number c that minimizes the square error function

    $$S(x) = (x - (-1))^2 + (x - 4)^2 + (x - 6)^2.$$

    (b) Find the number that is the least square approximation to the four numbers $-1, 0, 4, 6$.
    (c) Find the number that is the least square approximation to the n numbers $a_1, a_2, \ldots, a_n$; that is, find the number c that minimizes the square error function

    $$S(x) = (x - a_1)^2 + (x - a_2)^2 + \cdots + (x - a_n)^2.$$

92. Measures other than the least square approximation of the total error are important in statistics, too. For example, we could minimize the total of the absolute values of the errors.

  (a) Find the number that is the best total absolute error approximation to the three numbers -1, 4, 6; that is, find the number c that minimizes the error function

  $$A(x) = |x - (-1)| + |x - 4| + |x - 6|.$$

  (b) Find the number that is the best total absolute error approximation to the four numbers -1, 0, 4, 6.

  (c) Find the number that is the best total absolute error approximation to the n numbers $a_1, a_2, \ldots, a_n$; that is, find the number c that minimizes the error function

  $$A(x) = |x - a_1| + |x - a_2| + \cdots + |x - a_n|.$$

  (*Hint:* Sketch the graph of the functions.)

93. Material for the lateral surface and bottom of a cylinder with no top costs $0.09/in.^2 and welds around the circular bottom and up the side cost $0.06/in. The cylinder is to have volume 120 in.^{3}.

  (a) Express the total cost as a function of the radius of the cylinder.

  (b) Find the approximate dimensions of the cylinder that is least expensive to produce.

94. Material for the lateral surface, top, and bottom of a cylinder costs $0.15/in.^2 and welds around the circular top and bottom cost $0.04/in. The cylinder is to have volume 460 in.^{3}.

  (a) Express the total cost as a function of the radius of the cylinder.

  (b) Find the approximate dimensions of the cylinder that is least expensive to produce.

95. Explain why it is not adequate to define a function to be concave upward if the graph "holds water," and give an example.

96. If $y = ax + b$ is an oblique asymptote of f, show that $\lim\limits_{x \to \infty} \dfrac{f(x)}{ax + b} = 1$. Explain why this limit may be one, even though $y = ax + b$ is not an oblique asymptote of f.

EXTENDED APPLICATION

STAMPING CIRCULAR BLANKS
FROM A STEEL COIL

Allegheny Ludlum Steel Corporation produces steel from scrap metal at two locations in western Pennsylvania. During the manufacturing process, hot slabs of steel weighing approximately 10 tons are pressed by a series of rollers into thin strips over 1000 feet long and between 15 and 48 inches wide. The strips are called *coils,* because they are usually rolled up like a very long, narrow carpet while they are still hot.

One of Allegheny Ludlum's customers regularly orders large quantities of circular steel discs, called *blanks,* to be used in manufacturing kitchenware. The blanks, 12 inches in diameter, are stamped out of a coil. Two possible patterns are shown in Figure 1. Of course, the company chooses a pattern that minimizes the total amount of waste involved.

FIGURE 1

You are a summer employee at Allegheny Ludlum's Brackenridge plant. Your boss, having just learned that you aced your calculus course, has asked you to find a waste-minimizing pattern for stamping out blanks, in order to double-check the company's calculations. To keep you honest, he has refused to tell you the pattern presently employed. You are allowed to simplify the problem as follows:

- Ignore the waste at each end of the coil.
- Consider only coils between 15 and 24 inches wide.
- Consider only repeating patterns of two vertical columns of equally spaced blanks, as indicated in Figure 2. The width of the coil depends on the angle.

FIGURE 2

Tasks:

1. Consider a rectangular portion of the coil representing one "period" of the pattern, as in Figure 2. Explain why the total area occupied by blanks, or portions of blanks, within such a rectangle is independent of θ.
2. Express the area of the rectangle in terms of θ. (*Hint:* You will need one formula for $\theta \leq \pi/6$ and another one for $\theta > \pi/6$.)
3. Find the value of θ that minimizes the waste.

Writing Assignment:
Write a report to your boss, describing the optimal pattern and explaining how you determined that it minimizes waste. The boss took calculus years ago and remembers how to take derivatives. He was never very good at optimization problems, though, so provide plenty of detail.

5 The Definite Integral

Many physical problems involve a certain type of sum, called a Riemann sum. In this chapter we will see how these types of sums are related to problems of determining the area of regions in the plane and finding the volume of solids. We will also see how the sums are related to problems that involve motion of an object along a number line.

Riemann sums lead us to the concept of the definite integral. A definite integral is a form of a limit of Riemann sums. We develop the concept of the definite integral in a way that is intended to emphasize the relation of Riemann sums to applications of the integral.

We will study the fundamental relation between the concepts of differentiation and integration. This fundamental relation will allow us to develop formulas for the relatively easy evaluation of many types of definite integrals. Finally, we will investigate techniques for obtaining approximate values of definite integrals and compare the accuracy of some representative methods.

Connections

Differentiation formulas, 3.1–4.

5.1 INDEFINITE INTEGRALS

Before we begin our study of Riemann sums, it is convenient to introduce a problem that seems to have no connection with sums. In subsequent sections we will see that the study of Riemann sums and definite integrals is closely related to this problem, which is of interest in itself.

Problem: *For a given function f, can we find a function (or functions) F such that $F'(x) = f(x)$ for x in an interval?*

Let us see that the problem does not have a simple answer. We first observe that if F is any function that satisfies $F'(x) = f(x)$, then for any constant C, the function $G(x) = F(x) + C$ also has derivative $G'(x) = F'(x) = f(x)$. Hence, if we can find one solution $F(x)$, then each of the functions $G(x) = F(x) + C$ is also a solution. This leads us to the question: Are there any other, different types of solutions? Fortunately, the answer to this question is no. That is, if $F'(x) = f(x)$ and $G'(x) = f(x)$ on an interval, then $F'(x) = G'(x)$ on the interval. Theorem 4 of Section 3.8 then implies there is a constant C such that $G(x) = F(x) + C$ on the interval. The following theorem summarizes these results.

THEOREM 1

Suppose $F'(x) = f(x)$ on an interval I.

(a) If $G(x) = F(x) + C$ for some constant C, then $G'(x) = f(x)$ on I.
(b) If $G'(x) = f(x)$ on I, then there is some constant C such that
$G(x) = F(x) + C$ on I.

Notation

We have seen that the solution to the problem, if there is a solution, consists of a collection of functions G such that $G'(x) = f(x)$ on an interval. This collection of functions is denoted by the *symbol*

$$\int f(x)\,dx.$$

This symbol is called the **indefinite integral** of f. The function f is called the **integrand** of the indefinite integral. At this time $\int$ and dx are merely parts of a symbol and have no individual meaning.

A function F such that $F'(x) = f(x)$ on an interval is called an **antiderivative** of f. If F is a known antiderivative of f, the collection of functions G such that $G'(x) = f(x)$ on an interval can be described as $F(x) + C$. Thus, we can write

$$\int f(x)\,dx = F(x) + C.$$

A statement of this form is called an **integration formula**. Finding a particular antiderivative F of f is called evaluating the indefinite integral $\int f(x)\,dx$. The mathematical statement

$$\int f(x)\,dx = F(x) + C \text{ means } F'(x) = f(x) \text{ on an interval.}$$

Verifying Integration Formulas

An integration formula $\int f(x)\,dx = F(x) + C$ can be verified by showing that $F'(x) = f(x)$ on an interval.

EXAMPLE 1

Verify the following integration formulas by differentiation.

(a) $\int x\sqrt{x^2 + 1}\,dx = \dfrac{1}{3}(x^2 + 1)^{3/2} + C.$

(b) $\int \cos^2 x\,dx = \dfrac{\cos x \cdot \sin x}{2} + \dfrac{x}{2} + C.$

(c) $\int x \sin x\,dx = -x \cos x + \sin x + C.$

(d) $\int \sin x \cdot \cos x\,dx = \dfrac{1}{2}\sin^2 x + C.$

(e) $\int \sin x \cdot \cos x\,dx = -\dfrac{1}{2}\cos^2 x + C.$

SOLUTION   In each case, we must check that the derivative of the right-hand side equals the integrand. (The derivative of the constant C is zero.)

(a) $\dfrac{d}{dx}\left(\dfrac{1}{3}(x^2 + 1)^{3/2}\right) = \left(\dfrac{1}{3}\right)\left(\dfrac{3}{2}\right)(x^2 + 1)^{1/2}(2x) = x\sqrt{x^2 + 1}$, so

$$\int x\sqrt{x^2 + 1}\,dx = \dfrac{1}{3}(x^2 + 1)^{3/2} + C.$$

(b) $\dfrac{d}{dx}\left(\dfrac{\cos x \cdot \sin x}{2} + \dfrac{x}{2}\right) = \dfrac{1}{2}((\cos x)(\cos x) + (\sin x)(-\sin x) + 1)$

$$= \dfrac{1}{2}(\cos^2 x - \sin^2 x + 1)$$

$$= \dfrac{1}{2}(\cos^2 x - (1 - \cos^2 x) + 1)$$

$$= \cos^2 x, \text{ so}$$

$$\int \cos^2 x\,dx = \dfrac{\cos x \cdot \sin x}{2} + \dfrac{x}{2} + C.$$

(We used the fact that $\sin^2 x + \cos^2 x = 1$ to write the derivative in the desired form.)

(c) $\dfrac{d}{dx}(-x \cos x + \sin x) = -((x)(-\sin x) + (\cos x)(1)) + \cos x$

$$= x \sin x, \text{ so}$$

$$\int x \sin x\,dx = -x \cos x + \sin x + C.$$

(d) $\dfrac{d}{dx}\left(\dfrac{1}{2}\sin^2 x\right) = \dfrac{1}{2}(2)(\sin x)(\cos x)$

$$= \sin x \cdot \cos x, \text{ so}$$

$$\int \sin x \cdot \cos x\,dx = \dfrac{1}{2}\sin^2 x + C.$$

(e) $\dfrac{d}{dx}\left(-\dfrac{1}{2}\cos^2 x\right) = -\dfrac{1}{2}(2)(\cos x)(-\sin x)$

$$= \sin x \cdot \cos x, \text{ so}$$

$$\int \sin x \cdot \cos x \, dx = -\dfrac{1}{2}\cos^2 x + C. \quad \blacksquare$$

Note that

$$\dfrac{1}{2}\sin^2 x - \left(-\dfrac{1}{2}\cos^2 x\right) = \dfrac{1}{2}(\sin^2 x + \cos^2 x) = \dfrac{1}{2}.$$

That is, the two different antiderivatives of $\sin x \cdot \cos x$ that were verified in Examples 1(d) and 1(e) differ by a constant, as Theorem 1 says they must.

EXAMPLE 2

Find all k for which $\int \sin 2x \, dx = k \cos 2x + C$.

SOLUTION   The expression $\int \sin 2x \, dx = k \cos 2x + C$ means

$$\dfrac{d}{dx}(k \cos 2x) = \sin 2x.$$

Using the Chain Rule to carry out the differentiation, we obtain the equation

$$k(-\sin 2x)(2) = \sin 2x,$$
$$-2k = 1,$$
$$k = -\dfrac{1}{2}.$$

The integration formula is true for $k = -1/2$. $\quad \blacksquare$

Note that

$$\int F'(x)\,dx = F(x) + C. \tag{1}$$

Statement 1 is true because the derivative of $F(x)$ on the right is equal to $F'(x)$, the integrand of the indefinite integral.

Some Basic Integration Formulas

We have still not addressed the problem of how to evaluate an indefinite integral. Much effort will be given to this question in a later chapter. For now, we will see how to solve the problem for a few basic functions. The idea is to transform a known differentiation formula $F'(x) = f(x)$ into the corresponding integration formula $\int f(x)\,dx = F(x) + C$. For example, we know that if r is a rational number and $r \neq -1$, then

$$\dfrac{d}{dx}\left(\dfrac{x^{r+1}}{r+1}\right) = x^r, \text{ so } \int x^r \, dx = \dfrac{x^{r+1}}{r+1} + C.$$

We also know that

$$\dfrac{d}{dx}(\sin x) = \cos x, \text{ so } \int \cos x \, dx = \sin x + C.$$

The integration formulas that correspond to the formulas for the derivatives of power functions and the basic trigonometric functions are listed below:

$$\int x^r \, dx = \frac{x^{r+1}}{r+1} + C, \quad r \text{ rational}, r \neq -1,$$

$$\int \cos x \, dx = \sin x + C,$$

$$\int \sin x \, dx = -\cos x + C,$$

$$\int \sec^2 x \, dx = \tan x + C,$$

$$\int \csc^2 x \, dx = -\cot x + C,$$

$$\int \sec x \cdot \tan x \, dx = \sec x + C,$$

$$\int \csc x \cdot \cot x \, dx = -\csc x + C.$$

These formulas will occur often enough in our work that it is worthwhile to memorize them, especially the first three. You may also find it helpful to verify these formulas by differentiation each time they are used. You should be very familiar with the corresponding differentiation formulas, so the differentiation can be carried out without any additional writing. Be especially careful with the signs in the trigonometric formulas.

EXAMPLE 3

Using the first formula above we have

(a) $\displaystyle\int x^2 \, dx = \frac{x^3}{3} + C,$

(b) $\displaystyle\int \sqrt{x} \, dx = \int x^{1/2} \, dx = \frac{x^{(1/2)+1}}{(1/2)+1} + C = \frac{2}{3} x^{3/2} + C,$

(c) $\displaystyle\int \frac{1}{\sqrt{x}} \, dx = \int x^{-1/2} \, dx = \frac{x^{1/2}}{1/2} + C = 2x^{1/2} + C,$

(d) $\displaystyle\int 1 \, dx = \int x^0 \, dx = \frac{x^{0+1}}{0+1} + C = x + C.$ ■

Since $(aF + bG)' = aF' + bG'$, we have

$$\int (af(x) + bg(x)) \, dx = a \int f(x) \, dx + b \int g(x) \, dx. \tag{2}$$

This says that we can evaluate the indefinite integral of a sum by evaluating the indefinite integral of each term. Also, the indefinite integral of a constant multiple of a function is the constant times the indefinite integral of the function. A constant C should be added on the right-hand side when the indefinite integrals are evaluated.

EXAMPLE 4

Using (2) we have

(a) $\displaystyle\int (x^2 + 3x - 1)\,dx = \frac{x^3}{3} + 3\frac{x^2}{2} - x + C,$

(b) $\displaystyle\int (\cos x + \sin x)\,dx = \sin x - \cos x + C,$

(c) $\displaystyle\int (2\sec^2 x - \sec x \cdot \tan x)\,dx = 2\tan x - \sec x + C,$

(d) $\displaystyle\int x(6x - 4)\,dx = \int (6x^2 - 4x)\,dx = 6\frac{x^3}{3} - 4\frac{x^2}{2} + C.$

Note that it was necessary to carry out the multiplication before we could use (2) to evaluate the indefinite integral in (d).

$$\textbf{(e)} \quad \int \frac{x + 1}{x^{1/3}}\,dx = \int (x^{2/3} + x^{-1/3})\,dx$$
$$= \frac{x^{5/3}}{5/3} + \frac{x^{2/3}}{2/3} + C$$
$$= \frac{3}{5}x^{5/3} + \frac{3}{2}x^{2/3} + C.$$

It was necessary to carry out the division before we could use (2) to evaluate the indefinite integral in (e). ∎

Finding Particular Antiderivatives

If G is an antiderivative of f and

$$\int f(x)\,dx = F(x) + C,$$

then

$$G(x) = F(x) + C \text{ for some constant } C.$$

We can determine a particular antiderivative G by using additional information to evaluate the constant C.

EXAMPLE 5

Find a function G such that $G'(x) = \sin x$ and $G(0) = 0$.

SOLUTION   We know from (1) that G is an antiderivative of $G'(x) = \sin x$. We have

$$\int \sin x\,dx = -\cos x + C,$$

so $G(x) = -\cos x + C$, for some constant C. The constant C is evaluated by solving the equation $G(0) = 0$ for C. We have

$$G(0) = 0,$$
$$-\cos 0 + C = 0,$$
$$-1 + C = 0,$$
$$C = 1.$$

The desired function is $G(x) = -\cos x + 1$.

The graphs of some of the functions $-\cos x + C$ for different values of C are sketched in Figure 5.1.1. The graph of only one of these functions passes through the point $(0, 0)$. That is the one we found. ∎

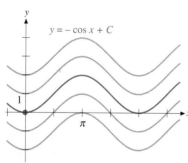

$y = -\cos x + C$

FIGURE 5.1.1   The graph of $-\cos x + C$ passes through the point $(0, 0)$ when $C = 1$.

Rectilinear Motion

Let us see how the ideas of an indefinite integral can be used in pr
involve motion along a number line. If the position of an object alon
line at time t is given by $s(t)$, we know the velocity is $v(t) = s'(t)$ and the
acceleration is $a(t) = v'(t)$; that is, $s(t)$ is an antiderivative of $v(t)$, so

$$s(t) \text{ is given by } \int v(t)\,dt.$$

Since $v(t)$ is an antiderivative of $a(t)$,

$$v(t) \text{ is given by } \int a(t)\,dt.$$

If we know the formula for the acceleration $a(t)$ and the velocity at some time
t_0, we can obtain the formula for the velocity $v(t)$ by using the method of
Example 5. Similarly, we can obtain the formula for the position $s(t)$ if we
know the formula for the velocity $v(t)$ and the position at some time.

Acceleration due to the gravitational attraction of the earth is essentially
constant at points near the surface of the earth. Depending on the units of length
used, we will use either

$$g = 32 \text{ ft/s}^2 \text{ or } g = 9.8 \text{ m/s}^2$$

for the magnitude of this acceleration. For problems that involve vertical motion
near the surface of the earth, it is generally convenient to choose a vertical
coordinate line with positive direction up and origin at ground level. If other
forces such as air resistance are neglected, we then have

$$a(t) = -g.$$

The negative value of the acceleration indicates that the acceleration is downward, in the negative direction. We then have

$$\int a(t)\,dt = \int -g\,dt = -gt + C, \text{ so}$$
$$v(t) = -gt + C.$$

It is easy to see that $C = v(0)$. The value $v(0)$ is customarily denoted v_0 and is
called the **initial velocity**. Thus,

$$v(t) = -gt + v_0.$$

We then have

$$\int v(t)\,dt = \int (-gt + v_0)\,dt = -\frac{1}{2}gt^2 + v_0 t + C, \text{ so}$$
$$s(t) = -\frac{1}{2}gt^2 + v_0 t + C.$$

The value of the above constant C is $s(0)$. The value $s(0)$ is denoted s_0 and
called the **initial position**. We then have the following formula for the height
above ground level of an object that is thrown vertically from an initial height
s_0 with initial velocity v_0.

$$s(t) = -\frac{1}{2}gt^2 + v_0 t + s_0. \tag{3}$$

(Although this formula is easily derived from the formula $a(t) = -g$, some students may just memorize the result. In any case, we are concerned with how to interpret and use the formula.)

EXAMPLE 6

A ball is thrown vertically upward from a height of 48 ft above ground level at a speed of 32 ft/s. **(a)** How high above ground will it get? **(b)** How long after it is thrown will it hit the ground? **(c)** At what speed will it hit the ground? **(d)** At what speed will a ball hit the ground if it is thrown downward from a height of 48 ft at 32 ft/s?

SOLUTION

(a) We choose $t = 0$ to be the time the ball is thrown. We are told that the initial position is $s_0 = 48$ and the initial velocity is $v_0 = +32$. (The initial velocity is upward, so v_0 is positive.) Since the unit of length is feet, we use the value $g = 32$ ft/s^2. Formula 3 then gives

$$s(t) = -16t^2 + 32t + 48.$$

The velocity is

$$v(t) = s'(t) = -32t + 32.$$

The formulas for $s(t)$ and $v(t)$ are valid only from the time the ball is thrown ($t = 0$) until it returns to ground level.

The highest above ground that the ball reaches is the maximum value of the function $s(t)$. This occurs when $s'(t) = v(t) = 0$. (We know that the ball will go upward [positive velocity] to its highest point and then start falling [negative velocity]. At its highest point the velocity changes from positive to negative, and is zero.) Solving the equation $v(t) = 0$ for t, we have

$$v(t) = 0,$$
$$-32t + 32 = 0,$$
$$-32t = -32,$$
$$t = 1.$$

When $t = 1$, the height is

$$s(1) = -16(1)^2 + 32(1) + 48 = 64.$$

The ball reaches a maximum height of 64 ft above ground.

(b) The ball is at ground level when $s(t) = 0$. Using the formula for $s(t)$ from (a), we can solve this equation for t. We have

$$s(t) = 0,$$
$$-16t^2 + 32t + 48 = 0,$$
$$-16(t^2 - 2t - 3) = 0,$$
$$-16(t - 3)(t + 1) = 0.$$

The solutions are $t = 3$ and $t = -1$. The ball will hit the ground 3 s after it is thrown.

(c) The speed is the absolute value of the velocity. Using the formula from (a), we see that when $t = 3$, the velocity is

$$v(3) = -32(3) + 32 = -64.$$

The ball hits the ground at a speed of 64 ft/s. The negative velocity tells us the ball is falling when it hits the ground. See Figure 5.1.2.

FIGURE 5.1.2 The height in feet of a ball t seconds after it has been thrown vertically upward from 48 ft above ground level at speed 32 ft/s is given by $s(t) = -16t^2 + 32t + 48$.

(d) If a ball is thrown downward from a height of 48 ft at 32 ft/s, it has initial position $s_0 = 48$ and initial velocity $v_0 = -32$. (The initial velocity is negative, because the motion of the ball is downward, in the negative direction of our coordinate line.) The position at time t is then

$$s(t) = -16t^2 - 32t + 48.$$

The velocity is

$$v(t) = s'(t) = -32t - 32.$$

The ball will hit the ground when $s(t) = 0$. Solving this equation for t, we have

$$\begin{aligned} s(t) &= 0, \\ -16t^2 - 32t + 48 &= 0, \\ -16(t^2 + 2t - 3) &= 0, \\ -16(t + 3)(t - 1) &= 0, \\ t = 1, \quad t &= -3. \end{aligned}$$

The ball hits the ground 1 s after it is thrown. When $t = 1$, the velocity is

$$v(1) = -32(1) - 32 = -64.$$

It hits the ground at 64 ft/s. See Figure 5.1.3. ∎

FIGURE 5.1.3 The height in feet of a ball t seconds after it has been thrown vertically downward from 48 ft above ground level at speed 32 ft/s is given by $s(t) = -16t^2 - 32t + 48$.

Comparing the answers in Example 6c and 6d, we see that a ball that is thrown directly upward from a height of 48 ft at 32 ft/s hits the ground with the same speed as a ball that is thrown directly downward from the same height at the same speed.

EXERCISES 5.1

Verify the integration formulas in Exercises 1–8 by differentiation.

1. $\displaystyle\int \frac{x}{\sqrt{x^2 + 4}}\,dx = \sqrt{x^2 + 4} + C$

2. $\displaystyle\int x\sqrt{x^2 + 9}\,dx = \frac{1}{3}(x^2 + 9)^{3/2} + C$

3. $\displaystyle\int \frac{1}{\sqrt{x}(1 + \sqrt{x})^2}\,dx = -\frac{2}{1 + \sqrt{x}} + C$

4. $\displaystyle\int \frac{(1 + x^{1/3})^3}{x^{2/3}}\,dx = \frac{3}{4}(1 + x^{1/3})^4 + C$

5. $\displaystyle\int x \sin x\,dx = \sin x - x \cos x + C$

6. $\displaystyle\int x \cos x\,dx = \cos x + x \sin x + C$

7. $\displaystyle\int \cos^3 x\,dx = -\frac{1}{3}\sin^3 x + \sin x + C$

8. $\displaystyle\int \sin^3 x\,dx = \frac{1}{3}\cos^3 x - \cos x + C$

Find all k for which the integration formulas in Exercises 9–20 are valid.

9. $\displaystyle\int \sqrt{2x + 3}\,dx = k(2x + 3)^{3/2} + C$

10. $\displaystyle\int \sqrt{1 - x}\,dx = k(1 - x)^{3/2} + C$

11. $\displaystyle\int (3x - 1)^{-1/3}\,dx = k(3x - 1)^{2/3} + C$

12. $\displaystyle\int x\sqrt{x^2 + 1}\,dx = k(x^2 + 1)^{3/2} + C$

13. $\displaystyle\int x^2(3x^3 + 1)^3\,dx = k(3x^3 + 1)^4 + C$

14. $\displaystyle\int \frac{x^2}{(x^3 + 1)^3}\,dx = \frac{k}{(x^3 + 1)^2} + C$

15. $\displaystyle\int \cos(x/2)\,dx = k \sin(x/2) + C$

16. $\displaystyle\int \sec(3x)\tan(3x)\,dx = k \sec(3x) + C$

17. $\displaystyle\int \cos 3x\,dx = k \sin 3x + C$

18. $\int \cos \frac{x}{3} dx = k \sin \frac{x}{3} + C$

19. $\int \cos(\pi - x) dx = k \sin(\pi - x) + C$

20. $\int \sin\left(\frac{\pi}{4} - x\right) dx = k \cos\left(\frac{\pi}{4} - x\right) + C$

Evaluate the indefinite integrals in Exercises 21–40.

21. $\int (x^2 - 2x + 3) dx$

22. $\int (3x^3 + 2x + 1) dx$

23. $\int (x^{1/2} + x^{-1/2}) dx$

24. $\int (x^{1/3} + x^{-1/3}) dx$

25. $\int (\sin x + \cos x) dx$

26. $\int (\sec^2 x + \sec x \cdot \tan x) dx$

27. $\int \csc x \cot x \, dx$

28. $\int \csc^2 x \, dx$

29. $\int (x + 1)(x - 2) dx$

30. $\int x(x^2 + 2) dx$

31. $\int (2x - 1)(x + 3) dx$

32. $\int x(x - 2)(x + 2) dx$

33. $\int \frac{x^2 + 1}{x^2} dx$

34. $\int \frac{x - 1}{\sqrt{x}} dx$

35. $\int \frac{\tan x + 1}{\sec x} dx$

36. $\int \frac{1 + \sin x}{\cos^2 x} dx$

37. $\int (\sec^2 x + 1) dx$

38. $\int (\tan^2 x + 1) dx$

39. $\int \cos(x - c) dx$

40. $\int \sin(c - x) dx$

In Exercises 41–50, find a function $G(x)$ that satisfies the given conditions.

41. $G'(x) = 6x^2 + 2x - 3, G(0) = 5$

42. $G'(x) = 3x^2 + 1, G(0) = -2$

43. $G'(x) = \sqrt{x}, G(1) = 2$

44. $G'(x) = 1/\sqrt{x}, G(4) = 0$

45. $G'(x) = \sin x, G(0) = 2$

46. $G'(x) = \cos x, G(\pi) = -1$

47. $G''(x) = 12x, G'(0) = 2, G(0) = 3$

48. $G''(x) = 6x + 4, G'(0) = -1, G(0) = 4$

49. $G''(x) = \sin x + \cos x, G'(0) = 1, G(0) = 1$

50. $G''(x) = \sin x + \cos x, G'(0) = 0, G(0) = 0$

51. A ball is thrown vertically upward from ground level at a speed of 32 ft/s. **(a)** How high above ground will it get? **(b)** How long will it take the ball to return to ground level? **(c)** What will the velocity of the ball be when it hits the ground?

52. A ball is thrown vertically upward from ground level at a speed of 64 ft/s. **(a)** How high above ground will it get? **(b)** How long will it take the ball to return to ground level? **(c)** What will the velocity of the ball be when it hits the ground?

53. A rock is thrown directly downward at a speed of 48 ft/s from the top of a building 64 ft high. **(a)** How long will it take to hit the ground? **(b)** At what speed will it hit the ground?

54. A rock is thrown directly downward at a speed of 64 ft/s from the top of a building 80 ft high. **(a)** How long will it take to hit the ground? **(b)** At what speed will it hit the ground?

55. A hot air balloon is rising vertically at a rate of 0.7 m/s. A sandbag is released from the balloon when it is at a height of 42 m. **(a)** How long will it take the sandbag to hit the ground? **(b)** At what speed will it hit the ground?

56. A rocket is rising vertically at a rate of 4.9 m/s at a height of 147 m when it releases a spent fuel tank. **(a)** How long will it take the tank to hit the ground? **(b)** At what speed will the tank hit the ground?

57. A gull flying parallel to the surface of the water drops a clam, which takes 3 seconds to hit the water. How high was the gull at the instant it dropped the clam? (**Calculus Explorer & Tutor II, Introduction to Integration, 9.**)

58. A rocket is accelerating from a position of rest at a rate of $6t$ m/s². How far will it travel in the interval of time from **(a)** $t = 0$ to $t = 2$? **(b)** $t = 2$ to $t = 4$?

59. A racer starts from rest and accelerates at a constant rate of 4 m/s². How long will it take to travel 400 m?

60. Show that $v^2 = v_0^2 - \frac{2k}{R}\left(1 - \frac{R}{s}\right)$ implies the acceleration is $a = -\frac{k}{s^2}$.

61. The **deflection** of a uniformly loaded beam that is fixed at one end satisfies

$$\frac{EI}{\omega} \frac{d^2 y}{dx^2} = \frac{1}{2}(\ell - x)^2, \quad y(0) = y'(0) = 0,$$

where E, I, ω, and ℓ are constants. See figure. **(a)** Find a formula for $y(x)$. **(b)** What is the maximum deflection of the beam?

5.2   RIEMANN SUMS

Connections

Intuitive ideas of area, volume, and velocity.

In this section we will introduce Riemann sums and see how they relate to problems that involve area, volume, and position along a line.

Definition and Notation

In simple terms, a Riemann sum is a sum of products. Each term is the product of the length of an interval and the value of a function at a point in the interval. We describe Riemann sums in precise mathematical terms to establish the notation and provide a common frame of reference for the examples that follow. Do not expect to understand Riemann sums by reading the mathematical description. It is the set of *examples* that gives us an understanding of the idea and how it is used.

DEFINITION

A **Riemann sum** of a function f over an interval $[a, b]$ is any sum of the form

$$f(x_1^*)\Delta x_1 + f(x_2^*)\Delta x_2 + \cdots + f(x_n^*)\Delta x_n, \tag{1}$$

where the points $x_0, x_1, \ldots, x_n$ with

$$a = x_0 < x_1 < \cdots < x_n = b$$

divide the interval $[a, b]$ into n subintervals,

$$[x_0, x_1], \quad [x_1, x_2], \quad \ldots, \quad [x_{n-1}, x_n],$$

the length of the subinterval $[x_{j-1}, x_j]$ is

$$\Delta x_j = x_j - x_{j-1},$$

and x_j^* is any point in the interval $[x_{j-1}, x_j]$, so

$$x_{j-1} \le x_j^* \le x_j, \quad j = 1, \ldots, n.$$

Points $x_0, x_1, \ldots, x_n$ with $a = x_0 < x_1 < \cdots < x_n = b$ are said to form a **partition** of $[a, b]$. The largest of the lengths Δx_j is denoted max Δx_j and called the **norm** of the partition.

Although it is not required by the definition, we will always divide the interval $[a, b]$ into n subintervals of equal length. Their common length is then

$$\Delta x_j = \frac{b - a}{n}, \quad j = 1, \ldots, n,$$

and the points of subdivision are given by the formula

$$x_j = a + (j)\left(\frac{b - a}{n}\right), \quad j = 0, 1, 2, \ldots, n. \tag{2}$$

A Riemann sum is a *number* that is a sum of products. Although we will calculate the value of some Riemann sums, calculation is only of secondary interest. Our primary goal is to see how Riemann sums are related to some physical problems. We will see that it is the *interpretation* of the products $f(x_j^*)\Delta x_j$ that leads to the many applications of the concept. This is analogous to the derivative, where it is the interpretation of the difference quotient that leads to the applications.

 We will use a calculator to evaluate Riemann sums.

Application of Riemann Sums to Area Under a Curve

Suppose the function f is nonnegative and continuous on the interval $a \le x \le b$. The region bounded by

$$y = f(x), \quad y = 0, \quad x = a, \quad \text{and } x = b$$

is called the **region under the curve** $y = f(x), a \le x \le b$. See Figure 5.2.1a. To find an approximate value of the area of the region, let us slice the region into n thin vertical strips of width Δx_j, $j = 1, \dots, n$, as indicated in Figure 5.2.1b. Since f is continuous, the values of f do not change very much as x varies over the width of a thin strip, so the area of each vertical strip is approximately the area of a rectangle with length $f(x_j^*)$ and width Δx_j, as indicated in Figure 5.2.1c. Also, the error in approximation of the area in each strip becomes *proportionately* smaller than the area in the strip as the width of the strip becomes smaller. This means that the total area of the rectangles becomes closer to the exact area of the region as the approximating rectangles become thinner. See Figure 5.2.1d. That is, if all Δx_j are small enough,

Area of region $\approx$ Sum of areas of rectangles.

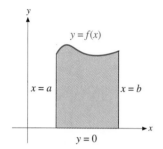

FIGURE 5.2.1a   If f is nonnegative and continuous on the interval $a \le x \le b$, the region bounded by $y = f(x), y = 0, x = a$, and $x = b$ is called the region under the curve.

FIGURE 5.2.1b   The region under the curve is sliced into n thin vertical strips of width $\Delta x_j, j = 1, \dots, n$.

FIGURE 5.2.1c   The area of the region inside each strip is approximately the area of a rectangle with length $f(x_j^*)$ and width Δx_j.

FIGURE 5.2.1d   The total area of the rectangles becomes closer to the exact area of the region as the approximating rectangles become thinner.

We evaluate the sum of the rectangles to obtain the following.

If f is nonnegative and continuous on the interval $a \le x \le b$, the region under the curve $y = f(x), a \le x \le b$ has

$$\text{Area} \approx f(x_1^*)\Delta x_1 + f(x_2^*)\Delta x_2 + \cdots + f(x_n^*)\Delta x_n.$$

The sum on the right is a Riemann sum of the function f. The product $f(x_j^*)\Delta x_j$ is interpreted as the area of a rectangle with length $f(x_j^*)$ and width Δx_j. The Riemann sum is simply the sum of the areas of these rectangles.

Note that it doesn't make sense to interpret the product $f(x_j^*)\Delta x_n$ as the area of a rectangle unless the values of f are nonnegative. We can form a Riemann sum of a function that has negative values, but we will not interpret such a sum as an area. Later in this section we will see an interpretation of Riemann sums that makes perfectly good sense for functions with positive and negative values.

EXAMPLE 1

(a) Use Riemann sums with $n = 8$ subintervals of equal length and $x_j^* = x_j$, the right endpoint of the subinterval $[x_{j-1}, x_j]$, $j = 1, \ldots, n$, to find an approximate value of the area under the curve $y = \sqrt{4 - x^2}$, $-2 \le x \le 2$.

(b) Repeat (a), except use $n = 16$ subintervals.

SOLUTION   The region is sketched in Figure 5.2.2.

(a) We want to evaluate the Riemann sum of $f(x) = \sqrt{4 - x^2}$ over the interval $[-2, 2]$, using $n = 8$ subintervals of equal length

$$\Delta x_j = \frac{b - a}{n} = \frac{2 - (-2)}{8} = 0.5,$$

and $x_j^* = x_j$, the right endpoint of each subinterval.

The points $x_0 = -2$, $x_1 = -1.5$, $x_2 = -1$, $x_3 = -0.5$, $x_4 = 0$, $x_5 = 0.5$, $x_6 = 1$, $x_7 = 1.5$, and $x_8 = 2$ divide the interval $[-2, 2]$ into eight subintervals of equal length. The desired Riemann sum is

$$\begin{aligned}
R_8 &= f(x_1)\Delta x_1 + f(x_2)\Delta x_2 + f(x_3)\Delta x_3 + f(x_4)\Delta x_4 \\
&\quad + f(x_5)\Delta x_5 + f(x_6)\Delta x_6 + f(x_7)\Delta x_7 + f(x_8)\Delta x_8 \\
&= \sqrt{4 - (-1.5)^2}(0.5) + \sqrt{4 - (-1)^2}(0.5) + \sqrt{4 - (-0.5)^2}(0.5) \\
&\quad + \sqrt{4 - (0)^2}(0.5) + \sqrt{4 - (0.5)^2}(0.5) + \sqrt{4 - (1)^2}(0.5) \\
&\quad + \sqrt{4 - (1.5)^2}(0.5) + \sqrt{4 - (2)^2}(0.5) \\
&\approx 5.9914181.
\end{aligned}$$

R_8 is the sum of the areas of the rectangles pictured in Figure 5.2.3a. There are eight rectangles if we count the "rectangle" on the right that has height and area zero.

(b) The points $x_0 = -2$, $x_1 = -1.75$, $x_2 = -1.50$, $x_3 = -1.25$, $\ldots$, $x_{15} = 1.75$, and $x_{16} = 2$ divide the interval $[-2, 2]$ into sixteen subintervals of equal length

$$\Delta x_j = \frac{b - a}{n} = \frac{2 - (-2)}{16} = 0.25.$$

The desired Riemann sum is then

$$\begin{aligned}
R_{16} &= f(x_1)\Delta x_1 + f(x_2)\Delta x_2 + \cdots + f(x_{16})\Delta x_{16} \\
&= \sqrt{4 - (-1.75)^2}(0.25) + \sqrt{4 - (-1.50)^2}(0.25) \\
&\quad + \sqrt{4 - (-1.25)^2}(0.25) + \sqrt{4 - (-1.00)^2}(0.25) \\
&\quad + \sqrt{4 - (-0.75)^2}(0.25) + \sqrt{4 - (-0.50)^2}(0.25) \\
&\quad + \sqrt{4 - (-0.25)^2}(0.25) + \sqrt{4 - (0)^2}(0.25) + \sqrt{4 - (0.25)^2}(0.25) \\
&\quad + \sqrt{4 - (0.50)^2}(0.25) + \sqrt{4 - (0.75)^2}(0.25) + \sqrt{4 - (1.00)^2}(0.25) \\
&\quad + \sqrt{4 - (1.25)^2}(0.25) + \sqrt{4 - (1.50)^2}(0.25) + \sqrt{4 - (1.75)^2}(0.25) \\
&\quad + \sqrt{4 - (2)^2}(0.25) \approx 6.1796383.
\end{aligned}$$

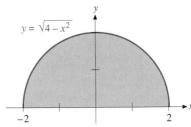

$y = \sqrt{4 - x^2}$

FIGURE 5.2.2   The graph of $y = \sqrt{4 - x^2}$ is the upper half of the circle $x^2 + y^2 = 4$.

FIGURE 5.2.3a R_8 is the sum of the areas of the eight rectangles, including the "rectangle" on the right that has height and area zero.

FIGURE 5.2.3b   The area under the curve is more accurately approximated when smaller subdivisions are used.

FIGURE 5.2.4 x_j^* is chosen to be the midpoint of the interval $[x_{j-1}, x_j]$.

FIGURE 5.2.5a x_j^* is chosen so that $f(x_j^*) = \dfrac{f(x_{j-1}) + f(x_j)}{2}$.

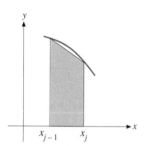

FIGURE 5.2.5b $f(x_j^*)\Delta x_j = \dfrac{f(x_{j-1}) + f(x_j)}{2}\Delta x_j$ is the area of the trapezoid.

R_{16} is the sum of the areas of the rectangles pictured in Figure 5.2.3b.

The region under the curve in this example is half of a circle of radius 2. The exact area of the region is $\dfrac{1}{2}\pi(2)^2 = 2\pi \approx 6.2831853$. We see the Riemann sum $R_{16} \approx 6.1796383$ is closer to the actual value than is $R_8 \approx 5.9914181$. ∎

Let us note that we are using Riemann sums with subintervals of equal length and evaluating the function at the right endpoint of the interval, so $x_j^* = x_j$, as a matter of convenience at this time. We could also approximate the area under the graph of a nonnegative function by using Riemann sums with a different choice of x_j^*. For example, we could choose $x_j^* = \dfrac{x_{j-1} + x_j}{2}$, the *midpoint* of the interval $[x_{j-1}, x_j]$. See Figure 5.2.4. We could also choose x_j^* so that $f(x_j^*) = \dfrac{f(x_{j-1}) + f(x_j)}{2}$, the *average* of the values of f at the endpoints of the interval $[x_{j-1}, x_j]$. See Figure 5.2.5a. The product $f(x_j^*)\Delta x_j = \dfrac{f(x_{j-1}) + f(x_j)}{2}\Delta x_j$ is then the *area of the trapezoid* with vertices at $(x_{j-1}, 0)$, $(x_{j-1}, f(x_{j-1}))$, $(x_j, f(x_j))$, and $(x_j, 0)$. See Figure 5.2.5b and note that we can determine the area of the trapezoid without determining the value of x_j^*. (If f is continuous, the Mean Value Theorem guarantees the existence of a point x_j^* such that $f(x_j^*) = \dfrac{f(x_{j-1}) + f(x_j)}{2}$.)

We could also approximate the area under a curve by using the graph to determine values of $f(x_j^*)$ for which the area of the rectangle with length $f(x_j^*)$ and width Δx_j is approximately the area under the curve between x_{j-1} and x_j. The value of the corresponding Riemann sum is then the sum of the areas of the approximating rectangles. For example, if we choose the values $f(x_j^*)$ as indicated in Figure 5.2.6, the corresponding Riemann sum has value

$$(0.9)(0.5) + (1.5)(0.5) + (1.8)(0.5) + (2.0)(0.5) + (2.0)(0.5)$$
$$+ (1.8)(0.5) + (1.5)(0.5) + (0.9)(0.5) = 6.2,$$

which is closer to the actual area than is R_{16}. Note that we need only to guess appropriate values $f(x_j^*)$, it is not necessary to determine values of the points x_j^*.

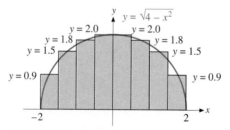

FIGURE 5.2.6 We can approximate the area under a curve by using the graph to determine values of $f(x_j^*)$ for which the area of the rectangle with length $f(x_j^*)$ and width Δx_j is approximately the area under the curve between x_{j-1} and x_j.

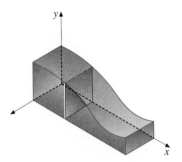

FIGURE 5.2.7a Cross-sections perpendicular to the x-axis have area $A(x)$.

FIGURE 5.2.7b To approximate the volume of the solid, we divide it into n slices of thickness $\Delta x_j, j = 1, \ldots, n$.

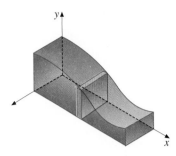

FIGURE 5.2.7c The volume of a slice is approximately $A(x_j^*)\Delta x_j$.

Application of Riemann Sums to Volume

Let us consider a solid with the property that variable cross-sections perpendicular to the x-axis have area $A(x)$, where A is a continuous function. Such a solid is illustrated in Figure 5.2.7a. To approximate the volume of the solid, we divide it into n slices of thickness Δx_j, $j = 1, \ldots, n$, by cutting perpendicular to the x-axis, as indicated in Figure 5.2.7b. If a slice has edges perpendicular to the faces, then cross-sections perpendicular to the faces have constant area, and we know

Volume of slice = (Area of cross section)(Thickness).

This is illustrated in Figure 5.2.7c. In case the area of the cross-sections of the slice is not constant, we can approximate the volume of the slice by the product

$$A(x_j^*)\Delta x_j,$$

where $A(x_j^*)$ is the area of the cross-section at some point within the slice. If A is continuous and the slice is thin enough, the error in the approximate value will be proportionately smaller than the volume of the slice. This means the sum of approximate volumes of the slices will be near the actual volume of the solid. That is, if all Δx_j are small enough,

Volume of solid $\approx$ sum of approximate volumes of slices,

Volume $\approx A(x_1^*)\Delta x_1 + A(x_2^*)\Delta x_2 + \cdots + A(x_n^*)\Delta x_n.$

The sum is a Riemann sum of the cross-sectional area function $A(x)$. The product $A(x_j^*)\Delta x_j$ is interpreted as the volume of a slice that has area of cross-section $A(x_j^*)$ and thickness Δx_j. The Riemann sum is the sum of the volumes of n slices.

EXAMPLE 2

Use Riemann sums with $n = 8$ subintervals of equal length and $x_j^* = x_j$, $j = 1, \ldots, n$, to find an approximate value of the volume of the prism pictured in Figure 5.2.8. Cross-sections perpendicular to the x-axis are isosceles right triangles with area

$$A(x) = \frac{1}{2}\left(\frac{4-x}{2}\right)^2.$$

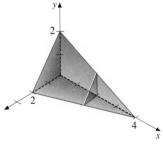

FIGURE 5.2.8 Cross-sections are right triangles with area $A(x) = \frac{1}{2}\left(\frac{4-x}{2}\right)^2$.

SOLUTION   The points $x_0 = 0$, $x_1 = 0.5$, $x_2 = 1.0$, $x_3 = 1.5$, $x_4 = 2.0$, $x_5 = 2.5$, $x_6 = 3.0$, $x_7 = 3.5$, and $x_8 = 4.0$ divide the interval $[0, 4]$ into eight subintervals of equal length

$$\Delta x_j = \frac{b-a}{n} = \frac{4-0}{8} = 0.5.$$

The desired Riemann sum is

$$
\begin{aligned}
R_8 = {} & A(0.5)\Delta x_1 + A(1.0)\Delta x_2 + A(1.5)\Delta x_3 + A(2.0)\Delta x_4 \\
& + A(2.5)\Delta x_5 + A(3.0)\Delta x_6 + A(3.5)\Delta x_7 + A(4.0)\Delta x_8 \\
= {} & \frac{1}{2}\left(\frac{4-(0.5)}{2}\right)^2 (0.5) + \frac{1}{2}\left(\frac{4-(1.0)}{2}\right)^2 (0.5) \\
& + \frac{1}{2}\left(\frac{4-(1.5)}{2}\right)^2 (0.5) + \frac{1}{2}\left(\frac{4-(2.0)}{2}\right)^2 (0.5) \\
& + \frac{1}{2}\left(\frac{4-(2.5)}{2}\right)^2 (0.5) + \frac{1}{2}\left(\frac{4-(3.0)}{2}\right)^2 (0.5) \\
& + \frac{1}{2}\left(\frac{4-(3.5)}{2}\right)^2 (0.5) + \frac{1}{2}\left(\frac{4-(4.0)}{2}\right)^2 (0.5) \\
= {} & 2.1875.
\end{aligned}
$$

This Riemann sum is the volume of the eight smaller prisms pictured in Figure 5.2.9, where we have counted the "prism" on the far right that has zero volume. ■

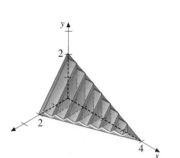

FIGURE 5.2.9   The volume of the prism is approximated by the sum of the volumes of the smaller prisms.

Application of Riemann Sums to Velocity and Position Along a Line

If we travel for 2 hours at a rate that varies between 50 and 60 mph, it is clear that we will cover between 100 and 120 miles. That is,

Distance $\geq$ (Minimum velocity)(Length of time) $= (50)(2) = 100$

and

Distance $\leq$ (Maximum velocity)(Length of time) $= (60)(2) = 120.$

We could improve the accuracy of our estimate of the distance traveled by breaking the 2 hours into smaller intervals of time over which the speed does not vary so much and then adding the results. This idea leads to a Riemann sum.

Suppose the position of an object along a number line at time t is given by $s(t)$. Also, assume that the function s is differentiable, so the (instantaneous) velocity of the object is the derivative $s'(t)$. Let us see how the difference in position of the object at times $t = a$ and $t = b$ is given by a Riemann sum of s' over the time interval $[a, b]$. That is, choose points

$$a = t_0 < t_1 < \cdots < t_{n-1} < t_n = b,$$

so the interval $[a, b]$ is partitioned into n subintervals. Then the difference in position can be written

$$
\begin{aligned}
s(b) - s(a) = {} & s(t_n) - s(t_0) \\
= {} & [s(t_n) - s(t_{n-1})] + [s(t_{n-1}) - s(t_{n-2})] \\
& + \cdots + [s(t_2) - s(t_1)] + [s(t_1) - s(t_0)].
\end{aligned}
$$

Applying the Mean Value Theorem to each of the above terms, we have

$$s(b) - s(a) = s'(t_n^*) \cdot (t_n - t_{n-1}) + s'(t_{n-1}^*) \cdot (t_{n-1} - t_{n-2})$$
$$+ \cdots + s'(t_2^*) \cdot (t_2 - t_1) + s'(t_1^*) \cdot (t_1 - t_0). \qquad (3)$$

The sum in (3) is a Riemann sum of s' over the interval $[a, b]$. (The terms are written in reverse order, but that doesn't matter.) In this application the terms $s'(t_j^*)\Delta t_j$ are interpreted as the product of velocity and a length of time. This product gives the distance and the direction traveled by an object that travels with average velocity $s'(t_j^*)$ over a length of time $\Delta t_j = t_j - t_{j-1}$. The sign of the velocity gives the direction traveled; s' may have positive and negative values. The algebraic sum of the terms gives the **net displacement** $s(b) - s(a)$.

We may not know the values of the t_j^*'s that give the average velocities over the intervals $[t_{j-1}, t_j]$, $j = 1, \ldots, n$. However, if each t_j^* in (3) is replaced by the t that gives the maximum value of $s'(t)$ in the interval $[t_{j-1}, t_j]$, the corresponding Riemann sum will be greater than or equal to $s(b) - s(a)$. Similarly, if each t_j^* is replaced by a value that gives the minimum of $s'(t)$ in the interval $[t_{j-1}, t_j]$, we obtain a sum that is less than or equal to $s(b) - s(a)$. We can obtain approximate values of $s(b) - s(a)$ by evaluating these two Riemann sums.

EXAMPLE 3

The position of an object along a number line at time t is given by $s(t)$. The formula for $s(t)$ is not known, but it is known that $s(0) = 2$ and $s'(t) = 0.75 - t$. Use Riemann sums of s' over the interval $[0, 1]$ with $n = 10$ subintervals of equal length to find upper and lower bounds of $s(1)$, the position at $t = 1$.

SOLUTION   The points $t_0 = 0$, $t_1 = 0.1$, $t_2 = 0.2, \ldots$, and $t_{10} = 1.0$ clearly divide $[0, 1]$ into ten subintervals of equal length 0.1.

Since the function $s'(t) = 0.75 - t$ is decreasing, the maximum occurs at the left endpoint of each subinterval t_{j-1}, and the minimum occurs at the right endpoint t_j. See Figure 5.2.10a. That is,

$$s'(t_j) \leq s'(t_j^*) \leq s'(t_{j-1}), \quad j = 1, \ldots, 10.$$

Therefore, $s(1) - s(0)$ is between the corresponding Riemann sums of s'.

The Riemann sum of $s'(t) = 0.75 - t$ over the interval $[0, 1]$ with $n = 10$ subintervals of equal length and $x_j^* = x_j$, the right endpoint of each subinterval, is (see Figure 5.2.10b)

$$s'(t_1)\Delta t_1 + s'(t_2)\Delta t_2 + \cdots + s'(t_{10})\Delta t_{10}$$
$$= (0.75 - 0.1)(0.1) + (0.75 - 0.2)(0.1) + (0.75 - 0.3)(0.1)$$
$$+ (0.75 - 0.4)(0.1) + (0.75 - 0.5)(0.1) + (0.75 - 0.6)(0.1)$$
$$+ (0.75 - 0.7)(0.1) + (0.75 - 0.8)(0.1) + (0.75 - 0.9)(0.1)$$
$$+ (0.75 - 1.0)(0.1) = 0.2.$$

Using $x_j^* = x_{j-1}$, the left endpoint of each subinterval, we obtain (see Figure 5.2.10c)

$$s'(t_0)\Delta t_1 + s'(t_1)\Delta t_2 + \cdots + s'(t_9)\Delta t_{10}$$
$$= (0.75 - 0)(0.1) + (0.75 - 0.1)(0.1) + (0.75 - 0.2)(0.1)$$
$$+ (0.75 - 0.3)(0.1) + (0.75 - 0.4)(0.1) + (0.75 - 0.5)(0.1)$$
$$+ (0.75 - 0.6)(0.1) + (0.75 - 0.7)(0.1) + (0.75 - 0.8)(0.1)$$
$$+ (0.75 - 0.9)(0.1) = 0.3.$$

FIGURE 5.2.10a The velocity is decreasing, so the maximum velocity in each subinterval of time occurs at the left endpoint of the subinterval, t_{j-1}, and the minimum velocity in the subinterval occurs at the right endpoint, t_j.

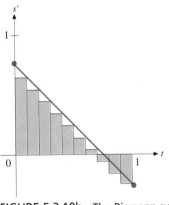

FIGURE 5.2.10b The Riemann sum with x_j^* chosen to give the minimum value of the velocity in each interval has value less than $s(b) - s(a)$. The product $s'(t_j^*)\Delta t_j$ is interpreted as the product of velocity and a length of time. The velocity may be positive or negative.

FIGURE 5.2.10c The Riemann sum with x_j^* chosen to give the maximum value of the velocity in each interval has value greater than $s(b) - s(a)$.

We conclude that

$$0.2 \leq s(1) - s(0) \leq 0.3,$$
$$0.2 \leq s(1) - 2 \leq 0.3,$$
$$2.2 \leq s(1) \leq 2.3. \quad \blacksquare$$

We can use the methods of Section 5.1 to determine that a function $s(t)$ with $s'(t) = 0.75 - t$ and $s(0) = 2$ must be $s(t) = 2 + 0.75t - 0.5t^2$. It follows that $s(1) = 2.25$. This is consistent with the range for $s(1)$ that we found in Example 4. The connection between the two methods is fundamental to the study of calculus. We will discuss this connection in Section 5.4.

EXERCISES 5.2

In Exercises 1–8, use Riemann sums with the given number n of subintervals of equal length and $x_j^* = x_j$, the right endpoint of each subinterval, to find approximate values of the area under the given curves.

1. $y = x, 0 \leq x \leq 1; n = 10$

2. $y = x^2, 0 \leq x \leq 2; n = 8$

3. $y = \sin x, 0 \leq x \leq \pi; n = 6$

4. $y = \cos x, 0 \leq x \leq \pi/2; n = 6$

5. $y = 1/x, 2 \leq x \leq 4; n = 4$

6. $y = 1/(x^2 + 1), -1 \leq x \leq 1; n = 4$

7. $y = x, -1 \leq x \leq 1; n = 5$

8. $y = x^2, 0 \leq x \leq 2; n = 4$

The shape of cross-sections perpendicular to the x-axis is indicated for each of the solids pictured in Exercises 9–16. Use Riemann sums with $n = 8$ subintervals of equal length and $x_j^* = x_j, j = 1, \ldots, 8$, to find approximate values of the volumes of the solids.

9. Circles, $A(x) = \pi/4, 0 \leq x \leq 2$

10. Circles, $A(x) = \pi/4, 0 \leq x \leq 2$

11. Circles, $A(x) = \pi \left(1 - \dfrac{x}{2}\right)^2, 0 \leq x \leq 2$

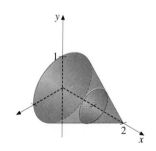

12. Circles, $A(x) = \pi x^2/4, 0 \le x \le 2$

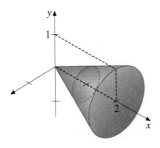

13. Isosceles right triangles, $A(x) = x^2/8, 0 \le x \le 2$

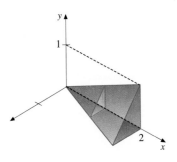

14. Rectangles, $A(x) = x/2, 0 \le x \le 2$

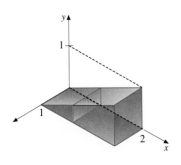

15. Circles (solid is a hemisphere), $A(x) = \pi(1 - x^2)$, $0 \le x \le 1$

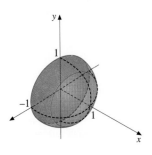

16. Squares, $A(x) = x^2/4, 0 \le x \le 2$

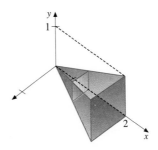

In Exercises 17–24, the initial position $s(0)$, a terminal time T, and the instantaneous velocity $s'(t)$ of an object on a number line are given. Use Riemann sums of s' over the interval $[0, T]$ with the given number n of subintervals of equal length to find upper and lower bounds of $s(T)$.

17. $s'(t) = 2t, s(0) = 0, T = 1, n = 4$

18. $s'(t) = 2t, s(0) = 0, T = 2, n = 8$

19. $s'(t) = t^2, s(0) = 4, T = 2, n = 8$

20. $s'(t) = t^2, s(0) = 4, T = 1, n = 4$

21. $s'(t) = 32 - 32t, s(0) = 0, T = 1, n = 4$

22. $s'(t) = 32 - 32t, s(0) = 0, T = 2, n = 8$

23. $s'(t) = \sqrt{1 + t^3}, s(0) = 1, T = 2, n = 8$

24. $s'(t) = \sqrt{1 + t^3}, s(0) = 1, T = 1, n = 4$

In Exercises 25–26, use the given graphs to determine where R_n, the Riemann sum with n subintervals of equal length and $x_j^* = x_j$, is greater than or less than the area under the curves.

25.

26.

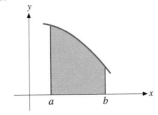

In Exercise 27–28, use the given graphs to determine whether R_n, the Riemann sum with n subintervals of equal length and $x_j^* = x_j$, is greater than or less than the volume of the solid pictured.

27.

28.

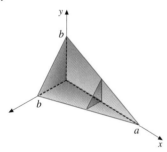

In Exercises 29–32, use Riemann sums with the given number n of subintervals of equal length and $x_j^* = \dfrac{x_{j-1} + x_j}{2}$, the midpoint of each interval, to find the approximate values of the areas under the curves.

29. $y = x^2, 0 \le x \le 1; n = 4$

30. $y = \sin x, 0 \le x \le \pi; n = 3$

31. $y = 1/x, 1 \le x \le 3; n = 4$

32. $y = 1/(1 + x^2), 0 \le x \le 1; n = 2$

In Exercises 33–36, use Riemann sums with the given number n of subintervals of equal length and $f(x_j^*) = \dfrac{f(x_{j-1}) + f(x_j)}{2}$ to find the approximate values of the areas under the given curves. (Hint: You do not need to determine x_j^*. The product $f(x_j^*)\Delta x_j = \dfrac{f(x_{j-1}) + f(x_j)}{2}\Delta x_j$ is the area of the trapezoid with vertices $(x_{j-1}, 0)$, $(x_{j-1}, f(x_{j-1}))$, $(x_j, f(x_j))$, and $(x_j, 0)$.)

33. $y = 2x, 0 \le x \le 2; n = 4$

34. $y = 3x^2, 0 \le x \le 2; n = 4$

35. $y = \sin x, 0 \le x \le \pi; n = 4$

36. $y = 1/x, 1 \le x \le 3; n = 4$

In Exercises 37–40, use Riemann sums with four subintervals of equal length and $f(x_j^*)$ chosen from the graph so that $f(x_j^*)\Delta x_j$ appears to be approximately the area under the curve between x_{j-1} and x_j to approximate the area under the curves.

37.

38.

39.

40.

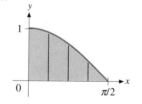

41. Use the Riemann sum with four subintervals of equal length and $x_j^* = x_{j-1}$, the left endpoint of each subinterval, to find an approximate value of the area under the curve $y = 1 + x^2, 0 \le x \le 1$. (**Calculus Explorer & Tutor II, Introduction to Integration, 10.**)

5.3 LIMITS OF RIEMANN SUMS; THE DEFINITE INTEGRAL

Connections

Bounds of functions, 2.2.

Limits at infinity, 4.2.

We will investigate the behavior of some Riemann sums that involve a large number of subintervals. To do this it is convenient to use $\sum$ (sigma) notation.

Sigma Notation

The symbol $\displaystyle\sum_{j=1}^{n} a_j$ is defined by the equation

$$\sum_{j=1}^{n} a_j = a_1 + a_2 + \cdots + a_n.$$

Similarly, $\displaystyle\sum_{j=m}^{n} a_j = a_m + a_{m+1} + \cdots + a_n$ for $m \leq n$.

The values of the terms a_j are often given by a formula. That is, a_j is given as a function of j, $a_j = a(j)$. We can then evaluate a sum $\displaystyle\sum_{j=m}^{n} a_j$ by substitution.

EXAMPLE 1

Evaluate

(a) $\displaystyle\sum_{j=1}^{4} 2j$,

(b) $\displaystyle\sum_{j=1}^{5} (2 - j)$,

(c) $\displaystyle\sum_{j=1}^{6} 4$,

(d) $\displaystyle\sum_{j=2}^{5} j^2$.

SOLUTION

(a) The terms of the sum $\displaystyle\sum_{j=1}^{4} 2j$ are given by the formula $a_j = 2j$, $1 \leq j \leq 4$. Hence,

$$\sum_{j=1}^{4} 2j = 2(1) + 2(2) + 2(3) + 2(4) = 20.$$

(b) The terms of the sum $\displaystyle\sum_{j=1}^{5} (2 - j)$ are given by the formula $a_j = 2 - j$, $1 \leq j \leq 5$, so

$$\sum_{j=1}^{5} (2 - j) = (2 - 1) + (2 - 2) + (2 - 3) + (2 - 4) + (2 - 5) = -5.$$

(c) The terms of the sum $\displaystyle\sum_{j=1}^{6} 4$ are given by the formula $a_j = 4, 1 \le j \le 6$, so the values of a_j are given by a constant function. Then

$$\sum_{j=1}^{6} 4 = 4 + 4 + 4 + 4 + 4 + 4 = 6(4) = 24.$$

(d) The terms of the sum $\displaystyle\sum_{j=2}^{5} j^2$ are given by the formula $a_j = j^2, 2 \le j \le 5$, so

$$\sum_{j=2}^{5} j^2 = 2^2 + 3^2 + 4^2 + 5^2 = 54. \quad \blacksquare$$

We will use the following property of sums.

$$\sum_{j=1}^{n} (ca_j + db_j) = c\left(\sum_{j=1}^{n} a_j\right) + d\left(\sum_{j=1}^{n} b_j\right). \tag{1}$$

Proof Equation 1 is verified easily by using familiar properties of arithmetic. We have

$$\sum_{j=1}^{n} (ca_j + db_j) = (ca_1 + db_1) + (ca_2 + db_2) + \cdots + (ca_n + db_n)$$

$$= (ca_1 + ca_2 + \cdots + ca_n) + (db_1 + db_2 + \cdots + db_n)$$

$$= c(a_1 + a_2 + \cdots + a_n) + d(b_1 + b_2 + \cdots + b_n)$$

$$= c\left(\sum_{j=1}^{n} a_j\right) + d\left(\sum_{j=1}^{n} b_j\right). \quad \blacksquare$$

The following formulas can be verified by Mathematical Induction:

$$\sum_{j=1}^{n} c = c + c + \cdots + c = nc, \tag{2}$$

$$\sum_{j=1}^{n} j = 1 + 2 + \cdots + n = \frac{n(n+1)}{2}, \tag{3}$$

$$\sum_{j=1}^{n} j^2 = 1^2 + 2^2 + \cdots + n^2 = \frac{n(n+1)(2n+1)}{6}, \tag{4}$$

$$\sum_{j=1}^{n} j^3 = 1^3 + 2^3 + \cdots + n^3 = \left(\frac{n(n+1)}{2}\right)^2. \tag{5}$$

EXAMPLE 2

Evaluate $\displaystyle\sum_{j=1}^{50}(j^2 - 5j + 3)$.

SOLUTION   Using Formula 1 and then Formulas 2, 3, and 4 with $n = 50$, we have

$$\sum_{j=1}^{50}(j^2 - 5j + 3) = \left(\sum_{j=1}^{50}j^2\right) - 5\left(\sum_{j=1}^{50}j\right) + \left(\sum_{j=1}^{50}3\right)$$

$$= \frac{50(50 + 1)(2(50) + 1)}{6} - 5\left(\frac{50(50 + 1)}{2}\right) + 50(3)$$

$$= 36,700. \quad\blacksquare$$

Limits of Riemann Sums

We can use Formulas 1–5 to investigate the behavior of some Riemann sums that involve a large number of subintervals.

EXAMPLE 3

(a) Find a simple formula for the Riemann sum of $f(x) = 2x$ over the interval $[0, 1]$, using n subintervals of equal length and $x_j^* = x_j$, $j = 1, \ldots, n$.

(b) Find the limit as n approaches infinity of the Riemann sum found in (a).

SOLUTION

(a) From Formula 2 of Section 5.2 we have

$$x_j = a + (j)\left(\frac{b - a}{n}\right) = 0 + (j)\left(\frac{1 - 0}{n}\right) = \frac{j}{n}, \quad j = 1, \ldots, n.$$

Also,

$$\Delta x_j = \frac{b - a}{n} = \frac{1}{n}, \quad j = 1, \ldots, n.$$

Using $\sum$ notation we can express the desired Riemann sum as

$$R_n = \sum_{j=1}^{n} f(x_j)\Delta x_j = \sum_{j=1}^{n} f\left(\frac{j}{n}\right) \cdot \left(\frac{1}{n}\right) = \sum_{j=1}^{n}\left(2\frac{j}{n}\right)\left(\frac{1}{n}\right).$$

Using Equation 1 and Formula 3 we have

$$R_n = \frac{2}{n^2}\left(\sum_{j=1}^{n} j\right) = \frac{2}{n^2}\left(\frac{n(n + 1)}{2}\right) = 1 + \frac{1}{n}.$$

(b) The limit of R_n as n approaches infinity is evaluated by using the techniques of Section 4.2 that were used to evaluate limits as a real variable x approaches infinity. We have

$$\lim_{n\to\infty} R_n = \lim_{n\to\infty}\left(1 + \frac{1}{n}\right) = 1. \quad\blacksquare$$

EXAMPLE 4

(a) Find a simple formula for the Riemann sum of $f(x) = 1 - x^2$ over the interval $[0, 1]$, using n subintervals of equal length and $x_j^* = x_j$, $j = 1, \ldots, n$.

(b) Find the limit as n approaches infinity of the Riemann sum found in (a).

SOLUTION

(a) As in Example 3 we have $x_j = j/n$ and $\Delta x_j = 1/n$, $j = 1, \ldots, n$. The desired Riemann sum is

$$R_n = \sum_{j=1}^{n} f(x_j)\Delta x_j = \sum_{j=1}^{n} f\left(\frac{j}{n}\right) \cdot \left(\frac{1}{n}\right) = \sum_{j=1}^{n} \left(1 - \left(\frac{j}{n}\right)^2\right)\left(\frac{1}{n}\right).$$

Using Equation 1 and Formulas 2 and 4, we have

$$R_n = \frac{1}{n}\left(\sum_{j=1}^{n} 1\right) - \frac{1}{n^3}\left(\sum_{j=1}^{n} j^2\right)$$

$$= \frac{1}{n}(n) - \frac{1}{n^3}\left(\frac{n(n+1)(2n+1)}{6}\right)$$

$$= 1 - \frac{2n^2 + 3n + 1}{6n^2}$$

$$= 1 - \frac{1}{3} - \frac{1}{2n} - \frac{1}{6n^2}$$

$$= \frac{2}{3} - \frac{1}{2n} - \frac{1}{6n^2}.$$

(b) $\displaystyle\lim_{\pi \to \infty} R_n = \lim_{n \to \infty}\left(\frac{2}{3} - \frac{1}{2n} - \frac{1}{6n^2}\right) = \frac{2}{3}.$ ∎

EXAMPLE 5

(a) Find a simple formula for the Riemann sum of $f(x) = x^3$ over the interval $[0, 2]$, using n subintervals of equal length and $x_j^* = x_j$, $j = 1, \ldots, n$.

(b) Find the limit as n approaches infinity of the Riemann sum found in (a).

SOLUTION

(a) From Formula 2 of Section 5.2 we have

$$x_j = a + (j)\left(\frac{b-a}{n}\right) = 0 + (j)\left(\frac{2-0}{n}\right) = \frac{2j}{n}, \quad j = 1, \ldots, n.$$

Also,

$$\Delta x_j = \frac{b-a}{n} = \frac{2}{n}, \quad j = 1, \ldots, n.$$

The desired Riemann sum is

$$R_n = \sum_{j=1}^{n} f(x_j)\Delta x_j = \sum_{j=1}^{n} f\left(\frac{2j}{n}\right) \cdot \left(\frac{2}{n}\right) = \sum_{j=1}^{n} \left(\frac{2j}{n}\right)^3 \left(\frac{2}{n}\right)$$

$$= \frac{16}{n^4}\left(\sum_{j=1}^{n} j^3\right).$$

Using Formula 5, we have

$$R_n = \frac{16}{n^4}\left(\frac{n(n+1)}{2}\right)^2 = \frac{16n^2 + 32n + 16}{4n^2} = 4 + \frac{8}{n} + \frac{4}{n^2}.$$

(b) $\displaystyle\lim_{n\to\infty} R_n = \lim_{n\to\infty}\left(4 + \frac{8}{n} + \frac{4}{n^2}\right) = 4.$ ■

The Definite Integral

Examples 3, 4, and 5 illustrate the idea of a limit of Riemann sums. Let us express this concept in terms of a formal definition.

DEFINITION

A function f is said to be **integrable** over the interval $[a, b]$ if there is a number, denoted $\displaystyle\int_a^b f(x)\,dx$, such that, for each $\epsilon > 0$, there is a positive number δ such that

$$\left|\sum_{j=1}^n f(x_j^*)\Delta x_j - \int_a^b f(x)\,dx\right| \leq \epsilon$$

whenever $\displaystyle\sum_{j=1}^n f(x_j^*)\Delta x_j$ is any Riemann sum over $[a, b]$ that has partition norm max $\Delta x_j < \delta$. If it exists, $\displaystyle\int_a^b f(x)\,dx$ is called the **definite integral** of f from a to b. The function f is called the **integrand** and the numbers a and b are called **limits of integration.**

The value of a definite integral does not depend on the variable used. For example, $\displaystyle\int_a^b f(x)\,dx$ and $\displaystyle\int_a^b f(t)\,dt$ represent the same number. The variable of integration is sometimes called a **dummy variable.**

The number $\displaystyle\int_a^b f(x)\,dx$, if it exists, is a complicated form of a limit of Riemann sums. The notation is suggestive of the Riemann sum $\displaystyle\sum_{j=1}^n f(x_j^*)\Delta x_j$. That is, the symbol $\int$ corresponds to the summation sign $\sum$; $f(x)$ corresponds to $f(x_j^*)$; and dx corresponds to the length Δx_j. In applications of the definite integral, we will think of $\displaystyle\int_a^b f(x)\,dx$ as the "sum" of the "products" $f(x) \cdot dx$; $f(x)$ and dx will be interpreted as physical quantities, as though they were components of a Riemann sum.

If f is nonnegative for $a \leq x \leq b$, we have seen in Section 5.2 that the Riemann sums

$$f(x_1)\Delta x_1 + f(x_2)\Delta x_2 + \cdots + f(x_n)\Delta x_n$$

FIGURE 5.3.1   If f is nonnegative, $\int_a^b f(x)\,dx$ can be interpreted geometrically as the area under the curve $y = f(x)$, $a \le x \le b$.

give approximate values of the area under the curve. It should then seem reasonable that the definite integral $\int_a^b f(x)\,dx$ can be interpreted *geometrically* as the area under the curve. See Figure 5.3.1.

The condition of the definition requires that *all* Riemann sums of f over $[a, b]$ that have all subintervals of small enough length be near the number $\int_a^b f(x)\,dx$. As in the case of the limit of a function at a point, this implies there is at most one number that can satisfy the condition of the definition. Also, as in the case of limits of functions, we are not interested in verifying the condition of the definition. We are interested in using theory to evaluate definite integrals.

The following theorem guarantees the existence of the definite integral for an important class of functions.

THEOREM

If f is continuous on the closed interval $[a, b]$, then the definite integral of f from a to b exists.

We will not verify this theorem, although we note that the continuity of f implies that values $f(x_j^*)$ are nearly equal for all x_j^* in the interval $[x_{j-1}, x_j]$ if $\Delta x_j = x_j - x_{j-1}$ is small. This implies that the largest and smallest values of $f(x_j^*)$ for x_j^* in the interval $[x_{j-1}, x_j]$ are nearly the same, so there is little difference between the largest and smallest Riemann sums that correspond to a particular subdivision of $[a, b]$.

If we know $\int_a^b f(x)\,dx$ exists, then we can use the condition of the definition to our advantage. For example, we can then find an approximate value of $\int_a^b f(x)\,dx$ by using *any* convenient type of Riemann sum. We are not restricted to any particular choice of points of subdivision or of the points x_j^* in the interval $[x_{j-1}, x_j]$. The Riemann sum used will be a good approximation of the definite integral as long as $[a, b]$ is partitioned into small enough subintervals. In particular, if n is large enough, the Riemann sum R_n with n subintervals of equal length and $x_j^* = x_j$ must be near the number $\int_a^b f(x)\,dx$.

As n approaches infinity, the Riemann sums R_n must approach $\int_a^b f(x)\,dx$. That is, if f is continuous on the interval $[a, b]$, then we have

$$\int_a^b f(x)\,dx = \lim_{n \to \infty} \sum_{j=1}^n f(x_j)\Delta x_j, \tag{6}$$

where $x_j = a + (j)\left(\dfrac{b-a}{n}\right)$ and $\Delta x_j = \dfrac{b-a}{n}$.

Let us review the results of Example 3 in terms of Formula 6. In Example 3 we showed that the Riemann sum R_n of the function $f(x) = 2x$ over the interval $[0, 1]$ with n subintervals of equal length and $x_j^* = x_j$ satisfies $\lim_{n \to \infty} R_n = 1$.

y

2

$y = 2x$

0   1   x

FIGURE 5.3.2 $\int_0^1 2x\,dx = 1$ is the area of the triangular region under the curve $y = 2x$, $0 \le x \le 1$.

The function $f(x) = 2x$ is continuous on the interval $[0, 1]$, so Formula 6 then implies

$$\int_0^1 2x\,dx = 1.$$

(The value of 1 is the area under the curve $y = 2x$, $0 \le x \le 1$. See Figure 5.3.2.) Similarly, the results of Examples 4 and 5 tell us

$$\int_0^1 (1 - x^2)\,dx = \frac{2}{3} \text{ and } \int_0^2 x^3\,dx = 4, \text{ respectively.}$$

The method used in Examples 3–5 is not the most efficient way to evaluate a definite integral. These examples are intended to illustrate the idea that a *definite integral is a limit of Riemann sums*. We will learn better ways to evaluate definite integrals, but we must remember what a definite integral is. The idea of a definite integral as a "sum" is basic to understanding how it is used in applications.

Properties of Definite Integrals

Many properties of the definite integral follow from the corresponding properties of sums.

$$\int_a^b (cf(x) + dg(x))\,dx = c\int_a^b f(x)\,dx + d\int_a^b g(x)\,dx. \tag{7}$$

Proof We assume the three integrals exist. The value of each of them can then be approximated by using the *same subintervals* of $[a, b]$ and the *same points* x_j^*. This is true because the definition requires only that $[a, b]$ be partitioned into small enough subintervals. Using Equation 1 we then have

$$\int_a^b (cf(x) + dg(x))\,dx \approx \sum_{j=1}^n (cf(x_j^*) + dg(x_j^*))\Delta x_j$$

$$= c\left(\sum_{j=1}^n f(x_j^*)\Delta x_j\right) + d\left(\sum_{j=1}^n g(x_j^*)\Delta x_j\right)$$

$$\approx c \cdot \int_a^b f(x)\,dx + d \cdot \int_a^b g(x)\,dx.$$

Equation 7 follows from the fact that the errors in the above approximations can be made less than any prescribed positive number. ∎

The following formulas define $\int_a^b f(x)\,dx$ in case $b \le a$.

$$\int_a^b f(x)\,dx = -\int_b^a f(x)\,dx \text{ and } \int_a^a f(x)\,dx = 0. \tag{8}$$

The formulas in (8) may be considered to be properties of the definite integral. Another important property is

$$\int_a^b f(x)\,dx = \int_a^c f(x)\,dx + \int_c^b f(x)\,dx. \tag{9}$$

Proof We first assume $a < c < b$. We then choose a partition of $[a, b]$ that includes c as a point of subdivision. Let us say $c = x_m$. Separating the terms of an approximating Riemann sum into those involving subintervals to the left of c and those involving subintervals to the right of c, we have

$$\int_a^b f(x)\,dx \approx \sum_{j=1}^n f(x_j^*)\Delta x_j$$

$$= \sum_{j=1}^m f(x_j^*)\Delta x_j + \sum_{j=m+1}^n f(x_j^*)\Delta x_j$$

$$\approx \int_a^c f(x)\,dx + \int_c^b f(x)\,dx.$$

Other cases follow from the case $a < c < b$. For example, if $a < b < c$, we know from the above argument that

$$\int_a^c f(x)\,dx = \int_a^b f(x)\,dx + \int_b^c f(x)\,dx, \text{ so}$$

$$\int_a^b f(x)\,dx = \int_a^c f(x)\,dx - \int_b^c f(x)\,dx.$$

Property 8 then implies

$$\int_a^b f(x)\,dx = \int_a^c f(x)\,dx - \left[-\int_c^b f(x)\,dx\right]$$

$$= \int_a^c f(x)\,dx + \int_c^b f(x)\,dx. \ \blacksquare$$

FIGURE 5.3.3 The area under the curve $y = f(x)$ between $x = a$ and $x = b$ is equal to the sum of the area between $x = a$ and $x = c$ and the area between $x = c$ and $x = b$.

Property 9 is illustrated geometrically in Figure 5.3.3. The area under the curve $y = f(x)$ from $x = a$ to $x = b$ is equal to the sum of the area from $x = a$ to $x = c$ and the area from $x = c$ to $x = b$.

EXAMPLE 6

Given that $\int_0^7 f(x)\,dx = 3$, $\int_0^4 f(x)\,dx = 5$, and $\int_0^4 g(x)\,dx = 7$, use properties of the definite integral to evaluate

(a) $\int_4^7 f(x)\,dx$ and **(b)** $\int_0^4 (6f(x) - 4g(x))\,dx$.

SOLUTION

(a) Using Property 9, we have

$$\int_4^7 f(x)\,dx = \int_0^7 f(x)\,dx - \int_0^4 f(x)\,dx = 3 - 5 = -2.$$

(b) Property 7 implies

$$\int_0^4 (6f(x) - 4g(x))\,dx = 6\left(\int_0^4 f(x)\,dx\right) - 4\left(\int_0^4 g(x)\,dx\right)$$

$$= 6(5) - 4(7) = 2. \ \blacksquare$$

Finding Bounds of Definite Integrals

The following four results are useful for estimating values of definite integrals:

$m \leq f(x) \leq M$ for $a \leq x \leq b$ implies

$$m(b-a) \leq \int_a^b f(x)dx \leq M(b-a), \tag{10}$$

$f(x) \geq 0$ for $a \leq x \leq b$ implies $\int_a^b f(x)dx \geq 0,$ (11)

$f(x) \leq g(x)$ for $a \leq x \leq b$ implies $\int_a^b f(x)dx \leq \int_a^b g(x)dx,$ (12)

$$\left| \int_a^b f(x)dx \right| \leq \int_a^b |f(x)|dx. \tag{13}$$

Proof Let us verify the right-hand inequality in (10). Verification of the left-hand inequality is similar. We have

$$\int_a^b f(x)dx \approx \sum_{j=1}^n f(x_j^*)\Delta x_j$$

$$\leq \sum_{j=1}^n M\Delta x_j = M \sum_{j=1}^n \Delta x_j = M(b-a).$$

The above inequality is true because $f(x_j^*) \leq M$ for every choice of x_j^* and because $\Delta x_j > 0$. Also, we have used the fact that $\sum_{j=1}^n \Delta x_j = b - a$. That is, the sum of the lengths of the subintervals of $[a, b]$ is the length of $[a, b]$.

Property 11 follows from (10) with $m = 0$.

Property 12 follows from the fact that both integrals can be approximated by using the same partition, so

$$\int_a^b f(x)dx \approx \sum_{j=1}^n f(x_j^*)\Delta x_j$$

$$\leq \sum_{j=1}^n g(x_j^*)\Delta x_j \approx \int_a^b g(x)dx.$$

We should think of (13) as saying the absolute value of a sum is less than or equal to the sum of absolute values. Using the Triangle Inequality to verify (13), we have

$$\left| \int_a^b f(x)dx \right| \approx \left| \sum_{j=1}^n f(x_j^*)\Delta x_j \right| \leq \sum_{j=1}^n |f(x_j^*)|\Delta x_j \approx \int_a^b |f(x)|dx. \quad \blacksquare$$

EXAMPLE 7

Show that

$$\frac{36}{13} \leq \int_0^3 \frac{12}{x^2+4}dx \leq 9.$$

SOLUTION We need to find upper and lower bounds of the integrand for $0 \leq x \leq 3$. We have $x^2 \geq 0$ so

$$\frac{12}{x^2+4} \leq \frac{12}{0+4} = 3.$$

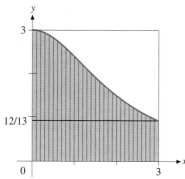

FIGURE 5.3.4 The area of the region under the curve $y = \dfrac{12}{x^2 + 4}$ between $x = 0$ and $x = 3$ is greater than the area of the rectangle contained in the region and less than the area of the rectangle that contains the region.

Also, $0 \le x \le 3$ implies $x^2 + 4 \le 3^2 + 4 = 13$, so

$$\frac{12}{x^2 + 4} \ge \frac{12}{13}.$$

It then follows from (10) that

$$\int_0^3 \frac{12}{x^2 + 4}\, dx \le 3(3 - 0) = 9$$

and

$$\int_0^3 \frac{12}{x^2 + 4}\, dx \ge \frac{12}{13}(3 - 0) = \frac{36}{13}. \quad \blacksquare$$

Example 7 is illustrated geometrically in Figure 5.3.4. The area under the curve $y = 12/(x^2 + 4)$ between $x = 0$ and $x = 3$ is greater than the area of the rectangle with base 3 and height $12/13$, and less than the area of the rectangle with base 3 and height 3.

Mean Value Theorem for Definite Integrals

The definite integral can be used to define the average value of a function $f(x)$, $a \le x \le b$. To do this, we divide the interval into n subintervals of equal length $\Delta x = (b - a)/n$. Let x_j^* denote any point in the jth subinterval, $j = 1, n$. The average value of the numbers $f(x_1^*), \ldots, f(x_n^*)$ is

$$\frac{f(x_1^*) + \cdots + f(x_n^*)}{n} = \frac{1}{b - a}(f(x_1^*) + \cdots + f(x_n^*))\Delta x.$$

If the interval is divided into small enough subintervals, the Riemann sum $(f(x_1^*) + \cdots + f(x_n^*))\Delta x$ is approximately $\displaystyle\int_a^b f(x)\,dx$. The following definition should then seem reasonable.

DEFINITION

The **average value of** f over the interval $a \le x \le b$ is

$$f_{avg} = \frac{1}{b - a}\int_a^b f(x)\,dx.$$

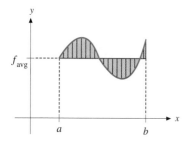

FIGURE 5.3.5 The area of the region above the line $y = f_{avg}$ and below $y = f(x)$ is equal to the area of the region below $y = f_{avg}$ and above $y = f(x)$.

Geometrically, the area of the region above the line $y = f_{avg}$ and below $y = f(x)$ is equal to the area of the region below $y = f_{avg}$ and above $y = f(x)$. See Figure 5.3.5.

We conclude this section with the following.

MEAN VALUE THEOREM FOR DEFINITE INTEGRALS

If f is continuous on $[a, b]$, there is a number z such that $a < z < b$ and

$$\int_a^b f(x)\,dx = f(z)(b - a).$$

Proof   Since f is continuous on an interval that contains both endpoints, we know from the Extreme Value Theorem of Section 3.8 that f has a minimum value $f(c')$ and maximum value $f(c'')$ on $[a, b]$. Then $f(c') \le f(x) \le f(c'')$ on $[a, b]$, so (10) implies

$$f(c')(b-a) \le \int_a^b f(x)\,dx \le f(c'')(b-a), \text{ or}$$

$$f(c') \le \frac{1}{b-a} \int_a^b f(x)\,dx \le f(c'').$$

Since

$$\frac{1}{b-a} \int_a^b f(x)\,dx$$

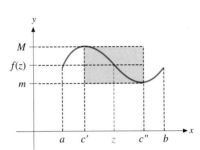

FIGURE 5.3.6   Since $\dfrac{1}{b-a}\displaystyle\int_a^b f(x)\,dx$ is between the values $f(c')$ and $f(c'')$, the Intermediate Value Theorem implies there is at least one number z between c' and c'' with $f(z) = \dfrac{1}{b-a}\displaystyle\int_a^b f(x)\,dx$.

is between the values $f(c')$ and $f(c'')$, the Intermediate Value Theorem of Section 3.7 tells us there is at least one number z that is between c' and c'' with

$$f(z) = \frac{1}{b-a} \int_a^b f(x)\,dx, \text{ so } \int_a^b f(x)\,dx = f(z)(b-a).$$

Since c' and c'' are in the interval $[a, b]$, so is z. See Figure 5.3.6.   ∎

The Mean Value Theorem for Integrals tells us that if f is continuous on an interval I, then there is a point z in I such that $f(z)$ is equal to the average value of f on I.

EXERCISES 5.3

Evaluate the sums in Exercises 1–10 either directly or by using Formulas 1–5.

1. $\displaystyle\sum_{j=1}^{4} 3j$

2. $\displaystyle\sum_{j=1}^{4} 2j$

3. $\displaystyle\sum_{j=1}^{4} (1 - 2j)$

4. $\displaystyle\sum_{j=1}^{4} (3 - 2j)$

5. $\displaystyle\sum_{j=3}^{6} (j^2 - j)$

6. $\displaystyle\sum_{j=3}^{6} (j^2 - 2j)$

7. $\displaystyle\sum_{j=1}^{12} (j^3 - 6j)$

8. $\displaystyle\sum_{j=1}^{10} (j^3 - j^2)$

9. $\displaystyle\sum_{j=1}^{100} (2j^2 + 4j - 3)$

10. $\displaystyle\sum_{j=1}^{120} (3j^2 - 4j - 5)$

In Exercises 11–16, (a) use the method of Examples 3–5 to find a simple formula for the Riemann sum corresponding to the given integrals, using n subintervals of equal length and $x_j^* = x_j$, $j = 1, \ldots, n$. (b) Evaluate the integrals by finding the limit as n approaches infinity of the Riemann sums found in (a).

11. $\displaystyle\int_0^3 (2x - 4)\,dx$

12. $\displaystyle\int_0^2 (4x - 2)\,dx$

13. $\displaystyle\int_0^4 (3x^2 - 8x - 4)\,dx$

14. $\displaystyle\int_0^3 (x^2 - 6x - 1)\,dx$

15. $\displaystyle\int_0^2 (6x^3 - 6x^2 - 1)\,dx$

16. $\displaystyle\int_0^4 (x^3 - 6x^2 + 5)\,dx$

Use the method of Examples 3–5 to evaluate the area under the curves given in Exercises 17–18.

17. $y = 4x + 6$, $0 \le x \le 3$ **18.** $y = 3x^2 + 2$, $0 \le x \le 2$

Use the method of Examples 3–5 to evaluate the average values of the functions given in Exercises 19–20 over the indicated intervals.

19. $f(x) = 6x - 4$, $0 \le x \le 4$

20. $f(x) = 1 - x^2$, $0 \le x \le 3$

Use the method of Examples 3–5 to verify the formulas in Exercises 21–24. You may assume that $b > 0$.

21. $\displaystyle\int_0^b c\,dx = cb$, c a constant

22. $\displaystyle\int_0^b x\,dx = \frac{b^2}{2}$

23. $\displaystyle\int_0^b x^2\,dx = \frac{b^3}{3}$

24. $\displaystyle\int_0^b x^3\,dx = \frac{b^4}{4}$

Use the results of Exercises 21–24 and properties of integrals to verify the formulas in Exercises 25–28. You may assume that $0 < a < b$.

25. $\displaystyle\int_a^b c\,dx = c(b-a)$, c a constant

26. $\displaystyle\int_a^b x\,dx = \frac{b^2}{2} - \frac{a^2}{2}$ **27.** $\displaystyle\int_a^b x^2\,dx = \frac{b^3}{3} - \frac{a^3}{3}$

28. $\displaystyle\int_a^b x^3\,dx = \frac{b^4}{4} - \frac{a^4}{4}$

Use the properties of the definite integral to evaluate the integrals in Exercises 29–36, given that $\displaystyle\int_5^7 f(x)\,dx = 3$, $\displaystyle\int_1^5 f(x)\,dx = 4$, $\displaystyle\int_1^7 g(x)\,dx = 5$, and $\displaystyle\int_5^7 g(x)\,dx = 6$.

29. $\displaystyle\int_5^1 f(x)\,dx$ **30.** $\displaystyle\int_7^5 g(x)\,dx$

31. $\displaystyle\int_1^7 f(x)\,dx$ **32.** $\displaystyle\int_1^5 g(x)\,dx$

33. $\displaystyle\int_5^7 [4f(x) + 3g(x)]\,dx$ **34.** $\displaystyle\int_5^7 [2f(x) - 4g(x)]\,dx$

35. $\displaystyle\int_1^5 [3f(x) - 5g(x)]\,dx$ **36.** $\displaystyle\int_1^5 [5f(x) - 2g(x)]\,dx$

37. Show that $0 \le \displaystyle\int_0^\pi \sin x\,dx \le \pi$.

38. Show that $\left| \displaystyle\int_0^{2\pi} f(x)\sin x\,dx \right| \le \displaystyle\int_0^{2\pi} |f(x)|\,dx$.

39. If $|f(x)| \le B$ for $a \le x \le b$, show that $\displaystyle\int_a^b [f(x)]^2\,dx \le B\displaystyle\int_a^b |f(x)|\,dx$.

40. Show that $\left| \displaystyle\int_0^1 x^k f(x)\,dx \right| \le \displaystyle\int_0^1 |f(x)|\,dx$ for all nonnegative integers k.

41. (a) Use Property (10) to find upper and lower bounds of
$$\int_0^{1/2} \frac{4}{1+x^2}\,dx \text{ and } \int_{1/2}^1 \frac{4}{1+x^2}\,dx.$$
(b) Add the results of (a) to show that
$$2.6 \le \int_0^1 \frac{4}{1+x^2}\,dx \le 3.6.$$

42. (a) Use Property (10) to find upper and lower bounds of
$$\int_{2^{j-1}}^{2^j} \frac{1}{x}\,dx.$$
(b) Add the results of (a) for $j = 1, \ldots, n$ to show that
$$\frac{n}{2} \le \int_1^{2^n} \frac{1}{x}\,dx \le n.$$

43. What is the average value of the function $f(x) = 2x - 1$ over the interval $[0, 2]$? (*Hint:* A sketch of the graph might help.)

44. If f is a linear function, at what point c in the interval $[a, b]$ is $f(c) = f_{avg}$?

45. What is the average value of an odd function over an interval $[-a, a]$?

46. If $m \le f(x) \le M$ for $a \le x \le b$, show that $m \le f_{avg} \le M$.

47. If f and g are continuous with $g > 0$ on the interval $I = [a, b]$, show that there is a point z in I such that
$$\int_a^b f(x)g(x)\,dx = f(z)\int_a^b g(x)\,dx.$$

48. (a) Use the formula
$$\sum_{j=1}^n \sin hj = \frac{\cos\left(\dfrac{h}{2}\right) - \cos\left(\left(n + \dfrac{1}{2}\right)h\right)}{2\sin\left(\dfrac{h}{2}\right)},$$
to evaluate the Riemann sum of $\sin x$ over the interval $[0, \pi/2]$ with n subintervals of equal length and $x_j^* = x_j$.
(b) Evaluate the limit of the Riemann sum in (a) as n approaches infinity to obtain the value of $\displaystyle\int_0^{\pi/2} \sin x\,dx$.

49. If f is unbounded on the interval $[a, b]$, then $\displaystyle\int_a^b f(x)\,dx$ does not exist as a limit of Riemann sums. For example, if
$$f(x) = \begin{cases} 1/x, & 0 < x \le 1, \\ 0, & x = 0, \end{cases}$$
show that for each number $B \ge 1$ we can choose $0 < x_1^* < 1/n$ so that the Riemann sum of f over $[0, 1]$ with n subintervals of equal length satisfies
$$\sum_{j=1}^n f(x_j^*)\Delta x_j > f(x_1^*) \cdot \frac{1}{n} > B.$$

5.4 THE FUNDAMENTAL THEOREM OF CALCULUS

Connections

Derivative at a point, 2.6.

Differentiation formulas, 3.1–4.

Indefinite integrals, 5.1.

Many (but not all) of the definite integrals we deal with can be evaluated very simply, without calculation of Riemann sums. The idea utilizes the fundamental connection between the concepts of the derivative and the integral.

The Fundamental Theorem

THE FUNDAMENTAL THEOREM OF CALCULUS

If f and F are continuous on the interval $a \leq x \leq b$ and $F'(x) = f(x)$ for $a < x < b$, then

$$\int_a^b f(x)\,dx = F(b) - F(a).$$

Proof   The proof of this result that we present emphasizes that the definite integral is a limit of Riemann sums. To begin, let us suppose the points $a = x_0 < x_1 < \cdots < x_n = b$ divide the interval $[a, b]$ into n subintervals. Then

$$
\begin{aligned}
F(b) - F(a) &= F(x_n) - F(x_0) \\
&= [F(x_n) - F(x_{n-1})] + [F(x_{n-1}) - F(x_{n-2})] \\
&\quad + \cdots + [F(x_1) - F(x_0)] \\
&= \sum_{j=1}^n [F(x_j) - F(x_{j-1})].
\end{aligned}
$$

Applying the Mean Value Theorem on each of the subintervals, we obtain $F(x_j) - F(x_{j-1}) = F'(x_j^*)(x_j - x_{j-1})$ for some x_j^* between x_j and x_{j-1}. Since $x_j - x_{j-1} = \Delta x_j$ and $F'(x_j^*) = f(x_j^*)$, $j = 1, \ldots, n$, we obtain

$$F(b) - F(a) = \sum_{j=1}^n f(x_j^*)\Delta x_j.$$

We then see that the number $F(b) - F(a)$ can be written as a Riemann sum of $\int_a^b f(x)\,dx$. But all such Riemann sums that have all subintervals small enough must be near $\int_a^b f(x)\,dx$. It follows that the number $F(b) - F(a)$ cannot be different from the number $\int_a^b f(x)\,dx$, so

$$\int_a^b f(x)\,dx = F(b) - F(a). \quad \blacksquare$$

EXAMPLE 1

(a) We know that $f(x) = 2x$ is the derivative of $F(x) = x^2$, so Theorem 1 tells us

$$\int_3^5 2x\,dx = \int_3^5 f(x)\,dx = F(5) - F(3) = 5^2 - 3^2 = 16.$$

(b) The function $f(x) = 2 \sin x \cos x$ is the derivative of $F(x) = \sin^2 x$, so Theorem 1 tells us

$$\int_0^{\pi/2} 2 \sin x \cos x \, dx = \int_0^{\pi/2} f(x) \, dx = F\left(\frac{\pi}{2}\right) - F(0)$$

$$= \sin^2 \frac{\pi}{2} - \sin^2 0 = (1)^2 - (0)^2 = 1. \quad \blacksquare$$

Using the Fundamental Theorem

To use the Fundamental Theorem to evaluate the definite integral $\int_a^b f(x) \, dx$, we need to know a function F with $F'(x) = f(x)$. The problem of finding a function F with $F' = f$ is the same as the problem of evaluating an indefinite integral, as was studied in Section 5.1. Recall that, for certain functions f, we were able to use known differentiation formulas to find functions F such that $F'(x) = f(x)$. Let us recall the basic integration formulas from Section 5.1.

$$\int x^r \, dx = \frac{x^{r+1}}{r+1}, r \text{ rational}, r \neq -1,$$

$$\int \cos x \, dx = \sin x + C,$$

$$\int \sin x \, dx = -\cos x + C,$$

$$\int \sec^2 x \, dx = \tan x + C,$$

$$\int \csc^2 x \, dx = -\cot x + C,$$

$$\int \sec x \cdot \tan x \, dx = \sec x + C,$$

$$\int \csc x \cdot \cot x \, dx = -\csc x + C,$$

$$\int (cf(x) + dg(x)) \, dx = c \int f(x) \, dx + d \int g(x) \, dx.$$

The methods we will use to find a function F such that $F'(x) = f(x)$ lead naturally to formulas that involve x. That is, we first find a formula for $F(x)$ and then use substitution to find the value of $F(b) - F(a)$. It is convenient to introduce the notation

$$F(x) \Big]_a^b = F(b) - F(a).$$

Recall from Section 5.1 that $\int f(x) \, dx = F(x) + C$ means $F'(x) = f(x)$. Hence, if we assume f and F are continuous on $[a, b]$, Theorem 1 tells us

$$\int f(x) \, dx = F(x) + C \text{ implies } \int_a^b f(x) \, dx = F(x) \Big]_a^b = F(b) - F(a).$$

We know that we may have $\int f(x) \, dx = F(x) + C$ and $\int f(x) \, dx = G(x) + C$ for different functions F and G. In this case we have $F'(x) =$

$f(x) = G'(x)$, so there is some constant C_0 such that $G(x) = F(x) + C_0$. Then

$$G(b) - G(a) = [F(b) + C_0] - [F(a) + C_0] = F(b) - F(a),$$

so we may use either F or G in the Fundamental Theorem. Generally, we use the most convenient function.

EXAMPLE 2
Evaluate the following definite integrals.

(a) $\int_1^4 x^{-1/2} dx$,

(b) $\int_0^4 (x^3 - 4x - 3) dx$,

(c) $\int_0^\pi \sin t \, dt$,

(d) $\int_{-\pi/4}^{\pi/4} \sec^2 \theta \, d\theta$.

SOLUTION   In each case we will use the basic integration formulas as in Section 5.1 to find a function F such that $\int_a^b f(x) dx = F(x)]_a^b$. Also, recall that the variable of integration is a dummy variable; we may use any real variable in place of x.

(a) $\int_1^4 x^{-1/2} dx = \dfrac{x^{1/2}}{1/2} \Big]_1^4 = 2x^{1/2} \Big]_1^4$

$$= 2(4)^{1/2} - 2(1)^{1/2} = 2(2) - 2(1) = 2.$$

(b) $\int_0^4 (x^3 - 4x - 3) dx = \dfrac{x^4}{4} - 4\dfrac{x^2}{2} - 3x \Big]_0^4$

$$= \left(\dfrac{(4)^4}{4} - 2(4)^2 - 3(4) \right) - \left(\dfrac{(0)^4}{4} - 2(0)^2 - 3(0) \right)$$

$$= 20.$$

(c) $\int_0^\pi \sin t \, dt = -\cos t \Big]_0^\pi = (-\cos \pi) - (-\cos 0) = -(-1) + 1 = 2.$

(d) $\int_{-\pi/4}^{\pi/4} \sec^2 \theta \, d\theta = \tan \theta \Big]_{-\pi/4}^{\pi/4} = \tan \dfrac{\pi}{4} - \tan \left(-\dfrac{\pi}{4} \right)$

$$= (1) - (-1) = 2. \quad \blacksquare$$

A word of caution concerning the Fundamental Theorem is in order. We must be sure that $F'(x) = f(x)$ for all x in the open interval (a, b) and that F and f are continuous on the closed interval $[a, b]$. For example, we know that

$$\int \dfrac{1}{x^2} dx = -\dfrac{1}{x} + C, \text{ because } \dfrac{d}{dx} \left(-\dfrac{1}{x} \right) = \dfrac{1}{x^2}, \quad x \neq 0.$$

Since $(-1/x)' \neq 1/x^2$ for $x = 0$, we cannot use the theorem to evaluate $\int_a^b (1/x^2) dx$ if the interval $[a, b]$ contains the point $x = 0$. In particular,

$$\int_{-1}^1 \dfrac{1}{x^2} dx \underset{(?)}{=} -\dfrac{1}{x} \Big]_{-1}^1 = \left(-\dfrac{1}{1} \right) - \left(-\dfrac{1}{-1} \right) = -2 \text{ is nonsense.}$$

The theorem does not apply. (The integral of the positive function $1/x^2$ over the interval $[-1, 1]$ would not be negative, if it made sense at all.) Even though $F(x) = -1/x$ and $f(x) = 1/x^2$ satisfy $F'(x) = f(x)$ on the interval $0 < x < 1$, we cannot use the Fundamental Theorem to evaluate

$$\int_0^1 \frac{1}{x^2}\,dx$$

because $f(x) = 1/x^2$ and $F(x) = -1/x$ are not continuous on the interval $[0, 1]$. (The fact that $f(0)$ and $F(0)$ are undefined can be corrected by assigning them any values. However, it is impossible to assign values at $x = 0$ that make the functions *continuous* at $x = 0$.)

Rectilinear Motion

Suppose the position of an object along a number line at time t is given by $s(t)$. If s' is continuous on the interval $a \leq t \leq b$, Theorem 1 implies

$$s(b) - s(a) = \int_a^b s'(t)\,dt.$$

This formula is consistent with the fact that the **net displacement** of the object over the time interval $[a, b]$ can be approximated by Riemann sums of the velocity. In Section 5.2 we used the formula

$$s(b) - s(a) \approx s'(t_1^*)\Delta t_1 + s'(t_2^*)\Delta t_2 + \cdots + s'(t_n^*)\Delta t_n.$$

EXAMPLE 3

The instantaneous velocity of an object on a number line is given by $s'(t) = 64 - 32t$, $0 \leq t \leq 3$. Find the net displacement of the object.

SOLUTION   The net displacement over the time interval from $t = 0$ to $t = 3$ is

$$s(3) - s(0) = \int_0^3 s'(t)\,dt = \int_0^3 (64 - 32t)\,dt$$

$$= 64t - 16t^2 \Big]_0^3$$

$$= (64(3) - 16(3)^2) - (64(0) - 16(0)^2)$$

$$= 48.$$

FIGURE 5.4.1   The object is traveling in the positive direction for $0 \leq t \leq 2$, when the velocity is positive. It is traveling in the negative direction for $2 \leq t \leq 3$, when the velocity is negative.

The net displacement is 48 units in the positive direction. See Figure 5.4.1. ■

The Second Fundamental Theorem

Let us now investigate another important relation between the concepts of differentiation and integration.

If f is continuous on the interval $[a, b]$, then the function

$$G(x) = \int_a^x f(t)\,dt$$

is defined for $a \leq x \leq b$. The following theorem tells us that $G(x)$ is differentiable for $a < x < b$, with derivative $f(x)$. The function $\int_a^x f(t)\,dt$ is called an **indefinite integral of** f, although it is actually a definite integral with a variable limit of integration. (The fact that we defined an indefinite integral to be something else in Section 5.1 should cause no difficulty.)

THE SECOND FUNDAMENTAL THEOREM OF CALCULUS

If f is continuous on the interval $[a, b]$ and $G(x) = \displaystyle\int_a^x f(t)\,dt$, then
$G'(x) = f(x), a < x < b.$

Proof   The proof uses the definition of the derivative and properties of the integral. We will show that $G'(c) = f(c)$, c any point in the interval (a, b).

For $x \neq c$, we have

$$\frac{G(x) - G(c)}{x - c} = \frac{1}{x - c}\left(\int_a^x f(t)\,dt - \int_a^c f(t)\,dt\right) = \frac{1}{x - c}\int_c^x f(t)\,dt.$$

The Mean Value Theorem for definite integrals that was given in Section 5.3 implies there is a number z between c and x such that

$$\int_c^x f(t)\,dt = f(z)(x - c).$$

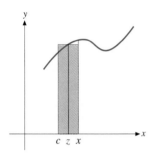

FIGURE 5.4.2   If f is continuous, then there is a number z between c and x such that $\displaystyle\int_c^x f(t)\,dt = f(z)(x - c).$

See Figure 5.4.2. Substitution then gives

$$\frac{G(x) - G(c)}{x - c} = \frac{1}{x - c}(f(z)(x - c)) = f(z).$$

Since z is between x and c, z must approach c as x approaches c. But f is continuous at c, so $f(z)$ must approach $f(c)$ as z approaches c. It follows that

$$G'(c) = \lim_{x \to c} \frac{G(x) - G(c)}{x - c} = f(c). \quad\blacksquare$$

If we have a function F such that $F'(x) = f(x)$ on $[a, b]$, we can use the Fundamental Theorem to express the function G of the Second Fundamental Theorem in terms of F. That is,

$$G(x) = \int_a^x f(t)\,dt = F(t)\Big]_a^x = F(x) - F(a).$$

It is then clear that $G'(x) = F'(x) = f(x)$. (The derivative of the constant $F(a)$ is zero.) The Second Fundamental Theorem tells us how to evaluate the derivative of $\displaystyle\int_a^x f(t)\,dt$ *without* finding a function F with $F'(x) = f(x)$. The derivative is simply the value of the integrand at x, which is $f(x)$.

Using the Second Fundamental Theorem

Let us restate the conclusion of the Second Fundamental Theorem in the form

$$\frac{d}{dx}\left(\int_a^x f(t)\,dt\right) = f(x). \tag{1}$$

Note that the upper limit of integration of the indefinite integral is identical to the variable of differentiation and the lower limit of integration is a constant. The value of the derivative of the indefinite integral is then given by the integrand, evaluated at the variable of differentiation.

EXAMPLE 4

From Formula 1 we have

(a) $\dfrac{d}{dx}\left(\displaystyle\int_{\pi}^{x}\sin t\,dt\right)=\sin x,$

(b) $\dfrac{d}{dx}\left(\displaystyle\int_{1}^{x}\dfrac{1}{t}\,dt\right)=\dfrac{1}{x},\,x>0,$

(c) $\dfrac{d}{dx}\left(\displaystyle\int_{0}^{x}\sin t^2\,dt\right)=\sin x^2.$ ∎

In Example 4a we can write $\displaystyle\int_{\pi}^{x}\sin t\,dt=-\cos x-1$, and we then differentiate to obtain the derivative $\sin x$. At this time we cannot evaluate the integral in Example 4b because we do not know a function F with $F'(x)=1/x$. Such a function will be introduced in Chapter 7, so we will be able to evaluate the integral in Example 4b. There is no combination of elementary functions that has derivative $\sin x^2$, so we must use the Second Fundamental Theorem in Example 4c.

EXAMPLE 5

Evaluate

$$\frac{d}{dx}\left(\int_{x}^{0}\frac{1}{4+t^2}\,dt\right).$$

SOLUTION   To use Formula 1, the indefinite integral must have the variable of differentiation as the upper limit of integration and a constant as the lower limit of integration. The given integral can be written in this form by interchanging the limits of integration. Since interchanging the limits of integration changes the sign of the integral, we have

$$\frac{d}{dx}\left(\int_{x}^{0}\frac{1}{4+t^2}\,dt\right)=\frac{d}{dx}\left(-\int_{0}^{x}\frac{1}{4+t^2}\,dt\right)=-\frac{1}{4+x^2}.\quad∎$$

If the upper limit of integration is a function of the variable of differentiation, we can use the Chain Rule to find the derivative. That is, if

$$y=\int_{a}^{u}f(t)\,dt \text{ and } u=u(x),$$

then y is a function of x and the Chain Rule gives

$$\frac{dy}{dx}=\frac{dy}{du}\frac{du}{dx}=f(u)\cdot\frac{du}{dx}.$$

We have used the fact that Formula 1 implies

$$\frac{d}{du}\left(\int_{a}^{u}f(t)\,dt\right)=f(u),$$

since the upper limit of integration is the variable of differentiation.

We can restate the Chain Rule version of Theorem 2 in the form

$$\frac{d}{dx}\left(\int_a^{u(x)} f(t)\,dt\right) = f(u(x))\frac{du}{dx}. \tag{2}$$

Note that in this formula the upper limit of integration is a function of the variable of integration. If the upper limit is exactly the limit of integration, this formula is not needed. If the lower limit of integration is not a constant, we must modify the indefinite integral to fit the formula.

EXAMPLE 6

Evaluate the derivatives.

(a) $\dfrac{d}{dx}\left(\displaystyle\int_1^{x^2} \frac{1}{t}\,dt\right),$

(b) $\dfrac{d}{dx}\left(\displaystyle\int_0^{\sqrt{x}} \sin t^2\,dt\right),$

(c) $\dfrac{d}{dx}\left(\displaystyle\int_{x^2}^{x^3} \frac{1}{t}\,dt\right), x > 0.$

SOLUTION

(a) Formula 2 implies

$$\frac{d}{dx}\left(\int_1^{x^2} \frac{1}{t}\,dt\right) = \left(\frac{1}{x^2}\right)(2x) = \frac{2}{x}.$$

(b) From Formula 2, we have

$$\frac{d}{dx}\left(\int_0^{\sqrt{x}} \sin t^2\,dt\right) = (\sin(\sqrt{x})^2)\left(\frac{1}{2\sqrt{x}}\right) = \frac{\sin x}{2\sqrt{x}}.$$

(c) To use Formula 2, we express the function to be differentiated as a sum of integrals that have a variable expression as the upper limit of integration and a constant as the lower limit. We can then write the derivative of the sum as above. We could use any positive number in place of 1 below:

$$\begin{aligned}
\frac{d}{dx}\left(\int_{x^2}^{x^3} \frac{1}{t}\,dt\right) &= \frac{d}{dx}\left(\int_{x^2}^{1} \frac{1}{t}\,dt + \int_1^{x^3} \frac{1}{t}\,dt\right) \\
&= \frac{d}{dx}\left(-\int_1^{x^2} \frac{1}{t}\,dt + \int_1^{x^3} \frac{1}{t}\,dt\right) \\
&= -\left(\frac{1}{x^2}\right)(2x) + \left(\frac{1}{x^3}\right)(3x^2) \\
&= -\frac{2}{x} + \frac{3}{x} = \frac{1}{x}. \quad\blacksquare
\end{aligned}$$

EXERCISES 5.4

Use Theorem 1 to evaluate the definite integrals given in Exercises 1–24.

1. $\int_1^4 \sqrt{x}\,dx$

2. $\int_1^8 x^{1/3}\,dx$

3. $\int_1^8 x^{-1/3}\,dx$

4. $\int_1^9 x^{-1/2}\,dx$

5. $\int_1^2 \frac{1}{x^3}\,dx$

6. $\int_1^3 \frac{1}{x^4}\,dx$

7. $\int_0^1 (x^2 + x + 1)\,dx$

8. $\int_0^1 (x^2 - x - 1)\,dx$

9. $\int_1^3 (x^2 - 6x + 2)\,dx$

10. $\int_{-1}^3 (2x^2 + 4x + 5)\,dx$ **(Calculus Explorer & Tutor II, Introduction to Integration, 1.)**

11. $\int_0^{\pi/2} \cos x\,dx$

12. $\int_0^{\pi/2} \sin x\,dx$

13. $\int_0^{\pi/4} \sec^2 x\,dx$

14. $\int_{\pi/6}^{\pi/2} \csc^2 x\,dx$

15. $\int_{\pi/6}^{\pi/3} \csc x \cdot \cot x\,dx$

16. $\int_{\pi/6}^{\pi/3} \sec x \cdot \tan x\,dx$

17. $\int_0^{\pi} \sin x\,dx$

18. $\int_0^{\pi} \cos x\,dx$

19. $\int_0^{2\pi} \cos x\,dx$

20. $\int_0^{2\pi} \sin x\,dx$

21. $\int_{\pi/6}^{\pi/3} \csc^2 x\,dx$

22. $\int_{\pi/6}^{\pi/3} \sec^2 x\,dx$

23. $\int_{-\pi/4}^{\pi/4} \sec x \tan x\,dx$

24. $\int_{\pi/4}^{\pi/2} \csc x \cot x\,dx$

In Exercises 25–30, the instantaneous velocity of an object on a number line is given. Find the net displacement of the object over the given time interval.

25. $s'(t) = 32 - 32t,\ 0 \le t \le 3$

26. $s'(t) = 29.4 - 9.8t,\ 0 \le t \le 5$

27. $s'(t) = t - t^3,\ 0 \le t \le 2$

28. $s'(t) = t^2 - 3t + 2,\ 0 \le t \le 3$

29. $s'(t) = \sin t,\ 0 \le t \le 2\pi$

30. $s'(t) = \sin t - \cos t,\ 0 \le t \le \pi$

Use Theorem 2 to evaluate the derivatives in Exercises 31–48.

31. $\dfrac{d}{dx}\left(\int_1^x \cos t^2\,dt\right)$

32. $\dfrac{d}{dx}\left(\int_2^x \cos t^2\,dt\right)$

33. $\dfrac{d}{dx}\left(\int_0^x \sec t^2\,dt\right),\ |x| < \sqrt{\pi/2}$

34. $\dfrac{d}{dx}\left(\int_0^x \tan t\,dt\right),\ |x| < \dfrac{\pi}{2}$

35. $\dfrac{d}{dx}\left(\int_x^1 (1 - t)^5\,dt\right)$

36. $\dfrac{d}{dx}\left(\int_x^8 (10 - t)^{-2}\,dt\right),\ x < 10$

37. $\dfrac{d}{dx}\left(\int_0^{x+1} (t - 1)^4\,dt\right)$ **38.** $\dfrac{d}{dx}\left(\int_0^{x-1} (t + 1)^5\,dt\right)$

39. $\dfrac{d}{dx}\left(\int_1^{2x} \dfrac{1}{t}\,dt\right),\ x > 0$

40. $\dfrac{d}{dx}\left(\int_x^{x^2} \tan t\,dt\right),\ |x| < \sqrt{\pi/2}$

41. $\dfrac{d}{dx}\left(\int_x^{x^2} \sec t\,dt\right),\ |x| < \sqrt{\pi/2}$

42. $\dfrac{d}{dx}\left(\int_{2x}^{3x} \dfrac{1}{t}\,dt\right),\ x > 0$

43. $\dfrac{d}{dx}\left(\int_1^{\sec x} \dfrac{1}{t}\,dt\right),\ |x| < \pi/2$

44. $\dfrac{d}{dx}\left(\int_1^{\sec x + \tan x} \dfrac{1}{t}\,dt\right),\ 0 < x < \pi/2$

45. $\dfrac{d}{dx}\left(\int_{2x}^{4x} \sin^2 t\,dt\right)$ **46.** $\dfrac{d}{dx}\left(\int_{x/2}^x \cos^2 t\,dt\right)$

47. $\dfrac{d}{dx}\left(\int_{\cos x}^{\sin x} \dfrac{1}{1 + t^2}\,dt\right)$ **48.** $\dfrac{d}{dx}\left(\int_{1/x}^x \dfrac{1}{1 + t^2}\,dt\right)$

49. $G(x) = \int_{-1}^x \dfrac{1}{t^2 + 1}\,dt$. Find $G''(2)$.

50. $G(x) = \int_0^x \sqrt{25 - t^2}\,dt,\ -5 \le x \le 5$. Find $G''(3)$.

51. Find the equation of the line tangent to the graph of $y = \int_2^x \dfrac{t^2}{t^4 + 16}\,dt$ at the point where $x = 2$.

52. Find the equation of the line tangent to the graph of $y = 1 + \int_0^x \cos(t^2)\,dt$ at the point where $x = 0$.

In Exercises 53–56, sketch the graph of f near the point where $x = c$. Indicate the slope and concavity.

53. $f(x) = \int_1^x \dfrac{1}{t}\,dt,\ c = 1$

54. $f(x) = \int_0^x \dfrac{1}{t^2 + 1}\,dt,\ c = 0$

55. $f(x) = \int_0^x \dfrac{t}{(t + 1)^{1/3}}\,dt,\ c = 0$

56. $f(x) = 1 + \int_0^x \dfrac{1}{\sqrt{1 + t^2}}\,dt,\ c = 0$

Verify Theorem 2 for the functions in Exercises 57–62 by **(a)** evaluating each indefinite integral, **(b)** differentiating the result of (a), and **(c)** using Theorem 2 directly to evaluate the derivative of the given indefinite integral and comparing the result with the result of (b).

57. $\int_0^x \sin t\, dt$

58. $\int_a^x t^2\, dt$

59. $\int_x^1 \frac{1}{t^2}\, dt$

60. $\int_x^b \frac{1}{\sqrt{t}}\, dt$, b a constant

61. $\int_{\sqrt{c}}^{\sqrt{x}} \cos t\, dt$, c a constant

62. $\int_{x^2}^{c^2} \cos t\, dt$, c a constant

5.5 SUBSTITUTION AND CHANGE OF VARIABLE

Connections

Chain Rule, 3.4.

Integration formulas, 5.4.

We have seen in Section 5.4 that an integration formula $\int f(x)\,dx = F(x) + C$ greatly facilitates evaluation of the definite integral $\int_a^b f(x)\,dx$. However, we do not yet have a systematic method of evaluating indefinite integrals, except for our basic formulas. In this section we introduce a method that extends the class of functions for which we can evaluate indefinite integrals by using the Fundamental Theorem. The method is called **substitution**. Substitution is a type of *algebra* that is used to change variables and evaluate integrals.

Even if you are using a computer with a symbolic calculus program to evaluate integrals, it is important that you understand how to change variables in integrals.

The Change of Variable Formula

The change of variable formula for integrals corresponds to the Chain Rule for differentiation.
If

$$\frac{d}{du}(F(u)) = f(u) \text{ for } u \text{ in an interval,}$$

we have

$$\int f(u)\,du = F(u) + C. \tag{1}$$

If $u = u(x)$ is a differentiable function of x, the Chain Rule gives

$$\frac{d}{dx}(F(u(x))) = \frac{dF}{du}\frac{du}{dx} = f(u(x)) \cdot u'(x).$$

The integral form of this statement is

$$\int f(u(x))u'(x)\,dx = F(u(x)) + C. \tag{2}$$

Combining Formulas 1 and 2, we obtain the following formula for change of variables in indefinite integrals.

If $u = u(x)$, then

$$\int f(u(x))u'(x)\,dx = \int f(u)\,du. \tag{3}$$

Note that you can obtain the left side of (3) by substituting $u(x)$ for u and $u'(x)\,dx$ for du on the right. Recall from Section 3.6 that $du = u'(x)\,dx$ is the differential of $u(x)$.

Using Substitution to Evaluate Indefinite Integrals

We can use the change of variable formula to simplify and evaluate integrals. The idea is to replace $u(x)$ by u and $u'(x)\,dx$ by du on the left side of (3). This substitution reduces the left side of (3) to the simpler form of the right side of (3). If we can evaluate the integral on the right in terms of the variable u, we can substitute $u(x)$ for u to obtain a solution in terms of the original variable x.

Solution form for using substitution to evaluate indefinite integrals

$$\int f(u(x))u'(x)\,dx = \int f(u)\,du \qquad \text{(Substitute)}$$
$$= F(u) + C \qquad \text{(Evaluate)}$$
$$= F(u(x)) + C \qquad \text{(Substitute back).}$$

Of course, to successfully use this method, we must be able to evaluate the simplified integral $\int f(u)\,du$. For now, this means that $\int f(u)\,du$ must be one of the following types, or a *constant* multiple of one of them, or a sum of such types.

$$\int u^r\,du = \frac{u^{r+1}}{r+1} + C, \qquad r \text{ rational}, r \neq -1, \qquad (4)$$

$$\int \cos u\,du = \sin u + C, \qquad (5)$$

$$\int \sin u\,du = -\cos u + C, \qquad (6)$$

$$\int \sec^2 u\,du = \tan u + C, \qquad (7)$$

$$\int \csc^2 u\,du = -\cot u + C, \qquad (8)$$

$$\int \sec u \cdot \tan u\,du = \sec u + C, \qquad (9)$$

$$\int \csc u \cdot \cot u\,du = -\csc u + C, \qquad (10)$$

The above formulas of integration are those that we used in Sections 5.1 and 5.4, except we are now using u for the variable of integration.

Of course, we can use the formula

$$\int (cf(u) + dg(u))\,du = c\int f(u)\,du + d\int g(u)\,du$$

to evaluate linear sums of the above types.

> ## Strategy to evaluate an indefinite integral
>
> - Recognize which of the above formulas applies.
> - Choose $u(x)$.
> - Evaluate the differential $du = u'(x)dx$.
> - Substitute u for $u(x)$ and du for $u'(x)dx$.
> - Evaluate the simplified integral in terms of u. (If we cannot reduce a given integral to a linear sum of the above types, then we cannot evaluate it at this time.)
> - Substitute $u(x)$ for u.

EXAMPLE 1

Determine which of the integrals reduce to a constant multiple of one of the types (4)–(10). Evaluate those that do.

(a) $\displaystyle \int 2x\sqrt{x^2 + 1}\,dx,$

(b) $\displaystyle \int x\sqrt{x^2 + 1}\,dx,$

(c) $\displaystyle \int \sqrt{x^2 + 1}\,dx,$

(d) $\displaystyle \int x^2\sqrt{x^2 + 1}\,dx,$

(e) $\displaystyle \int \frac{1}{(1 + \sqrt{x})^2}\frac{1}{\sqrt{x}}\,dx,$

(f) $\displaystyle \int \frac{1}{(1 + \sqrt{x})^2 x}\,dx,$

(g) $\displaystyle \int x^2 \sin x^3\,dx,$

(h) $\displaystyle \int \cos x^2\,dx,$

(i) $\displaystyle \int \frac{\cos x}{\sqrt{\sin x + 1}}\,dx,$

(j) $\displaystyle \int (\cos x)^{1/3}\,dx,$

(k) $\displaystyle \int \frac{\sin x}{\cos x}\,dx,$

(l) $\displaystyle \int (2x - 1)^{12}\,dx.$

SOLUTION

(a) The integral $\int 2x\sqrt{x^2 + 1}\,dx$ involves the square root of a function. Let us see if we can reduce the integral to $\int u^{1/2}\,du$. If it is of this form, we must have $u = x^2 + 1$. Let us try

$$u = x^2 + 1, \text{ so } du = 2x\,dx.$$

Noting that the integrand contains the factor $2x$, we substitute to obtain

$$\int 2x\sqrt{x^2 + 1}\,dx = \int \underbrace{\sqrt{x^2 + 1}}_{\sqrt{u}}\,\underbrace{2x\,dx}_{du} = \int \sqrt{u}\,du = \int u^{1/2}\,du.$$

Using (4) to evaluate the simplified integral, we obtain

$$\int u^{1/2}\,du = \frac{u^{3/2}}{3/2} + C.$$

Substituting $x^2 + 1$ for u, and combining results, we have

$$\int 2x\sqrt{x^2+1}\,dx = \frac{2}{3}(x^2+1)^{3/2} + C.$$

Recall that we can verify an integration formula $\int f(x)\,dx = F(x) + C$ by showing $F'(x) = f(x)$. Using the Chain Rule to verify the formula obtained above, we have

$$\frac{d}{dx}\left(\frac{2}{3}(x^2+1)^{3/2}\right) = \left(\frac{2}{3}\right)\left(\frac{3}{2}\right)(x^2+1)^{1/2}(2x) = 2x\sqrt{x^2+1},$$

so the formula is correct. (Since we used the integral form of the Chain Rule to evaluate this integral, we should expect that the Chain Rule would be used to verify the formula by differentiation.)

(b) The integral $\int x\sqrt{x^2+1}\,dx$ involves the square root of a function. As in (a), we try to reduce the integral to $\int u^{1/2}\,du$. Let us try

$$u = x^2 + 1, \qquad du = 2x\,dx.$$

We can then write

$$\int x\sqrt{x^2+1}\,dx = \int \underbrace{\sqrt{x^2+1}}_{\sqrt{u}}\left(\frac{1}{2}\right)\underbrace{(2x)\,dx}_{du}.$$

We have introduced the factor

$$\left(\frac{1}{2}\right)(2)$$

in the integrand so the exact expression $du = 2x\,dx$ appears in the integrand. For the substitution $u = x^2 + 1$ to simplify the integral, the integrand must contain the *essential factor* x, since $u'(x) = (\text{constant})x$. We can modify the essential factor only by a constant. Substitution now gives

$$\int \sqrt{x^2+1}\left(\frac{1}{2}\right)(2x)\,dx = \int u^{1/2}\left(\frac{1}{2}\right)du.$$

This integral is of the form of (4), except for the constant factor $1/2$. We evaluate to obtain

$$\int u^{1/2}\left(\frac{1}{2}\right)du = \left(\frac{1}{2}\right)\left(\frac{u^{3/2}}{3/2}\right) + C = \frac{1}{3}u^{3/2} + C.$$

Substituting $x^2 + 1$ for u, we then obtain

$$\int x\sqrt{x^2+1}\,dx = \frac{1}{3}(x^2+1)^{3/2} + C.$$

An alternate method of evaluating the above integral avoids introduction of the factor

$$\left(\frac{1}{2}\right)(2).$$

That is, noting that the integrand contains the factor $x\,dx$, we solve $du = 2x\,dx$ for $x\,dx$. We obtain

$$x\,dx = \frac{1}{2}du.$$

Direct substitution of $(1/2)\,du$ for $x\,dx$ gives

$$\int x\sqrt{x^2 + 1}\,dx = \int \sqrt{u}\,\frac{1}{2}\,du.$$

We then continue as before.

    (c) The integral $\int \sqrt{x^2 + 1}\,dx$ involves the square root of a function. We try $u = x^2 + 1$, so $du = 2x\,dx$. This substitution will not reduce the integral to the simplified form $\int (\text{constant})u^{1/2}\,du$, because the integrand does not contain the essential factor x. Attempts to substitute will lead to either meaningless expressions that involve both u and x or complicated expressions of the variable u if we try to eliminate x by solving $u = x^2 + 1$ for x. For example, suppose we try to modify the integrand by the factor

$$\left(\frac{1}{2x}\right)(2x).$$

We would obtain

$$\int \sqrt{x^2 + 1}\,dx = \int \sqrt{x^2 + 1}\left(\frac{1}{2x}\right)(2x)\,dx.$$

We could substitute $u^{1/2}$ for $\sqrt{x^2 + 1}$ and du for $2x\,dx$, but we would still have a factor of x in the denominator.

NOTE   *We cannot simply move a variable factor of the integrand outside of the integral sign, as we can with a constant.*

We could solve the equation $u = x^2 + 1$ for x, to obtain $x = \pm(u - 1)^{1/2}$. Substitution of this expression for x gives

$$\int \frac{u^{1/2}}{\pm 2(u - 1)^{1/2}}\,du,$$

which is more complicated than the integral we started with. We cannot reduce this integral to one of the types (4)–(10).

    (d) The integral $\int x^2\sqrt{x^2 + 1}\,dx$ is similar to that in (c). The substitution $u = x^2 + 1$ will not reduce $\int x^2\sqrt{x^2 + 1}\,dx$ to the form $\int (\text{constant})u^{1/2}\,du$, because the integrand contains an extra factor of x. We cannot reduce this integral to one of the types (4)–(10).

    (e) The integral

$$\int \frac{1}{(1 + \sqrt{x})^2}\frac{1}{\sqrt{x}}\,dx$$

involves a function to the negative two power. Let us try to reduce the integral to a constant multiple of $\int u^{-2}\,du$. We try

$$u = 1 + \sqrt{x}, \text{ so } du = \frac{1}{2\sqrt{x}}\,dx.$$

We note that the integrand contains the essential factor $1/\sqrt{x}$. We can then write

$$\int \frac{1}{(1+\sqrt{x})^2} \frac{1}{\sqrt{x}} dx = \int \frac{1}{(1+\sqrt{x})^2} (2) \frac{1}{2\sqrt{x}} dx$$

$$= \int u^{-2}(2) \, du$$

$$= 2\frac{u^{-1}}{-1} + C$$

$$= -2(1+\sqrt{x})^{-1} + C.$$

(**f**) The integral $\displaystyle\int \frac{1}{(1+\sqrt{x})^2 x} dx$ involves a function to the negative two power. Let us try the substitution

$$u = 1+\sqrt{x}, \text{ so } du = \frac{1}{2\sqrt{x}} dx.$$

We can then write

$$\int \frac{1}{(1+\sqrt{x})^2 x} dx = \int \frac{1}{(1+\sqrt{x})^2} \frac{2}{\sqrt{x}} \frac{1}{2\sqrt{x}} dx.$$

We can solve $u = 1+\sqrt{x}$ for $\sqrt{x}$ to obtain $\sqrt{x} = u - 1$. Substitution of u for $1+\sqrt{x}$, $u - 1$ for $\sqrt{x}$, and du for $\dfrac{1}{2\sqrt{x}} dx$ then gives

$$\int \frac{2}{u^2(u - 1)} \, du.$$

This integral does not reduce to one of the types (4)–(10).

(**g**) The integral $\int x^2 \sin x^3 \, dx$ involves the sine of a function. Let us set $u = x^3$, so $du = 3x^2 \, dx$. Noting that the integrand contains the essential factor x^2, we write

$$\int x^2 \sin x^3 \, dx = \int (\sin x^3) \cdot \left(\frac{1}{3}\right)(3x^2) \, dx$$

$$= \int (\sin u) \left(\frac{1}{3}\right) du$$

$$= \frac{1}{3}(-\cos u) + C$$

$$= -\frac{1}{3}\cos x^3 + C.$$

(**h**) The substitution $u = x^2$ does not reduce $\int \cos x^2 \, dx$ to $\int \cos u \, du$, because the integrand does not contain the essential factor x; $u = x^2$ implies $du = 2x \, dx$. We cannot reduce this integral to one of the types (4)–(10).

(**i**) The integral

$$\int \frac{\cos x}{\sqrt{\sin x + 1}} dx$$

involves the negative one-half power of a function. Let us set

$$u = \sin x + 1, \text{ so } du = \cos x \, dx.$$

The integrand contains the essential factor $\cos x$. We have

$$\int \frac{\cos x}{\sqrt{\sin x + 1}} dx = \int (\sin x + 1)^{-1/2} \cos x \, dx$$

$$= \int u^{-1/2} du$$

$$= \frac{u^{1/2}}{1/2} + C$$

$$= 2\sqrt{\sin x + 1} + C.$$

(**j**) The integral $\int (\cos x)^{1/3} dx$ does not reduce to one of the desired types. The substitution $u = \cos x$ requires the essential factor $\sin x$, since $du = -\sin x \, dx$. We cannot reduce this integral to one of the types (4)–(10).

(**k**) The integrand of

$$\int \frac{\sin x}{\cos x} dx$$

contains the factor $\sin x$. This leads us to try

$$u = \cos x, \quad \text{so } du = -\sin x \, dx.$$

Substitution then gives

$$\int \frac{\sin x}{\cos x} dx = \int (\cos x)^{-1}(-1)(-\sin x) dx = -\int u^{-1} du.$$

We cannot evaluate this integral at this time. Formula 4 does not apply because the exponent is -1.

(**l**) The integral $\int (2x - 1)^{12} dx$ involves the twelfth power of a function. We set

$$u = 2x - 1, \quad \text{so } du = 2dx.$$

Substitution then gives

$$\int (2x - 1)^{12} dx = \int (2x - 1)^{12} \left(\frac{1}{2}\right)(2) dx$$

$$= \int u^{12} \left(\frac{1}{2}\right) du$$

$$= \frac{1}{2} \frac{u^{13}}{13} + C$$

$$= \frac{1}{26}(2x - 1)^{13} + C. \quad \blacksquare$$

Using Substitution to Evaluate Definite Integrals

We can evaluate a definite integral by using a substitution to evaluate the corresponding indefinite integral and then using the resulting integration formula to evaluate the definite integral.

EXAMPLE 2

Evaluate $\int_0^3 \frac{x}{\sqrt{x^2 + 16}}\,dx$.

SOLUTION Let us first use the method of Example 1 to evaluate the corresponding indefinite integral. We set

$$u = x^2 + 16, \text{ so } du = 2x\,dx.$$

Then

$$\int \frac{x}{\sqrt{x^2 + 16}}\,dx = \int (x^2 + 16)^{-1/2}\left(\frac{1}{2}\right)(2x)\,dx$$

$$= \int u^{-1/2}\left(\frac{1}{2}\right)du$$

$$= \left(\frac{1}{2}\right)\left(\frac{u^{1/2}}{1/2}\right) + C$$

$$= \sqrt{x^2 + 16} + C.$$

It follows from the formula for the indefinite integral that

$$\int_0^3 \frac{x}{\sqrt{x^2 + 16}}\,dx = \sqrt{x^2 + 16}\,\Big]_0^3 = \sqrt{3^2 + 16} - \sqrt{0^2 + 16}$$

$$= 5 - 4 = 1. \quad \blacksquare$$

The Change of Variable Formula for Definite Integrals

In some problems it is desirable to change the limits of integration to correspond to a change of the variable of integration. To see how this is done we assume $F'(x) = f(x)$. The Fundamental Theorem of Calculus in Section 5.4 and Formula 2 then imply

$$\int_a^b f(u(x))u'(x)\,dx = F(u(x))\,\Big]_a^b = F(u(b)) - F(u(a)),$$

and Formula 1 gives us

$$\int_{u(a)}^{u(b)} f(u)\,du = F(u)\,\Big]_{u(a)}^{u(b)} = F(u(b)) - F(u(a)).$$

Assuming appropriate differentiability, so that both formulas hold, we conclude that

$$\int_a^b f(u(x))u'(x)\,dx = \int_{u(a)}^{u(b)} f(u)\,du. \qquad (11)$$

In Formula 11, the limits of integration of the variable u are obtained from the formula for $u(x)$; $x = a$ corresponds to $u = u(a)$; $x = b$ corresponds to $u = u(b)$.

EXAMPLE 3

Express $\displaystyle\int_0^3 \frac{x}{\sqrt{x^2+16}}\,dx$ as a definite integral in terms of the variable $u = x^2 + 16$ and evaluate.

SOLUTION   We have $u = x^2 + 16$, so $du = 2x\,dx$,

$$u(0) = (0)^2 + 16 = 16, \text{ and } u(3) = (3)^2 + 16 = 25.$$

Formula 11 then gives

$$\int_0^3 \frac{x}{\sqrt{x^2+16}}\,dx = \int_0^3 (x^2+16)^{-1/2}\left(\frac{1}{2}\right)(2x)\,dx$$

$$= \int_{16}^{25} u^{-1/2}\left(\frac{1}{2}\right)du.$$

Note that the limits of integration were changed to correspond to the variable u. We now evaluate the integral on the right in terms of u, without substituting back to the variable x. We have

$$\int_{16}^{25} u^{-1/2}\left(\frac{1}{2}\right)du = \left(\frac{1}{2}\right)\left(\frac{u^{1/2}}{1/2}\right)\Bigg]_{16}^{25} = \sqrt{25} - \sqrt{16} = 1.$$

This agrees with the result of Example 2. ∎

NOTE   *Be sure that the limits of integration correspond to the variable of integration.*

For example, it is incorrect to write

$$\int_0^3 \frac{x}{\sqrt{x^2+16}}\,dx \underset{(?)}{=} \int_0^3 u^{-1/2}\left(\frac{1}{2}\right)du$$

$$\underset{(?)}{=} \left(\frac{1}{2}\right)\left(\frac{u^{1/2}}{1/2}\right)\Bigg]_0^3$$

$$\underset{(?)}{=} \sqrt{x^2+16}\,\Bigg]_0^3 = 1.$$

The first equality above is invalid because the limits of integration 0 and 3 of the integral on the right are not the values of u that correspond to the values $x = 0$ and $x = 3$. The second equality is invalid because the Fundamental Theorem of Calculus does not apply to $\displaystyle\int_0^3 u^{-1/2}\,du$ since $u^{-1/2}$ is not continuous on the interval $[0, 3]$. The third equality is invalid because

$$\left(\frac{1}{2}\right)\left(\frac{u^{1/2}}{1/2}\right)\Bigg]_0^3 = \sqrt{3} \neq 1.$$

EXERCISES 5.5

Determine which of the integrals in Exercises 1–24 reduce to a constant multiple of one of the types in Formulas 4–10. Evaluate those that do.

1. $\displaystyle\int (1-x)^{10}\,dx$

2. $\displaystyle\int (6x-5)^3\,dx$ **(Calculus Explorer & Tutor II, Introduction to Integration, 2.)**

3. $\displaystyle\int x(x^2+1)^{1/3}\,dx$

4. $\displaystyle\int \frac{x}{\sqrt{x^2+2}}\,dx$ **(Calculus Explorer & Tutor II, Introduction to Integration, 3.)**

5. $\displaystyle\int \frac{1}{1+\sqrt{x}}\frac{1}{\sqrt{x}}\,dx$

6. $\displaystyle\int \frac{1}{1+\sqrt{x}}\,dx$

7. $\displaystyle\int \frac{\sqrt{1+\sqrt{x}}}{\sqrt{x}}\,dx$

8. $\displaystyle\int x^2(1+x^3)^{1/3}\,dx$

9. $\displaystyle\int \frac{x+1}{(x^2+2x+3)^2}\,dx$

10. $\displaystyle\int \frac{x+2}{(x^2+2x+4)^2}\,dx$

11. $\displaystyle\int \frac{\cos\sqrt{x}}{\sqrt{x}}\,dx$

12. $\displaystyle\int \frac{\sin\sqrt{x}}{\sqrt{x}}\,dx$

13. $\displaystyle\int \sin\sqrt{x}\,dx$

14. $\displaystyle\int \cos x^2\,dx$

15. $\displaystyle\int x\sec^2(x^2)\,dx$

16. $\displaystyle\int \sec(x^2)\,dx$

17. $\displaystyle\int \sec 2x \cdot \tan 2x\,dx$

18. $\displaystyle\int \sec^2(2x+1)\,dx$

19. $\displaystyle\int \sqrt{1+\sin x}\,\cos x\,dx$

20. $\displaystyle\int \frac{\cos x}{\sqrt{1+\sin x}}\,dx$

21. $\displaystyle\int 2\sin x \cdot \cos x\,dx$

22. $\displaystyle\int \cos^2 x\,dx$

23. $\displaystyle\int \tan^6 x \sec^2 x\,dx$

24. $\displaystyle\int \tan^4 x \sec^2 x\,dx$

Evaluate the definite integrals in Exercises 25–36.

25. $\displaystyle\int_0^2 \sqrt{1+4x}\,dx$

26. $\displaystyle\int_0^1 \frac{1}{\sqrt{1+4x}}\,dx$

27. $\displaystyle\int_{\pi^3/8}^{\pi^3} \frac{\sin x^{1/3}}{x^{2/3}}\,dx$

28. $\displaystyle\int_0^{\sqrt{\pi}} x\sin x^2\,dx$

29. $\displaystyle\int_3^4 \frac{x}{\sqrt{25-x^2}}\,dx$

30. $\displaystyle\int_3^4 x\sqrt{25-x^2}\,dx$

31. $\displaystyle\int_1^8 \frac{(1+x^{1/3})^3}{x^{2/3}}\,dx$

32. $\displaystyle\int_1^8 \frac{(1+x^{2/3})^3}{x^{1/3}}\,dx$

33. $\displaystyle\int_{-\pi/4}^{\pi/4} \sin 2x \cos 2x\,dx$

34. $\displaystyle\int_{-\pi/6}^{\pi/6} \sin 3x \cos 3x\,dx$

35. $\displaystyle\int_0^{\pi/3} \sqrt{4\sec x+1}\,\sec x\tan x\,dx$

36. $\displaystyle\int_0^{\pi/4} \sqrt{\tan x}\,\sec^2 x\,dx$

Express each integral in Exercises 37–42 as a definite integral in terms of the indicated variable u.

37. $\displaystyle\int_0^{\pi/2} \frac{\cos x}{1+\sin x}\,dx,\ u=1+\sin x$

38. $\displaystyle\int_{-1}^1 \frac{1}{1+9x^2}\,dx,\ u=3x$

39. $\displaystyle\int_0^1 \frac{1}{(1+\sqrt{x})\sqrt{x}}\,dx,\ u=1+\sqrt{x}$

40. $\displaystyle\int_1^8 \frac{(1+x^{2/3})^6}{x^{1/3}}\,dx,\ u=1+x^{2/3}$

41. $\displaystyle\int_0^3 \frac{x}{\sqrt{x+1}}\,dx,\ u=\sqrt{x+1}$

42. $\displaystyle\int_0^7 x(x+1)^{1/3}\,dx,\ u=(x+1)^{1/3}$

43. Show that f even implies $\displaystyle\int_{-a}^a f(x)\,dx = 2\int_0^a f(x)\,dx$.

44. Show that f odd implies $\displaystyle\int_{-a}^a f(x)\,dx = 0$.

45. Show that $\displaystyle\int_0^a f(x)\,dx = \int_0^a f(a-x)\,dx$.

46. Show that $f(a-x)=f(x)$ for $0\le x\le a$ implies $\displaystyle\int_0^a f(x)\,dx = 2\int_0^{a/2} f(x)\,dx$.

47. Show that $\displaystyle\int_1^{1/x} \frac{1}{t}\,dt = -\int_1^x \frac{1}{t}\,dt,\ x>0$.

48. Show that $\displaystyle\int_1^{x^2} \frac{1}{t}\,dt = 2\int_1^x \frac{1}{t}\,dt,\ x>0$.

5.6 APPROXIMATE VALUES OF DEFINITE INTEGRALS; NUMERICAL METHODS

Connections

Bounds of functions, 2.2.

Sigma notation, 5.3.

There are many ways that a calculator or computer can be programmed to calculate approximate values of definite integrals. In this section we will introduce some of these approximation methods and investigate the efficiency of each method.

We assume f is continuous on the closed interval $[a, b]$, so the definite integral $\int_a^b f(x)\,dx$ exists.

The Right Endpoint Rule

The first step in our systematic construction of Riemann sums of f over $[a, b]$ is to choose points of subdivision of $[a, b]$. It seems reasonable to divide $[a, b]$ into n subintervals of equal length. From Statement 2 of Section 5.2 we then have

$$x_j = a + (j)\left(\frac{b-a}{n}\right), \qquad j = 0, 1, \ldots, n.$$

Also,

$$\Delta x_j = \frac{b-a}{n}, \qquad j = 1, \ldots, n.$$

The next step in constructing our Riemann sums is to choose points x_j^* in each of the intervals $[x_{j-1}, x_j]$, $j = 1, \ldots, n$. One simple way to do this is to choose $x_j^* = x_j$, so x_j^* is the right endpoint of the subinterval $[x_{j-1}, x_j]$, $j = 1, \ldots, n$. Let R_n denote the corresponding Riemann sum. That is,

$$R_n = \left(\frac{b-a}{n}\right)(f(x_1) + f(x_2) + \cdots + f(x_n)).$$

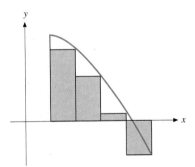

FIGURE 5.6.1 The Riemann sum R_n has subintervals of equal length $\Delta x_j = \dfrac{b-a}{n}$ and $x_j^* = x_j$, the right endpoint of each subinterval.

R_n is illustrated geometrically in Figure 5.6.1. Terms of the Riemann sum that correspond to rectangles below the x-axis give the *negative* of the area of the corresponding rectangle.

Writing

$$R_n = \left(\frac{b-a}{n}\right)(0 \cdot f(x_0) + 1 \cdot f(x_1) + 1 \cdot f(x_2) + \cdots + 1 \cdot f(x_n)),$$

we see that R_n is given by the following formula.

Right Endpoint Rule:

$$R_n = B\left(\sum_{j=0}^{n} c_j f(x_j)\right), \tag{1}$$

where $B = \dfrac{b-a}{n}$, $c_0 = 0$, *and* $c_1 = c_2 = \cdots = c_n = 1$.

We will study other types of Riemann sums. They will all be of the form

$$B\left(\sum_{j=0}^{n} c_j f(x_j)\right), \text{ where } B, c_0, c_1, \ldots, c_n \text{ are fixed constants.} \qquad (2)$$

Different types of Riemann sums give different constants $B, c_0, c_1, \ldots, c_n$. It is quite easy to program a computer to carry out the calculation of sums and products of the form of (2). We will use a calculator and present the calculations in tabular form.

It is useful to know how accurately a particular type of Riemann sum approximates the value of $\int_a^b f(x)\,dx$. The following result gives a bound of the error associated with Riemann sums of type R_n.

Error bound for Right Endpoint Rule:
If $|f'(x)| \le M$ for x in $[a, b]$, then

$$\left|\int_a^b f(x)\,dx - R_n\right| \le \frac{M(b-a)^2}{2n}. \qquad (3)$$

Statement 3 is verified at the end of this section. For now, let us see how to use the result.

EXAMPLE 1

(a) Use R_{10} to find an approximate value of $\int_1^2 (1/x)\,dx$.

(b) How large should n be to insure that $\left|\int_1^2 (1/x)\,dx - R_n\right| \le 0.0001$?

SOLUTION

(a) The calculations are given in Table 5.6.1. The j column runs from 0 to $n = 10$. The x_j column goes from $a = 1$ to $b = 2$ in steps of $(b-a)/n = 0.1$. The c_j column reflects that we are choosing x_j^* to be the right endpoint of each subinterval and the value of f at x_0 does not contribute. In the $f(x_j)$ column, we have substituted the value of x_j into the formula for $f(x)$, but we have not evaluated each term separately. We can then use a calculator to evaluate the expression $\sum c_j f(x_j)$ directly, without making any preliminary calculations. (This is one way to avoid excessive rounding in preliminary calculations.) The value of $\sum c_j f(x_j)$ is then multiplied by B to obtain the value of the desired Riemann sum. We see that $R_{10} = 0.66877$.

(b) We have $f(x) = x^{-1}$, so $f'(x) = -x^{-2} = -1/x^2$. We need a bound of $\frac{1}{x^2}$ for $1 \le x \le 2$. It is clear that the values of $\frac{1}{|x|^2}$ for x in the interval $1 \le x \le 2$ are smaller than the value at one, the left endpoint of the interval. We then have

$$|f'(x)| = \frac{1}{|x|^2} \le \frac{1}{1^2} = 1, \qquad 1 \le x \le 2,$$

TABLE 5.6.1

j	x_j	c_j	$f(x_j) = 1/x_j$
0	1	0	1
1	1.1	1	1/1.1
2	1.2	1	1/1.2
3	1.3	1	1/1.3
4	1.4	1	1/1.4
5	1.5	1	1/1.5
6	1.6	1	1/1.6
7	1.7	1	1/1.7
8	1.8	1	1/1.8
9	1.9	1	1/1.9
10	2	1	1/2

$\sum c_j f(x_j) = 6.68771$

$B = (b-a)/n = 0.1$

$R_{10} = B(\sum c_j f(x_j)) = 0.66877$

so we can use $M = 1$ in (3). Substituting into (3), we then obtain

$$\left| \int_1^2 \frac{1}{x} dx - R_n \right| \leq \frac{M(b-a)^2}{2n} = \frac{(1)(2-1)^2}{2n} = \frac{1}{2n}.$$

It follows that

$$\left| \int_1^2 \frac{1}{x} dx - R_n \right| < 0.0001 \text{ whenever } \frac{1}{2n} < 0.0001$$

$$\text{or } n > \frac{1}{0.0002} = 5000.$$

(If we actually used more than 5000 terms to find an approximate value of the integral, we would need to worry about round-off error in each of the terms. A small error in each of 5000 terms could add up to a significant error.) ∎

The Trapezoid Rule

We can improve the accuracy of approximate values of a definite integral by choosing the points x_j^* so that $f(x_j^*)$ is between the values $f(x_{j-1})$ and $f(x_j)$. In particular, since f is continuous and the average $(f(x_{j-1}) + f(x_j))/2$ is between the values $f(x_{j-1})$ and $f(x_j)$, the Intermediate Value Theorem tells us there is some point x_j^* between x_{j-1} and x_j such that $f(x_j^*) = (f(x_{j-1}) + f(x_j))/2$. See Figure 5.6.2a. The Riemann sums that correspond to this choice of x_j^* are

$$\sum_{j=1}^n f(x_j^*) \Delta x_j = \sum_{j=1}^n \left(\frac{f(x_{j-1}) + f(x_j)}{2} \right) \left(\frac{b-a}{n} \right)$$

$$= \left(\frac{b-a}{2n} \right) \left(\sum_{j=1}^n (f(x_{j-1}) + f(x_j)) \right)$$

$$= \left(\frac{b-a}{2n} \right) ((f(x_0) + f(x_1))$$
$$+ (f(x_1) + f(x_2)) + \cdots + (f(x_{n-1}) + f(x_n)))$$

$$= \left(\frac{b-a}{2n} \right) (1 \cdot f(x_0) + 2 \cdot f(x_1) + 2 \cdot f(x_2)$$
$$+ \cdots + 2 \cdot f(x_{n-1}) + 1 \cdot f(x_n)).$$

This Riemann sum is called the nth **Trapezoidal approximation** of the definite integral $\int_a^b f(x) \, dx$. (If $f(x) \geq 0$ for $a \leq x \leq b$, the product

$$\left(\frac{f(x_{j-1}) + f(x_j)}{2} \right) \Delta x_j$$

may be interpreted as the area of the trapezoid pictured in Figure 5.6.2b.) We will denote the nth Trapezoidal approximation by T_n. The sum T_n involves only values of f at our points of subdivision of $[a, b]$. We can express T_n in the form of (2). In particular, we have the following.

Trapezoidal Rule:

$$T_n = B \left(\sum_{j=0}^n c_j f(x_j) \right), \quad \text{where } B = \frac{b-a}{2n}, \tag{4}$$
$$c_0 = c_n = 1, \quad \text{and } c_1 = c_2 = \cdots = c_{n-1} = 2.$$

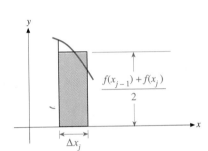

FIGURE 5.6.2a The point x_j^* is chosen so that $f(x_j^*) = \dfrac{f(x_{j-1}) + f(x_j)}{2}$.

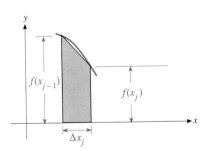

FIGURE 5.6.2b The product $\left(\dfrac{f(x_{j-1}) + f(x_j)}{2} \right) \Delta x_j$ may be interpreted as the area of a trapezoid.

TABLE 5.6.2

j	x_j	c_j	$f(x_j) = 1/x_j$
0	1	1	1
1	1.1	2	1/1.1
2	1.2	2	1/1.2
3	1.3	2	1/1.3
4	1.4	2	1/1.4
5	1.5	2	1/1.5
6	1.6	2	1/1.6
7	1.7	2	1/1.7
8	1.8	2	1/1.8
9	1.9	2	1/1.9
10	2	1	1/2

$\sum c_j f(x_j) = 13.8754$

$B = (b-a)/2n = 0.05$

$T_{10} = 0.69377$

Corresponding to (3) we have the following result for the Trapezoidal Rule.

Error bound for Trapezoidal Rule:
If $|f''(x)| \le M$ for $a \le x \le b$, then

$$\left| \int_a^b f(x)\,dx - T_n \right| \le \frac{M(b-a)^3}{12n^2}. \tag{5}$$

We will not verify (5). The ideas are similar to those involved in (3).

EXAMPLE 2

(a) Use T_{10} to find an approximate value of $\int_1^2 (1/x)\,dx$.

(b) How large should n be to insure that $\left| \int_1^2 (1/x)\,dx - T_n \right| < 0.0001$?

SOLUTION

(a) The calculations are carried out in Table 5.6.2. The constants $c_0, \ldots, c_{10}$ and B for the Trapezoidal Rule are given by (4). We see that $T_{10} = 0.69377$.

(b) We can evaluate the second derivative of $f(x) = 1/x$ to obtain

$$f''(x) = \frac{2}{x^3}.$$

We can determine as in Example 1 that

$$\frac{1}{|x|^3} \le \frac{1}{1^3} = 1, \qquad 1 \le x \le 2,$$

so we have

$$|f''(x)| = 2\left(\frac{1}{|x|^3}\right) \le 2(1) = 2.$$

This shows that we can use $M = 2$ in (5). Then

$$\left| \int_1^2 \frac{1}{x}\,dx - T_n \right| \le \frac{M(b-a)^3}{12n^2} = \frac{(2)(2-1)^3}{12n^2} = \frac{1}{6n^2}.$$

It follows that

$$\left| \int_1^2 \frac{1}{x}\,dx - T_n \right| < 0.0001 \text{ whenever } \frac{1}{6n^2} < 0.0001$$

or $n > \sqrt{1/0.0006} \approx 40.8$. Since n must be an integer, we should use $n \ge 41$. (Recall from Example 1b that $n > 5000$ is required to insure that R_n is within 0.0001 of the exact value of the integral.) ∎

Simpson's Rule

The value of T_n may be interpreted as $\int_a^b f_n(x)\,dx$, where f_n is the function that has the same values as f at the points $x_0, x_1, \ldots, x_n$ and is linear between these points. See Figure 5.6.3. From this point of view it seems that we could obtain a more accurate approximate value of $\int_a^b f(x)\,dx$ if we would choose a function g_n that is closer to f than the trapezoidal function f_n. Our experience from Section 3.9 indicates that second-order approximations give more accurate results than linear, first-order approximations. Hence, we could try to approximate the values of $f(x)$ for x between the x_j's by polynomials of second order. Let us describe such a method, called **Simpson's Rule**.

Let us note that a linear function is determined by its value at two points, so it is natural to approximate the graph of a function f by a linear function between two points on the graph, as in Figure 5.6.3. Since a quadratic function is determined by its value at three points, we will approximate the graph of f by a quadratic function that agrees with the values of f at three points on the graph of f. See Figure 5.6.4.

We begin by defining a function g_n for $x_0 \le x \le x_2$ by the formula $g_n(x) = c_0 + c_1(x - x_1) + c_2(x - x_1)^2$, where the numbers c_0, c_1, and c_2 are chosen so the graph of g_n passes through the points $(x_0, f(x_0))$, $(x_1, f(x_1))$, and $(x_2, f(x_2))$. The condition $g_n(x_1) = f(x_1)$ clearly implies $c_0 = f(x_1)$. The other two conditions then give us the equations

$$f(x_0) = f(x_1) + c_1(x_0 - x_1) + c_2(x_0 - x_1)^2 \text{ and}$$
$$f(x_2) = f(x_1) + c_1(x_2 - x_1) + c_2(x_2 - x_1)^2.$$

Noting that $x_2 - x_1 = x_1 - x_0 = (b - a)/n$, we can solve the above system of two equations for the unknown values of c_1 and c_2. We obtain

$$c_1 = \frac{f(x_2) - f(x_0)}{2(x_1 - x_0)} = \frac{(f(x_2) - f(x_0))n}{2(b - a)} \text{ and}$$
$$c_2 = \frac{f(x_0) - 2f(x_1) + f(x_2)}{2(x_1 - x_0)^2} = \frac{(f(x_0) - 2f(x_1) + f(x_2))n^2}{2(b - a)^2}.$$

Using these values of c_0, c_1, and c_2 in the expression for $g_n(x)$, we can show that

$$\int_{x_0}^{x_2} g_n(x)\,dx = \frac{b - a}{3n}(f(x_0) + 4f(x_1) + f(x_2)).$$

We define g_n similarly in each of the intervals $[x_2, x_4]$, $[x_4, x_6]$, ..., $[x_{n-2}, x_n]$. This works out evenly only if *n is an even integer.* Evaluating the integral of g_n over each of the above integrals, we obtain

$$\int_a^b g_n(x)\,dx = \int_{x_0}^{x_2} g_n(x)\,dx + \int_{x_2}^{x_4} g_n(x)\,dx + \cdots + \int_{x_{n-2}}^{x_n} g_n(x)\,dx$$
$$= \left(\frac{b - a}{3n}\right)((f(x_0) + 4f(x_1) + f(x_2))$$
$$+ (f(x_2) + 4f(x_3) + f(x_4))$$
$$+ \cdots + (f(x_{n-2}) + 4f(x_{n-1}) + f(x_n))).$$

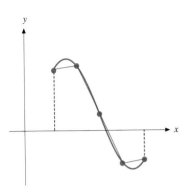

FIGURE 5.6.3 T_n may be interpreted as $\int_a^b f_n(x)\,dx$, where f_n has the same values as f at $x_0, x_1, \ldots, x_n$ and is linear between these points.

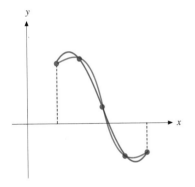

FIGURE 5.6.4 The graph of f is approximated by quadratic functions that agree with the values of f at three points on the graph of f.

This expression is denoted S_n. Grouping like terms, we obtain the following.

Simpson's Rule:

$$S_n = B\left(\sum_{j=0}^{n} c_j f(x_j)\right),\tag{6}$$

where $B = (b-a)/3n$, n is an even integer, $c_0 = c_n = 1$, $c_1 = c_3 = \cdots = c_{n-1} = 4$ (odd-numbered c's), and $c_2 = c_4 = \cdots = c_{n-2} = 2$ (even-numbered c's, except for c_0 and c_n).

A bound of the error in using S_n to find an approximate value of $\int_a^b f(x)\,dx$ is given by the following formula, which we will not verify.

Error bound for Simpson's Rule:
If $|f^{(4)}(x)| \le M$ for $a \le x \le b$, then

$$\left|\int_a^b f(x)\,dx - S_n\right| \le \frac{M(b-a)^5}{180n^4}.\tag{7}$$

EXAMPLE 3

(a) Use S_{10} to find an approximate value of $\int_1^2 (1/x)\,dx$.

(b) How large should n be to insure that $\left|\int_1^2 (1/x)\,dx - S_n\right| < 0.0001$?

SOLUTION

(a) The calculations are carried out in Table 5.6.3. The constants $c_0, \ldots, c_{10}$ and B for Simpson's Rule are given by (6). We see that $S_{10} = 0.69315$.

(b) As in Examples 1 and 2, we can determine that

$$|f^{(4)}(x)| = \frac{24}{|x|^5} \le 24, \quad 1 \le x \le 2,$$

so we can use $M = 24$ in (7). Then

$$\left|\int_1^2 \frac{1}{x}dx - S_n\right| \le \frac{M(b-a)^5}{180n^4} = \frac{(24)(2-1)^5}{180n^4} = \frac{(24)(2-1)^5}{180n^4} = \frac{2}{15n^4}.$$

It follows that

$$\left|\int_1^2 \frac{1}{x}dx - S_n\right| < 0.0001 \text{ whenever } \frac{2}{15n^4} < 0.0001$$

or $n > (2/0.0015)^{1/4} \approx 6.04$. Since n must be an even integer for Simpson's Rule, we should use $n \ge 8$. (Recall that the Trapezoidal Rule required $n \ge 41$ to insure an approximate value of T_n within 0.0001 of the integral.) ∎

TABLE 5.6.3

j	x_j	c_j	$f(x_j) = 1/x_j$
0	1	1	1
1	1.1	4	1/1.1
2	1.2	2	1/1.2
3	1.3	4	1/1.3
4	1.4	2	1/1.4
5	1.5	4	1/1.5
6	1.6	2	1/1.6
7	1.7	4	1/1.7
8	1.8	2	1/1.8
9	1.9	4	1/1.9
10	2	1	1/2

$\sum c_j f(x_j) = 20.79451$
$B = (b-a)/(3n) = 0.03333$
$S_{10} = 0.69315$

Proof of Error Bound for R_n

Let us conclude this section by proving (3), which gives a bound of the error associated with R_n. The proof uses many of the facts we have learned.

Proof We will show that if $|f'(x)| \leq M$ for x in $[a, b]$, then

$$\left| \int_a^b f(x)\,dx - R_n \right| \leq \frac{M(b-a)^2}{2n}.$$

We begin by using the fact that the integral over $[a, b]$ is equal to the sum of the integrals over the subintervals. This gives

$$\left| \int_a^b f(x)\,dx - R_n \right| = \left| \sum_{j=1}^n \int_{x_{j-1}}^{x_j} f(x)\,dx - \sum_{j=1}^n f(x_j)\Delta x_j \right|.$$

Since $f(x_j)$ is a fixed number, we have

$$f(x_j)\Delta x_j = f(x_j)(x_j - x_{j-1}) = \int_{x_{j-1}}^{x_j} f(x_j)\,dx.$$

Substitution of this integral for $f(x_j)\Delta x_j$ gives us

$$\left| \sum_{j=1}^n \int_{x_{j-1}}^{x_j} f(x)\,dx - \sum_{j=1}^n f(x_j)\Delta x_j \right|$$

$$= \left| \sum_{j=1}^n \int_{x_{j-1}}^{x_j} f(x)\,dx - \sum_{j=1}^n \int_{x_{j-1}}^{x_j} f(x_j)\,dx \right|.$$

Using properties of sums and integrals we have

$$\left| \sum_{j=1}^n \int_{x_{j-1}}^{x_j} f(x)\,dx - \sum_{j=1}^n \int_{x_{j-1}}^{x_j} f(x_j)\,dx \right|$$

$$= \left| \sum_{j=1}^n \left(\int_{x_{j-1}}^{x_j} f(x)\,dx - \int_{x_{j-1}}^{x_j} f(x_j)\,dx \right) \right|$$

$$= \left| \sum_{j=1}^n \int_{x_{j-1}}^{x_j} (f(x) - f(x_j))\,dx \right|$$

$$\leq \sum_{j=1}^n \left| \int_{x_{j-1}}^{x_j} (f(x) - f(x_j))\,dx \right|$$

$$\leq \sum_{j=1}^n \int_{x_{j-1}}^{x_j} |f(x) - f(x_j)|\,dx.$$

The Mean Value Theorem implies $|f(x) - f(x_j)| = |f'(z_j)(x - x_j)|$ for some z_j between x and x_j. Since $|f'(z_j)| \leq M$ and $x_j - x > 0$, we have $|f(x) - f(x_j)| \leq M(x_j - x)$. We then have

$$\sum_{j=1}^n \int_{x_{j-1}}^{x_j} |f(x) - f(x_j)|\,dx \leq \sum_{j=1}^n \int_{x_{j-1}}^{x_j} M(x_j - x)\,dx$$

$$= \sum_{j=1}^n -M\frac{(x_j - x)^2}{2}\Bigg]_{x_{j-1}}^{x_j}$$

$$= \frac{M}{2} \sum_{j=1}^{n} (x_j - x_{j-1})^2$$

$$= \frac{M}{2} \sum_{j=1}^{n} \left(\frac{b-a}{n}\right)^2 = \frac{M}{2} \left(\frac{b-a}{n}\right)^2 (n) = \frac{M(b-a)}{2n}$$

Collecting results, we have

$$\left| \int_a^b f(x)\,dx - R_n \right| \le \frac{M(b-a)^2}{2n}. \quad \blacksquare$$

EXERCISES 5.6

In Exercises 1–8, calculate **(a)** R_n, **(b)** T_n, and **(c)** S_n for the given functions f, the intervals $[a, b]$, and the integers n. **(d)** Use Formula 3 to find a bound for the error in (a). **(e)** Use Formula 5 to find a bound for the error in (b). **(f)** Use Formula 7 to find a bound for the error in (c). **(g)** Find values of n that guarantee $\left| \int_a^b f(x)\,dx - R_n \right| < 0.0001$. **(h)** Find values of n that guarantee $\left| \int_a^b f(x)\,dx - T_n \right| < 0.0001$. **(i)** Find values of n that guarantee $\left| \int_a^b f(x)\,dx - S_n \right| < 0.0001$.

1. $f(x) = 1 - x^2$, $[0, 1]$, $n = 8$
2. $f(x) = x - x^2$, $[0, 1]$, $n = 8$
3. $f(x) = x^2$, $[-1, 1]$, $n = 10$
4. $f(x) = x^3$, $[-1, 1]$, $n = 10$
5. $f(x) = \sin x$, $[0, \pi]$, $n = 4$
6. $f(x) = \cos x$, $[0, \pi]$, $n = 4$
7. $f(x) = x \sin x$, $[0, \pi]$, $n = 6$
8. $f(x) = x \cos x$, $[0, \pi]$, $n = 6$

In Exercises 9–10, use Simpson's Rule with $n = 8$ to find approximate values of the given integrals.

9. **(a)** $\int_1^2 \frac{1}{x}\,dx$; **(b)** $\int_2^4 \frac{1}{x}\,dx$
10. **(a)** $\int_4^8 \frac{1}{x}\,dx$; **(b)** $\int_8^{16} \frac{1}{x}\,dx$

In Exercises 11–16, use Simpson's Rule with $n = 4$ to find approximate values of the given integrals.

11. $\int_0^1 \frac{4}{x^2+1}\,dx$
12. $\int_0^{1/\sqrt{2}} 8(\sqrt{1-x^2} - x)\,dx$
13. $\int_{-2}^2 2\sqrt{4-x^2}\,dx$
14. $\int_0^4 \sqrt{x}\,dx$

15. $\int_0^{\sqrt{\pi}} \sin(x^2)\,dx$

16. $\int_0^\pi \frac{\sin x}{x}\,dx$. (Hint: What value would make $\frac{\sin x}{x}$ continuous at zero?)

17. Explain why we cannot use Formula 7 to estimate the error in Exercise 13.

18. Explain why we cannot use Formula 7 to estimate the error in Exercise 14.

In Exercises 19–20, the velocity (ft/s) of an object is given at intervals of 1 s over an interval of 10 s. Use Simpson's Rule to find an approximate value of the integral $\int_0^{10} s'(t)\,dt$. (The exact value of this integral is the displacement of the object over the interval of time.)

19.

t	0	1	2	3	4	5	6	7	8	9	10
$s'(t)$	0	30	65	95	130	160	190	225	255	290	320

20.

t	0	1	2	3	4	5	6	7	8	9	10
$s'(t)$	0	1	5	10	15	25	35	50	65	80	100

21. Use Simpson's Rule to find an approximate value that is within 0.05 of the actual value of the area under the curve $y = x^5$, $0 \le x \le 2$.

22. Use Simpson's Rule to find an approximate value that is within 0.0000001 of the actual value of the area under the curve $y = x^3$, $0 \le x \le 2$.

23. In Chapter 7 we will see that $\int_1^3 \frac{1}{x}\,dx = \ln 3$. Use Formula 5 to find n for which $\left| \int_1^3 \frac{1}{x}\,dx - T_n \right| < 0.0005$.

24. In Chapter 7 we will see that $\int_1^3 \frac{1}{x}\,dx = \ln 3$. Use Formula 7 to find n for which $\left| \int_1^3 \frac{1}{x}\,dx - S_n \right| < 0.0005$.

In Exercise 25–28, use Simpson's Rule with $n = 4$ and the values $f(x_j)$ determined from the given graphs to approximate the area under the given curves.

25.

26.

27.

28.

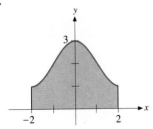

Use the given graphs of f in Exercises 29–32 to determine

(a) whether $R_n \leq \int_a^b f(x)\,dx$ or $R_n \geq \int_a^b f(x)\,dx$ and

(b) whether $T_n \leq \int_a^b f(x)\,dx$ or $T_n \geq \int_a^b f(x)\,dx$.

29.

30.

31.

32.

33. Use Simpson's Rule with $n = 4$ to approximate the average temperature on a very pleasant day in St. Louis from the temperatures given below.

Time	12	3	6	9	12	3	6	9	12
Temperature	65	61	59	68	75	79	81	76	71

34. Use Simpson's Rule with $n = 8$ to approximate the average temperature for the day for the temperatures given above.

35. Show that the approximation of $\dfrac{1}{b - a} \int_a^b f(x)\,dx$ obtained by using T_n to approximate the integral is the average of the n numbers

$$f(x_1), f(x_2), \ldots, f(x_{n-1}), \frac{f(x_0) + f(x_n)}{2}.$$

36. Show that the Trapezoidal Rule gives the exact value of $\int_a^b (c_1 x + c_0)\,dx$. (*Hint:* What is the error?)

37. If $f(x) = c_3 x^3 + c_2 x^2 + c_1 x + c_0$, verify that Simpson's Rule with $n = 2$ gives the exact value of $\int_a^b f(x)\,dx$. (*Hint:* What is the error?)

38. (a) Calculate R_n for $\int_0^1 x\, dx$.

(b) Evaluate $\left| \int_0^1 x\, dx - R_n \right|$.

(c) Use Formula 3 to obtain a bound of $\left| \int_0^1 x\, dx - R_n \right|$.

(d) Compare the results of (b) and (c). **This shows that the actual error may be as large as that given by (3).**

39. (a) Calculate T_n for $\int_0^1 x^2\, dx$.

(b) Evaluate $\left| \int_0^1 x^2\, dx - T_n \right|$.

(c) Use Formula 5 to obtain a bound of $\left| \int_0^1 x^2\, dx - T_n \right|$.

(d) Compare the results of (b) and (c). **This shows that the actual error may be as large as that given by (5).**

REVIEW OF CHAPTER 5 CONCEPTS

- The definite integral is a limit of Riemann sums. Think of a definite integral as a sum.
- We have seen that Riemann sums may represent area, volume, or displacement.
- If $F'(x) = f(x)$ for x in an interval, then F is called an antiderivative of f. If, also, $G'(x) = f(x)$ on the same interval, then there is a constant C such that $G(x) = F(x) + C$ on the interval.
- The integration formula $\int f(x)\, dx = F(x) + C$ means $F'(x) = f(x)$ on an interval. Every differentiation formula $F'(x) = f(x)$ has a corresponding integration formula $\int f(x)\, dx = F(x) + C$.
- Relations between the concepts of the derivative and the definite integral are expressed by the Fundamental Theorems of Calculus.
- If f and F are continuous on the interval $a \le x \le b$ and $F'(x) = f(x)$ for $a < x < b$, the Fundamental Theorem of Calculus implies $\int_a^b f(x)\, dx = F(b) - F(a)$.
- The Fundamental Theorem can be expressed as $\int_a^b F'(x)\, dx = F(b) - F(a)$.

- If $\int f(x)\, dx = F(x) + C$, then $\int_a^b f(x)\, dx = F(b) - F(a)$, so we can use an integration formula and the Fundamental Theorem to evaluate a definite integral.
- If f is continuous, the Second Fundamental Theorem implies

$$\frac{d}{dx}\left(\int_a^x f(t)\, dt \right) = f(x).$$

You should know the Chain Rule version of this formula.
- If $u = u(x)$, then $\int f(u(x))u'(x)\, dx = \int f(u)\, du$. This is the formula for change of variables in indefinite integrals.

$$\int_a^b f(u(x))u'(x)\, dx = \int_{u(a)}^{u(b)} f(u)\, du.$$

- Numerical methods such as the Trapezoidal Rule and Simpson's Rule can be used to approximate values of definite integrals. Some methods of approximation are more efficient than others.

CHAPTER 5 REVIEW EXERCISES

Find all k for which the integration formulas in Exercises 1–4 are valid.

1. $\displaystyle\int \sin 2x\, dx = k \cos 2x + C$

2. $\displaystyle\int \sec^2 \frac{x}{3}\, dx = k \tan \frac{x}{3} + C$

3. $\displaystyle\int \sqrt{5 - 2x}\, dx = k(5 - 2x)^{3/2} + C$

4. $\displaystyle\int \frac{x}{\sqrt{x^2 + 1}}\, dx = k\sqrt{x^2 + 1} + C$

5. Find a function $G(x)$ that satisfies $G'(x) = 6x^2 + 4x - 1$ and $G(1) = 2$.

6. Find a function $G(x)$ that satisfies $G''(x) = 12x^2$, $G'(1) = 3$, and $G(1) = 4$.

7. A ball is thrown vertically upward from ground level at a speed of 64 ft/s. How high above ground will it get? How long will it take to return to ground level?

8. A hot-air balloon is rising vertically at a rate of 2 ft/s. A sandbag is released from the balloon when it is at a height of 1580 ft. How long will it take the sandbag to hit the ground? At what speed will it hit the ground?

Use R_4 to find an approximate value of the area under the curves given in Exercises 9–12.

9. $y = x^2$, $0 \le x \le 2$

10. $y = x^2$, $-2 \le x \le 2$

11. $y = 1/(x + 1)$, $0 \le x \le 2$

12. $y = x/(x + 1)$, $0 \le x \le 2$

In Exercises 13–14, approximate the area under the curve by using the Riemann sum with $n = 4$ subintervals of equal length and x_j^* chosen from the graph so that $f(x_j^*)\Delta x_j$ appears to be approximately the area under the curve between x_{j-1} and x_j.

13.

14.

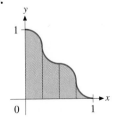

In Exercises 15–18, use R_4 to find an approximate value of the volume of the solids pictured.

15. Cross-sections perpendicular to the x-axis are squares,
$$A(x) = \frac{x^2}{4}, \qquad 0 \le x \le 4.$$

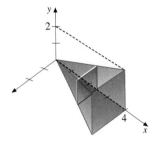

16. Cross-sections perpendicular to the x-axis are squares,
$$A(x) = \frac{1}{4}(4 - x)^2, \qquad 0 \le x \le 4.$$

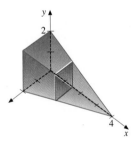

17. Cross-sections are quarter-circles,
$$A(x) = \frac{\pi}{36}(3 - x)^2, \qquad 0 \le x \le 3.$$

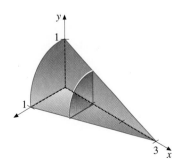

18. Cross-sections are quarter-circles, $A(x) = \dfrac{\pi x^2}{36}$, $0 \le x \le 3$.

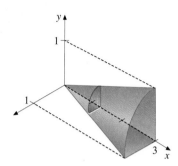

In Exercises 19–20, the velocity of an object on a number line is given by the formula $v(t)$, $0 \le t \le 1$. Use Riemann sums of $v(t)$ over the interval $[0, 1]$ with six subintervals of equal length to obtain upper and lower bounds for $s(1) - s(0)$, the change in position of the object. (Use the maximum value of $v(t)$ in each subinterval to determine the upper bound; use the minimum value in each subinterval to determine the lower bound.)

19. $v(t) = 2 \sin \pi t$ **20.** $v(t) = 2 \cos \pi t$

Evaluate the sums in Exercises 21–26.

21. $\displaystyle\sum_{j=1}^{3}(2j - 1)$ **22.** $\displaystyle\sum_{j=3}^{5}(3j^2 - j)$

23. $\displaystyle\sum_{j=1}^{16}(j^3 + 2j)$ **24.** $\displaystyle\sum_{j=1}^{100}(2j^2 - 3j + 4)$

25. $\displaystyle\sum_{j=1}^{40}(j^2 - j^3)$ **26.** $\displaystyle\sum_{j=1}^{60}(4 + j^2)$

Find a simple formula for the sums in Exercises 27–30.

27. $\displaystyle\sum_{j=1}^{n}(6j^2 - 1)$ **28.** $\displaystyle\sum_{j=1}^{n}(j^3 + 2j)$

29. $\displaystyle\sum_{j=1}^{2n} j$ **30.** $\displaystyle\sum_{j=1}^{2n} j^2$

In Exercises 31–34, (a) find a simple formula for R_n that corresponds to the given integral, (b) evaluate $\lim_{n \to \infty} R_n$, and (c) use the Fundamental Theorem to evaluate the integral.

31. $\displaystyle\int_0^2 (3x^2 - 4x)\,dx$ **32.** $\displaystyle\int_0^1 (4x^3 + 3)\,dx$

33. $\displaystyle\int_0^3 (12x^3 - 5)\,dx$ **34.** $\displaystyle\int_0^1 (2x - 3x^2)\,dx$

Evaluate the integrals in Exercises 35–36, given that
$$\int_0^3 f(x)\,dx = 7, \quad \int_0^9 f(x)\,dx = 4, \quad and \quad \int_0^3 g(x)\,dx = 5.$$

35. $\displaystyle\int_0^3 [4g(x) - f(x)]\,dx - \int_3^9 2f(x)\,dx$

36. $\displaystyle\int_0^3 [3f(x) - 2g(x)]\,dx - \int_3^9 f(x)\,dx$

37. Show that $\left| \displaystyle\int_0^{2\pi} x \sin x\,dx \right| \le 2\pi^2$.

38. Show that $\left| \displaystyle\int_0^{\pi} x^2 \sin x\,dx \right| \le \dfrac{\pi^3}{3}$.

39. The **variance of a random variable** X with distribution $f(x)$ is
$$Var(X) = \int_{-c}^{c} x^2 f(x)\,dx - \left[\int_{-c}^{c} x f(x)\,dx \right]^2$$
and the variance of $aX + b$ is
$$Var(aX + b)$$
$$= \int_{-c}^{c} (ax + b)^2 f(x)\,dx - \left[\int_{-c}^{c} (ax + b) f(x)\,dx \right]^2,$$
where a, b and c are constants and $\displaystyle\int_{-c}^{c} f(x)\,dx = 1$. Show that $Var(aX + b) = a^2 \, Var(X)$.

40. Show that $\displaystyle\int_1^{\sqrt{c}} \dfrac{1}{x}\,dx = \int_{\sqrt{c}}^{c} \dfrac{1}{x}\,dx$.

Evaluate the derivatives in Exercises 41–44.

41. $\dfrac{d}{dx}\left(\displaystyle\int_0^x \sin t^2\,dt \right)$ **42.** $\dfrac{d}{dx}\left(\displaystyle\int_x^1 \dfrac{1}{t}\,dt \right)$, $x > 0$

43. $\dfrac{d}{dx}\left(\displaystyle\int_1^{1/x} \dfrac{1}{t}\,dt \right)$ **44.** $\dfrac{d}{dx}\left(\displaystyle\int_x^{x^3} \dfrac{1}{t}\,dt \right)$, $x > 0$

Use l'Hôpital's Rule to evaluate the limits in Exercises 45–46.

45. $\displaystyle\lim_{x \to 0} \dfrac{1}{x^2} \int_0^x \dfrac{t}{t^2 + 1}\,dt$ **46.** $\displaystyle\lim_{x \to 0} \dfrac{1}{x^3} \int_0^x \sin(t^2)\,dt$

Find the equation of the line tangent to the given graph at the given point in Exercises 47–48.

47. $y = 1 - \displaystyle\int_0^x (t^2 + 1)^{10}\,dt$, $(0, 1)$

48. $y = x + \displaystyle\int_0^x \sec t\,dt$, $(0, 0)$

Evaluate the indefinite integrals in Exercises 49–64.

49. $\displaystyle\int (x^3 - x + 2)\,dx$ **50.** $\displaystyle\int (x^{2/3} - x^{-1/3})\,dx$

51. $\displaystyle\int (\sin x - \cos x)\,dx$ **52.** $\displaystyle\int (\csc x)(\csc x + \cot x)\,dx$

53. $\displaystyle\int \sqrt{x}(x+1)\,dx$

54. $\displaystyle\int \frac{x^2+1}{x^2}\,dx$

55. $\displaystyle\int \sec \pi x \tan \pi x\,dx$

56. $\displaystyle\int \csc^2(x/2)\,dx$

57. $\displaystyle\int \sqrt{2x+1}\,dx$

58. $\displaystyle\int \frac{1}{\sqrt{1-x}}\,dx$

59. $\displaystyle\int x\sqrt{x^2+4}\,dx$

60. $\displaystyle\int \frac{x}{\sqrt{1-x^2}}\,dx$

61. $\displaystyle\int \frac{(1+\sqrt{x})^3}{\sqrt{x}}\,dx$

62. $\displaystyle\int \frac{(1+x^{2/3})^2}{x^{1/3}}\,dx$

63. $\displaystyle\int \sin^2 x \cos x\,dx$

64. $\displaystyle\int \tan^2 x \sec^2 x\,dx$

Evaluate the definite integrals in Exercises 65–70.

65. $\displaystyle\int_1^2 (6x^2-4x+1)\,dx$

66. $\displaystyle\int_{-\pi/2}^{\pi/2} \sin x\,dx$

67. $\displaystyle\int_0^1 (2x+1)^{-4}\,dx$

68. $\displaystyle\int_{-\pi/2}^{\pi/2} \sec^2 \frac{x}{2}\,dx$

69. $\displaystyle\int_0^4 \frac{x}{(x^2+9)^2}\,dx$

70. $\displaystyle\int_0^{\pi/2} \sin x \cos x\,dx$

Express the integrals in Exercises 71–74 as a definite integral in terms of the indicated variable.

71. $\displaystyle\int_0^{\pi/2} \frac{\cos x}{3\sin^2 x + 4\cos^2 x}\,dx, \quad u = \sin x$

72. $\displaystyle\int_0^{\pi/4} \frac{\sin x}{1+4\cos^2 x}\,dx, \quad u = 2\cos x$

73. $\displaystyle\int_1^{64} \frac{1}{x^{1/2}+x^{1/3}}\,dx, \quad u = x^{1/6}$

74. $\displaystyle\int_0^2 \frac{x}{1+4x^2}\,dx, \quad u = 1+4x^2$

75. The study of **fully developed laminar flow between infinite parallel plates** leads to the equation

$$v = k(ay - y^2),$$

where v is the velocity of the flow at distance y from the bottom plate, a is the distance between the plates, and k is a constant. Find the average velocity. Find the maximum velocity in terms of the average velocity.

76. Show that $V = \displaystyle\int_a^b [f(x)-c]^2\,dx$ is minimum when c is the average value of f over the interval $a \le x \le b$.

*In Exercises 77–80, **(a)** use T_4 to find an approximate value of the given integral. **(b)** Use S_4 to find an approximate value of the given integral. **(c)** Find a bound of the error in (a). **(d)** Find a bound of the error in (b). **(e)** Find values of n that guarantee T_n is within 0.005 of the value of the integral. **(f)** Find values of n that guarantee S_n is within 0.005 of the value of the integral.*

77. $\displaystyle\int_{-\pi/2}^{\pi/2} \cos x\,dx$

78. $\displaystyle\int_0^2 x^4\,dx$

79. $\displaystyle\int_0^2 \frac{x}{x+1}\,dx$

80. $\displaystyle\int_0^4 (2x+1)^{-1}\,dx$

81. In Chapter 7 we will see that $\displaystyle\int_{-1}^1 \frac{1}{x^2+1}\,dx = \frac{\pi}{2}$.

(a) Use Formula 5 of Section 5.6 to find n for which

$$\left| \int_{-1}^1 \frac{1}{x^2+1}\,dx - T_n \right| < 0.0005.$$

(b) Use Formula 7 of Section 5.6 to find n for which

$$\left| \int_{-1}^1 \frac{1}{x^2+1}\,dx - S_n \right| < 0.0005.$$

82. **(a)** Can a right triangle with base 8 in. and height 3 in. be covered with strips perpendicular to the base, as indicated below by using *rectangular* strips cut from 13 in. of 1-in. tape? **(b)** Will rectangular strips cut from 26 in. of 1/2-in. tape cover the triangle? Explain.

Use the graph of the velocity of particle moving along a number line given in Exercises 83–84 to determine the first time the particle returns to its initial position.

83.

84.

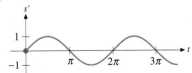

85. Write a definite integral that represents the area between the curves $y = f_1(x)$ and $y = f_2(x)$, $a \le x \le b$, where $f_1(x) \le f_2(x)$ for $a \le x \le b$.

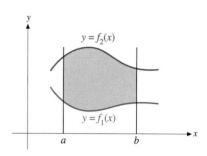

86. Write a definite integral that represents the area between the curve $y = f(x)$ and the x-axis, $a \le x \le b$, where $f(x) \le 0$ for $a \le x \le b$.

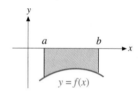

87. Use the figure below to explain why $\sum_{j=1}^{n} j = n(n+1)/2$.

88. Find a function f for which the Trapezoidal approximations of $\int_{a}^{b} f(x)\,dx$ are more accurate than the corresponding Simpson's Rule approximations. Explain.

EXTENDED APPLICATION

CATCHING A TOXIC WASTE DUMPER

You have been employed by the Environmental Protection Agency to analyze water samples from rivers and streams in western Pennsylvania. Recent samples from the Allegheny River indicate that toxic waste is being dumped into the river just above Pittsburgh. You have located the probable source, and the agency's lawyers are preparing a case against the suspected polluter. To strengthen their case, they would like to know how much waste is entering the river in a typical week.

In order to do your calculation, you must know the amount of water flowing past the checkpoint where your samples were taken, as well as the concentration of waste in the water. Figure 1 shows a cross-section of the river at that point.

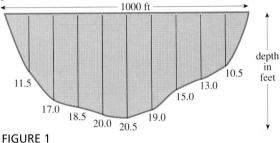

FIGURE 1

Tasks:

1. Use a numerical integration technique to estimate the area of the cross-section shown. Write a few sentences to explain why you chose the method you used.
2. During the week when the samples were taken, the current was 1.5 miles per hour. How many cubic feet of water flowed past the checkpoint each day during that week? How many gallons of water flowed past the checkpoint each day? (One gallon of water occupies 0.1337 cubic foot.)
3. The following table lists the concentration of waste, in gallons of waste per gallon of water, in samples you took on eight consecutive days. Estimate the rate at which waste was flowing past the checkpoint at the time of each sample, in gallons of waste per day.

Time	Concentration		Time	Concentration	
2 P.M. Mon.	6.235	10^9	2 P.M. Fri.	7.916	10^9
2 P.M. Tue.	6.928	10^9	2 P.M. Sat.	0.857	10^9
2 P.M. Wed.	7.470	10^9	2 P.M. Sun.	0.113	10^9
2 P.M. Thu.	8.044	10^9	2 P.M. Mon.	6.481	10^9

4. Use your results from Task 3 to sketch a rough graph of the rate of flow of waste past your checkpoint as a function of time.
5. Use a numerical integration technique to estimate the total number of gallons of waste entering the river during the week. Write a few sentences to explain why you chose the method you used.

Writing Assignment:

Submit two reports. The first, to the agency's lawyers, should present your findings, along with a nontechnical summary of your methods for the benefit of a jury. It can contain phrases such as "Using the methods of calculus, I determined that ...." The second report, for the benefit of the expert witnesses on both sides, should present your calculations, supported by clear explanations in the language of calculus.

6 Applications of the Definite Integral

Applications of the definite integral to a variety of physical problems are illustrated. The underlying principle involved is the interpretation of the definite integral as a sum of products, with each component of the integral representing a physical quantity.

Each component of a definite integral that represents a physical quantity must be expressed in terms of the variable of integration. We cannot set up such an integral properly unless we can describe physical quantities in terms of a variable. This basic analytic geometry skill is studied in the first section of this chapter.

6.1 ANALYTIC GEOMETRY FOR APPLICATIONS OF THE DEFINITE INTEGRAL

Connections

Variable points on a graph, length and midpoint formulas, 1.3.

In this section we will illustrate how analytic geometry can be used to describe selected geometric quantities such as length and area in terms of a variable. The skills developed here are necessary in order to apply the concept of the definite integral to physical problems. The examples in this section are presented exactly as they will be used in later sections as parts of calculus problems.

Note that a computer can be used to evaluate a definite integral, but the computer will evaluate only the integral you give it. It is up to you to set up a definite integral that represents the desired physical quantity.

For simplicity of exposition we will use the usual variables x and y with their usual orientation of the y-axis "vertical" and the x-axis "horizontal." There should be no difficulty in applying the concepts to other variables.

Lengths of Variable Vertical Line Segments

Let us begin by considering the length of a **variable vertical line segment**. Recall from Section 1.3 that the coordinates of a variable point on the graph of $y = f(x)$ can be expressed in terms of the variable x as $(x, f(x))$. For example, (x, x^2) is a variable point on the graph of $y = x^2$ and $(x, x+2)$ is a variable point on the graph of $y = x + 2$. Two points are on the same vertical line if they have the same x-coordinate. For example, the points (x, x^2) and $(x, x+2)$ are on the same vertical line for any choice of the variable x; the line segment between these points is a variable vertical line segment. See Figure 6.1.1.

Use x as the independent variable in problems that deal with variable vertical line segments.

It is not necessary to use the distance formula to obtain the length of a vertical line segment. We should label the coordinates of the endpoints and then use

(Length of vertical line segment)

$\qquad = (y\text{-coordinate of top point}) - (y\text{-coordinate of bottom point}).$

EXAMPLE 1

Find the length of a variable vertical line segment between $y = x^2$ and the x-axis.

SOLUTION   The length of the line segment depends on its position. That is, we will determine length as a function of x.

The first step is to make a sketch and label coordinates in terms of the variable x. This is done in Figure 6.1.2. The point (x, x^2) is a variable point on the graph of $y = x^2$. The x-axis has equation $y = 0$, so $(x, 0)$ is a variable point on the x-axis.

The length of a variable vertical line segment is given by the y-coordinate of the top point minus the y-coordinate of the bottom point. From our sketch we see that (x, x^2) is the top of the line segment and $(x, 0)$ is the bottom, for all values of x. We can then write

$$\text{Length} = x^2 - 0 \text{ or } L(x) = x^2. \quad \blacksquare$$

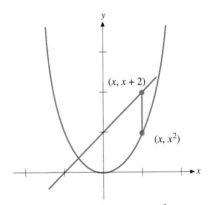

FIGURE 6.1.1 The point (x, x^2) is a variable point on the graph of $y = x^2$ and $(x, x+2)$ is a variable point on the graph of $y = x + 2$. The line segment between these points is a variable vertical line segment between the graphs.

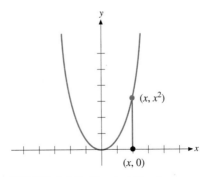

FIGURE 6.1.2 The length of a variable vertical line segment between $y = x^2$ and the x-axis is $L(x) = x^2$.

Let us note that the distance formula gives the same formula obtained in Example 1. That is,

$$L(x) = \sqrt{(x - x)^2 + (x^2 - 0)^2} = \sqrt{(x^2)^2} = |x^2| = x^2.$$

EXAMPLE 2

Find the length of a variable vertical line segment between points on the graph of $x + 1 - (y - 1)^2 = 0$.

SOLUTION We wish to express length as a function of x. To do this, we need to solve the given equation for y in terms of x:

$$x + 1 - (y - 1)^2 = 0,$$
$$(y - 1)^2 = x + 1,$$
$$y - 1 = \pm\sqrt{x + 1},$$
$$y = 1 + \sqrt{x + 1}, \qquad y = 1 - \sqrt{x + 1}.$$

We can now sketch the parabola and label coordinates. This is done in Figure 6.1.3.

We use the sketch to write

$$\text{Length} = (1 + \sqrt{x + 1}) - (1 - \sqrt{x + 1}),$$
$$L(x) = 2\sqrt{x + 1}.$$

The domain of this function is $x \geq -1$. When $x = -1$, the "line segment" has length zero. The vertical line $x = a$ does not intersect the parabola when $a < -1$. ∎

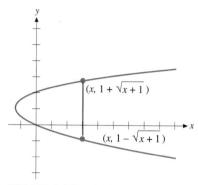

FIGURE 6.1.3 The length of a variable vertical line segment between points on the graph of $x + 1 - (y - 1)^2 = 0$ is $L(x) = 2\sqrt{x + 1}$.

Midpoint of a Variable Line Segment

Recall from Section 1.3 that the **midpoint** of the line segment between the points (x_1, y_1) and (x_2, y_2) is

$$\left(\frac{x_1 + x_2}{2}, \frac{y_1 + y_2}{2} \right).$$

EXAMPLE 3

Find the length and midpoint of a variable vertical line segment between $y = x^2$ and $x + y = 2$.

SOLUTION The graphs are sketched and labeled in Figure 6.1.4. We have used the fact that $x + y = 2$ implies $y = 2 - x$.

The line is above the parabola between their points of intersection, and the parabola is above the line outside the points of intersection.

There is not a single variable vertical line segment whose labeling of top and bottom holds for all x.

There are two different types of vertical segments with different labeling. This will be reflected in the fact that two different formulas will be used to express the length—one the negative of the other. To determine intervals where each formula holds, we need the x-coordinates of the points of intersection. If a point (x, y) is in the intersection of the graphs, its coordinates will satisfy both equations. This leads to the simultaneous equations

$$y = x^2, \qquad y = 2 - x.$$

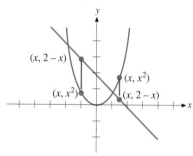

FIGURE 6.1.4 The length of a variable vertical line segment between $y = x^2$ and $x + y = 2$ is $(2 - x) - x^2$ between their points of intersection and $x^2 - (2 - x)$ outside their points of intersection.

We do not need to find pairs (x, y) that satisfy both equations, but only the x-coordinates of the solutions.

We eliminate the undesired variable.

This is done easily by equating the two expressions for y. We obtain

$$x^2 = 2 - x.$$

Solving for x, we have

$$x^2 + x - 2 = 0,$$
$$(x + 2)(x - 1) = 0,$$
$$x = -2, \qquad x = 1.$$

The length of a variable vertical line segment is

$$L(x) = \begin{cases} (2 - x) - x^2, & -2 \le x \le 1, \\ x^2 - (2 - x), & x < -2 \text{ or } x > 1. \end{cases}$$

The midpoint is

$$\left(\frac{x + x}{2}, \frac{x^2 + (2 - x)}{2} \right) = \left(x, \frac{x^2 - x + 2}{2} \right).$$

The two different types of vertical line segments give the same formula for the coordinates of their midpoints. ■

If the distance formula were used to determine the length of the line segment in Example 3, we would have

$$L(x) = \sqrt{(x - x)^2 + \big((2 - x) - x^2 \big)^2}$$
$$= |(2 - x) - x^2|$$
$$= \begin{cases} (2 - x) - x^2, & (2 - x) - x^2 \ge 0, \\ x^2 - (2 - x), & (2 - x) - x^2 < 0. \end{cases}$$

We would then need to solve the above inequalities to determine intervals of x for which each formula for $L(x)$ is valid. The result would be the same as that obtained in Example 3. (Absolute value signs are not convenient for the calculations we will be making in applications of the definite integral.)

EXAMPLE 4

Find the length of a variable vertical line segment between $x - y = 0$ and $x^2 - y + 1 = 0$.

SOLUTION   The graphs are sketched and labeled in Figure 6.1.5.

It appears from our sketch that the parabola is always above the line. To be sure of this, we attempt to find the x-coordinates of points of intersection. This leads to the equations

$$y = x^2 + 1, \qquad y = x.$$

Equating the two expressions for y and then solving for x, we obtain

$$x^2 + 1 = x,$$
$$x^2 - x + 1 = 0.$$

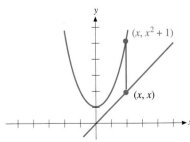

FIGURE 6.1.5 The length of a variable vertical line segment between $x - y = 0$ and $x^2 - y + 1 = 0$ is $L(x) = (x^2 + 1) - x$.

The Quadratic Formula gives the solutions

$$x = \frac{-(-1) \pm \sqrt{(-1)^2 - 4(1)(1)}}{2(1)}.$$

Since $b^2 - 4ac = (-1)^2 - 4(1)(1) = -3 < 0$, there is no real number solution of this equation. This means the graphs do not intersect. We have length

$$L(x) = (x^2 + 1) - x. \quad \blacksquare$$

Variable Horizontal Line Segments

The same ideas that we used for vertical line segments also apply to horizontal line segments. Points on a horizontal line have a common y-coordinate.

Use y as the independent variable in problems that deal with variable horizontal line segments.

In this case, we must label the coordinates on our graphs in terms of their y-coordinate. Also,

(Length of horizontal line segment)

= (x-coordinate of right endpoint) − (x-coordinate of left endpoint).

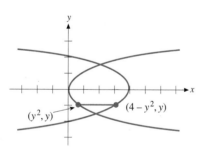

FIGURE 6.1.6 The length of a variable horizontal line segment between $x = y^2$ and $x = 4 - y^2$ is $L(y) = (4 - y^2) - y^2$ between their points of intersection.

EXAMPLE 5

Find the length of a variable horizontal line segment between $x = y^2$ and $x = 4 - y^2$ for values of y that are between the y-coordinates of the points of intersection of the graphs.

SOLUTION   The graphs are sketched and labeled in Figure 6.1.6.

We need the y-coordinates of the points of intersection of the graphs. In this case, we eliminate the variable x from the simultaneous equations

$$x = y^2, \qquad x = 4 - y^2$$

to obtain

$$y^2 = 4 - y^2.$$

Solving for y, we have

$$2y^2 = 4,$$
$$y^2 = 2,$$
$$y = \sqrt{2}, \qquad y = -\sqrt{2}.$$

For values of y between the y-coordinates of the points of intersection, the parabola $x = 4 - y^2$ is to the right of the parabola $x = y^2$. This means the length is

$$L(y) = (4 - y^2) - y^2, \qquad -\sqrt{2} \le y \le \sqrt{2}. \quad \blacksquare$$

Variable Cross-Sections

We will need to be able to find the area of a variable cross-section of certain three-dimensional solids.

Use x as the independent variable in problems that deal with variable cross-sections perpendicular to the x-axis.

EXAMPLE 6

Cross-sections perpendicular to the x-axis of the prism pictured in Figure 6.1.7a are squares. Find the area of a variable cross-section.

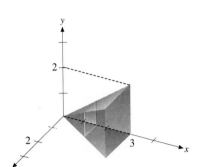

SOLUTION We see from Figure 6.1.7a that the length of one side of the square formed by a variable cross-section is the length of a variable (vertical) line segment between the x-axis and the line in the xy-plane through the origin and the point $(3, 2)$. The slope of this line is

$$\frac{2 - 0}{3 - 0} = \frac{2}{3}.$$

FIGURE 6.1.7a Cross-sections are squares with one side of the square formed by a variable line segment between the x-axis and the line in the xy-plane through the origin and the point $(3, 2)$.

The equation of the line through the origin with slope $2/3$ is

$$y = \frac{2}{3}x.$$

This line and a variable vertical line segment between it and the x-axis are sketched and labeled in Figure 6.1.7b. We see that the length of one side of the square is

$$\text{Length } = \frac{2}{3}x.$$

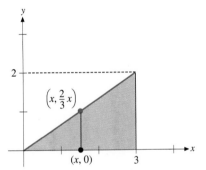

The area of a variable square cross-section is

$$\text{Area} = (\text{Length})^2,$$

$$A(x) = \left(\frac{2}{3}x\right)^2,$$

$$A(x) = \frac{4}{9}x^2, \qquad 0 \le x \le 3. \quad \blacksquare$$

FIGURE 6.1.7b The length of a variable vertical line segment between $y = \frac{2}{3}x$ and the x-axis is $\frac{2}{3}x$.

Use y as the independent variable in problems that deal with variable cross-sections perpendicular to the y-axis.

EXAMPLE 7

Cross-sections perpendicular to the y-axis of the right circular cone pictured in Figure 6.1.8a are circles. Find the area of a variable cross-section.

SOLUTION We need to determine the radius of a variable circular cross-section as a function of y. This is the length of a variable (horizontal) line segment between the y-axis and the line in the xy-plane through the points $(0, 3)$ and $(1, 0)$. The slope of this line is -3, and the equation is

$$y - 0 = -3(x - 1).$$

FIGURE 6.1.8a   Cross-sections are circles with radius the length of a variable line segment between the *y*-axis and the line in the *xy*-plane through (0, 3) and (1, 0).

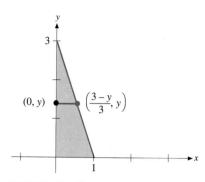

FIGURE 6.1.8b   The length of a variable horizontal line segment between $3x = 3 - y$ and the *y*-axis is $\dfrac{3-y}{3}$.

Solving for x we have

$$y = -3x + 3,$$
$$3x = 3 - y,$$
$$x = \frac{3 - y}{3}.$$

The line and a variable horizontal line segment are sketched and labeled in Figure 6.1.8b. We see that

$$\text{Radius} = \frac{3 - y}{3}.$$

The area of a circular cross-section is

$$\text{Area} = \pi (\text{Radius})^2,$$
$$A(y) = \pi \left(\frac{3 - y}{3} \right)^2, \qquad 0 \le y \le 3. \quad \blacksquare$$

Revolving Points and Line Segments About a Line

A point that is revolved about a line traces out a circle. Figure 6.1.9a illustrates a circle obtained by revolving a point about a line that is parallel to the *x*-axis. Figure 6.1.9b illustrates a circle obtained by revolving a point about a line that is parallel to the *y*-axis. These circles in three-dimensional space are drawn as ellipses on the two-dimensional sketch plane in order to represent an oblique view of the circles.

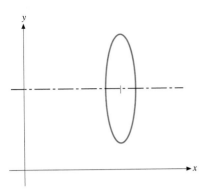

FIGURE 6.1.9a A point that is revolved about a line traces out a circle.

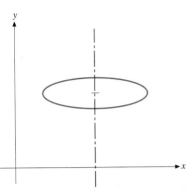

FIGURE 6.1.9b Circles in three-dimensional space that are not parallel to the sketch plane are drawn as ellipses.

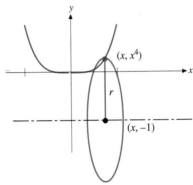

FIGURE 6.1.10 The circle obtained by revolving a variable point on the graph of $y = x^4$ about the line $y = -1$ has radius $r = x^4 + 1$.

EXAMPLE 8

Find the **(a)** area and **(b)** circumference of the circle obtained by revolving a variable point on the graph of $y = x^4$ about the line $y = -1$, in terms of x.

SOLUTION   The graph is sketched in Figure 6.1.10. We see that the radius of the circle is $r = x^4 - (-1)$.

(a) The area is $A = \pi r^2 = \pi(x^4 + 1)^2$.
(b) The circumference is $C = 2\pi r = 2\pi(x^4 + 1)$. ∎

When a line segment is revolved through space, a surface is traced out. To sketch the surface, first sketch the circles obtained by revolving each endpoint of the segment about the line. This should indicate the type of surface obtained.

If a line segment is revolved about a perpendicular line, the resulting surface will be a disk if the segment touches the line (Figure 6.1.11a) and a washer if the segment does not touch the line (Figure 6.1.11b). A line segment that is revolved about a parallel line generates a cylindrical surface (Figure 6.1.11c).

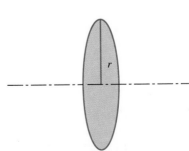

FIGURE 6.1.11a If a line segment is revolved about a perpendicular line through an endpoint of the segment, the resulting surface is a disk.

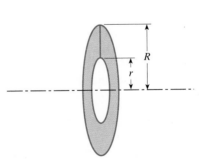

FIGURE 6.1.11b If a line segment is revolved about a perpendicular line that does not intersect the segment, the resulting surface is a washer.

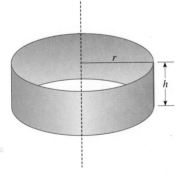

FIGURE 6.1.11c A line segment that is revolved about a parallel line generates a cylindrical surface.

FIGURE 6.1.12 The area of a cylindrical surface is the area of the rectangle obtained by cutting and unrolling the cylinder, so $A = 2\pi rh$.

The areas of these surfaces are given by the formulas

$$\text{Area of disk } = \pi r^2,$$
$$\text{Area of washer } = \pi R^2 - \pi r^2,$$
$$\text{Area of cylindrical surface } = 2\pi rh.$$

Note that the area of the cylindrical surface is the area of the rectangle obtained by cutting and unrolling the cylinder. See Figure 6.1.12.

EXAMPLE 9

Find the area of the disk obtained by revolving a variable vertical line segment between $y = \sqrt{x}$ and the x-axis about the x-axis.

SOLUTION In Figure 6.1.13 we have sketched the curve, sketched and labeled the endpoints of a variable vertical line segment, and then sketched the circle obtained by revolving the upper endpoint about the x-axis. The lower endpoint is on the x-axis so that the point remains stationary as the line segment is revolved about the x-axis. The radius of the resulting disk is

$$r = \sqrt{x} - 0.$$

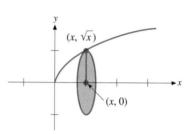

FIGURE 6.1.13 If a variable vertical line segment between $y = \sqrt{x}$ and the x-axis is revolved about the x-axis, a disk with radius $\sqrt{x}$ is generated.

The area of the disk is

$$\text{Area } = \pi r^2,$$
$$A(x) = \pi(\sqrt{x})^2,$$
$$A(x) = \pi x, \qquad x \geq 0. \quad \blacksquare$$

EXAMPLE 10

A variable vertical line segment between $y = \sqrt{x}$ and the x-axis is revolved about the line $y = -1$. Find the area of the resulting washer.

SOLUTION The first step is to sketch the curve, draw a variable vertical line segment, and sketch the line $y = -1$. The circles obtained by revolving each endpoint of the segment about the axis of revolution are then sketched, indicating a washer. We then label both endpoints of the variable line segment and the center of the washer in terms of x. This is done in Figure 6.1.14. From the sketch, we see that the washer has

$$\text{Outer radius: } R = \sqrt{x} - (-1) = \sqrt{x} + 1,$$
$$\text{Inner radius: } r = 0 - (-1) = 1.$$

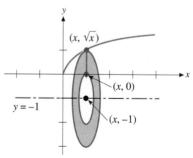

FIGURE 6.1.14 If a variable vertical line segment between $y = \sqrt{x}$ and the x-axis is revolved about the line $y = -1$, a washer with outer radius $\sqrt{x} + 1$ and inner radius 1 is generated.

The area is

$$\text{Area } = \pi R^2 - \pi r^2,$$
$$A(x) = \pi(\sqrt{x} + 1)^2 - \pi(1)^2, \qquad x \geq 0. \quad \blacksquare$$

EXAMPLE **11**

Find the area of the cylindrical surface obtained by revolving a variable vertical line segment between $y = \sqrt{x}$ and the x-axis about the y-axis.

SOLUTION   We first sketch the curve and a variable vertical line segment. Each endpoint of the segment is then revolved about the y-axis, forming a circle. Since the vertical segment is parallel to the axis of revolution, the radii of these two circles are equal. These two circles are then connected by a vertical line segment that indicates the edge of the cylinder generated by revolving the segment. The final step is to label the endpoints of the segment and a point on the axis of revolution. This is done in Figure 6.1.15. From the sketch, we see the cylinder has

$$\text{Radius: } r = x - 0,$$
$$\text{Height: } h = \sqrt{x} - 0.$$

The area of the cylindrical surface is

$$\text{Area } = 2\pi rh$$
$$A(x) = 2\pi (x)(\sqrt{x}), \qquad x \geq 0. \quad \blacksquare$$

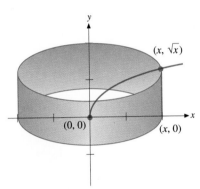

FIGURE 6.1.15   If a variable vertical line segment between $y = \sqrt{x}$ and the x-axis is revolved about the y-axis, a cylindrical surface with radius x and height $\sqrt{x}$ is generated.

EXERCISES 6.1

Sketch, label, and find the length of the variable line segments described in Exercises 1–12. Use x as the variable in problems that involve vertical line segments; use y as the variable in problems that involve horizontal line segments. Do not use absolute value signs in formulas for length.

1. Vertical, between $y = x^2$ and $y = -2$

2. Vertical, between $y = -x^2$ and $y = 1$

3. Vertical, between $y = x^2 + 1$ and $y + x^2 = 0$

4. Vertical, between $y = 1 - x^2$ and $x + y = 2$

5. Horizontal, between $x = y^2$ and $x + y = -1$

6. Horizontal, between $x = 1 - y^2$ and $x = 2$

7. Vertical, between points on $(x - 2)^2 + y^2 = 4$

8. Horizontal, between points on $(x - 2)^2 + y^2 = 4$

9. Horizontal, between points on $y = (x - 1)^2 - 1$

10. Vertical, between points on $x = (y + 1)^2 - 1$

11. Vertical, between $y = 2x$ and the x-axis

12. Horizontal, between $y = 3x$ and $y = 2x$

Sketch, label, and find the length and midpoint of the line segments described in Exercises 13–20. Use x as the variable in problems that involve vertical line segments; use y as the variable in problems that involve horizontal line segments. Do not use absolute value signs in formulas for length.

13. Vertical, between $y = 2 - x^2$ and $y = x$

14. Vertical, between $y = x^2$ and $y = 4 - x^2$

15. Horizontal, between $x = y^2$ and $2x = 1 + y^2$

16. Horizontal, between $x + y^2 = 1$ and $x - y + 1 = 0$

17. Vertical, between $y = x^2$ and $y = x, 0 \le x \le 1$

18. Horizontal, between $y = x^3$ and $y = x, 0 \le y \le 1$

19. Horizontal, between $y = x^3$ and $y = x^2, 0 \le y \le 1$

20. Vertical, between $y = x^3$ and $y = x^2, 0 \le x \le 1$

The shape of cross-sections perpendicular to the x-axis is indicated for each of the solids in Exercises 21–26. Express the area of a variable cross-section as a function of x.

21. Circles

22. Circles

23. Squares

24. Squares

25. Isosceles right triangles

26. Rectangles

The shape of cross-sections perpendicular to the y-axis is indicated for each of the solids in Exercises 27–30. Express the area of a variable cross-section as a function of y.

27. Squares

28. Squares

29. Rectangles

30. Isosceles right triangles

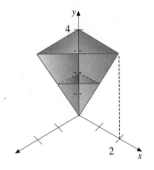

*In Exercises 31–42, a variable point on the graph of the given equation is revolved about the given line. Find the **(a)** area and **(b)** circumference of the resulting circle in terms of the indicated variable.*

31. $y = x^2$, about the x-axis, in terms of x

32. $y = \sqrt{x}$, about the x-axis, in terms of x

33. $y = \sin x$, about the line $y = 1$, in terms of x

34. $y = \cos x$, about the line $y = -1$, in terms of x

35. $y = \sqrt{x}$, about the y-axis, in terms of y

36. $y = x^3$, $x \geq 0$, about the y-axis, in terms of y

37. $x = y^2$, about the line $x = -1$, in terms of y

38. $y = \sqrt{1 - x}$, about the line $x = 2$, in terms of y

39. $y = x^2$, about the y-axis, in terms of x

40. $y = \sqrt{x}$, about the y-axis, in terms of x

41. $y = \sqrt{x}$, about the x-axis, in terms of y

42. $y = x^2$, about the x-axis, in terms of y

In Exercises 43–56, revolving the given line segment about the indicated line generates either a disk, washer, or cylinder surface. Sketch and label whichever one is generated in each exercise. Find the area of each surface obtained. Use x as the variable in problems that involve vertical line segments; use y as the variable in problems that involve horizontal line segments. Do not use absolute value signs in formulas for length.

43. Vertical, between $y = x^2 + 1$ and the x-axis; about the x-axis

44. Vertical, between $y = x^2 - 1$ and $y = 1$, $-\sqrt{2} \leq x \leq \sqrt{2}$; about $y = 1$

45. Vertical, between $y = x^2 + 1$ and the x-axis; about $y = -1$

46. Vertical, between $y = x^2 + 1$ and $y = 1$; about the x-axis

47. Vertical, between $hy = rx$ and the x-axis, $0 \leq x \leq h$; about the x-axis; assume h and r are positive

48. Horizontal, between $ry = hx$ and $x = r$, $0 \leq y \leq h$; about the y-axis; assume h and r are positive

49. Vertical, between $(x/r) + (y/h) = 1$ and the x-axis, $0 \leq x \leq r$; about the y-axis; assume h and r are positive

50. Horizontal, between $(x/r) + (y/h) = 1$ and the y-axis, $0 \leq y \leq h$; about the y-axis; assume h and r are positive

51. Horizontal, between $(x/h) + (y/r) = 1$ and the y-axis, $0 \leq y \leq r$; about the x-axis; assume h and r are positive

52. Vertical, between points on $x = y^2 + 1$; about the y-axis

53. Horizontal, between points on $y = x^2$; about the x-axis

54. Horizontal, between $x = \sqrt{r^2 - y^2}$ and the y-axis; about the y-axis

55. Horizontal, between $x = y^2$ and $x - y = 2$, $-1 \leq y \leq 2$; about $x = 4$

56. Vertical, between points on $x^2 + y^2 = r^2$; about $x = R$, where $0 < r < R$

6.2   AREA OF REGIONS IN THE PLANE

Connections

Area as a limit of Riemann sums, 5.2.

Length of variable line segments, 6.1.

In Chapter 5 we saw how the definite integral is related to the area of the region under the graph of a nonnegative function. In this section we will see how definite integrals can be used to find the area of more general regions.

Using Vertical Area Strips

Let us consider the region bounded by the curves

$$y = f(x), \quad y = g(x), \quad x = a, \quad \text{and } x = b.$$

The region is sketched in Figure 6.2.1a. We assume that f and g are continuous and that $g(x) \leq f(x)$ for $a \leq x \leq b$. It is clear intuitively that the "area" of the region is related to a sum of the areas of rectangles, as pictured in Figure 6.2.1b. The typical rectangle sketched in Figure 6.2.1c has vertical length $f(x_j^*) - g(x_j^*)$ and width Δx_j, so its area is given by the product $(f(x_j^*) - g(x_j^*))\Delta x_j$. The "area" of the entire region is related to a sum of the areas of such typical rectangles. That is,

Area of region $\approx$ Sum of areas of rectangles,

$$\text{``Area''} \approx \sum_{j=1}^{n}(f(x_j^*) - g(x_j^*))\Delta x_j. \tag{1}$$

We now note that the difference between the area of the rectangle and the "area" bounded by $y = f(x)$, $y = g(x)$, $x = x_{j-1}$, and $x = x_j$ is at most a small proportion of the area of the rectangle if Δx_j is small enough. It follows that the sum in (1) is very near the "area" of the region when all Δx_j's are small enough. On the other hand, since the sum is a Riemann sum of $f - g$ over the interval $[a, b]$, we know such sums are near the number $\int_a^b [f(x) - g(x)]dx$. It should then seem reasonable to *define* the area of the region by the following formula.

$$\text{Area} = \int_a^b (f(x) - g(x))dx. \tag{2}$$

This definition of the area of the region is consistent with our intuitive notion of area and gives familiar formulas when applied to simple geometric shapes. When $f(x) \geq 0$ and $g(x) = 0$ for $a \leq x \leq b$, the integral in Formula 2 becomes $\int_a^b f(x)dx$, which represents the area under the curve $y = f(x)$.

We should not think of (2) as a formula to be memorized. The formula merely reflects the fact that the area is given as a limit of sums of areas of rectangles. We should think of the components of the integral as representing specific physical quantities. That is,

$$\text{Area} = \int_a^b \underbrace{\underbrace{(f(x) - g(x))}_{\text{Length}} \underbrace{dx}_{\text{Width}}}_{\text{Area of rectangle}}.$$

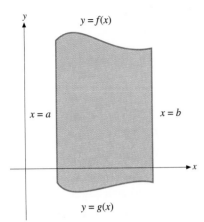

FIGURE 6.2.1a The region is bounded on the top by $y = f(x)$, on the bottom by $y = g(x)$, and on the sides by $x = a$ and $x = b$.

FIGURE 6.2.1b The area of the region is approximately the sum of the areas of the rectangles.

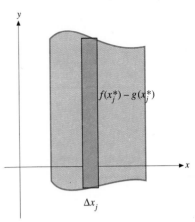

FIGURE 6.2.1c A typical rectangle has vertical length $f(x_j^*) - g(x_j^*)$ and width Δx_j.

To find the area of a region by using vertical strips

- Sketch the region.
- Draw a variable vertical rectangular area strip.
- Label coordinates of endpoints of the strip in terms of x.
- Write "area" of strip as (Length)(Width).
- Write integral that gives "sum" of areas of strips.
- Evaluate the definite integral.

EXAMPLE 1

Set up and evaluate a definite integral that gives the area of the region bounded by

$$y = \frac{x}{2}, \quad y = -x, \quad x = 0, \text{ and } x = 1.$$

SOLUTION The region is sketched and a variable vertical area strip is drawn and labeled in Figure 6.2.2. The variable rectangular strip is drawn as a vertical line segment. The coordinates of the top and bottom of the strip have been labeled in terms of the variable x. The strip is typical because the same labeling applies as x varies between $x = 0$ and $x = 1$. We now think of our typical strip as a rectangle with

$$\text{Width} = dx,$$

and

$$\text{Length} = (y\text{-coordinate of top}) - (y\text{-coordinate of bottom})$$

$$= \left(\frac{x}{2}\right) - (-x).$$

(It is clear from Figure 6.2.2 that $x/2 > -x$ for $0 \le x \le 1$.)

The area of the region is a limit of the sum of the areas of the rectangles. This is given by the definite integral

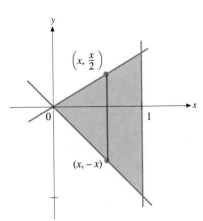

FIGURE 6.2.2 The vertical length of a variable rectangular strip is $\left(\frac{x}{2}\right) - (-x), 0 \le x \le 1.$

$$\text{Area} = \int_0^1 \left(\left(\frac{x}{2} \right) - (-x) \right) dx.$$

$$\underbrace{\underbrace{}_{\text{Sum}} \underbrace{\underbrace{}_{\text{Length}} \underbrace{}_{\text{Width}}}_{\text{Area of rectangle}}}$$

Evaluating the integral, we have

$$\text{Area} = \int_0^1 \frac{3}{2} x \, dx = \left(\frac{3}{2} \right) \left(\frac{x^2}{2} \right) \Bigg]_0^1 = \left(\frac{3}{2} \right) \left(\frac{1^2}{2} \right) - \left(\frac{3}{2} \right) \left(\frac{0^2}{2} \right) = \frac{3}{4}.$$

In this example the region is a triangle with (vertical) base 3/2 and (horizontal) height 1. The usual formula for the area of a triangle gives

$$\text{Area} = \frac{1}{2}(\text{Base})(\text{Height}) = \left(\frac{1}{2} \right) \left(\frac{3}{2} \right) (1) = \frac{3}{4}.$$

This value agrees with that obtained by using the definite integral. ∎

Of course, we are interested in using the integral method to find the area of regions for which area formulas are not available. We will concentrate on setting up area integrals. Evaluation is a separate step. We will not carry out the details of evaluating the integrals for the remaining examples in this section.

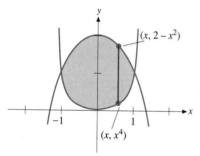

FIGURE 6.2.3 The vertical length is $(2 - x^2) - (x^4)$ between points of intersection of the curves.

EXAMPLE 2

Set up a definite integral that gives the area of the region bounded by $y = x^4$ and $y = 2 - x^2$.

SOLUTION We have sketched the region, drawn a typical rectangular strip, and labeled coordinates in Figure 6.2.3.

The variable x varies between the x-coordinates of the points of intersection of the curves $y = x^4$ and $y = 2 - x^2$. These coordinates are found by solving the two equations simultaneously for x. That is,

$$y = x^4 \text{ and } y = 2 - x^2$$

imply

$$x^4 = 2 - x^2,$$
$$x^4 + x^2 - 2 = 0,$$
$$(x^2 + 2)(x^2 - 1) = 0,$$
$$(x^2 + 2)(x - 1)(x + 1) = 0,$$

so the real number solutions are $x = -1$ and $x = 1$.

We can now write the definite integral that gives the limit of the sum of the areas of the rectangles. We have

$$\text{Area} = \int_{-1}^1 \underbrace{[(2 - x^2) - (x^4)]}_{\text{Length}} \, \underbrace{dx}_{\text{Width}}.$$

$$\underbrace{}_{\text{Sum}} \quad \underbrace{}_{\text{Area of rectangle}}$$

(It is not difficult to verify that the above integral has value 44/15.) ∎

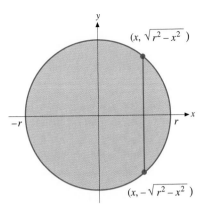

FIGURE 6.2.4 Vertical strips have length $2\sqrt{r^2 - x^2}$.

EXAMPLE 3

Set up a definite integral that gives the area of the region bounded by the circle $x^2 + y^2 = r^2$.

SOLUTION   The equation $x^2 + y^2 = r^2$ implies $y = \sqrt{r^2 - x^2}$ or $y = -\sqrt{r^2 - x^2}$. The graph of $y = \sqrt{r^2 - x^2}$ is the upper half of the circle, while $y = -\sqrt{r^2 - x^2}$ gives the lower half. The circle is sketched and a variable vertical rectangular area strip has been drawn and labeled in Figure 6.2.4.

The length of the typical rectangular strip is

$$\sqrt{r^2 - x^2} - (-\sqrt{r^2 - x^2}) = 2\sqrt{r^2 - x^2}.$$

The variable x clearly varies between $x = -r$ and $x = r$. The area of the circle is given by the definite integral

$$\text{Area} = \int_{-r}^{r} 2\sqrt{r^2 - x^2}\,dx.$$

(Later we will learn a technique for evaluating the above integral. Of course, we will discover that its value is πr^2, the area of a circle of radius r.) ∎

Using More Than One Integral

EXAMPLE 4

Set up an integral or integrals that give the area of the region bounded by $y = 4x$ and $y = x(x + 1)(x - 2)$.

SOLUTION   The region is sketched in Figure 6.2.5.

We see that the region is composed of two parts, one where the cubic curve is above the line and another where the line is above the cubic curve. This means that two formulas are required to describe the length of a variable vertical line segment between the graphs.

We need the x-coordinates of the points of intersection of $y = 4x$ and $y = x(x + 1)(x - 2)$. Solving these two equations for values of x, we have

$$x(x + 1)(x - 2) = 4x,$$
$$x^3 - x^2 - 2x = 4x,$$
$$x^3 - x^2 - 6x = 0,$$
$$x(x^2 - x - 6) = 0,$$
$$x(x - 3)(x + 2) = 0.$$

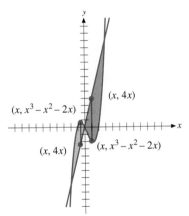

FIGURE 6.2.5 Vertical strips have length $(x^3 - x^2 - 2x) - (4x)$ for $-2 \le x \le 0$ and have length $(4x) - (x^3 - x^2 - 2x)$ for $0 \le x \le 3$.

The solutions are $x = 0, 3, -2$.

We then see from Figure 6.2.5 that the length of a variable vertical line segment between the graphs is

$$L(x) = \begin{cases} (x^3 - x^2 - 2x) - (4x), & \text{if } -2 \le x \le 0, \\ (4x) - (x^3 - x^2 - 2x), & \text{if } 0 \le x \le 3. \end{cases}$$

The area is then the sum of the areas of the two parts.

$$\text{Area} = \int_{-2}^{0} ((x^3 - x^2 - 2x) - (4x))\,dx + \int_{0}^{3} ((4x) - (x^3 - x^2 - 2x))\,dx.$$

(The sum of the two integrals is $253/12$.) ∎

In Example 4 we could have written

$$\text{Area} = \int_{-2}^{3} |(x^3 - x^2 - 2x) - 4x| \, dx,$$

but we could not use our formulas for evaluating definite integrals without determining intervals where the expression inside the absolute value is of constant sign. We have used the sketch in Figure 6.2.5 to help determine the sign of $(x^3 - x^2 - 2x) - 4x$.

A computer program that uses a numerical method such as those illustrated in Section 5.6 could be used to give an approximate value of an integral that involves absolute values.

Using Horizontal Area Strips

It is often convenient to divide a region into horizontal area strips.

EXAMPLE 5

Set up an integral or integrals that give the area of the triangular region bounded by $x = 2y$, $y + 6 = 2x$, and $y = 0$.

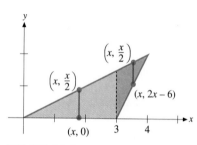

FIGURE 6.2.6a Vertical strips have length $\dfrac{x}{2}$ for $0 \le x \le 3$ and have length $\dfrac{x}{2} - (2x - 6)$ for $3 \le x \le 4$.

SOLUTION The region is sketched in Figure 6.2.6a. We see that there is no single variable vertical area strip that is typical for the entire region. Some vertical strips run from $y = 0$ to $y = x/2$ and other vertical strips run from $y = 2x - 6$ to $y = x/2$. This means that two formulas are required to describe the length of a variable vertical area strip, so we must express the area of the region as a sum of two integrals with respect to the variable x.

The limits of integration are determined by finding the x-coordinates of the points of intersection of the boundary lines. It is easy to see that

$$x = 2y \text{ and } y = 0 \text{ imply } x = 0;$$
$$y + 6 = 2x \text{ and } y = 0 \text{ imply } x = 3; \text{ and}$$
$$x = 2y \text{ and } y + 6 = 2x \text{ imply } x = 4.$$

We then have

$$\text{Area} = \int_{0}^{3} \left(\frac{x}{2} - 0\right) dx + \int_{3}^{4} \left(\frac{x}{2} - (2x - 6)\right) dx.$$

(Each of the above integrals represents the area of a triangular portion of the region.)

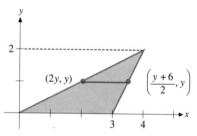

FIGURE 6.2.6b Horizontal strips have length $\left(\dfrac{y+6}{2}\right) - (2y)$ for $0 \le y \le 2$.

An alternate solution for this problem is suggested in Figure 6.2.6b. We see that there is a single variable horizontal area strip that is typical for the entire region. In the case of horizontal strips, the variable is y and each component of the corresponding definite integral must be expressed in terms of y. We see that y varies between $y = 0$ and the y-coordinate of the point of intersection of the lines $x = 2y$ and $x = (y + 6)/2$, which is $y = 2$. The length of a horizontal line segment is

$$\text{Length} = (x\text{-coordinate of right endpoint}) - (x\text{-coordinate of left endpoint})$$
$$= \frac{y+6}{2} - 2y.$$

The area is

$$\text{Area} = \int_0^2 \left(\frac{y+6}{2} - 2y \right) dy.$$

It is easy to check that both of the above methods give the value of 3 for the area of the triangle. This agrees with the value obtained from the formula

$$\text{Area of triangle} = \frac{1}{2}(\text{Base})(\text{Height}) = \frac{1}{2}(3)(2) = 3. \quad \blacksquare$$

You should be prepared to divide a region into either vertical or horizontal area strips, whichever is more convenient.

EXAMPLE 6

Set up an integral or integrals that give the area of the region bounded by $x - y = 2$ and $x = y^2 - 1$.

SOLUTION   The region is sketched in Figure 6.2.7. There is a single variable horizontal strip that is typical for the entire region. Two integrals would be required if x were chosen as the variable. We should use the more convenient variable strips.

The variable y varies between the y-coordinates of the points of intersection of $x = y + 2$ and $x = y^2 - 1$. These values are given by the equation

$$y^2 - 1 = y + 2 \text{ or } y^2 - y - 3 = 0.$$

The Quadratic Formula gives the solutions

$$y = \frac{-(-1) \pm \sqrt{(-1)^2 - 4(1)(-3)}}{2(1)}.$$

That is,

$$y = \frac{1 + \sqrt{13}}{2} \text{ and } y = \frac{1 - \sqrt{13}}{2}.$$

We have

$$\text{Area} = \int_{(1-\sqrt{13})/2}^{(1+\sqrt{13})/2} ((y + 2) - (y^2 - 1)) \, dy.$$

(The value of this integral is $13\sqrt{13}/6 \approx 7.81$.) $\blacksquare$

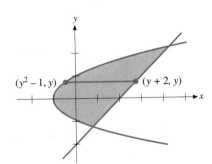

FIGURE 6.2.7   Horizontal strips have length $(y + 2) - (y^2 - 1)$ between points of intersection.

EXERCISES 6.2

Sketch and label the regions bounded by the given lines in Exercises 1–10. Set up and evaluate integrals that give the areas of the regions. Verify the results by using formulas from geometry to find the areas of the regions.

1. $y = 2x, y = 0, x = 3$

2. $x + y = 4, y = 0, x = 0$

3. $3x + 2y = 6, x - y = 2, x = 0$

4. $y = x, y = -2x, x = 2$

5. $y = 0, x - y + 2 = 0, 2y + x = 4$

6. $y = 0, x + y + 2 = 0, 2y - x + 4 = 0$

7. $y = x, y = x + 3, x = 0, x = 2$

8. $y = x + 2, y = 0, x = 0, x = 3$

9. $y = x, y = 2, y = 0, x + y = 5$

10. $y = x, y = 2, y = 0, y = x - 5$

In Exercises 11–24, sketch and label the regions bounded by the given curves. Set up and evaluate integrals that give the areas of the regions.

11. $y = x^2, y = 0, x = 2$

12. $y = 4 - x^2, y = 0$

13. $y = x^2, y = x$

14. $y = x^2, y = \sqrt{x}, x = 1, x = 4$ (**Calculus Explorer & Tutor II, Applications of Integration, 1.**)

15. $y = \sec^2 x, y = 0, x = 0, x = \pi/4$

16. $y = \cos x, y = 0, x = -\pi/2, x = \pi/2$

17. $x = y^2, x + y = 2$

18. $x = y^2, x - y = 2$

19. $y = \sin x, y = 0, x = -\pi, x = \pi$

20. $y = x, y = 0, x = -3, x = 2$ (**Calculus Explorer & Tutor II, Introduction to Integration, 8.**)

21. $y = \sin x, y = 1/2, x = 0 \ (x \geq 0)$

22. $y = \cos x, y = 1/2, x = 0 \ (x \geq 0)$

23. $y = x^3/3, y = x^2$

24. $y = 32x^{1/3}, y = x^2$

In Exercises 25–30, sketch and label the regions bounded by the given curves. Set up integrals that give the areas of the regions.

25. Inside $x^2 + y^2 = 4$ and above $x + y = 2$

26. Inside $x^2 + y^2 = 4$ and above $y = 1$

27. Inside $x^2 + y^2 = 4$ and above $y = -1$

28. Inside $x^2 + y^2 = 4$ and above $2x + y = 2$

29. $y = \sin x, y = x, x = \pi$

30. $y = \tan x, y = x, x = \pi/4$

31. Find c so that the vertical line $x = c$ divides the region bounded by $y = \sqrt{x}, x = 1$, and the x-axis into two regions of equal area.

32. Find c so that the vertical line $x = c$ divides the region bounded by $y = \cos x$ and the x-axis between $x = 0$ and $x = \pi/2$ into two regions of equal area.

33. Find c so that the horizontal line $y = c$ divides the region bounded by $y = x^3, y = 1$, and the y-axis into two regions of equal area.

34. Find c so that the horizontal line $y = c$ divides the region bounded by $y = kx, y = k \ (k > 0)$, and the y-axis into two regions of equal area.

35. A lawn mower cuts strips 18 in. wide. If turning is not considered, which pattern requires more walking, cutting

parallel to the sides or at an angle? See figure. Explain in terms of calculus.

36. Find the area of the region enclosed by the graph of $|x|^{1/2} + |y|^{1/2} = a^{1/2}$.

37. Find the area of the region enclosed by the loop formed by the graph of $y^2 = x(4 - x)^2$. See figure.

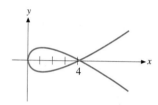

38. Find the area of the region enclosed by the two loops of the graph of $y^2 = x^2(4 - x^2)$. See figure.

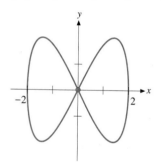

39. Sketch the region bounded by
$$y = 1, \quad y = 0, \quad x = 0, \quad x = 2\pi$$
and the region bounded by
$$y = 1 + \sin^2 x, \quad y = \sin^2 x, \quad x = 0, \quad x = 2\pi.$$
Show that the two regions have equal areas.

40. Sketch the region bounded by
$$x = \sec y, \quad x = 0, \quad y = -\frac{\pi}{4}, \quad y = \frac{\pi}{4}$$
and the region bounded by
$$x = \frac{1}{2}\sec y, \quad x = -\frac{1}{2}\sec y, \quad y = -\frac{\pi}{4}, \quad y = \frac{\pi}{4}.$$
Show that the two regions have equal areas.

41. Show that the length of a variable line segment between two sides of a triangle and parallel to the base depends only on the length of the base and height of the triangle, and on the distance of the segment from the base. Use an area integral to show that triangles that have equal bases and heights have the same area, even though they may not be the same shape. See figure.

(a)   (b)

42. Show that the length of a variable line segment between sides of a parallelogram and parallel to the base depends only on the length of the base. Use an area integral to show that parallelograms that have equal bases and heights have the same area, even though they may not be the same shape. See figure.

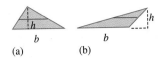

(a)   (b)

In Exercises 43–44, use the absolute value function to set up a single integral that expresses the area of the indicated regions as a single definite integral. Use a computer to find an approximate value of the integral.

43. Between $y = 3\cos x$ and $y = 1 - x$

44. Between $y = \ln(x + 1)$ and $y = 8(-x^3 + 3x^2 - 2x)$

In Exercises 45–48, use the functions

$$\max(f, g) = \frac{f + g + |f - g|}{2} \quad \text{and}$$

$$\min(f, g) = \frac{f + g - |f - g|}{2}$$

to set up a single definite integral that expresses the area of the regions indicated as a limit of sums of the areas of vertical area strips. Use a computer to find an approximate value of the integral.

45.

46.

47.

48.

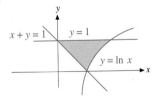

6.3 VOLUMES BY SLICING

Connections

Volume as a limit of Riemann sums, 5.2.

Area of variable cross-sections, 6.1.

The main objective of this section is to strengthen understanding of how the idea of the definite integral as a sum is used in physical problems. The particular problem that we consider involves the volume of three-dimensional solids.

Cross-Sections Perpendicular to the x-Axis

The idea of obtaining the volume of a solid by slicing and adding the volumes of the slices was introduced in Section 5.2 as an application of Riemann sums. Let us review the idea.

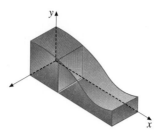

FIGURE 6.3.1a Cross-sections perpendicular to the x-axis have area A(x).

FIGURE 6.3.1b The volume of the solid is the sum of the volumes of the slices. The volume of a variable slice is approximately $A(x_j^*)\Delta x_j$.

Consider a solid object such as is pictured in Figure 6.3.1a. We assume that the area of cross-sections perpendicular to the x-axis is given by a continuous function $A(x)$, $a \le x \le b$. Let us now slice the solid as indicated in Figure 6.3.1b. If the slices are thin enough, the volume of that part of the solid between $x = x_{j-1}$ and $x = x_j$ is approximately $A(x_j^*)\Delta x_j$, where x_j^* is any point between x_j and x_{j-1}. An approximate value of the volume of the solid is then given by the sum of the approximate volumes of the parts. That is,

$$\text{Volume of solid} \approx \text{Sum of approximate volumes of slices,}$$

$$\text{Volume} \approx \sum_{j=1}^{n} A(x_j^*)\Delta x_j.$$

If all slices are thin enough, the above Riemann sum is a good approximation of both the volume of the solid and the value of the definite integral $\int_a^b A(x)\,dx$. Hence, it should seem reasonable that

$$\text{Volume} = \int_a^b A(x)\,dx.$$

Each component of this integral should have a definite physical meaning.

$$\text{Volume} = \int_a^b \underbrace{\underbrace{A(x)}_{\left(\substack{\text{Area of} \\ \text{cross section}}\right)} \underbrace{dx}_{\left(\substack{\text{Thickness} \\ \text{of slice}}\right)}}_{\substack{\text{Sum} \quad \text{Volume of slice}}}.$$

The limits of integration indicate that the solid consists of slices between $x = a$ and $x = b$.

To use the method of slicing to find the volume of a solid

- Sketch the solid.
- Draw a variable cross-section perpendicular to the x-axis.
- Find the area of the variable cross-section in terms of x.
- Write the "volume" of a slice as (area)(thickness).
- Write the integral that gives the "sum" of the slice volumes.
- Evaluate the definite integral.

Let us now use a definite integral to find the exact volume of the prism considered in Example 2 of Section 5.2. We can now use the technique of Section 6.1 to express the area of the cross-section as a function of x.

EXAMPLE 1

Set up and evaluate a definite integral that gives the volume of the prism sketched in Figure 6.3.2. Cross-sections perpendicular to the x-axis are isosceles right triangles.

SOLUTION   In Figure 6.3.2 we have used b to indicate the equal sides of a variable right triangular cross-section. The area of the cross-section can then be expressed as

$$A = \frac{1}{2}b^2,$$

where b depends on the position along the x-axis. We see that b is the length of the vertical line segment between the x-axis and the line through the points $(0, 2)$ and $(4, 0)$ in the xy-plane. To express b in terms of the variable x, we need the equation of this line. This line has slope $-1/2$. The equation of the line through the point $(4, 0)$ with slope $-1/2$ is

$$y - 0 = -\frac{1}{2}(x - 4).$$

The height (and base) of the triangle at position x are given by the y-coordinate of the point on the line. Solving the equation for y, we obtain

$$y = \frac{4 - x}{2},$$

so

$$b = \frac{4 - x}{2}.$$

The area of a variable cross-section is then

$$A(x) = \frac{1}{2}\left(\frac{4 - x}{2}\right)^2, \qquad 0 \le x \le 4.$$

The volume of the prism is given by the sum of the volumes of the slices between $x = 0$ and $x = 4$. We have

$$\text{Volume} = \int_0^4 \underbrace{\underbrace{\frac{1}{2}\left(\frac{4 - x}{2}\right)^2}_{\substack{\text{Area of} \\ \text{cross section}}} \underbrace{dx}_{\substack{\text{Thickness} \\ \text{of slice}}}}.$$

Sum   Volume of slice

(We are thinking of the integral as a sum of the volumes of a finite number of thin slices. The definite integral gives the limit of such sums.)

Evaluating the definite integral, we obtain

$$\text{Volume} = \int_0^4 \frac{1}{8}(16 - 8x + x^2)\,dx$$

$$= \frac{1}{8}\left(16x - 4x^2 + \frac{1}{3}x^3\right)\Bigg]_0^4$$

$$= \frac{1}{8}\left(16(4) - 4(4)^2 + \frac{1}{3}(4)^3\right) = \frac{8}{3}. \quad \blacksquare$$

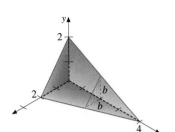

FIGURE 6.3.2   Cross-sections are isosceles right triangles with area $\frac{1}{2}b^2$.

Cross-Sections Perpendicular to the y-Axis

The same ideas can be used to find the volume of a solid by slicing perpendicular to the y-axis. In this case we use y as the variable of integration, so the area of cross-section must be expressed as a function of y.

EXAMPLE 2

Set up an integral that gives the volume of the solid that intersects the xy-plane in the region bounded by $y = x^2$ and $y = 1$, if cross-sections perpendicular to the y-axis are disks with centers on the y-axis.

SOLUTION   The solid is pictured in Figure 6.3.3. A variable cross-section perpendicular to the y-axis has been drawn and labeled. The position of a variable slice is given by the variable y, so we have labeled the sketch in terms of y.

The cross-sections perpendicular to the y-axis are disks and have area πr^2, where the radius r depends on the position of the slice. From the sketch we see that the radius r, in terms of the variable y, is given by $r = \sqrt{y}$. The volume of the solid is

$$\text{Volume} = \int_0^1 \underbrace{\pi(\sqrt{y})^2}_{\substack{\left(\begin{array}{c}\text{Area of}\\ \text{cross section}\end{array}\right)}} \underbrace{dy.}_{\left(\begin{array}{c}\text{Thickness}\\ \text{of slice}\end{array}\right)}$$

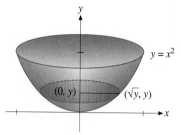

FIGURE 6.3.3   Cross-sections are circles of radius $\sqrt{y}$.

(The integral has value $\pi/2$.)   ∎

Solids with Given Base and Cross-Section Type

EXAMPLE 3

The base of a solid consists of the region bounded by

$$2x + 3y = 6, \quad y = 0, \quad \text{and} \quad x = 0.$$

Set up an integral that gives the volume of the solid if cross-sections perpendicular to the x-axis are semicircles with diameters on the base.

SOLUTION   The solid is sketched in Figure 6.3.4a.

Let b denote the length of the base of a cross-section at the variable position x along the x-axis. The cross-section is a semicircle with diameter on the base, so b is the length of the diameter of the semicircle. From the sketch of Figure 6.3.4b we see that the area of the cross-section is one-half the area of a circle of radius $b/2$, or

$$A = \frac{1}{2}\pi \left(\frac{b}{2}\right)^2 = \frac{\pi}{8}b^2.$$

We need to express b in terms of the variable x. From Figure 6.2.4a, we see that b is given by the y-coordinate of the point on the line $2x + 3y = 6$. Solving the equation for y in terms of the variable x, we have

$$y = \frac{6 - 2x}{3}, \quad \text{so } b = \frac{6 - 2x}{3}.$$

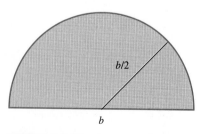

FIGURE 6.3.4a   Cross-sections are semicircles with diameters on the base.

FIGURE 6.3.4b   The area of the cross-section is one-half the area of a circle of radius $b/2$.

The volume is

$$\text{Volume} = \int_0^3 \frac{\pi}{8} \left(\frac{6 - 2x}{3} \right)^2 dx.$$

(This integral has value $\pi/2$.) ■

EXAMPLE 4

The base of a solid consists of the region inside the circle $x^2 + y^2 = 4$. Set up an integral that gives the volume of the solid if cross-sections perpendicular to the x-axis are isosceles right triangles with hypotenuses on the base.

SOLUTION   The solid is sketched in Figure 6.3.5a. The area of a variable cross-section is given by

$$a = \frac{1}{2}bh,$$

where b is the base (hypotenuse) and h is the height. Since the equal angles in an isosceles right triangle are $45°$, we have

$$h = \frac{b}{2}.$$

See Figure 6.3.5b. Then

$$A = \frac{1}{2}bh = \frac{1}{2}b\left(\frac{b}{2}\right) = \frac{b^2}{4}.$$

We need to express b in terms of the variable x. The equation $x^2 + y^2 = 4$ implies $y = \sqrt{4 - x^2}$ and $y = -\sqrt{4 - x^2}$, so we see that $b = \sqrt{4 - x^2} - (-\sqrt{4 - x^2}) = 2\sqrt{4 - x^2}$. The volume is

$$\text{Volume} = \int_{-2}^{2} \frac{1}{4}(2\sqrt{4 - x^2})^2 dx.$$

(This integral has value $32/3$.) ■

 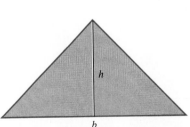

FIGURE 6.3.5a   Cross-sections are isosceles right triangles with hypotenuses on the base.

FIGURE 6.3.5b   The area of the cross-section is $\frac{1}{2}bh$. Since the equal angles in an isosceles right triangle are $45°$, $h = b/2$.

Intersection of Cylinders

EXAMPLE 5

Find the volume of the solid that is bounded by two right circular cylinders of radius r, if the axes of the cylinders meet at right angles.

FIGURE 6.3.6a The solid is bounded by two right circular cylinders of radius r whose axes meet at right angles.

FIGURE 6.3.6b We can see the cross-section of the intersection of the two cylinders if we slice the cylinders parallel to their axes and remove the upper portion.

SOLUTION The solid is sketched in Figure 6.3.6a. In Figure 6.3.6b we have sketched two solid right circular cylinders with equal radii that intersect at right angles, sliced along a plane that is parallel to their axes, with the upper portion removed. This shows that the cross-section of the intersection of the cylinders in this plane is a square. Using a two-dimensional coordinate system with axes through the center of one of the cylinders and perpendicular to the axis of the cylinder, with y directed upward, we can determine that the length of each side of the square cross-section at level y is $2\sqrt{r^2 - y^2}$. The area of the cross-section of the intersection of the cylinders at level y is then

$$A(y) = (2\sqrt{r^2 - y^2})^2 = 4(r^2 - y^2).$$

As y varies from $-r$ to r, these cross-sections generate the entire intersection of the cylinders. We then have

$$\begin{aligned}
\text{Volume} &= \int_{-r}^{r} 4(r^2 - y^2)\, dy \\
&= 4r^2 y - 4\frac{y^3}{3} \Bigg]_{-r}^{r} \\
&= \left(4r^2(r) - 4\frac{(r)^3}{3}\right) - \left(4r^2(-r) - 4\frac{(-r)^3}{3}\right) = \frac{16}{3}r^3. \quad \blacksquare
\end{aligned}$$

FIGURE 6.3.7 The two solids have corresponding cross-sections of equal area, so Cavalieri's Theorem implies they have equal volume.

Cavalieri's Theorem

It should be clear from the method of slicing that the volumes of two solids are equal if their corresponding cross-sections have equal area. This general result is called **Cavalieri's Theorem**.

EXAMPLE 6

Cavalieri's Theorem implies that the volumes of the solids pictured in Figure 6.3.7 are equal, because corresponding cross-sections have equal area. ■

EXERCISES 6.3

Set up and evaluate integrals that give the volume of the solids indicated in Exercises 1–12.

1. Cross-sections are rectangles.

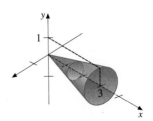

2. Cross-sections are right triangles.

3. Cross-sections are circles.

4. Cross-sections are squares.

5. Cross-sections are squares.

6. Cross-sections are circles.

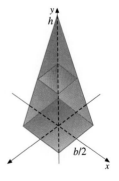

7. Intersection with the xy-plane is the region bounded by

$$x^2 - y^2 = 1, \quad y = 1, \text{ and } y = -1.$$

Cross-sections perpendicular to the y-axis are circles with centers on the y-axis.

8. Intersection with the xy-plane is the region bounded by

$$y = 4 - x^2 \text{ and } y = 0.$$

Cross-sections perpendicular to the y-axis are circles with centers on the y-axis.

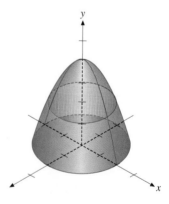

9. The base is the region in the xy-plane bounded by

$$x + 3y = 3, \quad x - 3y = 3, \text{ and } x = 0.$$

Cross-sections perpendicular to the x-axis are **(a)** squares, **(b)** isosceles right triangles with hypotenuses on the base, **(c)** isosceles right triangles with one side on the base.

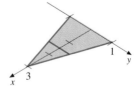

10. The base consists of the region in the xy-plane inside the circle $x^2 + y^2 = 4$. Cross-sections perpendicular to the

x-axis are **(a)** squares, **(b)** semicircles with diameters on the base, **(c)** equilateral triangles with one side on the base.

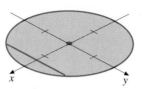

11. The base is the region in the xy-plane bounded by

$$x = 4 - y^2 \text{ and } x = 0.$$

Cross-sections perpendicular to the x-axis are **(a)** semicircles with diameters on the base, **(b)** equilateral triangles with one side on the base, **(c)** rectangles with height half the base.

12. The base is the region in the xy-plane bounded by

$$y^2 - x^2 = 1, \quad x = 1, \text{ and } x = -1.$$

Cross-sections perpendicular to the x-axis are **(a)** semicircles with diameters on the base, **(b)** isosceles right triangles with hypotenuses on the base, **(c)** rectangles with height half the base.

13. A cylindrical hole of radius r is drilled along the axis of a right circular cone with radius of base $2r$ and height $4r$. Set up and evaluate an integral that gives the volume of the resulting solid.

14. A cylindrical hole of radius r is drilled along the diameter of a sphere of radius $2r$. Set up and evaluate an integral that gives the volume of the resulting solid.

15. A wedge is formed from a right circular cylinder of radius 2 in. by cutting the cylinder perpendicular to the axis and at an angle of $30°$ from the perpendicular plane, in such a way that the edge of the wedge is a diameter of the cylinder. See figure. Set up and evaluate an integral that gives the volume of the solid.

16. A solid is formed by slicing a right circular cylinder of radius 20 cm by two parallel planes that are 7 cm apart, each at an angle θ with the axis of the cylinder. See figure. Set up and evaluate an integral that gives the volume of the solid. (Note that $\int_{-20}^{20} 2\sqrt{20^2 - x^2}\,dx$ is the area of a circle of radius 20.)

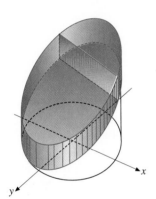

17. A solid intersects the xy-plane in the region bounded by

$$y = \sin x, \quad y = -\sin x, \quad x = 0, \text{ and } x = \pi.$$

Cross-sections perpendicular to the x-axis are circles with centers on the x-axis. See figure. Set up an integral that gives the volume of the solid.

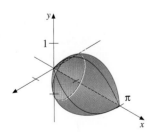

18. A solid intersects the xy-plane in the region bounded by

$$y = \cos x, \quad y = -\cos x, \quad x = 0, \text{ and } x = \frac{\pi}{2}.$$

Cross-sections perpendicular to the x-axis are circles with centers on the x-axis. See figure. Set up an integral that gives the volume of the solid.

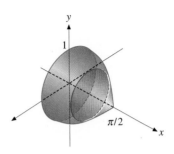

19. Find the volume of the solid that is bounded by two right circular cylinders of radius r if their axes meet at angle θ. See figure.

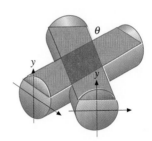

20. The axes of a right circular cylinder of radius r and a right circular cylinder of radius $2r$ meet at a right angle. See figure. Set up an integral that gives the volume of the solid bounded by the cylinders.

21. The axes of a right circular cylinder of radius r and a square beam with length of sides r meet at a right angle. See figure.

Set up an integral that gives the volume of that part of the beam that is inside the cylinder.

22. A square beam with length of sides r intersects a right circular cylinder of radius r at a right angle and so that the bottom of the beam meets the axis of the cylinder. See figure. Set up an integral that gives the volume of that part of the beam that is inside the cylinder.

23. Find the volume of the solid that has cross-sections perpendicular to the x-axis that are squares with endpoints of a diagonal on the curves $y = x^2/4$ and $y = -x^2/4$, $0 \le x \le 2$. See figure.

24. Find the volume of the solid that has cross-sections perpendicular to the x-axis that are circles with endpoints of a diameter on the curve $y = x^2/4$ and the x-axis, $0 \le x \le 2$.

25. Find the volume of the solid that has cross-sections perpendicular to the y-axis that are circles with endpoints of a diameter on the curves $y = \sqrt{x}$ and $y = 3\sqrt{x}/2$, $0 \le y \le 1$.

26. Find the volume of the solid that has cross-sections perpendicular to the y-axis that are squares with endpoints of a diagonal on the curves $y = \sqrt{x}$ and $y = x^2$, $0 \le y \le 1$.

27. A square with length of sides one lies in a plane perpendicular to the y-axis, with one corner of the square on the y-axis. Find the volume of the solid generated as the square revolves uniformly once about the y-axis as it moves along the y-axis from $y = 0$ to $y = 2\pi$. See figure.

28. Find the volume of the solid that has cross-sections perpendicular to the x-axis that are squares with endpoints of a diagonal on the curve $|x|^{1/2} + |y|^{1/2} = 1$.

29. Use integral calculus to explain why a stack of coins has the same total volume whether or not the coins are stacked directly on top of one another. See figure.

30. Show that a hemisphere of radius r has the same volume as the solid obtained by removing a right circular cone with radius r and height r from a right circular cylinder of radius r and height r.

31. The circumferences of a football at points equally spaced along its axis are given in the figure. Approximate the volume of the football by using Simpson's Rule to approximate the integral that gives its volume.

Circumference

45 cm   53 cm   45 cm

28 cm

32. The circumferences of a bowling pin at points equally spaced along its axis are given in the figure.

Circumference

12.5 cm
18.5 cm
14.7 cm
29.0 cm
38.4 cm
33.1 cm
19.2 cm

38 cm

Approximate the volume of the bowling pin by using Simpson's Rule to approximate the integral that gives its volume.

Connections

Revolving line segments about a line, 6.1.

6.4   VOLUMES OF SOLIDS OF REVOLUTION

In this section we consider the problem of finding the volume of a solid that is obtained by revolving a region in the plane about a line that is in the plane and does not pass through the region.

Solution for Thin Rectangular Regions

We begin by studying the case where the region is a thin rectangular strip and the axis of revolution is either perpendicular or parallel to the longer side of the strip.

In the case where the axis of revolution is perpendicular to the strip, the solid generated is either a disk or a washer, as illustrated in Figure 6.4.1. We get a disk if the strip touches the axis of revolution. If the strip does not touch the axis of revolution we get a disk with a hole, called a washer. The volumes are easily calculated:

$$\text{Volume of disk} = \pi(\text{Radius})^2(\text{Thickness})$$

and

$$\text{Volume of washer} = (\pi(\text{Outer radius})^2 - \pi(\text{Inner radius})^2)(\text{Thickness}).$$

In the case where the axis of revolution is parallel to the longer side of the strip, the solid obtained will be a thin cylindrical shell, as in Figure 6.4.2. The volume is

$$V = \pi(r + \Delta r)^2 h - \pi r^2 h$$
$$= \pi r^2 h + 2\pi r h \Delta r + \pi h (\Delta r)^2 - \pi r^2 h$$
$$= 2\pi r h \Delta r + \pi h (\Delta r)^2.$$

If Δr is small, then $\pi h (\Delta r)^2$ is proportionately much smaller than $2\pi r h \Delta r$. We use the formula

FIGURE 6.4.1a   The volume of the disk is $V = \pi r^2 \Delta h$.

FIGURE 6.4.1b   The volume of the washer is $V = \pi R^2 \Delta h - \pi r^2 \Delta h$.

FIGURE 6.4.2   The volume of the thin cylindrical shell is approximately the volume of the rectangular slab obtained by cutting and unrolling the shell, so $V \approx 2\pi r h \Delta r$.

> Approximate volume of a thin cylindrical shell
> $= 2\pi(\text{Radius})(\text{Height})(\text{Thickness}).$

The approximate volume of the cylindrical shell may be thought of as the volume of the rectangular slab obtained by cutting and unrolling the shell. The shell unrolls to a length of $2\pi(\text{Radius})$, which is the circumference of a circle.

NOTE   *We are distinguishing between washers and thin cylindrical shells, even though they are of the same geometric type. The distinction, as we are using the terms, depends on whether the height or difference in radii is small. A washer has the shape of a washer used with nuts and bolts. A thin cylindrical shell has the shape of a tin can with both ends removed.*

More General Regions

By using integral calculus we can find the volumes of solids obtained by revolving more general plane regions about a line in the plane. The idea is to divide the area into thin rectangular strips and use the integral to find the limit of the sums of the volumes of the solids obtained from each of the strips. See Figure 6.4.3.

FIGURE 6.4.3a   If a thin rectangular strip is revolved about a perpendicular line that touches one end of the strip, the solid generated is a disk.

FIGURE 6.4.3b   If a thin rectangular strip is revolved about a perpendicular line that does not touch the strip, the resulting solid is a washer.

FIGURE 6.4.3c   If a thin rectangular strip is revolved about a parallel line, the resulting solid is a thin cylindrical shell.

> ### Strategy for finding volume of a solid of revolution
>
> - Sketch region and axis of revolution.
> - Choose a convenient variable rectangular area strip.
> - Sketch the circular path of the endpoints of the strip and identify the solid obtained by revolving the strip as either a disk, washer, or cylindrical shell.
> - Label appropriate coordinates in terms of the variable that gives the position of the variable area strip (x for vertical strips and y for horizontal strips).
> - Write an integral that gives the limit of sums of the volumes of the disks or washers, or the approximate volumes of the cylindrical shells.
> - Evaluate the integral.

Recall from Section 6.1 that circles that are not parallel to the sketching plane appear as ellipses. See Figure 6.4.4. Some care should be used to sketch your circle through a point that is equidistant and on the opposite side of the axis of revolution from the point being revolved.

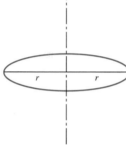

FIGURE 6.4.4a Circles that are not parallel to the sketching plane appear as ellipses.

FIGURE 6.4.4b Care should be used to sketch a circle through a point that is equidistant and on the opposite side of the axis of revolution from the point being revolved.

EXAMPLE 1

Consider the region bounded by

$$y = \frac{x}{2}, \quad y = 0, \text{ and } x = 2.$$

Set up an integral that gives the volume of the solid obtained by revolving the region about the **(a)** x-axis, **(b)** line $y = -1$, **(c)** y-axis, **(d)** line $x = 3$.

SOLUTION   We could use either a vertical strip or a horizontal strip, but let us use a vertical strip for each of the four parts of the problem. This means the components of our integrals will be expressed in terms of the variable x.

(a) The region, a variable vertical area strip, and the disk obtained by revolving the strip about the x-axis are sketched in Figure 6.4.5a. We have labeled coordinates of the endpoints of the area strip in terms of the variable x. The lower endpoint of the variable vertical strip is on the axis of revolution and

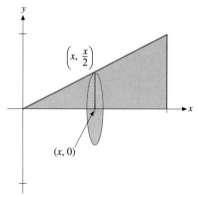

FIGURE 6.4.5a Revolving a variable vertical strip between $y = \dfrac{x}{2}$ and the x-axis about the x-axis generates a disk with radius $\dfrac{x}{2}$.

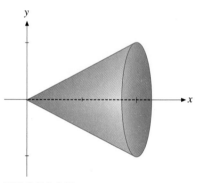

FIGURE 6.4.5b The complete solid is a right circular cone.

is the center of the disk obtained by revolving the strip. The radius of the disk is the length of the vertical line segment between the points $(x, x/2)$ and $(x, 0)$. The length of a vertical line segment is the difference in the y-coordinates of its endpoints. The equation

$$\text{Length} = (y\text{-coordinate of top}) - (y\text{-coordinate of bottom})$$

gives nonnegative length.

The volume is

$$V = \int_0^2 \underbrace{\pi}_{\substack{\pi \\ \text{Sum}}} \underbrace{\left(\frac{x}{2}\right)^2}_{(\text{Radius})^2} \underbrace{dx}_{(\text{Thickness})}.$$

$$\underbrace{\phantom{V = \int_0^2 \pi \left(\frac{x}{2}\right)^2 dx}}_{\text{Volume of disk}}$$

(The value of this integral is $2\pi/3$.)

We have drawn only one typical disk in Figure 6.4.5a. Sketches that are too detailed tend to hide the essential features. The complete solid is sketched in Figure 6.4.5b.

(b) From Figure 6.4.6a we see that revolving a variable vertical area strip about the line $y = -1$ results in a washer. We have labeled the endpoints of the area strip and the center of the washer. Labeling coordinates allows us to read the lengths of the outer radius and inner radius from our sketch.

The volume is

$$V = \int_0^2 \left(\underbrace{\pi}_{(\pi} \underbrace{\left(\frac{x}{2} - (-1)\right)^2}_{(\text{Outer radius})^2} \underbrace{-\pi}_{-\pi} \underbrace{(0 - (-1))^2}_{(\text{Inner radius})^2} \right) \underbrace{dx}_{(\text{Thickness})}.$$

$$\underbrace{\phantom{V = \int_0^2 \left(\pi \left(\frac{x}{2} - (-1)\right)^2 -\pi (0 - (-1))^2 \right) dx}}_{\substack{\text{Sum} \\ \text{Volume of washer}}}$$

(The value of the integral is $8\pi/3$.)

The complete solid is sketched in Figure 6.4.6b.

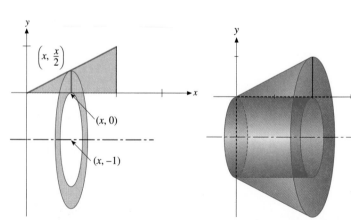

FIGURE 6.4.6a Revolving a variable vertical strip between $y = \dfrac{x}{2}$ and the x-axis about the line $y = -1$ generates a washer with outer radius $\left(\dfrac{x}{2}\right) - (-1)$ and inner radius $0 - (-1)$.

FIGURE 6.4.6b The solid generated is a right circular cone with a coaxial cylinder removed.

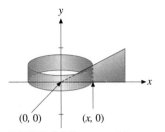

FIGURE 6.4.7a Revolving a variable vertical strip between $y = \dfrac{x}{2}$ and the x-axis about the y-axis generates a thin cylindrical shell with radius x and height $\dfrac{x}{2}$.

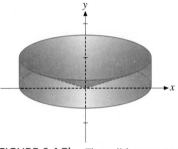

FIGURE 6.4.7b The solid generated is a right circular cylinder with a coaxial cone removed.

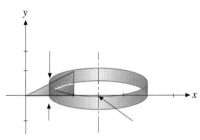

FIGURE 6.4.8a Revolving a variable vertical strip between $y = \dfrac{x}{2}$ and the x-axis about the line $x = 3$ generates a thin cylindrical shell with radius $3 - x$ and height $\dfrac{x}{2}$.

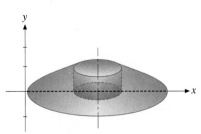

FIGURE 6.4.8b The solid generated is a right circular cone with a coaxial cylinder removed.

(c) Since the y-axis is parallel to the variable vertical area strip, we obtain a cylindrical shell. We have labeled the endpoints of the strip and a corresponding point on the axis of revolution in Figure 6.4.7a. The variable is x. The height of the cylindrical shell is the length of the variable vertical line segment between $(x, x/2)$ and $(x, 0)$. The radius is the length of the variable horizontal line segment between $(0, 0)$ and $(x, 0)$. The length of a horizontal line segment is the difference of the x-coordinates of its endpoints. That is,

Length = (x-coordinate of right endpoint) − (x-coordinate of left endpoint).

We then have

$$V = \int_0^2 \underbrace{2\pi}_{\substack{2\pi}} \underbrace{(x - 0)}_{(\text{Radius})} \underbrace{\left(\frac{x}{2} - 0\right)}_{(\text{Height})} \underbrace{dx}_{(\text{Thickness})}.$$

$$\underbrace{}_{\text{Sum}} \qquad \underbrace{\phantom{2\pi (x-0)\left(\frac{x}{2}-0\right)}}_{\substack{\text{Approximate volume} \\ \text{of cylindrical shell}}}$$

(This integral has value $8\pi/3$.)

The complete solid is indicated in Figure 6.4.7b.

(d) Revolving a vertical area strip about the vertical line $x = 3$ gives a cylindrical shell. A typical shell is sketched and labeled in Figure 6.4.8a. The volume is

$$V = \int_0^2 \underbrace{2\pi}_{\substack{2\pi}} \underbrace{(3 - x)}_{(\text{Radius})} \underbrace{\left(\frac{x}{2} - 0\right)}_{(\text{Height})} \underbrace{dx}_{(\text{Thickness})}.$$

$$\underbrace{}_{\text{Sum}} \qquad \underbrace{\phantom{2\pi (3-x)\left(\frac{x}{2}-0\right)}}_{\substack{\text{Approximate volume} \\ \text{of cylindrical shell}}}$$

(The value of the integral is $10\pi/3$.)

The complete solid is indicated in Figure 6.4.8b. ■

The method of solution we are using does not require that we memorize a different formula for each case that might occur. The only formulas used are that of the area of a circle for disks and washers, and the circumference of a circle for cylindrical shells. Since both of these are well-known formulas, it should not matter whether revolving a variable area strip results in a disk, washer, or cylindrical shell. Just choose the most convenient area strip and use whatever you get. Labeling key coordinates in terms of the appropriate variable allows us to read radii and heights from our sketches. As in all applications of the integral, it is most important to think of the definite integral as a sum. Each component of the integral should be thought of in terms of a specific physical quantity.

EXAMPLE 2

Consider the region bounded by

$$x = (y - 1)^2 \text{ and } x = y + 1.$$

Set up an integral that gives the volume of the solid obtained by revolving the region about the **(a)** x-axis, **(b)** line $y = 3$, **(c)** y-axis, **(d)** line $x = 4$.

SOLUTION In this problem it is clearly more convenient to use horizontal area strips, so y is the variable. (Note that there is not a single vertical area strip that is typical for the region. One type of vertical strip would go from the bottom to the top of the parabola, while another type of vertical strip would go from the line to the top of the parabola. Two integrals would be required to write the volume as integrals with respect to x.)

Appropriate sketches are given in Figures 6.4.9–6.4.12. Note that sketches of the complete solids would be very complicated. The coordinates are labeled in terms of the variable y. In each case, the variable y ranges between the y-coordinates of the points of intersection of the parabola and the line.

$$x = (y - 1)^2 \text{ and } x = y + 1 \text{ imply}$$
$$(y - 1)^2 = y + 1,$$

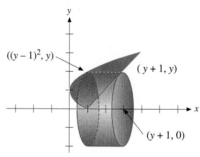

FIGURE 6.4.9 Revolving a variable horizontal strip between $x = (y - 1)^2$ and $x = y + 1$ about the x-axis generates a thin cylindrical shell with radius y and height $(y + 1) - (y - 1)^2$.

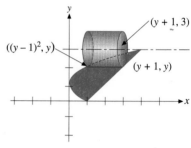

FIGURE 6.4.10 Revolving a variable horizontal strip between $x = (y - 1)^2$ and $x = y + 1$ about the line $y = 3$ generates a thin cylindrical shell with radius $3 - y$ and height $(y + 1) - (y - 1)^2$.

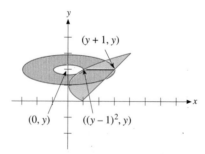

FIGURE 6.4.11 Revolving a variable horizontal strip between $x = (y - 1)^2$ and $x = y + 1$ about the y-axis generates a washer with outer radius $y + 1$ and inner radius $(y - 1)^2$.

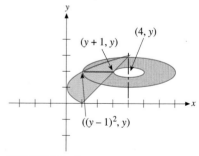

FIGURE 6.4.12 Revolving a variable horizontal strip between $x = (y - 1)^2$ and $x = y + 1$ about the line $x = 4$ generates a washer with outer radius $4 - (y - 1)^2$ and inner radius $4 - (y + 1)$.

$$y^2 - 2y + 1 = y + 1,$$
$$y^2 - 3y = 0,$$
$$y(y - 3) = 0.$$

The desired coordinates are $y = 0$ and $y = 3$.

Integrals that give the desired volumes are given below. Use Figures 6.4.9–6.4.12 to identify and check the components of each integral.

(a) Revolving horizontal strips about the x-axis gives thin cylindrical shells (Figure 6.4.9) with total volume

$$V = \int_0^3 2\pi(y)((y+1) - (y-1)^2)\,dy \quad (= 27\pi/2).$$

(b) Revolving horizontal strips about the line $y = 3$ gives thin cylindrical shells (Figure 6.4.10) with total volume

$$V = \int_0^3 2\pi(3 - y)((y+1) - (y-1)^2)\,dy \quad (= 27\pi/2).$$

(c) Revolving horizontal strips about the y-axis gives washers (Figure 6.4.11) with total volume

$$V = \int_0^3 (\pi(y+1)^2 - \pi((y-1)^2)^2)\,dy \quad (= 72\pi/5).$$

(d) Revolving horizontal strips about the line $x = 4$ gives washers (Figure 6.4.12) with total volume

$$V = \int_0^3 (\pi(4 - (y-1)^2)^2 - \pi(4 - (y+1))^2)\,dy \quad (= 108\pi/5). \quad \blacksquare$$

EXERCISES 6.4

In Exercises 1–30, set up and evaluate integrals that give the volume of the solids obtained by revolving the regions bounded by the given curves about the given lines.

1. $y = x^2$, $y = 0$, $x = 1$; about **(a)** x-axis, **(b)** $y = 1$, **(c)** y-axis, **(d)** $x = 1$

2. $y = x^3$, $y = 0$, $x = 1$; about **(a)** x-axis, **(b)** $y = 1$, **(c)** y-axis, **(d)** $x = 1$

3. $y = x^2$, $y = x$; about **(a)** x-axis, **(b)** $y = 1$, **(c)** y-axis, **(d)** $x = 1$

4. $y = 1 - x^2$, $x + y = 1$; about **(a)** x-axis, **(b)** $y = 1$, **(c)** y-axis, **(d)** $x = 1$

5. $y = x + x^2$, $y = 0$, $x = 1$, $x = 3$; about the x-axis **(Calculus Explorer & Tutor II, Applications of Integration, 2.)**

6. $y = 4x - x^2$, $x = y$; about the x-axis

7. $y = x^2$, $y = 1$, $x = 0$, $x = 1$; about the y-axis

8. $y = 2x$, $y = x$, $x = 2$, $x = 4$; about the x-axis **(Calculus Explorer & Tutor II, Applications of Integration, 4.)**

9. $y = 4x + x^2$, $x + y = 0$; about the y-axis

10. $y = \sqrt{x^2 + 1}$, $y = x^2/8$, $x = 3$, $x = 8$; about the y-axis **(Calculus Explorer & Tutor II, Applications of Integration, 3.)**

11. $x = y$, $x + 2y = 3$, $y = 0$; about **(a)** x-axis, **(b)** y-axis

12. $x + y = 1$, $x - 2y = 1$, $y = 1$; about **(a)** x-axis, **(b)** y-axis

13. $y = x^2$, $x + y = 2$, $x = 0 (x \geq 0)$; about the y-axis

14. $y = 2 - x^2$, $y = 0$, $x = 0 (x \geq 0)$; about the y-axis

15. $x = 2 - y^2$, $x = 0$; about the y-axis

16. $x = y^2 + 1$, $x = 0$, $y = 1$, $y = -1$; about the y-axis

17. $y = 1/x$, $y = 0$, $x = 1$, $x = 2$; about **(a)** x-axis, **(b)** y-axis

18. $y = 1/x^3$, $y = 0$, $x = 1$, $x = 2$; about **(a)** x-axis, **(b)** y-axis

19. $y = 2 - x^2$, $x + y = 2$; about **(a)** x-axis , **(b)** y-axis

20. $y = x^2$, $x + y = 2$, $y = 0$; about **(a)** x-axis, **(b)** y-axis

21. $y = 1/x^3$, $x = 0$, $y = 1$, $y = 8$; about **(a)** x-axis, **(b)** y-axis

22. $y = 1/x$, $x = 0$, $y = 1$, $y = 2$; about **(a)** x-axis , **(b)** y-axis

23. $y = \sec x$, $y = 1$, $x = 0$, $x = \pi/4$; about the x-axis

24. $y = 2 \sec x$, $y = 1$, $x = 0$, $x = \pi/4$; about the x-axis

25. $y = \sin x^2$, $y = 0$, $x = 0$, $x = \sqrt{\pi}$; about the y-axis

26. $y = \cos x^2$, $y = 0$, $x = 0$, $x = \sqrt{\pi/2}$; about the y-axis

27. $\dfrac{x}{r} + \dfrac{y}{h} = 1$, $y = 0$, $x = 0$; about the y-axis. (This gives the volume of a right circular cone with height h and radius of base r.)

28. $x = \sqrt{r^2 - y^2}$, $x = 0$; about the y-axis. (This gives the volume of a sphere of radius r.)

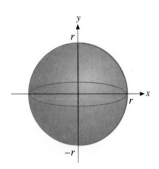

29. Inside $x^2 + y^2 = r^2$, above $y = r - h$, and to the right of $x = 0$; about the y-axis. (This gives the volume of a segment of a sphere.)

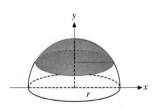

30. $(r - R)y = h(x - R)$, $y = h$, $y = 0$, $x = 0$;   above   the y-axis. (This gives the volume of a truncated cone.)

In Exercises 31–38, set up integrals that give the volume of the solids obtained by revolving the given regions about the given lines.

31. Inside $(x - 2)^2 + y^2 = 1$; about the y-axis

32. Inside $x^2 + y^2 = 1$; about $x = 2$

33. $y = \sin x$, $y = 0$, $x = 0$, $x = \pi$; about **(a)** x-axis, **(b)** y-axis

34. $y = \cos x$, $y = 0$, $x = 0$, $x = \pi/2$; about **(a)** x-axis, **(b)** y-axis

35. $y = \cos x$, $y = 1$, $x = 0$, $x = \pi/2$; about **(a)** x-axis, **(b)** y-axis

36. $y = \sin x$, $y = x$, $x = 0$, $x = \pi$;   about   **(a)** x-axis,   **(b)** y-axis

37. $y = \tan x$, $y = 0$, $x = 0$, $x = \pi/4$;   about   **(a)** x-axis,   **(b)** $x = \pi/4$

38. $y = \cot x$, $y = 0$, $x = \pi/4$, $x = \pi/2$; about **(a)** x-axis, **(b)** $x = \pi/4$

39. The region bounded by $y = kx$, $y = c$, $x = 0$, and $x = 1$ is revolved about $y = c$. Find c, $0 \le c \le 1$, such that the resulting solid has minimum volume. See figure.

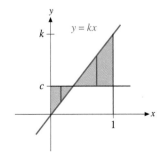

40. The region bounded by $y = 1 - x^2$
$x = 1$ is revolved about $y = c$. Find c
the resulting solid has minimum volu

6.5 ARC LENGTH AND SURFACE AREA

Connections

Length of a line segment, 1.3.

Revolving line segments about a line, 6.1.

We are familiar with formulas for the length of a line segment and for the circumference of a circle. We may also know formulas for the area of the surface of a sphere and the lateral surface area of right circular cylinders and cones. In this section we will use integral calculus to find the length of more general curves and the area of more general surfaces.

Length of Smooth Curves

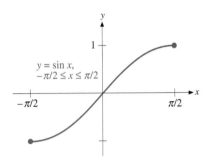

FIGURE 6.5.1 The curve $y = \sin x$, $-\pi/2 \le x \le \pi/2$, is smooth.

> **DEFINITION**
>
> The graph of a continuous function $y = f(x)$, $a \le x \le b$, is called a **curve**. If the derivative f' is continuous on the interval $[a, b]$, the graph is said to be a **smooth curve**.

A smooth curve has a tangent line at each point and the slope of the tangent line varies continuously as x varies. A curve is not smooth if it contains a sharp corner. For example, the graph of $y = \sin x$, $-\pi/2 \le x \le \pi/2$, is a smooth curve. See Figure 6.5.1. The graph of $y = |x|$, $-1 \le x \le 1$, is not a smooth curve. See Figure 6.5.2.

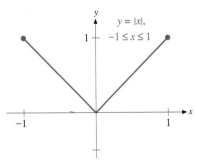

FIGURE 6.5.2 The curve $y = |x|$, $-1 \le x \le 1$ is not smooth because it contains a sharp corner at the origin.

Let us consider the **arc length** of a smooth curve that is the graph of $y = f(x)$, $a \le x \le b$. To calculate an approximate value of the arc length of the curve, let us choose points

$$a = x_0 < x_1 < \cdots < x_{n-1} < x_n = b$$

that divide $[a, b]$ into n subintervals. We can then use the length of the line segment between $(x_{j-1}, f(x_{j-1}))$ and $(x_j, f(x_j))$ as an approximate value of the "arc length" of the curve between these points, $j = 1, 2, \ldots, n - 1, n$. See Figure 6.5.3. The distance between $(x_{j-1}, f(x_{j-1}))$ and $(x_j, f(x_j))$ is

$$\Delta s_j = \sqrt{(x_j - x_{j-1})^2 + (f(x_j) - f(x_{j-1}))^2}$$
$$= \sqrt{(\Delta x_j)^2 + (\Delta y_j)^2}.$$

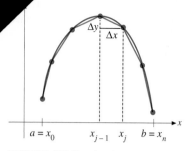

FIGURE 6.5.3 The length of a smooth curve can be approximated by a sum of the lengths of line segments between points on the curve.

Since the curve is smooth, the actual arc length of the curve will be very near the sum of the lengths of these line segments, if $[a, b]$ is divided into small enough subintervals. That is,

$$\text{Arc length} \approx \sum_{j=1}^{n} \Delta s_j = \sum_{j=1}^{n} \sqrt{(\Delta x_j)^2 + (\Delta y_j)^2}.$$

To relate the arc length of the curve to a definite integral, we need to express the expression we have for the approximate arc length as a Riemann sum. To do this, let us use the Mean Value Theorem to obtain

$$\Delta y_j = f(x_j) - f(x_{j-1}) = f'(x_j^*)(x_j - x_{j-1}) = f'(x_j^*)\Delta x_j$$

for some x_j^* between x_j and x_{j-1}. Then

$$\Delta s_j = \sqrt{(\Delta x_j)^2 + (\Delta y_j)^2}$$

$$= \sqrt{(\Delta x_j)^2 + (f'(x_j^*)\Delta x_j)^2}$$

$$= \sqrt{(1 + (f'(x_j^*))^2)(\Delta x_j)^2}$$

$$= \sqrt{1 + (f'(x_j^*))^2}\,\Delta x_j.$$

We then have

$$\text{Arc length} \approx \sum_{j=1}^{n} \Delta s_j = \sum_{j=1}^{n} \sqrt{1 + (f'(x_j^*))^2}\,\Delta x_j.$$

We then see that an approximate value of the arc length of the curve is given by a Riemann sum of the continuous function $\sqrt{1 + (f'(x))^2}$ over the interval $[a, b]$. The formula

$$\boxed{\text{Arc length} = \int_a^b \sqrt{1 + (f'(x))^2}\,dx}$$

should satisfy our intuitive notion of arc length of the smooth curve $y = f(x)$, $a \leq x \leq b$. The formula may be considered to be the definition of the arc length of a smooth curve. (It is possible to define arc length in terms of sums of lengths of the line segments we used, without expressing the sums as Riemann sums. Such a definition could be used to calculate the arc length of curves that are not smooth.)

It is useful to think of

$$ds = \sqrt{1 + (f'(x))^2}\,dx$$

as the arc length of a small piece of the smooth curve, located at the point $(x, f(x))$, $a \leq x \leq b$. The integral

$$\int_a^b \sqrt{1 + (f'(x))^2}\,dx$$

then represents the "sum" of the arc lengths of the pieces of the curve. The formula for ds is suggested by the sketch in Figure 6.5.4. That is:

If $y = f(x)$, then $dy = f'(x)\,dx$ and

$$ds = \sqrt{(dx)^2 + (dy)^2} = \sqrt{(dx)^2 + (f'(x)dx)^2} = \sqrt{1 + (f'(x))^2}\,dx.$$

We call ds the **differential arc length** of the curve.

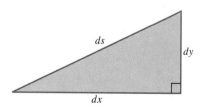

FIGURE 6.5.4 The Pythagorean Theorem suggests the formula $ds = \sqrt{(dx)^2 + (dy)^2} = \sqrt{1 + \left(\dfrac{dy}{dx}\right)^2}\,dx.$

Arc-Length Integrals

Let us see that the integral formula for arc length agrees with the distance formula for the length of a line segment.

EXAMPLE 1

Set up and evaluate an integral that gives the arc length of the curve $y = 2x - 1$, $0 \le x \le 3$.

SOLUTION   The curve is the line segment between the points $(0, -1)$ and $(3, 5)$. See Figure 6.5.5. The distance formula tells us the length of the segment is

$$s = \sqrt{(3 - 0)^2 + (5 - (-1))^2} = \sqrt{45} = 3\sqrt{5}.$$

To use the integral formula for arc length, we need the derivative of y with respect to x. We have

$$y = 2x - 1, \text{ so } \frac{dy}{dx} = 2.$$

Then

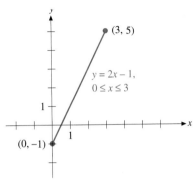

FIGURE 6.5.5   The curve $y = 2x - 1$, $0 \le x \le 3$, is the line segment from $(0, -1)$ to $(3, 5)$.

$$ds = \sqrt{1 + \left(\frac{dy}{dx}\right)^2}\, dx = \sqrt{1 + (2)^2}\, dx = \sqrt{5}\, dx.$$

The arc length is

$$s = \int_0^3 \sqrt{5}\, dx = 3\sqrt{5}.$$

This agrees with the value obtained from the distance formula. ∎

EXAMPLE 2

Set up and evaluate an integral that gives the arc length of the curve $y = x^{3/2}$, $0 \le x \le 4$.

SOLUTION   We have

$$y = x^{3/2}, \text{ so } \frac{dy}{dx} = \frac{3}{2} x^{1/2}.$$

Then

$$ds = \sqrt{1 + \left(\frac{dy}{dx}\right)^2}\, dx$$

$$= \sqrt{1 + \left(\frac{3}{2} x^{1/2}\right)^2}\, dx$$

$$= \sqrt{1 + \frac{9}{4} x}\, dx.$$

The arc length is

$$s = \int_0^4 \sqrt{1 + \frac{9}{4} x}\, dx.$$

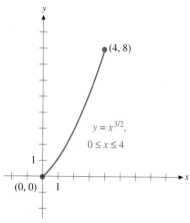

FIGURE 6.5.6   The curve $y = x^{3/2}, 0 \le x \le 4$, extends from $(0, 0)$ to $(4, 8)$.

The definite integral may be evaluated by using the substitution

$$u = 1 + \frac{9}{4}x, \text{ so } du = \frac{9}{4}dx, u(0) = 1, \text{ and } u(4) = 10.$$

Then

$$s = \int_0^4 \left(1 + \frac{9}{4}x\right)^{1/2} \left(\frac{4}{9}\right) \left(\frac{9}{4}\right) dx$$

$$= \int_1^{10} u^{1/2} \left(\frac{4}{9}\right) du = \left(\frac{4}{9}\right) \left(\frac{u^{3/2}}{3/2}\right)\Bigg]_1^{10}$$

$$= \frac{8}{27}(10)^{3/2} - \frac{8}{27}(1)^{3/2} \approx 9.0734.$$

The curve is sketched in Figure 6.5.6. ■

EXAMPLE 3
Find the arc length of the curve

$$y = \frac{x^3}{3} + \frac{1}{4x}, \qquad 1 \le x \le 2.$$

SOLUTION   We have

$$y = \frac{x^3}{3} + \frac{1}{4x}, \text{ so } \frac{dy}{dx} = x^2 - \frac{1}{4x^2}.$$

Then

$$ds = \sqrt{1 + \left(\frac{dy}{dx}\right)^2}\, dx = \sqrt{1 + \left(x^2 - \frac{1}{4x^2}\right)^2}\, dx$$

$$= \sqrt{1 + \left(x^4 - 2x^2\frac{1}{4x^2} + \frac{1}{16x^4}\right)}\, dx$$

$$= \sqrt{1 + \left(x^4 - \frac{1}{2} + \frac{1}{16x^4}\right)}\, dx$$

$$= \sqrt{x^4 + \frac{1}{2} + \frac{1}{16x^4}}\, dx$$

$$= \sqrt{\left(x^2 + \frac{1}{4x^2}\right)^2}\, dx$$

$$= \left(x^2 + \frac{1}{4x^2}\right) dx.$$

In this example the expression inside the square root turned out to be a perfect square. This greatly simplifies the evaluation of the arc length integral. We have

$$s = \int_1^2 \left(x^2 + \frac{1}{4x^2}\right) dx$$

$$= \left(\frac{x^3}{3} - \frac{1}{4x}\right)\Bigg]_1^2 = \left(\frac{8}{3} - \frac{1}{8}\right) - \left(\frac{1}{3} - \frac{1}{4}\right) = \frac{59}{24}.$$

The curve is sketched in Figure 6.5.7. ■

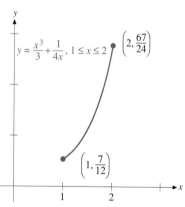

FIGURE 6.5.7   The curve $y = \frac{x^3}{3} + \frac{1}{4x}, 1 \le x \le 2$, extends from $(1, 7/12)$ to $(2, 67/24)$.

FIGURE 6.5.8a Rotating a small piece of a curve about a line generates a ribbonlike band.

FIGURE 6.5.8b We can cut and unroll the band to obtain a "rectangle" with one side of length $2\pi r$ and the other side of length ds, so the area is approximately $dS = 2\pi r\,ds$.

Area of Surfaces of Revolution

A problem closely associated with arc length is that of the area of a surface of revolution. Let us consider the surface obtained by revolving the smooth curve

$$y = f(x), \qquad a \leq x \leq b,$$

about a line. As before, we divide the curve into small pieces that are very nearly line segments of length Δs_j, $j = 1, \ldots, n$. The result of revolving one of the small pieces about the axis of revolution is a ribbonlike band that is very nearly part of the surface of a right circular cone. We can obtain an approximate value of the area of this band by cutting and unrolling it. We can then think of the band as a rectangle with one length the circumference of the circle formed by the band and the other dimension the *slant height* of the band. See Figure 6.5.8. Noting that the slant height is the arc length of the small piece, we obtain that the approximate area of the band is

$$dS = 2\pi r\,ds.$$

NOTE *We are using capital S for surface area and lowercase s for arc length. Be careful not to confuse them.*

It should then seem reasonable that the surface area is given by the following integral:

$$\text{Surface area} = S = \int_a^b 2\pi r \sqrt{1 + (f'(x))^2}\,dx,$$

where r must be expressed in terms of the variable x.

When setting up an integral to determine the area of a surface of revolution, it is useful to draw a typical small piece of the curve and sketch the band obtained by revolving the piece about the axis of revolution. Labeling the location of the piece as $(x, f(x))$ and labeling the center of the band allows us to read the radius of the band in terms of x. Don't forget to use ds, the slant height of the band.

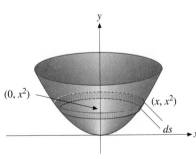

FIGURE 6.5.9 Revolving a piece of the curve $y = x^2$, $0 \leq x \leq 1$, about the y-axis generates a circular band of radius x.

EXAMPLE 4
Find the area of the surface obtained by revolving the curve $y = x^2$, $0 \leq x \leq 1$, about the y-axis.

SOLUTION A typical piece of the surface is sketched and labeled in Figure 6.5.9. We see that the radius of the band is $r = x$.

$$y = x^2, \text{ so } \frac{dy}{dx} = 2x.$$

Then

$$ds = \sqrt{1 + \left(\frac{dy}{dx}\right)^2}\,dx = \sqrt{1 + 4x^2}\,dx.$$

The surface area is given by

$$S = \int_0^1 \underbrace{\underbrace{2\pi x}_{\text{(Circumference)}} \underbrace{\sqrt{1 + 4x^2}\,dx}_{\text{(Slant height, } ds\text{)}}}_{\text{Area of band}}.$$

Sum

This integral involves the square root of a function, so we can try to reduce it to $\int u^{1/2}\,du$. We set $u = 1 + 4x^2$, so $du = 8x\,dx$, $u(0) = 1$, and $u(1) = 5$. Noting that the integrand contains the essential factor x, we have

$$\begin{aligned}
S &= \int_0^1 2\pi x \sqrt{1 + 4x^2}\,dx \\
&= \int_0^1 2\pi (1 + 4x^2)^{1/2}\left(\frac{1}{8}\right)(8x)\,dx \\
&= \int_1^5 2\pi u^{1/2}\left(\frac{1}{8}\right)du \\
&= 2\pi \left(\frac{1}{8}\right)\left(\frac{u^{3/2}}{3/2}\right)\Bigg]_1^5 \\
&= \frac{\pi}{6}(5^{3/2} - 1) \approx 5.3304. \quad \blacksquare
\end{aligned}$$

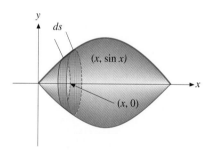

FIGURE 6.5.10 Revolving a piece of the curve $y = \sin x$, $0 \le x \le \pi$, about the x-axis generates a circular band of radius $\sin x$.

EXAMPLE 5

Set up an integral that gives the area of the surface obtained by revolving the curve $y = \sin x$, $0 \le x \le \pi$, about the x-axis.

SOLUTION   We have $y = \sin x$, so $\dfrac{dy}{dx} = \cos x$. Then

$$ds = \sqrt{1 + \left(\frac{dy}{dx}\right)^2}\,dx = \sqrt{1 + \cos^2 x}\,dx.$$

From Figure 6.5.10 we see that the radius of a typical band is $r = \sin x$. The surface area is

$$S = \int_0^1 \underbrace{\underbrace{2\pi \sin x}_{\text{(Circumference)}} \underbrace{\sqrt{1 + \cos^2 x}\,dx}_{\text{(Slant height, } ds\text{)}}}_{\text{Area of band}}.$$

Sum

(This integral can be evaluated by using techniques we will learn in Chapter 8. It has approximate value 14.4236.) $\blacksquare$

There are very few types of curves for which either the arc-length integral or the integral for the corresponding surface of revolution is convenient to evaluate by using the Fundamental Theorem of Calculus. Of course, an approximate value of any definite integral can be obtained by using an appropriate computer program.

EXERCISES 6.5

Find the arc lengths of the curves given in Exercises 1–4.

1. $y = \frac{4}{3}(x + 1)^{3/2}, 0 \le x \le 3$

2. $y = \frac{x^4}{8} + \frac{1}{4x^2}, \frac{1}{2} \le x \le 2$ **(Calculus Explorer & Tutor II, Applications of Integration, 6.)**

3. $y = 2x^{2/3}, 1 \le x \le 8$

4. $y = [4 - (x - 8)^{2/3}]^{3/2}, 0 \le x \le 7$

In Exercises 5–8, find the areas of the surfaces obtained by revolving the given curves about the given lines.

5. $y = \frac{1}{3}x^3, 0 \le x \le 1$; about the x-axis

6. $y = \frac{1}{2}x^2 + 1, 0 \le x \le 1$; about the y-axis

7. $y = \sqrt{25 - x^2}, 0 \le x \le 3$; about the y-axis

8. $y = \sqrt{25 - x^2}, 0 \le x \le 3$; about the x-axis

In Exercises 9–14, find **(a)** *the arc lengths of the given curves and* **(b)** *the areas of the surfaces obtained by revolving the curves about the given line.*

9. $y = 4 - 2x, 0 \le x \le 2$; about the y-axis

10. $y = 4 - 2x, 0 \le x \le 2$; about the x-axis

11. $y = \frac{1}{6}x^3 + \frac{1}{2x}, 1 \le x \le 2$; about the x-axis

12. $y = \frac{2x^6 + 1}{8x^2}, 1 \le x \le 3$; about the x-axis

13. $y = \frac{1}{24}(x^2 + 16)^{3/2}, 0 \le x \le 3$; about the y-axis

14. $y = \frac{2}{27}(x^2 + 9)^{3/2}, 0 \le x \le 4$; about the y-axis

15. Find the lateral surface area of a right circular cone with height h and radius of base r.

16. Find the lateral surface area of a right circular cylinder with height h and radius r.

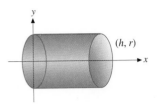

17. **(a)** Find the area of the surface obtained by revolving the curve $y = \sqrt{r^2 - x^2}, 0 \le x \le x_0$, about the y-axis, where $0 < x_0 < r$. **(b)** Find the surface area of a hemisphere of radius r by taking the limit as x_0 approaches r from the left of the area found in (a).

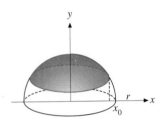

18. **(a)** Find the area of the surface obtained by revolving the curve $y = \sqrt{r^2 - x^2}, 0 \le x \le x_0$, about the x-axis, where $0 < x_0 < r$. **(b)** Find the surface area of a hemisphere of radius r by taking the limit as x_0 approaches r from the left of the area found in (a).

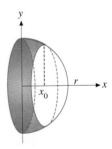

19. A cable is hanging in the form of a parabola from two points that are of the same height and a distance $2a$ apart horizontally. At its lowest point midway between the supports, the cable hangs a distance h below its supports. Set up an integral that gives the length of the cable.

20. A cable is hanging in the shape of the parabola $y = x^2/320$, $-40 \le x \le 40$. Use Simpson's Rule with $\Delta x_j = 10$ to find the approximate length of the cable.

Coordinates of a variable point on a curve can be expressed in terms of the arc length from a fixed point on the curve. In Exercises 21–24: **(a)** *Express the length s of the given curve from the given point to a variable point* (x, y) *on the curve as a function of x.* **(b)** *Solve the equation in (a) for x in terms of s and express the coordinates of a variable point on the curve in terms of the arc length s.*

21. $y = 2x - 1,\ (0, -1)$

22. $y = mx + b,\ (0, b)$

23. $y = \dfrac{2}{3}x^{3/2},\ (0, 0)$

24. $x^{2/3} + y^{2/3} = 1,\ (1, 0)$, for $0 < x \le 1$ and $0 \le y < 1$

25. **(a)** Express the length s of the arc subtended on the circle $x^2 + y^2 = r^2$ by the angle θ in standard position in terms of θ. **(b)** Express the (x, y) coordinates of a variable point on the circle in terms of s.

26. Use Δx_j's, Δy_j's, and the Mean Value Theorem to show that the arc length of a curve $x = g(y)$, $a \le y \le b$, can be expressed as $L = \displaystyle\int_a^b \sqrt{(g'(y))^2 + 1}\, dy$.

Use the formula given in Exercise 26 to set up an integral with respect to y that gives the length of the curves given in Exercises 27–28.

27. $y^3 = x^2$, between $(1, 1)$ and $(64, 16)$

28. $x = \sin y$, between $(0, 0)$ and $(1, \pi/2)$

In Exercises 29–33, find approximate values of the arc length of the given curves by: **(a)** *adding the lengths of the six line segments that connect points on the graph with x-coordinates equally spaced;* **(b)** *using Simpson's Rule with* $n = 6$ *(or a computer program) to find an approximate value of the arc length integral.*

29. $y = x^2,\ 0 \le x \le 3$

30. $y = x^3,\ 0 \le x \le 3$

31. $y = \sin x,\ 0 \le x \le \pi$

32. $y = \sin 2x,\ 0 \le x \le \pi$

33. $y = \sqrt{1 - x^2},\ 0 \le x \le 1/\sqrt{2}$. (Note that the arc length is one-eighth the circumference of a circle with radius 1.)

34. A sheet of metal is to be corrugated so that its cross-section is given by the curve $y = 0.5 \sin(\pi x/2)$, where the units are inches. Use Simpson's Rule with $\Delta x_j = 1$ (or a computer program) to approximate the width of the original flat sheet if it is to be 48 in. wide after it is corrugated. (*Hint:* You need to approximate the arc-length integrals only as x varies between 0 and 2.)

In Exercises 35–38, use Simpson's Rule with the indicated value of n (or a computer program) to approximate the integral that gives the area of the surface obtained by revolving the given curve about the given line.

35. $y = x^2,\ 0 \le x \le 1$, about the x-axis, $n = 4$

36. $y = \sin x,\ 0 \le x \le \pi$, about the x-axis, $n = 6$

37. $y = x^3/3,\ 0 \le x \le 1$, about the y-axis, $n = 4$

38. $y = \cos x,\ 0 \le x \le \pi/2$, about the y-axis, $n = 4$

39. If $f'(t)$ is continuous for $a \le t \le b$, then the arc-length function $s(x) = \displaystyle\int_a^x \sqrt{1 + (f'(t))^2}\, dt$ is defined for $a \le x \le b$. Show that the differential of s, $ds = s'(x)\,dx$, is what we have called the differential arc length, $ds = \sqrt{1 + (f'(x))^2}\, dx$. Show that $ds = \sec \gamma\, dx$, where γ is the acute angle between the y-axis and the line normal to $y = f(x)$ at $(x, f(x))$.

6.6   WORK

Connections

Concept of definite integrals as limits of Riemann sums, 5.2–3.

Work is a quantity associated with a force acting on an object as it moves from one position to another.

Force has magnitude and direction. In this section we will consider the work done when an object moves along a line, subject to a force in a direction along the line. Work done by a *constant force* is then given by

$$\text{Work} = (\text{Constant force}) \cdot (\text{Distance}).$$

In this formula, distance should be interpreted as directed distance, with direction indicated by sign. Also, the direction along the line in which a force acts is indicated by its sign. Work done by a force is positive if the direction of the force is in the direction of motion, while the work is negative if the force is opposite to the motion. Units of work in English and metric systems are indicated in Table 6.6.1.

TABLE 6.6.1

System	Distance	Force	Work
English	feet (ft)	pound (lb)	ft · lb
mks (SI)	meter (m)	newton (N)	N · m = joule (J)
cgs	centimeter (cm)	dyne	dyne · cm = erg

Gravitational Force

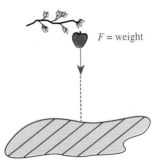

F = weight

FIGURE 6.6.1 For problems that involve motion over relatively small distances near the surface of the earth, we assume the weight of an object does not vary and the force on the object due to gravity is directed downward.

The magnitude of the force on an object due to the gravitational attraction of the earth is the weight of the object. The force is directed toward the center of the earth. The magnitude of the force depends on the distance of the object from the center of the earth, so objects weigh slightly more at sea level than they do at the top of a high mountain. For problems that involve motion near the surface of the earth we assume that the weight of an object does not vary. If the distances involved are relatively small compared to the diameter of the earth, we will consider the earth to be flat and the force downward. See Figure 6.6.1.

EXAMPLE 1
A 500-lb ball is to be raised 15 ft. The work required is $W = (500) \cdot (15) = 7500$ ft·lb. ■

In Example 1 we calculated the work required to overcome the force due to gravity. This requires an upward force, in the direction of motion, so the work required is positive. The work done by gravity is negative, since the direction of motion is opposite to the force due to gravity. If an object is lowered, the work done by gravity is positive, and the work required to overcome gravity is negative.

Work associated with the force of gravity has the property that it depends only on the initial and terminal heights of the object being moved. For example, if an object is raised and then lowered back to its initial position, the net work done is zero. This is because the work required to overcome gravity is negative when the direction of motion is downward, so it cancels the work required to lift the object. Gravity does not directly offer any resistance to horizontal motion, so no work is required to overcome gravity for horizontal motion. Of course, the force of friction must be overcome for horizontal motion. The force due to friction of a moving object is always in the direction opposite to the motion. Hence, if an object is moved horizontally and then returned to its original position, the work done to overcome moving friction is positive.

Adding Components of Work

To calculate work done in procedures more complicated than that illustrated in Example 1, we use the following basic fact.

Total work is the sum of the work of each component.

This allows us to break down a complicated procedure into component parts. Each component must consist of a (nearly) constant force applied over a distance along a straight line. We can then use the basic formula Work = (Constant force)·(Distance) to approximate the component work and add results to determine the total work associated with the procedure. We are particularly

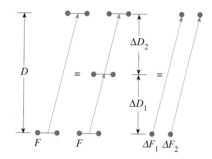

FIGURE 6.6.2 Total work done is the sum of each component. For lifting an object, either the distance or the object being lifted may be divided into components.

interested in cases where a definite integral can be used to evaluate the limit of the sums of component work.

Either the distance or the object being moved may be divided into components. Figure 6.6.2 illustrates three procedures for lifting a weight. The entire weight may be lifted the entire distance ($W = F \cdot D$); the weight may be lifted to an intermediate level and then the rest of the way ($W = F \cdot \Delta D_1 + F \cdot \Delta D_2$); or components of the weight may be lifted separately ($W = \Delta F_1 \cdot D + \Delta F_2 \cdot D$).

If $D = \Delta D_1 + \Delta D_2$ and $F = \Delta F_1 + \Delta F_2$, then

$$W = F \cdot D = F \cdot \Delta D_1 + F \cdot \Delta D_2 = \Delta F_1 \cdot D + \Delta F_2 \cdot D.$$

Fluid Pumping Problems

EXAMPLE 2

The conical tank pictured in Figure 6.6.3 is 8 ft deep and 8 ft across the top. It contains water to a depth of 6 ft. Find the work required to pump the water to the level of the top of the tank. Water weighs $62.4\,\text{lb/ft}^3$.

SOLUTION   In this problem we need to determine the work required to overcome the force of gravity. This means that the work required depends only on the initial and terminal height of each portion of water. Each thin layer of water at level y must be pumped to the level of the top of the tank. The variable y varies from 0 to 6, since these are the levels of water that we wish to remove. The volume of a thin layer of water at level y and the distance that the water at level y is to be moved depend on the variable y.

The cross-section of water at level y is a disk. From Figure 6.6.3 we see that the radius of the disk is the x-coordinate of the point on the line through the points $(0, 0)$ and $(4, 8)$. This line has slope 2 and equation $y = 2x$, so $x = (1/2)y$. The coordinates of the endpoints of a typical radius have been labeled in Figure 6.6.3 in terms of the variable y. We can express the work required as

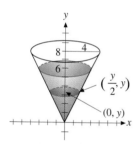

FIGURE 6.6.3   A thin layer of water at level y forms a disk with radius $\dfrac{y}{2}$ and thickness dy.

$$W = \int_0^6 \underbrace{(8-y)}_{\substack{\text{(Distance)}}} \quad \underbrace{(62.4)}_{\substack{\text{(Weight/volume)}}} \quad \underbrace{\pi \left(\frac{y}{2}\right)^2}_{\substack{\text{Area of}\\ \text{cross-section}}} \quad \underbrace{dy}_{\text{(Thickness)}}.$$

(Force, weight of disk)

Sum   Work

We then have

$$W = \int_0^6 \frac{(62.4)\pi}{4}(8y^2 - y^3)\,dy$$

$$= \frac{(62.4)\pi}{4}\left(8\frac{y^3}{3} - \frac{y^4}{4}\right)\Bigg]_0^6$$

$$= \frac{(62.4)\pi}{4}\left(8\frac{6^3}{3} - \frac{6^4}{4}\right)$$

$$= (62.4)(\pi)(63) \approx 12,350\,\text{ft·lb.}\quad\blacksquare$$

Chain-Lifting Problems

EXAMPLE 3

A chain 15 ft long and weighing 3 lb/ft is hanging vertically from the top of a building. Find the work required to raise 10 ft of the chain to the level of the top of the building, so 5 ft remain hanging.

SOLUTION   The work required to overcome gravity depends only on the initial and terminal heights of the components of the chain. Let us choose a vertical coordinate system as in Figure 6.6.4.

We can calculate the work by using the integral that corresponds to the procedure of lifting the chain a little at a time. This procedure is illustrated in Figure 6.6.4. The force is then the weight of the chain left hanging each time it is lifted. The variable y represents the position of the bottom of the chain; y varies from 0 to 10. The work done is

FIGURE 6.6.4   A chain of length $15 - y$ is lifted a distance of dy.

$$W = \int_0^{10} \underbrace{\underbrace{3}_{\text{(Weight/length)}} \underbrace{(15 - y)}_{\text{(Length)}}}_{\underbrace{}_{\text{(Force, weight)}}} \underbrace{dy}_{\text{(Distance)}}.$$

$$\underbrace{\phantom{W = \int_0^{10} 3(15-y)dy}}_{\text{Sum}} \quad \underbrace{}_{\text{Work}}$$

Evaluating this integral gives

$$W = \left(3(15)y - 3\frac{y^2}{2} \right) \Bigg]_0^{10} = (450 - 150) - (0) = 300 \text{ ft·lb}.$$

We can also calculate the work by using an integral that corresponds to the procedure of cutting the chain into small pieces and lifting each piece separately. This is illustrated in Figure 6.6.5. The variable y represents the level of the small piece that is being lifted. Each part of the chain is lifted some distance, so y varies from 0 to 15. For $0 \leq y \leq 5$, the piece at height y is lifted 10 ft. For $5 \leq y \leq 15$ the piece is lifted from height y to height 15, a distance of $15 - y$. We can express the work as the sum of two integrals.

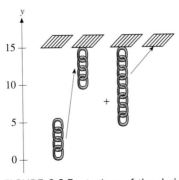

FIGURE 6.6.5   A piece of the chain of length dy at level y is lifted 10 ft if $0 \leq y \leq 5$ and it is lifted a distance $15 - y$ if $5 \leq x \leq 15$.

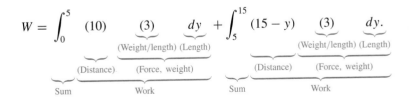

$$W = \int_0^5 \underbrace{(10)}_{\text{(Distance)}} \underbrace{(3)}_{\text{(Weight/length)}} \underbrace{dy}_{\text{(Length)}} + \int_5^{15} \underbrace{(15 - y)}_{\text{(Distance)}} \underbrace{(3)}_{\text{(Weight/length)}} \underbrace{dy}_{\text{(Length)}}.$$

Then

$$W = (10)(3)y \Bigg]_0^5 + \left(45y - 3\frac{y^2}{2} \right) \Bigg]_5^{15} = 150 + 150 = 300 \text{ ft·lb}.$$

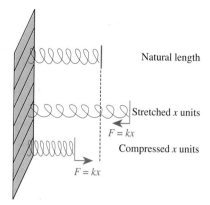

FIGURE 6.6.6 Hooke's Law says the force due to the displacement of a spring from its natural length is directly proportional to the displacement.

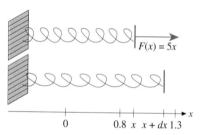

FIGURE 6.6.7 The force on a spring is nearly constant as the displacement varies over a small interval.

The first integral gives the work done in lifting a $(3)(5)$-lb piece of chain a distance of 10 ft. That is, we could have thought of lifting the bottom 5 ft of the chain as a single unit.

Note that both methods of calculating the work give the same value, as they should for a problem that involves the work required to overcome gravity. ∎

Hooke's Law

Hooke's Law says the force due to the displacement of a spring from its natural position is directly proportional to the displacement.

$$F = kx, \qquad k \text{ a constant.}$$

The force is directed toward the natural position of the spring. See Figure 6.6.6.

EXAMPLE 4

A 4-lb weight stretches a certain spring 0.8 ft beyond its natural length. How much work is required to stretch the spring from 0.8 ft to 1.3 ft beyond its natural length?

SOLUTION We must first determine the force on the spring. From Hooke's Law we know $F(x) = kx$, where x is the distance the spring is stretched beyond its natural length and k is a constant. We need the value of k. Since $x = 0.8$ when $F = 4$, substitution in the equation $F(x) = kx$ gives $4 = k(0.8)$, so $k = 4/0.8 = 5$ and $F(x) = 5x$.

The force $F(x) = 5x$ is not constant as x varies from 0.8 to 1.3. However, the force is nearly constant as x ranges over very small intervals. We can then calculate the work done over each small subinterval of $[0.8, 1.3]$ and use a definite integral to evaluate the limit of the sums. See Figure 6.6.7. We have

$$W = \int_{0.8}^{1.3} \underbrace{\underbrace{5x}_{\text{(Force)} \cdot \text{(Distance)}} \underbrace{dx}_{}}_{\substack{\text{Sum} \qquad \text{Work}}}.$$

Evaluating this integral, we have

$$W = \frac{5x^2}{2} \bigg]_{0.8}^{1.3} = \frac{5}{2}(1.3)^2 - \frac{5}{2}(0.8)^2 = 2.625 \text{ ft·lb.} \quad ∎$$

The same reasoning as was used in Example 4 shows that the work associated with any continuously varying force $F(x)$ in a direction along the line of motion between a and b is given by

$$W = \int_a^b F(x)\,dx.$$

This integral represents the sum of the components of work given by a force of $F(x)$ over a distance dx.

Work Done by an Ideal Gas

Let us consider a gas that is contained in a cylinder with a movable piston that allows the volume of the gas to vary. See Figure 6.6.8a. The **state** of the gas is determined by its pressure (P), volume (V), and temperature (T). We assume

FIGURE 6.6.8a The volume of a gas varies as a piston moves.

that the pressure and temperature of the gas are the same at every point inside the container, so the gas is in equilibrium. The force due to gas pressure on the piston is

$$F = P \cdot A,$$

where A is the area of the piston cross-section. If the force due to gas pressure is different from the exterior force on the piston, the piston will move. If the piston is moved a small distance dy, the change in volume of the gas is

$$dV = A \, dy,$$

and the work done by the gas is

$$dW = F \cdot dy = (P \cdot A) dy = P \cdot (A \, dy) = P \, dV.$$

See Figure 6.6.8b. If the state of the gas changes relatively slowly, so the gas is essentially at equilibrium at each intermediate state, the work done by the gas is given by the sum of the components $dW = P \, dV$. If the volume of the gas is initially V_i and terminally V_t, the total work can be expressed as an integral,

$$W = \int_{V_i}^{V_t} P \, dV.$$

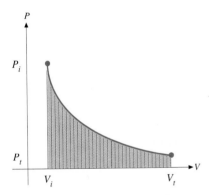

FIGURE 6.6.8b If the piston moves a small distance dy, the change in volume of the gas is $dV = A \, dy$, where A is the area of the piston cross-section.

The work done by a gas depends not only on the initial and terminal states of the gas; it also depends on the intermediate states. An **ideal gas** satisfies the condition

$$\frac{PV}{T} = \text{Constant}$$

(T is the absolute temperature in degrees Kelvin), so the state of an ideal gas can be determined from its pressure and volume. Changes in the state of a gas can be indicated by a **PV diagram**. It is customary to interpret the work done by a gas as the area under the curve in the PV diagram. See Figure 6.6.9.

A process in which an ideal gas changes state without exchanging heat with its surrounding is called an **adiabatic process**. In any adiabatic process, the volume and pressure of an ideal gas are related by an equation of the form

$$PV^{\gamma} = k,$$

where γ and k are constants.

FIGURE 6.6.9 The work done by a gas is the area under the curve in the PV diagram of the process.

EXAMPLE 5

Air in the cylinder of a diesel engine expands from 1.2 in.^{3} to 12 in.^{3}. The initial pressure is 600 lb/in.^{2}. Find the work done by the gas if the air behaves as an ideal gas and the expansion is adiabatic with $PV^{1.4} = k$.

SOLUTION We need to determine the value of the constant k. Initially, $P_i = 600$ lb/in.^{2} and $V_i = 1.2$ in.^{3}, so

$$k = (600)(1.2)^{1.4} \approx 774.$$

We then have $PV^{1.4} = 774$, so

$$P = 774 \, V^{-1.4}.$$

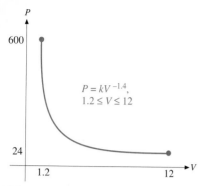

The terminal volume is $V_t = 12\,\text{in.}^3$ and the work done by the gas is

$$W = \int_{V_i}^{V_t} P\,dV = \int_{1.2}^{12} 774\,V^{-1.4}\,dV$$

$$= 774\,\frac{V^{-0.4}}{-0.4}\Bigg]_{1.2}^{12} = -1935\big((12)^{-0.4} - (1.2)^{-0.4}\big)$$

$$\approx 1083\,\text{in.·lb}.$$

The PV diagram for this adiabatic process is given in Figure 6.6.10. ■

FIGURE 6.6.10 $P = kV^{-1.4}$, $1.2 \le V \le 12$. The constant k is determined by the initial condition $P_i = 600$ and $V_i = 12$.

EXERCISES 6.6

1. An elevator carries two passengers from the first to the second floor, where an additional passenger enters. Two people get out at the third floor and the remaining person rides to the fourth floor. Find the work required to lift the passengers if each passenger weighs 150 lb and the floors are 12 ft apart.

2. Eight boxes 1 ft high and weighing 40 lb each are at floor level. Find the work required to stack them one on top of the other.

The containers pictured in Exercises 3–4 are full of water. Find the work required to pump all the water to (a) the level of the top of the containers and (b) a level 2 ft above the top. (Weight = 62.4 lb/ft³.)

3.

4.

The containers pictured in Exercises 5–8 contain water. (Weight = 62.4 lb/ft³.) Find the work required to pump the water to the level of the top of the containers if (a) they are full, (b) they contain water to a depth of 1 ft.

5.

6.

7.

8.

9. A 10-ft chain that weighs 0.8 lb/ft is hanging vertically. **(a)** Find the work required to raise the entire chain to the level of the top of the chain. **(b)** Find the work required to raise the chain as in (a) if there is a 20-lb weight attached to the bottom of the chain.

10. A 30-ft chain that weighs 2 lb/ft is hanging vertically from a crane. **(a)** Find the work required to roll up 20 ft of the chain, so the bottom is raised 20 ft. **(b)** Find the work required to raise the chain as in (a) if there is a 200-lb weight attached to the bottom of the chain.

11. A uniform chain 12 ft long and weighing 4 lb is lying at ground level. Find the work required to raise the chain so it hangs vertically with the bottom of the chain **(a)** at ground level and **(b)** 8 ft above ground level.

12. A uniform chain 10 ft long and weighing 32 lb is lying at ground level. Find the work required to raise the entire chain to a platform that is 4 ft above the ground.

13. A 2-lb weight stretches a certain spring 1.5 ft. Find the work required to stretch the spring from its natural length to **(a)** 1.5 ft and **(b)** 3 ft beyond its natural length.

14. A spring has natural length of 4 meters. A force of 2 newtons compresses the spring to a length of 3 meters. How much work is done in compressing the spring from a length of 4 meters to a length of 2 meters? (**Calculus Explorer & Tutor II, Applications of Integration, 9.**)

15. Two foot-pounds of work are required to stretch a certain spring 1.2 ft beyond its natural length. How far beyond its natural length will a 4-lb weight stretch the spring?

16. Three foot-pounds of work are required to stretch a certain spring from 2 ft to 4 ft beyond its natural length. How far beyond its natural length will a 5-lb weight stretch the spring?

17. For large distances above the surface of the earth, the force due to gravitational attraction of the earth on an object changes according to the law

$$F = 4 \times 10^8 \frac{m}{r^2}.$$

F is the force in newtons, m is the mass in kilograms, and r is the distance from the center of the earth in kilometers. (The radius of the earth is about 6400 kilometers.) How much work does it take to move a 200-kilogram satellite from the surface of the earth to an orbit 1000 kilometers above the earth's surface?

18. The wind exerts a constant force of 6 lb on a canoe. Determine the work required to paddle the canoe **(a)** 1000 ft against the wind and **(b)** 1000 ft with the wind in a lake of still water if a force of 10 lb is required to move the canoe through still water with no wind.

19. A sandbag is being lifted 10 ft at the rate of 2 ft/s. Find the work done if the sandbag initially weighs 80 lb and loses sand at a uniform rate of 3 lb/s as it is being lifted.

20. A leaky bucket that weighs 3 lb is to be lifted 24 ft at a rate of 2 ft/s by a rope that weighs 0.2 lb/ft. The bucket initially contains 20 lb of water and loses 0.1 lb/s as it is being lifted. What is the total work required to lift the bucket, water, and rope?

21. The magnitude of the force due to friction on an object that is moving along a level surface is $F = \mu w$, where μ is the **coefficient of kinetic friction** and w is the weight of the object. The direction of the force is opposite to the direction of movement. Find the work required to move a 40-lb object 30 ft along a level surface and then return it to its original position if the coefficient of friction is $\mu = 0.5$.

22. The coefficient of kinetic friction of rubber on concrete is $\mu = 0.xxx$; the coefficient of rolling friction is $\mu = 0.yyy$. Find the work required to overcome friction if a 30-pound rubber wheel is **(a)** slid without rolling and **(b)** rolled for 100 ft along a level concrete surface.

23. Air in the cylinder of a diesel engine expands from 2 in.³ to 16 in³. The initial pressure is 500 lb/in.². Find the work done by the gas if the air behaves as an ideal gas and the expansion is adiabatic with $PV^{1.4} = k$.

24. An ideal gas expands from volume 20 in.³ at pressure 15 lb/in.² to volume 30 in.³ at 30 lb/in.². Sketch the PV diagram and find the work done by the gas **(a)** if the gas expands with pressure fixed and then pressure increases with fixed volume and **(b)** if pressure increases with fixed volume and the gas then expands with pressure fixed.

Connections

Concept of definite integrals as limits of Riemann sums, 5.2–3.

Midpoint of variable line segment, 6.1.

6.7 MASS, MOMENTS, AND CENTERS OF MASS

The idea of a center of mass, or balance point, of an object or system of masses involves the mathematical concept of moments.

Systems of Point Masses

Moments of a system of masses are defined in terms of mass. Let us distinguish between the mass and the weight of an object. The weight of an object is the force due to the gravitational attraction of the earth. **Newton's Second Law of Motion** tells us that

$$w = ma,$$

where w is the weight, m is the mass, and a is the acceleration due to gravity. The weight and acceleration due to gravity depend on the distance of the object from the center of the earth. The mass depends on the molecular structure and not on the position of the object. For problems that involve motion near the surface of the earth, we assume that acceleration due to gravity has a constant value g. We then have

$$w = mg,$$

so the weight of an object is proportional to its mass. We will use either $g = 32 \, \text{ft/s}^2$ or $g = 9.8 \, \text{m/s}^2$. Units of force, mass, and acceleration in English and metric systems are given in Table 6.7.1.

TABLE 6.7.1

System	Mass	Acceleration	Force
English	slug	ft/s^2	slug $\cdot$ ft/s^2 = lb
mks (SI)	kilogram (kg)	m/s^2	kg $\cdot$ m/s^2 = newton (N)
cgs	gram (g)	cm/s^2	g $\cdot$ cm/s^2 = dyne

English and metric units are related by the following equations.

$$1 \, \text{slug} = 14.57 \, \text{kg} \quad \text{and} \quad 1 \, \text{N} = 0.225 \, \text{lb}.$$

Thus, the *weight* in pounds of a 1-kg *mass* near the surface of the earth is

$$w = mg = (1 \, \text{kg})(9.8 \, \text{m/s}^2) = 9.8 \, \text{N} = (9.8 \, \text{N}) \left(\frac{0.225 \, \text{lb}}{1 \, \text{N}} \right) = 2.205 \, \text{lb}.$$

DEFINITION

The **moment about** $x = a$ of a system of masses $m_1, m_2, \ldots, m_n$ located at points $x_1, x_2, \ldots, x_n$, respectively, on a number line is

$$M_{x=a} = \sum_{j=1}^{n} (x_j - a)m_j.$$

We may think of $M_{x=a}$ as a measure of the tendency of the system to rotate about the point $x = a$ due to the force of gravity acting downward on each mass of the system.

FIGURE 6.7.1a Equal masses at equal distances on opposite sides of the fulcrum balance.

FIGURE 6.7.1b If unequal masses are equal distances from the fulcrum, the lever rotates toward the larger mass.

FIGURE 6.7.1c If equal masses are unequal distances from the fulcrum, the lever will rotate toward the farther mass.

The idea of moments and balance points shoul
experience. For example, we know that two equal ma
are placed equal distances on opposite sides of the
ure 6.7.1a). If one mass is either increased (Figure 6
from the fulcrum (Figure 6.7.1c), the lever will rotate
mass and up on the opposite side.

It is a rule of physics that a system of masses wi
$x = a$ if $M_{x=a} = 0$. We can use the definition of $M_{x=a}$ to solve the equation for the balance point. Thus, if $x = a$ is the balance point,

$$M_{x=a} = 0,$$

$$\sum_{j=1}^{n}(x_j - a)m_j = 0,$$

$$\sum_{j=1}^{n} x_j m_j - a \sum_{j=1}^{n} m_j = 0,$$

$$a = \frac{\displaystyle\sum_{j=1}^{n} x_j m_j}{\displaystyle\sum_{j=1}^{n} m_j}.$$

We have $\displaystyle\sum_{j=1}^{n} x_j m_j = M_{x=0}$; $M = \displaystyle\sum_{j=1}^{n} m_j$ is the total mass of the system. It is customary to denote the balance point by $\overline{x}$. Then

$$\overline{x} = \frac{\displaystyle\sum_{j=1}^{n} x_j m_j}{\displaystyle\sum_{j=1}^{n} m_j} = \frac{M_{x=0}}{M}.$$

The balance point $\overline{x}$ is called the **center of mass** of the system.

EXAMPLE 1

Find the center of mass of the system of masses $m_1 = 5$, $m_2 = 3$, $m_3 = 2$, located at points $x_1 = -4$, $x_2 = 2$, $x_3 = 8$, respectively.

SOLUTION The total mass is

$$M = m_1 + m_2 + m_3 = 5 + 3 + 2 = 10.$$

The moment about $x = 0$ of the system is

$$M_{x=0} = x_1 m_1 + x_2 m_2 + x_3 m_3 = (-4)(5) + (2)(3) + (8)(2) = 2.$$

The center of mass is

$$\overline{x} = \frac{M_{x=0}}{M} = \frac{2}{10} = 0.2.$$

FIGURE 6.7.2 The center of mass of the system of three masses is $\overline{x} = 0.2$.

See Figure 6.7.2. ■

The same ideas of moments and centers of mass apply to a system of masses $m_1, m_2, \ldots, m_n$ located at points $(x_1, y_1), (x_2, y_2), \ldots, (x_n, y_n)$ in the plane.

$$M_{x=0} = \sum_{j=1}^{n} x_j m_j \text{ and } M_{y=0} = \sum_{j=1}^{n} y_j m_j$$

represent the moments about the *lines* $x = 0$ and $y = 0$, respectively. The center of mass of the system is $(\bar{x}, \bar{y})$, where

$$\bar{x} = \frac{\sum_{j=1}^{n} x_j m_j}{\sum_{j=1}^{n} m_j} = \frac{M_{x=0}}{M} \text{ and } \bar{y} = \frac{\sum_{j=1}^{n} y_j m_j}{\sum_{j=1}^{n} m_j} = \frac{M_{y=0}}{M}.$$

It is common to designate $M_{x=0}$ by M_y and $M_{y=0}$ by M_x. That is,

$$M_{x=0} = M_{y-\text{axis}} = M_y \text{ and } M_{y=0} = M_{x-\text{axis}} = M_x.$$

EXAMPLE 2

Find the center of mass of the system of masses $m_1 = 6$, $m_2 = 3$, $m_3 = 2$ located at points $(x_1, y_1) = (-4, 2)$, $(x_2, y_2) = (2, 1)$, $(x_3, y_3) = (7, -5)$, respectively.

SOLUTION We have

$$M = m_1 + m_2 + m_3 = 6 + 3 + 2 = 11,$$
$$M_{x=0} = x_1 m_1 + x_2 m_2 + x_3 m_3 = (-4)(6) + (2)(3) + (7)(2) = -4,$$

and

$$M_{y=0} = y_1 m_1 + y_2 m_2 + y_3 m_3 = (2)(6) + (1)(3) + (-5)(2) = 5.$$

Hence,

$$\bar{x} = \frac{M_{x=0}}{M} = \frac{-4}{11} \text{ and } \bar{y} = \frac{M_{y=0}}{M} = \frac{5}{11}.$$

FIGURE 6.7.3 The center of mass of the system of three masses is $(\bar{x}, \bar{y}) = (-4/11, 5/11)$.

The center of mass is located at $(\bar{x}, \bar{y}) = (-4/11, 5/11)$. See Figure 6.7.3. ■

It is not difficult to see that a system will balance about any line through its center of mass. For example, suppose that $(\bar{x}, \bar{y}) = (0, 0)$, so

$$\sum_{j=1}^{n} x_j m_j = \sum_{j=1}^{n} y_j m_j = 0.$$

From Figure 6.7.4 we see that the directed distance of a point (x_j, y_j) from the line ℓ through the origin at an angle θ with the x-axis is given by

$$y_j' = (y_j - x_j \tan \theta)(\cos \theta) = y_j \cos \theta - x_j \sin \theta.$$

Hence, the moment about ℓ of the system is

$$M_\ell = \sum_{j=1}^{n} y_j' m_j = \cos \theta \sum_{j=1}^{n} y_j m_j - \sin \theta \sum_{j=1}^{n} x_j m_j = 0.$$

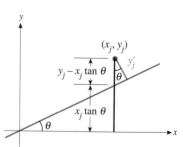

FIGURE 6.7.4 The directed distance of the point (x_j, y_j) from the line ℓ is y_j'.

Actually, the system will balance if supported only at the point that is the center of mass.

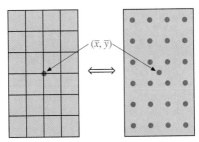

FIGURE 6.7.5 We can consider a homogeneous rectangular lamina to be a system of point masses that are evenly distributed over the lamina.

FIGURE 6.7.6 A homogeneous circular lamina balances at its center.

Homogeneous Laminas

Let us now extend the ideas of mass, moment, and center of mass from systems of point masses to solid objects. In this section we will consider thin, flat objects whose shapes are described by regions in the plane. Such an object is called a **lamina**. The **density** of a lamina is its mass per unit area. **Homogeneous** laminas have uniform density, given by a constant ρ. The mass is then the product of the density and the area of the region that describes the object:

$$M = \rho A.$$

We will use the following fact.

The center of mass of a homogeneous rectangular lamina is located at the geometric center of the rectangle. (1)

Statement 1 is intuitively clear from the interpretation of center of mass as the balance point. It is worthwhile to use the ideas of calculus to relate the problem to that of a system of point masses. For example, we could divide the rectangle into small pieces of equal size, as illustrated in Figure 6.7.5. Each small piece could then be considered as a point mass and we could determine the center of mass of the system of point masses. Such a procedure would lead to (1).

Similar reasoning gives the following:

The center of mass of a homogeneous circular lamina is located at the center of the circle.

See Figure 6.7.6. We can also determine the center of mass of a homogeneous triangular lamina.

The center of mass of a homogeneous triangular lamina is located at the intersection of the medians of the triangle.

Statement 1 implies that each thin rectangular area strip in Figure 6.7.7 balances about a point very near its center, so the balance point of the triangle must be very near the line formed by the centers of the rectangular strips. The line formed by the centers of the strips is a median of the triangle. Since the strips can be drawn parallel to any side of the triangle, we conclude that the intersection of the medians of the triangular lamina is the balance point. It is useful to recall that the intersection of the medians of a triangle is located at one-third the height from each side. See Figure 6.7.8.

FIGURE 6.7.7 Each thin strip balances at a point near its center, so a homogeneous triangular lamina balances along each median.

FIGURE 6.7.8 The intersection of the medians of a triangle is located one-third the height from each side.

We can rewrite the definition of coordinates of the center of mass of a system in the form

$$M_{x=0} = \bar{x}M \text{ and } M_{y=0} = \bar{y}M. \tag{2}$$

We then observe that $\bar{x}M$ and $\bar{y}M$ can be interpreted as the moments about $x = 0$ and $y = 0$, respectively, of a system that consists of a single point mass of M located at the point $(\bar{x}, \bar{y})$. Statement 2 then gives the following fact.

The moments about $x = 0$ and $y = 0$ of a system of point masses with total mass M are equivalent to the corresponding moments of the system that consists of a single point mass of M located at the center of mass of the original system.

When calculating moments we can replace a system or an object by a point mass located at the center of mass. We can also replace any part of a system by a corresponding point mass at the center of mass of the part.

EXAMPLE 3

Find the center of mass of the homogeneous lamina pictured in Figure 6.7.9.

SOLUTION   Let us obtain an equivalent system of three point masses by replacing the circle, the rectangle, and the triangle by point masses located at their respective centers of mass. The mass of each part is determined by the formula

$$\text{Mass} = (\text{Density})(\text{Area})$$

and each center of mass is determined geometrically.

The mass of the circle is $m_1 = \rho\pi(1)^2$, where ρ is the constant density. The center of mass is $(x_1, y_1) = (1, 4)$.

The rectangle has mass $m_2 = \rho(2)(3)$ and center of mass $(x_2, y_2) = (1, 3/2)$.

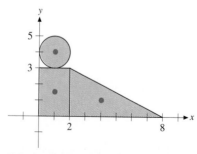

FIGURE 6.7.9   We replace the circle, rectangle, and triangle by point masses located at their respective centers of mass.

The triangle has mass $m_3 = \rho(6)(3)/2$ and center of mass $(x_3, y_3) = (4, 1)$.

For the system of three masses, we have

$$M = \pi\rho + 6\rho + 9\rho = (\pi + 15)\rho,$$
$$M_{x=0} = (1)(\pi\rho) + (1)(6\rho) + (4)(9\rho) = (\pi + 42)\rho,$$

and

$$M_{y=0} = (4)(\pi\rho) + \left(\frac{3}{2}\right)(6\rho) + (1)(9\rho) = (4\pi + 18)\rho.$$

Hence,

$$\overline{x} = \frac{M_{x=0}}{M} = \frac{(\pi + 42)\rho}{(\pi + 15)\rho} = \frac{\pi + 42}{\pi + 15} \approx 2.488,$$
$$\overline{y} = \frac{M_{y=0}}{M} = \frac{(4\pi + 18)\rho}{(\pi + 15)\rho} = \frac{4\pi + 18}{\pi + 15} \approx 1.685.$$

From Figure 6.7.9 we see that $(\overline{x}, \overline{y}) \approx (2.488, 1.685)$ seems to be a reasonable position for the balance point. This serves as a rough check on our calculations. ∎

Using Definite Integrals

We can use calculus to determine the center of mass of a homogeneous lamina whose shape is given by equations in x and y. See Figure 6.7.10.

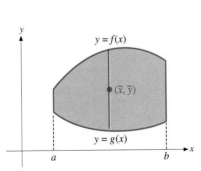

FIGURE 6.7.10a Homogeneous laminas may be divided into thin area strips.

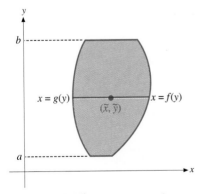

FIGURE 6.7.10b The center of mass of a homogeneous area strip is located at the midpoint of the strip.

To find the center of mass of a homogeneous lamina

- Sketch the region and draw a variable rectangular area strip.
- Label the center of the mass of the variable area strip in terms of the appropriate variable. Denote the center of mass of the rectangle by $(\tilde{x}, \tilde{y})$. (The formula for the midpoint of a line segment can be used to determine $(\tilde{x}, \tilde{y})$.)
- Use the integral as a sum to calculate the total mass of the lamina. The mass of a variable rectangle is

$$dM = \rho\, dA,$$

where ρ is the density and dA represents the area of the rectangle. The total mass is the limit of sums,

$$M = \int dM.$$

- Use the integral as a sum to calculate the moments about $x = 0$ and $y = 0$:

$$M_{x=0} = \int \tilde{x}\, dM \text{ and } M_{y=0} = \int \tilde{y}\, dM.$$

(Note that $\tilde{x}\, dM$ and $\tilde{y}\, dM$ represent the moments about $x = 0$ and $y = 0$, respectively, of the variable rectangle. The integrals give the limits of the sums.)
- The center of mass is given by

$$\overline{x} = \frac{\int \tilde{x}\, dM}{\int dM} = \frac{M_{x=0}}{M} \text{ and } \overline{y} = \frac{\int \tilde{y}\, dM}{\int dM} = \frac{M_{y=0}}{M}.$$

Let us use the integral formulas to find the center of mass of a homogeneous triangular lamina.

EXAMPLE 4

Find the center of mass of the homogeneous triangular lamina bounded by $y = 3x$, $y = 0$, and $x = 2$.

SOLUTION   The lamina is sketched in Figure 6.7.11. We have drawn a vertical area strip, so we will use x as our variable. The strips range between $x = 0$ and $x = 2$, so the limits of integration are 0 and 2. The endpoints and midpoint of the strip are labeled in terms of the variable x. We have used the formula for the midpoint of the line segment between $(x, 3x)$ and $(x, 0)$ to determine that

$$(\tilde{x}, \tilde{y}) = \left(\frac{x + x}{2}, \frac{3x + 0}{2}\right) = \left(x, \frac{3x}{2}\right).$$

The variable area strip has

$$\text{Length} = 3x - 0 = 3x,$$
$$\text{Width} = dx,$$
$$\text{Area} = dA = 3x\, dx,$$
$$\text{Mass} = dM = \rho\, dA = \rho 3x\, dx.$$

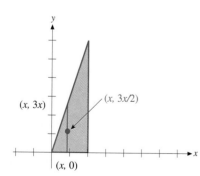

FIGURE 6.7.11 The homogeneous vertical area strip between $y = 3x$ and the x-axis has center of mass $(x, 3x/2)$.

It follows that

$$M = \int dM = \int_0^2 \rho 3x \, dx = 6\rho,$$

$$M_{x=0} = \int \tilde{x} \, dM = \int_0^2 x\rho 3x \, dx = \int_0^2 \rho 3x^2 \, dx = 8\rho, \quad \text{and}$$

$$M_{y=0} = \int \tilde{y} \, dM = \int_0^2 \left(\frac{3x}{2}\right)\rho 3x \, dx = 12\rho, \quad \text{so}$$

$$\bar{x} = \frac{M_{x=0}}{M} = \frac{8\rho}{6\rho} = \frac{4}{3} \text{ and } \bar{y} = \frac{M_{y=0}}{M} = \frac{12\rho}{6\rho} = 2.$$

We see that the point

$$(\bar{x}, \bar{y}) = \left(\frac{4}{3}, 2\right)$$

is the intersection of the medians of the triangle as we knew it should be. ■

EXAMPLE 5

Find the center of mass of the homogeneous lamina bounded by $y = x^2$ and $y = x$.

SOLUTION The region is sketched in Figure 6.7.12. We choose a vertical area strip, so the variable is x. Note that x ranges between $x = 0$ and $x = 1$, the x-coordinates of the points of intersection of $y = x^2$ and $y = x$.

Using the formula for the midpoint of the line segment between (x, x^2) and (x, x), we have

$$(\tilde{x}, \tilde{y}) = \left(\frac{x + x}{2}, \frac{x + x^2}{2}\right) = \left(x, \frac{x + x^2}{2}\right).$$

A variable rectangle has

$$\begin{aligned}
\text{Length} &= x - x^2, \\
\text{Width} &= dx, \\
\text{Area} = dA &= (x - x^2)dx, \\
\text{Mass} = dM = \rho \, dA &= \rho(x - x^2)dx.
\end{aligned}$$

It follows that

$$M = \int dM = \int_0^1 \rho(x - x^2)dx = \frac{\rho}{6},$$

$$M_{x=0} = \int \tilde{x} \, dM = \int_0^1 x\rho(x - x^2)dx = \frac{\rho}{12},$$

and

$$M_{y=0} = \int \tilde{y} \, dM = \int_0^1 \left(\frac{x + x^2}{2}\right)\rho(x - x^2)dx = \frac{\rho}{2}\int_0^1 (x^2 - x^4)dx = \frac{\rho}{15},$$

so

$$\bar{x} = \frac{M_{x=0}}{M} = \frac{\rho/12}{\rho/6} = \frac{1}{2} \text{ and } \bar{y} = \frac{M_{y=0}}{M} = \frac{\rho/15}{\rho/6} = \frac{2}{5}.$$

The center of mass is located at $(\bar{x}, \bar{y}) = (1/2, 2/5)$. ■

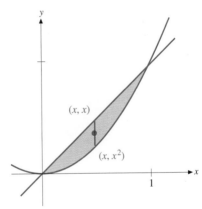

FIGURE 6.7.12 The homogeneous vertical area strip between $y = x^2$ and $y = x$ has center of mass $(x, \frac{x + x^2}{2})$.

You may notice that the density ρ of a homogeneous lamina cancels out of the calculation of the center of mass of the lamina. It is a good idea to use ρ in the calculations for homogeneous laminas, even though it does cancel. It is absolutely necessary that ρ be included in calculations that involve variable density.

Variable Density

If the density of a lamina is given by a continuous function of either x or y, but not both, we can use definite integrals to evaluate the mass and moments of the lamina. If $\rho = \rho(x)$, then thin vertical rectangles are essentially homogeneous and have center of mass very near their geometric center. We can then calculate mass and moments by dividing the lamina into vertical area strips and using x as the variable of integration. See Figure 6.7.13a. If $\rho = \rho(y)$, then thin horizontal rectangles are essentially homogeneous and we can calculate mass and moments by dividing the lamina into horizontal area strips and using y as the variable of integration. See Figure 6.7.13b.

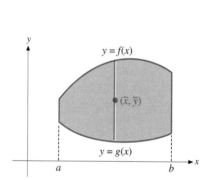

FIGURE 6.7.13a If $\rho = \rho(x)$, then thin vertical strips are essentially homogeneous and have center of mass near their geometric center.

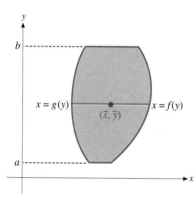

FIGURE 6.7.13b If $\rho = \rho(y)$, then thin horizontal strips are essentially homogeneous and have center of mass near their geometric center.

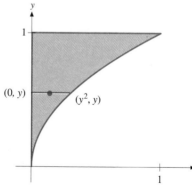

FIGURE 6.7.14 Since the density does not depend on x, thin horizontal area strips are essentially homogeneous.

EXAMPLE 6

Find the center of mass of the lamina that is bounded by $y = \sqrt{x}$, $y = 1$, and the y-axis, if the density is given by $\rho = y$.

SOLUTION The region is sketched in Figure 6.7.14. Since the density is a function of y, we must use y as the variable of integration. Thin horizontal rectangles are essentially homogeneous. We have labeled a typical horizontal area strip in terms of the variable y. Note that $y = \sqrt{x}$ implies $x = y^2$. A variable horizontal area strip has

$$\text{Area} = dA = y^2\,dy,$$
$$\text{Mass} = dM = \rho\,dA = (y)(y^2\,dy) = y^3\,dy,$$

and midpoint

$$(\tilde{x}, \tilde{y}) = \left(\frac{y^2}{2}, y\right).$$

The total mass is

$$M = \int dM = \int_0^1 y^3 \, dy = \frac{1}{4}.$$

The moments are

$$M_{x=0} = \int \tilde{x} \, dM = \int_0^1 \left(\frac{y^2}{2}\right) y^3 \, dy = \int_0^1 \frac{y^5}{2} \, dy = \frac{1}{12}$$

and

$$M_{y=0} = \int \tilde{y} \, dM = \int_0^1 (y) y^3 \, dy = \int_0^1 y^4 \, dy = \frac{1}{5}, \text{ so}$$

$$\overline{x} = \frac{M_{x=0}}{M} = \frac{1/12}{1/4} = \frac{1}{3} \text{ and } \overline{y} = \frac{M_{y=0}}{M} = \frac{1/5}{1/4} = \frac{4}{5}.$$

The center of mass is located at

$$(\overline{x}, \overline{y}) = \left(\frac{1}{3}, \frac{4}{5}\right). \quad \blacksquare$$

In the case of a homogeneous lamina defined by a region in the plane, it is customary to refer to the center of mass of the lamina as the **centroid** of the region.

Solids, Curves, and Surfaces of Revolution

Definite integrals can be used to determine the centers of mass of various types of solids.

• If a cross-section perpendicular to the y-axis of a solid has area $A(y)$, then a slice of the solid at level y has differential mass $dM = \rho \, dV = \rho A(y) \, dy$, where the density ρ may depend on y. See Figure 6.7.15. The y-component of the center of mass of a slice of the solid at level y is $\tilde{y} = y$, so

$$\overline{y} = \frac{\int y \, dM}{\int dM}.$$

• For objects that may be thought of as lying along a curve, we can speak of density per unit length. If the curve is $y = f(x)$, the differential mass is $dM = \rho \, ds = \rho \sqrt{1 + (f'(x))^2} \, dx$, where the density at a point on the curve may depend on both x and $y = f(x)$. The center of mass of a small segment of the curve at a point (x, y) on the curve is $(\tilde{x}, \tilde{y}) = (x, f(x))$. See Figure 6.7.16. We have

$$\overline{x} = \frac{\int x \, dM}{\int dM} \quad \text{and} \quad \overline{y} = \frac{\int f(x) \, dM}{\int dM}.$$

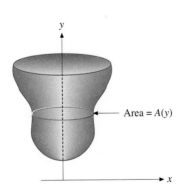

FIGURE 6.7.15 $dM = \rho \, dV = \rho A(y) \, dy$ where the density ρ may depend on y.

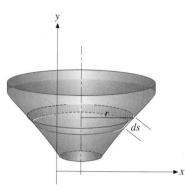

FIGURE 6.7.16 $dM = \rho\, ds =$ $\rho\sqrt{1 + (f'(x))^2}\, dx$, where the density may depend on both x and $y = f(x)$.

FIGURE 6.7.17 $dM = \rho\, dS =$ $\rho 2\pi r\, ds = \rho 2\pi r\sqrt{1 + (f'(x))^2}\, dx$, where the density may depend on both x and $y = f(x)$.

• The differential mass at level y of the surface obtained by revolving a curve $y = f(x)$ about a vertical line is

$$dM = \rho\, dS = \rho 2\pi r\, ds = \rho 2\pi r\sqrt{1 + (f'(x))^2}\, dx,$$

where the density ρ may be variable. See Figure 6.7.17. The y-coordinate of the center of mass of a ribbon of the surface at level y is $\tilde{y} = y$, so

$$\overline{y} = \frac{\int y\, dM}{\int dM}.$$

EXERCISES 6.7

In Exercises 1–4, find $\overline{x}$, the balance point of the given system of masses on a coordinate line.

1. $m_1 = 4, m_2 = 3; x_1 = -2, x_2 = 4$

2. $m_1 = 2, m_2 = 5; x_1 = -4, x_2 = 3$

3. $m_1 = 2, m_2 = 3, m_3 = 5; x_1 = -6, x_2 = -2, x_3 = 4$

4. $m_1 = 6, m_2 = 4, m_3 = 3; x_1 = -4, x_2 = 2, x_3 = 5$

In Exercises 5–8, find $(\overline{x}, \overline{y})$, the center of mass of the given system of masses in the plane.

5. $m_1 = 1, m_2 = 2, m_3 = 3; (x_1, y_1) = (-4, 5),$ $(x_2, y_2) = (-1, 2), (x_3, y_3) = (2, -4)$

6. $m_1 = 2, m_2 = 3, m_3 = 4; (x_1, y_1) = (-6, 5),$ $(x_2, y_2) = (-2, 2), (x_3, y_3) = (6, 4)$

7. $m_1 = 2, m_2 = 3, m_3 = 3, m_4 = 4; (x_1, y_1) = (-4, -2),$ $(x_2, y_2) = (3, 4), (x_3, y_3) = (1, -4), (x_4, y_4) = (2, -5)$

8. $m_1 = 1, m_2 = 2, m_3 = 3, m_4 = 4; (x_1, y_1) = (5, 4),$ $(x_2, y_2) = (2, 4), (x_3, y_3) = (1, -4), (x_4, y_4) = (-2, 1)$

Find the center of mass of each homogeneous lamina pictured in Exercises 9–14.

9.

10.

11.

12.

13.

14.

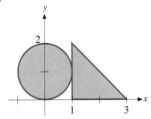

In Exercises 15–24, find the center of mass of the homogeneous lamina bounded by the given curves.

15. $y = x^2$, $y = 4$ (**Calculus Explorer & Tutor II, Applications of Integration, 7.**)

16. $y = x^2$, $y = 4x$, $x = 0$, $x = 1$

17. $y = x^2$, $y = x^2 + 1$, $x = -1$, $x = 1$

18. $y = x^2 + 2$, $y = 1 - x^2$, $x = -1$, $x = 1$

19. $y = x^2$, $y = x + 2$

20. $y = 2 - x^2$, $y = x$

21. $x = 2y^2$, $x = y^2 + 1$

22. $x = y^2 + 1$, $x = 0$, $y = -1$, $y = 1$

23. $x = 2y$, $x = y^2 + 1$, $y = 0$

24. $x = 1 - y^2$, $x = 2 - 2y$, $y = 0$

25. Verify that the integration method gives the center of mass of the homogeneous triangular lamina pictured to be $(b/3, h/3)$.

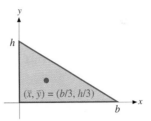

26. Verify that the integration method gives the center of mass of a homogeneous circular lamina to be the center of the circle. ($M = \rho A = \rho \pi r^2$.)

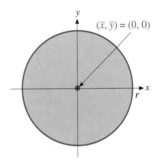

In Exercises 27–34, find the center of mass of the lamina that is bounded by the given curves and has the given density function.

27. $y = 1$, $y = 0$, $x = 0$, $x = 1$; $\rho = x$

28. $y = x$, $y = 0$, $x = 1$; $\rho = x^2$

29. $y = x^3$, $y = 0$, $x = 0$, $x = 1$; $\rho = 2x$

30. $y = x$, $y = x^2$; $\rho = x$

31. $x = \sqrt{y}$, $y = 1$, $x = 0$; $\rho = y$

32. $y = x$, $y = 1$, $x = 0$; $\rho = y$

33. $x + y = 1$, $y = 0$, $x = 0$; $\rho = y^2$

34. $x = 1 - y^2$, $x = 0$; $\rho = y^2$

35. Find the center of mass of a solid right circular cone with radius r and height h.

36. Find the center of mass of a solid hemisphere of radius r.

37. Find the center of mass of a thin homogeneous wire positioned along the curve $y = \sqrt{r^2 - x^2}$, $-\sqrt{3}r/2 \le x \le \sqrt{3}r/2$. ($M = \rho 2\pi r/3$.)

38. Find the center of mass of two thin homogeneous bars of length ℓ that have their ends welded together so the angle between them is $2\pi/3$. (*Hint:* You don't need calculus to find the balance point of this system.)

39. Find the center of mass of the surface obtained by revolving the line $y = hx/r$, $0 \le x \le r$, about the y-axis.

40. Find the center of mass of a thin hemispherical shell of radius r.

41. (a) Find the center of mass of a homogeneous lamina that is between $y = x^n$ and the x-axis, $0 \le x \le 1$. (b) What point do the centers of mass in (a) approach as n approaches infinity?

42. Find the height above its vertex of the center of mass of an inverted right circular cone if the density at each point is proportional to the square root of the height of the point above the vertex. (**Calculus Explorer & Tutor II, Applications of Integration, 8.**)

43. Let R be the region bounded by $y = f(x)$, $y = g(x)$, $x = a$, and $x = b$. Assume f and g are continuous with $g(x) \le f(x)$ for $a \le x \le b$. Let $\bar{x}$ be the x-coordinate of the centroid of R. Let A denote the area of R, and let V denote the volume of the solid obtained by revolving R about a vertical line $x = x_0$, $x_0 \le a$. Show that $V = 2\pi(\bar{x} - x_0)A$, so that the volume of the solid is the distance traveled by the centroid of the region times the area of the region. This illustrates a general result called **Pappus' Theorem**.

In Exercises 44–46, use the result of Exercise 43 to find the volume of the solids obtained by revolving the regions bounded by the given curves about the given lines.

44. $(x - 1)^2 + y^2 = 1$; about the y-axis

45. $y = x$, $y = 0$, $x = 3$; about the y-axis

46. $\dfrac{x}{r} + \dfrac{y}{h} = 1$, $y = 0$, $x = 0$ $(r, h > 0)$; about the y-axis

---

6.8 FORCE DUE TO FLUID PRESSURE

Connections

Concept of center of mass, 6.7.

Fluid exerts force due to fluid pressure on submerged surfaces. We will study forces on flat surfaces. We will see that calculation of force due to fluid pressure on such surfaces involves moment integrals.

Fluid Pressure

Let us begin by considering some basic principles of fluid pressure. Pressure is force per unit area. We will consider problems in which the force is due to the weight of the fluid. We assume the fluid is located near the surface of the earth, so weight is a constant multiple of the mass, $w = mg$. Fluid pressure at a point in a standing body of fluid is then directly proportional to the depth of the point below the level of the fluid surface. **Pascal's Principle** asserts that the pressure is equal in each direction.

If a flat surface is submerged horizontally, then the fluid pressure is equal at each point on the surface. The magnitude of the force due to fluid pressure is then given by

$$F = \rho g h A,$$

where ρ is the density of the fluid, h is the depth of the surface, and A is the area of the surface. (The density ρ is mass per unit volume and ρg is the weight per unit volume of the fluid.) The force acts in a direction perpendicular to the surface. Note that hA is the volume of a generalized right cylinder with height h and area of base A; ρhA is the mass and $\rho g h A$ is the weight of that amount of fluid. See Figure 6.8.1a.

It is a consequence of Pascal's Principle that the force due to fluid pressure depends on the depth of the surface below the uppermost fluid level. It does not matter how much fluid is directly above the surface. Thus, if the containers pictured in Figure 6.8.1a and 6.8.1b contain the same type fluid at the same uppermost level and have bases of equal area, then the forces on their bottom surfaces will be equal. This is because the fluid exerts an upward force on the horizontal surface below the uppermost fluid level in Figure 6.8.1b, and

FIGURE 6.8.1a The force on the bottom surface is equal to the weight of the fluid in the container.

FIGURE 6.8.1b If the uppermost fluid levels are the same, the force on the bottom surface is the same as that in Figure 6.8.1a, even though this container has less fluid.

FIGURE 6.8.1c The horizontal surface must exert a force to keep the fluid in the container.

FIGURE 6.8.1d If a thin straw were inserted, the fluid in the straw would rise to the uppermost fluid level, so the force exerted by the horizontal surface compensates exactly for the missing fluid.

that surface must exert an equal downward force to contain the fluid. See Figure 6.8.1c. The downward force exerted by the container is transmitted to the total force on the bottom of the container. The downward force exerted by the container is equal to the force exerted by a column of fluid with height at the uppermost fluid level. For example, we know that if a thin straw were inserted as in Figure 6.8.1d, the fluid would be forced up the straw until it reached the uppermost level of the surface.

Using Definite Integrals

Let us now consider the force on a submerged flat surface that is not necessarily horizontal. We divide the surface into strips so that points on the same strip are at approximately the same level. The force on a strip of area dA, located at a depth of approximately h, is approximately

$$dF = \rho g h \, dA.$$

See Figure 6.8.2. It is important that each point on the strip be at approximately the same level and

Depth of strip = (Uppermost fluid level) − (Level of strip).

The force acts in the direction perpendicular to the surface. (Recall that Pascal's Principle tells us the pressure at a point is equal in every direction.) Since we have assumed that the surface is flat, the forces on different strips act in the same direction. The total force on the surface is the sum of the forces on its parts, which we write as an integral:

$$F = \int \rho g h \, dA.$$

Uppermost fluid level

h

FIGURE 6.8.2 A submerged flat surface can be divided into strips so that points on the same strip are at approximately the same level.

EXAMPLE 1

A cylindrical tank is lying on its side. The radius of each circular end is 3 ft. Set up an integral that gives the force on one circular end if the tank is **(a)** completely full of a fluid with weight $\rho g = 62.4$ lb/ft^3 and **(b)** half full.

SOLUTION   Since the tank is lying on its side, the circular ends are vertical. We choose a coordinate system with origin at the center of one end, so that end is described by the equation $x^2 + y^2 = 3^2$. See Figure 6.8.3. The variable y represents the level of a variable area strip. We need to calculate the force on each strip as y varies between the smallest and largest values of the y-coordinates of the submerged surface.

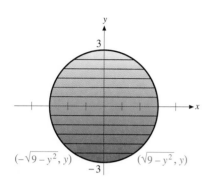

FIGURE 6.8.3a If the tank is completely full, the uppermost fluid level is $y = 3$, so the depth of a variable strip at level y is $3 - y$.

(a) If the tank is full, the uppermost fluid level is $y = 3$, so the depth of a variable strip is

$$h = \text{(Uppermost fluid level)} - \text{(Level of strip)} = 3 - y.$$

The variable y ranges from -3 to 3. See Figure 6.8.3a. A variable horizontal strip runs from $(-\sqrt{9 - y^2}, y)$ to $(\sqrt{9 - y^2}, y)$. The area of a variable strip is $dA = 2\sqrt{9 - y^2}\, dy$. The total force is

$$F = \int_{-3}^{3} (62.4)(3 - y)2\sqrt{9 - y^2}\, dy.$$

(The above integral has value $(62.4)(27\pi) \approx 5293$.)

(b) If the tank is only half full, the uppermost fluid level is $y = 0$, so $h = 0 - y = -y$. The submerged surface is between levels $y = -3$ and $y = 0$. These values give the limits of integration. See Figure 6.8.3b. The total force is

$$F = \int_{-3}^{0} (62.4)(0 - y)2\sqrt{9 - y^2}\, dy.$$

(This integral has value $(62.4)(18) = 1123.3$.) ∎

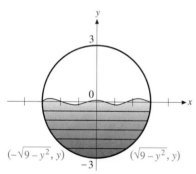

FIGURE 6.8.3b If the tank is half full, the uppermost fluid level is $y = 0$, so the depth of a variable strip at level y is $0 - y$.

Using Centroids

Let us now see how *force due to fluid pressure is related to a moment integral.* We merely note that

$$F = \int \rho g h\, dA = g\left(\int h\rho\, dA\right).$$

The last integral may be interpreted as the moment integral about the uppermost level of the fluid of the system of masses $\rho\, dA$ at level h below the uppermost fluid level. This observation allows us to use what we know about moment integrals to evaluate forces on submerged flat surfaces. In particular, since the moment of a system of masses is the same as the moment of the system with total mass located at the center of mass, we have

> *Force on submerged flat surface*
> $= (Depth\ of\ centroid\ of\ surface)\ (\rho g)(Area\ of\ surface).$

We can use the above formula to evaluate the integral in Example 1(a). The centroid of the circular surface is located at its center, 3 ft below the uppermost fluid level. Thus,

$$F = \bar{y}\rho g A = (3)(62.4)(\pi 3^2) \approx 5293 \text{ lb.}$$

It is not convenient to use this method to evaluate the integral in Example 1(b), because the location of the centroid of a semicircular flat surface is not geometrically obvious.

EXAMPLE 2

The swimming pool pictured in Figure 6.8.4 is full of water. Find the force due to fluid pressure on **(a)** the vertical rectangular surface at the deep end, **(b)** one triangular side, and **(c)** the slanted bottom of the pool. ($\rho g = 62.4\,\text{lb/ft}^3$.)

SOLUTION

(a) The centroid of the rectangle is located 6 ft below the fluid level. We have

$$F = (\text{Depth of centroid})(\rho g)(\text{Area})$$
$$= (6)(62.4)(20)(12) = 89,856\,\text{lb}.$$

(b) The centroid of the triangle is located one-third of the distance down from the horizontal side at the top of the triangle, 4 ft below the fluid level. The force is

$$F = (4)(62.4)\left(\frac{1}{2}\right)(30)(12) = 44,928\,\text{lb}.$$

(c) The center of the slanted, rectangular bottom of the pool is 6 ft deep. The length of the slanted side is $\sqrt{30^2 + 12^2} = \sqrt{1044}$, so

$$F = (6)(62.4)(20)(\sqrt{1044}) \approx 241,945\,\text{lb}. \quad \blacksquare$$

FIGURE 6.8.4 We can use the area and depth of the centroid of each surface to calculate the force on that surface.

EXERCISES 6.8

1. The containers pictured are filled with a fluid of weight density $\rho g = 60\,\text{lb/ft}^3$. Find the force due to fluid pressure on the circular bottom surface of each container.

2. The rectangular tanks pictured are filled with a fluid of weight density $\rho g = 60\,\text{lb/ft}^3$. Find the force due to fluid pressure on the indicated surface of each tank.

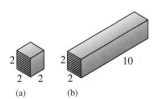

In Exercises 3–6, use the given coordinate system to set up an integral that gives the force due to fluid pressure on the vertically submerged flat surfaces pictured.

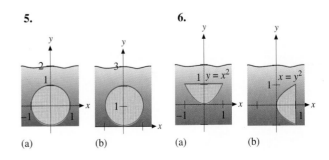

In Exercises 7–10, use the given coordinate system to set up an integral that gives the force due to fluid pressure on the indicated vertical surfaces if the containers have fluid to the level indicated.

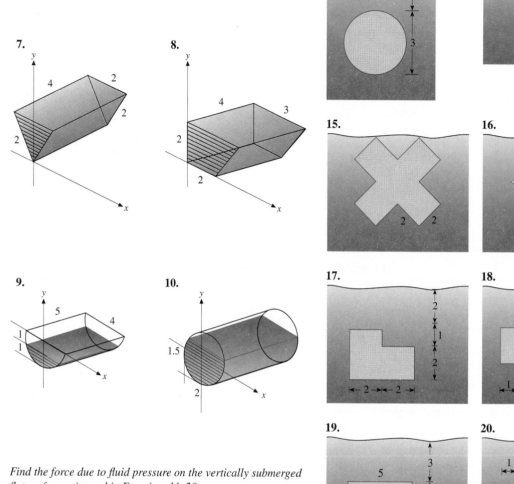

7.

8.

9.

10.

Find the force due to fluid pressure on the vertically submerged flat surfaces pictured in Exercises 11–20.

11.

12.

13.

14.

15.

16.

17.

18.

19.

20.

In Exercises 21–24, the containers are filled with fluid. Find the force due to fluid pressure on the indicated flat surfaces.

21.

22.

23.

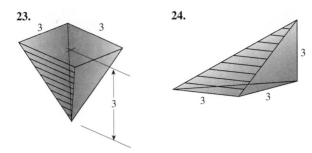

24.

25. A rectangular block of density ρ_b is submerged in a fluid of density ρ_f as pictured. **(a)** Find the force due to fluid pressure on the top surface of the block. **(b)** Find the force due to fluid pressure on the bottom surface of the block. **(c)** Find the force on the block due to gravity. **(d)** Find the net vertical force on the block due to the forces in (a), (b), and (c). What is the direction of the force if $\rho_b > \rho_f$?

26. A rectangular block of density ρ_b is submerged in a fluid of density ρ_f as pictured. Assume $\rho_b < \rho_f$. **(a)** Find the force due to fluid pressure on the bottom surface of the block. **(b)** Find the force on the block due to gravity. **(c)** Find d so that the net vertical force on the block due to the forces in (a) and (b) is zero. **(d)** For the value of d found in (c), show that the weight of fluid displaced by the submerged part of the block equals the total weight of the block.

27. (a) Disregarding other forces, show that the vertical force due to fluid pressure on the completely submerged rectangular block in Exercise 25 is equal to the weight of the fluid displaced by the block. **(b)** Show that the vertical force due to fluid pressure on the partially submerged rectangular block in Exercise 26 is equal to the weight of the fluid displaced by the block.

28. The vertical force due to fluid pressure on the submerged flat surface indicated below is $F\cos\theta$, where F is the force due to fluid pressure on the surface. Show that the vertical force on the indicated surface is independent of θ, as long as the center of the surface remains at depth h. It then follows from Exercise 27 that the vertical force due to fluid pressure on a submerged block is equal to the weight of the fluid displaced, even if the top and bottom of the block are slanted. Since any shaped object could be considered to be composed of thin vertical blocks with tops and bottoms that may be slanted, and the total vertical force on the object is the sum of the vertical forces on the individual parts, we can conclude that **the vertical force due to fluid pressure on any shaped submerged object is equal to the weight of the fluid displaced**.

29. What is the approximate force due to fluid pressure on a door of a car that is submerged upright in water to the top of the door if the door is approximately a rectangle that is 36 in. wide and 42 in. high? This force can be considered to act at the center of the door, so the door could be opened with half that force applied at the edge opposite the hinges. See figure.

30. An aquarium is 16 inches deep and has trapezoidal ends, each with upper edge 10 inches long and lower edge 12 inches long. The aquarium is completely filled with water. Find the total force on one end of the aquarium due to the pressure of the water. (**Calculus Explorer & Tutor II, Applications of Integration, 10.**)

REVIEW OF CHAPTER 6 CONCEPTS

- Applications of definite integrals depend upon the interpretation of the definite integral as a sum of products, with each component of the integral representing a physical quantity.
- Geometric applications require that analytic geometry be used to describe quantities such as length, area, and volume in terms of a variable.
- If a region is divided into thin strips, the area of each strip is approximately the area of a thin rectangle. A definite integral gives the area of the region as a limit of the sums of the areas of the approximating rectangles.
- If a solid is cut into thin slices, the volume of each slice is approximately the area of its cross-section times its thickness. A definite integral gives the volume of the solid as a limit of the sums of the volumes of the slices.
- If a thin rectangle is revolved about a perpendicular line, the result is either a disk or washer. If a thin rectangle is revolved about a parallel line, the result is a thin cylindrical shell.
- The volume of a solid of revolution obtained by revolving a region about a line is approximately the sum of the volumes of the solids obtained by revolving thin area strips of the region about the line. Use area strips that are convenient for the region and then determine if the resulting solid is a disk, washer, or thin cylindrical shell and then write the appropriate definite integral that gives the sum of the volumes of the components.
- The arc length of a smooth curve can be thought of as the sum of the lengths of small, nearly straight pieces of the curve with length $ds = \sqrt{(dx)^2 + (dy)^2}$.
- Total work is the sum of the work associated with each component of a process. Either the object or the distance can be divided into small increments. Choose a particular process and then write the corresponding integral.
- Moments of a solid are given by the sum of the moments of each of the parts of the solid. We can use a definite integral to evaluate a moment of a homogeneous lamina by replacing thin strips of the lamina by point masses at the center of mass of each strip.
- Centers of mass give balance points of solids.
- Force due to fluid pressure can be calculated by using either a definite integral or known centroids.

CHAPTER 6 REVIEW EXERCISES

1. Sketch, label, and find the length and midpoint of a variable vertical line segment between $y = 4 - x^2$ and $x + y = 2$, between points of intersection of the parabola and the line.

2. Sketch, label, and find the length and midpoint of a variable horizontal line segment between $x = y^2 - 3$ and $x = 5 - y^2$, between points of intersection of the parabolas.

Find the area of a variable cross-section of the solids pictured in Exercises 3–4.

3. Cross-sections perpendicular to the x-axis are semicircles.

4. Cross-sections perpendicular to the x-axis are circles.

5. A variable point on the graph of $y = 1/x^2$ is revolved about the y-axis. Find the **(a)** area and **(b)** circumference of the resulting circle in terms of x.

6. A variable point on the graph of $y = 1/x^2$ is revolved about the x-axis. Find the **(a)** area and **(b)** circumference of the resulting circle in terms of x.

In Exercises 7–8, sketch and label either the disk, washer, or cylindrical cylinder, whichever one is generated by revolving the given line segment about the indicated line. Find the area of each surface obtained.

7. The variable vertical line segment between $y = \sin x$ and the x-axis, $0 \leq x \leq \pi$; **(a)** about the x-axis, **(b)** about the y-axis, **(c)** about the line $y = -1$

8. The variable horizontal line segment between $x - y = 1$ and the y-axis, $0 \leq y \leq 1$; **(a)** about the x-axis, **(b)** about the y-axis, **(c)** about the line $x = 2$

Set up integrals that give the area of the region bounded by the curves given in Exercises 9–16.

9. $y = x^4$, $y = 8x$

10. $y = x(x - 3)$, $y = x$

11. $y = x$, $x = y(y - 2)$

12. $y = 2x$, $2x + 3y = 6$, x-axis

13. $y = 2x$, $y = x/2$, $x + y = 3$

14. $y = x(x - 1)(x + 2)$, $y = 4x$

15. $y = \cos x$, $y = \sin x$, one loop

16. $y = \sec^2 x$, $y = 2$, one loop

Set up an integral that gives the volume of each solid pictured in Exercises 17–20.

17. Cross-sections perpendicular to the x-axis are isosceles right triangles.

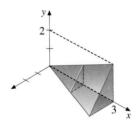

18. Cross-sections perpendicular to the y-axis are isosceles right triangles.

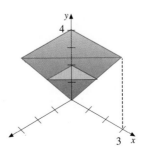

19. Cross-sections perpendicular to the x-axis are squares from which quarter-circles that have centers on the corners and radii half the side of the square have been removed.

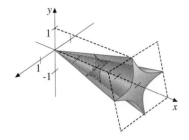

20. Cross-sections perpendicular to the y-axis are circles with endpoints of a diameter on the curves $y = x$ and $y = x^2$, $0 \leq x \leq 1$.

21. Set up an integral that gives the volume of the solid with base the region in the xy-plane bounded by $x^2 + y^2 = 1$, if cross-sections perpendicular to the x-axis are squares.

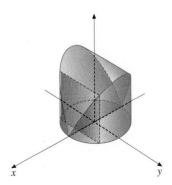

22. Set up an integral that gives the volume of the solid with base the region in the xy-plane bounded by $x = 4 - y^2$ and the y-axis, if cross-sections perpendicular to the x-axis are isosceles right triangles with hypotenuses on the base.

23. Set up an integral that gives the volume of the solid obtained by revolving the region bounded by $y = x^2$ and $y = x$ **(a)** about the x-axis, **(b)** about the y-axis.

24. Set up an integral that gives the volume of the solid obtained by revolving the region bounded by $y = \sin x$, the x-axis, $x = 0$, and $x = \pi$ **(a)** about the x-axis, **(b)** about the y-axis, **(c)** about the line $y = 1$.

25. Set up an integral that gives the volume of the solid obtained by revolving the region bounded by $y = x^3$, $y = 0$, and $x = 1$ **(a)** about the x-axis, **(b)** about the y-axis, and **(c)** about $y = 1$.

26. Set up an integral that gives the volume of the solid obtained by revolving the region bounded by $y = \cos x$, $y = 0$, and $x = 0$ $(x \geq 0)$, **(a)** about the x-axis, **(b)** about the y-axis, and **(c)** about $x = \pi/2$.

27. A solid is obtained by revolving the region in the first quadrant bounded by $y = x^2$, $y = 1$, and the y-axis, about the y-axis. Find c so a slice perpendicular to the y-axis at $y = c$ divides the solid into two pieces with equal volumes.

28. A solid is obtained by revolving the region in the first quadrant bounded by $y = \sqrt{x}$, $y = 1$, and the y-axis, about the y-axis. Find c so a slice perpendicular to the y-axis at $y = c$ divides the solid into two pieces with equal volumes.

29. **(a)** Set up an integral that gives the arc length of the curve $y = \sin x$, $0 \leq x \leq \pi$. Set up integrals that give the area of the surface obtained by revolving the curve about **(b)** the x-axis and **(c)** the y-axis.

30. **(a)** Set up an integral that gives the arc length of the curve $y = x^2$, $0 \leq x \leq 2$. Set up integrals that give the area of the surface obtained by revolving the curve about **(b)** the x-axis and **(c)** the y-axis.

31. **(a)** Set up an integral that gives the arc length of the curve $y = 1/x^2$, $1 \leq x \leq 2$. Set up integrals that give the area of the surface obtained by revolving the curve about **(b)** the x-axis, **(c)** the y-axis, and **(d)** the line $x = 1$.

32. **(a)** Set up an integral that gives the arc length of the curve $y = \dfrac{x}{x+1}$, $0 \leq x \leq 1$. Set up integrals that give the area of the surface obtained by revolving the curve about **(b)** the x-axis, **(c)** the y-axis, and **(d)** the line $y = 1$.

33. **(a)** Set up an integral that gives the arc length of the curve $y = \displaystyle\int_1^x \frac{1}{t}\, dt$, $1 \leq x \leq 4$. Set up integrals that give the area of the surface obtained by revolving the curve about **(b)** the y-axis and **(c)** the line $x = 4$.

34. **(a)** Set up an integral that gives the arc length of the curve $y = \displaystyle\int_0^x \tan t\, dt$, $0 \leq x \leq \pi/4$. Set up integrals that give the area of the surface obtained by revolving the curve about **(b)** the y-axis and **(c)** the line $x = \pi/4$.

35. A 2-lb weight stretches a certain spring 2 ft beyond its natural length. Find the work required to stretch the spring from 1 ft to 4 ft beyond its natural length.

36. A 4-lb weight stretches a certain spring 0.5 ft beyond its natural length. Find the work required to stretch the spring from 1 ft to 1.5 ft beyond its natural length.

37. A conical tank is 6 m high and the radius of the circular top is 2 m. The vertex of the cone is at the bottom and the tank contains water to a depth of 4 m at the deepest part. Find the work required to pump all the water to the level of the top of the tank. (Density $= 1000\,\text{kg/m}^3$, $g = 9.8\,\text{m/s}^2$, $\text{kg}\cdot\text{m/s}^2 = \text{N}$, $\text{N}\cdot\text{m} = \text{J}$.)

38. A spherical tank has radius 8 m. The tank contains water to a depth of 5 m at the deepest part. Find the work required to pump all the water to the level of the top of the tank. (Density $= 1000\,\text{kg/m}^3$, $g = 9.8\,\text{m/s}^2$, $\text{kg}\cdot\text{m/s}^2 = \text{N}$, $\text{N}\cdot\text{m} = \text{J}$.)

39. A 50-ft chain that weighs 4 lb/ft is hanging vertically from a building. Find the work required to raise 30 ft of the chain to the level of the top of the building, so 20 ft of chain remains hanging.

40. A 15-ft chain that weighs 2 lb/ft is lying at ground level. Find the work required to raise the chain so it hangs vertically with the bottom of the chain 5 ft above ground level.

41. A cylindrical barrel with diameter 2.5 feet and height 3.5 feet is floating freely in water with its circular bottom surface 1.5 feet below the surface. Find the work required to push the barrel down vertically from its floating depth until the top is at water level. Water weighs 62.4 lb/ft^3.

42. A beach ball with diameter 16 in. is floating with 1 in. under water. Use the fact that the force due to fluid pressure is equal to the weight of the fluid displaced to determine the work required to push the ball completely under water. Water weighs 62.4 lb/ft^3 and the volume under water of a sphere of radius r submerged to depth h is $\pi h^2 \left(r - \dfrac{h}{3} \right)$.

In Exercises 43–44, find the centers of mass of the homogeneous laminas pictured.

43.   **44.**

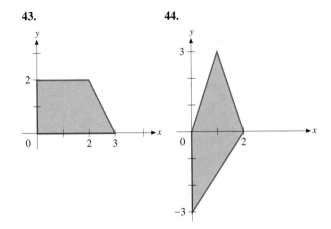

45. Find the center of mass of the homogeneous lamina inside $x^2 + y^2 = 1$ and above the x-axis. (Use a formula for the area of a semicircle.)

46. Find the center of mass of the homogeneous lamina bounded by $y = x(x - 2)$ and $y = x$.

47. A lamina is bounded by $y = x$, $x = 1$ and the x-axis. The density per unit area at a point (x, y) is $\rho = y$. **(a)** Find the mass of the lamina. **(b)** Find the center of mass of the lamina.

48. A lamina is bounded by $y = x$, $x = 1$, and the x-axis. The density per unit area at a point (x, y) is given by $\rho = x$. **(a)** Find the mass of the lamina. **(b)** Find the center of mass of the lamina.

In Exercises 49–50, find the centers of mass of the homogeneous solids pictured.

49.   **50.**

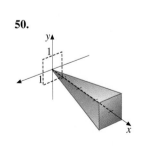

51. Find the mass of a beam of length π if the density per unit length of the beam at a point x units from one end of the beam is given by $\rho(x) = \sin x$.

52. Set up an integral that gives the mass of a rod if the rod is positioned along the curve $y = x^2$, $0 \le x \le 1$, and the density per unit length of the rod at points along the rod is given by $\rho = xy$.

53. Find the center of mass of the homogeneous surface obtained by revolving the line segment $y = 2 - x$, $0 \le x \le 1$, about the x-axis.

54. Find an approximate value of the y–coordinate of the center of mass of the homogeneous surface obtained by revolving the curve $y = x^2/2$, $0 \le x \le 1$, about the x-axis by using Simpson's Rule with $n = 4$ (or a computer program) to approximate the appropriate integrals.

55. Show that moments of a system of masses located at points satisfy $M_{x=a} = M_{x=b} + (b - a)M$.

56. Show that $M_{x=\bar{x}} = 0$.

Find the force due to fluid pressure on the vertically submerged surfaces pictured in Exercises 57–60.

57.   **58.**

59.   **60.**

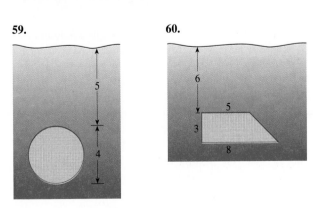

61. The speed of an object traveling along a number line is the absolute value of its velocity. If the speed is constant, the total distance traveled is the product of the speed times the length of time. Use calculus to find the distance traveled if the velocity is $v(t) = 1 - t^2, 0 \le t \le 2$.

62. The velocity of an object t seconds after it has been thrown vertically at 48 ft/s is $v = 48 - 32t$. What is the total distance traveled between $t = 0$ and $t = 2$? (See Exercise 61.)

63. A body of fluid has horizontal cross-section area $A(y)$ for $a \le y \le b$.
  (a) Set up an integral that gives the work required to pump the fluid to level $y = h$.
  (b) Set up a definite integral that gives the mass of the body of fluid.
  (c) Set up a definite integral that gives the moment of the mass of fluid about $y = 0$.
  (d) Show that the work required is the weight of the water times the distance of the center of mass from the level $y = h$.

64. The **moment of inertia** about the x-axis of a system of masses $m_1, \ldots, m_n$ located at points $(x_1, y_1), \ldots, (x_n, y_n)$, respectively, in a coordinate plane, is defined to be
$$I_{y=0} = \sum y_j^2 m_j.$$
Use an integral to find the moment of inertia about the x-axis of the homogeneous lamina bounded by $x = 0$, $x = h$, $y = 0$, $y = k$.

65. Show that the moment of inertia, $I_{y=a} = \sum (y_j - a)^2 m_j$, satisfies $I_{y=a} = I_{y=0} - \bar{y}^2 M + (\bar{y} - a)^2 M$.

66. Show that $I_{y=a} = I_{y=\bar{y}} + (\bar{y} - a)^2 M$.

67. The **period of motion** of a compound pendulum is
$$T = 2\pi \sqrt{\frac{I}{mgd}},$$
where m is the mass of the pendulum, g is the acceleration due to gravity, d is the distance of the center of mass of the pendulum to the pivot, and I is the moment of inertia about an axis through the pivot perpendicular to the plane of oscillation. Find the period of a **simple pendulum** with mass m a distance L from the pivot on a rod of negligible mass.

68. **(a)** Find the moment of inertia of a thin uniform rod of length L and mass m about an axis through one end of the rod, perpendicular to the rod.
  (b) Find the period of motion of a uniform rod of length L about one end of the rod. See Exercise 67.

69. **(a)** Find the moment of inertia of a thin circular ring of radius R and mass m about a pin on which it is hanging. $\left(\text{Hint: } I = \int d^2 \dfrac{m}{2\pi R} ds. \right.$ See figure. The

Law of Cosines can be used to express the distance d in terms of θ and the differential arc length of the ring is $ds = R d\theta$. $\left. \int_0^{2\pi} \cos^2 \theta \, d\theta = \pi. \right)$

  (b) Find the period of motion of the thin circular ring as it pivots about a pin. See Exercise 67.

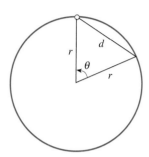

70. A certain random variable X has **density function**
$$f(x) = \begin{cases} 1/(b-a), & a \le x \le b, \\ 0, & \text{otherwise}. \end{cases}$$
The probability density is
$$P[X \le x] = \int_a^x f(t) \, dt,$$
which represents the area under the density curve from negative infinity to x.
  (a) Show that
$$P[X \le x] = \begin{cases} 0, & x < a, \\ \dfrac{x-a}{b-a}, & a \le x \le b, \\ 1, & x > b. \end{cases}$$
  (b) Evaluate the expectation, $E(X) = \displaystyle\int_a^b x f(x) \, dx$.
  (c) Evaluate the variance, $Var(X) = \displaystyle\int_a^b x^2 f(x) \, dx - \left[\displaystyle\int_a^b x f(x) \, dx \right]^2$.

71. Write a description for a noncalculus student of how to approximate the area inside a circle. Use a sketch to illustrate your description.

72. Write a description for a noncalculus student of how to approximate the circumference of a circle. Use a sketch to illustrate your description.

73. Find the length of the n-step curve from $(0, 0)$ to $(1, 1)$ illustrated below. As n approaches infinity, the points on the n-step curve approach points on the line segment between $(0, 0)$ and $(1, 1)$. Do the lengths of the n-step curves approach the length of the line segment? Explain.

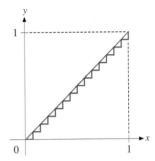

EXTENDED APPLICATION

SPACE STATION DESIGN

Your company is submitting a bid to build a compartment of a space station for NASA. The compartment you design must meet the following specifications:

- It must have a circular base 5 meters in radius.
- Its height at the center must be 10 meters.
- It must hold a payload with a volume of at least 500 cubic meters.
- Because part of the construction will be completed by robots, the compartment must be a solid of revolution, obtained by revolving the graph of an elementary function $y = f(x)$ about the y-axis, as in Figure 1.

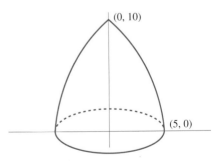

FIGURE 1

Furthermore, because the materials used in constructing the compartment are expensive, the surface area should be small. The president of your company is confident that you can win the contract if you can specify an elementary function $f(x)$ that will produce a surface area smaller than those which your competitors will submit.

There are several techniques for creating elementary functions $y = f(x)$ with $f(0) = 10$ and $f(5) = 0$. Here are two:

- Begin with a function $g(x)$ having $g(0) = p$ and $g(q) = 0$. Let

$$f(x) = \frac{10}{p} g\left(\frac{qx}{5}\right).$$

- Begin with a function $g(x)$ having $g(0) = p$ and $g(5) = r$. Let

$$f(x) = \frac{10[g(x) - r]}{p - r}.$$

Tasks:

1. Choose several elementary functions $y = f(x)$ with $f(0) = 10$ and $f(5) = 0$. Include a variety of types, such as polynomials, square roots of polynomials, rational functions, and trigonometric functions. It may help to experiment with a graphing utility.
2. For each function, calculate the volume and surface area of the solid obtained by revolving the arc $y = f(x)$, $0 \le x \le 5$, about the y-axis. Obtain exact values if you can. Numerical approximations are acceptable if you can prove that they are accurate to at least two decimal places.

Writing Assignment:

Submit a recommended design to the company president. Include your calculations for each elementary function you created, accompanied by a verbal summary making a convincing argument that your design is likely to produce a smaller surface area than any submitted by your competitors.

Source: Computer Modules, Copyright ©1991 by D. Damiano, M. Freije, J. Little III, P. Perkins, and M. Tews, College of the Holy Cross.

7 Selected Transcendental Functions

Functions that can be expressed in terms of sums, products, quotients, powers, and roots of polynomials are called **algebraic** functions; all other functions are called **transcendental** functions. The trigonometric functions are examples of transcendental functions. In this chapter, we will study some of the transcendental functions that occur in physical problems. The development includes differentiation formulas for each new function. The integration formulas that correspond to the new differentiation formulas allow us to use the Fundamental Theorem of Calculus to evaluate a wider variety of integrals.

7.1 INVERSE FUNCTIONS; INVERSE TRIGONOMETRIC FUNCTIONS

Connections

Functions and graphs of functions, 2.1.

Intuitive idea of continuity, 2.3.

The concept of inverse functions is essential for the development of the transcendental functions that are introduced in this chapter.

In Section 1.6 we reviewed how to use calculator values of inverse trigonometric functions to solve some basic trigonometric equations. In this section we are interested in developing the inverse trigonometric functions as functions and using them to illustrate how the relation between a function and its inverse allows us to determine properties of the inverse function from corresponding properties of the function.

Characterization of Inverse Functions

A function g is an inverse of a function f if g reverses the action of f in the sense that $g(f(x)) = x$ for all x in the domain of f.

Figure 7.1.1a illustrates a graphic process for determining the value $y = f(x)$ from a value x in the domain of f. Figure 7.1.1b illustrates the reverse process of determining the value $x = g(y)$ of an inverse function from the value $y = f(x)$ in the range of f. Note that the process cannot be reversed unless each y in the range of f corresponds to exactly one x in the domain of f. A function that has this property is said to be **one-to-one**. See Figure 7.1.1c. We conclude:

A function has an inverse if and only if each horizontal line that intersects the graph of the function intersects it at exactly one point.

FIGURE 7.1.1a The graph of f determines a value $y = f(x)$ for each x in the domain of f.

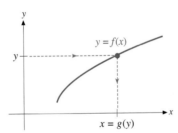

FIGURE 7.1.1b The reverse process determines a value $x = g(y)$ for each y in the range of f.

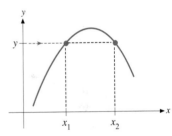

FIGURE 7.1.1c The process is not reversible if some horizontal line intersects the graph of f at more than one point.

The condition that $g(f(x)) = x$ for each x in the domain of f determines the value of an inverse function for each $y = f(x)$ in the range of f. It follows that a function f can have at most one inverse function that has domain equal to the range of f. If it exists, this inverse function is denoted by f^{-1}, read "f inverse."

NOTE   *The* -1 *in the symbol* f^{-1} *denotes an inverse function and is not an exponent. The negative first power of* $f(x)$ *can be denoted by either* $(f(x))^{-1}$ *or* $1/f(x)$.

It follows from the reversibility of the process that f is the inverse of f^{-1}. We then have

$$f^{-1}(f(x)) = x \quad \text{for all } x \text{ in the domain of } f$$
$$\text{and} \tag{1}$$
$$f(f^{-1}(y)) = y \quad \text{for all } y \text{ in the domain of } f^{-1}.$$

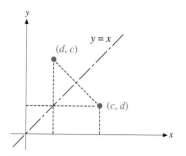

FIGURE 7.1.2 The points (c, d) and (d, c) are symmetric with respect to the line $y = x$.

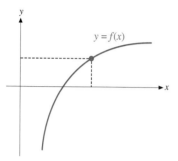

FIGURE 7.1.3a Each horizontal line that intersects the graph of f intersects it at exactly one point, so f has an inverse function.

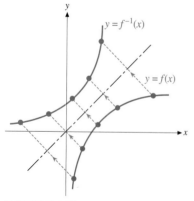

FIGURE 7.1.3b The graphs of $y = f(x)$ and $y = f^{-1}(x)$ are symmetric with respect to the line $y = x$.

Statement 1 characterizes inverse functions in terms of composite functions.

If $y = f(x)$, then (1) implies $f^{-1}(y) = f^{-1}(f(x)) = x$. Similarly, $x = f^{-1}(y)$ implies $f(x) = f(f^{-1}(y)) = y$. This gives a characterization of inverse functions in terms of equivalent equations:

$$y = f(x) \quad \text{if and only if} \quad x = f^{-1}(y). \tag{2}$$

Statement 2 holds for all x in the domain of f (equal to the range of f^{-1}) and all y in the domain of f^{-1} (equal to the range of f).

Graphs of a Function and Its Inverse

If (c, d) is on the graph of $y = f(x)$, then $d = f(c)$. Condition 2 then implies that $c = f^{-1}(d)$, so (d, c) is on the graph of $y = f^{-1}(x)$. From Figure 7.1.2 we see that the points (c, d) and (d, c) are **symmetric with respect to the line** $y = x$. (This means that the line segment between the points is perpendicular to and is bisected by the line $y = x$.) We then conclude:

The graphs of $y = f(x)$ and $y = f^{-1}(x)$ are symmetric with respect to the line $y = x$.

If f has an inverse function, we can use the fact that the graphs of f and f^{-1} are symmetric with respect to the line $y = x$ to construct the graph of f^{-1}.

Using the graph of f to sketch the graph of f^{-1}

- First, use the horizontal line test to determine that f has an inverse function.
- Lightly sketch the line $y = x$.
- Select several points on the graph of f and plot corresponding points that are symmetric with respect to the line $y = x$.
- Sketch the graph of f^{-1} through the points that have been plotted in such a way that the graphs of f and f^{-1} appear symmetric with respect to the line $y = x$.

EXAMPLE 1

The graph of a function is given in Figure 7.1.3a. We see that each horizontal line that intersects the graph of f intersects it at exactly one point, so f has an inverse function f^{-1}. In Figure 7.1.3b we have lightly sketched the line $y = x$, selected several point on the graph of f and plotted corresponding points that are symmetric with respect to the line $y = x$, and then sketched the graph of f^{-1}. ∎

We can think of the two equations in (2) as different forms of the same equation. In some cases we can find a formula for values of f^{-1} by using the usual rules of algebra to solve the equation $y = f(x)$ for x to obtain $x = f^{-1}(y)$. This gives us a formula for $f^{-1}(y)$. To express values of the inverse function f^{-1} in terms of the variable x, we simply substitute x for y in the formula for $f^{-1}(y)$.

EXAMPLE 2

Sketch the graph of f and determine whether f has an inverse f^{-1}. If f^{-1} exists, find a formula for $f^{-1}(x)$, sketch the graph of $y = f^{-1}(x)$, and verify that $f^{-1}(f(x)) = x$ and $f(f^{-1}(y)) = y$.

(a) $f(x) = (6 - 2x)/3$,
(b) $f(x) = x^3 + 1$,
(c) $f(x) = x^2$,
(d) $f(x) = x^2, x \geq 0$.

SOLUTION

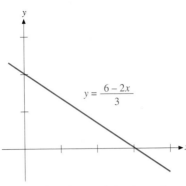

(a) From Figure 7.1.4a we see that the function $f(x) = (6 - 2x)/3$ has an inverse function, since each horizontal line intersects the graph at exactly one point.

Substituting into the equation $y = f(x)$ and then solving for x, we have

$$y = f(x),$$

$$y = \frac{6 - 2x}{3},$$

$$3y = 6 - 2x,$$

$$2x = 6 - 3y,$$

$$x = \frac{6 - 3y}{2}.$$

FIGURE 7.1.4a Each horizontal line intersects the graph of $f(x) = (6 - 2x)/3$ at exactly one point, so f has an inverse function.

Since $y = f(x)$ is equivalent to $x = f^{-1}(y)$, this means

$$f^{-1}(y) = \frac{6 - 3y}{2}.$$

Substituting x for y in this equation, we obtain

$$f^{-1}(x) = \frac{6 - 3x}{2}.$$

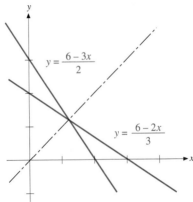

In Figure 7.1.4b we have used the graph of f to sketch the graph of f^{-1}.

We have

$$f^{-1}(f(x)) = f^{-1}\left(\frac{6 - 2x}{3}\right) = \frac{6 - 3\left(\dfrac{6 - 2x}{3}\right)}{2} = \frac{6 - (6 - 2x)}{2} = x$$

FIGURE 7.1.4b The graphs of $f(x) = (6 - 2x)/3$ and $f^{-1}(x) = (6 - 3x)/2$ are symmetric with respect to the line $y = x$.

and

$$f(f^{-1}(y)) = f\left(\frac{6 - 3y}{2}\right) = \frac{6 - 2\left(\dfrac{6 - 3y}{2}\right)}{3} = \frac{6 - (6 - 3y)}{3} = y.$$

(b) From Figure 7.1.5a we see that $f(x) = x^3 + 1$ has an inverse function. Solving $y = x^3 + 1$ for x, we have

$$y = x^3 + 1,$$

$$x^3 = y - 1,$$

$$x = (y - 1)^{1/3}.$$

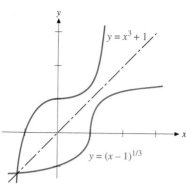

FIGURE 7.1.5a The function $f(x) = x^3 + 1$ has an inverse function.

FIGURE 7.1.5b The graphs of $f(x) = x^3 + 1$ and $f^{-1}(x) = (x-1)^{1/3}$ are symmetric with respect to the line $y = x$.

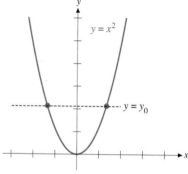

FIGURE 7.1.6 The function $f(x) = x^2$ does not have an inverse function, since there are horizontal lines that intersect the graph at more than one point.

This means $f^{-1}(y) = (y - 1)^{1/3}$, so $f^{-1}(x) = (x - 1)^{1/3}$. The graph of f^{-1} is sketched in Figure 7.1.5b.

We have

$$f^{-1}(f(x)) = f^{-1}(x^3 + 1) = ((x^3 + 1) - 1)^{1/3} = (x^3)^{1/3} = x$$

and

$$f(f^{-1}(y)) = f((y - 1)^{1/3}) = ((y - 1)^{1/3})^3 + 1 = (y - 1) + 1 = y.$$

(c) $f(x) = x^2$ does not have an inverse function. For any $y_0 > 0$, the horizontal line $y = y_0$ intersects the graph at two points. See Figure 7.1.6. The equation $y = x^2$ does not have a unique solution for x in terms of y. For $y > 0$ the equation has two solutions, $x = \sqrt{y}$ and $x = -\sqrt{y}$.

(d) Figure 7.1.7a indicates that $f(x) = x^2, x \geq 0$, has an inverse function.

The equation $y = x^2$ has the *single* solution $x = \sqrt{y}$ if x is restricted to nonnegative values, since the solution $x = -\sqrt{y}$ is negative if $y > 0$. We have $f^{-1}(y) = \sqrt{y}$, so $f^{-1}(x) = \sqrt{x}$. The graph of f^{-1} is given in Figure 7.1.7b.

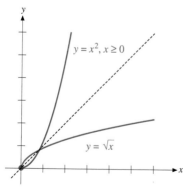

FIGURE 7.1.7a The function $f(x) = x^2, x \geq 0$ has an inverse function.

FIGURE 7.1.7b The graphs of $f(x) = x^2, x \geq 0$, and $f^{-1}(x) = \sqrt{x}$ are symmetric with respect to the line $y = x$.

If $x \geq 0$, we have

$$f^{-1}(f(x)) = f^{-1}(x^2) = \sqrt{x^2} = |x| = x.$$

If $y \geq 0$, we have

$$f(f^{-1}(y)) = f(\sqrt{y}) = (\sqrt{y})^2 = y. \quad \blacksquare$$

Let us compare Example 2(c) and 2(d). The function $f(x) = x^2$ of Example 2(c) does not have an inverse function, but the *different function* obtained in Example 2(d) by restricting the domain to nonnegative values of x does have an inverse.

Inverse Trigonometric Functions

We can use the idea of restricting the domain of a function to define the inverse trigonometric functions.

The function $\sin x$ does not have an inverse function, because horizontal lines that intersect its graph intersect it infinitely often, which is more than once. See Figure 7.1.8. However, we can obtain an inverse function by restricting the domain. We want an interval on which $\sin x$ assumes each value between -1 and 1 exactly once. The interval $-\pi/2 \leq x \leq \pi/2$ is most convenient. Thus, the function

$$f(x) = \sin x, \quad -\frac{\pi}{2} \leq x \leq \frac{\pi}{2},$$

has an inverse function, which we denote either arcsin x or $\sin^{-1} x$ (read " sine inverse of x"). See Figure 7.1.9a. The graph of $y = \sin^{-1} x$ is obtained from the graph of $\sin x$, $-\pi/2 \leq x \leq \pi/2$, by reflection about the line $y = x$. See Figure 7.1.9b. The function $\sin^{-1} x$ is defined for $-1 \leq x \leq 1$ and has values $-\pi/2 \leq y \leq \pi/2$.

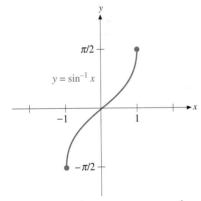

FIGURE 7.1.8 The function $\sin x$ does not have an inverse function, because horizontal lines that intersect its graph intersect it at more than one point.

FIGURE 7.1.9a The function $f(x) = \sin x$, $-\pi/2 \leq x \leq \pi/2$, has inverse function $\sin^{-1} x$.

FIGURE 7.1.9b The function $\sin^{-1} x$ is defined for $-1 \leq x \leq 1$ and has values $-\pi/2 \leq y \leq \pi/2$.

NOTE *The -1 in the notation $\sin^{-1} x$ corresponds to the -1 in the notation f^{-1}. In particular, this -1 is not intended to be an exponent.*

Corresponding to Statement 1, we have

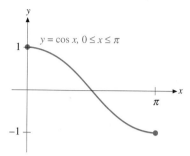

FIGURE 7.1.10a $f(x) = \cos x, 0 \leq x \leq \pi$, has inverse function $\cos^{-1} x$.

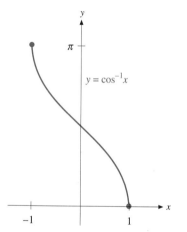

FIGURE 7.1.10b   The function $\cos^{-1} x$ is defined for $-1 \leq x \leq 1$ and has values $0 \leq y \leq \pi$.

$$\sin^{-1}(\sin x) = x \qquad \text{for } -\frac{\pi}{2} \leq x \leq \frac{\pi}{2},$$
$$\text{and} \tag{3}$$
$$\sin(\sin^{-1} y) = y \qquad \text{for } -1 \leq y \leq 1.$$

Note that $\sin^{-1}(\sin x) = x$ only if $-\pi/2 \leq x \leq \pi/2$. For example:

$$\sin^{-1}\left(\sin \frac{5\pi}{6}\right) = \sin^{-1}\left(\frac{1}{2}\right) = \frac{\pi}{6} \neq \frac{5\pi}{6}.$$

The sine and inverse sine functions are real-valued functions of a real variable, but it may be convenient for some applications to think of $\sin^{-1} y$ as an angle. This is analogous to thinking of $\sin x$ as the sine of an angle x.

We can obtain other inverse trigonometric functions by restricting domains of the corresponding trigonometric functions, as we did with the sine function.

The function $f(x) = \cos x, 0 \leq x \leq \pi$, has an inverse function, denoted either $\arccos x$ or $\cos^{-1} x$. See Figure 7.1.10a. The graph of $y = \cos^{-1} x$ is given in Figure 7.1.10b. The function $\cos^{-1} x$ is defined for $-1 \leq x \leq 1$ and has values $0 \leq y \leq \pi$. We have

$$\cos^{-1}(\cos x) = x \qquad \text{for } 0 \leq x \leq \pi,$$
$$\text{and} \tag{4}$$
$$\cos(\cos^{-1} y) = y \qquad \text{for } -1 \leq y \leq 1.$$

The function $f(x) = \tan x, -\pi/2 < x < \pi/2$, has an inverse function, denoted either $\arctan x$ or $\tan^{-1} x$. See Figure 7.1.11a. The graph of $y = \tan^{-1} x$ is given in Figure 7.1.11b. The function $\tan^{-1} x$ is defined for all x and has values $-\pi/2 < y < \pi/2$. We have

$$\tan^{-1}(\tan x) = x \qquad \text{for } -\pi/2 < x < \pi/2,$$
$$\text{and} \tag{5}$$
$$\tan(\tan^{-1} y) = y \qquad \text{for all } y.$$

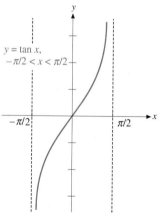

FIGURE 7.1.11a $f(x) = \tan x, -\pi/2 < x < \pi/2$, has inverse function $\tan^{-1} x$.

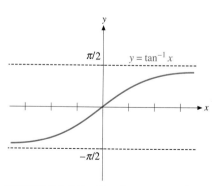

FIGURE 7.1.11b   The function $\tan^{-1} x$ is defined for all x and has values $-\pi/2 < y < \pi/2$.

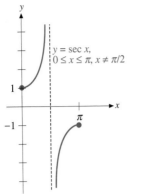

FIGURE 7.1.12a $f(x) = \sec x, 0 \leq x \leq \pi, x \neq \pi/2$, has inverse function $\sec^{-1} x$.

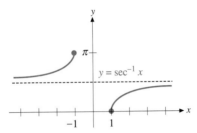

FIGURE 7.1.12b   The function $\sec^{-1} x$ is defined for $|x| \geq 1$ and has values $0 \leq y \leq \pi, y \neq \pi/2$.

The function $f(x) = \sec x, 0 \leq x \leq \pi, x \neq \pi/2$, has an inverse function, denoted either arcsec x or $\sec^{-1} x$. See Figure 7.1.12a. The graph of $y = \sec^{-1} x$ is given in Figure 7.1.12b. The function $\sec^{-1} x$ is defined for $|x| \geq 1$ and has values $0 \leq y \leq \pi, y \neq \pi/2$. We have

$$\sec^{-1}(\sec x) = x \qquad \text{for } 0 \leq x \leq \pi, x \neq \pi/2,$$
$$\text{and} \tag{6}$$
$$\sec(\sec^{-1} y) = y \qquad \text{for } |y| \geq 1.$$

Obtaining Properties of f^{-1} from Properties of f

The symmetry with respect to the line $y = x$ of the graphs of a function and its inverse implies a relation between asymptotes of the graphs.

Vertical asymptotes of the graph of $y = f(x)$ correspond to horizontal asymptotes of the graph of $y = f^{-1}(x)$.

Horizontal asymptotes of the graph of $y = f(x)$ correspond to vertical asymptotes of the graph of $y = f^{-1}(x)$.

For example, compare the graphs of $\tan x$ and $\tan^{-1} x$ in Figure 7.1.11 and the graphs of $\sec x$ and $\sec^{-1} x$ in Figure 7.1.12. Figure 7.1.11b also shows that

$$\lim_{x \to \infty} \tan^{-1} x = \frac{\pi}{2} \text{ and } \lim_{x \to -\infty} \tan^{-1} x = -\frac{\pi}{2}, \tag{7}$$

and Figure 7.1.12b shows that

$$\lim_{x \to \infty} \sec^{-1} x = \frac{\pi}{2} \text{ and } \lim_{x \to -\infty} \sec^{-1} x = \frac{\pi}{2}. \tag{8}$$

The following formula relates values of the inverse secant function and the inverse cosine function.

$$\sec^{-1} x = \cos^{-1} \frac{1}{x}. \tag{9}$$

Proof   Using (4), we have

$$\sec\left(\cos^{-1}\frac{1}{x}\right) = \frac{1}{\cos\left(\cos^{-1}\frac{1}{x}\right)} = \frac{1}{1/x} = x.$$

We know that $0 \leq \cos^{-1}(1/x) \leq \pi$. Also, $\frac{1}{x} \neq 0$ implies $\cos^{-1}(1/x) \neq \pi/2$. We can then apply (6) to obtain

$$\sec^{-1} x = \sec^{-1} \sec\left(\cos^{-1}\frac{1}{x}\right) = \cos^{-1}\frac{1}{x}.$$

(Note that $\sec^{-1}(\sec y) \neq y$, unless $0 \leq y \leq \pi, y \neq \pi/2$.) ∎

 Formula 9 allows us to evaluate the inverse secant function by using the inverse cosine function on a calculator.

Simplifying Inverse Trigonometric Expressions

Certain combinations of trigonometric functions and inverse trigonometric functions simplify to algebraic expressions. A sketch is often useful for this type of simplification.

EXAMPLE 3

Write $\cos(\sin^{-1} x)$ as an algebraic expression.

SOLUTION   We have $\cos(\sin^{-1} x) = \cos\theta$, where $\theta = \sin^{-1} x$. Thus, θ satisfies $-\pi/2 \leq \theta \leq \pi/2$ and $\sin\theta = x$. Representative triangles with $\sin\theta = x$ and $-\pi/2 \leq \theta \leq \pi/2$ are sketched in Figure 7.1.13. We have used $\sin\theta = x/1$ to label one side and the hypotenuse of each triangle and then used the Pythagorean Theorem to determine that the other side is $\sqrt{1 - x^2}$. We can read from our sketch that $\cos(\sin^{-1} x) = \cos\theta = \sqrt{1 - x^2}$. (Note that $\sqrt{1 - x^2} \geq 0$ and $-\pi/2 \leq \theta \leq \pi/2$ implies $\cos\theta \geq 0$.) ■

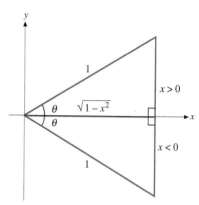

FIGURE 7.1.13 $\sin\theta = x, -\pi/2 \leq \theta \leq \pi/2$, implies $\cos\theta = \sqrt{1 - x^2}$.

Inverse Functions and Continuity

We most often deal with functions that are continuous on an interval. We first note that any function that is defined and monotonic on an interval has an inverse. See Figure 7.1.14. If a *continuous* function is not monotonic on an interval, then the Intermediate Value Theorem can be used to show that some horizontal line intersects its graph at more than one point. See Figure 7.1.15. We conclude:

If f is defined and continuous on an interval, then f has an inverse if and only f is monotonic on the interval.

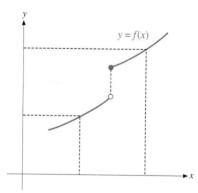

FIGURE 7.1.14   A function that is defined and monotonic on an interval has an inverse.

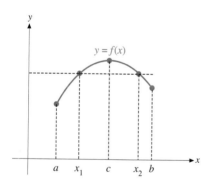

FIGURE 7.1.15   If a continuous function is not monotonic on an interval, then the Intermediate Value Theorem can be used to show that some horizontal line intersects its graph at more than one point.

If f is differentiable and its derivative is either positive at each point in an interval or negative at each point in an interval, then f is continuous and monotonic on the interval, and so has an inverse function. Since the graphs of a function and its inverse are symmetric with respect to the line $y = x$, we should expect the inverse of a continuous function to be continuous.

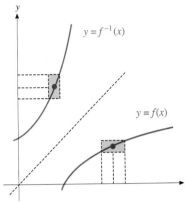

THEOREM 1

If f is differentiable on an interval I with either $f'(x) > 0$ on I or $f'(x) < 0$ on I, then f^{-1} is defined and continuous at points $d = f(c)$, c in I.

Formal verification of Theorem 1 follows from the fact that $f^{-1}(x)$ is between $f^{-1}(d) - \epsilon$ and $f^{-1}(d) + \epsilon$ whenever x is between $f(c - \epsilon)$ and $f(c + \epsilon)$. See Figure 7.1.16.

FIGURE 7.1.16 If either $f'(x) > 0$ on an interval I or $f'(x) < 0$ on I, then f is continuous and monotonic, so f has an inverse. The inverse is continuous.

EXERCISES 7.1

Use the graphs of f given in Exercises 1–4 to determine whether f has an inverse function. If f^{-1} exists, sketch its graph.

1.

2.

3.

4.

In Exercises 5–16, sketch the graph of f and determine if f has an inverse function f^{-1}. If f^{-1} exists, find a formula for $f^{-1}(x)$, sketch the graph of $y = f^{-1}(x)$, and verify that $f^{-1}(f(x)) = x$ and $f(f^{-1}(y)) = y$.

5. $f(x) = 2x + 1$ **6.** $f(x) = 1 - \dfrac{x}{2}$

7. $f(x) = 2(x - 1)^{1/3}$ **8.** $f(x) = 1 - x^3$

9. $f(x) = x^4 - 1$ **10.** $f(x) = 2x - x^2$

11. $f(x) = \sqrt{4 - x^2}$,   **12.** $f(x) = 2\sqrt{1 - x^2}$,
     $0 \le x \le 2$ $0 \le x \le 1$

13. $f(x) = \sqrt{x + 1}$ **14.** $f(x) = 1 - \sqrt{1 - x}$

15. $f(x) = \dfrac{1}{x-1}$ **16.** $f(x) = \dfrac{x}{x+1}$

Evaluate the expressions in Exercises 17–28.

17. $\sin^{-1}(1/2)$ **18.** $\cos^{-1}(-1)$

19. $\tan^{-1}(1)$ **20.** $\sec^{-1}(\sqrt{2})$

21. $\sin^{-1}(0.01)$ **22.** $\cos^{-1}(-0.2)$

23. $\tan^{-1}(3)$ **24.** $\sec^{-1}(2.5)$

25. $\cos[\sin^{-1}(-1/\sqrt{2})]$ **26.** $\sin[\cos^{-1}(\sqrt{3}/2)]$

27. $\sin^{-1}[\sin(3\pi/4)]$ **28.** $\tan^{-1}[\tan(4\pi/3)]$

   Solve for x in Exercises 29–36.

29. $\sin^{-1}(2x) = 0.23$ **30.** $\sin^{-1}(x/2) = -0.14$

31. $\sin(2x) = 0.25, -\pi/4 \le x \le \pi/4$

32. $\tan(2x-1) = -0.5, -\pi/2 < 2x - 1 < \pi/2$

33. $\cos x = 1/3, 0 \le x \le 2\pi$

34. $\sin x = -1/4, 0 \le x \le 2\pi$

35. $\cot(2x) = -4, 0 \le x \le \pi$

36. $\cot(2x) = -3, 0 \le x \le \pi$

Write the expressions in Exercises 37–42 as algebraic expressions.

37. $\tan[\sin^{-1} x]$ **38.** $\sin[\tan^{-1} x]$

39. $\cos[\sin^{-1}(x/2)]$ **40.** $\cos[\tan^{-1}(x/3)]$

41. $\sin[\sec^{-1}(x/3)]$ **42.** $\tan[\sec^{-1}(x/2)]$

43. We define the **inverse cotangent function** to be the inverse of the function $f(x) = \cot x, 0 < x < \pi$. **(a)** Sketch the graphs of f and f^{-1}. **(b)** What is the domain of $\cot^{-1} x$? **(c)** What is the range of $\cot^{-1} x$?

44. We define the **inverse cosecant function** to be the inverse of the function $f(x) = \csc x, -\pi/2 \le x \le \pi/2, x \ne 0$. **(a)** Sketch the graphs of f and f^{-1}. **(b)** What is the domain of $\csc^{-1} x$? **(c)** What is the range of $\csc^{-1} x$?

Use the graphs of f in Exercises 45–48 to determine the largest intervals on which f, restricted to the intervals, has an inverse function.

45.

46.

47.

48.

*In Exercises 49–54, **(a)** find the largest interval that contains the given point x_0 and has the property that the given function f, restricted to the interval, has an inverse function. (Hint: Sketch the graph of f.) **(b)** Find a formula for $f^{-1}(x)$. (Hint: Each of the equations $y = f(x)$ has more than one solution for x. Choose the solution that gives the value x_0.)*

49. $f(x) = x^2 - 1, x_0 = -1$ **50.** $f(x) = x^2 - 2x, x_0 = 0$

51. $f(x) = \tan x, x_0 = \pi$ **52.** $f(x) = \cos x, x_0 = -\pi/2$

53. $f(x) = \sin x, x_0 = \pi$ **54.** $f(x) = \sin x, x_0 = 2\pi$

55. A sign 16 ft high is painted on a vertical wall. The bottom of the sign is 9 ft above eye level. Express the angle of sight between the top and bottom of the sign as a function of the distance from the wall.

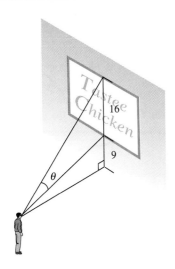

56. Show that the inverse of a linear function, $f(x) = ax + b$, $a \neq 0$, is a linear function.

Use the definitions as in the proof of Statement 9 to verify the identities in Exercises 57–60.

57. $\sin^{-1}(-x) = -\sin^{-1} x$

58. $\cos^{-1}(-x) = \pi - \cos^{-1} x$

59. $\cos^{-1} x = \dfrac{\pi}{2} - \sin^{-1} x$

60. $\tan^{-1} x + \tan^{-1}(1/x) = \pi/2, x > 0$

61. Verify that f increasing on $[a, b]$ implies f is one-to-one on $[a, b]$.

62. Verify that f continuous and increasing on $[a, b]$ implies f^{-1} increasing on $[f(a), f(b)]$.

63. Find a function f that is one-to-one on $[a, b]$ but is neither increasing on $[a, b]$ nor decreasing on $[a, b]$.

64. Find a function f that is increasing on $[a, b]$, but f^{-1} is not defined at some points of $[f(a), f(b)]$.

7.2 DIFFERENTIATION OF INVERSE FUNCTIONS

Connections

Geometric interpretation of derivative, 2.6.

Chain Rule, 3.4.

Inverse functions, 7.1.

To apply the concept of inverse functions to calculus problems, we need to establish differentiation and integration formulas. In this section, we will illustrate how to obtain formulas for the derivatives of inverse functions by developing formulas for the derivatives of the inverse trigonometric functions. The corresponding integration formulas allow us to evaluate some new types of integrals.

We conclude the section by briefly discussing some general conditions under which an inverse function is differentiable and how the derivative of a function is related to the derivative of its inverse.

Differentiation Formulas

Let us illustrate how implicit differentiation can be used to find a formula for the derivative of an inverse function. The method requires that the inverse function be differentiable.

If f is differentiable on an open interval I with either $f'(x) > 0$ on I or $f'(x) < 0$ on I, then Theorem 1 of Section 7.1 tells us that f has an inverse. At the end of this section we will show that the inverse function is differentiable. For now we assume this is true, so we can use the Method of Implicit Differentiation to find a formula for the derivative of the inverse function.

The function $\sin x$ is differentiable with derivative $\cos x > 0$ for $-\pi/2 < x < \pi/2$. It follows that the inverse function $\sin^{-1} x$ exists and is differentiable for $-1 < x < 1$. To find a formula for the derivative of $\sin^{-1} x$, we set

$$y = \sin^{-1} x, \text{ so } -\pi/2 \leq y \leq \pi/2 \text{ and } \sin y = \sin(\sin^{-1} x) = x.$$

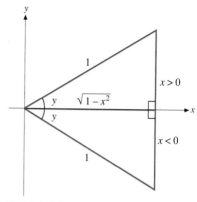

FIGURE 7.2.1 $\sin y = x, -\pi/2 \le y \le \pi/2$, implies $\cos y = \sqrt{1 - x^2}$.

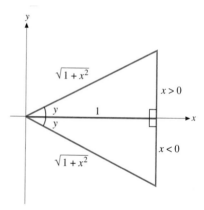

FIGURE 7.2.2 $\tan y = x, -\pi/2 < y < \pi/2$, implies $\sec y = \sqrt{1 + y^2}$.

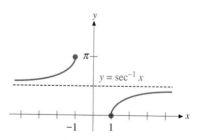

FIGURE 7.2.3   The derivative of $\sec^{-1} x$ is positive at all points where it exists.

Differentiating the equation $\sin y = x$ with respect to x, and using the Chain Rule to differentiate the sine of the function y, we have

$$\cos y \frac{dy}{dx} = 1, \text{ so } \frac{dy}{dx} = \frac{1}{\cos y}.$$

Since y is a function of x, it would be convenient to have a formula for the derivative of y in terms of x. We need to express $\cos y$ in terms of x. We know $\sin y = x$ and $-\pi/2 \le y \le \pi/2$. Representative right triangles with "angle" y are sketched in Figure 7.2.1. We can determine from the sketch that $\sin y = x, -\pi/2 \le y \le \pi/2$, implies $\cos y = \sqrt{1 - x^2}$. Substitution then gives the formula

$$\frac{d}{dx}(\sin^{-1} x) = \frac{1}{\sqrt{1 - x^2}}.$$

The function $\tan x$ has a positive derivative for $-\pi/2 < x < \pi/2$, so the inverse function $\tan^{-1} x$ exists and is differentiable. Let

$$y = \tan^{-1} x, \text{ so } -\pi/2 < y < \pi/2 \text{ and } \tan y = \tan(\tan^{-1} x) = x.$$

Differentiating the equation $\tan y = x$ with respect to x, and using the Chain Rule, we obtain

$$\sec^2 y \frac{dy}{dx} = 1, \text{ or } \frac{dy}{dx} = \frac{1}{\sec^2 y}.$$

From Figure 7.2.2 we see that $\tan y = x, -\pi/2 < y < \pi/2$, implies $\sec y = \sqrt{1 + x^2}$. Substitution then gives

$$\frac{d}{dx}(\tan^{-1} x) = \frac{1}{1 + x^2}.$$

We can use the technique illustrated above and the formula

$$\frac{d}{dx}(\sec x) = \sec x \tan x$$

to verify that

$$\frac{d}{dx}(\sec^{-1} x) = \frac{1}{|x|\sqrt{x^2 - 1}}.$$

We see from the graph in Figure 7.2.3 that the derivative of $\sec^{-1} x$ is positive at all points where it exists.

From Exercise 59 of Section 7.1 we know that

$$\cos^{-1} x = \frac{\pi}{2} - \sin^{-1} x, \text{ so } \frac{d}{dx}(\cos^{-1} x) = -\frac{d}{dx}(\sin^{-1} x), \text{ or}$$

$$\frac{d}{dx}(\cos^{-1} x) = -\frac{1}{\sqrt{1 - x^2}}.$$

Using the Chain Rule

We must know the Chain Rule version of each of the new differentiation formulas. For example,

$$\frac{d}{dx}(\sin^{-1}(\text{Function})) = \frac{1}{\sqrt{1 - (\text{Function})^2}} \frac{d}{dx}(\text{Function}).$$

EXAMPLE 1

(a) $\dfrac{d}{dx}(\sin^{-1}(x^2)) = \dfrac{1}{\sqrt{1 - (x^2)^2}}(2x).$

(b) $\dfrac{d}{dx}\left(\tan^{-1}\dfrac{x}{2}\right) = \left(\dfrac{1}{1 + (x/2)^2}\right)\left(\dfrac{1}{2}\right) = \dfrac{2}{4 + x^2}.$

(c) $\dfrac{d}{dx}(\sec^{-1}(|x|)) = \dfrac{1}{|(|x|)|\sqrt{(|x|)^2 - 1}} \dfrac{d}{dx}(|x|).$

If $x > 1$, then $|(|x|)| = x$ and $\dfrac{d}{dx}(|x|) = 1$. If $x < -1$, then $|(|x|)| = -x$ and $\dfrac{d}{dx}(|x|) = -1$. In either case the result in (c) simplifies to

$$\frac{d}{dx}(\sec^{-1}(|x|)) = \frac{1}{x\sqrt{x^2 - 1}}. \quad \blacksquare$$

Let us restate the result of Example 1(c):

$$\frac{d}{dx}(\sec^{-1}(|x|)) = \frac{1}{x\sqrt{x^2 - 1}}.$$

EXAMPLE 2

A boy is sitting 30 ft from a road, watching cars go by. Find the rate of change of his angle of sight as a car going 30 mph passes the point on the road nearest him.

SOLUTION A sketch is given in Figure 7.2.4. The position of the car along the vertical line $x = 30$ is indicated by the variable y and the angle of sight has been labeled θ. It is important that y and θ are considered to be variables. Note that y and θ are negative and increasing as the car approaches the point nearest the boy. The units of y should be feet, the units used to indicate the distance of the boy from the street; θ is measured in radians.

Following the procedure of Section 4.6, we express what we want and what we know in mathematical terms and display the information clearly for future reference.

$$\textit{We want: } \frac{d\theta}{dt} \text{ when } y = 0.$$

Let us change 30 mph to units of ft/s.

$$30 \text{ mph} = \left(\frac{30 \text{ mi}}{\text{h}}\right)\left(\frac{1 \text{ h}}{60 \text{ min}}\right)\left(\frac{1 \text{ min}}{60 \text{ s}}\right)\left(\frac{5280 \text{ ft}}{1 \text{ mi}}\right) = 44 \text{ ft/s}.$$

$$\textit{We know: } \frac{dy}{dt} = 44 \text{ ft/s}.$$

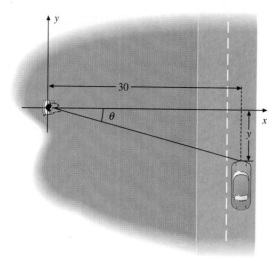

FIGURE 7.2.4   We see that $\tan\theta = y/30$.

From the sketch, we see that the variables are related by the equation

$$\tan\theta = \frac{y}{30}, \text{ with } -\frac{\pi}{2} < \theta < \frac{\pi}{2}, \text{ so } \theta = \tan^{-1}\left(\frac{y}{30}\right).$$

Differentiating the latter equation with respect to t, we have

$$\frac{d\theta}{dt} = \frac{1}{1+\left(\dfrac{y}{30}\right)^2}\left(\frac{1}{30}\frac{dy}{dt}\right).$$

Substitution of $y = 0$ and $\dfrac{dy}{dt} = 44$ into the above formula gives

$$\frac{d\theta}{dt} = \frac{1}{30}(44) \approx 1.467,$$

so the angle of sight is changing at a rate of approximately 1.467 rad/s. ■

Integration Formulas

The following integration formulas may be verified by differentiation. Recall that $\int f(x)dx = F(x) + C$ means $F' = f$ on some interval. We assume $a > 0$.

$$\int \frac{1}{\sqrt{a^2 - u^2}}du = \sin^{-1}\left(\frac{u}{a}\right) + C,$$

$$\int \frac{1}{a^2 + u^2}du = \frac{1}{a}\tan^{-1}\left(\frac{u}{a}\right) + C,$$

$$\int \frac{1}{u\sqrt{u^2 - a^2}}du = \frac{1}{a}\sec^{-1}\left(\frac{|u|}{a}\right) + C.$$

Since $\sec^{-1} x = \cos^{-1}(1/x)$, we also have

$$\int \frac{1}{u\sqrt{u^2 - a^2}}\,du = \frac{1}{a}\cos^{-1}\left(\frac{a}{|u|}\right) + C.$$

This form may be more convenient for evaluating definite integrals if your calculator does not have an inverse secant key.

Note that the integral that gives an inverse sine involves the square root of

$$(\text{Constant})^2 - (\text{Variable})^2.$$

The integral that gives either an inverse secant or the above form of the inverse cosine involves the square root of

$$(\text{Variable})^2 - (\text{Constant})^2.$$

EXAMPLE 3

Evaluate $\displaystyle\int \frac{1}{\sqrt{1 - 4x^2}}\,dx$.

SOLUTION The integral involves the square root of $(\text{Constant})^2 - (\text{Variable})^2$, so it may reduce to a $\sin^{-1}$ integral. We have $4x^2 = (2x)^2$ in place of u^2. Let us set $u = 2x$, so that $du = 2dx$. Then

$$\int \frac{1}{\sqrt{1 - 4x^2}}\,dx = \int \frac{(1/2)2}{\sqrt{1 - (2x)^2}}\,dx$$

$$= \int \frac{1/2}{\sqrt{1 - u^2}}\,du$$

$$= \frac{1}{2}\sin^{-1}u + C$$

$$= \frac{1}{2}\sin^{-1}2x + C. \quad \blacksquare$$

EXAMPLE 4

Evaluate $\displaystyle\int \frac{x}{x^4 + 16}\,dx$.

SOLUTION The integral involves $x^4 + 16 = (x^2)^2 + (4)^2$, so it may reduce to a $\tan^{-1}$ integral. Let us try the substitution $u = x^2$, so $du = 2x\,dx$. Then

$$\int \frac{x}{x^4 + 16}\,dx = \int \frac{(1/2)2x}{(x^2)^2 + 4^2}\,dx$$

$$= \int \frac{1/2}{u^2 + 4^2}\,du$$

$$= \frac{1}{2}\left(\frac{1}{4}\tan^{-1}\left(\frac{u}{4}\right)\right) + C$$

$$= \frac{1}{8}\tan^{-1}\left(\frac{x^2}{4}\right) + C. \quad \blacksquare$$

EXAMPLE 5

Evaluate $\displaystyle\int \frac{1}{x\sqrt{x^4 - 4}}dx$.

SOLUTION The integral involves the square root of $(\text{Variable})^2 - (\text{Constant})^2$, so it may reduce to a $\sec^{-1}$ or $\cos^{-1}$ integral. Setting $u = x^2$, we have $du = 2x\,dx$, so

$$\int \frac{1}{x\sqrt{x^4 - 4}}dx = \int \frac{(1/2)2x}{x^2\sqrt{(x^2)^2 - 2^2}}dx$$

$$= \int \frac{1/2}{u\sqrt{u^2 - 2^2}}du$$

$$= \frac{1}{2}\left(\frac{1}{2}\sec^{-1}\frac{|u|}{2}\right) + C$$

$$= \frac{1}{4}\sec^{-1}\frac{x^2}{2} + C.$$

We could also express the integral as

$$\int \frac{1}{x\sqrt{x^4 - 4}}dx = \frac{1}{4}\cos^{-1}\frac{2}{x^2} + C. \quad\blacksquare$$

EXAMPLE 6

Evaluate $\displaystyle\int \frac{\tan^{-1} x}{1 + x^2}dx$.

SOLUTION In this integral we recognize $1/(1 + x^2)$ as the derivative of $\tan^{-1} x$. This suggests the substitution

$$u = \tan^{-1} x, \ \text{so}\ du = \frac{1}{1 + x^2}dx.$$

Then

$$\int \frac{\tan^{-1} x}{1 + x^2}dx = \int u\,du = \frac{u^2}{2} + C = \frac{1}{2}(\tan^{-1} x)^2 + C. \quad\blacksquare$$

EXAMPLE 7

Find the volume of the solid obtained by revolving the region bounded by

$$y = \frac{3}{\sqrt{x^2 + 9}}, \qquad y = 0, \quad x = 0, \quad x = 3,$$

about the **(a)** x-axis, **(b)** y-axis.

SOLUTION

    (a) A typical vertical area strip generates a disk when revolved about the x-axis. From the sketch in Figure 7.2.5a we see that the radius of the disk is $3/\sqrt{x^2 + 9}$. The volume is

$$V = \int_0^3 \pi \left(\frac{3}{\sqrt{x^2 + 9}}\right)^2 dx = \int_0^3 \frac{9\pi}{x^2 + 3^2}dx = 9\pi\frac{1}{3}\tan^{-1}\frac{x}{3}\Big]_0^3$$

$$= 3\pi\tan^{-1}(1) - 3\pi\tan^{-1}(0) = 3\pi\frac{\pi}{4} = \frac{3\pi^2}{4} \approx 7.40.$$

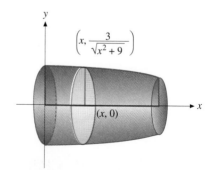

FIGURE 7.2.5a A typical disk has radius $3/\sqrt{x^2 + 9}$ and thickness dx.

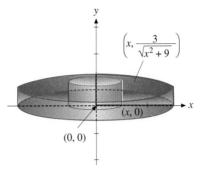

(b) In this case a typical vertical area strip generates a thin cylindrical shell with radius x and height $3/\sqrt{x^2 + 9}$. See Figure 7.2.5b. The volume is

$$V = \int_0^3 2\pi x \left(\frac{3}{\sqrt{x^2 + 9}} \right) dx = \int_0^3 3\pi (x^2 + 9)^{-1/2} 2x \, dx.$$

Let $u = x^2 + 9$, so $du = 2x \, dx$, $u(0) = 9$, and $u(3) = 18$. Substitution then gives

$$V = \int_9^{18} 3\pi u^{-1/2} du = 3\pi \frac{u^{1/2}}{1/2} \bigg]_9^{18} = 6\pi \sqrt{18} - 6\pi \sqrt{9} \approx 23.42. \quad \blacksquare$$

FIGURE 7.2.5b A typical thin cylindrical shell has radius x, height $3/\sqrt{x^2 + 9}$, and thickness of cylindrical wall dx.

The Differentiation Theorem

We know that the value of the derivative of a function is given by the slope of a line tangent to its graph. It follows from the fact that the graphs of f and f^{-1} are symmetric with respect to the line $y = x$ that the lines tangent to the graphs at corresponding points are also symmetric with respect to $y = x$. See Figure 7.2.6. From Figure 7.2.7 we see that two lines that are symmetric with respect to the line $y = x$ have slopes that are reciprocal. It follows that we should expect that the derivatives of f and f^{-1} at corresponding points are reciprocal. Let us state this important result as a theorem and give a proof in terms of the definition of the derivative.

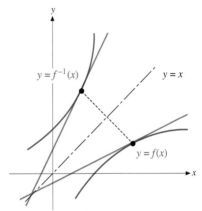

FIGURE 7.2.6 Since the graphs of f and f^{-1} are symmetric with respect to the line $y = x$, so are corresponding tangent lines.

THEOREM 1

Suppose f is differentiable on an open interval I with either $f'(x) > 0$ on I or $f'(x) < 0$ on I. Then f has an inverse function f^{-1}, and for each c in I, f^{-1} is differentiable at $d = f(c)$ with

$$(f^{-1})'(d) = \frac{1}{f'(c)}.$$

Proof It follows from Theorem 1 of Section 7.1 that f^{-1} exists and is continuous. To show that f^{-1} is differentiable and verify the differentiation formula, we look at the definition of the derivative of f^{-1} at the point $d = f(c)$. We have

$$(f^{-1})'(d) = \lim_{y \to d} \frac{f^{-1}(y) - f^{-1}(d)}{y - d}.$$

We know that $d = f(c)$ implies $f^{-1}(d) = c$. Setting $x = f^{-1}(y)$, so $y = f(x)$, and noting that $y \neq d$ implies $x = f^{-1}(y) \neq f^{-1}(d) = c$, we can rewrite the above quotient as

$$\frac{x - c}{f(x) - f(c)} \quad \text{or} \quad \frac{1}{\dfrac{f(x) - f(c)}{x - c}}.$$

Since f^{-1} is continuous, $x = f^{-1}(y) \to f^{-1}(d) = c$ as $y \to d$. This means

$$\frac{f(x) - f(c)}{x - c} \to f'(c) \text{ as } y \to d.$$

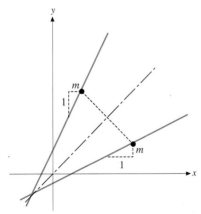

FIGURE 7.2.7 Two lines that are symmetric with respect to the line $y = x$ have slopes that are reciprocal.

Combining results, we obtain the theorem. ■

EXAMPLE 8

Suppose it is known that f has a differentiable inverse function, $f(3) = 2$, and $f'(3) = 4$. What can you conclude about values of f^{-1} and $(f^{-1})'$?

SOLUTION The basic formula that relates a function and its inverse can be used to determine the value of f^{-1} that corresponds to the value $f(3) = 2$. That is,

$$f(3) = 2 \text{ implies } f^{-1}(2) = 3.$$

Theorem 1 implies that f^{-1} is differentiable at 2 and that

$$(f^{-1})'(2) = \frac{1}{f'(3)} = \frac{1}{4}. \quad ■$$

EXERCISES 7.2

Evaluate the derivative in Exercises 1–12.

1. $\dfrac{d}{dx}\left(\sin^{-1}\dfrac{x}{3}\right)$

2. $\dfrac{d}{dx}(\sin^{-1} 5x)$ **(Calculus Explorer & Tutor II, Transcendental Functions, 1.)**

3. $\dfrac{d}{dx}\left(\cos^{-1}\dfrac{x}{2}\right)$ **4.** $\dfrac{d}{dx}\left(\cos^{-1}\dfrac{1}{x}\right)$

5. $\dfrac{d}{dx}\left(\tan^{-1}\dfrac{1}{x}\right)$ **6.** $\dfrac{d}{dx}\left(\tan^{-1}\dfrac{x}{a}\right)$

7. $\dfrac{d}{dx}(\sec^{-1}(x^2 - 4x))$ **(Calculus Explorer & Tutor II, Transcendental Functions, 3.)**

8. $\dfrac{d}{dx}(\sec^{-1} 2x)$ **9.** $\dfrac{d}{dx}\left(\sin^{-1}\dfrac{x}{a}\right)$, $a > 0$

10. $\dfrac{d}{dx}\left(\sec^{-1}\dfrac{|x|}{a}\right)$, $a > 0$ **11.** $\dfrac{d}{dx}\left(\cos^{-1}\dfrac{x}{a}\right)$, $a > 0$

12. $\dfrac{d}{dx}\left(x \tan^{-1} x\right)$ **(Calculus Explorer & Tutor II, Transcendental Functions, 2.)**

Evaluate the integrals in Exercises 13–26.

13. $\displaystyle\int \frac{1}{\sqrt{9 - x^2}}\,dx$ **14.** $\displaystyle\int \frac{1}{\sqrt{4 - 9x^2}}\,dx$

15. $\displaystyle\int \frac{1}{9 + 4x^2}\,dx$

16. $\displaystyle\int \frac{1}{4 + x^2}\,dx$ **(Calculus Explorer & Tutor II, Transcendental Functions, 4.)**

17. $\displaystyle\int \frac{1}{x\sqrt{4x^2 - 1}}\,dx$ **18.** $\displaystyle\int \frac{x}{\sqrt{1 - x^4}}\,dx$

19. $\displaystyle\int \frac{x^3}{\sqrt{x^4 - 1}}\,dx$ **20.** $\displaystyle\int \frac{x}{\sqrt{x^2 - 4}}\,dx$

21. $\displaystyle\int \frac{\sin^{-1} x}{\sqrt{1 - x^2}}\,dx$ **22.** $\displaystyle\int \frac{\sqrt{\tan^{-1} x}}{1 + x^2}\,dx$

23. $\displaystyle\int \frac{1}{x\sqrt{x^4 - 1}}\,dx$ **24.** $\displaystyle\int \frac{1}{x^2 + 3}\,dx$

25. $\displaystyle\int \frac{1}{\sqrt{3 - x^2}}\,dx$ **26.** $\displaystyle\int \frac{1}{\sqrt{x}\sqrt{1 - x}}\,dx$

Evaluate the definite integrals in Exercises 27–30.

27. $\displaystyle\int_0^1 \frac{1}{\sqrt{4 - x^2}}\,dx$ **28.** $\displaystyle\int_{-1}^1 \frac{1}{\sqrt{2 - x^2}}\,dx$

29. $\displaystyle\int_0^3 \frac{1}{9 + x^2}\,dx$ **30.** $\displaystyle\int_4^8 \frac{1}{x\sqrt{x^2 - 4}}\,dx$

31. Find an equation of the line tangent to the graph of $y = x^2 \tan^{-1} x$ at the point where $x = 1$.

32. Find an equation of the line tangent to the graph of $y = x \sin^{-1} x$ at the point where $x = 1/2$.

33. A balloon is rising vertically at a rate of 6 ft/s from a point on the ground that is 30 ft from an observer. Find the rate of change of the angle of elevation from the observer to the balloon when the balloon is 90 ft above ground.

34. A baseball is thrown by the pitcher toward home plate at a rate of 110 ft/s. At what rate does the batter need to rotate his line of sight to follow the ball as it passes home plate at a point 2 ft from him?

35. Find the area of the region bounded by $y = 1/\sqrt{4 - x^2}$, $y = 0$, $x = -1$, $x = 1$.

36. Find the area of the region bounded by $y = 1/(x^2 + 4)$, $y = 0$, $x = -2$, $x = 2$.

37. Find the volume of the solid obtained by revolving the region bounded by $y = 4/\sqrt{x^2 + 16}$, $y = 0$, $x = 0$, $x = 4\sqrt{3}$ about the **(a)** x-axis, **(b)** y-axis.

38. Find the volume of the solid obtained by revolving the region bounded by $y = 1/(16 - x^2)^{1/4}$, $y = 0$, $x = 0$, $x = 2\sqrt{3}$ about the **(a)** x-axis, **(b)** y-axis.

In Exercises 39–42, suppose that f has a differentiable inverse function and that f and f' have the given values. What can you conclude about values of f^{-1} and $(f^{-1})'$?

39. $f(1) = 2$, $f'(1) = 3$ **40.** $f(2) = 3$, $f'(2) = 5$

41. $f(3) = 5/2$, $f'(3) = 2/3$

42. $f(1/2) = -1$, $f'(1/2) = 1/3$

Evaluate the limits in Exercises 43–46.

43. $\displaystyle \lim_{x \to \infty} \int_0^x \frac{1}{1 + t^2} dt$ **44.** $\displaystyle \lim_{x \to 1^-} \int_0^x \frac{1}{\sqrt{1 - t^2}} dt$

45. $\displaystyle \lim_{x \to \infty} x \left(\frac{\pi}{2} - \tan^{-1}(2x) \right)$ **46.** $\displaystyle \lim_{x \to \infty} \frac{\frac{\pi}{2} - \tan^{-1} x}{\frac{\pi}{2} - \sec^{-1} x}$

47. A sign 12 ft high is painted on a vertical wall. The bottom of the sign is 8 ft above eye level. How far from the wall should one stand to obtain the maximum angle of sight between the top and bottom of the sign? See figure.

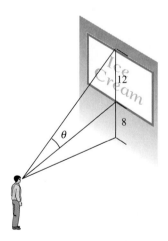

48. A triangle, not necessarily a right triangle, has a side of length a opposite an angle θ and another side of length b, where $a < b$. The remaining side has variable length x.
   (a) Express θ in terms of x.

(b) Show that θ is maximum when $x = \sqrt{b^2 - a^2}$.

(c) Show that the triangle with maximum θ is a right triangle.

(d) Find the maximum value of θ.

49. Show that $\theta = \cos^{-1}\left(\dfrac{\sqrt{b^2 - a^2}}{b} \right)$, $0 < a < b$ implies $\sin \theta = a/b$.

50. **(a)** Evaluate the length, s, of the curve $y = \sqrt{r^2 - x^2}$, $0 \le x < r$, between the points $(0, r)$ and $(x, \sqrt{r^2 - x^2})$ on the curve. See figure.
   (b) Solve the equation in (a) for x in terms of s.
   (c) Use (b) to express the coordinates of a point on the curve in terms of s.

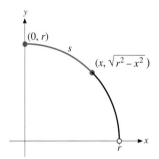

51. Find the area of the surface obtained by revolving the curve $y = \sqrt{r^2 - x^2}$, $c \le x \le c + h$, about the x-axis, where $-r \le c < c + h \le r$. See figure. (Your result should show that the area of that part of the surface of a sphere between two planes that are a fixed distance apart does not depend on where the planes slice the sphere, as long as both planes intersect the sphere.)

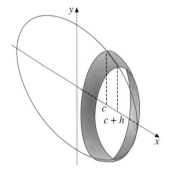

52. Verify that $\dfrac{d}{dx}(\sec^{-1} x) = \dfrac{1}{|x|\sqrt{x^2 - 1}}$ by using implicit differentiation of the equation $\sec y = x$ with respect to x.

7.3 THE NATURAL LOGARITHMIC FUNCTION

Connections

Chain Rule, 3.4.

Graphing, asymptotes, 4.1–4.

The Fundamental Theorem of Calculus, 5.4.

Up to this point in our development of calculus, we still cannot use the Fundamental Theorem to evaluate $\int \frac{1}{x} dx$, because we do not know of a function that has derivative $\frac{1}{x}$. In this section we will define and study a function that has this property. This important function happens to be of a type you studied in algebra courses. Namely, it is a logarithmic function.

Formal Definition and Properties

We can use an integral to define a function that has derivative $1/x$.

DEFINITION

The **natural logarithmic function** is

$$\ln x = \int_1^x \frac{1}{t} dt, \qquad x > 0.$$

Note that $x > 0$ implies that the integrand $1/t$ is continuous on the closed interval with endpoints 1 and x, so the integral that defines $\ln x$ exists. The integral does not exist for $x \leq 0$ and $\ln x$ is undefined for $x \leq 0$.

We can use the definition of $\ln x$ to determine properties of the function. It is particularly easy to use the integral form of the definition to evaluate the derivative of $\ln x$. That is, the Fundamental Theorem of Calculus implies

$$\frac{d}{dx}(\ln x) = \frac{d}{dx}\left(\int_1^x \frac{1}{t} dt\right) = \frac{1}{x}.$$

This shows that we have defined a function with derivative $1/x$, as was desired. We can also use the integral definition to show that $\ln x$ satisfies the following properties, which are characteristic of logarithmic functions.

$$\ln(ab) = \ln a + \ln b. \tag{1}$$

$$\ln(a^r) = r \ln a, \quad r \text{ a rational number.} \tag{2}$$

$$\ln\left(\frac{1}{b}\right) = -\ln b. \tag{3}$$

$$\ln\left(\frac{a}{b}\right) = \ln a - \ln b. \tag{4}$$

Proof To verify (1) we use a property of integrals to write

$$\ln(ab) = \int_1^{ab} \frac{1}{t} dt = \int_1^a \frac{1}{t} dt + \int_a^{ab} \frac{1}{t} dt.$$

In the integral on the right above, we use the substitution $u = t/a$, so $du = dt/a$, $u(a) = 1$, and $u(ab) = b$, to obtain

$$\int_a^{ab} \frac{1}{t} dt = \int_a^{ab} \frac{1}{t/a} \frac{1}{a} dt = \int_1^b \frac{1}{u} du.$$

Combining results then gives

$$\ln(ab) = \int_1^a \frac{1}{t}\,dt + \int_1^b \frac{1}{u}\,du = \ln a + \ln b.$$

Similarly, the substitution $u = t^{1/r}$ can be used to verify (2). Since $1/x = x^{-1}$, (3) is a special case of (2). Equation (4) follows from (1) and (3). ■

Using the Chain Rule and Logarithmic Differentiation

The Chain Rule version of the differentiation formula for $\ln x$ is

$$\frac{d}{dx}(\ln(\text{Function})) = \frac{1}{\text{Function}}\frac{d}{dx}(\text{Function}).$$

EXAMPLE 1

(a) $\dfrac{d}{dx}\left(\ln(x^2 + 1)\right) = \dfrac{1}{x^2 + 1}(2x),$

(b) $\dfrac{d}{dx}(\ln(\sec x)) = \dfrac{1}{\sec x}(\sec x \tan x) = \tan x.$ ■

It is helpful to use the properties of logarithms to simplify differentiation of the natural logarithmic function of products, quotients, and powers.

EXAMPLE 2

$$\frac{d}{dx}\left(\ln\left(x\sqrt{x^2 + 1}\right)\right) = \frac{d}{dx}\left(\ln x + \frac{1}{2}\ln(x^2 + 1)\right)$$

$$= \frac{1}{x} + \frac{1}{2}\left(\frac{1}{x^2 + 1}\right)(2x). \quad ■$$

It is useful to have a formula for the differentiation of the function $\ln |x|$. We have

$$\ln |x| = \begin{cases} \ln x, & x > 0, \\ \ln(-x), & x < 0, \end{cases}$$

so

$$\frac{d}{dx}(\ln |x|) = \begin{cases} \dfrac{1}{x}, & x > 0, \\ \dfrac{1}{-x}(-1) = \dfrac{1}{x}, & x < 0. \end{cases}$$

This gives

$$\frac{d}{dx}(\ln |x|) = \frac{1}{x}, \quad x \neq 0. \tag{5}$$

We can use Formula 5 to simplify finding the derivative of functions that contain several factors. The procedure is called **logarithmic differentiation.** We illustrate the procedure with an example.

EXAMPLE 3

$$y = \frac{x^2(x - 1)^3}{x + 1}.$$

Use the method of logarithmic differentiation to find dy/dx.

SOLUTION   We first write the absolute value of y in terms of the absolute values of the factors and then take logarithms. (The use of the absolute values insures that the logarithm of each factor exists at each point where it is nonzero.) Using properties of logarithms to simplify by expanding, we then obtain

$$|y| = \frac{|x|^2|x-1|^3}{|x+1|},$$
$$\ln|y| = 2\ln|x| + 3\ln|x-1| - \ln|x+1|.$$

Using the Chain Rule to differentiate the logarithm of each function, we obtain

$$\frac{1}{y}\frac{dy}{dx} = \frac{2}{x} + \frac{3}{x-1} - \frac{1}{x+1},$$

$$\frac{dy}{dx} = y\left(\frac{2(x-1)(x+1) + 3(x)(x+1) - (x)(x-1)}{x(x-1)(x+1)}\right)$$
$$= \frac{x^2(x-1)^3}{x+1}\left(\frac{4x^2+4x-2}{x(x-1)(x+1)}\right) = \frac{2x(x-1)^2(2x^2+2x-1)}{(x+1)^2}. \quad\blacksquare$$

The Graph of ln x

We know that

$$\frac{d}{dx}(\ln x) = \frac{1}{x}, \quad x > 0.$$

It follows that $\ln x$ is continuous and increasing on the interval $x > 0$. The second derivative is $-1/x^2$, which is negative, so $\ln x$ is concave downward. We have $\ln(1) = 0$, so the graph passes through the point $(1,0)$. In Example 3 of Section 5.6, we used Simpson's Rule to show that

$$\ln 2 = \int_1^2 \frac{1}{t}dt \approx 0.69315,$$

with error less than 0.0001. This allows us to plot the point $(2, \ln 2)$. From (2) we have

$$\ln 2^n = n\ln 2.$$

Thus, if n is increased by one, the value of $\ln 2^n$ is increased by $\ln 2$. That is,

$$\ln 2^{n+1} = (n+1)\ln 2 = n\ln 2 + \ln 2 = \ln 2^n + \ln 2.$$

Since $\ln x$ is increasing, this gives

$$\lim_{x\to\infty} \ln x = \infty. \tag{6}$$

Similarly, if n is decreased by one, $\ln 2^n$ is decreased by $\ln 2$. Since $2^n \to 0$ as $n \to -\infty$, this implies $\ln x$ must have vertical asymptote $x = 0$ and

$$\lim_{x\to 0^+} \ln x = -\infty. \tag{7}$$

Statements 6 and 7 and the continuity of $\ln x$ imply that the range of $\ln x$ is all real numbers. The graph of $\ln x$ is given in Figure 7.3.1.

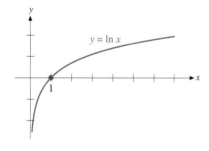

FIGURE 7.3.1 $\ln x$ is continuous, increasing, and concave down on the interval $x > 0$, with $\ln 1 = 0$.

Asymptotic Behavior

It is useful to know the relative sizes of $\ln x$ and x^r.

If x is large, the values of x^r, $r > 0$, are much larger than corresponding values of $\ln x$.

In fact,

$$\lim_{x \to \infty} \frac{\ln x}{x^r} = 0, \qquad r > 0. \tag{8}$$

We also have:

For small, nonzero x, the values of $|x|^{-r}$, $r > 0$, are much larger than the corresponding values of $\ln x$.

This means

$$\lim_{x \to 0} \frac{\ln |x|}{|x|^{-r}} = \lim_{x \to 0} |x|^r \ln |x| = 0, \qquad r > 0. \tag{9}$$

Limits of the types in (8) and (9) can be verified by using l'Hôpital's Rule. For example, if $r = 1/2$ in (8), we have

$$\lim_{x \to \infty} \frac{\ln x}{x^{1/2}} \underset{(\infty/\infty, \text{l'H})}{=} \lim_{x \to \infty} \frac{1/x}{1/(2x^{1/2})} = \lim_{x \to \infty} 2x^{-1/2} = 0. \tag{10}$$

If you remember the relative sizes of x^r and $\ln x$, then it is not necessary to apply l'Hôpital's Rule each time it is desired to evaluate a limit of these types.

The Number e

Since $\ln 2 \approx 0.69315 < 1$ and $\ln 4 = \ln 2^2 = 2 \ln 2 > 1$, the Intermediate Value Theorem for Continuous Functions tells us there is a number e with $2 < e < 4$ and

$$\ln e = 1.$$

See Figure 7.3.2. It can be shown that the number e is an irrational number with approximate value

$$e \approx 2.7182818.$$

(You should verify with a calculator that $\ln(2.7182818) \approx 1$.)

From (2) we have $\ln(e^r) = r \ln e = r$, so e^r is the unique solution of the equation $\ln x = r$. That is,

$$\ln x = r \text{ if and only if } x = e^r, r \text{ a rational number.} \tag{11}$$

We can use (11) to solve the equation $\ln x = r$ for x.

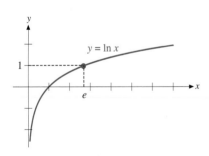

FIGURE 7.3.2 The Intermediate Value Theorem for Continuous Functions tells us there is a number e with $\ln e = 1$.

EXAMPLE 4

Sketch the graph of $f(x) = x \ln x$.

SOLUTION   We first note that $f(x)$ is defined only for $x > 0$. We see that $f(1) = 0$, $f(x)$ is positive for $x > 1$, and $f(x)$ is negative for $0 < x < 1$.

Using the product rule, we have

$$f'(x) = \frac{d}{dx}(x \ln x)$$

$$= (x)\left(\frac{1}{x}\right) + (\ln x)(1)$$

$$= 1 + \ln x.$$

The equation $f'(x) = 0$ implies $1 + \ln x = 0$, so $\ln x = -1$. Statement 11 then tells us $x = e^{-1}$. We see that $f(e^{-1}) = e^{-1}\ln(e^{-1}) = -e^{-1}$. The graph has a horizontal tangent at $(e^{-1}, -e^{-1})$. (Using a calculator, we have $e^{-1} \approx 0.37$.) The first derivative $f'(x)$ is negative and the function is decreasing for $0 < x < e^{-1}$; $f'(x)$ is positive and the function is increasing for $x > e^{-1}$; $f(e^{-1}) = -e^{-1}$ is the minimum value of $f(x)$.

We have

$$f''(x) = \frac{1}{x}, \qquad x > 0.$$

The second derivative $f''(x)$ is positive and the graph is concave up for $x > 0$.

We can determine the behavior of $f(x) = x \ln x$ as x approaches zero from the right either by remembering that the values of $1/x$ are much larger than the corresponding values of $\ln x$, or by applying (9) with $r = 1$, or by using l'Hôpital's Rule. In either case, we obtain

$$\lim_{x \to 0^+} f(x) = \lim_{x \to 0^+} x \ln x = 0.$$

This tells us that the graph approaches the point $(0, 0)$ as x approaches zero from the right. If we note that $f'(x) = 1 + \ln x$ approaches negative infinity as x approaches zero from the right, we could conclude that the graph becomes vertical at the origin. The graph is sketched in Figure 7.3.3. ■

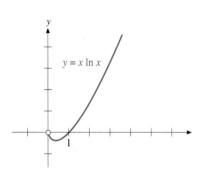

FIGURE 7.3.3 $f(x) = x \ln x$ has a local minimum at $(e^{-1}, -e^{-1})$. The graph approaches the origin as x approaches zero from the right.

EXERCISES 7.3

Evaluate the derivatives in Exercises 1–20.

1. $\frac{d}{dx}(\ln(2x+1))$

2. $\frac{d}{dx}(\ln(3x-2))$

3. $\frac{d}{dx}(\ln(\sec x + \tan x))$

4. $\frac{d}{dx}(\ln(\csc x))$

5. $\frac{d}{dx}\left(\ln \frac{1}{x^2}\right)$

6. $\frac{d}{dx}\left(\ln \frac{1}{1+x^2}\right)$

7. $\frac{d}{dx}(\ln(x\sqrt{x^2+1}))$

8. $\frac{d}{dx}(\ln\sqrt{x(x^2+1)})$

9. $\frac{d}{dx}\left(\ln \frac{2x+1}{3x-5}\right)$

10. $\frac{d}{dx}\left(\ln \frac{x}{x+2}\right)$

11. $\frac{d}{dx}\left(\ln \frac{x}{x^2+3}\right)$

12. $\frac{d}{dx}(\ln(\ln x))$

13. $\frac{d}{dx}((\ln x)^2)$

14. $\frac{d}{dx}(\sqrt{\ln x})$

15. $\frac{d}{dx}(x^3 \ln x)$

16. $\frac{d}{dx}\left(\frac{x^2}{\ln x}\right)$ **(Calculus Explorer & Tutor II, Transcendental Functions, 7.)**

17. $\frac{d}{dx}(\ln|\sec x|)$

18. $\frac{d}{dx}(\tan^{-1}(\ln 3x))$ **(Calculus Explorer & Tutor II, Transcendental Functions, 8.)**

19. $\frac{d}{dx}\left(\int_1^{x^2} \ln t \, dt\right)$ **20.** $\frac{d}{dx}\left(\int_x^{2x} \ln t \, dt\right)$

Use logarithmic differentiation to find dy/dx in Exercises 21–24.

21. $y = x(x-1)^{1/3}(x+1)^{2/3}$ **22.** $y = \frac{\sqrt{x}(x+1)}{x-1}$

23. $y = \dfrac{x(x-1)^{1/3}}{(x+1)^{2/3}}$ 　　**24.** $y = \dfrac{x\sqrt{x^2+1}}{(2x-1)^2}$

25. Find an equation of the line tangent to the graph of $y = x\ln x$ at the point where $x = e^2$.

26. (a) Find an equation of the line tangent to the graph of $y = \ln x$ at the point where $x = c$. (b) For what value of c does the tangent line contain the origin?

Evaluate the limits in Exercises 27–32.

27. $\lim\limits_{x\to 0^+} x\ln x$ 　　**28.** $\lim\limits_{x\to\infty} \dfrac{\ln x}{x}$

29. $\lim\limits_{x\to\infty} \dfrac{\ln x}{\sqrt{x^2+1}}$ 　　**30.** $\lim\limits_{x\to 0^+} \dfrac{\ln(2x)}{\ln x}$

31. $\lim\limits_{x\to\infty}(\sqrt{x}-\ln x)$ 　　**32.** $\lim\limits_{x\to 0^+}\left(\dfrac{1}{x}+\ln x\right)$

Sketch the graphs of the functions given in Exercises 33–38. Indicate relative extrema, points of inflection, the behavior as x approaches zero from the right, and the behavior as x approaches infinity.

33. $y = x^2\ln x$ 　　**34.** $y = \sqrt{x}\ln x$

35. $y = \dfrac{\ln x}{x}$ 　　**36.** $y = \dfrac{\ln x}{\sqrt{x}}$

37. $y = \dfrac{1}{x}+\ln x$ 　　**38.** $y = x-\ln x$

39. (a) Use Simpson's Rule with $n = 6$ to approximate $\ln 4$. (b) Find a bound of the error in (a).

40. (a) Use Simpson's Rule with $n = 4$ to approximate $\int_1^3 \ln x\,dx$. (b) Find a bound of the error in (a).

41. Use the definition of the logarithmic function as an integral and a change of variables to verify that $\ln x^r = r\ln x$, r a rational number.

42. Verify that $\ln(x/y) = \ln(x)-\ln(y)$.

7.4　$\int u^{-1}du$

Connections

Substitution and change of variable, 5.5.

In the previous section we obtained the formula

$$\frac{d}{dx}(\ln|x|) = \frac{1}{x}.$$

The corresponding integration formula is

$$\int \frac{1}{u}\,du = \ln|u| + C.$$

This formula is used to integrate a quotient whose numerator is a constant multiple of the derivative of its denominator. We will use the method of substitution that was introduced in Section 5.5.

Substitution

EXAMPLE 1

Evaluate $\int \dfrac{1}{1-2x}\,dx$.

SOLUTION　We choose $u = 1-2x$, so $du = -2dx$. Then

$$\int \frac{1}{1-2x}\,dx = \int \frac{1}{1-2x}\left(-\frac{1}{2}\right)(-2)\,dx$$
$$= \int \frac{1}{u}\left(-\frac{1}{2}\right)du$$
$$= -\frac{1}{2}\ln|u| + C = -\frac{1}{2}\ln|1-2x| + C. \quad\blacksquare$$

EXAMPLE 2

Evaluate $\int \dfrac{\cos x}{2 + \sin x}\, dx$.

SOLUTION   Setting $u = 2 + \sin x$, so $du = \cos x\, dx$, we obtain

$$\int \frac{\cos x}{2 + \sin x}\, dx = \int \frac{1}{u}\, du$$
$$= \ln |u| + C$$
$$= \ln(2 + \sin x) + C. \quad \blacksquare$$

Rewriting the Integrand

EXAMPLE 3

Evaluate $\int \dfrac{2x - 1}{x + 1}\, dx$.

SOLUTION   The integrand is a rational function that has the degree of its numerator equal to the degree of its denominator. Let us carry out the division in the integrand to obtain the sum of a constant and a rational function that has constant numerator and denominator $x + 1$. We have

$$
\begin{array}{r}
2 \\
x + 1 \overline{)2x - 1} \\
\underline{2x + 2} \\
-3
\end{array}
$$

Then

$$\int \frac{2x - 1}{x + 1}\, dx = \int \left(2 - \frac{3}{x + 1} \right) dx$$
$$= 2x - 3 \int \frac{1}{x + 1}\, dx.$$

Using the substitution $u = x + 1$, $du = dx$, to evaluate $\int \dfrac{1}{x + 1}\, dx$, we obtain

$$\int \frac{2x - 1}{x + 1}\, dx = 2x - 3 \ln |x + 1| + C. \quad \blacksquare$$

EXAMPLE 4

Evaluate $\int \tan x\, dx$.

SOLUTION   In this example, we write

$$\tan x = \frac{\sin x}{\cos x}$$

and then recognize that the numerator, $\sin x$, is a constant multiple of the derivative of the denominator, $\cos x$. We set $u = \cos x$, so $du = -\sin x \, dx$. Substitution then gives

$$\int \tan x \, dx = \int \frac{\sin x}{\cos x} \, dx = \int \frac{1}{\cos x}(-1)(-\sin x) \, dx$$
$$= \int \frac{1}{u}(-1) \, du$$
$$= -\ln |u| + C$$
$$= -\ln |\cos x| + C$$
$$= \ln |\cos x|^{-1} + C$$
$$= \ln |\sec x| + C. \quad \blacksquare$$

Recognizing Other Forms

EXAMPLE 5

Evaluate $\displaystyle\int \frac{x}{x^2 + 4} \, dx$.

SOLUTION Let us set $u = x^2 + 4$, so $du = 2x \, dx$. Noting that the integrand contains the essential factor x, we obtain

$$\int \frac{x}{x^2 + 4} \, dx = \int \frac{1}{x^2 + 4} \left(\frac{1}{2}\right)(2x) \, dx$$
$$= \int \frac{1}{u} \left(\frac{1}{2}\right) du$$
$$= \frac{1}{2} \ln |u| + C$$
$$= \frac{1}{2} \ln(x^2 + 4) + C. \quad \blacksquare$$

EXAMPLE 6

Evaluate $\displaystyle\int \frac{1}{x^2 + 4} \, dx$.

SOLUTION Even though the integrand has a factor in the denominator, this integral is not of the form $\int \frac{1}{u} \, du$. This is a basic inverse tangent integral. We have

$$\int \frac{1}{x^2 + 4} \, dx = \frac{1}{2} \tan^{-1} \left(\frac{x}{2}\right) + C. \quad \blacksquare$$

EXAMPLE 7

Evaluate $\displaystyle\int \frac{(\ln x)^2}{x} \, dx$.

SOLUTION This example is not of the form $\int u^{-1} du$, but we recognize that $1/x$ is the derivative of $\ln x$. We then choose

$$u = \ln x, \text{ so } du = \frac{1}{x} dx.$$

Substitution gives

$$\int \frac{(\ln x)^2}{x} dx = \int u^2 du = \frac{u^3}{3} + C = \frac{1}{3}(\ln x)^3 + C. \quad \blacksquare$$

EXAMPLE 8

Evaluate $\displaystyle\int \frac{2x}{\sqrt{x^2+1}} dx$.

SOLUTION   In this example, the numerator, $2x$, is not a constant multiple of the derivative of the denominator, $\sqrt{x^2+1}$. The formula $\int u^{-1} du$ does not apply. The substitution $u = x^2 + 1$, $du = 2x\, dx$, gives

$$\int \frac{2x}{\sqrt{x^2+1}} dx = \int \frac{1}{\sqrt{u}} du = \int u^{-1/2} du$$

$$= \frac{u^{1/2}}{1/2} + C$$

$$= 2\sqrt{x^2+1} + c. \quad \blacksquare$$

NOTE   *Do not confuse*

$$\int \frac{1}{u} du \quad and \quad \int \frac{1}{u^r} du, r \neq 1.$$

For $r \neq 1$, we have

$$\int \frac{1}{u^r} du = \int u^{-r} du = \frac{u^{-r+1}}{-r+1} + C.$$

This formula does not make sense for $r = 1$. When $r = 1$, we have

$$\int \frac{1}{u} du = \ln|u| + C.$$

We cannot use the formula

$$\int \frac{1}{u} du = \ln|u| + C$$

to evaluate a definite integral

$$\int_a^b \frac{1}{u} du$$

if $a \leq 0 \leq b$. The Fundamental Theorem does not apply in this case. (The Fundamental Theorem requires that the integrand be continuous on the interval $a \leq u \leq b$. The function $1/u$ becomes unbounded as u approaches zero, so it cannot be assigned any value at zero that makes it continuous on an interval that contains zero.)

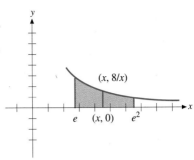

FIGURE 7.4.1 A variable vertical area strip has length $8/x$ and width dx.

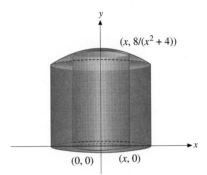

FIGURE 7.4.2 A typical thin cylindrical shell has radius x, height $8/(x^2+4)$, and thickness of cylindrical wall dx.

Applications

EXAMPLE 9

Find the area of the region bounded by $y = 8/x$, $y = 0$, $x = e$, and $x = e^2$. (Recall that $\ln e = 1$.)

SOLUTION The region is sketched in Figure 7.4.1. We have labeled the end-points of a variable vertical area strip in terms of the variable x. We see that a typical strip has height $8/x$ and width dx. The integral that gives the "sum" of the areas of the rectangles is

$$\text{Area} = \int_e^{e^2} \frac{8}{x}dx = 8 \ln|x| \Big]_e^{e^2}$$
$$= 8(\ln e^2 - \ln e) = 8(2 \ln e - \ln e) = 8(2 - 1) = 8. \quad \blacksquare$$

EXAMPLE 10

Find the volume of the solid obtained by revolving the region bounded by $y = 8/(x^2+4)$, $y = 0$, $x = 0$, and $x = 1$, about the y-axis.

SOLUTION The region is sketched in Figure 7.4.2. A variable vertical area strip generates a thin cylindrical shell when revolved about the y-axis. From the labeling of the sketch we see that the cylinder has a radius $= x$, height $= 8/(x^2+4)$, and thickness dx. We then have

$$\text{Volume} = \int_0^1 2\pi x \left(\frac{8}{x^2+4}\right) dx.$$

Using the substitution $u = x^2 + 4$, so $du = 2x\,dx$, $u(0) = 4$, and $u(1) = 5$, we obtain

$$\text{Volume} = \int_0^1 \frac{8\pi}{x^2+4} 2x\,dx$$
$$= \int_4^5 \frac{8\pi}{u} du$$
$$= 8\pi \ln|u| \Big]_4^5$$
$$= 8\pi \ln 5 - 8\pi \ln 4$$
$$= 8\pi \ln \frac{5}{4} \approx 5.6082. \quad \blacksquare$$

EXERCISES 7.4

Evaluate the integrals in Exercises 1–22.

1. $\displaystyle\int \frac{1}{2x+1}dx$

2. $\displaystyle\int \frac{1}{3x+2}dx$

3. $\displaystyle\int \frac{x}{x^2+1}dx$

4. $\displaystyle\int \frac{x^2}{x^3+1}dx$

5. $\displaystyle\int \frac{x}{x-3}dx$

6. $\displaystyle\int \frac{x^2}{x+1}dx$

7. $\displaystyle\int \frac{\sin x}{1+\cos x}dx$

8. $\displaystyle\int \frac{\sin x - \cos x}{\sin x + \cos x}dx$

9. $\displaystyle\int \frac{\sec x \tan x + \sec^2 x}{\sec x + \tan x}dx$

10. $\displaystyle\int \cot x\,dx$

11. $\displaystyle\int \frac{1}{x \ln x}\, dx$

12. $\displaystyle\int \frac{1}{x(\ln x)^2}\, dx$

13. $\displaystyle\int \frac{x^2}{x^2 + 1}\, dx$

14. $\displaystyle\int \frac{x^3}{x^2 + 1}\, dx$

15. $\displaystyle\int \frac{1}{x^{1/3}(1 + x^{2/3})}\, dx$

16. $\displaystyle\int \frac{1}{\sqrt{x}(1 + \sqrt{x})}\, dx$

17. $\displaystyle\int \frac{1}{x^2 + 1}\, dx$

18. $\displaystyle\int \frac{x}{(x^2 + 1)^2}\, dx$

19. $\displaystyle\int \frac{1}{\sqrt{4 - x^2}}\, dx$

20. $\displaystyle\int \frac{x}{\sqrt{4 - x^2}}\, dx$

21. $\displaystyle\int \frac{x - 1}{x^2 + 4}\, dx$

22. $\displaystyle\int \frac{2x + 1}{x^2 + 1}\, dx$

Evaluate the definite integrals in Exercises 23–26.

23. $\displaystyle\int_1^3 \frac{1}{x}\, dx$

24. $\displaystyle\int_0^1 \frac{1}{2x + 1}\, dx$

25. $\displaystyle\int_0^4 \frac{x}{x^2 + 9}\, dx$

26. $\displaystyle\int_0^{\pi/8} \tan 2x\, dx$

27. Find the area of the region bounded by

$$y = \frac{1}{x}, \quad y = \frac{1}{x + 1}, \quad x = 1, \quad x = 2.$$

28. Find the area of the region bounded by

$$y = \frac{1}{x}, \quad y = 0, \quad x = a, \quad x = 2a,$$

where a is a positive constant.

29. Find the volume of the solid generated by revolving the region bounded by

$$y = \frac{2}{x + 1}, \quad y = 0, \quad x = 0, \quad x = 1,$$

about the **(a)** x-axis, **(b)** y-axis.

30. Find the volume of the solid generated by revolving the region bounded by

$$y = \frac{1}{x}, \quad y = 0, \quad x = 1, \quad x = 2,$$

about the vertical line **(a)** $x = 1$, **(b)** $x = 2$.

31. For what value of c does the vertical line $x = c$ divide the region under the curve $y = 1/x$ between $x = 1$ and $x = 9$ into equal areas?

32. A gas within a piston-cylinder device has initial volume $0.020\,\text{m}^3$ and pressure $10^5\,\text{N/m}^2$. Determine the work done as the gas expands to a volume of $0.040\,\text{m}^3$ with **(a)** the pressure constant and **(b)** $PV = $ constant. $(1\,\text{J} = 1\,\text{N} \cdot \text{m}.)$

7.5 EXPONENTIAL FUNCTIONS; INDETERMINATE FORMS 0^0, 1^∞, ∞^0

Connections

Limits, 2.5.

Graphing, asymptotes, 4.1–4.

Concept of inverse function, 7.1.

Properties of logarithmic functions, 7.3.

We use the natural logarithmic function to define exponential functions and establish their properties.

Definition and Properties

We know that the derivative of $\ln x$ is positive for $x > 0$, so $\ln x$ has an inverse function. Let us denote the inverse function of $\ln x$ by $\exp x$. The range of $\ln x$ is all real numbers, so the domain of $\exp x$ is all real numbers. The domain of $\ln x$ is all positive real numbers, so the range of $\exp x$ is all positive real numbers. That is, $\exp x$ is defined and positive for all real x. We have

$$\exp(\ln x) = x \quad \text{for all } x > 0$$

and

$$\ln(\exp y) = y \quad \text{for all } y.$$

Properties of the **exponential function** $y = \exp x$ can be obtained from what we know of the logarithmic function. For example:

$$\ln 1 = 0 \text{ implies } \exp 0 = \exp(\ln 1) = 1.$$
$$\ln e = 1 \text{ implies } \exp 1 = \exp(\ln e) = e.$$
$$\lim_{x \to 0^+} \ln x = -\infty \text{ implies } \lim_{x \to -\infty} \exp x = 0.$$

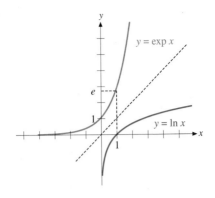

FIGURE 7.5.1 The exponential function, exp x, is the inverse of the natural logarithmic function, ln x.

$$\lim_{x \to \infty} \ln x = \infty \text{ implies } \lim_{x \to \infty} \exp x = \infty.$$

See Figure 7.5.1, where the graphs of ln x and exp x are sketched.

For $a > 0$ and r rational, the expression a^r has been defined in terms of powers and roots of a; a^x has not been defined for irrational x. We will see that the exponential function exp x can be used to define a^x for all x. We first consider the case with base $a = e$, where $\ln e = 1$.

If r is rational, we know that $\ln e^r = r \ln e = r$. It follows that

$$\exp r = \exp(\ln e^r) = e^r.$$

Since the value of e^x agrees with that of exp x for rational x, we *define* e^x to be exp x for every value of x and use the simpler notation e^x in place of exp x. The relations between ln x and its inverse function can then be written

$$\ln e^x = x \quad \text{for all } x,$$
$$e^{\ln x} = x, \quad x > 0,$$
$$\text{and}$$
$$y = \ln x \text{ if and only if } x = e^y.$$

For $a > 0$ and any number x, we define a^x by the equation

$$\boxed{a^x = e^{x \ln a}.}$$

If x is rational, this definition agrees with the previous definition of a^x in terms of powers and roots. That is, from Section 7.3 we know that $\ln a^x = x \ln a$, so $a^x = e^{x \ln a}$.

When we showed that $\ln a^r = r \ln a$ in Section 7.3, it was necessary that r be a rational number because a^r was defined only for rational numbers r. We now have

$$\ln a^b = \ln e^{b \ln a} = b \ln a \quad \text{for all values of } b.$$

The following properties of e^x hold for all real number exponents:

$$e^a e^b = e^{a+b}$$
$$(e^a)^b = e^{ab}.$$
$$e^{-a} = \frac{1}{e^a}.$$
$$\frac{e^a}{e^b} = e^{a-b}.$$

These are the familiar rules of exponents. The only difference is that the expressions are now defined for all real values of the exponents; the rules are no longer restricted to rational exponents. These properties follow from corresponding properties of the logarithmic function. For example,

$$\ln(e^a e^b) = \ln e^a + \ln e^b = a \ln e + b \ln e = a + b,$$

so

$$e^{a+b} = e^{\ln(e^a e^b)} = e^a e^b.$$

The Derivative of e^x

We know that e^x is differentiable, since e^x is the inverse function of a function that has positive derivative. Let us obtain a formula for the derivative. We set

$$y = e^x, \text{ so } \ln y = x.$$

Using the Chain Rule to differentiate the latter equation with respect to x, we obtain

$$\left(\frac{1}{y}\right)\left(\frac{dy}{dx}\right) = 1, \text{ so } \frac{dy}{dx} = y.$$

Since $y = e^x$, substitution gives the formula

$$\frac{d}{dx}(e^x) = e^x.$$

That is, the function e^x has the extraordinary property of being its own derivative! The Chain Rule version of the formula for the differentiation of the exponential function is

$$\frac{d}{dx}(e^{\text{Function}}) = e^{\text{Function}}\frac{d}{dx}(\text{Function}).$$

EXAMPLE 1

(a) $\dfrac{d}{dx}(e^{3x}) = e^{3x}(3),$

(b) $\dfrac{d}{dx}(e^{x/2}) = e^{x/2}\left(\dfrac{1}{2}\right),$

(c) $\dfrac{d}{dx}(e^{x^2}) = e^{x^2}(2x),$

(d) $\dfrac{d}{dx}(xe^x - e^x) = ((x)(e^x) + (e^x)(1)) - e^x = xe^x.$ ∎

The usual formula for differentiation of a power function holds for all *constant* powers of the variable of differentiation. It is no longer necessary to require that the power be a rational number.

$$\frac{d}{dx}(x^\alpha) = \alpha x^{\alpha-1}.$$

Proof

$$\frac{d}{dx}(x^\alpha) = \frac{d}{dx}(e^{\alpha \ln x}) = e^{\alpha \ln x}\left(\alpha\frac{1}{x}\right) = \alpha x^\alpha x^{-1} = \alpha x^{\alpha-1}. \quad ∎$$

The Chain Rule version of the power rule is

$$\frac{d}{dx}((\text{Function})^\alpha) = \alpha(\text{Function})^{\alpha-1}\frac{d}{dx}(\text{Function}), \quad \alpha \text{ a constant.}$$

Differentiating f^g

We can use the Chain Rule version of the formula $\frac{d}{dx}(x^\alpha) = \alpha x^{\alpha-1}$ to differentiate a function raised to a *constant* power. We cannot use this formula to differentiate a function raised to a *variable* power. If f and g are functions, we could differentiate f^g by using logarithmic differentiation. However, the following procedure is more direct.

To differentiate (FUNCTION)^{Function}

- Write

$$(\text{FUNCTION})^{\text{Function}} = e^{(\text{Function})(\ln \text{FUNCTION})}.$$

- Use the formula $\frac{d}{dx}(e^u) = e^u \frac{du}{dx}$ to differentiate the exponential expression obtained. (Note that the Product Rule and the Chain Rule are needed to evaluate the product (Function)($\ln$ FUNCTION).)

EXAMPLE 2

$$\frac{d}{dx}(x^x) = \frac{d}{dx}(e^{x \ln x})$$

$$= e^{x \ln x}\left((x)\left(\frac{1}{x}\right) + (\ln x)(1)\right)$$

$$= x^x(1 + \ln x). \quad \blacksquare$$

Integration Formula

It follows from the differentiation formula $\frac{d}{dx}(e^x) = e^x$ that

$$\int e^u \, du = e^u + C.$$

EXAMPLE 3

Evaluate $\int e^{3x} \, dx$.

SOLUTION Setting $u = 3x$, so $du = 3dx$, we have

$$\int e^{3x} \, dx = \int e^{3x}\left(\frac{1}{3}\right) 3dx$$

$$= \int e^u \left(\frac{1}{3}\right) du$$

$$= \frac{1}{3}e^u + C$$

$$= \frac{1}{3}e^{3x} + C. \quad \blacksquare$$

Integrals of the type of Example 3 can also be evaluated by the *guess and check* method. That is, we know that $\int e^{3x} \, dx = ke^{3x} + C$ for some constant

k. We can *guess* a value of k and then *check* by differentiating mentally. Thus, if we guess $k = \dfrac{1}{3}$, we have

$$\int e^{3x}\,dx = \frac{1}{3}e^{3x} + C, \text{ because } \frac{d}{dx}\left(\frac{1}{3}e^{3x}\right) = \frac{1}{3}e^{3x}(3) = e^{3x}.$$

If differentiation does not give the integrand, we try a different value of k.

EXAMPLE 4

Evaluate $\int x e^{-x^2}\,dx$.

SOLUTION Using $u = -x^2$, so $du = -2x\,dx$, we have

$$\int x e^{-x^2}\,dx = \int e^{-x^2}\left(-\frac{1}{2}\right)(-2x)\,dx$$
$$= \int e^{u}\left(-\frac{1}{2}\right)du$$
$$= -\frac{1}{2}e^{u} + C$$
$$= -\frac{1}{2}e^{-x^2} + C. \quad \blacksquare$$

EXAMPLE 5

Evaluate $\int \cos x\, e^{\sin x}\,dx$.

SOLUTION Set $u = \sin x$, so $du = \cos x\,dx$. Then

$$\int \cos x\, e^{\sin x}\,dx = \int e^{u}\,du$$
$$= e^{u} + C$$
$$= e^{\sin x} + C. \quad \blacksquare$$

EXAMPLE 6

Evaluate $\displaystyle\int \frac{e^{x}}{1 + e^{x}}\,dx$.

SOLUTION In this example, we recognize e^{x} as the derivative of $1 + e^{x}$. We then set $u = 1 + e^{x}$, so $du = e^{x}\,dx$ and

$$\int \frac{e^{x}}{1 + e^{x}}\,dx = \int \frac{1}{u}\,du$$
$$= \ln|u| + C$$
$$= \ln(1 + e^{x}) + C. \quad \blacksquare$$

Asymptotic Behavior

We have seen in Section 7.3 that $\ln x$ is dominated by any function x^α, $\alpha > 0$, as x approaches infinity. Similarly, x^α is dominated by e^x as x approaches infinity. We say:

If x is large, the values of e^x are much larger than the corresponding values of x^α, $\alpha > 0$.

That is,

$$\lim_{x \to \infty} \frac{x^\alpha}{e^x} = 0, \quad \alpha > 0. \tag{1}$$

Also:

If x is large, the values $e^{-|x|}$ are much smaller than the corresponding values of $|x|^{-\alpha}$, $\alpha > 0$.

This means

$$\lim_{x \to -\infty} \frac{e^x}{|x|^{-\alpha}} = \lim_{x \to -\infty} |x|^\alpha e^x = 0, \quad \alpha > 0. \tag{2}$$

Formulas 1 and 2 can be verified formally by using l'Hôpital's Rule. For example, in case $\alpha = 2$ in Formula 1, we have

$$\lim_{x \to \infty} \frac{x^2}{e^x} \underset{(\infty/\infty,\,\text{l'H})}{=} \lim_{x \to \infty} \frac{2x}{e^x} \underset{(\infty/\infty,\,\text{l'H})}{=} \lim_{x \to \infty} \frac{2}{e^x} = 0.$$

Note that it is not necessary to use l'Hôpital's Rule to evaluate limits of types (1) and (2) if you remember the relative size of the functions involved.

EXAMPLE 7

Sketch the graph of $y = xe^{-x}$.

SOLUTION   We will investigate the sign of y and the first two derivatives.

$$y = xe^{-x},$$
$$y' = (x)(e^{-x})(-1) + (e^{-x})(1) = e^{-x}(1 - x),$$
$$y'' = (e^{-x})(-1) + (1 - x)(e^{-x})(-1) = e^{-x}(x - 2).$$

We know that $e^{-x} > 0$ for all x. Hence, $y = 0$ implies $x = 0$. The only intercept is $(0, 0)$. We see that $y > 0$ (graph above x-axis) whenever $x > 0$ and $y < 0$ (graph below x-axis) whenever $x < 0$.

The equation $y' = 0$ implies $x = 1$ and $y = (1)e^{-(1)} = 1/e$. The graph has a horizontal tangent at $(1, 1/e)$. We see that $y' > 0$ (y increasing) whenever $x < 1$; we also see that $y' < 0$ (y decreasing) whenever $x > 1$. The graph has a local maximum at $(1, 1/e)$.

The equation $y'' = 0$ implies $x = 2$ and $y = (2)e^{-(2)} = 2/e^2$. We see that $y'' > 0$ (y concave upward) whenever $x > 2$; we also see that $y'' < 0$ (y concave downward) whenever $x < 2$. The point $(2, 2/e^2)$ is a point of inflection.

We can determine the behavior of xe^{-x} as x approaches infinity by writing xe^{-x} as $\dfrac{x}{e^x}$ and then either remembering that the values of e^x are much larger

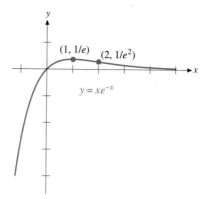

FIGURE 7.5.2 $f(x) = xe^{-x}$ has a local maximum at $(1, 1/e)$ and point of inflection $(2, 1/e^2)$. The graph approaches the horizontal asymptote $y = 0$ as x approaches infinity.

than the corresponding values of x, or by applying (1) with $\alpha = 1$, or by using l'Hôpital's Rule to evaluate the limit. In either case, we obtain

$$\lim_{x \to \infty} xe^{-x} = \lim_{x \to \infty} \frac{x}{e^x} = 0.$$

This shows that the graph approaches the horizontal asymptote $y = 0$ as x approaches infinity. As x approaches negative infinity, xe^{-x} approaches negative infinity. The graph is sketched in Figure 7.5.2. ∎

Indeterminate Forms 0^0, 1^∞, ∞^0

Depending on the values of the limits of f and g at a point, the function f^g is said to have **indeterminate form 0^0, 1^∞, or ∞^0** at the point. Generally, $\ln f^g = g \ln f$ then has indeterminate form $0 \cdot \infty$ at the point. For example, if $\lim f(x) = \lim g(x) = 0$, then we have $\ln f^g = g \ln f$, where $\ln f$ approaches negative infinity. Thus, $g \ln f$ has indeterminate form $0 \cdot \infty$.

NOTE 1^∞ *doesn't look like an indeterminate form, because 1 raised to any real number power is 1. However, you can verify on your calculator that x^y can have either very small or very large values as x approaches one, $x \neq 1$, and y approaches infinity.*

We can evaluate the limit of an indeterminate form f^g by writing f^g in the form $e^{g \ln f}$ and then evaluating the limit of the exponent. Since the exponential function is continuous, we have

$$\lim f^g = \lim e^{g \ln f} = e^{\lim(g \ln f)}. \qquad (3)$$

L'Hôpital's Rule may be needed to evaluate the indeterminate form $0 \cdot \infty$ in the exponent.

EXAMPLE 8

Evaluate $\lim_{x \to \infty} \left(1 + \dfrac{1}{x}\right)^x$.

SOLUTION The expression $\left(1 + \dfrac{1}{x}\right)^x$ has indeterminate form 1^∞ at infinity. Following the form in (3), we have

$$\lim_{x \to \infty} \left(1 + \frac{1}{x}\right)^x = \lim_{x \to \infty} e^{x \ln\left(1 + \frac{1}{x}\right)} = e^{\lim_{x \to \infty} x \ln\left(1 + \frac{1}{x}\right)}.$$

We now apply l'Hôpital's Rule to evaluate the limit in the exponent. We obtain

$$\lim_{x \to \infty} x \ln\left(1 + \frac{1}{x}\right) \underset{\text{(Rewriting)}}{=} \lim_{x \to \infty} \frac{\ln\left(1 + \dfrac{1}{x}\right)}{\dfrac{1}{x}}$$

$$\underset{(0/0, \text{l'H})}{=} \lim_{x \to \infty} \frac{\dfrac{1}{1 + (1/x)}\left(-\dfrac{1}{x^2}\right)}{-\dfrac{1}{x^2}}$$

$$\underset{\text{(Simplifying)}}{=} \lim_{x \to \infty} \frac{1}{1 + \dfrac{1}{x}} = 1,$$

so the value of the limit in the exponent is one. Combining results, we have

$$\lim_{x \to \infty} \left(1 + \frac{1}{x}\right)^x = e^{\lim_{x \to \infty} x \ln\left(1 + \frac{1}{x}\right)} = e^1 = e. \quad \blacksquare$$

Let us restate and emphasize the result of Example 8.

$$\lim_{x \to \infty} \left(1 + \frac{1}{x}\right)^x = e. \tag{4}$$

Statement 4 gives us a way to evaluate approximate values of e, the base of the natural logarithmic function. You can use your calculator to verify that

$$\left(1 + \frac{1}{x}\right)^x \approx 2.7182818 \quad \text{for large values of } x.$$

Bases Other than e

For some applications it is convenient to use logarithms and exponentials to a base other than e. We have already defined exponentials a^x. Namely,

$$a^x = e^{x \ln a}, \quad a > 0.$$

For $a > 0$, $a \neq 1$, the function a^x has derivative

$$\frac{d}{dx}(a^x) = \frac{d}{dx}(e^{x \ln a}) = e^{x \ln a}(\ln a) = (\ln a)a^x.$$

We see that the derivative is positive if $a > 1$ and negative if $0 < a < 1$. In either case, a^x has an inverse function, which we denote by $\log_a x$. Thus,

$$y = a^x \text{ if and only if } x = \log_a y.$$

The function $\log_a x$ is the familiar function that is discussed in algebra classes. The function $\log_{10} x$ is called the **common logarithm** of x; $\log_e x = \ln x$ is the **natural logarithm** of x. Logarithms to any base are related to natural logarithms by the formula

$$\log_a x = \frac{\ln x}{\ln a}.$$

Proof Since

$$a^{(\ln x)/(\ln a)} = e^{[(\ln x)/(\ln a)] \ln a} = e^{\ln x} = x,$$

we have

$$\log_a x = \log_a a^{(\ln x)/(\ln a)} = \frac{\ln x}{\ln a}. \quad \blacksquare$$

We have

$$\frac{d}{dx}(\log_a x) = \left(\frac{1}{\ln a}\right)\left(\frac{1}{x}\right).$$

The differentiation formula

$$\frac{d}{dx}(a^x) = (\ln a)a^x$$

corresponds to the integration formula

$$\int a^x dx = \frac{a^x}{\ln a} + C.$$

Note that the factor $\ln a$ appears in the above differentiation and integration formulas. The formulas are simpler for the base e, because $\ln e = 1$. That is why the base e is "natural" for calculus.

EXERCISES 7.5

Evaluate the derivatives in Exercises 1–20.

1. $\dfrac{d}{dx}(e^{-x^3})$ **(Calculus Explorer & Tutor II, Transcendental Functions, 6.)**

2. $\dfrac{d}{dx}(e^{\sqrt{x}})$

3. $\dfrac{d}{dx}(e^{2x-1})$

4. $\dfrac{d}{dx}(e^{1-3x})$

5. $\dfrac{d}{dx}(e^{\tan x})$

6. $\dfrac{d}{dx}(e^{\sin x})$ **(Calculus Explorer & Tutor II, Transcendental Functions, 5.)**

7. $\dfrac{d}{dx}(xe^{-x})$

8. $\dfrac{d}{dx}((\sin x)(e^{-x}))$

9. $\dfrac{d}{dx}\left(\dfrac{e^{-x}}{x}\right)$

10. $\dfrac{d}{dx}\left(\dfrac{e^{2x}}{x^2}\right)$

11. $\dfrac{d}{dx}(\ln(1+e^x))$

12. $\dfrac{d}{dx}(\tan^{-1}(e^x))$

13. $\dfrac{d}{dx}(x^{\sqrt{x}})$ **(Calculus Explorer & Tutor II, Transcendental Functions, 11.)**

14. $\dfrac{d}{dx}(x^{\ln x})$

15. $\dfrac{d}{dx}(10^x)$

16. $\dfrac{d}{dx}(2^{5x})$ **(Calculus Explorer & Tutor II, Transcendental Functions, 10.)**

17. $\dfrac{d}{dx}(10^{-\sqrt{x}})$

18. $\dfrac{d}{dx}(3^{0.02x})$

19. $\dfrac{d}{dx}(\log_{10} x)$

20. $\dfrac{d}{dx}(\log_{10}\sqrt{x})$

Evaluate the integrals in Exercises 21–34.

21. $\displaystyle\int e^{-x}\, dx$

22. $\displaystyle\int e^{2x}\, dx$

23. $\displaystyle\int \frac{e^{\sqrt{x}}}{\sqrt{x}}\, dx$

24. $\displaystyle\int e^{x^{1/3}} x^{-2/3}\, dx$

25. $\displaystyle\int e^x\sqrt{e^x+1}\, dx$

26. $\displaystyle\int \frac{e^x}{\sqrt{1-e^{2x}}}\, dx$

27. $\displaystyle\int \frac{e^x}{1+e^{2x}}\, dx$ **(Calculus Explorer & Tutor II, Transcendental Functions, 15.)**

28. $\displaystyle\int \frac{e^x}{(1+e^x)^2}\, dx$

29. $\displaystyle\int \frac{e^{2x}}{1+e^{2x}}\, dx$

30. $\displaystyle\int e^x e^{e^x}\, dx$

31. $\displaystyle\int 2^{-x}\, dx$

32. $\displaystyle\int 10^x\, dx$

33. $\displaystyle\int x6^{-x^2}\, dx$ **(Calculus Explorer & Tutor II, Transcendental Functions, 13.)**

34. $\displaystyle\int \frac{\log_{10} x}{x}\, dx$ **(Calculus Explorer & Tutor II, Transcendental Functions, 14.)**

Evaluate the definite integrals in Exercises 35–38.

35. $\displaystyle\int_0^2 e^{x/2}\, dx$

36. $\displaystyle\int_0^{1/3} e^{3x}\, dx$

37. $\displaystyle\int_0^1 2^x\, dx$

38. $\displaystyle\int_0^1 \frac{e^x+e^{-x}}{2}\, dx$

Evaluate the limits in Exercises 39–54.

39. $\displaystyle\lim_{x\to\infty} \frac{\ln x}{\sqrt{x}}$

40. $\displaystyle\lim_{x\to\infty} \frac{x^{10}}{e^{x/2}}$

41. $\displaystyle\lim_{x\to 0^+} \sqrt{x}\ln x$

42. $\displaystyle\lim_{x\to-\infty} x^2 e^x$

43. $\displaystyle\lim_{x\to 0} \frac{e^{2x}-1-2x}{x^2}$

44. $\displaystyle\lim_{x\to\infty} x\ln(1+e^{-x})$

45. $\displaystyle\lim_{x\to 0^+} x^x$

46. $\displaystyle\lim_{x\to 0^+} x^{1/\ln x}$

47. $\displaystyle\lim_{x\to 0^+} (1-\sqrt{x})^{1/x}$

48. $\displaystyle\lim_{x\to 0^+} (1+2x)^{1/x}$

49. $\displaystyle\lim_{x\to 0^+} x^{1/x}$

50. $\displaystyle\lim_{x\to\infty} x^x$

51. $\displaystyle\lim_{x\to\infty} x^{1/x}$

52. $\displaystyle\lim_{x\to\infty} (1+2x)^{1/x}$

53. $\displaystyle\lim_{x\to\infty} x^{-1/\ln x}$

54. $\displaystyle\lim_{x\to\infty} \left(\frac{x}{x+1}\right)^{\ln x}$

Sketch the graphs of the functions in Exercises 55–60. Indicate asymptotes, local extrema, and points of inflection.

55. $y = xe^x$

56. $y = x^2 e^x$

57. $y = \dfrac{e^x - e^{-x}}{2}$

58. $y = \dfrac{e^x + e^{-x}}{2}$

59. $y = e^x/x$

60. $y = 1/(e^x - 1)$

61. Find the area of the region bounded by $y = e^{-x}$, $y = 0$, $x = 0$, and $x = 1$.

62. Find the volume of the solid obtained by revolving the region bounded by $y = e^x$, $y = 0$, $x = 0$, and $x = 1$, about the x-axis.

 Evaluate the expressions in Exercises 63–66.

63. $\log_{10} 0.02$

64. $\log_{10} 20$

65. $\log_5 2$

66. $\log_2 5$

67. Express $3 \cdot 2^{-5}$ as a power of e.

68. Express $2 \cdot 10^{35}$ as a power of e. $\longrightarrow$ $\ln y = \ln\left(2\cdot 10^{35}\right)$

69. Express $e^{-\frac{\ln 2}{4}t}$ as a power of 2. $\ln y = \ln 2 + 35 \ln 10$

70. Express $e^{(2\ln 2)t}$ as a power of 2. $\ln 2 + 35 \ln 10$

71. Express e^{-3t} as a power of 10. $y = e^{\underline{\ln 2 + 35 \ln 10}}$

72. Express e^{2t} as a power of 10.

Find the fourth-order Taylor polynomials about $c = 0$ of the functions given in Exercises 73–74. Taylor polynomials were defined in Section 3.10.

73. $f(x) = e^x$

74. $f(x) = \ln(1 - x)^{-1}$

Use the corresponding property of $\ln x$ to verify the equations in Exercises 75–76.

75. $(e^x)^y = e^{xy}$

76. $\dfrac{e^x}{e^y} = e^{x-y}$

77. $f(x) = \dfrac{1}{1 + e^{1/x}}$.

 (a) Show that $f(x) + f(-x) = 1$.

 (b) Evaluate $\displaystyle\lim_{x\to 0^+} f(x)$.

 (c) Evaluate $\displaystyle\lim_{x\to\infty} f(x)$.

 (d) Use the above information to sketch the graph of f.

78. The **Henderson-Hasselbach equation** for computation of the pH of human blood in terms of the ratio $r = [HCO_3^-]/[H_2CO_3]$ of bicarbonate and carbonic acid is

$$\text{pH} = 6.1 + \log_{10} r.$$

The normal pH of blood for adults ranges from 7.35 to 7.45. A pH of over 7.0 is alkaline, so blood is slightly alkaline. A pH of either less than 7.0 or over 9.9 is usually fatal. **(a)** What ratio r corresponds to a pH of 7.4? **(b)** What is the rate of change of pH with respect to r when the pH is 7.4?

In Exercises 79–80, evaluate $\displaystyle\lim_{x\to\infty} f(x)$ and use your calculator to evaluate $f(10), f(100), f(1000),$ and $f(10,000)$.

79. $f(x) = \left(1 + \dfrac{1}{x}\right)^x$

80. $f(x) = \left(1 + \dfrac{1}{x^2}\right)^x$

7.6 EXPONENTIAL GROWTH AND DECAY

Connections

Exponential function, 7.5.

In many natural processes the rate of change of a physical quantity is proportional to the current amount of the quantity. Thus, we are led to find a function y that satisfies the equation

$$y' = ky, \quad k \text{ a constant.}$$

This equation is called a **differential equation.** A function y that satisfies the equation in some interval is called a **solution** of the differential equation. In this section we will solve this differential equation and see how it is related to several scientific problems.

The Differential Equation $y' = ky$

Let us see that we can solve the differential equation $y' = ky$ by integration. First, we note that any solution y must be continuous, since it is differentiable. Since y must be continuous and $y' = ky$, y' must also be continuous.

If y is a solution of $y' = ky$, then $y'(t) = ky(t)$ for t in an interval. In case there is some point t_0 in the interval for which $y(t_0)$ is not zero, then the continuous function y must be nonzero on some interval, so we can write

$$\frac{y'(t)}{y(t)} = k.$$

Integrating with respect to t, we obtain

$$\int \frac{y'(t)}{y(t)}\, dt = \int k\, dt.$$

Using the substitution $y = y(t)$, so $dy = y'(t)\, dt$, to evaluate the integral on the left, we have

$$\int \frac{1}{y}\, dy = \int k\, dt,$$

$$\ln|y| = kt + c, \quad c \text{ a constant.}$$

Substitution of $y(t)$ for y then gives

$$\ln|y(t)| = kt + c,$$
$$|y(t)| = e^{kt+c},$$
$$|y(t)| = e^c e^{kt}.$$

This shows that $y(t)$ is never zero. Since continuous functions cannot change sign without becoming zero, we must have

$$y(t) = Ce^{kt},$$

where either $C = e^c$ or $C = -e^c$. This shows that if there is any point t_0 in the interval with $y(t_0) \neq 0$, then any solution of $y' = ky$ must be of the form $y(t) = Ce^{kt}$. If there are no points on the interval with $y(t) \neq 0$, we have $y(t) = Ce^{kt}$ with $C = 0$. Thus, any solution of $y' = ky$ must be of the form $y(t) = Ce^{kt}$. It is easy to check that any function of this form is a solution of $y' = ky$. That is, $y(t) = Ce^{kt}$ implies $y'(t) = Ce^{kt}(k)$, so $y'(t) = ky(t)$ for all t.

We have seen that for each constant C, $y(t) = Ce^{kt}$ is a solution of the differential equation $y' = ky$, and that all solutions are of this form. Thus, Ce^{kt} represents all solutions of the differential equation. The collection of functions Ce^{kt} is called the **general solution** of the differential equation. We can find a **particular solution** by using additional information to determine a value of the constant C. In particular, if it is required that a solution of the differential equation $y' = ky$ satisfy the condition $y(0) = y_0$, substitution into the equation $y(t) = Ce^{kt}$ gives

$$y(0) = Ce^{k(0)},$$
$$y_0 = C, \text{ so } y(t) = y_0 e^{kt}.$$

The problem of finding a function y that satisfies the differential equation $y' = ky$ and the **initial condition** $y(0) = y_0$ is called an **initial value problem**. It is convenient to memorize the solution of the above initial value problem, which we restate.

$$y' = ky, \quad y(0) = y_0 \quad \text{has solution} \quad y(t) = y_0 e^{kt}. \tag{1}$$

EXAMPLE 1

The initial value problem

$$y' = -2y, \quad y(0) = 3$$

has solution

$$y(t) = 3e^{-2t}. \quad \blacksquare$$

Let us now see how the equation $y' = ky$ and its solution, $y = Ce^{kt}$, occur in some physical problems.

Exponential Growth

Under certain conditions, such as when they are allowed enough food and space, bacteria colonies grow at a rate that is directly proportional to the population of the colony. If $P(t)$ denotes the population of the colony at a time t, this means $P' = kP$, so Statement 1 implies

$$P(t) = P_0 e^{kt}. \tag{2}$$

The initial population P_0 and the constant k can be determined if we know the population of the colony at two different times.

EXAMPLE 2

The population of a certain colony of bacteria is known to increase at a rate that is directly proportional to its size. Initially, when $t = 0$, the population was 120. Three hours later the population was 200. What was the population when $t = 2$, two hours after the initial time?

SOLUTION   We know from Formula 2 that $P(t) = P_0 e^{kt}$ for some constants P_0 and k. We are given the initial population $P_0 = 120$. Also, we know that $P(3) = 200$. That is,

$$120 e^{k(3)} = 200.$$

Solving this equation for k, we have

$$e^{3k} = 5/3,$$
$$3k = \ln(5/3),$$
$$k = \frac{\ln(5/3)}{3}.$$

We then have

$$P(t) = 120 e^{\{[\ln(5/3)]/3\}t}.$$

We were asked to find the population when $t = 2$. This is

$$P(2) = 120 e^{\{[\ln(5/3)]/3\}(2)} \approx 169.$$

(We have used a calculator to find the approximate value.) The graph of $P(t)$ is given in Figure 7.6.1. $\blacksquare$

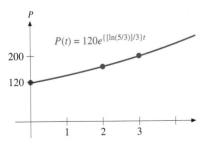

FIGURE 7.6.1   The population grows exponentially.

We can use properties of the logarithm to express the formula for the population in Example 2 in a different form. That is,

$$
\begin{aligned}
P(t) &= 120e^{\{[\ln(5/3)]/3\}t} \\
&= 120e^{(t/3)\ln(5/3)} \\
&= 120e^{\ln[(5/3)^{t/3}]} \\
&= 120\left(\frac{5}{3}\right)^{t/3}.
\end{aligned}
$$

Radioactive Decay

Radioactive material decays at a rate that is directly proportional to the current amount of material. This means $A' = kA$, where $A(t)$ is the amount of material at time t. We then know from Statement 1 that $A(t) = A_0 e^{kt}$ for some constant k. (Since the amount is decreasing, k will be negative.)

Suppose that the amount of a particular radioactive substance at time t is given by the equation $A(t) = A_0 e^{kt}$. After some time t_1, half of the initial amount will have decayed, so half will remain. This means

$$
A(t_1) = \frac{1}{2}A_0.
$$

Solving this equation for t_1, we have

$$
\begin{aligned}
A_0 e^{kt_1} &= \tfrac{1}{2}A_0, \\
e^{kt_1} &= 2^{-1}, \\
kt_1 &= \ln(2^{-1}), \\
t_1 &= -\frac{\ln 2}{k}.
\end{aligned}
$$

This shows that the length of time it takes for half of the substance to decay does not depend on the initial amount A_0, but only on the constant k. The value of k depends on the type of radioactive substance. The length of time it takes for half of a radioactive substance to decay is called the **half-life** of the substance.

If the amount of a radioactive substance at time t is given by $A(t) = A_0 e^{kt}$, we have shown that the half-life is given by the formula

$$
\text{Half-life} = -\frac{\ln 2}{k}, \text{ so } k = -\frac{\ln 2}{\text{Half-life}}.
$$

We can then write

$$
A(t) = A_0 e^{-[(\ln 2)/(\text{Half-life})]t}. \tag{3}
$$

This formula is convenient when dealing with radioactive substances for which the half-life is known.

EXAMPLE 3

The carbon isotope ^{11}C has a half-life of 20 min. How long will it take for 90% of a sample of ^{11}C to decay?

SOLUTION From Formula 3 with half-life $= 20$ min, we have $A(t) = A_0 e^{-[(\ln 2)/20]t}$, where t is measured in minutes. When 90% of the sample

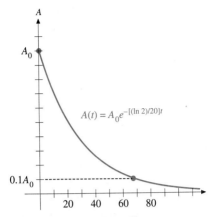

FIGURE 7.6.2 Carbon ^{11}C decays exponentially. When 90% has decayed, there will be 10% remaining.

has decayed, there will be 10% remaining. Thus, we need to determine t such that $A(t) = 0.10A_0$. Solving this equation for t we have

$$A_0 e^{-[(\ln 2)/20]t} = 0.10A_0,$$

$$-\left(\frac{\ln 2}{20}\right)t = \ln 0.10,$$

$$t = \frac{-\ln 0.10}{\dfrac{\ln 2}{20}} = \frac{\ln\left(\dfrac{1}{10}\right)^{-1}}{\ln 2}(20) = \frac{\ln 10}{\ln 2}(20) \approx 66.4.$$

Ninety percent of the sample will decay in approximately 66.4 min. The graph of $A(t)$ is sketched in Figure 7.6.2. ■

Carbon Dating

Radiocarbon ^{14}C is a radioactive form of carbon that is present in the earth's atmosphere. All living plants and animals absorb radiocarbon as a part of their life processes. Thus, the radiocarbon that decays in a living thing is replaced, so the level remains essentially constant. After the living thing dies, the ^{14}C is not replaced and decays with half-life 5750 years. The time of death of an ancient living thing can be determined by measuring the proportion of radiocarbon remaining in a sample.

EXAMPLE 4

Find the age of a sample that has 70% of the radiocarbon it had while living.

SOLUTION   Using Formula 3 with half-life $= 5750$ years, we have

$$A(t) = A_0 e^{-[(\ln 2)/5750]t}.$$

We want to find t so that

$$A(t) = 0.70A_0.$$

Substitution gives

$$A_0 e^{-[(\ln 2)/5750]t} = 0.70A_0,$$

$$-\left(\frac{\ln 2}{5750}\right)t = \ln 0.70,$$

$$t = -\frac{5750}{\ln 2}\ln 0.70 \approx 2959.$$

The sample is approximately 2960 years old. ■

Compound Interest

If simple annual interest rate r is paid on a principal of P_0 dollars for t years, the earned interest is $I = rtP_0$ and the total value of the principal and interest is given by the linear function

$$P_s(t) = (1 + rt)P_0. \tag{4}$$

EXAMPLE 5

What is the value of an investment of $1000 at 8% simple interest after one year?

SOLUTION Formula 4 with $P_0 = 1000$, $r = 0.08$, and $t = 1$ gives

$$P_s(1) = (1 + (0.08)(1))(1000) = \$1080.00.$$

The investment is worth $1,080.00 after one year. ∎

In the case of simple interest, interest is paid only on the initial investment, P_0, even though the value of the account increases. In a different type of savings or loan plan, the interest is periodically compounded, or added to the account, and interest is then paid on the new value of the account. For example, suppose the interest is compounded k times a year, so $k = 1$ corresponds to compounding annually, $k = 12$ to compounding monthly, and $k = 365$ to compounding daily. Let us determine $P_k(t)$, the value of the account t years after an initial investment of P_0. Until $t = 1/k$, when interest is first credited to the account, interest is paid on the initial amount and $P_k(t)$ is given by the linear function

$$P_k(t) = (1 + rt)P_0, \qquad 0 \le t \le \frac{1}{k},$$

so

$$P_k\left(\frac{1}{k}\right) = \left(1 + \frac{r}{k}\right) P_0.$$

At time $t = 1/k$, the interest from the first period is credited to the account and interest is paid on the amount $P_k(1/k)$ for t between $1/k$ and $2/k$. At the end of the second period of $1/k$ years, we have

$$P_k\left(\frac{2}{k}\right) = \left(1 + \frac{r}{k}\right)\left(P_k\left(\frac{1}{k}\right)\right) = \left(1 + \frac{r}{k}\right)^2 P_0.$$

Similarly, after n periods of $1/k$ years the value is

$$P_k\left(\frac{n}{k}\right) = \left(1 + \frac{r}{k}\right)^n P_0.$$

Letting $t = n/k$ denote the time in years, we can rewrite the above as

$$P_k(t) = \left(1 + \frac{r}{k}\right)^{kt} P_0, \tag{5}$$

where t is the time in years, r is the interest rate and the interest is compounded k times a year. Formula 5 holds for $t = n/k$, $n = 0, 1, \ldots$, and the function P_k is linear between these values of t. See Figure 7.6.3.

FIGURE 7.6.3 The function P_k is linear between times when the interest is compounded.

$$P_k(t) = \left(1 + \frac{r}{k}\right)^{kt} P_0,$$

$$t = n/k, \ n = 0, 1, 2, \ldots$$

EXAMPLE 6

What is the value of an investment of $1000 at 8% after one year if the interest is compounded monthly?

SOLUTION Formula 5 with $P_0 = 1000$, $r = 0.08$, $k = 12$, and $t = 1$ gives

$$P_{12}(1) = \left(1 + \frac{0.08}{12}\right)^{12(1)} (1000) \approx 1,083.00.$$

The investment is worth $1,083.00 after one year. ∎

We see from Examples 5 and 6 that compounding the interest monthly increased the value over that obtained from simple interest. This leads to the question, what if the interest is compounded even more often? To answer this question, let us consider the limit of $P_k(t)$ as k approaches infinity. We can use Formula 5 and the techniques of Section 7.5 to determine that

$$P(t) = \lim_{k \to \infty} P_k(t) = \lim_{k \to \infty} \left(1 + \frac{r}{k}\right)^{kt} P_0 = \boxed{P_0 e^{rt}.} \qquad (6)$$

This formula is said to correspond to interest being **compounded continuously.**

EXAMPLE 7

What is the value of an investment of $1000 at 8% after one year if the interest is compounded continuously?

SOLUTION   Formula 6 with $P_0 = 1000$, $r = 0.08$, and $t = 1$ gives

$$P(1) = 1000e^{0.08(1)} \approx 1,083.29.$$

The investment is worth $1,083.29 after one year. ∎

We see from Example 7 that compounding continuously does give a larger value than compounding monthly, but the change is not dramatic.

It is worthwhile to compare the rate of change of the amount for different plans. For simple interest, the rate of change is

$$P_s'(t) = r P_0,$$

so the rate of change is always proportional to the initial amount. Since $P_k(t)$ increases according to the formula for simple interest between times when the interest is compounded, the rate of change of P_k is r times its value at the last time interest was credited to the account. If k is large, so the interest is credited often, we have

$$P_k'(t) \approx r P_k(t)$$

at each point where P_k is differentiable. Of course, the function $P(t) = P_0 e^{rt}$ satisfies the differential equation

$$P' = r P.$$

This means that the rate of change of the value is proportional to the value when the interest is compounded continuously.

Electric Circuits

Consider a simple series circuit with resistance R (ohms), capacitance C (farads), and impressed voltage $E(t)$ (volts), as indicated in Figure 7.6.4. Let $i(t)$ (amperes) denote the current at time t. One of Kirchhoff's Laws implies that $i(t)$ is a solution of the differential equation

$$Ri' + \frac{1}{C}i = E'(t).$$

FIGURE 7.6.4 The circuit has resistance R, capacitance C, and impressed voltage E(t).

If the impressed voltage is constant, we have $E'(t) = 0$, so the equation reduces to

$$i' = -\frac{1}{RC}i.$$

We know from (1) that this equation has solution

$$i(t) = i_0 e^{-t/(RC)}. \tag{7}$$

From Equation 7 we see that $i(t)$ approaches zero as t approaches infinity. That is, the capacitor builds up a charge that approaches the constant value of the impressed voltage and the current decreases to zero over long intervals of time.

EXAMPLE 8

A simple series circuit consists of a charged capacitance, a 200-Ω resistor, and a constant impressed voltage of 6 V. Find the capacitance if the current in the circuit decreases from 10 A to 5 A in 0.3 s.

SOLUTION   The impressed voltage is constant, $E(t) = 6$, so we know the current is given by Formula 7. The resistance is $R = 200$, and the capacitance C is unknown. We take the initial current to be $i_0 = 10$, so

$$i(t) = 10e^{-t/(200C)}.$$

We also know that $i(0.3) = 5$. Substitution then gives

$$5 = 10e^{-0.3/(200C)}.$$

Solving for C, we obtain

$$e^{0.3/(200C)} = 2,$$

$$\frac{0.3}{200C} = \ln 2,$$

$$C = \frac{0.3}{200 \ln 2} \approx 0.00216404.$$

The capacitance is approximately 0.002 F (farads). ∎

EXERCISES 7.6

In Exercises 1–8, find the solution of the given differential equation that satisfies the indicated condition.

1. $y' = -y,\ y(0) = 2$

2. $y' = y,\ y(0) = 3$

3. $y' = -0.1y,\ y(0) = 0.8$

4. $y' = 1.4y,\ y(0) = 3.2$

5. $y' = 2y,\ y(1) = 1$

6. $y' = 3y,\ y(1/3) = 5$

7. $y' = -4y,\ y(1) = e$

8. $y' = -y/2,\ y(2) = e$

9. A certain bacteria colony increases at a rate proportional to the population. The initial population was 200. After 2 h the population was 400. What was the population 4 h after the initial time?

10. A bacteria colony is observed to increase from 100 to 300 in 2 h. Assuming exponential growth, how long will it take the colony to grow from 300 to 500?

11. Each year the population of a certain town increases by 5%, so $P(t + 1) = (1.05)P(t)$, t in years. Assuming exponential growth, how long will it take to grow from 10,000 to 25,000?

12. The population of a colony of bacteria increases by 2% each day, so $P(t + 1) = (1.02)P(t)$, t in days. Assuming exponential growth, what will the population of the colony be one week after the population is 900?

13. Five percent of a certain radioactive substance decays in one day. What is the half-life of the substance?

14. Carbon ^{11}C has half-life 20 min. What percentage of a sample of ^{11}C remains after 12 minutes?

15. Find the age of an ancient skull that is measured to have 40% of the ^{14}C it contained while living. (^{14}C has half-life 5750 years.)

16. Find the age of a sample of wood that contains 25% of the ^{14}C it contained while a living tree. (^{14}C has half-life 5750 years.)

17. What is the value of an investment of $2000 at 6% after five years if the interest is compounded (a) quarterly, (b) monthly, and (c) continuously?

18. What is the value of an investment of $3000 at $7\frac{1}{2}\%$ after two years if the interest is compounded (a) quarterly, (b) monthly, and (c) continuously?

19. At what interest rate will an investment of $10,000 grow to $11,200 after being compounded (a) monthly and (b) continuously for one year?

20. At what interest rate will an investment of $4000 grow to $5000 after being compounded (a) monthly and (b) continuously for three years?

21. How long does it take an investment at 8% simple interest to triple if the interest is not compounded?

22. How long does it take an investment at 8% to triple if the interest is compounded continuously?

23. Interest on a **zero coupon bond** is compounded continuously for the period of the bond. What is the initial cost of a zero coupon bond that pays $7\frac{1}{2}\%$ and is worth $10,000 at the end of twelve years?

24. Interest on **federal savings bonds** is compounded continuously for the period of the bond. What is the initial cost of a bond that pays 6% interest and is worth $100 at the end of ten years?

25. A simple series circuit consists of a charged 0.02-F capacitance, a 200-Ω resistor, and no impressed voltage. The initial current is 10 A. Find the current after 0.2 s.

26. A simple series circuit consists of a charged 0.0004-F capacitance, a resistor, and impressed charge of 6 V. Find the resistance if the current in the circuit decreases from 8 A to 2 A in 0.1 s.

27. A simple RL circuit satisfies the differential equation

$$\frac{di}{dt} + \frac{R}{L}i = 0.$$

See figure.

(a) Find the solution that satisfies the initial condition $i(0) = i_0$.

(b) Plot the graph of $i(t)/i_0$. This is the **response curve** of the circuit.

(c) Find the **time constant** for the circuit, the time required for the response curve to drop to zero if it would decay at a constant rate that is equal to its initial rate of decay.

28. The voltage of a certain **critically damped RLC circuit** is given by

$$v(t) = 400te^{-2.5t}.$$

See figure.

(a) What is the maximum voltage?

(b) Determine an approximate value of the **settling time** of the circuit, when the voltage is 1% of the maximum voltage.

The rate of change of the temprature of an object is directly proportional to the difference between its temperature and the temperature of its surroundings. (We assume that all parts of the object are the same temperature and that the temperature of the object does not significantly affect the temperature of the surroundings. These assumptions are reasonable if the object is small and is composed of material that is a good conductor of heat.) If $T(t)$ denotes the temperature of the object and K is the constant temperature of the surrounding region, then **Newton's Law of Cooling** *can be expressed as*

$$T' = k(T - K).$$

Since the constant K has derivative zero, it follows that $T(t) - K$ is a solution of the differential equation $y' = ky$, so

$$T(t) - K = Ce^{kt}.$$

29. A small object had temperature 200°F when it was dropped into a large pool of 80°F water. Five minutes after the object was put into the water, the temperature of the object was 100°F. What was the temperature of the water 10 minutes after the object was put into the water?

30. A thermometer reads 90° when it is moved to a 70° room. Ten minutes later the thermometer reads 80°. What will the thermometer read after an additional 10 minutes?

31. A small beef roast at a temperature of 50° is put into a 350° oven. After one hour the temperature of the roast is 90°. How much longer will it take the roast to reach 150°?

32. A thermometer reads 70° when it is first moved outdoors. After being outside one minute, the thermometer reads 60°. One minute later it reads 55°. Find the outdoor temperature.

33. Newton's Law of Cooling leads to the formula

$$T(t) = K + (T(0) + K)e^{kt}.$$

Physical considerations imply that $k < 0$ if either $T(0) - K > 0$ or $T(0) - K < 0$. Evaluate $\lim_{t \to \infty} T(t)$ and give a physical interpretation of the meaning of the limit.

34. If a piston has orifices, its motion through oil in a cylinder is governed by the equation $\dfrac{dv}{dt} = -kv$, where v is the velocity and k is a positive constant. See figure.

Piston with orifices

(a) Use the formula $v(t) = v_0 e^{-kt}$ to show that $x(t) = \dfrac{v_0}{k}(1 - e^{-kt})$, where x is the distance the piston has moved in time t. ($v = \dfrac{dx}{dt}$, so x is an antiderivative of v.)

(b) What is the range of distances over which the piston can move?

7.7 HYPERBOLIC AND INVERSE HYPERBOLIC FUNCTIONS

Connections

Relation between differentiation and integration formulas, 5.1.

Concept of inverse function, 7.1.

Differentiation of inverse functions, 7.2.

Exponential function, 7.5.

Certain combinations of the functions e^x and e^{-x} occur often enough in applications that it is convenient to give them a name.

The Hyperbolic Functions

The **hyperbolic sine** is

$$\sinh x = \frac{e^x - e^{-x}}{2}.$$

The **hyperbolic cosine** is

$$\cosh x = \frac{e^x + e^{-x}}{2}.$$

We can use the usual methods to graph $y = \sinh x$. We know that $e^0 = 1$, $e^x > 1 > e^{-x}$ for $x > 0$, and $e^x < 1 < e^{-x}$ for $x < 0$. It follows that $\sinh 0 = 0$, $\sinh x > 0$ for $x > 0$, and $\sinh x < 0$ for $x < 0$.

$$\frac{d}{dx}(\sinh x) = \frac{d}{dx}\left(\frac{e^x - e^{-x}}{2}\right) = \frac{e^x - e^{-x}(-1)}{2} = \frac{e^x + e^{-x}}{2}.$$

The derivative is positive for all x, so $\sinh x$ is increasing. The second derivative can be used to show that the graph is concave down for $x < 0$ and concave up for $x > 0$. $(0, 0)$ is a point of inflection. The graph is sketched in Figure 7.7.1.

We can also use the usual methods to analyze the function $\cosh x$. The derivative tells us that $\cosh x$ is decreasing for $x < 0$ and increasing for $x > 0$. The minimum value is $\cosh 0 = 1$. The second derivative tells us that $\cosh x$ is concave up. The graph is sketched in Figure 7.7.2.

It can be shown that a homogeneous flexible cable that is suspended as illustrated in Figure 7.7.3 will hang in the shape of the curve $y = a \cosh(x/a)$.

$y = \sinh x$

FIGURE 7.7.1 $\sinh x$ is increasing with $\sinh(0) = 0$.

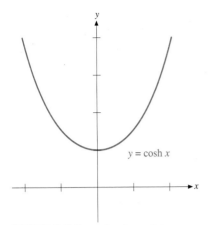

FIGURE 7.7.2 $\cosh x$ has minimum value $\cosh(0) = 1$. The graph is increasing on the interval $x \geq 0$.

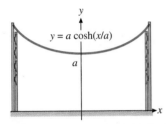

FIGURE 7.7.3 A homogeneous flexible cable that is suspended from both ends hangs in the shape of a catenary, $y = a\cosh(x/a)$.

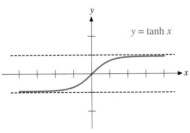

FIGURE 7.7.4 $\tanh x$ is increasing with values $-1 < y < 1$.

This curve is called a **catenary**. Note that the weight of a homogeneous hanging cable is not distributed uniformly along the x-axis, since there is more cable per unit length along the x-axis where the curve is steeper. A cable that is loaded uniformly along the x-axis, such as a bridge support, hangs in the shape of a parabola.

The functions $\sinh x$ and $\cosh x$ are related by the following identity:

$$\cosh^2 x - \sinh^2 x = 1. \tag{1}$$

Proof   The identity is verified by substitution of exponential expressions for $\sinh x$ and $\cosh x$, and then simplifying. We have

$$\cosh^2 x - \sinh^2 x = \left(\frac{e^x + e^{-x}}{2}\right)^2 - \left(\frac{e^x - e^{-x}}{2}\right)^2$$

$$= \frac{(e^x)^2 + 2e^x e^{-x} + (e^{-x})^2}{4} - \frac{(e^x)^2 - 2e^x e^{-x} + (e^{-x})^2}{4}$$

$$= \frac{e^{2x} + 2e^{x-x} + e^{-2x}}{4} - \frac{e^{2x} - 2e^{x-x} + e^{-2x}}{4}$$

$$= \frac{e^{2x} + 2 + e^{-2x} - e^{2x} + 2 - e^{-2x}}{4} = \frac{4}{4} = 1. \quad \blacksquare$$

From (1) we see that the coordinates $x = \cosh t$, $y = \sinh t$ satisfy the equation of the hyperbola $x^2 - y^2 = 1$. This suggests the name of the hyperbolic functions. Analogously, the coordinates $x = \sin t$, $y = \cos t$ satisfy the equation of the circle $x^2 + y^2 = 1$. The trigonometric functions are sometimes called the circular functions. We will see that, except for possible changes in sign, many formulas for trigonometric functions also hold for hyperbolic functions.

Corresponding to the trigonometric functions, we define the following:

$$\tanh x = \frac{\sinh x}{\cosh x} = \frac{e^x - e^{-x}}{e^x + e^{-x}},$$

$$\coth x = \frac{\cosh x}{\sinh x} = \frac{e^x + e^{-x}}{e^x - e^{-x}},$$

$$\text{sech } x = \frac{1}{\cosh x} = \frac{2}{e^x + e^{-x}},$$

$$\text{csch } x = \frac{1}{\sinh x} = \frac{2}{e^x - e^{-x}}.$$

The graphs of these functions are given in Figures 7.7.4–7.7.7.

Hyperbolic analogues of many trigonometric identities are easily verified by using the definitions. For example, dividing each side of equation (1) by $\cosh^2 x$, we obtain

$$1 - \tanh^2 x = \text{sech}^2 x. \tag{2}$$

Some other identities are given in the exercises.

Differentiation Formulas

It is convenient to express the derivatives of the hyperbolic functions in terms of hyperbolic functions. We have the following:

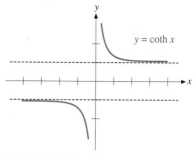

FIGURE 7.7.5 coth x is decreasing on each of the intervals $x > 0$ and $x < 0$ with values $|y| \geq 1$.

$$\frac{d}{dx}(\sinh x) = \cosh x,$$

$$\frac{d}{dx}(\cosh x) = \sinh x,$$

$$\frac{d}{dx}(\tanh x) = \operatorname{sech}^2 x,$$

$$\frac{d}{dx}(\coth x) = -\operatorname{csch}^2 x,$$

$$\frac{d}{dx}(\operatorname{sech} x) = -\operatorname{sech} x \tanh x,$$

$$\frac{d}{dx}(\operatorname{csch} x) = -\operatorname{csch} x \coth x.$$

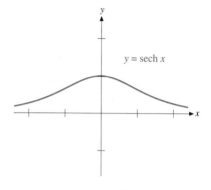

FIGURE 7.7.6 sech x is positive with maximum value sech(0) = 1. The graph is decreasing on the interval $x \geq 0$.

The first two of these formulas are easily verified by using the definitions of $\sinh x$ and $\cosh x$ in terms of e^x and e^{-x}. These two formulas can then be used to derive the others. For example,

$$\frac{d}{dx}(\operatorname{sech} x) = \frac{d}{dx}((\cosh x)^{-1}) = (-1)(\cosh x)^{-2}(\sinh x)$$

$$= -\left(\frac{1}{\cosh x}\right)\left(\frac{\sinh x}{\cosh x}\right) = -\operatorname{sech} x \tanh x.$$

The differentiation formulas for the hyperbolic functions correspond to those of the trigonometric functions, except for the differences in sign.

We should be able to use the Chain Rule version of the differentiation formulas.

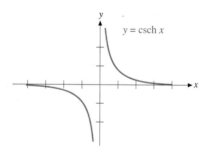

FIGURE 7.7.7 csch x is decreasing on each of the intervals $x > 0$ and $x < 0$ with values $y \neq 0$.

EXAMPLE 1

use chain rule

(a) $\dfrac{d}{dx}\left(a \cosh\left(\dfrac{x}{a}\right)\right) = a \sinh\left(\dfrac{x}{a}\right)\left(\dfrac{1}{a}\right) = \sinh\left(\dfrac{x}{a}\right),$

(b) $\dfrac{d}{dx}(\tanh(x^2)) = \operatorname{sech}^2(x^2)(2x).$ ∎

Integration Formulas

The following integration formulas correspond to the above differentiation formulas:

$$\int \cosh u \, du = \sinh u + C,$$

$$\int \sinh u \, du = \cosh u + C,$$

$$\int \operatorname{sech}^2 u \, du = \tanh u + C,$$

$$\int \operatorname{csch}^2 u \, du = -\coth u + C,$$

$$\int \operatorname{sech} u \tanh u \, du = -\operatorname{sech} u + C,$$

$$\int \operatorname{csch} u \coth u \, du = -\operatorname{csch} u + C.$$

EXAMPLE 2

Evaluate

$$\int_0^3 \sinh \frac{x}{3} \, dx.$$

SOLUTION   Using $u = x/3$, so $du = (1/3)dx$, $u(0) = 0$, and $u(3) = 1$, we have

$$\int_0^3 \sinh \frac{x}{3} \, dx = \int_0^3 \sinh \frac{x}{3}(3)\left(\frac{1}{3}\right) dx$$

$$= \int_0^1 \sinh u \, (3) \, du$$

$$= 3 \cosh u \Big]_0^1$$

$$= 3 \cosh(1) - 3 \cosh(0)$$

$$= 3\left(\frac{e + e^{-1}}{2}\right) - 3(1) \approx 1.63. \quad \blacksquare$$

Inverse Hyperbolic Functions

We know that $\sinh x$ has positive derivative for all x. It follows that $\sinh x$ has a differentiable inverse function, $\sinh^{-1} x$. The defining relation is

$$y = \sinh^{-1} x \text{ if and only if } x = \sinh y.$$

We can obtain a formula for $\sinh^{-1} x$ by solving the equation $x = \sinh y$ for y. That is,

$$x = \sinh y,$$

$$x = \frac{e^y - e^{-y}}{2},$$

$$e^y - 2x - e^{-y} = 0.$$

Multiplying the above equation by e^y, we obtain

$$(e^y)^2 - 2x(e^y) - 1 = 0.$$

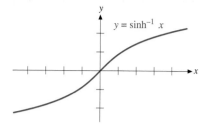

FIGURE 7.7.8 $\sinh x$ has inverse function $\sinh^{-1} x$.

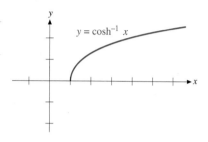

FIGURE 7.7.9 $\cosh x, x \geq 0$, has inverse function $\cosh^{-1} x$.

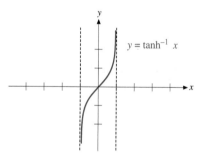

FIGURE 7.7.10 $\tanh x$ has inverse function $\tanh^{-1} x$.

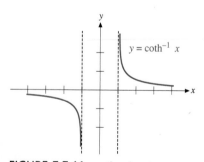

FIGURE 7.7.11 $\coth x$ has inverse function $\coth^{-1} x$.

We then note that we have an equation that is quadratic in e^y. The Quadratic Formula gives the solution

$$e^y = \frac{-(-2x) \pm \sqrt{(-2x)^2 - 4(1)(-1)}}{2(1)}.$$

Since $e^y > 0$, we discard the negative solution and simplify to obtain

$$e^y = x + \sqrt{x^2 + 1},$$
$$y = \ln(x + \sqrt{x^2 + 1}).$$

The formula for $\sinh^{-1} x$ is

$$\sinh^{-1} x = \ln(x + \sqrt{x^2 + 1}).$$

We can use the same method to obtain formulas for other inverse hyperbolic functions. In some cases it is necessary to restrict the domain to obtain an inverse function. Let us list the formulas.

- The inverse function of $\sinh x$ is

$$\sinh^{-1} x = \ln(x + \sqrt{x^2 + 1}). \qquad \text{(Figure 7.7.8)}$$

- The inverse function of $\cosh x, x \geq 0$, is

$$\cosh^{-1} x = \ln(x + \sqrt{x^2 - 1}). \qquad \text{(Figure 7.7.9)}$$

- The inverse function of $\tanh x$ is

$$\tanh^{-1} x = \frac{1}{2} \ln \frac{1 + x}{1 - x}. \qquad \text{(Figure 7.7.10)}$$

- The inverse function of $\coth x$ is

$$\coth^{-1} x = \frac{1}{2} \ln \frac{x + 1}{x - 1}. \qquad \text{(Figure 7.7.11)}$$

- The inverse function of $\operatorname{sech} x, x \geq 0$ is

$$\operatorname{sech}^{-1} x = \ln \frac{1 + \sqrt{1 - x^2}}{x}. \qquad \text{(Figure 7.7.12)}$$

- The inverse function of $\operatorname{csch} x$ is

$$\operatorname{csch}^{-1} x = \ln \left(\frac{1}{x} + \frac{\sqrt{1 + x^2}}{|x|} \right). \qquad \text{(Figure 7.7.13)}$$

We can use the Chain Rule to obtain the following differentiation formulas:

FIGURE 7.7.12 $\operatorname{sech} x, x \geq 0$, has inverse function $\operatorname{sech}^{-1}x$.

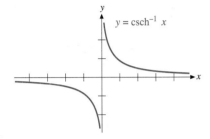

FIGURE 7.7.13 $\operatorname{csch} x$ has inverse function $\operatorname{csch}^{-1}x$.

$$\frac{d}{dh}(\sinh^{-1} x) = \frac{1}{\sqrt{1+x^2}},$$

$$\frac{d}{dx}(\cosh^{-1} x) = \frac{1}{\sqrt{x^2-1}}, \qquad x > 1,$$

$$\frac{d}{dx}(\tanh^{-1} x) = \frac{1}{1-x^2}, \qquad |x| < 1,$$

$$\frac{d}{dx}(\coth^{-1} x) = \frac{1}{1-x^2}, \qquad |x| > 1,$$

$$\frac{d}{dx}(\operatorname{sech}^{-1} x) = -\frac{1}{x\sqrt{1-x^2}}, \qquad 0 < x < 1,$$

$$\frac{d}{dx}(\operatorname{csch}^{-1} x) = -\frac{1}{|x|\sqrt{1+x^2}}, \qquad x \neq 0.$$

The above differentiation formulas can be used to verify the following integration formulas. We assume that $a > 0$.

$$\int \frac{1}{\sqrt{a^2+u^2}}\,dx = \sinh^{-1}\frac{u}{a} + C = \ln(u + \sqrt{u^2+a^2}) + C,$$

$$\int \frac{1}{\sqrt{u^2-a^2}}\,dx = \cosh^{-1}\frac{u}{a} + C = \ln(u + \sqrt{u^2-a^2}) + C,$$

$$\int \frac{1}{a^2-u^2}\,du = \frac{1}{a}\tanh^{-1}\frac{u}{a} + C = \frac{1}{2a}\ln\frac{a+u}{a-u} + C, |u| < a,$$

$$\int \frac{1}{a^2-u^2}\,du = \frac{1}{a}\coth^{-1}\frac{u}{a} + C = \frac{1}{2a}\ln\frac{u+a}{u-a} + C, |u| > a,$$

$$\int \frac{1}{u\sqrt{a^2-u^2}}\,du = -\frac{1}{a}\operatorname{sech}^{-1}\frac{|u|}{a} + C = -\frac{1}{a}\ln\frac{a+\sqrt{a^2-u^2}}{|u|} + C,$$

$$\int \frac{1}{u\sqrt{a^2+u^2}}\,du = -\frac{1}{a}\operatorname{csch}^{-1}\frac{|u|}{a} + C = -\frac{1}{a}\ln\frac{a+\sqrt{a^2+u^2}}{|u|} + C.$$

EXAMPLE 3
Evaluate

$$\int_0^3 \frac{1}{\sqrt{9+x^2}}\,dx.$$

SOLUTION The integration formula gives

$$\int_0^3 \frac{1}{\sqrt{9+x^2}}\,dx = \sinh^{-1}\frac{x}{3}\Big]_0^3 = \sinh^{-1}(1) - \sinh^{-1}(0) = \sinh^{-1}(1)$$

$$= \ln(1 + \sqrt{(1)^2+1}) = \ln(1 + \sqrt{2}) \approx 0.88. \quad \blacksquare$$

EXERCISES 7.7

Verify the identities in Exercises 1–10.

1. $\sinh(-x) = -\sinh x$

2. $\cosh(-x) = \cosh x$

3. $\sinh 2x = 2\sinh x \cosh x$

4. $\cosh 2x = \cosh^2 x + \sinh^2 x$

5. $\cosh(x + y) = \cosh x \cosh y + \sinh x \sinh y$

6. $\sinh(x + y) = \sinh x \cosh y + \cosh x \sinh y$

7. $\tanh(x + y) = \dfrac{\tanh x + \tanh y}{1 + \tanh x \tanh y}$

8. $\tanh 2x = \dfrac{2\tanh x}{1 + \tanh^2 x}$

9. $\cosh^2 \dfrac{x}{2} = \dfrac{1 + \cosh x}{2}$

10. $\tanh \dfrac{x}{2} = \dfrac{\sinh x}{1 + \cosh x}$

Evaluate the derivatives in Exercises 11–20.

11. $\dfrac{d}{dx}(\sinh x^2)$

12. $\dfrac{d}{dx}(\sinh \sqrt{x})$

13. $\dfrac{d}{dx}(\cosh^2 x)$

14. $\dfrac{d}{dx}(\cosh x^2)$

15. $\dfrac{d}{dx}(\ln(\cosh x))$

16. $\dfrac{d}{dx}(\tan^{-1}(\sinh x))$

17. $\dfrac{d}{dx}(\sinh^{-1}(2x))$

18. $\dfrac{d}{dx}\left(\cosh^{-1}\dfrac{x}{3}\right)$

19. $\dfrac{d}{dx}(\tanh^{-1}(\sin x))$

20. $\dfrac{d}{dx}(\text{sech}^{-1}(\cos x))$

Evaluate the integrals in Exercises 21–30.

21. $\displaystyle\int \sinh 3x \, dx$

22. $\displaystyle\int \sinh \dfrac{x}{2} \, dx$

23. $\displaystyle\int x \, \text{sech}^2(x^2) \, dx$

24. $\displaystyle\int \dfrac{\cosh \sqrt{x}}{\sqrt{x}} \, dx$

25. $\displaystyle\int \tanh x \, dx$

26. $\displaystyle\int \sinh x \cosh x \, dx$

27. $\displaystyle\int \dfrac{1}{\sqrt{9 + 4x^2}} \, dx$

28. $\displaystyle\int \dfrac{1}{\sqrt{9x^2 - 4}} \, dx$

29. $\displaystyle\int \dfrac{e^x}{\sqrt{e^{2x} - 1}} \, dx$

30. $\displaystyle\int_4^8 \dfrac{x}{x^4 - 16} \, dx$

Evaluate the definite integrals in Exercises 31–36.

31. $\displaystyle\int_0^2 \dfrac{1}{9 - x^2} \, dx$

32. $\displaystyle\int_3^5 \dfrac{1}{x^2 - 4} \, dx$

33. $\displaystyle\int_0^2 \dfrac{1}{\sqrt{4 + x^2}} \, dx$

34. $\displaystyle\int_3^5 \dfrac{1}{\sqrt{x^2 - 4}} \, dx$

35. $\displaystyle\int_0^1 \dfrac{1}{9 - x^2} \, dx$

36. $\displaystyle\int_3^2 \dfrac{1}{1 - x^2} \, dx$

37. Use the definitions in terms of the exponential function to verify that $\dfrac{d}{dx}(\sinh x) = \cosh x$.

38. Use the definitions in terms of the exponential function to verify that $\dfrac{d}{dx}(\cosh x) = \sinh x$.

39. Use formulas for the derivatives of $\sinh x$ and $\cosh x$ to verify that $\dfrac{d}{dx}(\tanh x) = \text{sech}^2 x$.

40. Use the formula for the derivative of $\sinh x$ to verify that $\dfrac{d}{dx}(\text{csch } x) = -\text{csch } x \coth x$.

41. Verify that $\text{sech}^{-1} x = \ln \dfrac{1 + \sqrt{1 - x^2}}{x}$ by solving the equation $x = \dfrac{2}{e^y + e^{-y}}$ for y, $y \geq 0$.

42. Verify that $\tanh^{-1} x = \dfrac{1}{2} \ln \dfrac{1 + x}{1 - x}$ by solving the equation $x = \dfrac{e^y - e^{-y}}{e^y + e^{-y}}$ for y.

43. Verify that $\dfrac{d}{dx}(\sinh^{-1} x) = \dfrac{1}{\sqrt{1 + x^2}}$ by **(a)** differentiating the logarithmic expression for $\sinh^{-1} x$ and then simplifying; **(b)** implicit differentiation of the equation $\sinh y = x$ and then using $\cosh^2 y - \sinh^2 y = 1$ to express y in terms of x.

44. Verify that $\dfrac{d}{dx}(\tanh^{-1} x) = \dfrac{1}{1 - x^2}$ by **(a)** differentiating the logarithmic expression for $\tanh^{-1} x$ and then simplifying; **(b)** implicit differentiation of the equation $\tanh y = x$ and then using $1 - \tanh^2 y = \text{sech}^2 y$ to express y' in terms of x.

45. Find the fifth-order Taylor polynomial of $f(x) = \sinh x$ about zero.

46. Find the sixth-order Taylor polynomial of $f(x) = \cosh x$ about zero.

47. Find the length of the catenary $y = a\cosh(x/a)$, $-a \leq x \leq a$.

REVIEW OF CHAPTER 7 CONCEPTS

- A function has an inverse if every horizontal line that intersects the graph of the function intersects it at exactly one point. A function that has this property is said to be one-to-one.
- The graphs of a function and its inverse are symmetric with respect to the line $y = x$.
- Properties of a function can be derived from properties of its inverse.
- It is necessary to restrict the domain of trigonometric functions in order that they have inverse functions. The restricted domain becomes the range of the inverse.
- Derivatives of a function and its inverse, evaluated at corresponding points, are reciprocal. The derivative of an inverse can be determined from the derivative of the function by using implicit differentiation.

$$\frac{d}{dx}(\sin^{-1} x) = \frac{1}{\sqrt{1 - x^2}} \text{ and}$$

$$\int \frac{1}{\sqrt{a^2 - x^2}} dx = \sin^{-1} \frac{x}{a} + C.$$

$$\frac{d}{dx}(\tan^{-1} x) = \frac{1}{1 + x^2} \text{ and}$$

$$\int \frac{1}{a^2 + x^2} dx = -\frac{1}{a} \tan \frac{x}{a} + C.$$

$$\frac{d}{dx}(\sec^{-1} x) = \frac{1}{|x|\sqrt{x^2 - 1}} \text{ and}$$

$$\int \frac{1}{x\sqrt{x^2 - a^2}} dx = \frac{1}{a} \sec^{-1} \frac{|x|}{a} + C$$

$$= \frac{1}{a} \cos^{-1} \frac{a}{|x|} + C.$$

- The natural logarithmic function x and the exponential function e^x are inverses of each other.
- As x approaches infinity, each of the functions e^x, $x^\alpha (\alpha > 0)$, and $\ln x$ approaches infinity. For x large, the values of e^x are much larger than corresponding values of x^α and the values of x^α are much larger than corresponding values of $\ln x$.
- As x approaches infinity, both of the functions $e^{-|x|}$ and $|x|^{-\alpha}(\alpha > 0)$ approach zero. For x large, the values of $e^{-|x|}$ are much smaller than corresponding values of $|x|^{-\alpha}$.
- As x approaches zero, both of the functions $\ln |x|$ and $|x|^{-\alpha}(\alpha > 0)$ approach infinity. For small, nonzero x, the values of $|x|^{-\alpha}$ are much larger than corresponding values of $\ln |x|$.
- The formulas $a^x = e^{x \ln a}$ and $\log_a x = \frac{\ln x}{\ln a}$ allow us to express exponential and logarithmic expressions with bases other than e in terms of the natural exponential and logarithmic functions.
- The formula $f^g = e^{g \ln f}$ can be used to evaluate limits and derivatives of functions raised to variable powers.
- Limits of type $1^\infty, 0^0$, and ∞^0 are of indeterminate form and are evaluated by considering their logarithms.
- The differential equation $y' = ky$ indicates that the instantaneous rate of change of y at time t is proportional to the value of y at t. This occurs in many natural problems.
- The initial value problem $y' = ky$, $y(0) = y_0$, has solution $y(t) = y_0 e^{kt}$.
- Hyperbolic functions are defined in terms of e^x and e^{-x}.

CHAPTER 7 REVIEW EXERCISES

In Exercises 1–4, use the given graph of f to sketch the graph of f^{-1}.

1.

2.

3.

4.

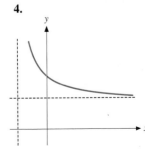

Find the inverse function of the functions given in Exercises 5–8.

5. $f(x) = 3x - 2$

6. $f(x) = x^2 - 2x, x \geq 1$

7. $f(x) = \tan(2x) - 1, -\pi/4 < x < \pi/4$

8. $f(x) = 2\sin(x/2), -\pi \leq x \leq \pi$

9. Evaluate $\sin^{-1}\left(-\dfrac{\sqrt{3}}{2}\right)$.

10. Evaluate $\cos^{-1}\left(\cos\left(-\dfrac{\pi}{4}\right)\right)$.

11. Write $\sin\left(\cos^{-1}\dfrac{x}{2}\right)$ as an algebraic expression.

12. Write $\sec\left(\tan^{-1}\dfrac{x}{4}\right)$ as an algebraic expression.

Find the derivatives in Exercises 13–32.

13. $\dfrac{d}{dx}(\sin^{-1} x^2)$

14. $\dfrac{d}{dx}(\sin^{-1}\sqrt{x})$

15. $\dfrac{d}{dx}\left(\sec^{-1}\dfrac{1}{x}\right)$

16. $\dfrac{d}{dx}\left(\cos^{-1}\dfrac{x}{3}\right)$

17. $\dfrac{d}{dx}\left(\tan^{-1}\dfrac{x}{2}\right)$

18. $\dfrac{d}{dx}\left(\tan^{-1}\left(\dfrac{2}{x}\right)\right)$

19. $\dfrac{d}{dx}\left(\ln\left(\dfrac{x^2}{x^2+1}\right)\right)$

20. $\dfrac{d}{dx}\left(\ln(x\sqrt{x^2+1})\right)$

21. $\dfrac{d}{dx}(\ln|\sec x|)$

22. $\dfrac{d}{dx}(\ln|\sec x + \tan x|)$

23. $\dfrac{d}{dx}(\ln\sqrt{2 - \sin^2 x})$

24. $\dfrac{d}{dx}(e^{-x^2/2})$

25. $\dfrac{d}{dx}(2xe^{2x} - e^{2x})$

26. $\dfrac{d}{dx}(xe^{-x} + e^{-x})$

27. $\dfrac{d}{dx}(3^{x^{1/3}})$

28. $\dfrac{d}{dx}(x2^{-x})$

29. $\dfrac{d}{dx}(\log_{10}(x^2+1))$

30. $\dfrac{d}{dx}(\log_3 x^{1/3})$

31. $\dfrac{d}{dx}((\ln x)^x)$

32. $\dfrac{d}{dx}(x^{\ln x})$

Evaluate the indefinite integrals in Exercises 33–50.

33. $\displaystyle\int \dfrac{x}{9+x^2}\, dx$

34. $\displaystyle\int \dfrac{x}{\sqrt{9-x^2}}\, dx$

35. $\displaystyle\int \dfrac{1}{9+x^2}\, dx$

36. $\displaystyle\int \dfrac{1}{x\sqrt{x^2-9}}\, dx$

37. $\displaystyle\int \dfrac{1}{\sqrt{9-x^2}}\, dx$

38. $\displaystyle\int \tan 2x\, dx$

39. $\displaystyle\int e^{-x/2}\, dx$

40. $\displaystyle\int e^{2x}\, dx$

41. $\displaystyle\int \dfrac{\log_{10} x}{x}\, dx$

42. $\displaystyle\int 2^{-x}\, dx$

43. $\displaystyle\int \sec^2 x\, e^{\tan x}\, dx$

44. $\displaystyle\int \dfrac{(\ln x)^3}{x}\, dx$

45. $\displaystyle\int \dfrac{(\sin^{-1} x)^2}{\sqrt{1-x^2}}\, dx$

46. $\displaystyle\int \dfrac{e^x - e^{-x}}{e^x + e^{-x}}\, dx$

47. $\displaystyle\int \dfrac{1}{4+(x-1)^2}\, dx$

48. $\displaystyle\int \dfrac{1}{\sqrt{4-(x-1)^2}}\, dx$

49. $\displaystyle\int \cosh(2x)\, dx$

50. $\displaystyle\int \operatorname{sech}^2(x/2)\, dx$

Evaluate the definite integrals in Exercises 51–54.

51. $\displaystyle\int_0^2 \dfrac{1}{4+x^2}\, dx$

52. $\displaystyle\int_0^1 \dfrac{1}{\sqrt{4-x^2}}\, dx$

53. $\displaystyle\int_0^1 \dfrac{x}{4+x^2}\, dx$

54. $\displaystyle\int_{2\sqrt{2}}^4 \dfrac{1}{x\sqrt{x^2-4}}\, dx$

55. Use the equation $x = 2\tan\theta$ to express $\displaystyle\int_0^2 \sqrt{x^2+4}\, dx$ as a definite integral with respect to θ.

56. Use the equation $x = \sin\theta$ to express $\displaystyle\int_0^1 \sqrt{1-x^2}\, dx$ as a definite integral with respect to θ.

Evaluate the limits in Exercises 57–68.

57. $\displaystyle\lim_{x\to 0} \dfrac{1 - \cos 4x}{x^2}$

58. $\displaystyle\lim_{x\to 0} \dfrac{\sin^{-1} 2x}{x}$

59. $\displaystyle\lim_{x\to -\infty} xe^{x/2}$

60. $\displaystyle\lim_{x\to\infty} xe^{-\sqrt{x}}$

61. $\displaystyle\lim_{x\to 0^+} x^{1/3} \ln x$

62. $\displaystyle\lim_{x\to\infty} x^{-1/3} \ln x$

63. $\displaystyle\lim_{x\to 0^+} \tan^{-1}\left(\dfrac{1}{x}\right)$

64. $\displaystyle\lim_{x\to\pi} \sin^{-1}(\cos x)$

65. $\displaystyle\lim_{x\to 0}(1 - 2x)^{1/x}$

66. $\displaystyle\lim_{x\to 0}(1 - x)^{1/x}$

67. $\displaystyle\lim_{x\to\infty}\left(1 + \dfrac{3}{x}\right)^x$

68. $\displaystyle\lim_{x\to\infty}(1 + 3x)^{1/(\ln x)}$

Sketch the graphs of the equations given in Exercises 69–72. Indicate asymptotes, local extrema, and points of inflection.

69. $y = xe^{-x/2}$

70. $y = e^{-x^2}$

71. $y = x\ln x$

72. $y = x - \ln x$

73. A bacteria colony is observed to increase from 200 to 600 in 3 hours. Assuming exponential growth, how long will it take the colony to grow from 600 to 1000?

74. Determine the age of a sample of wood that contains 30% of the ^{14}C it contained while a living tree. (Radiocarbon ^{14}C has a half-life of 5750 years.)

75. A circuit consists of a 0.02-F capacitor connected in series to a 60-Ω resistor, with an impressed voltage of 12 V.

Kirchhoff's Second Law can be used to show that the current in amperes is a solution of the differential equation

$$60i' + \frac{1}{0.02}i = \frac{d}{dt}(1).$$

Find the current at time t if the initial current is 8 A.

76. Find the volume of the solid generated when the region bounded by $y = 1/x^2$ and the x-axis between $x = 1$ and $x = 2$ is revolved about the y-axis.

77. Find the volume of the solid generated when the region bounded by $y = \ln x$, the x-axis, and $x = e$ is revolved about the y-axis. (*Hint:* Use the horizontal area strips.)

78. (a) Explain in words why, if f and g are one-to-one, then $f \circ g$ is one-to-one. (b) Explain in words why, if $f \circ g$ is one-to-one, then g is one-to-one, but f may not be. Give an example.

79. Explain how the usual rules of logarithms would lead to contradictions if the formula $\ln(-x) = -\ln x$ were used to extend the definition of the logarithmic function to all nonzero real numbers.

EXTENDED APPLICATION

DETERMINING BLOOD ALCOHOL LEVELS

In recent years the problem of drunken driving has received increased attention in the United States. For the purpose of framing and enforcing laws, it has been necessary to establish objective methods for measuring levels of intoxication. The measure most commonly used is a person's Blood Alcohol Count (BAC), a number that represents the percentage of alcohol in the bloodstream. Thus, for example, a BAC of 0.07 means that alcohol occupies 0.07% of the bloodstream. Although a given BAC affects each person differently, it is roughly true that 0.05 is high enough to produce possible impairment of one's driving abilities. It is illegal to drive in some states with a BAC of 0.08, and in all states with a BAC of 0.10.

If you have been drinking, your BAC depends on several factors, including your weight, your gender, your usual drinking habits, and the amount of food in your stomach. For now we will consider only your weight and the amount of alcohol in your system. We will use the following variables:

t = time, in hours, since you began drinking,

t_0 = time at which you stop drinking,

$A(t)$ = number of ounces of alcohol in your bloodstream at time t,

A_0 = number of ounces of alcohol in your bloodstream at the time you stop drinking,

$B(t)$ = your Blood Alcohol Count at time t,

B_0 = your Blood Alcohol Count at the time you stop drinking,

$C(t)$ = your rate of alcohol intake, in ounces per hour, while drinking,

W = your weight, in pounds.

It is approximately true that $B(t) = 7.2A(t)/W$. In creating formulas for $A(t)$ and $B(t)$, we will assume that your body eliminates alcohol from the bloodstream at a rate proportional to the amount present at any time. We will also assume that your rate of alcohol intake $C(t)$ is a constant C.

Tasks:

1. Explain why the functions $A(t)$ and $B(t)$ should have the following properties:

   a. $A(0) = B(0) = 0$.

   b. For $t > t_0$, $A'(t) = -kA(t)$ and $B'(t) = -kB(t)$ for some $k > 0$.

   c. For $0 < t < t_0$, $A'(t) = C - kA(t)$ and $B'(t) = \dfrac{7.2C}{W} - kB(t)$. (*Hint:* While you are drinking, $A'(t)$ is the difference between the rate at which alcohol enters your system and the rate at which you eliminate it.)

2. Demonstrate that the functions

$$A(t) = \begin{cases} 4C(1 - e^{-t/4}), & \text{if } 0 \le t < t_0, \\ A_0 e^{-(t-t_0)/4}, & \text{if } t \ge t_0 \end{cases}$$

and

$$B(t) = \begin{cases} \dfrac{28.8C}{W}(1 - e^{-t/4}), & \text{if } 0 \leq t < t_0, \\ B_0 e^{-(t-t_0)/4}, & \text{if } t \geq t_0 \end{cases}$$

are continuous for $t \geq 0$ and differentiable except at t_0, if $A_0 = 4C(1 - e^{-t_0/4})$ and $B_0 = \dfrac{28.8C}{W}(1 - e^{-t_0/4})$. Also show that they have the properties listed in Task 1.

3. Paula plans to go bar-hopping tonight. She usually drinks one glass of wine, with an alcoholic content of 0.55 oz, per hour. She does not want her BAC to go above 0.06 at any time, and she does not want to drive with a BAC higher than 0.04. Paula weighs 120 pounds. According to the model in Task 2, how long can she drink before stopping? How long after that should she wait before driving?

4. Put yourself in Paula's place, and repeat Task 3. Use your own weight, and choose a type of drink from the following list:

Type of drink	Alcoholic content
light beer (12 oz)	0.50 oz
table wine (5 oz)	0.55 oz
gin and tonic	0.60 oz
margarita	0.70 oz
dry martini	1.00 oz
frozen daiquiri	1.20 oz

5. Several models for determining the rate of elimination of alcohol from the bloodstream are in general use. According to the Traffic Institute at Northwestern University, the rate of elimination is 0.5 oz of alcohol per hour. Heublein, Inc., estimates that the rate of elimination will reduce the BAC by 0.01 per hour.

   a. For each model, write a formula for $B(t)$, $t \geq t_0$, assuming that $B(t_0) = B_0$ and no alcohol is consumed thereafter.

   b. Three drinkers, weighing 100, 150, and 200 pounds, respectively, each have $B_0 = 0.08$. According to the model in Task 2, what will each person's BAC be after two hours? After seven hours? Repeat your calculations using each of the models from part a.

Writing Assignment:

Your local Chief of Police must sometimes estimate a driver's BAC at the time of an accident by using a reading taken later. She has read the studies conducted by the Traffic Institute and by Heublein and has seen your work from Task 2. She would like to know how nearly compatible the three models are and how the predicted rates of elimination compare. Organize your results from Task 5b into tabular form, with an accompanying nontechnical explanation, to make the comparison as clear as possible.

8 Techniques of Integration

We have seen that the solution of many problems that arise in science and engineering requires the evaluation of an integral. We have also seen that the Fundamental Theorem of Calculus can be used to evaluate an integral, if the appropriate integration formula either is known or can be found in a table of integrals. Computers and some calculators can also now be used to evaluate integrals.

It is important to understand how the form of integrals can be manipulated algebraically, even if integration tables or computers are used to evaluate integrals. In this chapter, we see how substitution, integration by parts, and partial fractions are used to change the form of integrals. We will use these methods to develop formulas for a table of integrals and illustrate the use of tables throughout the chapter. In later chapters, we will refer to a table of integrals for the evaluation of integrals.

8.1   REVIEW OF KNOWN FORMULAS

Connections

Integration formulas, 5.5, 7.2, 7.4–5.

Substitution and change of variable, 5.5.

Let us review the integration formulas that we know:

$$\int u^\alpha \, du = \frac{u^{\alpha+1}}{\alpha+1} + C, \qquad \alpha \neq -1.$$

$$\int \frac{1}{u} \, du = \ln|u| + C.$$

$$\int e^u \, du = e^u + C.$$

$$\int \cos u \, du = \sin u + C.$$

$$\int \sin u \, du = -\cos u + C.$$

$$\int \sec^2 u \, du = \tan u + C.$$

$$\int \csc^2 u \, du = -\cot u + C.$$

$$\int \sec u \tan u \, du = \sec u + C.$$

$$\int \csc u \cot u \, du = -\csc u + C.$$

$$\int \frac{1}{\sqrt{a^2 - u^2}} \, du = \sin^{-1} \frac{u}{a} + C.$$

$$\int \frac{1}{a^2 + u^2} \, du = \frac{1}{a} \tan^{-1} \frac{u}{a} + C.$$

$$\int \frac{1}{u\sqrt{u^2 - a^2}} \, du = \frac{1}{a} \sec^{-1} \frac{|u|}{a} + C = \frac{1}{a} \cos^{-1} \frac{a}{|u|} + C.$$

A simple substitution will often transform a given integral into one of the above basic types. In later sections we will learn more sophisticated methods for evaluating integrals, but we should always look first for a simple solution.

Recall that an integration formula can be checked by differentiation. That is, we can verify the formula $\int f(x)\,dx = F(x) + C$ by checking that $F' = f$. In many integration problems it is very easy to differentiate the answer mentally, and we should do so. In practice, as the answers become more complicated, we will not check every problem, but the possibility is still there.

Linear Substitutions

The simplest substitutions are linear substitutions $u = ax + b$, where a and b are constants and $a \neq 0$. The differential is then $du = a\,dx$. We can easily modify the integrand without changing its value by introducing the unit factor $\left(\frac{1}{a}\right)(a)$, so the integrand contains the derivative $u'(x) = a$.

EXAMPLE 1

Evaluate $\int e^{x/3}\,dx$.

SOLUTION The integrand is e with exponent the linear function $x/3$. Let us try to simplify the integral by setting $u = x/3$, so $du = \dfrac{1}{3}\,dx$. We will introduce the factor (3) $\left(\dfrac{1}{3}\right)$ in the integrand so it contains the exact expression $du = \dfrac{1}{3}\,dx$. We have

$$
\begin{aligned}
\int e^{x/3}\,dx &= \int e^{x/3}(3)\left(\frac{1}{3}\right)dx \\
&= \int e^{u}(3)\,du \\
&= 3e^{u} + C \\
&= 3e^{x/3} + C.
\end{aligned}
$$

This result is easily checked. Using the Chain Rule, we have

$$
\frac{d}{dx}(3e^{x/3}) = 3e^{x/3}\left(\frac{1}{3}\right) = e^{x/3},
$$

so the integration result is correct. ∎

EXAMPLE 2

Evaluate $\int \sin(2x + 1)\,dx$.

SOLUTION The integrand is the sine of the linear function $2x + 1$. We try to simplify the integral by setting $u = 2x + 1$, so $du = 2\,dx$. We then have

$$
\begin{aligned}
\int \sin(2x + 1)\,dx &= \int \sin(2x + 1)\left(\frac{1}{2}\right)(2)\,dx \\
&= \int \sin u \left(\frac{1}{2}\right)du \\
&= \frac{1}{2}(-\cos u) + C \\
&= -\frac{1}{2}\cos(2x + 1) + C.
\end{aligned}
$$

Using the Chain Rule to check our work, we have

$$
\frac{d}{dx}\left(-\frac{1}{2}\cos(2x + 1)\right) = -\frac{1}{2}(-\sin(2x + 1))(2) = \sin(2x + 1). \quad ∎
$$

Generally, it is a good idea to write each step of a substitution, even if it is a simple substitution. However, as you gain experience and confidence, you may carry out simple substitutions without writing the details. When you do this, you should be sure to check the result by differentiating, at least mentally. For example, in Example 1 you may use what you know about the derivative of the exponential function to conclude that $\int e^{x/3}\,dx$ is some constant multiple of $e^{x/3}$. You may guess that

$$
\int e^{x/3}\,dx = 3e^{x/3} + C.
$$

You can then check your guess by differentiation. If your guess is incorrect, then the check should indicate the correct antiderivative. That is, if you guessed

$$\int e^{x/3}\,dx \underset{(?)}{=} e^{x/3},$$

your check,

$$\frac{d}{dx}(e^{x/3}) = e^{x/3}\left(\frac{1}{3}\right),$$

would indicate that you need a factor of 3 to cancel the $1/3$ and give the integrand. Note that the Chain Rule must be used when checking this type of indefinite integral by differentiation.

In Example 2, you may guess that

$$\int \sin(2x + 1)\,dx = -\frac{1}{2}\cos(2x + 1) + C,$$

and then check by differentiating.

Note that **guess and check** is a legitimate method of evaluating some *simple* indefinite integrals. However, guessing without checking by differentiating is not recommended.

NOTE   *If you do not write the details of a substitution in an integral, always check your work by differentiation.*

The differentiation may be done mentally, but don't forget to do it.

It is not worthwhile to try to evaluate any but the simplest, most familiar types of integrals by guessing. Also, checking by differentiation could be more difficult than using an integration method to evaluate the integral.

Choosing a Simplifying Substitution

The change of variable formula,

$$\int f(u(x))u'(x)\,dx = \int f(u)\,du,$$

requires that the integrand of the original integral contain a factor that is a constant multiple of $u'(x)$, the derivative of $u(x)$. When considering a simplifying substitution $u = u(x)$, we should check that the integrand contains the *essential*, nonconstant factors of the derivative $u'(x)$.

EXAMPLE 3
Evaluate $\int x\sqrt{x^2 + 1}\,dx$.

SOLUTION   This integral involves the square root of a function. Let us try to reduce the integral to $\int u^{1/2}\,du$. We see that the substitution $u = x^2 + 1$, $du = 2x\,dx$ will work because the integrand contains the *essential factor x*, the nonconstant factor of the derivative of $x^2 + 1$. We have

$$\int x\sqrt{x^2 + 1}\,dx = \int \left(\frac{1}{2}\right)2x(x^2 + 1)^{1/2}\,dx$$

$$= \int \frac{1}{2}u^{1/2}\,du$$

$$= \frac{1}{2}\left(\frac{u^{3/2}}{3/2}\right) + C$$

$$= \frac{1}{3}(x^2 + 1)^{3/2} + C.$$

Checking our work, we have

$$\frac{d}{dx}\left(\frac{1}{3}(x^2 + 1)^{3/2}\right) = \frac{1}{3}\left(\frac{3}{2}\right)(x^2 + 1)^{1/2}(2x) = x(x^2 + 1)^{1/2}. \quad \blacksquare$$

EXAMPLE 4

Evaluate $\displaystyle\int \frac{x^3}{x^4 + 1}\,dx$.

SOLUTION   We see that the numerator contains the essential factor x^3 necessary for the simplifying substitution $u = x^4 + 1$, $du = 4x^3 dx$.

$$\int \frac{x^3}{x^4 + 1}\,dx = \int \frac{(1/4)4x^3}{x^4 + 1}\,dx$$

$$= \int \frac{1}{4}u^{-1}\,du$$

$$= \frac{1}{4}\ln|u| + C$$

$$= \frac{1}{4}\ln(x^4 + 1) + C.$$

Checking this result, we have

$$\frac{d}{dx}\left(\frac{1}{4}\ln(x^4 + 1)\right) = \frac{1}{4}\left(\frac{1}{x^4 + 1}\right)(4x^3) = \frac{x^3}{x^4 + 1}. \quad \blacksquare$$

Some Simple Trigonometric Integrals

Since the simplifying substitution $u = u(x)$ requires that the integrand contain nonconstant factors of the derivative $u'(x)$, the factors of the integrand can suggest possible substitutions. That is, if we recognize a factor of the integrand as the derivative of a function $u(x)$, we may want to try the simplifying substitution $u = u(x)$. Let us see how this idea works with some integrals that have integrands that are powers of a trigonometric function times the derivative of that function.

EXAMPLE 5

Evaluate $\int \sin^2 x \cos x\,dx$.

SOLUTION   We first note that this integral does not match up with any of the formulas we have for the integral of trigonometric functions. The integrand involves powers of $\sin x$ and contains the factor $\cos x$, which we recognize as the derivative of $\sin x$. This means we can try the substitution $u = \sin x$, $du = \cos x\,dx$, which simplifies the integral.

$$\int \sin^2 x \cos x\,dx = \int u^2\,du$$

$$= \frac{u^3}{3} + C$$

$$= \frac{1}{3}(\sin x)^3 + C.$$

This result is easy enough to check. We have

$$\frac{d}{dx}\left(\frac{1}{3}(\sin x)^3\right) = \frac{1}{3}(3)(\sin x)^2(\cos x) = \sin^2 x \cos x. \quad \blacksquare$$

It should be coming clear that you must know the formulas for differentiation of the basic functions forwards and backwards. That is, it isn't enough to know that the derivative of $\sin x$ is $\cos x$. When you see $\cos x$, you must recognize that it is the derivative of $\sin x$.

EXAMPLE 6
Evaluate $\int \tan^4 x \sec^2 x \, dx$.

SOLUTION   The integrand involves powers of $\tan x$ and it contains the factor $\sec^2 x$, which we recognize as the derivative of $\tan x$. This suggests the substitution $u = \tan x$, so $du = \sec^2 x \, dx$.

$$\int \tan^4 x \sec^2 x \, dx = \int u^4 \, du$$

$$= \frac{u^5}{5} + C$$

$$= \frac{1}{5} \tan^5 x + C. \quad \blacksquare$$

EXAMPLE 7
Evaluate $\int \sec^3 x \tan x \, dx$.

SOLUTION   The integrand involves powers of $\sec x$, and it contains the factor $\sec x \tan x$, which is the derivative of $\sec x$. This suggests that we try $u = \sec x$, so $du = \sec x \tan x \, dx$. We have

$$\int \sec^3 x \tan x \, dx = \int \sec^2 x \sec x \tan x \, dx$$

$$= \int u^2 \, du$$

$$= \frac{u^3}{3} + C$$

$$= \frac{1}{3} \sec^3 x + C. \quad \blacksquare$$

Rewriting Integrands

EXAMPLE 8

Evaluate $\displaystyle\int \frac{1}{x\sqrt{4x^2-9}}\,dx$.

SOLUTION   We can put this integral into the exact form of a known integral by factoring the coefficient of the quadratic term. We have

$$\int \frac{1}{x\sqrt{4x^2-9}}\,dx = \int \frac{1}{x\sqrt{4(x^2-9/4)}}\,dx$$

$$= \frac{1}{2}\int \frac{1}{x\sqrt{x^2-(3/2)^2}}\,dx$$

$$= \frac{1}{2}\frac{1}{3/2}\cos^{-1}\frac{3/2}{|x|}+C$$

$$= \frac{1}{3}\cos^{-1}\frac{3}{2|x|}+C. \quad \blacksquare$$

You should be alert to the possibility of writing an integral as a sum and evaluating each part separately.

EXAMPLE 9

Evaluate $\displaystyle\int \frac{x+2}{\sqrt{4-x^2}}\,dx$.

SOLUTION   Writing the integral as a sum, we have

$$\int \frac{x+2}{\sqrt{4-x^2}}\,dx = \int \frac{x}{\sqrt{4-x^2}}\,dx + \int \frac{2}{\sqrt{4-x^2}}\,dx.$$

We see that the first integral on the right contains the factor x, which is the essential factor of the derivative of $4-x^2$. We use the substitution $u = 4-x^2$, so $du = -2x\,dx$, to evaluate this integral. The second integral on the right is already of the form of one of our integration formulas. We then have

$$\int \frac{x+2}{\sqrt{4-x^2}}\,dx = \int \frac{x}{\sqrt{4-x^2}}\,dx + \int \frac{2}{\sqrt{4-x^2}}\,dx$$

$$= \int (4-x^2)^{-1/2}\left(-\frac{1}{2}\right)(-2x)\,dx + 2\int \frac{1}{\sqrt{2^2-x^2}}\,dx$$

$$= \int u^{-1/2}\left(-\frac{1}{2}\right)du + 2\int \frac{1}{\sqrt{2^2-x^2}}\,dx$$

$$= -\frac{1}{2}\frac{u^{1/2}}{1/2} + 2\sin^{-1}\frac{x}{2} + C$$

$$= -\sqrt{4-x^2} + 2\sin^{-1}\frac{x}{2} + C. \quad \blacksquare$$

Definite Integrals

Recall that the formula for change of variables in a definite integral is

$$\int_a^b f(u(x))u'(x)\,dx = \int_{u(a)}^{u(b)} f(u)\,du.$$

It is important that the limits of integration correspond to the variable of integration.

EXAMPLE 10

Evaluate $\displaystyle\int_1^3 \frac{1}{1+x}\frac{1}{\sqrt{x}}\,dx$.

SOLUTION We see that the integrand contains the factor $1/\sqrt{x}$. This suggests that we may try the substitution $u = \sqrt{x}$, so $du = \dfrac{1}{2\sqrt{x}}\,dx$. Then $u(1) = \sqrt{1} = 1$ and $u(3) = \sqrt{3}$, so we have

$$\int_1^3 \frac{1}{1+x}\frac{1}{\sqrt{x}}\,dx = \int_1^3 \frac{1}{1+(\sqrt{x})^2}(2)\left(\frac{1}{2\sqrt{x}}\right)dx$$

$$= \int_1^{\sqrt{3}} \frac{1}{1+u^2}(2)\,du \qquad \text{[Note that we have changed the limits of integration to correspond to the new variable of integration.]}$$

$$= 2\tan^{-1} u\,\Big]_1^{\sqrt{3}}$$

$$= 2\tan^{-1}\sqrt{3} - 2\tan^{-1} 1 \approx 0.5236.$$

(You may recognize that

$$2\tan^{-1}\sqrt{3} - 2\tan^{-1} 1 = 2\left(\frac{\pi}{3}\right) - 2\left(\frac{\pi}{4}\right) = \frac{\pi}{6},$$

so the exact answer can be expressed in simplified form.) ∎

EXERCISES 8.1

Evaluate the integrals in Exercises 1–36.

1. $\displaystyle\int \frac{x}{\sqrt{x^2+4}}\,dx$

2. $\displaystyle\int x\sqrt{1-x^2}\,dx$ **(Calculus Explorer & Tutor II, Techniques of Integration, 1.)**

3. $\displaystyle\int \frac{1}{\sqrt{1-9x^2}}\,dx$

4. $\displaystyle\int \frac{x}{\sqrt{1-9x^2}}\,dx$

5. $\displaystyle\int \frac{\cos x}{2-\sin x}\,dx$

6. $\displaystyle\int \frac{e^x - e^{-x}}{e^x + e^{-x}}\,dx$

7. $\displaystyle\int xe^{-x^2}\,dx$

8. $\displaystyle\int \frac{(\ln x)^3}{x}\,dx$

9. $\displaystyle\int \frac{1}{x\ln x}\,dx$

10. $\displaystyle\int e^{\sin x}\cos x\,dx$

11. $\displaystyle\int \sin(7x-3)\,dx$ **(Calculus Explorer & Tutor II, Introduction to Integration, 7.)**

12. $\displaystyle\int \cos(x/2)\,dx$

13. $\displaystyle\int \frac{\sec(x^{2/3})\tan(x^{2/3})}{x^{1/3}}\,dx$

14. $\displaystyle\int x\sec^2(x^2)\,dx$

15. $\displaystyle\int e^{3x-1}\,dx$

16. $\displaystyle\int \frac{1}{1-2x}\,dx$

17. $\displaystyle\int \frac{1}{(3x-1)^2}\,dx$

18. $\displaystyle\int \frac{x}{(x^2+2)^2}\,dx$

19. $\displaystyle\int \frac{1}{2+9x^2}\,dx$

20. $\displaystyle\int \frac{1}{\sqrt{2-9x^2}}\,dx$

21. $\displaystyle\int \csc^2(2x)\,dx$

22. $\displaystyle\int \frac{x\sin\sqrt{2x^2+1}}{\sqrt{2x^2+1}}\,dx$ **(Calculus Explorer & Tutor II, Introduction to Integration, 5.)**

23. $\displaystyle\int \sin^4 x \cos x \, dx$

24. $\displaystyle\int \frac{\sin x}{\cos^2 x} \, dx$

25. $\displaystyle\int \tan^5 x \sec^2 x \, dx$

26. $\displaystyle\int \frac{\sec^2 x}{\tan x} \, dx$

27. $\displaystyle\int \sec^5 x \tan x \, dx$

28. $\displaystyle\int \frac{\tan x}{\sec^3 x} \, dx$

29. $\displaystyle\int \frac{e^x}{\sqrt{1 - e^{2x}}} \, dx$

30. $\displaystyle\int \frac{e^x}{\sqrt{9 - e^{2x}}} \, dx$

31. $\displaystyle\int \frac{1}{\sqrt{4x - x^2}} \, dx$

32. $\displaystyle\int \frac{1}{\sqrt{9x - x^2}} \, dx$

33. $\displaystyle\int \frac{1}{x^2 + 2x + 5} \, dx$

34. $\displaystyle\int \frac{1}{x^2 - 4x + 6} \, dx$

35. $\displaystyle\int \frac{x}{x^2 - 2x + 5} \, dx$

36. $\displaystyle\int \frac{x}{x^2 + 6x + 10} \, dx$

Evaluate the definite integrals in Exercises 37–46.

37. $\displaystyle\int_{-1}^{1} e^{2x} \, dx$

38. $\displaystyle\int_{0}^{\pi/2} \sin(2x) \, dx$

39. $\displaystyle\int_{0}^{1} \frac{1}{\sqrt{4 - x^2}} \, dx$

40. $\displaystyle\int_{0}^{2} \frac{1}{x^2 + 4} \, dx$

41. $\displaystyle\int_{0}^{1} \frac{x}{\sqrt{4 - x^2}} \, dx$

42. $\displaystyle\int_{0}^{2} \frac{x}{x^2 + 4} \, dx$

43. $\displaystyle\int_{0}^{\pi/4} \sin 2x \cos 2x \, dx$

44. $\displaystyle\int_{0}^{\pi/2} \tan (x/2) \sec^2 (x/2) \, dx$

45. $\displaystyle\int_{0}^{1} \frac{2x + 1}{x^2 + 1} \, dx$

46. $\displaystyle\int_{0}^{\sqrt{3}/2} \frac{1 - x}{\sqrt{1 - x^2}} \, dx$

8.2   INTEGRATION BY PARTS

Connections

Product Rule, 3.2.
Fundamental Theorem, 5.4.

We know that the formula for the differentiation of a product is

$$(uv)'(x) = u(x)v'(x) + v(x)u'(x).$$

The corresponding integration formula is

$$\int (u(x)v'(x) + v(x)u'(x)) \, dx = u(x)v(x) + C.$$

Since $\int 0 \, dx = C$, we can rewrite this equation as

$$\int (u(x)v'(x) + v(x)u'(x)) \, dx = u(x)v(x) + \int 0 \, dx.$$

We now subtract $\int v(x)u'(x) \, dx$ from each side of the above equation and combine the integrals on each side to obtain

$$\int u(x)v'(x) \, dx = u(x)v(x) - \int v(x)u'(x) \, dx. \qquad (1)$$

This formula is called the **Integration by Parts Formula.**

It is convenient to express the Integration by Parts Formula in terms of differentials. Substituting $du = u'(x) \, dx$ and $dv = v'(x) \, dx$ into Formula 1, we obtain

$$\int u \, dv = uv - \int v \, du. \qquad (2)$$

This is the form of the Integration by Parts Formula that we will use. Note that the integrand of the integral on the left contains a factor that is the derivative of v and the integrand of the integral on the right contains a factor that is the derivative of u, as in Formula 1.

Integration by parts can be used to change the form of an integral. In many cases it is easier to evaluate $\int v \, du$ than it is to evaluate $\int u \, dv$. Depending on the interpretation of the variables, the formula for integration by parts can give useful relations between physical quantities.

When you integrate by parts

- Choose u to be a factor of the integrand that has a derivative that is "simpler" than the function u. (The functions x^n, $\ln x$, and the inverse trigonometric functions are good choices for u.)
- The choice of u determines what $dv = v'\,dx$ must be, since the integrand must be the product uv'. (We must be able to find an antiderivative of the function v'. The function v can be any antiderivative of the function v'. Different antiderivatives differ by a constant.)
- The function v should be "no worse" than the function v' that is a factor of the original integral. (Powers of x, exponentials, and the sine and cosine functions are good candidates for v'.)

NOTE *Do not try to use integration by parts to evaluate all integrals. Always look for a simple substitution before trying to integrate by parts.*

Eliminating Powers of x

Some integrals can be simplified by eliminating a power of x from the integrand. Note that $x^0 = 1$ is a "simpler" factor than other powers of x.

EXAMPLE 1
Evaluate $\int x \sin x\,dx$.

SOLUTION We first note that a simple substitution does not reduce this integral to one of the types of Section 8.1. We then see that we could easily evaluate the integral, if we could replace the factor x in the integrand with the simpler factor one. This suggests that we try integration by parts with $u = x$, since differentiation of x gives the constant function 1. That is, $u = x$ implies $du = dx$, so du does not contain the factor x that we were trying to eliminate.

Since $u\,dv = x \sin x\,dx$ and we have chosen $u = x$, we must have $dv = \sin x\,dx$. Antiderivatives of $\sin x$ are all of the form $-\cos x + C$. We choose $v = -\cos x$, the most convenient of these antiderivatives. Note that $-\cos x$ is "no worse" to integrate than the factor $\sin x$ in the given integral.

We have

$$u = x, \qquad dv = \sin x\,dx,$$
$$du = dx, \qquad v = -\cos x.$$

Then

$$\int x \sin x\,dx = \int u\,dv = uv - \int v\,du$$
$$= (x)(-\cos x) - \int (-\cos x)\,dx$$
$$= -x \cos x + \sin x + C.$$

The formula for the differentiation of a product is required to check this integral. We have

$$\frac{d}{dx}(-x \cos x + \sin x) = -((x)(-\sin x) + (\cos x)(1)) + \cos x$$
$$= x \sin x. \quad \blacksquare$$

Note what would happen in Example 1 if we had chosen $u = \sin x$ and $dv = x\, dx$. We would then have $du = \cos x\, dx$ and $v = x^2/2$. The Integration by Parts Formula would then give

$$\int x \sin x\, dx = \int u\, dv = uv - \int v\, du$$

$$= (\sin x)\left(\frac{x^2}{2}\right) - \int \left(\frac{x^2}{2}\right)(\cos x)\, dx.$$

We see that we would obtain an integral that is not as simple as the original one.

NOTE   *If integration by parts seems to result in a more complicated integral, try a different choice of u and dv.*

We may need to integrate by parts more than once to evaluate an integral. In that case we should check that we are simplifying the integral at each step.

EXAMPLE 2
Evaluate $\int x^2 e^{2x}\, dx$.

SOLUTION   We would like to eliminate the factor x^2 in this integral. We can at least lower the power of x by one if we integrate by parts with $u = x^2$. Then

$$u = x^2, \qquad dv = e^{2x}\, dx,$$

$$du = 2x\, dx, \qquad v = \frac{1}{2}e^{2x}.$$

(To find a function v that satisfies $dv = e^{2x}\, dx$, we need to evaluate the integral $\int e^{2x}\, dx$. This involves the substitution $u = 2x$, $du = 2\, dx$, which can be carried out mentally. You can easily check that $v = e^{2x}/2$ satisfies $dv = e^{2x}\, dx$ by using the Chain Rule to differentiate v mentally.) Then

$$\int x^2 e^{2x}\, dx = \int u\, dv = uv - \int v\, du$$

$$= (x^2)\left(\frac{1}{2}e^{2x}\right) - \int \frac{1}{2}e^{2x} 2x\, dx$$

$$= \frac{1}{2}x^2 e^{2x} - \int e^{2x} x\, dx.$$

We integrate by parts once again to eliminate the remaining power of x. We use

$$u = x, \qquad dv = e^{2x}\, dx,$$

$$du = dx, \qquad v = \frac{1}{2}e^{2x}.$$

We then have

$$\int x^2 e^{2x}\, dx = \frac{1}{2}x^2 e^{2x} - \left(\int u\, dv\right)$$

$$= \frac{1}{2}x^2 e^{2x} - \left(uv - \int v\, du\right)$$

$$= \frac{1}{2}x^2 e^{2x} - \left((x)\left(\frac{1}{2}e^{2x} \right) - \int \frac{1}{2}e^{2x}\,dx \right)$$

$$= \frac{1}{2}x^2 e^{2x} - \frac{1}{2}xe^{2x} + \frac{1}{2}\int e^{2x}\,dx$$

$$= \frac{1}{2}x^2 e^{2x} - \frac{1}{2}xe^{2x} + \frac{1}{4}e^{2x} + C. \quad \blacksquare$$

Choosing *u* so *du* Involves Powers of *x*

The functions $\ln x$ and the inverse trigonometric functions have derivatives that involve only power functions. This means that we can use Integration by Parts to transform some integrals that involve either $\ln x$ or inverse trigonometric functions into integrals that involve only power functions.

EXAMPLE 3
Evaluate $\int x \ln x\,dx$.

SOLUTION   It wouldn't do much good to eliminate x in this example, because it isn't particularly easy to integrate $\ln x$. Let us try $u = \ln x$. Then du is a power of x and powers of x are generally not too bad to work with. Also, $dv = x\,dx$ gives a v that is a power of x. We have

$$u = \ln x, \qquad dv = x\,dx,$$
$$du = \frac{1}{x}\,dx, \qquad v = \frac{x^2}{2}.$$

Then

$$\int x \ln x\,dx = \int u\,dv = uv - \int v\,du$$

$$= (\ln x)\left(\frac{x^2}{2} \right) - \int \left(\frac{x^2}{2} \right)\left(\frac{1}{x} \right) dx$$

$$= (\ln x)\left(\frac{x^2}{2} \right) - \int \frac{x}{2}\,dx$$

$$= \frac{1}{2}x^2 \ln x - \frac{1}{4}x^2 + C. \quad \blacksquare$$

EXAMPLE 4
Evaluate $\int \sin^{-1} x\,dx$.

SOLUTION   The derivative of $\sin^{-1} x$ involves powers of x, so let us try to integrate by parts with $u = \sin^{-1} x$. We must then have $dv = dx$. We have

$$u = \sin^{-1} x, \qquad dv = dx,$$
$$du = \frac{1}{\sqrt{1 - x^2}}\,dx, \qquad v = x.$$

Then

$$\int \sin^{-1} x\,dx = \int u\,dv$$

$$= uv - \int v \, du$$

$$= (\sin^{-1} x)(x) - \int x \frac{1}{\sqrt{1 - x^2}} \, dx.$$

The integral above on the right is evaluated by using the substitution $u = 1 - x^2$, so $du = -2x \, dx$. We have

$$\int \sin^{-1} x \, dx = x \sin^{-1} x - \int (1 - x^2)^{-1/2} \left(-\frac{1}{2}\right)(-2x) \, dx$$

$$= x \sin^{-1} x - \left(-\frac{1}{2}\right) \left(\frac{(1 - x^2)^{1/2}}{1/2}\right) + C$$

$$= x \sin^{-1} x + \sqrt{1 - x^2} + C. \quad \blacksquare$$

Integrals That Involve Exponential and Trigonometric Functions

The following formulas are useful for evaluating some integrals that will occur in later problems. These formulas are contained in the Table of Integrals found in the inside back cover of the text and need not be memorized.

$$\int e^{ax} \cos bx \, dx = \frac{e^{ax}}{a^2 + b^2} (b \sin bx + a \cos bx) + C. \qquad (3)$$

$$\int e^{ax} \sin bx \, dx = \frac{e^{ax}}{a^2 + b^2} (a \sin bx - b \cos bx) + C. \qquad (4)$$

Proof   Let us derive (3) in the special case $a = b = 1$. The technique involves integrating by parts twice. The method works because integrating (or differentiating) $\cos x$ twice gives us a negative multiple of $\cos x$, while differentiating (or integrating) e^x twice gives us a positive multiple of e^x. The result is that the original integral appears on the opposite side of the equation, multiplied by a factor that is not one. We have

$$u = e^x, \qquad dv = \cos x \, dx,$$
$$du = e^x dx, \qquad v = \sin x.$$

Then

$$\int e^x \cos x \, dx = \int u \, dv = uv - \int v \, du$$

$$= e^x \sin x - \int e^x \sin x \, dx.$$

We now use

$$u = e^x, \qquad dv = \sin x \, dx,$$
$$du = e^x dx, \qquad v = -\cos x.$$

We then have

$$\int e^x \cos x \, dx = e^x \sin x - \left(\int u \, dv\right)$$

$$= e^x \sin x - \left(uv - \int v \, du\right)$$

$$= e^x \sin x - \left(e^x(-\cos x) - \int e^x(-\cos x)dx \right)$$

$$= e^x \sin x + e^x \cos x - \int e^x \cos x\, dx.$$

We now have

$$\int e^x \cos x\, dx = e^x \sin x + e^x \cos x - \int e^x \cos x\, dx.$$

We now add $\int e^x \cos x\, dx$ to each side of the above equation and combine integrals on each side to obtain

$$2 \int e^x \cos x\, dx = e^x \sin x + e^x \cos x + \int 0\, dx,$$

$$\int e^x \cos x\, dx = \frac{e^x}{2}(\sin x + \cos x) + \int \frac{0}{2} dx,$$

$$\int e^x \cos x\, dx = \frac{e^x}{2}(\sin x + \cos x) + C. \quad \blacksquare$$

It doesn't matter which of the two factors of the integrand were chosen for u and dv in the first integration by parts above. However, we must be very careful to choose u and dv in the second integration by parts so that one function is differentiated twice while the other function is integrated twice. We would simply undo the previous integration by parts if we integrated the function we just obtained by differentiation and differentiated the function we just obtained by integration.

EXAMPLE 5

Evaluate $\displaystyle\int_0^{4\pi} e^{-x} \sin 2x\, dx$.

SOLUTION We recognize that this integral is of the form of Formula 4 with $a = -1$ and $b = 2$. Substitution into Formula 4 then gives

$$\int_0^{4\pi} e^{-x} \sin 2x\, dx = \frac{e^{-x}}{(-1)^2 + (2)^2} \left((-1) \sin 2x - 2 \cos 2x \right) \Bigg]_0^{4\pi}$$

$$= \frac{e^{-4\pi}}{5}(-\sin 8\pi - 2\cos 8\pi) - \frac{e^0}{5}(-\sin 0 - 2\cos 0)$$

$$= \frac{e^{-4\pi}}{5}(-2) - \frac{1}{5}(-2) = \frac{2}{5}(1 - e^{-4\pi}) \approx 0.3999986. \quad \blacksquare$$

EXERCISES 8.2

Use integration by parts to evaluate the integrals in Exercises 1–18.

1. $\displaystyle\int xe^{2x}\, dx$

2. $\displaystyle\int xe^{-x}\, dx$

3. $\displaystyle\int x \cos(x/2)\, dx$

4. $\displaystyle\int x \sin 2x\, dx$ **(Calculus Explorer & Tutor II, Techniques of Integration, 2.)**

5. $\int x \dfrac{2x}{(1-x^2)^{3/2}} \, dx$

6. $\int x^2 \dfrac{2x}{(x^2+1)^2} \, dx$

7. $\int \sqrt{x} \dfrac{\sin\sqrt{x}}{\sqrt{x}} \, dx$

8. $\int x^2 \cos(x/2) \, dx$

9. $\int x^3 e^{-x} \, dx$

10. $\int x^3 \sin x \, dx$

11. $\int \sqrt{x} \ln x \, dx$

12. $\int x^3 \ln x \, dx$

13. $\int \tan^{-1} x \, dx$

14. $\int \ln x \, dx$ (**Calculus Explorer & Tutor II, Techniques of Integration, 3.**)

15. $\int x \tan^{-1} x \, dx$

16. $\int x \sec^{-1} |x| \, dx$

17. $\int e^{\sqrt{x}} \, dx$

18. $\int \sin\sqrt{x} \, dx$

Evaluate the definite integrals in Exercises 19–22.

19. $\int_0^1 x e^x \, dx$

20. $\int_0^{\pi/2} x \cos x \, dx$

21. $\int_1^e \ln x \, dx$

22. $\int_0^{\pi} e^{-x} \sin x \, dx$

23. Find the volume of the solid obtained by revolving the region bounded by $y = \ln x$, $y = 0$, $1 \le x \le e$, about the x-axis.

24. Find the volume of the solid obtained by revolving the region bounded by $y = \cos x$, $y = 0$, $0 \le x \le \pi/2$, about the y-axis.

Use the technique that was used to verify the special case of Formula 3 to evaluate the integrals in Exercises 25–28.

25. $\int e^{2x} \cos x \, dx$

26. $\int e^x \sin x \, dx$

27. $\int \sin(\ln x) \, dx$

28. $\int \cos(\ln x) \, dx$

29. Use integration by parts to derive the formula

$$\int x^n \ln x \, dx = \frac{x^{n+1} \ln x}{n+1} - \frac{x^{n+1}}{(n+1)^2} + C.$$

30. Use integration by parts to derive the formula

$$\int (\ln x)^n \, dx = x(\ln x)^n - n \int (\ln x)^{n-1} \, dx.$$

Use Formulas 3–4 and the results of Exercises 29 and 30 to evaluate the integrals in Exercises 31–38.

31. $\int e^{-x} \sin(x/2) \, dx$

32. $\int e^{2x} \sin 3x \, dx$

33. $\int e^{-3x} \cos 4x \, dx$

34. $\int e^{-2x} \cos 3x \, dx$

35. $\int (\ln x)^2 \, dx$

36. $\int (\ln x)^3 \, dx$

37. $\int x^3 \ln x \, dx$

38. $\int x^2 \ln x \, dx$

39. Use integration by parts to derive the formula

$$\int e^{ax} \cos bx \, dx = \frac{e^{ax}}{a^2+b^2} (b \sin bx + a \cos bx) + C.$$

8.3 TRIGONOMETRIC INTEGRALS; REDUCTION FORMULAS

Connections

Simple trigonometric formulas, 1.6.

In this section, we will use integration by parts to develop formulas to evaluate powers of the sine and cosine functions. We will also see that some trigonometric integrals can be evaluated by a simplifying substitution or by using trigonometric identities.

Simple Integrals of Powers of $\sin x$ and $\cos x$

Generally, we will say an integral is *simple* to integrate if a simplifying substitution reduces it to one of the fundamental formulas that we know. Integrals of the form $\int \sin^m x \cos x \, dx$ are *simple* to evaluate by using the simplifying substitution $u = \sin x$, so $du = \cos x \, dx$. Similarly, $\int \cos^n x \sin x \, dx$ is *simple* to integrate if we use the substitution $u = \cos x$, so $du = -\sin x \, dx$.

EXAMPLE 1

Evaluate $\int \sin^3 x \cos x \, dx$.

SOLUTION This is of the *simple* type, since the integrand involves powers of $\sin x$ and contains the essential factor $\cos x$. Setting $u = \sin x$, so $du = \cos x \, dx$, we have

$$\int \sin^3 x \cos x \, dx = \int u^3 du$$

$$= \frac{u^4}{4} + C$$

$$= \frac{\sin^4 x}{4} + C. \quad \blacksquare$$

If n is an odd integer, $\int \sin^m x \cos^n x \, dx$ can be simplified by using the substitution $u = \sin x$, so $du = \cos x \, dx$. To do this we save one of the odd number of factors of $\cos x$ to form du and use the identity $\sin^2 x + \cos^2 x = 1$ to express the other even number of factors of $\cos x$ in terms of $\sin x$. An analogous technique can be used if m is an odd integer.

EXAMPLE 2

Evaluate $\int \sin^3 x \cos^3 x \, dx$.

SOLUTION We see that the power of $\cos x$ in the integrand is an odd integer. We then save one factor of $\cos x$ and use $\sin^2 x + \cos^2 x = 1$ to express the remaining even number of factors of $\cos x$ in terms of $\sin x$. The substitution $u = \sin x$, $du = \cos x \, dx$, then simplifies the integral. We have

$$\int \sin^3 x \cos^3 x \, dx = \int \sin^3 x \cos^2 x \cos x \, dx$$

$$= \int \sin^3 x (1 - \sin^2 x) \cos x \, dx$$

$$= \int u^3 (1 - u^2) \, du$$

$$= \int (u^3 - u^5) \, du$$

$$= \frac{u^4}{4} - \frac{u^6}{6} + C$$

$$= \frac{\sin^4 x}{4} - \frac{\sin^6 x}{6} + C.$$

Since the power of $\sin x$ in the integrand is odd, we could also evaluate $\int \sin^3 x \cos^3 x \, dx$ by saving one factor of $\sin x$, using $\sin^2 x + \cos^2 x = 1$ to express the other factors of $\sin x$ in terms of $\cos x$, and then using the substitution $u = \cos x$, $du = -\sin x \, dx$. This gives

$$\int \sin^3 x \cos^3 x \, dx = \int \sin^2 x \cos^3 x \sin x \, dx$$

$$= \int (1 - \cos^2 x) \cos^3 x (-1)(-\sin x) \, dx$$

$$= \int (1 - u^2) u^3 (-1) \, du$$

$$= \int (u^5 - u^3)\,du$$

$$= \frac{u^6}{6} - \frac{u^4}{4} + C$$

$$= \frac{\cos^6 x}{6} - \frac{\cos^4 x}{4} + C.$$

The two answers are different, but we can check by differentiation that they are both correct. That is,

$$\frac{d}{dx}\left(\frac{\sin^4 x}{4} - \frac{\sin^6 x}{6}\right) = \sin^3 x \cos x - \sin^5 x \cos x$$

$$= \sin^3 x (1 - \sin^2 x) \cos x = \sin^3 x \cos^3 x$$

and

$$\frac{d}{dx}\left(\frac{\cos^6 x}{6} - \frac{\cos^4 x}{4}\right) = \cos^5 x(-\sin x) - \cos^3 x(-\sin x)$$

$$= \sin x(-\cos^2 x + 1)\cos^3 x = \sin^3 x \cos^3 x. \quad \blacksquare$$

Reduction Formulas for Powers of $\sin x$ and $\cos x$

If m and n are both even integers, we cannot simplify $\int \sin^m x \cos^n x\,dx$ as we can if either m or n is an odd integer. If m and n are both even, we use the following **reduction formulas** to reduce powers of the trigonometric functions in the integrands. These formulas are contained in the Table of Integrals given in the inside back cover of the text and need not be memorized.

$$\int \sin^n x\,dx = -\frac{\sin^{n-1} x \cos x}{n} + \frac{n-1}{n}\int \sin^{n-2} x\,dx. \qquad (1)$$

$$\int \cos^n x\,dx = \frac{\cos^{n-1} x \sin x}{n} + \frac{n-1}{n}\int \cos^{n-2} x\,dx. \qquad (2)$$

$$\int \sin^2 x\,dx = -\frac{\sin x \cos x}{2} + \frac{x}{2} + C. \qquad (3)$$

$$\int \cos^2 x\,dx = \frac{\sin x \cos x}{2} + \frac{x}{2} + C. \qquad (4)$$

$$\int \sin^m x \cos^n x\,dx = -\frac{\sin^{m-1} x \cos^{n+1} x}{m+n} + \frac{m-1}{m+n}\int \sin^{m-2} x \cos^n x\,dx. \qquad (5)$$

$$\int \sin^m x \cos^n x\,dx = \frac{\sin^{m+1} x \cos^{n-1} x}{m+n} + \frac{n-1}{m+n}\int \sin^m x \cos^{n-2} x\,dx. \qquad (6)$$

Let us see how to derive (2). We begin by writing $\cos^n x$ as a product and integrating by parts. We then use the identity $\sin^2 x + \cos^2 x = 1$. We have

$$u = \cos^{n-1} x, \qquad dv = \cos x \, dx,$$
$$du = (n-1)(\cos^{n-2} x)(-\sin x)dx, \qquad v = \sin x.$$

Then

$$\int \cos^n x \, dx = \int \cos^{n-1} x \cos x \, dx = \int u \, dv = uv - \int v \, du$$

$$= (\cos^{n-1} x)(\sin x) - \int (\sin x)(n-1)(\cos^{n-2} x)(-\sin x)dx$$

$$= \cos^{n-1} x \sin x + (n-1) \int \cos^{n-2} x \sin^2 x \, dx$$

$$= \cos^{n-1} x \sin x + (n-1) \int \cos^{n-2} x (1 - \cos^2 x)dx$$

$$= \cos^{n-1} x \sin x + (n-1) \int \cos^{n-2} x \, dx - (n-1) \int \cos^n x \, dx.$$

We now have

$$\int \cos^n x \, dx = \cos^{n-1} x \sin x + (n-1) \int \cos^{n-2} x \, dx - (n-1) \int \cos^n x \, dx.$$

Adding $(n-1) \int \cos^n x \, dx$ to each side of the above equation gives

$$n \int \cos^n x \, dx = \cos^{n-1} x \sin x + (n-1) \int \cos^{n-2} x \, dx,$$

so

$$\int \cos^n x \, dx = \frac{\cos^{n-1} x \sin x}{n} + \frac{n-1}{n} \int \cos^{n-2} x \, dx.$$

This gives Formula 2. Formulas 1, 5, and 6 can be derived similarly. Formulas 3 and 4 can be obtained by setting $n = 2$ in Formulas 1 and 2, respectively. For example, when $n = 2$ in Formula 2, we have

$$\int \cos^2 x \, dx = \frac{\sin x \cos^{2-1} x}{2} + \frac{2-1}{2} \int \cos^{2-2} x \, dx$$

$$= \frac{\sin x \cos x}{2} + \frac{1}{2} \int 1 \, dx \qquad [\cos^{2-2} x = (\cos x)^0 = 1]$$

$$= \frac{\sin x \cos x}{2} + \frac{x}{2} + C.$$

It is also possible to evaluate $\int \cos^2 x \, dx$ by using the trigonometric identity

$$\cos^2 x = \frac{1 + \cos 2x}{2}.$$

That is,

$$\int \cos^2 x \, dx = \int \frac{1 + \cos 2x}{2} dx = \frac{x}{2} + \left(\frac{1}{2}\right)\left(\frac{\sin 2x}{2}\right) + C.$$

The identity $\sin 2x = 2 \sin x \cos x$ can then be used to obtain the same expression obtained by using Formula 2. Similarly, we can evaluate $\int \sin^2 x \, dx$ by using either Formula 1 or the identity

$$\sin^2 x = \frac{1 - \cos 2x}{2}.$$

In any case, we now consider (3) and (4) to be established formulas that can be used in future applications.

EXAMPLE 3
Evaluate $\int \sin^2 x \cos^2 x \, dx$.

SOLUTION   This integral is not one of the *simple* types, so we will use the reduction formulas.

$$\int \sin^2 x \cos^2 x \, dx \underset{(5)}{=} -\frac{\sin x \cos^3 x}{4} + \frac{1}{4} \int \cos^2 x \, dx$$

$$\underset{(4)}{=} -\frac{\sin x \cos^3 x}{4} + \frac{1}{4} \left(\frac{\sin x \cos x}{2} + \frac{x}{2} \right) + C$$

$$= -\frac{\sin x \cos^3 x}{4} + \frac{\sin x \cos x}{8} + \frac{x}{8} + C. \quad \blacksquare$$

We could also evaluate the integral in Example 3 by applying (6) to reduce the power of $\cos x$, and then using (3). In that case we would obtain an answer that contains different powers of $\sin x$ and $\cos x$. Recall that two correct answers can have different forms.

Integrals of even powers of $\sin x$ and $\cos x$ can also be evaluated by using the trigonometric formulas

$$\sin^2 x = \frac{1 - \cos 2x}{2} \quad \text{and} \quad \cos^2 x = \frac{1 + \cos 2x}{2}$$

to reduce the powers. This method results in answers that involve the sine and cosine of multiples of x, unless you use additional trigonometric formulas to switch back to $\sin x$ and $\cos x$. This is not convenient for the work we will be doing in Section 8.4, so we will systematically use the reduction formulas.

In some cases we have a choice of writing an integral as a sum of integrals of *simple* type or using a reduction formula. For example, we can write $\int \cos^3 x \, dx = \int (1 - \sin^2 x) \cos x \, dx$ or we can use (2). Which to use is a matter of personal preference.

Changing Variables Before Using an Integration Formula

It is important in Formulas 1–6 that the trigonometric functions be functions of the variable of integration. If they are not, we must change variables before using the formulas.

EXAMPLE 4
Evaluate $\int \cos^2 2x \, dx$.

SOLUTION   We see that the integrand involves the cosine of $2x$, which is not identical to the cosine of the variable of integration. We use the substitution $u = 2x$, so $du = 2 \, dx$ and then apply Formula 2.

$$\int \cos^2 2x \, dx = \int \cos^2 2x \left(\frac{1}{2} \right) (2) \, dx$$

$$= \int \cos^2 u \left(\frac{1}{2}\right) du$$

$$= \frac{1}{2} \left(\int \cos^2 u \, du \right)$$

$$= \frac{1}{2} \left(\frac{\sin u \cos u}{2} + \frac{u}{2} \right) + C$$

$$= \frac{1}{2} \left(\frac{\sin 2x \cos 2x}{2} + \frac{2x}{2} \right) + C$$

$$= \frac{1}{4} \sin 2x \cos 2x + \frac{1}{2}x + C. \quad \blacksquare$$

Simple Integrals of Powers of sec x and tan x

Integrals that involve powers of sec x and tan x can be evaluated by using the same general principles that we used for powers of sin x and cos x. Namely:

Strategy for evaluating trigonometric integrals

- Look for a simple substitution.
- If you don't see a simple substitution, use a reduction formula.

Integrals of the form $\int \tan^n x \sec^2 x \, dx$ are *simple* to evaluate. We use $u = \tan x$, so $du = \sec^2 x \, dx$. Similarly, $\int \sec^{m-1} x \sec x \tan x \, dx$ is *simple* to evaluate. Use $u = \sec x$, so $du = \sec x \tan x \, dx$. If either m is even or n is odd, we can write $\int \sec^m x \tan^n x \, dx$ as sum of integrals of *simple* type. To do this we use $\tan^2 x + 1 = \sec^2 x$ to replace extra even powers.

EXAMPLE 5

Evaluate $\int \sec^7 x \tan x \, dx$.

SOLUTION The integrand can be written as a power of sec x times sec x tan x. We can then use the substitution $u = \sec x$, $du = \sec x \tan x \, dx$ to write the integral in the form of a *simple* type.

$$\int \sec^7 x \tan x \, dx = \int \sec^6 x \sec x \tan x \, dx$$

$$= \int u^6 du$$

$$= \frac{u^7}{7} + C$$

$$= \frac{1}{7} \sec^7 x + C. \quad \blacksquare$$

EXAMPLE 6

Evaluate $\int \tan^2 x \sec^4 x \, dx$.

SOLUTION We save $\sec^2 x$ and use $\sec^2 x = \tan^2 x + 1$ to express the remaining even power of sec x in terms of tan x. The substitution $u = \tan x$, $du = \sec^2 x \, dx$ then gives

$$\int \tan^2 x \sec^4 x \, dx = \int \tan^2 x (\sec^2 x) \sec^2 x \, dx$$

$$= \int \tan^2 x (\tan^2 x + 1) \sec^2 x \, dx$$

$$= \int u^2 (u^2 + 1) du$$

$$= \int (u^4 + u^2) du$$

$$= \frac{u^5}{5} + \frac{u^3}{3} + C$$

$$= \frac{1}{5} \tan^5 x + \frac{1}{3} \tan^3 x + C. \quad \blacksquare$$

Reduction Formulas for Powers of sec *x* and tan *x*

If m is odd and n is even, $\int \sec^m x \tan^n x \, dx$ is not a simple type. In this case we need the following results:

$$\int \tan x \, dx = \ln |\sec x| + C. \tag{7}$$

$$\int \sec x \, dx = \ln |\sec x + \tan x| + C. \tag{8}$$

Formula 7 was derived in Example 4 of Section 7.4 by writing $\tan x = \sin x / \cos x$ and then using the substitution $u = \cos x$. Formula 8 may be verified by differentiation. It can be derived by a clever trick:

$$\int \sec x \, dx = \int \sec x \frac{\tan x + \sec x}{\tan x + \sec x} \, dx = \int \frac{\sec x \tan x + \sec^2 x}{\sec x + \tan x} \, dx.$$

The substitution $u = \sec x + \tan x$ then gives (8).

Higher powers of $\sec x$ can be reduced by using

$$\int \sec^m x \, dx = \frac{1}{m-1} \sec^{m-2} x \tan x + \frac{m-2}{m-1} \int \sec^{m-2} x \, dx. \tag{9}$$

Formula 9 can be derived by using a method similar to that used to derive Formula 2. That is, we write $\sec^m x = \sec^{m-2} x \sec^2 x$, integrate by parts with $u = \sec^{m-2} x$ and $dv = \sec^2 x \, dx$, and use the trigonometric formula $\tan^2 x + 1 = \sec^2 x$. Let us derive (9) in the special case $m = 3$. We have

$$u = \sec x, \qquad dv = \sec^2 x \, dx,$$
$$du = \sec x \tan x \, dx, \qquad v = \tan x.$$

Then

$$\int \sec^3 x \, dx = \int \sec x \sec^2 x \, dx = \int u \, dv = uv - \int v \, du$$

$$= (\sec x)(\tan x) - \int (\tan x)(\sec x \tan x) dx$$

$$= \sec x \tan x - \int \sec x \tan^2 x \, dx$$

$$= \sec x \tan x - \int \sec x (\sec^2 x - 1) dx$$

$$= \sec x \tan x - \int \sec^3 x \, dx + \int \sec x \, dx.$$

We now have

$$\int \sec^3 x \, dx = \sec x \tan x - \int \sec^3 x \, dx + \int \sec x \, dx, \quad \text{so}$$

$$2 \int \sec^3 x \, dx = \sec x \tan x + \int \sec x \, dx,$$

$$\int \sec^3 x \, dx = \frac{1}{2} \sec x \tan x + \frac{1}{2} \int \sec x \, dx.$$

This agrees with (9) when $m = 3$.

The following formula can be used to reduce powers of $\tan x$:

$$\int \tan^n x \, dx = \frac{1}{n-1} \tan^{n-1} x - \int \tan^{n-2} x \, dx. \qquad (10)$$

Formula 10 is not difficult to derive. If $n \neq 1$, we have

$$\int \tan^n x \, dx = \int \tan^{n-2} x \tan^2 x \, dx$$

$$= \int \tan^{n-2} x (\sec^2 x - 1) dx$$

$$= \int \tan^{n-2} x \sec^2 x \, dx - \int \tan^{n-2} x \, dx.$$

The first integral above can be evaluated by using the substitution $u = \tan x$, so $du = \sec^2 x \, dx$. We then have

$$\int \tan^n x \, dx = \int u^{n-2} du - \int \tan^{n-2} x \, dx$$

$$= \frac{u^{n-1}}{n-1} - \int \tan^{n-2} x \, dx$$

$$= \frac{1}{n-1} \tan^{n-1} x - \int \tan^{n-2} x \, dx.$$

If m is odd and n is even (not one of the *simple* cases), we can use $\tan^2 x + 1 = \sec^2 x$ to express $\sec^m x \tan^n x$ as a sum of powers of $\sec x$, and then reduce the power by using (9). Continue reducing the powers until the integrand is $\sec x$, then use (8).

EXAMPLE 7

Evaluate $\int \tan^2 x \sec^3 x\, dx$.

SOLUTION   This cannot be written as a *simple* type, so we will use a reduction formula. We first use $\tan^2 x + 1 = \sec^2 x$ to express the integral as a sum of integrals of powers of $\sec x$. We will then use a reduction formula to reduce the integral with the *highest* power of $\sec x$. We will then combine integrals of like powers of $\sec x$ before using another reduction formula.

$$\int \tan^2 x \sec^3 x\, dx$$

$$= \int (\sec^2 x - 1) \sec^3 x\, dx$$

$$= \left(\int \sec^5 x\, dx \right) - \int \sec^3 x\, dx$$

$$\underset{(9)}{=} \left(\frac{1}{4} \sec^3 x \tan x + \frac{3}{4} \int \sec^3 x\, dx \right) - \int \sec^3 x\, dx$$

$$= \frac{1}{4} \sec^3 x \tan x - \frac{1}{4} \left(\int \sec^3 x\, dx \right)$$

$$\underset{(9)}{=} \frac{1}{4} \sec^3 x \tan x - \frac{1}{4} \left(\frac{1}{2} \sec x \tan x + \frac{1}{2} \int \sec x\, dx \right)$$

$$= \frac{1}{4} \sec^3 x \tan x - \frac{1}{8} \sec x \tan x - \frac{1}{8} \left(\int \sec x\, dx \right)$$

$$\underset{(8)}{=} \frac{1}{4} \sec^3 x \tan x - \frac{1}{8} \sec x \tan x - \frac{1}{8} \ln |\sec x + \tan x| + C. \quad \blacksquare$$

The integrals that we have discussed so far in this section involve many cases of odd and even powers. Rather than trying to memorize all cases, it is recommended that you be familiar with the derivatives and always *look for a simple substitution*. If you can't find a *simple* substitution, then try a *reduction formula*.

Using Trigonometric Product Formulas

The following trigonometric identities can be used either to evaluate integrals or to derive integration formulas. The formulas are included in the list inside the front cover of the text.

$$\sin mx \sin nx = \frac{1}{2}(\cos((m - n)x) - \cos((m + n)x)),$$

$$\sin mx \cos nx = \frac{1}{2}(\sin((m + n)x) + \sin((m - n)x)),$$

$$\cos mx \cos nx = \frac{1}{2}(\cos((m + n)x) + \cos((m - n)x)).$$

EXAMPLE 8

Evaluate $\int \sin 2x \sin x\, dx$.

SOLUTION   We use the formula for $\sin mx \sin nx$ with $m = 2$ and $n = 1$.

$$\int \sin 2x \sin x \, dx = \int \frac{1}{2}(\cos x - \cos 3x)dx = \frac{1}{2}\left(\sin x - \frac{1}{3}\sin 3x\right) + C$$

$$= \frac{1}{2}\sin x - \frac{1}{6}\sin 3x + C. \quad \blacksquare$$

EXERCISES 8.3

Evaluate the integrals in Exercises 1–30.

1. $\int \cos^2 x \sin x \, dx$ **(Calculus Explorer & Tutor II, Introduction to Integration, 4.)**

2. $\int \sin x \cos x \, dx$

3. $\int \tan x \sec^2 x \, dx$ **(Calculus Explorer & Tutor II, Introduction to Integration, 6.)**

4. $\int \tan^2 x \sec^2 x \, dx$
5. $\int \sec^5 x \tan x \, dx$

6. $\int \sec^3 x \tan^3 x \, dx$
7. $\int \sin^2 2x \, dx$

8. $\int \tan^2(x/2)dx$
9. $\int \cos^3 x \, dx$

10. $\int \tan^4 x \, dx$ **(Calculus Explorer & Tutor II, Techniques of Integration, 6.)**

11. $\int \sec^3 2x \, dx$
12. $\int \cos^4(x/3)dx$

13. $\int \tan^6 x \sec^4 x \, dx$
14. $\int \sec^6 x \, dx$

15. $\int \sin^2 x \cos^3 x \, dx$ **(Calculus Explorer & Tutor II, Techniques of Integration, 7.)**

16. $\int \cos^2 x \sin^5 x \, dx$
17. $\int \frac{\sin^3 x}{\cos^2 x}dx$

18. $\int \sec^4 x \, dx$
19. $\int \cos^4 x \sin^2 x \, dx$

20. $\int \sec^5 x \, dx$
21. $\int \frac{1}{\sec^2 x}dx$

22. $\int \frac{1}{\sec^4 x}dx$
23. $\int \tan^3 x \, dx$

24. $\int \cos^5 x \, dx$
25. $\int \sin^3 x \cos^2 x \, dx$

26. $\int \sin^6 x \cos^4 x \, dx$
27. $\int \sin 3x \sin 2x \, dx$

28. $\int \sin 5x \cos 3x \, dx$
29. $\int \cos 4x \sin x \, dx$

30. $\int \cos 5x \cos 2x \, dx$

Evaluate the definite integrals in Exercises 31–36.

31. $\int_0^\pi \sin^2 x \, dx$
32. $\int_0^{\pi/2} \cos^4 x \, dx$

33. $\int_0^{\pi/6} \cos^3 2x \, dx$
34. $\int_0^{\pi/3} \sin^3(x/2)dx$

35. $\int_0^{\pi/4} \tan x \sec^2 x \, dx$
36. $\int_0^{\pi/4} \tan^2 x \sec x \, dx$

37. If m and n are positive integers, show that

$$\int_0^{2\pi} \sin mx \sin nx \, dx = \begin{cases} \pi, & m = n, \\ 0, & m \neq n. \end{cases}$$

38. If m and n are positive integers, show that

$$\int_0^{2\pi} \sin mx \cos nx \, dx = 0.$$

39. Use integration by parts to derive the formula

$$\int \sin^n x \, dx = -\frac{\sin^{n-1} x \cos x}{n} + \frac{n-1}{n}\int \sin^{n-2} x \, dx.$$

40. Study of **two-dimensional heat flow along a flat plate** leads to the integral

$$\int_0^1 \sin \frac{\pi}{2}\eta \left(1 - \sin \frac{\pi}{2}\eta\right)d\eta.$$

Evaluate this integral.

41. Find c so that the solid generated by revolving the region between $y = \sin x$ and the line $y = c$, $0 \leq x \leq \pi$ about the line $y = c$ has minimum volume. See figure.

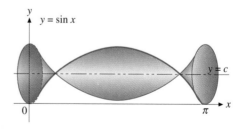

42. The base of a solid is the region in the xy-plane bounded by $y = \cos x$, $y = 0$, $x = 0$, and $x = \pi/2$. Cross-sections perpendicular to the x-axis are semicircles with diameters on the base. Find the volume of the solid. **(Calculus Explorer & Tutor II, Applications of Integration, 5.)**

43. Find the length of the curve $y = \ln(\cos x)$, $0 \leq x \leq \pi/4$.

44. The **root mean square** of a sinusoidal current $i = I\sin(\omega t)$ is

$$I_{rms} = \sqrt{(i^2)_{avg}} = \sqrt{\frac{\omega}{2\pi}\int_0^{2\pi/\omega} I^2\sin^2(\omega t)\,dt}.$$

Show that $I_{rms} = I/\sqrt{2}$. (115 V ac means the rms voltage is 115 V. The peak voltage is then $V = \sqrt{2}\,V_{rms}$.)

8.4 TRIGONOMETRIC SUBSTITUTIONS; MORE INTEGRATION FORMULAS

Connections

Substitution and change of variable, 5.5.

Trigonometric integrals, reduction formulas, 8.3.

We will study some substitutions that can be used to transform certain integrals that involve quadratic expressions into trigonometric integrals. The methods of Section 8.3 can then be used to evaluate the integrals. We will also illustrate the use of a table of integrals to evaluate integrals of these types.

The idea of the trigonometric substitutions is to replace a quadratic expression with an expression that is the square of a trigonometric function. The method is based on the trigonometric formulas $\sin^2\theta + \cos^2\theta = 1$ and $\tan^2\theta + 1 = \sec^2\theta$.

Trigonometric substitutions

- To simplify $a^2 - x^2$, set $x = a\sin\theta$, so $dx = a\cos\theta\,d\theta$ and
  $$a^2 - x^2 = a^2 - a^2\sin^2\theta = a^2(1 - \sin^2\theta) = a^2\cos^2\theta.$$
- To simplify $a^2 + x^2$, set $x = a\tan\theta$, so $dx = a\sec^2\theta\,d\theta$ and
  $$a^2 + x^2 = a^2 + a^2\tan^2\theta = a^2(1 + \tan^2\theta) = a^2\sec^2\theta.$$
- To simplify $x^2 - a^2$, set $x = a\sec\theta$, so $dx = a\sec\theta\tan\theta\,d\theta$ and
  $$x^2 - a^2 = a^2\sec^2\theta - a^2 = a^2(\sec^2 - 1) = a^2\tan^2\theta.$$

The choice of which substitution to use depends on the form of the quadratic expression that you want to simplify. If your choice does not transform the quadratic expression into a positive perfect square, try a different choice!

Trigonometric substitutions are particularly useful for integrals that involve the square root of a quadratic expression. The fact that the substitution replaces the quadratic expression with the square of a trigonometric function allows us the eliminate the square root.

Using Trigonometric Substitutions

We will use the following procedure.

To use a trigonometric substitution

- Choose the trigonometric substitution that simplifies the quadratic expression in the integrand.
- Evaluate the differential dx.
- Simplify the quadratic expression that is in the integrand.
- Substitute for all x in the integrand and for the differential dx.
- Evaluate the trigonometric integral.
- Use a sketch of a representative right triangle with angle θ to transform functions of θ back to functions of x. See Figures 8.4.1–8.4.3.

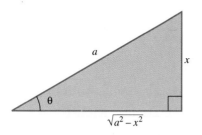

FIGURE 8.4.1 $x = a\sin\theta$, $dx = a\cos\theta\,d\theta$, $\sqrt{a^2 - x^2} = a\cos\theta$, $\theta = \sin^{-1}(x/a)$.

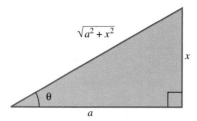

FIGURE 8.4.2 $x = a\tan\theta$, $dx = a\sec^2\theta\,d\theta$, $\sqrt{a^2 + x^2} = a\sec\theta$, $\theta = \tan^{-1}(x/a)$.

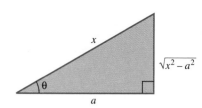

FIGURE 8.4.3 $x = a\sec\theta$, $dx = a\sec\theta\tan\theta\,d\theta$, $\sqrt{x^2 - a^2} = a\tan\theta$, $\theta = \sec^{-1}(x/a)$.

EXAMPLE 1

Evaluate $\displaystyle\int \frac{1}{(1 - x^2)^{3/2}}\,dx$.

SOLUTION The integrand contains the quadratic expression $1 - x^2$, so we use the substitution $x = \sin\theta$. Then $dx = \cos\theta\,d\theta$, and $1 - x^2 = 1 - \sin^2\theta = \cos^2\theta$. Substitution then gives

$$\int \frac{1}{(1 - x^2)^{3/2}}\,dx = \int \frac{1}{(\cos^2\theta)^{3/2}}\cos\theta\,d\theta$$

$$= \int \frac{1}{\cos^2\theta}\,d\theta$$

$$= \int \sec^2\theta\,d\theta = \tan\theta + C.$$

A representative right triangle with angle θ that satisfies $\sin\theta = x$ is sketched in Figure 8.4.4. The sketch can be used to express $\tan\theta$ in terms of x. We obtain

$$\int \frac{1}{(1 - x^2)^{3/2}}\,dx = \frac{x}{\sqrt{1 - x^2}} + C. \quad \blacksquare$$

FIGURE 8.4.4 $\sin\theta = x$, so $\tan\theta = \dfrac{x}{\sqrt{1 - x^2}}$.

EXAMPLE 2

Evaluate $\displaystyle\int \frac{1}{(9 + x^2)^2}\,dx$.

SOLUTION The integrand contains the quadratic expression $9 + x^2$, so we use the substitution $x = 3\tan\theta$. Then $dx = 3\sec^2\theta\,d\theta$, and $9 + x^2 = 9 + 9\tan^2\theta = 9\sec^2\theta$. We substitute to obtain

$$\int \frac{1}{(9 + x^2)^2}\,dx = \int \frac{1}{(9\sec^2\theta)^2}\,3\sec^2\theta\,d\theta$$

$$= \int \frac{1}{27\sec^2\theta}\,d\theta$$

$$= \frac{1}{27}\int \cos^2\theta\,d\theta$$

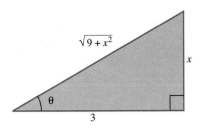

FIGURE 8.4.5 $\tan\theta = x/3$, so $\sin\theta = \dfrac{x}{\sqrt{9+x^2}}$, $\cos\theta = \dfrac{3}{\sqrt{9+x^2}}$, and $\theta = \tan^{-1}\dfrac{x}{3}$.

$$= \frac{1}{27}\left(\frac{\sin\theta\cos\theta}{2} + \frac{\theta}{2}\right) + C \qquad \text{[From Table of Integrals]}$$

$$= \frac{1}{54}\left(\frac{x}{\sqrt{9+x^2}}\right)\left(\frac{3}{\sqrt{9+x^2}}\right) + \frac{1}{54}\tan^{-1}\frac{x}{3} + C$$

[From Figure 8.4.5]

$$= \frac{x}{18(9+x^2)} + \frac{1}{54}\tan^{-1}\frac{x}{3} + C. \quad \blacksquare$$

EXAMPLE 3

Evaluate $\displaystyle\int \frac{1}{\sqrt{x^2-4}}\,dx$.

SOLUTION   The integrand contains the factor $x^2 - 4$, so we use the substitution $x = 2\sec\theta$. Then $dx = 2\sec\theta\tan\theta\,dx$ and $x^2 - 4 = \sec^2\theta - 4 = 4\tan^2\theta$. Substitution yields

$$\int \frac{1}{\sqrt{x^2-4}}\,dx = \int \frac{1}{\sqrt{4\tan^2\theta}}2\sec\theta\tan\theta\,d\theta$$

$$= \int \frac{1}{2\tan\theta}2\sec\theta\tan\theta\,d\theta$$

$$= \int \sec\theta\,d\theta$$

$$= \ln|\sec\theta + \tan\theta| + C \qquad \text{[From Table of Integrals]}$$

$$= \ln\left|\frac{x}{2} + \frac{\sqrt{x^2-4}}{2}\right| + C. \qquad \text{[From Figure 8.4.6]}$$

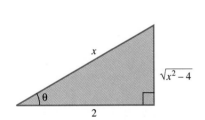

FIGURE 8.4.6 $\sec\theta = x/2$, so $\sec\theta + \tan\theta = \dfrac{x}{2} + \dfrac{\sqrt{x^2-4}}{2}$.

The answer can also be expressed as $\ln|x + \sqrt{x^2-4}| + C$, since

$$\ln\left|\frac{x}{2} + \frac{\sqrt{x^2-4}}{2}\right| = \ln|x + \sqrt{x^2-4}| - \ln 2,$$

so the two functions differ by the constant $\ln 2$ and, hence, have the same derivative. $\blacksquare$

Using Integration Formulas

Trigonometric substitutions can be used to derive the formulas that involve $a^2 + u^2$, $a^2 - u^2$, and $u^2 - a^2$ in the Table of Integrals. We will use these formulas instead of deriving them each time they occur.

It is important that we understand how to perform the algebra involved in using a trigonometric substitution to change the form of an integral. However, integrals of the types that can be evaluated by trigonometric substitutions can also be evaluated by using a table of integrals or an appropriate computer program.

It may be of some interest that we can now evaluate an integral that represents the area of a circle.

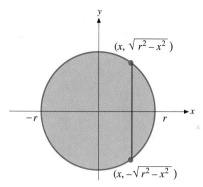

FIGURE 8.4.7 A variable vertical area strip has length $2\sqrt{r^2 - x^2}$ and width dx.

EXAMPLE 4

Find the area of a circle of radius r.

SOLUTION   Let us consider the circle $x^2 + y^2 = r^2$. A sketch of the circle is given in Figure 8.4.7. We see that a variable vertical area strip has length $2\sqrt{r^2 - x^2}$. The area of the circle is

$$A = \int_{-r}^{r} 2\sqrt{r^2 - x^2}\,dx.$$

Let us evaluate this integral by using Formula 37 from the Table of Integrals given inside the back cover of the text. We have

$$A = \int_{-r}^{r} 2\sqrt{r^2 - x^2}\,dx = 2\left[\frac{x}{2}\sqrt{r^2 - x^2} + \frac{r^2}{2}\sin^{-1}\frac{x}{r}\right]_{-r}^{r}$$

$$= \left[r\sqrt{r^2 - r^2} + r^2\sin^{-1}\left(\frac{r}{r}\right)\right] - \left[-r\sqrt{r^2 - (-r)^2} + r^2\sin^{-1}\left(\frac{-r}{r}\right)\right]$$

$$= r^2\left(\frac{\pi}{2}\right) - r^2\left(-\frac{\pi}{2}\right) = \pi r^2. \quad \blacksquare$$

We can use the substitution $u = a\tan\theta$ and then the reduction formula for powers of $\sec\theta$ to derive the reduction formula

$$\int \frac{1}{(u^2 + a^2)^n}\,du$$

$$= \frac{u}{a^2(2n-2)(u^2 + a^2)^{n-1}} + \frac{2n-3}{a^2(2n-2)}\int \frac{1}{(u^2 + a^2)^{n-1}}\,du. \qquad (1)$$

This formula is useful for the type of integrals we will consider in the next section. The formula is included in our Table of Integrals.

EXAMPLE 5

Evaluate $\displaystyle\int \frac{1}{(x^2 + 4)^2}\,dx$.

SOLUTION   Using Formula 1 with $n = 2$ and $a^2 = 4$, we have

$$\int \frac{1}{(x^2 + 4)^2}\,dx = \frac{x}{4(2)(x^2 + 4)} + \frac{1}{4(2)}\int \frac{1}{x^2 + 4}\,dx$$

$$= \frac{x}{8(x^2 + 4)} + \frac{1}{16}\tan^{-1}\frac{x}{2} + C. \quad \blacksquare$$

Completing a Square

If the quadratic expression contains a linear term x, we must complete a square before making a trigonometric substitution or using an integration formula.

EXAMPLE 6

Evaluate $\displaystyle\int \frac{1}{(4x + x^2)^{3/2}}\,dx$.

SOLUTION   Completing the square, we have

$$4x + x^2 = x^2 + 4x + [4] - [4] = (x + 2)^2 - 2^2.$$

We evaluate this integral by first setting $u = x + 2$, so $du = dx$, and then using Formula 40 from the Table of Integrals. We have

$$\int \frac{1}{(4x + x^2)^{3/2}}\, dx = \int \frac{1}{((x+2)^2 - 2^2)^{3/2}}\, dx$$

$$= \int \frac{1}{(u^2 - 2^2)^{3/2}}\, du$$

$$= -\frac{u}{2^2\sqrt{u^2 - 2^2}} + C$$

$$= -\frac{x + 2}{4\sqrt{(x+2)^2 - 4}} + C$$

$$= -\frac{x + 2}{4\sqrt{x^2 + 4x}} + C. \quad \blacksquare$$

NOTE   *Don't forget to look for an easier method before trying a trigonometric substitution.*

For example, the substitution $x = 2\tan\theta$ could be used to show that

$$\int \frac{1}{x^2 + 4}\, dx = \frac{1}{2}\tan^{-1}\frac{x}{2} + C,$$

but we should recognize this as an integral for which we already know the formula. Also, we could use $x = \sin\theta$ to show

$$\int x\sqrt{1 - x^2}\, dx = -\frac{1}{3}(1 - x^2)^{3/2} + C,$$

but it is much easier to use the substitution $u = 1 - x^2$.

EXERCISES 8.4

*In Exercises 1–6, (**a**) use a trigonometric substitution to evaluate each integral. (**b**) Verify the results of (a) by using a different method to evaluate each integral.*

1. $\displaystyle\int \frac{1}{\sqrt{4 - x^2}}\, dx$

2. $\displaystyle\int \frac{x}{\sqrt{4 - x^2}}\, dx$

3. $\displaystyle\int \frac{1}{x^2 + 4}\, dx$

4. $\displaystyle\int \frac{x}{x^2 + 4}\, dx$

5. $\displaystyle\int x\sqrt{x^2 - 4}\, dx$

6. $\displaystyle\int \frac{1}{x\sqrt{x^2 - 4}}\, dx$

Evaluate the integrals in Exercises 7–28.

7. $\displaystyle\int \sqrt{4 - x^2}\, dx$

8. $\displaystyle\int \frac{x^2}{\sqrt{4 - x^2}}\, dx$

9. $\displaystyle\int \frac{1}{(2 - x^2)^{3/2}}\, dx$

10. $\displaystyle\int \frac{x^3}{\sqrt{9 - x^2}}\, dx$

11. $\displaystyle\int \frac{1}{(1 + 9x^2)^2}\, dx$

12. $\displaystyle\int \frac{1}{x\sqrt{1 + x^2}}\, dx$ (**Calculus Explorer & Tutor II, Techniques of Integration, 10.**)

13. $\displaystyle\int x^3\sqrt{1 - x^2}\, dx$

14. $\displaystyle\int x\sqrt{x^2 - 4}\, dx$

15. $\displaystyle\int \frac{1}{\sqrt{x^2 - 3}}\, dx$

16. $\displaystyle\int \frac{1}{\sqrt{4x^2 - 1}}\, dx$

17. $\displaystyle\int \frac{1}{x^2\sqrt{x^2 - 4}}\, dx$ (**Calculus Explorer & Tutor II, Techniques of Integration, 9.**)

18. $\displaystyle\int \frac{x^2}{\sqrt{x^2 + 4}}\, dx$

19. $\displaystyle\int x^2\sqrt{9 - x^2}\, dx$

20. $\displaystyle\int \frac{x^2}{(1 - x^2)^{3/2}}\, dx$

21. $\displaystyle\int \frac{x}{4 + x^2}\, dx$

22. $\displaystyle\int \sqrt{x^2 - 4}\, dx$

23. $\displaystyle\int \frac{1}{\sqrt{x^2 + 4x + 5}}\, dx$

24. $\int \dfrac{1}{\sqrt{x^2 + 4x + 3}}\, dx$ **25.** $\int \dfrac{1}{9 - x^2}\, dx$

26. $\int \dfrac{1}{\sqrt{x^2 - 4x}}\, dx$ **(Calculus Explorer & Tutor II, Techniques of Integration, 11.)**

27. $\int \dfrac{1}{(x^2 + 9)^2}\, dx$ **28.** $\int \dfrac{1}{(x^2 + 9)^3}\, dx$

Evaluate the definite integrals in Exercises 29–34. (Integration formulas are given inside the back cover of the text.)

29. $\int_{-1}^{1} \dfrac{1}{(4 - x^2)^{3/2}}\, dx$ **30.** $\int_{-2}^{2} \sqrt{4 - x^2}\, dx$

31. $\int_{-2}^{2} \dfrac{1}{(x^2 + 4)^2}\, dx$ **32.** $\int_{1}^{2} \dfrac{1}{x\sqrt{x^2 + 4}}\, dx$

33. $\int_{3}^{5} \dfrac{1}{(x^2 - 4)^{3/2}}\, dx$ **34.** $\int_{3}^{5} \dfrac{1}{\sqrt{x^2 - 4}}\, dx$

35. Find the area of the region bounded by the ellipse $\dfrac{x^2}{a^2} + \dfrac{y^2}{b^2} = 1$.

36. Find the area of the region bounded by the hyperbola

$\dfrac{x^2}{a^2} - \dfrac{y^2}{b^2} = 1$ and the line $x = \sqrt{a^2 + b^2}$.

37. Find the volume of the solid that is obtained by revolving the region bounded by $y = x/\sqrt{x^2 + 1}$, $y = 0$, and $x = 1$, about the y-axis.

38. Find the volume of the solid that is obtained by revolving the region bounded by $y = x/\sqrt{x^2 + 1}$, $y = 0$, and $x = 1$, about the x-axis.

39. Find the arc length of the curve $y = x^2$ between $(0, 0)$ and $(2, 4)$.

40. Find the area of the surface obtained by revolving the curve $y = x^2$, $0 \le x \le 2$, about the x-axis.

41. Use the substitution $u = a \tan \theta$ and then the formula for the reduction of powers of $\cos \theta$ (Formula 2 of Section 8.3) to derive the formula

$$\int \frac{1}{(u^2 + a^2)^n}\, du$$
$$= \frac{u}{a^2(2n - 2)(u^2 + a^2)^{n-1}}$$
$$+ \frac{2n - 3}{a^2(2n - 2)} \int \frac{1}{(u^2 + a^2)^{n-1}}\, du.$$

8.5 PARTIAL FRACTIONS

We will illustrate a technique that can be used to evaluate integrals of the form

$$\int \frac{P(x)}{Q(x)}\, dx,$$

where P and Q are polynomials. The idea is to use algebra to express the rational function P/Q as a sum of terms that can be integrated by using techniques we have previously developed. The particular expression we will describe is called the **partial fraction expansion** of P/Q.

Expanding a Rational Function

The partial fraction expansion of a rational function P/Q consists of the sum of a polynomial and terms that have a power of a factor of Q in the denominators. Let us illustrate the general case with a specific example.

Let us consider the identity

$$\frac{x^6 + x^3 + 4}{x^5 - x^4 + x^3 - x^2} = x + 1 + \frac{-4}{x} + \frac{-4}{x^2} + \frac{3}{x - 1} + \frac{x + 2}{x^2 + 1}. \quad (1)$$

We first note that we know how to integrate each term of the sum on the right side of (1), so we can obtain the integral of the left side by integrating the terms on the right side. Also, we know how to combine terms on the right side to obtain the left side. The question we now address is, given the left side of (1), how can we expand the expression to obtain the right side? We want to establish some rules that allow us to do this in a systematic way. The expansion obtained is called the partial fraction expansion of the rational function.

Strategy for the partial fraction expansion of a rational function

- Check the degrees of the numerator and denominator. If $\deg P \geq \deg Q$, divide P by Q to obtain polynomials S and R with $\deg R < \deg Q$ and

$$\frac{P(x)}{Q(x)} = S(x) + \frac{R(x)}{Q(x)}.$$

If $\deg P < \deg Q$, then $S(x) = 0$, $R(x) = P(x)$, and it is not necessary to divide.

For example, the first step in expanding the left side of (1) is to note that the degree of the numerator is greater than the degree of the denominator, so we must divide. We carry out the division of $P(x) = x^6 + x^3 + 4$ by $Q(x) = x^5 - x^4 + x^3 - x^2$ to obtain

$$\frac{x^6 + x^3 + 4}{x^5 - x^4 + x^3 - x^2} = x + 1 + \frac{x^3 + x^2 + 4}{x^5 - x^4 + x^3 - x^2}, \tag{2}$$

so $S(x) = x + 1$ and $R(x) = x^3 + x^2 + 4$.

- Factor the denominator into powers of distinct linear terms and powers of irreducible quadratic terms. This factorization determines the form of the expansion of R/Q.

A theorem of algebra guarantees that any polynomial with real number coefficients can be factored into powers of linear terms and powers of **irreducible** quadratic terms. The linear factors can be written in the form $x - r$, where r is a zero of the denominator. The quadratic $ax^2 + bx + c$ is irreducible if it has no real zeros. That is, if $b^2 - 4ac < 0$.

In (1), we have

$$x^5 - x^4 + x^3 - x^2 = x^2(x^3 - x^2 + x - 1) = x^2(x - 1)(x^2 + 1), \tag{3}$$

so the denominator contains the second power of the linear factor x, the first power of the linear factor $x - 1$ and the first power of the irreducible quadratic factor $x^2 + 1$.

- If $Q(x)$ contains exactly m identical linear factors $x - r$, the expansion of R/Q contains a sum of the form

$$\frac{A_1}{x - r} + \frac{A_2}{(x - r)^2} + \cdots + \frac{A_m}{(x - r)^m},$$

where $A_1, A_2, \ldots, A_m$ are unknown constants.
- If $Q(x)$ contains exactly n identical irreducible quadratic factors $ax^2 + bx + c$, the expansion of R/Q contains a sum of the form

$$\frac{B_1 x + C_1}{ax^2 + bx + c} + \frac{B_2 x + C_2}{(ax^2 + bx + c)^2} + \cdots + \frac{B_n x + C_n}{(ax^2 + bx + c)^n},$$

where $B_1, C_1, B_2, C_2, \ldots, B_n, C_n$ are unknown constants.

Each distinct factor of Q contributes a term or terms of the type indicated to the expansion of R/Q. In each case, the denominators are powers of the factors, where the powers run from the first power up to the power of the factor in Q. Terms that have powers of linear factors in the denominator have numerators that are numbers. Terms that have powers of irreducible quadratic factors in the denominator have numerators of the form (number)x + (number).

For example, we use the factorization in (3) to expand the expression R/Q on the right side of (2). That is, we write

$$\frac{x^3 + x^2 + 4}{x^5 - x^4 + x^3 - x^2} = \frac{A}{x} + \frac{B}{x^2} + \frac{C}{x - 1} + \frac{Dx + E}{x^2 + 1}, \tag{4}$$

where A, B, C, D, and E are unknown constants. Note that the second power of the linear factor x^2 in the factorization of $Q(x)$ corresponds to the terms

$$\frac{A}{x} + \frac{B}{x^2},$$

the first power of $x - 1$ in the factorization of $Q(x)$ corresponds to single term

$$\frac{C}{x - 1},$$

and the first power of the irreducible factor $x^2 + 1$ in the factorization of $Q(x)$ corresponds to the term

$$\frac{Dx + E}{x^2 + 1}.$$

- Complete the partial fraction expansion by solving for the values of the unknown constants and substituting the expansion of R/Q into the expression $\dfrac{P(x)}{Q(x)} = S(x) + \dfrac{R(x)}{Q(x)}$.

A theorem of algebra guarantees a unique solution for the values. Techniques for solving for the unknown constants will be illustrated below.

We could use (4) to determine that $A = -4$, $B = -4$, $C = 3$, $D = 1$, and $E = 2$. Replacing $\dfrac{x^3 + x^2 + 4}{x^5 - x^4 + x^3 - x^2}$ by its expansion in equation (2) then completes the expansion of the left side of the equation in (1) to the expression on the right side of the equation.

Distinct Linear Factors

EXAMPLE 1

Evaluate $\displaystyle\int \frac{x - 3}{x^2 - 3x + 2}\,dx$.

SOLUTION We first note that

$$\deg(x - 3) = 1 < 2 = \deg(x^2 - 3x + 2),$$

so it is not necessary to divide.

Factoring, we have $x^2 - 3x + 2 = (x - 1)(x - 2)$. We see there are two distinct linear factors, each to the first power. We then know that the partial fraction expansion is of the form

$$\frac{x - 3}{(x - 1)(x - 2)} = \frac{A}{x - 1} + \frac{B}{x - 2}.$$

Multiplying both sides of the above equation by $(x - 1)(x - 2)$ to clear fractions, we obtain

$$x - 3 = A(x - 2) + B(x - 1). \qquad (*)$$

There are unique numbers A and B so that equation $(*)$ is true for all x. We must determine these unknown numbers. We will illustrate two methods for doing this.

Method 1: Combine like terms in equation $(*)$ to obtain an equation of the form $0 = $ (polynomial). We then obtain a system of linear equations in the unknowns A and B by setting all coefficients and the constant term of the polynomial equal to zero. Solution for the unknowns then gives the desired identity. From $(*)$ we have

$$x - 3 = Ax - 2A + Bx - B,$$
$$0 = (A + B - 1)x + (-2A - B + 3).$$

Then

$$\begin{cases} A + B - 1 = 0, & \text{[Coefficient of } x\text{]} \\ -2A - B + 3 = 0. & \text{[Constant term]} \end{cases}$$

This system of linear equations can be solved by using any of the techniques you learned in your algebra course. For example, we can eliminate B by adding the two equations to obtain

$$-A + 2 = 0, \text{ so } A = 2.$$

Substitution into the first equation then gives

$$(2) + B - 1 = 0, \text{ so } B = -1.$$

Method 2: We determine the values of the unknown constants by substituting specific values of x into equation $(*)$. Values of x for which $Q(x) = 0$ give simple equations and these should be chosen before other values of x.

Substitution of $x = 1$ into $(*)$ gives

$$1 - 3 = A(1 - 2) + B(1 - 1), \text{ so } A = 2.$$

Substitution of $x = 2$ into $(*)$ gives

$$2 - 3 = A(2 - 2) + B(2 - 1), \text{ so } B = -1.$$

Method 1 will work in any case and is easily programmed so the work can be done by a computer. Method 2 is a convenient shortcut in the case of distinct linear factors raised to the first power.

By either method, we obtain

$$\frac{x - 3}{x^2 - 3x + 2} = \frac{2}{x - 1} - \frac{1}{x - 2}.$$

(The above equation can be checked by combining the terms on the right.) Then

$$\int \frac{x-3}{x^2-3x+2}\,dx = \int \frac{2}{x-1}\,dx - \int \frac{1}{x-2}\,dx$$

$$= 2\ln|x-1| - \ln|x-2| + C. \quad \blacksquare$$

EXAMPLE 2

Evaluate $\displaystyle\int \frac{x^3 - x - 3\sqrt{2}}{x^2 - 2}\,dx$.

SOLUTION   We have $\deg(x^3 - x - 3\sqrt{2}) = 3 \neq 2 = \deg(x^2 - 2)$, so we must divide:

$$
\begin{array}{r}
x \\
x^2 - 2 \overline{)\; x^3 - x - 3\sqrt{2}} \\
\underline{x^3 - 2x } \\
x - 3\sqrt{2}
\end{array}
$$

It follows that

$$\frac{x^3 - x - 3\sqrt{2}}{x^2 - 2} = x + \frac{x - 3\sqrt{2}}{x^2 - 2}.$$

We need the partial fraction expansion of the term $\dfrac{x - 3\sqrt{2}}{x^2 - 2}$.

The quadratic $x^2 - 2$ does not have rational zeros, but it is zero when $x = \pm\sqrt{2}$; $x^2 - 2$ is not an irreducible quadratic. Factoring, we have

$$x^2 - 2 = (x - \sqrt{2})(x + \sqrt{2}).$$

We see that there are two distinct linear factors, each to the first power. We then know that the desired partial fraction expansion is of the form

$$\frac{x - 3\sqrt{2}}{x^2 - 2} = \frac{A}{x - \sqrt{2}} + \frac{B}{x + \sqrt{2}}.$$

Clearing fractions, we obtain

$$x - 3\sqrt{2} = A(x + \sqrt{2}) + B(x - \sqrt{2}). \qquad (*)$$

We need to find the unique numbers A and B that make equation $(*)$ true for all x.

Method 1:

$$x - 3\sqrt{2} = Ax + \sqrt{2}A + Bx - \sqrt{2}B,$$
$$0 = (A + B - 1)x + (\sqrt{2}A - \sqrt{2}B + 3\sqrt{2}).$$
$$\begin{cases} A + B - 1 = 0, & \text{[Coefficient of } x] \\ \sqrt{2}A - \sqrt{2}B + 3\sqrt{2} = 0. & \text{[Constant term]} \end{cases}$$

We can solve this system by dividing the second equation by $\sqrt{2}$ and then eliminating B by adding the new second equation to the first. We obtain

$$2A + 2 = 0, \text{ so } A = -1.$$

Substitution into the first equation then gives

$$(-1) + B - 1 = 0, \text{ so } B = 2.$$

Method 2: Substitution of $x = \sqrt{2}$ into $(*)$ gives

$$\sqrt{2} - 3\sqrt{2} = A(2\sqrt{2}) + B(0), \text{ so } A = -1.$$

Substitution of $x = -\sqrt{2}$ into $(*)$ gives

$$-\sqrt{2} - 3\sqrt{2} = A(0) + B(-2\sqrt{2}), \text{ so } B = 2.$$

Both methods give

$$\frac{x - 3\sqrt{2}}{x^2 - 2} = \frac{-1}{x - \sqrt{2}} + \frac{2}{x + \sqrt{2}}.$$

Combining this identity with the result of our division, we obtain

$$\frac{x^3 - x - 3\sqrt{2}}{x^2 - 2} = x - \frac{1}{x - \sqrt{2}} + \frac{2}{x + \sqrt{2}}.$$

(This equation can be checked by simplifying the expression on the right.) We then have

$$\int \frac{x^3 - x - 3\sqrt{2}}{x^2 - 2} \, dx = \int \left(x - \frac{1}{x - \sqrt{2}} + \frac{2}{x + \sqrt{2}} \right) dx$$

$$= \frac{x^2}{2} - \ln|x - \sqrt{2}| + 2\ln|x + \sqrt{2}| + C. \quad \blacksquare$$

If you forgot to divide in Example 2 and simply wrote

$$\frac{x^3 - x - 3\sqrt{2}}{x^2 - 2} \underset{(?)}{=} \frac{A}{x - \sqrt{2}} + \frac{B}{x + \sqrt{2}},$$

this incorrect expression could not be an identity for any choice of the unknown numbers A and B. Method 1 would lead to an inconsistent system of linear equations. Method 2 would give values of A and B, but the resulting equation would be incorrect. Be very careful when using Method 2.

The Quadratic Formula can be used to find the zeros and "factor" a quadratic expression. For example, the Quadratic Formula implies the zeros of $x^2 + 2x - 1$ are

$$x = \frac{-(2) \pm \sqrt{(2)^2 - 4(1)(-1)}}{2(1)} = -1 \pm \sqrt{2}.$$

We then have the factorization

$$x^2 + 2x - 1 = \left(x - (-1 + \sqrt{2}) \right) \left(x - (-1 - \sqrt{2}) \right).$$

Of course, if $b^2 - 4ac < 0$, then $ax^2 + bx + c$ has no real zeros and is irreducible.

Powers of Linear Factors

EXAMPLE 3

Evaluate $\displaystyle\int \frac{12}{x^4 - x^3 - 2x^2}\, dx$.

SOLUTION We have $\deg(12) = 0 < 4 = \deg(x^4 - x^3 - 2x^2)$, so we don't need to divide.

Factoring, we have

$$x^4 - x^3 - 2x^2 = x^2(x - 2)(x + 1).$$

In this example we have three distinct linear factors, one squared and the others to the first power. Note that x^2 is not an irreducible quadratic. The partial fraction expansion is of the form

$$\frac{12}{x^4 - x^3 - 2x^2} = \frac{A}{x} + \frac{B}{x^2} + \frac{C}{x - 2} + \frac{D}{x + 1},$$

$$12 = A(x)(x - 2)(x + 1) + B(x - 2)(x + 1)$$
$$+ C(x^2)(x + 1) + D(x^2)(x - 2).$$

$(*)$

We will use Method 1 for this example.

$$12 = A(x)(x^2 - x - 2) + B(x^2 - x - 2) + C(x^2)(x + 1) + D(x^2)(x - 2),$$
$$12 = Ax^3 - Ax^2 - 2Ax + Bx^2 - Bx - 2B + Cx^3 + Cx^2 + Dx^3 - 2Dx^2,$$
$$0 = (A + C + D)x^3 + (-A + B + C - 2D)x^2 + (-2A - B)x$$
$$+ (-2B - 12)$$

$$\begin{cases} A + C + D = 0, & \text{[Coefficient of } x^3\text{]} \\ -A + B + C - 2D = 0, & \text{[Coefficient of } x^2\text{]} \\ -2A - B = 0, & \text{[Coefficient of } x\text{]} \\ - 2B - 12 = 0, & \text{[Constant term]} \end{cases}$$

This system is solved easily. The fourth equation implies $B = -6$. The third equation then implies $A = 3$. Substitution of $A = 3$ and $B = -6$ into the first two equations then gives two linear equations in the variables C and D, which can be solved as before to obtain $C = 1$ and $D = -4$. We then have

$$\frac{12}{x^4 - x^3 - 2x^2} = \frac{3}{x} + \frac{-6}{x^2} + \frac{1}{x - 2} + \frac{-4}{x + 1}.$$

Hence,

$$\int \frac{12}{x^4 - x^3 - 2x^2}\, dx = \int \left(\frac{3}{x} + \frac{-6}{x^2} + \frac{1}{x - 2} + \frac{-4}{x + 1} \right) dx$$

$$= 3\ln|x| + \frac{6}{x} + \ln|x - 2| - 4\ln|x + 1| + C. \quad \blacksquare$$

Irreducible Quadratic Factors

If $ax^2 + bx + c$ is an irreducible quadratic, we can evaluate

$$\int \frac{Bx + C}{(ax^2 + bx + c)^n}\, dx$$

by using the formulas

$$\int \frac{u}{u^2 + d^2} \, du = \frac{1}{2} \ln(u^2 + d^2) + C, \tag{5}$$

$$\int \frac{u}{(u^2 + d^2)^n} \, du = \frac{-1}{2(n-1)(u^2 + d^2)^{n-1}} + C, \quad n > 1, \tag{6}$$

$$\int \frac{1}{u^2 + d^2} \, du = \frac{1}{d} \tan^{-1} \frac{u}{d} + C, \tag{7}$$

and the reduction formula

$$\int \frac{1}{(u^2 + d^2)^n} \, du \tag{8}$$

$$= \frac{u}{d^2(2n-2)(u^2+d^2)^{n-1}} + \frac{2n-3}{d^2(2n-2)} \int \frac{1}{(u^2+d^2)^{n-1}} \, du, \quad n > 1.$$

If $b \neq 0$, we need to complete the square to obtain $ax^2 + bx + c = a\left((x-e)^2 + d^2\right)$ and then use the linear change of variables $u = x - e$ before using the above integration formulas.

EXAMPLE 4

Evaluate $\displaystyle\int \frac{x}{(x^2 + x + 1)(x - 1)} \, dx.$

SOLUTION   We do not need to divide in this example.

We have one linear term and one irreducible quadratic in the denominator. Note that $(1)^2 - 4(1)(1) = -3 < 0$, so $x^2 + x + 1$ is irreducible. The numerator of the term that corresponds to the factor $x^2 + x + 1$ is (Unknown number)x + (Unknown number). We write

$$\frac{x}{(x^2 + x + 1)(x - 1)} = \frac{A}{x - 1} + \frac{Bx + C}{x^2 + x + 1},$$

$$x = A(x^2 + x + 1) + (Bx + C)(x - 1), \tag{*}$$

$$x = Ax^2 + Ax + A + Bx^2 - Bx + Cx - C,$$

$$0 = (A + B)x^2 + (A - B + C - 1)x + (A - C).$$

$$\begin{cases} A + B & = 0, \quad \text{[Coefficient of } x^2\text{]} \\ A - B + C - 1 = 0, \quad \text{[Coefficient of } x\text{]} \\ A \quad\;\; - C \quad = 0. \quad \text{[Constant term]} \end{cases}$$

The solution is $(A, B, C) = (1/3, -1/3, 1/3)$. We obtain

$$\frac{x}{(x^2 + x + 1)(x - 1)} = \frac{1/3}{x - 1} + \frac{(-1/3)x + (1/3)}{x^2 + x + 1}$$

$$= \frac{1}{3}\frac{1}{x - 1} + \frac{1}{3}\frac{1 - x}{x^2 + x + 1}.$$

The integral of the first term on the right is easy. Let us work on evaluating the integral of the second term. We first complete the square.

$$\int \frac{1 - x}{x^2 + x + 1} \, dx = \int \frac{1 - x}{(x^2 + x + (1/4)) - (1/4) + 1} \, dx$$

$$= \int \frac{1-x}{(x+(1/2))^2 + (\sqrt{3}/2)^2} \, dx.$$

Using $u = x + (1/2)$, so that $du = dx$ and $x = u - (1/2)$, we see that the above integral is

$$\int \frac{1 - (u - (1/2))}{u^2 + (\sqrt{3}/2)^2} \, du$$

$$= \frac{3}{2} \int \frac{1}{u^2 + (\sqrt{3}/2)^2} \, du - \int \frac{1}{u^2 + (\sqrt{3}/2)^2} \left(\frac{1}{2} \right) (2u) \, du$$

$$= \frac{3}{2} \left(\frac{1}{\sqrt{3}/2} \right) \tan^{-1} \left(\frac{u}{\sqrt{3}/2} \right) - \frac{1}{2} \ln \left(u^2 + (\sqrt{3}/2)^2 \right) + C$$

$$= \sqrt{3} \tan^{-1} \left(\frac{2(x + (1/2))}{\sqrt{3}} \right) - \frac{1}{2} \ln \left((x + (1/2))^2 + (\sqrt{3}/2)^2 \right) + C$$

$$= \sqrt{3} \tan^{-1} \left(\frac{2x+1}{\sqrt{3}} \right) - \frac{1}{2} \ln(x^2 + x + 1) + C.$$

We then write

$$\int \frac{x}{(x^2 + x + 1)(x - 1)} \, dx$$

$$= \frac{1}{3} \int \frac{1}{x - 1} \, dx + \frac{1}{3} \int \frac{1 - x}{x^2 + x + 1} \, dx$$

$$= \frac{1}{3} \ln|x - 1| + \frac{\sqrt{3}}{3} \tan^{-1} \frac{2x + 1}{\sqrt{3}} - \frac{1}{6} \ln(x^2 + x + 1) + C. \quad \blacksquare$$

EXAMPLE 5

Evaluate $\displaystyle\int \frac{x^3 + 1}{(x^2 + 1)^2} \, dx$.

SOLUTION We have $\deg(x^3 + 1) = 3 < 4 = \deg((x^2 + 1)^2)$, so we don't need to divide. The term $x^2 + 1$ is irreducible. We have

$$\frac{x^3 + 1}{(x^2 + 1)^2} = \frac{Ax + B}{x^2 + 1} + \frac{Cx + D}{(x^2 + 1)^2}.$$

Clearing fractions and using Method 1 as before, we can determine the solution is $(A, B, C, D) = (1, 0, -1, 1)$, so

$$\frac{x^3 + 1}{(x^2 + 1)^2} = \frac{x + 0}{x^2 + 1} + \frac{-x + 1}{(x^2 + 1)^2}$$

$$= \frac{x}{x^2 + 1} - \frac{x}{(x^2 + 1)^2} + \frac{1}{(x^2 + 1)^2}.$$

Using Formulas 5, 6, 7, and 8, we then have

$$\int \frac{x^3 + 1}{(x^2 + 1)^2} \, dx = \int \frac{x}{x^2 + 1} \, dx - \int \frac{x}{(x^2 + 1)^2} \, dx + \int \frac{1}{(x^2 + 1)^2} \, dx$$

$$= \frac{1}{2}\ln(x^2+1) - \frac{-1}{2(x^2+1)} + \frac{x}{2(x^2+1)} + \frac{1}{2}\int \frac{1}{x^2+1}\,dx$$

$$= \frac{1}{2}\ln(x^2+1) + \frac{1}{2}\frac{1}{x^2+1} + \frac{1}{2}\frac{x}{x^2+1} + \frac{1}{2}\tan^{-1}x + C. \quad \blacksquare$$

It is important that the denominator be factored correctly. For example, we have

$$(x^2-1)(x-1) = (x-1)(x+1)(x-1) = (x-1)^2(x+1),$$

so the correct form of the partial fraction expansion of $1/[(x^2-1)(x-1)]$ is

$$\frac{1}{(x^2-1)(x-1)} = \frac{A}{x-1} + \frac{B}{(x-1)^2} + \frac{C}{x+1}.$$

If we incorrectly considered x^2-1 to be an irreducible quadratic and wrote

$$\frac{1}{(x^2-1)(x-1)} \underset{(?)}{=} \frac{A}{x-1} + \frac{Bx+C}{x^2-1},$$

we could not obtain an identity for any values of A, B, or C. Method 1 would result in an inconsistent system of linear equations. Method 2 would give values of A, B, and C, but an incorrect equation.

EXERCISES 8.5

Evaluate the integrals in Exercises 1–38.

1. $\int \frac{1}{x^2-2x-3}\,dx$ (Calculus Explorer & Tutor II, Techniques of Integration, 12.)

2. $\int \frac{2x-1}{x^2-x-2}\,dx$

3. $\int \frac{x-6}{x^2-3x}\,dx$

4. $\int \frac{x-5}{2x^2+x-1}\,dx$

5. $\int \frac{2x^3-9x+10}{x^2-4}\,dx$

6. $\int \frac{3x^2-33}{x^2-9}\,dx$

7. $\int \frac{2x-3}{(x-1)^2}\,dx$

8. $\int \frac{3x+4}{(x+2)^2}\,dx$

9. $\int \frac{x^2+8x-4}{x^3-4x}\,dx$

10. $\int \frac{3x^2-3x-6}{x^3-x}\,dx$

11. $\int \frac{4x^2-11x+4}{x^3-3x^2+2x}\,dx$

12. $\int \frac{x+5}{x^3-4x^2+3x}\,dx$

13. $\int \frac{x}{(x-1)^3}\,dx$

14. $\int \frac{x^2}{(x+1)^3}\,dx$

15. $\int \frac{x^4-x^3+x^2+2x-2}{x^3-x^2}\,dx$

16. $\int \frac{x^5-3}{x^3-x^2}\,dx$

17. $\int \frac{x^2+2x+2}{x^4+x^3}\,dx$

18. $\int \frac{4}{x^4-x^3}\,dx$

19. $\int \frac{1}{x(x+1)^2}\,dx$ (Calculus Explorer & Tutor II, Techniques of Integration, 13.)

20. $\int \frac{4x-4}{x^2(x-2)^2}\,dx$

21. $\int \frac{x^2}{x^2-3}\,dx$

22. $\int \frac{x^4}{x^2-2}\,dx$

23. $\int \frac{3x^2+2}{x(x^2+1)}\,dx$

24. $\int \frac{x^2-x-4}{x(x^2+4)}\,dx$

25. $\int \frac{4}{x^2-1}\,dx$

26. $\int \frac{32}{x^2-16}\,dx$

27. $\int \frac{8}{(x^2+1)(x^2+9)}\,dx$

28. $\int \frac{3x}{(x^2+1)(x^2+4)}\,dx$

29. $\int \frac{x^3+4x-2}{x^3-8}\,dx$

30. $\int \frac{x^3+x+2}{x^3-1}\,dx$

31. $\int \frac{x^2}{(x^2+9)^2}\,dx$

32. $\int \frac{x^3-20x+4}{(x^2+4)^2}\,dx$

33. $\int \frac{5}{x(x^2+2x+5)}\,dx$

34. $\displaystyle\int \frac{5}{x(x^2 + 4x + 5)}\,dx$ **35.** $\displaystyle\int \frac{2x}{x^2 + 2x - 1}\,dx$

36. $\displaystyle\int \frac{2x}{x^2 - 2x - 1}\,dx$ **37.** $\displaystyle\int \frac{x}{(2-x)^4}\,dx$

38. $\displaystyle\int \frac{x^2}{(1-2x)^5}\,dx$

Evaluate the definite integrals in Exercises 39–44.

39. $\displaystyle\int_1^2 \frac{1}{(x+1)(x+2)}\,dx$ **40.** $\displaystyle\int_1^2 \frac{1}{x^2(x+1)}\,dx$

41. $\displaystyle\int_{-1}^1 \frac{1}{(x-2)(x+2)}\,dx$ **42.** $\displaystyle\int_0^2 \frac{x}{(x+2)^2}\,dx$

43. $\displaystyle\int_1^3 \frac{1}{x(x^2+4)}\,dx$ **44.** $\displaystyle\int_1^4 \frac{1}{x^2(x^2+9)}\,dx$

45. Analysis of **frictional adiabatic flow** leads to the integral

$$\int_{M_0}^1 \frac{2(1-M^2)}{kM^3\left(1 + \dfrac{k-1}{2}M^2\right)}\,dM,$$

where k is a constant. Use partial fractions to evaluate this integral.

8.6   RATIONALIZING SUBSTITUTIONS

Connections

Substitution and change of variable, 5.5.

We will illustrate some substitutions that can be used to transform certain types of integrals into integrals of rational functions, which can then be evaluated by the method of partial fractions.

$u^n = ax + b$

The substitution $u^n = ax + b$ can sometimes be used to simplify integrals that contain expressions of the form $(ax + b)^{1/n}$.

EXAMPLE 1

Evaluate $\int x\sqrt{3-x}\,dx$.

SOLUTION   The equation $u^2 = 3 - x,\, u \geq 0$, implies $2u\,du = -dx,\, x = 3 - u^2$, and $\sqrt{3-x} = u$, so substitution gives

$$\int x\sqrt{3-x}\,dx = \int (3-u^2)(u)(-2u)\,du = \int (2u^4 - 6u^2)\,du$$

$$= 2\frac{u^5}{5} - 6\frac{u^3}{3} + C$$

$$= \frac{2}{5}(\sqrt{3-x})^5 - 2(\sqrt{3-x})^3 + C$$

$$= \frac{2}{5}(3-x)^{5/2} - 2(3-x)^{3/2} + C. \quad\blacksquare$$

In the previous example, we could also use the substitution $u = 3 - x$. This would result in an integrand that is a linear sum of fractional powers of u.

A substitution of the form $u = ax + b$ can be used to express integrals that have powers of $ax + b$ in the denominator as a sum of powers of u.

EXAMPLE 2

Evaluate $\displaystyle\int \frac{x}{(2x-1)^6}\,dx$.

SOLUTION   We could use the method of partial fractions to evaluate this integral. However, let us see that the substitution $u = 2x - 1$ simplifies the integral. The equation $u = 2x - 1$ implies $du = 2\,dx$ and $x = \dfrac{u+1}{2}$, so

$$
\begin{aligned}
\int \frac{x}{(2x-1)^6}\,dx &= \int \frac{(1/2)(u+1)}{u^6}\,\frac{1}{2}\,du \\[2mm]
&= \int \left(\frac{1}{4}u^{-5} + \frac{1}{4}u^{-6}\right) du \\[2mm]
&= \frac{1}{4}\left(\frac{u^{-4}}{-4}\right) + \frac{1}{4}\left(\frac{u^{-5}}{-5}\right) + C \\[2mm]
&= -\frac{1}{16(2x-1)^4} - \frac{1}{20(2x-1)^5} + C \\[2mm]
&= \frac{1-10x}{80(2x-1)^5} + C. \quad\blacksquare
\end{aligned}
$$

We can remove fractional powers of x from an integrand by using the substitution $u^n = x$, where n is the least common multiple of the denominators of the fractional powers of x.

EXAMPLE 3

Evaluate $\displaystyle\int \frac{x^{1/2}}{x^{1/3}+1}\,dx$.

SOLUTION   The equation $u^6 = x$ implies $6u^5\,du = dx$, $x^{1/2} = u^3$, and $x^{1/3} = u^2$, so

$$
\begin{aligned}
\int \frac{x^{1/2}}{x^{1/3}+1}\,dx &= \int \frac{u^3}{u^2+1}(6u^5)\,du = \int 6\frac{u^8}{u^2+1}\,du \\[2mm]
&= \int 6\left(u^6 - u^4 + u^2 - 1 + \frac{1}{u^2+1}\right) du \\[2mm]
&= 6\left(\frac{u^7}{7} - \frac{u^5}{5} + \frac{u^3}{3} - u + \tan^{-1}u\right) + C \qquad {\scriptstyle[u=x^{1/6}]} \\[2mm]
&= \frac{6}{7}x^{7/6} - \frac{6}{5}x^{5/6} + 2x^{1/2} - 6x^{1/6} + 6\tan^{-1}(x^{1/6}) + C. \quad\blacksquare
\end{aligned}
$$

$u = e^{ax}$

The substitution $u = e^{ax}$ can be used to transform some integrals that involve e^{ax} into integrals that involve powers of u.

EXAMPLE 4

Evaluate $\displaystyle\int \frac{1}{e^x + 1}\,dx$.

SOLUTION The equation $u = e^x$ implies $du = e^x\,dx$, so

$$\int \frac{1}{e^x + 1}\,dx = \int \frac{1}{e^x + 1}\frac{1}{e^x}e^x\,dx$$

$$= \int \frac{1}{u + 1}\frac{1}{u}\,du$$

$$= \int \left(\frac{1}{u} - \frac{1}{u + 1}\right)\,du$$

$$= \ln|u| - \ln|u + 1| + C$$

$$= \ln(e^x) - \ln(e^x + 1) + C = x - \ln(e^x + 1) + C. \quad \blacksquare$$

$u = \tan(x/2)$

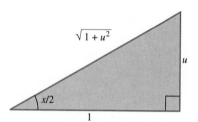

Rational expressions of $\sin x$ and $\cos x$ can be transformed to rational expressions of u by the change of variables $u = \tan(x/2)$. From the representative triangle sketched in Figure 8.6.1 we see that

$$\sin \frac{x}{2} = \frac{u}{\sqrt{1 + u^2}} \text{ and } \cos \frac{x}{2} = \frac{1}{\sqrt{1 + u^2}},$$

FIGURE 8.6.1 $\tan(x/2) = u$, so $\sin\dfrac{x}{2} = \dfrac{u}{\sqrt{1+u^2}}$ and $\cos\dfrac{x}{2} = \dfrac{1}{\sqrt{1+u^2}}$.

so

$$\cos x = 1 - 2\sin^2\frac{x}{2} = 1 - 2\left(\frac{u}{\sqrt{1 + u^2}}\right)^2 = \frac{1 - u^2}{1 + u^2}$$

and

$$\sin x = 2\sin \frac{x}{2}\cos \frac{x}{2} = 2\left(\frac{u}{\sqrt{1 + u^2}}\right)\left(\frac{1}{\sqrt{1 + u^2}}\right) = \frac{2u}{1 + u^2}.$$

The equation $u = \tan(x/2)$, $-\pi < x < \pi$, implies $\tan^{-1} u = x/2$, so $x = 2\tan^{-1} u$. Then

$$dx = \frac{2}{1 + u^2}\,du.$$

Let us summarize:

✳ The substitution

$$u = \tan \frac{x}{2}, \qquad -\pi < x < \pi,$$

gives the equations

$$\cos x = \frac{1 - u^2}{1 + u^2}, \quad \sin x = \frac{2u}{1 + u^2}, \text{ and } dx = \frac{2}{1 + u^2}\,du.$$

These equations can be used to transform an integral of a rational function of $\sin x$ and $\cos x$ into a rational function of u. After evaluating the integral of this rational function, we can use $u = \tan(x/2)$ to express the indefinite integral as a function of x. If we wish to express the answer in terms of $\sin x$ and $\cos x$, we can use the half-angle formula

$$\tan \frac{x}{2} = \frac{\sin x}{1 + \cos x}.$$

EXAMPLE 5

Evaluate $\displaystyle\int \frac{1}{1 + \sin x} \, dx$.

SOLUTION Using the formulas for the substitution $u = \tan(x/2)$, we have

$$\int \frac{1}{1 + \sin x} \, dx = \int \frac{1}{1 + \dfrac{2u}{1 + u^2}} \frac{2}{1 + u^2} \, du$$

$$= \int \frac{2}{(1 + u^2) + (2u)} \, du$$

$$= \int \frac{2}{u^2 + 2u + 1} \, du$$

$$= \int \frac{2}{(u + 1)^2} \, du$$

$$= -\frac{2}{u + 1} + C$$

$$= -\frac{2}{\tan(x/2) + 1} + C. \quad\blacksquare$$

EXERCISES 8.6

Evaluate the integrals in Exercises 1–26.

1. $\displaystyle\int \frac{x}{\sqrt{2x - 1}} \, dx$

2. $\displaystyle\int x^2 \sqrt{x + 2} \, dx$

3. $\displaystyle\int \frac{x}{(3x + 2)^{1/3}} \, dx$

4. $\displaystyle\int \frac{x}{(3x - 1)^{2/3}} \, dx$

5. $\displaystyle\int \frac{1}{1 + \sqrt{x}} \, dx \quad u^2 = x$

6. $\displaystyle\int \frac{x}{1 + \sqrt{x}} \, dx$

7. $\displaystyle\int \frac{1 - \sqrt{x}}{1 + \sqrt{x}} \, dx$

8. $\displaystyle\int x(x + 1)^{1/3} \, dx$

9. $\displaystyle\int \frac{1}{x^{1/2} + x^{1/3}} \, dx$

10. $\displaystyle\int \frac{1}{x^{1/2} + x^{1/4}} \, dx$

11. $\displaystyle\int \frac{x^{1/3}}{1 + x^{2/3}} \, dx$

12. $\displaystyle\int \frac{x^{2/3}}{1 + x^{1/3}} \, dx$

13. $\displaystyle\int \frac{x}{(2x + 5)^{5/2}} \, dx$

14. $\displaystyle\int \frac{x}{(3x - 2)^{7/3}} \, dx$

15. $\displaystyle\int \frac{x}{(2 - x)^4} \, dx$

16. $\displaystyle\int \frac{x^2}{(1 - 2x)^5} \, dx$

17. $\displaystyle\int \frac{e^{2x}}{e^x + 1} \, dx$

18. $\displaystyle\int \frac{1}{e^{2x} + 1} \, dx$

19. $\displaystyle\int e^{2x} \sqrt{1 + e^x} \, dx$

20. $\displaystyle\int \frac{e^{2x}}{\sqrt{1 + e^x}} \, dx$

21. $\displaystyle\int \frac{1}{1 - \sin x} \, dx$

22. $\displaystyle\int \frac{1}{1 - \cos x} \, dx$

23. $\displaystyle\int \frac{1}{1 + \cos x} \, dx$

24. $\displaystyle\int \frac{1}{1 + \sin x + \cos x} \, dx$

25. $\displaystyle\int \frac{1}{3 \sin x + 4 \cos x} \, dx$

26. $\displaystyle\int \frac{1}{3 + 5 \cos x} \, dx$

Evaluate the definite integrals in Exercises 27–32.

27. $\displaystyle\int_0^1 \frac{\sqrt{x}}{x + 1} \, dx$

28. $\displaystyle\int_0^2 x\sqrt{4x + 1} \, dx$

29. $\displaystyle\int_0^{\ln 3} \sqrt{1 + e^x} \, dx$

30. $\displaystyle\int_0^4 \sqrt{x} \, e^{\sqrt{x}} \, dx$

31. $\displaystyle\int_0^{\pi/2} \frac{1}{5 + 3 \cos x} \, dx$

32. $\displaystyle\int_0^{\pi/2} \frac{1}{5 + 3 \sin x} \, dx$

8.7 IMPROPER INTEGRALS

Connections

Evaluation of limits, 2.5, 4.2.

Recall from Section 5.3 that the definite integral of a function over a finite interval was defined as a limit of Riemann sums. In this section we will see how to assign values to certain definite integrals, even though the integrals do not exist as limits of Riemann sums.

Definitions and Notation

The definition of definite integral requires that the integrand be defined at each point of the interval of integration. However, an integrand that is undefined at only a *finite* number of points in the interval of integration can be "fixed" by assigning it any values at the points where it is undefined. The choice of values affects neither the existence nor the value of the resulting integral. (See Exercise 39.)

If f is unbounded on the interval of integration, we can form Riemann sums, but it is a fact that $\int_a^b f(x)\,dx$ does not exist as a limit of Riemann sums unless f is bounded on $[a, b]$. (See Exercise 40.)

We have not defined Riemann sums of a function over an infinite interval of integration. It is not convenient to define a definite integral over an infinite interval as a limit of Riemann sums.

The above facts lead us to the following definition.

DEFINITION

The integral $\int_a^b f(x)\,dx$ is called **improper** if either:

  (i) f is not bounded on $[a, b]$ or
  (ii) the interval of integration is infinite.

- The integral $\int_0^1 \dfrac{\sin x}{x}\,dx$ is not considered to be improper even though the integrand is undefined at the left endpoint of the interval of integration. We know that $\lim\limits_{x \to 0} \dfrac{\sin x}{x} = 1$, so the integrand is bounded. Assigning the integrand the value one at zero gives a continuous integrand, so we know the integral exists as a limit of Riemann sums.

- The integral $\int_0^1 \dfrac{1}{\sqrt{1 - x^2}}\,dx$ is improper because $1/\sqrt{1 - x^2}$ becomes unbounded as x approaches one, the right endpoint of the interval of integration.

- The integral $\int_0^1 \dfrac{1}{x}\,dx$ is improper because $1/x$ becomes unbounded as x approaches zero, the left endpoint of the interval of integration.

- The integral $\int_1^\infty \dfrac{1}{x^2}\,dx$ is improper because $[1, \infty)$ is an infinite interval of integration.

- The integral $\int_{-1}^1 \dfrac{1}{x^2}\,dx$ is improper because $1/x^2$ becomes unbounded as x approaches zero, a point inside the interval of integration.

- The integral $\int_0^\infty x^{-2/3}\,dx$ is improper because $x^{-2/3}$ becomes unbounded as x approaches zero and because $[0, \infty)$ is an infinite interval of integration.

So far, an improper integral is only a symbol with no meaning. We will now describe some cases for which certain types of improper integrals can be assigned a numerical value.

Suppose $\int_a^b f(x)\,dx$ is improper only because of the behavior at one endpoint of the interval of integration. For example, suppose that either f becomes unbounded as x approaches b or $b = \infty$, but also that $\int_a^\beta f(x)\,dx$ exists for every $a < \beta < b$. (This means that we can apply the usual techniques of integration to evaluate $\int_a^\beta f(x)\,dx$.) We then say that $\int_a^b f(x)\,dx$ is **convergent** and define its value by

$$\int_a^b f(x)\,dx = \lim_{\beta \to b^-} \int_a^\beta f(x)\,dx,$$

provided the limit exists and is finite. Otherwise, we say that $\int_a^b f(x)\,dx$ is **divergent** and no numerical value is assigned to the symbol. Similarly, if $\int_a^b f(x)\,dx$ is improper only at a, we say it is convergent with value

$$\int_a^b f(x)\,dx = \lim_{\alpha \to a^+} \int_\alpha^b f(x)\,dx,$$

provided the limit exists and is finite.

If $f(x)$ is *positive* on the interval of integration, we can interpret an improper integral as area under a curve. For example, let us suppose $f(x)$ is positive and $\int_a^b f(x)\,dx$ is improper only because of the behavior near the right endpoint b. If $a < \beta < b$, then we know that $\int_a^\beta f(x)\,dx$ represents the area under the curve $y = f(x)$, $a \le x \le \beta$. It should then seem reasonable to define the area under the curve for $a \le x \le b$ to be the value of the improper integral. Convergent improper integrals of positive functions correspond to finite area under the curve; divergent improper integrals of positive functions correspond to infinite area under the curve. See Figure 8.7.1.

Integrands That Become Unbounded at an Endpoint

EXAMPLE 1

Evaluate $\int_0^1 \dfrac{1}{\sqrt{1 - x^2}}\,dx$.

SOLUTION The integral is improper only at the right endpoint $x = 1$. For $0 < \beta < 1$, we have

$$\int_0^\beta \frac{1}{\sqrt{1 - x^2}}\,dx = \sin^{-1} x \Big]_0^\beta = \sin^{-1}\beta - \sin^{-1} 0 = \sin^{-1}\beta;$$

$$\lim_{\beta \to 1^-} \int_0^\beta \frac{1}{\sqrt{1 - x^2}}\,dx = \lim_{\beta \to 1^-} \sin^{-1}\beta = \sin^{-1} 1 = \frac{\pi}{2}.$$

The integral is convergent with

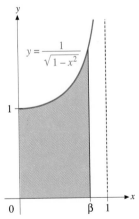

FIGURE 8.7.1 Convergent improper integrals of positive functions correspond to finite area under the curve; divergent improper integrals of positive functions correspond to infinite area under the curve.

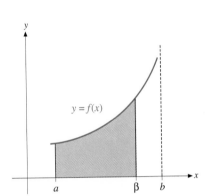

FIGURE 8.7.2 The area under the curve $y = \dfrac{1}{\sqrt{1 - x^2}}$, $0 \le x \le \beta$, approaches $\pi/2$ as β approaches 1 from the left. $\int_0^1 \dfrac{1}{\sqrt{1 - x^2}}\,dx = \dfrac{\pi}{2}$.

$$\int_0^1 \frac{1}{\sqrt{1-x^2}}\,dx = \frac{\pi}{2}.$$

See Figure 8.7.2. ■

Note that the Fundamental Theorem of Calculus does not apply directly to

$$\int_0^1 \frac{1}{\sqrt{1-x^2}}\,dx,$$

because the integrand cannot be assigned any value at 1 that would make it continuous on the interval [0, 1].

EXAMPLE 2

Evaluate $\displaystyle\int_0^1 \frac{1}{x}\,dx$.

SOLUTION   The integral is improper only at the left endpoint $x = 0$. For $0 < \alpha < 1$, we have

$$\int_\alpha^1 \frac{1}{x}\,dx = \ln x\Big]_\alpha^1 = \ln 1 - \ln \alpha = -\ln \alpha.$$

Then

$$\lim_{\alpha \to 0^+} \int_\alpha^1 \frac{1}{x}\,dx = \lim_{\alpha \to 0^+} (-\ln \alpha) = +\infty.$$

Since the limit does not converge to a finite number, we say that

$$\int_0^1 \frac{1}{x}\,dx$$

is divergent. This integral is not assigned a numerical value. See Figure 8.7.3. ■

Infinite Limits of Integration

EXAMPLE 3

Evaluate $\displaystyle\int_1^\infty \frac{1}{x^2}\,dx$.

SOLUTION   The integral is improper only because the right endpoint is infinite. For $1 < \beta < \infty$, we have

$$\int_1^\beta \frac{1}{x^2}\,dx = -\frac{1}{x}\Big]_1^\beta = -\frac{1}{\beta} + \frac{1}{1};$$

$$\lim_{\beta \to \infty} \int_1^\beta \frac{1}{x^2}\,dx = \lim_{\beta \to \infty} \left(-\frac{1}{\beta} + \frac{1}{1}\right) = 1.$$

Hence, the integral is convergent with

$$\int_1^\infty \frac{1}{x^2}\,dx = 1.$$

See Figure 8.7.4. ■

Note that we cannot apply the Fundamental Theorem of Calculus directly to the integral in Example 3, because the interval of integration does not contain its right endpoint. Infinity is not a real number.

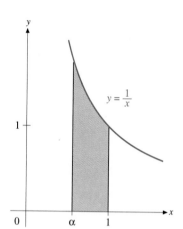

FIGURE 8.7.3   The area under the curve $y = \dfrac{1}{x}, \alpha \le x \le 1$, approaches infinity as α approaches 0 from the right. $\displaystyle\int_0^1 \frac{1}{x}\,dx$ is divergent.

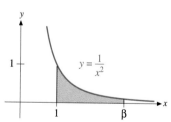

FIGURE 8.7.4   The area under the curve $y = \dfrac{1}{x^2}, 1 \le x \le \beta$, approaches 1 as β approaches infinity. $\displaystyle\int_1^\infty \frac{1}{x^2}\,dx = 1$.

Integrals Improper at Other Than One Endpoint

Suppose $\int_a^b f(x)dx$ is improper at other than one endpoint, but we can partition the interval of integration into a finite number of subintervals with the property that the integral over each subinterval is improper at only one endpoint. If *every* integral in the sum is convergent, we say $\int_a^b f(x)dx$ is convergent and has a value equal to the sum of the values of the integrals that form the sum. If *any* of the integrals in the sum diverge, we say $\int_a^b f(x)dx$ is divergent and no numerical value is assigned to the symbol. The convergence or divergence of an improper integral does not depend on how the interval of integration is divided, as long as we write the integral as a sum of integrals that are improper at only one endpoint.

Remember that the integral over every one of the subintervals must be convergent for the original integral to be convergent. If the integral over any one of the subintervals diverges, then the integral over the original interval is divergent.

EXAMPLE 4

Evaluate $\int_{-1}^1 x^{-2}dx$.

SOLUTION   This integral is improper at the point $x = 0$, a point in the interval other than an endpoint. We write

$$\int_{-1}^1 x^{-2}dx = \int_{-1}^0 x^{-2}dx + \int_0^1 x^{-2}dx.$$

Each of the two integrals in the sum is improper only at one endpoint. We must check that both converge.

Let us check the convergence of

$$\int_{-1}^0 x^{-2}dx.$$

We have

$$\lim_{\beta \to 0^-} \int_{-1}^{\beta} x^{-2}dx = \lim_{\beta \to 0^-} \left(-\frac{1}{x}\right)\Bigg]_{-1}^{\beta} = \lim_{\beta \to 0^-} \left(-\frac{1}{\beta} - 1\right) = \infty.$$

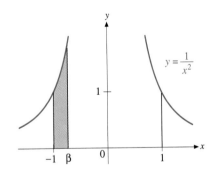

FIGURE 8.7.5   The area under the curve $y = \dfrac{1}{x^2}$, $-1 \le x \le \beta$, approaches infinity as β approaches zero from the left. Since the integral over one subinterval is divergent, $\int_{-1}^1 \dfrac{1}{x^2}\,dx$ is divergent.

Since this integral is divergent, we know that $\int_{-1}^1 x^{-2}dx$ is divergent. It is not necessary to check the convergence of the integral over the interval between zero and one. See Figure 8.7.5. ∎

If you did not notice that the integral in Example 4 is improper, you might write

$$\int_{-1}^1 x^{-2}dx \underset{(?)}{=} -x^{-1}\Bigg]_{-1}^1 = -(1) - (-(-1)) = -2,$$

which is an incorrect answer. The Fundamental Theorem of Calculus does not apply to this integral.

EXAMPLE 5

Evaluate $\displaystyle\int_0^\infty x^{-2/3}\,dx$.

SOLUTION   The integral is improper at $x = 0$ and ∞. Let us subdivide the interval of integration at $x = 1$, a convenient number, and write

$$\int_0^\infty x^{-2/3}\,dx = \int_0^1 x^{-2/3}\,dx + \int_1^\infty x^{-2/3}\,dx.$$

The first integral on the right is improper only at the endpoint $x = 0$, and the second is improper only at ∞. We must check the convergence of each separately:

$$\lim_{\alpha \to 0^+}\int_\alpha^1 x^{-2/3}\,dx = \lim_{\alpha \to 0^+} 3x^{1/3}\Big]_\alpha^1 = \lim_{\alpha \to 0^+}(3 - 3\alpha^{1/3}) = 3,$$

so the first integral converges;

$$\lim_{\beta \to \infty}\int_1^\beta x^{-2/3}\,dx = \lim_{\beta \to \infty} 3x^{1/3}\Big]_1^\beta = \lim_{\beta \to \infty}(3\beta^{1/3} - 3) = \infty,$$

so the second integral diverges. Since one of the parts diverges, $\displaystyle\int_0^\infty x^{-2/3}\,dx$ is divergent. See Figure 8.7.6.   ∎

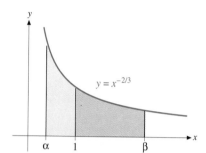

FIGURE 8.7.6 $\int_0^\infty x^{-2/3}\,dx$ is divergent. The area under the curve $y = x^{-2/3}$, $1 \le x \le \beta$, approaches infinity as β approaches infinity.

EXAMPLE 6

Evaluate $\displaystyle\int_0^\infty \frac{1}{2x^2 + 3x + 1}\,dx$.

SOLUTION   The integral is improper at infinity. We have $2x^2 + 3x + 1 = (2x + 1)(x + 1)$, so the zeros of the denominator are not contained in the interval of integration. The integrand $1/(2x^2 + 3x + 1)$ does not become unbounded on the interval of integration. For $0 < \beta < \infty$, we have

$$\int_0^\beta \frac{1}{2x^2 + 3x + 1}\,dx = \int_0^\beta \frac{1}{(2x + 1)(x + 1)}\,dx.$$

The partial fraction expansion of the integrand is

$$\frac{1}{(2x + 1)(x + 1)} = \frac{A}{2x + 1} + \frac{B}{x + 1}.$$

It is easy to determine that $A = 2$ and $B = -1$. We then have

$$\begin{aligned}
\int_0^\beta \frac{1}{2x^2 + 3x + 1}\,dx &= \int_0^\beta \frac{1}{(2x + 1)(x + 1)}\,dx\\
&= \int_0^\beta \left(\frac{2}{2x + 1} - \frac{1}{x + 1}\right)dx\\
&= \ln|2x + 1| - \ln|x + 1|\Big]_0^\beta\\
&= \ln\left|\frac{2x + 1}{x + 1}\right|\,\Big]_0^\beta
\end{aligned}$$

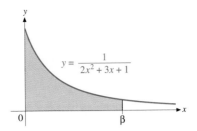

FIGURE 8.7.7 The area under the curve $y = \dfrac{1}{2x^2 + 3x + 1}$, $0 \le x \le \beta$, approaches $\ln 2$ as β approaches infinity. $\displaystyle\int_0^\infty \frac{1}{2x^2 + 3x + 1}\,dx = \ln 2.$

$$= \ln \frac{2\beta + 1}{\beta + 1} - \ln 1$$
$$= \ln \frac{2\beta + 1}{\beta + 1},$$

so

$$\int_0^\infty \frac{1}{2x^2 + 3x + 1}\,dx = \lim_{\beta \to \infty} \ln \frac{2\beta + 1}{\beta + 1} = \ln 2.$$

The integral is convergent with value $\ln 2$. See Figure 8.7.7. ∎

In Example 6, we cannot evaluate $\displaystyle\lim_{\beta \to \infty} (\ln(2\beta + 1) - \ln(\beta + 1))$ as the difference of limits, since each term approaches infinity as β approaches infinity. We need to combine terms to obtain

$$\lim_{\beta \to \infty} \ln \frac{2\beta + 1}{\beta + 2}.$$

Determining Convergence or Divergence by Comparison

In Section 4.2 we investigated limits at infinity of functions that involved roots and powers by factoring the highest power of x from each of the numerator and denominator. Let us now see how we can use this technique to determine the convergence or divergence of some improper integrals without evaluating a definite integral over a subinterval and taking a limit. It is important that we consider only improper integrals that have positive integrands, so the convergence or divergence corresponds to area under a curve.

We will need the following result, which is verified easily.

$$\int_1^\infty \frac{1}{x^p}\,dx \text{ is convergent if } p > 1 \text{ and divergent if } p \le 1. \tag{1}$$

See Figure 8.7.8, parts a and b.

We can use Statement 1 to obtain information about the convergence or divergence of other integrals that are improper at infinity.

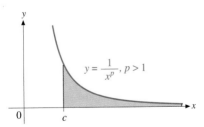

FIGURE 8.7.8a The area under the curve $y = \dfrac{1}{x^p}$, $x \ge c > 0$, is finite if $p > 1$.

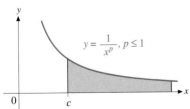

FIGURE 8.7.8b The area under the curve $y = \dfrac{1}{x^p}$, $x \ge c > 0$, is infinite if $p \le 1$.

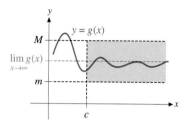

FIGURE 8.7.9 If $0 < \lim\limits_{x \to \infty} g(x) < \infty$, then there are numbers m, M, and c such that $0 < m < g(x) < M < \infty$ whenever $x \geq c$.

THEOREM 1

If g is positive and continuous on $[a, \infty)$, $a > 0$, and the limit at infinity of g is a finite, nonzero number, then

$$\int_1^\infty \frac{g(x)}{x^p}\, dx \text{ is convergent if } p > 1 \text{ and divergent if } p \leq 1.$$

Proof If $0 < \lim\limits_{x \to \infty} g(x) < \infty$, then there are numbers m, M, and $c \geq a$ such that $0 < m < g(x) < M < \infty$ for $x \geq c$. See Figure 8.7.9. It follows that

$$0 \leq \frac{m}{x^p} < \frac{g(x)}{x^p} < \frac{M}{x^p} \quad \text{for } x \geq c.$$

If $p > 1$, then $g(x)/x^p < M/x^p$, $x \geq c$, implies the area under the curve $y = g(x)/x^p$ is less than the area under the curve $y = M/x^p$. Since Statement 1 implies the area under the curve $y = M/x^p$ is finite, the area under the curve $y = g(x)/x^p$ is also finite. This implies $\int_c^\infty g(x)/x^p\, dx$ is convergent. Since $\int_a^c g(x)/x^p\, dx$ is not improper, we conclude that $\int_a^\infty g(x)/x^p\, dx$ is convergent. See Figure 8.7.10a. If $p \leq 1$, then the area under the curve $y = m/x^p$ is infinite. Then $m/x^p < g(x)/x^p$ implies the area under the curve $y = g(x)/x^p$ is also infinite, so $\int_a^\infty g(x)/x^p\, dx$ is divergent. See Figure 8.7.10b. ∎

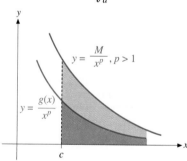

FIGURE 8.7.10a If $p > 1$, the area under the curve $y = M/x^p$, $x \geq c$ is finite. Then $0 \leq g(x)/x^p < M/x^p$ implies the area under the curve $y = g(x)/x^p$ is also finite.

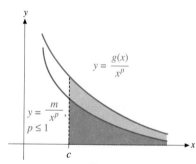

FIGURE 8.7.10b If $p \leq 1$, the area under the curve $y = m/x^p$, $x \geq c$ is infinite. Then $g(x)/x^p > m/x^p \geq 0$ implies the area under the curve $y = g(x)/x^p$ is also infinite.

EXAMPLE 7

Determine whether the improper integrals are convergent or divergent.

(a) $\displaystyle\int_1^\infty \frac{x+1}{3x^3 + 2x + 1}$,

(b) $\displaystyle\int_1^\infty \frac{2x}{\sqrt{x^4 + 1}}\, dx$.

SOLUTION

(a) We factor the highest power of x from each of the numerator and denominator to obtain

$$\frac{x+1}{3x^3+2x+1} = \frac{x\left(1+\dfrac{1}{x}\right)}{x^3\left(3+\dfrac{2}{x^2}+\dfrac{1}{x^3}\right)} = \frac{1}{x^2}\left(\frac{1+\dfrac{1}{x}}{3+\dfrac{2}{x^2}+\dfrac{1}{x^3}}\right).$$

The expression

$$\frac{1+\dfrac{1}{x}}{3+\dfrac{2}{x^2}+\dfrac{1}{x^3}}$$

is positive and continuous on $[1, \infty)$ and has finite, nonzero limit $1/3$ as $x \to \infty$. Since the power of x in the denominator is $p = 2$, so $p > 1$, Theorem 1 tells us that $\displaystyle\int_1^\infty \frac{x+1}{3x^3+2x+1}\,dx$ is convergent.

(b) We factor the highest power of x from the denominator to obtain

$$\frac{2x}{\sqrt{x^4+1}} = \frac{2x}{x^2\sqrt{1+\dfrac{1}{x^4}}} = \frac{1}{x}\frac{2}{\sqrt{1+\dfrac{1}{x^4}}}.$$

The expression

$$\frac{1}{\sqrt{1+\dfrac{1}{x^4}}}$$

is positive and continuous on $[1, \infty)$ and has a finite, nonzero limit as x approaches infinity. Theorem 1 with $p = 1$, so $p \le 1$, then implies $\displaystyle\int_1^\infty \frac{x}{\sqrt{x^4+1}}\,dx$ is divergent. ∎

The following results for integrals that are improper at one point of a finite interval correspond to Statement 1 and Theorem 1.

If $a \le c \le b$, then

$$\int_a^b \frac{1}{|x-c|^\gamma}\,dx \text{ is convergent if } \gamma < 1 \text{ and divergent if } \gamma \ge 1.$$

See Figure 8.7.11, parts a and b.

THEOREM 2

If g is nonnegative and continuous on the finite interval $[a, b]$, $a \le c \le b$, and $g(c) > 0$, then

$$\int_a^b \frac{g(x)}{|x-c|^\gamma}\,dx \text{ is convergent if } \gamma < 1 \text{ and divergent if } \gamma \ge 1.$$

Proof The integral is improper only at $x = c$. If $g(x)$ is continuous on $[a, b]$ and $g(c) > 0$, then there are numbers m and M such that $0 < m < g(x) < M < \infty$ for all x that are within some positive distance of c. See Figure 8.7.12. The proof is then similar to that of Theorem 1. ∎

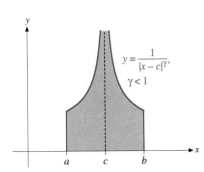

FIGURE 8.7.11a If $a \le c \le b$, the area under the curve $y = \dfrac{1}{|x-c|^\gamma}$, $a \le x \le b$, is finite if $\gamma < 1$.

FIGURE 8.7.11b If $a \le c \le b$, the area under the curve $y = \dfrac{1}{|x-c|^\gamma}$, $a \le x \le b$, is infinite if $\gamma \ge 1$.

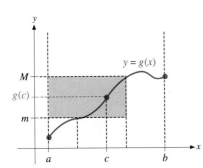

FIGURE 8.7.12 If g is continuous at c and $g(c) > 0$, then there are numbers m and M such that $0 < m < g(x) < M < \infty$ for all x that are within some positive distance of c.

EXAMPLE 8

Determine whether the improper integrals are convergent or divergent.

(a) $\displaystyle\int_1^2 \frac{x}{(x-1)^{3/2}}\, dx,$ (b) $\displaystyle\int_0^1 \frac{x^2}{\sqrt{1-x^2}}\, dx.$

SOLUTION

(a) The integral $\displaystyle\int_1^2 \frac{x}{(x-1)^{3/2}}\, dx$ is improper only at $x = 1$. The factor x is positive and continuous on $[1, 2]$. Theorem 2 with $\gamma = 3/2$, so $\gamma \geq 1$, then implies divergence of $\displaystyle\int_1^2 \frac{x}{(x-1)^{3/2}}\, dx.$

(b) The integral $\displaystyle\int_0^1 \frac{x^2}{\sqrt{1-x^2}}\, dx$ is improper only at $x = 1$. We have

$$\frac{x^2}{\sqrt{1-x^2}} = \frac{x^2}{\sqrt{(1-x)(1+x)}} = \frac{1}{(1-x)^{1/2}}\left(\frac{x^2}{\sqrt{1+x}}\right).$$

The factor $\dfrac{x^2}{\sqrt{1+x}}$ is nonnegative and continuous on $[0, 1]$, and is not zero at $x = 1$. Theorem 2 with $\gamma = 1/2$, so $\gamma < 1$, then implies $\displaystyle\int_0^1 \frac{x^2}{\sqrt{1-x^2}}\, dx$ is convergent. ■

EXERCISES 8.7

Evaluate or indicate divergence for each of the integrals in Exercises 1–32.

1. $\displaystyle\int_0^1 \frac{1}{1-x}\, dx$

2. $\displaystyle\int_0^2 \frac{1}{x-2}\, dx$

3. $\displaystyle\int_0^1 \frac{1}{\sqrt{1-x}}\, dx$

4. $\displaystyle\int_0^2 \frac{1}{(2-x)^{1/3}}\, dx$

5. $\displaystyle\int_0^{\pi/2} \sec^2 x\, dx$

6. $\displaystyle\int_0^{\pi/2} \tan x\, dx$

7. $\displaystyle\int_0^{\pi/2} \frac{\sin x}{1-\cos x}\, dx$

8. $\displaystyle\int_0^8 x^{-1/3}\, dx$ (Calculus Explorer & Tutor II, Techniques of Integration, 14.)

9. $\displaystyle\int_{1/2}^1 \frac{1}{x\ln x}\, dx$

10. $\displaystyle\int_e^\infty \frac{1}{x(\ln x)^2}\, dx$ (Calculus Explorer & Tutor II, Techniques of Integration, 15.)

11. $\displaystyle\int_0^\infty xe^{-x^2}\, dx$

12. $\displaystyle\int_0^\infty \frac{1}{4+x^2}\, dx$

13. $\displaystyle\int_1^\infty \frac{1}{x-1}\, dx$

14. $\displaystyle\int_0^\infty \sin x\, dx$

15. $\displaystyle\int_{-1}^1 \frac{1}{\sqrt{1-x^2}}\, dx$

16. $\displaystyle\int_1^\infty \frac{1}{x\sqrt{x^2-1}}\, dx$

17. $\displaystyle\int_0^\pi \tan x\, dx$

18. $\displaystyle\int_0^\pi \sec x\, dx$

19. $\displaystyle\int_0^\infty \frac{e^{-\sqrt{x}}}{\sqrt{x}}\, dx$

20. $\displaystyle\int_0^\infty \frac{1}{x+1}\, dx$

21. $\displaystyle\int_{-1}^1 x^{-1/3}\, dx$

22. $\displaystyle\int_{-1}^1 x^{-4/3}\, dx$

23. $\displaystyle\int_{-\infty}^\infty \frac{1}{x^2+4x+5}\, dx$

24. $\displaystyle\int_{-\infty}^\infty \frac{x}{x^2+16}\, dx$

25. $\displaystyle\int_{-\infty}^\infty \frac{1}{x^2+4x+3}\, dx$

26. $\displaystyle\int_1^\infty \frac{1}{x^2+x}\, dx$

27. $\displaystyle\int_0^\infty xe^{-x/2}\, dx$

28. $\displaystyle\int_0^1 \ln x\, dx$

29. $\displaystyle\int_1^\infty \frac{1}{x(x^2+1)}\, dx$

30. $\displaystyle\int_0^\infty x^2 e^{-x}\, dx$

31. $\displaystyle\int_0^\infty \frac{1}{(x^2+4)^{3/2}}\, dx$

32. $\displaystyle\int_0^\infty e^{-x}\sin x\, dx$

33. Show that $\displaystyle\int_1^\infty (1/x^p)\, dx$ is convergent for $p > 1$ and divergent for $p \leq 1$.

34. Show that $\int_0^1 (1/x^\gamma)dx$ is convergent for $\gamma < 1$ and divergent for $\gamma \geq 1$.

35. The velocity at time t of an object that is moving along a straight line is given by the formula

$$v(t) = \frac{1}{1+t^2}, \qquad t \geq 0.$$

What is the total distance the object will travel?

36. The velocity at time t of an object that is moving along a straight line is given by the formula

$$v(t) = \frac{1}{1+t}, \qquad t \geq 0.$$

What is the total distance the object will travel?

37. Show that the arc length of the curve $y = 3x^{2/3}, 0 \leq x \leq 8$, is given by an improper integral and evaluate the integral.

38. Show that the integral for the surface area of the hemisphere of radius r obtained by revolving the curve $y = \sqrt{r^2 - x^2}, 0 \leq x \leq r$, about the y-axis is improper and evaluate the integral.

39. Let

$$f(x) = \begin{cases} a, & x = 0, \\ x, & 0 < x \leq 1. \end{cases}$$

Show that the difference of corresponding Riemann sums of $\int_0^1 f(x)dx$ and $\int_0^1 x\,dx$ is no greater than $|a|\Delta x_1$, which approaches zero as Δx_1 approaches zero. Since we know that $\int_0^1 x\,dx = \frac{1}{2}$, it follows that the Riemann sums of f must converge to $1/2$ for any choice of a. A similar argument can be used to show that **neither the existence nor the values of a definite integral are affected by the values of the integrand at a finite number of points of the interval of integration.**

40. Let

$$f(x) = \begin{cases} 0, & x = 0, \\ 1/\sqrt{x}, & 0 < x \leq 1. \end{cases}$$

Show that for any given number B and for any positive integer n, there is a Riemann sum R of f over the interval $[0, 1]$ with n subdivisions of equal length and $R > B$. This shows that $\int_0^1 1/\sqrt{x}\,dx$ cannot be the limit of Riemann sums of f over $[0, 1]$. (*Hint:* $R \geq f(x_1^*)/n$ for $0 \leq x_1^* \leq 1/n$). The argument is similar to show that **any function that is unbounded on the interval of integration is not integrable.**

Determine whether the improper integrals in Exercises 41–54 are convergent or divergent. You do not need to evaluate the integrals.

41. $\int_1^\infty \frac{x}{x^2 + 1}\,dx$

42. $\int_1^\infty \frac{x}{x^3 + 1}\,dx$

43. $\int_0^\infty \frac{1}{(4 + x^2)^{3/2}}\,dx$

44. $\int_0^\infty \frac{1}{\sqrt{x}(1 + x)}\,dx$

45. $\int_1^\infty \frac{1}{x^{1/2} + x^{3/2}}\,dx$

46. $\int_1^\infty \frac{1}{x + x^2}\,dx$

47. $\int_0^2 \frac{1}{\sqrt{4 - x^2}}\,dx$

48. $\int_0^1 \frac{1}{(1 - x^2)^{3/2}}\,dx$

49. $\int_0^1 \frac{1}{x(x + 1)}\,dx$

50. $\int_1^3 \frac{x}{\sqrt{x - 1}}\,dx$

51. $\int_0^\infty \frac{1}{x(x + 1)}\,dx$

52. $\int_0^1 \frac{1}{1 - \sqrt{x}}\,dx$

53. $\int_0^\infty \left(\frac{\pi}{2} - \tan^{-1} x\right) dx$

54. $\int_0^\infty \left(\frac{\pi}{2} - \tan^{-1} x\right)^2 dx$

55. It has been established at a radar telemetry tracking station that the distance X in inches between tape-surface flaws of high-quality magnetic tapes has probability density function

$$f(x) = \begin{cases} 0.01e^{-0.01x}, & x \geq 0, \\ 0, & x < 0. \end{cases}$$

The probability that an additional flaw will be found within the next n inches of a flaw is

$$P[0 \leq X \leq n] = \int_0^n f(x)dx.$$

(a) Find the probability that an additional flaw will be found within the next 100 inches of a flaw.

(b) Find the probability that at least 100 inches will occur before the next flaw occurs,

$$P[100 \leq X < \infty] = \int_{100}^\infty f(x)dx.$$

(c) Evaluate $E(X) = \int_0^\infty xf(x)dx$, the **expectation** of X. This is the average number of inches of tape between each pair of consecutive flaws encountered.

(d) The **variance** in distance between flaws is

$$\text{Var}(X) = \int_0^\infty x^2 f(x)dx - \left[\int_0^\infty xf(x)dx\right]^2.$$

Find the **standard deviation**, the square root of the variance.

56. Show that $\int_0^1 \frac{\sin x}{x^{3/2}}dx$ is convergent.

57. Consider the infinite funnel obtained by revolving the curve $y = 1/x, 1 \leq x < \infty$, about the x-axis.

(a) Show that the volume of the funnel is finite.

(b) Show that the surface area of the funnel is infinite. (This shows you can fill the funnel with a finite amount of paint, but it takes an infinite amount of paint to paint the funnel with a *uniform* coat of paint.)

58. The **gamma function** is

$$\Gamma(r) = \int_0^\infty x^{r-1}e^{-x}\,dx, \quad r > 0.$$

(a) Show that $\Gamma(1) = 1$.

(b) Show that $\Gamma(r+1) = r\,\Gamma(r)$.
(It follows from (a) and (b) that $\Gamma(k+1) = k!$ for every nonnegative integer k.)

REVIEW OF CHAPTER 8 CONCEPTS

- Substitution can be used to simplify integrals. It is a good idea to write the details for all but the simplest substitutions.
- Always look for a simple substitution before trying to integrate by parts.
- The Integration by Parts Formula is $\int u\,dv = uv - \int v\,du$. The functions x^n, $\ln x$, and the inverse trigonometric functions are good choices for u. Powers of x, exponentials, the sine and cosine functions are good choices for v'. If integration by parts seems to result in a more complicated integral, try a different choice of u and v.
- To integrate powers of either sines and cosines or tangents and secants, look for a simple substitution. If you don't see a simple substitution, use a reduction formula.
- Trigonometric substitutions can be used to transform certain integrals that contain square roots into trigonometric integrals without square roots. You should be able to carry out these types of substitutions. Integrals that can be evaluated by trigonometric substitution can usually be found in a table of integrals.

- The technique of partial fractions can be used to integrate any rational function. This technique is required for many practical problems.
- An integral is improper if either the integrand is not bounded or the interval of integration is infinite.
- Integrals that are improper because of the behavior at one endpoint of the interval of convergence can be evaluated by evaluating the integral over a subinterval and then taking a limit.
- An improper integral should be written as a sum of integrals that are improper at exactly one endpoint of the interval of integration. The original integral is then convergent if every improper integral in the sum is convergent; it is divergent if any integral in the sum is divergent.
- You should be able to determine convergence or divergence of improper integrals of functions that involve powers and roots by observation, without evaluating the integral over subintervals and taking a limit.

CHAPTER 8 REVIEW EXERCISES

Evaluate the integrals in Exercises 1–38.

1. $\int x \sin 2x\,dx$

2. $\int x \sin x^2\,dx$

3. $\int \ln x\,dx$

4. $\int \frac{\ln x}{x}\,dx$

5. $\int x^2 e^{-x}\,dx$

6. $\int \sin^4 x \cos x\,dx$

7. $\int \sin^6 x \cos^3 x\,dx$

8. $\int \sin^2 x\,dx$

9. $\int \cos^4 x\,dx$

10. $\int \sec^2 x \tan^8 x\,dx$

11. $\int \sec^4 x \tan^8 x\,dx$

12. $\int \sec^2 x \tan x\,dx$

13. $\int \tan^2 x\,dx$

14. $\int \frac{1}{(9-x^2)^{3/2}}\,dx$

15. $\int x\sqrt{1-x^2}\,dx$

16. $\int \sqrt{4+x^2}\,dx$

17. $\int \frac{1}{\sqrt{9+x^2}}\,dx$

18. $\int \sqrt{x^2-9}\,dx$

19. $\int \frac{1}{(x^2+1)^2}\,dx$

20. $\int \frac{1}{x^2-4}\,dx$

21. $\int \frac{x^3}{x^2-1}\,dx$

22. $\int \frac{x^2}{x^2+2}\,dx$

23. $\int \frac{1}{x^2-3x+2}\,dx$

24. $\int \frac{x}{x^2+6x+10}\,dx$

25. $\int \frac{x}{x^2-4x+8}\,dx$

26. $\int \frac{8x}{(x^2-4)(x+2)}\,dx$

27. $\displaystyle\int \frac{1}{x^3 + x^2}\,dx$

28. $\displaystyle\int \frac{x+1}{x^3 + x}\,dx$

29. $\displaystyle\int x\sqrt{x+1}\,dx$

30. $\displaystyle\int x^2\sqrt{1-x}\,dx$

31. $\displaystyle\int \frac{\cos x}{(1+\cos x)^2}\,dx$

32. $\displaystyle\int \frac{\sin x}{(1+\sin x)^2}\,dx$

33. $\displaystyle\int \frac{1}{1+\sqrt{x}}\,dx$

34. $\displaystyle\int \frac{1}{1+e^{2x}}\,dx$

35. $\displaystyle\int \frac{1}{e^x + e^{-x}}\,dx$

36. $\displaystyle\int \frac{\sqrt{x}}{\sqrt{x}+1}\,dx$

37. $\displaystyle\int \frac{1}{\sqrt{1+\sqrt{x}}}\,dx$

38. $\displaystyle\int \sqrt{1+\sqrt{x}}\,dx$

Evaluate the definite integrals in Exercises 39–44.

39. $\displaystyle\int_1^2 (x^2 - 1)^{3/2}\,dx$

40. $\displaystyle\int_2^3 x^2\sqrt{x^2 - 3}\,dx$

41. $\displaystyle\int_{-1}^1 \frac{1}{(x^2 + 1)^2}\,dx$

42. $\displaystyle\int_1^2 \frac{1}{x\sqrt{x^2 + 4}}\,dx$

43. $\displaystyle\int_{-1}^1 \sqrt{4 - x^2}\,dx$

44. $\displaystyle\int_0^1 (1 - x^2)^{3/2}\,dx$

Evaluate or indicate divergence for each of the integrals in Exercises 45–52.

45. $\displaystyle\int_0^2 \frac{1}{\sqrt{4 - x^2}}\,dx$

46. $\displaystyle\int_0^\infty x e^{-x^2}\,dx$

47. $\displaystyle\int_0^4 \frac{1}{x - 1}\,dx$

48. $\displaystyle\int_0^2 \frac{1}{(x - 1)^{4/3}}\,dx$

49. $\displaystyle\int_1^\infty x e^{-x}\,dx$

50. $\displaystyle\int_0^1 x \ln(1/x)\,dx$

51. $\displaystyle\int_0^\infty \frac{1}{(x^2 + 9)^{3/2}}\,dx$

52. $\displaystyle\int_1^\infty \frac{1}{x^2 + x}\,dx$

Determine whether the integrals in Exercises 53–58 are convergent or divergent. You do not need to evaluate the integrals.

53. $\displaystyle\int_1^\infty \frac{1}{\sqrt{x^2 + 1}}\,dx$

54. $\displaystyle\int_0^\infty \frac{x}{(1 + x^2)^{3/2}}\,dx$

55. $\displaystyle\int_0^\infty \frac{1}{x^{3/2} + x^{2/3}}\,dx$

56. $\displaystyle\int_1^\infty \frac{1}{x(x+1)}\,dx$

57. $\displaystyle\int_1^2 \frac{1}{(x-1)^{3/2}(2-x)^{2/3}}\,dx$

58. $\displaystyle\int_0^1 \frac{1}{\sqrt{x - x^2}}\,dx$

59. Find the volume of the solid generated when the region under the curve $y = \cos x$, $0 \le x \le \pi/2$, is revolved about the x-axis.

60. Find the volume of the solid generated when the region under the curve $y = \sin x$, $0 \le x \le \pi$, is revolved about the y-axis.

61. Find the volume of the solid generated when the region under the curve $y = \sec^2 x$, $0 \le x \le \pi/4$, is revolved about the x-axis.

62. Find the volume of the solid generated when the region under the curve $y = \sec^2 x$, $0 \le x \le \pi/4$, is revolved about the y-axis.

63. Find the length of the curve $y = x^2$, $0 \le x \le 1$.

64. Find the area of the surface generated when the curve $y = x^2$, $0 \le x \le 1$, is revolved about the x-axis.

65. The **magnetic field** at a point P due to an electric current in a straight wire is

$$B = \int_{-\infty}^\infty \frac{\mu_0 I}{4\pi}\frac{a}{(x^2 + a^2)^{3/2}}\,dx.$$

See figure. Evaluate this integral.

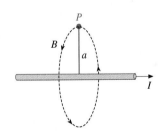

66. The **magnetic field** at the center of a **solenoid** of length ℓ and radius R containing N closely spaced turns and carrying a steady current I is

$$B = \int_{-\ell/2}^{\ell/2} \frac{\mu_0 R^2}{2(x^2 + R^2)^{3/2}} I\left(\frac{N}{\ell}\right)\,dx.$$

See figure. Evaluate this integral.

67. The probability that a certain random variable X with gamma distribution satisfies $X > \alpha$ is

$$P[X > \alpha] = \int_0^\alpha \lambda^2 x e^{-\lambda x}\, dx.$$

Evaluate this integral.

68. Gauss' probability integral is

$$I_0(\alpha) = \int_0^\infty e^{-\alpha x^2}\, dx = \frac{1}{2}\sqrt{\frac{\pi}{\alpha}}.$$

Evaluate

(a) $I_1(\alpha) = \displaystyle\int_0^\infty x e^{-\alpha x^2}\, dx,$

(b) $I_2(\alpha) = \displaystyle\int_0^\infty x^2 e^{-\alpha x^2}\, dx.$

(*Hint:* Use integration by parts to show that $I_2(\alpha) = \dfrac{1}{2\alpha} I_0(\alpha).$)

69. A random variable X has **gamma distribution**

$$f(x) = \begin{cases} \dfrac{\lambda^r}{\Gamma(r)} x^{r-1} e^{-\lambda x}, & x \geq 0, \\ 0, & x < 0, \end{cases}$$

where $\Gamma(r) = \displaystyle\int_0^\infty x^{r-1} e^{-x}\, dx.$ Show   that $E(X) = \displaystyle\int_0^\infty x f(x)\, dx = r/\lambda.$ (*Hint:* Use a change of variables and integration by parts to express $\displaystyle\int_0^\infty x^r e^{-\lambda x}\, dx$ in terms of $\Gamma(r)$.)

70. Two of your classmates obtained different answers when they evaluated $\int \tan x \sec^2 x\, dx$. Explain how this might have happened and why they may both have correct answers.

71. Explain   what   is   wrong   with   saying   "$\displaystyle\int_0^1 1/x\, dx = -\int_{-1}^0 1/x\, dx,$ so $\displaystyle\int_{-1}^1 1/x\, dx = 0$."

9 INFINITE SERIES

One of the main objectives of this chapter is to study the approximation of functions by Taylor polynomials. Recall that Taylor polynomials were briefly introduced in Section 3.9, where we studied the approximation of functions by their first- and second-order Taylor polynomials. In this chapter we will reintroduce Taylor polynomials and see that higher-order Taylor polynomials can be used to obtain approximate values that are as accurate as we desire for a large class of functions. We will illustrate the technique by approximating values of trigonometric functions, inverse trigonometric functions, logarithmic and exponential functions, and other transcendental functions.

Taylor polynomials of successive orders are obtained by adding a term to the previous sum. The study of Taylor polynomials of higher order then leads to the investigation of the behavior of successive sums of real numbers as the number of terms in the sums approaches infinity. We begin our study by carefully considering the behavior of such sums of real numbers, called infinite series.

9.1 SEQUENCES

Connections

Limits at infinity, 4.2.

The investigation of infinite series depends on the concept of limits of sequences.

Definitions and Notation

DEFINITION

A **sequence**, denoted either $\{a_n\}_{n \geq 1}$ or $a_1, a_2, a_3, \ldots$, is a function whose domain is the set of all positive integers n; the value a_n is called the **nth term** of the sequence.

We are primarily interested in sequences of numbers. Usually, the numbers a_n are given by a formula. For example, the formula

$$a_n = \frac{1}{n}, \qquad n \geq 1,$$

defines the sequence we express as either

$$\left\{\frac{1}{n}\right\}_{n \geq 1} \qquad \text{or} \qquad 1, \frac{1}{2}, \frac{1}{3}, \frac{1}{4}, \frac{1}{5}, \frac{1}{6}, \ldots.$$

The sequence $\{a\}_{n \geq 1}$ is the constant sequence $a, a, a, a, \ldots$. Do not confuse the constant *sequence* $\{a\}_{n \geq 1}$ with the *set* that consists of the single element a. The sequence $\{a\}_{n \geq 1}$ consists of an infinite number of terms a, one for each $n \geq 1$.

The important characteristic of a sequence is that there is a first term, second term, third term, and so on. It is not necessary that the first term of a sequence correspond to $n = 1$ in the formula that defines the terms of the sequence. The *formula* that defines the terms may have domain $n \geq n_0$ for any integer n_0. Thus,

$$\{a_n\}_{n \geq n_0} \text{ denotes the sequence } a_{n_0}, a_{n_0+1}, a_{n_0+2}, \ldots,$$

which has first term a_{n_0}, second term a_{n_0+1}, etc. For example,

$$\{2n\}_{n \geq 3} \text{ is the sequence } 6, 8, 10, 12, \ldots.$$

The formula

$$a_n = \frac{n}{n+1}, \qquad n \geq 0,$$

gives the sequence

$$0, \frac{1}{2}, \frac{2}{3}, \frac{3}{4}, \ldots.$$

Note that it is theoretically unsound to try to determine a formula for a_n from a list of only finitely many terms. Either no formula or many formulas that give the terms listed may be evident. In spite of this, it is customary to indicate some simple sequences by a partial list of terms. For example,

$$1, 2, 3, 4, \ldots \quad \text{indicates the sequence } \{n\}_{n \geq 1},$$

$$\frac{1}{2}, \frac{2}{3}, \frac{3}{4}, \frac{4}{5}, \ldots \quad \text{indicates the sequence } \left\{\frac{n}{n+1}\right\}_{n \geq 1}, \quad \text{and}$$

$$1^2, 2^2, 3^2, 4^2, \ldots \quad \text{indicates the sequence } \{n^2\}_{n \geq 1}.$$

A sequence may be defined inductively, or recursively. For example, in Example 3 of Section 3.7 we used Newton's Method to obtain the formula

$$a_{n+1} = \frac{a_n}{2} + \frac{1}{a_n}$$

for successive approximations of the positive solution of the equation $x^2 - 2 = 0$. If we choose our first approximation to be $a_1 = 2$, the equations

$$a_1 = 2, \quad a_{n+1} = \frac{a_n}{2} + \frac{1}{a_n}, \quad n \geq 1,$$

define a sequence $\{a_n\}_{n \geq 1}$. The equation $a_1 = 2$ implies

$$a_2 = \frac{a_1}{2} + \frac{1}{a_1} = \frac{(2)}{2} + \frac{1}{(2)} = 1.5.$$

Then $a_2 = 1.5$ implies

$$a_3 = \frac{a_2}{2} + \frac{1}{a_2} = \frac{(1.5)}{2} + \frac{1}{(1.5)} \approx 1.4166667.$$

Then a_3 determines $a_4 \approx 1.4142157$ and a_4 determines $a_5 \approx 1.4142136$. Generally, each term determines the succeeding term. (As n increases, a_n approaches $\sqrt{2}$, the positive solution of $x^2 - 2 = 0$.)

Limits

The idea of a limit of a sequence is important for many problems. Let us state the formal definition.

DEFINITION

We say that the real number L is the **limit** of the sequence $\{a_n\}_{n \geq 1}$, written

$$\lim_{n \to \infty} a_n = L,$$

if, for each $\varepsilon > 0$, there is a number N such that $|a_n - L| < \varepsilon$ whenever $n \geq N$. If $\lim_{n \to \infty} a_n = L$, we say the sequence $\{a_n\}_{n \geq 1}$ is **convergent**. If the limit does not exist, we say the sequence is **divergent**. A sequence that approaches infinity is considered divergent.

The condition $|a_n - L| < \varepsilon$ whenever $n \geq N$ means that the points (n, a_n) are between the horizontal lines $y = L + \varepsilon$ and $y = L - \varepsilon$ for all $n \geq N$. (The points may or may not be within the lines for $n < N$.) The statement

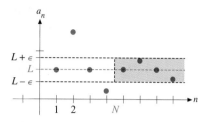

FIGURE 9.1.1 For each $\epsilon > 0$, there is a number N such that $|a_n - L| < \epsilon$ whenever $n \geq N$. The points (n, a_n) approach the horizontal line $y = L$ as n approaches infinity.

FIGURE 9.1.2 If $\lim\limits_{n \to \infty} a_n = L$, then the points a_n on a number line approach the point L as n approaches infinity.

$\lim\limits_{n \to \infty} a_n = L$ implies the points (n, a_n) approach the horizontal line $y = L$ as n approaches infinity. See Figure 9.1.1.

It is sometimes useful to plot the terms a_n of a sequence along a number line. The statement $\lim\limits_{n \to \infty} a_n = L$ then corresponds to the points a_n approaching the point L on the line. See Figure 9.1.2.

As is the case with other limits we have studied, if a sequence has a limit, the limit is unique. A sequence cannot have more than one limit.

The usual rules for limits of functions apply to limits of sequences. In particular, we have the following.

LIMIT THEOREM

If $\lim\limits_{n \to \infty} a_n$ and $\lim\limits_{n \to \infty} b_n$ both exist and c and d are constants, then $\lim\limits_{n \to \infty} (ca_n + db_n)$ exists and

$$\lim_{n \to \infty} (ca_n + db_n) = c(\lim_{n \to \infty} a_n) + d(\lim_{n \to \infty} b_n).$$

The definition of the limit of a sequence corresponds to that of the limit as x approaches infinity of a function f of a real variable x. The following is an immediate consequence of the definitions.

THEOREM RELATING LIMITS OF SEQUENCES AND LIMITS OF FUNCTIONS AT INFINITY

If $f(x)$ is defined for all real numbers $x \geq 1$, $a_n = f(n)$ for all positive integers n, and $\lim\limits_{x \to \infty} f(x)$ exists, then $\lim\limits_{n \to \infty} a_n$ exists and

$$\lim_{n \to \infty} a_n = \lim_{x \to \infty} f(x).$$

This means we can evaluate limits of sequences by treating the formula for a_n as a formula for the values of a function of a real variable.

We can then use the results for evaluating limits of functions to evaluate limits of sequences.

The following basic examples are useful in determining the limit of sequences that are more complicated, but of similar type.

If $\alpha > 0$, then $\lim\limits_{n \to \infty} \dfrac{1}{n^\alpha} = 0$, and $\lim\limits_{n \to \infty} n^\alpha$ is divergent.

$\lim\limits_{n \to \infty} r^n = 0$ *if $|r| < 1$;* $\lim\limits_{n \to \infty} r^n$ *is divergent if $|r| > 1$.*

(See Exercise 41.)

Examples

The ideas and techniques illustrated in the following examples should seem familiar.

FIGURE 9.1.3 $\lim\limits_{n\to\infty} \dfrac{1}{\sqrt{n}} = 0.$

FIGURE 9.1.4 $\lim\limits_{n\to\infty} 2 = 2.$

FIGURE 9.1.5 $\lim\limits_{n\to\infty}\left(2 + \dfrac{1}{n}\right) = 2.$

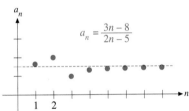

FIGURE 9.1.6 $\lim\limits_{n\to\infty} \dfrac{3n-8}{2n-5} = \dfrac{3}{2}.$

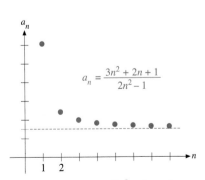

FIGURE 9.1.7 $\lim\limits_{n\to\infty} \dfrac{3n^2 + 2n + 1}{2n^2 - 1} = \dfrac{3}{2}.$

EXAMPLE 1

$$\lim_{n\to\infty} \frac{1}{\sqrt{n}} = 0.$$

(See Figure 9.1.3. Note that it appears from the graph that the terms $1/\sqrt{n}$ do not approach zero very quickly.)   ∎

EXAMPLE 2

$$\lim_{n\to\infty} 2 = 2.$$

See Figure 9.1.4.   ∎

EXAMPLE 3

$$\lim_{n\to\infty}\left(2 + \frac{1}{n}\right) = 2.$$

See Figure 9.1.5.   ∎

We can determine the limit as n approaches infinity of an algebraic expression of n by factoring the highest power of n from each of the numerator and denominator, as we did in Section 4.2.

EXAMPLE 4

$$\lim_{n\to\infty} \frac{3n-8}{2n-5} = \lim_{n\to\infty} \frac{n\left(3 - \dfrac{8}{n}\right)}{n\left(2 - \dfrac{5}{n}\right)} = \lim_{n\to\infty} \frac{3 - \dfrac{8}{n}}{2 - \dfrac{5}{n}} = \frac{3-0}{2-0} = \frac{3}{2}.$$

See Figure 9.1.6.   ∎

EXAMPLE 5

$$\lim_{n\to\infty} \frac{3n^2 + 2n + 1}{2n^2 - 1} = \lim_{n\to\infty} \frac{n^2\left(3 + \dfrac{2}{n} + \dfrac{1}{n^2}\right)}{n^2\left(2 - \dfrac{1}{n^2}\right)} = \lim_{n\to\infty} \frac{3 + \dfrac{2}{n} + \dfrac{1}{n^2}}{2 - \dfrac{1}{n^2}} = \frac{3}{2}.$$

See Figure 9.1.7.   ∎

EXAMPLE 6

$$\lim_{n\to\infty} \frac{\sqrt{4n^2 + 9}}{n} = \lim_{n\to\infty} \frac{\sqrt{n^2\left(4 + \dfrac{9}{n^2}\right)}}{n}$$

$$= \lim_{n\to\infty} \frac{n\sqrt{4 + \dfrac{9}{n^2}}}{n}$$

$$= \lim_{n\to\infty} \sqrt{4 + \frac{9}{n^2}}$$

$$= \sqrt{4} = 2.$$

See Figure 9.1.8.   ∎

FIGURE 9.1.8 $\displaystyle\lim_{n\to\infty}\frac{\sqrt{4n^2+9}}{n}=2$.

FIGURE 9.1.9 $\displaystyle\lim_{n\to\infty}\sin\frac{n\pi}{2}$ is divergent.

FIGURE 9.1.10 $\displaystyle\lim_{n\to\infty}\frac{\cos n\pi}{\sqrt{n}}=0$.

FIGURE 9.1.11 $\displaystyle\lim_{n\to\infty}\left(-\frac{2}{3}\right)^n=0$.

FIGURE 9.1.12 $\displaystyle\lim_{n\to\infty}ne^{-n}=0$.

EXAMPLE 7

$\displaystyle\lim_{n\to\infty}\sin\frac{n\pi}{2}$ is divergent.

The sequence $\{\sin(n\pi/2)\}_{n\ge 1}$ corresponds to $1, 0, -1, 0, 1, 0, -1, 0, \ldots$. Thus, we see that the terms repeat the three values $-1, 0, 1$ infinitely often; the terms do not approach a *single* number L as n approaches infinity. See Figure 9.1.9. ∎

EXAMPLE 8

$$\lim_{n\to\infty}\frac{\cos n\pi}{\sqrt{n}}=0.$$

Note that $(\cos n\pi)/\sqrt{n}$ is of the form $a_n b_n$, where $\displaystyle\lim_{n\to\infty} a_n = \lim_{n\to\infty} 1/\sqrt{n} = 0$ and $|b_n| = |\cos n\pi| \le 1$ for all n, so $\{b_n\}_{n\ge 1}$ is a bounded sequence. It follows that the product $a_n b_n$ approaches zero as n approaches infinity, as in Statement 3 of Section 2.5. See Figure 9.1.10. ∎

EXAMPLE 9

$$\lim_{n\to\infty}\left(-\frac{2}{3}\right)^n=0.$$

This is a consequence of the basic fact that $\displaystyle\lim_{n\to\infty} r^n = 0$ for $|r| < 1$. We have $r = -2/3$ in this example. See Figure 9.1.11. ∎

EXAMPLE 10

$$\lim_{n\to\infty}ne^{-n}=\lim_{n\to\infty}\frac{n}{e^n}\underset{(\infty/\infty,\text{l'H})}{=}\lim_{n\to\infty}\frac{1}{e^n}=0.$$

We have used the theorem relating limits of sequences and limits of functions of a real variable to treat n as a real variable and then applied l'Hôpital's Rule as in Section 7.5. See Figure 9.1.12. ∎

EXAMPLE 11

Evaluate $\displaystyle\lim_{n\to\infty}\frac{2^n}{n!}$.

SOLUTION

$$\frac{2^n}{n!}=\frac{2\cdot 2\cdot 2\cdot 2\cdots 2}{1\cdot 2\cdot 3\cdot 4\cdots n}$$

$$=\left(\frac{2}{1}\right)\left(\frac{2}{2}\right)\left(\frac{2}{3}\right)\left(\frac{2}{4}\right)\cdots\left(\frac{2}{n}\right)$$

$$\le\left(\frac{2}{1}\right)\left(\frac{2}{2}\right)\left(\frac{2}{3}\right)\left(\frac{1}{2}\right)^{n-3},$$

for $n \ge 3$. Since $2^n/n! > 0$ and

$$\lim_{n\to\infty}\frac{4}{3}\left(\frac{1}{2}\right)^{n-3}=0,$$

FIGURE 9.1.13 $\displaystyle\lim_{n\to\infty}\frac{2^n}{n!}=0.$

we can use either Statement 1 of Section 2.5 or the Squeeze Theorem at the end of that section to conclude that

$$\lim_{n\to\infty}\frac{2^n}{n!}=0.$$

See Figure 9.1.13. ■

It is useful to note that

$$(n+1)!=(1)(2)\cdots(n)(n+1)=n!\,(n+1).$$

Some General Results

The following fact about limits of sequences should be clear.

Changing the values of a finite number of terms of a sequence does not affect the convergence or divergence of a sequence, nor the limit of the sequence, if it is convergent. (1)

Statement 1 reflects the fact that the limit of a sequence depends only on a_n for n large, $n \geq N$.

The following result is of fundamental importance.

MONOTONE CONVERGENCE THEOREM

If there is a real number B such that either $a_n \leq a_{n+1} \leq B$ for all n or $a_n \geq a_{n+1} \geq B$ for all n, then $\displaystyle\lim_{n\to\infty} a_n$ exists.

We will not prove the Monotone Convergence Theorem, but let us see how the Bisection Method can be used to locate a candidate L for the limit of a sequence that satisfies the condition $a_n \leq a_{n+1} \leq B$ for all n. We first note that this condition implies

$$a_1 \leq a_2 \leq a_3 \leq a_4 \leq \cdots \leq B.$$

This means that all terms a_n, $n \geq 1$, are contained in the interval between a_1 and B. It follows that L must also be in this interval. We then consider the point $B_1 = (a_1 + B)/2$. If there is no $a_n > B_1$, then L must be between a_1 and B_1. If there is some point $a_n > B_1$, then

$$B_1 \leq a_n \leq a_{n+2} \leq a_{n+3} \leq a_{n+4} \leq \cdots \leq B,$$

so L must be between B_1 and B. We continue to bisect as above indefinitely. This gives a sequence of intervals that are nested, each one inside the previous one, and the length of the intervals approaches zero. The intervals shrink down to a single point L and it can be shown that L is the limit of the sequence. See Figure 9.1.14.

A sequence that satisfies either $a_n \leq a_{n+1}$ for all n or $a_n \geq a_{n+1}$ for all n is called a **monotone** sequence.

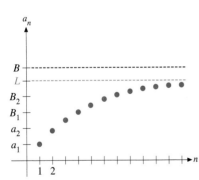

FIGURE 9.1.14 If $a_n \leq a_{n+1} \leq B$ for all n, then $\displaystyle\lim_{n\to\infty} a_n$ exists. The limit is greater than or equal to any term a_n and is less than or equal to any upper bound B.

The Monotone Convergence Theorem says that a bounded, monotone sequence is convergent.

In many important cases we can use a derivative to show that a given sequence is monotone. The proof is based on the facts that f is increasing on intervals where $f'(x)$ is positive and f is decreasing on intervals where $f'(x)$ is negative.

THE DERIVATIVE TEST FOR MONOTONICITY OF A SEQUENCE

Suppose $f(x)$ is defined and differentiable for $x \geq n_0$ and that $a_n = f(n)$ for all positive integers $n \geq n_0$.
If $f'(x) \geq 0$ for $x \geq n_0$, then $a_{n+1} \geq a_n$ for all $n \geq n_0$.
If $f'(x) \leq 0$ for $x \geq n_0$, then $a_{n+1} \leq a_n$ for all $n \geq n_0$.

For example, if

$$f(x) = \frac{10x}{x^2 + 5},$$

then

$$f'(x) = \frac{(x^2 + 5)(10) - (10x)(2x)}{(x^2 + 5)^2} = \frac{10(5 - x^2)}{(x^2 + 5)^2}.$$

Since $f'(x) \leq 0$ for $x \geq \sqrt{5}$, we can conclude from the theorem that $a_n = \frac{10n}{n^2 + 5}$ satisfies $a_{n+1} \leq a_n$ for all $n \geq 3$. From Figure 9.1.15 it appears that $\left\{\dfrac{10n}{n^2 + 5}\right\}_{n \geq 2}$ is monotone, but that the sequence $\left\{\dfrac{10n}{n^2 + 5}\right\}_{n \geq 1}$ is not monotone. These facts can be verified by showing that $f(2) > f(3)$ and that $f(1) < f(2)$, respectively.

The following result is sometimes useful for dealing with sequences that are defined recursively.

FIGURE 9.1.15 The function $f(x) = \dfrac{10x}{x^2 + 5}$ satisfies $f'(x) \leq 0$ for $x \geq \sqrt{5}$, so the sequence $\left\{\dfrac{10n}{n^2 + 5}\right\}_{n \geq 3}$ is monotone.

THEOREM

Suppose $\lim\limits_{n \to \infty} a_n = L$, where a_n is defined recursively by $a_{n+1} = f(a_n)$, $n \geq 1$. If the function f is continuous at L, then L must satisfy the equation $L = f(L)$.

This result follows from the continuity of f at L and the fact that $\lim\limits_{n \to \infty} a_n = L$ implies $\lim\limits_{n \to \infty} a_{n+1} = L$.

EXAMPLE 12

Show that $\lim\limits_{n \to \infty} a_n = \sqrt{2}$, where $a_1 = 2$ and $a_{n+1} = \dfrac{a_n}{2} + \dfrac{1}{a_n}$, $n \geq 1$.

SOLUTION Let us first show that the sequence is monotone and bounded, so we know that the limit of the sequence exists.

We can solve the inequality

$$\frac{a_n}{2} + \frac{1}{a_n} > \sqrt{2}$$

to determine that $a_n > \sqrt{2}$ implies $a_{n+1} > \sqrt{2}$. Mathematical Induction can then be used to show $a_n > \sqrt{2}$ for all $n \geq 1$.

It is not difficult to show that

$$\frac{a_n}{2} + \frac{1}{a_n} < a_n \quad \text{for } a_n > \sqrt{2}.$$

Since we know that $a_n > \sqrt{2}$ for all $n \geq 1$, we conclude that $a_{n+1} < a_n$ for $n \geq 1$, so the sequence is monotone. The sequence is also bounded, since $\sqrt{2} < a_n \leq a_1 = 2$ for all $n \geq 1$.

If $\lim_{n \to \infty} a_n = L$, then we can take limits as n approaches infinity in the equation

$$a_{n+1} = \frac{a_n}{2} + \frac{1}{a_n}$$

to obtain

$$L = \frac{L}{2} + \frac{1}{L},$$
$$\frac{L}{2} = \frac{1}{L},$$
$$L^2 = 2,$$
$$L = \sqrt{2},$$

where we have used the fact that $L > 0$. ∎

EXERCISES 9.1

Evaluate the limits in Exercises 1–40.

1. $\displaystyle\lim_{n \to \infty} \frac{3n - 1}{2n + 1}$

2. $\displaystyle\lim_{n \to \infty} \frac{\sqrt{n}}{2n - 3}$

3. $\displaystyle\lim_{n \to \infty} \frac{n}{n^2 + 1}$

4. $\displaystyle\lim_{n \to \infty} \frac{1 - n^2}{1 + n^2}$

5. $\displaystyle\lim_{n \to \infty} \frac{(2n - 1)^2}{3n^2 - n + 7}$

6. $\displaystyle\lim_{n \to \infty} \frac{2n^2 + 3n + 1}{(2n + 3)^2}$

7. $\displaystyle\lim_{n \to \infty} \frac{n}{\sqrt{4n^2 + 9}}$

8. $\displaystyle\lim_{n \to \infty} \frac{n^2}{\sqrt{9n^2 + 4}}$

9. $\displaystyle\lim_{n \to \infty} (-1)^n$

10. $\displaystyle\lim_{n \to \infty} \frac{1 + (-1)^n}{2}$

11. $\displaystyle\lim_{n \to \infty} \sin(\pi/n)$

12. $\displaystyle\lim_{n \to \infty} \cos(\pi/n)$

13. $\displaystyle\lim_{n \to \infty} n \cos n\pi$

14. $\displaystyle\lim_{n \to \infty} e^{-n} \sin(n\pi/2)$

15. $\displaystyle\lim_{n \to \infty} (-1)^n e^{1/n}$

16. $\displaystyle\lim_{n \to \infty} \frac{\cos n\pi}{n}$

17. $\displaystyle\lim_{n \to \infty} 2^{-n} \cos(n\pi)$

18. $\displaystyle\lim_{n \to \infty} \frac{(-1)^n}{\ln n}$

19. $\displaystyle\lim_{n \to \infty} \frac{e^{1/n}}{n}$

20. $\displaystyle\lim_{n \to \infty} \frac{n(-1)^n}{n + 1}$

21. $\displaystyle\lim_{n \to \infty} n 2^{-n}$

22. $\displaystyle\lim_{n \to \infty} n \sin(1/n)$ **(Calculus Explorer & Tutor II, Sequences and Series, 1.)**

23. $\displaystyle\lim_{n \to \infty} \frac{2^n}{3^n + 1}$

24. $\displaystyle\lim_{n \to \infty} \frac{3^n}{2^n + 1}$

25. $\displaystyle\lim_{n \to \infty} \tan(1/n)$

26. $\displaystyle\lim_{n \to \infty} \cot(1/n)$

27. $\displaystyle\lim_{n \to \infty} \frac{e^n - e^{-n}}{e^n + e^{-n}}$

28. $\displaystyle\lim_{n \to \infty} \frac{2}{e^n + e^{-n}}$

29. $\displaystyle\lim_{n \to \infty} \frac{\ln n}{n}$

30. $\displaystyle\lim_{n \to \infty} n^2 e^{-n}$

31. $\displaystyle\lim_{n \to \infty} \frac{e^n}{n!}$

32. $\displaystyle\lim_{n \to \infty} \frac{n!}{(n + 1)!}$

33. $\displaystyle\lim_{n \to \infty} \frac{n^2 n!}{(n + 2)!}$

34. $\displaystyle\lim_{n \to \infty} \frac{(n!)^2}{(2n)!}$

35. $\displaystyle\lim_{n \to \infty} \frac{n!}{1 \cdot 3 \cdot 5 \cdots (2n - 1)}$

36. $\displaystyle\lim_{n \to \infty} \frac{2 \cdot 4 \cdot 6 \cdots (2n)}{4 \cdot 7 \cdot 10 \cdots (3n + 1)}$

37. $\displaystyle\lim_{n \to \infty} (2 + 3n)^{1/n}$

38. $\displaystyle\lim_{n \to \infty} \left(\frac{e^n + e^{-n}}{2} \right)^{1/n}$

39. $\displaystyle\lim_{n\to\infty}\left(1-\frac{2}{n}\right)^n$

40. $\displaystyle\lim_{n\to\infty}\left(\frac{n}{n+1}\right)^n$

41. Use properties of the exponential function and the fact that $|r|^n = e^{n\ln|r|}$ to show $\displaystyle\lim_{n\to\infty} r^n = 0$ if $|r| < 1$, and $\displaystyle\lim_{n\to\infty} r^n$ is divergent if $|r| > 1$.

42. Show that $\displaystyle\lim_{r\to\infty} r^n = 1$ if $r = 1$, and that $\displaystyle\lim_{r\to\infty} r^n$ does not exist if $r = -1$.

Sketch the graph of the indicated function in each of Exercises 43–44. What is the domain of each function? At what points is each function continuous?

43. $f(x) = \displaystyle\lim_{n\to\infty}(1 - |x|)^n$

44. $f(x) = \displaystyle\lim_{n\to\infty}\frac{1}{|x|^n + 1}$

45. (a) Express the perimeter of a regular polygon with n sides that is inscribed in a circle of radius r in terms of n. See figure.
   (b) Find the limit as n approaches infinity of the perimeter of the polygons in (a).

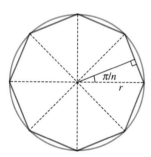

46. Find an example of a function f that is continuous on $x \geq 1$, the sequence $\{f(n)\}_{n\geq 1}$ is convergent, but $\displaystyle\lim_{x\to\infty} f(x)$ does not exist.

47. Find an example of sequences such that $\{a_n\}_{n\geq 1}$ is convergent, $\{b_n\}_{n\geq 1}$ is bounded, and $\{a_n b_n\}_{n\geq 1}$ is divergent.

48. Find an example of sequences such that $\{a_n\}_{n\geq 1}$ is divergent, $\{b_n\}_{n\geq 1}$ is divergent, and $\{a_n + b_n\}_{n\geq 1}$ is convergent.

49. Use the definition of limit to prove that, if $\{a_n\}_{n\geq 1}$ and $\{b_n\}_{n\geq 1}$ are convergent, then $\{a_n + b_n\}_{n\geq 1}$ is convergent.

50. Suppose $\displaystyle\lim_{n\to\infty} a_n = A$ and $\displaystyle\lim_{n\to\infty} b_n = B$.

   (a) If $A = B$, show that the sequence $a_1, b_1, a_2, b_2, a_3, b_3, a_4, b_4, \ldots$ has limit $A = B$.
   (b) If $A \neq B$, show that the sequence $a_1, b_1, a_2, b_2, a_3, b_3, a_4, b_4, \ldots$ is divergent.

51. Show that $\displaystyle\lim_{n\to\infty} a_n$ exists and find its value, where $a_1 = 1$ and $a_{n+1} = \sqrt{2 + a_n}, n \geq 1$. (*Hint:* Use Mathematical Induction to show that $a_n < 2$ for all $n \geq 1$ and $a_{n+1} > a_n$ for all $n \geq 1$.)

52. Show that $\displaystyle\lim_{n\to\infty} a_n$ exists and find its value, where $a_1 = 3$ and $a_{n+1} = \sqrt{1 + 2a_n}, n \geq 1$. (*Hint:* Use Mathematical Induction to show that $a_n > 1 + \sqrt{2}$ for all $n \geq 1$ and $a_{n+1} < a_n$ for all $n \geq 1$.)

53. If ϵ is positive and less than one, for what values of n is
$$\left|\frac{n}{n+1} - 1\right| < \epsilon?$$

54. If ϵ is positive and less than one, for what values of n is
$$\left|e^{-1/n} - 1\right| < \epsilon?$$

55. For what values of n is $ne^{-n} < 0.005$?

56. For what values of n is $\left|n\sin\dfrac{1}{n} - 1\right| < 0.005$?

9.2 INFINITE SERIES

Connections

Sigma notation, 5.3.

Let us consider the following question:

If a ball always bounces back to half its previous height, will it bounce forever?

We know from Formula 3 of Section 5.1 that the height of a ball t seconds after it is dropped from an initial height of s_0 feet is given by

$$s(t) = s_0 - 16t^2.$$

Hence, it will first hit ground level after t_0 seconds, where

$$s(t_0) = 0,$$
$$s_0 - 16t_0^2 = 0,$$
$$t_0 = \frac{\sqrt{s_0}}{4}.$$

After the first bounce, the ball will bounce to a height $s_1 = s_0/2$ and return to ground level after an additional time interval of length

$$t_1 = 2\frac{\sqrt{s_1}}{4} = \frac{\sqrt{s_0/2}}{2}.$$

It will then bounce to a height $s_2 = s_1/2 = s_0/2^2$ and return to ground level after an additional time interval of length

$$t_2 = 2\frac{\sqrt{s_2}}{4} = \frac{\sqrt{s_0/2^2}}{2}.$$

Generally, after the nth bounce, the ball will bounce to a height of $s_n = s_0/2^n$ and return to ground level in a time interval of length

$$t_n = \frac{\sqrt{s_0/2^n}}{2}.$$

Thus, we see that the ball bounces an infinite number of times and the total time that the ball bounces should be the "sum"

$$T = t_0 + t_1 + t_2 + t_3 + \cdots$$
$$= \frac{\sqrt{s_0}}{4} + \frac{\sqrt{s_0/2}}{2} + \frac{\sqrt{s_0/2^2}}{2} + \frac{\sqrt{s_0/2^3}}{2} + \cdots.$$

In this section we will investigate the meaning of such "sums" of an infinite number of terms and then return to answer the question about the bouncing ball.

Definitions and Notation

> **DEFINITION**
>
> The symbol
>
> $$\sum_{n=1}^{\infty} a_n \text{ or } a_1 + a_2 + a_3 + \cdots$$
>
> denotes an **infinite series.** Infinite series are sometimes called infinite sums or, simply, series. The **nth partial sum** of a series is the number
>
> $$S_n = \sum_{j=1}^{n} a_j = a_1 + a_2 + \cdots + a_n,$$
>
> the sum of the first n terms of the series. If the sequence of partial sums $\{S_n\}_{n\geq 1}$ converges to a number S,
>
> $$\lim_{n\to\infty} S_n = S,$$
>
> we say the series is **convergent** with **sum** S and write
>
> $$\sum_{n=1}^{\infty} a_n = S.$$
>
> If the sequence of partial sums is divergent, we say the series is **divergent** and it is not assigned a value.

We will also denote infinite series by the symbol

$$\sum_{n=n_0}^{\infty} a_n \quad \text{or} \quad a_{n_0} + a_{n_0+1} + a_{n_0+2} + \cdots,$$

where n_0 is any integer. The nth partial sum of the series is still the sum of the first n terms, even if these terms do not correspond to the integers $1, 2, \ldots, n$.

Do not confuse the series $\sum_{n=1}^{\infty} a_n$ with the sequence $\{a_n\}_{n\geq 1}$. The terms *series* and *sequence* are similar in everyday usage, but they are not the same in mathematics. Think of a sequence as a list of terms and think of a series as a sum of terms. A sequence is convergent if the terms a_n approach a finite limit. However, convergence of a series requires that the sequence of partial sums $S_n = a_1 + a_2 + \cdots + a_n$ approach a finite limit. For example, the *sequence* $\left\{2 - \dfrac{1}{n}\right\}_{n\geq 1}$ is convergent with

$$\lim_{n\to\infty}\left(2 - \frac{1}{n}\right) = 2,$$

but the *series* $\sum_{n=1}^{\infty}\left(2 - \dfrac{1}{n}\right)$ is divergent, since

$$S_n = \left(2 - \frac{1}{1}\right) + \left(2 - \frac{1}{2}\right) + \left(2 - \frac{1}{3}\right) + \cdots + \left(2 - \frac{1}{n}\right)$$
$$\geq \underbrace{1 + 1 + 1 + \cdots + 1}_{n \text{ terms}} = n \to \infty \text{ as } n \to \infty.$$

The fact that a series corresponds to a sum is suggested by the notation $a_1 + a_2 + a_3 + \cdots$. The Greek letter Σ (sigma) is used as an abbreviation of sum in the notation $\sum_{n=1}^{\infty} a_n$.

The statement $a_1 + a_2 + a_3 + \cdots = S$ does not mean that S is the algebraic sum of an infinite number of terms, although the notation seems to suggest this. We cannot perform the operation of adding an infinite number of terms. We can add any finite number of terms and, hence, determine S_n for any given value of n. The statement $a_1 + a_2 + a_3 + \cdots = S$ means that S is the limit of the partial sums S_n.

It is usually difficult to evaluate the sum of a convergent series, because we usually do not have a closed formula for the values of the partial sums. We will use the partial sums S_n as approximate values of the sum of a *convergent* series.

EXAMPLE 1

Find the first four partial sums of the series

$$\sum_{n=1}^{\infty} \frac{(-1)^{n+1}}{(2n-1)!}(0.5)^{2n-1}.$$

Compare with a calculator value of $\sin 0.5$. (In later sections we will see that the partial sums of the series are Taylor polynomials of $\sin x$ about $c = 0$, evaluated at $x = 0.5$, and that the series is convergent with sum $\sin 0.5$. Since the Taylor polynomials involve derivatives of trigonometric functions, we must use radian measure.)

SOLUTION   We see that

$$a_n = \frac{(-1)^{n+1}}{(2n-1)!}(0.5)^{2n-1}, \quad n \geq 1.$$

Recall that $n! = (1)(2)(3)\cdots(n)$. Then

$$S_1 = a_1 = \frac{1}{1!}(0.5) = 0.5,$$

$$S_2 = a_1 + a_2 = \frac{1}{1!}(0.5) - \frac{1}{3!}(0.5)^3 = 0.47916667,$$

$$S_3 = a_1 + a_2 + a_3 = \frac{1}{1!}(0.5) - \frac{1}{3!}(0.5)^3 + \frac{1}{5!}(0.5)^5 = 0.47942708,$$

$$S_4 = a_1 + a_2 + a_3 + a_4$$
$$= \frac{1}{1!}(0.5) - \frac{1}{3!}(0.5)^3 + \frac{1}{5!}(0.5)^5 - \frac{1}{7!}(0.5)^7 = 0.47942553.$$

A calculator value of the sine of 0.5 *radian* is $\sin(0.5) \approx 0.47942554$, which is very close to the value S_4. ■

The following formula relates the nth term and the nth partial sum of a series.

$$S_n = S_{n-1} + a_n, \qquad n > 1. \tag{1}$$

That is,

$$S_{n-1} + a_n = (a_1 + a_2 + \cdots + a_{n-1}) + a_n = S_n.$$

Mathematical Induction can be used to prove the statement formally.

Convergent series have some of the properties of finite sums. The following result is a consequence of the corresponding property of finite sums and the Limit Theorem of Section 9.1.

THEOREM

If $\displaystyle\sum_{n=1}^{\infty} a_n$ and $\displaystyle\sum_{n=1}^{\infty} b_n$ are convergent and c and d are constants, then $\displaystyle\sum_{n=1}^{\infty}(ca_n + db_n)$ is convergent and

$$\sum_{n=1}^{\infty}(ca_n + db_n) = c\sum_{n=1}^{\infty} a_n + d\sum_{n=1}^{\infty} b_n.$$

This theorem allows us to add corresponding terms of convergent sequences and factor common multiples from terms of a convergent sequence. For example, if the series are convergent for a particular value of x, we have

$$(1 + x + x^2 + x^3 + x^4 + x^5 + \cdots) - (1 - x + x^2 - x^3 + x^4 - x^5 + \cdots)$$
$$= (1 - 1) + (1 + 1)x + (1 - 1)x^2 + (1 + 1)x^3$$
$$+ (1 - 1)x^4 + (1 + 1)x^5 + \cdots$$
$$= 2x + 2x^3 + 2x^5 + \cdots$$
$$= 2x(1 + x^2 + x^4 + \cdots).$$

Geometric Series

We now consider a very important type of series for which we can obtain a closed formula for partial sums.

The series

$$\sum_{n=0}^{\infty} r^n = 1 + r + r^2 + r^3 + r^4 + \cdots$$

is called a **geometric series**. The nth partial sum is

$$S_n = 1 + r + r^2 + \cdots + r^{n-1}.$$

Multiplying the above equation by r, we obtain

$$r S_n = r + r^2 + \cdots + r^{n-1} + r^n.$$

Then

$$S_n - r S_n = 1 + r + r^2 + \cdots + r^{n-1}$$
$$-r - r^2 - \cdots - r^{n-1} - r^n$$
$$= 1 - r^n,$$

so

$$(1 - r)S_n = 1 - r^n.$$

If $r \neq 1$, we then divide by $1 - r$ to obtain the formula

$$S_n = 1 + r + r^2 + \cdots + r^{n-1} = \frac{1 - r^n}{1 - r}, \qquad r \neq 1. \tag{2}$$

If $|r| < 1$, we know that $\lim_{n \to \infty} r^n = 0$, so

$$\lim_{n \to \infty} S_n = \lim_{n \to \infty} \frac{1 - r^n}{1 - r} = \frac{1}{1 - r}.$$

This means that the geometric series is convergent with sum $1/(1 - r)$ if $|r| < 1$. If $|r| \geq 1$, the partial sums of the geometric series do not converge. Note that $S_n = n$ if $r = 1$. Summarizing, we have

$$\sum_{n=0}^{\infty} r^n = 1 + r + r^2 + r^3 + \cdots = \frac{1}{1 - r}, \qquad |r| < 1. \tag{3}$$

The series is divergent if $|r| \geq 1$.

Formula 3 is very important. You should know this formula and how to use it.

Series of the form $\sum_{n=n_0}^{\infty} ar^n$ are also geometric series. We can obtain partial sums and the sum (if $|r| < 1$) of such series by factoring the common first term from each term of the series and then using Formulas 2 and 3.

EXAMPLE 2

Find the formula for the nth partial sum S_n and evaluate $\displaystyle\sum_{n=1}^{\infty} 8\left(-\frac{1}{3}\right)^n$.

SOLUTION   We recognize that this is a geometric series with $r = -1/3$. The series converges because $r = -1/3$ satisfies $|r| < 1$. Let us write out the first few terms and then factor the common first term. We have

$$\sum_{n=1}^{\infty} 8\left(-\frac{1}{3}\right)^n = 8\left(-\frac{1}{3}\right) + 8\left(-\frac{1}{3}\right)^2 + 8\left(-\frac{1}{3}\right)^3 + \cdots$$

$$= 8\left(-\frac{1}{3}\right)\left(1 + \left(-\frac{1}{3}\right) + \left(-\frac{1}{3}\right)^2 + \cdots\right).$$

We can now use Formula 2 to obtain a formula for S_n. In particular, the nth partial sum of the above series is the product of the factor $8(-1/3)$ and the nth partial sum of a series that is of the form $1 + r + r^2 + r^3 + \cdots$. Formula 2 tells us that the latter sum has nth partial sum

$$\frac{1 - r^n}{1 - r}, \quad \text{with } r = -\frac{1}{3}$$

We then obtain

$$S_n = 8\left(-\frac{1}{3}\right)\left(1 + \left(-\frac{1}{3}\right) + \cdots + \left(-\frac{1}{3}\right)^{n-1}\right) \qquad \text{[Formula 2 with } r = -1/3]$$

$$= 8\left(-\frac{1}{3}\right)\left(\frac{1 - \left(-\frac{1}{3}\right)^n}{1 - \left(-\frac{1}{3}\right)}\right)$$

$$= -2\left(1 - \left(-\frac{1}{3}\right)^n\right).$$

We can then obtain the sum as the limit of the partial sums. That is,

$$S = \lim_{n \to \infty} S_n = \lim_{n \to \infty} \left(-2\left(1 - \left(-\frac{1}{3}\right)^n\right)\right) = -2. \quad\blacksquare$$

EXAMPLE 3

Express the infinitely repeating decimal number $0.77777\ldots$ as a ratio of two integers.

SOLUTION   We have

$$0.77777\ldots = 0.7 + 0.07 + 0.007 + 0.0007 + 0.00007 + \cdots$$

$$= 0.7(1 + 0.1 + 0.01 + 0.001 + 0.0001 + \cdots)$$

$$= 0.7(1 + 0.1 + (0.1)^2 + (0.1)^3 + (0.1)^4 + \cdots).$$

The expression inside the parentheses is a convergent geometric series with $r = 0.1$. Substitution into Formula 3 then gives

$$0.77777\ldots = 0.7\left(\frac{1}{1 - 0.1}\right) = 0.7\left(\frac{1}{0.9}\right) = \frac{7}{9}. \quad\blacksquare$$

Let us now return to the question of the bouncing ball. We determined that the total time that the ball bounces is

$$T = \frac{\sqrt{s_0}}{4} + \frac{\sqrt{s_0/2}}{2} + \frac{\sqrt{s_0/2^2}}{2} + \frac{\sqrt{s_0/2^3}}{2} + \frac{\sqrt{s_0/2^4}}{2} + \cdots$$

$$= \frac{\sqrt{s_0}}{4} + \frac{\sqrt{s_0/2}}{2} \left(1 + \sqrt{1/2} + \sqrt{1/2^2} + \sqrt{1/2^3} + \cdots \right)$$

$$= \frac{\sqrt{s_0}}{4} + \frac{\sqrt{s_0/2}}{2} \left(1 + \sqrt{1/2} + (\sqrt{1/2})^2 + (\sqrt{1/2})^3 + \cdots \right).$$

We see that the series inside the parentheses is a convergent geometric series with $r = \sqrt{1/2}$, so the ball will not bounce forever, even though it bounces an infinite number of times. We can use Formula 3 to determine the total time. We have

$$T = \frac{\sqrt{s_0}}{4} + \frac{\sqrt{s_0/2}}{2} \left(\frac{1}{1 - \sqrt{1/2}} \right) \approx 1.457 \sqrt{s_0}.$$

Telescoping Series

We can find simple formulas for the nth partial sums of certain types of series because of cancellation between the terms. Let us illustrate with some examples.

EXAMPLE 4

Find a formula for the nth partial sum S_n and evaluate the series

$$\sum_{n=1}^{\infty} \left(\frac{1}{n} - \frac{1}{n+1} \right).$$

SOLUTION   Let us write a few partial sums and see if a pattern can be determined.

$$S_1 = a_1 = \frac{1}{1} - \frac{1}{1+1} = 1 - \frac{1}{2},$$

$$S_2 = S_1 + a_2 = \left(1 - \frac{1}{2} \right) + \left(\frac{1}{2} - \frac{1}{2+1} \right) = 1 + \left(-\frac{1}{2} + \frac{1}{2} \right) - \frac{1}{3}$$

$$= 1 - \frac{1}{3},$$

$$S_3 = S_2 + a_3 = \left(1 - \frac{1}{3} \right) + \left(\frac{1}{3} - \frac{1}{4} \right) = 1 + \left(-\frac{1}{3} + \frac{1}{3} \right) - \frac{1}{4} = 1 - \frac{1}{4}.$$

Generally, we have

$$S_n = 1 - \frac{1}{n+1}.$$

(We have verified this formula for $n = 1$, 2, and 3. It can be proved formally for all positive integers n by using Mathematical Induction.)

The sum of the series is the limit of the partial sums. We have

$$S = \lim_{n \to \infty} S_n = \lim_{n \to \infty} \left(1 - \frac{1}{n+1} \right) = 1.$$

The series is convergent with

$$\sum_{n=1}^{\infty} \left(\frac{1}{n} - \frac{1}{n+1} \right) = 1. \quad \blacksquare$$

EXAMPLE 5

Find a formula for the nth partial sum S_n and evaluate the series

$$\sum_{n=1}^{\infty} \left(\frac{n}{n+1} - \frac{n+2}{n+3} \right).$$

SOLUTION   We have

$$S_1 = a_1 = \frac{1}{1+1} - \frac{1+2}{1+3} = \frac{1}{2} - \frac{3}{4},$$

$$S_2 = S_1 + a_2 = \left(\frac{1}{2} - \frac{3}{4} \right) + \left(\frac{2}{3} - \frac{4}{5} \right) = \frac{1}{2} + \frac{2}{3} - \frac{3}{4} - \frac{4}{5},$$

$$S_3 = S_2 + a_3 = \left(\frac{1}{2} + \frac{2}{3} - \frac{3}{4} - \frac{4}{5} \right) + \left(\frac{3}{4} - \frac{5}{6} \right) = \frac{1}{2} + \frac{2}{3} - \frac{4}{5} - \frac{5}{6},$$

$$S_4 = S_3 + a_4 = \left(\frac{1}{2} + \frac{2}{3} - \frac{4}{5} - \frac{5}{6} \right) + \left(\frac{4}{5} - \frac{6}{7} \right) = \frac{1}{2} + \frac{2}{3} - \frac{5}{6} - \frac{6}{7},$$

$$S_5 = S_4 + a_5 = \left(\frac{1}{2} + \frac{2}{3} - \frac{5}{6} - \frac{6}{7} \right) + \left(\frac{5}{6} - \frac{7}{8} \right) = \frac{1}{2} + \frac{2}{3} - \frac{6}{7} - \frac{7}{8}.$$

Generally, we have

$$S_n = \frac{1}{2} + \frac{2}{3} - \frac{n+1}{n+2} - \frac{n+2}{n+3}, \qquad n \geq 2.$$

(This formula can be proved formally by Mathematical Induction.) The series is convergent with sum

$$S = \lim_{n \to \infty} S_n = \lim_{n \to \infty} \left(\frac{1}{2} + \frac{2}{3} - \frac{n+1}{n+2} - \frac{n+2}{n+3} \right) = \frac{1}{2} + \frac{2}{3} - 1 - 1 = -\frac{5}{6}.$$

$$\blacksquare$$

Series of the form

$$\sum_{n=1}^{\infty} (b_n - b_{n+k})$$

are called **telescoping series**. The series

$$\sum_{n=1}^{\infty} \left(\frac{1}{n} - \frac{1}{n+1} \right)$$

of Example 4 is a telescoping series with $b_n = 1/n$ and $k = 1$. The series

$$\sum_{n=1}^{\infty} \left(\frac{n}{n+1} - \frac{n+2}{n+3} \right)$$

of Example 5 is a telescoping series with $b_n = n/(n+1)$ and $k = 2$.

We can obtain a formula for the partial sums of a telescoping series by writing a few partial sums and observing the pattern of cancellation, as we did in Examples 4 and 5.

Let us note that Examples 2, 4, and 5 are intended to reinforce the fact that *the sum of a series is the limit of the partial sums.* In most practical problems it is not necessary to find a formula for the partial sums of a series. Numerical values of a finite number of the partial sums are calculated and used as approximate values of the sum, as in Example 1. However, it is important that we know that the limit of partial sums exists. It doesn't make sense to try to approximate the limit of the sequence of partial sums if this sequence does not have a limit.

EXERCISES 9.2

Find the first four partial sums of the series and compare with a calculator value of the given sum in Exercises 1–10.

1. $\displaystyle\sum_{n=1}^{\infty} \frac{1}{n}(0.5)^n = \ln 2$

2. $\displaystyle\sum_{n=1}^{\infty} \frac{1}{2n-1}(0.6)^{2n-1} = \ln 2$

3. $\displaystyle\sum_{n=1}^{\infty} \frac{1}{2n-1}(0.8)^{2n-1} = \ln 3$

4. $\displaystyle\sum_{n=1}^{\infty} \frac{1}{n}\left(\frac{2}{3}\right)^n = \ln 3$

5. $\displaystyle\sum_{n=1}^{\infty} \frac{1}{(n-1)!}2^{n-1} = e^2 \quad (0! = 1)$

6. $\displaystyle\sum_{n=1}^{\infty} \frac{1}{(n-1)!}(-1)^{n-1} = e^{-1} \quad (0! = 1)$

7. $\displaystyle\sum_{n=1}^{\infty} \frac{(-1)^{n+1}}{2n-1}(0.5)^{2n-1} = \tan^{-1} 0.5$

8. $\displaystyle\sum_{n=1}^{\infty} \frac{(-1)^{n+1}}{(2n-2)!}(0.4)^{2n-2} = \cos 0.4 \text{ (cosine of 0.4 rad)}$

9. $\displaystyle\sum_{n=0}^{\infty}(-1)^n \frac{(0.3)^{2n+1}}{(2n+1)!} = \sin 0.3 \text{ (sine of 0.3 rad)}$

10. $\displaystyle(0.35) + \frac{(0.35)^3}{6} + \frac{3(0.35)^5}{40} + \frac{5(0.35)^7}{112} + \cdots = \sin^{-1} 0.35$

Find a formula for the nth partial sum S_n and evaluate the series that are convergent in Exercises 11–26.

11. $\displaystyle\sum_{n=0}^{\infty} 8(0.2)^n$

12. $\displaystyle\sum_{n=0}^{\infty} 6(-0.2)^n$

13. $\displaystyle\sum_{n=1}^{\infty}(-1.1)^n$

14. $\displaystyle\sum_{n=1}^{\infty}(9.9)^n$

15. $\displaystyle\sum_{n=0}^{\infty} \frac{3}{2^n}$

16. $\displaystyle\sum_{n=0}^{\infty} \frac{(-1)^n}{3^n}$

17. $\displaystyle\sum_{n=1}^{\infty}\left(\frac{n}{n+1} - \frac{n+1}{n+2}\right)$

18. $\displaystyle\sum_{n=1}^{\infty}\left(\frac{n}{2n-1} - \frac{n+1}{2n+1}\right)$

19. $\displaystyle\sum_{n=1}^{\infty}\left(\frac{n}{2^n} - \frac{n+1}{2^{n+1}}\right)$

20. $\displaystyle\sum_{n=1}^{\infty}\left(\frac{2n}{3n-2} - \frac{2n+4}{3n+4}\right)$

21. $\displaystyle\sum_{n=1}^{\infty}\left(\frac{1}{n} - \frac{1}{n+2}\right)$

22. $\displaystyle\sum_{n=1}^{\infty}\left(\frac{n^2}{n+1} - \frac{(n+2)^2}{n+3}\right)$

23. $\displaystyle\sum_{n=1}^{\infty} \frac{1}{n^2+4n+3}$. (*Hint:* Use partial fractions.)

24. $\displaystyle\sum_{n=1}^{\infty} \frac{1}{(3n+1)(3n+4)}$. (**Calculus Explorer & Tutor II, Sequences and Series, 2.**)

25. $\displaystyle\sum_{n=1}^{\infty} \ln \frac{n}{n+1}$

26. $\displaystyle\sum_{n=1}^{\infty}\left(\frac{1}{2^n} - \frac{1}{2^{n+1}}\right)$

Find the sum of each series in Exercises 27–34.

27. $\displaystyle\sum_{n=2}^{\infty} 3(-2)^{-n}$

28. $\displaystyle\sum_{n=-3}^{\infty} 2\left(\frac{1}{3}\right)^n$

29. $\displaystyle\sum_{n=2}^{\infty} \frac{2^n}{3^n}$ (**Calculus Explorer & Tutor II, Sequences and Series, 3.**)

30. $\displaystyle\sum_{n=1}^{\infty} \frac{5^n}{3^{2n}}$

31. $\displaystyle\sum_{n=1}^{\infty}\left(\frac{2}{n} - \frac{2}{n+1}\right)$

32. $\displaystyle\sum_{n=1}^{\infty}\left(\frac{n}{n+2} - \frac{n+1}{n+3}\right)$

33. $\displaystyle\sum_{n=1}^{\infty}\left(\frac{n}{3n+1} - \frac{n-1}{3n-2}\right)$

34. $\displaystyle\sum_{n=1}^{\infty}\left(\frac{1}{n+1} - \frac{1}{n+3}\right)$

35. Express the infinitely repeating decimal number 0.55555... as a geometric series and then use Formula 3 to express the number as a ratio of two integers.

36. Express the infinitely repeating decimal number 0.252525... as a geometric series and then use Formula 3 to express the number as a ratio of two integers.

37. A handball bounces back to two-thirds the height from which it is dropped. Find the total distance traveled by a handball that is dropped from a height of 6 ft.

38. How long after the handball in Exercise 37 is dropped will it stop bouncing?

39. Show that the telescoping series $\sum_{n=1}^{\infty}(b_n - b_{n+2})$ has partial sums
$$S_n = (b_1 + b_2) - (b_{n+1} + b_{n+2}), n \geq 2.$$

40. Show that the telescoping series $\sum_{n=1}^{\infty}(b_n - b_{n+k})$ has partial sums
$$S_n = (b_1 + b_2 + \cdots + b_k) - (b_{n+1} + b_{n+2} + \cdots + b_{n+k}),$$
$$n \geq k.$$

41. Show that $\sum_{n=0}^{\infty} r^n$ diverges for $r = 1$ and $r = -1$.

42. Find a formula for the sum of the series
$$1 + x - x^2 - x^3 + x^4 + x^5 - x^6 - x^7 + x^8 + \cdots$$
and determine the values of x for which the formula is valid.

43. Find a formula for the sum of the series $\sum_{n=0}^{\infty}\left(\dfrac{x}{x+1}\right)^n$
and determine the values of x for which the formula is valid.

44. Find the total length of the zigzag line segments in the figure.

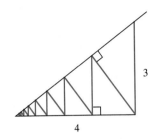

For Exercises 45–46, consider the sequence of disjoint subintervals of $[0, 1]$ obtained by first selecting the middle third of $[0, 1]$, then selecting the middle third of each of the two subintervals of $[0, 1]$ on each side of the first selected interval, and continuing to repeat the process of selecting the middle third of each of the 2^n subintervals left after the nth step.

45. Show that the (infinite) sum of the lengths of the selected intervals is one.

46. What is the (infinite) sum of the areas of the equilateral triangles that have bases on the selected subintervals? See figure.

47. For which values of n is
$$\left|\sum_{k=0}^{n-1}\left(\frac{2}{3}\right)^k - 3\right| < 0.005?$$

48. For which values of n is
$$\left|\sum_{k=1}^{n}\left(\frac{1}{k} - \frac{1}{k+1}\right) - 1\right| < 0.005?$$

Connections

Convergence or divergence of improper integrals by comparison, 8.7.

Limits of sequences, 9.1.

9.3 COMPARISON TESTS

In this section we will develop some tests that are used to answer the question:

Is $\sum_{n=1}^{\infty} a_n$ *convergent or divergent?*

This is a reasonable question to ask *before* we start trying to find approximate values of the series. If we know the series is convergent, then the partial sums of the series can be used for approximate values of the sum. If the series is divergent, it does not have a value and it does not make sense to try to find approximate values.

Preliminary Results

The following preliminary observations should be kept in mind.

Changing the value of any finite number of terms of a series does not affect convergence or divergence of the series. (1)

Of course, changing the value of finitely many terms could change the value of the sum of a convergent series.

If $c \neq 0$, *then* $\sum_{n=1}^{\infty} a_n$ *and* $\sum_{n=1}^{\infty} ca_n$ *are either both convergent or both divergent.* (2)

We will use the following fact about the terms of a convergent series:

$$\text{If } \sum_{n=1}^{\infty} a_n \text{ is convergent, then } \lim_{n \to \infty} a_n = 0. \tag{3}$$

Proof   If $\lim_{n \to \infty} S_n = S$, then

$$\lim_{n \to \infty} a_n = \lim_{n \to \infty} (S_n - S_{n-1}) = \lim_{n \to \infty} S_n - \lim_{n \to \infty} S_{n-1} = S - S = 0. \quad \blacksquare$$

Statement 3 cannot be used as a test for convergence, but we can restate it to obtain a logically equivalent *test for divergence.*

THE nTH TERM TEST FOR DIVERGENCE

$$\text{If } \lim_{n \to \infty} a_n \neq 0, \text{ then } \sum_{n=1}^{\infty} a_n \text{ is divergent.} \tag{4}$$

EXAMPLE 1

We see that $\displaystyle\sum_{n=1}^{\infty} \frac{n}{n+1}$ is divergent, because $\displaystyle\lim_{n \to \infty} \frac{n}{n+1} = 1 \neq 0$. The nth term does not approach zero. $\blacksquare$

Note that Statement 4 is only a test for divergence of a series.

NOTE   *Statement 4 cannot be used as a test for convergence.*

It is very important to note that Statement 3 does not say that a series is convergent if its nth term approaches zero as n approaches infinity. A series may diverge, even if its nth term approaches zero. For example, $\lim_{n \to \infty} (1/n) = 0$, but

$$\sum_{n=1}^{\infty} \frac{1}{n} \text{ is divergent.} \tag{5}$$

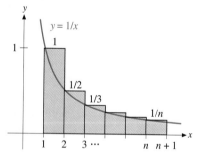

FIGURE 9.3.1 S_n is the sum of the areas of the rectangles, so $S_n > \displaystyle\int_1^{n+1} \frac{1}{x} dx = \ln(x+1) \to \infty$ as $n \to \infty$.

Proof   We can verify Statement 5 by showing that the partial sums of the series become unbounded as n approaches infinity. To do this, we first note that any positive number a_j can be interpreted as the area of a rectangle with height a_j and base one. Using Figure 9.3.1 to compare the sum of the areas of the rectangles with the corresponding area under the curve $y = 1/x$, we see that

$$S_n = 1 + \frac{1}{2} + \frac{1}{3} + \frac{1}{4} + \cdots + \frac{1}{n}$$

$$> \int_1^{n+1} \frac{1}{x} dx = \ln(n+1) \to \infty \quad \text{as } n \to \infty. \quad \blacksquare$$

Let us compare the amount of effort required to show divergence of the series in Example 1 and in Statement 5. In Example 1 we used available theory in the form of Statement 4. We could then determine divergence by applying a rather simple test that involved looking at just the nth term of the series. In Statement 5 we did not have the appropriate theory available, so we needed to estimate the size of S_n to show divergence. It should seem much easier to use a known general result to determine convergence or divergence of a series.

Series with Nonnegative Terms

We will now extended our theoretical results concerning the convergence and divergence of series. For the remainder of this section we will consider only series that have nonnegative terms. The following result is the basis of the theory of convergence of series with nonnegative terms.

THEOREM 1

Let S_n denote the nth partial sum of a series $\displaystyle\sum_{n=1}^{\infty} a_n$ with nonnegative terms. Then, either S_n diverges to infinity and $\displaystyle\sum_{n=1}^{\infty} a_n$ is divergent or there is some number B such that $S_n \leq B$ for all n and $\displaystyle\sum_{n=1}^{\infty} a_n$ is convergent.

Proof We first note that the partial sums of a series with *nonnegative* terms form a nondecreasing sequence. That is, $a_{n+1} \geq 0$ for all n, so $S_{n+1} = S_n + a_{n+1} \geq S_n$ for all n. It follows that either the sequence of partial sums S_n increases to infinity or there is some number B such that $S_n \leq B$ for all n. See Figure 9.3.2. Of course, the series is divergent if S_n increases to infinity. If the partial sums are bounded from above, convergence follows from the Monotone Convergence Theorem in Section 9.1, which tells us that a *bounded, monotone* sequence is convergent. ∎

Let us see how Theorem 1 can be used to verify some important results that we will use later.

$$\sum_{n=1}^{\infty} \frac{1}{n^p} \text{ is convergent if } p > 1. \qquad (6)$$

Proof According to Theorem 1, it is enough to show there is a number B such that $S_n \leq B$ for all n. To do this, we use Figure 9.3.3 to compare the second through nth terms of the partial sum S_n with the corresponding areas under the curve $y = x^{-p}$. We see that

$$S_n = \frac{1}{1^p} + \frac{1}{2^p} + \frac{1}{3^p} + \cdots + \frac{1}{n^p}$$

$$\leq 1 + \int_1^2 \frac{1}{x^p}\, dx + \int_2^3 \frac{1}{x^p}\, dx + \cdots + \int_{n-1}^n \frac{1}{x^p}\, dx$$

$$= 1 + \int_1^n \frac{1}{x^p}\, dx$$

$$= 1 + \left[\frac{1}{1-p} x^{1-p} \right]_1^n$$

$$= 1 + \frac{n^{1-p}}{1-p} - \frac{1}{1-p}$$

$$= 1 + \frac{1}{p-1} - \frac{n^{1-p}}{p-1}.$$

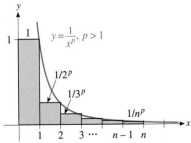

FIGURE 9.3.2 The partial sums S_n of a series with *nonnegative* terms satisfy $S_{n+1} \geq S_n$. If there is a number B such that $S_n \leq B$ for all n, then the monotone sequence of partial sums converges, so $\displaystyle\sum_{n=1}^{\infty} a_n$ is convergent.

FIGURE 9.3.3 S_n is the sum of the areas of the rectangles, so $S_n < 1 + \int_1^n \frac{1}{x^p}\, dx = 1 + \frac{1}{p-1} - \frac{n^{1-p}}{p-1} < \frac{p}{p-1}$ if $p > 1$.

Since $p > 1$ implies

$$-\frac{n^{1-p}}{p-1} < 0,$$

we see that

$$S_n < 1 + \frac{1}{p-1} = \frac{p}{p-1}.$$

We have shown that

$$S_n \leq \frac{p}{p-1} \text{ for all } n,$$

so Theorem 1 tells us the series is convergent. ∎

We also have the following:

$$\sum_{n=1}^{\infty} \frac{1}{n^p} \text{ is divergent if } p < 1. \tag{7}$$

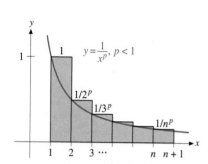

FIGURE 9.3.4 S_n is the sum of the areas of the rectangles, so $S_n > \int_1^{n+1} \frac{1}{x^p} dx = \frac{1}{1-p}\left((n+1)^{1-p} - 1\right) \to \infty$ as $n \to \infty$ if $p < 1$.

Proof To verify Statement 7, we use Figure 9.3.4 to obtain

$$S_n = \frac{1}{1^p} + \frac{1}{2^p} + \frac{1}{3^p} + \cdots + \frac{1}{n^p}$$

$$\geq \int_1^{n+1} \frac{1}{x^p} dx = \frac{1}{1-p}((n+1)^{1-p} - 1).$$

The condition $p < 1$ implies the right-hand term diverges to infinity, so S_n must also diverge to infinity. Theorem 1 then implies the series is divergent. ∎

At this point let us repeat the warning given in Section 9.2 not to confuse the *series* $\sum_{n=1}^{\infty} a_n$ with the *sequence* $\{a_n\}_{n\geq 1}$. In particular, the *sequences*

$$\left\{\frac{1}{n}\right\}_{n\geq 1} \text{ and } \left\{\frac{1}{n^2}\right\}_{n\geq 1}$$

are both convergent with limit zero. However,

the *series* $\sum_{n=1}^{\infty} \frac{1}{n}$ is divergent

and

the *series* $\sum_{n=1}^{\infty} \frac{1}{n^2}$ is convergent.

That is, we have

$$1 + \frac{1}{2} + \frac{1}{3} + \frac{1}{4} + \cdots + \frac{1}{n} > \int_1^{n+1} \frac{1}{x} dx = \ln(n+1) \to \infty \text{ as } n \to \infty.$$

Thus, even though the nth term of the series $\displaystyle\sum_{n=1}^{\infty}(1/n)$ approaches zero as n approaches infinity, $1/n$ does not approach zero *fast enough* to keep the partial sums from becoming unbounded. On the other hand, we have

$$1 + \frac{1}{2^2} + \frac{1}{3^2} + \cdots + \frac{1}{n^2} < 1 + \int_1^n \frac{1}{x^2}\,dx = 2 - \frac{1}{n} < 2.$$

Thus, $1/n^2$ approaches zero *fast enough* so that the partial sums of the series $\displaystyle\sum_{n=1}^{\infty}(1/n^2)$ remain bounded and the series is convergent.

p-Series

Let us summarize Statements 5, 6, and 7.

The p-series $\displaystyle\sum_{n=1}^{\infty}\frac{1}{n^p}$ is convergent if $p > 1$ and divergent if $p \leq 1$.

We can then use Statement 2 to determine convergence or divergence of constant multiples of p-series.

EXAMPLE 2

Determine convergence or divergence of the series $\displaystyle\sum_{n=1}^{\infty}\frac{1}{3\sqrt{n}}$.

SOLUTION   We have

$$a_n = \frac{1}{3\sqrt{n}} = \frac{1}{3}\frac{1}{n^{1/2}} = c\frac{1}{n^p}, \text{ where } c = \frac{1}{3} \neq 0 \text{ and } p = \frac{1}{2} < 1.$$

A nonzero multiple of a divergent p-series is divergent, so

$$\sum_{n=1}^{\infty}\frac{1}{3\sqrt{n}}$$

is divergent.   ■

EXAMPLE 3

Determine convergence or divergence of the series $\displaystyle\sum_{n=1}^{\infty}\frac{100}{n^{4/3}}$.

SOLUTION   We have

$$a_n = \frac{100}{n^{4/3}} = c\frac{1}{n^p}, \text{ where } c = 100 \neq 0 \text{ and } p = \frac{4}{3} > 1.$$

A multiple of a convergent p-series is convergent, so

$$\sum_{n=1}^{\infty}\frac{100}{n^{4/3}}$$

is convergent.   ■

We could use the partial sums of the convergent series in Example 3 to obtain approximate values of its sum. It does not make sense to use the partial sums of the divergent series in Example 2 to approximate its sum, since divergent series do not have a sum.

Comparison Tests

The following theorem can be used to determine convergence or divergence of a series $\sum a_n$ with nonnegative terms by comparing the size of its terms to the terms of a series $\sum b_n$ with nonnegative terms for which convergence or divergence is known. Note that we now know which p-series and which geometric series converge and which diverge.

COMPARISON TEST FOR SERIES WITH NONNEGATIVE TERMS

If $0 \le a_n \le b_n$ for all n, then $\sum_{n=1}^{\infty} b_n$ convergent implies $\sum_{n=1}^{\infty} a_n$ convergent.

If $a_n \ge b_n \ge 0$ for all n, then $\sum_{n=1}^{\infty} b_n$ divergent implies $\sum_{n=1}^{\infty} a_n$ divergent.

Proof The Comparison Test is an immediate consequence of Theorem 1. That is, $a_n \le b_n$ for all n implies

$$a_1 + a_2 + \cdots + a_n \le b_1 + b_2 + \cdots + b_n.$$

If $\sum_{n=1}^{\infty} b_n$ is convergent, the partial sums on the right remain bounded. It follows

that the partial sums on the left also are bounded, so Theorem 1 implies $\sum_{n=1}^{\infty} a_n$ is convergent. (Note that we are dealing with series that have nonnegative terms, so Theorem 1 applies.) Similarly, $a_n \ge b_n$ for all n implies

$$a_1 + a_2 + \cdots + a_n \ge b_1 + b_2 + \cdots + b_n.$$

If $\sum_{n=1}^{\infty} b_n$ is divergent, then the partial sums on the right become unbounded.

This implies the partial sums on the left also become unbounded, so $\sum_{n=1}^{\infty} a_n$ is divergent. ■

The Comparison Test for Series with Nonnegative Terms tells us that if the nonnegative terms b_n approach zero *fast enough* to ensure that $\sum_{n=1}^{\infty} b_n$ is convergent, then any smaller nonnegative terms a_n also approach zero fast enough to ensure $\sum_{n=1}^{\infty} a_n$ is convergent. If the terms b_n do not approach zero fast enough, so $\sum_{n=1}^{\infty} b_n$ is divergent, then any larger terms a_n do not approach zero fast enough for $\sum_{n=1}^{\infty} a_n$ to be convergent.

The following is a much more useful version of the Comparison Test.

LIMIT VERSION OF THE COMPARISON TEST

If a_n and b_n are both positive for all n and $0 < \lim\limits_{n\to\infty} \dfrac{a_n}{b_n} < \infty$, then $\sum\limits_{n=1}^{\infty} a_n$

and $\sum\limits_{n=1}^{\infty} b_n$ are either both convergent or both divergent.

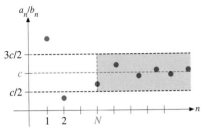

FIGURE 9.3.5 If $\lim\limits_{n\to\infty} a_n/b_n = c$, $0 < c < \infty$, then there is a number N such that $\dfrac{c}{2} \le \dfrac{a_n}{b_n} \le \dfrac{3c}{2}$ whenever $n \ge N$.

Proof   If $\lim\limits_{n\to\infty} \dfrac{a_n}{b_n} = c$, where $0 < c < \infty$, then there is a number N such that

$$\frac{c}{2} \le \frac{a_n}{b_n} \le \frac{3c}{2} \quad \text{for all } n \ge N.$$

See Figure 9.3.5.

If $\sum\limits_{n=1}^{\infty} b_n$ is convergent, then Statements 1 and 2 imply that $\sum\limits_{n=N}^{\infty} \dfrac{3c}{2} b_n$ is also

convergent, since $3c/2 \ne 0$. Since $0 \le a_n \le \dfrac{3c}{2} b_n$ for all $n \ge N$, the Comparison Test then shows that $\sum\limits_{n=N}^{\infty} a_n$ is convergent. It follows from Statement 1 that

$\sum\limits_{n=1}^{\infty} a_n$ is also convergent.

Similarly, if $\sum\limits_{n=1}^{\infty} b_n$ is divergent, we can use the inequality $a_n \ge \dfrac{c}{2} b_n \ge 0$

to show that $\sum\limits_{n=1}^{\infty} a_n$ must also be divergent. ∎

We can extend the Limit Version of the Comparison Test to include the cases where the limit of the ratio of terms is either zero or infinity.

EXTENSION OF THE LIMIT VERSIONS OF THE COMPARISON TEST

If a_n and b_n are both positive for all n and $\lim\limits_{n\to\infty} \dfrac{a_n}{b_n} = 0$, then $\sum\limits_{n=1}^{\infty} b_n$

convergent implies $\sum\limits_{n=1}^{\infty} a_n$ convergent.

If a_n and b_n are both positive for all n and $\lim\limits_{n\to\infty} \dfrac{a_n}{b_n} = \infty$, then $\sum\limits_{n=1}^{\infty} b_n$

divergent implies $\sum\limits_{n=1}^{\infty} a_n$ divergent.

Proof   If $\lim\limits_{n\to\infty} \dfrac{a_n}{b_n} = 0$, then there is a number N such that $a_n \le b_n$ for all $n \ge N$. We can then argue as in the proof of the Limit Version of the Comparison Test to show that $\sum\limits_{n=1}^{\infty} b_n$ convergent implies $\sum\limits_{n=1}^{\infty} a_n$ convergent.

If $\lim\limits_{n\to\infty} \dfrac{a_n}{b_n} = \infty$, then there is a number N such that $a_n \geq b_n$ for all $n \geq N$.

It follows that $\sum\limits_{n=1}^{\infty} b_n$ divergent implies $\sum\limits_{n=1}^{\infty} a_n$ divergent. ■

If $0 < \lim\limits_{n\to\infty} \dfrac{a_n}{b_n} < \infty$, we should think of a_n and b_n as essentially the same size, so the series $\sum\limits_{n=1}^{\infty} a_n$ and $\sum\limits_{n=1}^{\infty} b_n$ either both converge or both diverge.

If $\lim\limits_{n\to\infty} \dfrac{a_n}{b_n} = 0$, we should think of a_n as being smaller than b_n, so the series $\sum\limits_{n=1}^{\infty} a_n$ must be convergent if $\sum\limits_{n=1}^{\infty} b_n$ is convergent.

If $\lim\limits_{n\to\infty} \dfrac{a_n}{b_n} = \infty$, we should think of a_n as being larger than b_n, so the series $\sum\limits_{n=1}^{\infty} a_n$ must be divergent if $\sum\limits_{n=1}^{\infty} b_n$ is divergent.

Using the Limit Version of the Comparison Test

We can now use the results for the convergence or divergence of p-series, along with the Limit Version of the Comparison Test, to determine *easily* the convergence or divergence of series of positive terms that are algebraic expressions of n. To determine which p-series to use for comparison, we factor the highest power of n from each of the numerator and the denominator.

EXAMPLE 4

Determine convergence or divergence of the series $\sum\limits_{n=1}^{\infty} \dfrac{n}{3n^2 + 1}$.

SOLUTION   Since the terms of this series are algebraic expressions of n, we should compare its terms to a p-series. We have

$$a_n = \frac{n}{3n^2 + 1} = \frac{n}{n^2(3 + 1/n^2)} = \frac{1}{n}\left(\frac{1}{3 + 1/n^2}\right).$$

Since for large n, $1/(3 + 1/n^2) \approx 1/3$, this suggests we compare with the series

$$\sum_{n=1}^{\infty} \frac{1}{n}.$$

We have

$$\lim_{n\to\infty} \frac{a_n}{1/n} = \lim_{n\to\infty} \frac{1}{3 + (1/n^2)} = \frac{1}{3}.$$

Since

$$\sum_{n=1}^{\infty} \frac{1}{n}$$

is divergent, the Limit Version of the Comparison Test tells us that so is

$$\sum_{n=1}^{\infty} \frac{n}{3n^2 + 1}. \quad \blacksquare$$

EXAMPLE 5

Determine convergence or divergence of the series $\sum_{n=1}^{\infty} \frac{3\sqrt{n}}{2n^2 - 1}$.

SOLUTION   The terms are algebraic expressions of n, so we compare with a p-series. We have

$$a_n = \frac{3\sqrt{n}}{2n^2 - 1} = \frac{3n^{1/2}}{n^2(2 - (1/n^2))} = \frac{1}{n^{3/2}} \left(\frac{3}{2 - 1/n^2} \right).$$

This suggests we compare with

$$\sum_{n=1}^{\infty} \frac{1}{n^{3/2}}.$$

We have

$$\lim_{n \to \infty} \frac{a_n}{1/n^{3/2}} = \lim_{n \to \infty} \frac{3}{2 - (1/n^2)} = \frac{3}{2}.$$

Since

$$\sum_{n=1}^{\infty} \frac{1}{n^{3/2}}$$

is convergent, so is

$$\sum_{n=1}^{\infty} \frac{3\sqrt{n}}{2n^2 - 1}. \quad \blacksquare$$

It is useful for the purpose of comparison to recall from Chapter 7 the following relations between the sizes of the logarithmic, exponential, and power functions.

$$\lim_{n \to \infty} \frac{\ln n}{n^\alpha} = 0, \qquad \alpha > 0,$$

$$\lim_{n \to \infty} \frac{n^\alpha}{e^n} = 0.$$

The first limit tells us that $\ln n$ becomes smaller than any positive power of n as n approaches infinity; the second limit tells us that any power of n becomes smaller than e^n as n approaches infinity.

EXAMPLE 6

Determine convergence or divergence of the series $\sum_{n=1}^{\infty} \frac{\ln n}{n^2}$.

SOLUTION   Let us compare the terms

$$a_n = \frac{\ln n}{n^2}$$

with those of a p-series. Our first choice might be $p = 2$. We have

$$\lim_{n \to \infty} \frac{(\ln n)/n^2}{1/n^2} = \lim_{n \to \infty} \left(\frac{\ln n}{n^2} \right) \left(\frac{n^2}{1} \right) = \lim_{n \to \infty} \ln n = \infty.$$

Since $\sum_{n=1}^{\infty} 1/n^2$ is convergent and the ratio $((\ln n)/n^2)/(1/n^2)$ approaches infinity, the Limit Version of the Comparison Test tells us nothing about the convergence or divergence of the series $\sum_{n=1}^{\infty} (\ln n)/n^2$.

We might next try $p = 1$. Then

$$\lim_{n \to \infty} \frac{(\ln n)/n^2}{1/n} = \lim_{n \to \infty} \left(\frac{\ln n}{n^2} \right) \left(\frac{n}{1} \right) = \lim_{n \to \infty} \frac{\ln n}{n} = 0.$$

Since $\sum_{n=1}^{\infty} 1/n$ is divergent and the ratio $((\ln n)/n^2)/(1/n)$ approaches zero, the Limit Version of the Comparison Test tells us nothing about the convergence or divergence of the series $\sum_{n=1}^{\infty} (\ln n)/n^2$.

Let us try $p = 1.5$. Then

$$\lim_{n \to \infty} \frac{(\ln n)/n^2}{1/n^{1.5}} = \lim_{n \to \infty} \left(\frac{\ln n}{n^2} \right) \left(\frac{n^{1.5}}{1} \right) = \lim_{n \to \infty} \frac{\ln n}{n^{0.5}} = 0.$$

Since $\sum_{n=1}^{\infty} 1/n^{1.5}$ is convergent and the ratio $((\ln n)/n^2)/(1/n^{1.5})$ approaches zero, the Limit Version of the Comparison Test tells us that the series $\sum_{n=1}^{\infty} (\ln n) / n^2$ is also convergent. (We could reach the same conclusion by comparison with any p-series, $1 < p < 2$.) ■

Series with positive terms that have exponential factors may be compared to either a geometric series or a p-series.

EXAMPLE 7

Determine convergence or divergence of the series $\sum_{n=1}^{\infty} n^2 2^{-n}$.

SOLUTION   It does not help to compare $a_n = n^2 2^{-n}$ to the convergent geometric series $\sum_{n=1}^{\infty} 2^{-n}$, because $n^2 2^{-n}$ is larger than 2^{-n}. However, we have

$$a_n = n^2 2^{-n} = n^2 (2^{-n/2} 2^{-n/2}) = (n^2 2^{-n/2}) 2^{-n/2}.$$

We can then compare the series $\sum_{n=1}^{\infty} a_n$ with the convergent geometric series $\sum_{n=1}^{\infty} 2^{-n/2}$. We have

$$\lim_{n \to \infty} \frac{a_n}{2^{-n/2}} = \lim_{n \to \infty} n^2 2^{-n/2} = 0.$$

The Limit Version of the Comparison Test then implies the series $\sum_{n=1}^{\infty} n^2 2^{-n}$ is convergent.

Note that

$$\lim_{n \to \infty} \frac{n^2 2^{-n}}{1/n^p} = 0$$

for all values of p. It follows that we could determine the convergence of the series $\sum_{n=1}^{\infty} n^2 2^{-n}$ by comparison with any convergent p-series. ■

EXAMPLE 8

Determine convergence or divergence of the series $\sum_{n=1}^{\infty} \frac{1}{n!}$.

SOLUTION   We have

$$\frac{1}{n!} = \frac{1}{1 \cdot 2 \cdots (n-1)n} \le \frac{1}{(n-1)n} \le \frac{2}{n^2}, \qquad n \ge 2.$$

Since $\sum_{n=1}^{\infty} 2/n^2$ is a convergent p-series, the Comparison Test shows that $\sum_{n=1}^{\infty} 1/n!$ is also convergent. ■

EXERCISES 9.3

Determine convergence or divergence for each of the series in Exercises 1–36.

1. $\displaystyle\sum_{n=1}^{\infty} \frac{7}{n^{2/3}}$

2. $\displaystyle\sum_{n=1}^{\infty} \frac{3}{n^{3/2}}$

3. $\displaystyle\sum_{n=1}^{\infty} \frac{1}{2n-1}$

4. $\displaystyle\sum_{n=1}^{\infty} \frac{n}{n^2+4}$

5. $\displaystyle\sum_{n=1}^{\infty} n^{-1.01}$

6. $\displaystyle\sum_{n=1}^{\infty} n^{-0.99}$

7. $\displaystyle\sum_{n=1}^{\infty} \frac{n}{\sqrt{n^2+1}}$

8. $\displaystyle\sum_{n=1}^{\infty} \frac{1}{n\sqrt{n^2+4}}$

9. $\displaystyle\sum_{n=1}^{\infty} \frac{2n}{3n+1}$

10. $\displaystyle\sum_{n=2}^{\infty} \frac{1}{n^3-1}$

11. $\displaystyle\sum_{n=1}^{\infty} \frac{1}{\sqrt{2n-1}}$

12. $\displaystyle\sum_{n=1}^{\infty} \frac{1}{\sqrt{n^4+16}}$

13. $\displaystyle\sum_{n=1}^{\infty} \sin(1/n^2)$

14. $\displaystyle\sum_{n=1}^{\infty} \cos(1/n^2)$

15. $\displaystyle\sum_{n=1}^{\infty} \frac{\ln n}{n^3}$

16. $\displaystyle\sum_{n=1}^{\infty} \frac{1}{1+\sqrt{n}}$

17. $\displaystyle\sum_{n=1}^{\infty} \frac{2n+5}{n^2+3n+2}$

18. $\displaystyle\sum_{n=1}^{\infty} \frac{2n^2+3}{n^4+2n^2+4}$

19. $\displaystyle\sum_{n=1}^{\infty} e^{-n}$

20. $\displaystyle\sum_{n=1}^{\infty} \left(1+\frac{2}{n}\right)^n$ (Calculus Explorer & Tutor II, Sequences and Series, 4.)

21. $\displaystyle\sum_{n=1}^{\infty} \frac{1}{n(n+1)(n+2)}$

22. $\displaystyle\sum_{n=1}^{\infty} ne^{-n}$

23. $\displaystyle\sum_{n=1}^{\infty} \frac{e^n}{1+e^{2n}}$

24. $\displaystyle\sum_{n=1}^{\infty} \frac{n}{(n+1)e^n}$

25. $\displaystyle\sum_{n=1}^{\infty} \frac{2^n}{n3^n}$ (Calculus Explorer & Tutor II, Sequences and Series, 5.)

26. $\displaystyle\sum_{n=1}^{\infty} n^3 e^{-n}$

27. $\displaystyle\sum_{n=1}^{\infty} \frac{\cos(1/n)}{n^{1.004}}$ (Calculus Explorer & Tutor II, Sequences and Series, 9.)

28. $\sum_{n=1}^{\infty} \dfrac{n + |\sin(\pi \sqrt{n})|}{n\sqrt{n}}$ (Calculus Explorer & Tutor II, Sequences and Series, 8.)

29. $\sum_{n=1}^{\infty} \dfrac{\ln n}{n}$

30. $\sum_{n=1}^{\infty} \dfrac{(\ln n)^2}{n^2}$

31. $\sum_{n=1}^{\infty} \dfrac{\sqrt{2n + 1}}{(n^4 + 1)^{1/3}}$

32. $\sum_{n=1}^{\infty} \dfrac{\sqrt{3n - 1}}{(n^5 + 5)^{1/3}}$

33. $\sum_{n=1}^{\infty} \dfrac{1}{2n!}$

34. $\sum_{n=1}^{\infty} \dfrac{e^n}{n!}$

35. $\sum_{n=1}^{\infty} \dfrac{n!}{1 \cdot 3 \cdot 5 \cdots (2n - 1)}$

36. $\sum_{n=1}^{\infty} \dfrac{2 \cdot 4 \cdot 6 \cdots (2n)}{4 \cdot 7 \cdot 10 \cdots (3n + 1)}$

37. Find an example such that $\sum_{n=1}^{\infty} a_n$ is divergent, but $\sum_{n=1}^{\infty} a_n^2$ is convergent.

38. Find an example of a series that is divergent, but its sequence of partial sums is bounded.

39. If a_n and b_n are positive for all n, $\sum_{n=1}^{\infty} a_n$ is convergent, and $\{b_n\}_{n \geq 1}$ is bounded, show that $\sum_{n=1}^{\infty} a_n b_n$ is convergent.

40. If $a_n > 0$ and $\sum_{n=1}^{\infty} a_n$ is convergent, show that $\sum_{n=1}^{\infty} a_n^2$ is convergent.

41. Give another proof that $\sum_{n=1}^{\infty} \dfrac{1}{n}$ diverges by showing

$$\sum_{n=2^{k-1}+1}^{2^k} \frac{1}{n} \geq \frac{1}{2}, \quad k \geq 1, \text{ so } \sum_{n=1}^{2^N} \frac{1}{n} \geq 1 + \frac{N}{2}.$$

$$\left(Hint: 1 + \underbrace{\frac{1}{2}}_{=1/2} + \underbrace{\frac{1}{3} + \frac{1}{4}}_{>1/2} + \underbrace{\frac{1}{5} + \frac{1}{6} + \frac{1}{7} + \frac{1}{8}}_{>1/2} + \cdots \right.$$

$$\left. + \underbrace{\frac{1}{2^{N-1} + 1} + \cdots + \frac{1}{2^N}}_{>1/2}. \right)$$

9.4 THE INTEGRAL TEST; ACCURACY OF APPROXIMATION

Connections

Bounds of functions, 2.2.

Comparison tests, 9.3.

In the previous section we used integrals to determine convergence or divergence of p-series. In this section we will use the same ideas to establish a test for convergence or divergence of other series with positive terms.

The Integral Test

We first note that, if f is positive and continuous on the interval $x \geq N$, then the integrals $\int_N^n f(x)\,dx$ exist for each $n \geq N$ and *increase* as n approaches infinity. It can then be shown, as in the Monotone Convergence Theorem of Section 9.1, that the improper integral $\int_N^{\infty} f(x)\,dx$ is convergent if there is some real number B such that $\int_N^n f(x)\,dx \leq B$ for all $n \geq N$; $\int_N^{\infty} f(x)\,dx$ is divergent if the integrals $\int_N^n f(x)\,dx$ approach infinity as n approaches infinity.

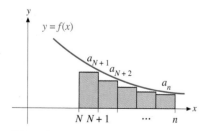

FIGURE 9.4.1a   The sum of the areas of the rectangles is less than the area under the curve between N and n.

THE INTEGRAL TEST

If f is positive, decreasing, and continuous on the interval $x \geq N$, with $a_n = f(n)$ for $n \geq N$, then the series $\sum_{n=1}^{\infty} a_n$ and the improper integral $\int_N^{\infty} f(x)\,dx$ are either both convergent or both divergent.

Proof   From Figure 9.4.1a, we see that the fact that f is decreasing with

$f(n) = a_n$ implies

$$S_n = S_N + a_{N+1} + a_{N+2} + \cdots + a_n \leq S_N + \int_N^n f(x)\,dx.$$

If the improper integral is convergent, then the numbers $\int_N^n f(x)\,dx$ *increase* to the real number $\int_N^\infty f(x)\,dx$ as n approaches infinity. We then have

$$S_n \leq S_N + \int_N^n f(x)\,dx \leq S_N + \int_N^\infty f(x)\,dx,$$

so the partial sums of the series $\sum_{n=1}^\infty a_n$ satisfy $S_n \leq S_N + \int_N^\infty f(x)\,dx$ for all $n \geq N$. Since the terms of the series are positive, this implies the series is convergent.

From Figure 9.4.1b we see that

$$S_n = S_{N-1} + a_N + a_{N+1} + \cdots + a_n \geq S_{N-1} + \int_N^{n+1} f(x)\,dx.$$

If the improper integral is divergent, then $\int_N^{n+1} f(x)\,dx$ approaches infinity as n approaches infinity. The partial sums of the series then approach infinity as n approaches infinity, so the series is also divergent. ∎

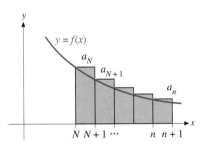

FIGURE 9.4.1b The sum of the areas of the rectangles is greater than the area under the curve between N and $n+1$.

EXAMPLE 1

Determine convergence or divergence of the series $\displaystyle\sum_{n=2}^\infty \frac{1}{n \ln n}$.

SOLUTION We first note that we cannot determine convergence or divergence of the series by comparison with a p-series. That is,

$$\frac{1}{n \ln n} < \frac{1}{n} \quad \text{for } n \geq 3,$$

but it doesn't help to know that the terms are less than those of the divergent series $\displaystyle\sum_{n=1}^\infty \frac{1}{n}$. On the other hand, if $p > 1$, we have

$$\frac{1}{n \ln n} > \frac{1}{n^p} \quad \text{for all sufficiently large } n.$$

However, it doesn't help to know that the terms are greater than those of a convergent series.

We note that the nth term of the series is the value at $x = n$ of the function $f(x) = \dfrac{1}{x \ln x}$, which is positive, decreasing, and continuous on the interval $x \geq 2$. The Integral Test then tells us that the series and the improper integral $\int_2^\infty f(x)\,dx$ are either both convergent or both divergent.

Using the substitution $u = \ln x$, so $du = \dfrac{1}{x}\,dx$, we have

$$\int \frac{1}{x \ln x}\,dx = \int \frac{1}{u}\,du = \ln |u| + C = \ln |\ln x| + C.$$

Then

$$\lim_{b \to \infty} \int_2^b \frac{1}{x \ln x} \, dx = \lim_{b \to \infty} (\ln |\ln b| - \ln |\ln 2|) = \infty,$$

so the improper integral is divergent. It follows that the corresponding series is also divergent. ■

Combining the Comparison and Integral Tests

We cannot apply the Integral Test unless we can integrate the corresponding function f. However, we can sometimes use the Limit Version of the Comparison Test to replace the given series with a series for which the corresponding function is easier to integrate.

EXAMPLE 2

Determine convergence or divergence of the series $\displaystyle\sum_{n=2}^{\infty} \frac{n}{(n^2 - 5)(\ln n)^2}$.

SOLUTION   The first step is to determine a simpler series that converges or diverges as does the given series. To do this, we factor n^2 from the quadratic expression in the denominator to obtain

$$a_n = \frac{n}{(n^2 - 5)(\ln n)^2} = \frac{n}{n^2(1 - (5/n^2))(\ln n)^2} = \frac{1}{n(1 - (5/n^2))(\ln n)^2}.$$

This indicates that we could compare $\sum a_n$ with the series $\sum b_n$, where

$$b_n = \frac{1}{n(\ln n)^2}.$$

With this choice of b_n,

$$\lim_{n \to \infty} \frac{a_n}{b_n} = 1,$$

so the sequences are either both convergent or both divergent. We can use the integral test to determine the convergence or divergence of $\sum b_n$. We have

$$b_n = f(n), \quad \text{where } f(x) = \frac{1}{x(\ln x)^2}.$$

The function f is positive, decreasing, and continuous on the interval $x \geq 2$, so the conditions of the Integral Test are satisfied. Using the substitution $u = \ln x$, so $du = \frac{1}{x} dx$, $u(2) = \ln 2$, and $u(n) = \ln n$, we have

$$\int_2^n \frac{1}{x(\ln x)^2} \, dx = \int_{\ln 2}^{\ln n} u^{-2} \, du$$

$$= -\frac{1}{u}\Bigg]_{\ln 2}^{\ln n} = \frac{1}{\ln 2} - \frac{1}{\ln n} \to \frac{1}{\ln 2} \text{ as } n \to \infty,$$

so the improper integral is convergent. It follows that $\displaystyle\sum_{n=1}^{\infty} \frac{1}{n(\ln n)^2}$ and, hence,

$$\sum_{n=1}^{\infty} \frac{n}{(n^2 - 5)(\ln n)^2} \text{ are convergent. } ■$$

Accuracy of Approximation

Let us now see how to use the Integral Test to obtain bounds of error in using S_n to approximate the sum of a series $\sum\limits_{n=1}^{\infty} a_n$. We assume the Integral Test applies, so there is a function f that is nonnegative, nonincreasing, and continuous on an interval $x \geq N$, with $a_n = f(n)$ for $n \geq N$. We also assume the improper integral $\int_{N}^{\infty} f(x)\,dx$ is convergent, so the series is convergent.

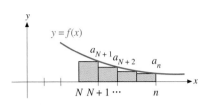

FIGURE 9.4.2 $S_n - S_N$ is the sum of the areas of the rectangles, so $S_n - S_N \leq \int_{N}^{n} f(x)\,dx$, $n > N$.

From Figure 9.4.2, we see that

$$0 \leq S_n - S_N = a_{N+1} + a_{N+2} + \cdots + a_n \leq \int_{N}^{n} f(x)\,dx, \quad n > N.$$

Then

$$0 \leq S - S_N = \lim_{n \to \infty} (S_n - S_N) \leq \lim_{n \to \infty} \int_{N}^{n} f(x)\,dx = \int_{N}^{\infty} f(x)\,dx.$$

We conclude that if the conditions of the Integral Test are satisfied, then

$$0 \leq S - S_N \leq \int_{N}^{\infty} f(x)\,dx. \tag{1}$$

This inequality gives a bound of the error in using S_N to approximate the sum S.

EXAMPLE 3

Approximate $S = \sum\limits_{n=1}^{\infty} e^{-n}$ by a partial sum S_N with $0 \leq S - S_N \leq 0.005$.

SOLUTION   The series is a convergent geometric series with ratio $r = e^{-1}$. Also, the Integral Test applies with $f(x) = e^{-x}$. Then

$$0 \leq S - S_N \leq \int_{N}^{\infty} e^{-x}\,dx = \lim_{n \to \infty} \int_{N}^{n} e^{-x}\,dx = \lim_{n \to \infty} \left(e^{-N} - e^{-n}\right) = e^{-N}.$$

We then see that

$$0 \leq S - S_N \leq 0.005 \text{ whenever } e^{-N} < 0.005, \text{ or } N > -\ln 0.005 \approx 5.3.$$

Choosing $N = 6$, the smallest integer that satisfies $N > 5.3$, we obtain the approximate value

$$S_6 = e^{-1} + e^{-2} + e^{-3} + e^{-4} + e^{-5} + e^{-6} \approx 0.5805.$$

In this example we know that the geometric series has sum

$$\sum_{n=1}^{\infty} e^{-n} = e^{-1} + e^{-2} + e^{-3} + e^{-4} + e^{-5} + e^{-6} + \cdots$$

$$= e^{-1}\left(1 + e^{-1} + e^{-2} + e^{-3} + e^{-4} + e^{-5} + \cdots\right)$$

$$= e^{-1}\left(1 + \left(e^{-1}\right) + \left(e^{-1}\right)^2 + \left(e^{-1}\right)^3 + \left(e^{-1}\right)^4 + \left(e^{-1}\right)^5 + \cdots\right)$$

$$= \frac{1}{e}\left(\frac{1}{1-e^{-1}}\right) = \frac{1}{e-1} \approx 0.581976706,$$

so we see that our approximate value is within 0.005 of the exact value. ■

EXAMPLE 4

Approximate $S = \sum_{n=1}^{\infty} \frac{1}{n^3}$ by a partial sum S_N with $0 \leq S - S_N \leq 0.01$.

SOLUTION The series is a convergent p-series, so we know that the sum S is a real number. We will use the integral test to determine the accuracy of using S_N as an approximate value of S. The conditions of the Integral Test apply with $f(x) = x^{-3}$. We then have

$$0 \leq S - S_N \leq \int_N^{\infty} x^{-3}dx = \lim_{n\to\infty} \int_N^n x^{-3}dx = \lim_{n\to\infty}\left(\frac{1}{2N^2} - \frac{1}{2n^2}\right) = \frac{1}{2N^2}.$$

It follows that

$$0 \leq S - S_N \leq 0.01 \text{ whenever } \frac{1}{2N^2} < 0.01 \text{ or } N^2 > 50.$$

The smallest integer N to satisfy this condition is $N = 8$. The desired approximate value of S is

$$S_8 = \frac{1}{1^3} + \frac{1}{2^3} + \frac{1}{3^3} + \frac{1}{4^3} + \frac{1}{5^3} + \frac{1}{6^3} + \frac{1}{7^3} + \frac{1}{8^3} \approx 1.195.$$

(We have used a calculator to evaluate S_8. Note that we need to be concerned about round-off error if N is large.) ■

EXAMPLE 5

Determine values of N for which the partial sum S_N of $S = \sum_{n=1}^{\infty} \frac{1}{n^2+1}$ is guaranteed by Inequality 1 to satisfy $0 \leq S - S_N \leq 0.005$.

SOLUTION The Comparison Test can be used to show that this series is convergent. The Integral Test applies with $f(x) = \frac{1}{x^2+1}$. Then

$$0 \leq S - S_N \leq \int_N^{\infty} \frac{1}{x^2+1}dx = \lim_{n\to\infty} \int_N^n \frac{1}{x^2+1}dx$$

$$= \lim_{n\to\infty}\left(\tan^{-1} n - \tan^{-1} N\right) = \frac{\pi}{2} - \tan^{-1} N.$$

Inequality 1 then guarantees the desired accuracy whenever

$$\frac{\pi}{2} - \tan^{-1} N < 0.005.$$

Solving this inequality for N, we obtain

$$N > \tan\left(\frac{\pi}{2} - 0.005\right) \approx 199.998.$$

We are sure to have $0 \leq S - S_N \leq 0.005$ if $N \geq 200$. ■

EXERCISES 9.4

Use either a Comparison Test, the Integral Test, or a combination of the tests to determine convergence or divergence of each of the series in Exercises 1–18.

1. $\displaystyle\sum_{n=1}^{\infty} \frac{n^{1/2}}{n^{17/12} + 1}$

2. $\displaystyle\sum_{n=1}^{\infty} \frac{n^{1/3}}{n^{17/12} + 1}$

3. $\displaystyle\sum_{n=1}^{\infty} \frac{n}{n^2 + 1}$

4. $\displaystyle\sum_{n=1}^{\infty} \frac{n^2}{n^3 + 1}$

5. $\displaystyle\sum_{n=1}^{\infty} \frac{\ln n}{n^2}$

6. $\displaystyle\sum_{n=1}^{\infty} \frac{\ln n}{2n - 1}$

7. $\displaystyle\sum_{n=1}^{\infty} \sin \frac{1}{n}$

8. $\displaystyle\sum_{n=1}^{\infty} \cos \frac{1}{n}$

9. $\displaystyle\sum_{n=2}^{\infty} \frac{1}{n\sqrt{\ln n}}$

10. $\displaystyle\sum_{n=2}^{\infty} \frac{1}{n(\ln n)^2}$ (**Calculus Explorer & Tutor II, Sequences and Series, 7.**)

11. $\displaystyle\sum_{n=3}^{\infty} \frac{1}{n(\ln n)(\ln \ln n)^2}$

12. $\displaystyle\sum_{n=3}^{\infty} \frac{1}{n(\ln n)(\ln \ln n)}$

13. $\displaystyle\sum_{n=1}^{\infty} \frac{1}{n \ln (n + 1)}$

14. $\displaystyle\sum_{n=1}^{\infty} \frac{n}{(n^2 + 1) \ln (2n + 1)}$

15. $\displaystyle\sum_{n=1}^{\infty} n e^{-n}$

16. $\displaystyle\sum_{n=1}^{\infty} n^2 e^{-n}$

17. $\displaystyle\sum_{n=1}^{\infty} n e^{-n^2}$

18. $\displaystyle\sum_{n=1}^{\infty} n e^{-n^3}$

19. Determine values of p for which $\displaystyle\sum_{n=3}^{\infty} \frac{1}{n(\ln n)(\ln \ln n)^p}$ is convergent.

20. Determine values of p for which $\displaystyle\sum_{n=2}^{\infty} \frac{1}{n(\ln n)^p}$ is convergent.

21. Approximate $S = \displaystyle\sum_{n=1}^{\infty} n^{-4}$ by a partial sum S_N with $0 \le S - S_N \le 0.01$.

22. Approximate $S = \displaystyle\sum_{n=1}^{\infty} e^{-n}$ by a partial sum S_N with $0 \le S - S_N \le 0.01$.

23. Approximate $S = \displaystyle\sum_{n=1}^{\infty} n e^{-n^2}$ by a partial sum S_N with $0 \le S - S_N \le 0.01$.

24. Approximate $S = \displaystyle\sum_{n=1}^{\infty} n e^{-n}$ by a partial sum S_N with $0 \le S - S_N \le 0.01$.

25. Determine values of N for which the partial sum S_N of $S = \displaystyle\sum_{n=1}^{\infty} \frac{1}{n^2 + 1}$ is guaranteed by Inequality 1 to satisfy $0 \le S - S_N \le 0.001$.

26. Determine values of N for which the partial sum S_N of $S = \displaystyle\sum_{n=1}^{\infty} \frac{1}{(n + 1)[\ln(n + 1)]^2}$ is guaranteed by Inequality 1 to satisfy $0 \le S - S_N \le 0.1$.

27. Use an appropriate integral to obtain an upper bound of the error $S - S_N$ for the series $\displaystyle\sum_{n=1}^{\infty} \frac{5}{(2n + 1)(3n - 1)}$. Evaluate the error bound for $N = 10, 100$, and 1000. Note that $S - S_N \approx \dfrac{C}{N}$.

28. Use an appropriate integral to obtain an upper bound of the error $S - S_N$ for the series $\displaystyle\sum_{n=1}^{\infty} \frac{2}{n(n + 1)(n + 2)}$. Evaluate the error bound for $N = 10, 100$, and 1000. Note that $S - S_N \approx \dfrac{C}{N^2}$.

29. The function $f(x) = 1 - \cos 2\pi x + \dfrac{1}{x^2}$ satisfies $f(n) = \dfrac{1}{n^2}$ for each positive integer n. We know that $\displaystyle\sum_{n=1}^{\infty} \frac{1}{n^2}$ is convergent. Show that the improper integral $\displaystyle\int_{1}^{\infty} f(x)\, dx$ is divergent. What condition of the Integral Test is not satisfied by f?

30. If $0 < r < 1$, we know that $S = \displaystyle\sum_{n=1}^{\infty} r^n = \frac{r}{1 - r}$ and $S_N = r + r^2 + r^3 + \cdots + r^N = \dfrac{r - r^{N+1}}{1 - r}$, so $S - S_N = \dfrac{r^{N+1}}{1 - r}$. What is the bound of $S - S_N$ given by the Integral Test? $\left(\text{Hint: } \displaystyle\int r^x\, dx = \frac{r^x}{\ln r} + C.\right)$

31. A basketball arena is to be constructed at a large midwestern university. The rows are to be one meter apart horizontally and tiered so the line of sight from each row to midcourt clears the point δ meters above the eyes of the person in the next lower row. Let (n, y_n) represent the point of sight from the nth row, $n \ge 1$. Suppose the center of the court is located at the point $(-d, 0)$, and that the line of sight from $(1, y_1)$

to $(-d, 0)$ passes through the point $(0, \delta)$, as indicated in the figure.

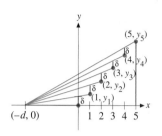

(a) Show that

$$y_n = \delta(n + d)\left(\frac{1}{d} + \frac{1}{d+1} + \frac{1}{d+2} + \cdots + \frac{1}{d+n-1}\right),$$
$$n \geq 1.$$

(b) Show that the approximate height of the rows is given by the function

$$f(x) = \delta(x + d)\ln\left(\frac{x + d}{d}\right).$$

9.5 RATIO AND ROOT TESTS; SUMMARY

Connections

Indeterminate forms, 7.5.

We develop additional tests that aid us in determining the convergence or divergence of infinite series that have positive terms.

The Ratio Test

THE RATIO TEST

If $a_n > 0$ for all n, then

$$\sum_{n=1}^{\infty} a_n \text{ is convergent if } \lim_{n \to \infty} \frac{a_{n+1}}{a_n} < 1;$$

$$\sum_{n=1}^{\infty} a_n \text{ is divergent if } \lim_{n \to \infty} \frac{a_{n+1}}{a_n} > 1.$$

Proof The proof of the Ratio Test involves comparison with a geometric series. For example,

$$\lim_{n \to \infty} \frac{a_{n+1}}{a_n} = \rho < 1$$

implies that the ratio approaches ρ as n approaches infinity. This means there are numbers r and N such that

$$\rho < r < 1 \text{ and } \frac{a_{n+1}}{a_n} < r \quad \text{for } n \geq N.$$

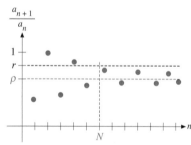

FIGURE 9.5.1 If $\lim_{n \to \infty} \frac{a_{n+1}}{a_n} = \rho < 1$, then for fixed r, $\rho < r < 1$, there is a number N such that $\frac{a_{n+1}}{a_n} < r$ whenever $n \geq N$.

See Figure 9.5.1. We then have $a_{n+1} < ra_n$ for all $n \geq N$, so

$$a_{N+1} < a_N r,$$
$$a_{N+2} < a_{N+1}r < a_N r^2,$$
$$a_{N+3} < a_{N+2}r < a_N r^3,$$
$$\vdots$$
$$a_{N+k} < a_N r^k.$$

Since $a_{N+k} < a_N r^k$ for $k \geq 1$ and $\displaystyle\sum_{k=1}^{\infty} a_N r^k$ is a convergent geometric series

for $0 < r < 1$, the Comparison Test implies $\displaystyle\sum_{k=1}^{\infty} a_{N+k}$ is convergent, so

$$\sum_{n=1}^{\infty} a_n = a_1 + a_2 + \cdots + a_N + \sum_{k=1}^{\infty} a_{N+k}$$

is also convergent.

If $a_n > 0$ for all n, then $\displaystyle\lim_{n \to \infty} \frac{a_{n+1}}{a_n} > 1$ implies a_n does not approach zero

as n approaches infinity, so the series $\displaystyle\sum_{n=1}^{\infty} a_n$ must be divergent. ∎

The Ratio Test gives no information if $\displaystyle\lim_{n \to \infty} \frac{a_{n+1}}{a_n} = 1.$ *The series* $\displaystyle\sum_{n=1}^{\infty} a_n$ *may*

either converge or diverge in this case.

For example, both of the series

$$\sum_{n=1}^{\infty} n^{-2} \text{ and } \sum_{n=1}^{\infty} n^{-1/2} \text{ have } \lim_{n \to \infty} \frac{a_{n+1}}{a_n} = 1.$$

The first series is convergent and the second is divergent.

Generally, the Ratio Test is not useful for series with terms that involve only algebraic expressions of n.

Series with Factorials and/or *n*th Powers of Constants

The Ratio Test is very useful for series with terms that contain factorials and/or nth powers of constants.

EXAMPLE 1

Determine convergence or divergence of the series $\displaystyle\sum_{n=1}^{\infty} \frac{2^n}{n!}$.

SOLUTION   We see that the terms of this series involves factorials and *n*th powers of two. This suggests that we use the Ratio Test. We have

$$a_n = \frac{2^n}{n!}, \text{ so } a_{n+1} = \frac{2^{n+1}}{(n+1)!}.$$

Since

$$(n+1)! = (1)(2) \cdots (n)(n+1) = n!(n+1),$$

we have

$$\frac{a_{n+1}}{a_n} = a_{n+1}\frac{1}{a_n} = \frac{2^{n+1}}{(n+1)!}\frac{n!}{2^n} = \frac{2^{n+1}}{n!(n+1)}\frac{n!}{2^n}$$

$$= \frac{2}{n+1} \to 0 < 1 \text{ as } n \to \infty.$$

The Ratio Test then implies convergence. ∎

EXAMPLE 2

Determine convergence or divergence of the series $\sum_{n=1}^{\infty} n^2 e^{-n}$.

SOLUTION   The terms of this series involve nth powers of e^{-1}, which suggests we try the Ratio Test. We have

$$a_n = n^2 e^{-n}, \text{ so } a_{n+1} = (n+1)^2 e^{-(n+1)}.$$

Then

$$\frac{a_{n+1}}{a_n} = \frac{(n+1)^2 e^{-(n+1)}}{n^2 e^{-n}} = \left(\frac{n+1}{n}\right)^2 \frac{1}{e} \to \frac{1}{e} < 1 \text{ as } n \to \infty.$$

Convergence follows from the Ratio Test.   ■

In Section 9.3 we showed that the series in Example 2 can be shown to be convergent by comparing it to either the convergent geometric series $\sum_{n=1}^{\infty} 2^{-n/2}$ or a convergent p-series. The use of the Ratio Test in Example 2 should seem much more straightforward than using the Comparison Test for the series.

EXAMPLE 3

Determine convergence or divergence of the series $\sum_{n=1}^{\infty} \frac{5^n}{(2n-1)!}$.

SOLUTION   The nth power of five and the factorial suggest the Ratio Test. We have

$$a_n = \frac{5^n}{(2n-1)!}, \text{ so } a_{n+1} = \frac{5^{(n+1)}}{(2(n+1)-1)!} = \frac{5^{n+1}}{(2n+1)!}.$$

We have

$$(2n+1)! = (1)(2) \cdots (2n-1)(2n)(2n+1) = (2n-1)!(2n)(2n+1).$$

Then

$$\frac{a_{n+1}}{a_n} = \frac{5^{n+1}}{(2n+1)!} \frac{(2n-1)!}{5^n} = \frac{5^{n+1}}{(2n-1)!(2n)(2n+1)} \frac{(2n-1)!}{5^n}$$

$$= \frac{5}{2n(2n+1)} \to 0 < 1 \text{ as } n \to \infty.$$

The Ratio Test then implies convergence.   ■

EXAMPLE 4

Determine convergence or divergence of the series $\sum_{n=1}^{\infty} \frac{2^n}{n^2}$.

SOLUTION   The factor 2^n suggests the Ratio Test. We have

$$a_n = \frac{2^n}{n^2}, \text{ so } a_{n+1} = \frac{2^{n+1}}{(n+1)^2}.$$

Then

$$\frac{a_{n+1}}{a_n} = \frac{2^{n+1}}{(n+1)^2} \frac{n^2}{2^n} \to 2 > 1 \text{ as } n \to \infty.$$

The Ratio Test implies divergence.   ■

The Root Test

THE ROOT TEST

If $a_n \geq 0$ for all n, then

$$\sum_{n=1}^{\infty} a_n \text{ is convergent if } \lim_{n \to \infty} (a_n)^{1/n} < 1;$$

$$\sum_{n=1}^{\infty} a_n \text{ is divergent if } \lim_{n \to \infty} (a_n)^{1/n} > 1.$$

The proof of the Root Test is similar to that of the Ratio Test. Note that $0 \leq (a_n)^{1/n} < r$ implies $a_n < r^n$. As in the case of the Ratio Test, we can conclude nothing from the Root Test when the limit of the nth root of a_n is one.

The Root Test is useful for series with terms that involve variable expressions raised to the nth power.

EXAMPLE 5

Determine convergence or divergence of the series $\displaystyle\sum_{n=1}^{\infty} \frac{(\ln n)^{2n}}{n^n}$.

SOLUTION   The term of this series are nth powers, which suggests the Root Test. We have

$$a_n = \frac{(\ln n)^{2n}}{n^n}, \text{ so } (a_n)^{1/n} = \frac{(\ln n)^2}{n}.$$

Since $\ln n$ is dominated by any positive power of n as n approaches infinity, we have

$$\lim_{n \to \infty} (a_n)^{1/n} = \lim_{n \to \infty} \frac{(\ln n)^2}{n} = 0 < 1.$$

The Root Test then implies convergence. ∎

Summary of Convergence Tests

Strategy for determining convergence or divergence of series with positive terms

- If a_n does not approach zero, then $\displaystyle\sum_{n=1}^{\infty} a_n$ diverges. This condition may be discovered while using either the Comparison, Ratio, or Root Test. Remember that a series may diverge even though its nth term does approach zero. Showing that a_n approaches zero is no help in determining if the series is convergent or divergent.
- If the terms are either less than those of a known convergent series or greater than those of a known divergent series, use the Comparison Test.

- If the terms involve only powers and roots, use the Limit Version of the Comparison Test. Factor the highest power of n from each of the numerator and denominator to obtain $a_n = \dfrac{1}{n^p} b_n$, where b_n has finite, nonzero limit as n approaches infinity. The p-series $\displaystyle\sum_{n=1}^{\infty} \dfrac{1}{n^p}$ converges if $p > 1$ and diverges if $p \leq 1$.
- If the terms contain factorials and/or nth powers of constants, use the Ratio Test. $\displaystyle\sum_{n=1}^{\infty} a_n$ converges if $\displaystyle\lim_{n\to\infty} \dfrac{a_{n+1}}{a_n} < 1$ and diverges if $\displaystyle\lim_{n\to\infty} \dfrac{a_{n+1}}{a_n} > 1$. The Ratio Test gives no information if the limit of a_{n+1}/a_n is equal to one.
- If the terms involve variable expressions raised to the nth power, the Root Test could be helpful. $\displaystyle\sum_{n=1}^{\infty} a_n$ converges if $\displaystyle\lim_{n\to\infty} (a_n)^{1/n} < 1$ and diverges if $\displaystyle\lim_{n\to\infty} (a_n)^{1/n} > 1$. The Root Test gives no information if the limit of $(a_n)^{1/n}$ is equal to one. It is useful to note that $\displaystyle\lim_{n\to\infty} (an + b)^{1/n} = 1$, unless a and b are both zero.
- If we can integrate the function f with $f(n) = a_n$, we might consider the Integral Test. We can use the Limit Version of the Comparison Test to replace the given series with a series for which the corresponding function is easier to integrate.

EXERCISES 9.5

Determine convergence or divergence for each of the series in Exercises 1–32.

1. $\displaystyle\sum_{n=1}^{\infty} \dfrac{n}{2^n}$

2. $\displaystyle\sum_{n=1}^{\infty} \dfrac{3^n}{n!}$

3. $\displaystyle\sum_{n=1}^{\infty} \dfrac{e^n}{n}$

4. $\displaystyle\sum_{n=1}^{\infty} \dfrac{1}{n^n}$

5. $\displaystyle\sum_{n=1}^{\infty} \dfrac{n^2 2^n}{(2n-1)!}$

6. $\displaystyle\sum_{n=1}^{\infty} \dfrac{5^{4n}}{(n!)^2}$ **(Calculus Explorer & Tutor II, Sequences and Series, 6.)**

7. $\displaystyle\sum_{n=2}^{\infty} \dfrac{1}{(\ln n)^n}$

8. $\displaystyle\sum_{n=1}^{\infty} \left(\dfrac{n}{2n-1}\right)^{2n}$

9. $\displaystyle\sum_{n=1}^{\infty} \dfrac{n!}{2^n}$

10. $\displaystyle\sum_{n=1}^{\infty} \dfrac{2n+1}{3n^3+n}$

11. $\displaystyle\sum_{n=1}^{\infty} \dfrac{2n}{n^2+1}$

12. $\displaystyle\sum_{n=2}^{\infty} \dfrac{1}{\sqrt{n}\ln n}$

13. $\displaystyle\sum_{n=2}^{\infty} \dfrac{\ln n}{n}$

14. $\displaystyle\sum_{n=1}^{\infty} \dfrac{e^n}{(n+1)^n}$

15. $\displaystyle\sum_{n=1}^{\infty} \dfrac{(1)(3)\cdots(2n-1)}{(3)(6)\cdots(3n)}$

16. $\displaystyle\sum_{n=1}^{\infty} \dfrac{(2)(4)\cdots(2n)}{(2)(5)\cdots(3n-1)}$

17. $\displaystyle\sum_{n=1}^{\infty} \dfrac{3^n}{[1+\cos(1/n)]^n}$

18. $\displaystyle\sum_{n=1}^{\infty} \dfrac{2n}{n+1}$

19. $\displaystyle\sum_{n=1}^{\infty} \dfrac{n!n^n}{(2n)!}$

20. $\displaystyle\sum_{n=1}^{\infty} \dfrac{n!n^{2n}}{(2n)!}$

21. $\displaystyle\sum_{n=1}^{\infty} \dfrac{\sqrt{n}}{n+4}$

22. $\displaystyle\sum_{n=1}^{\infty} \dfrac{\sqrt{n}}{n^2+4}$

23. $\displaystyle\sum_{n=1}^{\infty} \left(1-\cos\dfrac{1}{n}\right)$

24. $\displaystyle\sum_{n=1}^{\infty} (e^{1/n}-1)$

25. $\displaystyle\sum_{n=1}^{\infty} (0.5)^n$

26. $\displaystyle\sum_{n=1}^{\infty} (1.02)^n$

27. $\displaystyle\sum_{n=1}^{\infty} \left(1-\dfrac{1}{n}\right)^n$

28. $\displaystyle\sum_{n=1}^{\infty} \left(1-\dfrac{1}{n}\right)^{n^2}$

29. $\displaystyle\sum_{n=3}^{\infty} \dfrac{1}{n(\ln n)(\ln\ln n)^3}$

30. $\displaystyle\sum_{n=3}^{\infty} \dfrac{1}{n(\ln n)\sqrt{\ln\ln n}}$

31. $\displaystyle\sum_{n=1}^{\infty} \dfrac{2}{e^n+e^{-n}}$

32. $\displaystyle\sum_{n=1}^{\infty} \dfrac{3^n}{2^n+1}$

33. Determine values of p for which $\displaystyle\sum_{n=1}^{\infty} n^p e^{-n}$ is convergent.

9.6 ALTERNATING SERIES; ABSOLUTE AND CONDITIONAL CONVERGENCE

Connections

Summary for series with positive terms, 9.5.

In this section we will study series for which successive terms are of opposite sign. We will establish a test for convergence and a simple method of determining a bound of error in using partial sums to approximate certain series of this type. We will see that some series may be convergent, although the series formed by the absolute values of the terms may be divergent.

The Alternating Series Test

> **DEFINITION**
>
> A series for which successive terms are of opposite sign is called an **alternating series**.

For example,

- $\displaystyle\sum_{n=1}^{\infty} \frac{(-1)^n}{n} = -1 + \frac{1}{2} - \frac{1}{3} + \frac{1}{4} - \frac{1}{5} + \frac{1}{6} - \cdots$ is an alternating series.

- $\displaystyle\sum_{n=1}^{\infty} \frac{1}{n} = 1 + \frac{1}{2} + \frac{1}{3} + \frac{1}{4} + \frac{1}{5} + \frac{1}{6} + \cdots$ is not an alternating series.

- $1 + \dfrac{1}{2} - \dfrac{1}{3} - \dfrac{1}{4} + \dfrac{1}{5} + \dfrac{1}{6} - \cdots$ is not an alternating series, although the *different* series obtained by grouping successive positive terms and successive negative terms is an alternating series. That is,

- $\left(1 + \dfrac{1}{2}\right) - \left(\dfrac{1}{3} + \dfrac{1}{4}\right) + \left(\dfrac{1}{5} + \dfrac{1}{6}\right) - \cdots$ is an alternating series.

It is very important to recognize alternating series, since this is an important factor with regard to the convergence or divergence of the series, as the following theorem indicates.

> **THE ALTERNATING SERIES TEST**
>
> If there is a number n_0 such that $\displaystyle\sum_{n=n_0}^{\infty} a_n$ is an alternating series that satisfies the conditions
>
> (i) $\displaystyle\lim_{n\to\infty} |a_n| = 0$ and
>
> (ii) $|a_{n+1}| < |a_n|$ for all $n \geq n_0$,
>
> then the series $\displaystyle\sum_{n=1}^{\infty} a_n$ is convergent. Moreover, if $\displaystyle\sum_{n=1}^{\infty} a_n = S$, then
>
> $$|S - S_N| < |a_{N+1}|, \quad \text{for all } N \geq n_0.$$

Let us see why this theorem is true. The conditions of the Alternating Series Test imply that S_{N+k} is between S_N and $S_{N+1} = S_N + a_{N+1}$ for all $N \geq n_0$ and all $k > 1$. That is, adding a_{N+1} to S_N gives S_{N+1}. Since a_{N+2} has sign opposite to that of a_{N+1} and $|a_{N+2}| < |a_{N+1}|$, adding a_{N+2} to S_{N+1} gives a value of

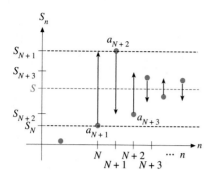

FIGURE 9.6.1 If $\sum_{n=1}^{\infty} a_n$ is an alternating series with (i) $\lim_{n\to\infty} |a_n| = 0$ and (ii) $|a_{n+1}| < |a_n|$ for $n \geq N$, then the series converges to a value S that is between S_N and S_{N+1}, so $|S_N - S| < |a_{N+1}|$.

S_{N+2} that is between S_N and S_{N+1}. Similarly, adding a_{N+3} to S_{N+2} gives a value of S_{N+3} that is between S_{N+1} and S_{N+2}, and so forth. See Figure 9.6.1. It is geometrically evident that $\lim_{n\to\infty} |a_n| = 0$ then implies S_n converges to a point S that is within $|a_{N+1}|$ of S_N.

Using the Alternating Series Test

Later in this section we will see how to use the condition $|S - S_N| < |a_{N+1}|$ to determine a bound of the error in using S_N to approximate the sum of a series that satisfies the conditions of the Alternating Series Test. For now, let us see how to use the test to determine convergence or divergence.

To apply the Alternating Series Test

- First check that successive terms of the series are of opposite sign, at least for all terms a_n with n greater than or equal to some number n_0.
- Verify (i) $\lim_{n\to\infty} |a_n| = 0$.
- Finally, verify (ii) $|a_{n+1}| < |a_n|$ for all n greater than or equal to some number n_0.

It is useful to note that we can use a derivative to establish Condition (ii):

Condition (ii), $|a_{n+1}| < |a_n|$ for all $n \geq n_0$, is satisfied if there is a function f with $f(n) = |a_n|$ for all $n \geq n_0$ and $f'(x) < 0$ for $x \geq n_0$.

EXAMPLE 1

Determine convergence or divergence of the series $\sum_{n=1}^{\infty} \dfrac{(-1)^n}{n^{1/2}}$.

SOLUTION We first note that successive terms have opposite signs, so the series is alternating. The values $|a_n| = n^{-1/2}$ satisfy

(i) $\lim_{n\to\infty} n^{-1/2} = 0$ and

(ii) $(n+1)^{-1/2} < n^{-1/2}$.

The Alternating Series Test then implies convergence. ∎

EXAMPLE 2

Determine convergence or divergence of the series $\sum_{n=1}^{\infty} \dfrac{(-1)^n (\ln n)}{n}$.

SOLUTION This is an alternating series. We have

$$|a_n| = \frac{\ln n}{n}.$$

(i) $\lim_{n\to\infty} \dfrac{\ln n}{n} \underset{(\infty/\infty, \text{l'H})}{=} \lim_{n\to\infty} \dfrac{1/n}{1} = 0$. To check condition (ii), we set

$$f(x) = \frac{\ln x}{x}.$$

Let us use the derivative of f to determine intervals on which f is decreasing. We have

$$f'(x) = \frac{(x)(1/x) - (\ln x)(1)}{x^2}.$$

We see that $f'(x) < 0$, so f is decreasing, for $1 - \ln x < 0$, or $x > e$. It follows that

(ii) $\dfrac{\ln(n+1)}{n+1} < \dfrac{\ln n}{n}$ for $n \geq 3 > e$, which is enough.

The Alternating Series Test implies convergence. ∎

EXAMPLE 3

Determine convergence or divergence of the series $\displaystyle\sum_{n=1}^{\infty} \frac{n(-1)^n}{n+1}$.

SOLUTION   This is an alternating series. We have

$$|a_n| = \frac{n}{n+1} \to 1 \neq 0 \text{ as } n \to \infty.$$

Since the terms do not approach zero, the series must diverge. ∎

Accuracy of Approximation

Let us now see how to use the inequality $|S - S_N| < |a_{N+1}|$ to determine a bound of the error in using S_N to approximate the sum of a series that satisfies the conditions of the Alternating Series Test.

EXAMPLE 4

Approximate $\displaystyle\sum_{n=0}^{\infty} \frac{(-1)^n}{n!}$ with error less than 0.005.

SOLUTION   We first note that the Alternating Series Test applies and implies convergence. We then write out some terms of the series,

$$S = 1 - 1 + \frac{1}{2!} - \frac{1}{3!} + \frac{1}{4!} - \frac{1}{5!} + \frac{1}{6!} - \frac{1}{7!} + \cdots.$$

Checking the absolute values of successive terms of the series, we see that $1/6! \approx 0.001388888$ is the first term that is less than the acceptable error of 0.005. We have $|S - S_N| < |a_{N+1}| < 0.05$, so

$$S \approx S_6 = 1 - 1 + \frac{1}{2!} - \frac{1}{3!} + \frac{1}{4!} - \frac{1}{5!} \approx 0.3667$$

with error less than 0.005. (We will see in Section 9.9 that the series has sum $e^{-1} \approx 0.367879441$.) ∎

EXAMPLE 5

Determine values of N for which the partial sum S_N of $S = \displaystyle\sum_{n=1}^{\infty} \frac{(-1)^n}{n}$ is guaranteed by the Alternating Series Test to be within 0.05 of S.

SOLUTION   The Alternating Series Test applies and implies convergence. We have $|a_n| = \dfrac{1}{n}$, so

$$|S - S_N| < |a_{N+1}| = \frac{1}{N+1} \leq 0.05 \text{ whenever } N + 1 \geq \frac{1}{0.05} \text{ or } N \geq 19.$$

If $N \geq 19$, then the Alternating Series Test guarantees that S_N is within 0.05 of S. (We will see in Section 9.8, Exercise 41, that $\displaystyle\sum_{n=1}^{\infty} \frac{(-1)^n}{n} = -\ln 2$.) ∎

Absolute and Conditional Convergence

Let us now see how the investigation of the series formed by the absolute values of the terms of a series can give useful information about the original series.

DEFINITION

A series $\sum_{n=1}^{\infty} a_n$ is said to be **absolutely convergent** if the series $\sum_{n=1}^{\infty} |a_n|$ is convergent.

EXAMPLE 6

Show that $\sum_{n=1}^{\infty} (-1)^n n^{-2}$ is absolutely convergent.

SOLUTION We have $a_n = (-1)^n n^{-2}$, so

$$|a_n| = n^{-2} = \frac{1}{n^p}, \quad \text{where } p = 2 > 1.$$

We see that $\sum_{n=1}^{\infty} |a_n|$ is a convergent p-series, so $\sum_{n=1}^{\infty} a_n$ is absolutely convergent. ∎

Note that the Alternating Series Test shows that the series in Example 6 is convergent. The following general result is true.

THEOREM

If $\sum_{n=1}^{\infty} a_n$ is absolutely convergent, then the series is convergent.

Proof Let $b_n = |a_n| - a_n$. Note that $b_n = 0$ if $a_n \geq 0$, and $b_n = 2|a_n|$ if $a_n < 0$. In either case we have $0 \leq b_n \leq 2|a_n|$. The Comparison Test for series with nonnegative terms then shows that $\sum_{n=1}^{\infty} b_n$ is convergent, since $\sum_{n=1}^{\infty} |a_n|$ is convergent. It follows that

$$\sum_{n=1}^{\infty} a_n = \sum_{n=1}^{\infty} (|a_n| - b_n)$$

is convergent. ∎

Let us consider the above theorem from a different point of view. If S_n denotes the nth partial sum of the series $\sum_{n=1}^{\infty} a_n$, then the Triangle Inequality implies that a difference of partial sums satisfies

$$|S_n - S_N| = |a_{N+1} + a_{N+2} + \cdots + a_n|$$
$$\leq |a_{N+1}| + |a_{N+2}| + \cdots + |a_n|, \quad n > N.$$

The sum on the right is the corresponding difference of partial sums of the series $\sum_{n=1}^{\infty} |a_n|$. Thus, the partial sums of $\sum_{n=1}^{\infty} a_n$ differ most if the terms are all of the same sign. Differences of sign tend to make the partial sums closer to each other, and make convergence more likely.

To determine convergence of a series, we should first check for absolute convergence. We can use the tests we have developed for convergence of series with nonnegative terms to test absolute convergence. If it is absolutely convergent, we know it is also convergent. Absolute convergence tells more about a series than does convergence. A series may be convergent, even though it is not absolutely convergent.

DEFINITION

We say $\sum_{n=1}^{\infty} a_n$ is **conditionally convergent** if the series is convergent, but not absolutely convergent.

(The expression "conditionally convergent" indicates that the convergence of the series depends on the order of the terms. It is true that a series that is absolutely convergent is convergent and has the same sum for every arrangement of its terms. We will not use or verify this fact. See Exercises 49 and 50.)

We will use the following strategy.

Strategy for testing absolute convergence, conditional convergence, or divergence

- If you notice that the nth term of the series does not approach zero, then the series must be divergent and you are done.
- Test for absolute convergence by using tests for series with positive terms to determine convergence or divergence of $\sum_{n=1}^{\infty} |a_n|$. If the series is absolutely convergent, you are done.
- If either the Ratio Test or Root Test gives a limit that is greater than one, then the nth term of the series does not approach zero, so the series must be divergent and you are done.
- If the series is not absolutely convergent and the nth term approaches zero, check convergence of the series $\sum_{n=1}^{\infty} a_n$. Remember that a series may be divergent even though its nth term approaches zero. If the series is convergent but not absolutely convergent, it is conditionally convergent and you are done.

Note that it is convenient to test for convergence of series with positive and negative terms only if the series is alternating, so you can use the Alternating Series Test. There are series for which convergence or divergence cannot be determined by using only the tests we have developed.

EXAMPLE 7

Determine whether the series $\displaystyle\sum_{n=1}^{\infty}(-1)^n n^{-1/2}$ is absolutely convergent, conditionally convergent, or divergent.

SOLUTION We have $a_n = (-1)^n n^{-1/2}$, so $|a_n| = n^{-1/2}$. Then $\displaystyle\sum_{n=1}^{\infty}|a_n|$ is a divergent p-series, so $\displaystyle\sum_{n=1}^{\infty}a_n$ is not absolutely convergent. We then note that the terms a_n alternate sign, $|a_n| \to 0$ as $n \to \infty$, and $|a_{n+1}| < |a_n|$. The Alternating Series Test then shows that $\displaystyle\sum_{n=1}^{\infty}a_n$ is convergent. Since the series is convergent, but not absolutely convergent, it is conditionally convergent. ∎

EXERCISES 9.6

Determine convergence or divergence of the series in Exercises 1–20.

1. $\displaystyle\sum_{n=1}^{\infty}\frac{(-1)^n}{n^2+4}$

2. $\displaystyle\sum_{n=1}^{\infty}\frac{(-1)^n}{n^{2/3}}$

3. $\displaystyle\sum_{n=1}^{\infty}(-1)^n \cos\frac{1}{n}$

4. $\displaystyle\sum_{n=1}^{\infty}(-1)^n \sin\frac{1}{n}$

5. $\displaystyle\sum_{n=2}^{\infty}(-1)^n \ln\left(1-\frac{1}{n}\right)$

6. $\displaystyle\sum_{n=2}^{\infty}\frac{(-1)^n}{\sqrt{\ln n}}$

7. $\displaystyle\sum_{n=1}^{\infty}\frac{(-2)^n}{3^n+3^{-n}}$

8. $\displaystyle\sum_{n=1}^{\infty}\frac{(-3)^n}{2^n+2^{-n}}$

9. $\displaystyle\sum_{n=1}^{\infty}(-1)^n\left(1-\frac{1}{n}\right)^n$

10. $\displaystyle\sum_{n=1}^{\infty}(-1)^n\left(\frac{\pi}{2}-\tan^{-1}n\right)$

11. $\displaystyle\sum_{n=1}^{\infty}\frac{(-1)^n}{n+1}$

12. $\displaystyle\sum_{n=1}^{\infty}\frac{(-1)^n n}{n+1}$

13. $\displaystyle\sum_{n=1}^{\infty}\frac{(-1)^n}{n^{1/4}}$ (Calculus Explorer & Tutor II, Sequences and Series, 10.)

14. $\displaystyle\sum_{n=2}^{\infty}\frac{(-1)^n}{\ln n}$

15. $\displaystyle\sum_{n=1}^{\infty}\frac{(-1)^n}{\sqrt{1+\ln n}}$

16. $\displaystyle\sum_{n=1}^{\infty}\frac{(-1)^n}{\sqrt{n+1}}$

17. $\displaystyle\sum_{n=1}^{\infty}\frac{(-e)^n}{1+e^n}$

18. $\displaystyle\sum_{n=1}^{\infty}\cos n\pi \sin(\pi/n)$

19. $\displaystyle\sum_{n=1}^{\infty}\frac{\cos n\pi}{1+\sqrt{n}}$

20. $\displaystyle\sum_{n=1}^{\infty}\frac{(-1)^n n}{e^n}$

Approximate the values of the series in Exercises 21–24 with error less than 0.01.

21. $\displaystyle\sum_{n=1}^{\infty}\frac{(-1)^n}{n!}$

22. $\displaystyle\sum_{n=1}^{\infty}\frac{(-1/2)^n}{n}$

23. $\displaystyle\sum_{n=1}^{\infty}\frac{(-1)^{n+1}}{(2n-1)!}\left(\frac{\pi}{4}\right)^{2n-1}$

24. $\displaystyle\sum_{n=1}^{\infty}\frac{(-1)^{n+1}}{(2n-2)!}\left(\frac{\pi}{3}\right)^{2n-2}$

Determine either absolute convergence, conditional convergence, or divergence for each of the series in Exercises 25–38.

25. $\displaystyle\sum_{n=0}^{\infty}\frac{(-1)^n}{1+n^2}$

26. $\displaystyle\sum_{n=0}^{\infty}\frac{(-1)^n n}{1+n^2}$

27. $\displaystyle\sum_{n=0}^{\infty}\frac{(-1)^n n^2}{1+n^2}$

28. $\displaystyle\sum_{n=0}^{\infty}\frac{n(-2)^n}{1+3^n}$

29. $\displaystyle\sum_{n=1}^{\infty}\frac{(-1)^n}{n(1+2^n)}$

30. $\displaystyle\sum_{n=1}^{\infty}\frac{(-3)^n}{n(1+2^n)}$

31. $\displaystyle\sum_{n=1}^{\infty} \frac{(-1)^n \ln n}{n^2}$

32. $\displaystyle\sum_{n=1}^{\infty} \frac{(-1)^n \ln n}{n}$

33. $\displaystyle\sum_{n=2}^{\infty} \frac{(-1)^n}{\ln n}$

34. $\displaystyle\sum_{n=2}^{\infty} \frac{(-1)^n}{n(\ln n)^2}$

35. $\displaystyle\sum_{n=0}^{\infty} \frac{(-5)^n}{n!}$

36. $\displaystyle\sum_{n=0}^{\infty} \frac{(-1)^n (n! + 1)}{(n+1)!}$

37. $\displaystyle\sum_{n=1}^{\infty} \frac{(-1)^n}{n + n^2}$

38. $\displaystyle\sum_{n=1}^{\infty} \tan \frac{(-1)^n}{n}$

39. Determine values of N for which the partial sum S_N of $S = \displaystyle\sum_{n=1}^{\infty} \frac{(-1)^n}{n^2 + 1}$ is guaranteed by the Alternating Series Test to be within 0.001 of S.

40. Determine values of N for which the partial sum S_N of $S = \displaystyle\sum_{n=1}^{\infty} \frac{(-1)^n}{(n+1)[\ln(n+1)]^2}$ is guaranteed by the Alternating Series Test to be within 0.1 of S.

Consider the behavior of the partial sums of the series in Exercises 41–45 to determine whether they are convergent or divergent. (The Alternating Series Test does not apply directly to these series.)

41. $1 + \dfrac{1}{2} - \dfrac{1}{3} - \dfrac{1}{4} + \dfrac{1}{5} + \dfrac{1}{6} - \dfrac{1}{7} - \dfrac{1}{8} + \cdots$

42. $1 - \dfrac{1}{2} - \dfrac{1}{2} + \dfrac{1}{3} + \dfrac{1}{3} + \dfrac{1}{3} - \dfrac{1}{4} - \dfrac{1}{4} - \dfrac{1}{4} - \dfrac{1}{4} + \cdots$

43. $1 - \dfrac{1}{2^2} + \dfrac{1}{3} - \dfrac{1}{4^2} + \dfrac{1}{5} - \dfrac{1}{6^2} + \dfrac{1}{7} - \dfrac{1}{8^2} + \dfrac{1}{9} - \dfrac{1}{10^2} + \cdots$

44. $1 - \dfrac{1}{2^2} + \dfrac{1}{3^3} - \dfrac{1}{4^2} + \dfrac{1}{5^3} - \dfrac{1}{6^2} + \dfrac{1}{7^3} - \dfrac{1}{8^2} + \dfrac{1}{9^3} - \dfrac{1}{10^2} + \cdots$

45. $1 - \dfrac{1}{2} - \dfrac{1}{4} + \dfrac{1}{3} - \dfrac{1}{6} - \dfrac{1}{8} + \dfrac{1}{5} - \dfrac{1}{10} - \cdots$

46. If $\displaystyle\sum_{n=1}^{\infty}(a_{2n-1} + a_{2n})$ is convergent and $\lim_{n \to \infty} a_n = 0$, show that $\displaystyle\sum_{n=1}^{\infty} a_n$ is convergent.

47. Find an example of a sequence such that $\displaystyle\sum_{n=1}^{\infty}(a_{2n-1} + a_{2n})$ is convergent, but $\displaystyle\sum_{n=1}^{\infty} a_n$ is divergent.

48. If $\displaystyle\sum_{n=1}^{\infty} a_n$ and $\displaystyle\sum_{n=1}^{\infty} b_n$ are both absolutely convergent, show that $\displaystyle\sum_{n=1}^{\infty}(a_n + b_n)$ is absolutely convergent.

49. Suppose a series is obtained by **rearranging** the terms of the series $\displaystyle\sum_{n=1}^{\infty} \frac{(-1)^{n+1}}{n}$ as follows. The first term is 1 and the second term is $-\dfrac{1}{2}$. We next take a string of positive terms, in original order, until the sum of the string is greater than 0.5, and then take the next remaining negative term, which is $-1/4$. We then take another string of positive terms that have sum greater than 0.5, and then the next negative term. This process is repeated indefinitely. Does the resulting series converge? What is the limit of the partial sums?

$\Bigg($ Hint:

$$1 - \frac{1}{2} + \underbrace{\frac{1}{3} + \frac{1}{5}}_{>0.5} - \frac{1}{4} + \underbrace{\frac{1}{7} + \frac{1}{9} + \frac{1}{11} + \frac{1}{13} + \frac{1}{15} + \frac{1}{17}}_{>0.5}$$

$$-\frac{1}{6} + \frac{1}{19} + \frac{1}{21} + \cdots \Bigg)$$

50. Suppose a series is obtained by **rearranging** the terms of the series $\displaystyle\sum_{n=1}^{\infty} \frac{(-1)^{n+1}}{n}$ as follows. The first term is 1. We then take a string of negative terms, in original order, until the first time we obtain a negative partial sum. We then take a string of positive terms, in original order, until we obtain a positive partial sum, and then take another string of negative terms. We repeat indefinitely the process of taking strings of positive terms until the partial sum becomes positive followed by a string of negative terms until the partial sum becomes negative. Does the resulting series converge? What is the limit of the partial sums?

(*Hint:*

$$1 \underbrace{- \frac{1}{2} - \frac{1}{4} - \frac{1}{6} - \frac{1}{8}}_{<0} \underbrace{+ \frac{1}{3} - \frac{1}{10} - \frac{1}{12} - \frac{1}{14} - \frac{1}{16}}_{>0} \underbrace{+ \frac{1}{5} - \frac{1}{18} \cdots}_{<0} \cdots)$$

9.7 CONVERGENCE SETS OF POWER SERIES

Connections

Inequalities with absolute values, 1.1.

Ratio test, 9.5.

Aternating series test, 9.6.

We will investigate the convergence of series that have terms that are constant multiples of powers of $x - c$, where x is a variable and c represents a fixed number. In later sections we will see how such series can be used in various approximation problems.

Definition and Notation

DEFINITION

A series of the form

$$\sum_{n=0}^{\infty} a_n (x - c)^n = a_0 + a_1(x - c) + a_2(x - c)^2 + a_3(x - c)^3 + \cdots$$

is called a **power series** about c.

For each value of the variable x, the power series $\sum_{n=0}^{\infty} a_n (x - c)^n$ is a series of numbers. We can then determine convergence or divergence by using any of the tests we have previously developed.

A power series is a function of x, with domain the set of all x for which the series converges. We first note that a power series about c always converges for $x = c$, since all partial sums are equal to a_0 when $x = c$. We will illustrate by example and then prove that the convergence set of a power series about c is always an interval with center c. The series may converge at either both endpoints, one endpoint, or neither endpoint of the interval of convergence.

Examples

EXAMPLE 1

Find the convergence set of $\sum_{n=1}^{\infty} nx^n$.

SOLUTION   The series is a power series about zero with $a_n = n$. Let us use the Ratio Test to test for absolute convergence of the series. We set $u_n = n|x|^n$, so $u_{n+1} = (n + 1)|x|^{n+1}$. Then

$$\frac{u_{n+1}}{u_n} = \frac{(n + 1)|x|^{n+1}}{n|x|^n} = \frac{n + 1}{n}|x| \to |x| \text{ as } n \to \infty.$$

The limit is less than one whenever $|x| < 1$. The Ratio Test then shows $\sum_{n=1}^{\infty} u_n$ is convergent for $|x| < 1$, so $\sum_{n=1}^{\infty} nx^n$ is absolutely convergent for $|x| < 1$. Since absolute convergence implies convergence, we can conclude that $\sum_{n=1}^{\infty} nx^n$ is convergent for $|x| < 1$. For $|x| > 1$ the Ratio Test shows that $u_n = |nx^n|$ approaches infinity as n approaches infinity. This implies that $\sum_{n=1}^{\infty} nx^n$ is divergent for $|x| > 1$.

We know that the series is convergent for $-1 < x < 1$. We need to check convergence at the endpoints of this interval. When $x = 1$, we have $\sum_{n=1}^{\infty} n$. Since the nth term of this series does not approach zero as n approaches infinity, we know this series must diverge. Similarly, the series is divergent when $x = -1$. The convergence set is $-1 < x < 1$. ■

EXAMPLE 2

Find the convergence set of $\sum_{n=1}^{\infty} \dfrac{x^n}{n}$.

SOLUTION   This is a power series about zero. Let us use the Ratio Test to determine absolute convergence. We set

$$u_n = \frac{|x|^n}{n}, \text{ so } u_{n+1} = \frac{|x|^{n+1}}{n+1}.$$

Then

$$\frac{u_{n+1}}{u_n} = \frac{|x|^{n+1}}{n+1}\frac{n}{|x|^n} = \frac{n}{n+1}|x| \to |x| \text{ as } n \to \infty.$$

The limit is less than one whenever $|x| < 1$ and greater than one whenever $|x| > 1$. The Ratio Test then implies $\sum_{n=1}^{\infty} \dfrac{x^n}{n}$ is absolutely convergent, so convergent, for $|x| < 1$ and divergent for $|x| > 1$.

We need to check convergence at the endpoints of the interval $-1 < x < 1$. When $x = 1$, we have

$$\sum_{n=1}^{\infty} \frac{1}{n}.$$

This is a divergent p-series. When $x = -1$, we have

$$\sum_{n=1}^{\infty} \frac{(-1)^n}{n}.$$

The Alternating Series Test shows that this series is convergent. The convergence set is $-1 \le x < 1$. ■

EXAMPLE 3

Find the convergence set of $\sum_{n=1}^{\infty} \dfrac{x^n}{n^2}$.

SOLUTION   As before, we set

$$u_n = \frac{|x|^n}{n^2}, \text{ so } u_{n+1} = \frac{|x|^{n+1}}{(n+1)^2}.$$

Then

$$\frac{u_{n+1}}{u_n} = \frac{|x|^{n+1}}{(n+1)^2}\frac{n^2}{|x|^n} = \frac{n^2}{(n+1)^2}|x| \to |x| \text{ as } n \to \infty.$$

The Ratio Test then implies that this series is absolutely convergent for $|x| < 1$ and divergent for $|x| > 1$. It is easy to see that the series is absolutely convergent and, hence, convergent at both endpoints. The convergence set is $-1 \le x \le 1$. ∎

EXAMPLE 4

Find the convergence set of $\sum_{n=1}^{\infty} n!x^n$.

SOLUTION   Let

$$u_n = n!|x|^n, \text{ so } u_{n+1} = (n+1)!|x|^{n+1}.$$

Then

$$\frac{u_{n+1}}{u_n} = \frac{(n+1)!|x|^{n+1}}{n!|x|^n} = \frac{n!(n+1)|x|^{n+1}}{n!|x|^n} = (n+1)|x| \to \infty \text{ as } n \to \infty$$

for every $x \neq 0$. The Ratio Test then implies that the series is divergent for $x \neq 0$. We know that a power series about zero is convergent for $x = 0$. This series converges only for $x = 0$. ∎

EXAMPLE 5

Find the convergence set of $\sum_{n=1}^{\infty} \frac{x^n}{n!}$.

SOLUTION   Let

$$u_n = \frac{|x|^n}{n!}, \text{ so } u_{n+1} = \frac{|x|^{n+1}}{(n+1)!}.$$

Then

$$\frac{u_{n+1}}{u_n} = \frac{|x|^{n+1}}{(n+1)!}\frac{n!}{|x|^n} = \frac{|x|}{n+1} \to 0 \text{ as } n \to \infty$$

for each value of x. The Ratio Test then implies that this series converges absolutely for all x. ∎

Since the nth term of a convergent series approaches zero as n approaches infinity, we can conclude from Example 5 that

$$\lim_{n\to\infty}\frac{x^n}{n!} = 0 \quad \text{for all } x. \tag{1}$$

We will need to use (1) in Section 9.9.

EXAMPLE 6

Find the convergence set of $\sum_{n=1}^{\infty}\frac{(-1)^n(x-2)^n}{n4^n}$.

SOLUTION   Let

$$u_n = \frac{|x-2|^n}{n4^n}, \text{ so } u_{n+1} = \frac{|x-2|^{n+1}}{(n+1)4^{n+1}}.$$

Then

$$\frac{u_{n+1}}{u_n} = \frac{|x - 2|^{n+1}}{(n+1)4^{n+1}} \frac{n4^n}{|x - 2|^n} = \frac{n|x - 2|}{(n+1)4} \to \frac{|x - 2|}{4} \text{ as } n \to \infty.$$

The Ratio Test gives convergence for $\dfrac{|x - 2|}{4} < 1$ and divergence for $\dfrac{|x - 2|}{4} > 1$.

Solving the first inequality above, we have

$$|x - 2| < 4,$$
$$-4 < x - 2 < 4,$$
$$-2 < x < 6.$$

When $x = -2$, the series is

$$\sum_{n=1}^{\infty} \frac{(-1)^n(-2 - 2)^n}{n4^n} = \sum_{n=1}^{\infty} \frac{1}{n},$$

which is a divergent p-series. When $x = 6$, we have

$$\sum_{n=1}^{\infty} \frac{(-1)^n(6 - 2)^n}{n4^n} = \sum_{n=1}^{\infty} \frac{(-1)^n}{n}.$$

The Alternating Series Test shows that this series is convergent. The convergence set is $-2 < x \le 6$. ■

EXAMPLE 7

Find the convergence set of $\displaystyle\sum_{n=1}^{\infty} \frac{(x - 2)^{2n}}{4^n}$.

SOLUTION   This is a power series about $c = 2$ with $a_n = 0$ for n odd. We can investigate convergence by applying the Ratio Test to the nonzero terms. Let

$$u_n = \frac{|x - 2|^{2n}}{4^n}, \text{ so } u_{n+1} = \frac{|x - 2|^{2(n+1)}}{4^{n+1}}.$$

$$\frac{u_{n+1}}{u_n} = \frac{|x - 2|^{2(n+1)}}{4^{n+1}} \frac{4^n}{|x - 2|^{2n}}$$
$$= \frac{|x - 2|^2}{4} \to \frac{|x - 2|^2}{4} \text{ as } n \to \infty.$$

The Ratio Test gives convergence for $\dfrac{|x - 2|^2}{4} < 1$ and divergence for $\dfrac{|x - 2|^2}{4} > 1$.

Solving the first inequality above, we have

$$|x - 2|^2 < 4,$$
$$|x - 2| < 2,$$
$$-2 < x - 2 < 2,$$
$$0 < x < 4.$$

When either $x = 0$ or $x = 4$, the series is $\displaystyle\sum_{n=1}^{\infty} 1$, which is clearly divergent. The convergence set is $0 < x < 4$. ■

Example 8

Find the convergence set of $\displaystyle\sum_{n=1}^{\infty} \frac{n!}{(1)(4)(7)\cdots(3n-2)} x^n$.

Solution   Let

$$u_n = \frac{n!}{(1)(4)(7)\cdots(3n-2)} |x|^n, \text{ so}$$

$$u_{n+1} = \frac{(n+1)!}{(1)(4)(7)\cdots(3n-2)(3(n+1)-2)} |x|^{n+1}$$

$$= \frac{(n+1)!}{(1)(4)(7)\cdots(3n-2)(3n+1)} |x|^{n+1}.$$

$$\frac{u_{n+1}}{u_n} = \frac{(n+1)!|x|^{n+1}}{(1)(4)(7)\cdots(3n-2)(3n+1)} \frac{(1)(4)(7)\cdots(3n-2)}{n!|x|^n}$$

$$= \frac{n+1}{3n+1}|x| \rightarrow \frac{|x|}{3} \text{ as } n \rightarrow \infty.$$

We obtain absolute convergence for $|x| < 3$ and divergence for $|x| > 3$. To check convergence at the points $x = 3$ and $x = -3$, we note that

$$\frac{n!3^n}{(1)(4)(7)\cdots(3n-2)} = \frac{(1)(2)(3)\cdots(n)3^n}{(1)(4)(7)\cdots(3n-2)}$$

$$= \frac{(3)(6)(9)\cdots(3n)}{(1)(4)(7)\cdots(3n-2)}$$

$$= \left(\frac{3}{1}\right)\left(\frac{6}{4}\right)\left(\frac{9}{7}\right)\cdots\left(\frac{3n}{3n-2}\right) > 1.$$

This shows that the nth term of the series with either $x = 3$ or $x = -3$ does not approach zero, so the series must diverge for these values. The convergence set is $-3 < x < 3$. ∎

The General Case

We have seen that each power series in Examples 1–8 has a convergence set that is an interval. Let us now show that this is true for all power series.

The series

$$\sum_{n=0}^{\infty} a_n(x-c)^n$$

converges to a_0 when $x = c$, because the partial sums

$$S_n = a_0 + a_1(c-c)^2 + a_2(c-c)^2 + \cdots + a_{n-1}(c-c)^{n-1} = a_0$$

for all n. It is possible that the series converges only for $x = c$. If the series converges at a point $x_1 \neq c$, we can use the following fact to obtain additional information about the convergence set.

If a power series $\displaystyle\sum_{n=0}^{\infty} a_n(x-c)^n$ converges at a point $x = x_1$, then the series is absolutely convergent for values of x that satisfy $|x - c| < |x_1 - c|$.   (2)

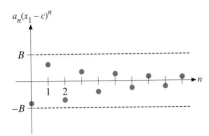

FIGURE 9.7.1 If $\lim\limits_{n\to\infty} a_n(x_1 - c)^n = 0$, then the sequence $\{a_n(x_1 - c)^n\}_{n\geq 0}$ is bounded.

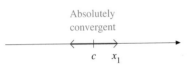

FIGURE 9.7.2 If the power series $\sum\limits_{n=0}^{\infty} a_n(x - c)^n$ converges at a point x_1, then it converges absolutely for $|x - c| < |x_1 - c|$.

Proof   We first note that convergence of the series $\sum\limits_{n=0}^{\infty} a_n(x_1 - c)^n$ implies

$$\lim_{n\to\infty} a_n(x_1 - c)^n = 0.$$

It follows as in Statement 4 of Section 2.5 that the sequence $\{a_n(x_1 - c)^n\}_{n\geq 0}$ is bounded, so there is a number B such that $|a_n(x_1 - c)^n| \leq B$ for all n. See Figure 9.7.1 We then have

$$|a_n(x - c)^n| = |a_n(x - c)^n| \left(\frac{|x_1 - c|}{|x_1 - c|} \right)^n$$

$$= |a_n(x_1 - c)^n| \left(\frac{|x - c|}{|x_1 - c|} \right)^n < B \left(\frac{|x - c|}{|x_1 - c|} \right)^n \quad \text{for } n \geq 0.$$

For a fixed value of x with $|x - c| < |x_1 - c|$, we have

$$\frac{|x - c|}{|x_1 - c|} = r < 1,$$

and the absolute convergence of $\sum\limits_{n=0}^{\infty} a_n(x - c)^n$ is a consequence of comparison of $\sum\limits_{n=0}^{\infty} |a_n(x - c)^n|$ with the convergent geometric series $\sum\limits_{n=0}^{\infty} r^n$. See Figure 9.7.2. ∎

If a power series $\sum\limits_{n=0}^{\infty} a_n(x - c)^n$ is not absolutely convergent at a point $x = x_2$, then the series cannot be convergent for any x that satisfies $|x - c| > |x_2 - c|$. That is, if the series is convergent at some x that satisfies $|x - c| > |x_2 - c|$ then Statement 2 would imply that it is absolutely convergent at $x = x_2$. This gives the following result.

If a power series $\sum\limits_{n=0}^{\infty} a_n(x - c)^n$ does not converge absolutely at a point $x = x_2$, then the series is divergent for values of x that satisfy $|x - c| > |x_2 - c|$. (3)

Statements 2 and 3 can be used to verify a fundamental fact of power series.

CONVERGENCE SET OF A POWER SERIES

For each power series $\sum\limits_{n=0}^{\infty} a_n(x - c)^n$, either

(i) the series converges only for $x = c$,

(ii) the series converges absolutely for all x, or

(iii) there is a number R such that the series converges absolutely for $|x - c| < R$ and diverges for $|x - c| > R$.

Proof   If (i) is not true, then there is a point $x_1 \neq c$ such that $\sum\limits_{n=0}^{\infty} a_n(x_1 - c)^n$ is convergent. If (ii) is also not true, there is a point x_2 such that $\sum\limits_{n=0}^{\infty} a_n(x_2 - c)^n$

FIGURE 9.7.3 If the power series $\sum_{n=0}^{\infty} a_n(x - c)^n$ converges at a point x_1 and is not absolutely convergent at x_2, then it converges absolutely for $|x - c| < |x_1 - c|$ and diverges for $|x - c| > |x_2 - c|$.

is not absolutely convergent. From Statements 2 and 3 we can then conclude that $\sum_{n=0}^{\infty} a_n(x - c)^n$ is absolutely convergent for all x with $|x - c| < |x_1 - c|$ and divergent for all x with $|x - c| > |x_2 - c|$. It follows that $|x_1 - c| \le |x_2 - c|$. Now choose x_3 with $|x_3 - c| = (|x_1 - c| + |x_2 - c|)/2$. See Figure 9.7.3. If $\sum_{n=1}^{\infty} a_n(x - c)^n$ converges for $x = x_3$, then the series is absolutely convergent for $|x - c| < |x - x_3|$; if the series is divergent at $x = x_3$, then $\sum_{n=1}^{\infty} a_n(x_3 - c)^n$ is not absolutely convergent, so the series is divergent for $|x - c| > |x_3 - c|$. Continuing to bisect the gap between the interval of absolute convergence and the intervals of divergence, we obtain a sequence $\{x_n\}_{n=1}$ that converges to a point that determines the number R of case (iii). ▪

The number R in case (iii) of the previous theorem is called the **radius of convergence** of the power series. In case (i), we say the radius of convergence is zero. In case (ii), we say the radius of convergence is infinity. The convergence set of a power series is then called the **interval of convergence**. The interval of convergence contains all x that satisfy $|x - c| < R$ and, as we have seen, possibly one or both endpoints of that interval.

EXERCISES 9.7

Find the interval of convergence and the radius of convergence of the series in Exercises 1–24.

1. $\sum_{n=0}^{\infty} (x - 1)^n$

2. $\sum_{n=0}^{\infty} n(x - 1)^{n-1}$

3. $\sum_{n=0}^{\infty} \frac{(x - 1)^{n+1}}{n + 1}$

4. $\sum_{n=1}^{\infty} \frac{(x + 1)^{n+2}}{(n + 1)(n + 2)}$

5. $\sum_{n=0}^{\infty} \frac{x^n}{1 + 2^n}$

6. $\sum_{n=0}^{\infty} \frac{x^{2n}}{1 + 2^n}$

7. $\sum_{n=2}^{\infty} (\ln n) x^n$

8. $\sum_{n=2}^{\infty} \frac{x^n}{\ln n}$

9. $\sum_{n=1}^{\infty} \frac{n(x - 2)^n}{2^n}$

10. $\sum_{n=1}^{\infty} \frac{(x + 4)^n}{n3^n}$ **(Calculus Explorer & Tutor II, Sequences and Series, 12.)**

11. $\sum_{n=1}^{\infty} \frac{(-1)^n (x - 3)^n}{n^2 3^n}$

12. $\sum_{n=0}^{\infty} \frac{(-1)^n (x + 2)^n}{4^n}$

13. $\sum_{n=0}^{\infty} \frac{n!}{2^n} x^n$

14. $\sum_{n=1}^{\infty} n^n x^n$

15. $\sum_{n=0}^{\infty} \frac{(-1)^n x^{2n}}{(2n)!}$

16. $\sum_{n=0}^{\infty} \frac{(-1)^n x^{2n+1}}{(2n + 1)!}$

17. $\sum_{n=0}^{\infty} \frac{(x - 2)^{2n}}{4^n}$

18. $\sum_{n=0}^{\infty} 8(x - 1)^{3n}$

19. $\sum_{n=1}^{\infty} \frac{x^n}{2^{2n}}$

20. $\sum_{n=1}^{\infty} \frac{(-1)^n x^n}{n}$

21. $\sum_{n=1}^{\infty} \frac{(1)(3) \cdots (2n - 1)}{(n + 2)!} x^n$

22. $\sum_{n=1}^{\infty} \frac{(2)(5) \cdots (3n - 1)}{(n + 2)!} x^n$

23. $\sum_{n=1}^{\infty} \frac{(3)(5) \cdots (2n + 1)}{n!} x^n$

24. $\sum_{n=1}^{\infty} \frac{(4)(7) \cdots (3n + 1)}{n!} x^n$

25. If $\sum_{n=0}^{\infty} a_n x^n$ is convergent for $x = 3$, is it convergent for $x = 2$? Must it be convergent for $x = 4$? Explain your answer.

26. If $\sum_{n=0}^{\infty} a_n(x - 3)^n$ is convergent for $x = 1$, is it convergent for $x = 2$? Must it be convergent for $x = 0$? Explain your answer.

9.8 POWER SERIES REPRESENTATION OF FUNCTIONS

Connections

Geometric series, 9.2.

Concept of power series, 9.7.

We will see that functions that can be represented as power series can be treated as polynomials. This means, within their interval of convergence, we can add, subtract, multiply, differentiate, and integrate term by term.

A function f is said to be **represented by a power series** if

$$f(x) = \sum_{n=0}^{\infty} a_n(x-c)^n \quad \text{for all } x \text{ in some interval.}$$

Of course, a function can be represented by a particular power series only within the interval of convergence of the series. A function may be represented by different power series in different intervals.

We already know of one function that can be represented as a power series. Namely, in Section 9.2 we showed that the geometric series

$$\sum_{n=0}^{\infty} r^n \text{ is convergent with sum } \frac{1}{1-r} \text{ if } |r| < 1.$$

This gives the following power series representation of the function $1/(1-x)$.

$$\frac{1}{1-x} = 1 + x + x^2 + x^3 + \cdots, \qquad |x| < 1. \tag{1}$$

We can obtain power series representations of other functions by *substitution.*

Substitution into Known Formulas

Substitution into known formulas to obtain new formulas is an extremely important technique.

EXAMPLE 1
Show that

$$\frac{1}{1+x} = 1 - x + x^2 - x^3 + \cdots, \qquad |x| < 1.$$

SOLUTION   Substituting $-x$ for x in (1), we obtain

$$\frac{1}{1+x} = \frac{1}{1-(-x)} = 1 + (-x) + (-x)^2 + (-x)^3 + \cdots$$

$$= 1 - x + x^2 - x^3 + \cdots, \qquad |-x| < 1. \blacksquare$$

Substitution of $-x^2$ for x in (1) gives

$$\frac{1}{1+x^2} = 1 - x^2 + x^4 - x^6 + \cdots, \qquad |x| < 1. \tag{2}$$

Arithmetic Operations

The following theorem tells us how to perform the operations of addition, subtraction, and multiplication by a constant.

ADDITION, SUBTRACTION, AND MULTIPLICATION OF POWER SERIES BY A CONSTANT

Suppose $\sum_{n=0}^{\infty} a_n(x - c)^n$ has radius of convergence R_1 and $\sum_{n=0}^{\infty} b_n(x - c)^n$ has radius of convergence R_2. Let $R = \text{minimum}\,(R_1, R_2)$. If d and e are constants, then

$$d\left(\sum_{n=0}^{\infty} a_n(x - c)^n\right) + e\left(\sum_{n=0}^{\infty} b_n(x - c)^n\right) = \sum_{n=0}^{\infty}(da_n + eb_n)(x - c)^n,$$

$$|x - c| < R.$$

This means we can carry out the operations on the left by combining terms with like powers of $x - c$, as if the series were finite sums. Replacing each of the two series on the left with its nth partial sum, and then combining terms with like powers of $x - c$, gives the nth partial sum of the series on the right. The conclusion of the theorem is then an immediate consequence of the Limit Theorem.

MULTIPLICATION OF TWO POWER SERIES

Suppose $\sum_{n=0}^{\infty} a_n(x - c)^n$ has radius of convergence R_1 and $\sum_{n=0}^{\infty} b_n(x - c)^n$ has radius of convergence R_2. Let $R = \text{minimum}\,(R_1, R_2)$.
  If $c_n = a_0 b_n + a_1 b_{n-1} + a_2 b_{n-2} + \cdots + a_{n-1} b_1 + a_n b_0$, then

$$\left(\sum_{n=0}^{\infty} a_n(x - c)^n\right)\left(\sum_{n=0}^{\infty} b_n(x - c)^n\right) = \sum_{n=0}^{\infty} c_n(x - c)^n, \qquad |x - c| < R.$$

This result tells us that, within both intervals of convergence, we can multiply two power series as if they were finite sums. Note that multiplication of the partial sums

$$a_0 + a_1(x - c) + a_2(x - c)^2 + \cdots + a_N(x - c)^N \text{ and}$$
$$b_0 + b_1(x - c) + b_2(x - c)^2 + \cdots + b_N(x - c)^N$$

and then combining terms with like powers of $x - c$ gives a polynomial in $x - c$ with coefficients c_n for $0 \le n \le N$, plus some higher-degree terms. (See Exercise 39.) The fact that the sum of the extra, higher-degree, terms approaches zero as N approaches infinity depends on the absolute convergence of power series at points interior to their interval of convergence. We will not verify this.

EXAMPLE 2

Show that

$$\frac{1+x}{2-x} = \frac{1}{2} + \frac{3x}{4} + \frac{3x^2}{8} + \frac{3x^3}{16} + \cdots, \qquad |x| < 1.$$

SOLUTION   Rewriting, using Formula 1 and then Theorems 1 and 2, we have

$$\frac{1+x}{2-x} = (1+x)\left(\frac{1}{2}\right)\left(\frac{1}{1-(x/2)}\right)$$

$$= \frac{1}{2}(1+x)\left(1 + \left(\frac{x}{2}\right) + \left(\frac{x}{2}\right)^2 + \left(\frac{x}{2}\right)^3 + \cdots\right)$$

$$= \frac{1}{2}\left(1 + \frac{x}{2} + \frac{x^2}{4} + \frac{x^3}{8} + \cdots\right) + \frac{x}{2}\left(1 + \frac{x}{2} + \frac{x^2}{4} + \frac{x^3}{8} + \cdots\right)$$

$$= \left(\frac{1}{2} + \frac{x}{4} + \frac{x^2}{8} + \frac{x^3}{16} + \cdots\right) + \left(\frac{x}{2} + \frac{x^2}{4} + \frac{x^3}{8} + \cdots\right)$$

$$= \frac{1}{2} + \frac{3x}{4} + \frac{3x^2}{8} + \frac{3x^3}{16} + \cdots, \qquad |x| < 1. \quad \blacksquare$$

Generally, the radius of convergence of a power series that is obtained by division cannot be described in terms of the radii of convergence of the dividend and the divisor. It seems reasonable that we would need to avoid values of x for which the denominator is zero and the numerator is nonzero.

Differentiation and Integration

The next result tells us that derivatives and integrals of functions that have a power series representation can be obtained by differentiating and integrating the power series term by term.

DIFFERENTIATION AND INTEGRATION OF POWER SERIES

Let R be the radius of convergence of $\displaystyle\sum_{n=0}^{\infty} a_n(x-c)^n$. Then $|x-c| < R$ implies

$$\frac{d}{dx}\left(a_0 + a_1(x-c) + a_2(x-c)^2 + a_3(x-c)^3 + a_4(x-c)^4 + \cdots\right)$$

$$= a_1 + 2a_2(x-c) + 3a_3(x-c)^2 + 4a_4(x-c)^3 + \cdots$$

and

$$\int \left(a_0 + a_1(x-c) + a_2(x-c)^2 + a_3(x-c)^3 + a_4(x-c)^4 + \cdots\right)dx$$

$$= C + a_0(x-c) + \frac{a_1}{2}(x-c)^2 + \frac{a_2}{3}(x-c)^3 + \frac{a_3}{4}(x-c)^4$$

$$+ \frac{a_4}{5}(x-c)^5 + \cdots .$$

We will not verify this result.

EXAMPLE 3

Evaluate the third derivative at zero of the function $f(x) = \sum_{n=1}^{\infty} \frac{1}{n} x^n$.

SOLUTION   Let us write out a few terms of the series. We have

$$f(x) = x + \frac{1}{2}x^2 + \frac{1}{3}x^3 + \frac{1}{4}x^4 + \frac{1}{5}x^5 + \cdots .$$

We then differentiate term by term to obtain

$$f'(x) = 1 + x + x^2 + x^3 + x^4 + \cdots ,$$
$$f''(x) = 1 + 2x + 3x^2 + 4x^3 + \cdots ,$$
$$f'''(x) = 2 + 6x + 12x^2 + \cdots .$$

Substitution then gives

$$f'''(0) = 2 + 6(0) + 12(0)^2 + \cdots = 2. \quad \blacksquare$$

Let us see how to use the theorem on differentiation and integration of power series to obtain power series representations of additional types of functions.

EXAMPLE 4

Find a power series representation about $c = 0$ of

(a) $\dfrac{-2x}{(1+x^2)^2} = \dfrac{d}{dx}\left(\dfrac{1}{1+x^2}\right),$

(b) $\ln\left(\dfrac{1}{1-x}\right) = \displaystyle\int_0^x \dfrac{1}{1-t}\,dt.$

SOLUTION

(a) We use (2) and then differentiate the series term by term to obtain

$$\frac{-2x}{(1+x^2)^2} = \frac{d}{dx}\left(\frac{1}{1+x^2}\right) = \frac{d}{dx}(1 - x^2 + x^4 - x^6 + \cdots)$$
$$= -2x + 4x^3 - 6x^5 + \cdots , \qquad |x| < 1.$$

(b) Using (1) and then integrating term by term, we have

$$\ln\frac{1}{1-x} = \int_0^x \frac{1}{1-t}\,dt$$
$$= \int_0^x (1 + t + t^2 + t^3 + \cdots)\,dt$$
$$= x + \frac{x^2}{2} + \frac{x^3}{3} + \frac{x^4}{4} + \cdots , \qquad |x| < 1. \quad \blacksquare$$

Let us restate the result of Example 4b for future reference.

$$\ln\frac{1}{1-x} = x + \frac{1}{2}x^2 + \frac{1}{3}x^3 + \frac{1}{4}x^4 + \cdots , \qquad |x| < 1. \qquad (3)$$

Approximating Function Values

If a function f can be expressed as a power series, then the partial sums of the power series can be used as approximate values of $f(x)$. Let us illustrate this important idea.

EXAMPLE 5

Use partial sums of the power series expansion of

$$\ln \frac{1}{1-x}$$

with one, two, three, and four nonzero terms to obtain approximate values of $\ln 2$. Compare with a calculator value of $\ln 2$.

SOLUTION From (3) we see that the partial sums of the power series for

$$\ln \frac{1}{1-x}$$

that have one, two, three, and four nonzero terms are

$$P_1(x) = x,$$

$$P_2(x) = x + \frac{1}{2}x^2,$$

$$P_3(x) = x + \frac{1}{2}x^2 + \frac{1}{3}x^3, \text{ and}$$

$$P_4(x) = x + \frac{1}{2}x^2 + \frac{1}{3}x^3 + \frac{1}{4}x^4, \text{ respectively.}$$

To use these formulas, we need to find x such that

$$\frac{1}{1-x} = 2.$$

Solving for x, we have

$$1 = 2(1 - x),$$
$$1 = 2 - 2x,$$
$$2x = 1,$$
$$x = 0.5.$$

We then obtain

$$P_1(0.5) = 0.5,$$

$$P_2(0.5) = 0.5 + \frac{1}{2}(0.5)^2 = 0.625,$$

$$P_3(0.5) = 0.5 + \frac{1}{2}(0.5)^2 + \frac{1}{3}(0.5)^3 = 0.66666667,$$

$$P_4(0.5) = 0.5 + \frac{1}{2}(0.5)^2 + \frac{1}{3}(0.5)^3 + \frac{1}{4}(0.5)^4 = 0.68229167.$$

A calculator gives $\ln 2 \approx 0.69314718$.

The graphs of

$$\ln \frac{1}{1-x}, \qquad |x| < 1,$$

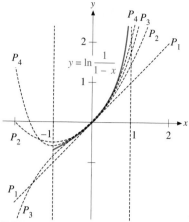

FIGURE 9.8.1 The power series expansion of $\ln \dfrac{1}{1-x}$ converges to the function for $-1 \le x < 1$.

and the approximating partial sums of the series are given in Figure 9.8.1. Note that the series for $\ln(1/(1-x))$ converges for $x = -1$, by the Alternating Series Test. It appears from the graph that the limit is

$$\ln \frac{1}{1-(-1)} = \ln 0.5.$$

This is true. (See Exercise 41.) ■

We can use the power series of

$$\ln \frac{1}{1-x}$$

to obtain approximate values of $\ln u$ for

$$u = \frac{1}{1-x}, \qquad |x| < 1.$$

Solving the inequality $|x| < 1$ for the expression that gives u, we obtain

$$-1 < x < 1,$$
$$1 > -x > -1,$$
$$2 > 1 - x > 0,$$
$$\frac{1}{1-x} > \frac{1}{2}.$$

It follows that we can use this formula to approximate values of $\ln u$ only for $u > 0.5$. However, if $0 < u \le 0.5$, then $\dfrac{1}{u} \ge 2 > 0.5$, so we can write $\ln u = -\ln \dfrac{1}{u}$ and apply the formula to approximate $\ln \dfrac{1}{u}$.

The following result can be used to calculate approximate values of $\ln u$ for all $u > 0$. This series tends to converge more quickly than the series given in (3).

$$\ln \sqrt{\frac{1+x}{1-x}} = x + \frac{x^3}{3} + \frac{x^5}{5} + \frac{x^7}{7} + \frac{x^9}{9} + \cdots, \qquad |x| < 1. \qquad (4)$$

Proof We have

$$\ln \sqrt{\frac{1+x}{1-x}} = \frac{1}{2} \left(\ln \frac{1}{1-x} - \ln \frac{1}{1-(-x)} \right).$$

The formula in (3) gives a power series representation of $\ln \dfrac{1}{1-x}$. Formula 3, with $-x$ in place of x, also gives a power series representation of $\ln \dfrac{1}{1-(-x)}$. We then obtain

$$\ln \sqrt{\frac{1+x}{1-x}} = \frac{1}{2} \left(\left(x + \frac{x^2}{2} + \frac{x^3}{3} + \frac{x^4}{4} + \frac{x^5}{5} + \cdots \right) \right.$$
$$\left. - \left(-x + \frac{x^2}{2} - \frac{x^3}{3} + \frac{x^4}{4} - \frac{x^5}{5} + \cdots \right) \right)$$
$$= x + \frac{x^3}{3} + \frac{x^5}{5} + \cdots, \qquad |x| < 1. \ ■$$

To use (4) to find an approximate value of $\ln u$, we solve

$$u = \sqrt{\frac{1+x}{1-x}}$$

for x. We obtain

$$u^2 = \frac{1+x}{1-x},$$
$$u^2(1-x) = 1+x,$$
$$u^2 - u^2 x = 1+x,$$
$$-u^2 x - x = -u^2 + 1,$$
$$-(u^2+1)x = -(u^2-1),$$
$$x = \frac{u^2-1}{u^2+1}.$$

It is not difficult to verify that each positive number u corresponds to a value of x that satisfies $|x| < 1$.

EXAMPLE 6
Use partial sums of the series expansion of

$$\ln \sqrt{\frac{1+x}{1-x}}$$

with one, two, three, and four nonzero terms to approximate values of $\ln 2$. Compare with a calculator value of $\ln 2$.

SOLUTION We want

$$\sqrt{\frac{1+x}{1-x}} = 2, \text{ so } x = \frac{2^2-1}{2^2+1} = 0.6.$$

The partial sum of the series in (4) that contains n nonzero terms is

$$P_{2n}(x) = x + \frac{x^3}{3} + \frac{x^5}{5} + \frac{x^7}{7} + \cdots + \frac{x^{2n-1}}{2n-1}, \qquad n \geq 1.$$

(This notation is consistent with that used in the next section. Note that $P_{2n}(x)$ contains all terms of the series that have degree less than or equal to $2n$.) The desired partial sums are $P_{2(1)} = P_2$, $P_{2(2)} = P_4$, $P_2 = P_6$, and $P_{2(4)} = P_8$, evaluated at $x = 0.6$. We have

$$P_2(0.6) = 0.6,$$
$$P_4(0.6) = 0.6 + \frac{(0.6)^3}{3} = 0.672,$$
$$P_6(0.6) = 0.6 + \frac{(0.6)^3}{3} + \frac{(0.6)^5}{5} = 0.687552,$$
$$P_8(0.6) = 0.6 + \frac{(0.6)^3}{3} + \frac{(0.6)^5}{5} + \frac{(0.6)^7}{7} = 0.69155109.$$

A calculator gives $\ln 2 = 0.69314718$.

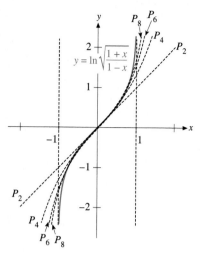

FIGURE 9.8.2 The power series expansion of $\ln\sqrt{\dfrac{1+x}{1-x}}$ converges to the function for $-1 < x < 1$.

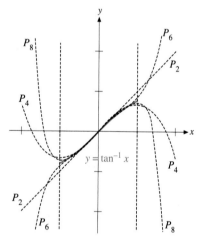

FIGURE 9.8.3 The power series expansion of $\tan^{-1} x$ converges to the function for $-1 \le x \le 1$.

Note that the approximate value obtained by using four nonzero terms in Example 6 is closer to the exact value of $\ln 2$ than the corresponding approximate value in Example 5. The graphs of

$$\ln\sqrt{\frac{1+x}{1-x}}$$

and some of the approximating partial sums are given in Figure 9.8.2. ∎

$$\tan^{-1} x = x - \frac{x^3}{3} + \frac{x^5}{5} - \frac{x^7}{7} + \cdots, \qquad |x| < 1. \tag{5}$$

Proof To prove (5), we integrate the power series of $\dfrac{1}{1+x^2}$, which is given in (2). That is,

$$
\begin{aligned}
\tan^{-1} x &= \int_0^x \frac{1}{1+t^2}\,dt \\
&= \int_0^x (1 - t^2 + t^4 - t^6 + \cdots)\,dt \\
&= x - \frac{x^3}{3} + \frac{x^5}{5} - \frac{x^7}{7} + \cdots, \qquad |x| < 1.
\end{aligned}
$$

The graph of $\tan^{-1} x$ and some of the approximating partial sums are given in Figure 9.8.3. It can be shown that the series also converges to $\tan^{-1} x$ for $x = \pm 1$. See Exercise 42. ∎

EXAMPLE 7
Find a power series representation about $c = 0$ of

$$
f(x) = \begin{cases} \dfrac{\tan^{-1} x}{x}, & x \ne 0, \\ 1, & x = 0. \end{cases}
$$

SOLUTION From (5), we have

$$
\begin{aligned}
f(x) &= \frac{x - \dfrac{x^3}{3} + \dfrac{x^5}{5} - \dfrac{x^7}{7} + \cdots}{x} \\
&= \frac{x\left(1 - \dfrac{x^2}{3} + \dfrac{x^4}{5} - \dfrac{x^6}{7} + \cdots\right)}{x} \\
&= 1 - \frac{x^2}{3} + \frac{x^4}{5} - \frac{x^6}{7} + \cdots,
\end{aligned}
$$

for $0 < |x| < 1$. The value of the power series also agrees with that of f when $x = 0$. ∎

Approximating Values of Definite Integrals

EXAMPLE 8

Approximate $\int_{0}^{0.5} \frac{\tan^{-1} x}{x} dx$ with error less than 0.001.

SOLUTION Using the power series representation we obtained in Example 7, we have

$$\int_{0}^{0.5} \frac{\tan^{-1} x}{x} dx = \int_{0}^{0.5} \left(1 - \frac{x^2}{3} + \frac{x^4}{5} - \frac{x^6}{7} + \cdots \right) dx$$

$$= 0.5 - \frac{1}{9}(0.5)^3 + \frac{1}{25}(0.5)^5 - \frac{1}{49}(0.5)^7 + \cdots .$$

We see that the series alternates sign, and that the absolute values of the terms *decrease* to zero. The Alternating Series Test then tells us that the difference in the exact value of the series and a partial sum is less than the absolute value of the first term omitted from the partial sum. It follows that

$$\int_{0}^{0.5} \frac{\tan^{-1} x}{x} dx \approx 0.5 - \frac{1}{9}(0.5)^3 + \frac{1}{25}(0.5)^5 \approx 0.48736111$$

with error less than

$$\frac{1}{49}(0.5)^7 \approx 0.00015944 < 0.001. \quad \blacksquare$$

Summary of Formulas

Let us restate some of the formulas that were developed in this section.

$$\frac{1}{1 - x} = \sum_{n=0}^{\infty} x^n = 1 + x + x^2 + x^3 + x^4 + \cdots, \ |x| < 1,$$

$$\frac{1}{1 + x^2} = \sum_{n=0}^{\infty} (-1)^n x^{2n} = 1 - x^2 + x^4 - x^6 + x^8 + \cdots, \ |x| < 1,$$

$$\ln \frac{1}{1 - x} = \sum_{n=1}^{\infty} \frac{x^n}{n} = x + \frac{x^2}{2} + \frac{x^3}{3} + \frac{x^4}{4} + \cdots, \ |x| < 1,$$

$$\ln \sqrt{\frac{1 + x}{1 - x}} = \sum_{n=0}^{\infty} \frac{x^{2n+1}}{2n + 1} = x + \frac{x^3}{3} + \frac{x^5}{5} + \frac{x^7}{7} + \frac{x^9}{9} + \cdots, \ |x| < 1,$$

$$\tan^{-1} x = \sum_{n=0}^{\infty} (-1)^n \frac{x^{2n+1}}{2n + 1} = x - \frac{x^3}{3} + \frac{x^5}{5} - \frac{x^7}{7} + \frac{x^9}{9} - \cdots, \ |x| < 1.$$

These formulas can be used to obtain other formulas by using substitution, addition, subtraction, multiplication, differentiation, and integration.

EXERCISES 9.8

Use the formulas in the above summary and techniques developed in this section to find power series representations about $c = 0$ for the functions in Exercises 1–16.

1. (a) $\dfrac{1}{1-x^2}$, (b) $\dfrac{x}{1-x^2}$,

  (c) $\dfrac{2x}{(1-x^2)^2} = \dfrac{d}{dx}\left(\dfrac{1}{1-x^2}\right)$

2. (a) $\dfrac{1}{2x-3}$, (b) $\dfrac{x}{2x-3}$,

  (c) $\dfrac{1}{(2x-3)^2} = -\dfrac{1}{2}\dfrac{d}{dx}\left(\dfrac{1}{2x-3}\right)$

3. (a) $\dfrac{1}{2-x}$, (b) $\dfrac{1}{(2-x)^2} = \dfrac{d}{dx}\left(\dfrac{1}{2-x}\right)$,

  (c) $\ln(2-x) = \ln 2 - \displaystyle\int_0^x \dfrac{1}{2-t}\,dt$

4. (a) $\dfrac{3}{3-x}$, (b) $\dfrac{3}{(3-x)^2} = \dfrac{d}{dx}\left(\dfrac{3}{3-x}\right)$,

  (c) $3\ln(3-x) = 3\ln 3 - \displaystyle\int_0^x \dfrac{3}{3-t}\,dt$

5. (a) $\ln(1+x)$, (b) $x\ln(1+x)$, (c) $\dfrac{\ln(1+2x)}{x}$

6. (a) $\tan^{-1} 2x$, (b) $\dfrac{\tan^{-1} 2x}{x}$, (c) $\dfrac{\tan^{-1} 2x - 2x}{x^3}$

7. $\dfrac{1}{x-1} + \dfrac{1}{x+1}$ **8.** $\dfrac{1}{x+2} - \dfrac{1}{x-2}$

9. (a) $\dfrac{1}{x^2+2x-3}$, (b) $\dfrac{x^2}{x^2+2x-3}$

10. (a) $\dfrac{1}{2x^2-x-3}$, (b) $\dfrac{x}{2x^2-x-3}$

11. (a) $\dfrac{1+x}{1-x}$, (b) $\dfrac{x^2+1}{x-1}$

12. (a) $(x-1)\tan^{-1}x$, (b) $(x^2-1)\tan^{-1}x$

13. $\dfrac{1}{1-x} - \dfrac{1}{1-x^2}$ **14.** $\ln\dfrac{1}{1-x} - \tan^{-1}x$

15. $\dfrac{\tan^{-1}x}{1-x}$ **16.** $\dfrac{\ln(1-x)}{1-x}$

Use partial sums of the series in Formula 4 with one, two, three, and four terms to find approximate values of the numbers in Exercises 17–20. Compare with calculator values.

17. $\ln 0.25$ **18.** $\ln 0.2$

19. $\ln 5$ **20.** $\ln 4$

Use partial sums of the series in Formula 5 with one, two, three,

and four terms to find approximate values of the numbers in Exercises 21–24. Compare with calculator values.

21. $\tan^{-1} 0.05$ **22.** $\tan^{-1} 0.2$

23. $\tan^{-1} 0.4$ **24.** $\tan^{-1} 0.8$

Approximate the integrals in Exercises 25–32 with errors less than 0.001.

25. $\displaystyle\int_0^{0.2} \ln(1+x)\,dx$ **26.** $\displaystyle\int_0^{0.3} \tan^{-1}x\,dx$

27. $\displaystyle\int_0^{0.25} x\tan^{-1}x\,dx$ **28.** $\displaystyle\int_0^{0.5} x^2\ln(1+x)\,dx$

29. $\displaystyle\int_0^{0.5} \dfrac{x}{1+x^2}\,dx$ **30.** $\displaystyle\int_0^{0.2} \dfrac{x^2}{1+x^2}\,dx$

31. $\displaystyle\int_0^{0.2} \dfrac{x}{1+x}\,dx$ **32.** $\displaystyle\int_0^{0.3} \dfrac{\tan^{-1}x}{x}\,dx$

33. $f(x) = \displaystyle\sum_{n=0}^{\infty} 2^n x^n$. Find $f''(0)$.

34. $f(x) = \displaystyle\sum_{n=1}^{\infty} nx^n$. Find $f''(0)$.

35. $f(x) = \displaystyle\sum_{n=0}^{\infty} \dfrac{1}{n!}x^{2n}$. Find $f'''(0)$.

36. $f(x) = \displaystyle\sum_{n=0}^{\infty} \dfrac{1}{(2n+1)!}x^{2n+1}$. Find $f'''(0)$.

37. $f(x) = \displaystyle\sum_{n=0}^{\infty} \dfrac{x^n}{n!}$. Show that $f'(x) = f(x)$.

38. $f(x) = 1 + kx + \dfrac{k(k-1)}{2!}x^2 + \dfrac{k(k-1)(k-2)}{3!}x^3 +$
$\dfrac{k(k-1)(k-2)(k-3)}{4!}x^4 + \cdots$, $|x| < 1$. Show that
$(1+x)f'(x) = kf(x)$.

39. If $c_n = a_0 b_n + a_1 b_{n-1} + a_2 b_{n-2} + \cdots + a_{n-1}b_1 + a_n b_0$, verify that

$(a_0 + a_1 x + a_2 x^2)(b_0 + b_1 x + b_2 x^2)$
$= c_0 + c_1 x + c_2 x^2 +$ (Terms of degree higher than two).

40. Suppose $f(x) = \displaystyle\sum_{n=1}^{\infty} a_n x^n$, for all x and that $a_1 > 0$. Then $f(0) = 0$. Show that there is an open interval I that contains zero such that zero is the only point in I for which $f(x) = 0$.

41. Show that $\displaystyle\sum_{n=1}^{\infty} \dfrac{(-1)^n}{n} = -\ln 2$ by showing that

$\left| -\ln 2 - \displaystyle\sum_{n=1}^{N} \dfrac{(-1)^n}{n} \right| \to 0$ as $N \to \infty$.

(*Hint:* We have

$$\ln \frac{1}{1-x} = \int_0^x \frac{1}{1-t}\,dt$$

$$= \int_0^x \left(1 + t + t^2 + \cdots + t^{N-1} + \frac{t^N}{1-t}\right) dt$$

$$= x + \frac{x^2}{2} + \frac{x^3}{3} + \cdots + \frac{x^N}{N} + \int_0^x \frac{t^N}{1-t}\,dt,$$

$$x < 1.$$

Show that

$$\int_0^x \frac{t^N}{1-t}\,dt \to 0 \text{ as } N \to \infty \text{ when } x = -1.)$$

42. Show that $\displaystyle\sum_{n=0}^{\infty} \frac{(-1)^n}{2n+1} = \frac{\pi}{4}$ by showing that

$$\left| \frac{\pi}{4} - \sum_{n=0}^{N} \frac{(-1)^n}{2n+1} \right| \to 0 \text{ as } N \to \infty.$$

(*Hint:* We have

$$\tan^{-1} x$$

$$= \int_0^x \frac{1}{1+t^2}\,dt$$

$$= \int_0^x \left(1 - t^2 + t^4 + \cdots + (-t^2)^N + \frac{(-t^2)^{N+1}}{1+t^2}\right) dt$$

$$= x - \frac{x^3}{3} + \frac{x^5}{5} + \cdots + \frac{(-1)^N x^{2N+1}}{2N+1} + \int_0^x \frac{(-t^2)^{N+1}}{1+t^2}\,dt,$$

$$-\infty < x < \infty.$$

Show that

$$\int_0^x \frac{t^{2N+2}}{1+t^2}\,dt \to 0 \text{ as } N \to \infty \text{ when } x = 1.)$$

9.9 TAYLOR POLYNOMIALS WITH REMAINDER; TAYLOR SERIES

Connections

Bounds of functions, 2.2.

Taylor polynomials, accuracy of approximation, 3.9.

Substitution into power series formulas, 9.8.

In the two previous sections we studied properties of power series and found power series representations of selected functions. Let us now consider the question:

Given a function f, can we find a power series representation of f?

Theorem 1 provides us with some important information about the answer to this question.

THEOREM 1

If $f(x) = \displaystyle\sum_{n=0}^{\infty} a_n (x-c)^n,\ |x-c| < R,\ R > 0$, then

$$a_n = \frac{f^{(n)}(c)}{n!}, \qquad n \geq 0.$$

Proof To verify this result, we evaluate the derivatives of f at the point c. We will illustrate the idea by carrying out the calculations for the first few terms. Differentiating the series term by term, we have

$$f(x) = a_0 + a_1(x-c) + a_2(x-c)^2 + a_3(x-c)^3 + a_4(x-c)^4 + \cdots,$$
$$f'(x) = a_1 + a_2(2)(x-c) + a_3(3)(x-c)^2 + a_4(4)(x-c)^3 + \cdots,$$
$$f''(x) = a_2(2) + a_3(3)(2)(x-c) + a_4(4)(3)(x-c)^2 + \cdots,$$
$$f'''(x) = a_3(3)(2) + a_4(4)(3)(2)(x-c) + \cdots.$$

Setting $x = c$ in the first of the above formulas, we obtain

$$f(c) = a_0, \text{ so } a_0 = \frac{f^{(0)}(c)}{0!},$$

since $f^{(0)}(c) = f(c)$ and $0! = 1$. Then, setting $x = c$ in each of the other three formulas above we obtain

$$f'(c) = a_1, \text{ so } a_1 = \frac{f^{(1)}(c)}{1!},$$

$$f''(c) = a_2(2), \text{ so } a_2 = \frac{f^{(2)}(c)}{2!},$$

$$f'''(c) = a_3(3)(2), \text{ so } a_3 = \frac{f^{(3)}(c)}{3!}.$$

This verifies the formula for a_n, $0 \le n \le 3$. Mathematical Induction could be used to verify the formula for all n. ∎

Taylor Series

Recall from Section 3.9 that the polynomial

$$P_n(x) = f(c) + f'(c)(x - c) + \frac{f''(c)}{2!}(x - c)^2 + \cdots + \frac{f^{(n)}(c)}{n!}(x - c)^n$$

is called the nth-order **Taylor polynomial** of f about c. The series

$$\sum_{n=0}^{\infty} \frac{f^{(n)}(c)}{n!}(x - c)^n$$

is called the **Taylor series** of f about c. If $c = 0$, the series is also called the **Maclaurin series** of f. Theorem 1 tells us, if f does have a power series representation about c, then the series must be the Taylor series of f about c. For example, the power series representations

$$\frac{1}{1 - x} = 1 + x + x^2 + x^3 + \cdots, \quad |x| < 1,$$

$$\ln\left(\frac{1}{1 - x}\right) = x + \frac{x^2}{2} + \frac{x^3}{3} + \frac{x^4}{4} + \cdots, \quad |x| < 1,$$

$$\ln\sqrt{\frac{1 + x}{1 - x}} = x + \frac{x^3}{3} + \frac{x^5}{5} + \frac{x^7}{7} + \cdots, \quad |x| < 1,$$

$$\tan^{-1} x = x - \frac{x^3}{3} + \frac{x^5}{5} - \frac{x^7}{7} + \cdots, \quad |x| < 1, \text{ and}$$

$$\frac{\tan^{-1} x}{x} = 1 - \frac{x^2}{3} + \frac{x^4}{5} - \frac{x^6}{7} + \cdots, \quad |x| < 1,$$

which were established in Section 9.8, are all Taylor series.

Let us now return to the original question. For a given function f, if f has a power series representation, it must be the Taylor series of f. If f has derivatives of all orders at c, we can form the Taylor series of f,

$$\sum_{n=0}^{\infty} \frac{f^{(n)}(c)}{n!}(x - c)^n.$$

We know that the series is convergent for $x = c$ and that

$$\sum_{n=0}^{\infty} \frac{f^{(n)}(c)}{n!}(c - c)^n = f(c).$$

It is possible that the series is convergent only for $x = c$. In this case, f does not have a power series representation about c, since any power series representation of f about c must be the Taylor series of f about c. If the series is convergent in an interval I of positive length with center c, then either

(i) $f(x) = \displaystyle\sum_{n=0}^{\infty} \frac{f^{(n)}(c)}{n!}(x - c)^n$ for all x in I,

so the Taylor series of f is a power series representation of f, or

(ii) $f(x) \neq \displaystyle\sum_{n=0}^{\infty} \frac{f^{(n)}(c)}{n!}(x - c)^n$ for some x in I.

In the latter case, f does not have a power series representation on I, since any power series representation of f on I must be the Taylor series of f.

An example of case (ii) is the function

$y = \begin{cases} e^{-1/x^2}, & x \neq 0 \\ 0, & x = 0 \end{cases}$

$$f(x) = \begin{cases} e^{-1/x^2}, & x \neq 0, \\ 0, & x = 0. \end{cases}$$

We can use the definition of the derivative as the limit of a difference quotient to show that $f^{(n)}(0) = 0$ for all $n \geq 1$. (See Exercise 31.) Hence, the Maclaurin series of f converges to zero for all x. This is not equal to $f(x)$, except for $x = 0$. The graph of f is indicated in Figure 9.9.1. Note that $f(x)$ is positive for $x \neq 0$.

FIGURE 9.9.1 The Maclaurin series of the function is zero for all x, but the function is zero only for $x = 0$.

Taylor Polynomials with Remainder

To investigate if the Taylor series of a function converges to the function, we define $R_n(x)$ to be the **remainder** when $P_n(x)$ is subtracted from $f(x)$. That is,

$$R_n(x) = f(x) - P_n(x)$$

$$= f(x) - \left(f(c) + f'(c)(x - c) + \frac{f''(c)}{2!}(x - c)^2 + \cdots \right.$$

$$\left. + \frac{f^{(n)}(c)}{n!}(x - c)^n \right).$$

Since $R_n(x)$ is the difference between $f(x)$ and $P_n(x)$, $P_n(x)$ approaches $f(x)$ as n approaches infinity if and only if $R_n(x)$ approaches zero as n approaches infinity. That is,

$$f(x) = \sum_{n=0}^{\infty} \frac{f^{(n)}(c)}{n!}(x - c)^n$$

at exactly those points x for which $R_n(x) \to 0$ as $n \to \infty$.

We can find a simple formula for $R_n(x)$ in case $f(x) = 1/(1 - x)$. The nth-order Taylor polynomial of $1/(1 - x)$ about $c = 0$ is

$$P_n(x) = 1 + x + x^2 + x^3 + \cdots + x^n.$$

We can then use Formula 2 of Section 9.2 to obtain

$$\frac{1}{1 - x} - (1 + x + x^2 + x^3 + \cdots + x^n) = \frac{1}{1 - x} - \frac{1 - x^{n+1}}{1 - x} = \frac{x^{n+1}}{1 - x}.$$

It follows that the nth remainder is

$$R_n(x) = f(x) - P_n(x) = \frac{x^{n+1}}{1-x}.$$

Thus, we see that $R_n(x) \to 0$ as $n \to \infty$ for $|x| < 1$, as expected, since

$$\frac{1}{1-x} = 1 + x + x^2 + x^3 + \cdots, \quad |x| < 1.$$

If f is any function that has derivatives of all orders in some open interval that contains c, the following theorem provides us with a concise formula for the remainder term $R_n(x)$.

TAYLOR'S FORMULA WITH REMAINDER

If f, f', f'', ..., $f^{(n)}$ and $f^{(n+1)}$ are continuous on the closed interval between x and c, then there is some point z between x and c such that

$$f(x) = f(c) + f'(c)(x-c) + \frac{f''(c)}{2!}(x-c)^2 + \cdots$$
$$+ \frac{f^{(n)}(c)}{n!}(x-c)^n + \frac{f^{(n+1)}(z)}{(n+1)!}(x-c)^{(n+1)}.$$

Hence,

$$R_n(x) = \frac{f^{(n+1)}(z)}{(n+1)!}(x-c)^{n+1} \text{ for some } z \text{ between } x \text{ and } c.$$

The remainder can also be expressed in integral form as

$$R_n(x) = \int_c^x \frac{f^{(n+1)}(t)}{n!}(x-t)^n dt.$$

Proof   The theorem can be verified by applying Rolle's Theorem to an auxiliary function g. This is the same idea that was used to verify the Mean Value Theorem in Section 3.8 and Taylor's Formulas in Section 3.9. We will only outline the steps here.

For x and c fixed and values of the variable t between x and c, we set

$$g(t) = f(x) - f(t) - f'(t)(x-t) - \frac{f''(t)}{2!}(x-t)^2 - \cdots$$
$$- \frac{f^{(n)}(t)}{n!}(x-t)^n - R_n(x)\frac{(x-t)^{n+1}}{(x-c)^{n+1}},$$

where $R_n(x) = f(x) - P_n(x)$. It is clear that $g(x) = 0$. Also, $g(c) = 0$. Differentiating $g(t)$ with respect to t, and noting the cancellation of terms that occurs, we obtain

$$g'(t) = -\frac{f^{(n+1)}(t)}{n!}(x-t)^n + R_n(x)(n+1)\frac{(x-t)^n}{(x-c)^{n+1}}.$$

We can then apply Rolle's Theorem to obtain a point z between x and c such that $g'(z) = 0$. Substituting the formula for g' in the equation $g'(z) = 0$, and solving for $R_n(x)$, we obtain the first form of the remainder. To obtain the integral form of the remainder, we note that $\displaystyle\int_c^x g'(t)dt = g(x) - g(c) = 0$. The integral form is obtained by substituting the formula for g' in the equation $\displaystyle\int_c^x g'(t)dt = 0$. ■

Examples

EXAMPLE 1

Show that

$$e^x = 1 + x + \frac{x^2}{2!} + \frac{x^3}{3!} + \cdots \text{ for all } x.$$

SOLUTION   We first calculate the coefficients of the Maclaurin series of $f(x) = e^x$. We have $f^{(n)}(x) = e^x$, so $f^{(n)}(0) = 1$ for all $n \geq 0$. This shows that

$$\frac{f^{(n)}(c)}{n!} = \frac{1}{n!},$$

so the Maclaurin series of e^x is

$$1 + x + \frac{x^2}{2!} + \frac{x^3}{3!} + \cdots.$$

We must show that this series converges to e^x for all x. We will show that $R_n(x) \to 0$ as $n \to \infty$ for all x. We have

$$R_n(x) = \frac{f^{(n+1)}(z)}{(n+1)!}x^{n+1} = \frac{e^z}{(n+1)!}x^{n+1}$$

for some z between x and 0. If $x \geq 0$, then $0 \leq z \leq x$, so $e^z \leq e^x$. If $x < 0$, then $x < z < 0$, so $e^z < e^0 = 1$. We then have

$$|R_n(x)| \leq \begin{cases} \dfrac{e^x}{(n+1)!}|x|^{n+1}, & x \geq 0, \\[2ex] \dfrac{1}{(n+1)!}|x|^{n+1}, & x < 0. \end{cases}$$

Each of the expressions on the right above approaches zero as n approaches infinity for every fixed value of x. This is a consequence, for example, of the fact that

$$\sum_{n=1}^{\infty} \frac{|x|^{n+1}}{(n+1)!}$$

converges, as was noted in Statement 1 of Section 9.7. It follows that $R_n(x)$ approaches zero as n approaches infinity, so the Maclaurin series of e^x converges to e^x for all x. The graph of e^x and some of the approximating Taylor polynomials are sketched in Figure 9.9.2. ■

EXAMPLE 2

Show that

$$\sin x = x - \frac{x^3}{3!} + \frac{x^5}{5!} - \frac{x^7}{7!} + \cdots \text{ for all } x.$$

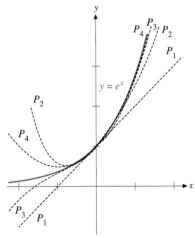

FIGURE 9.9.2   The Maclaurin series of e^x converges to e^x for all x.

SOLUTION Calculations for determining the Maclaurin series of $\sin x$ are given in Table 9.9.1.

TABLE 9.9.1

n	$f^{(n)}(x)$	$f^{(n)}(0)$	$f^{(n)}(0)/n!$
0	$\sin x$	0	0
1	$\cos x$	1	$\dfrac{1}{1!}$
2	$-\sin x$	0	0
3	$-\cos x$	-1	$-\dfrac{1}{3!}$
4	$\sin x$	0	0
5	$\cos x$	1	$\dfrac{1}{5!}$
$\vdots$			

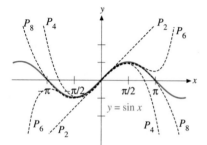

FIGURE 9.9.3 The Maclaurin series of $\sin x$ converges to $\sin x$ for all x.

The series will converge to $\sin x$ if $R_n(x)$ approaches zero as n approaches infinity. We see that $f^{(n+1)}(z)$ is either $\pm \sin z$ or $\pm \cos z$. Since the sine and cosine functions are both bounded by one, we have

$$|R_n(x)| = \left| \frac{f^{(n+1)}(z)}{(n+1)!} x^{n+1} \right| \leq \frac{|x|^{n+1}}{(n+1)!}.$$

The expression on the right above and, hence, $R_n(x)$ do approach zero as n approaches infinity for any fixed x, as was indicated in Example 1. This means that the Maclaurin series of $\sin x$ converges to $\sin x$ for all x. The graph of $\sin x$ and some of the approximating Taylor polynomials are sketched in Figure 9.9.3. ■

EXAMPLE 3

Find the Taylor series of $\sin x$ about $\pi/4$ and show that it converges to $\sin x$ for all x.

SOLUTION Calculations for determining the Taylor series of $\sin x$ about $\pi/4$ are given in Table 9.9.2.

TABLE 9.9.2

n	$f^{(n)}(x)$	$f^{(n)}(\pi/4)$	$f^{(n)}(\pi/4)/n!$
0	$\sin x$	$\dfrac{1}{\sqrt{2}}$	$\dfrac{1}{\sqrt{2}}$
1	$\cos x$	$\dfrac{1}{\sqrt{2}}$	$\dfrac{1}{\sqrt{2}}$
2	$-\sin x$	$-\dfrac{1}{\sqrt{2}}$	$-\dfrac{1}{2\sqrt{2}}$
3	$-\cos x$	$-\dfrac{1}{\sqrt{2}}$	$-\dfrac{1}{6\sqrt{2}}$
4	$\sin x$	$\dfrac{1}{\sqrt{2}}$	$\dfrac{1}{24\sqrt{2}}$
5	$\cos x$	$\dfrac{1}{\sqrt{2}}$	$\dfrac{1}{120\sqrt{2}}$
$\vdots$			

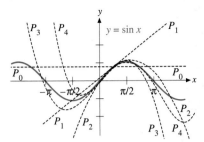

FIGURE 9.9.4 The Taylor series of $\sin x$ about $\pi/4$ converges to $\sin x$ for all x.

The Taylor series of $\sin x$ about $\pi/4$ is

$$\frac{1}{\sqrt{2}} + \frac{1}{\sqrt{2}}\left(x - \frac{\pi}{4}\right) - \frac{1}{2\sqrt{2}}\left(x - \frac{\pi}{4}\right)^2 - \frac{1}{6\sqrt{2}}\left(x - \frac{\pi}{4}\right)^3$$
$$+ \frac{1}{24\sqrt{2}}\left(x - \frac{\pi}{4}\right)^4 + \cdots.$$

The series will converge to $\sin x$ if $R_n(x)$ approaches zero as n approaches infinity. Since $f^{(n+1)}(z)$ is either $\pm \sin z$ or $\pm \cos z$, we can verify, as in Example 2, that $R_n(x)$ approaches zero as n approaches infinity for all x. The function $\sin x$ and some of the Taylor polynomials about $\pi/4$ are sketched in Figure 9.9.4. ■

It is not always necessary to evaluate the derivatives $f^{(n)}(c)$ in order to find a Taylor series. Theorem 1 tells us that if we arrive at a power series representation of f by any (legitimate) method, the power series is the Taylor series of f.

EXAMPLE 4
Show that

$$\cos x = 1 - \frac{x^2}{2!} + \frac{x^4}{4!} - \frac{x^6}{6!} + \cdots \qquad \text{for all } x.$$

SOLUTION We have

$$\cos x = \frac{d}{dx}(\sin x) = \frac{d}{dx}\left(x - \frac{x^3}{3!} + \frac{x^5}{5!} + \cdots\right)$$
$$= 1 - \frac{3x^2}{3!} + \frac{5x^4}{5!} - \frac{7x^6}{7!} + \cdots$$
$$= 1 - \frac{x^2}{2!} + \frac{x^4}{4!} - \frac{x^6}{6!} + \cdots.$$

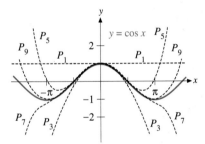

FIGURE 9.9.5 The Maclaurin series of $\cos x$ converges to $\cos x$ for all x.

This is the Maclaurin series of $\cos x$. We can see from Figure 9.9.5 how the Taylor polynomials P_n begin to approach $\cos x$ as n increases. ■

EXAMPLE 5
Show that

$$\frac{\sin x}{x} = 1 - \frac{x^2}{3!} + \frac{x^4}{5!} - \frac{x^6}{7!} + \cdots.$$

SOLUTION We assume that $(\sin x)/x$ is defined to be 1 at $x = 0$. (Since $\lim_{x \to 0}(\sin x)/x = 1$, this choice makes $(\sin x)/x$ continuous.) For $x \neq 0$, Example 2 gives

$$\frac{\sin x}{x} = \frac{x - \dfrac{x^3}{3!} + \dfrac{x^5}{5!} - \dfrac{x^7}{7!} + \cdots}{x}$$
$$= \frac{x\left(1 - \dfrac{x^2}{3!} + \dfrac{x^4}{5!} - \dfrac{x^6}{7!} + \cdots\right)}{x}$$
$$= 1 - \frac{x^2}{3!} + \frac{x^4}{5!} - \frac{x^6}{7!} + \cdots.$$

Since the value of the series agrees with the (assigned) value of $(\sin x)/x$ at $x = 0$, the above formula holds for all x. The series is the Maclaurin series of the function. ∎

Using the Alternating Series Test to Determine Accuracy

EXAMPLE 6

Approximate $\displaystyle\int_0^{0.5} \sin x^2\, dx$ with error less than 0.001.

SOLUTION   Substituting x^2 for x in the series established in Example 2, we have

$$\int_0^{0.5} \sin x^2\, dx = \int_0^{0.5} \left((x^2) - \frac{(x^2)^3}{3!} + \frac{(x^2)^5}{5!} - \frac{(x^2)^7}{7!} + \frac{(x^2)^9}{9!} - \cdots \right) dx$$

$$= \int_0^{0.5} \left(x^2 - \frac{x^6}{6} + \frac{x^{10}}{120} - \cdots \right) dx$$

$$= \frac{(0.5)^3}{3} - \frac{(0.5)^7}{42} + \frac{(0.5)^{11}}{1320} - \cdots .$$

The Alternating Series Test applies and tells us that the error in using a partial sum to approximate the sum is less than the absolute value of the first term omitted. Since

$$\frac{(0.5)^7}{42} = 0.00018601 < 0.001,$$

we have

$$\int_0^{0.5} \sin x^2\, dx \approx \frac{(0.5)^3}{3} \approx 0.0417,$$

with error less than 0.001. ∎

Using the Remainder Term to Determine Accuracy

Many calculations that involve power series involve alternating series and we can use the Alternating Series Test to determine the accuracy of the calculations. In case we do not have an alternating series, we can use estimates of the remainder term to determine accuracy.

In Section 7.3 we showed that $2 < e < 4$. We can now use Taylor series to greatly improve this preliminary estimate of the value of e.

EXAMPLE 7

Assuming that you know that $e < 4$, use the Maclaurin series of e^x to approximate the value of e with error less than 0.0001.

SOLUTION   Note that $e = e^1$. We use partial sums of the Maclaurin series of e^x with $x = 1$ to obtain approximate values of e. We have

$$e^x = 1 + x + \frac{1}{2!}x^2 + \frac{1}{3!}x^3 + \frac{1}{4!}x^4 + \cdots .$$

The partial sum that contains n nonzero terms is

$$P_{n-1}(x) = 1 + x + \frac{1}{2!}x^2 + \frac{1}{3!}x^3 + \frac{1}{4!}x^4 + \cdots + \frac{1}{(n-1)!}x^{n-1}.$$

We know from the solution of Example 1 that

$$|R_{n-1}(x)| \leq \frac{e^x}{n!}|x|^n, \quad x \geq 0.$$

Using $x = 1$ and the fact that $e < 4$, we have

$$|R_{n-1}(1)| \leq \frac{4^1}{n!}|1|^n = \frac{4}{n!}.$$

In Table 9.9.3 we have tabulated values of $4/n!$ in order to find an integer n that gives the desired accuracy.

TABLE 9.9.3

n	$4/n!$
2	2
3	0.66666667
4	0.16666667
5	0.03333333
6	0.00555556
7	0.00079365
8	0.00009921 < 0.0001

We see that we need $n = 8$ nonzero terms of the Maclaurin series of e^x with $x = 1$. We have

$$e \approx P_7(1) = 1 + \frac{1}{1!} + \frac{1}{2!} + \frac{1}{3!} + \frac{1}{4!} + \frac{1}{5!} + \frac{1}{6!} + \frac{1}{7!} = 2.718254.$$

This gives an approximate value of e with error less than $0.00009921 < 0.0001$. (A calculator gives $e \approx 2.7182818$.) ■

We can use bounds of $f^{(n+1)}$ to determine a set of values of x for which $P_n(x)$ is within a prescribed positive distance ϵ of $f(x)$. That is, if B is a bound of $f^{(n+1)}(z)$ for z between x and c, then Taylor's Formula with Remainder implies

$$|f(x) - P_n(x)| = |R_n(x)| = \frac{|f^{n+1}(z)|}{(n+1)!}|x - c|^{n+1} \leq \frac{B}{(n+1)!}|x - c|^{n+1}.$$

It follows that

$$|f(x) - P_n(x)| < \epsilon \text{ if } x \text{ satisfies } |x - c| < \left(\frac{(n+1)!\epsilon}{B}\right)^{1/(n+1)}.$$

EXAMPLE 8

Use a bound of the third derivative of e^x for $|x| < 0.5$ to determine a set of values of x for which

$$\left| e^x - \left(1 + x + \frac{x^2}{2}\right) \right| < 0.005.$$

SOLUTION The expression $1 + x + \dfrac{x^2}{2}$ is the second-order Taylor polynomial of $f(x) = e^x$ about zero. Taylor's Formula then implies

$$\left| e^x - \left(1 + x + \frac{x^2}{2} \right) \right| = |R_2| = \left| \frac{f'''(z)}{3!} x^3 \right| = \frac{e^z}{6} |x|^3 \le \frac{e^{|x|}}{6} |x|^3,$$

since z between x and 0 implies that $e^z \le e^{|x|}$. Using the preliminary assumption that $|x| \le 0.5$ to find a bound of the factor $e^{|x|}$, we then have

$$\left| e^x - \left(1 + x + \frac{x^2}{2} \right) \right| \le \frac{e^{|x|}}{6} |x|^3 \le \frac{e^{0.5}}{6} |x|^3.$$

The expression $\left| e^x - \left(1 + x + \dfrac{x^2}{2} \right) \right|$ is then less than 0.005 if $|x| \le 0.5$ and

$$\frac{e^{0.5}}{6} |x|^3 < 0.005 \qquad \text{or} \qquad |x| < \left(\frac{6(0.005)}{e^{0.5}} \right)^{1/3} \approx 0.26.$$

Noting that our preliminary assumption is satisfied if $|x| < 0.26$, we then conclude that

$$\left| e^x - \left(1 + x + \frac{x^2}{2} \right) \right| < 0.005 \text{ if } |x| < 0.26.$$

Note that it is possible (and true) that $\left| e^x - \left(1 + x + \dfrac{x^2}{2} \right) \right| < 0.005$ for values of x with $|x| \ge 0.26$. ∎

Summary of Formulas

Let us restate the basic formulas that were developed in this section. These formulas will be needed for the exercises.

$$e^x = \sum_{n=0}^{\infty} \frac{x^n}{n!} = 1 + x + \frac{x^2}{2!} + \frac{x^3}{3!} + \frac{x^4}{4!} + \frac{x^5}{5!} + \cdots,$$

$$\sin x = \sum_{n=0}^{\infty} (-1)^n \frac{x^{2n+1}}{(2n+1)!} = x - \frac{x^3}{3!} + \frac{x^5}{5!} - \frac{x^7}{7!} + \frac{x^9}{9!} - \cdots,$$

$$\cos x = \sum_{n=0}^{\infty} (-1)^n \frac{x^{2n}}{(2n)!} = 1 - \frac{x^2}{2!} + \frac{x^4}{4!} - \frac{x^6}{6!} + \frac{x^7}{7!} - \cdots.$$

EXERCISES 9.9

Use values of the derivatives at c to find the Taylor series about c of the functions in Exercises 1–6.

1. $f(x) = x^3 + 3x - 1, c = 1$

2. $f(x) = x^4 - x^2 + 1, c = -1$

3. $f(x) = (x - 2)^3, c = 0$

4. $f(x) = x^3 - 9x^2 + 27x - 27, c = 3$

5. $f(x) = \sinh x, c = 0$

6. $f(x) = \cosh x, c = 0$

Use established formulas to find the Maclaurin series of the functions in Exercises 7–14.

7. $f(x) = \sin 2x$

8. $f(x) = \cos(x/2)$

9. $f(x) = e^{-x^2}$

10. $f(x) = xe^{-x}$

11. $f(x) = \cos^2 x$

12. $f(x) = \cos x^2$

13. $f(x) = \dfrac{1 - \cos x}{x^2}$

14. $f(x) = \dfrac{1 - e^x}{x}$

If $f(x) = \displaystyle\sum_{n=0}^{\infty} a_n x^n$ for x in some open interval that contains zero, then Theorem 1 implies $a_n = f^{(n)}(0)/n!$. Use this fact to evaluate the derivatives in Exercises 15–18 at $x = 0$.

15. $\dfrac{d^5}{dx^5}(x^2 \sin 2x)$

16. $\dfrac{d^5}{dx^5}\left(\dfrac{x - \sin x}{x^3}\right)$

17. $\dfrac{d^{100}}{dx^{100}}(x \cos x)$

18. $\dfrac{d^{99}}{dx^{99}}(e^{x^2})$

Use Taylor series to approximate the expressions in Exercises 19–26 with error less than 0.001.

19. $\sin 0.125$

20. $\cos 0.25$

21. $1/\sqrt{e}$

22. e^{-2}

23. $\displaystyle\int_0^{0.5} \dfrac{\sin x}{x}\, dx$

24. $\displaystyle\int_0^{0.2} \dfrac{1 - \cos x}{x^2}\, dx$

25. $\displaystyle\int_0^{0.2} e^{-x^2}\, dx$

26. $\displaystyle\int_0^{0.3} x \sin x\, dx$

27. Use a bound of the third derivative of $\sin x$ to determine a set of values of x for which $|\sin x - x| < 0.001$. Note that $x = P_1 = P_2$, so we can use R_2.

28. Use a bound of the fourth derivative of $\cos x$ to determine a set of values of x for which $\left|\cos x - \left(1 - \dfrac{x^2}{2}\right)\right| < 0.005$.

Note that $1 - \dfrac{x^2}{2} = P_2 = P_3$, so we can use R_3.

29. Use a bound of the third derivative of e^{-x} for $x \geq 0$ to determine a set of nonnegative values of x for which $\left|e^{-x} - \left(1 - x + \dfrac{x^2}{2}\right)\right| < 0.005$.

30. Use two terms of the Maclaurin series of $\sin x^2$ to approximate $\displaystyle\int_0^1 \sin t^2\, dt$. Use a bound of $\displaystyle\int_0^1 R_4(t^2)\, dt$, where $R_4(x) = \dfrac{\sin^{(5)} z}{5!} x^5$, to obtain a bound of the error. **(Calculus Explorer & Tutor II, Sequences and Series, 15.)**

31. $f(x) = \begin{cases} e^{-1/x^2}, & x \neq 0, \\ 0, & x = 0. \end{cases}$

 (a) Show that $\displaystyle\lim_{x \to 0} \dfrac{f(x) - f(0)}{x - 0} = 0$. This shows that $f'(0) = 0$.

 (b) Show that $\displaystyle\lim_{x \to 0} \dfrac{f'(x) - f'(0)}{x - 0} = 0$. This shows that $f''(0) = 0$. (Mathematical Induction could be used to show that $f^{(n)}(0) = 0$ for all n.)

32. $f(x) = \begin{cases} 1 - x, & x \leq 1, \\ 0, & x > 1. \end{cases}$ Find the Maclaurin series of f.

For what values of x does the series converge? For what values of x does the series converge to $f(x)$?

33. Show that

$$\frac{1}{x} = 1 - (x - 1) + (x - 1)^2 - (x - 1)^3 + (x - 1)^4 - \cdots,$$
$$|x - 1| < 1.$$

$$\left(\text{Hint: } \frac{1}{x} = \frac{1}{1 + (x - 1)}.\right)$$

Use the result of Exercise 33 to find power series representations about $c = 1$ for the functions in Exercises 34–35.

34. $\ln x$

35. $1/x^2$

36. Find the Maclaurin series of $\sinh x$ by using the Maclaurin series of e^x and e^{-x}.

9.10 BINOMIAL SERIES

Connections

Substitution into power series formulas, 9.8–9.

Remainder term in Taylor's formula, 9.9.

The Maclaurin series of $(1 + x)^k$ is called the **binomial series.** We have

$$(1 + x)^k = 1 + kx + \frac{k(k - 1)}{2!}x^2 + \frac{k(k - 1)(k - 2)}{3!}x^3 \qquad (1)$$
$$+ \frac{k(k - 1)(k - 2)(k - 3)}{4!}x^4 + \cdots, \qquad |x| < 1.$$

To verify Formula 1, we first evaluate the coefficients of the Maclaurin series of $(1 + x)^k$. This is done in Table 9.10.1. We see that $f^{(n)}(0)/n! = k(k - 1)(k - 2) \cdots (k - n + 1)/n!$, $n > 0$. This shows that the series is the Maclaurin series of $(1 + x)^k$. The Ratio Test can be used to determine that the series is convergent for $|x| < 1$, but we must verify that the series has sum $(1 + x)^k$. This can be accomplished by showing that the integral form of the remainder $R_n(x)$ approaches zero as n approaches infinity. Later in this section we will illustrate how this is done by considering the special case $k = -1/2$. For now, let us see how the formula is used.

TABLE 9.10.1

n	$f^{(n)}(x)$	$f^{(n)}(0)$	$f^{(n)}(0)/n!$
0	$(1+x)^k$	1	1
1	$k(1+x)^{k-1}$	k	k
2	$k(k-1)(1+x)^{k-2}$	$k(k-1)$	$k(k-1)/2!$
3	$k(k-1)(k-2)(1+x)^{k-3}$	$k(k-1)(k-2)$	$k(k-1)(k-2)/3!$
$\vdots$			

If k is a positive integer, Formula 1 contains only a finite number of nonzero terms. In this case (1) reduces to the usual binomial formula and is true for all x. For example, substitution of $k = 2$ and $k = 3$ in Formula 1 gives

$$(1+x)^2 = 1 + 2x + x^2 + 0x^3 + 0x^3 + 0x^4 + \cdots = 1 + 2x + x^2$$

and

$$(1+x)^3 = 1 + 3x + 3x^2 + x^3 + 0x^4 + 0x^5 + \cdots = 1 + 3x + 3x^2 + x^3,$$

respectively.

EXAMPLE 1
Show that

$$(1+x)^{-2} = 1 - 2x + 3x^2 - 4x^3 + \cdots, \quad |x| < 1.$$

SOLUTION This is immediate from (1) with $k = -2$. ∎

EXAMPLE 2
Show that

$$\sqrt{1-x} = 1 - \frac{1}{2}x - \frac{1}{8}x^2 - \frac{1}{16}x^3 - \cdots, \quad |x| < 1.$$

SOLUTION This follows from (1) with $k = 1/2$ and x replaced by $-x$. ∎

EXAMPLE 3
Show that

$$\sqrt{x} = 1 + \frac{1}{2}(x-1) - \frac{1}{8}(x-1)^2 + \frac{1}{16}(x-1)^3 - \cdots, \quad |x-1| < 1.$$

SOLUTION Write $\sqrt{x} = \sqrt{1+(x-1)}$ and then substitute in (1) with $k = 1/2$ and $x-1$ in place of x. ∎

EXAMPLE 4
Show that

$$\frac{1}{\sqrt{1-x^2}} = 1 + \frac{1}{2}x^2 + \frac{3}{8}x^4 + \frac{5}{16}x^6 + \cdots, \quad |x| < 1.$$

SOLUTION Use (1) with $k = -1/2$ and x replaced by $-x^2$. ∎

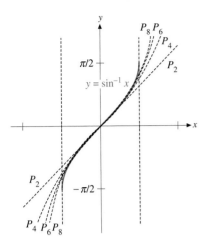

FIGURE 9.10.1 The Maclaurin series of $\sin^{-1} x$ converges to $\sin^{-1} x$ for $-1 < x < 1$.

EXAMPLE 5

Show that

$$\sin^{-1} x = x + \frac{1}{6}x^3 + \frac{3}{40}x^5 + \frac{5}{112}x^7 + \cdots, \quad |x| < 1.$$

SOLUTION   Using the results of Example 4, we have

$$\sin^{-1} x = \int_0^x \frac{1}{\sqrt{1 - t^2}} dt$$

$$= \int_0^x \left(1 + \frac{1}{2}t^2 + \frac{3}{8}t^4 + \frac{5}{16}t^6 + \cdots \right) dt$$

$$= x + \frac{1}{6}x^3 + \frac{3}{40}x^5 + \frac{5}{112}x^7 + \cdots, \quad |x| < 1.$$

The graph of $\sin^{-1} x$ and some of the approximating Taylor polynomials are given in Figure 9.10.1. ■

Let us restate the result of Example 5:

$$\sin^{-1} x = x + \frac{1}{6}x^3 + \frac{3}{40}x^5 + \frac{5}{112}x^7 + \cdots, \quad |x| < 1.$$

Approximating Function Values

EXAMPLE 6

Use the first four nonzero terms of the Maclaurin series of $\sin^{-1} x$ to find an approximate value of $\sin^{-1} 0.3$. Compare with the calculator value of $\sin^{-1} 0.3$.

SOLUTION   From Example 5 we see that the desired Taylor polynomial of $\sin^{-1} x$ is

$$P_8(x) = x + \frac{1}{6}x^3 + \frac{3}{40}x^5 + \frac{5}{112}x^7.$$

Substituting 0.3 for x, we obtain

$$P_8(0.3) = 0.3 + \frac{1}{6}(0.3)^3 + \frac{3}{40}(0.3)^5 + \frac{5}{112}(0.3)^7 = 0.30469201.$$

A calculator gives $\sin^{-1} 0.3 \approx 0.30469265$ rad. ■

Verification of Special Case of Binomial Series

Let us now return to the verification of Formula 1 for the binomial series in the case $k = -1/2$. We will show that the integral form of the remainder approaches zero as n approaches infinity for $|x| < 1$. Recall from Taylor's Formula with Remainder in Section 9.9 that the integral form of the remainder is

$$R_n(x) = \int_0^x \frac{f^{(n+1)}(t)}{n!}(x - t)^n dt.$$

For $f(t) = (1 + t)^{-1/2}$, we have

$$f^{(n+1)}(t)$$

$$= \left(-\frac{1}{2}\right)\left(-\frac{3}{2}\right)\left(-\frac{5}{2}\right)\cdots\left(-\frac{2n-1}{2}\right)\left(-\frac{2n+1}{2}\right)(1+t)^{-(2n+3)/2}$$

$$= (-1)^{n+1}\frac{(1)(3)(5)\cdots(2n-1)(2n+1)}{2^{n+1}}(1+t)^{-n-3/2}.$$

Substitution then gives

$$R_n(x) = \int_0^x (-1)^{n+1}\frac{(1)(3)(5)\cdots(2n-1)(2n+1)}{2^{n+1}n!}(1+t)^{-n-3/2}(x-t)^n\,dt$$

$$= \int_0^x (-1)^{n+1}\frac{(1)(3)(5)\cdots(2n-1)(2n+1)}{(2)(4)(6)\cdots(2n)(2)}(1+t)^{-3/2}\left(\frac{x-t}{1+t}\right)^n\,dt.$$

We can now obtain a bound of $R_n(x)$ by finding a bound of each of the factors in the integrand. We have

$$\left|(-1)^{n+1}\frac{(1)(3)(5)\cdots(2n-1)(2n+1)}{(2)(4)(6)\cdots(2n)(2)}\right|$$

$$= \left(\frac{1}{2}\right)\left(\frac{3}{4}\right)\left(\frac{5}{6}\right)\cdots\left(\frac{2n-1}{2n}\right)\left(\frac{2n+1}{2}\right)$$

$$< (1)(1)(1)\cdots(1)\left(\frac{2n+1}{2}\right) = \frac{2n+1}{2}.$$

The conditions $|x| < 1$, t between x and 0, imply

$$|1+t|^{-3/2} \leq \begin{cases} 1, & 0 \leq x < 1, \\ (1-|x|)^{-3/2}, & -1 < x < 0 \end{cases}$$

(the negative power of $|1 + t|$ is largest when $1 + t$ is nearest zero) and

$$\left|\frac{x-t}{1+t}\right|^n \leq |x|^n$$

(find the maximum value of $|(x - t)/(1 + t)|$ as a function of t for each of the cases $0 \leq t \leq x < 1$ and $-1 < x \leq t \leq 0$).

We then have

$$|R_n(x)| \leq \int_0^{|x|}\frac{2n+1}{2}(1-|x|)^{-3/2}|x|^n\,dt$$

$$= \frac{2n+1}{2}(1-|x|)^{-3/2}|x|^{n+1},$$

where the factor $(1 - |x|)^{-3/2}$ is not needed if x is positive. We can show $(2n + 1)|x|^{n+1}$ approaches zero as n approaches infinity for $|x| < 1$. It follows that $R_n(x)$ also approaches zero for $|x| < 1$.

EXERCISES 9.10

Use established series to find the Maclaurin series of the functions in Exercises 1–12.

1. $f(x) = \sqrt{1 + x}$

2. $f(x) = \dfrac{1}{\sqrt{1 + x}}$

3. $f(x) = \dfrac{1}{(1 - x)^{1/3}}$

4. $f(x) = (1 + x)^4$

5. $f(x) = \sqrt{4 + x^2}$

6. $f(x) = \dfrac{1}{\sqrt{9 + x^2}}$

7. $f(x) = \dfrac{1}{(x + 2)^3}$

8. $f(x) = \dfrac{x}{(x - 2)^2}$

9. $f(x) = \sqrt{2x + 1}$

10. $f(x) = (3x - 1)^{1/3}$

11. $f(x) = \dfrac{\sin^{-1} x}{x}$

12. $f(x) = \sin^{-1} 2x$

Use the first four nonzero terms of the Maclaurin series to approximate the expressions in Exercises 13–18.

13. $\sin^{-1} 0.15$

14. $\sin^{-1} 0.25$

15. $\displaystyle\int_0^{0.5} \dfrac{x^2}{\sqrt{1 + x^4}}\, dx$

16. $\displaystyle\int_0^{0.3} \dfrac{1}{\sqrt{1 + x^4}}\, dx$

17. $\displaystyle\int_0^{0.2} \sin^{-1} x\, dx$

18. $\displaystyle\int_0^{1} \sqrt{4 + x^2}\, dx$

REVIEW OF CHAPTER 9 CONCEPTS

- Limits of sequences of real numbers correspond to the limit at infinity of functions of a real variable. We can use the methods of real variable functions to evaluate limits of sequences.
- Bounded, monotone sequences are convergent.
- Be sure to distinguish between a sequence and a series. Think of a sequence as a list of terms and a series as a sum of terms.
- A convergent infinite series has value that is the limit of its partial sums. An infinite series is not assigned a value if the sequence of its partial sums is divergent.
- Partial sums of geometric series and telescopic series can be expressed by simple formulas from which convergence or divergence and the sum of the convergent series can be determined.
- The nth term of a convergent series must approach zero, but a series may be divergent even though its nth term approaches zero.
- Series with nonnegative terms are convergent if and only if the sequence of their partial sums is bounded.
- If the Integral Test applies to a convergent series, then the inequality $|S - S_N| < \displaystyle\int_N^{\infty} f(x)\, dx$ can be used to obtain a bound of the error in using S_N as an approximate value of the series.
- For series with positive terms the following strategy may be used for determining convergence or divergence:

If a_n does not approach zero, then $\displaystyle\sum_{n=1}^{\infty} a_n$ diverges.

This condition may be discovered while using either the Comparison, Ratio, or Root Test. Remember that a series may diverge even though its nth term does approach zero. Showing that a_n approaches zero is no help in determining if the series is convergent or divergent.

If the terms are either less than those of a known convergent series or greater than those of a known divergent series, use the Comparison Test.

If the terms involve only powers and roots, use the Limit Version of the Comparison Test. Factor the highest power of n from each of the numerator and denominator to obtain $a_n = \dfrac{1}{n^p} b_n$, where b_n has finite, nonzero limit as n approaches infinity. The p-series $\displaystyle\sum_{n=1}^{\infty} \dfrac{1}{n^p}$ converges if $p > 1$ and diverges if $p \le 1$.

If the terms contain factorials and/or nth powers of constants, use the Ratio Test. $\displaystyle\sum_{n=1}^{\infty} a_n$ converges if $\displaystyle\lim_{n \to \infty} \dfrac{a_{n+1}}{a_n} < 1$ and diverges if $\displaystyle\lim_{n \to \infty} \dfrac{a_{n+1}}{a_n} > 1$. The Ratio Test gives no information if the limit of a_{n+1}/a_n is equal to one.

If the terms involve variable expressions raised to the nth power, the Root Test could be helpful. $\displaystyle\sum_{n=1}^{\infty} a_n$ converges if $\displaystyle\lim_{n \to \infty} (a_n)^{1/n} < 1$ and diverges if $\displaystyle\lim_{n \to \infty} (a_n)^{1/n} > 1$. The Root Test gives no information if the limit of $(a_n)^{1/n}$ is equal to one. It is useful to note that $\displaystyle\lim_{n \to \infty} (an + b)^{1/n} = 1$.

If the conditions of the Integral Test hold and we can integrate the function f with $f(n) = a_n$, we can use the Integral Test. We can use the Limit Version of the Comparison Test to replace the given series with a series for which the corresponding function is easier to integrate.

- A series for which successive terms are of opposite sign is called an alternating series.
- If the conditions of the Alternating Series Test hold, the formula $|S - S_N| < |a_{N+1}|$ tells us that the error in approximating the series is less than the absolute value of the first term that is excluded from the approximating sum.
- A series is absolutely convergent if the series formed by the absolute values of its terms is convergent. Absolutely convergent series are convergent.
- A series that is convergent, but not absolutely convergent, is called conditionally convergent.
- Power series $\sum_{n=0}^{\infty} a_n(x - c)^n$ converge at all points of some interval with center c. The series may or may not converge at each endpoint of the interval.
- Within their radius of convergence, we can add, subtract, multiply, differentiate, and integrate power series term by term.
- We can substitute into known formulas to obtain new formulas for power series representation of functions.
- We can use partial sums of a power series representation of a function to obtain approximate values of a function and definite integrals of the function.
- If a function is represented by a power series, then the series must be a Taylor series of the function.
- The difference between a function and its nth order Taylor polynomial is denoted $R_n(x)$ and is called the nth remainder term. Taylor's Formula with Remainder gives a simple form of the remainder in terms of an unknown point that is between x and c.
- The Taylor series of a function converges to the function for a particular value of x if and only if the value at x of the remainder given by Taylor's Formula approaches zero.

- Some of the Taylor series that have been developed in this chapter are listed below, for reference.

$$\frac{1}{1 - x} = \sum_{n=0}^{\infty} x^n$$
$$= 1 + x + x^2 + x^3 + x^4 + \cdots, \quad |x| < 1,$$

$$\frac{1}{1 + x^2} = \sum_{n=0}^{\infty} (-1)^n x^{2n}$$
$$= 1 - x^2 + x^4 - x^6 + x^8 + \cdots, \quad |x| < 1,$$

$$\ln \frac{1}{1 - x} = x + \frac{x^2}{2} + \frac{x^3}{3} + \frac{x^4}{4} + \cdots, \quad |x| < 1,$$

$$\ln \sqrt{\frac{1 + x}{1 - x}} = \sum_{n=0}^{\infty} \frac{x^{2n+1}}{2n + 1}$$
$$= x + \frac{x^3}{3} + \frac{x^5}{5} + \frac{x^7}{7} + \frac{x^9}{9} + \cdots, \quad |x| < 1,$$

$$\tan^{-1} x = \sum_{n=0}^{\infty} (-1)^n \frac{x^{2n+1}}{2n + 1}$$
$$= x - \frac{x^3}{3} + \frac{x^5}{5} - \frac{x^7}{7} + \frac{x^9}{9} - \cdots, \quad |x| < 1,$$

$$e^x = \sum_{n=0}^{\infty} \frac{x^n}{n!}$$
$$= 1 + x + \frac{x^2}{2!} + \frac{x^3}{3!} + \frac{x^4}{4!} + \frac{x^5}{5!} + \cdots,$$

$$\sin x = \sum_{n=0}^{\infty} (-1)^n \frac{x^{2n+1}}{(2n + 1)!}$$
$$= x - \frac{x^3}{3!} + \frac{x^5}{5!} - \frac{x^7}{7!} + \frac{x^9}{9!} - \cdots,$$

$$\cos x = \sum_{n=0}^{\infty} (-1)^n \frac{x^{2n}}{(2n)!}$$
$$= 1 - \frac{x^2}{2!} + \frac{x^4}{4!} - \frac{x^6}{6!} + \frac{x^7}{7!} - \cdots,$$

$$(1 + x)^k = 1 + kx + \frac{k(k - 1)}{2!}x^2 + \frac{k(k - 1)(k - 2)}{3!}x^3$$
$$+ \frac{k(k - 1)(k - 2)(k - 3)}{4!}x^4 + \cdots, \quad |x| < 1.$$

CHAPTER 9 REVIEW EXERCISES

Evaluate or indicate "does not exist" for the limits in Exercises 1–8.

1. $\lim\limits_{n \to \infty} \dfrac{3n + 5}{2n - 1}$

2. $\lim\limits_{n \to \infty} \dfrac{\sqrt{9n^2 - 4n + 16}}{2n}$

3. $\lim\limits_{n \to \infty} \dfrac{6n^2 + n - 3}{2n^2 - 3n + 4}$

4. $\lim\limits_{n \to \infty} \dfrac{(-1)^n n}{2n + 1}$

5. $\lim\limits_{n \to \infty} \dfrac{\cos n\pi}{n}$

6. $\lim\limits_{n \to \infty} \dfrac{n^2}{n + 2}$

7. $\lim\limits_{n \to \infty} n e^{-\sqrt{n}}$

8. $\lim\limits_{n \to \infty} \dfrac{\ln n}{n}$

Find a simple formula for the nth partial sum S_n and evaluate the sum S for the series in Exercises 9–14.

9. $\sum_{n=1}^{\infty} \left(\dfrac{2n}{n+1} - \dfrac{2n+4}{n+3} \right)$ **10.** $\sum_{n=1}^{\infty} 2 \left(\dfrac{1}{3} \right)^n$

11. $\sum_{n=1}^{\infty} \dfrac{2}{n(n+2)}$ **12.** $\sum_{n=1}^{\infty} \left(\dfrac{n}{2n-1} - \dfrac{n+1}{2n+1} \right)$

13. $\sum_{n=1}^{\infty} 4 \left(\dfrac{3}{4} \right)^n$ **14.** $\sum_{n=0}^{\infty} 9 \left(-\dfrac{1}{2} \right)^n$

Indicate convergent or divergent for the series in Exercises 15–34.

15. $\sum_{n=1}^{\infty} \dfrac{n}{n^2+3}$ **16.** $\sum_{n=1}^{\infty} \dfrac{n}{2n+1}$

17. $\sum_{n=1}^{\infty} (2n+7)e^{-n^2}$ **18.** $\sum_{n=1}^{\infty} \dfrac{2^n}{n^2}$

19. $\sum_{n=1}^{\infty} \dfrac{3^n}{n^n}$ **20.** $\sum_{n=1}^{\infty} \dfrac{(-1)^n}{n^{1/3}}$

21. $\sum_{n=1}^{\infty} \dfrac{\sqrt{n^3+2n}}{n^3}$ **22.** $\sum_{n=1}^{\infty} \dfrac{1}{n^2 \ln(n+1)}$

23. $\sum_{n=1}^{\infty} \dfrac{1}{n(\ln(n+1))^2}$ **24.** $\sum_{n=1}^{\infty} \left(\dfrac{n}{2n+1} \right)^n$

25. $\sum_{n=1}^{\infty} \dfrac{2n^2+7}{n^3+2}$ **26.** $\sum_{n=1}^{\infty} \dfrac{(-1)^n n}{2n+3}$

27. $\sum_{n=1}^{\infty} \dfrac{(1)(3)\cdots(2n-1)}{(3)(6)\cdots(3n)}$ **28.** $\sum_{n=1}^{\infty} \dfrac{(-1)^n n}{n^2+1}$

29. $\sum_{n=1}^{\infty} \dfrac{1}{\sqrt{1+n^2}}$ **30.** $\sum_{n=1}^{\infty} \dfrac{n!}{(1)(3)(5)\cdots(2n-1)}$

31. $\sum_{n=1}^{\infty} \dfrac{2^n}{3^n-1}$ **32.** $\sum_{n=1}^{\infty} \dfrac{\ln n}{n^2+\ln n}$

33. $\sum_{n=1}^{\infty} \sin^2(1/n)$ **34.** $\sum_{n=1}^{\infty} \cos^2(1/n)$

Indicate absolutely convergent, conditionally convergent, or divergent for the series in Exercises 35–42.

35. $\sum_{n=1}^{\infty} \dfrac{(-1)^n}{n^{2/3}}$ **36.** $\sum_{n=1}^{\infty} \dfrac{(-1)^n}{n^{3/2}}$

37. $\sum_{n=1}^{\infty} \dfrac{(-1)^n}{n^{1/3}}$ **38.** $\sum_{n=1}^{\infty} \dfrac{(-1)^n n}{\sqrt{n^2+1}}$

39. $\sum_{n=1}^{\infty} \dfrac{(-1)^n}{n+n^2}$ **40.** $\sum_{n=1}^{\infty} \dfrac{(-1)^n}{n+\sqrt{n}}$

41. $\sum_{n=1}^{\infty} \dfrac{\cos n\pi}{n}$ **42.** $\sum_{n=1}^{\infty} \dfrac{n \cos n\pi}{n+1}$

Find the interval of convergence of the series in Exercises 43–48.

43. $\sum_{n=1}^{\infty} \dfrac{1}{n}(x-1)^n$ **44.** $\sum_{n=1}^{\infty} \dfrac{1}{n^2} x^n$

45. $\sum_{n=1}^{\infty} n(x-2)^n$ **46.** $\sum_{n=1}^{\infty} \dfrac{(-1)^n n}{n^2+1}(x-1)^n$

47. $\sum_{n=1}^{\infty} \dfrac{nx^{2n}}{4^n}$ **48.** $\sum_{n=1}^{\infty} \dfrac{(-1)^n (x-1)^{2n}}{4^n}$

Use an estimate of $|S - S_n|$ to find all n so that S_n is within 0.001 of S in Exercises 49–52.

49. $\sum_{n=1}^{\infty} n^{-3}$ **50.** $\sum_{n=1}^{\infty} n^{-4}$

51. $\sum_{n=1}^{\infty} \dfrac{(-1)^n}{n^2}$ **52.** $\sum_{n=1}^{\infty} \dfrac{(-1)^n}{n^3+1}$

Find the first three nonzero terms of the Maclaurin series of the functions in Exercises 53–70.

53. $\dfrac{\cos x}{1+x}$ **54.** $\sin 2x$

55. $\dfrac{1}{x-2}$ **56.** $\dfrac{1-\cos x}{x^2}$

57. $\dfrac{x}{1+x^2}$ **58.** $\ln(1+x^2)$

59. $\dfrac{1}{1+x^2}$ **60.** $\dfrac{-2x}{(1+x^2)^2}$

61. $(1-x)^{3/2}$ **62.** $(3+x)^{1/3}$

63. $(2-x)^{-3/2}$ **64.** $\sin^2 x$

65. $\tan x$ **66.** $\sec x$

67. $e^x \sin x$ **68.** $x \cos x$

69. $\sin x + \cos x$ **70.** $e^x - \sin x$

Approximate the values of the integrals in Exercises 71–74 with error less than 0.0001.

71. $\displaystyle\int_0^{0.3} \cos x^2 \, dx$ **72.** $\displaystyle\int_0^{0.2} e^{-x^2} \, dx$

73. $\displaystyle\int_0^{0.2} \dfrac{\sin x}{x} \, dx$ **74.** $\displaystyle\int_0^{0.1} \dfrac{1-\cos 2x}{x^2} \, dx$

75. Find all N so that

$$\left| \sin 0.2 - \sum_{n=0}^{N} (-1)^n \dfrac{(0.2)^{2n+1}}{(2n+1)!} \right| < 0.005.$$

   (*Hint:* Use a bound of R_{2N+2}.)

76. Find all N so that

$$\left| e^{-0.3} - \sum_{n=0}^{N} \dfrac{(-0.3)^n}{n!} \right| < 0.005.$$

77. Use a bound of the second derivative of e^x for $|x| \le 0.5$ to determine a set of values of x for which $|e^x - (1+x)| < 0.005$.

78. Use a bound of the second derivative of $\ln(1+x)$ for $|x| \leq 0.5$ to determine a set of values of x for which $|\ln(1+x) - x| < 0.005$.

79. If $\displaystyle\sum_{n=1}^{\infty} a_n$ and $\displaystyle\sum_{n=1}^{\infty} b_n$ are convergent series with positive terms, show that $\displaystyle\sum_{n=1}^{\infty} a_n b_n$ is convergent.

80. Find an example of convergent series $\displaystyle\sum_{n=1}^{\infty} a_n$ and $\displaystyle\sum_{n=1}^{\infty} b_n$ for which $\displaystyle\sum_{n=1}^{\infty} a_n b_n$ is divergent.

81. If one pair of newborn rabbits is present initially and one pair is born each successive month to each pair not less than two months old and none die, then the **Fibonacci Sequence** $\{a_n\}_{n \geq 1}$ determined by the number a_n of rabbits after n months satisfies the equation $a_{n+2} = a_n + a_{n+1}$. **(a)** Explain how this formula is derived. **(b)** Show that the limit of the ratio of two consecutive terms of this sequence is $(-1 + \sqrt{5})/2$.

82. List several ways to approximate the number e.

83. A version of **Zeno's paradox** states that in order for a hare to catch tortoise, the hare must cover first half the distance between them, then half of the remaining distance, and so on. Thus, the hare must perform an infinite number of steps before catching the tortoise. Explain why this is possible.

EXTENDED APPLICATION

FUNDING AN ENDOWED SCHOLARSHIP

A wealthy benefactor recently made a $100,000 gift to your college, in order to establish an endowed scholarship fund. The school's financial aid officer wants to be sure that the amount of money provided by the fund will increase each year, in order to cover rising tuition costs. She will hire a consultant to set up the fund, but she already has two alternative plans in mind. Since you are a work-study student in her office and the star of your calculus class, she has asked you to do a preliminary analysis of her plans.

Scholarships will be provided from interest earned on investments made with the benefactor's gift. Both plans call for an initial award to be made during the current year. The award will then increase each year, either by a fixed dollar amount or by a fixed percentage. She would like you to compare the awards provided by each plan in several specific cases, in terms of both initial size and long-term growth.

In your analysis, you will need to know that if P dollars is invested for t years at an annual rate of $100r\%$, compounded continuously, the return is A dollars, where $A = Pe^{rt}$. Although the rate of return on the college's investments will vary from year to year, you may assume a constant rate of 8%, compounded continuously.

Tasks:

1. Show that in order to get a return of A dollars at the end of t years at 8%, one must initially invest P dollars, where $P = Ae^{-.08t}$.

2. The scholarship fund will award $f(n)$ dollars at the end of the nth year. If the fund is to remain solvent for N years, tell why the initial investment must be at least

$$\sum_{n=0}^{N} f(n)e^{-.08n} \text{ dollars.}$$

3. a. Tell why the sum in Task 2 can be approximated by

$$\int_0^{N+1} f(t)e^{-.08t}\, dt.$$

b. Tell why the fund will remain solvent indefinitely if

$$\int_0^{\infty} f(t)e^{-.08t}\, dt = 100,000.$$

4. Let S denote the dollar amount of the initial award. The financial aid officer's first plan has $f(t) = S + mt$ for some $m > 0$.

a. From the equation in Task 3b, write an equation relating S and m.

b. For the specific cases $S = 2500$, $S = 5000$, and $S = 7500$, find the corresponding value of m, and determine the size of the award to be given after 2 years, 10 years, and 50 years.

5. Repeat Task 4 for the other plan, where $f(t) = Se^{mt}$ for some $m > 0$.

Writing Assignment:

Write two reports. The first, to the financial aid officer, should contain a nontechnical summary of your findings. Include your results in the form of a table organized for easy comparisons. The second report, to the consultant, should include your calculations, along with an explanation, in the language of calculus, showing why your methods are appropriate.

10 Conic Sections; Polar Coordinates; Three-Dimensional Space

The chapter begins with a study of parabolas, ellipses, and hyperbolas. We will see how to obtain information about these curves from their equations, and how to determine the equation of a parabola, ellipse, or hyperbola that has prescribed properties. These curves occur in many scientific and engineering applications. For example, natural laws result in nearly elliptical orbits of planets and satellites; parabolas and hyperbolas are used in the design of a variety of light and sound reflectors.

The chapter continues with a study of the polar coordinate system. We will see how to use polar coordinates to sketch curves and to solve calculus problems. Finally, rectangular coordinates for three-dimensional space are introduced. Hints are given for sketching curves and surfaces in three-dimensional space. Sketching three-dimensional figures is a skill that is very useful for understanding calculus of several variables.

10.1   PARABOLAS

We will investigate how to determine information about a parabola from its equation and how to determine an equation of a parabola that has prescribed characteristics. We will also illustrate a reflection property that is important for many applications of parabolas.

Standard Equations of Parabolas

We begin with some definitions.

> **DEFINITION**
>
> A **parabola** is the set of all points P in a plane that are equidistant from a fixed line (**directrix**) and a fixed point (**focus**) not on the line.

Connections

Completing a square.

Graphs of parabolas, 1.5.

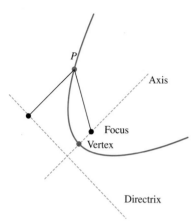

FIGURE 10.1.1 Points P on a parabola are equidistant from the focus and the directrix.

The line through the focus, perpendicular to the directrix, is called the **axis** of the parabola. The point on the axis that is midway between the focus and the directrix is called the **vertex** of the parabola. The vertex is the intersection of the parabola and the axis. The parabola is symmetric with respect to its axis. It curves around the focus, away from the directrix. See Figure 10.1.1.

We can use the geometric properties of the definition of parabola to find an equation that describes a parabola. Let us find an equation satisfied by the parabola with vertex (h, k) and vertical axis $x = h$. The parabola will have focus on the axis, say at $(h, k + p)$, and the directrix will then be the horizontal line $y = k - p$. See Figure 10.1.2. The parabola opens upward, as in Figure 10.1.2, if $p > 0$; the parabola opens downward if $p < 0$. We see from Figure 10.1.2 that (x, y) is on the parabola if and only if

$$d_1 = d_2,$$
$$\sqrt{(x - h)^2 + (y - (k + p))^2} = |y - (k - p)|,$$
$$\sqrt{(x - h)^2 + ((y - k) - p)^2} = |(y - k) + p|,$$
$$(x - h)^2 + (y - k)^2 - 2p(y - k) + p^2 = (y - k)^2 + 2p(y - k) + p^2,$$
$$(x - h)^2 = 4p(y - k).$$

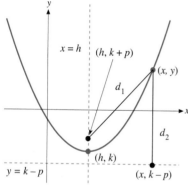

FIGURE 10.1.2 The standard form of the equation of the parabola with vertex (h, k), axis $x = h$, directrix $y = k - p$, and focus $(h, k + p)$ is $(x - h)^2 = 4p(y - k)$.

The equation

$$(x - h)^2 = 4p(y - k)$$

is the **standard form** of the equation of the parabola with vertex (h, k), axis $x = h$, directrix $y = k - p$, and focus $(h, k + p)$. The parabola opens upward if $p > 0$ and downward if $p < 0$. Solving the standard equation for y, we see that the equation of the parabola can also be expressed in the form

$$y = ax^2 + bx + c, \text{ where } a, b, \text{ and } c \text{ are constants and } a \neq 0.$$

Similarly,

$$(y - k)^2 = 4p(x - h)$$

is the **standard form** of the equation of the parabola with vertex (h, k), axis $y = k$, directrix $x = h - p$, and focus $(h + p, k)$. The parabola opens to the right if $p > 0$ and to the left if $p < 0$. See Figure 10.1.3. This equation can also be expressed in the form

$$x = ay^2 + by + c, \text{ where } a, b, \text{ and } c \text{ are constants and } a \neq 0.$$

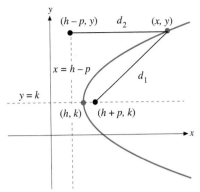

FIGURE 10.1.3 The standard form of the equation of the parabola with vertex (h, k), axis $y = k$, directrix $x = h - p$, and focus $(h + p, k)$ is $(x - h)^2 = 4p(y - k)$.

Locating Key Points and Sketching Parabolas

The standard form of the equation of a parabola can be used to determine important points and sketch the graph of the parabola.

> **To locate important points and sketch the graph of either $(x - h)^2 = 4p(y - k)$ or $(y - k)^2 = 4p(x - h)$**
>
> - Plot the vertex, (h, k), and draw the axis through the vertex. The axis will be vertical if the equation contains a term $(x - h)^2$, and the axis will be horizontal if the equation contains a term $(y - k)^2$. See Figure 10.1.4a.
> - Plot the focus by moving a directed distance of p units from the vertex along the axis. Move in the positive direction if $p > 0$ and move in the negative direction if $p < 0$. See Figure 10.1.4b.
> - Draw the directrix perpendicular to the axis, $|p|$ units from the vertex, on the side opposite the focus. See Figure 10.1.4c.
> - Draw the parabola through the vertex, wrapping around the focus so the distance from the focus is approximately equal to the distance from the directrix. The graph should be symmetric with respect to the axis of the parabola. See Figure 10.1.4d.

FIGURE 10.1.4a Plot the vertex and draw the axis through the vertex.

FIGURE 10.1.4b Plot the focus by moving a directed distance of p units from the vertex along the axis.

FIGURE 10.1.4c Draw the directrix perpendicular the the axis, $|p|$ units from the vertex, on the side opposite the focus.

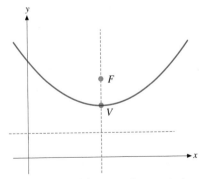

FIGURE 10.1.4d Draw the parabola through the vertex, wrapping around the focus so the distance from the focus is approximately equal to the distance from the directrix.

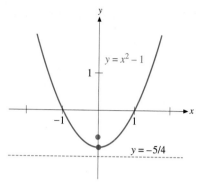

FIGURE 10.1.5 The standard form $(x - 0)^2 = 4(1/4)(y - (-1))$ shows that the vertex is $(0, -1)$, $p = 1/4$, and the axis is vertical, so the focus is $(0, -1 + 1/4)$ and the parabola opens upward.

EXAMPLE 1

Find the vertex, focus, and directrix, and then sketch the parabola $y = x^2 - 1$.

SOLUTION Rewriting the equation in standard form, we have

$$x^2 = y + 1 \quad \text{or}$$
$$(x - 0)^2 = 4\left(\frac{1}{4}\right)(y - (-1)).$$

Comparing this to the standard form

$$(x - h)^2 = 4p(y - k),$$

we see that $h = 0$, $k = -1$, and $p = 1/4$. It follows that the parabola has vertex $(0, -1)$, focus $(0, -1 + 1/4) = (0, -3/4)$, and directrix $y = -1 - 1/4$, or $y = -5/4$. See Figure 10.1.5. ∎

In some cases it is necessary to complete a square to write the equation of a parabola in standard form.

EXAMPLE 2

Find the vertex, focus, and directrix, and then sketch the parabola $y^2 + 6y = 8x - 1$.

SOLUTION Rewriting the equation in standard form, we have

$$y^2 + 6y = 8x - 1,$$
$$y^2 + 6y + (9) = 8x - 1 + (9),$$
$$(y + 3)^2 = 4(2)(x + 1).$$

Comparing this to the standard form

$$(y - k)^2 = 4p(x - h),$$

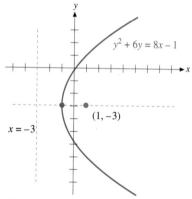

FIGURE 10.1.6 The standard form $(y + 3)^2 = 4(2)(x + 1)$ shows that the vertex is $(-1, -3)$, $p = 2$, and the axis is horizontal, so the focus is $(-1 + 2, -3)$ and the parabola opens to the right.

we see that $h = -1$, $k = -3$, and $p = 2$. The parabola has vertex $(-1, -3)$, focus $(-1 + 2, -3) = (1, -3)$, and directrix $x = -1 - 2$, or $x = -3$. See Figure 10.1.6. ∎

Finding Equations of Parabolas

If we are given any two of the focus, directrix, or vertex of a parabola that has either vertical or horizontal axis, we can use a sketch to determine the values needed to write the standard form of the equation of the parabola.

EXAMPLE 3

Find an equation of the parabola with focus $(2, 1)$ and directrix $y = 3$.

SOLUTION The parabola is sketched in Figure 10.1.7. The axis of the parabola is the vertical line through the focus, $x = 2$. The axis intersects the directrix at the point $(2, 3)$. The vertex is at the midpoint of the line segment between $(2, 1)$ and $(2, 3)$, so $(h, k) = (2, 2)$. We know that $|p|$ is the distance between

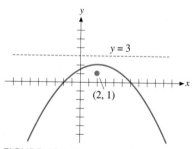

FIGURE 10.1.7 The axis is vertical, so the parabola has standard form $(x - h)^2 = 4p(y - k)$. The distance between the focus and the directrix is $2|p| = 2$. The focus is below the vertex, so p must be negative.

the focus and the vertex, so $|p| = 1$. Since the parabola opens downward, we have $p = -1$. The standard equation of the parabola is then

$$(x - 2)^2 = -4(y - 2). \quad \blacksquare$$

EXAMPLE 4

Find an equation of the parabola that has horizontal axis, vertex $(2, 4)$, and passes through the origin.

SOLUTION   Since the parabola has horizontal axis, we know the standard form is

$$(y - k)^2 = 4p(x - h).$$

We are given that the vertex is $(h, k) = (2, 4)$, so we have

$$(y - 4)^2 = 4p(x - 2).$$

To determine the value of the unknown number p, we use the fact that a point is on the graph of an equation if the coordinates of the point satisfy the equation. If the parabola passes through the origin, the coordinates $(0, 0)$ must satisfy the equation of the parabola. Substitution then gives

$$(0 - 4)^2 = 4p(0 - 2),$$
$$16 = -8p,$$
$$p = -2.$$

We can then write the equation in standard form as

$$(y - 4)^2 = -8(x - 2).$$

The graph is sketched in Figure 10.1.8. $\quad \blacksquare$

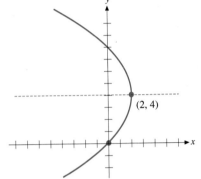

(2, 4)

FIGURE 10.1.8 A parabola with horizontal axis and vertex $(2, 4)$ has standard form $(y - 4)^2 = 4p(x - 2)$. We can determine the value of p by substituting the coordinates of a point that is on the parabola into the equation.

There is exactly one parabola with vertical axis that contains three given points that are not collinear and have distinct x-coordinates. We can determine the equation of the parabola by substituting the coordinates of each of the given points for (x, y) into the equation

$$y = ax^2 + bx + c.$$

This gives us three linear equations in the three unknowns a, b, and c, which we solve. The equations have a unique solution if the three given points have distinct x-coordinates, and $a \neq 0$ if the points are not collinear. Whenever $a \neq 0$, the graph of $y = ax^2 + bx + c$ is a parabola. To find the equation of the parabola with horizontal axis that contains three given points, we follow the same procedure, but use the equation

$$x = ay^2 + by + c.$$

EXAMPLE 5

Find the equation of the parabola that has vertical axis and contains the points $(0, 0)$, $(2, 0)$, and $(3, 6)$.

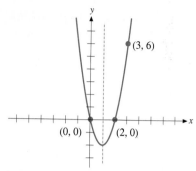

FIGURE 10.1.9 A parabola with vertical axis can be written in the form $y = ax^2 + bx + c$. We can determine the values of a, b, and c by substituting the coordinates of three points that are on the parabola onto the equation.

SOLUTION The equation of the parabola is of the form $y = ax^2 + bx + c$, where a, b, and c are unknown numbers. Substitution of the (x, y) coordinates of each of the three given points into the equation gives

$$\begin{cases} 0 = a(0)^2 + b(0) + c, & [(x, y) = (0,0)] \\ 0 = a(2)^2 + b(2) + c, & [(x, y) = (2,0)] \\ 6 = a(3)^2 + b(3) + c. & [(x, y) = (3,6)] \end{cases}$$

The resulting system of linear equations,

$$\begin{cases} 0 = c, \\ 0 = 4a + 2b + c, \\ 6 = 9a + 3b + c, \end{cases}$$

has solution $(a, b, c) = (2, -4, 0)$. The equation is then

$$y = 2x^2 - 4x.$$

The graph is sketched in Figure 10.1.9. ∎

Reflection Property of Parabolas

Parabolas have the property that lines parallel to the axis reflect through the focus, as illustrated in Figure 10.1.10. (See Exercises 25 and 26.)

The reflection property of parabolas is used in applications. For example, a light source that is positioned at the focus of a parabolic reflector produces a beam of parallel light rays, or a spotlight. Parabolic reflectors are also used in the listening devices you may have seen being used during television broadcasts. Sound waves from the direction in which the reflector is aimed are concentrated at the focus, while undesired noise from other directions is deflected away from the focus. Parabolas are also used in the design of radio telescopes, reflecting optical telescopes, and satellite dishes.

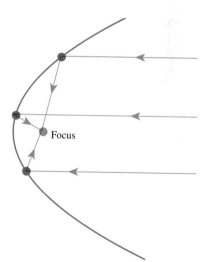

FIGURE 10.1.10 All rays parallel to the axis reflect through the focus of a parabola.

EXERCISES 10.1

Find the vertex, focus, and directrix, and then sketch each of the parabolas in Exercises 1–12.

1. $y = x^2$

2. $x = y^2/4$

3. $4y = 4 - x^2$

4. $8y = 16 - x^2$

5. $8x = 16 - y^2$

6. $4x = 4 - y^2$

7. $y = x^2 - 3x$

8. $2y = 2x - x^2$

9. $8x = 3y^2 + 6y - 5$

10. $8x = 13 - 6y - 3y^2$

11. $4y = 3x^2 - 6x + 7$

12. $5y = 4x^2 + 8x - 11$

Find an equation of the parabola that satisfies the conditions given in each of Exercises 13–24.

13. Focus $(1, 0)$; directrix $x = -1$

14. Focus $(0, -2)$; directrix $y = 2$

15. Focus $(0, 0)$; directrix $x = -2$

16. Focus $(0, 0)$; directrix $y = 4$

17. Focus $(4, 0)$; vertex $(4, 2)$

18. Focus $(-2, 2)$; vertex $(0, 2)$

19. Vertex $(0, -1)$; directrix $y = -3$

20. Vertex $(1, 1)$; directrix the y–axis

21. Vertex $(6, -4)$; contains $(0, 0)$; horizontal axis

22. Vertex $(2, -1)$; contains $(0, -4)$; vertical axis

23. Contains $(-1, 0)$, $(0, 3)$, and $(3, 0)$; vertical axis

24. Contains $(0, 3)$, $(0, 1)$, and $(3, 0)$; horizontal axis

25. The parabola $4y = x^2 - 4$ is sketched below.

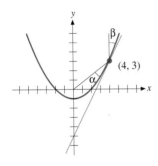

(a) Find the equation of the line tangent to the parabola at the point $(4, 3)$.

(b) Find the y-intercept of the tangent line of (a).

(c) Find the focus of the parabola.

(d) Find the distance between the focus and $(4, 3)$ and the distance between the focus and the y-intercept of (b).

(e) Find a relation between the angles α and β in the figure. (This verifies the reflection property for this parabola at this point on the parabola.)

26. The parabola $x^2 = 4p(y + p)$, $p > 0$, is sketched below.

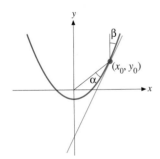

(a) Find the equation of the line tangent to the parabola at a point (x_0, y_0) on the parabola, $x_0 \neq 0$.

(b) Find the y-intercept of the tangent line of (a).

(c) Find the focus of the parabola.

(d) Find the distance between the focus and (x_0, y_0) and the distance between the focus and the y-intercept of (b).

(e) Find a relation between the angles α and β in the figure. (This verifies the reflection property for parabolas.)

27. A cable supporting a suspension bridge hangs from supports that are at equal heights and 160 ft apart. The low point of the cable is 60 ft below the top. See figure. The uniform horizontal distribution of the load on the cable causes it to hang in the shape of a parabola. Find the angle between the cable and the horizontal at the supports.

28. A parabola has focus $(p, 0)$ and directrix $x = -p$. (a) What is the length of the chord through the focus, perpendicular to the axis? (b) Find an equation of the line tangent to the graph at the point where $x = p$ and y is positive.

29. Find equations of the family of parabolas that have vertical axes and contain the points $(0, 0)$ and $(4, 0)$.

30. Find an equation satisfied by all points (x, y) for which the distance between (x, y) and the origin is equal to the distance of (x, y) to the line $x = -d$, $d \neq 0$.

31. The position at time t of an object that is thrown near the surface of the earth at an angle θ from the horizontal with initial speed v_0 is given by the coordinate equations $x(t) = v_0(\cos\theta)t$ and $y(t) = v_0(\sin\theta)t - \dfrac{g}{2}t^2$. Show that the path of the object is a parabola.

10.2 ELLIPSES; HYPERBOLAS

We will investigate properties of ellipses and hyperbolas, which are similar in many ways.

Connections

Graphs of ellipses, 1.5.

Graphs with asymptotes, 1.5, 4.2.

Standard Equations of Ellipses

We begin with the definition of terms.

> **DEFINITION**
>
> An **ellipse** is the set of all points P in a plane such that the sum of the distances between P and two distinct fixed points (**foci**) is constant and greater than the distance between the fixed points.

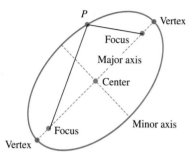

FIGURE 10.2.1 The sum of the distances between a point P on an ellipse and two distinct foci is constant.

The midpoint of the line segment between the two foci is called the **center** of the ellipse. The two points of intersection of the ellipse with the line through the foci are called **vertices**. The line segment between the vertices is called the **major axis**. The line segment through the center, perpendicular to the major axis, with endpoints on the ellipse is called the **minor axis**. The ellipse is symmetric with respect to its center and each of its axes. See Figure 10.2.1.

Let us determine an equation of the ellipse with center (h, k) and horizontal axis. The foci will then be $(h + c, k)$ and $(h - c, k)$, and the vertices will be $(h + a, k)$ and $(h - a, k)$, where $a > c > 0$. The sum of the distances from one vertex to each of the foci is $2a$. (Note that we would not obtain an ellipse unless $a > c$.) The sum of the distances from any point on the ellipse to each of the foci must also be $2a$. We then see from Figure 10.2.2 that a point (x, y) is on the ellipse if and only if

$$d_1 + d_2 = 2a,$$

FIGURE 10.2.2 The standard form of the ellipse with center (h, k), vertices $(h \pm a, k)$, and foci $(h \pm c, k)$ is

$$\frac{(x - h)^2}{a^2} + \frac{(y - k)^2}{b^2} = 1,$$

where $b^2 = a^2 - c^2$.

$$\sqrt{(x - (h - c))^2 + (y - k)^2} + \sqrt{(x - (h + c))^2 + (y - k)^2} = 2a,$$

$$\sqrt{((x - h) + c)^2 + (y - k)^2} = 2a - \sqrt{((x - h) - c)^2 + (y - k)^2},$$

$$((x - h) + c)^2 + (y - k)^2$$
$$= 4a^2 - 4a\sqrt{((x - h) - c)^2 + (y - k)^2} + ((x - h) - c)^2 + (y - k)^2,$$

$$(x - h)^2 + 2c(x - h) + c^2 + (y - k)^2$$
$$= 4a^2 - 4a\sqrt{((x - h) - c)^2 + (y - k)^2}$$
$$+ (x - h)^2 - 2c(x - h) + c^2 + (y - k)^2,$$

$$a\sqrt{((x - h) - c)^2 + (y - k)^2} = a^2 - c(x - h),$$

$$a^2[((x - h) - c)^2 + (y - k)^2] = (a^2 - c(x - h))^2,$$

$$a^2(x - h)^2 - 2a^2c(x - h) + a^2c^2 + a^2(y - k)^2$$
$$= a^4 - 2a^2c(x - h) + c^2(x - h)^2,$$

$$(a^2 - c^2)(x - h)^2 + a^2(y - k)^2 = a^2(a^2 - c^2),$$

$$\frac{(x - h)^2}{a^2} + \frac{(y - k)^2}{b^2} = 1,$$

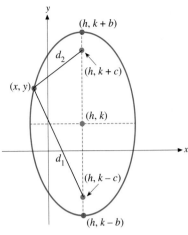

FIGURE 10.2.3 The standard form of the ellipse with center (h, k), vertices $(h, k \pm b)$, and foci $(h, k \pm c)$ is

$$\frac{(x-h)^2}{a^2} + \frac{(y-k)^2}{b^2} = 1,$$

where $a^2 = b^2 - c^2$.

where $b^2 = a^2 - c^2$, $b > 0$. (Note that $a > b$.) Summarizing, we say that the **standard form** of the ellipse with horizontal major axis, center (h, k), vertices $(h - a, k)$ and $(h + a, k)$, and foci $(h - c, k)$, and $(h + c, k)$ is

$$\frac{(x-h)^2}{a^2} + \frac{(y-k)^2}{b^2} = 1,$$

where $a > c > 0$, $a > b > 0$, and $b^2 = a^2 - c^2$.

We can use Figure 10.2.3 and a calculation as above to obtain an equation of an ellipse with vertical major axis. That is, the **standard form** of the ellipse with vertical major axis, center (h, k), vertices $(h, k - b)$ and $(h, k + b)$, and foci $(h, k - c)$ and $(h, k + c)$ is

$$\frac{(x-h)^2}{a^2} + \frac{(y-k)^2}{b^2} = 1,$$

where $b > c > 0$, $b > a > 0$, and $a^2 = b^2 - c^2$.

Locating Key Points and Sketching Ellipses

The standard form of the equation of an ellipse can be used to determine important points and sketch the graph of the ellipse.

To locate important points and sketch the graph of
$$\frac{(x-h)^2}{a^2} + \frac{(y-k)^2}{b^2} = 1, \ a \neq b \ ,$$

- Plot the center, (h, k), and draw horizontal and vertical line segments through the center. See Figure 10.2.4a.
- Plot the extreme points, located by moving horizontally left and right a units, and vertically up and down b units from the center. See Figure 10.2.4b.
- Draw an ellipse through the four extreme points that have been plotted. See Figure 10.2.4c.
- The foci are located on the longer axis, c units from the center, where the square of c is the square of the larger of a and b minus the square of the smaller of a and b. See Figure 10.2.4d.

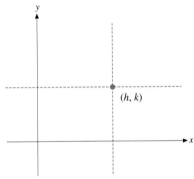

FIGURE 10.2.4a Plot the center and draw horizontal and vertical line segments through the center.

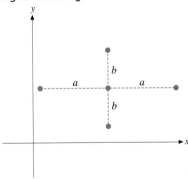

FIGURE 10.2.4b Plot the extreme points by moving horizontally left and right a units and vertically up and down b units from the center.

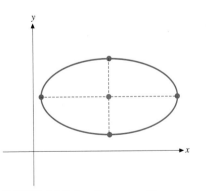

FIGURE 10.2.4c Draw an ellipse through the four extreme points.

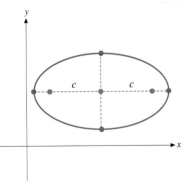

FIGURE 10.2.4d The foci are located on the longer axis, c units from the center, where the square of c is the square of the larger of a and b minus the square of the smaller of a and b.

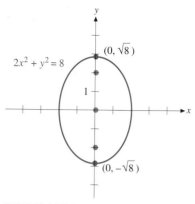

$2x^2 + y^2 = 8$

$(0, \sqrt{8})$

$(0, -\sqrt{8})$

FIGURE 10.2.5 The standard form

$$\frac{x^2}{2^2} + \frac{y^2}{(\sqrt{8})^2} = 1$$

shows that the center is $(0, 0)$. The extreme points are $(\pm 2, 0)$ and $(0, \pm\sqrt{8})$. The longer axis is vertical, so $c^2 = (\sqrt{8})^2 - 2^2 = 4$, and the foci are $(0, \pm 2)$.

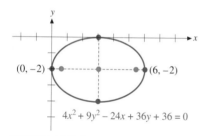

$(0, -2)$ $(6, -2)$

$4x^2 + 9y^2 - 24x + 36y + 36 = 0$

FIGURE 10.2.6 The standard form

$$\frac{(x-3)^2}{3^2} + \frac{(y+2)^2}{2^2} = 1$$

shows that the center is $(3, -2)$. The extreme points are $(3 \pm 3, -2)$ and $(3, -2 \pm 2)$. The longer axis is horizontal, so $c^2 = 3^2 - 2^2 = 5$ and the foci are $(3 \pm \sqrt{5}, -2)$.

EXAMPLE 1

Find the center, vertices, and foci of the ellipse $2x^2 + y^2 = 8$. Sketch the graph.

SOLUTION   Writing the equation in standard form, we have

$$2x^2 + y^2 = 8,$$
$$\frac{x^2}{4} + \frac{y^2}{8} = 1,$$
$$\frac{x^2}{2^2} + \frac{y^2}{(\sqrt{8})^2} = 1.$$

We see that the ellipse has center $(h, k) = (0, 0)$, $a = 2$, and $b = \sqrt{8}$. Since $b > a$, the ellipse has vertical axis. The vertices are $(0, \sqrt{8})$ and $(0, -\sqrt{8})$. We have $c^2 = 8 - 4 = 4$, so $c = \pm 2$. The foci are $(0, 2)$ and $(0, -2)$. The graph is sketched in Figure 10.2.5. ∎

EXAMPLE 2

Find the center, vertices, and foci of the ellipse

$$4x^2 + 9y^2 - 24x + 36y + 36 = 0.$$

Sketch the graph.

SOLUTION   We first complete the squares.

$$4x^2 - 24x + 9y^2 + 36y = -36,$$
$$4(x^2 - 6x + [9]) + 9(y^2 + 4y + [4]) = -36 + 4[9] + 9[4],$$
$$4(x-3)^2 + 9(y+2)^2 = 36,$$
$$\frac{(x-3)^2}{9} + \frac{(y+2)^2}{4} = 1,$$
$$\frac{(x-3)^2}{3^2} + \frac{(y+2)^2}{2^2} = 1.$$

We see that the center is $(h, k) = (3, -2)$. Since $3^2 > 2^2$, the ellipse has horizontal major axis and vertices $(0, -2)$ and $(6, -2)$. We have $c^2 = 3^2 - 2^2 = 5$, so $c = \sqrt{5}$. The foci are $(3 - \sqrt{5}, -2)$ and $(3 + \sqrt{5}, -2)$. The graph is sketched in Figure 10.2.6. ∎

Finding Equations of Ellipses

We can use properties of symmetry and the relation between a, b, and c to determine the equation of ellipses with prescribed characteristics.

EXAMPLE 3

Find an equation of the ellipse that has vertices at $(0, 0)$ and $(0, 10)$ and a focus at $(0, 1)$.

SOLUTION   The ellipse is sketched in Figure 10.2.7.

The center is the midpoint of the line segment between the vertices $(0, 0)$ and $(0, 10)$, so $(h, k) = (0, 5)$. The distance between the center and each focus is $c = 4$, and the distance between the center and each vertex is $b = 5$. We then

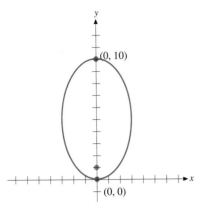

FIGURE 10.2.7 The center is the midpoint of the line segment between the vertices, so $(h, k) = (0, 5)$. The distance between the center and each focus is $c = 4$ and the distance between the center and each vertex is $b = 5$, so $a^2 = 5^2 - 4^2 = 9$.

obtain $a^2 = 5^2 - 4^2 = 9$, so $a = 3$. We then write the equation in standard form as

$$\frac{(x - 0)^2}{3^2} + \frac{(y - 5)^2}{5^2} = 1. \quad \blacksquare$$

> **DEFINITION**
>
> *FOCAL DISTANCE*
> *½ MAJOR AXIS*
>
> The **eccentricity of an ellipse** is
>
> $$e = \frac{\text{Distance between center and a focus}}{\text{Distance between center and a vertex}}.$$

The distance between the center and each vertex is the larger of a and b, and $0 < e < 1$. The ellipse is long and thin if the eccentricity is near one. The shape approaches that of a circle as the eccentricity approaches zero.

EXAMPLE 4

Sketch the graphs of the ellipses with center at the origin, vertices $(-5, 0)$ and $(5, 0)$, and eccentricity (a) 0.6, (b) 0.8.

SOLUTION

(a) The distance between the center and each vertex is $a = 5$. Then $e = c/a$ implies $0.6 = c/5$, so $c = 3$. Then $b^2 = 5^2 - 3^2 = 16$, so $b = 4$. The equation is

$$\frac{x^2}{5^2} + \frac{y^2}{4^2} = 1.$$

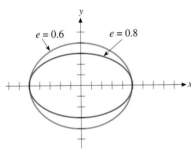

FIGURE 10.2.8 Ellipses with eccentricity near one are elongated. The ellipses become more circular as the eccentricity approaches zero.

(b) We have $a = 5$. Then $e = c/a$ implies $0.8 = c/5$, so $c = 4$. We have $b^2 = 5^2 - 4^2 = 9$, so $b = 3$. The equation is

$$\frac{x^2}{5^2} + \frac{y^2}{3^2} = 1.$$

The graphs are sketched in Figure 10.2.8. $\blacksquare$

Reflection Property of Ellipses

Sound waves that are emitted in all directions from one focus of an ellipse will reflect off the ellipse toward the other focus, as illustrated in Figure 10.2.9. (See Exercises 25 and 26.) The waves will travel the same time along each path. In a room that has an ellipsoidal ceiling, words whispered at one focus can be heard quite distinctly by a listener at the other focus, but not by listeners at other points of the room.

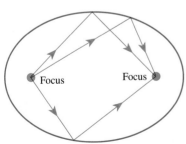

FIGURE 10.2.9 All rays emanating from one focus of an ellipse reflect through the other focus. All such paths have the same length.

Standard Equations of Hyperbolas

We begin by defining terms.

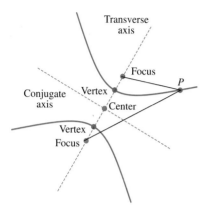

FIGURE 10.2.10 The positive difference between the distances between a point P on a hyperbola and two distinct foci is constant.

> **DEFINITION**
>
> A **hyperbola** is the set of all points P in a plane such that the positive difference between distances from P and two distinct fixed points (**foci**) is constant and less than the distance between the fixed points.

The midpoint of the line segment between the two foci is called the **center** of the hyperbola. The line through the foci is called the **transverse axis.** The line through the center, perpendicular to the transverse axis, is called the **conjugate axis.** The two points of intersection of the transverse axis and the hyperbola are called **vertices.** A hyperbola consists of two separate parts called **branches.** The hyperbola is symmetric with respect to its center and each of its axes. See Figure 10.2.10.

In Figure 10.2.11 we have labeled the center (h, k), vertices $(h - a, k)$ and $(h + a, k)$, and foci $(h - c, k)$ and $(h + c, k)$ of a hyperbola with horizontal transverse axis $y = k$, where $c > a > 0$. The positive difference of the distances between either vertex and each of the foci is $2a$. (Note that we would not obtain a hyperbola unless $a < c$.) A point (x, y) is on the hyperbola if and only if

$$d_2 - d_1 = \pm 2a.$$

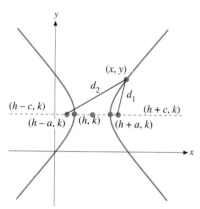

FIGURE 10.2.11 The standard form of the hyperbola with center (h, k), vertices $(h \pm a, k)$, and foci $(h \pm c, k)$ is

$$\frac{(x - h)^2}{a^2} - \frac{(y - k)^2}{b^2} = 1,$$

where $c^2 = a^2 + b^2$.

We can now follow the procedure that was used to find the equation of an ellipse to find the equation of the hyperbola. (Be sure to isolate one radical on one side of the equation before squaring. You will see that the term $\pm a$ appears squared in the final formula, and $(\pm a)^2 = a^2$. Since $c > a$, $a^2 - c^2$ is negative. We set $-b^2 = a^2 - c^2$, $b > 0$.) We conclude that the **standard form** of the equation of a hyperbola with center (h, k), vertices $(h - a, k)$ and $(h + a, k)$, and foci $(h - c, k)$ and $(h + c, k)$ is

$$\frac{(x - h)^2}{a^2} - \frac{(y - k)^2}{b^2} = 1,$$

where $c > a > 0$, $c > b > 0$, and $c^2 = a^2 + b^2$.

We can use Figure 10.2.12 to establish the equation of a hyperbola that has vertical transverse axis. The **standard form** of the equation of a hyperbola with center (h, k), vertices $(h, k - b)$ and $(h, k + b)$, and foci $(h, k - c)$ and $(h, k + c)$ is

$$-\frac{(x - h)^2}{a^2} + \frac{(y - k)^2}{b^2} = 1,$$

where $c > a > 0$, $c > b > 0$, and $c^2 = a^2 + b^2$.

The hyperbolas given by each of the above standard forms have **asymptotes**

$$\frac{y - k}{b} = \frac{x - h}{a} \quad \text{and} \quad \frac{y - k}{b} = -\frac{x - h}{a}.$$

This means that the graphs of the hyperbolas approach the asymptotic lines as $|x|$ and $|y|$ approach infinity. For example, in Figure 10.2.13 the distance

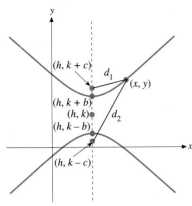

FIGURE 10.2.12 The standard form of the hyperbola with center (h, k), vertices $(h, k \pm b)$, and foci $(h, k \pm c)$ is

$$-\frac{(x-h)^2}{a^2} + \frac{(y-k)^2}{b^2} = 1,$$

where $c^2 = a^2 + b^2$.

between the asymptotic line $y = k + (b/a)(x - h)$ and a corresponding point on the upper-right portion of the hyperbola

$$\frac{(x-h)^2}{a^2} - \frac{(y-k)^2}{b^2} = 1$$

satisfies

$$\lim_{x \to \infty} \left(\left(k + \frac{b}{a}(x - h) \right) - \left(k + \frac{b}{a}\sqrt{(x-h)^2 - a^2} \right) \right)$$

$$= \lim_{x \to \infty} \frac{b}{a} \left((x - h) - \sqrt{(x-h)^2 - a^2} \right)$$

$$= \lim_{x \to \infty} \frac{b}{a} \left((x - h) - \sqrt{(x-h)^2 - a^2} \right) \left(\frac{(x-h) + \sqrt{(x-h)^2 - a^2}}{(x-h) + \sqrt{(x-h)^2 - a^2}} \right)$$

$$= \lim_{x \to \infty} \frac{b}{a} \frac{a^2}{(x - h) + \sqrt{(x-h)^2 - a^2}} = 0.$$

Note that the asymptotes of a hyperbola can be obtained by replacing the 1 in the general equation by 0. For example,

$$\frac{(x-h)^2}{a^2} - \frac{(y-k)^2}{b^2} = 0 \text{ implies } \frac{y-k}{b} = \pm \frac{x-h}{a}.$$

Locating Key Points and Sketching Hyperbolas

The standard form of the equation of a hyperbola can be used to determine important points and sketch the graph of the hyperbola.

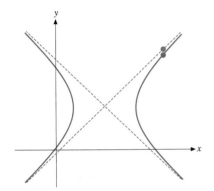

FIGURE 10.2.13 The graph of a hyperbola approaches asymptotic lines as $|x|$ and $|y|$ approach infinity.

> **To locate important points and sketch the graph of either** $\dfrac{(x-h)^2}{a^2} - \dfrac{(y-k)^2}{b^2} = 1$ **or** $-\dfrac{(x-h)^2}{a^2} + \dfrac{(y-k)^2}{b^2} = 1$
>
> - Plot the center, (h, k), and draw horizontal and vertical line segments through the center. See Figure 10.2.14a.
> - Locate the pair of points that are a units horizontally left and right of the center and the pair of points that are b units vertically up and down from the center, and lightly sketch a rectangle, with sides parallel to the coordinate axis, through these points. See Figure 10.2.14b.
> - Lightly sketch lines through opposite vertices of the rectangle. These are the asymptotes of the hyperbola. See Figure 10.2.14c.
> - Determine which two points on the rectangle are on the hyperbola, either the pair horizontally from the center or the pair vertically from the center. (The coordinates of these points satisfy the equation.) Plot these vertices. See Figure 10.2.14d.
> - Draw both branches of the hyperbola, through the vertices and approaching the asymptotes. See Figure 10.2.14e.
> - The foci are located on the axis through the center and vertices, c units from the center, where $c^2 = a^2 + b^2$. See Figure 10.2.14f.

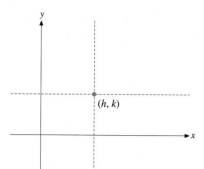

FIGURE 10.2.14a Plot the center and draw horizontal and vertical line segments through the center.

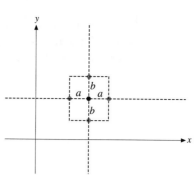

FIGURE 10.2.14b Plot the pair of points that are *a* units horizontally left and right of the center and the pair of points that are *b* units vertically up and down from the center, and lightly sketch a rectangle, with sides parallel to the coordinate axes, through these points.

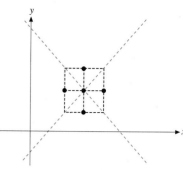

FIGURE 10.2.14c Lightly sketch the asymptotes through opposite corners of the rectangle.

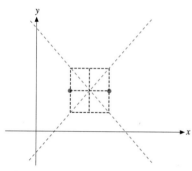

FIGURE 10.2.14d Determine which two points satisfy the equation of the hyperbola and plot these points.

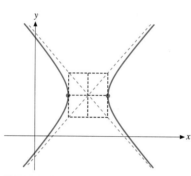

FIGURE 10.2.14e Draw both branches of the hyperbola, through the vertices and approaching the asymptotes.

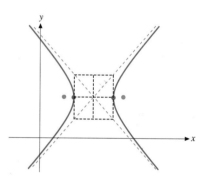

FIGURE 10.2.14f The foci are located on the axis through the center and vertices, *c* units from the center, where $c^2 = a^2 + b^2$.

EXAMPLE 5

Find the center, vertices, foci, and asymptotes of the hyperbola $x^2 - 4y^2 = 4$. Sketch the graph.

SOLUTION We have

$$\frac{x^2}{2^2} - \frac{y^2}{1^2} = 1.$$

The center is $(0, 0)$. We see that $a = 2$ and $b = 1$. The four points through which we draw a rectangle are $(-2, 0)$, $(2, 0)$, $(0, -1)$, and $(0, 1)$. The coordinates $(-2, 0)$ and $(2, 0)$ satisfy the equation, so the vertices are $(-2, 0)$ and $(2, 0)$. We have $c^2 = 2^2 + 1^2 = 5$, so $c = \pm\sqrt{5}$. The foci are $(-\sqrt{5}, 0)$ and $(\sqrt{5}, 0)$. The asymptotes are

$$\frac{x^2}{2^2} - \frac{y^2}{1^2} = 0 \text{ or } \frac{x}{2} = \pm y.$$

The graph is sketched in Figure 10.2.15. ∎

FIGURE 10.2.15 The standard form $\frac{x^2}{2^2} - \frac{y^2}{1^2} = 1$ shows that the center is $(0, 0)$. The vertices are $(\pm 2, 0)$ and the foci are $(\pm\sqrt{5}, 0)$.

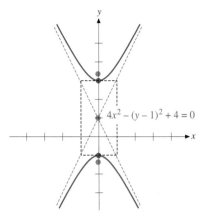

FIGURE 10.2.16 The standard form
$$-\frac{x^2}{1^2} + \frac{(y-1)^2}{2^2} = 1$$
shows that the center is $(0, 1)$. The vertices are $(0, 1 \pm 2)$ and the foci are $(0, 1 \pm \sqrt{5})$.

EXAMPLE 6

Find the center, vertices, foci, and asymptotes of the hyperbola $4x^2 - (y-1)^2 + 4 = 0$. Sketch the graph.

SOLUTION   The equation $4x^2 - (y-1)^2 + 4 = 0$ implies

$$-\frac{x^2}{1^2} + \frac{(y-1)^2}{2^2} = 1.$$

The center is $(0, 1)$. We see that $a = 1$ and $b = 2$. The four points through which we draw a rectangle are $(-1, 1)$, $(1, 1)$, $(0, -1)$, and $(0, 3)$. The coordinates of the points $(0, -1)$ and $(0, 3)$ satisfy the equation and give the vertices of the hyperbola. We have $c^2 = 1^2 + 2^2 = 5$, so $c = \pm\sqrt{5}$. The foci are $(0, 1 - \sqrt{5})$ and $(0, 1 + \sqrt{5})$. The asymptotes are

$$-\frac{x^2}{1^2} + \frac{(y-1)^2}{2^2} = 0 \text{ or } \frac{y-1}{2} = \pm x.$$

The graph is sketched in Figure 10.2.16. ∎

Finding Equations of Hyperbolas

EXAMPLE 7

Find an equation of the hyperbola that has foci $(0, -5)$ and $(0, 5)$ and one vertex $(0, 4)$.

SOLUTION   We know that the center is the midpoint of the line segment between the foci, so $(h, k) = (0, 0)$. We have $c = 5$ and $b = 4$, so $5^2 = a^2 + 4^2$ and $a = 3$. We then have

$$-\frac{x^2}{3^2} + \frac{y^2}{4^2} = 1.$$

A sketch is given in Figure 10.2.17. ∎

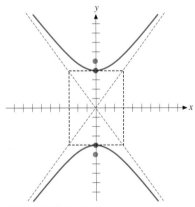

FIGURE 10.2.17 The center is the midpoint of the line segment between the foci, so $(h, k) = (0, 0)$. The distance between the center and each focus is $c = 5$ and the distance between the center and each vertex is $b = 4$, so $a^2 = 5^2 - 4^2 = 9$.

DEFINITION

The **eccentricity of a hyperbola** is

$$e = \frac{\text{Distance between center and a focus}}{\text{Distance between center and a vertex}}.$$

Note that the eccentricity of a hyperbola is greater than one.

Reflection Property of Hyperbolas

Hyperbolas have the reflection property illustrated in Figure 10.2.18. (See Exercises 51 and 52.) Light rays emitted from one focus reflect off a hyperbolic reflector in the direction of the line from the other focus through the point of reflection. This spreads the light and produces a floodlight effect.

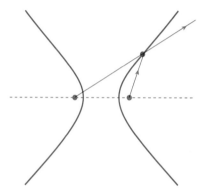

FIGURE 10.2.18 All rays emanating from one focus of a hyperbola reflect in the direction of the line from the other focus through the point of reflection.

Conic Sections

Parabolas, ellipses, and hyperbolas are called **conic sections**. This reflects the fact that they can be generated as the intersection of a double-napped cone and a plane. From Figure 10.2.19a we see that the intersection is a parabola if the plane intersects the axis of the cone at an angle that is equal to the angle between the axis and a generating line of the cone. We include a single line as a **degenerate case** of a parabola, as illustrated in Figure 10.2.19b. If the plane intersects the cone at an angle that is greater than the angle between the axis and a generating line of the cone, the intersection is an ellipse, as illustrated in Figure 10.2.20a, or one of the degenerate cases of either a circle, as in Figure 10.2.20b, or a point, as in Figure 10.2.20c. If the plane intersects the axis of the cone at an angle that is less than the angle between the axis and a generating line, the intersection is a hyperbola, as illustrated in Figure 10.2.21a, or the degenerate case of two intersecting lines, as in Figure 10.2.21b.

FIGURE 10.2.19a The intersection of a cone and a plane that is parallel to a generating line of the cone is a parabola.

FIGURE 10.2.19b A single line is a degenerate case of a parabola.

FIGURE 10.2.20a The intersection of a cone and a plane that intersects the axis of the cone at an angle greater than the angle between the axis and a generating line of the cone is an ellipse.

FIGURE 10.2.20b A circle is a degenerate case of an ellipse.

FIGURE 10.2.20c A single point is a degenerate case of an ellipse.

FIGURE 10.2.21a The intersection of a double-napped cone and a plane that intersects the axis of the cone at an angle that is less than the angle between the axis and a generating line of the cone is a hyperbola.

FIGURE 10.2.21b Two intersecting lines are a degenerate case of a hyperbola.

Parabolas, ellipses, and hyperbolas can be expressed in terms of a **focus-directrix equation**,

$$|PF| = e|PD|,$$

where $|PF|$ is the distance from a point P on the curve to the focus and $|PD|$ is the distance of P to the directrix. See Figure 10.2.22. The graph is an ellipse with eccentricity e if $e < 1$. The graph is a hyperbola with eccentricity e if $e > 1$. If $e = 1$ the graph is a parabola. We then define the **eccentricity of a parabola** to be one.

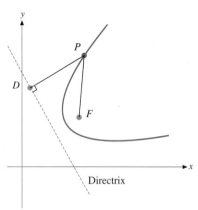

FIGURE 10.2.22 The focus-directrix form of the equation of a conic section is $|PF| = e|PD|$, where e is the eccentricity.

EXERCISES 10.2

Find the center, vertices, and foci, and then sketch each of the ellipses in Exercises 1–10.

1. $25x^2 + 9y^2 = 225$

2. $25x^2 + 169y^2 = 4225$

3. $x^2 - 4x + 4y^2 = 0$

4. $4x^2 + y^2 + 4y = 0$

5. $100x^2 + 36y^2 - 180y = 0$

6. $4x^2 - 12x + 36y^2 = 0$

7. $4x^2 - 12x + 9y^2 - 18y = -9$

8. $4x^2 - 8x + y^2 - 4y = -7$

9. $3x^2 - 6x + y^2 - 2y = 0$

10. $x^2 + 4x + 5y^2 + 10y = 0$

Find an equation of the ellipse that satisfies the conditions given in each of Exercises 11–18.

11. Center $(0, 0)$; vertex $(0, 13)$; focus $(0, 12)$

12. Center $(0, 0)$; vertex $(5, 0)$; focus $(3, 0)$

13. Vertices $(-5, 3)$ and $(5, 3)$; focus $(4, 3)$

14. Vertex $(0, 0)$; foci $(0, 2)$ and $(0, 8)$

15. Vertices $(1, -4)$ and $(1, 4)$; contains $(2, 0)$

16. Vertices $(0, 2)$ and $(6, 2)$; contains $(3, 0)$

17. Center $(0, 0)$; vertex $(4, 0)$; contains $(2, 1)$

18. Center $(0, 0)$; vertex $(0, 2)$; contains $(1, 1)$

19. Find equations of all ellipses that have a focus at the origin, vertex $(5, 0)$, and eccentricity 0.5.

20. Find equations of all ellipses that have a focus at the origin, vertex $(0, 2)$, and eccentricity 0.6.

21. Describe the set of all (x, y) for which the sum of the distances between (x, y) and each of the points $(-c, 0)$ and $(c, 0)$ is $2c$, $c > 0$.

22. Describe the set of all (x, y) for which the sum of the distances between (x, y) and each of the points $(-c, 0)$ and $(c, 0)$ is $2d$, $c > d > 0$.

23. Find an equation satisfied by all (x, y) for which the distance between (x, y) and the origin is one-half the distance of (x, y) from the line $x = -3$.

24. Find an equation satisfied by all (x, y) for which the distance between (x, y) and the origin is e times the distance of (x, y) from the line $x = d$, $0 < e < 1$, $d \neq 0$.

25. The ellipse $16x^2 + 25y^2 = 400$ is sketched below.

(a) Find the slope of the line tangent to the ellipse at the point $(4, 12/5)$.
(b) Find the foci of the ellipse.
(c) Find the slopes of the lines through $(4, 12/5)$ and each focus.
(d) Find the $\tan \alpha$ and $\tan \beta$. (Note that the tangent of the angle between lines that have slopes m_1 and m_2 is $(m_2 - m_1)/(1 + m_1 m_2)$.)
(e) Find a relation between the angles α and β in the figure. (This verifies the reflection property of this ellipse at this point on the ellipse.)

26. The ellipse $b^2 x^2 + a^2 y^2 = a^2 b^2$, $a > b > 0$, is sketched below.

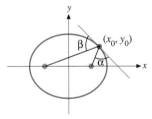

(a) Find the slope of the line tangent to the ellipse at a point (x_0, y_0) on the ellipse, $y_0 \neq 0$.
(b) Find the foci of the ellipse.
(c) Find the slopes of the lines through (x_0, y_0) and each focus.
(d) Find the $\tan \alpha$ and $\tan \beta$. (Note that the tangent of the angle between lines that have slopes m_1 and m_2 is $(m_2 - m_1)/(1 + m_1 m_2)$.)
(e) Find a relation between the angles α and β in the figure. (This verifies the reflection property for ellipses.)

Find the center, vertices, foci, and asymptotes, and then sketch each of the hyperbolas in Exercises 27–36.

27. $25x^2 - 144y^2 = 3600$
28. $25x^2 - 144y^2 + 3600 = 0$
29. $9x^2 - 16y^2 + 144 = 0$
30. $9x^2 - 16y^2 = 144$
31. $x^2 - 2x - y^2 = 0$
32. $x^2 - 4y^2 - 16y = 12$
33. $4x^2 + 8x - y^2 + 4y = 1$
34. $x^2 - 2x - y^2 - 2y + 1 = 0$
35. $9x^2 - 36x - 16y^2 + 32y = 16$
36. $16x^2 + 48x - 9y^2 - 36y = 36$

Find an equation of the hyperbola that satisfies the conditions given in each of Exercises 37–44.

37. Center $(0, 0)$; vertex $(0, 12)$; focus $(0, 13)$
38. Center $(0, 0)$; vertex $(4, 0)$; focus $(5, 0)$
39. Vertices $(-4, 3)$ and $(4, 3)$; focus $(5, 3)$
40. Vertex $(0, 0)$; foci $(0, -2)$ and $(0, 8)$
41. Vertices $(1, -3)$ and $(1, 3)$; contains $(0, 5)$
42. Vertices $(0, 2)$ and $(6, 2)$; contains $(-2, 0)$
43. Center $(0, 0)$; vertex $(0, 4)$; contains $(1, 6)$
44. Center $(0, 0)$; vertex $(3, 0)$; contains $(4, 5)$
45. Find equations of all hyperbolas that have a focus at the origin, vertex $(2, 0)$, and eccentricity 4.
46. Find equations of all ellipses that have a focus at the origin, vertex $(0, 3)$, and eccentricity 3.
47. Describe the set of all (x, y) for which the positive difference of the distances between (x, y) and each of the points $(-c, 0)$ and $(c, 0)$ is $2c$, $c > 0$.
48. Describe the set of all (x, y) for which the positive difference of the distances between (x, y) and each of the points $(-c, 0)$ and $(c, 0)$ is $2d$, $d > c > 0$.
49. Find an equation satisfied by all (x, y) for which the distance between (x, y) and the origin is twice the distance of (x, y) from the line $x = -3$.
50. Find an equation satisfied by all (x, y) for which the distance between (x, y) and the origin is e times the distance of (x, y) from the line $x = d$, $e > 1$, $d \neq 0$.
51. The hyperbola $16x^2 - 9y^2 = 144$ is sketched below.

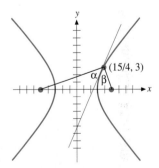

(a) Find the slope of the line tangent to the hyperbola at point $(15/4, 3)$.
(b) Find the foci of the hyperbola.
(c) Find the slopes of the lines through $(15/4, 3)$ and each focus.

(d) Find $\tan \alpha$ and $\tan \beta$. (Note that the tangent of the angle between lines that have slopes m_1 and m_2 is $(m_2 - m_1)/(1 + m_1 m_2)$.)

(e) Find a relation between the angles α and β in the figure. (This verifies the reflection property of this hyperbola at this point of the hyperbola.)

52. The hyperbola $b^2 x^2 - a^2 y^2 = a^2 b^2$, $a > 0$, $b > 0$, is sketched below.

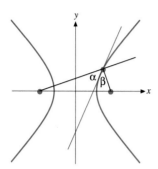

(a) Find the slopes of the lines tangent to the hyperbola at point (x_0, y_0) on the hyperbola, $y_0 \neq 0$.

(b) Find the foci of the hyperbola.

(c) Find the slopes of the lines through (x_0, y_0) and each focus.

(d) Find $\tan \alpha$ and $\tan \beta$. (Note that the tangent of the angle between lines that have slopes m_1 and m_2 is $(m_2 - m_1)/(1 + m_1 m_2)$.)

(e) Find a relation between the angles α and β in the figure. (This verifies the reflection property for hyperbolas.)

53. Use Figure 10.2.11 to verify that the equation $d_2 - d_1 = \pm 2a$ leads to the standard form of the equation of a hyperbola.

54. Find the area of the surface obtained by revolving the curve $y = \sqrt{x^2 + 2}$, $-1 \leq x \leq 1$, about the x-axis.

55. Find the area of the surface obtained by revolving the curve $y = 2\sqrt{1 - x^2}$, $-1 \leq x \leq 1$ about the x-axis.

56. Find the volume of the solid obtained by revolving the region bounded by $y = \sqrt{x^2 + 1}$, $y = 2$, and the y-axis ($x \geq 0$) about the x-axis.

57. Find the volume of the solid obtained by revolving the region bounded by $y = \sqrt{x^2 + 1}$, $y = 2$, and the y-axis ($x \geq 0$) about the y-axis.

58. Two families are building houses in the country, along a straight road, 300 yd apart. A well is to be dug 400 yd from the main road, next to an access road that is perpedicular to the main road and intersects the main road at a point between the houses, 200 yd from one of the houses. They plan to buy 1" pipe to lay from the well to a junction point and use 400 yd of 3/4" pipe that they already have to go from the junction to each house. See figure. Where should they locate the junction to minimize the amount of 1" pipe they need to buy?

10.3 TRANSLATION AND ROTATION OF AXES

A change of coordinates often facilitates the solution of a physical problem. A famous example of this involves planetary motion. Astronomers tried for years to understand the motion of planets using the earth as the center of the coordinate system, but it wasn't until they used the sun as the center of the coordinate system that understanding of planetary motion was achieved.

In this section we will see two different ways to change from a given coordinate system to a new coordinate system. The new coordinates will be used to simplify and analyze equations given in terms of the original coordinates.

Translation

Let us consider a rectangular coordinate system (x', y') with origin at the point $(x, y) = (h, k)$ of the xy-plane, x'-axis parallel to the x-axis, and y'-axis parallel to the y-axis, as illustrated in Figure 10.3.1a. Such a system is said to be obtained by a **translation of axes**. We see from Figure 10.3.1b that the relations between coordinates of the two systems are given by the equations

$$\begin{cases} x = x' + h, \\ y = y' + k. \end{cases}$$

These equations of translation give the original (x, y) coordinates in terms of the new coordinates (x', y'). This form is convenient for substitution in an equation that involves x and y in order to obtain an equation in x' and y'. The translated coordinates (x', y') can be expressed in terms of the original coordinates by the equivalent system of equations

$$\begin{cases} x' = x - h, \\ y' = y - k. \end{cases}$$

(Note that we can interpret these equations as corresponding to a translation of the $x'y'$-axes back to the xy-axes.)

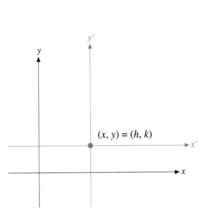

FIGURE 10.3.1a The translated coordinate system (x', y') has origin at the point $(x, y) = (h, k)$ and axes parallel to the axes of the (x, y) coordinate system.

FIGURE 10.3.1b The equations
$$\begin{cases} x = x' + h, \\ y = y' + k \end{cases}$$
relate (x, y) coordinates of a point and the (x', y') coordinates of the translated system that has origin at the point $(x, y) = (h, k)$.

EXAMPLE 1

Express the equation $3x^2 - 12x - 8y = 0$ in terms of the (x', y') coordinates given by the translation

$$\begin{cases} x = x' + 2, \\ y = y' - \dfrac{3}{2}. \end{cases}$$

Sketch the graph and both sets of coordinate axes.

SOLUTION   Substitution of the equations of translation gives

$$3x^2 - 12x - 8y = 0,$$

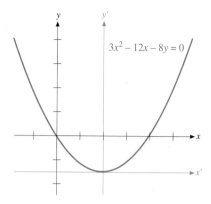

FIGURE 10.3.2   The equations

$$x = x' + 2, y = y' - \frac{3}{2}$$

translate the axes to the vertex of the parabola.

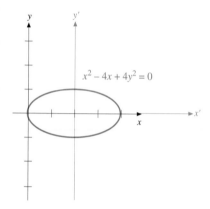

FIGURE 10.3.3   The equations

$$x = x' + 2, y = y'$$

translate the axes to the center of the ellipse.

$$3(x' + 2)^2 - 12(x' + 2) - 8\left(y' - \frac{3}{2}\right) = 0,$$

$$3x'^2 + 12x' + 12 - 12x' - 24 - 8y' + 12 = 0,$$

$$3x'^2 - 8y' = 0.$$

We see that the graph is a parabola with vertex at $(x', y') = (0, 0)$. The equations of translation tell us that $(x', y') = (0, 0)$ corresponds to $(x, y) = (2, -3/2)$. The (x', y') coordinate axes have been drawn with origin at the point $(x, y) = (2, -3/2)$, and the graph is sketched in Figure 10.3.2. ∎

EXAMPLE 2

Find the equations of translation for which the origin of the $x'y'$-plane is at the center of the ellipse $x^2 - 4x + 4y^2 = 0$. Express the equation in terms of the translated coordinates (x', y'). Sketch both sets of coordinate axes and the graph of the ellipse.

SOLUTION   The center of the ellipse is found by completing the square and writing the equation in standard form. We have

$$x^2 - 4x + [4] + 4y^2 = [4],$$

$$(x - 2)^2 + 4y^2 = 4,$$

$$\frac{(x - 2)^2}{2^2} + \frac{y^2}{1^2} = 1.$$

We see that the ellipse has center $(h, k) = (2, 0)$. The desired equations of translation are then

$$\begin{cases} x = x' + 2, \\ y = y'. \end{cases}$$

(It is a good idea to check that $(x', y') = (0, 0)$ corresponds to the proper (x, y) coordinates, $(x, y) = (2, 0)$ in this case.) In terms of the (x', y') coordinates, the equation is

$$\frac{x'^2}{2^2} + \frac{y'^2}{1^2} = 1.$$

The (x', y') coordinate axes have been drawn with origin at $(x, y) = (2, 0)$, and the graph sketched in Figure 10.3.3. ∎

Rotation

We want to establish a new coordinate system by **rotation of axes** about the origin by an angle θ. This is illustrated in Figure 10.3.4.

A point P in the plane has coordinates (x, y) in the original coordinate system, and coordinates (x', y') in the new, rotated system. We want to determine the relation between the two coordinate systems. From Figure 10.3.4, we have

$$x' = d \cos \alpha \text{ and } y' = d \sin \alpha.$$

Also,

$$x = d \cos(\alpha + \theta)$$

$$= d(\cos \alpha \cos \theta - \sin \alpha \sin \theta)$$

$$= x' \cos \theta - y' \sin \theta,$$

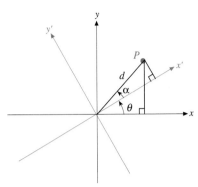

FIGURE 10.3.4 The equations

$$x = x' \cos\theta - y' \sin\theta,$$
$$y = x' \sin\theta + y' \cos\theta$$

relate (x, y) coordinates of a point and a rectangular coordinate system obtained by rotating the x- and y-axes about the origin by an angle θ.

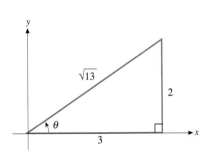

FIGURE 10.3.5 A representative sketch of $\theta = \tan^{-1}(2/3)$ can be used to determine $\sin\theta$ and $\cos\theta$.

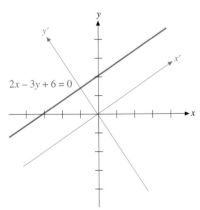

FIGURE 10.3.6 If the axes are rotated $\tan^{-1}(2/3)$, the line is parallel to the x'-axis and has equation $y' = 6\sqrt{13}/13$.

and

$$
\begin{aligned}
y &= d\sin(\alpha + \theta) \\
&= d(\cos\alpha\sin\theta + \sin\alpha\cos\theta) \\
&= x'\sin\theta + y'\cos\theta.
\end{aligned}
$$

Summarizing, we have

$$
\begin{cases}
x = x'\cos\theta - y'\sin\theta, \\
y = x'\sin\theta + y'\cos\theta.
\end{cases}
$$

Substitution of the above values of x and y in an equation gives an equivalent equation in terms of the new coordinates that are obtained by rotation of the (x, y) axes by an angle of θ.

The equations of rotation can be solved for x' and y' in terms of x and y. We obtain

$$
\begin{cases}
x' = x\cos\theta + y\sin\theta, \\
y' = -x\sin\theta + y\cos\theta.
\end{cases}
$$

(Note that we can interpret these equations as corresponding to a rotation of the (x', y') axes by an angle of $-\theta$.)

EXAMPLE 3

Express the equation $2x - 3y + 6 = 0$ in terms of the (x', y') coordinates given by a rotation of $\theta = \tan^{-1}(2/3)$. Sketch the graph and both sets of coordinate axes.

SOLUTION   A representative sketch of the angle $\theta = \tan^{-1}(2/3)$ is given in Figure 10.3.5. We see that $\sin\theta = 2/\sqrt{13}$ and $\cos\theta = 3/\sqrt{13}$. Substitution into the equations of rotation then gives

$$x = x'\cos\theta - y'\sin\theta = \frac{3x' - 2y'}{\sqrt{13}},$$

$$y = x'\sin\theta + y'\cos\theta = \frac{2x' + 3y'}{\sqrt{13}}.$$

We then have

$$2x - 3y + 6 = 0,$$

$$2\left(\frac{3x' - 2y'}{\sqrt{13}}\right) - 3\left(\frac{2x' + 3y'}{\sqrt{13}}\right) + 6 = 0,$$

$$6x' - 4y' - 6x' - 9y' + 6\sqrt{13} = 0,$$

$$y' = \frac{6\sqrt{13}}{13}.$$

The graph and both sets of axes are sketched in Figure 10.3.6. Note that the line is parallel to the x'-axis. This is true because the angle of rotation θ is equal to the angle between the line and the x-axis. ∎

EXAMPLE 4

Express the equation $xy - x + y = 0$ in terms of coordinates (x', y') that correspond to a rotation of θ. Find $0 < \theta < \pi/2$ such that the new equation contains no $x'y'$ term. Sketch the graph and both sets of axes.

SOLUTION   Substitution gives

$$xy - x + y = 0,$$
$$(x'\cos\theta - y'\sin\theta)(x'\sin\theta + y'\cos\theta)$$
$$- (x'\cos\theta - y'\sin\theta) + (x'\sin\theta + y'\cos\theta) = 0,$$
$$(\cos\theta\sin\theta)x'^2 + (\cos^2\theta - \sin^2\theta)x'y' - (\sin\theta\cos\theta)y'^2$$
$$+ (\sin\theta - \cos\theta)x' + (\sin\theta + \cos\theta)y' = 0.$$

We see that the new equation will contain no $x'y'$ term if θ is chosen so that

$$\cos^2\theta - \sin^2\theta = 0.$$

This gives $\tan\theta = \pm 1$. This equation and the condition that $0 < \theta < \pi/2$ are satisfied by the angle $\theta = \pi/4$. Since $\sin(\pi/4) = \cos(\pi/4) = 1/\sqrt{2}$, substitution of $\theta = \pi/4$ into the above equation in (x', y') gives

$$\left(\frac{1}{\sqrt{2}}\frac{1}{\sqrt{2}}\right)x'^2 + \left(\frac{1}{2} - \frac{1}{2}\right)x'y' - \left(\frac{1}{\sqrt{2}}\frac{1}{\sqrt{2}}\right)y'^2 + \left(\frac{1}{\sqrt{2}} - \frac{1}{\sqrt{2}}\right)x'$$
$$+ \left(\frac{1}{\sqrt{2}} + \frac{1}{\sqrt{2}}\right)y' = 0,$$
$$x'^2 - y'^2 + \frac{4}{\sqrt{2}}y' = 0,$$
$$x'^2 - y'^2 + 2\sqrt{2}y' = 0.$$

Completing the square to write this equation in standard form, we obtain

$$x'^2 - (y'^2 - 2\sqrt{2}y' + [2]) = -([2]),$$
$$x'^2 - (y' - \sqrt{2})^2 = -2,$$
$$-\frac{x'^2}{(\sqrt{2})^2} + \frac{(y' - \sqrt{2})^2}{(\sqrt{2})^2} = 1.$$

We see that this is the equation of a hyperbola with center $(x', y') = (0, \sqrt{2})$. The vertices are $(x', y') = (0, 0)$ and $(x', y') = (0, 2\sqrt{2})$. The graph and both sets of axes are sketched in Figure 10.3.7. ∎

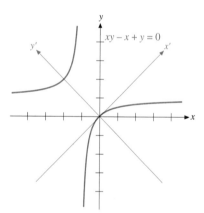

FIGURE 10.3.7   Rotation of axes by $\pi/4$ gives the equation

$$-\frac{x'^2}{(\sqrt{2})^2} + \frac{(y' - \sqrt{2})^2}{(\sqrt{2})^2} = 1.$$

This is the equation of a hyperbola.

EXAMPLE 5

Find the (x, y) coordinates of the center, vertices, and foci of the hyperbola $xy - x + y = 0$ of Example 4. Find the equations of the asymptotes in terms of (x, y).

SOLUTION   From the equation obtained in Example 4,

$$-\frac{x'^2}{(\sqrt{2})^2} + \frac{(y' - \sqrt{2})^2}{(\sqrt{2})^2} = 1,$$

we see that the center is $(x', y') = (0, \sqrt{2})$. Since the angle of rotation is $\theta = \pi/4$, the corresponding (x, y) coordinates are

$$x = x' \cos\theta - y' \sin\theta = (0)\left(\frac{1}{\sqrt{2}}\right) - (\sqrt{2})\left(\frac{1}{\sqrt{2}}\right) = -1,$$

$$y = x' \sin\theta + y' \cos\theta = (0)\left(\frac{1}{\sqrt{2}}\right) + (\sqrt{2})\left(\frac{1}{\sqrt{2}}\right) = 1.$$

The center is $(x, y) = (-1, 1)$. Similarly, the vertices $(x', y') = (0, 0)$ and $(x', y') = (0, 2\sqrt{2})$ correspond to $(x, y) = (0, 0)$ and $(x, y) = (-2, 2)$, respectively.

We need the (x', y') coordinates of the foci. From the (x', y') equation of the hyperbola, we see that $c^2 = a^2 + b^2 = 2 + 2 = 4$, so $c = 2$. The foci are $(x', y') = (0, \sqrt{2} - 2)$ and $(x', y') = (0, \sqrt{2} + 2)$. These correspond to $(x, y) = (\sqrt{2} - 1, -\sqrt{2} + 1)$ and $(x, y) = (-\sqrt{2} - 1, \sqrt{2} + 1)$, respectively.

The equations of the asymptotes are

$$-\frac{x'^2}{(\sqrt{2})^2} + \frac{(y' - \sqrt{2})^2}{(\sqrt{2})^2} = 0,$$

or $y' = x' + \sqrt{2}$ and $y' = -x' + \sqrt{2}$. Using the equations $x' = x \cos\theta + y \sin\theta$, $y' = -x \sin\theta + y \cos\theta$, $\theta = \pi/4$, and substituting, we have

$$y' = x' + \sqrt{2},$$
$$(-x \sin\theta + y \cos\theta) = (x \cos\theta + y \sin\theta) + \sqrt{2},$$
$$-\frac{x}{\sqrt{2}} + \frac{y}{\sqrt{2}} = \frac{x}{\sqrt{2}} + \frac{y}{\sqrt{2}} + \sqrt{2},$$
$$-\frac{2x}{\sqrt{2}} = \sqrt{2},$$
$$x = -1.$$

Similarly, the equation $y' = -x' + \sqrt{2}$ can be shown to have (x, y) equation $y = 1$. ∎

General Second-Degree Equations

The equation

$$Ax^2 + Bxy + Cy^2 + Dx + Ey + F = 0$$

(A, B, C not all three zero) is called the **general second-degree equation**. Rotation of axes by an angle θ transforms this equation to the equation

$$A'x'^2 + B'x'y' + C'y'^2 + D'x' + E'y' + F' = 0,$$

where

$$A' = A \cos^2\theta + B \sin\theta \cos\theta + C \sin^2\theta,$$
$$B' = B(\cos^2\theta - \sin^2\theta) + 2(C - A) \sin\theta \cos\theta,$$
$$C' = A \sin^2\theta - B \sin\theta \cos\theta + C \cos^2\theta,$$
$$D' = D \cos\theta + E \sin\theta,$$
$$E' = -D \sin\theta + E \cos\theta,$$
$$F' = F.$$

In particular, we see that we can obtain an equation that has no $x'y'$ term if we choose θ such that

$$B(\cos^2 \theta - \sin^2 \theta) + 2(C - A) \sin \theta \cos \theta = 0.$$

We can use trigonometric identities for double angles to solve this equation for θ. That is,

$$B \cos 2\theta + (C - A) \sin 2\theta = 0,$$

$$\cot 2\theta = \frac{A - C}{B}.$$

If $B \neq 0$, this equation will be satisfied by an angle $0 < \theta < \pi/2$. If $B = 0$, the original equation has no xy term.

It is possible to determine the character of the graph of the equation

$$Ax^2 + Bxy + Cy^2 + Dx + Ey + F = 0$$

from the sign of the expression $B^2 - 4AC$, called the **discriminant** of the equation. To see how this is done, we first note that rotation of axes does not change the value of the discriminant. That is:

If the equation $A'x'^2 + B'x'y' + C'y'^2 + D'x' + E'y' + F' = 0$ is obtained from the equation $Ax^2 + Bxy + Cy^2 + Dx + Ey + F = 0$ by any rotation of axes, then $B'^2 - 4A'C' = B^2 - 4AC$.

(See Exercise 24.) If the rotation is chosen so $B' = 0$ and the new equation has no $x'y'$ term, then $B^2 - 4AC = -4A'C'$. Then $B^2 - 4AC < 0$ implies $-4A'C' < 0$, so A' and C' have the same sign and the graph of $A'x'^2 + C'y'^2 + D'x' + E'y' + F' = 0$ and, hence, the graph of $Ax^2 + Bxy + Cy^2 + Dx + Ey + F = 0$ is an ellipse (or a degenerate case of either a circle or a single point). If $B^2 - 4AC > 0$, then $-4A'C' > 0$, so A' and C' have opposite signs and the graph is a hyperbola (or the degenerate case of intersecting lines). If $B^2 - 4AC = 0$, then $-4A'C' = 0$, so one of A' and C' is zero and the graph is a parabola (or the degenerate case of a line). Summarizing:

THE DISCRIMINANT TEST

The graph of the equation $Ax^2 + Bxy + Cy^2 + Dx + Ey + F = 0$ is

an ellipse if $B^2 - 4AC < 0$,

a hyperbola if $B^2 - 4AC > 0$, and

a parabola if $B^2 - 4AC = 0$.

(Each case includes degenerate cases.)

EXERCISES 10.3

Express the equations in Exercises 1–4 in terms of the (x', y') coordinates of the given translations. Sketch the graphs and both sets of axes.

1. $x = y^2 - 1; x = x' - 1, y = y'$

2. $x + y^2 + 1 = 0; x = x' - 1, y = y'$

3. $y = 2x - x^2; x = x' + 1, y = y' + 1$

4. $y = x^2 + 4x; x = x' - 2, y = y' - 4$

Find the equations of translation that transform the origin to the center of the graphs of Exercises 5–8. Sketch the graphs and both sets of axes.

5. $x^2 - 4x + 4y^2 - 8y + 7 = 0$

6. $x^2 + 9y^2 - 18y + 8 = 0$

7. $x^2 - y^2 + 2y = 0$

8. $4x^2 - 12x - 9y^2 + 18y + 9 = 0$

Express the equations in Exercises 9–12 in terms of the (x', y') coordinates obtained by rotation by the given angle θ. Sketch the graphs and both sets of axes.

9. $x - y + 2 = 0; \theta = \pi/4$

10. $\sqrt{3}x + y = 3; \theta = \pi/6$

11. $2x + y = 2; \theta = \tan^{-1}(1/2)$

12. $3x - 4y + 12 = 0; \theta = \tan^{-1}(3/4)$

In Exercises 13–16, find $-\pi/2 < \theta < \pi/2$ so rotation by θ gives an equation with no x' term. Express the equation in terms of the (x', y') coordinates obtained by that rotation. Sketch the graphs and both sets of axes.

13. $x + y = \sqrt{2}$

14. $y = \sqrt{3}(x - 1)$

15. $\sqrt{3}y = x + \sqrt{3}$

16. $4x + 3y = 12$

In Exercises 17–22, find $0 < \theta < \pi/2$ so rotation by θ gives an equation with no $x'y'$ term. Express the equations in terms

of the (x', y') coordinates obtained by that rotation. Sketch the graphs and both sets of axes.

17. $xy + 1 = 0$

18. $xy = x + 1$

19. $x^2 - 2xy + y^2 - \sqrt{2}x - \sqrt{2}y = 0$

20. $x^2 + 2\sqrt{3}xy + 3y^2 + 2\sqrt{3}x - 2y = 0$

21. $7x^2 - 6\sqrt{3}xy + 13y^2 = 16$

22. $31x^2 + 10\sqrt{3}xy + 21y^2 = 144$

23. Find $-\pi/2 < \theta < \pi/2$ so rotation by θ changes the equation $ax + by = c$, $b \neq 0$, into an equation with no x' term. Express the equation in terms of the (x', y') coordinates obtained by that rotation.

24. Verify that rotation by any angle θ changes the equation

$$Ax^2 + Bxy + Cy^2 + Dx + Ey + F = 0$$

to an equation

$$A'x'^2 + B'x'y' + C'y'^2 + D'x' + E'y' + F' = 0$$

with $B'^2 - 4A'C' = B^2 - 4AC$.

Find the vertices and foci of the conics in Exercises 25–28.

25. $xy = 1$

26. $x^2 - 10\sqrt{3}xy + 11y^2 + 16 = 0$

27. $x^2 + 2\sqrt{3}xy + 3y^2 + 8\sqrt{3}x - 8y = 16$

28. $5x^2 - 6xy + 5y^2 = 8$

Use the discriminant test to identify the graphs of the equations in Exercises 29–38.

29. $x^2 - y = 0$

30. $y^2 + x = 1$

31. $x^2 + 4y^2 = 4$

32. $x^2 - y^2 = 1$

33. $x^2 - 4xy + 3y^2 = 8$

34. $x^2 - 4xy + 4y^2 = 8$

35. $x^2 - 4xy + 5y^2 = 8$

36. $2x^2 + 3xy + y^2 = 4$

37. $3x^2 + 6xy + 3y^2 = 16$

38. $-2x^2 + 5xy + 4y^2 = 8$

10.4 POLAR COORDINATE SYSTEM

Connections

Sign of expressions, 1.2.

Angles in standard position, 1.6.

Values of trigonometric functions of common angles, 1.6.

Graphs of conic sections, 10.1–2.

Many problems can be described most easily in terms of a direction and distance from a fixed reference point. For example, the motion of the planets is most clear when described in terms of a direction and distance from the sun. The polar coordinate system is convenient for such problems.

In this section we will introduce the polar coordinate system and investigate polar equations of some familiar curves.

Polar Coordinates

Consider a plane with a fixed number line (**polar axis**) with origin (**pole**) O. We will choose the polar axis to be horizontal with positive direction to the

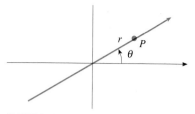

FIGURE 10.4.1 The point P with polar coordinates (r, θ) is located by rotating the polar axis an angle θ and then moving r units along the rotated number line.

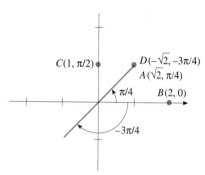

FIGURE 10.4.2 The points A and D are equal, even though their polar coordinates are different.

right, as the x-axis is usually oriented. To find the point P in the plane that has **polar coordinates** (r, θ), we first rotate the polar axis an angle θ and then move r units along the rotated number line. See Figure 10.4.1. The polar coordinates r and θ can be any real numbers, positive, negative, or zero. If r is positive, we move in the positive direction of the rotated number line. If r is negative, we move in the negative direction of the rotated number line. As usual, positive angles θ correspond to counterclockwise rotations and negative angles θ correspond to clockwise rotations.

EXAMPLE 1

Points with polar coordinates $A(\sqrt{2}, \pi/4)$, $B(2, 0)$, $C(1, \pi/2)$, and $D(-\sqrt{2}, -3\pi/4)$ are plotted in Figure 10.4.2. ■

We see in Example 1 that the points A and D have different polar coordinates, but correspond to the same point. Adding an integer multiple of 2π to the polar angle clearly does not change the corresponding point, so

$$P(r, \theta) = P(r, \theta + 2\pi n), \quad n \text{ any integer.}$$

See Figure 10.4.3. Also, rotating the polar axis an angle $\theta + \pi$ and then moving $-r$ units along the rotated number line brings us to the point that has polar coordinates (r, θ), so $P(r, \theta) = P(-r, \theta + \pi)$, or

$$P(r, \theta) = P(-r, \theta + \pi + 2\pi n), \quad n \text{ any integer.}$$

See Figure 10.4.4. Finally, we note that the pole corresponds to $r = 0$ and any value of θ, so

$$O(0, \theta) = O(0, \phi) \quad \text{for all values of } \theta \text{ and } \phi.$$

The above three formulas show that a single point can be represented by different polar coordinates. However, polar coordinates that represent the same point must be related by one of these equations.

Polar coordinates determine rectangular coordinates. From Figure 10.4.5 we see that if the polar axis and the x-axis coincide, then

$$x = r\cos\theta, \quad y = r\sin\theta.$$

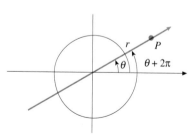

FIGURE 10.4.3 Adding an integer multiple of 2π to the polar angle does not change the corresponding point, so $P(r, \theta) = P(r, \theta + 2\pi n)$, n any integer.

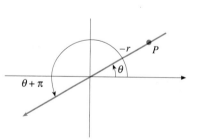

FIGURE 10.4.4 Rotating the polar axis an angle of $\theta + \pi$ and then moving $-r$ units along the rotated number line brings us to the point with polar coordinates (r, θ), so $P(r, \theta) = P(-r, \theta + \pi)$.

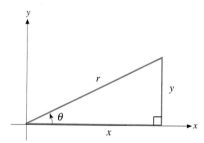

FIGURE 10.4.5 Polar and rectangular equations are related by the equations

$x = r\cos\theta$ and $y = r\sin\theta$.

EXAMPLE 2

Find rectangular coordinates of the points with polar coordinates $A(\sqrt{2}, \pi/4)$, $B(2, 0)$, $C(1, \pi/2)$, and $D(-\sqrt{2}, -3\pi/4)$.

SOLUTION   Using $x = r\cos\theta$ and $y = r\sin\theta$, we have

$$A(x, y) = \left(\sqrt{2}\cos\frac{\pi}{4}, \sqrt{2}\sin\frac{\pi}{4}\right) = (1, 1),$$
$$B(x, y) = (2\cos 0, 2\sin 0) = (2, 0),$$
$$C(x, y) = \left(1\cos\frac{\pi}{2}, 1\sin\frac{\pi}{2}\right) = (0, 1), \text{ and}$$
$$D(x, y) = \left(-\sqrt{2}\cos\left(-\frac{3\pi}{4}\right), -\sqrt{2}\sin\left(-\frac{3\pi}{4}\right)\right) = (1, 1).$$

These points were plotted by using polar coordinates in Figure 10.4.2. ■

To change rectangular coordinates to polar coordinates, we use a sketch and the equations

$$r^2 = x^2 + y^2,$$
$$\tan\theta = \frac{y}{x}.$$

These equations follow easily from the equations $x = r\cos\theta$ and $y = r\sin\theta$.

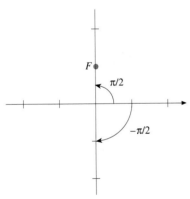

FIGURE 10.4.6 $E(x, y) = (-1, \sqrt{3})$ has polar coordinates $E(r, \theta) = (2, 2\pi/3)$ and $(-2, -\pi/3)$.

EXAMPLE 3

Find two sets of polar coordinates for each of the points $E(x, y) = (-1, \sqrt{3})$ and $F(x, y) = (0, 1)$.

SOLUTION   We see that $E(x, y) = (-1, \sqrt{3})$ has $r^2 = x^2 + y^2 = 4$, so $r = \pm 2$. We also have $\tan\theta = y/x = -\sqrt{3}/1$. This equation is satisfied by $\theta = -\pi/3 + n\pi$, n any integer. Note that not all polar coordinates $(r, \theta) = (\pm 2, -\pi/3 + n\pi)$, n an integer, correspond to the point E. From the sketch in Figure 10.4.6 we see that two sets of polar coordinates for E are $E(r, \theta) = (2, 2\pi/3)$ and $E(r, \theta) = (-2, -\pi/3)$.

$F(x, y) = (0, 1)$ has $r^2 = 1$ and $\theta = \pm\pi/2$. From Figure 10.4.7 we see that two sets of polar coordinates for F are $F(r, \theta) = (1, \pi/2)$ and $F(r, \theta) = (-1, -\pi/2)$. ■

FIGURE 10.4.7 $F(x, y) = (0, 1)$ has polar coordinates $F(r, \theta) = (1, \pi/2)$ and $(-1, -\pi/2)$.

Polar Graphs

A point P is on the graph of a polar equation in r and θ if and only if any of the (infinitely many) polar coordinates (r, θ) of P satisfy the equation.

The graph of a polar equation in r and θ is the set of all points P that correspond to polar coordinates (r, θ) that satisfy the equation.

EXAMPLE 4

The graph of the polar equation $r = 2$ is given in Figure 10.4.8. Note that $(r, \theta) = (2, \theta)$ satisfies the equation for any choice of θ. Also, note that the polar equation $r = -2$ has the same graph as $r = 2$, since $P(2, \theta) = P(-2, \theta + \pi)$. Both polar equations $r = 2$ and $r = -2$ satisfy $r^2 = 4$. The corresponding equation in rectangular coordinates is $x^2 + y^2 = 4$. ■

FIGURE 10.4.8   The graph of $r = r_0$ is a circle with center at the origin and radius $|r_0|$.

The graph of $r = r_0$ is a circle with center at the origin and radius $|r_0|$.

FIGURE 10.4.9 The graph of $\theta = \theta_0$ is a line through the pole at angle θ_0.

EXAMPLE 5

The graph of $\theta = \pi/4$ is given in Figure 10.4.9. Note that $P(r, \pi/4)$ is on the graph for all choices of r. ■

The graph of $\theta = \theta_0$ is a line through the pole at angle θ_0.

EXAMPLE 6

The graph of

$$r = \frac{1}{\cos\theta + \sin\theta}$$

is sketched in Figure 10.4.10. We could graph this equation either by plotting lots of points or by noticing that

$$r = \frac{1}{\cos\theta + \sin\theta}$$

implies $r\cos\theta + r\sin\theta = 1$. The polar equation $r\cos\theta + r\sin\theta = 1$ corresponds to the rectangular equation $x + y = 1$, which is the equation of a line. If we recognize the original polar equation is a line, we can determine the graph by plotting any two polar points. For example, $\theta = 0$ implies

FIGURE 10.4.10 The graph of $r = \dfrac{c}{a\cos\theta + b\sin\theta}$ is a line. The graph of $r = \dfrac{1}{\cos\theta + \sin\theta}$ has intercepts $(r, \theta) = (1, 0)$ and $(1, \pi/2)$.

$$r = \frac{1}{\cos 0 + \sin 0} = 1,$$

so $(r, \theta) = (1, 0)$ is a polar point on the graph. Similarly, we can determine that the polar point $(r, \theta) = (1, \pi/2)$ is on the graph. It is not necessary to change from polar to rectangular coordinates to sketch the graph. ■

The graph of $r = \dfrac{c}{a\cos\theta + b\sin\theta}$ is a line.

We see that $r \to \pm\infty$ as $a\cos\theta + b\sin\theta \to 0$ and that there is no point on the graph with $a\cos\theta + b\sin\theta = 0$, or $\tan\theta = -a/b$. The slope of the line is $-a/b$. See Figure 10.4.11.

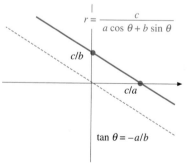

FIGURE 10.4.11 The line $r = \dfrac{c}{a\cos\theta + b\sin\theta}$ has intercepts $(r, \theta) = (c/a, 0)$ and $(c/b, \pi/2)$. There is no point on the graph with $a\cos\theta + b\sin\theta = 0$. The slope is $\tan\theta = -a/b$.

FIGURE 10.4.12 We can rewrite $r = 2\sec\theta$ as $r = \dfrac{2}{\cos\theta}$. This is the polar form of the equation of a line.

EXAMPLE 7

The graph of $r = 2\sec\theta$ is given in Figure 10.4.12. We may see that $r = 2\sec\theta$ implies $r\cos\theta = 2$, so the rectangular form of the equation is $x = 2$. Also, the polar equation can be written as

$$r = \frac{2}{1\cos\theta + 0\sin\theta},$$

which we know is the polar equation of a line. ■

EXAMPLE 8

The graph of $r = 2\sin\theta$ is given in Figure 10.4.13. We could plot points or recognize that $r = 2\sin\theta$ implies $r^2 = 2r\sin\theta$. Changing to rectangular coordinates, we then obtain $x^2 + y^2 = 2y$. Completing the square then gives $x^2 + (y-1)^2 = 1$. ■

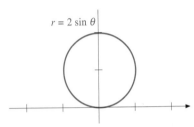

FIGURE 10.4.13 The graphs of $r = 2a\cos\theta$ and $r = 2a\sin\theta$ are circles through the pole.

The graph of $r = 2a\cos\theta$ is a circle through the pole with center $(r, \theta) = (a, 0)$. The graph of $r = 2a\sin\theta$ is a circle through the pole with center $(r, \theta) = (a, \pi/2)$.

Using Key Points to Sketch Graphs

Let us reconsider Example 8 and see how to obtain the graph of $r = 2\sin\theta$ without changing to rectangular coordinates. The idea is to plot a few key points and then draw a smooth curve through the points, in the direction of increasing θ. Coordinates of some key points are given in Table 10.4.1.

Each of the points determined in the table is plotted. We then note that $r = 0$ when $\theta = 0$, and the value of r increases to 2 as θ increases from 0 to $\pi/2$. We use this information to sketch that part of the curve that corresponds to $0 \le \theta \le \pi/2$. The curve should be traced in the direction of increasing θ, from the pole to the point $(r, \theta) = (2, \pi/2)$. This step is illustrated in Figure 10.4.14a.

TABLE 10.4.1

θ	0	$\pi/2$	π	$3\pi/2$	2π
$r = 2\sin\theta$	0	2	0	-2	0

FIGURE 10.4.14a As θ increases from 0 to $\pi/2$, $r = 2\cos\theta$ increases from zero to two.

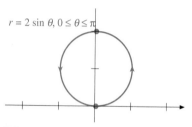

FIGURE 10.4.14b As θ increases from $\pi/2$ to π, $r = 2\sin\theta$ decreases from two to zero. As θ increases from π to 2π, $r = 2\sin\theta$ is negative and the graph retraces the circle.

Continuing, we note that r decreases from 2 to zero as θ ranges from $\pi/2$ to π. This means that the graph goes from the point $(r, \theta) = (2, \pi/2)$ to the pole as θ goes from $\pi/2$ to π. That part of the curve that corresponds to $\pi/2 \leq \theta \leq \pi$ has been added in Figure 10.4.14b. Again, the curve is traced in the direction of increasing θ. As θ varies from π to 2π, r is negative and the graph retraces the circle. Finally, we note that if we know the graph of $r = 2\sin\theta$ is a circle through the pole, we can use the above method to obtain a very accurate sketch. Also, note that the point $(r, \theta) = (2, \pi/2)$ then determines the orientation and scale of the circle.

As usual, we will try to sketch graphs by using general characteristics and key points. Key points of polar graphs should include points with $\theta = 0, \pi/2, \pi, 3\pi/2$, and 2π. Values of θ for which either $r = 0$ or r is undefined should also be noted. The sign of r is an important characteristic. Use values of $r(\theta)$ to *trace the curve in the direction of increasing θ*, between the key points that have been plotted. Of course, sketching any graph is easier if you know the general shape in advance, so you need only determine orientation and scale. Always check the equation to see if it is a familiar type.

EXAMPLE 9

Sketch the graphs of the polar equations

(a) $r = \dfrac{6}{2 - \sin\theta}$,

(b) $r = \dfrac{2}{1 + \cos\theta}$,

(c) $r = \dfrac{3}{1 - 2\sin\theta}$.

SOLUTION

TABLE 10.4.2

θ	0	$\pi/2$	π	$3\pi/2$	2π
$r = 6/(2 - \sin\theta)$	3	6	3	2	3

(a) Table 10.4.2 contains some key values of r. The corresponding points are plotted and a smooth curve is then drawn through them, as in Figure 10.4.15. Starting at the point $(r, \theta) = (3, 0)$, the curve is traced in the direction of increasing θ, as indicated by the arrows. We have used the fact that r is always positive.

TABLE 10.4.3

θ	0	$\pi/2$	π	$3\pi/2$	2π
$r = 2/(1 + \cos\theta)$	1	2	*	2	1

(b) Key values of r are given in Table 10.4.3. The key points are plotted and that part of the graph that corresponds to $0 \leq \theta < \pi$ is sketched in Figure 10.4.16a. Note that r approaches infinity as θ approaches π from below; r is undefined when $\theta = \pi$. As θ increases from π to 2π, the values of r decrease from infinity to one. That part of the graph has been added in Figure 10.4.16b.

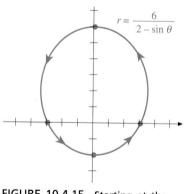

FIGURE 10.4.15 Starting at the point $(r, \theta) = (3, 0)$, the curve is traced in the direction of increasing θ, through the key points that have been plotted. The values of $r = \dfrac{6}{2 - \sin\theta}$ are always positive. The graph appears to be an ellipse.

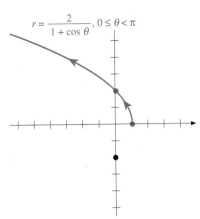

$r = \dfrac{2}{1 + \cos \theta}, 0 \leq \theta < \pi$

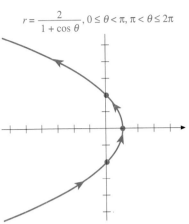

$r = \dfrac{2}{1 + \cos \theta}, 0 \leq \theta < \pi, \pi < \theta \leq 2\pi$

FIGURE 10.4.16a As θ increases from 0 to π, $r = \dfrac{2}{1 + \cos \theta}$ increases from one to infinity.

FIGURE 10.4.16b As θ increases from π to 2π, $r = \dfrac{2}{1 + \cos \theta}$ decreases from infinity to one. The graph appears to be a parabola.

(c) In this example, we see that $r = 3/(1 - 2\sin\theta)$ becomes unbounded as θ approaches $\pi/6$ and $5\pi/6$, where $1 - 2\sin\theta = 0$; r is undefined at these values of θ. We add these values of θ to our list of key points in Table 10.4.4.

TABLE 10.4.4

θ	0	$\pi/6$	$\pi/2$	$5\pi/6$	π	$3\pi/2$	2π
$r = 3/(1 - 2\sin\theta)$	3	*	-3	*	3	1	3

As θ varies from 0 to $\pi/6$, r increases from 3 to infinity. This part of the graph is sketched in Figure 10.4.17a. As θ varies from $\pi/6$ to $5\pi/6$, we see that r is negative. This means that the part of the graph that corresponds to $\pi/6 < \theta < 5\pi/6$ will be drawn in the sector opposite to that between $\pi/6$ and $5\pi/6$. As θ increases from $\pi/6$, r increases from negative infinity to -3 when $\theta = \pi/2$, and then decreases to negative infinity as θ approaches $5\pi/6$ from below. This gives the lower part of the graph that has been added in Figure 10.4.17b. As θ increases from $5\pi/6$, r decreases from infinity to 3 when $\theta = \pi$, continues decreasing to 1 when $\theta = 3\pi/2$, and then increases to 3 as θ approaches 2π. This part of the graph has been added in Figure 10.4.17c. ∎

You may suspect from looking at the sketches that the graphs of the equations in Example 9 are conics. They are. For example,

$$r = \frac{1}{1 + \cos\theta} \text{ implies}$$
$$r + r\cos\theta = 2,$$
$$r = 2 - r\cos\theta,$$
$$r^2 = (2 - r\cos\theta)^2,$$
$$x^2 + y^2 = (2 - x)^2,$$
$$x^2 + y^2 = 4 - 4x + x^2,$$
$$y^2 = 4 - 4x, \text{ a parabola.}$$

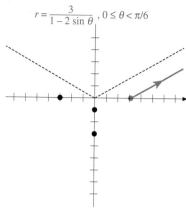

$r = \dfrac{3}{1 - 2\sin\theta}$, $0 \le \theta < \pi/6$

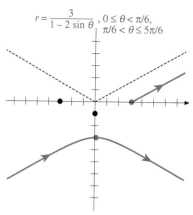

$r = \dfrac{3}{1 - 2\sin\theta}$, $0 \le \theta < \pi/6$, $\pi/6 < \theta \le 5\pi/6$

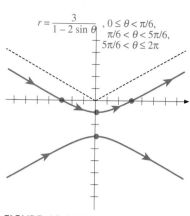

$r = \dfrac{3}{1 - 2\sin\theta}$, $0 \le \theta < \pi/6$, $\pi/6 < \theta < 5\pi/6$, $5\pi/6 < \theta \le 2\pi$

FIGURE 10.4.17a As θ increases from 0 to $\pi/6$, $r = \dfrac{3}{1 - 2\sin\theta}$ increases from three to infinity.

FIGURE 10.4.17b As θ increases from $\pi/6$ to $5\pi/6$, $r = \dfrac{3}{1 - 2\sin\theta}$ increases from negative infinity to negative three when $\theta = \pi/2$ and then decreases to negative infinity as θ increases to $5\pi/6$. Since r is negative, the graph is in the sector opposite $\pi/6 < \theta < 5\pi/6$ as θ ranges over these values.

FIGURE 10.4.17c As θ increases from $5\pi/6$ to 2π, $r = \dfrac{3}{1 - 2\sin\theta}$ decreases from infinity to one when $\theta = 3\pi/2$ and then increases to three as θ approaches 2π. The graph appears to be a hyperbola.

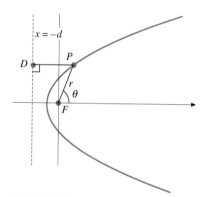

FIGURE 10.4.18 The focus-directrix equation of a conic with focus at the pole and axis along a coordinate axis is $|PF| = e|PD|$, where e is the eccentricity. This simplifies to either $r = \dfrac{c}{1 \pm e\sin\theta}$ or $r = \dfrac{c}{1 \pm e\cos\theta}$.

Polar Equations of Conics

Conics that have a focus at the pole and vertices along a coordinate axis can be expressed conveniently as polar equations. For example, the focus-directrix equation of the conic with focus at the origin, directrix $x = -d$, and eccentricity e is

$$|PF| = e|PD|.$$

See Figure 10.4.18. Expressing this equation in polar coordinates, we have

$$r = e(r\cos\theta + d),$$
$$(1 - e\cos\theta)r = ed,$$
$$r = \frac{ed}{1 - e\cos\theta}.$$

Considering all conics with focus at the origin and vertices on a coordinate axis gives the following.

The graphs of

$$r = \frac{c}{1 \pm e\sin\theta} \quad and \quad r = \frac{c}{1 \pm e\cos\theta}$$

are conics with a focus at the pole and axis either $\theta = 0$ or $\theta = \pi/2$.

The graph is an ellipse with eccentricity e if $0 < e < 1$, a parabola if $e = 1$, and a hyperbola with eccentricity e if $e > 1$. It is a circle if $e = 0$.

If we know in advance that the graph is a conic, then the pattern made by the points on the coordinate axes tells which conic it is, as well as gives the orientation and scale.

We have seen how to change polar equations of lines and certain circles and conics to rectangular equations. The general idea in changing from polar to rectangular coordinates is to manipulate the polar equation so that it contains only variable expressions of the form r^2, $r \cos \theta$, and $r \sin \theta$. The procedure differed in each of the cases studied. We should not expect to be able to change every polar equation to a rectangular equation, although this may be possible with great effort. On the other hand, the equations $x = r \cos \theta$ and $y = r \sin \theta$ can be used to change easily any rectangular equation to a polar equation.

EXERCISES 10.4

Plot the points with the given polar coordinates in Exercises 1–2.

1. $A(1, 0)$, $B(2, \pi/2)$, $C(-\sqrt{2}, \pi/4)$, $D(-2, 3\pi/2)$, $E(1, 1)$

2. $A(2, 0)$, $B(1, \pi/2)$, $C(-2, \pi)$, $D(2, 7\pi/6)$, $E(\pi/2, 2)$

Find the rectangular coordinates of the points with the given polar coordinates in Exercises 3–4.

3. $A(-1, \pi)$, $B(2, \pi/3)$, $C(3, \pi/2)$, $D(2, 3\pi/4)$

4. $A(-2, 0)$, $B(\sqrt{2}, \pi/4)$, $C(2, -\pi/6)$, $D(-3, \pi/2)$

Find two sets of polar coordinates of the points with the rectangular coordinates in Exercises 5–6.

5. $A(1, 0)$, $B(0, 2)$, $C(-1, \sqrt{3})$, $D(2, -2)$

6. $A(2, 0)$, $B(0, 1)$, $C(-\sqrt{2}, \sqrt{2})$, $D(-\sqrt{3}, 1)$

Change the polar equations in Exercises 7–14 to equivalent equations in rectangular coordinates.

7. $r = 2$

8. $\theta = \dfrac{\pi}{3}$

9. $r = \sec \theta$

10. $r = \dfrac{6}{2 \cos \theta + 3 \sin \theta}$

11. $r = 4 \cos \theta$

12. $r = \dfrac{6}{2 + \cos \theta}$

13. $r = \dfrac{2}{1 - \sin \theta}$

14. $r = \dfrac{3}{1 + 2 \cos \theta}$

Change the rectangular equations given in Exercises 15–22 to polar equations.

15. $x^2 + y^2 = 9$

16. $y + x = 0$

17. $y = 3$

18. $y = 2x - 1$

19. $x^2 + 6x + y^2 = 0$

20. $x^2 + y^2 - 4y = 0$

21. $y^2 = 4x + 4$

22. $16x^2 - 96x + 25y^2 = 256$

Identify and sketch the graph of the polar equations in Exercises 23–40.

23. $r = 3$

24. $r = -2$

25. $\theta = \pi/2$

26. $\theta = \pi/3$

27. $r = \dfrac{2}{\cos \theta + 2 \sin \theta}$

28. $r = \dfrac{2}{2 \cos \theta - \sin \theta}$

29. $r = \csc \theta$

30. $r = -\sec \theta$

31. $r = \sin \theta$

32. $r = \cos \theta$

33. $r = -3 \cos \theta$

34. $r = -2 \sin \theta$

35. $r = \dfrac{3}{1 + \sin \theta}$

36. $r = \dfrac{6}{2 - \sin \theta}$

37. $r = \dfrac{6}{2 + \cos \theta}$

38. $r = \dfrac{3}{1 - \cos \theta}$

39. $r = \dfrac{2}{1 - 2 \sin \theta}$

40. $r = \dfrac{2}{1 + 2 \cos \theta}$

41. Find a formula for the distance between the points with polar coordinates (r_0, θ_0) and (r_1, θ_1).

42. Show that the equation $r = a \sin \theta + b \cos \theta$ is the polar equation of a circle if $(a, b) \neq (0, 0)$.

43. Find a polar equation satisfied by all points for which the distance of the point from the pole is equal to the distance of the point from the line $x = -d$, $d \neq 0$.

44. Find a polar equation satisfied by all points for which the distance of the point from the pole is twice the distance of the point from the line $x = -d$, $d \neq 0$.

45. Find a polar equation satisfied by all points for which the distance of the point from the pole is half the distance of the point from the line $x = -d$, $d \neq 0$.

46. Verify that
$$r = \frac{c}{1 + e \sin \theta}$$
is an ellipse with one focus at the pole and eccentricity e if $0 < e < 1$.

47. Verify that
$$r = \frac{c}{1 - e \cos \theta}$$
is a hyperbola with one focus at the pole and eccentricity e if $e > 1$.

48. Verify that
$$r = \frac{c}{1 - \sin \theta}$$
is a parabola with focus at the pole.

10.5 GRAPHS OF POLAR EQUATIONS; TANGENT LINES

Connections

Geometric interpretation of derivative, 2.6.

Polar graphing strategy, 10.4.

Section 10.4 was intended to increase confidence in handling polar equations by providing practice with familiar curves. In this section we will look at some curves that are more convenient to study in polar coordinates than in rectangular coordinates. We will study graphs of polar equations of the form $r = r(\theta)$, where r is a continuous function of θ. We will first use the sign of $r(\theta)$ to analyze the graph. This corresponds to graphing a rectangular equation $y = f(x)$ by using the sign of $f(x)$, as was done in Section 4.1. Finally, we will see how the derivative of $r(\theta)$ can be used to study polar graphs.

Tangents at the Pole and Loops

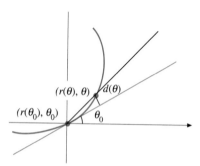

FIGURE 10.5.1a Under usual conditions, if $r(\theta_0) = 0$, the graph of $r = r(\theta)$ approaches and leaves the pole at angle $\theta = \theta_0$ as θ passes through θ_0.

> **THEOREM 1**
>
> Suppose that r is a continuous function of θ and $r(\theta_0) = 0$, and there is some $\delta > 0$ such that $r(\theta) \neq 0$ for $0 < |\theta - \theta_0| < \delta$. Then the graph of the polar equation $r = r(\theta)$ approaches and leaves the pole at the angle of the line $\theta = \theta_0$ as θ passes through θ_0. That is, the curve is tangent to the line $\theta = \theta_0$ at the pole.

Proof We first note that $r(\theta) \to r(\theta_0) = 0$ as $\theta \to \theta_0$ because $r(\theta)$ is continuous. This implies the point $(r(\theta), \theta)$ approaches the pole as θ approaches θ_0. In Figure 10.5.1a we have set $d(\theta)$ equal to the distance from the point $(r(\theta), \theta)$ to the line $\theta = \theta_0$. The distance of the point from the pole is $r(\theta)$. To show that the point approaches the line tangentially, we must verify that the ratio $d(\theta)/r(\theta) \to 0$ as $\theta \to \theta_0$. From Figure 10.5.1a we see that

$$\frac{d(\theta)}{r(\theta)} = \sin(\theta - \theta_0).$$

Since sine is a continuous function, $\sin(\theta - \theta_0) \to 0$ as $\theta \to \theta_0$. We conclude that

$$\lim_{\theta \to \theta_0} \frac{d(\theta)}{r(\theta)} = 0,$$

so either the graph is tangent to the line $\theta = \theta_0$ as θ approaches θ_0, as indicated in Figure 10.5.1a, or the graph forms a cusp at angle θ_0, as indicated in Figure 10.5.1b. ∎

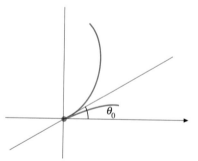

FIGURE 10.5.1b The graph may either pass smoothly through the pole or have a cusp at angle θ_0 if $r(\theta_0) = 0$.

A continuous function $r(\theta)$ does not change sign between successive zeros. The graph of $r = r(\theta)$ will make a **positive loop** where $r(\theta)$ is positive between successive zeros of r. This is illustrated in Figure 10.5.2a. The graph will make a **negative loop** where $r(\theta)$ is negative between successive zeros of r. See Figure 10.5.2b. Note that a negative loop is drawn in the sector opposite to that of the corresponding θ.

When sketching the graph of a polar equation, we should determine those values of θ_0 for which $r(\theta_0) = 0$ and lightly draw the lines $\theta = \theta_0$ through the pole. This marks the angles at which the graph approaches the pole. By determining the sign of $r(\theta)$ between successive zeros, we can determine whether to sketch a positive loop or a negative loop.

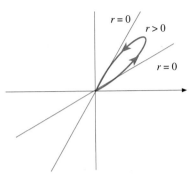

FIGURE 10.5.2a The graph of $r = r(\theta)$ will make a positive loop between successive zeros of r where $r(\theta)$ is positive.

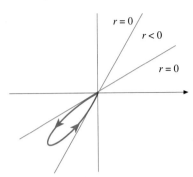

FIGURE 10.5.2b The graph makes a negative loop between successive zeros of r where $r(\theta)$ is negative. A negative loop is drawn in the sector opposite to that of the corresponding θ.

EXAMPLE 1

Sketch the graph of the polar equation $r = 1 - \cos\theta$.

SOLUTION We determine key values of $r(\theta)$ from key values of $\cos\theta$. See Table 10.5.1.

TABLE 10.5.1

θ	0	$\pi/2$	π	$3\pi/2$	2π
r	0	1	2	1	0

We see that $r(0) = r(2\pi) = 0$ and that $r(\theta)$ is positive for $0 < \theta < 2\pi$. The graph makes a positive loop between $\theta = 0$ and $\theta = 2\pi$. Since $r(0) = 0$, we know that the graph must leave the pole at the angle of the polar axis as θ increases from 0. The equation $r(2\pi) = 0$ implies that the graph must approach the pole tangent to the line $\theta = 2\pi$ as θ increases toward 2π. The graph is sketched in Figure 10.5.3. ∎

$r = 1 - \cos\theta$

FIGURE 10.5.3 The graph makes a positive loop between $\theta = 0$ and $\theta = 2\pi$.

EXAMPLE 2

Sketch the graph of the polar equation $r = 1 + 2\sin\theta$.

SOLUTION The equation $r = 0$ implies $1 + 2\sin\theta = 0$, or $\sin\theta = -1/2$. The solutions $\theta = 7\pi/6$ and $\theta = 11\pi/6$ are included in our table of key values. See Table 10.5.2.

TABLE 10.5.2

θ	0	$\pi/2$	π	$7\pi/6$	$3\pi/2$	$11\pi/6$	2π
r	1	3	1	0	-1	0	1

$r = 1 + 2\sin\theta$

FIGURE 10.5.4 The graph makes a negative loop between $\theta = 7\pi/6$ and $\theta = 11\pi/6$. The graph makes a positive loop between $\theta = -\pi/6$ and $\theta = 7\pi/6$.

The graph is sketched in Figure 10.5.4. Note that the graph passes through the pole at the angle of the lines $\theta = 7\pi/6$ and $\theta = 11\pi/6$. The graph makes a negative loop between $\theta = 7\pi/6$ and $\theta = 11\pi/6$. This is the smaller, inner loop. Note that the line $\theta = 11\pi/6$ is the same as the line $\theta = -\pi/6$. We can then describe the larger, outer loop as a positive loop between $\theta = -\pi/6$ and $\theta = 7\pi/6$. It is important for the integration problems that we will do in later sections that loops be described in terms of increasing θ. ∎

The graph of an equation of the form either

$$r = a + b\sin\theta \ or \ r = a + b\cos\theta$$

is called a **limaçon**. *If* $|a| = |b|$, *the graph is also called a* **cardioid**.

The graph will have an inner loop if $|b| > |a|$. If $|a| > |b|$, the graph will not intersect the pole. We will use the derivative to obtain more information about this case later in this section.

EXAMPLE 3

Sketch the graph of the polar equation $r = 2\sin 2\theta$.

SOLUTION   Table 10.5.3 includes the values of θ between 0 and 2π where $\sin 2\theta$ is 0, 1, and -1. We see that values of $r(\theta)$ repeat in intervals of length π.

$r = 2\sin 2\theta, 0 \le \theta \le \pi$

FIGURE 10.5.5a   The graph makes a positive loop between $\theta = 0$ and $\theta = \pi/2$. The graph makes a negative loop between $\theta = \pi/2$ and $\theta = \pi$.

TABLE 10.5.3

θ	0	$\pi/4$	$\pi/2$	$3\pi/4$	π	$5\pi/4$	$3\pi/2$	$7\pi/4$	2π
r	0	2	0	-2	0	2	0	-2	0

The graph approaches the pole at each of the angles $0, \pi/2, \pi, 3\pi/2$, and 2π. The graph makes a positive loop between $\theta = 0$ and $\theta = \pi/2$, a negative loop between $\theta = \pi/2$ and $\theta = \pi$, a positive loop between $\theta = \pi$ and $\theta = 3\pi/2$, and then a negative loop between $\theta = 3\pi/2$ and $\theta = 2\pi$. In Figure 10.5.5a we have sketched the graph as θ varies between 0 and π. As θ varies between $\pi/2$ and π, the corresponding negative loop is drawn in the opposite sector, the fourth quadrant. The complete graph is sketched in Figure 10.5.5b. ∎

$r = 2\sin 2\theta, 0 \le \theta \le 2\pi$

FIGURE 10.5.5b   The graph makes another positive loop and then another negative loop as θ increases from π to 2π.

EXAMPLE 4

Sketch the graph of the polar equation $r = \cos 3\theta$.

SOLUTION   Values of $r(\theta)$ repeat in intervals of length $2\pi/3$. Table 10.5.4 includes values where $r = 0, 1$, or -1, for θ between 0 and π.

TABLE 10.5.4

θ	0	$\pi/6$	$\pi/3$	$\pi/2$	$2\pi/3$	$5\pi/6$	π
r	1	0	-1	0	1	0	-1

The graph completes part of a positive loop as θ varies between 0 and $\pi/6$, and makes a negative loop between $\theta = \pi/6$ and $\theta = \pi/2$. This part of the graph is sketched in Figure 10.5.6a. The graph makes a positive loop between $\theta = \pi/2$ and $\theta = 5\pi/6$, and makes half of a negative loop between $\theta = 5\pi/6$ and $\theta = \pi$. Note that this completes the graph, as illustrated in Figure 10.5.6b. The graph is retraced as θ varies between π and 2π. ∎

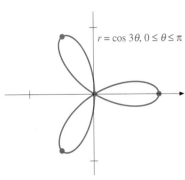

FIGURE 10.5.6a The graph makes half of a positive loop between $\theta = 0$ and $\theta = \pi/6$. The graph makes a negative loop between $\theta = \pi/6$ and $\theta = \pi/2$.

FIGURE 10.5.6b The graph makes a positive loop between $\theta = \pi/2$ and $\theta = 5\pi/6$. The graph makes half of a negative loop as θ increases from $5\pi/6$ to π. The graph is retraced as θ increases from π to 2π.

Graphs of equations of the form either

$$r = a \sin n\theta \ or \ r = a \cos n\theta,$$

*n an integer, are called **rose curves**.*

The zeros divide $[0, 2\pi]$ into $2n$ intervals of length π/n. The graph alternates positive and negative loops as θ varies over these intervals. There will be $2n$ **leaves** if n is even, and each leaf is traversed once as θ increases from 0 to 2π. If n is odd, there will be n leaves, and each leaf is transversed twice as θ increases from 0 to 2π.

EXAMPLE 5

Sketch the graph of the polar equation $r^2 = 2 \sin 2\theta$.

SOLUTION   Note that $r^2 = 2 \sin 2\theta$ implies either

$$r = +\sqrt{2 \sin 2\theta} \ or \ r = -\sqrt{2 \sin 2\theta}.$$

Since these two polar equations have the same graphs, let us restrict our attention to $r = \sqrt{2 \sin 2\theta}$. Values of $r(\theta)$ repeat in intervals of π. Table 10.5.5 contains key values of $r(\theta)$ for θ between 0 and π.

TABLE 10.5.5

θ	0	$\pi/4$	$\pi/2$	$3\pi/4$	π
r	0	$\sqrt{2}$	0	*	0

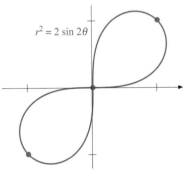

FIGURE 10.5.7 Since $2 \sin 2\theta$ is negative for $\pi/2 < \theta < \pi$ and $3\pi/2 < \theta < 2\pi$, no coordinates (r, θ) satisfy the equation $r^2 = 2 \sin 2\theta$ for θ in these intervals.

The graph is sketched in Figure 10.5.7. The loop in the first quadrant is formed by $r = \sqrt{2 \sin 2\theta}$ as θ varies between 0 and $\pi/2$. The loop in the third quadrant is formed by $r = \sqrt{2 \sin 2\theta}$ as θ varies between π and $3\pi/2$. (The loops would simply be interchanged if we had used $r = -\sqrt{2 \sin 2\theta}$.) There are no points on the graph that correspond to values of θ between either $\pi/2$ and π or $3\pi/2$ and 2π, because $\sin 2\theta$ is negative for those values of θ and there is no real number r that satisfies $r^2 < 0$. ∎

The graph of an equation of the form either

$$r^2 = a \sin 2\theta \ or \ r^2 = a \cos 2\theta$$

*is called a **lemniscate**.*

FIGURE 10.5.8 The graph of $r = a\theta$ is a spiral.

EXAMPLE 6

Sketch the graph of the polar equation $r = \theta/2$, $\theta \geq 0$.

SOLUTION Some key points of the graph are given in Table 10.5.6.

TABLE 10.5.6

θ	0	$\pi/2$	π	$3\pi/2$	2π
r	0	$\pi/4$	$\pi/2$	$3\pi/4$	π

We see that the graph leaves the pole tangent to the line $\theta = 0$ and that $r(\theta)$ increases as θ increases. The graph is sketched in Figure 10.5.8. ◼

Graphs of equations of the form

$$r = a\theta$$

are called **spirals**.

Tangents to Polar Curves

We have seen that the line $\theta = \theta_0$ is tangent to the graph of the polar equation $r = r(\theta)$ at the pole if $r(\theta_0) = 0$. Let us now use the derivative to investigate tangents to polar curves at points other than the pole.

If $r(\theta)$ is differentiable at θ_0 and $r(\theta_0) \neq 0$, it should seem reasonable that the graph of $r = r(\theta)$ has a tangent line at the point $(r(\theta_0), \theta_0)$. Let us verify this fact and develop a formula for the slope of the tangent line. We will see that it is convenient to express the slope of a polar curve in terms of the angle β illustrated in Figure 10.5.9; β is the angle from the radial line $\theta = \theta_0$ to the line tangent to the graph at $(r(\theta_0), \theta_0)$.

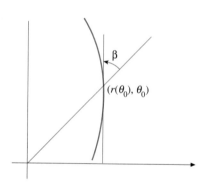

FIGURE 10.5.9 The angle β is the angle from the radial line $\theta = \theta_0$ to the tangent to the graph at $(r(\theta_0), \theta_0)$.

Let β' denote the angle from the radial line $\theta = \theta_0$ to the line through the points $(r(\theta_0), \theta_0)$ and $(r(\theta), \theta)$, as illustrated in Figure 10.5.10. Note that the circular arc from $(r(\theta), \theta)$ to $(r(\theta), \theta_0)$ intersects the radial line $\theta = \theta_0$ at a right angle. Also, if θ is near θ_0, $\theta \neq \theta_0$, the arc is nearly straight, so β' can be thought of as being an angle of a right triangle that has length of the opposite side approximately $r(\theta)(\theta - \theta_0)$ and length of the side adjacent approximately $r(\theta) - r(\theta_0)$. It follows that

$$\tan \beta' \approx \frac{r(\theta)(\theta - \theta_0)}{r(\theta) - r(\theta_0)} = \frac{r(\theta)}{\dfrac{r(\theta) - r(\theta_0)}{\theta - \theta_0}} \approx \frac{r(\theta_0)}{r'(\theta_0)},$$

where we have assumed that r is differentiable at θ_0 with $r(\theta_0) \neq 0$ and $r'(\theta_0) \neq 0$. As θ approaches θ_0, the angles β' approach the value β. We obtain the formula

$$\tan \beta = \frac{r(\theta_0)}{r'(\theta_0)},$$

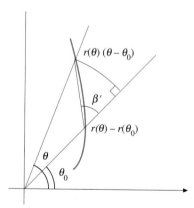

FIGURE 10.5.10 The angle β' approaches β as θ approaches θ_0. We have

$$\tan \beta' \approx \frac{r(\theta)(\theta - \theta_0)}{r(\theta) - r(\theta_0)} \approx \frac{r(\theta_0)}{r'(\theta_0)}.$$

where β is the angle from the radial line $\theta = \theta_0$ to the line tangent to the graph of $r = r(\theta)$ at $(r(\theta_0), \theta_0)$.

If $r(\theta_0) \neq 0$ and $r'(\theta_0) = 0$, we choose $\beta = \pm \pi/2$. In case $r(\theta_0) = 0$, we know the graph approaches and leaves the pole tangent to the line $\theta = \theta_0$, so we choose $\beta = 0$.

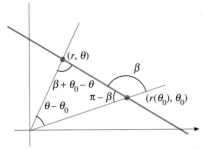

FIGURE 10.5.11 The Law of Sines implies

$$\frac{r}{\sin(\pi - \beta)} = \frac{r(\theta_0)}{\sin(\beta + \theta_0 - \theta)}.$$

It is not difficult to obtain the polar equation of a line tangent to a polar curve. To do this, let β denote the angle associated with the line tangent to the curve $r = r(\theta)$ at the point $(r(\theta_0), \theta_0)$, so $\tan \beta = r(\theta_0)/r'(\theta_0)$. A point (r, θ) is on the tangent line whenever the angle from the radial line $\theta = \theta_0$ to the line through (r, θ) and $(r(\theta_0), \theta_0)$ is β. We can then apply the Law of Sines to the triangle in Figure 10.5.11 to obtain

$$\frac{r}{\sin(\pi - \beta)} = \frac{r(\theta_0)}{\sin(\beta + \theta_0 - \theta)}.$$

Using the trigonometric identity for the sine of a difference to simplify, we obtain the equation

$$r = \frac{r(\theta_0) \sin \beta}{\sin(\beta + \theta_0) \cos \theta - \cos(\beta + \theta_0) \sin \theta}.$$

This is the polar equation of the line tangent to the graph of the polar curve $r = r(\theta)$ at the point $(r(\theta_0), \theta_0)$.

EXAMPLE 7

Find a polar equation of the line tangent to $r = 2\cos\theta$ at the point where $\theta = \pi/3$.

SOLUTION We need to determine the angle β at the point where $\theta_0 = \pi/3$. We have $r = 2\cos\theta$ and $r' = -2\sin\theta$. When $\theta = \pi/3$, $r = 2(1/2) = 1$ and $r' = -2(\sqrt{3}/2) = -\sqrt{3}$. Then $\tan\beta = r/r' = -1/\sqrt{3}$, so $\beta = -\pi/6$ and $\theta_0 + \beta = \pi/6$. Substitution then gives

FIGURE 10.5.12 The line tangent to the graph of $r = 2\cos\theta$ at the point where $\theta = \pi/3$ is

$$r = \frac{-1}{\cos\theta - \sqrt{3}\sin\theta}.$$

$$r = \frac{r(\theta_0) \sin \beta}{\sin(\beta + \theta_0) \cos \theta - \cos(\beta + \theta_0) \sin \theta},$$

$$r = \frac{(1)(\sin(-\pi/6))}{\sin(\pi/6)\cos\theta - \cos(\pi/6)\sin\theta},$$

$$r = \frac{-1/2}{(1/2)\cos\theta - (\sqrt{3}/2)\sin\theta},$$

$$r = \frac{-1}{\cos\theta - \sqrt{3}\sin\theta}.$$

The graph is sketched in Figure 10.5.12. ∎

The slope of the line tangent to the polar curve $r = r(\theta)$ at a point $(r(\theta), \theta)$ is $\tan\alpha$, as illustrated in Figure 10.5.13. Since $\alpha = \theta + \beta$, we have

$$\tan\alpha = \tan(\theta + \beta) = \frac{\tan\theta + \tan\beta}{1 - \tan\theta\tan\beta}$$

$$= \frac{\dfrac{\sin\theta}{\cos\theta} + \dfrac{r}{r'}}{1 - \left(\dfrac{\sin\theta}{\cos\theta}\right)\left(\dfrac{r}{r'}\right)} \left(\frac{r'\cos\theta}{r'\cos\theta}\right)$$

$$= \frac{r'\sin\theta + r\cos\theta}{r'\cos\theta - r\sin\theta}.$$

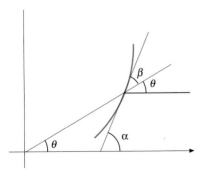

FIGURE 10.5.13 The slope of the line tangent to the graph of $r = r(\theta)$ is $\tan\alpha = \tan(\theta + \beta)$.

Summarizing, we have

$$\text{Slope} = \frac{r'\sin\theta + r\cos\theta}{r'\cos\theta - r\sin\theta}.$$

There are horizontal tangents where $r' \sin\theta + r\cos\theta = 0$, and vertical tangents where $r'\cos\theta - r\sin\theta = 0$, unless both expressions are zero. Note that both expressions are zero only if $r = r' = 0$.

EXAMPLE 8

Sketch the graph of the polar curve $r = 3 - 2\cos\theta$. Find all points on the graph that have vertical tangent lines.

SOLUTION The graph will have vertical tangents whenever

$$r'\cos\theta - r\sin\theta = 0.$$

We have $r = 3 - 2\cos\theta$, so $r' = 2\sin\theta$. Substitution then gives

$$(2\sin\theta)(\cos\theta) - (3 - 2\cos\theta)(\sin\theta) = 0,$$
$$(\sin\theta)(4\cos\theta - 3) = 0.$$

We see then that $\sin\theta = 0$ gives the values $\theta = 0$ and $\theta = \pi$, and that $4\cos\theta - 3 = 0$ has solution $\cos^{-1}(3/4)$ in the first quadrant and solution $2\pi - \cos^{-1}(3/4)$ in the fourth quadrant. These values of θ are included in Table 10.5.7. The graph is sketched in Figure 10.5.14. ■

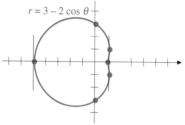

FIGURE 10.5.14 The graph of $r = 3 - 2\cos\theta$ has four points with vertical tangents. The graph has an indentation at $(r, \theta) = (1, 0)$.

TABLE 10.5.7

θ	0	$\cos^{-1}(3/4)$	$\pi/2$	π	$3\pi/2$	$2\pi - \cos^{-1}(3/4)$	2π
r	1	3/2	3	5	3	3/2	1

EXAMPLE 9

Sketch the graph of the polar curve $r = 3 - \cos\theta$. Find all points on the graph that have vertical tangent lines.

SOLUTION The graph will have vertical tangents whenever

$$r'\cos\theta - r\sin\theta = 0.$$

Since $r' = \sin\theta$, substitution gives

$$(\sin\theta)(\cos\theta) - (3 - \cos\theta)(\sin\theta) = 0,$$
$$(\sin\theta)(2\cos\theta - 3) = 0.$$

The equation $\sin\theta = 0$ gives the values $\theta = 0$ and $\theta = \pi$, and $2\cos\theta - 3 = 0$ implies $\cos\theta = 3/2$. The latter equation has no solution, since the cosine function is bounded by one. Some key values are given in Table 10.5.8. The graph is sketched in Figure 10.5.15. ■

FIGURE 10.5.15 The graph of $r = 3 - \cos\theta$ has only two points with vertical tangents. The graph does not have an indentation.

TABLE 10.5.8

θ	0	$\pi/2$	π	$3\pi/2$	2π
r	2	3	4	3	2

Let us compare the graphs in Figure 10.5.3, 10.5.14, and 10.5.15. Each is the graph of a limaçon, with equation

$$r = a + b\cos\theta, \qquad |a| \geq |b|.$$

It can be shown that the graph has a cusp (sharp point) at the pole when $|a| = |b|$ (Figure 10.5.3). The graph has a smooth indentation at $(a + b, 0)$ for $|b| < |a| < 2|b|$ (Figure 10.5.15). The indentation disappears for $|a| \geq 2|b|$ (Figure 10.5.15). The limaçon has an inner loop if $|a| < |b|$, as we have seen in Figure 10.5.4.

EXERCISES 10.5

Sketch the graphs of the polar equations given in Exercises 1–28.

1. $r = 1 + \cos\theta$ **2.** $r = 2 - 2\cos\theta$

3. $r = 2 - 2\sin\theta$ **4.** $r = 1 + \sin\theta$

5. $r = 1 - 2\cos\theta$ **6.** $r = 1 + 2\cos\theta$

7. $r = 1 - \sqrt{2}\sin\theta$ **8.** $r = \sqrt{3} + 2\sin\theta$

9. $r = 2\cos 2\theta$ **10.** $r = \sin 2\theta$

11. $r = -\cos 3\theta$ **12.** $r = -\sin 3\theta$

13. $r = \sin 4\theta$ **14.** $r = \cos 4\theta$

15. $r^2 = 4\cos 2\theta$ **16.** $r^2 = \sin 2\theta$

17. $r^2 = -9\sin 2\theta$ **18.** $r^2 = \cos 2\theta$

19. $r = \theta/\pi, \theta \geq 0$ **20.** $r = \theta/4, \theta \geq 0$

21. $r^2 = \sin 3\theta$ **22.** $r^2 = \sin 4\theta$

23. $r = \sin(\theta/2)$ **24.** $r = \sin(\theta/3)$

25. $r^2 = \theta$ **26.** $r = \cos^2\theta$

27. $r = \begin{cases} \sin(4\theta/3), & 0 \leq \theta \leq 3\pi/4, \\ \sin(4\theta), & 3\pi/4 \leq \theta \leq \pi, \\ \sin(4\theta/3 - 4\pi/3), & \pi \leq \theta \leq 7\pi/4, \\ \sin(4\theta), & 7\pi/4 \leq \theta \leq 2\pi \end{cases}$

28. $r = \ln\theta$

Determine the angle β between the graphs given in Exercises 29–34 and each of the radial lines $\theta = 0, \pi/2, \pi, 3\pi/2,$ and 2π. Sketch the graphs.

29. $r = 1 + \sin\theta$ **30.** $r = 1 + 2\sin\theta$

31. $r = \dfrac{1}{\cos\theta + 2\sin\theta}$ **32.** $r = \dfrac{6}{2 + \sin\theta}$

33. $r = \dfrac{2}{1 + \sin\theta}$ **34.** $r = \dfrac{3}{1 - 2\sin\theta}$

Find the polar equation of the line tangent to the given graphs at the point with the given value of θ_0 in Exercises 35–38. Sketch the graphs.

35. $r = 2, \theta_0 = \pi/3$

36. $r = 2\cos\theta, \theta_0 = \pi/6$

37. $r = \sin 3\theta, \theta_0 = \pi/2$

38. $r = 2\cos 2\theta, \theta = 0$

Sketch the graphs in Exercises 39–42. Find all points on the graphs that have vertical tangent lines.

39. $r = 1 + \cos\theta$ **40.** $r = \sqrt{2} + \cos\theta$

41. $r = 2 + \cos\theta$ **42.** $r = 3 + \cos\theta$

Sketch the graphs in Exercises 43–46. Find all points on the graph that have horizontal tangent lines.

43. $r = 1 + \cos\theta$ **44.** $r = 2\cos\theta$

45. $r = \dfrac{6}{2 + \sin\theta}$ **46.** $r = \dfrac{2}{1 + \sin\theta}$

47. $r = a - b\cos\theta, a > b > 0$.

  (a) Show that the graph has four distinct points with vertical tangent lines whenever $a < 2b$. Sketch the graph.

  (b) Show that the graph has only two distinct points with vertical tangent lines whenever $a \geq 2b$. Sketch the graph.

48. Find polar coordinates of the point nearest the origin where the graph of $r = \ln\theta$ intersects itself.

49. Sketch the graph of

$$r(\theta) = \begin{cases} 1 - \sqrt{2}\cos\theta, & \pi/4 \leq \theta \leq 7\pi/4, \\ 0, & -\pi/4 < \theta < \pi/4. \end{cases}$$

Is the graph tangent to the line $\theta = 0$ at the pole? Why doesn't the theorem about tangent lines at the pole apply?

50. Use the formula $d(fg) = f\,dg + g\,df$ from Section 3.6 to express dx and dy in terms of $r, \theta, dr,$ and $d\theta$. Evaluate the ratio $\dfrac{dy}{dx}$ and compare with the formula given for the slope of a polar curve.

10.6 AREA OF REGIONS BOUNDED BY POLAR CURVES

Connections

Polar graphs, 10.5.

Concept of area as an integral, 6.2.

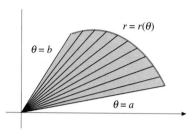

FIGURE 10.6.1 The region bounded by the polar curves $r = r(\theta)$, $\theta = a$, and $\theta = b$ is divided into thin sectors by slicing along radial lines.

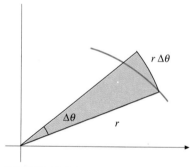

FIGURE 10.6.2 The area of the region in each sector is approximately $\frac{1}{2}r^2\Delta\theta$, the area of the sector of a circle with radius r and central angle $\Delta\theta$.

FIGURE 10.6.3 The sectors run from $r = 0$ to $r = 2\sin\theta$, $0 \le \theta \le \pi$.

Consider the region in the plane bounded by the polar curves

$$r = r(\theta), \quad \theta = a, \quad \text{and} \quad \theta = b.$$

We assume that r is continuous for $a \le \theta \le b$ and that $b - a \le 2\pi$. The values of r can be either positive, negative, or zero. To find the area of the region, we divide it into sectors by slicing it along radial lines $\theta = \theta_j$ from the pole to the curve $r = r(\theta)$, as illustrated in Figure 10.6.1. Note that the condition $b - a \le 2\pi$ keeps the sectors from overlapping as θ increases from a to b. The area of each sector is approximately $\frac{1}{2}r^2\Delta\theta$, the area of a sector of a circle with radius r and central angle $\Delta\theta$. See Figure 10.6.2. The area of the region should be the limit of sums of the areas of the sectors. That is,

$$\text{Area} = \int_a^b \frac{1}{2}(r(\theta))^2\,d\theta.$$

Be careful to use the above area formula only for regions to which it applies. The integral represents a limit of sums of the areas of sectors. For it to apply to a region, radial lines must divide the region into sectors. For any fixed θ between a and a *larger* number b, radial lines at angle θ must intersect the region from the pole to the point $(r(\theta), \theta)$ on the boundary curve.

Regions Between a Polar Curve and the Pole

EXAMPLE 1
Find the area of the region enclosed by $r = 2\sin\theta$.

SOLUTION The region and a typical area sector are sketched in Figure 10.6.3. The sector runs from the pole to the curve $r = 2\sin\theta$. These sectors will cover the region once as θ ranges from $\theta = 0$ to $\theta = \pi$. The area is

$$\begin{aligned}
\text{Area} &= \int_0^\pi \frac{1}{2}(2\sin\theta)^2\,d\theta \\
&= 2\int_0^\pi \sin^2\theta\,d\theta \\
&= 2\left[-\frac{\sin\theta\cos\theta}{2} + \frac{\theta}{2}\right]_0^\pi \quad \text{[Table of Integrals]} \\
&= (0 + \pi) - (0 + 0) = \pi.
\end{aligned}$$

Note that the graph of $r = 2\sin\theta$ is a circle of radius 1, so the integral formula gives the correct area. ∎

EXAMPLE 2
Find the area of the region bounded by the lines $\theta = 0$ and $\theta = 2\pi/3$ and that part of the spiral $r = e^\theta$ with $0 \le \theta \le 2\pi/3$.

SOLUTION The region and a typical area sector are pictured in Figure 10.6.4. The sectors run from the pole to $r = e^\theta$. The region is covered once as the sectors range from $\theta = 0$ to $\theta = 2\pi/3$. The area is

$$\text{Area} = \int_0^{2\pi/3} \frac{1}{2}(e^\theta)^2 \, d\theta = \frac{1}{2}\int_0^{2\pi/3} e^{2\theta} \, d\theta$$

$$= \frac{1}{2}\frac{e^{2\theta}}{2}\Bigg]_0^{2\pi/3} = \frac{1}{4}\left(e^{4\pi/3} - 1\right). \ \blacksquare$$

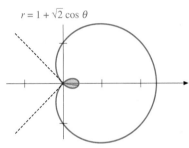

$r = e^\theta, 0 \le \theta \le 2\pi/3$

FIGURE 10.6.4 The sectors run from $r = 0$ to $r = e^\theta, 0 \le \theta \le 2\pi/3$.

$r = 1 + \sqrt{2}\cos\theta$

FIGURE 10.6.5 The sectors run from $r = 0$ to $r = 1 + \sqrt{2}\cos\theta$. The inner loop is generated as θ increases from $\theta = 3\pi/4$ to $\theta = 5\pi/4$.

EXAMPLE 3

Find the area of the region bounded by the inner loop of $r = 1 + \sqrt{2}\cos\theta$.

SOLUTION The graph and a typical area sector are sketched in Figure 10.6.5. Typical area sectors go from the pole to the inner loop. The limits of integration of the area integral are given by those values of θ that give the initial and terminal angles of the inner, negative loop. Loops begin and end at angles for which $r = 0$. In this case, $r = 0$ implies $\cos\theta = -1/\sqrt{2}$, so $\theta = 3\pi/4$ and $\theta = 5\pi/4$. Note that the smaller loop is generated as θ ranges from $3\pi/4$ to $5\pi/4$. It doesn't matter that r is negative in this interval. The area is

$$\text{Area} = \int_{3\pi/4}^{5\pi/4} \frac{1}{2}(1 + \sqrt{2}\cos\theta)^2 \, d\theta$$

$$= \int_{3\pi/4}^{5\pi/4} \frac{1}{2}(1 + 2\sqrt{2}\cos\theta + 2\cos^2\theta) \, d\theta$$

$$= \left[\frac{\theta}{2} + \sqrt{2}\sin\theta + \frac{\sin\theta\cos\theta}{2} + \frac{\theta}{2}\right]_{3\pi/4}^{5\pi/4} \quad \text{[Table of Integrals]}$$

$$= \left[\theta + \sqrt{2}\sin\theta + \frac{1}{2}\sin\theta\cos\theta\right]_{3\pi/4}^{5\pi/4}$$

$$= \left[\frac{5\pi}{4} + \sqrt{2}\left(-\frac{1}{\sqrt{2}}\right) + \left(\frac{1}{2}\right)\left(-\frac{1}{\sqrt{2}}\right)\left(-\frac{1}{\sqrt{2}}\right)\right]$$

$$\quad - \left[\frac{3\pi}{4} + \sqrt{2}\left(\frac{1}{\sqrt{2}}\right) + \left(\frac{1}{2}\right)\left(\frac{1}{\sqrt{2}}\right)\left(-\frac{1}{\sqrt{2}}\right)\right]$$

$$= \frac{\pi}{2} - \frac{3}{2}. \ \blacksquare$$

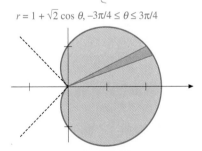

$r = 1 + \sqrt{2}\cos\theta, -3\pi/4 \le \theta \le 3\pi/4$

FIGURE 10.6.6 The sectors run from $r = 0$ to $r = 1 + \sqrt{2}\cos\theta$. The outer loop is generated as θ increases from $\theta = -3\pi/4$ to $\theta = 3\pi/4$.

EXAMPLE 4

Find the area of the region bounded by the outer loop of $r = 1 + \sqrt{2}\cos\theta$.

SOLUTION The graph and typical area sector are sketched in Figure 10.6.6. Typical area sectors go from the pole to the outer loop. We need values of θ that give the initial and terminal angles of the outer loop. As in Example 3, $r = 0$ implies $\cos\theta = -1\sqrt{2}$, but this time we choose the solutions $\theta = -3\pi/4$ and $\theta = 3\pi/4$. The larger loop is generated as θ increases from $-3\pi/4$ to the *larger* value $3\pi/4$. The area is

$$\text{Area} = \int_{-3\pi/4}^{3\pi/4} \frac{1}{2}(1 + \sqrt{2}\cos\theta)^2 \, d\theta$$

$$= \left[\theta + \sqrt{2}\sin\theta + \frac{1}{2}\sin\theta\cos\theta\right]_{-3\pi/4}^{3\pi/4} \quad \text{[As in Example 3]}$$

$$= \left[\frac{3\pi}{4} + \sqrt{2}\left(\frac{1}{\sqrt{2}}\right) + \left(\frac{1}{2}\right)\left(\frac{1}{\sqrt{2}}\right)\left(-\frac{1}{\sqrt{2}}\right) \right]$$

$$- \left[-\frac{3\pi}{4} + \sqrt{2}\left(-\frac{1}{\sqrt{2}}\right) + \left(\frac{1}{2}\right)\left(-\frac{1}{\sqrt{2}}\right)\left(-\frac{1}{\sqrt{2}}\right) \right]$$

$$= \frac{3\pi}{2} + \frac{3}{2}. \quad \blacksquare$$

Note that we cannot describe the outer loop of Example 4 by saying that θ ranges (backward) from $5\pi/4$ to $3\pi/4$. Evaluation of the corresponding area integral,

$$\int_{5\pi/4}^{3\pi/4} \frac{1}{2}[1 + \sqrt{2}\cos\theta]^2 \, d\theta = -\int_{3\pi/4}^{5\pi/4} \frac{1}{2}[1 + \sqrt{2}\cos\theta]^2 \, d\theta,$$

would give the negative of the area of the smaller loop.

NOTE *It is very important that the limits of integration go from one number to a larger number, and that they describe those area sectors that cover the region exactly once.*

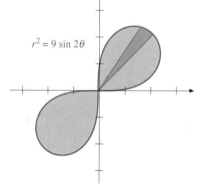

$r^2 = 9\sin 2\theta$

FIGURE 10.6.7 The sectors run from $r = 0$ to $r = \sqrt{9\sin 2\theta}$. One loop is generated as θ increases from $\theta = 0$ to $\theta = \pi/2$.

EXAMPLE 5

Find the area of the region enclosed by $r^2 = 9\sin 2\theta$.

SOLUTION Let us first note that the equation $r^2 = 9\sin 2\theta$ is not really of the form $r = r(\theta)$ that is required to apply our area formula. However, $r^2 = 9\sin 2\theta$ implies $r = \sqrt{9\sin 2\theta}$ or $r = -\sqrt{9\sin 2\theta}$, and each of the latter equations has the same graph as the original equation. Let us work with the equation $r = \sqrt{9\sin 2\theta}$, so $r(\theta) \geq 0$ whenever θ is in either the first or third quadrant. We see that $r(\theta)$ is undefined for θ in either the second or fourth quadrant. We have $[r(\theta)]^2 = 9\sin 2\theta$ whenever $r(\theta)$ is defined. The region is sketched in Figure 10.6.7. *Using symmetry,* we see that the entire area is twice the area of one loop. The region in the first quadrant is covered by sectors from the pole to the curve, as θ varies between 0 and $\pi/2$. Note that 0 and $\pi/2$ are successive zeros of r. We have

$$\text{Area} = 2(\text{Area of loop in first quadrant})$$

$$\text{Area} = 2\int_0^{\pi/2} \frac{1}{2}9\sin 2\theta \, d\theta = -\frac{9}{2}\cos 2\theta \Big]_0^{\pi/2} = -\frac{9}{2}(\cos\pi - \cos 0)$$

$$= -\frac{9}{2}(-1 - 1) = 9. \quad \blacksquare$$

Note that we cannot express the area of the region in Example 5 as $\int_0^{2\pi} (1/2)9\sin 2\theta \, d\theta$. Evaluation of this integral gives zero. Even though there are no points on the graph of $r = \sqrt{9\sin 2\theta}$ in the second and fourth quadrants, the expression $9\sin 2\theta$ is well defined and can be integrated over the corresponding values of θ. The reason the integral does not give the proper area is that $9\sin 2\theta \neq (\sqrt{9\sin 2\theta})^2$ for θ in the second and fourth quadrants, because we cannot take the positive square root of a negative number. Thus, the integral does not represent the area bounded by $r = \sqrt{9\sin 2\theta}$, $\theta = 0$, and $\theta = 2\pi$.

Regions Between Two Polar Curves

We can use the area formula we have developed to find the area of a region between two polar curves. As in the case of rectangular coordinates, this involves finding the points of intersection of the curves.

The problem of finding all points of intersection of two polar curves deserves a word of caution. The difficulty arises because different coordinate pairs (r, θ) can represent the same point. If a point has a coordinate pair (r, θ) that satisfies both equations, it will correspond to a point of intersection. However, there may be points of intersection for which no coordinate pair of the point satisfies both of the given equations. It may be necessary to replace the given equations by different equations that have the same graph. For example, from Figure 10.6.8 we see that the graphs of $r = 2$ and $\theta = \pi/3$ have two points of intersection. The point of intersection in the first quadrant has a coordinate pair $(2, \pi/3)$ that satisfies both equations. The coordinates of this point are given by the simultaneous solution of the equations $r = 2$, $\theta = \pi/3$. The point of intersection in the third quadrant does not have a coordinate pair that satisfies both given equations. The coordinates of this point are not given by a simultaneous solution of the equations $r = 2$, $\theta = \pi/3$. However, if we replace the equation $r = 2$ by the equivalent equation $r = -2$, we can obtain the point of intersection in the third quadrant as the simultaneous solution of the equations $r = -2$, $\theta = \pi/3$. Note that the polar equations $r = 2$ and $r = -2$ have the same graph.

FIGURE 10.6.8 The coordinates of points of intersection of polar curves may not satisfy the given equations of the curves. Use a sketch of both curves to roughly locate and determine the number of points of intersection of two polar curves.

To find all points of intersection of two polar curves

- Sketch both curves and use the sketch to roughly locate and determine the number of points of intersection.
- Use simultaneous solutions of the two given equations to find coordinate pairs that satisfy both equations.
- If this does not give all points of intersection, substitute equivalent equations for the given equations.

Let us now return to the problem of area. We will see that the area of regions between two polar curves can sometimes be determined as a difference of areas of regions of the type to which our area formula applies.

EXAMPLE 6

Find the area of the region inside $r = 1 + \sin \theta$ and outside $r = 1$.

SOLUTION The region is sketched in Figure 10.6.9. The points of intersection of the graphs are easily seen to be given by the values $\theta = 0$ and $\theta = \pi$. Radial lines at angles θ, for θ between the points of intersection, intersect the region from $(1, \theta)$ to $(1 + \sin \theta, \theta)$. The area of the region is the positive difference of the areas of the regions bounded by $r = 1 + \sin \theta$ and $r = 1$, between the values of θ. That is,

$$\text{Area} = \int_0^\pi \frac{1}{2}(1 + \sin \theta)^2 \, d\theta - \int_0^\pi \frac{1}{2}(1)^2 \, d\theta$$

$$= \int_0^\pi \frac{1}{2}\left((1 + \sin \theta)^2 - (1)^2\right) \, d\theta$$

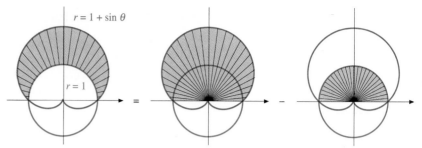

FIGURE 10.6.9 The area of the region inside $r = 1 + \sin \theta$ and outside $r = 1$ is equal to the area of the sectors from $r = 0$ to $r = 1 + \sin \theta$, $0 \leq \theta \leq \pi$, minus the area of the sectors between $r = 0$ and $r = 1$, $0 \leq \theta \leq \pi$.

$$= \int_0^\pi \frac{1}{2}(1 + 2\sin\theta + \sin^2\theta - 1)\,d\theta = \int_0^\pi \left(\sin\theta + \frac{1}{2}\sin^2\theta \right) d\theta$$

$$= -\cos\theta + \frac{1}{2}\left(-\frac{\sin\theta\cos\theta}{2} + \frac{\theta}{2} \right) \Big]_0^\pi \qquad \text{[Table of Integrals]}$$

$$= \left(1 + \frac{\pi}{4} \right) - (-1) = 2 + \frac{\pi}{4}. \quad \blacksquare$$

EXAMPLE 7

Find the area of the region inside $r = 2\cos\theta$ and outside $r = 1$.

SOLUTION   The region is sketched in Figure 10.6.10. Radial lines at angles θ, for θ between the points of intersection, intersect the region from $(1, \theta)$ to $(2\cos\theta, \theta)$. We need the points of intersection of the curves. The equations $r = 2\cos\theta$ and $r = 1$ imply $\cos\theta = 1/2$, so $\theta = -\pi/3$ and $\theta = \pi/3$ are the desired coordinates. The area of the region is the difference of the areas of the regions bounded by $r = 2\cos\theta$ and $r = 1$, between these values of θ. We have

$$\text{Area} = \int_{-\pi/3}^{\pi/3} \frac{1}{2}(2\cos\theta)^2\,d\theta - \int_{-\pi/3}^{\pi/3} \frac{1}{2}(1)^2\,d\theta = \int_{-\pi/3}^{\pi/3} \left(2\cos^2\theta - \frac{1}{2} \right) d\theta$$

$$= 2\left(\frac{\sin\theta\cos\theta}{2} + \frac{\theta}{2} \right) - \frac{\theta}{2} \Big]_{-\pi/3}^{\pi/3} \qquad \text{[Table of Integrals]}$$

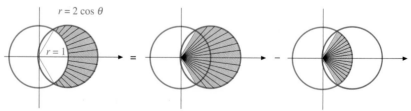

FIGURE 10.6.10 The area of the region inside $r = 2\cos\theta$ and outside $r = 1$ is equal to the area of the sectors from $r = 0$ to $r = 2\cos\theta$, $-\pi/3 \leq \theta \leq \pi/3$, minus the area of the sectors between $r = 0$ and $r = 1$, $-\pi/3 \leq \theta \leq \pi/3$.

$$= \sin\theta\cos\theta + \frac{\theta}{2}\Big]_{-\pi/3}^{\pi/3}$$

$$= \left(\left(\frac{\sqrt{3}}{2}\right)\left(\frac{1}{2}\right) + \frac{\pi}{6}\right) - \left(\left(-\frac{\sqrt{3}}{2}\right)\left(\frac{1}{2}\right) - \frac{\pi}{6}\right) = \frac{\sqrt{3}}{2} + \frac{\pi}{3}. \quad \blacksquare$$

EXAMPLE 8

Find the area of the region between the inner and outer loops of $r = 1 + \sqrt{2}\cos\theta$.

SOLUTION   The region is sketched in Figure 10.6.11. The desired area is the area of the outer loop minus the area of the inner loop. Using the results of Examples 3 and 4, we have

Area $=$ (Area of outer loop) $-$ (Area of inner loop)

$$= \left(\int_{-3\pi/4}^{3\pi/4} \frac{1}{2}[1 + \sqrt{2}\cos\theta]^2 d\theta\right) - \left(\int_{3\pi/4}^{5\pi/4} \frac{1}{2}[1 + \sqrt{2}\cos\theta]^2 d\theta\right)$$

$$= \left(\frac{3\pi}{2} + \frac{3}{2}\right) - \left(\frac{\pi}{2} - \frac{3}{2}\right) = \pi + 3.$$

Note that the integrals that give the areas of these two loops have different limits of integration. We cannot combine the integrals in this example. $\quad \blacksquare$

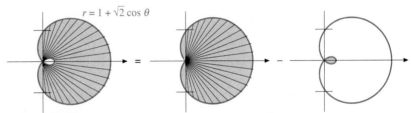

FIGURE 10.6.11   The area between the inner and outer loops of $r = 1 + \sqrt{2}\cos\theta$ is equal to the area inside the outer loop minus the area inside the inner loop.

EXAMPLE 9

Find the area of the region in the first quadrant that is inside both $r = 1$ and $r = 2\cos\theta$.

SOLUTION   The region is sketched in Figure 10.6.12. We see that there are two distinct types of radial lines associated with the region. One type runs from the pole to $r = 1$, and the other runs from the pole to $r = 2\cos\theta$. This means

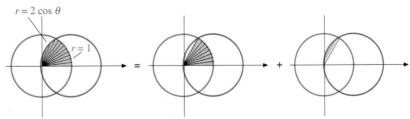

FIGURE 10.6.12   The area in the first quadrant that is inside both $r = 1$ and $r = 2\cos\theta$ is equal to the area of the sectors from $r = 0$ to $r = 1, 0 \le \theta \le \pi/3$, plus the area of the sectors from $r = 0$ to $r = 2\cos\theta, \pi/3 \le \theta \le \pi/2$.

that we can express the area as a sum of the areas of two regions to which our formula applies. We need the point of intersection. The equations $r = 1$ and $r = 2\cos\theta$ imply $2\cos\theta = 1$. The solution in the first quadrant is $\theta = \pi/3$. The area is

$$\text{Area} = \int_0^{\pi/3} \frac{1}{2}(1)^2 d\theta + \int_{\pi/3}^{\pi/2} \frac{1}{2}(2\cos\theta)^2 d\theta$$

$$= \int_0^{\pi/3} \frac{1}{2}d\theta + \int_{\pi/3}^{\pi/2} 2\cos^2\theta\, d\theta = \frac{\pi}{6} + \left[\sin\theta\cos\theta + \theta\right]_{\pi/3}^{\pi/2}$$

[Table of Integrals]

$$= \frac{\pi}{6} + \left[\left((1)(0) + \frac{\pi}{2}\right) - \left(\left(\frac{\sqrt{3}}{2}\right)\left(\frac{1}{2}\right) + \frac{\pi}{3}\right)\right] = \frac{\pi}{3} - \frac{\sqrt{3}}{4}. \quad \blacksquare$$

EXERCISES 10.6

Find the area of the regions described in Exercises 1–26.

1. Bounded by $r = a$, $\theta = 0$, $\theta = 2$
2. Bounded by $r = a$, $\theta = 0$, $\theta = \alpha$ $(0 \le \alpha \le 2\pi)$
3. Bounded by $r = 2\sec\theta$, $\theta = 0$, $\theta = \pi/3$
4. Bounded by $r = 2\csc\theta$, $\theta = \pi/4$, $\theta = \pi/2$
5. Bounded by $r = \theta$, $\theta = 0$, $\theta = \pi$
6. Bounded by $r = e^{-\theta/2}$, $\theta = 0$, $\theta = \pi$
7. Enclosed by $r = 2 - 2\sin\theta$
8. Enclosed by $r = 3 - 2\sin\theta$
9. Enclosed by $r^2 = \sin\theta$
10. Enclosed by $r^2 = \cos 2\theta$
11. Enclosed by $r = \sin 2\theta$
12. Enclosed by $r = \cos 3\theta$
13. The smaller loop of $r = 1 - 2\cos\theta$
14. The larger loop of $r = 1 - 2\cos\theta$
15. Inside $r = 2$ and to the right of $r = \sec\theta$
16. Inside $r = 1$ and outside $r = 1 - \cos\theta$
17. Bounded by $r = 2\theta$, $r = \theta$, $\theta = 0$, $\theta = \pi$
18. Inside $r = 2\cos\theta$ and outside $r = \cos\theta$
19. Inside $r = 2\sin 2\theta$ and outside $r = 1$
20. Inside $r^2 = 2\cos 2\theta$ and outside $r = 1$
21. Between loops of $r = 2\cos\theta - 1$
22. Between loops of $r = 2\sin\theta + \sqrt{3}$
23. Inside $r = 2$ and above $r = -\csc\theta$
24. In the first quadrant, inside both $r = 1$ and $r^2 = 2\cos\theta$
25. Inside both $r^2 = \sin\theta$ and $r = \sqrt{2}\sin\theta$
26. Inside both $r = \cos\theta$ and $r = \sqrt{3}\sin\theta$

*In Section 6.7 we evaluated **centers of mass of laminas** by using the formulas*

$$\bar{x} = \frac{\int \tilde{x}\,dM}{\int dM} \quad and \quad \bar{y} = \frac{\int \tilde{y}\,dM}{\int dM},$$

where $(\tilde{x}, \tilde{y})$ is the center of mass of a part of the lamina that has mass dM and $(\bar{x}, \bar{y})$ are the rectangular coordinates of the center of mass of the lamina. We can use polar coordinates to evaluate these integrals if each component of the integrals can be expressed in terms of the polar angle θ. For example, if a homogeneous lamina is bounded by the polar curve $r = r(\theta)$, $a \le \theta \le b$, we can express the mass of a thin sector as

$$dM = \rho\,dA = \rho\frac{1}{2}[r(\theta)]^2\,d\theta,$$

where ρ is the constant density. Also, since a thin sector is essentially a triangle, the rectangular coordinates of the center of mass of a thin sector are

$$(\tilde{x}, \tilde{y}) = \left(\frac{2}{3}r(\theta)\cos\theta, \frac{2}{3}r(\theta)\sin\theta\right).$$

See figure. If desired, we can change the rectangular coordinates of the center of mass to polar coordinates. Use this technique to find polar coordinates of the center of mass of the homogeneous lamina bounded by the polar curves given in each of Exercises 27–30.

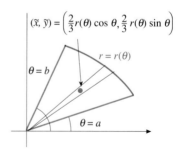

27. $r = a$, $\theta = 0$, and $\theta = \pi$
28. $r = a$, $\theta = -\alpha$, and $\theta = \alpha$ $(0 < \alpha < \pi)$
29. $r = 1 + \cos\theta$
30. $r = 2 + \sin\theta$

10.7 THREE-DIMENSIONAL RECTANGULAR COORDINATE SYSTEM; SKETCHING

Connections

Sketching lines, 1.4.

Sketching conic sections, 10.1–2.

FIGURE 10.7.1 A three-dimensional rectangular coordinate system can be envisioned by looking at the corner of a rectangular room.

The basic concept of a **three-dimensional rectangular coordinate system** can be illustrated by viewing a rectangular room. We choose a corner where two walls intersect the floor to be the origin. The intersection of the floor with the wall to the left as you face the corner forms the x-axis, the intersection of the floor with the wall to the right forms the y-axis, and the intersection of the two walls forms the z-axis. Any point in the room can be located by specifying its coordinates with respect to these three coordinate axes. That is, the ordered triple of coordinates (x, y, z) is associated with the point found by moving from the origin x units parallel to the x-axis, y units parallel to the y-axis, and z units parallel to the z-axis. See Figure 10.7.1, where we have superimposed the coordinate system on a photograph of the corner of a room.

Plotting Points

To sketch objects in three-dimensional space we need to establish a point from which the object is viewed. We will ordinarily orient the coordinate axes as viewed from a point that is an equal distance from each axis, as illustrated in Figure 10.7.1. This view is symmetric with respect to the three coordinate axes and is the most common view used in computer graphics and other technical drawings. We can represent it pictorially by drawing the axes at angles of $120°$ and using the same scale on each axis. The z-axis is drawn with positive direction upward, the positive x-axis is drawn to the left and downward, while the positive y-axis is drawn to the right and downward. See Figure 10.7.2.

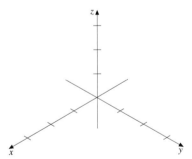

FIGURE 10.7.2 We will ordinarily draw the coordinate axes at angles of $120°$ and use the same scale on each axis.

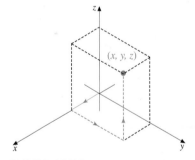

FIGURE 10.7.3 The point with coordinates (x, y, z) is plotted by moving from the origin x units parallel to the x-axis, y units parallel to the y-axis, and z units parallel to the z-axis.

FIGURE 10.7.4 The axes are oriented so that, if the z-axis is grasped with the right hand so the fingers wrap from the direction of the positive x-axis to the direction of the positive y-axis, then the thumb will point in the direction of the positive z-axis.

The point with coordinates (x, y, z) is plotted by moving from the origin x units parallel to the x-axis, y units parallel to the y-axis, and z units parallel to the z-axis. See Figure 10.7.3. The axes are oriented to form a **right-hand rectangular coordinate system**. This means that, if the z-axis is grasped with the *right hand* so the fingers wrap from the direction of the positive x-axis to the direction of the positive y-axis, the thumb will point in the direction of the positive z-axis. See Figure 10.7.4.

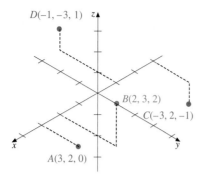

FIGURE 10.7.5 The first octant consists of all points in three-dimensional space that have all three coordinates positive.

The coordinate planes divide three-dimensional space into eight parts, called **octants**. The **first octant** is that portion that has all three coordinates positive.

EXAMPLE 1

Plot the points $A(3, 2, 0)$, $B(2, 3, 2)$, $C(-3, 2, -1)$, and $D(-1, -3, 1)$. Which points are in the first octant?

SOLUTION   The points are plotted in Figure 10.7.5. Only B is in the first octant. ■

Distance and Midpoint Formulas

The **distance** between two points (x_1, y_1, z_1) and (x_2, y_2, z_2) is

$$d = \sqrt{(x_2 - x_1)^2 + (y_2 - y_1)^2 + (z_2 - z_1)^2}.$$

The distance formula can be verified by two applications of the Pythagorean Theorem. See Figure 10.7.6. The **midpoint** of the line segment between the points is

$$M = \left(\frac{x_1 + x_2}{2}, \frac{y_1 + y_2}{2}, \frac{z_1 + z_2}{2} \right).$$

Each coordinate of the midpoint is the average of the corresponding coordinates of the endpoints of the segment. See Figure 10.7.7.

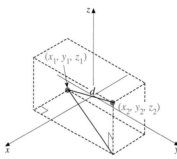

FIGURE 10.7.6 The distance formula can be verified by two applications of the Pythagorean Theorem.

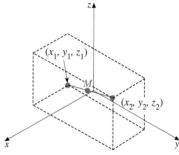

FIGURE 10.7.7 Each coordinate of the midpoint of a line segment is the average of the corresponding coordinates of the endpoints of the segment.

EXAMPLE 2

Sketch the line segment between the points $(1, -1, 0)$ and $(-1, 3, 4)$. Find the length and midpoint of the segment.

SOLUTION   The points are plotted and the segment is sketched in Figure 10.7.8. The length is

$$d = \sqrt{((-1) - 1)^2 + (3 - (-1))^2 + (4 - 0)^2}$$
$$= \sqrt{4 + 16 + 16} = \sqrt{36} = 6.$$

The midpoint is

$$M = \left(\frac{1 - 1}{2}, \frac{-1 + 3}{2}, \frac{0 + 4}{2} \right) = (0, 1, 2). \quad ■$$

FIGURE 10.7.8 The line segment between the points $(1, -1, 0)$ and $(-1, 3, 4)$ has length $d = 6$ and midpoint $M = (0, 1, 2)$.

A **sphere** with center (h, k, ℓ) and radius r is the set of all points (x, y, z) in three-dimensional space that are at a distance r from (h, k, ℓ). It follows from the distance formula that the equation of the sphere with center (h, k, ℓ) and radius r is

$$(x - h)^2 + (y - k)^2 + (z - \ell)^2 = r^2.$$

This is called the **standard equation** of the sphere.

EXAMPLE 3

Find the standard equation of the sphere that passes through the points $(2, 3, -1)$ and $(-2, 1, 3)$ and has center at the midpoint of the line segment between the points.

SOLUTION   We need the coordinates of the center and the length of the radius to write the equation of the sphere.

The midpoint of the line segment between the points is

$$M = \left(\frac{2 - 2}{2}, \frac{3 + 1}{2}, \frac{-1 + 3}{2} \right) = (0, 2, 1).$$

This is the center of the sphere.

The radius is the distance between the center and any point on the sphere. We know that $(2, 3, -1)$ is on the sphere, so

$$r = \sqrt{(2 - 0)^2 + (3 - 2)^2 + (-1 - 1)^2} = 3.$$

The standard equation of the sphere is

$$x^2 + (y - 2)^2 + (z - 1)^2 = 3^2.$$

The sphere is sketched in Figure 10.7.9. ■

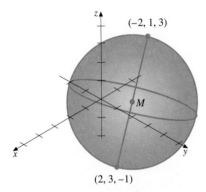

FIGURE 10.7.9 The center of a sphere is the midpoint of any diameter and the radius is the distance between the center and any point on the sphere.

An equation of the form

$$x^2 + y^2 + z^2 + ax + by + cz + d = 0$$

can be written in the form

$$(x - h)^2 + (y - k)^2 + (z - \ell)^2 = D$$

by completing the squares. If $D > 0$, this is the standard form of the equation of a sphere. If $D = 0$, the graph will be a single point; the equation will have no points on its graph if $D < 0$.

EXAMPLE 4

Find the center and radius of the sphere

$$x^2 + y^2 + z^2 + 4x - 6z = 0.$$

SOLUTION   The center and radius are easily read from the standard form of the sphere. Completing the squares, we have

$$x^2 + y^2 + z^2 + 4x - 6z = 0,$$
$$x^2 + 4x + [4] + y^2 + z^2 - 6z + [9] = 0 + [4] + [9],$$
$$(x + 2)^2 + y^2 + (z - 3)^2 = (\sqrt{13})^2.$$

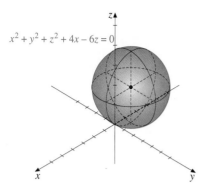

$x^2 + y^2 + z^2 + 4x - 6z = 0$

FIGURE 10.7.10 The center and radius of a sphere are easily read from the standard form of the sphere.

The center is $(-2, 0, 3)$ and the radius is $\sqrt{13}$. The sphere is sketched in Figure 10.7.10. ∎

EXAMPLE 5

Find the distance from the point $(-3, 2, -1)$ to the z-axis. Find the distance from the point $(-3, 2, -1)$ to the xy-plane.

SOLUTION The distance from the point $(-3, 2, -1)$ to the z-axis is the distance between $(-3, 2, -1)$ and the point on the z-axis that has the same z-coordinate, $(0, 0, -1)$. This is the point on the z-axis that is closest to $(-3, 2, -1)$. See Figure 10.7.11. We have

$$\sqrt{(-3 - 0)^2 + (2 - 0)^2 + (-1 - (-1))^2} = \sqrt{13}.$$

The distance from the point $(-3, 2, -1)$ to the xy-plane is the distance between $(-3, 2, -1)$ and $(-3, 2, 0)$. See Figure 10.7.11. This distance is one. ∎

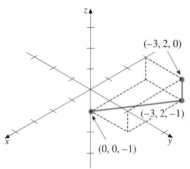

FIGURE 10.7.11 The distance from the point $(-3, 2, -1)$ to the z-axis is the distance between $(-3, 2, -1)$ and the point on the z-axis that has the same z-coordinate, $(0, 0, -1)$.

Sketching Curves

For the remainder of this section we will concentrate on sketching three-dimensional curves and surfaces. The ability to sketch at least a few types of three-dimensional figures is very useful for the understanding of calculus of several variables.

The most common types of drawings that are used to represent objects in three-dimensional space in computer graphics and other technical drawings are **axonometric drawings**. In axonometric drawings, lines that are parallel to an axis are drawn parallel to and with the same scale as the axis. The direction and scale at which each axis is drawn depend on the point from which an object is viewed. Computers can be programmed to calculate appropriate directions and scales for any view and change views in a way that makes it appear that an object is rotating slowly.

The easiest type of axonometric drawing to make manually, and the type most commonly used in technical drawing, is an **isometric drawing**. In isometric drawings, axes are drawn at angles of $120°$, and equal scales are used on the three axes. This corresponds to viewing the origin from a point that is an equal distance from each of the coordinate axes. An isometric drawing of a cube with one corner at the origin and edges along the axes is given in Figure 10.7.12.

As a general rule, if a sketch does not look good in the usual isometric sketch, then we should change the view. For example, we can represent the view from a point that is lower than usual by drawing the x-axis and the y-axis at an angle greater than $120°$ and using a relatively large scale on the z-axis, as is illustrated in Figure 10.7.13a. A view from a point that is higher than usual is obtained by drawing the x-axis and y-axis at an angle less than $120°$ and using a relatively small scale on the z-axis, as in Figure 10.7.13b. In Figure 10.7.13c we have obtained a different view by decreasing the angle between the z-axis and the y-axis, increasing the angle between the z-axis and the x-axis, and using a relatively small scale on the x-axis.

We are primarily interested in the graphs of equations. Generally speaking, we should expect the graph of one equation in three variables to be a surface in three-dimensional space. The graph of two simultaneous equations in three

FIGURE 10.7.12 An isometric view of a cube with edges on the coordinate axes gives a view in the direction of a diagonal of the cube.

FIGURE 10.7.13a We can represent a view from a point that is lower than usual by drawing the *x*-axis and the *y*-axis at an angle greater than 120° and using a relatively large scale on the *z*-axis.

FIGURE 10.7.13b A view from a point that is higher than usual is obtained by drawing the *x*-axis and the *y*-axis at an angle less than 120° and using a relatively small scale on the *z*-axis.

FIGURE 10.7.13c A view from closer to the *x*-axis is obtained by decreasing the angle between the *z*-axis and the *y*-axis, increasing the angle between the *z*-axis and the *x*-axis, and using a relatively small scale on the *x*-axis.

variables is the intersection of two surfaces and is generally a curve in three-dimensional space. For example, a line (curve) in space may be considered to be the intersection of two planes (surfaces).

We begin our study of sketching by noting the following:

- $x = x_0$ is the equation of the plane that contains $(x_0, 0, 0)$ and is parallel to the yz-plane. See Figure 10.7.14a.
- $y = y_0$ is the equation of the plane that contains $(0, y_0, 0)$ and is parallel to the xz-plane. See Figure 10.7.14b.
- $z = z_0$ is the equation of the plane that contains $(0, 0, z_0)$ and is parallel to the xy-plane. See Figure 10.7.14c.

Let us see how to sketch curves that are contained in a plane that is parallel to one of the coordinate planes. This is a basic skill that is necessary for sketching three-dimensional figures.

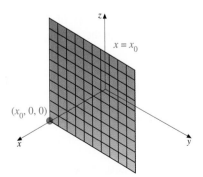

FIGURE 10.7.14a $x = x_0$ is the equation of the plane that contains $(x_0, 0, 0)$ and is parallel to the yz-plane.

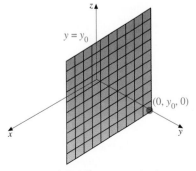

FIGURE 10.7.14b $y = y_0$ is the equation of the plane that contains $(0, y_0, 0)$ and is parallel to the xz-plane.

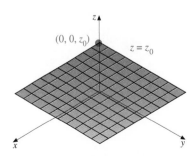

FIGURE 10.7.14c $z = z_0$ is the equation of the plane that contains $(0, 0, z_0)$ and is parallel to the xy-plane.

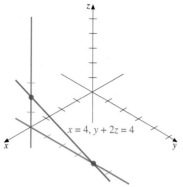

FIGURE 10.7.15 The graph is a line in the plane $x = 4$.

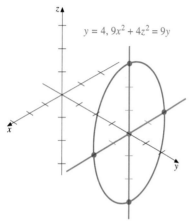

FIGURE 10.7.16 The graph is an ellipse in the plane $y = 4$.

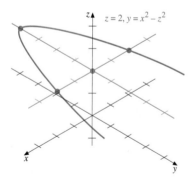

FIGURE 10.7.17 The graph is a parabola in the plane $z = 2$.

EXAMPLE 6

Sketch the curve $x = 4$, $y + 2z = 4$.

SOLUTION The curve consists of all points with coordinates (x, y, z) that satisfy both equations. The equation $x = 4$ tells us that the graph is in the plane $x = 4$. We begin by sketching scaled lines through $(4, 0, 0)$, parallel to each of the y-axis and the z-axis. This gives us a coordinate system for the y and z variables in the plane $x = 4$. We then graph the equation $y + 2z = 4$, considered to be a function of the two variables y and z, in this coordinate plane. The graph is a line in the plane $x = 4$. It is easily sketched by plotting the intercepts with the coordinate axes we have drawn in the plane $x = 4$. See Figure 10.7.15. ∎

EXAMPLE 7

Sketch the curve $y = 4$, $9x^2 + 4z^2 = 9y$.

SOLUTION The equation $y = 4$ tells us the curve is contained in the plane $y = 4$. We establish a coordinate system for the variables x and z in this plane by drawing scaled lines through the point $(0, 4, 0)$, parallel to each of the x-axis and the z-axis. We then plot the graph of the equation $9x^2 + 4z^2 = 9y$ in this plane. Since $y = 4$, we can substitute 4 for y in the equation $9x^2 + 4z^2 = 9y$ to obtain $9x^2 + 4z^2 = 9(4)$, or

$$\frac{x^2}{2^2} + \frac{z^2}{3^2} = 1.$$

We then see that the graph is an ellipse in the plane $y = 4$. It is sketched, as usual, by plotting and then sketching through the four points on the axes, except we use the coordinate plane we have established in the plane $y = 4$. See Figure 10.7.16. ∎

EXAMPLE 8

Sketch the curve $z = 2$, $y = x^2 - z^2$.

SOLUTION The equation $z = 2$ indicates the curve is in the plane $z = 2$. A coordinate system for the variables x and y in this plane is established by drawing scaled lines through $(0, 0, 2)$, parallel to each of the x-axis and the y-axis. We want to sketch the graph of $y = x^2 - z^2$ in this plane. Since $z = 2$, we can substitute 2 for z to obtain an equation in the variables x and y. We obtain $y = x^2 - 2^2$, or $y = x^2 - 4$. The graph is a parabola in the plane $z = 2$. It is sketched by finding its intercepts on the coordinate axes we have drawn in the plane $z = 2$, and then sketching the parabola through the intercepts. See Figure 10.7.17. ∎

Let us now turn to the problem of sketching surfaces. Generally, it is difficult to represent a surface in three-dimensional space by a sketch on a two-dimensional surface such as a piece of paper. As with all sketches, the idea is to illustrate the general character. A standard method for representing a surface is to draw a number of curves that are the intersection of the surface with planes that are parallel to a coordinate plane. The number of such curves that are drawn depends on the complexity of the surface. An example of a surface that is represented by many curves is given in Figure 10.7.18. This is a familiar type of computer-generated illustration, much more detailed than the sketches we will be doing.

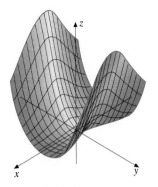

FIGURE 10.7.18 Computers can be used to represent a surface by drawing many curves that are the intersection of the surface with planes that are parallel to a coordinate plane.

The intersection of a surface with a plane is called the **trace** of the surface in the plane. We will use a selection of traces in planes parallel to the coordinate planes to represent three-dimensional surfaces. Generally, we should sketch the traces in each coordinate plane. However, the sketches should be kept as simple as possible. If you draw too many curves, the general character of the surface may be lost in the complexity of your sketch. Many surfaces can be represented very nicely by drawing only a trace in one coordinate plane and a typical trace in a perpendicular plane. We will use traces to give our sketches boundary edges, rather than letting the surfaces just fade away.

Points of intersection of a surface with each of the coordinate axes are called **intercepts**. Intercepts are often useful in determining the traces of the surface in the coordinate planes.

Sketching Planes

The simplest surfaces in three-dimensional space are planes. The **general equation** of a plane in three-dimensional space is

$$ax + by + cz + d = 0, \quad \text{where } (a, b, c) \neq (0, 0, 0).$$

We will derive this equation and study the character of planes in the next chapter. At this time we are interested only in sketching the graph. To sketch a plane, we should determine and plot the intercepts. The intercepts are then used to sketch the coordinate traces. The coordinate traces of a plane are lines through the intercepts.

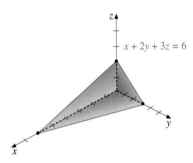

FIGURE 10.7.19 The coordinate traces of a plane can be determined from the intercepts.

EXAMPLE 9

Sketch the plane $x + 2y + 3z = 6$.

SOLUTION The x-intercept is found by setting $y = z = 0$ in the equation. This gives $x + 2(0) + 3(0) = 6$, so $(6, 0, 0)$ is the x-intercept. The other intercepts are easily seen to be $(0, 3, 0)$ and $(0, 0, 2)$. The three intercepts are plotted and used to sketch the traces. The traces are the lines through the intercepts in each of the coordinate planes. The segments of the traces between the intercepts are used to form the boundary edges of our sketch. See Figure 10.7.19. ■

EXAMPLE 10

Sketch the graph of $x + y = 2$ (in three-dimensional space).

SOLUTION We recognize that the graph is a plane (not a line), because we know we are dealing with three-dimensional space. The xy-trace is the line $z = 0$, $x + y = 2$, which is drawn easily by plotting the intercepts on the x-axis and the y-axis. The yz-trace is given by the equations $x = 0, x + y = 2$. Substituting 0 for x in the equation $x + y = 2$, we obtain $y = 2$. The graph is a vertical line in the yz-plane, through the intercept $(0, 2, 0)$. Similarly, we determine that the xz-trace is the vertical line through the intercept $(2, 0, 0)$. The coordinate traces are sketched in Figure 10.7.20a.

In Figure 10.7.20b we have drawn the boundary curve $z = 3$, $x + y = 2$ to give the sketch an upper boundary edge. This line is drawn between the points on our sketch where the coordinate traces of the plane $z = 3$ intersect the coordinate traces of the plane $x + y = 2$. ■

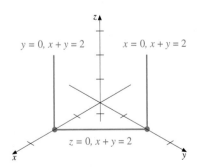

FIGURE 10.7.20a The xy-trace is obtained by connecting the intercepts. The yz-trace and the xz-trace are lines through the intercepts, parallel to the z-axis.

Note that the equation $x + y = 2$ of Example 10 does not involve the variable z. This means that whenever the coordinates of a point (x_0, y_0, z_0)

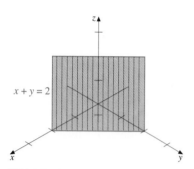

FIGURE 10.7.20b The sketch is completed by drawing an upper boundary curve and shading.

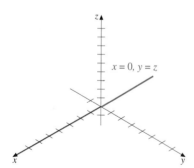

FIGURE 10.7.21 Since the equation $x + y = 2$ does not involve the variable z, the graph is generated by lines through the xy-trace, parallel to the z-axis.

satisfy the equation $x + y = 2$, then the coordinates of the point (x_0, y_0, z) satisfy the equation for every value of z. In particular, if a point $(x_0, y_0, 0)$ is on the xy-trace, then the point (x_0, y_0, z) is also on the graph for every value of z. As z varies over all real numbers, the points (x_0, y_0, z) form a line through $(x_0, y_0, 0)$, parallel to the z-axis. The entire plane is generated by these lines as $(x_0, y_0, 0)$ varies over the xy-trace. See Figure 10.7.21. These ideas allow us to use the xy-trace to sketch the yz-trace and the xz-trace, without dealing directly with the equations of the latter traces. That is, the yz-trace is obtained by drawing a line through the y-intercept $(0, 2, 0)$, parallel to the z-axis; the xz-trace is drawn through the x-intercept $(2, 0, 0)$, parallel to the z-axis.

EXAMPLE 11

Sketch the plane $y = z$.

SOLUTION The graph is a plane in three-dimensional space. The yz-trace is the line $x = 0$, $y = z$. Since the equation $y = z$ does not involve the variable x, we can use the sketch of the yz-trace to determine the other coordinate traces. That is, whenever a point $(0, y_0, z_0)$ is on the yz-trace, then the point (x, y_0, z_0) is on the graph for every value of x. These points generate a line through $(0, y_0, z_0)$ parallel to the x-axis. In particular, since the yz-trace contains the point $(0, 0, 0)$, we know that the graph contains the line through $(0, 0, 0)$, parallel to the x-axis. This line is the x-axis, which is the trace of the plane in the xy-plane and in the xz-plane. (When a coordinate trace contains a coordinate axis, be sure to mark the axis in a way that will remind you that it is a trace.) The coordinate traces are sketched in Figure 10.7.22a. We see that the traces form a single line, so the isometric view of the plane is parallel to the plane and the plane appears as a line. In Figure 10.7.22b we have redrawn the traces as viewed from a point that is higher than usual, so the x-axis and the y-axis are drawn at an angle less than $120°$ and a relatively small scale is used on the z-axis. In Figure 10.7.22c we have used the intersection of the plane $y = z$ with each of the arbitrarily chosen planes $x = 6$ and $y = 6$ to give the sketch boundary edges and have indicated that the plane $y = z$ is generated by lines through the yz-trace, parallel to the x-axis. ∎

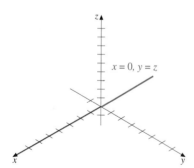

FIGURE 10.7.22a The isometric view of the plane is parallel to the plane and the plane appears as a straight line.

FIGURE 10.7.22b We redraw the traces as viewed from a higher point of view.

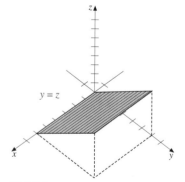

FIGURE 10.7.22c Since the equation $y = z$ does not involve the variable x, the plane is generated by lines through the yz-trace, parallel to the x-axis.

Sketching Generalized Cylinders

A surface in three-dimensional space that is generated by a collection of parallel lines is called a **generalized cylinder**. The graph in three-dimensional space of an equation that has one (or more) of the three variables missing will be a generalized cylinder with generating lines parallel to the axis of the missing variable(s). For example, the plane $x + y = 2$ of Example 10 is a generalized cylinder with generating lines parallel to the z-axis. The plane $y = z$ of Example 11 is a generalized cylinder with generating lines parallel to the x-axis.

A generalized cylinder can be sketched by drawing a few of the generating lines between two traces that are parallel to each other and perpendicular to the generating lines.

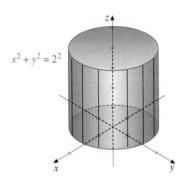

$x^2 + y^2 = 2^2$

FIGURE 10.7.23 Since the equation $x^2 + y^2 = 2^2$ does not involve the variable z, the surface is generated by lines through the xy-trace, parallel to the z-axis.

EXAMPLE 12
Sketch the graph of $x^2 + y^2 = 2^2$ (in three-dimensional space).

SOLUTION   The graph is sketched in Figure 10.7.23. The graph is a (generalized) cylinder in three-dimensional space. The generating lines are parallel to the z-axis and perpendicular to the xy-plane. The xy-trace is the circle $z = 0$, $x^2 + y^2 = 2^2$. We have sketched the xy-trace and the parallel trace $z = 4$, $x^2 + y^2 = 4$, and drawn some of the generating lines between these traces. ■

EXAMPLE 13
Sketch the graph of $z = x^2$.

SOLUTION   We assume that we are interested in a graph in three-dimensional space, so the graph is a cylinder with generating lines parallel to the y-axis. See Figure 10.7.24. We have drawn the trace $y = 0$, $z = x^2$ (a parabola), the parallel trace $y = 4$, $z = x^2$, and some of the generating lines between these traces. ■

$z = x^2$

FIGURE 10.7.24 Since the equation $z = x^2$ does not involve the variable y, the surface is generated by lines through the xz-trace, parallel to the y-axis.

EXAMPLE 14
Sketch the graph of $z = \sin y$.

SOLUTION   The graph is sketched in Figure 10.7.25. The surface is a generalized cylinder with generating lines parallel to the x-axis. We have drawn the trace $x = 0$, $z = \sin y$, the parallel trace $x = 6$, $z = \sin y$, and some of the generating lines between these traces. ■

$z = \sin y$

FIGURE 10.7.25 Since the equation $z = \sin y$ does not involve the variable x, the surface is generated by lines through the yz-trace, parallel to the x-axis.

EXERCISES 10.7

1. Plot the points $A(2, 2, 0)$, $B(1, 2, 1)$, $C(-1, -2, 1)$, and $D(2, -1, 1)$. Which points are in the first octant?

2. Plot the points $A(1, 3, 1)$, $B(0, 1, 2)$, $C(-3, -3, -1)$, and $D(3, -2, 1)$. Which points are in the first octant?

3. Sketch the line segment between the points $(1, -1, 0)$ and $(-3, 3, -2)$. Find the length and midpoint of the segment.

4. Sketch the line segment between the points $(2, 0, 3)$ and $(0, 4, 1)$. Find the length and midpoint of the segment.

5. Sketch the points $A(2, 0, 0)$, $B(0, 0, 3)$, $C(1, 2, 2)$, and $D(3, 2, -1)$. Show that the midpoint of the line segment between A and C coincides with the midpoint of the line segment between B and D, so the four points form a parallelogram.

6. Sketch the points $A(2, 1, 0)$, $B(0, 0, 3)$, and $C(3, 1, 3)$ and show that they form an isosceles triangle. Find the midpoint of its longest side.

7. Find all values of a so the points $A(0, 0, a)$, $B(3, 0, 0)$, and $C(1, 2, 1)$ form a right triangle with right angle at C. (If the lengths of the sides of a triangle satisfy $d_1^2 = d_2^2 + d_3^2$, the Law of Cosines implies the angle opposite the side of length d_1 is a right angle.)

8. Find an equation satisfied by all points (x, y, z) that are equidistant from the point $(1, 3, 2)$ and the origin.

Find the standard form of the equation of the sphere described in each of Exercises 9–14.

9. The points $(2, 0, 0)$ and $(0, 4, 2)$ form the endpoints of a diameter.

10. Center $(1, 2, 0)$; contains $(0, 0, 2)$

11. Center $(4, 3, 12)$; contains the origin

12. Center $(1, 2, 3)$; tangent to the xy-plane $(x-1)^2 + (y-2)^2 + (z-3)^2 = 9$

13. Center $(1, -2, 1)$; tangent to the z-axis

14. Radius 2; center in the first octant; tangent to all three coordinate planes

15. Find the center and radius of the sphere
$$x^2 + y^2 + z^2 - 2x + 4y = 0.$$

16. Find the center and radius of the sphere
$$x^2 + y^2 + z^2 + 6y + 2z + 7 = 0.$$

17. Find the distance from the point $(2, 3, 2)$ to the z-axis.

18. Find the distance from the point $(-1, -2, 1)$ to the x-axis.

19. Find the distance from the point $(2, -4, 3)$ to the xy-plane.

20. Find the distance from the point $(-3, 4, 12)$ to the yz-plane.

Sketch the curves with equations given in Exercises 21–30.

21. $x = -2, 3z - 2y = 6$

22. $x = 3, z = 1 - y^2$

23. $y = 2, x^2 + z^2 = 4$

24. $y = 2, 2x + 3z = 6$

25. $z = 2, x^2 + y^2 = 2z$

26. $z = 4, z = x + 2y$

27. $y = 2, z = y^2 - x^2$

28. $z = 0, 4x^2 + 4z^2 = y^2$

29. $z = 1, z = y^2 - x^2$

30. $y = 0, x^2 - y^2 - z^2 = 1$

Sketch the surfaces with equations given in Exercises 31–50.

31. $2x + y + 3z = 6$

32. $2x + 3y + 3z = 6$

33. $3x + 2z = 6$

34. $3y + 4z = 12$

35. $z = x + 2y$

36. $z = 2x + 2y$

37. $2x - 2y + z = 4$

38. $3x + 2y - 2z = 6$

39. $y = 2x$

40. $z = 2x$

41. $x^2 + y^2 - 2y = 0$

42. $y^2 + z^2 = 1$

43. $4x^2 + z^2 = 4$

44. $4x^2 + y^2 - 8x = 0$

45. $z = 4 - y^2$

46. $z = y^2 - 1$

47. $y = \sin x$

48. $z = \cos y$

49. $z = |y|$

50. $z = -y/|y|$

10.8   QUADRIC SURFACES

Connections

Three-dimensional sketching, 10.7.

We will be dealing with surfaces that have equations of the form

$$Ax^2 + By^2 + Cz^2 + Dx + Ey + Fz + G = 0,$$

where A, B, and C are not all equal to zero. Surfaces of this type are called **quadric surfaces**. Many of these surfaces can be sketched by drawing only one coordinate trace and a typical trace in a perpendicular plane. We should look particularly for typical traces that are circles or ellipses. These occur quite often in quadric surfaces, they are relatively easy to draw, and they give a good indication of the character of the surface.

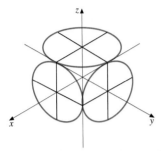

FIGURE 10.8.1 Circles that are not parallel to the sketch plane appear as ellipses. Be sure you can draw circles in three-dimensional space.

Using Circles and Traces to Sketch Quadric Surfaces

Be sure you can draw circles in three-dimensional space. Figure 10.8.1 contains sketches of circles parallel to each of the coordinate planes. Each of these circles appears as an ellipse. Circles are sketched easily by locating the center, drawing diameters that are parallel to coordinate axes, marking off the length of the radius along the diameters that have been drawn, and then sketching an ellipse through the marks. Tangents to the circles at the points we have marked are drawn parallel to coordinate axes.

EXAMPLE 1
Sketch the graph of $z = x^2 + y^2$.

SOLUTION   We begin our sketch by looking at the yz-trace. This is given by the equations

$$x = 0, \qquad z = x^2 + y^2.$$

Substituting 0 for x in the second equation, we obtain $z = y^2$. The yz-trace is then seen to be a parabola, which we sketch. See Figure 10.8.2a. We then notice that traces of the surface in planes $z = z_0$ have equations

$$z = z_0, \qquad z_0 = x^2 + y^2.$$

If $z_0 > 0$, these are circles that are parallel to the xy-plane, with centers on the z-axis. We sketch the trace in the plane $z = 4$ to indicate a typical circle and to give our sketch a top edge. The radius of the circle is determined by the intersection of the coordinate traces of the plane $z = 4$ and the parabola in the yz-plane. See Figure 10.8.2b. The sketch is given side edges and shaded to complete the picture. The side edges of the sketch do not coincide with either the yz-trace or the xz-trace. See Figure 10.8.2c.   ∎

*Surfaces of the type illustrated in Figure 10.8.2 are called **paraboloids**.*

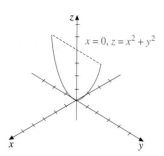

FIGURE 10.8.2a The trace in the yz-plane is a parabola.

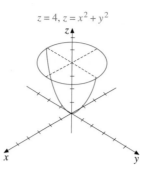

FIGURE 10.8.2b Traces parallel to the xy-plane are circles with centers on the z-axis and radii determined by the trace we have drawn in the yz-plane.

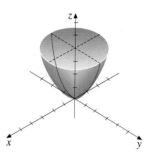

FIGURE 10.8.2c The sketch is completed by drawing side edges and shading. The graph is a paraboloid.

EXAMPLE 2

Sketch the graph of $4x^2 - y^2 + 4z^2 = 0$.

SOLUTION The first step is to determine the yz-trace. The yz-trace is given by the equations

$$x = 0, \qquad 4x^2 - y^2 + 4z^2 = 0.$$

Simplifying by substituting 0 for x in the second equation, we obtain $-y^2 + 4z^2 = 0$, so $y = 2z$ or $y = -2z$. The graph of the yz-trace is then seen to be two lines, which we sketch. See Figure 10.8.3a. The intersection of the surface with the plane $y = y_0$, y_0 a constant, is given by the equations

$$y = y_0, \qquad x^2 + z^2 = \frac{y_0^2}{4}.$$

These curves are circles that are parallel to the xz-plane, with centers on the y-axis. We draw the circles in the planes $y = -4$ and $y = 4$ to represent typical traces and give our sketch side edges. The yz-trace we have drawn determines the radius of these circles. See Figure 10.8.3b. The sketch is completed by drawing top and bottom edges and shading. See Figure 10.8.3c. ∎

The graph in Figure 10.8.3 is a **cone**.

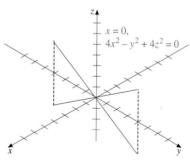

FIGURE 10.8.3a The trace in the yz-plane is a pair of intersecting lines.

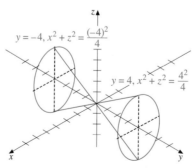

FIGURE 10.8.3b Traces parallel to the xz-plane are circles with center on the y-axis and radii determined by the yz-trace we have sketched.

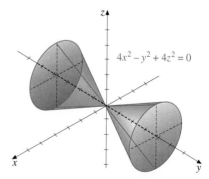

FIGURE 10.8.3c Side edges and shading complete the sketch. The graph is a cone.

EXAMPLE 3

Sketch the graph of $16x^2 + 9y^2 + 36z^2 = 144$.

SOLUTION   It is easy to determine the equations of the three coordinate traces. We obtain

$$x = 0, \quad \frac{y^2}{4^2} + \frac{z^2}{2^2} = 1,$$

$$y = 0, \quad \frac{x^2}{3^2} + \frac{z^2}{2^2} = 1,$$

$$z = 0, \quad \frac{x^2}{3^2} + \frac{y^2}{4^2} = 1.$$

The coordinate traces are ellipses, which we draw. See Figure 10.8.4a. The sketch is then given an outer edge and shaded. See Figure 10.8.4b. ■

*The graph in Figure 10.8.4 is called an **ellipsoid**.*

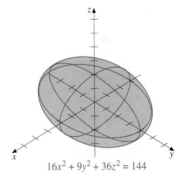

$16x^2 + 9y^2 + 36z^2 = 144$

FIGURE 10.8.4a   The coordinate traces are ellipses.

FIGURE 10.8.4b   The sketch is given an outer edge and shaded. The graph is an ellipsoid.

EXAMPLE 4

Sketch the graph of $4x^2 - y^2 + 4z^2 + 4 = 0$.

SOLUTION   The yz-trace is given by the equations

$$x = 0, \quad \frac{y^2}{2^2} - \frac{z^2}{1^2} = 1.$$

This is a hyperbola, which we draw. See Figure 10.8.5a.
   If $y_0^2 > 4$, the traces

$$y = y_0, \quad x^2 + z^2 = \frac{y_0^2 - 4}{4}$$

are circles that are parallel to the xz-plane, with centers on the y-axis. The traces are single points for $y_0 = \pm 2$, and empty for $-2 < y_0 < 2$. We have added the circles in the planes $y = 5$ and $y = -5$ in Figure 10.8.5b.
   The top and bottom edges have been added, and the figure shaded in Figure 10.8.5c. ■

*The surface in Figure 10.8.5 is called a **hyperboloid of two sheets**.*

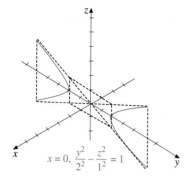

FIGURE 10.8.5a The *yz*-trace is a hyperbola.

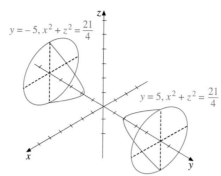

FIGURE 10.8.5b Traces parallel to the *xz*-plane are circles with centers on the *y*-axis and radii determined by the *yz*-trace we have drawn.

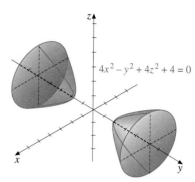

FIGURE 10.8.5c Side edges and shading complete the sketch. The graph is a hyperboloid of two sheets.

EXAMPLE 5

Sketch the graph of $4x^2 + 4y^2 - z^2 = 4$.

SOLUTION   The *yz*-trace is given by the equations

$$x = 0, \quad \frac{y^2}{1^2} - \frac{z^2}{2^2} = 1.$$

This is a hyperbola, which we draw. See Figure 10.8.6a. The traces

$$z = z_0, \quad x^2 + y^2 = 1 + \frac{z_0^2}{4}$$

are circles that are parallel to the xy-plane, with centers on the z-axis. We have drawn the circles in the planes $z = 4$ and $z = -4$ in Figure 10.8.6b. The side edges have been added and the figure shaded in Figure 10.8.6c. ■

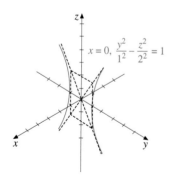

FIGURE 10.8.6a The *yz*-trace is a hyperbola.

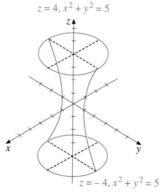

FIGURE 10.8.6b Traces parallel to the *xy*-plane are circles with centers on the *z*-axis and radii determined by the *yz*-trace we have sketched.

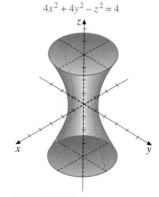

FIGURE 10.8.6c The sketch is given side edges and shaded. The graph is a hyperboloid of one sheet.

The surface in Figure 10.8.6 is called a **hyperboloid of one sheet.**

It is not necessary to memorize the form of the equations to distinguish between a hyperboloid of one sheet and a hyperboloid of two sheets. In either case, we first sketch a hyperbolic trace, determine the coordinate plane to which the circular traces are parallel, and then sketch the circular traces. If these traces are drawn correctly, the completed sketch will have the correct number of sheets.

Hyperbolic Paraboloids

EXAMPLE 6

Sketch the graph of $4z = y^2 - 4x^2$.

SOLUTION   The yz-trace is given by

$$x = 0, \quad 4z = y^2,$$

which is a parabola that opens upward. This is drawn in Figure 10.8.7a. The xz-trace is given by the equations

$$y = 0, \quad 4z = -4x^2,$$

a parabola that opens downward. This trace is added in Figure 10.8.7b. The traces given by

$$y = y_0, \quad 4z = y_0^2 - 4x^2$$

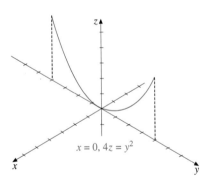

FIGURE 10.8.7a The yz-trace is a parabola that opens upward.

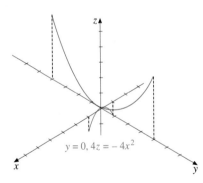

FIGURE 10.8.7b The xz-trace is a parabola that opens downward.

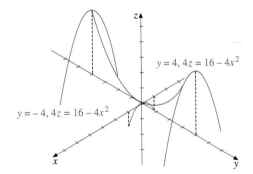

FIGURE 10.8.7c Traces parallel to the xz-plane are parabolas that open downward. We can think of the surface as being generated by hanging the parabolas $y = y_0, 4z = y_0^2 - 4x^2$ from the parabola $x = 0, 4z = y^2$.

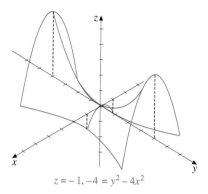

FIGURE 10.8.7d Traces parallel to the xy-plane are hyperbolas. We use the hyperbola $z = -1, -4 = y^2 - 4x^2$ to give our sketch a bottom edge.

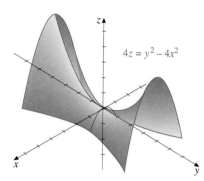

FIGURE 10.8.7e The sketch is given a top edge and shaded. The graph is a hyperbolic paraboloid.

are also parabolas. We sketch the parabolas in the planes $y = 4$ and $y = -4$ for side edges of our sketch. See Figure 10.8.7c. Note that we can think of the surface as being generated by hanging the parabolas $y = y_0$, $4z = y_0^2 - 4x^2$ from the parabola $x = 0$, $4z = y^2$. The traces given by

$$z = z_0, \quad 4z_0 = y^2 - 4x^2$$

are hyperbolas for $z_0 \neq 0$. When $z_0 = 0$, the xy-trace is a pair of lines that intersect at the origin. We sketch the hyperbola in the plane $z = -1$ to give our sketch a bottom edge. See Figure 10.8.7d. Finally, we fill in the top edge and shade the figure. See Figure 10.8.7e. ∎

*The surface in Figure 10.8.7, which resembles a saddle, is called a **hyperbolic paraboloid**.*

These are difficult to draw because no trace is a circle or ellipse.

Changing the Orientation of the Axes

EXAMPLE 7
Sketch the graph of $x = \sqrt{y^2 + z^2}$.

SOLUTION The yz-trace consists of the single point at the origin. The xz-trace is given by

$$y = 0, \quad x = \sqrt{z^2}.$$

Since $\sqrt{z^2} = |z|$, the second equation can be written $x = |z|$. We have sketched the xz-trace in Figure 10.8.8a. The traces of the surface in planes $x = x_0$, $x_0 > 0$, are given by

$$x = x_0, \quad y^2 + z^2 = x_0^2,$$

which are circles that are parallel to the yz-plane, with centers on the x-axis. We have drawn the circle in the plane $x = 4$ and added the edge lines in Figure 10.8.8b. At this point we notice that the sketch does not seem to indicate the general character of the surface very well. The graph should look like a single-napped cone. We can obtain a better picture by moving the point of view toward the yz-plane, so the y-axis and the z-axis are drawn at an angle greater than $120°$ and a relatively large scale is used on the x-axis. This is done in Figure 10.8.8c. ■

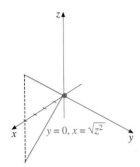

FIGURE 10.8.8a The yz-trace is a single point at the origin. The xz-trace consists of two rays from the origin.

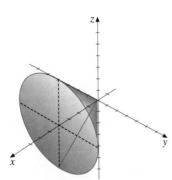

FIGURE 10.8.8b Traces parallel to the yz-plane are circles with centers on the x-axis and radii determined by the xz-trace we have sketched.

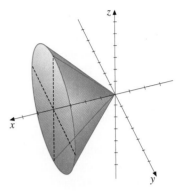

FIGURE 10.8.8c We obtain a better sketch by moving the point of view toward the yz-plane.

EXERCISES 10.8

Sketch the graphs of the equations given in Exercises 1–34.

1. $x^2 + z^2 = y$
2. $x^2 + y^2 + z = 4$
3. $4x^2 + 4y^2 + z^2 = 4$
4. $x^2 + y^2 + z^2 = 9$
5. $x^2 - 9y^2 - 9z^2 = 0$
6. $4x^2 + 4y^2 = z^2$
7. $z = \sqrt{3x^2 + 3y^2}$
8. $z = -2\sqrt{x^2 + y^2}$
9. $z = 4 - 2\sqrt{x^2 + y^2}$
10. $3y^2 + 3z^3 = x$
11. $4x^2 + 4y^2 - z^2 + 4 = 0$
12. $x^2 - 4y^2 - 4z^2 = 4$
13. $9x^2 - y^2 + 9z^2 = 9$
14. $x^2 - 9y^2 - 9z^2 + 9 = 0$
15. $8z = x^2 - 8y^2$
16. $z = y^2 - 2x^2$
17. $x = 4y^2 + 2z^2$
18. $16x^2 - y^2 + 4z^2 + 4 = 0$

19. $16x^2 - y^2 + 4z^2 - 4 = 0$
20. $z = 4 - \sqrt{4x^2 + y^2}$
21. $x^2 + (z - y/2 - 1)^2 = 1 \ (0 \le y \le 4)$
22. $x^2 + (z - y^2)^2 = 1 \ (-1 \le y \le 2)$
23. $xyz = 1 \ (1 \le x \le 5, 1 \le y \le 5)$
24. $z = xy/4 \ (0 \le x \le 1, 0 \le y \le 1)$
25. $z = 1/(x + y) \ (0 \le x \le 2, 0 \le y \le 2, z \le 4)$
26. $xy + xz + yz = 1 \ (0 \le x \le 4, 0 \le y \le 4, 0 \le z \le 4)$
27. $2x + 2y + 4z = 4$
28. $3x + y + z = 3$
29. $2x + z = 2$
30. $y + 2z = 2$
31. $z = x + 2y$
32. $y = 2x + 4z$
33. $y = 2z$
34. $z = 2x$

REVIEW OF CHAPTER 10 CONCEPTS

CONIC SECTIONS

- The standard forms $y - k = 4p(x - h)^2$ and $x - h = 4p(y - k)^2$ can be used to determine the vertices, foci, directrix, and graph of the corresponding parabolas. The number p gives the directed distance from the vertex to the focus. The distance between the vertex and the directrix is $|p|$.

- The graph of $y = ax^2 + bx + c$, where a, b, and c are constants with $a \neq 0$, is a parabola with vertical axis. This form is convenient for finding the equation of the parabola that has vertical axis and contains three given points that are not collinear and have distinct x-coordinates. Similarly, $x = ay^2 + by + c$, $a \neq 0$, corresponds to a parabola that has horizontal axis.

- The standard form $\dfrac{(x - h)^2}{a^2} + \dfrac{(y - k)^2}{b^2} = 1$ can be used to determine the center, vertices, foci, and graph of the corresponding ellipse. The vertices are the points on the ellipse that are farthest from the center. The foci are inside the ellipse on the line through the vertices.

- The standard forms

$$\frac{(x - h)^2}{a^2} - \frac{(y - k)^2}{b^2} = 1 \quad \text{and}$$

$$-\frac{(x - h)^2}{a^2} + \frac{(y - k)^2}{b^2} = 1$$

can be used to determine the center, vertices, foci, asymptotes, and graph of the corresponding hyperbola. The distance between the center and a focus is $c = \sqrt{a^2 + b^2}$.

- The focus-directrix equation of a conic is

  (Distance from focus) $= e$ (Distance from directrix),

  where e is the eccentricity of the conic. The graph is an ellipse if $e < 1$, a hyperbola if $e > 1$, and a parabola if $e = 1$.

- The equations $x = x' + h$, $y = y' + k$ relate rectangular coordinates (x, y) to coordinates with respect to new axes obtained by translation of the origin of the (x, y) axes to the point $(x, y) = (h, k)$.

- The equations $x = x' \cos\theta - y' \sin\theta$, $y = x' \sin\theta + y' \cos\theta$ relate rectangular coordinates (x, y) to coordinates with respect to new axes obtained by rotation of the (x, y) axes by an angle θ about the origin.

- The equation $Ax^2 + Bxy + Cy^2 + Dx + Ey + F = 0$, A, B, C not all three zero, is the general second-degree equation. The graph is an ellipse if $B^2 - 4AC < 0$, a hyperbola if $B^2 - 4AC > 0$, and a parabola if $B^2 - 4AC = 0$. Rotation of axes by an angle θ that satisfies $\cot 2\theta = \dfrac{A - C}{B}$ gives a new equation in (x', y') that has no $x'y'$ term.

POLAR COORDINATES

- Different polar coordinates may correspond to the same point. This is true exactly when the coordinates are related by either

$$P(r, \theta) = P(r, \theta + 2\pi n), \quad n \text{ any integer,}$$
$$P(r, \theta) = P(-r, \theta + \pi + 2\pi n), \quad n \text{ any integer, or}$$
$$O(0, \theta) = O(0, \phi), \quad \theta \text{ and } \phi \text{ any values.}$$

- Rectangular and polar coordinates are related by the basic equations

$$x = r\cos\theta, \ y = r\sin\theta, \ x^2 + y^2 = r^2, \ \tan\theta = \frac{y}{x}.$$

- A conic section with focus at the pole and axis on one of the coordinate axes has polar equation of the form either

$$r = \frac{c}{1 \pm e\sin\theta} \text{ or } r = \frac{c}{1 \pm e\cos\theta}.$$

- Graphs of polar equations $r = r(\theta)$ can be determined by finding values θ_0 for which $r(\theta_0) = 0$; the graph approaches and leaves the pole at the angle of the line $\theta = \theta_0$ as θ passes through these values θ_0. The graph forms either a positive loop or a negative loop as θ ranges between successive zeros of $r(\theta)$.

- The area of the region in the plane bounded by the polar curves $r = r(\theta)$, $\theta = a$, and $\theta = b$, where r is continuous for $a \leq \theta \leq b$ and $b - a \leq 2\pi$, is given by $\displaystyle\int_a^b \frac{1}{2}(r(\theta))^2 \, d\theta$.

- To find all points of intersection of two polar curves, sketch both curves and use the sketch to determine the number and approximate location of the points of intersection. Simultaneous solutions of the two given equations can be used to find coordinate pairs that satisfy both equations. It may be necessary to substitute equivalent equations for the given equations in order to determine coordinates of all points of intersection.

THREE-DIMENSIONAL SPACE

- Three coordinates, (x, y, z), determine a point in rectangular three-dimensional space.

- The most common type of manual three-dimensional drawing is an isometric drawing, where the axes are drawn at angles of $120°$ and equal scales are used on each of the three axes.

- Computer drawings of a surface in three-dimensional space consist of a net formed by traces of the surface in planes parallel to a coordinate plane. Sketching curves in planes that are parallel to a coordinate axis is an important basic skill for sketching three-dimensional surfaces.

- The graph of $ax + by + cz + d = 0$, a, b, c not all three zero, is a plane in three-dimensional space.

- The graph in three-dimensional space of an equation with one of the three variables missing is a generalized cylinder with generating lines parallel to the axis of the missing variable.

- The graph of an equation of the form $Ax^2 + By^2 + Cz^2 + Dx + Ey + Fz + G = 0$ is a quadric surface. Many of these surfaces can be sketched by drawing only one coordinate trace and a typical trace in a perpendicular plane. Your sketch can then be completed by drawing the edges of the surface. You should look for traces that are circles or ellipses, since circles and ellipses are relatively easy to sketch and can be recognized in your sketch.

CHAPTER 10 REVIEW EXERCISES

1. Find an equation of the parabola with directrix $x = -2$ and focus $(2, 0)$.

2. Find an equation of the parabola with vertical axis that has vertex $(0, -2)$ and contains the point $(-1, -1)$.

3. Find an equation of the ellipse that has center $(4, 0)$, one focus $(0, 0)$, and contains $(4, 3)$.

4. Find an equation of the ellipse that has foci $(-3, 0)$ and $(3, 0)$ and contains $(0, 2)$.

5. Find an equation of the hyperbola with foci $(-1, 0)$ and $(9, 0)$ and eccentricity 1.25.

6. Find an equation of the hyperbola with vertices $(0, 0)$ and $(-2, 0)$ and one focus $(1, 0)$.

Sketch the graphs of the equations in Exercises 7–12.

7. $x^2 = 16 - 4y^2$

8. $4x^2 + y^2 - 4y = 0$

9. $y^2 = 2 + x$

10. $x^2 = 8 - 8y$

11. $\dfrac{y^2}{4} - \dfrac{x^2}{1} = 1$

12. $4x^2 - y^2 - 8x = 0$

In Exercises 13–14, find $-\pi/2 < \theta < \pi/2$ so the equations $x = x' \cos\theta - y' \sin\theta$, $y = x' \sin\theta + y' \cos\theta$ transform the given equation to an equation in (x', y') with no x' term. Express the equation in terms of the corresponding (x', y') coordinates. Sketch both sets of axes and the graph of the equation.

13. $y = \sqrt{3}(x + 1)$

14. $x + y = 2$

In Exercises 15–16, find $0 < \theta < \pi/2$ so the equations $x = x' \cos\theta - y' \sin\theta$, $y = x' \sin\theta + y' \cos\theta$ transform the given equation to an equation in (x', y') with no $x'y'$ term. Express the equation in terms of the corresponding (x', y') coordinates. Sketch both sets of axes and the graph of the equation.

15. $xy = y + 1$

16. $2x^2 + \sqrt{3}xy + y^2 + 2x - 2\sqrt{3}y = 0$

17. Find the coordinates of the vertices of the ellipse $3x^2 - 2xy + 3y^2 = 4$.

18. Find the coordinates of the vertices of the hyperbola $3x^2 + 10xy + 3y^2 = 8$.

Find the equations of translation that translate the origin to the center of the graphs in Exercises 19–22. Sketch the graphs and both sets of axes.

19. $x^2 + 4y^2 - 8y = 0$

20. $4x^2 + y^2 - 16x - 2y + 1 = 0$

21. $x^2 - y^2 + 4x + 2y = 1$

22. $x^2 - y^2 - 2x = 0$

Use the discriminant test to identify the graphs in Exercises 23–26.

23. $x^2 + 3xy + 2y^2 = 6$

24. $2x^2 + 5xy + 3y^2 = 7$

25. $x^2 - 2xy + y^2 + 2x = 1$

26. $3x^2 - 6xy + 4y^2 = 12$

Sketch the graphs of the polar equations in Exercises 27–42.

27. $r = 2\cos\theta$

28. $r = -2$

29. $r = \dfrac{2}{\cos\theta + \sin\theta}$

30. $r = \dfrac{2}{2\sin\theta - \cos\theta}$

31. $r = \dfrac{1}{1 - \sin\theta}$

32. $r = \dfrac{2}{1 + \cos\theta}$

33. $r = \dfrac{3}{1 - 2\cos\theta}$

34. $r = \dfrac{3}{2 - \sin\theta}$

35. $r = \sin 2\theta$

36. $r = \cos 3\theta$

37. $r^2 = \sin 2\theta$

38. $r = e^\theta$

39. $r = 1 + 2\sin\theta$

40. $r = 1 - \cos\theta$

41. $r = 2 + \cos\theta$

42. $r = 3 + 2\sin\theta$

Sketch the graphs of the polar equations in Exercises 43–44. Find all points on the graphs that have vertical tangent lines.

43. $r = 5 - 3\cos\theta$ **44.** $r = 5 - 2\cos\theta$

Sketch the graphs of the polar equations in Exercises 45–46. Find all points on the graphs that have horizontal tangent lines.

45. $r = 3 - \sin\theta$

46. $r = 3 - 2\sin\theta$

47. Find the angle between the line $\theta = \pi/2$ and the line tangent to the polar curve $r = 2 - 3\cos\theta$ at the point $(2, \pi/2)$.

48. Find the angle between the line $\theta = \pi/2$ and the line tangent to the polar curve $r = e^{2\theta}$ at the point $(e^\pi, \pi/2)$.

49. Find the angle between the lines tangent to the polar curves $r = 1 - \cos\theta$ and $r = -3\cos\theta$ at the point $(3/2, 2\pi/3)$.

50. Find the angle between lines tangent to the polar curves $r = 2\cos\theta$ and $r = 2\sin\theta$ at each point of intersection.

Set up integrals that give the areas of the regions indicated in Exercises 51–54.

51. Inside $r = 2 - 2\cos\theta$

52. Inside $r^2 = \sin 2\theta$

53. Inside $r = 2 - 2\cos\theta$ and outside $r = 3$

54. Inside $r = 2\sin\theta$ and outside $r = 1$

55. Sketch the line segment between the points $(3, -2, 1)$ and $(-1, 2, 3)$. Find the length and midpoint of the segment.

56. Sketch the line segment between the points $(2, 3, -1)$ and $(-3, -2, 2)$. Find the length and midpoint of the segment.

57. Find the standard equation of the sphere that has center $(2, 0, 1)$ and contains the point $(0, 2, 2)$.

58. Find the standard equation of the sphere that has center $(3, 2, 4)$ and is tangent to the xy-plane.

59. Find the center and radius of the sphere
$$x^2 + y^2 + z^2 + 4x - 6y = 0.$$

60. Find the center and radius of the sphere
$$x^2 + y^2 + z^2 + 2x - y - 3z = 0.$$

Sketch the curves with equations given in Exercises 61–66.

61. $x = 2, 2z + 3y = 6$

62. $x = -1, 4y^2 + 9z^2 = 36$

63. $y = -2, z = x^2 - y^2$

64. $y = 1, x^2 - z^2 = y$

65. $z = -1, x^2 + 4y^2 = 4z^2$

66. $z = -2, x + 2y + z + 2 = 0$

Sketch the surfaces with equations given in Exercises 67–82.

67. $3x + 2y + 6z = 6$ **68.** $3y + 2z = 6$

69. $z = 2x + 2y$ **70.** $z = 2y$

71. $x^2 + y^2 - 4x = 0$ **72.** $x^2 + 4z^2 = 4$

73. $z = y^2$ **74.** $z^2 = 4y^2$

75. $z = x^2 + y^2$ **76.** $z = 4 - x^2 - y^2$

77. $y^2 = 4x^2 + 4z^2$ **78.** $9x^2 + 4y^2 + 36z^2 = 36$

79. $z^2 = 4x^2 + 4y^2 + 4$ **80.** $y^2 = 4x^2 + 4z^2 - 4$

81. $z = y^2 - x^2$ **82.** $z = x^2 - y^2$

83. Find the volume of the paraboloid of revolution that has altitude h and radius of base r.

84. Find the surface area of the ellipsoid $x^2 + 2y^2 + z^2 = 2$.

85. An ideal gas satisfies the equation $Pv = RT$, where P is the pressure, v is the volume, T is the absolute temperature, and R is a positive constant. Sketch the graph of the surface $Pv = RT$ in the first octant of a rectangular TvP-coordinate system. Indicate constant-temperature curves, called **isotherms**, and curves of constant volume on the surface. Identify these two types of traces.

86. An outfielder throws a baseball to a catcher 250 ft away. What is the maximum height of the ball if it is thrown at an angle of $20°$ with the horizontal, with initial and terminal height 6 ft above ground, and it follows a parabolic path?

87. Find an equation satisfied by all points (x, y, z) with their distance from $(p, 0, 0)$ equal to their distance from the plane $x = -p$. What is the graph?

88. Find an equation satisfied by all points (x, y, z) that have the sum of their distance from $(0, c, 0)$ and their distance from $(0, -c, 0)$ equal to $2a$, where $a > c > 0$. What is the graph?

89. Find an equation satisfied by all points (x, y, z) that have the positive difference of their distance from $(0, 0, c)$ and their distance from $(0, 0, -c)$ equal to $2a$, where $c > a > 0$. What is the graph?

90. If the function r has a continuous derivative, we see from the figure that a small arc of the polar curve $r = r(\theta)$ is approximately the hypotenuse of a right triangle that has sides of lengths Δr and $r\,\Delta\theta$.

    (a) Use the Pythagorean Theorem to find a formula for an approximate value of Δs in terms of Δr and $\Delta\theta$.

    (b) If $\Delta\theta$ is small, then $\dfrac{\Delta r}{\Delta\theta} \approx r'$. Substitute r' for $\dfrac{\Delta r}{\Delta\theta}$ in the formula from (a) to express Δs in terms of r' and $\Delta\theta$.

(c) Use the formula in (b) to determine a formula for the length of the polar curve $r = r(\theta)$, $\theta_0 \leq \theta \leq \theta_1$, as an integral with respect to θ.

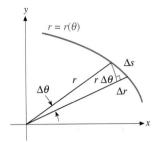

91. Describe in words the relation between the graphs of the equations $y = f(x)$ and $y - k = f(x - h)$.

92. Discuss symmetry with respect to the polar axis, the line $\theta = \pi/2$ and the pole of the graph of a polar equation $r = r(\theta)$.

93. Three-dimensional sketches can be made by drawing the y-axis horizontal, the z-axis vertical, and the x-axis at $45°$ from the horizontal, with equal scales on the y-axis and z-axis and 2/3 scale on the x-axis, as indicated below. Show that spheres do not appear round in this type drawing and explain why this type drawing does not give a true view of a cube.

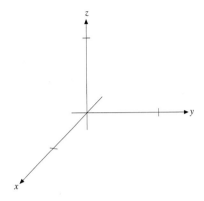

EXTENDED APPLICATION

FEASIBLE SPACE SHUTTLE TRANSFER ORBITS

You have been included in a team of mathematicians, scientists, and engineers whose assignment is to plan the next space shuttle mission. You have planned for the shuttle to spend some time in a circular orbit (called a *parking orbit*) with a radius of 120 miles, while taking detailed photographs of the earth. Afterward it will move to another parking orbit with a radius of 300 miles, in order to minimize atmospheric scattering while making deep-space observations. You have been assigned to investigate possible *transfer orbits* from the first parking orbit to the second.

It can be shown that every transfer orbit must be an arc of a conic with the earth at a focus. To begin your investigation, you will consider only elliptic orbits, although parabolic and hyperbolic arcs can also occur. Two elliptic orbits are shown in Figure 1. The one in Figure 1b, which is tangent to both parking orbits, is called a *Hohmann transfer orbit* (after Walter Hohmann, who discovered in 1925 that such an orbit minimizes the energy needed to transfer between two given circular parking orbits).

If a polar coordinate system is introduced with the earth at the pole and the major axis of the transfer orbit along the polar axis, then the transfer orbit can be described by the polar equation $r = \dfrac{\rho}{1 - e \cos \theta}$. In this equation e is the eccentricity of the orbit and $\rho = a(1 - e^2)$, where a is the semimajor axis.

Tasks:

1. Explain why every feasible elliptical transfer orbit must satisfy each of the following conditions:

   a. $\dfrac{\rho}{1 - e \cos \theta} \leq 120$ for some θ.

   b. $\dfrac{\rho}{1 - e \cos \theta} \geq 300$ for some θ.

   c. $0 < e < 1$.

2. Write and graph a system of inequalities in the (e, ρ)-plane to show those combinations of e and ρ that satisfy the conditions from Task 1.

3. On your graph from Task 2, indicate those orbits that are tangent to the inner parking orbit and those that are tangent to the outer parking orbit. Also indicate the Hohmann orbit.

4. Find both polar and rectangular equations for the Hohmann orbit.

Writing Assignment:

Write a report to your colleagues at NASA for the purpose of presenting and explaining your graph. In particular, explain your graphical representation of the Hohmann orbit and those orbits that are tangent to each of the parking orbits.

Source: Marvin Keener and Jeanne Agnew, Oklahoma State University, *Industry Related Problems for Mathematics Students.*

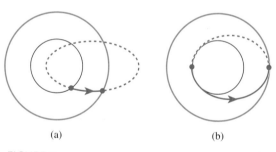

(a)   (b)

FIGURE 1

11 Vectors and Parametric Curves

We will introduce parametric equations of curves and see how to find parametric equations for several types of curves. We will see how to use parametric equations to determine the slope, concavity, length, and sharpness of turning of curves. In Chapter 14 we will see that parametric equations of curves play an important role in several problems that involve measurement of physical quantities along a curve.

Vectors are the mathematical objects used to study problems that involve magnitude and direction. We will introduce two-dimensional and three-dimensional vectors and illustrate how they are used in applications. We will use vectors to study planes and lines in three-dimensional space. Finally, we will see how vectors are related to the study of curves and motion of objects along curves.

11.1 PARAMETRIC EQUATIONS

In this section we will introduce a way of describing curves that is convenient for the study of both static properties of curves (such as length and sharpness of turns) and dynamic properties associated with motion along curves (such as velocity and acceleration).

If x and y are continuous functions of t for t in an interval I, we can describe a curve C in the xy-plane by writing

$$C : x = x(t), \qquad y = y(t), \qquad t \text{ in } I.$$

The equations $x = x(t)$, $y = y(t)$ are called **parametric equations** and the real variable t is called a **parameter** of the curve. The **graph** of the **parametric curve** C is the set of all points $(x(t), y(t))$, t in I. Note that each value of the parameter t in I determines a point $(x(t), y(t))$ on the graph of the curve; the curve is traced as the parameter varies over I. The **direction** of the curve is along the graph, in the direction of increasing t. If the interval I contains its left endpoint a, the point $(x(a), y(a))$ is called the **initial point** of the curve; if I contains its right endpoint b, $(x(b), y(b))$ is called the **terminal point** of the curve. See Figure 11.1.1.

Parametric equations can be used to obtain a dynamic description of a moving particle. In this application the parameter represents time and the parametric equations give the position of the particle at time t. See Figure 11.1.2.

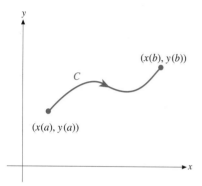

FIGURE 11.1.1   This curve has initial point $(x(a), y(a))$ and terminal point $(x(b), y(b))$.

Identifying and Sketching Parametric Curves

A basic method of sketching a parametric curve is to draw line segments between points on the curve, where the points are ordered according to increasing values of the parameter.

EXAMPLE 1
Sketch the curve

$$C : x = \cos t - \cos 2t, \quad y = \sin t - \sin 2t, \quad 0 \le t \le 2\pi,$$

by connecting points that correspond to increments of $\Delta t = \pi/4$ with line segments. Sketch a smooth curve through the points that have been plotted.

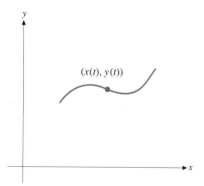

FIGURE 11.1.2   Parametric equations can be used to give the position of a particle at time t.

SOLUTION   The values of the coordinates that correspond to values of t in increments of $\Delta t = \pi/4$ from $t = 0$ to $t = 2\pi$ are given in Table 11.1.1. The

TABLE 11.1.1

t	x	y
0	0	0
$\pi/4$	$1/\sqrt{2} \approx 0.7$	$1/\sqrt{2} - 1 \approx -0.3$
$\pi/2$	1	1
$3\pi/4$	$-1/\sqrt{2} \approx -0.7$	$1/\sqrt{2} + 1 \approx 1.7$
π	-2	0
$5\pi/4$	$-1/\sqrt{2} \approx -0.7$	$-1/\sqrt{2} - 1 \approx -1.7$
$3\pi/2$	1	-1
$7\pi/4$	$1/\sqrt{2} \approx 0.7$	$-1/\sqrt{2} + 1 \approx 0.3$
2π	0	0

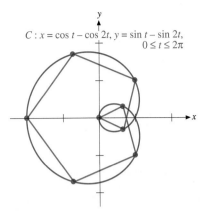

$C : x = \cos t - \cos 2t, \ y = \sin t - \sin 2t,$
$0 \le t \le 2\pi$

FIGURE 11.1.3 Parametric curves can be sketched by plotting some points on the curve and then connecting the points in the order of increasing values of the parameter.

graph is sketched by first plotting the initial point and then repeating the process of plotting the next point and drawing a line segment from the previous point to the new point. This process is continued until the terminal point is reached. The sketch is given in Figure 11.1.3, where we have superimposed a smooth curve. ■

The graphing technique of Example 1 is well suited for computer graphics. It is easy to program a computer to repeat the process of calculating coordinates of a point from parametric equations and then drawing a line segment from the previous point to the new point. The figures in this text were computer generated in this way. The curves appear quite smooth if the increment Δt is small enough, as indicated by the smooth curve in Figure 11.1.3.

In some cases we can identify the graph of a plane parametric curve by eliminating the parameter to obtain a familiar equation in x and y that is satisfied by points on the curve. The curve can then be described without reference to a parameter by indicating its initial and terminal points on the graph of the equation in x and y.

NOTE *The graph of the parametric curve may not include all points on the graph of the equation obtained by eliminating the parameter.*

EXAMPLE 2
Identify, sketch, and describe the curve

$$C : x = 2t - 1, \qquad y = t + 1, \qquad 0 \le t \le 2.$$

SOLUTION In this example we can eliminate the parameter by solving the second equation for t in terms of y and then substituting into the first equation. That is,

$$y = t + 1 \text{ implies } t = y - 1$$

and then

$$x = 2t - 1 \text{ implies } x = 2(y - 1) - 1.$$

Points of the curve satisfy the equation

$$x = 2(y - 1) - 1,$$
$$x = 2y - 2 - 1,$$
$$x - 2y + 3 = 0.$$

This is the equation of a line. The curve consists of the line segment from the initial point $(-1, 1)$, where $t = 0$, to the terminal point $(3, 3)$, where $t = 2$. See Figure 11.1.4. ■

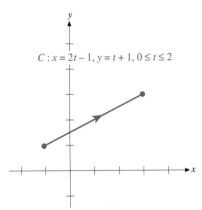

$C : x = 2t - 1, y = t + 1, 0 \le t \le 2$

FIGURE 11.1.4 The curve goes along the line segment from $(-1, 1)$ to $(3, 3)$.

EXAMPLE 3
Identify, sketch, and describe the curve

$$C : x = 2t^3 + 1, \qquad y = t^3 + 2, \qquad -1 \le t \le 1.$$

SOLUTION We see that

$$y = t^3 + 2 \text{ implies } t^3 = y - 2,$$

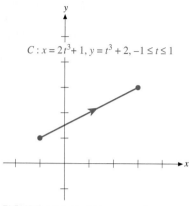

FIGURE 11.1.5 This curve also goes along the line segment from $(-1, 1)$ to $(3, 3)$.

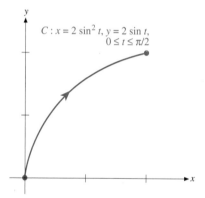

FIGURE 11.1.6 The curve goes along the parabola from $(0, 0)$ to $(2, 2)$.

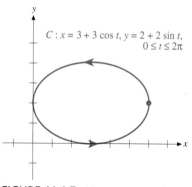

FIGURE 11.1.7 The curve goes from the point $(6, 2)$, counterclockwise once around the ellipse.

and then

$$x = 2t^3 + 1 \text{ implies } x = 2(y - 2) + 1, \text{ so}$$
$$x = 2y - 4 + 1,$$
$$x - 2y + 3 = 0.$$

The curve consists of the segment of this line from the initial point $(-1, 1)$, where $t = -1$, to the terminal point $(3, 3)$, where $t = 1$. See Figure 11.1.5. ■

The parametric descriptions of the curves in Examples 2 and 3 are different, but the curves have the same graph and the same direction. The different parametric descriptions may represent the motion of different particles that follow the same path in the same direction, but have different speeds.

EXAMPLE 4

Identify, sketch, and describe the curve

$$C : x = 2\sin^2 t, \qquad y = 2\sin t, \qquad 0 \le t \le \pi/2.$$

SOLUTION We eliminate t by noting that $\sin t = y/2$, so

$$x = 2(\sin t)^2,$$
$$x = 2\left(\frac{y}{2}\right)^2,$$
$$x = \frac{y^2}{2}.$$

This is the equation of a parabola. The curve consists of that portion of the parabola from the initial point $(0, 0)$ to the terminal point $(2, 2)$. The graph is sketched in Figure 11.1.6. ■

EXAMPLE 5

Identify, sketch, and describe the curve

$$C : x = 3 + 3\cos t, \qquad y = 2 + 2\sin t, \qquad 0 \le t \le 2\pi.$$

SOLUTION In this case we eliminate t by using the identity $\sin^2 t + \cos^2 t = 1$. We have

$$\sin t = \frac{y - 2}{2} \text{ and } \cos t = \frac{x - 3}{3},$$

so

$$\frac{(y - 2)^2}{2^2} + \frac{(x - 3)^2}{3^2} = 1.$$

This is the equation of an ellipse. The curve goes from the initial point $(6, 2)$, counterclockwise once around the ellipse, and returns to the initial point. The sketch is given in Figure 11.1.7. ■

Choosing Parametric Equations

To solve problems that involve the measurement of physical quantities at points along a curve, we will need to find parametric equations for a collection of curves that are described geometrically or in terms of equations in x and y. The following rules are useful.

The curve along the graph of $y = f(x)$ in the direction of increasing x from $(a, f(a))$ to $(b, f(b))$ can be expressed as

$$C : x = t, \qquad y = f(t), \qquad a \leq t \leq b. \tag{1}$$

See Figure 11.1.8.

The curve along the graph of $x = f(y)$ in the direction of increasing y from $(f(a), a)$ to $(f(b), b)$ can be expressed as

$$C : x = f(t), \qquad y = t, \qquad a \leq t \leq b. \tag{2}$$

See Figure 11.1.9.

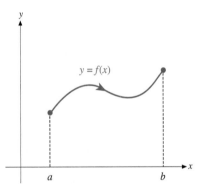

FIGURE 11.1.8 We can parametrize the graph of $y = f(x)$ in the direction of increasing x by setting the parameter equal to x.

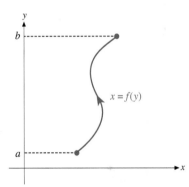

FIGURE 11.1.9 We can parametrize the graph of $x = f(y)$ in the direction of increasing y by setting the parameter equal to y.

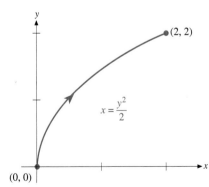

FIGURE 11.1.10 Since the equation $x = y^2/2$ gives x as a function of y and the direction is that of increasing y, we can set the parameter equal to y.

EXAMPLE 6

Find parametric equations of the curve from $(0, 0)$ to $(2, 2)$ along the parabola $x = y^2/2$.

SOLUTION   The equation $x = y^2/2$ gives x as a function of y, and the direction of the curve is in the direction of increasing y. See Figure 11.1.10. Formula 2 then gives the parametric representation

$$C : x = \frac{t^2}{2}, \qquad y = t, \qquad 0 \leq t \leq 2. \quad \blacksquare$$

The curves in Examples 4 and 6 have the same graph and the same direction, but the parametric equations that describe the curves are different. Examples 2 and 3 also illustrated that different parametric equations can describe the

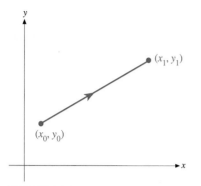

FIGURE 11.1.11 The line segment from (x_0, y_0) to (x_1, y_1) is given by $C : x = x_0 + (x_1 - x_0)t$, $y = y_0 + (y_1 - y_0)t, 0 \le t \le 1$.

same curve. There is not a unique way to parametrize a curve. We should try to choose the parameter in a systematic way that gives simple equations. Note that the parametric equations in Example 6 appear more natural than the equations used in Example 4.

We can find parametric equations of a line segment in the xy-plane by using the following formula, which is easily verified. See Figure 11.1.11.

The line segment from (x_0, y_0) to (x_1, y_1) is given by

$$C : x = x_0 + (x_1 - x_0)t, \quad y = y_0 + (y_1 - y_0)t, \quad 0 \le t \le 1. \quad (3)$$

Example 7

Find parametric equations of the line segment from $(1, -2)$ to $(-3, 1)$.

Solution From Formula 3, we obtain

$$C : x = 1 + (-3 - 1)t, \quad y = -2 + (1 - (-2))t, \quad 0 \le t \le 1.$$

Simplifying, we have

$$C : x = 1 - 4t, \quad y = -2 + 3t, \quad 0 \le t \le 1.$$

The graph is sketched in Figure 11.1.12. ∎

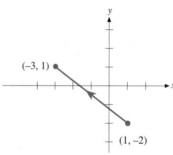

FIGURE 11.1.12 Formula 3 gives parametric equations of the line segment.

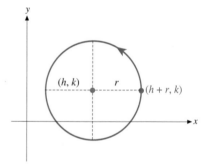

FIGURE 11.1.13 The curve that goes from $(h + r, k)$, counterclockwise once around the circle, is $C : x = h + r \cos t, y = k + r \sin t, 0 \le t \le 2\pi$.

Parametric equations of the curve that goes from the point $(h + r, k)$, counterclockwise once around the circle $(x - h)^2 + (y - k)^2 = r^2$, are

$$C : x = h + r \cos t, \quad y = k + r \sin t, \quad 0 \le t \le 2\pi. \quad (4)$$

See Figure 11.1.13.

Example 8

Find parametric equations of the curve that goes from the point $(2, 0)$, counterclockwise once around the circle with center at the origin and radius 2.

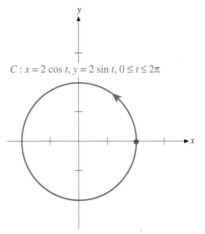

$C : x = 2 \cos t, y = 2 \sin t, 0 \le t \le 2\pi$

FIGURE 11.1.14 Formula 4 gives parametric equations of the circle.

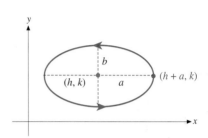

FIGURE 11.1.15 The curve that goes from $(h + a, k)$, counterclockwise once around the ellipse, is

$$C : x = h + a \cos t, y = k + b \sin t,$$

$$0 \le t \le 2\pi.$$

SOLUTION   Using Formula 4 with $(h, k) = (0, 0)$ and $r = 2$, we obtain

$$C : x = 2 \cos t, \quad y = 2 \sin t, \quad 0 \le t \le 2\pi.$$

See Figure 11.1.14.   ■

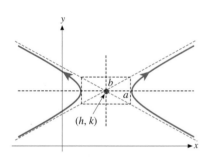

FIGURE 11.1.16 Parametric equations of the hyperbola are $C : x = h + a \sec t, y = k + b \tan t$.

Parametric equations of the curve that goes from the point $(h + a, k)$, counterclockwise once around the ellipse $\dfrac{(x - h)^2}{a^2} + \dfrac{(y - k)^2}{b^2} = 1$, are

$$C : x = h + a \cos t, \quad y = k + b \sin t, \quad 0 \le t \le 2\pi. \tag{5}$$

See Figure 11.1.15.

Parametric equations of the hyperbola $\dfrac{(x - h)^2}{a^2} - \dfrac{(y - k)^2}{b^2} = 1$ are

$$C : x = h + a \sec t, \quad y = k + b \tan t. \tag{6}$$

One branch is traced for $-\pi/2 < t < \pi/2$; the other branch is traced for $\pi/2 < t < 3\pi/2$. See Figure 11.1.16.

Parametric equations of the hyperbola $-\dfrac{(x - h)^2}{a^2} + \dfrac{(y - k)^2}{b^2} = 1$ are

$$C : x = h + a \tan t, \quad y = k + b \sec t. \tag{7}$$

One branch is traced for $-\pi/2 < t < \pi/2$; the other branch is traced for $\pi/2 < t < 3\pi/2$. See Figure 11.1.17.

In some cases we can use geometry and trigonometry to determine parametric equations of a curve.

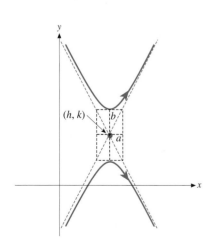

FIGURE 11.1.17 Parametric equations of the hyperbola are $C : x = h + a \tan t, y = k + b \sec t$.

EXAMPLE 9

Find parametric equations of the curve traced by a point on the circumference of a wheel of radius r as the wheel rolls along a level surface.

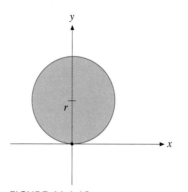

FIGURE 11.1.18a We choose a coordinate system so the initial position of the point on the circumference is $(0,0)$.

FIGURE 11.1.18b The horizontal distance t is equal to the length $r\theta$ along the circumference.

SOLUTION We choose a coordinate system as indicated in Figure 11.1.18a, so that the initial position of the point on the circumference is $(0,0)$. As the wheel rolls, the position of the point depends upon the change in position of the center (t) and the angle of rotation about the center (θ). See Figure 11.1.18b. We see that

$$x = t - r \sin\theta \quad \text{and} \quad y = r - r\cos\theta.$$

We can obtain a relation between t and θ by noting that as the wheel rolls, the horizontal distance t is equal to the length along the circumference subtended by θ. This gives

$$r\theta = t.$$

We can solve this equation for θ in terms of t and then substitute into the above equations for x and y to obtain

$$C : x = t - r\sin\frac{t}{r}, \quad y = r - r\cos\frac{t}{r}.$$

This curve is called a **cycloid**. A sketch is given in Figure 11.1.19. ■

FIGURE 11.1.19 The point on the circumference traces a curve called a cycloid.

The Negative of a Curve

DEFINITION

The negative of a curve C is the curve, denoted $-C$, that has the same graph as C, but opposite direction.

The negative of the curve

$$C : x = f(t), \quad y = g(t), \quad a \le t \le b,$$

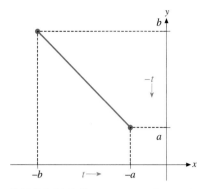

FIGURE 11.1.20 As t varies from $-b$ to $-a$, the expression $-t$ varies from b to a.

can be expressed as

$$-C : x = f(-t), \quad y = g(-t), \quad -b \le t \le -a. \tag{8}$$

That is, we can obtain a parametric representation of $-C$ by substituting $-t$ for t in the parametric representation of C. Note that the equation $-b \le t \le -a$ is equivalent to $a \le -t \le b$. It follows that each point $(f(-t), g(-t))$, $-b \le t \le -a$, on the graph of $-C$ corresponds to a point on the graph of C. As t varies from $-b$ to $-a$, the expression $-t$ varies from b to a. See Figure 11.1.20. This shows that the directions of C and $-C$ are opposite.

EXAMPLE 10

Find parametric equations of the curve $-C$, where

$$C : x = \cos t, \quad y = \sin t, \quad 0 \le t \le 2\pi.$$

SOLUTION Substituting $-t$ for t in the parametric representation of C, we obtain

$$-C : x = \cos(-t), \quad y = \sin(-t), \quad 0 \le -t \le 2\pi.$$

Using the trigonometric identities $\cos(-t) = \cos t$ and $\sin(-t) = -\sin t$, and multiplying the inequality $0 \le -t \le 2\pi$ by -1, we obtain

$$-C : x = \cos t, \quad y = -\sin t, \quad -2\pi \le t \le 0.$$

The curve $-C$ goes clockwise from $(1, 0)$ once around the circle $x^2 + y^2 = 1$. See Figure 11.1.21. ■

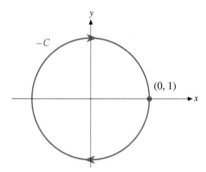

FIGURE 11.1.21 The curve $-C$ goes from $(1, 0)$, clockwise once around the circle.

For the applications of parametric curves that we will study in Chapter 14, we can use parametric equations of either C or $-C$. Thus, if it is more convenient, we find parametric equations of $-C$ instead of C. We need only to indicate which of C or $-C$ we have parametrized.

EXAMPLE 11

The curve C goes from $(2, 4)$ to $(0, 0)$ along the parabola $y = x^2$. Find parametric equations of either C or $-C$, whichever seems more convenient.

SOLUTION The curve C is sketched in Figure 11.1.22. Note that both x and y are decreasing, so Formulas 1 and 2 do not apply directly to C. However, both formulas apply to $-C$, since both x and y are increasing in the direction of $-C$. We then use Formula 1 to obtain

$$-C : x = t, \quad y = t^2, \quad 0 \le t \le 2.$$

Since $C = -(-C)$, we could now use Formula 8 to determine parametric equations of C, but that is not necessary. ■

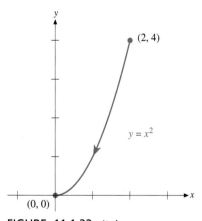

FIGURE 11.1.22 It is more convenient to parametrize $-C$ than C, since the direction of C is in the direction of negative x and negative y.

Polar Curves

The (x, y) coordinates of a point in the plane that has polar coordinates (r, θ) are given by the equations

$$x = r \cos \theta, \quad y = r \sin \theta.$$

See Figure 11.1.23. It follows that

The polar curve $r = r(\theta)$ in the direction of increasing θ can be expressed as

$$C : x = r(t) \cos t, \quad y = r(t) \sin t. \tag{9}$$

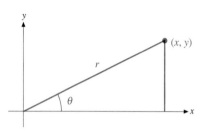

FIGURE 11.1.23 Polar coordinates and rectangular coordinates are related by the formulas $x = r \cos \theta, y = r \sin \theta$.

EXAMPLE 12

Find parametric equations of the curve that goes in the direction of increasing θ from the point $(r, \theta) = (1, 0)$ once around the polar curve $r = \cos 3\theta$.

SOLUTION   The graph is sketched in Figure 11.1.24. Using the polar graphing techniques of Section 10.5, we can determine that the curve is traced once as θ varies from 0 to π. Also, the given initial point corresponds to $\theta = 0$. From Formula 9 we then obtain

$$C : x = (\cos 3t)(\cos t), \quad y = (\cos 3t)(\sin t), \quad 0 \le t \le \pi. \quad \blacksquare$$

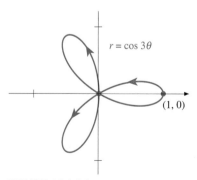

FIGURE 11.1.24 The polar curve is given by $C : x = (\cos 3t)(\cos t), y = (\cos 3t)(\sin t), 0 \le t \le \pi$.

EXERCISES 11.1

Sketch the curves in Exercises 1–8 by connecting points that correspond to the given increments with line segments. Sketch a smooth curve through the points.

1. $C : x = \cos t,\ y = \sin t,\ 0 \le t \le 2\pi,\ \Delta t = \pi/4$

2. $C : x = 2\cos t,\ y = \sin t,\ 0 \le t \le 2\pi,\ \Delta t = \pi/4$

3. $C : x = 2\cos^2 t,\ y = \sin 2t,\ -\pi/2 \le t \le \pi/2,\ \Delta t = \pi/6$

4. $C : x = \sin 2t,\ y = \sin^2 t,\ 0 \le t \le \pi,\ \Delta t = \pi/6$

5. $C : x = \dfrac{2\cos t}{\cos t + \sin t},\ y = \dfrac{2\sin t}{\cos t + \sin t},\ -\pi/6 \le t \le 2\pi/3,$
$\Delta t = \pi/6$

6. $C : x = 2,\ y = 2\tan t,\ -\pi/3 \le t \le \pi/3,\ \Delta t = \pi/6$

7. $C : x = t,\ y = t^2,\ -2 \le t \le 2,\ \Delta t = 1$

8. $C : x = t^2,\ y = t,\ -2 \le t \le 2,\ \Delta t = 1$

Identify, sketch, and describe the curves given below in Exercises 9–20.

9. $C : x = -3 + 6t,\ y = -1 + 4t,\ 0 \le t \le 1$

10. $C : x = 3 - 3t,\ y = -2t,\ 0 \le t \le 1$

11. $C : x = t,\ y = 2t - t^2,\ 0 \le t \le 2$

12. $C : x = t^2,\ y = t,\ 0 \le t \le 2$

13. $C : x = t,\ y = \cos t,\ 0 \le t \le 2\pi$

14. $C : x = \cos t,\ y = t,\ 0 \le t \le \pi$

15. $C : x = \sin t,\ y = t,\ -\pi/2 \le t \le \pi/2$

16. $C : x = t,\ y = \sin t,\ -\pi/2 \le t \le \pi/2$

17. $C : x = \cos t,\ y = \sin t,\ 0 \le t \le \pi$

18. $C : x = \cos t,\ y = -\sin t,\ 0 \le t \le \pi$

19. $C : x = \dfrac{2t}{1 + t^2},\ y = \dfrac{1 - t^2}{1 + t^2},\ -\infty < t < \infty$

20. $C : x = \tan t,\ y = \sec t,\ -\pi/2 < t < \pi/2$

For the curves C described in Exercises 21–40, find parametric equations at either C or −C, whichever seems more convenient.

21. Along $y = x^3$, from $(-1, -1)$ to $(2, 8)$

22. Along $y = x^2 - 2x$, from $(0, 0)$ to $(3, 3)$

23. Along $x = y^2$, from $(1, -1)$ to $(4, 2)$

24. Along $x = y^3 - y$, from $(-6, -2)$ to $(6, 2)$

25. Along the line segment from $(0, 0)$ to $(3, 2)$

26. Along the line segment from $(0, 3)$ to $(2, 0)$

27. Along the line segment from $(2, 1)$ to $(-2, 0)$

28. Along the line segment from $(1, 3)$ to $(-1, 1)$

29. Once around the circle $x^2 + y^2 = 4$, counterclockwise from $(2, 0)$

30. Once around the circle $x^2 + (y - 1)^2 = 1$, counterclockwise from $(1, 1)$

31. Once around the ellipse $9x^2 + 4y^2 = 36$, counterclockwise from $(2, 0)$

32. Once around the ellipse $x^2 + 4y^2 = 4$, counterclockwise from $(2, 0)$

33. Along the branch of the hyperbola $x^2 - 4y^2 = 4$ with $x > 0$, in the direction of increasing y

34. Along the branch of the hyperbola $x^2 - 4y^2 = 4$ with $x < 0$, in the direction of increasing y

35. Along the branch of the hyperbola $4x^2 - 9y^2 + 36 = 0$ with $y > 0$, in the direction of increasing x

36. Along the branch of the hyperbola $4x^2 - 9y^2 + 36 = 0$ with $y < 0$, in the direction of increasing x

37. Along $y = x^2$, from $(2, 4)$ to $(-2, 4)$

38. Along $y = x^3$, from $(2, 8)$ to $(0, 0)$

39. Once around the circle $x^2 + y^2 = 1$, clockwise from $(1, 0)$

40. Along $x = 1 - y^2$, from $(0, 1)$ to $(-3, -2)$

Find parametric equations of the polar curves given in Exercises 41–44.

41. $r = 1 - \cos\theta,\ 0 \le \theta \le 2\pi$

42. $r = 1 + 2\cos\theta,\ 0 \le \theta \le 2\pi$

43. $r = \theta/2,\ 0 \le \theta \le 2\pi$

44. $r = \sin 3\theta,\ 0 \le \theta \le \pi$

45. Find parametric equations of the curve traced by a point P that is a distance d from the center of a wheel of radius r as the wheel rolls along a level surface. Assume the wheel rolls along the positive x-axis with the initial position of P at the point $(0, r - d)$.

46. Assume that f is continuous on the interval $a \le x \le b$, and that $a \le c < d \le b$ implies $f(c) < f(d)$. This implies that f has an inverse function f^{-1}. Use values of f to determine parametric equations that describe the graph of $y = f^{-1}(x)$ in the direction of increasing x.

We can find parametric equations for the general second-degree curve

$$Ax^2 + Bxy + Cy^2 + Dx + Ey + F = 0$$

by using the equations of rotation

$$x = x'\cos\theta - y'\sin\theta, \quad y = x'\sin\theta + y'\cos\theta$$

with $\cot 2\theta = (A - C)/B$ *to obtain an equation that contains no $x'y'$ term. We then use the methods of this section to parametrize the $x'y'$ curve. The equations of rotation then give parametric equations for the variables x and y.*

47. Find parametric equations of $3x^2 + 2xy + 3y^2 = 4$.

48. Find parametric equations of $x^2 + 4xy + y^2 = 3$.

49. One end of a 36-ft rope is attached to the side of a circular silo with a radius of 12 ft, and the other end is attached to a goat. The region where the goat can graze is indicated in the figure. The boundary of the region in the first and fourth quadrants is a semicircular arc. In the second and third quadrants the boundary is generated by the endpoint of the rope as it wraps partway around the silo and then follows a straight tangential path from the silo.

(a) Express the coordinates of that part of the boundary curve in the second quadrant in terms of the angle θ subtended by the arc of the silo between the point where the rope is attached and the point where it is tangent to the silo.

(b) Use the fact that the area in the second quadrant where the goat can graze is a limit of sums of the areas of the "triangles" indicated to express that area as a definite integral with respect to θ.

(c) Find the total area of the region where the goat can graze.

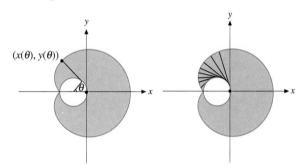

50. Parametric equations of a line are $C : x = a + bt, \ y = c + dt$.

(a) Express the distance of a point on the line from the origin in terms of t.

(b) Find the minimum distance of the line from the origin.

51. Use the figure to find parametric equations of the curve traced by a point P on the circumference of a circle as the circle rolls around another circle of equal radius r_0. (The curve is a cardioid.)

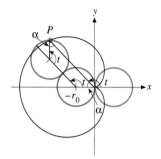

52. Sketch the curve

$$C : x(t) = \begin{cases} t + 2, & -3 \le t \le -1, \\ 1 - \dfrac{1}{\sqrt{3}} \tan \dfrac{\pi}{3}(1 - t^2), & -1 < t \le 0, \\ -1 + \dfrac{1}{\sqrt{3}} \tan \dfrac{\pi}{3}(1 - t^2), & 0 < t \le 1, \end{cases}$$

$$y(t) = \begin{cases} 0, & -3 \le t \le -1, \\ \dfrac{1}{\sqrt{3}} \tan \dfrac{\pi}{3}(1 - t^2), & -1 < t \le 1. \end{cases}$$

11.2   SLOPE; CONCAVITY; ARC LENGTH; CHANGE OF PARAMETER

Connections

Geometric interpretation of f' and f'', 4.3–4.

Arc length, 6.5.

In this section we will see how parametric equations of a curve can be used to obtain information about the curve.

$\dfrac{dy}{dx}$ and the Slope of a Parametric Curve

If y is a function of x, we know that

$$\frac{dy}{dx} \quad \text{and} \quad \frac{d^2y}{dx^2}$$

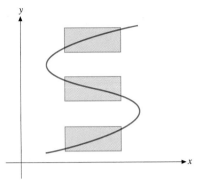

FIGURE 11.2.1   That part of the graph inside each rectangle defines *y* as a function of *x*.

give information about the slope and concavity of the graph of *y*. If the parametric equations of a curve determine *y* as a function of *x*, we can use what we know about the geometric interpretation of the first and second derivatives of a function to obtain information about the parametric curve.

As in the case of functions that are defined implicitly by an equation in *x* and *y*, parametric equations of a curve may define *y* as a function of *x* only for various parts of the curve. See Figure 11.2.1.

Let us consider the curve

$$C : x = x(t), \quad y = y(t), \quad a \le t \le b,$$

where *x* and *y* are differentiable functions of *t*. We first note that, if $dx/dt \ne 0$, the sign of dx/dt determines the direction of *C*:

If $\dfrac{dx}{dt} > 0$ for t in a subinterval I of $[a, b]$, then x is increasing as t increases, so the curve is in the direction of increasing x for t in I.

See Figure 11.2.2a.

If $\dfrac{dx}{dt} < 0$ for t in I, the curve is in the direction of decreasing x for t in I.

See Figure 11.2.2b.

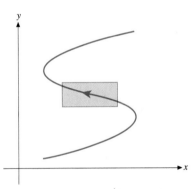

FIGURE 11.2.2a   If $\dfrac{dx}{dt} > 0$, the direction of the curve is the direction of increasing *x*.

FIGURE 11.2.2b   If $\dfrac{dx}{dt} < 0$, the direction of the curve is the direction of decreasing *x*.

We assume that dx/dt is either always positive or always negative as *t* varies over a subinterval *I* of $[a, b]$. This condition guarantees that the function $x(t)$, *t* in *I*, has an inverse function. From Theorem 1 of Section 7.2 we know that the equation $x = x(t)$, *t* in *I*, defines $t = t(x)$ as a differentiable function of *x* with

$$\frac{dt}{dx} = \frac{1}{\dfrac{dx}{dt}}.$$

For *t* in *I*, the equations

$$y = y(t) \text{ and } t = t(x)$$

define *y* as a differentiable function of *x* and the Chain Rule gives

$$\frac{dy}{dx} = \frac{dy}{dt}\frac{dt}{dx}.$$

Combining results, we obtain

$$\frac{dy}{dx} = \frac{\dfrac{dy}{dt}}{\dfrac{dx}{dt}}. \tag{1}$$

This formula gives dy/dx as a function of the parameter t. Its value is the slope of the line tangent to the curve at the point $(x(t), y(t))$.

EXAMPLE 1

Sketch the curve

$$C : x = t^2 - 1, \quad y = t, \quad -2 \le t \le 3.$$

Find an equation of the line tangent to the curve at the point where $t = 2$.

SOLUTION   Substituting y for t in the first equation, we obtain

$$x = y^2 - 1.$$

This is the equation of a parabola. The curve consists of that portion of the parabola from $(3, -2)$ to $(8, 3)$. See Figure 11.2.3.

From the equations $x = t^2 - 1$ and $y = t$, we obtain

$$\frac{dx}{dt} = 2t \text{ and } \frac{dy}{dt} = 1.$$

Formula 1 then implies

$$\frac{dy}{dx} = \frac{\dfrac{dy}{dt}}{\dfrac{dx}{dt}} = \frac{1}{2t}.$$

When $t = 2$, we have

$$x = 2^2 - 1 = 3, \quad y = 2, \text{ and } \frac{dy}{dx} = \frac{1}{4}.$$

An equation of the line through the point $(3, 2)$ with slope $1/4$ is

$$y - 2 = \tfrac{1}{4}(x - 3),$$
$$4y - 8 = x - 3,$$
$$x - 4y + 5 = 0. \quad \blacksquare$$

In Example 1 we have $dx/dt = 2t$, so $dx/dt = 0$ when $t = 0$. Formula 1 does not apply when $t = 0$. From Figure 11.2.3 we see that the curve does not determine y as a function of x as the curve passes through the point $(-1, 0)$, which corresponds to $t = 0$.

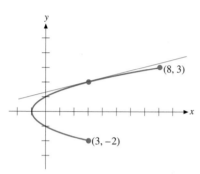

FIGURE 11.2.3 The slope of the tangent line is the value of

$$\frac{dy}{dx} = \frac{\dfrac{dy}{dt}}{\dfrac{dx}{dt}}.$$

$\dfrac{d^2y}{dx^2}$ and Concavity

Note that Formula 1 can be used to differentiate any function of t with respect to x. In particular, if we express $\dfrac{dy}{dx}$ as a function of t, we can use Formula 1 to find its derivative with respect to x. Applying Formula 1 with y replaced by $\dfrac{dy}{dx}$, we obtain

$$\frac{d^2y}{dx^2} = \frac{d}{dx}\left(\frac{dy}{dx}\right) = \frac{\dfrac{d}{dt}\left(\dfrac{dy}{dx}\right)}{\dfrac{dx}{dt}}. \tag{2}$$

This formula gives d^2y/dx^2 in terms of the parameter t. The second derivative gives us information about the concavity of the graph.

EXAMPLE 2

Evaluate dy/dx and d^2y/dx^2 at the point on the curve

$$C : x = e^t \cos t, \quad y = e^t \sin t,$$

where $t = \pi/2$. Sketch an arc of the curve as it passes through that point. Indicate slope, concavity, and the direction of the curve as it passes through the point.

SOLUTION   We have

$$x = e^t \cos t,$$
$$y = e^t \sin t,$$
$$\frac{dx}{dt} = (e^t)(-\sin t) + (\cos t)(e^t) = e^t(\cos t - \sin t), \text{ and}$$
$$\frac{dy}{dt} = (e^t)(\cos t) + (\sin t)(e^t) = e^t(\cos t + \sin t).$$

It then follows from Formula 1 that

$$\frac{dy}{dx} = \frac{\dfrac{dy}{dt}}{\dfrac{dx}{dt}} = \frac{\cos t + \sin t}{\cos t - \sin t}, \text{ so}$$

$$\frac{d}{dt}\left(\frac{dy}{dx}\right) = \frac{(\cos t - \sin t)(-\sin t + \cos t) - (\cos t + \sin t)(-\sin t - \cos t)}{(\cos t - \sin t)^2}.$$

Substitution of $t = \pi/2$ then gives $x = 0$, $y = e^{\pi/2}$, and $dy/dx = -1$. When $t = \pi/2$, we also have

$$\frac{dx}{dt} = -e^{\pi/2} \text{ and } \frac{d}{dt}\left(\frac{dy}{dx}\right) = 2,$$

so Formula 2 implies

$$\frac{d^2y}{dx^2} = \frac{\dfrac{d}{dt}\left(\dfrac{dy}{dx}\right)}{\dfrac{dx}{dt}} = \frac{2}{-e^{\pi/2}} = -2e^{-\pi/2}.$$

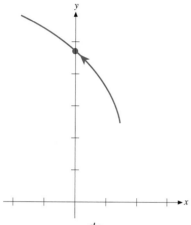

FIGURE 11.2.4 $\dfrac{dy}{dx} < 0$, so the graph is decreasing. $\dfrac{d^2y}{dx^2} < 0$, so the graph is concave downward. $\dfrac{dx}{dt} < 0$, so the direction is that of decreasing x.

FIGURE 11.2.5 The length of the curve is approximately the length of the polygonal path through points on the curve determined by small increments.

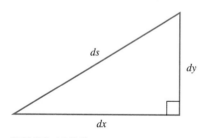

FIGURE 11.2.6 We can think of the differential arc length as the hypotenuse of a right triangle with length of sides dx and dy, so

$$ds = \sqrt{(dx)^2 + (dy)^2}.$$

We see that the graph of the curve has slope -1 as it passes through the point $(0, e^{\pi/2})$ and it is concave down near that point. Also, since dx/dt is negative, the x-coordinates of points on the curve are decreasing as t increases. These features are indicated in the segment of the graph sketched in Figure 11.2.4. (The entire graph is a spiral that spirals counterclockwise away from the origin as t increases.) ■

Arc Length

Let us consider the length of the parametric curve

$$C : x = x(t), \quad y = y(t), \quad a \leq t \leq b.$$

We assume the derivatives dx/dt and dy/dt are continuous. We divide the interval $[a, b]$ into subintervals of length Δt and draw a polygonal path through points on the curve that correspond to the points of subdivision. See Figure 11.2.5.

If Δt is small enough, the length of the curve should be approximately

$$\sum \sqrt{(\Delta x)^2 + (\Delta y)^2} = \sum \sqrt{\left(\frac{\Delta x}{\Delta t}\right)^2 + \left(\frac{\Delta y}{\Delta t}\right)^2} \, \Delta t.$$

The following definition should then seem reasonable.

DEFINITION

The **arc length** of the curve $C : x = x(t)$, $y = y(t)$, $a \leq t \leq b$, is

$$L = \int_a^b \sqrt{\left(\frac{dx}{dt}\right)^2 + \left(\frac{dy}{dt}\right)^2} \, dt.$$

It is convenient to think of the differential expression

$$ds = \sqrt{(dx)^2 + (dy)^2} = \sqrt{\left(\frac{dx}{dt}\right)^2 + \left(\frac{dy}{dt}\right)^2} \, dt$$

as the length of a small segment of the curve and the integral

$$\int_a^b \sqrt{\left(\frac{dx}{dt}\right)^2 + \left(\frac{dy}{dt}\right)^2} \, dt$$

as the "sum" of the lengths of the segments. This reinforces the fact that the integral is a limit of Riemann sums. See Figure 11.2.6.

EXAMPLE 3

Find the length of the curve

$$C : x = 3\cos t, \quad y = 3\sin t, \quad 0 \leq t \leq 2\pi.$$

SOLUTION The curve traces the circle $x^2 + y^2 = 3^2$ once, counterclockwise from the point $(3, 0)$. See Figure 11.2.7. We have

$$\frac{dx}{dt} = -3\sin t \text{ and } \frac{dy}{dt} = 3\cos t,$$

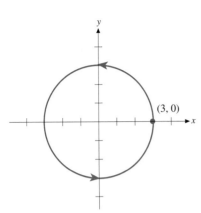

FIGURE 11.2.7 The arc length integral gives the circumference of the circle.

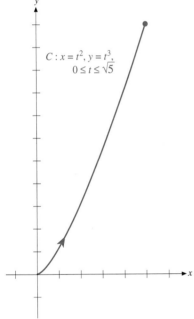

$$C : x = t^2, y = t^3,$$
$$0 \leq t \leq \sqrt{5}$$

FIGURE 11.2.8 The graph shows that 335/27 ≈ 12.4 is a reasonable answer for the length of this curve.

so

$$ds = \sqrt{(-3\sin t)^2 + (3\cos t)^2}\, dt = \sqrt{9(\sin^2 t + \cos^2 t)}\, dt = 3\, dt.$$

The arc length is

$$L = \int_0^{2\pi} 3\, dt = 6\pi.$$

This is the circumference of a circle of radius 3, as we know. ∎

EXAMPLE 4

Find the length of the curve

$$C : x = t^2, \quad y = t^3, \quad 0 \leq t \leq \sqrt{5}.$$

SOLUTION   We have

$$\frac{dx}{dt} = 2t \text{ and } \frac{dy}{dt} = 3t^2.$$

Hence,

$$ds = \sqrt{(2t)^2 + (3t^2)^2}\, dt = \sqrt{4t^2 + 9t^4}\, dt = t\sqrt{4 + 9t^2}\, dt, \quad t \geq 0.$$

$$L = \int_0^{\sqrt{5}} t\sqrt{4 + 9t^2}\, dt.$$

Let us express this integral in terms of the variable $u = 4 + 9t^2$. Then $du = 18t\, dt$, $u(0) = 4$, and $u(\sqrt{5}) = 49$. The arc length is then

$$L = \int_4^{49} \frac{1}{18} u^{1/2}\, du$$

$$= \frac{1}{18} \frac{u^{3/2}}{3/2} \Big]_4^{49}$$

$$= \frac{1}{27}\left((49)^{3/2} - (4)^{3/2}\right)$$

$$= \frac{1}{27}(343 - 8) = \frac{335}{27}.$$

The curve is sketched in Figure 11.2.8. ∎

If a curve is the graph of a differentiable function, $y = f(x)$, $a \leq x \leq b$, we can write

$$C : x = x, \quad y = f(x), \quad a \leq x \leq b.$$

In the above equations we are using x as the parameter that determines the coordinates (x, y) of points on the curve. The differential arc length is then

$$ds = \sqrt{(dx)^2 + (dy)^2} = \sqrt{\left(\frac{dx}{dx}\right)^2 + \left(\frac{dy}{dx}\right)^2}\, dx = \sqrt{1 + (f'(x))^2}\, dx.$$

This is the same formula we derived in Section 6.5.

We can find the arc length of a polar curve $r = r(\theta)$ by writing

$$C : x = r(\theta)\cos\theta, \quad y = r(\theta)\sin\theta, \quad a \leq \theta \leq b.$$

Here we are using θ as the parameter that determines the rectangular coordinates (x, y) of points on the curve. Then

$$\left(\frac{dx}{d\theta}\right)^2 + \left(\frac{dy}{d\theta}\right)^2 = (-r\sin\theta + r'\cos\theta)^2 + (r\cos\theta + r'\sin\theta)^2$$

$$= r^2\sin^2\theta - 2rr'\sin\theta\cos\theta + r'^2\cos^2\theta$$
$$+ r^2\cos^2\theta + 2rr'\sin\theta\cos\theta + r'^2\sin^2\theta$$
$$= r^2 + (r')^2.$$

Hence, we obtain the formula

$$\text{Arc length} = \int_a^b \sqrt{r^2 + \left(\frac{dr}{d\theta}\right)^2}\, d\theta.$$

NOTE   *This formula applies to curves that are given in terms of polar coordinates and it is not necessary to write the curve in parametric form in order to use the formula. If the rectangular coordinates of a curve are given by parametric equations, we use the formula $ds = \sqrt{(dx)^2 + (dy)^2}$, even if the parameter happens to be θ.*

EXAMPLE 5

Find the length of the polar spiral

$$r = \theta^2, \qquad 0 \le \theta \le \pi.$$

SOLUTION   We use the formula for the differential arc length of a polar curve. We have

$$r = \theta^2, \text{ so } \frac{dr}{d\theta} = 2\theta.$$

Then

$$ds = \sqrt{r^2 + \left(\frac{dr}{d\theta}\right)^2}\, d\theta = \sqrt{(\theta^2)^2 + (2\theta)^2}\, d\theta = \theta\sqrt{\theta^2 + 4}\, d\theta, \quad \theta \ge 0.$$

$$L = \int_0^\pi \theta\sqrt{\theta^2 + 4}\, d\theta$$

$$= \frac{1}{3}(\theta^2 + 4)^{3/2}\Big]_0^\pi$$

$$= \frac{1}{3}((\pi^2 + 4)^{3/2} - 8).$$

The curve is sketched in Figure 11.2.9. ∎

$r = \theta^2, 0 \le \theta \le \pi$

FIGURE 11.2.9 For polar curves, $r = r(\theta)$, we use the formula

$$ds = \sqrt{r^2 + \left(\frac{dr}{d\theta}\right)^2}.$$

Area of Surfaces of Revolution

Recall from Section 6.5 that differential arc length ds is used to determine the area of a surface obtained by revolving a plane curve about a line in the plane. See Figure 11.2.10. We have

$$\text{Surface area} = \int_0^L 2\pi r\, ds,$$

where r is the radius of the band obtained when a small variable piece of the curve of length ds is revolved about the line and L is the length of the curve.

FIGURE 11.2.10 The area of the ribbon-like band obtained by revolving a differential piece of a curve about a line is $dS = 2\pi(\text{radius})\, ds$, where ds is differential arc length.

If the curve is given in parametric form, we can evaluate the surface area by expressing r, ds, and the limits of integration in terms of the parameter and then evaluating the integral.

EXAMPLE 6

Set up an integral that gives the area of the surface obtained by revolving the half-ellipse

$$C : x = 3\cos t, \quad y = 2\sin t, \quad 0 \le t \le \pi,$$

about the x-axis.

SOLUTION   The curve is sketched in Figure 11.2.11. We have also sketched the band that is obtained when a typical piece of arc is revolved about the x-axis. The area of a typical band is

$$dS = 2\pi(\text{Radius})(\text{Slant height}) = 2\pi y\, ds.$$

We need to express y and the differential arc length ds in terms of the parameter t, and then use a definite integral to give us the limit of the sums of areas of the bands. We have $x = 3\cos t$ and $y = 2\sin t$, so

$$\frac{dx}{dt} = -3\sin t \text{ and } \frac{dy}{dt} = 2\cos t.$$

Then

$$ds = \sqrt{(-3\sin t)^2 + (2\cos t)^2}\, dt = \sqrt{9\sin^2 t + 4\cos^2 t}\, dt.$$

Since $y = 2\sin t$, we have

$$\text{Surface area} = \overbrace{\int_0^\pi}^{\text{Sum}} \overbrace{2\pi 2(\sin t)}^{2\pi r} \overbrace{\sqrt{9\sin^2 t + 4\cos^2 t}\, dt}^{ds}.$$

(This integral could be evaluated by writing $9\sin^2 t + 4\cos^2 t = 9 - 5\cos^2 t$ and then using the substitution $u = (\sqrt{5}/3)\cos t$. We would then need a trigonometric substitution or an integration formula. Also, the methods of Section 5.6 could be used to obtain an approximate value of the integral.) ∎

Change of Parameter

We have seen that a curve that is described geometrically or in terms of an equation in x and y can be parametrized in different ways. Let us now systematically study how to change the parametrization of a curve

$$C : x = f(t), \quad y = g(t), \quad a \le t \le b.$$

We assume that $u = \phi(t)$, where ϕ' exists and is positive for $a \le t \le b$, so ϕ has an inverse function ϕ^{-1} with

$$u = \phi(t) \text{ if and only if } t = \phi^{-1}(u).$$

See Figure 11.2.12. As u varies from $\phi(a)$ to $\phi(b)$, the variable $t = \phi^{-1}(u)$ assumes each value from a to b; these values of t are assumed exactly once and no other values of t are assumed for $\phi(a) \le u \le \phi(b)$. It follows that the curve

$$C_\phi : x = f(\phi^{-1}(u)), \quad y = g(\phi^{-1}(u)), \quad \phi(a) \le u \le \phi(b),$$

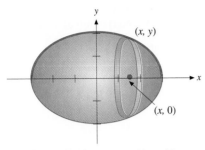

FIGURE 11.2.11   A typical band has radius $y = 2\sin t$.

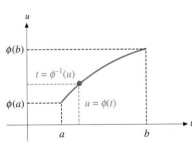

FIGURE 11.2.12   If ϕ' exists and is positive, then ϕ has an inverse function.

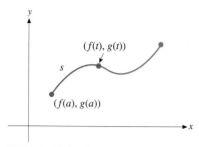

FIGURE 11.2.13 The variable $s = \phi(t)$ represents the length of the curve between $(f(a), g(a))$ and $(f(t), g(t))$.

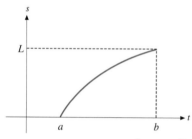

FIGURE 11.2.14 If $\left(\dfrac{dx}{dt}\right)^2 + \left(\dfrac{dy}{dt}\right)^2 \neq 0$, the arc length function $\phi(t)$ increases from zero at $t = a$ to the length of the curve at $t = b$.

has the same graph and the same direction as C. In this sense, we consider C and C_ϕ to be different parametrizations of the same curve.

Let us see how to use arc length to obtain a parametrization of a curve

$$C : x = f(t), \quad y = g(t), \quad a \leq t \leq b.$$

We assume that f and g have continuous derivatives and that

$$(x')^2 + (y')^2 \neq 0 \text{ on the interval } a \leq t \leq b.$$

We then define the **arc length function**

$$\phi(t) = \int_a^t \sqrt{(x'(\tau))^2 + (y'(\tau))^2}\, d\tau, \quad a \leq t \leq b,$$

where we have used τ for the variable of integration in order that the variable of integration not be confused with the upper limit of integration, t. The variable $s = \phi(t)$ represents the length of that portion of the curve between $(f(a), g(a))$ and $(f(t), g(t))$. See Figure 11.2.13. The Fundamental Theorem of Calculus tells us that ϕ is differentiable with

$$\phi' = \sqrt{(x')^2 + (y')^2} > 0.$$

See Figure 11.2.14. Note that $\phi(a) = 0$ and $\phi(b) = L$, where L is the total arc length of the curve. The function ϕ has an inverse function ϕ^{-1} with

$$s = \phi(t) \text{ if and only if } t = \phi^{-1}(s).$$

The curve

$$C_\phi : x = f(\phi^{-1}(s)), \quad y = g(\phi^{-1}(s)), \quad 0 \leq s \leq L,$$

is called the **parametrization of C in terms of arc length**. Parametrization in terms of arc length plays an important role in the vector analysis of curves, a topic we will cover later in this chapter, after we have introduced vectors.

In order to parametrize a curve in terms of arc length, we need to determine the arc length function and then find its inverse. This is theoretically possible for any curve as described above, but in practice we can carry out the details and obtain formulas in only a few cases.

EXAMPLE 7

Find the parametrization in terms of arc length of the curve

$$C : x = 4 - 4t, \quad y = 3t, \quad 0 \leq t \leq 1.$$

SOLUTION This curve is the line segment from $(4, 0)$ to $(0, 3)$. See Figure 11.2.15. We see that

$$x'(t) = -4 \text{ and } y'(t) = 3,$$

so the arc length function is

$$s = \int_0^t \sqrt{(x'(\tau))^2 + (y'(\tau))^2}\, d\tau = \int_0^t \sqrt{(-4)^2 + (3)^2}\, d\tau = 5t.$$

This gives the relation

$$s = 5t.$$

Solving this equation for t, we obtain t in terms of the arc length s,

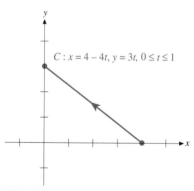

FIGURE 11.2.15 The parametrization of this line segment in terms of arc length is $C : x = 4 - 0.8s, y = 0.6s, 0 \leq s \leq 5$.

$$t = \frac{s}{5}.$$

Substitution into the given parametric equations of C gives

$$x = 4 - 4\left(\frac{s}{5}\right) = 4 - 0.8s \text{ and } y = 3\left(\frac{s}{5}\right) = 0.6s.$$

The total length of the curve is given by setting $t = 1$ in the equation $s = 5t$, so $L = 5$. The parametrization of C in terms of arc length is

$$C : x = 4 - 0.8s, \quad y = 0.6s, \quad 0 \le s \le 5. \quad \blacksquare$$

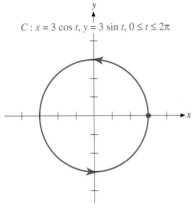

$C : x = 3 \cos t, y = 3 \sin t, 0 \le t \le 2\pi$

FIGURE 11.2.16 The parametrization of this circle in terms of arc length is $C : x = 3 \cos \frac{s}{3}, y = 3 \sin \frac{s}{3}, 0 \le s \le 6\pi$.

EXAMPLE 8

Find the parametrization in terms of arc length of the curve

$$C : x = 3 \cos t, \quad y = 3 \sin t, \quad 0 \le t \le 2\pi.$$

SOLUTION The curve is a circle with center $(0, 0)$ and radius 3. See Figure 11.2.16. We have

$$x'(t) = -3 \sin t \text{ and } y'(t) = 3 \cos t.$$

Expressing these derivatives in terms of the variable of integration τ, we have

$$s = \int_0^t \sqrt{(x'(\tau))^2 + (y'(\tau))^2} \, d\tau$$

$$= \int_0^t \sqrt{(-3 \sin \tau)^2 + (3 \cos \tau)^2} \, d\tau$$

$$= \int_0^t \sqrt{(3)^2(\sin^2 \tau + \cos^2 \tau)} \, d\tau$$

$$= \int_0^t 3 \, d\tau = 3t.$$

We now have

$$s = 3t.$$

Solving this equation for t, we obtain t in terms of the arc length s,

$$t = \frac{s}{3}.$$

When $t = 2\pi$, the equation $s = 3t$ gives the total arc length $L = 6\pi$. The parametrization of C in terms of arc length is

$$C : x = 3 \cos \frac{s}{3}, \quad y = 3 \sin \frac{s}{3}, \quad 0 \le s \le 6\pi. \quad \blacksquare$$

EXAMPLE 9

Find the parametrization in terms of arc length of the curve

$$C : x = 3t, \quad y = 2t^{3/2}, \quad 0 \le t \le 3.$$

SOLUTION We have

$$x'(t) = 3 \text{ and } y'(t) = 3t^{1/2},$$

so the arc length function is

$$s = \int_0^t \sqrt{(3)^2 + (3\tau^{1/2})^2}\, d\tau$$

$$= \int_0^t \sqrt{9 + 9\tau}\, d\tau$$

$$= \int_0^t 3(1 + \tau)^{1/2}\, d\tau$$

$$= 3\frac{(1+\tau)^{3/2}}{3/2}\Bigg]_0^t$$

$$= 2(1 + t)^{3/2} - 2.$$

We now have

$$s = 2(1 + t)^{3/2} - 2.$$

When $t = 3$, this equation gives the total arc length,

$$L = 2(1 + 3)^{3/2} - 2 = 14.$$

Solving the equation for t in terms of the arc length s, we have

$$s = 2(1 + t)^{3/2} - 2,$$
$$s + 2 = 2(1 + t)^{3/2},$$
$$(1 + t)^{3/2} = \frac{s + 2}{2},$$
$$1 + t = \left(\frac{s + 2}{2}\right)^{2/3},$$
$$t = \left(\frac{s + 2}{2}\right)^{2/3} - 1.$$

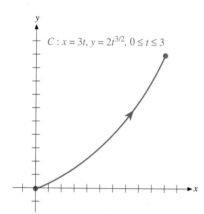

$C : x = 3t,\ y = 2t^{3/2},\ 0 \le t \le 3$

FIGURE 11.2.17 The arc length function of this curve is $s = 2(1 + t)^{3/2} - 2$. To parametrize in terms of arc length, we solve this equation for t and substitute into the given parametric equations.

The parametrization in terms of arc length is

$$C : x = 3\left[\left(\frac{s + 2}{2}\right)^{2/3} - 1\right], \quad y = 2\left[\left(\frac{s + 2}{2}\right)^{2/3} - 1\right]^{3/2}, \quad 0 \le s \le 14.$$

The graph is sketched in Figure 11.2.17. ■

EXERCISES 11.2

In each of Exercises 1–4, find the equation of the line tangent to the curve at the indicated point, and sketch the curve and the tangent line.

1. $C : x = t^2,\ y = t^3,\ -1 \le t \le 2;\ t = 1$

2. $C : x = t^3,\ y = t^2,\ -1 \le t \le 2;\ t = 1$

3. $C : x = \sin t,\ y = t,\ -\pi/2 \le t \le \pi/2;\ t = \pi/6$

4. $C : x = \cos t,\ y = t,\ 0 \le t \le \pi;\ t = \pi/4$

In Exercises 5–10, evaluate dy/dx and d^2y/dx^2 at the indicated point. Sketch an arc of the curve as it passes through that point. Indicate slope, concavity, and the direction of the curve as it passes through the point.

5. $C : x = t^2,\ y = t^3;\ t = 1$

6. $C : x = 1 + \cos t,\ y = \sin t;\ t = \pi/6$

7. $C : x = t^2 + t,\ y = t^2 + 2t - 1;\ t = 1$

8. $C : x = t - t^2, y = t^2 + t; t = 1$

9. $C : x = \ln t, y = t^2; t = 1$

10. $C : x = t \cos t, y = t \sin t; t = 3\pi/2$

Find the length of the curves in Exercises 11–20.

11. $C : x = 4t, y = 3 - 3t, 0 \le t \le 1$

12. $C : x = 12t, y = 5t, 0 \le t \le 1$

13. $C : x = t^3, y = t^2, 0 \le t \le 2$

14. $C : x = t^2/2, y = t^3/3, 0 \le t \le 1$

15. $C : x = \cos t, y = 1 + \sin t, 0 \le t \le 2\pi$

16. $C : x = 2 \cos t, y = 2 \sin t, 0 \le t \le \pi$

17. $C : x = t^2 \cos t, y = t^2 \sin t, 0 \le t \le 2\pi$

18. $C : x = e^{2t} \cos t, y = e^{2t} \sin t, 0 \le t \le 2\pi$

19. $C : x = 2 \ln t, y = t + (1/t), 1 \le t \le 4$

20. $C : x = 2e^t, y = (1/3)e^{3t} + e^{-t}, 0 \le t \le 1$

Find the length of the polar curves in Exercises 21–26.

21. $r = 4 \cos \theta$

22. $r = 1 + \cos \theta$. (*Hint:* $1 + \cos \theta = 2 \cos^2(\theta/2)$.)

23. $r = e^{\theta/2}, 0 \le \theta \le 2\pi$

24. $r = \sin^2(\theta/2)$

25. $r = \theta^2, 0 \le \theta \le 2\pi$

26. $r = \tan^2 \theta \sec \theta, 0 \le \theta \le \pi/6$

In Exercises 27–34, find the area of the surface obtained when the given parametric curve is revolved about the indicated line.

27. $C : x = 3t, y = 4t, 0 \le t \le 1$; about the y-axis

28. $C : x = rt, y = h - ht, 0 \le t \le 1$; about the y-axis

29. $C : x = t^2, y = t, 0 \le t \le 2$; about the x-axis

30. $C : x = t, y = t^3/3, 0 \le t \le 3$; about the x-axis

31. $C : x = 2 \cos t, y = 2 \sin t, 0 \le t \le \pi$; about the x-axis

32. $C : x = r \cos t, y = r \sin t, 0 \le t \le \pi$; about the x-axis

33. $C : x = 3 + \cos t, y = \sin t, 0 \le t \le 2\pi$; about the y-axis

34. $C : x = R + r \cos t, \ \ y = r \sin t, \ \ 0 \le t \le 2\pi$; about the y-axis; $R > r > 0$

Find the surface area of the surface of revolution obtained by revolving the polar curves given in Exercises 35–38 about the polar axis.

35. $r = \dfrac{2}{1 + \cos \theta}, 0 \le \theta \le \pi/2$

36. $r = e^{-\theta/2}, 0 \le \theta \le \pi$

37. $r^2 = \sin 2\theta$

38. $r = \sin^2(\theta/2)$ (*Hint:* $2 \sin(\theta/2) \cos(\theta/2) = \sin \theta$.)

Find the surface area of the surface of revolution obtained by revolving the polar curves given in Exercises 39–42 about the line $\theta = \pi/2$.

39. $r = 2 \cos \theta$

40. $r = 1 + \sin \theta$

41. $r = e^{-\theta/2}, -\pi/2 \le \theta \le \pi/2$

42. $r = r_0, -\pi/2 \le \theta \le \pi/2$

Find the parametrization in terms of arc length of the curves in Exercises 43–48.

43. $C : x = 4t, y = 3t, 0 \le t \le 1$

44. $C : x = at, y = b - bt, 0 \le t \le 1$

45. $C : x = 3 \cos 2t, y = 3 \sin 2t, 0 \le t \le \pi$

46. $C : x = r \cos \omega t, y = r \sin \omega t, 0 \le t \le 2\pi/\omega$

47. $C : x = 2t^3/3, y = t^2, 0 \le t \le 3$

48. $C : x = t^2/2, y = t^3/3, 0 \le t \le 1$

Find the parametrization in terms of arc length of the polar curves in Exercises 49–50.

49. $r = e^\theta, 0 \le \theta \le 2\pi$

50. $r = \theta^2, 0 \le \theta \le 2\pi$

Use Simpson's Rule with $n = 6$ or a computer program to approximate the integrals that give the length of the curves in Exercises 51–52.

51. $C : x = t^2 + 1, y = t^2 + t, 0 \le t \le 3$

52. $C : x = \sin t, y = \sin 2t, 0 \le t \le \pi$.

In Exercises 53–54, (a) set up a definite integral that gives the length of the polar curve. (b) Use Simpson's Rule with the indicated value of n or a computer program to find an approximate value of the length.

53. $r = 2 + \cos \theta, 0 \le \theta \le \pi; n = 6$

54. One loop of $r = \sin 2\theta; n = 4$

55. If $\phi(t)$ has a continuous, positive derivative for $a \le t \le b$, use a change of variables in an integral to show that the smooth curve

$$C : x = f(t), \quad y = g(t), \quad a \le t \le b,$$

and the reparametrization

$$C_\phi : x = f(\phi^{-1}(u)), \ y = g(\phi^{-1}(u)), \ \phi(a) \le u \le \phi(b),$$

have equal lengths.

11.3 VECTORS IN THE PLANE

Connections

Rectangular coordinate systems, 1.3, 10.7.
Work, 6.6.

Many problems involve magnitude and direction. For example, force has magnitude and is applied in a direction. Motion involves the direction and magnitude (speed) of the motion.

In this section we will define a mathematical system that embodies the concepts of magnitude and direction as well as the arithmetical operations we will need for applications.

Definitions and Basic Properties

DEFINITION

An ordered pair of real numbers (a, b) is called a **two-dimensional vector**. The numbers a and b are called **components** of the vector. Two vectors are **equal** if and only if corresponding components are equal. That is,

$$(a, b) = (c, d) \text{ means } a = c \text{ and } b = d.$$

The **sum** of two vectors is

$$(a, b) + (c, d) = (a + c, b + d),$$

and the **product** of a real number and a vector is

$$c(a, b) = (ca, cb).$$

Addition of two vectors and multiplying a vector by a real number involve the operations of addition and multiplication of real numbers, carried out component by component. The sum of two vectors is a vector, and the product of a real number and a vector is a vector.

Within the context of vectors, a real number is called a **scalar**. We will use boldface to distinguish vectors from scalars. For example, $c\mathbf{U}$ denotes the product of the scalar c and the vector $\mathbf{U}$. Another way to distinguish vectors is to draw an arrow over them, as in the expression $c\overrightarrow{U}$.

The **zero vector** is

$$\mathbf{0} = (0, 0).$$

The **negative** of the vector $\mathbf{U} = (a, b)$ is

$$-\mathbf{U} = -(a, b) = (-a, -b).$$

The *vector* $\mathbf{U} = (a, b)$ can be represented geometrically by an **arrow** from the origin to the *point* (a, b), or by any other arrow that has the same length and direction. See Figure 11.3.1. Note that the vector (a, b) determines the length and direction of corresponding arrows, but not their position. For some applications it is convenient to position an arrow with its initial point at a point other than the origin. All arrows that have the same length and direction are considered to represent the same vector. We will not necessarily distinguish between a vector and a representative arrow of the vector. Thus, we may speak of the position of a two-dimensional vector when we mean the position of a representative arrow.

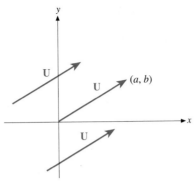

FIGURE 11.3.1 The vector (a, b) can be represented geometrically by an arrow from the origin to the point (a, b), or any other arrow that has the same length and direction.

Addition of vectors and multiplication of a vector by a scalar have geometric interpretations.

The sum of **U** and **V** can be determined geometrically by moving the initial point of **V** to the terminal point of **U**. The sum **U** + **V** is then the vector from the initial point of **U** to the terminal point of **V**. See Figure 11.3.2.

If $c > 0$, c**U** is a vector in the same direction as **U** with length c times the length of **U**. If $c < 0$, c**U** is in the direction opposite to **U** with length $|c|$ times the length of **U**. The vector $-$**V** $= (-1)$**V** is in the direction opposite to **V** with the same length as **V**. See Figure 11.3.3.

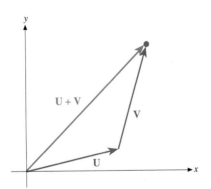

FIGURE 11.3.2 If the initial point of **V** is positioned at the terminal point of **U**, then **U** + **V** is the vector from the initial point of **U** to the terminal point of **V**.

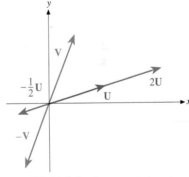

FIGURE 11.3.3 If $c > 0$, c**U** has the same direction as **U** with length c times the length of **U**. If $c < 0$, c**U** has direction opposite to **U** and length $|c|$ times the length of **U**.

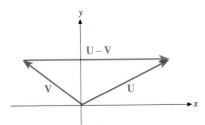

FIGURE 11.3.4 If **U** and **V** have the same initial point, then **U** − **V** is the vector from the terminial point of **V** to the terminal point of **U**.

The vector **U** − **V** can be determined geometrically. If the vectors **U** and **V** are drawn with the same initial point, then **U** − **V** is the vector from the terminal point of **V** to the terminal point of **U**. Be careful to draw the difference vector in the correct direction. See Figure 11.3.4.

Operations on vectors have the following properties, which correspond to properties of addition and multiplication of real numbers.

FIGURE 11.3.5 The vector **i** is represented by the arrow of length one in the direction of the positive *x*-axis. The vector **j** is represented by the arrow of length one in the direction of the positive *y*-axis.

$$U + (V + W) = (U + V) + W,$$
$$U + V = V + U,$$
$$U + 0 = 0 + U = U,$$
$$U + (-U) = (-U) + U = 0,$$
$$c(U + V) = cU + cV,$$
$$(c + d)U = cU + dU,$$
$$(cd)U = c(dU),$$
$$1U = U.$$

These properties are shared by many mathematical systems. They are characteristic properties of a **vector space**.

It is convenient to define two special two-dimensional vectors,

$$\mathbf{i} = (1, 0) \text{ and } \mathbf{j} = (0, 1).$$

See Figure 11.3.5.

The vector (a, b) can be expressed in terms of the vectors **i** and **j**. We have

$$(a, b) = a\mathbf{i} + b\mathbf{j}. \tag{1}$$

Proof

$$(a, b) = (a, 0) + (0, b)$$
$$= a(1, 0) + b(0, 1)$$
$$= a\mathbf{i} + b\mathbf{j}. \ \blacksquare$$

Addition and scalar multiplication of vectors can be expressed in terms of the **i** and **j** notation as follows:

$$(a\mathbf{i} + b\mathbf{j}) + (c\mathbf{i} + d\mathbf{j}) = (a + c)\mathbf{i} + (b + d)\mathbf{j} \tag{2}$$

and

$$c(a\mathbf{i} + b\mathbf{j}) = ca\mathbf{i} + cb\mathbf{j}. \tag{3}$$

This means we can perform the operations of vector addition and scalar multiplication by using the usual rules of algebra and combining "like" terms. The vectors **i** and **j** act as place holders for the components.

EXAMPLE 1

(a) $3(2\mathbf{i} - \mathbf{j}) = 6\mathbf{i} - 3\mathbf{j}.$
(b) $(2\mathbf{i} + 3\mathbf{j}) + (\mathbf{i} - 2\mathbf{j}) = 2\mathbf{i} + 3\mathbf{j} + \mathbf{i} - 2\mathbf{j} = 3\mathbf{i} + \mathbf{j}.$
(c) $5(\mathbf{i} + 2\mathbf{j}) - 2(2\mathbf{i} + 7\mathbf{j}) = 5\mathbf{i} + 10\mathbf{j} - 4\mathbf{i} - 14\mathbf{j} = \mathbf{i} - 4\mathbf{j}.$ $\blacksquare$

The length of a two-dimensional vector is defined to agree with the length of any of its representative arrows.

DEFINITION

The **length** or **magnitude** of a two-dimensional vector is

$$|a\mathbf{i} + b\mathbf{j}| = \sqrt{a^2 + b^2}.$$

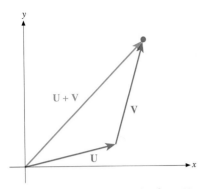

FIGURE 11.3.6 The length of $U + V$ is less than or equal to the sum of the lengths of U and V.

The length of vectors has the following two properties:

$$|cU| = |c|\,|U|,$$
$$|U + V| \leq |U| + |V|. \qquad \text{(The Triangle Inequality)}$$

These properties correspond to properties of the absolute value of real numbers. (Note that $|c|$ is the absolute value of the real number c.) The first property can be verified directly from the definitions. The second property is evident from the geometric interpretation of the sum of two vectors. That is, the length of one side of a triangle is less than the sum of the lengths of the other two sides. See Figure 11.3.6. Equality occurs if and only if U and V are in the same direction.

EXAMPLE 2

$$|3i - 4j| = \sqrt{3^2 + (-4)^2} = 5.$$

See Figure 11.3.7. ∎

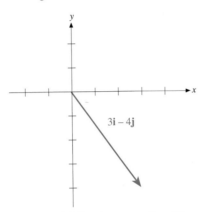

FIGURE 11.3.7 The vector $3i - 4j$ has length 5.

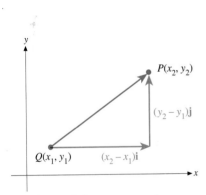

FIGURE 11.3.8 The vector from $Q(x_1, y_1)$ to $P(x_2, y_2)$ is $QP = (x_2 - x_1)i + (y_2 - y_1)j$.

Although arrows are convenient for describing many geometric problems, it is more convenient for the solution of the problems to express the arrows in the analytic form of vectors. We can use the following two observations to do this.

If $Q(x_1, y_1)$ and $P(x_2, y_2)$ are points in a coordinate plane, the vector from Q to P is

$$QP = (x_2 - x_1)i + (y_2 - y_1)j.$$

See Figure 11.3.8.

If $r > 0$, the vector

$$r \cos\theta i + r \sin\theta j$$

has length r and direction at an angle θ with the positive x-axis.

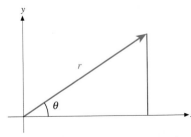

FIGURE 11.3.9 The vector $r \cos\theta i + r \sin\theta j$ has length r and direction θ.

See Figure 11.3.9.

Vector Representation of Force

An important application of vectors is to problems that involve force. That is, a force has magnitude and direction and can be associated with a vector. If two or more forces are applied simultaneously to an object, the **resultant force** is the vector sum of the corresponding vectors.

The zero vector is the resultant of two forces that are of equal magnitude and opposite direction. The effect of several forces can be counteracted by a single force with magnitude that of the resultant force and direction opposite to the resultant force.

EXAMPLE 3

A person with a broken femur is in **Russell's traction**, as indicated in Figure 11.3.10a. The magnitude of each of the three forces acting on the leg due to the pulley system is 5 lb, the same as the weight at the end of the rope. Find the magnitude and direction of the resultant force on the femur due to the traction.

SOLUTION   The vertical force on the sling at the knee can be represented by the vector $5\mathbf{j}$. Each of the two horizontal forces at the foot can be represented by the vector $5\mathbf{i}$. The resultant force is then

$$(5\mathbf{j}) + (5\mathbf{i}) + (5\mathbf{i}) = 10\mathbf{i} + 5\mathbf{j}.$$

The resultant vector has length

$$\sqrt{10^2 + 5^2} = \sqrt{125} = 5\sqrt{5}.$$

The resultant force has magnitude $5\sqrt{5}$ lb.

FIGURE 11.3.10a   Vectors can be used to determine the force on the leg due to the pulley system.

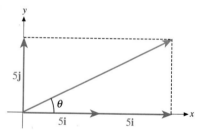

FIGURE 11.3.10b   The resultant force has magnitude $5\sqrt{5}$ lb and direction $\theta = \tan^{-1}(5/10) \approx 26.6°$.

From Figure 11.3.10b we see that the angle that the resultant vector makes with the horizontal is $\theta = \tan^{-1}(5/10) \approx 26.6°$. This is the direction of the resultant force. ∎

Dot Product and Angle Between Vectors

We can use the triangle formed by the vectors $\mathbf{U}$, $\mathbf{V}$, and $\mathbf{U} - \mathbf{V}$ to express the angle between $\mathbf{U}$ and $\mathbf{V}$ in terms of the components of $\mathbf{U}$ and $\mathbf{V}$. See Figure 11.3.11. If the angle θ between two nonzero vectors $\mathbf{U} = a\mathbf{i} + b\mathbf{j}$ and $\mathbf{V} = c\mathbf{i} + d\mathbf{j}$ satisfies $0 < \theta < \pi$, the Law of Cosines (Section 1.6) gives

$$|\mathbf{U} - \mathbf{V}|^2 = |\mathbf{U}|^2 + |\mathbf{V}|^2 - 2|\mathbf{U}||\mathbf{V}|\cos\theta.$$

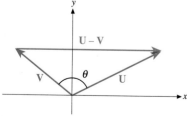

FIGURE 11.3.11   The angle θ between the vectors $\mathbf{U}$ and $\mathbf{V}$ can be determined by applying the Law of Cosines to the triangle formed by the vectors $\mathbf{U}$, $\mathbf{V}$, and $\mathbf{U} - \mathbf{V}$.

This equality is also true in case either $\mathbf{U} = \mathbf{0}$, $\mathbf{V} = \mathbf{0}$, $\theta = 0$, or $\theta = \pi$. Expressing the squares of the lengths of the vectors in the above equality in terms of their components and then simplifying, we have

$$(a - c)^2 + (b - d)^2 = a^2 + b^2 + c^2 + d^2 - 2|\mathbf{U}||\mathbf{V}|\cos\theta,$$
$$a^2 - 2ac + c^2 + b^2 - 2bd + d^2 = a^2 + b^2 + c^2 + d^2 - 2|\mathbf{U}||\mathbf{V}|\cos\theta,$$
$$2|\mathbf{U}||\mathbf{V}|\cos\theta = 2ac + 2bd,$$
$$|\mathbf{U}||\mathbf{V}|\cos\theta = ac + bd. \tag{4}$$

We see that the formula for the angle between the vectors $a\mathbf{i} + b\mathbf{j}$ and $c\mathbf{i} + d\mathbf{j}$ involves the expression $ac + bd$. This suggests the following definition.

DEFINITION

The **dot product** of two-dimensional vectors is

$$(a\mathbf{i} + b\mathbf{j}) \cdot (c\mathbf{i} + d\mathbf{j}) = ac + bd.$$

The dot product is the sum of the products of corresponding components. The dot product of two vectors is a *scalar.*

We can express Equation 4 in terms of the dot product as

$$|\mathbf{U}||\mathbf{V}|\cos\theta = \mathbf{U} \cdot \mathbf{V}.$$

This gives the following.

The angle between two nonzero vectors $\mathbf{U}$ and $\mathbf{V}$ is

$$\theta = \cos^{-1}\left(\frac{\mathbf{U} \cdot \mathbf{V}}{|\mathbf{U}||\mathbf{V}|}\right).$$

See Figure 11.3.11. Two vectors are said to be **orthogonal** if their dot product is zero. Two nonzero vectors in the plane are orthogonal if and only if they are perpendicular. The zero vector is orthogonal to every vector.

Two nonzero vectors are **parallel** if one of the vectors is a scalar multiple of the other. The vectors may have the same direction ($\theta = 0$) or opposite directions ($\theta = \pi$).

Since $|\cos\theta| \le 1$, $\mathbf{U} \cdot \mathbf{V} = |\mathbf{U}||\mathbf{V}|\cos\theta$ implies

$$|\mathbf{U} \cdot \mathbf{V}| \le |\mathbf{U}||\mathbf{V}|. \quad \text{(The Cauchy-Schwarz Inequality)}$$

EXAMPLE 4

Sketch the vectors and find the angle between the vectors.

(a) $12\mathbf{i} - 3\mathbf{j}$, $2\mathbf{i} + 8\mathbf{j}$,
(b) $\mathbf{i} - \mathbf{j}$, $\mathbf{j}$,
(c) $2\mathbf{i} + \mathbf{j}$, $-4\mathbf{i} - 2\mathbf{j}$.

SOLUTION

(a) The vectors are sketched in Figure 11.3.12.

$$(12\mathbf{i} - 3\mathbf{j}) \cdot (2\mathbf{i} + 8\mathbf{j}) = (12)(2) + (-3)(8) = 0.$$

Hence, $\cos\theta = 0$, so $\theta = \pi/2$. The vectors are orthogonal.

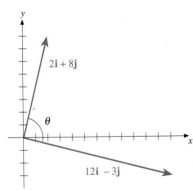

FIGURE 11.3.12 $(12\mathbf{i} - 3\mathbf{j}) \cdot (2\mathbf{i} + 8\mathbf{j}) = 0$, so the vectors are orthogonal.

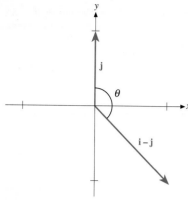

FIGURE 11.3.13 $\cos\theta = -1/\sqrt{2}$, so $\theta = 3\pi/4$.

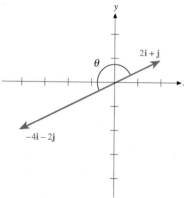

FIGURE 11.3.14 $\cos\theta = -1$, so $\theta = \pi$.

(b) The vectors are sketched in Figure 11.3.13.

$$(\mathbf{i} - \mathbf{j}) \cdot \mathbf{j} = (\mathbf{i} - \mathbf{j}) \cdot (0\mathbf{i} + \mathbf{j}) = (1)(0) + (-1)(1) = -1.$$

We have $|\mathbf{i} - \mathbf{j}| = \sqrt{2}$ and $|\mathbf{j}| = 1$.   Hence,

$$\cos\theta = \frac{(\mathbf{i} - \mathbf{j}) \cdot \mathbf{j}}{|\mathbf{i} - \mathbf{j}||\mathbf{j}|} = \frac{-1}{(\sqrt{2})(1)}, \text{ so } \theta = \frac{3\pi}{4}.$$

(c) The vectors are sketched in Figure 11.3.14.

$$(2\mathbf{i} + \mathbf{j}) \cdot (-4\mathbf{i} - 2\mathbf{j}) = (2)(-4) + (1)(-2) = -10.$$

We have $|2\mathbf{i} + \mathbf{j}| = \sqrt{5}$ and $|-4\mathbf{i} - 2\mathbf{j}| = \sqrt{20}$. Hence,

$$\cos\theta = \frac{(2\mathbf{i} + \mathbf{j}) \cdot (-4\mathbf{i} - 2\mathbf{j})}{|2\mathbf{i} + \mathbf{j}||-4\mathbf{i} - 2\mathbf{j}|} = \frac{-10}{(\sqrt{5})(\sqrt{20})} = -1, \text{ so } \theta = \pi.$$

The vectors are parallel and in opposite directions.   ∎

The dot product has the following properties.

$$(\mathbf{U} + \mathbf{V}) \cdot \mathbf{W} = \mathbf{U} \cdot \mathbf{W} + \mathbf{V} \cdot \mathbf{W},$$
$$(c\mathbf{U}) \cdot \mathbf{V} = c(\mathbf{U} \cdot \mathbf{V}) = \mathbf{U} \cdot (c\mathbf{V}),$$
$$\mathbf{U} \cdot \mathbf{V} = \mathbf{V} \cdot \mathbf{U},$$
$$\mathbf{U} \cdot \mathbf{U} > 0 \text{ unless } \mathbf{U} = \mathbf{0}.$$

These properties correspond to those of a product of real numbers. We have

$$(a\mathbf{i} + b\mathbf{j}) \cdot (a\mathbf{i} + b\mathbf{j}) = a^2 + b^2 = |a\mathbf{i} + b\mathbf{j}|^2.$$

That is,

$$\mathbf{U} \cdot \mathbf{U} = |\mathbf{U}|^2. \tag{5}$$

It follows from the definition that

$$\mathbf{i} \cdot \mathbf{j} = (1\mathbf{i} + 0\mathbf{j}) \cdot (0\mathbf{i} + 1\mathbf{j}) = (1)(0) + (0)(1) = 0.$$

In the same way, we can verify that

$$\mathbf{i} \cdot \mathbf{j} = \mathbf{j} \cdot \mathbf{i} = 0 \text{ and } \mathbf{i} \cdot \mathbf{i} = \mathbf{j} \cdot \mathbf{j} = 1. \tag{6}$$

We can treat each vector factor $a\mathbf{i} + b\mathbf{j}$ in a dot product as a sum and use the properties of the dot product to remove parentheses. We can then use the formulas in (6) to evaluate the dot product.

EXAMPLE 5

Evaluate

(a) $(\mathbf{i} + 2\mathbf{j}) \cdot (-3\mathbf{j})$,
(b) $(2\mathbf{i} - \mathbf{j}) \cdot (\mathbf{i} + 2\mathbf{j})$.

SOLUTION   From the definition of dot product, we have

(a) $(\mathbf{i} + 2\mathbf{j}) \cdot (-3\mathbf{j}) = (1\mathbf{i} + 2\mathbf{j}) \cdot (0\mathbf{i} - 3\mathbf{j}) = (1)(0) + (2)(-3) = -6$ and
(b) $(2\mathbf{i} - \mathbf{j}) \cdot (\mathbf{i} + 2\mathbf{j}) = (2)(1) + (-1)(2) = 0.$

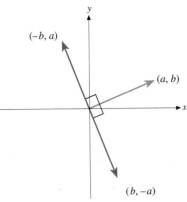

FIGURE 11.3.15 Both of the vectors $b\mathbf{i} - a\mathbf{j}$ and $-b\mathbf{i} + a\mathbf{j}$ are orthogonal to $a\mathbf{i} + b\mathbf{j}$.

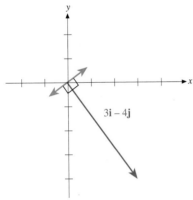

FIGURE 11.3.16 There are two unit vectors that are orthogonal to $3\mathbf{i} - 4\mathbf{j}$.

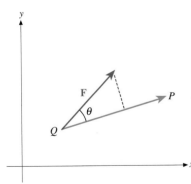

FIGURE 11.3.17 The work associated with a force $\mathbf{F}$ that is applied as an object moves in a straight line from a point Q to a point P is $W = \mathbf{F} \cdot \mathbf{QP} = |\mathbf{F}||\mathbf{QP}| \cos\theta$.

Considering these expressions as dot products of sums and using (6), we have

(a) $(\mathbf{i} + 2\mathbf{j}) \cdot (-3\mathbf{j}) = -3\mathbf{i} \cdot \mathbf{j} - 6\mathbf{j} \cdot \mathbf{j} = -3(0) - 6(1) = -6$ and
(b) $(2\mathbf{i} - \mathbf{j}) \cdot (\mathbf{i} + 2\mathbf{j}) = 2\mathbf{i} \cdot \mathbf{i} + 4\mathbf{i} \cdot \mathbf{j} - \mathbf{j} \cdot \mathbf{i} - 2\mathbf{j} \cdot \mathbf{j} = 2 + 0 - 0 - 2 = 0.$ ∎

The vector $a\mathbf{i} + b\mathbf{j}$ is called a **unit vector** if it has length one. That is, $a\mathbf{i} + b\mathbf{j}$ is a unit vector if $|a\mathbf{i} + b\mathbf{j}| = \sqrt{a^2 + b^2} = 1$.

If $\mathbf{V} \neq \mathbf{0}$, $\mathbf{V}/|\mathbf{V}|$ is the unit vector in the direction of $\mathbf{V}$.

In some problems it is desired to find a vector that is orthogonal to a given vector $a\mathbf{i} + b\mathbf{j}$. We have

$$(a\mathbf{i} + b\mathbf{j}) \cdot (b\mathbf{i} - a\mathbf{j}) = (a)(b) + (b)(-a) = 0,$$

so the vectors $a\mathbf{i} + b\mathbf{j}$ and $b\mathbf{i} - a\mathbf{j}$ are orthogonal. Any scalar multiple of $b\mathbf{i} - a\mathbf{j}$ is also orthogonal to $a\mathbf{i} + b\mathbf{j}$. In particular, the vector $-b\mathbf{i} + a\mathbf{j}$ is also orthogonal to $a\mathbf{i} + b\mathbf{j}$. See Figure 11.3.15.

EXAMPLE 6
Find two unit vectors that are orthogonal to $3\mathbf{i} - 4\mathbf{j}$.

SOLUTION Interchanging components and changing the sign of one component, we obtain

$$4\mathbf{i} + 3\mathbf{j} \text{ and } -4\mathbf{i} - 3\mathbf{j}.$$

Both of these vectors are orthogonal to $3\mathbf{i} - 4\mathbf{j}$. Both vectors have length $\sqrt{4^2 + 3^2} = 5$. Unit vectors in the direction of each of the two vectors are obtained by dividing the vectors by their length. This gives the vectors

$$0.8\mathbf{i} + 0.6\mathbf{j} \text{ and } -0.8\mathbf{i} - 0.6\mathbf{j}.$$

These two unit vectors are orthogonal to $3\mathbf{i} - 4\mathbf{j}$. See Figure 11.3.16. ∎

Work

DEFINITION

The **work** associated with a constant force $\mathbf{F}$ that is applied to an object as the object moves in a straight line from a point Q to a point P is

$$W = \mathbf{F} \cdot \mathbf{QP}.$$

We have

$$W = \mathbf{F} \cdot \mathbf{QP} = |\mathbf{F}||\mathbf{QP}| \cos\theta,$$

where θ is the angle between the force and the direction of motion. See Figure 11.3.17. If the force is parallel to the direction of motion, then $\cos\theta = \pm 1$ and the work is plus or minus the product of the magnitude of the force and the distance moved. This agrees with the definition of work associated with a force that acts in the direction of motion, as defined in Section 6.6.

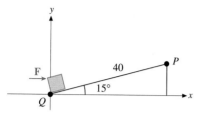

FIGURE 11.3.18 The horizontal force can be represented by the vector 30i. The motion is represented by the vector $\mathbf{QP} = (40\cos 15°)\mathbf{i} + (40\sin 15°)\mathbf{j}$.

EXAMPLE 7

A horizontal force of 30 lb is applied to a box as the box moves a distance of 40 ft along an incline of $15°$. Find the work associated with the force.

SOLUTION   A sketch is given in Figure 11.3.18. We can represent the force by the vector $\mathbf{F} = 30\mathbf{i}$. The motion is represented by the vector

$$\mathbf{QP} = (40\cos 15°)\mathbf{i} + (40\sin 15°)\mathbf{j}.$$

The work is then

$$
\begin{aligned}
W = \mathbf{F} \cdot \mathbf{QP} &= (30\mathbf{i}) \cdot [(40\cos 15°)\mathbf{i} + (40\sin 15°)\mathbf{i}] \\
&= (30)(40\cos 15°) + (0)(40\sin 15°) \\
&= 1200\cos 15° \approx 1159 \text{ ft-lb.} \quad \blacksquare
\end{aligned}
$$

Projection of Vectors

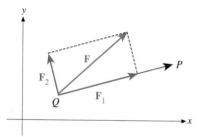

FIGURE 11.3.19 The work associated with the force vector **F** as an object moves from Q to P is the same as that associated with $\mathbf{F}_1$. The work associated with the force $\mathbf{F}_2$ that is orthogonal to the direction of motion is zero.

Suppose that a force **F** is applied as an object moves in a straight line from a point Q to a point P. Further, suppose that we can write $\mathbf{F} = \mathbf{F}_1 + \mathbf{F}_2$, where $\mathbf{F}_1$ is parallel to $\mathbf{QP}$ and $\mathbf{F}_2$ is orthogonal to $\mathbf{QP}$. See Figure 11.3.19. The work associated with **F** is then

$$W = \mathbf{F} \cdot \mathbf{QP} = (\mathbf{F}_1 + \mathbf{F}_2) \cdot \mathbf{QP} = \mathbf{F}_1 \cdot \mathbf{QP} + \mathbf{F}_2 \cdot \mathbf{PQ} = \mathbf{F}_1 \cdot \mathbf{QP},$$

since $\mathbf{F}_2 \cdot \mathbf{QP} = 0$. It follows that work associated with **F** is the same as the work associated with $\mathbf{F}_1$, which is in the direction of motion. Zero work is associated with $\mathbf{F}_2$, which is orthogonal to the direction of motion. This leads us to consider the following problem.

Given a vector **U** *and a nonzero vector* **V**, *can we find vectors* $\mathbf{U}_1$ *and* $\mathbf{U}_2$ *such that* $\mathbf{U} = \mathbf{U}_1 + \mathbf{U}_2$, $\mathbf{U}_1$ *is parallel to* **V**, *and* $\mathbf{U}_2$ *is orthogonal to* **V**?

To solve this problem, we note that $\mathbf{U}_1$ parallel to **V** means $\mathbf{U}_1 = c\mathbf{V}$ for some scalar c; $\mathbf{U}_2$ orthogonal to **V** means $\mathbf{U}_2 \cdot \mathbf{V} = 0$. If

$$\mathbf{U} = \mathbf{U}_1 + \mathbf{U}_2, \quad \mathbf{U}_1 = c\mathbf{V}, \text{ and } \mathbf{U}_2 \cdot \mathbf{V} = 0,$$

then

$$
\begin{aligned}
\mathbf{U} \cdot \mathbf{V} &= (\mathbf{U}_1 + \mathbf{U}_2) \cdot \mathbf{V} \\
&= (c\mathbf{V} + \mathbf{U}_2) \cdot \mathbf{V} \\
&= (c\mathbf{V}) \cdot \mathbf{V} + \mathbf{U}_2 \cdot \mathbf{V} \\
&= c(\mathbf{V} \cdot \mathbf{V}) + 0.
\end{aligned}
$$

It follows that

$$c = \frac{\mathbf{U} \cdot \mathbf{V}}{\mathbf{V} \cdot \mathbf{V}}.$$

If we set

$$\mathbf{U}_1 = \frac{\mathbf{U} \cdot \mathbf{V}}{\mathbf{V} \cdot \mathbf{V}}\mathbf{V} \text{ and } \mathbf{U}_2 = \mathbf{U} - \mathbf{U}_1 = \mathbf{U} - \frac{\mathbf{U} \cdot \mathbf{V}}{\mathbf{V} \cdot \mathbf{V}}\mathbf{V},$$

then $\mathbf{U} = \mathbf{U}_1 + \mathbf{U}_2$, $\mathbf{U}_1$ is parallel to **V**, and

$$\mathbf{U}_2 \cdot \mathbf{V} = \left(\mathbf{U} - \frac{\mathbf{U} \cdot \mathbf{V}}{\mathbf{V} \cdot \mathbf{V}}\mathbf{V}\right) \cdot \mathbf{V} = \mathbf{U} \cdot \mathbf{V} - \frac{\mathbf{U} \cdot \mathbf{V}}{\mathbf{V} \cdot \mathbf{V}}\mathbf{V} \cdot \mathbf{V} = 0,$$

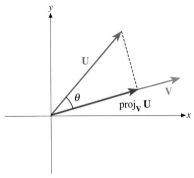

FIGURE 11.3.20 The vector $\text{proj}_V\mathbf{U}$ is in the direction of **V** if the angle between **U** and **V** satisfies $0 \leq \theta < \pi/2$.

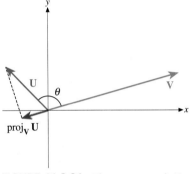

FIGURE 11.3.21 The vector $\text{proj}_V\mathbf{U}$ is in the direction opposite to that of **V** if the angle between **U** and **V** satisfies $\pi/2 < \theta \leq \pi$.

so $\mathbf{U}_2$ is orthogonal to **V**. The vectors $\mathbf{U}_1$ and $\mathbf{U}_2$ then solve our problem.

The vector

$$\text{proj}_V\mathbf{U} = \frac{\mathbf{U} \cdot \mathbf{V}}{\mathbf{V} \cdot \mathbf{V}}\mathbf{V}$$

is called the **projection of U in the direction of V**. See Figures 11.3.20 and 11.3.21. We have

$$(\text{proj}_V\mathbf{U}) \cdot \mathbf{V} = \left(\frac{\mathbf{U} \cdot \mathbf{V}}{\mathbf{V} \cdot \mathbf{V}}\mathbf{V}\right) \cdot \mathbf{V} = \frac{\mathbf{U} \cdot \mathbf{V}}{\mathbf{V} \cdot \mathbf{V}}(\mathbf{V} \cdot \mathbf{V}) = \mathbf{U} \cdot \mathbf{V}. \tag{7}$$

Also,

$$\mathbf{U} - \text{proj}_V\mathbf{U} \; \textit{is orthogonal to } \mathbf{V}. \tag{8}$$

EXAMPLE 8

Find the projection of $\mathbf{U} = \mathbf{i} + 2\mathbf{j}$ in the direction of $\mathbf{V} = 3\mathbf{i} - 4\mathbf{j}$. Find vectors $\mathbf{U}_1$ and $\mathbf{U}_2$ such that $\mathbf{U} = \mathbf{U}_1 + \mathbf{U}_2$, $\mathbf{U}_1$ is parallel to **V**, and $\mathbf{U}_2$ is orthogonal to **V**.

SOLUTION We have

$$\mathbf{U} \cdot \mathbf{V} = (\mathbf{i} + 2\mathbf{j}) \cdot (3\mathbf{i} - 4\mathbf{j}) = (1)(3) + (2)(-4) = 3 - 8 = -5 \text{ and}$$
$$\mathbf{V} \cdot \mathbf{V} = (3\mathbf{i} - 4\mathbf{j}) \cdot (3\mathbf{i} - 4\mathbf{j}) = (3)(3) + (-4)(-4) = 9 + 16 = 25,$$

so

$$\text{proj}_V\mathbf{U} = \frac{\mathbf{U} \cdot \mathbf{V}}{\mathbf{V} \cdot \mathbf{V}}\mathbf{V} = \frac{-5}{25}(3\mathbf{i} - 4\mathbf{j}) = -0.6\mathbf{i} + 0.8\mathbf{j}.$$

The vector $\mathbf{U}_1 = \text{proj}_V\mathbf{U} = -0.6\mathbf{i} + 0.8\mathbf{j}$ is parallel to **V**, $\mathbf{U}_2 = \mathbf{U} - \text{proj}_V\mathbf{U} = (\mathbf{i} + 2\mathbf{j}) - (-0.6\mathbf{i} + 0.8\mathbf{j}) = 1.6\mathbf{i} + 1.2\mathbf{j}$ is orthogonal to **V**, and $\mathbf{U} = \mathbf{U}_1 + \mathbf{U}_2$. See Figure 11.3.22. ∎

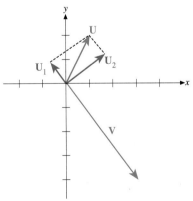

FIGURE 11.3.22 $\mathbf{U} = \mathbf{U}_1 + \mathbf{U}_2$, where $\mathbf{U}_1$ is parallel to **V** and $\mathbf{U}_2$ is orthogonal to **V**.

EXERCISES 11.3

Evaluate the expressions in Exercises 1–12.

1. $-(2\mathbf{i} - 3\mathbf{j})$

2. $3(3\mathbf{i} - 2\mathbf{j})$

3. $2(\mathbf{i} + 2\mathbf{j}) - 3(2\mathbf{i} - 3\mathbf{j})$

4. $4(3\mathbf{i} + \mathbf{j}) - 5(2\mathbf{i} + 2\mathbf{j})$

5. $|12\mathbf{i} - 5\mathbf{j}|$

6. $|4\mathbf{i} - 3\mathbf{j}|$

7. $|-2\mathbf{i} + 3\mathbf{j}|$

8. $|4\mathbf{i} - \mathbf{j}|$

9. $(2\mathbf{i} + 3\mathbf{j}) \cdot (4\mathbf{i})$

10. $(4\mathbf{i} - 3\mathbf{j}) \cdot (2\mathbf{i} + 2\mathbf{j})$

11. $(5\mathbf{i} - 3\mathbf{j}) \cdot (2\mathbf{i} + 3\mathbf{j})$

12. $(4\mathbf{i} - 3\mathbf{j}) \cdot (4\mathbf{i} - 3\mathbf{j})$

Evaluate $\mathbf{U} + \mathbf{V}$, $\mathbf{U} - \mathbf{V}$, $2\mathbf{U}$ *and* $-\mathbf{V}$ *for the vectors given in Exercises 13–16. Sketch each vector.*

13. $\mathbf{U} = \mathbf{i} + \mathbf{j}$, $\mathbf{V} = \mathbf{i} - \mathbf{j}$

14. $\mathbf{U} = \mathbf{i} + 3\mathbf{i}$, $\mathbf{V} = 2\mathbf{i} - \mathbf{j}$

15. $\mathbf{U} = 2\mathbf{i} - 3\mathbf{j}$, $\mathbf{V} = \mathbf{i} + 2\mathbf{j}$

16. $\mathbf{U} = 3\mathbf{i} - 2\mathbf{j}$, $\mathbf{V} = 2\mathbf{i} + 3\mathbf{j}$

Find the angle between the vectors in Exercises 17–24.

17. $3\mathbf{i} + 4\mathbf{j}$, $4\mathbf{i} - 3\mathbf{j}$

18. $\mathbf{i} + \mathbf{j}$, $\mathbf{i} - \mathbf{j}$

19. $2\mathbf{i} - \mathbf{j}$, $4\mathbf{i} - 2\mathbf{j}$

20. $\mathbf{i} - 2\mathbf{j}$, $-\mathbf{i} + 2\mathbf{j}$

21. $\sqrt{3}\mathbf{i} + \mathbf{j}$, $-\sqrt{3}\mathbf{i} - \mathbf{j}$

22. $-\mathbf{i} + \mathbf{j}$, $-\mathbf{i} - \mathbf{j}$

23. $\mathbf{i} + 2\mathbf{j}$, $\mathbf{i} + 3\mathbf{j}$

24. $2\mathbf{i} - 3\mathbf{j}$, $\mathbf{i} + 2\mathbf{j}$

Find two unit vectors that are orthogonal to the vectors given in Exercises 25–28.

25. $5\mathbf{i} - 12\mathbf{j}$

26. $2\mathbf{i} + \mathbf{j}$

27. $3\mathbf{i}$

28. $4\mathbf{i} - 3\mathbf{j}$

In Exercises 29–32, find the projection of $\mathbf{U}$ *in the direction of* $\mathbf{V}$. *Find vectors* $\mathbf{U}_1$ *and* $\mathbf{U}_2$ *such that* $\mathbf{U} = \mathbf{U}_1 + \mathbf{U}_2$, $\mathbf{U}_1$ *is parallel to* $\mathbf{V}$, *and* $\mathbf{U}_2$ *is orthogonal to* $\mathbf{V}$.

29. $\mathbf{U} = 2\mathbf{i} - \mathbf{j}$, $\mathbf{V} = 3\mathbf{i} + 4\mathbf{j}$

30. $\mathbf{U} = -\mathbf{i} + 2\mathbf{j}$, $\mathbf{V} = 12\mathbf{i} + 5\mathbf{j}$

31. $\mathbf{U} = 3\mathbf{i} + 2\mathbf{j}$, $\mathbf{V} = -\mathbf{i} + \mathbf{j}$

32. $\mathbf{U} = 2\mathbf{i} + 3\mathbf{j}$, $\mathbf{V} = 3\mathbf{i} - 2\mathbf{j}$

Find the resultant of the forces given in Exercises 33–34.

33. $\mathbf{F}_1 = -2\mathbf{j}$, $\mathbf{F}_2 = \mathbf{i} + 3\mathbf{j}$

34. $\mathbf{F}_1 = -\mathbf{i} - 2\mathbf{j}$, $\mathbf{F}_2 = -\mathbf{i} - \mathbf{j}$

Find a single force that counteracts the resultant of the forces given in each of Exercises 35–36.

35. $\mathbf{F}_1 = 2\mathbf{i}$, $\mathbf{F}_2 = \mathbf{i} + 3\mathbf{j}$

36. $\mathbf{F}_1 = 5\mathbf{i} - 4\mathbf{j}$, $\mathbf{F}_2 = 3\mathbf{i} + 4\mathbf{j}$

Find the projection of the force vector $\mathbf{F}$ *in the direction of* $\mathbf{V}$ *in Exercises 37–38. Write* $\mathbf{F}$ *as a sum of a force in the direction of* $\mathbf{V}$ *and a force that is orthogonal to* $\mathbf{V}$.

37. $\mathbf{F} = \mathbf{i} + \mathbf{j}$, $\mathbf{V} = -\mathbf{i} + 2\mathbf{j}$

38. $\mathbf{F} = 4\mathbf{i}$, $\mathbf{V} = \mathbf{i} + \mathbf{j}$

39. A force of 18 lb is applied at an angle of $30°$ from the horizontal to pull a wagon 80 ft along a flat road. Find the work associated with the force.

40. A boat is pulled 50 ft along a $20°$ incline by two men who are pulling with a force of 60 lb each on a rope that is parallel to the incline, while two women are lifting the boat with a vertical force of 45 lb each. Find the total work done.

41. A person is in **Russell's traction**, as indicated in the figure. The magnitude of each of the three forces along the rope is 5 lb, the same as the weight. Find the magnitude and direction of the resultant force on the femur.

42. If an object is moving along a flat surface, the force due to **kinetic friction** $\mathbf{F}_k$ acts opposite to the direction of motion and has magnitude that is proportional to the magnitude of the force on the object in the direction orthogonal to the level surface. That is, $|\mathbf{F}_k| = \mu_k|\mathbf{N}|$. See part (a) of the figure below. The number μ_k is called the **coefficient of kinetic friction**. Find the work required to overcome kinetic friction if a horizontal force of 40 lb is applied to a 50-lb box as the box moves 10 ft up an incline of $12°$, if the coefficient of kinetic friction is 0.1. See part (b) of the figure below. What is the work required to overcome kinetic friction if the 40-lb force is applied parallel to the incline? See part (c) of the figure below.

(a)

(b)

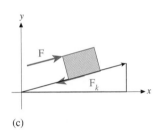

(c)

43. A man can row 1.2 m/s in still water. He wishes to row directly across a stream that is 45 m wide. At what angle upstream from directly across the stream should he head if the current of the stream is 0.3 m/s? How long will it take him to cross the stream? See the figure below.

44. A 120-lb person with a broken femur is in **Buck's traction**, as indicated in the figure below. At what angle θ should the bed be tilted so that the force due to the person's weight has a component parallel to the bed that counteracts the force from the traction, if the force due to the 5-lb weight has a

magnitude of 5 lb and acts in the direction parallel to the bed?

45. A boy and a girl are pulling a wagon along a straight line. The girl is pulling with a force of 8 lb on a horizontal rope that makes an angle of 30° with the direction of motion of the wagon. The boy is pulling on a horizontal rope that makes an angle of 45° with the direction of motion. What is the magnitude of the force applied by the boy?

46. A barge is being towed as indicated in the figure. Find the angle α so the force required from A is minimum, if the resultant of the forces from A and B has magnitude 6000 lb.

Use the technique of the proof of Equation 1 to prove the formulas in Exercises 47–48.

47. $(a\mathbf{i} + b\mathbf{j}) + (c\mathbf{i} + d\mathbf{j}) = (a + c)\mathbf{i} + (b + d)\mathbf{j}$

48. $c(a\mathbf{i} + b\mathbf{j}) = ca\mathbf{i} + cb\mathbf{j}$

49. Use the fact that the sides of an equilateral triangle satisfy $|\mathbf{U}| = |\mathbf{V}| = |\mathbf{U} - \mathbf{V}|$ to show that $2\mathbf{U} \cdot \mathbf{V} = |\mathbf{U}||\mathbf{V}|$, so the angles of an equilateral triangle are $\pi/3$.

50. Verify that $|\mathbf{U} + \mathbf{V}|^2 + |\mathbf{U} - \mathbf{V}|^2 = 2|\mathbf{U}|^2 + 2|\mathbf{V}|^2$ and relate the equality to the length of sides and diagonals of a parallelogram.

51. Show that the diagonals of a rhombus are perpendicular and bisect each other. (*Hint:* Express the diagonals as a vector sum and a vector difference of the vectors that represent the sides.)

52. For a fixed point $P_0(x_0, y_0)$ and a fixed nonzero vector $\mathbf{N} = a\mathbf{i} + b\mathbf{j}$, find an equation satisfied by all points $P(x, y)$ such that $\mathbf{P_0P}$ is orthogonal to $\mathbf{N}$.

11.4   VECTORS IN THREE-DIMENSIONAL SPACE; CROSS PRODUCT

Connections

Properties of two-dimensional vectors, 11.3.

The definitions and basic properties of three-dimensional vectors are completely analogous to those of two-dimensional vectors.

Definitions and Basic Properties

Three-dimensional vectors are defined to be ordered triples of real numbers (a, b, c). The equation

$$(a_1, b_1, c_1) = (a_2, b_2, c_2) \text{ means } a_1 = a_2, \quad b_1 = b_2, \text{ and } c_1 = c_2.$$

We define the **sum** of two vectors by

$$(a_1, b_1, c_1) + (a_2, b_2, c_2) = (a_1 + a_2, b_1 + b_2, c_1 + c_2).$$

Multiplication by a scalar is defined by

$$d(a, b, c) = (da, db, dc).$$

The **zero vector** is

$$\mathbf{0} = (0, 0, 0),$$

and the **negative** of $\mathbf{U} = (a, b, c)$ is

$$-\mathbf{U} = -(a, b, c) = (-a, -b, -c).$$

These definitions can be used to verify that three-dimensional vectors satisfy the characteristic properties of a vector space given in Section 11.3.

We define the special three-dimensional vectors

$$\mathbf{i} = (1, 0, 0), \quad \mathbf{j} = (0, 1, 0), \text{ and } \mathbf{k} = (0, 0, 1).$$

See Figure 11.4.1.

We can use the definitions to verify that

$$(a, b, c) = a\mathbf{i} + b\mathbf{j} + c\mathbf{k}.$$

Sums of vectors and multiplication of vectors by scalars are then evaluated as in the case of two-dimensional vectors.

FIGURE 11.4.1   Three-dimensional vectors can be expressed as a linear combination of the vectors **i**, **j**, and **k**.

EXAMPLE 1

$$2(4\mathbf{i} - 5\mathbf{k}) - 3(\mathbf{i} + 2\mathbf{j} - 4\mathbf{k}) = 8\mathbf{i} - 10\mathbf{k} - 3\mathbf{i} - 6\mathbf{j} + 12\mathbf{k}$$
$$= 5\mathbf{i} - 6\mathbf{j} + 2\mathbf{k}. \quad \blacksquare$$

The **dot product** of the two three-dimensional vectors is

$$(a_1\mathbf{i} + b_1\mathbf{j} + c_1\mathbf{k}) \cdot (a_2\mathbf{i} + b_2\mathbf{j} + c_2\mathbf{k}) = a_1 a_2 + b_1 b_2 + c_1 c_2.$$

The dot product of three-dimensional vectors has the same properties as the dot product of two-dimensional vectors. Recall that these properties are analogous to those of the product of real numbers. We have

$$\mathbf{i} \cdot \mathbf{j} = \mathbf{j} \cdot \mathbf{i} = \mathbf{i} \cdot \mathbf{k} = \mathbf{k} \cdot \mathbf{i} = \mathbf{j} \cdot \mathbf{k} = \mathbf{k} \cdot \mathbf{j} = 0 \text{ and } \mathbf{i} \cdot \mathbf{i} = \mathbf{j} \cdot \mathbf{j} = \mathbf{k} \cdot \mathbf{k} = 1.$$

The **length** of a three-dimensional vector is

$$|a\mathbf{i} + b\mathbf{j} + c\mathbf{k}| = \sqrt{a^2 + b^2 + c^2}.$$

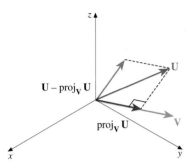

FIGURE 11.4.2 proj$_V$U is parallel to **V**. **U** − proj$_V$**U** is orthogonal to **V**.

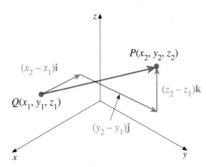

FIGURE 11.4.3 The arrow from the point $Q(x_1, y_1, z_1)$ to the point $P(x_2, y_2, z_2)$ is associated with the vector $(x_2 - x_1)\mathbf{i} + (y_2 - y_1)\mathbf{j} + (z_2 - z_1)\mathbf{k}$.

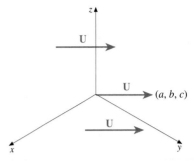

FIGURE 11.4.4 The vector $a\mathbf{i} + b\mathbf{j} + c\mathbf{k}$ is associated with the arrow from the origin to the point (a, b, c), and all other arrows that have the same length and direction.

The **angle between two nonzero vectors** is

$$\theta = \cos^{-1} \frac{\mathbf{U} \cdot \mathbf{V}}{|\mathbf{U}| |\mathbf{V}|}.$$

The vectors **U** and **V** are **orthogonal** if $\mathbf{U} \cdot \mathbf{V} = 0$. If $\mathbf{V} \neq \mathbf{0}$, the **projection of U in the direction of V** is the vector

$$\text{proj}_\mathbf{V} \mathbf{U} = \frac{\mathbf{U} \cdot \mathbf{V}}{\mathbf{V} \cdot \mathbf{V}} \mathbf{V}.$$

$\mathbf{U} - \text{proj}_\mathbf{V} \mathbf{U}$ is orthogonal to **V**. See Figure 11.4.2.

The **arrow** from a point $Q(x_1, y_1, z_1)$ to a point $P(x_2, y_2, z_2)$ in three-dimensional space is associated with the vector

$$\mathbf{QP} = (x_2 - x_1)\mathbf{i} + (y_2 - y_1)\mathbf{j} + (z_2 - z_1)\mathbf{k}.$$

See Figure 11.4.3. The vector $\mathbf{U} = a\mathbf{i} + b\mathbf{j} + c\mathbf{k}$ is associated with the arrow from the origin to the point (a, b, c), and all other arrows that have the same length and direction. See Figure 11.4.4.

Three-dimensional vectors can be added, subtracted, or multiplied by a scalar geometrically, in a manner similar to that of two-dimensional vectors. Length of a vector and the angle between vectors correspond to the geometric interpretation.

EXAMPLE 2

Sketch the vectors $\mathbf{U} = 2\mathbf{i} + \mathbf{j} + 2\mathbf{k}$ and $\mathbf{V} = 3\mathbf{j} + 4\mathbf{k}$. Find the angle between **U** and **V**. Find the projection of **U** in the direction of **V**. Find vectors $\mathbf{U}_1$ and $\mathbf{U}_2$ such that $\mathbf{U} = \mathbf{U}_1 + \mathbf{U}_2$, $\mathbf{U}_1$ is parallel to **V**, and $\mathbf{U}_2$ is orthogonal to **V**.

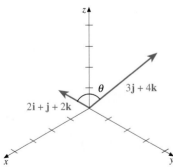

FIGURE 11.4.5 The angle between the vectors is $\theta = \cos^{-1} \dfrac{\mathbf{U} \cdot \mathbf{V}}{|\mathbf{U}||\mathbf{V}|} =$ $\cos^{-1}(11/15) \approx 0.7476.$

SOLUTION The vectors **U** and **V** are sketched in Figure 11.4.5. We have

$$\mathbf{U} \cdot \mathbf{V} = (2\mathbf{i} + \mathbf{j} + 2\mathbf{k}) \cdot (3\mathbf{j} + 4\mathbf{k}) = (2)(0) + (1)(3) + (2)(4) = 11,$$
$$\mathbf{U} \cdot \mathbf{U} = |\mathbf{U}|^2 = |2\mathbf{i} + \mathbf{j} + 2\mathbf{k}|^2 = 2^2 + 1^2 + 2^2 = 9,$$
$$\mathbf{V} \cdot \mathbf{V} = |\mathbf{V}|^2 = |3\mathbf{j} + 4\mathbf{k}|^2 = 0^2 + 3^2 + 4^2 = 25.$$

The angle θ between **U** and **V** satisfies

$$\cos \theta = \frac{\mathbf{U} \cdot \mathbf{V}}{|\mathbf{U}||\mathbf{V}|} = \frac{11}{(3)(5)},$$

so the angle between the vectors is $\theta = \cos^{-1}(11/15) \approx 0.7476$ rad.

The projection of **U** in the direction of **V** is the vector

$$\text{proj}_{\mathbf{v}}\mathbf{U} = \frac{\mathbf{U} \cdot \mathbf{V}}{\mathbf{V} \cdot \mathbf{V}}\mathbf{V} = \left(\frac{11}{25}\right)(3\mathbf{j} + 4\mathbf{k}) = \frac{33}{25}\mathbf{j} + \frac{44}{25}\mathbf{k}.$$

$\mathbf{U}_1 = \text{proj}_{\mathbf{v}}\mathbf{U} = \dfrac{33}{25}\mathbf{j} + \dfrac{44}{25}\mathbf{k}$ is parallel to **V**,

$$\mathbf{U}_2 = \mathbf{U} - \text{proj}_{\mathbf{v}}\mathbf{U}$$

$$= (2\mathbf{i} + \mathbf{j} + 2\mathbf{k}) - \left(\frac{33}{25}\mathbf{j} + \frac{44}{25}\mathbf{k}\right)$$

$$= 2\mathbf{i} - \frac{8}{25}\mathbf{j} + \frac{6}{25}\mathbf{k}$$

is orthogonal to **V**, and $\mathbf{U} = \mathbf{U}_1 + \mathbf{U}_2$. ∎

Direction Cosines

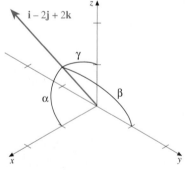

FIGURE 11.4.6 The direction cosines of a vector **U** are the cosines of the angles between **U** and the positive coordinate axes.

The angles α, β, γ between a vector **U** and the vectors **i**, **j**, **k**, respectively, are called **direction angles** of **U**. See Figure 11.4.6. The cosines of the angles are called **direction cosines** of U. If $\mathbf{U} = a\mathbf{i} + b\mathbf{j} + c\mathbf{k}$, then

$$\cos \alpha = \frac{\mathbf{U} \cdot \mathbf{i}}{|\mathbf{U}||\mathbf{i}|} = \frac{a}{|\mathbf{U}|},$$

$$\cos \beta = \frac{\mathbf{U} \cdot \mathbf{j}}{|\mathbf{U}||\mathbf{j}|} = \frac{b}{|\mathbf{U}|}, \text{ and}$$

$$\cos \gamma = \frac{\mathbf{U} \cdot \mathbf{k}}{|\mathbf{U}||\mathbf{k}|} = \frac{c}{|\mathbf{U}|}.$$

We see that $\cos \alpha\,\mathbf{i} + \cos \beta\,\mathbf{j} + \cos \gamma\,\mathbf{k} = \mathbf{U}/|\mathbf{U}|$, so $\cos \alpha\,\mathbf{i} + \cos \beta\,\mathbf{j} + \cos \gamma\,\mathbf{k}$ is the unit vector in the direction of **U**.

EXAMPLE 3

Find the direction cosines of the vector $\mathbf{i} - 2\mathbf{j} + 2\mathbf{k}$.

SOLUTION We have

$$|\mathbf{i} - 2\mathbf{j} + 2\mathbf{k}| = \sqrt{1^2 + (-2)^2 + 2^2} = 3,$$

so the unit vector in the direction of $\mathbf{i} - 2\mathbf{j} + 2\mathbf{k}$ is

$$\frac{\mathbf{i} - 2\mathbf{j} + 2\mathbf{k}}{3} = \frac{1}{3}\mathbf{i} - \frac{2}{3}\mathbf{j} + \frac{2}{3}\mathbf{k}.$$

It follows that $\cos \alpha = 1/3, \cos \beta = -2/3$, and $\cos \gamma = 2/3$. See Figure 11.4.7. ∎

FIGURE 11.4.7 The direction cosines of **U** are given by the components of the vector **U**/|**U**|.

Definition and Basic Properties of Cross Products

Many physical problems involve a particular type of product of three-dimensional vectors, called a **vector product** or **cross product**. For example, torque, angular momentum, and the force on a charged particle as it moves through a magnetic field can be described in terms of a cross product of vectors.

DEFINITION

The **cross product** of two three-dimensional vectors is

$$(a_1\mathbf{i} + b_1\mathbf{j} + c_1\mathbf{k}) \times (a_2\mathbf{i} + b_2\mathbf{j} + c_2\mathbf{k})$$
$$= (b_1c_2 - b_2c_1)\mathbf{i} - (a_1c_2 - a_2c_1)\mathbf{j} + (a_1b_2 - a_2b_1)\mathbf{k}.$$

FIGURE 11.4.8a The vectors **U**, **V**, and **U** × **V** (in that order) form a right-hand system. That is, if the right hand is placed on the plane formed by **U** and **V** so the fingers curl from **U** toward **V** in the direction that the angle from **U** to **V** is smaller, then the thumb points in the direction of **U** × **V**.

FIGURE 11.4.8b The direction of **V** × **U** is opposite to that of **U** × **V**.

The cross product of two three-dimensional vectors is a three-dimensional vector.

$$\mathbf{U} \times \mathbf{V} \text{ is orthogonal to both } \mathbf{U} \text{ and } \mathbf{V}. \tag{1}$$

Proof We need to show that $\mathbf{U} \cdot (\mathbf{U} \times \mathbf{V}) = 0$ and $\mathbf{V} \cdot (\mathbf{U} \times \mathbf{V}) = 0$. If

$$\mathbf{U} = a_1\mathbf{i} + b_1\mathbf{j} + c_1\mathbf{k} \text{ and } \mathbf{V} = a_2\mathbf{i} + b_2\mathbf{j} + c_2\mathbf{k},$$

then

$$\begin{aligned}
\mathbf{U} \cdot (\mathbf{U} \times \mathbf{V}) &= (a_1)(b_1c_2 - b_2c_1) - (b_1)(a_1c_2 - a_2c_1) \\
&\quad + (c_1)(a_1b_2 - a_2b_1) \\
&= a_1b_1c_2 - a_1b_2c_1 - a_1b_1c_2 + a_2b_1c_1 + a_1b_2c_1 - a_2b_1c_1 \\
&= 0.
\end{aligned}$$

Verification that $\mathbf{V} \cdot (\mathbf{U} \times \mathbf{V}) = 0$ is similar. ■

We know from (1) that $\mathbf{U} \times \mathbf{V}$ is orthogonal to both **U** and **V**. It is also true that the vectors **U**, **V**, and $\mathbf{U} \times \mathbf{V}$ (in that order) form a right-hand system. That is, if the right hand is placed on the plane formed by **U** and **V** in such a way that the fingers curl from **U** toward **V** in the direction in which the angle from **U** to **V** is smaller, between 0 and π, then the thumb indicates the direction of $\mathbf{U} \times \mathbf{V}$. See Figure 11.4.8a. Note that the direction of $\mathbf{V} \times \mathbf{U}$ is obtained by wrapping the fingers of the right hand from **V** toward **U** in the direction in which the angle from **V** to **U** is smaller; this angle is in the direction opposite to the smaller angle from **U** to **V**. This indicates the direction of $\mathbf{V} \times \mathbf{U}$ is opposite to that of $\mathbf{U} \times \mathbf{V}$. See Figure 11.4.8b.

We can verify the following equation by using the method of the proof of (1).

$$|\mathbf{U} \times \mathbf{V}|^2 = |\mathbf{U}|^2|\mathbf{V}|^2 - (\mathbf{U} \cdot \mathbf{V})^2. \tag{2}$$

If θ is the angle between **U** and **V**, we can use (2) and the fact that $\mathbf{U} \cdot \mathbf{V} = |\mathbf{U}||\mathbf{V}|\cos\theta$ to express the length of $\mathbf{U} \times \mathbf{V}$ in terms of θ and the lengths of **U** and **V**. We have

$$\begin{aligned}
|\mathbf{U} \times \mathbf{V}|^2 &= |\mathbf{U}|^2|\mathbf{V}|^2 - (\mathbf{U} \cdot \mathbf{V})^2 \\
&= |\mathbf{U}|^2|\mathbf{V}|^2 - (|\mathbf{U}||\mathbf{V}|\cos\theta)^2 \\
&= |\mathbf{U}|^2|\mathbf{V}|^2(1 - \cos^2\theta) \\
&= |\mathbf{U}|^2|\mathbf{V}|^2\sin^2\theta.
\end{aligned}$$

Taking the square root of each side of the above equation gives

$$|\mathbf{U} \times \mathbf{V}| = |\mathbf{U}||\mathbf{V}|\sin\theta. \tag{3}$$

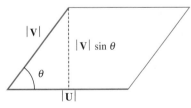

FIGURE 11.4.9 The area of the parallelogram is

$$(|\mathbf{U}|)(|\mathbf{V}|\sin\theta) = |\mathbf{U} \times \mathbf{V}|.$$

From Figure 11.4.9 we see that the area of the parallelogram formed by two nonzero vectors $\mathbf{U}$ and $\mathbf{V}$ is

$$(\text{Base})(\text{Height}) = (|\mathbf{U}|)(|\mathbf{V}| \sin \theta).$$

Collecting results, we have

$$(\text{Area of parallelogram formed by } \mathbf{U} \text{ and } \mathbf{V}) = |\mathbf{U} \times \mathbf{V}|. \tag{4}$$

It is straightforward, but tedious, to use the definition of cross product to verify the following properties.

$$\mathbf{U} \times (\mathbf{V} + \mathbf{W}) = \mathbf{U} \times \mathbf{V} + \mathbf{U} \times \mathbf{W},$$
$$(\mathbf{U} + \mathbf{V}) \times \mathbf{W} = \mathbf{U} \times \mathbf{W} + \mathbf{V} \times \mathbf{W},$$
$$(c\mathbf{U}) \times \mathbf{V} = c(\mathbf{U} \times \mathbf{V}) = \mathbf{U} \times (c\mathbf{V}),$$
$$\mathbf{0} \times \mathbf{U} = \mathbf{U} \times \mathbf{0} = \mathbf{0},$$
$$\mathbf{U} \times \mathbf{U} = \mathbf{0},$$
$$\mathbf{U} \times \mathbf{V} = -\mathbf{V} \times \mathbf{U}.$$

The first four properties above correspond to usual properties of multiplication. The properties $\mathbf{U} \times \mathbf{U} = \mathbf{0}$ and $\mathbf{U} \times \mathbf{V} = -\mathbf{V} \times \mathbf{U}$ do not correspond to multiplication of real numbers.

NOTE   *Since the Commutative Law does not hold for cross products, we must be careful with the order of the factors in a cross product of three-dimensional vectors.*

Since the area of the parallelogram formed by parallel vectors is zero, we can conclude from Formula 4 that

$$\mathbf{i} \times \mathbf{i} = \mathbf{j} \times \mathbf{j} = \mathbf{k} \times \mathbf{k} = \mathbf{0}.$$

Since the area of the parallelogram formed by the unit vectors $\mathbf{i}$ and $\mathbf{j}$ is one, and since the vectors $\mathbf{i}$, $\mathbf{j}$, and $\mathbf{i} \times \mathbf{j}$ form a right-hand system, we must have $\mathbf{i} \times \mathbf{j} = \mathbf{k}$. Similarly, we can determine that $\mathbf{j} \times \mathbf{i} = -\mathbf{k}$. Each of the following formulas may be derived in this manner.

$$\mathbf{i} \times \mathbf{j} = \mathbf{k}, \quad \mathbf{j} \times \mathbf{i} = -\mathbf{k},$$
$$\mathbf{j} \times \mathbf{k} = \mathbf{i}, \quad \mathbf{k} \times \mathbf{j} = -\mathbf{i},$$
$$\mathbf{k} \times \mathbf{i} = \mathbf{j}, \quad \mathbf{i} \times \mathbf{k} = -\mathbf{j}.$$

See Figure 11.4.10. The above formulas can be used with the Distributive Law to evaluate cross products. For example,

$$(\mathbf{i} + \mathbf{j}) \times (2\mathbf{i} - \mathbf{k}) = 2\mathbf{i} \times \mathbf{i} - \mathbf{i} \times \mathbf{k} + 2\mathbf{j} \times \mathbf{i} - \mathbf{j} \times \mathbf{k}$$
$$= \mathbf{0} + \mathbf{j} - 2\mathbf{k} - \mathbf{i} = -\mathbf{i} + \mathbf{j} - 2\mathbf{k}.$$

The Cross Product as the Determinant of a Matrix

There is a relatively easy way to remember the definition of cross product. This involves the idea of the **determinant** of a **matrix**. A matrix is simply a rectangular array of numbers.

DEFINITION

The **determinant** of a 2×2 matrix is

$$\begin{vmatrix} a & b \\ c & d \end{vmatrix} = ad - cb.$$

FIGURE 11.4.10a $\mathbf{i} \times \mathbf{j} = \mathbf{k}$.

FIGURE 11.4.10b $\mathbf{j} \times \mathbf{k} = \mathbf{i}$.

FIGURE 11.4.10c $\mathbf{k} \times \mathbf{i} = \mathbf{j}$.

The determinant of a 3×3 matrix can be expressed in terms of determinants of 2×2 matrices.

DEFINITION

The **determinant** of a 3×3 matrix is

$$\begin{vmatrix} a_0 & b_0 & c_0 \\ a_1 & b_1 & c_1 \\ a_2 & b_2 & c_2 \end{vmatrix} = \begin{vmatrix} b_1 & c_1 \\ b_2 & c_2 \end{vmatrix} a_0 - \begin{vmatrix} a_1 & c_1 \\ a_2 & c_2 \end{vmatrix} b_0 + \begin{vmatrix} a_1 & b_1 \\ a_2 & b_2 \end{vmatrix} c_0.$$

The expression on the right above is called a **row expansion** of the 3×3 determinant on the left. Note that the terms a_0, b_0, and c_0 on the right side of the above equation are the elements of the first *row* of the determinant on the left. The 2×2 determinant that is multiplied by a_0 on the right is obtained by deleting the first row and first column of the determinant on the left, which are the row and column that contain a_0. That is,

$$\begin{vmatrix} \not{a_0} & \not{b_0} & \not{c_0} \\ \not{a_1} & b_1 & c_1 \\ \not{a_2} & b_2 & c_2 \end{vmatrix} = \begin{vmatrix} b_1 & c_1 \\ b_2 & c_2 \end{vmatrix}.$$

The determinant coefficient of b_0 on the right is the *negative* of the 2×2 determinant obtained by deleting the first row and second column of the determinant on the left. These are the row and column that contain b_0. Deletion of the first row and the third column on the left, the row and column that contain c_0, gives the determinant coefficient of c_0 on the right.

We can use the idea of determinants to express cross products as

$$(a_1\mathbf{i} + b_1\mathbf{j} + c_1\mathbf{k}) \times (a_2\mathbf{i} + b_2\mathbf{j} + c_2\mathbf{k})$$
$$= \begin{vmatrix} \mathbf{i} & \mathbf{j} & \mathbf{k} \\ a_1 & b_1 & c_1 \\ a_2 & b_2 & c_2 \end{vmatrix} = \begin{vmatrix} b_1 & c_1 \\ b_2 & c_2 \end{vmatrix} \mathbf{i} - \begin{vmatrix} a_1 & c_1 \\ a_2 & c_2 \end{vmatrix} \mathbf{j} + \begin{vmatrix} a_1 & b_1 \\ a_2 & b_2 \end{vmatrix} \mathbf{k}.$$

Note that the above "3×3 determinant" is not really a determinant, because the terms are not numbers. However, the same rules apply formally.

Remember that the order of the factors is important in a cross product. Be sure to use the correct order when using the above formula to evaluate a cross product. In particular, note that the components of the first factor form the middle row of the 3×3 determinant, and the components of the second factor form the bottom row. Also, don't forget the negative sign that precedes the $\mathbf{j}$ term in the expansion. Finally, note that it is very helpful to use parentheses when evaluating a cross product.

EXAMPLE 4

$$\mathbf{U} = \mathbf{i} - \mathbf{j} \text{ and } \mathbf{V} = \mathbf{j} + 2\mathbf{k}.$$

Evaluate

    (a) $\mathbf{U} \times \mathbf{V}$,
    (b) $\mathbf{V} \times \mathbf{U}$,
    (c) $\mathbf{U} \cdot (\mathbf{U} \times \mathbf{V})$ and $\mathbf{V} \cdot (\mathbf{U} \times \mathbf{V})$,

(d) $|\mathbf{U} \times \mathbf{V}|$,

(e) the angle θ between $\mathbf{U}$ and $\mathbf{V}$, and $|\mathbf{U}|\,|\mathbf{V}|\sin\theta$.

SOLUTION

(a) $\mathbf{U} \times \mathbf{V} = (\mathbf{i} - \mathbf{j}) \times (\mathbf{j} + 2\mathbf{k})$

$$= \begin{vmatrix} \mathbf{i} & \mathbf{j} & \mathbf{k} \\ 1 & -1 & 0 \\ 0 & 1 & 2 \end{vmatrix} = \begin{vmatrix} -1 & 0 \\ 1 & 2 \end{vmatrix}\mathbf{i} - \begin{vmatrix} 1 & 0 \\ 0 & 2 \end{vmatrix}\mathbf{j} + \begin{vmatrix} 1 & -1 \\ 0 & 1 \end{vmatrix}\mathbf{k}$$

$$= [(-1)(2) - (1)(0)]\mathbf{i} - [(1)(2) - (0)(0)]\mathbf{j} + [(1)(1) - (0)(-1)]\mathbf{k}$$

$$= -2\mathbf{i} - 2\mathbf{j} + \mathbf{k}.$$

(b) $\mathbf{V} \times \mathbf{U} = (\mathbf{j} + 2\mathbf{k}) \times (\mathbf{i} - \mathbf{j})$

$$= \begin{vmatrix} \mathbf{i} & \mathbf{j} & \mathbf{k} \\ 0 & 1 & 2 \\ 1 & -1 & 0 \end{vmatrix} = \begin{vmatrix} 1 & 2 \\ -1 & 0 \end{vmatrix}\mathbf{i} - \begin{vmatrix} 0 & 2 \\ 1 & 0 \end{vmatrix}\mathbf{j} + \begin{vmatrix} 0 & 1 \\ 1 & -1 \end{vmatrix}\mathbf{k}$$

$$= [(1)(0) - (-1)(2)]\mathbf{i} - [(0)(0) - (1)(2)]\mathbf{j} + [(0)(-1) - (1)(1)]\mathbf{k}$$

$$= 2\mathbf{i} + 2\mathbf{j} - \mathbf{k}.$$

$(\mathbf{U} \times \mathbf{V} = -\mathbf{V} \times \mathbf{U}$, as is true in general.)

(c) $\mathbf{U} \cdot (\mathbf{U} \times \mathbf{V}) = (\mathbf{i} - \mathbf{j}) \cdot (-2\mathbf{i} - 2\mathbf{j} + \mathbf{k})$

$$= (1)(-2) + (-1)(-2) + (0)(1) = 0$$

and

$$\mathbf{V} \cdot (\mathbf{U} \times \mathbf{V}) = (\mathbf{j} + 2\mathbf{k}) \cdot (-2\mathbf{i} - 2\mathbf{j} + \mathbf{k}) = (0)(-2) + (1)(-2) + (2)(1) = 0.$$

$(\mathbf{U} \times \mathbf{V}$ is orthogonal to both $\mathbf{U}$ and $\mathbf{V}$, as is true in general.)

(d) $|\mathbf{U} \times \mathbf{V}| = |-2\mathbf{i} - 2\mathbf{j} + \mathbf{k}| = \sqrt{(-2)^2 + (-2)^2 + (1)^2} = 3.$

(e) We have

$$\mathbf{U} \cdot \mathbf{V} = (\mathbf{i} - \mathbf{j}) \cdot (\mathbf{j} + 2\mathbf{k}) = (1)(0) + (-1)(1) + (0)(2) = -1,$$

$$|\mathbf{U}| = |\mathbf{i} - \mathbf{j}| = \sqrt{(1)^2 + (-1)^2} = \sqrt{2},$$

$$|\mathbf{V}| = |\mathbf{j} + 2\mathbf{k}| = \sqrt{(1)^2 + (2)^2} = \sqrt{5}.$$

The angle between $\mathbf{U}$ and $\mathbf{V}$ is

$$\theta = \cos^{-1}\left(\frac{\mathbf{U} \cdot \mathbf{V}}{|\mathbf{U}|\,|\mathbf{V}|}\right) = \cos^{-1}\left(\frac{-1}{\sqrt{10}}\right) \approx 1.8925 \text{ radians.}$$

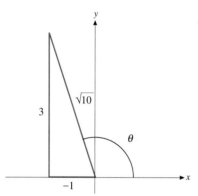

FIGURE 11.4.11 A representative sketch of $\theta = \cos^{-1}\left(-1/\sqrt{10}\right)$ can be used to determine that $\sin\theta = 3/\sqrt{10}$.

From Figure 11.4.11 we see that $\sin\theta = 3/\sqrt{10}$, so

$$|\mathbf{U}|\,|\mathbf{V}|\sin\theta = (\sqrt{2})(\sqrt{5})\left(\frac{3}{\sqrt{10}}\right) = 3.$$

$(|\mathbf{U} \times \mathbf{V}| = |\mathbf{U}|\,|\mathbf{V}|\sin\theta$, as is true in general.) ∎

EXAMPLE 5

Evaluate $\mathbf{i} \times (\mathbf{i} \times \mathbf{j})$ and $(\mathbf{i} \times \mathbf{i}) \times \mathbf{j}$.

SOLUTION We begin the evaluation of $\mathbf{i} \times (\mathbf{i} \times \mathbf{j})$ by evaluating the vector inside the parentheses. We have

$$\mathbf{i} \times \mathbf{j} = \begin{vmatrix} \mathbf{i} & \mathbf{j} & \mathbf{k} \\ 1 & 0 & 0 \\ 0 & 1 & 0 \end{vmatrix} = \begin{vmatrix} 0 & 0 \\ 1 & 0 \end{vmatrix}\mathbf{i} - \begin{vmatrix} 1 & 0 \\ 0 & 0 \end{vmatrix}\mathbf{j} + \begin{vmatrix} 1 & 0 \\ 0 & 1 \end{vmatrix}\mathbf{k}$$

$$= [(0)(0) - (1)(0)]\mathbf{i} - [(1)(0) - (0)(0)]\mathbf{j} + [(1)(1) - (0)(0)]\mathbf{k}$$

$$= \mathbf{k}.$$

Then

$$\mathbf{i} \times (\mathbf{i} \times \mathbf{j}) = \mathbf{i} \times (\mathbf{k}) = \begin{vmatrix} \mathbf{i} & \mathbf{j} & \mathbf{k} \\ 1 & 0 & 0 \\ 0 & 0 & 1 \end{vmatrix}$$

$$= \begin{vmatrix} 0 & 0 \\ 0 & 1 \end{vmatrix} \mathbf{i} - \begin{vmatrix} 1 & 0 \\ 0 & 1 \end{vmatrix} \mathbf{j} + \begin{vmatrix} 1 & 0 \\ 0 & 0 \end{vmatrix} \mathbf{k}$$

$$= [(0)(1) - (0)(0)]\mathbf{i} - [(1)(1) - (0)(0)]\mathbf{j} + [(1)(0) - (0)(0)]\mathbf{k}$$

$$= -\mathbf{j}.$$

In order to evaluate $(\mathbf{i} \times \mathbf{i}) \times \mathbf{j}$, we again evaluate the vector inside the parentheses first. Since the cross product of any vector with itself is the zero vector, and the cross product of the zero vector with any vector is the zero vector, we have

$$(\mathbf{i} \times \mathbf{i}) \times \mathbf{j} = (\mathbf{0}) \times \mathbf{j} = \mathbf{0}. \quad \blacksquare$$

From Example 5 we see that $\mathbf{U} \times (\mathbf{V} \times \mathbf{W})$ is generally *not* equal to $(\mathbf{U} \times \mathbf{V}) \times \mathbf{W}$.

NOTE   *Since the Associative Law does not hold for cross products, we must be very careful with parentheses (as well as order) when dealing with expressions that involve cross products.*

In particular, we cannot express either $\mathbf{U} \times (\mathbf{V} \times \mathbf{W})$ or $(\mathbf{U} \times \mathbf{V}) \times \mathbf{W}$ as $\mathbf{U} \times \mathbf{V} \times \mathbf{W}$; $\mathbf{U} \times \mathbf{V} \times \mathbf{W}$ is undefined.

Finding a Vector That Is Orthogonal to Two Given Vectors

One use of the cross product is to find a vector that is orthogonal to two given vectors.

EXAMPLE 6

Find two unit vectors that are orthogonal to both $2\mathbf{i} - \mathbf{j} + 2\mathbf{k}$ and $\mathbf{i} + 2\mathbf{j} - 3\mathbf{k}$.

SOLUTION   The cross product of the two given vectors (in either order) is orthogonal to both vectors. The unit vector in the direction of the cross product and the negative of that unit vector are both unit vectors that are orthogonal to the given vectors. See Figure 11.4.12. We have

$$(2\mathbf{i} - \mathbf{j} + 2\mathbf{k}) \times (\mathbf{i} + 2\mathbf{j} - 3\mathbf{k})$$

$$= \begin{vmatrix} \mathbf{i} & \mathbf{j} & \mathbf{k} \\ 2 & -1 & 2 \\ 1 & 2 & -3 \end{vmatrix} = \begin{vmatrix} -1 & 2 \\ 2 & -3 \end{vmatrix} \mathbf{i} - \begin{vmatrix} 2 & 2 \\ 1 & -3 \end{vmatrix} \mathbf{j} + \begin{vmatrix} 2 & -1 \\ 1 & 2 \end{vmatrix} \mathbf{k}$$

$$= [(-1)(-3) - (2)(2)]\mathbf{i} - [(2)(-3) - (1)(2)]\mathbf{j} + [(2)(2) - (1)(-1)]\mathbf{k},$$

$$= -\mathbf{i} + 8\mathbf{j} + 5\mathbf{k}.$$

The length of the cross product is

$$|-\mathbf{i} + 8\mathbf{j} + 5\mathbf{k}| = \sqrt{(-1)^2 + 8^2 + 5^2} = \sqrt{90}.$$

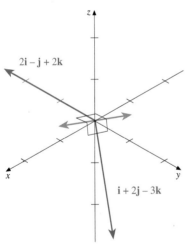

$2\mathbf{i} - \mathbf{j} + 2\mathbf{k}$

$\mathbf{i} + 2\mathbf{j} - 3\mathbf{k}$

FIGURE 11.4.12   The cross product $\mathbf{U} \times \mathbf{V}$ is orthogonal to both $\mathbf{U}$ and $\mathbf{V}$, so $\dfrac{\mathbf{U} \times \mathbf{V}}{|\mathbf{U} \times \mathbf{V}|}$ and $-\dfrac{\mathbf{U} \times \mathbf{V}}{|\mathbf{U} \times \mathbf{V}|}$ are two unit vectors that are orthogonal to both $\mathbf{U}$ and $\mathbf{V}$.

The unit vector in the direction of the cross product and the negative of that unit vector are

$$\frac{-\mathbf{i} + 8\mathbf{j} + 5\mathbf{k}}{\sqrt{90}} \text{ and } \frac{\mathbf{i} - 8\mathbf{j} - 5\mathbf{k}}{\sqrt{90}}.$$

These are the two unit vectors that are orthogonal to the given vectors. (Note that if we had evaluated the cross product of the given vectors in the opposite order, we would obtain the same two unit vectors.) ■

Geometric Applications

We have seen in Equation 4 that the area of parallelogram formed by **U** and **V** is given by $|\mathbf{U} \times \mathbf{V}|$. It follows that

$$\text{(Area of triangle formed by } \mathbf{U} \text{ and } \mathbf{V}) = \frac{1}{2}|\mathbf{U} \times \mathbf{V}|. \quad (5)$$

See Figure 11.4.13.

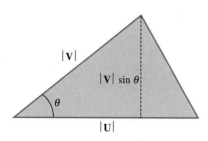

FIGURE 11.4.13 The triangle has area $\frac{1}{2}(|\mathbf{U}|)(|\mathbf{V}|\sin\theta) = \frac{1}{2}|\mathbf{U} \times \mathbf{V}|$.

EXAMPLE 7
Find the area of the triangle with vertices at the points $A(0, -1, 1)$, $B(3, 1, 0)$, and $C(-2, 0, 2)$.

SOLUTION The triangle is sketched in Figure 11.4.14. We see that the triangle is formed by the *vectors* **AB** and **AC**. (We could consider the triangle to be formed by vectors that correspond to any two sides of the triangle.) We have

$$\mathbf{AB} = (3 - 0)\mathbf{i} + (1 - (-1))\mathbf{j} + (0 - 1)\mathbf{k} = 3\mathbf{i} + 2\mathbf{j} - \mathbf{k}$$

and

$$\mathbf{AC} = (-2 - 0)\mathbf{i} + (0 - (-1))\mathbf{j} + (2 - 1)\mathbf{k} = -2\mathbf{i} + \mathbf{j} + \mathbf{k}.$$

Then

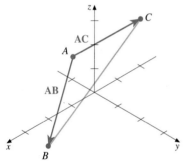

FIGURE 11.4.14 The triangle formed by the points A, B, and C is formed by the vectors **AB** and **AC**.

$$\mathbf{AB} \times \mathbf{AC} = \begin{vmatrix} \mathbf{i} & \mathbf{j} & \mathbf{k} \\ 3 & 2 & -1 \\ -2 & 1 & 1 \end{vmatrix}$$

$$= \begin{vmatrix} 2 & -1 \\ 1 & 1 \end{vmatrix}\mathbf{i} - \begin{vmatrix} 3 & -1 \\ -2 & 1 \end{vmatrix}\mathbf{j} + \begin{vmatrix} 3 & 2 \\ -2 & 1 \end{vmatrix}\mathbf{k}$$

$$= (2 + 1)\mathbf{i} - (3 - 2)\mathbf{j} + (3 + 4)\mathbf{k}$$

$$= 3\mathbf{i} - \mathbf{j} + 7\mathbf{k}.$$

Formula 5 then gives

$$\text{Area of triangle} = \frac{1}{2}|3\mathbf{i} - \mathbf{j} + 7\mathbf{k}| = \frac{1}{2}\sqrt{3^2 + (-1)^2 + 7^2} = \frac{\sqrt{59}}{2}. \quad ■$$

FIGURE 11.4.15 The volume of the parallelepiped is $(|\mathbf{U} \times \mathbf{V}|)(|\mathbf{W}||\cos\phi|) = |(\mathbf{U} \times \mathbf{V}) \cdot \mathbf{W}|$.

From Figure 11.4.15 we see that the vectors **U**, **V**, and **W** form a parallelepiped with area of base $|\mathbf{U} \times \mathbf{V}|$ and height $|\mathbf{W}||\cos\phi|$, since $\mathbf{U} \times \mathbf{V}$ is perpendicular to the base formed by **U** and **V**. The volume of the parallelepiped is then

$$\text{(Area of base)(Height)} = (|\mathbf{U} \times \mathbf{V}|)(|\mathbf{W}||\cos\phi|)$$

$$= |(\mathbf{U} \times \mathbf{V}) \cdot \mathbf{W}|.$$

That is,

(Volume of parallelepiped formed by $\mathbf{U}$, $\mathbf{V}$, and $\mathbf{W}$) $= |(\mathbf{U} \times \mathbf{V}) \cdot \mathbf{W}|$.

The expression on the right above can be evaluated by using determinants. That is, let

$$\mathbf{U} = a_1\mathbf{i} + b_2\mathbf{j} + c_1\mathbf{k}, \quad \mathbf{V} = a_2\mathbf{i} + b_2\mathbf{j} + c_2\mathbf{k}, \text{ and } \mathbf{W} = a_0\mathbf{i} + b_0\mathbf{j} + c_0\mathbf{k}.$$

Then

$$
\begin{aligned}
(\mathbf{U} \times \mathbf{V}) \cdot \mathbf{W} &= (b_1c_2 - b_2c_1)a_0 - (a_1c_2 - a_2c_1)b_0 + (a_1b_2 - a_2b_1)c_0 \\
&= \begin{vmatrix} b_1 & c_1 \\ b_2 & c_2 \end{vmatrix} a_0 - \begin{vmatrix} a_1 & c_1 \\ a_2 & c_2 \end{vmatrix} b_0 + \begin{vmatrix} a_1 & b_1 \\ a_2 & b_2 \end{vmatrix} c_0 \\
&= \begin{vmatrix} a_0 & b_0 & c_0 \\ a_1 & b_1 & c_1 \\ a_2 & b_2 & c_2 \end{vmatrix}.
\end{aligned}
$$

Finally, note that the volume of the parallelepiped does not depend on which two vectors are considered to form the base. The scalar

$$(\mathbf{U} \times \mathbf{V}) \cdot \mathbf{W}$$

is called the **triple product** of $\mathbf{U}$, $\mathbf{V}$, and $\mathbf{W}$. We have

$$(\mathbf{U} \times \mathbf{V}) \cdot \mathbf{W} = \mathbf{U} \cdot (\mathbf{V} \times \mathbf{W}).$$

EXAMPLE 8

Find the volume of the parallelepiped formed by $\mathbf{i} + 2\mathbf{j}$, $3\mathbf{j}$, and $\mathbf{i} + 2\mathbf{j} + 3\mathbf{k}$.

SOLUTION We begin by evaluating the determinant that corresponds to the triple product of the three given vectors. We have

$$
\begin{vmatrix} 1 & 2 & 0 \\ 0 & 3 & 0 \\ 1 & 2 & 3 \end{vmatrix} = \begin{vmatrix} 3 & 0 \\ 2 & 3 \end{vmatrix}(1) - \begin{vmatrix} 0 & 0 \\ 1 & 3 \end{vmatrix}(2) + \begin{vmatrix} 0 & 3 \\ 1 & 2 \end{vmatrix}(0)
$$

$$= (9)(1) - (0)(2) + (-3)(0) = 9.$$

The volume is then $|9| = 9$. The parallelepiped is sketched in Figure 11.4.16. (In this example, one face of the parallelepiped is contained in the xy-plane and has area 3. The height from that side is 3, so we can determine geometrically that the volume is 9.) ∎

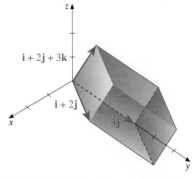

FIGURE 11.4.16 The volume of the parallelepiped is given by the triple product of $\mathbf{i} + 2\mathbf{j}$, $3\mathbf{j}$, and $\mathbf{i} + 2\mathbf{j} + 3\mathbf{k}$.

EXERCISES 11.4

Evaluate the expressions in Examples 1–20.

1. $-2(3\mathbf{i} - 3\mathbf{j} + \mathbf{k})$
2. $(\mathbf{i} - 2\mathbf{j} + 3\mathbf{k}) + (\mathbf{i} + 3\mathbf{j} - 2\mathbf{k})$
3. $2(3\mathbf{i} + \mathbf{j} - 2\mathbf{k}) - 3(2\mathbf{i} + \mathbf{k})$
4. $3(4\mathbf{i} - 2\mathbf{j} + \mathbf{k}) - 2(6\mathbf{i} + \mathbf{j} + \mathbf{k})$
5. $|3\mathbf{i} - 4\mathbf{j} + 5\mathbf{k}|$
6. $|12\mathbf{i} - 4\mathbf{j} - 3\mathbf{k}|$
7. $|\mathbf{i} - 2\mathbf{j} + 2\mathbf{k}|$
8. $|2\mathbf{i} + \mathbf{j} - \mathbf{k}|$
9. $(2\mathbf{i} - 3\mathbf{j} + \mathbf{k}) \cdot (\mathbf{i} - \mathbf{j} - \mathbf{k})$
10. $(2\mathbf{i} + 3\mathbf{j} + 2\mathbf{k}) \cdot (\mathbf{i} + \mathbf{j} + \mathbf{k})$
11. $(\mathbf{i} + 4\mathbf{j}) \cdot (3\mathbf{i} - \mathbf{k})$
12. $(3\mathbf{i} + 4\mathbf{j}) \cdot (3\mathbf{j} + 4\mathbf{k})$
13. $(\mathbf{i} + 2\mathbf{j} + 3\mathbf{k}) \times \mathbf{i}$
14. $(4\mathbf{i} - 5\mathbf{j} - 6\mathbf{k}) \times (\mathbf{i} + \mathbf{j})$
15. $(3\mathbf{i} + 4\mathbf{j} - 5\mathbf{k}) \times (\mathbf{i} + \mathbf{j} + \mathbf{k})$
16. $(5\mathbf{i} + 2\mathbf{j} + 3\mathbf{k}) \times (\mathbf{i} - \mathbf{j} - \mathbf{k})$
17. $\mathbf{i} \times (\mathbf{i} + \mathbf{k})$
18. $(\mathbf{i} + \mathbf{k}) \times \mathbf{i}$
19. $[(\mathbf{i} - \mathbf{j}) \times (\mathbf{i} + \mathbf{k})] \cdot \mathbf{i}$
20. $[(2\mathbf{i} + \mathbf{j}) \times (\mathbf{i} + 2\mathbf{k})] \cdot \mathbf{k}$

In Exercises 21–24, sketch the vectors **U** *and* **V**, *find the angle between* **U** *and* **V**, *and find the projection of* **U** *in the direction of* **V**.

21. $\mathbf{U} = \mathbf{i} + \mathbf{j} + \mathbf{k}, \mathbf{V} = \mathbf{i} + \mathbf{j}$
22. $\mathbf{U} = 2\mathbf{i} + 2\mathbf{j} + \mathbf{k}, \mathbf{V} = 2\mathbf{i} + 2\mathbf{j} - \mathbf{k}$
23. $\mathbf{U} = 2\mathbf{i} - 3\mathbf{j}, \mathbf{V} = 3\mathbf{i} + 4\mathbf{k}$
24. $\mathbf{U} = 2\mathbf{i} - \mathbf{j} + 2\mathbf{k}, \mathbf{V} = 2\mathbf{i} + 2\mathbf{j} - \mathbf{k}$

Find the direction cosines of the vectors in Exercises 25–28.

25. $4\mathbf{i} - 5\mathbf{k}$
26. $\mathbf{i} - 2\mathbf{j} - 2\mathbf{k}$
27. $3\mathbf{i} + 2\mathbf{j} + 3\mathbf{k}$
28. $3\mathbf{i} - 12\mathbf{j} + 4\mathbf{k}$

Find the two unit vectors that are orthogonal to the vectors given in each of Exercises 29–32.

29. $\mathbf{i} + \mathbf{j}, \mathbf{j} + \mathbf{k}$
30. $\mathbf{i} + 2\mathbf{j} + 2\mathbf{k}, \mathbf{i} + \mathbf{j} - \mathbf{k}$
31. $2\mathbf{i} - 2\mathbf{j} + 3\mathbf{k}, \mathbf{i} + \mathbf{j}$
32. $3\mathbf{i} + 4\mathbf{j} + 2\mathbf{k}, 2\mathbf{i} + 3\mathbf{j} - 4\mathbf{k}$

Find the area of the parallelogram formed by the vectors given in Exercises 33–36.

33. $\mathbf{i}, \mathbf{j} + \mathbf{k}$
34. $6\mathbf{i} + 4\mathbf{j}, \mathbf{i} - \mathbf{k}$
35. $2\mathbf{i} + 3\mathbf{j} + \mathbf{k}, \mathbf{j} + \mathbf{k}$
36. $3\mathbf{i} + 4\mathbf{j}, 3\mathbf{j} + 4\mathbf{k}$

Find the volume of the parallelepiped formed by the vectors given in Exercises 37–40.

37. $\mathbf{i}, \mathbf{i} + \mathbf{j}, 2\mathbf{k}$
38. $\mathbf{i} + \mathbf{j}, \mathbf{i} - \mathbf{j}, \mathbf{i} + \mathbf{j} + 2\mathbf{k}$
39. $2\mathbf{i} - 2\mathbf{j} + \mathbf{k}, 2\mathbf{i} + 2\mathbf{j} + \mathbf{k}, 3\mathbf{i}$
40. $3\mathbf{i} - 4\mathbf{j} + 2\mathbf{k}, 4\mathbf{i} + 3\mathbf{j} + 2\mathbf{k}, \mathbf{i} + \mathbf{j}$

Find the area of the triangle formed by the three points given in Exercises 41–44.

41. $(0, 0, 0), (1, 2, 0), (-1, 3, 2)$
42. $(0, 0, 0), (-1, 1, 1), (1, 2, 1)$
43. $(4, 0, 0), (0, 3, 0), (0, 0, 4)$
44. $(1, 1, 1), (0, 3, 0), (2, 3, 1)$

In Exercises 45–48, express **F** *as a sum of two forces, one parallel to* **V** *and one orthogonal to* **V**.

45. $\mathbf{F} = 3\mathbf{k}, \mathbf{V} = \mathbf{i} + 2\mathbf{j} + 2\mathbf{k}$
46. $\mathbf{F} = 4\mathbf{i} - \mathbf{k}, \mathbf{V} = 2\mathbf{i} + 3\mathbf{j}$
47. $\mathbf{F} = 2\mathbf{i} - \mathbf{j} + \mathbf{k}, \mathbf{V} = \mathbf{i}$
48. $\mathbf{F} = -\mathbf{i} + \mathbf{j} + 2\mathbf{k}, \mathbf{V} = \mathbf{j}$

In Exercises 49–50, find the angle $\angle ABC$.

49. $A(1, 3, 2), B(2, 1, -1), C(-2, 2, 3)$
50. $A(4, -1, 2), B(-2, 2, 1), C(3, -2 - 1)$
51. Show that $\begin{vmatrix} a_0 & b_0 & c_0 \\ a_1 & b_1 & c_1 \\ a_2 & b_2 & c_2 \end{vmatrix} = \begin{vmatrix} a_0 & a_1 & a_2 \\ b_0 & b_1 & b_2 \\ c_0 & c_1 & c_2 \end{vmatrix}$.
52. Show that the area of the parallelogram formed by vectors $\mathbf{U} = u_1\mathbf{i} + u_2\mathbf{j}$ and $\mathbf{V} = v_1\mathbf{i} + v_2\mathbf{j}$ is the absolute value of the determinant

$$\begin{vmatrix} u_1 & u_2 \\ v_1 & v_2 \end{vmatrix}.$$

Connections

Angle between vectors, projection of vectors, 11.4.

FIGURE 11.5.1a A point is in the plane of the floor if and only if the vector from the corner of the floor to the point is perpendicular to the vertical.

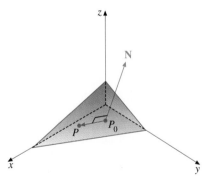

FIGURE 11.5.1b A point P is in the plane that contains the point P_0 and has normal vector **N** if and only if **P₀P** is orthogonal to **N**.

11.5 PLANES

Consider the plane of the floor in a rectangular room. The intersection of two walls determines a vertical vector that is perpendicular to the floor. Any vector in the plane of the floor is horizontal and perpendicular to the vertical. A point is in the plane of the floor if and only if the vector from a corner of the floor to the point is perpendicular to the vertical. See Figure 11.5.1a. Similarly, any **plane** is determined by one point on the plane and a nonzero vector **N** that is perpendicular to the plane. A vector that is perpendicular to all vectors in a plane is said to be a **normal vector** of the plane. A variable point $P(x, y, z)$ is in the plane that contains the fixed point $P_0(x_0, y_0, z_0)$ and has normal vector $\mathbf{N} = a\mathbf{i} + b\mathbf{j} + c\mathbf{k}$ if and only if the vector $\mathbf{P_0P} = (x - x_0)\mathbf{i} + (y - y_0)\mathbf{j} + (z - z_0)\mathbf{k}$ is orthogonal to **N**. See Figure 11.5.1b. Expressing the orthogonality in terms of the dot product, we obtain the **vector form** of an equation of the plane,

$$\mathbf{N} \cdot \mathbf{P_0P} = 0.$$

Evaluating the dot product, we have

$$(a\mathbf{i} + b\mathbf{j} + c\mathbf{k}) \cdot ((x - x_0)\mathbf{i} + (y - y_0)\mathbf{j} + (z - z_0)\mathbf{k}) = 0,$$

$$a(x - x_0) + b(y - y_0) + c(z - z_0) = 0. \tag{1}$$

Equation 1 is used to *write* an equation of the plane that contains the point $P_0(x_0, y_0, z_0)$ and has normal vector $N = a\mathbf{i} + b\mathbf{j} + c\mathbf{k}$. For example, an equation of the plane that contains the point $(2, -1, 3)$ and has normal vector $\mathbf{N} = \mathbf{i} + 2\mathbf{j} - 5\mathbf{k}$ is

$$(1)(x - 2) + (2)(y - (-1)) + (-5)(z - 3) = 0.$$

Let us simplify the equation we obtained for the plane that contains the point $P_0(x_0, y_0, z_0)$ and has normal vector $\mathbf{N} = a\mathbf{i} + b\mathbf{j} + c\mathbf{k}$. We have

$$a(x - x_0) + b(y - y_0) + c(z - z_0) = 0,$$
$$ax - ax_0 + by - by_0 + cz - cz_0 = 0,$$
$$ax + by + cz + (-ax_0 - by_0 - cz_0) = 0,$$
$$ax + by + cz + d = 0,$$

where $d = -ax_0 - by_0 - cz_0$. On the other hand, suppose we start with an equation of the form

$$ax + by + cz + d = 0,$$

where $a, b,$ and c are not all equal to zero. We then find some point $P_0(x_0, y_0, z_0)$ that satisfies the equation, so $ax_0 + by_0 + cz_0 + d = 0$. We can then reverse the above steps to obtain

$$a(x - x_0) + b(y - y_0) + c(z - z_0) = 0.$$

We conclude that

$$ax + by + cz + d = 0 \tag{2}$$

is an equation of a plane that has normal vector $\mathbf{N} = a\mathbf{i} + b\mathbf{j} + c\mathbf{k}$, if $a, b,$ and c are not all equal to zero. Equation 2 is called the **general equation of a plane**. Note that we can *read* the components of a normal vector from

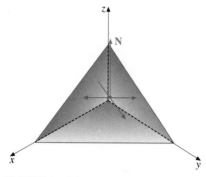

FIGURE 11.5.2 A normal vector of a plane is orthogonal to any vector in the plane.

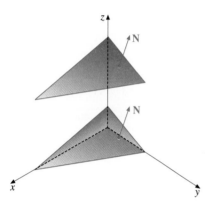

FIGURE 11.5.3 Parallel planes have parallel normal vectors.

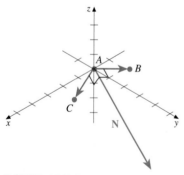

FIGURE 11.5.4 The plane that contains the points A, B, and C has normal vector that is orthogonal to both of the vectors **AB** and **AC**.

the general equation of a plane. For example, The plane $x - 3y + 2z = 0$ has normal vector $\mathbf{N} = \mathbf{i} - 3\mathbf{j} + 2\mathbf{k}$.

Any nonzero multiple of a normal vector of a plane is also a normal vector of the plane. Also, a normal vector of a plane is orthogonal to any vector in the plane. See Figure 11.5.2.

Parallel planes have parallel normal vectors.

See Figure 11.5.3.

Finding Equations of Planes

EXAMPLE 1

Find an equation of the plane that contains the point $(1, -2, 3)$ and is parallel to the plane $z = 2x + 3y - 4$.

SOLUTION To write an equation of a plane, we need a point on the plane and a normal vector $\mathbf{N}$. We have the point $(1, -2, 3)$. To determine a normal vector, we use the fact that a normal vector of a plane is also a normal vector of all parallel planes. This means we can choose $\mathbf{N}$ to be a vector that is normal to the plane $z = 2x + 3y - 4$. Writing this equation in standard form $-2x - 3y + z + 4 = 0$, we see that the vector $-2\mathbf{i} - 3\mathbf{j} + \mathbf{k}$ is a normal vector. An equation of the plane that contains the point $(1, -2, 3)$ and has normal vector $\mathbf{N} = -2\mathbf{i} - 3\mathbf{j} + \mathbf{k}$ is

$$-2(x - 1) - 3(y + 2) + (z - 3) = 0.$$

This equation simplifies to

$$-2x - 3y + z - 7 = 0 \text{ or } z = 2x + 3y + 7. \quad \blacksquare$$

EXAMPLE 2

Find an equation of the plane that contains the three points $A(1, 1, 1)$, $B(2, 4, 3)$, and $C(3, 2, 1)$.

SOLUTION We need a point P_0 on the plane and a normal vector $\mathbf{N}$. We are given three points on the plane. Let us choose the point $A(1, 1, 1)$ to be P_0. To find a normal vector, we first note that vectors from any one of the points A, B, C to any other of these points are in the plane that contains the points. Let us consider

$$\mathbf{AB} = (2 - 1)\mathbf{i} + (4 - 1)\mathbf{j} + (3 - 1)\mathbf{k} = \mathbf{i} + 3\mathbf{j} + 2\mathbf{k}$$

and

$$\mathbf{AC} = (3 - 1)\mathbf{i} + (2 - 1)\mathbf{j} + (1 - 1)\mathbf{k} = 2\mathbf{i} + \mathbf{j}.$$

See Figure 11.5.4. Since a normal vector of a plane is orthogonal to any vector in the plane, we want a normal vector that is orthogonal to both of the vectors **AB** and **AC**. Since the cross product of two vectors is orthogonal to both factors, this condition will be satisfied by the vector

$$\mathbf{N} = \mathbf{AB} \times \mathbf{AC} = \begin{vmatrix} \mathbf{i} & \mathbf{j} & \mathbf{k} \\ 1 & 3 & 2 \\ 2 & 1 & 0 \end{vmatrix} = \begin{vmatrix} 3 & 2 \\ 1 & 0 \end{vmatrix} \mathbf{i} - \begin{vmatrix} 1 & 2 \\ 2 & 0 \end{vmatrix} \mathbf{j} + \begin{vmatrix} 1 & 3 \\ 2 & 1 \end{vmatrix} \mathbf{k}$$

$$= -2\mathbf{i} + 4\mathbf{j} - 5\mathbf{k}.$$

An equation of the plane that contains the point $(1, 1, 1)$ and has normal vector $\mathbf{N} = -2\mathbf{i} + 4\mathbf{j} - 5\mathbf{k}$ is

$$-2(x - 1) + 4(y - 1) - 5(z - 1) = 0$$
$$-2x + 2 + 4y - 4 - 5z + 5 = 0,$$
$$-2x + 4y - 5z + 3 = 0.$$

We would get an equivalent equation if any of the points A, B, C were used as the point P_0 and any two vectors that connect two different pairs of points A, B, C were used to determine a normal vector. ∎

Angle Between Planes

The **angle of intersection** of two intersecting planes with normal vectors $\mathbf{N}_1$ and $\mathbf{N}_2$ is defined to be

$$\theta = \cos^{-1}\left(\frac{|\mathbf{N}_1 \cdot \mathbf{N}_2|}{|\mathbf{N}_1||\mathbf{N}_2|}\right).$$

This is the angle between the normal vectors, if the normal vectors are chosen so the angle between them is acute. Taking the absolute value of the dot product of $\mathbf{N}_1$ and $\mathbf{N}_2$ ensures $0 \le \theta \le \pi/2$. Geometrically, the angle between two planes is the smaller of the two angles between the lines formed by a cross section that is perpendicular to their line of intersection. See Figure 11.5.5.

FIGURE 11.5.5 The angle between two planes is the acute angle determined by their normal vectors.

EXAMPLE 3

Find the angle of intersection of the plane $x + y - z = 0$ and the plane $x - 3y + z - 1 = 0$.

SOLUTION The plane $x + y - z = 0$ has normal vector $\mathbf{N}_1 = \mathbf{i} + \mathbf{j} - \mathbf{k}$, and the plane $x - 3y + z - 1 = 0$ has normal vector $\mathbf{N}_2 = \mathbf{i} - 3\mathbf{j} + \mathbf{k}$. Then

$$\frac{|\mathbf{N}_1 \cdot \mathbf{N}_2|}{|\mathbf{N}_1||\mathbf{N}_2|} = \frac{|(1)(1) + (1)(-3) + (-1)(1)|}{\sqrt{1^2 + 1^2 + (-1)^2}\sqrt{1^2 + (-3)^2 + 1^2}} = \frac{|-3|}{\sqrt{3}\sqrt{11}} = \frac{\sqrt{3}}{\sqrt{11}}.$$

The angle between the planes is

$$\theta = \cos^{-1}\left(\frac{\sqrt{3}}{\sqrt{11}}\right) \approx 1.0213 \text{ rad.} \quad ∎$$

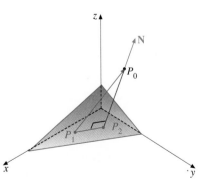

FIGURE 11.5.6 The distance from a point P_0 to the plane is the distance between P_0 and the point P_2 in the plane that is closest to P_0, so $\mathbf{P}_2\mathbf{P}_0$ is perpendicular to the plane. If P_1 is any point in the plane then $|\mathbf{P}_2\mathbf{P}_0| = |\text{proj}_\mathbf{N}\mathbf{P}_1\mathbf{P}_0|$.

Distance Between a Point and a Plane

The distance from a point P_0 to a plane is the distance from P_0 to the point in the plane that is closest to P_0. If P_2 is the point on the plane with the property that $\mathbf{P}_2\mathbf{P}_0$ is perpendicular to the plane, then the distance between P_2 and P_0 is smaller than the distance between P_0 and any other point on the plane, so

(Distance of P_0 to the plane) $= |\mathbf{P}_2\mathbf{P}_0|$.

See Figure 11.5.6. Let us see how to use the equation of the plane to evaluate the distance $|\mathbf{P}_2\mathbf{P}_0|$. The equation of the plane determines a normal vector $\mathbf{N}$.

We can also use the equation of the plane to find the coordinates of a point P_1 on the plane by choosing any convenient values (such as zero) for two of the variables x, y, z, and then solving the equation of the plane for the other variable. From Figure 11.5.6 we see that the distance between P_0 and P_2 is the length of the projection of $\mathbf{P_1P_0}$ in the direction of $\mathbf{P_2P_0}$. Since $\mathbf{P_2P_0}$ is perpendicular to the plane, $\mathbf{P_2P_0}$ is parallel to $\mathbf{N}$, so the projection of $\mathbf{P_1P_0}$ in the direction of $\mathbf{P_2P_0}$ is equal to the projection of $\mathbf{P_1P_0}$ in the direction of $\mathbf{N}$. We then have

(Distance of P_0 to plane) $= |\mathbf{P_2P_0}|$

$$= |\text{proj}_\mathbf{N}\mathbf{P_1P_0}| = \left| \frac{\mathbf{P_1P_0} \cdot \mathbf{N}}{\mathbf{N} \cdot \mathbf{N}} \mathbf{N} \right| = \frac{|\mathbf{P_1P_0} \cdot \mathbf{N}|}{\mathbf{N} \cdot \mathbf{N}} |\mathbf{N}|$$

$$= \frac{|\mathbf{P_1P_0} \cdot \mathbf{N}|}{|\mathbf{N}|^2} |\mathbf{N}| = \frac{|\mathbf{P_1P_0} \cdot \mathbf{N}|}{|\mathbf{N}|}.$$

If the plane has equation $ax + by + cz + d = 0$, we take $\mathbf{N} = a\mathbf{i} + b\mathbf{j} + c\mathbf{k}$. If $P_1(x_1, y_1, z_1)$ is on the plane, then we have $ax_1 + by_1 + cz_1 + d = 0$. Then

$$\mathbf{P_1P_0} \cdot \mathbf{N} = a(x_0 - x_1) + b(y_0 - y_1) + c(z_0 - z_1)$$
$$= ax_0 + by_0 + cz_0 - (ax_1 + by_1 + cz_1) = ax_0 + by_0 + cz_0 + d.$$

We then obtain the simplified formula

$$(\text{Distance of } P_0 \text{ to plane}) = \frac{|ax_0 + by_0 + cz_0 + d|}{\sqrt{a^2 + b^2 + c^2}}.$$

Also,

$$\text{proj}_\mathbf{N}\,\mathbf{P_1P_0} = \frac{ax_0 + by_0 + cz_0 + d}{a^2 + b^2 + c^2}\mathbf{N},$$

where P_1 is any point on the plane. If desired, we can find the coordinates of the point P_2 in the plane that is closest to the point P_0 by subtracting the components of $\text{proj}_\mathbf{N}\,\mathbf{P_1P_0}$ from the coordinates of P_0.

EXAMPLE 4

Find the distance from the point $P_0(2, 5, 4)$ to the plane $x + 2y + 2z = 2$. Find the point on the plane that is closest to P_0.

SOLUTION   We first write the equation of the plane as $x + 2y + 2z - 2 = 0$, with zero on one side of the equation. The simplified formula for the distance of a point to a plane then gives

$$\text{Distance of point to plane} = \frac{|(1)(2) + (2)(5) + (2)(4) - 2|}{\sqrt{1^2 + 2^2 + 2^2}} = \frac{18}{\sqrt{9}} = 6.$$

If P_1 is any point on the plane, the projection of $\mathbf{P_1P_0}$ in the direction of $\mathbf{N}$ is

$$\text{proj}_\mathbf{N}\,\mathbf{P_1P_0} = \frac{(1)(2) + (2)(5) + (2)(4) - 2}{1^2 + 2^2 + 2^2}(\mathbf{i} + 2\mathbf{j} + 2\mathbf{k})$$

$$= \frac{18}{9}(\mathbf{i} + 2\mathbf{j} + 2\mathbf{k}) = 2\mathbf{i} + 4\mathbf{j} + 4\mathbf{k}.$$

Let $\mathbf{OP_0}$ denote the vector from the origin to the point P_0. Then

$$\mathbf{OP_0} - \text{proj}_\mathbf{N}\,\mathbf{P_1P_0} = (2\mathbf{i} + 5\mathbf{j} + 4\mathbf{k}) - (2\mathbf{i} + 4\mathbf{j} + 4\mathbf{k}) = \mathbf{j},$$

so the point on the plane that is closest to P_0 is $(0, 1, 0)$. ∎

Intersection of Surfaces

FIGURE 11.5.7a Coordinate traces of each plane are sketched.

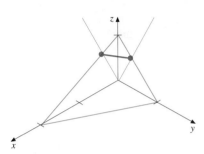

FIGURE 11.5.7b The line of intersection of the planes is drawn by connecting the points of intersection of the coordinate traces.

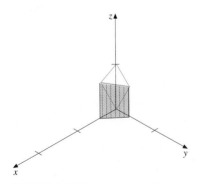

FIGURE 11.5.7c The vertical plane that contains the line of intersection is indicated by drawing vertical lines between the intersection and the xy-plane.

EXAMPLE 5

Find a vertical plane that contains the line of intersection of the planes $x + 2y + 2z = 2$ and $2x + 2y - z = 0$. Sketch all three planes and the line of intersection.

SOLUTION   Sketches of the planes are made by determining their coordinate traces. Coordinate traces of the plane $x + 2y + 2z = 2$ are sketched by determining the intercepts and then drawing the lines between the intercepts. The plane $2x + 2y - z = 0$ is indicated by sketching the trace $x = 0$, $z = 2y$, and the trace $y = 0$, $z = 2x$. See Figure 11.5.7a.

To sketch the line of intersection of the two planes, we note that the intersection of their traces in the yz-plane and the intersection of their traces in the xz-plane give two points on the line of intersection. Since the intersection of two planes is a straight line, we can sketch the line of intersection by drawing a straight line through the points we have determined from the two coordinate traces. See Figure 11.5.7b.

To find a vertical plane that contains the line of intersection, we note that points in the line of intersection satisfy both of the equations

$$x + 2y + 2z = 2 \text{ and } 2x + 2y - z = 0.$$

If these equations are algebraically combined to obtain a new equation, then points on the intersection will satisfy the new equation. This means that the graph of the new equation contains the line of intersection. We want the new equation to be the equation of a vertical plane. A plane is vertical if its equation contains no z term. That is, we need to algebraically combine the equations

$$x + 2y + 2z = 2 \text{ and } 2x + 2y - z = 0,$$

to obtain a single equation that contains no z term. One way to do this is to solve one equation for z and then substitute for z in the other equation. Solving the second equation for z, we obtain $z = 2x + 2y$. Substitution into the first equation then gives

$$x + 2y + 2(2x + 2y) = 2,$$
$$x + 2y + 4x + 4y = 2,$$
$$5x + 6y = 2.$$

This is the equation of a vertical plane that contains the line of intersection of the two given planes. We have indicated this plane in Figure 11.5.7c by drawing vertical lines between the line of intersection and the xy-plane. ∎

The intersection of the xy-plane with the vertical generalized cylinder that contains a curve is called the **projection** of the curve onto the xy-plane.

In Chapter 13 we will learn a procedure to describe certain types of regions in three-dimensional space. An important part of this procedure involves finding a vertical generalized cylinder that contains the intersection of two given surfaces. We will use the procedure introduced in Example 5. Namely,

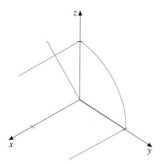

FIGURE 11.5.8a Coordinate traces of the cylinder and plane are sketched.

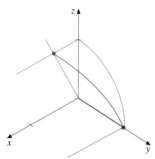

FIGURE 11.5.8b Points of intersection of the coordinate traces are connected by an elliptical arc.

FIGURE 11.5.8c The vertical generalized cylinder that contains the intersection is indicated by vertical lines. The projection of the intersection onto the *xy*-plane is the intersection of the vertical generalized cylinder with the *xy*-plane.

> ### To sketch the intersection of two surfaces
> - Sketch the coordinate traces of each surface.
> - Find points of intersection of the coordinate traces.
> - Connect the points of intersection you have found with a continuous curve. (We assume that the surfaces intersect in a single continuous curve.)

To find an equation of a vertical generalized cylinder that contains the intersection of two given surfaces, we combine the equations of the surfaces in such a way that z is eliminated from the new equation.

If the equations of two given surfaces are combined algebraically, the graph of the new equation is a surface that contains the intersection of the two surfaces. If the new equation contains no z term, its graph is a vertical generalized cylinder. That is, the graph is a generalized cylinder with generating lines parallel to the z-axis. The intersection of this generalized cylinder with the xy-plane is the projection of the curve of intersection onto the xy-plane.

EXAMPLE 6

Sketch the part in the first octant of the cylinder $y^2 + z^2 = 1$, the plane $z = 2x$, and their curve of intersection. Find a vertical generalized cylinder that contains the curve of intersection. Sketch the generalized cylinder and the projection of the curve of intersection onto the xy-plane.

SOLUTION The surface $y^2 + z^2 = 1$ is a cylinder with generating lines parallel to the x-axis. The yz-trace is a circle, which we sketch. The other traces are then obtained by drawing lines parallel to the x-axis from the intersections of the yz-trace with each of the y-axis and z-axis.

The yz-trace of the plane $z = 2x$ is the y-axis. The xz-trace is the line $y = 0, z = 2x$.

The traces of the cylinder and plane are sketched in Figure 11.5.8a. In Figure 11.5.8b, we have indicated intersections of the traces in the yz-plane and the xz-plane, and then drawn the curve of intersection between these two points of intersection.

To find a vertical cylinder that contains the curve of intersection, we combine the equations

$$y^2 + z^2 = 1, \quad z = 2x,$$

by substituting $2x$ for z in the first equation to obtain

$$y^2 + 4x^2 = 1.$$

This is the equation of a vertical elliptical cylinder. This cylinder and the projection of the curve of intersection onto the xy-plane have been added to the sketch in Figure 11.5.8c. ∎

EXERCISES 11.5

Find an equation of the plane that satisfies the conditions given in Exercises 1–12.

1. Contains $(2, 3, 1)$, normal vector $4\mathbf{i} + 3\mathbf{j} - 2\mathbf{k}$

2. Contains $(3, -1, 2)$, normal vector $\mathbf{i} - \mathbf{j} + 2\mathbf{k}$

3. Contains $(1, -1, 0)$, parallel to $x - 3y + 2z = 0$

4. Contains $(4, 3, -2)$, parallel to $2x - y + 3z = 2$

5. Contains $(5, 1, 3)$, parallel to xy-plane

6. Contains $(2, -2, 1)$, parallel to yz-plane

7. Contains $(1, -2, 1)$, $(2, 2, 2)$, and $(0, 0, 0)$

8. Contains $(3, 2, 0)$, $(5, 2, 0)$, and $(4, 0, 0)$

9. Contains $(4, 6, 5)$, $(3, 2, 7)$, and $(2, 1, 4)$

10. Contains $(1, 0, -1)$, $(-1, 1, 0)$, and $(0, 1, 1)$

11. Contains $(2, 1, -2)$, contains vectors in the direction of $\mathbf{i} + 2\mathbf{k}$ and $\mathbf{j} - 3\mathbf{k}$

12. Contains $(3, -1, 2)$, contains vectors in the direction of $\mathbf{i} - 2\mathbf{k}$ and $\mathbf{j}$

Find the angle of intersection of the planes in Exercises 13–16.

13. $x - y + z = 1$, $2x + y - z = 3$

14. $x = z$, $x + y + z = 0$

15. $2x + y - z = 2$, $3x + y - z = 3$

16. $3x - 2y = z$, $2x + 3y = z$

Find the distance of the given point from the plane, and the point in the plane that is closest to the given point in each of Exercises 17–20.

17. $(0, 0, 0)$, $2x + 3y + 2z = 6$

18. $(1, 2, 0)$, $3x - y + 2z = 6$

19. $(-1, 1, 2)$, $x + y = 2$

20. $(3, -2, 2)$, $2x + 2y + z = 6$

Find the distance between the parallel planes given in Exercises 21–22.

21. $2x - 3y + z = 6$, $2x - 3y + z = 2$

22. $x + 2y - z = 0$, $x + 2y - z = 4$

In Exercises 23–32, sketch the part in the first octant of the given surfaces and their curve of intersection. Find a vertical generalized cylinder that contains the intersection of the surfaces. Sketch the generalized cylinder and the projection of the curve of intersection onto the xy-plane.

23. $x + y + z = 3$, $z = 2x$

24. $x + z = 3$, $y + z = 3$

25. $x + 2y + 2z = 4$, $y = 2z$

26. $z = 2x + 2y$, $z = 3$

27. $z = 2x + 2y$, $x + z = 3$

28. $z = 3x$, $x + z = 3$

29. $z = 3x + 3y$, $z = 4 - x^2 - y^2$

30. $2y = z$, $x^2 + z^2 = 9$

31. $x + 2y + 2z = 6$, $x^2 + y^2 = 4$

32. $x^2 + y^2 = 9$, $x^2 + z^2 = 9$

In Exercises 33–37, find extreme values of z for points on the given planes with x- and y-coordinates that satisfy the given conditions. (Hint: A sketch should help you solve the problem without using calculus.)

33. $x - 2y + 2z = 2$; $0 \le x \le 1$, $0 \le y \le 1$

34. $2x + 3y + 3z = 6$; $0 \le x \le 1$, $0 \le y \le 1$

35. $2y - z + 1 = 0$; $x^2 + y^2 \le 1$

36. $x + y + 2z = 1$; $x^2 + y^2 \le 1$

11.6   LINES IN THREE-DIMENSIONAL SPACE

Connections

Parametric equations of lines in the plane, 11.1.

Angles between vectors and projection, 11.4.

Planes, 11.5.

Let us consider a vertical line through the corner of a rectangular room. We can describe such a line as the intersection of the two planes determined by the walls. The line can also be characterized as the set of points for which the vector from the corner of the room to the point is vertical. See Figure 11.6.1a. We will see that each of these types of descriptions is useful.

As in the case of a vertical line through the corner of a rectangular room, any **line** in three-dimensional space is determined by a point on the line and a nonzero direction vector. A point $P(x, y, z)$ is on the line that contains the point $P_0(x_0, y_0, z_0)$ and has direction vector $\mathbf{V} = a\mathbf{i} + b\mathbf{j} + c\mathbf{k}$ if and only if the vector $\mathbf{P_0P} = (x - x_0)\mathbf{i} + (y - y_0)\mathbf{j} + (z - z_0)\mathbf{k}$ is a multiple of $\mathbf{V}$. See Figure 11.6.1b. That is,

$$\mathbf{P_0P} = t\mathbf{V}, \quad t \text{ any real number.}$$

FIGURE 11.6.1a The line of intersection of the walls can be characterized as the set of points for which the vector from the corner to the point is vertical.

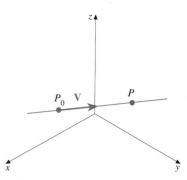

FIGURE 11.6.1b A point P is on the line that contains the point P_0 and has direction vector $\mathbf{V}$ if and only if $\mathbf{P_0P}$ is parallel to $\mathbf{V}$.

This is the **vector form** of the equation of the line. Let us express this equation in terms of the components of the vectors. This gives

$$(x - x_0)\mathbf{i} + (y - y_0)\mathbf{j} + (z - z_0)\mathbf{k} = t(a\mathbf{i} + b\mathbf{j} + c\mathbf{k}),$$

$$(x - x_0)\mathbf{i} + (y - y_0)\mathbf{j} + (z - z_0)\mathbf{k} = at\mathbf{i} + bt\mathbf{j} + ct\mathbf{k}.$$

Equating corresponding components of the vectors on each side of the above equation, we obtain the equations

$$x - x_0 = at, \quad y - y_0 = bt, \quad z - z_0 = ct,$$

or

$$x = x_0 + at, \quad y = y_0 + bt, \quad z = z_0 + ct.$$

These equations are **parametric equations** of the line that contains the point (x_0, y_0, z_0) and has direction vector $\mathbf{V} = a\mathbf{i} + b\mathbf{j} + c\mathbf{k}$. The parameter t ranges over all real numbers.

We can obtain a different form of the equations of a line by eliminating the parameter from the parametric equations of the line. If none of a, b, or c is zero, we can solve each of the three parametric equations of a line for the parameter t. Thus,

$$t = \frac{x - x_0}{a}, \quad t = \frac{y - y_0}{b}, \quad t = \frac{z - z_0}{c}.$$

We can then express the equations of the line as

$$\frac{x - x_0}{a} = \frac{y - y_0}{b} = \frac{z - z_0}{c}.$$

This is called the **symmetric form** of the equations of the line that contains the point (x_0, y_0, z_0) and has direction vector $\mathbf{V} = a\mathbf{i} + b\mathbf{j} + c\mathbf{k}$. If one or two of a, b, or c are zero, the corresponding variable is constant. We should replace the equation that involves the corresponding variable by the equation that expresses the constant value. For example, the equations of the line through the point $(1, 3, 2)$ with direction vector $\mathbf{V} = \mathbf{j} - 2\mathbf{k}$ should be written as

$$x = 1, \quad \frac{y - 3}{1} = \frac{z - 2}{-2}, \quad \text{instead of} \quad \frac{x - 1}{0} = \frac{y - 3}{1} = \frac{z - 2}{-2}.$$

Symmetric equations of a line can be considered as two equations. For example, the symmetric equations

$$\frac{x - x_0}{a} = \frac{y - y_0}{b} = \frac{z - z_0}{c}$$

give the two equations

$$\frac{x - x_0}{a} = \frac{y - y_0}{b} \quad \text{and} \quad \frac{y - y_0}{b} = \frac{z - z_0}{c}.$$

The third equation, which relates x and z, is a consequence of these two equations. These equations are both equations of planes. Thus, we can think of a line in three-dimensional space as the intersection of two planes. Also, the equation

$$\frac{x - x_0}{a} = \frac{y - y_0}{b}$$

is a vertical generalized cylinder that contains the line. This plane gives the projection of the line onto the xy-plane. Similarly, the equation

$$\frac{y - y_0}{b} = \frac{z - z_0}{c}$$

gives the projection of the line onto the yz-plane. See Figure 11.6.2.

Equations of a line that correspond to different points P_0 on the line and different (but parallel) direction vectors $\mathbf{V}$ may appear very different, even though they represent the same line.

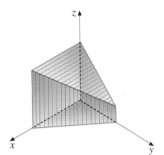

FIGURE 11.6.2 The line is the intersection of the vertical plane $\dfrac{x - x_0}{a} = \dfrac{y - y_0}{b}$ and the plane $\dfrac{y - y_0}{b} = \dfrac{z - z_0}{c}$.

To show that two different sets of equations correspond to the same line

- Check that the coordinates of some point on one line satisfy the equations of the other line.
- Check that the direction vectors of the two lines are parallel. That is, check that $\mathbf{V}_1 = k\mathbf{V}_2$ for some constant k.

We can *read* a direction vector from either the parametric or symmetric equations of a line.

- The line with symmetric equations

$$\frac{x - 1}{2} = \frac{y + 3}{-1} = \frac{z}{3}$$

has direction vector $2\mathbf{i} - \mathbf{j} + 3\mathbf{k}$.
- The line with parametric equations

$$x = -1 + 2t, \quad y = 3 + 5t, \quad z = -1 - 4t,$$

has direction vector $2\mathbf{i} + 5\mathbf{j} - 4\mathbf{k}$.

We also note:

Parallel lines have parallel direction vectors.

Finding Equations of Lines

EXAMPLE 1

Find symmetric equations and parametric equations of the line that contains the point $(3, 5, 2)$ and is parallel to the line

$$\frac{x}{-2} = \frac{y}{1} = \frac{z}{3}.$$

SOLUTION We need a point on the line and a direction vector $\mathbf{V}$. We have the point $(3, 5, 2)$ and we know that parallel lines have parallel direction vectors. A direction vector of the given parallel line is $\mathbf{V} = -2\mathbf{i} + \mathbf{j} + 3\mathbf{k}$. Using $\mathbf{V}$ for the direction vector, we obtain the symmetric equations

$$\frac{x-3}{-2} = \frac{y-5}{1} = \frac{z-2}{3}.$$

Parametric equations of the line are

$$x = 3 - 2t, \quad y = 5 + t, \quad z = 2 + 3t.$$

See Figure 11.6.3. ■

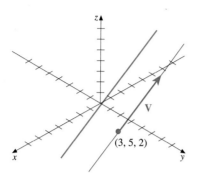

FIGURE 11.6.3 A line parallel to the line $\frac{x}{-2} = \frac{y}{1} = \frac{z}{3}$ has direction vector $-2\mathbf{i} + \mathbf{j} + 3\mathbf{k}$.

EXAMPLE 2

Find symmetric equations and parametric equations of the line that contains the point $(3, 2, 1)$ and is perpendicular to the plane

$$x + 2y - 2x = 2.$$

SOLUTION We have a point on the line. We need a direction vector. If the line is to be perpendicular to the plane, then any direction vector of the line must be parallel to any normal vector of the plane. See Figure 11.6.4. This means we can use a normal vector of the plane as a direction vector of the line. A vector normal to the plane $x + 2y - 2z = 2$ is $\mathbf{N} = \mathbf{i} + 2\mathbf{j} - 2\mathbf{k}$. Symmetric equations of the line that contains the point $(3, 2, 1)$ and has direction vector $\mathbf{i} + 2\mathbf{j} - 2\mathbf{k}$ are

$$\frac{x-3}{1} = \frac{y-2}{2} = \frac{z-1}{-2}.$$

Parametric equations of the line are

$$x = 3 + t, \quad y = 2 + 2t, \quad z = 1 - 2t. \quad ■$$

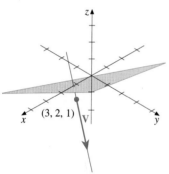

FIGURE 11.6.4 A line that is perpendicular to the plane $x + 2y - 2z = 2$ has direction vector $\mathbf{i} + 2\mathbf{j} - 2\mathbf{k}$.

EXAMPLE 3

Find symmetric equations and parametric equations of the line that contains the two points $A(1, -2, 3)$ and $B(3, 4, 2)$.

SOLUTION We need a point on the line and a direction vector. Let us use the point $A(1, -2, 3)$ as the fixed point on the line. We need a direction vector that is parallel to the vector $\mathbf{AB}$. See Figure 11.6.5. Choose

$$\mathbf{V} = \mathbf{AB} = (3-1)\mathbf{i} + (4+2)\mathbf{j} + (2-3)\mathbf{k} = 2\mathbf{i} + 6\mathbf{j} - \mathbf{k}.$$

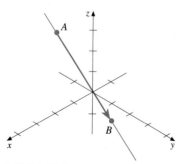

FIGURE 11.6.5 The line that contains the points A and B has direction vector $\mathbf{AB}$.

The symmetric equations are

$$\frac{x-1}{2} = \frac{y+2}{6} = \frac{z-3}{-1}.$$

Parametric equations of the line are

$$x = 1 + 2t, \quad y = -2 + 6t, \quad z = 3 - t. \quad \blacksquare$$

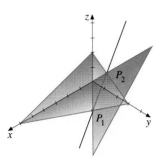

FIGURE 11.6.6 The line of intersection of two planes has direction vector $\mathbf{P_1P_2}$, where P_1 and P_2 are any two points on the intersection.

EXAMPLE 4

Find symmetric equations and parametric equations of the line of intersection of the two planes $x - y + z = 0$ and $x + 2y + 3z = 6$.

SOLUTION We need a point on the line and a direction vector. We also note that any two points on the line can be used to determine a direction vector. See Figure 11.6.6.

Coordinates of points on the line of intersection of the planes are solutions of the simultaneous equations

$$\begin{cases} x - y + z = 0, \\ x + 2y + 3z = 6. \end{cases}$$

We can find a solution of these equations by choosing any convenient value for one of the variables and then solving the equations for the other two variables. For example, let $z = 0$. We then have

$$\begin{cases} x - y = 0, \\ x + 2y = 6. \end{cases}$$

The solutions are $x = 2$ and $y = 2$. The point $P_1 = (2, 2, 0)$ is on the line of intersection. Setting $x = 0$ in the equations of the planes, we obtain

$$\begin{cases} -y + z = 0, \\ 2y + 3z = 6. \end{cases}$$

Solution of these equations gives $y = z = 6/5$. The point $P_2 = (0, 6/5, 6/5)$ is on the line.

We can use the vector between the two points we have found as a direction vector of the line. We have

$$\mathbf{P_1P_2} = (0 - 2)\mathbf{i} + \left(\frac{6}{5} - 2\right)\mathbf{j} + \left(\frac{6}{5} - 0\right)\mathbf{k}$$

$$= -2\mathbf{i} - \frac{4}{5}\mathbf{j} + \frac{6}{5}\mathbf{k}.$$

Since any nonzero multiple of a direction vector of a line is also a direction vector of the line, we can choose our direction vector to be any nonzero multiple of $\mathbf{P_1P_2}$. We multiply by $5/2$ in order to obtain a simplified vector with integer components. We have

$$\mathbf{V} = \frac{5}{2}\mathbf{P_1P_2} = \frac{5}{2}\left(-2\mathbf{i} - \frac{4}{5}\mathbf{j} + \frac{6}{5}\mathbf{k}\right) = -5\mathbf{i} - 2\mathbf{j} + 3\mathbf{k}.$$

Thus, symmetric equations of the line through $P_1(2, 2, 0)$ with direction vector $-5\mathbf{i} - 2\mathbf{j} + 3\mathbf{k}$ are

$$\frac{x-2}{-5} = \frac{y-2}{-2} = \frac{z-0}{3}.$$

Parametric equations of the line are

$$x = 2 - 5t, \quad y = 2 - 2t, \quad z = 3t. \quad \blacksquare$$

Since the line of intersection of two planes is contained in each plane, any direction vector of the line is orthogonal to normal vectors of each plane. This means that we can find a direction vector of the line of intersection of two planes by taking a cross product of normal vectors of the planes.

For example, the planes in Example 4 have normal vectors $\mathbf{N}_1 = \mathbf{i} - \mathbf{j} + \mathbf{k}$ and $\mathbf{N}_2 = \mathbf{i} + 2\mathbf{j} + 3\mathbf{k}$. See Figure 11.6.7. You can verify that $\mathbf{N}_1 \times \mathbf{N}_2 = -5\mathbf{i} - 2\mathbf{j} + 3\mathbf{k}$, so $\mathbf{N}_1 \times \mathbf{N}_2$ is a direction vector of the line of intersection of the planes.

FIGURE 11.6.7 The cross product of normal vectors of two planes gives the direction of the line of intersection of the planes.

Intersection of a Line and a Plane

Parametric equations of a line are convenient for finding the point of intersection of the line and a plane.

EXAMPLE 5

Find the point of intersection of the line

$$\frac{x-1}{3} = \frac{y+1}{1} = \frac{z-2}{-2}$$

and the plane $2x - 3y + z = 6$.

SOLUTION Parametric equations of the line are

$$x = 1 + 3t, \quad y = -1 + t, \quad z = 2 - 2t.$$

The point (x, y, z) is on the plane if and only if the coordinates satisfy the equation of the plane. Substitution into the equation of the plane gives

$$2x - 3y + z = 6,$$
$$2(1 + 3t) - 3(-1 + t) + (2 - 2t) = 6,$$
$$2 + 6t + 3 - 3t + 2 - 2t = 6,$$
$$t = -1.$$

It follows that the point on the line that corresponds to $t = -1$ is also on the plane. Substituting $t = -1$ into the parametric equations of the line gives $(x, y, z) = (-2, -2, 4)$. This is the point of intersection of the line and the plane. $\blacksquare$

NOTE   *If a line is parallel to a plane, the technique of Example 5 will give either an equation (such as $0 = 1$) that is not satisfied for any values of the parameter if the line does not intersect the plane or an equation (such as $0 = 0$) that is satisfied for all values of the parameter if the line is contained in the plane.*

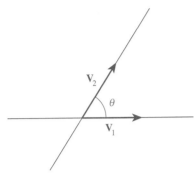

FIGURE 11.6.8a The angle of intersection of two intersecting lines is

$$\theta = \cos^{-1} \frac{|\mathbf{V}_1 \cdot \mathbf{V}_2|}{|\mathbf{V}_1||\mathbf{V}_2|}$$

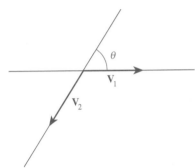

FIGURE 11.6.8b The absolute value of the dot product in the formula insures that $0 \le \theta \le \pi/2$, so the angle between the lines is the smaller of the angles formed by the lines.

Intersecting Lines

The **angle of intersection** of two intersecting lines with direction vectors $\mathbf{V}_1$ and $\mathbf{V}_2$ satisfies

$$\cos \theta = \frac{|\mathbf{V}_1 \cdot \mathbf{V}_2|}{|\mathbf{V}_1||\mathbf{V}_2|}.$$

See Figure 11.6.8.

The intersection of two lines can be found by using parametric equations of the lines, but *different parameters* must be used for the lines. We can then express each of the three coordinates of the point of intersection in terms of each of the two parameters. This gives three equations in the two parametric variables.

NOTE *Be sure to check that the coordinates you find for the point of intersection of two lines satisfy the equations of both lines.*

EXAMPLE 6

Find the point of intersection and the angle between the lines

$$\frac{x+1}{2} = \frac{y-2}{1} = \frac{z-4}{-2} \quad \text{and} \quad \frac{x-3}{-1} = \frac{y-1}{1} = \frac{z-0}{1}.$$

SOLUTION Using different parameters to express the lines in parametric form, we obtain

$$x = -1 + 2t, \quad y = 2 + t, \quad z = 4 - 2t,$$

and

$$x = 3 - s, \quad y = 1 + s, \quad z = s.$$

The coordinates of the point of intersection, if the lines actually intersect, must satisfy

$$-1 + 2t = x = 3 - s,$$
$$2 + t = y = 1 + s,$$
$$4 - 2t = z = s,$$

so we obtain the system of linear equations

$$\begin{cases} -1 + 2t &= 3 - s, \\ 2 + t &= 1 + s, \\ 4 - 2t &= s. \end{cases}$$

It is not difficult to determine that this system has solution $t = 1, s = 2$. Substitution of these values into the parametric equations gives

$$x = -1 + 2(1) = 1, \quad y = 2 + (1) = 3, \quad z = 4 - 2(1) = 2,$$

and

$$x = 3 - (2) = 1, \quad y = 1 + (2) = 3, \quad z = (2) = 2.$$

This shows that the point $(1, 3, 2)$ is on both lines.

The lines have direction vectors

$$\mathbf{V}_1 = 2\mathbf{i} + \mathbf{j} - 2\mathbf{k} \text{ and } \mathbf{V}_2 = -\mathbf{i} + \mathbf{j} + \mathbf{k}.$$

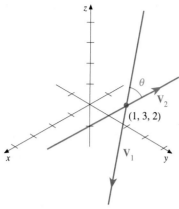

FIGURE 11.6.9 The lines intersect at the point $(1, 3, 2)$. The angle between the lines is approximately $\theta = 0.9553$ rad.

The angle between the lines satisfies

$$
\begin{aligned}
\cos \theta &= \frac{|\mathbf{V}_1 \cdot \mathbf{V}_2|}{|\mathbf{V}_1||\mathbf{V}_2|} \\
&= \frac{|(2\mathbf{i} + \mathbf{j} - 2\mathbf{k}) \cdot (-\mathbf{i} + \mathbf{j} + \mathbf{k})|}{|2\mathbf{i} + \mathbf{j} - 2\mathbf{k}||-\mathbf{i} + \mathbf{j} + \mathbf{k}|} \\
&= \frac{|-2 + 1 - 2|}{\sqrt{4 + 1 + 4}\sqrt{1 + 1 + 1}} \\
&= \frac{3}{3\sqrt{3}} = \frac{1}{\sqrt{3}}.
\end{aligned}
$$

Then $\theta = \cos^{-1}(1/\sqrt{3}) \approx 0.9553$ rad. See Figure 11.6.9. ■

We can also use symmetric equations of lines to find the intersection of the lines. Symmetric equations of a line give two equations in the variables x, y, and z. Thus, the coordinates of the point of intersection of two lines must satisfy four equations. When finding the point of intersection of two lines, be sure to check that the coordinates you find actually satisfy symmetric equations of both lines.

EXAMPLE 7

Find the equation of the plane that contains the line

$$
\frac{x - 1}{1} = \frac{y - 4}{-2} = \frac{z - 5}{-1}
$$

and is parallel to the line

$$
\frac{x}{1} = \frac{y}{2} = \frac{z}{-1}.
$$

Show that the two lines do not intersect.

SOLUTION To write the equation of a plane, we need a point on the plane and a normal vector. Since the plane is to contain the line

$$
\frac{x - 1}{1} = \frac{y - 4}{-2} = \frac{z - 5}{-1},
$$

the plane must contain the point $(1, 4, 5)$, which is on the line, and the normal of the plane must be orthogonal to the direction vector of the line, $\mathbf{V}_1 = \mathbf{i} - 2\mathbf{j} - \mathbf{k}$. Since the plane is to be parallel to the line

$$
\frac{x}{1} = \frac{y}{2} = \frac{z}{-1},
$$

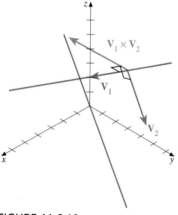

FIGURE 11.6.10 $\mathbf{V}_1 \times \mathbf{V}_2$ is a normal vector of the plane that contains the line with direction $\mathbf{V}_1$ and is parallel to the line with direction $\mathbf{V}_2$.

the normal to the plane must also be orthogonal to a direction vector of that line, $\mathbf{V}_2 = \mathbf{i} + 2\mathbf{j} - \mathbf{k}$. See Figure 11.6.10. Since the cross product is orthogonal to both vectors, we can use $\mathbf{V}_1 \times \mathbf{V}_2$ as a normal vector of the plane.

We have

$$
\mathbf{V}_1 \times \mathbf{V}_2 = \begin{vmatrix} \mathbf{i} & \mathbf{j} & \mathbf{k} \\ 1 & -2 & -1 \\ 1 & 2 & -1 \end{vmatrix} = \begin{vmatrix} -2 & -1 \\ 2 & -1 \end{vmatrix} \mathbf{i} - \begin{vmatrix} 1 & -1 \\ 1 & -1 \end{vmatrix} \mathbf{j} + \begin{vmatrix} 1 & -2 \\ 1 & 2 \end{vmatrix} \mathbf{k}
$$

$$
= 4\mathbf{i} + 4\mathbf{k}.
$$

The equation of the plane that contains the point $(1, 4, 5)$ with normal vector $4\mathbf{i} + 4\mathbf{k}$ is

$$4(x - 1) + 0(y - 4) + 4(z - 5) = 0,$$
$$4x - 4 + 4z - 20 = 0,$$
$$4x + 4z = 24,$$
$$x + z = 6.$$

Let us now use parametric equations of the lines to verify that the two lines do not intersect. Recall that the lines must be expressed in terms of different parameters. Parametric equations of the lines are

$$x = 1 + t, \quad y = 4 - 2t, \quad z = 5 - t$$

and

$$x = s, \quad y = 2s, \quad z = -s.$$

Setting corresponding coordinates of points on each line equal, we obtain the system of equations

$$\begin{cases} 1 + t = s, \\ 4 - 2t = 2s, \\ 5 - t = -s. \end{cases}$$

The first two equations have solution $(t, s) = (1/2, 3/2)$. Since these values do not satisfy the third equation, the system of equations has no solution. Since any point of intersection of the lines must satisfy this system of equations, we conclude that the lines do not intersect. (Note that the values $t = 1/2$ and $s = 3/2$ give points on the lines that have the same x- and y-coordinates, but different z-coordinates.) ■

We can use the ideas of Example 7 to find parallel planes that contain two nonparallel lines. Suppose the directions of two lines are given by the nonparallel vectors $\mathbf{V}_1$ and $\mathbf{V}_2$. See Figure 11.6.11. A plane that contains the line with direction $\mathbf{V}_1$ has normal vector that is orthogonal to $\mathbf{V}_1$. A plane that contains the line with direction $\mathbf{V}_2$ has normal vector orthogonal to $\mathbf{V}_2$. Parallel planes have parallel normal vectors, so we can choose both normal vectors to be any vector that is orthogonal to both $\mathbf{V}_1$ and $\mathbf{V}_2$. The cross product $\mathbf{N} = \mathbf{V}_1 \times \mathbf{V}_2$ is such a vector.

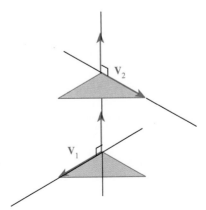

FIGURE 11.6.11 The cross product $\mathbf{V}_1 \times \mathbf{V}_2$ gives a vector that is normal to planes that are parallel to the lines with direction vectors $\mathbf{V}_1$ and $\mathbf{V}_2$.

FIGURE 11.6.12 The distance of P_0 to the line is $|\mathbf{P}_1\mathbf{P}_0|\sin\theta$, where P_1 is any point on the line.

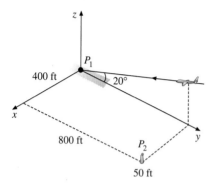

FIGURE 11.6.13 The shortest distance between the airplane and the light is the distance between the line of descent of the plane and the point where the light is located.

Distance Between a Point and a Line

Let us see how to find the distance of a point P_0 from a line in three-dimensional space. We first choose any point P_1 that is on the line. From Figure 11.6.12 we see that the distance of the point from the line is

$$d = |\mathbf{P}_1\mathbf{P}_0|\sin\theta,$$

where θ is the angle between the vector $\mathbf{P}_1\mathbf{P}_0$ and a direction vector $\mathbf{V}$ of the line. From Section 11.4 we know that

$$|\mathbf{P}_1\mathbf{P}_0 \times \mathbf{V}| = |\mathbf{P}_1\mathbf{P}_0||\mathbf{V}|\sin\theta.$$

Substitution into the formula for d then gives

$$(\text{Distance of } \mathbf{P}_0 \text{ to line}) = |\mathbf{P}_1\mathbf{P}_0|\frac{|\mathbf{P}_1\mathbf{P}_0 \times \mathbf{V}|}{|\mathbf{P}_1\mathbf{P}_0||\mathbf{V}|} = \frac{|\mathbf{P}_1\mathbf{P}_0 \times \mathbf{V}|}{|\mathbf{V}|}.$$

EXAMPLE 8

An airplane approaches the end of a runway from the east with an angle of descent of $20°$. A tower that is 50 ft high is located 400 ft south and 800 ft east of the end of the runway. How close does the airplane come to a warning light that is at the top of the tower?

SOLUTION   A coordinate system is chosen as indicated in Figure 11.6.13. We want the distance of the point $P_0(400, 800, 50)$ from the line of descent of the airplane. The point $P_1(0, 0, 0)$ is on the line and the line has direction vector

$$\mathbf{V} = \cos 20°\mathbf{j} + \sin 20°\mathbf{k}.$$

We then have

$$\mathbf{P}_1\mathbf{P}_0 = 400\mathbf{i} + 800\mathbf{j} + 50\mathbf{k}$$

and

$$\mathbf{P}_1\mathbf{P}_0 \times \mathbf{V} = \begin{vmatrix} \mathbf{i} & \mathbf{j} & \mathbf{k} \\ 400 & 800 & 50 \\ 0 & \cos 20° & \sin 20° \end{vmatrix}$$

$$= \begin{vmatrix} 800 & 50 \\ \cos 20° & \sin 20° \end{vmatrix}\mathbf{i} - \begin{vmatrix} 400 & 50 \\ 0 & \sin 20° \end{vmatrix}\mathbf{j} + \begin{vmatrix} 400 & 800 \\ 0 & \cos 20° \end{vmatrix}\mathbf{k}$$

$$= (800\sin 20° - 50\cos 20°)\mathbf{i} - 400\sin 20°\mathbf{j} + 400\cos 20°\mathbf{k}.$$

The distance of P_0 from the line is

$$d = \frac{|\mathbf{P}_1\mathbf{P}_0 \times \mathbf{V}|}{|\mathbf{V}|}$$

$$= \frac{|(800\sin 20° - 50\cos 20°)\mathbf{i} - 400\sin 20°\mathbf{j} + 400\cos 20°\mathbf{k}|}{|\cos 20°\mathbf{j} + \sin 20°\mathbf{k}|}$$

$$= \frac{\sqrt{(800\sin 20° - 50\cos 20°)^2 + (-400\sin 20°)^2 + (400\cos 20°)^2}}{\sqrt{\cos^2 20° + \sin^2 20°}}$$

$$\approx 459.74105.$$

The airplane is never closer than approximately 460 ft from the warning light.

EXERCISES 11.6

Find symmetric equations and parametric equations of the lines that satisfy the conditions given in Exercises 1–14.

1. Contains $(1, -1, 2)$, direction vector $2\mathbf{i} - 2\mathbf{j} + 3\mathbf{k}$

2. Contains $(2, 3, -1)$, direction vector $3\mathbf{i} + \mathbf{j} + 2\mathbf{k}$

3. Contains $(2, 3, -1)$, parallel to $\dfrac{x-1}{2} = \dfrac{y-3}{-1} = \dfrac{z+1}{3}$

4. Contains $(3, 1, -2)$, parallel to $\dfrac{x+1}{3} = \dfrac{y-2}{2} = \dfrac{z-4}{-2}$

5. Contains $(2, 4, -3)$, parallel to x-axis

6. Contains $(-3, 2, -1)$, parallel to y-axis

7. Contains $(1, -2, 3)$, perpendicular to $2x - 3y + z = 6$

8. Contains $(3, 2, -2)$, perpendicular to $x + 2y = 2$

9. Contains $(2, 2, 1)$ and $(1, 1, 3)$

10. Contains $(1, -2, 1)$ and $(3, 1, 2)$

11. Contains $(5, 3, 7)$ and $(2, 4, 3)$

12. Contains $(3, 2, 5)$ and $(3, 2, 1)$

13. Intersection of $x + y - z = 0$ and $2x + 2y - z = 2$

14. Intersection of $3x + y + 2z = 6$ and $2x - 2y + 3z = 6$

Find the point of intersection of the line and plane given in each of Exercises 15–18.

15. $3x - 3y + 2z = 6$, $\dfrac{x-1}{2} = \dfrac{y-1}{2} = \dfrac{z+1}{-1}$

16. $2x + 2y + z = 4$, $\dfrac{x}{2} = \dfrac{y+2}{1} = \dfrac{z-1}{1}$

17. $x + y + z = 1$, $\dfrac{x-2}{3} = \dfrac{y+1}{-2} = \dfrac{z-3}{2}$

18. $3x + y + 2z = 13$, $\dfrac{x}{2} = \dfrac{y}{1} = \dfrac{z}{3}$

Determine if the lines in Exercises 19–22 intersect. If they do, find the point of intersection and the angle of intersection.

19. $\dfrac{x-2}{2} = \dfrac{y-3}{2} = \dfrac{z-3}{1}, \dfrac{x+2}{2} = \dfrac{y-2}{-1} = \dfrac{z}{2}$

20. $\dfrac{x}{1} = \dfrac{y}{1} = \dfrac{z}{2}, \dfrac{x+2}{2} = \dfrac{y+2}{2} = \dfrac{z-2}{1}$

21. $\dfrac{x}{1} = \dfrac{y+1}{1} = \dfrac{z+2}{2}, \dfrac{x-1}{1} = \dfrac{y-1}{1} = \dfrac{z-2}{2}$

22. $\dfrac{x-2}{2} = \dfrac{y-2}{2} = \dfrac{z+1}{-1}, \dfrac{x-4}{2} = \dfrac{y+4}{-2} = \dfrac{z}{1}$

In Exercises 23–26, find the distance between the point and the line.

23. $(0, 0, 0)$, $\dfrac{x-1}{3} = \dfrac{y-3}{2} = \dfrac{z-2}{4}$

24. $(1, -2, -1)$, $\dfrac{x-2}{2} = \dfrac{y+2}{-3} = \dfrac{z-1}{1}$

25. $(2, 3, -1)$, $\dfrac{x+1}{1} = \dfrac{y+2}{1} = \dfrac{z-2}{-1}$

26. $(3, -2, 2)$, $\dfrac{x+2}{3} = \dfrac{y-1}{-1} = \dfrac{z+1}{2}$

27. Find an equation of the plane that contains
$$\dfrac{x-1}{2} = \dfrac{y-1}{2} = \dfrac{z-2}{-1}$$
and is parallel to
$$\dfrac{x}{1} = \dfrac{y}{2} = \dfrac{x}{1}.$$

28. Find an equation of the plane that contains
$$\dfrac{x+1}{2} = \dfrac{y-2}{-2} = \dfrac{z-3}{1}$$
and is parallel to
$$\dfrac{x}{3} = \dfrac{y}{2} = \dfrac{x}{4}.$$

In each of Exercises 29–30, use the projection of a vector between points on the two given lines in the direction of a vector that is orthogonal to both lines to find the (shortest) distance between points on the planes. This is the distance between the lines.

29. $\dfrac{x-1}{2} = \dfrac{y-1}{1} = \dfrac{z-1}{1}$ and $\dfrac{x-2}{1} = \dfrac{y-1}{-2} = \dfrac{z+1}{2}$

30. $\dfrac{x-2}{1} = \dfrac{y+1}{1} = \dfrac{z+2}{-2}$ and $\dfrac{x-1}{2} = \dfrac{y-2}{-2} = \dfrac{z+1}{1}$

In Exercises 31–36, determine if there is a plane that contains the two given lines. If there is such a plane, find its equation.

31. $\dfrac{x-2}{1} = \dfrac{y-0}{3} = \dfrac{z-1}{-1}, \dfrac{x-1}{3} = \dfrac{y-1}{1} = \dfrac{z-1}{2}$

32. $\dfrac{x-2}{1} = \dfrac{y-0}{3} = \dfrac{z-1}{-1}, \dfrac{x-3}{3} = \dfrac{y-3}{1} = \dfrac{z-0}{2}$

33. $\dfrac{x+1}{2} = \dfrac{y-1}{-1} = \dfrac{z+2}{2}, \dfrac{x-1}{-2} = \dfrac{y-0}{3} = \dfrac{z-0}{1}$

34. $\dfrac{x+1}{2} = \dfrac{y-1}{-1} = \dfrac{z+2}{2}, \dfrac{x-1}{-2} = \dfrac{y-2}{3} = \dfrac{z-0}{1}$

35. $\dfrac{x-2}{1} = \dfrac{y-0}{3} = \dfrac{z-1}{-1}, \dfrac{x-1}{1} = \dfrac{y-1}{3} = \dfrac{z+1}{-1}$

36. $\dfrac{x+1}{2} = \dfrac{y-1}{-1} = \dfrac{z+2}{2}, \dfrac{x-1}{2} = \dfrac{y-2}{-1} = \dfrac{z-0}{2}$

37. A 5-ft-square target is leaning with its bottom edge on the floor. The bottom edge is 3 ft from the west wall and one side is against the north wall. An arrow is shot to the center of the target from a point that is 21.5 ft from the west wall, 8.5 ft from the north wall, and 4 ft above the floor. Although the path of the arrow is parabolic, we may assume it is a straight line over short distances. Find the angle between the

arrow and a vector that is normal to the target. See figure below.

38. An airplane that is flying a straight line course at an altitude of 2000 ft passes over a point on the ground that is 8000 ft due south of a radio station at ground level and passes over another point that is 8000 ft due east of the station. What is the closest that the airplane comes to the station? See figure.

The parametric equations of the lines in each of Exercises 39–42 give the position of two particles at time t. Find points where the paths of the particles intersect and determine if the particles collide.

39. $x = t, y = 2t, z = 1 - t; x = 2 - t, y = 1 + t, z = -1 + t$

40. $x = -1 + t, y = t, z = 2 - t; x = -4 + t, y = 6 - t, z = -4 + t$

41. $x = 2t, y = 1 - t, z = t; x = 1 + t, y = 2 + t, z = 1 + t$

42. $x = t, y = 1 - t, z = 1 + t; x = 1 + t, y = t, z = 2 + t$

43. For what value of c is the line $\dfrac{x-1}{1} = \dfrac{y-2}{2} = \dfrac{z-3}{c}$ parallel to the plane $2x + 3y - z = 6$?

11.7   VECTOR-VALUED FUNCTIONS

Connections

Rectilinear motion, 5.1.

Vectors, 11.3–4.

In this section we will study **vector-valued functions** of a real variable t and see they can be used to study the motion of objects in a coordinate plane or three-dimensional space. Vector-valued simply means that the values $\mathbf{R}(t)$ are vectors. The vectors $\mathbf{R}(t)$ can be either two-dimensional or three-dimensional vectors.

A vector-valued function that has values that are two-dimensional vectors can be expressed in terms of two real-valued **component functions**, f and g. That is,

$$\mathbf{R}(t) = f(t)\mathbf{i} + g(t)\mathbf{j},$$

where $\mathbf{i}$ and $\mathbf{j}$ are the usual two-dimensional vectors. Similarly, a vector-valued function that has values that are three-dimensional vectors can be expressed in terms of three real-valued component functions,

$$\mathbf{R}(t) = f(t)\mathbf{i} + g(t)\mathbf{j} + h(t)\mathbf{k},$$

where $\mathbf{i}, \mathbf{j}$, and $\mathbf{k}$ are the usual three-dimensional vectors.

We can use vector functions to describe parametric curves. For example, we may express the parametric curve

$$C : x = x(t), \quad y = y(t)$$

in **vector form**

$$\mathbf{R}(t) = x(t)\mathbf{i} + y(t)\mathbf{j}.$$

Note that the components of the vector are simply the coordinates of points on the curve.

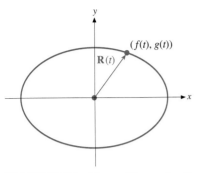

FIGURE 11.7.1   The position vector **R** = $f(t)\mathbf{i} + g(t)\mathbf{j}$, drawn from the origin, gives the coordinates $(f(t), g(t))$ of an object at time t.

We can represent the position of an object in a coordinate plane at time t by the **position vector**

$$\mathbf{R}(t) = f(t)\mathbf{i} + g(t)\mathbf{j}.$$

That is, the coordinates of the position at time t are $(f(t), g(t))$. See Figure 11.7.1. Similarly, we can represent the position of an object in three-dimensional space at time t by a three-dimensional vector-valued function

$$\mathbf{R}(t) = f(t)\mathbf{i} + g(t)\mathbf{j} + h(t)\mathbf{k}.$$

The position at time t then has coordinates $(f(t), g(t), h(t))$. See Figure 11.7.2.

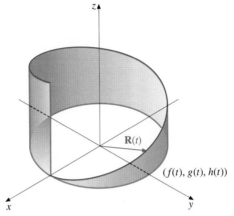

FIGURE 11.7.2   Three-dimensional position vectors give the coordinates of an object in three-dimensional space.

A position vector $\mathbf{R}(t)$ should be drawn as an arrow with initial point at the origin. The tip of the arrow then gives the position of the object at time t.

If the component functions are continuous, then the tip of the position arrow traces a curve as t varies over an interval of time. We will study properties of the curve and how properties of the curve relate to motion along the curve in Sections 11.8 and 11.9.

Calculus of Vector-Valued Functions

To study the motion of objects in a plane or three-dimensional space, we need to develop calculus of vector-valued functions. The development involves the same ideas as calculus of real-valued functions. As in the case of real-valued functions, the concepts of differentiation and integration of vector-valued functions depend on limits.

Much of the calculus of vector-valued functions is completely parallel to that of real-valued functions. We merely use the length of a vector in place of the absolute value of a real number. For example, a vector-valued function $\mathbf{R}(t)$ is said to be **bounded** on an interval I if there is a number B such that $|\mathbf{R}(t)| \le B$ for all t in I. $\mathbf{R}(t)$ is close to $\mathbf{L}$ (within ε) whenever $|\mathbf{R}(t) - \mathbf{L}|$ is small (less than ε). The definition of the limit of a vector-valued function can then be stated in the same terms as the limit of a real-valued function. The formal definition follows.

DEFINITION

We say that a vector $\mathbf{L}$ is the **limit of $\mathbf{R}(t)$ as t approaches c** (written $\lim\limits_{t \to c} \mathbf{R}(t) = \mathbf{L}$) if, for each $\varepsilon > 0$, there is a $\delta > 0$ such that

$$|\mathbf{R}(t) - \mathbf{L}| < \varepsilon \text{ whenever } 0 < |t - c| < \delta.$$

The properties

$$|c(t)\mathbf{R}(t)| = |c(t)||\mathbf{R}(t)| \text{ and } |\mathbf{U}(t) + \mathbf{V}(t)| \leq |\mathbf{U}(t)| + |\mathbf{V}(t)|$$

can be used to verify the Limit Theorem for vector-valued functions. That is, the limit of a sum of two vector-valued functions is the sum of the limits, and the limit of the product of a real-valued function and a vector-valued function is the product of the limits.

It is important that we understand the geometric meaning of the limit of a vector-valued function. Geometrically, $\lim\limits_{t \to c} \mathbf{R}(t) = \mathbf{L}$ means that the length of $\mathbf{R}(t)$ approaches the length of $\mathbf{L}$ and, if $\mathbf{L} \neq \mathbf{0}$, the direction of $\mathbf{R}(t)$ approaches the direction of $\mathbf{L}$ as t approaches c. See Figure 11.7.3.

To evaluate the limit of vector-valued functions, we use the fact that $\mathbf{R}(t)$ approaches $\mathbf{L}$ if and only if each component of $\mathbf{R}(t)$ approaches the corresponding component of $\mathbf{L}$. That is, for two-dimensional vector-valued functions,

$$\lim_{t \to c}(f(t)\mathbf{i} + g(t)\mathbf{j}) = \left(\lim_{t \to c} f(t)\right)\mathbf{i} + \left(\lim_{t \to c} g(t)\right)\mathbf{j}.$$

We can evaluate limits of vector-valued functions component by component.

This statement also holds for three-dimensional vector-valued functions.

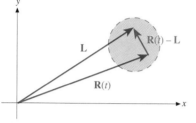

FIGURE 11.7.3 As $\mathbf{R}(t)$ approaches $\mathbf{L}$, the length of $\mathbf{R}(t)$ approaches the length of $\mathbf{L}$ and, if $\mathbf{L} \neq \mathbf{0}$, the direction of $\mathbf{R}(t)$ approaches the direction of $\mathbf{L}$.

EXAMPLE 1

$\mathbf{R}(t) = t^2\mathbf{i} - 3t\mathbf{j}$. Evaluate $\lim\limits_{t \to 2} \dfrac{\mathbf{R}(t) - \mathbf{R}(2)}{t - 2}$.

SOLUTION We first note that the expression

$$\frac{\mathbf{R}(t) - \mathbf{R}(2)}{t - 2}$$

is well defined for $t \neq 2$. That is, we can divide the vector $\mathbf{R}(t) - \mathbf{R}(2)$ by the real number $t - 2$, since this is equivalent to multiplying the vector by the real number $1/(t - 2)$. Let us now write the expression in terms of components and simplify. For $t \neq 2$, we have

$$\frac{\mathbf{R}(t) - \mathbf{R}(2)}{t - 2} = \frac{(t^2\mathbf{i} - 3t\mathbf{j}) - ((2)^2\mathbf{i} - 3(2)\mathbf{j})}{t - 2}$$

$$= \frac{t^2 - 2^2}{t - 2}\mathbf{i} + \frac{-3t + 3(2)}{t - 2}\mathbf{j}$$

$$= \frac{(t - 2)(t + 2)}{t - 2}\mathbf{i} - \frac{3(t - 2)}{t - 2}\mathbf{j}$$

$$= (t + 2)\mathbf{i} - 3\mathbf{j}.$$

Then

$$\lim_{t \to 2} \frac{\mathbf{R}(t) - \mathbf{R}(2)}{t - 2} = \lim_{t \to 2}((t + 2)\mathbf{i} - 3\mathbf{j}) = 4\mathbf{i} - 3\mathbf{j}. \quad\blacksquare$$

The **derivative** of a vector-valued function is defined to be

$$\mathbf{R}'(c) = \lim_{t \to c} \frac{\mathbf{R}(t) - \mathbf{R}(c)}{t - c}.$$

$\mathbf{R}'$ is a vector-valued function.

We can evaluate the derivative of vector-valued functions component by component.

That is,

$$(f\mathbf{i} + g\mathbf{j})' = f'\mathbf{i} + g'\mathbf{j}$$

and

$$(f\mathbf{i} + g\mathbf{j} + h\mathbf{k})' = f'\mathbf{i} + g'\mathbf{j} + h'\mathbf{k}.$$

Each component is a real-valued function and we can use all the formulas that we already know to find the derivatives. The vectors $\mathbf{i}$, $\mathbf{j}$, and $\mathbf{k}$ are *vector* constants. Don't forget that scalar constants have derivative zero. Vector constants have derivative equal to the *zero vector.*

EXAMPLE 2

$\mathbf{R}(t) = (1 - 2t)\mathbf{i} + (t^2 - t)\mathbf{j}$. Find $\mathbf{R}'(t)$ and $\mathbf{R}''(t)$.

SOLUTION

$$\mathbf{R}(t) = (1 - 2t)\mathbf{i} + (t^2 - t)\mathbf{j},$$
$$\mathbf{R}'(t) = -2\mathbf{i} + (2t - 1)\mathbf{j},$$
$$\mathbf{R}''(t) = \mathbf{0} + 2\mathbf{j} = 2\mathbf{j}. \quad \blacksquare$$

The following rules of differentiation hold for vector-valued functions $\mathbf{U}$ and $\mathbf{V}$. Here f denotes a scalar function.

$$(\mathbf{U} + \mathbf{V})' = \mathbf{U}' + \mathbf{V}', \tag{1}$$
$$(f\mathbf{V})' = f\mathbf{V}' + f'\mathbf{V}, \tag{2}$$
$$(\mathbf{U} \cdot \mathbf{V})' = \mathbf{U} \cdot \mathbf{V}' + \mathbf{U}' \cdot \mathbf{V}, \tag{3}$$
$$(\mathbf{U} \times \mathbf{V})' = \mathbf{U} \times \mathbf{V}' + \mathbf{U}' \times \mathbf{V}. \tag{4}$$

These formulas can be verified by expressing the functions in terms of their components and then differentiating component by component.

NOTE   *Note the order of the cross products in Formula 4. Don't forget that $\mathbf{U} \times \mathbf{V} = -\mathbf{V} \times \mathbf{U}$, so we must be careful with the order of a cross product.*

If $\mathbf{U}(t)$ is a vector with constant length, then $\mathbf{U}$ and $\mathbf{U}'$ are orthogonal.   (5)

Proof   Since $|\mathbf{U}|^2 = \mathbf{U} \cdot \mathbf{U}$, we can use Formula 3 to obtain

$$\frac{d}{dt}(|\mathbf{U}|^2) = \frac{d}{dt}(\mathbf{U} \cdot \mathbf{U}) = \mathbf{U} \cdot \mathbf{U}' + \mathbf{U}' \cdot \mathbf{U} = 2\mathbf{U} \cdot \mathbf{U}'.$$

On the other hand, if $|\mathbf{U}|$ is constant, then $|\mathbf{U}|^2$ is also constant, so

$$\frac{d}{dt}(|\mathbf{U}|^2) = 0.$$

Collecting results, we have $2\mathbf{U} \cdot \mathbf{U}' = 0$, so $\mathbf{U} \cdot \mathbf{U}' = 0$. This shows that $\mathbf{U}$ and $\mathbf{U}'$ are orthogonal. ■

If $\mathbf{R}(t)$ is defined for $a \leq t \leq b$, we can form a Riemann sum $\displaystyle\sum_{j=1}^{n} \mathbf{R}(t_j^*)\Delta t_j$.
Each term of the sum is the product of a scalar Δt_j and a vector $\mathbf{R}(t_j^*)$. Hence, the sum is a vector. In case $\mathbf{R}(t)$ is continuous, the Riemann sums that have all Δt_j's small enough will be close to a single vector $\displaystyle\int_a^b \mathbf{R}(t)\,dt$. We can evaluate vector-valued integrals component by component. In case of two-dimensional vectors, we have

$$\int_a^b (f(t)\mathbf{i} + g(t)\mathbf{j})\,dt = \left(\int_a^b f(t)\,dt\right)\mathbf{i} + \left(\int_a^b g(t)\,dt\right)\mathbf{j},$$

$$\int (f(t)\mathbf{i} + g(t)\mathbf{j})\,dt = \left(\int f(t)\,dt\right)\mathbf{i} + \left(\int g(t)\,dt\right)\mathbf{j} + \mathbf{C}.$$

Note that the constant of integration in the indefinite integral is a vector.

We can integrate vector-valued functions component by component.

EXAMPLE 3
Find $\mathbf{R}(t)$ if $\mathbf{R}'(t) = 6t\mathbf{i} + \mathbf{j}$ and $\mathbf{R}(0) = \mathbf{i} - \mathbf{j}$.

SOLUTION   We have

$$\int \mathbf{R}'(t)\,dt = \int (6t\mathbf{i} + \mathbf{j})\,dt = 3t^2\mathbf{i} + t\mathbf{j} + \mathbf{C},$$

so $\mathbf{R}(t) = 3t^2\mathbf{i} + t\mathbf{j} + \mathbf{C}$ for some constant vector $\mathbf{C}$. We have $\mathbf{R}(0) = \mathbf{i} - \mathbf{j}$, so substitution gives

$$\mathbf{i} - \mathbf{j} = \mathbf{0} + \mathbf{0} + \mathbf{C}.$$

Hence, $\mathbf{C} = \mathbf{i} - \mathbf{j}$ and

$$\mathbf{R}(t) = 3t^2\mathbf{i} + t\mathbf{j} + \mathbf{i} - \mathbf{j} = (3t^2 + 1)\mathbf{i} + (t - 1)\mathbf{j}. ■$$

Motion

Let us now see how we can use calculus of vector-valued functions to study motion of objects in a coordinate plane or three-dimensional space.

If $\mathbf{R}(t)$ is the position of an object at time t, the change in position over the interval of time t to $t + \Delta t$ is

$$\Delta \mathbf{R} = \mathbf{R}(t + \Delta t) - \mathbf{R}(t).$$

The change in position divided by the length of the time interval is

$$\frac{\Delta \mathbf{R}}{\Delta t} = \frac{\mathbf{R}(t + \Delta t) - \mathbf{R}(t)}{\Delta t}.$$

See Figure 11.7.4. The limit of the above ratio as $\Delta t \to 0$ is the derivative $\mathbf{R}'$, which is the **velocity** of the object,

$$\mathbf{V} = \mathbf{R}'.$$

FIGURE 11.7.4   The rate of change of position with respect to time is

$$\mathbf{R}' = \lim_{\Delta t \to 0} \frac{\mathbf{R}(t + \Delta t) - \mathbf{R}(t)}{\Delta t}.$$

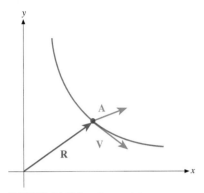

FIGURE 11.7.5 The position vector **R** is drawn from the origin, while the velocity **V** and acceleration **A** are drawn from the position of the object.

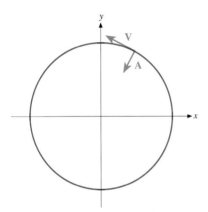

FIGURE 11.7.6 As an object moves around a circle with constant speed, its velocity is tangent to the circle and its acceleration is toward the origin.

FIGURE 11.7.7 The object is spiraling up the vertical cylinder $x^2 + y^2 = 1$.

Similarly, the **acceleration** is given by

$$\mathbf{A} = \mathbf{V}' = \mathbf{R}''.$$

The velocity and acceleration of an object that is moving either in a coordinate plane or in three-dimensional space are vectors.

The speed of the object is the scalar value $|\mathbf{V}|$.

Note that the position vector **R** is represented by an arrow from the origin, while the velocity **V** and acceleration **A** are drawn from the position of the object. This is done to indicate the motion of the object as it passes through the point. See Figure 11.7.5.

EXAMPLE 4
$\mathbf{R} = \cos \dfrac{t}{2}\mathbf{i} + \sin \dfrac{t}{2}\mathbf{j}$. Find the velocity, speed, and acceleration.

SOLUTION

$$\mathbf{R} = \cos \frac{t}{2}\mathbf{i} + \sin \frac{t}{2}\mathbf{j},$$

$$\mathbf{V} = \mathbf{R}' = -\frac{1}{2}\sin \frac{t}{2}\mathbf{i} + \frac{1}{2}\cos \frac{t}{2}\mathbf{j},$$

$$\mathbf{A} = \mathbf{V}' = \mathbf{R}'' = -\frac{1}{4}\cos \frac{t}{2}\mathbf{i} - \frac{1}{4}\sin \frac{t}{2}\mathbf{j},$$

$$\text{Speed} = |\mathbf{V}| = \sqrt{\left(-\frac{1}{2}\sin \frac{t}{2}\right)^2 + \left(\frac{1}{2}\cos \frac{t}{2}\right)^2} = \frac{1}{2}. \quad \blacksquare$$

Note that the (x, y) coordinates of the object in Example 4 satisfy $x^2 + y^2 = 1$. This means that the path of the object is along the circle with center at the origin and radius one. Representative velocity and acceleration vectors are sketched in Figure 11.7.6. Note that the velocity vector is tangent to the circle and the acceleration vector always points toward the origin. Since **V** is of constant length, we know from (5) that **V** and $\mathbf{A} = \mathbf{V}'$ must be orthogonal.

EXAMPLE 5
$\mathbf{R} = \cos t\mathbf{i} + \sin t\mathbf{j} + t\mathbf{k}$. Find the velocity and acceleration.

SOLUTION

$$\mathbf{R} = \cos t\mathbf{i} + \sin t\mathbf{j} + t\mathbf{k},$$
$$\mathbf{V} = \mathbf{R}' = -\sin t\mathbf{i} + \cos t\mathbf{j} + \mathbf{k},$$
$$\mathbf{A} = \mathbf{V}' = \mathbf{R}'' = -\cos t\mathbf{i} - \sin t\mathbf{j}.$$

The path of the object and representative velocity and acceleration vectors are sketched in Figure 11.7.7. $\quad \blacksquare$

Since $\mathbf{V}' = \mathbf{A}$ and $\mathbf{R}' = \mathbf{V}$, we can determine the velocity and the position from the acceleration by integration, if initial values are known.

Motion of a Projectile

Let us determine the equation of motion of an object that is thrown near the surface of the earth. We assume that the distances involved are small relative to the radius of the earth. We choose a two-dimensional coordinate system with y-axis vertical and origin at ground level. We use the vector-valued function

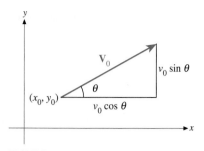

FIGURE 11.7.8 The initial position is $\mathbf{R}(0) = x_0\mathbf{i} + y_0\mathbf{j}$. The initial velocity is $\mathbf{V}_0 = v_0 \cos\theta\,\mathbf{i} + v_0 \sin\theta\,\mathbf{j}$.

$\mathbf{R}(t)$ to represent the position of the object at time t. The initial position of the object is $\mathbf{R}(0) = x_0\mathbf{i} + y_0\mathbf{j}$. The initial velocity is $\mathbf{V}_0 = v_0 \cos\theta\,\mathbf{i} + v_0 \sin\theta\,\mathbf{j}$, where θ is the angle of elevation of the initial velocity and v_0 is the initial speed. See Figure 11.7.8. We neglect air resistance, so the only force on the object is due to gravity. The acceleration due to gravity has constant magnitude and is directed downward. That is, $\mathbf{A} = -g\mathbf{j}$, where $g = 32\ \text{ft/s}^2$ or $9.8\ \text{m/s}^2$. We can integrate the acceleration vector twice and substitute the initial values of the velocity and position (as in Example 3) to obtain

$$\mathbf{R}(t) = (x_0 + v_0(\cos\theta)t)\mathbf{i} + \left(y_0 + v_0(\sin\theta)t - \frac{g}{2}t^2\right)\mathbf{j}. \tag{6}$$

The components of the position vector $\mathbf{R}(t)$ give the coordinates of the object at time t. We have

$$x = x_0 + v_0(\cos\theta)t \text{ and } y = y_0 + v_0(\sin\theta)t - \frac{g}{2}t^2.$$

By solving the first of the above equations for t, substituting into the second equation, and completing the square, we obtain the equation

$$y - \left(y_0 + \frac{v_0^2 \sin^2\theta}{2g}\right) = \frac{-g}{2v_0^2 \cos^2\theta}\left(x - \left(x_0 + \frac{v_0^2 \sin\theta \cos\theta}{g}\right)\right)^2.$$

This is the equation of a parabola with vertex

$$\left(x_0 + \frac{v_0^2 \sin\theta \cos\theta}{g}, \quad y_0 + \frac{v_0^2 \sin^2\theta}{2g}\right).$$

A representative path is sketched in Figure 11.7.9. Note that the y-coordinate of the vertex gives the maximum height of the object. This is the value of y when

$$\frac{dy}{dt} = 0.$$

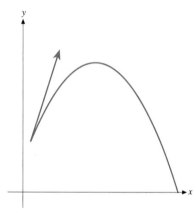

FIGURE 11.7.9 The path of the object that is thrown near the surface of the earth is a parabola, if we assume that gravity is the only force acting on the object.

EXAMPLE 6

A ball is thrown with a speed of 64 ft/s from a height of 48 ft at an angle of 30° from horizontal. How long will it take the ball to hit the ground? Where will the ball hit the ground? At what speed will the ball hit the ground? What is the maximum height the ball will attain?

SOLUTION A coordinate system is sketched in Figure 11.7.10. The initial position is $(x_0, y_0) = (0, 48)$.

We are given that the initial speed is $v_0 = 64$ and that $\theta = 30°$. The units are feet and seconds, so we use $g = 32\ \text{ft/s}^2$. Substitution of the data into the equation of motion, Equation 6, then gives

$$\mathbf{R}(t) = (0 + 64 \cos 30° t)\mathbf{i} + (48 + 64(\sin 30°)t - 16t^2)\mathbf{j}$$
$$= (32\sqrt{3}t)\mathbf{i} + (48 + 32t - 16t^2)\mathbf{j}.$$

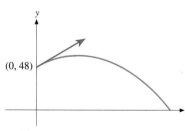

FIGURE 11.7.10 The initial position is $(x_0, y_0) = (0, 48)$ and the initial velocity is $\mathbf{V}_0 = 64 \cos 30°\,\mathbf{i} + 64 \sin 30°\,\mathbf{j}$.

The ball will hit the ground when the **j** component of the position vector is 0. This corresponds to $y = 0$. That is, when

$$48 + 32t - 16t^2 = 0,$$
$$-16(t^2 - 2t - 3) = 0,$$
$$-16(t - 3)(t + 1) = 0.$$

This implies $t = 3$, since we want $t > 0$. When $t = 3$, the position vector is

$$\mathbf{R}(3) = (32\sqrt{3})(3)\mathbf{i} + (0)\mathbf{j} = 96\sqrt{3}\mathbf{i}.$$

The ball will hit the ground at a point that is $96\sqrt{3} \approx 166.3$ ft from the point on the ground that is directly below where it was thrown.

We have

$$\mathbf{R}(t) = (32\sqrt{3}t)\mathbf{i} + (48 + 32t - 16t^2)\mathbf{j},$$
$$\mathbf{V}(t) = \mathbf{R}'(t) = 32\sqrt{3}\mathbf{i} + (32 - 32t)\mathbf{j}.$$

When $t = 3$, we have

$$\mathbf{V}(3) = 32\sqrt{3}\mathbf{i} - 64\mathbf{j},$$

so the speed when the ball hits the ground is

$$|\mathbf{V}(3)| = \sqrt{(32\sqrt{3})^2 + (-64)^2} = \sqrt{7168} \approx 84.66 \text{ ft/s}.$$

The maximum height is the maximum value of the y-coordinate, $y = 48 + 32t - 16t^2$. We can determine the time when the maximum height is obtained by solving $y'(t) = 0$ for t. We have $y'(t) = 32 - 32t$, so $y'(t) = 0$ implies $t = 1$. The maximum height is then $y(1) = 48 + 32(1) - 16(1)^2 = 64$ ft. ∎

Maximum Horizontal Distance of a Projectile

Suppose that a projectile is launched at ground level with fixed initial speed v_0. Let us determine the angle of elevation that gives the maximum horizontal distance L the projectile travels before returning to ground level. See Figure 11.7.11. The initial position is taken to be $x_0 = y_0 = 0$. The equation of motion for a particular choice of θ is then

$$\mathbf{R}(t) = (v_0(\cos\theta)t)\mathbf{i} + \left(v_0(\sin\theta)t - \frac{g}{2}t^2\right)\mathbf{j}.$$

We see from the sketch that the horizontal distance L occurs at that time when the y-coordinate of the position vector is zero. Thus, at some time t we must have

$$v_0(\cos\theta)t = L \text{ and}$$

$$v_0(\sin\theta)t - \frac{g}{2}t^2 = 0.$$

The distance L and the hitting time t both depend on the angle of elevation θ. We can eliminate t from the two equations above and obtain an equation that relates L and θ. That is, the second equation gives

$$t = \frac{2v_0\sin\theta}{g}.$$

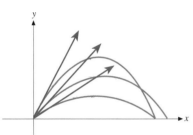

FIGURE 11.7.11 For a fixed initial speed, the horizontal distance the projectile travels depends on the angle of elevation.

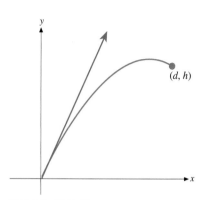

FIGURE 11.7.12 We want the projectile to hit a point that is d units horizontally and h units above the launch point.

Substitution into the first equation then gives

$$L = v_0 \cos \theta \left(\frac{2v_0 \sin \theta}{g} \right), \quad \text{or}$$

$$L = \frac{2v_0^2}{g} \sin \theta \cos \theta, \quad 0 \le \theta \le \frac{\pi}{2}.$$

We now have the length L as a continuous function of the elevation angle θ. We can then determine the maximum distance and the corresponding angle θ as usual. Namely, check values where the derivative is zero and at endpoints. We find that the maximum occurs for $\theta = \pi/4$.

Hitting a Fixed Target

A projectile is to hit a point that is d units horizontally and h units above the launch point. See Figure 11.7.12. Let us determine the launch angle that gives the minimum initial speed.

We choose initial position $x_0 = y_0 = 0$. The position vector is then

$$\mathbf{R}(t) = (v_0 (\cos \theta)t)\mathbf{i} + \left(v_0 (\sin \theta)t - \frac{g}{2}t^2 \right)\mathbf{j}.$$

The path of the projectile is required to pass through the point (d, h). This means, for some value of t, we must have

$$d = v_0 (\cos \theta)t \quad \text{and}$$

$$h = v_0 (\sin \theta)t - \frac{g}{2}t^2.$$

Note that t, v_0, and θ are unknown variables in the above two equations. We can obtain an equation that relates θ and v_0 by solving the first equation for t and then substituting into the second equation. We have

$$t = \frac{d}{v_0 \cos \theta},$$

and then

$$h = v_0 \sin \theta \left(\frac{d}{v_0 \cos \theta} \right) - \frac{g}{2} \left(\frac{d}{v_0 \cos \theta} \right)^2,$$

$$2hv_0^2 \cos^2 \theta = 2dv_0^2 \sin \theta \cos \theta - gd^2,$$

$$2 \cos \theta (h \cos \theta - d \sin \theta)v_0^2 = -gd^2,$$

$$v_0^2 = \frac{gd^2}{2 \cos \theta (d \sin \theta - h \cos \theta)}.$$

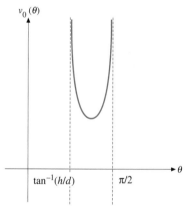

FIGURE 11.7.13 The target can be hit with any launch angle between $\tan^{-1}(h/d)$ and $\pi/2$. The angle

$$\theta = \tan^{-1} \left(\frac{h + \sqrt{h^2 + d^2}}{d} \right)$$

gives the minimum initial speed with which the target can be hit.

A representative sketch of the graph of $v_0(\theta)$, $\tan^{-1}(h/d) < \theta < \pi/2$, is given in Figure 11.7.13. (Note that $g > 0$. Since v_0^2 must be positive, we must have $d \sin \theta - h \cos \theta > 0$. This implies $\tan \theta > h/d$.) The minimum value of v_0 occurs where

$$\frac{dv_0}{d\theta} = 0.$$

With some work, we can see that this condition gives the equation

$$\tan \theta = \frac{h + \sqrt{h^2 + d^2}}{d}.$$

There is exactly one value of θ in the interval $\tan^{-1}(h/d) < \theta < \pi/2$ that satisfies this equation. This is the launch angle that gives the minimum initial speed.

It is possible to verify that the x-coordinate of the vertex of the parabolic path of a projectile that is launched at the above angle θ with the corresponding initial speed v_0 is less than d. Hence, the projectile is falling as it approaches the point (d, h), as was indicated in Figure 11.7.12.

Momentum and Torque

Suppose the position of an object of mass m at time t is given by $\mathbf{r}(t)$ and its velocity is $\mathbf{v}(t)$. The **linear momentum** of the object is

$$\mathbf{p} = m\mathbf{v}.$$

The **angular momentum about the origin** is

$$\mathbf{L} = \mathbf{r} \times \mathbf{p}.$$

If the motion of an object is due to a resultant exterior force $\mathbf{F}$, then Newton's Second Law says

$$\mathbf{F} = \frac{d\mathbf{p}}{dt}.$$

That is, the exterior force is equal to the rate of change of the linear momentum of the object. The rate of change of the angular momentum about the origin of the object is

$$\frac{d\mathbf{L}}{dt} = \frac{d}{dt}(\mathbf{r} \times \mathbf{p}).$$

Using Property 4, we obtain

$$\frac{d\mathbf{L}}{dt} = \mathbf{r} \times \frac{d\mathbf{p}}{dt} + \frac{d\mathbf{r}}{dt} \times \mathbf{p}$$
$$= \mathbf{r} \times \mathbf{F} + \mathbf{v} \times (m\mathbf{v})$$
$$= \mathbf{r} \times \mathbf{F},$$

since $\mathbf{v} \times (m\mathbf{v}) = m(\mathbf{v} \times \mathbf{v}) = m\mathbf{0} = \mathbf{0}$. For any vector function $\mathbf{F}$, we define the **torque about the origin** to be

$$\tau = \mathbf{r} \times \mathbf{F}.$$

(The units of torque are length times force. Torque is not a force.) We have shown that the rate of change of the angular momentum about the origin of an object is the torque about the origin of the resultant exterior force on the object.

EXAMPLE 7

One end of a string is attached to a weight and the other end passes through a small hole in a flat table top. The weight is revolving with constant speed v_0 in a circle of radius r_0 when the string is pulled through the hole until the radius

FIGURE 11.7.14a A weight tied to a string is revolving with constant speed v_0 in a circle of radius r_0.

FIGURE 11.7.14b The string is pulled through the hole in the table until the radius of rotation is $r_0/2$.

of rotation is $r_0/2$. See Figure 11.7.14. What is the speed of the object when the radius of rotation is $r_0/2$, if forces due to friction are negligible?

SOLUTION The force **F** on the weight due to pulling of the string is toward the center of rotation, so the rate of change of the angular momentum is $\mathbf{r} \times \mathbf{F} = \mathbf{0}$. Since the force due to gravity is counteracted by a force from the table top and other external forces (such as friction or a small force to overcome friction) on the weight are negligible, this means that the angular momentum about the center of rotation is essentially constant. That is, we may assume that $\mathbf{L} = \mathbf{r} \times (m\mathbf{v})$ is constant. When the weight is moving in a circle of constant radius, the velocity is perpendicular to the radius vector $\mathbf{r}$, so $|\mathbf{r} \times (m\mathbf{v})| = m|\mathbf{r}||\mathbf{v}|$. When $|\mathbf{r}| = r_0$, we have

$$|\mathbf{r} \times (m\mathbf{v})| = m|\mathbf{r}||\mathbf{v}| = mr_0v_0,$$

and when the radius is $r_0/2$ we have

$$|\mathbf{r} \times (m\mathbf{v})| = m|\mathbf{r}||\mathbf{v}| = m(r_0/2)v_1,$$

where v_1 is the speed at radius $r_0/2$. Since the angular momentum is constant, we must have

$$mr_0v_0 = mr_0v_1/2, \text{ so } v_1 = 2v_0.$$

The speed at radius $r_0/2$ is twice the speed at radius r_0. ∎

The **angular speed** ω of an object rotating in a circle of radius r at speed v is

$$\omega = \frac{v}{r}.$$

Thus, when the radius is halved and the speed is doubled, the angular speed increases by a factor of four. The weight in Example 7 then makes four times as many revolutions per unit time at radius $r_0/2$ as at radius r_0.

EXERCISES 11.7

Evaluate the expressions in Exercises 1–18.

1. $\lim\limits_{t \to 2} (t^2\mathbf{i} - 3t\mathbf{j})$

2. $\lim\limits_{t \to \pi/2} (\sin t\mathbf{i} + \cos t\mathbf{j})$

3. $\lim\limits_{t \to 3} \left(\dfrac{t^2 - 9}{t - 3}\mathbf{i} + \dfrac{2t - 6}{t - 3}\mathbf{j} \right)$

4. $\lim\limits_{t \to 0} \left(\dfrac{\sin 2t}{t}\mathbf{i} + \dfrac{1 - \cos^2 2t}{t}\mathbf{j} \right)$

5. $\dfrac{d}{dt}(t^2\mathbf{i} - 3t\mathbf{j})$

6. $\dfrac{d}{dt}(t\mathbf{i} + t^2\mathbf{j} + t^3\mathbf{k})$

7. $\dfrac{d}{dt}(t\mathbf{i} + (e^{-t}\sin t)\mathbf{j} + (e^{-t}\cos t)\mathbf{k})$

8. $\dfrac{d}{dt}(te^{-t}\mathbf{i} + e^{-3t}\mathbf{j} + e^{-t}\mathbf{k})$

9. $\dfrac{d}{dt}(\tan t\mathbf{i} + \sec t\mathbf{j})$

10. $\dfrac{d}{dt}(\tan^{-1} t\mathbf{i} + \ln t\mathbf{j})$

11. $\displaystyle\int (2t\mathbf{i} + 6t^2\mathbf{j})\,dt$

12. $\displaystyle\int (t^2\mathbf{i} + t^3\mathbf{j} + t^4\mathbf{k})\,dt$

13. $\displaystyle\int (\sin t\mathbf{i} + \cos t\mathbf{j})\,dt$

14. $\displaystyle\int (2\sin 2t\mathbf{i} + 2\cos 2t\mathbf{j})\,dt$

15. $\displaystyle\int (\sec^2 2t\mathbf{i} + (\sec 2t \tan 2t)\mathbf{j})\,dt$

16. $\displaystyle\int (\sin^2 t\mathbf{i} + \cos^2 t\mathbf{j})\,dt$

17. $\displaystyle\int \left(\dfrac{t}{t^2 + 4}\mathbf{i} + \dfrac{1}{t^2 + 4}\mathbf{j} \right)\,dt$

18. $\int \left(\frac{t}{t^2 - 4}\mathbf{i} + \frac{1}{t^2 - 4}\mathbf{j} \right) dt$

In Exercises 19–22, find the velocity, acceleration, and speed of an object with the given position vector.

19. $\mathbf{R}(t) = t\mathbf{i} + t^2\mathbf{j}$

20. $\mathbf{R}(t) = 2t^2\mathbf{i} - 3t^2\mathbf{j}$

21. $\mathbf{R}(t) = \cos 2t\mathbf{i} + \sin 2t\mathbf{j}$

22. $\mathbf{R}(t) = \cos 3t\mathbf{i} + \sin 3t\mathbf{j} + 3t\mathbf{k}$

In Exercises 23–26, find the position vector for an object with the given acceleration, initial position, and initial velocity.

23. $\mathbf{A}(t) = -32\mathbf{j}$, $\mathbf{R}_0 = 0$, $\mathbf{V}_0 = \mathbf{i} + 2\mathbf{j}$

24. $\mathbf{A}(t) = -32\mathbf{k}$, $\mathbf{R}_0 = 0$, $\mathbf{V}_0 = \mathbf{i} + \mathbf{j} + \mathbf{k}$

25. $\mathbf{A}(t) = 12t\mathbf{j}$, $\mathbf{R}_0 = 3\mathbf{j}$, $\mathbf{V}_0 = 4\mathbf{i}$

26. $\mathbf{A}(t) = 12t^2\mathbf{i} + 6t\mathbf{j} + \mathbf{k}$, $\mathbf{R}_0 = 2\mathbf{k}$, $\mathbf{V}_0 = \mathbf{i} + 2\mathbf{j}$

27. A projectile is launched with initial speed 64 ft/s from ground level at an angle of 60° from the horizontal. How high is it after 3 s? What is the maximum height the projectile will reach?

28. A projectile is launched with initial speed 64 ft/s from ground level at an angle of 30° from the horizontal. How far will it travel horizontally?

29. A projectile is launched horizontally from a point 64 ft above ground level at an initial speed of 80 ft/s. How long will it take to reach ground level? How long would it take for an object that was dropped from a height of 64 ft to reach the ground?

30. A projectile is to be launched from ground level at an angle of 45° from the horizontal. Find the initial speed needed for it to reach a maximum height of 144 ft.

31. A projectile is to pass through a point that is 120 ft horizontally from and 90 ft above the launch point. What is the minimum initial velocity that must be used?

32. Let $\mathbf{U}(t) = f_1(t)\mathbf{i} + g_1(t)\mathbf{j}$ and $\mathbf{V}(t) = f_2(t)\mathbf{i} + g_2(t)\mathbf{j}$.
   (a) Evaluate $(\mathbf{U} \cdot \mathbf{V})'$ by expressing the dot product in terms of the components and then differentiating.
   (b) Express $\mathbf{U} \cdot \mathbf{V}' + \mathbf{U}' \cdot \mathbf{V}$ in terms of components.
   (c) Show that the expressions in (a) and (b) are equal. This verifies the formula $(\mathbf{U} \cdot \mathbf{V})' = \mathbf{U} \cdot \mathbf{V}' + \mathbf{U}' \cdot \mathbf{V}$.

33. If the variables v_0 and θ are related by the equation

$$v_0^2 = \frac{gd^2}{2\cos\theta(d\sin\theta - h\cos\theta)}, \quad \tan^{-1}(h/d) < \theta < \pi/2,$$

show that

$$\frac{dv_0}{d\theta} = 0 \text{ implies } \tan\theta = \frac{h + \sqrt{h^2 + d^2}}{d}.$$

34. If

$$v_0^2 = \frac{gd^2}{2\cos\theta(d\sin\theta - h\cos\theta)}, \quad \tan^{-1}(h/d) < \theta < \pi/2,$$

and

$$\tan\theta = \frac{h + \sqrt{h^2 + d^2}}{d},$$

show that the x-coordinate of the vertex of the parabolic path of a projectile with initial position 0 is less than d.

35. A ball is thrown at an angle of θ with the horizontal from ground level on the side of a hill in a direction where the hill makes an angle of α with the horizontal. See figure. (Take α to be positive if the ball is thrown up the hill and negative if the ball is thrown down the hill.) For a fixed initial speed v_0, show that the maximum distance along the hill that the ball travels before hitting the ground is obtained when the ball is thrown at an angle θ that satisfies $(\tan 2\theta)(\tan\alpha) = -1$.

36. A projectile is launched with initial speed v_0. Express the maximum height the projectile can be when it has gone a distance d_0 horizontally, in terms of v_0 and d_0.

37. Show that local extreme values of the distance between points on a curve $\mathbf{R}(t)$ and the origin are obtained where $\mathbf{R} \cdot \mathbf{R}' = 0$.

38. Set the torque about the hinge point of the lower jaw equal to zero to determine the force on a piece of hard candy if the **masseter muscle** exerts a force of 200 N. See figure.

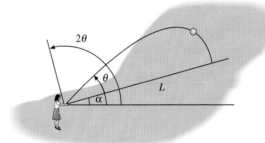

$F_m = 200$ N

2.9 cm

5.0 cm

$\mathbf{F}_{\text{chew reaction}}$

39. Set the torque about the elbow equal to zero to determine the force the **bicep muscle** must exert to hold up the forearm and calculus textbook as indicated in the figure. (Force, N = (Mass, kg)(9.8 m/s^2).)

40. Planetary motion is determined by the force $\mathbf{F} = \dfrac{k}{|\mathbf{r}|^3}\mathbf{r}$ due to the gravitational attraction of the sun, where $\mathbf{r}$ is the vector from the sun to the planet. Show that the rate of change of the angular momentum of the planet about the sun is zero, so $\mathbf{L} = \mathbf{L}_0$, a constant vector. Since $\mathbf{r} \times \mathbf{p} = \mathbf{L}_0$, we have $\mathbf{r} \cdot \mathbf{L}_0 = 0$. It follows that $\mathbf{r}$ is always in the plane that contains the sun and has normal vector $\mathbf{L}_0$ so the motion of the planet is in this plane. See figure.

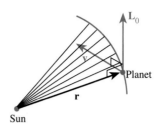

41. Show that the time for a ball to roll along a straight line path from the top of a vertical circle to any other point on the circle does not depend on the angle of the straight line path. See figure. (*Hint:* What is the component of the acceleration due to gravity in the direction of the path at angle θ from the vertical?)

11.8 UNIT TANGENT AND NORMAL VECTORS; CURVATURE

Connections

Vector-valued functions, 11.7.

Given a curve in parametric form,

$$C : x = x(t), \quad y = y(t), \quad a \le t \le b,$$

we saw in Section 11.7 that we can express the curve in **vector form** by writing

$$\mathbf{R}(t) = x(t)\mathbf{i} + y(t)\mathbf{j}, \quad a \le t \le b.$$

In this section we will see that the vector representation of C is convenient for studying properties of the curve.

Unit Tangent and Normal Vectors

Let us think of the vector function

$$\mathbf{R}(t) = x(t)\mathbf{i} + y(t)\mathbf{j}, \quad a \le t \le b,$$

as representing the position of an object along a curve. Recall from Section 11.7 that the position vector is drawn from the origin, so the point of the arrow gives the position of the object. See Figure 11.8.1. We assume that x and y are differentiable and $(x')^2 + (y')^2 \ne 0$ for $a \le t \le b$. This means the derivative

$$\mathbf{R}'(t) = x'(t)\mathbf{i} + y'(t)\mathbf{j}$$

FIGURE 11.8.1 $\mathbf{R}(t) = x(t)\mathbf{i} + y(t)\mathbf{j}$ is the vector form of the parametric curve.

exists and is nonzero. For $\Delta t \neq 0$, the secant vector from $\mathbf{R}(t)$ to $\mathbf{R}(t + \Delta t)$ is $\mathbf{R}(t + \Delta t) - \mathbf{R}(t)$. The vector

$$\frac{\mathbf{R}(t + \Delta t) - \mathbf{R}(t)}{\Delta t}$$

has direction of a secant line between two points on the curve. The curve has a tangent at the point $\mathbf{R}(t)$ exactly when the secant lines approach a single line through the point as Δt approaches zero. But

$$\lim_{\Delta t \to 0} \frac{\mathbf{R}(t + \Delta t) - \mathbf{R}(t)}{\Delta t} = \mathbf{R}'(t),$$

so the direction of the secant vectors approaches the direction of the *nonzero* vector $\mathbf{R}'(t)$ as Δt approaches zero. This means that $\mathbf{R}'(t)$ is tangent to the curve at position $\mathbf{R}(t)$. Also, $\mathbf{R}'(t)$ points in the direction of increasing t. See Figure 11.8.2a. The unit vector in the direction of $\mathbf{R}'(t)$ is

$$\mathbf{T}(t) = \frac{\mathbf{R}'(t)}{|\mathbf{R}'(t)|}.$$

Since $\mathbf{T}$ is a positive multiple of $\mathbf{R}'$, $\mathbf{T}$ is also a tangent to the curve and points in the direction of increasing t. The vector $\mathbf{T}$ is called the **unit tangent vector** of the curve. The unit tangent vector at a point on a curve should be drawn from the point on the curve, so it indicates the direction of the curve at that point. See Figure 11.8.2b.

FIGURE 11.8.2a   If $\mathbf{R}'(t) \neq 0$, the vector $\dfrac{\mathbf{R}(t + \Delta t) - \mathbf{R}(t)}{\Delta t}$ approaches the direction of the nonzero vector $\mathbf{R}'(t)$ as Δt approaches zero, so $\mathbf{R}'(t)$ is tangent to the curve and points in the direction of increasing t.

FIGURE 11.8.2b   The unit tangent vector at a point on the curve is drawn from the point on the curve, so it indicates the direction of the curve at that point.

Since $\mathbf{T}(t)$ is a vector with constant length one, it follows from Statement 5 of Section 11.7 that $\mathbf{T}$ and $\mathbf{T}'$ are orthogonal. If $\mathbf{T}' \neq \mathbf{0}$, we define the (principal) **unit normal vector** to the curve to be

$$\mathbf{N} = \frac{\mathbf{T}'}{|\mathbf{T}'|}.$$

Since **N** is a multiple of **T′**,

T and N are orthogonal.

The unit normal vector at a point on the curve should be drawn from the point on the curve. We will see in Section 11.9 that the unit normal vector **N** does not depend on the direction the curve is traced; **N** is always orthogonal to the curve and points in the direction the curve is bending. See Figure 11.8.3.

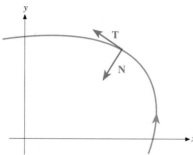

FIGURE 11.8.3a The unit normal vector **N** at a point on a curve is drawn from the point on the curve. **N** is always orthogonal to the curve and points in the direction the curve is bending.

FIGURE 11.8.3b The unit normal vector does not depend on the direction of the curve.

EXAMPLE 1

Find the unit tangent and normal vectors of the curve

$$C : x = 3\cos 2t, \quad y = 3\sin 2t, \quad 0 \le t \le 2\pi.$$

SOLUTION   We express the curve in vector form by writing

$$\mathbf{R}(t) = (3\cos 2t)\mathbf{i} + (3\sin 2t)\mathbf{j}.$$

The derivative is

$$\mathbf{R}'(t) = (-6\sin 2t)\mathbf{i} + (6\cos 2t)\mathbf{j},$$

and the length of the derivative is

$$|\mathbf{R}'(t)| = \sqrt{(-6\sin 2t)^2 + (6\cos 2t)^2}$$
$$= \sqrt{36(\sin^2 2t + \cos^2 2t)} = 6.$$

The unit tangent vector is

$$\mathbf{T}(t) = \frac{\mathbf{R}'(t)}{|\mathbf{R}'(t)|} = \frac{(-6\sin 2t)\mathbf{i} + (6\cos 2t)\mathbf{j}}{6} = (-\sin 2t)\mathbf{i} + (\cos 2t)\mathbf{j}.$$

We have

$$\mathbf{T}'(t) = (-2\cos 2t)\mathbf{i} + (-2\sin 2t)\mathbf{j},$$

so

$$|\mathbf{T}'(t)| = \sqrt{(-2\cos 2t)^2 + (-2\sin 2t)^2} = 2 \text{ and}$$
$$\mathbf{N} = \frac{\mathbf{T}'}{|\mathbf{T}'|} = (-\cos 2t)\mathbf{i} + (-\sin 2t)\mathbf{j}.$$

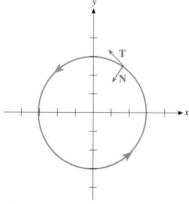

FIGURE 11.8.4 The unit normal vector to a circle points toward the center of the circle.

The graph of the curve and typical unit tangent and normal vectors are sketched in Figure 11.8.4. Note that the curve is a circle and that the normal vector points toward the center of the circle. ■

Smooth Curves

The curve

$$C : x = x(t), \quad y = y(t), \quad a \le t \le b,$$

is called **smooth** if x and y have continuous derivatives and

$$\left(\frac{dx}{dt}\right)^2 + \left(\frac{dy}{dt}\right)^2 \ne 0 \text{ for } a \le t \le b.$$

If C is a smooth curve, the unit tangent vector **T** exists at every point on the curve and the direction of **T** varies continuously. This means the graph of a smooth curve cannot have any sharp corners for $a < t < b$, so the curve in Figure 11.8.5 cannot be the graph of a smooth curve. If the initial and terminal points coincide, the graph will have a sharp corner at that point unless $\mathbf{T}(a) = \mathbf{T}(b)$. Thus, the graph of a smooth curve may have one corner, if that corner is both the initial point and the terminal point of the curve. See Figure 11.8.6.

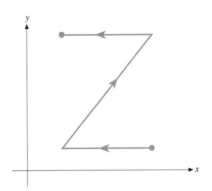

FIGURE 11.8.5 The graph of a smooth curve cannot have any sharp corners for $a < t < b$, so this cannot be the graph of a smooth curve.

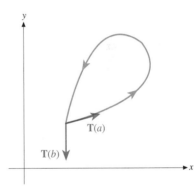

FIGURE 11.8.6 The graph of a smooth curve may have one corner, if that corner is both the initial point and the terminal point of the curve.

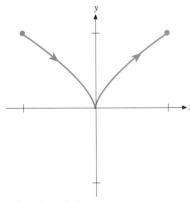

FIGURE 11.8.7 The graph may have a sharp corner at a point where $\left(\frac{dx}{dt}\right)^2 + \left(\frac{dy}{dt}\right)^2 = 0$, even if the coordinate functions have continuous derivatives.

If the graph of a curve appears smooth, then it is theoretically possible to use a parametrization in terms of arc length to obtain a smooth curve with the same graph. As we know, it is usually difficult to carry out the calculations of the arc length parametrization in specific cases.

The graph of a curve can have a sharp corner, even if the coordinate functions have continuous derivatives. If the coordinate functions have continuous derivatives, the velocity of a corresponding moving object will be continuous, so the motion will be "smooth." This implies that the object must stop at each corner along its path. For example, consider the curve

$$C : x = t^3, \quad y = t^2, \quad -1 \le t \le 1.$$

The derivatives dx/dt and dy/dt are continuous, but the graph has a corner at the point $(0, 0)$ where $t = 0$. Note that $(dx/dt)^2 + (dy/dt)^2 = 0$ at that point. See Figure 11.8.7.

The condition $(dx/dt)^2 + (dy/dt)^2 \ne 0$ ensures that a smooth curve cannot abruptly reverse direction and double back on itself. For example, consider

$$C : x = \sin t, \quad y = 0, \quad 0 \le t \le 3\pi/2.$$

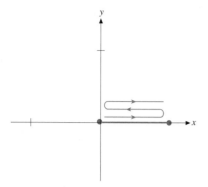

FIGURE 11.8.8 A smooth curve cannot abruptly reverse direction and double back on itself.

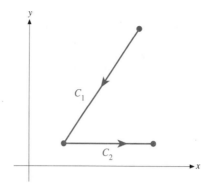

FIGURE 11.8.9 If the terminal point of C_1 coincides with the initial point of C_2, we can combine the curves to form the curve $C_1 + C_2$.

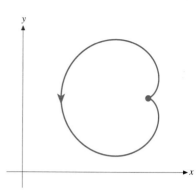

FIGURE 11.8.10 A curve is closed if its initial point coincides with its terminal point.

The graph of the curve is the line segment between $(0, 0)$ and $(1, 0)$, which appears "smooth." See Figure 11.8.8. However, the curve goes from $(0, 0)$ to $(1, 0)$ as t varies from 0 to $\pi/2$, doubles back on itself to $(0, 0)$ as t varies from $\pi/2$ to π, and then returns to $(1, 0)$ as t varies from π to $3\pi/2$. We have

$$\left(\frac{dx}{dt}\right)^2 + \left(\frac{dy}{dt}\right)^2 = 0 \text{ for } t = \pi/2 \text{ and } t = 3\pi/2,$$

so the curve is not smooth. The curve

$$C' : x = t, \quad y = 0, \quad 0 \leq t \leq 1,$$

has the same graph as C and is a smooth curve. C' goes directly from the point $(0, 0)$ to the point $(1, 0)$.

If the terminal point of a curve C_1 coincides with the initial point of a curve C_2, we denote by $C_1 + C_2$ the curve that goes from the initial point of C_1, along C_1 to the terminal point of C_1, which is the initial point of C_2, and then along C_2 to the terminal point of C_2. See Figure 11.8.9. A curve that is the sum of a finite number of smooth curves is called **piecewise smooth**.

Although a smooth curve cannot abruptly reverse direction and double back on itself, it can intersect itself. If the initial point of a curve coincides with its terminal point, we say the curve is **closed**. See Figure 11.8.10. If a curve does not intersect itself except possibly at its end points, the curve is said to be **simple**. See Figure 11.8.11. A simple curve may or may not be closed; a closed curve may or may not be simple.

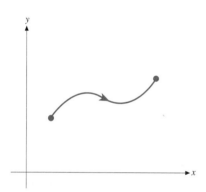

FIGURE 11.8.11 A curve is called simple if it does not intersect itself except possibly at its endpoints.

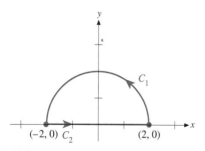

FIGURE 11.8.12 This curve is simple, closed, and piecewise smooth.

EXAMPLE 2

Consider the curve C that goes from the point $(2, 0)$ counterclockwise around the circle $x^2 + y^2 = 4$ to the point $(-2, 0)$ and then follows the x-axis to $(2, 0)$. Determine if C is either simple or closed. Determine if C can be expressed as either a smooth or a piecewise smooth parametric curve, and find a parametric representation.

SOLUTION The curve is sketched in Figure 11.8.12. C is simple because it does not intersect itself except at its endpoints. C is closed because its initial point coincides with its terminal point. C cannot be expressed as a smooth parametric curve because of the corner at $(-2, 0)$. We have $C = C_1 + C_2$, where

$$C_1 : x = 2\cos t, \quad y = 2\sin t, \quad 0 \le t \le \pi, \text{ and}$$
$$C_2 : x = t, \quad y = 0, \quad -2 \le t \le 2.$$

C_1 and C_2 are smooth, so C is piecewise smooth. ∎

Curvature

Given a smooth curve

$$C : x = x(t), \quad y = y(t), \quad a \le t \le b,$$

we have seen in Section 11.2 that it is theoretically possible to express the curve in terms of arc length s, where

$$s = \int_a^t \sqrt{(x'(\tau))^2 + (y'(\tau))^2} \, d\tau \text{ and } \frac{ds}{dt} = \sqrt{(x')^2 + (y')^2}.$$

The vector representation of C is

$$\mathbf{R} = x\mathbf{i} + y\mathbf{j}, \text{ so}$$
$$\mathbf{R}' = x'\mathbf{i} + y'\mathbf{j}, \text{ and}$$
$$|\mathbf{R}'| = \sqrt{(x')^2 + (y')^2} = \frac{ds}{dt}.$$

It follows that

$$\mathbf{T} = \frac{\mathbf{R}'}{|\mathbf{R}'|} = \frac{x'}{|\mathbf{R}'|}\mathbf{i} + \frac{y'}{|\mathbf{R}'|}\mathbf{j} = \frac{\dfrac{dx}{dt}}{\dfrac{ds}{dt}}\mathbf{i} + \frac{\dfrac{dy}{dt}}{\dfrac{ds}{dt}}\mathbf{j} = \frac{dx}{ds}\mathbf{i} + \frac{dy}{ds}\mathbf{j}.$$

That is,

$$\mathbf{T} = \frac{d\mathbf{R}}{ds},$$

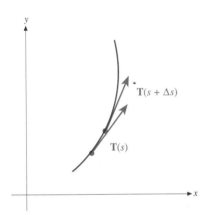

FIGURE 11.8.13a $|d\mathbf{T}/ds|$ measures the rate of change of the direction of **T**.

so the unit tangent vector is the rate of change of the position with respect to arc length. Let us see that $|d\mathbf{T}/ds|$ measures the change of the direction of **T**. In Figure 11.8.13a we have sketched unit tangent vectors at points on the curve that correspond to s and $s + \Delta s$, $\Delta s \ne 0$. From Figure 11.8.13b we see that, since $\mathbf{T}(s + \Delta s)$ and $\mathbf{T}(s)$ are unit vectors, the length of the vector $\mathbf{T}(s + \Delta s) - \mathbf{T}(s)$ is approximately the radian measure of the angle between the vectors, for Δs small. That is,

$$\left| \frac{\mathbf{T}(s + \Delta s) - \mathbf{T}(s)}{\Delta s} \right| \approx \left| \frac{\Delta \theta}{\Delta s} \right|.$$

It follows that $|d\mathbf{T}/ds|$ is the instantaneous rate of change of the direction of **T** with respect to arc length. The expression

$$\kappa = \left| \frac{d\mathbf{T}}{ds} \right|$$

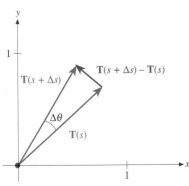

FIGURE 11.8.13b   Since $\mathbf{T}(s + \Delta s)$ and $\mathbf{T}(s)$ are unit vectors, we have $|\mathbf{T}(s + \Delta s) - \mathbf{T}(s)| \approx |\mathbf{T}(s)||\Delta \theta| = |\Delta \theta|.$

is called the **curvature** of the curve. We can calculate the curvature from given parametric equations of the curve by using the formula

$$\kappa = \left| \frac{d\mathbf{T}}{ds} \right| = \frac{\left| \dfrac{d\mathbf{T}}{dt} \right|}{\dfrac{ds}{dt}} = \frac{|\mathbf{T}'|}{|\mathbf{R}'|}.$$

EXAMPLE 3

Find the curvature of the circle $(x - h)^2 + (y - k)^2 = r^2$.

SOLUTION   We can express the circle in vector form as

$$\mathbf{R}(t) = (h + r \cos t)\mathbf{i} + (k + r \sin t)\mathbf{j}.$$

It follows that

$$\mathbf{R}'(t) = (-r \sin t)\mathbf{i} + (r \cos t)\mathbf{j},$$
$$|\mathbf{R}'(t)| = r,$$
$$\mathbf{T}(t) = (-\sin t)\mathbf{i} + (\cos t)\mathbf{i},$$
$$\mathbf{T}'(t) = (-\cos t)\mathbf{i} + (-\sin t)\mathbf{i}, \text{ and}$$
$$|\mathbf{T}'(t)| = 1.$$

The curvature is then

$$\kappa = \frac{|\mathbf{T}'|}{|\mathbf{R}'|} = \frac{1}{r}. \quad \blacksquare$$

We define the **radius of curvature** at a point on a curve where the curvature is nonzero to be

$$\rho = \frac{1}{\kappa}.$$

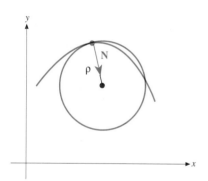

FIGURE 11.8.14 The circle of curvature has radius $\rho = 1/\kappa$ and center at a distance of ρ along the normal vector **N** from the point on the curve. The circle of curvature approximates the graph of the curve.

The **circle of curvature** is the circle that has radius ρ and center at a distance of ρ along the normal vector **N** from the point on the curve. The center of the circle of curvature is called the **center of curvature**. See Figure 11.8.14. It follows from Example 3 that the circle of curvature of a circle coincides with the circle. A curve with curvature κ at a point is bending the same amount as a circle with radius $1/\kappa$ as the curve passes through the point.

EXAMPLE 4

Find the unit tangent vector, the unit normal vector, and the curvature of the curve $y = \sin x$. Find the radius of curvature and the center of curvature at the point $(\pi/2, 1)$.

SOLUTION   We have

$$\mathbf{R} = t\mathbf{i} + (\sin t)\mathbf{j},$$
$$\mathbf{R}' = \mathbf{i} + (\cos t)\mathbf{j},$$
$$|\mathbf{R}'| = \sqrt{1 + \cos^2 t},$$
$$\mathbf{T} = \frac{1}{(1 + \cos^2 t)^{1/2}}\mathbf{i} + \frac{\cos t}{(1 + \cos^2 t)^{1/2}}\mathbf{j},$$

$$\mathbf{T}' = -\frac{1}{2}(1 + \cos^2 t)^{-3/2}(2\cos t)(-\sin t)\mathbf{i}$$
$$+ \frac{(1 + \cos^2 t)^{1/2}(-\sin t) - (\cos t)(1/2)(1 + \cos^2 t)^{-1/2}(2\cos t)(-\sin t)}{1 + \cos^2 t}\mathbf{j}$$

$$= \frac{\sin t \cos t}{(1 + \cos^2 t)^{3/2}}\mathbf{i} - \frac{\sin t}{(1 + \cos^2 t)^{3/2}}\mathbf{j},$$

$$|\mathbf{T}'| = \frac{|\sin t|}{1 + \cos^2 t}.$$

(The absolute value comes from taking the nonnegative square root of $\sin^2 t$.)

$$\mathbf{N} = \left(\frac{\sin t}{|\sin t|}\right)\left(\frac{\cos t}{\sqrt{1 + \cos^2 t}}\mathbf{i} - \frac{1}{\sqrt{1 + \cos^2 t}}\mathbf{j}\right),$$

$$\kappa = \frac{|\mathbf{T}'|}{|\mathbf{R}'|} = \frac{|\sin t|}{(1 + \cos^2 t)^{3/2}}.$$

The point $(\pi/2, 1)$ corresponds to $t = \pi/2$. When $t = \pi/2$, the radius of curvature is

$$\rho = \frac{1}{\kappa} = 1.$$

The center of curvature is located at a distance of $\rho = 1$ from $(\pi/2, 1)$ in the direction of the unit normal vector $\mathbf{N}(\pi/2) = -\mathbf{j}$. The graph of $y = \sin x$ and the circle of curvature at $(\pi/2, 1)$ are sketched in Figure 11.8.15. Notice that the direction of the unit normal vector $\mathbf{N}$ corresponds to the concavity of the curve. The unit normal is undefined where $y = 0$. These points are points of inflection of the curve. ∎

FIGURE 11.8.15 The unit normal points in the direction the curve is bending. The unit normal vector does not exist at points of inflection of the curve.

Formulas for Curvature

Let us develop a formula for the curvature in terms of the parametric equations of a curve,

$$C : x = x(t), \quad y = y(t), \quad a \le t \le b.$$

The vector form of the curve is

$$\mathbf{R} = x\mathbf{i} + y\mathbf{j}.$$

The curvature is $\kappa = |\mathbf{T}'|/|\mathbf{R}'|$. We have

$$\mathbf{R}' = x'\mathbf{i} + y'\mathbf{j} \text{ and } |\mathbf{R}'| = \sqrt{x'^2 + y'^2},$$

$$\mathbf{T}' = \frac{d}{dt}(\mathbf{T}) = \frac{d}{dt}\left(\frac{\mathbf{R}'}{|\mathbf{R}'|}\right) = \frac{d}{dt}\left(\frac{x'}{\sqrt{x'^2 + y'^2}}\mathbf{i} + \frac{y'}{\sqrt{x'^2 + y'^2}}\mathbf{j}\right).$$

Carrying out the differentiation and then simplifying, we obtain

$$\mathbf{T}' = \frac{y'(y'x'' - x'y'')}{(x'^2 + y'^2)^{3/2}}\mathbf{i} + \frac{x'(x'y'' - y'x'')}{(x'^2 + y'^2)^{3/2}}\mathbf{j}.$$

Calculating the length of this vector, we obtain

$$|\mathbf{T}'| = \frac{|x'y'' - y'x''|}{x'^2 + y'^2}.$$

Dividing by $|\mathbf{R}'|$, we obtain the formula

$$\kappa = \frac{|x'y'' - y'x''|}{(x'^2 + y'^2)^{3/2}}.$$

This formula is convenient for finding the curvature of a curve when we do not need to calculate the unit normal vector.

EXAMPLE 5

Find a formula for the curvature of the ellipse

$$C : x = 3\cos t, \quad y = 2\sin t, \quad 0 \le t \le 2\pi.$$

Find maximum and minimum values of the curvature and find points on the ellipse where these values occur.

SOLUTION   We have $x' = -3\sin t$, $x'' = -3\cos t$, $y' = 2\cos t$, and $y'' = -2\sin t$. From the formula we then obtain

$$\kappa = \frac{|(-3\sin t)(-2\sin t) - (2\cos t)(-3\cos t)|}{((-3\sin t)^2 + (2\cos t)^2)^{3/2}}$$

$$= \frac{6(\sin^2 t + \cos^2 t)}{(9\sin^2 t + 4\cos^2 t)^{3/2}}$$

$$= 6(9\sin^2 t + 4\cos^2 t)^{-3/2}.$$

Let us use the derivative of κ to find the maximum and minimum values. We have

$$\frac{d\kappa}{dt} = 6\left(-\frac{3}{2}\right)(9\sin^2 t + 4\cos^2 t)^{-5/2}((18)(\sin t)(\cos t) + (8)(\cos t)(-\sin t))$$

$$= \frac{-90(\sin t)(\cos t)}{(9\sin^2 t + 4\cos^2 t)^{5/2}}.$$

The derivative is zero whenever $(\sin t)(\cos t) = 0$—that is, whenever $t = 0$, $\pi/2$, π, $3\pi/2$, and 2π. The maximum curvature is $\kappa(0) = \kappa(\pi) = \kappa(2\pi) = 6(4)^{-3/2} = 3/4$, at the vertices of the ellipse. The minimum curvature is $\kappa(\pi/2) = \kappa(3\pi/2) = 6(9)^{-3/2} = 2/9$, at the endpoints of the minor axis of the ellipse.

In this example, we could have determined the maximum and minimum value of the curvature without using a derivative. That is, we know that $\sin^2 t + \cos^2 t = 1$, so

$$\kappa = 6(9\sin^2 t + 4\cos^2 t)^{-3/2}$$

$$= 6(9\sin^2 t + 4(1 - \sin^2 t))^{-3/2}$$

$$= 6(5\sin^2 t + 4)^{-3/2}.$$

Since $4 \le 5\sin^2 t + 4 \le 9$, we see that

$$6(9)^{-3/2} \le \kappa \le 6(4)^{-3/2}.$$

The ellipse and the circles of curvature at these points are sketched in Figure 11.8.16. ∎

FIGURE 11.8.16 The maximum curvature of the ellipse occurs at its vertices.

If a curve is the graph of a function $y = y(x)$, we can parametrize the curve by writing

$$C : x = x, \quad y = y(x).$$

In the above equations we are using x as the parameter. We have $x' = 1, x'' = 0$, so the formula for curvature reduces to

$$\kappa = \frac{|y''|}{(1 + y'^2)^{3/2}}, \text{ if } y = y(x).$$

EXERCISES 11.8

Determine which of the curves in Exercises 1–8 are simple and which are closed. Determine which can be expressed as either smooth or piecewise smooth parametric curves, and find parametric representations.

1. From $(-1, 1)$, along $y = x^{2/3}$ to $(1, 1)$

2. From $(-1, -1)$, along $y = x^{1/3}$ to $(1, 1)$

3. From $(-1, -1)$, along $y = x^3$ to $(1, 1)$, along $y = x$ to $(-1, -1)$

4. From $(0, 0)$, along $y = x^2$ to $(1, 1)$, along $y = x$ to $(0, 0)$

5. From $(1, 0)$, counterclockwise twice around $x^2 + y^2 = 1$

6. From $(1, 0)$, counterclockwise one-and-one-half times around $x^2 + y^2 = 1$

7. From $(1, 0)$, counterclockwise once around $x^2 + y^2 = 1$

8. From $(-1, 0)$, along the x-axis to $(1, 0)$, counterclockwise once around $x^2 + y^2 = 1$

9. A curve goes once around the polar curve $r = 1 - \cos \theta$. For which initial points does the curve have a smooth parametrization?

10. A smooth curve that is given by a function $f(x)$ can have a unit tangent vector that is vertical. Can f be differentiable at such a point?

Find the unit tangent vector, the unit normal vector, and the curvature of the curves given in Exercises 11–16.

11. $C : x = 2 \cos t, y = 2 \sin t$

12. $C : x = 1 - 2t, y = 2 + 2t$

13. $C : x = t, y = \ln t$

14. $C : x = t^2/2, y = t^3/3, t > 0$

15. $C : x = e^t \cos t, y = e^t \sin t$

16. $C : x = 2t, y = e^t + e^{-t}$

For the curves given in Exercises 17–20, find the unit tangent vector, the unit normal vector, the curvature, the radius of curvature, and the center of curvature at the indicated point. Sketch the graph of the curve and sketch the circle of curvature at the indicated point.

17. $C : x = t, y = \cos t; t = 0$

18. $C : x = e^t, y = t; t = 0$

19. $C : x = 4 \cos t, y = 3 \sin t; t = 0, \pi/2$

20. $C : x = t \cos t, y = t \sin t; t = 0$

Find a formula for the curvature of the curves given in Exercises 21–28.

21. $y = 1/x, x > 0$

22. $y = 1/x^2$

23. $y = \ln(\sec x), -\pi/2 < x < \pi/2$

24. $y = \ln(\sec x + \tan x), 0 \le x < \pi/2$

25. $C : x = 2 \cos t, y = \sin t$

26. $C : x = e^t + e^{-t}, y = e^t - e^{-t}$

27. $C : x = \sec t, y = \tan t$

28. $C : x = t - r \cos(t/r), y = r - r \sin(t/r)$

29. Find the maximum curvature of the curve $y = \ln x$, and find the point on the curve where the curvature is maximum.

30. Find the maximum curvature of the curve $y = e^x$, and find the point on the curve where the curvature is maximum.

31. Find the maximum curvature of the curve $C : x = 2t$, $y = e^t + e^{-t}$, and find the point on the curve where the curvature is maximum.

32. Show that the curvature of a polar curve $r = r(\theta)$ is given by the formula

$$\kappa = \frac{|r^2 + 2(r')^2 - rr''|}{(r^2 + (r')^2)^{3/2}},$$

where the derivatives are with respect to θ.

Use the formula of Exercise 32 to find a formula for the curvature of the polar curves given in Exercises 33–35.

33. $r = 2 - \cos\theta$

34. $r = 1 + \sin\theta$

35. $r = 1 - \sin\theta$

36. Use the formula of Exercise 32 to show that the curvature of a polar curve $r = r(\theta)$ at the pole is given by

$$\kappa = \frac{2}{|r'|}.$$

Use the formula of Exercise 36 to find the curvature at the pole of the polar curves given in Exercises 37–39.

37. $r = 2\cos 2\theta$

38. $r = 2\sin 3\theta$

39. $r = \sin 4\theta$

40. Find the maximum curvature of the polar curve $r = \sin 3\theta$.

41. Find the minimum curvature of the polar curve $r = 1 - \cos\theta$.

42. The curve from $(-1, -1)$, along $y = -\sqrt{1 - (x + 1)^2}$ to $(0, 0)$, and then along $y = \sqrt{4 - (x - 2)^2}$ to $(2, 2)$ is

smooth, because the unit tangent vector varies continuously. Use the fact that the curve consists of two circular arcs to discuss the curvature. (Abrupt changes in curvature along the cross-section of an airfoil could seriously affect its aerodynamic properties.)

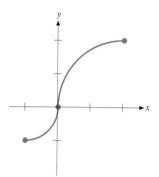

43. A particle is forced along a curve in such a way that its position at time t is $\mathbf{R}(t) = t^2\mathbf{i} + 2t\mathbf{j}$, $0 \le t \le 2$. At $t = 2$, the particle flies off on a tangent to the curve and moves free of any force. What is the position of the particle at time $t = 3$?

11.9 CURVES IN THREE-DIMENSIONAL SPACE; COMPONENTS OF ACCELERATION

Connections

Parametric equations and graphs in the plane, 11.1.

*Motion vectors, **R**, **V**, **A**, 11.7.*

*__**T**, **N**, κ, 11.8.__*

The concepts we developed for curves in the plane also apply to space curves. The parametric form of a curve in three-dimensional space is

$$C : x = x(t), \quad y = y(t), \quad z = z(t),$$

where x, y, and z are continuous functions of t.

Sketching Parametric Curves in Three-Dimensional Space

We can sketch parametric curves in three-dimensional space by connecting an ordered set of points on the curve. To illustrate the three-dimensional character of the curve, it is useful to indicate the relation between the curve and its projection onto one of the coordinate planes. For example, if we combine the equations for x and y in such a way that t is eliminated, we will obtain the equation of the vertical generalized cylinder that contains C. The intersection of this cylinder with the xy-plane is the projection of C onto the xy-plane. Drawing the projection and some of the vertical lines between the projection and the curve helps orient the curve in the sketch.

EXAMPLE 1
Sketch the graph of the curve

$$C : x = 3\cos t, \quad y = 3\sin t, \quad z = t, \quad 0 \le t \le 2\pi.$$

Find the equation of the vertical generalized cylinder that contains C; sketch the projection of C onto the xy-plane and draw some of the vertical lines between the curve and its projection.

TABLE 11.9.1

t	x	y	z
0	3	0	0
$\dfrac{\pi}{2}$	0	3	$\dfrac{\pi}{2}$
π	-3	0	π
$\dfrac{3\pi}{2}$	0	-3	$\dfrac{3\pi}{2}$
2π	3	0	2π

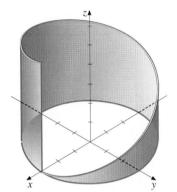

FIGURE 11.9.1 The curve spirals up the vertical cylinder $x^2 + y^2 = 3^2$, which gives the projection of the curve onto the xy-plane.

SOLUTION   Values of the coordinates of points on the curve that correspond to increments of $\Delta t = \pi/2$ are given in Table 11.9.1. From the equations $x = 3\cos t$ and $y = 3\sin t$, we see that

$$x^2 + y^2 = 3^2.$$

This is the vertical generalized cylinder that contains the curve; it is a right circular cylinder with radius 3 and axis along the z-axis. The projection onto the xy-plane is the circle with center at the origin and radius 3. The z-coordinate of a point on the curve increases as the curve winds counterclockwise around the cylinder. The curve is called a **helix**. The graph is sketched in Figure 11.9.1. ■

In Section 11.6 we observed that parametric equations of the line in three-dimensional space that contains the point (x_0, y_0, z_0) and has direction vector $\mathbf{V} = a\mathbf{i} + b\mathbf{j} + c\mathbf{k}$ are

$$x = x_0 + at, \quad y = y_0 + bt, \quad z = z_0 + ct.$$

We also know that the line segment from (x_0, y_0, z_0) to (x_1, y_1, z_1) has direction $\mathbf{V} = (x_1 - x_0)\mathbf{i} + (y_1 - y_0)\mathbf{j} + (z_1 - z_0)\mathbf{k}$. See Figure 11.9.2. It follows that

The line segment from (x_0, y_0, z_0) to (x_1, y_1, z_1) is given by

$$C : x = x_0 + (x_1 - x_0)t, \quad y = y_0 + (y_1 - y_0)t,$$
$$z = z_0 + (z_1 - z_0)t, \quad 0 \leq t \leq 1.$$

This formula corresponds to the formula for the parametric equations of the line segment between two points in the plane.

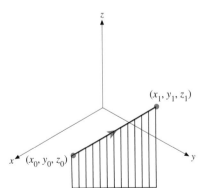

FIGURE 11.9.2 The line segment has parametric equations

$$C : x = x_0 + (x_1 - x_0)t,$$
$$y = y_0 + (y_1 - y_0)t,$$
$$z = z_0 + (z_1 - z_0)t, 0 \leq t \leq 1$$

Intersection of a Quadric Surface and a Plane

We can find parametric equations for the curve of intersection of a quadric surface

$$Ax^2 + By^2 + Cz^2 + Dx + Ey + Fz + G = 0$$

and a plane $ax + by + cz + d = 0$ by using the equation of the plane to eliminate one variable from the quadric equation, finding parametric equations for the resulting second-order equation in two variables, and then using the linear equation to obtain the other variable in terms of the parameter.

EXAMPLE 2

Find parametric equations of the curve of intersection of the ellipsoid $4x^2 + y^2 + 4z^2 = 4$ and the plane $x + y = 1$. Sketch the curve.

SOLUTION   Let us solve the linear equation for y and then substitute in the quadric equation to eliminate y.

$$x + y = 1 \text{ implies } y = 1 - x.$$

Substitution then gives

$$4x^2 + y^2 + 4z^2 = 4,$$
$$4x^2 + (1 - x)^2 + 4z^2 = 4,$$
$$4x^2 + 1 - 2x + x^2 + 4z^2 = 4,$$
$$5x^2 - 2x + 4z^2 = 3.$$

This is the equation of an ellipse in the variables x and z. We can find parametric equations by completing the square and expressing the equation in the standard form of an ellipse. We have

$$5\left(x^2 - \frac{2}{5}x + \frac{1}{25}\right) + 4z^2 = 3 + 5\left(\frac{1}{25}\right),$$

$$5\left(x - \frac{1}{5}\right)^2 + 4z^2 = \frac{16}{5},$$

$$\frac{\left(x - \frac{1}{5}\right)^2}{\left(\frac{4}{5}\right)^2} + \frac{z^2}{\left(\frac{2}{\sqrt{5}}\right)^2} = 1.$$

From Formula 5 of Section 11.1, we then obtain the parametric equations

$$x = \frac{1}{5} + \frac{4}{5}\cos t, \quad z = \frac{2}{\sqrt{5}}\sin t, \quad 0 \le t \le 2\pi.$$

Since $y = 1 - x$, we also have

$$y = 1 - \left(\frac{1}{5} + \frac{4}{5}\cos t\right) = \frac{4}{5} - \frac{4}{5}\cos t.$$

Combining results, we obtain the following parametric representation:

$$C : x = \frac{1}{5} + \frac{4}{5}\cos t, \quad y = \frac{4}{5} - \frac{4}{5}\cos t, \quad z = \frac{2}{\sqrt{5}}\sin t, \quad 0 \le t \le 2\pi.$$

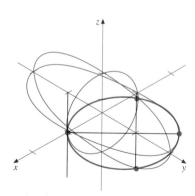

FIGURE 11.9.3 The curve of intersection is sketched by drawing an ellipse through the points of intersection of the coordinate traces.

To sketch the curve, we first sketch the traces of the ellipsoid and the plane in each coordinate plane, determine the points of intersection of the traces in each coordinate plane, and then draw the curve of intersection through these points. This is done in Figure 11.9.3. Note that you are looking for points of intersection of traces that are in the same coordinate plane. You must be very careful to recognize which traces are in which coordinate plane. It is also a great help to have some idea of the general character of the curve of intersection of an ellipsoid and a plane. ∎

Arc Length

If the coordinate functions have continuous derivatives, we define the **differential arc length** of the curve

$$C : x = x(t), \quad y = y(t), \quad z = z(t), \quad a \leq t \leq b,$$

to be

$$ds = \sqrt{(dx)^2 + (dy)^2 + (dz)^2} = \sqrt{\left(\frac{dx}{dt}\right)^2 + \left(\frac{dy}{dt}\right)^2 + \left(\frac{dz}{dt}\right)^2}\, dt.$$

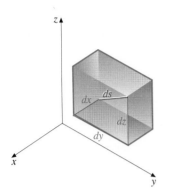

FIGURE 11.9.4 The differential, arc length is $ds = \sqrt{(dx)^2 + (dy)^2 + (dz)^2}$.

See Figure 11.9.4. The **length** of the curve is

$$L = \int_a^b \sqrt{\left(\frac{dx}{dt}\right)^2 + \left(\frac{dy}{dt}\right)^2 + \left(\frac{dz}{dt}\right)^2}\, dt.$$

The **arc length function** is

$$s = \int_a^t \sqrt{(x'(\tau))^2 + (y'(\tau))^2 + (z'(\tau))^2}\, d\tau.$$

If the coordinate functions are continuous with

$$\left(\frac{dx}{dt}\right)^2 + \left(\frac{dy}{dt}\right)^2 + \left(\frac{dz}{dt}\right)^2 \neq 0 \text{ for } a \leq t \leq b,$$

so the curve is smooth, we can use the inverse of the arc length function to parametrize the curve in terms of arc length.

EXAMPLE 3

Find the length of the helix

$$C : x = 3\cos t, \quad y = 3\sin t, \quad z = t, \quad 0 \leq t \leq 2\pi.$$

Find the parametrization of the curve in terms of arc length.

SOLUTION We have

$$x'(t) = -3\sin t, \quad y'(t) = 3\cos t, \text{ and } z'(t) = 1,$$

so

$$\left(x'(t)\right)^2 + \left(y'(t)\right)^2 + \left(z'(t)\right)^2 = (-3\sin t)^2 + (3\cos t)^2 + (1)^2$$
$$= 9(\sin^2 t + \cos^2 t) + 1 = 9 + 1 = 10.$$

It follows that the total length of the curve is

$$L = \int_0^{2\pi} \sqrt{10}\, d\tau = 2\pi\sqrt{10}$$

and the arc length function is

$$s = \int_0^t \sqrt{10}\, d\tau = t\sqrt{10}, \quad 0 \leq t \leq 2\pi.$$

Solving the equation $s = t\sqrt{10}$ for t, we obtain

$$t = \frac{s}{\sqrt{10}}.$$

Substitution then gives the parametric representation of C in terms of arc length,

$$C : x = 3\cos\frac{s}{\sqrt{10}}, \quad y = 3\sin\frac{s}{\sqrt{10}}, \quad z = \frac{s}{\sqrt{10}}, \quad 0 \le s \le 2\pi\sqrt{10}. \quad \blacksquare$$

T, N, B, and κ

The space curve

$$C : x = x(t), \quad y = y(t), \quad z = z(t)$$

can be expressed in **vector form** as

$$\mathbf{R} = x(t)\mathbf{i} + y(t)\mathbf{j} + z(t)\mathbf{k}.$$

We then have

$$|\mathbf{R}'| = \sqrt{\left(\frac{dx}{dt}\right)^2 + \left(\frac{dy}{dt}\right)^2 + \left(\frac{dz}{dt}\right)^2} = \frac{ds}{dt}.$$

If C is a smooth curve, we can define the unit tangent vector, the unit normal vector, and the curvature in terms of the arc length parametrization of C. The definitions correspond to those of a curve in the plane. As before, we can carry out the calculations in terms of any given parametrization.

The unit tangent vector is $\mathbf{T} = \dfrac{d\mathbf{R}}{ds} = \dfrac{\mathbf{R}'}{|\mathbf{R}'|}.$

The principal unit normal vector is $\mathbf{N} = \dfrac{\dfrac{d\mathbf{T}}{ds}}{\left|\dfrac{d\mathbf{T}}{ds}\right|} = \dfrac{\mathbf{T}'}{|\mathbf{T}'|}.$

(Note that the principal unit normal vector is not defined if $\mathbf{T}' = 0$.)

The curvature is $\kappa = \left|\dfrac{d\mathbf{T}}{ds}\right| = \dfrac{\left|\dfrac{d\mathbf{T}}{dt}\right|}{\dfrac{ds}{dt}} = \dfrac{|\mathbf{T}'|}{|\mathbf{R}'|}.$

For curves in three-dimensional space we define the **unit binormal** to be

$$\mathbf{B} = \mathbf{T} \times \mathbf{N}.$$

The vector $\mathbf{B}$ is a unit vector, since

$$|\mathbf{B}| = |\mathbf{T} \times \mathbf{N}| = |\mathbf{B}||\mathbf{T}|\sin\theta = 1,$$

where $\theta = \pi/2$ is the angle between the unit vectors $\mathbf{T}$ and $\mathbf{N}$. The unit binormal vector at a point on a curve is drawn from the point on the curve, as are the unit tangent and unit normal vectors. The vectors $\mathbf{T}$, $\mathbf{N}$, and $\mathbf{B}$ (in that order) form a right-handed rectangular coordinate system at the corresponding point on the curve. See Figure 11.9.5.

FIGURE 11.9.5   The vectors **T**, **N**, and **B** (in that order) form a right-hand rectangular coordinate system at the corresponding point on the curve.

EXAMPLE 4

Find the unit tangent vector, the principal unit normal vector, the unit binormal, and the curvature of the helix

$$C : x = 3\cos t, \quad y = 3\sin t, \quad z = t, \quad 0 \le t \le 2\pi.$$

SOLUTION   Using the vector form of the curve, we have

$$\mathbf{R} = (3\cos t)\mathbf{i} + (3\sin t)\mathbf{j} + t\mathbf{k}, \text{ so}$$
$$\mathbf{R}' = (-3\sin t)\mathbf{i} + (3\cos t)\mathbf{j} + \mathbf{k} \text{ and}$$
$$|\mathbf{R}'| = \sqrt{(-3\sin t)^2 + (3\cos t)^2 + (1)^2} = \sqrt{10}.$$

The unit tangent vector is

$$\mathbf{T} = \frac{\mathbf{R}'}{|\mathbf{R}'|} = \left(\frac{-3\sin t}{\sqrt{10}}\right)\mathbf{i} + \left(\frac{3\cos t}{\sqrt{10}}\right)\mathbf{j} + \left(\frac{1}{\sqrt{10}}\right)\mathbf{k}.$$

We have

$$\mathbf{T}' = \left(\frac{-3\cos t}{\sqrt{10}}\right)\mathbf{i} + \left(\frac{-3\sin t}{\sqrt{10}}\right)\mathbf{j}, \text{ so}$$
$$|\mathbf{T}'| = \frac{3}{\sqrt{10}}.$$

The principal unit normal vector is

$$\mathbf{N} = \frac{\mathbf{T}'}{|\mathbf{T}'|} = (-\cos t)\mathbf{i} + (-\sin t)\mathbf{j}.$$

The binormal is

$$\mathbf{B} = \mathbf{T} \times \mathbf{N} = \begin{vmatrix} \mathbf{i} & \mathbf{j} & \mathbf{k} \\ \dfrac{-3\sin t}{\sqrt{10}} & \dfrac{3\cos t}{\sqrt{10}} & \dfrac{1}{\sqrt{10}} \\ -\cos t & -\sin t & 0 \end{vmatrix}$$

$$= \begin{vmatrix} \dfrac{3\cos t}{\sqrt{10}} & \dfrac{1}{\sqrt{10}} \\ -\sin t & 0 \end{vmatrix}\mathbf{i} - \begin{vmatrix} \dfrac{-3\sin t}{\sqrt{10}} & \dfrac{1}{\sqrt{10}} \\ -\cos t & 0 \end{vmatrix}\mathbf{j} + \begin{vmatrix} \dfrac{-3\sin t}{\sqrt{10}} & \dfrac{3\cos t}{\sqrt{10}} \\ -\cos t & -\sin t \end{vmatrix}\mathbf{k}$$

$$= \frac{\sin t}{\sqrt{10}}\mathbf{i} - \frac{\cos t}{\sqrt{10}}\mathbf{j} + \frac{3}{\sqrt{10}}\mathbf{k}.$$

The curvature is

$$\kappa = \frac{|\mathbf{T}'|}{|\mathbf{R}'|} = \frac{\dfrac{3}{\sqrt{10}}}{\sqrt{10}} = 0.3.$$

The curve and representative vectors **T**, **N**, and **B** are sketched in Figure 11.9.6. ▪

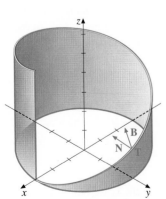

FIGURE 11.9.6 The unit normal vector of the helix is horizontal and points toward the z-axis.

Motion

Let us now see how to use the properties of a curve to study the motion of an object along the curve. In particular, let

$$\mathbf{R}(t) = x(t)\mathbf{i} + y(t)\mathbf{j} + z(t)\mathbf{k}$$

represent the position vector of an object at time t. We can also think of $\mathbf{R}$ as the vector form of the curve traced by the object. We want to relate the velocity and acceleration of the object to the unit tangent and normal vectors of the curve.

The velocity of the object is $\mathbf{V} = \mathbf{R}'$ and the speed is

$$|\mathbf{V}| = |\mathbf{R}'| = \sqrt{\left(\frac{dx}{dt}\right)^2 + \left(\frac{dy}{dt}\right)^2 + \left(\frac{dz}{dt}\right)^2} = \frac{ds}{dt}.$$

We have

$$\mathbf{V} = \mathbf{R}' = |\mathbf{R}'|\mathbf{T} = \frac{ds}{dt}\mathbf{T},$$

so the velocity is in the direction of the unit tangent vector. Also, the acceleration is

$$\mathbf{A} = \mathbf{V}' = \frac{d}{dt}(\mathbf{V}) = \frac{d}{dt}\left(\frac{ds}{dt}\mathbf{T}\right) = \frac{d^2s}{dt^2}\mathbf{T} + \frac{ds}{dt}\mathbf{T}' = \frac{d^2s}{dt^2}\mathbf{T} + \kappa\left(\frac{ds}{dt}\right)^2\mathbf{N},$$

where we have used

$$\mathbf{T}' = |\mathbf{T}'|\mathbf{N} \text{ and } |\mathbf{T}'| = \kappa\frac{ds}{dt}, \text{ so } \mathbf{T}' = \kappa\frac{ds}{dt}\mathbf{N}.$$

This shows that the acceleration of the object is in the plane determined by the vectors $\mathbf{T}$ and $\mathbf{N}$.

$a_{\mathbf{T}} = \dfrac{d^2s}{dt^2}$ is called the **tangential component of acceleration.**

$a_{\mathbf{N}} = \kappa\left(\dfrac{ds}{dt}\right)^2$ is called the **normal component of acceleration.**

The normal component of acceleration is nonnegative, so the *principal normal vector points in the direction the curve is bending.* Also, the magnitude of the force associated with the normal component of acceleration varies jointly as the curvature and the square of the speed. This suggests that a car should slow down in order to navigate a sharp turn, where the curvature is large and the turning radius is small.

The tangential and normal components of acceleration can be calculated from $\mathbf{R}'$ and $\mathbf{R}''$, without calculating $\mathbf{T}$ and $\mathbf{N}$. Since $\mathbf{A} = \mathbf{R}''$, we have

$$\mathbf{R}'' = \frac{d^2s}{dt^2}\mathbf{T} + \kappa\left(\frac{ds}{dt}\right)^2\mathbf{N}.$$

Taking the dot product of each side of this equation with $\mathbf{T}$, we obtain

$$\mathbf{T} \cdot \mathbf{R}'' = \mathbf{T} \cdot \left(\frac{d^2s}{dt^2}\mathbf{T} + \kappa\left(\frac{ds}{dt}\right)^2\mathbf{N}\right) = \frac{d^2s}{dt^2}\mathbf{T} \cdot \mathbf{T} + \kappa\left(\frac{ds}{dt}\right)^2\mathbf{T} \cdot \mathbf{N} = \frac{d^2s}{dt^2}.$$

We also have

$$\mathbf{T} \cdot \mathbf{R}'' = \left(\frac{\mathbf{R}'}{|\mathbf{R}'|} \right) \cdot \mathbf{R}'' = \frac{\mathbf{R}' \cdot \mathbf{R}''}{|\mathbf{R}'|}.$$

We conclude that

$$a_{\mathbf{T}} = \frac{d^2 s}{dt^2} = \frac{\mathbf{R}' \cdot \mathbf{R}''}{|\mathbf{R}'|}.$$

Similarly, we have

$$\mathbf{T} \times \mathbf{R}'' = \mathbf{T} \times \left(\frac{d^2 s}{dt^2} \mathbf{T} + \kappa \left(\frac{ds}{dt} \right)^2 \mathbf{N} \right) = \frac{d^2 s}{d^2 t} \mathbf{T} \times \mathbf{T} + \kappa \left(\frac{ds}{dt} \right)^2 \mathbf{T} \times \mathbf{N}$$

$$= \kappa \left(\frac{ds}{dt} \right)^2 \mathbf{B}.$$

Since

$$\mathbf{T} \times \mathbf{R}'' = \left(\frac{\mathbf{R}'}{|\mathbf{R}'|} \right) \times \mathbf{R}'' = \frac{\mathbf{R}' \times \mathbf{R}''}{|\mathbf{R}'|} \text{ and } |\mathbf{B}| = 1,$$

we conclude that

$$a_{\mathbf{N}} = \kappa \left(\frac{ds}{dt} \right)^2 = \frac{|\mathbf{R}' \times \mathbf{R}''|}{|\mathbf{R}'|}.$$

Note that the above formula for $a_{\mathbf{N}}$ involves the *length* of the vectors $\mathbf{R}' \times \mathbf{R}''$ and $\mathbf{R}'$. The components of acceleration are *numbers*, not vectors.

EXAMPLE 5

The position of an object at time t is given by

$$\mathbf{R} = t^2 \mathbf{i} + t \mathbf{j} + \frac{t^2}{2} \mathbf{k}.$$

Find the tangential and normal components of acceleration.

SOLUTION
$$\mathbf{R} = t^2 \mathbf{i} + t \mathbf{j} + \frac{t^2}{2} \mathbf{k},$$
$$\mathbf{R}' = 2t \mathbf{i} + \mathbf{j} + t \mathbf{k},$$
$$\mathbf{R}'' = 2 \mathbf{i} + 0 \mathbf{j} + \mathbf{k},$$
$$|\mathbf{R}'| = \sqrt{(2t)^2 + (1)^2 + (t)^2} = \sqrt{5t^2 + 1},$$
$$\mathbf{R}' \cdot \mathbf{R}'' = (2t)(2) + (1)(0) + (t)(1) = 5t,$$

$$\mathbf{R}' \times \mathbf{R}'' = \begin{vmatrix} \mathbf{i} & \mathbf{j} & \mathbf{k} \\ 2t & 1 & t \\ 2 & 0 & 1 \end{vmatrix}$$

$$= \begin{vmatrix} 1 & t \\ 0 & 1 \end{vmatrix} \mathbf{i} - \begin{vmatrix} 2t & t \\ 2 & 1 \end{vmatrix} \mathbf{j} + \begin{vmatrix} 2t & 1 \\ 2 & 0 \end{vmatrix} \mathbf{k}$$

$$= \mathbf{i} - 0\mathbf{j} - 2\mathbf{k},$$
$$|\mathbf{R}' \times \mathbf{R}''| = \sqrt{(1)^2 + (0)^2 + (-2)^2} = \sqrt{5}.$$

The tangential component of acceleration is

$$a_{\mathbf{T}} = \frac{\mathbf{R}' \cdot \mathbf{R}''}{|\mathbf{R}'|} = \frac{5t}{\sqrt{5t^2 + 1}}.$$

The normal component of acceleration is

$$a_{\mathbf{N}} = \frac{|\mathbf{R}' \times \mathbf{R}''|}{|\mathbf{R}'|} = \frac{\sqrt{5}}{\sqrt{5t^2 + 1}}.$$

The graph of the curve $\mathbf{R} = t^2\mathbf{i} + t\mathbf{j} + (t^2/2)\mathbf{k}$ is sketched in Figure 11.9.7. The curve is the intersection of the parabolic cylinder $x = y^2$ and the plane $x = 2z$. ∎

We have

$$\begin{aligned}
|\mathbf{R}''|^2 &= \mathbf{R}'' \cdot \mathbf{R}'' \\
&= (a_{\mathbf{T}}\,\mathbf{T} + a_{\mathbf{N}}\,\mathbf{N}) \cdot (a_{\mathbf{T}}\,\mathbf{T} + a_{\mathbf{N}}\,\mathbf{N}) \\
&= a_{\mathbf{T}}^2\,\mathbf{T} \cdot \mathbf{T} + 2a_{\mathbf{T}}\,a_{\mathbf{N}}\,\mathbf{T} \cdot \mathbf{N} + a_{\mathbf{N}}^2\,\mathbf{N} \cdot \mathbf{N} \\
&= a_{\mathbf{T}}^2 + a_{\mathbf{N}}^2.
\end{aligned}$$

This gives the formula

$$a_{\mathbf{N}} = \sqrt{|\mathbf{R}''|^2 - a_{\mathbf{T}}^2}.$$

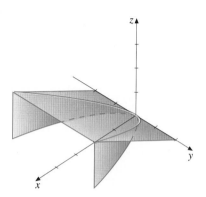

FIGURE 11.9.7 This curve is the intersection of the parabolic cylinder $x = y^2$ and the plane $x = 2z$.

We can use this formula to calculate the normal component of acceleration without evaluating the cross product $\mathbf{R}' \times \mathbf{R}''$. For example, in Example 5 we have

$$|\mathbf{R}''|^2 = |2\mathbf{i} + 0\mathbf{j} + \mathbf{k}|^2 = (2)^2 + (0)^2 + (1)^2 = 5$$

and

$$a_{\mathbf{T}} = \frac{5t}{\sqrt{5t^2 + 1}}.$$

It follows that

$$\begin{aligned}
a_{\mathbf{N}} &= \sqrt{|\mathbf{R}''|^2 - a_{\mathbf{T}}^2} \\
&= \sqrt{5 - \left(\frac{5t}{\sqrt{5t^2 + 1}}\right)^2} \\
&= \sqrt{\frac{5(5t^2 + 1) - 5^2 t^2}{5t^2 + 1}} \\
&= \frac{\sqrt{5}}{\sqrt{5t^2 + 1}},
\end{aligned}$$

as we have seen.

Since

$$a_{\mathbf{N}} = \kappa\left(\frac{ds}{dt}\right)^2 = \frac{|\mathbf{R}' \times \mathbf{R}''|}{|\mathbf{R}'|} \quad \text{and} \quad \frac{ds}{dt} = |\mathbf{R}'|,$$

we have the formula

$$\kappa = \frac{|\mathbf{R}' \times \mathbf{R}''|}{|\mathbf{R}'|^3}. \qquad (1)$$

This formula gives the curvature in terms of $\mathbf{R}'$ and $\mathbf{R}''$.

The formulas that we derived for space curves may also be applied to plane curves, since we may think of a plane curve as a space curve with z-coordinate zero.

If

$$C : x = x(t), \quad y = y(t), \quad z = 0$$

is a curve in the xy-plane, Formula 1 becomes

$$\kappa = \frac{|x'y'' - y'x''|}{(x'^2 + y'^2)^{3/2}}.$$

This is the same formula we used in Section 11.8.

EXERCISES 11.9

Sketch the graphs of the parametric curves given in Exercises 1–8.

1. $C : x = t, y = t, z = t^2$

2. $C : x = t^2, y = t, z = t^2$

3. $C : x = \cos t, y = \sin t, z = t/2$

4. $C : x = \cos t, y = -\cos t, z = \sin t$

5. $C : x = t^2, y = t, z = t^3$

6. $C : x = -t^3, y = t, z = t^3$

7. $C : x = \dfrac{\cos t}{t+1}, y = t, z = \dfrac{\sin t}{t+1}, t \geq 0$

8. $C : x = t^2 \cos t, y = t^2 \sin t, z = t$

Find parametric equations of the curves described in Exercises 9–14 and sketch the curves.

9. Line segment from $(2, 1, -1)$ to $(5, -2, 3)$

10. Line segment from $(0, 2, 1)$ to $(4, 3, 5)$

11. Intersection of $x^2 + y^2 = 1$ and $x + z = 1$

12. Intersection of $x^2 + z^2 = 4$ and $x + y = 0$

13. Intersection of $x^2 + y^2 + z^2 = 4$ and $x + z = 0$

14. Intersection of $z = x^2 + y^2$ and $2x + 2y + z = 2$

Find the length of each curve in Exercises 15–18.

15. $C : x = 1 - t, y = 2t, z = 2 + t, 0 \leq t \leq 1$

16. $C : x = 3t^3, y = 3t^2, z = 2t, 0 \leq t \leq 1$

17. $C : x = t, y = \sqrt{3}t, z = t^2/2, 0 \leq t \leq 1$

18. $C : x = \sin^2 t, y = \cos^2 t, z = \sqrt{8} \sin t, 0 \leq t \leq \pi/2$

Find the parametrization in terms of arc length of the curves in Exercises 19–20.

19. $C : x = t, y = \cos 2t, z = \sin 2t, 0 \leq t \leq \pi/2$

20. $C : x = 2t, y = 1 - t, z = 2 + 2t, 0 \leq t \leq 1$

Find the unit tangent vector, the principal unit normal vector, the unit binormal, and the curvature of the curves in Exercises 21–26.

21. $C : x = e^t \cos t, y = e^t \sin t, z = e^t$

22. $C : x = \cos t, y = \sqrt{2} \sin t, z = \cos t$

23. $C : x = t, y = t^2/2, z = t$; where $t = 0$

24. $C : x = e^t + e^{-t}, y = \sqrt{2}t, z = \sqrt{2}t$; where $t = 0$

25. $C : x = \sqrt{2} \ln t, y = 1/t, z = t$; where $t = 1$

26. $C : x = e^t, y = e^{-t}, z = \sqrt{2}t$; where $t = 0$

Find the tangential and normal components of acceleration for the motion described by the position vectors given in Exercises 27–34.

27. $\mathbf{R} = 3 \cos 2t \mathbf{i} + 3 \sin 2t \mathbf{j}$

28. $\mathbf{R} = 2 \cos t \mathbf{i} + 3 \sin t \mathbf{j}$

29. $\mathbf{R} = t \mathbf{i} + t^2 \mathbf{j}$

30. $\mathbf{R} = t^3 \mathbf{i} + t^2 \mathbf{j}$

31. $\mathbf{R} = \cos t \mathbf{i} + \sin t \mathbf{j} + t \mathbf{k}$

32. $\mathbf{R} = \ln t \mathbf{i} + t \mathbf{j} + \ln t \mathbf{k}$

33. $\mathbf{R} = e^{-t} \cos t \mathbf{i} + e^{-t} \sin t \mathbf{j} + t \mathbf{k}$; where $t = 0$

34. $\mathbf{R} = t \cos t \mathbf{i} + t \sin t \mathbf{j} + t \mathbf{k}$; where $t = 0$

Find the curvature of the curves given in Exercises 35–36.

35. $C : x = t, y = t^2, z = t^3$; where $t = 1$

36. $C : x = t^2, y = t^2, z = t^3$; where $t = 1$

37. The position of an object at time t is given by

$$\mathbf{R} = t^2\mathbf{i} + t^3\mathbf{j} + t\mathbf{k}.$$

Find the equation of the plane determined by the vectors $\mathbf{T}$ and $\mathbf{N}$ at the point on the curve where $t = 1$. (*Hint:* The vectors $\mathbf{R}'$ and $\mathbf{R}''$ are in the plane.)

38. Find parametric equations of the line tangent to the curve of

intersection of the sphere $x^2 + y^2 + z^2 = 9$ and the plane $y = 2x$ at the point $(1, 2, 2)$.

39. A particle moving along a curve has velocity $\mathbf{V} = 2\mathbf{i} + \mathbf{j}$ and acceleration $\mathbf{A} = \mathbf{i} - \mathbf{k}$ at a point on the curve. Find the unit tangent vector $\mathbf{T}$ and the unit normal vector $\mathbf{N}$ at the point.

40. Express the unit tangent vector $\mathbf{T}$ and the unit normal vector $\mathbf{N}$ in terms of the velocity $\mathbf{V}$ and the acceleration $\mathbf{A}$.

REVIEW OF CHAPTER 11 CONCEPTS

PARAMETRIC CURVES

- You should be able to identify and sketch some common parametric curves in the plane.
- You should be able to find parametric equations for line segments, circles, conics, curves of the form either $y = f(x)$ or $x = f(y)$, and polar curves.
- If y and x are each functions of t, the formula $\dfrac{dy}{dx} = \dfrac{\dfrac{dy}{dt}}{\dfrac{dx}{dt}}$

  shows how to find the derivative of y with respect to x. We can apply the formula with $\dfrac{dy}{dt}$ in place of y to

  obtain a formula for $\dfrac{d^2y}{dx^2}$.
- The arc length of the curve $C : x = x(t), y = y(t), a \le t \le b$, is

$$L = \int_a^b \sqrt{\left(\frac{dx}{dt}\right)^2 + \left(\frac{dy}{dt}\right)^2}\, dt.$$

  The arc length function is

$$\phi(t) = \int_a^t \sqrt{(x'(\tau))^2 + (y'(\tau))^2}\, d\tau, \quad a \le t \le b.$$

  Theoretically, if $\left(x'\right)^2 + \left(y'\right)^2 \ne 0$ we can solve $s = \phi(t)$ for t in terms of the variable s to obtain $t = \phi^{-1}(s)$. The curve

$$C_\phi : x = x(\phi^{-1}(s)), y = y(\phi^{-1}(s)), 0 \le s \le L$$

  is the parametrization of C in terms of arc length.
- Parametrization in terms of arc length is convenient for theory, but it is difficult to actually evaluate the arc length function, except for a few simple cases.

TWO- AND THREE-DIMENSIONAL VECTORS

- Vectors in the plane can be expressed in terms of the two-dimensional vectors $\mathbf{i}$ and $\mathbf{j}$. Vectors in three-

dimensional space can be expressed in terms of the three-dimensional vectors $\mathbf{i}$, $\mathbf{j}$, and $\mathbf{k}$.
- You should know the geometric interpretation of two-dimensional and three-dimensional vectors and the operations of addition, subtraction, and scalar multiplication.
- You should know how to evaluate the length of a vector and the dot product of two vectors. Unit vectors have length one. Two vectors are orthogonal if their dot product is zero.
- The smaller angle between two vectors is $\theta = \cos^{-1}\left(\dfrac{\mathbf{U} \cdot \mathbf{V}}{|\mathbf{U}||\mathbf{V}|}\right)$.
- If $\mathbf{V}$ is a nonzero vector, we can write any vector $\mathbf{U}$ as a sum, $\mathbf{U} = \mathbf{U}_1 + \mathbf{U}_2$, where

$$\mathbf{U}_1 = \text{proj}_\mathbf{V}\mathbf{U} = \frac{\mathbf{U} \cdot \mathbf{V}}{\mathbf{V} \cdot \mathbf{V}}\mathbf{V} \text{ is parallel to } \mathbf{V}, \text{ and}$$

$$\mathbf{U}_2 = \mathbf{U} - \text{proj}_\mathbf{V}\mathbf{U} \text{ is orthogonal to } \mathbf{V}.$$

- The cross product of two three-dimensional vectors is orthogonal to each of the vectors and has length equal to the area of the parallelogram formed by the two vectors. The vectors $\mathbf{U}$, $\mathbf{V}$, and $\mathbf{U} \times \mathbf{V}$ (in that order) form a right-hand system.
- The cross product satisfies $\mathbf{U} \times \mathbf{V} = -\mathbf{V} \times \mathbf{U}$, so the order of the factors is important.
- The absolute value of the triple product $(\mathbf{U} \times \mathbf{V}) \cdot \mathbf{W}$ is the volume of the parallelepiped formed by $\mathbf{U}$, $\mathbf{V}$, and $\mathbf{W}$.

LINES AND PLANES

- An equation of the plane that contains the point $P_0(x_0, y_0, z_0)$ with normal vector $\mathbf{N} = a\mathbf{i} + b\mathbf{j} + c\mathbf{k}$ is

$$a(x - x_0) + b(y - y_0) + c(z - z_0) = 0.$$

- The equation $ax + by + cz + d = 0$ is the general equation of a plane with normal vector $\mathbf{N} = a\mathbf{i} + b\mathbf{j} + c\mathbf{k}$.

- The line that contains the point $P(x_0, y_0, z_0)$ with direction vector $\mathbf{V} = a\mathbf{i} + b\mathbf{j} + c\mathbf{k}$ has parametric form

$$x = x_0 + at, \quad y = y_0 + bt, \quad z = z_0 + ct$$

and symmetric equations

$$\frac{x - x_0}{a} = \frac{y - y_0}{b} = \frac{z - z_0}{c}.$$

- A projection vector can be used to determine the distance between a point and a plane, a point and a line, two planes, and two lines.

VECTOR-VALUED FUNCTIONS

- Vector-valued functions are simply functions with values that are vectors. Vector-valued functions can be expressed in terms of their component functions.
- Limits, derivatives, and integrals of vector-valued functions of a real variable can be evaluated component by component.
- If the position at time t of an object is given by the vector-valued function $\mathbf{R}(t)$, the velocity of the object is $\mathbf{V} = \mathbf{R}'$ and the acceleration is $\mathbf{A} = \mathbf{R}''$. The speed is the scalar quantity $|\mathbf{V}|$.
- If we assume that gravity is the only force acting on an object that is thrown near the surface of the earth

with initial position $\mathbf{R}_0 = (x_0, y_0)$ and initial velocity $\mathbf{V}_0 = v_0 \cos\theta \mathbf{i} + v_0 \sin\theta \mathbf{j}$, its position at time t is

$$\mathbf{R}(t) = (x_0 + v_0(\cos\theta)t)\mathbf{i} + \left(y_0 + v_0(\sin\theta)t - \frac{g}{2}t^2\right)\mathbf{j},$$

where $g = 32$ ft/s^2 or 9.8 m/s^2.
- The curve that is expressed in vector form as $\mathbf{R}(t)$ has unit tangent vector $\mathbf{T} = \dfrac{\mathbf{R}'}{|\mathbf{R}'|}$, unit normal vector $\mathbf{N} = \dfrac{\mathbf{T}'}{|\mathbf{T}'|}$, and curvature $\kappa = \dfrac{|\mathbf{T}'|}{|\mathbf{R}'|}$.
- If the position at time t of an object is given by the vector-valued function $\mathbf{R}(t)$, the acceleration can be expressed as $\mathbf{A} = a_{\mathbf{T}}\mathbf{T} + a_{\mathbf{N}}\mathbf{N}$, where the tangential component of acceleration is $a_{\mathbf{T}} = \dfrac{d^2s}{dt^2} = \dfrac{\mathbf{R}' \cdot \mathbf{R}''}{|\mathbf{R}'|}$ and the normal component of acceleration is $a_{\mathbf{N}} = \kappa\left(\dfrac{ds}{dt}\right)^2 = \sqrt{|\mathbf{R}''|^2 - a_{\mathbf{T}}^2}$.
- The fact that the normal component of acceleration is nonnegative indicates that the normal vector $\mathbf{N}$ always points in the direction the curve is bending.

CHAPTER 11 REVIEW EXERCISES

In Exercises 1–4, find an xy-equation of each given curve. Sketch the graph of each curve; show initial point, terminal point, and direction.

1. $C : x = -\cos t, y = \sin t, 0 \le t \le \pi$
2. $C : x = 1/\sqrt{t-1}, y = 1/(t-1), 5/4 \le t \le 5$
3. $C : x = t + 1, y = 3 - 2t, 0 \le t \le 1$
4. $C : x = 2\sin t, y = 3\cos t, 0 \le t \le 2\pi$

Sketch the curves in Exercises 5–6.

5. $C : x = 2\cos t, y = 2\sin t, z = t/2, 0 \le t \le 3\pi$
6. $C : x = t, y = t^2, z = t/4, -2 \le t \le 2$

In Exercises 7–12, find parametric equations of the indicated curves.

7. Along the line segment from $(2, -1, 3)$ to $(-1, 3, 1)$
8. From $(2, 0)$, counterclockwise around $9x^2 + 4y^2 = 36$
9. Along $x = y^2$ from $(4, 2)$ to $(0, 0)$
10. The intersection of the ellipsoid $4x^2 + 36y^2 + 9z^2 = 36$ and the plane $z + y = 1$

11. From the origin, counterclockwise once around the polar curve $r = 1 + \cos\theta$
12. From the origin, counterclockwise once around the polar curve $r = \sin\theta$
13. Find parametric equations of the line tangent to the curve $C : x = te^t, y = e^t$, at the point where $t = 1$.
14. Find parametric equations of the line tangent to the curve $C : x = 2t^2, y = t^3, z = t^2/2$, at the point where $t = 1$.

Evaluate dy/dx and d^2y/dx^2 at the indicated point for the curves given in Exercises 15–16. Sketch an arc of each curve as it passes through the points; show slope, concavity, and direction.

15. $C : x = 1 - t^2, y = t^4; t = 1$
16. $C : x = 4\cos t, y = 2 - 2\sin t; t = 3\pi/4$
17. Find the length of the curve $C : x = t^3, y = t^2, 0 \le t \le 1$.
18. Set up an integral that gives the length of the curve $C : x = 2\cos t, y = 3\sin t, 0 \le t \le 2\pi$.

In Exercises 19–20 set up an integral that gives the area of the surface obtained when the given curve is rotated about the indicated line.

19. $C : x = t + \sin t, y = 1 - \cos t, 0 \le t \le 2\pi$; rotated about the x-axis

20. $C : x = e^t + e^{-t}, \ y = e^t - e^{-t}, \ 0 \le t \le 1$; rotated about the y-axis

Set up integrals that give the length of the polar curves given in Exercises 21–22.

21. $r = 2 - \sin \theta$

22. $r = \theta, 0 \le \theta \le 2\pi$

In Exercises 23–24, find the area of the surface obtained by revolving the given polar curve about the given line.

23. $r^2 = \cos 2\theta$, about the polar axis, $\theta = 0$

24. $r = \sin \theta$, about the line $\theta = \pi/2$

25. Find the length of the arc of a circle of radius a subtended by an angle α that has vertex on the circle.

26. (a) Express the differential arc length of the polar curve $\theta = \theta_0, 0 \le r \le r_0$, in terms of the polar coordinate r. (b) Set up a definite integral in terms of r that gives the area of the surface obtained by revolving the curve about the line $\theta = \pi/2$. (c) Evaluate the integral.

Find the parametrization in terms of arc length of the curves given in Exercises 27–30.

27. $C : x = 1 - 3t, y = 3 + 4t, 0 \le t \le 1$

28. $C : x = -2 \cos 3t, y = 2 - 2 \sin 3t, 0 \le t \le \pi/3$

29. $C : x = t^3/3, y = t^2/2, 0 \le t \le 2$

30. The polar curve $r = e^{2\theta}, 0 \le \theta \le \pi$

Determine which of the curves in Exercises 31–32 are simple and which are closed. Determine which can be expressed as either smooth or piecewise smooth parametric curves, and find parametric representations.

31. From $(0, 0)$, along $y = x^2$ to $(1, 1)$, then along $y = x$ to $(-1, -1)$

32. From $(1, 0)$, along the x-axis to $(-1, 0)$, then along the circle $x^2 + y^2 = 1$ through $(0, -1)$ to $(1, 0)$

33. $C : x = 2 \cos t, y = 3 \sin t$. Find the unit tangent vector, the unit normal vector, the curvature, the radius of curvature, and the center of curvature at the point where $t = 0$. Sketch the curve and the circle of curvature.

34. $\mathbf{R} = \sqrt{5}t\mathbf{i} + [(t^2/2) + 2t]\mathbf{j}$. Find the unit tangent vector, the unit normal vector, and the curvature at the point where $t = 0$.

35. $\mathbf{R} = t\mathbf{i} + (2/3)t^{3/2}\mathbf{j}$. Find $\mathbf{T}$, $\mathbf{N}$, and κ at the point where $t = 3$.

36. $\mathbf{R} = \cos t\mathbf{i} + \sin t\mathbf{j} + t\mathbf{k}$. Find $\mathbf{T}$, $\mathbf{N}$, $\mathbf{B}$, and κ at the point where $t = 0$.

Find formulas for the curvature of the curves given in Exercises 37–40.

37. $C : x = t^2, y = t^3/3$

38. $C : x = t^2, y = t, z = t^3/3$

39. $y = \cos x$

40. $y = x^3 - 1$

Find the components of acceleration and the curvature for the curves given in Exercises 41–42.

41. $\mathbf{R} = t^2\mathbf{i} + t\mathbf{j} + t^2\mathbf{k}$

42. $\mathbf{R} = e^{2t}\mathbf{i} + t^2\mathbf{j}$

43. $\mathbf{U} = 2\mathbf{i} + 6\mathbf{j}, \mathbf{V} = \mathbf{i} - 3\mathbf{j}$. Find
   (a) $\mathbf{U} - 2\mathbf{V}$
   (b) $|\mathbf{U}|$
   (c) $|\mathbf{V}|$
   (d) $\mathbf{U} \cdot \mathbf{V}$
   (e) the angle between $\mathbf{U}$ and $\mathbf{V}$
   (f) a unit vector in the direction of $\mathbf{V}$
   (g) $\text{proj}_{\mathbf{V}}\mathbf{U}$

44. $\mathbf{U} = \mathbf{i} + 2\mathbf{j} + 2\mathbf{k}, \mathbf{V} = 3\mathbf{i} + 4\mathbf{j} - 12\mathbf{k}$. Find
   (a) $2\mathbf{U} - \mathbf{V}$
   (b) $|\mathbf{U}|$
   (c) $|\mathbf{V}|$
   (d) $\mathbf{U} \cdot \mathbf{V}$
   (e) the angle between $\mathbf{U}$ and $\mathbf{V}$
   (f) direction cosines of $\mathbf{V}$
   (g) $\mathbf{U} \times \mathbf{V}$

45. Find the projection of the force $\mathbf{F} = 5\mathbf{i} + 2\mathbf{j}$ in the direction of the vector $\mathbf{V} = 3\mathbf{i} + 4\mathbf{j}$.

46. Write $\mathbf{F} = 2\mathbf{i} + \mathbf{j}$ as a sum of vectors $\mathbf{F}_1$ and $\mathbf{F}_2$, where $\mathbf{F}_1$ is parallel to $\mathbf{V} = \mathbf{i} + 3\mathbf{j}$ and $\mathbf{F}_2$ is orthogonal to $\mathbf{V}$.

47. Find the area of the parallelogram formed by the vectors $2\mathbf{i} + \mathbf{j}$ and $-\mathbf{i} + \mathbf{j}$.

48. Find the volume of the parallelepiped formed by the vectors $2\mathbf{i} + 2\mathbf{j} - \mathbf{k}, -\mathbf{i} + 2\mathbf{j}$, and $\mathbf{i} + \mathbf{j} + \mathbf{k}$.

49. Find the area of the triangle formed by the three points $(6, 4, 3), (3, 1, 1)$, and $(4, 2, 2)$.

50. Find the area of the parallelogram formed by the points $(0, 0), (1, 2), (3, -1)$, and $(4, 1)$.

In Exercises 51–54, find an equation of the plane that satisfies the given conditions.

51. Contains the three points $(3, 0, 3), (5, 4, 1), (1, 1, 1)$

52. Contains $(1, -2, 2)$ and is parallel to the plane $x - 2y + 3z = 4$

53. Contains $(-1, 2, 3)$ and is perpendicular to the line

$$\frac{x}{3} = \frac{y}{4} = \frac{z}{-5}$$

54. Contains the line

$$\frac{x}{3} = \frac{y}{-1} = \frac{z}{2}$$

and is parallel to the line

$$\frac{x-4}{5} = \frac{y+2}{3} = \frac{z-1}{-4}$$

In Exercises 55–58, find symmetric equations and parametric equations of the line that satisfies the given conditions.

55. Contains $(7, 5, 3)$ and $(4, 1, 2)$

56. Contains $(3, 5, -2)$ and is parallel to the line

$$\frac{x}{-2} = \frac{y}{1} = \frac{z}{3}$$

57. Contains $(1, -1, 2)$ and is perpendicular to the plane $2x - y + 3z = 1$

58. Line of intersection of the planes $x + y = 2$ and $x + z = 2$

59. Find the angle between the lines

$$\frac{x}{2} = \frac{y}{-2} = \frac{z}{1} \text{ and } \frac{x}{-1} = \frac{y}{-2} = \frac{z}{2}.$$

60. Find the angle between the planes $3x + 4y = 0$ and $2x - y - 2z = 0$.

61. Find the point of intersection of the line

$$\frac{x}{2} = \frac{y}{-3} = \frac{z}{1}$$

and the plane $2x + 3y + z = 6$.

62. Find the point of intersection of the line

$$\frac{x-2}{3} = \frac{y+1}{4} = \frac{z+3}{-2}$$

and the plane $z = x + 2y$.

63. Find the distance between the point $(3, 1, -2)$ and the plane $z = 2x - 3y$.

64. Find the distance between the point $(1, 3, -2)$ and the line

$$\frac{x}{4} = \frac{y}{-2} = \frac{z}{3}.$$

In Exercises 65–66, the position of an object at time t is given by **R**. *Find the velocity, speed, and acceleration at the time indicated.*

65. $\mathbf{R} = 2\cos t\mathbf{i} + 3\sin t\mathbf{j}$, when $t = \pi/3$

66. $\mathbf{R} = \sin^2 t\mathbf{i} + \cos t\mathbf{j}$, when $t = \pi/4$

67. Find $\mathbf{R}$ if $\mathbf{R}' = 16t^3\mathbf{i} - 15t^2\mathbf{j}$, $\mathbf{R}(0) = 6\mathbf{i}$.

68. Find $\mathbf{R}$ if $\mathbf{R}' = 3t^{1/2}\mathbf{i} + t^{-1/2}\mathbf{j}$, $\mathbf{R}(0) = \mathbf{i} + \mathbf{j}$.

69. A golf ball is hit from ground level at 90 ft/s at an angle of $40°$ from the horizontal. How far will it travel horizontally before returning to ground level?

70. A ball is thrown at an angle of $45°$ to a point that is 100 ft horizontally and 40 ft vertically away. Determine the initial speed.

71. Find parametric equations of the curve traced by a point on the circumference of a wheel of radius r as the wheel rolls along a flat surface that makes an angle of α with the horizontal. See figure. (*Hint:* Find equations for the (x', y') coordinates and then use equations of rotation to find equations for the (x, y) coordinates.)

72. Water flies off a spinning wheel tangentially, as indicated in the figure. Find the maximum height of the water.

73. Find the curvature of the curve $x = y - y^2$ where $y = 0$.

74. Find a formula for the curvature of a curve $x = g(y)$ in terms of g, g' and g''.

75. A particle moving along a curve has velocity $\mathbf{V} = 3\mathbf{i}$ and acceleration $\mathbf{A} = 2\mathbf{i} + \mathbf{j}$ at a point on the curve. Find the unit tangent vector $\mathbf{T}$ and the unit normal vector $\mathbf{N}$ at the point.

76. The **magnetic force** in newtons on a particle with charge q coulombs moving with velocity $\mathbf{v}$ meters per second through a magnetic field of $\mathbf{B}$ tesla is $\mathbf{F} = q\mathbf{v} \times \mathbf{B}$. Find the force on an electron with charge 1.6×10^{-19} coulomb that is moving at 8×10^6 meters per second in the direction of the positive x-axis through a uniform magnetic field of $\mathbf{B} = 2\mathbf{i} + \mathbf{j} - 2\mathbf{k}$ tesla.

77. The magnetic field due to an electric current along the z-axis is $\mathbf{B} = k\left(\dfrac{-y\mathbf{i} + x\mathbf{j}}{x^2 + y^2}\right)$. See figure. Find the force,

$\mathbf{F} = q\mathbf{v} \times \mathbf{B}$, on a charged particle at the point $(2, -1, 3)$ moving with velocity $\mathbf{v} = -\mathbf{i} + 2\mathbf{j} + 2\mathbf{k}$.

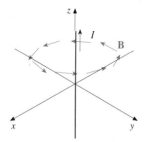

78. The force on a current-carrying straight wire from the point A to the point C in a uniform magnetic field $\mathbf{B}$ is $\mathbf{F} = I\mathbf{AC} \times \mathbf{B}$, where I is the current. Find the force on a wire from $(1, 1, 2)$ to $(2, 3, 1)$ with current I in the uniform field $\mathbf{B} = 2\mathbf{i}$.

79. Let $C_{m,n} : x = \sin mt, \quad y = \sin nt, \quad t \geq 0$.
  (a) Sketch the graph of the curve $C_{2,3}$.
  (b) If m and n are integers, for what values of t does $C_{m,n}$ first start retracing its graph?
  (c) If m and n are integers, how many times does the graph of $C_{m,n}$ touch the line $x = 1$ before it starts retracing the graph? How many times does it touch the line $y = 1$ before it starts retracing the graph?

80. Sketch the graph of the curve
$$C_{\alpha,\beta} : x = \sin \alpha t, \quad y = \sin \beta t, \quad t \geq 0,$$
for various values of α and β, including irrational values.

81. A fly creeps from the center of a long-playing record to the rim, as the record is spinning. Find parametric equations of the path of the fly.

82. Explain in words how to derive the equation of a plane that **(a)** contains three non-collinear points, **(b)** contains a line and a point not on the line, and **(c)** contains two parallel lines.

83. Explain in words how to find all three-dimensional vectors that are perpendicular to a given nonzero vector $a\mathbf{i} + b\mathbf{j} + c\mathbf{k}$.

84. The position of a particle at a time t is given by $\mathbf{R} = 2 \cos t\mathbf{i} + 3 \sin t\mathbf{j}$. Discuss what happens to the speed of the particle as it moves around the ellipse and explain why the speed of the particle changes.

85. Explain what happens to the binomial vector $\mathbf{B}$ of a curve $C : x = x, y = f(x), z = 0, a \leq x \leq b$, at points where the curve, considered as a curve in the xy-plane, is concave downward, concave upward, and at a point of inflection.

EXTENDED APPLICATION

OPTIMAL LAUNCH TIME FOR A SPACE PROBE

Although the human race has made truly amazing progress in space exploration during the twentieth century, we do not yet have the capability to launch a spacecraft into deep space without assistance from the motion of the earth in its orbit. Launch times must therefore be chosen so that the earth lines up with the spacecraft's destination in a favorable way. In Figure 1a, the path of the spacecraft lies in a plane nearly perpendicular to the plane of the earth's orbit, and the earth's velocity vector contributes little to the desired initial velocity vector of the spacecraft. A more favorable arrangement is shown in Figure 1b, where the spacecraft's path is more nearly parallel to the plane of the earth orbit, called the *ecliptic plane*.

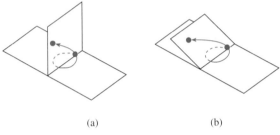

(a)   (b)

FIGURE 1

You are a member of a team of scientists and mathematicians employed by NASA to plan a space probe to Mars. At the present time you are attempting to determine the launch time. In order to analyze the problem, you have constructed a three-dimensional coordinate system in which the ecliptic plane is the xy-plane and the sun is at the origin.

For the purpose of conducting a quick preliminary analysis, you have represented the elliptical orbits of the earth and Mars as circles of radius 1 and 1.520, respectively. The path of the spacecraft is in the plane determined by its initial and terminal points and the sun. The orbit of Mars is in a plane inclined to the xy-plane at an angle of about 6.4°.

Other considerations have led you to consider two possible launch times:

a. Earth at $(1, 0, 0)$ at launch time; Mars at $(-1.519, 0, 0.050)$ at arrival time.

b. Earth at $(0, -1, 0)$ at launch time; Mars at $(1.075, -1.075, 0.080)$ at arrival time.

In each case, the path of the probe will lie in a plane that includes the sun.

Tasks:

1. Let O, A, and B denote, respectively, the positions of the sun, the earth at launch time, and Mars at arrival time. Explain why the vector $\mathbf{N} = \mathbf{OA} \times \mathbf{OB}$ is normal to the plane of the probe's path.

2. Let $\theta = \cos^{-1} \dfrac{\mathbf{N} \cdot \mathbf{k}}{|\mathbf{N}|}$. The launch will be assisted most by the motion of the earth's orbit when $\theta = 0°$ or $180°$. Explain why this is true.

3. Calculate the vector $\mathbf{N}$ and the angle θ for each of the two launch times under consideration.

Writing Assignment:

Write a report to your colleagues, comparing the merits of the two launch times. Make a recommendation, and use your explanations and calculations to justify it.

Source: Marvin Keener and Jeanne Agnew, Oklahoma State University, *Industry Related Problems for Mathematics Students.*

12 Differential Calculus of Functions of Several Variables

Physical quantities that depend on more than one real variable lead to the study of functions of several variables. As in the case of functions of one real variable, the calculus of functions of several variables depends on the concepts of limits, continuity, and the derivative. We will extend these concepts to functions of several variables and then study approximation and extreme value problems.

12.1 FUNCTIONS OF SEVERAL VARIABLES

Connections

Concept of function, 2.1

Three-dimensional sketching, 10.7–8.

The concepts and notation of real-valued functions of several variables are similar to those of real-valued functions of one variable. For example, $f(x, y)$ signifies that f is a function of the variables x and y. Similarly, $g(x, y, z)$ indicates that g is a function of the three variables x, y, and z. As in the case of one variable, the values of functions in several variables will often be given by a formula. For example, $f(x, y) = x^2 + y^2$ or $g(x, y, z) = y + 1$. Unless stated otherwise, the domain of a function that is given by a formula includes all points where the formula gives a real number. This is the same convention that was used for functions of one variable.

Domain and Values

EXAMPLE 1

Find the domain of the function

$$f(x, y) = \frac{x^2 + y^2}{x - y}.$$

Evaluate $f(2, 0)$, $f(0, -3)$, and $f(2a, a)$ for $a \neq 0$.

SOLUTION   The function is defined for all (x, y) except for (x, y) on the line $y = x$ where the denominator is zero. See Figure 12.1.1. Substitution into the formula gives

$$f(2, 0) = \frac{(2)^2 + (0)^2}{(2) - (0)} = 2,$$

$$f(0, -3) = \frac{(0)^2 + (-3)^2}{(0) - (-3)} = 3.$$

If $a \neq 0$, then $(2a, a)$ is not on the line $y = x$, so $f(2a, a)$ is defined. We then have

$$f(2a, a) = \frac{(2a)^2 + (a)^2}{(2a) - (a)} = 5a, \qquad a \neq 0. \quad \blacksquare$$

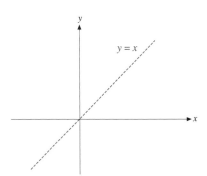

FIGURE 12.1.1 The function $f(x, y) = \dfrac{x^2 + y^2}{x - y}$ is undefined on the line $y = x$.

EXAMPLE 2

Find the domain of the function $f(x, y) = \sqrt{4 - x^2 - 4y^2}$. Evaluate $f(1, 1/2)$ and $f(1, 0)$.

SOLUTION   The function is defined for all (x, y) that satisfy $4 - x^2 - 4y^2 \geq 0$, or $x^2 + 4y^2 \leq 4$. The domain consists of all points on or inside the ellipse

$$\frac{x^2}{2^2} + \frac{y^2}{1^2} = 1.$$

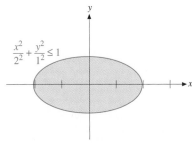

FIGURE 12.1.2 The function $f(x, y) = \sqrt{4 - x^2 - 4y^2}$ is defined inside and on the ellipse $\dfrac{x^2}{2^2} + \dfrac{y^2}{1^2} = 1$.

See Figure 12.1.2. Substitution into the formula gives

$$f\left(1, \frac{1}{2}\right) = \sqrt{4 - (1)^2 - 4\left(\frac{1}{2}\right)^2} = \sqrt{2},$$

$$f(1, 0) = \sqrt{4 - (1)^2 - 4(0)^2} = \sqrt{3}. \quad \blacksquare$$

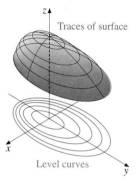

FIGURE 12.1.3 The level curve $f(x, y) = z_0$ is the projection of the trace $z = z_0, z = f(x, y)$ onto the xy-plane.

Graph and Level Curves

The **graph of a function** f of two variables x and y is the set of all points (x, y, z) in three-dimensional space such that $z = f(x, y)$. Generally, we would expect the graph of a function of two variables to be a surface in three-dimensional space. Graphing surfaces in three-dimensional space was discussed in Chapter 10. It may be useful to review sections 10.7 and 10.8 at this time.

For a fixed value of z_0, the set of all points (x, y) in the plane that satisfy the equation $f(x, y) = z_0$ is called a **level curve** of f. Sketching and labeling a selection of level curves is one way to indicate the values and general character of a function of two variables.

The graph of $z = f(x, y)$ is a surface in three-dimensional space. The level curve $f(x, y) = z_0$ is a curve in the xy-plane. The level curve $f(x, y) = z_0$ is related to the curve in three-dimensional space with equations $z = z_0, z = f(x, y)$, which is the trace of the surface $z = f(x, y)$ in the plane $z = z_0$. The level curve $f(x, y) = z_0$ is the projection of the trace $z = z_0, z = f(x, y)$ onto the xy-plane. The graph of $z = f(x, y)$ can be constructed from level curves by using the level curves $f(x, y) = z_0$ for several values of z_0 to sketch the traces $z = z_0, z = f(x, y)$. That is, each level curve $f(x, y) = z_0$ is simply "lifted" from the xy-plane to the plane $z = z_0$. See Figure 12.1.3.

Level curves are used on topographic maps to indicate the function whose value at a point is the elevation above sea level of the point. These level curves are called **contour curves**. All points on a particular contour curve are at the same elevation. The elevation changes more rapidly—the hill is steeper—where the contour curves are closer together. See Figure 12.1.4.

FIGURE 12.1.4a This is the sandtrap and first green of the Purdue South Golf Course.

FIGURE 12.1.4b The closeness of the contour lines indicates a steep grade near the edge of the green.

EXAMPLE 3

Characterize and sketch several level curves of the function $f(x, y) = 1 - \ln(x^2 + y^2)$. Sketch the graph of the function.

SOLUTION   The level curve that corresponds to a number z_0 is the set of all (x, y) that satisfy the equation

$$f(x, y) = z_0, \text{ or}$$
$$1 - \ln(x^2 + y^2) = z_0.$$

Rewriting the equation, we obtain

$$\ln(x^2 + y^2) = 1 - z_0,$$
$$x^2 + y^2 = e^{(1-z_0)}.$$

This is the equation of a circle with center at the origin and radius $e^{(1-z_0)/2}$. The level curves that correspond to $z_0 = -1, 0, 1$, and 2 are sketched in Figure 12.1.5a. (These level curves become closer as they move toward the origin. Since the z-increments between the level curves are equal, this indicates that the values of the function are changing more rapidly at (x, y) as (x, y) moves toward the origin. This should be compared with the fact that a hill is steeper where the contour curves are closer together.) The graph of f is sketched in Figure 12.1.5b, where the traces that correspond to $z_0 = -1, 0, 1$, and 2 have been indicated. Note that the values $f(x, y)$ increase more rapidly as (x, y) moves toward the origin. ■

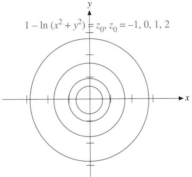

FIGURE 12.1.5a   The level curves of the function $f(x, y) = 1 - \ln(x^2 + y^2)$ are circles with centers at the origin.

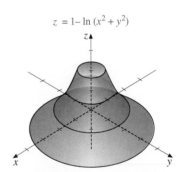

FIGURE 12.1.5b   The values of the function change more rapidly near the origin, where the level curves are closer together.

EXAMPLE 4

Characterize and sketch several level curves of the function

$$f(x, y) = \frac{y^2 - x^2}{2}.$$

Sketch the graph of the function.

SOLUTION   The level curves that correspond to $f(x, y) = z_0$ are

$$y^2 - x^2 = 2z_0.$$

For $z_0 \neq 0$, this is the equation of a hyperbola with center at the origin and asymptotes $y = \pm x$; the vertices are $(\pm\sqrt{-2z_0}, 0)$ if $z_0 < 0$ and $(0, \pm\sqrt{2z_0})$ if $z_0 > 0$. The level curve that corresponds to $z_0 = 0$ consists of the two lines $y = x$ and $y = -x$. Level curves that correspond to $z_0 = -1, 0, 1$, and 2 are given in Figure 12.1.6a. The graph of f is given in Figure 12.1.6b, where the traces that correspond to $z_0 = -1, 0, 1$, and 2 have been indicated. ■

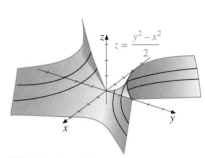

FIGURE 12.1.6a The level curves of the function $f(x, y) = \dfrac{y^2 - x^2}{2}$ are hyperbolas with centers at the origin.

FIGURE 12.1.6b The graph is a hyperbolic paraboloid.

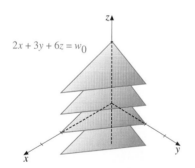

FIGURE 12.1.7 Level surfaces of $f(x, y, z) = 2x + 3y + 6z$ are parallel planes.

Level Surfaces

The graph of a function $f(x, y, z)$ of three variables is the set of all points (x, y, z, w) such that $w = f(x, y, z)$. It doesn't seem likely that we would be able to effectively sketch a graph of points in four-dimensional space. It is possible, however, to indicate the **level surfaces** $f(x, y, z) = w_0$ of a function of three variables as surfaces in three-dimensional space.

EXAMPLE 5

Characterize and sketch several level surfaces of the function $f(x, y, z) = 2x + 3y + 6z$.

SOLUTION   The level surface that corresponds to $f(x, y, z) = w_0$ is

$$2x + 3y + 6z = w_0.$$

This is the equation of a plane. Several of the level surfaces are sketched in Figure 12.1.7. ■

EXAMPLE 6

Characterize and sketch several level surfaces of the function $f(x, y, z) = y - 2x^2 - 2z^2$.

SOLUTION   The level surface that corresponds to $f(x, y, z) = w_0$ is

$$y - 2x^2 - 2z^2 = w_0.$$

This is the equation of a paraboloid with vertex at the point $(0, w_0, 0)$ and axis along the y-axis. Several of the level surfaces are sketched in Figure 12.1.8. ■

FIGURE 12.1.8 Level surfaces of $f(x, y, z) = y - 2x^2 - 2z^2$ are paraboloids with axes along the y-axis.

Topological Properties of Regions

Let us conclude this section with a brief discussion of some topological concepts that are useful for characterizing the domain of a function of several variables.

DEFINITION

A point is called an **interior point** of a set S if there is some $\delta > 0$ such that all points that are within δ of the point are in S. A set S is called **open** if all points in S are interior points of S.

EXAMPLE 7

The set S of points (x, y) that satisfy $x^2 + y^2 < 1$ is an open set. For any point in S we may choose the positive number δ in the definition of interior point to be less than or equal to the distance of the point to the circle $x^2 + y^2 = 1$. See Figure 12.1.9. ■

DEFINITION

A point is called a **boundary point** of a set S if, for each $\delta > 0$, there is a point in S within δ of the point and a point not in S that is within δ of the point. A set S is called **closed** if it contains all its boundary points.

EXAMPLE 8

The set S of points (x, y) that satisfy $x^2 + y^2 \leq 1$ is a closed set. The boundary of S consists of those points on the circle $x^2 + y^2 = 1$, and S contains this circle. The interior of S is the set of points that are inside the circle. See Figure 12.1.10. ■

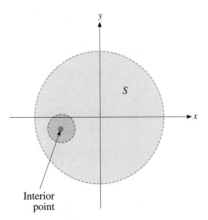

FIGURE 12.1.9 The set S of points (x, y) that satisfy $x^2 + y^2 < 1$ is an open set, because each point of S is the center of a disk that is contained in S.

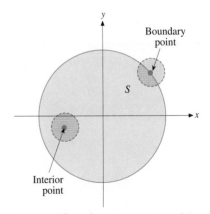

FIGURE 12.1.10 The boundary of the set S of points (x, y) that satisfy $x^2 + y^2 \leq 1$ consists of the circle $x^2 + y^2 = 1$. Each disk with center on the circle contains both points in S and points not in S.

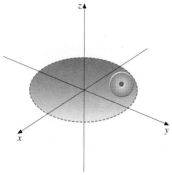

FIGURE 12.1.11 The set of points (x, y, z) that satisfy $x^2 + y^2 < 1$ and $z = 0$ is neither open nor closed. The boundary consists of all points that satisfy $x^2 + y^2 \leq 1$ and $z = 0$, since spheres with centers at these points contain both points in S and points not in S.

Every point of a set S must be either an interior point of S or a boundary point of S; no point of S can be both an interior point and a boundary point of S.

A set S is not open if S contains some boundary point of S. S is not closed if some boundary point of S is not in S.

EXAMPLE 9

The set S of points (x, y, z) that satisfy $x^2 + y^2 < 1$ and $z = 0$ is neither open nor closed. That is, for any point $(x, y, 0)$ in S and any $\delta > 0$, there are points that are within δ of $(x, y, 0)$ that have nonzero z-coordinate and, hence, are not in S. It follows that none of the points of S are interior points, so S is not open. The boundary of S consists of all points in S and all points on the circle $z = 0$, $x^2 + y^2 = 1$. Since those boundary points of S on the circle are not contained in S, S is not closed. See Figure 12.1.11. ◼

EXERCISES 12.1

Find the domain and indicated values of the functions in Exercises 1–6.

1. $f(x, y) = \dfrac{\sin(x + y)}{x}$; $f(2, -2)$, $f(\pi, \pi/2)$

2. $f(x, y) = \dfrac{2x - y}{x + 2y}$; $f(2, 3)$, $f(3, 2)$

3. $f(x, y) = \dfrac{x + y}{x^2 + y^2}$; $f(0, 1)$, $f(2, 0)$, $f(a, a)$, $a \neq 0$

4. $f(x, y) = \dfrac{x + y}{x^2 - y^2}$; $f(0, 1)$, $f(0, 2)$, $f(1, a)$, $a \neq \pm 1$

5. $f(x, y) = \sqrt{9 - x^2 - y^2}$; $f(2, 1)$, $f(-2, 1)$

6. $f(x, y, z) = \dfrac{1}{\sqrt{36 - x^2 - 4y^2 - 9z^2}}$; $f(0, 2, 1)$, $f(4, -1, 1)$

Characterize and sketch several level curves of the functions given in Exercises 7–16. Sketch the graphs of the functions.

7. $f(x, y) = 3 - x - y$

8. $f(x, y) = 2x + 2y$

9. $f(x, y) = e^{-x^2 - y^2}$

10. $f(x, y) = \sqrt{9 - x^2 - y^2}$

11. $f(x, y) = x^2 + y^2$

12. $f(x, y) = 2\sqrt{x^2 + y^2}$

13. $f(x, y) = \dfrac{\sqrt{x^2 - y^2}}{2}$

14. $f(x, y) = \dfrac{\sqrt{y^2 - x^2}}{2}$

15. $f(x, y) = \dfrac{x^2 + y^2}{2x}$

16. $f(x, y) = \dfrac{1 - x^2}{y^2}$

Characterize and sketch several level surfaces of the functions given in Exercises 17–24.

17. $f(x, y, z) = 6x + 3y + 2z$

18. $f(x, y, z) = 2x + 3y$

19. $f(x, y, z) = x^2 + y^2 + z$

20. $f(x, y, z) = x^2 + y^2 + z^2$

21. $f(x, y, z) = z - 2\sqrt{x^2 + y^2}$

22. $f(x, y, z) = z - x^2 - y^2$

23. $f(x, y, z) = \dfrac{x^2 + y^2}{z^2}$

24. $f(x, y, z) = \sqrt{\dfrac{x^2 + y^2}{z}}$

Find the equation of the level curve that contains the given point in Exercises 25–26.

25. $f(x, y) = x^2 + 2x + y^2$; $(-1, 2)$

26. $f(x, y) = x^2 - 4x - y^2 + 2y$; $(2, -3)$

27. A rectangular box has one corner at the origin and the opposite corner at the point (x, y, z) on the plane $3x + 2y + 2z = 6$.
  (a) Express the volume of the box as a function of (x, y). Indicate the domain of the boundary function.
  (b) What is the volume of the box at a boundary point of the domain?

28. A rectangular box has one corner at the origin and the opposite corner at the point (x, y, z) on the plane $3x + 2y + 2z = 6$.

   (a) Express the total area of the six sides of the box as a function of (x, y). Indicate the domain of the function.

   (b) What is the surface area of the box at boundary points of the domain? (*Hint:* The surface is a function of one variable on each segment of the boundary.)

Sketch the sets indicated in Exercises 29–38. Describe the boundary of each set and determine if the set is open, closed, or neither.

29. $(x, y) : x > 2y + 1$

30. $(x, y) : x \leq 3y$

31. $(x, y) : x \leq 2y - 1, y \leq 2$

32. $(x, y) : x \geq y^2, x \leq 4$

33. $(x, y) : x + y > 3, y \leq 5$

34. $(x, y) : x + 2y \leq 4, x > 0$

35. $(x, y) : 1 < x^2 + y^2 \leq 4$

36. $(x, y) : x^2 + 4y^2 \leq 9$

37. $(x, y) : x^2 + y^2 = \dfrac{1}{n^2}, n = 1, 2, 3, \ldots$

38. $(x, y) : x$ and y integers

Describe the boundary of each set in Exercises 39–44 and determine if the set is open, closed, or neither.

39. $(x, y, z) : x + y + z > 5$

40. $(x, y, z) : x^2 + y^2 \geq z$

41. $(x, y, z) : x^2 + y^2 \leq 1, z > 0$

42. $(x, y, z) : y^2 + z^2 < 1, x = 1$

43. $(x, y, z) : x^2 + y^2 + z^2 = 1$

44. $(x, y, z) : 0 < z < 1$

12.2 LIMITS AND CONTINUITY

Connections

Bounds of functions, 2.2.

Concept of limit, 2.3.

The ideas of limits and continuity are important for the development of calculus. In this section we will briefly discuss these ideas for functions of several variables.

Definitions and Basic Results

We begin with the formal definition of the limit of a function of two variables.

> **DEFINITION**
>
> We say $f(x,y)$ **has limit** L **as** (x,y) **approaches** (a,b) (written $\lim\limits_{(x,y)\to(a,b)} f(x, y) = L$) if, for each $\varepsilon > 0$, there is some $\delta > 0$ such that $|f(x, y) - L| < \varepsilon$ whenever $0 < \sqrt{(x - a)^2 + (y - b)^2} < \delta$.

The definition of $\lim\limits_{(x,y)\to(a,b)} f(x, y) = L$ requires that for each $\varepsilon > 0$, there is a positive number δ such that the values $f(x, y)$ are within ε of L for all (x, y) that are within δ of (a, b), excluding the point (a, b). This means that the values $f(x, y)$ must approach a value or values that are within ε of L as (x, y) approaches (a, b) along any curve that does not include (a, b). The fact that ε can be any positive number excludes the possibility that $f(x, y)$ approaches any number other than L. It follows that $f(x, y)$ must approach the single number L as (x, y) approaches (a, b) along any path that does not include (a, b). See Figure 12.2.1a.

Geometrically, $\lim\limits_{(x,y)\to(a,b)} f(x, y) = L$ means that points $(x, y, f(x, y))$ on the graph of f approach the point (a, b, L) as (x, y) approaches (a, b). See Figure 12.2.1b.

The formal definition of the limit of a function of two variables corresponds to the definition of the limit of a function of one variable, except that here the distance of (x, y) from (a, b) is given by $\sqrt{(x - a)^2 + (y - b)^2}$, while the

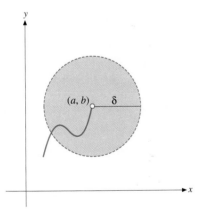

FIGURE 12.2.1a If $\lim\limits_{(x,y)\to(a,b)} f(x, y) = L$, then the values $f(x, y)$ must approach L as (x, y) approaches (a, b) along any curve.

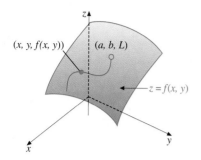

FIGURE 12.2.1b The points $(x, y, f(x, y))$ approach the point (a, b, L) as (x, y) approaches (a, b).

distance between numbers x and a on the line is given by $|x - a|$. Since $|x - a| = \sqrt{(x - a)^2}$, the definitions are completely analogous. The definition of limit for functions of three variables is similar. We merely use the formula for the distance between two points in three-dimensional space. For functions of more than three variables, we use the obvious extensions of the formulas for distance. All of the theory we developed for the evaluation of functions of one variable also holds for functions of several variables. In particular, we could prove the following.

THE LIMIT THEOREM

Results for the evaluation of the limit of sums, products, and quotients of functions of a single variable also hold for functions of several variables.

We also have the following result for the composition of functions. This result is proved in the same way as the corresponding one-variable result was proved earlier in Section 2.4.

If $\lim\limits_{(x,y) \to (a,b)} g(x, y) = M$ *and* $f(t)$ *is continuous at* M, *then*

$$\lim_{(x,y) \to (a,b)} f(g(x, y)) = f(\lim_{(x,y) \to (a,b)} g(x, y)) = f(M).$$

We say $f(x, y)$ is **continuous** at a point (a, b) if

$$\lim_{(x,y) \to (a,b)} f(x, y) = f(a, b).$$

Continuity of functions of more than two variables is defined similarly.

The theory of limits can be used to determine the continuity of many functions. We first use the definition of limit to prove that

$$\lim_{(x,y) \to (a,b)} x = a \text{ and } \lim_{(x,y) \to (a,b)} y = b.$$

(See Exercises 37 and 38.) This shows that the functions $f(x, y) = x$ and $g(x, y) = y$ are continuous at every point. The Limit Theorem then tells us that sums, products, and quotients of x and y are continuous, except at points where the denominator of a quotient is zero. The above-stated result for the limit of composite functions tells us that the composition $f \circ g$ of a continuous function f of a real variable and a continuous function g of (x, y) is a continuous function of (x, y).

We can evaluate the limit of a function that is known to be continuous by substitution.

EXAMPLE 1

Evaluate $\lim\limits_{(x,y) \to (0,0)} \cos xy$.

SOLUTION We know the product xy is a continuous function that approaches zero as (x, y) approaches $(0, 0)$. Also, the cosine function is a continuous function of one variable, so

$$\lim_{(x,y) \to (0,0)} \cos xy = \cos 0 = 1.$$

The graph is sketched in Figure 12.2.2. ∎

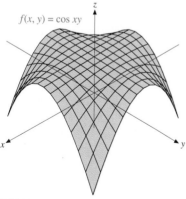

$f(x, y) = \cos xy$

FIGURE 12.2.2 The function $f(x, y) = \cos xy$ is continuous at every point.

Some Limits That Do Not Exist

The definition of limit of $f(x, y)$ at (a, b) requires that $f(x, y)$ must approach a *single* number as (x, y) approaches (a, b) along any curve. In particular:

If $f(x, y)$ approaches different numbers as (x, y) approaches (a, b) along different curves, then $\lim\limits_{(x,y)\to(a,b)} f(x, y)$ *does not exist.*

We can show that a limit does not exist at a point by showing that $f(x, y)$ approaches different values as (x, y) approaches the point along two different curves.

EXAMPLE 2

Show that $\lim\limits_{(x,y)\to(0,0)} \dfrac{x^2 - y^2}{x^2 + y^2}$ does not exist.

SOLUTION The expression

$$\frac{x^2 - y^2}{x^2 + y^2}$$

has value 1 at all points $(x, 0)$ on the x-axis except the origin, so the values approach 1 as $(x, y) = (x, 0)$ approaches the origin along the x-axis. Similarly, the values approach -1 as $(x, y) = (0, y)$ approaches the origin along the y-axis. Since the values approach different numbers as (x, y) approaches the origin along these curves, the limit at the origin does not exist. See Figure 12.2.3. ∎

If there is a path to (a, b) for which the values of $f(x, y)$ do not approach a limit, then $\lim\limits_{(x,y)\to(a,b)} f(x, y)$ *does not exist.*

EXAMPLE 3

Show that

$$\lim_{(x,y)\to(0,0)} \frac{y}{x^2 + y^2}$$

does not exist.

SOLUTION When $(x, y) = (0, y)$, the expression

$$\frac{y}{x^2 + y^2}$$

reduces to $1/y$. It follows that the values approach infinity as (x, y) approaches the origin along the positive y-axis, so the limit does not exist. See Figure 12.2.4, where we have sketched the graph. ∎

We cannot conclude that a limit exists just because the values approach the same limit along a few test paths. It is possible that the limit may not exist even if $f(x, y)$ approaches the same value along every straight line path to the point.

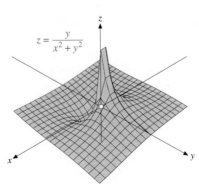

FIGURE 12.2.3 $\dfrac{x^2 - y^2}{x^2 + y^2}$ approaches 1 as (x, y) approaches $(0, 0)$ along the x-axis. The values approach -1 as (x, y) approaches $(0, 0)$ along the y-axis.

$$\lim_{(x,y)\to(0,0)} \frac{x^2 - y^2}{x^2 + y^2} \text{ does not exist.}$$

FIGURE 12.2.4 $\dfrac{y}{x^2 + y^2}$ approaches infinity as (x, y) approaches $(0, 0)$ along the positive y-axis, so

$$\lim_{(x,y)\to(0,0)} \frac{y}{x^2 + y^2} \text{ does not exist.}$$

EXAMPLE 4

Suppose

$$f(x, y) = \begin{cases} \dfrac{xy^2}{x^2 + y^4}, & (x, y) \neq (0, 0), \\ 0, & (x, y) = (0, 0). \end{cases}$$

(a) Show that $f(x, y) \to 0$ as $(x, y) \to (0, 0)$ along any line through the origin. **(b)** Show that $\lim\limits_{(x,y)\to(0,0)} f(x, y)$ does not exist.

SOLUTION

(a) For each fixed α we have

$$f(x, \alpha x) = \frac{x(\alpha x)^2}{x^2 + (\alpha x)^4} = \frac{\alpha^2 x}{1 + \alpha^4 x^2} \to 0 \text{ as } (x, \alpha x) \to (0, 0),$$

so $f(x, y) \to 0$ as $(x, y) \to (0, 0)$ along any nonvertical line $y = \alpha x$. Also, $f(x, y) \to 0$ as $(x, y) \to (0, 0)$ along the y-axis, so $f(x, y) \to 0$ as $(x, y) \to (0, 0)$ along any line through the origin.

(b) For (x, y) on the curve $y = \sqrt{x}, x > 0$, we have

$$f(x, \sqrt{x}) = \frac{x(x)}{x^2 + x^2} = \frac{1}{2},$$

so $f(x, y) \to 1/2$ as $(x, y) \to (0, 0)$ along the curve $y = \sqrt{x}$. Since $f(x, y)$ approaches more than one value as $(x, y) \to (0, 0)$ along different curves, $\lim\limits_{(x,y)\to(0,0)} f(x, y)$ does not exist. The graph of f is given in Figure 12.2.5. ∎

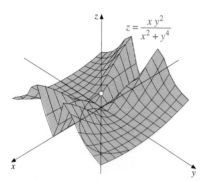

FIGURE 12.2.5 $f(x, y)$ approaches zero as (x, y) approaches $(0, 0)$ along any line, but $\lim\limits_{(x,y)\to(0,0)} f(x, y)$ does not exist.

Using Inequalities to Verify Limits

We need to be able to establish the existence of limits of functions that are not known to be continuous. As in Chapter 2, we can verify that $\lim\limits_{(x,y)\to(a,b)} f(x, y) = L$ by showing that

$$|f(x, y) - L| \leq z(x, y), \quad (x, y) \neq (a, b),$$

where $z(x, y)$ is a function that is known to approach zero as (x, y) approaches (a, b). We will illustrate the technique by considering functions with a factor $x^2 + y^2$ in the denominator. Such functions occur often in applications of vector calculus, as we shall see in Chapter 14.

We will need the following inequalities, which are geometrically evident from Figure 12.2.6.

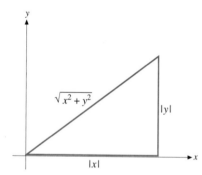

FIGURE 12.2.6 The length of the hypotenuse of a right triangle is greater than the length of either side.

$$|x| \leq \sqrt{x^2 + y^2},$$
$$|y| \leq \sqrt{x^2 + y^2}.$$

EXAMPLE 5

Show that $\lim\limits_{(x,y)\to(0,0)} \dfrac{y^4}{x^2 + y^2} = 0.$

SOLUTION   The expression

$$\frac{y^4}{x^2 + y^2}$$

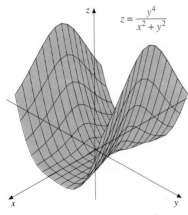

$$z = \frac{y^4}{x^2 + y^2}$$

FIGURE 12.2.7 $\displaystyle\lim_{(x,y)\to(0,0)} \frac{y^4}{x^2 + y^2}$
$= 0.$

is defined for all (x, y), except for $(x, y) = (0, 0)$. Since $y^4 \le (\sqrt{x^2 + y^2})^4 = (x^2 + y^2)^2$, we have

$$\left| \frac{y^4}{x^2 + y^2} \right| \le \frac{(x^2 + y^2)^2}{x^2 + y^2} = x^2 + y^2 \to 0 \text{ as } (x, y) \to (0, 0).$$

The graph is sketched in Figure 12.2.7. ∎

We have seen that the investigation of limits of functions of several variables can be complicated. Let us organize what we have observed into a procedure.

Procedure for investigating $\displaystyle\lim_{(x,y)\to(a,b)} f(x, y)$

- Determine whether or not $f(x, y)$ is a function that is known to be continuous at (a, b). If $f(x, y)$ is known to be continuous at (a, b), the limit exists and can be evaluated by substitution. If $f(x, y)$ is not noticeably continuous at (a, b), we continue.
- Evaluate the limit of $f(x, y)$ along several paths to the point (a, b). If either the limit along one of these paths does not exist or two paths give different limits, the limit does not exist and we are done.
- If all test paths give the same number L, then try to show

$$|f(x, y) - L| \le z(x, y), \ (x, y) \ne (a, b),$$

where $z(x, y) \to 0$ as $(x, y) \to (a, b)$. This shows the limit exists and has value L.

EXAMPLE 6

Either evaluate and verify the existence of

$$\lim_{(x,y)\to(2,1)} \frac{x^2}{x^2 + y^2},$$

or show the limit does not exist.

SOLUTION   The numerator and the denominator are each continuous functions of (x, y), and the value of the denominator at $(2, 1)$ is nonzero. It follows that we can evaluate the limit by substitution, so

$$\lim_{(x,y)\to(2,1)} \frac{x^2}{x^2 + y^2} = \frac{(2)^2}{(2)^2 + (1)^2} = 0.8.$$

The graph is sketched in Figure 12.2.8. ∎

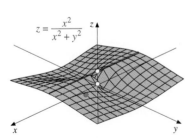

$$z = \frac{x^2}{x^2 + y^2}$$

FIGURE 12.2.8 $\dfrac{x^2}{x^2 + y^2}$ is continuous

at every point except the origin.

$$\lim_{(x,y)\to(2,1)} \frac{x^2}{x^2 + y^2} = \frac{(2)^2}{(2)^2 + (1)^2} = 0.8.$$

EXAMPLE 7

Either evaluate and verify the existence of

$$\lim_{(x,y)\to(0,0)} \frac{xy}{x^2 + y^2}$$

or show the limit does not exist.

SOLUTION   We cannot evaluate this limit by substitution, because the denominator is zero at the origin. Let us evaluate limits along some simple paths to the origin.

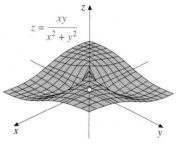

FIGURE 12.2.9 $\dfrac{xy}{x^2+y^2}$ approaches zero as (x, y) approaches the origin along the coordinate axes, but the values approach 1/2 as (x, y) approaches the origin along the line $y = x$. $\displaystyle\lim_{(x,y)\to(0,0)}\dfrac{xy}{x^2+y^2}$ does not exist.

The values of $\dfrac{xy}{x^2+y^2}$ are zero on both coordinate axes, except at the origin, so the values approach zero as (x, y) approaches zero along either axis. However, when $y = x$, the values of $\dfrac{xy}{x^2+y^2}$ are 1/2, so the values approach 1/2 as (x, y) approaches the origin along the line $y = x$. This shows that the values do not approach the same value along all paths to the origin, so the limit does not exist. See Figure 12.2.9. ∎

EXAMPLE 8

Either evaluate and verify the existence of

$$\lim_{(x,y)\to(0,0)}\frac{3x^2 - x^3 + 3y^2 - y^3}{x^2 + y^2}$$

or show the limit does not exist.

SOLUTION We cannot evaluate this limit by substitution, so we try to evaluate limits along some simple paths to the origin.

When $x = 0$ and $y \neq 0$, the expression $\dfrac{3x^2 - x^3 + 3y^2 - y^3}{x^2 + y^2}$ reduces to $3 - y$, so the values approach 3 as (x, y) approaches the origin along the y-axis. The values also approach 3 along the x-axis. In fact, if $y = \alpha x$ and $x \neq 0$, we have

$$\frac{3x^2 - x^3 + 3y^2 - y^3}{x^2 + y^2} = \frac{3x^2 - x^3 + 3(\alpha x)^2 - (\alpha x)^3}{x^2 + (\alpha x)^2}$$

$$= \frac{x^2(3 - x + 3\alpha^2 - \alpha^3 x)}{x^2(1 + \alpha^2)}$$

$$= \frac{3(1 + \alpha^2) - (1 + \alpha^3)x}{1 + \alpha^2}$$

$$= 3 - \frac{1 + \alpha^3}{1 + \alpha^2}x \to 3 \text{ as } (x, y) \to (0, 0).$$

It follows that the values approach 3 along every straight line path to the origin. We know that this does not, in itself, imply the limit is 3, but it is enough to indicate that we should try to verify that the limit is 3.

For $(x, y) \neq (0, 0)$, we have

$$\left|\frac{3x^2 - x^3 + 3y^2 - y^3}{x^2 + y^2} - 3\right| = \left|\frac{-x^3 - y^3}{x^2 + y^2}\right|$$

$$\leq \frac{|x|^3}{x^2 + y^2} + \frac{|y|^3}{x^2 + y^2}$$

$$= \frac{(\sqrt{x^2 + y^2})^3}{x^2 + y^2} + \frac{(\sqrt{x^2 + y^2})^3}{x^2 + y^2}$$

$$= 2\sqrt{x^2 + y^2} \to 0 \text{ as } (x, y) \to (0, 0).$$

This shows that the limit exists and has value 3. See Figure 12.2.10. ∎

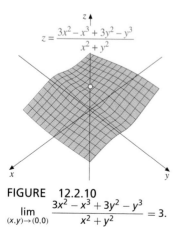

FIGURE 12.2.10 $\displaystyle\lim_{(x,y)\to(0,0)}\dfrac{3x^2 - x^3 + 3y^2 - y^3}{x^2 + y^2} = 3.$

Restricting the Variables

Recall that one-sided limits at c of a function of one variable were obtained by restricting the variable to one side of c in the definition of limit. One-sided limits were then used to investigate limits of functions that were defined by more than one formula and to define continuity on an interval. We can obtain analogous concepts for functions of several variables by restricting the variables in the definition of limit to a particular set S. We use the notation

$$\lim_{\substack{(x,y)\to(a,b)\\(x,y)\ \text{in}\ S}} f(x, y) = L$$

to indicate that only values $f(x, y)$ for (x, y) in S are considered in the definition of limit. We assume that, for each $\delta > 0$, there are points in S other than (a, b) that are within δ of (a, b).

If (a, b) is an *interior* point of S, then

$$\lim_{(x,y)\to(a,b)} f(x, y) \quad \text{and} \quad \lim_{\substack{(x,y)\to(a,b)\\(x,y)\ \text{in}\ S}} f(x, y)$$

are identical, since all paths to (a, b) eventually involve only points of S. Restricting the points (x, y) to S allows us to consider limits at boundary points of S. See Figure 12.2.11.

FIGURE 12.2.11 Restricting the points (x, y) to S allows us to consider limits at boundary points of S.

If a function $f(x, y)$ is defined by different formulas for different values of (x, y), we can use each formula to evaluate the limit of $f(x, y)$ at a point in the *interior* of the set where the formula holds. Each formula also can be used to determine limits of $f(x, y)$ with (x, y) restricted to the set where the formula holds. If a point can be approached from sets where different formulas hold, then $f(x, y)$ will have a limit at the point if and only if the limits restricted to each of the sets have the same value.

EXAMPLE 9

Find all points where the function

$$f(x, y) = \begin{cases} x, & x \ge 0, \\ y, & x < 0 \end{cases}$$

is continuous.

SOLUTION The interior of the set $x \ge 0$ is the set $x > 0$. If $a > 0$, we have

$$\lim_{(x,y)\to(a,b)} f(x, y) = \lim_{\substack{(x,y)\to(a,b)\\x\ge 0}} f(x, y) = \lim_{\substack{(x,y)\to(a,b)\\x\ge 0}} x = a = f(a, b),$$

so $f(x, y)$ is continuous at (a, b) if $a > 0$. Similarly, if $a < 0$,

$$\lim_{(x,y)\to(a,b)} f(x, y) = \lim_{\substack{(x,y)\to(a,b)\\x< 0}} f(x, y) = \lim_{\substack{(x,y)\to(a,b)\\x< 0}} y = b = f(a, b),$$

so $f(x, y)$ is also continuous at (a, b) if $a < 0$. Points $(0, b)$ on the y-axis can be approached from the set $x \ge 0$ and from the set $x < 0$, so we must consider limits from within each of these sets. We have

$$\lim_{\substack{(x,y)\to(0,b)\\x\ge 0}} f(x, y) = \lim_{\substack{(x,y)\to(0,b)\\x\ge 0}} x = 0$$

and

$$\lim_{\substack{(x,y)\to(0,b)\\x< 0}} f(x, y) = \lim_{\substack{(x,y)\to(0,b)\\x< 0}} y = b.$$

These two limits are equal only when $b = 0$, so $(0, b) = (0, 0)$ is the only point on the x-axis where $f(x, y)$ is continuous. Collecting results, we conclude that $f(x, y)$ is continuous at each point except those on either the positive or the negative y-axis. See Figure 12.2.12. ■

We say that f is **continuous on a set** S if

$$\lim_{\substack{(x,y)\to(a,b)\\(x,y)\ \text{in}\ S}} f(x, y) = f(a, b) \text{ for every } (a, b) \text{ in } S.$$

This allows us to say, for example, that $f(x, y) = \sqrt{4 - x^4 - y^2}$ is continuous on its domain. See Figure 12.2.13.

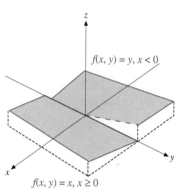

FIGURE 12.2.12 $f(x, y) =$
$\begin{cases} x, & x \ge 0, \\ y, & x < 0 \end{cases}$ is discontinuous at
all points on the y-axis except the origin.

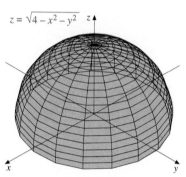

FIGURE 12.2.13 $f(x, y) =$
$\sqrt{4 - x^2 - y^2}$ is continuous on its
domain, the set of points (x, y) that
satisfy $x^2 + y^2 \le 4$.

EXERCISES 12.2

For each of the limits in Exercises 1–28, evaluate the limit or show the limit does not exist.

1. $\lim_{(x,y)\to(4,2)} (2x - 3y + 5)$

2. $\lim_{(x,y)\to(3,\pi/2)} \sin xy$

3. $\lim_{(x,y)\to(0,0)} e^{x^2+y^2}$

4. $\lim_{(x,y)\to(3,-2)} \ln(x + y)$

5. $\lim_{(x,y)\to(0,0)} \cot(x^2 + y^2)$

6. $\lim_{(x,y)\to(3,-2)} \ln \sqrt{x^2 + y^2}$

7. $\lim_{(x,y)\to(1,-1)} \dfrac{x^2 - 2xy + y^2}{x - y}$

8. $\lim_{(x,y)\to(2,3)} \dfrac{x^2 - y^2}{x - y}$

9. $\lim_{(x,y)\to(3,1)} \dfrac{x^2 - y^2}{x - y}$

10. $\lim_{(x,y)\to(0,-2)} \dfrac{x^2 - y^2}{x - y}$

11. $\lim_{(x,y)\to(0,0)} \dfrac{x^2}{x^2 + y^2}$

12. $\lim_{(x,y)\to(0,0)} \dfrac{y^3}{x^2 + y^2}$

13. $\lim_{(x,y)\to(0,0)} \dfrac{xy^2}{x^2 + y^2}$

14. $\lim_{(x,y)\to(0,0)} \dfrac{x(x + y)}{x^2 + y^2}$

15. $\lim_{(x,y)\to(0,0)} \dfrac{x^2 + 2y^2}{x^2 + y^2}$

16. $\lim_{(x,y)\to(0,0)} \dfrac{x^3 + y^3}{x^2 + y^2}$

17. $\lim_{(x,y)\to(0,0)} \dfrac{x^3 + y}{x^2 + y^2}$

18. $\lim_{(x,y)\to(0,0)} \dfrac{x^3 + 2x^2y}{x^2 + y^2}$

19. $\lim_{(x,y)\to(0,0)} \ln(x^2 + y^2)$

20. $\lim_{(x,y)\to(0,0)} \dfrac{x}{x^2 + y^2}$

21. $\lim_{(x,y)\to(0,0)} \dfrac{x}{\sqrt{x^2 + y^2}}$

22. $\lim_{(x,y)\to(0,0)} \dfrac{xy}{\sqrt{x^2 + y^2}}$

23. $\lim_{(x,y,z)\to(0,0,0)} \dfrac{xyz}{x^2 + y^2 + z^2}$

24. $\lim_{(x,y,z)\to(0,0,0)} \dfrac{x^2}{x^2 + y^2 + z^2}$

25. $\lim_{(x,y)\to(0,0)} f(x, y)$, where $f(x, y) = \begin{cases} x, & y = x^2 \\ 0, & y \ne x^2 \end{cases}$

26. $\lim_{(x,y)\to(0,0)} f(x, y)$, where $f(x, y) = \begin{cases} x, & y = x^3 \\ y, & y \neq x^3 \end{cases}$

27. $\lim_{(x,y)\to(0,0)} f(x, y)$, where $f(x, y) = \begin{cases} 1, & y = x^2 + 1 \\ 0, & y \neq x^2 + 1 \end{cases}$

28. $\lim_{(x,y)\to(0,0)} f(x, y)$, where $f(x, y)$

$$= \begin{cases} 1 - x^2 - y^2, & x^2 + y^2 \leq 1 \\ 0, & x^2 + y^2 > 1 \end{cases}$$

29. Determine the set where the function $f(x, y) = \sqrt{1 - x^2 - y^2}$ is continuous.

30. Determine the set where the function $f(x, y) = \ln(1 - x^2 - y^2)$ is continuous.

31. Determine the set where the function $f(x, y) = \ln(x^2 + y^2)$ is continuous.

32. Determine the set where the function $f(x, y) = \tan^{-1}\frac{y}{x}$ is continuous.

33. Determine the set where the function in Exercise 25 is continuous.

34. Determine the set where the function in Exercise 26 is continuous.

35. Determine the set where the function in Exercise 27 is continuous.

36. Determine the set where the function in Exercise 28 is continuous.

37. Use the ϵ-δ definition of limit to show
$$\lim_{(x,y)\to(a,b)} x = a.$$

38. Use the ϵ-δ definition of limit to show
$$\lim_{(x,y)\to(a,b)} y = b.$$

39. Use the ϵ-δ definition of limit to show
$$\lim_{(x,y)\to(a,b)} \frac{x^2}{\sqrt{x^2 + y^2}} = 0.$$

40. Use the ϵ-δ definition of limit to show
$$\lim_{(x,y)\to(a,b)} \frac{x^3 + x^2 + y^2 + y^3}{x^2 + y^2} = 1.$$

In Exercises 41–44, either evaluate the limit or show that the limit does not exist.

41. $\lim_{\substack{(x,y)\to(2,2) \\ x \neq y}} \frac{x^2 - y^2}{x - y}$

42. $\lim_{\substack{(x,y)\to(2,1) \\ x \neq 2y}} \frac{x^2 - 4y^2}{x - 2y}$

43. $\lim_{\substack{(x,y)\to(0,0) \\ x > y}} \frac{|x - y|}{x - y}$

44. $\lim_{\substack{(x,y)\to(0,0) \\ x < y}} \frac{|x - y|}{x - y}$

45. $f(x, y) = \begin{cases} \dfrac{xy^2}{x^2 + y^6}, & (x, y) \neq (0, 0), \\ 0, & (x, y) = (0, 0). \end{cases}$

(a) Show $f(x, y) \to 0$ as $(x, y) \to (0, 0)$ along every line through the origin.

(b) Show $\lim_{(x,y)\to(0,0)} f(x, y)$ does not exist.

12.3 PARTIAL DERIVATIVES

Connections

Differentiation formulas, 3.1–4.

We begin our study of differentiation of functions of several variables by defining **partial derivatives**. At this point we are mostly concerned with the notation and the evaluation of partial derivatives. In later sections we will see how they are used in applications.

Definition and Notation

DEFINITION

The **partial derivative of $f(x, y)$ with respect to x** at the point (a, b) is
$$\frac{\partial f}{\partial x} = \lim_{x \to a} \frac{f(x, b) - f(a, b)}{x - a}, \quad \text{if the limit exists.}$$

The **partial derivative with respect to y** at (a, b) is
$$\frac{\partial f}{\partial y} = \lim_{y \to b} \frac{f(a, y) - f(a, b)}{y - b}.$$

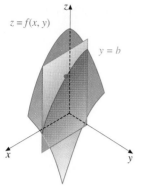

FIGURE 12.3.1a The partial derivative of $f(x, y)$ with respect to x at (a, b) gives the slope of the curve of intersection of the surface $z = f(x, y)$ and the plane $y = b$.

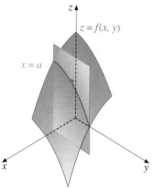

FIGURE 12.3.1b The partial derivative of $f(x, y)$ with respect to y at (a, b) gives the slope of the curve of intersection of the surface $z = f(x, y)$ and the plane $x = a$.

The partial derivatives of $f(x, y)$ with respect to x and y at (a, b) may be interpreted geometrically as the slopes of the curves of intersection of the surface $z = f(x, y)$ with the planes $y = b$ and $x = a$, respectively. See Figure 12.3.1.

Generally, a function has a partial derivative that corresponds to each of its variables. For example, a function $f(x, y, z)$ of three variables has three partial derivatives, $\partial f / \partial x$, $\partial f / \partial y$, and $\partial f / \partial z$, defined by

$$\frac{\partial f}{\partial x} = \lim_{x \to a} \frac{f(x, b, c) - f(a, b, c)}{x - a},$$

$$\frac{\partial f}{\partial y} = \lim_{y \to b} \frac{f(a, y, c) - f(a, b, c)}{y - b}, \text{ and}$$

$$\frac{\partial f}{\partial z} = \lim_{z \to c} \frac{f(a, b, z) - f(a, b, c)}{z - c}.$$

The notation f_x is sometimes used to denote the partial derivative of f with respect to the variable x. Also, f_1 is used to denote the partial derivative of f with respect to the first variable, f_2 denotes the partial derivative with respect to the second variable, and so forth. If $w = f(x, \ldots)$, we may also write $\partial w / \partial x$ in place of $\partial f / \partial x$ or w_x in place of f_x.

As in the case of the derivative of a function of one real variable, we have defined partial derivatives at a *point* in the domain of the function. As this point varies we obtain a *function*. The partial derivatives of f are functions of the same variables as the original function. The domain of each partial derivative consists of those points in the domain of f where the partial derivative exists.

Evaluating Partial Derivatives

Each partial derivative is defined in terms of a limit of a single real variable. Partial derivatives are ordinary derivatives of the functions of one variable that are obtained by fixing the values of all but one variable of the function. This means that we can use all of the rules of differentiation we developed for functions of one variable to evaluate partial derivatives. It is only necessary to treat all variables, except the variable of partial differentiation, as constants. If the function is given by a formula, the rules of differentiation give us formulas for the partial derivatives. The value of a partial derivative at a point can then be determined by substitution.

EXAMPLE 1

$f(x, y) = x^2 y^3$. Evaluate $\dfrac{\partial f}{\partial x}(3, 5)$ and $\dfrac{\partial f}{\partial y}(3, 5)$.

SOLUTION   We first find formulas for $\dfrac{\partial f}{\partial x}(x, y)$ and $\dfrac{\partial f}{\partial y}(x, y)$. To find $\dfrac{\partial f}{\partial x}(x, y)$ we consider y to be constant and differentiate $f(x, y) = x^2 y^3$ with respect to x. The expression $x^2 y^3$ is then read as x^2 times the "constant" y^3. The derivative with respect to x is then the derivative of x^2 times the "constant" y^3. We obtain

$$\frac{\partial f}{\partial x}(x, y) = (2x)(y^3) = 2xy^3.$$

The partial derivative with respect to y is found by considering x to be a constant and differentiating with respect to y. In this case, $x^2 y^3$ is read as the "constant" x^2 times y^3. The derivative with respect to y is the "constant" x^2 times the derivative of y^3. This gives

$$\frac{\partial f}{\partial y}(x, y) = (x^2)(3y^2) = 3x^2y^2.$$

Substitution into the above formulas gives

$$\frac{\partial f}{\partial x}(3, 5) = 2(3)(5)^3 = 750 \text{ and}$$

$$\frac{\partial f}{\partial y}(3, 5) = 3(3)^2(5)^2 = 675. \quad \blacksquare$$

EXAMPLE 2

$f(x, y) = x^2 - 2xy + 3y^2$. Evaluate $\frac{\partial f}{\partial x}(x, y)$ and $\frac{\partial f}{\partial y}(x, y)$.

SOLUTION To evaluate the partial derivative with respect to x, we consider y to be a constant and differentiate with respect to the variable x. We have

$$f(x, y) = x^2 - 2xy + 3y^2, \text{ so}$$

$$\frac{\partial f}{\partial x}(x, y) = 2x - 2y + 0 = 2x - 2y.$$

The partial derivative with respect to y is found by considering x to be a constant and differentiating with respect to y. We have

$$f(x, y) = x^2 - 2xy + 3y^2, \text{ so}$$

$$\frac{\partial f}{\partial y}(x, y) = 0 - 2x + 3(2y) = -2x + 6y. \quad \blacksquare$$

EXAMPLE 3

$f(x, y) = xe^{xy}$. Find $f_x(x, y)$ and $f_y(x, y)$.

SOLUTION If $f(x, y) = xe^{xy}$, then

$$f_x(x, y) = (x)(e^{xy}y) + (e^{xy})(1) = e^{xy}(xy + 1) \text{ and}$$
$$f_y(x, y) = x(e^{xy}x) = x^2e^{xy}.$$

Note that the Product Rule was used to find f_x and the Chain Rule was used to differentiate $e^{(\text{Function})}$ in both formulas. $\quad \blacksquare$

EXAMPLE 4

$f(x, y, z) = ze^{x^2y^3}$. Find $f_x(x, y, z)$, $f_y(x, y, z)$, and $f_z(x, y, z)$.

SOLUTION The partial derivative with respect to x is found by considering the variables y and z to be constants and then differentiating $f(x, y, z) = ze^{x^2y^3}$ with respect to x. We obtain

$$f_x(x, y, z) = z(e^{x^2y^3})(2xy^3).$$

If x and z are considered to be constants, differentiation of $f(x, y, z) = ze^{x^2y^3}$ with respect to y gives

$$f_y(x, y, z) = z(e^{x^2y^3})(3x^2y^2).$$

The partial derivative with respect to z is obtained by considering x and y to be constants while we differentiate $f(x, y, z) = ze^{x^2y^3}$ with respect to z. This gives

$$f_z(x, y, z) = e^{x^2y^3}. \quad \blacksquare$$

EXAMPLE 5

$f(x, y, z) = \sin(x - y)$. Find $f_x(x, y, z)$, $f_y(x, y, z)$, and $f_z(x, y, z)$.

SOLUTION   If $f(x, y, z) = \sin(x - y)$, then

$$f_x(x, y, z) = (\cos(x - y))(1 - 0) = \cos(x - y),$$
$$f_y(x, y, z) = (\cos(x - y))(0 - 1) = -\cos(x - y),$$
$$f_z(x, y, z) = 0. \quad \blacksquare$$

Higher-Order Partial Derivatives

Since partial derivatives are functions, we may take partial derivatives of the partial derivatives. Two different notations for such higher-order partial derivatives are indicated below.

$$\frac{\partial}{\partial x}\left(\frac{\partial f}{\partial x}\right) = \frac{\partial^2 f}{\partial x^2}, \qquad (f_x)_x = f_{xx},$$

$$\frac{\partial}{\partial y}\left(\frac{\partial f}{\partial x}\right) = \frac{\partial^2 f}{\partial y \partial x}, \qquad (f_x)_y = f_{xy},$$

$$\frac{\partial}{\partial x}\left(\frac{\partial f}{\partial y}\right) = \frac{\partial^2 f}{\partial x \partial y}, \qquad (f_y)_x = f_{yx},$$

$$\frac{\partial}{\partial y}\left(\frac{\partial f}{\partial y}\right) = \frac{\partial^2 f}{\partial y^2}, \qquad (f_y)_y = f_{yy}.$$

Note the order of differentiation indicated by the two different notations. The partial derivatives

$$\frac{\partial^2 f}{\partial y \partial x} = \frac{\partial}{\partial y}\left(\frac{\partial f}{\partial x}\right) \text{ and } f_{xy} = (f_x)_y$$

are evaluated by first taking the partial derivative with respect to x and then taking the partial derivative with respect to y. The order is reversed to evaluate

$$\frac{\partial^2 f}{\partial x \partial y} = \frac{\partial}{\partial x}\left(\frac{\partial f}{\partial y}\right) \text{ and } f_{yx} = (f_y)_x.$$

The functions f_x and f_y are called **first-order** partial derivatives of f; f_{xx}, f_{xy}, f_{yx} and f_{yy} are called **second-order** partial derivatives of f.

EXAMPLE 6

Find all first- and second-order partial derivatives of the function $f(x, y) = x \sin y$.

SOLUTION   Since f is a function of two variables, there are two first-order derivatives. We have $f(x, y) = x \sin y$, so

$$f_x = \frac{\partial}{\partial x}(x \sin y) = \sin y,$$

$$f_y = \frac{\partial}{\partial y}(x \sin y) = x \cos y.$$

The second-order derivatives are obtained by taking partial derivatives of each of the first-order partials. We obtain

$$f_{xx} = (f_x)_x = \frac{\partial}{\partial x}(\sin y) = 0,$$

$$f_{xy} = (f_x)_y = \frac{\partial}{\partial y}(\sin y) = \cos y,$$

$$f_{yx} = (f_y)_x = \frac{\partial}{\partial x}(x \cos y) = \cos y,$$

$$f_{yy} = (f_y)_y = \frac{\partial}{\partial y}(x \cos y) = -x \sin y. \quad \blacksquare$$

For most of the functions we deal with, it is true that corresponding **mixed partials** such as f_{xy} and f_{yx} are equal, even though the order of differentiation is different. The following theorem tells us that the order in which the partial derivatives are taken does not matter if all of the partial derivatives involved are continuous. If the partial derivatives are not continuous, it is possible that $f_{xy} \neq f_{yx}$. (See Exercise 43.)

THEOREM

If $f(x, y)$ is defined in an open region R, and the functions f, f_x, f_y, f_{xy}, and f_{yx} are all continuous in R, then $f_{xy} = f_{yx}$ in R.

We will not prove this theorem.

EXERCISES 12.3

1. $f(x, y) = x^2 - y^2$. Evaluate $f_x(2, 3)$ and $f_y(2, 3)$.

2. $f(x, y) = x^2 + 2xy + y^2$. Evaluate $f_x(1, 2)$ and $f_y(1, 2)$.

3. $f(x, y) = x^3 y^2 - xy^2$. Evaluate $f_x(-1, 3)$ and $f_y(-1, 3)$.

4. $f(x, y) = x^3 - 3x^2 y^2 + 3xy^2 - y^3$. Evaluate $f_x(3, 2)$ and $f_y(3, 2)$.

Find all first-order partial derivatives of the functions in Exercises 5–24.

5. $f(x, y) = 2x^2 y - xy^3 + 3x - y + 7$

6. $f(x, y) = x^3 y^2 - 4xy^4 - x^2 + 2y - 5$

7. $f(x, y) = \dfrac{x}{x + y}$

8. $f(x, y) = \dfrac{2x - y}{3x + 5y}$

9. $f(x, y) = x\sqrt{x^2 + y^2}$

10. $f(x, y) = y \sin(x + y)$

11. $f(x, y) = \ln(x^2 + y^2)$

12. $f(x, y) = e^{-x^2 - y^2}$

13. $f(x, y) = \tan^{-1}(y/x)$

14. $f(x, y) = \sin^{-1} xy$

15. $f(x, y) = \tan(2x - y)$

16. $f(x, y) = \cos 2xy$

17. $f(x, y) = \sin^2(x^2 y^2)$

18. $f(x, y) = 1/\sqrt{x^2 + y^2}$

19. $f(x, y, z) = xy + yz + xz$

20. $f(x, y, z) = \sqrt{x^2 + y^2 + z^2}$

21. $f(x, y, z) = \dfrac{x - y}{x + z}$

22. $f(x, y, z) = 2\sqrt{xyz}$

23. $f(x, y, z) = e^{2x-y}$

24. $f(x, y, z) = z \ln yz$

Find all first- and second-order partial derivatives of the functions in Exercises 25–30.

25. $f(x, y) = x^3 + x^2y - 3xy^2 + y^3$

26. $f(x, y) = 3x^2 - 2xy + 5y^2 - x + 4y + 1$

27. $f(x, y) = x \ln y$

28. $f(x, y) = e^{xy}$

29. $f(x, y) = \sin xy$

30. $f(x, y) = 1 - \sqrt{x^2 + y^2}$

31. $x = r \cos\theta$. Find $\dfrac{\partial x}{\partial r}$ and $\dfrac{\partial x}{\partial \theta}$.

32. $x = \rho \sin\phi \cos\theta$. Find $\dfrac{\partial x}{\partial \rho}$, $\dfrac{\partial x}{\partial \phi}$ and $\dfrac{\partial x}{\partial \theta}$.

Show that the components of each vector function $M(x, y)\mathbf{i} + N(x, y)\mathbf{j}$ given in Exercises 33–36 satisfies the indicated equation.

33. $x\mathbf{i} + y\mathbf{j}$; $M_y = N_x$

34. $y\mathbf{i} - x\mathbf{j}$; $M_x + N_y = 0$

35. $\dfrac{x}{x^2 + y^2}\mathbf{i} + \dfrac{y}{x^2 + y^2}\mathbf{j}$; $M_x + N_y = 0$, $x^2 + y^2 \neq 0$

36. $\dfrac{2xy}{(x^2 + y^2)^2}\mathbf{i} + \dfrac{y^2 - x^2}{(x^2 + y^2)^2}\mathbf{j}$; $M_y = N_x$, $x^2 + y^2 \neq 0$

37. Suppose z is a differentiable function of (x, y) and $x^2 + y^2 + z^2 = 1$. Express z_x and z_y in terms of x, y, and z.

38. Suppose z is a differentiable function of (x, y) and $z^3 = xyz + 1$. Express z_x and z_y in terms of x, y, and z.

*Show that the functions in Exercises 39–40 satisfy the **Laplace equation**, $f_{xx} + f_{yy} = 0$.*

39. $f(x, y) = \ln \sqrt{x^2 + y^2}$

40. $f(x, y) = \tan^{-1} \dfrac{y}{x}$, $x > 0$

*Show that the functions in Exercises 41–42 satisfy the **heat equation**, $f_t = f_{xx}$.*

41. $f(x, t) = e^{-a^2 t} \sin ax$, a a constant

42. $f(x, t) = \displaystyle\int_0^{x/\sqrt{2t}} e^{-\tau^2/2} d\tau$, $x > 0, t > 0$

43. $f(x, y) = \begin{cases} \dfrac{xy^3 - x^3y}{x^2 + y^2}, & (x, y) \neq (0, 0), \\ 0, & (x, y) = (0, 0). \end{cases}$ Show that $f_{xy}(0, 0) \neq f_{yx}(0, 0)$. (*Hint:* Show that $f_x(0, y) = y$, so $f_{xy}(0, 0) = 1$. Show that $f_y(x, 0) = -x$, so $f_{yx}(0, 0) = -1$.)

12.4 LINEAR APPROXIMATION; DIFFERENTIABILITY

Connections

Differentiability, 2.6.

Approximation by tangent lines, 3.6.

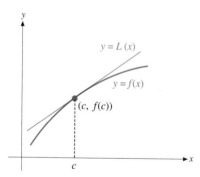

FIGURE 12.4.1a The line $y = L(x)$ is tangent to the graph of $y = f(x)$ at the point $(c, f(c))$.

In this section we will begin the study of linear approximation of functions of several variables by introducing differentials and seeing how they are used to approximate error. We will also introduce the concept of differentiability for functions of several variables. In later sections we will see how differentiability relates to the Chain Rule for derivatives, rates of change, and planes tangent to surfaces.

If a function f of a single variable is differentiable at c, we have seen in Section 3.6 that the linear function

$$L(x) = f(c) + f'(c)(x - c)$$

can be used to find approximate values of $f(x)$ for x near c. Geometrically, if f is differentiable at c, then the graph of $y = L(x)$ is the line tangent to the graph of $y = f(x)$ at the point $(c, f(c))$. See Figure 12.4.1a. The function L can be characterized as the linear function with the property that its value and the value of its derivative at c agree with those of f. This suggests that, for functions of f of two variables, we consider the linear function

$$L(x, y) = f(a, b) + f_x(a, b)(x - a) + f_y(a, b)(y - b).$$

The value of this function and the values of its two first partial derivatives at (a, b) agree with those of f. The graph of $z = L(x, y)$ is a plane that contains the point $(a, b, f(a, b))$ and has normal vector $f_x(a, b)\mathbf{i} + f_y(a, b)\mathbf{j} - \mathbf{k}$.

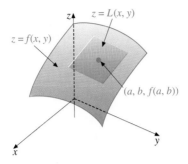

FIGURE 12.4.1b The plane $z = L(x, y)$ is tangent to the graph of $z = f(x, y)$ at the point $(a, b, f(a, b))$.

Later in this section we will see that the existence of the partial derivatives of f at (a, b) is not enough to guarantee that $L(x, y)$ is a good approximation of $f(x, y)$ for (x, y) near (a, b). We will also investigate a condition that does guarantee this result. Functions that satisfy this condition are called **differentiable**. Fortunately, most of the functions we encounter are differentiable at points where their partial derivatives are defined. If f is differentiable at (a, b), then the graph of $z = L(x, y)$ is the plane "tangent" to the graph of $z = f(x, y)$ at the point $(a, b, f(a, b))$. See Figure 12.4.1b.

Differentials

As in the case of a single variable, it is convenient to describe linear approximation in terms of differentials. If f has partial derivatives $f_x(a, b)$ and $f_y(a, b)$, we define the **differential** of f at (a, b) to be

$$df(a, b, dx, dy) = f_x(a, b)dx + f_y(a, b)dy,$$

where dx and dy are real-valued variables. The differential of a function of any number of variables is defined similarly.

EXAMPLE 1
If $f(x, y) = x^2 y$, then

$$df(x, y, dx, dy) = f_x(x, y)dx + f_y(x, y)dy = 2xy\,dx + x^2\,dy. \quad \blacksquare$$

EXAMPLE 2
If $f(x, y, z) = z \sin(x^2 + y^2)$, then

$$\begin{aligned} df(x, y, z, dx, dy, dz) &= f_x(x, y, z)dx + f_y(x, y, z)dy + f_z(x, y, z)dz \\ &= z\cos(x^2 + y^2)(2x)dx + z\cos(x^2 + y^2)(2y)dy \\ &\quad + \sin(x^2 + y^2)dz. \quad \blacksquare \end{aligned}$$

If f is differentiable at (a, b) and (x, y) is near (a, b), we have

$$f(x, y) \approx L(x, y),$$
$$f(x, y) \approx f(a, b) + f_x(a, b)(x - a) + f_y(a, b)(y - b),$$
$$f(x, y) - f(a, b) \approx f_x(a, b)(x - a) + f_y(a, b)(y - b).$$

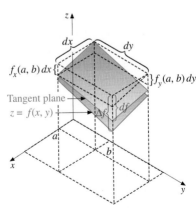

FIGURE 12.4.2 If f is differentiable at (a, b), then $\Delta f \approx df$ for (x, y) near (a, b).

That is, if

$$\Delta f = f(x, y) - f(a, b)$$

is the **change in f corresponding to a change in (x, y)** and

$$df = df(a, b, x - a, y - b) = f_x(a, b)(x - a) + f_y(a, b)(y - b)$$

is the **differential approximation of Δf**, we have

$$\Delta f \approx df.$$

Figure 12.4.2 illustrates the relation between Δf and df. The ideas are the same for functions of any number of variables.

EXAMPLE 3

$f(x, y) = 1/\sqrt{x^2 + y^2}$. Calculate $f(3, 4)$, $f(3.2, 3.9)$, $\Delta f = f(3.2, 3.9) - f(3, 4)$, and the differential approximation of Δf.

SOLUTION

$$f(3, 4) = \frac{1}{\sqrt{3^2 + 4^2}} = \frac{1}{5} = 0.2,$$

$$f(3.2, 3.9) = \frac{1}{\sqrt{(3.2)^2 + (3.9)^2}} \approx 0.19822394 \quad \text{(calculator value)},$$

$$\Delta f = f(3.2, 3.9) - f(3, 4) \approx -0.00177606.$$

The differential approximation of Δf is

$$df(3, 4, 3.2 - 3, 3.9 - 4) = f_x(3, 4)(3.2 - 3) + f_y(3, 4)(3.9 - 4).$$

We have

$$f_x(x, y) = -\frac{1}{2}(x^2 + y^2)^{-3/2}(2x) = -\frac{x}{(x^2 + y^2)^{3/2}} \quad \text{and}$$

$$f_y(x, y) = -\frac{1}{2}(x^2 + y^2)^{-3/2}(2y) = -\frac{y}{(x^2 + y^2)^{3/2}},$$

so substitution gives $f_x(3, 4) = -0.024$ and $f_y(3, 4) = -0.032$. Then

$$df(3, 4, 3.2 - 3, 3.9 - 4) = (-0.024)(0.2) + (-0.032)(-0.1) = -0.0016.$$

∎

Approximating Error

The differential df is sometimes interpreted as the **approximate error** in f due to **possible errors** of dx and dy in the variables x and y, respectively.

EXAMPLE 4

Two adjacent sides and the included angle of a triangle are measured to be 4 ft, 5 ft, and 30°, respectively. Find the area of the triangle if the measurements are assumed to be exact. Use differentials to find the approximate maximum error in the area of the triangle due to possible errors of 0.2 ft in the measurements of the sides and a possible error of 5° in the measurement of the angle.

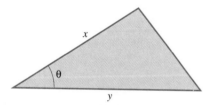

FIGURE 12.4.3 The lengths of two sides and the included angle of a triangle are considered to be variables in order to study the effect of errors in their measurement on the area of the triangle.

SOLUTION A sketch of the triangle is given in Figure 12.4.3. The variables x and y represent the lengths of the sides in feet and the variable θ represents the angle. Note that we cannot study the effect of errors in measurements of the sides and angle on the area unless we consider the values of the measurements to be variables.

The area of the triangle in square feet is

$$A(x, y, \theta) = \frac{1}{2}xy \sin \theta.$$

Since we are going to be differentiating a trigonometric function in this problem, we measure angles in *radians*. (Recall that the formulas for the differentiation of the trigonometric functions were derived for radian measure.) We have

$$30° = \frac{\pi}{6} \text{ rad} \quad \text{and} \quad 5° = 5° \left(\frac{\pi \text{ rad}}{180°} \right) = \frac{\pi}{36} \text{ rad}.$$

If the measurements are exact, the area is

$$A(4, 5, \pi/6) = \frac{1}{2}(4)(5)\left(\frac{1}{2}\right) = 5 \text{ ft}^2.$$

The desired approximate value of the error in area is given by the differential dA, with the partial derivatives evaluated at $(x, y, \theta) = (4, 5, \pi/6)$, $dx = \pm 0.2$, $dy = \pm 0.2$, and $d\theta = \pm \pi/36$. We have

$$A_x(x, y, \theta) = \frac{1}{2}y \sin \theta, \text{ so } A_x(4, 5, \pi/6) = 1.25,$$

$$A_y(x, y, \theta) = \frac{1}{2}x \sin \theta, \text{ so } A_y(4, 5, \pi/6) = 1,$$

$$A_\theta(x, y, \theta) = \frac{1}{2}xy \cos \theta, \text{ so } A_\theta(4, 5, \pi/6) = 5\sqrt{3}.$$

Then

$$dA = A_x dx + A_y dy + A_\theta d\theta$$
$$= (1.25)(\pm 0.2) + (1)(\pm 0.2) + (5\sqrt{3})(\pm \pi/36).$$

The *maximum* possible error would occur when the errors that correspond to each variable have the same sign. The approximate value of the maximum possible error is the sum of the absolute values of each of the terms in the differential. Thus, the approximate maximum error is

$$|(1.25)(0.2)| + |(1)(0.2)| + |(5\sqrt{3})(\pi/36)| \approx 1.2057 \text{ ft}^2. \quad \blacksquare$$

EXAMPLE 5
The magnitude of the force due to gravitational attraction between two masses m_1 and m_2 that are a distance D apart is

$$F = k\frac{m_1 m_2}{D^2},$$

where k is a constant. Find the approximate maximum percent error in F due to possible errors of 5 percent in measurement of the masses and a possible error of 4 percent in the measurement in the distance.

SOLUTION   The magnitude of the force is a function of the three variables m_1, m_2, and D. We have

$$F_{m_1} = \frac{km_2}{D^2}, \quad F_{m_2} = \frac{km_1}{D^2}, \text{ and } F_D = -\frac{2km_1 m_2}{D^3}.$$

The errors in measurements are $dm_1 = \pm 0.05m_1$, $dm_2 = \pm 0.05m_2$, and $dD = \pm 0.04D$. The differential gives the approximate error in F. We have

$$dF = F_{m_1}dm_1 + F_{m_2}dm_2 + F_D dD$$
$$= \left(\frac{km_2}{D^2}\right)(\pm 0.05m_1) + \left(\frac{km_1}{D^2}\right)(\pm 0.05m_2) + \left(\frac{-2km_1 m_2}{D^3}\right)(\pm 0.04D)$$
$$= \pm 0.05F \pm 0.05F \mp 0.08F.$$

The approximate maximum error is the sum of the absolute values of the terms of the differential,

$$0.05F + 0.05F + 0.08F = 0.18F, \text{ or } 18 \text{ percent of } F. \quad \blacksquare$$

It is important that we realize that the idea of differentials, or linear approximation, is to obtain information about a function from its value and the values of its partial derivatives at a point. Thus, we may find approximate values of a function for which we may not have a formula, but for which values of the partial derivatives have been determined experimentally.

EXAMPLE 6

We desire to know the effect on the total resistance of a complicated electric circuit due to changes in the resistance of two resistors in the circuit. It is determined experimentally that keeping the second resistor fixed and changing the first by ΔR_1 results in a change of $0.2\Delta R_1$ in the resistance of the circuit. Changing the second resistor by ΔR_2 while the first is kept fixed gives a change of $0.15\Delta R_2$ in the resistance of the circuit. What is the approximate change in the resistance of the circuit if the first resistor is increased by 10 ohms while the second resistor is decreased by 4 ohms?

SOLUTION   The resistance of the circuit R is given by an unknown function of the resistance of the two variable resistors R_1 and R_2,

$$R = R(R_1, R_2).$$

For fixed values of R_2, the rate of change of R with respect to R_1 is the partial derivative of R with respect to R_1. The experimental data indicate that

$$\frac{\partial R}{\partial R_1} \approx \frac{\Delta R}{\Delta R_1} = \frac{0.2\Delta R_1}{\Delta R_1} = 0.2,$$

so we take $\partial R / \partial R_1 = 0.2$. Similarly,

$$\frac{\partial R}{\partial R_2} \approx \frac{\Delta R}{\Delta R_2} = \frac{0.15\Delta R_2}{\Delta R_2} = 0.15,$$

so we take $\partial R / \partial R_1 = 0.15$. Using $dR_1 = +10$ and $dR_2 = -4$, we have

$$dR = \frac{\partial R}{\partial R_1}dR_1 + \frac{\partial R}{\partial R_2}dR_2 = (0.2)(10) + (0.15)(-4) = 1.4.$$

The resistance of the circuit will increase by approximately 1.4 ohms. $\quad \blacksquare$

Differentiability

The problem of determining when the linear function $L(x, y)$ gives good approximate values of $f(x, y)$ is more complicated than in the case of functions of a single variable. First, we should note that the partial derivatives $f_x(a, b)$ and $f_y(a, b)$ depend only on the behavior of $f(x, y)$ for (x, y) on the two lines $x = a$ and $y = b$ in the xy-plane. The values of $f(x, y)$ at other points do not affect L. Figure 12.4.4 contains a sketch of the graph of the function

$$f(x, y) = \begin{cases} \dfrac{xy}{x^2 + y^2}, & (x, y) \neq (0, 0), \\ 0, & (x, y) = (0, 0). \end{cases}$$

$$f(x, y) = \begin{cases} \dfrac{xy}{x^2 + y^2}, & (x, y) \neq (0, 0) \\ 0, & (x, y) = (0, 0) \end{cases}$$

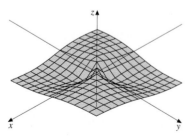

FIGURE 12.4.4 Both partial derivatives of f exist at the origin, but f is not continuous at the origin. The linear function $L(x, y) = f(0, 0) + f_x(0, 0)x + f_y(0, 0)y = 0$ is not a good approximation of $f(x, y)$ for (x, y) near $(0, 0)$.

We have $f(0, 0) = f_x(0, 0) = f_y(0, 0) = 0$, so the linear "approximation" of $f(x, y)$ near the origin is $L(x, y) = 0$. We do have $L(x, y) = f(x, y) = 0$ for (x, y) on either the x-axis or the y-axis, but zero is not a good approximation of $f(x, y)$ for points that are not on the axes. For example, $f(x, x) = 1/2$ for $x \neq 0$, so the value of $f(x, y)$ is $1/2$ as (x, y) approaches the origin along the line $y = x$. This example also shows that existence of the partial derivatives at a point does not even imply that the function is continuous at the point.

The following theorem gives us conditions that guarantee that the values of $L(x, y)$ and $f(x, y)$ are close to each other for (x, y) near (a, b).

THEOREM

If $f(x, y)$ has partial derivatives f_x and f_y in a disk with center (a, b), and f_x and f_y are continuous at (a, b), then there are functions ε_1 and ε_2 of x and y such that $\varepsilon_1(x, y) \to 0$ and $\varepsilon_2(x, y) \to 0$ as $(x, y) \to (a, b)$, and

$$f(x, y) - f(a, b) = f_x(a, b)(x - a) + f_y(a, b)(y - b)$$
$$+ \varepsilon_1(x, y) \cdot (x - a) + \varepsilon_2(x, y) \cdot (y - b).$$

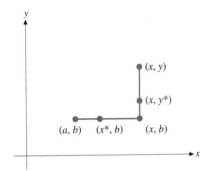

FIGURE 12.4.5 The path from (a, b) to (x, y) is divided into a segment where only x varies and a segment where only y varies. We can then apply the Mean Value Theorem for functions of one variable to f on each segment.

Proof Let us write

$$f(x, y) - f(a, b) = f(x, y) - f(x, b) + f(x, b) - f(a, b).$$

Since f_y is the derivative of the function of y that we obtain by fixing the value of x in $f(x, y)$, the Mean Value Theorem applied to that function of y gives $f(x, y) - f(x, b) = f_y(x, y^*)(y - b)$ for some y^* between y and b. Similarly, $f(x, b) - f(a, b) = f_x(x^*, b)(x - a)$ for some x^* between x and a. See Figure 12.4.5. Combining results, we obtain

$$f(x, y) - f(a, b) = f_y(x, y^*)(y - b) + f_x(x^*, b)(x - a)$$
$$= f_x(a, b)(x - a) + f_y(a, b)(y - b)$$
$$+ [f_x(x^*, b) - f_x(a, b)](x - a)$$
$$+ [f_y(x, y^*) - f_y(a, b)](y - b).$$

The conclusion of the theorem is satisfied if we choose $\varepsilon_1(x, y) = f_x(x^*, b) - f_x(a, b)$ and $\varepsilon_2(x, y) = f_y(x, y^*) - f_y(a, b)$. The continuity of f_x and f_y at (a, b) implies that $\varepsilon_1(x, y)$ and $\varepsilon_2(x, y)$ approach zero as (x, y) approaches (a, b). ∎

Let us rewrite the conclusion of the above theorem in the form

$$f(x, y) - [f(a, b) + f_x(a, b)(x - a) + f_y(a, b)(y - b)]$$
$$= \varepsilon_1(x, y) \cdot (x - a) + \varepsilon_2(x, y) \cdot (y - b), \text{ or}$$
$$f(x, y) - L(x, y) = \varepsilon_1(x, y) \cdot (x - a) + \varepsilon_2(x, y) \cdot (y - b).$$

This shows that the values of $f(x, y)$ and $L(x, y)$ are close whenever (x, y) is close to (a, b). In fact, we have

$$\left| \frac{f(x, y) - L(x, y)}{\sqrt{(x - a)^2 + (y - b)^2}} \right| = \left| \frac{\varepsilon_1(x, y) \cdot (x - a)}{\sqrt{(x - a)^2 + (y - b)^2}} + \frac{\varepsilon_2(x, y) \cdot (y - b)}{\sqrt{(x - a)^2 + (y - b)^2}} \right|$$
$$\leq |\varepsilon_1(x, y)| + |\varepsilon_2(x, y)| \to 0 \text{ as } (x, y) \to (a, b),$$

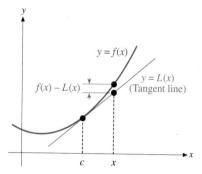

FIGURE 12.4.6 The graph of the linear function $L(x) = f(c) + f'(c)(x - c)$ is the line tangent to the graph of $f(x)$ where $x = c$. We have seen that $\dfrac{f(x) - L(x)}{x - c} \to 0$ as $x \to 0$.

so the ratio of the difference $f(x, y) - L(x, y)$ and the distance between (x, y) and (a, b) approaches zero as (x, y) approaches (a, b). This corresponds to the behavior of a function of one variable near a point where the function is differentiable. That is, $y = L(x)$ is the line tangent to the graph of $f(x)$ at $(c, f(c))$. The differentiability of $f(x)$ at c implies that the ratio of the difference $f(x) - L(x)$ and the distance between x and c approaches zero as x approaches c. See Figure 12.4.6.

For functions of two variables, we take the condition of the conclusion of the above theorem to be the definition of differentiable. It is convenient to express the condition in terms of delta notation. For a fixed point (a, b), let

$$\Delta x = x - a, \quad \Delta y = y - b, \text{ and } \Delta f = f(x, y) - f(a, b).$$

DEFINITION

A function $f(x, y)$ is said to be **differentiable** at (a, b) if the partial derivatives $f_x(a, b)$ and $f_y(a, b)$ exist, and if there are functions ε_1 and ε_2 such that $\varepsilon_1(\Delta x, \Delta y)$ and $\varepsilon_2(\Delta x, \Delta y)$ approach zero as $(\Delta x, \Delta y)$ approaches $(0, 0)$, and

$$\Delta f = f_x(a, b)\Delta x + f_y(a, b)\Delta y + \varepsilon_1(\Delta x, \Delta y)\Delta x + \varepsilon_2(\Delta x, \Delta y)\Delta y.$$

If f is differentiable at (a, b), the plane

$$z = f(a, b) + f_x(a, b)(x - a) + f_y(a, b)(y - b)$$

is said to be the **plane tangent to the surface $z = f(x, y)$** at the point $(a, b, f(a, b))$.

We shall see that the above form of the definition of differentiability is convenient for *theory*. In particular, we will use the definition to develop formulas of differentiation. The definition is not convenient for determining if a given function is differentiable. We will not need to verify the condition of the definition for individual functions. The previous theorem tells us that a *function that has first partial derivatives at all points near (a, b) is differentiable at (a, b) if the partials are continuous there.*

Most of the functions we encounter are easily seen to be continuous and have continuous partial derivatives wherever they are defined, and so are differentiable.

EXAMPLE 7

Find an equation of the plane tangent to the graph of $f(x, y) = 6 - x^2 - y^2$ at the point $(1, 2, 1)$.

SOLUTION   The function f has continuous partial derivatives, so it is differentiable. Note that $f(1, 2) = 1$, so the point $(1, 2, 1)$ is on the graph of f. We have

$$f(x, y) = 6 - x^2 - y^2,$$
$$f_x(x, y) = -2x,$$
$$f_y(x, y) = -2y,$$

so

$$f_x(1, 2) = -2,$$
$$f_y(1, 2) = -4.$$

The plane tangent to the graph of f at $(1, 2, 1)$ is then

$$z = 1 - 2(x - 2) - 4(y - 1),$$
$$z = 9 - 2x - 4y.$$

See Figure 12.4.7. ∎

FIGURE 12.4.7 The plane $z = 9 - 2x - 4y$ is tangent to the paraboloid $z = 6 - x^2 - y^2$ at the point $(1, 2, 1)$.

The *form* of the condition of differentiability extends easily to functions of any number of variables. The only difference is in the geometric interpretation of the condition. The theory is the same. For example, a function $f(x, y, z)$ is defined to be **differentiable** at (a, b, c) if all first partial derivatives of f exist at (a, b, c), and there are functions ε_1, ε_2, and ε_3 such that $\varepsilon_1(\Delta x, \Delta y, \Delta z)$, $\varepsilon_2(\Delta x, \Delta y, \Delta z)$, and $\varepsilon_3(\Delta x, \Delta y, \Delta z)$ approach zero as $(\Delta x, \Delta y, \Delta z)$ approaches $(0, 0, 0)$, and

$$\Delta f = f_x(a, b, c)\Delta x + f_y(a, b, c)\Delta y + f_z(a, b, c)\Delta z + \varepsilon_1(\Delta x, \Delta y, \Delta z)\Delta x$$
$$+ \varepsilon_2(\Delta x, \Delta y, \Delta z)\Delta y + \varepsilon_3(\Delta x, \Delta y, \Delta z)\Delta z.$$

A function of three variables is differentiable at (a, b, c) if it has first partial derivatives at all points near (a, b, c) and these derivatives are continuous at (a, b, c).

We say a function is **differentiable on a set** if it is differentiable at each point of the set. This follows from the definition of differentiability.

A function that is differentiable at a point is continuous at the point.

Continuity requires only that $\Delta f \to 0$ as $(\Delta x_1, \dots, \Delta x_n) \to (0, \dots, 0)$.

EXERCISES 12.4

Evaluate the differential $df(x, y, dx, dy)$ for each of the functions given in Exercises 1–6.

1. $f(x, y) = 3x^2 - 5xy + y^2$

2. $f(x, y) = xy^2 - x + 2y - 7$

3. $f(x, y) = \sin^{-1}\left(\dfrac{y}{\sqrt{x^2 + y^2}}\right)$

4. $f(x, y) = \ln(x^2 + y^2)$

5. $f(x, y) = e^{xy}$

6. $f(x, y) = e^{-2x} \cos y$

Evaluate the differential $df(x, y, z, dx, dy, dz)$ for each of the functions given in Exercises 7–12.

7. $f(x, y, z) = x^2 y^3 z$

8. $f(x, y, z) = xy^2/z$

9. $f(x, y, z) = xy + xz + yz$

10. $f(x, y, z) = xyz$

11. $f(x, y, z) = \dfrac{1}{x^2 + y^2 + z^2}$

12. $f(x, y, z) = \sqrt{x^2 + y^2 + z^2}$

13. $f(x, y) = \sqrt{x^2 + y^2}$. Calculate $f(4, 3)$, $f(3.9, 3.1)$, and $\Delta f = f(3.9, 3.1) - f(4, 3)$. Find df, the differential approximation of Δf.

14. $f(x, y) = x^2 y$. Calculate $f(2, 3)$, $f(1.9, 3.1)$, and $\Delta f = f(1.9, 3.1) - f(2, 3)$. Find df, the differential approximation of Δf.

15. $f(x, y, z) = xyz$. Calculate $f(2, 3, 1)$, $f(2.1, 3.1, 0.9)$, and $\Delta f = f(2.1, 3.1, 0.9) - f(2, 3, 1)$. Find df, the differential approximation of Δf.

16. $f(x, y, z) = 2x + y - z$. Calculate $f(1, 2, 3)$, $f(1.2, 1.9, 3.1)$, and $\Delta f = f(1.2, 1.9, 3.1) - f(1, 2, 3)$. Find df, the differential approximation of Δf.

17. The angle of elevation to the top of a vertical flagpole from a point measured to be 100 ft horizontally from the base of the flagpole is measured to be 30°. Find the height of the flagpole if the measurements are assumed to be exact. Use differentials to find the approximate maximum error in the height due to possible errors of 2 ft in the measurement of the distance and 1° in the measurement of the angle.

18. Two adjacent sides and the included angle of a triangle are measured to be 2 ft, 3 ft, and 60°. Find the area of the triangle if the measurements are assumed to be exact. Use differentials to find the approximate maximum error in the area due to possible errors of 0.15 ft in the measurement of the sides and an error of 2° in the measurement of the angle.

19. Find the total surface area of a right circular cylinder that has radius 2 m and height 3 m. Use differentials to find the approximate maximum error in the surface area due to possible errors of 0.1 m in the measurement of the radius and 0.15 m in the measurement of the height. ($S = 2\pi r^2 + 2\pi rh$.)

20. Find the volume of a right circular cone that has radius of base 3 m and height 6 m. Use differentials to find the approximate maximum error in the volume due to possible errors of 0.2 m in the measurement of the radius and 0.4 m in the measurement of the height. ($V = \pi r^2 h/3$.)

21. Find the approximate maximum percent error in the volume of a rectangular box due to possible errors of 3 percent in the measurement of each of its length, width, and height.

22. Find the approximate maximum percent error in the volume of a rectangular box with square base due to possible errors of 5 percent in the measurement of the length of each side of its square base and its height.

23. Find the approximate maximum percent error in the total surface area of the six sides of the rectangular box of Exercise 21.

24. Find the approximate maximum percent error in the total surface area of the six sides of the rectangular box of Exercise 22.

25. It has been determined experimentally that changing one resistor in a circuit by ΔR_1 with other variables in the circuit fixed results in a change of $0.3\Delta R_1$ in the resistance of the circuit. Changing a second resistor by ΔR_2 with other variables fixed gives a change of $0.5\Delta R_2$ in the resistance of the circuit. What is the approximate change in the resistance of the circuit if the first resistor is increased by 25 Ω while the second resistor is decreased by 12 Ω?

26. The temperature in degrees centigrade at three coordinate points on a flat surface is measured to be $T(0, 0) = 124$, $T(0, 0.42) = 122$, and $T(0.24, 0) = 125$. Find an approximate value of the temperature at the point $(0.30, 0.40)$. On what assumptions is your solution based?

In Exercises 27–32, find an equation of the plane tangent to the graph of the given function at the indicated point on the graph.

27. $f(x, y) = x^2 - y^3$; $(3, 2, 1)$
28. $f(x, y) = (2x + 3y)^2$; $(2, -1, 1)$
29. $f(x, y) = \sqrt{x^2 + y^2}$; $(3, 4, 5)$
30. $f(x, y) = \dfrac{2}{5}(x^2 + y^2)$; $(4, 3, 10)$

31. $f(x, y) = \sqrt{1 - x^2 - y^2}$; $(0, 0, 1)$
32. $f(x, y) = x^2 + y^2 - 8$; $(1, -2, -3)$

In Exercises 33–36, find an equation of the plane tangent to the graph of the given equation at the indicated point on the graph.

33. $x^2 + y^2 + z^2 = 169$; $(3, 4, 12)$
34. $z^2 = x^2 + y^2$; $(3, 2, 5)$
35. $z^2 = xyz - 8$; $(-2, 3 - 4)$
36. $z^3 = xz + yz + xy$; $(1, 2, 2)$

In Exercises 37–40, let $g(t) = f(a + ht, b + kt)$, where f has continuous second partial derivatives in some disk with center (a, b), so f, f_x, and f_y are differentiable and $f_{xy} = f_{yx}$ in the disk.

37. Use the formula for Δf in the definition of the differentiability of f to show
$$g'(t) = f_x(a + ht, b + kt)h + f_y(a + ht, b + kt)k.$$

38. Apply the differentiation formula of Exercises 37 to f_x and f_y to show
$$g''(t) = f_{xx}(a + ht, b + kt)h^2 + f_{xy}(a + ht, b + kt)hk$$
$$+ f_{yy}(a + ht, b + kt)k^2.$$

39. Show that evaluation at $t = 1$ of the first-order Taylor polynomial of g about zero gives the linear approximation of $f(a + h, b + k)$ for $(a + h, b + k)$ near (a, b).

40. Obtain a **second-order approximation** of $f(a + h, b + k)$ for $(a + h, b + k)$ near (a, b) by evaluating the second-order Taylor polynomial of g about zero at $t = 1$.

We can apply Taylor's Formula from Section 3.9 to the function g of the previous exercises to show that there is some number $0 < z < 1$ such that
$$f(a + h, b + k) = f(a, b) + f_x(a, b)h + f_y(a, b)k$$
$$+ \frac{1}{2}(f_{xx}(a + zh, b + zk)h^2$$
$$+ 2f_{xy}(a + zh, b + zk)hk$$
$$+ f_{yy}(a + zh, b + zk)k^2).$$

If $|f_{xx}(x, y)| \le M$, $|f_{xy}(x, y)| \le M$, and $|f_{yy}(x, y)| \le M$ for (x, y) in a region that contains (a, b), then
$$\left| \frac{1}{2}(f_{xx}(a + zh, b + zk)h^2 + 2f_{xy}(a + zh, b + zk)hk \right.$$
$$\left. + f_{yy}(a + zh, b + zk)k^2) \right|$$
$$\le \frac{1}{2}(|f_{xx}(a + zh, b + zk)|h^2$$
$$+ 2|f_{xy}(a + zh, b + zk)||hk|$$
$$+ |f(yy(a + zh, b + zk)|k^2)$$

$$\leq \frac{M}{2}(h^2 + 2|hk| + k^2) = \frac{M}{2}(|h| + |k|)^2$$

whenever $(a + h, b + k)$ *is in the region. Setting* $(x, y) = (a + h, b + k)$, *so* $h = x - a$ *and* $k = b - y$, *we then have*

$$f(x, y) \approx f(a, b) + f_x(a, b)(x - a) + f_y(a, b)(y - b)$$

with error bounded by

$$\frac{M}{2}(|x - a| + |y - b|)^2.$$

In Exercises 41–46, **(a)** *find the linear approximation of the given function for* (x, y) *near the given point* (a, b). **(b)** *Use the above formula to determine a bound of the error in using the linear approximation for* (x, y) *in the indicated region.*

41. $f(x, y) = x^2 + y^2$; $(1, 0)$; $|x - 1| \leq 0.05$, $|y| \leq 0.05$

42. $f(x, y) = x^3 + 2y^3$; $(1, -1)$; $|x - 1| \leq 0.05$, $|y + 1| \leq 0.05$

43. $f(x, y) = xy$; $(0, 0)$; $|x| \leq 0.05$, $|y| \leq 0.05$

44. $f(x, y) = x \cos y$; $(0, 0)$; $|x| \leq 0.005$, $0 \leq y \leq 2\pi$

45. $f(x, y) = \ln \sqrt{x^2 + y^2}$; $(3, 4)$; $|x - 3| \leq 0.05$, $|y - 4| \leq 0.05$

46. $f(x, y) = \sqrt{x^2 + y^2}$; $(3, 4)$; $|x - 3| \leq 0.05$, $|y - 4| \leq 0.05$

12.5 THE CHAIN RULE

Connections

Chain Rule, 3.4.

The Chain Rule is used to differentiate compositions of functions.

The Derivative of $f(x(t), y(t))$

If $z = f(x, y)$ is a function of x and y, and if x and y are functions of t, then $z = f(x(t), y(t))$ is a function of t. We wish to develop a formula for the derivative of z with respect to t. We assume that f is a differentiable function of (x, y) at $(x(c), y(c))$ and that x and y are differentiable functions of t at c.

The derivative of z with respect to t at a point c is

$$\frac{dz}{dt} = \lim_{t \to c} \frac{f(x(t), y(t)) - f(x(c), y(c))}{t - c}.$$

Since $f(x, y)$ is differentiable at $(x(c), y(c))$, we know there are functions ε_1 and ε_2 such that $\varepsilon_1(x, y)$ and $\varepsilon_2(x, y)$ approach zero as (x, y) approaches $(x(c), y(c))$ and such that

$$
\begin{aligned}
f(x(t), y(t)) &- f(x(c), y(c)) \\
&= f_x(x(c), y(c)) \cdot (x(t) - x(c)) + f_y(x(c), y(c)) \cdot (y(t) - y(c)) \\
&\quad + \varepsilon_1(x(t), y(t)) \cdot (x(t) - x(c)) + \varepsilon_2(x(t), y(t)) \cdot (y(t) - y(c)).
\end{aligned}
$$

Dividing the above equation by $t - c$, taking the limit as t approaches c, and noting that $(x(t), y(t)) \to (x(c), y(c))$ as $t \to c$, we see that

$$\frac{dz}{dt} = f_x(x(c), y(c))\frac{dx}{dt}(c) + f_y(x(c), y(c))\frac{dy}{dt}(c).$$

It is useful to write this result in the following form.

If $z = z(x, y)$, $x = x(t)$, and $y = y(t)$, then (assuming appropriate differentiability)

$$\frac{dz}{dt} = \frac{\partial z}{\partial x}\frac{dx}{dt} + \frac{\partial z}{\partial y}\frac{dy}{dt}.$$

EXAMPLE 1

$z = \ln(y/x)$, $x = \cos t$, $y = \sin t$. Find dz/dt.

SOLUTION   We first write the formula for dz/dt and then carefully substitute for each term. (Note that $z = \ln(y/x) = \ln y - \ln x$, so the partial derivatives are easy to evaluate.)

$$\frac{dz}{dt} = \frac{\partial z}{\partial x}\frac{dx}{dt} + \frac{\partial z}{\partial y}\frac{dy}{dt}$$

$$= \left(-\frac{1}{x}\right)(-\sin t) + \left(\frac{1}{y}\right)(\cos t). \quad \blacksquare$$

If we wanted to express the derivative in Example 1 in terms of t, we could substitute to obtain

$$\frac{dz}{dt} = \left(-\frac{1}{\cos t}\right)(-\sin t) + \left(\frac{1}{\sin t}\right)(\cos t) = \tan t + \cot t.$$

We could also substitute before differentiating to obtain

$$z = \ln\left(\frac{\sin t}{\cos t}\right) = \ln(\sin t) - \ln(\cos t).$$

The Chain Rule for functions of one variable would then give us

$$\frac{dz}{dt} = \left(\frac{1}{\sin t}\right)(\cos t) - \left(\frac{1}{\cos t}\right)(-\sin t).$$

This agrees with what we obtained previously.

Partial Derivatives of $z(x(s, t), y(s, t))$

Let us now consider the case $z = z(x, y)$, where x and y are functions of the two variables s and t. Then $z = z(x(s, t), y(s, t))$ is a function of s and t, so we may consider formulas for the two partial derivatives $\partial z/\partial s$ and $\partial z/\partial t$. Since partial derivatives are ordinary derivatives of the function obtained by holding all variables but the variable of partial differentiation constant, we can use the formula that we derived previously. Thus, we have the following results.

If $z = z(x, y)$, $x = x(s, t)$, and $y = y(s, t)$, then (assuming appropriate differentiability)

$$\frac{\partial z}{\partial s} = \frac{\partial z}{\partial x}\frac{\partial x}{\partial s} + \frac{\partial z}{\partial y}\frac{\partial y}{\partial s} \quad \text{and} \quad \frac{\partial z}{\partial t} = \frac{\partial z}{\partial x}\frac{\partial x}{\partial t} + \frac{\partial z}{\partial y}\frac{\partial y}{\partial t}.$$

We have used ∂ to indicate partial derivatives of a function of more than one variable; d indicates differentiation of a function of a single variable.

EXAMPLE 2

$w = x^2 - y^2$, $x = r\cos\theta$, $y = r\sin\theta$. Find $\partial w/\partial r$ and $\partial w/\partial\theta$.

SOLUTION   We write the appropriate formula and substitute for each term.

$$\frac{\partial w}{\partial r} = \frac{\partial w}{\partial x}\frac{\partial x}{\partial r} + \frac{\partial w}{\partial y}\frac{\partial y}{\partial r}$$

$$= (2x)(\cos\theta) + (-2y)(\sin\theta).$$

Similarly,

$$\frac{\partial w}{\partial \theta} = \frac{\partial w}{\partial x}\frac{\partial x}{\partial \theta} + \frac{\partial w}{\partial y}\frac{\partial y}{\partial \theta}$$
$$= (2x)(-r\sin\theta) + (-2y)(r\cos\theta). \quad \blacksquare$$

The General Case

The above examples should serve as a hint that there are many versions of the Chain Rule for functions of several variables. At this point it is worthwhile to consider the general case, so that we will have a common frame of reference for the discussion of individual cases.

Suppose $z = z(x_1, \ldots, x_m)$, where

$$x_1 = x_1(t_1, \ldots, t_n), \ldots, x_m = x_m(t_1, \ldots, t_n).$$

Then z is a function of $(t_1, \ldots, t_n)$. Assuming appropriate differentiability, we have

$$\frac{\partial z}{\partial t_1} = \frac{\partial z}{\partial x_1}\frac{\partial x_1}{\partial t_1} + \cdots + \frac{\partial z}{\partial x_m}\frac{\partial x_m}{\partial t_1},$$
$$\vdots$$
$$\frac{\partial z}{\partial t_n} = \frac{\partial z}{\partial x_1}\frac{\partial x_1}{\partial t_n} + \cdots + \frac{\partial z}{\partial x_m}\frac{\partial x_m}{\partial t_n}.$$

We can say that z is a function of m *dependent* variables $x_1, \ldots, x_m$, and that each dependent variable is a function of n *independent* variables $t_1, \ldots, t_n$. The Chain Rule then gives a formula for the partial derivative of z with respect to each of the n independent variables. Each formula contains m terms, one for each dependent variable. Each term contains two factors. The first factor in each term is a partial derivative of z with respect to a dependent variable, and the second factor is the partial derivative of that dependent variable with respect to the variable of partial differentiation, an independent variable.

Strategy for using the Chain Rule

- Write the function in terms of the dependent variables.
- Write each dependent variable in terms of the independent variables.
- Write out the formula for the derivative you want. There is a Chain Rule formula for the partial derivative with respect to each of the independent variables. Each formula contains a term corresponding to each dependent variable.
- Write each derivative in its proper place in the formula you have written.

EXAMPLE 3

$w = x^2 + y^2 + z^2$, $x = t + 1$, $y = 2t + 1$, $z = 3t + 1$. Find dw/dt.

SOLUTION   We see that w is a function of three dependent variables x, y, and z. Each of x, y, and z is a function of the single independent variable t. Thus, the Chain Rule gives us one formula, for the derivative of w with respect to t.

The formula has three terms, one for each of the three dependent variables. We write the appropriate formula and then substitute for each term.

$$\frac{dw}{dt} = \frac{\partial w}{\partial x}\frac{dx}{dt} + \frac{\partial w}{\partial y}\frac{dy}{dt} + \frac{\partial w}{\partial z}\frac{dz}{dt}$$
$$= (2x)(1) + (2y)(2) + (2z)(3).$$

Note that we used d's in place of ∂'s to indicate when we were taking the derivative of a function of a single variable, not a partial derivative of a function of more than one variable. ∎

We can use the Chain Rule even if we do not have a formula for each function involved. It is only necessary that we know the appropriate values.

EXAMPLE 4

Suppose $w = F(t^2 + 2t + 1, t^3 - t + 2)$, where $F(u, v)$ is differentiable with $F_u(1, 2) = 3$ and $F_v(1, 2) = 4$. Find dw/dt when $t = 0$.

SOLUTION   We first write the functions in the form that we have been using for the Chain Rule. We have

$$w = F(u, v), \quad u = t^2 + 2t + 1, \quad v = t^3 - t + 2.$$

The Chain Rule gives

$$\frac{dw}{dt} = \frac{\partial F}{\partial u}\frac{du}{dt} + \frac{\partial F}{\partial v}\frac{dv}{dt}.$$

When $t = 0$, we have

$$u = (0)^2 + 2(0) + 1 = 1 \text{ and } v = (0)^3 - (0) + 2 = 2.$$

The partial derivatives of F are evaluated at $(u, v) = (1, 2)$ and we are given that $F_u(1, 2) = 3$ and $F_v(1, 2) = 4$. The derivatives of u and v are evaluated at $t = 0$. We have

$$\frac{du}{dt} = 2t + 2 \text{ and } \frac{dv}{dt} = 3t^2 - 1.$$

When $t = 0$,

$$\frac{du}{dt} = 2 \text{ and } \frac{dv}{dt} = -1.$$

Substitution of the values into the formula for dw/dt gives

$$\frac{dw}{dt} = (3)(2) + (4)(-1) = 2. \quad ∎$$

EXAMPLE 5

Suppose $w = x^2 - 3xy + y^2$, where $x(t)$ and $y(t)$ are differentiable with

$$x(2) = -1, \quad y(2) = 2, \quad \frac{dx}{dt}(2) = 3, \text{ and } \frac{dy}{dt}(2) = -2.$$

Find dw/dt when $t = 2$.

SOLUTION We have

$$w = x^2 - 3xy + y^2, \quad x = x(t), \quad y = y(t).$$

The Chain Rule gives

$$\frac{dw}{dt} = \frac{\partial w}{\partial x}\frac{dx}{dt} + \frac{\partial w}{\partial y}\frac{dy}{dt}$$

$$= (2x - 3y)\left(\frac{dx}{dt}\right) + (-3x + 2y)\left(\frac{dy}{dt}\right).$$

When $t = 2$, we are given that $x = -1$, $y = 2$, $dx/dt = 3$, and $dy/dt = -2$. Substitution then gives

$$\frac{dw}{dt} = (2(-1) - 3(2))(3) + (-3(-1) + 2(2))(-2) = -38. \quad \blacksquare$$

Suppose $z = f(x, y)$ and $y = y(x)$, so z is a function of x. Let us find a formula for dz/dx. We can fit this problem into our general scheme by writing

$$z = f(x, y), \quad x = x, \quad y = y(x).$$

We then consider z to be a function of two dependent variables x and y. These variables are functions of the single independent variable x. Thus, we have one formula that has two terms. Namely,

$$\frac{dz}{dx} = \frac{\partial f}{\partial x}\frac{dx}{dx} + \frac{\partial f}{\partial y}\frac{dy}{dx}.$$

Since $dx/dx = 1$, we obtain the formula

$$\frac{dz}{dx} = \frac{\partial f}{\partial x} + \frac{\partial f}{\partial y}\frac{dy}{dx}, \quad \text{where } z = f(x, y) \text{ and } y = y(x). \tag{1}$$

$F(x, y) = d, \ y = y(x)$

Let us suppose the equation $F(x, y) = d$, d a constant, defines y (implicitly) as a differentiable function of x. If F is a differentiable function of (x, y), we can obtain a formula for the derivative of y with respect to x in terms of partial derivatives of F. Differentiating each side of the equation $F(x, y) = d$ with respect to x, and using Formula 1, we obtain

$$\frac{\partial F}{\partial x} + \frac{\partial F}{\partial y}\frac{dy}{dx} = 0.$$

If $\partial F/\partial y \neq 0$, we then have

$$\frac{dy}{dx} = -\frac{\dfrac{\partial F}{\partial x}}{\dfrac{\partial F}{\partial y}} = -\frac{F_x}{F_y}, \quad \text{where } F(x, y) = d \text{ and } y = y(x). \tag{2}$$

If y is given explicitly by an equation $y = f(x)$, we can write the equation in the form $F(x, y) = 0$, where $F(x, y) = f(x) - y$. Equation 2 then gives

$$\frac{dy}{dx} = -\frac{F_x}{F_y} = -\frac{f'(x)}{-1} = f'(x),$$

as it should.

EXAMPLE 6

Assume the equation $x^3 + y^3 = x - y$ defines y as a differentiable function of x. Find dy/dx.

SOLUTION   Let us write $F(x, y) = x^3 + y^3 - x + y$. Our equation can then be expressed as $F(x, y) = 0$. We then use the formula

$$\frac{dy}{dx} = -\frac{F_x}{F_y}$$

to obtain

$$\frac{dy}{dx} = -\frac{3x^2 - 1}{3y^2 + 1}.$$

(This result could also be obtained by the one-variable method of implicit differentiation. That is, we could consider y to be a function of x, differentiate the equation $x^3 + y^3 = x - y$ with respect to x, and then solve the resulting equation for dy/dx. Formula 2 avoids the algebra required for the one-variable method.) ∎

$F(x, y, z) = d,\ z = z(x, y)$

If F is differentiable and the equation $F(x, y, z) = d$, d a constant, defines z as a differentiable function of x and y, we can express the partial derivatives of z with respect to x and y in terms of partial derivatives of F. We consider $F(x, y, z)$ to be a function of the three dependent variables x, y, and z. These variables are functions of the two independent variables x and y. That is, $x = x$, $y = y$, and $z = z(x, y)$. There are then two formulas for partial derivatives of F. Each formula has three terms. Using the Chain Rule, we take partial derivatives of both sides of the equation $F(x, y, z) = d$ to obtain

$$\frac{\partial F}{\partial x}\frac{\partial x}{\partial x} + \frac{\partial F}{\partial y}\frac{\partial y}{\partial x} + \frac{\partial F}{\partial z}\frac{\partial z}{\partial x} = 0 \text{ and}$$

$$\frac{\partial F}{\partial x}\frac{\partial x}{\partial y} + \frac{\partial F}{\partial y}\frac{\partial y}{\partial y} + \frac{\partial F}{\partial z}\frac{\partial z}{\partial y} = 0.$$

The variable y is constant while we are taking the partial derivative with respect to x, so $\partial y/\partial x = 0$. Since $\partial x/\partial x = 1$, the first formula above gives us

$$\frac{\partial F}{\partial x} + \frac{\partial F}{\partial z}\frac{\partial z}{\partial x} = 0.$$

Similarly, the second formula above gives us

$$\frac{\partial F}{\partial y} + \frac{\partial F}{\partial z}\frac{\partial z}{\partial y} = 0.$$

Summarizing, if $\partial F/\partial z \neq 0$, we have

$$\frac{\partial z}{\partial x} = -\frac{\dfrac{\partial F}{\partial x}}{\dfrac{\partial F}{\partial z}} = -\frac{F_x}{F_z}, \qquad \frac{\partial z}{\partial y} = -\frac{\dfrac{\partial F}{\partial y}}{\dfrac{\partial F}{\partial z}} = -\frac{F_y}{F_z}, \tag{3}$$

where $F(x, y, z) = d, \quad z = z(x, y)$.

EXAMPLE 7

$yz + xz + xy = 1$. Find $\partial z/\partial x$, $\partial z/\partial y$, and $\partial^2 z/\partial x^2$.

SOLUTION Our equation is of the form $F(x, y, z) = 1$, where $F(x, y, z) = yz + xz + xy$. We then have

$$\frac{\partial z}{\partial x} = -\frac{F_x}{F_z} = -\frac{z+y}{y+x},$$

$$\frac{\partial z}{\partial y} = -\frac{F_y}{F_z} = -\frac{z+x}{y+x}.$$

The second partial $\partial^2 z/\partial x^2$ is obtained by taking the partial derivative of $\partial z/\partial x$ with respect to x. To do this, we must remember that y is considered to be constant and that z is then considered a function of x. Using the Quotient Rule, we have

$$\frac{\partial^2 z}{\partial x^2} = \frac{\partial}{\partial x}\left(\frac{\partial z}{\partial x}\right) = \frac{\partial}{\partial x}\left(-\frac{z+y}{y+x}\right)$$

$$= -\frac{(y+x)(z_x + 0) - (z+y)(0+1)}{(y+x)^2}.$$

We now substitute the formula for z_x that we obtained above to obtain

$$\frac{\partial^2 z}{\partial x^2} = -\frac{(y+x)\left(-\dfrac{z+y}{y+x}\right) - (z+y)}{(y+x)^2}$$

$$= -\frac{-z-y-z-y}{(y+x)^2} = \frac{2z+2y}{(y+x)^2}. \quad \blacksquare$$

EXERCISES 12.5

Use the appropriate version of the Chain Rule to evaluate the derivatives in Exercises 1–18. (Do not eliminate variables by substitution.)

1. $z = x^2 y^3$, $x = 2t - 1$, $y = 3t + 4$. Find dz/dt.

2. $z = \sqrt{xy}$, $x = t^2 + 1$, $y = t^4 + 1$. Find dz/dt.

3. $z = \ln(x^2 + y^2)$, $x = \cos t$, $y = \sin t$. Find dz/dt.

4. $z = e^{2xy}$, $x = \ln t$, $y = t$. Find dz/dt.

5. $z = x^2 + y^2$, $x = r\cos\theta$, $y = r\sin\theta$. Find $\partial z/\partial r$ and $\partial z/\partial\theta$.

6. $z = x^2 y$, $x = r\cos\theta$, $y = r\sin\theta$. Find $\partial z/\partial r$ and $\partial z/\partial\theta$.

7. $z = r^2 \cos\theta$, $r = \sqrt{x^2 + y^2}$, $\theta = \tan^{-1}(y/x)$. Find $\partial z/\partial x$ and $\partial z/\partial y$.

8. $z = r\sin\theta$, $r = \sqrt{x^2 + y^2}$, $\theta = \tan^{-1}(y/x)$. Find $\partial z/\partial x$ and $\partial z/\partial y$.

9. $w = xy^2/z$, $x = t - 1$, $y = 3t + 2$, $z = 2t + 3$. Find dw/dt.

10. $w = xy + xz + yz$, $x = t^2$, $y = t^3$, $z = 2t$. Find dw/dt.

11. $w = u/v$, $u = x^2 - y^2$, $v = 2xy$. Find $\partial w/\partial x$ and $\partial w/\partial y$.

12. $w = \ln(u + v)$, $u = x^2 + y^2$, $v = 2xy$. Find $\partial w/\partial x$ and $\partial w/\partial y$.

13. $w = x^2 + xy + y^2$, $x = t^2 + 1$, $y = 2t - 1$. Find $\dfrac{dw}{dt}$ when $t = 2$.

14. $w = x^3 + y^3$, $x = \sin t$, $y = \cos t$. Find $\dfrac{dw}{dt}$ when $t = \pi/3$.

15. $w = 3x^2 - 2xy + y^2$, $x = 2r - s$, $y = 3r + 2s$. Find $\dfrac{\partial w}{\partial r}$ and $\dfrac{\partial w}{\partial s}$ when $r = 2$ and $s = -1$.

16. $w = x^2 + 3xy - y^2$, $x = r^2 + s^2$, $y = r^2 - s^2$. Find $\dfrac{\partial w}{\partial r}$ and $\dfrac{\partial w}{\partial s}$ when $r = 3$ and $s = 2$.

17. $w = r \cos \theta$, $r = \sqrt{x^2 + y^2}$, $\theta = \tan^{-1} \frac{y}{x}$. Find $\frac{\partial w}{\partial x}$ and $\frac{\partial w}{\partial y}$ when $x = 3$ and $y = 4$.

18. $w = r \sin \theta$, $r = \sqrt{x^2 + y^2}$, $\theta = \tan^{-1} \frac{y}{x}$. Find $\frac{\partial w}{\partial x}$ and $\frac{\partial w}{\partial y}$ when $x = 3$ and $y = 4$.

19. $x^3 + xy + y^3 = 0$. Find dy/dx.

20. $x^2 + xy^2 + y^4 = 0$. Find dy/dx.

21. $e^{xy} + e^{-xy} = 4$. Find dy/dx.

22. $\ln(x^2 + y^2) = 2xy$. Find dy/dx.

23. $e^{xyz} - x^2 - y^2 - z^2 = 1$. Find $\partial z/\partial x$ and $\partial z/\partial y$.

24. $yx^2 + y^2z + xz^2 = 1$. Find $\partial z/\partial x$ and $\partial z/\partial y$.

25. $x + yz + xz^2 = 0$. Find $\partial z/\partial x$, $\partial z/\partial y$, and $\partial^2 z/\partial x^2$.

26. $xy - yz + xz = 1$. Find $\partial z/\partial x$, $\partial z/\partial y$, and $\partial^2 z/\partial x \partial y$.

27. $w = F(t^2, t^3)$, where $F(u, v)$ is differentiable with $F_u(1, 1) = 5$ and $F_v(1, 1) = 7$. Find dw/dt when $t = 1$.

28. $w = F(3t + 2, 2t + 1)$, where $F(u, v)$ is differentiable with $F_u(2, 1) = -3$ and $F_v(2, 1) = 5$. Find dw/dt when $t = 0$.

29. $w = u^2/v$, where $u(t)$ and $v(t)$ are differentiable with $u(1) = 3$, $v(1) = 2$, $u'(1) = 5$, and $v'(1) = -4$. Find dw/dt when $t = 1$.

30. $w = u^2 v^3$, where $u(t)$ and $v(t)$ are differentiable with $u(2) = 2$, $v(2) = 3$, $u'(2) = -1$, and $v'(2) = 5$. Find dw/dt when $t = 2$.

31. If $y = uv$, $u = u(x)$, $v = v(x)$, show that the formula for dy/dx given by the Chain Rule is the Product Rule.

32. If $y = u/v$, $u = u(x)$, $v = v(x)$, show that the formula for dy/dx given by the Chain Rule is the Quotient Rule.

33. The length, width, and height of a certain rectangular box are functions of time.
(a) Use the Chain Rule for functions of several variables to find the rate of change of the volume of the box in terms of its length, width, height, and their rates of change.
(b) What is the rate of change of the volume of the box when the length is 3 ft, the width is 2 ft, and the height

is 1 ft, if the length and width are increasing at a rate of 1 ft/s while the height is decreasing at a rate of 1 ft/s?

34. Two sides and the included angle of a certain triangle are functions of time.
(a) Use the Chain Rule for functions of several variables to find the rate of change of the area of the triangle in terms of the lengths of the two sides, the included angle, and their rates of change.
(b) If the two sides of the triangle are each increasing at a rate of 2 cm/s while the included angle is decreasing at a rate of 5°/s, what is the rate of change of the area of the triangle when one of the sides is 30 cm, the other side is 20 cm, and the included angle is 30°?

Suppose the equations in Exercises 35–38 define u and v as differentiable functions of (x, y). Find formulas for $\frac{\partial u}{\partial x}$, $\frac{\partial v}{\partial x}$, $\frac{\partial u}{\partial y}$, and $\frac{\partial v}{\partial y}$.

35. $x + y + 2u + 3v = 4$, $2x - y + u - 3v = 3$

36. $2x - 3y + u + 2v = 3$, $3x + 2y - u + 2v = 1$

37. $x^2 + y^2 + u^2 + v^2 = 4$, $x + y + u + v = 1$

38. $x^2 + y^2 + u^2 + v^2 = 4$, $2x + 3y - 3u - 2v = 0$

39. Suppose the functions $u = u(x, y)$ and $v = v(x, y)$ satisfy the equation $F(x, y, u, v) = 0$. Find a formula that relates partial derivatives of F and the partial derivatives of u and v with respect to **(a)** x and **(b)** y.

40. Suppose the equations $F(x, y, u, v) = 0$ and $G(x, y, u, v) = 0$ define u and v as differential functions of (x, y) and that $F_u G_v - G_u F_v \neq 0$.
(a) Find formulas for $\frac{\partial u}{\partial x}$ and $\frac{\partial v}{\partial x}$ in terms of partial derivatives of F and G by solving the system of linear equations obtained by taking the partial derivative with respect to x of each of the two equations.
(b) Find formulas for $\frac{\partial u}{\partial y}$ and $\frac{\partial v}{\partial y}$ in terms of partial derivatives of F and G.

12.6 DIRECTIONAL DERIVATIVES AND THE GRADIENT

Connections

Vectors, 11.3–4.

Partial derivatives, 12.3.

The first partial derivatives $f_x(a, b)$ and $f_y(a, b)$ may be considered to be the rate of change of $f(x, y)$ at the point (a, b) in the direction of the x-axis and y-axis, respectively. We will extend this idea to include the rate of change of f in any direction.

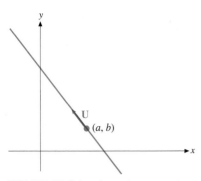

FIGURE 12.6.1 The unit vector **U** determines the direction and scale of a number line through the point (a, b).

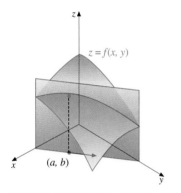

FIGURE 12.6.2 The directional derivative of $f(x, y)$ in the direction of the unit vector **U** at (a, b) gives the slope of the curve of intersection of the surface $z = f(x, y)$ and the vertical plane that contains the vector **U** with its initial point at the point (a, b) in the xy-plane.

Directional Derivatives

Suppose $\mathbf{U} = u_1\mathbf{i} + u_2\mathbf{j}$ is a unit vector, so $u_1^2 + u_2^2 = 1$. We use the vector **U** to determine the direction and scale of a line through (a, b). This line has parametric equations

$$x = a + u_1 t, \quad y = b + u_2 t.$$

The point on this line where $t = 0$ has coordinates (a, b). The positive direction of the line is given by the direction of **U**. The fact that **U** is a unit vector implies that distance along the line in units of t agrees with distance in terms of (x, y) coordinates. That is, this is a parametrization of the line in terms of arc length. See Figure 12.6.1.

We now consider the function of the real variable t defined by the equations

$$z = f(x, y), \quad x = a + u_1 t, \quad y = b + u_2 t.$$

The derivative of z with respect to t at $t = 0$, if it exists, is called the **directional derivative** of f at (a, b) in the direction of the unit vector **U**, and is denoted $D_{\mathbf{U}} f(a, b)$. $D_{\mathbf{U}} f(a, b)$ may be interpreted as the rate of change of f at (a, b) in the direction of **U**. See Figure 12.6.2.

If f is a differentiable function of (x, y) and

$$z = f(x, y), \quad x = a + u_1 t, \quad y = b + u_2 t,$$

then z is a differentiable function of t and the Chain Rule gives

$$\frac{dz}{dt} = \frac{\partial f}{\partial x}\frac{dx}{dt} + \frac{\partial f}{\partial y}\frac{dy}{dt}$$
$$= f_x(x, y)u_1 + f_y(x, y)u_2.$$

When $t = 0$, we have $x = a$, $y = b$, and

$$D_{\mathbf{U}}(a, b) = f_x(a, b)u_1 + f_y(a, b)u_2. \tag{1}$$

If $\mathbf{U} = \mathbf{i}$, then Formula 1 with $u_1 = 1$ and $u_2 = 0$ gives $D_{\mathbf{i}} f = f_x$. Similarly, $D_{\mathbf{j}} f = f_y$, so the directional derivatives of f in the directions of the coordinate axes are partial derivatives of f.

The Gradient

The vector

$$\nabla f = f_x\mathbf{i} + f_y\mathbf{j}$$

is called the **gradient** of $f(x, y)$. The gradient of $f(x, y)$ is a vector-valued function of (x, y). **Grad** f is also used to denote the gradient.

It follows from (1) that the directional derivative of $f(x, y)$ in the direction of a *unit* vector **U** can be expressed as

$$D_{\mathbf{U}} f = \nabla f \cdot \mathbf{U}. \tag{2}$$

Ordinarily, we will use Formula 2 to evaluate directional derivatives.

EXAMPLE 1

$f(x, y) = x^2 + y^2$. Evaluate the directional derivative of f at $(3, 2)$ in the direction of the vector $4\mathbf{i} - 3\mathbf{j}$.

SOLUTION   We have

$$\nabla f = f_x \mathbf{i} + f_y \mathbf{j} = 2x\mathbf{i} + 2y\mathbf{j}.$$

When $(x, y) = (3, 2)$,

$$\nabla f = 2(3)\mathbf{i} + 2(2)\mathbf{j} = 6\mathbf{i} + 4\mathbf{j}.$$

We need a *unit* vector in the direction of $4\mathbf{i} - 3\mathbf{j}$. We have

$$|4\mathbf{i} - 3\mathbf{j}| = \sqrt{(4)^2 + (-3)^2} = 5,$$

so we take

$$\mathbf{U} = \frac{4\mathbf{i} - 3\mathbf{j}}{|4\mathbf{i} - 3\mathbf{j}|} = \frac{4}{5}\mathbf{i} - \frac{3}{5}\mathbf{j}.$$

Then

$$\begin{aligned}
D_\mathbf{U} f(3, 2) &= \nabla f \cdot \mathbf{U} \\
&= (6\mathbf{i} + 4\mathbf{j}) \cdot \left(\frac{4}{5}\mathbf{i} - \frac{3}{5}\mathbf{j}\right) \\
&= (6)\left(\frac{4}{5}\right) + (4)\left(-\frac{3}{5}\right) = \frac{12}{5}. \quad \blacksquare
\end{aligned}$$

Maximum and Minimum Rates of Change

We know that the dot product of two vectors can be expressed in terms of the angle between the vectors. We can use this fact to obtain information about the directional derivatives of a function $f(x, y)$. If $\mathbf{U}$ is a unit vector, we have

$$D_\mathbf{U} f = \nabla f \cdot \mathbf{U} = |\nabla f||\mathbf{U}| \cos\theta = |\nabla f| \cos\theta,$$

where θ is the angle between ∇f and $\mathbf{U}$. Since $-1 \le \cos\theta \le 1$, we have

$$-|\nabla f| \le D_\mathbf{U} f \le |\nabla f|.$$

The maximum value of the directional derivative is $|\nabla f|$, obtained when $\theta = 0$. When $\theta = 0$ the angle between ∇f and $\mathbf{U}$ is zero, so $\mathbf{U}$ is a unit vector in the direction of the gradient. The minimum value of the directional derivative is $-|\nabla f|$, obtained when $\mathbf{U}$ is a unit vector in the direction opposite to that of the gradient. This gives the following.

THEOREM 1

If f is a differentiable function of (x, y), then the directional derivative has maximum value $|\nabla f|$ in the direction of ∇f and the directional derivative has minimum value $-|\nabla f|$ in the direction of $-\nabla f$.

Note that, at points where $\nabla f = \mathbf{0}$, ∇f does not have a direction and the directional derivatives at the point are zero in all directions.

EXAMPLE 2

$f(x, y) = xy^2 - 9y$. Find a unit vector $\mathbf{U}$ such that the rate of change of f at $(3, 2)$ in the direction of $\mathbf{U}$ is maximum. Evaluate the rate of change of f at $(3, 2)$ in the direction of that vector $\mathbf{U}$.

SOLUTION Theorem 1 tells us that the maximum rate of change of f is in the direction of ∇f, and that the maximum rate of change is $|\nabla f|$. We have

$$f(x, y) = xy^2 - 9y, \text{ so}$$
$$\nabla f = f_x\mathbf{i} + f_y\mathbf{j} = y^2\mathbf{i} + (2xy - 9)\mathbf{j}.$$

When $(x, y) = (3, 2)$,

$$\nabla f = (2)^2\mathbf{i} + (2(3)(2) - 9)\mathbf{j} = 4\mathbf{i} + 3\mathbf{j} \text{ and}$$
$$|\nabla f| = \sqrt{4^2 + 3^2} = 5.$$

The unit vector in the direction of ∇f is

$$\mathbf{U} = \frac{\nabla f}{|\nabla f|} = \frac{4\mathbf{i} + 3\mathbf{j}}{5} = 0.8\mathbf{i} + 0.6\mathbf{j}.$$

The rate of change of f at $(3, 2)$ is maximum in the direction of $\mathbf{U}$, and the maximum rate of change is

$$D_{\mathbf{U}} f = |\nabla f| = 5. \quad \blacksquare$$

EXAMPLE 3

A marble is set on the surface $z = e^{-2x-3y}$ at the point $(0, 0, 1)$. Along what direction in the xy-plane will the marble roll?

SOLUTION The surface is pictured in Figure 12.6.3. We assume that the marble will roll in the direction in which the surface is falling away at the fastest rate. That is, the marble will roll in the direction in which the rate of change of z is minimum. This is the direction opposite to ∇z, namely, in the direction of $-\nabla z$. We have

$$z_x = e^{-2x-3y}(-2) \text{ and } z_y = e^{-2x-3y}(-3).$$

When $x = 0$ and $y = 0$, the gradient of z is

$$\nabla z = z_x\mathbf{i} + z_y\mathbf{i} = -2\mathbf{i} - 3\mathbf{j}.$$

The marble will roll along the surface in the direction in the xy-plane of the vector

$$-\nabla z = 2\mathbf{i} + 3\mathbf{j}. \quad \blacksquare$$

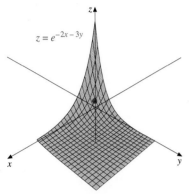

$z = e^{-2x-3y}$

FIGURE 12.6.3 The marble will roll in the direction in which the surface is falling away at the fastest rate. This is the direction in the xy-plane for which the directional derivative of $f(x, y)$ is minimum.

The ideas of gradient and directional derivatives extend to any number of variables. For example, we define the gradient of a function F of three variables to be the vector

$$\nabla F(x, y, z) = F_x(x, y, z)\mathbf{i} + F_y(x, y, z)\mathbf{j} + F_z(x, y, z)\mathbf{k}.$$

Generally, the gradient of a function of n variables is a vector that has n components, each component being a partial derivative of F. The gradient of F is a vector-valued function of the same variables as is F.

If F is a differentiable function of any number of variables, the directional derivative of F in the direction of a unit vector $\mathbf{U}$ is given by the dot product

$$D_{\mathbf{U}} F = \nabla F \cdot \mathbf{U}, \qquad |\mathbf{U}| = 1.$$

$\mathbf{U}$ must have the same number of components as the gradient of F.

The analogue of Theorem 1 is true for any number of variables.

THEOREM 2

If F is a differentiable function of $(x_1, \ldots, x_n)$, the directional derivative has maximum value $|\nabla F|$ in the direction of ∇F; the directional derivative has minimum value $-|\nabla F|$ in the direction of $-\nabla F$.

EXAMPLE 4

$F(x, y, z) = x^2 + y^2 + xyz$. Find a unit vector $\mathbf{U}$ such that the rate of change of F at $(2, 3, -1)$ in the direction of $\mathbf{U}$ is maximum. Evaluate $D_{\mathbf{U}} F$ for that vector $\mathbf{U}$.

SOLUTION   We know that the value of the directional derivative is maximum in the direction of the gradient. We have

$$\nabla F = F_x \mathbf{i} + F_y \mathbf{j} + F_z \mathbf{k} = (2x + yz)\mathbf{i} + (2y + xz)\mathbf{j} + (xy)\mathbf{k}.$$

When $(x, y, z) = (2, 3, -1)$, the gradient is

$$\nabla F = \mathbf{i} + 4\mathbf{j} + 6\mathbf{k}.$$

A unit vector in the direction of the gradient at $(2, 3, -1)$ is

$$\mathbf{U} = \frac{\nabla F}{|\nabla F|} = \frac{\mathbf{i} + 4\mathbf{j} + 6\mathbf{k}}{\sqrt{53}}.$$

The maximum rate of increase of F at $(2, 3, -1)$ is

$$D_{\mathbf{U}} F = |\nabla F| = \sqrt{53}. \quad \blacksquare$$

In case $z = f(x)$ is a function of one variable, the gradient of f is a vector with one component. Namely, $\nabla f = f'\mathbf{i}$. This vector points in the positive direction when f' is positive and in the negative direction when f' is negative. In each case, the gradient points in the direction in which f is increasing, as indicated in Theorem 2. See Figure 12.6.4.

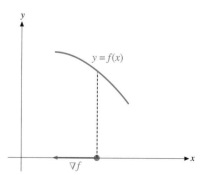

FIGURE 12.6.4a   If $f(x)$ is a function of one variable, then $\nabla f = f'\mathbf{i}$. If $f'(x)$ is positive, the values $f(x)$ increase as x increases and the gradient points in the direction of increasing x.

FIGURE 12.6.4b   If $f'(x)$ is negative, the values $f(x)$ increase as x decreases and the gradient points in the direction of decreasing x.

EXERCISES 12.6

In Exercises 1–6, evaluate the directional derivative of the given function at the given point in the direction of the indicated vector.

1. $f(x, y) = xy^2$; $(3, 2)$; $4\mathbf{i} + 3\mathbf{j}$
2. $f(x, y) = x^2 - 4xy + y^2$; $(-1, 2)$; $3\mathbf{i} - 4\mathbf{j}$
3. $f(x, y) = e^{xy}$; $(0, 2)$; $\mathbf{i}$
4. $f(x, y) = \ln(2x + +3y)$; $(-1, 1)$; $3\mathbf{i} + 2\mathbf{j}$
5. $f(x, y, z) = x^2 y^3 z$; $(2, -1, 3)$; $\mathbf{i} - 2\mathbf{j} - 2\mathbf{k}$
6. $f(x, y, z) = x^2 + y^2 + z^2$; $(1, -1, 2)$; $2\mathbf{i} - 2\mathbf{j} + \mathbf{k}$

In Exercises 7–14, find a unit vector $\mathbf{U}$ such that the rate of change of f in the direction of $\mathbf{U}$ at the given point is maximum and evaluate $D_{\mathbf{U}} f$ at that point.

7. $f(x, y) = 2x - x^2$; $(3, 1)$
8. $f(x, y) = 8 - y^3$; $(2, -1)$
9. $f(x, y) = x^2 - 3xy - y^2$; $(1, 2)$
10. $f(x, y) = x^2 y + 4xy$; $(1, -1)$
11. $f(x, y, z) = e^{xyz}$; $(2, 3, 0)$
12. $f(x, y, z) = \sin xyz$; $(3, 5, \pi)$
13. $f(x, y, z) = x^2 + y^2 + z^2 - xyz$; $(2, -1, 3)$
14. $f(x, y, z) = x^2 yz + xy^2 z + xyz^2$; $(-1, 2, 3)$

In Exercises 15–18, find two unit vectors $\mathbf{U}$ such that the rate of change of f in the direction of $\mathbf{U}$ at the indicated point is zero.

15. $f(x, y) = 3x - 4y$; $(2, -3)$
16. $f(x, y) = x^2 - 3xy + y^2$; $(3, 2)$
17. $f(x, y) = x^2 y^3$; $(2, -1)$
18. $f(x, y) = \sqrt{x^2 + y^2}$; $(3, 4)$
19. If $D_{(\mathbf{i}-\mathbf{j})/\sqrt{2}} f(a, b) = 1$ and $D_{(\mathbf{i}+\mathbf{j})/\sqrt{2}} f(a, b) = 3$, what is $\nabla f(a, b)$?
20. If $D_{(\mathbf{i}+\mathbf{j})/\sqrt{2}} f(a, b) = 2/\sqrt{2}$ and $D_{(3\mathbf{i}+4\mathbf{j})/5} f(a, b) = 2$, what is $\nabla f(a, b)$?

21. A rectangle has length $\ell = 4$ and width $w = 3$. **(a)** Which will change the area more, a small change in the length or the same change in the width? **(b)** For what ratio $\dfrac{\Delta \ell}{\Delta w}$ will the area be approximately unchanged?

22. A right circular cylinder has radius $r = 3$ and height $h = 2$. **(a)** Which will change the volume of the cylinder more, a small change in the radius or the same change in the height? **(b)** For what ratio $\dfrac{\Delta r}{\Delta h}$ will the volume be approximately unchanged?

23. Suppose (x, y) changes a distance δ in the direction of the unit vector $\mathbf{U}$, so $dx\mathbf{i} + dy\mathbf{j} = \delta \mathbf{U}$. Show that $df = \delta \cdot D_{\mathbf{U}} f$, so the expression $\delta \cdot D_{\mathbf{U}} f$ gives the approximate change of f.

24. If $D_{\mathbf{U}} f = 3.2$, what is the approximate change in $f(x, y)$ if (x, y) changes a distance of 0.15 in the direction of the unit vector $\mathbf{U}$?

In Exercises 25–28, find the directional derivative of the given function $f(x, y)$ in the direction of the unit tangent vector of the given curve at the given point.

25. $f(x, y) = x^2 y$; $C : x = 2\cos t$, $y = 2\sin t$; $t = \pi/3$
26. $f(x, y) = x^2 + y^2$; $x = \cos t$, $y = \sin t$; $t = 5\pi/6$
27. $f(x, y) = y^2 - x^2$; $C : x = t$, $y = t^2$; $t = \sqrt{2}$
28. $f(x, y) = x^2 - xy + y^2$; $C : x = t^2 + t$, $y = t^2 - t$; $t = 2$
29. Find the rate of change of $f(x, y) = xy^2 - yx^2$ at (x, y) in the direction of the origin.
30. Find the rate of change of $f(x, y) = x^2 y + x - 2y + xy^2$ as (x, y) moves directly away from the origin.
31. Show that $\nabla(kf) = k\nabla f$, k a constant.
32. Show that $\nabla(fg) = f\nabla g + g\nabla f$.

12.7 NORMAL LINES AND TANGENT PLANES

Connections

Planes, 11.5.

Lines in space, 11.6.

Chain Rule, 12.5.

In this section we will see how the gradient can be used to find lines normal to plane curves and surfaces. The results are particularly useful for curves and surfaces that are defined implicitly.

Lines Normal to Plane Curves

The values of a function are constant along any level curve of the function, so we should expect the directional derivative in the direction of a level curve to be zero. Theorem 1 is the technical tool that allows us to verify that this is true.

> **THEOREM 1**
>
> Suppose that $f(x, y)$ is differentiable at (a, b),
>
> $$C: \quad x = x(t), \quad y = y(t)$$
>
> is a curve with $(x(t_0), y(t_0)) = (a, b)$, and x and y are differentiable at t_0 with $(x'(t_0))^2 + (y'(t_0))^2 = 1$, so the unit tangent vector to C at (a, b) is $\mathbf{T} = x'(t_0)\mathbf{i} + y'(t_0)\mathbf{j}$. Then
>
> $$\lim_{t \to t_0} \frac{f(x(t), y(t)) - f(a, b)}{t - t_0} = D_{\mathbf{T}} f(a, b).$$

Proof   We first note that

$$\lim_{t \to t_0} \frac{f(x(t), y(t)) - f(a, b)}{t - t_0} = \frac{dz}{dt}(t_0),$$

where

$$z = f(x, y), \quad x = x(t), \quad y = y(t).$$

The Chain Rule then gives

$$\frac{dz}{dt}(t_0) = f_x(x(t_0), y(t_0))\frac{dx}{dt}(t_0) + f_y(x(t_0), y(t_0))\frac{dy}{dt}(t_0)$$

$$= f_x(a, b)x'(t_0) + f_y(a, b)y'(t_0)$$

$$= \nabla f(a, b) \cdot \mathbf{T} = D_{\mathbf{T}} f(a, b). \quad \blacksquare$$

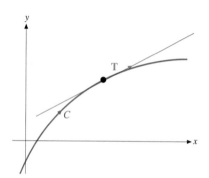

FIGURE 12.7.1   The values of f along any curve with unit tangent vector $\mathbf{T}$ at (a, b) determine the value of the directional derivative $D_{\mathbf{T}} f(a, b)$.

Note that $D_{\mathbf{T}} f(a, b)$ depends only on the values $f(x, y)$ for (x, y) on the *line* through (a, b) in the direction of $\mathbf{T}$. Theorem 1 tells us that we can evaluate the directional derivative from values of $f(x, y)$ for (x, y) on any curve with unit tangent vector $\mathbf{T}$ at (a, b). See Figure 12.7.1.

If the curve C in Theorem 1 is a level curve of f, so $f(x(t), y(t))$ is constant $(= f(a, b))$ for all values of t near t_0, then the limit in Theorem 1 is zero. This implies $D_{\mathbf{T}} f = \nabla f \cdot \mathbf{T} = 0$. This gives the following results. See Figure 12.7.2.

> **THEOREM 2**
>
> If $f(x, y)$ is differentiable at (a, b) and the level curve of f through (a, b) has a unit tangent vector at (a, b), then:
> **(a)** The directional derivative of f at (a, b) is zero in the direction of the level curve of $f(x, y)$ through (a, b);
> **(b)** The gradient $\nabla f(a, b)$ is orthogonal to the level curve of $f(x, y)$ through (a, b).

FIGURE 12.7.2   The values $f(x, y)$ are constant along level curves of f, so we should expect the directional derivative of f in the direction of a level curve to be zero. Since $D_{\mathbf{T}} f = \nabla f \cdot \mathbf{T} = 0$, the gradient is orthogonal to the level curve of f at (a, b).

EXAMPLE 1

$f(x, y) = xy$. Sketch the level curves $f(x, y) = z_0$, for $z_0 = 1, 2, 3$. Evaluate $\nabla f(1, 2)$ and sketch that vector with initial position at $(1, 2)$. Evaluate the directional derivative of f at $(1, 2)$ in the direction of a unit vector in the direction of ∇f. Sketch the unit vector $\mathbf{U} = (1/\sqrt{5})\mathbf{i} + (-2/\sqrt{5})\mathbf{j}$ with initial position at $(1, 2)$ and evaluate $D_{\mathbf{U}} f(1, 2)$.

SOLUTION   We have $f(x, y) = xy$, so $f_x = y$ and $f_y = x$. The gradient is $\nabla f = y\mathbf{i} + x\mathbf{j}$. When $(x, y) = (1, 2)$, we have $\nabla f(1, 2) = 2\mathbf{i} + \mathbf{j}$.

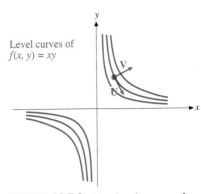

FIGURE 12.7.3 The level curves of $f(x, y) = xy$ are hyperbolas. The unit vector **U** is tangent to the level curve $xy = 2$, so $D_{\mathbf{U}}f = 0$. The gradient points in the direction of greatest increase of f.

A unit vector in the direction of $\nabla f(1, 2)$ is given by

$$\mathbf{V} = \frac{\nabla f}{|\nabla f|} = \frac{2\mathbf{i} + \mathbf{j}}{\sqrt{5}} = \frac{2}{\sqrt{5}}\mathbf{i} + \frac{1}{\sqrt{5}}\mathbf{j}.$$

The directional derivative of f in the direction of a unit vector in the direction of ∇f at $(1, 2)$ is

$$D_{\mathbf{V}}f = \nabla f \cdot \mathbf{V} = |\nabla f| = \sqrt{5}.$$

The directional derivative of f in the direction of the unit vector $\mathbf{U} = (1/\sqrt{5})\mathbf{i} - (2/\sqrt{5})\mathbf{j}$ is

$$D_{\mathbf{U}}f = \nabla f \cdot \mathbf{U} = (2)\left(\frac{1}{\sqrt{5}}\right) + (1)\left(\frac{-2}{\sqrt{5}}\right) = 0.$$

The three level curves and the vectors **V** and **U** are sketched in Figure 12.7.3. The level curves are hyperbolas. We see that the vector **U** is tangent to the level curve $xy = 2$ at $(1, 2)$; the directional derivative of f in the direction of the level curve is zero. We also see that the gradient of f is perpendicular to the level curve $xy = 2$ at $(1, 2)$. Also, the gradient points toward the level curve $xy = 3$, where the values of $f(x, y)$ are larger than the value $f(1, 2) = 2$. We know from Section 12.6 that ∇f points in the direction of greatest increase of f. ■

Since we can think of a curve that is defined by an equation $f(x, y) = d$ as a level curve of $f(x, y)$, Theorem 2b implies the following.

THEOREM 3

If $f(x, y)$ is differentiable and the equation $f(x, y) = d$ defines a curve that has a unit tangent vector at (a, b), then the gradient $\nabla f(a, b)$ is orthogonal to the curve $f(x, y) = d$ at (a, b).

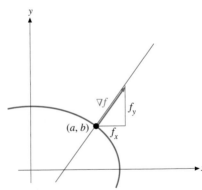

FIGURE 12.7.4 The line through (a, b) in the direction of the vector $\nabla f(a, b)$ has slope $f_y(a, b)/f_x(a, b)$.

Note that the line through (a, b) with nonzero direction vector $\nabla f(a, b) = f_x(a, b)\mathbf{i} + f_y(a, b)\mathbf{j}$ can be expressed in parametric form as

$$C: \quad x = a + f_x(a, b)t, \quad y = b + f_y(a, b)t.$$

If $f_x(a, b) \neq 0$, the line has slope $f_y(a, b)/f_x(a, b)$. If $f_x(a, b) = 0$, the line is vertical. See Figure 12.7.4.

Theorem 3 tells us that the line normal to a curve $f(x, y) = d$ has slope f_y/f_x. The slope of the line tangent to the curve then has slope that is the negative reciprocal, $-f_x/f_y$. We have seen in Formula 2 of Section 12.5 that if the equation $f(x, y) = d$ defines y as a differentiable function of x, then $dy/dx = -f_x/f_y$, so the slope of the tangent line given by Theorem 3 agrees with that given in Section 12.5 in case $f_y \neq 0$. Theorem 3 includes cases where $f_y = 0$, but $\nabla f \neq 0$.

EXAMPLE 2

Find an equation of the line that is normal to the curve $x^2 = y^2 + 5$ at the point $(3, 2)$.

SOLUTION We write the equation of the curve in the form $f(x, y) = 5$, where

$$f(x, y) = x^2 - y^2.$$

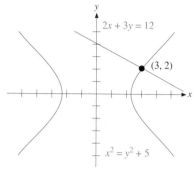

FIGURE 12.7.5 The line $2x + 3y = 12$ is normal to the curve $x^2 = y^2 + 5$ at the point $(3, 2)$.

We then know from Theorem 3 that the gradient of f at $(3, 2)$ is orthogonal to the curve at that point. We have

$$\nabla f = f_x \mathbf{i} + f_y \mathbf{j} = 2x\mathbf{i} - 2y\mathbf{j}.$$

When $(x, y) = (3, 2)$,

$$\nabla f = 2(3)\mathbf{i} - 2(2)\mathbf{j} = 6\mathbf{i} - 4\mathbf{j}.$$

The direction of the gradient gives the slope of the line that is normal to the curve at $(3, 2)$. This slope is

$$\frac{\text{Rise}}{\text{Run}} = \frac{\mathbf{j}\text{-component of } \nabla f}{\mathbf{i}\text{-component of } \nabla f} = \frac{-4}{6} = -\frac{2}{3}.$$

An equation of the line through the point $(3, 2)$ with slope $-2/3$ is

$$\begin{aligned}
y - 2 &= -\tfrac{2}{3}(x - 3), \\
3(y - 2) &= -2(x - 3), \\
3y - 6 &= -2x + 6, \\
2x + 3y &= 12.
\end{aligned}$$

This is an equation of the line that is normal to the curve at the point $(3, 2)$. See Figure 12.7.5. ∎

Planes Tangent to Surfaces $F(x, y, z) = d$

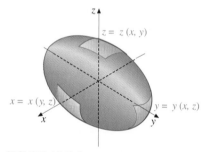

FIGURE 12.7.6 Various parts of a surface may define one variable as a function of the other two variables.

FIGURE 12.7.7 The surface $x^2 + y^2 - z^2 = 0$ does not define any one variable as a function of the other two in any open rectangular box with center the origin.

Various parts of a surface in three-dimensional space may define one variable as a function of the other two variables in a manner analogous to the way parts of a curve in the plane may define one variable as a function of the other variable. For example, a surface S defines z as a function of (x, y) in a rectangular box with center at a point on the surface if each line parallel to the z-axis that intersects the box intersects the surface at exactly one point. See Figure 12.7.6.

If F has continuous first partial derivatives and $\nabla F(a, b, c) \neq \mathbf{0}$, then it is true that the equation $F(x, y, z) = d$ defines one of the three variables x, y, z as a differentiable function of the other two for (x, y, z) in some open rectangular box with center (a, b, c). We can then define the **plane tangent to the surface** $F(x, y, z) = d$ as the plane tangent to the graph of a differentiable function of two variables, as was done in Section 12.4.

Note that a surface $F(x, y, z) = d$ may not define any one of the variables as a function of the other two in any rectangle with center a point on the surface where $\nabla F = \mathbf{0}$. For example, the function $F(x, y, z) = x^2 + y^2 - z^2$ has continuous partial derivatives, but the surface $x^2 + y^2 - z^2 = 0$ does not define any one of the variables as a function of the other two in any open rectangular box with center the origin, where $\nabla F = \mathbf{0}$. See Figure 12.7.7.

The analogue of Theorem 2 holds for any number of variables and gives the following result for functions of three variables.

THEOREM 4

If $F(x, y, z)$ is differentiable at (a, b, c) and the level surface of F through (a, b, c) has a tangent plane at (a, b, c), then:

(a) The directional derivative of F at (a, b, c) is zero in the direction of any vector in the plane tangent to the level surface of F at (a, b, c);

(b) The gradient $\nabla F(a, b, c)$ is orthogonal to the level surface of $F(x, y, z)$ through (a, b, c).

Since we can think of the surface $F(x, y, z) = d$ as a level surface of $F(x, y, z)$, Theorem 4 gives the following result.

THEOREM 5

If $F(x, y, z)$ is differentiable at (a, b, c), the gradient $\nabla F(a, b, c)$ is orthogonal to the surface $F(x, y, z) = d$ at each point (a, b, c) on the surface that has a tangent plane. An equation of the plane tangent to the surface at (a, b, c) is

$$F_x(a, b, c)(x - a) + F_y(a, b, c)(y - b) + F_z(a, b, c)(z - a) = 0.$$

If F is differentiable and the surface $F(x, y, z) = d$ defines z as a differentiable function of (x, y), we know from Section 12.4 that the plane tangent to the surface $z = z(x, y)$ at $(a, b, z(a, b))$ has equation

$$z = z(a, b) + z_x(a, b)(x - a) + z_y(a, b)(y - b).$$

If $F_z(a, b, c) \neq 0$, we know from Formula 3 of Section 12.5 that

$$z_x = -\frac{F_x}{F_z} \quad \text{and} \quad z_y = -\frac{F_y}{F_z}.$$

Combining results, we obtain the equation

$$z = z(a, b) - \frac{F_x(a, b)}{F_z(a, b)}(x - a) - \frac{F_y(a, b)}{F_z(a, b)}(y - b),$$

$$\frac{F_x(a, b)}{F_z(a, b)}(x - a) + \frac{F_y(a, b)}{F_z(a, b)}(y - b) + z - z(a, b) = 0,$$

$$F_x(a, b)(x - a) + F_y(a, b)(y - b) + F_z(a, b)(z - c) = 0,$$

which agrees with the result of Theorem 5 in this case. Theorem 5 gives an equation of the tangent plane in case any one of the variables is a differentiable function of the other two.

EXAMPLE 3

Find an equation of the plane tangent to the surface $z^2 = x^2 + y^2$ at the point $(3, 4, 5)$.

SOLUTION   Let us express the surface as $F(x, y, z) = 0$, where

$$F(x, y, z) = z^2 - x^2 - y^2.$$

The surface has normal vector

$$\nabla F = F_x \mathbf{i} + F_y \mathbf{j} + F_z \mathbf{k} = -2x\mathbf{i} - 2y\mathbf{j} + 2z\mathbf{k}.$$

When $(x, y, z) = (3, 4, 5)$, we have

$$\nabla F = -6\mathbf{i} - 8\mathbf{j} + 10\mathbf{k}.$$

The fact that ∇F is normal to the surface at the point $(3, 4, 5)$ means that it is normal to the tangent plane at that point. It follows that an equation of the tangent plane is

$$(-6)(x - 3) + (-8)(y - 4) + (10)(z - 5) = 0.$$

Simplifying the equation, we obtain

$$
\begin{aligned}
-3(x - 3) - 4(y - 4) + 5(z - 5) &= 0, \\
-3x + 9 - 4y + 16 + 5z - 25 &= 0, \\
3x + 4y - 5z &= 0. \quad \blacksquare
\end{aligned}
$$

EXAMPLE 4

Find symmetric equations and parametric equations of the line that is normal to the surface $z^2 = x^2 + y^2 + 4$ at the point $(2, -1, 3)$.

SOLUTION   The equation of the surface can be written as $F(x, y, z) = 4$, where

$$F(x, y, z) = z^2 - x^2 - y^2.$$

The gradient of F is then a vector that is normal to the surface, so the gradient is in the direction of the normal line. We have

$$\nabla F = -2x\mathbf{i} - 2y\mathbf{j} + 2z\mathbf{k}.$$

When $(x, y, z) = (2, -1, 3)$, we have

$$\nabla F = -4\mathbf{i} + 2\mathbf{j} + 6\mathbf{k}.$$

Symmetric equations of the normal line are

$$\frac{x - 2}{-4} = \frac{y + 1}{2} = \frac{z - 3}{6}.$$

We can obtain parametric equations of the line by setting each term in the symmetric equations equal to a parameter t. This gives

$$x = 2 - 4t, \quad y = -1 + 2t, \quad z = 3 + 6t. \quad \blacksquare$$

Planes Tangent to Surfaces $z = f(x, y)$

We know from Section 12.4 that the plane tangent to the surface $z = f(x, y)$ at the point $(a, b, f(a, b))$ has equation

$$
\begin{aligned}
z &= f(a, b) + f_x(a, b)(x - a) + f_y(a, b)(y - b), \\
-f_x(a, b)(x - a) &- f_y(a, b)(x - b) + (z - f(a, b)) = 0.
\end{aligned}
$$

It follows that:

The plane tangent to the surface $z = f(x, y)$ at $(a, b, f(a, b))$ has normal vector $-f_x(a, b)\mathbf{i} - f_y(z, b)\mathbf{j} + \mathbf{k}$.

This fact can be related to Theorem 5 by writing the equation $z = f(x, y)$ in the form $F(x, y, z) = 0$, with $F(x, y, z) = z - f(x, y)$. Theorem 5 then tells us the surface has normal vector $\nabla F = -f_x\mathbf{i} - f_y\mathbf{j} + \mathbf{k}$.

NOTE   *Do not confuse the gradient of $f(x, y)$ with a vector normal to the surface $z = f(x, y)$.*

The gradient $\nabla f = f_x\mathbf{i} + f_y\mathbf{j}$ is a vector in the xy-plane that is orthogonal to the level curves of $f(x, y)$. The vector $-f_x\mathbf{i} - f_y\mathbf{j} + \mathbf{k}$ that is normal to the surface $z = f(x, y)$ is a vector in three-dimensional space.

EXAMPLE 5

Find an equation of the plane that is tangent to the surface $z = x^3 + y^3$ at the point $(2, -1, 7)$.

SOLUTION   The equation of the surface is of the form $z = f(x, y)$, where $f(x, y) = x^3 + y^3$. The tangent plane has normal vector

$$\mathbf{N} = -f_x\mathbf{i} - f_y\mathbf{j} + \mathbf{k} = -(3x^2)\mathbf{i} - (3y^2)\mathbf{j} + \mathbf{k}.$$

When $(x, y, z) = (2, -1, 7)$,

$$\mathbf{N} = -3(2)^2\mathbf{i} - 3(-1)^2\mathbf{j} + \mathbf{k} = -12\mathbf{i} - 3\mathbf{j} + \mathbf{k}.$$

An equation of the plane through the point $(2, -1, 7)$ with normal vector $-12\mathbf{i} - 3\mathbf{j} + \mathbf{k}$ is

$$(-12)(x - 2) + (-3)(y + 1) + (1)(z - 7) = 0,$$
$$-12x + 24 - 3y - 3 + z - 7 = 0,$$
$$z = 12x + 3y - 14. \quad \blacksquare$$

EXERCISES 12.7

1. At what points on the surface $x^2 + y^2 + z^2 = 1$ does the surface define **(a)** x, **(b)** y, **(c)** z, as a function of the other two variables in some rectangular box with center at the point?

2. At what points on the surface $x^2 + y^2 - z^2 = 0$ does the surface define **(a)** x, **(b)** y, **(c)** z, as a function of the other two variables in some rectangular box with center at the point?

In Exercises 3–10, sketch the level curves $f(x, y) = z_0$; evaluate ∇f and $D_{\mathbf{U}}f$ at the indicated point; sketch ∇f and $\mathbf{U}$ with initial position at the point indicated.

3. $f(x, y) = y/x^2$; $z_0 = 1, 2, 3$; $\mathbf{U} = (\mathbf{i} + 4\mathbf{j})/\sqrt{17}$; $(1, 2)$

4. $f(x, y) = (4 - x)/y^2$; $z_0 = 1, 2, 3$; $\mathbf{U} = (4\mathbf{i} + \mathbf{j})/\sqrt{17}$; $(2, -1)$

5. $f(x, y) = \ln(x^2 + y^2)$; $z_0 = -1, 0, 1$; $\mathbf{U} = (-\mathbf{i} + \mathbf{j})/\sqrt{2}$; $(1/\sqrt{2}, 1/\sqrt{2})$

6. $f(x, y) = \ln xy$; $z_0 = -1, 0, 1$; $\mathbf{U} = (\mathbf{i} - \mathbf{j})/\sqrt{2}$; $(1, 1)$

7. $f(x, y) = x^2 + y^2$; $z_0 = 1, 2, 3$; $\mathbf{U} = (-\mathbf{i} + \mathbf{j})/\sqrt{2}$; $(1, 1)$

8. $f(x, y) = y - x^2$; $z_0 = -1, 0, 1$; $\mathbf{U} = (\mathbf{i} + 2\mathbf{j})/\sqrt{5}$; $(1, 1)$

9. $f(x, y) = x^2 - y^2$; $z_0 = -1, 0, 1$; $\mathbf{U} = -(\mathbf{i} + \mathbf{j})/\sqrt{2}$; $(1, -1)$

10. $f(x, y) = y^2/x$; $z_0 = 1, 2, 3$; $\mathbf{U} = (\mathbf{i} - 2\mathbf{j})/\sqrt{5}$; $(2, 2)$

In Exercises 11–18, find an equation of the line that is normal to given curves at the indicated point.

11. $x^2 + y^2 = 25$; $(3, -4)$

12. $xy = 8$; $(2, 4)$

13. $x^2 - 3xy + 2y^2 = 0$; $(2, 1)$

14. $x = ye^y$; $(0, 0)$

15. $x = \sin y$; $(1/2, \pi/6)$

16. $x^2 + 2xy - 2y^2 = 4$; $(-6, 2)$

17. $x^2 - 4y^2 = 9$; $(5, -2)$

18. $x^2y = 12$; $(2, 3)$

In Exercises 19–24, find an equation of the plane that is tangent to the given surface at the indicated point.

19. $z^2 = x^2 + y^2$; $(3, -4, -5)$

20. $z^2 = xyz + 3$; $(-1, 2, -3)$

21. $ze^z = x^2 + y^2$; $(e, -e, 2)$

22. $z^2 + z - 2xy^2 = 0$; $(3, 1, 2)$

23. $z = xy$; $(-2, 3, -6)$

24. $z = \cos(x^2 + y^2)$; $(0, \sqrt{\pi}, -1)$

In Exercises 25–30, find symmetric equations and parametric equations of the line that is normal to the given surface at the indicated point.

25. $z = x^2 + y^2$; $(-1, 2, 5)$

26. $z^2 = 2x - 3y$; $(3, -1, -3)$

27. $e^{xyz} = 2x + y$; $(2, -3, 0)$

28. $6e^{xz} = yx - yz$; $(2, 3, 0)$

29. $z = xe^y$; $(0, 0, 0)$

30. $z = \sin x \cos y$; $(0, 0, 0)$

31. Find a vector that is tangent to the curve of intersection of the surfaces $z = xy$ and $x^2 + y^2 + z^2 = 9$ at the point $(1, 2, 2)$.

32. Find a vector that is tangent to the curve of intersection of the cylinders $y^2 + z^2 = 1$ and $x^2 + z^2 = 1$ at the point $(1/2, 1/2, \sqrt{3}/2)$.

33. Find a vector that is tangent to the curve of intersection of the plane $z = 2x + 3y$ and the sphere $x^2 + y^2 + z^2 = 6$ at the point $(2, -1, 1)$.

34. Find a vector that is tangent to the curve of intersection of the plane $2x + 3y + 4z = 1$ and the surface $xyz = 6$ at the point $(3, -1, -2)$.

35. What is the rate of change of $f(x, y, z) = xy^2z^3$ in the direction of the upward normal vector (with positive **k** component) to the surface $z = x^2y + xy^2$ at the point $(1, 2, 6)$?

36. What is the rate of change of $f(x, y, z) = 3x^2 + 2xy - y^2$ in the direction of the upward normal vector (with positive **k** component) to the surface $z = x^2 + y^2$ at the point $(2, 3, 13)$?

37. At what point does the line normal to the paraboloid $z = x^2 + y^2$ at the point $(3, 4, 25)$ intersect the xy-plane?

38. At what point does the line normal to the sphere $x^2 + y^2 + (z - 3)^2 = 9$ at the point $(2, 2, 2)$ intersect the xy-plane?

12.8 EXTREMA

Connections

Extreme value theorem, 3.8.

Applications, 4.7.

Partial derivatives, 12.3.

Chain Rule, 12.5.

The theory of extrema for functions of several variables is similar to the theory for functions of one variable. For example, we say that $f(c_1, \ldots, c_n)$ is the (absolute) **maximum** value of f in a region R if $(c_1, \ldots, c_n)$ is in R and $f(c_1, \ldots, c_n) \geq f(x_1, \ldots, x_n)$ for all $(x_1, \ldots, x_n)$ in R; $f(c_1, \ldots, c_n)$ is the (absolute) **minimum** value of f in R if $(c_1, \ldots, c_n)$ is in R and $f(c_1, \ldots, c_n) \leq f(x_1, \ldots, x_n)$ for all $(x_1, \ldots, x_n)$ in R. The value $f(c_1, \ldots, c_n)$ is a **local** maximum or minimum value of f if there is some $\delta > 0$ such that $f(c_1, \ldots, c_n)$ is a maximum or minimum, respectively, of f in the set $(x_1 - c_1)^2 + \cdots + (x_n - c_n)^2 < \delta^2$.

Critical Points and Local Extrema

A function $f(x_1, \ldots, x_n)$ cannot have a local extreme value at a point where f is either increasing or decreasing in any direction. It follows that f can have a local extreme value only at points where either

> **(i)** all first partial derivatives of f are zero, or
> **(ii)** some first partial derivative of f does not exist.

Points $(c_1, \ldots, c_n)$ in the interior of the domain of f where either (i) or (ii) holds are called **critical points** of f. As in the case of one variable, values at critical points are not necessarily local extrema. We can sometimes use the graph of f to determine if the value at a critical point is a local maximum, a local minimum, or neither of these.

EXAMPLE 1

Find critical points and local extrema of

$$f(x, y) = x^2 - 6x + y^2 - 4y.$$

SOLUTION   This function is defined with first partial derivatives for all (x, y). The first partial derivatives are

$$f_x(x, y) = 2x - 6,$$
$$f_y(x, y) = 2y - 4.$$

Setting each first partial derivative equal to zero, we obtain the following system of two equations:

$$\begin{cases} 2x - 6 = 0, \\ 2y - 4 = 0. \end{cases}$$

The solution of this system is $(x, y) = (3, 2)$. This is the only critical point of the function. The corresponding value is

$$f(3, 2) = (3)^2 - 6(3) + (2)^2 - 4(2) = -13.$$

In this example we can complete squares to write

$$f(x, y) = x^2 - 6x + (9) + y^2 - 4y + (4) - (9) - (4)$$
$$= (x - 3)^2 + (y - 2)^2 - 13.$$

We then see that the graph is a paraboloid with vertex at $(3, 2, -13)$ and that $f(3, 2) = -13$ is a local minimum which is also the absolute minimum value of the function. See Figure 12.8.1. ■

$$z = x^2 - 6x + y^2 - 4y$$

FIGURE 12.8.1   The minimum value of f is obtained at the vertex of the paraboloid, where $f_x = f_y = 0$.

EXAMPLE 2

Find critical points and extrema of

$$f(x, y) = 4 - 2\sqrt{x^2 + y^2}.$$

SOLUTION   For $(x, y) \neq (0, 0)$, we have

$$f_x(x, y) = -2\left(\frac{1}{2}\right)(x^2 + y^2)^{-1/2}(2x),$$
$$f_y(x, y) = -2\left(\frac{1}{2}\right)(x^2 + y^2)^{-1/2}(2y).$$

These partial derivatives are not *simultaneously* zero for $(x, y) \neq (0, 0)$. Since

$$f(x, 0) = 4 - 2|x| \text{ and } f(0, y) = 4 - 2|y|,$$

neither first partial derivative of f exists at $(0, 0)$, so $(0, 0)$ is a critical point. The corresponding value is

$$f(0, 0) = 4.$$

The graph is a cone with vertex at the point $(0, 0, 4)$. The value $f(0, 0) = 4$ is a local maximum which is also the absolute maximum of the function. See Figure 12.8.2. ■

$$z = 4 - 2\sqrt{x^2 + y^2}$$

FIGURE 12.8.2   The maximum value of f is at the vertex of the cone, where f_x and f_y do not exist.

EXAMPLE 3

Find critical points and local extrema of

$$f(x, y) = \frac{x^2 - y^2}{4}.$$

SOLUTION The function is defined with partial derivatives for all (x, y). We have

$$f_x(x, y) = \frac{2x}{4},$$

$$f_y(x, y) = \frac{-2y}{4}.$$

Setting each first partial derivative equal to zero, we obtain

$$x = 0 \text{ and } y = 0.$$

The point $(0, 0)$ is the only critical point. The corresponding value is

$$f(0, 0) = 0.$$

The graph of f is a hyperbolic paraboloid. The value $f(0, 0) = 0$ is neither a maximum nor a minimum value; the graph is concave up at the origin in the direction of the x-axis; the graph is concave down at the origin in the direction of the y-axis. See Figure 12.8.3. ∎

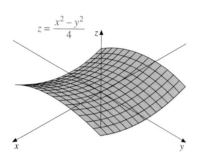

$$z = \frac{x^2 - y^2}{4}$$

FIGURE 12.8.3 Both partial derivatives of f are zero at the origin, but f does not have a local extreme value at the origin. The graph is concave upward at the origin along the x-axis and concave downward at the origin along the y-axis.

The Second Derivative Test

If the value of a function at a critical point is neither a maximum nor a minimum value, the corresponding point on the graph is called a **saddle point**.

For functions of a single real variable, we can sometimes use the Second Derivative Test to determine if a local extreme value is a local maximum or a local minimum. Let us see what the corresponding result is for functions of two variables. First, we note that it is not enough to require that both second partial derivatives at a critical point be nonzero with the same sign. For example, it is possible that $f(x, y)$ is continuous with $f_x(0, 0) = f_y(0, 0) = 0$, $f_{xx}(0, 0) > 0$, $f_{yy}(0, 0) > 0$, and $f(0, 0)$ not a local minimum. This is illustrated in Figure 12.8.4. The second partial derivatives being positive implies that the traces of the surface in the planes $x = 0$ and $y = 0$ are concave up at the origin. We would seem to have a better chance of having a local minimum if the graph were concave up in *every* direction at $(0, 0)$. This is illustrated in Figure 12.8.5.

The idea illustrated in Figure 12.8.5 can be used to develop a version of the Second Derivative Test for Local Extrema of functions of two variables. That is, we calculate the second derivative with respect to t of the function

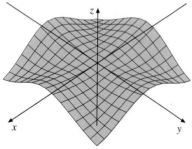

FIGURE 12.8.4 The graph is concave upward at the origin along both coordinate axes, but it is concave downward at the origin in some other directions.

$$z = f(x, y), \quad x = a + u_1 t, \quad y = b + u_2 t,$$

where $u_1^2 + u_2^2 = 1$. The sign of $z''(t)$ gives the concavity of f at (a, b) in

FIGURE 12.8.5 This graph is concave upward at the origin in every direction. The function has a local minimum at the origin.

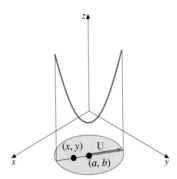

FIGURE 12.8.6 We can determine the concavity of f in the direction of the unit vector $\mathbf{U} = u_1\mathbf{i} + u_2\mathbf{j}$ by investigating the sign of the second derivative of the function $z = f(x, y), x = a + u_1t, y = b + u_2t$.

the direction of the vector $u_1\mathbf{i} + u_2\mathbf{j}$. See Figure 12.8.6. Assuming appropriate differentiability, we obtain

$$z'(t) = f_x(x, y)u_1 + f_y(x, y)u_2 \text{ and}$$
$$z''(t) = (f_{xx}(x, y)u_1 + f_{xy}(x, y)u_2)u_1 + (f_{yx}(x, y)u_1 + f_{yy}(x, y)u_2)u_2.$$

Assuming the second partial derivatives are continuous, so that $f_{xy} = f_{yx}$, we have

$$z''(t) = f_{xx}(x, y)u_1^2 + 2f_{xy}(x, y)u_1u_2 + f_{yy}(x, y)u_2^2.$$

We want $z''(t)$ to be of constant sign for all values of t, u_1, and u_2 for which $(x, y) = (a + u_1t, b + u_2t)$ is in a disk with center (a, b). If (a, b) is a critical point of f, so that $f_x(a, b) = f_y(a, b) = 0$, we then have

$$z'(0) = f_x(a, b)u_1 + f_y(a, b)u_2 = 0$$

and two applications of the Mean Value Theorem give

$$\begin{aligned} f(x, y) - f(a, b) &= z(t) - z(0) \\ &= z'(t_1)(t - 0) \\ &= (z'(t_1) - z'(0))t \\ &= z''(t_2)t_1t, \end{aligned}$$

where either $0 < t_2 < t_1 < t$ or $t < t_1 < t_2 < 0$. It follows that $f(x, y) - f(a, b)$ is of constant sign as (x, y) varies over a disk with center (a, b), so $f(a, b)$ is either a local maximum or a local minimum, depending on the constant sign of z''. To determine when $z''(t)$ is of constant sign, we note that a quadratic expression $Ax^2 + Bx + C$ is never zero if the discriminant $B^2 - 4AC$ is negative. We can use this fact to conclude that $z''(t)$ is not zero for any choice of (u_1, u_2) with $u_1^2 + u_2^2 = 1$ if

$$[2f_{xy}(x, y)]^2 - 4f_{xx}(x, y)f_{yy}(x, y) < 0, \text{ or}$$
$$f_{xx}(x, y)f_{yy}(x, y) - [f_{xy}(x, y)]^2 > 0.$$

Since we have assumed that the second partial derivatives of f are continuous, if we have

$$f_{xx}(a, b)f_{yy}(a, b) - [f_{xy}(a, b)]^2 > 0,$$

then we also have

$$f_{xx}(x, y)f_{yy}(x, y) - [f_{xy}(x, y)]^2 > 0$$

for all (x, y) in a disk with center (a, b). The condition

$$f_{xx}(a, b)f_{yy}(a, b) - [f_{xy}(a, b)]^2 > 0$$

implies that $f_{xx}(a, b)$ and $f_{yy}(a, b)$ are of the same sign and that the sign of $z''(t)$ agrees with them both. Also, if

$$f_{xx}(a, b)f_{yy}(a, b) - [f_{xy}(a, b)]^2 < 0,$$

then the concavity of f is upward in one direction and downward in another direction, so $f(a, b)$ must be a saddle point. This gives the following theorem.

THEOREM 1

If $f(x, y)$ has second partial derivatives that are continuous in a disk with center (a, b), and $f_x(a, b) = f_y(a, b) = 0$, then:

(**i**) f has a local maximum at (a, b) if

$$f_{xx}(a, b)f_{yy}(a, b) - [f_{xy}(a, b)]^2 > 0 \text{ and } f_{xx}(a, b) < 0.$$

(**ii**) f has a local minimum at (a, b) if

$$f_{xx}(a, b)f_{yy}(a, b) - [f_{xy}(a, b)]^2 > 0 \text{ and } f_{xx}(a, b) > 0.$$

(**iii**) f has a saddle point at (a, b) if

$$f_{xx}(a, b)f_{yy}(a, b) - [f_{xy}(a, b)]^2 < 0.$$

The theorem does not give information for all possible cases. If $f_{xx}f_{yy} - (f_{xy})^2 = 0$, other tests are required to determine if a value at a critical point is either a local maximum, a local minimum, or a saddle point.

If $f_{xy} = f_{yx}$, the expression $f_{xx}f_{yy} - (f_{xy})^2$ is the value of the determinant

$$\begin{vmatrix} f_{xx} & f_{xy} \\ f_{yx} & f_{yy} \end{vmatrix}.$$

This is called the **Hessian** of the function f. The extension of the above theorem to functions of more than two variables involves corresponding determinants. We leave this to more advanced texts.

EXAMPLE 4

Find all critical points of the function

$$f(x, y) = \frac{1}{3}x^3 - x + xy^2.$$

Find the value at each critical point and characterize it as a local maximum, local minimum, or saddle point.

SOLUTION We have

$$f_x(x, y) = x^2 - 1 + y^2,$$
$$f_y(x, y) = 2xy.$$

Setting the first partial derivatives equal to zero, we obtain the system

$$\begin{cases} x^2 - 1 + y^2 = 0, & (1) \\ 2xy = 0. & (2) \end{cases}$$

Equation 2 implies either $x = 0$ or $y = 0$. If $x = 0$, substitution into Equation 1 gives

$$(0)^2 - 1 + y^2 = 0,$$
$$y^2 = 1,$$
$$y = 1, -1.$$

This gives the critical points $(0, 1)$ and $(0, -1)$. Similarly, if $y = 0$, Equation 1 implies $x = 1, -1$ and we obtain the critical points $(1, 0)$ and $(-1, 0)$.

Let us use the second derivatives to determine if each critical point corresponds to a local maximum, a local minimum, or a saddle point. We have

$$f_x = x^2 - 1 + y^2, \quad f_y = 2xy,$$
$$f_{xx} = 2x, \quad f_{yy} = 2x,$$
$$f_{xy} = 2y,$$
$$f_{xx}f_{yy} - (f_{xy})^2 = (2x)(2x) - (2y)^2 = 4x^2 - 4y^2.$$

The values of f, $f_{xx}f_{yy} - (f_{xy})^2$, and f_{xx} at each critical point are given in Table 12.8.1. We see that $(0, 1)$ and $(0, -1)$ correspond to saddle points; $f(1, 0) = -2/3$ is a local minimum, and $f(-1, 0) = 2/3$ is a local maximum. The graph is sketched in Figure 12.8.7. ∎

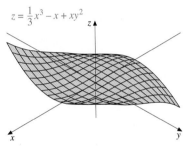

$z = \frac{1}{3}x^3 - x + xy^2$

FIGURE 12.8.7 The signs of $f_{xx}f_{yy} - [f_{xy}]^2$ and f_{xx} can be used to determine if f has a local extreme or saddle point at a critical point.

TABLE 12.8.1

(x, y)	f	$f_{xx}f_{yy} - (f_{xy})^2$	f_{xx}	Type
$(0, 1)$	0	$-4 < 0$	—	Saddle point
$(0, -1)$	0	$-4 < 0$	—	Saddle point
$(1, 0)$	$-2/3$	$4 > 0$	$2 > 0$	Local minimun
$(-1, 0)$	$2/3$	$4 > 0$	$-2 < 0$	Local maximum

Extrema Subject to a Constraint

Many problems involve finding maximum or minimum values of a function of several variables subject to an equation of constraint that relates the variables.

EXAMPLE 5

Find the minimum value of $f(x, y) = x^2 - 6x + y^2 - 4y$ subject to the condition that (x, y) is on the line $x + y = 7$.

SOLUTION The values of $f(x, y)$ that we want to consider are indicated in Figure 12.8.8a.

We can use one-variable techniques to find the desired maximum value. That is, we use the condition $x + y = 7$ to express y in terms of x. In particular, $y = 7 - x$. We can then substitute for y in the expression for $f(x, y)$ to obtain a function of x. We have

$$
\begin{aligned}
z &= f(x, 7 - x) \\
&= x^2 - 6x + (7 - x)^2 - 4(7 - x) \\
&= x^2 - 6x + 49 - 14x + x^2 - 28 + 4x \\
&= 2x^2 - 16x + 21.
\end{aligned}
$$

This is a quadratic function of x that has a minimum at the point where $dz/dx = 0$. We have

$$
\frac{dz}{dx} = 4x - 16,
$$

$\dfrac{dz}{dx} = 0$ implies $4x - 16 = 0$, so $x = 4$.

When $x = 4$, we have $y = 7 - x = 3$. The desired minimum value is

$$
f(4, 3) = (4)^2 - 6(4) + (3)^2 - 4(3) = -11. \quad \blacksquare
$$

In Figure 12.8.8b we have sketched some level curves of the function $f(x, y) = x^2 - 6x + y^2 - 4y$ in Example 5. Note that a level curve of f is tangent to the line $x + y = 7$ at the point $(4, 3)$, where the function assumes its minimum value on the line.

FIGURE 12.8.8a The graph of $f(x, y) = x^2 - 6x + y^2 - 4y$ subject to the constraint $x + y = 7$ is the curve of intersection of the surface $z = f(x, y)$ and the vertical plane that contains the line.

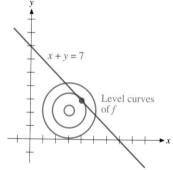

FIGURE 12.8.8b The extreme value of $f(x, y)$ for (x, y) restricted to the line occurs at a point where a level curve of f is tangent to the line.

Extrema in Closed, Bounded Regions

We know from the Extreme Value Theorem in Section 3.8 that a function of a single variable that is continuous on a closed and bounded interval has maximum and minimum values on the interval, and that these values occur at either critical points or endpoints of the interval. Analogous results are true for functions of several variables. Recall from Section 12.1 that a region R in

n-dimensional space is **closed** if it contains all its boundary points. The region R is said to be **bounded** if there is a number B such that $x_1^2 + \cdots + x_n^2 < B^2$ for all $(x_1, \ldots, x_n)$ in R. We then have the following result for functions of several variables.

THEOREM 2

If f is continuous on a closed, bounded region R, then f has a maximum value and a minimum value on the region. These extrema occur either
(**i**) at an interior point of R where all first partial derivatives of f are zero;
(**ii**) at an interior point of R where some first partial derivative of f does not exist; or
(**iii**) on the boundary of the region.

We will not verify this result. Note that in order to find boundary extrema we must find extrema subject to the constraint of the equation that determines the boundary. Parts of the boundary that are given by different equations must be treated separately.

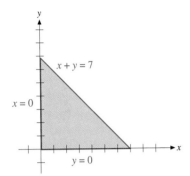

FIGURE 12.8.9a   The values of $f(x, y)$ are restricted to the closed region bounded by the x-axis, the y-axis, and the line $x + y = 7$.

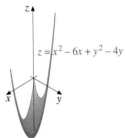

FIGURE 12.8.9b   Extreme values of f occur either at critical points in the interior of the region or on the boundary. We must consider parts of the boundary with different equations separately, with the equation of the boundary as a constraint.

EXAMPLE 6
Find maximum and minimum values of

$$f(x, y) = x^2 - 6x + y^2 - 4y$$

on the closed region bounded by the x-axis, the y-axis, and the line $x + y = 7$.

SOLUTION   The region bounded by the x-axis, the y-axis, and the line $x + y = 7$ is sketched in Figure 12.8.9a. The graph of f is sketched in Figure 12.8.9b. The continuous function has maximum and minimum values on the closed and bounded region.

In Example 1 we saw that the first partial derivatives of f are both zero at the point $(3, 2)$. This is an interior point of the domain. There are no points where the first partial derivatives of the function do not exist.

The boundary of the region consists of segments of the lines $y = 0$, $x = 0$, and $x + y = 7$. Each part must be treated separately.

We need extrema of $f(x, y)$ for (x, y) on the line $x + y = 7$, between the points $(0, 7)$ and $(7, 0)$. As in Example 5, we obtain a function of x by substituting $7 - x$ for y. However, we must now restrict the domain to those values of x that correspond to points on the line segment between $(0, 7)$ and $(7, 0)$. We obtain

$$f(x, 7 - x) = 2x^2 - 16x + 21, \quad 0 \le x \le 7.$$

Setting the derivative with respect to x equal to zero gives the point $(4, 3)$. We must also consider the endpoints $(0, 7)$ and $(7, 0)$.

Consideration of extrema on the boundary of the region along $y = 0$ leads to the function

$$f(x, 0) = x^2 - 6x, \quad 0 \le x \le 7.$$

Setting the derivative with respect to x equal to zero gives the point $(3, 0)$. The endpoints are $(0, 0)$ and $(7, 0)$.

TABLE 12.8.2

(x, y)	$f(x, y)$
$(3, 2)$	-13
$(4, 3)$	-11
$(3, 0)$	-9
$(0, 2)$	-4
$(0, 0)$	0
$(7, 0)$	7
$(0, 7)$	21

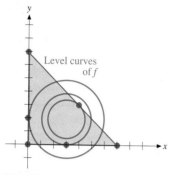

FIGURE 12.8.10 The candidates for extrema on the boundary are exactly corner points of the boundary and points on the boundary where a level curve of *f* is tangent to the boundary.

Consideration of extrema on the boundary of the region along $x = 0$ leads to the function

$$f(0, y) = y^2 - 4y, \quad 0 \le y \le 7.$$

Setting the derivative with respect to y equal to zero gives the point $(0, 2)$. The endpoints are $(0, 0)$ and $(0, 7)$.

All candidates for the maximum and minimum values are listed in Table 12.8.2. We see that the minimum value is $f(3, 2) = -13$; the maximum value is $f(0, 7) = 21$. ∎

In Figure 12.8.10 we have sketched the domain and some level curves of the function of Example 6. Note that the candidates for extrema on the boundary are exactly corner points of the boundary and points on the boundary where a level curve of f is tangent to the boundary.

Minimum Distance

EXAMPLE 7

Find the minimum distance between the point $(1, 2, 0)$ and the cone $z = 2\sqrt{x^2 + y^2}$.

SOLUTION A sketch is given in Figure 12.8.11. It is more convenient to work with the square of the distance between two points than with the distance. The minimum value of the square of the distance corresponds to the minimum value of the distance. The square of the distance between the points $(1, 2, 0)$ and (x, y, z) is

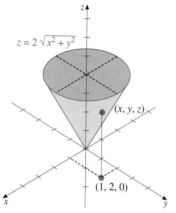

FIGURE 12.8.11 The distance between the point $(1, 2, 0)$ and a point (x, y, z) on the cone is $\sqrt{(x-1)^2 + (y-2)^2 + z^2}$.

$$D = (x - 1)^2 + (y - 2)^2 + z^2.$$

A point (x, y, z) is on the cone if the coordinates of the point satisfy the equation of the cone. Thus, the equation of the cone is a constraint for this problem. Using the equation of the cone to eliminate z from the expression for D, we obtain

$$D(x, y) = (x - 1)^2 + (y - 2)^2 + 4(x^2 + y^2).$$

Setting the two partial derivatives of $D(x, y)$ equal to zero, we obtain the system

$$\begin{cases} 2(x - 1) + 4(2x) = 0, \\ 2(y - 2) + 4(2y) = 0. \end{cases}$$

The solution of the system is $x = 0.2$, $y = 0.4$. The corresponding value of z is $z = 2\sqrt{(0.2)^2 + (0.4)^2} = 2\sqrt{0.2} \approx 0.8944$. It is clear from the geometry that these values give the minimum distance. We have

$$D(0.2, 0.4) = (0.2 - 1)^2 + (0.4 - 2)^2 + 4((0.2)^2 + (0.4)^2) = 4.$$

The minimum distance is then $\sqrt{D} = 2$. ∎

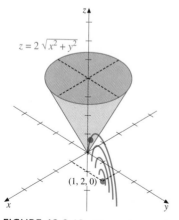

FIGURE 12.8.12 The point on the cone that is closest to the point $(1, 2, 0)$ is the point where a level surface of $D(x, y, z)$ is tangent to the cone.

Note that the level surfaces of the function

$$D = (x - 1)^2 + (y - 2)^2 + z^2$$

are spheres with center $(1, 2, 0)$. It seems that the point on the cone $z = 2\sqrt{x^2 + y^2}$ that is closest to $(1, 2, 0)$ should be the first point on the cone that is touched as the level surfaces expand from $(1, 2, 0)$, and that the sphere should be tangent to the cone at that point. This is true. See Figure 12.8.12.

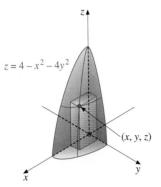

$z = 4 - x^2 - 4y^2$

(x, y, z)

FIGURE 12.8.13 The rectanglar box has one corner at the origin and the opposite corner on the paraboloid $x = 4 - x^2 - 4y^2$.

Maximum Volume

EXAMPLE 8

Find the dimensions and volume of the rectangular box of maximum volume that is in the first octant with one corner at the origin and the opposite corner on the paraboloid $z = 4 - x^2 - 4y^2$.

SOLUTION   The box is sketched in Figure 12.8.13. The volume of the box is

$$V = xyz.$$

The point (x, y, z) is on the paraboloid if the coordinates of the point satisfy the equation of the paraboloid. The equation of the paraboloid is a constraint for the problem. We use the equation of constraint to eliminate z from the expression for the volume. We obtain

$$V(x, y) = xy(4 - x^2 - 4y^2)$$
$$= 4xy - x^3y - 4xy^3.$$

The domain of $V(x, y)$ is the closed region in the first quadrant bounded by $x = 0$, $y = 0$, and $x^2 + 4y^2 = 4$. These correspond to each of x, y, and z being zero. Since $V(x, y)$ is continuous on a closed and bounded region, we know there is a maximum value. The volume is positive in the interior of the region and zero on the boundary, so the maximum must occur at a critical point. Setting the partial derivatives of $V(x, y)$ equal to zero, we obtain the system

$$\begin{cases} 4y - 3x^2y - 4y^3 = 0, \\ 4x - x^3 - 12xy^2 = 0, \end{cases}$$

or

$$\begin{cases} y(4 - 3x^2 - 4y^2) = 0, \\ x(4 - x^2 - 12y^2) = 0. \end{cases}$$

We may disregard cases where either $x = 0$ or $y = 0$, since these values correspond to boundary points. The system then simplifies to

$$\begin{cases} 4 - 3x^2 - 4y^2 = 0, \\ 4 - x^2 - 12y^2 = 0. \end{cases}$$

Subtracting the second equation above from the first, we obtain

$$-2x^2 + 8y^2 = 0, \text{ so } x = 2y,$$

where we have used the fact that x and y are nonnegative. Substitution then gives

$$4 - 3(2y)^2 - 4y^2 = 0, \text{ so } y = \frac{1}{2}.$$

Then $x = 2y = 2(1/2) = 1$ and $z = 4 - (1)^2 - 4(1/2)^2 = 2$. The box of maximum volume has dimensions $1 \times 1/2 \times 2$ and volume

$$V\left(1, \frac{1}{2}\right) = 4(1)\left(\frac{1}{2}\right) - (1)^3\left(\frac{1}{2}\right) - 4(1)\left(\frac{1}{2}\right)^3 = 1. \quad \blacksquare$$

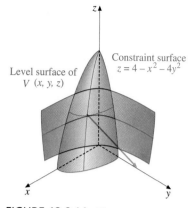

Level surface of $V(x, y, z)$

Constraint surface $z = 4 - x^2 - 4y^2$

FIGURE 12.8.14 The extreme value of $V(x, y, z)$ subject to the constraint $z = 4 - x^2 - 4y^2$ occurs at a point where a level surface of V is tangent to the paraboloid.

In Example 8 we found the maximum value of $V = xyz$ subject to the constraint $z = 4 - x^2 - 4y^2$. Let us see that the level surface of $V = xyz$ is tangent to the paraboloid $z = 4 - x^2 - 4y^2$ at the point $(1, 1/2, 2)$ that gives the maximum value. In particular, the gradient of $V = xyz$ at $(1, 1/2, 2)$ is

$$\nabla V = V_x \mathbf{i} + V_y \mathbf{j} + V_z \mathbf{k} = yz\mathbf{i} + xz\mathbf{j} + xy\mathbf{k} = \mathbf{i} + 2\mathbf{j} + \frac{1}{2}\mathbf{k}.$$

The gradient of V is normal to the level surface of V. Let us write the equation of constraint as $g(x, y, z) = 4$, where $g(x, y, z) = x^2 + 4y^2 + z$. The gradient of g is then normal to the paraboloid. When $(x, y, z) = (1, 1/2, 2)$, we have

$$\nabla g = g_x \mathbf{i} + g_y \mathbf{j} + g_z \mathbf{k} = 2x\mathbf{i} + 8y\mathbf{j} + \mathbf{k} = 2\mathbf{i} + 4\mathbf{j} + \mathbf{k}.$$

Since $\nabla V = \nabla g/2$, the normal vectors of the surfaces are parallel and the surfaces are tangent. See Figure 12.8.14.

EXERCISES 12.8

Find all critical points of the functions given in Exercises 1–10. Find the value at each critical point and characterize it as a local maximum, local minimum, or saddle point.

1. $f(x, y) = x^3 - 3x + y^2$

2. $f(x, y) = x^3 - 12x + y^2 + 8y$

3. $f(x, y) = x^3 - 3x + y^3 - 3y$

4. $f(x, y) = x^3 + 3x^2 + y^3 - 3y^2$

5. $f(x, y) = 2x^3 + 6xy + 3y^2$

6. $f(x, y) = 2x^3 - 6xy + y^2$

7. $f(x, y) = x^2 + xy + y^2 + x + 2y$

8. $f(x, y) = x^3 + y^3 - 3xy$

9. $f(x, y) = \sqrt{x^2 + y^2}$

10. $f(x, y) = \sqrt{x^2 + y^2 - 2x + 4y + 5}$

In Exercises 11–14, find extrema of the given functions subject to the indicated conditions; determine if each value is a maximum or minimum value subject to the condition.

11. $f(x, y) = xy, y = 6 - 3x$

12. $f(x, y) = xy, y = x - 4$

13. $f(x, y) = x^2 - y^2, y = 3 - 2x$

14. $f(x, y) = x^2 - y^2, 2y = x + 3$

In Exercises 15–20, find the maximum and minimum values of the given functions in the indicated regions.

15. $f(x, y) = x^3 - 3x + y^2, x^2 - 2x + y^2 \le 0$

16. $f(x, y) = xy + 3, x^2 + 4y^2 \le 8$

17. $f(x, y) = x^2 - 3y + y^3, x^2 + 4y^2 \le 36$

18. $f(x, y) = 9x^2 + y^3 - 12y, x^2 + y^2 - 3y \le 0$

19. $f(x, y) = x^2 + 2x + y^2 - 2y$, region bounded by $x = 0$, $x = 3, y = 0$, and $y = 3$

20. $f(x, y) = 4x - x^2 + 4y - y^2$, region bounded by $x = 0$, $x = 3, y = 0$, and $y = 1$

21. Find the point on the line $y = 2x + 3$ that is nearest the origin.

22. Find the point or points on the parabola $2y = x^2 - 1$ that are nearest the origin.

23. Find the point or points on the parabola $y = x^2 - 1$ that are nearest the origin.

24. Find the point or points on the paraboloid $z = x^2 + (y^2/4) - 1$ that are nearest the origin.

25. Find the dimensions and volume of the rectangular box of maximum volume that is in the first octant with one corner at the origin and the opposite corner on the plane $x + 2y + z = 3$.

26. Find the dimensions and volume of the rectangular box of maximum volume that is in the first octant with one corner at the origin and the opposite corner on the paraboloid $x^2 + 2y^2 + z = 8$.

27. The volume of an open-top container, constructed by cutting a wedge from each corner of a square piece of cardboard and folding as indicated in the figure, is given by the function

$$V(x, y) = \frac{1}{6}(x^2 + xy + y^2)\sqrt{(a-x)^2 - (y-x)^2},$$

$$0 \le x \le y \le a.$$

Show that the container that has maximum volume is neither a rectangular box ($x = y$) nor a square pyramid ($x = 0$). (*Hint:* Investigate the signs of the partial derivatives $V_y(y, y)$ and $V_x(0, y)$.)

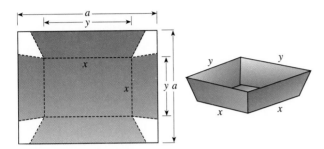

28. (a) Find the maximum and minimum values of $f(x, y) = x + y$ for (x, y) on the circle $C : x = \cos t$, $y = \sin t$, $0 \le t \le 2\pi$. **(b)** Sketch the circle C and the level curves of f that correspond to the extreme values of f. **(c)** What is the relation between the circle and the level curves?

29. (a) Find the minimum value of $f(x, y) = y - 2x$ for (x, y) on the parabola $y = x^2$. **(b)** Sketch the parabola and the level curve of f that corresponds to the minimum value of f. **(c)** What is the relation between the parabola and the level curve?

30. It is assumed that experimentally determined points (x_1, y_1), $(x_2, y_2), \ldots, (x_n, y_n)$ lie near some line $y = \alpha x + \beta$, and it is desired to find the line that "best" fits the data. See figure.

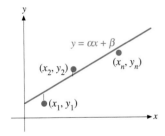

One way to do this is to choose the line that minimizes the sum of the squares of the vertical distances of the data points from the line. That is, choose α and β such that

$$(\alpha x_1 + \beta - y_1)^2 + (\alpha x_2 + \beta - y_2)^2 + \cdots + (\alpha x_n + \beta - y_n)^2$$

is a minimum. This line is called the **least squares regression line** for the data. Show that

$$\alpha = \frac{\left(\sum_{j=1}^{n} x_j\right)\left(\sum_{j=1}^{n} y_j\right) - n\left(\sum_{j=1}^{n} x_j y_j\right)}{\left(\sum_{j=1}^{n} x_j\right)^2 - n\left(\sum_{j=1}^{n} x_j^2\right)}$$

and

$$\beta = \frac{\left(\sum_{j=1}^{n} x_j\right)\left(\sum_{j=1}^{n} x_j y_j\right) - \left(\sum_{j=1}^{n} x_j^2\right)\left(\sum_{j=1}^{n} y_j\right)}{\left(\sum_{j=1}^{n} x_j\right)^2 - n\left(\sum_{j=1}^{n} x_j^2\right)}.$$

Use the formulas from the previous exercise to find the least squares regression line for the data points given in Exercises 31–34. (It is convenient to organize the data in columns of x_j, y_j, $x_j y_j$, and x_j^2, and sum each column from $j = 1$ to n.)

31. $(1, 3), (2, 3), (4, 1)$

32. $(1, 1), (2, 2), (4, 3)$

33. $(1, 2), (2, 3), (3, 3), (4, 4)$

34. $(1, 4), (2, 3), (3, 3), (4, 4)$

If a level curve $f(x, y) = z_0$ intersects a closed set S, then z_0 is a value of $f(x, y)$ for some (x, y) in S. It follows that the extreme values of $f(x, y)$ for (x, y) in S are the extreme values of z_0 for which the level curves $f(x, y) = z_0$ intersect S. In simple cases we can determine extreme values of z_0 for which the level curves $f(x, y) = z_0$ intersect S by sketching S and some level curves of f. See figure. Use this method in Exercises 35–40 to find the extreme values of the given functions on the indicated sets.

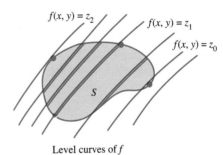

Level curves of f

35. $f(x, y) = y - x$, (x, y) on the triangular path from $(0, 0)$ to $(2, 0)$, to $(0, 1)$, and back to $(0, 0)$

36. $f(x, y) = x + y$, (x, y) on the triangular path from $(0, 0)$ to $(2, 0)$, to $(0, 1)$, and back to $(0, 0)$

37. $f(x, y) = x + 3y$, (x, y) on the triangular path from $(0, 0)$ to $(2, 0)$, to $(0, 1)$, and back to $(0, 0)$

38. $f(x, y) = x^2 + y^2$, (x, y) on the triangular path from $(0, 0)$ to $(2, 0)$, to $(0, 1)$, and back to $(0, 0)$

39. $f(x, y) = x^2 + y^2$, (x, y) on the line $x + y = 2$

40. $f(x, y) = x - y$, (x, y) on the circle $x^2 + y^2 = 1$

12.9 LAGRANGE MULTIPLIERS

Connections

Extreme values, 12.8.

Directional derivatives, 12.6.

The method of substitution used in Section 12.8 for finding extrema subject to constraints requires that certain variables be chosen as independent variables and other variables eliminated. In many problems it is difficult or impossible to solve the equations of constraint to obtain explicit formulas for the variables that are to be eliminated. Also, we must determine the domain of the chosen independent variables. In this section we will discuss a different method of finding extrema of a function subject to constraints. The method allows us to treat all of the original independent variables in a similar manner, without choosing some to be considered as independent variables and eliminating others.

Extrema of $f(x, y)$ Subject to a Constraint $g(x, y) = d$

Let us consider the problem of finding (x, y) that give extrema of $f(x, y)$ subject to the constraint $g(x, y) = d$, where d is a constant. That is, we want the extrema of $f(x, y)$ for (x, y) on the curve given by the equation $g(x, y) = d$. If the directional derivative of f in the direction tangent to the curve $g(x, y) = d$ is not zero at a point on the curve, then the values of $f(x, y)$ are either increasing or decreasing as (x, y) moves along the curve through the point. This means that extreme values of $f(x, y)$ on the curve $g(x, y) = d$ can occur only at points where the directional derivative of f in the direction of the curve is zero, so the gradient of f must be orthogonal to the curve at those points. Since the plane vectors ∇f and ∇g are both orthogonal to the curve $g(x, y) = d$ at such points on the curve, ∇f must be parallel to ∇g. If $\nabla g \neq \mathbf{0}$, this means that ∇f must be a scalar multiple of ∇g. We can also conclude that the level curves of $f(x, y)$ are tangent to the curve $g(x, y) = d$ at the points on the curve where the extrema of $f(x, y)$ on the curve occur. See Figure 12.9.1.

Let us summarize the above results in the following form.

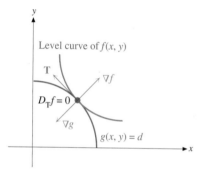

FIGURE 12.9.1 Extreme values of $f(x, y)$ subject to the constraint $g(x, y) = d$ occur at points where a level curve of f is tangent to the curve $g(x, y) = d$.

LAGRANGE MULTIPLIER THEOREM

If f and g are differentiable functions of (x, y) and $\nabla g \neq \mathbf{0}$ at points where $g(x, y) = d$, then extrema of $f(x, y)$ subject to the constraint $g(x, y) = d$ can occur only at points where there is a number λ such that $\nabla f = \lambda \nabla g$.

The Greek letter λ (lambda) introduced above is called a **Lagrange multiplier**. To use the Lagrange multiplier, we note that the vector equation $\nabla f = \lambda \nabla g$ and the equation of constraint give the system of three equations

$$\begin{cases} f_x(x, y) = \lambda g_x(x, y), \\ f_y(x, y) = \lambda g_y(x, y), \\ g(x, y) = d. \end{cases}$$

Solutions of this system for the variables x and y give candidates for extrema of $f(x, y)$ subject to the constraint $g(x, y) = d$. This system is generally not linear and there is not a fixed method of solution that works nicely for all cases. We will illustrate some methods that are convenient for certain cases. However, any legitimate method you choose is fine.

EXAMPLE 1

Find extrema of $f(x, y) = 4x + 3y$ for (x, y) on the circle $x^2 + y^2 = 25$.

SOLUTION We have

$$f(x, y) = 4x + 3y, \text{ so } \nabla f = 4\mathbf{i} + 3\mathbf{j}.$$

The constraint is of the form $g(x, y) = d$, where $g(x, y) = x^2 + y^2$, so

$$\nabla g = 2x\mathbf{i} + 2y\mathbf{j}.$$

We note that f and g are differentiable and $\nabla g \neq \mathbf{0}$ on the curve $g(x, y) = 25$, so the Lagrange Multiplier Theorem applies.

Substitution into the vector equation $\nabla f = \lambda \nabla g$ gives

$$4\mathbf{i} + 3\mathbf{j} = \lambda(2x\mathbf{i} + 2y\mathbf{j}).$$

This vector equation and the equation of constraint then give the system of equations

$$\begin{cases} 4 = \lambda(2x), \\ 3 = \lambda(2y), \\ x^2 + y^2 = 25. \end{cases}$$

We want the (x, y) values of the solution of this system. In this example, it is convenient to solve the first and second equations for λ. Noting that the first equation implies $x \neq 0$ and the second equation implies $y \neq 0$, we obtain

$$\frac{4}{2x} = \lambda = \frac{3}{2y}, \text{ so}$$

$$y = \frac{3x}{4}.$$

We now substitute into the third equation and solve for x:

$$x^2 + y^2 = 25,$$
$$x^2 + \left(\frac{3x}{4}\right)^2 = 25,$$
$$16x^2 + 9x^2 = 25(16),$$
$$25x^2 = 25(16),$$
$$x^2 = 16,$$
$$x = 4, -4.$$

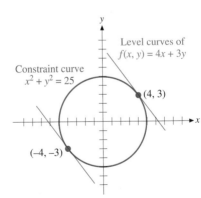

Constraint curve
$x^2 + y^2 = 25$

Level curves of
$f(x, y) = 4x + 3y$

$(4, 3)$

$(-4, -3)$

FIGURE 12.9.2 The level curves of $f(x, y) = 4x + 3y$ are tangent to the constraint curve $x^2 + y^2 = 25$ at points where f has extreme values subject to the constraint.

When $x = 4$, $y = \dfrac{3(4)}{4} = 3$. When $x = -4$, $y = \dfrac{3(-4)}{4} = -3$. The (x, y) values that give solutions of our system of equations are $(4, 3)$ and $(-4, -3)$. These are the only candidates to give extrema of $f(x, y)$ on the circle. Since f is continuous on the circle, which is a closed and bounded set, f has a maximum value and a minimum value on the circle. The value $f(4, 3) = 25$ is the maximum value and the value $f(-4, -3) = -25$ is the minimum value. From Figure 12.9.2 we see that the level curves of $f(x, y) = 4x + 3y$ are tangent to the circle $x^2 + y^2 = 25$ at the points $(4, 3)$ and $(-4, -3)$. ∎

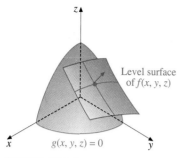

FIGURE 12.9.3 Extreme values of $f(x, y, z)$ subject to the constraint $g(x, y, z) = d$ occur at points where a level surface of f is tangent to the constraint surface.

Extrema of $f(x, y, z)$ Subject to a Constraint $g(x, y, z) = d$

Let us now consider extrema of $f(x, y, z)$, subject to the constraint $g(x, y, z) = d$, where d is a constant.

Extrema can occur only at points (x, y, z) on the surface $g(x, y, z) = d$ where the directional derivative of f is zero in every direction of the plane tangent to the surface. This means that the gradient of f is orthogonal to the surface $g(x, y, z) = d$, so the gradient of f is parallel to the gradient of g, and the level surface of f is tangent to the surface $g(x, y, z) = d$. See Figure 12.9.3. The analogue of the Lagrange Multiplier Theorem holds for functions of three variables, and the procedure for determining the extrema is similar. The vector equation $\nabla f = \lambda \nabla g$ gives three equations, and the equation of constraint gives the fourth equation of the system.

EXAMPLE 2

Material for the top of a rectangular box costs $3/ft^2. Material for the bottom and four sides costs $5/ft^2. What is the largest volume the box can have if the total cost of material for the box is $96?

SOLUTION   The box is sketched in Figure 12.9.4. The variables are the length, width, height, and the volume of the box. The volume is the max-min variable. We first express the max-min variable in terms of the other variables in the problem.

$$V(x, y, z) = xyz.$$

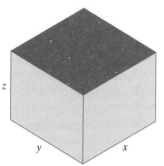

FIGURE 12.9.4 The length, width, and height of the rectangular box are variables. The max-min variable is the volume of the box.

The total cost gives a constraint. We have

Total cost $= $ (Cost of top) $+$ (Cost of bottom and sides)
$\qquad = 3$(Area of top) $+$ 5(Area of bottom plus area of sides)
$\qquad = 3xy + 5(xy + 2xz + 2yz)$.

If the total cost is to be $96, we obtain the equation of constraint

$$96 = 3xy + 5(xy + 2xz + 2yz),$$

which we rewrite as

$$8xy + 10xz + 10yz = 96,$$
$$4xy + 5xz + 5yz = 48.$$

This is of the form $g(x, y, z) = d$, where $g(x, y, z) = 4xy + 5xz + 5yz$, so

$$\nabla g = (4y + 5z)\mathbf{i} + (4x + 5z)\mathbf{j} + (5x + 5y)\mathbf{k}.$$

It is not difficult to check that $\nabla g = \mathbf{0}$ implies $x = y = z = 0$, so $\nabla g \neq \mathbf{0}$ on the surface $g(x, y, z) = 48$. The functions f and g are differentiable, so we can apply the Lagrange Multiplier Theorem to find an extreme value of the volume subject to the cost constraint. Since the variables are not restricted to a bounded region, we have no theoretical guarantee that there is a box of maximum volume. However, it can be verified that the extreme value obtained by the Lagrange Multiplier method is a maximum value. (See Exercise 35.)

We have $V = xyz$, so $\nabla V = yz\mathbf{i} + xz\mathbf{j} + zy\mathbf{k}$. The vector equation $\nabla V = \lambda\nabla g$ and the equation of constraint then give the system of equations

$$\begin{cases} yz = \lambda(4y + 5z), \\ xz = \lambda(4x + 5z), \\ xy = \lambda(5x + 5y), \\ 4xy + 5xz + 5yz = 48. \end{cases}$$

This system could be solved by solving each of the first three equations for λ and using these equations as in Example 1 to obtain equations that relate x, y, and z. Let us illustrate a different solution method that utilizes the symmetry of the variables in this example. We multiply the first three equations by x, y, and z, respectively. This gives the system

$$\begin{cases} xyz = \lambda(4xy + 5xz), \\ xyz = \lambda(4xy + 5yz), \\ xyz = \lambda(5xz + 5yz), \\ 4xy + 5xz + 5yz = 48. \end{cases}$$

Subtracting the second equation from the first, we obtain $5\lambda(x - y)z = 0$. This implies that either $\lambda = 0$, $x - y = 0$, or $z = 0$. The equation $z = 0$ clearly does not give the maximum volume, so we disregard that case. Similarly, if $\lambda = 0$, then one of x, y, or z must be zero, so we disregard that case. We conclude that the maximum value occurs when $x - y = 0$, so $y = x$. Subtracting the third equation from the second, we obtain $\lambda(4y - 5z)x = 0$. As before, we conclude that $z = \dfrac{4y}{5}$. We now have x and z in terms of y. Substituting these expressions for x and z into the fourth equation above, we obtain

$$4(y)y + 5(y)\left(\frac{4y}{5}\right) + 5y\left(\frac{4y}{5}\right) = 48,$$
$$12y^2 = 48,$$
$$y^2 = 4,$$
$$y = 2 \qquad \text{(we disregard } y = -2\text{)}.$$
$$y = 2 \text{ implies } x = 2 \text{ and } z = \frac{4(2)}{5} = \frac{8}{5}.$$

The maximum volume is

$$V\left(2, 2, \frac{8}{5}\right) = (2)(2)\left(\frac{8}{5}\right) = \frac{32}{5} = 6.4 \text{ ft}^3$$

when the lengths of the sides of the base are 2 and the height is 8/5. ∎

EXAMPLE 3

A rectangular box with no top is to have volume 8 ft^3. Material for the bottom of the box costs twice as much as material for the sides. Find the dimensions of the box that has minimum total cost of materials.

SOLUTION The box is pictured in Figure 12.9.5. The variables are x, y, z, and C, the total cost of material.

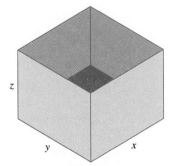

FIGURE 12.9.5 The length, width, and height of the rectangular box with no top are variables. The max-min variable is the cost of materials of the box.

Let k denote the cost per unit area of material for the sides of the box. The cost per unit area of material for the bottom of the box is then $2k$. The total cost of material is

$$C(x, y, z) = 2kxy + k(2yz + 2xz).$$

The cost is subject to the constraint that the volume is 8. That is,

$$xyz = 8.$$

This is of the form $g(x, y, z) = d$, where $g(x, y, z) = xyz$, so

$$\nabla g = yz\mathbf{i} + xz\mathbf{j} + xy\mathbf{k}.$$

There are no points on the surface $xyz = 8$ where $\nabla g = \mathbf{0}$, since $\nabla g = \mathbf{0}$ implies $xyz = 0$. We can use the Lagrange Multiplier method to find an extreme value of the cost subject to the volume constraint. We assume the extreme value obtained is a minimum value. (See Exercise 36.)

We have $\nabla C = (2ky + 2kz)\mathbf{i} + (2kx + 2kz)\mathbf{j} + (2ky + 2kx)\mathbf{k}$. The vector equation $\nabla V = \lambda \nabla g$ and the equation of constraint then give the system

$$\begin{cases} 2ky + 2kz = \lambda yz, \\ 2kx + 2kz = \lambda xz, \\ 2ky + 2kx = \lambda xy, \\ xyz = 8. \end{cases}$$

We can use the method of Example 2 to solve this system. Multiplying the first three equations by x, y, and z, respectively, we obtain the system

$$\begin{cases} 2kxy + 2kxz = \lambda xyz, \\ 2kxy + 2kyz = \lambda xyz, \\ 2kyz + 2kxz = \lambda xyz, \\ xyz = 8. \end{cases}$$

Subtracting the second equation from the first gives

$$2kz(x - y) = 0, \quad \text{so } x = y. \qquad \text{(We disregard } z = 0.)$$

Subtracting the third equation from the second gives

$$2kx(y - z) = 0, \quad \text{so } z = y. \qquad \text{(We disregard } x = 0.)$$

Since $x = y = z$, the fourth equation gives $x^3 = 8$, so $x = 2$. The least expensive box has length of each edge 2. ■

Problems with More than One Constraint

Let us consider the problem of finding extrema of $f(x, y, z)$ subject to $g_1(x, y, z) = d_1$ and $g_2(x, y, z) = d_2$, where d_1 and d_2 are constants. That is, we want extreme values of $f(x, y, z)$ for (x, y, z) on the curve of intersection of the surfaces $g_1(x, y, z) = d_1$ and $g_2(x, y, z) = d_2$. Extrema can occur only where the directional derivative of f in the direction of the curve is zero. If $\nabla g_1 \neq \mathbf{0}$ and $\nabla g_2 \neq \mathbf{0}$, this means that the gradient of f must be in the plane determined by ∇g_1 and ∇g_2. See Figure 12.9.6. It follows that there are numbers λ_1 and λ_2 such that

$$\nabla f = \lambda_1 \nabla g_1 + \lambda_2 \nabla g_2.$$

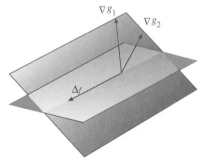

FIGURE 12.9.6 If ∇f is orthogonal to the intersection of the surfaces $g_1(x, y, z) = d_1$ and $g_2(x, y, z) = d_2$, then ∇f must be in the plane determined by ∇g_1 and ∇g_2, so there are constants λ_1 and λ_2 such that $\nabla f = \lambda_1 \nabla g_1 + \lambda_2 \nabla g_2$.

This vector equation and the two equations of constraint form the system of equations that gives the extrema of f.

The same ideas can be used for problems that involve any number of variables and any number of constraints. If f, $g_1, \ldots, g_m$ are differentiable functions of $(x_1, \ldots, x_n)$ and $\nabla g_j \neq \mathbf{0}$ at points where $g_j(x_1, \ldots, x_n) = d_j$, $j = 1, \ldots, m$, then extrema of the function

$$f(x_1, \ldots, x_n)$$

subject to the constraints

$$g_1(x_1, \ldots, x_n) = d_1, \ldots, g_m(x_1, \ldots, x_n) = d_m$$

occur at points that are solutions of the system of $n + m$ equations that consists of the n equations derived from the vector equation $\nabla f = \lambda_1 \nabla g_1 + \cdots + \lambda_m \nabla g_m$ and the m equations of constraint. Note that we introduce one Lagrange multiplier for each equation of constraint.

For the special case of extrema of $f(x, y, z)$ subject to two constraints $g_1(x, y, z) = d_1$, $g_2(x, y, z) = d_2$, we can simplify the system of equations. That is, we note that $\nabla g_1 \times \nabla g_2$ is orthogonal to both ∇g_1 and ∇g_2. Then

$$\nabla f = \lambda_1 \nabla g_1 + \lambda_2 \nabla g_2 \text{ implies}$$
$$\nabla f \cdot (\nabla g_1 \times \nabla g_2) = \lambda_1 \nabla g_1 \cdot (\nabla g_1 \times \nabla g_2) + \lambda_2 \nabla g_2 \cdot (\nabla g_1 \times \nabla g_2) = 0.$$

This means that we can replace the three component equations of the vector equation $\nabla f = \lambda_1 \nabla g_1 + \lambda_2 \nabla g_2$ by the single equation $\nabla f \cdot (\nabla g_1 \times \nabla g_2) = 0$, which does not involve the multipliers λ_1 and λ_2. The extreme values of $f(x, y, z)$ occur at points (x, y, z) that are solutions of the system of three equations

$$\begin{cases} \nabla f \cdot (\nabla g_1 \times \nabla g_2) = 0, \\ g_1(x, y, z) = d_1, \\ g_2(x, y, z) = d_2. \end{cases}$$

FIGURE 12.9.7 The curve of intersection of the cone and the plane can be sketched by connecting points of intersection of the traces in each coordinate plane. The curve is elliptical in shape.

EXAMPLE 4

Find points on the curve of intersection of the cone $z^2 = 4x^2 + 4y^2$ and the plane $4x + 3y + 4z = 7$ that are nearest to and farthest from the origin.

SOLUTION The curve is sketched in Figure 12.9.7. Note that the curve of intersection of the cone with the plane is a closed set. Since the distance from the origin is a continuous function, we know that there is a point on the curve that gives the maximum distance and a point on the curve that gives the minimum distance. We want the extrema of

$$D(x, y, z) = x^2 + y^2 + z^2$$

subject to the constraints

$$4x^2 + 4y^2 - z^2 = 0 \text{ and } 4x + 3y + 4z = 7.$$

Note that $D(x, y, z)$ is the *square* of the function that gives the distance of the point (x, y, z) from the origin. This function is easier to work with and has extrema at the same points as the distance function.

The extreme values of D occur at points (x, y, z) that are solutions of the system

$$\begin{cases} \nabla D \cdot (\nabla(4x^2 + 4y^2 - z^2) \times \nabla(4x + 3y + 4z)) = 0, & (1) \\ 4x^2 + 4y^2 - z^2 = 0, & (2) \\ 4x + 3y + 4z = 7. & (3) \end{cases}$$

We have

$$\nabla D = 2x\mathbf{i} + 2y\mathbf{j} + 2z\mathbf{k},$$
$$\nabla(4x^2 + 4y^2 - z^2) = 8x\mathbf{i} + 8y\mathbf{j} - 2z\mathbf{k},$$
$$\nabla(4x + 3y + 4z) = 4\mathbf{i} + 3\mathbf{j} + 4\mathbf{k},$$
$$\nabla(4x^2 + 4y^2 - z^2) \times \nabla(4x + 3y + 4z)$$
$$= \begin{vmatrix} \mathbf{i} & \mathbf{j} & \mathbf{k} \\ 8x & 8y & -2z \\ 4 & 3 & 4 \end{vmatrix}$$
$$= \begin{vmatrix} 8y & -2z \\ 3 & 4 \end{vmatrix} \mathbf{i} - \begin{vmatrix} 8x & -2z \\ 4 & 4 \end{vmatrix} \mathbf{j} + \begin{vmatrix} 8x & 8y \\ 4 & 3 \end{vmatrix} \mathbf{k}$$
$$= (32y + 6z)\mathbf{i} - (32x + 8z)\mathbf{j} + (24x - 32y)\mathbf{k}.$$

Equation 1 then implies

$$(2x)(32y + 6z) - (2y)(32x + 8z) + (2z)(24x - 32y) = 0,$$
$$64xy + 12xz - 64xy - 16yz + 48xz - 64yz = 0,$$
$$60xz - 80yz = 0,$$
$$20z(3x - 4y) = 0.$$

This means either $z = 0$ or $3x - 4y = 0$.

If $z = 0$, Equation 2 implies $x = y = 0$, but $(x, y, z) = (0, 0, 0)$ does not satisfy Equation 3, so the system does not have a solution with $z = 0$.

If $3x - 4y = 0$, substitution into Equation 2 implies either $z = 5x/2$ or $z = -5x/2$. Substitution of $y = 3x/4$ and $z = 5x/2$ into Equation 3 then implies $x = 28/65$, so $(28/65, 21/65, 70/65)$ is a solution of the system. Similarly, $z = -5x/2$ gives the solution $(-28/15, -21/15, 70/15)$.

The maximum and minimum values of the distance between points on the curve and the origin can occur only at

$$\left(\frac{28}{65}, \frac{21}{65}, \frac{70}{65}\right) \quad \text{or} \quad \left(-\frac{28}{15}, -\frac{21}{15}, \frac{70}{15}\right).$$

The distance of the point $(28/65, 21/65, 70/65)$ from the origin is

$$\sqrt{\left(\frac{28}{65}\right)^2 + \left(\frac{21}{65}\right)^2 + \left(\frac{70}{65}\right)^2} \approx 1.204;$$

this is the point on the curve that is nearest to the origin.

The distance of the point $(-28/15, -21/15, 70/15)$ from the origin is

$$\sqrt{\left(-\frac{28}{15}\right)^2 + \left(-\frac{21}{15}\right)^2 + \left(\frac{70}{15}\right)^2} \approx 5.217;$$

this is the point on the curve that is farthest from the origin. ∎

Extrema in Closed, Bounded Regions

EXAMPLE 5

Find the maximum and minimum values of

$$f(x, y, z) = x^2 - 2x + y^2 - 4y + z^2 - 4z$$

in the region $x^2 + y^2 + z^2 \leq 36$.

SOLUTION Since f is continuous on the closed and bounded region, we know that there are maximum and minimum values of f in the region. The maximum and minimum values can occur either at critical points in the interior of the region or on the boundary of the region.

Let us begin by determining the critical points of f. The first partial derivatives of f are

$$f_x = 2x - 2, \quad f_y = 2y - 4, \quad f_z = 2z - 4.$$

Setting each partial derivative of f equal to zero, we obtain the system

$$\begin{cases} 2x - 2 = 0, \\ 2y - 4 = 0, \\ 2z - 4 = 0. \end{cases}$$

This system of equations has solution $(1, 2, 2)$, which is a point inside the sphere $x^2 + y^2 + z^2 = 36$.

We also need to determine extrema of f on the boundary of the sphere. That is, we need extrema of

$$x^2 - 2x + y^2 - 4y + z^2 - 4z$$

subject to the constraint

$$x^2 + y^2 + z^2 = 36.$$

The vector equation $\nabla(x^2 - 2x + y^2 - 4y + z^2 - 4z) = \lambda \nabla(x^2 + y^2 + z^2)$ and the equation of constraint give the system of equations

$$\begin{cases} 2x - 2 = \lambda(2x), \\ 2y - 4 = \lambda(2y), \\ 2z - 4 = \lambda(2z), \\ x^2 + y^2 + z^2 = 36. \end{cases}$$

We can solve this system by solving the first three equations for λ.

$$2x - 2 = \lambda(2x) \text{ implies } x \neq 0 \text{ and } \lambda = \frac{x - 1}{x}.$$

$$2y - 4 = \lambda(2y) \text{ implies } y \neq 0 \text{ and } \lambda = \frac{y - 2}{y}.$$

It follows that

$$\frac{y - 2}{y} = \frac{x - 1}{x},$$
$$x(y - 2) = (x - 1)y,$$
$$xy - 2x = xy - y,$$
$$y = 2x.$$

Similarly,

$$2z - 4 = \lambda(2z) \text{ implies } z \neq 0, \quad \lambda = \frac{z-2}{z}, \text{ and } z = 2x.$$

Substitution into the fourth equation of our system then gives

$$x^2 + (2x)^2 + (2x)^2 = 36,$$
$$9x^2 = 36,$$
$$x^2 = 4,$$
$$x = 2, -2.$$

TABLE 12.9.1

(x, y, z)	$f(x, y, z)$
$(1, 2, 2)$	-9
$(2, 4, 4)$	0
$(-2, -4, -4)$	72

When $x = 2$, $y = 4$ and $z = 4$. When $x = -2$, $y = -4$ and $z = -4$.

The values of $f(x, y)$ at each of the candidates for extrema are given in Table 12.9.1. We see that $f(1, 2, 2) = -9$ is the minimum value and $f(-2, -4, -4) = 72$ is the maximum value. ■

EXERCISES 12.9

In Exercises 1–14, find all extrema of the given functions subject to the indicated constraints.

1. $f(x, y) = 3x - 4y$, $x^2 + y^2 = 25$

2. $f(x, y) = 12x + 5y$, $x^2 + y^2 = 169$

3. $f(x, y) = xy$, $x^2 + 3y^2 = 6$

4. $f(x, y) = (x^2/4) - y^2$, $x^2 + y^2 = 1$

5. $f(x, y, z) = x + 2y - 2z$, $x^2 + y^2 + z^2 = 9$

6. $f(x, y, z) = 3x + 4y + 12z$, $x^2 + y^2 + z^2 = 169$

7. $f(x, y, z) = xyz$, $xy + xz + 3yz = 4$ (one local maximum and one local minimum)

8. $f(x, y, z) = xyz$, $xy + 3xz + 4yz = 9$ (one local maximum and one local minimum)

9. $f(x, y, z) = 2xy + 4xz + yz$, $xyz = 1$ (one local minimum)

10. $f(x, y, z) = 9xy + 3xz + yz$, $xyz = 1$ (one local minimum)

11. $f(x, y, z) = xyz$, $x^2 + y^2 + z^2 = 4$, $x + y = 2$

12. $f(x, y, z) = x + y + z$, $x^2 + y^2 + z^2 = 1$, $x + 2y - 2z = 0$

13. $f(x, y, z) = 2x - 3y + z$, $x^2 + y^2 = 1$, $y^2 + z^2 = 1$

14. $f(x, y, z) = xyz$, $x^2 + y^2 = 1$, $y^2 + z^2 = 1$

In Exercises 15–20, find the maximum and minimum values of the given functions in the indicated regions.

15. $f(x, y) = -x^2 + 4x - y^2 + 2y$, $x^2 + y^2 \leq 20$

16. $f(x, y) = x^2 - 12x + y^2 - 9y$, $x^2 + y^2 \leq 25$

17. $f(x, y, z) = x^2 - 4x + y^2 + 2y + z^2 - 6z$, $x^2 + y^2 + z^2 \leq 56$

18. $f(x, y, z) = x^2 - 2x + y^2 - 4y + z^2 - 4z$, $x^2 + y^2 + z^2 \leq 9$

19. $f(x, y, z) = 2xy$, $x^2 + y^2 + z^2 \leq 1$

20. $f(x, y, z) = 2xy + 2z^2$, $x^2 + y^2 + z^2 \leq 1$

21. A rectangular box with square base and no top is to have volume 32. Find the dimensions and material required for the box that requires the least material for the four sides and bottom.

22. A rectangular box with no top is to have volume 12 ft^3. Material for the bottom of the box costs \$3/ft^2. Material for the four sides costs \$1/ft^2. Find the dimensions and cost of the box that has minimum total cost of materials.

23. Material for the top of a rectangular box costs \$2/ft^2. Material for the bottom and four sides costs \$1/ft^2. Find the dimensions and volume of the box with largest volume that can be made with \$9 for cost of material.

24. Find the point on the plane $x + 2y - z = 12$ that is nearest the origin.

25. Show that the volume of a rectangular box with fixed total surface area of the six sides is largest for a cube.

26. Show that the area of a triangle with fixed perimeter is largest for an equilateral triangle.

27. A trough is to be made from a rectangular piece of metal that is 12 in. wide. Equal lengths from each side along the length are to be folded up the same angle to form slanted sides. See figure. Find the length that should be folded from

each end and the angle that the sides should be bent so the cross-sectional area of the trough is maximum.

28. A trough with cross-sectional area 4 ft² is to be made from a rectangular piece of metal by bending equal lengths from each side along the length. Each side is to be folded up the same angle. Find the length that should be folded from each end and the angle that the sides should be bent so the material required is smallest.

In Exercises 29–32, find the extrema of the given functions in the indicated regions by sketching the regions and level curves of the functions.

29. $f(x, y) = x - y, x^2 + y^2 \leq 1$

30. $f(x, y) = 2x - y, x^2 + y^2 \leq 25$

31. $f(x, y) = x^2 + y^2, 1 \leq x + y \leq 2, x \geq 0, y \geq 0$

32. $f(x, y) = xy, x^2 + y^2 \leq 4$

33. Find the minimum of $d_1^2 + d_2^2 + d_3^2$, the sum of the squares of the distances from a point inside a triangle to each side of the triangle. See figure. (*Hint:* Obtain a linear equation of constraint by noting that the area of the triangle can be expressed as $\frac{1}{2}(ad_1 + bd_2 + cd_3) = A$.)

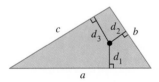

34. **All points determined by the Lagrange Multiplier method are not extrema of the function subject to the constraint.** Verify this by showing that the Lagrange Multiplier method applied to the function $f(x, y) = xy^2$ subject to the constraint $x + y = 0$ gives the point $(0, 0)$, but $f(0, 0)$ is not an extreme value of $f(x, y)$ subject to the constraint.

35. Verify that the volume $V = xyz$ subject to the constraint $4xy + 5xz + 5yz = 48$ has a maximum value for $x, y, z \geq 0$ as indicated:
(a) If $4xy + 5xz + 5yz = 48$ and $x, y, z > 0$, then
$$y = \frac{48 - 5xz}{4x + 5z} \leq \frac{48}{4x + 5z} \leq \frac{48}{4x} = \frac{12}{x}.$$

Show that $z \leq \dfrac{48}{5x}$.

(b) Use the results of (a) to show that $V = xyz \leq \dfrac{12(48)}{5x}$.

(c) Use the method of (a) and (b) to show that $4xy + 5xz + 5yz = 48$, $x, y, z > 0$, implies $V \leq \dfrac{12(48)}{5y}$ and $V \leq \dfrac{(48)(48)}{25z}$.

(d) Find value x_0, y_0, z_0 for which $V(x, y, z) < 32/5$ if either $x > x_0, y > y_0$, or $z > z_0$. Since $V(2, 2, 8/5) = 32/5$, it follows that no value of $V(x, y, z)$ with either $x > x_0, y > y_0$, or $z > z_0$ can be a maximum value of V subject to the constraint.

(e) Explain why V has a maximum value on the surface $4xy + 5xz + 5yz = 48, 0 \leq x \leq x_0, 0 \leq y \leq y_0, 0 \leq z \leq z_0$. It then follows from (d) that this maximum value is the maximum value of V subject to the constraint.

36. Verify that the cost $C = 2kxy + k(2yz + 2xz)$ subject to the volume constraint $xyz = 8$ has a minimum value for $x, y, z > 0$ as indicated:
(a) If $xyz = 8$ and $x, y, z > 0$, then
$$C \geq 2kyz = \frac{16k}{x}.$$

Show that $C \geq \dfrac{16k}{y}$ and $C \geq \dfrac{16k}{z}$.

(b) Find value $x_0 > 0, y_0 > 0, z_0 > 0$ for which $C(x, y, z) > 24k$ if either $0 < x < x_0, 0 < y < y_0$, or $0 < z < z_0$. Since $V(2, 2, 2) = 24k$, it follows that no value of $C(x, y, z)$ with either $0 < x < x_0, 0 < y < y_0$, or $0 < z < z_0$ can be a minimum value of C subject to the constraint.

(c) Explain why V has a minimum value on the surface $xyz = 8, x \geq x_0, y \geq y_0, z \geq z_0$. (*Hint:* $x \geq x_0, y \geq y_0, z \geq z_0$, and $xyz = 8$ imply $x \leq \dfrac{8}{y_0 z_0}, y \leq \dfrac{8}{x_0 z_0}$, and $z \leq \dfrac{8}{x_0 y_0}$.) It then follows from (b) that this minimum value is the minimum value of C subject to the constraint.

REVIEW OF CHAPTER 12 CONCEPTS

- In this chapter we have studied limits, continuity, and differentiability of functions of several variables. Limits and differentiability of functions of more than one variable are complicated by the fact that there are an infinite number of ways to approach a particular point.
- In order that $\lim\limits_{(x,y)\to(a,b)} f(x, y) = L$, the values $f(x, y)$ must approach L along every path that approaches (a, b). One way to show that $\lim\limits_{(x,y)\to(a,b)} f(x, y)$ does not exist is to show that there are two different paths to (a, b) along which $f(x, y)$ approaches different numbers.
- We can use the Limit Theorem and the Composition Theorem to show that many of the functions with which we deal are continuous at each point where they are defined.
- Partial derivatives of f give the rate of change of f in the directions of the coordinate axes. Directional derivatives give the rate of change in any direction.
- Partial derivatives are evaluated by using the rules of differentiation of functions of a single variable. Directional derivatives are evaluated by using the gradient. The gradient of a function gives the direction of the maximum rate of change of the function.
- Linear approximation and differentiability depend on the fact that there is a linear function that approximates the function in each direction. The definition of differentiability is chosen to be convenient for the

theory. Functions that have continuous first partial derivatives are differentiable.
- The Chain Rule for functions of several variables has many forms that depend on the number of dependent variables and the number of independent variables.
- The gradient $\nabla f(x, y)$ is orthogonal to the level curves of $f(x, y)$. Similarly, the gradient $\nabla F(x, y, z)$ is orthogonal to the level surfaces of $F(x, y, z)$.
- The gradient ∇F is orthogonal to the surface $F(x, y, z) = d$. The three-dimensional vector $-f_x(x, y)\mathbf{i} - f_y(x, y)\mathbf{j} + \mathbf{k}$ is orthogonal to the surface $z = f(x, y)$. Do not confuse the vector normal to the surface $z = f(x, y)$ with the gradient of f; $\nabla f = f_x\mathbf{i} + f_y\mathbf{j}$ is a two-dimensional vector.
- Local extrema can occur only at critical points, where either all first partial derivatives are zero or some first partial derivative does not exist. Extreme values of a continuous function must occur at either a critical point or a point on the boundary of a region.
- Extrema of $f(x, y)$ subject to the constraint $g(x, y) = 0$ can occur only at points where the level curves of f are tangent to the constraint curve. This implies that ∇f and ∇g are parallel, so there is a number λ such that $\nabla f = \lambda \nabla g$. Similar results hold for functions of more than two variables and lead to the Lagrange method for finding extrema subject to constraints.

CHAPTER 12 REVIEW EXERCISES

Find the domain of the functions in Exercises 1–2.

1. $f(x, y) = \dfrac{x^2 - y^2}{x - y}$

2. $f(x, y) = \sqrt{1 - \ln(x^2 + y^2)}$

Characterize and sketch several level curves of the functions given in Exercises 3–4. Sketch the graphs of the functions.

3. $f(x, y) = x^2 + 2y$

4. $f(x, y) = \dfrac{x^2 + y^2}{y}$

Characterize and sketch several level surfaces of the functions in Exercises 5–6.

5. $f(x, y, z) = 2\sqrt{x^2 + z^2} - y$

6. $f(x, y, z) = z + x^2 + y^2$

For each of the limits in Exercises 7–10, either evaluate and verify the existence of the limit or show that the limit does not exist.

7. $\lim\limits_{(x,y)\to(0,0)} \dfrac{x^2}{x^2 + y^2}$

8. $\lim\limits_{(x,y)\to(0,0)} \dfrac{2xy}{\sqrt{x^2 + y^2}}$

9. $\lim\limits_{(x,y)\to(0,0)} \dfrac{2x^2 - y^2}{x^2 + y^2}$

10. $\lim\limits_{(x,y)\to(0,0)} \dfrac{x^3}{x^2 + y^2}$

11. Sketch the set of points (x, y) such that $y > x^2$. **(a)** What is the boundary of the set? **(b)** Is the set open, closed, or neither?

12. Sketch the set of points (x, y) such that $x^2 + y^2 < 4$ and $y > x$. **(a)** What is the boundary of the set? **(b)** Is the set open, closed, or neither?

13. Sketch the set of points (x, y) such that $x \geq 0$, $y \geq 0$, and $x + 2y \leq 2$. **(a)** What is the boundary of the set? **(b)** Is the set open, closed, or neither?

14. Sketch the set of points (x, y) such that $x^2 + y^2 \leq 4$ and $x > 0$. **(a)** What is the boundary of the set? **(b)** Is the set open, closed, or neither?

Find all points of continuity of the functions in Exercises 15–18.

15. $f(x, y) = \begin{cases} x^2 + y^2, & x^2 + y^2 \leq 1, \\ 1, & x^2 + y^2 > 1 \end{cases}$

16. $f(x, y) = \begin{cases} y, & y \geq x^2, \\ x, & y < x^2 \end{cases}$

17. $f(x, y) = \begin{cases} \dfrac{x^2 - y^2}{x - y}, & x \neq y, \\ 2, & x = y \end{cases}$

18. $f(x, y) = \begin{cases} \dfrac{|x - y|}{x - y}, & x \neq y, \\ 0, & x = y \end{cases}$

Find all first- and second-order partial derivatives of the functions in Exercises 19–24.

19. $f(x, y) = x^2 - 2xy + 3y^2$

20. $f(x, y) = \sin xy$

21. $f(x, y) = \ln \sqrt{x^2 + y^2}$

22. $f(x, y) = \tan^{-1}(x/y)$

23. $f(x, y, z) = xy^2z^3$

24. $f(x, y, z) = 3xy + 2xz + yz$

25. Express the differential of $f(x, y) = \dfrac{\sqrt{x}}{y}$ in terms of the variables x, y, dx, and dy.

26. Express the differential of $f(x, y) = e^{2x-y}$ in terms of the variables x, y, dx, and dy.

27. The magnitude of the force due to gravity of an object with mass m at a distance d from the center of the earth is $F = km/d^2$, where k is a constant. Find the approximate maximum percent error in F due to possible errors of 0.2 percent in measurement of the mass and 1.5 percent in measurement of the distance.

28. Find the approximate maximum percent error in the current I due to possible errors of 5 percent in measurement of the voltage V and 3 percent in measurement of the resistance R. $(I = V/R.)$

29. $z = x^2/y$, $x = \tan t$, $y = \cos^2 t$. Find dz/dt.

30. $z = \cos(x^2 - y^2)$, $x = t^2 - 3t + 4$, $y = t^3 - 1$. Find dz/dt.

31. $z = x^2 - y^2$, $x = r \cos \theta$, $y = r \sin \theta$. Find $\partial z/\partial r$ and $\partial z/\partial \theta$.

32. $z = u^2v - 2uv$, $u = 2x - 3y$, $v = 3x + 2y$. Find $\partial z/\partial x$ and $\partial z/\partial y$.

33. $x^2y + y^2z + z^2x = 0$. Find $\partial z/\partial x$ and $\partial z/\partial y$.

34. $xz^2 + yz = 1$. Find $\partial z/\partial x$ and $\partial z/\partial y$.

35. $z^2 = x^2 + y^2$. Find $\dfrac{\partial^2 z}{\partial y \partial x}$.

36. $x^2 + y^2 + z^2 = 4$. Find $\dfrac{\partial^2 z}{\partial y \partial x}$.

Suppose the equations in Exercises 37–38 define u and v as differentiable functions of (x, y). Find formulas for $\dfrac{\partial u}{\partial x}$, $\dfrac{\partial v}{\partial x}$, $\dfrac{\partial u}{\partial y}$ and $\dfrac{\partial v}{\partial y}$.

37. $2x - y + u - 2v = 2$, $x - 3y - 2u + v = 3$

38. $x^2 + y^2 + 2u^2 + v^2 = 6$, $x - y + u - 2v = 1$

39. $z = F(x, y)$, where $F_x(2, 3) = 2$ and $F_y(2, 3) = 3$. Find the rate of change of z when $x = 2$ and $y = 3$ if x is increasing at a rate of 5 units/s and y is decreasing at a rate of 4 units/s at that time.

40. The radius of a right circular cylinder is increasing at a rate of 2 ft/s while the height is decreasing at a rate of 3 ft/s. Find the rate of change of **(a)** the volume and **(b)** the total surface area of the cylinder when the radius is 6 ft and the height is 4 ft.

41. Find the rate of change of $f(x, y, z) = x^2 - xz + yz + y^2$ at the point $(2, -1, 3)$ in the direction of $\mathbf{i} - 2\mathbf{j} + 2\mathbf{k}$.

42. Find the rate of change of the function $f(x, y, z) = x^2 + y^2 + z^2 - 4xyz$ at the point (x, y, z) in the direction of the origin.

43. Find the rate of change of $f(x, y) = 3x^2 + 4xy - y^3$ in the direction of the unit tangent vector of the curve
$$C : x = 3 \cos t, \qquad y = 2 \sin t$$
at the point where $t = 4\pi/3$.

44. Find the rate of change of $f(x, y) = \ln(x^2 + y^2)$ in the direction of the unit tangent vector of the curve
$$C : x = 1 + t^2, \qquad y = 1 - t$$
at the point where $t = 2$.

45. Find the rate of change of $f(x, y, z) = x^2 + y^2 + z^2$ in the direction of the upward normal vector (with positive $\mathbf{k}$) to the surface $z = 3x^2 + 2y^2$ at the point $(2, 3, 30)$.

46. Find the rate of change of $f(x, y, z) = xyz$ in the direction of the upward normal vector (with positive $\mathbf{k}$) to the surface $z = \sqrt{9 - x^2 - y^2}$ at the point $(2, 1, 2)$.

For the functions given in Exercises 47–50, sketch the level curves $f(x, y) = 2, 3,$ and 4. Evaluate ∇f at the point $(2, 1)$. Find two unit vectors U such that $D_U f(2, 1) = 0$. Sketch ∇f and U at $(2, 1)$.

47. $f(x, y) = x + y$

48. $f(x, y) = x^2 - y^2$

49. $f(x, y) = x^2 - y$

50. $f(x, y) = 2x - y$

In Exercises 51–52, find a unit vector U such that the rate of change of the given function at $(2, 3)$ in the direction of U is maximum and evaluate that maximum value.

51. $f(x, y) = xy^2 - yx^2$

52. $f(x, y) = x^2 y^2 - xy$

Find an equation of the plane tangent to the given surface at the indicated point in each of Exercises 53–56.

53. $z = x^2 + y^2$, at $(2, 3, 13)$

54. $z = xe^{-y}$, at $(1, 0, 1)$

55. $z^2 = x^2 + y^2$, at $(3, 4, -5)$

56. $xy + xz + yz = 1$, at $(2, -1, 3)$

Find symmetric equations and parametric equations of the line normal to the given surface at the indicated point in each of Exercises 57–60.

57. $z^2 = x^2 - y^2$, at $(5, -4, 3)$

58. $z = xy$, at $(3, 2, 6)$

59. $z = x^2 + y^2$, at $(3, 4, 25)$

60. $z^2 = x^2 + y^2$, at $(3, 4, -5)$

Find all extrema of the functions given in Exercises 61–66.

61. $f(x, y) = x^2 + 2xy + 3y^2$

62. $f(x, y) = x^3 + 3xy - y^3$

63. $f(x, y) = x^4 - 4xy + y^4$

64. $f(x, y) = \dfrac{y}{x^2 + y^2 + 1}$

65. $f(x, y, z) = x^2 + y^2 + z^2, x - y + z = 1$ (minimum)

66. $f(x, y, z) = x^2 + y^2 + z^2, x - y = 1$ (minimum)

67. A rectangular box with no top has volume 32 ft³. Find the dimensions and surface area of the box that has surface area a minimum.

68. A rectangular box with no top has surface area 48 ft². Find the dimensions and volume of the box that has volume a maximum.

In Exercises 69–72, find the maximum and minimum values of the given functions on the indicated regions.

69. $f(x, y) = x^2 y, x^2 + y^2 \leq 1$

70. $f(x, y) = x^2 y^2, x^2 + y^2 \leq 1$

71. $f(x, y) = xy^2$, region bounded by $x = 0, y = 0$, and $x + y = 1$

72. $f(x, y) = xy + y - x^2 - y^2$, region bounded by $x = 0, y = 0,$ and $x + y = 2$

In Exercises 73–76, use a sketch of the region and some level curves of f to find the maximum and minimum values of the given functions on the indicated regions.

73. $f(x, y) = x + y, x^2 + y^2 \leq 1$

74. $f(x, y) = x + y, x \geq 0, y \geq 0, x + 2y \leq 2$

75. $f(x, y) = x^2 + y^2, x \geq 0, y \geq 0, x + 2y \leq 2, x + 2y \geq 4$

76. $f(x, y) = x^2 + y^2, 0 \leq x \leq 1, x + y \leq 2, x + y \geq 1$

77. At which point(s) on the surface $z = \dfrac{1}{3}(x^3 + y^3)$ is the tangent plane parallel to the plane $x - y - z = 4$?

78. A **van der Waals fluid** obeys the equation of state $\left(p + \dfrac{a}{v^2}\right)(v - b) = RT$, where $a, b,$ and R are constants. Assuming the equation defines p as a function of (v, T), find **(a)** $\dfrac{\partial p}{\partial v}$ and **(b)** $\dfrac{\partial p}{\partial T}$.

79. Assume the van der Waals equation,
$$\left(p + \frac{a}{v^2}\right)(v - b) = RT,$$
defines p as a function of (v, T). Show that $\dfrac{\partial^2 p}{\partial T^2} = 0$.

80. Assume the van der Waals equation,
$$\left(p + \frac{a}{v^2}\right)(v - b) = RT,$$
defines v as a function of (p, T). Show that
$$\frac{\partial^2 v}{\partial T^2} = \frac{R^2(2a/v^3)[(3b/v) - 1]}{[p - (a/v^2) + (2ab/v^3)]^3}.$$

81. Explain in words some of the things that can cause the limit at a point of a function of one real variable not to exist. What can cause the limit of a function of two variables not to exist?

82. Explain how you would use the Lagrange Multiplier method to find extreme values of $f(x, y)$ for (x, y) restricted to line segments that form the boundary of a triangle. Explain why the Lagrange method might not give extreme values that occur at a corner of the triangle.

EXTENDED APPLICATION

LAGRANGE MULTIPLIERS AND HYDROELECTRIC POWER

The Great Northern Paper Company operates several hydroelectric generating stations, including one on the West Branch of the Penobscot River in central Maine. Water enters the station through a pipe from a dam and can be directed toward any or all of the three turbines. The first two turbines can each accept a volume of up to 1110 ft^3/sec, and the third can accept up to 1225 ft^3/sec. The rate of flow through each should be at least 250 ft^3/sec. The efficiency of each turbine in converting water power to electric power is a function of the rate of flow, and the efficiency is greatest about midway between the minimum and maximum rates. The level of efficiency is different for each turbine, and the rate of flow through each is constrained by the total rate of flow through the station.

As a company engineer, you are to devise a plan for distributing the total flow to the three turbines in a way that will produce the maximum amount of power at any level of flow. You know that the power generated by each turbine is given by Bernoulli's equation

$$W_i = (\Delta P + \Delta V + \Delta z - L(Q_T))h_i(Q_i),$$

where

W_i = power generated by turbine i (kilowatts)

ΔP = change in pressure between headwater (above the station) and tailwater (below the station)

ΔV = change in fluid velocity between headwater and tailwater

Δz = change in elevation between headwater and tailwater

$L(Q_T)$ = mechanical energy loss due to friction, a function of the total rate of flow Q_T (ft^3/sec)

$h_i(Q_i)$ = turbine efficiency, a function of the rate of flow Q_i (ft^3/sec) through turbine i

For this power station, $\Delta P = \Delta V = 0$ and $\Delta z = 170$ feet. Based on past observations, you have estimated that

$$L(Q_T) = 1.6 \times 10^{-6}Q_T^2$$

and that within the normal operating capacities of the turbines, the functions $h_i(Q_i)$ can be modeled by quadratic polynomials, as follows:

$$h_1(Q_1) = -18.89 + 0.1277Q_1 - (0.408 \times 10^{-4})Q_1^2,$$
$$250 \le Q_1 \le 1110$$
$$h_2(Q_2) = -24.51 + 0.1358Q_2 - (0.469 \times 10^{-4})Q_2^2,$$
$$250 \le Q_2 \le 1110$$
$$h_3(Q_3) = -27.02 + 0.1380Q_3 - (0.384 \times 10^{-4})Q_3^2,$$
$$250 \le Q_3 \le 1225$$

Tasks:

1. In terms of Q_T, find the values of Q_1, Q_2, and Q_3 that will maximize $W_1 + W_2 + W_3$ subject to the constraint $Q_1 + Q_2 + Q_3 = Q_T$.
2. Give an interval of values for Q_T for which your solution is valid. (For smaller values of Q_T, one or more turbines must be shut down. For larger values, one or more turbines must operate at full capacity. Devising a plan for these cases is a separate problem.)

Writing Assignment:

Write a report to your supervisor, giving your solution and explaining the methods used to obtain it. Your explanation should use language related to the power station, rather than abstract mathematics. Include your calculations on a separate sheet.

Source: Marvin Keener and Jeanne Agnew, Oklahoma State University, *Industry Related Problems for Mathematics Students.*

13 Multiple Integrals

We have seen that definite integrals of functions of one real variable have applications to many physical problems. In this chapter we will study analogous results for functions of several variables.

13.1 REGIONS IN THE PLANE AND IN THREE-DIMENSIONAL SPACE

Connections

Sketching, 10.7–8.

In later sections we will study integrals over certain types of regions in the plane and in three-dimensional space. In order to determine limits of integration of these integrals, we must be able to describe the regions in the way that we develop in this section.

We consider only regions that are bounded.

Vertically Simple Plane Regions

Let us consider the region in the plane bounded by

$$y = g_1(x), \quad y = g_2(x), \quad x = a, \quad \text{and} \quad x = b,$$

where g_1 and g_2 are continuous with $g_1(x) \leq g_2(x)$ for $a \leq x \leq b$. A typical such region of this type is sketched in Figure 13.1.1a. These regions have the property that vertical lines that intersect the region intersect it in a line segment. See Figure 13.1.1b. An arrow shot vertically from below the region would follow a trajectory with x fixed, enter the region at a point on the curve $y = g_1(x)$, and leave the region at a point on the curve $y = g_2(x)$. See Figure 13.1.1c. We call such regions **vertically simple**, or **simple in the direction of the y-axis.**

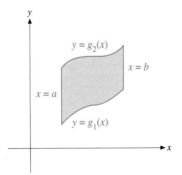

FIGURE 13.1.1a The region is bounded on the bottom by the curve $y = g_1(x)$ and on the top by the curve $y = g_2(x)$. It is bounded on the sides by the vertical lines $x = a$ and $x = b$.

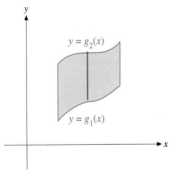

FIGURE 13.1.1b Vertical lines that intersect the region intersect it in a line segment.

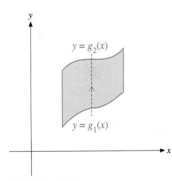

FIGURE 13.1.1c An arrow shot vertically from below the region would enter the region at a point on the curve $y = g_1(x)$ and leave the region at a point on the curve $y = g_2(x)$. The region is vertically simple.

The region pictured in Figure 13.1.2a is not vertically simple since some vertical lines intersect the region in two disjoint segments. This region can be divided into two vertically simple regions, as indicated in Figure 13.1.2b.

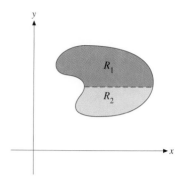

FIGURE 13.1.2a Some vertical lines intersect this region in two disjoint segments. The region is not vertically simple.

FIGURE 13.1.2b The region can be divided into two vertically simple regions.

We can **describe** a vertically simple region by the statements

For x fixed, y varies from $y = g_1(x)$ to $y = g_2(x)$.
x varies from $x = a$ to $x = b$.

For x fixed, y varies from the *lower* curve $y = g_1(x)$ to the *upper* curve $y = g_2(x)$; x varies from the *smaller* number $x = a$ to the *larger* number $x = b$.

The description of the region determines the region and can be used to sketch the region. To determine the region from the description it is better to start with the last statement of the description. The statement "x varies from $x = a$ to $x = b$" tells us that the region is bounded on the sides by the two vertical lines $x = a$ and $x = b$. These vertical lines should be sketched as in Figure 13.1.3a. The statement "For x fixed, y varies from $y = g_1(x)$ to $y = g_2(x)$" tells us that the region is bounded on the bottom by $y = g_1(x)$ and on the top by $y = g_2(x)$. A sketch of the region can then be completed by adding the top and bottom curves between the vertical lines we have drawn, as in Figure 13.1.3b.

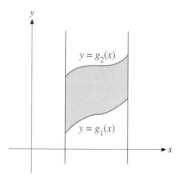

FIGURE 13.1.3a The statement "x varies from $x = a$ to $x = b$" tells us that the region is bounded on the sides by the two vertical lines $x = a$ and $x = b$.

FIGURE 13.1.3b The statement "For x fixed, y varies from $y = g_1(x)$ to $y = g_2(x)$" tells us that the region is bounded on the bottom by $y = g_1(x)$ and on the top by $y = g_2(x)$.

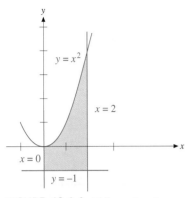

FIGURE 13.1.4 This region is bounded on the bottom by the line $y = -1$ and on the top by the curve $y = x^2$. It is bounded on the sides by the vertical lines $x = 0$ and $x = 2$.

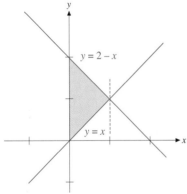

FIGURE 13.1.5 This region is bounded on the bottom by the curve $y = x$ and on the top by the curve $y = 2 - x$. It is bounded on the left side by the vertical line $x = 0$. The region is bounded on the right side by the vertical line through the intersection of the top and bottom curves.

EXAMPLE 1

Sketch and describe the region in the plane bounded by $y = x^2$, $y = -1$, $x = 0$, and $x = 2$.

SOLUTION   Each of the four curves is sketched and labeled in Figure 13.1.4. We see that the region bounded by the curves is vertically simple. From the sketch, we see that

For x fixed, y varies from $y = -1$ to $y = x^2$.
x varies from $x = 0$ to $x = 2$.   ∎

If the upper and lower boundary curves of a vertically simple region intersect at a point that forms a side boundary of the region, the equation of that side ($x = $ a constant) is generally not given. We must then determine the equation of that side by finding the x-coordinate of the point of intersection of the upper and lower boundary curves.

EXAMPLE 2

Sketch and describe the region in the plane bounded by $y = x$, $x + y = 2$, and the y-axis.

SOLUTION   The region is sketched and labeled in Figure 13.1.5. Note that $x + y = 2$ implies $y = 2 - x$. We see from the sketch that the region is vertically simple. Also,

For x fixed, y varies from $y = x$ to $y = 2 - x$.

Since the top and bottom intersect to form the right-side boundary, we must combine the equations of the top and bottom to find the x-coordinate of the intersection:

$$y = x \text{ and } y = 2 - x \text{ imply } x = 2 - x, \text{ so } x = 1.$$

The region is bounded on the right by the vertical line $x = 1$. The region is bounded on the left by the y-axis, which has equation $x = 0$. Hence,

x varies from $x = 0$ to $x = 1$.   ∎

Horizontally Simple Plane Regions

A region in the plane bounded by

$$x = g_1(y), \quad x = g_2(y), \quad y = a, \quad \text{and} \quad y = b,$$

where g_1 and g_2 are continuous with $g_1(y) \le g_2(y)$ for $a \le y \le b$, is called **horizontally simple** or **simple in the direction of the x-axis**. See Figure 13.1.6a. These regions have the property that horizontal lines that intersect the region intersect it in a line segment. See Figure 13.1.6b. An arrow shot horizontally from the left of the region would follow a trajectory with y fixed, enter the region at a point on the curve $x = g_1(y)$, and leave the region at a point on the curve $x = g_2(y)$. See Figure 13.1.6c.

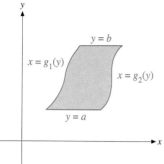

FIGURE 13.1.6a This region is horizontally simple.

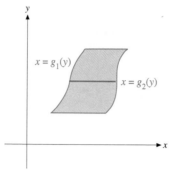

FIGURE 13.1.6b Horizontal lines that intersect the region intersect it in a line segment.

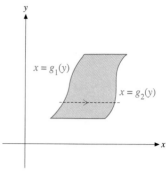

FIGURE 13.1.6c An arrow shot horizontally from the left of the region would enter the region at a point on the curve $x = g_1(y)$ and leave the region at a point on the curve $x = g_2(y)$.

We can describe a horizontally simple region by writing

For y fixed, x varies from $x = g_1(y)$ to $x = g_2(y)$.
y varies from $y = a$ to $y = b$.

Regions that are both vertically simple and horizontally simple may be described in either way, although one of the descriptions may be more convenient than the other. Also, it may be convenient to divide a region into subregions and describe each subregion separately.

EXAMPLE 3

Describe the region bounded by $x = y^2 - 2y$ and $y = x$ **(a)** as one or more horizontally simple regions and **(b)** as one or more vertically simple regions.

SOLUTION

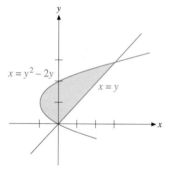

FIGURE 13.1.7a It is convenient to describe this region as a horizontally simple region.

(a) The region is sketched in Figure 13.1.7a. We see that the region is horizontally simple. Also,

For y fixed, x varies from $x = y^2 - 2y$ to $x = y$.

We need the y-coordinate of the intersection of the line and the parabola. The equations $x = y$ and $x = y^2 - 2y$ imply

$$y^2 - 2y = y,$$
$$y^2 - 3y = 0,$$
$$y(y - 3) = 0,$$
$$y = 0, \quad y = 3.$$

We conclude that

y varies from $y = 0$ to $y = 3$.

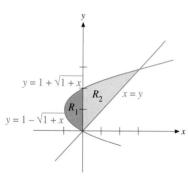

FIGURE 13.1.7b It is convenient to describe this region as two vertically simple regions.

(b) It is convenient to divide this region into two subregions that are vertically simple, as indicated in Figure 13.1.7b. We see that the top of each subregion is formed by the upper part of the parabola. The bottom of the left subregion is formed by the lower part of the parabola, and the bottom of the right subregion is formed by the line.

We need to express the upper and lower parts of the parabola as functions of x. To do this we use the Quadratic Formula to solve the equation $x = y^2 - 2y$ for y in terms of x. The equation $y^2 - 2y - x = 0$ implies

$$y = \frac{-(-2) \pm \sqrt{(-2)^2 - 4(1)(-x)}}{2(1)}$$

$$= \frac{2 \pm \sqrt{4 + 4x}}{2}$$

$$= 1 \pm \sqrt{1 + x}.$$

The upper part of the parabola is $y = 1 + \sqrt{1 + x}$; the lower part is $y = 1 - \sqrt{1 + x}$.

$y = 1 + \sqrt{1 + x}$ and $y = 1 - \sqrt{1 + x}$ intersect where $x = -1$.
$y = 1 - \sqrt{1 + x}$ and $y = x$ intersect where $x = 0$.
$y = 1 + \sqrt{1 + x}$ and $y = x$ intersect where $x = 3$.

The left subregion is then described by the statements

For x fixed, y varies from $y = 1 - \sqrt{1 + x}$ to $y = 1 + \sqrt{1 + x}$.
x varies from $x = -1$ to $x = 0$.

The right subregion is described by the statements

For x fixed, y varies from $y = x$ to $y = 1 + \sqrt{1 + x}$.
x varies from $x = 0$ to $x = 3$.   ∎

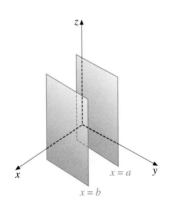

FIGURE 13.1.8a The statement "x varies from $x = a$ to $x = b$" tells us that the region is bounded on the "back" and "front" by the two planes, $x = a$ and $x = b$.

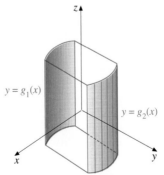

FIGURE 13.1.8b The statement "For x fixed, y varies from $y = g_1(x)$ to $y = g_2(x)$" tells us that the region is bounded on the "left side" by the vertical generalized cylinder $y = g_1(x)$ and on the "right side" by the vertical generalized cylinder $y = g_2(x)$.

Regions in Three-Dimensional Space

The same ideas that were used to describe regions in the plane can be used to describe regions in three-dimensional space. We will deal with regions in three-dimensional space that can be described by writing

For x and y fixed, z varies from $z = h_1(x, y)$ to $z = h_2(x, y)$.
For x fixed, y varies from $y = g_1(x)$ to $y = g_2(x)$.
x varies from $x = a$ to $x = b$.

The statement "x varies from $x = a$ to $x = b$" tells us the region is bounded in the **front** by the plane $x = b$ and is bounded in the **back** by the plane $x = a$. See Figure 13.1.8a. "For x fixed, y varies from $y = g_1(x)$ to $y = g_2(x)$" tells us the region is bounded on the **right side** by the vertical generalized cylinder $y = g_2(x)$ and is bounded on the **left side** by the vertical generalized cylinder $y = g_1(x)$. See Figure 13.1.8b. The statement "For x and y fixed, z varies from $z = h_1(x, y)$ to $z = h_2(x, y)$" tells us that the region is bounded on the **top** by the surface $z = h_2(x, y)$ and is bounded on the **bottom** by the surface $z = h_1(x, y)$. An arrow shot vertically from below the region would follow a trajectory with x and y fixed, enter the region at a point on the bottom surface $z = h_1(x, y)$, travel parallel to the vertical surfaces $y = g_1(x)$,

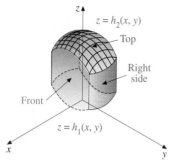

FIGURE 13.1.8c The statement "For x and y fixed, z varies from $z = h_1(x, y)$ to $z = h_2(x, y)$" tells us that the region is bounded on the "bottom" by the surface $z = h_1(x, y)$ and on the "top" by the surface $z = h_2(x, y)$.

$y = g_2(x)$, $x = a$, and $x = b$, and leave the region at a point on the top surface $z = h_2(x, y)$. See Figure 13.1.8c.

The statements "For x fixed, y varies from $y = g_1(x)$ to $y = g_2(x)$" and "x varies from $x = a$ to $x = b$," from the description of a region in three-dimensional space, correspond to the description of a region in the xy-plane that is simple in the direction of the y-axis. This region in the xy-plane is the **projection** of the region in three-dimensional space onto the xy-plane. See Figure 13.1.9.

As in the case of regions in the plane, it may be necessary to combine given equations to find all the formulas needed to describe a region in three-dimensional space. We will use the following procedure.

FIGURE 13.1.9 The projection of this region in three-dimensional space onto the xy-plane is simple in the direction of the y-axis.

To sketch and describe a region in three-dimensional space

- Sketch coordinate traces of each surface.
- Find points of intersection of traces in each coordinate plane.
- Sketch region; draw all boundary curves.
- Identify top and bottom surfaces. The equations of the top and bottom must involve the variable z.
- Identify side cylinders. The equations of the side cylinders must involve y and cannot involve z. If top and bottom surfaces intersect to form a side edge (Figure 13.1.10a), use the equations of the top and bottom to eliminate z. This gives the equation of a cylinder that contains the side edge and is used as the side cylinder.
- Identify front and back planes. If either the sides intersect (Figure 13.1.10b), the top and bottom intersect (Figure 13.1.10c), or both (Figure 13.1.10d), there will be either a front or back edge or a front or back point. The equations can then be used to find the x-coordinate of the intersection. This gives the front or back plane.

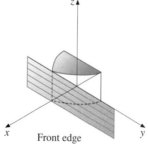

FIGURE 13.1.10a The top and bottom surfaces may intersect to form a side edge.

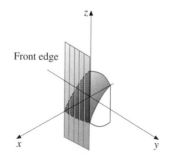

FIGURE 13.1.10b The side cylinders may intersect to form a front edge.

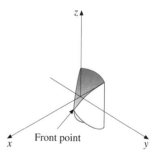

FIGURE 13.1.10c The top and bottom surfaces may intersect to form a front edge.

FIGURE 13.1.10d The top and bottom surfaces and the side cylinders may all intersect at a front point.

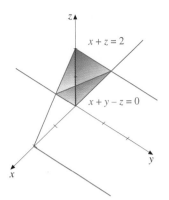

FIGURE 13.1.11 The region is sketched by drawing the coordinate traces of each plane and then connecting points of intersection of the traces with a line segment.

EXAMPLE 4

Sketch and describe the region in the first octant bounded by $x + z = 2$ and $x + y - z = 0$.

SOLUTION The region is sketched in Figure 13.1.11. The points of intersection of the planes in the yz-plane and the xz-plane have been connected to indicate a boundary edge. The fact that the region is in the first octant tells us that $x = 0$, $y = 0$, and $z = 0$ are candidates for boundary surfaces of the region.

For x and y fixed, z varies from $z = x + y$ to $z = 2 - x$.

The top and bottom intersect to form a right-side edge. We see that

$$z = x + y \text{ and } z = 2 - x \text{ imply } x + y = 2 - x, \text{ so } y = 2 - 2x.$$

This equation gives the right-side cylinder.

For x fixed, y varies from $y = 0$ to $y = 2 - 2x$.

The sides, top, and bottom intersect to form a front point. We use the equations of the sides to find the x-coordinate of the intersection:

$$y = 0 \text{ and } y = 2 - 2x \text{ imply } 0 = 2 - 2x, \text{ so } x = 1.$$

x varies from $x = 0$ to $x = 1$.

Figure 13.1.12 illustrates how the description of the region determines the region. Note that the right-side cylinder $y = 2 - 2x$ contains the line of intersection of the top and the bottom planes and gives the projection of that line onto the xy-plane. ■

FIGURE 13.1.12a The region is bounded in back by the plane $x = 0$ and in front by the plane $x = 1$.

FIGURE 13.1.12b The region is bounded on the sides by the vertical planes $y = 0$ and $y = 2 - 2x$.

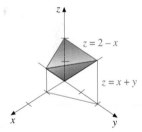

FIGURE 13.1.12c The region is bounded on the bottom by the surface $z = x + y$ and on the top by the surface $z = 2 - x$.

EXAMPLE 5

Sketch and describe the region that is inside the cylinder $x^2 + y^2 = 1$ and between the planes $z = 0$ and $y + z = 2$.

SOLUTION The region is sketched in Figure 13.1.13.

For x and y fixed, z varies from $z = 0$ to $z = 2 - y$.

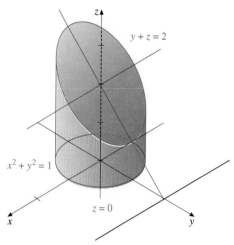

FIGURE 13.1.13 The vertical right circular cylinder $x^2 + y^2 = 1$ is cut off at the bottom by the plane $z = 0$ and at the top by the plane $y + z = 2$.

The side cylinders are given by the equation $x^2 + y^2 = 1$, which gives $y = \sqrt{1 - x^2}$ and $y = -\sqrt{1 - x^2}$.

For x fixed, y varies from $y = -\sqrt{1 - x^2}$ to $y = \sqrt{1 - x^2}$.

The sides intersect to form front and back edges. It is clear from the sketch that the x-coordinates of the intersections of the front and back edges are 1 and -1, respectively. Algebraically, $y = \sqrt{1 - x^2}$ and $y = -\sqrt{1 - x^2}$ imply $1 - x^2 = 0$, so $x = -1, 1$.

x varies from $x = -1$ to $x = 1$.

Figure 13.1.14 illustrates how the description of the region determines the region. The projection of the region onto the xy-plane is given by the equation $x^2 + y^2 = 1$. ∎

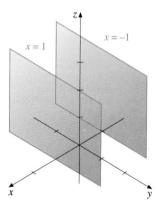

FIGURE 13.1.14a The region is bounded in back by the plane $x = -1$ and in front by the plane $x = 1$.

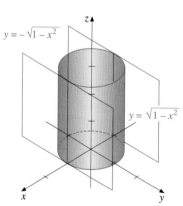

FIGURE 13.1.14b The region is bounded on the sides by the vertical half-cylinders $y = -\sqrt{1 - x^2}$ and $y = \sqrt{1 - x^2}$.

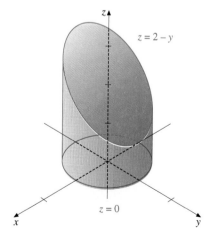

FIGURE 13.1.14c The region is bounded on the bottom by the plane $z = 0$ and on the top by the plane $z = 2 - y$.

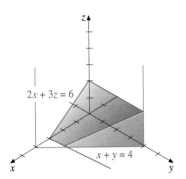

FIGURE 13.1.15 The top and bottom surfaces intersect to form a front edge. We need the x-coordinate of the intersection to describe the front of the region.

EXAMPLE 6

Sketch and describe the region in the first octant bounded by the planes $2x + 3z = 6$ and $x + y = 4$.

SOLUTION   The region is sketched in Figure 13.1.15.

For x and y fixed, z varies from $z = 0$ to $z = \frac{2}{3}(3 - x)$.

For x fixed, y varies from $y = 0$ to $y = 4 - x$.

The top and bottom intersect to form a front edge. We need the x-coordinate of that edge:

$$z = \frac{2}{3}(3 - x) \text{ and } z = 0 \text{ imply } \frac{2}{3}(3 - x) = 0, \text{ so } x = 3.$$

x varies from $x = 0$ to $x = 3$.

See Figure 13.1.16. The projection of the region onto the xy-plane is given by the equations $y = 0$, $y = 4 - x$, $x = 0$, and $x = 3$. ∎

We know that certain regions in the plane can be described with either x or y fixed in the first step. Similarly, certain regions in three-dimensional space can be described by fixing the variables x, y, and z in one or more of six possible orders. We will deal with such regions in Section 13.5.

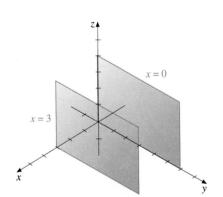

FIGURE 13.1.16a The region is bounded in back by the plane $x = 0$ and in front by the plane $x = 3$.

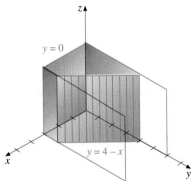

FIGURE 13.1.16b The region is bounded on the sides by the coordinate plane $y = 0$ and the vertical plane $y = 4 - x$.

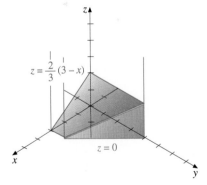

FIGURE 13.1.16c The region is bounded on the bottom by the coordinate plane $z = 0$ and on the top by the plane $z = \frac{2}{3}(3 - x)$.

EXERCISES 13.1

In Exercises 1–14, sketch and describe the regions in the plane bounded by the given equalities (a) as one or more vertically simple regions and (b) as one or more horizontally simple regions.

1. $y = x$, $y = 1$, $x = 0$
2. $y = x$, $y = 0$, $x = 1$
3. $x + 2y = 2$, $y = 0$, $x = 0$
4. $y = 4 - x^2$, $y = 0$, $x = 0$ $(x \geq 0)$
5. $x^2 + y^2 = 4$, $y = 0$ $(y \geq 0)$
6. $x^2 + y^2 = 4$, $x = 0$ $(x \geq 0)$
7. $(x - 1)^2 + y^2 = 1$, $x = 1$ $(x \geq 1)$
8. $x^2 + (y - 1)^2 = 1$, $y = 1$ $(y \leq 1)$
9. $y = x^3$, $y = -1$, $x = 1$
10. $y = \cos x$, $y = 0$, $x = 0$ $(x \geq 0)$
11. $y = \cos^{-1} x$, $y = \pi$, $x = 1$
12. $y = \sin^{-1} x$, $y = 0$, $x = 1$
13. $y = 2x$, $y = x + 1$, $x = 0$
14. $x + y = 3$, $x = 2y$, $x = 0$

In Exercises 15–20, sketch and describe the regions in the plane bounded by the given equations in any convenient way.

15. $y = \sin x$, $y = \cos x$, $x = 0$ $(x \geq 0)$
16. $y = x^2 + 1$, $y = x$, $x = 0$, $x = 1$
17. $y = x$, $x = 2 - y^2$
18. $y = 1$, $y = 0$, $x + y^2 = 0$, $x = 1 + y^2$
19. $y = 1$, $y = 0$, $x = 0$, $y = \ln x$
20. $y = 1$, $y = 0$, $y = e^x - 1$, $x = 2$

In Exercises 21–38, sketch and describe the regions in three-dimensional space bounded by the given equations.

21. Above $z = x + y$, below $x + y + z = 2$, in the first octant
22. $y + z = 4$, $z = 2x + 2y$, in the first octant
23. Below $x + y + z = 3$, above $z = x$, in the first octant
24. Below $x + 2y + z = 6$, above $z = 2y$, in the first octant
25. Inside $x^2 + y^2 = 1$, below $x + y + z = 2$, above $z = 0$
26. Inside $x^2 + y^2 + z^2 = 4$, above $z = 1$
27. $y + z = 3$, $x + y = 2$, in the first octant
28. Inside $z = \sqrt{2x^2 + 2y^2}$, below $x + z = 1$, in the first octant
29. Inside $x^2 + y^2 = 4$ and $y^2 + z^2 = 4$, in the first octant
30. Inside $x^2 + y^2 + z^2 = 4$, above $z = \sqrt{3}$, in the first octant
31. Inside $x^2 + y^2 = 1$, outside $z^2 = 4x^2 + 4y^2$
32. $x^2 + z^2 = 1$, $x + y = 3$, in the first octant
33. $y = 3$, $z = 2$, $z = 2x$, in the first octant

34. $y = 3$, $2x + 3z = 6$, in the first octant

35. (a) $x = 2$, $y = 2$, $z = 2$, in the first octant
    (b) $x = 2$, $y + 2z = 4$, between $y = 0$ and $y = 2$, in the first octant
    (c) $x = 2$, above $y = 2z$, below $y + 2z = 4$, in the first octant
    (d) $x + y = 2$, above $y = 2z$, below $y + 2z = 4$, in the first octant

36. (a) $x = 2$, $y = 2$, $z = 6$, $x = 0$, $y = 0$, $z = 0$
    (b) $x = 2$, $y = 2$, below $x + z = 6$, $x = 0$, $y = 0$, $z = 0$
    (c) $y = 2$, below $x + z = 6$, above $z = 2x$, $x = 0$, $y = 0$
    (d) $x + y = 2$, below $x + z = 6$, above $z = 2x$, $x = 0$, $y = 0$

37. (a) $x + 3z = 3$, $y = 3$, $x = 0$, $y = 0$, $z = 0$
    (b) $x + 3z = 3$, $x + y = 3$, $x = 0$, $y = 0$, $z = 0$
    (c) $x + 3z = 3$, $x + y = 3$, $x = 2$, $x = 0$, $y = 0$, $z = 0$

38. (a) above $z = x + y$, below $z = 3$, $x = 0$, $y = 0$
    (b) above $z = x + y$, below $z = 3$, $x = 2$, $x = 0$, $y = 0$

Describe the regions pictured in Exercises 39–44.

39.

40.

41.

42.

43.   **44.**

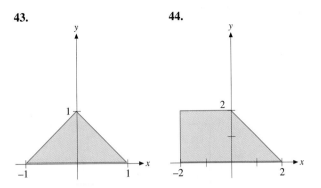

13.2 DOUBLE INTEGRALS

Connections

Describing regions, 13.1.

We will discuss the concept of the double integral of a bounded function over a region in a rectangular coordinate plane and see how double integrals relate to the area of regions in the plane.

Definition and Basic Properties

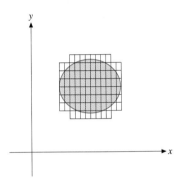

FIGURE 13.2.1 We cover the region with rectangles R_j and form a sum $\sum f(x_j, y_j)\Delta A_j$, where (x_j, y_j) is in R_j and ΔA_j is the area of R_j.

Let us consider a function f that is defined and bounded for (x, y) in a region R. We assume that R is a bounded region. That is, R is contained in some disk. To define the double integral of f over R, we need to consider some analogue of the Riemann sums that were used to define the integral of a function of one variable over an interval. The idea is to cover the region with rectangles R_j that have sides parallel to the coordinate axes, choose a point (x_j, y_j) in each rectangle, and take the sum of the products of $f(x_j, y_j)$ times the area of the corresponding rectangle R_j. See Figure 13.2.1. One difficulty is what to do with rectangles that are part in and part out of the region. A nice way out of this difficulty is to consider only regions for which it doesn't matter what we do with the "boundary rectangles." Thus, we assume that R satisfies the following condition.

The region R can be completely covered with rectangles in such a way that the total area of all rectangles that are part in and part out of the region is smaller than any prescribed positive number. (1)

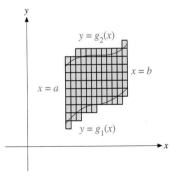

FIGURE 13.2.2 If the region is either horizontally or vertically simple, we can choose the rectangles small enough to make the total area of the rectangles that intersect the boundary less than any prescribed positive number. This means we can neglect these rectangles in a Riemann sum of a bounded function.

It can be proved that regions that are bounded by a finite number of smooth curves satisfy this condition. Experience with using the definite integral to calculate the area of vertically simple and horizontally simple regions in the plane should lead us to believe that regions of these types satisfy condition (1). This is illustrated in Figure 13.2.2. If the top and bottom of the region are given by functions that are continuous for $a \le x \le b$, it seems reasonable that the total area of the "boundary rectangles" approaches zero as the width and height of the rectangles approach zero. The sum of the products of the area of each of these rectangles times the value of a *bounded* function $f(x, y)$ also approaches zero.

For regions R that have property (1), a **Riemann sum** of a bounded function $f(x, y)$ over R is any sum of the form

$$\sum f(x_j, y_j)\Delta A_j,$$

where R is covered by a grid of rectangles R_j that have sides parallel to the axes, (x_j, y_j) is any point in R_j, ΔA_j is the area of R_j, and the sum is over all rectangles that are inside R.

We are interested in the case when the Riemann sums of f over R approach a single number as the lengths of sides of the rectangles approach zero. This idea is expressed by the following formal definition.

DEFINITION

A bounded function f is **integrable** over the bounded region R if there is a number (denoted $\iint_R f(x, y)\, dA$) such that, for each $\varepsilon > 0$, there is a positive number δ such that

$$\left| \sum f(x_j, y_j) \Delta A_j - \iint_R f(x, y)\, dA \right| < \varepsilon$$

whenever the lengths of the sides of all rectangles in the Riemann sum are less than δ.

There can be at most one number $\iint_R f(x, y)\, dA$ that satisfies the condition of the definition, since a Riemann sum could not be within ε of two different numbers if those numbers differed by more than 2ε. If it exists, that number is called the **double integral** of f over R. It can be proved that the integral exists if R satisfies condition (1) and f is continuous on the set that consists of R and all boundary points of R. In particular, the integral exists if R is bounded by a finite number of smooth curves and f is continuous.

As in the case of a single variable, we will establish methods of evaluating double integrals without dealing directly with Riemann sums, but the idea of the integral as a limit of Riemann sums is very important for understanding applications of the concept. When setting up a double integral to represent a particular physical quantity, we should think in terms of a corresponding Riemann sum.

The usual properties of the integral are true for double integrals of *integrable* functions. For example, if all of the functions involved are integrable over the appropriate regions, then

$$\iint_R (f + g)\, dA = \iint_R f\, dA + \iint_R g\, dA,$$

$$\iint_R kf\, dA = k \left(\iint_R f\, dA \right), \quad k \text{ a constant,}$$

and, if R is the union of the disjoint regions R' and R'',

$$\iint_R f\, dA = \iint_{R'} f\, dA + \iint_{R''} f\, dA.$$

See Figure 13.2.3.

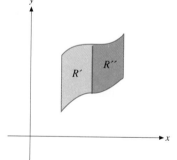

FIGURE 13.2.3   A double integral over a set is equal to the sum of the integrals over its disjoint parts.

Iterated Double Integrals

Let us consider a vertically simple region that can be described by the following statements:

For x fixed, y varies from $y = g_1(x)$ to $y = g_2(x)$.
x varies from $x = a$ to $x = b$.

We assume the boundary functions are continuous with $g_1(x) \le g_2(x)$ for $a \le x \le b$. See Figure 13.2.4a. As y varies from $y = g_1(x)$ to $y = g_2(x)$ with x fixed, the points (x, y) form a vertical line segment between the top and bottom of the region. See Figure 13.2.4b. As x varies from $x = a$ to $x = b$, these line segments sweep out the entire region. See Figure 13.2.4c.

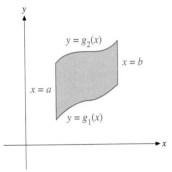

FIGURE 13.2.4a   This is a typical vertically simple region.

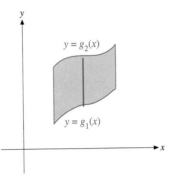

FIGURE 13.2.4b   As y varies from $y = g_1(x)$ to $y = g_2(x)$ with x fixed, the points (x, y) form a vertical line segment between the top and bottom of the region.

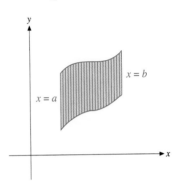

FIGURE 13.2.4c   As x varies from $x = a$ to $x = b$, these line segments sweep out the entire region.

We need to develop a formula for the evaluation of the double integral of an integrable function f over a vertically simple region R as described above. We begin by covering the region with a grid of rectangles R_{ij} that have sides parallel to the coordinate axes and area $\Delta A_{ij} = \Delta x_i \, \Delta y_j$. This is illustrated in Figure 13.2.5a. We evaluate f at the lower left-hand corner of each rectangle that is contained in the region to obtain the Riemann sum

$$\sum_{i,j} f(x_i, y_j) \, \Delta A_{ij}.$$

The terms that involve the same x_i correspond to a column of rectangles. See Figure 13.2.5b. We group and add the terms that involve the same x_i to obtain

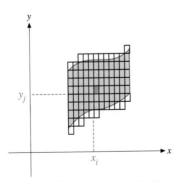

FIGURE 13.2.5a   The vertically simple region is covered by rectangles that have sides parallel to the coordinate axes.

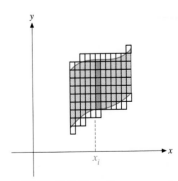

FIGURE 13.2.5b   We group and add the terms of the Riemann sum that correspond to a column of rectangles.

$$\sum_i \left[\sum_j f(x_i, y_j) \Delta y_j \right] \Delta x_i.$$

We see that the sum inside the brackets above is a Riemann sum with respect to y of the function of one variable $f(x_i, y)$ obtained by fixing the value x_i; for x_i fixed, y varies from essentially $y = g_1(x_i)$ to $y = g_2(x_i)$. If every Δy_j is small enough, the sum inside the brackets is approximately

$$\int_{g_1(x_i)}^{g_2(x_i)} f(x_i, y) \, dy,$$

so our double Riemann sum is approximately

$$\sum_i \left[\int_{g_1(x_i)}^{g_2(x_i)} f(x_i, y) \, dy \right] \Delta x_i.$$

This is a Riemann sum with respect to x of the function of x defined by

$$\int_{g_1(x)}^{g_2(x)} f(x, y) \, dy.$$

The variable x varies from $x = a$ to $x = b$. If every Δx_i is small enough, this Riemann sum is approximately

$$\int_a^b \left[\int_{g_1(x)}^{g_2(x)} f(x, y) \, dy \right] dx.$$

If f is continuous on R, it can be shown that the total error in the above approximations can be made less than any prescribed positive number. It follows that the double integral of f over a vertically simple region R that can be described by the statements "For x fixed, y varies from $y = g_1(x)$ to $y = g_2(x)$" and "x varies from $x = a$ to $x = b$" satisfies

$$\iint_R f(x, y) \, dA = \int_a^b \left[\int_{g_1(x)}^{g_2(x)} f(x, y) \, dy \right] dx.$$

The integral on the right is called an **iterated integral**. It is evaluated by considering x to be constant (fixed) while we integrate the inner integral with respect to y from $g_1(x)$ to $g_2(x)$. This gives a function of x, which we then integrate with respect to x from a to b. Note how *the description of the region corresponds to the limits of integration and the evaluation of the iterated integral.* In particular, the statement "For x fixed, y varies from $y = g_1(x)$ to $y = g_2(x)$" gives the limits of integration of the inner integral and suggests that x is fixed while that integral is evaluated. The statement "x varies from $x = a$ to $x = b$" gives the limits of integration of the outer integral.

It is common to omit the brackets around the inner integral of an iterated integral.

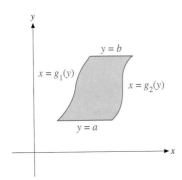

FIGURE 13.2.6a This is a typical horizontally simple region.

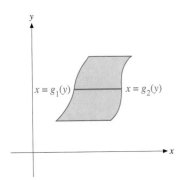

FIGURE 13.2.6b As x varies from $x = g_1(y)$ to $x = g_2(y)$ with y fixed, the points (x, y) form a horizontal line segment between the sides of the region.

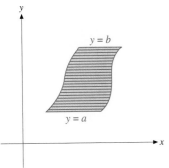

FIGURE 13.2.6c As y varies from $y = a$ to $y = b$, these line segments sweep out the entire region.

EXAMPLE 1

Evaluate $\displaystyle\int_0^1 \int_{\sqrt{x}}^1 24x^2 y \, dy \, dx$.

SOLUTION   We begin by evaluating the inner integral. The variable x is considered to be constant while we integrate with respect to y. We obtain

$$\int_0^1 \int_{\sqrt{x}}^1 24x^2 y \, dy \, dx = \int_0^1 \left[\int_{\sqrt{x}}^1 24x^2 y \, dy\right] dx$$

$$= \int_0^1 \left[24x^2 \frac{y^2}{2}\right]_{y=\sqrt{x}}^{y=1} dx.$$

(We have written $y = 1$ and $y = \sqrt{x}$ to remind us that y is the variable of integration and we must substitute for y in the expression $24x^2(y^2/2)$.) We then obtain

$$\int_0^1 \int_{\sqrt{x}}^1 24x^2 y \, dy \, dx = \int_0^1 [12x^2(1)^2 - 12x^2(\sqrt{x})^2] dx$$

$$= \int_0^1 [12x^2 - 12x^3] dx$$

$$= 12\frac{x^3}{3} - 12\frac{x^4}{4}\Big]_0^1$$

$$= [4(1)^3 - 3(1)^4] - [4(0)^3 - 3(0)^4] = 1. \quad\blacksquare$$

Let us consider a horizontally simple region R that can be described by the following statements:

For y fixed, x varies from $x = g_1(y)$ to $x = g_2(y)$.
y varies from $y = a$ to $y = b$.

See Figure 13.2.6a. As x varies from $x = g_1(y)$ to $x = g_2(y)$ with y fixed, the points (x, y) form a horizontal line segment between the sides of the region. See Figure 13.2.6b. As y varies from $y = a$ to $y = b$, these line segments sweep out the entire region. See Figure 13.2.6c.

In a manner that is completely analogous to the case of a vertically simple region, we then obtain

$$\iint_R f(x, y) \, dA = \int_a^b \left[\int_{g_1(y)}^{g_2(y)} f(x, y) \, dx\right] dy.$$

Iterated integrals of this type are evaluated by treating y as a constant while evaluating the inner integral with respect to x, and then integrating with respect to y.

EXAMPLE 2

Evaluate $\displaystyle\int_0^\pi \int_0^y \sin x \, dx \, dy$.

SOLUTION   The inner integral is with respect to x. We have

$$
\begin{aligned}
\int_0^\pi \int_0^y \sin x \, dx \, dy &= \int_0^\pi \left[\int_0^y \sin x \, dx \right] dy \\
&= \int_0^\pi \left[-\cos x \right]_{x=0}^{x=y} dy \\
&= \int_0^\pi [(-\cos y) - (-\cos 0)] dy \\
&= \int_0^\pi [1 - \cos y] dy \\
&= y - \sin y \Big]_0^\pi \\
&= (\pi - \sin \pi) - (0 - \sin 0) = \pi. \quad \blacksquare
\end{aligned}
$$

NOTE   *Remember that the inner integral of an iterated double integral is evaluated first. Be sure to distinguish the variable of integration from the variable that is considered constant during this first integration.*

Evaluating Double Integrals

If R is either a vertically simple region or a horizontally simple region, we can use the corresponding iterated integral to evaluate the double integral over R of an integrable function. If R is both vertically simple and horizontally simple, we can use an iterated integral with either of the two orders of integration. The two iterated integrals will have the same value, since they represent two different ways of evaluating the same double integral.

We have seen how the limits of integration of an iterated integral correspond to the description of the region. We can determine the limits of integration of an iterated integral over a region by exactly the same method that we used in Section 13.1 to describe the region.

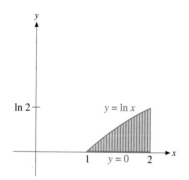

FIGURE 13.2.7a This region is vertically simple.

EXAMPLE 3

Evaluate $\iint_R 4x \, dA$, where R is the region bounded by $y = \ln x$, $y = 0$, and $x = 2$.

SOLUTION   The region is sketched in Figure 13.2.7a. We see that the region is vertically simple and can be described by the following statements:

For x fixed, y varies from y = 0 to y = ln x.
x varies from x = 1 to x = 2.

The corresponding iterated integral is

$$
\int_1^2 \int_0^{\ln x} 4x \, dy \, dx = \int_1^2 4xy \Big]_{y=0}^{y=\ln x} dx
$$
$$
= \int_1^2 4x \ln x \, dx.
$$

Using the Table of Integrals (or integrating by parts), we obtain

$$\int_1^2 \int_0^{\ln x} 4x\,dy\,dx = 2x^2 \ln x - x^2\Big]_1^2$$
$$= (8\ln 2 - 4) - (0 - 1)$$
$$= 8\ln 2 - 3.$$

From the sketch in Figure 13.2.7b we see that the region is also horizontally simple. We can express the equation $y = \ln x$ as $x = e^y$. Also:

$$x = e^y \text{ and } x = 2 \text{ imply } e^y = 2, \text{ so } y = \ln 2.$$

We can then describe the region by the following statements:

For y fixed, x varies from $x = e^y$ to $x = 2$.
y varies from $y = 0$ to $y = \ln 2$.

The corresponding iterated integral is

$$\int_0^{\ln 2} \int_{e^y}^2 4x\,dx\,dy = \int_0^{\ln 2} 2x^2\Big]_{x=e^y}^{x=2} dy$$
$$= \int_0^{\ln 2} [2(2)^2 - 2(e^y)^2]dy$$
$$= \int_0^{\ln 2} (8 - 2e^{2y})dy$$
$$= 8y - e^{2y}\Big]_0^{\ln 2}$$
$$= (8\ln 2 - e^{2\ln 2}) - (0 - e^0)$$
$$= 8\ln 2 - 4 + 1 = 8\ln 2 - 3.$$

We see that the integration is more elementary in the latter iterated integral than in the integral with the opposite order of integration. ■

Area

Let us see how to use a double integral to find the area of a region. We cover the region R with a grid of rectangles that have sides parallel to the axes and form the sum of the areas of the rectangles contained in R. If the total area of the rectangles that touch the boundary of R is small, then the sum of the areas of the rectangles inside R is approximately the area of R. See Figure 13.2.8. This suggests that the area of the region be defined by a double integral. That is,

$$\text{Area} = \iint_R dA.$$

EXAMPLE 4

Use double integrals to find the area of the region bounded by $x = y - y^2$ and $x + y = 0$.

SOLUTION The region is sketched in Figure 13.2.9. From the sketch, we see that the region is vertically simple and horizontally simple. However, it is not convenient to describe the region as a vertically simple region. We can describe the region by writing:

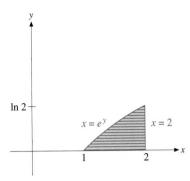

FIGURE 13.2.7b The region is also horizontally simple.

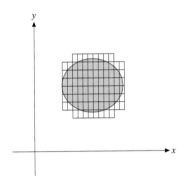

FIGURE 13.2.8 If the total sum of the areas of the rectangles that intersect the boundary is small, then the sum of the areas of the rectangles inside the region is approximately the area of the region.

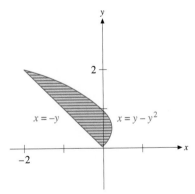

FIGURE 13.2.9 It is convenient to describe this region as a horizontally simple region.

For y fixed, x varies from x = −y to x = y − y².

We see that y varies between the y-coordinates of the intersection of the two boundary curves. The equations

$$x = y - y^2 \text{ and } x = -y \text{ imply } -y = y - y^2, \text{ so } y^2 - 2y = 0.$$

The solutions are $y = 0$ and $y = 2$, so

y varies from y = 0 to y = 2.

We then have

$$
\begin{aligned}
\text{Area} &= \int_0^2 \int_{-y}^{y-y^2} dx\,dy \\
&= \int_0^2 x \Big]_{x=-y}^{x=y-y^2} dy \\
&= \int_0^2 [(y - y^2) - (-y)]dy \\
&= \int_0^2 (2y - y^2)dy \\
&= y^2 - \frac{y^3}{3}\Big]_0^2 \\
&= \left(4 - \frac{8}{3}\right) - (0) = \frac{4}{3}. \quad \blacksquare
\end{aligned}
$$

[This integral represents the sum of the areas of rectangles with horizontal length $(y - y^2) - (-y)$ and width dy. This corresponds to the integral obtained by the single-variable method we used to determine area.]

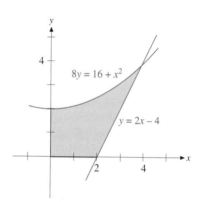

FIGURE 13.2.10a It is not convenient to describe this region as either a vertically simple region or a horizontally simple region.

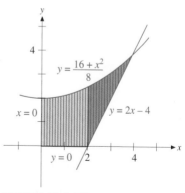

FIGURE 13.2.10b We can divide the region into two vertically simple subregions that are described easily.

If a region does not have a simple description, it may require more than one iterated integral to evaluate a double integral over the region.

EXAMPLE 5

Use double integrals to find the area of the region in the first quadrant bounded by $y = 2x - 4$ and $8y = 16 + x^2$.

SOLUTION From the sketch in Figure 13.2.10a we see that the region does not seem to have a simple description. We can express the region as the union of two disjoint vertically simple regions that each have a simple description, as indicated in Figure 13.2.10b. (The region could also be expressed as a union of disjoint horizontal simple regions that each have a simple description.) We need the x-coordinate of the point of intersection of $y = 2x - 4$ and $8y = 16 + x^2$.

$$
\begin{aligned}
y = 2x - 4 \text{ and } 8y = 16 + x^2 \text{ imply } 8(2x - 4) &= 16 + x^2, \text{ so} \\
16x - 32 &= 16 + x^2, \\
x^2 - 16x + 48 &= 0, \\
(x - 4)(x - 12) &= 0, \\
x = 4, \quad x &= 12.
\end{aligned}
$$

The solution we want is $x = 4$.

The total area is then the sum of the areas of the parts. We have

$$\text{Area} = \int_0^2 \int_0^{(16+x^2)/8} dy\,dx + \int_2^4 \int_{2x-4}^{(16+x^2)/8} dy\,dx$$

$$= \int_0^2 \left(2 + \frac{x^2}{8}\right) dx + \int_2^4 \left[\left(2 + \frac{x^2}{8}\right) - (2x - 4)\right] dx$$

$$= \int_0^2 \left(2 + \frac{x^2}{8}\right) dx + \int_2^4 \left(\frac{x^2}{8} - 2x + 6\right) dx$$

$$= \left[2x + \frac{x^3}{24}\right]_0^2 + \left[\frac{x^3}{24} - x^2 + 6x\right]_2^4$$

$$= \left[\left(4 + \frac{1}{3}\right) - (0)\right] + \left[\left(\frac{8}{3} - 16 + 24\right) - \left(\frac{1}{3} - 4 + 12\right)\right] = \frac{20}{3}.$$

Changing the Order of Integration

It is useful to be able to change the order of integration of an iterated integral. For example, a scientific problem may lead to an iterated double integral that is easier to evaluate with the opposite order of integration. We will use the following steps to change the order of integration.

To change the order of integration of an iterated double integral

- Use the given limits of integration to determine the equations of the boundary of the corresponding region.
- Sketch the region.
- Use the sketch and boundary equations to determine the desired limits of integration.

NOTE   *Do not try to go directly from the limits of integration for one order of integration to the limits of integration for the opposite order. Always go from the limits of integration to a sketch of the region and then to the other limits of integration.*

EXAMPLE 6
Express

$$\int_0^4 \int_{x/2}^2 e^{y^2}\,dy\,dx$$

as an iterated integral with the opposite order of integration and evaluate.

SOLUTION   From the given limits of integration we see that

For x fixed, y varies from y = x/2 to y = 2.
x varies from x = 0 to x = 4.

The corresponding region is sketched in Figure 13.2.11. From the sketch, we see that the region can also be described by the following statements:

For y fixed, x varies from x = 0 to x = 2y.
y varies from y = 0 to y = 2.

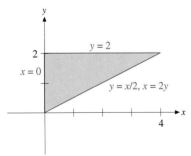

FIGURE 13.2.11 The limits of integration of the iterated integral give the equations of the boundary of the region of integration.

The corresponding iterated integral is

$$\int_0^2 \int_0^{2y} e^{y^2} dx\, dy.$$

This integral has the same value as the given integral, since both iterated integrals represent the same double integral.

We cannot use the Fundamental Theorem of Calculus to evaluate the original iterated integral, because the indefinite integral $\int e^{y^2} dy$ cannot be expressed in terms of any of the usual functions. However, we do have

$$\int_0^2 \int_0^{2y} e^{y^2} dx\, dy = \int_0^2 xe^{y^2}\Big]_{x=0}^{x=2y} dy$$

$$= \int_0^2 2y e^{y^2} dy$$

$$= e^{y^2}\Big]_0^2 = e^4 - 1. \quad \blacksquare$$

It is worthwhile to consider different methods of evaluating a double integral, since some methods may be significantly easier than others.

EXERCISES 13.2

Evaluate the integrals in Exercises 1–16.

1. $\displaystyle\int_0^1 \int_0^x xy\, dy\, dx$

2. $\displaystyle\int_0^1 \int_0^{\sqrt{x}} xy\, dy\, dx$

3. $\displaystyle\int_0^1 \int_{\sqrt{x}}^1 y\, dy\, dx$

4. $\displaystyle\int_0^1 \int_x^1 x\, dy\, dx$

5. $\displaystyle\int_0^{\pi/2} \int_0^y \cos x \sin y\, dx\, dy$

6. $\displaystyle\int_0^{\pi/2} \int_0^{\sin y} x \sin y\, dx\, dy$

7. $\displaystyle\int_0^{\pi/4} \int_0^y \sec x \tan x\, dx\, dy$

8. $\displaystyle\int_0^{\pi/4} \int_0^y \sec^2 x\, dx\, dy$

9. $\displaystyle\iint_R x^2\, dA$, where R is the region bounded by $x + y = 4$, $y = 0$, $x = 0$, and $x = 2$

10. $\displaystyle\iint_R x^2\, dA$, where R is the region bounded by $y = 3 - x^2$ and $y = 0$

11. $\displaystyle\iint_R \sqrt{y}\, dA$, where R is the region bounded by $y = x^2$, $y = 4$, and $x = 0 (x \geq 0)$

12. $\displaystyle\iint_R \cos y\, dA$, where R is the region bounded by $y = x/2$, $y = 0$, and $x = \pi/2$

13. $\displaystyle\iint_R e^{x/2}\, dA$, where R is the region bounded by $x + y = 4$, $y = 0$, $y = 2$, and $x = 0$

14. $\displaystyle\iint_R 2xy\, dA$, where R is the region bounded by $x + y = 0$, $y = x/2$, and $x = 2$

15. $\displaystyle\iint_R 2xy\, dA$, where R is the region bounded by $y = 2x$, $y = 2$, and $x = 0$

16. $\displaystyle\iint_R x^2\, dA$, where R is the region bounded by $y = \sin^{-1} x$, $y = \pi/2$, and $x = 0$

Use double integrals to find the area of the regions bounded by the equations given in Exercises 17–24.

17. $y = \cos x$, $y = 1/2 \ (-\pi/3 \leq x \leq \pi/3)$

18. $y = \sin x$, $y = 2x/\pi \ (0 \leq x \leq \pi/2)$

19. $x = 4 - y^2$, $x = 0$

20. $x = y^2$, $x - y = 2$

21. $y = x$, $y = x - 1$, $y = 0$, $x = 2$

22. $y - 1 = x/2$, $y + 2 = 2x$, $x = 0$, $y = 0$

23. $y = x(x - 2)(x + 1)$, $y = 0$

24. $y = \cos x$, $y = 1 - (2x/\pi)$, $x = 0$, $x = 2\pi/3$

Express the integrals in Exercises 25–36 as iterated double integrals with order of integration opposite to that given and evaluate.

25. $\displaystyle\int_0^2 \int_{x^2}^{2x} xy\,dy\,dx$

26. $\displaystyle\int_0^3 \int_{2(3-x)/3}^2 (x+y)\,dy\,dx$

27. $\displaystyle\int_0^{\pi/2} \int_{\sin x}^1 \cos x\,dy\,dx$

28. $\displaystyle\int_0^{\pi/2} \int_0^{\cos x} \sin x\,dy\,dx$

29. $\displaystyle\int_{1/2}^1 \int_{\sin^{-1} x}^{\pi/2} \cot y\,dy\,dx$

30. $\displaystyle\int_0^1 \int_{\tan^{-1} x}^{\pi/4} \cos y\,dy\,dx$

31. $\displaystyle\int_1^e \int_{\ln x}^1 x\,dy\,dx$

32. $\displaystyle\int_0^1 \int_x^1 e^{y^2}\,dy\,dx$

33. $\displaystyle\int_0^1 \int_{x^3}^1 3x^2\,dy\,dx$

34. $\displaystyle\int_0^1 \int_0^x 2xy\,dy\,dx$

35. $\displaystyle\int_{-1}^1 \int_0^{\sqrt{1-x^2}} 2y\,dy\,dx$

36. $\displaystyle\int_0^1 \int_{-\sqrt{x}}^{\sqrt{x}} 2x\,dy\,dx$

In Exercises 37–40, set up an iterated double integral of $f(x, y)$ over the indicated regions.

37.

38.

39.

40.

41. Why does the double integral of e^{xy} over R exist, where R is the region that consists of points (x, y) with $0 \le x \le 1$ and $0 \le y \le 1$? We have

$$\iint_R e^{xy}\,dA = \int_0^1 \int_0^1 e^{xy}\,dy\,dx.$$

Evaluate $\displaystyle\int_0^1 e^{xy}\,dy$ for $0 < x \le 1$ and for $x = 0$. Is $\displaystyle\int_0^1 e^{xy}\,dy$ a continuous function of x?

42. Do you think we would obtain the same value of a double integral if we used either parallelograms or triangles in place of rectangles in Riemann sums?

13.3 APPLICATIONS OF DOUBLE INTEGRALS

Connections

Volume, 6.3–4.

Surface area, 6.5.

Mass and moments, 6.7.

Describing regions, 13.1.

We will see how to use double integrals to find both the volume of regions and the area of surfaces in three-dimensional space. We will also see how to find the mass, moments, and centers of mass of laminas that have variable density.

Volume

Consider a region in three-dimensional space that is generated by vertical line segments between the surfaces $z = h_1(x, y)$ and $z = h_2(x, y)$ as (x, y) varies over a region R in the xy-plane. We assume that h_1 and h_2 are continuous with $h_1(x, y) \le h_2(x, y)$ for (x, y) in R. R is the projection of the region in space onto the xy-plane. See Figure 13.3.1a. Let us divide the region R in the xy-plane into rectangles with area ΔA. That part of the region in three-dimensional space that is above each rectangle is essentially a rectangular column with volume

$$\Delta V \approx (\text{Height})(\text{Area of base}) \approx (h_2(x, y) - h_1(x, y))\Delta A.$$

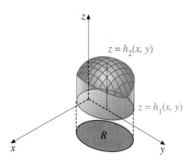

FIGURE 13.3.1a The region is generated by vertical line segments between the surfaces $z = h_1(x, y)$ and $z = h_2(x, y)$ as (x, y) varies over the region R in the xy-plane.

See Figure 13.3.1b. The height of the column depends on the (x, y) coordinates of its position. The total volume of the region is the sum of the volumes of its parts, so

$$\text{Volume} = \sum (\text{Volume of columns}) \approx \sum (h_2(x, y) - h_1(x, y))\Delta A.$$

The latter sum is a Riemann sum of the height function $h_2 - h_1$ over the projection of the region in space onto the xy-plane. It should seem reasonable that the volume of the region in space should be defined by the corresponding double integral,

$$\text{Volume} = \int \int_R (h_2 - h_1) \, dA.$$

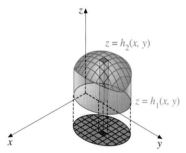

FIGURE 13.3.1b That part of the region that is above a rectangle in the xy-plane is essentially a rectangular column.

If the region R is simple in the direction of either the x-axis or the y-axis, we can express this volume as an iterated double integral.

Suppose a region in three-dimensional space can be described by the statements

For x and y fixed, z varies from $z = h_1(x, y)$ to $z = h_2(x, y)$.
For x fixed, y varies from $y = g_1(x)$ to $y = g_2(x)$.
x varies from $x = a$ to $x = b$.

The surfaces $z = h_1(x, y)$ and $z = h_2(x, y)$ form the bottom and top of the region. The surfaces $y = g_1(x)$, $y = g_2(x)$, $x = a$, and $x = b$ form the vertical sides of the region and give the projection of the region onto the xy-plane. The projection of the region onto the xy-plane is a region R of the xy-plane that is simple in the direction of the y-axis. See Figure 13.3.2a. As z varies from $z = h_1(x, y)$ to $z = h_2(x, y)$ with x and y fixed, the points (x, y, z) form a vertical line segment between the top and bottom of the region. See Figure 13.3.2b. As (x, y) varies over the projection R, these line segments generate the entire region in space. Since the double integral over R can be expressed as an iterated integral, we have

$$\text{Volume} = \int \int_R (h_2 - h_1) \, dA = \int_a^b \int_{g_1(x)}^{g_1(x)} (h_2(x, y) - h_1(x, y)) \, dy \, dx.$$

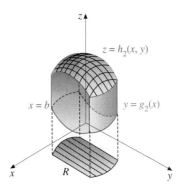

FIGURE 13.3.2a The projection of this region onto the xy-plane is simple in the direction of the y-axis.

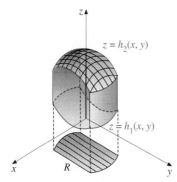

FIGURE 13.3.2b The region is generated by vertical line segments between the surfaces $z = h_1(x, y)$ and $z = h_2(x, y)$ as (x, y) varies over the projection of the region onto the xy-plane.

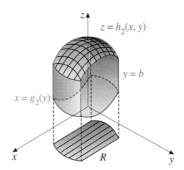

FIGURE 13.3.3 The projection of this region onto the xy-plane is simple in the direction of the x-axis.

Note that the equations used in the description of the region give the integrand and the limits of integration of the volume integral. Ordinarily, the equations of all six boundary surfaces are not given. We will use the procedure of Section 13.1 to determine missing equations. The same ideas apply if the roles of x and y are interchanged, so the projection of the region onto the xy-plane is a region of the xy-plane that is simple in the direction of the x-axis. See Figure 13.3.3.

EXAMPLE 1

Use a double integral to find the volume of the region in the first octant bounded by $x + 2y + 3z = 6$.

SOLUTION   The region is sketched in Figure 13.3.4a. Let us use the method of Section 13.1 to describe the region. We see that

For x and y fixed, z varies from $z = 0$ to $z = \dfrac{6 - x - 2y}{3}$.

The top and bottom intersect to form a right-side edge. The equation of the vertical generalized cylinder that contains this edge is obtained by eliminating z from the equations of the top and bottom.

$$z = 0 \text{ and } z = \frac{6 - x - 2y}{3} \text{ imply } 6 - x - 2y = 0, \text{ so } y = \frac{6 - x}{2}.$$

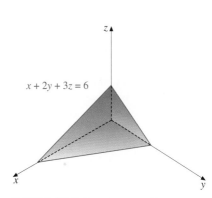

FIGURE 13.3.4a The coordinate traces of the plane are used to sketch the region.

For x fixed, y varies from $y = 0$ to $y = \dfrac{6 - x}{2}$.

The sides intersect to form a front point.

$$y = 0 \text{ and } y = \frac{6 - x}{2} \text{ imply } 0 = 6 - x, \text{ so } x = 6.$$

x varies from $x = 0$ to $x = 6$.

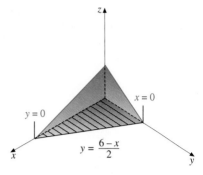

FIGURE 13.3.4b The projection of the region onto the *xy*-plane is simple in the direction of the *y*-axis.

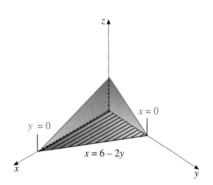

FIGURE 13.3.5 The projection of the region onto the *xy*-plane is also simple in the direction of the *x*-axis.

The volume of the region is the double integral of the height function over the projection of the region onto the *xy*-plane. We can now use the description of the region to set up an iterated double integral that gives the volume of the region. The height of a variable vertical column is given by the equations of the top and bottom of the region. We have

$$\text{Height} = (\text{Top}) - (\text{Bottom}) = \frac{6 - x - 2y}{3}.$$

The equations that describe the two sides and the front and back boundaries of the region in three-dimensional space give a description of the projection of the region in space onto the *xy*-plane. This description corresponds to a region of the *xy*-plane that is simple in the direction of the *y*-axis. See Figure 13.3.4b. We then have

$$\begin{aligned}
\text{Volume} &= \int_0^6 \int_0^{(6-x)/2} \frac{6 - x - 2y}{3}\, dy\, dx \\
&= \int_0^6 \int_0^{(6-x)/2} \left(2 - \frac{x}{3} - \frac{2y}{3}\right) dy\, dx \\
&= \int_0^6 \left[2y - \frac{xy}{3} - \frac{y^2}{3}\right]_{y=0}^{y=(6-x)/2} dx \\
&= \int_0^6 \left[(6 - x) - \frac{x(6 - x)}{6} - \frac{(6 - x)^2}{12}\right] dx \\
&= \int_0^6 \left(6 - x - x + \frac{x^2}{6} - 3 + x - \frac{x^2}{12}\right) dx \\
&= \int_0^6 \left(\frac{x^2}{12} - x + 3\right) dx \quad \text{\small[This integral represents the sum of} \\
&\qquad\qquad\qquad\qquad\qquad\;\; \text{\small the volumes of slices of the region} \\
&= \frac{x^3}{36} - \frac{x^2}{2} + 3x\Big]_0^6 \quad \text{\small perpendicular to the } x\text{-axis that have} \\
&\qquad\qquad\qquad\qquad\qquad\;\; \text{\small cross-sectional areas } \frac{x^2}{12} - x + 3. \text{\small This} \blacksquare \\
&= 6 - 18 + 18 = 6. \quad \text{\small corresponds to the integral used to} \\
&\qquad\qquad\qquad\qquad\qquad\;\; \text{\small obtain the volume by slicing.]}
\end{aligned}$$

The projection of the region of Example 1 onto the *xy*-plane is a region of the *xy*-plane that is simple in the direction of the *x*-axis as well as simple in the direction of the *y*-axis. See Figure 13.3.5. The projection can be described by the following statements:

For y fixed, x varies from $x = 0$ to $x = 6 - 2y$.
y varies from $y = 0$ to $y = 3$.

The volume of the region can then be expressed as

$$\text{Volume} = \int_0^3 \int_0^{6-2y} \frac{6 - x - 2y}{3}\, dx\, dy.$$

This integral has value 6.

Surface Area

Consider a surface given by

$$z = f(x, y), \qquad (x, y) \text{ in the region } R.$$

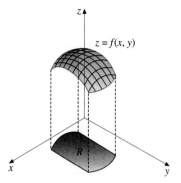

FIGURE 13.3.6 The surface is given by $z = f(x, y)$, (x, y) in R.

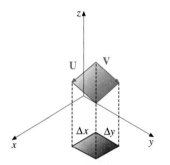

FIGURE 13.3.7 We can use vectors to approximate the area of that part of the surface that is above a small rectangle in the xy-plane.

See Figure 13.3.6. We assume that f has continuous first partial derivatives for (x, y) in R and on the boundary of R, so f is differentiable and the surface has a tangent plane at each point. The area of that part of the surface above a small rectangle R_j with one corner at (x_j, y_j) and opposite corner at $(x_j + \Delta x, y_j + \Delta y)$ is approximately the area of the part above R_j of the plane tangent to the surface at (x_j, y_j). Let us use vectors to find the area of that part of the tangent plane.

The vector

$$\mathbf{R} = x\mathbf{i} + y\mathbf{j} + f(x, y)\mathbf{k}$$

gives the position for the point on the surface $z = f(x, y)$ that is above the point (x, y) in the xy-plane. The change in position due to changes of Δx in x and Δy in y is

$$\Delta \mathbf{R} = \Delta x \mathbf{i} + \Delta y \mathbf{j} + \Delta f \mathbf{k}$$
$$\approx \Delta x \mathbf{i} + \Delta y \mathbf{j} + \left(\frac{\partial f}{\partial x} \Delta x + \frac{\partial f}{\partial y} \Delta y \right) \mathbf{k}.$$

The vector

$$\mathbf{U} = \Delta x \mathbf{i} + f_x(x_j, y_j)(\Delta x)\mathbf{k}$$

corresponds to a change in the xy-plane from (x_j, y_j) to $(x_j + \Delta x, y_j)$. The vector

$$\mathbf{V} = \Delta y \mathbf{j} + f_y(x_j, y_j)(\Delta y)\mathbf{k}$$

corresponds to a change in the xy-plane from (x_j, y_j) to $(x_j, y_j + \Delta y)$. See Figure 13.3.7. The area of that part of the surface $z = f(x, y)$ above the rectangle R_j is essentially the area of the parallelogram formed by $\mathbf{U}$ and $\mathbf{V}$. From Section 11.4 we know that this area is given by $|\mathbf{U} \times \mathbf{V}|$. We have

$$\mathbf{U} \times \mathbf{V} = \begin{vmatrix} \mathbf{i} & \mathbf{j} & \mathbf{k} \\ \Delta x & 0 & f_x(x_j, y_j)(\Delta x) \\ 0 & \Delta y & f_y(x_j, y_j)(\Delta y) \end{vmatrix}$$

$$= \begin{vmatrix} 0 & f_x(x_j, y_j)(\Delta x) \\ \Delta y & f_y(x_j, y_j)(\Delta y) \end{vmatrix} \mathbf{i} - \begin{vmatrix} \Delta x & f_x(x_j, y_j)(\Delta x) \\ 0 & f_y(x_j, y_j)(\Delta y) \end{vmatrix} \mathbf{j} + \begin{vmatrix} \Delta x & 0 \\ 0 & \Delta y \end{vmatrix} \mathbf{k}$$

$$= -f_x(x_j, y_j)(\Delta x)(\Delta y)\mathbf{i} - f_y(x_j, y_j)(\Delta x)(\Delta y)\mathbf{j} + (\Delta x)(\Delta y)\mathbf{k}.$$

The area of that part of the tangent plane above the rectangle R_j is then

$$|\mathbf{U} \times \mathbf{V}| = \sqrt{(f_x(x_j, y_j))^2 + (f_y(x_j, y_j))^2 + 1}\, \Delta x \, \Delta y.$$

This area is approximately the area of the surface above R_j. That is,

$$\Delta S_j \approx |\mathbf{U} \times \mathbf{V}| = \sqrt{(f_x(x_j, y_j))^2 + (f_y(x_j, y_j))^2 + 1}\, \Delta A_j,$$

where $\Delta A_j = \Delta x \, \Delta y$ is the area of the rectangle R_j in the xy-plane. The error in approximation is proportionately smaller than ΔA_j, if Δx and Δy are small. It follows that the area of the surface over the region R is

$$S = \sum \Delta S_j \approx \sum \sqrt{(f_x(x_j, y_j))^2 + (f_y(x_j, y_j))^2 + 1}\, \Delta A_j,$$

where the sum is taken over all rectangles in R. This is a Riemann sum, so it should seem reasonable to define

$$S = \int \int_R \sqrt{(f_x(x, y))^2 + (f_y(x, y))^2 + 1}\, dA.$$

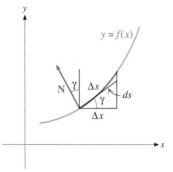

FIGURE 13.3.8 We see that
$\sec\gamma = \dfrac{ds}{dx}$, where γ is the angle
between a normal to the curve and
the vertical.

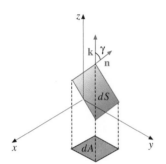

FIGURE 13.3.9 γ is the angle
between a normal to the surface
and the vertical.

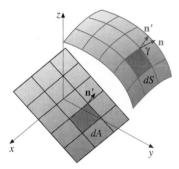

FIGURE 13.3.10 We have $dS =$
$\sec\gamma\, dA$, where γ is the acute angle
between a normal vector to the
surface and a vector that is normal to
the projection plane.

The formula for surface area given above should be compared with the
formula

$$s = \int_a^b \sqrt{(f'(x))^2 + 1}\, dx$$

for the arc length of the curve $y = f(x), a \le x \le b$. From Figure 13.3.8 we see
that, as the lengths approach zero, the ratio of the arc length of a small piece of
the curve $y = f(x)$ and the corresponding length along the x-axis approaches

$$\sec\gamma = \frac{ds}{dx} = \sqrt{(f')^2 + 1},$$

where γ is the acute angle between a normal to the curve and the vertical. Let
us see that, as the areas approach zero, the ratio of the area of a small piece of
the surface $z = f(x, y)$ and the corresponding area in the xy-plane approaches

$$\sec\gamma = \sqrt{(f_x)^2 + (f_y)^2 + 1},$$

where γ is the acute angle between a normal to the surface and the vertical.
See Figure 13.3.9. That is, the vector

$$\mathbf{n} = -f_x\mathbf{i} - f_y\mathbf{j} + \mathbf{k}$$

is normal to the surface $z = f(x, y)$. ($\mathbf{n} = \nabla F$, where $F(x, y, z) =
z - f(x, y)$.) If γ is the acute angle between $\mathbf{n}$ and the vertical vector $\mathbf{k}$,
we have

$$\cos\gamma = \frac{\mathbf{n} \cdot \mathbf{k}}{|\mathbf{n}|\,|\mathbf{k}|} = \frac{1}{\sqrt{(f_x)^2 + (f_y)^2 + 1}}, \quad \text{so}$$
$$\sec\gamma = \sqrt{(f_x)^2 + (f_y)^2 + 1}.$$

We use the symbolism

$$dS = \sec\gamma\, dA \qquad (1)$$

to indicate the relation between the area of a small rectangle in the xy-plane
and the area of the corresponding part of the surface $z = f(x, y)$. Generally,
this relation holds whenever a surface is projected orthogonally onto a plane
and γ is the acute angle between a normal vector to the surface and a vector
that is normal to the plane. See Figure 13.3.10. For example, we know that the
surface $F(x, y, z) = 0$ has normal vector $\mathbf{n} = \nabla F$. If the plane onto which the
surface is projected orthogonally has unit normal $\mathbf{n}'$, we have

$$\sec\gamma = \frac{1}{\cos\gamma} = \frac{|\nabla F|\,|\mathbf{n}'|}{|\nabla F \cdot \mathbf{n}'|} = \frac{|\nabla F|}{|\nabla F \cdot \mathbf{n}'|}.$$

This gives the formula

$$dS = \sec\gamma\, dA = \frac{|\nabla \mathbf{F}|}{|\nabla \mathbf{F} \cdot \mathbf{n}'|}\, dA, \qquad (2)$$

where the surface S is given by $F(x, y, z) = 0$, dA represents area in the
projection plane, and $\mathbf{n}'$ is a unit vector that is normal to the projection plane.

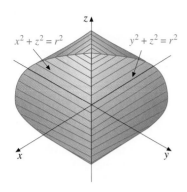

$x^2 + z^2 = r^2$ $y^2 + z^2 = r^2$

FIGURE 13.3.11a The surface consists of parts of two right circular cylinders.

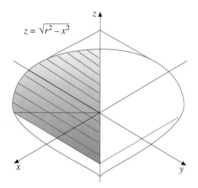

$z = \sqrt{r^2 - x^2}$

FIGURE 13.3.11b That part of the surface that is above the region in the xy-plane bounded by $y = -x$, $y = x$, and $x = r$ is one-eighth of the total surface area.

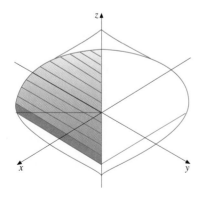

FIGURE 13.3.11c The area of that part of the surface above the strip in the xy-plane between $x = 1 - b$ and $x = 1$ approaches zero as b approaches one.

In cases where a surface is given by $z = z(x, y)$ and the projection plane is the xy-plane, we have $\nabla(z - z(x, y)) = -z_x \mathbf{i} - x_y \mathbf{j} + \mathbf{k}$, and $\mathbf{n}' = \mathbf{k}$ is a unit vector to the projection plane, so

$$\sec \gamma = \frac{|\nabla F|}{|\nabla F \cdot \mathbf{n}'|} = \sqrt{(z_x)^2 + (z_y)^2 + 1},$$

as we have seen previously. We will apply (2) only when the projection plane is a coordinate plane, so $\mathbf{n}'$ is either $\mathbf{i}$, $\mathbf{j}$, or $\mathbf{k}$.

EXAMPLE 2

Find the surface area of the solid that is bounded by the cylinders $x^2 + z^2 = r^2$ and $y^2 + z^2 = r^2$.

SOLUTION   The solid is sketched in Figure 13.3.11a. We see that the total surface area is eight times the area of that part of the surface indicated in Figure 13.3.11b. We need equations of the region in the xy-plane below that part of the surface. To find these equations, we note that coordinates of points on the intersection of the cylinders satisfy both

$$x^2 + z^2 = r^2 \text{ and } y^2 + z^2 = r^2, \text{ so } x^2 = y^2.$$

The total surface area is eight times that part of the surface $z = \sqrt{r^2 - x^2}$ above the region of the xy-plane bounded by $y = -x$, $y = x$, and $x = r$.

$$z = \sqrt{r^2 - x^2} \text{ implies } z_x = \frac{-x}{\sqrt{r^2 - x^2}} \text{ and } z_y = 0, \text{ so}$$

$$\sqrt{z_x^2 + z_y^2 + 1} = \sqrt{\frac{x^2}{r^2 - x^2} + 1} = \frac{r}{\sqrt{r^2 - x^2}}.$$

The surface area should be

$$S = 8 \int_0^r \int_{-x}^x \frac{r}{\sqrt{r^2 - x^2}} \, dy \, dx.$$

However, we see that the integrand becomes unbounded as (x, y) approaches the line $x = r$, so the formula we developed does not apply directly. We can overcome this difficulty by treating the integral as an improper integral. That is, we evaluate the integral with the limit of integration r replaced by a number b with $0 < b < r$ and then take the limit as b approaches r from the left. This means we omit part of the surface in the evaluation of the surface area; the surface area of the part omitted approaches 0 as b approaches r. See Figure 13.3.11c. We have

$$8 \int_0^b \int_{-x}^x \frac{r}{\sqrt{r^2 - x^2}} \, dy \, dx = 8 \int_0^b \frac{r2x}{\sqrt{r^2 - x^2}} \, dx$$

$$= -16r\sqrt{r^2 - x^2} \Big]_0^b$$

$$= -16r\sqrt{r^2 - b^2} + 16r^2.$$

We then have

$$\lim_{b \to r^-} \left(-16r\sqrt{r^2 - b^2} + 16r^2\right) = 16r^2,$$

so the surface area is $16r^2$. ∎

Mass, Moments, and Center of Mass

The **mass** of an object that has constant density is given by

$$\text{Mass} = (\text{Density})(\text{Volume}).$$

If the density varies continuously, we can divide the object into small parts with nearly constant density. Then

$$\text{Total mass} = \sum (\text{Mass of parts}).$$

If an object is thin and flat (i.e., a lamina), we may think of the object as occupying a region in a plane, rather than a region in space. We may then describe the density of the lamina as density per unit area. Dividing the region into small parts and adding the masses leads to a double integral. That is, let

$$\rho(x, y) = \text{Density per unit area at } (x, y).$$

The mass of a rectangular piece of the lamina located at (x_i, y_j) is approximately the density per unit area at the point (x_i, y_j) times the area of the rectangle. That is,

$$\Delta M \approx \rho(x_i, y_j) \, \Delta A_{ij}.$$

See Figure 13.3.12. This suggests the notation

$$dM = \rho(x, y) \, dA.$$

The total mass is then

$$M = \int \int_R dM = \int \int_R \rho(x, y) \, dA.$$

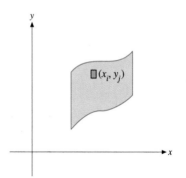

FIGURE 13.3.12 If the density varies continuously, then the density is essentially constant over small rectangles.

The **moment** of a point mass in a coordinate plane about one of the coordinate axes in the plane is the product of the mass times the directed distance of the mass from the axis. The total moment of a system of point masses is the sum of the moments of each mass. This leads to the following formulas for moments of a lamina about the axes:

$$M_{y=0} = \int \int_R y \, dM, \qquad M_{x=0} = \int \int_R x \, dM.$$

These moments are related to the tendency of the lamina to rotate about each coordinate axis.

The **center of mass** (balance point) of a lamina is $(\overline{x}, \overline{y})$, where

$$\overline{x} = \frac{M_{x=0}}{M} = \frac{\int \int_R x \, dM}{\int \int_R dM} \text{ and } \overline{y} = \frac{M_{y=0}}{M} = \frac{\int \int_R y \, dM}{\int \int_R dM}.$$

EXAMPLE 3

Find the center of mass of the lamina bounded by $y = x^2$ and $y = 2x$ if the density is given by $\rho(x, y) = 840xy$.

SOLUTION   The lamina is sketched in Figure 13.3.13. We need to determine the limits of integration for iterated integrals over the region. We see that

For x fixed, y goes from $y = x^2$ to $y = 2x$.

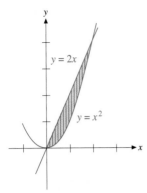

FIGURE 13.3.13 This region is vertically simple.

We need the x-coordinate of each point of intersection of the top and bottom curves. The equations $y = x^2$ and $y = 2x$ imply $x^2 = 2x$, so

$$x^2 - 2x = 0,$$
$$x(x - 2) = 0,$$
$$x = 0, \quad x = 2.$$

Hence,

x goes from x = 0 to x = 2.

The mass of a small piece of the lamina is represented by

$$dM = \rho \, dA = 840xy \, dy \, dx.$$

We then have

$$M = \int \int_R dM = \int_0^2 \int_{x^2}^{2x} 840xy \, dy \, dx$$

$$= \int_0^2 420xy^2 \Big]_{y=x^2}^{y=2x} dx$$

$$= \int_0^2 (1680x^3 - 420x^5) \, dx$$

$$= 420x^4 - 70x^6 \Big]_0^2$$

$$= (6720 - 4480) - (0) = 2240,$$

$$M_{x=0} = \int \int_R x \, dM = \int_0^2 \int_{x^2}^{2x} 840x^2 y \, dy \, dx = 3072,$$

$$M_{y=0} = \int \int_R y \, dM = \int_0^2 \int_{x^2}^{2x} 840xy^2 \, dy \, dx = 5376,$$

$$\bar{x} = \frac{M_{x=0}}{M} = \frac{3072}{2240} = \frac{48}{35} \approx 1.37 \text{ and } \bar{y} = \frac{M_{y=0}}{M} = \frac{5376}{2240} = \frac{12}{5} = 2.4.$$

The center of mass is at the point $(48/35, 12/5)$. ∎

EXERCISES 13.3

Use double integrals to find the volume of the regions bounded by the surfaces given in Exercises 1–6.

1. $x + z = 3$, $y = 3$, in the first octant

2. $x + 2y + 3z = 6$, in the first octant

3. Below $2x + 3y + 3z = 6$, above $z = x + y$, in the first octant

4. Below $x + z = 4$, above $z = x + 2y$, in the first octant

5. $z = 4 - x^2 - y^2$, $z = 0$

6. Inside $x^2 + y^2 = 1$, below $z = 1 - xy$, in the first octant

Find the area of each surface given in Exercises 7–12.

7. $z = 2\sqrt{x^2 + y^2}$; $0 \le x \le 2, 0 \le y \le 3$

8. $x + 3y + 2z = 6$; in the first octant

9. $z = x + (2/3)y^{3/2}$; $0 \le x \le 1, 0 \le y \le 1$

10. $z = \sqrt{2xy}$; $0 \le x \le 1, 0 \le y \le 1$

11. $x^2 + z^2 = r^2$; $y = 0$, $y = h$, in the first octant

12. A sphere of radius r

Find the mass of each lamina indicated in Exercises 13–18.

13. $y + 2x = 4$, $y = 0$, $x = 0$; $\rho(x, y) = x$

14. $y = 4 - x^2$, $y = 0$; $\rho(x, y) = y$

15. $x + y = 2$, $x = 0$ between $y = 0$ and $y = 1$; $\rho(x, y) = 2 - y$

16. $x = y^2$, $x + y = 2$, $y = 0$, in the first quadrant; $\rho(x, y) = xy$

17. $y = \sin x$, $y = 0$, between $x = 0$ and $x = \pi$; $\rho(x, y) = y$

18. $y = \cos x$, $y = 0$, $x = 0 (x \ge 0)$; $\rho(x, y) = x$

Find the center of mass of each lamina indicated in Exercises 19–24.

19. $y = 0$, $y = 4$, $x = 0$, $x = 2$; $\rho(x, y) = 2y$

20. $x + y = 3$, $y = 0$, $x = 0$; $\rho(x, y) = 6xy$

21. $x = \sqrt{1 - y^2}$, $x = 0$; $\rho(x, y) = x$

22. $x^2 + y^2 = 4$, in the first quadrant; $\rho(x, y) = xy$

23. $x^2 + y^2 = 4$, $y = 0$, $y = x$, in the first quadrant; $\rho(x, y) = xy$

24. $x^2 + y^2 = 4$, in the first quadrant; $\rho(x, y) = x^2 + y^2$

25. Set up an iterated double integral in the variables x and z that gives the volume of the region in the first octant below both $x + z = 1$ and $x + y + z = 2$.

26. Set up an iterated double integral in the variables x and z that gives the volume of the region in the first octant below both $x + y + z = 3$ and $y = 2z$.

27. Set up an iterated double integral in the variables x and z that gives the area of the surface $y = 1 - x^2 - z^2$, $y \ge 0$.

28. Set up an iterated double integral in the variables y and z that gives the area of the surface $x = 2\sqrt{y^2 + z^2}$, $x \le 2$.

Moments of inertia *of a lamina about the coordinate axes are* $I_{x\text{-}axis} = \int_R y^2 dM$ *and* $I_{y\text{-}axis} = \int_R x^2 dM$. *Use these formulas to find the indicated moments of inertia of the laminas described in Exercises 29–32.*

29. $2x + y = 2$, $x = 0$, $y = 0$; $\rho(x, y) = 1$; about **(a)** the x-axis and **(b)** the y-axis

30. $bx + ay = ab$, $x = 0$, $y = 0$ (a and b positive); $\rho(x, y) = 1$; about **(a)** the x-axis and **(b)** the y-axis

31. $y = \sqrt{1 - x^2}$, $y = 0$; $\rho(x, y) = y$; about the x-axis

32. A circular homogeneous lamina of radius r; about a diameter

Is $I_{x\text{-}axis}$ or $I_{y\text{-}axis}$ larger for the homogeneous laminas indicated in Exercises 33–34?

33.

34.

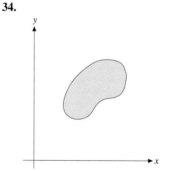

35. **Fully developed laminar flow** of a fluid in a horizontal pipe of radius R at a distance r from the center of the pipe is given by the equation $V = k(R^2 - r^2)$, where k is a constant. Find the volume flow through the pipe, $Q = \int \int_A V \, dA$. Show that the maximum velocity of the fluid is twice the average velocity, Q/A.

13.4   DOUBLE INTEGRALS IN POLAR COORDINATES

Many problems involve regions and functions that can be conveniently described in terms of polar coordinates. We develop double integrals in terms of polar coordinates for these problems.

The Basic Formulas

For a given plane region R and bounded function f, let us use a grid of circular arcs $r =$ constant and radial lines $\theta =$ constant to divide R into "polar rectangles" R_j with area ΔA_j, as indicated in Figure 13.4.1. We then choose (r_j, θ_j) in R_j and form the sum

$$\sum_j f(r_j, \theta_j)\,\Delta A_j.$$

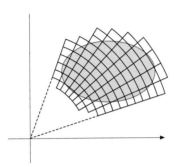

FIGURE 13.4.1   The region is divided into "polar rectangles" bounded by radial lines $\theta =$ constant and circular arcs $r =$ constant.

If f is integrable over R, we should be able to approximate $\iint_R f\,dA$ by such sums. (Recall that we defined the double integral in terms of rectangles with sides parallel to the axes. To be completely rigorous, it would be necessary to show that we arrive at the same value of the double integral by using polar rectangles.)

In order to use polar coordinates to evaluate a double integral, we need to express the area of a polar rectangular in terms of r and θ. The formula for the area of a sector of a circle can be used to show that the area of the typical polar rectangle pictured in Figure 13.4.2 is

$$\frac{1}{2}(r + \Delta r)^2 \Delta\theta - \frac{1}{2}r^2 \Delta\theta = \left(r + \frac{\Delta r}{2}\right)\Delta r\,\Delta\theta,$$

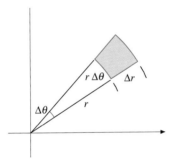

FIGURE 13.4.2   The area of the polar rectangle is $\Delta A \approx r\,\Delta r\,\Delta\theta$.

which is essentially $r\,\Delta r\,\Delta\theta$ when Δr and $\Delta\theta$ are small. In particular, there is a small *percentage* error in using $r\,\Delta r\,\Delta\theta$ as an approximate value of the actual area. Note that the sides of the polar rectangle are orthogonal. For small values of Δr and $\Delta\theta$, the polar rectangle is very nearly an actual rectangle with sides of lengths Δr and $r\Delta\theta$; we can think of $r\,\Delta r\,\Delta\theta$ as the area of this rectangle. We then obtain

$$\Delta A \approx r\,\Delta r\,\Delta\theta.$$

Using the notation of differentials, we write

$$dA = r\,dr\,d\theta.$$

Let us now suppose that the region R can be described in terms of polar coordinates by the following statements:

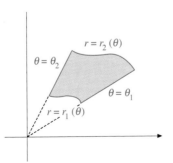

FIGURE 13.4.3a   This region can be described conveniently in terms of polar coordinates.

For θ fixed, r goes from $r = r_1(\theta)$ to $r = r_2(\theta)$.
θ goes from $\theta = \theta_1$ to $\theta = \theta_2$.

(We assume $0 \le r_1(\theta) \le r_2(\theta)$, $0 \le \theta_2 - \theta_1 \le 2\pi$.) Such a region is sketched in Figure 13.4.3a. An arrow that is shot from the pole follows a trajectory with θ fixed, enters the region at a point on the curve $r = r_1(\theta)$, and leaves the region at a point on $r = r_2(\theta)$. As r varies from $r = r_1(\theta)$ to $r = r_2(\theta)$ with θ fixed, the points (r, θ) form a line segment along the ray from the origin at angle θ. See Figure 13.4.3b. As θ varies from θ_1 to θ_2, these line segments sweep out the

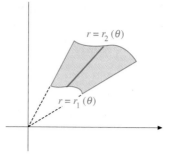

FIGURE 13.4.3b As r varies from $r = r_1(\theta)$ to $r = r_2(\theta)$ with θ fixed, the points (r, θ) form a line segment along the ray from the pole at angle θ.

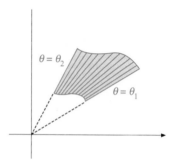

FIGURE 13.4.3c These line segments sweep out the region as θ varies from $\theta = \theta_1$ to $\theta = \theta_2$.

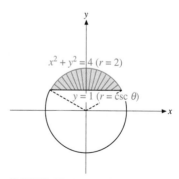

FIGURE 13.4.4 An arrow shot from the pole enters the region at a point on the (polar) line $r = \csc\theta$ and leaves the region at a point on the polar curve $r = 2$.

entire region. See Figure 13.4.3c. We cover R with a grid of polar rectangles R_{ij} bounded by circular arcs $r = r_i$ and radial lines $\theta = \theta_j$. We then have

$$
\begin{aligned}
\iint_R f\, dA &\approx \sum_{i,j} f(r_i, \theta_j)\, \Delta A_{ij} \\
&\approx \sum_{i,j} f(r_i, \theta_j)\, r_i\, \Delta r_i\, \Delta \theta_j \\
&= \sum_j \left(\sum_i f(r_i, \theta_j)\, r_i\, \Delta r_i \right) \Delta \theta_j \\
&\approx \sum_j \left(\int_{r_1(\theta_j)}^{r_2(\theta_j)} f(r, \theta_j)\, r\, dr \right) \Delta \theta_j \\
&\approx \int_{\theta_1}^{\theta_2} \left(\int_{r_1(\theta)}^{r_2(\theta)} f(r, \theta)\, r\, dr \right) d\theta.
\end{aligned}
$$

If f is continuous, the total error in the above approximations can be made less than any prescribed positive number. We conclude that

$$
\iint_R f\, dA = \int_{\theta_1}^{\theta_1} \int_{r_1(\theta)}^{r_2(\theta)} f(r, \theta)\, r\, dr\, d\theta.
$$

To express $\iint_R f\, dA$ in terms of polar coordinates

- Sketch and describe the region R in terms of polar coordinates.
- Use the description of the region to determine the limits of integration.
- Express the integrand f in terms of polar coordinates.
- Express dA as $r\, dr\, d\theta$.

NOTE   *Don't forget the factor r when expressing dA in terms of polar coordinates.*

Recall that we can use the equations

$$
x = r \cos\theta \text{ and } y = r \sin\theta
$$

to change from rectangular to polar coordinates. We also have

$$
x^2 + y^2 = r^2.
$$

Applications

EXAMPLE 1

Find the area of the region inside the circle $x^2 + y^2 = 4$ and above the line $y = 1$.

SOLUTION   The region is sketched in Figure 13.4.4.

We know that the circle $x^2 + y^2 = 4$ can be expressed by the very simple polar equation $r = 2$. This suggests that we might want to try to use polar coordinates in this problem. The line $y = 1$ can be expressed as a polar equation by writing

$$
r \sin\theta = 1, \text{ so } r = \csc\theta.
$$

From Figure 13.4.3 we see that an arrow that is shot from the pole follows a trajectory with θ fixed, enters the region at a point on the line $r = \csc\theta$, and leaves at a point on the circle $r = 2$. The intersection of the line and the circle satisfies

$$r = 2 \text{ and } r = \csc\theta, \text{ so } \csc\theta = 2, \text{ or } \sin\theta = \frac{1}{2}.$$

The θ-coordinates of the points of intersection are $\theta = \pi/6$ and $\theta = 5\pi/6$. The area is then given by

$$
\begin{aligned}
A &= \iint_R dA \\
&= \int_{\pi/6}^{5\pi/6} \int_{\csc\theta}^{2} r\,dr\,d\theta \\
&= \int_{\pi/6}^{5\pi/6} \frac{r^2}{2}\bigg]_{r=\csc\theta}^{r=2} d\theta \\
&= \int_{\pi/6}^{5\pi/6} \frac{1}{2}(4 - \csc^2\theta)\,d\theta \\
&= \frac{1}{2}(4\theta + \cot\theta)\bigg]_{\pi/6}^{5\pi/6} \\
&= \frac{1}{2}\left(4\left(\frac{5\pi}{6}\right) - \sqrt{3}\right) - \frac{1}{2}\left(4\left(\frac{\pi}{6}\right) + \sqrt{3}\right) \\
&= \frac{4\pi}{3} - \sqrt{3}. \quad\blacksquare
\end{aligned}
$$

We can use an iterated polar integral to evaluate any double integral that occurs in applications, as long as the region can be described in terms of polar coordinates. We should be alert for regions and integrands that can be conveniently expressed in terms of polar coordinates.

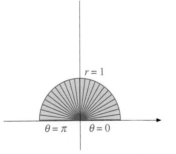

FIGURE 13.4.5　Polar coordinates are convenient for this region.

EXAMPLE 2

Find the center of mass of a homogeneous half-disk of radius 1.

SOLUTION　We need to choose a coordinate system. It is a good idea to center the disk at the origin and to use polar coordinates. The half-disk is sketched in Figure 13.4.5. It occupies the region inside the (polar) circle $r = 1$ for $0 \le \theta \le \pi$. An arrow that is shot from the pole follows a trajectory with fixed θ, enters the region at the pole $r = 0$, and leaves at a point on the circle $r = 1$.

Let k denote the density of the lamina. We then set up the integrals for the mass, $\bar{x}$ and $\bar{y}$ as usual, except we use iterated polar integrals to evaluate the integrals. (Don't forget that we must use $dA = r\,dr\,d\theta$ when using polar coordinates to evaluate $\iint_R f\,dA$.) The mass is

$$M = \iint_R k\,dA = \int_0^{\pi} \int_0^{1} kr\,dr\,d\theta = \int_0^{\pi} \frac{kr^2}{2}\bigg]_{r=0}^{r=1} d\theta = \int_0^{\pi} \frac{k}{2}\,d\theta = \frac{k\pi}{2}.$$

The moment about the x-axis is

$$M_{y=0} = \iint_R yk\,dA = \int_0^\pi \int_0^1 r\sin\theta\, kr\,dr\,d\theta$$

[We used $y = r\sin\theta$ to express y in terms of polar coordinates.]

$$= \int_0^\pi \int_0^1 kr^2 \sin\theta\, dr\, d\theta$$

$$= \int_0^\pi \left[k\left(\frac{r^3}{3}\right)\sin\theta \right]_{r=0}^{r=1} d\theta$$

$$= \int_0^\pi \frac{k}{3}\sin\theta\, d\theta$$

$$= \frac{k}{3}(-\cos\theta)\Big]_{\theta=0}^{\theta=\pi} = \frac{2k}{3}.$$

We then have

$$\overline{y} = \frac{M_{y=0}}{M} = \frac{2k/3}{k\pi/2} = \frac{4}{3\pi}.$$

By symmetry, we know that the center of mass is on the y-axis, so $\overline{x} = 0$. The center of mass is at the point with xy-coordinates $(0, 4/(3\pi))$. ∎

EXAMPLE 3

Find the surface area of the part of the paraboloid $z = x^2 + y^2$ that is below the plane $z = 1$.

SOLUTION The surface is sketched in Figure 13.4.6a. The intersection of the paraboloid and the plane is contained in the cylinder $x^2 + y^2 = 1$. The intersection of this cylinder with the xy-plane gives the region R in the xy-plane over which we integrate. See Figure 13.4.6b. The surface area of the part of the paraboloid $z = x^2 + y^2$ above the region R is

$$S = \iint_R \sqrt{z_x^2 + z_y^2 + 1}\,dA$$

$$= \iint_R \sqrt{(2x)^2 + (2y)^2 + 1}\,dA.$$

At this point we notice that both the region of integration and the function that is being integrated are very well suited for polar coordinates. The region is bounded by the (polar) circle $r = 1$. The expression $x^2 + y^2$ in the integrand can be replaced by the polar expression r^2. We then have

$$S = \int_0^{2\pi} \int_0^1 \sqrt{4r^2 + 1}\, r\,dr\,d\theta$$

$$= \int_0^{2\pi} \frac{1}{12}(4r^2 + 1)^{3/2} \Big]_{r=0}^{r=1} d\theta$$

$$= \int_0^{2\pi} \frac{1}{12}(5^{3/2} - 1)d\theta = \frac{\pi}{6}(5^{3/2} - 1) \approx 5.3304. \ ∎$$

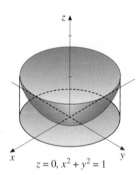

FIGURE 13.4.6a We want the area of that part of the paraboloid that is below the plane.

FIGURE 13.4.6b The projection of the surface onto the xy-plane is bounded by the circle $x^2 + y^2 = 1$. This region of integration and the function being integrated are well suited for polar coordinates.

A Different Order of Integration

Let us now consider a region R that can be described in terms of polar coordinates by the statements

For r fixed, θ goes from θ = θ₁(r) to θ = θ₂(r).
r goes from r = r₁ to r = r₂.

(We assume $0 \le \theta_2(r) - \theta_1(r) \le 2\pi, 0 \le r_1 \le r_2$.) Such a region is sketched in Figure 13.4.7a. Each counterclockwise circular trajectory that intersects the region has r fixed, enters the region at a point on the curve $\theta = \theta_1(r)$, and leaves the region at a point on $\theta = \theta_2(r)$. As θ varies from $\theta = \theta_1(r)$ to $\theta = \theta_2(r)$ with r fixed, the points (r, θ) form an arc of a circle with center at the pole and radius r. See Figure 13.4.7b. As r varies from $r = r_1$ to $r = r_2$, these arcs sweep out the entire region. See Figure 13.4.7c. As before, we have

$$\iint_R f \, dA = \int_{r_1}^{r_1} \int_{\theta_1(r)}^{\theta_2(r)} f(r, \theta) r \, d\theta \, dr.$$

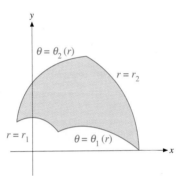

FIGURE 13.4.7a   This region can be described conveniently in terms of polar coordinates.

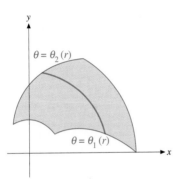

FIGURE 13.4.7b   As θ varies from $\theta = \theta_1(r)$ to $\theta = \theta_2(r)$ with r fixed, the points (r, θ) form an arc of a circle with center at the pole and radius r.

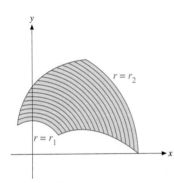

FIGURE 13.4.7c   These circular arcs sweep out the region as r varies from $r = r_1$ to $r = r_2$.

EXAMPLE 4

Find the area of the region bounded by the polar curves $r = e^\theta \, (0 \le \theta \le 2)$, $r = 1 \, (0 \le \theta \le \pi)$, $r = e^2 \, (2 \le \theta \le \pi)$, and the polar axis.

SOLUTION   The region is sketched in Figure 13.4.8. We see that an arrow shot from the pole could leave the region at a point on either of the curves $r = e^\theta$ or $r = e^2$. It is not convenient to express the area of the region as a single iterated integral with the r-integration inside. However, counterclockwise trajectories with fixed r enter the region at a point on $r = e^\theta$ and leave at a point on $\theta = \pi$. Since

$$r = e^\theta \text{ implies } \theta = \ln r,$$

we can describe the region with the following statements:

For r fixed, θ varies from θ = ln r to θ = π.
r varies from r = 1 to r = e².

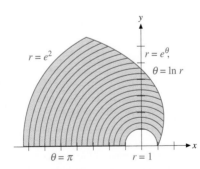

FIGURE 13.4.8   It is not convenient to express the area of the region as a single iterated polar integral with the r-integration inside. It is convenient to express the area as a single iterated polar integral with the θ-integration on the inside.

The area of the region is given by the corresponding iterated polar integral. We have

$$\text{Area} = \int_1^{e^2} \int_{\ln r}^{\pi} r \, d\theta \, dr = \int_1^{e^2} (\pi r - r \ln r) \, dr.$$

Using the Table of Integrals (or integration by parts) to evaluate the indefinite integral of $r \ln r$, we obtain

$$\text{Area} = \frac{\pi r^2}{2} - \frac{r^2 \ln r}{2} + \frac{r^2}{4} \Bigg]_1^{e^2}$$

$$= \left(\frac{\pi e^4}{2} - \frac{e^4 \ln e^2}{2} + \frac{e^4}{4} \right) - \left(\frac{\pi}{2} - \frac{\ln 1}{2} + \frac{1}{4} \right)$$

$$= \frac{\pi e^4}{2} - \frac{3e^4}{4} - \frac{\pi}{2} - \frac{1}{4} \approx 42.993. \quad \blacksquare$$

EXERCISES 13.4

Find the area of the regions indicated in Exercises 1–10.

1. $x^2 + y^2 = 2$, above $y = 1$

2. Between $x^2 + y^2 = 9$ and $x^2 + y^2 = 4$

3. $x^2 + y^2 = 9$, $2y = x$, x-axis, in the first quadrant

4. Inside $(x - 1)^2 + y^2 = 1$ and outside $x^2 + y^2 = 1$

5. Inside $x^2 + (y - 1)^2 = 1$ and above $x + y = 0$

6. Inside both $x^2 + y^2 = 1$ and $(x - 1)^2 + y^2 = 1$

7. Bounded by the polar curve $r = \sin 3\theta$

8. Bounded by the polar curve $r = \sin 4\theta$

9. Bounded by the polar curve $r^2 = \sin 2\theta$

10. Bounded by the polar curve $r = 1 - \cos \theta$

11. Find the volume of a sphere of radius R.

12. Find the volume of a right circular cone with height H and radius of base R.

13. Find the volume of the region inside both $x^2 + y^2 = 1$ and $x^2 + y^2 + z^2 = 4$.

14. Find the volume of the region bounded by $x^2 + y^2 = 1$, $x + z = 2$, and $z = 0$.

15. Find the mass of a circular lamina with radius R and density at a point proportional to the square of the distance of the point to the center.

16. Find the mass of a circular lamina with radius R and density at a point proportional to the distance of the point to the center.

Find the center of mass of each lamina described below in Exercises 17–20.

17. $y = \sqrt{1 - x^2}$, $y = 0$; $\rho(x, y) = ky$, k a constant

18. $x^2 + y^2 = 4$, in the first quadrant; $\rho(x, y) = \sqrt{x^2 + y^2}$

19. Between $x^2 + y^2 = 4$ and $x^2 + y^2 = 1$, in the first quadrant; $\rho(x, y) = k$

20. Inside $(x - 1)^2 + y^2 = 1$, between $y = -x$ and $y = x$; homogeneous

21. Find the area of the curved surface of a right circular cone with height H and radius of base R.

22. Find the area of that part of a plane that is inside a right circular cylinder with radius R if the plane intersects the cylinder at an angle ϕ_0, so the angle between the axis of the cylinder and the normal to the plane is $\pi/2 - \phi_0$.

23. Find the area of that part of the surface of a sphere of radius R that is inside a right circular cone with vertex at the center of the sphere if each line along the cone through its vertex makes an angle of ϕ_0 with the axis of the cone.

24. Find the surface area of a sphere of radius R.

Find the area of the regions bounded by the polar curves given in Exercises 25–28.

25. $r = \theta \, (0 \le \theta \le 2)$, $r = 2\theta \, (0 \le \theta \le 1)$, $r = 2 \, (1 \le \theta \le 2)$

26. $r = e^\theta \, (0 \le \theta \le 2)$, $r = e^{2\theta} \, (0 \le \theta \le 1)$, $r = e^2 \, (1 \le \theta \le 2)$

27. $\theta = \pi/(1 + r^2) \, (\pi/4 \le \theta \le \pi)$, $r = \sqrt{3} \, (0 \le \theta \le \pi/4)$, and the polar axis

28. $r = \ln(\pi/(2\theta)) \, (\pi/6 \le \theta \le \pi/2)$, $r = \ln 3 \, (0 \le \theta < \pi/6)$, and the polar axis

29. The **moment of inertia** of a lamina in the xy-plane about the z-axis is $I_{z\text{-axis}} = \iint_R (x^2 + y^2) dM$.
   (a) Find the moment of inertia about the z-axis of a homogeneous lamina in the xy-plane bounded by $x^2 + y^2 = r_0^2$, in terms of the mass of the lamina.
   (b) Find the moment of inertia about the z-axis of a homogeneous lamina in the xy-plane bounded by $x^2 + y^2 = r_0^2$ and $x^2 + y^2 = (4r_0/5)^2$, in terms of the mass of the lamina.
   (c) If the laminas in (a) and (b) have equal mass, which has the larger moment of inertia?

30. (a) Find polar coordinates of the center of mass of a homogeneous lamina bounded by the polar curves $r = r_0$, $\theta = -\theta_0$, and $\theta = \theta_0$.

(b) What are the polar coordinates of the point that is the limit as θ_0 approaches zero of the points in (a)?

31. An important formula in statistics involves the integral

$$I = \int_0^\infty e^{-x^2} dx.$$

In this exercise we will illustrate a way of evaluating this integral. (The Fundamental Theorem of Calculus cannot be used to evaluate $\int e^{-x^2} dx$.)

(a) Use polar coordinates to evaluate

$$\int\int_{D_R} e^{-x^2-y^2} dA,$$

where D_R is the region in the first quadrant bounded by $x^2 + y^2 = R^2$.

(b) Show that

$$I_R^2 = \int\int_{S_R} e^{-x^2-y^2} dA,$$

where $I_R = \int_0^R e^{-x^2} dx = \int_0^R e^{-y^2} dy$ and S_R is the region given by $0 \le x \le R$ and $0 \le y \le R$.

(c) Show that

$$\int\int_{D_R} e^{-x^2-y^2} dA \le I_R^2 \le \int\int_{D_{\sqrt{2}R}} e^{-x^2-y^2} dA.$$

(d) Evaluate

$$I = \lim_{R\to\infty} I_R.$$

13.5 TRIPLE INTEGRALS; APPLICATIONS

Connections

Describing regions, 13.1.

Concept of iterated integrals, 13.2.

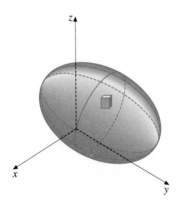

FIGURE 13.5.1 We divide the region into small rectangular prisms R_j and form a sum $\sum f(x_j, y_j, z_j)\Delta V_j$, where (x_j, y_j, z_j) is in R_j and ΔV_j is the volume of R_j.

Triple integrals involve ideas analogous to those for double integrals. A region R in three-dimensional space is divided into small rectangular prisms with sides parallel to the coordinate planes. See Figure 13.5.1. A Riemann sum over R of a bounded function f is the sum of products of f evaluated at a point in each prism times the volume of the prism,

$$\sum f(x_j, y_j, z_j)\Delta V_j.$$

The sum is taken over all prisms that are inside the region. As we did in the case of double integrals, we must avoid regions that do not have a nice enough boundary. In particular, we require that we can cover the region in such a way that the total volume of all rectangular prisms that touch the boundary is less than any prescribed positive number. If f is continuous at all points of R and the boundary of R, then the Riemann sums of f over R approach a single number, denoted

$$\int\int\int_R f(x, y, z)\,dV,$$

as the maximum of the lengths of sides of the rectangular prisms approach zero. As usual, applications of the idea depend on the interpretation of the factors in a corresponding Riemann sum as physical quantities.

Iterated Triple Integrals

Let us consider a region R that can be described by the following statements:

For x and y fixed, z varies from $z = h_1(x, y)$ to $z = h_2(x, y)$.
For x fixed, y varies from $y = g_1(x)$ to $y = g_2(x)$.
x varies from $x = a$ to $x = b$.

We assume the boundary functions are continuous with $h_1(x, y) \le h_2(x, y)$ and $g_1(x) \le g_2(x)$. The surfaces $z = h_1(x, y)$ and $z = h_2(x, y)$ form the bottom and top of the region, the vertical generalized cylinders $y = g_1(x)$ and $y =$

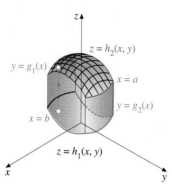

FIGURE 13.5.2a An arrow shot vertically from below the region would enter the region at a point on the surface $z = h_1(x, y)$ and leave the region at a point on the surface $z = h_2(x, y)$.

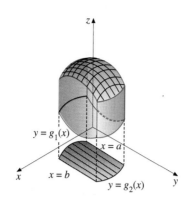

FIGURE 13.5.2b The projection of the region onto the xy-plane is simple in the direction of the y-axis.

$g_2(x)$ form the sides of the region, and the vertical planes $x = a$ and $x = b$ form the back and front boundaries of the region. See Figure 13.5.2a. An arrow shot vertically from below the region would follow a trajectory with x and y fixed, enter the region at a point on the surface $z = h_1(x, y)$, travel parallel to the vertical surfaces $y = g_1(x)$, $y = g_2(x)$, $x = a$, and $x = b$, and leave the region at a point on the surface $z = h_2(x, y)$. The intersection of the surfaces $y = g_1(x)$, $y = g_2(x)$, $x = a$, and $x = b$ with the xy-plane is the projection of the region onto the xy-plane, which is a region of the xy-plane that is simple in the direction of the y-axis. See Figure 13.5.2b. As z varies from $z = h_1(x, y)$ to $z = h_2(x, y)$ with x and y fixed, the points (x, y, z) form a vertical line segment between the top and the bottom of the region. See Figure 13.5.3a. As y then varies from $y = g_1(x)$ to $y = g_2(x)$ with x fixed, these line segments generate a cross-section of the region. See Figure 13.5.3b. As x varies from $x = a$ to $x = b$, the cross-sections sweep out the entire region. See Figure 13.5.3c. In a manner similar to that used for double integrals, we can express a triple integral

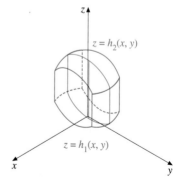

FIGURE 13.5.3a As z varies from $z = h_1(x, y)$ to $z = h_2(x, y)$ with x and y fixed, the points (x, y, z) form a line segment between the top and bottom surfaces of the region.

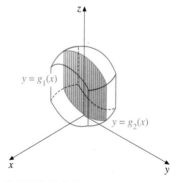

FIGURE 13.5.3b As y varies from $y = g_1(x)$ to $y = g_2(x)$ with x fixed, these line segments generate a cross-section of the region.

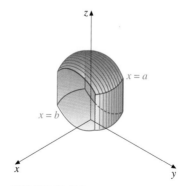

FIGURE 13.5.3c As x varies from $x = a$ to $x = b$, these cross-sections sweep out the entire region.

over R as an **iterated triple integral**. We obtain

$$\iiint_R f(x, y, z)\,dV = \int_a^b \int_{g_1(x)}^{g_2(x)} \int_{h_1(x,y)}^{h_2(x,y)} f(x, y, z)\,dz\,dy\,dx.$$

Note how the limits of integration of the iterated triple integral correspond to the description of the region. We will use the method of Section 13.1 to determine the limits of integration. It is convenient to use the notation of differentials to express the volume of a small rectangular prism as

$$dV = dz\,dy\,dx.$$

The iterated triple integral

$$\int_a^b \int_{g_1(x)}^{g_2(x)} \int_{h_1(x,y)}^{h_2(x,y)} f(x, y, z)\,dz\,dy\,dx$$

is evaluated by first integrating the inner integral with respect to z while x and y are treated as constants. The next integration is carried out by treating x as a constant while we integrate with respect to y. We then integrate with respect to x.

The volume of a region R is expressed as

$$\text{Volume} = \iiint_R dV.$$

EXAMPLE 1

Use a triple integral to find the volume of the region in the first octant that is below the plane $x + y + z = 2$ and above the plane $z = x + y$.

SOLUTION   The region is sketched in Figure 13.5.4a. Note that the points of intersection of the planes in the xz-and yz-planes have been connected with a straight line. An arrow shot parallel to the z-axis from below the region would enter the region at a point on the surface $z = x + y$ and leave the region at a point on the surface $z = 2 - x - y$. The projection of the region onto the xy-plane is given by the vertical planes $x = 0$, $y = 0$ and the vertical plane that contains the line of intersection of the top and bottom of the region. See Figure 13.5.4b. We have

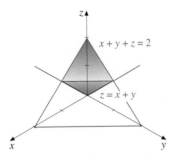

FIGURE 13.5.4a Points of intersection of the coordinate traces are connected by a line segment.

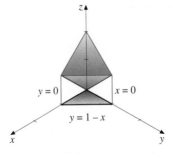

FIGURE 13.5.4b The projection of the region onto the xy-plane is given by the planes $x = 0$, $y = 0$ and the vertical plane that contains the intersection of the top and bottom surfaces of the region.

For x and y fixed, z goes from z = x + y to z = 2 − x − y.

The top and bottom intersect to form a right-side edge. $z = 2 - x - y$ and $z = x + y$ imply $x + y = 2 - x - y$, so $2x + 2y = 2$, or $y = 1 - x$.

For x fixed, y goes from y = 0 to y = 1 − x.

The sides intersect to form a front point. The equations $y = 1 - x$ and $y = 0$ imply $0 = 1 - x$, so $x = 1$.

x goes from x = 0 to x = 1.

The volume is

$$
\begin{aligned}
V &= \int_0^1 \int_0^{1-x} \int_{x+y}^{2-x-y} dz\, dy\, dz \\
&= \int_0^1 \int_0^{1-x} [(2 - x - y) - (x + y)]dy\, dx \\
&= \int_0^1 \int_0^{1-x} (2 - 2x - 2y)dy\, dx \\
&= \int_0^1 \left[2(1 - x)y - y^2 \right]_{y=0}^{y=1-x} dx \\
&= \int_0^1 [2(1 - x)^2 - (1 - x)^2]dx \\
&= \int_0^1 (1 - x)^2 dx \\
&= -\frac{(1 - x)^3}{3} \bigg]_0^1 = \frac{1}{3}. \quad \blacksquare
\end{aligned}
$$

Different Orders of Integration

Triple integrals can be expressed as iterated integrals with six possible orders of integration, if the region is appropriate. Triple integrals over some regions cannot be expressed as a single iterated integral for some of the possible orders. To express a triple integral as a single iterated integral with a particular variable on the inside, it is necessary that the region satisfy the condition that arrows that follow a trajectory parallel to the axis of that variable must all enter the region through the same surface and must all leave the region through the same surface. Also, it must be possible to describe the projection of the region onto the plane of the other two variables in terms of these two variables.

EXAMPLE 2

Express the volume of the region in the first octant that is below both of the planes $2x + 3y + 3z = 6$ and $x + 2z = 2$ as a single iterated triple integral for every order of integration for which this is convenient.

SOLUTION   The region is sketched in Figure 13.5.5a.

From Figure 13.5.5b we see that an arrow shot parallel to the z-axis from below the region could leave the region at a point on either the plane

$2x + 3y + 3z = 6$ or the plane $x + 2z = 2$. Hence, there is not a single equation that describes the top of the region. It is not convenient to express the volume of this region as a single iterated triple integral that has the integration with respect to z on the inside, to be done first.

Similarly, from Figure 13.5.5c we see that an arrow shot parallel to the x-axis from behind the region could leave the region at a point on either of two planes. It is not convenient to express the volume as a single iterated triple integral that has the integration with respect to x on the inside.

An arrow that is shot parallel to the y-axis from the left of the region would enter the region at a point on the plane $y = 0$; travel parallel to the planes $x = 0$, $z = 0$, and $x + 2z = 2$; and leave the region at a point on the plane $2x + 3y + 3z = 6$. See Figure 13.5.5d. We can say:

For x and z fixed, y varies from $y = 0$ to $y = \dfrac{6 - 2x - 3z}{3}$.

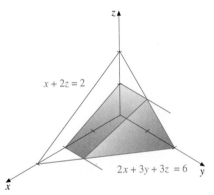

FIGURE 13.5.5a The points of intersection of coordinate traces are connected by a line segment.

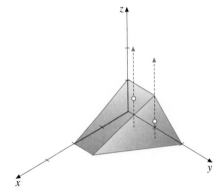

FIGURE 13.5.5b It is not convenient to describe this region as simple in the direction of the z-axis, since arrows parallel to the z-axis could leave the region on either of two different planes.

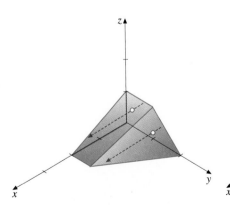

FIGURE 13.5.5c It is not convenient to describe this region as simple in the direction of the x-axis, since arrows parallel to the x-axis could leave the region on either of two different planes.

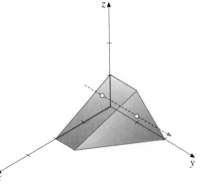

FIGURE 13.5.5d The region is described easily as simple in the direction of the y-axis, since arrows parallel to the y-axis all enter the region on the same plane and all leave the region on the same plane.

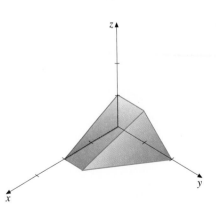

FIGURE 13.5.5e The projection of the region onto the xz-plane is described easily as simple in the direction of the x-axis and as simple in the direction of the y-axis.

The projection of the region onto the xz-plane is given by the equations

$$x = 0, \quad z = 0, \text{ and } x + 2z = 2.$$

See Figure 13.5.5e. This projection is simple in the direction of both the x-axis and the z-axis. We can then say either of the following:

For x fixed, z varies from $z = 0$ to $z = (2 - x)/2$.
x varies from $x = 0$ to $x = 2$.

For z fixed, x varies from $x = 0$ to $x = 2 - 2z$.
z varies from $z = 0$ to $z = 1$.

The volume of the region is then given by the corresponding two integrals,

$$\int_0^2 \int_0^{(2-x)/2} \int_0^{(6-2x-3z)/3} dy\, dz\, dx$$

and

$$\int_0^1 \int_0^{2-2z} \int_0^{(6-2x-3z)/3} dy\, dx\, dz. \quad \blacksquare$$

Mass and Moments

Triple integrals are particularly well suited for the evaluation of the mass, moments, and center of mass of solid objects.

The relation between the mass and volume of a small solid with density ρ is expressed in differential notation as

$$dM = \rho\, dV.$$

The **mass** of an object that occupies a region R in three-dimensional space and has variable density given by a continuous function ρ is

$$M = \iiint_R dM = \iiint_R \rho\, dV.$$

First **moments** of an object about the coordinate planes are

$$M_{x=0} = \iiint_R x\, dM, \quad M_{y=0} = \iiint_R y\, dM, \text{ and } M_{z=0} = \iiint_R z\, dM.$$

Coordinates of the **center of mass** are

$$\bar{x} = \frac{M_{x=0}}{M} = \frac{\iiint_R x\, dM}{\iiint_R dM},$$

$$\bar{y} = \frac{M_{y=0}}{M} = \frac{\iiint_R y\, dM}{\iiint_R dM}, \text{ and}$$

$$\bar{z} = \frac{M_{z=0}}{M} = \frac{\iiint_R z\, dM}{\iiint_R dM}.$$

$z = 4 - x^2 - y^2$

FIGURE 13.5.6 Symmetry implies that the center of mass of this homogeneous solid must be on the z-axis.

EXAMPLE 3

Find the center of mass of a homogeneous solid that is bounded by $z = 4 - x^2 - y^2$, $z = 0$, $y = -1$, $y = 1$, $x = -1$, and $x = 1$.

SOLUTION   The solid is sketched in Figure 13.5.6. The density is constant, $\rho(x, y, z) = k$. The mass is

$$M = \iiint_R dM = \iiint_R \rho \, dV$$

$$= \int_{-1}^{1} \int_{-1}^{1} \int_{0}^{4-x^2-y^2} k \, dz \, dy \, dx$$

$$= \int_{-1}^{1} \int_{-1}^{1} k(4 - x^2 - y^2) \, dy \, dx$$

$$= \int_{-1}^{1} k \left[4y - x^2 y - \frac{1}{3} y^3 \right]_{y=-1}^{y=1} dx$$

$$= \int_{-1}^{1} k \left[\left(4 - x^2 - \frac{1}{3} \right) - \left(-4 + x^2 + \frac{1}{3} \right) \right] dx$$

$$= \int_{-1}^{1} k \left(\frac{22}{3} - 2x^2 \right) dx$$

$$= k \left[\frac{22}{3} x - 2 \frac{x^3}{3} \right]_{-1}^{1}$$

$$= k \left[\left(\frac{22}{3} - \frac{2}{3} \right) - \left(-\frac{22}{3} + \frac{2}{3} \right) \right] = \frac{40}{3} k.$$

The moment about the xy-plane is

$$M_{z=0} = \iiint_R z \, dM = \iiint_R z \rho \, dV$$

$$= \int_{-1}^{1} \int_{-1}^{1} \int_{0}^{4-x^2-y^2} kz \, dz \, dy \, dx = \frac{1016}{45} k.$$

Then

$$\overline{z} = \frac{M_{z=0}}{M} = \frac{\frac{1016}{45} k}{\frac{40}{3} k} = \left(\frac{1016}{45} \right) \left(\frac{3}{40} \right) = \frac{127}{75} \approx 1.693.$$

Symmetry implies that the center of mass is on the z-axis, so $\overline{x} = 0$ and $\overline{y} = 0$. The center of mass is at the point

$$(\overline{x}, \overline{y}, \overline{z}) = \left(0, 0, \frac{127}{75} \right). \quad \blacksquare$$

Let us see how the study of **kinetic energy** (K.E.) leads to a type of moment integral that is different from first moments. The kinetic energy of an object with mass m that is moving with speed v is

$$\text{K.E.} = \frac{1}{2} m v^2.$$

Suppose the object is moving in a circular path with radius of rotation R and constant angular velocity ω. The speed is then

$$v = \omega R$$

and the kinetic energy is

$$\text{K.E.} = \frac{1}{2}m(\omega R)^2 = \frac{\omega^2}{2}R^2 m.$$

If the object is rotating about the z-axis, the radius of rotation is

$$R = \sqrt{x^2 + y^2}$$

and the kinetic energy is

$$\text{K.E.} = \frac{\omega^2}{2}(x^2 + y^2)m.$$

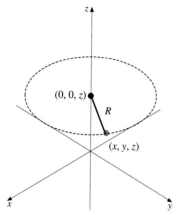

FIGURE 13.5.7   The object is moving in a circular path about the z-axis, so the radius of rotation is $R = \sqrt{x^2 + y^2}$.

See Figure 13.5.7. Let us define the **moment of inertia** about the z-axis of the object to be

$$I_{z\text{-axis}} = (x^2 + y^2)m.$$

The moment of inertia about the z-axis of the object is then proportional to the kinetic energy of the object as it rotates about the z-axis with constant angular speed.

The moment of inertia about the z-axis of a system of masses is the sum of the moments of inertia of each part. The moment of inertia about the z-axis of a solid object that occupies a region R in space and has density function $\rho(x, y, z)$ is defined to be the triple integral

$$I_{z\text{-axis}} = \iiint_R (x^2 + y^2)\, dM = \iiint_R (x^2 + y^2)\rho(x, y, z)\, dV.$$

The moments of inertia of a solid about the other coordinate axes are

$$I_{x\text{-axis}} = \iiint_R (y^2 + z^2)\rho\, dV \quad \text{and}$$

$$I_{y\text{-axis}} = \iiint_R (x^2 + z^2)\rho\, dV.$$

Moments of inertia are also called **second moments**.

EXAMPLE 4

A homogeneous beam has length ℓ and cross-sections squares with length of sides s. Find the moment of inertia of the beam about its axis. Express the answer in terms of the mass and dimensions of the beam.

FIGURE 13.5.8   The beam is positioned with axis along the z-axis and sides parallel to coordinate planes. We then want the moment of inertia about the z-axis.

SOLUTION   The first step is to choose a coordinate system. Let us orient the beam with axis along the z-axis and sides parallel to coordinate planes as indicated in Figure 13.5.8. The desired moment of inertia is then $I_{z\text{-axis}}$. The density of a homogeneous object is constant. Let k denote the constant density. We then have

$$
\begin{aligned}
I_{z\text{-axis}} &= \iiint_R (x^2 + y^2)\,dM = \iiint_R (x^2 + y^2)\rho\,dV \\
&= \int_{-s/2}^{s/2} \int_{-s/2}^{s/2} \int_0^\ell k(x^2 + y^2)\,dz\,dy\,dx \\
&= \int_{-s/2}^{s/2} \int_{-s/2}^{s/2} k\ell(x^2 + y^2)\,dy\,dx \\
&= \int_{-s/2}^{s/2} k\ell \left[x^2 y + \frac{y^3}{3} \right]_{y=-s/2}^{y=s/2} dx \\
&= \int_{-s/2}^{s/2} k\ell \left[sx^2 + \frac{s^3}{12} \right] dx \\
&= k\ell \left[s\frac{x^3}{3} + \frac{s^3}{12}x \right]_{-s/2}^{s/2} \\
&= k\ell \left[\frac{s^4}{12} + \frac{s^4}{12} \right] = k\frac{\ell s^4}{6}.
\end{aligned}
$$

We can express the moment of inertia of the beam in terms of its mass and dimensions by noting that the mass of the beam is

$$
m = k(\text{Volume}) = ks^2\ell, \text{ so } k = \frac{m}{s^2\ell}.
$$

Substitution then gives

$$
I_{z\text{-axis}} = \left(\frac{m}{s^2\ell} \right) \frac{\ell s^4}{6} = \frac{ms^2}{6}. \quad \blacksquare
$$

Let us see how moments of inertia can be related to the **angular momentum** of a solid. We assume that an object is rotating about the z-axis with constant angular velocity ω. The angular momentum about the origin of a small piece of the object with mass dm is

$$
d\mathbf{L} = \mathbf{r} \times \mathbf{v}\,dm,
$$

where $\mathbf{r} = x\mathbf{i} + y\mathbf{j} + z\mathbf{k}$ is the position vector of the object and $\mathbf{v}$ is its velocity. If the object is rotating counterclockwise about the z-axis, we have

$$
\mathbf{v} = -\omega y\mathbf{i} + \omega x\mathbf{j}.
$$

Note that the speed $|\mathbf{v}|$ is ω times the radius of rotation and $\mathbf{v}$ is tangent to the circle of rotation. See Figure 13.5.9. Then

$$
\mathbf{r} \times \mathbf{v} = (x\mathbf{i} + y\mathbf{j} + z\mathbf{k}) \times (-\omega y\mathbf{i} + \omega x\mathbf{j}) = -\omega xz\mathbf{i} - \omega yz\mathbf{j} + \omega(x^2 + y^2)\mathbf{k}.
$$

The total angular momentum is the vector sum of the angular momentum of each of the parts,

$$
\mathbf{L} = \iiint d\mathbf{L} = \iiint \mathbf{r} \times \mathbf{v}\,dm.
$$

If the object is *symmetric about the z-axis*, the $\mathbf{i}$ and $\mathbf{j}$ components of $\mathbf{r} \times \mathbf{v}$ of diametric points will cancel each other and the angular momentum will become

$$
\mathbf{L} = \iiint \omega(x^2 + y^2)\mathbf{k}\,dm = \omega I_{z\text{-axis}}\mathbf{k}.
$$

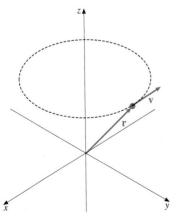

FIGURE 13.5.9 The object is at position **r**, rotating about the z-axis with constant angular velocity ω.

See Figure 13.5.10. We know from Section 11.7 that $d\mathbf{L}/dt = \mathbf{0}$ if there is no external force acting on the object, so $\mathbf{L}$ is constant. This means that

$$\omega I_{z\text{-axis}} = \text{constant.}$$

It follows that the angular velocity ω will increase if the moment of inertia is made smaller. For example, a figure skater can increase the angular velocity of a spin by drawing his or her arms and legs closer to the axis of rotation.

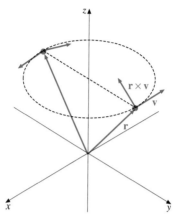

FIGURE 13.5.10 If an object is symmetric about the z-axis, the $\mathbf{i}$ and $\mathbf{j}$ components of $\mathbf{r} \times \mathbf{v}$ of diametric points will cancel each other.

EXERCISES 13.5

Evaluate the integrals in Exercises 1–10.

1. $\displaystyle\int_0^3 \int_0^2 \int_0^1 xyz\,dz\,dy\,dx$

2. $\displaystyle\int_0^1 \int_0^1 \int_0^1 x^2 y e^{xyz}\,dz\,dy\,dx$

3. $\displaystyle\int_0^1 \int_0^x \int_0^y xyz\,dz\,dy\,dx$

4. $\displaystyle\int_0^1 \int_0^x \int_0^{xy} z\,dz\,dy\,dx$

5. $\displaystyle\int_0^\pi \int_0^x \int_0^{xy} x^2 \sin z\,dz\,dy\,dx$

6. $\displaystyle\int_0^r \int_0^{\sqrt{r^2-x^2}} \int_0^{\sqrt{r^2-x^2-y^2}} xy\,dz\,dy\,dx$

7. $\displaystyle\int_0^1 \int_0^z \int_0^y z e^{y^2}\,dx\,dy\,dz$

8. $\displaystyle\int_0^1 \int_0^{\sqrt{z}} \int_0^{\sqrt{yz}} x\,dx\,dy\,dz$

9. $\displaystyle\int_1^e \int_0^y \int_0^{1/y} x^2 z\,dx\,dz\,dy$

10. $\displaystyle\int_1^2 \int_0^{y^2} \int_0^{z^2} yz\,dx\,dz\,dy$

Use triple integrals to find the volume of each region described in Exercises 11–16.

11. $2x + y + 2z = 4$, in the first octant

12. $z = 2x + 2y$, $z = 4$, in the first octant

13. Below $x + 2y + 2z = 4$, above $z = 2x$, in the first octant

14. Below $x + z = 2$, above $z = x + y$, in the first octant

15. Inside both $x^2 + y^2 = r^2$ and $x^2 + z^2 = r^2$

16. Inside $x^2 + y^2 = 4$, below $x + y + z = 4$, in the first octant

Set up an iterated triple integral that gives the volume of each region indicated in Exercises 17–20.

17. Inside both $x^2 + y^2 = r^2$ and $x^2 + z^2 = R^2$, $0 < r < R$

18. Inside both $x^2 + y^2 = r^2$ and $x^2 + y^2 + z^2 = R^2$, $0 < r < R$

19. Inside both $x^2 + y^2 = 1$ and $(y + 4)^2 = 4x^2 + 4z^2$

20. Inside $x^2 + y^2 = 1$, between $z = 4 - \sqrt{x^2 + y^2}$ and $z = 0$

In Exercises 21–26, express the volume of each of the given regions as a single iterated triple integral for every order of integration for which this is convenient.

21. Below both $z = 2x$ and $x + 2y + 3z = 6$, in the first octant

22. $x + z = 4$, $z = 2$, $y = 4$, $x = 0$, $y = 0$, $z = 0$

23. $x + y + z = 4$, $z = 1$, $x = 0$, $y = 0$, $z = 0$

24. $x + z = 3$, $y + z = 3$, $z = 1$, $x = 0$, $y = 0$, $z = 0$

25. Below both $x + y + z = 6$ and $x + z = 4$, above $z = 2x$, in the first octant

26. Below both $y + z = 4$ and $y = 3z$, $y = x$, $x = 0$, $z = 0$, in the first octant

In Exercises 27–32, express the given integrals as a single iterated triple integral for every other order of integration for which this is convenient.

27. $\displaystyle\int_0^4 \int_0^{4-x} \int_{x+y}^4 f(x, y, z)\,dz\,dy\,dx$

28. $\displaystyle\int_0^2 \int_0^{4-2x} \int_{x+y}^{4-x} f(x, y, z)\,dz\,dy\,dx$

29. $\displaystyle\int_0^2 \int_0^{2-x} \int_{x+y}^{4-x-y} f(x, y, z)\,dz\,dy\,dx$

30. $\displaystyle\int_0^1 \int_0^{4-x} \int_{x+y}^4 f(x, y, z)\,dz\,dy\,dx$

31. $\displaystyle\int_0^1 \int_0^{4-2x} \int_{x+y}^{4-x} f(x, y, z)\,dz\,dy\,dx$

32. $\displaystyle\int_0^1 \int_0^{2-x} \int_{x+y}^{4-x-y} f(x, y, z)\,dz\,dy\,dx$

Find the mass of the solids described in Exercises 33–36.

33. Bounded by $z = 2 - y$, $x = 3$, in the first octant; $\rho(x, y, z) = z$

34. Bounded by $x + 2y + 2z = 6$, in the first octant; $\rho(x, y, z) = 6 - x$

35. Bounded by $x = 4$, $y = 6$, $z = 3$, in the first octant; $\rho(x, y, z) = x^2 + y^2 + z^2$

36. Bounded by $x^2 + y^2 = 4$, $z = 1$, $z = 0$; $\rho(x, y, z) = z$

Find the center of mass of each solid described in Exercises 37–40.

37. Bounded by $z = 2x + 2y$, $x = 1$, $y = 1$, in the first octant; $\rho(x, y, z) = 6$

38. Bounded by $x + y + z = 1$, in the first octant; $\rho(x, y, z) = 1$

39. Bounded by $x = 3$, $y = 2$, $z = 1$, in the first octant; $\rho(x, y, z) = 1 + xyz$

40. $x^2 + y^2 = 1$, $z = 4$, $z = 0$; $\rho(x, y, z) = z^2$

41. Find the moment of inertia of a homogeneous right circular cylinder with radius R and height H about its axis. Express the answer in terms of the mass and dimensions of the cylinder.

42. Find the moment of inertia of a homogeneous cube with length of each edge S about a line through the center of opposite sides. Express the answer in terms of the mass and dimensions of the cube.

43. Find the moment of inertia of a homogeneous cube with length of each edge S about a line along one edge. Express the answer in terms of the mass and dimensions of the cube.

44. Set up an iterated triple integral that gives the moment of inertia of a homogeneous solid cone with radius of base r and height h, about its axis. ($\rho(x, y, z) = 3m/(\pi r^2 h)$.)

45. Find the moment of inertia of a homogeneous wedge bounded by $y = 2x$, $y = 4$, $z = 4$, $z = 0$, $x = 0$, about the z-axis. Express the answer in terms of the mass and dimensions of the wedge.

46. The moment of inertia of a homogeneous right circular cylinder with radius r and height h about its axis is $I = k\dfrac{\pi h r^4}{2}$. What is the moment of inertia of a homogeneous cylindrical shell with inner radius r_1, outer radius r_2, and height h, about its axis?

47. The moment of inertia of a homogeneous sphere with radius r about a diameter is $I = k\dfrac{8\pi r^5}{15}$. What is the moment of inertia of a spherical shell with inner radius r_1 and outer radius r_2, about a diameter?

48. Two homogeneous spheres have the same mass, but unequal radii. For which sphere is the moment of inertia about a diameter larger?

The moment of inertia of an object about any line in three-dimensional space is defined to be the integral of the product of the density function at each point of the object and the square of the distance of the point from the line. From Section 11.6 we know that the distance d of a point P from a line that contains the point P_0 and has direction vector $\mathbf{V}$ is given by the formula

$$d = \frac{|\mathbf{PP_0} \times \mathbf{V}|}{|\mathbf{V}|}.$$

49. Find the moment of inertia of a homogeneous cube with length of edge S about a line through opposite corners. Express the answer in terms of the mass of the cube. (*Hint:* Position the cube in the first octant with one corner at the origin. Evaluate the moment of inertia about the line through the origin with direction vector $\mathbf{i} + \mathbf{j} + \mathbf{k}$.)

50. Set up an iterated triple integral that gives the moment of inertia of the homogeneous solid cone with radius of base r and height h, about a line through its vertex along its curved surface. ($\rho(x, y, z) = 3m/(\pi r^2 h)$.) (*Hint:* Position the cone with vertex at the origin and axis along the positive z-axis. Set up the integral that gives the moment of inertia about the line through the origin with direction vector $r\mathbf{j} + h\mathbf{k}$.)

51. Let $(\bar{x}, \bar{y}, \bar{z})$ be the center of mass of a solid. The first moments of the solid about $x = \bar{x}$ and $y = \bar{y}$ then satisfy $M_{x=\bar{x}} = M_{y=\bar{y}} = 0$. Let ℓ denote the vertical line through the center of mass of the solid, and let M denote the mass of the solid. Show that $I_{z\text{-axis}} = I_\ell + (\bar{x}^2 + \bar{y}^2)M$, so that the second moment about the z-axis is the second moment about the parallel line through the center of mass, plus the square of the distance between the lines times the mass. This illustrates the **Parallel Axis Theorem** for second moments.

52. The moment of inertia of a homogeneous right sphere with radius r about a diameter is $I = k\dfrac{8\pi r^5}{15}$. Use the formula

in Exercise 51 to find the moment of inertia of the sphere about a line that is $2r$ from its center.

53. The moment of inertia of a homogeneous right circular cylinder with radius r and height h about its axis is $I = k\dfrac{\pi h r^4}{2}$. Use the formula in Exercise 51 to find the moment of inertia of the cylinder about a line along the side of the cylinder, parallel to its axis.

54. A figure skater is spinning at a rate of three revolutions per second. Assuming that there are no external forces acting on the skater, find the rate of spin after she decreases the moment of inertia of her body about the axis of rotation by 20 percent by drawing her arms and legs closer to the axis of rotation.

13.6 CYLINDRICAL AND SPHERICAL COORDINATES

Connections

Polar coordinates, 10.4.

Concept of describing regions, 13.1.

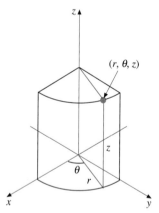

FIGURE 13.6.1 It is important to know the geometric interpretation of each component of cylindrical coordinates.

We have seen that it is convenient to express some double integrals in terms of polar coordinates. In Sections 13.7 and 13.8 we will see that many triple integrals are considerably easier to evaluate if they are expressed in coordinates other than rectangular. In this section we will introduce cylindrical and spherical coordinate systems and use them to sketch and describe regions in three-dimensional space. As in the case of rectangular coordinates, the description of the regions will be used to determine limits of integration of iterated triple integrals.

Cylindrical Coordinates (r, θ, z)

The point in three-dimensional space that is associated with the cylindrical coordinates (r, θ, z) is illustrated in Figure 13.6.1. The cylindrical z-coordinate is the same as the z-coordinate in rectangular coordinates. The cylindrical coordinates r and θ correspond to polar coordinates of the projection of the point onto the xy-plane, although the cylindrical r-coordinate is generally taken to be nonnegative. Cylindrical and rectangular coordinates are related by the equations

$$x = r\cos\theta,$$
$$y = r\sin\theta,$$
$$z = z.$$

We also have

$$x^2 + y^2 = r^2.$$

The three basic surfaces associated with cylindrical coordinates are

- $r = r_0$, a right circular cylinder (Figure 13.6.2a).
- $\theta = \theta_0 \, (r \geq 0)$, a half-plane with edge along the z-axis (Figure 13.6.2b).
- $z = z_0$, a horizontal plane (Figure 13.6.2c).

FIGURE 13.6.2a The graph of the cylindrical equation $r = r_0$ is a right circular cylinder with axis the z-axis.

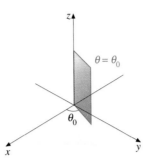

FIGURE 13.6.2b The graph of the cylindrical equation $\theta = \theta_0$ is a half-plane with edge along the z-axis.

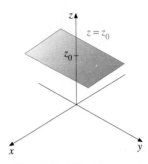

FIGURE 13.6.2c The graph of the cylindrical equation $z = z_0$ is a horizontal plane.

Let us consider a region in three-dimensional space that can be described by the following statements:

For r and θ fixed, z varies from $z = z_1(r, \theta)$ to $z = z_2(r, \theta)$.
For θ fixed, r varies from $r = r_1(\theta)$ to $r = r_2(\theta)$.
θ varies from $\theta = \theta_1$ to $\theta = \theta_2$.

We assume the boundary functions are continuous with $z_1(r, \theta) \leq z_2(r, \theta)$, $0 \leq r_1(\theta) \leq r_2(\theta)$, and $0 \leq \theta_2 - \theta_1 \leq 2\pi$. See Figure 13.6.3a. The vertical generalized cylinders $r = r_1(\theta), r = r_2(\theta), \theta = \theta_1$, and $\theta = \theta_2$ form the vertical side boundaries of the region and give the projection of the region onto the xy-plane. See Figure 13.6.3b. The surfaces $z = z_1(r, \theta)$ and $z = z_2(r, \theta)$ form the bottom and top of the region. An arrow shot vertically from below the region would follow a trajectory with r and θ fixed; enter the region at a point on the bottom surface $z = z_1(r, \theta)$; travel parallel to the vertical surfaces $r = r_1(\theta)$, $r = r_2(\theta), \theta = \theta_1$, and $\theta = \theta_2$; and leave the region at a point on the top surface $z = z_2(r, \theta)$. See Figure 13.6.3c.

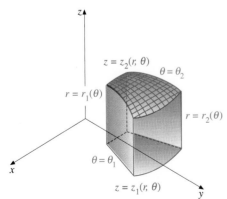

FIGURE 13.6.3a This region can be described in terms of cylindrical coordinates.

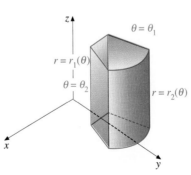

FIGURE 13.6.3b The region is bounded on the sides by the vertical generalized cylinders $r = r_1(\theta)$ and $r = r_2(\theta)$ and by the vertical planes $\theta = \theta_1$ and $\theta = \theta_2$.

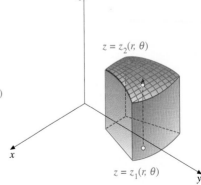

FIGURE 13.6.3c An arrow shot vertically from below the region would enter the region at a point on the surface $z = z_1(r, \theta)$ and leave the region at a point on the surface $z = z_2(r, \theta)$.

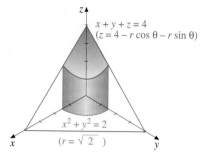

FIGURE 13.6.4 Equations of the cylinder and plane can be expressed easily in terms of polar coordinates.

EXAMPLE 1

Sketch and use cylindrical coordinates to describe the region in the first octant that is inside the cylinder $x^2 + y^2 = 2$ and below the plane $x + y + z = 4$.

SOLUTION The region is sketched in Figure 13.6.4. An arrow shot vertically from below the region would follow a trajectory with r and θ fixed, enter the region at a point on the plane $z = 0$, and leave the region at a point on the plane $x + y + z = 4$. We need to express the given equations in terms of cylindrical coordinates.

$$x^2 + y^2 = 2 \text{ implies } (r\cos\theta)^2 + (r\sin\theta)^2 = 2, \text{ so } r = \sqrt{2}.$$
$$x + y + z = 4 \text{ implies } r\cos\theta + r\sin\theta + z = 4, \text{ so } z = 4 - r\cos\theta - r\sin\theta.$$

For r and θ fixed, z varies from $z = 0$ to $z = 4 - r\cos\theta - r\sin\theta$.
For θ fixed, r varies from $r = 0$ to $r = \sqrt{2}$.
θ varies from $\theta = 0$ to $\theta = \pi/2$. ■

EXAMPLE 2

Sketch and use cylindrical coordinates to describe the region that is inside the cone $z = 2\sqrt{x^2 + y^2}$ and below the plane $z = 2$.

SOLUTION The region is sketched in Figure 13.6.5a. An arrow shot vertically would enter the region at a point on the cone and leave the region at a point on the plane $z = 2$. The cylindrical equation of the cone is $z = 2r$. The intersection of the surfaces $z = 2r$ and $z = 2$ is contained in the cylinder $r = 1$. See Figure 13.6.5b.

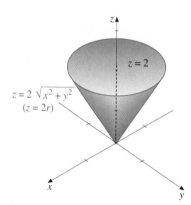

FIGURE 13.6.5a The cylindrical equation of the cone is $z = 2r$.

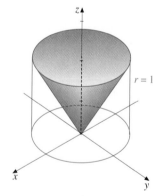

FIGURE 13.6.5b The cylinder $r = 1$ contains the intersection of the top and bottom surfaces.

For r and θ fixed, z varies from $z = 2r$ to $z = 2$.
For θ fixed, r varies from $r = 0$ to $r = 1$.
θ varies from $\theta = 0$ to $\theta = 2\pi$. ■

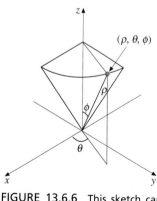

FIGURE 13.6.6 This sketch can be used to determine the relation between rectangular and spherical coordinates.

Spherical Coordinates (ρ, θ, ϕ)

The point in three-dimensional space that is associated with the spherical coordinates (ρ, θ, ϕ) is illustrated in Figure 13.6.6. The spherical coordinate ρ is the distance of the point from the origin, so $\rho \geq 0$. The spherical coordinate

θ corresponds to θ in cylindrical coordinates. The spherical coordinate ϕ is the smaller angle between the positive z-axis and the ray from the origin to the point, so $0 \le \phi \le \pi$; $\phi = 0$ is in the direction of the positive z-axis and $\phi = \pi$ is in the direction of the negative z-axis. Spherical and rectangular coordinates are related by the equations

$$x = \rho \sin \phi \cos \theta,$$
$$y = \rho \sin \phi \sin \theta,$$
$$z = \rho \cos \phi.$$

We also have

$$x^2 + y^2 + z^2 = \rho^2.$$

The three basic surfaces associated with spherical coordinates are

- $\rho = \rho_0$, a sphere (Figure 13.6.7a).
- $\theta = \theta_0$, a half-plane with edge along the z-axis (Figure 13.6.7b).
- $\phi = \phi_0$, a cone (Figure 13.6.7c). (The graph of $\phi = \pi/2$ is the xy-plane.)

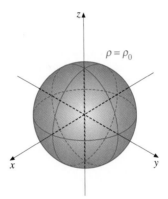

FIGURE 13.6.7a   The graph of the spherical equation $\rho = \rho_0$ is a sphere.

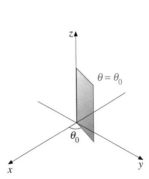

FIGURE 13.6.7b   The graph of the spherical equation $\theta = \theta_0$ is a half-plane with edge along the z-axis.

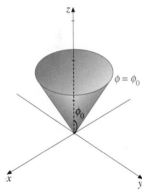

FIGURE 13.6.7c   The graph of the spherical equation $\phi = \phi_0$ is a single-napped cone if $\phi \ne \pi/2$. The graph of $\phi = \pi/2$ is the xy-plane.

A region of the type illustrated in Figure 13.6.8a may be described by the following statements:

For θ and ϕ fixed, ρ varies from $\rho = \rho_1(\theta, \phi)$ to $\rho = \rho_2(\theta, \phi)$.
For θ fixed, ϕ varies from $\phi = \phi_1(\theta)$ to $\phi = \phi_2(\theta)$.
θ varies from $\theta = \theta_1$ to $\theta = \theta_2$.

We assume the boundary functions are continuous with $0 \le \rho_1(\theta, \phi) \le \rho_2(\theta, \phi)$, $0 \le \phi_1(\theta) \le \phi_2(\theta) \le \pi$, and $0 \le \theta_2 - \theta_1 \le 2\pi$. The boundary surfaces $\phi = \phi_1(\theta)$, $\phi = \phi_2(\theta)$, $\theta = \theta_1$, and $\theta = \theta_2$ are generated by rays from the origin. See Figure 13.6.8b. An arrow shot from the origin along a trajectory with fixed θ and ϕ would enter the region at a point on the surface $\rho = \rho_1(\theta, \phi)$ and leave the region at a point on the surface $\rho = \rho_2(\theta, \phi)$. See Figure 13.6.8c.

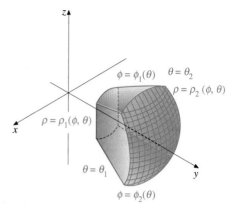

FIGURE 13.6.8a This region can be described in terms of spherical coordinates.

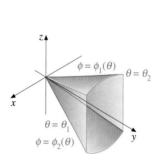

FIGURE 13.6.8b The boundary surfaces $\phi = \phi_1(\theta), \phi = \phi_2(\theta), \theta = \theta_1$, and $\theta = \theta_2$ are generated by rays from the origin.

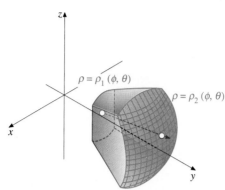

FIGURE 13.6.8c An arrow shot from the origin would enter the region at a point on the surface $\rho = \rho_1(\theta, \phi)$ and leave the region at a point on the surface $\rho = \rho_2(\theta, \phi)$.

EXAMPLE 3

Sketch and use spherical coordinates to describe the region that is inside both the cone $z = \sqrt{x^2 + y^2}$ and the sphere $x^2 + y^2 + z^2 = 4$.

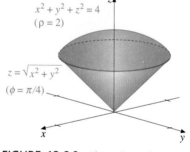

FIGURE 13.6.9 The sphere has spherical equation $\rho = 2$, and the cone has spherical equation $\phi = \pi/4$.

SOLUTION The region is sketched in Figure 13.6.9. An arrow that is shot from the origin enters the region at the origin and leaves the region at a point on the sphere. The spherical equation of the sphere is $\rho = 2$ and the spherical equation of the cone is $\phi = \pi/4$. That is,

$$x^2 + y^2 + z^2 = 4 \text{ implies } \rho^2 \sin^2 \phi \cos^2 \theta + \rho^2 \sin^2 \phi \sin^2 \theta + \rho^2 \cos^2 \phi = 4,$$
$$\rho^2 \sin^2 \phi (\cos^2 \theta + \sin^2 \theta) + \rho^2 \cos^2 \phi = 4,$$
$$\rho^2 \sin^2 \phi + \rho^2 \cos^2 \phi = 4,$$
$$\rho^2 = 4,$$
$$\rho = 2;$$
$$z^2 = x^2 + y^2 \text{ implies } \rho^2 \cos^2 \phi = \rho^2 \sin^2 \phi \cos^2 \theta + \rho^2 \sin^2 \phi \sin^2 \theta,$$
$$\cos^2 \phi = \sin^2 \phi,$$
$$\tan^2 \phi = 1,$$
$$\tan \phi = 1, \qquad \text{[} \tan \phi > 0 \text{ in this example,}$$
$$\qquad\qquad\qquad \text{since } z > 0.\text{]}$$
$$\phi = \frac{\pi}{4}.$$

For θ and ϕ fixed, ρ varies from $\rho = 0$ to $\rho = 2$.
For θ fixed, ϕ varies from $\phi = 0$ to $\phi = \pi/4$.
θ varies from $\theta = 0$ to $\theta = 2\pi$. ∎

EXAMPLE 4

Sketch and use spherical coordinates to describe the region in the first octant that is between the planes $y = 0$ and $y = x$ and inside the sphere $x^2 + y^2 + z^2 = 1$.

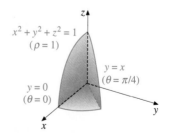

FIGURE 13.6.10 An arrow shot from the origin would enter the region at the origin and leave the region at a point on the sphere.

SOLUTION The region is sketched in Figure 13.6.10. An arrow shot from the origin would enter the region at the origin and leave the region at a point on the sphere. The spherical equation of the half-plane $y = 0$ $(x \geq 0)$ is $\theta = 0$.

$$y = x \text{ implies } \rho \sin \phi \sin \theta = \rho \sin \phi \cos \theta,$$

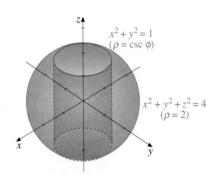

FIGURE 13.6.11a An arrow shot from the origin would enter the region at a point on the cylinder and leave the region at a point on the sphere.

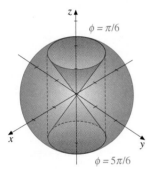

FIGURE 13.6.11b The region is between the two cones $\phi = \pi/6$ and $\phi = 5\pi/6$.

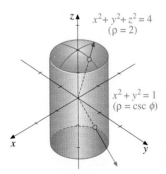

FIGURE 13.6.12 This region cannot be described simply in terms of spherical coordinates, because an arrow shot from the origin could leave the region at a point on either the sphere or the cylinder.

$$\sin\theta = \cos\theta,$$
$$\tan\theta = 1,$$
$$\theta = \frac{\pi}{4}.$$

The xy-plane has spherical equation $\phi = \pi/2$.

For θ and ϕ fixed, ρ varies from $\rho = 0$ to $\rho = 1$.
For θ fixed, ϕ varies from $\phi = 0$ to $\phi = \pi/2$.
θ varies from $\theta = 0$ to $\theta = \pi/4$. ∎

EXAMPLE 5
Sketch and use spherical coordinates to describe the region that is inside the sphere $x^2 + y^2 + z^2 = 4$ and outside the cylinder $x^2 + y^2 = 1$.

SOLUTION   The region is sketched in Figure 13.6.11a. An arrow shot from the origin would enter the region at a point on the cylinder and leave the region at a point on the sphere. We need spherical equations of the cylinder.

$$x^2 + y^2 = 1 \text{ implies } \rho^2 \sin^2\phi \cos^2\theta + \rho^2 \sin^2\phi \sin^2\theta = 1,$$
$$\rho^2 \sin^2\phi(\cos^2\theta + \sin^2\theta) = 1,$$
$$\rho^2 \sin^2\phi = 1,$$
$$\rho^2 = \csc^2\phi,$$
$$\rho = \csc\phi.$$

Points of intersection of the sphere $\rho = 2$ and the cylinder $\rho = \csc\phi$ satisfy $\csc\phi = 2$. It follows that the intersection is contained between the two cones, $\phi = \pi/6$ and $\phi = 5\pi/6$. See Figure 13.6.11b.

For θ and ϕ fixed, ρ varies from $\rho = \csc\phi$ to $\rho = 2$.
For θ fixed, ϕ varies from $\phi = \pi/6$ to $\phi = 5\pi/6$.
θ varies from $\theta = 0$ to $\theta = 2\pi$. ∎

The region that is inside both the sphere $x^2 + y^2 + z^2 = 4$ and the cylinder $x^2 + y^2 = 1$ is not the type of region that we can describe conveniently in terms of spherical coordinates with one set of equations, because an arrow shot from the origin could leave the region at a point on either the sphere or the cylinder. See Figure 13.6.12. The region is easy to describe in terms of cylindrical coordinates.

NOTE   *Be careful when using the inverse tangent function to find the spherical equation of a cone $z = -h\sqrt{x^2 + y^2}/r$, where h and r are positive constants.*

See Figure 13.6.13a. Substitution gives

$$\rho\cos\phi = -\frac{h(\rho\sin\phi)}{r}, \text{ so } \tan\phi = -\frac{r}{h}.$$

Since $\tan\phi < 0$, the value of $\tan^{-1}(-r/h)$ is between $-\pi/2$ and 0 and ϕ is between $\pi/2$ and π. We then have

$$\phi = \tan^{-1}\left(-\frac{r}{h}\right) + \pi.$$

See Figure 13.6.13b.

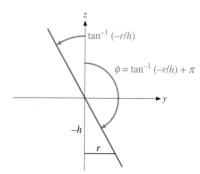

FIGURE 13.6.13a If h and r are positive, the cone $z = -h\sqrt{x^2 + y^2}/r$ opens downward, so ϕ is between $\pi/2$ and π.

FIGURE 13.6.13b Since $\tan^{-1}(-r/h)$ is between zero and $-\pi/2$ and ϕ is between $\pi/2$ and π, the equation $\tan\phi = -\dfrac{r}{h}$ implies $\phi = \tan^{-1}\left(-\dfrac{r}{h}\right) + \pi$.

EXERCISES 13.6

Plot the points with the given cylindrical coordinates (r, θ, z) in Exercises 1–2.

1. $A(1, 0, 0)$, $B(2, \pi/4, 2)$, $C(3, \pi, 1)$, $D(3, -\pi/2, -1)$

2. $A(2, \pi/2, 0)$, $B(2, \pi/3, 0)$, $C(2, 0, -1)$, $D(3, -\pi/4, 2)$

Plot the points with the given spherical coordinates (ρ, θ, ϕ) in Exercises 3–4.

3. $A(2, 0, \pi)$, $B(2, \pi/2, \pi/4)$, $C(3, -\pi/4, \pi/2)$, $D(3, \pi/2, 3\pi/4)$

4. $A(3, 0, 0)$, $B(3, \pi/2, \pi/6)$, $C(2, \pi/4, \pi/2)$, $D(3, -\pi/2, 5\pi/6)$

Find the rectangular coordinates of the points with the given cylindrical coordinates (r, θ, z) in Exercises 5–6.

5. $A(0, \pi, 2)$, $B(2, \pi/4, -1)$, $C(2, \pi/3, 3)$

6. $A(\pi, 0, \pi)$, $B(4, -\pi/4, 2)$, $C(2, 2\pi/3, 1)$

Find rectangular coordinates of the points with the given spherical coordinates (ρ, θ, ϕ) in Exercises 7–8.

7. $A(8, \pi/2, \pi/4)$, $B(1, \pi/4, \pi/2)$, $C(2, \pi/3, 2\pi/3)$

8. $A(4, \pi/2, \pi/3)$, $B(3, \pi, \pi/2)$, $C(2, \pi/4, 3\pi/4)$

Find cylindrical coordinates of the points with the given rectangular coordinates (x, y, z) in Exercises 9–10.

9. $A(2, 2, 2)$, $B(0, 0, -1)$, $C(4, -4\sqrt{3}, 1)$

10. $A(1, -1, 1)$, $B(0, 2, -1)$, $C(-2, 0, 2)$

Find spherical coordinates of the points with the given rectangular coordinates (x, y, z) in Exercises 11–12.

11. $A(0, 0, -2)$, $B(2, 2, 0)$, $C(0, 2, 2)$

12. $A(0, 0, 1)$, $B(0, 2, -2)$, $C(2, 0, 2)$

Sketch and use cylindrical coordinates to describe the regions given in Exercises 13–24.

13. Inside $x^2 + y^2 = 4$, below $z = 2x + 2y$, in the first octant

14. Inside both $x^2 + y^2 = 1$ and $x^2 + y^2 + z^2 = 4$

15. Inside $x^2 + y^2 = 9$, below $z = \sqrt{x^2 + y^2}$, above $z = 0$

16. Inside $x^2 + y^2 = 4$, below $x + z = 4$, in the first octant

17. $z = 4 - x^2 - y^2$, $z = 0$

18. $z = 4 - 2\sqrt{x^2 + y^2}$, $z = 0$

19. $z = 2 - e^{x^2 + y^2}$, $z = 0$

20. $z = \cos\sqrt{x^2 + y^2}$, $z = 0$ $[x^2 + y^2 \le (\pi/2)^2]$

21. Inside $x^2 + y^2 + z^2 = 16$, $x + y = 4$ $(x + y \le 4)$, in the first octant

22. $x + y + z = 3$, in the first octant

23. $x^2 + y^2 = 2x$, $z = 4$, $z = 0$

24. $x^2 + y^2 = 4y$, $y + z = 4$, $z = 0$

Sketch and use spherical coordinates to describe the regions given in Exercises 25–36.

25. $z = \sqrt{9 - x^2 - y^2}$, $z = 0$

26. Inside $x^2 + y^2 + z^2 = 4$, above $z = -\sqrt{x^2 + y^2}$

27. Inside $x^2 + y^2 + z^2 = 1$, in the first octant

28. Above $z = \sqrt{x^2 + y^2}$, below $z = 4$

29. Inside $x^2 + y^2 = 16$, below $z = \sqrt{x^2 + y^2}$, above $z = 0$

30. Inside $z = 4 - 2\sqrt{x^2 + y^2}$, outside $x^2 + y^2 = 1$, above $z = 0$

31. Inside $x^2 + y^2 + z^2 = 25$, above $z = 3$

32. Inside $x^2 + y^2 + z^2 = 25$, below $z = -3$

33. $x + y + z = 3$, in the first octant

34. Inside $x^2 + y^2 + z^2 = 1$, between $y = \sqrt{3}x$ and $y = x/\sqrt{3}$, in the first octant

35. Inside $z = 2\sqrt{x^2 + y^2}$, below $x + z = 2$

36. Inside $z = -3\sqrt{x^2 + y^2}$, above $z = -4$

37. Express the length of the segment OA in the figure in terms of **(a)** rectangular, **(b)** cylindrical, and **(c)** spherical coordinates.

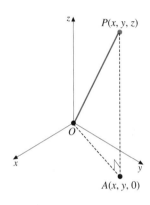

38. Express the length of the segment AP in the figure for Exercise 37 in terms of **(a)** rectangular, **(b)** cylindrical, and **(c)** spherical coordinates.

In Exercises 39–42, use the given descriptions of the regions in cylindrical coordinates to describe the regions in rectangular coordinates.

39. *For r and θ fixed, z varies from $z = 0$ to $z = \sqrt{4 - r^2}$. For θ fixed, r varies from $r = 0$ to $r = 1$. θ varies from $\theta = 0$ to $\theta = 2\pi$.*

40. *For r and θ fixed, z varies from $z = 2r$ to $z = 4$. For θ fixed, r varies from $r = 0$ to $r = 2$. θ varies from $\theta = 0$ to $\theta = 2\pi$.*

41. *For r and θ fixed, z varies from $z = 0$ to $z = 4 - 2r$. For θ fixed, r varies from $r = 0$ to $r = 2$. θ varies from $\theta = 0$ to $\theta = 2\pi$.*

42. *For r and θ fixed, z varies from $z = 0$ to $z = 2r$. For θ fixed, r varies from $r = 0$ to $r = 1$. θ varies from $\theta = 0$ to $\theta = \pi/2$.*

In Exercises 43–46, use the given descriptions of the regions in spherical coordinates to describe the regions in rectangular coordinates.

43. *For θ and ϕ fixed, ρ varies from $\rho = 0$ to $\rho = 2$. For θ fixed, ϕ varies from $\phi = 0$ to $\phi = \pi/2$. θ varies from $\theta = 0$ to $\theta = 2\pi$.*

44. *For θ and ϕ fixed, ρ varies from $\rho = 0$ to $\rho = 2\sqrt{5}$. For θ fixed, ϕ varies from $\phi = 0$ to $\phi = \tan^{-1}(1/2)$. θ varies from $\theta = 0$ to $\theta = 2\pi$.*

45. *For θ and ϕ fixed, ρ varies from $\rho = 0$ to $\rho = \csc\phi$. For θ fixed, ϕ varies from $\phi = \pi/4$ to $\phi = \pi/2$. θ varies from $\theta = 0$ to $\theta = 2\pi$.*

46. *For θ and ϕ fixed, ρ varies from $\rho = 0$ to $\rho = 4\sec\phi$. For θ fixed, ϕ varies from $\phi = 0$ to $\phi = \pi/6$. θ varies from $\theta = 0$ to $\theta = \pi/2$.*

13.7 MULTIPLE INTEGRALS IN CYLINDRICAL AND SPHERICAL COORDINATES

Cylindrical Coordinates

Connections

Cylindrical and spherical coordinates, 13.6.

Concept of iterated integrals, 13.5.

Let us see how to evaluate a triple integral

$$\iiint_R f\, dV$$

in terms of cylindrical coordinates. We begin by dividing the region R into cylindrical prisms R_j that are bounded by surfaces

$$r = \text{constant}, \quad \theta = \text{constant, and } z = \text{constant}.$$

See Figure 13.7.1a. We then choose a point in R_j with cylindrical coordinates (r_j, θ_j, z_j) and form the Riemann sum

$$\sum_j f(r_j, \theta_j, z_j)\Delta V_j,$$

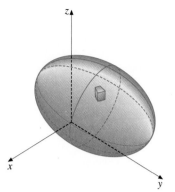

FIGURE 13.7.1a We divide the region into small cylindrical prisms R_j and form a sum $\sum f(x_j, y_j, z_j) \Delta V_j$, where (x_j, y_j, z_j) is in R_j and ΔV_j is the volume of R_j.

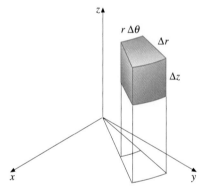

FIGURE 13.7.1b The volume of the cylindrical prism is $\Delta V \approx r \Delta z \Delta r \Delta \theta$.

where ΔV_j is the volume of the cylindrical prism R_j. If f is integrable over the region R, these sums approximate $\iiint_R f \, dV$ whenever all of the cylindrical prisms are small enough.

The next step in evaluating $\iiint_R f \, dV$ in terms of cylindrical coordinates is to find a formula for an approximate value of the volume of the small cylindrical prisms. To do this we note that each of these prisms has eight edges that are line segments and four edges that are arcs of circles. See Figure 13.7.1b. The edges are orthogonal and, for small increments, the four edges that are arcs of circles are nearly straight. This means that the prisms are nearly rectangular. We can then use the formula for the volume of a rectangle to obtain approximate values of the prisms. This gives

$$\Delta V \approx r \, \Delta z \, \Delta r \, \Delta \theta. \tag{1}$$

This approximation becomes proportionately more accurate as the increments approach zero. Formula 1 suggests the differential notation

$$dV = r \, dz \, dr \, d\theta.$$

Suppose now that the region R can be described in terms of cylindrical coordinates by the statements

For r and θ fixed, z varies from $z = z_1(r, \theta)$ to $z = z_2(r, \theta)$.
For θ fixed, r varies from $r = r_1(\theta)$ to $r = r_2(\theta)$.
θ varies from $\theta = \theta_1$ to $\theta = \theta_2$.

The surfaces $z = z_1(r, \theta)$ and $z = z_2(r, \theta)$ form the bottom and top of the region. The vertical surfaces $r = r_1(\theta), r = r_2(\theta), \theta = \theta_1$, and $\theta = \theta_2$ form the side boundaries of the region. See Figure 13.7.2a. As z varies from $z = z_1(r, \theta)$ to $z = z_2(r, \theta)$, the points (r, θ, z) form a vertical line segment between the top and bottom of the region. See Figure 13.7.2b. As r varies from $r = r_1(\theta)$ to $r = r_2(\theta)$ with θ fixed, these line segments generate a cross-section of the region in the half-plane with edge along the z-axis and in the direction of θ. See Figure 13.7.2c. As θ varies from $\theta = \theta_1$ to $\theta = \theta_2$, the cross-sections sweep out the entire region. See Figure 13.7.2d. We can then group the terms of a Riemann sum as we have done in Sections 13.2, 13.4, and 13.5 and use Formula 1 to show that

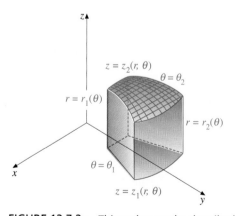

FIGURE 13.7.2a This region can be described in terms of cylindrical coordinates.

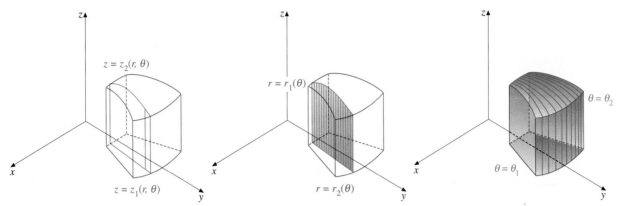

FIGURE 13.7.2b As z varies from $z = z_1(r, \theta)$ to $z = z_2(r, \theta)$ with r and θ fixed, the points (r, θ, z) form a vertical line segment from the bottom surface to the top surface.

FIGURE 13.7.2c As r varies from $r = r_1(\theta)$ to $r = r_2(\theta)$ with θ fixed, these line segments generate a cross-section of the region in the plane of θ.

FIGURE 13.7.2d As θ varies from $\theta = \theta_1$ to $\theta = \theta_2$, these sections sweep out the entire region.

$$\iiint_R f\, dV = \int_{\theta_1}^{\theta_2} \int_{r_1(\theta)}^{r_2(\theta)} \int_{z_1(r,\theta)}^{z_2(r,\theta)} f(r, \theta, z)\, r\, dz\, dr\, d\theta.$$

(If f is integrable over R, it can be shown that the total error of the approximations used in the development of the above formula can be made less than any prescribed positive number.)

In order to express a triple integral as an iterated integral with respect to cylindrical coordinates, we use the description of the region in terms of cylindrical coordinates to determine the limits of integration, express the integrand in terms of cylindrical coordinates, and use $r\, dz\, dr\, d\theta$ in place of dV.

Recall that we can use the equations

$$x = r \cos \theta,$$
$$y = r \sin \theta,$$
$$z = z$$

to change from rectangular coordinates to cylindrical coordinates. It is also useful to remember that

$$x^2 + y^2 = r^2.$$

We may use cylindrical coordinates to evaluate a triple integral whenever the region and the integrand are appropriate. The appearance of the expression $x^2 + y^2$ either in an equation of the boundary of the region or in the integrand is an indication that cylindrical coordinates might be appropriate.

EXAMPLE 1

Find the volume of a cone that has height H and radius of base R.

SOLUTION Let us position the cone as indicated in Figure 13.7.3a. The equation of the cone in rectangular coordinates is then $z = H\sqrt{x^2 + y^2}/R$. We want the volume of that part of the cone that is below $z = H$.

$$z = H \text{ and } z = H\sqrt{x^2 + y^2}/R \text{ imply } x^2 + y^2 = R^2.$$

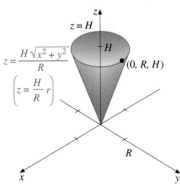

FIGURE 13.7.3a The cone is positioned with vertex at the origin and axis along the z-axis. The equation of the cone is then derived from the fact that $z = H$ when $\sqrt{x^2 + y^2} = R$.

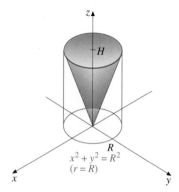

FIGURE 13.7.3b The cylinder $x^2 + y^2 = R^2$ contains the intersection of the cone and the plane.

See Figure 13.7.3b. We see that the region can be described easily in terms of cylindrical coordinates. The cone has cylindrical equation $z = Hr/R$. Using $dV = r\, dz\, dr\, d\theta$, we then have

$$V = \iiint_R dV = \int_0^{2\pi} \int_0^R \int_{Hr/R}^H r\, dz\, dr\, d\theta$$

$$= \int_0^{2\pi} \int_0^R \left(Hr - \frac{Hr^2}{R} \right) dr\, d\theta$$

$$= \int_0^{2\pi} \left(\frac{HR^2}{2} - \frac{HR^3}{3R} \right) d\theta = \left(\frac{HR^2}{6} \right)(2\pi) = \frac{\pi R^2 H}{3}. \quad \blacksquare$$

EXAMPLE 2

Find the moment of inertia of a homogeneous right circular cylinder with radius R and height H about the axis of the cylinder. Express the answer in terms of the mass and dimensions of the cylinder.

SOLUTION The first step is to choose a coordinate system. Let us choose the axis of the cylinder along the z-axis so the desired moment of inertia is $I_{z\text{-axis}}$. See Figure 13.7.4. The density of a homogeneous object is constant. Let k denote the constant density. We then have

$$I_{z\text{-axis}} = \iiint_R (x^2 + y^2)\, dM = \iiint_R (x^2 + y^2)\rho\, dV$$

$$= \int_0^{2\pi} \int_0^R \int_0^H k(r^2)r\, dz\, dr\, d\theta$$

$$= \int_0^{2\pi} \int_0^R kHr^3\, dr\, d\theta$$

$$= \int_0^{2\pi} kH\frac{R^4}{4}\, d\theta$$

$$= k\frac{\pi H R^4}{2}.$$

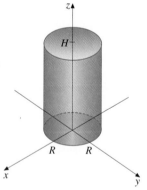

FIGURE 13.7.4 The cylinder is positioned with axis along the z-axis, and we calculate the moment of inertia about the z-axis.

We can express the moment of inertia of the cylinder in terms of its radius, height, and mass by noting that the total mass of the cylinder is

$$M = k(\text{Volume}) = k\pi R^2 H, \text{ so } k = \frac{M}{\pi R^2 H}.$$

We can then express the density in terms of M, R, and H. That is, substitution gives

$$I_{z\text{-axis}} = \frac{M}{\pi R^2 H} \frac{\pi H R^4}{2} = \frac{MR^2}{2}. \quad \blacksquare$$

Spherical Coordinates

To express a triple integral in terms of spherical coordinates, we first divide the region R along surfaces

$$\rho = \text{constant}, \phi = \text{constant}, \text{ and } \theta = \text{constant}$$

and form a Riemann sum as before. See Figure 13.7.5a. The sum of the products of f evaluated at a point in each prism times the volume of the prism should then be approximately $\iiint_R f\,dV$.

We need an approximate value of small spherical prisms. We note that the edges of the prisms meet at right angles and are very nearly straight for small increments, so the prisms are essentially rectangles with lengths of edges $\Delta\rho$, $\rho\,\Delta\phi$, $\rho\sin\phi\,\Delta\theta$. See Figure 13.7.5b. The volume is then approximately

$$\Delta V \approx \rho^2 \sin\phi\,\Delta\rho\,\Delta\phi\,\Delta\theta. \tag{2}$$

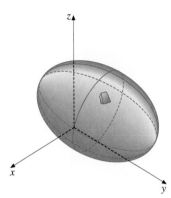

FIGURE 13.7.5a We divide the region into small prisms R_j and form a sum $\sum f(x_j, y_j, z_j)\Delta V_j$, where (x_j, y_j, z_j) is in R_j and ΔV_j is the volume of R_j.

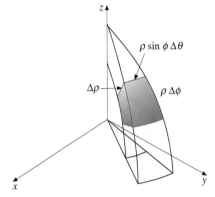

FIGURE 13.7.5b The volume of the spherical prism is $\Delta V \approx \rho^2 \sin\phi\,\Delta\rho\,\Delta\phi\,\Delta\theta$.

This approximation becomes proportionately more accurate as the increments approach zero. We express this by writing

$$dV = \rho^2 \sin\phi\,d\rho\,d\phi\,d\theta.$$

Suppose the region R can be described in terms of spherical coordinates by the following statements:

For θ and ϕ fixed, ρ varies from $\rho = \rho_1(\theta, \phi)$ to $\rho = \rho_2(\theta, \phi)$.
For θ fixed, ϕ varies from $\phi = \phi_1(\theta)$ to $\phi = \phi_2(\theta)$.
θ varies from $\theta = \theta_1$ to $\theta = \theta_2$.

The surfaces $\phi = \phi_1(\theta)$, $\phi = \phi_2(\theta)$, $\theta = \theta_1$, and $\theta = \theta_2$ are generated by rays from the origin. An arrow shot from the origin would follow a trajectory with θ and ϕ fixed, enter the region at a point on the surface $\rho = \rho_1(\theta, \phi)$, and leave the region at a point on the surface $\rho = \rho_2(\theta, \phi)$. See Figure 13.7.6a. As ρ varies from $\rho = \rho_1(\theta, \phi)$ to $\rho = \rho_2(\theta, \phi)$ with ϕ and θ fixed, the points (ρ, θ, ϕ) form a line segment along a ray from the origin. See Figure 13.7.6b. As ϕ varies from $\phi = \phi_1(\theta)$ to $\phi = \phi_2(\theta)$ with θ fixed, these line segments generate a cross-section of the region in the half-plane with edge along the z-axis and in the direction of θ. See Figure 13.7.6c. As θ varies from $\theta = \theta_1$ to

$\theta = \theta_2$, the cross-sections sweep out the entire region. See Figure 13.7.6d. As before, we can then obtain the formula

$$\iiint_R f \, dV = \int_{\theta_1}^{\theta_2} \int_{\phi_1(\theta)}^{\phi_2(\theta)} \int_{\rho_1(\theta,\phi)}^{\rho_2(\theta,\phi)} f(\rho, \theta, \phi) \rho^2 \sin\phi \, d\rho \, d\phi \, d\theta.$$

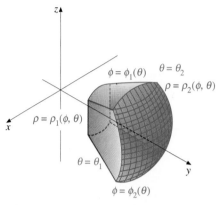

FIGURE 13.7.6a This region can be described in terms of spherical coordinates.

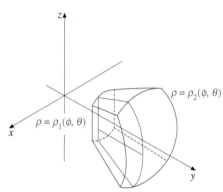

FIGURE 13.7.6b As ρ varies from $\rho = \rho_1(\phi, \theta)$ to $\rho = \rho_2(\phi, \theta)$ with ϕ and θ fixed, the points (ρ, ϕ, θ) form a line segment along a ray from the origin.

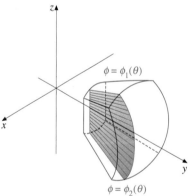

FIGURE 13.7.6c As ϕ varies from $\phi = \phi_1(\theta)$ to $\phi = \phi_2(\theta)$ with θ fixed, these line segments form a cross-section of the region in the plane of θ.

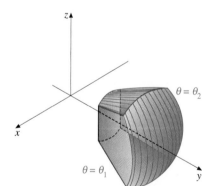

FIGURE 13.7.6d As θ varies from $\theta = \theta_1$ to $\theta = \theta_2$, these sections sweep out the entire region.

Note that the integrand must be expressed in terms of spherical coordinates and we must use the formula

$$dV = \rho^2 \sin\phi \, d\rho \, d\phi \, d\theta.$$

Recall that we can use the equations

$$x = \rho \sin\phi \cos\theta,$$
$$y = \rho \sin\phi \sin\theta,$$
$$z = \rho \cos\phi$$

to change from rectangular to spherical coordinates. Also,

$$x^2 + y^2 + z^2 = \rho^2.$$

We should consider using spherical coordinates to evaluate a triple integral if the integrand is a function of $x^2 + y^2 + z^2$ or if the boundary of the region consists of spheres and/or cones. Recall that the spherical equation $\phi = \phi_0$ is a cone, except when $\phi_0 = \pi/2$ it is the xy-plane. See Figure 13.7.7.

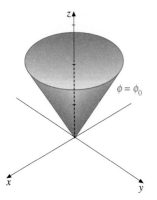

FIGURE 13.7.7 The graph of the spherical equation $\phi = \phi_0$ is a cone, except when $\theta_0 = \pi/2$.

EXAMPLE 3

Find the volume of a hemisphere of radius R.

SOLUTION Let us position the hemisphere with center at the origin and base in the xy-plane. See Figure 13.7.8. The hemisphere is then bounded by $x^2 + y^2 + z^2 = R^2$ and the xy-plane. These surfaces are described easily in terms of spherical coordinates. The sphere has spherical equation $\rho = R$ and the xy-plane has spherical equation $\phi = \pi/2$. Using $dV = \rho^2 \sin\phi \, d\rho \, d\phi \, d\theta$, we have

$$V = \iiint dV = \int_0^{2\pi} \int_0^{\pi/2} \int_0^R \rho^2 \sin\phi \, d\rho \, d\phi \, d\theta$$

$$= \int_0^{2\pi} \int_0^{\pi/2} \frac{R^3}{3} \sin\phi \, d\phi \, d\theta$$

$$= \int_0^{2\pi} \frac{R^3}{3} \left(-\cos\frac{\pi}{2} + \cos 0 \right) d\theta$$

$$= \frac{R^3}{3}(1)(2\pi) = \frac{2\pi R^3}{3}. \quad \blacksquare$$

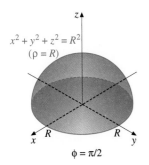

FIGURE 13.7.8 The xy-plane has spherical equation $\phi = \pi/2$.

EXAMPLE 4

Find the center of mass of the solid inside both the sphere $x^2 + y^2 + z^2 = 1$ and the cone $z = \sqrt{x^2 + y^2}$ if the density at a point is equal to the distance of the point from the origin.

SOLUTION The solid is pictured in Figure 13.7.9. The spherical equation of the sphere is $\rho = 1$. We can obtain the spherical equation of the cone either by noting that the cone has vertex angle $\phi = \pi/4$ or by arriving at this conclusion by substituting into the rectangular equation. That is,

$$z = \sqrt{x^2 + y^2},$$
$$\rho \cos\phi = \sqrt{(\rho \sin\phi \cos\theta)^2 + (\rho \sin\phi \sin\theta)^2},$$
$$\rho \cos\phi = \sqrt{\rho^2 \sin^2\phi (\cos^2\theta + \sin^2\theta)},$$
$$\rho \cos\phi = |\rho \sin\phi|,$$
$$\rho \cos\phi = \rho \sin\phi, \quad \text{[$\sin\phi \geq 0$, because $0 \leq \phi \leq \pi$]}$$
$$1 = \tan\phi,$$
$$\phi = \frac{\pi}{4}.$$

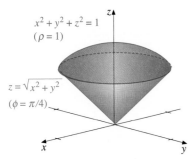

FIGURE 13.7.9 An arrow shot from the origin would enter the region at the origin and leave the region at a point on the sphere.

The density is equal to the spherical coordinate ρ in this problem. The mass of the solid is

$$
\begin{aligned}
M &= \iiint dM = \iiint \rho\, dV \\
&= \int_0^{2\pi} \int_0^{\pi/4} \int_0^1 \rho \rho^2 \sin\phi\, d\rho\, d\phi\, d\theta \\
&= \int_0^{2\pi} \int_0^{\pi/4} \int_0^1 \rho^3 \sin\phi\, d\rho\, d\phi\, d\theta \\
&= \int_0^{2\pi} \int_0^{\pi/4} \left[\frac{\rho^4}{4} \sin\phi \right]_{\rho=0}^{\rho=1} d\phi\, d\theta \\
&= \int_0^{2\pi} \int_0^{\pi/4} \frac{1}{4} \sin\phi\, d\phi\, d\theta \\
&= \int_0^{2\pi} \frac{1}{4} \left[-\cos\phi \right]_{\phi=0}^{\phi=\pi/4} d\theta \\
&= \int_0^{2\pi} \frac{1}{4} \left(-\frac{\sqrt{2}}{2} + 1 \right) d\theta \\
&= \left(\frac{1}{4} \right) \left(\frac{2 - \sqrt{2}}{2} \right) (2\pi) = \frac{(2 - \sqrt{2})\pi}{4}.
\end{aligned}
$$

We also have

$$
\begin{aligned}
M_{z=0} &= \iiint_R z\, dM = \iiint_R z\rho\, dV \\
&= \int_0^{2\pi} \int_0^{\pi/4} \int_0^1 \rho\cos\phi\, \rho\rho^2 \sin\phi\, d\rho\, d\phi\, d\theta \\
&= \int_0^{2\pi} \int_0^{\pi/4} \int_0^1 \rho^4 \cos\phi \sin\phi\, d\rho\, d\phi\, d\theta \\
&= \int_0^{2\pi} \int_0^{\pi/4} \left[\frac{\rho^5}{5} \cos\phi \sin\phi \right]_{\rho=0}^{\rho=1} d\phi\, d\theta \\
&= \int_0^{2\pi} \int_0^{\pi/4} \frac{1}{5} \cos\phi \sin\phi\, d\phi\, d\theta \qquad {\scriptstyle [u = \sin\phi,\, du = \cos\phi\, d\phi]} \\
&= \int_0^{2\pi} \frac{1}{5} \left[\frac{\sin^2 \phi}{2} \right]_{\phi=0}^{\phi=\pi/4} d\theta \\
&= \int_0^{2\pi} \frac{1}{20} d\theta = \frac{\pi}{10}.
\end{aligned}
$$

$$
\bar{z} = \frac{M_{z=0}}{M} = \frac{\dfrac{\pi}{10}}{\dfrac{(2 - \sqrt{2})\pi}{4}} = \frac{2}{5(2 - \sqrt{2})} \approx 0.6828.
$$

It follows from the symmetry of the region and the density function that $\bar{x} = \bar{y} = 0$. The center of mass is at the point with rectangular coordinates

$$
\left(0, 0, \frac{2}{5(2 - \sqrt{2})} \right). \qquad \blacksquare
$$

EXERCISES 13.7

Find the volume of the regions indicated in Exercises 1–10.

1. Bounded by $z = 4 - x^2 - y^2$, $z = 0$

2. Bounded by $z = H - H\sqrt{x^2 + y^2}/R$, $z = 0$ ($H > 0$, $R > 0$)

3. Inside both $x^2 + y^2 + z^2 = 16$ and $x^2 + y^2 = 4$

4. Inside $x^2 + (y - 1)^2 = 1$, between $y + z = 4$ and $z = 0$

5. A hemisphere of radius R

6. Inside both $x^2 + y^2 + z^2 = 4$ and $z = \sqrt{3x^2 + 3y^2}$

7. Inside $x^2 + y^2 + z^2 = 4$, above $z = 1$

8. Inside $z = x^2 + y^2$, below $z = 1$

9. Inside the "apple" that has spherical equation $\rho = 1 - \cos\phi$

10. Inside the "teardrop" that has spherical equation $\rho^2 = \cos 2\phi$, $0 \le \phi \le \pi/4$. (*Hint:* $\cos 2\phi = 2\cos^2\phi - 1$. Let $u = \sqrt{2}\cos\phi$ and then use the Table of Integrals.)

11. Find the center of mass of a homogeneous right circular cone with height H and radius of base R.

12. Find the center of mass of a homogeneous hemisphere with radius R.

13. Find the center of mass of the cylinder bounded by $x^2 + y^2 = R^2$, $z = H$, $z = 0$, if the density is $\rho(x, y, z) = 1 + z\sqrt{x^2 + y^2}$.

14. Find the center of mass of the hemisphere bounded by $z = \sqrt{R^2 - x^2 - y^2}$ and $z = 0$, if the density is $\rho(x, y, z) = 1 + \sqrt{x^2 + y^2 + z^2}$.

15. Find the moment of inertia of a homogeneous spherical shell with outer radius R and inner radius r, about a diameter. Express your answer in terms of the mass and dimensions of the shell.

16. Find the moment of inertia of a homogeneous cylindrical shell with outer radius R, inner radius r, and height H, about its axis. Express your answer in terms of mass and dimensions of the shell.

17. Find the moment of inertia about the z-axis of a homogeneous solid that is bounded by $z = x^2 + y^2$ and $z = 1$. Express your answer in terms of the mass and dimensions of the solid.

18. Find the moment of inertia of a wedge cut from a homogeneous sphere with radius R by two planes that intersect at an angle of θ_0 along a diameter, about the diameter. Express your answer in terms of the mass and dimensions of the wedge.

19. (a) Find the moment of inertia of a homogeneous right circular cylinder with radius R and height H, about a line that is parallel to the axis of the cylinder and at a distance of D from the axis. Express your answer in terms of the mass and dimensions of the cylinder.

(b) For what value of D is the moment of inertia in (a) minimum?

20. (a) Find the moment of inertia of a homogeneous sphere with radius R, about a line that is at a distance of D from the center of the sphere. Express your answer in terms of the mass and dimensions of the sphere.

(b) For what value of D is the moment of inertia in (a) minimum?

21. Find the mass of a pile of snow that is in the shape of the cone $z = 2 - \sqrt{x^2 + y^2}$ if the density of the snow is given by $\rho(x, y, z) = 2 - z$.

22. Find the moment of inertia of a homogeneous hemisphere of radius r about a line that is perpendicular to its circular base at the center of the base. Express the answer in terms of the mass and radius of the hemisphere.

23. Find the moment of inertia of a homogeneous right circular cylinder with radius R and height H about a line along its curved surface, parallel to its axis. Express the answer in terms of the mass and dimensions of the cylinder.

24. Find the moment of inertia of a homogeneous right circular cylinder with radius R and height H about a line that is perpendicular to the axis of the cylinder and through the axis at one end of the cylinder. Express the answer in terms of the mass and dimensions of the cylinder.

25. If a cylinder of liquid with radius R and height H is rotated with constant angular velocity ω about its axis, then the cylindrical coordinates of its points on its free surface satisfy the cylindrical equation $z = H_1 + (\omega r)^2/2g$, $0 \le r \le R$. See figure. Calculate the volume of the rotating liquid in terms of H_1, R, and ω, and then use the fact that the volume is $\pi R^2 H$ to express H_1 in terms of R, H, and ω.

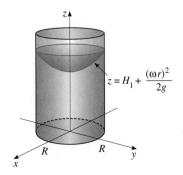

26. Find the center of mass of the truncated homogeneous cone pictured in the figure. (*Hint:* Use cylindrical coordinates with integration with respect to r first.)

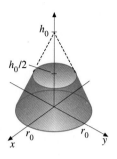

Evaluate the integrals in Exercises 27–30 by changing to either cylindrical or spherical coordinates.

27. $\displaystyle\int_{-1}^{1}\int_{-\sqrt{1-x^2}}^{\sqrt{1-x^2}}\int_{0}^{1}(x^2+y^2)^{3/2}\,dz\,dy\,dx$

28. $\displaystyle\int_{-1}^{1}\int_{-\sqrt{1-x^2}}^{\sqrt{1-x^2}}\int_{0}^{1-x^2-y^2}\sqrt{x^2+y^2}\,dz\,dy\,dx$

29. $\displaystyle\int_{-1}^{1}\int_{-\sqrt{1-x^2}}^{\sqrt{1-x^2}}\int_{0}^{\sqrt{1-x^2-y^2}}\sqrt{x^2+y^2+z^2}\,dz\,dy\,dx$

30. $\displaystyle\int_{-\sqrt{2}}^{\sqrt{2}}\int_{-\sqrt{2-x^2}}^{\sqrt{2-x^2}}\int_{\sqrt{x^2+y^2}}^{\sqrt{4-x^2-y^2}}\sqrt{x^2+y^2+z^2}\,dz\,dy\,dx$

13.8 GENERAL CHANGE OF VARIABLES IN MULTIPLE INTEGRALS

Connections

Integrals in polar coordinates, 13.4.

Integrals in cylindrical and spherical coordinates, 13.7.

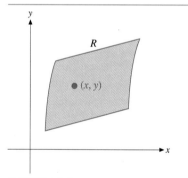

FIGURE 13.8.1a Points (x, y) in the region R are associated with points (u, v) in the region R'.

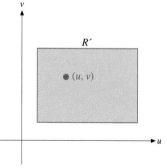

FIGURE 13.8.1b We are interested in the case that the points (x, y) cover R exactly once as the points (u, v) vary over R'.

We have seen in previous sections that it is useful to express certain integrals in terms of either polar, cylindrical, or spherical coordinates. In this section we will extend the concept to include general change of variables in multiple integrals and see how the results can be used to facilitate evaluation of certain integrals.

Transformation of Regions

Let us consider coordinates (u, v) that are related to coordinates (x, y) by the equations

$$x = x(u, v), \quad y = y(u, v). \tag{1}$$

These equations associate a point (u, v) in the uv-coordinate plane with a point (x, y) in the xy-coordinate plane. We can think of the point (u, v) as being transformed to the point (x, y). A region R' in the uv-plane is transformed to a region R in the xy-plane. See Figure 13.8.1. We are interested in the case that the points (x, y) cover the region R exactly once as the points (u, v) vary over the region R'.

Let us assume the functions $x(u, v)$ and $y(u, v)$ in (1) have continuous first partial derivatives with respect to u and v, so x and y are differentiable. The equations in (1) then imply

$$dx = \frac{\partial x}{\partial u}du + \frac{\partial x}{\partial v}dv, \quad dy = \frac{dy}{\partial u}du + \frac{\partial y}{\partial v}dv. \tag{2}$$

We also assume

$$\frac{\partial x}{\partial u}\frac{\partial y}{\partial v} - \frac{\partial y}{\partial u}\frac{\partial x}{\partial v} \neq 0. \tag{3}$$

If condition (3) is satisfied, the system of equations in (2) can be solved for (du, dv) in terms of (dx, dy), so small changes in x and y are related to specific changes in u and v. It can then be proved that the system of equations in (1) can theoretically be solved for u and v in terms of x and y, at least for all (u, v) in R' that are sufficiently close to any predetermined point of R'.

The Jacobian for Double Integrals

We want to establish a formula for expressing the double integral of a function f over a region R in the xy-plane in terms of the coordinates (u, v). To do this, let R' be a region in the uv-plane that has the property that every point (x, y) in R corresponds to exactly one point (u, v) in R'. We then divide R' into rectangles with vertical lines $u = u_0$ and horizontal lines $v = v_0$ in the uv-plane. The corresponding curves $C(u = u_0)$ and $C(v = v_0)$ divide R into curvilinear rectangles. See Figure 13.8.2. If the increments are small, these curvilinear rectangles are essentially parallelograms. See Figure 13.8.3. We need to express the area $\Delta A(x, y)$ of a curvilinear rectangle in terms of the area $\Delta A(u, v) = \Delta u \, \Delta v$ of the corresponding rectangle in the uv-plane. Let us use vectors to do this.

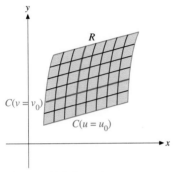

FIGURE 13.8.2a The curves $C(v = v_0)$ and $C(u = u_0)$ divide the region R into curvilinear rectangles.

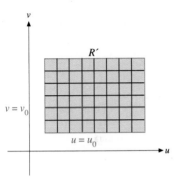

FIGURE 13.8.2b The lines $v = v_0$ and $u = u_0$ divide R' into rectangles.

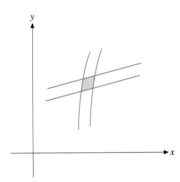

FIGURE 13.8.3a If the increments are small, the curvilinear rectangles are essentially parallelograms.

FIGURE 13.8.3b The area of the rectangle is $\Delta A(u, v) = \Delta u \, \Delta v$.

The vector

$$\mathbf{R} = x(u, v)\mathbf{i} + y(u, v)\mathbf{j}$$

gives the position of the point (x, y) in the xy-plane that corresponds to the point (u, v) in the uv-plane. From Formula 2 we see that the change in position corresponding to changes of Δu in u and Δv in v is

$$\Delta \mathbf{R} = \Delta x\mathbf{i} + \Delta y\mathbf{j} \approx \left(\frac{\partial x}{\partial u}\Delta u + \frac{\partial x}{\partial v}\Delta v\right)\mathbf{i} + \left(\frac{\partial y}{\partial u}\Delta u + \frac{\partial y}{\partial v}\Delta v\right)\mathbf{j}.$$

FIGURE 13.8.4 We can use vectors to approximate the area of the curvilinear rectangle.

The vector

$$\mathbf{U} = \left(\frac{\partial x}{\partial u}\Delta u\right)\mathbf{i} + \left(\frac{\partial y}{\partial u}\Delta u\right)\mathbf{j},$$

obtained by setting $\Delta v = 0$ in the formula for $\Delta \mathbf{R}$, corresponds to a change of Δu in u with v fixed; $\mathbf{U}$ is tangent to the curve $C(v = v_0)$. The vector

$$\mathbf{V} = \left(\frac{\partial x}{\partial v}\Delta v\right)\mathbf{i} + \left(\frac{\partial y}{\partial v}\Delta v\right)\mathbf{j},$$

obtained by setting $\Delta u = 0$ in the formula for $\Delta \mathbf{R}$, corresponds to a change of Δv in v with u fixed; $\mathbf{V}$ is tangent to the curve $C(u = u_0)$. The area of the parallelogram formed by $\mathbf{U}$ and $\mathbf{V}$ is an approximation of the value of the area of the curvilinear rectangle. See Figure 13.8.4. From Statement 4 of Section 11.4 we know that the area of the parallelogram formed by $\mathbf{U}$ and $\mathbf{V}$ is the length of the vector $\mathbf{U} \times \mathbf{V}$. We have

$$\mathbf{U} \times \mathbf{V} = \begin{vmatrix} \mathbf{i} & \mathbf{j} & \mathbf{k} \\ \dfrac{\partial x}{\partial u}\Delta u & \dfrac{\partial y}{\partial u}\Delta u & 0 \\ \dfrac{\partial x}{\partial v}\Delta v & \dfrac{\partial y}{\partial v}\Delta v & 0 \end{vmatrix} = \left(\frac{\partial x}{\partial u}\frac{\partial y}{\partial v} - \frac{\partial x}{\partial v}\frac{\partial y}{\partial u}\right)\Delta u\,\Delta v\,\mathbf{k}.$$

Since $\Delta u\,\Delta v = \Delta A(u, v)$, we then have

$$\Delta A(x, y) \approx |\mathbf{U} \times \mathbf{V}| = \left|\frac{\partial x}{\partial u}\frac{\partial y}{\partial v} - \frac{\partial x}{\partial v}\frac{\partial y}{\partial u}\right|\Delta A(u, v).$$

Let us define

$$\frac{\partial(x, y)}{\partial(u, v)} = \begin{vmatrix} \dfrac{\partial x}{\partial u} & \dfrac{\partial x}{\partial v} \\ \dfrac{\partial y}{\partial u} & \dfrac{\partial y}{\partial v} \end{vmatrix} = \frac{\partial x}{\partial u}\frac{\partial y}{\partial v} - \frac{\partial y}{\partial u}\frac{\partial x}{\partial v}.$$

This is called the **Jacobian** of x and y with respect to u and v. (Condition (3) says that the Jacobian is not zero.) We can express the relation between the area of a small curvilinear rectangle in the xy-plane and the area of the corresponding rectangle in the uv-plane in differential notation as

$$dA(x, y) = \left|\frac{\partial(x, y)}{\partial(u, v)}\right| dA(u, v). \tag{4}$$

It should then seem reasonable that

$$\int\int_R f(x, y)\,dA(x, y) = \int\int_{R'} f(x(u, v), y(u, v))\left|\frac{\partial(x, y)}{\partial(u, v)}\right| dA(u, v). \tag{5}$$

This is the formula that relates double integrals that correspond to the change of variables $x = x(u, v)$, $y = y(u, v)$.

Let us see how Formulas 4 and 5 relate to polar coordinates. We know that polar coordinates (r, θ) in the xy-plane are related to rectangular coordinates (x, y) by the equations

$$x = r\cos\theta, \quad y = r\sin\theta, \quad r > 0, \quad 0 \le \theta < 2\pi.$$

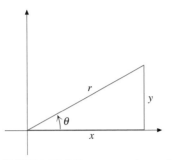

FIGURE 13.8.5 Rectangular and polar coordinates are related by the equations $x = r \cos \theta, y = r \sin \theta$.

See Figure 13.8.5. The line $r = r_0$ in the *rectangular $r\theta$-plane* corresponds to the circle $x^2 + y^2 = r_0^2$ in the xy-plane; the line $\theta = \theta_0$ in the *rectangular $r\theta$-plane* corresponds to the ray in the xy-plane from the origin that makes an angle of θ_0 with the positive x-axis. See Figure 13.8.6. (We have avoided the origin of the xy-plane and points in the $r\theta$-plane with $r = 0$ in order to satisfy the condition that each point in R corresponds to exactly one point in R.) The Jacobian is

$$\frac{\partial(x, y)}{\partial(r, \theta)} = \begin{vmatrix} \dfrac{\partial x}{\partial r} & \dfrac{\partial x}{\partial \theta} \\ \dfrac{\partial y}{\partial r} & \dfrac{\partial y}{\partial \theta} \end{vmatrix} = \begin{vmatrix} \cos \theta & -r \sin \theta \\ \sin \theta & r \cos \theta \end{vmatrix} = r.$$

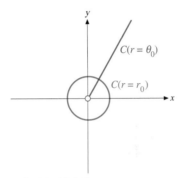

FIGURE 13.8.6a Equations $\theta = \theta_0$ correspond to rays from the origin in the *xy*-plane. Equations $r = r_0$ correspond to circles with centers at the origin.

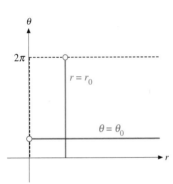

FIGURE 13.8.6b The graph of $\theta = \theta_0$ is a horizontal line in the rectangular $r\theta$-plane. Equations $r = r_0$ are vertical lines in the rectangular $r\theta$-plane.

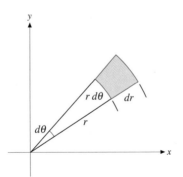

FIGURE 13.8.7 The area of the curvilinear rectangle is approximately $dA = r \, dr \, d\theta$.

Formula 4 then gives

$$dA(x, y) = \left| \frac{\partial(x, y)}{\partial(r, \theta)} \right| dA(r, \theta) = r \, dr \, d\theta,$$

which is the same formula we derived in Section 13.4. This should not be surprising, since the procedure to derive Formula 4 corresponds to that used in Section 13.4. See Figure 13.8.7. Also, note that we used this change of variables in cases where the region R in the xy-plane could be conveniently described in terms of polar coordinates. Generally, we apply Formula 5 to regions R that can be described easily in terms of the variables (u, v).

To use transformation equations $x = x(u, v)$, $y = y(u, v)$ to express an integral of (x, y) over a region R as an integral of (u, v) over a region R'

- Use substitution to transform each boundary curve of R to the corresponding boundary curve of R'. This determines the region R'.
- Calculate the Jacobian.
- Express the integrand in terms of (u, v) and substitute $\left| \dfrac{\partial(x, y)}{\partial(u, v)} \right| dA(u, v)$ for $dA(x, y)$.

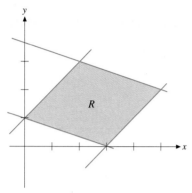

FIGURE 13.8.8a This region is not convenient for integration.

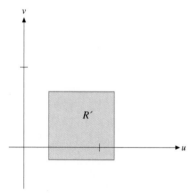

FIGURE 13.8.8b The transformed region is simple.

EXAMPLE 1

Use the linear transformation

$$x = 2u + 3v, \quad y = 2u - v$$

to evaluate $\int\int_R (x + 3y)\,dA$, where R is the parallelogram bounded by the lines

$$y = x - 3, \quad y = x + 1, \quad x + 3y = 3, \text{ and } x + 3y = 11.$$

SOLUTION The region R in the xy-plane is sketched in Figure 13.8.8a. We determine the boundary equations of the corresponding region in the uv-plane by substitution. From the equation $y = x - 3$, we obtain the corresponding equation

$$(2u - v) = (2u + 3v) - 3,$$
$$v = \frac{3}{4}.$$

This is a horizontal line in the uv-plane. The line $y = x + 1$, which is parallel to $y = x - 3$, corresponds to the horizontal line $v = -1/4$.

The equation $x + 3y = 3$ corresponds to

$$(2u + 3v) + 3(2u - v) = 3,$$
$$8u = 3,$$
$$u = \frac{3}{8}.$$

This is a vertical line in the uv-plane. The line $x + 3y = 11$ corresponds to the vertical line $u = 11/8$.

The region R' in the uv-plane that corresponds to the region R in the xy-plane is sketched in Figure 13.8.8b.

The Jacobian of the transformation is

$$\frac{\partial(x, y)}{\partial(u, v)} = \begin{vmatrix} \dfrac{\partial x}{\partial u} & \dfrac{\partial x}{\partial v} \\ \dfrac{\partial y}{\partial u} & \dfrac{\partial y}{\partial v} \end{vmatrix} = \begin{vmatrix} 2 & 3 \\ 2 & -1 \end{vmatrix} = (2)(-1) - (2)(3) = -8.$$

We then have

$$\int\int_T (x + 3y)\,dA(x, y) = \int\int_{R'} [(2u + 3v) + 3(2u - v)]| - 8|\,dA(u, v)$$

$$= \int_{3/8}^{11/8} \int_{-1/4}^{3/4} 64u\,dv\,du$$

$$= \int_{3/8}^{11/8} 64u \left[\frac{3}{4} - \left(-\frac{1}{4} \right) \right] du$$

$$= 32u^2 \Big]_{3/8}^{11/8} = 32 \left(\frac{11}{8} \right)^2 - 32 \left(\frac{3}{8} \right)^2 = 56. \quad \blacksquare$$

EXAMPLE 2

Use the transformation

$$x = 3u, \quad y = 2v$$

to evaluate the area of the region R in the first quadrant that is between the ellipses

$$\frac{x^2}{3^2} + \frac{y^2}{2^2} = 1 \text{ and } \frac{x^2}{6^2} + \frac{y^2}{4^2} = 1.$$

SOLUTION   The region R is sketched in Figure 13.8.9a. The ellipse

$$\frac{x^2}{3^2} + \frac{y^2}{2^2} = 1$$

corresponds to the circle $u^2 + v^2 = 1$; the ellipse

$$\frac{x^2}{6^2} + \frac{y^2}{4^2} = 1$$

corresponds to the circle $u^2 + v^2 = 4$; the x-axis corresponds to the u-axis; and the y-axis corresponds to the v-axis. The region R' in the uv-plane that corresponds to the region R in the xy-plane is sketched in Figure 13.8.9b. The Jacobian of the transformation is

$$\frac{\partial(x, y)}{\partial(u, v)} = \begin{vmatrix} \dfrac{\partial x}{\partial u} & \dfrac{\partial x}{\partial v} \\ \dfrac{\partial y}{\partial u} & \dfrac{\partial y}{\partial v} \end{vmatrix} = \begin{vmatrix} 3 & 0 \\ 0 & 2 \end{vmatrix} = (3)(2) - (0)(0) = 6.$$

We then have

$$\text{Area} = \int\int_R dA(x, y) = \int\int_{R'} |6| \, dA(u, v).$$

Let us use polar coordinates in the uv-plane to evaluate the latter integral. We have

$$u = r \cos\theta, \quad v = r \sin\theta, \text{ and}$$
$$dA(u, v) = r \, dA(r, \theta) = r \, dr \, d\theta.$$

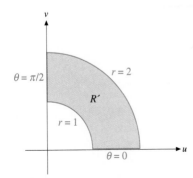

FIGURE 13.8.9a   This region is not convenient for integration with respect to *xy*-coordinates.

FIGURE 13.8.9b   We can use polar coordinates to evaluate the area of this region.

We then have

$$\text{Area} = \int\int_{R'} 6\,dA(u,v)$$

$$= \int_0^{\pi/2}\int_1^2 6r\,dr\,d\theta$$

$$= \int_0^{\pi/2} 3r^2\Big]_{r=1}^{r=2} d\theta$$

$$= \int_0^{\pi/2} 9\,d\theta = \frac{9\pi}{2}. \quad\blacksquare$$

Triple Integrals

The ideas involved in change of variables in triple integrals are analogous to those for double integrals. Equations

$$x = x(u,v,w), \quad y = y(u,v,w), \quad z = z(u,v,w)$$

relate points in a region R of xyz-space to points in a region R' of uvw-space. We assume that x, y, and z have continuous first partial derivatives. The **Jacobian** of x, y, and z with respect to u, v, and w is defined to be the determinant

$$\frac{\partial(x,y,z)}{\partial(u,v,w)} = \begin{vmatrix} \dfrac{\partial x}{\partial u} & \dfrac{\partial x}{\partial v} & \dfrac{\partial x}{\partial w} \\ \dfrac{\partial y}{\partial u} & \dfrac{\partial y}{\partial v} & \dfrac{\partial y}{\partial w} \\ \dfrac{\partial z}{\partial u} & \dfrac{\partial z}{\partial v} & \dfrac{\partial z}{\partial w} \end{vmatrix}.$$

We assume the Jacobian is nonzero, so small changes in x, y, and z correspond to specific changes in u, v, and w. We assume that every point (x,y,z) in R corresponds to exactly one point (u,v,w) in R'. Small rectangular prisms in R' with volume $\Delta V(u,v,w) = \Delta u\,\Delta v\,\Delta w$ correspond to curvilinear prisms in R that are essentially parallelepipeds. See Figure 13.8.10. The volume $\Delta V(x,y,z)$ of the curvilinear prisms is approximately the volume of the parallelepiped determined by the vectors

$$\mathbf{U} = \left(\frac{\partial x}{\partial u}\Delta u\right)\mathbf{i} + \left(\frac{\partial y}{\partial u}\Delta u\right)\mathbf{j} + \left(\frac{\partial z}{\partial u}\Delta u\right)\mathbf{k},$$

$$\mathbf{V} = \left(\frac{\partial x}{\partial v}\Delta v\right)\mathbf{i} + \left(\frac{\partial y}{\partial v}\Delta v\right)\mathbf{j} + \left(\frac{\partial z}{\partial v}\Delta v\right)\mathbf{k}, \text{ and}$$

$$\mathbf{W} = \left(\frac{\partial x}{\partial w}\Delta w\right)\mathbf{i} + \left(\frac{\partial y}{\partial w}\Delta w\right)\mathbf{j} + \left(\frac{\partial z}{\partial w}\Delta w\right)\mathbf{k}.$$

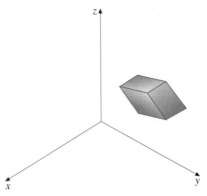

FIGURE 13.8.10a Rectangular prisms in *uvw*-space correspond to curvilinear prisms in *xyz*-space.

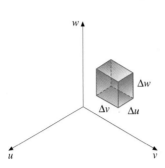

FIGURE 13.8.10b The rectangular prism has volume $\Delta V(u, v, w) = \Delta u \, \Delta v \, \Delta w$.

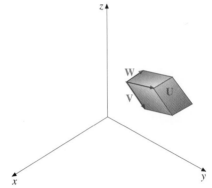

FIGURE 13.8.11 We can use the vectors **U**, **V**, and **W** to approximate the volume of the curvilinear prism.

See Figure 13.8.11. From Section 11.4 we know that the volume of the parallelepiped determined by **U**, **V**, and **W** is the absolute value of the determinant

$$
\begin{vmatrix}
\dfrac{\partial x}{\partial u} \Delta u & \dfrac{\partial x}{\partial v} \Delta v & \dfrac{\partial x}{\partial w} \Delta w \\[2ex]
\dfrac{\partial y}{\partial u} \Delta u & \dfrac{\partial y}{\partial v} \Delta v & \dfrac{\partial y}{\partial w} \Delta w \\[2ex]
\dfrac{\partial z}{\partial u} \Delta u & \dfrac{\partial z}{\partial v} \Delta v & \dfrac{\partial z}{\partial w} \Delta w
\end{vmatrix}.
$$

The value of this determinant can be expressed as

$$
\frac{\partial(x, y, z)}{\partial(u, v, w)} \Delta u \, \Delta v \, \Delta w, \quad \text{so}
$$

$$
\Delta V(x, y, z) \approx \left| \frac{\partial(x, y, z)}{\partial(u, v, w)} \right| \Delta V(u, v, w).
$$

Expressing this relation in differential notation, we have

$$
dV(x, y, z) = \left| \frac{\partial(x, y, z)}{\partial(u, v, w)} \right| dV(u, v, w).
$$

We then have

$$
\iiint_R f(x, y, z) \, dV(x, y, z)
$$

$$
= \iiint_{R'} f(x(u, v, w), y(u, v, w), z(u, v, w)) \left| \frac{\partial(x, y, z)}{\partial(u, v, w)} \right| dV(u, v, w).
$$

(6)

This is the formula that relates triple integrals that correspond to the change of variables

$$
x = x(u, v, w), \quad y = y(u, v, w), \quad z = z(u, v, w).
$$

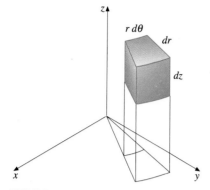

FIGURE 13.8.12 This curvilinear prism has volume $dV \approx r\,dr\,d\theta\,dz$.

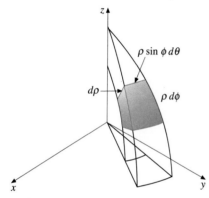

FIGURE 13.8.13 This curvilinear prism has volume $dV \approx \rho^2 \sin\theta\,d\rho\,d\phi\,d\theta$.

We know that rectangular coordinates (x, y, z) are related to cylindrical coordinates (r, θ, z) by the equations

$$x = r\cos\theta, \quad y = r\sin\theta, \quad z = z.$$

The Jacobian is

$$\frac{\partial(x, y, z)}{\partial(r, \theta, z)} = \begin{vmatrix} \dfrac{\partial x}{\partial r} & \dfrac{\partial x}{\partial \theta} & \dfrac{\partial x}{\partial z} \\[2mm] \dfrac{\partial y}{\partial r} & \dfrac{\partial y}{\partial \theta} & \dfrac{\partial y}{\partial z} \\[2mm] \dfrac{\partial z}{\partial r} & \dfrac{\partial z}{\partial \theta} & \dfrac{\partial z}{\partial z} \end{vmatrix} = \begin{vmatrix} \cos\theta & -r\sin\theta & 0 \\ \sin\theta & r\cos\theta & 0 \\ 0 & 0 & 1 \end{vmatrix} = r.$$

This corresponds to the formula

$$dV = r\,dr\,d\theta\,dz,$$

which we derived in Section 13.7. See Figure 13.8.12.

Rectangular coordinates (x, y, z) are related to spherical coordinates (ρ, θ, ϕ) by the equations

$$x = \rho\sin\phi\cos\theta, \quad y = \rho\sin\phi\sin\theta, \quad z = \rho\cos\phi.$$

As we may expect, the absolute value of the Jacobian is $\rho^2 \sin\phi$. This corresponds to the formula

$$dV = \rho^2 \sin\phi\,d\rho\,d\phi\,d\theta,$$

which we derived in Section 13.7. See Figure 13.8.13.

EXAMPLE 3

The region R is bounded by the planes $4x + y - 2z = 0$, $x - y + 2z = 0$, $y = 0$, and $x - y - 3z + 5 = 0$. Use the transformation $x = u - 2w$, $y = 2v$, $z = 2u + v + w$ to evaluate $\int\int\int_R y\,dV$.

SOLUTION The region R is sketched in Figure 13.8.14a. The surface in uvw-space that corresponds to the plane $4x + y - 2z = 0$ is

$$4(u - 2w) + (2v) - 2(2u + v + w) = 0,$$
$$-10w = 0,$$
$$w = 0.$$

Substitution into each of the other boundary planes of the region R gives the equations $u = 0$, $v = 0$, and $u + v + z = 1$. The region R' in uvw-space that corresponds to the region R in xyz-space is sketched in Figure 13.8.14b. The Jacobian of the transformation is

$$\begin{vmatrix} \dfrac{\partial x}{\partial u} & \dfrac{\partial x}{\partial v} & \dfrac{\partial x}{\partial w} \\[2mm] \dfrac{\partial y}{\partial u} & \dfrac{\partial y}{\partial v} & \dfrac{\partial y}{\partial w} \\[2mm] \dfrac{\partial z}{\partial u} & \dfrac{\partial z}{\partial v} & \dfrac{\partial z}{\partial w} \end{vmatrix} = \begin{vmatrix} 1 & 0 & -2 \\ 0 & 2 & 0 \\ 2 & 1 & 1 \end{vmatrix} = 10.$$

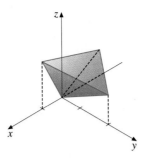

FIGURE 13.8.14a This region is not convenient for integration.

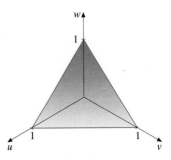

FIGURE 13.8.14b The transformed region is simple.

It follows that

$$
\begin{aligned}
\iiint_R y \, dV &= \iiint_{R'} (2v)|10| \, dV(u, v, w) \\
&= \int_0^1 \int_0^{1-u} \int_0^{1-u-v} 20v \, dw \, dv \, du \\
&= \int_0^1 \int_0^{1-u} (20v - 20uv - 20v^2) \, dv \, du \\
&= \int_0^1 \left(10(1-u)^2 - 10u(1-u)^2 - 20\frac{(1-u)^3}{3} \right) du \\
&= \int_0^1 \frac{10}{3}(1-u)^3 \, du \\
&= -\frac{10}{3}\frac{(1-u)^4}{4}\bigg]_0^1 = \frac{5}{6}. \quad \blacksquare
\end{aligned}
$$

Let us compare the formulas for change of variables in double and triple integrals with that for change of variables in definite integrals with respect to a single variable. We assume that $x = x(u)$ is differentiable and that dx/du is continuous and nonzero, so dx/du has constant sign. Then

$$
dx = \frac{dx}{du} \, du,
$$

$u = a$ corresponds to $x = x(a)$, and $u = b$ corresponds to $x = x(b)$, so

$$
\int_{x(a)}^{x(b)} f(x) \, dx = \int_a^b f(x(u)) \frac{dx}{du} \, du. \tag{7}
$$

We can rewrite (7) in the form

$$
\int_I f(x) \, dx = \int_{I'} f(x(u)) \left| \frac{dx}{du} \right| du,
$$

where $I' = [a, b]$ and I is either $[x(a), x(b)]$ or $[x(b), x(a)]$, depending on whether dx/du is positive or negative for $a \leq u \leq b$. We then see that the derivative dx/du in (7) corresponds to the Jacobian in (5) and (6). Also, we have expressed the limits of integration of the integral with respect to x in terms of the variable u, which is consistent with the way the formulas for double and triple integrals were expressed.

EXERCISES 13.8

1. Coordinates in the xy-plane are related to coordinates in the uv-plane by the equations $x = 2u - v$, $y = u + 2v$. Find the (x, y) coordinates that correspond to
(a) $(u, v) = (1, 0)$,   (b) $(u, v) = (0, 1)$.
Find the (u, v) coordinates that correspond to
(c) $(x, y) = (1, 0)$,   (d) $(x, y) = (0, 1)$.
(e) Solve the equations for (u, v) in terms of (x, y).

2. Coordinates in the xy-plane are related to coordinates in the uv-plane by the equations $x = 3u + 2v$, $y = u - 4v$. Find the (x, y) coordinates that correspond to
(a) $(u, v) = (1, 0)$,   (b) $(u, v) = (0, 1)$.
Find the (u, v) coordinates that correspond to
(c) $(x, y) = (1, 0)$,   (d) $(x, y) = (0, 1)$.
(e) Solve the equations for (u, v) in terms of (x, y).

3. Coordinates in xyz-space are related to coordinates in uvw-space by the equations $x = u - v$, $y = v - 2w$, $z = u + w$. Find the (x, y, z) coordinates that correspond to
(a) $(u, v, w) = (1, 0, 0)$,   (b) $(u, v, w) = (0, 1, 0)$,
(c) $(u, v, w) = (0, 0, 1)$.
Find the (u, v, w) coordinates that correspond to
(d) $(x, y, z) = (1, 0, 0)$,   (e) $(x, y, z) = (0, 1, 0)$,
(f) (f) $(x, y, z) = (0, 0, 1)$.
(g) Solve the equations for (u, v, w) in terms of (x, y, z).

4. Coordinates in xyz-space are related to coordinates in uvw-space by the equations $x = u + v + w$, $y = u + w$, $z = u + v$. Find the (x, y, z) coordinates that correspond to
(a) $(u, v, w) = (1, 0, 0)$,   (b) $(u, v, w) = (0, 1, 0)$,
(c) $(u, v, w) = (0, 0, 1)$.
Find the (u, v, w) coordinates that correspond to
(d) $(x, y, z) = (1, 0, 0)$,   (e) $(x, y, z) = (0, 1, 0)$,
(f) $(x, y, z) = (0, 0, 1)$.
(g) Solve the equations for (u, v, w) in terms of (x, y, z).

Coordinates in the xy-plane are related to coordinates in the uv-plane by the equations given in Exercises 5–10. Sketch the lines $u = u_0$ and $v = v_0$ in the uv-plane and sketch the corresponding curves in the xy-plane for the given values of u_0 and v_0.

5. $x = u + 2v$, $y = 2u - v$; $u_0 = -1, 0, 1$; $v_0 = -1, 0, 1$

6. $x = 3u + v$, $y = u - 2v$; $u_0 = -1, 0, 1$; $v_0 = -1, 0, 1$

7. $x = 2u$, $y = 3v$; $u_0 = -1, 0, 1$; $v_0 = -1, 0, 1$

8. $x = v \cos u$, $y = v \sin u$, $v > 0$, $0 \le u < 2\pi$; $u_0 = 0, \pi/4, 3\pi/4$; $v = 1, 2, 3$

9. $x = u^2 - v^2$, $y = u^2 + v^2$, u and v nonnegative; $u_0 = 0, 1, 2$; $v_0 = 0, 1, 2$

10. $x = u^2 - v^2$, $y = 2uv$, u and v positive; $u_0 = 1, 2, 3$; $v_0 = 1, 2, 3$

Coordinates in the xy-plane are related to coordinates in the uv-plane by the equations given in Exercises 11–16. Sketch the graph of the given xy-equation. Find the corresponding uv-equation and sketch its graph in the uv-plane.

11. $x = 3u + 3v$, $y = u + 2v$; $2x - 3y = 6$

12. $x = u - v$, $y = u + v$; $x - y = 1$

13. $x = u + 1$, $y = v - 2$; $x^2 + y^2 - 2x + 4y = 6$

14. $x = u - 2$, $y = v$; $x^2 - y^2 + 4x = 5$

15. $x = (u - v)/\sqrt{2}$, $y = (u + v)/\sqrt{2}$; $xy = 1$

16. $x = (u + v)/\sqrt{2}$, $y = (-u + v)/\sqrt{2}$; $xy = 1$

Coordinates in the xy-plane are related to coordinates in the uv-plane by the equations given in Exercises 17–22. Sketch the graph of the region in the xy-plane bounded by the given equations and sketch the graph of the corresponding region R' in the uv-plane.

17. $x = 2u$, $y = 3v$; $9x^2 + 4y^2 = 36$

18. $x = 4u$, $y = 3v$; $9x^2 + 16y^2 = 144$

19. $x = 3u + 2v$, $y = u + 3v$; $y = x/3$, $2y = 3x$, $2x + y = 7$

20. $x = 2u + v$, $y = u + 2v$; $y = x/2$, $y = 2x$, $x + y = 3$

21. $x = (u - v)/\sqrt{2}$, $y = (u + v)/\sqrt{2}$; $x + y = 2$, $x = 0$, $y = 0$

22. $x = (u + \sqrt{3}v)/2$, $y = (-\sqrt{3}u + v)/2$; $\sqrt{3}y = x$, $y = -\sqrt{3}x$, $x = 1$

Coordinates in xyz-space are related to coordinates in uvw-space by the equations given in Exercises 23–28. Sketch the graph of the region in xyz-space that is bounded by the given equations and sketch the graph of the corresponding region R' in uvw-space.

23. $x = 2u$, $y = 3v$, $z = 4w$; $36z = 9x^2 + 4y^2$, $z = 4$

24. $x = 2u$, $y = 6v$, $z = 3w$; $y = \sqrt{9x^2 + 4z^2}$, $y = 6$

25. $x = 4u$, $y = 3v + w$, $z = v + 3w$; $y = 3z$, $z = 3y$, $x = 0$, $x + y + z = 4$

26. $x = 3u + v$, $y = u + 3v$, $z = 4w$; $y = 3x$, $x = 3y$, $z = 0$, $x + y + z = 4$

27. $x = 2u + 3v$, $y = 2v + 3w$, $z = 3u + 2w$; $6y = 4x + 9z$, $6z = 9x + 4y$, $6x = 9y + 4z$, $x + y + z = 5$

28. $x = u + 2v$, $y = v + 2w$, $z = 2u + w$; $2y = x + 4z$, $2z = 4x + y$, $2x = 4y + z$, $x + y + z = 3$

29. Evaluate the Jacobian that corresponds to the translation of coordinates $x = u + h$, $y = v + k$, h and k constants.

30. Evaluate the Jacobian that corresponds to the rotation of coordinates $x = u \cos \theta - v \sin \theta$, $y = u \sin \theta + v \cos \theta$, θ constant.

31. Verify that the change from rectangular to cylindrical coordinates given by the equations $x = r\cos\theta$, $y = r\sin\theta$, $z = z$ has Jacobian r.

32. Verify that the change from rectangular to spherical coordinates given by the equations $x = \rho\sin\phi\cos\theta$, $y = \rho\sin\phi\sin\theta$, $z = \rho\cos\phi$ has Jacobian with absolute value $\rho^2\sin\phi$.

33. Show that a **linear transformation** $x = au + bv$, $y = cu + dv$ that has nonzero Jacobian transforms a line $Ax + By + C = 0$ in the xy-plane into a line in the uv-plane.

34. Show that a **linear transformation**

$$x = au + bv + cw, \quad y = du + ev + fw,$$
$$z = gu + hv + iw$$

that has nonzero Jacobian transforms a plane $Ax + By + Cz + D = 0$ in xyz-space into a plane in uvw-space.

In Exercises 35–50, express the given integrals as iterated integrals with respect to the indicated coordinates. Do not evaluate the integrals.

35. $\displaystyle\iint_R xy\,dA$, where R is the region bounded by the lines $y = 2x - 3$, $y = 2x$, $x = 2y - 3$, $x = 2y$; $x = 2u + v$, $y = u + 2v$

36. $\displaystyle\iint_R (x + y)\,dA$, where R is the region bounded by the lines $y = 2x - 6$, $x = 2y$, $y = x$; $x = 2u + v$, $y = u + 2v$

37. $\displaystyle\iint_R x\,dA$, where R is the region bounded by the lines $y = 2x$, $x = 2y$, $x + y = 6$; $x = 2u + v$, $y = u + 2v$

38. $\displaystyle\iint_R x\,dA$, where R is the region bounded by the lines $y = 2x$, $x = 2y$, $x + y = 6$; $x = 2u - v$, $y = u + v$

39. $\displaystyle\iint_R (2x + 3y)\,dA$, where R is the region bounded by the lines $4y = x - 3$, $4y = x + 2$, $2x + 3y = 6$, $2x + 3y = 17$; $x = 4u - 3v$, $y = u + 2v$

40. $\displaystyle\iint_R (x + y)\,dA$, where R is the region bounded by the lines $x + y = 1$, $x + y = 2$, $y = x - 1$, $y = x + 1$; $x = -u + v$, $y = u + v$

41. $\displaystyle\iint_R (x + y)\,dA$, where R is the region bounded by the lines $y = 2x$, $x = 2y$, $x + y = 6$; $x = (u - v)/\sqrt{2}$, $y = (u + v)/\sqrt{2}$

42. $\displaystyle\iint_R (x + y)\,dA$, where R is the region bounded by the lines $y = 2x$, $x = 2y$, $x + y = 6$; $x = (u + v)/\sqrt{2}$, $y = (-u + v)/\sqrt{2}$

43. $\displaystyle\iint_R (x^2 - y^2)^2\,dA$, where R is the region bounded by $xy = 1$, $y = x - 1$, $y = x + 1$; $x = (u + v)/\sqrt{2}$, $y = (-u + v)/\sqrt{2}$

44. $\displaystyle\iint_R (x^2 - y^2)^2\,dA$, where R is the region bounded by the parabola $x^2 - 2xy + y^2 + \sqrt{2}x + \sqrt{2}y = 12$ and the line $x + y = 2\sqrt{2}$; $x = (u + v)/\sqrt{2}$, $y = (-u + v)/\sqrt{2}$

45. $\displaystyle\iint_R e^{x^2 - y^2}\,dA$, where R is the region bounded by $x^2 - y^2 = 1$, $x^2 - y^2 = 16$, $x = 2y$, in the first quadrant; $x = r\sec\theta$, $y = r\tan\theta$

46. $\displaystyle\iint_R x\,dA$, where R is the region bounded by $y = x$, $4x^2 - y^2 = 16$, $4x^2 - y^2 = 4$, in the first quadrant; $x = r\sec\theta$, $y = 2r\tan\theta$

47. $\displaystyle\iiint_R (\sqrt{3}x + y)\,dV$, where R is the region bounded by the planes $y = \sqrt{3}x$, $y = \sqrt{3}x + 2\sqrt{3}$, $y = -\sqrt{3}x$, $y = -\sqrt{3}x + 2\sqrt{3}$, $z = 0$, $z = 2$; $x = u - v$, $y = \sqrt{3}u + \sqrt{3}v$, $z = 2w$

48. $\displaystyle\iiint_R dV$, where R is the region bounded by the planes $z = 0$, $2z = x$, $x - 2y - z = 0$, $x + 2y + z = 4$; $x = 2u + 2w$, $y = u + 2v + w/2$, $z = w$

49. $\displaystyle\iiint_R dV$, where R is the region bounded by $x = y^2 + z$, $z = x - 4$, $z = 0$, $z = 6$; $x = u + w$, $y = v$, $z = w$

50. $\displaystyle\iiint_R dV$, where R is the region bounded by $x^2 + y^2 + z^2 - 2yz = 1$, $z = 0$, $z = 4$; $x = u$, $y = v + w$, $z = w$

51. Find the area of the region in the first quadrant that is bounded by $x = 2y$, $x^2 + 4y^2 = 4$, and the x-axis. (*Hint:* Use a **change of scale transformation**, $x = au$, $y = bv$.)

52. Find the center of mass of a homogeneous lamina bounded by $3y = 2\sqrt{9 - x^2}$ and the x-axis. (*Hint:* Use a **change of scale transformation**, $x = au$, $y = bv$.)

53. Evaluate $\displaystyle\iint_R x\,dA$, where R is the region bounded by the lines $y = x$, $y = x - 4$, $y = -x$, $y = -x + 4$. (*Hint:* Use a **rotation transformation**, $x = (\cos\theta)u - (\sin\theta)v$, $y = (\sin\theta)u + (\cos\theta)v$.)

54. Evaluate $\displaystyle\iint_R (x + y)\,dA$, where R is the region bounded by the lines $y = x$, $y = x - 1$, $x = 2y$, $x = 2y - 1$. (*Hint:* Use a **linear transformation**, $x = au + bv$, $y = cu + dv$.)

55. Find the volume of the ellipsoid

$$\frac{x^2}{a^2} + \frac{y^2}{b^2} + \frac{z^2}{c^2} = 1.$$

56. Find the volume of the region bounded by the elliptical paraboloid $2z = 16 - 4x^2 - y^2$ and $z = 0$.

57. Find the volume of the region bounded by

$$\frac{x^2}{2} + \frac{y^2}{8} = z^2 + 1, \quad z = 1, \quad z = -1.$$

REVIEW OF CHAPTER 13 CONCEPTS

- The limits of integration of an iterated multiple integral correspond to a description of the region of integration. The problem of determining limits of integration is equivalent to that of describing the region.
- The technique for determining limits of integration of iterated triple integrals involves the same ideas as the technique for determining limits of integration of iterated double integrals. A sketch of the region is most helpful in any case.
- To change the order of integration of an iterated integral, use the given limits of integration to determine the corresponding region and then use a description of the region to obtain the limits of integration in the desired order. Do not try to go directly from the limits for one order of integration to the limits of integration for a different order without considering the region of integration.
- You should know the geometric interpretation of each coordinate in polar, cylindrical, and spherical coordinate systems.

- You should be prepared to use either polar, cylindrical, or spherical coordinates whenever the region and integrand are appropriate.
- Changing variables in a multiple integral involves the Jacobian of the transformation, which is simply the ratio of the areas (volumes) of corresponding differential area (volume) elements. For example, when changing from rectangular to polar coordinates we have $dA(x, y) = r\,dr\,d\theta$, where r is the absolute value of the Jacobian. For cylindrical coordinates we have $dV(x, y, z) = r\,dz\,dr\,d\theta$, where r is the absolute value of the Jacobian. For spherical coordinates we have $dV(x, y, z) = \rho^2 \sin\phi\,d\rho\,d\phi\,d\theta$, where $\rho^2 \sin\phi$ is the absolute value of the Jacobian.
- Do not forget the Jacobian when changing variables in a multiple integral.

CHAPTER 13 REVIEW EXERCISES

Evaluate the integrals in Exercises 1–12.

1. $\displaystyle\int_0^1 \int_0^x 6xy^2\,dy\,dx$

2. $\displaystyle\int_0^1 \int_x^1 (2y - 1)\,dy\,dx$

3. $\displaystyle\int_0^{\pi/2} \int_0^{\sin\theta} \cos\theta\, r\,dr\,d\theta$

4. $\displaystyle\int_0^{\pi/4} \int_0^{\sec\theta} \tan\theta\, r\,dr\,d\theta$

5. $\displaystyle\int_0^1 \int_0^{\sqrt{1-x^2}} \int_0^{\sqrt{1-x^2-y^2}} xy\,dz\,dy\,dx$

6. $\displaystyle\int_0^1 \int_0^x \int_0^y xyz\,dz\,dy\,dx$

7. $\displaystyle\iint_R y\,dA$, where R is the region bounded by $y = x^2$, $x + y = 2$, and the y-axis ($x \geq 0$)

8. $\displaystyle\iint_R x\,dA$, where R is the region bounded by $x = y^2$ and $x + y = 2$

9. $\displaystyle\iint_R x^3 y\,dA$, where R is the region bounded by $x = \sqrt{y}$, $y = 4$, and the y-axis

10. $\displaystyle\iint_R y\,dA$, where R is the region bounded by $y = x/2$, $x + y = 3$, and the x-axis

11. $\displaystyle\iint_R y^2\,dA$, where R is the region pictured in the figure

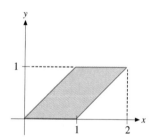

12. $\displaystyle\iint_R xy\,dA$, where R is the region pictured in the figure

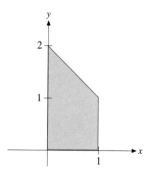

Express the integrals in Exercises 13–20 as an iterated double integral with the opposite order of integration.

13. $\displaystyle\int_0^1 \int_x^1 f(x, y)\,dy\,dx$

14. $\displaystyle\int_0^2 \int_0^{x/2} f(x, y)\,dy\,dx$

15. $\displaystyle\int_0^1 \int_1^{e^x} f(x, y)\,dy\,dx$

16. $\displaystyle\int_1^2 \int_0^{\ln x} f(x, y)\,dy\,dx$

17. $\displaystyle\int_0^{\sqrt{3}} \int_{\tan^{-1} x}^{\pi/3} f(x, y)\,dy\,dx$

18. $\displaystyle\int_0^1 \int_0^{\cos^{-1} x} f(x, y)\,dy\,dx$

19. $\displaystyle\int_0^1 \int_0^{2\sqrt{1-y^2}} f(x, y)\,dx\,dy$

20. $\displaystyle\int_0^2 \int_{y^2}^4 f(x, y)\,dx\,dy$

21. Find the surface area of that part of the surface $z = x^2 + y^2$ above the region in the xy-plane bounded by $x^2 + y^2 = 2$.

22. Find the surface area of that part of the surface $z = \sqrt{1 - x^2}$ that is inside the cylinder $x^2 + y^2 = 1$.

23. Find the center of mass of the lamina in the first quadrant bounded by $y = x$, $y = 0$, $x^2 + y^2 = 1$, if the density is given by $\rho(x, y) = \sqrt{x^2 + y^2}$.

24. Find the center of mass of the lamina in the first quadrant, between $x^2 + y^2 = 4$ and $x^2 + y^2 = 1$, if the density is proportional to the distance from the origin.

In Exercises 25–28, use each of rectangular, cylindrical, and spherical coordinates to set up iterated triple integrals that give the volume of the regions. Use one of the integrals to evaluate the volume.

25. Inside $x^2 + y^2 + z^2 = 4$ and above $z = 1$

26. Inside both $x^2 + y^2 + z^2 = 4$ and $z = \sqrt{x^2 + y^2}$

27. Inside $x^2 + y^2 = 1$, between $z = \sqrt{x^2 + y^2}$ and $z = 0$

28. In the first octant, below $x + y + z = 2$, and above $z = x + y$

In Exercises 29–30, use each of rectangular, cylindrical, and spherical coordinates to set up iterated triple integrals that give the moment of inertia about the z-axis of the solids. Do not evaluate the integrals.

29. The solid bounded by $z = \sqrt{x^2 + y^2}$ and $z = 2$, with density $\rho(x, y, z) = 1 + \sqrt{x^2 + y^2 + z^2}$

30. The solid bounded by $x^2 + y^2 + z^2 = 4$, with density $\rho(x, y, z) = x^2 + y^2$

Evaluate the integrals in Exercises 31–34 by changing to either cylindrical or spherical coordinates.

31. $\displaystyle\int_{-2}^2 \int_{-\sqrt{4-x^2}}^{\sqrt{4-x^2}} \int_{x^2+y^2}^4 \sqrt{z}\,dz\,dy\,dx$

32. $\displaystyle\int_0^1 \int_{-\sqrt{1-x^2}}^{\sqrt{1-x^2}} \int_{x^2+y^2}^{\sqrt{x^2+y^2}} \sqrt{z}\,dz\,dy\,dx$

33. $\displaystyle\int_{-1}^1 \int_{-\sqrt{1-x^2}}^{\sqrt{1-x^2}} \int_{\sqrt{x^2+y^2}}^{\sqrt{3x^2+3y^2}} \sqrt{z}\,dz\,dy\,dx$

34. $\displaystyle\int_0^1 \int_0^{\sqrt{1-x^2}} \int_0^{\sqrt{1-x^2-y^2}} \sqrt{z}\,dz\,dy\,dx$

In Exercises 35–40, express the volume of the indicated region as a single iterated triple integral with respect to rectangular coordinates for every order of integration for which that is convenient. Do not evaluate the integrals.

35. Bounded by $x + 2y + 3z = 6$, $x = 4$, $x = 0$, $y = 0$, $z = 0$

36. Bounded by $x + y = 4$, $y = x$, $x + 2z = 4$, $x = 0$, $z = 0$

37. Below both $3x + 2y + 3z = 6$ and $z = 1$, in the first octant

38. Below both $y + z = 2$ and $z = 1$, bounded by $x + y = 3$, in the first octant

39. Below both $x + z = 2$ and $z = 1$, bounded by $y = 2$, in the first octant

40. Below both $x + y + z = 3$ and $x + z = 1$, in the first octant

In Exercises 41–44, sketch the indicated region R, sketch the corresponding region R', and evaluate the Jacobian of the given transformation.

41. Region bounded by $x + y = 3$, $y = 2x$, and $x = 2y$; $x = 2u + v$, $y = u + 2v$

42. Region bounded by $x + y = 2$, $y = x$, and the x-axis; $x = u + 2v$, $y = u$

43. Region bounded by $y = x$, $y = x + 2$, $y = -x$, and $y = 2 - x$; $x = u - v$, $y = u + v$

44. Region bounded by $4x^2 + 9y^2 = 36$; $x = 3u$, $y = 2v$

In Exercises 45–50, express the given integrals as iterated integrals with respect to the indicated coordinates. Do not evaluate the integrals.

45. $\displaystyle\iint_R y\,dA$, where R is the region bounded by the lines $y = x$, $y = x + 2$, $x + y = 0$, $x + y = 2$; $x = (u - v)/\sqrt{2}$, $y = (u + v)/\sqrt{2}$

46. $\displaystyle\iint_R x^2\,dA$, where R is the region bounded by $x^2 - y^2 = 1$ and $x = 2/\sqrt{3}$ in the first quadrant; $x = r \sec\theta$, $y = r \tan\theta$

47. $\iint_R xy\,dA$, where R is the region in the first quadrant bounded by $x^2 + 4y^2 = 4$; $x = 2u$, $y = v$

48. $\iint_R (x - y)\,dA$, where R is the region bounded by $y = x$, $x = 2y$, and $x = 2$; $x = 2u + 2v$, $y = u + 2v$

49. $\iiint_R (y - x)\,dV$, where R is the region bounded by the planes $y = x$, $z = y$, $z = y - 2x$, $z = 1$; $x = -v + w$, $y = u - v + w$, $z = u + v + w$

50. $\iiint_R dV$, where R is the region bounded by the planes $2x - 4y + z = 0$, $4x - 2y - z = 0$, $z = 0$, and $x + y + z = 3$; $x = 2u + v + w$, $y = u + 2v + w$, $z = 2w$

51. Find the area of the region bounded by the ellipse $\dfrac{x^2}{2^2} + \dfrac{y^2}{4^2} = 1$.

52. Find the volume of the region bounded by the elliptical cone $z = 4\sqrt{x^2 + 4y^2}$ and the plane $z = 4$.

53. Find the moment of inertia of the homogeneous solid cone bounded by $z = H - \dfrac{H}{R}\sqrt{x^2 + y^2}$ and $z = 0$, with density $\rho(x, y, z) = k$, about its axis.

54. The moment of inertia of a homogeneous cube with length of edges x about a line through the centers of opposite sides is $I = k\dfrac{x^5}{6}$. Use the Parallel Axis Theorem (Section 13.5, Exercise 51) to find the moment of inertia of the cube about an edge.

55. The moment of inertia of a homogeneous sphere with radius r about a diameter is $I = k\dfrac{8\pi r^5}{15}$. Use the Parallel Axis Theorem (Section 13.5, Exercise 51) to find the moment of inertia of the sphere about a line that is tangent to the sphere.

56. Light intensity, light per unit area on a surface, is given by the formula $I = \dfrac{k\cos\theta}{d^2}$, where d is the distance of the light source from a point on the surface and θ is the angle the light makes with the normal to the surface at the point.

(a) Show that the total light on a sphere from a source at the center of the sphere is $4k\pi$.

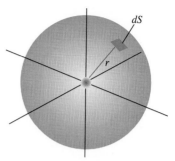

(b) Show that the total light on an infinite plane from a light source that is a distance d_0 from the plane is $2\pi k$.

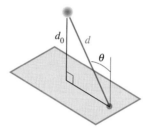

(c) Find the radius of the circle on the plane in part (b) that has center nearest the source and contains half the total light on the plane.

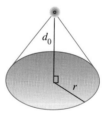

57. Describe conditions on R and $f(x, y)$ that guarantee that $\iint_R f(x, y)\,dA$ is four times the integral of f over that part of R that is in the first quadrant.

58. For each of the transformations **(a)** $x = v$, $y = u$, **(b)** $x = au$, $y = v$, and **(c)** $x = u$, $y = v + ku$, describe the region R in the xy-plane that corresponds to a rectangle in the uv-plane and relate the Jacobian of the transformation to the ratio of the areas of corresponding regions.

EXTENDED APPLICATION

THE PRESSURE-VOLUME-TEMPERATURE CELL

You have recently been employed by an oil company to work in a department whose responsibilities include the analysis of petroleum samples. One of the simplest tests applied to a sample is designed to determine the amount of vapor produced from a given quantity of liquid petroleum when it is subjected to different pressures. A possible technique is to use a pressure-volume-temperature cell. A carefully measured quantity of petroleum is introduced into a small right circular cylinder. A piston at one end of the cylinder is moved so as to increase or decrease the pressure in the cylinder. The other end has a glass plate marked off in graduated segments that allows the liquid level to be read. The volume of the sample is then calculated and compared to the original volume. The difference represents the portion of the sample's volume that has changed from a liquid to a gaseous state.

Since the sample may be quite small, it is useful to tilt the cell at an angle ϕ below the horizontal, so that the liquid level can be more easily read with the required degree of accuracy. Figure 1 shows a three-dimensional view of the cylinder, a lengthwise cross-section, and the face plate.

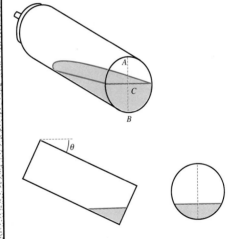

FIGURE 1

The volume of liquid in the cell depends on the values of several parameters:

ϕ = angle between the axis of the cylinder and the horizontal,

h = depth of the liquid at the center of the face plate,

L = distance from the face plate to the piston,

r = radius of the cell.

You have been asked to investigate the sensitivity of the cell to errors in measuring ϕ, h, L, and r. As a first step, you must find a formula for the volume of the liquid in terms of

ϕ, h, L, and r. You begin by creating a coordinate system in which the face plate lies in the xy-plane and the z-axis is the axis of the cylinder. In each of Figures 2a and 2b, the surface of the liquid is the plane DCE.

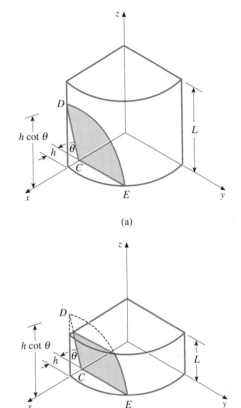

(a)

(b)

FIGURE 2

Tasks:

1. Determine an equation of the plane DCE in terms of ϕ, h, and r.
2. Find a formula for the volume of the cylinder in terms of ϕ, h, L, and r. (You will have to consider two separate cases, corresponding to Figures 2a and 2b.)

Writing Assignment:

Write a report to your supervisor, explaining how your formulas were obtained. In particular, explain why you needed two formulas and how your multiple integral represents the volume of liquid in each case.

Source: Marvin Keener and Jeanne Agnew, Oklahoma State University, *Industry Related Problems for Mathematics Students.*

14 Topics in Vector Calculus

In this chapter we will introduce integrals over curves and surfaces and establish some important results that relate integrals over a region to an integral over the boundary of the region. We will see how integrals of functions that measure components of forces along the unit tangent and normal vectors of curves and along the unit normal vector of surfaces relate to some physical problems.

14.1   LINE INTEGRALS

Connections

Concepts of arc length and surface area, 6.5.
Concept of mass, 6.7.
Parametric equations, 11.1, 11.9.
Arc length of parametric curves, 11.2, 11.9.

Line integrals are used to measure sums of physical quantities as the values vary along points on a curve. For example, we have see in Section 11.3 that the work associated with a force $\mathbf{F}$ as an object moves along a line from a point Q to a point P is $\mathbf{F} \cdot \mathbf{QP}$. Line integrals will allow us to calculate the work associated with $\mathbf{F}$ as the object follows any path from Q to P. As usual, the definition involves a limit of Riemann sums and applications of the concept depend on the physical interpretation of the terms of the sum.

Integrating with Respect to Arc Length

Suppose C is a curve that is in the plane, or in three-dimensional space, and f is defined at each point of C. We divide C into small pieces ΔC of length Δs, and form the Riemann sum

$$\sum f\,\Delta s,$$

where f is evaluated at a point in ΔC. See Figure 14.1.1. If the curve is "smooth enough" and f is continuous on C, theory guarantees that the Riemann sums converge to a single number $\int_C f\,ds$ as the lengths Δs approach zero. The number $\int_C f\,ds$ is called the **line integral** of f over the curve C.

$$\int_C ds \text{ is the arc length of } C.$$

FIGURE 14.1.1 We form a Riemann sum, $\sum f\,\Delta s$, where Δs is the length of a small segment of C and f is evaluated at a point on the segment.

If a smooth curve C is given in parametric form, we can express the line integral of a continuous function f over C as an integral with respect to the parameter. That is, let us consider the curve in the plane that is given by the parametric equations

$$C : x = x(t), \quad y = y(t), \quad a \le t \le b,$$

where $x(t)$ and $y(t)$ have continuous first partial derivatives and $(dx/dt)^2 + (dy/dt)^2 \ne 0$. If $[a, b]$ is divided into subintervals of length Δt, the points on the curve that correspond to the points of subdivision divide C into pieces of length Δs, where

$$\Delta s \approx \sqrt{(\Delta x)^2 + (\Delta y)^2} = \sqrt{\left(\frac{\Delta x}{\Delta t}\right)^2 + \left(\frac{\Delta y}{\Delta t}\right)^2}\,\Delta t.$$

It follows from the fact that the error in the above approximation is a small percentage of Δs for Δt small that

$$\sum f\,\Delta s \approx \sum f(x(t), y(t))\sqrt{\left(\frac{\Delta x}{\Delta t}\right)^2 + \left(\frac{\Delta y}{\Delta t}\right)^2}\,\Delta t.$$

We can then take the limits of the Riemann sums to obtain

$$\int_C f\,ds = \int_a^b f(x(t), y(t))\sqrt{\left(\frac{dx}{dt}\right)^2 + \left(\frac{dy}{dt}\right)^2}\,dt. \tag{1}$$

Formula 1 means that we can evaluate a line integral by substituting parametric values and then evaluating the resulting integral with respect to the real-valued parameter.

Line integrals over smooth curves in three-dimensional space can also be evaluated by substituting parametric values and using the appropriate formula for ds.

The differential expression

$$ds = \sqrt{\left(\frac{dx}{dt}\right)^2 + \left(\frac{dy}{dt}\right)^2}\, dt$$

that represents the arc length of a small piece of a parametric curve in the plane is the same as that derived in Section 11.2.

EXAMPLE 1

(a) $C : x = t, \ y = t^2, \ 1 \leq t \leq 2$. Evaluate $\int_C 3y/x\, ds$.

(b) $C' : x = \sqrt{u}, \ y = u, \ 1 \leq u \leq 4$, is a reparametrization of C. Evaluate $\int_{C'} 3y/x\, ds$.

(c) $-C : x = -v, \ y = v^2, \ -2 \leq v \leq -1$, has the same graph as C, but opposite direction. Evaluate $\int_{-C} 3y/x\, ds$.

SOLUTION

(a) We have $C : x = t, \ y = t^2, \ 1 \leq t \leq 2$, so

$$\frac{dx}{dt} = 1 \text{ and } \frac{dy}{dt} = 2t.$$

Substitution into Formula 1 then gives

$$\int_C \frac{3y}{x}\, ds = \int_1^2 \frac{3(t^2)}{(t)} \sqrt{(1)^2 + (2t)^2}\, dt$$

$$= \int_1^2 3t\sqrt{1 + 4t^2}\, dt$$

$$= \frac{3}{8} \frac{(1 + 4t^2)^{3/2}}{3/2} \Bigg]_1^2$$

$$= \frac{1}{4}(17)^{3/2} - \frac{1}{4}(5)^{3/2}.$$

(b) We have $C' : x = \sqrt{u}, \ y = u, \ 1 \leq u \leq 4$, so

$$\frac{dx}{du} = \frac{1}{2\sqrt{u}} \text{ and } \frac{dy}{du} = 1.$$

Substitution into Formula 1 then gives

$$\int_{C'} \frac{3y}{x}\, ds = \int_1^4 \frac{3u}{\sqrt{u}} \sqrt{\frac{1}{4u} + 1}\, du$$

$$= \int_1^4 \frac{3}{2}\sqrt{1 + 4u}\, du$$

$$= \frac{3}{8} \frac{(1 + 4u)^{3/2}}{3/2} \Bigg]_1^4$$

$$= \frac{1}{4}(17)^{3/2} - \frac{1}{4}(5)^{3/2}.$$

(c) We have $-C : x = -v,\ y = v^2,\ -2 \le v \le -1$, so

$$\frac{dx}{dv} = -1 \text{ and } \frac{dy}{dv} = 2v.$$

Substitution into Formula 1 then gives

$$\int_{-C} \frac{3y}{x}\, ds = \int_{-2}^{-1} \frac{3v^2}{-v} \sqrt{1 + 4v^2}\, dv$$

$$= -\int_{-2}^{-1} 3v\sqrt{1 + 4v^2}\, dv$$

$$= -\frac{3}{8} \frac{(1 + 4v^2)^{3/2}}{3/2} \bigg]_{-2}^{-1}$$

$$= \frac{1}{4}(17)^{3/2} - \frac{1}{4}(5)^{3/2}. \quad \blacksquare$$

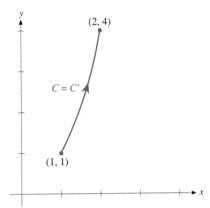

The curves C and C' in Example 1 consist of that portion of the parabola $y = x^2$ from the initial point $(1, 1)$ to the terminal point $(2, 4)$. See Figure 14.1.2a. The curve $-C$ goes along the parabola in the opposite direction, from $(2, 4)$ to $(1, 1)$. See Figure 14.1.2b. From Example 1 we see that

$$\int_{C} \frac{3y}{x}\, ds = \int_{C'} \frac{3y}{x}\, ds = \int_{-C} \frac{3y}{x}\, ds.$$

In fact, the equation $t = \sqrt{u}$ changes the parametrization of C to that of C' and the corresponding change of variables transforms the integral $\int_{1}^{4} \frac{3}{2}\sqrt{1 + 4u}\, du$

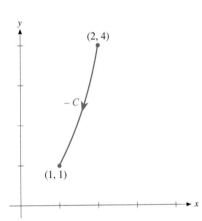

of (b) to the integral $\int_{1}^{2} 3t\sqrt{1 + 4t^2}\, dt$ of (a). The equation $t = -v$ trans-

forms the integral $-\int_{-2}^{-1} 3v\sqrt{1 + 4v^2}\, dv$ of (c) to that of (a). Different smooth

parametrizations of C give the same value of $\int_{C} f\, ds$. The direction does not

matter, because Δs represents length in the sum $\sum f\, \Delta s$, and length is non-negative. The length of a curve is the same in either direction.

We can use the formula for change of variable in a definite integral to verify the following general results.

If C' is a reparametrization of C, then

$$\int_{C'} f\, ds = \int_{C} f\, ds.$$

If $-C$ has the same graph, but opposite direction of C, then

$$\int_{-C} f\, ds = \int_{C} f\, ds.$$

The above formulas mean that we may choose the most convenient parametric representation of a curve or its negative to evaluate a line integral over the curve.

Choosing Parametric Equations to Evaluate Line Integrals

We need to be able to determine parametric equations for curves that are described geometrically or in terms of equations in x and y. Let us recall the following rules from Section 11.1.

The curve along the graph of $y = f(x)$ in the direction of increasing x from $(a, f(a))$ to $(b, f(b))$ can be expressed as

$$C : x = t, \quad y = f(t), \quad a \le t \le b. \tag{2}$$

The curve along the graph of $x = f(y)$ in the direction of increasing y from $(f(a), a)$ to $(f(b), b)$ can be expressed as

$$C : x = f(t), \quad y = t, \quad a \le t \le b. \tag{3}$$

The line segment from (x_0, y_0) to (x_1, y_1) is given by

$$C : x = x_0 + (x_1 - x_0)t, \quad y = y_0 + (y_1 - y_0)t, \quad 0 \le t \le 1. \tag{4}$$

Parametric equations of the curve that goes from the point $(h + r, k)$ counterclockwise once around the circle $(x - h)^2 + (y - k)^2 = r^2$ are

$$C : x = h + r\cos t, \quad y = k + r\sin t, \quad 0 \le t \le 2\pi. \tag{5}$$

Parametric equations of the curve that goes from the point $(h + a, k)$ counterclockwise once around the ellipse $\dfrac{(x - h)^2}{a^2} + \dfrac{(y - k)^2}{b^2} = 1$ are

$$C : x = h + a\cos t, \quad y = k + b\sin t, \quad 0 \le t \le 2\pi. \tag{6}$$

Let us emphasize:

If it is more convenient to find parametric equations of $-C$ than of C, use the formula

$$\int_C f\,ds = \int_{-C} f\,ds$$

to evaluate $\int_C f\,ds$.

Recall from Section 11.8 that the sum of a finite number of smooth curves is called **piecewise smooth**. We can evaluate line integrals over piecewise smooth curves by adding the integrals over each smooth part. For example, if C_1 and C_2 are smooth curves, we have

$$\int_{C_1 + C_2} f\,ds = \int_{C_1} f\,ds + \int_{C_1} f\,ds.$$

See Figure 14.1.3.

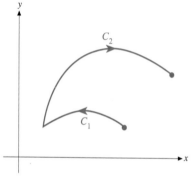

FIGURE 14.1.3 The integral over a piecewise smooth curve is the sum of the integrals over each smooth part.

EXAMPLE 2

Evaluate $\int_C y\,ds$, where C consists of the polygonal path from $(1, 0)$ to $(2, 3)$, to $(4, 3)$.

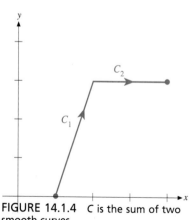

FIGURE 14.1.4 C is the sum of two smooth curves.

SOLUTION The curve is sketched in Figure 14.1.4. We need a parametric representation for each of the two smooth parts of the curve.

Formula 4 can be used to find parametric equations of the line segment from $(1, 0)$ to $(2, 3)$. We obtain

$$C_1 : x = 1 + t, \quad y = 3t, \quad 0 \leq t \leq 1.$$

$$\frac{dx}{dt} = 1, \quad \frac{dy}{dt} = 3, \text{ so } ds = \sqrt{(1)^2 + (3)^2}\, dt = \sqrt{10}\, dt.$$

The horizontal line segment from $(2, 3)$ to $(4, 3)$ can be described by the equation $y = 3$, $2 \leq x \leq 4$. Formula 2 then gives

$$C_2 : x = t, \quad y = 3, \quad 2 \leq t \leq 4.$$

$$\frac{dx}{dt} = 1, \quad \frac{dy}{dt} = 0, \text{ so } ds = \sqrt{(1)^2 + (0)^2}\, dt = dt.$$

Then

$$\int_C y\, ds = \int_{C_1} y\, ds + \int_{C_2} y\, ds$$

$$= \int_0^1 3t\sqrt{10}\, dt + \int_2^4 3\, dt$$

$$= \left[3\sqrt{10} \left(\frac{t^2}{2} \right) \right]_0^1 + \left[3t \right]_2^4$$

$$= \left[3\sqrt{10} \left(\frac{1}{2} \right) - (0) \right] + [3(4) - 3(2)] = \frac{3}{2}\sqrt{10} + 6. \quad \blacksquare$$

Any one of Formulas 2, 3, or 4 could be used to find parametric equations of the line segment from $(1, 0)$ to $(2, 3)$ in Example 2. For example, the equation of the line that contains $(1, 0)$ and $(2, 3)$ is $y = 3x - 3$. Formula 2 then gives

$$C_1 : x = t, \quad y = 3t - 3, \quad 1 \leq t \leq 2.$$

Since $y = 3x - 3$ implies $x = (y + 3)/3$, Formula 3 gives

$$C_1 : x = \frac{t + 3}{3}, \quad y = t, \quad 0 \leq t \leq 3.$$

The line integral $\int_{C_1} y\, ds$ has the same value for each of these parametrizations.

Lateral Surface Area

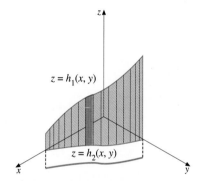

FIGURE 14.1.5 The lateral surface area of that part of the cylinder between the surfaces $z = h_1(x, y)$ and $z = h_2(x, y)$ above a segment of the curve of length ds is approximately $(h_2(x, y) - h_1(x, y))\, ds$.

The graph in three-dimensional space of an equation $F(x, y) = 0$ is a generalized cylinder with generating lines parallel to the z-axis. The **lateral surface area** of that part of the cylinder between the surfaces $z = h_1(x, y)$ and $z = h_2(x, y)$ is

$$\int_C (h_2 - h_1)\, ds,$$

where C is the curve given by the graph of $F(x, y) = 0$ in the xy-plane and $h_1 \leq h_2$ on C. See Figure 14.1.5.

EXAMPLE 3

Find the lateral surface area of that part of the cylinder $x^2 + y^2 = 4$ between the planes $z = 0$ and $x + 2y + z = 6$.

SOLUTION   The cylinder is sketched in Figure 14.1.6. We need a parametric

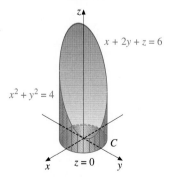

FIGURE 14.1.6 The lateral surface area of that part of the cylinder between $x + 2y + z = 6$ and $z = 0$ is $\int_C (6 - x - 2y)\, ds$.

representation of the circle $x^2 + y^2 = 4$ in the xy-plane. Formula 5 gives

$$C : x = 2\cos t, \quad y = 2\sin t, \quad 0 \le t \le 2\pi.$$

Then

$$\frac{dx}{dt} = -2\sin t \text{ and } \frac{dy}{dt} = 2\cos t.$$

It follows that

$$ds = \sqrt{(-2\sin t)^2 + (2\cos t)^2}\,dt = \sqrt{4(\sin^2 t + \cos^2 t)}\,dt = 2\,dt.$$

The lateral surface area is

$$S = \int_C (6 - x - 2y)\,ds = \int_0^{2\pi} [6 - 2\cos t - 2(2\sin t)](2)\,dt$$

$$= 12t - 4\sin t + 8\cos t \Big]_0^{2\pi} = 24\pi. \quad \blacksquare$$

Mass and Moments

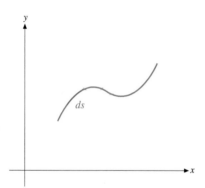

FIGURE 14.1.7 The mass of a differential segment of the wire is $dM = \rho\,ds$.

Suppose a wire is bent into the shape of a curve C and the density per unit length of the wire at a point on the curve is given by a function ρ. If ρ is continuous, the mass of a small piece of the wire is represented in differential form by

$$dM = \rho\,ds.$$

See Figure 14.1.7. The total mass of the wire is the sum of the masses of the small pieces. Taking the limit of the corresponding Riemann sums, we obtain

$$M = \int_C dM = \int_C \rho\,ds.$$

The moments of the wire are

$$M_{x=0} = \int_C x\rho\,ds, \quad M_{y=0} = \int_C y\rho\,ds, \text{ and } M_{z=0} = \int_C z\rho\,ds.$$

The center of mass is determined as usual by the mass and moments.

EXAMPLE 4

Find the point about which a homogeneous semicircle of radius r will balance horizontally.

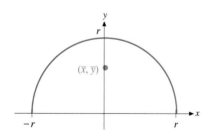

FIGURE 14.1.8 A homogeneous semicircle will balance horizontally from its center of mass.

SOLUTION   Let us position the semicircle in the xy-plane along the curve

$$C : x = r\cos t, \quad y = r\sin t, \quad 0 \le t \le \pi.$$

See Figure 14.1.8. The semicircle will balance horizontally about its center of mass.

Let $\rho(x, y) = k$, a constant. We have

$$ds = \sqrt{\left(\frac{dx}{dt}\right)^2 + \left(\frac{dy}{dt}\right)^2}\,dt = \sqrt{(-r\sin t)^2 + (r\cos t)^2}\,dt = r\,dt,$$

so

$$M = \int_C \rho\,ds = \int_0^\pi kr\,dt = k\pi r \text{ and}$$

$$M_{y=0} = \int_C y\rho\,ds = \int_0^\pi r\sin t \cdot kr\,dt = 2kr^2.$$

Then

$$\bar{y} = \frac{M_{y=0}}{M} = \frac{2kr^2}{k\pi r} = \frac{2r}{\pi}.$$

Symmetry implies $\bar{x} = 0$. The homogeneous semicircle would balance horizontally from a point $2r/\pi$ from its center, along its line of symmetry. ∎

EXERCISES 14.1

1. (a) $C: x = \cos t, y = \sin t, 0 \le t \le \pi/2$. Evaluate $\int_C xy\,ds$.
   (b) $C': x = \sqrt{1 - u^2}, y = u, 0 \le u \le 1$, is a reparametrization of C obtained by using the equation $u = \sin t$. Evaluate $\int_{C'} xy\,ds$.
   (c) $-C: x = \cos v, y = -\sin v, -\pi/2 \le v \le 0$, has the same graph as C, but opposite direction. Evaluate $\int_{-C} xy\,ds$.

2. (a) $C: x = t, y = t^2/2, 1 \le t \le 2$. Evaluate $\int_C x\,ds$.
   (b) $C': x = \sqrt{u - 1}, y = (u - 1)/2, 2 \le u \le 5$, is a reparametrization of C obtained by using the equation $u = t^2 + 1$. Evaluate $\int_{C'} x\,ds$.
   (c) $-C: x = -v, y = v^2/2, -2 \le v \le -1$, has the same graph as C, but opposite direction. Evaluate $\int_{-C} x\,ds$.

3. (a) $C: x = t, y = 0, 0 \le t \le 1$. Evaluate $\int_C \frac{x}{x^2 + 1}\,ds$.
   (b) $C': x = \sqrt{u - 1}, y = 0, 1 \le u \le 2$, is a reparametrization of C obtained by using the equation $u = t^2 + 1$. Evaluate $\int_{C'} \frac{x}{x^2 + 1}\,ds$.
   (c) $-C: x = -v, y = 0, -1 \le v \le 0$, has the same graph as C, but opposite direction. Evaluate $\int_{-C} \frac{x}{x^2 + 1}\,ds$.

4. (a) $C: x = 2t, y = t, 0 \le t \le 1$. Evaluate $\int_C xe^{y^2}\,ds$.
   (b) $C': x = 2\sqrt{u}, y = \sqrt{u}, 0 \le u \le 1$, is a reparametrization of C obtained by using the equation $u = t^2$. Evaluate $\int_{C'} xe^{y^2}\,ds$.
   (c) $-C: x = -2v, y = -v, -1 \le v \le 0$, has the same graph as C, but opposite direction. Evaluate $\int_{-C} xe^{y^2}\,ds$.

5. Evaluate $\int_C (2x + y)\,ds$, where C consists of the polygonal path from $(0, 0)$ to $(3, 0)$, to $(3, 2)$.

6. Evaluate $\int_C (2x + y)\,ds$, where C consists of the polygonal path from $(0, 0)$ to $(0, 2)$, to $(3, 2)$.

7. Evaluate $\int_C (x + y^2)\,ds$, where C consists of the path from $(2, 0)$, counterclockwise along the circle $x^2 + y^2 = 4$ to $(-2, 0)$, and then along the x-axis to $(2, 0)$.

8. Evaluate $\int_C (x + y^2)\,ds$, where C consists of the path from $(2, 0)$, counterclockwise once around the circle $x^2 + y^2 = 4$.

9. Find the lateral surface area of that part of the cylinder $x^2 + y^2 = 4$ between $z = 2 + xy$ and $z = 0$.

10. Find the lateral surface area of that part of the cylinder $x^2 + y^2 = 1$ between $z = x^2$ and $z = 0$.

11. Find the lateral surface area of that part of the vertical generalized cylinder $x = y^2/2, -1 \le y \le 1$, between $z = x$ and $z = 0$.

12. Find the lateral surface area of that part of the vertical generalized cylinder $y = \cos x, -\pi/2 \le x \le \pi/2$, between $z = y$ and $z = 0$.

In Exercises 13–16, a wire is bent in the shape of the given curve. Find the mass of the wire for the indicated density function.

13. $y = x^2, 0 \le x \le 2; \rho(x, y) = x$

14. $y = x^3, 0 \le x \le 2; \rho(x, y) = y$

15. $C: x = \cos t, y = \sin t, z = t, 0 \le t \le 2\pi; \rho(x, y, z) = z$

16. $C: x = t^2, y = t, z = t, 0 \le t \le 2; \rho(x, y, z) = z$

17. A wire with variable density is bent in the shape of the curve $C: x = \cos t, y = \sin t, 0 \le t \le \pi$, and its density is given by $\rho(x, y) = y$. Find its center of mass.

18. Find the center of mass of a homogeneous wire that is bent in the shape of the elliptical curve $C: x = 3\cos t, y = 2\sin t, 0 \le t \le \pi$.

19. Find the moment of inertia of a homogeneous ring with radius r about a diameter. Express your answer in terms of the mass and radius of the ring.

20. Find the moment of inertia of a homogeneous ring with radius r about a line through its center, orthogonal to the plane of the ring. Express your answer in terms of the mass and radius of the ring.

In Exercises 21–26, **(a)** *find the unit tangent vector* **T** *of the given curve C,* **(b)** *express* **F** · **T** *in terms of the parameter for the given vector function* **F**, *and* **(c)** *evaluate* $\int_C \mathbf{F} \cdot \mathbf{T}\, ds$.

21. $C: x = \cos t, y = \sin t, 0 \le t \le 2\pi; \mathbf{F} = x\mathbf{i} + y\mathbf{j}$

22. $C: x = \cos t, y = \sin t, 0 \le t \le 2\pi; \mathbf{F} = y\mathbf{i} - x\mathbf{j}$

23. $C: x = 2\cos t, y = \sin t, 0 \le t \le 2\pi; \mathbf{F} = y\mathbf{i} - x\mathbf{j}$

24. $C: x = 3\cos t, y = 2\sin t, 0 \le t \le 2\pi; \mathbf{F} = x\mathbf{j}$

25. $C: x = t, y = t^2, 0 \le t \le 2; \mathbf{F} = x\mathbf{i} + y\mathbf{j}$

26. $C: x = t^2, y = t^3, 0 \le t \le 1; \mathbf{F} = x^2\mathbf{i} + y\mathbf{j}$

27. Evaluate $\int_C \mathbf{F} \cdot \mathbf{n}\, ds$, where $C: x = \cos t, y = \sin t, 0 \le t \le 2\pi$, $\mathbf{F} = x\mathbf{i} + y\mathbf{j}$, and $\mathbf{n} = x\mathbf{i} + y\mathbf{j}$ is the outwardly directed unit normal vector of C.

28. Evaluate $\int_C \mathbf{F} \cdot \mathbf{n}\, ds$, where $C: x = \cos t, y = \sin t, 0 \le t \le 2\pi$, $\mathbf{F} = y\mathbf{i} - x\mathbf{j}$, and $\mathbf{n} = x\mathbf{i} + y\mathbf{j}$ is the outwardly directed unit normal vector of C.

29. Use a change of variables to show that $\int_{C_\phi} f\, ds = \int_C f\, ds$, where

$$C: x = x(t), \quad y = y(t), \quad a \le t \le b,$$
$$C_\phi: x = x(\phi^{-1}(u)), \quad y = y(\phi^{-1}(u)), \quad \phi(a) \le u \le \phi(b),$$
$$\text{and } \phi'(t) > 0 \text{ for } a \le t \le b.$$

30. Use a change of variables to show that $\int_{-C} f\, ds = \int_C f\, ds$, where

$$C: x = x(t), \quad y = y(t), \quad a \le t \le b, \text{ and}$$
$$-C: x = x(-u), \quad y = y(-u), \quad -b \le u \le -a.$$

31. Show that

$$\left| \int_C f\, ds \right| \le (\text{Maximum of } |f| \text{ on } C)(\text{Arc length of } C).$$

14.2 APPLICATIONS OF LINE INTEGRALS TO VECTOR FIELDS

The most important applications of line integrals involve vectors. We will introduce some of these applications in this section.

Vector Fields

A vector-valued function **F** that is defined on a region R is called a **vector field** on R. Each point P of R is associated with a vector $\mathbf{F}(P)$. R may be a region in the plane or in three-dimensional space.

If each vector value of **F** represents a force, the vector field is also called a **force field**. The forces may be due to gravitational or magnetic attraction, or an electric charge, to name a few. Vectors drawn at representative points can be used to indicate the direction and magnitude of the force at those points. The direction of vectors of a force field due to magnetic attraction is indicated in Figure 14.2.1.

Each vector **F** of a vector field may represent a velocity vector. A vector field that gives the velocity of water at the surface of a stream is indicated in Figure 14.2.2. The direction of the vector at a point indicates the direction of flow, and the length of the vector indicates the rate of flow. The sketch indicates that the stream is flowing more rapidly near the center than along the banks. Also, it indicates that the stream is flowing more rapidly where it is narrower.

If a function U is defined and has first partial derivatives in a region R, the vectors $\mathbf{F} = \nabla U$ form a vector field on R. Such vector fields are called **gradient fields** or **conservative fields**. Gradient fields are very common in applications.

Many of the applications of line integrals involve integrals of the form

$$\int_C \mathbf{F} \cdot \mathbf{T}\, ds,$$

where **F** is a vector field and **T** is the unit tangent vector of C. Let us develop a formula that allows us to evaluate this type of line integral without calculating

Connections

Parametric equations, 11.1, 11.9.

Work, 11.3.

Vector-valued functions, 11.7.

Line integrals, 14.1.

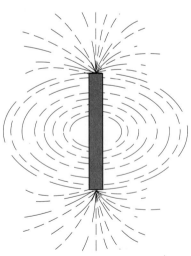

FIGURE 14.2.1 A magnet creates a force field.

FIGURE 14.2.2 The flow of water at the surface of a stream can be indicated by a velocity field.

the unit tangent vector and the differential form of the arc length of C in each case.

Suppose that $\mathbf{F}(x, y) = M(x, y)\mathbf{i} + N(x, y)\mathbf{j}$ and

$$C : x = x(t), \quad y = y(t), \quad a \le t \le b.$$

The vector form of the curve is then

$$\mathbf{R}(t) = x(t)\mathbf{i} + y(t)\mathbf{j}, \quad a \le t \le b.$$

We have

$$\mathbf{R}' = \frac{dx}{dt}\mathbf{i} + \frac{dy}{dt}\mathbf{j} \text{ and } |\mathbf{R}'| = \sqrt{\left(\frac{dx}{dt}\right)^2 + \left(\frac{dy}{dt}\right)^2}.$$

The unit tangent vector is

$$\mathbf{T} = \frac{\mathbf{R}'}{|\mathbf{R}'|} = \frac{\dfrac{dx}{dt}\mathbf{i} + \dfrac{dy}{dt}\mathbf{j}}{\sqrt{\left(\dfrac{dx}{dt}\right)^2 + \left(\dfrac{dy}{dt}\right)^2}}.$$

$$\mathbf{F} \cdot \mathbf{T} = (M\mathbf{i} + N\mathbf{j}) \cdot \left(\frac{\dfrac{dx}{dt}\mathbf{i} + \dfrac{dy}{dt}\mathbf{j}}{\sqrt{\left(\dfrac{dx}{dt}\right)^2 \left(\dfrac{dy}{dt}\right)^2}} \right) = \frac{M\dfrac{dx}{dt} + N\dfrac{dy}{dt}}{\sqrt{\left(\dfrac{dx}{dt}\right)^2 + \left(\dfrac{dy}{dt}\right)^2}}.$$

Then Formula 1 of Section 14.1 gives

$$\int_C \mathbf{F} \cdot \mathbf{T}\, ds = \int_a^b \frac{M\dfrac{dx}{dt} + N\dfrac{dy}{dt}}{\sqrt{\left(\dfrac{dx}{dt}\right)^2 + \left(\dfrac{dy}{dt}\right)^2}} \sqrt{\left(\frac{dx}{dt}\right)^2 + \left(\frac{dy}{dt}\right)^2}\, dt$$

$$= \int_a^b \left(M\frac{dx}{dt} + N\frac{dy}{dt} \right) dt.$$

Since

$$dx = \frac{dx}{dt}\, dt \text{ and } dy = \frac{dy}{dt}\, dt,$$

we can express the latter integral more simply as

$$\int_C M\, dx + N\, dy.$$

Summarizing:

If C is a plane curve and $\mathbf{F} = M\mathbf{i} + N\mathbf{j}$, then

$$\int_C \mathbf{F} \cdot \mathbf{T}\, ds = \int_C M\, dx + N\, dy.$$

Values of integrals of the form $\int_C M\, dx + N\, dy$ do not depend on the particular smooth parametrization of C used to evaluate the integrals, *except for direction.* The unit tangent vector of $-C$ is the negative of the unit tangent vector of C. See

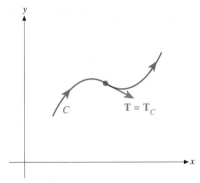

FIGURE 14.2.3a The unit tangent vector points in the direction of C.

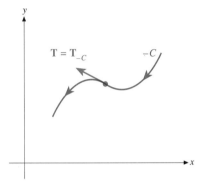

FIGURE 14.2.3b The unit tangent vector of $-C$ is the negative of the unit tangent vector of C.

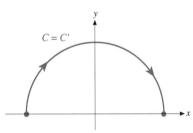

FIGURE 14.2.4a If C' has the same graph and direction as C, then $\int_{C'} \mathbf{F} \cdot \mathbf{T}\,ds = \int_C \mathbf{F} \cdot \mathbf{T}\,ds$.

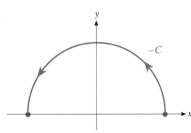

FIGURE 14.2.4b Since the unit tangent vector of $-C$ is the negative of the unit tangent vector of C, $\int_{-C} \mathbf{F} \cdot \mathbf{T}\,ds = -\int_C \mathbf{F} \cdot \mathbf{T}\,ds$.

Figure 14.2.3. Since the vector $\mathbf{T}$ in the integrand represents the unit tangent of the curve of integration, we can use the fact that $\int_{-C} f\,ds = \int_C f\,ds$ to obtain

$$
\begin{aligned}
\int_{-C} \mathbf{F} \cdot \mathbf{T}\,ds &= \int_{-C} \mathbf{F} \cdot \mathbf{T}_{-C}\,ds \\
&= \int_{-C} \mathbf{F} \cdot (-\mathbf{T}_C)\,ds \\
&= -\int_{-C} \mathbf{F} \cdot \mathbf{T}_C\,ds \\
&= -\int_{C} \mathbf{F} \cdot \mathbf{T}_C\,ds \\
&= -\int_{C} \mathbf{F} \cdot \mathbf{T}\,ds,
\end{aligned}
$$

so

$$
\int_{-C} M\,dx + N\,dy = -\left(\int_C M\,dx + N\,dy \right).
$$

Similarly, if C is a curve in three-dimensional space and $\mathbf{F} = M\mathbf{i} + N\mathbf{j} + P\mathbf{k}$, then

$$
\int_C \mathbf{F} \cdot \mathbf{T}\,ds = \int_C M\,dx + N\,dy + P\,dz \text{ and}
$$

$$
\int_{-C} M\,dx + N\,dy + P\,dz = -\left(\int_C M\,dx + N\,dy + P\,dz \right).
$$

We should use the above formulas for the integrals over $-C$ when it is more convenient to find parametric equations of $-C$ than of C.

Integrals of the form $\int_C M\,dx + N\,dy$ are evaluated by substitution of parametric values.

EXAMPLE 1

(a) $C\colon x = -\cos t,\, y = \sin t,\, 0 \le t \le \pi$. Evaluate $\displaystyle\int_C (2y - x)dx - y\,dy$.

(b) $C'\colon x = u,\, y = \sqrt{1 - u^2},\, -1 \le u \le 1$, is a reparametrization of C. Evaluate $\displaystyle\int_{C'} (2y - x)dx - y\,dy$.

(c) $-C\colon x = -\cos v,\, y = -\sin v,\, -\pi \le v \le 0$, has the same graph as C, but opposite direction. Evaluate $\displaystyle\int_{-C} (2y - x)dx - y\,dy$.

SOLUTION The curves are sketched in Figure 14.2.4.

(a) We have $C\colon x = -\cos t,\, y = \sin t,\, 0 \le t \le \pi$, so $dx = \sin t\,dt$, and $dy = \cos t\,dt$. Substitution then gives

$$
\begin{aligned}
\int_C (2y - x)dx - y\,dy &= \int_0^\pi \{[2(\sin t) - (-\cos t)](\sin t) - (\sin t)(\cos t)\}\,dt \\
&= \int_0^\pi 2\sin^2 t\,dt \\
&= -\sin t \cos t + t \Big]_0^\pi = \pi.
\end{aligned}
$$

(b) We have $C': x = u$, $y = \sqrt{1 - u^2}$, $-1 \le v \le 1$, so $dx = du$, $dy = \dfrac{-u}{\sqrt{1 - u^2}} du$. Then

$$\int_{C'} (2y - x)\,dx - y\,dy$$

$$= \int_{-1}^{1} \left[(2\sqrt{1 - u^2} - u) - (\sqrt{1 - u^2}) \left(\frac{-u}{\sqrt{1 - u^2}} \right) \right] du$$

$$= \int_{-1}^{1} 2\sqrt{1 - u^2}\,du$$

[The factor $-u/\sqrt{1 - u^2}$ becomes unbounded at ± 1, but the product with $\sqrt{1 - u^2}$ remains bounded and can be defined to be continuous at ± 1.]

$$= u\sqrt{1 - u^2} + \sin^{-1} u \bigg]_{-1}^{1}$$

$$= \sin^{-1}(1) - \sin^{-1}(-1) = \frac{\pi}{2} - \left(-\frac{\pi}{2} \right) = \pi.$$

(c) We have $-C: x = -\cos v$, $y = -\sin v$, $-\pi \le v \le 0$, so $dx = \sin v\,dv$, $dy = -\cos v\,dv$. Then

$$\int_{-C} (2y - x)\,dx - y\,dy$$

$$= \int_{-\pi}^{0} \{[2(-\sin v) - (-\cos v)](\sin v) - (-\sin v)(-\cos v)\}\,dv$$

$$= \int_{-\pi}^{0} -2\sin^2 v\,dv$$

$$= \sin v \cos v - v \bigg]_{-\pi}^{0} = -\pi.$$

We should expect the integrals in (a) and (b) to be equal, since C and C' have the same graph and the same direction. The integral in (c) is the negative of the integrals in (a) and (b), because $-C$ has opposite direction. ∎

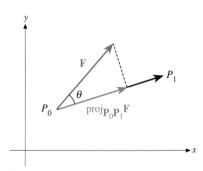

FIGURE 14.2.5 The force field **F** may vary in both magnitude and direction at different points on the curve.

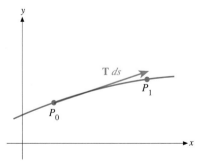

FIGURE 14.2.6a The work associated with a constant force **F** as an object moves in a straight line from P_0 to P_1 is $\mathbf{F} \cdot \mathbf{P_0 P_1}$.

Work

We now consider the problem of finding the work associated with a force field **F** as a small object moves along a curve C in the field. The force may vary in both magnitude and direction at different points on the curve. See Figure 14.2.5. Recall from Section 11.3 that the work associated with a *constant* force **F** as an object is moved in a *straight line* from points P_0 to P_1 on C is

$$\mathbf{F} \cdot (\mathbf{P_0 P_1}) = |\mathbf{F}||\mathbf{P_0 P_1}| \cos \theta.$$

See Figure 14.2.6a. If the curve is smooth, then small pieces of the curve are nearly straight and we have

$$\mathbf{P_0 P_1} \approx \mathbf{T}\,ds,$$

where **T** is the unit tangent vector of C at the point P_0 in the direction of motion and ds is the differential representation of the arc length of C between P_0 and P_1. See Figure 14.2.6b. If the force varies continuously, then **F** is nearly constant along small pieces of the curve. The work associated with moving a small distance along the curve is then approximately

$$dW = \mathbf{F} \cdot \mathbf{T}\,ds.$$

FIGURE 14.2.6b If the curve is smooth and ds is small, then $\mathbf{P_0 P_1} \approx \mathbf{T}\,ds$.

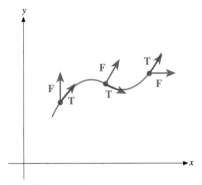

FIGURE 14.2.7a If the angle between **F** and **T** is between 0 and $\pi/2$ at all points of C, then the work associated with **F** as an object moves along C is positive.

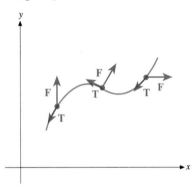

FIGURE 14.2.7b The work is negative if the angle between **F** and **T** is between $\pi/2$ and π at all points of C.

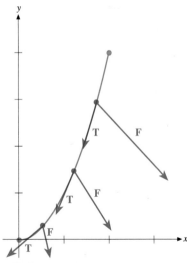

FIGURE 14.2.8 Sketches of representative vectors **F** and **T** at several points on C indicate the work associated with **F** is positive.

The total work is the sum of the work along each small part of the curve, so it should seem reasonable to define the total work associated with **F** along the curve C to be

$$W = \int_C \mathbf{F} \cdot \mathbf{T}\, ds.$$

The value of the work associated with a force field as an object moves along a curve depends on the direction of motion along the curve C. This corresponds to the formula

$$\int_{-C} \mathbf{F} \cdot \mathbf{T}\, ds = -\int_C \mathbf{F} \cdot \mathbf{T}\, ds.$$

If $\mathbf{F} \neq \mathbf{0}$ and the angle between **F** and **T** is less than $\pi/2$ at all points of C, as indicated in Figure 14.2.7a, then $\mathbf{F} \cdot \mathbf{T}$ is positive at every point of C and the work associated with **F** as an object moves along C will be positive. If the angle between **F** and **T** is between $\pi/2$ and π at all points of C, as indicated in Figure 14.2.7b, the work will be negative. If the angle between **F** and **T** varies between 0 and π, the net work may be either positive, negative, or zero.

EXAMPLE 2

Find the work associated with the force field $\mathbf{F} = (y/2)\mathbf{i} - x\mathbf{j}$ as an object moves along the curve $y = x^2$ from $(2, 4)$ to $(0, 0)$. Sketch the curve and representative vectors **F** and **T** at some points along the curve.

SOLUTION The curve and representative vectors **F** and **T** are sketched in Figure 14.2.8. Note that the angle between **F** and **T** appears to be less than $\pi/2$ at all points on the curve, so we should expect the work to be positive.

The work is $W = \int_C y/2\, dx - x\, dy$. We need parametric equations of the curve to evaluate the line integral. Since C is in the direction of decreasing x and decreasing y, it is more convenient to parametrize $-C$ than C. We can then use the formula

$$\int_C M\, dx + N\, dy = -\left(\int_{-C} M\, dx + N\, dy \right)$$

to evaluate the work. Formula 2 of Section 14.1 gives

$$-C : x = t, \quad y = t^2, \quad 0 \le t \le 2.$$

Then $dx = dt$ and $dy = 2t\, dt$. Substitution gives

$$W = \int_C \frac{y}{2}\, dx - x\, dy$$

$$= -\int_{-C} \frac{y}{2}\, dx - x\, dy$$

$$= -\int_0^2 \left[\frac{t^2}{2} - (t)(2t) \right] dt$$

$$= \int_0^2 \frac{3t^2}{2}\, dt = \frac{t^3}{2} \Big]_0^2 = 4.$$

As expected, the work is positive. ∎

Kinetic Energy

We can use the formula

$$W = \int_C \mathbf{F} \cdot \mathbf{T} \, ds$$

to calculate the work associated with the force $\mathbf{F}$ as an object moves along the curve C, even though there may be other forces acting on the object. In case $\mathbf{F}$ is an external force that is the only force acting on the object, then $\mathbf{F}$ determines the motion of the object. In this case, we can establish a relation between the work associated with $\mathbf{F}$ and the change in kinetic energy of the object. That is, if the position of the object at time t is given by the vector function $\mathbf{R}(t)$, then the velocity is $\mathbf{V} = \mathbf{R}'$ and the acceleration is $\mathbf{A} = \mathbf{V}'$. Newton's Second Law of Motion implies

$$\mathbf{F} = M\mathbf{A},$$

where M is the mass of the object and $\mathbf{A}$ is the acceleration vector. The unit tangent vector to C is

$$\mathbf{T} = \frac{\mathbf{R}'}{|\mathbf{R}'|} = \frac{\mathbf{V}}{|\mathbf{R}'|}.$$

Then

$$\mathbf{F} \cdot \mathbf{T} = (M\mathbf{A}) \cdot \left(\frac{\mathbf{V}}{|\mathbf{R}'|}\right)$$

$$= (M\mathbf{V}') \cdot \left(\frac{\mathbf{V}}{|\mathbf{R}'|}\right)$$

$$= \frac{M}{|\mathbf{R}'|}(\mathbf{V}' \cdot \mathbf{V}).$$

Since

$$\frac{d}{dt}(|\mathbf{V}|^2) = \frac{d}{dt}(\mathbf{V} \cdot \mathbf{V}) = \mathbf{V} \cdot \mathbf{V}' + \mathbf{V}' \cdot \mathbf{V} = 2\mathbf{V}' \cdot \mathbf{V},$$

we have

$$\mathbf{F} \cdot \mathbf{T} = \frac{M}{|\mathbf{R}'|}\left(\frac{1}{2}\frac{d}{dt}(|\mathbf{V}|^2)\right).$$

Since $ds = \sqrt{\left(\frac{dx}{dt}\right)^2 + \left(\frac{dy}{dt}\right)^2} \, dt = |\mathbf{R}'| dt$, it follows that

$$\int_C \mathbf{F} \cdot \mathbf{T} \, ds = \int_a^b \frac{M}{|\mathbf{R}'|}\left(\frac{1}{2}\frac{d}{dt}(|\mathbf{V}|^2)\right)|\mathbf{R}'| \, dt$$

$$= \int_a^b \frac{M}{2}\frac{d}{dt}(|\mathbf{V}|^2) \, dt$$

$$= \frac{M}{2}|\mathbf{V}(b)|^2 - \frac{M}{2}|\mathbf{V}(a)|^2,$$

where the last equality is a consequence of the Fundamental Theorem of Calculus. The expression $M|\mathbf{V}|^2/2$ is the **kinetic energy** of the object. We see that, if $\mathbf{F}$ is an external force that is the only force acting on an object, then the work associated with the force is equal to the change in kinetic energy.

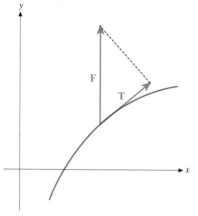

FIGURE 14.2.9 If **F** is a velocity field, then **F** · **T**ds is a measure of the flow along an arc of the curve of length ds.

Circulation

To obtain another interpretation of the line integral, let us consider a thin layer of fluid that is moving on the xy-plane and suppose that the velocity of the fluid at a point (x, y) is given by a vector function $\mathbf{F}(x, y)$. If C is a curve with unit tangent vector $\mathbf{T}$, then the inner product $\mathbf{F} \cdot \mathbf{T}$ is the component of the velocity $\mathbf{F}$ in the direction of $\mathbf{T}$ and $\mathbf{F} \cdot \mathbf{T}\,ds$ represents a measure of the flow of fluid along an arc of the curve of length ds. See Figure 14.2.9. The integral $\int_C \mathbf{F} \cdot \mathbf{T}\,ds$ is the net amount of flow along the curve. If C is a *simple closed curve*, the net flow is a measure of the circulation of the fluid around the curve. We may interpret the sign of the line integral as an indication of the net direction of rotation of the fluid along the curve. We say

$$\int_C \mathbf{F} \cdot \mathbf{T}\,ds \text{ is the \textbf{circulation} of } \mathbf{F} \text{ around the simple closed curve } C.$$

We use the term circulation, even though $\mathbf{F}$ may represent something other than velocity. For example, $\mathbf{F}$ may represent force due to gravity, an electric charge, or a magnetic force.

EXAMPLE 3

Evaluate the circulation of $\mathbf{F} = (x - 2y)\mathbf{i} + (2x + y)\mathbf{j}$ around the curve

$$C: x = \cos t, \quad y = \sin t, \quad 0 \le t \le 2\pi.$$

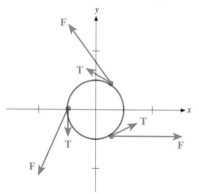

FIGURE 14.2.10 Components of **F** tangent to the curve indicate a counterclockwise circulation.

SOLUTION   The curve and some representative vectors of $\mathbf{F}$ along the curve are sketched in Figure 14.2.10. In this example, the vectors $\mathbf{F}$ indicate a counterclockwise rotation about the origin. Since the direction of C is also counterclockwise, we should expect the circulation around C to be positive.

The circulation is $\int_C \mathbf{F} \cdot \mathbf{T}\,ds$. We have

$$\mathbf{F} = (x - 2y)\mathbf{i} + (2x + y)\mathbf{j} = M\mathbf{i} + N\mathbf{j}, \text{ so}$$
$$M = x - 2y \text{ and } N = 2x + y.$$

Substitution then gives

$$\int_C \mathbf{F} \cdot \mathbf{T}\,ds = \int_C M\,dx + N\,dy$$
$$= \int_C (x - 2y)dx + (2x + y)dy.$$

From the parametric equations of C, we have

$$x = \cos t, \quad y = \sin t, \quad dx = -\sin t\,dt, \quad dy = \cos t\,dt.$$

The above integral then becomes

$$\int_0^{2\pi} [(\cos t - 2\sin t)(-\sin t) + (2\cos t + \sin t)(\cos t)]dt$$

$$= \int_0^{2\pi} 2\,dt = 4\pi. \quad \blacksquare$$

Flux

Let us return to the model of a fluid flowing along the xy-plane with velocity field $\mathbf{F}$. If C is a simple closed curve, the unit tangent vector of C is

$$\mathbf{T} = \frac{dx}{ds}\mathbf{i} + \frac{dy}{ds}\mathbf{j}.$$

The vector obtained by a clockwise rotation of $\mathbf{T}$ by $\pi/2$ is

$$\mathbf{n} = \frac{dy}{ds}\mathbf{i} - \frac{dx}{ds}\mathbf{j}.$$

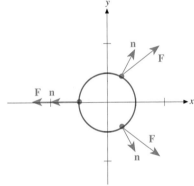

FIGURE 14.2.11 If the region enclosed by C is on the left as C is traced, then the unit vector $\mathbf{n} = \frac{dy}{ds}\mathbf{i} - \frac{dx}{ds}\mathbf{j}$ points outward.

If the region enclosed by C is on the left as C is traced, then $\mathbf{n}$ will point outward; $\mathbf{n}$ is then called the **outer unit normal** of the region enclosed by C. See Figure 14.2.11. The dot product $\mathbf{F} \cdot \mathbf{n}$ is the component of the velocity vector $\mathbf{F}$ in the direction of $\mathbf{n}$. The amount of fluid per unit time that flows in the direction of $\mathbf{n}$ across a small piece of the curve C with length ds is $\mathbf{F} \cdot \mathbf{n}\,ds$. The net flow across C is the sum of the flow over each small piece. The limit of these sums is

$$\int_C \mathbf{F} \cdot \mathbf{n}\,ds.$$

This integral is called the **outward flux** of $\mathbf{F}$ across C. If $\mathbf{F} = M\mathbf{i} + N\mathbf{j}$, then

$$\mathbf{F} \cdot \mathbf{n} = (M\mathbf{i} + N\mathbf{j}) \cdot \left(\frac{dy}{ds}\mathbf{i} - \frac{dx}{ds}\mathbf{j}\right) = M\frac{dy}{ds} - N\frac{dx}{ds}, \text{ and}$$

$$\int_C \mathbf{F} \cdot \mathbf{n}\,ds = \int_C M\,dy - N\,dx.$$

FIGURE 14.2.12 Components of $\mathbf{F}$ normal to the curve indicate an outward flow across C.

EXAMPLE 4

Evaluate the outward flux of $\mathbf{F} = 2x\mathbf{i} + y\mathbf{j}$ across the curve

$$C : x = \cos t, \quad y = \sin t, \quad 0 \le t \le 2\pi.$$

SOLUTION   The sketch in Figure 14.2.12 indicates an outward flow across C at each point of C, so we should expect the flux to be positive.

The flux is $\int_C \mathbf{F} \cdot \mathbf{n}\,ds$. We have

$$\mathbf{F} = 2x\mathbf{i} + y\mathbf{j} = M\mathbf{i} + N\mathbf{j}, \text{ so } M = 2x \text{ and } N = y.$$

Substitution into the formula for the outward flux of $\mathbf{F}$ across C then gives

$$\int_C \mathbf{F} \cdot \mathbf{n}\,ds = \int_C M\,dy - N\,dx$$

$$= \int_C 2x\,dy - y\,dx.$$

From the parametric equations of C, we have

$$x = \cos t, \quad y = \sin t, \quad dx = -\sin t\,dt, \quad dy = \cos t\,dt.$$

The integral that gives the flux then becomes

$$\int_0^{2\pi} [(2\cos t)(\cos t) - (\sin t)(-\sin t)]\,dt = \int_0^{2\pi} (\cos^2 t + \cos^2 t + \sin^2 t)\,dt$$

$$= \int_0^{2\pi} (\cos^2 t + 1)\,dt$$

$$= \left. \frac{\sin t \cos t}{2} + \frac{t}{2} + t \right]_0^{2\pi}$$

$$= \frac{2\pi}{2} + 2\pi = 3\pi. \quad \blacksquare$$

EXERCISES 14.2

Evaluate the line integrals in Exercises 1–6.

1. $\displaystyle\int_C 2x\,dx + xy\,dy;\ C: x = 3 - 3t,\ y = 3t,\ 0 \le t \le 1$

2. $\displaystyle\int_C 2x\,dx + xy\,dy;\ C: x = 3t,\ y = 3 - 3t,\ 0 \le t \le 1$

3. $\displaystyle\int_C y\,dx - x\,dy;\ C: x = 2\cos t,\ y = \sin t,\ 0 \le t \le 2\pi$

4. $\displaystyle\int_C x\,dx + y\,dy;\ C: x = \cos t,\ y = \sin t,\ 0 \le t \le 2\pi$

5. $\displaystyle\int_C yz\,dx + xz\,dy + xy\,dz;\ C: x = t,\ y = t^2,\ z = t^3,$
$0 \le t \le 2$

6. $\displaystyle\int_C -y\,dx + x\,dy + xy\,dz;\ C: x = \cos t,\ y = \sin t,\ z = 2t,$
$0 \le t \le 2\pi$

In Exercises 7–10, sketch the corresponding values of the vector-valued functions with initial points at the indicated points.

7. $\mathbf{F}(x, y) = x\mathbf{i} + y\mathbf{j};\ (1, 0),\ (0, 1),\ (-1/\sqrt{2}, 1/\sqrt{2}),$
$(-1/\sqrt{2}, -1/\sqrt{2}),\ (1/\sqrt{2}, -1/\sqrt{2})$

8. $\mathbf{F}(x, y) = -y\mathbf{i} + x\mathbf{j};\ (1, 0),\ (0, 1),\ (-1/\sqrt{2}, 1/\sqrt{2}),$
$(-1/\sqrt{2}, -1/\sqrt{2}),\ (1/\sqrt{2}, -1/\sqrt{2})$

9. $\mathbf{F}(x, y) = x\mathbf{i};\ (1, 0),\ (1/\sqrt{2}, 1/\sqrt{2}),\ (0, 1),\ (-1/\sqrt{2}, 1/\sqrt{2}),$
$(-1, 0),\ (-1/\sqrt{2}, -1/\sqrt{2}),\ (0, -1),\ (1/\sqrt{2}, -1/\sqrt{2})$

10. $\mathbf{F}(x, y) = (x - y)\mathbf{i} + (x + y)\mathbf{j};\ (1, 0),\ (0, 1),\ (-1/\sqrt{2}, 1/\sqrt{2}),$
$(-1/\sqrt{2}, -1/\sqrt{2}),\ (1/\sqrt{2}, -1/\sqrt{2})$

In Exercises 11–16, sketch the given curve and representative vectors $\mathbf{F}$ and $\mathbf{T}$ at some points on the curve. Find the work associated with the given force $\mathbf{F}$ as an object moves along the curve.

11. $C: x = 3t,\ y = 2t,\ 0 \le t \le 1;\ \mathbf{F} = \mathbf{i} - \mathbf{j}$

12. $C: x = 2t,\ y = 3t,\ 0 \le t \le 1;\ \mathbf{F} = x\mathbf{i} + y\mathbf{j}$

13. $C: x = \cos t,\ y = \sin t,\ 0 \le t \le 2\pi;\ \mathbf{F} = y\mathbf{i} - x\mathbf{j}$

14. $C: x = \cos t,\ y = \sin t,\ 0 \le t \le 2\pi;\ \mathbf{F} = x\mathbf{i} + y\mathbf{j}$

15. $C: x = \cos t,\ y = \sin t,\ z = t,\ 0 \le t \le 2\pi;\ \mathbf{F} = z\mathbf{k}$

16. $C: x = \cos t,\ y = \sin t,\ z = 1/t,\ 2\pi \le t \le 6\pi;$
$\mathbf{F} = (1/z^2)\mathbf{k}$

17. An object moves in a straight line from $(24, 10)$ to $(12, 5)$ due to an external force

$$\mathbf{F} = \frac{-x}{(x^2 + y^2)^{3/2}}\mathbf{i} + \frac{-y}{(x^2 + y^2)^{3/2}}\mathbf{j}.$$

Find the change in kinetic energy of the object.

18. An object moves in a straight line from $(3, 4)$ to $(6, 8)$ due to an external force

$$\mathbf{F} = \frac{x}{(x^2 + y^2)^{3/2}}\mathbf{i} + \frac{y}{(x^2 + y^2)^{3/2}}\mathbf{j}.$$

Find the change in kinetic energy of the object.

In Exercises 19–20, use the sketch of C and representative vectors $\mathbf{F}$ and $\mathbf{T}$ to determine whether the circulation of $\mathbf{F}$ around C is positive or negative.

19.

20.

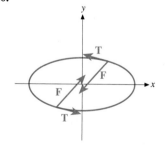

In Exercises 21–22, sketch the curve C and representative vectors $\mathbf{F}$ and $\mathbf{T}$. Find the circulation of $\mathbf{F}$ around C.

21. $\mathbf{F} = x\mathbf{i} + y\mathbf{j}$; C: $x = \cos t$, $y = \sin t$, $0 \le t \le 2\pi$

22. $\mathbf{F} = -y\mathbf{i} + x\mathbf{j}$; C: $x = \cos t$, $y = \sin t$, $0 \le t \le 2\pi$

In Exercises 23–26, find the circulation of the given vector function $\mathbf{F}$ around the indicated curve.

23. $\mathbf{F} = (x^2 - y)\mathbf{i} + (x - y^2)\mathbf{j}$; C: $x = \cos t$, $y = \sin t$,
$0 \le t \le 2\pi$

24. $\mathbf{F} = 2xy\mathbf{i} + (x^2 + y^2)\mathbf{j}$; C: $x = \cos t$, $y = \sin t$,
$0 \le t \le 2\pi$

25. $\mathbf{F} = xy^2\mathbf{i} + xy^3\mathbf{j}$; C is the polygonal path from $(0, 0)$ to $(1, 0)$, to $(0, 1)$, to $(0, 0)$

26. $\mathbf{F} = (x - 1)y\mathbf{i} + (x - 1)y^2\mathbf{j}$; C is the polygonal path from $(0, 0)$ to $(1, 1)$, to $(1, 0)$, to $(0, 0)$

In Exercises 27–28, use the sketch of representative vectors $\mathbf{F}$ and $\mathbf{n}$ to determine whether the outward flux of $\mathbf{F}$ around C is positive or negative.

27.

28.

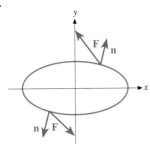

In Exercises 29–30, sketch the curve C and representative vectors $\mathbf{F}$ and $\mathbf{n}$. Find the outward flux of $\mathbf{F}$ around C.

29. $\mathbf{F} = x\mathbf{i} + y\mathbf{j}$; C: $x = \cos t$, $y = \sin t$, $0 \le t \le 2\pi$

30. $\mathbf{F} = -y\mathbf{i} + x\mathbf{j}$; C: $x = \cos t$, $y = \sin t$, $0 \le t \le 2\pi$

In Exercises 31–34, find the outward flux of the given vector function $\mathbf{F}$ across the indicated curve.

31. $\mathbf{F} = x\mathbf{i} + y\mathbf{j}$; C: $x = h + r\cos t$, $y = k + r\sin t$, $0 \le t \le 2\pi$

32. $\mathbf{F} = -y\mathbf{i} + x\mathbf{j}$; C: $x = h + r\cos t$, $y = k + r\sin t$, $0 \le t \le 2\pi$

33. $\mathbf{F} = (x^2 + y^2)\mathbf{i}$; C: $x = 2 + \cos t$, $y = \sin t$, $0 \le t \le 2\pi$

34. $\mathbf{F} = x\mathbf{i} + y\mathbf{j}$; C is the polygonal path from $(0, 0)$ to $(1, 0)$, to $(0, 1)$, to $(0, 0)$

35. $\mathbf{F} = x\mathbf{i} + y\mathbf{j}$
   (a) Evaluate $\int_{C_1} \mathbf{F} \cdot \mathbf{T}\,ds$, where C_1 goes along the x-axis from $(-2, 0)$ to $(2, 0)$.
   (b) Evaluate $\int_{C_2} \mathbf{F} \cdot \mathbf{T}\,ds$, where C_2 goes along the upper half of the ellipse $x^2 + 4y^2 = 4$ from $(-2, 0)$ to $(2, 0)$.

36. $\mathbf{F} = -y\mathbf{i} + x\mathbf{j}$
   (a) Evaluate $\int_{C_1} \mathbf{F} \cdot \mathbf{T}\,ds$, where C_1 goes along the x-axis from $(1, 0)$ to $(-1, 0)$.
   (b) Evaluate $\int_{C_2} \mathbf{F} \cdot \mathbf{T}\,ds$, where C_2 is the polygonal path from $(1, 0)$ to $(0, 1)$, to $(-1, 0)$.

37. Evaluate $\int_C x\,dy$, where C goes counterclockwise once around the triangle with vertices $(0, 0)$, $(1, 0)$, and $(0, 1)$. Compare with the area of the triangle.

38. Evaluate $\int_C -y\,dx$, where C goes counterclockwise once around the circle $x^2 + y^2 = r^2$. Compare with the area of the circle.

39. Show that
$$\left| \int_C \mathbf{F} \cdot \mathbf{T}\,ds \right| \le (\text{Maximum of } |\mathbf{F}| \text{ on } C)(\text{Length of } C).$$

40. An irrigation canal is 2 m wide. If an xy-coordinate system is positioned horizontally with the x-axis along the center of the canal, the water flow in m^3/min below a point (x, y) on the surface is given by the vector field $\mathbf{F} = (1 - y^4)\mathbf{i}$. See figure. Two vertical nets are placed across the canal, one along each of the curves C_1: $x = 2$, $y = t$, $-1 \le t \le 1$, and C_2: $x = t$, $y = t^3$, $-1 \le t \le 1$. Find the total flow (flux) past each net and compare the answers. Does the total flow depend on the configuration of the net?

14.3 GREEN'S THEOREM

In this section we will establish a relation between a double integral over a region and a line integral over the curve that bounds the region.

Development of the Formula

Suppose R is a vertically simple region, as illustrated in Figure 14.3.1. The curve $C = C_1 + C_2 + C_3 + C_4$ is the complete boundary of R, with direction chosen so the region is on the left as the boundary is traced. We assume that $M(x, y)$ and its first partial derivatives are continuous on R and on the boundary of R. We then have

$$\int_C M\,dx = \int_{C_1} M\,dx + \int_{C_2} M\,dx + \int_{C_3} M\,dx + \int_{C_4} M\,dx$$

$$= \int_{C_1} M\,dx + \int_{C_2} M\,dx - \int_{-C_3} M\,dx - \int_{-C_4} M\,dx.$$

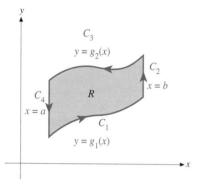

FIGURE 14.3.1 The vertically simple region is on the left as the boundary $C = C_1 + C_2 + C_3 + C_4$ is traced.

We have changed signs to obtain a curve $-C_3$ in the direction of increasing x and a curve $-C_4$ in the direction of increasing y. The above integrals can then be evaluated by using the parametric equations

$$C_1: x = x, \quad y = g_1(x), \quad a \le x \le b,$$
$$C_2: x = b, \quad y = y, \quad g_1(b) \le y \le g_2(b),$$
$$-C_3: x = x, \quad y = g_2(x), \quad a \le x \le b,$$
$$-C_4: x = a, \quad y = y, \quad g_1(a) \le y \le g_2(a).$$

Note that we have used x as the parameter for C_1 and $-C_3$; we have used y as the parameter for C_2 and $-C_4$. Since $dx = 0$ for C_2 and $-C_4$, we obtain

$$\int_C M\,dx = \int_{C_1} M\,dx - \int_{-C_3} M\,dx$$

$$= \int_a^b M(x, g_1(x))\,dx - \int_a^b M(x, g_2(x))\,dx$$

$$= \int_a^b [M(x, g_1(x)) - M(x, g_2(x))]\,dx.$$

For x fixed, $M(x, y)$ and $M_y(x, y)$ are continuous functions of y for $g_1(x, y) \le y \le g_2(x, y)$. Since $M_y(x, y)$ is the derivative of $M(x, y)$ with respect to y with x fixed, the Fundamental Theorem of Calculus tells us

$$M(x, g_1(x)) - M(x, g_2(x)) = \int_{g_1(x)}^{g_2(x)} -M_y(x, y)\,dy.$$

We then obtain

$$\int_C M\,dx = \int_a^b \int_{g_1(x)}^{g_2(x)} -M_y(x, y)\,dy\,dx$$

$$= \iint_R -M_y\,dA.$$

That is,

$$\int_C M\,dx = \iint_R -M_y\,dA. \tag{1}$$

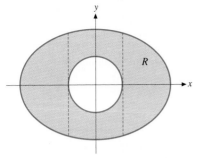

FIGURE 14.3.2 This region is finitely vertically simple, since it can be divided by vertical lines into a finite number (four) of vertically simple regions.

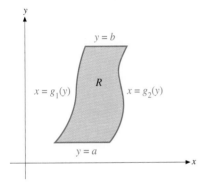

FIGURE 14.3.3 This region is horizontally simple.

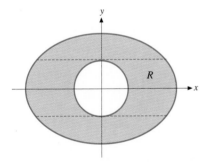

FIGURE 14.3.4 This region is finitely horizontally simple.

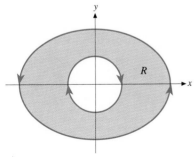

FIGURE 14.3.5 The region is on the left as the boundary is traced if outside boundary curves are traced counterclockwise and interior boundary curves are traced clockwise.

This result was obtained simply by using the Fundamental Theorem to evaluate the inner integral, with respect to y, of the iterated integral that corresponds to the double integral.

If a region is divided into subregions by cutting along vertical lines, the total boundary of the subregions consists of the boundary of the original region and the vertical cutting lines. Since the integral of $M\,dx$ over vertical lines is zero, the integral of $M\,dx$ over the original boundary is equal to the sum of the integrals over the boundaries of the subregions. The double integral of $-M_y$ over the region is also equal to the sum of the integrals over the subregions. It follows that (1) holds for regions that can be divided by vertical lines into a finite number of regions that are vertically simple. We call such regions **finitely vertically simple**. For example, the region R indicated in Figure 14.3.2 is finitely vertically simple; the two vertical dotted lines divide the region into four vertically simple regions.

If R is a horizontally simple region, as indicated in Figure 14.3.3, and $N(x, y)$ has continuous partial derivatives, we can verify that

$$\int_C N\,dy = \iint_R N_x\,dA.$$

This result holds if the region is **finitely horizontally simple**—that is, if the region can be divided by horizontal lines into a finite number of horizontally simple regions. For example, the region R in Figure 14.3.4 is finitely horizontally simple.

Let us say a region in the plane is **finitely simple** if it is both finitely vertically simple and finitely horizontally simple. Then both of the above results are true. We combine the results in the following theorem.

GREEN'S THEOREM

Suppose R is a finitely simple region in the plane and C is the complete boundary of R with the direction of each part chosen so R is on the left as the boundary is traced. If M, N, and their first partial derivatives are continuous on R and its boundary, then

$$\int_C M\,dx + N\,dy = \iint_R (N_x - M_y)\,dA.$$

If the region is to be on the left as the boundary is traced, then outside boundary curves are traced counterclockwise, while interior boundary curves are traced clockwise. See Figure 14.3.5.

Using the Formula

Green's Theorem relates a line integral along the complete boundary of a region to a double integral over the region.

EXAMPLE 1

Determine whether Green's Theorem can be applied directly to evaluate

$$\int_C y^2\,dx + 6xy\,dy,$$

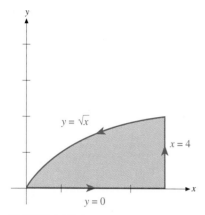

FIGURE 14.3.6 The region is on the left as the boundary curve is traced.

where C is the boundary curve of the region bounded by $y = \sqrt{x}$, $y = 0$, and $x = 4$, in the counterclockwise direction. Evaluate the line integral.

SOLUTION The region is sketched in Figure 14.3.6. We have

$$\int_C y^2 dx + 6xy\,dy = \int_C M\,dx + N\,dy,$$

so $M(x, y) = y^2$ and $N(x, y) = 6xy$. We then have $M_y = 2y$ and $N_x = 6y$. We see that M, N, and their first partial derivatives are continuous at all points of C and at all points of the region enclosed by C, so we can use Green's Theorem to evaluate the line integral.

We can either evaluate the line integral as the sum of the line integrals over each of the three smooth boundary curves of the region or use Green's Theorem. Let us use Green's Theorem. We note that the direction of the boundary curve is such that the region is on the left as the boundary curve is traced. We then have

$$\int_C y^2 dx + 6xy\,dy = \int_C M\,dx + N\,dy$$

$$= \iint_R (N_x - M_y)\,dA$$

$$= \iint_R (6y - 2y)\,dA$$

$$= \iint_R 4y\,dA$$

$$= \int_0^4 \int_0^{\sqrt{x}} 4y\,dy\,dx$$

$$= \int_0^4 2y^2 \Big]_{y=0}^{y=\sqrt{x}} dx$$

$$= \int_0^4 2x\,dx = x^2 \Big]_0^4 = 16. \quad \blacksquare$$

EXAMPLE 2

Determine whether Green's Theorem can be applied directly to evaluate

$$\int_C \frac{-y}{2 - x^2 - y^2}\,dx + \frac{x}{2 - x^2 - y^2}\,dy,$$

and evaluate the integral, where

(a) $C: x = \cos t,\ y = \sin t,\ 0 \le t \le 2\pi$,
(b) $C: x = \sqrt{3}\cos t,\ y = \sqrt{3}\sin t,\ 0 \le t \le 2\pi$.

SOLUTION We have

$$M = \frac{-y}{2 - x^2 - y^2} \quad \text{and} \quad N = \frac{x}{2 - x^2 - y^2}.$$

Using the Quotient Rule to differentiate, we then have

$$M_y = \frac{\partial}{\partial y}\left(\frac{-y}{2 - x^2 - y^2}\right) = \frac{(2 - x^2 - y^2)(-1) - (-y)(-2y)}{(2 - x^2 - y^2)^2}$$

$$= \frac{-2 + x^2 - y^2}{(2 - x^2 - y^2)^2}$$

and

$$N_x = \frac{2 + x^2 - y^2}{(2 - x^2 - y^2)^2}.$$

We see that M, N, and their first partial derivatives are continuous, except at points on the circle $x^2 + y^2 = 2$.

(a) C encloses points (x, y) with $x^2 + y^2 \le 1$. See Figure 14.3.7a. Since M and N have continuous first partial derivatives at these points, Green's Theorem can be used to evaluate the line integral. We could also use the parametric equations of C to evaluate the line integral. Let us use Green's Theorem, although that may not necessarily be easier than evaluating the line integral directly. We see that the region is on the left as the boundary curve is traced. We have

$$\int_C \frac{-y}{2 - x^2 - y^2} dx + \frac{x}{2 - x^2 - y^2} dy$$

$$= \int_C M\,dx + N\,dy$$

$$= \iint_R (N_x - M_y)\,dA$$

$$= \iint_R \left(\frac{2 + x^2 - y^2}{(2 - x^2 - y^2)^2} - \frac{-2 + x^2 - y^2}{(2 - x^2 - y^2)^2} \right) dA$$

$$= \iint_R \frac{4}{(2 - x^2 - y^2)^2}\,dA$$

$$= \int_0^{2\pi} \int_0^1 \frac{4}{(2 - r^2)^2}\,r\,dr\,d\theta \qquad \text{[in polar coordinates]}$$

$$= \int_0^{2\pi} \frac{2}{2 - r^2}\bigg]_{r=0}^{r=1} d\theta$$

$$= \int_0^{2\pi} (2 - 1)\,d\theta = 2\pi.$$

(b) C encloses points with $x^2 + y^2 \le 3$. See Figure 14.3.7b. Since this region includes points where the first partial derivatives of M and N do not exist, we cannot use Green's Theorem to evaluate the line integral. (The integral $\iint_R (N_x - M_y)\,dA$ would be a divergent improper integral in this example.) We can evaluate the line integral by using the parametric equations of C. We have

$$x = \sqrt{3}\cos t, \quad y = \sqrt{3}\sin t, \quad dx = -\sqrt{3}\sin t\,dt, \quad dy = \sqrt{3}\cos t\,dt.$$

Substitution then gives

$$\int_C \frac{-y}{2 - x^2 - y^2}\,dx + \frac{x}{2 - x^2 - y^2}\,dy$$

$$= \int_0^{2\pi} \left[\left(\frac{-\sqrt{3}\sin t}{2 - 3} \right)(-\sqrt{3}\sin t) + \left(\frac{\sqrt{3}\cos t}{2 - 3} \right)(\sqrt{3}\cos t) \right] dt$$

$$= \int_0^{2\pi} -3\,dt = -3(2\pi) = -6\pi. \quad \blacksquare$$

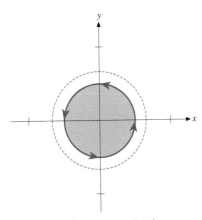

FIGURE 14.3.7a M and N have continuous first partial derivatives at all points on and enclosed by the boundary curve, so we can use Green's Theorem to evaluate the line integral over the complete boundary of the region.

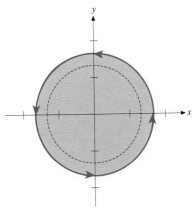

FIGURE 14.3.7b This region includes points where the first partials of M and N do not exist, so we cannot use Green's Theorem.

NOTE   *We cannot use Green's Theorem to evaluate the line integral if there is a point inside the region R where either* **F** *or one of its first partial derivatives is discontinuous, even if* **F** *and its first derivatives are continuous on C.*

If Green's Theorem applies, it gives us the *option* of evaluating a particular line integral by evaluating the corresponding double integral. It may not necessarily be easier to evaluate the double integral than it is to evaluate the line integral directly.

Area

Green's Theorem gives us the following formulas for the area of a region in terms of a line integral over the boundary curve of the region.

$$\text{Area} = \iint_R dA = \int_C x\,dy = \int_C -y\,dx = \frac{1}{2}\int_C x\,dy - y\,dx.$$

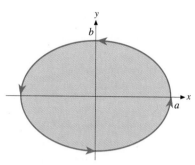

FIGURE 14.3.8 The boundary curve is parametrized so the region is to the left as the boundary is traced.

EXAMPLE 3

Find the area inside the ellipse $\dfrac{x^2}{a^2} + \dfrac{y^2}{b^2} = 1$.

SOLUTION   The ellipse is sketched in Figure 14.3.8. A parametric representation of the ellipse with direction so the region enclosed is on the left as the boundary is traced is given by

$$C: x = a\cos t, \quad y = b\sin t, \quad 0 \le t \le 2\pi.$$

We use the formula

$$\text{Area} = \frac{1}{2}\int_C x\,dy - y\,dx,$$

which turns out to give a very simple integral. We have

$$\begin{aligned}
\text{Area} &= \frac{1}{2}\int_C x\,dy - y\,dx \\
&= \frac{1}{2}\int_0^{2\pi} [(a\cos t)(b\cos t) - (b\sin t)(-a\sin t)]\,dt \\
&= \frac{1}{2}\int_0^{2\pi} ab\,dt = ab\pi. \quad \blacksquare
\end{aligned}$$

Scalar Curl and Circulation

We have seen in the previous section that, if $\mathbf{F} = M\mathbf{i} + N\mathbf{j}$, the circulation of **F** around a simple closed curve C is

$$\int_C \mathbf{F} \cdot \mathbf{T}\,ds = \int_C M\,dx + N\,dy. \tag{2}$$

If Green's Theorem applies to the function **F** and the region R bounded by a simple closed curve C, we can express the circulation of **F** around C as the integral of $N_x - M_y$ over R. That is,

$$\int_C \mathbf{F} \cdot \mathbf{T}\,ds = \int_C M\,dx + N\,dy = \iint_R (N_x - M_y)\,dA. \tag{3}$$

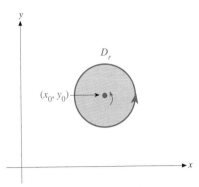

FIGURE 14.3.9 We consider the circulation around the boundary of a small disk with center at (x_0, y_0).

The expression $N_x - M_y$ can be related to the circulation of $\mathbf{F} = M\mathbf{i} + N\mathbf{j}$. To do this, let C_r denote the boundary of a disk D_r with center at a fixed point (x_0, y_0) and radius r, C_r directed counterclockwise around D_r. See Figure 14.3.9. Let us assume that $\mathbf{F}$ and its partial derivatives are continuous on some fixed disk with center (x_0, y_0). If r is sufficiently small, we then have

$$N_x(x, y) - M_y(x, y) \approx N_x(x_0, y_0) - M_y(x_0, y_0) \text{ for all } (x, y) \text{ on } C_r.$$

Then

$$\int_{C_r} \mathbf{F} \cdot \mathbf{T}\, ds = \iint_{D_r} (N_x - M_y)\, dA$$
$$\approx [N_x(x_0, y_0) - M_y(x_0, y_0)][\text{Area of } D_r].$$

This suggests that we think of the value of $N_x - M_y$ at a point as the circulation of $\mathbf{F}$ per unit area at the point. The expression $N_x - M_y$ is sometimes called the **scalar curl** of $\mathbf{F}$.

We may think of each point with nonzero curl as a vortex that contributes to the circulation around any curve that encircles the point. See Figure 14.3.10. Green's Theorem says the circulation around a curve C is the sum of the circulation around each point in the region enclosed by C. That is,

$$\underbrace{\int_C \mathbf{F} \cdot \mathbf{T}\, ds}_{\text{Circulation around } C} = \underbrace{\iint_R}_{\text{(Sum)}} \underbrace{\underbrace{(N_x - M_y)}_{\text{(Circulation per unit area)}} \underbrace{dA}_{\text{(Area)}}}_{\text{(Circulation around point)}}.$$

FIGURE 14.3.10a Circulation around a point contributes to the circulation around any curve that encircles the point.

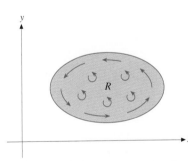

FIGURE 14.3.10b $\iint_R(N_x - M_y)\, dA$ represents the sum of the circulation around each point in the region R.

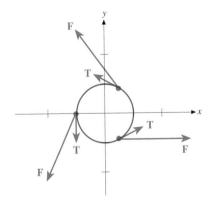

FIGURE 14.3.11 The representative vectors $\mathbf{F}$ and $\mathbf{T}$ indicate a counterclockwise, positive circulation.

EXAMPLE 4

Find the scalar curl of $\mathbf{F} = (x - 2y)\mathbf{i} + (2x + y)\mathbf{j}$. Evaluate the circulation of $\mathbf{F}$ around the curve

$$C: x = \cos t, \quad y = \sin t, \quad 0 \le t \le 2\pi.$$

SOLUTION The curve and some representative vectors $\mathbf{F}$ are sketched in Figure 14.3.11. We have

$$\mathbf{F} = (x - 2y)\mathbf{i} + (2x + y)\mathbf{j} = M\mathbf{i} + N\mathbf{j}, \text{ so}$$
$$M = x - 2y \text{ and } N = 2x + y.$$

The scalar curl is $N_x - M_y = 2 - (-2) = 4$, a constant function.

Since **F** and its first partial derivatives are continuous at all points, we can use Green's Theorem to evaluate the circulation of **F** around C. The region enclosed by C is on the left as C is traced, so

$$\int_C \mathbf{F} \cdot \mathbf{T}\, ds = \int_C M\, dx + N\, dy$$
$$= \iint_R (N_x - M_y)\, dA$$
$$= \iint_R 4\, dA$$
$$= 4 \left(\iint_R dA \right) \qquad \text{[since 4 is a \textit{constant}]}$$
$$= 4(\text{Area of } R) = 4\pi.$$

This agrees with the result of Example 3 of Section 14.2, where we calculated the circulation by using the parametric equations of C to evaluate the line integral. ∎

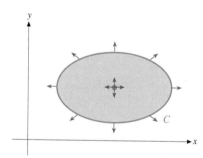

FIGURE 14.3.12 **F** · **n** measures the outward flux across C.

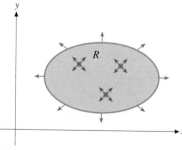

FIGURE 14.3.13a The divergence at a point contributes to the divergence across any curve that encircles the point.

Divergence and Flux

If C is the boundary curve of a region R, and if R is on the left as C is traced, then the outer unit normal is

$$\mathbf{n} = \frac{dy}{ds}\mathbf{i} - \frac{dx}{ds}\mathbf{j}.$$

See Figure 14.3.12. The **outward flux** of **F** across C is

$$\int_C \mathbf{F} \cdot \mathbf{n}\, ds = \int_C M\, dy - N\, dx. \qquad (4)$$

In cases where Green's Theorem applies, we have

$$\int_C \mathbf{F} \cdot \mathbf{n}\, ds = \int_C M\, dy - N\, dx = \iint_R (M_x + N_y)\, dA. \qquad (5)$$

The expression $M_x + N_y$ is called the **divergence** of **F**. We can think of the value of $M_x + N_y$ at a point as the flux per unit area at the point. Positive divergence at a point indicates that there is a net flow from the point, or a source at the point; negative divergence indicates a net flow into the point, or a sink at the point. See Figure 14.3.13a. Green's Theorem in the form of (5) then says that the outward flux of **F** over C is the sum of the flux per unit area (divergence) of **F** over the region enclosed by C. That is, the flow over the boundary of R is the net sum of the flow at each point of R. See Figure 14.3.13b. We have

$$\underbrace{\int_C \mathbf{F} \cdot \mathbf{n}\, ds}_{\substack{\text{Outward flux of } \mathbf{F} \text{ across } C}} = \iint_R \underbrace{(M_x + N_y)}_{\substack{\text{(Flux per unit area)} \\ \text{(Sum)}}} \underbrace{dA.}_{\substack{\text{(Area)} \\ \text{(Outward flux at point)}}}$$

FIGURE 14.3.13b $\iint_R (M_x + N_y)\, dA$ represents the sum of the divergence at each point in the region R.

NOTE *Do not use Formula 5 to evaluate the outward flux of* **F** *across* C *unless Green's Theorem applies.*

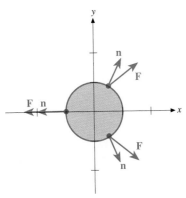

FIGURE 14.3.14 The representative vectors **F** and **n** indicate an outward flux of **F** across C.

EXAMPLE 5

Find the divergence of $\mathbf{F} = 2x\mathbf{i} + y\mathbf{j}$. Evaluate the outward flux of **F** across the curve

$$C: x = \cos t, \quad y = \sin t, \quad 0 \le t \le 2\pi.$$

SOLUTION The curve and some representative vectors **F** are sketched in Figure 14.3.14. We have

$$\mathbf{F} = 2x\mathbf{i} + y\mathbf{j} = M\mathbf{i} + N\mathbf{j}, \text{ so}$$
$$M = 2x, \quad N = y, \quad M_x = 2, \text{ and } N_y = 1.$$

The divergence is $M_x + N_y = 2 + 1 = 3$.

Since **F** and its first partial derivatives are continuous at all points, we can use Green's Theorem to evaluate the outward flux of **F** across C. Formula 5 implies

$$\int_C \mathbf{F} \cdot \mathbf{n}\, ds = \int_C M\, dy - N\, dx$$
$$= \int\int_R (M_x + N_y)\, dA$$
$$= \int\int_R 3\, dA$$
$$= 3\left(\int\int_R dA\right) \qquad \text{[since 3 is a \textit{constant}]}$$
$$= 3(\text{Area of } R) = 3\pi.$$

This agrees with the result of Example 4 of Section 14.2, where we calculated the flux by using the parametric equations of C to evaluate the line integral. ∎

EXAMPLE 6

Find the divergence of

$$\mathbf{F} = \frac{-x}{x^2 + y^2}\mathbf{i} + \frac{-y}{x^2 + y^2}\mathbf{j}.$$

Evaluate the outward flux of **F** across the curve

$$C_r: x = r\cos t, \quad y = r\sin t, \quad 0 \le t \le 2\pi.$$

SOLUTION The curve and some representative vectors of **F** are sketched in Figure 14.3.15.

We have

$$\mathbf{F} = \frac{-x}{x^2 + y^2}\mathbf{i} + \frac{-y}{x^2 + y^2}\mathbf{j} = M\mathbf{i} + N\mathbf{j}, \text{ so}$$
$$M = \frac{-x}{x^2 + y^2} \text{ and } N = \frac{-y}{x^2 + y^2}.$$

Using the Quotient Rule, we have

$$M_x = \frac{(x^2 + y^2)(-1) - (-x)(2x)}{(x^2 + y^2)^2} = \frac{x^2 - y^2}{(x^2 + y^2)^2}.$$

Similarly, we obtain

$$N_y = \frac{y^2 - x^2}{(x^2 + y^2)^2}.$$

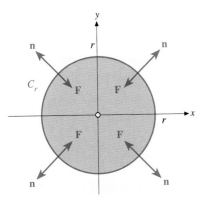

FIGURE 14.3.15 The representative vectors **F** and **n** indicate an inward, negative flux of **F** across C_r. We cannot apply Formula 5 because the region enclosed by C_r contains the origin, a point where **F** is not continuous.

The divergence is then

$$M_x + N_y = 0, \quad (x, y) \neq (0, 0).$$

Note that the function

$$\mathbf{F} = \frac{-x}{x^2 + y^2}\mathbf{i} + \frac{-y}{x^2 + y^2}\mathbf{j}$$

is discontinuous at $(0, 0)$, a point in the region enclosed by C. (Since the limit of $\mathbf{F}$ at $(0, 0)$ does not exist, no choice of $\mathbf{F}(0, 0)$ would give a function that is continuous at $(0, 0)$.) Since Green's Theorem requires that $\mathbf{F}$ and its first partial derivatives be continuous on C_r and in the region enclosed by C_r, we cannot apply Formula 5 to evaluate the line integral that gives the flux of $\mathbf{F}$ over the curve C_r. We must use Formula 4 to evaluate the flux directly as a line integral. We then have

$$\int_{C_r} \mathbf{F} \cdot \mathbf{n}\,ds = \int_{C_r} M\,dy - N\,dx$$

$$= \int_{C_r} \frac{-x}{x^2 + y^2}\,dy - \frac{-y}{x^2 + y^2}\,dx$$

$$= \int_0^{2\pi} \left[\left(\frac{-r\cos t}{(r\cos t)^2 + (r\sin t)^2} \right)(r\cos t) \right.$$

$$\left. - \left(\frac{-r\sin t}{(r\cos t)^2 + (r\sin t)^2} \right)(-r\sin t) \right] dt$$

$$= \int_0^{2\pi} (-1)\,dt = -2\pi.$$

We see that the flux is the same for all positive values of r. The negative flux indicates that there is a net flow into the region bounded by C_r. ∎

Even though $M_x + N_y = 0$ at every point except the origin in Example 6, $\int_C \mathbf{F} \cdot \mathbf{n}\,ds \neq \iint_R 0\,dA$ for curves C that enclose the origin. The flux across any simple closed curve that does not enclose the origin is zero.

EXERCISES 14.3

Determine whether Green's Theorem can be applied directly to evaluate the line integrals in Exercises 1–10. Evaluate the integrals.

1. $\int_C x\,dx + xy\,dy$; C is the boundary curve of the region in the first quadrant bounded by $2x + 3y = 6$, in the counterclockwise direction.

2. $\int_C x^2y\,dx + y\,dy$; C is the boundary curve of the region bounded by $y = x$, $y = 0$, and $x = 2$, in the counterclockwise direction.

3. $\int_C y\,dx + xy\,dy$; C is the boundary curve of the region bounded by $y = x^2$, $y = 0$, and $x = 2$, in the counterclockwise direction.

4. $\int_C xy^2\,dx + x^2y\,dy$; C is the boundary curve of the region bounded by $y = x^2$, $x + y = 2$, and $x = 0$, in the counterclockwise direction.

5. $\int_C (x^2 + y^2)\,dx + 2xy\,dy$; $C: x = \cos t, y = \sin t$, $0 \leq t \leq \pi$

6. $\int_C xy\,dx + dy$; $C: x = t, y = t^2, 0 \leq t \leq 2$

7. $\int_C -y\ln(x^2 + y^2)\,dx + x\ln(x^2 + y^2)\,dy$; $C: x = 2\cos t$, $y = 2\sin t, 0 \leq t \leq 2\pi$

8. $\int_C \frac{2xy}{(x^2 + y^2)^2}\,dx + \frac{y^2 - x^2}{(x^2 + y^2)^2}\,dy$; $C: x = \cos t, y = 2 + \sin t, 0 \leq t \leq 2\pi$

9. $\int_C \ln(x^2 + 1) \, dx - x \, dy$; C is the boundary curve of the region bounded by $y = \sqrt{4 - x^2}$ and $y = 0$, in the counterclockwise direction.

10. $\int_C (y + \sin x) \, dx - (x + \cos y) \, dy$; $C: x = \cos t, y = \sin t$, $0 \le t \le 2\pi$

In Exercises 11–12, verify the formulas for the coordinates of the center of mass of a homogeneous lamina in terms of line integrals. A denotes the area of the region occupied by the lamina.

11. $\bar{x} = \dfrac{1}{A} \int\!\!\int_R x \, dA = \dfrac{1}{2A} \int_C x^2 \, dy$

12. $\bar{y} = \dfrac{1}{A} \int\!\!\int_R y \, dA = \dfrac{-1}{2A} \int_C y^2 \, dx$

13. **(a)** Use the formula Area $= \int_C x \, dy$ to find the area of the region bounded by $y = \tan^{-1} x$, $y = -\tan^{-1} x$, and $x = 1$.

(b) Use the line integrals in Exercises 11 and 12 to find the center of mass of a homogeneous lamina that occupies the region in (a).

14. **(a)** Use the formula Area $= \int_C x \, dy$ to find the area of the region bounded by $y = \ln x$, $y = -\ln x$, and $x = e$.

(b) Use the line integrals in Exercises 11 and 12 to find the center of mass of a homogeneous lamina that occupies the region in (a).

In Exercises 15–20, find the scalar curl of the given vector function **F** *and the circulation of* **F** *around the indicated curve.*

15. $\mathbf{F} = x\mathbf{i} + y\mathbf{j}$; $C: x = h + r \cos t, y = k + r \sin t$, $0 \le t \le 2\pi$

16. $\mathbf{F} = -y\mathbf{i} + x\mathbf{j}$; $C: x = h + r \cos t, y = k + r \sin t$, $0 \le t \le 2\pi$

17. $\mathbf{F} = [1/(y^2 + 1)]\mathbf{i}$; C is the polygonal path from $(0, 0)$ to $(1, 0)$, to $(1, 1)$, to $(0, 1)$, to $(0, 0)$

18. $\mathbf{F} = [1/(y^2 + 1)]\mathbf{i}$; C is the polygonal path from $(-1, -1)$ to $(1, -1)$, to $(1, 1)$, to $(-1, 1)$, to $(-1, -1)$

19. $\mathbf{F} = e^x \sin y \mathbf{i} + e^x \cos y \mathbf{j}$; $C: x = \cos t, y = \sin t$, $0 \le t \le 2\pi$

20. $\mathbf{F}(x, y) = \begin{cases} y^2\mathbf{i}, & y \ge 0, \\ \mathbf{0}, & y < 0; \end{cases}$ $C: x = \cos t, y = \sin t$, $0 \le t \le 2\pi$

In Exercises 21–26, find the divergence of the given vector function **F** *and find the outward flux of* **F** *across the indicated curve.*

21. $\mathbf{F} = x\mathbf{i} + y\mathbf{j}$; $C: x = h + r \cos t, y = k + r \sin t$, $0 \le t \le 2\pi$

22. $\mathbf{F} = -y\mathbf{i} + x\mathbf{j}$; $C: x = h + r \cos t, y = k + r \sin t$, $0 \le t \le 2\pi$

23. $\mathbf{F} = (2x - y)\mathbf{i} + (x + 3y)\mathbf{j}$; $C: x = \cos t, y = \sin t$, $0 \le t \le 2\pi$

24. $\mathbf{F} = (3x + 2y)\mathbf{i} + (x - 2y)\mathbf{j}$; $C: x = \cos t, y = \sin t$, $0 \le t \le 2\pi$

25. $\mathbf{F} = \dfrac{x}{(x^2 + y^2)^{3/2}}\mathbf{i} + \dfrac{y}{(x^2 + y^2)^{3/2}}\mathbf{j}$; $C: x = 2\cos t, \ y = 2\sin t, \ 0 \le t \le 2\pi$

26. $\mathbf{F} = \dfrac{x^2 - y^2}{(x^2 + y^2)^2}\mathbf{i} + \dfrac{2xy}{(x^2 + y^2)^2}\mathbf{j}$; $C: x = \cos t, \ y = \sin t$, $0 \le t \le 2\pi$

27. Show that the circulation of a constant vector field around any simple closed curve is zero.

28. Show that the outward flux of a constant vector field across any simple closed curve is zero.

29. Show that the outward flux of

$$\mathbf{F} = \frac{-x}{x^2 + y^2}\mathbf{i} + \frac{-y}{x^2 + y^2}\mathbf{j}$$

across any simple closed curve that contains the origin in its interior is -2π. (*Hint:* Use the result of Example 7. Choose r so small that C_r is inside C, and then apply Green's Theorem to the region between C and C_r.)

30. Evaluate $\int_C \cos x \, dx + \sin y \, dy$, where $C: x = \cos t, y = \sin t, \ 0 \le t \le \pi$. (*Hint:* Apply Green's Theorem to the region enclosed by C and the line segment from $(-1, 0)$ to $(1, 0)$.)

14.4 INDEPENDENCE OF PATH OF LINE INTEGRALS IN THE PLANE

Connections

Partial derivatives, 12.3.

Green's theorem, 14.3.

We have seen that applications of line integrals lead to integrals of the form

$$\int_C M \, dx + N \, dy.$$

An integral of this type is evaluated by substitution of parametric values. We know that the value of the integral does not depend on the particular (smooth) parametric representation of C, except for the direction in which C is traced as the parameter increases. We will now investigate conditions under which the value of the integral depends only on the initial and terminal points of C and not on the path between these points. Many line integrals that occur in applications are of this type.

Conditions on *M, N,* and the Domain

For given functions M and N that are defined in a plane region R, it is desirable to know if the value of the line integral $\int_C M\,dx + N\,dy$ depends only on the initial and terminal points of C and not on the particular path taken within a region R. Such line integrals are said to be **independent of path** in R.

THEOREM 1

(a) Suppose C_1 and C_2 are curves that have the same initial and terminal points. If M, N, and their first partial derivatives are continuous with

$$M_y = N_x$$

at all points of C_1 and C_2 and the region between the curves, then

$$\int_{C_1} M\,dx + N\,dy = \int_{C_2} M\,dx + N\,dy.$$

(b) Suppose the plane region R has the property that whenever C' is a closed curve in R, the region enclosed by C' is also in R. If M, N, and their first partial derivatives are continuous with $M_y = N_x$ in R, then $\int_C M\,dx + N\,dy$ is independent of path in R.

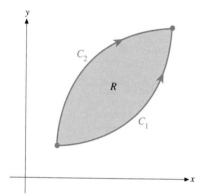

FIGURE 14.4.1 We can apply Green's Theorem to the region enclosed by $C_1 - C_2$.

Proof Statement (a) of Theorem 1 is a consequence of Green's Theorem applied to the region R bounded by $C = C_1 - C_2$, as illustrated in Figure 14.4.1. That is,

$$\int_{C_1} M\,dx + N\,dy - \int_{C_2} M\,dx + N\,dy = \int_C M\,dx + N\,dy$$
$$= \iint_R (N_x - M_y)\,dA = 0.$$

Statement (b) follows from part (a). Note that statement (b) does not apply to regions that have "holes," as illustrated in Figure 14.4.2. ∎

If a line integral is independent of path in a region, it may be easier to evaluate the integral by replacing the given curve with a more convenient path that has the same initial and terminal points. We also have the following result.

If $\int_C M\,dx + N\,dy$ is independent of path in R and C is a closed curve in R, then $\int_C M\,dx + N\,dy = 0$.

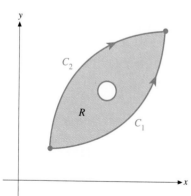

FIGURE 14.4.2 If a region R has holes, then $C_1 - C_2$ may not be the complete boundary of that part of R it encloses.

EXAMPLE 1

Evaluate $\int_C (x - 3y + 2)\,dx - (3x + 2y - 4)\,dy$, where C is

(a) C_1: The polygonal path from $(0, -1)$ to $(1, 0)$, to $(0, 1)$,

(b) C_2: The polygonal path from $(0, -1)$ to $(1, 0)$, to $(0, 1)$, to $(-1, 0)$, to $(0, -1)$.

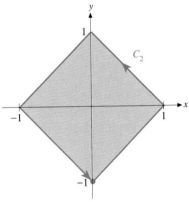

FIGURE 14.4.3a If an integral is independent of path, we may replace a given curve by a simpler curve.

FIGURE 14.4.3b If an integral is independent of path in *R*, then its integral over a closed path in *R* is zero.

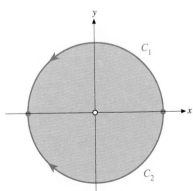

FIGURE 14.4.4 The region enclosed by $C_1 - C_2$ contains the origin, a point where M_y and N_x do not exist.

SOLUTION We have $M = x - 3y + 2$ and $N = -3x - 2y + 4$. This implies $M_y = -3 = N_x$ for all (x, y), so the line integrals are independent of path in the entire xy-plane.

(a) We replace C_1 by the path C_1' from $(0, -1)$ to $(0, 1)$ along the y-axis. See Figure 14.4.3a. This curve has parametric representation

$$C_1': x = 0, \quad y = t, \quad -1 \le t \le 1.$$

Note that $dx = 0$. Then

$$\int_{C_1} (x - 3y + 2)\,dx - (3x + 2y - 4)\,dy$$

$$= \int_{C_1'} (x - 3y + 2)\,dx - (3x + 2y - 4)\,dy$$

$$= 0 - \int_{-1}^{1} [3(0) + 2t - 4]\,dt$$

$$= -t^2 + 4t \bigg]_{-1}^{1} = 8.$$

(b) From Figure 14.4.3b, we see that C_2 is a closed curve, so

$$\int_{C_2} (x - 3y + 2)\,dx - (3x + 2y - 4)\,dy = 0. \quad \blacksquare$$

Theorem 1 does not guarantee that two line integrals over curves that have the same initial and terminal points have equal values unless $M_y = N_x$ for *every* point in the region enclosed by the curves.

EXAMPLE 2
Evaluate

$$\int_C \frac{-y}{x^2 + y^2}\,dx + \frac{x}{x^2 + y^2}\,dy,$$

where C is

(a) C_1: $x = \cos t, y = \sin t, 0 \le t \le \pi$,
(b) C_2: $x = \cos t, y = -\sin t, 0 \le t \le \pi$.

SOLUTION The curves are sketched in Figure 14.4.4. We see that the curves have the same initial and terminal points. Let us check the condition $M_y = N_x$ for independence of path. Using the Quotient Rule for differentiation, we have

$$M_y = \frac{(x^2 + y^2)(-1) - (-y)(2y)}{(x^2 + y^2)^2} = \frac{y^2 - x^2}{(x^2 + y^2)^2} \text{ and}$$

$$N_x = \frac{(x^2 + y^2)(1) - (x)(2x)}{(x^2 + y^2)^2} = \frac{y^2 - x^2}{(x^2 + y^2)^2}.$$

It follows that $M_y = N_x$ for $(x, y) \ne (0, 0)$. (Even if M and N were assigned values at the origin, M_y and N_x would not exist at the origin.) Since the region enclosed by C_1 and C_2 contains the point where M_y and N_x do not exist, we do not have $M_y = N_x$ for all points between the curves, so Theorem 1 does not guarantee that the values of the line integrals of **F** over these two curves will be the same.

(a) We have

$$\int_{C_1} M\,dx + N\,dy$$

$$= \int_0^\pi \left[\left(\frac{-\sin t}{\cos^2 t + \sin^2 t} \right)(-\sin t) + \left(\frac{\cos t}{\cos^2 t + \sin^2 t} \right)(\cos t) \right] dt$$

$$= \int_0^\pi 1\,dt = \pi.$$

(b) Similarly,

$$\int_{C_2} M\,dx + N\,dy$$

$$= \int_0^\pi \left[\left(\frac{-(-\sin t)}{\cos^2 t + \sin^2 t} \right)(-\sin t) + \left(\frac{\cos t}{\cos^2 t + \sin^2 t} \right)(-\cos t) \right] dt$$

$$= \int_0^\pi -1\,dt = -\pi. \ \blacksquare$$

The line integral in Example 2 is not independent of path in regions that contain the origin, a point where M_y and N_x do not exist. The integral is independent of path in regions that do not contain the origin and do not contain any holes. For example, the integral is independent of path in the region that consists of all points in the xy-plane except for $(x, 0)$ with $x \le 0$. See Figure 14.4.5. Closed curves in this region cannot encircle the origin.

FIGURE 14.4.5 The region that consists of all points in the xy-plane except for the origin and the negative x-axis contains no holes. No curve in this region can encircle the origin.

Conservative Vector Fields

We have seen that if M_y and N_x are continuous with $M_y = N_x$ in a region R that has no holes, then $\int_C M\,dx + N\,dy$ depends only on the initial and terminal points of C, as long as C is contained in R. The following theorem gives a different characterization of independence of path.

THEOREM 2

Suppose M and N are continuous in a region R.
(a) If there is a function U such that $\nabla U = M\mathbf{i} + N\mathbf{j}$ in R, then $\int_C M\,dx + N\,dy$ is independent of path in R. Moreover, if C is a smooth curve in R with initial point $(x(a), y(a))$ and terminal point $(x(b), y(b))$, then

$$\int_C M\,dx + N\,dy = U(x(b), y(b)) - U(x(a), y(a)).$$

(b) If $\int_C M\,dx + N\,dy$ is independent of path in R, then there is a function U such that $\nabla U = M\mathbf{i} + N\mathbf{j}$ in R.

Proof   To prove part (a) of this theorem, let us consider a smooth curve

$$C: x = x(t), \quad y = y(t), \quad a \le t \le b,$$

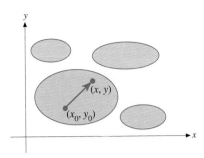

FIGURE 14.4.6   Each curve in R must remain in a single component of R. Components that cannot be connected by a path in R must be treated separately.

where C is contained in R. Then $U = U(x, y) = U(x(t), y(t))$ is a function of t, and the Chain Rule gives

$$\frac{dU}{dt} = \frac{\partial U}{\partial x}\frac{dx}{dt} + \frac{\partial U}{\partial y}\frac{dy}{dt}.$$

Since $\nabla U = (\partial U/\partial x)\mathbf{i} + (\partial U/\partial y)\mathbf{i} = M\mathbf{i} + N\mathbf{j}$, we have $M = \partial U/\partial x$ and $N = \partial U/\partial y$. Then

$$\int_C M\,dx + N\,dy = \int_a^b \left(\frac{\partial U}{\partial x}\frac{dx}{dt} + \frac{\partial U}{\partial y}\frac{dy}{dt} \right) dt$$

$$= \int_a^b \frac{dU}{dt}dt = U(x(b), y(b)) - U(x(a), y(a)),$$

where the last equality is a consequence of the Fundamental Theorem of Calculus.

To verify part (b) of the theorem, we assume that $\int_C M\,dx + N\,dy$ is independent of path in R. For a fixed point (x_0, y_0) in R, we define a function U by

$$U(x, y) = \int_{C(x,y)} M\,dx + N\,dy,$$

where $C(x, y)$ is any curve in R from (x_0, y_0) to (x, y). We are assuming that the line integral is independent of path, so the value $U(x, y)$ does not depend on the path of $C(x, y)$ from (x_0, y_0) to (x, y), as long as $C(x, y)$ is in R. If R has components that cannot be connected to (x_0, y_0) by a path in R, then we must treat each such component separately. Each curve in R must remain in a single component of R. See Figure 14.4.6.

Let us show that $U_x(c, d) = M(c, d)$. The partial derivative $U_x(c, d)$ depends only on the values of U on the horizontal line $y = d$. The values $U(x, d)$ can be determined by choosing $C(x, d)$ to be any path from (x_0, y_0) to a fixed point (x_1, d), and then along the line $y = d$ from (x_1, d) to (x, d). See Figure 14.4.7a.   Then

$$U(x, d) = \int_{C(x,d)} M\,dx + N\,dy$$

$$= \int_{C_1} M\,dx + N\,dy + \int_{C_2} M\,dx + N\,dy$$

$$= \int_{C_1} M\,dx + N\,dy + \int_{x_1}^x M(t, d)\,dt.$$

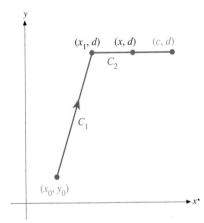

FIGURE 14.4.7a   We choose any path from (x_0, y_0) to a fixed point (x_1, d), and then approach (c, d) along the horizontal line from (x_1, d).

The first term above is a constant. The Fundamental Theorem of Calculus tells us the derivative of the second term with respect to x is $M(x, d)$. It follows that $U_x(x, d) = M(x, d)$, so $U_x(c, d) = M(c, d)$.

We can show that $U_y = N$ similarly, except we use the path indicated in Figure 14.4.7b.

We conclude that $M\mathbf{i} + N\mathbf{j} = U_x\mathbf{i} + U_y\mathbf{j} = \nabla U$.   ∎

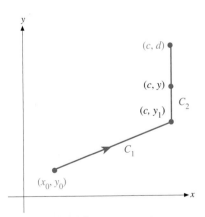

FIGURE 14.4.7b   We can also approach (c, d) along a path that is vertical near (c, d).

If there is a function U with $\nabla U = \mathbf{F}$, we say that $\mathbf{F}$ is a **conservative** vector field. The negative of U is called the **potential** of $\mathbf{F}$.

We know from Section 14.2 that if $\mathbf{F}$ is an external force that is the only force acting on an object, then the value of $\int_C \mathbf{F} \cdot \mathbf{T}\,ds$ is the change in kinetic energy of the object as it moves along C. That is, if C has initial point A and terminal point B, then

$$\int_C \mathbf{F} \cdot \mathbf{T}\,ds = \text{K.E.}(B) - \text{K.E.}(A),$$

where K.E. denotes kinetic energy. If $\mathbf{F}$ is *conservative* with potential function P, then Theorem 2 implies

$$\int_C \mathbf{F} \cdot \mathbf{T}\,ds = -P(B) + P(A).$$

It follows that the value of line integral is the change of **potential energy** of the object. Combining results, we have

$$\text{K.E.}(B) - \text{K.E.}(A) = -P(B) + P(A), \quad \text{so}$$
$$\text{K.E.}(B) + P(B) = \text{K.E.}(A) + P(A).$$

The last equality tells us that the sum of the kinetic energy and the potential energy is constant if $\mathbf{F}$ is a conservative force field. That is, energy is conserved in a conservative force field. Many electrical, magnetic, and gravitational fields are conservative.

If $\mathbf{F}$ is conservative in R, so $\int_C M\,dx + N\,dy$ is *independent of path* in R, we can use the notation

$$\int_A^B M\,dx + N\,dy$$

to denote the integral over any curve C in R that has initial point A and terminal point B. The integral can be evaluated either by choosing C to be any convenient curve with initial point A and terminal point B or by finding a function U with $\nabla U = \mathbf{F}$ and using the formula

$$\int_A^B M\,dx + N\,dy = U(B) - U(A).$$

Finding U such that $\nabla U = \mathbf{F}$

Theorem 2 tells us that $M\mathbf{i} + N\mathbf{j}$ is the gradient of a function U in R if and only if $\int_C M\,dx + N\,dy$ is independent of path in R. This raises two questions:

For a given vector function $\mathbf{F}$, how do we know when there is a function U with $\nabla U = \mathbf{F}$?

If such a function U exists, how do we find it?

We can combine Theorems 1 and 2 to answer the first question. That is, if M_y and N_x are continuous in a region R that has no holes and $M_y = N_x$ in R, then Theorem 1 tells us that $\int_C M\,dx + N\,dy$ is independent of path in R. Theorem 2 then implies there is a function U such that $\nabla U = \mathbf{F}$.

We might also note that if M_y and N_x are continuous in R, and if $\nabla U = \mathbf{F}$, then $M_y = N_x$. To see this, note that $U_x = M$ implies $U_{xy} = M_y$ and that $U_y = N$ implies $U_{yx} = N_x$. Since U_{xy} and U_{yx} are continuous, we know from Section 12.3 that $U_{xy} = U_{yx}$, so $M_y = N_x$.

Let us summarize the above remarks as a theorem.

THEOREM 3

If M_y and N_x are continuous in a region R that contains no holes, and $M_y = N_x$ in R, then there is a function U such that $\nabla U = \mathbf{F}$ in R, so $\mathbf{F}$ is conservative in R.

If M_y and N_x are continuous in R and there is a point of R where $M_y \neq N_x$, then there is no function U such that $\nabla U = \mathbf{F}$ in R, so $\mathbf{F}$ is not conservative in R.

Let us now turn to the second question above. Suppose that the vector function $\mathbf{F} = M\mathbf{i} + N\mathbf{j}$ has continuous first partial derivatives with $M_y = N_x$ in a region R that contains no holes. We then know that there is a function U such that the gradient of U is $\mathbf{F}$ in R. We want to determine a formula for $U(x, y)$. From the proof of Theorem 2b we know that, if $C(x, y)$ is any smooth curve in R with initial point (x_0, y_0) and terminal point (x, y), then we can take U to be

$$U(x, y) = \int_{C(x,y)} M\,dx + N\,dy. \tag{1}$$

This means that we can determine values of U by evaluating a line integral. We will illustrate a different technique for finding a function U with $\nabla U = \mathbf{F}$. If $\int_C M\,dx + N\,dy$ is not independent of path in R, the method we will illustrate does not determine a function U. If a particular path $C(x, y)$ to each point (x, y) were specified, then Formula 1 would determine a function U, but U would not satisfy $\nabla U = \mathbf{F}$ at all points of R. (See Exercise 40.)

EXAMPLE 3

(a) Show that the integral $\int_C -x\,dx + y\,dy$ is independent of path in the entire xy-plane.

(b) Find a function U such that $\nabla U = -x\mathbf{i} + y\mathbf{j}$.

(c) Evaluate $\displaystyle\int_{(1,0)}^{(3,4)} -x\,dx + y\,dy$.

SOLUTION

(a) We have $M = -x$ and $N = y$, so the functions M_y and N_x are continuous with $M_y = 0 = N_x$ in the entire xy-plane. The entire plane does not have any holes, so Theorem 1 implies $\int_C -x\,dx + y\,dy$ is independent of path.

(b) Theorem 3 implies that there is a function U such that $\nabla U = \mathbf{F}$ in the entire xy-plane. We need a formula for $U(x, y)$.

If $\nabla U = -x\mathbf{i} + y\mathbf{j}$, then $U_x = -x$. It follows that

$$U(x, y) = -\frac{x^2}{2} + h(y) \text{ for some function } h.$$

Note that the partial derivative U_x is found by holding y constant and differentiating with respect to x. Thus, we find U from U_x by holding y constant while we integrate U_x with respect to x. Also, since y is considered to be a constant, the "constant" of integration can depend on y. We have used $h(y)$ to denote the "constant" of integration.

From the above formula for U, we have

$$U_y = 0 + h'(y).$$

Also, $\nabla U = -x\mathbf{i} + y\mathbf{j}$ implies that $U_y = y$. Combining results, we obtain

$$h'(y) = y, \text{ so } h(y) = \frac{y^2}{2} + C.$$

Substitution of this expression for $h(y)$ in the previous formula for U gives us

$$U(x, y) = -\frac{x^2}{2} + \frac{y^2}{2} + C.$$

This function satisfies $\nabla U = -x\mathbf{i} + y\mathbf{j}$ for any choice of the constant C.

(c) Let us choose $C = 0$ in the above formula, so $U(x, y) = -\dfrac{x^2}{2} + \dfrac{y^2}{2}$. Then Theorem 2 implies

$$\int_{(1,0)}^{(3,4)} -x\,dy + y\,dy = U(3, 4) - U(1, 0)$$

$$= \left(-\frac{3^2}{2} + \frac{4^2}{2}\right) - \left(-\frac{1^2}{2} + \frac{0^2}{2}\right) = 4.$$

We could also evaluate the line integral by choosing a particular curve from $(1, 0)$ to $(3, 4)$. ∎

The method of Example 3 can be applied to the function

$$\mathbf{F} = \frac{-y}{x^2 + y^2}\mathbf{i} + \frac{x}{x^2 + y^2}\mathbf{j}, \quad y > 0$$

to obtain the function

$$U(x, y) = -\tan^{-1}\left(\frac{x}{y}\right) + C, \quad y > 0,$$

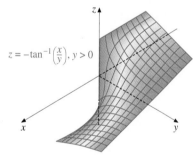

$z = -\tan^{-1}\left(\frac{x}{y}\right), y > 0$

FIGURE 14.4.8 This function cannot be extended to be continuous at the origin.

which satisfies $\nabla U = \mathbf{F}$, $y > 0$. The graph of U with $C = 0$ is sketched in Figure 14.4.8. We cannot find a function U such that $\nabla U = \mathbf{F}$ for all $(x, y) \neq (0, 0)$, since Example 2 shows $\int_C \mathbf{F} \cdot \mathbf{T}\,ds$ is not independent of path in that region.

It is worth noting what happens if we try to use the method illustrated above to find U with $\nabla U = \mathbf{F}$, for a vector field $\mathbf{F}$ with $M_y \neq N_x$. Let us try to find a function U such that $\nabla U = y\mathbf{i} - x\mathbf{j}$. Note that the functions $M = y$ and $N = -x$ do not satisfy $M_y = N_x$.

$$U_x = y \text{ implies } U(x, y) = yx + h(y), \text{ so } U_y = x + h'(y).$$

On the other hand, $U_y = -x$. We conclude that

$$x + h'(y) = -x, \text{ so } h'(y) = -2x.$$

The last equality does not make sense, because the left-hand side depends only on y and the right-hand side depends on x. The method does not determine a function U if the formula for $h'(y)$ depends on x. This indicates that the vector field is not conservative.

Summary of Techniques for Evaluating Line Integrals in the Plane

Let us now review the methods we have developed for evaluating line integrals.

> ### To evaluate $\int_C M\,dx + N\,dy$
>
> - Substitute parametric values of $\mathbf{F}$ and C. It may be necessary to write C as a sum of curves and parametrize each part separately.
> - Use the formula $\int_C M\,dx + N\,dy = -\int_{-C} M\,dx + N\,dy$ if it is easier to parametrize $-C$ than C.
> - We can use the formula $\int_C M\,dx + N\,dy = \iint_R (N_x - M_y)\,dA$ if Green's Theorem applies. M and N must have continuous first partial derivatives on and inside C, C must be the complete boundary of R, and C must be in the proper direction. It is not necessarily easier to use Green's Theorem than to evaluate the line integral directly.
> - If $\int_C M\,dx + N\,dy$ is independent of path in a region R that contains C, we may replace C with a path C' that has the same initial and terminal points as C. Check that $M_y = N_x$ at all points of C and C' and the region between the curves.
> - If $\nabla U = M\mathbf{i} + N\mathbf{j}$ in R and C is a curve in R with initial point $(x(a), y(a))$ and terminal point $(x(b), y(b))$, then $\int_C M\,dx + N\,dy = U(x(b), y(b)) - U(x(a), y(a))$. It may be easy to use the technique of Example 3 to find such a function U.

EXERCISES 14.4

Evaluate the line integrals in Exercises 1–16.

1. $\int_C xy\,dx + x^2\,dy$
  (a) $C: x = t^2,\ y = t,\ 0 \le t \le 2$
  (b) $C: x = 2t,\ y = t,\ 0 \le t \le 2$

2. $\int_C y^2\,dx + x^2\,dy$
  (a) $C: x = t,\ y = t^3,\ 0 \le t \le 2$
  (b) $C: x = t,\ y = 4t,\ 0 \le t \le 2$

3. $\int_C x^2\,dx + y^2\,dy$
  (a) $C: x = \cos t,\ y = \sin t,\ 0 \le t \le \pi$
  (b) $C: x = \cos t,\ y = -\sin t,\ 0 \le t \le \pi$

4. $\int_C x\,dx - y\,dy$
  (a) $C: x = \cos t,\ y = \sin t,\ 0 \le t \le \pi$
  (b) $C: x = \cos t,\ y = -\sin t,\ 0 \le t \le \pi$

5. $\int_C (x^2 + y^2)\,dx + 2xy\,dy$
  (a) $C: x = \cos t,\ y = \sin t,\ 0 \le t \le \pi$
  (b) $C: x = \cos t,\ y = -\sin t,\ 0 \le t \le \pi$

6. $\int_C (x^2 + y^2)\,dx + 2xy\,dy$
  (a) $C: x = -2\cos t,\ y = 2\sin t,\ 0 \le t \le \pi$
  (b) $C: x = t,\ y = 4 - t^2,\ -2 \le t \le 2$

7. $\int_C dx + x\,dy$
  (a) C: Polygonal path from $(0, 0)$ to $(1, 1)$, to $(0, 2)$
  (b) C: Along the y-axis from $(0, 0)$ to $(0, 2)$

8. $\int_C y\,dx - x\,dy$
  (a) C: Polygonal path from $(0, 0)$ to $(1, 1)$, to $(2, 0)$
  (b) C: Along the x-axis from $(0, 0)$ to $(2, 0)$

9. $\int_C (x^2 - y^2)\,dx + (y^2 - 2xy)\,dy$
  (a) C: Polygonal path from $(0, 0)$ to $(3, 2)$, to $(3, 0)$
  (b) C: Polygonal path from $(0, 0)$ to $(3, 2)$, to $(3, 0)$, to $(0, 0)$

10. $\int_C (x + y)\,dx + (x - y)\,dy$
  (a) C: Polygonal path from $(0, 0)$ to $(3, 0)$, to $(3, 3)$, to $(0, 3)$
  (b) C: Polygonal path from $(0, 0)$ to $(3, 0)$, to $(3, 3)$, to $(0, 3)$, to $(0, 0)$

11. $\int_C (y^2 + x)\,dx + (2xy + y)\,dy$
  (a) C: Polygonal path from $(0, 0)$ to $(4, 0)$, to $(4, 2)$, to $(0, 2)$
  (b) C: Polygonal path from $(0, 0)$ to $(4, 0)$, to $(4, 2)$, to $(0, 2)$, to $(0, 0)$

12. $\int_C (2xy + x^2)\,dx + (x^2 - y^2)\,dy$

    (a) C: Polygonal path from $(0, 0)$ to $(2, 0)$, to $(2, 2)$

    (b) C: Polygonal path from $(0, 0)$ to $(2, 0)$, to $(2, 2)$, to $(0, 0)$

13. $\int_C \dfrac{-y}{x^2 + y^2}\,dx + \dfrac{x}{x^2 + y^2}\,dy$

    (a) C: $x = 3 + \sin t$, $y = -2\cos t$, $0 \le t \le \pi$

    (b) C: $x = 3 + \sin t$, $y = -2\cos t$, $0 \le t \le 2\pi$

14. $\int_C \ln y\,dx + x/y\,dy$

    (a) C: $x = 2\cos t$, $y = 2 + \sin t$, $0 \le t \le \pi$

    (b) C: $x = 2\cos t$, $y = 2 + \sin t$, $0 \le t \le 2\pi$

15. $\int_C \dfrac{-y}{x^2 + y^2}\,dx + \dfrac{x}{x^2 + y^2}\,dy$; C: $x = \sqrt{3}\cos t$, $y = -1 + \sqrt{7}\sin t$, $0 \le t \le 2\pi$. (*Hint:* Apply Green's Theorem to the region inside C and outside $x^2 + y^2 = 1$.)

16. $\int_C \dfrac{y^3}{(x^2 + y^2)^2}\,dx - \dfrac{xy^2}{(x^2 + y^2)^2}\,dy$; C: $x = 2 + 7\cos t$, $y = 1 + 5\sin t$, $0 \le t \le 2\pi$. (*Hint:* Use Green's Theorem to replace C by a simpler curve that contains the origin.)

In Exercises 17–24, determine whether there is a function U such that ∇U is equal to the given vector function $\mathbf{F}$. If there is, find such a function U.

17. $\mathbf{F} = 2xy\mathbf{i} + x^2\mathbf{j}$

18. $\mathbf{F} = 2xy\mathbf{i} + (x^2 - y^2)\mathbf{j}$

19. $\mathbf{F} = (2x - y)\mathbf{i} + (x - 2y)\mathbf{j}$

20. $\mathbf{F} = (x - y)\mathbf{i} + (x + y)\mathbf{j}$

21. $\mathbf{F} = 2e^y\mathbf{i} + (2xe^y + y)\mathbf{j}$

22. $\mathbf{F} = (x^2 + xy)\mathbf{i} + \left(\dfrac{x^2}{2} + y^2\right)\mathbf{j}$

23. $\mathbf{F} = \dfrac{2x}{x^2 + y^2}\mathbf{i} + \dfrac{2y}{x^2 + y^2}\mathbf{j}$, $(x, y) \ne (0, 0)$

24. $\mathbf{F} = 3x\sqrt{x^2 + y^2}\,\mathbf{i} + 3y\sqrt{x^2 + y^2}\,\mathbf{j}$, $(x, y) \ne (0, 0)$

In Exercises 25–30, show that each integral given is independent of path in the entire xy-plane and find its value.

25. $\displaystyle\int_{(0,0)}^{(2,-1)} 2xy^2\,dx + 2x^2y\,dy$

26. $\displaystyle\int_{(0,0)}^{(2,3)} y\,dx + x\,dy$

27. $\displaystyle\int_{(-3,-4)}^{(4,3)} 2x\,dx + 2y\,dy$

28. $\displaystyle\int_{(1,-3)}^{(3,2)} 3x^2\,dx$

29. $\displaystyle\int_{(1,b)}^{(3,d)} 3x^2\,dx$, b and d any numbers

30. $\displaystyle\int_A^B 2x\,dx + 2y\,dy$, A and B any points on the circle $x^2 + y^2 = r^2$

31. Show that the line integral $\displaystyle\int_{(a,b)}^{(c,d)} f(x)\,dx$ is independent of path and that $\displaystyle\int_{(a,b)}^{(c,d)} f(x)\,dx = \int_a^c f(x)\,dx$ for continuous functions f.

Determine whether the integrals given in Exercises 32–34 are independent of path in the indicated regions.

32. $\displaystyle\int_C \dfrac{x}{x^2 + y^2}\,dx + \dfrac{y}{x^2 + y^2}\,dy$

    (a) R consists of all points (x, y) with $x > 0$

    (b) R consists of all points (x, y) with $x^2 + y^2 \ne 0$

33. $\displaystyle\int_C \dfrac{y}{x^2 - 2x + 1 + y^2}\,dx + \dfrac{1 - x}{x^2 - 2x + 1 + y^2}\,dy$

    (a) R consists of all points (x, y) with $y > 0$

    (b) R consists of all points (x, y) with $(x, y) \ne (1, 0)$

34. $\displaystyle\int_C \dfrac{2xy}{(x^2 + y^2)^2}\,dx + \dfrac{y^2 - x^2}{(x^2 + y^2)^2}\,dy$

    (a) R consists of all points (x, y) with $y > 0$

    (b) R consists of all points (x, y) with $x^2 + y^2 \ne 0$

35. If $\displaystyle\int_C M\,dx + N\,dy$ is independent of path in a rectangular region R, show that

$$\int_{(x_0,y_0)}^{(x,y)} M\,dx + N\,dy = \int_{x_0}^{x} M(u, y)\,du + \int_{y_0}^{y} N(x_0, v)\,dv$$

for (x, y) and (x_0, y_0) in R. (*Hint:* Consider the polygonal path from (x_0, y_0) to (x_0, y), to (x, y).)

36. Suppose $M_y = N_x$ in a rectangular region R. Show that Formula 1 becomes

$$U(x, y) = \int_{x_0}^{x} M(s, y)\,ds + \int_{y_0}^{y} N(x_0, t)\,dt,$$

if $C(x, y)$ is chosen to be the polygonal path from (x_0, y_0) to (x_0, y) to (x, y).

In Exercises 37–39, use the formula in Exercise 36 to find the function U that corresponds to the given vector function $\mathbf{F} = M\mathbf{i} + N\mathbf{j}$ and the given point (x_0, y_0).

37. $\mathbf{F} = (2xy - y^2)\mathbf{i} + (x^2 - 2xy)\mathbf{j}$; $(0, 0)$

38. $\mathbf{F} = (3x^2 - 2xy)\mathbf{i} + (2y - x^2)\mathbf{j}$; $(0, 0)$

39. $\mathbf{F} = ye^{xy}\mathbf{i} + (xe^{xy} + 1)\mathbf{j}$; $(0, 0)$

40. $\mathbf{F} = xy\mathbf{i} + x\mathbf{j}$.

    (a) Evaluate $U(x, y) = \displaystyle\int_{C(x,y)} \mathbf{F} \cdot \mathbf{T}\,ds$, where $C(x, y)$ is the polygonal path from $(0, 0)$ to $(0, y)$, to (x, y), and show that $U_x = M$, but $U_y \ne N$.

(b) Evaluate $U(x, y) = \displaystyle\int_{C(x,y)} \mathbf{F} \cdot \mathbf{T}\,ds$, where $C(x, y)$ is the polygonal path from $(0, 0)$ to $(x, 0)$, to (x, y), and show that $U_y = N$, but $U_x \neq M$.

(c) Evaluate $U(x, y) = \displaystyle\int_{C(x,y)} \mathbf{F} \cdot \mathbf{T}\,ds$, where $C(x, y)$ is the line segment from $(0, 0)$ to (x, y), and show that $U_x \neq M$ and $U_y \neq N$.

14.5 SURFACE INTEGRALS

Connections

Three-dimensional vectors, 11.4.

Surface area, 13.3.

Riemann sums of a function f over a surface S in three-dimensional space are obtained by dividing the surface into small pieces with surface areas ΔS. See Figure 14.5.1. We then form the sum

$$\sum f \, \Delta S,$$

where f is evaluated at a point on the corresponding small piece of the surface. The limit of these Riemann sums, in the usual sense, is called the **surface integral** of f over S and is denoted

$$\iint_S f \, dS,$$

where dS denotes **differential surface area.**

$$\iint_S dS \text{ is the \textbf{surface area} of } S.$$

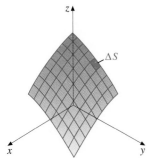

FIGURE 14.5.1 We form a Riemann sum, $\sum f \,\Delta S$, where ΔS is the area of a small piece of the surface and f is evaluated at a point on the piece.

Basic Formulas

In Section 13.3 we developed the formula

$$dS = \sec \gamma \, dA = \frac{|\nabla \mathbf{F}|}{|\nabla \mathbf{F} \cdot \mathbf{n}'|} \, dA,$$

where the surface S is given by $F(x, y, z) = 0$, dA is area in the projection plane, and $\mathbf{n}'$ is a unit vector that is normal to the projection plane. We can apply this formula to obtain formulas for surface integrals over surfaces that project onto a coordinate plane. We assume the functions involved are continuous, so the integrals exist.

If the surface S is given by $z = z(x, y)$, (x, y) in a region $R_{x,y}$ of the xy-plane (see Figure 14.5.2), then

$$\iint_S f \, dS = \iint_{R_{x,y}} f(x, y, z(x, y))\sqrt{(z_x)^2 + (z_y)^2 + 1}\, dA(x, y). \quad (1)$$

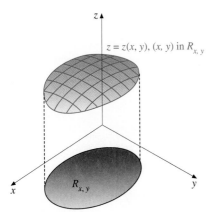

FIGURE 14.5.2 For surfaces $z = z(x, y)$, we have
$$dS = \sqrt{(z_x)^2 + (z_y)^2 + 1}\,dA,$$
where dA represents differential area in the xy-plane.

If the surface S is given by $y = y(x, z)$, (x, z) in a region $R_{x,z}$ of the xz-plane (see Figure 14.5.3), then

$$\iint_S f \, dS = \iint_{R_{x,z}} f(x, y(x, z), z)\sqrt{(y_x)^2 + 1 + (y_z)^2}\, dA(x, z). \quad (2)$$

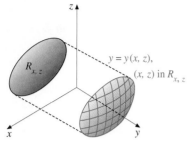

FIGURE 14.5.3 For surfaces $y = y(x, z)$, we have
$$dS = \sqrt{(y_x)^2 + 1 + (y_z)^2}\, dA,$$
where dA represents differential area in the xz-plane.

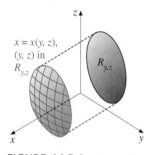

FIGURE 14.5.4 For surfaces $x = x(y, z)$, we have
$$dS = \sqrt{1 + (x_y)^2 + (x_z)^2}\, dA,$$
where dA represents differential area in the yz-plane.

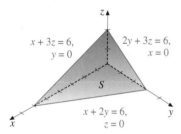

FIGURE 14.5.5a This surface has simple projections onto each coordinate plane.

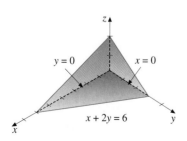

FIGURE 14.5.5b The projection onto the xy-plane is bounded by $x + 2y = 6$, $y = 0$, and $x = 0$.

If the surface S is given by $x = x(y, z)$, (y, z) in a region $R_{y,z}$ of the yz-plane (see Figure 14.5.4), then

$$\iint_S f\, dS = \iint_{R_{y,z}} f(x(y, z), y, z)\sqrt{1 + (x_y)^2 + (x_z)^2}\, dA(y, z). \quad (3)$$

EXAMPLE 1

Use each of Formulas 1, 2, and 3 to evaluate $\iint_S x\, dS$, where S is that part of the plane $x + 2y + 3z = 6$ that is in the first octant.

SOLUTION   The surface is sketched in Figure 14.5.5a.

(a) Let us express the surface as

$$z = \frac{6 - x - 2y}{3}, \quad \text{so}$$

$$z_x = -1/3, \quad z_y = -2/3, \quad \text{and}$$

$$\sqrt{(z_x)^2 + (z_y)^2 + 1} = \sqrt{\frac{1}{9} + \frac{4}{9} + 1} = \sqrt{\frac{14}{9}} = \frac{\sqrt{14}}{3}.$$

From Figure 14.5.5b, we see that we are interested in that part of the surface that is above the region in the xy-plane bounded by

$$x + 2y = 6, \quad y = 0, \quad \text{and } x = 0.$$

Substituting into Formula 1, we obtain

$$\iint_S x\, dS = \int_0^6 \int_0^{(6-x)/2} x\left(\frac{\sqrt{14}}{3}\right) dy\, dx$$

$$= \int_0^6 \frac{\sqrt{14}}{3} xy \Big]_{y=0}^{y=(6-x)/2} dx$$

$$= \int_0^6 \sqrt{14}\left(x - \frac{x^2}{6}\right) dx$$

$$= \sqrt{14}\left[\frac{x^2}{2} - \frac{x^3}{18}\right]_0^6 = 6\sqrt{14}.$$

(b) Let us express the surface as

$$y = \frac{6 - x - 3z}{2}, \quad \text{so}$$

$$y_x = -1/2, \quad y_z = -3/2, \quad \text{and}$$

$$\sqrt{(y_x)^2 + 1 + (y_z)^2} = \sqrt{\frac{1}{4} + 1 + \frac{9}{4}} = \sqrt{\frac{14}{4}} = \frac{\sqrt{14}}{2}.$$

In this case we see from Figure 14.5.5c that we integrate over the region in the xz-plane bounded by

$$x + 3z = 6, \quad z = 0, \quad \text{and } x = 0.$$

Substituting into Formula 2 and evaluating as before, we obtain

$$\iint_S x\, dS = \int_0^6 \int_0^{(6-x)/3} x\left(\frac{\sqrt{14}}{2}\right) dz\, dx = 6\sqrt{14}.$$

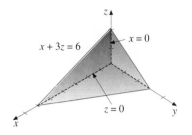

FIGURE 14.5.5c The projection onto the xz-plane is bounded by $x + 3z = 6$, $z = 0$, and $x = 0$.

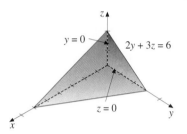

FIGURE 14.5.5d The projection onto the yz-plane is bounded by $2y + 3z = 6$, $y = 0$, and $z = 0$.

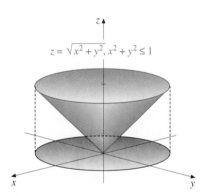

FIGURE 14.5.6 The equation of the cone is of the form $z = z(x, y)$, so we use the formula
$$dS = \sqrt{(z_x)^2 + (z_y)^2 + 1}\, dA,$$
where dA represents differential area in the xy-plane.

(c) If we express the surface as
$$x = 6 - 2y - 3z,$$
we have $x_y = -2$, $x_z = -3$, and
$$\sqrt{1 + (x_y)^2 + (x_z)^2} = \sqrt{1 + 4 + 9} = \sqrt{14}.$$
As indicated in Figure 14.5.5d, we integrate over the region in the yz-plane bounded by
$$2y + 3z = 6, \quad y = 0, \text{ and } z = 0.$$
Substituting into Formula 3 and evaluating, we obtain
$$\iint_S x\, dS = \int_0^3 \int_0^{(6-2y)/3} (6 - 2y - 3z)(\sqrt{14})\, dz\, dy = 6\sqrt{14}. \qquad \blacksquare$$

The three integrals in Example 1 have the same value, since they represent different ways of evaluating the same surface integral.

Mass and Moments

If a thin surface S has density per unit area at a point (x, y, z) on the surface given by a continuous function $\rho(x, y, z)$, the **differential mass** is
$$dM = \rho\, dS$$
and the **total mass** is the " sum"
$$M = \iint_S dM = \iint_S \rho\, dS.$$
The coordinates of the **center of mass** are given by the formulas
$$\bar{x} = \frac{\iint_S x\, dM}{\iint_S dM}, \quad \bar{y} = \frac{\iint_S y\, dM}{\iint_S dM}, \quad \bar{z} = \frac{\iint_S z\, dM}{\iint_S dM}.$$
The **moment of inertia about the z-axis** is
$$I_{z\text{-axis}} = \iint_S (x^2 + y^2)\, dM.$$

EXAMPLE 2
The surface that consists of that part of the cone $z = \sqrt{x^2 + y^2}$ with $x^2 + y^2 \le 1$ has density per unit area given by $\rho(x, y, z) = 1 - z$. Find its mass.

SOLUTION   The cone is sketched in Figure 14.5.6. The surface is given by the equation
$$z = \sqrt{x^2 + y^2}, \text{ so}$$
$$z_x = \frac{1}{2\sqrt{x^2 + y^2}}(2x) = \frac{x}{\sqrt{x^2 + y^2}} \text{ and}$$
$$z_y = \frac{1}{2\sqrt{x^2 + y^2}}(2y) = \frac{y}{\sqrt{x^2 + y^2}}.$$
The partial derivatives of z do not exist for $(x, y) = (0, 0)$. For $(x, y) \ne (0, 0)$, we have
$$(z_x)^2 + (z_y)^2 + 1 = \frac{x^2}{x^2 + y^2} + \frac{y^2}{x^2 + y^2} + 1 = \frac{x^2 + y^2}{x^2 + y^2} + 1 = 2.$$

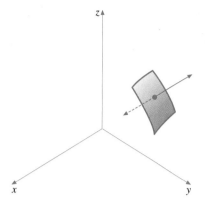

FIGURE 14.5.7 If a surface has a normal vector at a point on the surface, then it will have two unit normal vectors at the point, each the negative of the other.

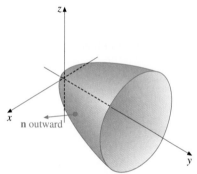

FIGURE 14.5.8 For surfaces that tend to enclose a region, we can describe the direction of a unit normal vector **n** as either inward or outward.

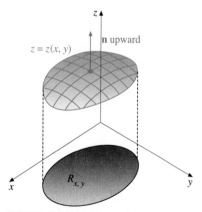

FIGURE 14.5.9 For surfaces of the form $z = z(x, y)$, we can describe the direction of a unit normal vector as either upward or downward. The unit normal vector **n** is upward if the **k**-component of **n** is positive; **n** is downward if its **k**-component is negative.

The region of integration $R_{x,y}$ in the xy-plane is bounded by $x^2 + y^2 = 1$. (The formulas for surface integrals that we are using require that the partial derivatives of $z(x, y)$ be continuous in the region $R_{x,y}$. Since the partial derivatives of z do not exist at the origin, we should integrate over the region defined by $a^2 \leq x^2 + y^2 \leq 1$ and then take the limit as a approaches zero from the right. Since the surface area of that part of the cone above $x^2 + y^2 \leq a^2$ approaches zero as a approaches zero, this would give the same result as that obtained by applying the formulas directly to the region $x^2 + y^2 \leq 1$, which we will do.) The mass is

$$
\begin{aligned}
M &= \iint_S dM = \iint_S \rho \, dS \\
&= \iint_S (1 - z) \, dS \\
&= \iint_{R_{x,y}} (1 - z)\sqrt{(z_x)^2 + (z_y)^2 + 1} \, dA \\
&= \iint_{R_{x,y}} (1 - \sqrt{x^2 + y^2})\sqrt{2} \, dA \\
&= \int_0^{2\pi} \int_0^1 (1 - r)\sqrt{2}\, r \, dr \, d\theta \qquad \text{[in polar coordinates]} \\
&= \int_0^{2\pi} \sqrt{2}\left[\frac{r^2}{2} - \frac{r^3}{3}\right]_{r=0}^{r=1} d\theta \\
&= \sqrt{2}\left(\frac{1}{2} - \frac{1}{3}\right)(2\pi) = \frac{\pi\sqrt{2}}{3}. \quad \blacksquare
\end{aligned}
$$

Flux Across a Surface

In Section 14.2 we used line integrals to measure the flux of a two-dimensional vector field across a plane curve. Let us now see how to use surface integrals to measure the flux of a three-dimensional vector field across a surface in three-dimensional space. We will measure the flux of a vector field **F** across a surface S in the direction of a unit normal vector **n**. We assume that the direction of **n** varies continuously over the surface.

If a surface has a normal vector at a point on the surface, then it will have two unit normal vectors, each the negative of the other, on opposite sides of the surface. See Figure 14.5.7. For a given surface S, we need to establish which of the two unit normal vectors at each point we are using to measure the flux of **F**. For surfaces that tend to enclose a region, we can describe the direction of a unit normal **n** as either *inward* or *outward*. Spheres, cones, paraboloids, and cylinders are of this type. If the direction of **n** is to vary continuously, then **n** must be chosen consistently—that is, either inward at each point or outward at each point. See Figure 14.5.8. For surfaces of the form $z = z(x, y)$, (x, y) in a region $R_{x,y}$ of the xy-plane, we can describe the direction of **n** as either *upward* or *downward*. The unit normal **n** is upward whenever the **k**-component of **n** is positive; **n** is downward whenever the **k**-component of **n** is negative. If the direction of **n** is to vary continuously, then **n** must be chosen consistently, either upward at each point or downward at each point. See Figure 14.5.9.

Let $\mathbf{n} = \cos \alpha \mathbf{i} + \cos \beta \mathbf{j} + \cos \gamma \mathbf{k}$ denote a unit vector that is normal to the surface S; the components are the direction cosines of $\mathbf{n}$. The **flux** of a vector function $\mathbf{F} = M\mathbf{i} + N\mathbf{j} + P\mathbf{k}$ across the surface S in the direction of $\mathbf{n}$ can be determined by evaluating $\mathbf{F} \cdot \mathbf{n}$ and then integrating $\mathbf{F} \cdot \mathbf{n}$ over the surface. That is,

$$\iint_S \mathbf{F} \cdot \mathbf{n} \, dS = \iint_S (M \cos \alpha + N \cos \beta + P \cos \gamma) \, dS. \tag{4}$$

If the surface is appropriate, we can use either Formula 1, 2, or 3 to evaluate each of the three terms of the surface integral.

If the surface is given by $z = z(x, y), (x, y)$ in a region $R_{x,y}$ of the xy-plane, the upward unit normal to the surface is

$$\mathbf{n} = \frac{-z_x \mathbf{i} - z_y \mathbf{j} + \mathbf{k}}{\sqrt{(z_x)^2 + (z_y)^2 + 1}}.$$

The direction cosines of $\mathbf{n}$ are

$$\cos \alpha = -\frac{z_x}{\sqrt{(z_x)^2 + (z_y)^2 + 1}},$$

$$\cos \beta = -\frac{z_y}{\sqrt{(z_x)^2 + (z_y)^2 + 1}},$$

$$\cos \gamma = \frac{1}{\sqrt{(z_x)^2 + (z_y)^2 + 1}}.$$

The differential surface area is

$$dS = \sqrt{(z_x)^2 + (z_y)^2 + 1} \, dA = \sec \gamma \, dA$$

and Formula 1 gives

$$\iint_S M \cos \alpha \, dS = \iint_{R_{x,y}} -M z_x \, dA, \tag{5}$$

$$\iint_S N \cos \beta \, dS = \iint_{R_{x,y}} -N z_y \, dA, \tag{6}$$

$$\iint_S P \cos \gamma \, dS = \iint_{R_{x,y}} P \, dA. \tag{7}$$

From Formula 4, we then see that the flux of a vector $\mathbf{F} = M\mathbf{i} + N\mathbf{j} + P\mathbf{k}$ across the surface $z = z(x, y), (x, y)$ in a region $R_{x,y}$ of the xy-plane, in the direction of the upward unit normal vector $\mathbf{n}$ can be expressed as

$$\iint_S \mathbf{F} \cdot \mathbf{n} \, dS = \iint_{R_{x,y}} (-M z_x - N z_y + P) \, dA. \tag{8}$$

Since the downward unit normal is the negative of the upward unit normal, the flux in the direction of the downward unit vector is given by the negative of the expression on the right side of the equation in (8).

EXAMPLE 3

Evaluate $\iint_S \mathbf{F} \cdot \mathbf{n} \, dS$, where $\mathbf{F} = x\mathbf{i} + y\mathbf{j} + z\mathbf{k}$, the surface S is that part of the paraboloid $z = 4 - x^2 - y^2$ with $z \geq 3$, and $\mathbf{n}$ is directed upward.

SOLUTION   The paraboloid is sketched in Figure 14.5.10.

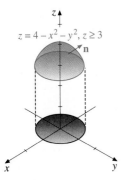

FIGURE 14.5.10 We want the flux across the paraboloid in the direction of the upward unit normal vector, so Formula 8 applies directly.

Let us use Formula 8 to evaluate the integral. We have

$$z = 4 - x^2 - y^2, \text{ so}$$
$$z_x = -2x \text{ and } z_y = -2y.$$

We have $\mathbf{F} = x\mathbf{i} + y\mathbf{j} + z\mathbf{k}$, so

$$M = x, \quad N = y, \text{ and } P = z = 4 - x^2 - y^2.$$

Note that z has been expressed in terms of x and y. That part of the paraboloid $z = 4 - x^2 - y^2$ with $z \geq 3$ satisfies $x^2 + y^2 \leq 1$. This inequality determines the region $R_{x,y}$. The unit normal vector that is given is upward, as in Formula 8. Substitution into Formula 8 then gives

$$\iint_S \mathbf{F} \cdot \mathbf{n} \, dS = \iint_{R_{x,y}} (-Mz_x - Nz_y + P) \, dA$$
$$= \iint_{R_{x,y}} [-(x)(-2x) - (y)(-2y) + (4 - x^2 - y^2)] \, dA$$
$$= \iint_{R_{x,y}} (4 + x^2 + y^2) \, dA.$$

At this point we notice that the double integral is well suited to polar coordinates. We then have

$$\iint_S \mathbf{F} \cdot \mathbf{n} \, dS = \int_0^{2\pi} \int_0^1 (4 + r^2) r \, dr \, d\theta$$
$$= \int_0^{2\pi} \int_0^1 (4r + r^3) \, dr \, d\theta$$
$$= \int_0^{2\pi} \left[2r^2 + \frac{r^4}{4} \right]_0^1 d\theta$$
$$= \int_0^{2\pi} \left(2 + \frac{1}{4} \right) d\theta = \left(\frac{9}{4} \right)(2\pi) = \frac{9\pi}{2}. \quad \blacksquare$$

FIGURE 14.5.11 We want the flux across the plane in the direction of the downward unit normal vector, so we must use the negative of the expression in Formula 8.

EXAMPLE 4

Evaluate the flux of $\mathbf{F} = x\mathbf{i} + y\mathbf{j} + z\mathbf{k}$ across that part of the surface $z = 3$ with $x^2 + y^2 \leq 1$, in the direction of the downward unit normal vector.

SOLUTION The surface S is sketched in Figure 14.5.11. The flux of $\mathbf{F}$ across the surface S in the direction of the unit normal $\mathbf{n}$ is

$$\iint_S \mathbf{F} \cdot \mathbf{n} \, dS.$$

The surface $z = 3$ is of the form $z = z(x, y)$. We are interested in that part of the surface above the region $R_{x,y}$ in the xy-plane bounded by $x^2 + y^2 = 1$. We can use Formula 8 to evaluate the integral that gives the flux, but since $\mathbf{n}$ is downward, we must change the sign of each term of the integral on the right of the equation.

$$z = 3 \text{ implies } z_x = z_y = 0.$$

$\mathbf{F} = x\mathbf{i} + y\mathbf{j} + z\mathbf{k}$ implies $M = x$, $N = y$, and $P = z = 3$. Note that z has constant value 3 on the surface of integration. From Formula 8, *with a change of sign,* we have

$$\iint_S \mathbf{F} \cdot \mathbf{n}\, dS = \iint_{R_{x,y}} (Mz_x + Nz_y - P)\, dA$$

$$= \iint_{R_{x,y}} [(x)(0) + (y)(0) - 3]\, dA$$

$$= -3(\text{Area of } R_{x,y}) = -3\pi. \ \blacksquare$$

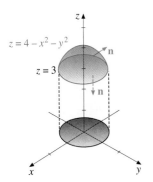

FIGURE 14.5.12 The outward unit normal of the region bounded by the paraboloid and the plane is upward on the paraboloid and downward on the plane.

$z = 4 - x^2 - y^2$

$z = 3$

The two surfaces in Examples 3 and 4 form the complete boundary of the region bounded by the paraboloid $z = 4 - x^2 - y^2$ and the plane $z = 3$. The normal is directed outward from the region on each of the boundary surfaces. See Figure 14.5.12. Example 3 tells us that the function $\mathbf{F} = x\mathbf{i} + y\mathbf{j} + z\mathbf{k}$ has an outward flux of $9\pi/2$ from the region across the paraboloid, and Example 4 tells us that $\mathbf{F}$ has an outward flux of -3π across the plane surface. The net flux out of the region is the algebraic sum of the outward flux across each boundary surface,

$$\left(\frac{9\pi}{2}\right) + (-3\pi) = \frac{3\pi}{2}.$$

EXAMPLE 5

Find the outward flux of

$$\mathbf{F} = \frac{x\mathbf{i} + y\mathbf{j} + z\mathbf{k}}{(x^2 + y^2 + z^2)^{3/2}}$$

across the sphere $x^2 + y^2 + z^2 = R^2$.

SOLUTION   The flux is the integral

$$\iint_S \mathbf{F} \cdot \mathbf{n}\, dS,$$

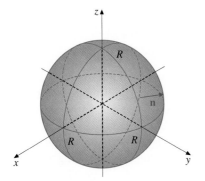

FIGURE 14.5.13 A normal vector to the sphere is given by ∇G, where $G(x, y, z) = x^2 + y^2 + z^2$.

where $\mathbf{n}$ is the outward unit normal to the sphere. See Figure 14.5.13. Let us evaluate the integral by finding $\mathbf{n}$, evaluating $\mathbf{F} \cdot \mathbf{n}$, and then integrating $\mathbf{F} \cdot \mathbf{n}$ over the surface, as in Formula 4.

The equation of the sphere is of the form $G(x, y, z) = R^2$, where $G(x, y, z) = x^2 + y^2 + z^2$. We then know that the gradient of G is normal to the surface. We have

$$\nabla G = G_x\mathbf{i} + G_y\mathbf{j} + G_z\mathbf{k} = 2x\mathbf{i} + 2y\mathbf{j} + 2z\mathbf{k}.$$

The two unit normal vectors that are normal to the sphere are

$$\pm\frac{\nabla G}{|\nabla G|} = \pm\frac{2x\mathbf{i} + 2y\mathbf{j} + 2z\mathbf{k}}{\sqrt{(2x)^2 + (2y)^2 + (2z)^2}} = \pm\frac{x\mathbf{i} + y\mathbf{j} + z\mathbf{k}}{\sqrt{x^2 + y^2 + z^2}}.$$

The outward unit normal to the sphere is

$$\mathbf{n} = \frac{x\mathbf{i} + y\mathbf{j} + z\mathbf{k}}{\sqrt{x^2 + y^2 + z^2}}.$$

We then see that

$$\mathbf{F} = \frac{x\mathbf{i} + y\mathbf{j} + z\mathbf{k}}{(x^2 + y^2 + z^2)^{3/2}} \text{ implies}$$

$$\mathbf{F} \cdot \mathbf{n} = \frac{(x)(x) + (y)(y) + (z)(z)}{(x^2 + y^2 + z^2)^{1/2}(x^2 + y^2 + z^2)^{3/2}}$$

$$= \frac{1}{x^2 + y^2 + z^2} = \frac{1}{R^2}$$

for all points (x, y, z) on the surface of the sphere. We then have

$$\iint_S \mathbf{F} \cdot \mathbf{n} \, dS = \iint_S \frac{1}{R^2} \, dS$$

$$= \frac{1}{R^2} \left(\iint_S dS \right) \qquad \text{[since } R \text{ is a constant]}$$

$$= \frac{1}{R^2} \text{ (Surface area of a sphere of radius } R)$$

$$= \frac{1}{R^2} (4\pi R^2) = 4\pi. \quad \blacksquare$$

We could express the surface of the sphere in Example 5 as the union of the two surfaces

$$z = \sqrt{R^2 - x^2 - y^2} \text{ (}\mathbf{n} \text{ upward) and}$$

$$z = -\sqrt{R^2 - x^2 - y^2} \text{ (}\mathbf{n} \text{ downward).}$$

We could then use Formula 8 with $R_{x,y}$ the region in the xy-plane bounded by $x^2 + y^2 = R^2$ to evaluate the outward flux over each of the two surfaces. (Since the first partial derivatives of the above functions z do not exist for $x^2 + y^2 = R^2$, the integral in Formula 8 would be *improper*. However, the improper integrals would be *convergent*. Generally, we can use Formula 8 to evaluate surface integrals over hemispheres, even though the resulting integrals are actually improper.)

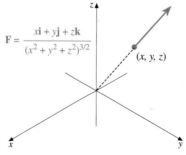

FIGURE 14.5.14 The magnitude of this inverse square field is inversely proportional to the square of the distance from the origin and the field is directed away from the origin.

The vector field $\mathbf{F} = (x\mathbf{i} + y\mathbf{j} + z\mathbf{k})/(x^2 + y^2 + z^2)^{3/2}$ of Example 5 is called an **inverse square field**. The magnitude is inversely proportional to the square of the distance from the origin and the direction of $\mathbf{F}$ is away from the origin. See Figure 14.5.14. Inverse square fields occur in several physical problems. For example, the force due to gravitational attraction between two objects has magnitude that is inversely proportional to the square of the distance between the objects and is in the direction between the objects. Example 5 shows that the outward flux of the inverse square field across a sphere with center at the origin and any radius is 4π.

EXERCISES 14.5

Find the area of the surfaces described in Exercises 1–6.

1. That part of the plane $2x + 3y + 6z = 12$ that is inside the cylinder $x^2 + y^2 = 4$

2. That part of the plane $3x + 3y + 2z = 6$ that is bounded by the planes $x = 0$, $x = 1$, $z = 0$, and $z = 1$

3. That part of the sphere $x^2 + y^2 + z^2 = 4$ that is inside the cone $x = \sqrt{y^2 + z^2}$

4. That part of the cylinder $x^2 + z^2 = 4$ that is above the plane $z = 1$ and between the planes $y = 0$ and $y = 1$

5. That part of the paraboloid $2y = x^2 + z^2$ cut off by the plane $y = 2$

6. That part of the paraboloid $x = 4 - y^2 - z^2$ cut off by $x = 0$

Evaluate the surface integrals in Exercises 7–12.

7. $\iint_S xy\,dS$; S is that part of the plane $z = 2x + 3y$ that is in the first octant and below $z = 6$.

8. $\iint_S z\,dS$; S is that part of the sphere $x^2 + y^2 + z^2 = 4$ above $z = 1$.

9. $\iint_S y\,dS$; S is that part of the cylinder $x^2 + y^2 = 4$ between $z = 0$ and $z = 3$ with $y \geq 1$.

10. $\iint_S x\,dS$; S is that part of the generalized cylinder $y = 1 - x^2$ between $z = 0$ and $z = 3$, in the first octant.

11. $\iint_S (y + z)\,dS$; S is that part of the cone $x = 2\sqrt{y^2 + z^2}$ with $x \leq 4$, in the first octant.

12. $\iint_S x\,dS$; S is that part of the generalized cylinder $x = \sin y$ with $0 \leq y \leq \pi$ and $0 \leq z \leq 2$.

*Find **(a)** the mass, **(b)** the center of mass, and **(c)** the moment of inertia about the z-axis of the surfaces S with density per unit area ρ given in Exercises 13–16.*

13. S is that part of the paraboloid $z = 9 - x^2 - y^2$ with $z \geq 0$; $\rho(x, y, z) = 1/\sqrt{4x^2 + 4y^2 + 1}$.

14. S is the hemisphere $z = \sqrt{1 - x^2 - y^2}$; $\rho(x, y, z) = z$.

15. S is that part of the cone $z = \sqrt{x^2 + y^2}$ with $z \leq 2$; $\rho(x, y, z) = x^2 + y^2$.

16. S is that part of the cone $z = \sqrt{x^2 + y^2}$ with $1 \leq z \leq 2$; $\rho(x, y, z) = 1$.

In Exercises 17–21, evaluate the flux of the given vector function across the given surface for the indicated normal vector.

17. $\mathbf{F} = x\mathbf{i} + y\mathbf{j} + z\mathbf{k}$; S is that part of the plane $x + y + z = 1$ in the first octant; $\mathbf{n}$ upward.

18. $\mathbf{F} = x\mathbf{i} + y\mathbf{j} + z\mathbf{k}$; S is that part of the paraboloid $z = 4 - x^2 - y^2$ with $0 \leq x \leq 1$ and $0 \leq y \leq 1$; $\mathbf{n}$ outward.

19. $\mathbf{F} = x\mathbf{i} + y\mathbf{j} + z\mathbf{k}$
  (a) S is that part of the cone $z = \sqrt{x^2 + y^2}$ with $z \leq 4$; $\mathbf{n}$ outward.
  (b) S is that part of the plane $z = 4$ with $x^2 + y^2 \leq 16$; $\mathbf{n}$ upward.

20. $\mathbf{F} = x\mathbf{i} + y\mathbf{j} + z\mathbf{k}$;
  (a) S is that part of the paraboloid $z = 9 - x^2 - y^2$ with $z \geq 0$; $\mathbf{n}$ outward.
  (b) S is that part of the plane $z = 0$ with $x^2 + y^2 \leq 9$; $\mathbf{n}$ downward.

21. $\mathbf{F} = x\mathbf{i} + y\mathbf{j} + z\mathbf{k}$; S is that part of the ellipsoid

$$\frac{x^2}{2^2} + \frac{y^2}{1^2} + \frac{z^2}{1^2} = 1$$

in the first octant; $\mathbf{n}$ outward. (*Hint:* Use $x(y, z)$.)

22. Find the outward flux of $\mathbf{F} = x\mathbf{i} + y\mathbf{j} + z\mathbf{k}$ across the sphere $x^2 + y^2 + z^2 = R^2$.

23. Find the outward flux of $\mathbf{F} = \dfrac{x\mathbf{i} + y\mathbf{j} + z\mathbf{k}}{x^2 + y^2 + z^2}$ across the sphere $x^2 + y^2 + z^2 = R^2$.

24. A surface is given by $z = f(r)$, $r = \sqrt{x^2 + y^2}$, $0 \leq r \leq r_0$.
  (a) Show that using Formula 1 to evaluate the area of the surface gives $S = \displaystyle\int_0^{r_0} 2\pi r \sqrt{1 + [f'(r)]^2}\,dr$.
  (b) Compare the formula in (a) with the formula for the area of the surface obtained by revolving the curve $y = f(x)$, $0 \leq x \leq x_0$, about the y-axis.

14.6 THE DIVERGENCE THEOREM

In this section we will develop a relation between a surface integral and a triple integral over the region bounded by the surface.

Development of the Theorem

Let D be the region in three-dimensional space bounded by the surfaces

$$z = h_2(x, y), \quad z = h_1(x, y), \quad y = g_2(x), \quad y = g_1(x), \quad x = b, \quad \text{and} \quad x = a,$$

and let $R_{x,y}$ be the region in the xy-plane bounded by the curves

$$y = g_2(x), \quad y = g_1(x), \quad x = b, \text{ and } x = a.$$

Let

$$\mathbf{n} = \cos\alpha\,\mathbf{i} + \cos\beta\,\mathbf{j} + \cos\gamma\,\mathbf{k}$$

denote the outwardly directed unit normal vector to the boundary surfaces of D. We have $\cos\gamma = 0$ on the vertical surfaces

$$y = g_2(x), \quad y = g_1(x), \quad x = b, \text{ and } x = a.$$

The unit vector $\mathbf{n}$ is directed upward on the top surface

$$S_2 : z = h_2(x, y), \quad (x, y), \text{ in } R_{x,y};$$

$\mathbf{n}$ is directed downward on the bottom surface

$$S_1 : z = h_1(x, y), \quad (x, y) \text{ in } R_{x,y}.$$

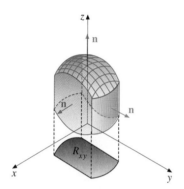

FIGURE 14.6.1 The outward unit normal vector is upward on the top surface, downward on the bottom surface, and horizontal on the four sides.

See Figure 14.6.1. If S is the complete boundary of D, we can then use Formula 7 of Section 14.5 to obtain

$$\iint_S P\cos\gamma\,dS = \iint_{S_2} P\cos\gamma\,dS + \iint_{S_1} P\cos\gamma\,dS$$

$$= \iint_{R_{x,y}} P(x, y, h_2(x, y))\,dA - \iint_{R_{x,y}} P(x, y, h_1(x, y))\,dA$$

$$= \int_a^b \int_{g_1(x)}^{g_2(x)} [P(x, y, h_2(x, y)) - P(x, y, h_1(x, y))]\,dy\,dx$$

$$= \int_a^b \int_{g_1(x)}^{g_2(x)} \left[\int_{h_1(x,y)}^{h_2(x,y)} \frac{\partial P}{\partial z}\,dz\right] dy\,dx$$

$$= \iiint_D \frac{\partial P}{\partial z}\,dV.$$

That is, we have

$$\iint_S P\cos\gamma\,dS = \iiint_D \frac{\partial P}{\partial z}\,dV. \tag{1}$$

If a region is divided into subregions by cutting along vertical generalized cylinders, the total of the boundaries of the subregions consists of the boundary of the original region and the vertical cutting surfaces. Since vertical surfaces do not contribute to the integral of $P\cos\gamma$ over a surface, the integral of $P\cos\gamma$ over the boundary of the original region is equal to the sum of the integrals over the boundaries of the subregions. The triple integral of $\partial P/\partial z$ over the region is also equal to the sum of the integrals over the subregions. It follows that Formula 1 holds for regions that can be divided by vertical surfaces into a

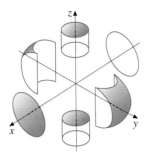

FIGURE 14.6.2 The region between two spheres can be divided by vertical surfaces into a finite number of regions of the type indicated in Figure 14.6.1.

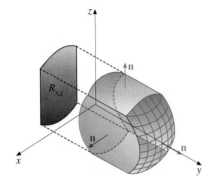

FIGURE 14.6.3 Triple integrals over regions of this type can be iterated with order $dy\,dz\,dx$.

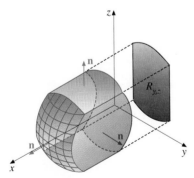

FIGURE 14.6.4 Triple integrals over regions of this type can be iterated with order $dx\,dz\,dy$.

finite number of regions of the type considered above. Let us call such regions **finitely simple in the direction of the z-axis**. For example, the region between the spheres $x^2 + y^2 + z^2 = 4$ and $x^2 + y^2 + z^2 = 1$ can be divided into six regions of the desired type by cutting along the vertical cylinder $x^2 + y^2 = 1$ and along the vertical planes $x = 1$ and $x = -1$, as indicated in Figure 14.6.2.

If the region D can be divided into a finite number of regions of the type illustrated in Figure 14.6.3 by cutting along generalized cylinders with generating lines parallel to the y-axis, we say the region is **finitely simple in the direction of the y-axis**. We then have

$$\iint_S N \cos \beta \, dS = \iiint_D \frac{\partial N}{\partial y} \, dV. \tag{2}$$

A region that can be divided into a finite number of regions of the type illustrated in Figure 14.6.4 by generalized cylinders with generating lines parallel to the x-axis is said to be **finitely simple in the direction of the x-axis**. We then have

$$\iint_S M \cos \alpha \, dS = \iiint_D \frac{\partial M}{\partial x} \, dV. \tag{3}$$

Let us say that a region is **finitely simple** if it is finitely simple in the direction of the three coordinate axes. We can then combine Formulas 1, 2, and 3. Let us express the result in vector form. We define the symbolic vector **del** to be

$$\nabla = \frac{\partial}{\partial x}\mathbf{i} + \frac{\partial}{\partial y}\mathbf{j} + \frac{\partial}{\partial z}\mathbf{k}.$$

If $\mathbf{F} = M\mathbf{i} + N\mathbf{j} + P\mathbf{k}$, we use the symbolism

$$\nabla \cdot \mathbf{F} = \frac{\partial M}{\partial x} + \frac{\partial N}{\partial y} + \frac{\partial P}{\partial z}.$$

The expression $\nabla \cdot \mathbf{F}$ is called the **divergence** of $\mathbf{F}$. If $\mathbf{n}$ is the outward unit normal to the boundary surface of a region D, then $\iint_S \mathbf{F} \cdot \mathbf{n}\,dS$ is the outward **flux** of $\mathbf{F}$ across the boundary of D. We then have the following.

DIVERGENCE THEOREM

Suppose D is a finitely simple region in three-dimensional space, S is the complete boundary of D, and $\mathbf{n}$ is the outward unit normal to the boundary surface. If $\mathbf{F}$ has continuous first partial derivatives in D and on the boundary of D, then

$$\iint_S \mathbf{F} \cdot \mathbf{n}\,dS = \iiint_D \nabla \cdot \mathbf{F}\,dV.$$

Note the correspondence between Green's Theorem in the form $\int_C \mathbf{F} \cdot \mathbf{n}\,ds = \int \int_R (M_x + N_y)\,dA$ and the Divergence Theorem.

The value of the divergence $\nabla \cdot \mathbf{F}$ at a point can be considered to be the flux per unit volume at the point. Positive divergence indicates a source at the point and negative divergence indicates a sink at the point.

The Divergence Theorem says that the flux across the boundary of a region is the net sum of the sources and sinks within the region.

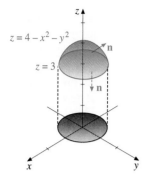

FIGURE 14.6.5 The outward unit normal vector is upward on the paraboloid and downward on the plane.

Using the Theorem

EXAMPLE 1

Evaluate $\iint_S \mathbf{F} \cdot \mathbf{n}\, dS$, where $\mathbf{F} = x\mathbf{i} + y\mathbf{j} + z\mathbf{k}$, S is the complete boundary of the region D bounded by the paraboloid $z = 4 - x^2 - y^2$ and the plane $z = 3$, and $\mathbf{n}$ is outward.

SOLUTION The region is sketched in Figure 14.6.5. $\mathbf{F}$ has continuous first partial derivatives, so we can use the Divergence Theorem to evaluate the surface integral. $\mathbf{F} = x\mathbf{i} + y\mathbf{j} + z\mathbf{k}$ implies $M = x$, $N = y$, and $P = z$. We then have

$$
\begin{aligned}
\iint_S \mathbf{F} \cdot \mathbf{n}\, dS &= \iiint_D \nabla \cdot \mathbf{F}\, dV \\
&= \iiint_D (M_x + N_y + P_z)\, dV \\
&= \iiint_D (1 + 1 + 1)\, dV \\
&= \int_0^{2\pi} \int_0^1 \int_3^{4-r^2} 3r\, dz\, dr\, d\theta \qquad \text{[in cylindrical coordinates]} \\
&= \int_0^{2\pi} \int_0^1 [(4 - r^2) - 3](3r)\, dr\, d\theta \\
&= \int_0^{2\pi} \int_0^1 (3r - 3r^3)\, dr\, d\theta \\
&= \int_0^{2\pi} \left(\frac{3}{2} - \frac{3}{4} \right) d\theta = \frac{3}{4}(2\pi) = \frac{3\pi}{2}. \quad \blacksquare
\end{aligned}
$$

In Examples 3 and 4 of Section 14.5 we calculated the outward flux across each of the boundary surfaces of the region in Example 1 above. We also noted that the net outward flux across the two surfaces was $3\pi/2$. This agrees with the calculation using the Divergence Theorem.

EXAMPLE 2

Evaluate the outward flux of $\mathbf{F} = x\mathbf{i} + z\mathbf{j} + y\mathbf{k}$ across the boundary of the region in the first octant inside the sphere $x^2 + y^2 + z^2 = 1$.

SOLUTION The region is sketched in Figure 14.6.6.

$\mathbf{F}$ has continuous first partial derivatives, so we can then apply the Divergence Theorem. The flux is

$$
\begin{aligned}
\iint_S \mathbf{F} \cdot \mathbf{n}\, dS &= \iiint_D \nabla \cdot \mathbf{F}\, dV \\
&= \iiint_D \left(\frac{\partial}{\partial x}(x) + \frac{\partial}{\partial y}(z) + \frac{\partial}{\partial z}(y) \right) dV \\
&= \iiint_D 1\, dV \\
&= \frac{1}{8} \, (\text{Volume of sphere with radius 1}) \\
&= \frac{1}{8} \left(\frac{4\pi(1)^3}{3} \right) = \frac{\pi}{6}. \quad \blacksquare
\end{aligned}
$$

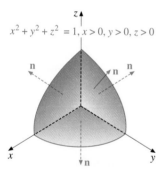

FIGURE 14.6.6 The region is bounded by the three coordinate planes and that part of the sphere that is in the first octant.

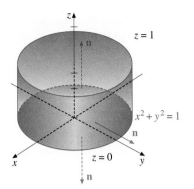

FIGURE 14.6.7 This region can be described easily in terms of cylindrical coordinates.

EXAMPLE 3

Evaluate $\iint_S \mathbf{F} \cdot \mathbf{n} \, dS$, where

$$\mathbf{F} = \tan^{-1}(y^2 + z^2)\mathbf{i} + \ln(1 + x^2 + z^2)\mathbf{j} + 2ze^{x^2+y^2}\mathbf{k},$$

S is the complete boundary of the region D that is inside the cylinder $x^2 + y^2 = 1$ and between the planes $z = 0$ and $z = 1$, and $\mathbf{n}$ is the outward unit normal.

SOLUTION The region is sketched in Figure 14.6.7.

We see $\mathbf{F}$ has continuous first partial derivatives and

$$\nabla \cdot \mathbf{F} = \frac{\partial}{\partial x}(\tan^{-1}(y^2 + z^2)) + \frac{\partial}{\partial y}(\ln(1 + x^2 + z^2)) + \frac{\partial}{\partial z}(2ze^{x^2+y^2})$$

$$= 2e^{x^2+y^2}.$$

The Divergence Theorem then gives

$$\iint_S \mathbf{F} \cdot \mathbf{n} \, dS = \iiint_D \nabla \cdot \mathbf{F} \, dV$$

$$= \iiint_D 2e^{x^2+y^2} \, dV.$$

We use cylindrical coordinates to evaluate the triple integral over the cylindrical region D. We have

$$\iiint_D 2e^{x^2+y^2} \, dV = \int_0^{2\pi} \int_0^1 \int_0^1 2e^{r^2} r \, dz \, dr \, d\theta$$

$$= \int_0^{2\pi} \int_0^1 e^{r^2} 2r \, dr \, d\theta$$

$$= \int_0^{2\pi} e^{r^2} \Big]_0^1 \, d\theta$$

$$= \int_0^{2\pi} (e - 1) \, d\theta = 2\pi(e - 1). \quad \blacksquare$$

Recall from Section 14.5 that the flux of a vector field $\mathbf{F} = M\mathbf{i} + N\mathbf{j} + P\mathbf{k}$ across the surface $z = z(x, y)$, (x, y) in a region $R_{x,y}$ of the xy-plane, in the direction of the upward unit normal vector $\mathbf{n}$ can be evaluated directly by using the formula

$$\iint_S \mathbf{F} \cdot \mathbf{n} \, dS = \iint_{R_{x,y}} (-Mz_x - Nz_y + P) \, dA.$$

This formula can be used in cases where the Divergence Theorem does not apply. Also, it may be easier to evaluate the surface integral directly, even though the Divergence Theorem does apply.

An Inverse Square Field

Let us consider the **inverse square field**

$$\mathbf{F} = \frac{x\mathbf{i} + y\mathbf{j} + z\mathbf{k}}{(x^2 + y^2 + z^2)^{3/2}}.$$

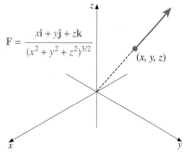

$$F = \frac{x\mathbf{i} + y\mathbf{j} + z\mathbf{k}}{(x^2 + y^2 + z^2)^{3/2}}$$

(x, y, z)

FIGURE 14.6.8 The inverse square field occurs in many natural problems.

The length of $\mathbf{F}$ is one over the square of the distance from the origin and the direction is directly away from the origin. See Figure 14.6.8. The divergence of $\mathbf{F}$ is

$$\nabla \cdot \mathbf{F} = \frac{\partial}{\partial x}\left(\frac{x}{(x^2 + y^2 + z^2)^{3/2}}\right) + \frac{\partial}{\partial y}\left(\frac{y}{(x^2 + y^2 + z^2)^{3/2}}\right)$$
$$+ \frac{\partial}{\partial z}\left(\frac{z}{(x^2 + y^2 + z^2)^{3/2}}\right).$$

Using the Quotient Rule to evaluate the first partial derivative above, we have

$$\frac{\partial}{\partial x}\left(\frac{x}{(x^2 + y^2 + z^2)^{3/2}}\right)$$
$$= \frac{(x^2 + y^2 + z^2)^{3/2}(1) - (x)(3/2)(x^2 + y^2 + z^2)^{1/2}(2x)}{(x^2 + y^2 + z^2)^{6/2}}$$
$$= \frac{(x^2 + y^2 + z^2) - 3x^2}{(x^2 + y^2 + z^2)^{5/2}}$$
$$= \frac{-2x^2 + y^2 + z^2}{(x^2 + y^2 + z^2)^{5/2}}.$$

The other two partial derivatives are evaluated similarly. We obtain

$$\nabla \cdot \mathbf{F} = \frac{-2x^2 + y^2 + z^2}{(x^2 + y^2 + z^2)^{5/2}} + \frac{x^2 - 2y^2 + z^2}{(x^2 + y^2 + z^2)^{5/2}} + \frac{x^2 + y^2 - 2z^2}{(x^2 + y^2 + z^2)^{5/2}}$$
$$= 0, \quad (x, y, z) \neq (0, 0, 0).$$

Let us state the above result as a theorem.

THEOREM 1

The inverse square vector field

$$\mathbf{F} = \frac{x\mathbf{i} + y\mathbf{j} + z\mathbf{k}}{(x^2 + y^2 + z^2)^{3/2}}$$

has divergence zero at every point of three-dimensional space except the origin.

If the divergence of $\mathbf{F}$ is zero at every point of a region D that has complete boundary S, the Divergence Theorem implies

$$\iint_S \mathbf{F} \cdot \mathbf{n}\, dS = \iiint_D \nabla \cdot \mathbf{F}\, dV = 0,$$

so the outward flux of $\mathbf{F}$ across S is zero. Theorem 1 then gives the following.

THEOREM 2

If D is any finitely simple region that does not contain the origin either in its interior or on its boundary, the outward flux of the inverse square field

$$\mathbf{F} = \frac{x\mathbf{i} + y\mathbf{j} + z\mathbf{k}}{(x^2 + y^2 + z^2)^{3/2}}$$

across the total boundary of D is zero.

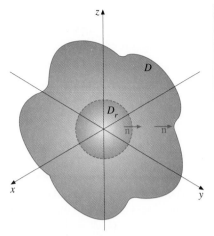

FIGURE 14.6.9 The outward unit normal vector to the region $D - D_r$ on the boundary of D_r is the negative of the outward unit normal to the sphere D_r.

THEOREM 3

If D is any finitely simple region that contains the origin in its interior, the outward flux of the inverse square field

$$\mathbf{F} = \frac{x\mathbf{i} + y\mathbf{j} + z\mathbf{k}}{(x^2 + y^2 + z^2)^{3/2}}$$

across the total boundary of D is 4π.

Proof We cannot apply the Divergence Theorem to the region D, because D contains a point where $\mathbf{F}$ does not have first partial derivatives. ($\mathbf{F}$ becomes unbounded at the origin, so it cannot have partial derivatives at that point.) However, we can apply the Divergence Theorem to the region $D - D_r$, where D_r is the sphere bounded by $x^2 + y^2 + z^2 = r^2$, and r is chosen small enough that D_r is contained in D. Let S denote the boundary of D, and let S_r denote the boundary of D_r. Let $\mathbf{n}$ denote the unit normal that is outward from the region D on S and outward from D_r on S_r. The outward flux from $D - D_r$ is then in the direction of $-\mathbf{n}$ on the surface S_r. See Figure 14.6.9.

The outward flux of F across the complete boundary of $D - D_r$ is

$$\iint_S \mathbf{F} \cdot \mathbf{n}\, dS - \iint_{S_r} \mathbf{F} \cdot \mathbf{n}\, dS = 0,$$

by Theorem 2. In Example 5 of Section 14.5, we determined that the outward flux of $\mathbf{F}$ across the surface S_r is 4π. It follows that

$$\iint_S \mathbf{F} \cdot \mathbf{n}\, dS = \iint_{S_r} \mathbf{F} \cdot \mathbf{n}\, dS = 4\pi. \quad \blacksquare$$

EXERCISES 14.6

In Exercises 1–18, find the outward flux of the given vector function across the boundary of the indicated region.

1. $\mathbf{F} = x^2\mathbf{i} + y^2\mathbf{j}$; D bounded by $x = 2$, $y = 3$, $z = 2$, in the first octant

2. $\mathbf{F} = xyz\mathbf{k}$; D bounded by $z = 2 - x - y$, in the first octant

3. $\mathbf{F} = x^2\mathbf{i} + y^2\mathbf{j} + z^2\mathbf{k}$; D bounded by $x^2 + y^2 = 1$, $z = 0$, $z = 4$

4. $\mathbf{F} = -y\mathbf{i} + x\mathbf{j}$; D bounded by $x^2 + y^2 = 4$, $z = 0$, $z = 3$

5. $\mathbf{F} = x^3\mathbf{i} + y^3\mathbf{j} + z^3\mathbf{k}$; D bounded by $z = \sqrt{x^2 + y^2}$ and $z = 1$

6. $\mathbf{F} = x^3\mathbf{i} + y^3\mathbf{j} + z^3\mathbf{k}$; D bounded by $z = \sqrt{1 - x^2 - y^2}$ and $z = 0$

7. $\mathbf{F} = xy\mathbf{i} + yz\mathbf{j} + zx\mathbf{k}$; D bounded by $z = 4 - x^2 - y^2$ and $z = 0$

8. $\mathbf{F} = x^2\mathbf{i} + y^2\mathbf{j} + z^2\mathbf{k}$; D bounded by $z = x^2 + y^2$ and $z = 4$

9. $\mathbf{F} = 2x\mathbf{i} + 3y\mathbf{j} + 5z\mathbf{k}$; D bounded by $z = 2x + 2y$ and $z = 2$, in the first octant

10. $\mathbf{F} = x\mathbf{i} + y\mathbf{j} + z\mathbf{k}$; D bounded by $z = 2y$, $z = 2$, and $x = 3$, in the first octant

11. $\mathbf{F} = x^2\mathbf{i}$; D in the first octant, below $x + y + z = 4$ and above $z = 2x$

12. $\mathbf{F} = x\mathbf{i} + y\mathbf{j} + z\mathbf{k}$; D bounded by $x + z = 1$ and $y = 2$, in the first octant

13. $\mathbf{F} = 2z\mathbf{k}$; D inside $x^2 + y^2 + z^2 = 4$ and above $z = 1$

14. $\mathbf{F} = x\mathbf{i} + y\mathbf{j} + z\mathbf{k}$; D inside both $x^2 + y^2 + z^2 = 8$ and $z = \sqrt{x^2 + y^2}$

15. $\mathbf{F} = (x\mathbf{i} + y\mathbf{j} + z\mathbf{k})/(x^2 + y^2 + z^2)$; D the sphere with center at the origin and radius r

16. $\mathbf{F} = (x\mathbf{i} + y\mathbf{j} + z\mathbf{k})/(x^2 + y^2 + z^2)^2$; D the sphere with center at the origin and radius r

17. $\mathbf{F} = (x\mathbf{i} + y\mathbf{j} + z\mathbf{k})/(x^2 + y^2 + z^2)^{3/2}$; D the sphere with center (h, k, ℓ) and radius r, $h^2 + k^2 + \ell^2 < r^2$

18. $\mathbf{F} = (x\mathbf{i} + y\mathbf{j} + z\mathbf{k})/(x^2 + y^2 + z^2)^{3/2}$; D the sphere with center (h, k, ℓ) and radius r, $h^2 + k^2 + \ell^2 > r^2$

19. The electric field due to a charge at the origin is given by

$$\mathbf{E} = \frac{x\mathbf{i} + y\mathbf{j} + z\mathbf{k}}{(x^2 + y^2 + z^2)^{3/2}}.$$

What is the outward flux across the boundary of a cubic box that has center at the origin and length of each edge 3 cm?

20. What is the outward flux of the electric field of Exercise 19 across the boundary of the ellipsoid $4x^2 + 4y^2 + 9z^2 = 36$?

21. Show that the flux of the electric field of Exercise 19 across that part of the sphere $x^2 + y^2 + z^2 = 25$ with $z \geq 3$ is equal to the flux across that part of the plane $z = 3$ with $x^2 + y^2 \leq 16$.

Connections

Cross products, 11.4.

Vector-valued functions, 11.7.

Line integrals, 14.2.

Surface integrals, 14.5.

14.7 STOKES' THEOREM

We have seen in Section 14.3 that Green's Theorem relates the circulation of a vector function $\mathbf{F} = M\mathbf{i} + N\mathbf{j}$ around a simple closed curve C in the xy-plane to the integral of the scalar curl $N_x - M_y$ over the region R in the xy-plane enclosed by C. In this section we will investigate a three-dimensional version of this result, called Stokes' Theorem.

Curl of a Vector Field

Let us define the **curl** of a vector field $\mathbf{F} = M\mathbf{i} + N\mathbf{j} + P\mathbf{k}$ to be

$$\text{curl } \mathbf{F} = \left(\frac{\partial P}{\partial y} - \frac{\partial N}{\partial z} \right)\mathbf{i} + \left(\frac{\partial M}{\partial z} - \frac{\partial P}{\partial x} \right)\mathbf{j} + \left(\frac{\partial N}{\partial x} - \frac{\partial M}{\partial y} \right)\mathbf{k}.$$

If $\mathbf{F} = M(x, y)\mathbf{i} + N(x, y)\mathbf{j}$, so M and N depend only on x and y and $P = 0$, we have $M_z = N_z = P_x = P_y = 0$, and

$$\text{curl } \mathbf{F} = \left(\frac{\partial N}{\partial x} - \frac{\partial M}{\partial y} \right)\mathbf{k};$$

the scalar curl

$$\frac{\partial N}{\partial x} - \frac{\partial M}{\partial y}$$

measures the circulation per unit area about the axis of rotation $\mathbf{k}$. See Figure 14.7.1a. Generally, the direction of the vector curl $\mathbf{F}$ at a point determines an axis of rotation of the circulation of $\mathbf{F}$ and the magnitude of curl $\mathbf{F}$ measures the magnitude of the net circulation about that axis at that point. See Figure 14.7.1b. The expression (curl $\mathbf{F}$) $\cdot \mathbf{n}$ measures the magnitude of the net circulation of $\mathbf{F}$ about the axis of rotation determined by the unit vector $\mathbf{n}$. See Figure 14.7.1c. If θ is the angle between curl $\mathbf{F}$ and the unit vector $\mathbf{n}$, we have

$$(\text{curl } \mathbf{F}) \cdot \mathbf{n} = |\text{curl } \mathbf{F}| \, |\mathbf{n}| \cos \theta = |\text{curl } \mathbf{F}| \cos \theta,$$

so we see that the circulation of $\mathbf{F}$ about $\mathbf{n}$ is greatest with magnitude $|\text{curl } \mathbf{F}|$ when $\mathbf{n}$ is the unit vector in the direction of curl $\mathbf{F}$.

It is convenient to express curl $\mathbf{F}$ in terms of the symbolic vector del,

$$\nabla = \frac{\partial}{\partial x}\mathbf{i} + \frac{\partial}{\partial y}\mathbf{j} + \frac{\partial}{\partial z}\mathbf{k}.$$

If $\mathbf{F} = M\mathbf{i} + N\mathbf{j} + P\mathbf{k}$, we use the symbolism

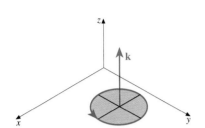

FIGURE 14.7.1a The scalar curl of a function of (x, y) measures the circulation per unit area about the .vertical axis **k**.

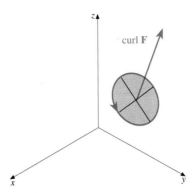

FIGURE 14.7.1b The direction of curl **F** determines an axis of rotation of the circulation of **F** and the magnitude of curl **F** measures the circulation about that axis.

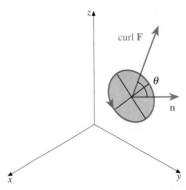

FIGURE 14.7.1c (curl **F**)·**n** measures the magnitude of the circulation about the axis of rotation determined by the unit vector **n**.

$$\nabla \times \mathbf{F} = \begin{vmatrix} \mathbf{i} & \mathbf{j} & \mathbf{k} \\ \dfrac{\partial}{\partial x} & \dfrac{\partial}{\partial y} & \dfrac{\partial}{\partial z} \\ M & N & P \end{vmatrix}$$

$$= \begin{vmatrix} \dfrac{\partial}{\partial y} & \dfrac{\partial}{\partial z} \\ N & P \end{vmatrix}\mathbf{i} - \begin{vmatrix} \dfrac{\partial}{\partial x} & \dfrac{\partial}{\partial z} \\ M & P \end{vmatrix}\mathbf{j} + \begin{vmatrix} \dfrac{\partial}{\partial x} & \dfrac{\partial}{\partial y} \\ M & N \end{vmatrix}\mathbf{k}$$

$$= \left(\dfrac{\partial P}{\partial y} - \dfrac{\partial N}{\partial z}\right)\mathbf{i} + \left(\dfrac{\partial M}{\partial z} - \dfrac{\partial P}{\partial x}\right)\mathbf{j} + \left(\dfrac{\partial N}{\partial x} - \dfrac{\partial M}{\partial y}\right)\mathbf{k},$$

so

$$\text{curl } \mathbf{F} = \nabla \times \mathbf{F}.$$

The cross product form is very convenient for remembering the definition of curl **F**. We can evaluate curl **F** either by substituting into the defining formula or by expressing the cross product as a determinant.

EXAMPLE 1
$\mathbf{F} = yz\mathbf{i} - xz\mathbf{j} + xy\mathbf{k}$.

    (a) Find curl **F**.
    (b) Evaluate $\iint_S (\text{curl } \mathbf{F}) \cdot \mathbf{n}\, dS$, where S is that part of the paraboloid $z = x^2 + y^2$ with $x^2 + y^2 \le 4$ and **n** is directed upward.
    (c) Evaluate $\int_S (\text{curl } \mathbf{F}) \cdot \mathbf{n}\, dS$, where S is that part of the plane $z = 4$ with $x^2 + y^2 \le 4$ and **n** is directed upward.
    (d) Evaluate $\int_C \mathbf{F} \cdot \mathbf{T}\, ds$, where

$$C: x = 2\cos t, \quad y = 2\sin t, \quad z = 4, \quad 0 \le t \le 2\pi.$$

SOLUTION

    (a) Using the determinant form of curl **F**, we have

$$\nabla \times \mathbf{F} = \begin{vmatrix} \mathbf{i} & \mathbf{j} & \mathbf{k} \\ \dfrac{\partial}{\partial x} & \dfrac{\partial}{\partial y} & \dfrac{\partial}{\partial z} \\ yz & -xz & xy \end{vmatrix}$$

$z = x^2 + y^2, x^2 + y^2 \leq 4$

n

FIGURE 14.7.2 The surface consists of that part of the paraboloid above the region in the xy-plane bounded by $x^2 + y^2 = 4$.

$$= \begin{vmatrix} \dfrac{\partial}{\partial y} & \dfrac{\partial}{\partial z} \\ -xz & xy \end{vmatrix} \mathbf{i} - \begin{vmatrix} \dfrac{\partial}{\partial x} & \dfrac{\partial}{\partial z} \\ yz & xy \end{vmatrix} \mathbf{j} + \begin{vmatrix} \dfrac{\partial}{\partial x} & \dfrac{\partial}{\partial y} \\ yz & -xz \end{vmatrix} \mathbf{k}$$

$$= (x + x)\mathbf{i} - (y - y)\mathbf{j} + (-z - z)\mathbf{k}$$

$$= 2x\mathbf{i} - 2z\mathbf{k}.$$

(b) We apply Formula 8 of Section 14.5 to the vector field curl $\mathbf{F} = 2x\mathbf{i} - 2z\mathbf{k}$. The surface is given by the formula $z = x^2 + y^2$, for (x, y) in the region $R_{x,y}$ of the xy-plane bounded by $x^2 + y^2 = 2^2$. See Figure 14.7.2. The components of curl $\mathbf{F} = \hat{M}\mathbf{i} + \hat{N}\mathbf{j} + \hat{P}\mathbf{k}$ are $\hat{M} = 2x$, $\hat{N} = 0$, and $\hat{P} = -2z = -2(x^2 + y^2)$. Then

$$\iint_S (\text{curl } \mathbf{F}) \cdot \mathbf{n} \, dS$$

$$= \iint_{R_{x,y}} (-\hat{M} z_x - \hat{N} z_y + \hat{P}) \, dA$$

$$= \iint_{R_{x,y}} \{-(2x)(2x) - (0)(2y) + [-2(x^2 + y^2)]\} \, dA$$

$$= \int_0^{2\pi} \int_0^2 (-4r^2 \cos^2 \theta - 2r^2) r \, dr \, d\theta \qquad \text{[in polar coordinates]}$$

$$= \int_0^{2\pi} \int_0^2 (-4r^3 \cos^2 \theta - 2r^3) \, dr \, d\theta$$

$$= \int_0^{2\pi} \left[-r^4 \cos^2 \theta - \frac{r^4}{2} \right]_{r=0}^{r=2} \, d\theta$$

$$= \int_0^{2\pi} (-16 \cos^2 \theta - 8) \, d\theta$$

$$= -8(\sin \theta \cos \theta + \theta) - 8\theta \Big]_0^{2\pi}$$

$$= -8(2\pi) - 8(2\pi) = -32\pi.$$

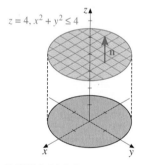

$z = 4, x^2 + y^2 \leq 4$

n

FIGURE 14.7.3 The surface consists of that part of the plane above the region in the xy-plane bounded by $x^2 + y^2 = 4$.

(c) We apply Formula 8 found in Section 14.5 to the vector field curl $\mathbf{F} = 2x\mathbf{i} - 2z\mathbf{k}$. The surface is given by $z = 4$, for (x, y) in the region $R_{x,y}$ of the xy-plane bounded by $x^2 + y^2 = 2^2$. See Figure 14.7.3. The components of curl $\mathbf{F} = \hat{M}\mathbf{i} + \hat{N}\mathbf{j} + \hat{P}\mathbf{k}$ are $\hat{M} = 2x$, $\hat{N} = 0$, and $\hat{P} = -2z = -2(4) = -8$. Then

$$\iint_S (\text{curl } \mathbf{F}) \cdot \mathbf{n} \, dS = \iint_{R_{x,y}} (-\hat{M} z_x - \hat{N} z_y + \hat{P}) \, dA$$

$$= \iint_{R_{x,y}} [-(2x)(0) - (0)(0) + (-8)] \, dA$$

$$= -8(\text{Area of } R_{x,y}) = -8(\pi 2^2) = -32\pi.$$

(d) The curve is sketched in Figure 14.7.4. Note that C is the boundary curve (or edge) of the surfaces in parts (b) and (c). We have $\mathbf{F} = yz\mathbf{i} - xz\mathbf{j} + xy\mathbf{k}$, so $M = yz$, $N = -xz$, and $P = xy$. Also, C: $x = 2 \cos t$, $y = 2 \sin t$, $z = 4$, $0 \leq t \leq 2\pi$. Then

$$\int_C \mathbf{F} \cdot \mathbf{T} \, ds = \int M \, dx + N \, dy + P \, dz$$

$$= \int_C yz \, dx - xz \, dy + xy \, dz$$

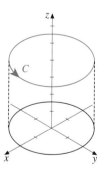

C

FIGURE 14.7.4 This curve is the boundary (or edge) of the surfaces in Figures 14.7.2 and 14.7.3.

$$= \int_0^{2\pi} [(2\sin t)(4)(-2\sin t) - (2\cos t)(4)(2\cos t)$$
$$+ (2\cos t)(2\sin t)(0)]dt$$
$$= \int_0^{2\pi} [-16\sin^2 t - 16\cos^2 t]dt$$
$$= \int_0^{2\pi} -16dt = -32\pi. \quad \blacksquare$$

The equality of the two surface integrals and the line integral in Example 1 is a consequence of a general result called Stokes' Theorem. To state this theorem, we need to establish a relation between the direction of the unit normal vector of a surface and the direction of the boundary curve of the surface.

Oriented Surfaces

We will consider surfaces S that are **oriented** in the sense that the surface has a unit normal vector **n** at each point and that the direction of **n** varies continuously over the surface. The boundary curve C (the edge) of an oriented surface S is said to be **n-directed** if the unit tangent vector **T** of C satisfies the condition that $\mathbf{T} \times \mathbf{n}$ points away from the surface. (Recall from Section 11.4 that the vectors $\mathbf{T}, \mathbf{n}$, and $\mathbf{T} \times \mathbf{n}$, in that order, form a right-hand system.) Informally, this means if you walked along the boundary of S in the direction of **T** with your head in the direction of **n**, the surface would be on your left. See Figure 14.7.5a. This is consistent with the relation between regions in the plane and the direction of their boundary curves that we used in Green's Theorem, if the unit normal vector **n** of the plane surface is chosen to be **k**. See Figure 14.7.5b.

If $f(x, y)$ has continuous first partial derivatives, the surface

$$z = f(x, y), \quad (x, y) \text{ in a region of the } xy\text{-plane}$$

is oriented; **n** may be chosen consistently to be either upward at every point or downward at every point. See Figure 14.7.6. Note the direction of the **n**-directed boundary curve in each case.

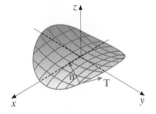

FIGURE 14.7.5a The boundary curve of a surface is **n**-directed if the surface would be to the left as you walked along the edge of the surface in the direction of **T** with your head in the direction of **n**.

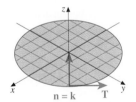

FIGURE 14.7.5b A region in the xy-plane is **n**-directed with **n** = **k** if the region is to the left as the boundary curve is traced.

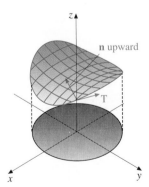

FIGURE 14.7.6a A surface of the form $z = f(x, y)$ is oriented. The unit normal vector may be chosen either upward at every point or downward at every point.

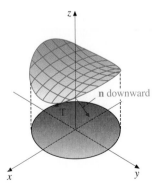

FIGURE 14.7.6b The direction of an **n**-oriented boundary curve depends on the choice of **n**.

The surface of a sphere is oriented. The unit normal vector may be chosen consistently to point either outward at every point or inward at every point. This surface has no boundary curve. See Figure 14.7.7.

The surface that consists of that part of the cylinder $y^2 + z^2 = 1$ between the planes $x = 0$ and $x = 4$ is oriented; the normal may be chosen consistently to point either outward at every point or inward at every point. The boundary consists of two circles. The directions of the **n**-directed boundary curves for **n** outward are indicated in Figure 14.7.8.

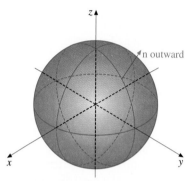

FIGURE 14.7.7 A sphere is an oriented surface.

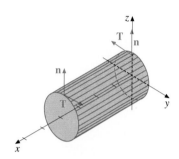

FIGURE 14.7.8 The cylinder is oriented. The **n**-oriented boundary curves have opposite directions.

FIGURE 14.7.9 A Möbius band can be constructed by turning over one end of a strip and then joining the ends.

A **Möbius band** is an example of a simple surface that is not oriented. A Möbius band can be constructed from a strip of paper by turning over one end and then joining the ends, as indicated in Figure 14.7.9. There is no choice of unit normal vectors that vary continuously over the surface. That is, if the unit normal vectors varied continuously around the Möbius band, then they would vary continuously over the flat surface from which the band was constructed. This means the normal vectors would all point toward the same side of the flat surface. The unit vectors would then be in the opposite direction on each side of the line where the ends are joined to form the Möbius band. The normal vector **n** cannot vary continuously at points where it abruptly changes direction, so it cannot vary continuously over the entire Möbius band.

Stokes' Theorem

STOKES' THEOREM

Suppose S is an oriented surface with unit normal vector **n**, C is the **n**-directed boundary curve of S, and **T** is the unit tangent vector of C. If **F** has continuous first partial derivatives, then

$$\int_C \mathbf{F} \cdot \mathbf{T}\,ds = \int\int_S (\text{curl } \mathbf{F}) \cdot \mathbf{n}\,dS.$$

Proof of some special cases If the surface S consists of a region in the xy-plane with $\mathbf{n} = \mathbf{k}$ and $\mathbf{F} = M(x, y)\mathbf{i} + N(x, y)\mathbf{j}$, then curl $\mathbf{F} = (N_x - M_y)\mathbf{k}$

and Stokes' Theorem reduces to Green's Theorem. Let us use Green's Theorem to prove Stokes' Theorem in the case where the surface S is of the form

$$z = z(x, y), \quad (x, y) \text{ in } R_{x,y}, \quad \mathbf{n} \text{ upward,}$$

where z has continuous second partial derivatives. In this case we can apply Formula 8 of Section 14.6 to the function

$$\text{curl } \mathbf{F} = (P_y - N_z)\mathbf{i} + (M_z - P_x)\mathbf{j} + (N_x - M_y)\mathbf{k}$$

to obtain

$$\iint_S (\text{curl } \mathbf{F}) \cdot \mathbf{n} \, dS = \iint_{R_{x,y}} (-(P_y - N_z)z_x - (M_z - P_x)z_y + (N_x - M_y)) \, dA.$$

On the other hand, let us suppose that the boundary curve of the region $R_{x,y}$ in the xy-plane is

$$C_{x,y}: x = x(t), \quad y = y(t), \quad a \le t \le b.$$

We can then use the equation of the surface to express the boundary curve C of S in parametric form as

$$C: x = x(t), \quad y = y(t), \quad z = z(x(t), y(t)), \quad a \le t \le b.$$

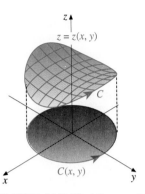

FIGURE 14.7.10 We can obtain a parametrization of C by adding the equation $z = z(x(t), y(t))$ to parametric equations of the curve $C(x, y)$ in the xy-plane.

See Figure 14.7.10. We then have

$$\int_C \mathbf{F} \cdot \mathbf{T} \, ds = \int_C M \, dx + N \, dy + P \, dz$$

$$= \int_a^b \left(M\frac{dx}{dt} + N\frac{dy}{dt} + P\left(z_x\frac{dx}{dt} + z_y\frac{dy}{dt}\right) \right) dt$$

$$= \int_{C_{x,y}} (M + Pz_x) \, dx + (N + Pz_y) \, dy,$$

which by Green's Theorem is

$$= \iint_{R_{x,y}} ((N + Pz_y)_x - (M + Pz_x)_y) \, dA.$$

Using the Chain Rule to evaluate the partial derivatives, we have

$$(N + Pz_y)_x = \frac{\partial}{\partial x}(N(x, y, z(x, y)) + P(x, y, z(x, y))z_y(x, y))$$

$$= (N_x + N_z z_x) + [(P)(z_{yx}) + (z_y)(P_x + P_z z_x)]$$

$$= N_x + N_z z_x + Pz_{yx} + P_x z_y + P_z z_x z_y$$

and, similarly,

$$(M + Pz_x)_y = M_y + M_z z_y + Pz_{xy} + P_y z_x + P_z z_x z_y.$$

Then

$$\iint_{R_{x,y}} ((N + Pz_y)_x - (M + Pz_x)_y) \, dA$$

$$= \iint_{R_{x,y}} ((N_x + N_z z_x + Pz_{yx} + P_x z_y + P_z z_x z_y)$$

$$- (M_y + M_z z_y + Pz_{xy} + P_y z_x + P_z z_x z_y)) \, dA$$

$$= \iint_{R_{x,y}} (-(P_y - N_z)z_x - (M_z - P_x)z_y + (N_x - M_y)) \, dA,$$

where we have used the fact that the continuity of the second partial derivatives of z implies that $z_{xy} = z_{yx}$. Combining results gives the conclusion of Stokes' Theorem in this special case.

We could prove Stokes' Theorem for surfaces given by either $y = y(x, z)$ or $x = x(y, z)$ in a manner analogous to that used in the case $z = z(x, y)$. We could then obtain Stokes' Theorem for oriented surfaces that can be divided into a finite number of the above type surfaces. ■

We may interpret (curl $\mathbf{F}$)·$\mathbf{n}$ as the circulation per unit surface area about a point on the surface in the plane tangent to the surface at the point. Each point on the surface contributes to the circulation around the boundary curve of the surface.

Stokes' Theorem says the circulation around the boundary of a surface is the sum of the contributions from all points on the surface.

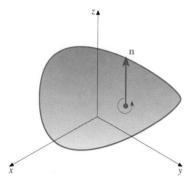

FIGURE 14.7.11a   (curl **F**)·**n** measures the circulation of **F** in the plane tangent to the surface at a point.

See Figure 14.7.11.

Using Stokes' Theorem

EXAMPLE 2

Use Stokes' Theorem to evaluate $\iint_S (\text{curl } \mathbf{F}) \cdot \mathbf{n}\,dS$, where $\mathbf{F} = z\mathbf{i} + x\mathbf{j} + y\mathbf{k}$, S is that part of the sphere $z = \sqrt{5 - x^2 - y^2} - 1$ with $z \geq 0$, and **n** is directed upward.

SOLUTION   The boundary of the surface is the intersection of the sphere and the plane $z = 0$. When $z = 0$, $z = \sqrt{5 - x^2 - y^2} - 1$ implies $1 = 5 - x^2 - y^2$, so $x^2 + y^2 = 2^2$. Since the unit normal is upward, we want to choose the direction of the boundary curve to be from the positive x-axis toward the positive y-axis. See Figure 14.7.12. We choose

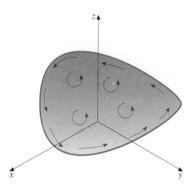

FIGURE 14.7.11b   Each point on a surface contributes to the circulation around the boundary curve of the surface.

$$C: x = 2\cos t, \quad y = 2\sin t, \quad z = 0, \quad 0 \leq t \leq 2\pi.$$

(It is a good idea to check the points on the curve that correspond to $t = 0$ and $t = \pi/2$ to determine if the parametric equations of the curve give the correct direction.)

We have $\mathbf{F} = z\mathbf{i} + x\mathbf{j} + y\mathbf{k}$, so $M = z$, $N = x$, and $P = y$. Stokes' Theorem implies

$$\iint_S (\text{curl } \mathbf{F}) \cdot \mathbf{n}\,dS = \int_C \mathbf{F} \cdot \mathbf{T}\,dS$$

$$= \int_C M\,dx + N\,dy + P\,dz$$

$$= \int_C z\,dx + x\,dy + y\,dz$$

$$= \int_0^{2\pi} [(0)(-2\sin t) + (2\cos t)(2\cos t)$$

$$+ (2\sin t)(0)]\,dt$$

$$= \int_0^{2\pi} 4\cos^2 t\,dt$$

$$= 2\cos t \sin t + 2t \Big]_0^{2\pi} = 4\pi. \quad ■$$

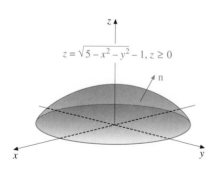

FIGURE 14.7.12   If we walked along the edge of the surface in the indicated direction with our head upward, the surface would be on the left.

$z = \sqrt{5 - x^2 - y^2} - 1, z \geq 0$

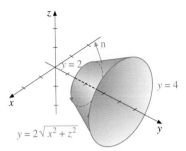

FIGURE 14.7.13 The two boundary curves have opposite directions.

EXAMPLE 3

Use Stokes' Theorem to evaluate $\iint_S (\text{curl } \mathbf{F}) \cdot \mathbf{n} \, dS$, where $\mathbf{F} = -yz\mathbf{i} + xy\mathbf{k}$, S is that part of the cone $y = 2\sqrt{x^2 + z^2}$ between $y = 2$ and $y = 4$, and $\mathbf{n}$ is outward.

SOLUTION   The surface is sketched in Figure 14.7.13. The surface has two boundary curves, one in the plane $y = 2$ and the other in the plane $y = 4$. Since $\mathbf{n}$ is outward, the right-side curve is in the direction from the positive x-axis toward the positive z-axis; the left-side curve is in the opposite direction. Let us choose

$$C_4: x = 2\cos t, \, y = 4, \, z = 2\sin t, 0 \le t \le 2\pi,$$
$$C_2: x = \sin t, \, y = 2, \, z = \cos t, 0 \le t \le 2\pi.$$

(Checking the points on the curves that correspond to $t = 0$ and $t = \pi/2$ shows that these curves are in the correct direction.) $\mathbf{F} = -yz\mathbf{i} + xy\mathbf{k}$ implies $M = -yz$, $N = 0$, and $P = xy$. Stokes' Theorem then gives

$$\iint_S (\text{curl } \mathbf{F}) \cdot \mathbf{n} \, dS = \int_C \mathbf{F} \cdot \mathbf{T} \, ds$$
$$= \int_C M \, dx + N \, dy + P \, dz$$
$$= \int_{C_4 + C_2} -yz \, dx + xy \, dz$$
$$= \int_{C_4} -yz \, dx + xy \, dz + \int_{C_2} -yz \, dx + xy \, dz$$
$$= \int_0^{2\pi} [-(4)(2\sin t)(-2\sin t) + (2\cos t)(4)(2\cos t)] \, dt$$
$$+ \int_0^{2\pi} [-(2)(\cos t)(\cos t) + (\sin t)(2)(-\sin t)] \, dt$$
$$= \int_0^{2\pi} [16\sin^2 t + 16\cos^2 t] \, dt$$
$$+ \int_0^{2\pi} [-2\cos^2 t - 2\sin^2 t] \, dt$$
$$= \int_0^{2\pi} 16 \, dt + \int_0^{2\pi} -2 \, dt = 32\pi - 4\pi = 28\pi. \quad \blacksquare$$

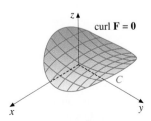

FIGURE 14.7.14 The net circulation around a surface that has no boundary curves is zero.

If S is a closed surface that has no boundary curves,

$$\iint_S (\text{curl } \mathbf{F}) \cdot \mathbf{n} \, dS = 0.$$

The net circulation of $\mathbf{F}$ around parts of the surface is zero, because of cancellation. See Figure 14.7.14.

If curl $\mathbf{F} = \mathbf{0}$ for all points in a region R in three-dimensional space, then the circulation of $\mathbf{F}$ around any smooth closed curve that is the boundary curve of a surface in R is zero. See Figure 14.7.15.

FIGURE 14.7.15 If curl $\mathbf{F} = \mathbf{0}$ for all points on a surface, then the circulation around the boundary of the surface is zero.

Independence of Path

If a region R in the plane has no holes, we know that $\int_C M\,dx + N\,dy$ is independent of path in R whenever M and N are continuous with scalar curl $N_x - M_y = 0$ in R. We also know that $\int M\,dx + N\,dy$ is independent of path in a region (with or without holes) if $\mathbf{F} = M\mathbf{i} + N\mathbf{j}$ is the gradient of a function U in the region. Analogous results hold for line integrals in three-dimensional space.

If C_1 and C_2 have the same initial point and the same terminal point, then $C_1 - C_2$ is a closed curve and

$$\int_{C_1} \mathbf{F} \cdot \mathbf{T}\,ds = \int_{C_2} \mathbf{F} \cdot \mathbf{T}\,ds \text{ if and only if } \int_{C_1 - C_2} \mathbf{F} \cdot \mathbf{T}\,ds = 0.$$

If $C_1 - C_2$ is the complete boundary of a surface S on which $\mathbf{F}$ has continuous first partial derivatives, then Stokes' Theorem gives

$$\int_{C_1 - C_2} \mathbf{F} \cdot \mathbf{T}\,ds = \iint_S \text{curl } \mathbf{F} \cdot \mathbf{n}\,ds.$$

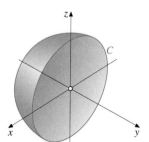

FIGURE 14.7.16   If curl $\mathbf{F} = \mathbf{0}$ for all points on a surface with complete boundary $C_1 - C_2$, then the circulation around $C_1 - C_2$ is zero.

See Figure 14.7.16. If curl $\mathbf{F} = \mathbf{0}$ at all points of S, it follows that $\int_{C_1} \mathbf{F} \cdot \mathbf{T}\,ds = \int_{C_2} \mathbf{F} \cdot \mathbf{T}\,ds$. This argument can be used to verify the following theorem.

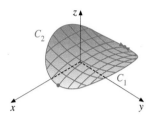

FIGURE 14.7.17   If a region R in the plane has holes, then some closed curves in the region are not the complete boundary of the region in R they enclose.

THEOREM 1

Suppose an open region R in three-dimensional space has the property that every smooth simple closed curve in R is the complete boundary of a bounded surface S in R. If $\mathbf{F}$ has continuous first partial derivatives and curl $\mathbf{F} = \mathbf{0}$ at all points of R, then $\int_C \mathbf{F} \cdot \mathbf{T}\,ds$ is independent of path in R.

FIGURE 14.7.18a   A curve in three-dimensional space can be the complete boundary of a surface in R even if R has interior holes.

The property that every smooth simple closed curve in the three-dimensional region R be the complete boundary of a bounded surface in R corresponds to the condition that a region in the plane not contain any holes, although the situation is more complicated in three-dimensional space. For example, a region in the plane that has interior holes does not have the property that every smooth simple closed curve in the region is the complete boundary of a surface that is *in the plane region*. See Figure 14.7.17. A curve in three-dimensional space can be the complete boundary of a surface in a three-dimensional region even if the region has interior holes. The surface can go around a hole in three-dimensional space, as indicated in Figure 14.7.18a. However, the region that consists of all of three-dimensional space except for a line does not have the desired property; a curve that encircles the line cannot be the complete boundary of a bounded surface that does not intersect the line. See Figure 14.7.18b.

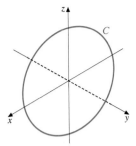

FIGURE 14.7.18b A curve that encircles a line cannot be the complete boundary of a bounded surface that does not intersect the line.

If $\int_C \mathbf{F} \cdot \mathbf{T}\, ds$ is independent of path in R, it may be easier to evaluate the integral by replacing the given curve with a more convenient path that has the same initial and terminal points.

If $\int_C \mathbf{F} \cdot \mathbf{T}\, ds$ is independent of path in R, then the line integral of $\mathbf{F} \cdot \mathbf{T}$ over a closed curve in R is zero.

Conservative Fields

The following theorem can be proved in a manner analogous to that of the corresponding result in the plane that was proved in Section 14.4.

THEOREM 2

Suppose M, N, and P are continuous in a region R.

(a) If there is a function U such that $\nabla U = M\mathbf{i} + N\mathbf{j} + P\mathbf{k}$ in R, then $\int_C M\, dx + N\, dy + P\, dz$ is independent of path in R. Moreover, if C is a smooth curve in R with initial point $(x(a), y(a), z(a))$ and terminal point $(x(b), y(b), z(b))$, then

$$\int_C M\, dx + N\, dy + P\, dz = U(x(b), y(b), z(b)) - U((x(a), y(a), z(a)).$$

(b) If $\int_C M\, dx + N\, dy + P\, dz$ is independent of path in R, then there is a function U such that $\nabla U = M\mathbf{i} + N\mathbf{j} + P\mathbf{k}$ in R.

As in the case of two-dimensional vector fields, a three-dimensional vector field $\mathbf{F}$ is called **conservative** in R if there is a function U such that $\mathbf{F} = \nabla U$ in R. The function $P = -U$ is then called the **potential** of $\mathbf{F}$.

The following theorem corresponds to Theorem 3 of Section 14.4.

THEOREM 3

If $\mathbf{F}$ has continuous first partial derivatives in a region R that has the property that every smooth simple closed curve in R is the complete boundary of a surface in R, and curl $\mathbf{F} = \mathbf{0}$ in R, then there is a function U such that $\nabla U = \mathbf{F}$ in R, so $\mathbf{F}$ is conservative in R.

If $\mathbf{F}$ has continuous first derivatives in R and there are points of R where curl $\mathbf{F} \neq \mathbf{0}$, then there is no function U such that $\nabla U = \mathbf{F}$ in R, so $\mathbf{F}$ is not conservative in R.

Proof The first statement of Theorem 3 follows from Theorems 1 and 2. To prove the second statement, we note that $\nabla U = M\mathbf{i} + N\mathbf{j} + P\mathbf{k}$ implies $U_x = M$, $U_y = N$, and $U_z = P$. Then $U_{xy} = M_y$, $U_{xz} = M_z$, $U_{yx} = N_x$, $U_{yz} = N_z$, $U_{zx} = P_x$, and $U_{zy} = P_y$. The continuity of the first partial derivatives of $\mathbf{F}$ implies the continuity of the second partials of U, so we have $U_{xy} = U_{yx}$, $U_{xz} = U_{zx}$, and $U_{yz} = U_{zy}$. It follows that curl $\mathbf{F} = (P_y - N_z)\mathbf{i} + (M_z - P_x)\mathbf{j} + (N_x - M_y)\mathbf{k} = \mathbf{0}$. ∎

If curl $\mathbf{F} = \mathbf{0}$ in an appropriate region R, we can find a function U such that $\nabla U = \mathbf{F}$ by using a method analogous to that used in Section 14.4.

EXERCISES 14.7

1. $\mathbf{F} = z\mathbf{i} + x\mathbf{j} + y\mathbf{k}$
 (a) Find curl $\mathbf{F}$.
 (b) Evaluate $\iint_S (\text{curl } \mathbf{F}) \cdot \mathbf{n}\, dS$, where S is that part of the paraboloid $z = x^2 + y^2$ with $x^2 + y^2 \le 1$, and $\mathbf{n}$ is downward.
 (c) Evaluate $\iint_S (\text{curl } \mathbf{F}) \cdot \mathbf{n}\, dS$, where S is that part of the plane $z = 1$ with $x^2 + y^2 \le 1$, and $\mathbf{n}$ is downward.
 (d) Evaluate $\int_C \mathbf{F} \cdot \mathbf{T}\, ds$, where $C: x = \sin t$, $y = \cos t$, $z = 1$, $0 \le t \le 2\pi$.

2. $\mathbf{F} = -yz\mathbf{i} + xz\mathbf{j} - xy\mathbf{k}$
 (a) Find curl $\mathbf{F}$.
 (b) Evaluate $\iint_S (\text{curl } \mathbf{F}) \cdot \mathbf{n}\, dS$, where S is that part of the cone $z = 2\sqrt{x^2 + y^2}$ with $x^2 + y^2 \le 4$, and $\mathbf{n}$ is upward.
 (c) Evaluate $\iint_S (\text{curl } \mathbf{F}) \cdot \mathbf{n}\, dS$, where S is that part of the plane $z = 4$ with $x^2 + y^2 \le 4$, and $\mathbf{n}$ is upward.
 (d) Evaluate $\int_C \mathbf{F} \cdot \mathbf{T}\, ds$, where $C: x = 2\cos t$, $y = 2\sin t$, $z = 4$, $0 \le t \le 2\pi$.

3. $\mathbf{F} = -y\mathbf{i} + x\mathbf{j} + \sqrt{x^2 + y^2}\,\mathbf{k}$
 (a) Find curl $\mathbf{F}$.
 (b) Evaluate $\iint_S (\text{curl } \mathbf{F}) \cdot \mathbf{n}\, dS$, where S is the hemisphere $z = \sqrt{1 - x^2 - y^2}$ and $\mathbf{n}$ is outward.
 (c) Evaluate $\iint_S (\text{curl } \mathbf{F}) \cdot \mathbf{n}\, dS$, where S is the hemisphere $z = -\sqrt{1 - x^2 - y^2}$ and $\mathbf{n}$ is outward.
 (d) Evaluate $\int_C \mathbf{F} \cdot \mathbf{T}\, ds$, where $C: x = \cos t$, $y = \sin t$, $z = 0$, $0 \le t \le 2\pi$.

4. $\mathbf{F} = -y^3\mathbf{i} + x^3\mathbf{j} + z^3\mathbf{k}$
 (a) Find curl $\mathbf{F}$.
 (b) Evaluate $\iint_S (\text{curl } \mathbf{F}) \cdot \mathbf{n}\, dS$, where S is that part of the cone $z = 2\sqrt{x^2 + y^2}$ between $z = 2$ and $z = 4$, and $\mathbf{n}$ is outward.
 (c) Evaluate $\int_C \mathbf{F} \cdot \mathbf{T}\, ds$, where $C: x = 2\cos t$, $y = 2\sin t$, $z = 4$, $0 \le t \le 2\pi$.
 (d) Evaluate $\int_C \mathbf{F} \cdot \mathbf{T}\, ds$, where $C: x = \sin t$, $y = \cos t$, $z = 2$, $0 \le t \le 2\pi$.

In Exercises 5–14, use Stokes' Theorem to evaluate $\iint_S (\text{curl } \mathbf{F}) \cdot \mathbf{n}\, dS$ for the given vector functions $\mathbf{F}$, the given surfaces S, and the indicated unit normal vector.

5. $\mathbf{F} = -y\mathbf{i} + x\mathbf{j} + z\mathbf{k}$; $z = 9 - x^2 - y^2$, $x^2 + y^2 \le 9$; $\mathbf{n}$ upward

6. $\mathbf{F} = x\mathbf{i} + y\mathbf{j} + z\mathbf{k}$; $z = 16 - x^2 - y^2$, $z \ge 12$; $\mathbf{n}$ upward

7. $\mathbf{F} = y\mathbf{i} - x\mathbf{j} - z\mathbf{k}$; $z = \sqrt{9 - x^2 - y^2}$; $\mathbf{n}$ upward

8. $\mathbf{F} = y\mathbf{i} - x\mathbf{j} - z\mathbf{k}$; $x^2 + y^2 + z^2 = 9$; $\mathbf{n}$ outward

9. $\mathbf{F} = x\mathbf{i} + y\mathbf{j} + z\mathbf{k}$; $z = \sqrt{x^2 + y^2}$, $z \le 2$; $\mathbf{n}$ outward

10. $\mathbf{F} = x\mathbf{i} + y\mathbf{j} + z\mathbf{k}$; $z = \sqrt{x^2 + y^2}$, $1 \le x \le 2$; $\mathbf{n}$ outward

11. $\mathbf{F} = -yz\mathbf{i} + xz\mathbf{j} - xy\mathbf{k}$; S is that part of the cylinder $x^2 + y^2 = 4$ with $0 \le z \le 2$; $\mathbf{n}$ outward

12. $\mathbf{F} = x^2\mathbf{i} + 2yz\mathbf{j} + z^2\mathbf{k}$; S is that part of the surface $z^2 = x^2 + y^2 - 1$ with $-1 \le z \le 1$; $\mathbf{n}$ outward

13. $\mathbf{F} = y\mathbf{k}$; $x = \sqrt{9 - y^2 - z^2}$; $\mathbf{n}$ outward

14. $\mathbf{F} = yz\mathbf{i} + xz\mathbf{j} - xy\mathbf{k}$; $y = \sqrt{9 - x^2 - z^2}$, $y \ge 2$; $\mathbf{n}$ outward

15. Evaluate
$$\int_C \sin(x^2)\, dx + (z + \cos(y^2))\, dy + \frac{1}{1 + z^4}\, dz,$$
where C is the polygonal path from $(1, 0, 0)$, to $(0, 1, 0)$, to $(0, 0, 1)$, and back to $(1, 0, 0)$. (*Hint: C is the boundary of a region in the plane $z = 1 - x - y$.*)

16. Evaluate
$$\int_C \left(z + \frac{1}{1 + x^2} \right) dx + \ln(1 + y^4)\, dy + e^{2z}\, dz,$$
where C is the polygonal path from $(1, 0, 0)$, to $(1, 1, 0)$, to $(0, 1, 1)$, to $(0, 0, 1)$, and back to $(1, 0, 0)$. (*Hint: C is the boundary of a region in the plane $z = 1 - x$.*)

17. Evaluate
$$\int_C y^2z^3\, dx + 2xyz^3\, dy + 3xy^2z^2\, dz,$$
where C is the polygonal path from $(1, 1, 1)$, to $(-1, 1, -1)$, to $(-1, -1, 1)$, to $(1, -1, -1)$, and back to $(1, 1, 1)$.

18. Evaluate
$$\int_C yz\, dx + xz\, dy + xy\, dz,$$
where C is the polygonal path from $(1, 0, 1)$, to $(0, 1, -1)$, to $(-1, 0, 1)$, to $(0, -1, -1)$, and back to $(1, 0, 1)$.

19. Evaluate $\int_C \mathbf{F} \cdot \mathbf{T}\, ds$, where
$$\mathbf{F} = \frac{x\mathbf{i} + y\mathbf{j} + z\mathbf{k}}{(x^2 + y^2 + z^2)^{3/2}}$$
and $C: x = \cos t$, $y = \sin t$, $z = z_0$, $0 \le t \le 2\pi$, z_0 a constant.

20. Evaluate $\int_C \mathbf{F} \cdot \mathbf{T}\, ds$, where
$$\mathbf{F} = \frac{x\mathbf{i} + y\mathbf{j} + z\mathbf{k}}{(x^2 + y^2 + z^2)^{3/2}}$$
and $C: x = \cos t$, $y = \sin t$, $z = z(t)$, $0 \le t \le 2\pi$, $z(t)$ any function that has a continuous derivative and $z(0) = z(2\pi)$.

21. Evaluate div curl $\mathbf{F}$, the divergence of curl $\mathbf{F}$. Assume the second partial derivatives are continuous.

22. Evaluate curl ∇U, the curl of the gradient of U. Assume the second partial derivatives are continuous.

Show that the line integrals in Exercises 23–26 are independent of path and find their value for the indicated curves.

23. $\displaystyle\int_C 2xyz\, dx + x^2z\, dy + x^2y\, dz$
 (a) C is the polygonal path from $(0, 0, 0)$, to $(2, 0, 0)$, to $(2, 3, 0)$, to $(2, 3, 4)$, to $(0, 3, 4)$
 (b) C is the polygonal path from $(0, 0, 0)$, to $(2, 0, 0)$, to $(2, 3, 0)$, to $(2, 3, 4)$, to $(0, 3, 4)$, to $(0, 0, 0)$

24. $\displaystyle\int_C 2xz^3\,dx + 3x^2z^2\,dz$
   (a) $C: x = \cos t,\ y = \sin t,\ z = t,\ 0 \le t \le 2\pi$
   (b) $C: x = \cos t,\ y = \sin t,\ z = \sin 2t,\ 0 \le t \le 2\pi$

25. $\displaystyle\int_C x\,dx + y\,dy + z\,dz$
   (a) $C: x = 2\cos t,\ y = 3\sin t,\ z = \cos t,\ 0 \le t \le \pi$
   (b) $C: x = 2\cos t,\ y = 3\sin t,\ z = \cos t,\ 0 \le t \le 2\pi$

26. $\displaystyle\int_C (x + y)\,dx + x\,dy + z\,dz$
   (a) $C: x = t - t^2,\ y = (t - t^2)^2,\ z = t^3,\ 0 \le t \le 1$
   (b) $C: x = t^2 - t,\ y = (t^2 - t)^2,\ z = t^3,\ 0 \le t \le 1$

In Exercises 27–30, find a function U such that ∇U is equal to the given vector function $\mathbf{F}$.

27. $\mathbf{F} = (2xy + z^2)\mathbf{i} + (x^2 + 2yz)\mathbf{j} + (y^2 + 2xz)\mathbf{k}$

28. $\mathbf{F} = (x^2 + y)\mathbf{i} + (x + z)\mathbf{j} + (y + 2z)\mathbf{k}$

29. $\mathbf{F} = \dfrac{x}{\sqrt{x^2 + y^2 + z^2}}\mathbf{i} + \dfrac{y}{\sqrt{x^2 + y^2 + z^2}}\mathbf{j} + \dfrac{z}{\sqrt{x^2 + y^2 + z^2}}\mathbf{k},$
   $(x, y, z) \ne (0, 0, 0)$

30. $\mathbf{F} = \dfrac{x}{x^2 + y^2 + z^2}\mathbf{i} + \dfrac{y}{x^2 + y^2 + z^2}\mathbf{j} + \dfrac{z}{x^2 + y^2 + z^2}\mathbf{k},$
   $(x, y, z) \ne (0, 0, 0)$

Show that the line integrals in Exercises 31–34 are independent of path and find their value.

31. $\displaystyle\int_{(0,0,0)}^{(1,2,-2)} x^2\,dx + y^2\,dy + z^2\,dz$

32. $\displaystyle\int_{(0,0,0)}^{(1,3,2)} (3x^2 + y^2)\,dx + 2xy\,dy + dz$

33. $\displaystyle\int_{(0,0,0)}^{(1,-1,2)} 3x^2\,dx + z^2\,dy + (2yz + 1)\,dz$

34. $\displaystyle\int_{(0,0,0)}^{(3,\pi/2,4)} yz\cos xy\,dx + xz\cos xy\,dy + \sin xy\,dz$

35. (a) Show that the inverse square field

$$\mathbf{F} = \frac{x\mathbf{i} + y\mathbf{j} + z\mathbf{k}}{(x^2 + y^2 + z^2)^{3/2}}$$

   satisfies curl $\mathbf{F} = \mathbf{0}$, $(x, y, z) \ne (0, 0, 0)$.
   (b) Find $U(x, y, z)$ such that $\nabla U = \mathbf{F}$, $(x, y, z) \ne (0, 0, 0)$.

36. Evaluate

$$\int_C x(x^2 + y^2 + z^2)^{-3/2}\,dx + y(x^2 + y^2 + z^2)^{-3/2}\,dy$$
$$+ z(x^2 + y^2 + z^2)^{-3/2}\,dz,$$

   where C is any smooth curve that is contained in the sphere $x^2 + y^2 + z^2 = r^2$.

37. The electrostatic force on a certain charged particle at the point (x, y, z) is

$$\mathbf{F} = \frac{-5(x\mathbf{i} + y\mathbf{j} + z\mathbf{k})}{(x^2 + y^2 + z^2)^{3/2}}.$$

   Find the work associated with $\mathbf{F}$ as the particle moves along the polygonal path from $(1, 0, 0)$ to $(0, 1, 0)$, to $(0, 1, 1)$.

REVIEW OF CHAPTER 14 CONCEPTS

- Line integrals $\int_C f\,ds$ are defined in terms of differential arc length. They are evaluated by substitution of parametric values.
- Many important applications of line integrals are of the form $\int_C M\,dx + N\,dy + P\,dz$.
- If $\mathbf{F} = M\mathbf{i} + N\mathbf{j}$, the circulation of $\mathbf{F}$ around a closed plane curve C is $\int_C \mathbf{F} \cdot \mathbf{T}\,ds = \int_C M\,dx + N\,dy$; the flux across C is $\int_C \mathbf{F} \cdot \mathbf{n}\,ds = \int_C M\,dy - N\,dx$.
- Green's Theorem states

$$\int_C M\,dx + N\,dy = \iint_R (N_x - M_y)\,dA.$$

   C must be the complete boundary of the region R and C must have the correct direction. The functions M and N must have continuous partial derivatives on C and the region enclosed by C.
- If Green's Theorem applies, the circulation of $\mathbf{F}$ around a closed plane curve C is

$$\int_C M\,dx + N\,dy = \iint_R (N_x - M_y)\,dA;$$

the divergence across C is

$$\int_C M\,dy - N\,dx = \iint_R (M_x + N_y)\,dA.$$

- You should know conditions that imply that a line integral is independent of path. We may choose any convenient path to evaluate a line integral that is independent of path.
- Surface integrals $\iint_S f\,dS$ are defined in terms of surface area.
- The flux of a function $\mathbf{F}$ across a surface S is $\iint_S \mathbf{F} \cdot \mathbf{n}\,dS$.
- The formula

$$\iint_S \mathbf{F} \cdot \mathbf{n}\,dS = \iint_{R_{x,y}} (-Mx_x - Nx_y + P)\,dA$$

applies to surfaces $z = z(x, y)$, (x, y) in $R_{x,y}$, with the unit normal vector $\mathbf{n}$ directed upward. Similar formulas hold for surfaces that are given as functions of either (y, z) or (x, z).
- The divergence of a function $\mathbf{F}$ is div $\mathbf{F} = \nabla \cdot \mathbf{F}$.

- The Divergence Theorem states

$$\iint_S \mathbf{F} \cdot \mathbf{n}\, dS = \iiint_D (M_x + N_y + P_z)\, dV.$$

  S must be the complete boundary of the region D.
- Green's Theorem in the form

$$\int_C \mathbf{F} \cdot \mathbf{n}\, ds = \iint_R (M_x + N_y)\, dA$$

corresponds to the Divergence Theorem.
- The curl of a function $\mathbf{F}$ is curl $\mathbf{F} = \nabla \times \mathbf{F}$.
- Stokes' Theorem states $\int_C \mathbf{F} \cdot \mathbf{T}\, ds = \iint_S (\text{curl } \mathbf{F}) \cdot \mathbf{n}\, dS$. The surface and boundary curve must be correctly oriented.
- Stokes' Theorem is a three-dimensional version of Green's Theorem.

CHAPTER 14 REVIEW EXERCISES

1. Set up and evaluate a line integral that gives the lateral surface area of that part of the cylinder $4x^2 + y^2 = 4$ in the first octant below $z = xy$.

2. Set up and evaluate a line integral that gives the lateral surface area of that part of the cylinder $y = x^2$, $0 \le x \le 1$, between $z = 0$ and $z = x$.

3. Set up and evaluate a line integral that gives the total mass of a wire bent in the shape of the curve

$$x = 2\cos t, \quad y = 3\sin t, \quad z = 4t, \quad 0 \le t \le \pi/2,$$

if the density at a point on the wire is given by the formula $\rho(x, y, z) = xy$.

4. Set up and evaluate a line integral that gives the total mass of a wire bent in the shape of the curve

$$C : x = t\cos t, \quad y = t\sin t, \quad z = t, \quad 0 \le t \le 2\pi,$$

if the density at a point on the wire is given by the formula $\rho(x, y, z) = \sqrt{x^2 + y^2 + z^2}$.

Evaluate the line integrals in Exercises 5–22.

5. $\int_C y\, ds$, C goes along $x = y^2$ from $(0, 0)$ to $(1, 1)$.

6. $\int_C y^2\, ds$, $C : x = \ln t$, $y = t^2$, $1 \le t \le 2$

7. $\int_C y\, dx - x\, dy$, $C : x = \cos t$, $y = \sin t$, $0 \le t \le 2\pi$

8. $\int_C y\, dx + dy$, C goes from $(0, 0)$, along $y = x^2$ to $(1, 1)$.

9. $\int_C (x^2 - y^2)\, dx - 2xy\, dy$, C goes from $(0, 0)$, along $y = \sin x$ to $(2\pi, 0)$.

10. $\int_C 2xy\, dx + (x^2 + y)\, dy$, C goes along $y = x^2$ from $(0, 0)$ to $(1, 1)$, then along $x + y = 2$ from $(1, 1)$ to $(0, 2)$.

11. $\int_C xy^2\, dx + x^2 y\, dy$, C is the polygonal path from $(0, 0)$, to $(1, 0)$, to $(2, 2)$, to $(0, 1)$, to $(0, 0)$.

12. $\int_C \frac{1}{2}\ln(x^2 + y^2)\, dx + \tan^{-1}(x/y)\, dy$, C goes from $(3, 2)$, counterclockwise once around $(x - 2)^2 + (y - 2)^2 = 1$.

13. $\int_C (y^2 - 2xy)\, dx + 2xy\, dy$, C is the polygonal path from $(0, 0)$, to $(1, 0)$, to $(1, 1)$, to $(0, 0)$.

14. $\int_C (y\sin x - y)\, dx + (y - \cos x)\, dy$, C is the polygonal path from $(0, 0)$, to $(1, 0)$, to $(1, 1)$, and then along $y = x^2$ from $(1, 1)$ to $(0, 0)$.

15. $\int_C x\, dx + y\, dy$, C goes along the hyperbola $x^2 - y^2 = 1$ from $(2, -\sqrt{3})$ to $(2, \sqrt{3})$.

16. $\int_C 2y\, dx + 3x\, dy$, C goes from $(3, 0)$, counterclockwise once around the ellipse $4x^2 + 9y^2 = 36$.

17. $\int_C (e^x - y^3)\, dx + (e^y + x^3)\, dx$, C goes from $(1, 0)$, counterclockwise once around $x^2 + y^2 = 1$.

18. $\int_C (\ln(2 - x) - y)\, dx + (\ln(2 - y) + x)\, dy$, C goes from $(1, 0)$, counterclockwise once around $x^2 + y^2 = 1$.

19. $\int_C x\, dx + y\, dy + z\, dz$; $C : x = 4\cos t$, $y = 4\sin t$, $z = \sin 2t$, $0 \le t \le 2\pi$

20. $\int_C y\, dx + x\, dy + z\, dz$; $C : x = e^{-t}\cos t$, $y = 2t$, $z = e^{-t}\sin t$, $0 \le t \le 2\pi$

21. $\int_C (x + y)\, dx + (x + z)\, dy + (y + z)\, dz$, C goes from $(1, 0, 0)$, counterclockwise once around the intersection of the sphere $x^2 + y^2 + z^2 = 1$ and the surface $z = xy$.

22. $\int_C x(x^2 + y^2 + z^2)^{-3/2}\, dx + y(x^2 + y^2 + z^2)^{-3/2}\, dy + z(x^2 + y^2 + z^2)^{-3/2}\, dz$; $C : x = \cos t$, $y = \sin t$, $z = \sin t + \cos t$, $0 \le t \le \pi/2$

In Exercises 23–24, an object is to be moved from $(0, 0)$ to $(3, 0)$, subject to the given force $\mathbf{F}$. Determine whether the work done is independent of path. If it is, evaluate the work.

23. $\mathbf{F} = x\mathbf{i} - y\mathbf{j}$

24. $\mathbf{F} = y\mathbf{i} - x\mathbf{j}$

Find the scalar curl of **F** *and the circulation around C in Exercises 25–28.*

25. $\mathbf{F} = y\mathbf{i} - x\mathbf{j}$, C goes from $(2, 0)$, counterclockwise once around $x^2 + y^2 = 4$.

26. $\mathbf{F} = x\mathbf{i} + y\mathbf{j}$, C goes from $(3, 0)$, counterclockwise once around $x^2 + y^2 = 9$.

27. $\mathbf{F} = \dfrac{x}{x^2 + y^2}\mathbf{i} + \dfrac{y}{x^2 + y^2}\mathbf{j}$, C goes from $(4, 0)$, counterclockwise once around $x^2 + y^2 = 16$.

28. $\mathbf{F} = \dfrac{-y}{x^2 + y^2}\mathbf{i} + \dfrac{x}{x^2 + y^2}\mathbf{j}$, C goes from $(4, 0)$, counterclockwise once around $x^2 + y^2 = 16$.

Find the divergence of **F** *and the flux across C in Exercises 29–32.*

29. $\mathbf{F} = y\mathbf{i} - x\mathbf{j}$, C goes from $(2, 0)$, counterclockwise once around $x^2 + y^2 = 4$.

30. $\mathbf{F} = x\mathbf{i} + y\mathbf{j}$, C goes from $(3, 0)$, counterclockwise once around $x^2 + y^2 = 9$.

31. $\mathbf{F} = \dfrac{x}{x^2 + y^2}\mathbf{i} + \dfrac{y}{x^2 + y^2}\mathbf{j}$, C goes from $(4, 0)$, counterclockwise once around $x^2 + y^2 = 16$.

32. $\mathbf{F} = \dfrac{-y}{x^2 + y^2}\mathbf{i} + \dfrac{x}{x^2 + y^2}\mathbf{j}$, C goes from $(4, 0)$ counterclockwise once around $x^2 + y^2 = 16$.

In Exercises 33–36, find a function $U(x, y)$ such that $\nabla U = \mathbf{F}$ for all (x, y) or indicate "not possible."

33. $\mathbf{F} = e^x \cos y\mathbf{i} - (e^x \sin y + 2y)\mathbf{j}$

34. $\mathbf{F} = (1 + 2xy)\mathbf{i} + (x^2 + y)\mathbf{j}$

35. $\mathbf{F} = y\mathbf{i} - x\mathbf{j}$

36. $\mathbf{F} = x\mathbf{i} - y\mathbf{j}$

In Exercises 37–38, find a function $U(x, y, z)$ such that $\nabla U = \mathbf{F}$ for all (x, y, z) or indicate "not possible."

37. $\mathbf{F} = (y^2 - z^2)\mathbf{i} + 2xy\mathbf{j} + (1 - 2xz)\mathbf{k}$

38. $\mathbf{F} = y\cos xy\mathbf{i} + (x\cos xy + z\sin yz)\mathbf{j} + y\sin yz\mathbf{k}$

39. Find the area of that part of the cone $z = \sqrt{x^2 + y^2}$ between $z = 1$ and $z = 2$.

40. Find the area of that part of the sphere $x^2 + y^2 + z^2 = 4$ that is inside the cylinder $x^2 + (y - 1)^2 = 1$ and satisfies $z > 0$.

41. Find the area of that part of the plane $x + y + z = 4$ that is inside the cylinder $(y - 1)^2 + z^2 = 1$.

42. Find the area of that part of the paraboloid $y = 4 - x^2 - z^2$ that is in the first octant.

43. Evaluate $\iint_S z\,dS$, where S is that part of the sphere $x^2 + y^2 + z^2 = 5$ above the plane $z = 1$.

44. Evaluate $\iint_S 1/z\,dS$, where S is that part of the sphere $x^2 + y^2 + z^2 = 5$ above the plane $z = 2$.

45. Evaluate $\iint_S(17 - 4x)\,dS$, where S is that part of the paraboloid $x = 4 - y^2 - z^2$ with $x \geq 0$.

46. Evaluate $\iint_S y^2\,dS$, where S is that part of the cone $y = 4 - 2\sqrt{x^2 + z^2}$ with $y \geq 0$.

Evaluate $\iint_S \mathbf{F} \cdot \mathbf{n}\,dS$ in Exercises 47–52.

47. $\mathbf{F} = x\mathbf{i} + y\mathbf{j} + z\mathbf{k}$; S is that part of the paraboloid $z = x^2 + y^2$ with $z \leq 1$; $\mathbf{n}$ downward.

48. $\mathbf{F} = x\mathbf{i} + y\mathbf{j} + z\mathbf{k}$; S is that part of the plane $z = 1$ with $x^2 + y^2 \leq 1$; $\mathbf{n}$ upward.

49. $\mathbf{F} = x\mathbf{i} + y\mathbf{j} + z\mathbf{k}$; S is the complete boundary of the region bounded by $z = x^2 + y^2$ and $z = 1$; $\mathbf{n}$ outward.

50. $\mathbf{F} = 2x\mathbf{i} + 2y\mathbf{j} + z\mathbf{k}$; S is the complete boundary of the region bounded by the cone $z = \sqrt{x^2 + y^2}$ and the plane $z = 4$; $\mathbf{n}$ outward.

51. $\mathbf{F} = 2x\mathbf{i} + 2y\mathbf{j} + z\mathbf{k}$; S is that part of the cone $z = \sqrt{x^2 + y^2}$ with $z \leq 4$; $\mathbf{n}$ downward.

52. $\mathbf{F} = 2x\mathbf{i} + 2y\mathbf{j} + z\mathbf{k}$; S is that part of the plane $z = 4$ with $x^2 + y^2 \leq 16$; $\mathbf{n}$ upward.

53. Evaluate $\iint_S \operatorname{curl} \mathbf{F} \cdot \mathbf{n}\,dS$, where $\mathbf{F} = -yz\mathbf{i} + xz\mathbf{j} + xy\mathbf{k}$, S is that part of the surface $z = x^2 + y^2$ below $z = 1$, and $\mathbf{n}$ is directed downward.

54. Evaluate $\iint_S \operatorname{curl} \mathbf{F} \cdot \mathbf{n}\,dS$, where $\mathbf{F} = -yz\mathbf{i} + xz\mathbf{j} + xy\mathbf{k}$, S is that part of the sphere $x^2 + y^2 + z^2 = 5$ below $z = 2$, and $\mathbf{n}$ is directed inward.

55. Evaluate $\iint_S \operatorname{curl} \mathbf{F} \cdot \mathbf{n}\,dS$, where $\mathbf{F} = -yz\mathbf{i} + xz\mathbf{j} + xy\mathbf{k}$, S is that part of the cylinder $x^2 + y^2 = 1$ between $z = 0$ and $z = 4$, and $\mathbf{n}$ is directed outward.

56. Evaluate $\iint_S \operatorname{curl} \mathbf{F} \cdot \mathbf{n}\,dS$, where $\mathbf{F} = -yz\mathbf{i} + xz\mathbf{j} + xy\mathbf{k}$, S is that part of the sphere $x^2 + y^2 + z^2 = 4$ that is between $y = 1$ and $y = -1$, and $\mathbf{n}$ is directed outward.

57. Find the unit vector $\mathbf{n}$ that is normal to the cone $x = \sqrt{y^2 + z^2}$ and directed outward from the cone.

58. Find the unit vector $\mathbf{n}$ that is normal to the paraboloid $2y = x^2 + z^2$ and directed outward from the paraboloid.

59. Show that $\operatorname{div}(f\mathbf{V}) = f(\operatorname{div}\mathbf{V}) + \nabla f \cdot \mathbf{V}$.

60. Show that $\operatorname{div}(\mathbf{U} \times \mathbf{V}) = \mathbf{V} \cdot \operatorname{curl}\mathbf{U} - \mathbf{U} \cdot \operatorname{curl}\mathbf{V}$.

61. Explain why $\int_{-C} f(x)\,ds = \int_C f(x, y)\,ds$, but

$$\int_{-C} M(x, y)\,dx + N(x, y)\,dy$$
$$= -\int_C M(x, y)\,dx + N(x, y)\,dy.$$

62. Discuss similarities of Green's Theorem and the Divergence Theorem.

EXTENDED APPLICATION

GREEN'S THEOREM IN LAND DEVELOPMENT

You have been hired as a computer consultant by an architectural engineering firm. One of your assignments is to write a program that will compute the area of each tract of land to be developed. The engineers need to know the area in order to determine lot sizes, as well as to ensure that the planned system of storm sewers is adequate to handle all rain runoff.

Information concerning the boundaries of each land tract is recorded on a blueprint, in a form similar to that shown in Figure 1. For this blueprint, a coordinate system has been chosen so that the x-axis is parallel to A_0A_1, and point A_0 has arbitrarily been assigned the coordinates (1000, 1000). All angles refer to interior angles. Your program must accept as input the lengths and angles from a blueprint of any polygonal tract.

$$
\begin{array}{ll}
A_0A_1 = 720 \text{ ft} & \angle A_1 = 51° \\
A_1A_2 = 195 \text{ ft} & \angle A_2 = 332° \\
A_2A_3 = 172 \text{ ft} & \angle A_3 = 81° \\
A_3A_4 = 172 \text{ ft} & \angle A_4 = 148° \\
A_4A_5 = 160 \text{ ft} & \angle A_5 = 90° \\
A_5A_6 = 860 \text{ ft} & \angle A_6 = 81° \\
A_6A_0 = 154.47 \text{ ft} & \angle A_0 = 117°
\end{array}
$$

Although the area of the tract in Figure 1 can be computed by subdividing the polygon into triangles, you realize quickly that a general scheme for finding a polygonal

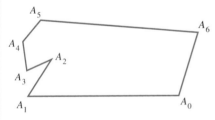

FIGURE 1

area through subdivision is difficult to program. You therefore consider the following alternative approach: As a consequence of Green's Theorem, the area of the tract is given by

$$
\frac{1}{2} \int_C (y\,dx - x\,dy),
$$

where C is the boundary of the tract, traversed in the clockwise direction. If the boundary is made up of n line segments, denote the length of the ith segment by L_i, its angle of inclination to the x-axis by α_i, and its initial point by (x_i, y_i). It can then be described by the parametric equations

$$
x = x_i + t\cos\alpha_i, \quad y = y_i + t\sin\alpha_i, \quad 0 \le t \le L_i.
$$

All values of L_i are given on the blueprint, as is the value of (x_0, y_0).

Tasks:

1. Develop a formula to express (x_{i+1}, y_{i+1}) in terms of (x_i, y_i), L_i, and α_i.
2. Develop a formula to express α_{i+1} in terms of α_i and A_{i+1}.

Using the values $x_0 = 1000$, $y_0 = 1000$, and $\alpha_0 = 180°$, your program can employ these formulas to find all remaining values of (x_i, y_i) and α_i.

3. For the tract shown in Figure 1, list all values of (x_i, y_i), L_i, and α_i.
4. Show that

$$
\frac{1}{2} \int_C (y\,dx - x\,dy) = \sum_{i=0}^{n-1} \frac{L_i}{2}(y_i\cos\alpha_i - x_i\sin\alpha_i).
$$

Use this fact to find the area of the tract shown in Figure 1.
5. (Optional) Write a program, in the language of your choice, to find the area of an arbitrary polygonal tract.

Writing Assignment:

Write a letter to the president of the engineering firm, explaining both how and why your program works. Use your calculations from Task 4 to illustrate.

Source: Marvin Keener and Jeanne Agnew, Oklahoma State University, *Industry Related Problems for Mathematics Students.*

15 Differential Equations

One of the most important applications of calculus is to the solution of differential equations. In Section 7.6 we saw some differential equations that are related to exponential growth and decay. In this chapter we will study a wider variety of differential equations and see how the concept relates to selected applications.

15.1 INTRODUCTION; SEPARABLE DIFFERENTIAL EQUATIONS

Connections

The derivative as a slope, 2.6.

Exponential growth and decay, 7.6.

A central idea in the application of mathematics to practical problems is to find a function that has some prescribed properties. In many problems the desired properties involve rates of change or derivatives of the unknown function. Thus, we are led to an equation that relates an unknown function and one or more of its derivatives to an independent variable. Such equations are called **differential equations.**

Some Basic Terminology

We begin with some basic definitions.

The **order** of a differential equation is the order of the highest-order derivative that occurs in the equation.

A function is said to be a **solution** of a differential equation if the function satisfies the equation for all values of the independent variable in some interval. The set of all solutions of a differential equation is called the **general solution** of the equation.

EXAMPLE 1

The equation $y' = 2y$ is a first-order differential equation. The unknown function y is considered to be a function of an independent variable x. We have seen in Section 7.6 that, for each constant C, the function $y(x) = Ce^{2x}$ is a solution. Also, we know that any solution must be of the form $y(x) = Ce^{2x}$ for some particular constant C. This means that the general solution of the differential equation $y' = 2y$ can be expressed as the collection of functions $y = Ce^{2x}$. ∎

EXAMPLE 2

The equation $y'' - 3y' + 2y = 0$ is a second-order differential equation. It is easy to check that, for any particular constants C_1 and C_2, the function $y(x) = C_1e^x + C_2e^{2x}$ is a solution. We will see in Section 15.5 that all solutions must be of this form, so the general solution can be expressed as $y = C_1e^x + C_2e^{2x}$. ∎

EXAMPLE 3

The equation $2xyy' = x^2 + y^2$ is a first-order differential equation. If C is any constant and y is a differentiable function that is defined implicitly by the equation $y^2 = x^2 + Cx$, then $2yy' = 2x + C$ and substitution shows that y is a solution. Later in this section we will see that all solutions are of this form, so the general solution of the differential equation $2xyy' = x^2 + y^2$ can be expressed by the equation $y^2 = x^2 + Cx$. ∎

Some methods of solving differential equations lead to solutions that are given implicitly by an equation in x and y, as in Example 3. Solutions in **implicit form** are considered satisfactory. It is not necessary to express y as an explicit function of x.

We should expect that the general solution of a first-order equation involves one arbitrary constant C and the general solution of a second-order equation involves two arbitrary constants. The general solution of an nth-order differential equation ordinarily involves n arbitrary constants. Determining the form of the general solution of a differential equation involves concepts that are not ordinarily considered a part of calculus. We will deal with this question only in very limited special cases.

The questions of **existence** and **uniqueness** of the solution of a differential equation are central to the theory. As with the question of the general solution, we will deal with these questions only in very limited special cases.

Generally speaking, in order to have a unique solution to an nth-order equation, it is necessary to impose n conditions on the solution. These conditions are used to determine the value of the n unknown constants in the general solution. One way is to prescribe the value of the solution and its first $n - 1$ derivatives at a fixed **initial point** x_0. That is, for fixed numbers $x_0, y_0, y_1, \ldots, y_{n-1}$, it is required that

$$y(x_0) = y_0, \quad y'(x_0) = y_1, \quad \ldots, \quad y^{(n-1)}(x_0) = y_{n-1}.$$

The problem of finding a solution of a differential equation for which the function and some of its derivatives have prescribed values at a point is called an **initial value problem**.

Separable Equations

Certain types of differential equations can be solved by integration.

We already know that $\int f(x)\,dx$ represents the collection of all functions y for which $y'(x) = f(x)$ on an interval. This means that:

The general solution of the first-order differential equation $y' = f(x)$ can be expressed as $y = \int f(x)\,dx$.

Note that we can think of obtaining the general solution by integration of the equation $y'(x) = f(x)$ with respect to x.

EXAMPLE 4

Find the general solution of the differential equation $y' = 2 \sin x + 3 \cos x$.

SOLUTION If y is any solution of the differential equation, then

$$y'(x) = 2 \sin x + 3 \cos x$$

for all x on some interval. Integrating with respect to x and evaluating then gives

$$\int y'(x)\,dx = \int (2 \sin x + 3 \cos x)\,dx,$$
$$y(x) = -2 \cos x + 3 \sin x + C,$$

where we have combined the constants of integration on the two sides of the equation as a single constant C. All solutions of the differential equation must be of this form and every function of this form is a solution of the differential equation. We can then express the general solution as

$$y = -2 \cos x + 3 \sin x + C. \quad \blacksquare$$

Let us now see how to use substitution and change of variables to solve differential equations of the form

$$M(x) + N(y)y' = 0.$$

Differential equations of this type are called **separable**.

Note that equations of the form $y' = f(x)$ are separable, since we can write the equation as $-f(x) + y' = 0$. This is of the form $M(x) + N(y)y' = 0$ with $M(x) = -f(x)$ and $N(y) = 1$.

If $y(x)$ is a solution of the separable differentiable equation $M(x) + N(y)y' = 0$, we have

$$M(x) + N(y(x))y'(x) = 0,$$
$$N(y(x))y'(x) = -M(x).$$

If M, N, and y' are continuous, we can integrate the above equation with respect to x to obtain

$$\int N(y(x))y'(x)\,dx = -\int M(x)\,dx.$$

Evaluation of the integral on the right gives a function of x. We can use the substitution $y = y(x)$, so that $dy = y'(x)\,dx$, to transform the integral on the left to $\int N(y)\,dy$. This integral can then be evaluated as a function of y. We then obtain an equation that gives the solution y as an implicit function of x. In some cases we can solve for y to obtain an explicit solution.

Curves

We will illustrate the solution of separable differential equations with some applications to curves.

We know that the slope of the line tangent to the curve $y = y(x)$ at a point (x, y) on the curve is given by $m = y'(x)$. A curve with prescribed slope m at each point is a solution of the differential equation $y' = m$. Generally, the prescribed slope m at a point (x, y) can depend on both x and y.

EXAMPLE 5

Find the equation of the curve that contains the point $(1, 2)$ and has slope $m = xy^2/2$ at each point (x, y) on the curve.

SOLUTION   The slope at a point (x, y) on the curve is the derivative of y with respect to x, so we need a solution of the initial value problem

$$y' = \frac{xy^2}{2}, \quad y(1) = 2.$$

If $y(x)$ is a solution of the differential equation, we have

$$y'(x) = \frac{x[y(x)]^2}{2},$$

$$[y(x)]^{-2}y'(x) = \frac{x}{2}.$$

Integrating with respect to x, we then obtain

$$\int [y(x)]^{-2}y'(x)\,dx = \int \frac{x}{2}\,dx.$$

The substitution $y = y(x)$, $dy = y'(x)\,dx$ in the integral on the left then gives

$$\int y^{-2}\,dy = \int \frac{x}{2}\,dx,$$

$$-\frac{1}{y} = \frac{x^2}{4} + C,$$

$$-\frac{1}{y(x)} = \frac{x^2}{4} + C.$$

Solving for $y(x)$, we obtain

$$y(x) = \frac{-4}{x^2 + 4C}.$$

The initial condition $y(1) = 2$ is then used to determine the value of C. Substituting $x = 1$ into the above equation and using the initial condition $y(1) = 2$, we obtain

$$2 = \frac{-4}{(1)^2 + 4C},$$

$$1 + 4C = -2,$$

$$C = -3/4.$$

The desired curve is then

$$y = \frac{-4}{x^2 - 3}.$$

See Figure 15.1.1. ∎

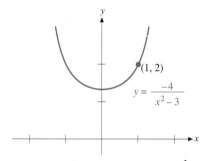

FIGURE 15.1.1   The curve $y = \dfrac{-4}{x^2 - 3}$ contains the point $(1, 2)$ and has slope $m = xy^2/2$ at each point (x, y) on the curve.

The collection of all curves that form the general solution of a first-order differential equation is called the **family of curves** associated with the differential equation.

The general solution of the separable differential equation

$$M(x) + N(y)y' = 0$$

is a family of curves with the property that the slope at a point (x, y) on any of the curves is

$$y' = -\frac{M(x)}{N(y)}.$$

Each curve of the family of curves that satisfy the separable differential equation

$$y' = \frac{N(y)}{M(x)}$$

has slope that is the negative reciprocal of the slope $y' = -M(x)/N(y)$. It follows that each time a curve from one of these families intersects a curve from the other family, *the curves intersect at right angles.* See Figure 15.1.2.

Two families of curves are called **orthogonal families** if each curve of one family intersects curves of the other family at right angles at every point of intersection.

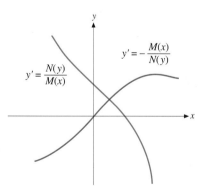

FIGURE 15.1.2   Curves with slopes $y' = -M(x)/N(y)$ intersect curves with slopes $y' = N(y)/M(x)$ at right angles whenever they intersect.

One example of orthogonal families of curves that occurs in applications involves steady-state heat flow along a thin, flat surface. The heat flows along a family of curves that is orthogonal to the family of isothermal curves where the temperature remains constant.

EXAMPLE 6

Determine the family of curves associated with the differential equation $x + 2yy' = 0$. Find an orthogonal family and sketch curves in each family.

SOLUTION   Let us first find the general solution of the equation

$$x + 2yy' = 0,$$
$$2yy' = -x.$$

Assuming that y is a function of x, we integrate with respect to x to obtain

$$\int 2yy'\,dx = -\int x\,dx.$$

Noting that $dy = y'dx$, we write

$$\int 2y\,dy = -\int x\,dx.$$

Evaluating the integrals then gives

$$y^2 = -\frac{x^2}{2} + C,$$
$$\frac{x^2}{2} + y^2 = C.$$

For $C > 0$, these curves are ellipses with center at the origin.

The family of curves associated with the equation $x + 2yy' = 0$ has slope $y' = -x/2y$. Any orthogonal family of curves has slope given by the negative reciprocal, so $y' = 2y/x$. Let us find the general solution of this equation. We have

$$\frac{1}{y}y' = \frac{2}{x}.$$

Integrating with respect to x gives

$$\int \frac{1}{y}y'\,dx = \int \frac{2}{x}\,dx,$$
$$\int \frac{1}{y}\,dy = \int \frac{2}{x}\,dx,$$
$$\ln|y| = 2\ln|x| + c,$$
$$|y| = e^{2\ln|x|+c},$$
$$|y| = e^{\ln x^2}e^c$$
$$y = Cx^2,$$

where $C = \pm e^c$. The curves in this family are parabolas.

Some curves from each family are sketched in Figure 15.1.3. ∎

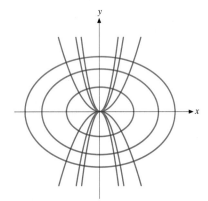

FIGURE 15.1.3 The ellipses $\frac{x^2}{2} + y^2 = C$ intersect the parabolas $y = Cx^2$ orthogonally at each point of intersection.

Homogeneous Equations

A first-order differential equation of the form $y' = f(y/x)$ is called a **homogeneous equation**.

A homogeneous equation can be transformed into a separable equation by setting $v = \dfrac{y}{x}$.

Then $y = xv$, so $y' = xv' + v$, and we have

$$y' = f\left(\frac{y}{x}\right),$$
$$xv' + v = f(v),$$
$$xv' = f(v) - v,$$
$$\frac{v'}{f(v) - v} = \frac{1}{x},$$

which is a separable differentiation in the unknown function v. If $v(x)$ is a solution of this equation, then $y(x) = xv(x)$ is a solution of the original equation $y' = f\left(\dfrac{y}{x}\right)$.

EXAMPLE 7

Find the general solution of the homogeneous differential equation $2xyy' = x^2 + y^2$.

SOLUTION   We have

$$2xyy' = x^2 + y^2,$$
$$y' = \frac{x^2 + y^2}{2xy},$$
$$y' = \frac{x}{2y} + \frac{y}{2x},$$
$$y' = \frac{1}{2y/x} + \frac{y/x}{2}.$$

We see that the right-hand side of the above equation is of the form $f(y/x)$, so we introduce the variable $v = y/x$. Then $y = xv$, so $y' = xv' + v$ and substitution gives

$$xv' + v = \frac{1}{2v} + \frac{v}{2},$$
$$xv' = \frac{1}{2v} - \frac{v}{2},$$
$$v' = -\frac{v^2 - 1}{2vx},$$
$$\frac{2v}{v^2 - 1}v' = -\frac{1}{x}.$$

Integrating with repect to x and using $v'\,dx = dv$, we obtain

$$\int \frac{2v}{v^2 - 1}\,dv = -\int \frac{1}{x}\,dx,$$

$$\ln|v^2 - 1| = -\ln|x| + c,$$
$$|v^2 - 1| = e^{\ln|x|^{-1} + c},$$
$$v^2 - 1 = \pm\frac{e^c}{x},$$
$$v^2 = 1 + \frac{C}{x},$$

where $C = \pm e^c$. Substitution of $v = \dfrac{y}{x}$ then gives the general solution,

$$\left(\frac{y}{x}\right)^2 = 1 + \frac{C}{x},$$
$$y^2 = x^2 + Cx.$$

This agrees with Example 3.   ■

EXERCISES 15.1

Show that the given functions are solutions of the differential equations in Exercises 1–6.

1. $y' = 2y$, $y(x) = Ce^{2x}$

2. $y'' - y' - 2y = 0$, $y(x) = C_1 e^{-x} + C_2 e^{2x}$

3. $y'' - 2y' + y = 0$, $y(x) = C_1 e^x + C_2 x e^x$

4. $y'' - 2y' + 2y = 0$, $y(x) = C_1 e^x \cos x + C_2 e^x \sin x$

5. $(1 + \ln y)y' = 2x$, $y \ln y = x^2 + C$

6. $yy' = x$, $y^2 - x^2 = C$

In Exercises 7–12, show that the given functions are solutions of the indicated initial value problems.

7. $y = 1 + e^{(1-x^2)/2}$; $y' + xy = x$, $y(1) = 2$

8. $y = \dfrac{\ln x}{x} + \dfrac{1}{x}$; $x^2 y' + xy = 1$, $y(1) = 1$

9. $y = 3e^x - e^{3x}$; $y'' - 4y' + 3y = 0$, $y(0) = 2$, $y'(0) = 0$

10. $y = e^{3x} - 4xe^{3x}$; $y'' - 6y' + 9y = 0$, $y(0) = 1$, $y'(0) = -1$

11. $y = \cos x - \sin x + e^{-x}$; $y'' + y = 2e^{-x}$, $y(0) = 2$, $y'(0) = -2$

12. $y = a \cosh\dfrac{x}{a}$; $y'' = \dfrac{1}{a}\sqrt{1 + (y')^2}$, $y(0) = a$, $y'(0) = 0$

Find the general solutions of the separable differential equations in Exercises 13–20.

13. $y' = 6x^2 - 4x + 3$

14. $y' = 6 \sin 2x - 12 \cos 3x$

15. $yy' = x$

16. $y^2 y' = 6$

17. $y^2 y' = x^2$

18. $\sqrt{y}\, y' = \sqrt{x}$

19. $\sqrt{x}\, y' = \sqrt{y}$

20. $yy' = (y^2 + 1)^2$

Solve the initial value problems in Exercises 21–34.

21. $y' + x = 0$, $y(1) = 2$

22. $y^2 y' = x^2$, $y(0) = 2$

23. $y' = 3y^2$, $y(0) = 1$

24. $y' = 3x^2 y^2$, $y(0) = 1$

25. $y' = \sec y$, $y(0) = 0$

26. $y' = \cos^2 y$, $y(0) = 0$

27. $y' = 1/\sqrt{x}$, $y(1) = 3$

28. $y' = \sin x$, $y(0) = 0$

29. $y'' = 6$, $y(0) = 2$, $y'(0) = 3$

30. $y'' = \sqrt{x}$, $y(1) = 1$, $y'(1) = 2$

31. $(y^2)' = 3x^2$, $y(0) = 1$

32. $\left(\dfrac{1}{y}\right)' = \dfrac{1}{x}$, $y(1) = 1$

33. $(xy)' = 2x$, $y(1) = 3$

34. $(x^2 y)' = \dfrac{1}{x}$, $y(1) = 1$

In Exercises 35–44, find the equation of the curve that contains the given point and has the given slope m at each point (x, y) on the curve.

35. $m = 4x - 1$, $(1, 3)$

36. $m = \sin(x/2)$, $(0, 1)$

37. $m = \dfrac{1}{1 - x}$, $(0, 1)$

38. $m = e^{2x}$, $(0, 0)$

39. $m = \dfrac{x}{y^2}$, $(3, 3)$

40. $m = \dfrac{2x^2}{y}$, $(1, 1)$

41. $m = 1 + y^2$, $(0, 0)$

42. $m = \sqrt{1 - y^2}$, $(0, 0)$

43. $m = y^2/x^2$, $(1, 2)$

44. $m = x/y$, $(3, 5)$

In Exercises 45–56, find the family of curves associated with the given differential equation and find an orthogonal family. Sketch three curves from each family.

45. $y' = e^x$

46. $y' = \dfrac{1}{1 + x^2}$

47. $y' = \sec^2 x$, $-\pi/2 < x < \pi/2$

48. $y' = 2y$

49. $y' = y^2$

50. $xy' = 3y$

51. $yy' + x = 0$

52. $y' = \sin x$, $0 < x < \pi$

53. $y' = y \tan x$, $-\pi/2 < x < \pi/2$

54. $y' = 2y \tan x$, $-\pi/2 < x < \pi/2$

55. $y' = \cot y$ (*Hint:* Sketch x as a function of y.)

56. $y' = y^2 + 1$ (*Hint:* Sketch x as a function of y.)

57. Find a differential equation satisfied by the family of curves $xy = C$ $(x > 0)$ and find a family orthogonal to this family.

58. The population at time t of a certain community is given approximately by the solution of the initial value problem

$$p' = 200p - 0.01p^2, \qquad p(0) = p_0.$$

Thus, the population increases when p is small and decreases when p is large. Find the solution for **(a)** $p(0) = 10{,}000$ and **(b)** $p(0) = 30{,}000$.

Solve the differential equations in Exercises 59–64.

59. $y' = \dfrac{y}{x} + 1$

60. $y' = \dfrac{y}{x} + \left(\dfrac{y}{x}\right)^2$

61. $xy' = 2y + x$

62. $2xyy' = y^2 - x^2$

63. $x^2 y' = x^2 + xy + y^2$

64. $xy' = y + \sqrt{y^2 - x^2}$, $x > 0$

15.2   EXACT DIFFERENTIAL EQUATIONS

Connections

Fundamental Theorem of Calculus, 5.4.

Finding U such that $\nabla U = \mathbf{F}$, 14.4.

In this section we will see how to use line integrals to solve certain differential equations of the form

$$M(x, y) + N(x, y)\frac{dy}{dx} = 0.$$

Exact Equations

If M and N have continuous first partial derivatives in an open region R, and $M_y = N_x$ in R, the differential equation

$$M(x, y) + N(x, y)\frac{dy}{dx} = 0 \qquad (1)$$

is said to be **exact** in R.

If the differential equation (1) is exact in a region that contains no holes, we know there is a differentiable function U of (x, y) such that

$$\nabla U = M\mathbf{i} + N\mathbf{j}, \quad \text{so } U_x = M \text{ and } U_y = N.$$

If $y = y(x)$ is differentiable with $(x, y(x))$ in R, the Chain Rule then gives

$$\frac{d}{dx}(U(x, y(x))) = U_x(x, y(x)) + U_y(x, y(x))\frac{dy}{dx}$$

$$= M(x, y(x)) + N(x, y(x))\frac{dy}{dx}.$$

It follows that, if $y = y(x)$ is any differentiable function that is defined implicitly by the equation

$$U(x, y) = K,$$

where K is a constant, then we can differentiate the equation with respect to x to obtain

$$M(x, y) + N(x, y)\frac{dy}{dx} = 0,$$

so $y = y(x)$ is a solution of the differential equation (1). On the other hand, if $y = y(x)$ is any solution of (1) with $(x, y(x))$ in R, we have

$$M(x, y(x)) + N(x, y(x))\frac{dy}{dx} = 0.$$

Then $U_x = M$ and $U_y = N$ imply

$$U_x(x, y(x)) + U_y(x, y(x))\frac{dy}{dx} = 0 \text{ so}$$

$$\frac{d}{dx}(U(x, y(x))) = 0, \text{ and}$$

$$U(x, y(x)) = K, \quad K \text{ a constant.}$$

This is, each solution of (1) is given implicitly by an equation

$$U(x, y) = K.$$

Let us summarize these results.

THEOREM 1

If the differential equation

$$M(x, y) + N(x, y)\frac{dy}{dx} = 0$$

is exact in a rectangular region R, then there is a differentiable function $U(x, y)$ such that $U_x = M$ and $U_y = N$ in R. If $y = y(x)$ is any differentiable function defined implicitly by the equation

$$U(x, y) = K,$$

then $y = y(x)$ is a solution of the differential equation. If $y = y(x)$ is any solution with $(x, y(x))$ in R, then y satisfies the equation $U(x, y) = K$ for some constant K.

If $U_x = M$ and $U_y = N$ in a rectangle R, then any solution of the exact differential equation

$$M + N\frac{dy}{dx} = 0$$

must satisfy the equation

$$U(x, y) = K, \quad K \text{ a constant.}$$

We can find a solution that satisfies an **initial condition**

$$y(x_0) = y_0$$

by substituting (x_0, y_0) into the equation $U(x, y) = K$ to obtain $K = U(x_0, y_0)$.

NOTE   *There may be points (x_0, y_0) in R for which the equation*

$$U(x, y) = U(x_0, y_0)$$

does not define y as a function of x. This means that the initial value problem may not have a solution $y = y(x)$ for every (x_0, y_0) in R. Even if there is a solution to the initial value problem, there may be a restriction on the values of x for which the solution is defined.

General Solutions

If the differential equation (1) is exact in a rectangle R, then $M_y = N_x$ in R and the function U of Theorem 1 satisfies $\nabla U = M\mathbf{i} + N\mathbf{j}$. It follows that we can use the method of Section 14.4 to find U.

EXAMPLE 1

Show that the differential equation

$$(2x - 2) + (8y - 16)\frac{dy}{dx} = 0$$

is exact and find the solution.

SOLUTION   We have $M = 2x - 2$ and $N = 8y - 16$, so $M_y = 0 = N_x$. This means the equation is exact in the entire xy-plane.

We now use the method of Section 14.4 to find a function U such that $U_x = 2x - 2$ and $U_y = 8y - 16$.

$U_x = 2x - 2$ implies $U = x^2 - 2x + f(y)$.
Then $U_y = 8y - 16$ implies $f'(y) = 8y - 16$, so $f(y) = 4y^2 - 16y + C$.

We conclude that the desired conditions are satisfied by the function

$$U(x, y) = x^2 - 2x + 4y^2 - 16y.$$

The general solution of the differential equation can be expressed in implicit form by the equation

$$x^2 - 2x + 4y^2 - 16y = K.$$

Graphs of the equation for several choices of the constant K are given in Figure 15.2.1a. The graphs are ellipses with center at the point $(1, 2)$.

None of the ellipses defines y as a function of x in a rectangle with center at a point with y-coordinate two; the differential equation does not have a solution that satisfies an initial condition $y(x_0) = 2$. If $y_0 > 2$, the graph of the solution of the differential equation that satisfies the initial condition $y(x_0) = y_0$ is the upper half of the ellipse that contains the point (x_0, y_0). If $y_0 < 2$, the graph of the solution that satisfies the initial condition $y(x_0) = y_0$ is the lower half of the ellipse that contains the point (x_0, y_0). For example, the graph of the solution that satisfies the initial condition $y(0) = 0$ is the lower half of the ellipse

$$x^2 - 2x + 4y^2 - 16y = 0.$$

See Figure 15.2.1b. There is no solution of the differential equation that is defined for all x. ∎

FIGURE 15.2.1a The differential equation has general solution given by the equation $x^2 - 2x + 4y^2 - 16y = K$. None of the ellipses defines y as a function of x in a rectangle with center at a point with y-coordinate two.

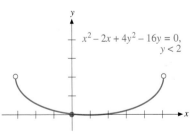

FIGURE 15.2.1b The solution that satisfies the initial condition $y(0) = 0$ is the lower half of the ellipse $x^2 - 2x + 4y^2 - 16y = 0$.

The differential equation in Example 1 is a separable differential equation. That is, it is of the form

$$M(x)\,dx + N(y)\,dy = 0.$$

All separable differential equations are exact, because

$$\frac{\partial}{\partial y}(M(x)) = 0 = \frac{\partial}{\partial x}(N(y)).$$

Let us now consider an exact differential equation that is not separable.

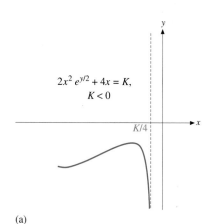

$2x^2\,e^{y/2} + 4x = K,$
$K < 0$

(a)

EXAMPLE 2

Show that the differential equation

$$(4xe^{y/2} + 4) + x^2e^{y/2}\frac{dy}{dx} = 0$$

is exact and find the solution.

SOLUTION   We have $M = 4xe^{y/2} + 4$ and $N = x^2e^{y/2}$, so $M_y = 4xe^{y/2}(1/2)$ and $N_x = 2xe^{y/2}$. Since $M_y = N_x$, the equation is exact in the entire xy-plane. We want U such that $U_x = 4xe^{y/2} + 4$ and $U_y = x^2e^{y/2}$.

$U_x = 4xe^{y/2} + 4$ implies $U = 2x^2e^{y/2} + 4x + f(y)$.
Then $U_y = x^2e^{y/2}$ implies $x^2e^{y/2} + f'(y) = x^2e^{y/2}$, so $f'(y) = 0$ and $f(y) = C$.

The desired conditions are satisfied by the function

$$U(x, y) = 2x^2e^{y/2} + 4x.$$

The general solution is given by the equation

$$2x^2e^{y/2} + 4x = K.$$

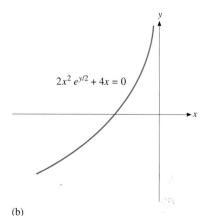

$2x^2\,e^{y/2} + 4x = 0$

(b)

Graphs of the equation for several choices of the constant K are given in Figure 15.2.2. Note that the equation does not define y as a function of x in any rectangle that has center with x-coordinate zero. ▪

It is common to write the exact differential equation

$$M(x, y) + N(x, y)\frac{dy}{dx} = 0$$

in the form

$$M\,dx + N\,dy = 0.$$

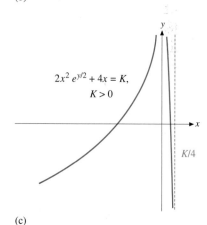

$2x^2\,e^{y/2} + 4x = K,$
$K > 0$

(c)

FIGURE 15.2.2 The character of the solution is different for $K < 0, K = 0$, and $K > 0$. The equation does not define y as a function of x in any rectangle that has center with x-coordinate zero.

This corresponds to the solution obtained by the line integral

$$\int_{(x_0, y_0)}^{(x, y)} M\,dx + N\,dy.$$

An Initial Value Problem

EXAMPLE 3

Find the solution of the initial value problem

$$x\,dx + y\,dy = 0, \quad y(3) = 4,$$

and sketch the graph of the solution.

SOLUTION   We have $M = x$ and $N = y$, so $M_y = 0 = N_x$. The differential equation is exact in the entire xy-plane.

We want U with $U_x = x$ and $U_y = y$.

$U_x = x$ implies $U = \dfrac{x^2}{2} + f(y)$.

Then $U_y = y$ implies $f'(y) = y$, so $f(y) = \dfrac{y^2}{2} + C$.

We can choose

$$U(x, y) = \frac{x^2}{2} + \frac{y^2}{2}.$$

The solution of the differential equation must satisfy the equation

$$\frac{x^2 + y^2}{2} = K$$

for some constant K. Substitution of the initial values $(3, 4)$ gives $K = 25/2$, so the solution of the initial value problem is given by

$$\frac{x^2 + y^2}{2} = \frac{25}{2} \quad \text{or} \quad x^2 + y^2 = 25.$$

The differentiable function $y = y(x)$ with $y(3) = 4$ defined by this equation is

$$y = \sqrt{25 - x^2},$$

which is defined and differentiable for $-5 < x < 5$. The graph of the solution is given in Figure 15.2.3. ∎

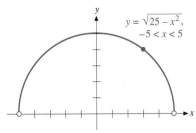

FIGURE 15.2.3   The solution of the initial value problem is the upper half of the circle $x^2 + y^2 = 25$.

EXERCISES 15.2

Show that the differential equations in Exercises 1–12 are exact and find their solutions.

1. $x + y\dfrac{dy}{dx} = 0$

2. $\dfrac{1}{y} - \dfrac{x}{y^2}\dfrac{dy}{dx} = 0,\ y > 0$

3. $\dfrac{1}{x} + \dfrac{1}{y}\dfrac{dy}{dx} = 0,\ x > 0,\ y > 0$

4. $\dfrac{y}{x} + \ln x\dfrac{dy}{dx} = 0$

5. $(2x + y - 1) + (x - 6y + 5)\dfrac{dy}{dx} = 0$

6. $(x + 2y + 3) + (2x - 4y + 1)\dfrac{dy}{dx} = 0$

7. $(3x^2 - y)dx + (2y - x)dy = 0$

8. $(e^x \sin y + x)dx + (e^x \cos y + y)dy = 0$

9. $(ye^x + x)dx + (e^x + y)dy = 0$

10. $(ye^{xy} + 1)dx + (xe^{xy} + 1)dy = 0$

11. $\cos x \cos^2 y\,dx - 2\sin x \sin y \cos y\,dy = 0$

12. $\tan y + x \sec^2 y\,dy = 0$

Solve the initial value problems in Exercises 13–16 and sketch the graphs of the solutions.

13. $y + x\dfrac{dy}{dx} = 0$
(a) $y(1) = 2$,
(b) $y(-1) = -2$

14. $x + y\dfrac{dy}{dx} = 0$
(a) $y(0) = 2$,
(b) $y(0) = -2$

15. $-y\,dx + x\,dy = 0,\ y(1) = \sqrt{3}$

16. $(2y - 2x)dx + 2x\,dy = 0$
(a) $y(1) = 2$,
(b) $y(-1) = 3$,
(c) $y(3) = 1$

15.3 FIRST-ORDER LINEAR DIFFERENTIAL EQUATIONS; INTEGRATING FACTORS

Connections

Separable differentiable equations, 15.1.

Equations of the form

$$y' + p(x)y = q(x)$$

are called **first-order linear** differential equations. If the functions p and q are continuous, we can solve an equation of this type by multiplying the equation by an **integrating factor** m, where m is chosen so that $m(y' + py) = (my)'$. If $\int^x p$ is any function in the collection of functions $\int p(x)\,dx$ that have derivative $p(x)$, we will see that

$e^{\int^x p}$ *is an integrating factor of the equation* $y' + p(x)y = q(x)$.

Let us illustrate how an integrating factor is used to solve the differential equation. Multiplying the equation by $e^{\int^x p}$, we obtain

$$e^{\int^x p}[y'(x) + p(x)y(x)] = e^{\int^x p}q(x). \tag{1}$$

Using the fact that

$$\frac{d}{dx}\left(e^{\int^x p}\right) = e^{\int^x p}\frac{d}{dx}\left(\int^x p\right) = e^{\int^x p}p(x),$$

it can be verified that the expression on the left above is the derivative of the product $e^{\int^x p}y$. Equation 1 can then be rewritten

$$\frac{d}{dx}\left(e^{\int^x p}y\right) = e^{\int^x p}q, \text{ so}$$

$$e^{\int^x p}y = \int e^{\int^x p}q(x)\,dx.$$

The formula for $y(x)$ is obtained by evaluating the integral on the right and then multiplying the resulting equation by $e^{-\int^x p}$.

It is not difficult to determine that an integrating factor of the equation $y' + py = q$ is of the form $\pm e^{\int^x p}$. That is, if $m(y' + py)$ is to be the derivative of the product my, we must have

$$(my)' = m(y' + py),$$
$$my' + ym' = my' + mpy,$$
$$ym' = mpy,$$
$$m' = mp.$$

Hence, m must satisfy the separable differential equation

$$\frac{m'}{m} = p.$$

Solving this equation for m by integrating, we obtain

$$\int \frac{m'}{m} dx = \int p\, dx,$$
$$\int \frac{1}{m} dm = \int p\, dx,$$
$$\ln |m| = \int p\, dx,$$
$$m = \pm e^{\int p\, dx},$$

which shows that any integrating factor of the equation $y' + py = q$ must be of the form $\pm e^{\int^x p}$.

Using Integrating Factors

Let us now see some examples of how to use integrating factors to solve first-order linear differential equations.

EXAMPLE 1

Find the solution of $y' + \dfrac{1}{x} y = \dfrac{1}{x^2}, \quad x > 0.$

SOLUTION We have $p(x) = x^{-1}$. Then

$$\int p(x)\, dx = \int x^{-1} dx = \ln x + C.$$

We choose $C = 0$ and use the integrating factor

$$e^{\int^x p} = e^{\ln x} = x.$$

Multiplying the differential equation by the integrating factor, we obtain

$$x\left(y' + \frac{1}{x} y\right) = x\frac{1}{x^2},$$
$$\frac{d}{dx}(xy) = x^{-1},$$
$$xy = \int x^{-1} dx,$$

$$xy = \ln x + C,$$
$$y = \frac{\ln x}{x} + \frac{C}{x}. \quad \blacksquare$$

EXAMPLE 2
Solve $y' + xy = x$.

SOLUTION In this example the coefficient of y is x, so
$$\int p(x)\,dx = \int x\,dx = \frac{x^2}{2} + C$$

and
$$e^{\int^x p} = e^{x^2/2}$$

is an integrating factor. Then
$$e^{x^2/2}(y' + xy) = e^{x^2/2}x,$$
$$\frac{d}{dx}(e^{x^2/2}y) = e^{x^2/2}x,$$
$$e^{x^2/2}y = \int e^{x^2/2}x\,dx.$$

The integral on the right is evaluated easily by using the substitution $u = x^2/2$, so $du = x\,dx$. We obtain
$$e^{x^2/2}y(x) = e^{x^2/2} + C,$$
$$y(x) = 1 + Ce^{-x^2/2}. \quad \blacksquare$$

EXAMPLE 3
Solve $y' + (\tan x)y = \sec x$, $-\pi/2 < x < \pi/2$.

SOLUTION We have
$$\int p(x)\,dx = \int \tan x\,dx = \ln \sec x + C,$$

so an integrating factor is
$$e^{\int^x p} = e^{\ln \sec x} = \sec x.$$

Then
$$(\sec x)[y' + (\tan x)y] = \sec^2 x,$$
$$\frac{d}{dx}((\sec x)y) = \sec^2 x,$$
$$(\sec x)y = \int \sec^2 x\,dx,$$
$$(\sec x)y = \tan x + C,$$
$$\frac{1}{\cos x}y = \frac{\sin x}{\cos x} + C,$$
$$y(x) = \sin x + C\cos x. \quad \blacksquare$$

FIGURE 15.3.1 A simple series circuit consists of a 0.025-F capacitor, a 20-Ω resistor, and an impressed voltage of $E(t) = 120 \sin t$.

An Electrical Circuit Problem

EXAMPLE 4

A simple series circuit consists of a 0.025-F capacitor, a 20-Ω resistor, and an impressed voltage of $E(t) = 120 \sin t$. See Figure 15.3.1. Kirchhoff's Second Law can be used to show that the current i in amperes is a solution of the differential equation

$$20i' + \frac{1}{0.025} i = \frac{d}{dt}(120 \sin t).$$

Find the current in the circuit at time t if the initial current in the circuit is zero.

SOLUTION The current is the solution of the initial value problem

$$20i' + \frac{1}{0.025} i = 120 \cos t, \quad i(0) = 0.$$

Simplifying the differential equation, we have

$$i' + 2i = 6 \cos t.$$

Since

$$\int p(t)\,dt = \int 2\,dt = 2t + C,$$

an appropriate integrating factor for this equation is e^{2t}. Then

$$e^{2t}(i' + 2i) = 6e^{2t} \cos t,$$

$$\frac{d}{dt}(e^{2t}i) = 6e^{2t} \cos t,$$

$$e^{2t}i = \int 6e^{2t} \cos t\,dt.$$

Using the Table of Integrals (or integrating by parts), we obtain

$$e^{2t}i = 6e^{2t}(0.4 \cos t + 0.2 \sin t) + C,$$
$$i(t) = 2.4 \cos t + 1.2 \sin t + Ce^{-2t}.$$

Since $i(0) = 0$, we have

$$0 = 2.4 + C, \text{ so } C = -2.4.$$

Then

$$i(t) = 2.4 \cos t + 1.2 \sin t - 2.4e^{-2t}.$$

The graph of $i(t)$ is given in Figure 15.3.2. For large values of t, we have

$$i(t) \approx 2.4 \cos t + 1.2 \sin t.$$

By using the formula

$$A \cos t + B \sin t = \sqrt{A^2 + B^2} \sin\left(t + \tan^{-1}\frac{A}{B}\right), \qquad A > 0, \ B > 0,$$

we can determine that

$$2.4 \cos t + 1.2 \sin t = 2.68 \sin(t + 1.107).$$

Thus, the current is essentially a sine curve that is 1.107 rad out of phase with the impressed voltage. ∎

FIGURE 15.3.2 The current is essentially a sine curve that is 1.107 rad out of phase with the impressed voltage.

EXERCISES 15.3

Solve the differential equations in Exercises 1–14.

1. $y' + \dfrac{2}{x}y = x, \; x > 0$

2. $y' + \dfrac{3}{x}y = 2, \; x > 0$

3. $y' - 2y = e^x$

4. $y' - 2y = e^{2x}$

5. $y' + \left(\dfrac{1}{x} + 1\right)y = \dfrac{1}{x}, \; x > 0$

6. $y' + y = \dfrac{1}{1 + e^x}$

7. $y' + \dfrac{2x}{1 + x^2}y = x$

8. $y' - \dfrac{2x}{1 + x^2}y = x$

9. $y' - \dfrac{1}{x}y = \dfrac{1}{x + 1}, \; x > 0$

10. $y' - \dfrac{1}{x + 1}y = \dfrac{1}{x}, \; x > 0$

11. $y' + (\tan x)y = \tan x, \; -\pi/2 < x < \pi/2$

12. $y' + (\sec x)y = \sec x, \; -\pi/2 < x < \pi/2$

13. $y' + (\sec x)y = 1, \; -\pi/2 < x < \pi/2$

14. $y' - (\tan x)y = \cos x, \; -\pi/2 < x < \pi/2$

15. A circuit consists of a 0.02-F capacitor connected in series to a 60-Ω resistor, with an expressed voltage of 12 V. Kirchhoff's Second Law can be used to show that the current in amperes is a solution of the differential equation

$$60i' + \frac{1}{0.02}i = \frac{d}{dt}(12).$$

Find the current at time t if the initial current is 8 A.

16. A circuit consists of a 0.1-F capacitor connected in series to a 60-Ω resistor, with an impressed voltage of $L(t) = 12\sin t$. Kirchhoff's Second Law can be used to show that the current in amperes is a solution of the differential equation

$$60i' + \frac{1}{0.1}i = \frac{d}{dt}(12\sin t).$$

Find the current at time t if the initial current is 8 A.

In Exercises 17–20, show that multiplying the given differential equation by the indicated **integrating factor** *gives an exact equation and find the solution of the equation.*

17. $1 - \dfrac{2x}{y}\dfrac{dy}{dx} = 0;$

   (a) $\dfrac{1}{y^2},$

   (b) $\dfrac{1}{x}$

18. $y + 2x\dfrac{dy}{dx} = 0;$

   (a) $\dfrac{1}{2\sqrt{x}},$

   (b) $\dfrac{1}{xy}$

19. $(1 + 2xy) - \dfrac{x}{y}\dfrac{dy}{dx} = 0; \; \dfrac{1}{y}$

20. $y + (x^2 y - x)\dfrac{dy}{dx} = 0; \; \dfrac{1}{x^2}$

15.4 SECOND-ORDER EQUATIONS THAT ARE REDUCIBLE TO FIRST ORDER

Connections

Solution of first-order equations, 15.1–3.

We will see that certain types of second-order differential equations can be reduced to first-order equations. Solution methods for first-order equations can then be used to solve these second-order equations. We will illustrate how equations of this type relate to hanging cables, parachutists, and rocket motion.

Equations with *y* Missing

If the dependent variable y is missing from a second-order differential equation, so the equation involves only y'', y', and the independent variable x, we can use the substitution

$$u = \frac{dy}{dx}. \tag{1}$$

We can then substitute $y' = u$ and $y'' = u'$ to obtain a first-order differential equation that involves u', u, and x. If we can solve this equation for $u = y'$ in terms of x, then we can obtain a solution for y by integration.

EXAMPLE 1

Find the solution of the initial value problem

$$y'' = \frac{1}{a}\sqrt{1 + (y')^2}, \quad y(0) = a, \quad y'(0) = 0.$$

SOLUTION   We see that the dependent variable y is missing from this second-order differential equation, so we can attempt to solve it by using the substitution $y' = u$. Then $y'' = u'$ and we obtain the first-order equation

$$u' = \frac{1}{a}\sqrt{1 + u^2}.$$

This is a separable differential equation, which we can solve by integration. Note that the unknown function u is considered to be a function of the independent variable x. We have

$$\int \frac{1}{\sqrt{1 + u^2}} \frac{du}{dx} dx = \int \frac{1}{a} dx,$$
$$\int \frac{1}{\sqrt{1 + u^2}} du = \int \frac{1}{a} dx.$$

Using the Table of Integrals to evaluate the integral on the left, we obtain

$$\sinh^{-1} u = \frac{x}{a} + C_1, \text{ so}$$
$$u = \sinh\left(\frac{x}{a} + C_1\right),$$

or

$$\frac{dy}{dx} = \sinh\left(\frac{x}{a} + C_1\right).$$

$y'(0) = 0$ implies $C_1 = 0$, so

$$\frac{dy}{dx} = \sinh\frac{x}{a}.$$

Integrating the above equation gives

$$y = a\cosh\frac{x}{a} + C_2.$$

$y(0) = a$ implies $C_2 = 0$, so we obtain the solution

$$y(x) = a\cosh\frac{x}{a}. \quad \blacksquare$$

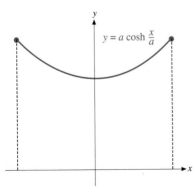

FIGURE 15.4.1a   A cable hangs in the shape of the catenary $y = a\cosh\frac{x}{a}$.

The solution of Example 1 is the **catenary** $y = a\cosh\frac{x}{a}$. This is the shape of a hanging cable, as illustrated in Figure 15.4.1a.

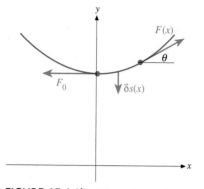

FIGURE 15.4.1b When the cable is hanging in a stable position, the total horizontal and vertical forces on the piece must be zero.

Let us see how to derive the differential equation that characterizes a hanging cable. We consider the forces on a piece of the curve from $(0, a)$ to (x, y) with arc length

$$s(x) = \int_0^x \sqrt{1 + (y'(t))^2}\, dt$$

and weight $w = \delta s(x)$. See Figure 15.4.1b. Forces due to tension in the rope must have direction in the direction tangent to the curve. The force due to tension at $(0, a)$ has magnitude F_0 and horizontal direction. The force due to tension at (x, y) has magnitude $F(x)$ and direction indicated by θ. The force on the piece due to its weight has magnitude $\delta s(x)$ and direction vertical. When the cable is hanging in a stable position, the total horizontal and vertical forces on the piece must be zero. This implies

$$F(x)\cos\theta = F_0 \quad \text{and} \quad F(x)\sin\theta = \delta s(x).$$

These equations imply

$$\tan\theta = \frac{\delta}{F_0} s(x).$$

Since $\tan\theta = \dfrac{dy}{dx}$, we obtain

$$\frac{dy}{dx} = \frac{\delta}{F_0} \int_0^x \sqrt{1 + (y'(t))^2}\, dt.$$

Differentiating with respect to x then gives

$$\frac{d^2y}{dx^2} = \frac{\delta}{F_0} \sqrt{1 + (y'(x))^2},$$

which is the differential equation of Example 1.

Equations with x Missing

If the independent variable x is missing from a second-order differential equation, we can attempt to solve the equation by using the substitution

$$u = \frac{dy}{dx}. \tag{2}$$

Then we can use the Chain Rule to obtain

$$\frac{d^2y}{dx^2} = \frac{d}{dx}\left(\frac{dy}{dx}\right) = \frac{du}{dx} = \frac{du}{dy}\frac{dy}{dx} = \frac{du}{dy}u. \tag{3}$$

Substitution then gives a first-order differential equation that involves $\dfrac{du}{dy}$, u, and y, but not x. If we can solve this first-order differential equation for the unknown function u as a function of y, we obtain an equation that relates $u = \dfrac{dy}{dx}$ and y. This is a first-order differential equation, which we can then attempt to solve for y in terms of x.

EXAMPLE 2

Find the general solution of the equation $(2y + 1)y'' + 2(y')^2 = 0$.

SOLUTION The independent variable x is missing from this second-order equation, so we can try the substitution $u = \dfrac{dy}{dx}$, as indicated in (2). Formula 3 then implies $\dfrac{d^2y}{dx^2} = \dfrac{du}{dy}u$. Substitution then gives

$$(2y + 1)\frac{du}{dy}u + 2(u)^2 = 0,$$

$$(2y + 1)\frac{du}{dy}u = -2(u)^2,$$

$$\frac{1}{u}\frac{du}{dy} = -\frac{2}{2y + 1}.$$

This is a separable equation, which we solve by integration. Note that we are considering the unknown function u to be a function of y. Integrating with respect to y then gives

$$\int \frac{1}{u}\frac{du}{dy}dy = -\int \frac{2}{2y + 1}dy,$$

$$\int \frac{1}{u}du = -\int \frac{2}{2y + 1}dy,$$

$$\ln|u| = -\ln|2y + 1| + c_1,$$

$$|u| = \frac{e^{c_1}}{|2y + 1|},$$

$$u = \frac{C_1}{2y + 1},$$

where $C_1 = \pm e^{c_1}$. Using the fact that $u = \dfrac{dy}{dx}$, we then obtain

$$\frac{dy}{dx} = \frac{C_1}{2y + 1},$$

$$(2y + 1)\frac{dy}{dx} = C_1,$$

$$\int (2y + 1)\frac{dy}{dx}dx = \int C_1\,dx,$$

$$y^2 + y = C_1 x + C_2.$$

This gives the general solution in implicit form. ∎

Mechanics

Let us consider the motion along a line of a particle with constant mass. We assume the particle is subject to a total external force that acts in a direction along the line. **Newton's Second Law of Motion** asserts that the force on the particle is proportional to mass times acceleration. For appropriate systems of units, the constant of proportionality is 1, so

$$F = ma.$$

Some common systems of units are given in Table 15.4.1.

TABLE 15.4.1

System	Length	Mass	Time	Force
English	feet	slugs	seconds	pounds
mks (SI)	meters	kilograms	seconds	newtons
cgs	centimeters	grams	seconds	dynes

The weight of an object is the force due to the gravitational attraction of the earth. Thus, Newton's Second Law tells us

$$w = ma,$$

where w is the weight, m is the mass, and a is the acceleration due to gravity. The weight and acceleration due to gravity depend on the distance of the object from the center of the earth. The mass depends on the molecular structure and not on the position of the object.

For problems that involve vertical motion near the surface of the earth, we assume that the magnitude of acceleration due to gravity has a constant value g. We will use either

$$g = 32 \text{ ft/s}^2 \text{ or } g = 9.8 \text{ m/s}^2,$$

depending on the units. These are approximate values of the acceleration due to gravity at sea level. The weight of an object near the surface of the earth and the mass of the object are then related by the equation

$$w = mg.$$

If s is the position of an object at time t, we know that the velocity is $v = s'$ and the acceleration is $a = v' = s''$. If the total force acting on an object is expressed in terms of either s, v, or t, then Newton's equation $F = ma$ gives us a differential equation. Recall that the net force due to two or more forces is the vector sum of the forces.

EXAMPLE 3

Determine the vertical velocity and height above ground of a 128-lb parachutist t seconds after jumping from an airplane that is flying slowly and horizontally at an altitude of 5000 ft. Assume that air resistance is eight times the speed and ignore horizontal motion.

SOLUTION   We choose a vertical coordinate line with origin at ground level and positive direction upward. The initial height of the parachutist is $s_0 = 5000$ ft. The initial vertical velocity is $v_0 = 0$. See Figure 15.4.2.

The (total) weight of the parachutist is 128 lb. The force due to gravity is then -128, where the negative sign indicates the force is directed downward. The mass is

$$m = \frac{w}{g} = \frac{128}{32} = 4 \text{ slugs.}$$

The force exerted by the parachute due to air resistance is opposite to the motion and has magnitude $8|v|$, so it is $-8v$. Note that the parachutist is falling, so the velocity is negative and the force $-8v$ is directed upward.

FIGURE 15.4.2   We choose a vertical coordinate system with origin at ground level and positive direction upward.

Newton's Second Law of Motion then gives

$$ma = F,$$
$$4s'' = -128 - 8s'.$$

This is a second-order differential equation that does not involve the independent variable s, so we can reduce the order by using the substitution $v = s'$. Then $s'' = v'$ and substitution gives

$$4v' = -128 - 8v,$$
$$v' = -2(16 + v).$$

This is a separable differential equation for the velocity as a function of time. Let us solve this equation to determine the velocity. We have

$$\frac{1}{16 + v}v' = -2.$$

Integrating with respect to t, we obtain

$$\int \frac{1}{16 + v}v'\,dt = \int -2\,dt,$$
$$\int \frac{1}{16 + v}\,dv = \int -2\,dt,$$
$$\ln|16 + v| = -2t + c,$$
$$|16 + v| = e^{-2t+c},$$
$$16 + v = Ce^{-2t},$$

where $C = \pm e^c$. Since the initial velocity is zero, we have

$$16 + 0 = C.$$

We then have the equation for the velocity

$$v(t) = -16 + 16e^{-2t}.$$

The height is a solution of the initial value problem

$$s' = -16 + 16e^{-2t}, \quad s(0) = 5000.$$

We have

$$s' = -16 + 16e^{-2t},$$
$$\int s'(t)\,dt = \int (-16 + 16e^{-2t})\,dt,$$
$$s = -16t - 8e^{-2t} + C.$$

Since the initial height is $s_0 = 5000$, we have $5000 = -16(0) - 8 + C$, so $C = 5008$. Thus,

$$s(t) = -16t - 8e^{-t} + 5008. \quad \blacksquare$$

It is interesting to compare the result of Example 3 with the result obtained if air resistance is neglected. This would be a somewhat reasonable assumption without the parachute. In that case we would have the familiar result

$$s(t) = -16t^2 + v_0 t + s_0 = -16t^2 + 5000.$$

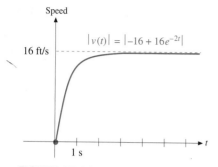

Speed

16 ft/s

$|v(t)| = |-16 + 16e^{-2t}|$

1 s

FIGURE 15.4.3   The parachutist is never falling as fast as the terminal velocity, 16 ft/s.

Thus, without a parachute, a person jumping from an altitude of 5000 ft would be expected to hit the ground approximately $\sqrt{5000/16} \approx 17.7$ s later at a speed of $32(17.7) \approx 566$ ft/s. With the greatly increased air resistance due to the parachute, the person in Example 3 is never falling as fast as 16 ft/s, as we can see from the equation

$$v(t) = -16 + 16e^{-2t}.$$

The limiting value of the velocity is called the **terminal velocity**. The speed the parachutist is falling at time t, $|v(t)|$, is indicated in Figure 15.4.3.

In many problems that involve motion, the external force depends on the position and/or the velocity of the object, but not on the independent variable time. We can then transform Newton's equation into a differential equation for the velocity as a function of the position. To do this, we note that the Chain Rule gives

$$v' = \frac{dv}{dt} = \frac{dv}{dx}\frac{dx}{dt} = \frac{dv}{dx}v.$$

Newton's equation can then be written

$$ms'' = F,$$
$$mv' = F,$$
$$mv\frac{dv}{dx} = F.$$

The latter form of Newton's equation is called the **phase plane** equation of the motion.

In a problem that involves the force of gravity acting on an object that does not remain near the surface of the earth, we cannot assume that the force due to gravity is constant. In that case we must use the fact that the force due to the gravitational pull of the earth is proportional to the mass of the object and inversely proportional to the square of the distance of the object from the center of the earth. We can express this force in terms of a vertical coordinate system with origin at the surface of the earth as

$$F = -\frac{R^2 gm}{(R+x)^2},$$

x

0

R

FIGURE 15.4.4   We choose a vertical coordinate system with origin at the surface of the earth and positive direction upward. The distance of an object from the center of the earth is $R + x$.

where R is the radius of the earth, g is the acceleration due to gravity at sea level, m is the fixed mass of the object, and x is the height of the object above sea level. See Figure 15.4.4.

The radius of the earth is approximately $R = 2.1 \times 10^7$ ft.

EXAMPLE 4

A rocket is launched from earth with initial velocity v_0. Find a formula for the velocity of the rocket in terms of its height. Assume that gravity is the only force acting on the rocket and that the mass of the rocket is constant. (The assumptions are reasonable if the fuel burns quickly, while the rocket is relatively near the surface of the earth.)

SOLUTION We choose a coordinate line as indicated in Figure 15.4.4. The phase plane equation of the motion is

$$mv\frac{dv}{dx} = -\frac{R^2 gm}{(R+x)^2}.$$

This is a separable differential equation. Canceling m from both sides of the equation and then integrating with respect to x, we obtain

$$\int v\frac{dv}{dx}\,dx = \int -\frac{R^2 g}{(R+x)^2}\,dx,$$

$$\int v\,dv = \int -\frac{R^2 g}{(R+x)^2}\,dx,$$

$$\frac{v^2}{2} = \frac{R^2 g}{R+x} + C.$$

Since $v = v_0$ when $x = 0$, we have

$$\frac{v_0^2}{2} = Rg + C, \text{ so } C = \frac{v_0^2}{2} - Rg.$$

The equation for the velocity in terms of the height is then

$$v^2 = \frac{2R^2 g}{R+x} + v_0^2 - 2Rg.$$

The above equation tells us that the velocity will never become zero if $v_0 \geq \sqrt{2Rg} \approx 36,660$ ft/s $\approx 25,000$ mph. This is the (approximate) initial velocity that a rocket would need to avoid being drawn back to the earth by the earth's gravity. Thus, $\sqrt{2Rg}$ is called the **escape velocity**. Representative graphs of the solution in the xv-phase plane for the cases $v_0 \geq \sqrt{2Rg}$ and $v_0 < \sqrt{2Rg}$ are given in Figure 15.4.5. ∎

FIGURE 15.4.5a If a rocket is launched from ground level with initial velocity v_0, where $v_0 \geq \sqrt{2Rg}$, the velocity is always positive and the rocket will not be drawn back to earth by the earth's gravity.

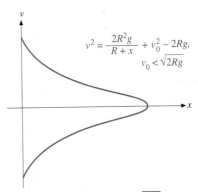

FIGURE 15.4.5b If $v_0 < \sqrt{2Rg}$, the rocket will be drawn back to earth by the earth's gravity.

EXERCISES 15.4

Solve the initial value problems in Exercises 1–10.

1. $y'' = (y')^2$, $y(0) = 3$, $y'(0) = 1$

2. $y'' = \sqrt{1 - (y')^2}$, $y(0) = 0$, $y'(0) = 0$

3. $y'' - y' = 1$, $y(0) = 3$, $y'(0) = 1$

4. $y'' - y' = x$, $y(0) = 3$, $y'(0) = 1$

5. $y'' + 2y' = e^{-2x}$, $y(0) = 0$, $y'(0) = 0$

6. $2y'' - 3y^2 = 0$, $y(0) = 4$, $y'(0) = 8$

7. $y'' + y = 0$
   (a) $y(0) = 0$, $y'(0) = 1$,
   (b) $y(0) = 1$, $y'(0) = 0$,
   (c) $y(0) = 3$, $y'(0) = 2$

8. $y'' = 2yy'$
   (a) $y(0) = y'(0) = 1$,
   (b) $y(0) = 0$, $y'(0) = 1$,
   (c) $y(0) = 2$, $y'(0) = 3$

9. $y'' - 4y = 0$, $y(0) = 0$, $y'(0) = 2$

10. $y'' - y = 0$, $y(0) = 1$, $y'(0) = 0$

Find the general solutions of the differential equations in Exercises 11–14.

11. $xy'' + y' = 0$

12. $(x + x^3)y'' = y'$

13. $y'' + 2y' = e^x$

14. $(y^2 + 1)y'' = 2y(y')^2$

15. Determine the velocity and height above ground of a 160-lb parachutist t seconds after jumping from an airplane that is flying slowly and horizontally at 3000 ft. Assume the air resistance is eight times the speed and ignore horizontal motion. Find the terminal velocity of the parachutist.

16. Determine the velocity and height above ground of a 2048-lb package t seconds after it is dropped by parachute from an airplane that is flying slowly and horizontally at 4000 ft. Assume the air resistance is eight times the speed and ignore horizontal motion. Find the terminal velocity of the package.

17. Determine the velocity and height above ground of a 160-lb skydiver t seconds after jumping from an airplane that is flying slowly and horizontally at 3000 ft. Assume the air resistance is four times the speed and ignore horizontal motion. Find the terminal velocity of the skydiver.

18. Determine the velocity and height above ground of a 160-lb skydiver t seconds after jumping from an airplane that is flying slowly and horizontally at 3000 ft. Assume the air resistance is twice the speed and ignore horizontal motion. Find the terminal velocity of the skydiver.

19. A rocket is launched vertically from earth. At the time the engines are turned off, the rocket has an altitude of $R = 2.1 \times 10^7$ ft and a speed of 20,000 ft/s. Assuming that gravity is the only force acting on the rocket, what is the maximum altitude the rocket will reach before it is drawn back toward the earth? R is the radius of the earth.

20. A rocket is launched vertically from earth. At the time the engines are turned off, the rocket has an altitude of $2R$ ft and a speed of 20,000 ft/s. Assuming that gravity is the only force acting on the rocket, what is the maximum altitude the rocket will reach before it is drawn back toward the earth? R is the radius of the earth.

21. What speed at an altitude of R ft would a rocket need to avoid being pulled by gravity back toward the earth? R is the radius of the earth.

22. What speed at an altitude of $2R$ ft would a rocket need to avoid being pulled by gravity back toward the earth? R is the radius of the earth.

23. Use Newton's Method or the solve function on a calculator to determine approximately how many seconds the parachutist in Example 3 falls.

24. A dog is chasing a rabbit as indicated in the figure. Initially, the rabbit is at the origin and the dog is at $(D, 0)$. The rabbit runs at constant speed v_r and the dog runs at constant speed v_d, where $v_d > v_r$. The rabbit runs along the y-axis and the dog always runs directly toward the rabbit.
   (a) Use the fact that the dog always runs directly toward the rabbit to show that points (x, y) on the path of the dog satisfy $x\dfrac{dy}{dx} = y - v_r t$.
   (b) Differentiate the equation in (a) with respect to x and use the fact that $\left(\dfrac{dx}{dt}\right)^2 + \left(\dfrac{dy}{dt}\right)^2 = v_d^2$ to obtain
   $$x\frac{d^2y}{dx^2} = \frac{v_r}{v_d}\sqrt{1 + \left(\frac{dy}{dx}\right)^2}.$$
   (c) Find the solution of the differential equation in (b) that satisfies the indicated initial conditions.
   (d) How long will it take the dog to catch the rabbit?

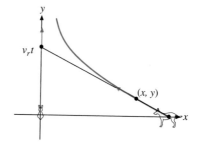

15.5 HOMOGENEOUS SECOND-ORDER LINEAR EQUATIONS

Connections

Complex numbers.

Solution of quadratic equations, 1.2.

We will briefly discuss some general properties of second-order linear differential equations and show how we can solve second-order linear equations with constant coefficients.

General Solutions

Equations of the form

$$y'' + p(x)y' + q(x)y = f(x)$$

are called **second-order linear differential equations**. They are called **homogeneous** if the function f is the zero function; they are called **nonhomogeneous** if f is not identically zero. They are said to have **constant coefficients** if the functions p and q are constants.

We have the following existence and uniqueness theorem, which we will not prove.

THEOREM 1

If p, q, and f are continuous on an open interval that contains the point x_0, then the initial value problem

$$y' + py' + qy = f, \quad y(x_0) = y_0, \quad y'(x_0) = y_1$$

has a unique solution on some open interval that contains x_0.

In this section we will study the solution of homogeneous second-order linear equations. In Sections 15.6 and 15.7 we will see how to use the general solution of a homogeneous equation to solve a corresponding nonhomogeneous equation.

We have the following result for the solution of homogeneous second-order linear differential equations.

THEOREM 2

If y_1 and y_2 are solutions of the homogeneous second-order linear equation

$$y'' + py' + qy = 0,$$

then, for any values of the constants C_1 and C_2,

$$y(x) = C_1 y_1(x) + C_2 y_2(x)$$

is also a solution.

Proof If $y = C_1 y_1 + C_2 y_2$, then

$$y' = C_1 y_1' + C_2 y_2' \text{ and}$$
$$y'' = C_1 y_1'' + C_2 y_2''.$$

Substitution then gives

$$\begin{aligned} y'' + py' + qy &= (C_1y_1'' + C_2y_2'') + p(C_1y_1' + C_2y_2') + q(C_1y_1 + C_2y_2) \\ &= C_1(y_1'' + py_1' + qy_1) + C_2(y_2'' + py_2' + qy_2) \\ &= C_1(0) + C_2(0) = 0, \end{aligned}$$

so $y = C_1y_1 + C_2y_2$ is a solution of the homogeneous equation. ■

Theorem 2 tells us that, if we have solutions y_1 and y_2 of a homogeneous second-order linear differential equation, then each function of the form $C_1y_1 + C_2y_2$ is also a solution. The next theorem gives us a condition on y_1 and y_2 that implies every solution of the equation is of this form, so $C_1y_1 + C_2y_2$ is the general solution.

THEOREM 3

Suppose y_1 and y_2 are solutions of the homogeneous second-order linear differential equation $y'' + py' + qy = 0$ on some open interval I and

$$y_1(x_0)y_2'(x_0) - y_1'(x_0)y_2(x_0) \neq 0 \text{ for some } x_0 \text{ in } I. \tag{1}$$

Then, if y is any solution on I, there are constants C_1 and C_2 such that

$$y(x) = C_1y_1(x) + C_2y_2(x) \text{ for } x \text{ in } I.$$

Proof Suppose that y is a solution of the homogeneous equation

$$y'' + py' + qy = 0.$$

Condition (1) implies that the linear system of equations

$$\begin{cases} y_1(x_0)u + y_2(x_0)v = y(x_0), \\ y_1'(x_0)u + y_2'(x_0)v = y'(x_0) \end{cases}$$

has a unique solution $(u, v) = (C_1, C_2)$, so

$$\begin{cases} y_1(x_0)C_1 + y_2(x_0)C_2 = y(x_0), \\ y_1'(x_0)C_1 + y_2'(x_0)C_2 = y'(x_0). \end{cases}$$

For the above choice of C_1 and C_2, the function

$$z(x) = y(x) - (C_1y_1(x) + C_2y_2(x))$$

is a solution of the initial value problem

$$y'' + py' + qy = 0, \quad y(x_0) = 0, \quad y'(x_0) = 0.$$

However, the zero function is also a solution of this initial value problem. Since Theorem 1 tells us that there is a *unique* solution of the initial value problem, we must have $z(x) = 0$ for all x in I. It follows that

$$y(x) = C_1y_1(x) + C_2y_2(x), \quad x \text{ in } I.$$

This shows that any solution of the homogeneous second-order linear differential equation must be of the form $C_1y_1 + C_2y_2$. ■

The expression $y_1(x_0)y_2'(x_0) - y_1'(x_0)y_2(x_0)$ is the value of the determinant

$$W(y_1, y_2; x_0) = \begin{vmatrix} y_1(x_0) & y_2(x_0) \\ y_1'(x_0) & y_2'(x_0) \end{vmatrix}$$

and is called the **Wronskian** of y_1 and y_2 at the point x_0. Theorem 3 tells us that if y_1 and y_2 are solutions of $y'' + py' + qy = 0$ with $W(y_1, y_2; x_0) \neq 0$, then the solution of the initial value problem

$$y'' + py' + qy = 0, \quad y(x_0) = y_0, \quad y'(x_0) = y_1$$

can be expressed in the form

$$y(x) = C_1 y_1(x) + C_2 y_2(x), \quad x \text{ in } I.$$

It is true that if y_1 and y_2 are solutions of the *same* linear differential equation $y'' + py' + qy = 0$ for x in some interval I and the Wronskian of y_1 and y_2 is nonzero at some point x_0 in I, then the Wronskian is nonzero for every x in I. (See Exercise 33.) If y_1 and y_2 are two functions that are not solutions of the same second-order linear differential equation in an interval, then the Wronskian can be zero at some points and nonzero at other points in the interval. (See Exercise 34.)

Reduction of Order

If y_1 is a solution of the homogeneous equation

$$y'' + py' + qy = 0,$$

we can use the substitution

$$y = vy_1$$

to find either a different solution of the homogeneous equation or a solution of the nonhomogeneous equation

$$y'' + py' + qy = f.$$

The function $y = vy_1$ is a solution of $y'' + py + qy = f$ exactly whenever the function v satisfies a related second-order equation that involves v'', v', and x, but not v. If we can solve this equation for v, then the formula $y = vy_1$ gives us the desired function y. Let us illustrate with an example.

EXAMPLE 1

(a) Show that $y_1(x) = e^{2x}$ is a solution of the equation $y'' - 4y' + 4y = 0$.
(b) Find a solution y_2 of the homogeneous equation in (a) that satisfies $W(y_1, y_2) \neq 0$.
(c) Find the general solution of the homogeneous equation in (a).

SOLUTION

(a) If $y(x) = e^{2x}$, then $y'(x) = 2e^{2x}$ and $y''(x) = 4e^{2x}$. We then have

$$y'' - 4y' + 4y = (4e^{2x}) - 4(2e^{2x}) + 4(e^{2x}) = 0,$$

so $y(x) = e^{2x}$ is a solution of $y'' - 4y' + 4y = 0$.

(b) Let us try the substitution $y_2(x) = v(x)e^{2x}$. Then

$$y_2'(x) = (v(x))(2e^{2x}) + (e^{2x})(v'(x))$$
$$= e^{2x}(2v(x) + v'(x))$$

and

$$y_2''(x) = (e^{2x})(2v'(x) + v''(x)) + (2v(x) + v'(x))(2e^{2x})$$
$$= e^{2x}(v''(x) + 4v'(x) + 4v(x)).$$

The function y_2 is a solution of

$$y'' - 4y' + 4y = 0$$

if and only if

$$y_2''(x) - 4y_2'(x) + 4y_2(x) = 0.$$

Replacing y_2, y_2', and y_2'' with the expressions determined above and simplifying, we have

$$e^{2x}(v'' + 4v' + 4v) - 4e^{2x}(2v + v') + 4(ve^{2x}) = 0,$$
$$v'' = 0.$$

Solving the above equation for v, we obtain

$$v'(x) = C_1 \text{ and then}$$
$$v(x) = C_1 x + C_2.$$

Then

$$y_2(x) = v(x)e^{2x} = C_1 x e^{2x} + C_2 e^{2x}$$

is a solution of

$$y'' - 4y' + 4y = 0$$

for any values of the constants C_1 and C_2. Note that $C_1 = 0$ implies $y_2(x) = C_2 e^{2x}$, a multiple of $y_1(x)$. It follows that if we would choose $C_1 = 0$, then we would obtain a solution y_2 of the homogeneous equation with $W(y_1, y_2) = 0$. Since we are looking for a solution of the homogeneous equation with $W(y_1, y_2) \neq 0$, we cannot choose $C_1 = 0$. For convenience we choose $C_1 = 1$ and $C_2 = 0$. This gives $v(x) = x$ and $y_2(x) = xe^{2x}$.

The Wronskian of $y_1(x) = e^{2x}$ and $y_2(x) = xe^{2x}$ is

$$W(y_1, y_2; x) = \begin{vmatrix} y_1 & y_2 \\ y_1' & y_2' \end{vmatrix} = \begin{vmatrix} e^{2x} & xe^{2x} \\ 2e^{2x} & e^{2x}(2x + 1) \end{vmatrix}$$
$$= e^{4x}(2x + 1) - 2xe^{4x} = e^{4x} \neq 0.$$

(c) Since $W(y_1, y_2) \neq 0$, the general solution of the homogeneous equation is

$$y = C_1 y_1(x) + C_2 y_2(x) = C_1 e^{2x} + C_2 x e^{2x}. \quad \blacksquare$$

Equations with Constant Coefficients

Let us consider solutions of

$$y'' + by' + cy = 0, \quad b \text{ and } c \text{ constants.} \tag{2}$$

If $y(x) = e^{rx}$, r a constant, then $y'(x) = re^{rx}$ and $y''(x) = r^2 e^{rx}$. If $y(x)$ is a solution of (2), then

$$r^2 e^{rx} + bre^{rx} + ce^{rx} = 0.$$

Since $e^{rx} \neq 0$, r is a solution of the equation

$$r^2 + br + c = 0. \tag{3}$$

On the other hand, if r is a solution of (3), then $y(x) = e^{rx}$ is a solution of (2). It follows that we can find solutions of (2) by solving Equation 3. Equation 3 is called the **characteristic equation** of the differential equation (2).

The quadratic equation

$$r^2 + br + c = 0$$

has solution

$$r = \frac{-b \pm \sqrt{b^2 - 4c}}{2}.$$

In case $b^2 - 4c > 0$, there are two distinct real roots. In case $b^2 - 4c = 0$, there is one double root. In case $b^2 - 4c < 0$, there are two complex, nonreal roots. We will discuss the solution of Equation 2 in each case.

Two Distinct Real Roots

If r_1 and r_2 are distinct real roots of the equation

$$r^2 + br + c = 0,$$

then $y_1(x) = e^{r_1 x}$ and $y_2(x) = e^{r_2 x}$ are both solutions of

$$y'' + by' + cy = 0.$$

Since $r_1 \neq r_2$,

$$W(y_1, y_2; x) = \begin{vmatrix} e^{r_1 x} & e^{r_2 x} \\ r_1 e^{r_1 x} & r_2 e^{r_2 x} \end{vmatrix} = e^{(r_1 + r_2)x}(r_2 - r_1) \neq 0,$$

so the general solution of the differential equation is

$$y(x) = C_1 e^{r_1 x} + C_2 e^{r_2 x}.$$

EXAMPLE 2

Find the general solution of

$$y'' - 3y' + 2y = 0.$$

SOLUTION   This second-order linear equation with constant coefficients has characteristic equation

$$r^2 - 3r + 2 = 0,$$
$$(r - 1)(r - 2) = 0,$$

so the solutions are $r_1 = 1$ and $r_2 = 2$. The corresponding solutions of the differential equation are

$$y_1(x) = e^x \text{ and } y_2(x) = e^{2x}.$$

The general solution is

$$y = C_1 e^x + C_2 e^{2x}. \quad \blacksquare$$

One Double Root

If $b^2 - 4c = 0$, so

$$r^2 + br + c = r^2 + br + \frac{b^2}{4} = \left(r + \frac{b}{2} \right)^2,$$

and $r^2 + rb + c = 0$ has one real root $r_1 = -b/2$ with multiplicity two, then we know that

$$y_1(x) = e^{r_1 x}$$

is a solution of

$$y'' + by' + cy = 0.$$

We can apply the method of reduction of order as in Example 1 to determine that another solution is

$$y_2(x) = xe^{r_1 x}.$$

Let us verify that y_2 is a solution. We have $r_1 = -b/2$, so

$$y_2(x) = xe^{-bx/2},$$

$$y_2'(x) = e^{-bx/2} \left(1 - \frac{b}{2}x \right),$$

$$y_2''(x) = e^{-bx/2} \left(\frac{b^2}{4}x - b \right).$$

We then have

$$y_2'' + by_2' + cy_2 = y_2'' + by_2' + \frac{b^2}{4} y_2$$

$$= e^{-bx/2} \left(\frac{b^2}{4}x - b \right) + be^{-bx/2} \left(1 - \frac{b}{2}x \right) + \frac{b^2}{4}xe^{-bx/2}$$

$$= e^{-bx/2} \left(\frac{b^2}{4}x - b + b - \frac{b^2}{2} + \frac{b^2}{4}x \right) = 0,$$

so y_2 is a solution of $y'' + by' + cy = 0$. We have

$$W(e^{r_1 x}, xe^{r_1 x}; x) = \begin{vmatrix} e^{r_1 x} & xe^{r_1 x} \\ r_1 e^{r_1 x} & e^{r_1 x}(r_1 x + 1) \end{vmatrix}$$

$$= e^{2r_1 x}(r_1 x + 1) - r_1 xe^{2r_1 x} = e^{2r_1 x} \neq 0,$$

so the general solution is

$$y = C_1 e^{r_1 x} + C_2 xe^{r_1 x}.$$

EXAMPLE 3

Find the general solution of

$$y'' - 6y' + 9y = 0.$$

SOLUTION The characteristic equation is

$$r^2 - 6t + 9 = 0,$$
$$(r - 3)^2 = 0,$$

so $r = 3$ is a double root. The general solution of the differential equation is then

$$y = C_1 e^{3x} + C_2 x e^{3x}. \quad \blacksquare$$

No Real Roots

In case $b^2 - 4c < 0$, the equation

$$r^2 + br + c = 0$$

has no real roots. However, the equation has the two distinct *complex roots*

$$r_1 = -\frac{b}{2} + \frac{\sqrt{b^2 - 4c}}{2} = \alpha + \beta i$$

and

$$r_2 = -\frac{b}{2} - \frac{\sqrt{b^2 - 4c}}{2} = \alpha - \beta i.$$

We now note that the exponential function can be defined for complex numbers $A + Bi$ and that

$$e^{A+Bi} = e^A (\cos B + i \sin B).$$

Based on the case of distinct real roots of the characteristic equation, we may suspect that

$$e^{(\alpha + \beta i)x} = e^{\alpha x}(\cos \beta x + i \sin \beta x)$$

and

$$e^{(\alpha - \beta i)x} = e^{\alpha x}(\cos \beta x - i \sin \beta x)$$

might be complex-valued solutions of

$$y'' + by' + cy = 0.$$

Since linear combinations of solutions of homogeneous linear differential equations are also solutions, we may then suspect that

$$\frac{e^{(\alpha + \beta i)x} + e^{(\alpha - \beta i)x}}{2} = e^{\alpha x} \cos \beta x$$

and

$$\frac{e^{(\alpha + \beta i)x} - e^{(\alpha - \beta i)x}}{2i} = e^{\alpha x} \sin \beta x$$

are solutions. It is not difficult to verify that

$$y_1(x) = e^{\alpha x} \cos \beta x$$

and

$$y_2(x) = e^{\alpha x} \sin \beta x$$

are, indeed, solutions of the differential equation. We have

$$W(y_1, y_2; x) = \begin{vmatrix} e^{\alpha x} \cos \beta x & e^{\alpha x} \sin \beta x \\ e^{\alpha x}(\alpha \cos \beta x - \beta \sin \beta x) & e^{\alpha x}(\alpha \sin \beta x + \beta \cos \beta x) \end{vmatrix}$$

$$= e^{2\alpha x}(\alpha \sin \beta x \cos \beta x + \beta \cos^2 \beta x)$$
$$- e^{2\alpha x}(\alpha \sin \beta x \cos \beta x - \beta \sin^2 \beta x)$$

$$= e^{2\alpha x} \beta (\cos^2 \beta x + \sin^2 \beta x) = \beta e^{2\alpha x} \neq 0,$$

since $\beta \neq 0$. The general solution is then

$$y = C_1 e^{\alpha x} \cos \beta x + C_2 e^{\alpha x} \sin \beta x.$$

EXAMPLE 4

Find the general solution of the equation

$$y'' + k^2 y = 0.$$

SOLUTION   This differential equation has characteristic equation

$$r^2 + k^2 = 0.$$

The equation has complex roots $r_1 = 0 + ki$ and $r_2 = 0 - ki$. The general solution of the differential equation is then

$$y = C_1 \sin kx + C_2 \cos kx. \quad \blacksquare$$

EXAMPLE 5

Find the general solution of

$$y'' + 4y' + 6y = 0.$$

SOLUTION   The characteristic equation is

$$r^2 + 4r + 6 = 0,$$

which has roots

$$r = \frac{-(4) \pm \sqrt{(4)^2 - 4(1)(6)}}{2(1)} = -2 \pm i\sqrt{2}.$$

The general solution is

$$y = C_1 e^{-2x} \cos \sqrt{2}x + C_2 e^{-2x} \sin \sqrt{2}x. \quad \blacksquare$$

EXERCISES 15.5

In Exercises 1–10: **(a)** *Show that the given function y_1 is a solution of the given equation.* **(b)** *Find a function v such that $y_2 = y_1 v$ is a solution of the given differential equation and $W(y_1, y_2) \neq 0$.* **(c)** *Find the general solution of the given equation.*

1. $y_1 = e^{2x};\ y'' - 3y' + 2y = 0$
2. $y_1 = e^{-x};\ y'' + 2y' + y = 0$
3. $y_1 = \sin x;\ y'' + y = 0$
4. $y_1 = \cos 2x;\ y'' + 4y = 0$
5. $y_1 = x;\ x^2 y'' + xy' - y = 0,\ x \neq 0$
6. $y_1 = x^2;\ x^2 y'' - 4xy' + 6y = 0,\ x \neq 0$
7. $y_1 = x;\ 3x^2 y'' - xy' + y = 0,\ x \neq 0$
8. $y_1 = \sqrt{x};\ 4x^2 y'' - 4xy' + 3y = 0,\ x > 0$
9. $y_1 = x;\ x^2 y'' - xy' + y = 0,\ x \neq 0$
10. $y_1 = x^2;\ x^2 y'' - 3xy' + 4y = 0,\ x \neq 0$

Find the general solutions of the equations in Exercises 11–26.

11. $y'' - 5y' + 6y = 0$
12. $y'' - y' - 6y = 0$
13. $y'' - 9y = 0$
14. $y'' - 4y' = 0$
15. $y'' - 2y' - y = 0$
16. $y'' + 4y' + y = 0$
17. $y'' - 2y' + y = 0$
18. $y'' + 2y' + y = 0$
19. $4y'' + 4y' + y = 0$
20. $y'' - 4y' + y = 0$
21. $y'' - 2y = 0$
22. $y'' - 3y = 0$
23. $y'' + 4y = 0$
24. $y'' + 3y = 0$
25. $y'' + 2y' + 2y = 0$
26. $y'' - 4y' + 5y = 0$

27. Show that $y_1(x) = e^x$ is a solution of the homogeneous equation $y'' - 3y' + 2y = 0$. Find a solution $y_2 = vy_1$ of the nonhomogeneous equation $y'' - 3y' + 2y = e^x$.

28. Show that $y_1(x) = e^x$ is a solution of the homogeneous equation $y'' - 3y' + 2y = 0$. Find a solution $y_2 = vy_1$ of the nonhomogeneous equation $y'' - 3y' + 2y = e^{2x}$.

29. The equation $a_2 x^2 y'' + a_1 xy' + a_0 y = 0$, where a_0, a_1, and a_2 are constants, is called a **Cauchy-Euler equation**. Show that $y = x^r$ is a solution if r satisfies the equation $a_2 r^2 + (a_1 - a_2)r + a_0 = 0$.

Use the result of Exercise 29 to find the general solutions of the equations in Exercises 30–32.

30. $x^2 y'' - 3xy' + 3y = 0$
31. $x^2 y'' + 4xy' + 2y = 0$
32. $4x^2 y'' + 4xy' - y = 0$

33. Suppose y_1 and y_2 are solutions of the homogeneous second-order linear differential equation $y'' + py' + qy = 0$, so (i) $y_1'' + py_1' + qy_1 = 0$ and (ii) $y_2'' + py_2' + qy_2 = 0$.
    (a) Multiply equation (i) by y_2, multiply equation (ii) by y_1, and subtract the first of the resulting equations from the second to obtain
    $$(y_1 y_2'' - y_1'' y_2) + p(y_1 y_2' - y_1' y_2) = 0.$$
    (b) Show that $W(y_1, y_2) = y_1 y_2' - y_1' y_2$ satisfies $W'(y_1, y_2) = y_1 y_2'' - y_1'' y_2$.
    (c) Conclude from (a) and (b) that $W(y_1, y_2)$ satisfies $W' + pW = 0$, so $W(y_1, y_2; x) = W(y_1, y_2; x_0)e^{-\int_{x_0}^x p}$.
    It follows from (c) that if y_1 and y_2 are both solutions of the same homogeneous second-order linear differential equation on an interval I and $W(y_1, y_2; x_0) \neq 0$ for some x_0 in I, then $W(y_1, y_2; x) \neq 0$ for all x in I.

34. Show that $W(x, x^2; x) = x^2$, so $W(x, x^2; 0) = 0$ and $W(x, x^2; x) \neq 0$ for $x \neq 0$. This shows that the Wronskian of two functions may be zero at a point without being identically zero. This cannot happen for two functions that are solutions of the same second-order linear differential equation.

15.6 VARIATION OF PARAMETERS

Connections

Solution of homogeneous equations, 15.5.

We will see that the general solution of the nonhomogeneous equation

$$y'' + py' + qy = f$$

can be expressed in terms of solutions of the homogeneous equation

$$y'' + py' + qy = 0$$

and a particular solution of the nonhomogeneous equation. Also, we will see how solutions of the homogeneous equation can be used to find a particular solution of the nonhomogeneous equation.

Nonhomogeneous Equations

Theorem 1 gives the form of the general solution of the nonhomogeneous equation.

THEOREM 1

If y_1 and y_2 are solutions of the homogeneous equation

$$y'' + py' + qy = 0$$

with $W(y_1, y_2; x_0) \neq 0$ and if y_p is a particular solution of the nonhomogeneous equation

$$y'' + py' + qy = f, \qquad (1)$$

then

$$y = C_1 y_1 + C_2 y_2 + y_p$$

is the general solution of (1).

Proof   It is easy to check that any function of the form $C_1 y_1 + C_2 y_2 + y_p$ is a solution of (1). We must show that any solution y is of the form $C_1 y_1 + C_2 y_2 + y_p$. To do this, we consider the two linear equations in the variables r and s,

$$y_1(x_0)r + y_2(x_0)s + y_p(x_0) = y(x_0)$$

and

$$y_1'(x_0)r + y_2'(x_0)s + y_p'(x_0) = y'(x_0).$$

Since $W(y_1, y_2; x_0) \neq 0$, this system of equations has a unique solution $r = C_1$ and $s = C_2$. For these choices of C_1 and C_2, the function

$$u = y - (C_1 y_1 + C_2 y_2 + y_p)$$

satisfies the initial value problem

$$y'' + py' + qy = 0, \quad y(x_0) = 0, \quad y'(x_0) = 0.$$

Since the zero function is also a solution of this initial value problem, it follows from Theorem 1 of Section 15.5 that $u(x) = 0$. It follows that

$$y(x) = C_1 y_1(x) + C_2 y_2 + y_p(x).$$

We conclude that this is the form of the general solution. ■

Variation of Parameters

We have seen that the general solution of the homogeneous equation and a particular solution of the nonhomogeneous equation give the general solution of the nonhomogeneous equation. Let us now see how we can use two solutions, y_1 and y_2, of the homogeneous equation

$$y'' + py' + qy = 0$$

with $W(y_1, y_2) \neq 0$ to find a particular solution of the nonhomogeneous equation

$$y'' + py' + qy = f.$$

The method is called **variation of parameters**. The idea is to find *functions* v_1 and v_2 such that

$$y_p = v_1 y_1 + v_2 y_2$$

is a particular solution of the nonhomogeneous equation.

If

$$y_p = v_1 y_1 + v_2 y_2,$$

then

$$y_p' = v_1 y_1' + v_1' y_1 + v_2 y_2' + v_2' y_2$$

and

$$y_p'' = v_1 y_1'' + 2v_1' y_1' + v_1'' y_1 + v_2 y_2'' + 2v_2' y_2' + v_2'' y_2.$$

The function y_p is a solution of

$$y'' + py' + qy = f$$

whenever

$$(v_1 y_1'' + 2v_1' y_1' + v_1'' y_1 + v_2 y_2'' + 2v_2' y_2' + v_2'' y_2)$$
$$+ p(v_1 y_1' + v_1' y_1 + v_2 y_2' + v_2' y_2) + q(v_1 y_1 + v_2 y_2) = f,$$
$$v_1(y_1'' + py_1' + qy_1) + v_2(y_2'' + py_2' + qy_2) + 2(v_1' y_1' + v_2' y_2')$$
$$+ p(v_1' y_1 + v_2' y_2) + (v_1'' y_1 + v_2'' y_2) = f.$$

Since y_1 and y_2 are solutions of the homogeneous equation, the above equation reduces to

$$2(v_1' y_1' + v_2' y_2') + p(v_1' y_1 + v_2' y_2) + (v_1'' y_1 + v_2'' y_2) = f. \qquad (2)$$

We want to find v_1 and v_2 so (2) is true. This can be done many ways. One way is to suppose that v_1 and v_2 can be chosen so that

$$v_1' y_1 + v_2' y_2 = 0. \qquad (3)$$

Note that this condition eliminates p from Equation 2. Condition (3) also gives

$$(v_1' y_1 + v_2' y_2)' = 0,$$
$$v_1' y_1' + v_1'' y_1 + v_2' y_2' + v_2'' y_2 = 0,$$
$$v_1'' y_1 + v_2'' y_2 = -(v_1' y_1' + v_2' y_2').$$

Equation 2 then simplifies to

$$v_1' y_1' + v_2' y_2' = f. \qquad (4)$$

Summarizing, we see that if y_1 and y_2 are solutions of

$$y'' + py' + qy = 0$$

with $W(y_1, y_2) \neq 0$, then

$$y_p = v_1 y_1 + v_2 y_2$$

is a solution of

$$y'' + py' + qy = f$$

whenever v_1 and v_2 satisfy

$$\begin{cases} v_1' y_1 + v_2' y_2 = 0, \\ v_1' y_1' + v_2' y_2' = f. \end{cases} \tag{5}$$

Since $W(y_1, y_2) \neq 0$, this system can be solved for v_1' and v_2' in terms of y_1, y_1', y_2, y_2', and f. We obtain

$$v_1' = \frac{-y_2 f}{y_1 y_2' - y_1' y_2} \qquad \text{and} \qquad v_2' = \frac{y_1 f}{y_1 y_2' - y_1' y_2}.$$

Integration then gives

$$v_1 = \int \frac{-y_2 f}{y_1 y_2' - y_1' y_2} \, dx \qquad \text{and} \qquad v_2 = \int \frac{y_1 f}{y_1 y_2' - y_1' y_2} \, dx. \tag{6}$$

Note that the denominators in the above integrands are the Wronskian, $W(y_1, y_2)$.

Particular Solutions

One way to implement the method of variation of parameters is to try a solution of the form $v_1 y_1 + v_2 y_2$, remember that we want to assume the simplifying condition (3), reduce the differential equation to Equation 4, and then solve for v_1' and v_2' in each case. We will just use the equations in (6).

EXAMPLE 1

Find a particular solution of

$$y'' - 3y' + 2y = e^{2x}.$$

SOLUTION   The homogeneous linear equation with constant coefficients

$$y'' - 3y' + 2y = 0$$

has characteristic equation

$$r^2 - 3r + 2 = 0,$$
$$(r - 1)(r - 2) = 0,$$

so we have solutions

$$y_1(x) = e^x \quad \text{and} \quad y_2(x) = e^{2x}$$

with

$$W(y_1, y_2) = \begin{vmatrix} e^x & e^{2x} \\ e^x & 2e^{2x} \end{vmatrix} = e^{3x} \neq 0.$$

Evaluating the integrals in (6), we have

$$v_1(x) = \int \frac{-y_2 f}{W(y_1, y_2)} dx = \int \frac{-e^{2x} e^{2x}}{e^{3x}} dx = \int -e^x dx = -e^x$$

and

$$v_2(x) = \int \frac{y_1 f}{W(y_1, y_2)} dx = \int \frac{e^x e^{2x}}{e^{3x}} dx = \int dx = x,$$

where we have chosen the constants of integration to be zero. A particular solution is then

$$
\begin{aligned}
y_p &= v_1 y_1 + v_2 y_2 \\
&= (-e^x)(e^x) + (x)(e^{2x}) \\
&= -e^{2x} + x e^{2x}. \quad \blacksquare
\end{aligned}
$$

EXAMPLE 2

Find a particular solution of

$$y'' - 4y' + 4y = e^{2x}.$$

SOLUTION   The homogeneous equation

$$y'' - 4y' + 4y = 0$$

has characteristic equation

$$
\begin{aligned}
r^2 - 4r + 4 &= 0, \\
(r - 2)^2 &= 0,
\end{aligned}
$$

so the homogeneous equation has solutions

$$y_1(x) = e^{2x} \text{ and } y_2(x) = x e^{2x}$$

with

$$W(y_1, y_2) = \begin{vmatrix} e^{2x} & x e^{2x} \\ 2e^x & e^{2x}(2x + 1) \end{vmatrix} = e^{4x} \neq 0.$$

We then have

$$v_1(x) = \int \frac{-y_2 f}{W(y_1, y_2)} dx = \int \frac{-x e^{2x} e^{2x}}{e^{4x}} dx = \int -x \, dx = -x^2/2$$

and

$$v_2(x) = \int \frac{y_1 f}{W(y_1, y_2)} dx = \int \frac{e^{2x} e^{2x}}{e^{4x}} dx = \int dx = x,$$

where we have chosen the constants of integration to be zero. Then

$$y_p = v_1 y_1 + v_2 y_2 = -\frac{1}{2} x^2 e^{2x} + x e^{2x}$$

is a particular solution. $\blacksquare$

EXAMPLE 3

Find a particular solution of

$$y'' - 3y' + 2y = e^x \sin x.$$

SOLUTION   As in Example 1, the homogeneous equation

$$y'' - 3y' + 2y = 0$$

has solutions

$$y_1(x) = e^x \text{ and } y_2(x) = e^{2x}$$

with

$$W(y_1, y_2) = e^{3x}.$$

We then have

$$v_1(x) = \int \frac{-y_2 f}{W(y_1, y_2)} dx = \int \frac{-(e^{2x})(e^x \sin x)}{e^{3x}} dx = \int -\sin x \, dx = \cos x$$

and

$$v_2(x) = \int \frac{y_1 f}{W(y_1, y_2)} dx = \int \frac{(e^x)(e^x \sin x)}{e^{3x}} dx = \int e^{-x} \sin x \, dx$$

$$= -\frac{e^{-x}}{2}(\sin x + \cos x),$$

where we have used the Table of Integrals to evaluate the latter integral. A particular solution is then

$$y_p = v_1 y_1 + v_2 y_2 = (\cos x)(e^x) + \left(-\frac{e^{-x}}{2}\right)(\sin x + \cos x)(e^{2x})$$

$$= \frac{e^x}{2}(\cos x - \sin x). \quad \blacksquare$$

General Solutions

EXAMPLE 4

Find the general solution of

$$y'' + y = e^{-x}.$$

Show that each solution is bounded on the interval $x \geq 0$.

SOLUTION   The corresponding homogeneous differential equation has characteristic equation

$$r^2 + 1 = 0$$

with complex roots $r_1 = i$ and $r_2 = -i$. The homogeneous equation then has solutions

$$y_1(x) = \cos x \text{ and } y_2(x) = \sin x$$

with

$$W(y_1, y_2) = \begin{vmatrix} \cos x & \sin x \\ -\sin x & \cos x \end{vmatrix} = 1.$$

We then have

$$v_1(x) = \int \frac{-y_2 f}{W(y_1, y_2)} dx = \int -e^{-x} \sin x \, dx = \frac{e^{-x}}{2}(\sin x + \cos x)$$

and

$$v_2(x) = \int \frac{y_1 f}{W(y_1, y_2)} dx = \int e^{-x} \cos x \, dx = \frac{e^{-x}}{2}(\sin x - \cos x),$$

where we have used the Table of Integrals to evaluate the integrals. The non-homogeneous equation then has a particular solution

$$y_p = v_1 y_1 + v_2 y_2$$

$$= \frac{e^{-x}}{2}(\sin x \cos x + \cos^2 x) + \frac{e^{-x}}{2}(\sin^2 x - \sin x \cos x)$$

$$= \frac{e^{-x}}{2}.$$

The general solution is

$$y = C_1 y_1 + C_2 y_2 + y_p = C_1 \cos x + C_2 \sin x + \frac{e^{-x}}{2}.$$

For $x \geq 0$, we have

$$\left| C_1 \cos x + C_2 \sin x + \frac{e^{-x}}{2} \right| \leq |C_1 \cos x| + |C_2 \sin x| + \left| \frac{e^{-x}}{2} \right|$$

$$\leq |C_1| + |C_2| + \frac{1}{2},$$

so each solution is bounded on $x \geq 0$. ∎

EXERCISES 15.6

Find a particular solution of the equations in Exercises 1–6.

1. $y'' - y' - 6y = e^{2x}$

2. $y'' - 5y' + 6y = e^{2x}$

3. $y'' - 2y' + y = e^x$

4. $y'' - 2y' + y = e^{-x}$

5. $y'' + 2y' - 8y = x$

6. $y'' - 4y' + 3y = xe^x$

Find the general solution of the equations in Exercises 7–10.

7. $y'' + y' - 6y = e^x$

8. $y'' - 6y' + 5y = e^{2x}$

9. $y'' + y = x$

10. $y'' + 4y = \sin 2x$

Solve the initial value problems in Exercises 11–14.

11. $y'' - y = \sin x$, $y(0) = 0$, $y'(0) = 1$

12. $y'' - y = xe^x$, $y(0) = 0$, $y'(0) = 0$

13. $y'' + y = e^x$, $y(0) = 0$, $y'(0) = 0$

14. $y'' + y = 1$, $y(0) = 0$, $y'(0) = 1$

15. (a) Show that $y_p(x) = \int_0^x \sin(x - t)b(t) \, dt$ is a particular solution of the equation $y'' + y = b(x)$, $x \geq 0$. (*Hint:* $\sin(A - B) = \sin A \cos B - \cos A \sin B$.)

    (b) Find the general solution of the equation in (a).

    (c) Show that each solution of the equation in (a) is bounded on the interval $x \geq 0$ if $\int_0^\infty |b(t)| \, dt = B < \infty$.

16. We know that $y_1 = e^{-\int p}$ is a solution of the homogeneous equation $y' + py = 0$. Show that $y = vy_1$ satisfies $y' + py = q$ whenever v satisfies $v' = qe^{\int p}$, so $y = vy_1 = e^{-\int p} \int qe^{\int p} \, dx$. This shows that the integrating factor method of solving first-order linear equations is related to the method of variation of parameters.

Connections

Solution of second-order linear equations, 15.5–6.

15.7 UNDETERMINED COEFFICIENTS

For certain types of nonhomogeneous equations it is fairly easy to obtain a particular solution without applying the method of variation of parameters. We will illustrate.

Differential Operators

It is convenient to associate the differential equation

$$y'' + py' + qy = f \qquad (1)$$

with the **differential operator** L that maps a function y into the function $L[y]$ defined by

$$L[y](x) = y''(x) + p(x)y'(x) + q(x)y(x).$$

The differential equation (1) can then be expressed as

$$L[y] = f.$$

It is not difficult to show that L is a **linear operator**. That is,

$$L[c_1 y_1 + c_2 y_2] = c_1 L[y_1] + c_2 L[y_2], \quad c_1 \text{ and } c_2 \text{ constants.}$$

It follows from the linearity of L that, if y_1 is a solution of $L[y] = f_1$ and y_2 is a solution of $L[y] = f_2$, then $y_1 + y_2$ is a solution of $L[y] = f_1 + f_2$. This allows **superposition** of solutions.

Familiarity of how the operator L transforms a few types of functions allows us to guess the form of the solutions of many differential equations that occur in applications.

EXAMPLE 1

$L[y] = y'' - 3y' + 2y$. Evaluate **(a)** $L[e^{3x}]$, **(b)** $L[e^{2x}]$, **(c)** $L[xe^{2x}]$, and **(d)** $L[\sin x]$.

SOLUTION

(a) We have

$$y = e^{3x},$$
$$y' = 3e^{3x},$$
$$y'' = 9e^{3x}.$$

Substitution then gives

$$L[e^{3x}] = 9e^{3x} - 3(3e^{3x}) + 2(e^{3x}) = 2e^{3x}.$$

(b) We have

$$y = e^{2x},$$
$$y' = 2e^{2x},$$
$$y'' = 4e^{2x}.$$

Substitution gives

$$L[e^{2x}] = 4e^{2x} - 3(2e^{2x}) + 2(e^{2x}) = 0.$$

(c) We have

$$y = xe^{2x},$$
$$y' = (x)(2e^{2x}) + (e^{2x})(1) = e^{2x}(2x + 1),$$
$$y'' = (e^{2x})(2) + (2x + 1)(2e^{2x}) = e^{2x}(4x + 4).$$

Substitution gives

$$L[xe^{2x}] = e^{2x}(4x + 4) - 3e^{2x}(2x + 1) + 2xe^{2x} = e^{2x}.$$

(d) We have

$$y = \sin 2x,$$
$$y' = 2\cos 2x,$$
$$y'' = -4\sin 2x.$$

Substitution gives

$$L[\sin 2x] = -4\sin 2x - 3(2\cos 2x) + 2(\sin 2x)$$
$$= -2\sin 2x - 6\cos 2x. \quad \blacksquare$$

EXAMPLE 2

$L[y] = y'' - 4y' + 4y$. Evaluate **(a)** $L[xe^{2x}]$ and **(b)** $L[x^2e^{2x}]$.

SOLUTION

(a) As in Example 1c, we have

$$y = xe^{2x},$$
$$y' = e^{2x}(2x + 1),$$
$$y'' = e^{2x}(4x + 4).$$

Then

$$L[xe^{2x}] = e^{2x}(4x + 4) - 4e^{2x}(2x + 1) + 4xe^{2x} = 0.$$

(b) We have

$$y = x^2e^{2x},$$
$$y' = (x^2)(2e^{2x}) + (e^{2x})(2x) = e^{2x}(2x^2 + 2x),$$
$$y'' = (e^{2x})(4x + 2) + (2x^2 + 2x)(2e^{2x}) = e^{2x}(4x^2 + 8x + 2).$$

Substitution gives

$$L[x^2e^{2x}] = e^{2x}(4x^2 + 8x + 2) - 4e^{2x}(2x^2 + 2x) + 4x^2e^{2x} = 2e^{2x}. \quad \blacksquare$$

If b and c are constants, the character of the function y can give us some information about $L[y] = y'' + by' + c$. For example:

- All derivatives of $e^{\alpha x}$ are constant multiples of $e^{\alpha x}$, so it is clear that $L[e^{\alpha x}]$ is a constant multiple of $e^{\alpha x}$.
- All derivatives of $x^n e^{\alpha x}$ are linear sums of powers of x times $e^{\alpha x}$, so $L[x^n e^{\alpha x}]$ must also be of this form.
- $L[\sin \beta x]$ and $L[\cos \beta x]$ must each be of the form $C_1 \sin \beta x + C_2 \cos \beta x$.
- $L[x^n]$ must be of the form $C_1 x^{n-2} + C_2 x^{n-1} + C_3 x^n$.

If the function f involves terms of either $e^{\alpha x}$, x^n, $x^n e^{\alpha x}$, $\sin \beta x$, or $\cos \beta x$, we can guess that the solution of $y'' + by' + c = f$ is a linear combination of corresponding terms. We can then substitute into the differential equation and solve for the unknown values of the constants in the linear combination to obtain a particular solution of the differential equation. This method is called the method of **undetermined coefficients**. If an incorrect general form were used, then we would not be able to solve for all of the constants and we would need to try again with the correct general form.

Equations That Involve $e^{\alpha x}$

In Example 1 we considered the operator $L[y] = y'' - 3y' + 2y$ with characteristic equation $r^2 - 3r + 2 = 0$. The characteristic equation has roots $r = 1$ and $r = 2$. Then $\alpha = 3$ is not a root of the characteristic equation and we saw in Example 1a that $L[e^{3x}] = 2e^{3x}$. This indicates that we should try $ce^{\alpha x}$ as a solution of $L[y] = de^{\alpha x}$ if α is not a solution of the characteristic equation.

The value $\alpha = 2$ is a root of the characteristic equation of the operator $L[y] = y'' - 3y' + 2y$. Example 1b shows $L[e^{2x}] = 0$ and Example 1c shows $L[xe^{2x}] = e^{2x}$. This indicates that a solution of $L[y] = de^{\alpha x}$ has form $cxe^{\alpha x}$ if α is a root of multiplicity one of the characteristic equation and that the solution has form $= cx^2 e^{\alpha x}$ if α is a double root of the characteristic equation.

> **To find a particular solution of the nonhomogeneous equation**
> $$y'' + by' + cy = de^{\alpha x}$$
>
> - If α is not a root of the characteristic equation, try $y = c_1 e^{\alpha x}$.
> - If α is root of multiplicity one, try $y = c_1 x e^{\alpha x}$.
> - If α is a double root, try $y = c_1 x^2 e^{\alpha x}$.

Note that if f is a sum of functions, the principle of superposition implies that then we can choose y to be a corresponding sum. We can think of each term of f as contributing to the particular solution.

EXAMPLE 3

Find a particular solution of the equation

$$y'' - 4y' + 3y = 4e^x - e^{2x}.$$

SOLUTION   The corresponding characteristic equation is

$$r^2 - 4r + 3 = 0.$$

The roots are $r = 1$ and $r = 3$.

Since $\alpha = 1$ is a solution of the characteristic equation, the term $4e^x$ on the right side of the differential equation contributes a term $c_1 x e^x$ to the particular solution. Since $\alpha = 2$ is not a root, the term e^{2x} contributes a term $c_2 e^{2x}$ to the particular solution. This leads us to try a particular solution of the form

$$y = c_1 x e^x + c_2 e^{2x}.$$

We then have

$$y' = c_1 (x + 1)e^x + 2c_2 e^{2x}$$

and

$$y'' = c_1 (x + 2)e^x + 4c_2 e^{2x}.$$

Substitution into the differential equation then gives

$$(c_1(x+2)e^x + 4c_2 e^{2x}) - 4(c_1(x+1)e^x + 2c_2 e^{2x}) + 3(c_1 x e^x + c_2 e^{2x})$$
$$= 4e^x - e^{2x}.$$

Collecting like terms and simplifying, we obtain

$$c_1 x e^x + 2c_1 e^x + 4c_2 e^{2x} - 4c_1 x e^x - 4c_1 e^x - 8c_2 e^{2x} + 3c_1 x e^x + 3c_2 e^{2x}$$
$$= 4e^x - e^{2x},$$
$$- 2c_1 e^x - c_2 e^{2x} = 4e^x - e^{2x}.$$

It is clear that the above equation has solution $c_1 = -2$ and $c_2 = 1$. A particular solution is then

$$y_p = c_1 x e^x + c_2 e^{2x} = e^{2x} - 2x e^x. \quad \blacksquare$$

Let us suppose that we had not noticed in Example 3 that $r = 1$ was a solution of the characteristic equation. We may have then tried a solution of the form

$$y = c_1 e^x + c_2 e^{2x}.$$

Substituting this y into the differential equation

$$y'' - 4y' + 3y = 4e^x - e^{2x}$$

and then simplifying as in Example 3 gives

$$-c_2 e^{2x} = 4e^x - e^{2x}.$$

There are no values of the constants c_1 and c_2 that satisfy this equation. There is no solution of the form $y = c_1 e^x + c_2 e^{2x}$.

EXAMPLE 4

Find a particular solution of the equation

$$y'' - 6y' + 9y = e^{3x}.$$

SOLUTION   The corresponding characteristic equation is

$$r^2 - 6r + 9 = 0,$$
$$(r - 3)^2 = 0,$$

which has the double root $r = 3$. The term e^{3x} on the right side of the differential equation then contributes a term $cx^2 e^{3x}$ to a particular solution. We have

$$y = cx^2 e^{3x},$$
$$y' = c(3x^2 + 2x)e^{3x},$$
$$y'' = c(9x^2 + 12x + 2)e^{3x}.$$

Substitution into the differential equation gives

$$(c(9x^2 + 12x + 2)e^{3x}) - 6(c(3x^2 + 2x)e^{3x}) + 9(cx^2 e^{3x}) = e^{3x},$$
$$2ce^{3x} = e^{3x},$$
$$2c = 1.$$

The above equation has solution $c = 1/2$. It follows that a particular solution is

$$y_p = \frac{1}{2}x^2 e^{3x}. \quad \blacksquare$$

Equations That Involve Polynomials

> **To find a particular solution of the nonhomogeneous equation**
> $$y'' + by' + cy = a_0 + a_1 x + \cdots + a_n x^n$$
>
> - If zero is not a root of the characteristic equation, try $y = c_0 + c_1 x + c_2 x^2 + \cdots + c_n x^n$.
> - If zero is a root of multiplicity one, try $y = c_0 x + c_1 x^2 + c_2 x^3 + \cdots + c_n x^{n+1}$.

EXAMPLE 5

Find a particular solution of the equation

$$y'' - 3y' + 2y = x^2.$$

SOLUTION   The corresponding characteristic equation is

$$r^2 - 3r + 2 = 0.$$

Zero is not a root, so we try

$$y = c_0 + c_1 x + c_2 x^2,$$
$$y' = c_1 + 2c_2 x,$$
$$y'' = 2c_2.$$

Substitution into the differential equation then gives

$$(2c_2) - 3(c_1 + 2c_2 x) + 2(c_0 + c_1 x + c_2 x^2) = x^2,$$
$$(2c_0 - 3c_1 + 2c_2) + (2c_1 - 6c_2)x + 2c_2 x^2 = x^2.$$

The solution for the values of the c's is given by the system of equations

$$\begin{cases} 2c_0 - 3c_1 + 2c_2 = 0, \\ 2c_1 - 6c_2 = 0, \\ 2c_2 = 1. \end{cases}$$

The system has solution $c_2 = 1/2$, $c_1 = 3/2$, $c_0 = 7/4$. A particular solution is

$$y_p = \frac{7}{4} + \frac{3}{2}x + \frac{1}{2}x^2. \quad \blacksquare$$

It is not correct to try a solution of the form $y(x) = cx^2$ in Example 5. Substitution into the equation

$$y'' - 3y' + 2y = x^2$$

and simplifying as in Example 4 would then give

$$2c - 6cx + 2cx^2 = x^2.$$

There is no value of the constant c that satisfies this equation. There is no solution of the equation of the form $y = cx^2$. We must include the lower-order terms c_0 and $c_1 x$ along with the term $c_2 x^2$ to find the correct solution.

Equations That Involve $\sin \beta x$ or $\cos \beta x$

Example 1d shows that the operator $L[y] = y'' - 4y' + 4y$ satisfies $L[\sin 2x] = -2 \sin 2x - 6 \cos 2x$. This result should be compared with the fact that, if $\alpha + \beta i$ is a root of the characteristic equation, we know that $L[e^{\alpha x} \sin \beta x] = L[e^{\alpha x} \cos \beta x] = 0$. In particular, if βi is a solution of the characteristic equation, then $L[\sin \beta x] = L[\cos \beta x] = 0$.

> **To find a particular solution of either the nonhomogeneous equation $y'' + by' + cy = d \sin \beta x$ or $y'' + by' + cy = d \cos \beta x$**
>
> - If βi is not a root of the characteristic equation, try $c_1 \sin \beta x + c_2 \cos \beta x$.
> - If βi is root, try $c_1 x \sin \beta x + c_2 x \cos \beta x$.

EXAMPLE 6

Find a particular solution of the equation

$$y'' + 2y' + 2y = \sin x.$$

SOLUTION   The characteristic equation is

$$r^2 + 2r + 2 = 0.$$

The complex number i is not a solution of the characteristic equation, so the term $\sin x$ on the right side of the differential equation contributes a term $c_1 \sin x + c_2 \cos x$ to the particular solution. Note that the particular solution we try involves both $\sin x$ and $\cos x$, even though the right-hand side of the differential equation involves only $\sin x$. We have

$$y = c_1 \sin x + c_2 \cos x,$$
$$y' = c_1 \cos x - c_2 \sin x,$$
$$y'' = -c_1 \sin x - c_2 \cos x.$$

Substitution gives

$$(-c_1 \sin x - c_2 \cos x) + 2(c_1 \cos x - c_2 \sin x) + 2(c_1 \sin x + c_2 \cos x) = \sin x,$$
$$(c_1 - 2c_2) \sin x + (2c_1 + c_2) \cos x = \sin x.$$

The above equation has solution for the constants c_1 and c_2 given by the system of equations

$$\begin{cases} c_1 - 2c_2 = 1, \\ 2c_1 + c_2 = 0. \end{cases}$$

The solution is $c_1 = 1/5$, $c_2 = -2/5$. A particular solution of the differential equation is

$$y_p = \frac{1}{5} \sin x - \frac{2}{5} \cos x. \quad \blacksquare$$

EXAMPLE 7

Solve the initial value problem

$$y'' + 9y = 10 \cos 2x - 12 \cos 3x, \quad y(0) = 3, \quad y'(0) = 1.$$

SOLUTION   In order to find the general solution of the nonhomogeneous equation we need the general solution of the corresponding homogeneous equation and a particular solution of the nonhomogeneous equation. The characteristic equation is

$$r^2 + 9 = 0,$$

which has roots $r = \pm 3i$. This implies that the homogeneous equation $y'' + 9y = 0$ has general solution

$$y_h = c_1 \sin 3x + c_2 \cos 3x.$$

We use the method of undetermined coefficients to find a particular solution of the nonhomogeneous equation.

Since $2i$ is not a solution of the characteristic equation, the term $10 \cos 2x$ of the differential equation contributes a term $c_3 \sin 2x + c_4 \cos 2x$ to the particular solution. Since $3i$ is a solution of the characteristic equation, the term $12 \cos 3x$ of the differential equation contributes a term $c_5 x \sin 3x + c_6 x \cos 3x$ to the particular solution.

For a particular solution of the nonhomogeneous equation we try

$$
\begin{aligned}
y_p &= c_3 \sin 2x + c_4 \cos 2x + c_5 x \sin 3x + c_6 x \cos 3x, \\
y_p' &= 2c_3 \cos 2x - 2c_4 \sin 2x + 3c_5 x \cos 3x + c_5 \sin 3x \\
&\quad - 3c_6 x \sin 3x + c_6 \cos 3x, \\
y_p'' &= -4c_3 \sin 2x - 4c_4 \cos 2x - 9c_5 x \sin 3x + 3c_5 \cos 3x + 3c_5 \cos 3x \\
&\quad - 9c_6 x \cos 3x - 3c_6 \sin 3x - 3c_6 \sin 3x.
\end{aligned}
$$

Substitution into the nonhomogeneous equation gives

$$
\begin{aligned}
[-4c_3 \sin 2x &- 4c_4 \cos 2x - 9c_5 x \sin 3x + 3c_5 \cos 3x + 3c_5 \cos 3x \\
&- 9c_6 x \cos 3x - 3c_6 \sin 3x - 3c_6 \sin 3x] \\
&+ 9[c_3 \sin 2x + c_4 \cos 2x + c_5 x \sin 3x + c_6 x \cos 3x] \\
&= 10 \cos 2x - 12 \cos 3x, \\
5c_3 \sin 2x + 5c_4 \cos 2x &+ 6c_5 \cos 3x - 6c_6 \sin 3x \\
&= 10 \cos 2x - 12 \cos 3x.
\end{aligned}
$$

The above equation has solution $c_3 = 0$, $c_4 = 2$, $c_5 = -2$, $c_6 = 0$. A particular solution of the nonhomogeneous equation is then $y_p = 2 \cos 2x - 2x \sin 3x$.

The general solution of the nonhomogeneous equation is

$$y = y_h + y_p = c_1 \sin 3x + c_2 \cos 3x + 2 \cos 2x - 2x \sin 3x.$$

We then have

$$y' = 3c_1 \cos 3x - 3c_2 \sin 3x - 4 \sin 2x - 6x \cos 3x - 2 \sin 3x.$$

$y(0) = 3$ implies $c_2 + 2 = 3$, so $c_2 = 1$.
$y'(0) = 1$ implies $3c_1 = 1$, so $c_1 = 1/3$.

The solution of the initial value problem is then

$$y = \frac{1}{3} \sin 3x + \cos 3x + 2 \cos 2x - 2x \cos 3x. \quad \blacksquare$$

EXERCISES 15.7

1. $L[y] = y'' - 5y' + 6y$. Evaluate
  (a) $L[e^{2x}]$,
  (b) $L[e^{3x}]$,
  (c) $L[e^{-x}]$,
  (d) $L[\sin 2x]$.

2. $L[y] = y'' - 3y' + 2y$. Evaluate
  (a) $L[e^x]$,
  (b) $L[e^{2x}]$,
  (c) $L[e^{-x}]$,
  (d) $L[\cos 2x]$.

3. $L[y] = y'' - 4y' + 3y$. Evaluate
  (a) $L[e^x]$,
  (b) $L[e^{3x}]$,
  (c) $L[xe^x]$,
  (d) $L[xe^{3x}]$.

4. $L[y] = y'' - y' - 6y$. Evaluate
  (a) $L[e^{3x}]$,
  (b) $L[e^{-2x}]$,
  (c) $L[xe^{3x}]$,
  (d) $L[xe^{-2x}]$.

5. $L[y] = y'' - 2y' + y$. Evaluate
  (a) $L[e^x]$,
  (b) $L[xe^x]$,
  (c) $L[x^2 e^x]$.

6. $L[y] = y'' + 2y' + y$. Evaluate
  (a) $L[e^{-x}]$,
  (b) $L[xe^{-x}]$,
  (c) $L[x^2 e^{-x}]$.

7. $L[y] = y'' + y$. Evaluate
  (a) $L[\sin x]$,
  (b) $L[\cos x]$,
  (c) $L[x \sin x]$,
  (d) $L[x \cos x]$.

8. $L[y] = y'' - 2y' + 2y$. Evaluate
  (a) $L[e^x \sin x]$,
  (b) $L[xe^x \cos x]$.

Find a particular solution of the equations in Exercises 9–22.

9. $y'' + 5y' + 6y = e^{-x}$

10. $y'' + 5y' - 6y = e^{3x}$

11. $y'' + 5y' - 6y = e^{-2x}$

12. $y'' + 2y' - 8y = e^{2x}$

13. $y'' - 2y' + y = e^x$

14. $4y'' - 4y' + y = e^{x/2}$

15. $y'' - 2y' + y = x^2$

16. $y'' + 2y' + y = 2x$

17. $y'' + 4y = \sin x$

18. $y'' + y = \cos 2x$

19. $y'' + 4y = \cos 2x$

20. $y'' + y = \sin x$

21. $y'' - 2y' + 2y = \sin x$

22. $y'' - 2y' + 2y = e^x \sin x$

Solve the initial value problems in Exercises 23–26.

23. $y'' - y' - 6y = e^x$, $y(0) = 0$, $y'(0) = 0$

24. $y'' - 5y' - 6y = e^{2x}$, $y(0) = 0$, $y'(0) = 0$

25. $y'' - 5y' - 6y = e^{-x}$, $y(0) = 0$, $y'(0) = 0$

26. $y'' - y' - 6y = e^{-2x}$, $y(0) = 0$, $y'(0) = 0$

Find the general solution of the equations in Exercises 27–30.

27. $y'' + 9y = x$

28. $y'' + y = e^x$

29. $y'' + 4y' + 5y = e^{-2x}$

30. $y'' - 2y' + 2y = \sin x$

31. $L[y] = y'' - 4y' + 5y$
  (a) Evaluate $L[xe^{2x}(C_1 \sin x + C_2 \cos x)]$.
  (b) Find a particular solution of $L[y] = e^{2x} \sin x$.

32. $L[y] = y'' - 2y' + 5y$
  (a) Evaluate $L[xe^x(C_1 \sin 2x + C_2 \cos 2x)]$.
  (b) Find a particular solution of $L[y] = e^x \cos 2x$.

15.8 POLYNOMIAL APPROXIMATIONS OF SOLUTIONS

Connections

Taylor series, 9.9.

Let us now see how infinite series can be used to find polynomial approximations of solutions of differential equations. The following theorem is the theoretical basis for doing this.

> **THEOREM 1**
>
> If $p_0, p_1, \ldots, p_n$, and q are polynomials, and $p_n(c) \neq 0$, then the initial value problem
>
> $$p_n y^{(n)} + \cdots + p_1 y' + p_0 y = q,$$
> $$y(c) = y_0, \ldots, y^{(n-1)}(c) = y_{n-1}$$
>
> has an infinite series solution
>
> $$y(x) = \sum_{n=0}^{\infty} \frac{y^{(n)}(c)}{n!}(x - c)^n$$
>
> in some open interval that contains c.

We will not prove this theorem.

Approximation of Solutions

We have seen that a small number of terms can sometimes give accurate approximate values of the series

$$\sum_{n=0}^{\infty} \frac{y^{(n)}(c)}{n!}(x - c)^n \text{ for } x \text{ near } c.$$

This suggests that the partial sums of an infinite series solution of an initial value problem can be used to give approximate values of the solution.

The terms of an infinite series solution of a differential equation are determined by the values of the derivatives of the solution at the initial point c. Setting $x = c$ in the differential equation and then using the given initial values gives the value at $x = c$ of the highest-order derivative in the equation. We can then differentiate the original differential equation to obtain a new equation that involves a higher-order derivative. Substitution of known values at $x = c$ then gives the value at $x = c$ of this higher-order derivative. We can continue to differentiate and substitute until we have determined the values at $x = c$ of as many derivatives as we please. Note that it is a good idea to write the highest-order derivative by itself on one side of the equation before differentiating.

We will illustrate the procedure by example.

EXAMPLE 1

Use the first four nonzero terms of the infinite series solution of the initial value problem

$$y' - 2y = 0, \quad y(0) = 3$$

to obtain a polynomial approximation of the solution.

SOLUTION   Note that the initial condition involves $c = 0$. Theory guarantees that there is a solution of the form

$$y(x) = \sum_{n=0}^{\infty} \frac{y^{(n)}(0)}{n!} x^n.$$

We begin the solution procedure by writing the differential equation in the form

$$y' = 2y.$$

Substitution of $y(0) = 3$ then gives

$$y'(0) = 2y(0) = 2(3) = 6.$$

Differentiating the equation for y' gives

$$y'' = 2y', \text{ so } y''(0) = 2y'(0) = 2(6) = 12.$$

Differentiating the equation for y'' gives

$$y''' = 2y'', \text{ so } y'''(0) = 2y''(0) = 2(12) = 24.$$

We now have four nonzero coefficients, so we stop. The desired polynomial solution is

$$P(x) = y(0) + y'(0)x + \frac{y''(0)}{2!}x^2 + \frac{y'''(0)}{3!}x^3$$

$$= 3 + 6x + \frac{12}{2}x^2 + \frac{24}{6}x^3$$

$$= 3 + 6x + 6x^2 + 4x^3.$$

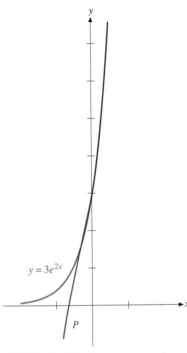

$y = 3e^{2x}$

P

FIGURE 15.8.1 The polynomial $P(x) = 3 + 6x + x^2 + 4x^3$ approximates $3e^{2x}$, the solution of the initial value problem.

We know from Section 7.6 that the solution of the initial value problem is $y(x) = 3e^{2x}$. It can be verified that we have obtained the first four nonzero terms of the Maclaurin series of $3e^{2x}$. The graphs of the solution $y(x)$ and the polynomial approximation $P(x)$ are given in Figure 15.8.1. ∎

EXAMPLE 2

Use the first four nonzero terms of the infinite series solution of the initial value problem

$$y'' + y = 0, \quad y(0) = 1, \quad y'(0) = 0$$

to obtain a polynomial approximation of the solution.

SOLUTION   Theory guarantees a solution of the form

$$y(x) = \sum_{n=0}^{\infty} \frac{y^{(n)}(0)}{n!}x^n.$$

Rewriting the equation with the second-order derivative on one side, we obtain

$$y'' = -y, \text{ so } y''(0) = -y(0) = -(1) = -1.$$

Differentiating the equation gives

$$y''' = -y', \text{ so } y'''(0) = -y'(0) = -(0) = 0.$$

Continuing, we obtain

$$y^{(4)} = -y'', \text{ so } y^{(4)}(0) = -y''(0) = -(-1) = 1,$$
$$y^{(5)} = -y''', \text{ so } y^{(5)}(0) = -y'''(0) = -(0) = 0,$$
$$y^{(6)} = -y^{(4)}, \text{ so } y^{(6)}(0) = -y^{(4)}(0) = -(1) = -1.$$

This gives four nonzero terms, so we stop. The desired approximating polynomial is

$$P(x) = y(0) + y'(0)x + \frac{y''(0)}{2!}x^2 + \frac{y'''(0)}{3!}x^3 + \frac{y^{(4)}(0)}{4!}x^4$$

$$+ \frac{y^{(5)}(0)}{5!}x^5 + \frac{y^{(6)}(0)}{6!}x^2$$

$$= 1 - \frac{1}{2}x^2 + \frac{1}{24}x^4 - \frac{1}{720}x^6.$$

It can be verified that $y(x) = \cos x$ is the solution of the given initial value problem. Note that we have obtained the first four nonzero terms of the Maclaurin series of $\cos x$. The graphs of the solution $y = \cos x$ and the polynomial approximation $P(x)$ are given in Figure 15.8.2. ■

$y = \cos x$

FIGURE 15.8.2 The polynomial $P(x) = 1 - \dfrac{x^2}{2} + \dfrac{x^4}{24} - \dfrac{x^6}{720}$ approximates $\cos x$, the solution of the initial value problem.

EXAMPLE 3

Use the first four nonzero terms of the infinite series solution of the initial value problem

$$y'' - 2y' + y = 0, \quad y(0) = 0, \quad y'(0) = 1$$

to obtain a polynomial approximation of the solution.

SOLUTION   Theory guarantees a solution of the form

$$y(x) = \sum_{n=0}^{\infty} \frac{y^{(n)}(0)}{n!}x^n.$$

We have

$$y'' = 2y' - y, \text{ so } y''(0) = 2y'(0) - y(0) = 2(1) - (0) = 2,$$
$$y''' = 2y'' - y', \text{ so } y'''(0) = 2y''(0) - y'(0) = 2(2) - (1) = 3,$$
$$y^{(4)} = 2y''' - y'', \text{ so } y^{(4)}(0) = 2y'''(0) - y''(0) = 2(3) - (2) = 4.$$

The desired polynomial approximation of the solution is

$$P(x) = y(0) + y'(0)x + \frac{y''(0)}{2!}x^2 + \frac{y'''(0)}{3!}x^3 + \frac{y^{(4)}(0)}{4!}x^4$$

$$= x + x^2 + \frac{1}{2}x^3 + \frac{1}{6}x^4.$$

It can be verified that $y(x) = xe^x$ is the solution of the initial value problem. We have obtained the first four nonzero terms of the Maclaurin series of xe^x. The solution and approximate solution are sketched in Figure 15.8.3. ■

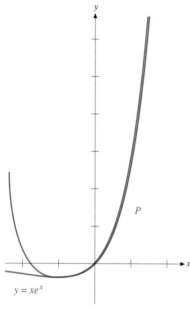

$y = xe^x$

FIGURE 15.8.3 The polynomial $P(x) = x + x^2 + \dfrac{x^3}{2} + \dfrac{x^4}{6}$ approximates xe^x, the solution of the initial value problem.

EXAMPLE 4

Use the first six nonzero terms of the infinite series solution of the initial value problem

$$y'' + (x^2 + 2)y = 4 - 3x, \quad y(0) = 1, \quad y'(0) = 2$$

to obtain a polynomial approximation of the solution.

SOLUTION   Theory guarantees a solution of the form

$$y(x) = \sum_{n=0}^{\infty} \frac{y^{(n)}(0)}{n!}x^n.$$

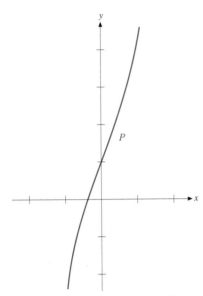

We have

$$y'' = 4 - 3x - (x^2 + 2)y, \text{ so}$$
$$y''(0) = 4 - 3(0) - ((0)^2 + 2)y(0) = 4 - 2(1) = 2,$$
$$y''' = -3 - (x^2 + 2)(y') - (y)(2x), \text{ so}$$
$$y'''(0) = -3 - ((0)^2 + 2)y'(0) - y(0)(2)(0) = -3 - (2)(2) = -7,$$
$$y^{(4)} = -(x^2 + 2)(y'') - (y')(2x) - (y)(2) - (2x)(y')$$
$$= -(x^2 + 2)y'' - 4xy' - 2y, \text{ so}$$
$$y^{(4)}(0) = -(2)(2) - 4(0)(2) - 2(1) = -6,$$
$$y^{(5)} = -(x^2 + 2)(y''') - (y'')(2x) - 4(x)(y'') - 4(y') - 2y'$$
$$= -(x^2 + 2)y''' - 6xy'' - 6y', \text{ so}$$
$$y^{(5)}(0) = -(2)(-7) - 6(0)(2) - 6(2) = 2.$$

The desired polynomial approximation of the solution is

$$P(x) = y(0) + y'(0)x + \frac{y''(0)}{2!}x^2 + \frac{y'''(0)}{3!}x^3 + \frac{y^{(4)}(0)}{4!}x^4 + \frac{y^{(5)}(0)}{5!}x^5$$

$$= 1 + 2x + x^2 - \frac{7}{6}x^3 - \frac{1}{4}x^4 + \frac{1}{60}x^5.$$

The graph of the polynomial approximation of the solution is given in Figure 15.8.4. The solution is not a familiar function. ∎

Recursive Formulas

We have seen in Examples 1 through 4 that n initial conditions determine the first n coefficients of an infinite series solution of an nth order initial value problem. Each successive coefficient is determined by previous coefficients, so it is possible to obtain a **recursive formula** that gives coefficients in terms of previous coefficients. The recursive formula could be used to write a computer program to generate partial sums of the infinite series solution.

FIGURE 15.8.4 The polynomial $P(x) = 1 + 2x + x^2 - \dfrac{7x^3}{6} - \dfrac{x^4}{4} + \dfrac{x^5}{60}$ approximates the solution of the initial value problem.

EXERCISES 15.8

Use the first four nonzero terms of the infinite series solutions of the initial value problems to obtain polynomial approximations of the solutions in Exercises 1–16.

1. $y' - y = 0,\ y(0) = 2$

2. $y' + y = 0,\ y(0) = 3$

3. $y'' - y = 0,\ y(0) = 1,\ y'(0) = 1$

4. $y'' - y = 0,\ y(0) = 1,\ y'(0) = -1$

5. $y'' + y = 0,\ y(0) = 0,\ y'(0) = 1$

6. $y'' + 4y = 0,\ y(0) = 3,\ y'(0) = 0$

7. $y'' - y = x^2,\ y(0) = 1,\ y'(0) = 2$

8. $y'' + y = x^2,\ y(0) = 1,\ y'(0) = 2$

9. $y'' - 3y' + 2y = 0,\ y(0) = 2,\ y'(0) = -1$

10. $y'' + 4y' + 4y = 0,\ y(0) = 3,\ y'(0) = 2$

11. $y'' + y' + 2y = 0,\ y(0) = 2,\ y'(0) = 1$

12. $y'' - y' + y = 0,\ y(0) = -2,\ y'(0) = 3$

13. $(1 - x)y' - y = 0,\ y(0) = 1$

14. $2(1 + x)y' - y = 0,\ y(0) = 1$

15. $y'' + y' + y = x + 2,\ y(0) = 3,\ y'(0) = 1$

16. $y'' - 2y' + y = 2x - 1,\ y(0) = 1,\ y'(0) = 4$

15.9 NUMERICAL SOLUTIONS OF FIRST-ORDER EQUATIONS

Connections

Calculator skills.

The derivative as a slope, 2.6.

Trapezoidal Rule, 5.6.

We will introduce two methods of generating numerical solutions of first-order differential equations. Numerical techniques are particularly well suited for use with computers.

We consider the initial value problem

$$y' = f(x, y), \quad y(x_0) = y_0,$$

where f is a function of the two variables x and y. We assume there is a unique solution of the initial value problem in some interval that contains x_0. Let us denote the exact solution of the problem by $\phi(x)$, so

$$\phi'(x) = f(x, \phi(x))$$

for all x in some interval that contains x_0. See Figure 15.9.1. For a fixed small value of h, we will use numerical methods to establish values y_n of an approximate solution $y(x)$ at points

$$x_n = x_0 + nh,$$

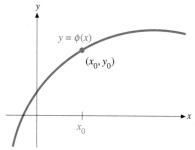

FIGURE 15.9.1   We assume that $y = \phi(x)$ is the exact solution of the initial value problem $y' = f(x, y), y(x_0) = y_0$, in some interval that contains x_0.

so

$$y(x_n) = y_n.$$

Of course, we choose the value of any approximate solution at the point x_0 to be y_0, so

$$y(x_0) = y_0.$$

Euler's Tangent Line Method

The first method we will use to determine the values of an approximate solution is called **Euler's tangent line method**. The idea is to use a simple linear approximation of the exact solution $\phi(x)$ between points x_n and x_{n+1}. Let us start at the point x_0 and see how to define our approximate solution at

$$x_1 = x_0 + h.$$

We have $y(x_0) = y_0$ and

$$\phi'(x_0) = f(x_0, \phi(x_0)).$$

Since $\phi(x_0) = y_0$, we have

$$\phi'(x_0) = f(x_0, y_0).$$

The line through the point (x_0, y_0) with slope $f(x_0, y_0)$ should be a good approximation of the graph of $\phi(x)$ for x near x_0. In particular, the point (x_1, y_1) on the line should be near the point $(x_1, \phi(x_1))$ if h is small. See Figure 15.9.2. The line has equation

$$y - y_0 = f(x_0, y_0)(x - x_0).$$

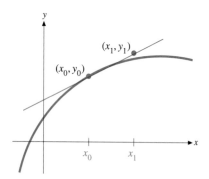

FIGURE 15.9.2   The line through the point (x_0, y_0) with slope $f(x_0, y_0)$ should be a good approximation of the graph of $\phi(x)$ for x near x_0.

If the point (x_1, y_1) is on the line, then

$$y_1 - y_0 = f(x_0, y_0)(x_1 - x_0), \text{ so}$$
$$y_1 = y_0 + f(x_0, y_0)(x_1 - x_0), \text{ or}$$
$$y_1 = y_0 + hf(x_0, y_0).$$

We choose the value of our approximate solution at the point x_1 to be y_1, so

$$y(x_1) = y_0 + hf(x_0, y_0).$$

We use the line through the point (x_1, y_1) with slope $f(x_1, y_1)$ to determine the value y_2 of our approximate solution at $x_2 = x_1 + h$. We obtain

$$y_2 = y_1 + hf(x_1, y_1).$$

The above typical step can be repeated to obtain additional points (x_n, y_n) that approximate the exact solution of the initial value problem. Let us summarize.

In order to use Euler's tangent line method to find points that approximate the exact solution of the initial value problem

$$y'(x) = f(x, y(x)), \quad y(x_0) = y_0,$$

choose a value of h and determine the coordinates of the approximating points by using the recursive formulas

$$x_{n+1} = x_n + h \quad \text{and} \quad y_{n+1} = y_n + hf(x_n, y_n). \tag{1}$$

EXAMPLE 1

For the initial value problem

$$y' = 2xy, \quad y(0) = 1,$$

(a) use Euler's tangent line method with $h = 0.1$ to determine the values of an approximate solution $y(x)$ at $x = 0.1, 0.2, 0.3$, and 0.4;

(b) use Euler's tangent line method with $h = 0.05$ to determine the values of an approximate solution $y(x)$ at $x = 0.1, 0.2, 0.3$, and 0.4;

(c) find the exact solution $\phi(x)$ and evaluate at $x = 0.1, 0.2, 0.3$, and 0.4.

SOLUTION

(a) We use Euler's tangent line recursive formulas

$$x_{n+1} = x_n + h \quad \text{and} \quad y_{n+1} = y_n + hf(x_n, y_n)$$

with $(x_0, y_0) = (0, 1)$, $h = 0.1$, and $f(x, y) = 2xy$. We have displayed the results of the calculator calculations in Table 15.9.1. Of course, it would be easy to program a computer to do the calculations.

(b) We use Formula 1 with $h = 0.05$. The results are given in Table 15.9.2.

TABLE 15.9.1

n	x_n	y_n	$y_{n+1} = y_n + (0.1)(2x_n y_n)$
0	0	1	1
1	0.1	1	1.02
2	0.2	1.02	1.0608
3	0.3	1.0608	1.124448
4	0.4	1.124448	

TABLE 15.9.2

n	x_n	y_n	$y_{n+1} = y_n + (0.05)(2x_n y_n)$
0	0	1	1
1	0.05	1	1.005
2	0.10	1.005	1.01505
3	0.15	1.01505	1.03027575
4	0.20	1.03027575	1.050881265
5	0.25	1.050881265	1.077153297
6	0.30	1.077153297	1.109467896
7	0.35	1.109467896	1.148299272
8	0.40	1.148299272	

(c) The differential equation is separable. If $\phi(x)$ is a solution, we have

$$\phi'(x) = 2x\phi(x),$$
$$\frac{\phi'(x)}{\phi(x)} = 2x,$$
$$\int \frac{\phi'(x)}{\phi(x)}\,dx = \int 2x\,dx,$$
$$\ln|\phi(x)| = x^2 + c,$$
$$|\phi(x)| = e^{x^2+c},$$
$$\phi(x) = Ce^{x^2},$$

where $C = \pm e^c$.

It is easy to see that the initial condition $\phi(0) = 1$ is satisfied when $C = 1$, so the exact solution of the initial value problem is

$$\phi(x) = e^{x^2}.$$

We have

$$\phi(0.1) = 1.010050167,$$
$$\phi(0.2) = 1.040810774,$$
$$\phi(0.3) = 1.094174284,$$
$$\phi(0.4) = 1.173510871. \quad \blacksquare$$

The results of Examples 1a, b, and c are compared in Table 15.9.3. We see that the results for $h = 0.05$ are more accurate than those for $h = 0.1$. This should not be surprising, since twice as many calculations were required for the smaller value of h.

TABLE 15.9.3

x_n	$y_n, h = 0.1$	$y_n, h = 0.05$	$\phi(x_n)$
0	1	1	1
0.1	1	1.005	1.010050167
0.2	1.02	1.03027575	1.040810774
0.3	1.0608	1.077153297	1.094174284
0.4	1.124448	1.148299272	1.173510871

A closer look at Table 15.9.3 reveals two more interesting facts. First, the error in the approximate values becomes larger with each successive step. Second, all of the approximate values are less than the corresponding exact values. Both of these facts can be explained geometrically. To do this we note from the solution of Example 1c that the general solution of the equation

$$y' = 2xy$$

is

$$y(x) = Ce^{x^2}.$$

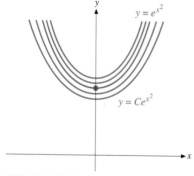

FIGURE 15.9.3a Each function $y = Ce^{x^2}$ is a solution of the differential equation $y' = 2xy$. The function $y = e^{x^2}$ satisfies the initial condition $y(0) = 1$.

Several of these solutions are sketched in Figure 15.9.3a. The solution of the initial value problem

$$y' = 2xy, \quad y(0) = 1$$

is the function ϕ with graph through the point $(0, 1)$. To determine the point (x_1, y_1) by using Euler's tangent line method, we follow the line tangent to

FIGURE 15.9.3b The tangent lines move to a different solution of $y' = 2xy$ at each step.

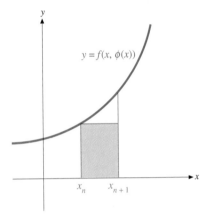

FIGURE 15.9.4a Euler's tangent line method corresponds to approximating $\int_{x_n}^{x_{n+1}} f(x, \phi(x))\, dx$ by substituting the constant $f(x_n, y_n)$ for the integrand.

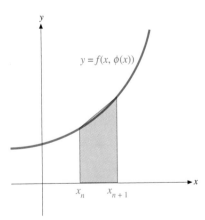

FIGURE 15.9.4b We should expect to improve the accuracy of Euler's method by using a trapezoidal-type approximation of the integral of $f(x, \phi(x))$.

the graph of ϕ at the point $(0, 1)$ to the point on the line with x-coordinate $x_1 = 0 + h$. Notice that ϕ is concave upward, so the tangent line is below the graph. Also, the point (x_1, y_1) is not on the graph of ϕ, but is on the graph of a *different* solution of the equation $y' = 2xy$. It is the tangent line of this new function that is used to determine the point (x_2, y_2). Again, this tangent line is below the graph of the function, so the point (x_2, y_2) moves to the graph of yet another solution of $y' = 2xy$, a solution that is farther away from the exact solution ϕ. See Figure 15.9.3b. This shows that the approximate values become less accurate with each successive step and that each error is in the same direction.

Improved Euler Method

In order to see how to improve the accuracy of the Euler tangent line method, let us look at the problem from a different point of view. If $\phi(x)$ is the exact solution of the initial value problem

$$y' = f(x, y), \quad y(x_0) = y_0,$$

then
$$\phi'(x) = f(x, \phi(x)).$$

We can then integrate the above differential equation with respect to x over the interval between x_n and x_{n+1} to obtain the **integral equation**

$$\phi(x_{n+1}) - \phi(x_n) = \int_{x_n}^{x_{n+1}} f(x, \phi(x))\, dx.$$

If we approximate the above integral by replacing $f(x, \phi(x))$ by the constant value $f(x_n, y_n)$, we obtain

$$\int_{x_n}^{x_{n+1}} f(x, \phi(x))\, dx \approx \int_{x_n}^{x_{n+1}} f(x_n, y_n)\, dx$$
$$= f(x_n, y_n)(x_{n+1} - x_n).$$

See Figure 15.9.4a. The expression $f(x_n, y_n)(x_{n+1} - x_n)$ is equal to $y_{n+1} - y_n$, the difference of the values of Euler's tangent line approximations of ϕ at the values x_n and x_{n+1}. This suggests that we might be able to improve the accuracy of Euler's method by using a trapezoidal-type approximation of the integral of $f(x, \phi(x))$. See Figure 15.9.4b. Let us illustrate a typical step.

We assume that x_n and y_n have already been calculated. To determine the approximate value y_{n+1} of $\phi(x_{n+1})$, we first use Euler's tangent line method to calculate a preliminary point

$$z_{n+1} = y_n + f(x_n, y_n)(x_{n+1} - x_n).$$

The first trapezoidal approximation of the integral gives

$$\int_{x_n}^{x_{n+1}} f(x, \phi(x))\, dx \approx \frac{f(x_n, \phi(x_n)) + f(x_{n+1}, \phi(x_{n+1}))}{2}(x_{n+1} - x_n).$$

Since we do not know the exact values of ϕ at the points x_n and x_{n+1}, we replace $\phi(x_n)$ and $\phi(x_{n+1})$ with the values y_n and z_{n+1}, respectively. Then

$$\phi(x_{n+1}) - \phi(x_n) = \int_{x_n}^{x_{n+1}} f(x, \phi(x))\, dx$$
$$\approx \frac{f(x_n, y_n) + f(x_{n+1}, z_{n+1})}{2}(x_{n+1} - x_n).$$

Using the expression on the right above as the difference between the approximate values y_n and y_{n+1}, we have

$$y_{n+1} = y_n + \frac{f(x_n, y_n) + f(x_{n+1}, z_{n+1})}{2}(x_{n+1} - x_n), \qquad (2)$$

where $z_{n+1} = y_n + f(x_n, y_n)(x_{n+1} - x_n)$.

Formula 2 gives the values of the **Improved Euler Method** for determining approximate values of the solution of the initial value problem

$$y' = f(x, y(x)), \quad y(x_0) = y_0.$$

The Improved Euler Method is a **predictor-corrector** method. Euler's tangent line method is used to *predict* z_{n+1} and then the formula for y_{n+1} is used to *correct* the prediction.

EXAMPLE 2

For the initial value problem

$$y' = 2xy, \quad y(0) = 1,$$

use the Improved Euler formula (2) with $h = 0.1$ to determine the values of an approximate solution $y(x)$ at $x = 0.1, 0.2, 0.3,$ and 0.4.

SOLUTION   We use Formula 2 with $(x_0, y_0) = (0, 1)$, $x_{n+1} - x_n = 0.1$, and $f(x, y) = 2xy$. Then

$$\begin{aligned} z_{n+1} &= y_n + f(x_n, y_n)(x_{n+1} - x_n) \\ &= y_n + (2x_n y_n)(0.1) \\ &= y_n + 0.2x_n y_n \end{aligned}$$

and

$$\begin{aligned} y_{n+1} &= y_n + \frac{f(x_n, y_n) + f(x_{n+1}, z_{n+1})}{2}(x_{n+1} - x_n) \\ &= y_n + (x_n y_n + x_{n+1} z_{n+1})(0.1). \end{aligned}$$

The calculations are arranged in Table 15.9.4. ■

In Table 15.9.5 we have added the results of Example 2 to the results of Example 1 that were given in Table 15.9.3. Note that the Improved Euler Method with $h = 0.1$ required about as many calculations as the Euler tangent line method with $h = 0.05$, but the improved method was much more accurate.

TABLE 15.9.4

n	x_n	y_n	z_{n+1}	y_{n+1}
0	0	1	1	1.01
1	0.1	1.01	1.0302	1.040704
2	0.2	1.040704	1.08233216	1.093988045
3	0.3	1.093988045	1.159627327	1.173192779
4	0.4	1.173192779		

TABLE 15.9.5

	Euler tangent line method		Improved Euler	
x_n	$y_n, h = 0.1$	$y_n, h = 0.05$	$y_n, h = 0.1$	$\phi(x_n)$
0	1	1	1	1
0.1	1	1.005	1.01	1.010050167
0.2	1.02	1.03027575	1.040704	1.040810774
0.3	1.0608	1.077153297	1.093988045	1.094174284
0.4	1.124448	1.148299272	1.173192779	1.173510871

Error Analysis

There are two types of errors to be considered in numerical solutions of differential equations. They are **formula error** and **round-off error**. Formula error occurs because of the approximation formula used to calculate y_{n+1} from y_n and the fact that y_n is not necessarily equal to the value $\phi(x_n)$ of the exact solution at x_n. **Local formula error** is the formula error in one step if we assume that $y_n = \phi(x_n)$.

Local formula error for the Euler tangent line method is proportional to h^2. The local formula error for the Improved Euler Method is proportional to h^3. Note that if h is chosen to be a small number, smaller than one, then h^3 is smaller than h^2.

Error can be controlled by using a sufficiently small value of h. However, if h is too small, then many calculations are required and there may be significant round-off error. Round-off error can be controlled by carrying extra digits in each calculation. One way to test the accuracy of a calculation is to recalculate using $h/2$ in place of h. If the new calculation does not differ from the previous one in a way that is significant for the desired accuracy, the calculations are likely to be sufficiently accurate.

EXERCISES 15.9

*For each of the initial value problems in Exercises 1–10, **(a)** use Euler's tangent line formula with $h = 0.1$ to determine values of an approximate solution $y(x)$ at $x = 0.1, 0.2, 0.3,$ and 0.4; **(b)** use the Improved Euler formula with $h = 0.1$ to determine values of an approximate solution $y(x)$ at $x = 0.1, 0.2, 0.3,$ and 0.4; **(c)** find the exact solution $\phi(x)$ and evaluate at $x = 0.1, 0.2, 0.3,$ and 0.4.*

1. $y' = 3x^2y^2,\ y(0) = 1$
2. $y' = 4x^3y^2,\ y(0) = 1$
3. $y' = -x/y,\ y(0) = 1$
4. $y' = x/y,\ y(0) = 1$
5. $y' = 1 + y^2,\ y(0) = 0$
6. $y' = \sqrt{1 - y^2},\ y(0) = 0$
7. $y' = y + 2,\ y(0) = 1$
8. $y' = 2y + 2,\ y(0) = 1$
9. $y' = e^{-x} - y,\ y(0) = 1$
10. $y' = 4x - 2y,\ y(0) = 1$

REVIEW OF CHAPTER 15 CONCEPTS

- Differential equations of the form $M(x) + N(y)y' = 0$ are called separable. Separable equations can be solved by integrating with respect to x.
- The collection of all curves that form the general solution of a first-order differential equation is called the family of curves associated with the differential equation.
- Any curve in the family of curves that satisfy $y' = f(x, y)$ intersects any curve in the family of curves that satisfy $y' = -1/f(x, y)$ orthogonally at each point of intersection.
- Differential equations of the form $M(x, y) + N(x, y)\dfrac{dy}{dx} = 0$ are called exact if $M_y = N_x$ in some open region R. Exact differential equations have solutions of the form $U(x, y) = K$, where K is a constant and U satisfies

$U_x = M$ and $U_y = N$. Such a function U can be found by using the technique of Section 14.4.
- Equations of the form $y' + p(x)y = q(x)$ can be solved by multiplying the equation by the integrating factor $e^{\int^x p}$ and then integrating.
- If a second-order differential equation does not involve the dependent variable y, the substitution $u = \dfrac{dy}{dx}$ transforms the equation to a first-order differential equation that relates u', u, and the independent variable x. If we can solve this equation for $u = y'$ in terms of x, then we can obtain a solution for y by integrating.
- If a second-order differential equation does not involve the independent variable x, the substitution $u = \dfrac{dy}{dx}$

transforms the equation to a first-order differential equation that relates $\dfrac{du}{dy}$, u, and y, but not x. We use the fact that

$$y'' = \frac{d}{dx}\left(\frac{dy}{dx}\right) = \frac{du}{dx} = \frac{du}{dy}\frac{dy}{dx} = \frac{du}{dy}u.$$

If we can solve this equation for $u = y'$ in terms of y, then we obtain an equation that relates $u = \dfrac{dy}{dx}$ and y. This is a first-order differential equation, which we can then attempt to solve for y in terms of x.

- Equations of the form $y'' + p(x)y' + q(x)y = f(x)$ are called second-order linear differential equations. The general solutions are of the form $y = C_1 y_1 + C_2 y_2 + y_p$, where y_1 and y_2 are solutions of the corresponding homogeneous equation $y'' + p(x)y' + q(x)y = 0$ with Wronskian $W(y_1, y_2) = y_1 y_2' - y_1' y_2 \neq 0$, and y_p is any particular solution of $y'' + p(x)y' + q(x)y = f(x)$.

- If y_1 and y_2 are solutions of the homogeneous equation $y'' + p(x)y' + q(x)y = 0$ with $W(y_1, y_2) \neq 0$, then $y_p = v_1 y_1 + v_2 y_2$ is a particular solution of the nonhomogeneous equation $y'' + p(x)y' + q(x)y = f(x)$ if the functions v_1 and v_2 satisfy

$$\begin{cases} v_1' y_1 + v_2' y_2 = 0, \\ v_1' y_1' + v_2' y_2' = f. \end{cases}$$

This solution method is called variation of parameters.

- If y_1 is a solution of the homogeneous equation $y'' + p(x)y' + q(x)y = 0$, we can use the substitution $y = vy_1$ to find either a different solution of the homogeneous equation or a solution of the corresponding nonhomogeneous equation $y'' + p(x)y' + q(x)y = f(x)$.

- The general solution of the homogeneous second-order linear equation with constant coefficients $y'' + by' + cy = 0$ can be determined from the roots of the characteristic equation $r^2 + br + c = 0$.

- For certain types of functions f, we should know that a particular solution of the equation $y'' + by' + cy = f$ is a linear sum of corresponding types of functions. We can then substitute into the equation to determine the values of the unknown constants in the sum. This solution method is called the method of undetermined coefficients.

- If $p_0, p_1, \ldots, p_n$, and q are polynomials, and $p_n(c) \neq 0$, then the initial value problem

$$p_n y^{(n)} + \cdots + p_1 y' + p_0 y = q,$$
$$y(c) = y_0, \ldots, y^{(n-1)}(c) = y_{n-1}$$

has an infinite series solution

$$y(x) = \sum_{n=0}^{\infty} \frac{y^{(n)}(c)}{n!}(x - c)^n$$

in some open interval that contains c. This means that we can approximate the solution with a partial sum of the series.

- Solution of initial value problems $y' = f(x, y)$, $y(x_0) = y_0$ can be approximated using numerical methods.

CHAPTER 15 REVIEW EXERCISES

Find the general solutions of the differential equations in Exercises 1–36.

1. $y' = x^3 y^2$

2. $y' = \sqrt{xy}$

3. $(1 + \sqrt{y})^2 y' = \sqrt{y}$

4. $(\sin y)y' = \cos x$

5. $y' = y^2(x + 1)$

6. $y' = y^2 - y$

7. $y' - 2y = 2$

8. $y' + 4y = e^x$

9. $y' + y = x$

10. $y' + y = \sin x$

11. $y' + \left(\dfrac{1}{x}\right)y = \dfrac{1}{x}, x > 0$

12. $y' + \dfrac{1}{x-1}y = 1, x > 1$

13. $y' + (\tan x)y = \cos x, -\pi/2 < x < \pi/2$

14. $y' + (\cos x)y = \cos x$

15. $y'' + 2y' - 3y = 0$

16. $y'' + y' - 6y = 0$

17. $y'' - 4y' + 4y = 0$

18. $9y'' - 6y' + y = 0$

19. $4y'' + y = 0$

20. $4y'' - y = 0$

21. $y'' + 2y' + 5y = 0$

22. $y'' + 4y' + 5y = 0$

23. $y'' - 3y' + 2y = e^{-x}$

24. $y'' - 3y' + 2y = \sin x$

25. $y'' + y = \sin x$

26. $y'' + y = e^x$

27. $y'' - 3y' + 2y = e^x$

28. $y'' - 3y' + 2y = x$

29. $4y'' - 4y' + y = e^x$

30. $4y'' - 4y' + y = e^{x/2}$

31. $y'' - 2y' + 2y = \sin x$

32. $y'' - 2y' + 2y = e^x \sin x$

33. $y'' - y = e^x + e^{-x}$

34. $y'' - y' - 2y = e^x + e^{-x}$

35. $y'' - 2y' + y = xe^x$

36. $y'' + y = x \sin x$

Solve the initial value problems in Exercises 37–62.

37. $y' = 2x - 1, y(0) = 2$

38. $y' = \sec^2 x$, $y(0) = 1$

39. $y'' = \sin x$, $y(0) = 0$, $y'(0) = 1$

40. $y'' = 6x + 1$, $y(0) = 2$, $y'(0) = 5$

41. $(xy)' = 3x^2$, $y(1) = 2$

42. $\left(\dfrac{y}{x}\right)' = \dfrac{1}{x}$, $y(1) = 1$

43. $(y \tan x)' = \sec^2 x$, $y(\pi/4) = 2$

44. $(y \cos x)' = \sin x$, $y(0) = 2$

45. $3(x^2 + 1)y^2 y' = x$, $y(0) = 2$

46. $xyy' = 1$, $y(1) = -2$

47. $yy' = -4x$, $y(0) = 2$

48. $(2y + 1)y' = 4x$, $y(0) = 2$

49. $2yy' = -1/x^2$, $y(1) = 2$

50. $y\,dx + (x - 1)\,dy = 0$, $y(2) = 6$

51. $(2xy - 6y)dx + (x - 3)^2 dy = 0$
  (a) $y(2) = 1$, **(b)** $y(4) = 1$

52. $y'' - y' - 2y = e^x$, $y(0) = 0$, $y'(0) = 0$

53. $y'' - y' - 2y = e^{-x}$, $y(0) = 0$, $y'(0) = 0$

54. $y'' - y' - 2y = 1$, $y(0) = 0$, $y'(0) = 1$

55. $y'' - y' - 2y = x$, $y(0) = 0$, $y'(0) = 1$

56. $y'' - y' - 2y = x^2$, $y(0) = 0$, $y'(0) = 1$

57. $y'' - y' - 2y = xe^x$, $y(0) = 0$, $y'(0) = 1$

58. $y'' - y' - 2y = xe^{-x}$, $y(0) = 0$, $y'(0) = 1$

59. $y'' - 2y' + y = e^x$, $y(0) = 0$, $y'(0) = 0$

60. $y'' - 2y' + y = e^{-x}$, $y(0) = 0$, $y'(0) = 0$

61. $y'' - 2y' + 2y = e^x$, $y(0) = 0$, $y'(0) = 0$

62. $y'' - 2y' + 2y = e^{-x}$, $y(0) = 0$, $y'(0) = 0$

Show that the differential equations in Exercises 63–64 are exact and find their solution.

63. $(2x + y) + (x + 2y)dy/dx = 0$

64. $2e^{2x} \cos y - e^{2x} \sin y\, dy/dx = 0$

In Exercises 65–66: **(a)** *Show that the given function y_1 is a solution of the given equation.* **(b)** *Find a function v such that $y_2 = y_1 v$ is a solution of the given differential equation and $W(y_1, y_2) \neq 0$.* **(c)** *Find the general solution of the given equation.*

65. $x^2 y'' - 3xy' + 3y = 0$; $y_1 = x$, $x > 0$

66. $x^2 y'' + xy' - 4y = 0$; $y_1 = x^2$, $x > 0$

In Exercises 67–70, find the equation of the curve that contains the given point and has slope $m(x, y)$ at each point (x, y) on the curve.

67. $(2, 1)$, $m = 2xy^2$

68. $(2, 0)$, $m = e^y$

69. $(0, 2)$, $m = y \sec x$

70. $(0, 0)$, $m = x \cos x$

In Exercises 71–72, find the family of curves associated with the given differential equation and find a mutually orthogonal family. Sketch three curves from each family.

71. $yy' = 1$ **72.** $yy' = x$

73. A simple series circuit consists of a 0.025-F capacitor, a 20-Ω resistor, and an impressed voltage of $E(t) = 120 \sin t$. See figure. Kirchhoff's Second Law can be used to show that the current in amperes is a solution of the differential equation

$$20i' + \frac{1}{0.025}i = \frac{d}{dt}(120 \sin t).$$

Find the current in the circuit at time t if the initial current in the circuit is zero.

74. A circuit consists of a 0.1-F capacitor connected in series to a 60-Ω resistor, with an impressed voltage of $E(t) = 12 \sin t$. Kirchhoff's Second Law can be used to show that the current in amperes is a solution of the differential equation

$$60i' + \frac{1}{0.1}i = \frac{d}{dt}(12 \sin t).$$

Find the current at time t if the initial current is 8 A.

75. A 256-lb parachutist jumps from an airplane that is flying slowly and horizontally at 3136 ft.
  (a) Determine the parachutist's velocity t seconds after jumping and the parachutist's terminal velocity if the air resistance is eight times the speed.
  (b) At what speed would the parachutist hit the ground if there were no air resistance? Ignore horizontal motion.

76. Determine the height above ground of a 160-lb skydiver t seconds after jumping from an airplane that is flying slowly and horizontally at 2400 ft. Assume the air resistance is ten times the speed and ignore horizontal motion.

77. A team of scientists wants to find the air resistance of a new parachute they have designed. In a test drop of a 200-lb weight from 5000 ft, the object fell for 240 s and hit the ground at approximately 25 ft/s. Assuming that the air resistance is proportional to the velocity, find the constant of proportionality.

78. The air resistance on a ball of mass $m = 200$ g is kv^2, where $k = 2 \times 10^{-4}$ and v is the velocity in meters per second. The motion of the ball in free fall is governed by the differential equation

$$kv^2 - mg = mv\frac{dv}{dy},$$

where y is the height of the ball above ground level. Find its terminal speed. Find the speed at which it hits the ground if it is dropped from 400 m.

79. A rocket is launched vertically from the earth. At the time the engines are turned off, the rocket has an altitude of $R/3$ and a speed of 10,000 ft/s, where the radius of the earth is $R = 2.1 \times 10^7$ ft. Assuming that gravity is the only force acting on the rocket, determine the maximum altitude it will reach. At an altitude of $R/3$ ft, what speed would the rocket need to avoid being pulled by gravity back toward the earth?

80. The cylindrical tank and the conical tank pictured are both full and contain the same amount of water. The flow of water through a hole in the bottom of the cylindrical tank and the conical tank pictured is governed by the differential equation

$$\frac{dV}{dt} = -k\sqrt{y},$$

where V is the volume of water in the tank when the depth of water is y and the constant k is the same for both tanks. See figure. Which tank do you think empties faster? Explain.

81. The flow of water through a hole in the bottom of a conical tank is governed by the differential equation

$$\frac{dV}{dt} = -k\sqrt{y},$$

where V is the volume of water in the tank when the depth of water is y and k is constant. See figure. How long will it take for a full tank to drain completely? (*Hint:* Express y in terms of V and then obtain a differential equation that relates dV/dt and V.)

82. The flow of water through a hole in the bottom of a cylindrical tank is governed by the differential equation

$$\frac{dV}{dt} = -k\sqrt{y},$$

where V is the volume of water in the tank when the depth of water is y and k is constant. See figure. How long will it take for a full tank to drain completely? (*Hint:* Express y in terms of V and then obtain a differential equation that relates dV/dt and V.)

83. Air flow rate at standard conditions in a flat duct is to be determined by installing pressure taps across a bend. The duct is 0.3 m deep and 0.1 m wide. The inner radius of the bend is 0.25 m. See figure. The change in pressure as the radius varies between the inner and outer radius of the bend is given by **Euler's equation**,

$$\frac{dp}{dr} = \frac{1.23V^2}{r}, \quad 0.25 \le r \le 0.35,$$

where the flow rate V is assumed to be constant. Find the flow rate if the pressure difference between the inner and outer surfaces of the bend is 400 N/m².

84. The maximum power output of an internal combustion engine decreases with altitude because the air density and hence the mass flow rate of air decreases. If we assume that the absolute temperature at altitude z is given by the linear equation $T = T_0 - m(z - z_0)$, the pressure and altitude are related by the differential equation

$$\frac{dp}{p} = -\frac{gd(z - z_0)}{R[T_0 - m(z - z_0)]}.$$

Obtain an equation that relates the pressure and altitude by finding the solution of this equation.

85. The motion of a **spring-mass system** is governed by the equation

$$\frac{d^2x}{dt^2} + 4x = 0,$$

where x is the displacement of the mass from a position of equilibrium. What is the equation of motion if the initial position is $x = 2$ and the initial velocity is zero?

86. The motion of a **spring-mass system** is governed by the equation

$$\frac{d^2x}{dt^2} + kx = 0,$$

where x is the displacement of the mass from a position of equilibrium and k is a *positive* constant. Determine the equation of motion. This type of motion is called **simple harmonic motion**.

*Exercises 87–90 involve **damped oscillation** of a spring-mass system. The motion is damped, for example, if the system is immersed in a fluid, so there is a friction force that is proportional to the velocity of the mass. Newton's equation can then be expressed as*

$$m\frac{d^2x}{dt^2} + c\frac{dx}{dt} + kx = 0,$$

where x is the displacement at time t, the constant m is the mass, c is the positive constant of proportionality of the damping force, and k is the positive spring constant.

87. Find the equation of motion of the spring-mass system governed by the initial value problem $x'' + 3x' + 2x = 0, x(0) = 2, x'(0) = 0$.

88. Find the equation of motion of the spring-mass system governed by the initial value problem $x'' + 2x' + x = 0, x(0) = 0, x'(0) = 2$.

89. Find the equation of motion of the spring-mass system governed by the initial value problem $x'' + 2x' + 2x = 0, x(0) = 2, x'(0) = 2$.

90. (a) A spring-mass system is said to be **critically damped** if $c^2 = 4mk$. Show that a critically damped spring-mass system approaches the equilibrium position $x = 0$, but does not oscillate, as indicated in the figure.

(b) A spring-mass system is called **overdamped** if $c^2 > 4mk$. Show that an overdamped spring-mass

system approaches the equilibrium position $x = 0$ after passing through the equilibrium position at most once, as indicated in the figure.

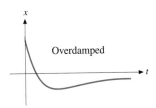

(c) A spring-mass system is called **underdamped** if $c^2 < 4mk$. Show that an underdamped spring-mass system passes through the equilibrium position periodically as it approaches the equilibrium position, as indicated in the figure.

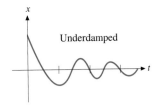

Use the first four nonzero terms of the infinite series solutions of the initial value problems to obtain polynomial approximations of the solutions in Exercises 91–94.

91. $y' - y = x, y(0) = 2$

92. $y' + 2y = 1 - 3x, y(0) = 3$

93. $y'' + xy' + x^2y = 0, y(0) = 1, y'(0) = 2$

94. $y'' - 2xy' + x^2y = x, y(0) = 2, y'(0) = 3$

For each of the initial value problems in Exercises 95–98, (a) use Euler's tangent line formula with $h = 0.1$ to determine values of an approximate solution $y(x)$ at $x = 0.1, 0.2, 0.3,$ and 0.4; (b) use the Improved Euler formula with $h = 0.1$ to determine values of an approximate solution $y(x)$ at $x = 0.1, 0.2, 0.3,$ and 0.4; (c) find the exact solution $\phi(x)$ and evaluate at $x = 0.1, 0.2, 0.3,$ and 0.4.

95. $y' = 2xy^2, y(0) = 1$ **96.** $y' = y^2 - 1, y(0) = 0$

97. $y' = e^x - y, y(0) = 1$ **98.** $y' = y + e^x, y(0) = 1$

99. Explain the terms of the phrase "separable first-order differential equation" in your own words.

100. Give some examples of first-order differential equations that are not linear and explain why they are not linear.

EXTENDED APPLICATION

CLEANING UP THE GREAT LAKES

You are part of a team of environmental scientists investigating pollution in the Great Lakes. After several weeks of extensive data collection, you are beginning to create a mathematical model that will allow you to judge the effectiveness of several proposed plans for reducing future pollution levels. As a starting point, you will create a simple model, using the following assumptions:

- Rainfall and evaporation balance each other, so the average rates of inflow and outflow are equal.
- The average rates of inflow and outflow do not vary much seasonally.
- Perfect mixing occurs when water enters the lakes, so the pollutants are uniformly distributed.
- Pollutants are removed from the lakes only by outflows, and not by decay, sedimentation, or any other means.
- Pollutants flow freely out of the lakes and are not retained.

The first two assumptions are valid, but the last three are not. For example, the pesticide DDT does not flow freely out of a lake. You will eventually have to refine your mathematical model to conform to the facts, but the basic model you are about to create will provide a foundation on which to build.

The model for each lake will be constructed in terms of the following parameters:

$V =$ volume of the lake, in millions of gallons,

$r =$ rate of inflow and outflow, in millions of gallons per year,

$t =$ time, in years,

$P_i(t) =$ pollution concentration in the inflow to the lake after t years, in gallons of pollutants per gallon of water,

$P_L(t) =$ pollution concentration in lake L after t years, in gallons of pollutants per gallon of water.

Tasks:

1. The rate of change in the quantity of pollution in a given lake is equal to the difference between the rate of inflow of pollutants and their rate of outflow. Use this fact to write a differential equation for $P_L(t)$ in terms of V, r, and $P_i(t)$.

2. Solve the differential equation under the assumption that a pollution control plan will keep the inflow pollution level $P_i(t)$ at a constant value of Q.

3. The ratio r/V has a value of 0.385 for Lake Erie and 0.00529 for Lake Superior.
   a. Write out the solutions $P_E(t)$ and $P_S(t)$ for Lakes Erie and Superior, respectively, in terms of the parameter Q.
   b. If a plan can achieve $Q = 0$, how many years will it take to reduce the pollution level in each lake to 50 percent of its present level? 30 percent? 10 percent?
   c. Repeat part b, assuming that the plan can achieve only $Q = 0.2P_L(0)$.
   d. If the goal is to reduce the pollution level by 90 percent over a fifty-year period, what must be the value of Q in terms of $P_L(0)$ for Lake Erie? For Lake Superior?

4. Lake Ontario receives 84 percent of its inflow from Lake Erie. Thus for Lake Ontario, $P_i(t)$ will not be constant at Q, but instead will be equal to $0.16Q + 0.84P_E(t)$, where $P_E(t)$ is your solution for Lake Erie from Task 3a.
   a. Assuming the same initial pollution level and the same value of Q in both lakes, write and solve the differential equation for $P_L(t)$. Use the value $r/V = 0.128$ for Lake Ontario.
   b. If $Q = 0$, how many years will it take to reduce the pollution level to 50 percent of its present level? What if $Q = 0.2P_L(0)$? (These questions can be answered most effectively with a graphing utility.)
   c. If the goal is to reduce the pollution level by 90 percent over a fifty-year period, what must be the value of Q in terms of $P_L(0)$ for Lake Ontario?

Writing Assignment:

Write a report to your colleagues, explaining how you constructed your mathematical model for each lake. Also communicate to them any insights you have gained into the problem and its solutions as a result of your explorations in Tasks 3 and 4.

Source: Margaret M. Lial, Charles D. Miller, Raymond M. Greenwell, *Calculus with Applications,* fifth edition, Copyright © 1993 by HarperCollins College Publishers, p. 515.

Answers to Odd-Numbered Exercises

CHAPTER ONE:
SOME PRELIMINARY TOPICS

EXERCISES 1.1

1. $x < 4$ **3.** $x \le -5/2$ **5.** $1 < x < 2$ **7.** $1 \le x \le 2$
9. $1 < x < 2$ **11.** $2/3 < x < 2$ **13.** $1 \le x \le 2$
15. $x > 5, x < 1$ **17.** $x \ge 3, x \le -2$ **19.** $1, 7$
21. $5, 5$ **23.** 1 **25.** 3 **27.** x^{-2} **29.** x^5 **31.** x^{-6}
33. $x^{-1/2}$ **35.** x **37.** $a^{1/2}b^{-1/2}$ **39.** $a^{-1}b$ **41.** 27
43. 2 **45.** 4 **47.** 4 **49.** 9 **51.** $-1/12$ **53.** $8/15$
55. 2.805855 **57.** 6.473008 **59.** 0.0078125
61. 3.659306 **63.** Undefined   **65.** 3.4027×10^4
67. 9.65×10^{-1} **69.** $73,200$ **71.** 0.0007361

73.

Interval notation	Inequality notation	Graph
$(0, 1)$	$0 < x < 1$	
$[1, 3]$	$1 \le x \le 3$	
$(-1, 1]$	$-1 < x \le 1$	
$(0, \infty)$	$x > 0$	
$(-\infty, 1]$	$x \le 1$	

75. $|x - 1| < 2$ **79.** 6.9×10^8 m/s   **81.** $653\frac{1}{3}$ ft

EXERCISES 1.2

1. $5 - 6x$ **3.** $(2x + 1)(30x + 1)$ **5.** $-\dfrac{x + 3}{2x^{5/2}}$

7. $\dfrac{x(7x + 6)}{3(x + 1)^{2/3}}$ **9.** $x + 2, x \ne 2$ **11.** $-\dfrac{x + 2}{4x^2}, x \ne 2$

13. $x^2 - 2x + 4, x \ne -2$ **15.** $x - 2, x \ne -1$
17. $-1, 6$ **19.** None   **21.** $0, 2$ **23.** $-2, -1/3$
25. $-1, 1$ **27.** $-1 \pm \sqrt{3}$ **29.** (a) 1,   (b) $-1/3, 7/3$
31. (a) 0,   (b) 2 **33.** (a) $x \le -1/2$,   (b) none
35. (a) $x < -2, x > 1$ (b) $-2 < x < 1$
37. (a) $-1 < x < 0, x > 2$,   (b) $x < -1, 0 < x < 2$
39. (a) $x > -1$,   (b) $x < -1$
41. (a) $x < 0, x > 1$,   (b) $0 < x < 1$
43. (a) $x > 0$,   (b) $-1 < x < 0$
45. (a) $x > 1/2$,   (b) $0 < x < 1/2$
47. $x < -1, 0 < x < 1$ **49.** $-2 < x < 1$
51. $0 < x < 1$ **53.** $0 < x < 4$
55. (a) -0.453,   (b) $x > -0.453$,   (c) $x < -0.453$
57. (a) $-2.115, 0.254, 1.861$,
(b) $-2.115 < x < 0.245, x > 1.861$,
(c) $x < -2.115, 0.254 < x < 1.861$

EXERCISES 1.3

1. $(1, -1)$ **3.** $(2, 1), (0, 1)$ **5.** $(-3, 2)$
7.   **9.**

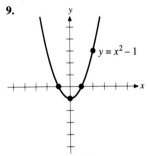

$3x + 2y = 6$

$y = x^2 - 1$

11.

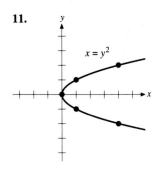

13. (a) $(x, 2x)$,   **(b)** $(y/2, y)$

15. (a) $(x, x^3 - 1)$,   **(b)** $((y + 1)^{1/3}, y)$

17. (a) $(x, \sqrt{x + 1}), (x, -\sqrt{x + 1})$,   **(b)** $(y^2 - 1, y)$

19. $5; -3/4; (1, 1/2)$ **21.** 5; undefined; $(1, 1/2)$

23. $\sqrt{5x^2 - 6x + 9}; \dfrac{2x}{x - 3}; \left(\dfrac{x + 3}{2}, x\right)$

25. $|x^2 + x - 2|$; undefined; $\left(x, \dfrac{x^2 - x}{2}\right)$

27. $|4 - y^2|; 0, y \neq \pm 2; \left(\dfrac{4 - y^2}{2}, y\right)$ **29.** $-2/3, x \neq -6$

31. $x, x \neq 0$ **33.** $x^2 + x + 1, x \neq 1$ **35.** $x - 1, x \neq 2$

37. $-1/x, x \neq 1$ **39.** $-\dfrac{x + 2}{3(x^2 - 1)}, x \neq 2$

41. $\left|\dfrac{x^3}{1 + x^2}\right|; \left(x, \dfrac{x^3 + 2x}{2(1 + x^2)}\right)$ **43.** Yes   **47.** No

49. $\sqrt{x^4 + x^2 - 2x + 1}; x \approx 0.59$ **51.** $\sqrt{x^4 - x^2 + 1}$; points on the graph near $(\pm 0.707, -0.5)$ are closer to the origin than is $(0, -1)$.   **53.** Slope decreases to two.

EXERCISES 1.4

1.

3.

5.

7.

9.

11.

$x = 0$

13.

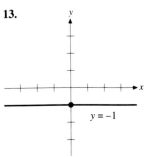

15. $x + y = 5$ **17.** $x - 3y + 7 = 0$ **19.** $x + y = 3$

21. $2x - y = 1$ **23.** $x - y + 2 = 0$ **25.** $y = 2$

27. $x = -1$ **29.** $6x + y + 9 = 0$ **31.** $y = -1$

33. $x + y = 1$ **35.** $y = 3x + 2$ **37.** $3x + 2y = 6$

39. $y = 2x$ **41.** $4x - 2y = 3$ **43.** $y = 1$ **45.** $2/\sqrt{5}$

49. $4/3$ **51.** 1

EXERCISES 1.5

1.

3.

5.

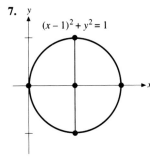

7.

$(x - 1)^2 + y^2 = 1$

9.

$x^2 + y^2 + 4y = 0$

11.

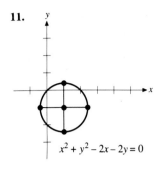

$x^2 + y^2 - 2x - 2y = 0$

25.

$y = -x^2 + 4$

27.

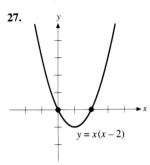

$y = x(x - 2)$

13.

$x^2 + y^2 - 3x - y + \dfrac{9}{4} = 0$

15.

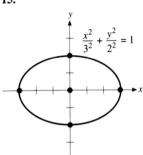

$\dfrac{x^2}{3^2} + \dfrac{y^2}{2^2} = 1$

29.

$x = y^2$

31.

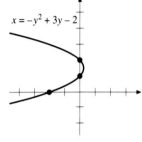

$x = -y^2 + 3y - 2$

17.

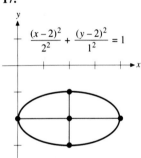

$\dfrac{(x-2)^2}{2^2} + \dfrac{(y-2)^2}{1^2} = 1$

19.

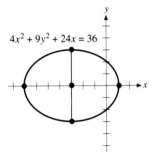

$4x^2 + 9y^2 + 24x = 36$

33.

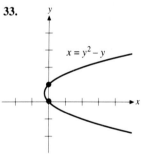

$x = y^2 - y$

35.

$y = \sqrt{x}$

21.

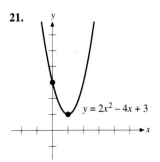

$y = 2x^2 - 4x + 3$

23.

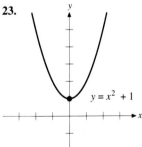

$y = x^2 + 1$

37.

$x = -\sqrt{y}$

39.

$y = \sqrt{x + 4}$

41.

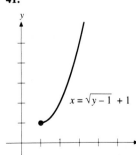

$x = \sqrt{y - 1} + 1$

43.

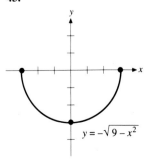

$y = -\sqrt{9 - x^2}$

45.

$x - \sqrt{2y - y^2} = 0$

47.

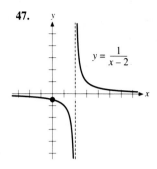

$y = \dfrac{1}{x-2}$

7. $b = 3\sin\theta$ **9.** $L = 2\cos\theta$
11. $h = 5\cot\theta$ **13. (a)** 0,   **(b)** -1,   **(c)** 0
15. (a) $1/2$,   **(b)** $-\sqrt{3}/2$,   **(c)** $-1/\sqrt{3}$
17. (a) $-1/\sqrt{2}$,   **(b)** $1/\sqrt{2}$,   **(c)** -1
19. (a) i,   **(b)** $-\sqrt{2}$,   **(c)** $-\sqrt{2}$
21. $4/5$ **23.**

$-2/\sqrt{5}$

49.

$y = \dfrac{1}{(x+1)^2}$

25. $\sqrt{1 - x^2}$

51. 36 ft   **53.** $(x+2)^2 + (y+2)^2 = (\sqrt{8})^2$; circle
55. $(1, 0)$; $x = 0$, $x = 2$,   **57.** $(1.73, 3.73)$; $(-1.73, 0.27)$
$y = 0$

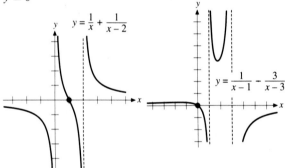

$y = \dfrac{1}{x} + \dfrac{1}{x-2}$

$y = \dfrac{1}{x-1} - \dfrac{3}{x-3}$

27. $\pi/6, 5\pi/6$ **29.** $\pi/3, 4\pi/3$ **31.** $0, \pi/2, \pi, 3\pi/2, 2\pi$
33. $\pi/6, \pi/2, 5\pi/6, 7\pi/6, 3\pi/2, 11\pi/6$
35. $0.3398, 2.8018$ **37.** $2.3603, 3.9229$
39. $2.8966, 6.0382$ **41.** $0, \pi/6, 5\pi/6, \pi, 2\pi$
43. $L^2 = 164 - 160\cos\theta$ **45.** $b = \csc\theta$
47. $16 = x^2(5 - 4\cos\theta)$ **49.** $26.4°, 117.3°, 36.3°$
51. $80°, 4.6, 2.9$

59. The ratio approaches zero.

EXERCISES 1.6

1. $4/5$

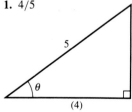

(4)

3. $\dfrac{x}{\sqrt{4 - x^2}}$

$\left(\sqrt{4 - x^2}\right)$

53.

$y = \sin x$,
$-\pi/2 \le x \le \pi/2$

55.

$y = \tan x$,
$-\pi/2 < x < \pi/2$

5. $\dfrac{x}{\sqrt{x^2 + 4}}$

$\left(\sqrt{x^2 + 4}\right)$

57.

$y = 1 - \sin x$

59. $y = \sin x + \cos x$

61.
$y = \cos\left(x - \dfrac{\pi}{2}\right)$

63.
$y = \sin 2x$

63.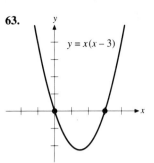
$y = x(x - 3)$

65.
$y = \sqrt{1 - x}$

65. $(1 - \sin^2 x)\cos x$ **67.** $(\cos^2 x - \cos^4 x)\sin x$

69. $(\tan^2 x + 1)\sec^2 x$ **71.** $(\sec^4 x - \sec^2 x)\sec x \tan x$

73. $\dfrac{3}{8} + \dfrac{1}{2}\cos 2x + \dfrac{1}{8}\cos 4x$ **75.** 11.3 ft   **77.** 3.27 rev

79. 33.7 mph

REVIEW EXERCISES

1. $x < -3$ **3.** $0 < x < 3$ **5.** $-1 < x < 2$

7. $x > -1, x < -3$ **9.** 1 **11.** x^4 **13.** $x^{1/4}$ **15.** $a^7 b^{-7}$

17. -54 **19.** 4.371×10^6 **21.** 0.002735 **23.** $\dfrac{ab}{a+b}$

25. $\dfrac{1}{(1-x^2)^{3/2}}$ **27.** $3/2, 13/18$ **29.** $\dfrac{3 \pm \sqrt{33}}{4}$

31. (a) -1,   (b) $0, -3/2$ **33.** (a) $1/2$,   (b) $3/8$

35. (a) $x < -2, x > 1$,   (b) $-2 < x < 1$

37. (a) $x < 0, 1 < x < 2, x > 2$,   (b) $0 < x < 1$

39. $\sqrt{y^2 - 6y + 13}; \dfrac{3-y}{2}; \left(1, \dfrac{y+3}{2}\right)$

41. $-x - 2, x \neq 2$ **43.** $\dfrac{1}{\sqrt{x}+2}, x \neq 4$

45. $(x+1)^2 - y^2 = 2^2$ **47.** $9x^2 + 25y^2 = 225$

49. $y = 3x - 5$ **51.** $4y = 3x - 3$ **53.** $2x + 3y = 6$

55.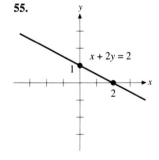
$x + 2y = 2$

57.
$2y = 3x$

59.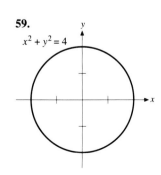
$x^2 + y^2 = 4$

61.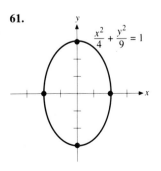
$\dfrac{x^2}{4} + \dfrac{y^2}{9} = 1$

67.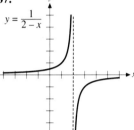
$y = \dfrac{1}{2 - x}$

69.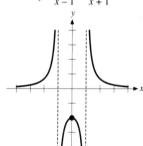
$y = \dfrac{1}{x-1} - \dfrac{1}{x+1}$

71.
$y = 1 - \cos x$

73.
$y = \cos(2x)$

75. $x = 2\sec\theta$ **77.** $x^2 = 50 - 50\cos\theta$ **79.** $3/4$

81. $-\sqrt{3}/2$ **83.** $2\pi/3, 4\pi/3$ **85.** $3\pi/4, 5\pi/4$

87. $0.6435, 2.4981$ **89.** $1.9823, 4.3009$

91. $1.2766, 4.4182$ **93.** $\pi/4, 3\pi/4, 5\pi/4, 7\pi/4$

103. $u = \dfrac{mv - Ft}{m}$ **105.** $m_2 = \dfrac{v + v_1}{v_2 - v}m_1$

107. 88.9 rev/min   **109.** 312 ft

CHAPTER TWO:
LIMITS, CONTINUITY, AND DIFFERENTIABILITY AT A POINT

EXERCISES 2.1

1. $15; 1$ **3.** $\dfrac{2-x}{x}; \dfrac{1}{2x - 1}$ **5.** $3x^2 - 6x + 1; 3x^2 - 3$

7. $x - 1, x \neq -2$ **9.** $3x^2; 5x^2 - x$ **11.** $12x^2 - 16x + 6$

13. Undefined, 0, 4

$y = x^2, x \geq 0$

15. Undefined, 0, $-1/2$

$y = \cos x, 0 \leq x \leq \pi$

17. 1, undefined, -1

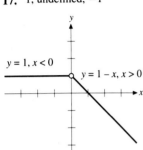

$y = 1, x < 0$

$y = 1 - x, x > 0$

19. Undefined, 2, 2

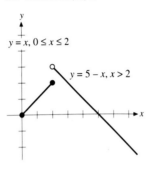

$y = x, 0 \leq x \leq 2$

$y = 5 - x, x > 2$

21. $x \neq -1, 1$ **23.** $x \leq 3/2$

25. $x \neq \dfrac{\pi}{2} + k\pi$, k an integer   **27.** $-2, x \neq 1$

29. $x + 1, x \neq 1$ **31.** $V(x) = x(2 - x)(3 - x), 0 \leq x \leq 2$

33. $S(x) = 20x - 3x^2, 0 \leq x \leq 5$

35. $C(s) = 2.5s^2 + 48s, s \geq 0$

37. (a) $A(x) = \pi x, x \geq 0$,   (b) $C(x) = 2\pi \sqrt{x}$

39. $L(x) = x^2$

41. Domain $-4 < x \leq 4$; range $-1 \leq y \leq 2$; undefined; 1; 2

43. Domain $0 \leq x < 4$; range $1 \leq y < 3$; undefined; 1; 2

45. $f(x) = \dfrac{-6 - 2x}{3}$; domain   **47.** y is not a function of x.

$-\infty < x < \infty$; range

$-\infty < y < \infty$

$2x + 3y + 6 = 0$

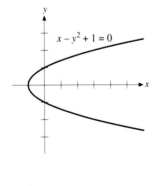

$x - y^2 + 1 = 0$

49. $f(x) = -\sqrt{1 - x}$; domain   **51.** $f(x) = 4/x^2$; domain $x \neq 0$;

$x \leq 1$; range $y \leq 0$ range $y > 0$

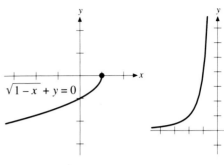

$\sqrt{1 - x} + y = 0$

$x^2 y = 4$

53. y is not a function of x.

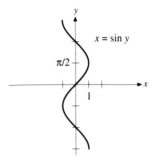

$x = \sin y$

55. Yes, $(2x + y)^2 = 0$ implies $y = -2x$.

57. $f(x) = \sqrt{4 - x^2}$ **59.** $f(x) = \begin{cases} x - 2, & x \geq 2, \\ 2 - x, & x < 2 \end{cases}$

61. $f(x) = \begin{cases} 1, & x > 2, \\ -1, & x < 2 \end{cases}$

65. (a) $f \circ g(x) = \sqrt{x^2 - 2}, |x| \geq \sqrt{2}$,

(b) $g \circ f(x) = x - 2, x \geq 1$

67. $f \circ g$ is defined only for $x \geq 0$, while h is

defined for all real x.   **69.** $g(x) = \dfrac{x + 1}{2}$

71. x-axis, y-axis, origin   **73.** y-axis   **75.** x-axis

77. Even   **79.** Neither even nor odd   **81.** Odd   **83.** Even

85. Even for n an even integer, odd for n odd   **87.** Even

89.

(a)

$y = f(-x)$

(b)

$y = -f(x)$

(c)

(d)

39.

41.

43.

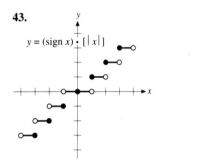

45. 0   **47.** 1/3

EXERCISES 2.2

1. $-1, 2$ **3.** $-1, 3$ **5.** $0, 2$ **7.** $0, 2\pi$
9. $3(2)^2 + 4(2) + 5 = 25$ **11.** $(2)^3 + 2(2) + 3 = 15$
13. Unbounded   **15.** $2(1) + 3(1) = 5$

17. $(1) + (1) = 2$ **19.** $\dfrac{1}{(2) + 3} = \dfrac{1}{5}$ **21.** Unbounded

23. $\dfrac{1}{3(4) - 2} = \dfrac{1}{10}$ **25.** $\dfrac{1}{1 - 2(0.3)} = 2.5$

27. $\dfrac{1}{(3)} = \dfrac{1}{3}$ **29.** $\dfrac{1}{0 + 4} = \dfrac{1}{4}$ **31.** $\dfrac{1}{0 + 5} = \dfrac{1}{5}$

33. $(3)(1) = 3$ **35.** $(\pi)^2(1)^2 = \pi^2$ **37.** $\dfrac{2}{|2(2) - 6|} = 1$

39. $\dfrac{1 + (0.2)}{1 - (0.2)} = 1.5$ **41.** $\dfrac{(5) + 4}{(3)^2} = 1$ **43.** $\dfrac{(2)^3}{0 + 4} = 2$

45. $\dfrac{(0.2)^3}{(1 - (0.2))^2} = 0.0125$ **47.** $\dfrac{(1) + (1)}{0 + 1} = 2$ **49.** 4
51. 1   **53.** 0.01   **63.** ≈ 0.4 **65.** ≈ 0.4 **67.** ≈ 0.4

EXERCISES 2.3

1. 0   **3.** Does not exist   **5.** Does not exist   **7.** 2
9. 3   **11.** (a) -2,   (b) does not exist,   (c) 0
13. Does not exist   **15.** 0   **17.** 1
19. (a) -1,   (b) 1,   (c) does not exist
21. (a) 1,   (b) 1,   (c) 1   **23.** (a) 0,   (b) 0,   (c) 0
25. Continuous at b, discontinuous at a and c
27. Discontinuous at a, b, and c
29. Continuous at zero when $a = 1$ **31.** Impossible
33. $a = -1/2, b = 2$
35. Continuous from left and   **37.** Continuous from left at zero
right at zero

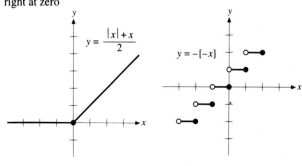

EXERCISES 2.4

1. $\epsilon/2$ **3.** $\epsilon^{1/3}$ **5.** ϵ^3 **7.** 0.0002   **9.** 0.012   **11.** ϵ
13. 0.001 m   **15.** 0.000999 m   **17.** 0.0008   **19.** 0.005 Ω
21. No, a tolerance of 2% is required.   **39.** (a) 3,   (b) 20
41. 4   **43.** $\epsilon > 1$

EXERCISES 2.5

1. 4   **3.** -10 **5.** 7   **7.** 3   **9.** 21/5   **11.** 1/2
13. $\sqrt{3}$ **15.** 0   **17.** 0   **19.** 8   **21.** 5   **23.** $-1/49$
25. 1/9   **27.** 1   **29.** 2   **31.** 5   **33.** 1/2   **35.** $\sqrt{2}$
37. -1 **39.** 1   **41.** $1/(2\sqrt{5})$ **43.** 0, 0, 0
45. $1, -1$, does not exist   **47.** $-1, 1$, does not exist
49. Does not exist, 0, does not exist   **51.** 0, 0, 0; 1, 1, 1
53. $-1, -1, -1$; 1, does not exist, does not exist   **55.** $3a^2$

EXERCISES 2.6

1. (a) $x - 2, x \neq 1$,   (b) -1,   (c) $x + y + 1 = 0$
3. (a) $3, x \neq -3$,   (b) 3,   (c) $y = 3x + 4$
5. (a) $x^2 + 2x + 4, x \neq 2$,   (b) 12,   (c) $y = 12x - 16$
7. (a) $-1/(4x), x \neq 4$,   (b) $-1/16$,   (c) $x + 16y = 8$
9. (a) $t + 2, t \neq 2$,   (b) 4   **11.** (a) $-2, t \neq 3$,   (b) -2
13. (a) $\dfrac{1}{\sqrt{t + 1} + 3}, t \neq 8$,   (b) 1/6
15. (a) $\dfrac{1}{4(t + 1)}, t \neq 3$,   (b) 1/16

17. $-16t + v_0$, $t \neq 0$; v_0 **19.** 6π **21.** $-4/5$ **23.** No
25. Yes   **27.** No   **29.** No   **31.** Yes   **33.** No   **35.** Yes
37. No   **39.** $1, 2$ **47. (a)** $2/3$,   **(b)** 0,   **(c)** $-2/3$
49. (a) 1,   **(b)** does not exist,   **(c)** -1
51. (a) 0,   **(b)** $-3/2$,   **(c)** 0
53. (a) 1,   **(b)** 2,   **(c)** $1/2$

REVIEW EXERCISES

1. (a) $2x^2 - 17x + 17$,   **(b)** $x^2 - 9x + 4$
3. (a) $3x + 4$, $x \neq 2$,   **(b)** $-32x + 24$
5. (a) $1 \leq x \leq 4$,   **(b)** $1 \leq x < 4$,   **(c)** $x \leq 3$
7. $x < 3/2$
9. $f(x) = \dfrac{3x - 12}{4}$, domain   **11.** y is not a function of x.
$-\infty < x < \infty$, range
$-\infty < y < \infty$

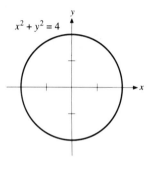

13. y is not a function of x.

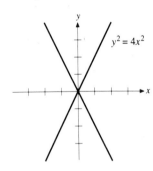

15. $V(x) = x(11 - 2x)(8 - 2x)$, $0 \leq x \leq 4$
17. (a) $A(x) = \pi x^4$,   **(b)** $C(x) = 2\pi x^2$
19. $L(x) = 1 + \cos x$

21. $f(x) = \begin{cases} x^2 - 4, & x \leq -2, \ x \geq 2, \\ 4 - x^2, & -2 < x < 2 \end{cases}$

23. $g(x) = \dfrac{x + 2}{3}$ **25.** $3(3)^2 + (3)(1) = 30$

27. $(2)(1) + (2)^2(1) = 6$ **29.** $\dfrac{1}{3(1) - 1} = \dfrac{1}{2}$

31. $\dfrac{(1)}{1 + 0} = 1$ **33.** $\dfrac{(1.2)}{2(0.8) - 0.8} = 1.5$

35. $\dfrac{(4)^3}{0 + 4} = 16$ **37.** Unbounded   **39.** 2 **41.** 1

43. (a) 1,   **(b)** 1,   **(c)** 1,   **(d)** -1,   **(e)** 1,   **(f)** -1
45. (a) No,   **(b)** 0.28,   **(c)** $5100, 23650, 50000$,   **(d)** no
47.
(a)

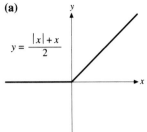

(b) domain $-\infty < x < \infty$, range $y \geq 0$,   **(c)** $0, 0, 1$,
(d) $0, 0$,   **(e)** yes, yes, yes

49.
(a)

(b) domain $-\pi/2 < x < \pi/2$, range $y > 0$,
(c) $-\sqrt{2}/2, 1, \sqrt{2}$,   **(d)** $1, 1$,   **(e)** yes, yes, yes
51.
(a)

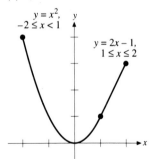

(b) domain $-2 \leq x \leq 2$, range $0 \leq y \leq 4$,   **(c)** $1, 1, 3$,
(d) yes,   **(e)** $2, 2$,   **(f)** yes

53.

(a)

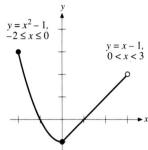

$y = x^2 - 1,$
$-2 \le x \le 0$

$y = x - 1,$
$0 < x < 3$

(b) domain $-2 \le x < 3$, range $-1 \le y \le 3$,
(c) 0, 1, undefined, **(d)** yes, **(e)** 0, 1, **(f)** no **55.** $\epsilon/3$
57. ϵ **59.** $\epsilon^{1/4}$ **61.** $\epsilon/2$ **63.** 3ϵ **65.** 0.0583 Ω
67. 0 **69.** $-16t, t \ne 2$; -32 **77.** y-axis **79.** x-axis
81. Even **83.** Neither even nor odd
85.
(a)

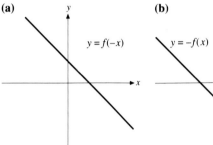

$y = f(-x)$

(b)

$y = -f(x)$

(c)

$y = f(|x|)$

(d)

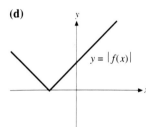

$y = |f(x)|$

CHAPTER THREE:
CONTINUITY AND DIFFERENTIABILITY ON AN INTERVAL

EXERCISES 3.1

1. $2x - 7$ **3.** $6x^2 - 1$ **5.** $2 - x^{-2}$
7. $\frac{1}{2}x^{-1/2} - \frac{1}{2}x^{-3/2}$ **9.** $4x^{-1/3} + 3x^{-4/3}$
11. $\frac{2^{4/3}}{3}x^{-1/3}$ **13.** $3x^{1/2} + \frac{1}{2}x^{-1/2}$ **15.** $\frac{1}{4} - \frac{1}{4}x^{-2}$

17. $12t^3 + 12t$ **19.** $-\frac{1}{2}t^{-2} - \frac{1}{4}t^{-3/2}$ **21.** $6x - 7$; 6; 0
23. $\frac{1}{3}x^{-2/3} - \frac{1}{3}x^{-4/3}$; $-\frac{2}{9}x^{-5/3} + \frac{4}{9}x^{-7/3}$;

$\frac{10}{27}x^{-8/3} - \frac{28}{27}x^{-10/3}$ **25.** $9x^8$; $72x^7$; $504x^6$
27. $y = 3x + 1$ **29.** $y = 4x - 6$ **31.** $x + 4y + 4 = 0$
33. $s'(t) = 50 - 32t$; $s(3) = 6$; $s'(3) = -46$
35. $s'(t) = 1 + 12t^2$; $s(1) = 5$; $s'(1) = 13$
37. $s'(t) = \frac{1}{2\sqrt{t}}$; $s(4) = 2$; $s'(4) = \frac{1}{4}$
39. **(a)** $v(t) = -32t + 32$, **(b)** $a(t) = -32$,
(c) $s(1) = 64$, **(d)** $v(-1) = 64$, $v(3) = -64$
41. **(a)** $v(t) = 1 + \frac{2}{t^3}$, **(b)** $a(t) = -\frac{6}{t^4}$,
(c) $s(-2^{1/3}) = -3 \cdot 2^{-2/3}$, **(d)** $v(1) = 3$
43. **(a)** $(3, -8)$, **(b)** none
45. **(a)** $(2, 4), (-2, -4)$, **(b)** none
47. **(a)** $(-1, -3)$, **(b)** $(0,0)$ **49.** **(a)** $(2, 16)$, **(b)** none
51. **(a)** $(1, 2), (-1, -2)$, **(b)** none
53. $(1, 0), (-1, 0)$ **55.** $(1, 1), (9, 81)$
57. $(-2, 4)$, 0.5404 rad; $(1, 1)$, 1.2490 rad

EXERCISES 3.2

1. $(x^2 + 1)(4x - 3) + (2x^2 - 3x + 1)(2x) =$
$8x^3 - 9x^2 + 6x - 3$
3. $(x + x^{-1})(1 + x^{-2}) + (x - x^{-1})(1 - x^{-2}) = 2x + 2x^{-3}$
5. $\frac{(x^2 + 1)(3) - (3x)(2x)}{(x^2 + 1)^2} = \frac{-3x^2 + 3}{(x^2 + 1)^2}$
7. $\frac{(4x + 1)(2) - (2x - 1)(4)}{(4x + 1)^2} = \frac{6}{(4x + 1)^2}$
9. $x + y = 1$ **11.** $5y = 8x - 7$
13. $2x + y = 7$ **15.** **(a)** None, **(b)** none
17. **(a)** $(1, 1/6), (-1, -1/6)$, **(b)** $(0,0)$
19. $2ff'$ **21.** $f'gh + fg'h + fgh'$
23. $-2g'/g^3$ **25.** $fg'' + 2f'g' + f''g$
27. 13 **29.** -1 **33.** **(a)** 0, **(b)** 0, **(c)** 2
35. **(a)** 0, **(b)** does not exist, **(c)** 1
37. **(a)** -2, **(b)** does not exist, **(c)** 2
39. **(a)** 4/3, **(b)** 0, **(c)** 4/3
41. **(a)** does not exist **(b)** does not exist, **(c)** -1

EXERCISES 3.3

1. $\cos x - \sin x$ **3.** $\sec x \tan x - 2\sec^2 x$
5. $(3x^{1/3})(\cos x) + (\sin x)(x^{-2/3}) = x^{-2/3}(3x \cos x + \sin x)$
7. $(2x)(\cos x) + (\sin x)(2) = 2x \cos x + 2\sin x$
9. $\frac{(x)(\cos x) - (\sin x)(1)}{(x)^2} = \frac{x \cos x - \sin x}{x^2}$
11. $\frac{(x^2 + 4)(-\sin x) - (\cos x)(2x)}{(x^2 + 4)^2} =$

$$\frac{-x^2 \sin x - 2x \cos x - 4 \sin x}{(x^2 + 4)^2}$$

13. $y = 2x$ **15.** $36y = 42x + 6\sqrt{3} - 7\pi$

17. $\left(\dfrac{2\pi}{3}, \dfrac{2\pi}{3} + \sqrt{3}\right), \left(\dfrac{4\pi}{3}, \dfrac{4\pi}{3} - \sqrt{3}\right)$

19. $\left(\dfrac{\pi}{3}, 2\right), \left(\dfrac{4\pi}{3}, -2\right)$ **21.** $0 \le t < \pi/3, \; 2\pi/3 < x \le 2\pi$

23. $0 \le t < \pi/4$ **31.** 1 **33.** 2 **35.** 9/2

37. (a) $f'(x) = \dfrac{2x \sin x - x^2 \cos x}{\sin^2 x}, \; x \ne 0$, **(b)** yes,

(c) yes, **(d)** yes **39.** ≈ 0.66

EXERCISES 3.4

1. $6(7x + 1)^5(7) = 42(7x + 1)^5$

3. $-\dfrac{2}{3}(2x^3 - 1)^{-5/3}(6x^2) = -4x^2(2x^3 - 1)^{-5/3}$

5. $(-1)(2 \sin x + 3)^{-2}(2 \cos x) = -\dfrac{2 \cos x}{(2 \sin x + 3)^2}$

7. $(\cos(x^3 - x))(3x^2 - 1) = (3x^2 - 1)\cos(x^3 - x)$

9. $\left(\dfrac{1}{2\sqrt{\sin x}}\right)(\cos x) = \dfrac{\cos x}{2\sqrt{\sin x}}$

11. $\left(\sec\dfrac{1}{x}\tan\dfrac{1}{x}\right)\left(-\dfrac{1}{x^2}\right) = -\dfrac{\sec\dfrac{1}{x}\tan\dfrac{1}{x}}{x^2}$

13. $(x)\left(\dfrac{1}{2\sqrt{x^2 + 1}}\right)(2x) + (\sqrt{x^2 + 1})(1) = \dfrac{2x^2 + 1}{\sqrt{x^2 + 1}}$

15. $7\left(\dfrac{x}{1 + x^2}\right)^6\left(\dfrac{(1 + x^2)(1) - (x)(2x)}{(1 + x^2)^2}\right) = \dfrac{7x^6(1 - x^2)}{(1 + x^2)^8}$

17. $\dfrac{(1 + \tan^2 x)(2 \sin x \cos x) - (\sin^2 x)(2 \tan x \sec^2 x)}{(1 + \tan^2 x)^2} =$

$2 \sin x \cos^3 x - 2 \sin^3 x \cos x$

19. $2(\tan(2x))(\sec^2(2x))(2) = 4 \tan(2x) \sec^2(2x)$

21. $\left(\dfrac{1}{2\sqrt{x^2 + \sin^2 x}}\right)(2x + (2 \sin x)(\cos x)) =$

$\dfrac{x + \sin x \cos x}{\sqrt{x^2 + \sin^2 x}}$

23. $\dfrac{1}{2}(x + (x + x^2)^{1/2})^{-1/2}\left(1 + \dfrac{1}{2}(x + x^2)^{-1/2}(1 + 2x)\right)$

25. $16y = 11x - 3$ **27.** $4x + 125y = 41$

29. $4x + y = \pi + 1$ **31.** $y = 3x - \pi$

33. (a) $(1, -1)$, **(b)** $(4/3, 0)$

35. (a) $(-2, -1/2)$, **(b)** none

37. (a) $(\pi/2, \pi/2 - 1)$, **(b)** none **39.** 20 **41.** 2/3

43. 0 **47.** $y - \dfrac{1}{\sqrt{2}} = \dfrac{5}{\sqrt{2}}\left(x - \dfrac{\pi}{4}\right)$ **49.** $2/x$

51. $\dfrac{2x}{x^2 + 1}$ **53.** Less when $r = 2$ than when $r = 3$

55. $10\pi/9$ ft/s **65. (a)** $f'(x) = 2x \sin\dfrac{1}{x} - \cos\dfrac{1}{x}, \; x \ne 0$,

(b) no, **(c)** yes, **(d)** no

EXERCISES 3.5

1. All points on graph except $(-1, 0)$ and $(1, 0)$

3. All points on graph **5.** All points on graph except $(0, 0)$

7. All points on graph except $(-1, 0)$, $(1, 0)$, and $(0, 0)$

9. x/y **11.** $1/(3y^2)$ **13.** $-\sqrt{y}/\sqrt{x}$

15. $\dfrac{1 - \cos(x + y)}{\cos(x + y)}$ **17.** $\dfrac{2xy - y^2}{2xy - x^2}$ **19.** $\dfrac{y - x^2}{y^2 - x}$

21. $\dfrac{\cos x \cos y}{1 + \sin x \sin y}$ **23.** $3x + y = 2$

25. $x + y = 2$ **27.** $46y = 9x + 120$

29. $32x + 4y = 15$ **31.** $6\sqrt{3}x + 12y = 3\sqrt{3} + 2\pi$

33. $8x + 4y = \pi + 8$ **37.** $(2\sqrt{6}/3)$ /ft

39. $\dfrac{dv}{du} = \dfrac{1}{2 + 6v}, \; \dfrac{du}{dv} = 2 + 6v$, they are reciprocals.

41. $-x/y; \; -4/y^3$ **43.** $\dfrac{y}{y - x}; \; \dfrac{-4}{(y - x)^3}$

45. $\cos^2 y; \; -2 \cos^3 y \sin y$

47.

(a)

(b) $y = x^3 + C$ implies $\dfrac{dy}{dx} = 3x^2$; $y = \dfrac{1}{3x} + C'$ implies

$\dfrac{dy}{dx} = -\dfrac{1}{3x^2}$

49.

(a)

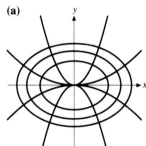

(b) $\dfrac{x^2}{2} + y^2 = C$ implies $\dfrac{dy}{dx} = -\dfrac{x}{2y}$; $y = C'x^2$ implies

$\dfrac{dy}{dx} = \dfrac{2y}{x}$ **53. (a)** 0, **(b)** $-1/2$

EXERCISES 3.6

1. 3.99   **3.** 1.51   **5.** 1   **7.** 1   **9.** 4.075
11. 0.885   **13.** 0.9   **15.** $du = 3\,dx$ **17.** $du = 2x\,dx$

19. $du = (3x^2 + 8x)\,dx$ **21.** $du = \dfrac{1}{2\sqrt{x}}\,dx$

23. $du = -3\sin(3x)\,dx$ **25.** $du = \sec^2 x\,dx$
27. $\Delta V = 2\pi r h_0\,dr + \pi h_0 (dr)^2,\, dV = 2\pi r h_0\,dr$
29. $\Delta A = 2s\,ds + (ds)^2,\, dA = 2s\,ds$
31. $V = 2\pi r^3,\, dV = \pm 9.048\ \text{m}^3,\, \pm 2.25\%$ **33.** 40 ± 1.4 ft
35. $dA = 3\pi r^2\,dr$ **37.** $dV = \pi r^2\,dh + 2\pi r h\,dr$

39. $dS = \pi r \dfrac{r\,dr + h\,dh}{\sqrt{r^2 + h^2}} + \pi\sqrt{r^2 + h^2}\,dr$

41. (a) $\delta = \dfrac{6w}{\pi D^3}$,   **(b)** 0.653,

(c) $d\delta = \dfrac{6}{\pi}\dfrac{D\,dw - 3w\,dD}{D^4}$,   **(d)** 0.0157 oz/in.^{3} **43.** 6%

45. $dh = \dfrac{x\,dx + y\,dy}{\sqrt{x^2 + y^2}}$

EXERCISES 3.7

1. $[-2, -1]$, $[0, 1]$, $[1, 2]$ **3.** $[-1, 0]$, $[0, 1]$, $[2, 3]$
5. $[1.2, 1.3]$ **7.** $[0.7, 0.8]$ **9.** $[1.8, 1.9]$ **11.** $[0.6, 0.7]$
15. f is undefined and discontinuous at zero.
17. 1.7321   **19.** -2.8284 **21.** 2.0801   **23.** 1.6818
25. 1.1593   **27.** 1.6180   **29.** 0.2638   **31.** 0.3222
33. 0.7391   **35.** 0.8241   **37.** $-1.879, 0.347, 1.532$
39. $-0.732, 1.000, 2.732$ **41.** $-2.422, 0.917, 4.505$
43. (a) $-\infty < x_1 < \infty$,   **(b)** $x_1 > 0$,   **(c)** $0 < x_1 < 4c$
45. (b) 0.306, 2.197   **47.** $0.26 < x_1 < 1.74$
49. $-1.39 < x_1 < 1.39$

EXERCISES 3.8

1. $f(3) = 1,\, f(0) = -2$ **3.** $f(0) = 1$, no minimum value

$y = x - 2,$
$0 \le x \le 3$

$y = 1 - x^2,$
$-2 < x < 1$

5. No maximum value, no minimum value   **7.** No maximum value, $f(0) = -1$

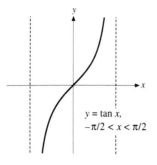

$y = \tan x,$
$-\pi/2 < x < \pi/2$

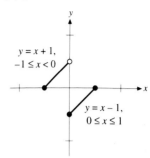

$y = x + 1,$
$-1 \le x < 0$

$y = x - 1,$
$0 \le x \le 1$

9. $f(0) = 3,\, f(2) = -5$
11. $f(0) = f(6) = 4,\, f(4) = -28$
13. $f(8) = 12,\, f(0) = 0$
15. $f(3/4) = 3(4)^{-4/3},\, f(2) = -2$
17. $f(1) = 1/2,\, f(0) = 0$
19. $f(5\pi/6) = 1,\, f(2\pi) = -0.58778525$
21. $f(\pi/4) = \sqrt{2},\, f(5\pi/4) = -\sqrt{2}$
23. $f(3/2) = 4,\, f(4) = -1$
25. 64 ft   **27.** 490 m   **29.** 9
31. $f(0.666) \approx 0.606$ **37. (a)** $\omega\sqrt{a^2 + b^2}$,   **(b)** ℓ
41. (a) Yes,   **(b)** 0   **43. (a)** No,   **(b)** none
45. (a) No,   **(b)** none   **47. (a)** No,   **(b)** $-1/8$
53. $[4.47, 4.53]$ **55.** $[2.2198, 2.2202]$
59. $y - f(a) = \left(\dfrac{f(b) - f(a)}{b - a}\right)(x - a)$ **65.** -1 **67.** No

EXERCISES 3.9

1. 0.4; 0.571; 0.171; 0.03078; 0.543794342
3. 2.0033333; 2.0033278; 0.00000556; 1.5432×10^{-8}; 2.00332779   **5.** 1; 0.9982; 0.0018; 0.000036; 0.998200539
7. 1.2; 1.24; 0.078125; 0.01953125; 1.25
9. 1.2; 1.22; 0.06; 0.004; 1.221402758
11. -0.25; -0.28125; 0.05555556; 0.01234568; -0.287682072
13. $P_1(x) = 4 - \dfrac{3}{4}(x - 3)$; $P_1(3.1) = 3.925$; $P_2(x) =$

$4 - \dfrac{3}{4}(x - 3) - \dfrac{25}{128}(x - 3)^2$; $P_2(3.1) =$

3.9230469; $\sqrt{25 - (3.1)^2} \approx 3.923009049$

15. $P_1(x) = \dfrac{\pi}{6} + \dfrac{2}{\sqrt{3}}\left(x - \dfrac{1}{2}\right)$;

$P_1(0.55) = 0.5813338$; $P_2(x) =$

$\dfrac{\pi}{6} + \dfrac{2}{\sqrt{3}}\left(x - \dfrac{1}{2}\right) + \dfrac{2}{3\sqrt{3}}\left(x - \dfrac{1}{2}\right)^2$; $P_2(0.55) =$

0.5822961; $\sin^{-1}(0.55) \approx 0.582364237$
23. $P_1(x) = 2(x - 1)$;
$P_1(1.2) = 0.4$; $P_2(x) = 2(x - 1) + 2(x - 1)^2$;
$P_2(1.2) = 0.48$; $P_3(x) = 2(x - 1) + 2(x - 1)^2 + (x - 1)^3$; $P_3(1.2) = 0.488$; $f(1.2) = 0.488$

25. $P_1(x) = -8 + 12x$; $P_1(-0.1) = -9.2$;
$P_2(x) = -8 + 12x - 6x^2$; $P_2(-0.1) = -9.26$;
$P_3(x) = -8 + 12x - 6x^2 + x^3$; $P_3(-0.1) = -9.261$;
$f(-0.1) = -9.261$
27. $P_1(x) = P_2(x) = x$; $P_1(0.2) = P_2(0.2) = 0.2$;
$P_3(x) = x + \dfrac{x^3}{3}$; $P_3(0.2) = 0.20266667$;
$f(0.2) \approx 0.20271003$
29. $P_1(x) = 1 + \dfrac{1}{2}(x-1)$; $P_1(0.98) = 0.99$;

$P_2(x) = 1 + \dfrac{1}{2}(x-1) - \dfrac{1}{8}(x-1)^2$; $P_2(0.98) = 0.98995$;

$P_3(x) = 1 + \dfrac{1}{2}(x-1) - \dfrac{1}{8}(x-1)^2 +$

$\dfrac{1}{16}(x-1)^3$; $P_3(0.98) = 0.9899495$; $f(0.98) \approx 0.98994949$

REVIEW EXERCISES

1. $3x^2 - 4x + 1$ **3.** $-2x^{-3}$ **5.** $12x^3 + 6x^2 - 6x$
7. $x^2 \sec^2 x + 2x \tan x$ **9.** $\sec^3 x + \sec x \tan^2 x$
11. $-\dfrac{2x \sin x + \cos x}{2x^{3/2}}$ **13.** $\dfrac{-2}{(1+x)^2}$

15. $\dfrac{2x \cos x + x^2 \sin x}{\cos^2 x}$ **17.** $-\dfrac{5}{3}(5x-1)^{-4/3}$

19. $-\dfrac{3 \sin 3x}{(1 - \cos 3x)^2}$ **21.** $2 \sin x \cos x$
23. $9 \tan^2 3x \sec^2 3x$ **25.** $4 \sec^2 2x \tan 2x$
27. $\csc (\cot x) \cot (\cot x) \csc^2 x$
29. $\dfrac{6x+1}{\sqrt{4x+1}}$ **31. (a)** 3,   **(b)** 0,   **(c)** 2

33. (a) 4,   **(b)** does not exist,   **(c)** -4
35. (a) 1/2,   **(b)** does not exist,   **(c)** 1/2
37. $2(1-2x)^{-2}$, $8(1-2x)^{-3}$, $48(1-2x)^{-4}$
39. $2 \sin x \cos x$, $-2 \sin^2 x + 2 \cos^2 x$, $-8 \sin x \cos x$
41. $4y = x + 4$ **43.** $12y + 6\sqrt{3}x = 2\pi + 3\sqrt{3}$
45. (a) $(-24, -48)$,   **(b)** $(-32, 0)$
47. $\left(\dfrac{7\pi}{6}, \dfrac{7\pi}{6} + \sqrt{3}\right)$, $\left(\dfrac{11\pi}{6}, \dfrac{11\pi}{6} - \sqrt{3}\right)$ **49.** $0 < t < 1$

51. $\dfrac{1}{\sqrt{a^2 - x^2}}$

53. $-3/\sqrt{15}, -\sqrt{3}, -9/\sqrt{7}, \left|\dfrac{dy}{dt}\right|$ increases without bound as x
approaches 12.
55.
(a)

$4x^2 + y^2 = 4$

(b) all except $(-1, 0)$ and $(1, 0)$,
(c) all except $(-1, 0)$ and $(1, 0)$,   **(d)** $-4x/y$
57.
(a)

$4x^2 - y^2 = 4$

(b) all except $(-1, 0)$ and $(1, 0)$,
(c) all except $(-1, 0)$ and $(1, 0)$,   **(d)** $4x/y$
59.
(a)

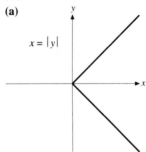

$x = |y|$

(b) all except $(0, 0)$,   **(c)** all except $(0, 0)$,   **(d)** $|y|/y$
61. x^2/y^2 **63.** $\dfrac{2x + 3y}{2y - 3x}$ **65.** $\dfrac{1}{2}\cos^2(2y)$
67. $-\csc y$ **69.** x/y; $-3/y^3$ **71.** $\cos y \cot y$;
$-\cos^2 y \cot y(\csc^2 y + 1) = -\cot^3 y(1 + \sin^2 y)$
73. $-\dfrac{\sin x}{\cos y}$; $\dfrac{\sin^2 x \sin y - \cos^2 y \cos x}{\cos^3 y}$
77. $du = 2x\,dx$ **79.** $du = \cos x\,dx$
81. $x\,dx + y\,dy = r\,dr$ **83.** $dx = -r \sin\theta\,d\theta + \cos\theta\,dr$
85. $dV = \pi r^2\,dh + 2\pi rh\,dr$ **87.** ± 0.12 m³; 3%
89. ± 2.64 **91.** [1.2, 1.3]   **93.** 0.4502   **95.** 0.666
97. $-2.361, -0.167, 2.529$ **99.** $0 < x_1 < 0.8$
101. $f(0) = 3$; no minimum value   **103.** $f(0) = 1$; no minimum value

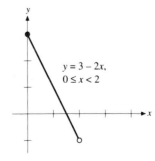

$y = 3 - 2x$, $0 \le x < 2$

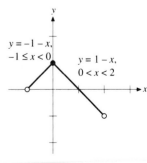

$y = -1 - x$, $-1 \le x < 0$
$y = 1 - x$, $0 < x < 2$

105. $f(2) = 2$; $f(-1) = -4$ **107.** $f(3) = 4$; $f(1/3) = 0$

109. 16   **115.** $P_1(x) = 1 + 16(x - 1)$;
$P_1(0.98) = 0.68$; $P_2(x) = 1 + 16(x - 1) + 120(x - 1)^2$;
$P_2(0.98) = 0.728$; 0.048; 0.00448; $(0.98)^{16} \approx 0.72379772$

117. $P_1(x) = 1 + 2\left(x - \dfrac{\pi}{4}\right)$; $P_1(0.75) = 0.929203673$;

$P_2(x) = 1 + 2\left(x - \dfrac{\pi}{4}\right) + 2\left(x - \dfrac{\pi}{4}\right)^2$;

$P_2(0.75) = 0.931709733$; 0.002506059; 0.000118279;

$\tan 0.75 \approx 0.931596459$ **119.** 0.987   **121.** $1 - 2x^2 + \dfrac{2}{3}x^4$

123. $2 + \dfrac{1}{4}(x - 4) - \dfrac{1}{64}(x - 4)^2 + \dfrac{1}{512}(x - 4)^3$

125. $1 + \dfrac{1}{2}x - \dfrac{1}{8}x^2 + \dfrac{1}{16}x^3$ **127.** f is undefined at zero.

9. $-\infty, -\infty$

11. $\infty, -\infty$

13. ∞, ∞

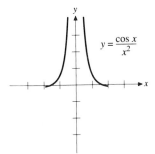

CHAPTER FOUR:
GRAPHS AND APPLICATIONS OF THE DERIVATIVE

EXERCISES 4.1

1. $-\infty, \infty$

3. $\infty, -\infty$

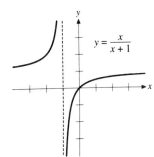

15. ∞ **17.** $-\infty$ **19.** ∞ **21.** ∞ **23.** 6
25. $\sqrt{3}/2$ **27.** 0   **29.** 2   **31.** $-1/6$ **33.** ∞
35. -2 **37.** 1   **39.** 1/2   **41.** 1   **43.** 1
45. 2   **55.** (a) $-\infty$,   (b) $-\infty$,   (c) $-\infty$
57. (a) 0,   (b) 0,   (c) 0

EXERCISES 4.2

1. 0   **3.** 1   **5.** 1   **7.** 1/2   **9.** ∞ **11.** $-\infty$ **13.** $-1/2$
15. 2   **17.** ∞ **19.** $-\infty$ **21.** $-\infty$ **23.** -2 **25.** 0
27. 1

5. $-\infty, \infty$

7. ∞, ∞

29.

31.

33.

$$y = \frac{x}{x^2 + x + 1}$$

35.

$$y = \frac{x - 2}{x^2 - 4x}$$

49. $f(x) = x + \dfrac{1}{x}$
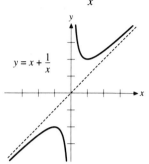
$$y = x + \frac{1}{x}$$

51. $f(x) = -x - 2 + \dfrac{5}{x + 2}$
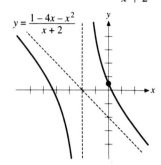
$$y = \frac{1 - 4x - x^2}{x + 2}$$

37.
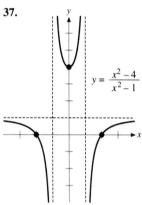
$$y = \frac{x^2 - 4}{x^2 - 1}$$

39.

$$y = \frac{x^2 - 1}{x^3}$$

53. $f(x) = x - 1 - \dfrac{1}{x - 1}$
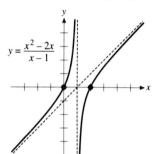
$$y = \frac{x^2 - 2x}{x - 1}$$

55. $f(x) = x + \dfrac{8}{x^2}$
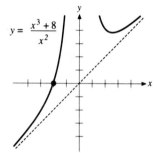
$$y = \frac{x^3 + 8}{x^2}$$

41.

$$y = \frac{1}{\sqrt{x^2 + 1}}$$

43.

$$y = \frac{|x|}{\sqrt{x^2 + 1}}$$

57. $f(x) = x^2 - \dfrac{1}{x}$

$$y = \frac{x^3 - 1}{x}$$

59. $f(x) = x^2 - \dfrac{1}{x^2}$

$$y = \frac{x^4 - 1}{x^2}$$

45.
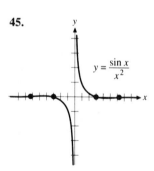
$$y = \frac{\sin x}{x^2}$$

47.
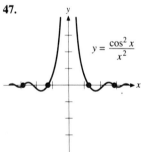
$$y = \frac{\cos^2 x}{x^2}$$

61. $f(x) = x^3 + \dfrac{1}{x}$
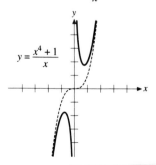
$$y = \frac{x^4 + 1}{x}$$

71. $x = -1, x = 0, x = 1;\ y = 0$ **73.** $y = \pi/4 \approx 0.78$

EXERCISES 4.3

1. Loc min $(0, 0)$

3. No local extrema

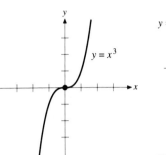

5. Loc max $(3, 19/6)$

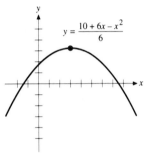

7. Loc min $(2, -2)$; loc max $(-2, 2)$

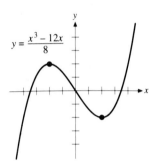

9. Loc min $(-1, 0)$; loc max $(1, 4)$

11. No local extrema

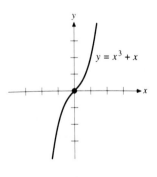

13. Loc min $(0, 0)$; loc max $(\sqrt{2}, 4)$, $(-\sqrt{2}, 4)$

15. Loc min $(0, 0)$

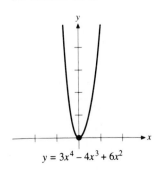

17. Loc min $(2, -2)$; loc max $(1, 7/8)$

19. Loc min $(1, -6)$; loc max $(-1, 6)$

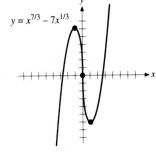

21. Loc min $(0, -1)$

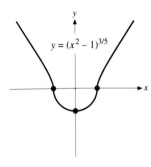

23. Loc min $(1, -2)$

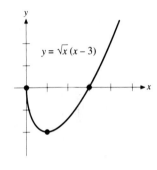

25. Loc min $(-2, -2)$; loc max $(2, 2)$

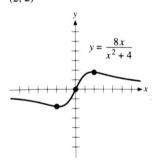

27. Loc min $(0, 0)$

29. Loc min $(-1, -1)$; loc max $(1, 1)$

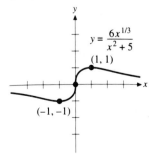

31. Loc min $(2, 4)$; loc max $(-2, -4)$

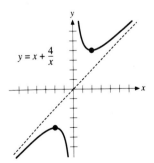

33. Loc min $(1, 2)$, $(-1, 2)$ **35.** Loc max $(0, -1)$

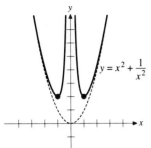

$y = x^2 + \dfrac{1}{x^2}$

$y = \dfrac{1}{x^2 - 1}$

49. Loc max $(\pi/6, -\sqrt{3})$ **51.** Loc min $(0, 1)$

$y = \tan x - 2 \sec x, \ -\pi/2 < x < \pi/2$

$y = \dfrac{1}{\sqrt{1 - x^2}}$

37. Loc min $(1, 4)$; loc max $(4, 1)$ **39.** Loc min $(1, 0)$, $(-1, 0)$

$y = \dfrac{8x - 20}{x^2 - 4}$

$y = \dfrac{(x^2 - 1)^2}{x^4}$

53. Loc max $(2, 2)$ **55.** No local extrema

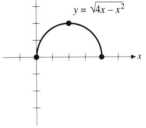

$y = \sqrt{4x - x^2}$

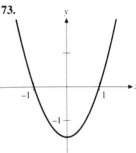

$y = \sqrt{x^2 - 1}$

41. No local extrema   **43.** No local extrema

$y = x + \sin x, \ 0 \le x \le 2\pi$

$y = x + \dfrac{1}{2} \sin 2x, \ 0 \le x \le 2\pi$

57. $x \le -2, \ x \ge 8$ **59.** $0 \le t < 2$

61. $0 \le t < \pi/2, \ 3\pi/2 < t \le 2\pi$

63. (a) $x = 1$,   (b) $x > 1$,   (c) $x < 1$

65. (a) $-1, 1$,   (b) $x < -1, \ x > 1$,   (c) $-1 < x < 1$

67. $g = f'$ **69.** $f = g'$

71.   **73.**

45. Loc min $(5\pi/4, -\sqrt{2})$; loc max $(\pi/4, \sqrt{2})$ **47.** No local extrema

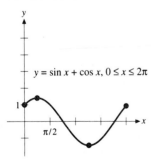

$y = \sin x + \cos x, \ 0 \le x \le 2\pi$

$y = \tan x - x, \ -\pi/2 < x < \pi/2$

75.   **77.**

79.

81.

13.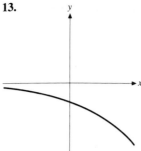

83. $f(1) = \dfrac{1}{2(1 + \gamma)}; \dfrac{\gamma - 1}{2\gamma^2}$

15. Impossible

17. Infl pt $(0, 1)$

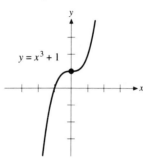

19. Loc min $(-1, -4)$; loc max $(1, 0)$; infl pt $(0, -2)$

EXERCISES 4.4

1.

3.

5. Impossible

7.

9.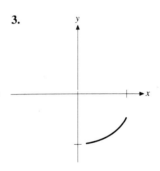

11. Impossible

21. Infl pt $(0, 0)$

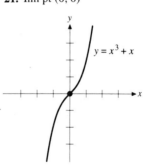

23. Loc min $(1/8, -1)$

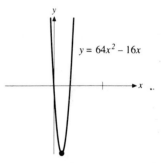

25. Loc min $(1, -2)$; loc max $(-1, 2)$; infl pt $(0, 0)$

27. Loc min $(1, -6)$; loc max $(-1, 6)$; infl pt $(0, 0)$

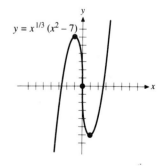

29. Loc min $(0, 0)$; loc max $(-16, 24 \cdot 2^{2/3})$; infl pt $(8, 48)$

$$y = \frac{x^{2/3}(x + 40)}{4}$$

31. Loc min $(-2, -2)$; loc max $(2, 2)$; infl pt $(0, 0)$, $(\sqrt{12}, \sqrt{3})$, $(-\sqrt{12}, -\sqrt{3})$

$$y = \frac{8x}{x^2 + 4}$$

41. Loc min $(5\pi/4, -\sqrt{2})$; loc max $(\pi/4, \sqrt{2})$; infl pt $(3\pi/4, 0)$, $(7\pi/4, 0)$

$y = \sin x + \cos x, \ 0 \le x \le 2\pi$

43.

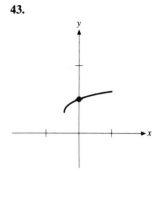

33. Infl pt $(0, 0)$

$$y = \frac{x}{x^2 - 1}$$

35. Loc max $(2, 1/4)$, infl pt $(3, 2/9)$

$$y = \frac{x - 1}{x^2}$$

45.

47.

49.

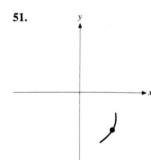

51.

37. Loc min $(1, 0)$, $(-1, 0)$; infl pt $(\sqrt{5/3}, 4/25)$, $(-\sqrt{5/3}, 4/25)$

$$y = \frac{(x^2 - 1)^2}{x^4}$$

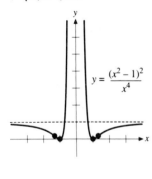

39. Loc min $(2\pi/3, 2\pi/3 - \sqrt{3}/2)$, $(5\pi/3, 5\pi/3 - \sqrt{3}/2)$; loc max $(\pi/3, \pi/3 + \sqrt{3}/2)$, $(4\pi/3, 4\pi/3 + \sqrt{3}/2)$; infl pt $(\pi/2, \pi/2)$, (π, π), $(3\pi/2, 3\pi/2)$

$y = x + \sin 2x, \ 0 \le x \le 2\pi$

53. (a) Loc min $(1, -1)$; loc max $(-1, 1)$,   **(b)** $(0, 0)$
55. (a) Loc max $(0, 1)$,   **(b)** $(-1.0, 0.75)$, $(1.0, 0.75)$
57. (a) Loc min $(1, -1)$,   **(b)** $(2.0, -0.7)$
59. (a) Loc min $(-1.9, -1.0)$; loc max $(1.9, 1.0)$,
(b) $(-2.9, -0.8)$, $(0, 0)$, $(2.9, 0.8)$ **61. (a)** 0,   **(b)** π

EXERCISES 4.5

1. (a) $2h + \ell = 11$,   **(b)** $2h + w = 8.5$,
(c) $V = (11 - 2h)(8.5 - 2h)(h)$

3. (a) $2w + \ell = 600$,   **(b)** $A = \ell\left(\dfrac{600 - \ell}{2}\right)$

5. (a) $h = \sqrt{3}\ell/2$,   **(b)** $A = \sqrt{3}\ell^2/4$

7. (a) $\ell w = 20,000$,   **(b)** $P = 2\ell + \dfrac{40,000}{\ell}$

9. (a) $\dfrac{4\pi r^3}{3} + \pi r^2 h = 600$,   **(b)** $S = \dfrac{4\pi r^2}{3} + \dfrac{1200}{r}$

11. $s^2 = x^2 + 150^2$ **13.** $L = 40 \sec\theta - 20 \tan\theta + 30$

15. (a) $4w = 12 - 3\ell$,   **(b)** $A = \ell\left(\dfrac{12 - 3\ell}{4}\right)$

17. $|AB| = W\csc\theta + L\sec\theta$ **19.** $A = 2h\sqrt{16 - h^2}$

21. $V = 9h^2$ **23.** $V = \dfrac{\pi h}{3}(b^2 + ab + a^2)$

25. $h = 50\tan\theta$ **27.** $F = 0.05x$

29. (a) $n = 110 - 0.1r$,   **(b)** $R = 110r - 0.1r^2$

31. $A = \dfrac{r^2\theta}{2} - \dfrac{r^2\sin\theta}{2}$ **33.** $A = \dfrac{d^2}{2}\sin\theta$

EXERCISES 4.6

1. (a) 4π ft/s,   **(b)** 24π ft²/s

3. (a) 720 cm²/s,   **(b)** 3600 cm³/s

5. $2/(9\pi)$ ft/min   **7.** $4/\sqrt{3}$ m/s   **9.** 5/18 ft/min

11. (a) 1/625 m/min,   **(b)** 1/1000 m/min

13. $16/(3\pi)$ m/min   **15. (a)** 3 ft/s,   **(b)** 3 ft/s

17. 30/13 ft/s   **19.** Decreasing at 18 mph   **21.** $10\sqrt{7}$ mph

23. $55\sqrt{2}$ mph   **25.** $80\pi/3$ m/s   **27.** 100π ft/s

29. (a) 0.4 rad/s,   **(b)** 0.1 rad/s   **31.** 0.075 in./s

33. Decreasing at 0.022 Ω/s   **35.** Decreasing at 15/16 m³/s

37. -0.12 N/s   **41.** $100\sqrt{3}$ m, $10\sqrt{3}$ m/s

43. $1/\sqrt{17}, 4/\sqrt{17}$ **45.** $\dfrac{dx}{dt} = -\dfrac{8}{\sqrt{73}}, \dfrac{dy}{dt} = \dfrac{3}{\sqrt{73}}$

47. $-\dfrac{4\sqrt{2}r^{-3/2}}{3\pi}$ **49.** 88 mph   **51.** 42.8 ft/s

EXERCISES 4.7

1. $A(\theta) = \dfrac{25}{2}\sin\theta\cos\theta, 0 \le \theta \le \pi/2; \pi/4, \pi/4,$

$\pi/2; A(\pi/4) = 25/4$ **3.** $A(x) = 450x - 3x^2/2,$
$0 \le x \le 300; 150\text{m} \times 225\text{m}; A(150) = 33,750\text{m}^2$

5. $A(y) = 10y - 2y^2, 0 \le y \le 5; 2.5\text{in.} \times 5\text{in.};$

$A(2.5) = 12.5\text{in.}^2$ **7. (a)** $F(x) = x + \dfrac{30,000}{x},$

$x > 0; 100\sqrt{3}\text{m} \times 50\sqrt{3}\text{m}; F(100\sqrt{3}) = 200\sqrt{3}\text{m},$

(b) $F(x) = x + \dfrac{30,000}{x}, 0 < x \le 150;$

$150\text{m} \times 100\text{m}; F(150) = 350\text{m}$

9. (a) $F(x) = \dfrac{32,000}{x} + 2x - 100, x \ge 100;$

$40\sqrt{10}\text{ft} \times 40\sqrt{10}\text{ft}; F(40\sqrt{10}) = 160\sqrt{10} - 100\text{ft},$

(b) $F(x) = \dfrac{32,000}{x} + 2x - 150, x \ge 150; 150\text{ft} \times 106\dfrac{2}{3}\text{ft};$

$F(150) = 363\dfrac{1}{3}\text{ft}$ **11. (a)** Minimum area is $\dfrac{9\sqrt{3}}{4(9 + 4\sqrt{3})}$ m²

when $\dfrac{12\sqrt{3}}{9 + 4\sqrt{3}}$ m is used for the square and $\dfrac{27}{9 + 4\sqrt{3}}$ m is

used for the triangle.   **(b)** No cut gives maximum area.
Maximum area is obtained when all the wire is used for the

square.   **13.** 5488 cm³   **15.** $C(x) = 8x + \dfrac{96,000}{x}, x > 0;$

$20\sqrt{30}\text{m} \times \dfrac{80\sqrt{30}}{3}\text{m}; C(20\sqrt{30}) = \$320\sqrt{30} \approx \$1753$

17. 4:46 A.M.   **19. (a)** $T(x) = \dfrac{\sqrt{x^2 + 300^2}}{2000} + \dfrac{200 - x}{4000},$

$0 \le x \le 200; T(100\sqrt{3}) \approx 0.18\text{hr} \approx 11\text{min},$

(b) $T(x) = \dfrac{\sqrt{x^2 + 300^2}}{2000} + \dfrac{160 - x}{4000},$

$0 \le x \le 160; T(160) \approx 0.17\text{hr} \approx 10.2\text{min},$

21. $T(x) = \dfrac{\sqrt{x^2 + 5^2}}{v_0} + \dfrac{\sqrt{(10 - x)^2 + 3^2}}{v_1}, 0 \le x \le 10$

23. c **25.** $(1/2, 1/\sqrt{2})$

27. $A(r) = 2r - r^2, 2/(1 + \pi) \le r \le 2; r = 1, \theta = 2$

31. $(W^{2/3} + L^{2/3})^{3/2}$ **33.** $(-1, -1)$

35. $R(r) = 80r - 0.1r^2, 400 \le r \le 800; \$400/\text{month}$

37. $\$6250$ per unit; $\$8,125,000$ **39.** $V(h) = \dfrac{\pi}{3}(36h - h^3),$

$0 \le h \le 6; 2\pi(1 - \sqrt{6}/3) \approx 1.15$ rad

41. (a) $A(\theta) = 36 - 18\csc\theta + 9\cot\theta, \pi/6 \le \theta \le \pi/2; \pi/3$

(b) $A(\theta) = 66 - 60.5\csc\theta + 30.25\cot\theta,$

$\sin^{-1}(5.5/6) \le \theta \le \pi/2; \sin^{-1}(5.5/6) \approx 1.16$ rad

43. $I(y) = \dfrac{ky}{(y^2 - 50y + 10,000)^{3/2}}, 0 \le y \le 200; 77.24$

45. $r = (V/2\pi)^{1/3}, h = 2(V/2\pi)^{1/3}$

47. $r = R/\sqrt{2}, h = \sqrt{2}R$ **49.** $32\pi r_0^3/81$

55. $\approx (1.392, 1.937)$

REVIEW EXERCISES

1. ∞ **3.** $-\infty$ **5.** 2   **7.** 0   **9.** 2   **11.** 3/2   **13.** -3

15. $-\infty$ **17.** 0

19. Loc min $(1, -2)$; loc max   **21.**
$(-1, 2)$; infl pt $(0, 0)$

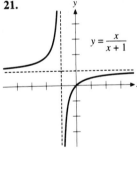

23. Loc max $(-2, -3)$ **25.**

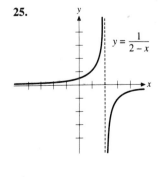

27. Loc min
$(2\pi + \tan^{-1}(-4/3), -5)$; loc
max $(\pi + \tan^{-1}(-4/3), 5)$;
infl pt $(\tan^{-1}(3/4), 0)$,
$(\pi + \tan^{-1}(3/4), 0)$

$y = 4\sin x - 3\cos x$,
$0 \le x \le 2\pi$

29. Loc min $(3\pi/2, 0)$; loc max
$(\pi/2, \sqrt{2})$

$y = \sqrt{1 + \sin x},\ 0 \le x \le 2\pi$

41. Loc min $(2, -2)$, loc max $(-2, 2)$

43. $f(x) = -2 + \dfrac{4}{x+2}$

45. $f(x) = x + 2 + \dfrac{1}{x}$

$y = \dfrac{-2x}{x+2}$

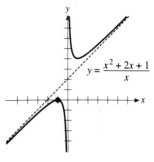

$y = \dfrac{x^2 + 2x + 1}{x}$

31. Infl pt
$\left(\sin^{-1}(\sqrt{3} - 2), \dfrac{2\sqrt{3} - 3}{\sqrt{4\sqrt{3} - 6}}\right)$

$y = \sec x + 2\tan x$,
$-\pi/2 < x < \pi/2$

33. Loc min
$(5\pi/3, -\sqrt{3} - 5\pi/3)$; loc
max $(\pi/3, \sqrt{3} - \pi/3)$; infl pt
$(\pi, -\pi)$

$y = 2\sin x - x,\ 0 \le x \le 2\pi$

47.

49.

51.

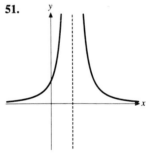

35. Loc min $(-1, 0)$, $(1, 0)$; loc
max $(0, 1)$; infl pt $(\sqrt{3}, 2^{2/3})$,
$(-\sqrt{3}, 2^{2/3})$

37.

$y = (x^2 - 1)^{2/3}$

53. **(a)** Loc max $(2, 2)$,   **(b)** $(0, 0)$, $(1.3, 1.2)$
55. **(a)** Loc min $(1, 0)$; loc max $(-1, 1.5)$,
(b) $(0.4, 0.5)$, $(-2.4, 1.0)$
57. $f(x) < 0$, $f'(x) > 0$, $f''(x) < 0$
59. $f(x) > 0$, $f'(x) < 0$, $f''(x) > 0$
61.

39.

63. Impossible   **65.** Impossible

67.

69. Volume is increasing at 144π in.^{3}/s. **71.** 0.01 rad/s
73. 2 cm/s **75.** 0.14 units/s **77.** $-k\sqrt{1.5}/4\pi$
79. (a) $A(y) = 2000y - 2y^2$, $0 \le y \le 1000$;
500 ft × 1000 ft; $A(500) = 500,000\,\text{ft}^2$,
(b) $A(y) = 2000y - 2y^2$, $0 \le y \le 400$; 400 ft × 1200 ft;
$A(400) = 480,000\,\text{ft}^2$ **81.** $r = \left(\dfrac{V_0}{3\pi}\right)^{1/3}$, $h = 3\left(\dfrac{V_0}{3\pi}\right)^{1/3}$
83. $h/h_0 = 1/3$, $r/r_0 = 2/3$ **85.** $(0,0)$
87. $A(b) = \dfrac{1}{2}\dfrac{b^2}{b-2}$, $b > 2$; $b = 4$, $h = 2$; $A(4) = 4$
89. (a) $y(\ell/2) = k\ell^4/16$, **(b)** $(1/2 \pm \sqrt{3}/6)\ell$
91. (a) 3, **(b)** 9/4, **(c)** $(a_1 + a_2 + \cdots + a_n)/n$
93. (a) $C(r) = 0.09\pi r^2 + 0.12\pi r + \dfrac{21.6}{r} + \dfrac{7.2}{\pi r^2}$, $r > 0$,
(b) $r \approx 3.23\,\text{in.}$, $h \approx 11.07\,\text{in.}$

CHAPTER FIVE:
THE DEFINITE INTEGRAL

EXERCISES 5.1

9. 1/3 **11.** 1/2 **13.** 1/36 **15.** 2 **17.** 1/3 **19.** -1
21. $\dfrac{1}{3}x^3 - x^2 + 3x + C$ **23.** $\dfrac{2}{3}x^{3/2} + 2x^{1/2} + C$
25. $-\cos x + \sin x + C$ **27.** $-\csc x + C$
29. $\dfrac{1}{3}x^3 - \dfrac{1}{2}x^2 - 2x + C$ **31.** $\dfrac{2}{3}x^3 + \dfrac{5}{2}x^2 - 3x + C$
33. $x - \dfrac{1}{x} + C$ **35.** $-\cos x + \sin x + C$
37. $\tan x + x + C$ **39.** $\sin(x - c) + C$
41. $G(x) = 2x^3 + x^2 - 3x + 5$ **43.** $G(x) = \dfrac{2}{3}x^{3/2} + \dfrac{4}{3}$
45. $G(x) = -\cos x + 3$ **47.** $G(x) = 2x^3 + 2x + 3$
49. $G(x) = -\sin x - \cos x + 2x + 2$
51. (a) 16 ft, **(b)** 2 s, **(c)** -32 ft/s
53. (a) 1 s, **(b)** 80 ft/s
55. (a) 3 s, **(b)** 28.7 m/s **57.** 144 ft **59.** $10\sqrt{2}$ s
61. (a) $y(x) = \dfrac{\omega}{24EI}(-\ell^4 + 4\ell^3 x + (\ell - x)^4)$, $0 \le x \le \ell$,
(b) $y(\ell) = \dfrac{\omega\ell^4}{8EI}$

EXERCISES 5.2

1. 0.55 **3.** $\pi(2 + \sqrt{3})/6$ **5.** 0.6345 **7.** 0.4
9. $\pi/2$ **11.** 1.7181 **13.** 0.3984 **15.** 1.8898
17. $0.75 \le s(1) \le 1.25$ **19.** $99/16 \le s(2) \le 115/16$
21. $12 \le s(1) \le 20$ **23.** $4.001744 \le s(2) \le 4.501744$
25. R_n is greater than area under curve.
27. R_n is greater than volume of solid. **29.** 0.3281
31. 1.0898 **33.** 4 **35.** $\pi(\sqrt{2} + 1)/4$ **37.** ≈ 1.0
39. ≈ 0.7 **41.** 1.21875

EXERCISES 5.3

1. 30 **3.** -16 **5.** 68 **7.** 5616
9. 696,600 **11. (a)** $-3 + 9/n$, **(b)** -3
13. (a) $-16 + \dfrac{32}{n} + \dfrac{32}{n^2}$, **(b)** -16
15. (a) $6 + \dfrac{24}{n} + \dfrac{16}{n^2}$, **(b)** 6 **17.** 36
19. 8 **29.** -4 **31.** 7 **33.** 30 **35.** 17
41. (a) $1.6 \le \displaystyle\int_0^{1/2} \dfrac{4}{1+x^2}\,dx \le 2$, $1 \le \displaystyle\int_{1/2}^1 \dfrac{4}{1+x^2}\,dx \le 1.6$
43. 1 **45.** 0

EXERCISES 5.4

1. 14/3 **3.** 9/2 **5.** 3/8 **7.** 11/6 **9.** $-34/3$
11. 1 **13.** 1 **15.** $-2/\sqrt{3} + 2$ **17.** 2 **19.** 0
21. $2/\sqrt{3}$ **23.** 0 **25.** -48 **27.** -2 **29.** 0
31. $\cos x^2$ **33.** $\sec x^2$ **35.** $-(1 - x)^5$ **37.** x^4
39. $1/x$ **41.** $2x \sec x^2 - \sec x$ **43.** $\tan x$
45. $4\sin^2(4x) - 2\sin^2(2x)$ **47.** $\dfrac{\cos x}{1 + \sin^2 x} + \dfrac{\sin x}{1 + \cos^2 x}$
49. $-4/25$ **51.** $8y = x - 2$

53. **55.**

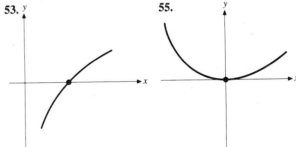

57. (a) $1 - \cos x$, **(b–c)** $\sin x$
59. (a) $\dfrac{1}{x} - 1$, **(b–c)** $-\dfrac{1}{x^2}$
61. (a) $\sin\sqrt{x} - \sin\sqrt{c}$, **(b–c)** $\dfrac{\cos\sqrt{x}}{2\sqrt{x}}$

EXERCISES 5.5

1. $-\dfrac{1}{11}(1 - x)^{11} + C$ **3.** $\dfrac{3}{8}(x^2 + 1)^{4/3} + C$

5. Not of type   **7.** $\frac{4}{3}(1 + \sqrt{x})^{3/2} + C$

9. $\dfrac{-1}{2(x^2 + 2x + 3)} + C$ **11.** $2 \sin \sqrt{x} + C$

13. Not of type   **15.** $\frac{1}{2} \tan x^2 + C$

17. $\frac{1}{2} \sec 2x + C$ **19.** $\frac{2}{3}(1 + \sin x)^{3/2} + C$

21. Either $\sin^2 x + C$ or $- \cos^2 x + C$ **23.** $\frac{1}{7} \tan^7 x + C$

25. 13/3   **27.** 3   **29.** 1   **31.** 195/4   **33.** 0

35. $(27 - 5^{3/2})/6$ **37.** $\displaystyle\int_1^2 \frac{1}{u}\,du$ **39.** $2\displaystyle\int_1^2 \frac{1}{u}\,du$

41. $2\displaystyle\int_1^2 (u^2 - 1)\,du$

EXERCISES 5.6

1. (a) 0.6015625,   **(b)** 0.6640625,   **(c)** 2/3,   **(d)** 0.125,
(e) 0.00260417,   **(f)** 0,   **(g)** $n \geq 10,001$,   **(h)** $n \geq 41$,
(i) $n \geq 2$ **3. (a)** 0.68,   **(b)** 0.68,   **(c)** 2/3,   **(d)** 0.4,
(e) 0.01333333,   **(f)** 0,   **(g)** $n \geq 40,001$,   **(h)** $n \geq 116$,
(i) $n \geq 2$ **5. (a)** 1.8961189,   **(b)** 1.8961189,
(c) 2.0045598,   **(d)** 1.2337005,   **(e)** 0.16149102,
(f) 0.00664105,   **(g)** $n \geq 49,349$,   **(h)** $n \geq 161$,
(i) $n \geq 12$ **7. (a)** 3.0694888,   **(b)** 3.0694888,
(c) 3.1429485,   **(d)** 3.4063234 (if $M = 1 + \pi$),
(e) 0.3690316 (if $M = 2 + \pi$),
(f) 0.0093684 (if $M = 4 + \pi$),
(g) $n \geq 204,380$,   **(h)** $n \geq 365$,   **(i)** $n \geq 20$
9. (a) 0.69315453,   **(b)** 0.693154533   **11.** 3.1415686
13. 11.904271   **15.** 0.903614   **19.** 1600
21. 10.683128   **23.** $n \geq 52$ **25.** ≈ 10.7 **27.** ≈ 2.7

29. (a) $R_n \geq \displaystyle\int_a^b f(x)\,dx$,   **(b)** $T_n \geq \displaystyle\int_a^b f(x)\,dx$

31. (a) $R_n \leq \displaystyle\int_a^b f(x)\,dx$,   **(b)** $T_n \geq \displaystyle\int_a^b f(x)\,dx$

33. 70.5° F   **39. (a)** $\dfrac{1}{3} + \dfrac{1}{6n^2}$,   **(b)** $\dfrac{1}{6n^2}$,   **(c)** $\dfrac{1}{6n^2}$,

(d) they are equal.

REVIEW EXERCISES

1. $-1/2$ **3.** $-1/3$ **5.** $G(x) = 2x^3 + 2x^2 - x - 1$
7. 64 ft; 4 s   **9.** 3.75   **11.** 0.95   **13.** 0.8   **15.** 7.5
17. 0.5154   **19.** $(1 + \sqrt{3})/3 \leq s(1) - s(0) \leq (3 + \sqrt{3})/3$
21. 9   **23.** 18,768   **25.** $-650,260$ **27.** $2n^3 + 3n^2$

29. $2n^2 + n$ **31. (a)** $\dfrac{4}{n} + \dfrac{4}{n^2}$,   **(b)** 0,   **(c)** 0

33. (a) $228 + \dfrac{486}{n} + \dfrac{243}{n^2}$,   **(b)** 228,   **(c)** 228   **35.** 19

41. $\sin x^2$ **43.** $-1/x$ **45.** 1/2   **47.** $x + y = 1$

49. $\dfrac{x^4}{4} - \dfrac{x^2}{2} + 2x + C$ **51.** $- \cos x - \sin x + C$

53. $\dfrac{2}{5} x^{5/2} + \dfrac{2}{3} x^{3/2} + C$ **55.** $\dfrac{1}{\pi} \sec \pi x + C$

57. $\dfrac{1}{3}(2x + 1)^{3/2} + C$ **59.** $\dfrac{1}{3}(x^2 + 4)^{3/2} + C$

61. $\dfrac{1}{2}(1 + \sqrt{x})^4 + C$ **63.** $\dfrac{1}{3} \sin^3 x + C$ **65.** 9

67. 13/81   **69.** 8/225   **71.** $\displaystyle\int_0^1 \frac{1}{4 - u^2}\,du$

73. $6\displaystyle\int_1^2 \frac{u^3}{u + 1}\,du$ **75.** $v_{ave} = ka^2/6$; $v_{max} = 3v_{ave}/2$

77. (a) 1.8961189,   **(b)** 2.0045598,   **(c)** 0.161491024,
(d) 0.006641052,   **(e)** $n \geq 23$,   **(f)** $n \geq 6$
79. (a) 0.8833333,   **(b)** 0.9,   **(c)** 0.083333,
(d) 0.016667,   **(e)** $n \geq 17$,   **(f)** $n \geq 6$
81. (a) $n \geq 104$ (if $M = 8$),   **(b)** $n \geq 20$ (if $M = 384$)
83. π

CHAPTER SIX:
APPLICATIONS OF THE DEFINITE INTEGRAL

EXERCISES 6.1

1. $L(x) = x^2 + 2$ **3.** $L(x) = 2x^2 + 1$

 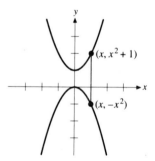

5. $L(y) = y^2 + y + 1$ **7.** $L(x) = 2\sqrt{4 - (x - 2)^2}$

9. $L(x) = 2\sqrt{y+1}$

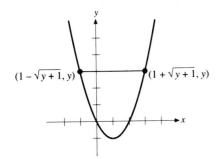

11. $L(x) = \begin{cases} -2x, & x < 0, \\ 2x, & x \geq 0 \end{cases}$

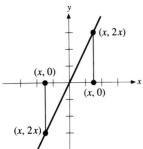

13. $L(x) = \begin{cases} 2 - x - x^2, & -2 \leq x \leq 1 \\ x^2 + x - 2, & x < -2, x > 1 \end{cases}$; $\left(x, \dfrac{2 + x - x^2}{2}\right)$

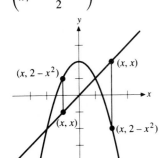

15. $L(y) = \begin{cases} (1 - y^2)/2, & -1 \leq y \leq 1 \\ (y^2 - 1)/2, & y < -1, y > 1 \end{cases}$; $\left(\dfrac{1 + 3y^2}{4}, y\right)$

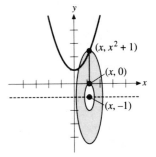

17. $L(x) = x - x^2, 0 \leq x \leq 1$; $\left(x, \dfrac{x + x^2}{2}\right)$

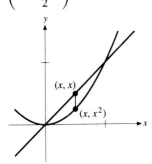

19. $L(y) = y^{1/3} - y^{1/2}$, $0 \leq y \leq 1$; $\left(\dfrac{y^{1/3} + y^{1/2}}{2}, y\right)$

21. $A(x) = \pi x^2/9, 0 \leq x \leq 3$

23. $A(x) = \left(2 - \dfrac{x}{2}\right)^2, 0 \leq x \leq 4$

25. $A(x) = x^2/8, 0 \leq x \leq 4$ **27.** $A(y) = y^2/4, 0 \leq y \leq 4$

29. $A(y) = 3(3 - y), 0 \leq y \leq 3$

31. (a) $A(x) = \pi x^4$,   (b) $C(x) = 2\pi x^2$

33. (a) $A(x) = \pi(1 - \sin x)^2$,   (b) $C(x) = 2\pi(1 - \sin x)$

35. (a) $A(y) = \pi y^4, y \geq 0$,   (b) $C(y) = 2\pi y^2, y \geq 0$

37. (a) $A(y) = \pi(y^2 + 1)^2$,   (b) $C(y) = 2\pi(y^2 + 1)$

39. (a) $A(x) = \pi x^2$,   (b) $C(x) = \begin{cases} -2\pi x, & x < 0 \\ 2\pi x, & x \geq 0 \end{cases}$

41. (a) $A(y) = \pi y^2, y \geq 0$,   (b) $C(y) = 2\pi y, y \geq 0$

43. $A(x) = \pi(x^2 + 1)^2$ **45.** $A(x) = \pi(x^2 + 2)^2 - \pi(1)^2$

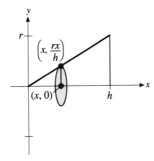

47. $A(x) = \pi\left(\dfrac{rx}{h}\right)^2$, $0 \leq x \leq h$

49. $S(x) = 2\pi xh\left(1 - \dfrac{x}{r}\right)$, $0 \leq x \leq r$

51. $S(y) = 2\pi yh\left(1 - \dfrac{y}{r}\right)$, $0 \leq y \leq r$

53. $S(y) = 4\pi y\sqrt{y}$

55. $A(y) = \pi(4 - y^2)^2 - \pi(2 - y)^2,$
$-1 \le y \le 2$

EXERCISES 6.2

1. $\int_0^3 2x\,dx = 9$

3. $\int_0^2 \left(5 - \frac{5}{2}x\right) dx = 5$

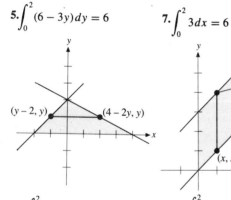

5. $\int_0^2 (6 - 3y)\,dy = 6$

7. $\int_0^2 3\,dx = 6$

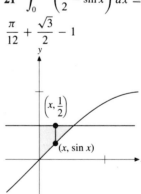

9. $\int_0^2 (5 - 2y)\,dy = 6$

11. $\int_0^2 x^2\,dx = \frac{8}{3}$

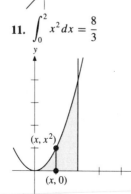

13. $\int_0^1 (x - x^2)\,dx = \frac{1}{6}$

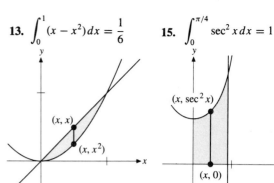

15. $\int_0^{\pi/4} \sec^2 x\,dx = 1$

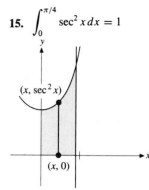

17. $\int_{-2}^1 (2 - y - y^2)\,dy = \frac{9}{2}$

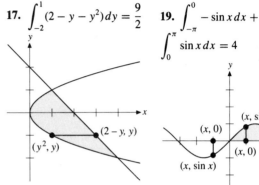

19. $\int_{-\pi}^0 -\sin x\,dx +$
$\int_0^\pi \sin x\,dx = 4$

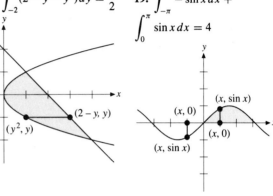

21 $\int_0^{\pi/6} \left(\frac{1}{2} - \sin x\right) dx =$
$\frac{\pi}{12} + \frac{\sqrt{3}}{2} - 1$

23. $\int_0^3 \left(x^2 - \frac{x^3}{3}\right) dx = \frac{9}{4}$

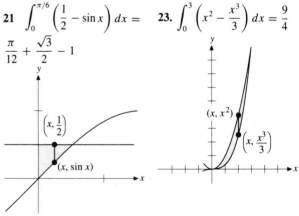

25. $\int_0^2 \left(\sqrt{4 - x^2} - 2 + x\right) dx$

27. $\int_{-1}^2 2\sqrt{4 - y^2}\,dy$

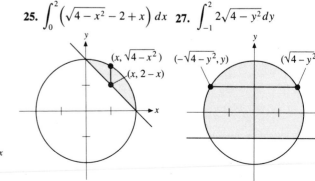

29. $\displaystyle\int_0^\pi (x - \sin x)\, dx$

31. $(1/2)^{2/3}$ **33.** $(1/2)^{3/4}$ **37.** $256/15$

39.

(a)

(b)

43. $\displaystyle\int_{-0.88947}^{3.63796} |3\cos x - (1 - x)|\, dx \approx 4.98489$

45. $\displaystyle\int_0^2 \frac{e^x + (2 - x) - |e^x - (2 - x)|}{2}\, dx \approx 1.76950$

47. $\displaystyle\int_0^2 \left(2 - \frac{e^{-x} + x + |e^{-x} - x|}{2}\right) dx \approx 1.72797$

EXERCISES 6.3

1. $\displaystyle\int_0^3 2\left(1 - \frac{x}{3}\right) dx = 3$ **3.** $\displaystyle\int_0^3 \pi\left(\frac{x}{3}\right)^2 dx = \pi$

5. $\displaystyle\int_0^h b^2\left(1 - \frac{y}{h}\right)^2 dy = \frac{b^2 h}{3}$ **7.** $\displaystyle\int_{-1}^1 \pi(1 + y^2)\, dy = \frac{8\pi}{3}$

9. (a) $\displaystyle\int_0^3 4\left(1 - \frac{x}{3}\right)^2 dx = 4$,   **(b)** $\displaystyle\int_0^3 \left(1 - \frac{x}{3}\right)^2 dx = 1$,

(c) $\displaystyle\int_0^3 2\left(1 - \frac{x}{3}\right)^2 dx = 2$ **11. (a)** $\displaystyle\int_0^4 \frac{\pi}{2}(4 - x)\, dx = 4\pi$,

(b) $\displaystyle\int_0^4 \sqrt{3}(4 - x)\, dx = 8\sqrt{3}$,   **(c)** $\displaystyle\int_0^4 2(4 - x)\, dx = 16$

13. $\displaystyle\int_0^{2r} \left(4\pi r^2\left(1 - \frac{y}{4r}\right)^2 - \pi r^2\right) dy = \frac{8\pi r^3}{3}$

15. $\displaystyle\int_{-2}^2 \frac{1}{2\sqrt{3}}(4 - x^2)\, dx = \frac{16}{3\sqrt{3}}$ in.^{3} **17.** $\displaystyle\int_0^\pi \pi \sin^2 x\, dx$

19. $\dfrac{16}{3} r^3 \csc\theta$ **21.** $\displaystyle\int_{-r/2}^{r/2} 2r\sqrt{r^2 - y^2}\, dy$ **23.** $4/5$

25. $5\pi/324$ **27.** 2π **31.** ≈ 4051 cm^3

EXERCISES 6.4

1. (a) $\displaystyle\int_0^1 \pi\left(x^2\right)^2 dx = \frac{\pi}{5}$,

(b) $\displaystyle\int_0^1 \left(\pi(1)^2 - \pi\left(1 - x^2\right)^2\right) dx = \frac{7\pi}{15}$,

(c) $\displaystyle\int_0^1 2\pi x\left(x^2\right) dx = \frac{\pi}{2}$,   **(d)** $\displaystyle\int_0^1 2\pi(1 - x)\left(x^2\right) dx = \frac{\pi}{6}$

3. (a) $\displaystyle\int_0^1 \left(\pi\,(x)^2 - \pi\left(x^2\right)^2\right) dx = \frac{2\pi}{15}$,

(b) $\displaystyle\int_0^1 \left(\pi\left(1 - x^2\right)^2 - \pi(1 - x)^2\right) dx = \frac{\pi}{5}$,

(c) $\displaystyle\int_0^1 2\pi x(x - x^2)\, dx = \frac{\pi}{6}$,

(d) $\displaystyle\int_0^1 2\pi(1 - x)(x - x^2)\, dx = \frac{\pi}{6}$

5. $\displaystyle\int_1^3 \pi\left(x + x^2\right)^2 dx = \frac{1456\pi}{15}$

7. $\displaystyle\int_0^1 2\pi x(1 - x^2)\, dx = \frac{\pi}{2}$

9. $\displaystyle\int_{-5}^0 2\pi(-x)(-5x - x^2)\, dx = \frac{625\pi}{6}$

11. (a) $\displaystyle\int_0^1 2\pi y(3 - 3y)\, dy = \pi$,

(b) $\displaystyle\int_0^1 \left(\pi(3 - 2y)^2 - \pi(y)^2\right) dy = 4\pi$

13. $\displaystyle\int_0^1 2\pi x(2 - x - x^2)\, dx = \frac{5\pi}{6}$

15. $\displaystyle\int_{-\sqrt{2}}^{\sqrt{2}} \pi(2 - y^2)^2\, dy = \frac{64\pi\sqrt{2}}{15}$

17. (a) $\displaystyle\int_1^2 \pi\left(\frac{1}{x}\right)^2 dx = \frac{\pi}{2}$,   **(b)** $\displaystyle\int_1^2 2\pi x\left(\frac{1}{x}\right) dx = 2\pi$

19. (a) $\displaystyle\int_0^1 \left(\pi\left(2 - x^2\right)^2 - \pi\,(2 - x)^2\right) dx = \frac{8\pi}{15}$,

(b) $\displaystyle\int_0^1 2\pi x(x - x^2)\, dx = \frac{\pi}{6}$

21. (a) $\displaystyle\int_1^8 2\pi y\left(y^{-1/3}\right) dy = \frac{186\pi}{5}$,

(b) $\displaystyle\int_1^8 \pi\left(y^{-1/3}\right)^2 dy = 3\pi$

23. $\displaystyle\int_0^{\pi/4} \left(\pi(\sec x)^2 - \pi(1)^2\right) dx = \pi - \frac{\pi^2}{4}$

25. $\displaystyle\int_0^{\sqrt{\pi}} 2\pi x(\sin x^2)\, dx = 2\pi$

27. $\displaystyle\int_0^r 2\pi x h\left(1 - \frac{x}{r}\right) dx = \int_0^h \pi r^2\left(1 - \frac{y}{h}\right)^2 dy = \frac{\pi r^2 h}{3}$

29. $\displaystyle\int_{r-h}^r \pi\left(r^2 - y^2\right) dy = \pi h^2\left(r - \frac{h}{3}\right)$

31. $\displaystyle\int_1^3 2\pi x\left(2\sqrt{1 - (x - 2)^2}\right) dx$

$\displaystyle = \int_{-1}^1 \left(\pi\left(2 + \sqrt{1 - y^2}\right)^2 - \pi\left(2 - \sqrt{1 - y^2}\right)^2\right) dy$

33. (a) $\int_0^\pi \pi(\sin x)^2\,dx$, **(b)** $\int_0^\pi 2\pi x(\sin x)\,dx$

35. (a) $\int_0^{\pi/2} \left(\pi(1)^2 - \pi(\cos x)^2\right)dx$,

(b) $\int_0^{\pi/2} 2\pi x(1 - \cos x)\,dx$

37. (a) $\int_0^{\pi/4} \pi(\tan x)^2\,dx$, **(b)** $\int_0^{\pi/4} 2\pi\left(\frac{\pi}{4} - x\right)(\tan x)\,dx$

39. $k/2$

EXERCISES 6.5

1. $(17^{3/2} - 5^{3/2})/6$ **3.** $(52^{3/2} - 125)/27$
5. $\pi(2^{3/2} - 1)/9$ **7.** 10π
9. (a) $2\sqrt{5}$, **(b)** $4\pi\sqrt{5}$ **11. (a)** $17/12$, **(b)** $47\pi/16$
13. (a) $33/8$, **(b)** $225\pi/16$ **15.** $\pi r\sqrt{r^2 + h^2}$
17. (a) $2\pi r^2 - 2\pi r\sqrt{r^2 - x_0^2}$, **(b)** $2\pi r^2$

19. $\int_{-a}^{a} \sqrt{1 + \left(\frac{2hx}{a^2}\right)^2}\,dx$

21. (a) $s = \sqrt{5}x$, **(b)** $\left(\frac{s}{\sqrt{5}}, \frac{2s}{\sqrt{5}} - 1\right)$

23. (a) $s = \frac{2}{3}(1 + x)^{3/2} - \frac{2}{3}$,

(b) $\left(\left(\frac{3}{2}s + 1\right)^{2/3} - 1, \frac{2}{3}\left(\left(\frac{3}{2}s + 1\right)^{2/3} - 1\right)^{3/2}\right)$

25. (a) $s = r\theta$, **(b)** $(r\cos(s/r), r\sin(s/r))$

27. $\int_1^{16} \sqrt{9y/4 + 1}\,dy$ **29. (a)** 9.7267, **(b)** $S_6 \approx 9.7505$
31. (a) 3.8066, **(b)** $S_6 \approx 3.8194$
33. (a) 0.7848, **(b)** $S_6 \approx 0.7855$ **35.** $S_4 \approx 3.8112$
37. $S_4 \approx 3.6071$

EXERCISES 6.6

1. 10,800 ft-lb **3. (a)** 249.6 ft-lb, **(b)** 1248 ft-lb
5. (a) 5324.8 ft-lb, **(b)** 832 ft-lb
7. (a) 196.04 ft-lb, **(b)** 183.78 ft-lb
9. (a) 40 ft-lb, **(b)** 240 ft-lb
11. (a) 24 ft-lb, **(b)** 56 ft-lb
13. (a) 1.5 ft-lb, **(b)** 6 ft-lb **15.** 1.44 ft
17. 1.69×10^9 J **19.** 725 ft-lb **21.** 1200 ft-lb
23. 1412 in.-lb

EXERCISES 6.7

1. 4/7 **3.** 1/5 **5.** $(0, -1/2)$ **7.** $(1, -2)$
9. $(11/7, 13/7)$ **11.** $(25/18, 19/18)$
13. $(0, (3\pi + 4)/(\pi + 4))$ **15.** $(0, 12/5)$
17. $(0, 5/6)$ **19.** $(1/2, 8/5)$ **21.** $(4/5, 0)$
23. $(4/5, 1/4)$ **27.** $(2/3, 1/2)$ **29.** $(5/6, 5/16)$
31. $(5/12, 5/7)$ **33.** $(1/5, 3/5)$
35. On axis, 1/4 the length of the height from the base

37. $(0, 3\sqrt{3}r/2\pi)$ **39.** $(0, 2h/3)$
41. (a) $\left(\frac{n+1}{n+2}, \frac{n+1}{4n+2}\right)$, **(b)** $\left(1, \frac{1}{4}\right)$ **45.** 18π

EXERCISES 6.8

1. (a) 1200π lb, **(b)** 1200π lb
3. (a) $\int_0^1 \rho g(1 - y)(y)\,dy$, **(b)** $\int_0^1 \rho g(2 - y)(y)\,dy$
5. (a) $\int_{-1}^1 \rho g(2 - y)(2\sqrt{1 - y^2})\,dy$,

(b) $\int_0^2 \rho g(3 - y)(2\sqrt{1 - (y - 1)^2})\,dy$

7. $\int_0^{\sqrt{3}} \rho g(\sqrt{3} - y)\left(\frac{2y}{\sqrt{3}}\right)dy$

9. $\int_{-2}^{-1} \rho g(-1 - y)(2\sqrt{4 - y^2})\,dy$ **11.** $45\rho g/2$
13. $81\pi \rho g/8$ **15.** $40\sqrt{2}\rho g$ **17.** $37\rho g$
19. $105\rho g/2$ **21.** $5\sqrt{3}\rho g$ **23.** $9\sqrt{5}\rho g/4$
25. (a) $-d\rho_f g b^2$, **(b)** $(d + h)\rho_f g b^2$, **(c)** $-\rho_b g h b^2$,
(d) $hb^2 g(\rho_f - \rho_b)$; downward **29.** 1147 lb

REVIEW EXERCISES

1. $L(x) = 2 + x - x^2$, $-1 \le x \le 2$;
$\left(x, \dfrac{6 - x - x^2}{2}\right)$

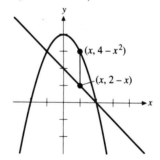

3. $A(x) = \pi(4 - x)^2/8, 0 \le x \le 4$

5. (a) $A(x) = \pi x^2$, **(b)** $C(x) = \begin{cases} -2\pi x, & x < 0 \\ 2\pi x, & x \ge 0 \end{cases}$

7. (a) $A(x) = \pi(\sin x)^2$, **(b)** $A(x) = 2\pi(x)(\sin x)$,
$0 \le x \le \pi$ $\qquad\qquad 0 \le x \le \pi$

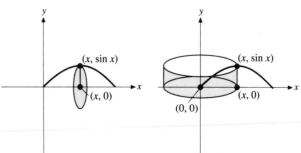

(c) $A(x) = \pi(\sin x + 1)^2 - \pi(1)^2,$
$0 \le x \le \pi$

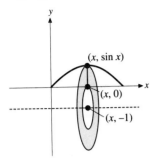

CHAPTER SEVEN:
SELECTED TRANSCENDENTAL FUNCTIONS

EXERCISES 7.1

1.

3.

9. $\displaystyle\int_0^2 (8x - x^4)\,dx$ **11.** $\displaystyle\int_0^3 (3y - y^2)\,dy$

13. $\displaystyle\int_0^1 \left(2x - \frac{x}{2}\right) dx + \int_1^2 \left((3 - x) - \frac{x}{2}\right) dx$

15. $\displaystyle\int_{\pi/4}^{5\pi/4} (\sin x - \cos x)\,dx$ **17.** $\displaystyle\int_0^3 \frac{1}{2}\left(\frac{2x}{3}\right)^2 dx$

19. $\displaystyle\int_0^4 \left(\left(\frac{x}{2}\right)^2 - \pi\left(\frac{x}{4}\right)^2\right) dx$ **21.** $\displaystyle\int_{-1}^1 4(1 - x^2)\,dx$

23. (a) $\displaystyle\int_0^1 \left(\pi(x)^2 - \pi(x^2)^2\right) dx$,   **(b)** $\displaystyle\int_0^1 2\pi x(x - x^2)\,dx$

25. (a) $\displaystyle\int_0^1 \pi\left(x^3\right)^2 dx$,   **(b)** $\displaystyle\int_0^1 2\pi x \cdot x^3\,dx$,

(c) $\displaystyle\int_0^1 \left(\pi(1)^2 - \pi\left(1 - x^3\right)^2\right) dx$ **27.** $1/\sqrt{2}$

29. (a) $\displaystyle\int_0^\pi \sqrt{1 + \cos^2 x}\,dx$,   **(b)** $\displaystyle\int_0^\pi 2\pi \sin x \sqrt{1 + \cos^2 x}\,dx$,

(c) $\displaystyle\int_0^\pi 2\pi x \sqrt{1 + \cos^2 x}\,dx$

31. (a) $\displaystyle\int_1^2 \frac{\sqrt{x^6 + 4}}{x^3}\,dx$,   **(b)** $\displaystyle\int_1^2 2\pi \left(\frac{1}{x^2}\right) \frac{\sqrt{x^6 + 4}}{x^3}\,dx$,

(c) $\displaystyle\int_1^2 2\pi x \frac{\sqrt{x^6 + 4}}{x^3}\,dx$,   **(d)** $\displaystyle\int_1^2 2\pi(x - 1)\frac{\sqrt{x^6 + 4}}{x^3}\,dx$

33. (a) $\displaystyle\int_1^4 \frac{\sqrt{x^2 + 1}}{x}\,dx$,   **(b)** $\displaystyle\int_1^4 2\pi x \frac{\sqrt{x^2 + 1}}{x}\,dx$,

(c) $\displaystyle\int_1^4 2\pi(4 - x)\frac{\sqrt{x^2 + 1}}{x}\,dx$

35. 15/2 ft-lb   **37.** 218,934 J   **39.** 4200 ft-lb
41. 613 ft-lb   **43.** (19/15, 14/15)
45. $(0, 4/3\pi)$ **47. (a)** 1/6,   **(b)** (3/4, 1/2)
49. $\left(0, \dfrac{h(R^2 + 2Rr + 3r^2)}{4(R^2 + Rr + r^2)}\right)$ **51.** 2   **53.** (4/9, 0)
57. $256\rho g/15$ **59.** $28\pi \rho g$ **61.** 2

63. (a) $W = \displaystyle\int_a^b \rho g(h - y)A(y)\,dy$,   **(b)** $M = \displaystyle\int_a^b \rho A(y)\,dy$,

(c) $M_{y=0} = \displaystyle\int_a^b y\rho A(y)\,dy$ **67.** $2\pi\sqrt{L/g}$
69. (a) $2mR^2$,   **(b)** $2\pi\sqrt{2R/g}$

5. $f^{-1}(x) = \dfrac{x - 1}{2}$ **7.** $f^{-1}(x) = \dfrac{x^3 + 8}{8}$

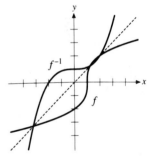

9. f^{-1} does not exist.   **11.** $f^{-1}(x) = \sqrt{4 - x^2}$, $0 \le x \le 2$

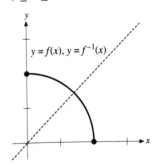

$y = f(x), y = f^{-1}(x)$

13. $f^{-1}(x) = x^2 - 1, x \geq 0$ **15.** $f^{-1}(x) = \dfrac{x+1}{x}$

17. $\pi/6$ **19.** $\pi/4$ **21.** 0.01000017 **23.** 1.24904577
25. $1/\sqrt{2}$ **27.** $\pi/4$ **29.** 0.11398876 **31.** 0.12634013
33. 1.23095942, 5.0522259 **35.** 1.44830700, 3.01910332

37. $\dfrac{x}{\sqrt{1-x^2}}$ **39.** $\dfrac{\sqrt{4-x^2}}{2}$ **41.** $\dfrac{\sqrt{x^2-9}}{|x|}$

43.
(a)

(b) $-\infty < x < \infty$, **(c)** $0 < y < \pi$
45. $(-\infty, 1], [1, \infty)$ **47.** $(-\infty, -1], [-1, 1], [1, \infty)$
49. (a) $(-\infty, 0]$, **(b)** $f^{-1}(x) = -\sqrt{x+1}$
51. (a) $(\pi/2, 3\pi/2)$, **(b)** $f^{-1}(x) = \tan^{-1} x + \pi$
53. (a) $[\pi/2, 3\pi/2]$, **(b)** $f^{-1}(x) = \pi - \sin^{-1} x$
55. $\theta(x) = \tan^{-1}(25/x) - \tan^{-1}(9/x), x > 0$

EXERCISES 7.2

1. $\dfrac{1}{\sqrt{9-x^2}}$ **3.** $\dfrac{-1}{\sqrt{4-x^2}}$ **5.** $\dfrac{-1}{x^2+1}$

7. $\dfrac{2x-4}{|x^2-4x|\sqrt{(x^2-4x)^2-1}}$ **9.** $\dfrac{1}{\sqrt{a^2-x^2}}$

11. $\dfrac{-1}{\sqrt{a^2-x^2}}$ **13.** $\sin^{-1}\dfrac{x}{3}+C$ **15.** $\dfrac{1}{6}\tan^{-1}\dfrac{2x}{3}+C$

17. $\sec^{-1}|2x|+C$ **19.** $\dfrac{1}{2}\sqrt{x^4-1}+C$

21. $\dfrac{1}{2}\left(\sin^{-1}x\right)^2+C$ **23.** $\dfrac{1}{2}\sec^{-1}(x^2)+C$

25. $\sin^{-1}\dfrac{x}{\sqrt{3}}+C$ **27.** $\pi/6$ **29.** $\pi/12$

31. $2(\pi+1)x - 4y = \pi + 2$ **33.** 0.02 rad/s

35. $\pi/3$ **37. (a)** $4\pi^2/3$, **(b)** 32π
39. $f^{-1}(2) = 1, \left(f^{-1}\right)'(2) = 1/3$
41. $f^{-1}(5/2) = 3, \left(f^{-1}\right)'(5/2) = 3/2$ **43.** $\pi/2$ **45.** $1/2$
47. $4\sqrt{10} \approx 12.6$ ft **51.** $2\pi rh$

EXERCISES 7.3

1. $\dfrac{2}{2x+1}$ **3.** $\sec x$ **5.** $-2/x$ **7.** $\dfrac{2x^2+1}{x(x^2+1)}$

9. $\dfrac{-13}{(2x+1)(3x-5)}$ **11.** $\dfrac{3-x^2}{x(x^2+3)}$ **13.** $\dfrac{2\ln x}{x}$

15. $x^2 + 3x^2 \ln x$ **17.** $\tan x$ **19.** $4x \ln x$

21. $\dfrac{6x^2-x-3}{3(x-1)^{2/3}(x+1)^{1/3}}$ **23.** $\dfrac{2x^2+3x-3}{3(x-1)^{2/3}(x+1)^{5/3}}$

25. $y = 3x - e^2$ **27.** 0 **29.** 0 **31.** ∞
33. Loc min $\left(e^{-1/2}, -e^{-1}/2\right)$; **35.** Loc max $\left(e, e^{-1}\right)$; infl pt
infl pt $\left(e^{-3/2}, -3e^{-3}/2\right)$ $\left(e^{3/2}, 3e^{-3/2}/2\right)$

37. Loc min $(1, 1)$; infl pt
$(2, 0.5 + \ln 2)$

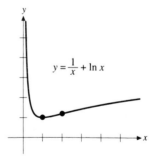

39. (a) 1.3877, **(b)** 0.025

EXERCISES 7.4

1. $\dfrac{1}{2}\ln|2x+1|+C$ **3.** $\dfrac{1}{2}\ln(x^2+1)+C$

5. $x + 3\ln|x-3|+C$ **7.** $-\ln(1+\cos x)+C$
9. $\ln|\sec x + \tan x|+C$ **11.** $\ln|\ln x|+C$

13. $x - \tan^{-1}x+C$ **15.** $\dfrac{3}{2}\ln\left(1+x^{2/3}\right)+C$

17. $\tan^{-1}x+C$ **19.** $\sin^{-1}\dfrac{x}{2}+C$

21. $\dfrac{1}{2}\ln(x^2+4)-\dfrac{1}{2}\tan^{-1}\dfrac{x}{2}+C$ **23.** $\ln 3$ **25.** $\ln(5/3)$

27. $\ln(4/3)$ **29.** (a) 2π,   (b) $4\pi-4\pi\ln 2$ **31.** 3

EXERCISES 7.5

1. $-3x^2 e^{-x^3}$ **3.** $2e^{2x-1}$ **5.** $e^{\tan x}\sec^2 x$ **7.** $e^{-x}(1-x)$

9. $\dfrac{-x-1}{x^2 e^x}$ **11.** $\dfrac{e^x}{1+e^x}$ **13.** $x^{\sqrt{x}}\dfrac{2+\ln x}{2\sqrt{x}}$ **15.** $10^x\ln 10$

17. $-\dfrac{10^{-\sqrt{x}}\ln 10}{2\sqrt{x}}$ **19.** $\dfrac{1}{x\ln 10}$ **21.** $-e^{-x}+C$

23. $2e^{\sqrt{x}}+C$ **25.** $\dfrac{2}{3}(e^x+1)^{3/2}+C$ **27.** $\tan^{-1}e^x+C$

29. $\dfrac{1}{2}\ln(1+e^{2x})+C$ **31.** $-\dfrac{2^{-x}}{\ln 2}+C$ **33.** $-\dfrac{6^{-x^2}}{2\ln 6}+C$

35. $2e-2$ **37.** $1/\ln 2$ **39.** 0   **41.** 0   **43.** 2   **45.** 1

47. 0   **49.** 0   **51.** 1   **53.** e^{-1}

55. Loc min $\left(-1,-e^{-1}\right)$; infl   **57.** Infl pt $(0,0)$
pt $\left(-2,-2e^{-2}\right)$

59. Loc min $(1,e)$

61. $1-e^{-1}$ **63.** -1.69897000
65. 0.43067656 **67.** $e^{\ln 3-5\ln 2}$ **69.** $2^{-t/4}$

71. $10^{-3t}\log_{10}e$ **73.** $1+x+\dfrac{x^2}{2!}+\dfrac{x^3}{3!}+\dfrac{x^4}{4!}$

77. (b) 0,   (c) 1/2,   (d)

79. e; 2.5937425, 2.70481383, 2.71692393, 2.71814593

EXERCISES 7.6

1. $y=2e^{-t}$ **3.** $y=0.8e^{-0.1t}$ **5.** $y=e^{-2}e^{2t}$
7. $y=e^5 e^{-4t}$ **9.** 800   **11.** 18.8 yr   **13.** 13.5 days
15. 7601 yr   **17.** (a) \$2694,   (b) \$2698,   (c) \$2700
19. (a) 11.4%,   (b) 11.3%   **21.** 25 yr   **23.** \$4066
25. 9.51 A
27. (a) $i=i_0 e^{-Rt/L}$,
(b)

(c) L/R
29. $83.3°$ F   **31.** 1 hour, 50 minutes   **33.** $\displaystyle\lim_{t\to\infty}T(t)=K$

EXERCISES 7.7

11. $2x\cosh(x^2)$ **13.** $2\cosh x\sinh x$ **15.** $\tanh x$

17. $\dfrac{2}{\sqrt{1+4x^2}}$ **19.** $\sec x$ **21.** $\dfrac{1}{3}\cosh 3x+C$

23. $\dfrac{1}{2}\tanh(x^2)+C$ **25.** $\ln\cosh x+C$

27. $\dfrac{1}{2}\sinh^{-1}\left(\dfrac{2x}{3}\right)+C$ **29.** $\cosh^{-1}e^x+C$

31. $\dfrac{1}{6}\ln 5\approx 0.26823965$ **33.** $\ln(1+\sqrt{2})\approx 0.88137359$

35. $\dfrac{1}{6}\ln 2\approx 0.11552453$ **45.** $x+\dfrac{x^3}{3!}+\dfrac{x^5}{5!}$

47. $2a\sinh 1=(e-e^{-1})a$

REVIEW EXERCISES

1. $f(x)=\sin x$,
$\dfrac{\pi}{2}\le x\le\dfrac{3\pi}{2}$
$f^{-1}(x)=\pi-\sin^{-1}(x)$

3. $f(x)=2-e^{-x}$ $f^{-1}(x)=-\ln(2-y)$

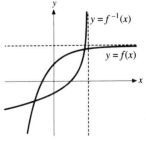

5. $f^{-1}(x) = \dfrac{x+2}{3}$ **7.** $f^{-1}(x) = \dfrac{1}{2}\tan^{-1}(x+1)$ **9.** $-\dfrac{\pi}{3}$

11. $\dfrac{\sqrt{4-x^2}}{2}$ **13.** $\dfrac{2x}{\sqrt{1-x^4}}$ **15.** $\dfrac{-1}{\sqrt{1-x^2}}$ **17.** $\dfrac{2}{4+x^2}$

19. $\dfrac{2}{x} - \dfrac{2x}{x^2+1}$ **21.** $\tan x$ **23.** $\dfrac{-\sin x \cos x}{2 - \sin^2 x}$

25. $4xe^{2x}$ **27.** $(\ln 3)3^{x^{1/3}-1}x^{-2/3}$ **29.** $\dfrac{2x}{\ln 10(x^2+1)}$

31. $(\ln x)^{x-1} + (\ln x)^x \ln(\ln x)$ **33.** $\dfrac{1}{2}\ln(9+x^2)+C$

35. $\dfrac{1}{3}\tan^{-1}\left(\dfrac{x}{3}\right)+C$ **37.** $\sin^{-1}\left(\dfrac{x}{3}\right)+C$

39. $-2e^{-x/2}+C$ **41.** $\dfrac{\ln 10}{2}\left(\log_{10}x\right)^2+C$ **43.** $e^{\tan x}+C$

45. $\dfrac{1}{3}\left(\sin^{-1}x\right)^3+C$ **47.** $\dfrac{1}{2}\tan^{-1}\left(\dfrac{x-1}{2}\right)+C$

49. $\dfrac{1}{2}\sinh(2x)+C$ **51.** $\dfrac{\pi}{8}$ **53.** $\dfrac{1}{2}\ln 1.25$

55. $\displaystyle\int_0^{\pi/4} 4\sec^3\theta\, d\theta$ **57.** 8   **59.** 0   **61.** 0   **63.** $\pi/2$

65. e^{-2} **67.** e^3

69. Loc max $(2, 2e^{-1})$; infl pt   **71.** Loc min $\left(e^{-1}, -e^{-1}\right)$
$(4, 4e^{-2})$

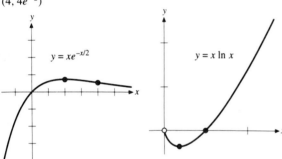

73. 1.39 hr   **75.** $i = 8e^{-t/1.2}$ **77.** $\dfrac{\pi e^2}{2} + \dfrac{\pi}{2}$

CHAPTER EIGHT:
TECHNIQUES OF INTEGRATION

EXERCISES 8.1

1. $\sqrt{x^2+4}+C$ **3.** $\dfrac{1}{3}\sin^{-1}3x+C$

5. $-\ln(2-\sin x)+C$ **7.** $-\dfrac{1}{2}e^{-x^2}+C$

9. $\ln|\ln x|+C$ **11.** $-\dfrac{1}{7}\cos(7x-3)+C$

13. $\dfrac{3}{2}\sec x^{2/3}+C$ **15.** $\dfrac{1}{3}e^{3x-1}+C$

17. $-\dfrac{1}{3}\dfrac{1}{3x-1}+C$ **19.** $\dfrac{1}{3\sqrt{2}}\tan^{-1}\dfrac{3x}{\sqrt{2}}+C$

21. $-\dfrac{1}{2}\cot 2x+C$ **23.** $\dfrac{1}{5}\sin^5 x+C$

25. $\dfrac{1}{6}\tan^6 x+C$ **27.** $\dfrac{1}{5}\sec^5 x+C$ **29.** $\sin^{-1}e^x+C$

31. $\sin^{-1}\dfrac{x-2}{2}+C$ **33.** $\dfrac{1}{2}\tan^{-1}\dfrac{x+1}{2}+C$

35. $\dfrac{1}{2}\ln(x^2-2x+5) + \dfrac{1}{2}\tan^{-1}\dfrac{x-1}{2}+C$

37. $\dfrac{1}{2}e^2 - \dfrac{1}{2}e^{-2}$ **39.** $\pi/6$ **41.** $2-\sqrt{3}$ **43.** 1/4

45. $\ln 2 + \pi/4$

EXERCISES 8.2

1. $\dfrac{1}{2}xe^{2x} - \dfrac{1}{4}e^{2x}+C$ **3.** $2x\sin\dfrac{x}{2} + 4\cos\dfrac{x}{2}+C$

5. $\dfrac{2x}{\sqrt{1-x^2}} - 2\sin^{-1}x+C$

7. $-2\sqrt{x}\cos\sqrt{x} + 2\sin\sqrt{x}+C$

9. $-e^{-x}\left(x^3+3x^2+6x+6\right)+C$

11. $\dfrac{2}{3}x^{3/2}\ln x - \dfrac{4}{9}x^{3/2}+C$

13. $x\tan^{-1}x - \dfrac{1}{2}\ln(1+x^2)+C$

15. $\dfrac{1}{2}x^2\tan^{-1}x - \dfrac{1}{2}x + \dfrac{1}{2}\tan^{-1}x+C$

17. $2\sqrt{x}e^{\sqrt{x}} - 2e^{\sqrt{x}}+C$ **19.** 1   **21.** 1

23. $\pi(e-2)$ **25.** $\dfrac{e^{2x}}{5}(\sin x + 2\cos x)+C$

27. $\dfrac{x}{2}\left(\sin(\ln x) - \cos(\ln x)\right)+C$

31. $\dfrac{4e^{-x}}{5}\left(-\sin\dfrac{x}{2} - \dfrac{1}{2}\cos\dfrac{x}{2}\right)+C$

33. $\dfrac{e^{-3x}}{25}(4\sin 4x - 3\cos 4x)+C$

35. $x(\ln x)^2 - 2x\ln x + 2x+C$ **37.** $\dfrac{1}{4}x^4\ln x - \dfrac{1}{16}x^4+C$

EXERCISES 8.3

1. $-\dfrac{1}{3}\cos^3 x+C$ **3.** $\dfrac{1}{2}\tan^2 x+C$ **5.** $\dfrac{1}{5}\sec^5 x+C$

7. $-\dfrac{1}{4}\sin 2x\cos 2x + \dfrac{1}{2}x+C$ **9.** $\sin x - \dfrac{1}{3}\sin^3 x+C$

11. $\dfrac{1}{4}\sec 2x\tan 2x + \dfrac{1}{4}\ln|\sec 2x + \tan 2x|+C$

13. $\dfrac{1}{9}\tan^9 x + \dfrac{1}{7}\tan^7 x+C$

15. $\dfrac{1}{3}\sin^3 x - \dfrac{1}{5}\sin^5 x+C$ **17.** $\cos x + \sec x+C$

19. $-\dfrac{1}{6}\sin x\cos^5 x + \dfrac{1}{24}\sin x\cos^3 x$
$+\dfrac{1}{16}\sin x\cos x + \dfrac{1}{16}x+C$ **21.** $\dfrac{\sin x\cos x}{2} + \dfrac{x}{2}+C$

23. $\frac{1}{2}\tan^2 x - \ln|\sec x| + C$ **25.** $\frac{1}{5}\cos^5 x - \frac{1}{3}\cos^3 x + C$

27. $\frac{1}{2}\sin x - \frac{1}{10}\sin 5x + C$

29. $-\frac{1}{10}\cos 5x + \frac{1}{6}\cos 3x + C$ **31.** $\pi/2$

33. $3\sqrt{3}/16$ **35.** $1/2$ **41.** $2/\pi$ **43.** $\ln(\sqrt{2}+1)$

EXERCISES 8.4

1. $\sin^{-1}\frac{x}{2} + C$ **3.** $\frac{1}{2}\tan^{-1}\frac{x}{2} + C$

5. $\frac{(x^2-4)^{3/2}}{3} + C$ **7.** $\frac{x\sqrt{4-x^2}}{2} + 2\sin^{-1}\frac{x}{2} + C$

9. $\frac{x}{2\sqrt{2-x^2}} + C$ **11.** $\frac{x}{2(1+9x^2)} + \frac{1}{6}\tan^{-1}3x + C$

13. $\frac{1}{5}(1-x^2)^{5/2} - \frac{1}{3}(1-x^2)^{3/2} + C$

15. $\ln|x + \sqrt{x^2-3}| + C$ **17.** $\frac{\sqrt{x^2-4}}{4x} + C$

19. $\frac{1}{4}x^3\sqrt{9-x^2} - \frac{9}{8}x\sqrt{9-x^2} + \frac{81}{8}\sin^{-1}\frac{x}{3} + C$

21. $\frac{1}{2}\ln(4+x^2) + C$ **23.** $\ln|\sqrt{x^2+4x+5} + x + 2| + C$

25. $\frac{1}{3}\ln\left|\frac{3}{\sqrt{9-x^2}} + \frac{x}{\sqrt{9-x^2}}\right| + C$

27. $\frac{x}{18(x^2+9)} + \frac{1}{54}\tan^{-1}\frac{x}{3} + C$ **29.** $\frac{1}{2\sqrt{3}}$

31. $\frac{1}{16} + \frac{\pi}{32}$ **33.** $\frac{3}{4\sqrt{5}} - \frac{5}{4\sqrt{21}}$ **35.** $ab\pi$

37. $\pi\left(\sqrt{2} - \ln(1+\sqrt{2})\right)$ **39.** $\sqrt{17} + \frac{1}{4}\ln(4+\sqrt{17})$

EXERCISES 8.5

1. $\frac{1}{4}\ln|x-3| - \frac{1}{4}\ln|x+1| + C$

3. $2\ln|x| - \ln|x-3| + C$

5. $x^2 + 2\ln|x-2| - 3\ln|x+2| + C$

7. $2\ln|x-1| + \frac{1}{x-1} + C$

9. $\ln|x| + 2\ln|x-2| - 2\ln|x+2| + C$

11. $2\ln|x| + 3\ln|x-1| - \ln|x-2| + C$

13. $-\frac{1}{x-1} - \frac{1}{2}\frac{1}{(x-1)^2} + C$

15. $\frac{x^2}{2} - \frac{2}{x} + \ln|x-1| + C$

17. $\ln|x| - \frac{1}{x^2} - \ln|x+1| + C$

19. $\ln|x| - \ln|x+1| + \frac{1}{x+1} + C$

21. $x + \frac{\sqrt{3}}{2}\ln|x - \sqrt{3}| - \frac{\sqrt{3}}{2}\ln|x + \sqrt{3}| + C$

23. $2\ln|x| + \frac{1}{2}\ln(x^2+1) + C$

25. $2\ln|x-1| - 2\ln|x+1| + C$

27. $\tan^{-1}x - \frac{1}{3}\tan^{-1}\frac{x}{3} + C$ **29.** $x + \frac{7}{6}\ln|x-2| -$

$\frac{7}{12}\ln(x^2+2x+4) + \frac{1}{2\sqrt{3}}\tan^{-1}\frac{x+1}{\sqrt{3}} + C$

31. $\frac{1}{6}\tan^{-1}\frac{x}{3} - \frac{1}{2}\frac{x}{x^2+9} + C$

33. $\ln|x| - \frac{1}{2}\ln(x^2+2x+5) - \frac{1}{2}\tan^{-1}\frac{x+1}{2} + C$

35. $\left(1+\frac{1}{\sqrt{2}}\right)\ln|x+1+\sqrt{2}| +$

$\left(1-\frac{1}{\sqrt{2}}\right)\ln|x+1-\sqrt{2}| + C$

37. $-\frac{1}{2}\frac{1}{(2-x)^2} + \frac{2}{3}\frac{1}{(2-x)^3} + C$

39. $\ln\frac{9}{8}$ **41.** $-\frac{1}{2}\ln 3$ **43.** $\frac{1}{8}\ln\frac{45}{13}$

45. $-\frac{1}{k} + \frac{k+1}{2k}\ln\left(\frac{k+1}{2}\right) + \frac{k+1}{k}\ln M_0 + \frac{1}{kM_0^2} -$

$\frac{k+1}{2k}\ln\left(1 + \frac{k-1}{2}M_0^2\right)$

EXERCISES 8.6

1. $\frac{1}{6}(2x-1)^{3/2} + \frac{1}{2}(2x-1)^{1/2} + C$

3. $\frac{1}{15}(3x+2)^{5/3} - \frac{1}{3}(3x+2)^{2/3} + C$

5. $2\sqrt{x} - 2\ln(1+\sqrt{x}) + C$

7. $-x + 4\sqrt{x} - 4\ln(\sqrt{x}+1) + C$

9. $2x^{1/2} - 3x^{1/3} + 6x^{1/6} - 6\ln\left(x^{1/6}+1\right) + C$

11. $\frac{3}{2}x^{2/3} - \frac{3}{2}\ln\left(x^{2/3}+1\right) + C$

13. $-\frac{1}{2}\frac{1}{(2x+5)^{1/2}} + \frac{5}{6}\frac{1}{(2x+5)^{3/2}} + C$

15. $-\frac{1}{2}\frac{1}{(2-x)^2} + \frac{2}{3}\frac{1}{(2-x)^3} + C$

17. $e^x - \ln(e^x+1) + C$

19. $\frac{2}{5}(1+e^x)^{5/2} - \frac{2}{3}(1+e^x)^{3/2} + C$

21. $\frac{2}{1-\tan(x/2)} + C$ **23.** $\tan\frac{x}{2} + C$

25. $\frac{1}{5}\ln\left|2\tan\frac{x}{2}+1\right| - \frac{1}{5}\ln\left|\tan\frac{x}{2}-2\right| + C$ **27.** $2 - \pi/2$

29. $4 - 2\sqrt{2} - \ln 3 + \ln(\sqrt{2}+1) - \ln(\sqrt{2}-1)$

31. $\frac{1}{2}\tan^{-1}\frac{1}{2}$

EXERCISES 8.7

1. Div **3.** 2 **5.** Div **7.** Div **9.** Div **11.** 1/2
13. Div **15.** π **17.** Div **19.** 2 **21.** 0 **23.** π
25. Div **27.** 4 **29.** $\ln 2/2$ **31.** 1/4 **35.** $\pi/2$
37. $\displaystyle\int_0^8 \sqrt{x^{2/3}+4}\,x^{-1/3}\,dx = 8^{3/2}-8$ **41.** Div **43.** Conv
45. Conv **47.** Conv **49.** Div **51.** Div **53.** Div
55. (a) $1-e^{-1}$, (b) e^{-1}, (c) 100, (d) 100

REVIEW EXERCISES

1. $-\dfrac{1}{2}x\cos 2x + \dfrac{1}{4}\sin 2x + C$ **3.** $x\ln x - x + C$

5. $-e^{-x}(x^2+2x+2)+C$ **7.** $\dfrac{1}{7}\sin^7 x - \dfrac{1}{9}\sin^9 x + C$

9. $\dfrac{1}{4}\cos^3 x \sin x + \dfrac{3}{8}\sin x \cos x + \dfrac{3}{8}x + C$

11. $\dfrac{1}{11}\tan^{11} x + \dfrac{1}{9}\tan^9 x + C$ **13.** $\tan x - x + C$

15. $-\dfrac{1}{3}(1-x^2)^{3/2}+C$ **17.** $\ln\left(\sqrt{9+x^2}+x\right)+C$

19. $\dfrac{1}{2}\dfrac{x}{x^2+1}+\dfrac{1}{2}\tan^{-1}x+C$

21. $\dfrac{1}{2}x^2 + \dfrac{1}{2}\ln|x-1| + \dfrac{1}{2}\ln|x+1| + C$

23. $\ln|x-2| - \ln|x-1| + C$

25. $\dfrac{1}{2}\ln(x^2-4x+8) + \tan^{-1}\dfrac{x-2}{2} + C$

27. $-\ln|x| - \dfrac{1}{x} + \ln|x+1| + C$

29. $\dfrac{2}{5}(x+1)^{5/2} - \dfrac{2}{3}(x+1)^{3/2} + C$

31. $\dfrac{1}{2}\tan\dfrac{x}{2} - \dfrac{1}{6}\tan^3\dfrac{x}{2} + C$ **33.** $2\sqrt{x} - 2\ln(1+\sqrt{x})+C$

35. $\tan^{-1}e^x + C$ **37.** $\dfrac{4}{3}(1+\sqrt{x})^{3/2} - 4(1+\sqrt{x})^{1/2} + C$

39. $\dfrac{3\sqrt{3}}{4} + \dfrac{3}{8}\ln(2+\sqrt{3})$ **41.** $\dfrac{1}{2}+\dfrac{\pi}{4}$ **43.** $\sqrt{3}+\dfrac{2\pi}{3}$

45. $\pi/2$ **47.** Div **49.** $2/e$ **51.** 1/9
53. Div **55.** Conv **57.** Div **59.** $\pi^2/4$

61. $4\pi/3$ **63.** $\dfrac{\sqrt{5}}{2}+\dfrac{1}{4}\ln(2+\sqrt{5})$ **65.** $\dfrac{\mu_0 I}{2\pi a}$

67. $1 - \alpha\lambda e^{-\lambda\alpha} - e^{-\lambda\alpha}$

CHAPTER NINE:
INFINITE SERIES

EXERCISES 9.1

1. 3/2 **3.** 0 **5.** 4/3 **7.** 1/2 **9.** Div **11.** 0 **13.** Div
15. Div **17.** 0 **19.** 0 **21.** 0 **23.** 0 **25.** 0 **27.** 1

29. 0 **31.** 0 **33.** 1 **35.** 0 **37.** 1 **39.** e^{-2}
43. Defined for $-2 < x < 2$;
continuous for $-2 < x < 0$
and $0 < x < 2$

45. (a) $P_n = 2nr\sin(\pi/n)$, (b) $2\pi r$
47. Need $\lim\limits_{n\to\infty} a_n \neq 0$ and $\{b_n\}_{n\geq 1}$ divergent **51.** 2
53. $n > 1/\epsilon - 1$ **55.** $n \geq 8$

EXERCISES 9.2

1. 0.5, 0.625, 0.66666667, 0.68229167; 0.69314718
3. 0.8, 0.97066667, 1.03620267, 1.06616198; 1.09861229
5. 1, 3, 5, 6.33333333; 7.389056099 **7.** 0.5,
0.458333333, 0.464583333, 0.463467261; 0.463647609
9. 0.3, 0.2955, 0.29552025, 0.295520206; 0.295520206
11. $10(1-(0.2)^n)$; 10 **13.** $-11(1-(-1.1)^n)/21$; div
15. $6(1-(1/2)^n)$; 6 **17.** $\dfrac{1}{2}-\dfrac{n+1}{n+2}$; $-\dfrac{1}{2}$
19. $\dfrac{1}{2}-\dfrac{n+1}{2^{n+1}}$; $\dfrac{1}{2}$ **21.** $1+\dfrac{1}{2}-\dfrac{1}{n+1}-\dfrac{1}{n+2}$; $\dfrac{3}{2}$
23. $\dfrac{1}{2}\left(\dfrac{1}{2}+\dfrac{1}{3}-\dfrac{1}{n+2}-\dfrac{1}{n+3}\right)$; $\dfrac{5}{12}$
25. $-\ln(n+1)$; div **27.** 1/2 **29.** 2 **31.** 2 **33.** 1/3
35. 5/9 **37.** 30 ft **43.** $x+1, x > -1/2$ **47.** $n \geq 16$

EXERCISES 9.3

1. Div **3.** Div **5.** Conv **7.** Div **9.** Div **11.** Div
13. Conv **15.** Conv **17.** Div **19.** Conv **21.** Conv
23. Conv **25.** Conv **27.** Conv **29.** Div **31.** Div
33. Conv **35.** Conv

EXERCISES 9.4

1. Div **3.** Div **5.** Conv **7.** Div **9.** Div
11. Conv **13.** Div **15.** Conv **17.** Conv **19.** $p > 1$
21. $S_4 \approx 1.079$ **23.** $S_2 \approx 0.4045$ **25.** $N \geq 1000$
27. $\ln\dfrac{6N+3}{6N-2}$; 0.0827, 0.0083, 0.0008

EXERCISES 9.5

1. Conv **3.** Div **5.** Conv **7.** Conv **9.** Div **11.** Div
13. Div **15.** Conv **17.** Div **19.** Conv **21.** Div
23. Conv **25.** Conv **27.** Div **29.** Conv **31.** Conv
33. $-\infty < p < \infty$

EXERCISES 9.6

1. Conv   **3.** Div   **5.** Conv   **7.** Conv   **9.** Div
11. Conv   **13.** Conv   **15.** Conv   **17.** Div
19. Conv   **21.** $S_4 = -0.625$ **23.** $S_2 \approx 0.70465265$
25. Abs conv   **27.** Div   **29.** Abs conv   **31.** Abs conv
33. Cond conv   **35.** Abs conv   **37.** Abs conv
39. $N \geq 31$ **41.** Conv   **43.** Div   **45.** Conv
49. The series is divergent. S_n approaches infinity.

EXERCISES 9.7

1. $0 < x < 2$; 1   **3.** $0 \leq x < 2$; 1   **5.** $-2 < x < 2$; 2
7. $-1 < x < 1$; 1   **9.** $0 < x < 4$; 2   **11.** $0 \leq x \leq 6$; 3
13. $x = 0$; 0   **15.** $-\infty < x < \infty$; ∞
17. $0 < x < 4$; 2   **19.** $-4 < x < 4$; 4
21. $-1/2 \leq x \leq 1/2$; 1/2   **23.** $-1/2 < x < 1/2$; 1/2
25. Conv for $x = 2$; may either converge or diverge for $x = 4$

EXERCISES 9.8

1. (a) $1 + x^2 + x^4 + x^6 + \cdots, |x| < 1,$
(b) $x + x^3 + x^5 + x^7 + \cdots, |x| < 1,$
(c) $2x + 4x^3 + 6x^5 + 8x^7 + \cdots, |x| < 1$
3. (a) $\dfrac{1}{2} + \dfrac{1}{4}x + \dfrac{1}{8}x^2 + \dfrac{1}{16}x^3 + \cdots, |x| < 2,$
(b) $\dfrac{1}{4} + \dfrac{1}{4}x + \dfrac{3}{16}x^2 + \dfrac{1}{8}x^3 + \cdots, |x| < 2,$
(c) $\ln 2 - \dfrac{1}{2}x - \dfrac{1}{8}x^2 - \dfrac{1}{24}x^3 - \dfrac{1}{64}x^4 - \cdots, |x| < 2$
5. (a) $x - \dfrac{x^2}{2} + \dfrac{x^3}{3} - \dfrac{x^4}{4} + \cdots, |x| < 1,$
(b) $x^2 - \dfrac{x^3}{2} + \dfrac{x^4}{3} - \dfrac{x^5}{4} + \cdots, |x| < 1,$
(c) $2 - 2x + \dfrac{8}{3}x^2 - 4x^3 + \cdots, |x| < \dfrac{1}{2}$
7. $-2x - 2x^3 - 2x^5 - 2x^7 - \cdots, |x| < 1$
9. (a) $-\dfrac{1}{3} - \dfrac{2}{9}x - \dfrac{7}{27}x^2 - \dfrac{20}{81}x^3 - \cdots, |x| < 1,$
(b) $-\dfrac{1}{3}x^2 - \dfrac{2}{9}x^3 - \dfrac{7}{27}x^4 - \dfrac{20}{81}x^5 - \cdots, |x| < 1$
11. (a) $1 + 2x + 2x^2 + 2x^3 + \cdots, |x| < 1,$
(b) $-1 - x - 2x^2 - 2x^3 - \cdots, |x| < 1$
13. $x + x^3 + x^5 + x^7 + \cdots, |x| < 1$
15. $x + x^2 + \dfrac{2}{3}x^3 + \dfrac{2}{3}x^4 + \dfrac{13}{15}x^5 + \cdots, |x| < 1$
17. $-0.88235294, -1.11133727, -1.21830227,$
$-1.27778601; -1.386294361$
19. $0.92307692, 1.18525262,$
$1.31928800, 1.40086491; 1.60943791$
21. $0.05, 0.04995833, 0.04995840, 0.04995840; 0.04995840$
23. $0.4, 0.37866667, 0.38071467, 0.38048061; 0.38050638$
25. 0.01866667 (two terms)   **27.** 0.00520833 (one term)

29. 0.11197917 (three terms)   **31.** 0.01733333 (two terms)
33. 8   **35.** 0

EXERCISES 9.9

1. $3 + 6(x - 1) + 3(x - 1)^2 + (x - 1)^3$
3. $-8 + 12x - 6x^2 + x^3$ **5.** $x + \dfrac{x^3}{3!} + \dfrac{x^5}{5!} + \dfrac{x^7}{7!} + \cdots$
7. $2x - \dfrac{2^3}{3!}x^3 + \dfrac{2^5}{5!}x^5 - \dfrac{2^7}{7!}x^7 + \cdots$
9. $1 - x^2 + \dfrac{x^4}{2!} - \dfrac{x^6}{3!} + \cdots$
11. $1 - x^2 + \dfrac{1}{3}x^4 - \dfrac{2}{45}x^6 + \cdots$
13. $\dfrac{1}{2!} - \dfrac{x^2}{4!} + \dfrac{x^4}{6!} - \dfrac{x^6}{8!} + \cdots$ **15.** -160 **17.** 0
19. 0.125 (one term)   **21.** 0.60677083 (five terms)
23. 0.49305556 (two terms)   **25.** 0.19733333 (two terms)
27. $|x| < (0.006)^{1/3}$ **29.** $0 \leq x < (0.03)^{1/3}$
35. $1 - 2(x - 1) + 3(x - 1)^2 - 4(x - 1)^3 + \cdots,$
$|x - 1| < 1$

EXERCISES 9.10

1. $1 + \dfrac{1}{2}x - \dfrac{1}{8}x^2 + \dfrac{1}{16}x^3 - \dfrac{5}{128}x^4 + \cdots$
3. $1 + \dfrac{1}{3}x + \dfrac{2}{9}x^2 + \dfrac{14}{81}x^3 + \dfrac{35}{243}x^4 + \cdots$
5. $2 + \dfrac{1}{4}x^2 - \dfrac{1}{64}x^4 + \dfrac{1}{512}x^6 - \cdots$
7. $\dfrac{1}{8} - \dfrac{3}{16}x + \dfrac{3}{16}x^2 - \dfrac{5}{32}x^3 + \cdots$
9. $1 + x - \dfrac{1}{2}x^2 + \dfrac{1}{2}x^3 - \dfrac{5}{8}x^4 + \cdots$
11. $1 + \dfrac{1}{6}x^2 + \dfrac{3}{40}x^4 + \dfrac{5}{112}x^6 + \cdots$ **13.** 0.15056827
15. 0.04112464   **17.** 0.02006748

REVIEW EXERCISES

1. 3/2   **3.** 3   **5.** 0   **7.** 0
9. $1 + \dfrac{4}{3} - \dfrac{2n + 2}{n + 2} - \dfrac{2n + 4}{n + 3}; -\dfrac{5}{3}$
11. $1 + \dfrac{1}{2} - \dfrac{1}{n + 1} - \dfrac{1}{n + 2}; \dfrac{3}{2}$ **13.** $12(1 - (3/4)^n)$; 12
15. Div   **17.** Conv   **19.** Conv   **21.** Conv
23. Conv   **25.** Div   **27.** Conv   **29.** Div
31. Conv   **33.** Conv   **35.** Cond conv
37. Cond conv   **39.** Abs conv   **41.** Cond conv
43. $0 \leq x < 2$ **45.** $1 < x < 3$ **47.** $-2 < x < 2$
49. $n \geq 23$ **51.** $n \geq 31$ **53.** $1 - x + \dfrac{x^2}{2} - \cdots$

55. $-\dfrac{1}{2} - \dfrac{x}{4} - \dfrac{x^2}{8} - \cdots$ **57.** $x - x^3 + x^5 - \cdots$

59. $1 - x^2 + x^4 - \cdots$ **61.** $1 - \dfrac{3}{2}x + \dfrac{3}{8}x^2 - \cdots$

63. $\dfrac{1}{2^{3/2}} + \dfrac{3}{2^{7/2}}x + \dfrac{15}{2^{13/2}}x^2 + \cdots$

65. $x + \dfrac{1}{3}x^3 + \dfrac{2}{15}x^5 + \cdots$ **67.** $x + x^2 + \dfrac{x^3}{3} + \cdots$

69. $1 + x - \dfrac{x^2}{2} - \cdots$ **71.** 0.299757 (two terms)

73. 0.19955556 (two terms) **75.** $N \geq 0$

77. $|x| < \sqrt{0.01/e^{0.5}}$

CHAPTER TEN:
CONIC SECTIONS;
POLAR COORDINATES;
THREE-DIMENSIONAL SPACE

EXERCISES 10.1

1. $(0,0)$; $(0, 1/4)$; $y = -1/4$ **3.** $(0,1)$; $(0,0)$; $y = 2$

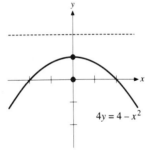

5. $(2,0)$; $(0,0)$; $x = 4$

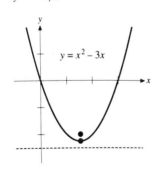

7. $(3/2, -9/4)$; $(3/2, -2)$; $y = -5/2$

9. $(-1,-1)$; $(-1/3, -1)$; $x = -5/3$

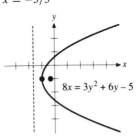

11. $(1,1)$; $(1, 4/3)$; $y = 2/3$

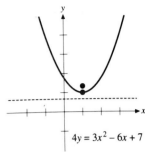

13. $y^2 = 4x$ **15.** $y^2 = 4(x+1)$

17. $(x-4)^2 = -8(y-2)$ **19.** $x^2 = 8(y+1)$

21. $(y+4)^2 = -\dfrac{8}{3}(x-6)$ **23.** $y = -x^2 + 2x + 3$

25. (a) $y = 2x - 5$, (b) $(0, -5)$, (c) $(0,0)$,

(d) $5, 5$, (e) $\alpha = \beta$ **27.** $\tan^{-1}(1.5) \approx 56°$

29. $y = a(x^2 - 4x), a \neq 0$

EXERCISES 10.2

1. $(0,0)$; $(0, \pm 5)$; $(0, \pm 4)$

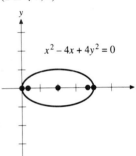

3. $(2,0)$; $(4,0)$; $(0,0)$; $(2 \pm \sqrt{3}, 0)$

5. $(0, 5/2)$; $(0,5)$; $(0,0)$; $(0, 9/2)$, $(0, 1/2)$

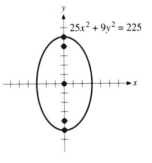

7. $(3/2, 1)$; $(3,1)$; $(0,1)$; $((3 \pm \sqrt{5})/2, 1)$

9. $(1, 1)$; $(1, 3)$, $(1, -1)$;
$(1, 1 \pm \sqrt{8/3})$

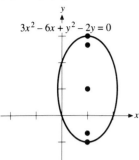

$3x^2 - 6x + y^2 - 2y = 0$

11. $\dfrac{x^2}{5^2} + \dfrac{y^2}{13^2} = 1$ **13.** $\dfrac{x^2}{5^2} + \dfrac{(y-3)^2}{3^2} = 1$

15. $\dfrac{(x-1)^2}{1^2} + \dfrac{y^2}{4^2} = 1$ **17.** $\dfrac{x^2}{4^2} + \dfrac{y^2}{(2/\sqrt{3})^2} = 1$

19. $\dfrac{(x+5)^2}{10^2} + \dfrac{y^2}{(5\sqrt{3})^2} = 1$, $\dfrac{(x-5/3)^2}{(10/3)^2} + \dfrac{y^2}{(5/\sqrt{3})^2} = 1$

21. $(x, 0)$, $-c \le x \le c$ **23.** $\dfrac{(x-1)^2}{2^2} + \dfrac{y^2}{(\sqrt{3})^2} = 1$

25. **(a)** $-16/15$,   **(b)** $(\pm 3, 0)$,   **(c)** $12/5$, $12/35$,
(d) $20/9$, $20/9$,   **(e)** $\alpha = \beta$

27. $(0, 0)$; $(\pm 12, 0)$; $(\pm 13, 0)$;   **29.** $(0, 0)$; $(0, \pm 3)$; $(0, \pm 5)$;
$y = \pm 5x/12$ $y = \pm 3x/4$

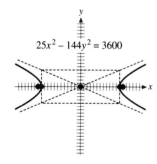

$25x^2 - 144y^2 = 3600$

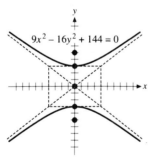

$9x^2 - 16y^2 + 144 = 0$

31. $(1, 0)$; $(2, 0)$, $(0, 0)$;   **33.** $(-1, 2)$; $(-3/2, 2)$,
$(1 \pm \sqrt{2}, 0)$; $y = \pm(x-1)$ $(-1/2, 2)$; $(-1 \pm \sqrt{5}/2, 2)$;
$y - 2 = \pm 2(x + 1)$

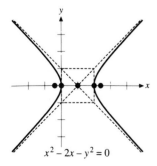

$x^2 - 2x - y^2 = 0$

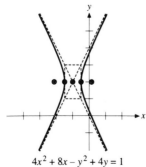

$4x^2 + 8x - y^2 + 4y = 1$

35. $(2, 1)$; $(4, 1)$, $(0, 1)$;
$(9/2, 1)$, $(-1/2, 1)$;
$x - 2 = \pm 4(y - 1)/3$

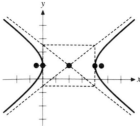

$9x^2 - 36x - 16y^2 + 32y = 16$

37. $-\dfrac{x^2}{5^2} + \dfrac{y^2}{12^2} = 1$ **39.** $\dfrac{x^2}{4^2} - \dfrac{(y-3)^2}{3^2} = 1$

41. $-\dfrac{(x-1)^2}{(3/4)^2} + \dfrac{y^2}{3^2} = 1$ **43.** $-\dfrac{x^2}{(2/\sqrt{5})^2} + \dfrac{y^2}{4^2} = 1$

45. $\dfrac{(x-8/3)^2}{(2/3)^2} - \dfrac{y^2}{(\sqrt{20/3})^2} = 1$;

$\dfrac{(x-8/5)^2}{(2/5)^2} - \dfrac{y^2}{(2\sqrt{15}/5)^2} = 1$

47. $(x, 0)$, $x \le -c$, $x \ge c$

49. $\dfrac{(x+4)^2}{2^2} - \dfrac{y^2}{(\sqrt{12})^2} = 1$

51. **(a)** $20/9$,   **(b)** $(\pm 5, 0)$,   **(c)** $12/35$, $-12/5$,
(d) $16/15$, $16/15$,   **(e)** $\alpha = \beta$

55. $8\pi + \dfrac{2\pi\sqrt{3}}{3} \ln \dfrac{2+\sqrt{3}}{2-\sqrt{3}}$ **57.** $4\pi/3$

EXERCISES 10.3

1. $x' = y'^2$

$x = y^2 - 1$

3. $y' = -x'^2$

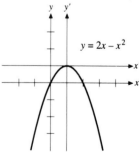

$y = 2x - x^2$

5. $x = x' + 2$, $y = y' + 1$ **7.** $x = x'$, $y = y' + 1$

$x^2 - 4x + 4y^2 - 8y + 7 = 0$

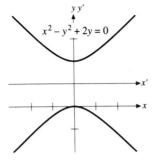

$x^2 - y^2 + 2y = 0$

9. $y' = \sqrt{2}$ **11.** $x' = 2/\sqrt{5}$

$x - y + 2 = 0$

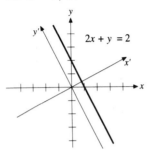

$2x + y = 2$

13. $-\pi/4$; $y' = 1$ **15.** $\pi/6$; $y' = \sqrt{3}/2$

$x + y = \sqrt{2}$

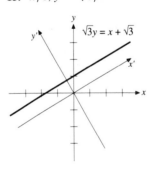

$\sqrt{3}y = x + \sqrt{3}$

17. $\pi/4$;

$$-\frac{x'^2}{(\sqrt{2})^2} + \frac{y'^2}{(\sqrt{2})^2} = 1$$

$xy + 1 = 0$

19. $\pi/4$; $x' = y'^2$

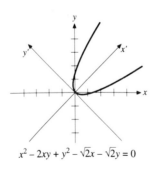

$x^2 - 2xy + y^2 - \sqrt{2}x - \sqrt{2}y = 0$

21. $\pi/6$; $\dfrac{x'^2}{2^2} + \dfrac{y'^2}{1^2} = 1$

$7x^2 - 6\sqrt{3}xy + 13y^2 = 16$

23. $\tan^{-1}(-a/b)$; $y' = \dfrac{b}{|b|}\dfrac{c}{\sqrt{a^2 + b^2}}$

25. $(1, 1)$, $(-1, -1)$; $(\sqrt{2}, \sqrt{2})$, $(-\sqrt{2}, -\sqrt{2})$

27. $(\sqrt{3}/2, -1/2)$; $(0, 0)$ **29.** Parabola   **31.** Ellipse

33. Hyperbola   **35.** Ellipse   **37.** Parabola

EXERCISES 10.4

1.

$B(2, \pi/2)$,
$D(-2, 3\pi/2)$

$E(1, 1)$

$A(1, 0)$

$C(-\sqrt{2}, \pi/4)$

3. $A(x, y) = (1, 0)$, $B(x, y) = (1, \sqrt{3})$, $C(x, y) = (0, 3)$,
$D(x, y) = (-\sqrt{2}, \sqrt{2})$ **5.** $A(r, \theta) = (1, 0)$, $(-1, \pi)$;
$B(r, \theta) = (2, \pi/2)$, $(-2, -\pi/2)$; $C(r, \theta) = (2, 2\pi/3)$,
$(-2, -\pi/3)$; $D(r, \theta) = (\sqrt{8}, -\pi/4)$, $(-\sqrt{8}, 3\pi/4)$

7. $x^2 + y^2 = 4$ **9.** $x = 1$ **11.** $x^2 + y^2 = 4x$

13. $x^2 = 4(y + 1)$ **15.** $r = 3$ **17.** $r \sin \theta = 3$

19. $r = -6 \cos \theta$ **21.** $r = \dfrac{2}{1 - \cos \theta}$

23. Circle   **25.** Line

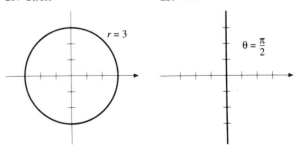

$r = 3$

$\theta = \dfrac{\pi}{2}$

27. Line

$$r = \frac{2}{\cos\theta + 2\sin\theta}$$

29. Line

$r = \csc\theta$

EXERCISES 10.5

1.

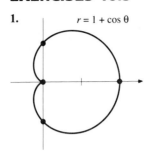

$r = 1 + \cos\theta$

3.

$r = 2 - 2\sin\theta$

31. Circle

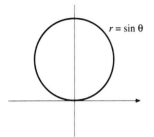

$r = \sin\theta$

33. Circle

$r = -3\cos\theta$

5. $r = 1 - 2\cos\theta$

7.

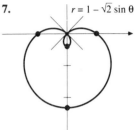

$r = 1 - \sqrt{2}\,\sin\theta$

35. Parabola

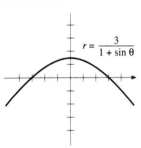

$$r = \frac{3}{1 + \sin\theta}$$

37. Ellipse

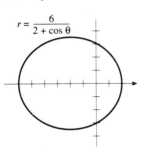

$$r = \frac{6}{2 + \cos\theta}$$

9.

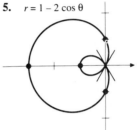

$r = 2\cos 2\theta$

11.

$r = -\cos 3\theta$

13.

$r = \sin 4\theta$

15.

$r^2 = 4\cos 2\theta$

39. Hyperbola

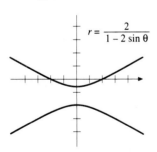

$$r = \frac{2}{1 - 2\sin\theta}$$

41. $d = \sqrt{r_0^2 + r_1^2 - 2r_0 r_1 \cos(\theta_1 - \theta_0)}$

43. $r = \dfrac{d}{1 - \cos\theta}$

45. $r = \dfrac{d}{2 - \cos\theta}$

17.

$r^2 = -9\sin 2\theta$

19.

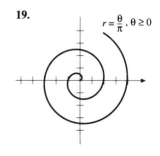

$r = \dfrac{\theta}{\pi},\ \theta \geq 0$

21.

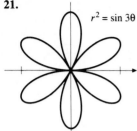

$r^2 = \sin 3\theta$

23.

$r = \sin \theta/2$

35.

$$r = \frac{4}{\cos\theta + \sqrt{3}\sin\theta}$$

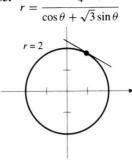

$r = 2$

37. $r = -\csc\theta$

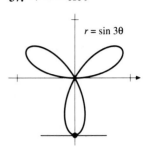

$r = \sin 3\theta$

25.

$r^2 = \theta$

27.

$r = \sin(4\theta)$, $3\pi/4 \le \theta \le \pi$

$r = \sin(4\theta/3)$, $0 \le \theta \le 3\pi/4$

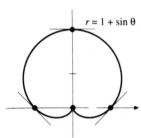

$r = \sin(4\theta)$, $7\pi/4 \le \theta \le 2\pi$

$r = \sin(4\theta/3 - 4\pi/3)$, $\pi \le \theta \le 7\pi/4$

39. $(2, 0)$, $(1/2, 2\pi/3)$, $(1/2, 4\pi/3)$

$r = 1 + \cos\theta$

41. $(3, 0)$, $(1, \pi)$

$r = 2 + \cos\theta$

29. $\pi/4$, $\pi/2$, $-\pi/4$, 0, $\pi/4$

$r = 1 + \sin\theta$

31. $\tan^{-1}(-1/2)$, $\tan^{-1} 2$, $\tan^{-1}(-1/2)$, $\tan^{-1} 2$, $\tan^{-1}(-1/2)$

$$r = \frac{1}{\cos\theta + 2\sin\theta}$$

43. $(3/2, \pi/3)$, $(0, \pi)$, $(3/2, 5\pi/3)$

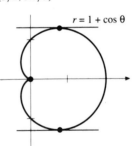

$r = 1 + \cos\theta$

45. $(2, \pi/2)$, $(6, 3\pi/2)$

$$r = \frac{6}{2 + \sin\theta}$$

33. $-\pi/4$, $\pi/2$, $\pi/4$, $*$, $-\pi/4$

$$r = \frac{2}{1 + \sin\theta}$$

47.

$r = a - b\cos\theta$, $0 < b < a < 2b$

$r = a - b\cos\theta$, $a \ge 2b > 0$

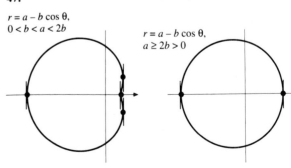

49. No, since $r(\theta) = 0$ for
$-\pi/4 < \theta < \pi/4$

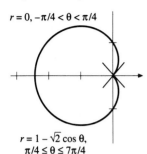

$r = 0, -\pi/4 < \theta < \pi/4$

$r = 1 - \sqrt{2}\cos\theta,$
$\pi/4 \le \theta \le 7\pi/4$

EXERCISES 10.6

1. a^2 **3.** $2\sqrt{3}$ **5.** $\pi^3/6$ **7.** 6π **9.** 2   **11.** $\pi/2$
13. $\pi - 3\sqrt{3}/2$ **15.** $4\pi/3 - \sqrt{3}$ **17.** $\pi^3/2$
19. $\sqrt{3} + 2\pi/3$ **21.** $3\sqrt{3} + \pi$ **23.** $8\pi/3 + \sqrt{3}$
25. $\pi/6 + \sqrt{3}/4$ **27.** $(4a/3\pi, \pi/2)$ **29.** $(5/6, 0)$

EXERCISES 10.7

1. B

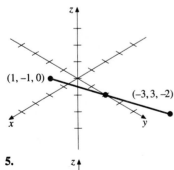

3. 6; $(-1, 1, -1)$

5.

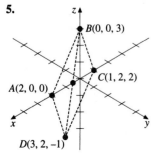

7. 3   **9.** $(x-1)^2 + (y-2)^2 + (z-1)^2 = (\sqrt{6})^2$
11. $(x-4)^2 + (y-3)^2 + (z-12)^2 = 13^2$
13. $(x-1)^2 + (y+2)^2 + (z-1)^2 = (\sqrt{5})^2$
15. $(1, -2, 0)$; $\sqrt{5}$ **17.** $\sqrt{13}$ **19.** 3

21.

23.

25.

27.

29.

31.

33.

35.

37.

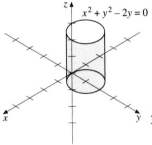

$2x - 2y + z = 4$

39.

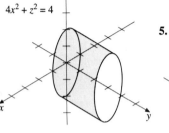

$y = 2x$

EXERCISES 10.8

1.

$x^2 + z^2 = y$

3.

$4x^2 + 4y^2 + z^2 = 4$

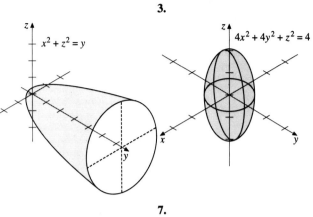

41.

$x^2 + y^2 - 2y = 0$

43.

$4x^2 + z^2 = 4$

5.

$x^2 - 9y^2 - 9z^2 = 0$

7.

$z = \sqrt{3x^2 + 3y^2}$

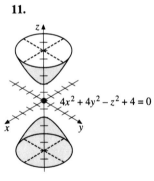

45.

$z = 4 - y^2$

47.

$y = \sin x$

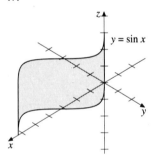

9.

$z = 4 - 2\sqrt{x^2 + y^2}$

11.

$4x^2 + 4y^2 - z^2 + 4 = 0$

49.

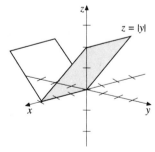

$z = |y|$

13.

$9x^2 - y^2 + 9z^2 = 9$

15.

$8z = x^2 - 8y^2$

17.

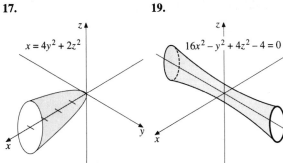

$x = 4y^2 + 2z^2$

19.

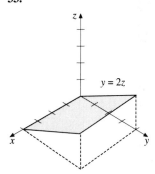

$16x^2 - y^2 + 4z^2 - 4 = 0$

33.

$y = 2z$

21.

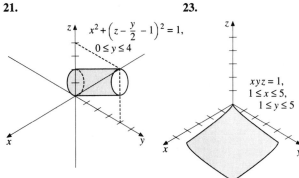

$x^2 + \left(z - \dfrac{y}{2} - 1\right)^2 = 1,$

$0 \le y \le 4$

23.

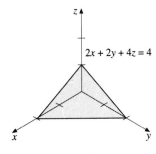

$xyz = 1,$
$1 \le x \le 5,$
$1 \le y \le 5$

REVIEW EXERCISES

1. $y^2 = 8x$ **3.** $\dfrac{(x-4)^2}{5^2} + \dfrac{y^2}{3^2} = 1$

5. $\dfrac{(x-4)^2}{4^2} - \dfrac{y^2}{3^2} = 1$

7.

$x^2 = 16 - 4y^2$

9.

$y^2 = 2 + x$

25.

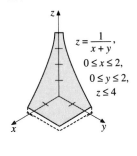

$z = \dfrac{1}{x+y},$

$0 \le x \le 2,$

$0 \le y \le 2,$

$z \le 4$

27.

$2x + 2y + 4z = 4$

11.

$\dfrac{y^2}{4} - \dfrac{x^2}{1} = 1$

13. $\pi/3;\ y' = \sqrt{3}/2$

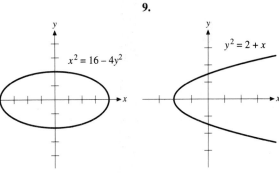

$y = \sqrt{3}\,(x + 1)$

29.

$2x + z = 2$

31.

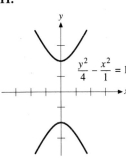

$z = x + 2y$

15. $\pi/4;\ \dfrac{(x' - 1/\sqrt{2})^2}{(\sqrt{2})^2} - \dfrac{(y' + 1/\sqrt{2})^2}{(\sqrt{2})^2} = 1$

$xy = y + 1$

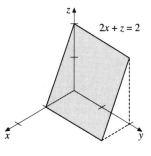

17. $(1, 1), (-1, -1)$

19. $x = x', y = y' + 1$

21. $x = x' - 2, y = y' + 1$

$x^2 + 4y^2 - 8y = 0$

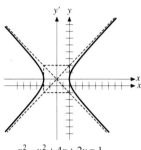

$x^2 - y^2 + 4x + 2y = 1$

23. Hyperbola   **25.** Parabola

27.

$r = 2 \cos \theta$

29.

$r = \dfrac{2}{\cos \theta + \sin \theta}$

31.

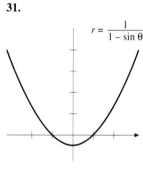

$r = \dfrac{1}{1 - \sin \theta}$

33.

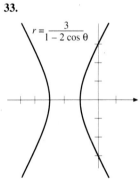

$r = \dfrac{3}{1 - 2 \cos \theta}$

35.

$r = \sin 2\theta$

37.

$r^2 = \sin 2\theta$

39.

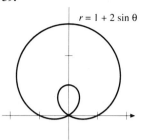

$r = 1 + 2 \sin \theta$

41.

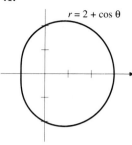

$r = 2 + \cos \theta$

43. $(2, 0)$,
$(5/2, \cos^{-1}(5/6)), (8, \pi)$,
$(5/2, 2\pi - \cos^{-1}(5/6))$

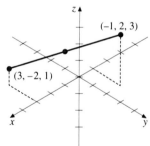

$r = 5 - 3 \cos \theta$

45. $(2, \pi/2), (4, 3\pi/2)$

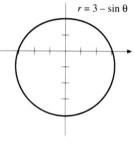

$r = 3 - \sin \theta$

47. $\tan^{-1}(2/3)$ **49.** $\pi/6$ **51.** $\displaystyle\int_0^{2\pi} \frac{1}{2}(2 - 2\cos\theta)^2 \, d\theta$

53. $\displaystyle\int_{2\pi/3}^{4\pi/3} \left(\frac{1}{2}(2 - 2\cos\theta)^2 - \frac{1}{2}(3)^2 \right) d\theta$

55. $6; (1, 0, 2)$

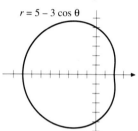

$(-1, 2, 3)$

$(3, -2, 1)$

57. $(x - 2)^2 + y^2 + (z - 1)^2 = 3^2$ **59.** $(-2, 3, 0); \sqrt{13}$

61.

$x = 2,\ 2z + 3y = 6$

63.

$y = -2,\ z = x^2 - y^2$

65.

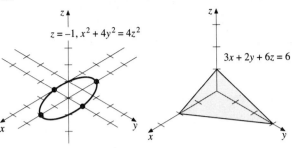

$$z = -1, \; x^2 + 4y^2 = 4z^2$$

67.

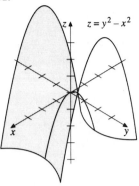

$$3x + 2y + 6z = 6$$

69.

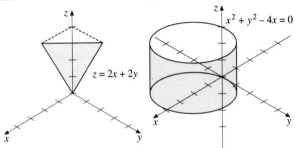

$$z = 2x + 2y$$

71.

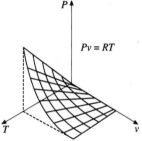

$$x^2 + y^2 - 4x = 0$$

73.

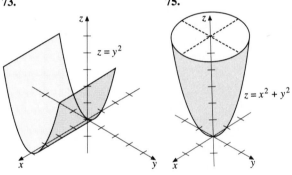

$$z = y^2$$

75.

$$z = x^2 + y^2$$

77.

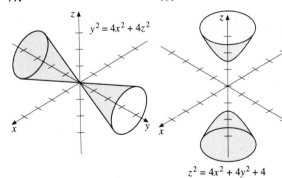

$$y^2 = 4x^2 + 4z^2$$

79.

$$z^2 = 4x^2 + 4y^2 + 4$$

81.

$$z = y^2 - x^2$$

83. $\pi r^2 h / 2$

85. Constant-temperature curves are hyperbolas. Constant-volume curves are lines.

$$Pv = RT$$

87. $4px = y^2 + z^2$, a paraboloid

89. $a^2 x^2 + a^2 y^2 + (a^2 - c^2) z^2 = a^4 - a^2 c^2$, a hyperboloid of two sheets

CHAPTER ELEVEN:
VECTORS AND PARAMETRIC CURVES

EXERCISES 11.1

1.

$$C : x = \cos t, \; y = \sin t,$$
$$0 \le t \le 2\pi$$

3.

$$C : x = 2\cos^2 t, \; y = \sin 2t,$$
$$-\pi/2 \le t \le \pi/2$$

5.

$$C: x = \frac{2\cos t}{\cos t + \sin t}, \, y = \frac{2\sin t}{\cos t + \sin t},$$
$$-\pi/6 \leq t \leq 2\pi/3$$

7.

$$C: x = t, \, y = t^2,$$
$$-2 \leq t \leq 2$$

17. Counterclockwise along $x^2 + y^2 = 1$ from $(1, 0)$ to $(-1, 0)$

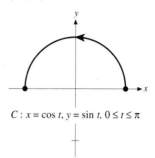

$$C: x = \cos t, \, y = \sin t, \, 0 \leq t \leq \pi$$

19. From $(0, -1)$, clockwise once around $x^2 + y^2 = 1$, excluding $(0, -1)$

$$C: x = \frac{2t}{1 + t^2}, \, y = \frac{1 - t^2}{1 + t^2},$$
$$-\infty < t < \infty$$

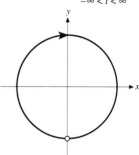

9. Along the line segment from $(-3, -1)$ to $(3, 3)$

$$C: x = -3 + 6t, \, y = -1 + 4t,$$
$$0 \leq t \leq 1$$

11. Along the parabola $y = 2x - x^2$ from $(0, 0)$ to $(2, 0)$

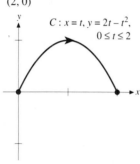

$$C: x = t, \, y = 2t - t^2,$$
$$0 \leq t \leq 2$$

21. $C: x = t, \, y = t^3, \, -1 \leq t \leq 2$
23. $C: x = t^2, \, y = t, \, -1 \leq t \leq 2$
25. $C: x = 3t, \, y = 2t, \, 0 \leq t \leq 1$
27. $C: x = 2 - 4t, \, y = 1 - t, \, 0 \leq t \leq 1$
29. $C: x = 2\cos t, \, y = 2\sin t, \, 0 \leq t \leq 2\pi$
31. $C: x = 2\cos t, \, y = 3\sin t, \, 0 \leq t \leq 2\pi$
33. $C: x = 2\sec t, \, y = \tan t, \, -\pi/2 < t < \pi/2$
35. $C: x = 3\tan t, \, y = 2\sec t, \, -\pi/2 < t < \pi/2$
37. $-C: x = t, \, y = t^2, \, -2 \leq t \leq 2$
39. $-C: x = \cos t, \, y = \sin t, \, 0 \leq t \leq 2\pi$
41. $C: x = (1 - \cos t)\cos t, \, y = (1 - \cos t)\sin t, \, 0 \leq t \leq 2\pi$
43. $C: x = (t\cos t)/2, \, y = (t\sin t)/2, \, 0 \leq t \leq 2\pi$
45. $C: x = t - d\sin(t/r), \, y = r - d\cos(t/r)$ **47.** $C:$
$x = (\cos t)/\sqrt{2} - \sin t, \, y = (\cos t)/\sqrt{2} + \sin t, \, 0 \leq t \leq 2\pi$
49. (a) $x = -12 + 12\cos\theta - (36 - 12\theta)\sin\theta,$
$y = 12\sin\theta + (36 - 12\theta)\cos\theta, \, 0 \leq \theta \leq 3,$
(b) $\displaystyle\int_0^3 \frac{1}{2}(36 - 12\theta)^2\,d\theta,$ **(c)** $648\pi + 1296 \text{ ft}^2$
51. $C: x = 2r_0(1 - \cos t)\cos t, \, y = 2r_0(1 - \cos t)\sin t$

13. Along $y = \cos x$ from $(0, 1)$ to $(2\pi, 1)$

$$C: x = t, \, y = \cos t,$$
$$0 \leq t \leq 2\pi$$

15. Along $x = \sin y$ from $(-1, -\pi/2)$ to $(1, \pi/2)$

$$C: x = \sin t, \, y = t,$$
$$-\pi/2 \leq t \leq \pi/2$$

EXERCISES 11.2

1. $2y = 3x - 1$

$$C: x = t^2, \, y = t^3, \, -1 \leq t \leq 2$$
$$2y = 3x - 1$$
$$(1, 1)$$

3. $6\sqrt{3}\,y = 12x - 6 + \pi\sqrt{3}$

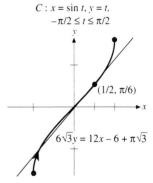

$$C: x = \sin t, \, y = t,$$
$$-\pi/2 \leq t \leq \pi/2$$
$$(1/2, \, \pi/6)$$
$$6\sqrt{3}\,y = 12x - 6 + \pi\sqrt{3}$$

5. 3/2; 3/4

$C: x = t^2, y = t^3$

7. 4/3; −2/27

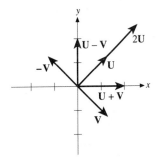

$C: x = t^2 + t, y = t^2 + 2t - 1$

9. 2; 4

$C: x = \ln t, y = t^2$

11. 5   **13.** $\left(40^{3/2} - 8\right)/27$ **15.** 2π

17. $8\left(\left(\pi^2 + 1\right)^{3/2} - 1\right)/3$ **19.** 15/4   **21.** 4π

23. $\sqrt{5}\,(e^\pi - 1)$ **25.** $8\left(\left(\pi^2 + 1\right)^{3/2} - 1\right)/3$

27. 15π **29.** $\pi\left(17^{3/2} - 1\right)/6$ **31.** 16π

33. $12\pi^2$ **35.** $8\pi(2\sqrt{2} - 1)/3$ **37.** 4π

39. $4\pi^2$ **41.** $\pi\sqrt{5}\left(e^{-\pi/2} + e^{\pi/2}\right)/2$

43. $C: x = 4s/5,\ y = 3s/5,\ 0 \le s \le 5$

45. $C: x = 3\cos(s/3),\ y = 3\sin(s/3),\ 0 \le s \le 6\pi$

47. $C: x = \dfrac{2}{3}\left(\left(\dfrac{3}{2}s + 1\right)^{2/3} - 1\right)^{3/2},$

$y = \left(\dfrac{3}{2}s + 1\right)^{2/3} - 1,\ 0 \le s \le \dfrac{2}{3}\left(10^{3/2} - 1\right)$

49. $C: x = \left(\dfrac{s}{\sqrt{2}} + 1\right)\cos\left(\ln\left(\dfrac{s}{\sqrt{2}} + 1\right)\right),$

$y = \left(\dfrac{s}{\sqrt{2}} + 1\right)\sin\left(\ln\left(\dfrac{s}{\sqrt{2}} + 1\right)\right),$

$0 \le s \le \sqrt{2}\left(e^{2\pi} - 1\right)$ **51.** 15.07

53. (a) $\displaystyle\int_0^\pi \sqrt{(2 + \cos\theta)^2 + (-\sin\theta)^2}\,d\theta,$ **(b)** 6.68

EXERCISES 11.3

1. $-2\mathbf{i} + 3\mathbf{j}$ **3.** $-4\mathbf{i} + 13\mathbf{j}$ **5.** 13   **7.** $\sqrt{13}$ **9.** 8

11. 1

13. $2\mathbf{i}; 2\mathbf{j}; 2\mathbf{i} + 2\mathbf{j}; -\mathbf{i} + \mathbf{j}$

15. $3\mathbf{i} - \mathbf{j}; \mathbf{i} - 5\mathbf{j}; 4\mathbf{i} - 6\mathbf{j};$
$-\mathbf{i} - 2\mathbf{j}$

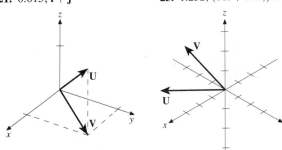

17. $\pi/2$ **19.** 0   **21.** π **23.** 0.1419

25. $\pm(12\mathbf{i} + 5\mathbf{j})/13$ **27.** $\pm\mathbf{j}$

29. $(6\mathbf{i} + 8\mathbf{j})/25;\ (6\mathbf{i} + 8\mathbf{j})/25;\ (44\mathbf{i} - 33\mathbf{j})/25$

31. $(\mathbf{i} - \mathbf{j})/2;\ (\mathbf{i} - \mathbf{j})/2;\ 5(\mathbf{i} + \mathbf{j})/2$

33. $\mathbf{i} + \mathbf{j}$ **35.** $-3\mathbf{i} - 3\mathbf{j}$

37. $(-\mathbf{i} + 2\mathbf{j})/5;\ \mathbf{F} = (-\mathbf{i} + 2\mathbf{j})/5 + (6\mathbf{i} + 3\mathbf{j})/5$

39. $720\sqrt{3}$ ft-lb   **41.** 10 lb, 30° from horizontal

43. $\approx 14°$; 39 s   **45.** $4\sqrt{2}$ lb

EXERCISES 11.4

1. $-6\mathbf{i} + 6\mathbf{j} - 2\mathbf{k}$ **3.** $2\mathbf{j} - 7\mathbf{k}$ **5.** $5\sqrt{2}$ **7.** 3   **9.** 4

11. 3   **13.** $3\mathbf{j} - 2\mathbf{k}$ **15.** $9\mathbf{i} - 8\mathbf{j} - \mathbf{k}$ **17.** $-\mathbf{j}$ **19.** -1

21. 0.615; $\mathbf{i} + \mathbf{j}$ **23.** 1.232; $(18\mathbf{i} + 24\mathbf{k})/25$

25. $4/\sqrt{41}, 0, -5/\sqrt{41}$ **27.** $3/\sqrt{22}, 2/\sqrt{22}, 3/\sqrt{22}$

29. $\pm(\mathbf{i} - \mathbf{j} + \mathbf{k})/\sqrt{3}$ **31.** $\pm(-3\mathbf{i} + 3\mathbf{j} + 4\mathbf{k})/\sqrt{34}$

33. $\sqrt{2}$ **35.** $2\sqrt{3}$ **37.** 2   **39.** 12   **41.** $\sqrt{45}/2$

43. $2\sqrt{34}$ **45.** $\mathbf{F} = 2(\mathbf{i} + 2\mathbf{j} + 2\mathbf{k})/3 + (-2\mathbf{i} - 4\mathbf{j} + 5\mathbf{k})/3$

47. $\mathbf{F} = (2\mathbf{i}) + (-\mathbf{j} + \mathbf{k})$ **49.** 0.578

EXERCISES 11.5

1. $4x + 3y - 2z = 15$ **3.** $x - 3y + 2z = 4$

5. $z = 3$ **7.** $z = x$ **9.** $14x - 5y - 3z = 11$

11. $-2x + 3y + z + 3 = 0$ **13.** $\pi/2$ **15.** 0.175

17. $6/\sqrt{17}$; $(12/17, 18/17, 12/17)$ **19.** $\sqrt{2}$; $(0, 2, 2)$

21. $4/\sqrt{14}$

23. $3x + y = 3$

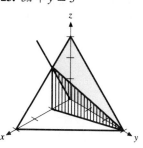

25. $x + 3y = 4$

27. $3x + 2y = 3$

29. $x^2 + y^2 + 3x + 3y = 4$

31. $x^2 + y^2 = 4$

33. $z = 1/2$ when $(x, y) = (1, 0)$; $z = 2$ when $(x, y) = (0, 1)$
35. $z = -1$ when $(x, y) = (0, -1)$; $z = 3$ when $(x, y) = (0, 1)$

EXERCISES 11.6

1. $\dfrac{x-1}{2} = \dfrac{y+1}{-2} = \dfrac{z-2}{3}$; $x = 1 + 2t$, $y = -1 - 2t$, $z = 2 + 3t$ **3.** $\dfrac{x-2}{2} = \dfrac{y-3}{-1} = \dfrac{z+1}{3}$; $x = 2 + 2t$, $y = 3 - t$, $z = -1 + 3t$
5. $y = 4$, $z = -3$; $x = 2 + t$, $y = 4$, $z = -3$
7. $\dfrac{x-1}{2} = \dfrac{y+2}{-3} = \dfrac{z-3}{1}$; $x = 1 + 2t$, $y = -2 - 3t$, $z = 3 + t$ **9.** $\dfrac{x-2}{1} = \dfrac{y-2}{1} = \dfrac{z-1}{-2}$; $x = 2 + t$, $y = 2 + t$, $z = 1 - 2t$ **11.** $\dfrac{x-5}{3} = \dfrac{y-3}{-1} = \dfrac{z-7}{4}$; $x = 5 + 3t$, $y = 3 - t$, $z = 7 + 4t$

13. $\dfrac{x-0}{1} = \dfrac{y-2}{-1}$, $z = 2$; $x = t$, $y = 2 - t$, $z = 2$
15. $(-7, -7, 3)$ **17.** $(-1, 1, 1)$
19. $(0, 1, 2)$; 1.110242335 **21.** Lines do not intersect.
23. $\sqrt{117/29}$ **25.** $\sqrt{8/3}$ **27.** $4x - 3y + 2z = 5$
29. $14/\sqrt{50}$ **31.** No such plane   **33.** $7x + 6y - 4z = 7$
35. $5x - 3y - 4z = 6$ **37.** 0.6094 **39.** Paths intersect at $(1, 2, 0)$; particles collide at $(1, 2, 0)$ when $t = 1$.
41. Paths intersect at $(0, 1, 0)$; particles do not collide.
43. 8

EXERCISES 11.7

1. $4\mathbf{i} - 6\mathbf{j}$ **3.** $6\mathbf{i} + 2\mathbf{j}$ **5.** $2t\mathbf{i} - 3\mathbf{j}$
7. $\mathbf{i} + e^{-t}(\cos t - \sin t)\mathbf{j} + e^{-t}(-\sin t - \cos t)\mathbf{k}$
9. $\sec^2 t\mathbf{i} + \sec t \tan t\mathbf{j}$ **11.** $t^2\mathbf{i} + 2t^3\mathbf{j} + \mathbf{C}$
13. $-\cos t\mathbf{i} + \sin t\mathbf{j} + \mathbf{C}$ **15.** $\dfrac{1}{2}\tan(2t)\mathbf{i} + \dfrac{1}{2}\sec(2t)\mathbf{j} + \mathbf{C}$
17. $\dfrac{1}{2}\ln(t^2 + 4)\mathbf{i} + \dfrac{1}{2}\tan^{-1}\dfrac{t}{2}\mathbf{j} + \mathbf{C}$
19. $\mathbf{i} + 2t\mathbf{j}$; $2\mathbf{j}$; $\sqrt{1 + 4t^2}$
21. $-2\sin(2t)\mathbf{i} + 2\cos(2t)\mathbf{j}$; $-4\cos(2t)\mathbf{i} - 4\sin(2t)\mathbf{j}$; 2
23. $t\mathbf{i} + (2t - 16t^2)\mathbf{j}$ **25.** $4t\mathbf{i} + (3 + 2t^3)\mathbf{j}$
27. 22.3 ft; 48 ft   **29.** 2 s; 2 s   **31.** 87.6 ft/s   **39.** 288 N

EXERCISES 11.8

1. Simple, not closed, not smooth, piecewise smooth; C_1: $x = -(-t)^{3/2}$, $y = -t$, $-1 \le t \le 0$; C_2: $x = t^{3/2}$, $y = t$, $0 \le t \le 1$ **3.** Not simple, closed, not smooth, piecewise smooth; C_1: $x = t$, $y = t^3$, $-1 \le t \le 1$; C_2: $x = -t$, $y = -t$, $-1 \le t \le 1$ **5.** Not simple, closed, smooth; C: $x = \cos t$, $y = \sin t$, $0 \le t \le 4\pi$ **7.** Simple, closed, smooth; C: $x = \cos t$, $y = \sin t$, $0 \le t \le 2\pi$
9. The pole   **11.** $-\sin t\mathbf{i} + \cos t\mathbf{j}$; $-\cos t\mathbf{i} - \sin t\mathbf{j}$; 1/2
13. $\dfrac{t}{\sqrt{t^2 + 1}}\mathbf{i} + \dfrac{1}{\sqrt{t^2 + 1}}\mathbf{j}$; $\dfrac{1}{\sqrt{t^2 + 1}}\mathbf{i} - \dfrac{t}{\sqrt{t^2 + 1}}\mathbf{j}$; $\dfrac{t}{(t^2 + 1)^{3/2}}$ **15.** $\left(\dfrac{-\sin t + \cos t}{\sqrt{2}}\right)\mathbf{i} + \left(\dfrac{\cos t + \sin t}{\sqrt{2}}\right)\mathbf{j}$; $\left(\dfrac{-\cos t - \sin t}{\sqrt{2}}\right)\mathbf{i} + \left(\dfrac{-\sin t + \cos t}{\sqrt{2}}\right)\mathbf{j}$; $\dfrac{1}{\sqrt{2}e^t}$
17. $\mathbf{i}$; $-\mathbf{j}$; 1; 1; $(0, 0)$ **19.** $\mathbf{j}$, $-\mathbf{i}$, 4/9, 9/4, $(7/4, 0)$; $-\mathbf{i}$, $-\mathbf{j}$, 3/16, 16/3, $(0, -7/3)$

$C: x = t$, $y = \cos t$

$C: x = 4\cos t$, $y = 3\sin t$

21. $\dfrac{2x^3}{\left(x^4 + 1\right)^{3/2}}$ **23.** $\cos x$ **25.** $\dfrac{2}{\left(4\sin^2 t + \cos^2 t\right)^{3/2}}$

27. $\dfrac{1}{\left(\tan^2 t + \sec^2 t\right)^{3/2}}$ **29.** $2/3^{3/2}$ at $(1/\sqrt{2}, \ln(1/\sqrt{2}))$

31. $1/2$ at $(0, 2)$ **33.** $\dfrac{6 - 6\cos\theta}{(5 - 4\cos\theta)^{3/2}}$ **35.** $\dfrac{3}{2^{3/2}\sqrt{1 - \sin\theta}}$

37. $1/2$ **39.** $1/2$ **41.** $3/4$ at $(r, \theta) = (2, \pi)$ **43.** $8\mathbf{i} + 6\mathbf{j}$

EXERCISES 11.9

1.

$C : x = t,\ y = t,\ z = t^2$

3.

$C : x = \cos t,\ y = \sin t,\ z = t/2$

5.

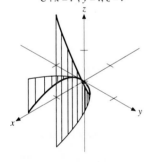

$C : x = t^2,\ y = t,\ z = t^3$

7.

$C : x = \dfrac{\cos t}{t + 1},\ y = t,\ z = \dfrac{\sin t}{t + 1},\ t \geq 0$

9. $C : x = 2 + 3t,$
$y = 1 - 3t,\ z = -1 + 4t,$
$0 \leq t \leq 1$

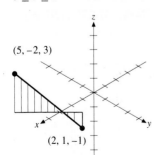

$(5, -2, 3)$

$(2, 1, -1)$

11. $C : x = \cos t,\ y = \sin t,$
$z = 1 - \cos t,\ 0 \leq t \leq 2\pi$

$x^2 + y^2 = 1,\ x + z = 1$

13. $C : x = \sqrt{2}\cos t,\ y = 2\sin t,\ z = -\sqrt{2}\cos t,\ 0 \leq t \leq 2\pi$

$x^2 + y^2 + z^2 = 4,$
$x + z = 0$

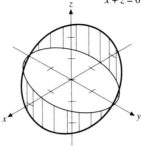

15. $\sqrt{6}$ **17.** $\sqrt{5}/2 + 2\ln\left((1 + \sqrt{5})/2\right)$

19. $C : x = s/\sqrt{5},\ y = \cos(2s/\sqrt{5}),\ z = \sin(2s/\sqrt{5}),$
$0 \leq s \leq \pi\sqrt{5}/2$

21. $\dfrac{-\sin t + \cos t}{\sqrt{3}}\mathbf{i} + \dfrac{\cos t + \sin t}{\sqrt{3}}\mathbf{j} + \dfrac{1}{\sqrt{3}}\mathbf{k};$

$\dfrac{-\cos t - \sin t}{\sqrt{2}}\mathbf{i} + \dfrac{-\sin t + \cos t}{\sqrt{2}}\mathbf{j};$

$\dfrac{\sin t - \cos t}{\sqrt{6}}\mathbf{i} + \dfrac{-\cos t - \sin t}{\sqrt{6}}\mathbf{j} + \dfrac{2}{\sqrt{6}}\mathbf{k};\ \dfrac{\sqrt{2}}{3e^t}$

23. $\dfrac{1}{\sqrt{2}}\mathbf{i} + \dfrac{1}{\sqrt{2}}\mathbf{k};\ \mathbf{j};\ -\dfrac{1}{\sqrt{2}}\mathbf{i} + \dfrac{1}{\sqrt{2}}\mathbf{k};\ \dfrac{1}{2}$

25. $\dfrac{1}{\sqrt{2}}\mathbf{i} - \dfrac{1}{2}\mathbf{j} + \dfrac{1}{2}\mathbf{k};\ \dfrac{1}{\sqrt{2}}\mathbf{j} + \dfrac{1}{\sqrt{2}}\mathbf{k};\ -\dfrac{1}{\sqrt{2}}\mathbf{i} - \dfrac{1}{2}\mathbf{j} + \dfrac{1}{2}\mathbf{k};\ \dfrac{1}{2\sqrt{2}}$

27. $0;\ 12$ **29.** $\dfrac{4t}{\sqrt{1 + 4t^2}};\ \dfrac{2}{\sqrt{1 + 4t^2}}$

31. $0;\ 1$ **33.** $-2/\sqrt{3};\ \sqrt{8/3}$

35. $\sqrt{19}/7\sqrt{14}$ **37.** $-3x + y + 3z = 1$

39. $\dfrac{2}{\sqrt{5}}\mathbf{i} + \dfrac{1}{\sqrt{5}}\mathbf{j};\ \dfrac{1}{\sqrt{30}}\mathbf{i} - \dfrac{2}{\sqrt{30}}\mathbf{j} - \dfrac{5}{\sqrt{30}}\mathbf{k}$

REVIEW EXERCISES

1. $x^2 + y^2 = 1$

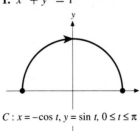

$C : x = -\cos t,\ y = \sin t,\ 0 \leq t \leq \pi$

3. $y = 5 - 2x$

$C : x = t + 1,\ y = 3 - 2t,$
$0 \leq t \leq 1$

5.

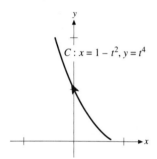

$$C: x = 2\cos t,$$
$$y = 2\sin t,$$
$$z = t, 0 \le t \le 3\pi$$

7. $C: x = 2 - 3t, y = -1 + 4t, z = 3 - 2t, 0 \le t \le 1$
9. $C: x = t^2, y = -t, -2 \le t \le 0$ **11.** C:
$x = (1 + \cos t)\cos t, y = (1 + \cos t)\sin t, -\pi \le t \le \pi$
13. $C: x = e + 2et, y = e + et$
15. $-2; 2$

$$C: x = 1 - t^2, y = t^4$$

17. $\left(13^{3/2} - 8\right)/27$
19. $\displaystyle\int_0^{2\pi} 2\pi\sqrt{2}(1 - \cos t)\sqrt{1 + \cos t}\,dt$
21. $\displaystyle\int_0^{2\pi} \sqrt{5 - 4\sin t}\,dt$ **23.** $2\pi(2 - \sqrt{2})$ **25.** $2a\alpha$
27. $C: x = 1 - 3s/5, y = 3 + 4s/5, 0 \le s \le 5$ **29.** C:
$x = \left((3s + 1)^{2/3} - 1\right)^{3/2}\big/3; y = \left((3s + 1)^{2/3} - 1\right)\big/2,$
$0 \le s \le \left(5^{3/2} - 1\right)\big/3$ **31.** Not simple, not closed, not
smooth, piecewise smooth; $C_1: x = t, y = t^2, 0 \le t \le 1; C_2$:
$x = -t, y = -t, -1 \le t \le 1$
33. $\mathbf{j}; -\mathbf{i}; 2/9; 9/2; (-5/2, 0)$

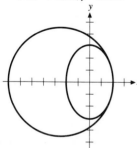

$$C: x = 2\cos t, y = 3\sin t$$

35. $\dfrac{1}{2}\mathbf{i} + \dfrac{\sqrt{3}}{2}\mathbf{j}; -\dfrac{\sqrt{3}}{2}\mathbf{i} + \dfrac{1}{2}\mathbf{j}; \dfrac{1}{16\sqrt{3}}$

37. $\dfrac{2}{|t|\left(4 + t^2\right)^{3/2}}$ **39.** $\dfrac{|\cos x|}{\left(1 + \sin^2 x\right)^{3/2}}$
41. $\dfrac{8t}{\sqrt{8t^2 + 1}}; \dfrac{2\sqrt{2}}{\sqrt{8t^2 + 1}}; \dfrac{2\sqrt{2}}{\left(8t^2 + 1\right)^{3/2}}$
43. (a) $12\mathbf{j}$,   **(b)** $\sqrt{40}$,   **(c)** $\sqrt{10}$,   **(d)** -16,
(e) 2.4981,   **(f)** $(\mathbf{i} - 3\mathbf{j})/\sqrt{10}$,   **(g)** $(-8\mathbf{i} + 24\mathbf{j})/5$
45. $(69\mathbf{i} + 92\mathbf{j})/25$ **47.** 3 **49.** $\sqrt{2}/2$
51. $3x - 4y - 5z + 6 = 0$ **53.** $3x + 4y - 5z + 10 = 0$
55. $\dfrac{x - 7}{3} = \dfrac{y - 5}{4} = \dfrac{z - 3}{1}; x = 7 + 3t, y = 5 + 4t,$
$z = 3 + t$ **57.** $\dfrac{x - 1}{2} = \dfrac{y + 1}{-1} = \dfrac{z - 2}{3}; x = 1 + 2t,$
$y = -1 - t, z = 2 + 3t$ **59.** 1.1102 **61.** $(-3, 9/2, -3/2)$
63. $5/\sqrt{14}$ **65.** $-\sqrt{3}\mathbf{i} + \dfrac{3}{2}\mathbf{j}; \dfrac{\sqrt{21}}{2}; -\mathbf{i} - \dfrac{3\sqrt{3}}{2}\mathbf{j}$
67. $(6 + 4t^4)\mathbf{i} - 5t^3\mathbf{j}$ **69.** 249 ft   **71.** C:
$x = \cos\alpha\,(t - r\sin(t/r)) - \sin\alpha\,(r - r\cos(t/r)),$
$y = \sin\alpha\,(t - r\sin(t/r)) + \cos\alpha\,(r - r\cos(t/r))$
73. $1/\sqrt{2}$ **75.** $\mathbf{i}; \mathbf{j}$
77. $-\dfrac{2qk}{5}(2\mathbf{i} - \mathbf{j} + 2\mathbf{k})$

79. (a)

$$C_{2,3}: x = \sin 2t, y = \sin 3t, t \ge 0$$

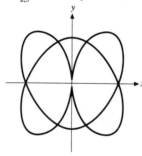

(b) $2\pi/\ell$, where ℓ is the largest common factor of m and n,
(c) $m/\ell; n/\ell$

CHAPTER TWELVE:
DIFFERENTIAL CALCULUS OF FUNCTIONS OF SEVERAL VARIABLES

EXERCISES 12.1

1. $x \ne 0; 0; -1/\pi$ **3.** $(x, y) \ne (0, 0); 1; 1/2; 1/a$
5. $x^2 + y^2 \le 9; 9; 2; 2$ **7.** Lines

(a) $3 - x - y = z_0$ **(b)** $z = 3 - x - y$

9. Circles

(a)

$e^{-x^2-y^2} = z_0$

(b)

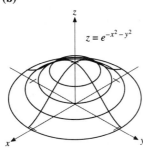

$z = e^{-x^2-y^2}$

15. Circles

(a)

$\dfrac{x^2+y^2}{2x} = z_0$

(b)

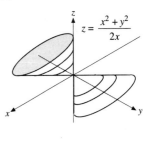

$z = \dfrac{x^2+y^2}{2x}$

11. Circles

(a)

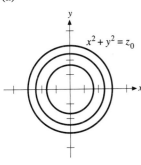

$x^2 + y^2 = z_0$

(b)

$z = x^2 + y^2$

17. Planes

$6x + 3y + 2z = w_0$

19. Paraboloids

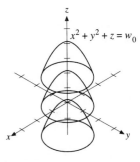

$x^2 + y^2 + z = w_0$

13. Hyperbolas

(a)

$\dfrac{\sqrt{x^2-y^2}}{2} = z_0$

(b)

$z = \dfrac{\sqrt{x^2-y^2}}{2}$

21. Single-napped cones

$z - 2\sqrt{x^2 - y^2} = w_0$

23. Cones

$\dfrac{x^2+y^2}{z^2} = w_0$

25. $x^2 + 2x + y^2 = 3$

27. (a) $V(x, y) = xy\left(\dfrac{6 - 3x - 2y}{2}\right)$, $x \geq 0$, $y \geq 0$,

$3x + 2y \leq 6$,   **(b)** 0

29. The boundary consists of the line $x = 2y + 1$, which is not in the set. The set is open.

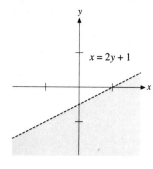

$x = 2y + 1$

31. The boundary consists of that part of the lines $x = 2y - 1$ and $y = 2$ with $x \leq 3$. The set contains its boundary, so it is closed.

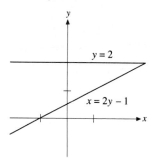

$y = 2$

$x = 2y - 1$

33. The boundary consists of that part of the lines $x + y = 3$ and $y = 5$ with $x > -2$. The set contains part, but not all of its boundary, so the set is neither open nor closed.

$y = 5$

$x + y = 3$

35. The boundary consists of the circles $x^2 + y^2 = 1$ and $x^2 + y^2 = 4$. The set contains part, but not all of its boundary, so the set is neither open nor closed.

$x^2 + y^2 = 4$

$x^2 + y^2 = 1$

37. The boundary consists of all points of the set and the point $(0, 0)$. The set contains part, but not all of its boundary, so the set is neither open nor closed.

$x^2 + y^2 = \dfrac{1}{n^2}$, $n = 1, 2, 3, \ldots$

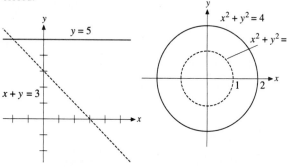

39. The boundary consists of the plane $x + y + z = 5$. The set is open. **41.** The boundary consists of the cylinder $x^2 + y^2 = 1, z > 0$, and the disk $x^2 + y^2 \leq 1, z = 0$. The set is neither open nor closed. **43.** The boundary consists of the sphere $x^2 + y^2 + z^2 = 1$. The set is closed.

EXERCISES 12.2

1. 7 **3.** 1 **5.** Does not exist **7.** 2 **9.** 4
11. Does not exist **13.** 0 **15.** Does not exist
17. Does not exist **19.** Does not exist **21.** Does not exist
23. 0 **25.** 0 **27.** 0 **29.** All (x, y) with $x^2 + y^2 \leq 1$
31. All $(x, y) \neq (0, 0)$ **33.** $(0, 0)$ and all (x, y) with $y \neq x^2$
35. All (x, y) with $y \neq x^2 + 1$ **41.** 4 **43.** 1

EXERCISES 12.3

1. $4; -6$ **3.** $18; 0$
5. $f_x = 4xy - y^3 + 3; \; f_y = 2x^2 - 3xy^2 - 1$
7. $f_x = \dfrac{y}{(x + y)^2}; \; f_y = -\dfrac{x}{(x + y)^2}$
9. $f_x = \dfrac{2x^2 + y^2}{\sqrt{x^2 + y^2}}; \; f_y = \dfrac{xy}{\sqrt{x^2 + y^2}}$
11. $f_x = \dfrac{2x}{x^2 + y^2}; \; f_y = \dfrac{2y}{x^2 + y^2}$
13. $f_x = \dfrac{-y}{x^2 + y^2}; \; f_y = \dfrac{x}{x^2 + y^2}$
15. $f_x = 2\sec^2(2x - y); \; f_y = -\sec^2(2x - y)$
17. $f_x = 4xy^2 \sin(x^2 y^2) \cos(x^2 y^2);$
$f_y = 4x^2 y \sin(x^2 y^2) \cos(x^2 y^2)$
19. $f_x = y + z; \; f_y = x + z; \; f_z = y + x$
21. $f_x = \dfrac{z + y}{(x + z)^2}; \; f_y = -\dfrac{1}{x + z}; \; f_z = \dfrac{y - x}{(x + z)^2}$
23. $f_x = 2e^{2x - y}; \; f_y = -e^{2x - y}; \; f_z = 0$
25. $f_x = 3x^2 + 2xy - 3y^2; \; f_y = x^2 - 6xy + 3y^2;$
$f_{xx} = 6x + 2y; \; f_{yy} = -6x + 6y;$
$f_{xy} = f_{yx} = 2x - 6y$ **27.** $f_x = \ln y; \; f_y = x/y;$
$f_{xx} = 0; \; f_{yy} = -x/y^2; \; f_{xy} = f_{yx} = 1/y$
29. $f_x = y \cos xy; \; f_y = x \cos xy; \; f_{xx} = -y^2 \sin xy;$
$f_{yy} = -x^2 \sin xy; \; f_{xy} = f_{yx} = -xy \sin xy + \cos xy$
31. $\dfrac{\partial x}{\partial r} = \cos \theta; \; \dfrac{\partial x}{\partial \theta} = -r \sin \theta$
37. $z_x = -x/z; \; z_y = -y/z$

EXERCISES 12.4

1. $(6x - 5y)\,dx + (-5x + 2y)\,dy$
3. $-\dfrac{xy}{|x|(x^2 + y^2)}\,dx + \dfrac{x^2}{|x|(x^2 + y^2)}\,dy$
5. $ye^{xy}\,dx + xe^{xy}\,dy$ **7.** $2xy^3z\,dx + 3x^2y^2z\,dy + x^2y^3\,dz$
9. $(y + z)\,dx + (x + z)\,dy + (x + y)\,dz$

11. $-\dfrac{2x}{\left(x^2 + y^2 + z^2\right)^2}\, dx - \dfrac{2y}{\left(x^2 + y^2 + z^2\right)^2}\, dy -$

$\dfrac{2z}{\left(x^2 + y^2 + z^2\right)^2}\, dz$ **13.** 5; 4.982; -0.018; -0.02

15. 6; 5.859; -0.141; -0.1 **17.** 57.7 ± 3.48 ft

19. 62.83 ± 6.28 m^2 **21.** 9%   **23.** 6%   **25.** 1.5 Ω

27. $z = 6x - 12y + 7$ **29.** $5z = 3x + 4y$ **31.** $z = 1$

33. $12z = 169 - 3x - 4y$ **35.** $z = 6x - 4y + 20$

41. (a) $1 + 2(x - 1)$,   **(b)** 0.01   **43. (a)** 0,   **(b)** 0.005

45. (a) $\ln 5 + 0.12(x - 3) + 0.16(y - 4)$,   **(b)** 0.0002

EXERCISES 12.5

1. $(2xy^3)(2) + (3x^2y^2)(3)$

3. $\left(\dfrac{2x}{x^2 + y^2}\right)(-\sin t) + \left(\dfrac{2y}{x^2 + y^2}\right)(\cos t)$

5. $\dfrac{\partial z}{\partial r} = (2x)(\cos\theta) + (2y)(\sin\theta)$;

$\dfrac{\partial z}{\partial\theta} = (2x)(-r\sin\theta) + (2y)(r\cos\theta)$

7. $\dfrac{\partial z}{\partial x} = (2r\cos\theta)\left(\dfrac{x}{\sqrt{x^2 + y^2}}\right) + (-r^2\sin\theta)\left(\dfrac{-y}{x^2 + y^2}\right)$;

$\dfrac{\partial z}{\partial y} = (2r\cos\theta)\left(\dfrac{y}{\sqrt{x^2 + y^2}}\right) + (-r^2\sin\theta)\left(\dfrac{x}{x^2 + y^2}\right)$

9. $\left(\dfrac{y^2}{z}\right)(1) + \left(\dfrac{2xy}{z}\right)(3) + \left(-\dfrac{xy^2}{z^2}\right)(2)$

11. $\dfrac{\partial w}{\partial x} = \left(\dfrac{1}{v}\right)(2x) + \left(-\dfrac{u}{v^2}\right)(2y)$;

$\dfrac{\partial w}{\partial y} = \left(\dfrac{1}{v}\right)(-2y) + \left(-\dfrac{u}{v^2}\right)(2x)$

13. 74   **15.** $\dfrac{\partial w}{\partial r} = 38$; $\dfrac{\partial w}{\partial s} = -26$

17. $\dfrac{\partial w}{\partial x} = 1$; $\dfrac{\partial w}{\partial y} = 0$ **19.** $-\dfrac{3x^2 + y}{x + 3y^2}$ **21.** $-y/x$

23. $\dfrac{\partial z}{\partial x} = -\dfrac{yze^{xyz} - 2x}{xye^{xyz} - 2z}$; $\dfrac{\partial z}{\partial y} = -\dfrac{xze^{xyz} - 2y}{xye^{xyz} - 2z}$

25. $\dfrac{\partial z}{\partial x} = -\dfrac{1 + z^2}{y + 2xz}$; $\dfrac{\partial z}{\partial y} = -\dfrac{z}{y + 2xz}$;

$\dfrac{\partial^2 z}{\partial x^2} = \dfrac{2(1 + z^2)(2yz + 3xz^2 - x)}{(y + 2xz)^3}$ **27.** 31   **29.** 24

33. (a) $\dfrac{dV}{dt} = wh\dfrac{d\ell}{dt} + \ell h\dfrac{dw}{dt} + \ell w\dfrac{dh}{dt}$,   **(b)** -1 ft^3/s

35. $\dfrac{\partial u}{\partial x} = -1$; $\dfrac{\partial v}{\partial x} = \dfrac{1}{3}$; $\dfrac{\partial u}{\partial y} = 0$; $\dfrac{\partial v}{\partial y} = -\dfrac{1}{3}$

37. $\dfrac{\partial u}{\partial x} = \dfrac{x - v}{v - u}$; $\dfrac{\partial v}{\partial x} = \dfrac{x - u}{u - v}$; $\dfrac{\partial u}{\partial y} = \dfrac{y - v}{v - u}$; $\dfrac{\partial v}{\partial y} = \dfrac{y - u}{u - v}$

39. (a) $F_x + F_u u_x + F_v v_x = 0$,   **(b)** $F_y + F_u u_y + F_v v_y = 0$

EXERCISES 12.6

1. 52/5   **3.** 2   **5.** $-76/3$ **7.** $-\mathbf{i}$; 4

9. $(-4\mathbf{i} - 7\mathbf{j})/\sqrt{65}$; $\sqrt{65}$ **11.** $\mathbf{k}$; 6

13. $(7\mathbf{i} - 8\mathbf{j} + 8\mathbf{k})/\sqrt{177}$; $\sqrt{177}$ **15.** $\pm(4\mathbf{i} + 3\mathbf{j})/5$

17. $\pm(3\mathbf{i} + \mathbf{j})/\sqrt{10}$ **19.** $2\sqrt{2}\mathbf{i} + \sqrt{2}\mathbf{j}$ **21. (a)** A small change in width changes the area more than an equal change in length.   **(b)** $-4/3$ **25.** $-5/2$ **27.** $2\sqrt{2}$

29. $\dfrac{3x^2y - 3xy^2}{\sqrt{x^2 + y^2}}$

EXERCISES 12.7

1. (a) All except the circle $x = 0$, $y^2 + z^2 = 1$;
(b) all except the circle $y = 0$, $x^2 + z^2 = 1$;
(c) all except the circle $z = 0$, $x^2 + y^2 = 1$
3. $-4\mathbf{i} + \mathbf{j}$; 0   **5.** $\sqrt{2}\mathbf{i} + \sqrt{2}\mathbf{j}$; 0

Level curves of
$f(x, y) = y/x^2$

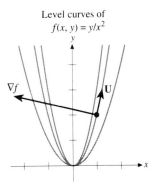

Level curves of
$f(x, y) = \ln(x^2 + y^2)$

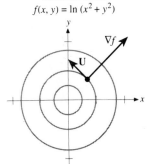

7. $2\mathbf{i} + 2\mathbf{j}$; 0   **9.** $2\mathbf{i} + 2\mathbf{j}$; $-2\sqrt{2}$

Level curves of
$f(x, y) = x^2 + y^2$

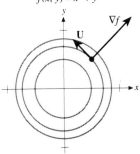

Level curves of
$f(x, y) = x^2 - y^2$

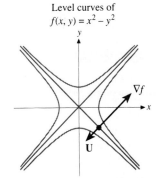

11. $4x + 3y = 0$ **13.** $2x + y = 5$

15. $6\sqrt{3}x + 12y = 2\pi + 3\sqrt{3}$ **17.** $5y = 8x - 50$

19. $-3x + 4y - 5z = 0$ **21.** $-2x + 2y + 3ez = 2e$

23. $-3x + 2y + z = 6$ **25.** $\dfrac{x + 1}{2} = \dfrac{y - 2}{-4} = \dfrac{z - 5}{1}$;

$x = -1 + 2t$, $y = 2 - 4t$, $z = 5 + t$

27. $\dfrac{x - 2}{-2} = \dfrac{y + 3}{-1} = \dfrac{z - 0}{-6}$; $x = 2 - 2t$, $y = -3 - t$,

$z = -6t$ **29.** $y = 0$, $\dfrac{x}{-1} = \dfrac{z}{1}$; $x = -t$, $y = 0$, $z = t$

31. $c(4\mathbf{i} - 5\mathbf{j} + 3\mathbf{k})$, $c \neq 0$ **33.** $c(\mathbf{i} - 2\mathbf{j} - 4\mathbf{k})$, $c \neq 0$
35. $-360\sqrt{10}$ **37.** $(153, 204, 0)$

EXERCISES 12.8

1. Loc min $f(1, 0) = -2$; saddle pt $f(-1, 0) = 2$
3. Loc min $f(1, 1) = -4$; saddle pt $f(1, -1) = 0$;
saddle pt $f(-1, 1) = 0$; loc max $f(-1, -1) = 4$
5. Saddle pt $f(0, 0) = 0$; loc min $f(1, -1) = -1$
7. Loc min $f(0, -1) = -1$ **9.** Loc min $f(0, 0) = 0$
11. Max $f(1, 3) = 3$ **13.** Max $f(2, -1) = 3$
15. Min $f(1, 0) = -2$; max $f(2, 0) = 2$
17. Min $f(0, -3) = -18$; max $f(\sqrt{320}/3, -1/3)$
$= f(-\sqrt{320}/3, -1/3) = 986/27$
19. Min $f(0, 1) = -1$; max $f(3, 3) = 18$ **21.** $(-6/5, 3/5)$
23. $(1/\sqrt{2}, -1/2)$, $(-1/\sqrt{2}, -1/2)$ **25.** $1 \times (1/2) \times 1$; $1/2$
29. (a) $f(1, 1) = -1$,
(b)

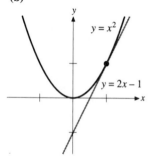

(c) The level curve is tangent to the parabola.
31. $y = (-5/7)x + 4$ **33.** $y = (3/5)x + 3/2$
35. Min $f(2, 0) = -2$; max $f(0, 1) = 1$
37. Min $f(0, 0) = 0$; max $f(0, 1) = 3$
39. Min $f(1, 1) = 2$

EXERCISES 12.9

1. Min $f(-3, 4) = -25$; max $f(3, -4) = 25$
3. Min $f(-\sqrt{3}, 1) = f(\sqrt{3}, -1) = -\sqrt{3}$;
max $f(\sqrt{3}, 1) = f(-\sqrt{3}, -1) = \sqrt{3}$
5. Min $f(-1, -2, 2) = -9$; max $f(1, 2, -2) = 9$
7. Loc min $f(-2, -2/3, -2/3) = -8/9$; loc max
$f(2, 2/3, 2/3) = 8/9$ **9.** Loc min $f(1/2, 2, 1) = 6$
11. Min $f(1, 1, -\sqrt{2}) = -\sqrt{2}$; max $f(1, 1, \sqrt{2}) = \sqrt{2}$
13. Min $f(-1/\sqrt{2}, 1/\sqrt{2}, -1/\sqrt{2}) = -3\sqrt{2}$;
max $f(1/\sqrt{2}, -1/\sqrt{2}, 1/\sqrt{2}) = 3\sqrt{2}$; loc min
$f(-1/\sqrt{10}, 3/\sqrt{10}, 1/\sqrt{10}) = -\sqrt{10}$; loc
max $f(1/\sqrt{10}, -3/\sqrt{10}, -1/\sqrt{10}) = \sqrt{10}$
15. Min $f(-4, -2) = -40$; max $f(2, 1) = 5$
17. Min $f(2, -1, 3) = -14$; max $f(-4, 2, -6) = 112$
19. Min $f(1/\sqrt{2}, -1/\sqrt{2}, 0) = f(-1/\sqrt{2}, 1/\sqrt{2}, 0) = -1$;
max $f(1/\sqrt{2}, 1/\sqrt{2}, 0) = f(-1/\sqrt{2}, -1/\sqrt{2}, 0) = 1$
21. $4 \times 4 \times 2$; 48 **23.** $(1 \text{ ft}) \times (1 \text{ ft}) \times (3/2 \text{ ft})$; $3/2 \text{ ft}^3$

27. 4 in.; $\pi/3$ **29.** Min $f(-1/\sqrt{2}, 1/\sqrt{2}) = -\sqrt{2}$;
max $f(1/\sqrt{2}, -1/\sqrt{2}) = \sqrt{2}$
31. Min $f(1/2, 1/2) = 1/2$; max $f(2, 0) = f(0, 2) = 4$
33. $\dfrac{4A^2}{a^2 + b^2 + c^2}$

REVIEW EXERCISES

1. $x \neq y$
3. Parabolas

(a)

(b)

5. Single-napped cones

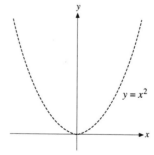

7. Does not exist   **9.** Does not exist
11. (a) $y = x^2$,   **(b)** open

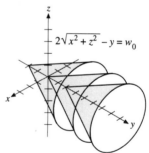

13. (a) The boundary consists of all points $(x, 0)$ with $0 \le x \le 2$, all points $(0, y)$ with $0 \le y \le 1$, and all points $(2 - 2y, y)$ with $0 \le y \le 1$.

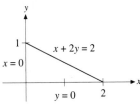

(b) The set is closed.   **15.** All (x, y)

17. $(1, 1)$ and all (x, y) with $x \ne y$ **19.** $f_x = 2x - 2y$; $f_y = -2x + 6y$; $f_{xx} = 2$; $f_{yy} = 6$; $f_{xy} = f_{yx} = -2$

21. $f_x = \dfrac{x}{x^2 + y^2}$; $f_y = \dfrac{y}{x^2 + y^2}$; $f_{xx} = \dfrac{y^2 - x^2}{\left(x^2 + y^2\right)^2}$;

$f_{yy} = \dfrac{x^2 - y^2}{\left(x^2 + y^2\right)^2}$; $f_{xy} = f_{yx} = -\dfrac{2xy}{\left(x^2 + y^2\right)^2}$

23. $f_x = y^2 z^3$; $f_y = 2xyz^3$; $f_z = 3xy^2z^2$; $f_{xx} = 0$; $f_{yy} = 2xz^3$; $f_{zz} = 6xy^2z$; $f_{xy} = f_{yx} = 2yz^3$; $f_{xz} = f_{zx} = 3y^2z^2$; $f_{yz} = f_{zy} = 6xyz^2$

25. $df = \dfrac{1}{2y\sqrt{x}}\,dx - \dfrac{\sqrt{x}}{y^2}\,dy$ **27.** 3.2%

29. $(2x/y)\left(\sec^2 t\right) + \left(-x^2/y^2\right)(2\cos t)(-\sin t)$

31. $\dfrac{\partial z}{\partial r} = (2x)(\cos\theta) + (-2y)(\sin\theta)$;

$\dfrac{\partial z}{\partial\theta} = (2x)(-r\sin\theta) + (-2y)(r\cos\theta)$

33. $\dfrac{\partial z}{\partial x} = -\dfrac{2xy + z^2}{y^2 + 2xz}$; $\dfrac{\partial z}{\partial y} = -\dfrac{x^2 + 2yz}{y^2 + 2xz}$

35. $\dfrac{\partial^2 z}{\partial y \partial x} = -\dfrac{xy}{z^3}$

37. $u_x = 4/3$; $v_x = 5/3$; $u_y = -7/3$; $v_y = -5/3$

39. Decreasing 2 unit/s   **41.** $-7/3$

43. $-(6 + 27\sqrt{3})/\sqrt{31}$ **45.** $-60/\sqrt{289}$

47. $\mathbf{i} + \mathbf{j}$; $\pm(\mathbf{i} - \mathbf{j})/\sqrt{2}$ **49.** $4\mathbf{i} - \mathbf{j}$; $\pm(\mathbf{i} + 4\mathbf{j})/\sqrt{17}$

Level curves of $f(x, y) = x^2 - y$

Level curves of $f(x, y) = x + y$

51. $(-3\mathbf{i} + 8\mathbf{j})/\sqrt{73}$; $\sqrt{73}$ **53.** $z = 4x + 6y - 13$

55. $3x + 4y + 5z = 0$ **57.** $\dfrac{x - 5}{-10} = \dfrac{y + 4}{-8} = \dfrac{z - 3}{6}$;

$x = 5 - 10t$, $y = -4 - 8t$, $z = 3 + 6t$

59. $\dfrac{x - 3}{-6} = \dfrac{y - 4}{-8} = \dfrac{z - 25}{1}$; $x = 3 - 6t$,

$y = 4 - 8t$, $z = 25 + t$ **61.** Loc min $f(0, 0) = 0$

63. Loc min $f(1, 1) = f(-1, -1) = -2$

65. Min $f(1/3, -1/3, 1/3) = 1/3$

67. (4 ft) $\times$ (4 ft) $\times$ (2 ft); 48 ft^2 **69.** Min

$f(-\sqrt{2}/\sqrt{3}, -1/\sqrt{3}) = f(\sqrt{2}/\sqrt{3}, -1/\sqrt{3}) = -2/3\sqrt{3}$;

max $f(\sqrt{2}/\sqrt{3}, 1/\sqrt{3}) = f(-\sqrt{2}/\sqrt{3}, 1/\sqrt{3}) = 2/3\sqrt{3}$

71. Min $f(x, 0) = 0$, $0 \le x \le 1$; $f(0, y) = 0$, $0 \le y \le 1$; max $f(1/3, 2/3) = 4/27$ **73.** Min

$f(-1/\sqrt{2}, -1/\sqrt{2}) = -\sqrt{2}$; max $f(1/\sqrt{2}, 1/\sqrt{2}) = \sqrt{2}$

75. Min $f(2/5, 4/5) = 4/5$; max $f(4, 0) = 16$ **77.** None

CHAPTER THIRTEEN:
MULTIPLE INTEGRALS

EXERCISES 13.1

1. (a)   For x fixed, y varies from $y = x$ to $y = 1$. x varies from $x = 0$ to $x = 1$.|**(b)** For y fixed, x varies from $x = 0$ to $x = y$. y varies from $y = 0$ to $y = 1$.

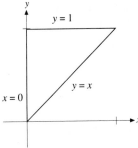

3. (a)   For x fixed, y varies from $y = 0$ to $y = (2 - x)/2$. x varies from $x = 0$ to $x = 2$.|**(b)** For y fixed, x varies from $x = 0$ to $x = 2 - 2y$. y varies from $y = 0$ to $y = 1$.

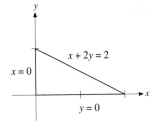

5. (a) For x fixed, y varies from $y = 0$ to $y = \sqrt{4 - x^2}$. x varies from $x = -2$ to $x = 2$. **(b)** For y fixed, x varies from $x = -\sqrt{4 - y^2}$ to $x = \sqrt{4 - y^2}$. y varies from $y = 0$ to $y = 2$.

7. (a) For x fixed, y varies from $y = -\sqrt{1 - (x - 1)^2}$ to $y = \sqrt{1 - (x - 1)^2}$. x varies from $x = 1$ to $x = 2$. **(b)** For y fixed, x varies from $x = 1$ to $x = 1 + \sqrt{1 - y^2}$. y varies from $y = -1$ to $y = 1$.

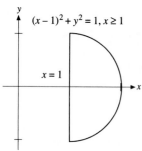

9. (a) For x fixed, y varies from $y = -1$ to $y = x^3$. x varies from $x = -1$ to $x = 1$. **(b)** For y fixed, x varies from $x = y^{1/3}$ to $x = 1$. y varies from $y = -1$ to $y = 1$.

11. (a) For x fixed, y varies from $y = \cos^{-1} x$ to $y = \pi$. x varies from $x = -1$ to $x = 1$. **(b)** For y fixed, x varies from $x = \cos y$ to $x = 1$. y varies from $y = 0$ to $y = \pi$.

13. (a) For x fixed, y varies from $y = 2x$ to $y = x + 1$. x varies from $x = 0$ to $x = 1$. **(b)** For y fixed, x varies from $x = 0$ to $x = y/2$. y varies from $y = 0$ to $y = 1$. For y fixed, x varies from $x = y - 1$ to $x = y/2$. y varies from $y = 1$ to $y = 2$.

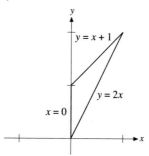

15. For x fixed, y varies from $y = \sin x$ to $y = \cos x$. x varies from $x = 0$ to $x = \pi/4$.

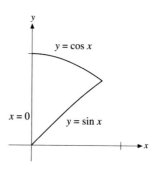

17. For y fixed, x varies from $x = y$ to $x = 2 - y^2$. y varies from $y = -2$ to $y = 1$.

19. For y fixed, x varies from $x = 0$ to $x = e^y$. y varies from $y = 0$ to $y = 1$.

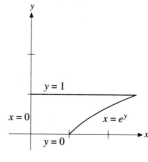

21. For x and y fixed, z varies from $z = x + y$ to $z = 2 - x - y$. For x fixed, y varies from $y = 0$ to $y = 1 - x$. x varies from $x = 0$ to $x = 1$.

$x = 0$ to $x = 1$.

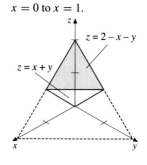

23. For x and y fixed, z varies from $z = x$ to $z = 3 - x - y$. For x fixed, y varies from $y = 0$ to $y = 3 - 2x$. x varies from $x = 0$ to $x = 3/2$.

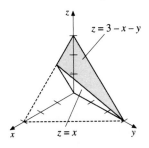

25. For x and y fixed, z varies from $z = 0$ to $z = 2 - x - y$. For x fixed, y varies from $y = -\sqrt{1 - x^2}$ to $y = \sqrt{1 - x^2}$. x varies from $x = -1$ to $x = 1$.

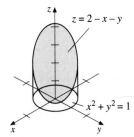

35.

(a) For x and y fixed, z varies from $z = 0$ to $z = 2$. For x fixed, y varies from $y = 0$ to $y = 2$. x varies from $x = 0$ to $x = 2$.

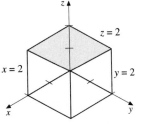

(b) For x and y fixed, z varies from $z = 0$ to $z = (4 - y)/2$. For x fixed, y varies from $y = 0$ to $y = 2$. x varies from $x = 0$ to $x = 2$.

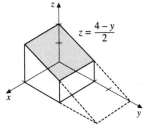

(c) For x and y fixed, z varies from $z = y/2$ to $z = (4 - y)/2$. For x fixed, y varies from $y = 0$ to $y = 2$. x varies from $x = 0$ to $x = 2$.

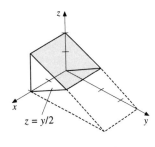

(d) For x and y fixed, z varies from $z = y/2$ to $z = (4 - y)/2$. For x fixed, y varies from $y = 0$ to $y = 2 - x$. x varies from $x = 0$ to $x = 2$.

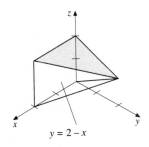

27. For x and y fixed, z varies from $z = 0$ to $z = 3 - y$. For x fixed, y varies from $y = 0$ to $y = 2 - x$. x varies from $x = 0$ to $x = 2$.

29. For x and y fixed, z varies from $z = 0$ to $z = \sqrt{4 - y^2}$. For x fixed, y varies from $y = 0$ to $y = \sqrt{4 - x^2}$. x varies from $x = 0$ to $x = 2$.

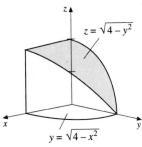

37.

(a) For x and y fixed, z varies from $z = 0$ to $z = (3 - x)/3$. For x fixed, y varies from $y = 0$ to $y = 3$. x varies from $x = 0$ to $x = 3$.

(b) For x and y fixed, z varies from $z = 0$ to $z = (3 - x)/3$. For x fixed, y varies from $y = 0$ to $y = 3 - x$. x varies from $x = 0$ to $x = 3$.

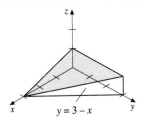

31. For x and y fixed, z varies from $z = -\sqrt{4x^2 + 4y^2}$ to $z = \sqrt{4x^2 + 4y^2}$. For x fixed, y varies from $y = -\sqrt{1 - x^2}$ to $y = \sqrt{1 - x^2}$. x varies from $x = -1$ to $x = 1$.

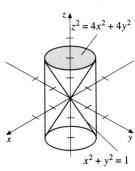

33. For x and y fixed, z varies from $z = 2x$ to $z = 2$. For x fixed, y varies from $y = 0$ to $y = 3$. x varies from $x = 0$ to $x = 1$.

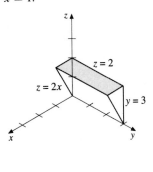

(c) For x and y fixed, z varies from $z = 0$ to $z = (3 - x)/3$. For x fixed, y varies from $y = 0$ to $y = 3 - x$. x varies from $x = 0$ to $x = 2$.

39. For x fixed, y varies from $y = 0$ to $y = (x + 4)/3$. x varies from $x = -1$ to $x = 2$.
41. For x fixed, y varies from $y = 0$ to $y = 2 - x$. x varies from $x = 0$ to $x = 1$.
43. For y fixed, x varies from $x = y - 1$ to $x = 1 - y$. y varies from $y = 0$ to $y = 1$.

EXERCISES 13.2

1. $1/8$ **3.** $1/4$ **5.** $\pi/4$ **7.** $\ln(\sqrt{2} + 1) - \pi/4$
9. $20/3$ **11.** 8 **13.** $-4e - 4 + 4e^2$
15. 1 **17.** $\sqrt{3} - \pi/3$ **19.** $32/3$ **21.** $3/2$
23. $37/12$ **25.** $\displaystyle\int_0^4 \int_{y/2}^{\sqrt{y}} xy\,dx\,dy = 8/3$
27. $\displaystyle\int_0^1 \int_0^{\sin^{-1} y} \cos x\,dx\,dy = 1/2$
29. $\displaystyle\int_{\pi/6}^{\pi/2} \int_{1/2}^{\sin y} \cot y\,dx\,dy = (1 - \ln 2)/2$
31. $\displaystyle\int_0^1 \int_1^{e^y} x\,dx\,dy = (e^2 - 3)/4$
33. $\displaystyle\int_0^1 \int_0^{y^{1/3}} 3x^2\,dx\,dy = 1/2$
35. $\displaystyle\int_0^1 \int_{-\sqrt{1-y^2}}^{\sqrt{1-y^2}} 2y\,dx\,dy = 4/3$
37. $\displaystyle\int_0^1 \int_0^{2-y} f(x, y)\,dx\,dy$ **39.** $\displaystyle\int_0^1 \int_x^{x+1} f(x, y)\,dy\,dx$
41. The function e^{xy} is continuous on the closed and bounded region $0 \le x \le 1, 0 \le y \le 1$.

$$\int_0^1 e^{xy}\,dy = \begin{cases} (e^x - 1)/x, & 0 < x \le 1, \\ 1, & x = 0. \end{cases}$$

Yes, it is a continuous function of x.

EXERCISES 13.3

1. $27/2$ **3.** $2/5$ **5.** 8π **7.** $6\sqrt{5}$ **9.** $2\left(3^{3/2} - 2^{3/2}\right)/3$
11. $\pi rh/2$ **13.** $8/3$ **15.** $7/3$ **17.** $\pi/4$ **19.** $(1, 8/3)$
21. $(3\pi/16, 0)$ **23.** $\left(8(4 - \sqrt{2})/15, 8\sqrt{2}/15\right)$

25. $\displaystyle\int_0^1 \int_0^{1-x} (2 - x - z)\,dz\,dx = \int_0^1 \int_0^{1-z} (2 - x - z)\,dx\,dz$
27. $\displaystyle\int_{-1}^1 \int_{-\sqrt{1-x^2}}^{\sqrt{1-x^2}} \sqrt{4x^2 + 1 + 4z^2}\,dz\,dx$
$$= \int_{-1}^1 \int_{-\sqrt{1-z^2}}^{\sqrt{1-z^2}} \sqrt{4x^2 + 1 + 4z^2}\,dx\,dz$$
29. (a) $2/3$,   **(b)** $1/6$ **31.** $4/15$
33. $I_{y-axis} > I_{x-axis}$

EXERCISES 13.4

1. $\pi/2 - 1$ **3.** $(9/2)\tan^{-1}(1/2)$ **5.** $(2 + 3\pi)/4$
7. $\pi/4$ **9.** 1 **11.** $4\pi R^3/3$ **13.** $4\pi\left(8 - 3^{3/2}\right)/3$
15. $k\pi R^4/2$ **17.** $(0, 3\pi/16)$ **19.** $(28/9\pi, 28/9\pi)$
21. $\pi R\sqrt{R^2 + H^2}$ **23.** $2\pi R^2(1 - \cos\phi_0)$
25. $4/3$ **27.** $\pi \ln 2$ **29. (a)** $Mr_0^2/2$,
(b) $41Mr_0^2/50$,   **(c)** The moment is larger in (b).
31. (a) $\pi\left(1 - e^{-R^2}\right)\big/4$,   **(d)** $\sqrt{\pi}/2$

EXERCISES 13.5

1. $9/2$ **3.** $1/48$ **5.** $\left(2\cos\pi^2 + \pi^4 - 2\right)/4$
7. $(e - 2)/4$ **9.** $1/6$ **11.** $8/3$ **13.** $8/15$
15. $16r^3/3$ **17.** $\displaystyle\int_{-r}^r \int_{-\sqrt{r^2-x^2}}^{\sqrt{r^2-x^2}} \int_{-\sqrt{R^2-x^2}}^{\sqrt{R^2-x^2}} dz\,dy\,dx$
19. $\displaystyle\int_{-1}^1 \int_{-\sqrt{1-x^2}}^{\sqrt{1-x^2}} \int_{-\sqrt{(y+4)^2-4x^2}/2}^{\sqrt{(y+4)^2-4x^2}/2} dz\,dy\,dx$
21. $\displaystyle\int_0^{12/7} \int_{z/2}^{6-3z} \int_0^{(6-x-3z)/2} dy\,dx\,dz,$
$\displaystyle\int_0^{12/7} \int_0^{(12-7z)/4} \int_{z/2}^{6-2y-3z} dx\,dy\,dz,$
$\displaystyle\int_0^3 \int_0^{(12-4y)/7} \int_{z/2}^{6-2y-3z} dx\,dz\,dy$
23. $\displaystyle\int_0^1 \int_0^{4-z} \int_0^{4-x-z} dy\,dx\,dz, \quad \int_0^1 \int_0^{4-z} \int_0^{4-y-z} dx\,dy\,dz$
25. $\displaystyle\int_0^{4/3} \int_{2x}^{4-x} \int_0^{6-x-z} dy\,dz\,dx$
27. $\displaystyle\int_0^4 \int_0^{4-y} \int_{x+y}^4 f(x, y, z)\,dz\,dx\,dy,$
$\displaystyle\int_0^4 \int_0^z \int_0^{z-y} f(x, y, z)\,dx\,dy\,dz,$
$\displaystyle\int_0^4 \int_y^4 \int_0^{z-y} f(x, y, z)\,dx\,dz\,dy,$
$\displaystyle\int_0^4 \int_0^z \int_0^{z-x} f(x, y, z)\,dy\,dx\,dz,$
$\displaystyle\int_0^4 \int_x^4 \int_0^{z-x} f(x, y, z)\,dy\,dz\,dx$
29. $\displaystyle\int_0^2 \int_0^{2-y} \int_{x+y}^{4-x-y} f(x, y, z)\,dz\,dx\,dy$

31. $\int_0^1 \int_x^{4-x} \int_0^{z-x} f(x, y, z)\,dy\,dz\,dx$ **33.** 4 **35.** 1464
37. $(7/12, 7/12, 7/6)$ **39.** $(12/7, 8/7, 4/7)$ **41.** $MR^2/2$
43. $2MS^2/3$ **45.** $26M/3$ **47.** $8k\pi\left(r_2^5 - r_1^5\right)\big/15$
49. $MS^2/6$ **53.** $3k\pi hr^4/2$

EXERCISES 13.6

1.

3.

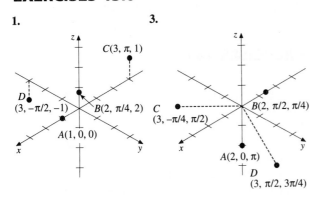

5. $A(x, y, z) = (0, 0, 2)$, $B(x, y, z) = (\sqrt{2}, \sqrt{2}, -1)$,
$C(x, y, z) = (1, \sqrt{3}, 3)$ **7.** $A(x, y, z) = (0, 4\sqrt{2}, 4\sqrt{2})$,
$B(x, y, z) = (\sqrt{2}/2, \sqrt{2}/2, 0)$, $C(x, y, z) = (\sqrt{3}/2, 3/2, -1)$
9. $A(r, \theta, z) = (2\sqrt{2}, \pi/4, 2)$, $B(r, \theta, z) = (0, 0, -1)$,
$C(r, \theta, z) = (8, -\pi/3, 1)$ **11.** $A(\rho, \theta, \phi) = (2, 0, \pi)$,
$B(\rho, \theta, \phi) = (2\sqrt{2}, \pi/4, \pi/2)$, $C(\rho, \theta, \phi) = (2\sqrt{2}, \pi/2, \pi/4)$

13. For r and θ fixed,
z varies from $z = 0$ to
$z = 2r(\cos\theta + \sin\theta)$. For θ
fixed, r varies from $r = 0$ to
$r = 2$. θ varies from $\theta = 0$ to
$\theta = \pi/2$.

15. For r and θ fixed, z varies
from $z = 0$ to $z = r$. For θ
fixed, r varies from $r = 0$ to
$r = 3$. θ varies from $\theta = 0$ to
$\theta = 2\pi$.

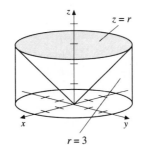

17. For r and θ fixed, z varies
from $z = 0$ to $z = 4 - r^2$. For
θ fixed, r varies from $r = 0$ to
$r = 2$. θ varies from $\theta = 0$ to
$\theta = 2\pi$.

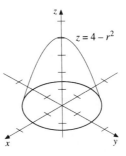

19. For r and θ fixed, z varies
from $z = 0$ to $z = 2 - e^{r^2}$.
For θ fixed, r varies from
$r = 0$ to $r = \sqrt{\ln 2}$. θ varies
from $\theta = 0$ to $\theta = 2\pi$.

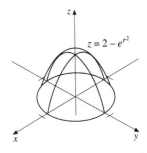

21. For r and θ fixed,
z varies from $z = 0$ to
$z = \sqrt{16 - r^2}$. For θ fixed,
r varies from $r = 0$ to
$r = 4/(\cos\theta + \sin\theta)$. θ
varies from $\theta = 0$ to $\theta = \pi/2$.

23. For r and θ fixed, z varies
from $z = 0$ to $z = 4$. For θ
fixed, r varies from $r = 0$ to
$r = 2\cos\theta$. θ varies from
$\theta = -\pi/2$ to $\theta = \pi/2$.

25. For θ and ϕ fixed, ρ
varies from $\rho = 0$ to $\rho = 3$.
For θ fixed, ϕ varies from
$\phi = 0$ to $\phi = \pi/2$. θ varies
from $\theta = 0$ to $\theta = 2\pi$.

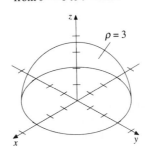

27. For θ and ϕ fixed, ρ
varies from $\rho = 0$ to $\rho = 1$.
For θ fixed, ϕ varies from
$\phi = 0$ to $\phi = \pi/2$. θ varies
from $\theta = 0$ to $\theta = \pi/2$.

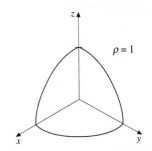

29. For θ and ϕ fixed, ρ varies from $\rho = 0$ to $\rho = 4 \csc \phi$. For θ fixed, ϕ varies from $\phi = \pi/4$ to $\phi = \pi/2$. θ varies from $\theta = 0$ to $\theta = 2\pi$.

31. For θ and ϕ fixed, ρ varies from $\rho = 3 \sec \phi$ to $\rho = 5$. For θ fixed, ϕ varies from $\phi = 0$ to $\phi = \cos^{-1}(3/5)$. θ varies from $\theta = 0$ to $\theta = 2\pi$.

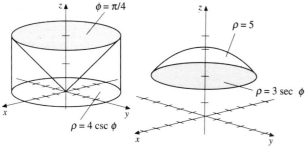

$\phi = \pi/4$

$\rho = 4 \csc \phi$

$\rho = 5$

$\rho = 3 \sec \phi$

EXERCISES 13.7

1. 8π **3.** $32\pi(8 - 3\sqrt{3})/3$ **5.** $2\pi R^3/3$
7. $5\pi/3$ **9.** $8\pi/3$ **11.** On axis, $H/4$ from base
13. $\left(0, 0, \dfrac{H(9 + 4HR)}{6(3 + HR)}\right)$ **15.** $\dfrac{2M(R^5 - r^5)}{5(R^3 - r^3)}$ **17.** $M/3$
19. (a) $M\left(R^2 + 2D^2\right)/2$, **(b)** 0 **21.** 4π **23.** $3MR^2/2$
25. $H_1 = H - \omega^2 R^2/4g$ **27.** $2\pi/5$ **29.** $\pi/2$

EXERCISES 13.8

1. (a) $(x, y) = (2, 1)$, **(b)** $(x, y) = (-1, 2)$,
(c) $(u, v) = (2/5, -1/5)$, **(d)** $(u, v) = (1/5, 2/5)$,
(e) $u = (2x + y)/5$, $v = (-x + 2y)/5$
3. (a) $(x, y, z) = (1, 0, 1)$,
(b) $(x, y, z) = (-1, 1, 0)$, **(c)** $(x, y, z) = (0, -2, 1)$,
(d) $(u, v, w) = (1/3, -2/3, -1/3)$,
(e) $(u, v, w) = (1/3, 1/3, -1/3)$,
(f) $(u, v, w) = (2/3, 2/3, 1/3)$, **(g)** $u = (x + y + 2z)/3$,
$v = (-2x + y + 2z)/3$, $w = (-x - y + z)/3$
5.
(a) **(b)**

33. For θ and ϕ fixed, ρ varies from $\rho = 0$ to $\rho = 3/(\sin\phi\cos\theta + \sin\phi\sin\theta + \cos\phi)$. For θ fixed, ϕ varies from $\phi = 0$ to $\phi = \pi/2$. θ varies from $\theta = 0$ to $\theta = \pi/2$.

35. For θ and ϕ fixed, ρ varies from $\rho = 0$ to $\rho = 2/(\sin\phi\cos\theta + \cos\phi)$. For θ fixed, ϕ varies from $\phi = 0$ to $\phi = \tan^{-1}(1/2)$. θ varies from $\theta = 0$ to $\theta = 2\pi$.

$$\rho = \frac{3}{\sin\phi\cos\theta + \sin\phi\sin\theta + \cos\theta}$$

$$\rho = \frac{2}{\sin\phi\cos\theta + \cos\phi}$$

$\phi = \tan^{-1}\dfrac{1}{2}$

7.
(a) **(b)**

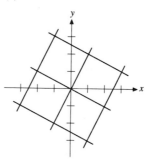

37. (a) $\sqrt{x^2 + y^2}$, **(b)** r, **(c)** $\rho \sin\phi$
39. For x and y fixed, z varies from $z = 0$ to $z = \sqrt{4 - x^2 - y^2}$. For x fixed, y varies from $y = -\sqrt{1 - x^2}$ to $y = \sqrt{1 - x^2}$. x varies from $x = -1$ to $x = 1$.
41. For x and y fixed, z varies from $z = 0$ to $z = 4 - 2\sqrt{x^2 + y^2}$. For x fixed, y varies from $y = -\sqrt{4 - x^2}$ to $y = \sqrt{4 - x^2}$. x varies from $x = -2$ to $x = 2$.
43. For x and y fixed, z varies from $z = 0$ to $z = \sqrt{4 - x^2 - y^2}$. For x fixed, y varies from $y = -\sqrt{4 - x^2}$ to $y = \sqrt{4 - x^2}$. x varies from $x = -2$ to $x = 2$.
45. For x and y fixed, z varies from $z = 0$ to $z = \sqrt{x^2 + y^2}$. For x fixed, y varies from $y = -\sqrt{1 - x^2}$ to $y = \sqrt{1 - x^2}$. x varies from $x = -1$ to $x = 1$.

9.

(a)   **(b)**

11. $u = 2$

(a)   **(b)**

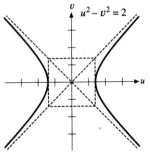

2x − 3y = 6   u = 2

13. $u^2 + v^2 = 11$

(a)   **(b)**

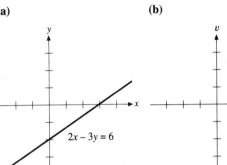

$x^2 + y^2 - 2x + 4y = 6$ $u^2 + v^2 = 11$

15. $u^2 - v^2 = 2$

(a)   **(b)**

$xy = 1$ $u^2 - v^2 = 2$

17.

(a)   **(b)**

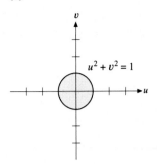

$9x^2 + 4y^2 = 36$ $u^2 + v^2 = 1$

19.

(a)   **(b)**

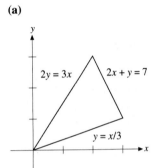

$2y = 3x$ $2x + y = 7$ $u = 0$ $u + v = 1$

$y = x/3$ $v = 0$

21.

(a)   **(b)**

$x + y = 2$ $v = u$

$u = \sqrt{2}$

$u = -v$

23.

(a)   **(b)**

$36z = 9x^2 + 4y^2$ $w = \dfrac{u^2 + v^2}{4}$

25.

(a)

(b)

$u + v + w = 1$

27.

(a)

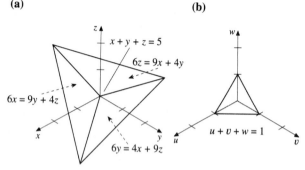

(b)

$u + v + w = 1$

29. 1   **35.** $\displaystyle\int_0^1 \int_0^1 (2u + v)(u + 2v)(3)\,dv\,du$

37. $\displaystyle\int_0^2 \int_0^{2-u} (2u + v)(3)\,dv\,du$

39. $\displaystyle\int_{6/11}^{17/11} \int_{-3/11}^{2/11} (11u)(11)\,dv\,du$

41. $\displaystyle\int_0^{3\sqrt{2}} \int_{-u/3}^{u/3} (\sqrt{2}u)(1)\,dv\,du$

43. $\displaystyle\int_{-1/\sqrt{2}}^{1/\sqrt{2}} \int_{-\sqrt{2+u^2}}^{\sqrt{2+u^2}} (4u^2v^2)(1)\,dv\,du$

45. $\displaystyle\int_0^{\pi/6} \int_1^4 e^{r^2} r \sec\theta\,dr\,d\theta$ **47.** $\displaystyle\int_0^1 \int_0^1 \int_0^1 24u\,dw\,dv\,du$

49. $\displaystyle\int_{-2}^2 \int_{v^2}^4 \int_0^6 (1)\,dw\,du\,dv$ **51.** $\pi/4$ **53.** 16

55. $4\pi abc/3$ **57.** $32\pi/3$

REVIEW EXERCISES

1. 2/5   **3.** 1/6   **5.** 1/15   **7.** 16/15

9. 16   **11.** 1/3   **13.** $\displaystyle\int_0^1 \int_0^y f(x, y)\,dx\,dy$

15. $\displaystyle\int_1^e \int_{\ln y}^1 f(x, y)\,dx\,dy$ **17.** $\displaystyle\int_0^{\pi/3} \int_0^{\tan y} f(x, y)\,dx\,dy$

19. $\displaystyle\int_0^2 \int_0^{\sqrt{4-x^2}/2} f(x, y)\,dy\,dx$

21. $13\pi/3$ **23.** $\left(\dfrac{3}{\pi\sqrt{2}}, \dfrac{3(\sqrt{2}-1)}{\pi\sqrt{2}}\right)$

25. $\displaystyle\int_{-\sqrt{3}}^{\sqrt{3}} \int_{-\sqrt{3-x^2}}^{\sqrt{3-x^2}} \int_1^{\sqrt{4-x^2-y^2}} dz\,dy\,dx,$

$\displaystyle\int_0^{2\pi} \int_0^{\sqrt{3}} \int_1^{\sqrt{4-r^2}} r\,dz\,dr\,d\theta,$

$\displaystyle\int_0^{2\pi} \int_0^{\pi/3} \int_{\sec\phi}^2 \rho^2 \sin\phi\,d\rho\,d\phi\,d\theta; \ 5\pi/3$

27. $\displaystyle\int_{-1}^1 \int_{-\sqrt{1-x^2}}^{\sqrt{1-x^2}} \int_0^{\sqrt{x^2+y^2}} dz\,dy\,dx, \int_0^{2\pi} \int_0^1 \int_0^r r\,dz\,dr\,d\theta,$

$\displaystyle\int_0^{2\pi} \int_{\pi/4}^{\pi/2} \int_0^{\csc\phi} \rho^2 \sin\phi\,d\rho\,d\phi\,d\theta; \ 2\pi/3$

29. $\displaystyle\int_{-2}^2 \int_{-\sqrt{4-x^2}}^{\sqrt{4-x^2}} \int_{\sqrt{x^2+y^2}}^2 (x^2 + y^2)$

$\left(1 + \sqrt{x^2 + y^2 + z^2}\right) dz\,dy\,dx,$

$\displaystyle\int_0^{2\pi} \int_0^2 \int_r^2 \left(1 + \sqrt{r^2 + z^2}\right) r^3\,dz\,dr\,d\theta,$

$\displaystyle\int_0^{2\pi} \int_0^{\pi/4} \int_0^{2\sec\phi} (1 + \rho)\rho^4 \sin^3\phi\,d\rho\,d\phi\,d\theta$

31. $64\pi/5$ **33.** $8\pi \left(3^{3/4} - 1\right)/21$

35. $\displaystyle\int_0^4 \int_0^{(6-x)/2} \int_0^{(6-x-2y)/3} dz\,dy\,dx,$

$\displaystyle\int_0^4 \int_0^{(6-x)/3} \int_0^{(6-x-3z)/2} dy\,dz\,dx$

37. $\displaystyle\int_0^1 \int_0^{2-z} \int_0^{(6-3x-3z)/2} dy\,dx\,dz,$

$\displaystyle\int_0^1 \int_0^{(6-3z)/2} \int_0^{(6-2y-3z)/3} dx\,dy\,dz$

39. $\displaystyle\int_0^1 \int_0^{2-z} \int_0^2 dy\,dx\,dz, \int_0^1 \int_0^2 \int_0^{2-z} dx\,dy\,dz,$

$\displaystyle\int_0^2 \int_0^1 \int_0^{2-z} dx\,dz\,dy$

41. 3

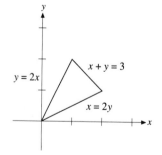

$y = 2x$ $x + y = 3$ $x = 2y$

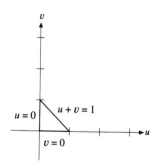

$u = 0$ $u + v = 1$ $v = 0$

43. 2

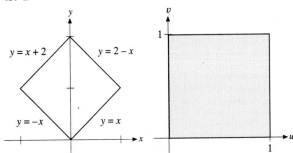

45. $\int_0^{\sqrt{2}} \int_0^{\sqrt{2}} \frac{u+v}{\sqrt{2}}\,dv\,du$ **47.** $\int_0^1 \int_0^{\sqrt{1-u^2}} (2uv)(2)\,dv\,du$

49. $\int_0^1 \int_0^{1-u} \int_0^{1-u-v} (u)(2)\,dw\,dv\,du$ **51.** 8π

53. $k\pi HR^4/10$ **55.** $28k\pi r^5/15$

CHAPTER FOURTEEN:
TOPICS IN VECTOR CALCULUS

EXERCISES 14.1

1. (a) $1/2$, **(b)** $1/2$, **(c)** $1/2$

3. (a) $\frac{1}{2}\ln 2$, **(b)** $\frac{1}{2}\ln 2$, **(c)** $\frac{1}{2}\ln 2$ **5.** 23

7. 4π **9.** 8π **11.** $\frac{3\sqrt{2}}{8} + \frac{1}{16}\ln\left(\frac{\sqrt{2}-1}{\sqrt{2}+1}\right)$

13. $\frac{1}{12}\left(17^{3/2} - 1\right)$ **15.** $2\sqrt{2}\pi^2$ **17.** $(0, \pi/4)$

19. $mr^2/2$ **21. (a)** $-\sin t\,\mathbf{i} + \cos t\,\mathbf{j}$, **(b)** 0, **(c)** 0

23. (a) $\frac{-2\sin t}{\sqrt{4\sin^2 t + \cos^2 t}}\mathbf{i} + \frac{\cos t}{\sqrt{4\sin^2 t + \cos^2 t}}\mathbf{j}$,

(b) $\frac{-2}{\sqrt{4\sin^2 t + \cos^2 t}}$, **(c)** -4π

25. (a) $\frac{1}{\sqrt{1+4t^2}}\mathbf{i} + \frac{2t}{\sqrt{1+4t^2}}\mathbf{j}$, **(b)** $\frac{t+2t^3}{\sqrt{1+4t^2}}$, **(c)** 10

27. 2π

EXERCISES 14.2

1. $-9/2$ **3.** -4π **5.** 64

7.

9.

11. 1

$C : x = 3t, y = 2t, 0 \le t \le 1$

13. -2π

$C : x = \cos t, y = \sin t, 0 \le t \le 2\pi$

15. $2\pi^2$

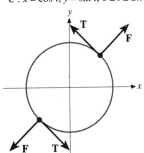

$C : x = \cos t, y = \sin t, z = t, 0 \le t \le 2\pi$

17. $1/26$ **19.** Positive

21. 0

$C : x = \cos t, y = \sin t, 0 \le t \le 2\pi$

23. 2π **25.** $-1/30$ **27.** Positive
29. 2π

$C : x = \cos t, y = \sin t, 0 \le t \le 2\pi$

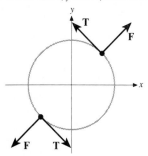

31. $2\pi r^2$ **33.** 4π **35. (a)** 0,   **(b)** 0
37. 1/2; equal to area of triangle

EXERCISES 14.3

1. Yes; 2   **3.** Yes; 8/15
5. Not directly, since C is not closed; $-2/3$
7. No; $8\pi \ln 4$ **9.** Yes; -2π
13. (a) $\pi/2 - \ln 2$,   **(b)** $\left(\dfrac{\pi - 2}{\pi - 2 \ln 2}, 0 \right)$ **15.** 0; 0
17. $\dfrac{2y}{(y^2 + 1)^2}; \dfrac{1}{2}$ **19.** 0, 0   **21.** 2; $2\pi r^2$ **23.** 5; 5π
25. $\dfrac{-1}{(x^2 + y^2)^{3/2}}; \pi$

EXERCISES 14.4

1. (a) 96/5,   **(b)** 64/3   **3. (a)** $-2/3$,   **(b)** $-2/3$
5. (a) $-2/3$,   **(b)** $-2/3$ **7. (a)** 1,   **(b)** 0
9. (a) 9,   **(b)** 0   **11. (a)** 2,   **(b)** 0
13. (a) $2 \tan^{-1}(2/3)$,   **(b)** 0   **15.** 2π **17.** $x^2y + C$
19. No such function   **21.** $2xe^y + y^2/2 + C$
23. $\ln(x^2 + y^2) + C$ **25.** 4   **27.** 0   **29.** 26
33. (a) Yes,   **(b)** no   **37.** $x^2y - xy^2$ **39.** $e^{xy} - 1 + y$

EXERCISES 14.5

1. $14\pi/3$ **3.** $2\pi(4 - 2\sqrt{2})$ **5.** $2\pi(5^{3/2} - 1)/3$
7. $3\sqrt{14}/2$ **9.** $12\sqrt{3}$ **11.** $16\sqrt{5}/3$
13. (a) 9π,   **(b)** $(0, 0, 9/2)$,   **(c)** $81\pi/2$
15. (a) $8\pi\sqrt{2}$,   **(b)** $(0, 0, 8/5)$,   **(c)** $64\pi\sqrt{2}/3$ **17.** 1/2
19. (a) 0,   **(b)** 64π **21.** π **23.** $4\pi R$

EXERCISES 14.6

1. 60   **3.** 16π **5.** $9\pi/10$ **7.** $32\pi/3$ **9.** 10/3
11. 64/27   **13.** $10\pi/3$ **15.** $4\pi r$ **17.** 4π **19.** 4π

EXERCISES 14.7

1. (a) $\mathbf{i} + \mathbf{j} + \mathbf{k}$,   **(b)** $-\pi$,   **(c)** $-\pi$,   **(d)** $-\pi$
3. (a) $\dfrac{y}{\sqrt{x^2 + y^2}}\mathbf{i} - \dfrac{x}{\sqrt{x^2 + y^2}}\mathbf{j} + 2\mathbf{k}$,   **(b)** 2π,
(c) -2π,   **(d)** 2π
5. 18π **7.** -18π **9.** 0   **11.** -16π **13.** 9π
15. $-1/2$ **17.** 0   **19.** 0   **21.** 0   **23. (a)** 0,   **(b)** 0
25. (a) 0,   **(b)** 0   **27.** $x^2y + xz^2 + y^2z + C$
29. $\sqrt{x^2 + y^2 + z^2} + C$ **31.** 1/3   **33.** -1
35. (b) $-1/\sqrt{x^2 + y^2 + z^2}$ **37.** $5/\sqrt{2} - 5$

REVIEW EXERCISES

1. $\displaystyle\int_0^{\pi/2} 2 \cos t \sin t \sqrt{\sin^2 t + 4\cos^2 t}\, dt = 14/9$
3. $\displaystyle\int_0^{\pi/2} (2\cos t)(3\sin t)\sqrt{4\sin^2 t + 9\cos^2 t + 16}\, dt =$
$50 - 16\sqrt{5}$
5. $(5^{3/2} - 1)/12$ **7.** -2π **9.** $8\pi^3/3$ **11.** 0   **13.** 2/3
15. 0   **17.** $3\pi/2$ **19.** 0   **21.** 0   **23.** Yes; 9/2
25. $-2; -8\pi$ **27.** 0, $(x, y) \ne (0, 0)$; 0   **29.** 0; 0
31. 0, $(x, y) \ne (0, 0)$; 2π **33.** $e^x \cos y - y^2 + C$
35. Not possible   **37.** $xy^2 - xz^2 + z + C$ **39.** $3\pi\sqrt{2}$
41. $\pi\sqrt{3}$ **43.** $4\pi\sqrt{5}$ **45.** $\pi\left(17^{5/2} - 1\right)/10$ **47.** $\pi/2$
49. $3\pi/2$ **51.** $128\pi/3$ **53.** -2π **55.** -8π
57. $\dfrac{-\sqrt{y^2 + z^2}\mathbf{i} + y\mathbf{j} + z\mathbf{k}}{\sqrt{2}\sqrt{y^2 + z^2}}$

CHAPTER FIFTEEN:
DIFFERENTIAL EQUATIONS

EXERCISES 15.1

13. $y = 2x^3 - 2x^2 + 3x + C$ **15.** $x^2 - y^2 = C$
17. $y^3 = x^3 + C$ **19.** $\sqrt{y} = \sqrt{x} + C$
21. $y = \left(-x^2 + 5\right)/2$ **23.** $y = 1/(1 - 3x)$
25. $y = \sin^{-1} x$ **27.** $y = 2\sqrt{x} + 1$ **29.** $y = 3x^2 + 3x + 2$
31. $y = \sqrt{x^3 + 1}$ **33.** $y = \left(x^2 + 2\right)/x$
35. $y = 2x^2 - x + 2$ **37.** $y = -\ln(1 - x) + 1, x < 1$
39. $y^3 = 3\left(x^2 + 9\right)/2$ **41.** $y = \tan x, -\pi/2 < x < \pi/2$
43. $y = 2x/(2 - x), x < 2$

45. $y = e^x + C$;
$y = e^{-x} + C$

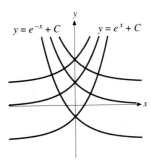

$y = e^{-x} + C$ $y = e^x + C$

47. $y = \tan x + C$; $y = -(\sin x \cos x + x)/2 + C$

$y = \dfrac{-\sin x \cos x + x}{2} + C$

$y = \tan x + C$

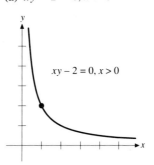

13.

(a) $xy - 2 = 0, x > 0$ **(b)** $xy - 2 = 0, x < 0$

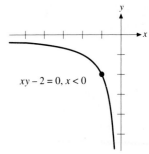

$xy - 2 = 0, x > 0$

$xy - 2 = 0, x < 0$

15. $y = \sqrt{3}x, x > 0$

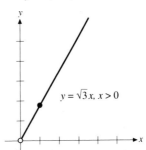

$y = \sqrt{3}x, x > 0$

49. $y = -1/(x + C)$;
$y^3/3 = -x + C$

$\dfrac{y^3}{3} = -x + C$ y

$y = -\dfrac{1}{x + C}$

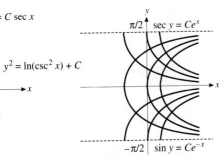

51. $x^2 + y^2 = C$; $y = Cx$

$x^2 + y^2 = C$

$y = Cx$

EXERCISES 15.3

1. $y = \dfrac{x^2}{4} + \dfrac{C}{x^2}$ **3.** $y = -e^x + Ce^{2x}$

5. $y = \dfrac{1}{x} + \dfrac{C}{xe^x}$ **7.** $y = \dfrac{2x^2 + x^4 + C}{4(1 + x^2)}$

9. $y = x \ln x - x \ln(x + 1) + Cx$ **11.** $y = 1 + C \cos x$

13. $y = \dfrac{\ln(\sec x + \tan x) + \ln \sec x + C}{\sec x + \tan x}$ **15.** $i(t) = 8e^{-5t/6}$

17. (a) $\dfrac{x}{y^2} = K$,   **(b)** $\ln x - \ln y^2 = K$ **19.** $\dfrac{x}{y} + x^2 = K$

53. $y = C \sec x$;
$y^2 = \ln(\csc^2 x) + C$

$y = C \sec x$

$y^2 = \ln(\csc^2 x) + C$

55. $\sec y = Ce^x$;
$\sin y = Ce^{-x}$

$\pi/2$ $\sec y = Ce^x$

$-\pi/2$ $\sin y = Ce^{-x}$

EXERCISES 15.4

1. $y = -\ln(1 - x) + 3, x < 1$ **3.** $y = 2e^x - x + 1$

5. $y = -e^{-2x}\left(\dfrac{1}{2}x + \dfrac{1}{4}\right) + \dfrac{1}{4}$ **7. (a)** $y = \sin x$,

(b) $y = \cos x$,   **(c)** $y = 2 \sin x + 3 \cos x$

9. $y = \sinh(2x)$ **11.** $y = C_1 \ln |x| + C_2$

13. $y = e^x/3 + C_1 e^{-2x} + C_2$ **15.** $v(t) = -20 + 20e^{-1.6t}$;
$s(t) = -20t - 12.5e^{-1.6t} + 3012.5$; 20 ft/s

17. $v(t) = -40 + 40e^{-0.8t}$; $s(t) = -40t - 50e^{-0.8t} + 3050$;
40 ft/s   **19.** 3.94R   **21.** $v_0 \geq \sqrt{Rg}$ **23.** 313 s

57. $y' = -y/x$; $y^2 = x^2 + C$ **59.** $y = x \ln |x| + Cx$
61. $y = Cx^2 - x$ **63.** $\tan^{-1}(y/x) = \ln |x| + C$

EXERCISES 15.2

1. $x^2 + y^2 = K$ **3.** $xy = K$
5. $x^2 + xy - x - 3y^2 + 5y = K$ **7.** $x^3 - xy + y^2 = K$
9. $2ye^x + x^2 + y^2 = K$ **11.** $\sin x \cos^2 y = K$

EXERCISES 15.5

1. (b) $v = e^{-x}$, **(c)** $y = C_1 e^{2x} + C_2 e^x$
3. (b) $v = \cot x$, **(c)** $y = C_1 \sin x + C_2 \cos x$
5. (b) $v = 1/x^2$, **(c)** $y = C_1 x + C_2/x$
7. (b) $v = x^{-2/3}$, **(c)** $y = C_1 x + C_2 x^{1/3}$
9. (b) $v = \ln|x|$, **(c)** $y = C_1 x + C_2 x \ln|x|$
11. $y = C_1 e^{2x} + C_2 e^{3x}$ **13.** $y = C_1 e^{3x} + C_2 e^{-3x}$
15. $y = C_1 e^{(1+\sqrt{2})x} + C_2 e^{(1-\sqrt{2})x}$
17. $y = C_1 e^x + C_2 x e^x$ **19.** $y = C_1 e^{-x/2} + C_2 x e^{-x/2}$
21. $y = C_1 e^{\sqrt{2}x} + C_2 e^{-\sqrt{2}x}$ **23.** $y = C_1 \cos 2x + C_2 \sin 2x$
25. $y = C_1 e^{-x} \cos x + C_2 e^{-x} \sin x$ **27.** $y_2 = -x e^x$
31. $y = C_1 x^{-1} + C_2 x^{-2}$

EXERCISES 15.6

1. $y_p = -e^{2x}/4$ **3.** $y_p = x^2 e^x/2$
5. $y_p = -(4x+1)/32$ **7.** $y = C_1 e^{-3x} + C_2 e^{2x} - e^x/4$
9. $y = C_1 \cos x + C_2 \sin x + x$
11. $y = \dfrac{3}{4} e^x - \dfrac{3}{4} e^{-x} - \dfrac{1}{2} \sin x$
13. $y = \dfrac{1}{2}(-\sin x - \cos x + e^x)$
15. (b) $y = C_1 \cos x + C_2 \sin x + y_p$

EXERCISES 15.7

1. (a) 0, **(b)** 0, **(c)** $12e^{-x}$, **(d)** $2\sin 2x - 10\cos 2x$
3. (a) 0, **(b)** 0, **(c)** $-2e^x$, **(d)** $2e^{3x}$
5. (a) 0, **(b)** 0, **(c)** $2e^x$
7. (a) 0, **(b)** 0, **(c)** $2\cos x$, **(d)** $-2\sin x$
9. $y_p = e^{-x}/2$ **11.** $y_p = -e^{-2x}/12$ **13.** $y_p = x^2 e^x/2$
15. $y_p = 6 + 4x + x^2$ **17.** $y_p = \dfrac{1}{3}\sin x$
19. $y_p = \dfrac{1}{4} x \sin 2x$ **21.** $y_p = \dfrac{1}{5}\sin x + \dfrac{2}{5}\cos x$
23. $y = \dfrac{1}{15} e^{-2x} + \dfrac{1}{10} e^{3x} - \dfrac{1}{6} e^x$
25. $\dfrac{1}{49} e^{6x} - \dfrac{1}{49} e^{-x} - \dfrac{1}{7} x e^{-x}$
27. $y = C_1 \sin 3x + C_2 \cos 3x + \dfrac{1}{9} x$
29. $y = C_1 e^{-2x}\cos x + C_2 e^{-2x}\sin 2x + e^{-2x}$
31. (a) $2C_1 e^{2x}\cos x - 2C_2 e^{2x}\sin x$,
(b) $y = -\dfrac{1}{2} x e^{2x}\cos x$

EXERCISES 15.8

1. $2 + 2x + x^2 + \dfrac{1}{3} x^3$ **3.** $1 + x + \dfrac{1}{2} x^2 + \dfrac{1}{6} x^3$
5. $x - \dfrac{1}{6} x^3 + \dfrac{1}{120} x^5 - \dfrac{1}{5040} x^7$ **7.** $1 + 2x + \dfrac{1}{2} x^2 + \dfrac{1}{3} x^3$

9. $2 - x - \dfrac{7}{2} x^2 - \dfrac{19}{6} x^3$ **11.** $2 + x - \dfrac{5}{2} x^2 + \dfrac{1}{2} x^3$
13. $1 + x + x^2 + x^3$ **15.** $3 + x - x^2 + \dfrac{1}{3} x^3$

EXERCISES 15.9

1.

x_n	y_n(tangent)	y_n(improved)	$y(x_n)$
0.000000000	1.000000000	1.000000000	1.000000000
0.100000000	1.000000000	1.001500000	1.001001001
0.200000000	1.003000000	1.009058733	1.008064516
0.300000000	1.015072108	1.029248525	1.027749229
0.400000000	1.042892135	1.070406956	1.068376068

3.

x_n	y_n(tangent)	y_n(improved)	$y(x_n)$
0.000000000	1.000000000	1.000000000	1.000000000
0.100000000	1.000000000	0.995000000	0.994987437
0.200000000	0.990000000	0.979822072	0.979795897
0.300000000	0.969797980	0.953981532	0.953939201
0.400000000	0.938863702	0.916578548	0.916515139

5.

x_n	y_n(tangent)	y_n(improved)	$y(x_n)$
0.000000000	0.000000000	0.000000000	0.000000000
0.100000000	0.100000000	0.100500000	0.100334672
0.200000000	0.201000000	0.203035327	0.202710036
0.300000000	0.305040100	0.309813786	0.309336250
0.400000000	0.414345046	0.423408346	0.422793219

7.

x_n	y_n(tangent)	y_n(improved)	$y(x_n)$
0.000000000	1.000000000	1.000000000	1.000000000
0.100000000	1.300000000	1.315000000	1.315512754
0.200000000	1.630000000	1.663075000	1.664208274
0.300000000	1.993000000	2.047697875	2.049576423
0.400000000	2.392300000	2.472706152	2.475474093

9.

x_n	y_n(tangent)	y_n(improved)	$y(x_n)$
0.000000000	1.000000000	1.000000000	1.000000000
0.100000000	1.000000000	0.995241871	0.995321160
0.200000000	0.990483742	0.982348115	0.982476904
0.300000000	0.973308443	0.962908839	0.963063687
0.400000000	0.950059421	0.938285321	0.938448064

REVIEW EXERCISES

1. $y = -4/(x^4 + C)$ **3.** $y = \left((3x/2 + C)^{1/3} - 1\right)^2$ or
$2\sqrt{y} + 2y + \dfrac{2y^{3/2}}{3} = x + C$
5. $y = -2/(x^2 + 2x + C)$
7. $y = -1 + Ce^{2x}$ **9.** $y = x - 1 + Ce^{-x}$
11. $y = 1 + C/x$ **13.** $y = x\cos x + C\cos x$
15. $y = C_1 e^x + C_2 e^{-3x}$ **17.** $y = C_1 e^{2x} + C_2 x e^{2x}$
19. $y = C_1 \sin(x/2) + C_2 \cos(x/2)$
21. $y = C_1 e^{-x}\sin 2x + C_2 e^{-x}\cos 2x$
23. $y = C_1 e^x + C_2 e^{2x} + e^{-x}/6$

25. $y = C_1 \sin x + C_2 \cos x - \dfrac{1}{2} x \cos x$

27. $y = C_1 e^x + C_2 e^{2x} - xe^x$

29. $y = C_1 e^{x/2} + C_2 x e^{x/2} + e^x$

31. $y = C_1 e^x \sin x + C_2 e^x \cos x + \dfrac{1}{5}\sin x + \dfrac{2}{5}\cos x$

33. $y = C_1 e^x + C_2 e^{-x} + \dfrac{1}{2}xe^x - \dfrac{1}{2}xe^{-x}$

35. $y = C_1 e^x + C_2 x e^x + \dfrac{1}{6}x^3 e^x$ **37.** $y = x^2 - x + 2$

39. $y = 2x - \sin x$ **41.** $y = x^2 + 1/x$ **43.** $y = 1 + \cot x$

45. $y = \left(\dfrac{1}{2}\ln\left(x^2 + 1\right) + 8\right)^{1/3}$ **47.** $y = \sqrt{4 - 4x^2}$

49. $y = \sqrt{1/x + 3}$

51. (a) $y = 1 \big/ \left(x^2 - 6x + 9\right),\ x < 3,$

(b) $y = 1 \big/ \left(x^2 - 6x + 9\right),\ x > 3$

53. $y = \dfrac{1}{9}e^{2x} - \dfrac{1}{9}e^{-x} - \dfrac{1}{3}xe^{-x}$

55. $y = \dfrac{5}{12}e^{2x} - \dfrac{2}{3}e^{-x} + \dfrac{1}{4} - \dfrac{1}{2}x$

57. $y = \dfrac{2}{3}e^{2x} - \dfrac{5}{12}e^{-x} - \dfrac{1}{2}xe^x - \dfrac{1}{4}e^x$ **59.** $y = \dfrac{1}{2}x^2 e^x$

61. $y = -e^x \cos x + e^x$ **63.** $x^2 + xy + y^2 = K$

65. (b) $y_2 = x^3$,   (c) $y = C_1 x + C_2 x^3$

67. $y = 1 \big/ \left(5 - x^2\right)$ **69.** $y = 2\sec x + 2\tan x$

71. $y^2/2 = x + C;\ y = Ce^{-x}$

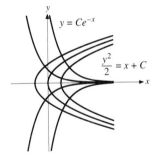

73. $i(t) = \dfrac{6}{5}\sin t + \dfrac{12}{5}\cos t - \dfrac{12}{5}e^{-2t}$

75. (a) $v(t) = -32 + 32e^{-t}$; 32 ft/s,   (b) 448 ft/s

77. 8   **79.** $\approx 10^7$ ft; $v_0 \geq \sqrt{3Rg/2}$

81. $27.7/k$ **83.** 31   **85.** $x = 2\cos 2t$

87. $x = 4e^{-t} - 2e^{-2t}$ **89.** $x = 4e^{-t}\sin t + 2e^{-t}\cos t$

91. $2 + 2x + \dfrac{3}{2}x^2 + \dfrac{1}{2}x^3$ **93.** $1 + 2x - \dfrac{1}{3}x^3 - \dfrac{1}{12}x^4$

95.

x_n	y_n(tangent)	y_n(improved)	$y(x_n)$
0.000000000	1.000000000	1.000000000	1.000000000
0.100000000	1.000000000	1.010000000	1.010101010
0.200000000	1.020000000	1.041435566	1.041666667
0.300000000	1.061616000	1.098432300	1.098901099
0.400000000	1.129237712	1.189462201	1.190476190

97.

x_n	y_n(tangent)	y_n(improved)	$y(x_n)$
0.000000000	1.000000000	1.000000000	1.000000000
0.100000000	1.000000000	1.005258546	1.005004168
0.200000000	1.010517092	1.020561813	1.020066756
0.300000000	1.031605658	1.046064506	1.045338514
0.400000000	1.063430973	1.082023259	1.081072372

Index

TABLE OF INTEGRALS

BASIC INTEGRALS

1. $\displaystyle\int u\,dv = uv - \int v\,du$

2. $\displaystyle\int u^\alpha\,du = \frac{u^{\alpha+1}}{\alpha+1} + C,\ \alpha \neq -1$

3. $\displaystyle\int \frac{1}{u}\,du = \ln|u| + C$

4. $\displaystyle\int e^u\,du = e^u + C$

5. $\displaystyle\int \cos u\,du = \sin u + C$

6. $\displaystyle\int \sin u\,du = -\cos u + C$

7. $\displaystyle\int \sec^2 u\,du = \tan u + C$

8. $\displaystyle\int \csc^2 u\,du = -\cot u + C$

9. $\displaystyle\int \sec u \tan u\,du = \sec u + C$

10. $\displaystyle\int \csc u \cot u\,du = -\csc u + C$

11. $\displaystyle\int \frac{1}{\sqrt{a^2 - u^2}}\,du = \sin^{-1}\frac{u}{a} + C$

12. $\displaystyle\int \frac{1}{a^2 + u^2}\,du = \frac{1}{a}\tan^{-1}\frac{u}{a} + C$

13. $\displaystyle\int \frac{1}{u\sqrt{u^2 - a^2}}\,du = \frac{1}{a}\sec^{-1}\frac{|u|}{a} + C$
$$= \frac{1}{a}\cos^{-1}\frac{a}{|u|} + C$$

INTEGRALS THAT INVOLVE TRIGONOMETRIC FUNCTIONS

14. $\displaystyle\int \tan u\,du = \ln|\sec u| + C$

15. $\displaystyle\int \cot u\,du = \ln|\sin u| + C$

16. $\displaystyle\int \sec u\,du = \ln|\sec u + \tan u| + C$

17. $\displaystyle\int \csc u\,du = \ln|\csc u - \cot u| + C$

18. $\displaystyle\int \sin^2 u\,du = -\frac{\sin u \cos u}{2} + \frac{u}{2} + C$

19. $\displaystyle\int \cos^2 u\,du = \frac{\sin u \cos u}{2} + \frac{u}{2} + C$

20. $\displaystyle\int \sec^3 u\,du = \frac{1}{2}\sec u \tan u + \frac{1}{2}\ln|\sec u + \tan u| + C$

21. $\displaystyle\int \sin^n u\,du = -\frac{\sin^{n-1} u \cos u}{n} + \frac{n-1}{n}\int \sin^{n-2} u\,du$

22. $\displaystyle\int \cos^n u\,du = \frac{\sin u \cos^{n-1} u}{n} + \frac{n-1}{n}\int \cos^{n-2} u\,du$

23. $\displaystyle\int \sin^m u \cos^n u\,du$
$$= -\frac{\sin^{m-1} u \cos^{n+1} u}{m+n} + \frac{m-1}{m+n}\int \sin^{m-2} u \cos^n u\,du$$

24. $\displaystyle\int \sin^m u \cos^n u\,du$
$$= \frac{\sin^{m+1} u \cos^{n-1} u}{m+n} + \frac{n-1}{m+n}\int \sin^m u \cos^{n-2} u\,du$$

25. $\displaystyle\int \tan^n u\,du = \frac{\tan^{n-1} u}{n-1} - \int \tan^{n-2} u\,du$

26. $\displaystyle\int \sec^n u\,du = \frac{\sec^{n-2} u \tan u}{n-1} + \frac{n-2}{n-1}\int \sec^{n-2} u\,du$

INTEGRALS THAT INVOLVE $a^2 + u^2$

27. $\displaystyle\int \frac{1}{(a^2 + u^2)^n}\,du$
$$= \frac{u}{a^2(2n-2)(a^2+u^2)^{n-1}} + \frac{2n-3}{a^2(2n-2)}\int \frac{1}{(a^2+u^2)^{n-1}}\,du$$

28. $\displaystyle\int \frac{1}{(a^2+u^2)^{3/2}}\,du = \frac{u}{a^2\sqrt{a^2+u^2}} + C$

29. $\displaystyle\int \frac{1}{u\sqrt{a^2+u^2}}\,du = -\frac{1}{a}\ln\left|\frac{a+\sqrt{a^2+u^2}}{u}\right| + C$

30. $\displaystyle\int \frac{1}{\sqrt{a^2+u^2}}\,du = \ln\left|u + \sqrt{a^2+u^2}\right| + C$

31. $\displaystyle\int \sqrt{a^2+u^2}\,du = \frac{u}{2}\sqrt{a^2+u^2} + \frac{a^2}{2}\ln\left|u+\sqrt{a^2+u^2}\right| + C$

32. $\displaystyle\int (a^2+u^2)^{3/2}\,du$
$$= \frac{u}{8}(2u^2+5a^2)\sqrt{a^2+u^2} + \frac{3a^4}{8}\ln\left|u+\sqrt{a^2+u^2}\right| + C$$

33. $\displaystyle\int u^2\sqrt{a^2+u^2}\,du$
$$= \frac{u}{8}(2u^2+a^2)\sqrt{a^2+u^2} - \frac{a^4}{8}\ln\left|u+\sqrt{a^2+u^2}\right| + C$$